LÜRSSEN

A TAILOR-MADE SOLUTION. WHATEVER THE CHALLENGE

From fast patrol boats and OPVs to frigates and fleet support vessels – Lürssen has over 135 years of experience building naval vessels of all types and sizes. We develop tailor-made maritime solutions to answer any of your requirements, whatever your international focus! And whenever the need arises, our logistic support services and spare parts supply are always there to help you. Anywhere in the world. All across the seven seas.

53° 10' N
8° 37' O

Fr. Lürssen Werft GmbH & Co. KG | Zum Alten Speicher 11 | 28759 Bremen | Germany
Tel.: +49 (0)421 6604 344 | Fax: +49 (0)421 6604 563 | Email: defence@luerssen.de | www.luerssen-defence.com

IHS™ Jane's® Fighting Ships

2014-2015

Commodore Stephen Saunders RN

ISBN - 978 0 7106 3101 5

© 2014 IHS. All rights reserved.
No part of this publication may be reproduced or transmitted, in any form or by any means, electronic, mechanical, photocopying, recording or otherwise, or be stored in any retrieval system of any nature, without prior written permission of IHS Global Limited. Applications for written permission should be directed to Christopher Bridge.

Any views or opinions expressed by contributors and third parties are personal to them and do not represent the views or opinions of IHS Global Limited, its affiliates or staff.

Disclaimer of liability
Whilst every effort has been made to ensure the quality and accuracy of the information contained in this publication at the time of going to press, IHS Global Limited, its affiliates, their officers, employees and agents assume no responsibility as to the accuracy or completeness of and, to the extent permitted by law, shall not be liable for any errors or omissions or any loss, damage or expense incurred by reliance on information or any statement contained in this publication.

Advertisement
Advertisers are solely responsible for the content of the advertising material which they submit to us and for ensuring that the material complies with applicable laws. IHS Global Limited is not responsible for any error, omission or inaccuracy in any advertisement. IHS Global Limited will not be liable for any damages arising from any use of products or services or any actions or omissions taken in reliance on information or any statement contained in advertising material. Inclusion of any advertisement is not intended to endorse any views expressed, nor products or services offered, nor the organisations sponsoring the advertisement.

Third party details and websites
Any third party details and websites are given for information and reference purposes only and IHS Global Limited does not control, approve or endorse these third parties or third party websites. Further, IHS Global Limited does not control or guarantee the accuracy, relevance, availability, timeliness or completeness of the information contained on any third party website. Inclusion of any third party details or websites is not intended to reflect their importance, nor is it intended to endorse any views expressed, products or services offered, nor the companies or organisations in question. You access any third party websites solely at your own risk.

Use of data
The company and personal data stated in any directory or database may be used for the limited purpose of enquiring about the products and services of the companies listed who have given permission for their data to be used for this purpose only. You may use the data only to the extent, and in such a manner, as is necessary for the authorised purpose. You must comply with the Data Protection Act 1998 and all other applicable data protection and privacy laws and regulations. In particular, you must not use the data (i) for any unlawful, harmful or offensive purpose; (ii) as a source for any kind of marketing or promotion activity; or (iii) for the purposes of compiling, confirming or amending your own database, directory or mailing list.

Trade Marks
IHS and Jane's are trade marks of IHS Global Limited.

This book was produced using FSC® certified paper
Printed and bound in the UK by Polestar Wheatons

thalesgroup.com/integratedmast

The advanced IM400 Integrated Mast

Everywhere it matters, we deliver

OPTIMAL PERFORMANCE
Increase situational awareness with unique 360° vision for all subsystems

COST-EFFICIENT SOLUTION
Gain savings through lower lifecycle costs, reduced and simplified maintenance

REDUCED RISK
Benefit from plug & play installation – tested, integrated and built in parallel with ship construction

LEAN MANNING OBJECTIVES
Promote reduced manning concept through easier operation

Critical decisions are made every day at sea by Naval Forces. Thales is already at the heart of this, with five IM400 systems on the Royal Netherlands Navy's patrol ships and joint support ship. The Thales IM400 Integrated Mast is a radically advanced engineering solution that resolves the electromagnetic conflicts and line-of-sight obstructions inherent in traditional topside antenna arrangements. Delivering remarkable advantages in terms of operational performance it is also cost effective and completely scalable to customer requirements. World-leading, the IM400 gives decision makers the information and control they need to make more effective responses. Everywhere, together with our customers, we are making a difference.

THALES
Together • Safer • Everywhere

Contents

Alphabetical list of advertisers ...[4]

Executive overview ...[10]

Acknowledgements ...[20]

Glossary ...[23]

Ranks and insignia of the world's navies ...[28]

Ensigns and flags of the world's navies ...[57]

Pennant list of major surface ships ...[67]

How to use ...[102]

World navies ...1

Albania ...2
Algeria ...3
Angola ...9
Anguilla ...10
Antigua and Barbuda ...10
Argentina ...11
Australia ...25
Azerbaijan ...40
Bahamas ...44
Bahrain ...46
Bangladesh ...51
Barbados ...62
Belgium ...62
Belize ...66
Benin ...66
Bermuda ...67
Bolivia ...68
Brazil ...69
British Indian Ocean Territory ...87
Brunei ...87
Bulgaria ...89
Cambodia ...94
Cameroon ...95
Canada ...98
Cape Verde ...113
Cayman Islands ...114
Chile ...115
China ...127
Colombia ...170
Comoros ...180
Democratic Republic of Congo ...181
Congo-Brazzaville ...182
Cook Islands ...182
Costa Rica ...183
Côte d'Ivoire ...184
Croatia ...184
Cuba ...189
Cyprus ...192
Denmark ...194
Djibouti ...203
Dominica ...204
Dominican Republic ...205
East Timor ...208
Ecuador ...209
Egypt ...217
El Salvador ...229
Equatorial Guinea ...231
Eritrea ...232
Estonia ...233
Falkland Islands ...237
Faroe Islands ...238
Fiji ...238
Finland ...239
France ...246
Gabon ...280
Gambia ...281
Georgia ...282
Germany ...284
Ghana ...300
Greece ...302
Grenada ...315
Guatemala ...316
Guinea ...317
Guinea-Bissau ...318
Guyana ...318
Haiti ...319
Honduras ...319
Hong Kong ...321
Hungary ...324
Iceland ...325
India ...326
Indonesia ...358
Iran ...378
Iraq ...391
Ireland ...393
Israel ...395
Italy ...400
Jamaica ...426
Japan ...428
Jordan ...460
Kazakhstan ...462
Kenya ...464
Kiribati ...465
Korea, North ...466
Korea, South ...473
Kuwait ...490
Latvia ...494
Lebanon ...497
Libya ...498
Lithuania ...501
Macedonia ...504
Madagascar ...504
Malawi ...505
Malaysia ...506
Maldives ...521
Malta ...522
Marshall Islands ...524
Mauritania ...524
Mauritius ...526
Mexico ...528
Federated States of Micronesia ...541
Montenegro ...542
Morocco ...543
Mozambique ...551
Myanmar ...552
Namibia ...559
NATO ...561
Netherlands ...561
New Zealand ...573
Nicaragua ...577
Nigeria ...578
Norway ...584
Oman ...593
Pakistan ...599
Palau ...611
Panama ...612
Papua New Guinea ...614
Paraguay ...615
Peru ...617
Philippines ...628
Poland ...637
Portugal ...649
Qatar ...656
Romania ...659
Russian Federation ...669
St Kitts and Nevis ...734
St Lucia ...734
St Vincent and the Grenadines ...735
Samoa ...736
Saudi Arabia ...736
Senegal ...743
Serbia ...746
Seychelles ...747
Sierra Leone ...749
Singapore ...749
Slovenia ...759
Solomon Islands ...760
South Africa ...760
Spain ...766
Sri Lanka ...786
Sudan ...792
Suriname ...793
Sweden ...794
Switzerland ...807
Syria ...808
Taiwan ...810
Tanzania ...825
Thailand ...827
Togo ...847
Tonga ...848
Trinidad and Tobago ...848
Tunisia ...850
Turkey ...854
Turkmenistan ...873
Tuvalu ...874
Ukraine ...874
United Arab Emirates ...883
United Kingdom ...894
United States ...927
Uruguay ...995
Vanuatu ...1000
Venezuela ...1001
Vietnam ...1008
Virgin Islands ...1016
Yemen ...1016
Zimbabwe ...1019

Indexes
Country abbreviations ...1023
Named ships ...1023
Named classes ...1041
Aircraft by countries ...1046

Alphabetical list of advertisers

Daewoo Shipbuilding & Marine Engineering Co Ltd
85 Da-dong, Jung-gu, Seoul, 100-180, Republic of Korea .... [7]

Fr. Lürssen Werft GmbH & Co Kg
Zum alten Speicher 11, D-28759 Bremen, Germany .... *Opposite title page*

Hanjin Heavy Industries & Construction Co Ltd
102-17, Garwol-dong, Yongsan-gu, Seoul, 140-700, Republic of Korea .... *Inside front cover*

Huntington Ingalls
2401 West Avenue, Newport News, VA-23607, US .... [5]

Hyundai Heavy Industries Co Ltd
1 Cheonha-dong, Dong-gu, Ulsan 682792, Republic of Korea .... [8]

Navantia, S.A
Viriato, 69, E-28010 Madrid, Spain .... [22]

Thales Nederland BV ta v. de
Postbus 42, NL-7550 GD Hengelo, Netherlands .... [2]

ThyssenKrupp Marine Systems GmbH
Werftstr. 112-114, D-24124 Kiel, Germany .... *Facing inside front cover*

Judy Howell, *Joiner* | Ingalls Shipbuilding, Pascagoula, Miss.

TOUGHER THAN STEEL
ONLY THE BEST SHIPBUILDERS IN THE WORLD CAN BUILD THE FINEST SHIPS AT SEA.

HuntingtonIngalls.com/**Judy** | #**TougherThanSteel**

Scan code to see more
Tougher Than Steel stories.

Hard Stuff Done Right™

EDITORIAL AND ADMINISTRATION

Managing Director: Blake Bartlett, e-mail: blake.bartlett@ihs.com
Group Publishing Director: Sean Howe, e-mail: sean.howe@ihs.com
Director IHS Jane's Reference and Data Transformation: Chris Bridge,
e-mail: chris.bridge@ihs.com
Director EMEA Editing and Design: Sara Morgan,
e-mail: sara.morgan@ihs.com
Product Manager Defence Equipment & Technology: Emma Cussell,
e-mail: emma.cussell@ihs.com
Compiler/Editor: Welcomes information and comments from users who should send material to:
Research and Information Services
IHS Jane's, IHS Global Limited, Sentinel House, 163 Brighton Road,
Coulsdon, Surrey CR5 2YH
e-mail: yearbook@ihs.com

SALES OFFICES

Europe/Middle East/Africa/Asia Pacific
Tel: (+44 0) 13 44 32 83 00 Fax: (+44 0) 13 44 32 80 05
e-mail: customer.support@ihs.com

North/Central/South America
Tel: Customer care to 1–800–IHS-CARE or 1–800–447–2273
e-mail: customercare@ihs.com

ADVERTISEMENT SALES OFFICES

UNITED KINGDOM
IHS Jane's, IHS Global Limited
Sentinel House, 163 Brighton Road,
Coulsdon, Surrey CR5 2YH, UK
Tel: (+44 20) 32 53 22 89 Fax: (+44 20) 32 53 21 03
e-mail: defadsales@ihs.com

Janine Boxall, Global Advertising Sales Director
Tel: (+44 20) 32 53 22 95 Fax: See UK
e-mail: janine.boxall@ihs.com

Richard West, Senior Key Accounts Manager
Tel: (+44 20) 32 53 22 92 Fax: See UK
e-mail: richard.west@ihs.com

Carly Litchfield, Advertising Sales Manager
Tel: (+44 20) 32 53 22 91 Fax: See UK
e-mail: carly.litchfield@ihs.com

Adam Smith, Advertising Sales Executive
Tel: (+44 20) 32 53 22 93 Fax: See UK
e-mail: adam.smith@ihs.com

UNITED STATES
IHS Jane's, IHS Global Inc.
110 N Royal Street, Suite 200,
Alexandria, Virginia 22314, US
Tel: (+1 703) 683 37 00 Fax: (+1 703) 836 55 37
e-mail: defadsales@ihs.com

Robert Sitch, US Advertising Sales Director,
Tel: (+1 703) 236 24 24 Fax: (+1 703) 836 55 37
e-mail: robert.sitch@ihs.com

Drucie DeVries, South and Southeast USA
Tel: (+1 703) 836 24 46 Fax: (+1 703) 836 55 37
e-mail: drucie.devries@ihs.com

Dave Dreyer, Northeastern USA
Tel: (+1 703) 438 78 38 Fax: (+1 703) 836 55 27
e-mail: dave.dreyer@ihs.com

Janet Murphy, Central USA
Tel: (+1 703) 836 31 39 Fax: (+1 703) 836 55 37
e-mail: janet.murphy@ihs.com

Richard L Ayer, Western USA and National Accounts
127 Avenida del Mar, Suite 2A, San Clemente, California 92672, US
Tel: (+1 949) 366 84 55 Fax: (+1 949) 366 92 89
e-mail: ayercomm@earthlink.com

REST OF THE WORLD
Australia: *Richard West* (UK Office)
Benelux: *Adam Smith* UK Office)
Brazil: *Drucie DeVries* (USA Office)

Canada: *Janet Murphy* (USA Office)
Eastern Europe (excl. Poland): MCW Media & Consulting Wehrstedt
Dr Uwe H Wehrstedt
Hagenbreite 9, D-06463 Ermsleben, Germany
Tel: (+49 03) 47 43/620 90 Fax: (+49 03) 47 43/620 91
e-mail: info@Wehrstedt.org

Germany and Austria: *MCW Media & Consulting Wehrstedt* (see Eastern Europe)
Greece: *Carly Litchfield* (UK Office)
Hong Kong: *Carly Litchfield* (UK Office)
India: *Carly Litchfield* (UK Office)
Israel: *Oreet International Media*
15 Kinneret Street, IL-51201 Bene Berak, Israel
Tel: (+972 3) 570 65 27 Fax: (+972 3) 570 65 27
e-mail: admin@oreet-marcom.com
Defence: Liat Heiblum
e-mail: liat_h@oreet-marcom.com

Italy and Switzerland: *Ediconsult Internazionale Srl*
Piazza Fontane Marose 3, I-16123 Genoa, Italy
Tel: (+39 010) 58 36 84 Fax: (+39 010) 56 65 78
e-mail: genova@ediconsult.com

Japan: *Carly Litchfield* (UK Office)
Middle East: *Adam Smith* (UK Office)
Pakistan: *Adam Smith* (UK Office)
Poland: *Adam Smith* (UK Office)

Russia: *Anatoly Tomashevich*
4-154, Teplichnyi Pereulok, Moscow, Russia, 123298
Tel/Fax: (+7 495) 942 04 65
e-mail: to-anatoly@tochka.ru

Scandinavia: *Falsten Partnership*
23, Walsingham Road, Hove, East Sussex BN41 2XA, UK
Tel: (+44 1273) 77 10 20 Fax: (+ 44 1273) 77 00 70
e-mail: sales@falsten.com

Singapore: *Richard West* (UK Office)
South Africa: *Richard West* (UK Office)
Spain: *Carly Litchfield* (UK Office)

ADVERTISING COPY
Sally Eason (UK Office)
Tel: (+44 20) 32 53 22 69 Fax: (+44 20) 87 00 38 59/37 44
e-mail: sally.eason@ihs.com

For North America, South America and Caribbean only:
Tel: (+1 703) 68 33 700 Fax: (+1 703) 83 65 5 37
e-mail: us.ads@ihs.com

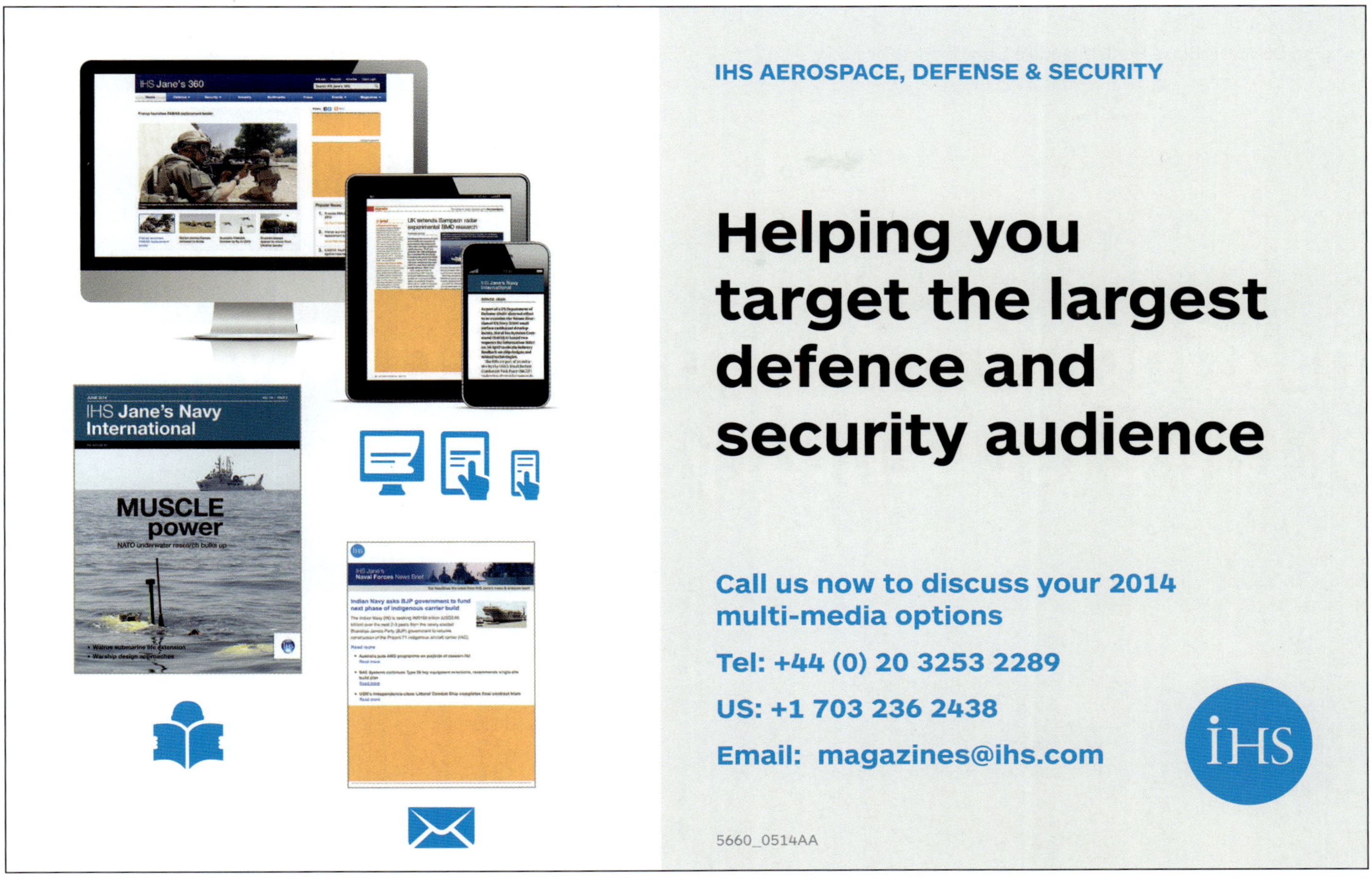

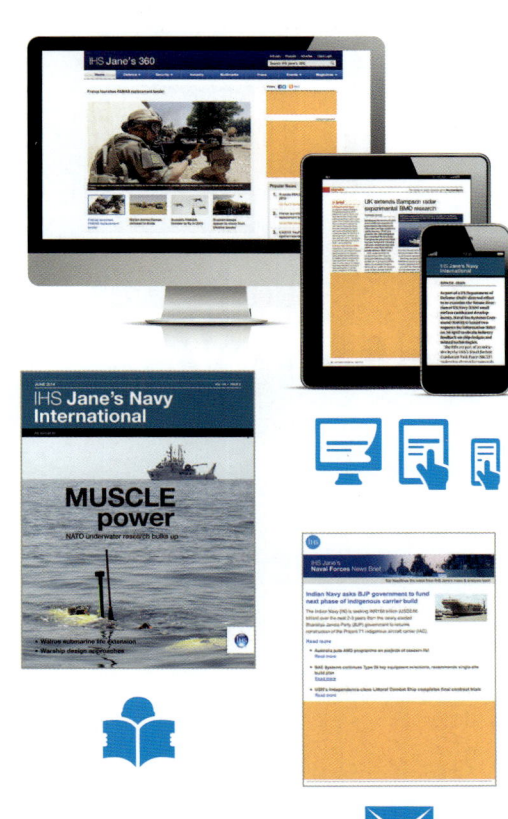

IHS AEROSPACE, DEFENSE & SECURITY

Helping you target the largest defence and security audience

Call us now to discuss your 2014 multi-media options

Tel: +44 (0) 20 3253 2289

US: +1 703 236 2438

Email: magazines@ihs.com

5660_0514AA

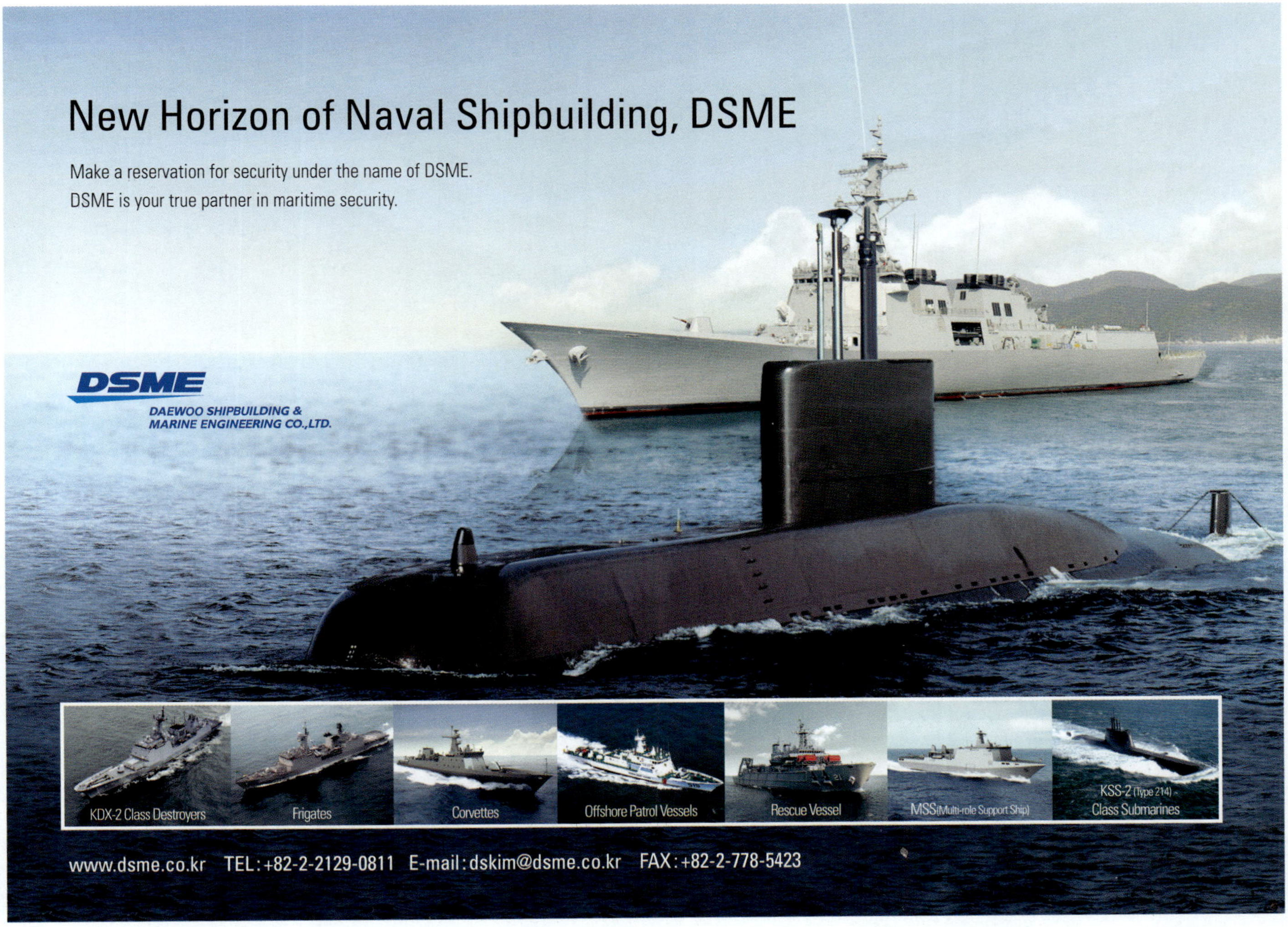

New Horizon of Naval Shipbuilding, DSME

Make a reservation for security under the name of DSME.
DSME is your true partner in maritime security.

DSME
DAEWOO SHIPBUILDING & MARINE ENGINEERING CO.,LTD.

KDX-2 Class Destroyers | Frigates | Corvettes | Offshore Patrol Vessels | Rescue Vessel | MSS (Multi-role Support Ship) | KSS-2 (Type 214) Class Submarines

www.dsme.co.kr   TEL:+82-2-2129-0811   E-mail:dskim@dsme.co.kr   FAX:+82-2-778-5423

Executive overview

Introduction

The Russian annexation of Crimea and the crisis in eastern Ukraine, which could lead to further territorial seizures, have triggered a significant change in international relations and the most dangerous threat to European security since the end of the Cold War. Whatever the rights and wrongs of the situation, it is clear from the actions and methods of the Russian government that a 25-year period of co-operation with the West, however uneasy it has been at times, is now over. While the new relationship does not yet amount to a new Cold War, it is hard to see how trust and good relations can easily be restored given the prevailing anti-western mind-set in Moscow.

The first result of this unexpected turn of events is that many of the policy assumptions that have underpinned western defence in recent years will have to be at least reviewed, if not changed. NATO, which has struggled at times to justify its existence, has become relevant again. Sweden and possibly Finland may consider joining the alliance and Ukraine itself, while unlikely in the near future to meet the criteria required to join a NATO Membership Action Plan, has taken a first westward step by signing an Association Agreement with the European Union. Meanwhile, relatively small-scale deployments of US and other allied forces to Poland, the Baltic States, and to the Black Sea have provided symbolic but nevertheless welcome reassurance to the countries concerned. NATO may now need to re-evaluate the balance between in-place and deployable forces; the steady progression over the past decade towards an expeditionary posture may have to be checked. In addition, individual states may wish to re-appraise the balance within their armed forces. For example, Turkey will now view the modernisation and expansion of the Russian Black Sea Fleet in a new light, while Baltic and Scandinavian nations will probably be dusting off their home-defence plans.

The prospect of a resurgent Russia also suggests that there is a need to re-consider the balance of investment across the spectrum of maritime capabilities. At the height of the Cold War, one of the most challenging technological and operational fields was anti-submarine warfare, which involved command and control of a complex panoply of sensors and weapons deployed on and below the surface of the sea, in the air, and on the seabed. Much of the equipment and many of the skills that were developed for this purpose have since been lost or at least allowed to 'wither on the vine' in favour of capabilities required for expeditionary operations. Funding priorities may now have to be adjusted. Meanwhile, increased urgency should now be injected into the NATO Ballistic Missile Defence (BMD) programme. Land-based BMD systems are to be based on a ballistic-missile defence radar at Kürecik, Turkey, and 'Aegis Ashore' capabilities at Deveselu airbase in Romania, and Redzikowo, Poland. At sea, USS *Donald Cook* became, in February 2014, the first of four Arleigh Burke-class guided missile destroyers to be stationed in Rota, Spain, while three additional BMD-capable ships are to join it by 2015. This US commitment should be matched by European nations. The Netherlands already plans to add an extended long-range missile search mode to the air-search radars of its De Zeven Provincien-class frigates and there is now a strong case for the navies of Denmark, France, Germany, Italy, and the United Kingdom, equipped with similar radars, to do the same.

United States and Canada

Against a background of an uncertain and dangerous world, the United States Defence Budget continues to be driven as much by domestic politics as by national defence needs. The 2013 Bipartisan Budget Act, approved in December 2013, brought a two-year relief from 'sequestration', the mechanism by which across-the-board spending cuts are imposed on the defence and other US federal budgets, but this indiscriminate process will return in 2016 if agreement cannot be reached. If this happens, "tough, tough choices are coming", warned Defense Secretary Chuck Hagel to a House of Representatives committee in March 2014. Unfortunately, options seem to focus primarily on cuts in manpower and major programmes rather than on reductions in infrastructure and support which, reports suggest, have considerable potential to yield funds for the front line.

Doubts about the future of the carrier *George Washington* (CVN-73) encapsulate the problems generated by the budget impasse. The ship is due to start a three-year Refueling and Complex Overhaul in 2016, having being replaced at its homeport of Yokosuka, Japan, by *Ronald Reagan* (CVN-76) in mid-2015. Despite the fact that the Navy is required by law to have 11 carriers, about which there seems to be little disagreement, there is a prospect that the ship will be defuelled and inactivated halfway through its service life if the necessary funding (USD7 billion) cannot be found. Speaking at a seminar on 6 April 2014 during the US Navy League's Sea-Air-Space exposition, General Amos, Commandant of the US Marine Corps, remarked that "we shouldn't be discussing refuelling", a clear indication of the frustration felt at all levels by this perplexing situation.

While it is widely believed or at least hoped that the enormity of the *George Washington* decision will concentrate Congressional minds, the Littoral Combat Ship (LCS) programme continues to be controversial. The Department of Defense's fiscal year 2015 (FY 2015) budget request included a 20-ship cut to the total number of LCS, thereby constraining production to 32 hulls. Ironically, the decision comes at a time when the programme is gathering pace; the next four years will see 16 ships enter service, the eight-month deployment of *Freedom* to Singapore during 2013 validated the 'three crews, two ships, one ship deployed' concept of operations, while progress is also being made towards the introduction of the three mission modules into service. Notwithstanding apparent truncation of the programme, LCS is officially on probation rather than cancelled; studies into alternative options for "a capable and lethal small surface combatant, generally consistent with the capabilities of a frigate" could yet uphold current plans. However, the consensus view is that it is more likely that the programme will end; indeed that it may not extend beyond the 24 ships currently in service and under contract. The main reason for a change of procurement strategy is that survivability of the ships and their suitability for open-ocean operations, particularly in the Pacific, continue to be questioned. While there is still a case for the shallow-

CHUNG-HOON

8/2013*, Chris Sattler / 1531202

water capabilities for which LCS was designed, a bigger, more durable vessel with longer legs is required to support strike group operations. Such a ship is likely to be equipped with proven, fixed-weapon systems, rather than through a modular approach. Whether a wholly new design will be required remains to be seen. However, the need to contain costs points to a development of a current US design; options include the US Coast Guard's national security cutter, the Oliver Hazard Perry class, and up-sized versions of the two LCS variants.

In contrast, the re-started DDG 51 destroyer programme is relatively stable and progressing well. Construction of the first four new ships (DDG 113-116) is in progress, while a multi-year contract for DDGs 117-126 was completed on 14 March 2014. The order for the latter 10 ships incorporates the transition from the Flight IIA configuration, equipped with SPY-1D air-defence radar, to Flight III ships fitted with Raytheon's Air and Missile Defense Radar. Under current plans, the S-band (NATO (E/F)) component of this radar is to be installed in the first 12 Flight III ships (probably DDGs 124-135), while the X-band (NATO I) component is to be provided by Northrop Grumman's SPQ-9B. Preliminary designs of the Flight III ships have been completed and, apart from a re-shaped stern to improve buoyancy, the ships' outward appearance is to look much the same as their predecessors; the weight margins are reported to be similar. As yet, it is unclear how long construction of the DDG-51 class variants (there has been mention of a Flight IV in the 2030s) will continue but, in any case, the Arleigh Burke design, first conceived in the 1980s, could still be in service in the 2060s. If so, it will prove to be the most successful design in the history of the US Navy.

While no US Navy programme could be considered to be 'ring-fenced', the Ohio Replacement Program comes close to falling into that category. The first patrol of the new class is to be conducted in 2031 which, although seemingly a long time in the future, necessitates a tight programme. Allowing for a seven-year construction phase and three years of trials and testing, construction of the first boat is scheduled to start in 2021. Many of the parameters of the new class have already been decided. There are to be 12 submarines, each armed with 16 tubes, based on the US/UK Common Missile Compartment, for the life-extended Trident D5 missile. Although there are eight fewer tubes than the Ohio class, overall dimensions are to be much the same; the new boats' modular design demands more space and their acoustic requirements dictate the length. The class is also to include a life-of-ship reactor core, that will power the class for its full 42-year service life, electric drive, and X-shaped stern control surfaces.

While the US Navy is going through turbulent times, the quest for innovation and new technology continues. Rail-guns, which use electro-magnetic energy to propel projectiles have long been, if not science-fiction, one of those technologies that remain forever just round the corner. However, the time has now come for the concept to undergo live testing at sea. A trial is to be conducted in Joint High Speed Vessel *Millinocket* (JHSV-3) in 2016. In the words of Admiral Jonathan Greenert, Chief of Naval Operations: "We're beyond lab coats, we're into engineering now." In a separate development, it was announced in April 2013 that former amphibious ship, *Ponce*, which has been converted into an interim Afloat Forward Staging Base, is to be used as a testbed to deploy a laser weapon system (LaWS) with the potential to combat drones, small boats, and other air or sea-borne threats. If successful, such weapons could become widely fitted in the future.

In Canada, two important projects gained momentum during 2013. Nearly a decade after the project was launched, it was confirmed in June 2013 that TKMS Canada's Berlin class had been selected as the design for the Joint Support Ship Project. To be known as the Queenston class, a construction contract is expected in 2016 for both ships to be built at Vancouver for entry into service in 2019 and 2020. Meanwhile, the Protecteur-class oilers are already 45 years old, although the life of *Protecteur* may already have ended following an engine-room fire in March 2014. In another major programme, a contract for the construction of up to eight Arctic/Offshore Patrol Ships is expected in 2014 with a view to work beginning at Halifax in 2015. The 6,400-tonne ships are to be built to Polar class 4/5 standards, with their principal role to conduct surveillance and sovereignty enforcement operations in Canada's maritime estate, particularly northern waters. The first ship is to enter service in 2018.

There can be few more protracted procurement projects than the CH-148 Cyclone helicopter project to replace the Sea King fleet. The contract for the acquisition of 28 aircraft was signed in 2004 with all aircraft to be delivered by 2012, but it may now be 2018 before the first fully operational aircraft is delivered. Wherever the faults may lie, this has not been a happy experience for the Canadian government but, given the money already invested, it could be an example of a project that is 'too big to fail'.

China

A near-collision between a Chinese amphibious ship and USS *Cowpens* took place in the South China Sea on 5 December 2013. The US warship was monitoring the operations of the Chinese carrier *Liaoning* in international waters when, presumably with the intention of deterring unwelcome attentions, a Chinese vessel closed to a distance of about 500 m ahead of the cruiser, forcing it to take evasive action. Fortunately, the event concluded without mishap but it is to be hoped that lessons have been learned on both sides. The Commander US Pacific Command, Admiral Samuel Locklear, suggested that Chinese operational inexperience and communication difficulties contributed to the confrontation, factors that the US Navy would need to allow for. For its part, the Chinese Navy certainly needs to accept that such surveillance operations in the vicinity of its capital ships are likely to continue but that they can be handled professionally and safely. The 1972 USA-USSR 'Prevention of Incidents On and Over the High Seas' (INCSea) agreement, which listed protocols for manoeuvring in company, collision-avoidance, and communications procedures, was developed as a result of similar incidents. Now might be the time to use it as a blue print for a similar USA-China bilateral agreement.

The near-collision occurred only two weeks after declaration of a Chinese Air Defence Identification Zone (ADIZ) covering a large portion of the East China Sea between South Korea and Taiwan. While the zone excludes most islands of the Japanese southern island chain, it includes the disputed Diaoyu/Senkaku islands. While there is nothing unusual about declaring an ADIZ (many nations have similar arrangements), the way in which the zone is applied will be watched with interest. The language of the official Chinese statement, which stated that "emergency defensive measures would be adopted" in response to aircraft not filing a flight plan, did not auger well. The principal concern is that China might wish to use the same approach to international airspace that it has applied towards the UN Convention on the Law of the Sea, in which it seeks to extend its rights in waters beyond its Territorial Seas. Therefore, while it was reassuring to hear Admiral Locklear state in February 2014 that Chinese forces had been acting professionally in the ADIZ, it was noteworthy that he also took

QINGDAO 10/2013*, Chris Sattler / 1531201

[12] Executive overview

the opportunity to reiterate US concerns about the zone. There is also the prospect that China might establish further ADIZs in the region, particularly over the South China Sea, a more complex area of territorial disputes involving Brunei, Malaysia, the Philippines, Taiwan, and Vietnam.

Reports that one or even two new aircraft carriers might be under construction at Dalian (where *Liaoning* was refitted 2002–12) and/or Jiangnan Shipyard on Changxin Island have been made on several occasions during the past year but, as yet, there has been no conclusive photographic evidence. Undoubtedly, aircraft carriers are part of China's future naval plans but, given the deliberate nature of Chinese naval construction over the past decade, it is unlikely that designs of a new ship would be finalised before all the lessons of bringing *Liaoning* into operational service have been fully absorbed. The Chinese will also take into account the fact that when the Kuznetsov class was designed in the 1970s, the Soviet Union had relatively limited experience of naval aviation on which to draw. It would be surprising if the Chinese Navy was to rush into such a major project.

While carrier stories tend to grab the headlines, the surface-ship building programme has continued apace. In 2013, China commissioned two Type 052C Luyang II-class destroyers, three Type 054A Jiangkai II-class frigates, nine Type 056 Jiangdao-class corvettes, two Fuchi-class replenishment ships, two Dalao-class submarine rescue ships, and two Wochi-class mine countermeasures vessels. This, by any measure, is an impressive feat. The destroyer programme looks set to continue with two further Luyang II and perhaps two Luyang III (incorporating modified 'Dragon Eye' phased-array radar) due to enter service in 2014. Overall, a class of 12 Luyang III may be built, although there have been unconfirmed reports that a larger 12,000-tonne Type 055 destroyer may be in the design phase. Meanwhile, a class of at least 20 Jiangkai II-class frigates is expected. A notable feature of surface ship programmes is that development of a modern mine countermeasures force is taking a long time. The reason for this is not clear. The capabilities of the new (predominantly Wochi class) specialist vessels are not fully understood and there has been no evidence that such vessels have been superseded by unmanned systems. Overall, it would appear that progress in this notoriously difficult warfare discipline is not proving easy and that, for the time being, mine countermeasures remains a weakness. This is a serious shortcoming for a nation that has an almost 8,000 n mile coastline to defend.

While 2013 was a relatively quiet year for submarine building programmes, 2014 could prove more interesting from an operational standpoint. Information about progress with the JL-2 ballistic missile has been scarce but, as some reports have suggested that operational status was achieved in 2013, it is possible that deterrent patrols by the Type 094 Jin class could soon start. The Chinese Navy will have thought long and hard about potential patrol areas but, assuming the missiles have a range in the order of 4,300 n miles, the deep waters of the Philippine Sea offer perhaps the best compromise between target range requirements and distance from base. The Jin class building programme still remains speculative but, according to the Pentagon's annual report to Congress *Military and Security Developments Involving the People's Republic of China 2013*, "up to five may enter service before China proceeds to its next-generation SSBN (Type 096) over the next decade". The report also stated that its force of nuclear-powered attack submarines (SSNs) is being expanded: "Two Shang class SSNs (Type 093) are already in service, and China is building four improved variants, which will replace the ageing Han-class SSNs (Type 091). In the next decade, China is likely to construct the Type 095 guided-missile attack submarine (SSGN), which may enable a submarine-based land-attack capability. In addition to better quieting technologies, the Type 095 will fulfil traditional anti-ship roles with the incorporation of torpedoes and anti-ship cruise missiles (ASCMs)".

The establishment of the new Chinese Coast Guard under the State Oceanic Administration in March 2013 was probably the final step in both consolidating the security activities of a variety of agencies and clarifying responsibilities for offshore constabulary duties. By 2010, five principal organisations had been established: China Marine Surveillance, Fisheries Law Enforcement Command, Maritime Safety Administration, China Coast Guard, and the Customs. However, it must soon have become obvious that it was both wasteful and operationally unsound to have several organisations with overlapping responsibilities and differing chains of command. The assets of the these agencies have now been subsumed within the new Coast Guard and only the Maritime Safety Administration, responsible for maritime safety and counter-pollution issues, remains a separate organisation, accountable to the Ministry of Transport.

United Kingdom

The announcement in November 2013 that three new offshore patrol vessels are to be built for the Royal Navy came as something as a surprise to most onlookers, even to the Royal Navy itself judging by its less than effusive reception of the news. The root cause of the Royal Navy's circumspection is that such ships, the first of which is to be delivered in 2017, provide unwelcome support to those who argue for a two-tier navy. Only two months before the decision, the First Sea Lord and Chief of Naval Staff, Admiral Sir George Zambellas, had remarked to a Royal United Services Institute conference that "a corvette designed for policing duties just doesn't pass the basic test of contingent flexibility". The dilemma for the service is that the ships are being built to sustain jobs in the UK's naval shipbuilding industry between the aircraft carrier and Type 26 frigate programmes rather than to meet a demonstrable operational need. On the face of it, the decision makes financial sense; the Ministry of Defence is apparently liable, under a 2009 agreement, to pay the costs of maintaining shipbuilding skills whether or not any warships are being built. While details of the ships have yet to be released, it is expected that they will be similar to the 90 m patrol vessels originally built for Trinidad and Tobago but subsequently purchased by Brazil in 2012. The ships are not frigates but, at approximately 2,300 tonnes and equipped with a helicopter deck and a small-calibre gun, they are not without utility. Rather than accept that the ships should replace the three smaller River-class fishery protection ships, that are only 10 years old, the new vessels provide an opportunity to take some of the strain out of an overstretched escort programme. The Caribbean guard ship role, in which counter-narcotic operations are combined with preserving ties with friendly nations and allies in the region, is an example of the sort of task that they might undertake. Their potential as training platforms, particularly for command, navigation, and seamanship skills, should also be exploited.

The UK's contribution of an SSN, *Tireless*, and a survey ship, *Echo*, to the search for the missing Malaysian Airlines Flight MH370 in the southern Indian Ocean, while generous, also served as an uncomfortable reminder that, should the UK be faced with a similar task, it is unable to call on a national airborne maritime surveillance capability. The Nimrod MRA4 programme was cancelled in 2010 but, nearly four years later, little progress has been made in finding an

DEFENDER

5/2013*, Michael Nitz / 1531200

alternative. Therefore, the news in April 2014 that a UK team is to be sent to train on Northrop Grumman's MQ-4C Triton unmanned aircraft system was welcome. The US Navy is to procure 68 of these aircraft to fulfil its Broad Area Maritime Surveillance requirement. However, while such a system can provide a real-time reconnaissance capability over large sea areas, it is not a suitable platform for more complex capabilities such as anti-submarine warfare. It is for this reason that some 20 UK personnel have been embedded in the US Navy's P-8 operations in order to retain core skills, while a further 12 are working with Australian, Canadian, and New Zealand P-3C teams. These initiatives point towards procurement of a mixture of manned and unmanned systems but while it is laudable that a small cadre of trained operators is being maintained, the problem will not be solved without the necessary funding.

The gap in Airborne Early Warning capability to be created by the retirement of the current fleet of Sea King ASaC.7 in 2016 is now to be shorter than once feared, following a decision in early 2014 to accelerate the Crowsnest replacement programme. It is now planned to align the project more closely with the new aircraft carrier, *Queen Elizabeth*; an interim operating capability of the new system is to be achieved in 2019 rather than 2021. Crowsnest is to be based on the procurement of 10 modular ASaC mission kits that can be fitted to any of the Merlin helicopters currently being upgraded to HM.2 standard. Each of these aircraft is to receive 'fit-to-receive' modifications to enable rapid installation. One of two competing systems is to be selected in 2016: Thales offers migration of the incumbent Sea King ASaC system, which includes the Cerberus tactical suite and Searchwater 2000AEW radar, while Lockheed Martin UK is proposing a pod-mounted active electronically scanned array (AESA) radar. While hastening of the Crowsnest programme has been a victory for commonsense, it has only been achieved at the expense of other operational capabilities. Of 40 Merlin HM.1 aircraft, with primary roles of anti-submarine (ASW) and anti-surface warfare (ASUW), only 30 are being upgraded to HM.2 standard. Of these, five Crowsnest-configured aircraft are likely to be embarked in a carrier, leaving only 25 optimised for ASW/ASUW, a 40% reduction of the original HM.1 force.

Europe and the Mediterranean
The failure of an M51.1 ballistic missile fired on 5 May 2013 from *Le Vigilant*, submerged off the coast of Brittany, was a set-back for the programme to replace the smaller-diameter M45 missile in France's 'Force de Dissuasion'. As a further success would have been confidently expected after five previous firings, the failure will have been troubling. The immediate ramifications are that confidence in the operational status of *Le Vigilant* and *Le Terrible*, already equipped with the missile, will have been reduced and that, with *Le Triomphant* in refit, full reliance can only be placed in the M45-equipped *Le Téméraire*. Meanwhile, studies have been commissioned to explore options for replacement of the force from about 2030, assuming a 35-year service life for the current boats. Progress in France's other major submarine programme, the Suffren class, has also been delayed, although by financial rather than technical issues. Whereas it had been planned to build the submarines to a 24-month 'drumbeat', the overall programme has been stretched by two years with the result that only one will have entered service by 2020 and that the last of the six-boat force, *Dupetit-Thouars*, will not join the fleet until 2029, some 30 years after studies for Project Barracuda were first launched.

The strength of the surface fleet has been reduced by the decommissioning of three destroyers: *De Grasse* and *Georges Leygues* were deleted in 2013, with *Dupleix* to follow in 2014. Meanwhile, the future destroyer programme has been further modified by financial constraints. The 2014–19 Defence Programming Law stated that a decision on whether to build the last three ships of the 11-ship FREMM programme had been suspended until 2016 and that the two air-defence (FREDA) variants were now to be the seventh and eighth hulls, rather than 10th and 11th as previously planned. These ships are to replace the two Standard SM-1 fitted Cassard class in 2022–23, which means that the last three FREMM, if they are built, will not be commissioned until the mid-2020s. While this falls well short of the expectations of the original 17-ship programme, there are plans to maintain the operational balance of the fleet by upgrading the La Fayette class or procuring a new surface combatant; this decision is also expected in 2016. On the credit side, the FREMM programme can at least claim an export success; the Moroccan frigate, *Mohammad VI*, was commissioned on 30 January 2014.

Plans to re-capitalise the patrol ship force were to have been based on the BATSIMAR (Batiments de Surveillance et d'intervention Maritime) programme to replace the inventory of ageing A69-class frigates and assorted offshore patrol ships with a single class of 20 ships. However, decisions on the way ahead have been postponed until 2019 so it will be the mid-2020s before the first ship enters service. As a result, the DCNS Gowind class demonstrator, *L'Adroit*, which has been on loan to the French Navy since 2012, is to be returned to its owners in 2014. One stop-gap measure has been to proceed with the construction of three B2M-class offshore patrol ships to replace the Batral-class landing ships stationed in French overseas territories. Based on an offshore supply ship design, the ships are to enter service from 2015. There has been little progress in procuring new auxiliaries. Replacement of the Durance-class tankers is the most pressing (the oldest ship, *Meuse*, is already 34 years old) but the time scale has yet to be confirmed. Finally, the BSAH (Bâtiments de Soutien et d'assistance Hauturiers) programme to replace the ocean-going tugs and some of the civilian-owned salvage and rescue vessels appears to have been suspended.

In Germany, two Batch 2 Type 212 submarines expected to enter service in the second half of 2014, bringing the total to six. The new boats are upgraded versions of the first four and include a non-penetrating optronic mast and a four-man lock-out chamber to facilitate the deployment of special forces. As there are no plans to procure further vessels, the successful German submarine industry will now have to rely entirely on exports to stay in business. An order in 2013 for two Type 218 boats from Singapore was timely.

Of the surface fleet, the once eight-strong Bremen-class frigates has now been reduced to four following the decommissioning of *Köln*, *Emden*, *Rheinland-Pfalz*, and *Bremen* in the past two years. The remaining ships are to be decommissioned by 2019. Of future ships, the first Type 125 frigate, *Baden-Württemberg*, was named on 12 December 2013 and is due to be commissioned in 2017. Plans to upgrade the Sachsen class with a BMD capability remain under consideration. This would involve modification of the SMART-L radar with an extended range mode. It is also intended to modernise the four Brandenburg frigates; principal improvements may (subject to funding) include the replacement of Exocet with, possibly, RBS15 and of NATO Sea Sparrow by Evolved Sea Sparrow. Looking ahead, the German Multi Role Combat Ship (MKS 180) programme is gaining momentum following completion of the analysis phase in March 2013. There are plans for four to six medium-sized surface combatants to replace the capabilities of the Type 122 frigates and Type 143A fast attack craft. Principal features are likely to include a medium-calibre gun, surface-to-surface missiles, a point-defence missile system, unmanned air and underwater systems, towed array sonar, up to two helicopters, and two special forces craft. Following a Development and Proposals Phase, decisions on the next phase are expected in 2015–16 and the vessels are planned to enter service 2020–24. Finally, the third Berlin-class combat support ship, *Bonn*, was commissioned on 13 September 2013.

In the Netherlands, construction of *Karel Doorman*, a new Joint Logistics Support Ship, has been the centrepiece of future naval plans since the contract was let in 2009. With primary tasks of maritime logistic support, strategic sealift, and support of land-based forces, the ship was to be able to act in concert with amphibious ships *Rotterdam* and *Johan de Witt*, or to be a significant platform in its own right. Therefore it was extremely surprising when it was announced in September 2013 that, due to budget cuts, the ship would not enter Royal Netherlands Navy service and even more extraordinary when, a month later, that decision was reversed. It was a somewhat bewildering episode. Among other programmes, the mid-life upgrade of the Walrus-class submarines began with *Zeeleeuw* in 2013. It is to be followed by the other boats by 2019. Principal upgrades include a new combat management system, upgrade of the sonar systems, and a new optronic mast. Even though the service-lives of these boats has been extended, a decision on their replacement will be required soon if a gap in capability is to be avoided. Modernisation of the two M-class frigates is also under way; work on *Van Speijk* was completed in 2012 and *Van Amstel* is to return to service in 2014. The most notable structural change is a new mainmast to incorporate the Thales Seastar and Gatekeeper systems. The surface fleet has also been enhanced by the commissioning of all four Holland-class patrol ships by late 2013. With their distinctive integrated masts, the frigate-sized ships are interesting newcomers to the naval scene.

The Italian Navy is emerging from a difficult financial period in a better state than might once have been expected, thanks to the Financial Stability Law approved on 23 December 2013. This legislation enabled funding of naval shipbuilding programmes to be drawn from national funds in addition to those of the defence budget. One of the most important projects to benefit from this scheme is procurement of a new amphibious assault ship, which is intended to become, with the carrier *Cavour*, the centrepiece of Italian power projection capabilities. Meanwhile, the former carrier *Giuseppe Garibaldi* is to continue in

VAN SPEIJK

service as a helicopter carrier as an interim measure, although full conversion to an amphibious assault role is now unlikely.

Carlo Bergamini, the first (general-purpose variant) of the FREMM-class frigates was commissioned in May 2013 and the second (ASW variant) ship, *Virginio Fasan*, followed late the same year. Both ships were lengthened by 3.6 m during 2013, a modification that is being made to the rest of the class in build. Future planned modifications include more powerful diesel engines for the seventh and eighth ships, to achieve a higher cruise speed, and an integrated mast in the eighth ship. The third ship, *Carlo Margottini*, is to enter service in 2014 and the remainder of the eight ships on order are to enter service by 2019. Decisions on whether there are to be ninth and tenth ships are awaited but it is possible these ships will be cancelled in favour of the UPAD (Unità Pattagliamento Altura Duale) programme for eight 5,000-tonne patrol ships, some of which are to be lightly armed offshore patrol ships (comparable to the Holland class) and a more heavily armed variant, equipped with missiles, guns, and a payload area for containerised capability packages. This programme, which has superseded the future corvette programme, has also been funded by the Financial Stability Law as has a new Logistic Support Ship, which is likely to be similar to the Deepak class built for India. Other programmes in the pipeline include a new 104 m submarine rescue ship, to replace *Anteo*, and two 60 m interception craft capable of 70 kt, probably for special forces.

In Spain, the Isaac Peral (S-80)-class submarine programme ran into serious problems in mid-2013 when it was discovered that the first of class was overweight, reportedly by a margin of some 70 tonnes. An error in design calculations may have been the cause. While US company Electric Boat has been contracted to assist in finding a solution, the remedy is unlikely to be easy or cheap. The main challenge is to improve the buoyancy of the boats, either by reducing the weight and/or by lengthening the hull. The latter option is likely to be preferred if all the capabilities of the boat, including air-independent propulsion (AIP), are to be retained. The effect of this setback is that the overall programme is to be delayed by a further two years; the first of class is now planned to be commissioned in 2018, 11 years after it was laid down.

Following the commissioning of the fifth and final air-defence destroyer, *Cristóbal Colón*, in 2012, consideration is now being given to the F 110 programme for up to five frigates to enter service from about 2022, when the Santa María-class frigates become 35 years old. The naval staff requirement, expected in mid-2014, is likely to identify the need for a multimission ship with particular emphasis on the littoral warfare capabilities. To this end, land-attack missiles, a medium-calibre gun, and the ability to insert special forces are expected to be key features. In addition, the ships are to be capable of operating the NH90 helicopter and will probably feature an integrated mast incorporating radars, communications, and electro-optic sensors. Finally, the potential for the ships to operate in polar waters is also being considered. This is unlikely to lead to a full ICE classification but the capability to operate systems in cold weather is a thought-provoking step for a southern European navy.

In Norway, a decision on a programme to replace the submarine flotilla from about 2020 is expected in 2014. A three-year project definition phase is then to follow. It is likely that a design will be selected from DCNS, Fincantieri, TKMS, or Daewoo which were sent requests for information in September 2012. However, as all the boats of the Ula class are of a similar age (they entered service 1989–92), it may be decided to life-extend at least some vessels in order to spread acquisition costs over a longer period. The Skjold-class fast patrol craft programme was finally completed in April 2013 when *Skjold*, the original pre-production vessel ordered in 1996, was commissioned after completing its upgrade to full production standard. An operational firing of the Naval Strike Missile was conducted from *Glimt* in October 2012. The most important programme now in progress is for a new logistics support ship. To be built by South Korea's Daewoo, the new vessel, based on the design of the UK's Tide-class tankers, is to complete the Royal Norwegian Navy's transition to an ocean-going capability when it enters service in 2016. Other naval programmes include construction of a new research ship to replace *Marjata* in 2015 and three new coastguard patrol ships to replace the Nordkapp class by 2020.

The Royal Danish Navy's newly commissioned frigate, *Iver Huitfeldt*, the first of a new class of three new ships, completed its maiden deployment to the Indian Ocean in 2013, thereby demonstrating the global role for which it was designed. The third ship, *Niels Juel*, started its sea trials at about the same time. The ships have yet to be equipped with the Standard SM-2 missile and so, initially, the ships will operate with an interim capability that includes Evolved Sea Sparrow. The ships will be further augmented by the MH-60R helicopter, which is to replace the Lynx fleet from 2016. The other significant naval programme is to construct a third offshore patrol vessel of the Knud Rasmussen class. Built to DNV ICE 1A standards, the principal role of these ships is to conduct sovereignty patrols in Arctic waters.

The Swedish A 26 submarine programme was thrown into confusion in February 2014, when it was announced by the Swedish procurement agency FMV that it had funded studies to investigate the ability of Saab to design and build the new submarine class. Having awarded a two-year design contract to Kockums in 2010, it had been expected that a contract for manufacture would be awarded to the same company in 2013. However, plans announced by the German parent company, TKMS, to make the Malmö yard the construction centre for small ships and submarines less than 1,000 tonnes, raised fears that the A 26 submarine, which is 1,860 tonnes, may not be built in Sweden. As of early May 2014, negotiations for the sale of all TKMS yards in Sweden to Saab had started, but until these industrial issues have been resolved, the new submarine programme is 'on hold', as is the Gotland class modernisation programme.

Among other Swedish programmes, the saga of the Visby-class corvette programme should be concluded in 2014 when the final vessel becomes operational. Both the Stockholm-class corvettes are to be decommissioned in 2015 but *Gävle* and *Sundsvall* are to be modernised (becoming the Gävle class in the process) while two decommissioned ships of the same class, *Kalmar* and *Göteborg*, are to be re-activated as patrol ships. Although the government announced plans in 2010 to build a new signals intelligence ship to replace *Orion*, no details of the new ship have been released. There have been no known developments in plans to build two logistic support ships and the project may have been postponed.

Russia

The annexation of Crimea by Russian Forces on 18 March 2014 was largely achieved by land forces but units of the Black Sea Fleet were also utilised to conduct a blockade of Ukrainian naval bases in the region and, in one of the more bizarre incidents of the conflict, the 'tactical scuttling' of the former Kara-class cruiser, *Admiral Ochakov*, to block the entrance to Lake Donuzlav. Naval forces may also have been involved in the seizure of Ukrainian naval bases at Sevastopol, Feodosia, Kerch, and Balaklava and naval units, including the recently re-commissioned submarine, *Zaporizya*. However, captured ships did not include the Krivak-class frigate, *Hetman Sagaidachny*, which diverted to Odessa on return from anti-piracy duties in the Indian Ocean.

The Black Sea Fleet is due to be boosted over the next few years. The Type 636 Kilo-class submarine building line was re-opened in 2010 and the first of these boats, *Novorossiysk*, was launched on 28 November 2013. It is expected to be commissioned in 2014 and, following trials and work-up, to join the fleet in 2015 as part of Russia's plans to boost naval forces in the Mediterranean. Ultimately, this is likely to include six submarines of which the fourth, *Krasnodar*, was laid down on 20 February 2014. Surface forces are also be strengthened by the addition of further frigates. *Admiral Grigorovich*, the first of six Project 11356-class ships, was floated out at Yantar Shipyard, Kaliningrad, on 14 March 2014. Four further ships are under construction and work on the sixth ship is expected to start in 2014. All six vessels are expected to be delivered by 2019, probably as replacements for the ageing Grisha class.

While the Ukrainian crisis has focused attention on the Black Sea, Russia's strategic interests in the Arctic region have been re-affirmed by the creation of a new Joint Strategic Command, which is to have similar status to the four existing joint commands that report directly to the General Staff and Defence Minister. The new headquarters, to be formed by the end of 2014, is to be based on the Northern Fleet but will also include other Arctic-based military units. The aim of the re-organisation is to protect Russia's mineral resources in the region and to co-ordinate the establishment of the infrastructure required to support the deployment of additional forces in the future. While this may take some years, it is confirmation of Russia's regional intent, following deployments of the nuclear-powered cruiser, *Pyotr Velikiy*, along the Northern Sea Route in both 2012 and 2013.

Meanwhile, other submarine programmes continue. The second unit of the Borey-class ballistic-missile submarines, *Alexander Nevsky*, was commissioned on 21 December 2013 and is likely to join the Pacific Fleet in 2014. Facilities at the submarine base at Rybachiy have reportedly been upgraded to receive the new vessels, which are to replace the three remaining Delta III boats, now more than 30 years old. The third vessel of the class, *Vladimir Monomach*, completed its first set of sea trials on 8 October 2013, while construction of the fourth boat, *Knyaz Vladimir*, to a modified Project 955A design, began in mid-2013. *Severodvinsk*, the first of a new class of SSNs, was finally commissioned on 30 December 2013, 20 years after being laid down. However, reported problems with propeller shaft bearings could delay its entry into operational service. Follow-on boats are built to a modified Project 885M design, the first of which, *Kazan*, is expected to be launched in 2014. The third boat, *Novosibirsk*, was laid down on 26 July 2013 and a fourth boat is likely to follow in 2014. Refits of older classes are also planned with Sierra I boats, *Karp* and *Kostroma*, due to be modernised at Zvedochka Shipyard by 2017. The Sierra IIs, *Pskov* and *Nizhny Novgorod*, will follow. Plans to build a new Kalina class of conventional submarine, equipped with AIP, have also been announced. As trials of AIP are to be conducted in the first Lada class, *Sankt Peterburg*, it is assumed that the second and third boats, *Kronshtadt* and *Sevastopol*, whose construction was halted for several years, will be the lead vessels of the new class. However, an entirely new design remains a possibility.

Among surface ship programmes, the expected refit of *Admiral Nakhimov*, third of the Kirov-class battle cruisers, was confirmed in mid-2013. Work on the ship is likely to take five years and to include major upgrades of weapon systems in addition to nuclear refuelling. The Admiral Gorshkov (Project 22350)-class frigate programme is the first new class of major surface combatant for the Russian Navy for many years, with its overdue sea trials likely to start in 2014. The Russian Navy has stated that it needs 20 to 30 of these ships, which are seen as replacements for the Krivak class and possibly for larger vessels such as the Udaloy and Sovremenny classes. It is unclear whether there is still a requirement for larger ships.

The Russian Navy's most controversial programme, construction of two Mistral-class amphibious ships at St Nazaire, France, took a step forward in March 2014 when first of class, *Vladivostok*, began sea trials. It is planned to be delivered to Russia in late 2014, to be equipped with Russian systems at Severnaya Verf Shipyard in St Petersburg, before being commissioned in the Pacific Fleet in 2015. The second ship, *Sevastopol*, is due to be floated out in October 2014. However, it remains to be seen whether the project will go ahead in light of Russian intervention in Ukraine. A decision is to be made by the French government in October 2014.

Indian Ocean and Gulf

The resignation of the Indian Navy Chief of Staff, Admiral D K Joshi, on 26 February 2014 was triggered by a fire on board the submarine *Sindhuratna* in which there were two fatalities. It was perhaps the final straw for an officer who no doubt felt he bore moral responsibility, not just for this accident but for an unfortunate succession of 10 safety-related incidents, including fires, collisions, and groundings, over the previous year. By far the most serious of these was the loss of the submarine, *Sindhurakshak*, in which 18 people died after an explosion in August 2013. The boat had only recently completed refit. While all navies experience periods of bad fortune from time to time, the first job of the new Chief Admiral R K Dhowan is to conduct an objective examination of this calamitous series of events and to put in place the measures necessary to restore morale and confidence. This may require some uncomfortable soul-searching because, while it is easy to cite lack of money, ageing equipment, and labyrinthine procurement procedures as contributory causes, it would be surprising if lapses in training, quality control, and professional standards were not also found to be significant factors.

Despite this unfortunate backdrop, the year of 2014 began on a high note with the arrival of the carrier, *Vikramaditya*, at its homeport of Karwar on 8 January, almost exactly 10 years after the contract was signed. Notwithstanding what has proved to be a frustrating and expensive project, the challenging task of working-up the ship to

STEREGUSHCHIY

7/2013*, M Declerck / 1531198

Executive overview

TRIKAND

operational status now lies ahead. This process is expected to take about a year, after which the 55-year-old *Viraat* will be retired. Progress has also been made with India's new home-built carrier, *Vikrant*, which was launched at Kochi on 12 August 2013 with a view to entering service in 2018. The surface fleet was also boosted by the emergence of the first Project 15A destroyer, *Kolkata*, on sea trials in 2013 and the commissioning of the sixth Talwar-class frigate, *Trikand*, in June 2013.

In contrast, the submarine flotilla is in some disarray after the loss of *Sindhurakshak* reduced the number of Kilo class to nine. Of these, *Sindhukirti* has been in refit at Hindustan Shipyard since 2006, while procurement of six new Scorpene-class boats continues to be beset by delays; launch of the first of class is not expected until late 2014/early 2015. The programme for six follow-on Project 75I submarines also remains undecided. Unsurprisingly, a stop-gap measure is required to maintain overall submarine numbers and life-extension refits for two (probably *Shalki* and *Shankul*) of the Shishumar (Type 209/1500) are now planned. Upgrades are to include the installation of anti-ship missiles. There are also question marks about the status of the Indian Navy's flagship project, the construction of its first indigenously built nuclear-powered ballistic missile submarine, *Arihant*. The submarine's reactor reportedly achieved criticality on 10 August 2013 but, although it was stated that harbour trials had been successfully completed in early 2014, sea trials had yet to start by early May.

The announcement by Pakistani officials in February 2014 that a contract to buy up to six submarines from China could be expected by the end of the year was confirmation that, following reports that western options would no longer be pursued, the government would turn to the country that has become its main naval supplier. The need for new boats has become increasingly pressing as the Hashmat (Agosta 70)-class boats are already about 35 years old. While the new class of submarines has yet to be confirmed, it is probable that the AIP-equipped Yuan class will be selected. Indeed, it is likely that the Pakistani Navy (PN), an experienced and respected submarine practitioner, has had an input into what could be a modified design. The choice of sensors and weapons will be of particular interest given the PN's experience with, and knowledge of, western systems. While the February announcement did not mention frigates, it would not be surprising if procurement of further Chinese-designed vessels was not also in the offing. The PN is faced with block obsolescence of the Tariq (Amazon) class of which *Badr* was the first to be decommissioned in 2013 and, following the commissioning at Karachi Shipyard of the fourth Sword class, *Aslat*, in April 2013, it would make sense for further vessels to be constructed at the same yard. Meanwhile, the shipyard is building *Dehshat*, the second of two Chinese-designed 63 m missile-armed patrol craft expected to be commissioned in 2014, while a new Turkish-designed 154 m double-hulled tanker was laid down in 2013. This ship is likely to replace *Moawin*.

The Chinese government is also playing a major part in the modernisation of the Bangladesh Navy, which is enjoying a period of significant expansion. The most important programme is for procurement of a submarine capability; it was confirmed in December 2013 that two Ming-class boats were to enter service from 2019 while construction of the necessary base facilities, probably at Kutubdia Island, is also to be initiated. It was also confirmed that two second-hand Jianghu III-class frigates, *Abu Bakr* and *Ali Haider*, are to replace two Leopard-class frigates of the same name in 2014 while two 64 m patrol vessels, *Durjoy* and *Nirmul*, were both commissioned in August 2013. These were in addition to five smaller 50 m patrol craft of Chinese design, built at Khulna Shipyard, which also joined the fleet in 2013. Surface forces have been further boosted by the acquisition of *Somudra Joy*, formerly the US Coast Guard cutter *Jarvis*, which was re-commissioned on 23 May 2013 while, looking ahead, two 90 m (possibly Type 056) corvettes are under construction at Wuhan and are expected to be commissioned in 2015. Another important asset in the pipeline is a new 80m tanker capable of replenishment at sea. Under construction at Ananda Shipyard, it is to be delivered in 2014. Aviation assets have also been boosted by two Dornier 228 maritime patrol aircraft, which were delivered in 2013. This follows the purchase in 2011 of two AgustaWestland AW109 helicopters to be operated from the frigate, *Bangabandhu*.

The most important development in Iran has been the unveiling of a new Fateh-class submarine in February 2014. Confirmation of the programme was first obtained in October 2013 when commercial imagery revealed the construction of two boats. The first was in the water at Bostanu Shipyard, some 15 n miles west of Bandar Abbas, and a second, surprisingly, at Bandar Anzali on the Caspian Sea. Formerly known as the Qaaem class, it had been assumed that the new vessels would be based on an existing design, probably the North Korean Sang-O class. However, the larger size of the boat and the distinctive design of the casing suggest that, as claimed in official announcements, the submarines have been both indigenously designed and built. If so, this would represent a significant achievement by Iran's naval industry. The size of the programme is not known, but about 10 boats are forecast. Looking ahead, the programme could well be a stepping-stone to production of larger submarines known as the Besat class. Replacement of the 1990s vintage Kilo class will be required from about 2025.

The Project Mowj frigate-building programme continues to make progress. The ships are modified versions of the three remaining UK-designed Alvand class built in the 1970s. Construction of the first Mowj unit (*Jamaran*) at Bandar Abbas was completed in 2010, while construction of a second (*Damavand*) at Bandar Anzali on the Caspian Sea and of a third unit (*Sahand*) at Bandar Abbas is in progress. Construction of a fourth at Bostanu Shipyard has also been reported. A notable feature of *Damavand* is an Iranian frequency-scanned 3D radar, Asr, which has replaced AWS 1 radar fitted in previous ships. It is likely that this will also be installed in further ships of the class. The future frigate programme has not been confirmed. While a fifth and sixth Mowj class may be built, reports have suggested that there is to be another (stretched) version of the Alvand class, known as Project Loghman. These 120 m ships incorporate a hangar and helicopter deck as well as surface-to-surface missiles and a medium-calibre gun. Development of the fast attack craft fleet has run in step with the frigate programme. The SINA project has involved the building of further vessels of the Kaman (Combattante) class, which entered service in 1977–81. The first new indigenously built craft was *Peykan*, launched in September 2003, followed by *Joshan* in 2006, and *Derafsh* in 2009. Four further SINA craft are under construction, one reportedly in the Caspian and the remainder at Bandar Abbas. Entry into service is expected in 2016.

Reports in late 2013 that agreement may be reached between Iran and the United Arab Emirates on the sovereignty of the islands of Abu Musa, and Greater and Lesser Tunb, were later denied in early 2014 by the Iranian government. The islands have been occupied by Iran since the British withdrawal in 1971 and it would have been surprising if a deal on the status of the strategically placed islands had been achieved. Meanwhile, the UAE continues to modernise its fleet. The third Baynunah-class corvette, *Al Dhafra*, was commissioned on 24 December 2013 while the sixth and final ship, *Al Hili*, was launched at Abu Dhabi Shipbuilding on 6 February 2014. The shipyard will have gained much experience as prime contractor of the Baynunah programme but there have been no indications so far as to follow-on programmes. Options include construction of two further Falaj 2-class patrol ships, the second of which, *Salahlah*, was commissioned in Italy in April 2013.

East Asia and Australasia
Australia's 2013 Fleet Review celebrated the 100th anniversary of the arrival of the new Australian Fleet Unit in Sydney. The importance of the event in 1913 was that, following the federation of the country in 1901 and the formal establishment of the Royal Australian Navy (RAN) in 1911, the entry into Sydney harbour of the Fleet Unit, led by the battle cruiser, *Australia*, was a visible expression of the new nation. One hundred years later, the modern RAN hierarchy would have been hoping to re-kindle some of that symbolism as 20 warships from 17 nations passed through the Sydney Heads. The aim would have been not only to impress the many thousands who watched from the shoreline but also a new Australian government, elected just a month beforehand. It was a fortuitous opportunity to familiarise new political bosses with naval matters before confronting the difficulties of formulating a new defence White Paper, to be published in early 2015. In the words of the Defence Minister David Johnston, the Liberal-National Party coalition had inherited "a complete mess" so while it has promised to increase defence spending from 1.59% of GDP to 2% in the next decade, some difficult decisions almost certainly lie ahead.

The principal challenge, as is the case in most countries, is to match defence capability aspirations with available resources. The strategic assessments of the earlier 2009 and 2013 papers, and the essentially maritime strategy that followed, are unlikely to change. If anything, they have been reinforced by experience over the last few years. Therefore, the requirements for an amphibious force, a submarine flotilla, capable surface units, and a small ship force are likely to endure. However, it would be surprising if all of the current programmes survive in their present form, particularly if the requirement to maintain a national shipbuilding industry precludes the option of buying off-the-shelf.

As it stands, the new amphibious force is taking shape (sea trials of the future HMAS *Canberra* started on 4 March 2014) and the first of a new generation of surface combatants, the destroyer *Hobart*, is expected to be launched by late 2014. However, looking ahead, the trickiest issue is the size and scope of the future submarine force. Against a background of enduring criticisms (operational availability, manning, and cost) of the current six-strong Collins-class force, the goal of achieving a future 12-boat force seems ambitious. Even if, as seems probable, the government opts for an evolved-Collins design, the first boat is unlikely to enter service before 2030. This could give rise to at least a dip in submarine capability. The timescale of the future frigate programme, which is due to come to fruition in about the same timescale, may also have to be adjusted. Original plans to procure a multipurpose hull to meet all future small ship force requirements appear to have been discarded in light of the more pressing need to maintain an effective patrol boat force. The 57 m Armidale class started entering service as recently as 2005 but their intensive work schedule has led to high maintenance demands, which in turn could necessitate early replacement. The Cape class being procured for the Customs service would be an obvious choice. As a result, the lives of the Huon-class mine countermeasures vessels are likely to be extended.

In summary, the RAN is in a state of transition and there are certainly grounds for optimism; getting to grips with new amphibious ships and air-warfare destroyers will be an exciting and rewarding experience. However, there are financial clouds on the horizon and the 2015 White Paper could prove to be an uncomfortable 'reality check'.

Japan's version of a defence White Paper, National Defense Program Guidelines (NDPG), was published on 17 December 2013, in tandem with National Security Strategy, the first document of its kind. The "Dynamic Defense Force" concept, first mentioned in the 2010 NDPG, represented at the time a marked change from the previous concept of passive deterrence and defence designed to counter the Soviet threat. There is now to be a further evolution into a "Dynamic Joint Defense Force", a subtle change perhaps, but one that seeks to improve operational integration between the service arms and to develop joint rapid response forces. The threats to security have not changed. On the one hand, Japan has to defend itself against a nuclear North Korea, led by its unpredictable rulers while, on the other, there is a need to respond to China's maritime assertiveness. While individual incidents have remained, thankfully, relatively low-key, it is clear that they are part of a co-ordinated plan to gain greater control over seas within the

CANBERRA

3/2014*, Chris Sattler / 1531227

[18] Executive overview

'first island chain' and (following the announcement of an ADIZ in the East China Sea) its associated airspace.

The implications for the Japanese Navy are that there is to be increased investment in intelligence and maritime domain awareness, particularly in the south-west (Nansei/Ryukyu) island chain, through procurement of high-altitude, long-endurance unmanned air systems and maritime patrol aircraft. By 2017, there are to be four helicopter carriers and their role as key platforms, from which rapid response forces can be deployed, was emphasised by the announcement in December 2013 of plans to procure 17 MV-22 Osprey tilt-rotor aircraft. This follows the successful landing of a US Marine Corps aircraft on board *Hyuga* on 14 June 2013. The destroyer force is to be increased from 47 to 54, including the addition of two further Aegis destroyers capable of BMD, while the submarine flotilla is to be increased from 16 to 22 boats, as previously announced.

Despite the gradual transition of Japan's defence policy into a more positive stance and a 3% increase in the defence budget for FY 2014, Japan's overall defence spending, about 1% of gross national product, remains low by international standards. Notwithstanding concerns about the potential for regional instability, Japan's approach remains cautious. This is, of course, driven by the constraints of the Japanese Constitution but is also underpinned by the 1952 Treaty of Mutual Cooperation and Security with the United States. While US policy to refocus on the Asia-Pacific region endorses the strength of these ties, budgetary realities may force Japan to play a bigger part in the future.

The announcement on 12 March 2013 that the Taiwan Ministry of National Defence had initiated a four-year study to explore the feasibility of building its own conventional submarines, suggests the government has accepted that there is little chance they might be built in the United States. Despite approval for the project in 2001 by then President George W Bush, there have been no indications of progress; the last conventional submarines to be built in the United States, the Barbel class, were commissioned in 1959. However, there is unlikely to be a simple national solution. While the industrial capability to build the boats may exist, the necessary design expertise and supporting technologies such as submarine command systems, sonars and weapons are unlikely to be found without foreign help. Overall, the project remains a significant challenge and, against this unpromising background, even the acquisition in late 2013 of 32 submarine-launched Harpoon missiles, to be fitted in the two Dutch-built Hai Lung class, will have been of little comfort.

In contrast, there has been progress in other capability areas. The first two of four Oliver Hazard Perry-class frigates to be acquired second-hand from the United States are to be transferred in 2015 and are to replace two of the ageing Knox class. The surface fleet has also been augmented by what is expected to be the first of 12 catamaran-hulled Hsun Hai fast attack craft of 60 m length. Larger than the Chinese Houbei class, with which it has been compared, the first of class is to enter service in 2015. In addition, a new 196 m combat support ship, *Panshih*, was named at Kaohsiung on 5 Nov 2013. Arguably, Taiwan's most significant maritime capability upgrade is the acquisition of 12 P-3C Orion maritime patrol aircraft from the United States. The first refurbished aircraft arrived at Pingtung-North airbase, in the south of the island, on 25 September 2013 and the remaining 11 are to be delivered by 2015. The new aircraft are to be operated by the air force, rather than the navy, as part of an anti-submarine warfare aviation group, which includes the Orions and 11 S-2T Trackers.

Indonesia's plans to acquire new submarines were confused by the announcement in December 2013 that procurement of new or refurbished missile-armed Kilo-class diesel-electric submarines was under consideration. While intentions to boost the submarine flotilla to a force of 10 are well known, it had been thought that this would be achieved by building Type 209/1400 submarines in partnership with Daewoo, South Korea. It was announced in December 2011 that two such boats were to be built in South Korea, while a third would be built under licence at PT PAL, Surabaya. Building of the necessary industrial facilities was to have been completed in 2016. While procurement of Russian boats was later rejected in March 2014, the fact that alternative solutions were apparently being explored suggests that the project may not be running smoothly. Technology transfer and industrial partnership are likely to be difficult issues to resolve.

Plans to boost the Indonesian surface fleet are proceeding more smoothly. Following the procurement of four 90 m Sigma-class corvettes from the Netherlands in 2007–08, the contract for the construction of two 105 m variants of the class was signed in June 2012. Modules of both ships are to be built both in the Netherlands and in Indonesia with final assembly to be undertaken at Surabaya. Agreement has also been reached on the acquisition of three Nakhoda Ragam corvettes, originally completed for Brunei in 2004, but not subsequently delivered. The ships are helicopter-capable (no hangar), armed with a 76 mm gun and it is reported that the original Seawolf system has been replaced by VL-MICA. Sea trials began in April 2014.

Singapore's acquisition of a submarine capability began in 1995 when the intention to acquire a training submarine from Sweden was first announced. Since then, three Challenger (Sjöormen)-class boats have entered operational service and, latterly, two Archer (Västergötland) class equipped with AIP. Although it had been expected that Singapore would maintain its links with Sweden by participating in the A 26 submarine programme, it was announced on 2 December 2013 a contract with TKMS for two new Type 218SG submarines had been signed. These are to replace the Challenger class from 2020. Details of the boats have not been released but the design is believed to be an adaptation of the Type 214 design in service with the navies of Greece and South Korea. Key features are to include air-independent propulsion and a combat system co-developed by Atlas Elektronik and ST Electronics. The other main development is that eight new 80 m Littoral Mission Vessels are to be procured to replace the 55 m Fearless class by 2016–20. A contract for design and build of the vessels was signed with Singapore Technologies Engineering in January 2013. Weapons are to include a medium-calibre gun and a point-defence system, while sensors are to include a surveillance radar and an electro-optic system. A stern ramp is to enable launch and recovery of two interception craft.

Latin America

Against a background of difficult economic times, there is little expectation that the Argentine Navy will emerge from its current parlous state in the near future. A recent blow to its fortunes has been

MARIATEGUI

that plans to upgrade 10 of its Super Étendard aircraft to SEM (Super Étendard Modernisé) 4/5 standard have been delayed by French plans to postpone retirement of its own aircraft until 2016; there is no indication of if, or when, the SEM kits, which include a new radar and upgraded avionics, will be delivered. As a result, only five of its fleet of 11 aircraft is believed to be operational. The Navy may also be encountering problems in upgrading its Exocet missiles, also supplied by France; it has been reported that the missiles fitted to the Almirante Brown-class destroyers are now to be modernised by the Argentine Institute for Scientific and Technological Research for Defense (CITEDEF). Plans for the procurement of new offshore patrol vessels under Project POM (Patrullero Oceánico Multipropósito) appear to have been reined back as procurement of 80 m vessels, probably based on a Fassmer design, have been abandoned in favour of smaller 35 m vessels. Perhaps for this reason it has been decided to convert the fast attack craft, *Indomita*, to a patrol craft role. While the vessel is still to be equipped with a 76 mm gun, the ship will now operate on two rather than four shafts giving a maximum speed of 22 kt. Its sister ship, *Intrépida*, continues in the missile-armed role having recently completed a refit at Ushuaia in mid-2013. Meanwhile, progress in refitting the icebreaker, *Almirante Irizar*, which was badly damaged by fire in 2007, continues to be slow; completion is forecast for 2015. The refit of the submarine, *San Juan*, has also been protracted; it is expected to return to service in 2014 after seven years in dockyard hands.

Despite a relatively stronger economy, Brazil's ambitious plans for the future seemed to have stalled over the last year. Whereas the conventional submarine programme for four Scorpene-class boats took a step forward when the forward section and sail of first of class *Riachuelo* was shipped from Cherbourg to Brazil in 2013, there has been no progress in the surface fleet's PROSUPER (Programa de Obtenção de Meios de Superfície) programme for five new frigates, five new offshore patrol vessels, and one logistics support ship. Indeed, reports that the navy may be considering life-extension of the Niteroi class or procurement of second-hand frigates as a stop-gap measure suggest that the project may have been scaled back. Account also needs to be taken of plans, announced in March 2014, to build a new aircraft carrier to enter service in about 2030. By that time, the current ship, *São Paulo*, will be almost 70 years old and modernisation, maintenance, and running costs are likely to prove increasingly costly.

The next step for the Chilean Navy is to procure a second amphibious ship following the acquisition of *Sargento Aldea* (ex-*Foudre*) in 2011. The most likely option is sister-ship, *Siroco*, which may be decommissioned from the French Navy earlier than expected. If so, a further 60 m CDIC-class landing craft will probably be part of the package. Until then, the principal ongoing programme is construction of a third 80 m Piloto Pardo-class offshore patrol vessel. To be named *Marinero Fuentealba*, it is planned to be launched in 2014.

In conclusion

The Ukraine crisis has concentrated western European minds but problems elsewhere have not gone away. US President Barack Obama's visit to Asian countries in early 2014 was intended to strengthen ties in the region. In Japan, he reassured his hosts that the uninhabited Diaoyu/Senkaku islands, claimed by both Tokyo and Beijing, "fall within the scope" of a US-Japanese security treaty, while in the Philippines a 10-year agreement to allow US forces to use bases in the Philippines, two decades after US personnel were invited to leave the country, raises the possibility of a return to Subic Bay naval base. Obama's visit to Seoul came at a time when North Korea seems to be preparing for another nuclear test, while his visit to Malaysia was the first by a US head of state for nearly 50 years.

In the Middle East, the Syrian civil war continues unabated and there seems to be no sign of an end to a tragedy that has already claimed more than 100,000 lives. At least some progress has been made in the programme, agreed in 2013, to destroy Syrian chemical weapons stocks. The most dangerous materials (including mustard gas) are being disposed of at sea by a specially adapted US Ready Reserve Force ship, *Cape Ray*. Meanwhile, the situation in neighbouring Egypt continues to be unstable, three years after the uprisings in the country began. Only in Iran, encouragingly, are there some grounds for optimism. If talks to reach a comprehensive deal on Tehran's nuclear programme are successful in 2014, as has been speculated, there is a chance that tensions in the Gulf region could be reduced.

However, the re-emergence of Russia as, if not a fully-fledged threat, a potentially awkward and disobliging nation for western governments to deal with, has been the most significant event of the past year. This may be what the Russian leadership wanted to achieve but, as conflicts over the past decade have demonstrated, events have an unfortunate habit of not turning out as first intended. In any case, there may be a price for Russia to pay. Western sanctions may have been apparently innocuous but their cumulative effect could yet prove damaging to the Russian economy. Efforts to reduce reliance on Russian energy supplies may take time but, ultimately, diminution or loss of energy export revenues could be very harmful. However, 'soft power' levers rarely have the desired effect if not backed up by military capability and political will. Perhaps the only good thing to have come out of the Ukraine crisis is that it has been a 'wake-up call' for European countries that had become over-complacent about their security. Following a quarter of a century of the 'Peace Dividend', the time has come to spend more on defence and adopt a firmer line. Cancellation by the French government of the sale of two Mistral-class amphibious ships to the Russian Navy would, in my opinion, be a good place to start.

Stephen Saunders   **May 2014**

Acknowledgements

The business of collecting information and recording change has always been a continuous process and, whilst *IHS Jane's Fighting Ships* in hard-copy remains annual, the online and offline versions are updated on a continuous basis and are ideal for those users more impatient for change as it happens. Amongst the many offerings in the online version are links to the latest IHS Jane's news and analysis, and many other developments are in the pipeline. Feedback on any aspect of the title in all its formats are always most welcome.

To the many anonymous people in government and industry who make data collection such a pleasure, my warmest thanks. We are not interested in secrets, but only in ensuring that open discussion on defence is based on reliable facts.

Thanks are due also to the many people who send colour photographs, whether every year or as the opportunity offers. While not every one can be published, any image, including those that are seemingly insignificant or of doubtful value, has the potential to be useful by corroborating other information about the ship(s) in question. Ideally, images should be at 300 dpi resolution although, exceptionally, lower quality images of rarely photographed ships will be considered for printing. Images should be sent by email or on a DVD or CD , preferably as soon as possible after they have been taken. For those who have not changed to the digital medium, colour prints are of course still gratefully received.

Ian Sturton's excellent scale line drawings have long been a major feature of the publication while changes to Ranks and Insignia have been further updated by Dr Nigel Thomas, an international expert. Similarly, updates to Ensigns and Flags are required each year and these have been provided by Graham Bartram, General Secretary of the Flag Institute, one of the world's main research and documentation centres for flags and vexicology. The importance of the US Navy in maritime affairs merits a special contributor in Tom Philpott who is the editor of *Military Update* in Washington DC.

Other individual contributors who are at the heart of the updating process, and who wish to be acknowledged include:

Captain M Annati, Señor A Campanera i Rovira, Mr M Carneiro, Herr H Carstens, Mr R Cheung, Mr J Ciślak, Lieutenant Commander M Condeno, Mr P Cornelis, Mr S da Costa, Mr G Davies, Mr M Declerck, Herr H Ehlers, Mr R Fildes, Herr F Findler, Mr D Fox, Dr Z Freivogel, Señor A E Galarce, Signor M Ghiglino, Signor G Ghiglione, Colonel W Globke, Rear Admiral J Goldrick, Mr E Hooton, Captain Shaun Jones, Mr Tohru Kizu, Mr P Körnefeldt, Mr A A de Kruijf, Colonel J Kürsener, Mr E Laursen, Mr M Laursen, Mr D Mahadzir, Mr M Mazumdar, Mr M Mokrus, Mr K Mommsen, Mr J Montes, Mr S Morison, Mr J Mortimer, Mr H Nakai, Mr L-G Nilsson, Herr M Nitz, Mr T Okano, Señor A Ortigueira Gil, Mr F Philips, Mr M Piché, Mr I J Plokker, Captain B Prézelin, Señor D Quevedo Carmona, Mr A J R Risseeuw, Monsieur J Y Robert, Mr A Roper, Mr F Sadek, Mr S San, Mr C Sattler, Monsieur A Sheldon-Duplaix, Mr D Swetnam, Captain T Tamura, Mr D Thomas, Mr G Toremans, Prof A Wessels, Herr M Winter, Mr J Wise, Mr C D Yaylali, Mr T Yüksel.

IHS Jane's staff at Coulsdon ease the production process and no praise can be high enough for Emma Donald (Senior Content Editor); Jack Brenchley (Senior Compositor) and the composition team at Amnet in Chennai, India; Jonathan Maynard (Copy Editing Senior Manager); Kevan Box, Wayne Sudbury, and Harriet Harding (Scanning Team); Jo Agius, Kate Whitehead, and Mike Johnson (Image Archivists); and Martyn Buchanan (Production Controller). Closer to home, my wife Ann is an indispensable member of the year-round editorial and administrative effort.

Cross referencing to other IHS Jane's publications is made easy by IHS Jane's online service which includes, inter alia: the *IHS Jane's All the World's Aircraft* series, *IHS Jane's Amphibious & Special Forces*, the *IHS Jane's Weapons* series, *IHS Jane's C4ISR & Mission Systems: Maritime*, *IHS Jane's International Defence Directory*, *IHS Jane's Unmanned Maritime Vehicles*, and *IHS Jane's World Air Forces*. *IHS Jane's Sentinel Security Assessments* are an excellent source of politico-military information while *IHS Jane's Defence Procurement - Military Ships*, is an online business tool for tracking and projecting military vessel upgrade and procurement programmes around the world. IHS Jane's magazines provide up to the minute reports on defence issues. These include *IHS Jane's Defence Weekly*, *IHS Jane's International Defence Review*, *IHS Jane's Intelligence Review*, *IHS Jane's Defence Industry*, *IHS Jane's Missiles and Rockets*, and *IHS Jane's Navy International*.

The focus of *IHS Jane's Fighting Ships* remains seagoing personnel, whether on the bridge or in the operations room. The aim is to provide the operational capabilities of a ship or navy in a consistent and concise format. Individual entries are composed so that there is no need to turn a page or cross-refer to other sections. It is always a pleasure to get feedback from those at sea.

All updating material should be sent to:
Commodore Stephen Saunders
IHS Jane's
Sentinel House
163 Brighton Road
Coulsdon
Surrey CR5 2YH
United Kingdom
Fax: (+44 20) 32 53 21 03
e-mail: yearbook@ihsjanes.com

Note: No illustration from this book may be reproduced without the publisher's permission, but the press may reproduce information and governmental photographs, provided that *IHS Jane's Fighting Ships* is acknowledged as the source. Photographs credited to other than official organisations must not be reproduced without permission from the originator.

Stephen Saunders

During a 32-year career in the Royal Navy, Stephen Saunders travelled extensively and worked with many different navies. A surface ship officer and anti-submarine warfare specialist, he served in most classes of warship from mine countermeasures vessels to aircraft carriers. He commanded the frigate HMS *Sirius* and, as Captain 1st Frigate Squadron, HMS *Coventry*; in the latter role he also commanded the Royal Navy's Armilla patrol when deployed to the Gulf. His broad staff experience included attachment to the NATO staff of Commander US 6th Fleet and several tours in the Ministry of Defence, London. Appointments in Naval Operational Requirements and Defence Concepts led to his final job as Director Force Development within the Defence Policy Division. He graduated from the National Defence College, Latimer, in 1982 and the Royal College of Defence Studies in 1994. Since leaving the Royal Navy in 1998, he has worked in the shipbuilding industry and as a defence consultant.

[24] Glossary

LF	low frequency
MAD	magnetic anomaly detector
MDF	Maritime Defence Force
Measurement	See *Tonnage*
MF	medium frequency
MFCS	Missile Fire-Control System
MG	machine gun
MIDAS	Mine and Ice Detection Avoidance System
MIRV	multiple, independently targetable re-entry vehicle
MPA	Maritime Patrol Aircraft
MSA	Japan Maritime Safety Agency
MSC	US Military Sealift Command
MW	megawatt
NBC	Nuclear, Biological and Chemical (warfare)
nt	net tonnage (see *Tonnage*)
n mile	nautical mile (mean value 1.8532 km)
NMRS	Near-term Mine Reconnaissance System
NTDS	Naval Tactical Direction System
oa	overall length
OTC	officer in tactical command
OTHT	over the horizon targeting
PAAMS	Principal Anti-Air Missile System
PAP	Poisson Auto Propulse
PDMS	Point Defence Missile System
PWR	pressurised water reactor
QRCC	quick reaction combat capability
RAIDS	Rapid Anti-ship missile Integrated Defence System
RAM	radar absorbent material
RAM	Rolling Airframe Missile
RAS	replenishment at sea
RAST	Recovery, Assist, Secure and Traverse System
RBU	anti-submarine rocket launcher
RCS	radar cross section
RIB	rigid inflatable boat
Ro-Ro	roll-on/roll-off
ROV	remote operated vehicle
rpm	revolutions per minute
SAM	surface-to-air missile
SAR	search and rescue
SATCOM	satellite communications
SAWCS	Submarine Acoustic Warfare Countermeasures System
SES	Surface Effect Ship
SHF	super high frequency
SINS	Ship's Inertial Navigation System
SLBM	submarine-launched ballistic missile
SLCM	ship-launched cruise missile
SLEP	Service Life Extension Programme
SMCS	Submarine Command System
SRBOC	Super Rapid Blooming Offboard Chaff
SSDE/SSE	submerged signal and decoy ejector
SSDS	Ship Self-Defence System
SSM	surface-to-surface missile
SSTDS	Surface Ship Torpedo Defence System
STIR	Surveillance Target Indicator Radar
STOBAR	short take-off and barrier arrested recovery
STOVL	short take-off and vertical landing
SUM	surface-to-underwater missile
SURTASS	Surface Towed Array Surveillance System
SWATH	small waterplane area twin hull
TACAN	tactical air navigation beacon
TACTASS	TACtical Towed Acoustic Sensor System
TAINS	Tercom Aided Inertial Navigation System
TAS	Target Acquisition System
TASM	Tomahawk anti-ship missile
TASS	Towed Array Surveillance System
TBMD	Theatre Ballistic Missile Defence
Tercom	terrain contour matching
TLAM	Tomahawk Land Attack Missile
Tonnage	Tonnage measurements are governed by an IMO Convention (International Convention on Tonnage Measurement of Ships, 1969 (London-Rules)) and apply to all ships built after July 1982.

(a) gt: gross tonnage - a function of the moulded volume of all enclosed spaces of the ship
(b) nt: net tonnage - a function of the moulded volume of all cargo spaces of the ship
(c) dwt: deadweight - is the displacement of a ship in tonnes at its load draft minus the lightship weight. It includes the crew, passengers, cargo, fuel, water, and stores.

Tonne	1,000 kilos = 2,204.6 lb
	Imperial (long) ton = 1.016 tonne or 2,240 lb
	US (short) ton = 0.9072 tonne or 2,000 lb
UAV	unmanned aerial vehicle
UCAV	unmanned combat aerial vehicle
UHF	ultra-high frequency
USM	underwater-to-surface missile
USV	unmanned surface vehicle
UUV	unmanned undersea vehicle
VDS	variable depth sonar, can be lowered to best listening depth. In helicopters called 'dunking sonar'
Vertrep	vertical replenishment
VLF	very low frequency radio
VLS	Vertical Launch System
VSTOL	vertical or short take-off/landing
VSV	very slender vessel
VTOL	vertical take-off/landing
wl	waterline length

NYKÖPING

9/2013*, Per Körnefeldt / 1531204

IHS Jane's Fighting Ships 2014-2015

© 2014 IHS

IHS Aerospace, Defence and Security products

With a legacy of over 100 years as Jane's, IHS is the leader in defence, security and transportation intelligence, delivering the most **reliable, comprehensive** and **up-to-date** open-source news, insight and analysis available.

Whether online, offline or in print, governments, defence organisations and businesses around the world rely on IHS **news, forecasting, reference** and **intelligence** products to support their critical plans, processes and decisions.

To learn more about how IHS defence, security and transportation intelligence can benefit your organisation, visit **www.ihs.com/defence**

IHS Jane's Defence Equipment & Technology Solutions

Intelligence

Defence Equipment & Technology Intelligence Centre

Reference

Aero Engines

All the World's Aircraft: Development & Production

All the World's Aircraft: In Service

All the World's Aircraft: Unmanned

C4ISR & Mission Systems: Air

C4ISR & Mission Systems: Joint & Common Equipment

C4ISR & Mission Systems: Land

C4ISR & Mission Systems: Maritime

EOD & CBRNE Defence Equipment

Flight Avionics

Fighting Ships

Land Warfare Platforms: Armoured Fighting Vehicles

Land Warfare Platforms: Artillery & Air Defence

Land Warfare Platforms: Logistics, Support & Unmanned

Land Warfare Platforms: System Upgrades

Mines & EOD Guide

Police & Homeland Security Equipment

Simulation & Training Systems

Space Systems & Industry

Unmanned Maritime Vehicles

Weapons: Air-Launched

Weapons: Ammunition

Weapons: Infantry

Weapons: Naval

Weapons: Strategic

News & Analysis

International Defence Review

Navy International

IHS Jane's Defence Industry Solutions

Intelligence

Defence Industry & Markets Intelligence Centre

Defence Procurement Intelligence Centre

Offsets Advisory Module

PEDS Complete

Forecasting

Defence Budgets

Defence Sector Budgets

DS Forecast

Reference

Aircraft Component Manufacturers

International ABC Aerospace Directory

International Defence Directory

World Defence Industry

News & Analysis

Defence Industry

Defence Weekly

IHS Jane's Security Intelligence Solutions

Intelligence

Chemical, Biological, Radiological and Nuclear Assessments Intelligence Centre

Military & Security Assessments Intelligence Centre

Sentinel Country Risk Assessments

Terrorism & Insurgency Centre

Terrorism Events Spatial Layer

World Insurgency & Terrorism

Reference

Amphibious & Special Forces

CBRN Response Handbook

World Air Forces

World Armies

World Navies

News & Analysis

Intelligence Review

IHS Jane's Transportation News & Reference

Reference

Air Traffic Control

Airports & Handling Agents

Airports, Equipment & Services

Urban Transport Systems

World Railways

News & Analysis

Airport Review

IHS Users' Charter

This publication is brought to you by IHS, a global company drawing on more than 100 years of history and an unrivalled reputation for impartiality, accuracy and authority.

Our collection and output of information and images is not dictated by any political or commercial affiliation. Our reportage is undertaken without fear of, or favour from, any government, alliance, state or corporation.

We publish information that is collected overtly from unclassified sources, although much could be regarded as extremely sensitive or not publicly accessible.

Our validation and analysis aims to eradicate misinformation or disinformation as well as factual errors; our objective is always to produce the most accurate and authoritative data.

In the event of any significant inaccuracies, we undertake to draw these to the readers' attention to preserve the highly valued relationship of trust and credibility with our customers worldwide.

If you believe that these policies have been breached by this title, or would like a copy of IHS's Code of Conduct for its editorial teams, you are invited to contact the Group Publishing Director.

www.ihs.com

TERMS AND CONDITIONS FOR THE SALE OF HARDCOPY PRODUCTS

All orders for the sale of hardcopy Products are subject to the following terms and conditions:

1. **DEFINITIONS**

    "**Client**" means the person, firm or company or any other entity that purchases the Products from IHS.

    "**Delivery Point**" where applicable, means the location as defined in the Order Confirmation where delivery of the Products is deemed to take place.

    "**Directory Products**" means IHS's proprietary database or any part thereof, including without limitation, details of particular company/organisation, key personnel, financial/statistical information, products/services description, organisational structure and any other information pertaining to such company(s)/organisation(s) operating in various industrial sectors.

    "**Fees**" means the money due and owing to IHS for Products supplied including any order processing charge and as set forth in the Order Confirmation. Fees are exclusive of taxes, which will be charged separately to the Client.

    "**Products**" means any publication, database, supplied to the Client in physical or electronic media, more specifically mentioned in the Order Confirmation. Products include Directory Products.

    "**Order Confirmation**" includes the order form or confirmation email or any other document which IHS sends to the Client to confirm that IHS has accepted the Client's order and which identifies the name of the Client, Product(s) being supplied, period of supply, delivery information, media of supply, Fees and any terms or conditions unique to the particular Product to be supplied hereunder.

2. Client will pay IHS the Fees as set forth in the Order Confirmation within 30 days from the date of the invoice. Any payments not received by IHS when due will be considered past due, and IHS may choose to accrue interest at the rate of five percent (5%) above the European Central Bank "Marginal lending facility" rate. Client has no right of set-off. Client will pay all the value-added, sales, use, import duties, customs or other taxes where applicable to the purchase of Products. IHS may request payment of the Fees before shipping the Products.

3. IHS grants to Client a nonexclusive, nontransferable license to use the Products for its internal business use only. Client may not copy, distribute, republish, transfer, sell, license, lease, give, disseminate in any form (including within its original cover), assign (whether directly or indirectly, by operation of law or otherwise), transmit, scan, publish on a network, or otherwise reproduce, disclose or make available to others, store in any retrieval system of any nature, create a database or create derivative works from the Product or any portion thereof, except as specifically authorized herein. Any information related to third party company and/or personal data included in the Directory Product(s), may be used by Client for the limited purpose of enquiring about the products and services of the companies/organisations listed therein and who have given permission for their data to be used for this purpose only. Client must comply with the UK Data Protection Act and all other applicable data protection and privacy laws and regulations. In particular, Client must not use such data (i) for any unlawful, harmful or offensive purpose; (ii) as a source for any kind of marketing or promotion activity; or (iii) for the purposes of compiling, confirming or amending its own database, directory or mailing list.

4. Client must not remove any proprietary legends or markings, including copyright notices, or any IHS-specific markings on the Products. Client acknowledges that all data, material and information contained in the Products are and will remain the copyright property and confidential information of IHS or any third party and are protected and that no rights in any of the data, material and information are transferred to Client. Client will take any and all actions that may reasonably be required by IHS to protect such proprietary rights as owned by IHS or any third party. Any unauthorised use may give rise to IHS bringing proceedings for copyright and/or database right infringement against the Client claiming an injunction, damages and costs.

5. Any dates specified in the Order Confirmation for delivery of the Products are intended to be an estimated time for delivery only and shall not be of the essence. IHS shall not be liable for any delay in the delivery of the Products. Unless otherwise agreed by the parties, packing and carriage charges are not included in the Fees and will be charged separately. The Products will be despatched and delivered to the Delivery Point as per Client's preferred method of delivery and as agreed by IHS. If special arrangements are required, then IHS reserves the right to additional charges. Except as provided hereunder, for all Products supplied hereunder, delivery is deemed to occur and risk of loss passes upon despatch of Products by IHS.

6. If for any reason IHS is unable to deliver the Products on time due to Client's failure to provide appropriate instructions, documents or authorisations etc; (i) any risk in the Products will pass to the Client; (ii) the Products will be deemed to have been delivered; and (iii) IHS may store the Products until delivery, whereupon the Client will be liable for all related costs and expenses.

7. Except as otherwise required by law, Client will not be entitled to object or to return or reject the Products or any part thereof unless the Products are damaged in transit. IHS's sole obligation and Clients' exclusive remedy for any claim with respect to such damaged Products will be to replace the damaged Products without any charge. No returns will be accepted by IHS without prior agreement and a returns number issued by IHS to accompany the Products to be returned. All return shipments are at the Client's risk and expense.

8. The possession and usage rights of the Products in accordance with clause 3 above will not pass to Client until IHS has received in full all sums due to it in respect of: (i) Fees; and (ii) all other sums which are or which become due to IHS from Client on any account. Until such rights have passed to Client, the Client will: (i) hold the Products in a fiduciary capacity; (ii) store the Products (at no cost to IHS) in such a way that they remain readily identifiable as IHS property; (iii) not destroy, deface or obscure any identifying mark or packaging on or relating to the Products; and (iv) maintain the Products in satisfactory condition and keep them insured on IHS' behalf for their full price against all risks to the reasonable satisfaction of IHS.

9. The quantity of any consignment of Products as recorded by IHS on despatch from IHS' place of business shall be conclusive evidence of the quantity received by the Client on delivery unless Client can provide conclusive evidence proving otherwise. IHS shall not be liable for any non-delivery of the Products (even if caused by IHS' negligence) unless Client provides conformed claims to IHS of the non-delivery. Any such conformed claim for non-receipt of the Products must be made in writing, quoting the account and Order Confirmation number to the IHS' Customer Service Department, within thirty (30) days of the estimated date of delivery as stated in the Order Confirmation.

10. The Products supplied herein are provided "AS IS" and "AS AVAILABLE". IHS does not warrant the completeness or accuracy of the data, material, third party advertisements or information as contained in the Product or that it will satisfy Client's requirements. IHS disclaims all other express or implied warranties, conditions and other terms, whether statutory, arising from course of dealing, or otherwise, including without limitation terms as to quality, merchantability, fitness for a particular purpose and noninfringement. To the extent permitted by law, IHS shall not be liable for any errors or omissions or any loss, damage or expense incurred by reliance on information, third party advertisements or any statement contained in the Products. Client assumes all risk in using the results of the Product(s).

11. If the Products supplied hereunder are subscription based, except as otherwise provided herein the period of supply will run for one calendar year from the start date as specified in the Order Confirmation and the Fees will cover the costs of supply of all issues of the Product published in that year. If Client attempts to cancel the Product subscription anytime during such period; (i) the Fees payable for that year will be invoiced by IHS in full; or (ii) where Client has already paid the Fees in advance, any Fees relating to the remaining period shall be forfeited. In addition to other rights and subject to the provisions of this clause, IHS in its sole discretion may discontinue the supply the Products in the event Client commits breach of any of the provision of these terms and conditions.

12. In the event of breach of any of the provision of these terms and conditions by IHS, IHS' total aggregate liability for any damages/losses incurred by the Client arising out of such breach shall not exceed at any time the Fees paid for the Product which is the subject matter of the claim. In no event shall IHS be liable for any indirect, special or consequential damages of any kind or nature whatsoever suffered by the Client including, without limitation, lost profits or any other economic loss arising out of or related to the subject matter of these terms and conditions. However, nothing in these terms and conditions shall limit or exclude IHS' liability for (i) death or personal injury caused by its negligence; (ii) fraud or fraudulent misrepresentation; or (iii) any breach of compelling consumer protection or other laws.

13. Client represents and warrants that it will not directly or indirectly engage in any acts that would constitute a violation of United States laws or regulations governing the export of United States products and technology.

14. The parties will comply with all applicable country laws relating to anti-corruption and anti-bribery, including the US Foreign Corrupt Practices Act and the UK Bribery Act. The parties represent and affirm that no bribes or corrupt actions have or will be offered, given, received or performed in relation to the procurement or performance of these terms and conditions. For the purposes of this clause, "bribes or corrupt actions" means any payment, gift, or gratuity, whether in cash or kind, intended to obtain or retain an advantage, or any other action deemed to be corrupt under the applicable country laws.

15. All Products supplied herein are subject to these terms and conditions only, to the exclusion of any other terms which would otherwise be implied by trade, custom, practice or course of dealing. Nothing contained in any Client-issued purchase order, Clients' acknowledgement, Clients' terms and conditions or invoice will in any way modify or add any additional terms to these terms and conditions. IHS reserves the right to amend these terms and conditions from time to time.

16. These terms and conditions and any dispute or claim arising out of or in connection with them or their subject matter shall be governed by and construed in accordance with the laws of England and Wales and shall be subject to the exclusive jurisdiction of the English Courts.

Ranks and insignia of the world's navies

Ranks and insignia of the world's navies

This section portrays the rank insignia worn by commissioned officers of the world's navies and coast guards on formal occasions. The rank titles are described in the language of the relevant country followed by the Royal Navy equivalent.

The traditional uniform pattern has been the very dark blue double-breasted service tunic introduced by the Royal Navy in the 19th Century, with rank insignia worn as thin, medium and wide gold braid rings, with a loop or 'curl' on the uppermost ring, around both cuffs. Other navies and coast guards have changed the tunic colour to black or a lighter shade of dark blue, varied the widths and order of the rings, or replaced the 'curl' with a star or other symbol. Some Middle East, African, Caribbean, or Pacific states wear Army insignia or Navy cuff rings on cloth shoulder-straps only, whilst others, especially from the former Warsaw Pact, wear rank insignia simultaneously on the shoulder-straps and cuffs of the tunic.

Similarly most states imitate the Royal Navy by wearing cuff rings on the shoulder-straps of greatcoats or white tropical dress tunics, with 'Flag-Officers' (admirals and sometimes commodores) having gold braid shoulder-straps with a national device above 5 – 1 silver wire Army-style stars. On shirts, pullovers, or camouflage field uniforms officers wear their cuff rings on cloth shoulder-loops slipped into cloth shoulder-straps or breast-loops, as metal badges on collars or rectangular breast-patches.

Albania (Forcat Detare Shqiptare)

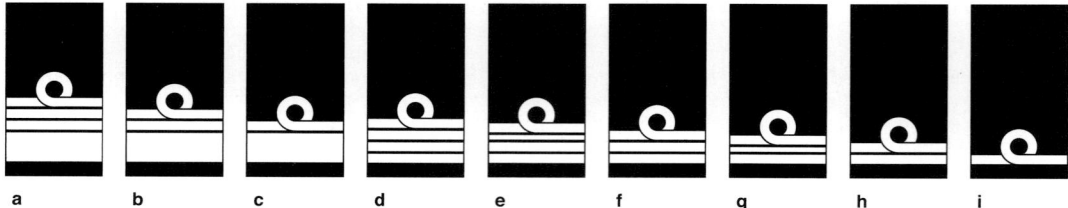

Gold braid rings with 'curl' on very dark blue cuffs. Rank titles are in Albanian. The Albanian Border Guard includes a Coast Guard (*Roja Bregdetare*).
a: *Admiral*, Vice Admiral (*rank not currently held*)   **b:** *Nënadmiral*, Rear Admiral (*rank not currently held*)   **c:** *Kundëradmiral*, Commodore (*Commander, Navy*)
d: *Kapiten i Rangut të I-rë (parë)*, Captain   **e:** *Kapiten i Rangut të II-rë (dytë)*, Commander   **f:** *Kapiten i Rangut të III-rë (tretë)*, Lieutenant Commander
g: *Kapiten Lejtnant*, Lieutenant   **h:** *Lejtnant*, Sub Lieutenant   **i:** *Nënlejtnant*, Acting Sub Lieutenant

Algeria (Al-Quwwat Al-Bahria Al-Djaza'eria)

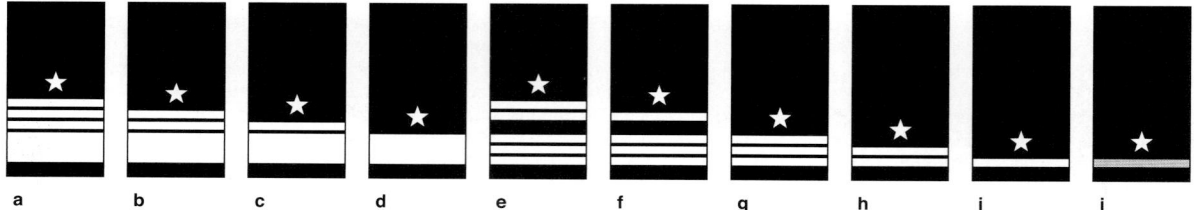

Gold braid stars and rings on mid-blue cloth cuffs; a Midshipman (j) has a silver braid star. Rank insignia is worn simultaneously on shoulder-straps. Algerian Army rank titles are used and written here in romanised Arabic. Algeria maintains a small Coast Guard.
a: *Farīq*, Vice Admiral (*rank not currently held*)   **b:** *Liwā'*, Rear Admiral (*Commander, Navy*)   **c:** *'Amid*, Commodore   **d:** *'Aqīd*, Captain
e: *Muqaddam*, Commander   **f:** *Rā'id*, Lieutenant Commander   **g:** *Naqīb*, Lieutenant   **h:** *Mulāzim Awwal*, Sub Lieutenant
i: *Mulāzim Thāni*, Acting Sub Lieutenant   **j:** *Murashshah*, Midshipman

Angola (Marinha de Guerra Angolana)

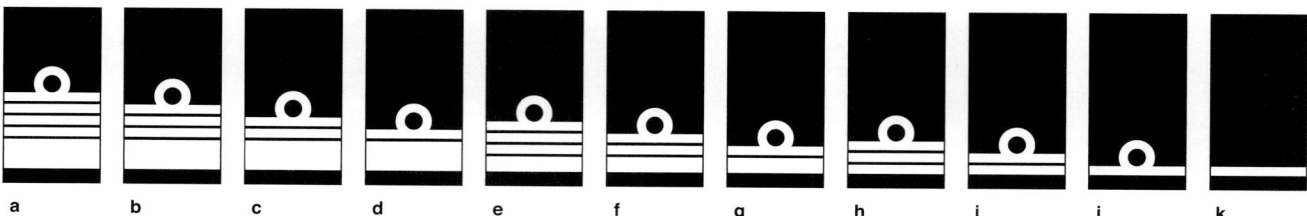

Gold braid rings with 'curl' on very dark blue cuffs. Rank titles are in Portuguese.
a: *Almirante da Armada*, Admiral (*rank not currently held*)   **b:** *Almirante*, Vice-Admiral (*Chief of Naval Staff*)   **c:** *Vice-Almirante*, Rear Admiral
d: *Contra-Almirante*, Commodore   **e:** *Capitão-de-Mar-e-Guerra*, Captain   **f:** *Capitão-de-Fragata*, Commander   **g:** *Capitão-de-Corveta*, Lieutenant Commander
h: *Tenente-de-Navio*, Lieutenant   **i:** *Tenente-de-Fragata*, Sub Lieutenant   **j:** *Tenente-de-Corveta*, Acting Sub Lieutenant   **k:** *Sub-Tenente*, Midshipman

Ranks and insignia of the world's navies [29]

Argentina (Armada Argentina)

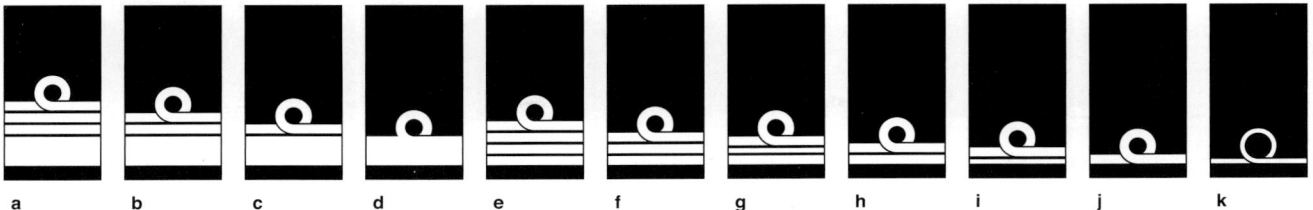

Gold braid rings with 'curl' on very dark blue cloth cuffs. Rank titles are in Spanish.
a: *Almirante*, Admiral **b:** *Vicealmirante*, Vice Admiral *(Chief of Naval Staff)* **c:** *Contraalmirante*, Rear Admiral **d:** *Comodoro de Marina*, Commodore
e: *Capitán de Navío*, Captain **f:** *Capitán de Fragata*, Commander **g:** *Capitán de Corbeta*, Lieutenant Commander **h:** *Teniente de Navío*, Lieutenant
i: *Teniente de Fragata*, (Senior) Sub Lieutenant **j:** *Teniente de Corbeta*, Sub Lieutenant **k:** *Guardiamarina*, Acting Sub Lieutenant

Argentina Coast Guard (Prefectura Naval Argentina)

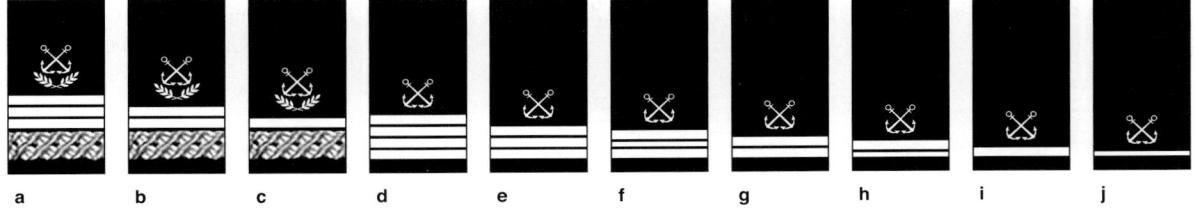

Gold wire anchors and wreaths, gold braid rings and twisted embroidery on very dark blue cuffs. Rank titles are in Spanish.
a: *Prefecto Nacional Naval*, Admiral *(Prefect-General, PNA)* **b:** *Subprefecto Nacional Naval*, Vice Admiral **c:** *Prefecto General*, Rear Admiral
d: *Prefecto Mayor*, Captain **e:** *Prefecto Principal*, Commander **f:** *Prefecto*, Lieutenant Commander **g:** *Subprefecto*, Lieutenant
h: *Oficial Principal*, (Senior) Sub Lieutenant **i:** *Oficial Auxiliar*, Sub Lieutenant **j:** *Oficial Ayudante*, Acting Sub Lieutenant

Australia (Royal Australian Navy)

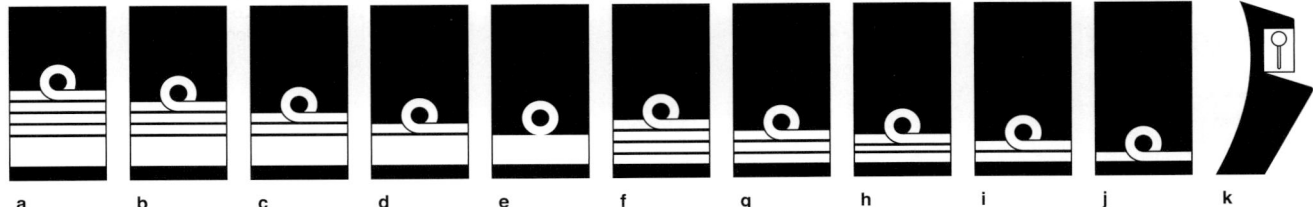

Gold braid rings with 'curl' on very dark blue cloth cuffs; brass button and white cord on white cloth collar-patch (k). British Royal Navy rank titles are used.
a: Admiral of the Fleet *(HRH Prince Philip, Duke of Edinburgh)* **b:** Admiral *(rank not currently held)* **c:** Vice Admiral *(Chief of Navy)* **d:** Rear Admiral
e: Commodore **f:** Captain **g:** Commander **h:** Lieutenant Commander **i:** Lieutenant **j:** Sub Lieutenant & Acting Sub Lieutenant **k:** Midshipman

Azerbaijan (Azerbycan Herbi Deniz Qüvveleri)

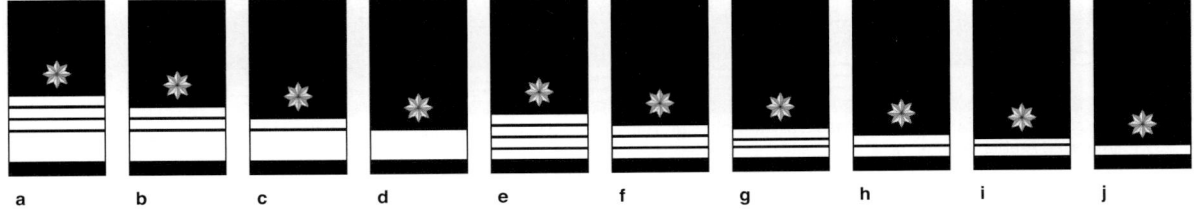

Gold wire stars and gold braid rings on black cloth cuffs; brass buttons. Rank insignia is worn simultaneously on shoulder-straps. Rank titles are in Azeri. The Azerbaijan Border Guard includes a small Coast Guard.
a: *Admiral*, Admiral *(rank not currently held)* **b:** *Vitse-admiral*, Vice Admiral *(Commander-in-Chief, Navy)* **c:** *Kontr-admiral*, Rear Admiral
d: *1 (Birinci) dereceli kapitan*, Captain **e:** *2 (Ikinci) dereceli kapitan*, Commander **f:** *3 (Üçüncü) dereceli kapitan*, Lieutenant Commander
g: *Kapitan-leytenant*, Lieutenant **h:** *Baş leytenant*, (Senior) Sub Lieutenant **i:** *Leytenant*, Sub Lieutenant **j:** *Kiçik leytenant*, Acting Sub Lieutenant

Bahamas (Royal Bahamas Defence Force)

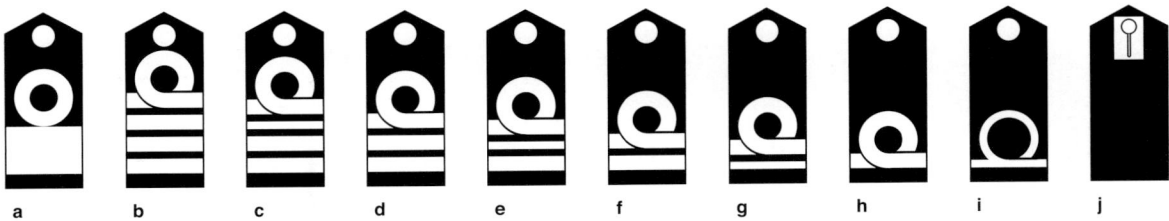

Gold braid rings with 'curl' on very dark blue cloth shoulder-straps with brass buttons; brass button and white cord on white rectangular cloth patch. Rank titles are in English.
a: *Commodore*, Commodore *(Commander, RBDF)* **b:** *Captain*, Captain *(Deputy Commander, RBDF)* **c:** *Senior Commander*, (Senior) Commander
d: *Commander*, Commander **e:** *Lieutenant Commander*, Lieutenant Commander **f:** *Senior Lieutenant*, Lieutenant **g:** *Lieutenant*, Sub Lieutenant
h: *Sub Lieutenant*, Acting Sub Lieutenant **i:** *Acting Sub Lieutenant*, (Junior) Acting Sub Lieutenant **j:** *Midshipman*, Midshipman

[30]   Ranks and insignia of the world's navies

Bahrain (Al-Quwwat Al-Bahria Al-Malakiyya Al-Bahrayn)

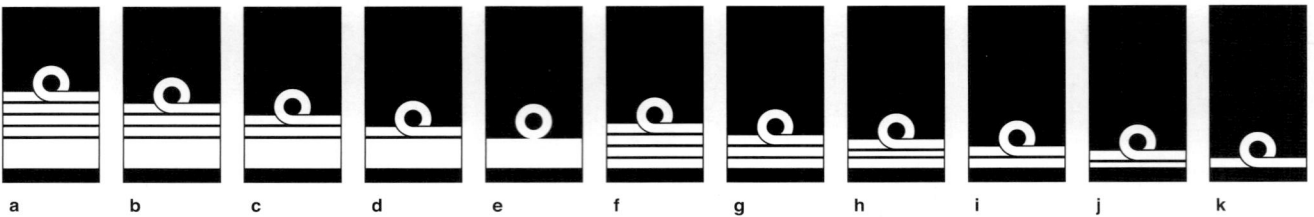

a   b   c   d   e   f   g   h   i   j   k

Gold braid rings with 'curl' on very dark blue cloth cuffs. Bahrain Army rank insignia is worn simultaneously on shoulder straps. Arabic Bahrain Army rank titles are used and written here in romanised script.
a: *Mushīr*, Admiral of the Fleet *(King of Bahrain)*   **b:** *Farīq Awwal*, Admiral *(rank not currently held)*   **c:** *Farīq*, Vice Admiral *(rank not currently held)*
d: *Liwā'*, Rear Admiral   **e:** *'Amid*, Commodore *(Commander, Navy)*   **f:** *'Aqīd*, Captain   **g:** *Muqaddam*, Commander   **h:** *Rā'id*, Lieutenant Commander
i: *Naqīb*, Lieutenant   **j:** *Mulāzim Awwal*, Sub Lieutenant   **k:** *Mulāzim Thāni*, Acting Sub Lieutenant

Bangladesh (Nou Bahini)

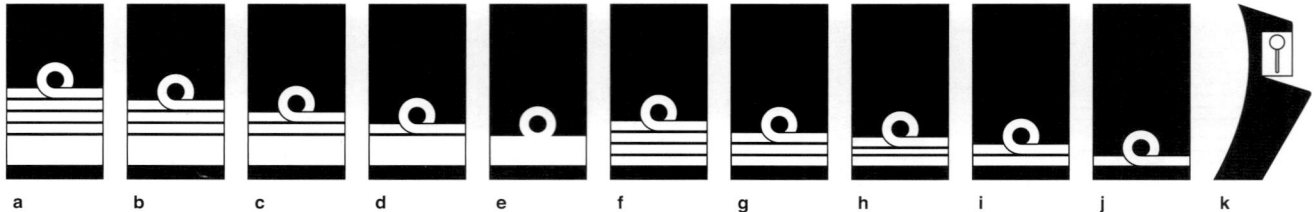

a   b   c   d   e   f   g   h   i   j   k

Gold braid rings with 'curl' on very dark blue cloth cuffs; brass button and white cord on white cloth collar-patch (k). Royal Navy rank titles are used. The Bangladesh Coast Guard forms part of the Ministry of Home Affairs. Personnel wear naval uniforms and insignia with a Commodore as the Director General.
a: Admiral of the Fleet *(rank not currently held)*   **b:** Admiral *(rank not currently held)*   **c:** Vice Admiral *(Chief of Naval staff)*   **d:** Rear Admiral   **e:** Commodore
f: Captain   **g:** Commander   **h:** Lieutenant Commander   **i:** Lieutenant   **j:** Sub Lieutenant & Acting Sub Lieutenant   **k:** Midshipman

Barbados Coast Guard

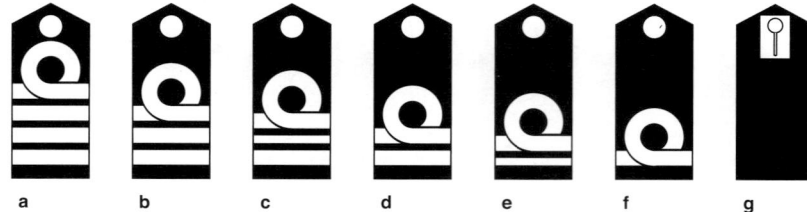

a   b   c   d   e   f   g

Gold braid rings with 'curl' on very dark blue cloth shoulder-straps with brass buttons; brass button and white cord on white rectangular cloth patch (g). Rank titles are in English.
a: *Captain*, Captain *(rank not currently held)*   **b:** *Commander*, Commander *(Commanding Officer, BCG)*   **c:** *Lieutenant Commander*, Lieutenant Commander
d: *Lieutenant*, Lieutenant   **e:** *Junior Lieutenant*, Sub Lieutenant   **f:** *Sub Lieutenant*, Acting Sub Lieutenant   **g:** *Midshipman*, Midshipman

Belgium (Naval Component) (Zeemacht/Force Navale)

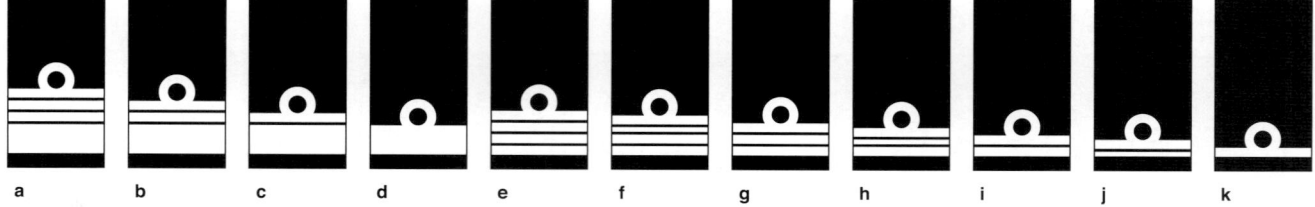

a   b   c   d   e   f   g   h   i   j   k

Gold braid rings on very dark blue cloth cuffs. Rank titles are in Flemish/French.
a: *Admiraal/Amiral*, Admiral *(King Philippe)*   **b:** *Vice-admiraal/Vice amiral*, Vice Admiral *(Chief of Court Protocol)*
c: *Divisie-admiraal/Amiral de division*, Rear Admiral *(Director of Defence Operations)*
d: *Flottieljeadmiraal/Amiral de flottille*, Commodore *(Commandant, Naval Component)*   **e:** *Kapitein-ter-zee/Capitaine de vaisseau*, Captain
f: *Fregatkapitein/Capitaine de frégate*, Commander   **g:** *Korvetkapitein/Capitaine de corvette*, Lieutenant Commander
h: *Luitenant-ter-zee 1ste (eerste) klasse/Lieutenant de vaisseau 1ère (première) classe*, (Senior) Lieutenant   **i:** *Luitenant-ter-zee/Lieutenant de vaisseau*, Lieutenant
j: *Vaandrig-ter-zee/Enseigne de vaisseau*, Sub Lieutenant
k: *Vaandrig-ter-zee 2de (tweede) klasse/Enseigne de vaisseau 2e (deuxième) classe*, Acting Sub Lieutenant

Benin (Forces Navales Béninoises)

a   b   c   d   e   f   g   h

Gold braid anchor, edging and rings and silver stars on very dark blue cloth shoulder-straps; a Commander (d) has silver second and fourth rings; brass buttons. Rank titles are in French.
a: *Vice-amiral*, Rear Admiral *(rank not currently held)*   **b:** *Contre-amiral*, Commodore *(rank not currently held)*
c: *Capitaine de vaisseau*, Captain *(rank not currently held)*   **d:** *Capitaine de frégate*, Commander *(Chief of Staff, Navy)*
e: *Capitaine de corvette*, Lieutenant Commander   **f:** *Lieutenant de vaisseau*, Lieutenant   **g:** *Enseigne de vaisseau de 1ère (première) classe*, Sub Lieutenant
h: *Enseigne de Vaisseau de 2e (deuxième) classe*, Acting Sub Lieutenant

IHS Jane's Fighting Ships 2014-2015   © 2014 IHS

Ranks and insignia of the world's navies [31]

Bolivia (Fuerza Naval Boliviana)

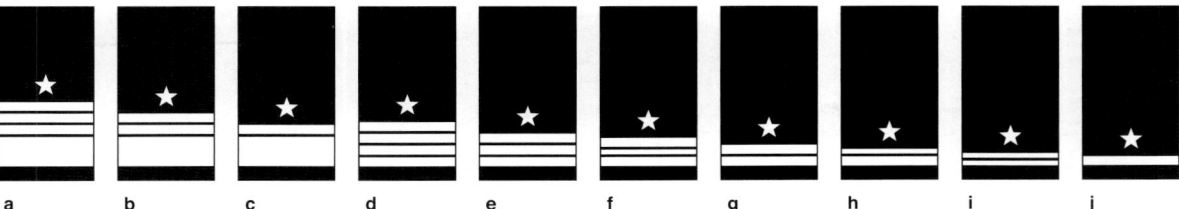

Gold wire stars and gold braid rings on very dark blue cloth cuffs. Rank titles are in Spanish.
a: *Almirante*, Admiral *(Commander-in-Chief, Navy)*   **b:** *Vicealmirante*, Vice Admiral   **c:** *Contralmirante*, Rear Admiral   **d:** *Capitán de Navío*, Captain
e: *Capitán de Fragata*, Commander   **f:** *Capitán de Corbeta*, Lieutenant Commander   **g:** *Teniente de Navío*, Lieutenant
h: *Teniente de Fragata*, (Senior) Sub Lieutenant   **i:** *Teniente de Corbeta*, Sub Lieutenant   **j:** *Alférez*, Acting Sub Lieutenant

Brazil (Marinha do Brasil)

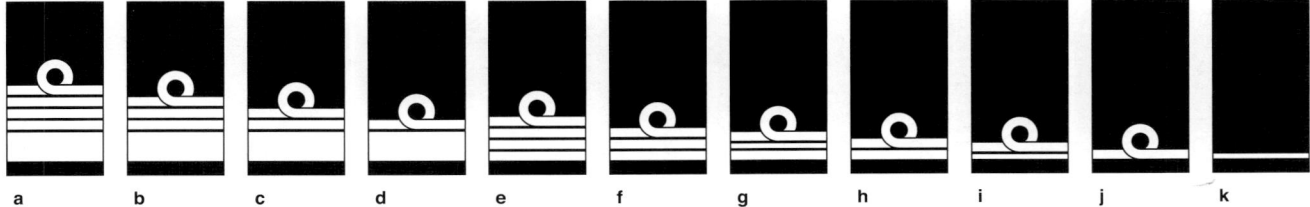

Gold braid stars and gold braid rings on very dark blue cloth cuffs. Rank titles are in Portuguese.
a: *Almirante*, Admiral *(rank only held in wartime)*   **b:** *Almirante-de-esquadra*, Admiral *(Commander of the Navy)*   **c:** *Vice-almirante*, Vice Admiral
d: *Contra-almirante*, Rear Admiral   **e:** *Capitão-de-mar-e-guerra*, Captain   **f:** *Capitão-de-fragata*, Commander   **g:** *Capitão-de-corveta*, Lieutenant Commander
h: *Capitão-tenente*, Lieutenant   **i:** *1° (Primeiro) Tenente*, Sub Lieutenant   **j:** *2° (Segundo) Tenente*, Acting Sub Lieutenant   **k:** *Guarda-marinha*, Midshipman

Brunei (Royal Brunei Navy) (Angkatan Tentera Laut Diraja Brunei)

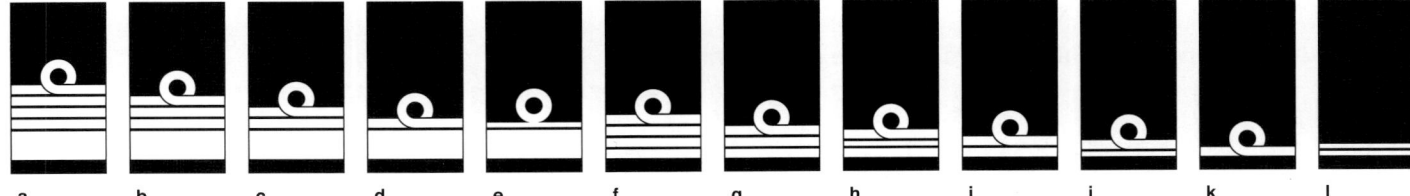

Gold braid rings with 'curl', two thin white rings (i), on very dark blue cuffs. Brunei Army rank titles are used and are written in Malay.
a: *Filed Marsyal (L)*, Admiral of the Fleet *(Sultan Haji Hassanal Bolkiah Mu'izzaddin Waddaulah)*   **b:** *Jeneral (L)*, Admiral *(rank not currently held)*
c: *Leftenan Jeneral (L)*, Vice Admiral *(rank not currently held)*   **d:** *Mejar Jeneral (L)*, Rear Admiral *(rank not currently held)*
e: *Brigedier Jeneral (L)*, Commodore/'First Admiral' *(Commander, Navy)*   **f:** *Kolonel (L)*, Captain   **g:** *Leftenan Kolonel (L)*, Commander
h: *Mejar (L)*, Lieutenant Commander   **i:** *Kapten (L)*, Lieutenant   **j:** *Leftenan (L)*, Sub Lieutenant   **k:** *Leftenan Muda (L)*, Acting Sub Lieutenant
l: *Pegawai Kadet (L)*, Midshipman

Bulgaria (Voennomorski sili na Balgariya)

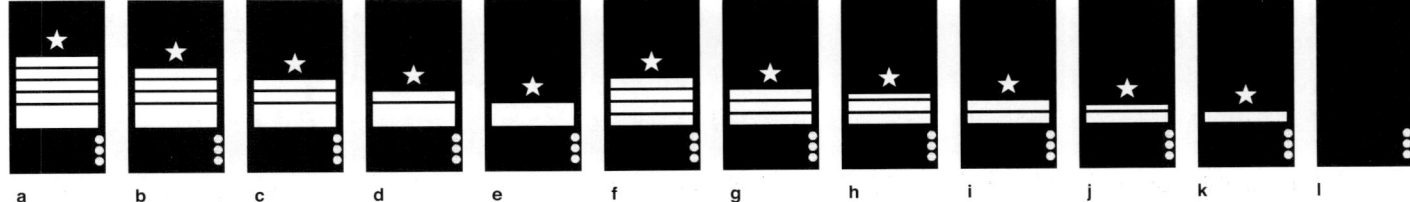

Gold wire stars and gold braid bars on black cuffs; brass buttons. Rank insignia is also worn simultaneously on gold braid shoulder-straps. Rank titles are in romanised Bulgarian. The Bulgarian Border Police *(Granichna Politsiya)* includes a small Coast Guard wearing dark-blue police uniforms and insignia.
a: *Admiral*, Admiral *(rank not currently held)*   **b:** *Vitseadmiral*, Vice Admiral *(Deputy Chief of Defence Staff)*
c: *Kontraadmiral*, Rear Admiral *(Commander-in-Chief, Navy)*   **d:** *Komodor*, Commodore   **e:** *Kapitan I (parvi) rang*, Captain   **f:** *Kapitan II (vtori) rang*, Commander
g: *Kapitan III (treti) rang*, Lieutenant Commander   **h:** *Kapitan-leytenant*, Lieutenant   **i:** *Starshi leytenant*, (Senior) Sub Lieutenant   **j:** *Leytenant*, Sub Lieutenant
k: *Mladshi leytenant*, Acting Sub Lieutenant   **l:** *Ofitserski kandidat*, Midshipman

Cambodia (Royal Cambodian Navy)

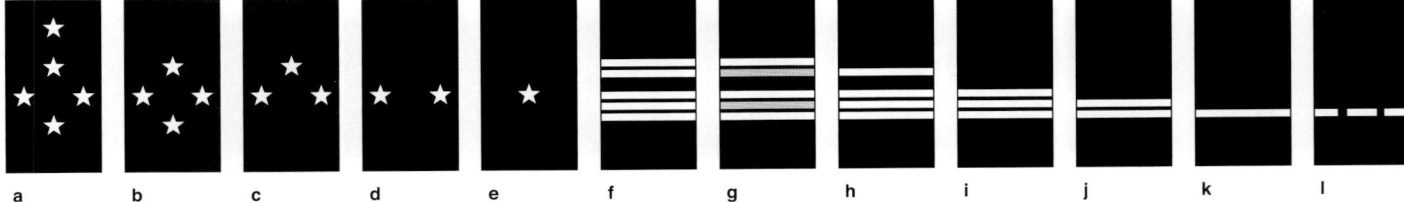

Gold metal stars and gold braid rings on dark blue cloth cuffs; a Commander (g) has silver braid second and fourth rings, a Midshipman (l) a gold braid ring with two mid-blue 'breaks'. Cambodian Army rank titles are used and written here in romanised Cambodian where available.
a: Admiral of the Fleet *(rank not currently held)*   **b:** Admiral *(rank not currently held)*   **c:** *Udon-Nearvey-Ek*, Vice Admiral *(Commander, Navy)*
d: *Udon-Nearvey-Tor*, Rear Admiral   **e:** *Udon-Nearvey-Trey*, Commodore   **f:** *Vorak-Nearvey-Ek*, Captain   **g:** *Vorak-Nearvey-Tor*, Commander
h: *Vorak-Nearvey-Trey*, Lieutenant Commander   **i:** *Aknouk-Nearvey-Ek*, Lieutenant   **j:** *Aknouk-Nearvey-Tor*, Sub Lieutenant
k: *Aknouk-Nearvey-Trey*, Acting Sub Lieutenant   **l:** *Niey Komnong*, Midshipman

[32] Ranks and insignia of the world's navies

Cameroon (Marine Nationale du Cameroun)

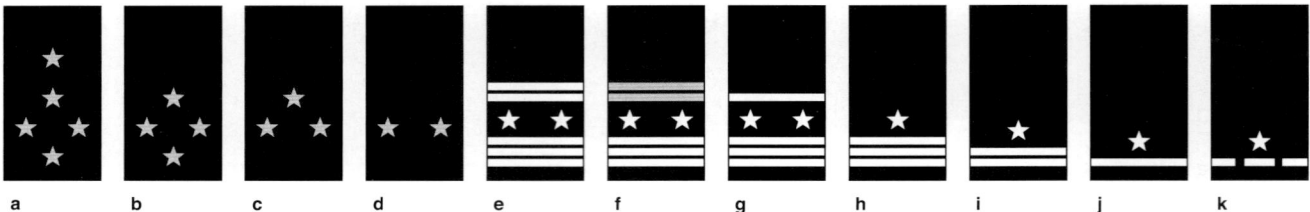

Silver metal stars (a-d), gold metal stars (e-k) and gold braid rings on dark blue cloth cuffs; a Commander (f) has silver braid first and second rings, a Midshipman (k) a gold braid ring with two mid-blue 'breaks'. Rank titles are in French.
a: *Amiral d'escadre*, Admiral *(rank not currently held)*   **b:** *Vice-amiral d'escadre*, Vice Admiral *(rank not currently held)*
c: *Vice-amiral*, Rear Admiral *(Chief of Naval Staff)*   **d:** *Contre-amiral*, Commodore   **e:** *Capitaine de vaisseau*, Captain   **f:** *Capitaine de frégate*, Commander
g: *Capitaine de corvette*, Lieutenant Commander   **h:** *Lieutenant de vaisseau*, Lieutenant   **i:** *Enseigne de vaisseau de 1ère (première) classe*, Sub Lieutenant
j: *Enseigne de vaisseau de 2e (deuxième) classe*, Acting Sub Lieutenant   **k:** *Aspirant*, Midshipman

Canada (Royal Canadian Navy/Marine Royale Canadienne)

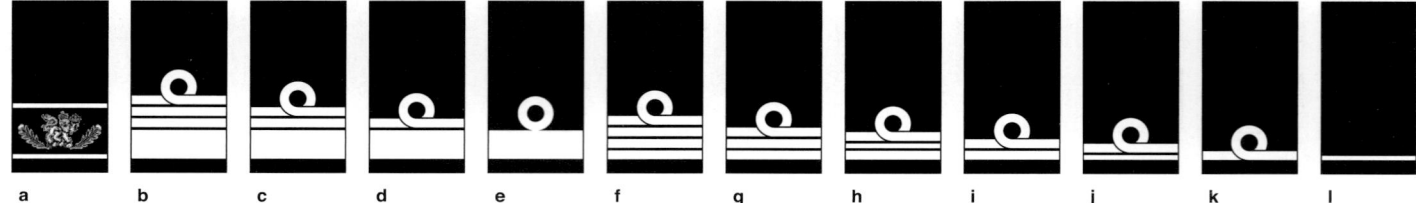

Gold braid rings with 'curl', gold wire oak leaves and gold, red, and white British lion on very dark blue cloth cuffs. Rank titles are in English/French.
a: *Commander-in-Chief/Commandant-en-chef*, Commander-in-Chief *(HM Queen Elizabeth II/Governor-General of Canada)*
b: *Admiral/Amiral*, Admiral *(rank not currently held)*   **c:** *Vice Admiral/Vice-amiral*, Vice Admiral *(Chief of Maritime Staff)*   **d:** *Rear Admiral/Contre-amiral*, Rear Admiral
e: *Commodore/Commodore*, Commodore   **f:** *Captain (N)/Capitaine de vaisseau*, Captain   **g:** *Commander/Capitaine de frégate*, Commander
h: *Lieutenant Commander/Capitaine de corvette*, Lieutenant Commander   **i:** *Lieutenant (N)/Lieutenant de vaisseau*, Lieutenant
j: *Sub Lieutenant/Enseigne de vaisseau de 1ère (première) classe*, Sub Lieutenant
k: *Acting Sub Lieutenant/Enseigne de vaisseau de 2e (deuxième) classe*, Acting Sub Lieutenant   **l:** *Naval Cadet/Aspirant de marine*, Midshipman

Canada (Canadian Coast Guard/Garde côtière canadienne)

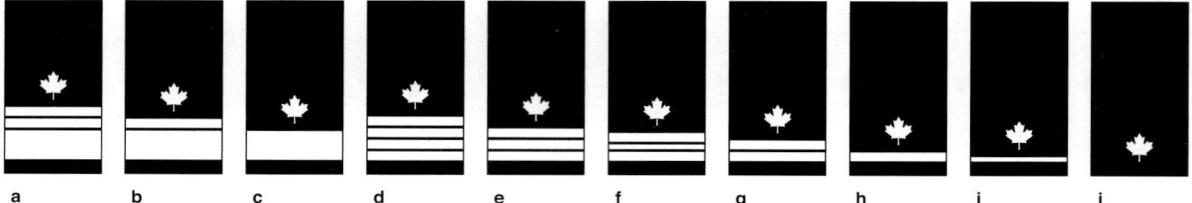

Gold wire maple-leaves and gold braid rings on very dark blue cloth cuffs. Rank titles are in English/French.
a: *Commissioner/Commissaire*, Vice Admiral *(Commissioner, CCG)*   **b:** *Deputy Commissioner/Sous-commissaire*, Rear Admiral
c: *Assistant Commissioner/Commissaire adjoint*   **d:** *Commanding Officer/Commandant*, Captain   **e:** *Chief Officer/Capitaine en 2e (Second)*, Commander
f: *1st (First) Officer/1er (Premier) Officier*, Lieutenant Commander   **g:** *2nd (Second) Officer/2e (Deuxième) Officier*, Lieutenant
h: *3rd (Third) Officer/3e (Troisième) Officier*, Sub Lieutenant   **i:** *4th (Fourth) Officer/4e (Quatrième) Officier*, Acting Sub Lieutenant
j: *Officer Cadet/Élève-officier*, Midshipman

Chile (Armada de Chile)

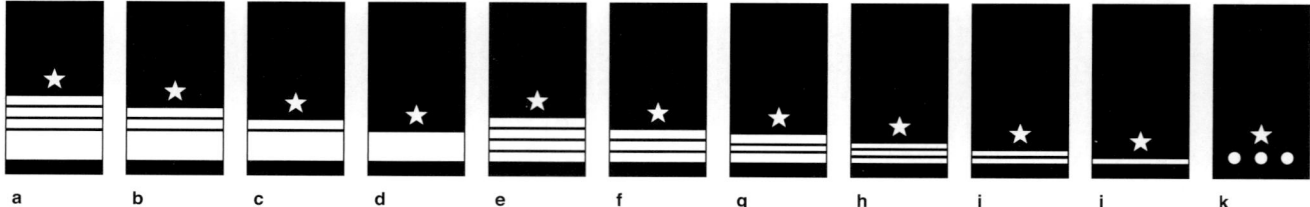

Gold wire, gold braid rings and brass buttons on very dark blue cloth cuffs. Rank titles are in Spanish.
a: *Almirante*, Admiral *(Commander-in Chief, Navy)*   **b:** *Vicealmirante*, Vice Admiral   **c:** *Contraalmirante*, Rear Admiral   **d:** *Comodoro*, Commodore
e: *Capitán de Navío*, Captain   **f:** *Capitán de Fragata*, Commander   **g:** *Capitán de Corbeta*, Lieutenant Commander   **h:** *Teniente 1° (Primero)*, Lieutenant
i: *Teniente 2° (Segundo)*, Sub Lieutenant   **j:** *Subteniente*, Acting Sub Lieutenant   **k:** *Guardiamarina*, Midshipman

China (People's Liberation Army Navy) (Zhōngguó Rénmín Jiěfàngjūn Hǎijūn)

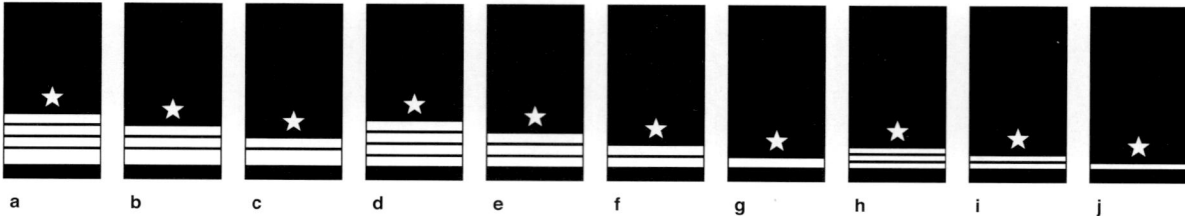

Gold wire stars and gold braid rings on dark blue cloth cuffs. Rank titles are in romanised Mandarin Chinese written in 'Hanyu Pinyin'. The Chinese Border Guard includes a Coast Guard.
a: *Hǎijūn Shangjiang*, Admiral *(Commander-in-Chief, Navy)*   **b:** *Hǎijūn Zhōngjiang*, Vice Admiral   **c:** *Hǎijūn Shaojiang*, Rear Admiral   **d:** *Hǎijūn Daxiao*, Commodore
e: *Hǎijūn Shangxiao*, Captain   **f:** *Hǎijūn Zhōngxiao*, Commander   **g:** *Hǎijūn Shaoxiao*, Lieutenant Commander   **h:** *Hǎijūn Shangwei*, Lieutenant
i: *Hǎijūn Zhōngwei*, Sub Lieutenant   **j:** *Hǎijūn Shaowei*, Acting Sub Lieutenant

Ranks and insignia of the world's navies [33]

Colombia (Armada Nacional de Colombia)

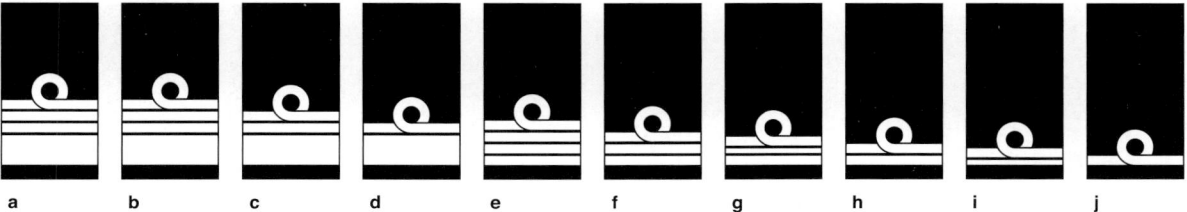

Gold braid rings with 'curl' on very dark blue cloth cuffs. Rank titles are in Spanish. The Colombian Navy maintains a small Coast Guard *(Guardacosta)*.
a: *Almirante*, Admiral *(rank not currently held)*   **b:** *Almirante de Esquadra*, Vice Admiral *(rank not currently held)*   **c:** *Vicealmirante*, Rear Admiral *(Commander, Navy)*
d: *Contraalmirante*, Commodore   **e:** *Capitán de Navío*, Captain   **f:** *Capitán de Fragata*, Commander   **g:** *Capitán de Corbeta*, Lieutenant Commander
h: *Teniente de Navío*, Lieutenant   **i:** *Teniente de Fragata*, Sub Lieutenant   **j:** *Teniente de Corbeta*, Acting Sub Lieutenant

Democratic Republic of Congo (Marine Nationale Congolaise)

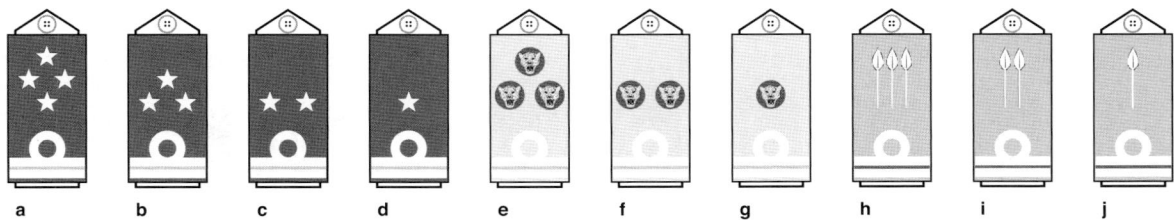

Gold metal stars, leopards' heads on red discs and spears above two medium gold braid rings with a 'curl' on the top ring on red cloth shoulder-loops with blue cloth between the rings (a-d), yellow cloth shoulder-loops with blue cloth between the rings (e-g) and blue cloth shoulder-loops with red cloth between the rings (h-j). Rank titles are in French.
a: *Grand Amiral*, Admiral   **b:** *Amiral*, Vice Admiral   **c:** *Vice-Amiral*, Rear Admiral   **d:** *Contre-Amiral*, Commodore   **e:** *Capitaine de Vaisseau*, Captain
f: *Capitaine de Frégate*, Commander   **g:** *Capitaine de Corvette*, Lieutenant Commander   **h:** *Lieutenant de Vaisseau*, Lieutenant
i: *Enseigne de vaisseau de 1ère (première) classe*, Sub Lieutenant   **j:** *Enseigne de vaisseau de 2e (deuxième) classe*, Acting Sub Lieutenant

Congo-Brazzaville (Marine Nationale du Congo)

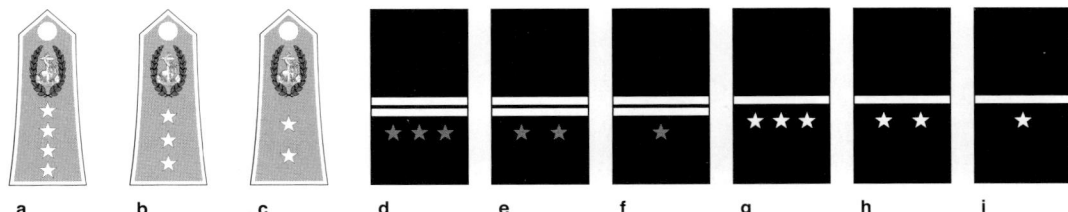

Flag-officers (a-c) wear the same gold wire cuff embroidery; also gold braid shoulder straps, rank stars, lion and anchor with green wreath and brass buttons. Other officers wear gold rings on very dark blue cuffs with red (d-f) and gold (g-i) stars and rank insignia simultaneously on very dark blue shoulder straps. Rank titles are in French.
a: *Vice-amiral d'escadre*, Vice Admiral *(rank not currently held)*   **b:** *Vice-amiral*, Rear Admiral *(rank not currently held)*
c: *Contre-amiral*, Commodore *(Commander, Navy)*   **d:** *Capitaine de vaisseau*, Captain   **e:** *Capitaine de frégate*, Commander
f: *Capitaine de corvette*, Lieutenant Commander   **g:** *Lieutenant de vaisseau*, Lieutenant   **h:** *Enseigne de vaisseau de 1ère (première) classe*, Sub Lieutenant
i: *Enseigne de vaisseau de 2e (deuxième) classe*, Acting Sub Lieutenant

Costa Rica (Coast Guard) (Servicio Nacional de Guardacostas)

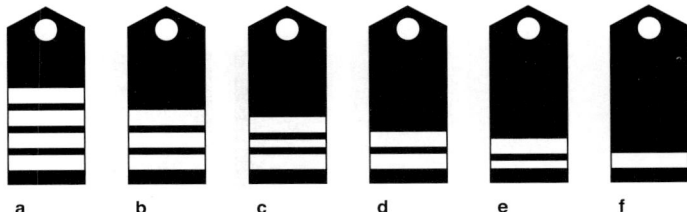

Gold braid rings on very dark blue shoulder-straps with brass buttons. Rank titles are in Spanish.
a: *Comisario*, Captain *(Director, Coast Guard)*   **b:** *Comisionado*, Commander   **c:** *Comandante*, Lieutenant Commander   **d:** *Capitán*, Lieutenant
e: *Intendente*, Sub Lieutenant   **f:** *Subintendente*, Acting Sub Lieutenant

Côte d'Ivoire (Marine Nationale de la Côte d'Ivoire)

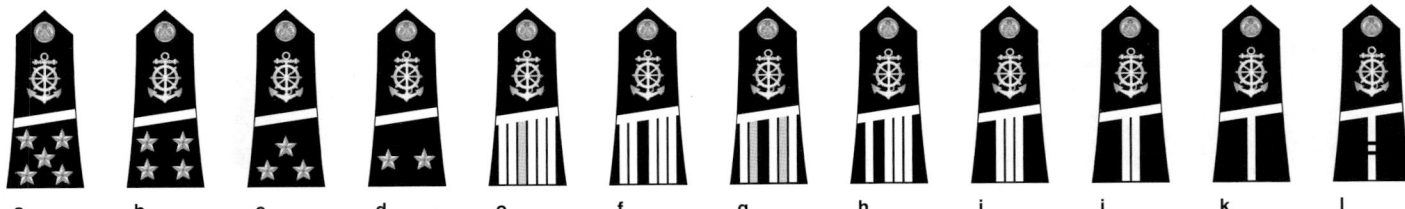

Silver lions, anchors and stars on very dark blue cloth shoulder-straps edged gold (a-d); gold anchors, stars and rings on very dark blue cloth shoulder-straps (e-k), brass buttons; a Senior Captain (e) has a silver braid third ring and a Commander (g) has silver braid first and second rings, a Midshipman (l) a gold braid bar with two mid-blue 'breaks'. Rank titles are in French.
a: *Amiral*, Admiral *(rank not currently held)*   **b:** *Vice-amiral d'escadre*, Vice Admiral *(rank not currently held)*   **c:** *Vice-amiral*, Rear Admiral *(rank not currently held)*
d: *Contre-amiral*, Commodore *(Commander, Navy)*   **e:** *Capitaine de vaisseau major*, (Senior) Captain   **f:** *Capitaine de vaisseau*, Captain
g: *Capitaine de frégate*, Commander   **h:** *Capitaine de corvette*, Lieutenant Commander   **i:** *Lieutenant de vaisseau*, Lieutenant
j: *Enseigne de vaisseau de 1ère (première) classe*, Sub Lieutenant   **k:** *Enseigne de vaisseau de 2e (deuxième) classe*, Acting Sub Lieutenant
l: *Aspirant*, Midshipman

[34] Ranks and insignia of the world's navies

Croatia (Hrvatska Ratna Mornarica)

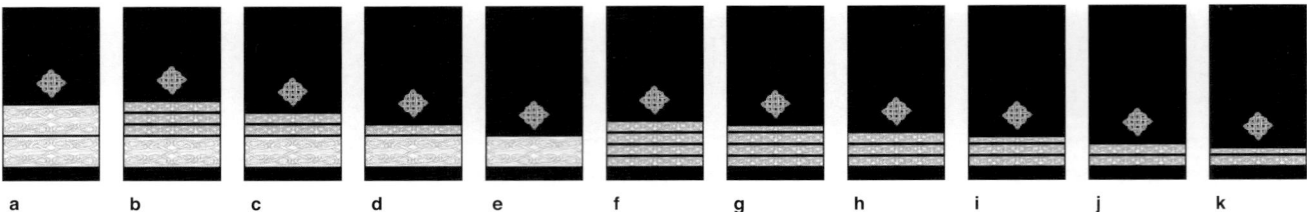

Gold wire rhomboids and gold braid rings on very dark blue cuffs. Rank titles are in Croatian. The Coast Guard *(Obalna straza)* forms part of the Navy, wears Navy uniforms and insignia and is commanded by a Captain seconded from the Navy.
a: *Stožerni admiral*, Admiral of the Fleet *(rank not currently held)*   **b:** *Admiral*, Admiral *(rank not currently held)*   **c:** *Viceadmiral*, Vice Admiral *(rank not currently held)*
d: *Kontraadmiral*, Rear Admiral *(rank not currently held)*   **e:** *Komodor*, Commodore *(Commander, Navy)*   **f:** *Kapetan bojnog broda*, Captain
g: *Kapetan fregate*, Commander   **h:** *Kapetan korvete*, Lieutenant Commander   **i:** *Poručnik bojnog broda*, Lieutenant   **j:** *Poručnik fregate*, Sub Lieutenant
k: *Poručnik korvete*, Acting Sub Lieutenant

Cuba (Marina de Guerra Revolucionaria)

Large gold stars with black anchors on red bosses and silver rays on black braid shoulder-loops (a–c); gold stars, chevrons and bars on black cloth shoulder-loops on white shoulder-straps; white bone buttons. Rank titles are in Spanish. The Cuban Border Guard includes a small Coast Guard.
a: *Almirante*, Vice Admiral *(rank not currently held)*   **b:** *Vicealmirante*, Rear Admiral *(Commander, Navy)*   **c:** *Contralmirante*, Commodore
d: *Capitán de Navío*, Captain   **e:** *Capitán de Fragata*, Commander   **f:** *Capitán de Corbeta*, Lieutenant Commander   **g:** *Teniente de Navío*, Lieutenant
h: *Teniente de Fragata*, (Senior) Sub Lieutenant   **i:** *Teniente de Corbeta*, Sub Lieutenant   **j:** *Alférez*, Acting Sub Lieutenant

Cyprus (Nautike Thiikissi Kiprou)

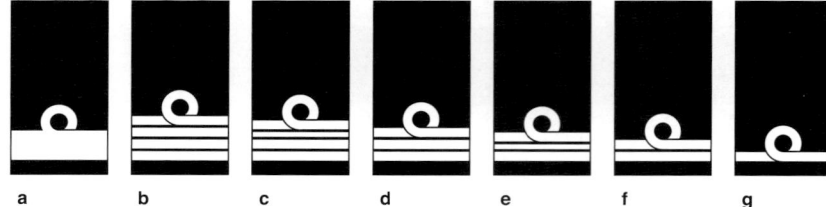

Gold braid rings with 'curl' on very dark blue cloth cuffs. Rank titles in romanised Greek. Cyprus also maintains a small Coast Guard entitled the Cyprus Port and Marine Police *(Limeniki & Nautiki Astinomia)* wearing dark blue Cyprus Police uniforms and insignia.
a: *Archiploiarchos*, Commodore *(rank not currently held)*   **b:** *Ploiarchos*, Captain *(Commander, Navy)*   **c:** *Antiploiarchos*, Commander
d: *Plotarchis*, Lieutenant Commander   **e:** *Ipoploiarchos*, Lieutenant   **f:** *Antipoploiarchos*, Sub Lieutenant   **g:** *Simaioforos*, Acting Sub Lieutenant

Denmark (Kongelige Danske Marine)

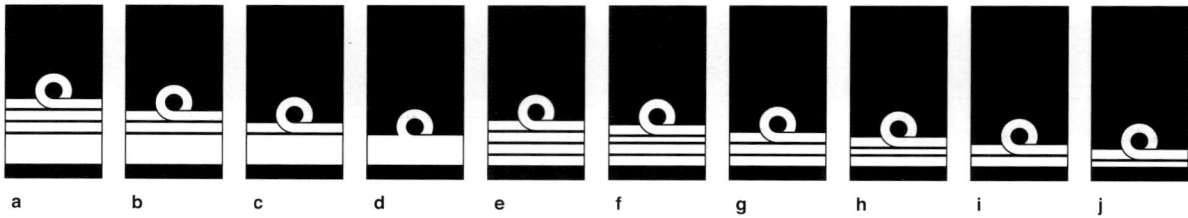

Gold braid rings with 'curl' on black cuffs. Rank titles are in Danish. The Danish Coast Guard *(Farlandsvæsenet)* is commanded by a Director-General.
a: *Admiral*, Admiral *(rank not currently held)*   **b:** *Viceadmiral*, Vice Admiral *(rank not currently held)*   **c:** *Kontreadmiral*, Rear Admiral *(Chief of the Navy)*
d: *Flotilleadmiral*, Commodore   **e:** *Kommandør*, Captain   **f:** *Kommandørkaptajn*, Commander   **g:** *Orlogskaptajn*, Lieutenant Commander
h: *Kaptajnløjtnant*, Lieutenant   **i:** *Premierløjtnant*, Sub Lieutenant   **j:** *Løjtnant*, Acting Sub Lieutenant

Djibouti (Marine Nationale Djiboutienne)

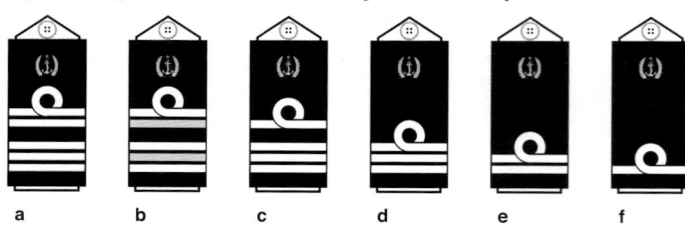

Gold wire fouled anchor and palm branches and red star above gold braid rings with 'curl' on very dark blue shoulder-loops on white shoulder-straps with white horn buttons; a Commander (b) has silver braid second and fourth rings. Army rank titles are used and are in French.
a: *Colonel*, Captain *(Commander, Navy)*   **b:** *Lieutenant-colonel*, Commander   **c:** *Commandant*, Lieutenant Commander   **d:** *Capitaine*, Lieutenant
e: *Lieutenant*, Sub-Lieutenant   **f:** *Sous-lieutenant*, Acting Sub-Lieutenant

IHS Jane's Fighting Ships 2014-2015

Dominican Republic (Marina de Guerra Dominicana)

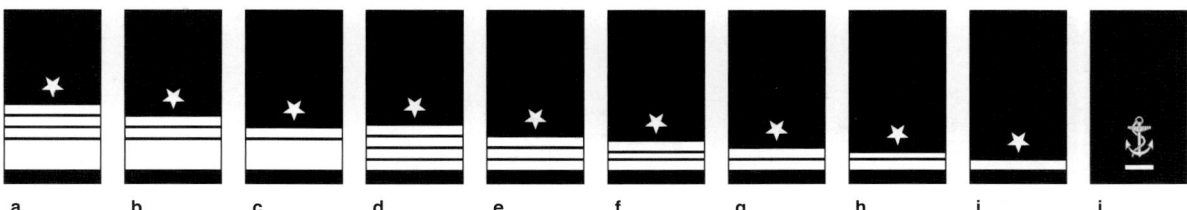

Gold wire stars and anchor and gold braid rings and bar on very dark blue cuffs. Rank titles are in Spanish.
a: *Almirante*, Admiral *(rank not currently held)*   **b:** *Vicealmirante*, Vice Admiral *(Commander, Navy)*   **c:** *Contralmirante*, Rear Admiral   **d:** *Capitán de Navío*, Captain
e: *Capitán de Fragata*, Commander   **f:** *Capitán de Corbeta*, Lieutenant Commander   **g:** *Teniente de Navío*, Lieutenant   **h:** *Alférez de Navío*, Sub Lieutenant
i: *Alférez de Fragata*, Acting Sub Lieutenant   **j:** *Guardiamarina*, Midshipman, 1st Year of training *(higher ranks 2–5 bars)*

East Timor Defence Force Naval Component (Contingente Naval, Forças de Defesa de Timor Leste)

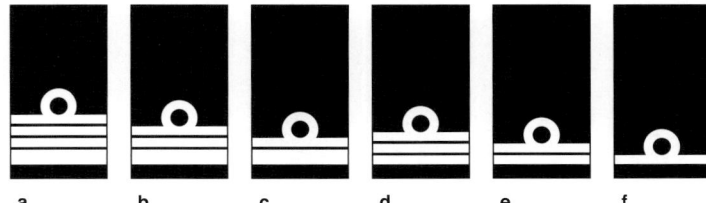

Gold braid rings with 'curl' on very dark blue cloth cuffs. Rank titles are in Portuguese.
a: *Capitão-de-mar-e-guerra*, Captain *(Commander, Naval Component)*   **b:** *Capitão-de-fragata*, Commander   **c:** *Capitão-tenente*, Lieutenant Commander
d: *Primeiro-tenente*, Lieutenant   **e:** *Segundo-tenente*, Sub Lieutenant   **f:** *Subtenente*, Acting Sub Lieutenant & *Guarda-marinha*, Midshipman

Ecuador (Armada del Ecuador)

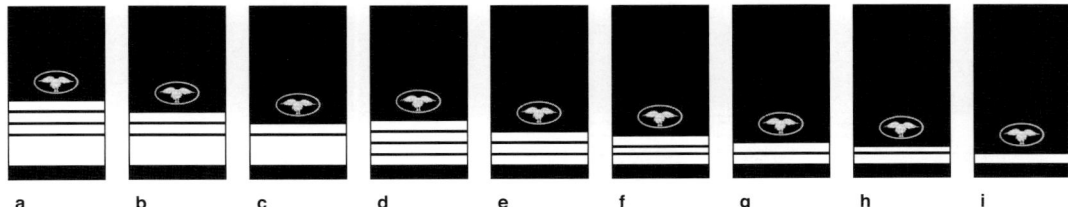

Gold wire condor in a very dark blue cloth oval edged in gold wire and gold braid rings on very dark blue cuffs. Rank titles are in Spanish.
a: *Almirante*, Admiral *(rank not currently held)*   **b:** *Vicealmirante*, Vice Admiral *(Commandant-General, Navy)*   **c:** *Contralmirante*, Rear Admiral
d: *Capitán de Navío*, Captain   **e:** *Capitán de Fragata*, Commander   **f:** *Capitán de Corbeta*, Lieutenant Commander   **g:** *Teniente de Navío*, Lieutenant
h: *Teniente de Fragata*, Sub Lieutenant   **i:** *Alférez de Fragata*, Acting Sub Lieutenant

Egypt (Egyptian Navy)

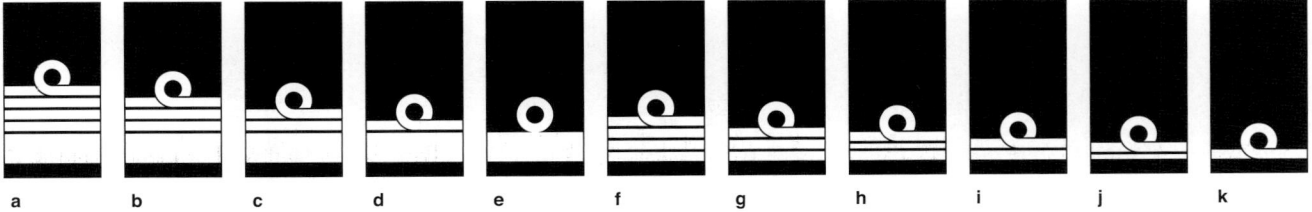

Gold braid rings with 'curl' on very dark blue cloth cuffs. Rank insignia is worn simultaneously on shoulder-straps. Arabic Egyptian Army rank titles are used and written here in romanised script.
a: *Mushīr*, Admiral of the Fleet *(rank not currently held)*   **b:** *Farīq Awwal*, Admiral *(rank not currently held)*   **c:** *Farīq*, Vice Admiral *(Commander, Navy)*
d: *Liwā'*, Rear Admiral   **e:** *'Amīd*, Commodore   **f:** *'Aqīd*, Captain   **g:** *Muqaddam*, Commander   **h:** *Rā'id*, Lieutenant Commander   **i:** *Naqīb*, Lieutenant
j: *Mulāzim Awwal*, Sub Lieutenant   **k:** *Mulāzim Thāni*, Acting Sub Lieutenant

El Salvador (Fuerza Naval de El Salvador)

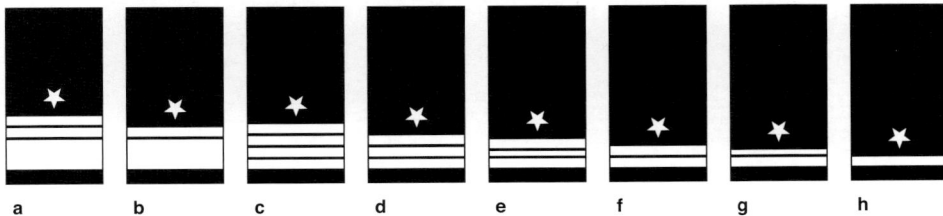

Gold wire stars and gold braid rings on very dark blue cuffs. Rank titles are in Spanish.
a: *Vice-Almirante*, Vice Admiral *(rank not currently held)*   **b:** *Contraalmirante*, Rear Admiral *(Inspector-General, Armed Forces)*
c: *Capitán de Navío*, Captain *(Chief of Naval Staff)*   **d:** *Capitán de Fragata*, Commander   **e:** *Capitán de Corbeta*, Lieutenant Commander
f: *Teniente de Navío*, Lieutenant   **g:** *Teniente de Fragata*, Sub Lieutenant   **h:** *Teniente de Corbeta*, Acting Sub Lieutenant

[36] Ranks and insignia of the world's navies

Equatorial Guinea (Marina de Guerra de Guinea Ecuatorial)

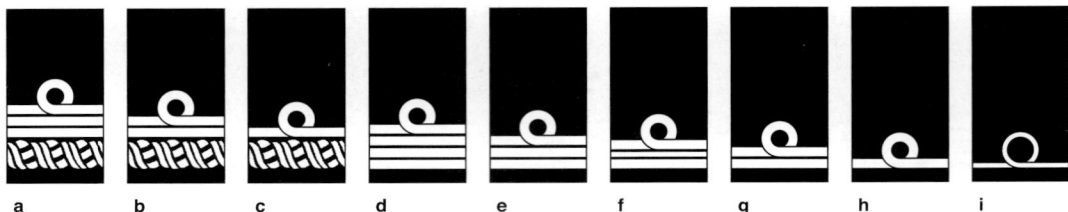

Gold gold braid rings with 'curl' and traditional Spanish generals' embroidery on very dark blue cloth cuffs. Rank titles are in Spanish.
a: *Almirante*, Vice Admiral *(President Obiong Nguema Mbasogo)* **b:** *Vicealmirante*, Rear Admiral *(Commander, Navy)* **c:** *Contraalmirante*, Commodore
d: *Capitán de Navío*, Captain **e:** *Capitán de Fragata*, Commander **f:** *Capitán de Corbeta*, Lieutenant Commander **g:** *Teniente de Navío*, Lieutenant
h: *Alférez de Navío*, Sub Lieutenant **i:** *Alférez de Fragata*, Acting Sub Lieutenant

Eritrea (Hayle Bahrie)

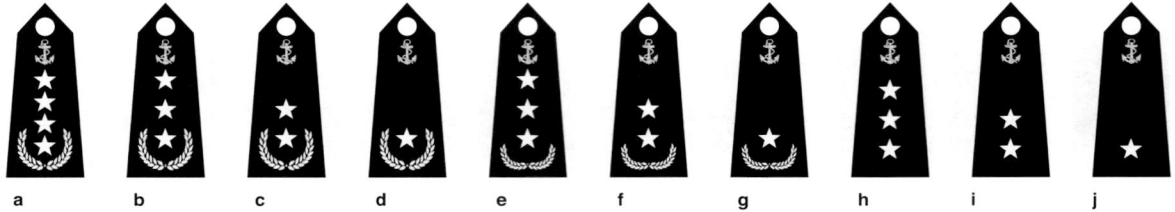

Gold metal buttons, fouled anchors and wreaths on very dark blue cloth shoulder-straps. Tigrinya rank-titles not available.
a: Admiral *(rank not currently held)* **b:** Vice Admiral *(rank not currently held)* **c:** Rear Admiral *(Commander, Navy)* **d:** Commodore **e:** Captain **f:** Commander
g: Lieutenant Commander **h:** Lieutenant **i:** Sub Lieutenant **j:** Acting Sub Lieutenant

Estonia (Eesti Merevägi)

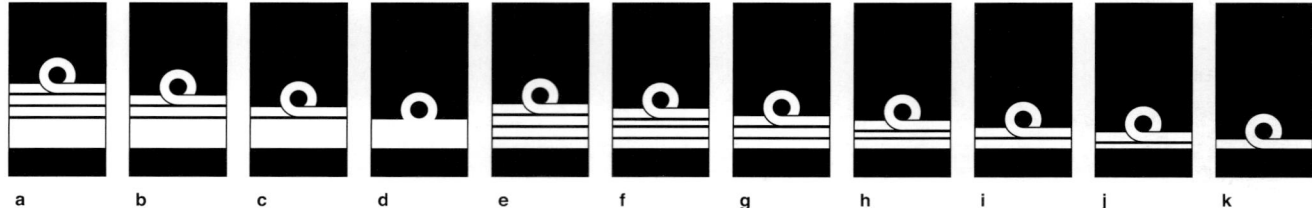

Gold braid rings with 'curl' on very dark blue cloth cuffs. Rank titles are in Estonian. The Estonian Border Guard includes a Coast Guard.
a: *Admiral*, Admiral *(rank not currently held)* **b:** *Viitseadmiral*, Vice Admiral *(rank not currently held)* **c:** *Kontradmiral*, Rear Admiral *(rank not currently held)*
d: *Kommodoor*, Commodore *(rank not currently held)* **e:** *Mereväekapten*, Captain *(Chief of Staff, Armed Forces; Commander, Navy)*
f: *Kaptenleitnant*, Commander **g:** *Kaptenmajor*, Lieutenant Commander **h:** *Vanemleitnant*, Lieutenant **i:** *Leitnant*, (Senior) Sub Lieutenant
j: *Nooremleitnant*, Sub Lieutenant **k:** *Lipnik*, Acting Sub Lieutenant

Fiji (Fijian Navy)

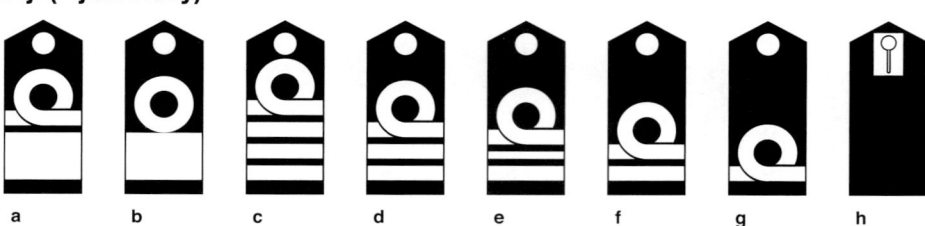

Gold braid rings with 'curl' on very dark blue cloth shoulder-straps; brass button and white cord on white cloth rectangular patch (h). Royal Navy rank titles are used.
a: Rear Admiral *(rank not currently held)* **b:** Commodore *(Commander, Royal Fiji Military Forces)* **c:** Captain *(Deputy Commander, RFMF)*
d: Commander *(Commanding Officer, Navy)* **e:** Lieutenant Commander **f:** Lieutenant **g:** Sub Lieutenant **h:** Midshipman

Finland (Suomen Merivoimat/Finska Marinen)

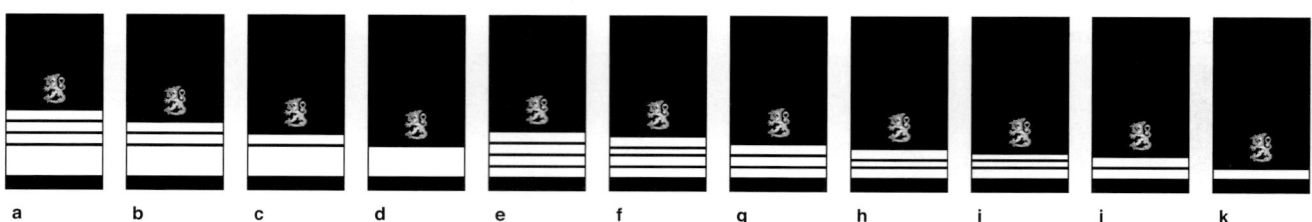

Gold wire Finnish heraldic lions and gold braid rings on very dark blue cuffs. Rank titles are in Finnish/Swedish. The Finnish Coast Guard *(Merivartiosto/Sjöbevakning)* forms part of the Interior Ministry. Personnel wear naval uniforms and insignia with Captain *(Kommodori/Kommodor)* as the highest rank.
a: *Amiraali/Amiral*, Admiral *(rank not currently held)* **b:** *Vara-amiraali/Viceamiral*, Vice Admiral *(Chief of Defence Command)*
c: *Kontra-amiraali/Konteramiral*, Rear Admiral *(Commander, Navy)* **d:** *Lippueamiraali/Flottiljamiral*, Commodore **e:** *Kommodori/Kommodor*, Captain
f: *Komentaja/Kommendör*, Commander **g:** *Komentajakapteeni/Kommendörkapten*, Lieutenant Commander **h:** *Kapteeniluutnantti/Kaptenlöjtnant*, Lieutenant
i: *Yliluutnantti/Premiärlöjtnant*, (Senior) Sub Lieutenant **j:** *Luutnantti/Löjtnant*, Sub Lieutenant **k:** *Aliluutnantti/Underlöjtnant*, Acting Sub Lieutenant

Ranks and insignia of the world's navies

France (Marine Nationale)

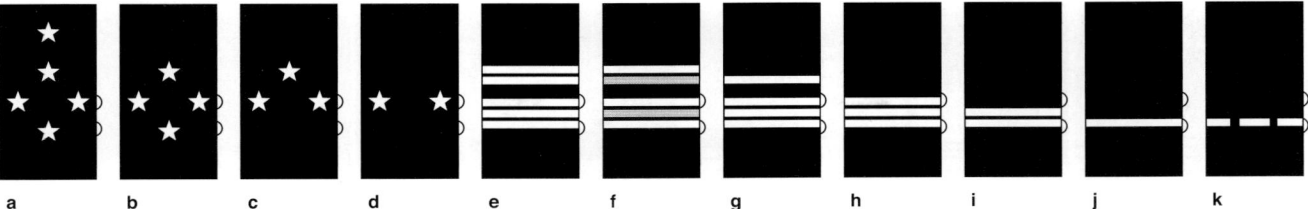

Silver metal stars and gold braid rings on very dark blue cloth cuffs, brass buttons; a Commander (f) has silver braid second and fourth rings, a Midshipman (k) a gold braid ring with two mid-blue 'breaks'. Rank titles are in French.
a: *Amiral*, Admiral *(Chief of Defence Staff, Chief of Naval Staff)* **b:** *Vice-amiral d'escadre*, Vice Admiral **c:** *Vice-amiral*, Rear Admiral
d: *Contre-amiral*, Commodore **e:** *Capitaine de vaisseau*, Captain **f:** *Capitaine de frégate*, Commander **g:** *Capitaine de corvette*, Lieutenant Commander
h: *Lieutenant de vaisseau*, Lieutenant **i:** *Enseigne de vaisseau de 1ère (première) classe*, Sub Lieutenant
j: *Enseigne de vaisseau de 2e (deuxième) classe*, Acting Sub Lieutenant **k:** *Aspirant*, Midshipman

Gabon (Marine Nationale Gabonaise)

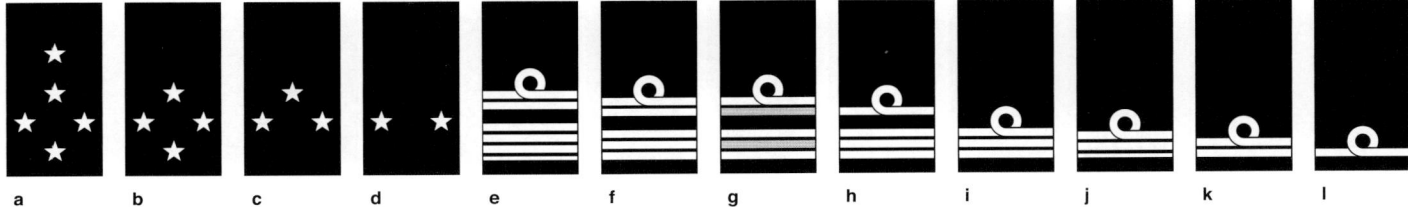

Gold metal stars and gold braid rings with 'curl' on very dark blue cloth cuffs; a Commander (f) has silver braid second and fourth rings. Senior officers in five ranks (f–i and k) wear a thin gold ring below their rank rings and their rank-titles are suffixed with 'major'; two ranks (e, j) are shown here as examples. Rank titles are in French.
a: *Amiral*, Admiral *(rank not currently held)* **b:** *Vice-amiral d'escadre*, Vice Admiral *(rank not currently held)* **c:** *Vice-amiral*, Rear Admiral *(Chief of Naval Staff)*
d: *Contre-amiral*, Commodore **e:** *Capitaine de vaisseau major*, (Senior) Captain **f:** *Capitaine de vaisseau*, Captain **g:** *Capitaine de frégate*, Commander
h: *Capitaine de corvette*, Lieutenant Commander **i:** *Lieutenant de vaisseau*, Lieutenant **j:** *Enseigne major*, (Senior) Sub Lieutenant
k: *Enseigne de vaisseau de 1ère (première) classe*, Sub Lieutenant **l:** *Enseigne de vaisseau de 2e (deuxième) classe*, Acting Sub Lieutenant

Gambia (Gambian Navy)

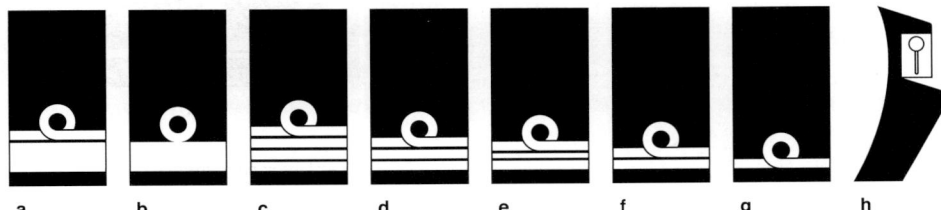

Gold braid rings with 'curl' on very dark blue cloth cuffs; brass button and white cord on white cloth collar-patch (h). British Royal Navy rank titles are used.
a: Rear Admiral **b:** Commodore *(Chief of Naval Staff)* **c:** Captain **d:** Commander **e:** Lieutenant Commander **f:** Lieutenant **g:** Sub Lieutenant
h: Midshipman

Georgia (Coast Guard)

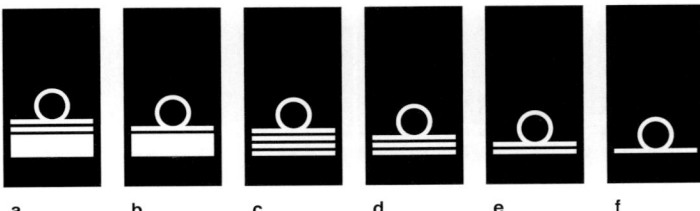

The Georgian Coast Guard forms part of the Border Guard Service under the Ministry of the Interior. Gold wire stars and bars with 'curl' on dark blue cloth cuffs. Rank titles are in Georgian written here in romanised script.
a: *Ge-2 (Meore) rangis kapitani*, Commander *(Commander, Coast Guard)* **b:** *Ge-3 (Mesame) rangis kapitani*, Lieutenant Commander
c: *Kapitan-leitenanti*, Lieutenant **d:** *Ufrosi leitenanti*, (Senior) Sub Lieutenant **e:** *Leitenanti*, Sub Lieutenant **f:** *Umcrosi leitenanti*, Acting Sub Lieutenant

Germany (Deutsche Marine)

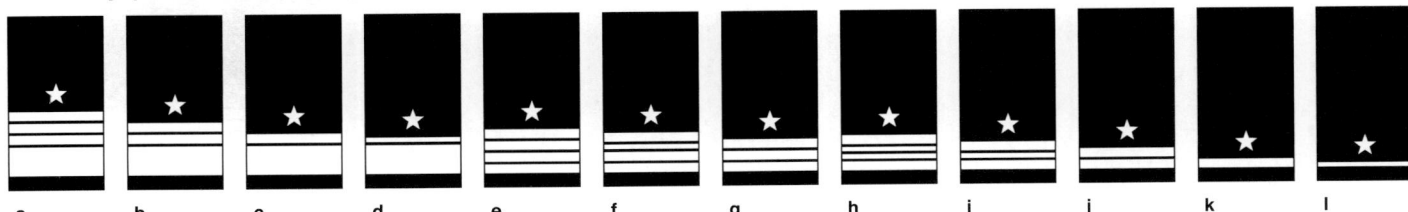

Gold wire stars and gold braid rings on very dark blue cloth cuffs. Rank titles are in German.
a: *Admiral*, Admiral *(rank not currently held)* **b:** *Vizeadmiral*, Vice Admiral *(Naval Inspector)* **c:** *Konteradmiral*, Rear Admiral **d:** *Flottillenadmiral*, Commodore
e: *Kapitän zur See*, Captain **f:** *Fregattenkapitän*, Commander **g:** *Korvettenkapitän*, Lieutenant Commander **h:** *Stabskapitänleutnant*, (Senior) Lieutenant
i: *Kapitänleutnant*, Lieutenant **j:** *Oberleutnant zur See*, Sub Lieutenant **k:** *Leutnant zur See*, Acting Sub Lieutenant **l:** *Oberfähnrich zur See*, Midshipman

[38] Ranks and insignia of the world's navies

Germany (Coast Guard) (Küstenwache des Bundes)

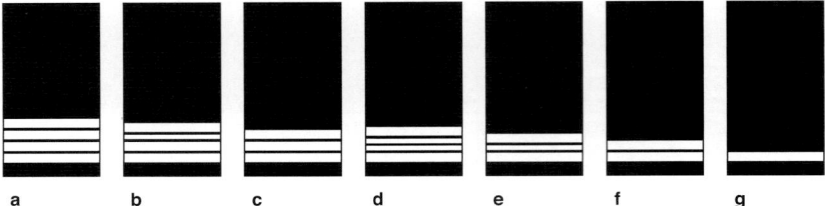

Gold braid rings on very dark blue cloth cuffs. Rank titles are in German.
a: *Polizeidirektor*, Captain *(Director, Coast Guard)* **b:** *Polizeioberrat*, Commander **c:** *Polizeirat*, Lieutenant Commander
d: *1. (Erster) Polizeihauptkommissar*, (Senior) Lieutenant **e:** *Polizeihauptkommissar*, Lieutenant **f:** *Polizeioberkommissar*, Sub Lieutenant
g: *Polizeikommissar*, Acting Sub Lieutenant

Ghana (Ghana Navy)

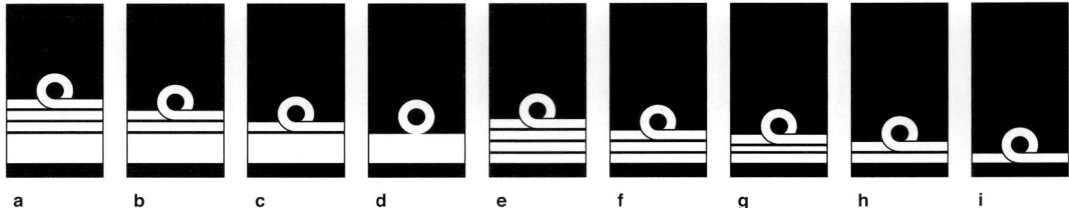

Gold braid rings with 'curl' on very dark blue cloth cuffs. British Royal Navy rank titles are used.
a: Admiral *(rank not currently held)* **b:** Vice Admiral *(rank not currently held)* **c:** Rear Admiral *(Chief of Naval Staff)* **d:** Commodore **e:** Captain **f:** Commander
g: Lieutenant Commander **h:** Lieutenant **i:** Sub Lieutenant & Acting Sub Lieutenant

Greece (Hellenic Navy) (Elliniko Polemiko Nautiko)

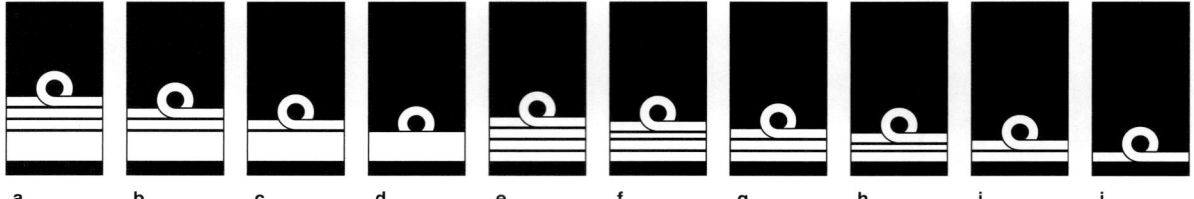

Gold braid rings with 'curl' on very dark blue cloth cuffs. Rank titles are in romanised Greek.
a: *Navarchos*, Admiral *(rank not currently held)* **b:** *Antinavarchos*, Vice Admiral *(Chief of Naval Staff)* **c:** *Yponavarchos*, Rear Admiral
d: *Archiploiarchos*, Commodore **e:** *Ploiarchos*, Captain **f:** *Antiploiarchos*, Commander **g:** *Plotarchis*, Lieutenant Commander **h:** *Ypoploiarchos*, Lieutenant
i: *Antipoploiarchos*, Sub Lieutenant **j:** *Simaioforos*, Acting Sub Lieutenant

Greece (Hellenic Coast Guard) (Limeniko Soma - Elliniki Aktofylaki)

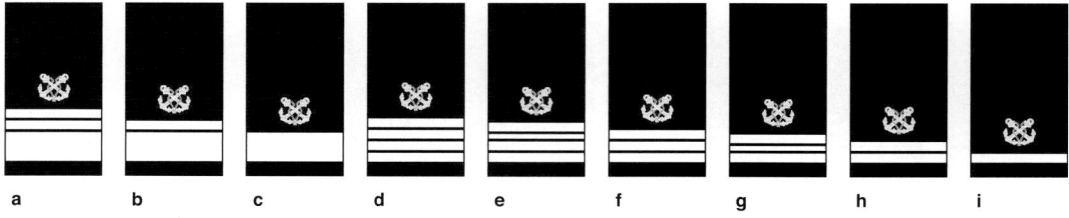

Gold wire crossed anchors and gold braid rings on very dark blue cloth cuffs. Hellenic Navy rank titles are used and written in romanised Greek.
a: *Antinavarchos*, Vice Admiral *(Commandant, Coast Guard)* **b:** *Yponavarchos*, Rear Admiral **c:** *Archiploiarchos*, Commodore **d:** *Ploiarchos*, Captain
e: *Antiploiarchos*, Commander **f:** *Plotarchis*, Lieutenant Commander **g:** *Ypoploiarchos*, Lieutenant **h:** *Antipoploiarchos*, Sub Lieutenant
i: *Simaioforos*, Acting Sub Lieutenant

Guatemala (Fuerza de Mar de Guatemala)

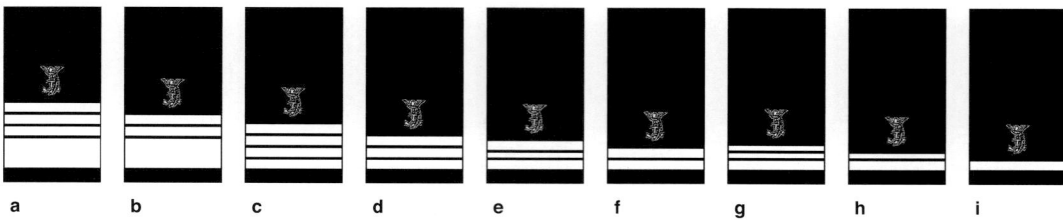

Gold wire Quetzal bird and anchor and gold braid rings on black cloth cuffs. Rank titles are in Spanish.
a: *Almirante*, Vice Admiral *(rank not currently held)* **b:** *Vicealmirante*, Rear Admiral *(Commanding Officer, Navy)* **c:** *Capitán de Navío*, Captain
d: *Capitán de Fragata*, Commander **e:** *Capitán de Corbeta*, Lieutenant Commander **f:** *Teniente de Navío*, (Senior) Lieutenant **g:** *Teniente de Fragata*, Lieutenant
h: *Alférez de Navío*, Sub Lieutenant **i:** *Alférez de Fragata*, Acting Sub Lieutenant

Ranks and insignia of the world's navies [39]

Guinea (Marine Nationale de Guinée)

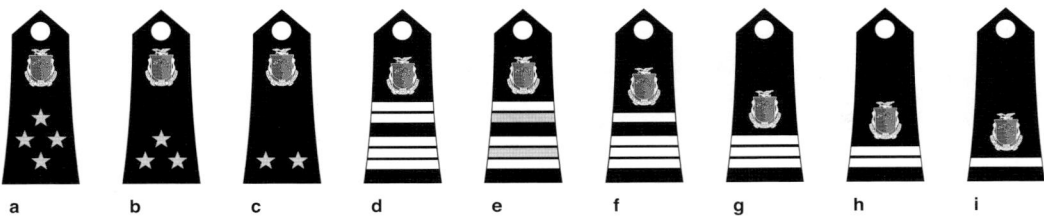

Silver metal stars, multicoloured metal coats of arms and gold braid rings on very dark blue cloth shoulder-straps; a Commander (e) has silver braid second and fourth rings; brass buttons. Rank titles are in French.
a: *Vice-amiral d'escadre*, Vice Admiral *(rank not currently held)* **b:** *Vice-amiral*, Rear Admiral *(rank not currently held)*
c: *Contre-amiral*, Commodore *(rank not currently held)* **d:** *Capitaine de vaisseau*, Captain **e:** *Capitaine de frégate*, Commander
f: *Capitaine de corvette*, Lieutenant Commander **g:** *Lieutenant de vaisseau*, Lieutenant **h:** *Enseigne de vaisseau 1ère (première) classe*, Sub Lieutenant
i: *Enseigne de vaisseau de 2e (deuxième) classe*, Acting Sub Lieutenant

Guinea-Bissau (Marinha de Guerra de Guiné-Bissau)

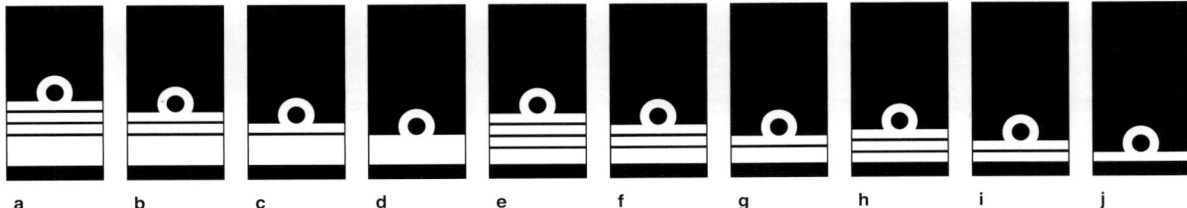

Gold braid rings with 'curl' on very dark blue cloth cuffs. Rank titles are in Portuguese.
a: *Almirante*, Admiral *(rank not currently held)* **b:** *Vice-almirante*, Vice Admiral *(rank not currently held)* **c:** *Contra-almirante*, Rear Admiral *(rank not currently held)*
d: *Comodoro*, Commodore *(rank not currently held)* **e:** *Capitão-de-mar-e-guerra*, Captain *(Acting Chief of Naval Staff)* **f:** *Capitão-de-fragata*, Commander
g: *Capitão-tenente*, Lieutenant Commander **h:** *Primeiro-tenente*, Lieutenant **i:** *Segundo-tenente*, Sub Lieutenant **j:** *Subtenente*, Acting Sub Lieutenant

Guyana (Guyana Defence Force Coast Guard)

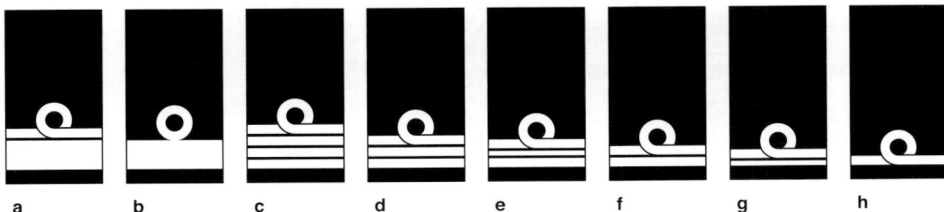

Gold braid rings with 'curl' on very dark blue cloth cuffs. British Royal Navy rank titles are used.
a: Rear Admiral *(rank not currently held)* **b:** Commodore *(rank not currently held)* **c:** Captain *(rank not currently held)* **d:** Commander *(Commander, Coast Guard)*
e: Lieutenant Commander **f:** Lieutenant **g:** Sub Lieutenant **h:** Acting Sub Lieutenant

Honduras (Fuerza Naval de Honduras)

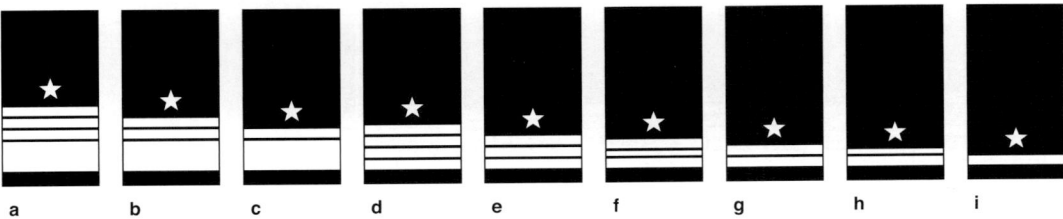

Gold wire stars and gold braid rings on very dark blue cuffs. Rank titles are in Spanish.
a: *Almirante*, Admiral *(rank not currently held)* **b:** *Vicealmirante*, Vice Admiral *(rank not currently held)* **c:** *Contralmirante*, Rear Admiral *(Commandant, Navy)*
d: *Capitán de Navío*, Captain **e:** *Capitán de Fragata*, Commander **f:** *Capitán de Corbeta*, Lieutenant Commander **g:** *Teniente de Navío*, Lieutenant
h: *Teniente de Fragata*, Sub Lieutenant **i:** *Alférez de Fragata*, Acting Sub Lieutenant

Hong Kong (Hong Kong Police Force Marine Region)

Silver-plated metal crossed tipstaves, wreaths, orchid-tree flowers in wreaths, Bath stars, bar, HKP shoulder-titles and buttons on dark blue cloth shoulder-straps. British Police Service rank titles are used.
a: *Assistant Commissioner*, Commodore *(C-in-C Marine Region)* **b:** *Chief Superintendent*, Captain **c:** *Senior Superintendent*, Commander
d: *Superintendent*, Lieutenant Commander **e:** *Chief Inspector*, Lieutenant **f:** *Senior Inspector*, (Senior) Sub Lieutenant **g:** *Inspector*, Sub Lieutenant
h: *Probationary Inspector*, Acting Sub Lieutenant

© 2014 IHS   IHS Jane's Fighting Ships 2014-2015

[40] Ranks and insignia of the world's navies

Iceland (Coast Guard) (Landhelgisgæsla Islands)

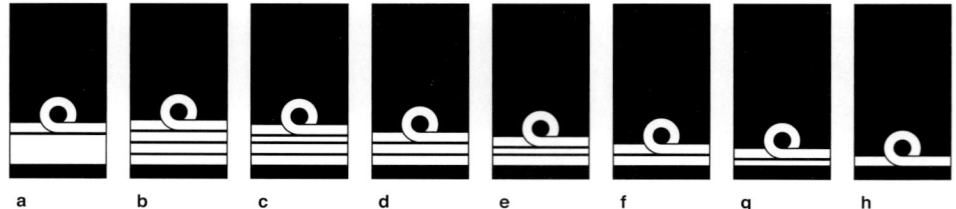

Gold braid rings with 'curl' on black cloth shoulder-straps. Rank titles are in Icelandic.
a: *Forstjóri Lanhelgisgæslu Islands*, Rear Admiral *(Director-General, Coast Guard)*   **b:** *Yfirmaður Gæsluframkvæmda*, Captain   **c:** *Skipherra 1°*, Commander
d: *Skipherra 2°*, Lieutenant Commander   **e:** *Yfirstýrimaður*, Lieutenant   **f:** *2. Stýrimaður*, Sub Lieutenant   **g:** *3. Stýrimaður*, (Senior) Acting Sub Lieutenant
h: *Foringjabyrjandi*, Acting Sub Lieutenant

India (Bharatiya Nau Sena)

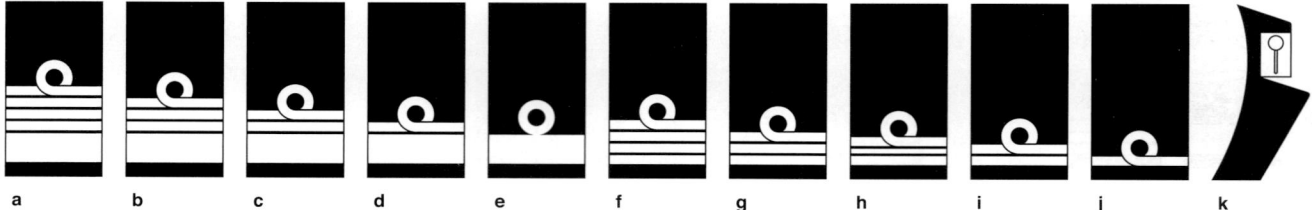

Gold braid rings with 'curl' on very dark blue cloth cuffs; brass button and white cord on white cloth collar-patch (k). British Royal Navy rank titles are used.
a: Admiral of the Fleet *(honorary rank not currently held)*   **b:** Admiral *(Chief of Naval Staff)*   **c:** Vice Admiral   **d:** Rear Admiral   **e:** Commodore   **f:** Captain
g: Commander   **h:** Lieutenant Commander   **i:** Lieutenant   **j:** Sub Lieutenant   **k:** Midshipman

India Coast Guard (Bharatiya Thatrakshak)

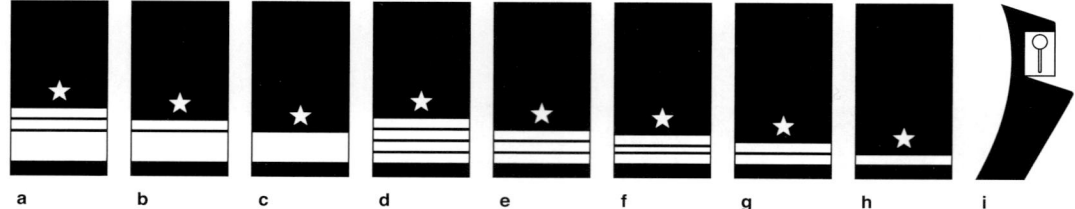

Gold wire stars and gold braid rings on very dark blue cuffs; brass button and white cord on white cloth collar-patch (i); brass buttons. The Director General is a serving
Vice Admiral of the Indian Navy. Rank titles are in English.
a: *Director General* and *Additional Director General*, Vice Admiral *(Director General, Coast Guard)*   **b:** *Inspector General*, Rear Admiral
c: *Deputy Inspector General*, Commodore   **d:** *Commandant*, Captain   **e:** *Commandant (Junior Grade)*, Commander
f: *Deputy Commandant*, Lieutenant Commander   **g:** *Assistant Commandant*, Lieutenant
h: *Assistant Commandant (under training after completion of Phase III afloat training)*, Sub Lieutenant
i: *Assistant Commandant (under training after completion of Phase II afloat training)*, Midshipman

Indonesia (Tentara Nasional Indonesia Angkatan Laut)

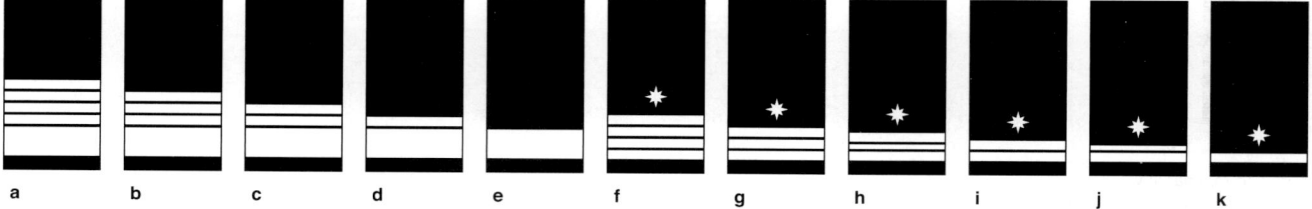

Gold wire stars and gold braid rings on very dark blue cuffs. Rank titles are in Indonesian. Indonesia maintains a 'Sea and Coast Guard' *(Kesatuan Penjaga Laut dan Pantai)*.
a: *Laksamana Besar*, Admiral of the Fleet *(wartime rank not currently held)*   **b:** *Laksamana*, Admiral *(Chief of Naval Staff)*   **c:** *Laksamana Madya*, Vice Admiral
d: *Laksamana Muda*, Rear Admiral   **e:** *Laksamana Pertama*, Commodore   **f:** *Kolonel*, Captain   **g:** *Letnan Kolonel*, Commander
h: *Mayor*, Lieutenant Commander   **i:** *Kapten*, Lieutenant   **j:** *Letnan Satu*, Sub Lieutenant   **k:** *Letnan Dua*, Acting Sub Lieutenant

Iran (Islamic Republic of Iran Navy)

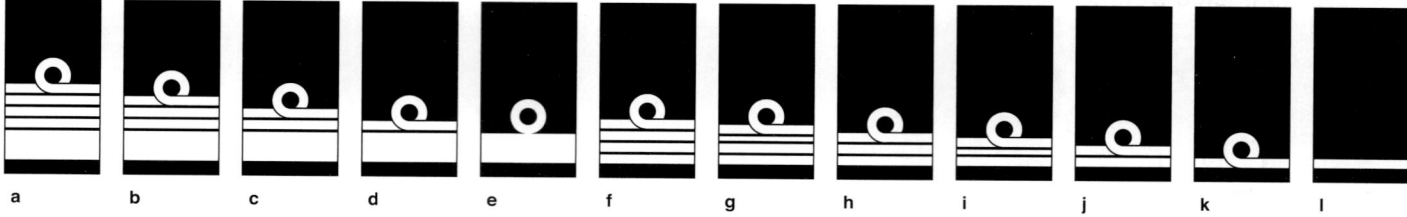

Gold braid rings with 'curl' on very dark blue cloth cuffs. Rank titles are in romanised Farsi. The Islamic Revolutionary Guard Corps *(Sepah-e Pasdaran-e Enqelab-e Eslami)* also maintains a Navy wearing Navy uniforms and shoulder-loop insignia, commanded by a Rear Admiral.
a: *Darybod*, Admiral *(rank not currently held)*   **b:** *Darysaklar*, Vice Admiral *(rank not currently held)*   **c:** *Daryban*, Rear Admiral *(Commander, Navy)*
d: *Darydar*, Commodore   **e:** *Darydar Dovom*, (Junior) Commodore   **f:** *Nakhoda Yekom*, Captain   **g:** *Nakhoda Dovom*, Commander
h: *Nakhoda Sevom*, Lieutenant Commander   **i:** *Navsarvan*, Lieutenant   **j:** *Navban Yekom*, Sub Lieutenant   **k:** *Navban Dovom*, Acting Sub Lieutenant
l: *Navban Sevom*, Midshipman

IHS Jane's Fighting Ships 2014-2015

Ranks and insignia of the world's navies [41]

Iraq (Al-Quwwat Al-Bahria Al-Iraq)

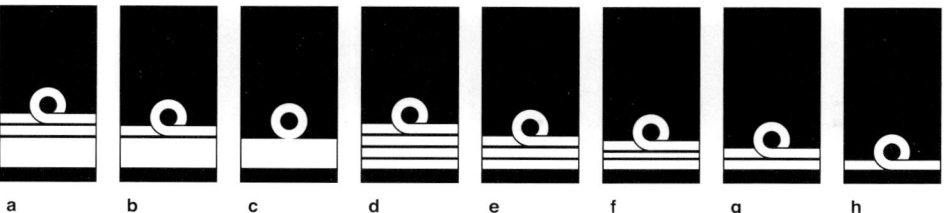

Gold braid rings with 'curl' on very dark blue cloth cuffs. Arabic Iraqi Army rank titles are used and written here in romanised script.
a: *Farīq'*, Vice Admiral *(Commanding Officer, Navy)*   **b:** *Liwā'*, Rear Admiral *(Commander, Navy)*   **c:** *'Amīd*, Commodore   **d:** *'Aqīd*, Captain
e: *Muqaddam*, Commander   **f:** *Rā'id*, Lieutenant Commander   **g:** *Naqīb*, Lieutenant   **h:** *Mulāzim Awwal*, Sub Lieutenant

Ireland (An Seirbhís Chabhlaigh na hÉireann)

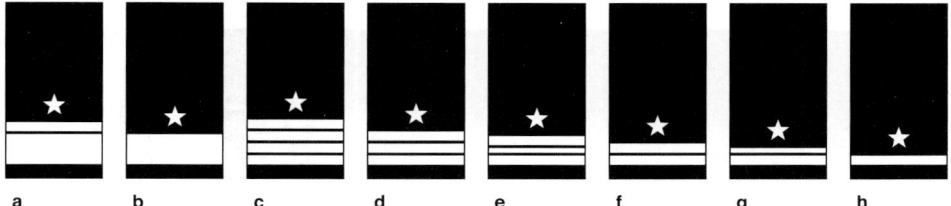

Gold wire stars and gold braid rings on very dark blue cuffs. Rank titles are in Irish/English. The Irish Coast Guard *(Garda Cósta na Éireann)*, commanded by a Director, is part of the Department of Transport; personnel wear naval style uniforms and rank insignia.
a: *Rear Admiral (Deputy Chief of Staff (Support))*   **b:** *Ceannasoir/Commodore*, Commodore *(Flag Officer Commanding Naval Service)*   **c:** *Captaen/Captain*, Captain
d: *Ceannasai/Commander*, Commander   **e:** *Lefteanant-Ceannasai/Lieutenant Commander*, Lieutenant Commander   **f:** *Lefteanant/Lieutenant*, Lieutenant
g: *Fo-Lefteanant/Sub Lieutenant*, Sub Lieutenant   **h:** *Meirgire/Ensign*, Acting Sub Lieutenant

Israel (Heil Ha Yam Ha Yisraeli)

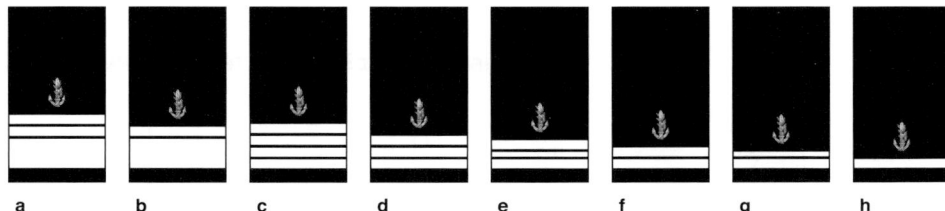

Gold wire anchor and leaf on very dark blue cloth cuffs. Israeli Army rank titles are used and the Hebrew is written here in romanised script.
a: *Alúf*, Vice Admiral *(Commander of the Navy)*   **b:** *Tat alúf*, Rear Admiral   **c:** *Alúf mishné*, Captain   **d:** *Sgan alúf*, Commander
e: *Rav séren*, Lieutenant Commander   **f:** *Séren*, Lieutenant   **g:** *Ségen*, Sub Lieutenant   **h:** *Ségen mishné*, Acting Sub Lieutenant

Italy (Marina Militare)

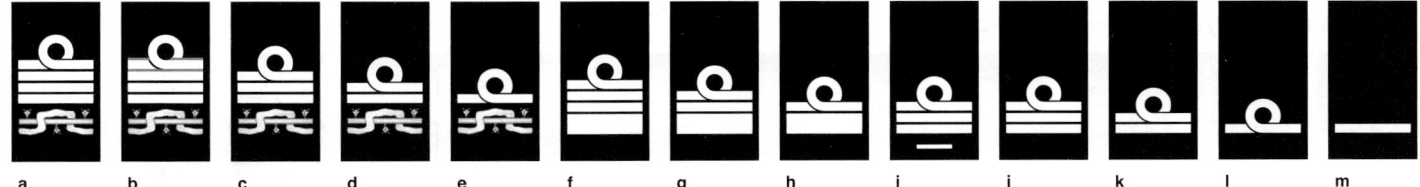

Gold braid bars and thin bar (i) with 'curl' and 'Greca' design on very dark blue cloth cuffs; upper and lower edges of top bar (but not the 'curl') edged in red cloth (b). Rank titles are in Italian. The Italian Coast Guard *(Capitanerie di Porto - Guardia Costiera)* forms part of the Navy. Personnel wear naval uniforms and insignia with the Commandant holding the rank of Vice-Admiral *(Ammiraglio di Squadra)*.
a: *Ammiraglio*, Admiral *(Chief of Defence Staff)*   **b:** *Ammiraglio di Squadra con Incarichi Speciali*, (Senior) Vice-Admiral *(Chief of Naval Staff)*
c: *Ammiraglio di Squadra*, Vice-Admiral   **d:** *Ammiraglio di Divisione*, Rear Admiral   **e:** *Contrammiraglio*, Commodore   **f:** *Capitano di Vascello*, Captain
g: *Capitano di Fregata*, Commander   **h:** *Capitano di Corvetta*, Lieutenant Commander   **i:** *1° (Primo) Tenente di Vascello*, (Senior) Lieutenant
j: *Tenente di Vascello*, Lieutenant   **k:** *Sottotenente di Vascello*, Sub Lieutenant   **l:** *Aspirante Guardiamarina*, Acting Sub Lieutenant
m: *Aspirante Guardiamarina*, Midshipman

Jamaica (Jamaica Defence Force Coast Guard)

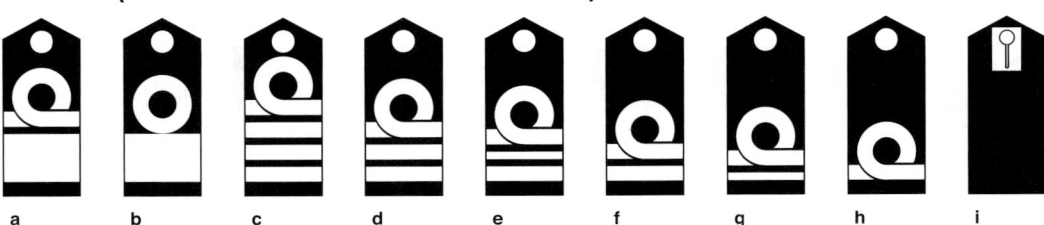

Gold braid rings with 'curl' on very dark blue cloth shoulder-straps with brass buttons; brass button and white cord on white rectangular cloth patch (i). Rank titles are in English.
a: *Rear Admiral*, Rear Admiral *(rank not currently held)*   **b:** *Commodore*, Commodore *(rank not currently held)*   **c:** *Captain (N)*, Captain *(rank not currently held)*
d: *Commander*, Commander *(Commander, Coast Guard)*   **e:** *Lieutenant Commander*, Lieutenant Commander   **f:** *Lieutenant senior grade*, Lieutenant
g: *Lieutenant junior grade*, Sub Lieutenant   **h:** *Sub Lieutenant*, Acting Sub Lieutenant   **i:** *Midshipman*, Midshipman

[48] Ranks and insignia of the world's navies

Oman (Al-Bahriyyat As-Sultaniyyat Al-'Umaniyyah)

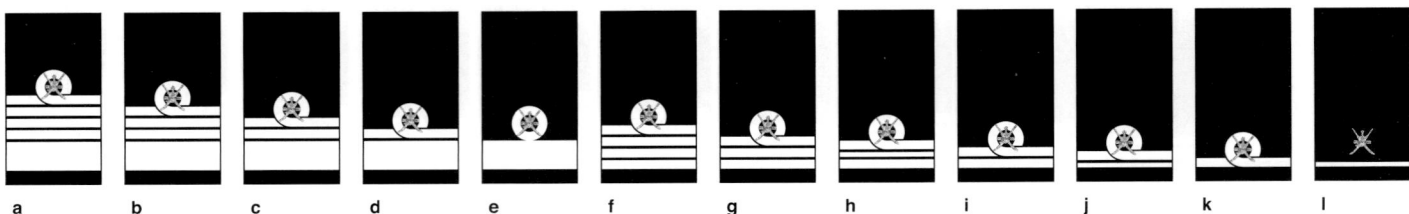

Gold wire dagger and crossed swords and gold braid rings with 'curl' on very dark blue cloth cuffs; a Midshipman (l) has a thin white ring. Arabic Oman Army rank titles are used and written here in romanised script. The Royal Oman Police Coast Guard is forms part of the Royal Oman Police wearing ROP uniforms and insignia.
a: *Mushīr*, Admiral of the Fleet *(Sultan Qabus ibn Sa'id)*   **b:** *Farīq Awwal*, Admiral *(rank not currently held)*   **c:** *Farīq*, Vice Admiral *(rank not currently held)*
d: *Liwā'*, Rear Admiral *(commander RNO)*   **e:** *'Amid*, Commodore   **f:** *'Aqīd*, Captain   **g:** *Muqaddam*, Commander   **h:** *Rā'id*, Lieutenant Commander
i: *Naqīb*, Lieutenant   **j:** *Mulāzim Awwal*, Sub Lieutenant   **k:** *Mulāzim Thāni*, Acting Sub Lieutenant   **l:** *Dābit Murashshah*, Midshipman

Pakistan (Pak Bahr'ya)

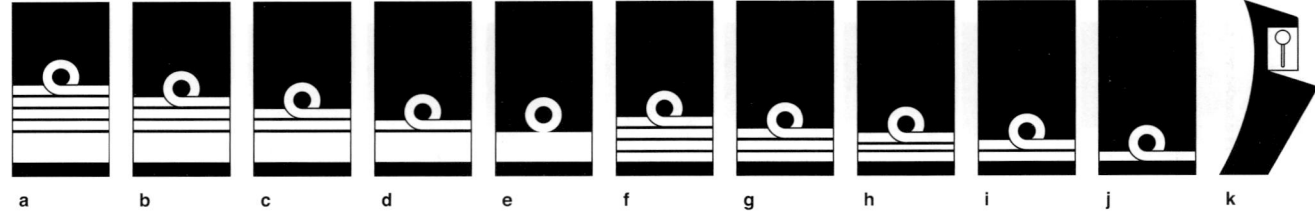

Gold braid rings with 'curl' on very dark blue cloth cuffs; brass button and white cord on white cloth collar-patch (k). British Royal Navy rank titles are used. The Pakistan Coast Guard is organised as a Pakistan Army reinforced brigade. Personnel wear Army uniforms and insignia with a seconded Army Brigadier as Director-General.
a: Admiral of the Fleet *(rank not currently held)*   **b:** Admiral *(Chief of the Naval Staff)*   **c:** Vice Admiral   **d:** Rear Admiral   **e:** Commodore   **f:** Captain
g: Commander   **h:** Lieutenant Commander   **i:** Lieutenant   **j:** Sub Lieutenant   **k:** Midshipman

Panama (Servicio Nacional Aéreonaval)

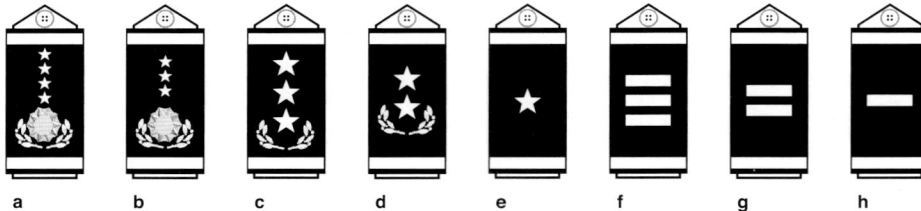

Gold braid stars, suns, wreaths and bars on black cloth shoulder-loops on white cloth shoulder-straps with white bone buttons. Panama National Police rank-titles are used and are in Spanish.
a: *Director General*, Rear Admiral *(Commander, SENAN)*   **b:** *Subdirector General*, Commodore   **c:** *Comisionado*, Captain   **d:** *Subcomisionado*, Commander
e: *Mayor*, Lieutenant Commander   **f:** *Capítan*, Lieutenant   **g:** *Teniente*, Sub Lieutenant   **h:** *Subteniente*, Acting Sub Lieutenant

Paraguay (Armada Nacional Paraguaya)

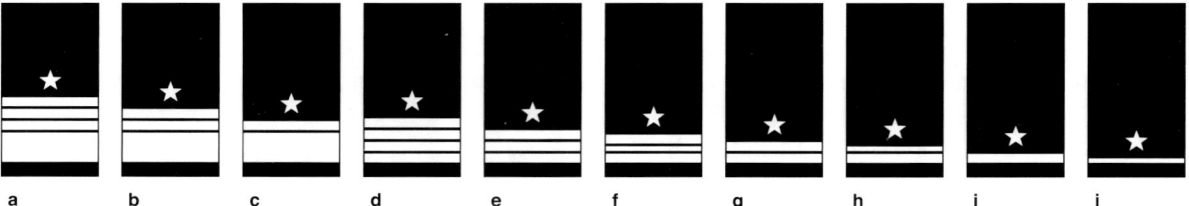

Gold wire stars and gold braid rings on very dark blue cuffs. Rank titles are in Spanish. The Paraguayan Coast Guard *(Prefectura General Naval)* forms part of the Navy and PGN personnel are commanded by serving naval officers.
a: *Almirante*, Admiral *(Commander of the Navy)*   **b:** *Vicealmirante*, Vice Admiral   **c:** *Contralmirante*, Rear Admiral   **d:** *Capitán de Navío*, Captain
e: *Capitán de Fragata*, Commander   **f:** *Capitán de Corbeta*, Lieutenant Commander   **g:** *Teniente de Navío*, Lieutenant
h: *Teniente de Fragata*, (Senior) Sub Lieutenant   **i:** *Teniente de Corbeta*, Sub Lieutenant   **j:** *Guardiamarina*, Acting Sub Lieutenant

Peru (Marina de Guerra del Perú)

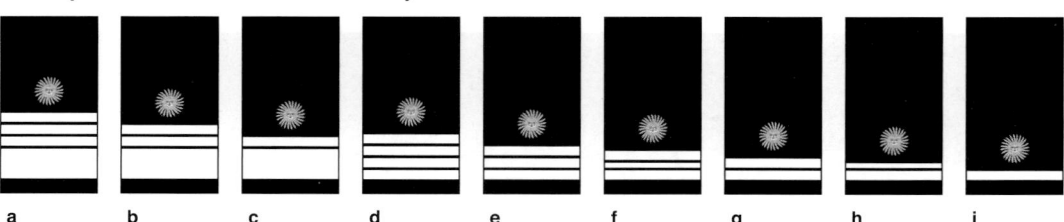

Gold wire suns and gold braid rings on very dark blue cuffs. Rank titles are in Spanish. There is a Peruvian Coast Guard *(Dirección General de Capitanías y Guardacostas)*.
a: *Almirante*, Admiral *(Commander, Navy)*   **b:** *Vicealmirante*, Vice Admiral   **c:** *Contralmirante*, Rear Admiral   **d:** *Capitán de Navío*, Captain
e: *Capitán de Fragata*, Commander   **f:** *Capitán de Corbeta*, Lieutenant Commander   **g:** *Teniente 1° (Primero)*, Lieutenant
h: *Teniente 2° (Segundo)*, Sub Lieutenant   **i:** *Alférez de Fragata*, Acting Sub Lieutenant

IHS Jane's Fighting Ships 2014-2015 © 2014 IHS

Ranks and insignia of the world's navies [49]

Philippines (Philippine Navy/Hukbong Dagat ng Pilipinas)

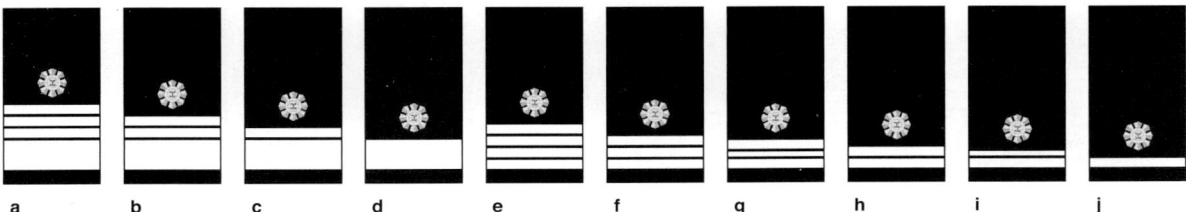

Gold wire suns and gold braid rings on very dark blue cuffs. The Navy title and rank titles are in English/Tagalog.
a: *Admiral/Admiral*, Admiral *(rank not currently held)*   **b:** *Vice Admiral/Bise Admiral*, Vice Admiral *(Flag-Officer-In-Command PN)*
c: *Rear Admiral/Rir Admiral*, Rear Admiral   **d:** *Commodore/Komodor*, Commodore   **e:** *Captain/Kapitan*, Captain   **f:** *Commander/Kumander*, Commander
g: *Lieutenant Commander/Tinyente Kumander*, Lieutenant Commander   **h:** *Lieutenant/Tinyente*, Lieutenant
i: *Lieutenant Junior Grade/Tinyente na Mababang Baitang*, Sub Lieutenant   **j:** *Ensign/Alferez*, Acting Sub Lieutenant

Philippines - Coast Guard/Tanurag Baybayin ng Pilipinas

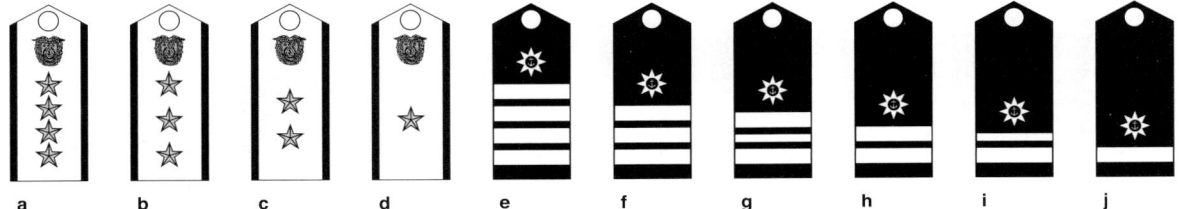

Silver wire coat of arms and silver metal stars on gold braid shoulder straps edged in very dark blue cloth (a-d). Gold wire suns with anchors and gold braid rings on very dark blue shoulder-straps (e-j). Titles are in English/Tagalog. The Philippine Coast Guard Auxiliary wears the PCGA badge above the rank stars/rings.
a: *Admiral/Admiral*, Admiral *(rank not currently held)*   **b:** *Vice Admiral/Bise Admiral*, Vice Admiral *(rank not currently held)*
c: *Rear Admiral/Rir Admiral*, Rear Admiral *(Commandant PCG)*   **d:** *Commodore/Komodor*, Commodore   **e:** *Captain/Kapitan*, Captain
f: *Commander/Kumander*, Commander   **g:** *Lieutenant Commander/Tinyente Kumander*, Lieutenant Commander   **h:** *Lieutenant/Tinyente*, Lieutenant
i: *Lieutenant Junior Grade/Tinyente na Mababang Baitang*, Sub Lieutenant   **j:** *Ensign/Alferez*, Acting Sub Lieutenant

Poland (Marynarka Wojenna)

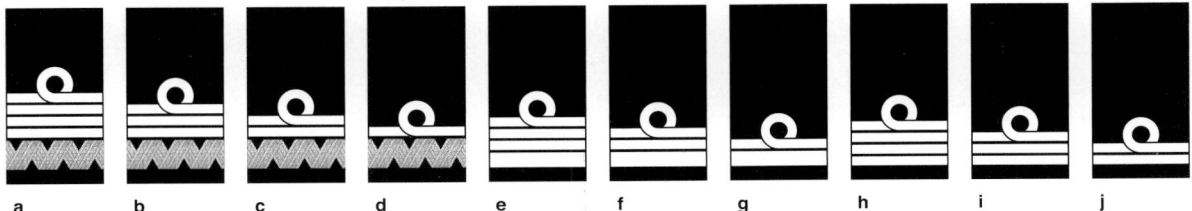

Gold braid rings with 'curl' and traditional Polish generals' embroidery on very dark blue cloth cuffs. Rank titles are in Polish. The Polish Border Guard *(Straż Graniczna)* includes a Marine Regional Unit, commander by a Rear Admiral, fulfilling coast guard functions and wearing Polish Navy uniforms and insignia.
a: *Admirał*, Admiral *(rank not currently held)*   **b:** *Admirał floty*, Vice Admiral *(Commander, Navy)*   **c:** *Wiceadmirał*, Rear Admiral   **d:** *Kontradmirał*, Commodore
e: *Komandor*, Captain   **f:** *Komandor porucznik*, Commander   **g:** *Komandor podporucznik*, Lieutenant Commander   **h:** *Kapitan marynarki*, Lieutenant
i: *Porucznik marynarki*, Sub Lieutenant   **j:** *Podporucznik marynarki*, Acting Sub Lieutenant

Portugal (Marinha de Guerra Portuguesa)

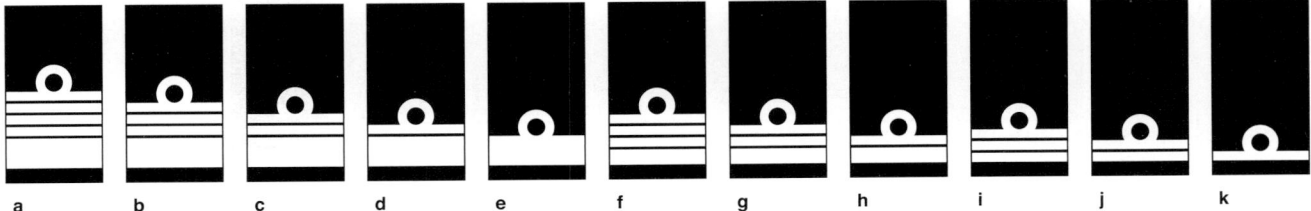

Gold braid rings with 'curl' on very dark blue cloth cuffs. Rank titles are in Portuguese.
a: *Almirante da Armada*, Admiral of the Fleet *(rank not currently held)*   **b:** *Almirante*, Admiral *(Chief of Naval Staff)*   **c:** *Vice-almirante*, Vice Admiral
d: *Contra-almirante*, Rear Admiral   **e:** *Comodoro*, Commodore   **f:** *Capitão-de-mar-e-guerra*, Captain   **g:** *Capitão-de-fragata*, Commander
h: *Capitão-tenente*, Lieutenant Commander   **i:** *Primeiro-tenente*, Lieutenant   **j:** *Segundo-tenente*, Sub Lieutenant
k: *Subtenente*, Acting Sub Lieutenant & *Guarda-marinha*, Midshipman

Qatar (Qatari Emiri Navy)

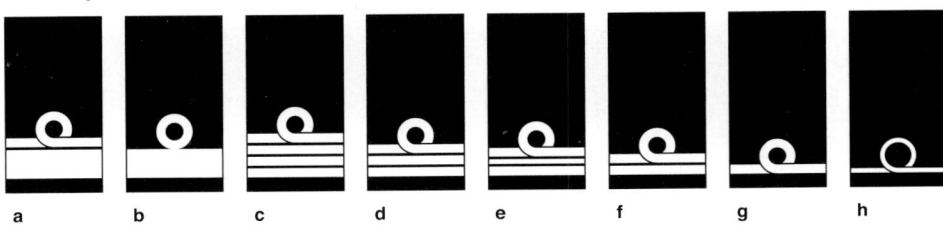

Gold braid rings with 'curl' on very dark blue cloth cuffs. Arabic Qatari Army rank titles are used and written here in romanised script. There is a Qatari Coast Guard wearing Qatari naval uniforms and commanded by a Captain.
a: *Liwā'*, Rear Admiral *(rank not currently held)*   **b:** *'Amīd*, Commodore *(Chief of Naval Staff)*   **c:** *'Aqīd*, Captain   **d:** *Muqaddam*, Commander
e: *Rā'id*, Lieutenant Commander   **f:** *Naqīb*, Lieutenant   **g:** *Mulāzim Awwal*, Sub Lieutenant   **h:** *Mulāzim Thāni*, Acting Sub Lieutenant

[50] Ranks and insignia of the world's navies

Romania (Forțele Navale Române)

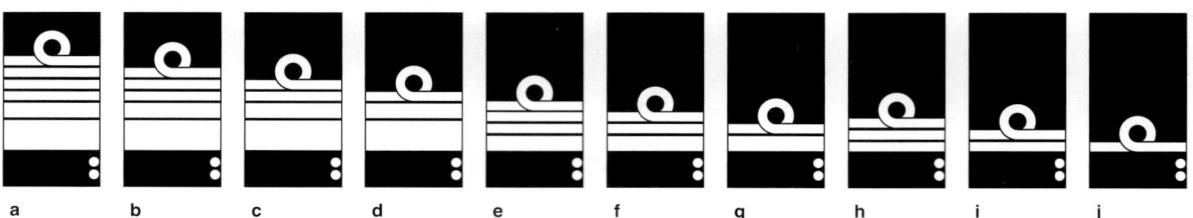

Gold braid rings with 'curl' on very dark blue cloth cuffs; brass buttons. Rank titles are in Romanian. The Romanian Border Guard, which forms part of the Interior Ministry, includes a small Coast Guard.
a: *Amiral*, Admiral *(rank not currently held)*   **b:** *Viceamiral*, Vice Admiral *(Chief of Naval Staff)*   **c:** *Contraamiral*, Rear Admiral   **d:** *Amiral de flotilă*, Commodore
e: *Comandor*, Captain   **f:** *Căpitan-comandor*, Commander   **g:** *Locotenent comandor*, Lieutenant Commander   **h:** *Căpitan*, Lieutenant
i: *Locotenent*, Sub Lieutenant   **j:** *Aspirant*, Acting Sub Lieutenant

Russian Federation (Voyenno-morsky Flot Rossiyskoy Federatsii)

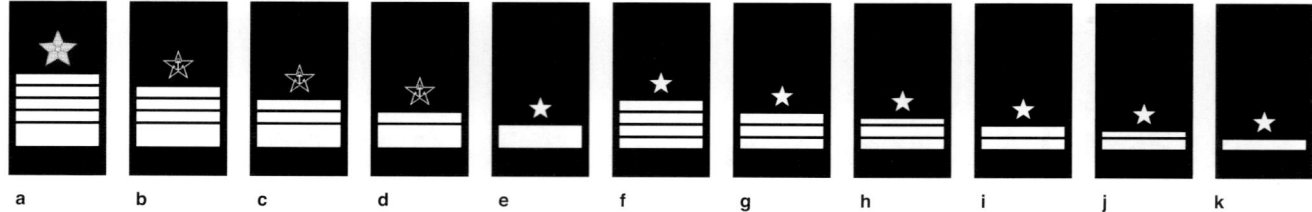

Gold wire large star (a) and small stars, black cloth stars edged in gold wire with a gold wire anchor (b-d) and gold braid bars on black cloth cuffs. Rank insignia is worn simultaneously on gold braid or black shoulder-straps. Rank titles are in romanised Russian. The Russian Federal Border Guard, which forms part of the Interior Ministry, includes a Coast Guard wearing naval-style uniforms.
a: *Admiral flota*, Admiral of the Fleet *(rank not currently held)*   **b:** *Admiral*, Admiral *(Commander-in-Chief, Navy)*   **c:** *Vitse-admiral*, Vice Admiral
d: *Kontr-admiral*, Rear Admiral   **e:** *Kapitan 1-go (pervogo) ranga*, Captain   **f:** *Kapitan 2-go (vtorogo) ranga*, Commander
g: *Kapitan 3-go (tretyego) ranga*, Lieutenant Commander   **h:** *Kapitan-leytenant*, Lieutenant   **i:** *Starshiy leytenant*, (Senior) Sub Lieutenant
j: *Leytenant*, Sub Lieutenant   **k:** *Mladshiy leytenant*, Acting Sub Lieutenant

Saudi Arabia (Al-Quwwat Al-Bahria Al-Malakiyya As-Su'udiyya)

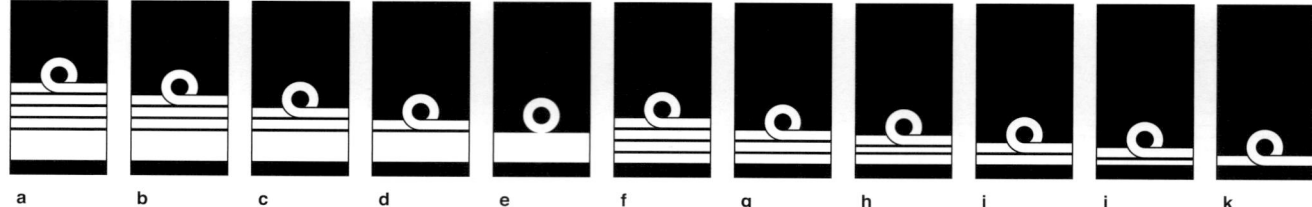

Gold braid rings with 'curl' on very dark blue cloth cuffs. Arabic Saudi Arabian Army rank titles are used and written here in romanised script. There is a Saudi Arabian Coast Guard, commanded by a Lieutenant-General.
a: *Mushīr*, Admiral of the Fleet   **b:** *Farīq Awwal*, Admiral *(rank not currently held)*   **c:** *Farīq*, Vice Admiral *(Commander, RNSF)*   **d:** *Liwā'*, Rear Admiral
e: *'Amīd*, Commodore   **f:** *'Aqīd*, Captain   **g:** *Muqaddam*, Commander   **h:** *Rā'id*, Lieutenant Commander   **i:** *Naqīb*, Lieutenant
j: *Mulāzim Awwal*, Sub Lieutenant   **k:** *Mulāzim*, Acting Sub Lieutenant

Senegal (Marine Nationale Sénégalaise)

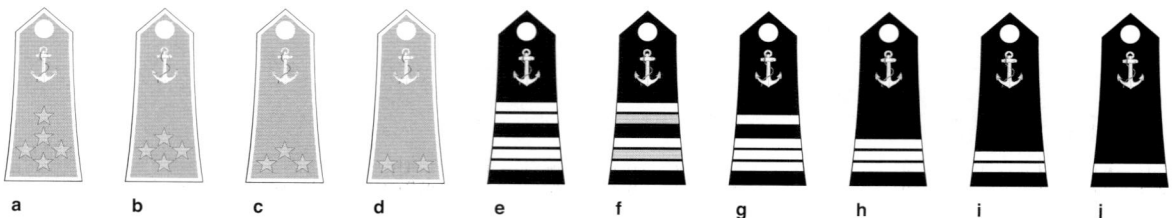

Gold wire anchor, edging, stars and rings on light blue (a-d) or very dark blue (e-j) cloth shoulder-straps; a Commander (f) has silver second and fourth rings; brass buttons. Rank titles are in French.
a: *Amiral*, Admiral *(rank not currently held)*   **b:** *Vice-amiral d'escadre*, Vice Admiral *(rank not currently held)*   **c:** *Vice-amiral*, Rear Admiral *(rank not currently held)*
d: *Contre-amiral*, Commodore *(Commander, Navy)*   **e:** *Capitaine de vaisseau*, Captain   **f:** *Capitaine de frégate*, Commander
g: *Capitaine de corvette*, Lieutenant Commander   **h:** *Lieutenant de vaisseau*, Lieutenant   **i:** *Enseigne de vaisseau de 1ère (première) classe*, Sub Lieutenant
j: *Enseigne de vaisseau de 2e (deuxième) classe*, Acting Sub Lieutenant

Seychelles (Seychelles Coast Guard)

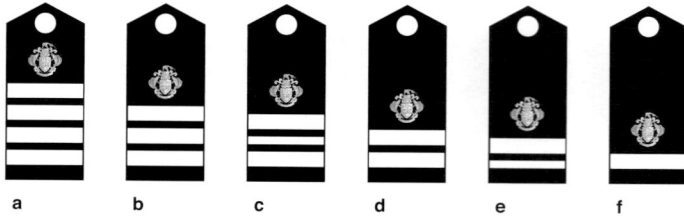

Gold wire Seychelles coats of arms and gold braid rings on very dark blue cloth shoulder-straps; brass buttons. Seychelles Army rank titles are used and written in English.
a: *Colonel*, Captain *(rank not currently held)*
b: *Lieutenant Colonel*, Commander *(Deputy Chief of Staff, Seychelles People's Defence Force; Commanding Officer, SCG)*   **c:** *Major*, Lieutenant Commander
d: *Captain*, Lieutenant   **e:** *First Lieutenant*, Sub Lieutenant   **f:** *Lieutenant*, Acting Sub Lieutenant

IHS Jane's Fighting Ships 2014-2015   © 2014 IHS

Ranks and insignia of the world's navies [51]

Sierra Leone (Republic of Sierra Leone Armed Forces Maritime Wing)

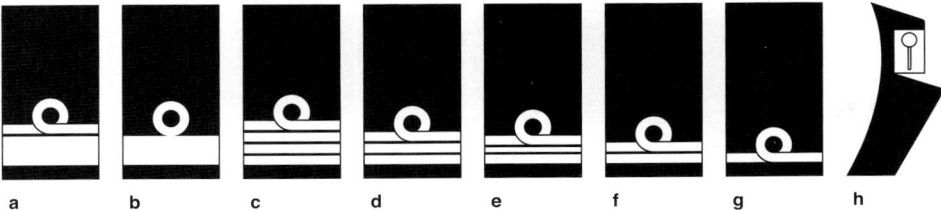

Gold braid rings with 'curl' on very dark blue cloth cuffs; brass button and white cord on white cloth collar-patch (h). British Royal Navy ranks are used.
a: Rear Admiral *(rank not currently held)* **b:** Commodore *(rank not currently held)* **c:** Captain (N) *(Commanding Officer, Maritime Wing)* **d:** Commander
e: Lieutenant Commander **f:** Lieutenant **g:** Sub Lieutenant **h:** Midshipman

Singapore (Republic of Singapore Navy/Angkastan Laut Republik Singapura)

Gold embroidered stars, crossed branches, coats of arms, bars and national titles on very dark blue cloth shoulder-straps; brass buttons. The Navy title is in English/Malayan, the rank titles in English only. The Singapore Police Coast Guard forms part of the Singapore Police. Personnel wear police uniforms and insignia with *Deputy Assistant Commissioner* (Commodore) as the highest rank.
a: *Vice Admiral*, Vice Admiral *(rank not currently held)* **b:** *Rear Admiral (2 stars)*, Rear Admiral *(Chief of Navy)* **c:** *Rear Admiral (1 star)*, Commodore
d: *Colonel*, Captain **e:** *Senior Lieutenant Colonel*, (Senior) Commander **f:** *Lieutenant Colonel*, Commander **g:** *Major*, Lieutenant Commander
h: *Captain*, Lieutenant **i:** *Lieutenant*, Sub Lieutenant **j:** *2nd (Second) Lieutenant*, Acting Sub Lieutenant

Slovenia (Slovenska Mornarica)

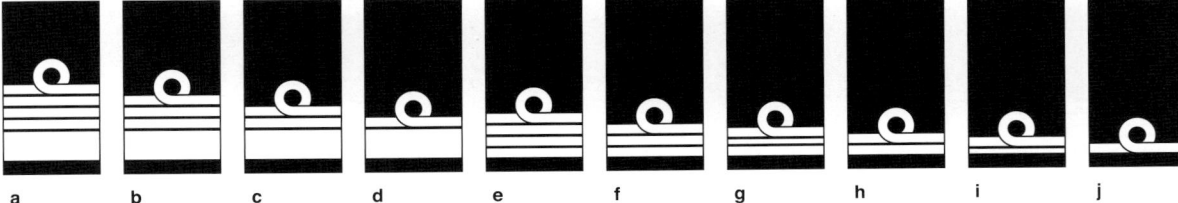

Gold braid rings with 'curl' on very dark blue cloth cuffs. Rank titles are in Slovene.
a: *Admiral*, Admiral *(rank not currently held)* **b:** *Viceadmiral*, Vice Admiral *(rank not currently held)* **c:** *Kontraadmiral*, Rear Admiral *(rank not currently held)*
d: *Kapitan*, Commodore *(rank not currently held)* **e:** *Kapitan bojne ladje*, Captain *(rank not currently held)* **f:** *Kapitan fregate*, Commander *(Chief of Naval Staff)*
g: *Kapitan korvete*, Lieutenant Commander **h:** *Poročnik bojne ladje*, Lieutenant **i:** *Poročnik fregate*, Sub Lieutenant **j:** *Poročnik korvete*, Acting Sub Lieutenant

South Africa

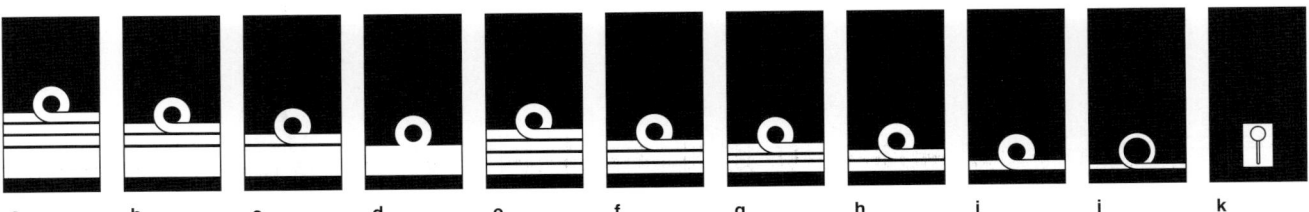

Gold braid rings on black cloth cuffs; brass button on white cord on white cloth cuff-patch (k); brass buttons. Rank titles are in English.
a: *Admiral*, Admiral *(rank not currently held)* **b:** *Vice Admiral*, Vice Admiral *(Chief of the Navy)* **c:** *Rear Admiral*, Rear Admiral
d: *Rear Admiral (Junior Grade)*, Commodore **e:** *Captain*, Captain **f:** *Commander*, Commander **g:** *Lieutenant Commander*, Lieutenant Commander
h: *Lieutenant*, Lieutenant **i:** *Sub-Lieutenant*, Sub Lieutenant **j:** *Ensign*, Acting Sub Lieutenant **k:** *Midshipman*, Midshipman

Spain (Armada Española)

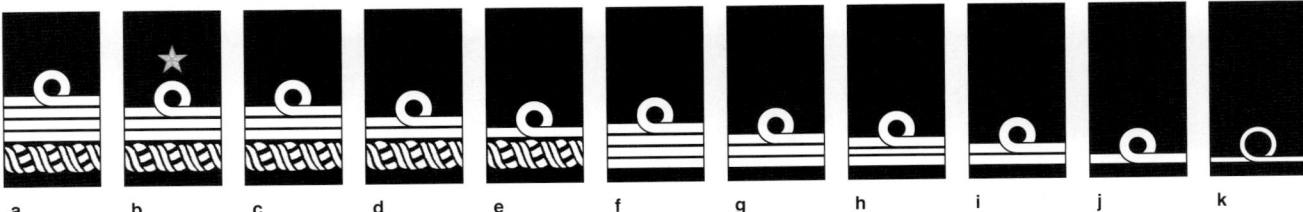

Gold wire star and gold braid rings with 'curl' and traditional Spanish generals' embroidery on very dark blue cloth cuffs. Rank titles are in Spanish. Coast Guard duties are carried out by the Maritime Service *(Servicio Maritimo)* of the Civil Guard *(Guardia Civil)* wearing green uniforms.
a: *Capitán General* (King Juan Carlos II) **b:** *Almirante General*, Admiral *(Chief of Naval Staff)* **c:** *Almirante*, Vice Admiral **d:** *Vicealmirante*, Rear Admiral
e: *Contraalmirante*, Commodore **f:** *Capitán de Navío*, Captain **g:** *Capitán de Fragata*, Commander **h:** *Capitán de Corbeta*, Lieutenant Commander
i: *Teniente de Navío*, Lieutenant **j:** *Alférez de Navío*, Sub Lieutenant **k:** *Alférez de Fragata*, Acting Sub Lieutenant

[52] Ranks and insignia of the world's navies

Sri Lanka (Sri Lanka Navy)

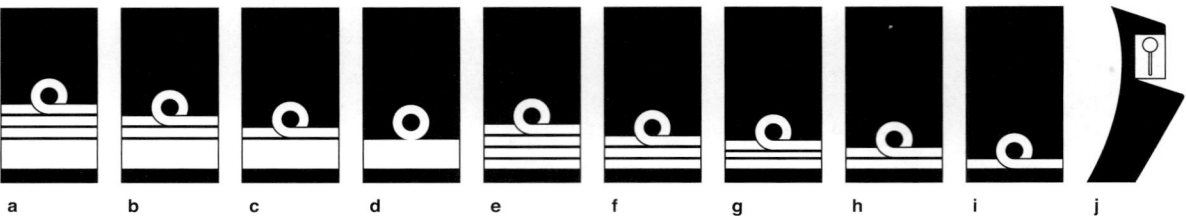

Gold braid rings with 'curl' on very dark blue cloth cuffs; brass button and white cord on white cloth collar-patch (j). British Royal Navy rank titles are used. The Sri Lanka Coast Guard, with a Rear-Admiral as Director-General, wear naval-style uniforms and shoulder-strap rank insignia.
a: Admiral *(Presidential National Security Adviser)* **b:** Vice Admiral *(Commander of the Navy)* **c:** Rear Admiral **d:** Commodore **e:** Captain **f:** Commander
g: Lieutenant Commander **h:** Lieutenant **i:** Sub Lieutenant and Acting Sub Lieutenant **j:** Midshipman

Sudan (Al-Quwwat Al-Bahria as-Sudaniya)

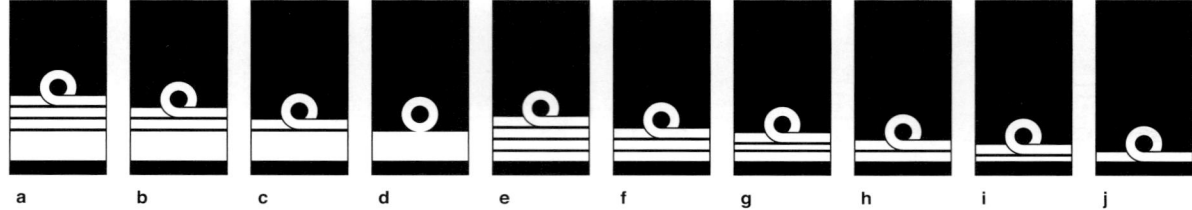

Gold wire stars and gold braid rings on very dark blue cloth cuffs. Arabic Sudan Army rank titles are used and written here in romanised script.
a: *Farīq Awwal*, Admiral *(rank not currently held)* **b:** *Farīq*, Vice Admiral *(Chief of Naval Staff)* **c:** *Liwā'*, Rear Admiral **d:** *'Amīd*, Commodore **e:** *'Aqīd*, Captain
f: *Muqaddam*, Commander **g:** *Rā'id*, Lieutenant Commander **h:** *Naqīb*, Lieutenant **i:** *Mulāzim Awwal*, Sub Lieutenant **j:** *Mulāzim Thāni*, Acting Sub Lieutenant

Suriname (Surinaamse Marine)

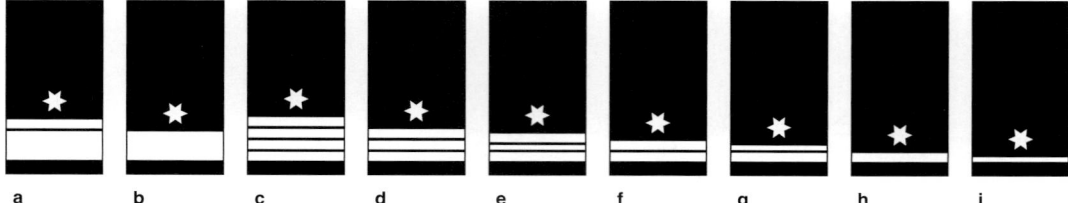

Gold wire stars and gold braid rings on very dark blue cloth cuffs. Suriname Army rank titles are used and written in Dutch.
a: *Generaal-majoor*, Rear Admiral *(rank not currently held)* **b:** *Brigade-generaal*, Commodore *(rank not currently held)* **c:** *Kolonel*, Captain *(rank not currently held)*
d: *Luitenant-kolonel*, Commander *(Commander, Navy)* **e:** *Majoor*, Lieutenant Commander **f:** *Kapitein*, Lieutenant **g:** *1e (Eerste) luitenant*, Sub Lieutenant
h: *2e (Tweede) luitenant*, Acting Sub Lieutenant **i:** *Vaandrig*, Warrant Officer

Sweden (Svenska Marinen)

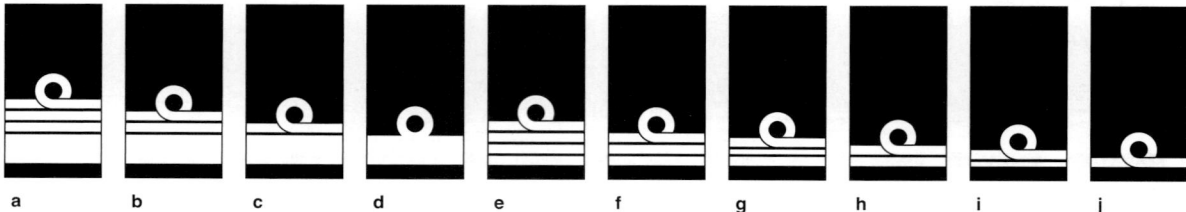

Gold braid rings with 'curl' on very dark blue cloth cuffs. Rank titles are in Swedish.
a: *Amiral*, Admiral *(King Carl Gustaf XVI)* **b:** *Viceamiral*, Vice Admiral *(rank not currently held)* **c:** *Konteramiral*, Rear Admiral *(Inspector, Navy)*
d: *Flottiljamiral*, Commodore **e:** *Kommendör*, Captain **f:** *Kommendörkapten*, Commander **g:** *Örlogskapten*, Lieutenant Commander **h:** *Kapten*, Lieutenant
i: *Löjnant*, Sub Lieutenant **j:** *Fänrik*, Acting Sub Lieutenant

Sweden Coast Guard (Kustbevakningen)

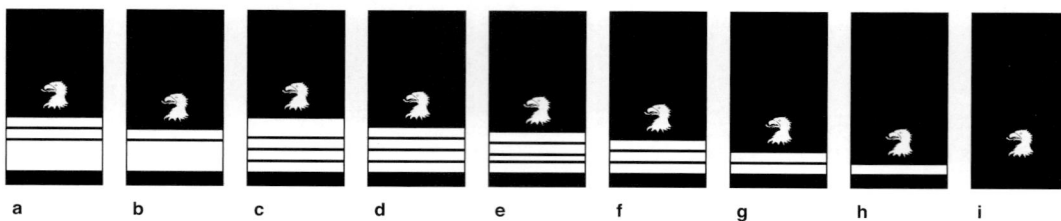

Gold metal eagles' heads and gold braid stripe and rings on very dark blue cloth cuffs. Rank titles are in Swedish.
a: *Generaldirektör*, Vice Admiral *(Director General, Coast Guard)* **b:** *Överdirektör*, Rear Admiral **c:** *Kustbevakningsdirektör*, Commodore
d: *Kustbevakningsöverinspektör*, Captain **e:** *1. (Förste) Kustbevakningsinspektör*, Commander **f:** *Kustbevakningsinspektör*, Lieutenant Commander
g: *Kustbevakningsassistent*, Lieutenant **h:** *Kustuppsyningsman*, Sub Lieutenant **i:** *Kustbevakningsaspirant*, Acting Sub Lieutenant

IHS Jane's Fighting Ships 2014-2015 © 2014 IHS

Ranks and insignia of the world's navies [53]

Syria (Al Quwwat Al-Bahria Al-Arabiya As-Souriya)

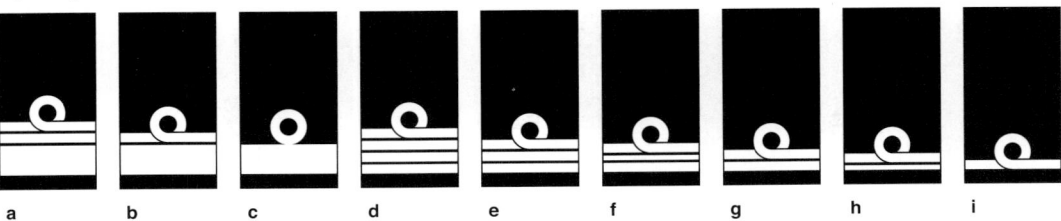

Gold braid rings with 'curl' on very dark blue cloth cuffs. Syrian Army rank titles are used and written here in romanised Arabic.
a: *Farīq*, Vice Admiral *(Chief of Naval Staff)*   **b:** *Liwā'*, Rear Admiral   **c:** *'Amid*, Commodore   **d:** *'Aqīd*, Captain   **e:** *Muqaddam*, Commander
f: *Rā'id*, Lieutenant Commander   **g:** *Naqīb*, Lieutenant   **h:** *Mulāzim Awwal*, Sub Lieutenant   **i:** *Mulāzim*, Acting Sub Lieutenant

Taiwan (Republic of China) (Zhōnghuá Mínguó Hǎijūn)

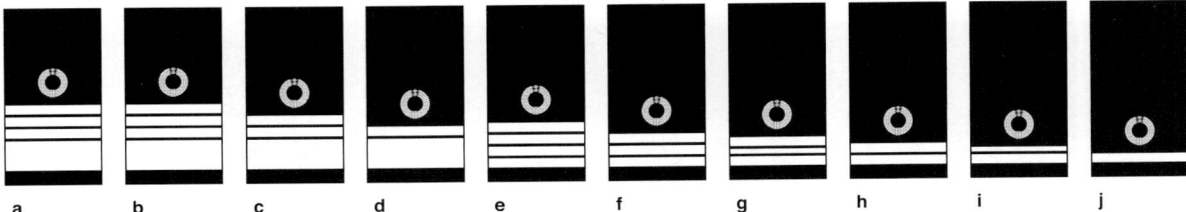

A gold wire cornsheaf on very dark blue cloth cuffs. Rank titles are in romanised Mandarin Chinese written in 'Hanyu Pinyin'. The Taiwanese Coast Guard is designated the 'Republic of China Coast Guard Administration'.
a: *Hǎijūn Yi-chi Shangjiang*, (Senior) Admiral *(4 shoulder-strap stars) (Commander-in-Chief of the Navy)*   **b:** *Hǎijūn Erh-chi Shangjiang*, Admiral *(3 shoulder-strap stars)*
c: *Hǎijūn Zhōngjiang*, Vice Admiral   **d:** *Hǎijūn Shaojiang*, Rear Admiral   **e:** *Hǎijūn Shangxiao*, Captain   **f:** *Hǎijūn Zhōngxiao*, Commander
g: *Hǎijūn Shaoxiao*, Lieutenant Commander   **h:** *Hǎijūn Shangwei*, Lieutenant   **i:** *Hǎijūn Zhōngwei*, Sub Lieutenant   **j:** *Hǎijūn Shaowei*, Acting Sub Lieutenant

Tanzania (Kamandi ya Jeshi la Majini)

Brass national coats of arms, crossed baton and scimitar, stars and national titles on dark blue shoulder-loops worn on light blue shirt shoulder-straps with white bone buttons. Tanzanian Army rank titles are used and written in Swahili.
a: *Meja Jenerali*, Rear Admiral *(Commander, Naval Command)*   **b:** *Brigedia Jenerali*, Commodore   **c:** *Kanali*, Captain   **d:** *Luteni Kanali*, Commander
e: *Meja*, Lieutenant Commander   **f:** *Kapteni*, Lieutenant   **g:** *Luteni*, Sub Lieutenant   **h:** *Luteni Usu*, Acting Sub Lieutenant

Thailand (Kongthap Ruea Thai)

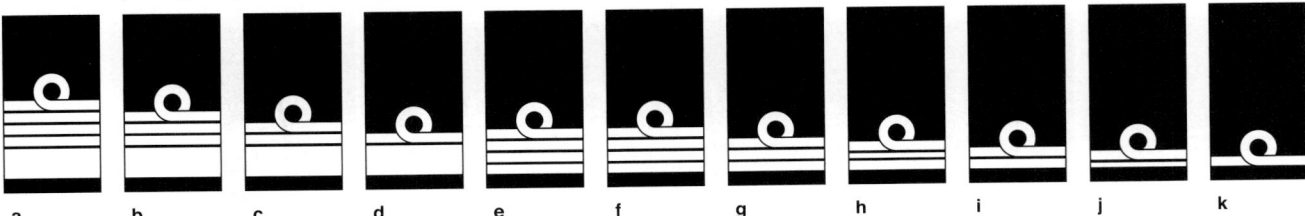

Gold braid rings with 'curl' on very dark blue cloth cuffs. Rank titles are in romanised Thai.
a: *Chom Phon Rua*, Admiral of the Fleet *(King Bhumibol Adulyadej)*   **b:** *Phon Rua Eg*, Admiral *(Commander-in-Chief, Navy)*   **c:** *Phon Rua Tho*, Vice Admiral
d: *Phon Rua Tri*, Rear Admiral   **e:** *Nawa Eg Phiset*, Commodore *(with flag-officer's cap peak embroidery)*   **f:** *Nawa Eg*, Captain   **g:** *Nawa Tho*, Commander
h: *Nawa Tri*, Lieutenant Commander   **i:** *Rua Eg*, Lieutenant   **j:** *Rua Tho*, Sub Lieutenant   **k:** *Rua Tri*, Acting Sub Lieutenant

Togo (Marine Nationale Togolaise)

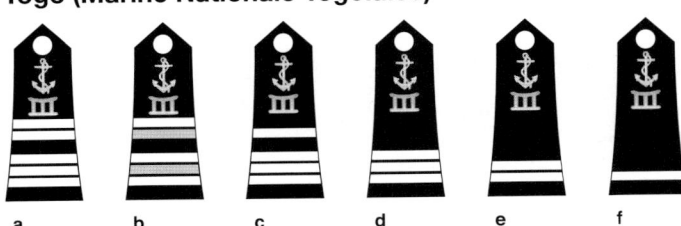

Gold wire anchors and monograms and gold braid rings on very dark blue cloth shoulder-straps; a Commander (b) has silver second and fourth rings; brass buttons. Rank titles are in French.
a: *Capitaine de vaisseau*, Captain *(Chief of Naval Staff)*   **b:** *Capitaine de frégate*, Commander   **c:** *Capitaine de corvette*, Lieutenant Commander
d: *Lieutenant de vaisseau*, Lieutenant   **e:** *Enseigne de vaisseau de 1ère (première) classe*, Sub Lieutenant
f: *Enseigne de vaisseau de 2e (deuxième) classe*, Acting Sub Lieutenant

[54] Ranks and insignia of the world's navies

Tonga (His Majesty's Armed forces' Maritime Force)

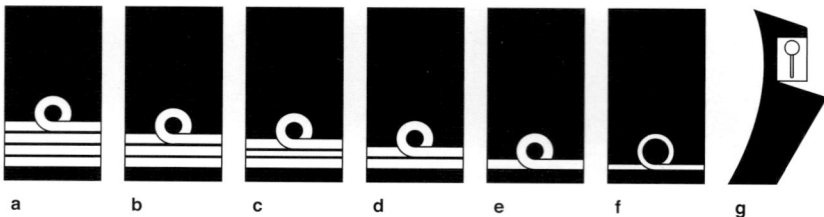

Gold braid rings with 'curl' on very dark blue cloth cuffs; brass button and white cord on white cloth collar-patch (g). Rank titles are in English.
a: *Captain*, Captain *(Acting Chief of the Defence Staff)* **b:** *Commander*, Commander *(Commander, Maritime Force)*
c: *Lieutenant Commander*, Lieutenant Commander **d:** *Lieutenant*, Lieutenant **e:** *Sub Lieutenant*, Sub Lieutenant **f:** *Ensign*, Acting Sub Lieutenant
g: *Midshipman*, Midshipman

Trinidad and Tobago Coast Guard

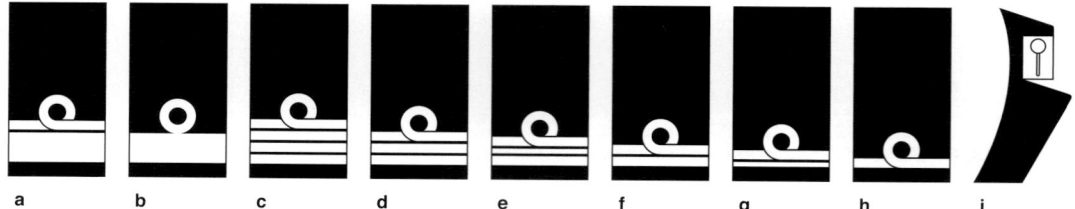

Gold braid rings with 'curl' on very dark blue cloth cuffs; brass button and white cord on white cloth collar-patch (h). Royal Naval rank titles are used.
a: Rear Admiral *(rank not currently held)* **b:** Commodore *(rank not currently held)* **c:** Captain (N) *(Commanding Officer, Coast Guard)* **d:** Commander
e: Lieutenant Commander **f:** Lieutenant **g:** Sub Lieutenant **h:** Acting Sub Lieutenant **i:** Midshipman

Tunisia (Al-Quwwat Al-Bahria at'Tunisia)

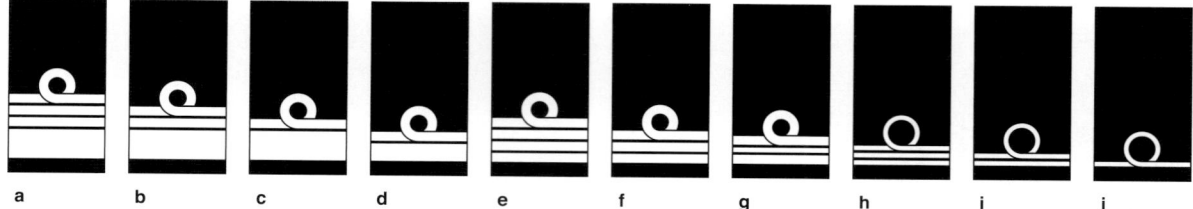

Gold braid rings with 'curl' on very dark blue cloth cuffs. Arabic Tunisian Army rank titles are used and written here in romanised Arabic. Tunisia maintains a small Coast Guard.
a: *Farīq*, Admiral *(rank not currently held)* **b:** *Liwā'*, Vice Admiral *(rank not currently held)* **c:** *'Amid*, Rear Admiral *(Chief of Naval Staff)*
d: *'Aqīd Awwal*, Commodore **e:** *'Aqīd*, Captain **f:** *Muqaddam*, Commander **g:** *Rā'id*, Lieutenant Commander **h:** *Naqīb*, Lieutenant
i: *Mulāzim Awwal*, Sub Lieutenant **j:** *Mulāzim*, Acting Sub Lieutenant

Turkey (Türk Deniz Kuvvetleri)

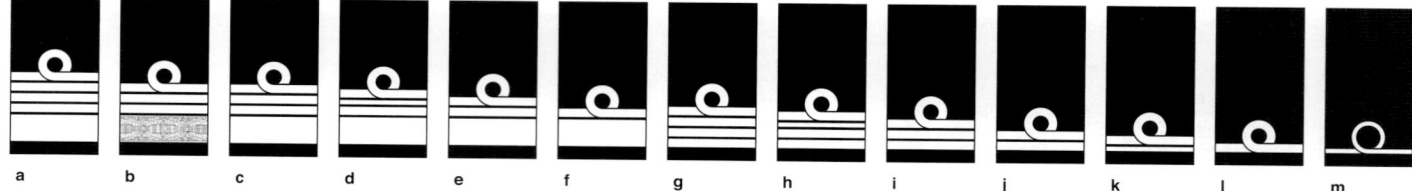

Gold braid rings with 'curl' on very dark blue cloth shoulder-straps; The Chief of General Staff (b) has a special pattern thick ring. Rank titles are in Turkish. The Turkish Coast Guard *(Sahil Güvenlik Kiliği)* is manned by seconded naval personnel and is commanded by a Rear Admiral *(Tümamiral)*. Personnel wear naval uniforms and insignia with the distinguishing shoulder-title 'Sahil Güvenlik'.
a: *Büyük amiral*, Admiral of the Fleet *(rank not currently held)* **b:** *Genelkumay Başkani*, Admiral or General *(Chief of Defence Staff)*
c: *Oramiral*, Admiral *(C-in-C Navy)* **d:** *Koramiral*, Vice Admiral **e:** *Tümamiral*, Rear Admiral **f:** *Tuğamiral*, Commodore **g:** *Albay*, Captain
h: *Yarbay*, Commander **i:** *Binbaşi*, Lieutenant Commander **j:** *Yüzbaşi*, Lieutenant **k:** *Üsteğmen*, Sub Lieutenant **l:** *Teğmen*, Acting Sub Lieutenant
m: *Asteğmen*, (Junior) Acting Sub Lieutenant

Turkmenistan

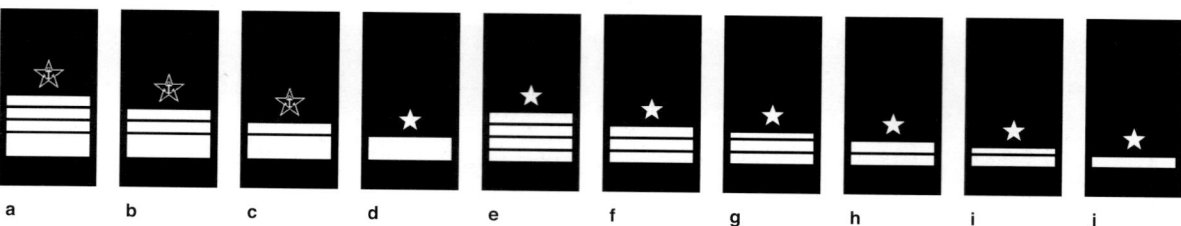

Black cloth stars edged in gold with a gold wire anchor. Gold stars and gold braid bars on black cloth cuffs. Rank insignia is worn simultaneously on shoulder-straps. Turkmen rank-titles not available.
a: Admiral *(rank not currently held)* **b:** Vice Admiral *(rank not currently held)* **c:** Rear Admiral *(Commander, Navy)* **d:** Captain **e:** Commander
f: Lieutenant Commander **g:** Lieutenant **h:** (Senior) Sub Lieutenant **i:** Sub Lieutenant **j:** Acting Sub Lieutenant

IHS Jane's Fighting Ships 2014-2015

Ranks and insignia of the world's navies [55]

Ukraine (Viys'kovo-Mors'ki Syly Ukrayiny)

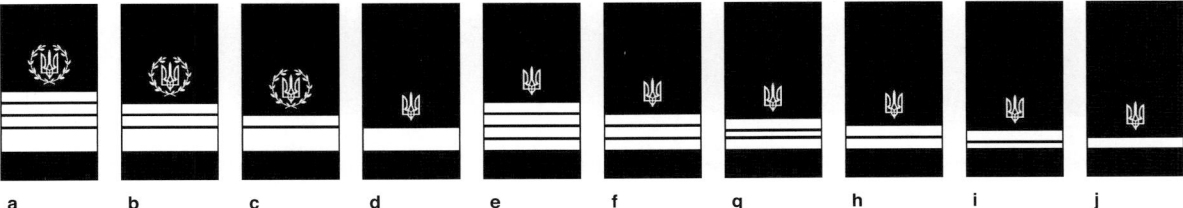

Gold wire tridents, wreaths and gold braid rings on black cloth cuffs with gold buttons. Rank titles are in romanised Ukrainian. The Ukrainian Border Guard, which forms part of the Interior Ministry, includes a Coast Guard *(Mors'ka Okhorona)* wearing naval-style uniforms.
a: *Admiral*, Admiral *(Commander-in-Chief, Navy)* **b:** *Vitse-admiral*, Vice Admiral **c:** *Kontr-admiral*, Rear Admiral **d:** *Kapitan 1-ho (pershoho) ranhu*, Captain
e: *Kapitan 2-ho (druhoho) ranhu*, Commander **f:** *Kapitan 3-ho (tret'oho) ranhu*, Lieutenant Commander **g:** *Kapitan-leytenant*, Lieutenant
h: *Starshiy-leytenant*, (Senior) Sub Lieutenant **i:** *Leytenant*, Sub Lieutenant **j:** *Molodshyi leytenant*, Acting Sub Lieutenant

United Arab Emirates (Al-Quwwat Al-Bahria Al-Dawlat Al-Imārāt Al-Arabiyya Al-Muttahida)

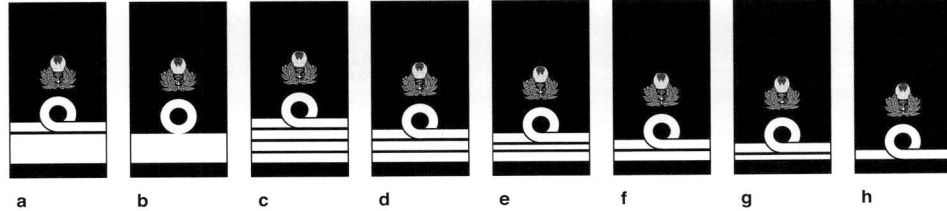

Gold wire cap-badge (gold eagle on red cloth, silver anchor, gold wreath) above gold braid rings with 'curl' on very dark blue cloth cuffs. Arabic UAE Army rank titles are used and written here in romanised script. The United Arab Emirates maintains a Coast Guard commanded by a Director-General.
a: *Liwā'*, Rear Admiral *(Commander, UAENF)* **b:** *'Amīd*, Commodore **c:** *'Aqīd*, Captain **d:** *Muqaddam*, Commander **e:** *Rā'id*, Lieutenant Commander
f: *Naqīb*, Lieutenant **g:** *Mulāzim Awwal*, Sub Lieutenant **h:** *Mulāzim*, Acting Sub Lieutenant

United Kingdom (Royal Navy)

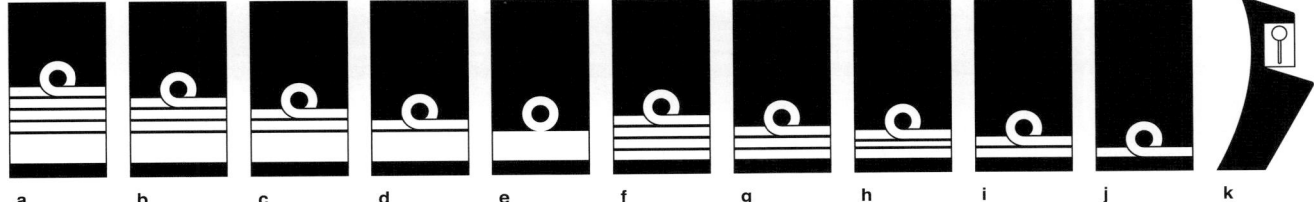

Gold braid rings with 'curl' on very dark blue cloth cuffs; brass button and white cord on white cloth collar-patch (k). Her Majesty's Coast Guard is a government agency. Personnel wear naval-style uniforms and insignia.
a: Admiral of the Fleet **b:** Admiral *(First Sea Lord & Chief of the Naval Staff)* **c:** Vice Admiral **d:** Rear Admiral **e:** Commodore **f:** Captain **g:** Commander
h: Lieutenant Commander **i:** Lieutenant **j:** Sub Lieutenant & Acting Sub Lieutenant **k:** Midshipman

United Kingdom (Royal Fleet Auxiliary Service)

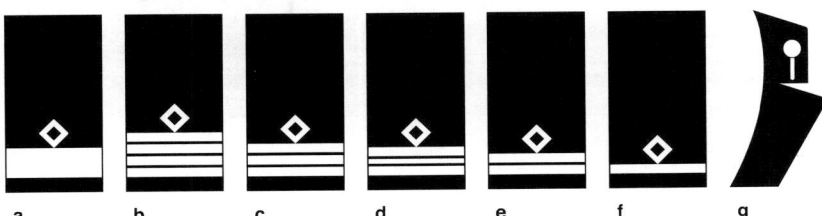

Gold braid rings and 'diamonds' on very dark blue cloth cuffs; brass button and white cord on collar (g). British Merchant Navy rank titles are used.
a: *Commodore*, Commodore *(Commanding Officer, RFA)* **b:** *Captain*, Captain **c:** *Chief Officer*, Commander **d:** *1st (First) Officer*, Lieutenant Commander
e: *2nd (Second) Officer*, Lieutenant **f:** *3rd (Third) Officer*, Sub Lieutenant **g:** *Deck Officer Cadet*, Midshipman

United States of America (United States Navy)

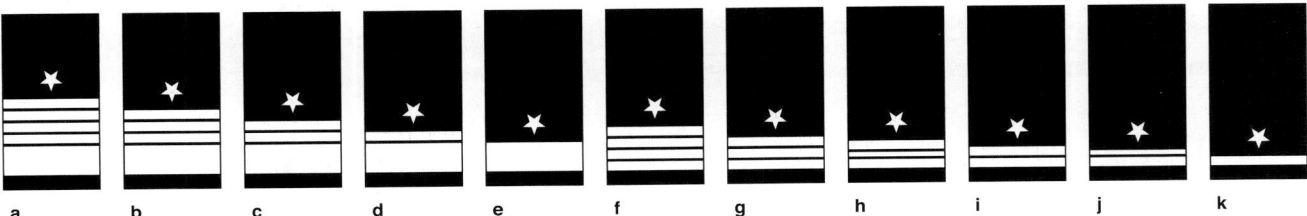

Gold wire stars and gold braid rings on very dark blue cloth cuffs. Rank titles are in English.
a: *Fleet Admiral*, Admiral of the Fleet *(rank not currently held)* **b:** *Admiral*, Admiral *(Chief of Naval Operations)* **c:** *Vice Admiral*, Vice Admiral
d: *Rear Admiral (upper half)*, Rear Admiral **e:** *Rear Admiral (lower half)*, Commodore **f:** *Captain*, Captain **g:** *Commander*, Commander
h: *Lieutenant Commander*, Lieutenant Commander **i:** *Lieutenant*, Lieutenant **j:** *Lieutenant (junior grade)*, Sub Lieutenant **k:** *Ensign*, Acting Sub Lieutenant

© 2014 IHS

IHS Jane's Fighting Ships 2014-2015

[56] Ranks and insignia of the world's navies

United States of America (United States Coast Guard)

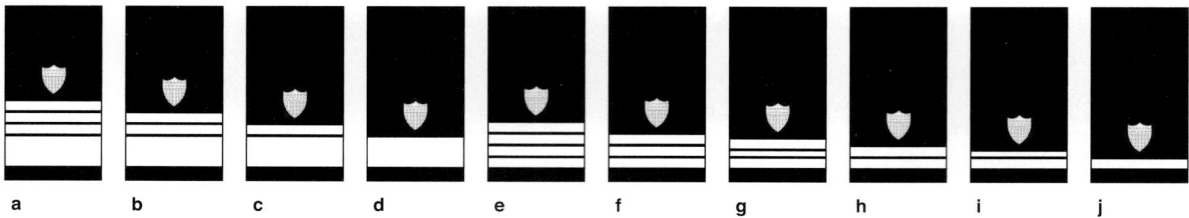

Gold wire shields and gold braid rings on blue cloth cuffs. Rank titles are in English.
a: *Admiral*, Admiral *(Commandant, USCG)*   **b:** *Vice Admiral*, Vice Admiral   **c:** *Rear Admiral (upper half)*, Rear Admiral   **d:** *Rear Admiral (lower half)*, Commodore
e: *Captain*, Captain   **f:** *Commander*, Commander   **g:** *Lieutenant Commander*, Lieutenant Commander   **h:** *Lieutenant*, Lieutenant
i: *Lieutenant (junior grade)*, Sub Lieutenant   **j:** *Ensign*, Acting Sub Lieutenant

Uruguay (Armada Nacional del Uruguay)

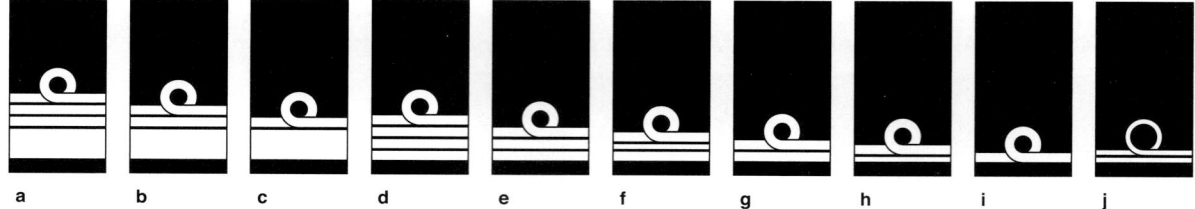

Gold braid rings with 'curl' on very dark blue cloth cuffs. Rank titles are in Spanish. The Uruguayan Coast Guard *(Prefectura Nacional Naval)* forms part of the Navy. Personnel wear naval uniforms and insignia with a Rear Admiral *(Contra Almirante)* as the commanding officer *(Prefecto Nacional Naval)*.
a: *Almirante*, Admiral *(Commander-in-Chief, Navy)*   **b:** *Vice Almirante*, Vice Admiral   **c:** *Contra Almirante*, Rear Admiral   **d:** *Capitán de Navío*, Captain
e: *Capitán de Fragata*, Commander   **f:** *Capitán de Corbeta*, Lieutenant Commander   **g:** *Teniente de Navio*, Lieutenant   **h:** *Alférez de Navio*, Sub Lieutenant
i: *Alférez de Fragata*, Acting Sub Lieutenant   **j:** *Guardiamarina*, Midshipman

Venezuela (Armada Nacional Bolivariana de Venezuela)

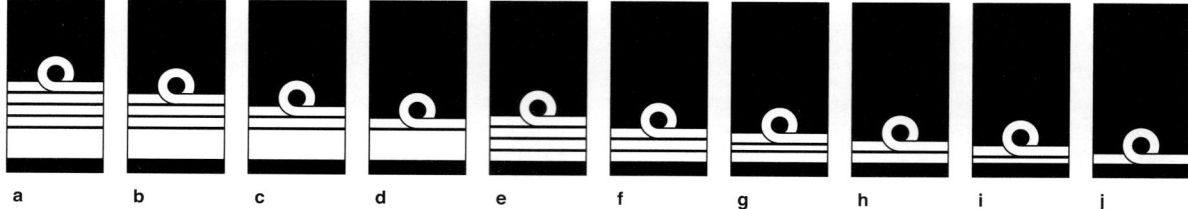

Gold braid rings with 'curl' on very dark blue cloth cuffs. Rank titles are in Spanish. The Coast Guard *(Comando de Guardacostas)* forms part of the Navy. Personnel wear naval uniforms and insignia with a Rear Admiral as the commanding officer.
a: *Almirante en Jefe*, Admiral of the Fleet *(Minister of People's Power for Defence)*   **b:** *Almirante*, Admiral *(Commandant-General, Navy)*
c: *Vicealmirante*, Vice Admiral   **d:** *Contralmirante*, Rear Admiral   **e:** *Capitán de Navío*, Captain   **f:** *Capitán de Fragata*, Commander
g: *Capitán de Corbeta*, Lieutenant Commander   **h:** *Teniente de Navío*, Lieutenant   **i:** *Teniente de Fragata*, Sub Lieutenant
j: *Alférez de Navío*, Acting Sub Lieutenant

Vietnam (Hai quan Nhan dan Viet Nam)

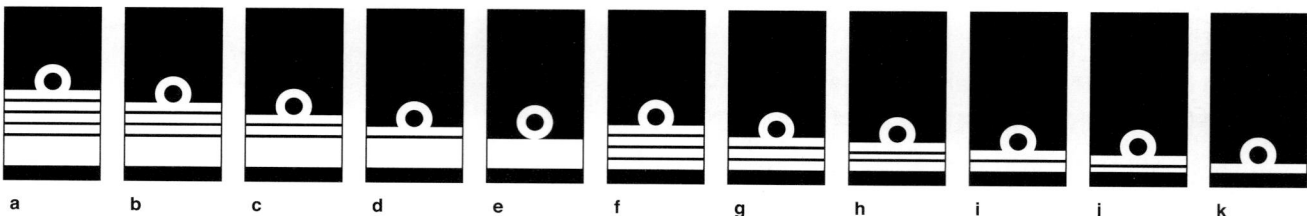

Gold braid rings with 'curl' on dark blue cuffs. Rank insignia is worn simultaneously on gold braid shoulder-straps piped black. Rank titles are in Vietnamese. There is also the Vietnam People's Coast Guard (Canh sat bien), commanded by a Major-General, wearing dark blue shoulder-straps piped yellow and using Army rank titles.
a: *Do Doc*, Admiral *(Commander, Navy)*   **b:** *Pho Do Doc*, Vice Admiral   **c:** *Chuan Do Doc*, Rear Admiral   **d:** *Da Ta*, Commodore   **e:** *Thuong Tai*, Captain
f: *Trung Ta*, Commander   **g:** *Thieu Ta*, Lieutenant Commander   **h:** *Da Uy*, (Senior) Lieutenant   **i:** *Thuong Uy*, Lieutenant   **j:** *Trung Uy*, Sub Lieutenant
k: *Thieu Uy*, Acting Sub Lieutenant

Yemen (Al Quwwat Al-Bahria Al-Yamaniya)

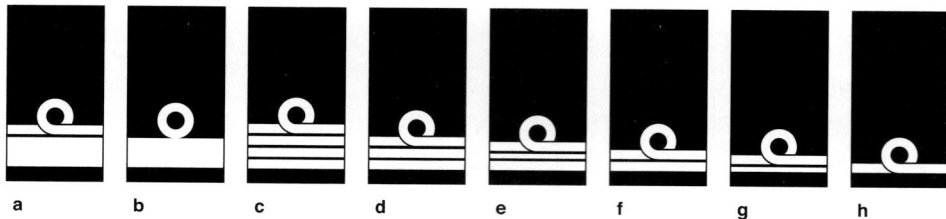

Gold braid rings with 'curl' on very dark blue cloth cuffs. Arabic Yemen Army rank titles are used and written here in romanised script. There is also a Coast Guard under the Ministry of Interior wearing naval uniforms and insignia and commanded by a Commodore.
a: *Liwā'*, Rear Admiral *(Commander, Navy)*   **b:** *'Amīd*, Commodore   **c:** *'Aqīd*, Captain   **d:** *Muqaddam*, Commander   **e:** *Rā'id*, Lieutenant Commander
f: *Naqīb*, Lieutenant   **g:** *Mulāzim Awwal*, Sub Lieutenant   **h:** *Mulāzim Thāni*, Acting Sub Lieutenant

Ensigns and flags of the world's navies

In cases where countries do not have ensigns their warships normally fly the national flag.

Albania
Ensign

Australia
Ensign

Belgium
Ensign

Algeria
Ensign

Azerbaijan
Ensign

Belize
National Flag and Ensign

Angola
National Flag and Ensign

Bahamas
Ensign

Benin
National Flag and Ensign

Anguilla
National Flag and Ensign

Bahrain
National Flag and Ensign

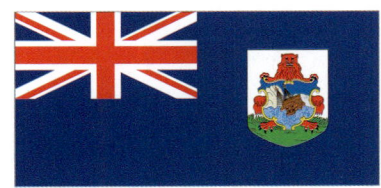
Bermuda
Ensign

Antigua and Barbuda
Ensign

Bangladesh
Ensign

Bolivia
Ensign

Argentina
National Flag and Ensign

Barbados
Ensign

Brazil
National Flag and Ensign

Flag Images courtesy of The Flag Institute, © 2014 Graham Bartram. All rights reserved.

© 2014 IHS

IHS Jane's Fighting Ships 2014-2015

[58] Ensigns and flags of the world's navies

British Indian Ocean Territory
National Flag and Ensign

Cayman Islands
National Flag and Ensign

Cook Islands
National Flag and Ensign

Brunei
Ensign

Chile
National Flag and Ensign

Costa Rica
State Flag and Ensign

Bulgaria
Ensign

China
National Flag and Ensign

Côte d'Ivoire
National Flag and Ensign

Cambodia
National Flag and Ensign

Colombia
Ensign

Croatia
Ensign

Cameroon
National Flag and Ensign

Comoros
National Flag and Ensign

Cuba
National Flag and Ensign

Canada
Ensign

Congo-Brazzaville
National Flag and Ensign

Cyprus
National Flag and Ensign

Cape Verde
National Flag

Democratic Republic of Congo
National Flag and Ensign

Cyprus, Turkish Republic of Northern
(Not recognised by United Nations)
National Flag and Ensign

Flag Images courtesy of The Flag Institute, © 2014 Graham Bartram. All rights reserved.

IHS Jane's Fighting Ships 2014-2015 © 2014 IHS

Ensigns and flags of the world's navies [59]

Denmark
Ensign

El Salvador
National Flag and Ensign

Fiji
Ensign

Djibouti
National Flag and Ensign

Equatorial Guinea
National Flag and Ensign

Finland
Ensign

Dominica
National Flag and Ensign

Eritrea
National Flag and Ensign

France
Ensign

Dominican Republic
State Flag and Ensign

Estonia
Ensign

Gabon
National Flag and Ensign

East Timor
National Flag and Ensign

European Union
Flag of the European Union

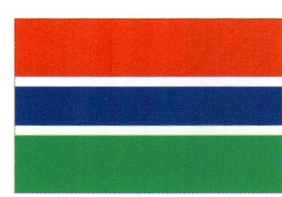
Gambia
National Flag and Ensign

Ecuador
State Flag and Ensign

Falkland Islands
Falkland Islands Flag and Ensign

Georgia
Ensign

Egypt
Ensign

Faroe Islands
Territory Flag and Ensign

Germany
Ensign

Flag Images courtesy of The Flag Institute, © 2014 Graham Bartram. All rights reserved.

[60]   **Ensigns and flags of the world's navies**

Ghana
Ensign

Haiti
State Flag and Ensign

Iran
National Flag and Ensign

Greece
National Flag and Ensign

Honduras
Ensign

Iraq
National Flag and Ensign

Grenada
Ensign

Hong Kong
Regional Flag and Ensign

Ireland
National Flag and Ensign

Guatemala
National Flag and Ensign

Hungary
National Flag and Ensign

Israel
Ensign

Guinea
National Flag and Ensign

Iceland
Ensign

Italy
Ensign

Guinea-Bissau
National Flag and Ensign

India
Ensign

Jamaica
Ensign

Guyana
National Flag and Ensign

Indonesia
National Flag and Ensign

Japan
Ensign

Flag Images courtesy of The Flag Institute, © 2014 Graham Bartram. All rights reserved.

IHS Jane's Fighting Ships 2014-2015   © 2014 IHS

Ensigns and flags of the world's navies [61]

Japan
Government Ensign

Kuwait
National Flag and Ensign

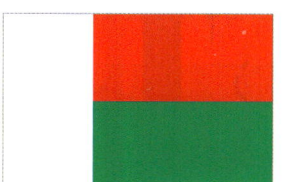
Madagascar
National Flag and Ensign

Jordan
Ensign

Latvia
Ensign

Malawi
National Flag and Ensign

Kazakhstan
Ensign

Lebanon
National Flag and Ensign

Malaysia
Ensign

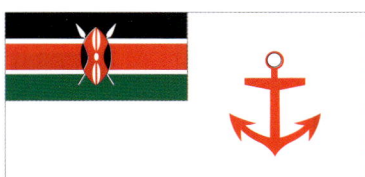
Kenya
Ensign

Liberia
National Flag and Ensign

Maldives
National Flag and Ensign

Kiribati
National Flag and Ensign

Libya
Ensign

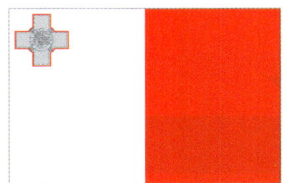
Malta
National Flag and Ensign

Korea, North
National Flag and Ensign

Lithuania
Ensign

Marshall Islands
National Flag and Ensign

Korea, South
National Flag and Ensign

Macedonia, Former Yugoslav Republic of
National Flag and Ensign

Mauritania
National Flag and Ensign

Flag Images courtesy of The Flag Institute, © 2014 Graham Bartram. All rights reserved.

IHS Jane's Defence Equipment & Technology

The IHS Defence Equipment & Technology portfolio is designed to provide business, defence and security customers with the highest quality defence reference content, grouped by subject and aligned to fit with their workflows, underpinned by structured content and overlaid with powerful search, manipulation and navigation tools.

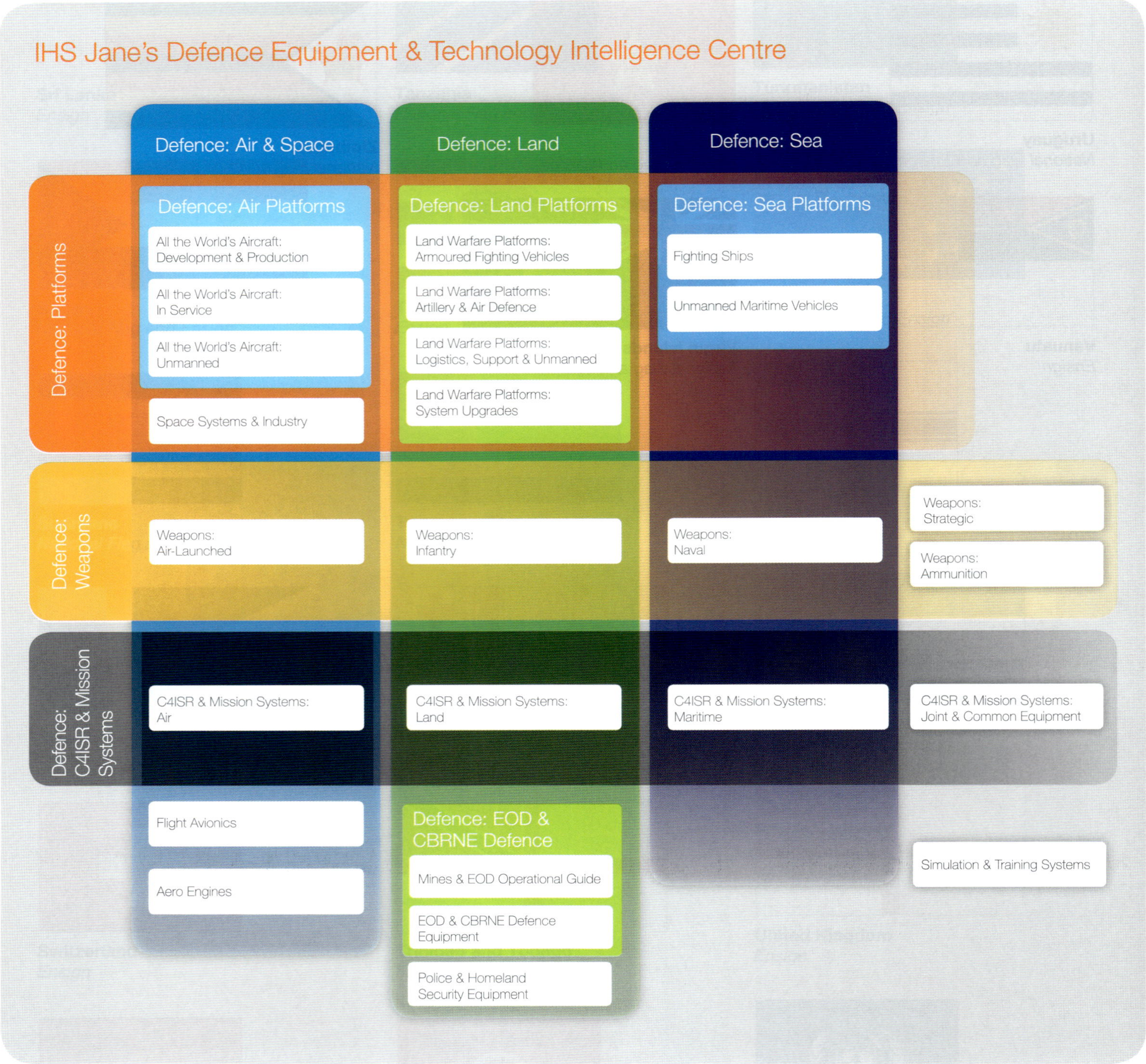

IHS Jane's Defence Equipment & Technology Intelligence Centre

Defence: Air & Space

Defence: Air Platforms
- All the World's Aircraft: Development & Production
- All the World's Aircraft: In Service
- All the World's Aircraft: Unmanned
- Space Systems & Industry

Defence: Land

Defence: Land Platforms
- Land Warfare Platforms: Armoured Fighting Vehicles
- Land Warfare Platforms: Artillery & Air Defence
- Land Warfare Platforms: Logistics, Support & Unmanned
- Land Warfare Platforms: System Upgrades

Defence: Sea

Defence: Sea Platforms
- Fighting Ships
- Unmanned Maritime Vehicles

Defence: Platforms

Defence: Weapons
- Weapons: Air-Launched
- Weapons: Infantry
- Weapons: Naval
- Weapons: Strategic
- Weapons: Ammunition

Defence: C4ISR & Mission Systems
- C4ISR & Mission Systems: Air
- C4ISR & Mission Systems: Land
- C4ISR & Mission Systems: Maritime
- C4ISR & Mission Systems: Joint & Common Equipment
- Flight Avionics
- Aero Engines
- Simulation & Training Systems

Defence: EOD & CBRNE Defence
- Mines & EOD Operational Guide
- EOD & CBRNE Defence Equipment
- Police & Homeland Security Equipment

Whether you're a researcher, an analyst, a planner, a strategist or a trainer, whether you're in the military, industry or academia, IHS will help you to understand the defence landscape and support you in your critical intelligence processes.

Contact us or visit
ihs.com/defence to find out more.

Pennant list of major surface ships

Ship type abbreviations

Notes: Designations specific to one nationality are followed by Country abbreviations.
The prefix W denotes a vessel of the Coastguard Service.
Suffixes to type indicators are as follows:
F denotes a vessel capable of speeds in excess of 35 kt.
G denotes a vessel with a force guided missile system, including SAM, USM, and SUM, usually with a range exceeding 20 miles.
H denotes a vessel equipped with a helicopter, or with a platform for operating one.
J denotes an air cushion or surface effect design.
K denotes a vessel equipped with hydrofoils.
M denotes a Combatant vessel with a close-range guided missile system.
N denotes a ship or submarine using nuclear propulsion

Submarines

AGSS	submarine, auxiliary, nuclear-powered (US)
DSRV	deep submergence rescue vehicle
DSV	deep submergence vehicle
SDV	swimmer delivery vehicle
SNA	submarine, attack, nuclear-powered (Fra)
SNLE	ballistic missile nuclear-powered submarine (Fra)
SS	submarine, general
SSA	submarine, auxiliary
SSA(N)	submarine, auxiliary, nuclear-powered
SSB	ballistic missile submarine (CPR)
SSBN	ballistic missile nuclear-powered submarine
SSC	submarine, coastal
SSGN	submarine, surface-to-surface missile, nuclear-powered
SSK	patrol submarine with ASW capability
SSW	submarine, midget
SSN	submarine, attack, nuclear-powered

Aircraft Carriers

CV (M)	aircraft carrier (guided missile system)
CVH (G)	helicopter carrier (guided missile system)
CVN (M)	aircraft carrier (nuclear-powered guided missile system)
PAN	aircraft carrier, nuclear-powered (Fra)

Cruisers

CG	guided missile cruiser
CGH	guided missile cruiser with helicopter
CGN	guided missile cruiser, nuclear-powered
CLM	guided missile cruiser (Per)

Destroyers

DD	destroyer
DDG (M)	guided missile destroyer
DDGH (M)	guided missile destroyer with helicopter, or helicopter platform
DDK	destroyer (Jpn)

Frigates

DE	destroyer escort (Jpn)
FF (L) (H)	frigate (light) (helicopter)
FFG (M)	guided missile frigate
FFGH (M)	guided missile frigate with helicopter, or helicopter platform
FS (G) (H) (M)	corvette (guided missile) (helicopter) (missile)

Patrol Forces

CF	river gunboat (Per)
CM	corvette (guided missile) (Per)
HSIC	high speed interception craft with speeds in excess of 55 kt
PB	coastal patrol vessel under 45 m without heavy armament
PB (F) (I) (R)	patrol boat (fast) (inshore) (river)
PBO (H)	offshore patrol vessel between 45 and 60 m (helicopter)
PC	vessel 35–55 m primarily for ASW role
PCK	As for PC but fitted with hydrofoils
PG	vessel 45–85 m equipped with at least 76 mm (3 in) gun
PGG	As for PG but with force guided missile system
PGGJ	As for PGG but air cushion or ground effect design
PGGK	As for PGG but fitted with hydrofoils
PSO (H)	offshore patrol vessel over 60 m (helicopter)
PTK	attack boat torpedo fitted with hydrofoils
PTGK	attack boat guided missile fitted with hydrofoils

SOC	special operations craft (US)

Landing Ships

AAAV	advanced amphibious assault vehicle
ACV	landing craft air cushion (Rus)
AGC	amphibious command ship (RoC)
ASDS	Advanced Swimmer-Seal Delivery System
EDCG	landing craft, utility (Brz)
LCA	landing craft, assault
LCAC	landing craft air cushion
LCC	amphibious command ship
LCH	landing craft, heavy (Aust)
LCM	landing craft, mechanised
LCP (L)	landing craft, personnel (large)
LCT	landing craft, tank
LCU	landing craft, utility
LCVP	landing craft, vehicle/personnel with bow ramp
LHA	amphibious assault ship general purpose with flooded well
LDW	swimmer delivery vehicle
LHD (M)	amphibious assault ship (multipurpose), can operate VSTOL aircraft and helicopters
LKA	amphibious cargo ship with own landing craft
LLP	assault ship, personnel
LPD	amphibious transport, dock with own LCMs and helicopter deck
LPH	amphibious assault ship, helicopter
LSD (H)	landing ship dock with own landing craft, helicopter
LSL (H)	landing ship logistic (Aust, UK, Sin), helicopter
LSM (H)	landing ship medium with bow doors and/or landing ramp, helicopter
LST (H)	landing ship tank with bow doors and/or landing ramp, helicopter
LSV	landing ship vehicle with bow doors and/or landing ramp
RCL	ramped craft, logistic (UK)
TCD	landing ship, dock (Fra)
UCAC	utility craft air cushion

Mine Warfare Ships

MCAC	mine clearance air cushion
MCD	mine countermeasures vessel, diving support
MCDV	maritime coast defence vessel (Can)
MCMV	mine countermeasures vessel
MCS	mine countermeasures support ship
MH (I) (C) (O)	minehunter (inshore) (coastal) (ocean)
MHCD	minehunter coastal with drone
MHSC	minehunter/sweeper coastal
ML (I) (C) (A)	minelayer (inshore) (coastal) (auxiliary)
MS (I) (C) (R)	minesweeper (inshore) (coastal) (river)
MSA (T)	minesweeper, auxiliary (tug)
MSB	minesweeper, boat
MSCD	coastal minesweeper capable of controlling drones
MSD	minesweeper, drone
MSO	minesweeper, ocean
SRMH	single role minehunter (UK)

Auxiliaries

ABU (H)	buoy tender (helicopter)
AD	destroyer tender
ADG	degaussing/deperming ship
AE (L)	ammunition ship capable of underway replenishment (small)
AEM	missile support ship
AET (L)	ammunition transport (small)
AF (L)	stores ship (small)
AFS	combat stores ship, capable of underway replenishment
AG (H)	auxiliary miscellaneous (helicopter)
AGB	icebreaker
AGDS	deep submergence support ship

AGE (H)	research ship (helicopter)
AGF (H)	auxiliary flag or command ship (helicopter)
AGI (H)	intelligence collection ship (helicopter)
AGM (H)	missile range instrumentation ship (helicopter)
AGOB	polar research ship
AGOR (H)	oceanographic research ship (helicopter)
AGOS (H)	ocean surveillance ship (helicopter)
AGP	patrol craft tender
AGS (C) (H)	surveying ship (coastal) (helicopter)
AH	hospital ship
AK (L) (R) (H)	cargo ship (light) (ro-ro) (helicopter)
AKE	armament stores carrier
AKR	roll on/roll off sealift ship
AKS (L) (H)	stores ship (light) (helicopter)
ANL	boom defence/cable/netlayer
AO	replenishment oiler (US)
AOE	fast combat support ship, primarily for POL replenishment
AOR (L) (H)	replenishment oiler (small) (helicopter)
AOT (L)	transport oiler (small)
AP (H)	personnel transport (helicopter)
APB	barracks ship
APCR	primary casualty receiving ship
AR (L)	repair ship (small)
ARC	submarine cable repair ship
ARS (D) (H)	salvage ship (heavy lift) (helicopter)
AS (L)	submarine tender (small)
ASE	research ship (Jpn)
ASR	submarine rescue ship
ATA	auxiliary ocean tug
ATF	fleet ocean tug and supply ship
ATR	fleet ocean tug (firefighting and rescue)
ATS	salvage and rescue ship
AVB	aviation support ship
AVM	aviation and missile support
AWT (L)	water tanker (small)
AX (L) (H)	training ship (small) (helicopter)
AXS	sail training ship
AXT	training tender
HSS	helicopter support ship
HSV	high speed logistic support vessel (catamaran)
TV	training ship (Jpn)

Service Craft

ASY	auxiliary yacht (Jpn)
SAR	search and rescue vessel
WFL	water/fuel lighter (Aust)
YAC	royal yacht
YAG	service craft, miscellaneous
YAGK	surface effect craft, experimental
YDG	degaussing vessel
YDT	diving tender
YE	ammunition lighter
YF	covered personnel transport under 40 m
YFB (H)	ferry (helicopter)
YFL	launch
YFRT	range safety vessel
YFU	former LCU used for cargo
YGS	survey launch
YH	ambulance boat
YM	dredging craft
YO (G)	fuel barge (gasolene)
YP	harbour patrol craft
YPB	floating barracks
YPC	oil pollution control vessel
YPT	torpedo recovery vessel
YT (B) (M) (L)	harbour tug (large) (medium) (small)
YTR	harbour fire/rescue craft with several monitors
YTT	torpedo trials craft
YW	water barge

© 2014 IHS

IHS Jane's Fighting Ships 2014-2015

[68] Pennant list

Pennant numbers of major surface ships in numerical order

Number	Ship's name	Type	Country	Page
001	President H I Remeliik	PB	Palau	612
001	San Juan	WPBO	Philippines	634
KBV 001	Poseidon	WPSO	Sweden	805
LRG 001	Constancia	PBR	Venezuela	1003
LSM 001	Punta Macolla	PB	Venezuela	1007
P 001	Pebane	PB	Mozambique	552
002	Edsa II	WPBO	Philippines	634
KBV 002	Triton	WPSO	Sweden	805
LRG 002	Perseverancia	PBR	Venezuela	1003
LSM 002	Farallón Centinela	PB	Venezuela	1007
003	Pampanga	WPBO	Philippines	634
AF 003	Amougna	PBR	Côte d'Ivoire	184
KBV 003	Amfitrite	WPSO	Sweden	805
LRG 003	Honestidad	PBR	Venezuela	1003
LSM 003	Charagato	PB	Venezuela	1007
004	Batangas	WPBO	Philippines	634
AF 004	Monsekela	PBR	Côte d'Ivoire	184
LRG 004	Tenacidad	PBR	Venezuela	1003
LSM 004	Bajo Brito	PB	Venezuela	1007
LRG 005	Integridad	PBR	Venezuela	1003
LSM 005	Bajo Araya	PB	Venezuela	1007
LRG 006	Lealtad	PBR	Venezuela	1003
LSM 006	Carecare	PB	Venezuela	1007
LSM 007	Vela De Cobo	PB	Venezuela	1007
LSM 008	Cayo Macereo	PB	Venezuela	1007
LSM 009	Chuspa	PB	Venezuela	1007
01	Rabaul	PB	Papua New Guinea	614
A 01	Contramaestre Casado	APH	Spain	780
A 01	Paluma	AGSC	Australia	34
A 01	Salema	WPB	Spain	782
ADR 01	Banderas	YM	Mexico	540
AMP 01	Huasteco	APH/AK/AH	Mexico	540
ARE 01	Otomi	ATF	Mexico	541
ATQ 01	Aguascalientes	YOG/YO	Mexico	541
BAL 01	Montes Azules	AK	Mexico	538
BE 01	Cuauhtémoc	AXS	Mexico	540
BI 01	Alejandro De Humboldt	AGOR	Mexico	539
BM 01	Mosoj Huayma	–	Bolivia	69
CSL 01	Wattle	YE	Australia	36
FL 01	Hiryu	FL/YTR	Japan	459
FM 01	Presidente Eloy Alfaro	FFGHM	Ecuador	210
FSM 01	Palikir	PB	Micronesia	541
G 01	Mazzei	PB/YXT	Italy	422
HL 01	Shoyo	AGS	Japan	459
KA 01	Kristaps	WPB	Latvia	495
L 01	Adelaide	LHD	Australia	31
L-01	Dourado	PB	Brazil	78
LC 01	Yopito	LCM	Venezuela	1004
LF 01	Cacine	PC	Guinea-Bissau	318
LP 01	Miguel Sotoa	PBR	Paraguay	616
LPH 01	Suboficial Rogelio Lesme	YGS	Paraguay	617
M 01	Al Hasbah	MHC	UAE	888
N-01	Tucunaré	PB	Brazil	78
NE-01	Barracuda	PB	Brazil	78
NGPWB 01	Bundeena	YFL/YDT	Australia	38
P 01	Bar	PB	Montenegro	543
P 01	Capitaine De Frégate Awore Paul	PB	Gabon	280
P 01	Capitán Cabral	PBR	Paraguay	616
P 01	Jarabakka	PB	Suriname	793
P 01	Liberta	PB	Antigua and Barbuda	11
P 01	Monte Sperone	PSO	Italy	423
P 01	Oryx	PBO	Namibia	559
P 01	Salamis	PBM	Cyprus	192
P 01	Trident	PB	Barbados	62
PB 01	Stalwart	PB	St Kitts and Nevis	734
PC 01	Matsunami	PC/PB	Japan	458
PL 01	Oki	PL/PSOH	Japan	452
PLH 01	Soya	PLH/PSOH	Japan	451
PS 01	Shinzan	PS/PBF	Japan	456
PT 01	Fatimah I	PB	Gambia	281
Q 01	Damsah	PGGF	Qatar	658
S 01	Gladan	AXS	Sweden	803
SE-01	Marlin	PB	Brazil	78
TNBH 01	Xavier Pinto Telleria	–	Bolivia	69
TNR 01	Jose Manuel Pando	YAG	Bolivia	69
U 01	Zaporizya	SS	Ukraine	875
02	Dreger	PB	Papua New Guinea	614
02	Hämeenmaa	ML	Finland	240
02	Tukoro	PB	Vanuatu	1000
A 02	Mermaid	AGSC	Australia	34
A 02	Rio Guadiaro	WPB	Spain	782
ADR 02	Magdalena	YM	Mexico	540
AMP 02	Zapoteco	APH/AK/AH	Mexico	540
ARE 02	Yaqui	ATF	Mexico	541
ATQ 02	Tlaxcala	YOG/YO	Mexico	541
BAL 02	Libertador	AK	Mexico	538
BI 02	Onjuku	AGS	Mexico	538
CSL 02	Boronia	YE	Australia	36
FM 02	Moran Valverde	FFGHM	Ecuador	210
FSM 02	Micronesia	PB	Micronesia	541
G 02	Vaccaro	PB/YXT	Italy	422
HL 02	Takuyo	AGS	Japan	459
KAL-IV-02	Baruna Jaya I	AGS/AGOR	Indonesia	371
L 02	Canberra	LHD	Australia	31
LF 02	Cacheu	PC	Guinea-Bissau	318
M 02	Al Murjan	MHC	UAE	888
MS 02	Saikai	YPC	Japan	460
NE-02	Anequim	PB	Brazil	78
NGPWB 02	Elouera	YFL/YDT	Australia	38
P 02	Defender	PB	St Lucia	735
P 02	Leonard C Banfield	PB	Barbados	62
P 02	Lieutenant De Vaisseau Mangoye Jean-Baptiste	PB	Gabon	280
P 02	Monte Cimone	PSO	Italy	423
P 02	Nanawa	PBR	Paraguay	616
P 02	Palmetto	PB	Antigua and Barbuda	11
P 02	Spari	PB	Suriname	793
PB 02	Ardent	PB	St Kitts and Nevis	734
PB 02	Levera	PB	Grenada	315
PL 02	Erimo	PL/PSOH	Japan	453
PLH 02	Tsugaru	PLH/PSOH	Japan	451
PM 02	Kitakami	PM/PSO	Japan	455
PS 02	Saroma	PS/PBF	Japan	456
PT 02	Sulayman Jun-Kung	PB	Gambia	281
Q 02	Al Ghariyah	PGGF	Qatar	658
S 02	Falken	AXS	Sweden	803
SE-02	Albacora	PB	Brazil	78
TNR 02	Nicolas Suarez	YAG	Bolivia	69
03	Lata	PB	Solomon Islands	760
03	Lomor	PB	Marshall Islands	524
03	Seeadler	PB	Papua New Guinea	614
03	Sydney	FFGHM	Australia	27
A 03	Rio Pisuerga	WPB	Spain	782
A 03	Shepparton	AGSC	Australia	34
ADR 03	Kino	YM	Mexico	540
ARE 03	Seri	ATF	Mexico	541
ATR 03	Tarasco	AK	Mexico	540
BI 03	Altair	AGOR	Mexico	539
CSL 03	Telopea	YE	Australia	36
G 03	Di Bartolo	PB/YXT	Italy	422
HL 03	Meiyo	AGS	Japan	459
KAL-IV-03	Baruna Jaya II	AGS/AGOR	Indonesia	371
MS 03	Katsuren	YPC	Japan	460
NGPWB 03	Shoalhaven	YFL/YDT	Australia	38
P 03	Enseigne De Vaisseau Betseng Adrien	PB	Gabon	280
P 03	Giovanni Denaro	PB	Italy	422
P 03	Gramorgu	PB	Suriname	793
P 03	Herceg Novi	PB	Montenegro	543
P 03	Lieutenant Commander Tsomakis	PB	Cyprus	193
P 03	Rudyard Lewis	PB	Barbados	62
P 03	Yellow Elder	PB	Bahamas	44
P 03	Zarrar	PGGF	Pakistan	607
PB 03	Rover I	PBF	St Kitts and Nevis	734
PL 03	Kudaka	PL/PSOH	Japan	453
PLH 03	Oosumi	PLH/PSOH	Japan	451
PS 03	Inasa	PS/PBF	Japan	456
Q 03	Rbigah	PGGF	Qatar	658
RK 03	Tiira	PB	Latvia	496
04	Auki	PB	Solomon Islands	760
04	Darwin	FFGHM	Australia	27
04	Moresby	PB	Papua New Guinea	614
A 04	Benalla	AGSC	Australia	34
A 04	Martín Posadillo	AKRH	Spain	780
A 04	Rio Nalon	WPB	Spain	782
ADR 04	Yavaros	YM	Mexico	540
ARE 04	Cora	ATF	Mexico	541
BI 04	Antares	AGOR	Mexico	539
G 04	Avallone	PB/YXT	Italy	422
HL 04	Tenyo	AGS	Japan	459
KAL-IV-04	Baruna Jaya III	AGS/AGOR	Indonesia	371
M 04	Imanta	MHC	Latvia	495
NGPWB 04	Sea Dragon	YFL/YDT	Australia	38
P 04	Carlskrona	AG	Sweden	803
P 04	Endeavour	PB	Barbados	62
P 04	Enseigne De Vaisseau Mendene	PB	Gabon	280
P 04	Karrar	PGGF	Pakistan	607
P 04	Lieutenant Commander Georgiu	PB	Cyprus	193
P 04	Protector	PB	St Lucia	735
P 04	Teniente Farina	PBR	Paraguay	616
PB 04	Hermitage Bay	PBF	St Kitts and Nevis	734
PL 04	Yahiko	PL/PSOH	Japan	453
PLH 04	Hayato	PLH/PSOH	Japan	451
PM 04	Tokachi	PM/PSO	Japan	455
PS 04	Kirishima	PS/PBF	Japan	456
Q 04	Barzan	PGGFM	Qatar	657
SVG 04	Hairoun	PB	St Vincent and the Grenadines	735
TNR 04	Max Paredes	YAG	Bolivia	69
05	Almirante Cochrane	FFGHM	Chile	119
05	Melbourne	FFGHM	Australia	27
05	Uusimaa	ML	Finland	240
A 05	Adventure	YGS	New Zealand	575
A 05	El Camino Español	AKR	Spain	780
ADR 05	Chamela	YM	Mexico	540
ARE 05	Iztaccihuatl	YTL	Mexico	541
BI 05	Rio Suchiate	AKS	Mexico	539
D 05	Ukale	PB	Dominica	205
FSM 05	Independence	PB	Micronesia	541
G 05	Oltramonti	PB/YXT	Italy	422
HL 05	Kaiyo	AGS	Japan	459
KAL-IV-05	Baruna Jaya IV	AGS/AGOR	Indonesia	371
L 05	President El Hadj Omar Bongo	LSTH	Gabon	281
M 05	Viesturs	MHC	Latvia	495
NGPWB 05	Ethel Joy	YFL/YDT	Australia	38
P 05	Enterprise	PB	Barbados	62
P 05	Itaipú	PBR	Paraguay	616
P 05	Skrunda	PB	Kuwait	494
PB 05	Oualie Bay	PBF	St Kitts and Nevis	734
PL 05	Dejima	PL/PSOH	Japan	453
PLH 05	Zao	PLH/PSOH	Japan	451
PM 05	Hitachi	PM/PSO	Japan	455
PS 05	Kamui	PS/PBF	Japan	456
Q05	Huwar	PGGFM	Qatar	657

Pennant list

Number	Ship's name	Type	Country	Page
TNR 05	Julio Olmos	YAG	Bolivia	69
06	Almirante Condell	FFGHM	Chile	119
06	Darussalam	PSO	Brunei	88
06	Newcastle	FFGHM	Australia	27
A 06	Rio Andarax	WPB	Spain	783
ADR 06	Tepoca	YM	Mexico	540
ARE 06	Popocateptl	YTL	Mexico	541
BI 06	Rio Hondo	AGS	Mexico	539
D 06	Dominique	PB	Dominica	205
G 06	Barbariso	PB/YXT	Italy	422
KA 06	Gaisma	WPB	Latvia	495
KAL-IV-06	Baruna Jaya VIII	AGOR	Indonesia	371
M 06	Talivaldis	MHC	Latvia	495
NGPWB 06	Reliance	YFL/YDT	Australia	38
P 06	Capitán Ortiz	PBF	Paraguay	616
P 06	Cēsis	PB	Latvia	494
P 06	Excellence	PB	Barbados	62
PL 06	Kurikoma	PL/PSOH	Japan	453
PLH 06	Okinawa	PLH/PSOH	Japan	451
PS 06	Raizan	PS/PBF	Japan	456
Q06	Al Udeid	PGGFM	Qatar	657
R 06	Illustrious	LPH	UK	909
TNBTL-06	Horacio Ugarteche	YAG	Bolivia	69
07	Almirante Lynch	FFGHM	Chile	119
07	Darulehsan	PSO	Brunei	88
A 07	Rio Guadalope	WPB	Spain	783
A 07	Takapu	YGS	New Zealand	575
ADR 07	Todos Santos	YM	Mexico	540
ARE 07	Citlaltepl	YTL	Mexico	541
BI 07	Moctezuma II	AGSC	Mexico	540
G 07	Paolini	PB/YXT	Italy	422
KA 07	Ausma	WPB	Latvia	495
M 07	Visvaldis	MHC	Latvia	495
NGPWB 07	Patonga	YFL/YDT	Australia	38
P 07	Général D'Armée Ba-Oumar	PBO	Gabon	280
P 07	Teniente Robles	PBF	Paraguay	616
P 07	Viesīte	PB	Latvia	494
PL 07	Satauma	PL/PSOH	Japan	453
PLH 07	Settsu	PLH/PSOH	Japan	451
PM 07	Isazu	PM/PSO	Japan	455
PS 07	Ashitaki	PS/PBF	Japan	456
Q07	Al Deebel	PGGFM	Qatar	657
TNR 07	Thames Crespo	YAG	Bolivia	69
08	Darulaman	PSO	Brunei	88
A 08	Rio Almanzora	WPB	Spain	783
A 08	Tarapunga	YGS	New Zealand	575
ADR 08	Asuncion	YM	Mexico	540
ARE 08	Xinantecatl	YTL	Mexico	541
BI 08	Alacran	YGS	Mexico	539
G 08	Greco	PB/YXT	Italy	422
KA 08	Saule	WPB	Latvia	495
L 08	Pono	LCU	Tanzania	826
M 08	Rūsiņš	MHC	Latvia	495
NGPWB 08	Bilgola	YFL/YDT	Australia	38
P 08	Colonel Djoue-Dabany	PBO	Gabon	280
P 08	Jelgava	PB	Latvia	494
P 08	Yhaguy	PBR	Paraguay	615
PL 08	Tosa	PL/PSOH	Japan	453
PLH 08	Echigo	PLH/PSOH	Japan	451
PM 08	Chitose	PM/PSO	Japan	455
PS 08	Kariba	PS/PBF	Japan	456
R 08	Queen Elizabeth	CV	UK	901
TNR 08	SO 2 Angel Orellano Vasquez	YAG	Bolivia	69
09	Daruttaqwa	PSO	Brunei	88
A 09	Manawanui	YDT	New Zealand	576
A 09	Rio Nervion	WPB	Spain	783
ADR 09	Almejas	YM	Mexico	540
ARE 09	Matlalcueye	YTL	Mexico	541
BI 09	Rizo	YGS	Mexico	539
G 09	Cinus	PB/YXT	Italy	422
KA 09	Klints	WPB	Latvia	495
L 09	Kibua	LCU	Tanzania	826
NGPWB 09	Sea Witch	YFL/YDT	Australia	38
P 09	Rēzekne	PB	Latvia	494
P 09	Tebicuary	PBR	Paraguay	615
PL 09	Kunigami	PL/PSOH	Japan	453
PLH 09	Ryukyu	PLH/PSOH	Japan	451
PM 09	Mabechi	PM/PSO	Japan	455
PS 09	Arase	PS/PBF	Japan	456
TNR 09	CNL Eduardo Avoroa Hidalgo	YAG	Bolivia	69
LSM 010	Isla De Medio	PB	Venezuela	1007
011	Chukotka	PBO	Russian Federation	726
011	PSKR 401	PBF	Russian Federation	729
011	Varyag	CGHM	Russian Federation	689
HQ 011	Dinh Tien Hoang	FFGM	Vietnam	1009
LSM 011	Morro De Barcelona	PB	Venezuela	1007
012	Astrakhan	PG	Russian Federation	703
012	Olenegorskiy Gorniak	LSTM	Russian Federation	704
012	Rais Hadj Mubarek	SSK	Algeria	4
HQ 012	Ly Thai To	FFGM	Vietnam	1009
LSM 012	Isla Papagallo	PB	Venezuela	1007
013	El Hadj Slimane	SSK	Algeria	4
013	PSKR 55	PBR	Russian Federation	731
014	Volgodonsk	PG	Russian Federation	703
015	Makhachkala	PG	Russian Federation	703
015	PSKR 632	PTF	Russian Federation	728
016	Georgiy Pobedonosets	LSTM	Russian Federation	704
016	Ural	PBO	Russian Federation	726
017	Podolsk	PGM	Russian Federation	728
018	Murmansk	PGH	Russian Federation	726
021	Akram Pacha	SSK	Algeria	4
021	Grad Sviyazhsk	PG	Russian Federation	703
021	Tolyatti	PCM	Russian Federation	727
022	Sidi Ahmed Rais	SSK	Algeria	4
022	Tver	PBO	Russian Federation	726
022	Uglich	PG	Russian Federation	703
023	Nakhodka	PCM	Russian Federation	727

Number	Ship's name	Type	Country	Page
023	Nevelsk	PGM	Russian Federation	728
024	Kaliningrad	PCM	Russian Federation	727
026	Yuzhno-Sakhalinsk	PGM	Russian Federation	728
027	Kondopoga	LSTM	Russian Federation	704
029	PSKR 635	PTF	Russian Federation	728
031	Alexander Otrakovskiy	LSTM	Russian Federation	704
031	Shapshal	PBF	Kazakhstan	463
031	Yaroslavl	PCM	Russian Federation	727
032	Batyr	PBF	Kazakhstan	463
033	Izet	PBF	Kazakhstan	463
035	PSKR 705	PTF	Russian Federation	728
035	Victor Kingisepp	PBO	Russian Federation	726
036	PSKR 709	PTF	Russian Federation	728
36-1	Puerto Quepos	PB	Costa Rica	183
037	Yastreb	PCM	Russian Federation	727
038	Zapolarye	PBO	Russian Federation	726
LP 039	Miguel Ela Edjodjomo	PB	Equatorial Guinea	231
040	Sarych	PCM	Russian Federation	727
041	Grif	PCM	Russian Federation	727
LP 041	Hipolito Micha	PB	Equatorial Guinea	231
042	Madeleine	PB	Lithuania	503
042	Orlan	PCM	Russian Federation	727
42-1	Primera Dama	PB	Costa Rica	183
043	Amur	PBO	Russian Federation	726
043	Novorossiysk	PCM	Russian Federation	727
044	AK 209	PGR	Russian Federation	732
044	Jamalyets	PGM	Russian Federation	728
044	Magadnets	PBO	Russian Federation	726
045	AK 223	PGR	Russian Federation	732
045	Karelia	PBO	Russian Federation	726
045	PSKR 719	PTF	Russian Federation	728
046	Brilliant	PSO	Russian Federation	730
046	PSKR 695	PTF	Russian Federation	728
047	AK 248	PGR	Russian Federation	732
047	Bata	PSO	Equatorial Guinea	232
050	Rubin	PSO	Russian Federation	730
052	Cheboksary	PCM	Russian Federation	727
052	Zemchuk	PSO	Russian Federation	730
053	Don	PBO	Russian Federation	726
053	Povorino	FFLM	Russian Federation	697
054	Eisk	FFLM	Russian Federation	697
054	Gomel	FSGM	Russian Federation	700
055	Admiral Nevelsky	LSTM	Russian Federation	704
055	Kasimov	FFLM	Russian Federation	697
055	Marshal Ustinov	CGHM	Russian Federation	689
056	Storm	PGR	Russian Federation	732
057	Groza	PGR	Russian Federation	732
058	Ladoga	PBO	Russian Federation	726
058	PSKR 57	PBR	Russian Federation	731
058	PSKR 645	PTF	Russian Federation	728
059	Aleksandrovets	FFLM	Russian Federation	697
060	Vladimirets	PGK	Russian Federation	703
061	Chang Bogo	SSK	Korea, South	474
061	PSKR 627	PTF	Russian Federation	728
062	60 Letiya Pogranvoysk	PGR	Russian Federation	732
062	Yi Chon	SSK	Korea, South	474
063	Admiral Kuznetsov	CVGM	Russian Federation	686
063	Choi Muson	SSK	Korea, South	474
063	Sokol	PCM	Russian Federation	727
064	Muromets	FFLM	Russian Federation	697
065	Briz	PGM	Russian Federation	728
065	Minsk	PCM	Russian Federation	727
065	Park Wi	SSK	Korea, South	474
065	PSKR 53	PBR	Russian Federation	731
066	Blagoveshchensk	PGR	Russian Federation	731
066	Lee Jongmu	SSK	Korea, South	474
066	Oslyabya	LSTM	Russian Federation	704
067	Jung Woon	SSK	Korea, South	474
068	Lee Sunsin	SSK	Korea, South	474
069	Na Daeyong	SSK	Korea, South	474
071	Lee Eokgi	SSK	Korea, South	474
071	Suzdalets	FFLM	Russian Federation	697
072	Sohn Won-il	SSK	Korea, South	474
073	Jeongji	SSK	Korea, South	474
075	Ahn Jung-Geun	SSK	Korea, South	474
076	Kim Jwa-Jin	SSK	Korea, South	474
077	Neptun	PGM	Russian Federation	728
077	Peresvet	LSTM	Russian Federation	704
078	Kobchik	PCM	Russian Federation	727
080	PSKR 706	PTF	Russian Federation	728
SSV 080	Pribaltika	AGIM	Russian Federation	712
081	Nikolay Vilkov	LSTM	Russian Federation	706
088	Cholmsk	PGM	Russian Federation	728
088	PSKA 295	PBO	Russian Federation	729
090	Nikolay Sipyagin	AK	Russian Federation	732
093	PSKR 56	PBR	Russian Federation	731
099	Pyotr Velikiy	CGHMN	Russian Federation	688
099	Siktivkar	PGM	Russian Federation	728
L 0203	Debundscha	LCU	Cameroon	97
0329	Wyatt Earp	YGS	Australia	34
1	Sea Based X-Band Radar-1	–	US	976
1	Uruguay	FF	Uruguay	995
AFDL 1	Hay Tan	YFD	Taiwan	820
AFSB(i)	Ponce	–	US	979
B 1	Patagonia	AORH	Argentina	20
C 1	Paraguay	PGR	Paraguay	615
DF 1	Endeavor	YFD	Dominican Republic	207
FPB 1	Taipei	PB	Gambia	281
FSF-1	Sea Fighter	AGE	US	970
HST 1	Guam	HSV	US	982
JHSV 1	Spearhead	TSV	US	980
LCC 1	Kao Hsiung	AGF	Taiwan	818
LCS 1	Freedom	–	US	950
LHD 1	Wasp	LHDM	US	964
LSV 1	Gen Frank S Besson Jr	LSV-ARMY	US	961
MCM 1	Avenger	MCM/MHSO	US	969
MOV 1	Dr Bernardo Houssay	WAXL/WAXS	Argentina	23
PB 1	Steve Obimpeh	PBF	Ghana	301
PCL 1	Ning Hai	PCF	Taiwan	817

[70] Pennant list

Number	Ship's name	Type	Country	Page
PRM-1	Falcon	–	US	972
PT 1	Amberjack	WPB	Singapore	758
RM 1	Guaroa	YTM/YTL	Dominican Republic	207
RSV 1	Safaga	MSI	Egypt	225
S 1	Shabab Oman	AXS	Oman	596
SAR 1	Dagma	PB	Denmark	197
T 1	Al Sultana	AKS	Oman	597
T-ACS 1	Keystone State	AK	US	982
T-AKE 1	Lewis And Clark	AKEH	US	974
T-MLP-1	Montford Point	AKR	US	978
Z 1	Al Bushra	PBO	Oman	595
Z1	Baltyk	AORL	Poland	644
2	Pedro Campbell	FF	Uruguay	995
2–508	Al Hirasa	FFL	Syria	808
82-2	Santamaria	PB	Costa Rica	183
A 2	Nasr Al Bahr	LSTH	Oman	596
A 2	Teniente Olivieri	PBO	Argentina	17
AFDL 2	Kim Men	YFD	Taiwan	820
DSV 2	Alvin	DSV	US	972
FPB 2	Kunta Kinteh	PB	Gambia	281
HD 2	Viken	YPT/YDT	Norway	591
HST 2	Puerto Rico	HSV	US	982
JHSV 2	Choctaw County	TSV	US	980
LCS 2	Independence	–	US	953
LHD 2	Essex	LHDM	US	964
LSV 2	CW 3 Harold C Clinger	LSV-ARMY	US	961
LSV-2	Cutthroat	DSV	US	972
MCM 2	Defender	MCM/MHSO	US	969
PB 2	Philemon Quaye	PBF	Ghana	301
PC 2	Tempest	PBFM	US	959
PCL 2	An Hai	PCF	Taiwan	817
PT 2	Salmon	WPB	Singapore	758
Q 2	Libertad	AXS	Argentina	19
R 2	Querandi	YTB/YTL	Argentina	21
RM 2	Guarionex	YTM/YTL	Dominican Republic	207
RSV 2	Abu El Ghoson	MSI	Egypt	225
SAR 2	Naja	PB	Denmark	197
T-ACS 2	Gem State	AK	US	982
T-AKE 2	Sacagawea	AKEH	US	974
T-MLP-2	John Glenn	AKR	US	978
Z 2	Al Mansoor	PBO	Oman	595
A 3	Francisco De Gurruchaga	PSO	Argentina	17
AFDL 3	Han Jih	YFD	Taiwan	820
B 3	Canal Beagle	AKS	Argentina	20
FPB 3	Kindy Camara	PB	Gambia	281
H3	Haras 3	PBF	Oman	598
JHSV 3	Millinocket	TSV	US	980
LCS 3	Fort Worth	–	US	950
LHD 3	Kearsarge	LHDM	US	964
LSV 3	Gen Brehon B Somervell	LSV-ARMY	US	961
MCM 3	Sentry	MCM/MHSO	US	969
PB 3	Joy Amedume	PBF	Ghana	301
PC 3	Hurricane	PBFM	US	959
PT 3	Tuna	WPB	Singapore	758
R 3	Tehuelche	YTB/YTL	Argentina	21
RM 3	Guarocuya	YTM/YTL	Dominican Republic	207
SI 3	Rodolfo D'Agostini	YTL/YTR	Argentina	24
T-ACS 3	Grand Canyon State	AK	US	982
T-AKE 3	Alan Shepard	AKEH	US	974
T-AVB 3	Wright	AVB	US	982
T-MLP-3	Lewis B Puller	AKR	US	978
YAC 3	Oriole	AXS	Canada	104
Z 3	Al Najah	PBO	Oman	595
4	General Artigas	ARL	Uruguay	999
4	Penyu	AT	Malaysia	516
B 4	Bahia San Blas	AKS	Argentina	20
D 4	Melville	PB	Dominica	204
FPB 4	Musa Maliya Jammeh	PB	Gambia	281
H4	Haras 4	PBF	Oman	598
HS 4	Sleipner	YPT/YDT	Norway	591
JHSV 4	Fall River	TSV	US	980
LCS 4	Coronado	–	US	953
LHD 4	Boxer	LHDM	US	964
LSV 4	LTG William B Bunker	LSV-ARMY	US	961
MCM 4	Champion	MCM/MHSO	US	969
PB 4	Dzang	PBF	Ghana	301
PC 4	Monsoon	PBFM	US	959
PT 4	Coral Cod	WPB	Singapore	758
RM 4	Magua	YTM/YTL	Dominican Republic	207
SB 4	Kodor	ARS/ATA	Russian Federation	723
T-ACS 4	Gopher State	AK	US	982
T-AKE 4	Richard E Byrd	AKEH	US	974
T-AVB 4	Curtiss	AVB	US	982
5	Bunga Mas Lima	AG	Malaysia	515
5	Uruguay	PBR	Uruguay	997
ARD 5	Fo Wu 5	YFD	Taiwan	820
ARDM 5	Arco	ARDM	US	973
B 5	Cabo De Hornos	AKS	Argentina	20
BP 5	Rettin	WPBR	Germany	298
H5	Haras 5	PBF	Oman	598
HS 5	Mjølner	YPT/YDT	Norway	591
JHSV 5	Trenton	TSV	US	980
LCS 5	Milwaukee	–	US	950
LHA 5	Peleliu	LHAM	US	963
LHD 5	Bataan	LHDM	US	964
LSV 5	MG Charles P Gross	LSV-ARMY	US	961
MSD 5	Hirsholm	MSD/AXL/AGSC	Denmark	198
PB 5	Owusu-Ansah	PBF	Ghana	301
PC 5	Typhoon	PBFM	US	959
PT 5	Cosby	WPB	Singapore	758
Q 5	Almirante Irizar	AGB/AGOB	Argentina	21
R 5	Angostura	YTM/YTL	Paraguay	617
R 5	Mocovi	YTB/YTL	Argentina	21
T-ACS 5	Flickertail State	AK	US	982
T-AKE 5	Robert E Peary	AKEH	US	974
6	Bunga Mas Enam	AG	Malaysia	515
A 6	Suboficial Castillo	PSO	Argentina	17
AFDL 6	Dynamic	AFDL	US	973
ARD 6	Fo Wu 6	YFD	Taiwan	820
CG 6	Cascadura	PB	Trinidad and Tobago	849
H 6	Haras 6	PB	Oman	598
JHSV 6	Brunswick	TSV	US	980
LCS 6	Jackson	–	US	953
LHA 6	America	LHA	US	966
LHD 6	Bonhomme Richard	LHDM	US	964
LSV 6	SP/4 James A Loux	LSV-ARMY	US	961
MCM 6	Devastator	MCM/MHSO	US	969
MSD 6	Saltholm	MSD/AXL/AGSC	Denmark	198
PB 6	Ohene-Kwapong	PBF	Ghana	301
PC 6	Sirocco	PBFM	US	959
PT 6	Dolphin	WPB	Singapore	758
R 6	Calchaqui	YTB/YTL	Argentina	21
SB 6	Moshchny	ARS/ATA	Russian Federation	723
T-ACS 6	Cornhusker State	AK	US	982
T-AKE 6	Amelia Earhart	AKEH	US	974
T-AOE 6	Supply	AOEH	US	974
A 7	Al Neemran	LCU	Oman	596
CG 7	Corozal Point	PB	Trinidad and Tobago	849
H 7	Haras 7	PB	Oman	598
HM 7	Kjeøy	YPT/YDT	Norway	591
JHSV 7	Carson City	TSV	US	980
LCS 7	Detroit	–	US	950
LHA 7	Tripoli	LHA	US	966
LHD 7	Iwo Jima	LHDM	US	964
LSV 7	SSGT Robert T Kuroda	LSV-ARMY	US	961
MCM 7	Patriot	MCM/MHSO	US	969
PB 7	Tom Annan	PBF	Ghana	301
PC 7	Squall	PBFM	US	959
PT 7	Leather Jacket	WPB	Singapore	758
R 7	Ona	YTB/YTL	Argentina	21
R 7	Esperanza	YTM/YTL	Paraguay	617
T-AKE 7	Carl M Brashear	AKEH	US	974
T-AOE 7	Rainier	AOEH	US	974
T-ARC 7	Zeus	ARC	US	978
V 7	Coco-Beach	PB	Gabon	280
YR 7	Esperanza	WAXL/WAXS	Argentina	23
8	Kepah	AT	Malaysia	515
A 8	Saba Al Bahr	LCM	Oman	596
CG 8	Crown Point	PB	Trinidad and Tobago	849
JHSV 8	Yuma	TSV	US	980
LCS 8	Montgomery	–	US	953
LHD 8	Makin Island	LHDM	US	964
LSV 8	Mg Robert Smalls	LSV-ARMY	US	961
MCM 8	Scout	MCM/MHSO	US	969
P 8	Paul Bogle	PB	Jamaica	427
PC 8	Zephyr	PBFM	US	959
PT 8	Mangrove Jack	WPB	Singapore	758
R 8	Toba	YTB/YTL	Argentina	21
SB 8	Canal Emilio Mitre	YTL/YTR	Argentina	24
T-AKE 8	Wally Schirra	AKEH	US	974
T-AOE 8	Arctic	AOEH	US	974
V 8	Port Gentil	PB	Gabon	280
A 9	Al Doghas	LCM	Oman	596
A 9	Alferez Sobral	PBO	Argentina	17
CG 9	Galera Point	PB	Trinidad and Tobago	849
H 9	Haras 9	PB	Oman	598
JHSV 9	Bismarck	TSV	US	980
LCS 9	Little Rock	–	US	950
LPD 9	Denver	LPD	US	966
MCM 9	Pioneer	MCM/MHSO	US	969
PC 9	Chinook	PBFM	US	959
PT 9	Oscar	WPB	Singapore	758
SB 9	Canal Costanero	YTL/YTR	Argentina	24
T-AKE 9	Matthew Perry	AKEH	US	974
V 9	Omboue	PB	Gabon	280
10	Al Riffa	PB	Bahrain	48
10	Colonia	PB	Uruguay	997
A 10	Al Temsah	LCM	Oman	596
A 10	Rio Guadalaviar	WPB	Spain	783
ACV 10	Roebuck Bay	PB	Australia	39
ADR 10	Chacagua	YM	Mexico	540
ARE 10	Tlaloc	YTL	Mexico	541
BI 10	Cabezo	YGS	Mexico	539
CG 10	Barcolet Point	PB	Trinidad and Tobago	849
D 10	Almirante Brown	DDGHM	Argentina	13
GED 10	Guarapari	EDCG/LCU	Brazil	80
H 10	Haras 10	PB	Oman	598
JHSV 10	Burlington	TSV	US	980
LCS 10	Gabrielle Giffords	–	US	953
MCM 10	Warrior	MCM/MHSO	US	969
NGPWB 10	Brutus	YFL/YDT	Australia	38
P 10	Général Nazaire Boulingui	PTM	Gabon	280
P 10	Piratini	PB	Brazil	80
PC 10	Firebolt	PBFM	US	959
PK 10	Sailfish	PBF	Singapore	758
PL 10	Motobu	PL/PSOH	Japan	453
PLH 10	Daisen	PLH/PSOH	Japan	451
PM 10	Sorachi	PM/PSO	Japan	455
PO 10	Veronica Adley	PBF	St Lucia	735
PS 10	Sanbe	PS/PBF	Japan	456
PT 10	Pari Burong	WPB	Singapore	758
R 10	Chulupi	YTB/YTL	Argentina	21
SVG 10	H K Tannis	PB	St Vincent and the Grenadines	736
T-AKE 10	Charles Drew	AKEH	US	974
T-AKR 10	Cape Island	AKR	US	982
T-AOE 10	Bridge	AOEH	US	974
U 10	Aspirante Nascimento	AXL	Brazil	85
V 10	Mayumba	PB	Gabon	280
WAGB 10	Polar Star	WAGBH	US	988
YTT 10	Battle Point	YTT	US	972
Z 10	Dhofar	PGGF	Oman	595
11	Capitán Prat	FFGM	Chile	118
11	Hawar	PB	Bahrain	48
11	Mahamiru	MHC	Malaysia	514
11	Rio Negro	PB	Uruguay	997
11	Smeli	FFLM	Bulgaria	89
11	Triglav	PBO	Slovenia	759
A 11	Endeavour	AORH	New Zealand	576
A 11	Rio Cabriel	WPB	Spain	783

Pennant list [71]

Number	Ship's name	Type	Country	Page
ADR 11	Coyuca	YM	Mexico	540
AGS 11	Sunjin	AGE	Korea, South	486
BE 11	Simón Bolívar	AXS	Venezuela	1006
BI 11	Anegagada De Adentro	YGS	Mexico	539
BO 11	Punta Brava	AGOR	Venezuela	1005
CF 11	Amazonas	CF/PGR	Peru	622
CG 11	Scarlet Ibis	PB	Trinidad and Tobago	850
CM 11	Esmeraldas	FSGHM	Ecuador	210
D 11	La Argentina	DDGHM	Argentina	13
F 11	Aung Zeya	FSG	Myanmar	553
GED 11	Tambaú	EDCG/LCU	Brazil	80
K 11	Felinto Perry	ASRH	Brazil	85
K 11	Stockholm	FSG	Sweden	799
LCS 11	Sioux City	–	US	950
LG 11	Los Taques	YAG	Venezuela	1006
LH 11	Gabriela	AGSC	Venezuela	1005
M 11	Styrsö	MHSDI/YDT	Sweden	802
MB 11	Alexander Piskunov	ATA	Russian Federation	723
MCM 11	Gladiator	MCM/MHSO	US	969
P 11	Brendan Simbwaye	PBO	Namibia	559
P 11	Elephant	PBO	Namibia	559
P 11	Pirajá	PB	Brazil	80
P 11	Zemaitis	PBO	Lithuania	502
PC 11	Constitución	PBG/PG	Venezuela	1003
PC 11	Hayanami	PC/PB/YTR	Japan	458
PC 11	Whirlwind	PBFM	US	959
PF 11	Rajah Humabon	FF	Philippines	629
PL 11	Dionysos	PB	Cyprus	193
PM 11	Yubari	PM/PSO	Japan	455
PO 11	Eusebe Lawrence	PBF	St Lucia	735
PO 11	Guaiqueri	PSOH	Venezuela	1004
PS 11	Mizuki	PS/PBF	Japan	456
PT 11	Piranha	WPB	Singapore	758
Q 11	Comodoro Rivadavia	AGOR	Argentina	19
RA 11	General Francisco De Miranda	ATA	Venezuela	1006
RTOP 11	Kralj Petar Kresimir IV	FSG	Croatia	186
SVK 11	Östhammar	PB	Sweden	800
T-AKE 11	Washington Chambers	AKEH	US	974
T-AKR 11	Cape Intrepid	AKR	US	982
U 11	Guarda Marinha Jansen	AXL	Brazil	85
YTT 11	Discovery Bay	YTT	US	972
Z 11	Al Sharqiyah	PGGF	Oman	595
12	Jerai	MHC	Malaysia	514
12	Paysandu	PB	Uruguay	997
12	Radhwa	YTB/YTM	Saudi Arabia	742
12-64	Al Whada	SAR	Morocco	551
12-65	Sebou	SAR	Morocco	551
A 12	Rio Cervantes	WPB	Spain	783
A 12	São Paulo	CVM	Brazil	72
ADR 12	Farrallon	YM	Mexico	540
BI 12	Rio Tuxpan	AGS	Mexico	539
CF 12	Loreto	CF/PGR	Peru	622
CG 12	Hibiscus	PB	Trinidad and Tobago	850
CM 12	Manabi	FSGHM	Ecuador	210
D 12	Heroina	DDGHM	Argentina	13
F 12	Kyan-Sit-Thar	FSG	Myanmar	553
GED 12	Camboriú	EDCG/LCU	Brazil	80
HARAS 12	Dheeb Al Bahar 2	PB	Oman	598
K 12	Malmö	FSG	Sweden	799
L 12	Ocean	LPH	UK	910
LCS 12	Omaha	–	US	953
LG 12	Los Cayos	YAG	Venezuela	1006
LH 12	Lely	AGSC	Venezuela	1005
M 12	Spårö	MHSDI/YDT	Sweden	802
MCM 12	Ardent	MCM/MHSO	US	969
P 12	Djukas	PBO	Lithuania	502
P 12	Pampeiro	PB	Brazil	80
PC 12	Federación	PBG/PG	Venezuela	1003
PC 12	Setogiri	PC/PB/YTR	Japan	458
PC 12	Thunderbolt	PBFM	US	959
PL 12	Kourion	PB	Cyprus	193
PM 12	Motoura	PM/PSO	Japan	455
PO 12	Warao	PSOH	Venezuela	1004
PS 12	Kouya	PS/PBF	Japan	456
PT 12	Porpoise	WPB	Singapore	758
R 12	Mataco	YTB/YTL	Argentina	21
RK 12	Valpas	WPBO	Latvia	497
RTOP 12	Kralj Dmitar Zvonimir	FSG	Croatia	186
T-AKE 12	William Mclean	AKEH	US	974
U 12	Guarda Marinha Brito	AXL	Brazil	85
Y 12	Şalopa 12	YC	Turkey	868
YDT 12	Granby	YDT	Canada	104
Z 12	Al Bat'nah	PGGF	Oman	595
13	Ledang	MHC	Malaysia	514
13	Reshitelni	FSM	Bulgaria	91
A 13	Rio Ara	WPB	Spain	783
A 13	Tunas Samudera	AXS	Malaysia	515
ADR 13	Chairel	YM	Mexico	540
B 13	Ingeniero Julio Krause	AKS/AOTL	Argentina	20
CF 13	Marañon	CF/PGR	Peru	622
CG 13	Humming Bird	PB	Trinidad and Tobago	850
CM 13	Los Rios	FSGHM	Ecuador	210
D 13	Sarandi	DDGHM	Argentina	13
GC 13	Delfin	WPSO	Argentina	22
LCS 13	Wichita	–	US	950
M 13	Skaftö	MHSDI/YDT	Sweden	802
MCM 13	Dextrous	MCM/MHSO	US	969
P 13	Parati	PB	Brazil	80
PC 13	Independencia	PBG/PG	Venezuela	1003
PC 13	Mizunami	PC/PB/YTR	Japan	458
PC 13	Shamal	PBFM	US	959
PL 13	Ilarion	PB	Cyprus	193
PM 13	Muromi	PM/PSO	Japan	455
PO 13	Yecuana	PSOH	Venezuela	1004
PS 13	Tsukuba	PS/PBF	Japan	456
PT 13	Hardyhead	WPB	Singapore	758
T-AKE 13	Medgar Evers	AKEH	US	974
14	Almirante Latorre	FFGM	Chile	118
14	Bodri	FSM	Bulgaria	91
14	Kinabalu	MHC	Malaysia	514
A 14	Patiño	AORH	Spain	780
A 14	Rio Adaja	WPB	Spain	783
ADR 14	San Andres	YM	Mexico	540
AGOR 14	Melville	AGOR	US	971
CF 14	Ucayali	CF/PGR	Peru	622
CG 14	Chaconia	PB	Trinidad and Tobago	850
CM 14	El Oro	FSGHM	Ecuador	210
J 14	Nirupak	AGSH	India	351
KA 14	Astra	WPB	Latvia	496
L 14	Albion	LPD	UK	911
LCS 14	Manchester	–	US	953
M 14	Sturkö	MHSDI/YDT	Sweden	802
MCM 14	Chief	MCM/MHSO	US	969
P 14	Aukstaitis	PBO	Lithuania	502
P 14	Bolong Kanta	PB	Gambia	281
P 14	Penedo	PB	Brazil	80
P 14	Perwira	PB	Brunei	88
PC 14	Iyonami	PC/PB/YTR	Japan	458
PC 14	Libertad	PBG/PG	Venezuela	1003
PC 14	Tornado	PBFM	US	959
PL 14	Karpasia	PB	Cyprus	193
PM 14	Tsurumi	PM/PSO	Japan	455
PO 14	Kariña	PSOH	Venezuela	1004
PS 14	Akagi	PS/PBF	Japan	456
R 14	Zbyszko	ARS	Poland	645
T-AKE 14	César Chávez	AKEH	US	974
U 14	Aspirante Moura	AGS	Brazil	83
Z 14	Mussandam	PGGF	Oman	595
15	Almirante Blanco Encalada	FFGHM	Chile	117
A 15	Cantabria	AORH	Spain	780
A 15	Nireekshak	ASR	India	352
A 15	Rio Duero	WPB	Spain	783
ADR 15	San Ignacio	YM	Mexico	540
AGOR 15	Knorr	AGOR	US	971
CF 15	Clavero	CF/PGR	Peru	622
CG 15	Poui	PB	Trinidad and Tobago	850
CM 15	Galapágos	FSGHM	Ecuador	210
F 15	Abu Bakr	FFG	Bangladesh	54
G 15	Paraguassú	AP	Brazil	85
HP 15	Hitra	YPT/YDT	Norway	591
J 15	Investigator	AGSH	India	351
L 15	Bulwark	LPD	UK	911
L 15	Kesari	LSTH	India	349
LCS 15	Billings	–	US	950
M 15	Aratu	MSC	Brazil	82
P 15	Pemburu	PB	Brunei	88
P 15	Poti	PB	Brazil	80
PC 15	Kurinami	PC/PB/YTR	Japan	458
PC 15	Patria	PBG/PG	Venezuela	1003
PF 15	Gregorio Del Pilar	FFH	Philippines	628
PL 15	Akamas	PB	Cyprus	193
PM 15	Teshio	PM/AGOB	Japan	455
PS 15	Bizan	PS/PBF	Japan	456
PT 15	Todak	WPB	Singapore	758
Q 15	Cormoran	AGSC	Argentina	19
R 15	Macko	ARS	Poland	645
SB 15	Tango	ARS	Argentina	24
U 15	Pará	YFB	Brazil	85
V 15	Imperial Marinheiro	PG/ATR	Brazil	79
16	Liaoning	CVGM	China	135
A 16	Rio Guadiana	WPB	Spain	783
ADR 16	Terminos	YM	Mexico	540
CF 16	Castilla	CF/PGR	Peru	622
CG 16	Teak	PB	Trinidad and Tobago	850
CM 16	Loja	FSGHM	Ecuador	210
F 16	Umar Farooq	FF	Bangladesh	59
J 16	Jamuna	AGSH	India	351
L 16	Absalon	AGF/AKR/AH	Denmark	200
L 16	Shardul	LSTH	India	349
LCS 16	Tulsa	–	US	953
M 16	Anhatomirim	MSC	Brazil	82
M 16	Almirante Diaz Pimienta	WPBF	Spain	783
P 16	Penyerang	PB	Brunei	88
PC 16	Hamanami	PC/PB/YTR	Japan	458
PC 16	Victoria	PBG/PG	Venezuela	1003
PF 16	Ramon A Alcaraz	FFH	Philippines	628
PS 16	Nobaru	PS/PBF	Japan	456
R 16	Capayán	YTB/YTL	Argentina	21
U 16	Doutor Montenegro	AH	Brazil	86
17	Ijhtihad	PB	Brunei	88
A 17	Rio Francoli	WPB	Spain	783
ADR 17	Teculapa	YM	Mexico	540
F 17	Ali Haider	FFG	Bangladesh	54
G 17	Potengi	AG	Brazil	86
J 17	Sutlej	AGSH	India	351
L 17	Esbern Snare	AGF/AKR/AH	Denmark	200
LCS 17	Indianapolis	–	US	950
LPD 17	San Antonio	LPDM	US	962
M 17	Atalaia	MSC	Brazil	82
M 17	Rio Arba	PB	Spain	783
PC 17	Shinonome	PC/PB/YTR	Japan	458
PS 17	Takachiho	PS/PBF	Japan	456
U 17	Parnaiba	PGRH	Brazil	80
18	Almirante Riveros	FFGHM	Chile	117
18	Berkat	PB	Brunei	88
A 18	Rio Guadalete	WPB	Spain	783
F 18	Osman	FFG	Bangladesh	53
H 18	Comandante Varella	ABU	Brazil	83
J 18	Sandhayak	AGSH	India	351
L 18	Cheetah	LSM/LSMH	India	350
LPD 18	New Orleans	LPDM	US	962
M 18	Araçatuba	MSC	Brazil	82
M 18	Rio Caudal	PB	Spain	783
P 18	Armatolos	PG	Greece	308
PC 18	Harunami	PC/PB/YTR	Japan	458
PS 18	Sanrei	PS/PBF	Japan	456
PT 18	Thread Fin	WPB	Singapore	758
R 18	Chiquillán	YTB/YTL	Argentina	21
U 18	Oswaldo Cruz	AHH	Brazil	86

[72] Pennant list

Number	Ship's name	Type	Country	Page
19	Almirante Williams	FFHM	Chile	118
19	Syafaat	PB	Brunei	88
A 19	Cabaleiro	WPB	Spain	783
DF 19	Parana	WAGH/AHH	Argentina	24
GS 19	Zhigulevsk	AGIM	Russian Federation	712
H 19	Tenente Castelo	ABU	Brazil	83
J 19	Nirdeshak	AGSH	India	351
L 19	Mahish	LSM/LSMH	India	350
LCC 19	Blue Ridge	LCCH/AGFH	US	961
LPD 19	Mesa Verde	LPDM	US	962
M 19	Abrolhos	MSC	Brazil	82
M 19	Rio Bernesga	PB	Spain	783
P 19	Navmachos	PG	Greece	308
P 19	Ngunguri	PB	Tanzania	826
PC 19	Kiyozuki	PC/PB/YTR	Japan	458
PS 19	Asaji	PS/PBF	Japan	456
PS 19	Miguel Malvar	FS	Philippines	630
PT 19	Dorado	WPB	Singapore	758
R 19	Morcoyán	YTB/YTL	Argentina	21
T-AGOS 19	Victorious	AGOS	US	976
T-AH 19	Mercy	AHH	US	980
U 19	Carlos Chagas	AHH	Brazil	86
V 19	Caboclo	PG/ATR	Brazil	79
20	Ahmad El Fateh	PGGF	Bahrain	48
20	Afiat	PB	Brunei	88
20	Thomson	SSK	Chile	117
20	Capitán Miranda	AXS	Uruguay	998
A 20	Moawin	AORH	Pakistan	609
A 20	Neptuno	ATF/AGDS	Spain	781
A 20	Rio Segre	PBO	Spain	783
ACV 20	Holdfast Bay	PB	Australia	39
CG 20	Nelson	PBO	Trinidad and Tobago	849
DF 20	Recalada	WAGH/AHH	Argentina	24
F 20	Godavari	FFGHM	India	339
H 20	Comandante Manhães	ABU	Brazil	83
HPB 20	Terrace Bay	PB	Namibia	559
L 20	Magar	LSTH	India	349
LCC 20	Mount Whitney	LCCH/AGFH	US	961
LPD 20	Green Bay	LPDM	US	962
M 20	Albardão	MSC	Brazil	82
M 20	Rio Martin	PB	Spain	783
P 20	Anthypoploiarchos Laskos	PGGF/PGG	Greece	307
P 20	Murature	AX	Argentina	20
P 20	Pedro Teixeira	PBR	Brazil	79
PC 20	Ayanami	PC/PB/YTR	Japan	458
PK 20	Spearfish	PBF	Singapore	758
PS 20	Magat Salamat	FS	Philippines	630
PT 20	Manta Ray	WPB	Singapore	757
PV 20	Poseidon	PBF	Cyprus	193
Q 20	Puerto Deseado	AGOB	Argentina	19
RK 20	Rānda	WPB	Latvia	496
T-AGOS 20	Able	AGOS	US	976
T-AH 20	Comfort	AHH	US	980
U 20	Cisne Branco	AXS	Brazil	85
WAGB 20	Healy	WAGBH	US	988
Z 20	Seeb	PB	Oman	595
21	Al Jabiri	PGGF	Bahrain	48
21	Cheong Hae Jin	ARS	Korea, South	486
21	Hejaz	LST	Iran	389
21	Mustaed	PB	Brunei	88
21	Simpson	SSK	Chile	117
21	Sirius	ABU	Uruguay	999
21	Sour	LCT	Lebanon	498
A 21	Kalmat	AOTL	Pakistan	609
BP 21	Bredstedt	WPSO	Germany	298
CG 21	Gaspar Grande	PBO	Trinidad and Tobago	850
CM 21	Velarde	CM/PGGFM	Peru	621
F 21	Gomati	FFGHM	India	339
F 21	Mahar Bandoola	FFG	Myanmar	553
F 21	Mariscal Sucre	FFGHM	Venezuela	1002
GC 21	Guaicamacuto	PSOH	Venezuela	1007
GC 21	Lynch	WPB	Argentina	22
H 21	Sirius	AGSH	Brazil	84
HPB 21	Möwe Bay	PB	Namibia	559
HPL 21	Ankaran	PBF	Slovenia	759
HS 21	Hamashio	YGS	Japan	459
J 21	Darshak	AGSH	India	351
L 21	Guldar	LSM/LSMH	India	350
LC 21	Curiapo	LCM	Venezuela	1004
LG 21	Polaris	PBF	Venezuela	1007
LM 21	Quito	–	Ecuador	212
LPD 21	New York	LPDM	US	962
M 21	Rio Guadalobon	PB	Spain	783
P 21	Bendeharu	PB	Brunei	89
P 21	King	AX	Argentina	20
P 21	Plotarchis Blessas	PGGF/PGG	Greece	307
P 21	Raposo Tavares	PBR	Brazil	79
PC 21	Tokinami	PC/PB/YTR	Japan	458
PF 21	Manaure	PBR	Venezuela	1004
PK 21	White Marlin	PBF	Singapore	758
PL 21	Kojima	PL/PSOH	Japan	452
PLH 21	Mizuho	PLH/PSOH	Japan	451
PM 21	Tokara	PM/PBO	Japan	456
PT 21	Angler Ray	WPB	Singapore	758
PV 21	Evagoras	PBF	Cyprus	193
R 21	Tritão	ATA	Brazil	87
RTOP 21	Sibenik	PTGF	Croatia	187
SSN 21	Seawolf	SSN	US	940
T-AGOS 21	Effective	AGOS	US	976
U 21	Soares De Meirelles	AH	Brazil	86
Z 21	Shinas	PB	Oman	595
22	Abdul Rahman Al Fadel	PGGF	Bahrain	48
22	Carrera	SSK	Chile	116
22	Damour	LCT	Lebanon	498
22	Karabala	LST	Iran	389
22	Oyarvide	AGS	Uruguay	998
A 22	Madadgar	AG	Pakistan	610
AM 22	Óbuda	MSR	Hungary	325
BP 22	Neustrelitz	WPBO	Germany	298
CG 22	Chacachacare	PBO	Trinidad and Tobago	850
CM 22	Santillana	CM/PGGFM	Peru	621
F 22	Almirante Brión	FFGHM	Venezuela	1002
F 22	Ganga	FFGHM	India	339
GC 22	Toll	WPB	Argentina	22
GC 22	Yavire	PSOH	Venezuela	1007
HS 22	Isoshi	YGS	Japan	459
J 22	Sarvekshak	AGSH	India	351
K 22	Gävle	FSG	Sweden	798
L 22	Kumbhir	LSM/LSMH	India	350
LG 22	Sirius	PBF	Venezuela	1007
LPD 22	San Diego	LPDM	US	962
M 22	Rio Cedena	PB	Spain	783
P 22	Aoife	PSO	Ireland	394
P 22	Kutaisi	PB	Georgia	283
P 22	Maharajalela	PB	Brunei	89
P 22	Tagomago	PB	Spain	775
P 22	Ypoploiarchos Mikonios	PGGF/PGG	Greece	307
PC 22	Hamagumo	PC/PB/YTR	Japan	458
PF 22	Mara	PBR	Venezuela	1004
PK 22	Silver Marlin	PBF	Singapore	758
PL 22	Miura	PL/PSOH	Japan	452
PLH 22	Yashima	PLH/PSOH	Japan	451
PM 22	Fukue	PM/PBO	Japan	456
PS 22	Sultan Kudarat	FS	Philippines	630
PV 22	Odysseus	PBF	Cyprus	193
R 22	Tridente	ATA	Brazil	87
R 22	Viraat	CVM	India	332
SSN 22	Connecticut	SSN	US	940
T-AGOS 22	Loyal	AGOS	US	976
Y 22	Şalopa 22	YC	Turkey	868
Z 22	Sadh	PB	Oman	595
23	Al Taweelah	PGGF	Bahrain	48
23	Kuha 23	MSI	Finland	240
23	Maldonado	PBO/AG	Uruguay	996
23	O'Higgins	SSK	Chile	116
A 23	Antares	AGS	Spain	779
A 23	Rasadgar	AG	Pakistan	610
AGOR-23	Thomas G Thompson	AGOR	US	971
BP 23	Bad Düben	WPBO	Germany	298
CM 23	De Los Heros	CM/PGGFM	Peru	621
F 23	General Urdaneta	FFGHM	Venezuela	1002
F 23	Mahar Thiha Thura	FFG	Myanmar	553
G 23	Almirante Gastão Motta	AOR	Brazil	86
GC 23	Naiguata	PSOH	Venezuela	1007
H 23	Lokys	PB	Lithuania	503
HS 23	Uzushio	YGS	Japan	459
L 23	Gharial	LSTH	India	349
LG 23	Rigel	PBF	Venezuela	1007
LM 23	Guayaquil	–	Ecuador	212
LPD 23	Anchorage	LPDM	US	962
M 23	Rio Ladra	PB	Spain	783
P 23	Aisling	PSO	Ireland	394
P 23	Kemaindera	PB	Brunei	89
P 23	Ypoploiarchos Troupakis	PGGF/PGG	Greece	307
PC 23	Ayuzuki	PC/PB/YTR	Japan	458
PF 23	Guaicaipuro	PBR	Venezuela	1004
PK 23	Striped Marlin	PBF	Singapore	758
PM 23	Oirase	PM/PBO	Japan	456
PT 23	Bull Ray	WPB	Singapore	758
PV 23	Thexas	PB	Cyprus	193
R 23	Triunfo	ATA	Brazil	87
SSN 23	Jimmy Carter	SSN	US	940
T-AGOS 23	Impeccable	AGOS	US	975
Y 23	Şalopa 23	YC	Turkey	868
Z 23	Khassab	PB	Oman	595
24	Farsi	LST	Iran	388
24	Kuha 24	MSI	Finland	240
24	Lieutenant Remus Lepri	MSC	Romania	665
A 24	Rah Naward	AXS	Pakistan	609
AGOR-24	Roger Revelle	AGOR	US	971
BP 24	Bad Bramstedt	WPSO	Germany	298
CM 24	Herrera	CM/PGGFM	Peru	621
F 24	General Soublette	FFGHM	Venezuela	1002
GC 24	Mantilla	WPSO	Argentina	21
GC 24	Tamanaco	PSOH	Venezuela	1007
HS 24	Okishio	YGS	Japan	459
K 24	Sundsvall	FSG	Sweden	798
L 24	Airavat	LSTH	India	349
L 24	Lina Maria	YAG	Panama	614
LG 24	Aldebaran	PBF	Venezuela	1007
LM 24	Cuenca	–	Ecuador	212
LPD 24	Arlington	LPDM	US	962
M 24	Rio Cervera	PB	Spain	783
P 24	Simeoforos Kavaloudis	PGGF/PGG	Greece	307
P 24	Sokhumi	PBF	Georgia	282
PC 24	Yufugiri	PC/PB/YTR	Japan	458
PF 24	Tamanaco	PBR	Venezuela	1004
PK 24	Black Marlin	PBF	Singapore	758
PM 24	Fuji	PM/PBO	Japan	456
PT 24	Butterfly Ray	WPB	Singapore	758
PV 24	Onisilos	PB	Cyprus	193
R 24	Almirante Guilhem	ATF	Brazil	87
T-AGM 24	Invincible	T-AGM	US	976
Y 24	Şalopa 24	YC	Turkey	868
25	Kasturi	FSGH	Malaysia	509
25	Kuha 25	MSI	Finland	240
25	Lieutenant Lupu Dunescu	MSC	Romania	665
25	Sardasht	LST	Iran	388
AGOR-25	Atlantis	AGOR	US	971
AT 25	Ang Pangulo	AP	Philippines	633
BP 25	Bayreuth	WPSO	Germany	298
CM 25	Larrea	CM/PGGFM	Peru	621
F 25	Bangabandhu	–	Bangladesh	52
F 25	General Salom	FFGHM	Venezuela	1002
G 25	Almirante Saboía	LSLH	Brazil	82
GC 25	Azopardo	WPSO	Argentina	21
H 25	Tenente Boanerges	ABU	Brazil	83
HS 25	Iseshio	YGS	Japan	459
LG 25	Antares	PBF	Venezuela	1007
LPD 25	Somerset	LPDM	US	962

Pennant list [73]

Number	Ship's name	Type	Country	Page
M 25	Rio Jucar	PB	Spain	783
PC 25	Tomonami	PC/PB/YTR	Japan	458
PK 25	Blue Marlin	PBF	Singapore	758
PM 25	Echizen	PM/PBO	Japan	456
PT 25	Cownose Ray	WPB	Singapore	758
R 25	Almirante Guillobel	ATF	Brazil	87
T-AGM 25	Howard O Lorenzen	AGM	US	976
26	Kuha 26	MSI	Finland	240
26	Lekir	FSGH	Malaysia	509
26	Sab Sahel	LST	Iran	388
26	Vanguardia	ARS	Uruguay	999
AGOR 26	Kilo Moana	AGOR	US	971
BP 26	Eschwege	WPSO	Germany	298
CM 26	Sanchez Carrión	CM/PGGFM	Peru	621
F 26	Almirante Garcia	FFGHM	Venezuela	1002
GC 26	Thompson	WPSO	Argentina	21
H 26	Faroleiro Mário Seixas	ABU	Brazil	82
HS 26	Hayashio	YGS	Japan	459
L 26	Isla Jicaron	YAG	Panama	614
LG 26	Canopus	PBF	Venezuela	1007
LPD 26	John P Murtha	LPDM	US	962
M 26	Rio Gallo	PB	Spain	783
P 26	Dzata	PBO	Ghana	301
P 26	Medas	PB	Spain	775
P 26	Ypoploiarchos Degiannis	PGGF/PGG	Greece	307
PK 26	Jumping Marlin	PBF	Singapore	758
PM 26	Kikuchi	PM/PBO	Japan	456
27	Banco Ortiz	YTB	Uruguay	1000
27	Pyong Taek	ATS	Korea, South	486
AGOR 27	Neil Armstrong	AGOR	US	971
CG 27	Plymouth	PB	Trinidad and Tobago	849
G 27	Marajo	AOR	Brazil	86
GC 27	Prefecto Fique	WPSO	Argentina	21
HS 27	Kurushima	YGS	Japan	459
LG 27	Altair	PBF	Venezuela	1007
LPD 27	Portland	LPDM	US	962
M 27	Rio Jiloca	PB	Spain	783
P 27	Inagua	PB	Bahamas	44
P 27	Sebo	PBO	Ghana	301
P 27	Simeoforos Xenos	PGGF/PGG	Greece	307
PM 27	Yoshino	PM/PBO	Japan	456
PT 27	Electric Ray	WPB	Singapore	758
U 27	Brasil	AXH	Brazil	84
Y 27	Şalopa 27	YC	Turkey	868
28	Kwang Yang	ATS	Korea, South	486
AGOR 28	Sally Ride	AGOR	US	971
CG 28	Caroni	PB	Trinidad and Tobago	849
F 28	Somudra Joy	FFH	Bangladesh	53
G 28	Mattoso Maia	LSTH	Brazil	81
GC 28	Prefecto Derbes	WPSO	Argentina	21
M 28	Rio Aragón	PB	Spain	783
P 28	Achimota	PG	Ghana	300
P 28	Simeoforos Simitzopoulos	PGGF/PGG	Greece	307
P 28	Tabarca	PB	Spain	775
PM 28	Isuzu	PM/PBO	Japan	456
PS 28	Cebu	FS	Philippines	630
PT 28	Flynose Ray	WPB	Singapore	758
S 28	Vanguard	SSBN	UK	896
SO 28	Pijao	SS	Colombia	171
U 28	Teniente Maximiano	AH	Brazil	86
29	Jebat	–	Malaysia	508
29	Lieutenant Dimitrie Nicolescu	MSC	Romania	665
CG 29	Galeota	PB	Trinidad and Tobago	849
G 29	Garcia D'Ávila	LSLH	Brazil	81
M 29	Rio Santa Eulalia	PB	Spain	783
P 29	Simeoforos Starakis	PGGF/PGG	Greece	307
P 29	Yogaga	PG	Ghana	300
PM 29	Yamakuni	PM/PBO	Japan	456
PT 29	Flying Ray	WPB	Singapore	758
S 29	Victorious	SSBN	UK	896
SO 29	Tayrona	SS	Colombia	171
U 29	Piraim	YFBH	Brazil	85
30	Al Jarim	PB	Bahrain	48
30	Lekiu	–	Malaysia	508
30	Sub Lieutenant Alexandru Axente	MSC	Romania	665
30	Vishwast	WPSOH	India	355
ACV 30	Botany Bay	PB	Australia	39
CG 30	Moruga	PB	Trinidad and Tobago	849
G 30	Ceará	LSDH	Brazil	81
LM 30	Casma	PGG	Chile	121
M 30	Ledbury	MHC	UK	913
M 30	Rio Ulla	PB	Spain	783
P 30	Anzone	PBO	Ghana	301
P 30	Roraima	PBR	Brazil	80
PK 30	Billfish	PBF	Singapore	758
PM 30	Kano	PM/PBO	Japan	456
PT 30	Eagle Ray	WPB	Singapore	757
Q 30	Al Mabrukah	FSH/AXL/AGS	Oman	594
RPB 30	Kozara	PBR	Serbia	747
S 30	Tupi	SSK	Brazil	71
S 30	Vigilant	SSBN	UK	896
V 30	Inhaúma	FSGH	Brazil	76
WLBB 30	Mackinaw	WLBB	US	988
Y 30	Şalopa 30	YC	Turkey	868
31	Al Jasrah	PB	Bahrain	48
31	Drummond	FFG	Argentina	14
31	Salamaua	LSM	Papua New Guinea	615
31	Temerario	MSC	Uruguay	998
31	Tongyeong	ARSH	Korea, South	486
31	Vijit	WPSOH	India	355
A 31	Malaspina	AGS	Spain	778
A 31	Ras El Hilal	YTB	Libya	501
AM 31	Dunaújváros	MSR	Hungary	325
BG 31	Bukovina	PCF	Ukraine	882
CG 31	Barracuda	PSOH	Mauritius	527
CG 31	Kairi	PB	Trinidad and Tobago	850
F 31	Brahmaputra	FFGHM	India	340
GS 31	Tchusovoy	AGE	Russian Federation	711
J 31	Makar	AGS	India	351
K 31	Visby	FSGH	Sweden	797
L 31	Damuan	YFU	Brunei	89
LD 31	Neyba	LCU	Dominican Republic	207
LG 31	Chichiriviche	PB	Venezuela	1007
LM 31	Chipana	PGG	Chile	121
M 31	Segura	MHC	Spain	778
M 31	Cattistock	MHC	UK	913
M 31	Corvo Marino	PB	Spain	783
P 31	Bonsu	PBO	Ghana	301
P 31	Eithne	PSOH	Ireland	393
P 31	Le Vigilant	PB	Seychelles	748
P 31	Rondônia	PBR	Brazil	80
P 31	Ureca	PB	Equatorial Guinea	231
PC 31	Kotonami	PC/PB	Japan	456
PF 31	Terepaima	PBR	Venezuela	1004
PG 31	Petrel	WPB	Venezuela	1006
PL 31	Izu	PL/PSOH	Japan	452
PLH 31	Shikishima	PLH/PSOH	Japan	451
PM 31	Abukuma	PM/PBO	Japan	456
PS 31	Pangasinan	FS	Philippines	630
PT 31	Giant Reef Ray	WPB	Singapore	758
Q 31	Qahir Al Amwaj	FSGMH	Oman	593
Q 31	That Assuari	WPBF	Qatar	658
S 31	Sábalo	SSK	Venezuela	1001
S 31	Salta	SSK	Argentina	13
S 31	Tamoio	SSK	Brazil	71
S 31	Vengeance	SSBN	UK	896
SS 31	Angamos	SSK	Peru	618
V 31	Jaceguai	FSGH	Brazil	76
Y 31	Şalopa 31	YC	Turkey	868
32	Buna	LSM	Papua New Guinea	615
32	Guerrico	FFG	Argentina	14
32	Tsibar	MHC	Bulgaria	91
32	Vaibhav	WPSOH	India	355
A 32	Al Ahweirif	YTB	Libya	501
A 32	Tofiño	AGS	Spain	778
AM 32	Dunafoldvar	MSR	Hungary	325
BG 32	Donbas	PCF	Ukraine	882
CG 32	Moriah	PB	Trinidad and Tobago	850
D 32	Daring	DDGHM	UK	902
K 32	Helsingborg	FSGH	Sweden	797
L 32	Puni	YFU	Brunei	89
LG 32	Caruanta	PB	Venezuela	1007
M 32	Rio Mijares	PB	Spain	783
M 32	Sella	MHC	Spain	778
P 32	Amapá	PBR	Brazil	80
P 32	David Hansen	PBI	Ghana	301
P 32	La Flèche	PB	Seychelles	748
P 32	Selis	PB	Lithuania	502
PC 32	Hatagumo	PC/PB	Japan	456
PG 32	Alcatraz	WPB	Venezuela	1006
PLH 32	Akitsushima	PLH/PSOH	Japan	451
PM 32	Minabe	PM/PBO	Japan	456
PS 32	Iloilo	FS	Philippines	630
PT 32	River Ray	WPB	Singapore	758
Q 32	Al Mua'zzar	FSGMH	Oman	593
S 32	Caribe	SSK	Venezuela	1001
S 32	Timbira	SSK	Brazil	71
SS 32	Antofagasta	SSK	Peru	618
V 32	Julio De Noronha	FSGH	Brazil	76
33	Fortuna	MSC	Uruguay	998
33	Granville	FFG	Argentina	14
A 33	Hespérides	AGOBH	Spain	778
AW 33	Lake Balusan	AWT	Philippines	634
CG 33	Matelot	PB	Trinidad and Tobago	850
D 33	Dauntless	DDGHM	UK	902
K 33	Härnösand	FSGH	Sweden	797
L 33	Serasa	LCU	Brunei	88
LG 33	Isla Santa Cruz	PB	Ecuador	216
M 33	Brocklesby	MHC	UK	913
M 33	Tambre	MHC	Spain	778
P 33	Abhay	FSM	India	341
P 33	Stephen Otu	PBF/PTF	Ghana	301
PF 33	Yaracuy	PBR	Venezuela	1004
PG 33	Albatros	WPB	Venezuela	1006
PM 33	Matsuura	PM/PBO	Japan	456
PT 33	Roughtail Ray	WPB	Singapore	758
R 33	Vikramaditya	CVGM	India	330
S 33	Tapajó	SSK	Brazil	71
SS 33	Pisagua	SSK	Peru	618
V 33	Frontin	FSGH	Brazil	76
Y 33	Şalopa 33	YC	Turkey	868
34	Audaz	MSC	Uruguay	998
AW 34	Lake Paoay	AWT	Philippines	634
D 34	Diamond	DDGHM	UK	902
H 34	Almirante Graça Aranha	ABUH	Brazil	82
K 34	Nyköping	FSGH	Sweden	797
LG-34	Isla San Cristóbal	WPB	Ecuador	215
LM 34	Angamos	PGG	Chile	121
M 34	Canal Bocayna	PB	Spain	782
M 34	Middleton	MHC	UK	913
M 34	Turia	MHC	Spain	778
P 34	Ajay	FSM	India	341
P 34	Blika	PBO	Ghana	302
PF 34	Sorocaima	PBR	Venezuela	1004
PG 34	Pelícano	WPB	Venezuela	1006
PM 34	Chikugo	PM/PBO	Japan	456
S 34	Tikuna	SSK	Brazil	70
SS 34	Chipana	SSK	Peru	618
V 34	Barroso	FSGH	Brunei	76
D 35	Dragon	DDGHM	UK	902
F 35	Bijoy	FSGH	Bangladesh	54
H 35	Amorim Do Valle	AGS	Brazil	83
K 35	Karlstad	FSGH	Sweden	797
LG 35	Isla Santa Rosa	WPB	Ecuador	214
M 35	Duero	MHC	Spain	778
M 35	Pico Del Teide	PB	Spain	782
MB 35	Evgeny Khorov	ARS/ATA	Russian Federation	723
P 35	Akshay	FSM	India	341

© 2014 IHS

IHS Jane's Fighting Ships 2014-2015

[74] Pennant list

Number	Ship's name	Type	Country	Page
P 35	Garinga	PBO	Ghana	302
PM 35	Kurose	PM/PBO	Japan	456
PS 35	Emilio Jacinto	FS	Philippines	629
PT 35	Shovelnose Ray	WPB	Singapore	758
SS 35	Islay	SSK	Peru	618
Y 35	Mesaha 1	YAG	Turkey	865
36	Varuna	WPSOH	India	355
D 36	Defender	DDGHM	UK	902
F 36	Dhaleshwari	FSGH	Bangladesh	54
H 36	Taurus	AGS	Brazil	83
LG 36	Isla Puná	WPB	Ecuador	214
M 36	Tajo	MHC	Spain	778
M 36	Rio Guadalquivir	PB	Spain	782
P 36	Agray	FSM	India	341
P 36	Chemle	PBO	Ghana	302
PM 36	Okitsu	PM/PBO	Japan	456
PS 36	Apolinario Mabini	FS	Philippines	629
PT 36	Spotted Ray	WPB	Singapore	758
RSRB 36	Sabac	YDG	Serbia	747
SS 36	Arica	SSK	Peru	618
Y 36	Mesaha 2	YAG	Turkey	865
37	Vajra	WPSOH	India	355
D 37	Duncan	DDGHM	UK	902
F 37	Beas	FFGHM	India	340
H 37	Garnier Sampaio	AGS	Brazil	83
LG 37	Isla De La Plata	WPBF	Ecuador	214
LM 37	Orella	PGG	Chile	120
M 37	Chiddingfold	MHC	UK	913
M 37	Rio Tordera	PB	Spain	782
P 37	Ehwor	PBO	Ghana	302
PM 37	Kunashiri	PM/PBO	Japan	456
PS 37	Artemio Ricarte	FS	Philippines	629
PT 37	Sting Ray	WPB	Singapore	758
38	Vivek	WPSOH	India	355
H 38	Cruzeiro Do Sul	AGS	Brazil	84
L 38	Galana	LCM	Kenya	465
LG 38	Isla Santa Clara	WPBF	Ecuador	214
LM 38	Serrano	PGG	Chile	120
M 38	Atherstone	MHC	UK	913
M 38	Rio Pas	PB	Spain	782
P 38	Yaa Asantewa	PB	Ghana	300
PM 38	Oumi	PM/PBO	Japan	456
PS 38	General Mariano Alvares	PB	Philippines	630
PT 38	Thomback Ray	WPB	Singapore	758
Y 38	Yüzbasi Nasit Öngören	YB	Turkey	868
39	Hobart	DDGHM	Australia	27
39	Vigraha	WPSOH	India	355
AS 39	Emory S Land	ASH	US	979
F 39	Betwa	FFGHM	India	340
GS 39	Syzran	AGIM	Russian Federation	712
L 39	Tana	LCM	Kenya	465
LG 39	Isla Fernandina	PBO	Ecuador	214
LM 39	Uribe	PGG	Chile	120
M 39	Hurworth	MHC	UK	913
M 39	Rio Guadalentin	PB	Spain	782
P 39	Naa Gbewaa	PB	Ghana	300
PM 39	Okushiri	PM/PBO	Japan	456
PT 39	Torpedo Ray	WPB	Singapore	758
WMEC 39	Alex Haley	PSOH/WMEC	US	986
Y 39	Binbasi Metin Sülüs	YB	Turkey	868
40	Al Zubara	LCU	Bahrain	49
40	Carlos Manuel De Cespedes	AXT	Cuba	191
40	Katanpää	MHSC	Finland	240
40	Varad	WPSOH	India	355
A 40	Attock	AOTL	Pakistan	609
ACV 40	Hervey Bay	PB	Australia	39
AS 40	Frank Cable	ASH	US	979
F 40	Niterói	FFGHM	Brazil	75
F 40	Talwar	FFGHM	India	337
FFG 40	Halyburton	FFH	US	952
H 40	Antares	AGS	Brazil	83
K 40	Veer	FSGM	India	344
LG 40	Isla Española	PBO	Ecuador	214
M 40	Rio Aller	PB	Spain	784
P 40	Grajaú	PBO	Brazil	79
PK 40	Swordfish	PBF	Singapore	758
PM 40	Natsui	PM/PBO	Japan	456
Q 40	Al-Shamikh	FSGHM	Oman	594
S 40	Riachuelo	SSK	Brazil	70
41	Aamchit	PB	Lebanon	497
41	Ajeera	YFU	Bahrain	48
41	Brisbane	DDGHM	Australia	27
41	Drazki	FFGM	Bulgaria	90
41	Espora	FFGH	Argentina	15
41	Purunpää	MHSC	Finland	240
41	Storm	YAG	UK	924
A 41	Behr Kusha	–	Pakistan	609
AP 41	Aquiles	APH	Chile	124
F 41	Defensora	FFGHM	Brazil	75
FFG 41	McClusky	FFH	US	952
H 41	Almirante Maximiano	AGOBH	Brazil	84
K 41	Nirbhik	FSGM	India	344
L 41	Jalashwa	LPD	India	349
LG 41	Isla San Salvador	PBO	Ecuador	214
LSD 41	Whidbey Island	LSD	US	967
M 41	Cabo De Gata	PB	Spain	784
M 41	Quorn	MHC	UK	913
P 41	Guaiba	PBO	Brazil	79
P 41	Meteoro	PSO	Spain	772
P 41	Orla	PSO	Ireland	394
PL 41	Aso	PL/PSO	Japan	453
Q 41	Al-Rahmani	FSGHM	Oman	594
RTOP 41	Vukovar	PTGM	Croatia	185
S 41	Humaita	SSK	Brazil	70
S 41	Santa Cruz	SSK	Argentina	12
42	Almirante Merino	AGP/ASH	Chile	124
42	Diablo	YAG	UK	924
42	Mashtan	LCU	Bahrain	48
42	Rosales	FFGH	Argentina	15
42	Samar	WPSOH	India	355
42	Sydney	DDGHM	Australia	27
42	Tabarja	PB	Lebanon	498
42	Vahterpää	MHSC	Finland	240
42	Verni	FFGM	Bulgaria	90
42	Yan Gyi Aung	FS	Myanmar	555
F 42	Constituição	FFGHM	Brazil	75
K 42	Nipat	FSGM	India	344
L 42	Umm Al Narr	LCU	UAE	889
LSD 42	Germantown	LSD	US	967
M 42	Rio Tormes	PB	Spain	784
N 42	Jotvingis	MCCS/AG	Lithuania	503
P 42	Ciara	PSO	Ireland	394
P 42	Graúna	PBO	Brazil	79
P 42	Rayo	PSO	Spain	772
PL 42	Dewa	PL/PSO	Japan	453
Q 42	Al-Rasikh	FSGHM	Oman	594
RTOP 42	Dubrovnik	PTGM	Croatia	185
S 42	San Juan	SSK	Argentina	12
S 42	Tonelero	SSK	Brazil	70
43	Al Kalamoun	PB	Lebanon	498
43	Gaspar Obiang Esono	PBR	Equatorial Guinea	231
43	Gordi	FFGM	Bulgaria	90
43	Mistral	YAG	UK	924
43	Mulnaya	FSGM	Bulgaria	90
43	Rubodh	LCU	Bahrain	48
43	Sangram	WPSOH	India	355
43	Spiro	FFGH	Argentina	15
BE 43	Esmeralda	AXS	Chile	123
F 43	Liberal	FFGHM	Brazil	75
F 43	Trishul	FFGHM	India	337
GC 43	Mandubi	WAX	Argentina	22
K 43	Nishank	FSGM	India	344
LSD 43	Fort McHenry	LSD	US	967
M 43	Rio Ter	PB	Spain	782
P 43	Goiana	PBO	Brazil	79
P 43	Relámpago	PSO	Spain	772
PL 43	Hakusan	PL/PSO	Japan	453
S 43	Angostura	SSK	Brazil	70
44	Parker	FFGH	Argentina	15
44	Sarang	WPSOH	India	355
44	Sirocco	YAG	UK	924
44	Suwad	LCU	Bahrain	48
A 44	Bholu	YTB	Pakistan	610
F 44	Independência	FFGHM	Brazil	75
F 44	Tabar	FFGHM	India	337
G 44	Mazzeo	–	Italy	422
H 44	Ary Rongel	AGOBH	Brazil	83
K 44	Nirghat	FSGM	India	344
L 44	Teraban	LCU	Brunei	88
LSD 44	Gunston Hall	LSD	US	967
M 44	Rio Genil	PB	Spain	782
P 44	Guajará	PBO	Brazil	79
P 44	Kirpan	FSGHM	India	343
P 44	Tornado	PSO	Spain	772
PO 44	Valle Del Cauca	–	Colombia	179
S 44	Shishumar	SSK	India	328
45	Fernando Nuara Engonda	PBR	Equatorial Guinea	231
45	Jaradah	LCU	Bahrain	48
45	Mikhail Kogalniceanu	PGR	Romania	664
45	Robinson	FFGH	Argentina	15
45	Sagar	WPSOH	India	355
A 45	Gama	YTB	Pakistan	610
F 45	Teg	FFGHM	India	337
F 45	União	FFGHM	Brazil	75
K 45	Vibhuti	FSGM	India	344
LSD 45	Comstock	LSD	US	967
M 45	Rio Gallego	PB	Spain	782
P 45	Guaporé	PBO	Brazil	79
PO 45	San Andres	PSO	Colombia	175
S 45	Shankush	SSK	India	328
T-AG 45	Waters	AGS	US	976
46	Gomez Roca	FFGH	Argentina	15
46	I C Bratianu	PGR	Romania	664
46	Sankalp	WPSOH	India	354
46	Tighatlib	YFL	Bahrain	49
AE 46	Cape Bojeador	ABU	Philippines	635
AP 46	Contre-Almirante Oscar Viel Toro	AGS/AGOBH	Chile	123
F 46	Greenhalgh	FFGHM	Brazil	74
F 46	Tarkash	FFGHM	India	337
G 46	Silanos	–	Italy	422
K 46	Vipul	FSGM	India	344
LSD 46	Tortuga	LSD	US	967
M 46	Rio Nacimiento	PB	Spain	782
P 46	Gurupá	PBO	Brazil	79
P 46	Kuthar	FSGHM	India	343
PO 46	20 De Julio	PSO	Colombia	174
S 46	Shalki	SSK	India	328
47	Dinar	LCU	Bahrain	49
47	Lascar Catargiu	PGR	Romania	664
47	Samrat	WPSOH	India	354
47	Sri Perlis	PB	Malaysia	512
A 47	Nasr	AORH	Pakistan	609
F 47	Shivalik	FFGHM	India	338
G 47	Ignesti	–	Italy	422
GC 47	Tonina	WARS	Argentina	22
K 47	Vinash	FSGM	India	344
LSD 47	Rushmore	LSD	US	967
M 47	Rio Nansa	PB	Spain	782
P 47	Gurupi	PBO	Brazil	79
P 47	Khanjar	FSGHM	India	343
PO 47	7 De Agosto	PSO	Colombia	174
S 47	Shankul	SSK	India	328
48	Al Hamra	LCU	Bahrain	49
F 48	Bosisio	FFGHM	Brazil	74
F 48	Satpura	FFGHM	India	338
FFG 48	Vandegrift	FFH	US	952
GC 48	Estrellemar	WPB	Argentina	23
K 48	Vidyut	FSGM	India	344

IHS Jane's Fighting Ships 2014-2015 © 2014 IHS

Pennant list

Number	Ship's name	Type	Country	Page
LSD 48	Ashland	LSD	US	967
M 48	Rio Tambre	PB	Spain	784
P 48	Guanabara	PBO	Brazil	79
SV 48	Behr Paima	AGS/AGOR	Pakistan	608
49	Sri Johor	PB	Malaysia	512
A 49	Gwadar	AOTL	Pakistan	609
F 49	Rademaker	FFGHM	Brazil	74
F 49	Sahyadri	FFGHM	India	338
GC 49	Remora	WPB	Argentina	23
LSD 49	Harpers Ferry	LSD	US	967
M 49	Rio Órbigo	PB	Spain	784
P 49	Guarujá	PBO	Brazil	79
P 49	Khukri	FSGHM	India	343
50	Al Manama	FSGH	Bahrain	47
A 50	Alster	AGI	Germany	295
A 50	Deepak	AORH	India	353
ACV 50	Corio Bay	PB	Australia	39
B 50	Sawahil	AGH	Kuwait	494
BG 50	Grigory Kuropiatnikov	PC	Ukraine	882
FFG 50	Taylor	FFH	US	952
GC 50	Congrio	WPB	Argentina	23
L 50	Tobruk	LSLH	Australia	32
LSD 50	Carter Hall	LSD	US	967
M 50	Rio Jándula	PB	Spain	784
N 50	Tyr	AGDS	Norway	591
P 50	Guaratuba	PBO	Brazil	79
P 50	Sukanya	PSOH	India	347
PH 50	Hammerhead Shark	WPB	Singapore	758
PK 50	Spikefish	PBF	Singapore	758
T-ARS 50	Safeguard	ARS	US	979
Y 50	Gölcük	YO	Turkey	867
51	Al Muharraq	FSGH	Bahrain	47
51	Iskar	MSB	Bulgaria	92
A 51	Gaj	YTM/YTL	India	354
A 51	Mahón	ATA	Spain	781
A 51	Temsah	YTM	UAE	890
BG 51	Poltava	PC	Ukraine	882
D 51	Rajput	DDGHM	India	336
DDG 51	Arleigh Burke	DDGHM	US	946
F 51	Trikand	FFGHM	India	337
FFG 51	Gary	FFH	US	952
FL 51	Almirante Padilla	FLGHM	Colombia	172
G 51	Fiore	–	Italy	422
GC 51	Mero	WPB	Argentina	23
L 51	Al Feyi	LCU	UAE	889
L 51	Galicia	LPD	Spain	776
LM 51	Korcula	MHI	Croatia	188
LSD 51	Oak Hill	LSD	US	967
P 51	Gravataí	PBO	Brazil	79
P 51	Roísín	PSO	Ireland	393
P 51	Subhadra	PSOH	India	347
PC 51	Yodo	PC/YTR	Japan	457
PG 51	Págalo	PB	Venezuela	1007
PH 51	Mako Shark	WPB	Singapore	758
PL 51	Hida	PL/PSO	Japan	452
T-ARS 51	Grasp	ARS	US	979
Y 51	Söndüren 1	YTB/YTM/YTL	Turkey	870
52	Almirante Montt	AOR	Chile	124
52	Dobrotich	MSB	Bulgaria	92
A 52	Las Palmas	AGOB	Spain	778
A 52	Oste	AGI	Germany	295
A 52	Ugaab	YTM	UAE	890
B 52	Hercules	LCC	Argentina	18
BG 52	Grigory Gnatenko	PC	Ukraine	882
CG 52	Bunker Hill	CGHM	US	944
D 52	Rana	DDGHM	India	336
DDG 52	Barry	DDGHM	US	946
FL 52	Caldas	FLGHM	Colombia	172
FM 52	Villavisencio	FFGHM	Peru	619
GC 52	Marsopa	WPB	Argentina	23
L 52	Castilla	LPD	Spain	776
L 52	Dayyinah	LCU	UAE	889
LSD 52	Pearl Harbor	LSD	US	967
M 52	Suduvis	MHC	Lithuania	502
P 52	Niamh	PSO	Ireland	393
P 52	Suvarna	PSOH	India	347
PC 52	Kotobiki	PC/YTR	Japan	457
PH 52	White Shark	WPB	Singapore	758
PL 52	Akaishi	PL/PSO	Japan	452
T-ARS 52	Salvor	ARS	US	979
Y 52	Doğanarslan	YTB/YTM/YTL	Turkey	870
53	Kapitan-Leytenant Kiril Minkov	MSB	Bulgaria	92
A 53	La Graña	ATA	Spain	781
A 53	Matanga	ATA/ATR	India	353
A 53	Oker	AGI	Germany	295
A 53	Virsaitis	MCCS/AG	Latvia	495
AO 53	Araucano	AOR	Chile	124
CG 53	Mobile Bay	CGHM	US	944
D 53	Ranjit	DDGHM	India	336
DDG 53	John Paul Jones	DDGHM	US	946
FL 53	Antioquia	FLGHM	Colombia	172
FM 53	Montero	FFGHM	Peru	619
GC 53	Petrel	WPB	Argentina	23
L 53	Jananah	LCU	UAE	889
M 53	Skalvis	MHC	Lithuania	502
P 53	Savitri	PSOH	India	347
PC 53	Nachi	PC/YTR	Japan	457
PH 53	Blue Shark	WPB	Singapore	758
PL 53	Kiso	PL/PSO	Japan	452
T-ARS 53	Grapple	ARS	US	979
Y 53	Kuvvet	YTB/YTM/YTL	Turkey	870
54	Balik	MSB	Bulgaria	92
CG 54	Antietam	CGHM	US	944
D 54	Ranvir	DDGHM	India	336
DDG 54	Curtis Wilbur	DDGHM	US	946
FFG 54	Ford	FFH	US	952
FL 54	Independiente	FLGHM	Colombia	172
FM 54	Mariategui	FFGHM	Peru	619
GC 54	Salmon	WPB	Argentina	23
M 54	Kuršis	MHC	Lithuania	502
P 54	Saryu	PSOH	India	347
PC 54	Nunobiki	PC/YTR	Japan	457
PH 54	Tiger Shark	WPB	Singapore	758
55	Kapitan-Leytenant Evstati Vinarov	MSB	Bulgaria	92
55	Orca	AXL	Canada	104
BG 55	Galichina	PCK	Ukraine	883
CG 55	Leyte Gulf	CGHM	US	944
D 55	Ranvijay	DDGHM	India	336
DDG 55	Stout	DDGHM	US	946
FFG 55	Elrod	FFH	US	952
FM 55	Aguirre	FFGHM	Peru	620
FV 55	Indaw	PBO	Myanmar	555
GC 55	Bigua	WPB	Argentina	23
P 55	Sharada	PSOH	India	347
P 55	Wellington	PBO	New Zealand	575
PC 55	Fudou	PC/YTR	Japan	457
PH 55	Basking Shark	WPB	Singapore	758
S 55	Sindhughosh	SSK	India	329
Y 55	Atil	YTB/YTM/YTL	Turkey	870
56	Kajava	AX	Finland	241
56	Kapitan Parvi Rang Dimitar Paskalev	MSB	Bulgaria	92
56	Raven	AXL	Canada	104
CG 56	San Jacinto	CGHM	US	944
DDG 56	John S McCain	DDGHM	US	946
FFG 56	Simpson	FFH	US	952
FM 56	Palacios	FFGHM	Peru	620
G 56	Tridenti	–	Italy	422
GC 56	Foca	WPB	Argentina	23
P 56	Sujata	PSOH	India	347
PC 56	Ryusei	PC/YTR	Japan	457
PH 56	Sandbar Shark	WPB	Singapore	758
S 56	Sindhudhvaj	SSK	India	329
Y 56	Pendik	YTB/YTM/YTL	Turkey	870
57	Caribou	AXL	Canada	104
57	Chun Jee	AORH	Korea, South	486
A 57	Shakti	AORH	India	353
BG 57	Mikolaiv	PCF	Ukraine	882
CG 57	Lake Champlain	CGHM	US	944
DDG 57	Mitscher	DDGHM	US	946
FM 57	Bolognesi	FFGHM	Peru	620
FV 57	Inya	PBO	Myanmar	555
GC 57	Tiburon	WPB	Argentina	23
P 57	Kasos	PG	Greece	308
P 57	Sunayna	PSOH	India	347
PC 57	Takataki	PC/YTR	Japan	457
PH 57	Thresher Shark	WPB	Singapore	758
S 57	Sindhuraj	SSK	India	329
Y 57	Aksaz	YTB/YTM/YTL	Turkey	870
58	Dae Chung	AORH	Korea, South	486
58	Renard	AXL	Canada	104
A 58	Jyoti	AORH	India	352
CG 58	Philippine Sea	CGHM	US	944
DDG 58	Laboon	DDGHM	US	946
FFG 58	Samuel B Roberts	FFH	US	952
FM 58	Quiñónes	FFGHM	Peru	620
G 58	Atzei	–	Italy	422
GC 58	Melva	WPB	Argentina	23
P 58	Sumedha	PSOH	India	347
PC 58	Aotaki	PC/YTR	Japan	457
PH 58	Whitetip Shark	WPB	Singapore	758
S 58	Sindhuvir	SSK	India	329
Y 58	Algarna 1	YD/YFDM	Turkey	869
59	Wolf	AXL	Canada	104
59	Hwa Chun	AORH	Korea, South	486
A 59	Aditya	AORH/AS	India	352
CG 59	Princeton	CGHM	US	944
DDG 59	Russell	DDGHM	US	946
FFG 59	Kauffman	FFH	US	952
GC 59	Lenguado	WPB	Argentina	23
P 59	Sumitra	PSOH	India	347
PC 59	Nachi	PC/YTR	Japan	457
PH 59	Blacktip Shark	WPB	Singapore	758
S 59	Sindhuratna	SSK	India	329
Y 59	Levent	YD/YFDM	Turkey	869
60	Grizzly	AXL	Canada	104
A 60	Gorch Fock	AXS	Germany	295
ACV 60	Arnhem Bay	PB	Australia	39
CG 60	Guardian	PB	Mauritius	526
CG 60	Normandy	CGHM	US	944
D 60	Mysore	DDGHM	India	334
DDG 60	Paul Hamilton	DDGHM	US	946
FFG 60	Rodney M Davis	FFH	US	952
G 60	Fidone	–	Italy	422
GC 60	Orca	WPB	Argentina	23
P 60	Bahamas	–	Bahamas	44
P 60	Bracui	PBO	Brazil	80
PC 60	Minoo	PC/YTR	Japan	457
S 60	Sindhukesari	SSK	India	329
T-AGS 60	Pathfinder	AGS	US	975
Y 60	Algarna 3	YD/YFDM	Turkey	869
61	Ashdod	LCT	Israel	399
61	Briz	MSC	Bulgaria	91
61	Cougar	AXL	Canada	104
AGS 61	Cabo De Hornos	AGOR	Chile	123
BG 61	Odessa	PCF	Ukraine	882
BP 61	Prignitz	WPB	Germany	299
CG 61	Monterey	CGHM	US	944
D 61	Delhi	DDGHM	India	334
DDG 61	Ramage	DDGHM	US	946
FFG 61	Ingraham	FFH	US	952
GC 61	Pinguino	WPB	Argentina	23
L 61	Juan Carlos I	LHD	Spain	777
M 61	Evniki	MHC	Greece	310
OB 61	Novigrad	PCM	Croatia	188
P 61	Baradero	PB	Argentina	17
P 61	Benevente	PBO	Brazil	80
P 61	Kora	FSGHM	India	342

© 2014 IHS

IHS Jane's Fighting Ships 2014-2015

[76] Pennant list

Number	Ship's name	Type	Country	Page
P 61	Nassau	–	Bahamas	44
P 61	Polemistis	PG	Greece	308
P 61	Samuel Buckett	PSO	Ireland	394
PL 61	Hateruma	PL/PSO	Japan	453
Q 61	Ciudad De Zarate	ABU	Argentina	20
S 61	Sindhukirti	SSK	India	329
T 61	Capana	LSTH	Venezuela	1005
T 61	Trinkat	PBO	India	348
T-AGS 61	Sumner	AGS	US	975
62	Moose	AXL	Canada	104
62	Shkval	MSC	Bulgaria	91
BG 62	Podilliya	PCF	Ukraine	882
BP 62	Uckermark	WPB	Germany	299
CG 62	Chancellorsville	CGHM	US	944
D 62	Mumbai	DDGHM	India	334
DDG 62	Fitzgerald	DDGHM	US	946
L 62	Al Khaznah	–	UAE	889
M 62	Evropi	MHS	Greece	310
OB 62	Solta	PCM	Croatia	188
P 62	Alboran	PSOH	Spain	775
P 62	Barranqueras	PB	Argentina	17
P 62	Bocaina	PBO	Brazil	80
P 62	James Joyce	PSO	Ireland	394
P 62	Kirch	FSGHM	India	342
PL 62	Ishigaki	PL/PSO	Japan	453
Q 62	Ciudad De Rosario	ABU	Argentina	20
S 62	Sindhuvijay	SSK	India	329
T 62	Esequibo	LSTH	Venezuela	1005
T-AGS 62	Bowditch	AGS	US	975
TR 62	Calicuchima	AETL	Ecuador	213
63	Priboy	MSC	Bulgaria	91
BG 63	Pavel Derzhavin	PCF	Ukraine	882
BP 63	Altmark	WPB	Germany	299
BRS 63	George Slight Marshall	ABU	Chile	123
CG 63	Cowpens	CGHM	US	944
D 63	Kolkata	DDGHM	India	333
DDG 63	Stethem	DDGHM	US	946
M 63	Kallisto	MHS	Greece	310
OB 63	Cavtat	PCM	Croatia	188
P 63	Arnomendi	PSOH	Spain	775
P 63	Babitonga	PBO	Brazil	80
P 63	Clorinda	PB	Argentina	17
P 63	Kulish	FSGHM	India	342
PL 63	Yonakuni	PL/PSO	Japan	453
PT 63	Black Ray	WPB	Singapore	758
Q 63	Punta Alta	ABU	Argentina	20
T 63	Goajira	LSTH	Venezuela	1005
T 63	Tarasa	PBO	India	348
T-AGS 63	Henson	AGS	US	975
TR 63	Atahualpa	AWT	Ecuador	213
BP 64	Börde	WPB	Germany	299
CG 64	Gettysburg	CGHM	US	944
D 64	Kochi	DDGHM	India	333
DDG 64	Carney	DDGHM	US	946
G 64	Darida	–	Italy	422
GC 64	Mar Del Plata	WPB	Argentina	22
M 64	Kalypso	MHC	Greece	310
OB 64	Hrvatska Kostajnica	PCM	Croatia	188
P 64	Concepción Del Uruguay	PB	Argentina	17
P 64	Karmuk	FSGHM	India	342
P 64	Tarifa	PSOH	Spain	775
PG 64	Palawan	PB	Philippines	635
PL 64	Shimokita	PL/PSO	Japan	453
PT 64	Jaguar Ray	WPB	Singapore	758
T 64	Los Llanos	LSTH	Venezuela	1005
T-AGS 64	Bruce C Heezen	AGS	US	975
TR 64	Quisquis	AWT	Ecuador	213
65-3	Cabo Blanco	PB	Costa Rica	183
65-4	Isla Burica	PB	Costa Rica	183
BP 65	Rhoen	WPB	Germany	299
CG 65	Chosin	CGHM	US	944
D 65	Chennai	DDGHM	India	333
DDG 65	Benfold	DDGHM	US	946
GC 65	Martin Garcia	WPB	Argentina	22
M 65	Alleppey	MSO	India	350
P 65	Punta Mogotes	PB	Argentina	18
PL 65	Shiretoko	PL/PSO	Japan	453
PT 65	Flower Ray	WPB	Singapore	758
S 65	Sindhushastra	SSK	India	329
T 65	Bangaram	PBO	India	348
T-AGS 65	Mary Sears	AGS	US	975
ATF 66	Galvarino	ATF	Chile	125
CG 66	Hue City	CGHM	US	944
DDG 66	Gonzalez	DDGHM	US	946
GC 66	Rio Lujan	WPB	Argentina	22
P 66	Rio Santiago	PB	Argentina	18
PL 66	Shikine	PL/PSO	Japan	453
PT 66	Fiddler Ray	WPB	Singapore	758
T 66	Bitra	PBO	India	348
T-AGS 66	Maury	AGS	US	975
ATF 67	Lautaro	ATF	Chile	125
CG 67	Shiloh	CGHM	US	944
DDG 67	Cole	DDGHM	US	946
GC 67	Rio Uruguay	WPB	Argentina	22
L 67	Zarkoh	LCT	UAE	889
M 67	Karwar	MSO	India	350
P 67	Mzizi	PB	Tanzania	825
P 67	Ypoploiarchos Roussen	PGG	Greece	306
PL 67	Amagi	PL/PSO	Japan	453
PT 67	Starfish Ray	WPB	Singapore	758
T 67	Batti Malv	PBO	India	348
68	Formidable	FFGHM	Singapore	752
CG 68	Anzio	CGHM	US	944
CVN 68	Nimitz	CVNM	US	942
DDG 68	The Sullivans	DDGHM	US	946
GC 68	Rio Paraguay	WPB	Argentina	22
M 68	Cannanore	MSO	India	350
P 68	Mzia	PB	Tanzania	825
P 68	Ypoploiarchos Daniolos	PGG	Greece	306
PL 68	Suzuka	PL/PSO	Japan	453
T 68	Baratang	PBO	India	348
69	Intrepid	FFGHM	Singapore	752
CG 69	Vicksburg	CGHM	US	944
CVN 69	Dwight D Eisenhower	CVNM	US	942
DDG 69	Milius	DDGHM	US	946
GC 69	Rio Parana	WPB	Argentina	22
M 69	Cuddalore	MSO	India	350
P 69	Ypoploiarchos Kristallidis	PGG	Greece	306
PL 69	Koshiki	PL/PSO	Japan	453
T 69	Car Nicobar	PBO	India	348
70	Rauma	PTGM	Finland	240
70	Steadfast	FFGHM	Singapore	752
ACV 70	Dame Roma Mitchell	PB	Australia	39
CG 70	Lake Erie	CGHM	US	944
CVN 70	Carl Vinson	CVNM	US	942
DDG 70	Hopper	DDGHM	US	946
GC 70	Rio De La Plata	WPB	Argentina	22
M 70	Kakinada	MSO	India	350
P 70	Macaé	PBO	Brazil	78
P 70	Ypoploiarchos Grigoropoulos	PGG	Greece	306
PS 70	Quezon	FS	Philippines	629
RA 70	Chimborazo	ATF	Ecuador	213
T 70	Chetlat	PBO	India	348
71	Alvand	FFG	Iran	380
71	Raahe	PTGM	Finland	240
71	Tara Bai	WPBO	India	356
71	Tenacious	FFGHM	Singapore	752
A 71	Juan Sebastián De Elcano	AXS	Spain	779
AT 71	Mangyan	ABU	Philippines	635
CG 71	Cape St George	CGHM	US	944
CG 71	Observer	PB	Mauritius	526
CVN 71	Theodore Roosevelt	CVNM	US	942
DDG 71	Ross	DDGHM	US	946
GC 71	La Plata	WPB	Argentina	22
M 71	Kozhikode	MSO	India	350
P 71	Anthypoploiarchos Ritsos	PGG	Greece	306
P 71	Macau	PBO	Brazil	78
P 71	Serviola	PSOH	Spain	774
PL 71	Iwami	PL/PSO	Japan	453
PSG 71	Micalvi	PB/AEM	Chile	121
S 71	Chakra	SSN	India	327
S 71	Galerna	SSK	Spain	767
T 71	Cora Divh	PBO	India	348
T 71	Margarita	LCU	Venezuela	1005
72	Ahalya Bai	WPBO	India	356
72	Alborz	FFG	Iran	380
72	Porvoo	PTGM	Finland	240
72	Stalwart	FFGHM	Singapore	752
A 72	Arosa	AXS	Spain	779
AF 72	Lake Taal	YO	Philippines	634
BS 72	Andrija Mohorovičić	AX	Croatia	188
CG 72	Vella Gulf	CGHM	US	944
CVN 72	Abraham Lincoln	CVNM	US	942
DDG 72	Mahan	DDGHM	US	946
GC 72	Buenos Aires	WPB	Argentina	22
M 72	Konkan	MSO	India	350
P 72	Centinela	PSOH	Spain	774
P 72	Maracaná	PBO	Brazil	78
P 72	Ypoploiarchos Votsis	PGFG	Greece	307
PC 72	Urayuki	PC/SAR	Japan	457
PL 72	Rebun	PL/PSO	Japan	453
PSG 72	Ortiz	PB/AEM	Chile	121
RB 72	Maranon	YTM/YTL	Ecuador	213
T 72	Cheriyam	PBO	India	348
T 72	La Orchila	LCU	Venezuela	1005
73	Collins	SSK	Australia	26
73	Lakshmi Bai	WPBO	India	356
73	Naantali	PTGM	Finland	240
73	Sabalan	FFG	Iran	380
73	Supreme	FFGHM	Singapore	752
A 73	Astravahini	YPT	India	353
A 73	Blanca	AXS	Spain	779
BS 73	Faust Vrančič	ASR	Croatia	189
CG 73	Port Royal	CGHM	US	944
CVN 73	George Washington	CVNM	US	942
DDG 73	Decatur	DDGHM	US	946
GC 73	Cabo Corrientes	WPB	Argentina	22
M 73	Koster	–	Sweden	802
P 73	Anthypoploiarchos Pezopoulos	PGFG	Greece	307
P 73	Mangaratiba	PBO	Brazil	78
P 73	Vigía	PSOH	Spain	774
PL 73	Kii	PL/PSO	Japan	453
PSG 73	Isaza	PB/AEM	Chile	121
RB 73	Cotopaxi	YTM/YTL	Ecuador	213
S 73	Arihant	SSBN/SSGN	India	327
S 73	Mistral	SSK	Spain	767
T 73	Cankarso	PBO	India	348
74	Farncomb	SSK	Australia	26
A 74	La Graciosa	AXS	Spain	779
A 74	Sagardhwani	–	India	350
CVN 74	John C Stennis	CVNM	US	942
DDG 74	McFaul	DDGHM	US	946
GC 74	Rio Quequen	WPB	Argentina	22
M 74	Kullen	–	Sweden	802
P 74	Atalaya	PSOH	Spain	774
P 74	Miramar	PBO	Brazil	78
PMD 74	Videla	PB/AEM	Chile	121
PS 74	Rizal	FS	Philippines	629
S 74	Tramontana	SSK	Spain	767
T 74	Kondul	PBO	India	348
75	Grigore Antipa	AGOR	Romania	665
75	Naiki Devi	WPBO	India	356
75	Waller	SSK	Australia	26
A 75	Sisargas	AXS	Spain	779
A 75	Tarangini	AXS	India	351
AU 75	Bessang Pass	PB	Philippines	635

IHS Jane's Fighting Ships 2014-2015

© 2014 IHS

Pennant list [77]

Number	Ship's name	Type	Country	Page	Number	Ship's name	Type	Country	Page
CVN 75	Harry S Truman	CVNM	US	942	F 82	Somerset	FFGHM	UK	904
DDG 75	Donald Cook	DDGHM	US	946	F 82	Victoria	FFGHM	Spain	768
GC 75	Bahia Blanca	WPB	Argentina	22	L 82	Sir Bu'nuer	–	UAE	889
M 75	Vinga	–	Sweden	802	OPV 82	Comandante Toro	PSO	Chile	121
P 75	Magé	PBO	Brazil	78	P 82	Formentor	PB	Spain	774
P 75	Plotarchis Maridakis	PGFG	Greece	307	PC 82	Awagiri	PC/SAR	Japan	457
RB 75	Iliniza	YTM/YTL	Ecuador	213	S 82	Narciso Monturiol	SSK	Spain	768
RM 75	Andagoya	YTL	Colombia	178	83	Armidale	PB	Australia	33
T 75	Kalpeni	PBO	India	348	83	Dristig	PBR	Sweden	800
76	Dechaineux	SSK	Australia	26	83	Hawkesbury	MHC	Australia	33
76	Hang Tuah	FFH/AX	Malaysia	514	83	Pori	PTGM	Finland	239
76	Jamaran	FFG	Iran	380	83	Rajtarang	PBO	India	357
A 76	Giralda	AXS	Spain	779	83	Unity	PCM/PGM	Singapore	754
CVN 76	Ronald Reagan	CVNM	US	942	A 83	Contramaestre Sánchez Fernández	AXL	Spain	779
DDG 76	Higgins	DDGHM	US	946	BG 83	Nizyn	PGR	Ukraine	883
GC 76	Ingeniero White	WPB	Argentina	22	DDG 83	Howard	–	US	948
M 76	Ven	–	Sweden	802	DF 83	Rio Orellana	YFD	Ecuador	213
P 76	Infanta Elena	PSOH/FSGM	Spain	773	F 83	Numancia	FFGHM	Spain	768
P 76	Maragogipe	PBO	Brazil	78	F 83	St Albans	FFGHM	UK	904
PC 76	Makigumo	PC/SAR	Japan	457	K 83	Nashak	FSGM	India	344
RB 76	Altar	YTM/YTL	Ecuador	213	OPV 83	Marinero Fuentealba	PSO	Chile	121
RB 76	Josué Alvarez	YTL	Colombia	178	PC 83	Shimagiri	PC/PB	Japan	457
T 76	Kabra	PBO	India	348	RF 83	Joves Fiallo	YTL	Colombia	178
77	Damavand	FFG	Iran	380	S 83	Cosme García	SSK	Spain	768
77	Rani Abbakka	PBO	India	355	84	Händig	PBR	Sweden	800
77	Sheean	SSK	Australia	26	84	Larrakia	PB	Australia	33
A 77	Almansa	AXS	Spain	779	84	Norman	MHC	Australia	33
CVN 77	George H W Bush	CVNM	US	942	84	Rajkiran	PBO	India	357
DDG 77	O'Kane	DDGHM	US	946	84	Sovereignty	PCM/PGM	Singapore	754
F 77	Te Kaha	FFHM	New Zealand	574	A 84	Contramaestre Antero	AXL	Spain	779
GC 77	Golfo San Matias	WPB	Argentina	22	DDG 84	Bulkeley	–	US	948
M 77	Ulvön	–	Sweden	802	F 84	Enymiri	FSM	Nigeria	579
P 77	Infanta Cristina	PSOH/FSGM	Spain	773	F 84	Reina Sofía	FFGHM	Spain	768
P 77	Kamorta	FFG	India	341	PC 84	Okinami	PC/PB	Japan	457
PC 77	Hamazuki	PC/SAR	Japan	457	S 84	Mateo García De Los Reyes	SSK	Spain	768
PSH 77	Cabrales	PB/AEM	Chile	121	85	Bathurst	PB	Australia	33
RB 77	Don Vizo	YTL	Colombia	178	85	Gascoyne	MHC	Australia	33
T 77	Koswari	PBO	India	348	85	Justice	PCM/PGM	Singapore	754
78	Rani Avantibai	PBO	India	355	85	Rajkamal	PBO	India	357
78	Rankin	SSK	Australia	26	85	Trygg	PBR	Sweden	800
A 78	Peregrina	AXS	Spain	779	A 85	Contramaestre Lamadrid	AXL	Spain	779
AF 78	Lake Buhi	YO	Philippines	634	DDG 85	McCampbell	–	US	948
CVN 78	Gerald R Ford	CVN	US	941	F 85	Navarra	FFGHM	Spain	768
DDG 78	Porter	DDGHM	US	946	P 85	Intrépida	PGGF	Argentina	18
F 78	Kent	FFGHM	UK	904	PC 85	Hayagiri	PC/PB	Japan	457
G 78	Ottonelli	PB	Italy	422	RF 85	Miguel Silva	YTL	Colombia	178
GC 78	Madryn	WPB	Argentina	22	86	Albany	PB	Australia	33
P 78	Cazadora	PSOH/FSGM	Spain	773	86	Diamantina	MHC	Australia	33
P 78	Kadmatt	FFG	India	341	86	Freedom	PCM/PGM	Singapore	754
PC 78	Isozuki	PC/SAR	Japan	457	86	Modig	PBR	Sweden	800
PSG 78	Sibbald	PB/AEM	Chile	121	86	Rajratan	PBO	India	357
RB 78	Portete	YTL	Colombia	178	A 86	Tir	AXH	India	351
RB 78	Quilotoa	YTM/YTL	Ecuador	213	DDG 86	Shoup	–	US	948
T 78	Karuva	PBO	India	348	F 86	Canarias	FFGHM	Spain	768
79	Rani Durgavati	PBO	India	355	H 86	Gleaner	YGS	UK	914
AE 79	Limasawa	ABU	Philippines	635	P 86	Indomita	PGGF	Argentina	18
CVN 79	John F Kennedy	CVN	US	941	PC 86	Natsugiri	PC/PB	Japan	457
DDG 79	Oscar Austin	–	US	948	87	Hurtig	PBR	Sweden	800
F 79	Portland	FFGHM	UK	904	87	Independence	PCM/PGM	Singapore	754
G 79	Barletta	PB	Italy	422	87	Pirie	PB	Australia	33
GC 79	Rio Deseado	WPB	Argentina	22	87	Rajdoot	PBO	India	357
P 79	Kiltan	FFG	India	341	87	Yarra	MHC	Australia	33
P 79	Vencedora	PSOH/FSGM	Spain	773	DDG 87	Mason	–	US	948
PC 79	Shimanami	PC/SAR	Japan	457	H 87	Echo	AGSH	UK	914
VTR 79	Dubnyak	AETL	Russian Federation	721	LT 87	South Cotabato	LST	Philippines	632
80	Hamina	PTGM	Finland	239	PC 87	Suganami	PC/PB	Japan	457
80	Rani Gaidinliu	PBO	India	355	RF 87	Capitán Vladimir Valek Moure	YTL	Colombia	178
ACV 80	Storm Bay	PB	Australia	39	88	Maitland	PB	Australia	33
BG 80	Dunai	AGF	Ukraine	883	88	Rajveer	PBO	India	357
CVN 80	Enterprise	CVN	US	941	88	Rapp	PBR	Sweden	800
DDG 80	Roosevelt	–	US	948	DDG 88	Preble	–	US	948
GC 80	Ushuaia	WPB	Argentina	22	G 88	La Malfa	PB	Italy	422
P 80	Kavaratti	FFG	India	341	GC 88	Medusa	WPB	Argentina	23
PC 80	Yuzuki	PC/SAR	Japan	457	H 88	Enterprise	AGSH	UK	914
RB 80	Cienaga De San Juan	YTL	Colombia	178	P 88	Victory	FSGM	Singapore	753
81	Bayandor	FS	Iran	382	RF 88	Teniente Luís Bernal Baquero	YTL	Colombia	178
81	Rani Rashmoni	PBO	India	355	S 88	Tireless	SSN	UK	898
81	Tornio	PTGM	Finland	239	89	Ararat	PB	Australia	33
81	Zhenghe	AXH	China	159	89	Rajdhwaj	PBO	India	357
BG 81	Lubny	PGR	Ukraine	883	89	Stolt	PBR	Sweden	800
CG 81	Rescuer	PB	Mauritius	526	AG 89	Kalinga	AKLH	Philippines	635
CLM 81	Almirante Grau	CG/CLM	Peru	618	DDG 89	Mustin	–	US	948
DBM 81	Cetina	LCT/ML	Croatia	187	F 89	Aradu	FFGHM	Nigeria	578
DDG 81	Winston S Churchill	–	US	948	G 89	Rosati	PB	Italy	422
F 81	Santa María	FFGHM	Spain	768	GC 89	Perca	WPB	Argentina	23
F 81	Sutherland	FFGHM	UK	904	P 89	Valour	FSGM	Singapore	753
GC 81	Canal De Beagle	WPB	Argentina	22	PM 89	Takatori	PM/PBO	Japan	455
OPV 81	Piloto Pardo	PSO	Chile	121	90	Ärlig	PBR	Sweden	800
P 81	Toralla	PB	Spain	774	90	Broome	PB	Australia	33
PC 81	Tamanami	PC/SAR	Japan	457	90	Sabha	FFGHM	Bahrain	46
PL 81	Taketomi	PL/PSOH	Japan	453	A 90	Varonis	AKS/AXL	Latvia	495
S 81	Isaac Peral	SSK	Spain	768	AC 90	Mactan	AK	Philippines	633
T 81	Ciudad Bolívar	AORH	Venezuela	1006	DDG 90	Chaffee	–	US	948
82	Djärv	PBR	Sweden	800	F 90	Thunder	FFH	Nigeria	579
82	Hanko	PTGM	Finland	239	GC 90	Calamar	WPB	Argentina	23
82	Huon	MHC	Australia	33	P 90	Vigilance	FSGM	Singapore	753
82	Naghdi	FS	Iran	382	S 90	Torbay	SSN	UK	898
82	Rajshree	PBO	India	357	91	Bundaberg	PB	Australia	33
82	Resilience	PCM/PGM	Singapore	754	91	Munter	PBR	Sweden	800
82	Shichang	HSS/AHH	China	160	91	Sargento Aldea	LSDH	Chile	122
82-3	Juan Rafael Mora Point Chico	PB	Costa Rica	183	A 91	Astrolabio	YGS	Spain	779
82-4	Pancha Carrasco Point Bridge	PB	Costa Rica	183	ASY 91	Hashidate	ASY/YAC	Japan	449
A 82	Contramaestre Navarrete	AXL	Spain	779	BE 91	Guayas	AXS	Ecuador	212
BG 82	Kaniv	PGR	Ukraine	883	BI 91	Orion	YGS	Ecuador	212
CG 82	Retriever	PB	Mauritius	526	DDG 91	Pinckney	–	US	948
DBV 82	Krka	LCT/ML	Croatia	187	GC 91	Hipocampo	WPB	Argentina	23
DDG 82	Lassen	–	US	948					
DF 82	Rio Napo	YFD	Ecuador	213					

© 2014 IHS

IHS Jane's Fighting Ships 2014-2015

[78] Pennant list

Number	Ship's name	Type	Country	Page
K 91	Pralaya	FSGM	India	344
M 91	Sagar	MSO	Bangladesh	58
P 91	Valiant	FSGM	Singapore	753
R 91	Charles De Gaulle	CVNM/PAN	France	252
S 91	Trenchant	SSN	UK	898
T 91	Los Frailes	AKL	Venezuela	1005
92	Orädd	PBR	Sweden	800
92	Rancagua	LSTH	Chile	122
92	Wollongong	PB	Australia	33
A 92	Escandallo	YGS	Spain	779
DDG 92	Momsen	–	US	948
G 92	Alberti	PBF	Italy	422
GC 92	Robaldo	WPB	Argentina	23
K 92	Prabal	FSGM	India	344
P 92	Vigour	FSGM	Singapore	753
S 92	Talent	SSN	UK	898
T 92	Los Testigos	AKL	Venezuela	1005
93	Childers	PB	Australia	33
DDG 93	Chung-Hoon	–	US	948
G 93	Angelini	PBF	Italy	422
GC 93	Camaron	WPB	Argentina	23
OB 93	Šokadija	PBR	Croatia	186
P 93	Vengeance	FSGM	Singapore	753
RF 93	Sejeri	YTL	Colombia	178
S 93	Triumph	SSN	UK	898
T 93	Los Roques	AKL	Venezuela	1005
94	Fearless	PCM/PGM	Singapore	754
94	Launceston	PB	Australia	33
DDG 94	Nitze	–	US	948
G 94	Cappelletti	PBF	Italy	422
GC 94	Gaviota	WPB	Argentina	23
PM 94	Kumano	PM/PBO	Japan	455
S 94	Astute	SSN	UK	900
T 94	Los Monjes	AKL	Venezuela	1005
95	Brave	PCM/PGM	Singapore	754
95	Chacabuco	LSTH	Chile	122
95	Maryborough	PB	Australia	33
DDG 95	James E Williams	–	US	948
G 95	Ciorlieri	PBF	Italy	422
GC 95	Abadejo	WPB	Argentina	23
M 95	Shapla	MHSC/PBO/AGS	Bangladesh	58
PM 95	Amami	PM/PBO	Japan	455
S 95	Artful	SSN	UK	900
Y 95	Torpido Tenderi	YPT	Turkey	868
96	Glenelg	PB	Australia	33
DDG 96	Bainbridge	–	US	948
G 96	D'Amato	PBF	Italy	422
LH 96	Sirius	AGSC	Ecuador	212
M 96	Saikat	MHSC/PBO/AGS	Bangladesh	58
PM 96	Kurokami	PM/PBO	Japan	455
S 96	Ambush	SSN	UK	900
97	Gallant	PCM/PGM	Singapore	754
DDG 97	Halsey	–	US	948
G 97	Fais	PBF	Italy	422
M 97	Surovi	MHSC/PBO/AGS	Bangladesh	58
PM 97	Ishikari	PM/PBO	Japan	455
S 97	Audacious	SSN	UK	900
98	Daring	PCM/PGM	Singapore	754
DDG 98	Forrest Sherman	–	US	948
G 98	Feliciani	PBF	Italy	422
M 98	Shaibal	MHSC/PBO/AGS	Bangladesh	58
PM 98	Horobetsu	PM/PBO	Japan	455
S 98	Anson	SSN	UK	900
Y 98	Takip 1	YPT	Turkey	868
99	Dauntless	PCM/PGM	Singapore	754
DDG 99	Farragut	–	US	948
G 99	Garzoni	PBF	Italy	422
S 99	Agamemnon	SSN	UK	900
Y 99	Takip 2	YPT	Turkey	868
100	Stavropol	PGM	Russian Federation	728
AU 100	Tirad Pass	PB	Philippines	635
BG 100	Sivas	PB	Ukraine	882
DDG 100	Kidd	–	US	948
G 100	Lippi	PBF	Italy	422
L 100	Choules	LSD	Australia	31
P 100	Mandume	PC	Angola	10
P 100	Andoni	PB	Nigeria	580
S 100	Ajax	SSN	UK	900
101	Al Hussein	PB	Jordan	460
101	Arkadash	PC	Turkmenistan	874
101	Fouque	LSL	Iran	389
101	General Matrosov	PBO	Russian Federation	726
101	Ho Hsing	WPSO	Taiwan	821
101	Levuka	PB	Fiji	239
101	Lilian	PB	Lithuania	504
101	Raif Denktas	WPBI	Turkey	872
101	Viteazul	ARS	Romania	667
A 101	Mar Caribe	ATF/AGDS	Spain	781
DD 101	Murasame	DDGHM	Japan	436
DDG 101	Gridley	–	US	948
F 101	Alvaro De Bazán	FFGHM	Spain	770
G 101	Lombardi	PBF	Italy	422
GC 101	Dorado	WPB	Argentina	22
GC 101	Aries	PB	Dominican Republic	206
P 101	Dorina	PB	Nigeria	582
P 101	Oecussi	PB	East Timor	208
P 101	Shet Gang	PB	Bangladesh	61
P 101	Tsotne Dadiani	WPB	Georgia	282
PC 101	Asogiri	PC/PB	Japan	458
PVL 101	Kindral Kurvits	PSO	Estonia	236
S 101	Manthatisi	SSK	South Africa	761
S 101	Shyri	SSK	Ecuador	209
SHV 101	Hårek	PB	Norway	589
TM 101	Almirante Grau	YFL	Bolivia	69
WTGB 101	Katmai Bay	WTGB	US	989
102	Al Hassan	PB	Jordan	460
102	Barkarar	PC	Turkmenistan	874
102	Derbent	PGM	Russian Federation	728
102	Kaliningrad	LSTM	Russian Federation	704
102	Kihu	PB	Lithuania	503
102	Lautoka	PB	Fiji	239
102	Sovetskiy Pogranichnik	AO	Russian Federation	733
102	Wei Hsung	WPSO	Taiwan	821
BG 102	Obolon	PB	Ukraine	882
DD 102	Harusame	DDGHM	Japan	436
DDG 102	Sampson	–	US	948
F 102	Almirante Juan De Borbón	FFGHM	Spain	770
G 102	Miccoli	PBF	Italy	422
GC 102	Anchoa	WPB	Argentina	23
LP 102	Manuel Trujillo	PBR	Paraguay	616
P 102	Atauro	PB	East Timor	208
P 102	General Mazniashvili	WPB	Georgia	282
P 102	Juan De La Barrera	PG/PGH	Mexico	535
P 102	Polar	PC	Angola	10
P 102	Porte Grande	PB	Bangladesh	61
PC 102	Murozuki	PC/PB	Japan	458
PG 102	Bagong Lakas	PB	Philippines	631
PM 102	Rafael Del Castillo Y Rada	PB	Colombia	174
S 102	Charlotte Maxeke	SSK	South Africa	761
S 102	Huancavilca	SSK	Ecuador	209
SHV 102	Hvasser	PB	Norway	589
WTGB 102	Bristol Bay	WTGB	US	989
103	Al Maks	ATA	Egypt	227
103	King Abdullah	PB	Jordan	460
BG 103	Darnitsya	PB	Ukraine	882
DD 103	Yuudachi	DDGHM	Japan	436
DDG 103	Truxtun	–	US	948
F 103	Blas De Lezo	FFGHM	Spain	770
G 103	Trezza	PBF	Italy	422
GC 103	Procion	PB	Dominican Republic	206
H 103	Guama	ABU	Cuba	191
P 103	Mariano Escobedo	PG/PGH	Mexico	535
PC 103	Urazuki	PC/PB	Japan	458
PL 103	Wakasa	PL/PSO	Japan	452
PM 103	José Maria Palas	PB	Colombia	175
PVL 103	Pikker	PB	Estonia	235
S 103	Queen Modjadji I	SSK	South Africa	761
SHV 103	Hekkingen	PB	Norway	589
TM 103	Comandante Arandia	YFL	Bolivia	69
WTGB 103	Mobile Bay	WTGB	US	989
DD 104	Kirisame	DDGHM	Japan	436
DDG 104	Sterett	–	US	948
F 104	Mendez Nuñez	FFGHM	Spain	770
G 104	Apruzzi	PBF	Italy	422
GC 104	Aldebarán	PB	Dominican Republic	206
P 104	Atlantico	PC	Angola	10
P 104	Bakassi	PBO	Cameroon	95
P 104	Manuel Doblado	PG/PGH	Mexico	535
PC 104	Kagayuki	PC/PB	Japan	458
PG 104	Bagong Silang	PB	Philippines	631
PL 104	Kii	PL/PSO	Japan	452
SHV 104	Kvitsøy	PB	Norway	589
WTGB 104	Biscayne Bay	WTGB	US	989
105	Al Agami	ATA	Egypt	227
105	Baykal	PBO	Russian Federation	726
105	Ivan Yevteyev	AK	Russian Federation	732
105	Mou Hsing	WPSO	Taiwan	822
105-1	Isla Del Coco	PB	Costa Rica	183
DD 105	Inazuma	DDGHM	Japan	436
DDG 105	Dewey	–	US	948
F 105	Cristóbal Colón	FFGHM	Spain	770
G 105	Ballali	PBF	Italy	422
GC 105	Antares	PB	Dominican Republic	206
L 105	Arromanches	RCL	UK	924
M 105	Bedok	MHS	Singapore	757
P 105	Jabanne	PB	Cameroon	96
P 105	Iveria	PB	Georgia	282
PB 105	Sir Milton	PB	Sierra Leone	749
PC 105	Hayagumo	PC/PBF	Japan	457
PM 105	Jaime Gómez Castro	PB	Colombia	179
SHV 105	Slotterøy	PB	Norway	589
WTGB 105	Neah Bay	WTGB	US	989
106	Abay	PB	Kazakhstan	462
106	Brest	PBO	Russian Federation	726
106	Capitan David Eyama Angue Osa	AK	Equatorial Guinea	232
106	Fu Hsing	WPSO	Taiwan	822
106	Nirolhu	PB	Maldives	521
106	Vyuga	PGR	Russian Federation	731
DD 106	Samidare	DDGHM	Japan	436
DDG 106	Stockdale	–	US	948
G 106	Bovienzo	PBF	Italy	422
GC 106	Bellatrix	PB	Dominican Republic	206
GC 106	Narval	WPB	Argentina	23
M 106	Kallang	MHS	Singapore	757
M 106	Penzance	MHC/SRMH	UK	912
P 106	Akwayafe	PB	Cameroon	96
P 106	Golfinho	PC	Angola	10
P 106	Mestia	PB	Georgia	282
P 106	Santos Degollado	PG/PGH	Mexico	535
PC 106	Murakumo	PC/PBF	Japan	457
PM 106	Juan Nepomuceno Peña	PB	Colombia	179
SHV 106	Halten	PB	Norway	589
WTGB 106	Morro Bay	WTGB	US	989
107	Al Antar	ATA	Egypt	227
107	Astrakhanets	MHSC/MHSCM	Russian Federation	708
DD 107	Ikazuchi	DDGHM	Japan	436
DDG 107	Gravely	–	US	948
G 107	Carreca	PBF	Italy	422
GC 107	Canopus	PB	Dominican Republic	206
L 107	Andalsnes	RCL	UK	924
M 107	Katong	MHS	Singapore	757
M 107	Pembroke	MHC/SRMH	UK	912
PC 107	Izunami	PC/PBF	Japan	457
PVL 107	Kou	PBO	Estonia	235
WTGB 107	Penobscot Bay	WTGB	US	989
DD 108	Akebono	DDGHM	Japan	436
DDG 108	Wayne E Meyer	–	US	948
G 108	Conversano	PBF	Italy	422
GC 108	Capella	PB	Dominican Republic	206

IHS Jane's Fighting Ships 2014-2015 © 2014 IHS

Pennant list [79]

Number	Ship's name	Type	Country	Page
M 108	Grimsby	MHC/SRMH	UK	912
M 108	Punggol	MHS	Singapore	757
P 108	Juan N Alvares	PG/PGH	Mexico	535
P 108	Poti	PBF	Georgia	283
PC 108	Yaegumo	PC/PBF	Japan	457
PS 108	Takatsuki	PS/PBF	Japan	456
WTGB 108	Thunder Bay	WTGB	US	989
109	Al Dekheila	ATA	Egypt	227
109	Kaifeng	DDG	China	136
109	Rasul Gamzatov	PGM	Russian Federation	728
DD 109	Ariake	DDGHM	Japan	436
DDG 109	Jason Dunham	–	US	948
G 109	Inzerilli	PBF	Italy	422
GC 109	Orion	PB	Dominican Republic	206
M 109	Bangor	MHC/SRMH	UK	912
P 109	Alibori	PB	Benin	67
P 109	Manuel Gutierrez Zamora	PG/PGH	Mexico	535
P 109	Pazisi	WPB	Georgia	282
PC 109	Natsugumo	PC/PBF	Japan	457
PS 109	Katsuragi	PS/PBF	Japan	456
WTGB 109	Sturgeon Bay	WTGB	US	989
110	Alexander Shabalin	LSTM	Russian Federation	704
110	Dalian	DDG	China	136
110	Sochi	PGM	Russian Federation	728
A 110	Orangeleaf	AOT	UK	915
ACV 110	Ashmore Guardian	PBO	Australia	39
BG 110	Ljubomir	PB	Ukraine	882
DD 110	Takanami	DDGHM	Japan	438
DDG 110	William P Lawrence	–	US	948
G 110	Letizia	PBF	Italy	422
GC 110	Sirius	PB	Dominican Republic	206
L 110	Aachen	RCL	UK	924
M 110	Ramsey	MHC/SRMH	UK	912
P 110	Oueme	PB	Benin	67
P 110	Valentin Gomez Farias	PG/PGH	Mexico	535
PC 110	Akigumo	PC/PBF	Japan	457
YFB 110	Meteoro	YFB	Chile	124
111	Al Iskandarani	ATA	Egypt	227
111	Dareen	YTB/YTM	Saudi Arabia	742
111	PSKR 609	PTF	Russian Federation	728
A 111	Alerta	AGI/AGOR	Spain	778
ACA 111	Caloyeras	YW/YO	Peru	624
D 111	Comodoro Manuel Azueta	FF/AX	Mexico	540
DD 111	Oonami	DDGHM	Japan	438
DDG 111	Spruance	–	US	948
F 111	Te Mana	FFHM	New Zealand	574
F 111	Marasesti	FFGH	Romania	661
G 111	Mazzarella	PBF	Italy	422
L 111	Arezzo	RCL	UK	924
M 111	Blyth	MHC/SRMH	UK	912
MS 111	Guns	PB	Pakistan	610
P 111	Ladse	PBF	Slovenia	759
P 111	Pabna	PBR	Bangladesh	61
P 111	Zou	PB	Benin	67
PC 111	Haginami	PC/PBF	Japan	457
PG 111	Bonny Serrano	PBF	Philippines	631
S 111	Nereus	SSK	Greece	303
112	Harbin	DDGHM	China	141
DD 112	Makinami	DDGHM	Japan	438
DDG 112	Michael Murphy	–	US	948
G 112	Nioi	PBF	Italy	422
GC 112	Altair	PB	Dominican Republic	206
LG 112	Rio Mataje	WPBR	Ecuador	215
M 112	Shoreham	MHC/SRMH	UK	912
MS 112	Sur	PB	Pakistan	610
P 112	Noakhali	PBR	Bangladesh	61
PC 112	Ikigumo	PC/PBF	Japan	457
PG 112	Bienvenido Salting	PBF	Philippines	631
PM 112	Quitasueño	PGF	Colombia	179
PVL 112	Valve	WPB	Estonia	235
S 112	Triton	SSK	Greece	303
T-AKR 112	Cape Texas	AKR	US	982
113	Al Areesh	ARS	Egypt	227
113	Berkut	PGM	Russian Federation	728
113	Qingdao	DDGHM	China	141
113	Tuwaig	YTB/YTM	Saudi Arabia	742
113	Yunga	FFLM	Russian Federation	697
DD 113	Sazanami	DDGHM	Japan	438
DDG 113	John Finn	–	US	948
G 113	Partipilo	PBF	Italy	422
L 113	Audemer	RCL	UK	924
MS 113	Malan	PB	Pakistan	610
P 113	Ignacio L Vallarta	PG/PGH	Mexico	535
P 113	Patuakhali	PBR	Bangladesh	61
PC 113	Natsuzuki	PC/PBF	Japan	457
PM 113	José Maria Garcia Y Toledo	PB	Colombia	179
S 113	Proteus	SSK	Greece	303
T-AKR 113	Cape Taylor	AKR	US	982
Y 113	Pinar 3	YW	Turkey	867
YFB 113	Buzo Sobenes	YFB	Chile	123
DD 114	Suzunami	DDGHM	Japan	438
DDG 114	Ralph Johnson	–	US	948
G 114	Puleo	PBF	Italy	422
GC 114	Arcturus	PB	Dominican Republic	206
GC 114	Don Lucas	WPB	Argentina	23
P 114	Jesus Gonzalez Ortega	PG/PGH	Mexico	535
P 114	Rangamati	PBR	Bangladesh	61
PC 114	Okigumo	PC/PBF	Japan	457
PG 114	Salvador Abcede	PBF	Philippines	631
PM 114	Juan Nepomuceno Eslava	PB	Colombia	179
Y 114	Pinar 4	YW	Turkey	867
YFB 114	Grumete Perez	YFB	Chile	124
115	Al Almain	ARS	Egypt	227
115	Burevi	PB	Maldives	522
115	Emil Racovita	AGS	Romania	665
115	Ivan Lednev	AK	Russian Federation	732
115	Shenyang	DDGHM	China	136
DD 115	Akizuki	DDHM	Japan	435
DDG 115	Rafael Peralta	–	US	948
G 115	Zannotti	PBF	Italy	422
GC 115	Talita II	WAXL/WAXS	Argentina	23
LG 115	Rio Daule	WPBR	Ecuador	215
P 115	Bogra	PBR	Bangladesh	61
PC 115	Awagumo	PC/PBF	Japan	457
PG 115	Ramon Aguirre	PBF	Philippines	631
PL 115	Noto	PL/PSO	Japan	452
PM 115	TECIM Jaime E Cárdenas Gomez	PB	Colombia	174
116	PSKR 670	PTF	Russian Federation	728
116	Shijiazhuang	DDGHM	China	136
116	Taipei	WPSO	Taiwan	823
DD 116	Teruzuki	DDHM	Japan	435
DDG 116	Thomas Hudner	–	US	948
G 116	Laganà	PB	Italy	422
LG 116	Rio Babahoyo	WPBR	Ecuador	215
PC 116	Shimagumo	PC/PBF	Japan	457
PG 116	Nicolas Mahusay	PBF	Philippines	631
S 116	Poseidon	SSK	Greece	303
Y 116	Pinar 6	YW	Turkey	867
117	PSKR 52	PBR	Russian Federation	731
117	Ras Al Fulaijah	MSO	Libya	501
117	Taichung	WPSO	Taiwan	822
117	Yinchuan	DDGHM	China	140
BG 117	Batutinets	PB	Ukraine	882
DD 117	Suzutsuki	DDHM	Japan	435
DDG 117	Paul Ignatius	–	US	948
G 117	Sanna	PB	Italy	422
GC 117	Adhara II	WAXL/WAXS	Argentina	23
LG 117	Rio Chone	WPBR	Ecuador	215
LG117	Rio Coca	WPB	Ecuador	216
P 117	Mariano Matamoros	PG/PGH	Mexico	535
PC 117	Yukigumo	PC/PBF	Japan	457
PG 117	Dionisio Ojeda	PBF	Philippines	631
PL 117	Suruga	PL/PSO	Japan	452
S 117	Amphitrite	SSK	Greece	303
118	Keelung	WPSO	Taiwan	822
118	Taiyuan	DDGHM	China	140
ACP 118	Noguera	YW/YO	Peru	624
BG 118	Arabat	PB	Ukraine	882
DD 118	Fuyuzuki	DDHM	Japan	435
DDG 118	Daniel Inouye	–	US	948
G 118	Inzucchi	PB	Italy	422
GC 118	Alumine	PB	Argentina	24
LG118	Rio Conambo	WPB	Ecuador	216
LG 118	Rio Zarumilla	WPBR	Ecuador	215
PC 118	Kitaguni	PC/PBF	Japan	457
PG 118	Emilio Liwanag	PBF	Philippines	631
S 118	Okeanos	SSK	Greece	303
119	Hualien	WPSO	Taiwan	822
119	Nikolay Starshinov	AK	Russian Federation	732
119	Ras Al Qula	MSO	Libya	501
ACP 119	Gauden	YW/YO	Peru	624
G 119	Vitali	PB	Italy	422
GC 119	Traful	PB	Argentina	24
LG119	Rio Cononaco	WPB	Ecuador	216
PC 119	Komayuki	PC/PBF	Japan	457
S 119	Pontos	SSK	Greece	303
120	Penhu	WPSO	Taiwan	822
G 120	Calabrese	PB	Italy	422
GC 120	Lacar	PB	Argentina	24
LG120	Rio Curaray	WPB	Ecuador	216
P 120	Amazonas	PSO	Brazil	79
PC 120	Kawagiri	PC/PBF	Japan	457
PL 120	Kunisaki	PL/PSO	Japan	452
S 120	Papanikolis	SSK	Greece	303
U 120	Skadovsk	PB	Ukraine	878
121	Moskva	CGHM	Russian Federation	689
A 121	Guardiamarina Barrutia	AXT	Spain	779
G 121	Kusar	FFL	Azerbaijan	40
G 121	Urso	PB	Italy	422
GC 121	Fontana	PB	Argentina	24
LG121	Rio Esmeraldas	WPBR	Ecuador	214
P 121	Apa	PSO	Brazil	79
P121	PGM 104	PB	Turkey	863
PC 121	Wakazuki	PC/PBF	Japan	457
PF 121	Diligente	PBR	Colombia	176
S 121	Pipinos	SSK	Greece	303
Y 121	Havuz 1	YD/YFDM	Turkey	869
122	Nantou	WPSO	Taiwan	823
A 122	Guardiamarina Chereguini	AXT	Spain	779
G 122	La Spina	PB	Italy	422
GC 122	Mascardi	PB	Argentina	24
LG 122	Rio Santiago	WPBR	Ecuador	214
P 122	Araguari	PSO	Brazil	79
P 122	PGM 105	PB	Turkey	863
P122	Teniente José Azueta	PSOH	Mexico	535
PC 122	Isonami	PC/PBF	Japan	457
PF 122	Juan Lucio	PBR	Colombia	176
PL 122	Iwami	PL/PSO	Japan	452
S 122	Matrozos	SSK	Greece	303
Y 122	Havuz 2	YD/YFDM	Turkey	869
123	Kinmen	WPSO	Taiwan	822
123	PSKR 58	PBR	Russian Federation	731
123	Ras Al Massad	MSO	Libya	501
A 123	Guardiamarina Rull	AXT	Spain	779
ARA 123	Guardian Rios	ATS	Peru	625
G 123	Salone	PB	Italy	422
GC 123	Viedna	PB	Argentina	24
NL 123	Sarucabey	LSTH/ML	Turkey	864
P 123	Capitán De Fragata Pedro Sáinz De Baranda	PSOH	Mexico	535
P123	PGM 106	PB	Turkey	863
PC 123	Nagozuki	PC/PBF	Japan	457
PF 123	Alfonso Vargas	PBR	Colombia	176
S 123	Katsonis	SSK	Greece	303
Y 123	Havuz 3	YD/YFDM	Turkey	869
A 124	Guardiamarina Salas	AXT	Spain	779
ARB 124	Selendán	YTL	Peru	625
G 124	Cavatorto	PB	Italy	422

© 2014 IHS

IHS Jane's Fighting Ships 2014-2015

[80] Pennant list

Number	Ship's name	Type	Country	Page
GC 124	San Martin	PB	Argentina	24
NL 124	Karamürselbey	LSTH/ML	Turkey	864
P 124	Comodoro Carlos Castillo Bretón	PSOH	Mexico	535
P124	PGM 107	PB	Turkey	863
PC 124	Yaezuki	PC/PBF	Japan	457
PF 124	Fritz Hagale	PBR	Colombia	176
Y 124	Havuz 4	YD/YFDM	Turkey	869
125	Lienchiang	WPSO	Taiwan	822
125	Ras Al Hani	MSO	Libya	501
ARB 125	Medina	YTL	Peru	625
G 125	Fusco	PB	Italy	422
GC 125	Buenos Aires	PB	Argentina	24
NL 125	Osman Gazi	LSTH/ML	Turkey	863
P 125	Vicealmirante Othón P Blanco	PSOH	Mexico	535
PF 125	Vengadora	PBR	Colombia	176
PL 125	Katori	PL/PSO	Japan	452
Y 125	Havuz 5	YD/YFDM	Turkey	869
126	Tainan	PSO	Taiwan	821
G 126	De Rosa	PB	Italy	422
GC 126	Musters	PB	Argentina	24
P 126	Contralmirante Angel Ortiz Monasterio	PSOH	Mexico	535
PL 126	Matsushima	PL/PSO	Japan	452
127	Hsin Bei	PSO	Taiwan	821
127	Minsk	LSTM	Russian Federation	704
DD 127	Isoyuki	DDGHM	Japan	440
G 127	Zaccola	PB	Italy	422
L 127	Brunei	LCH/LSM	Australia	32
P 127	AB 27	PC	Turkey	863
PL 127	Etomo	PL/PSO	Japan	452
CG 128	Yilan	PSO	Taiwan	824
DD 128	Haruyuki	DDGHM	Japan	440
G 128	Stanisci	PB	Italy	422
L 128	Labuan	LCH/LSM	Australia	32
PF 128	Carlos Galindo	PBR	Colombia	176
PL 128	Esan	PL/PSO	Japan	452
Y 128	Havuz 8	YD/YFDM	Turkey	869
CG 129	Kaohsiung	PSO	Taiwan	824
DD 129	Yamayuki	DDGHM	Japan	440
G 129	Sottile	PB	Italy	422
GC 129	Colhue	PB	Argentina	24
L 129	Tarakan	LCH/LSM	Australia	32
P 129	AB 29	PC	Turkey	863
PB-129	Merjen	WPB	Turkmenistan	874
Y 129	Havuz 9	YD/YFDM	Turkey	869
130	Korolev	LSTM	Russian Federation	704
DD 130	Matsuyuki	DDGHM	Japan	440
G 130	De Falco	PB	Italy	422
GC 130	Maria L Pendo	PB	Argentina	24
LG 130	Rio Yasuni	WPBR	Ecuador	216
U 130	Hetman Sagaidachny	FFHM	Ukraine	876
Y 130	Havuz 10	YD/YFDM	Turkey	869
131	Ingeniero Mery	YFD	Chile	123
GC 131	Roca	PB	Argentina	24
H 131	Scott	AGSH	UK	913
LG 131	Isla Isabela	WPB	Ecuador	215
P 131	Capitán De Navio Sebastian José Holzinger	PSOH	Mexico	533
P 131	Iliria	PB	Albania	3
SB 131	Nicolay Chiker	ATS	Russian Federation	722
132	Ibn Ouf	LSTH	Libya	500
132	Mutilla	YFD	Chile	123
132	PSKR 629	PTF	Russian Federation	728
A 132	Diligence	ARH	UK	918
DD 132	Asayuki	DDGHM	Japan	440
GC 132	Puelo	PB	Argentina	24
LG 132	Isla Seymour	WPB	Ecuador	214
P 132	Capitán De Navio Blas Godinez	PSOH	Mexico	533
P 132	Oriku	PB	Albania	3
133	Kaani	PB	Maldives	521
133	Talcauano	YFD	Chile	123
GC 133	Futalaufquen	PB	Argentina	24
P 133	Brigadier José Mariá De La Vega	PSOH	Mexico	533
P 133	Lisus	PB	Albania	3
134	Ibn Haritha	LSTH	Libya	500
134	Krechet	PCM	Russian Federation	727
134	PSKR 725	PTF	Russian Federation	728
F 134	Laksamana Hang Nadim	FSGM	Malaysia	511
GC 134	Falkner	PB	Argentina	24
P 134	Butrindi	PB	Albania	3
P 134	General Felipe B Berriozábal	PSOH	Mexico	533
Y 134	Havuz 11	YD/YFDM	Turkey	869
135	PSKR 110	PCK	Russian Federation	727
135	PSKR 500	PBF	Russian Federation	728
135	PSKR 712	PTF	Russian Federation	728
A 135	Argus	APCR	UK	918
F 135	Laksamana Tun Abdul Jamil	FSGM	Malaysia	511
GC 135	Hess	PB	Argentina	24
S 135	Hashmat	SSK	Pakistan	601
SB 135	Fotiy Krylov	ATS	Russian Federation	722
136	Hangzhou	DDGHM	China	137
136	PSKR 109	PCK	Russian Federation	727
136	PSKR 724	PTF	Russian Federation	728
136	Valentin Pikul	PGM	Russian Federation	728
F 136	Laksamana Muhammad Amin	FSGM	Malaysia	511
GC 136	Colhue Huapi	PB	Argentina	24
P 136	AB 36	PC	Turkey	863
PF 136	Leticia	PBR	Colombia	174
S 136	Hurmat	SSK	Pakistan	601
Y 136	Havuz 13	YD/YFDM	Turkey	869
137	Anatoly Korolev	PGM	Russian Federation	728
137	Fuzhou	DDGHM	China	137
137	Khabarovsk	PGR	Russian Federation	732
137	PSKR 400	PBF	Russian Federation	729
F 137	Laksamana Tan Pusmah	FSGM	Malaysia	511
GC 137	Cormoran	WPB	Argentina	23
PF 137	Arauca	PBR	Colombia	174
S 137	Khalid	SSK	Pakistan	600
138	Naryan-Mar	FFLM	Russian Federation	697
138	PSKR 501	PBF	Russian Federation	728
138	Shkval	PGR	Russian Federation	732
138	Taizhou	DDGHM	China	137
GC 138	Cisne	WPB	Argentina	23
S 138	Saad	SSK	Pakistan	600
139	Ishim	AO	Russian Federation	733
139	Kizljar	PGM	Russian Federation	728
139	Ningbo	DDGHM	China	137
GC 139	Pejerrey	WPB	Argentina	23
S 139	Hamza	SSK	Pakistan	600
140	IC 117	PB	India	357
GC 140	Yehuin	PB	Argentina	24
P 140	Rajshahi	PB	Pakistan	607
PG 140	Emilio Aguinaldo	PBO	Philippines	631
Y 140	H 500	YO	Turkey	867
141	IC 118	PB	India	357
141	Vyborg	PGM	Russian Federation	728
GC 141	Quillen	PB	Argentina	24
P 141	Justo Sierra Mendez	PSOH	Mexico	533
P 141	Mubarraz	PGGFM	UAE	887
P 141	Yunus 1	PBF	Turkey	863
PC 141	Cabo Corrientes Point Warde	PB	Colombia	179
PG 141	Antonio Luna	PBO	Philippines	631
Y 141	H 501	YO	Turkey	867
142	Bug	PBO	Russian Federation	726
142	IC 119	PB	India	357
142	Novocherkassk	LSTM	Russian Federation	704
GC 142	Surel	WPBF	Argentina	23
P 142	Makasib	PGGFM	UAE	887
P 142	Yunus 2	PBF	Turkey	863
PC 142	Cabo Manglares Point Wells	PB	Colombia	179
Y 142	H 502	YO	Turkey	867
143	AK 599	PGR	Russian Federation	732
143	Almaz	PGM	Russian Federation	728
143	IC 120	PB	India	357
143	PSKR 630	PTF	Russian Federation	728
143	Sergey Sudetsky	AK	Russian Federation	732
DDH 143	Shirane	DDHM	Japan	433
DT 143	Callao	LSTH	Peru	622
GC 143	Surubi	WPB	Argentina	23
P 143	Guillermo Prieto	PSOH	Mexico	533
PC 143	Cabo Tiburon	PB	Colombia	179
144	IC 121	PB	India	357
DDH 144	Kurama	DDHM	Japan	433
DT 144	Eten	LSTH	Peru	622
GC 144	Boga	WPB	Argentina	23
P 144	Matias Romero	PSOH	Mexico	533
PC 144	Cabo De La Vella	PB	Colombia	179
145	IC 122	PB	India	357
145	Dunay	PGH	Russian Federation	726
F 145	Amatola	FFGHM	South Africa	762
GC 145	Sabalo	WPB	Argentina	23
PC 145	11 De Noviembre	PBO	Colombia	176
146	IC 123	PB	India	357
146	PSKR 54	PBR	Russian Federation	731
F 146	Isandlwana	FFGHM	South Africa	762
GC 146	Huala	WPB	Argentina	23
147	AK 582	PGR	Russian Federation	732
147	IC 124	PB	India	357
F 147	Spioenkop	FFGHM	South Africa	762
GC 147	Pacu	WPB	Argentina	23
148	IC 125	PB	India	357
148	Orsk	LSTM	Russian Federation	706
F 148	Mendi	FFGHM	South Africa	762
GC 148	Manduruyu	WPB	Argentina	23
P 148	Otago	PBO	New Zealand	575
149	IC 126	PB	India	357
149	Kuban	PCM	Russian Federation	727
GC 149	Corvina	WPB	Argentina	23
150	Anzac	FFGHM	Australia	28
150	Changchun	DDGHM	China	139
150	Saratov	LSTM	Russian Federation	706
GC 150	Fagnano	PB	Argentina	24
V 150	Jägaren	PC	Sweden	800
151	Arunta	FFGHM	Australia	28
151	Azov	LSTM	Russian Federation	704
151	Midhili	PB	Maldives	521
151	Perantau	AGS	Malaysia	514
151	Zhengzhou	DDGHM	China	139
DD 151	Asagiri	DDGHM	Japan	439
GC 151	Nahuel Huapi	PB	Argentina	24
LR 151	Hamal	PB	Dominican Republic	207
P 151	Ban Yas	PGGF	UAE	887
P 151	Durango	PSOH	Mexico	534
P 151	Espadarte	PB	Cape Verde	114
152	Berkut	PCM	Russian Federation	727
152	Jinan	DDGHM	China	139
152	Nikolay Filchenkov	LSTM	Russian Federation	706
152	Warramunga	FFGHM	Australia	28
DD 152	Yamagiri	DDGHM	Japan	439
P 152	Marban	PGGF	UAE	887
P 152	Sonora	PSOH	Mexico	534
153	Stuart	FFGHM	Australia	28
153	Xian	DDGHM	China	139
BH 153	Quindio	–	Colombia	177
DD 153	Yuugiri	DDGHM	Japan	439
LR 153	Deneb	PB	Dominican Republic	207
P 153	Guanajuato	PSOH	Mexico	534
P 153	Rodqm	PGGF	UAE	887
U 153	Priluki	PGGK	Ukraine	877
154	Hefei	DDGHM	China	140

IHS Jane's Fighting Ships 2014-2015

© 2014 IHS

Pennant list [81]

Number	Ship's name	Type	Country	Page	Number	Ship's name	Type	Country	Page
154	Parramatta	FFGHM	Australia	28	172	Primorye	PBO	Russian Federation	726
ATP 154	Bayóvar	AOT	Peru	624	AGOR 172	Quest	AGORH	Canada	104
BO 154	Gorgona	AGSC	Colombia	177	AH 172	Stiglich	AGSC/AH	Peru	623
DD 154	Amagiri	DDGHM	Japan	439	DDG 172	Shimakaze	DDGHM	Japan	436
LR 154	Acamar	PB	Dominican Republic	207	LG-172	Rio Puyo	WPBR	Ecuador	216
P 154	Shaheen	PGGF	UAE	887	P 172	Al Hesen	PGGMH	UAE	886
P 154	Veracruz	PSOH	Mexico	534	T-ATF 172	Apache	ATF	US	978
U 154	Kahovka	PGGK	Ukraine	877	173	Anadyr	PGH	Russian Federation	726
155	AK 224	PGR	Russian Federation	732	173	Changsa	DDGHM	China	140
155	Ballarat	FFGHM	Australia	28	173	Perak	FSGHM	Malaysia	510
155	Nanjing	DDGHM	China	140	A 173	Protector	AGOBH	UK	908
ATP 155	Zorritos	AOT	Peru	624	DDG 173	Kongou	DDGHM	Japan	434
BO 155	Providencia	AGOR	Colombia	177	L 173	Chios	LSTH	Greece	309
DD 155	Hamagiri	DDGHM	Japan	439	LG-173	Rio Portoviejo	WPBR	Ecuador	216
LR 155	Pollux	PBF	Dominican Republic	206	P 173	Al Dhafra	PGGMH	UAE	886
P 155	Sagar	PGGF	UAE	887	P 173	Burutu	PB	Nigeria	581
U 155	Pridneprovye	FSGM	Ukraine	878	174	Terengganu	FSGHM	Malaysia	510
156	Orel	FFHM	Russian Federation	725	AEH 174	La Macha	AGSC/EH	Peru	623
156	Toowoomba	FFGHM	Australia	28	DDG 174	Kirishima	DDGHM	Japan	434
156	Yamal	LSTM	Russian Federation	704	L 174	Samos	LSTH	Greece	309
BO 156	Malpelo	AGOR	Colombia	177	LG-174	Rio Manta	WPBR	Ecuador	216
D 156	Nazim	DD	Pakistan	610	P 174	Mezyad	PGGMH	UAE	886
DD 156	Setogiri	DDGHM	Japan	439	P 174	Zaria	PB	Nigeria	581
LR 156	Castor	PBF	Dominican Republic	206	175	Guiyang	DDGHM	China	140
P 156	Tarif	PGGF	UAE	887	175	Kelantan	FSGHM	Malaysia	510
U 156	Kremenchuk	FSGM	Ukraine	878	AH 175	Carrillo	AGSC/EH	Peru	623
157	Perth	FFGHM	Australia	28	DDG 175	Myoukou	DDGHM	Japan	434
DD 157	Sawagiri	DDGHM	Japan	439	L 175	Ikaria	LSTH	Greece	309
LR 157	Shaula	PBF	Dominican Republic	206	P 175	Al Jahili	PGGMH	UAE	886
P 157	Larkana	PB	Pakistan	607	P 175	Okpoku	PB	Nigeria	582
158	Dzerzhinsky	FFHM	Russian Federation	725	SSV 175	Viktor Leonov	AGIM	Russian Federation	712
158	Tsesar Kunikov	LSTM	Russian Federation	704	176	Chengdu	DDGHM	China	140
158	Yung Chuan	MSC	Taiwan	819	176	Rahova	PGR	Romania	664
DD 158	Umigiri	DDGHM	Japan	439	176	Selangor	FSGHM	Malaysia	510
LR 158	Atria	PBF	Dominican Republic	206	176	Vyacheslav Denisov	AK	Russian Federation	732
160	Vorovsky	FFHM	Russian Federation	725	AH 176	Melo	AGSC/EH	Peru	623
BE 160	Gloria	AXS	Colombia	177	DDG 176	Choukai	DDGHM	Japan	434
S 160	Tridente	SSK	Portugal	649	L 176	Lesbos	LSTH	Greece	309
Y 160	Önder	YTB/YTM/YTL	Turkey	870	P 176	Bomadi	PB	Nigeria	582
161	Korshun	PCM	Russian Federation	727	P 176	Al Hili	PGGMH	UAE	886
BL 161	Cartagena De Indias	AGP	Colombia	178	177	Opanez	PGR	Romania	664
LG-161	Rio Coangos	WPBR	Ecuador	216	DDG 177	Atago	DDGHM	Japan	432
LI 161	Elnath	PBF	Dominican Republic	207	L 177	Rodos	LSTH	Greece	309
P 161	Oaxaca	PSOH	Mexico	534	P 177	Badagry	PB	Nigeria	582
P 161	Muray Jib	FSGHM	UAE	884	178	Smardan	PGR	Romania	664
S 161	Arpão	SSK	Portugal	649	DDG 178	Ashigara	DDGHM	Japan	432
Y 161	Öncü	YTB/YTM/YTL	Turkey	870	P 178	Ekpe	PGF	Nigeria	580
162	Yung Fu	MSC	Taiwan	819	179	Posada	PGR	Romania	664
BL 162	Buenaventura	AGP	Colombia	178	L 179	Paros	LCU	Greece	309
LG-162	Rio Muisne	WPBR	Ecuador	216	P 179	Damisa	PGF	Nigeria	580
LI 162	Polaris	PBF	Dominican Republic	207	180	Rovine	PGR	Romania	664
P 162	Baja California	PSOH	Mexico	534	L 180	Kefallinia	LCUJ	Greece	309
P 162	Das	FSGHM	UAE	884	LSP 180	Mantilla	YFL	Peru	625
Y 162	Özgen	YTB/YTM/YTL	Turkey	870	D 181	Tariq	FFHM/FFGH	Pakistan	604
163	Voron	PCM	Russian Federation	727	DDH 181	Hyuga	CVHG	Japan	431
LG-163	Rio Tangare	WPBR	Ecuador	216	L 181	Ithaki	LCUJ	Greece	309
LI 163	Nunki	PBF	Dominican Republic	207	LG-181	Rio Zamora	WPBR	Ecuador	216
M 163	Muhafiz	MHSC	Pakistan	608	LSP 181	Aguilar	YFL	Peru	625
P 163	Bicentenario De La Independencia	PSOH	Mexico	534	P 181	Siri	PGGF	Nigeria	581
P 163	Express	PB/AXL	UK	907	S 181	U 31	SSK	Germany	285
Y 163	Ödev	YTB/YTM/YTL	Turkey	870	D 182	Babur	FFHM/FFGH	Pakistan	604
164	Onega	FFLM	Russian Federation	697	DDH 182	Ise	CVHG	Japan	431
LG 164	Rio Taura	WPBR	Ecuador	216	HQ 182	Hanoi	SSK	Vietnam	1009
LI 164	Dubhe	PBF	Dominican Republic	207	L 182	Kerkira	LCUJ	Greece	309
M 164	Mujahid	MHSC	Pakistan	608	LG-182	Rio Palora	WPBR	Ecuador	216
P 164	Centenario De La Revolución	PSOH	Mexico	534	P 182	Ayam	PGGF	Nigeria	581
P 164	Explorer	PB/AXL	UK	907	S 182	U 32	SSK	Germany	285
Y 164	Özgür	YTB/YTM/YTL	Turkey	870	183	Volga	PGH	Russian Federation	726
165	Zhanjiang	DDG	China	136	D 183	Khaibar	FFHM/FFGH	Pakistan	604
LG 165	Rio Vinces	WPBR	Ecuador	216	DDH 183	Izumo	CVHG	Japan	432
LI 165	Regulus	PBF	Dominican Republic	207	HQ 183	Ho Chi Minh City	SSK	Vietnam	1009
MB 165	Serdity	ARS/ATA	Russian Federation	723	P 183	Ekun	PGGF	Nigeria	581
P 165	Example	PB/AXL	UK	907	S 183	U 33	SSK	Germany	285
166	Zhuhai	DDG	China	136	SFP 183	Akademik Seminikhin	AGS	Russian Federation	710
LG 166	Rio Bucay	WPBR	Ecuador	216	184	Mikhail Konovalov	AK	Russian Federation	732
LI 166	Denebola	PBF	Dominican Republic	207	HQ 184	Haiphong	SSK	Vietnam	1009
M 166	Munsif	MHSC	Pakistan	608	S 184	U 34	SSK	Germany	285
P 166	Yola	PB	Nigeria	582	185	Sakhalin	PBO	Russian Federation	726
167	Shenzhen	DDGHM	China	140	D 185	Tippu Sultan	FFHM/FFGH	Pakistan	604
167	Yung Ren	MSC	Taiwan	819	GC 185	Correa Falcon	PBO	Argentina	24
LG 167	Rio Jujan	WPBR	Ecuador	216	HQ 185	Da Nang	SSK	Vietnam	1009
LI 167	Acrux	PBF	Dominican Republic	207	S 185	U 35	SSK	Germany	285
P 167	Exploit	PB/AXL	UK	907	D 186	Shahjahan	FFHM/FFGH	Pakistan	604
P 167	Makurdi	PB	Nigeria	581	GC 186	Mariano Moreno	PBO	Argentina	23
168	Guangzhou	DDGHM	China	138	HQ 186	Khanh Hoa	SSK	Vietnam	1009
168	Yung Sui	MSC	Taiwan	819	S 186	U 36	SSK	Germany	285
P 168	Hadejia	PB	Nigeria	581	HQ 187	Vung Tau	SSK	Vietnam	1009
T-ATF 168	Catawba	ATF	US	978	T-AO 187	Henry J Kaiser	AOH	US	975
169	Wuhan	DDGHM	China	138	188	Zborul	FSG	Romania	664
L 169	Irakleia	LCU	Greece	309	T-AO 188	Joshua Humphries	AOH	US	975
MB 169	Pochetnyy	ARS/ATA	Russian Federation	723	189	Pescarusul	FSG	Romania	664
P 169	Brass	PB	Nigeria	582	189	PSKR 59	PBR	Russian Federation	731
T-ATF 169	Navajo	ATF	US	978	T-AO 189	John Lenthall	AOH	US	975
170	Lanzhou	DDGHM	China	139	190	Lastunul	FSG	Romania	664
170	Neva	PGH	Russian Federation	726	190	Monchegorsk	FFLM	Russian Federation	697
DF 170	Mayor Jaime Arias Arango	ASL	Colombia	178	LG 191	Rio Tiputini	PBF	Ecuador	215
L 170	Folegandros	LCU	Greece	309	P 191	Abu Dhabi	FSH	UAE	885
171	Haikou	DDGHM	China	139	LG 192	Rio Aguarico	PBF	Ecuador	215
171	Kedah	FSGHM	Malaysia	510	LSD 193	Shiu Hai	LSDH	Taiwan	817
AH 171	Carrasco	AGSC/EH	Peru	623	T-AO 193	Walter S Diehl	AOH	US	975
DDG 171	Hatakaze	DDGHM	Japan	436	T-AO 194	John Ericsson	AOH	US	975
LG-171	Rio Tena	WPBR	Ecuador	216	T-AO 195	Leroy Grumman	AOH	US	975
MB 171	Loksa	ARS/ATA	Russian Federation	723	196	Sneznogorsk	FFLM	Russian Federation	697
P 171	Baynunah	PGGMH	UAE	886	196	Zabaykalye	PBO	Russian Federation	726
T-ATF 171	Sioux	ATF	US	978	P 196	Andromeda	PB	Greece	307
172	Kunming	DDGHM	China	140	T-AO 196	Kanawha	AOH	US	975
172	Pahang	FSGHM	Malaysia	510	T-AO 197	Pecos	AOH	US	975
					198	Kamchatka	PBO	Russian Federation	726
					198	PSKR 322	PGR	Russian Federation	732
					P 198	Kyknos	PB	Greece	307

[82] Pennant list

Number	Ship's name	Type	Country	Page
T-AO 198	Big Horn	AOH	US	975
199	Brest	FFLM	Russian Federation	697
P 199	Pigasos	PB	Greece	307
T-AO 199	Tippecanoe	AOH	US	975
200	Perekop	–	Russian Federation	714
G 200	Buratti	PB	Italy	423
P 200	Zégbéla Togba Pivi	PB	Guinea	318
T-AO 200	Guadalupe	AOH	US	975
201	Kayvan	PB	Iran	385
201	Kujang	WPB	Indonesia	376
201	Kula	PB	Fiji	238
201	Samudra Prahari	WPSOH	India	354
201	Sardar	PB	Kazakhstan	463
A 201	Orion	AGIH	Sweden	802
F 201	Nicolas Bravo	FFH	Mexico	530
G 201	De Ianni	PB	Italy	423
GC 201	Rio Grande De Matagalpa	PB	Nicaragua	577
LP 201	Marina Pelayo Pratt Gill	PBR	Paraguay	616
P 201	5 Février 1979	PBO	Congo-Brazzaville	182
P 201	Cabo Fradera	PBR	Spain	774
P 201	Neiafu	PB	Tonga	848
P 201	Ruposhi Bangla	PB	Bangladesh	61
PO 201	Guardiamarina San Martin	PSO	Peru	625
PS 201	Tsuruugi	PS/PBOF	Japan	456
SSV 201	Priazove	AGIM	Russian Federation	712
T-AO 201	Patuxent	AOH	US	975
UAM 201	Creoula	AXS	Portugal	654
VS 201	Idabato	PB	Cameroon	96
WLB 201	Juniper	WLB/ABU	US	990
202	Kikau	PB	Fiji	238
202	Parang	WPB	Indonesia	376
202	Sakshi	PB	Kazakhstan	463
202	Samudra Pehredar	WPSOH	India	354
202	Smeul	PTF	Romania	664
202	Tiran	PB	Iran	385
F 202	Hermenegildo Galeana	FFH	Mexico	530
G 202	Salerno	PB	Italy	423
GC 202	Tendert	PB	Nicaragua	577
LM 202	Seiun	AKSL	Japan	460
P 202	31 Juillet 1968	PBO	Congo-Brazzaville	182
P 202	Joumhouria	PB	Tunisia	851
P 202	Pangai	PB	Tonga	848
PS 202	Hotaka	PS/PBOF	Japan	456
PVL 202	Kati	–	Estonia	235
T-AO 202	Yukon	AOH	US	975
VS 202	Isongo	PB	Cameroon	96
WLB 202	Willow	WLB/ABU	US	990
203	Balchik	AOTL	Bulgaria	92
203	Celurit	WPB	Indonesia	376
203	El Wacil	PB	Morocco	548
203	Kiro	PB	Fiji	238
203	Samudra Pavak	WPSOH	India	354
203	Zhenis	PB	Kazakhstan	463
G 203	Rossi	PB	Italy	423
LDG 203	Bacamarte	LCU	Portugal	654
LM 203	Sekiun	AKSL	Japan	460
P 203	15 Août 1960	PBO	Congo-Brazzaville	182
P 203	Savea	PB	Tonga	848
PM 203	Tortuguero	ABU	Dominican Republic	206
PS 203	Norikura	PS/PBOF	Japan	456
T-AO 203	Laramie	AOH	US	975
VS 203	Mouanco	PB	Cameroon	96
WLB 203	Kukui	WLB/ABU	US	990
204	Cundrik	WPB	Indonesia	376
204	El Jail	PB	Morocco	548
204	Mahan	PB	Iran	385
204	Semser	PB	Kazakhstan	463
204	Vijelia	PTF	Romania	664
BH 204	El Idrissi	AGS	Algeria	7
G 204	Garulli	PB	Italy	423
LM 204	Souun	AKSL	Japan	460
P 204	3 De Noviembre	PB	Panama	612
P 204	10 Juin 1991	PBO	Congo-Brazzaville	182
PM 204	Capotillo	ABU	Dominican Republic	206
PS 204	Kaimon	PS/PBOF	Japan	456
T-AO 204	Rappahannock	AOH	US	975
VS 204	Campo	PB	Cameroon	96
WLB 204	Elm	WLB/ABU	US	990
205	Belati	WPB	Indonesia	376
205	Chung Chien	LST	Taiwan	818
205	El Mikdam	PB	Morocco	548
G 205	Sanges	PB	Italy	423
GC 205	Rio Escondido	PB	Nicaragua	577
KM 205	Mamba	PB	Tanzania	826
LM 205	Reiun	AKSL	Japan	460
PS 205	Asama	PS/PBOF	Japan	456
T 205	Kassir	PBR	Kuwait	492
U 205	Lutsk	FFLM	Ukraine	877
WLB 205	Walnut	WLB/ABU	US	990
206	El Khafir	PB	Morocco	548
206	Golok	WSAR	Indonesia	376
206	Kapitan Parvi Rang Dimitri Dobrev	ADG/AX	Bulgaria	92
206	RT 249	MHC	Russian Federation	708
G 206	Corrias	PB	Italy	423
LM 206	Genun	AKSL	Japan	460
P 206	10 De Noviembre	PB	Panama	612
PS 206	Houou	PS/PBOF	Japan	456
U 206	Vinnitsa	FFLM	Ukraine	877
WLB 206	Spar	WLB/ABU	US	990
207	El Haris	PB	Morocco	548
207	Endurance	LPDM	Singapore	756
207	Panan	WSAR	Indonesia	376
F 207	Bremen	FFGHM	Germany	288
G 207	Cortile	PB	Italy	423
LM 207	Ayabane	AKSL	Japan	460
P 207	28 De Noviembre	PB	Panama	612
P 207	Utique	PB	Tunisia	851
U 207	Uzhgorod	PCM	Ukraine	878
WLB 207	Maple	WLB/ABU	US	990
208	Chung Shun	LST	Taiwan	818
208	El Essahir	PB	Morocco	548
208	Pedang	WSAR	Indonesia	376
208	Resolution	LPDM	Singapore	756
F 208	Niedersachsen	FFGHM	Germany	288
G 208	Casotti	PB	Italy	423
LM 208	Koun	AKSL	Japan	460
P 208	4 De Noviembre	PB	Panama	612
P 208	Jerba	PB	Tunisia	851
PC 208	Juan Antonio De La Fuente	PB	Mexico	536
SSV 208	Kurily	AGIM	Russian Federation	712
U 208	Khmelnitsky	PCM	Ukraine	878
WLB 208	Aspen	WLB/ABU	US	990
209	Erraid	WPB	Morocco	551
209	Kapak	WSAR	Indonesia	376
209	Persistence	LPDM	Singapore	756
209	Vulcanul	PTF	Romania	664
G 209	Prata	PB	Italy	423
P 209	5 De Noviembre	PB	Panama	612
P 209	Kuriat	PB	Tunisia	851
U 209	Ternopil	FFLM	Ukraine	877
WLB 209	Sycamore	WLB/ABU	US	990
210	Ayeda 4	AOTL/AWTL	Egypt	226
210	Endeavour	LPDM	Singapore	756
210	Erraced	WPB	Morocco	551
210	RT 273	MHC	Russian Federation	708
210	Smolny	–	Russian Federation	714
G 210	Marra	PB	Italy	423
PC 210	Ignacio Ramirez	PB	Mexico	536
T 210	Dastoor	PBR	Kuwait	492
WLB 210	Cypress	WLB/ABU	US	990
211	El Kaced	WPB	Morocco	551
211	Hadji Dimitar	ATS	Bulgaria	93
211	Maryut	AOTL/AWTL	Egypt	226
211	Parvin	PC	Iran	385
F 211	Ignacio Allende	FFHM	Mexico	529
G 211	Gottardi	PB	Italy	423
P 211	Meghna	PB	Bangladesh	57
PC 211	Ignacio Mariscal	PB	Mexico	536
WLB 211	Oak	WLB/ABU	US	990
212	Al Furat	AOTL/AWTL	Egypt	226
212	Al Qiaq	YFU	Saudi Arabia	741
212	Bahram	PC	Iran	385
212	Essaid	WPB	Morocco	551
212	Hristo Botev	ATS	Bulgaria	93
A 212	Ägir	YDT/AGF	Sweden	803
DCB 212	Máncora	PBR	Peru	626
F 212	Karlsruhe	FFGHM	Germany	288
F 212	Mariano Abasolo	FFHM	Mexico	529
G 212	La Piccirella	PB	Italy	423
P 212	Jaco	PB	East Timor	208
P 212	Jamuna	PB	Bangladesh	57
PF 212	Al Hani	FFGM	Libya	499
WLB 212	Hickory	WLB/ABU	US	990
213	Al Nil	AOTL/AWTL	Egypt	226
213	Nahid	PC	Iran	385
DCB 213	Huaura	PBR	Peru	626
F 213	Augsburg	FFGHM	Germany	288
F 213	Guadaloupe Victoria	FFHM	Mexico	529
G 213	Perissinotto	PB	Italy	423
PC 213	Parachique	PBR	Peru	627
WLB 213	Fir	WLB/ABU	US	990
214	Akdu	AOTL/AWTL	Egypt	226
214	Al Sulayel	YFU	Saudi Arabia	741
A 214	Belos III	ARSH	Sweden	804
DCB 214	Quilca	PBR	Peru	626
F 214	Lübeck	FFGHM	Germany	288
F 214	Francisco Javier Mina	FFHM	Mexico	529
G 214	Rocca	PB	Italy	423
PC 214	San Andreas	PBR	Peru	627
WLB 214	Hollyhock	WLB/ABU	US	990
215	Atbarah	AOTL/AWTL	Egypt	226
215	RT 234	MHC	Russian Federation	708
F 215	Brandenburg	FFGHM	Germany	286
G 215	Bertoldi	PB	Italy	423
P 215	Betano	PB	East Timor	208
T 215	Mahroos	PBR	Kuwait	492
WLB 215	Sequoia	WLB/ABU	US	990
216	Al Ula	YFU	Saudi Arabia	741
216	Ayeda 3	AOTL/AWTL	Egypt	226
216	Chung Kuang	LST	Taiwan	818
F 216	Schleswig-Holstein	FFGHM	Germany	286
G 216	Verdecchia	PB	Italy	423
PC 216	Chicama	PBR	Peru	626
PC 216	Francisco J Mugica	PB	Mexico	536
PC 216	Iseyuki	PC/PB	Japan	458
WLB 216	Alder	WLB/ABU	US	990
217	Chung Suo	LST	Taiwan	818
F 217	Bayern	FFGHM	Germany	286
G 217	De Santis	PB	Italy	423
PC 217	Huanchaco	PBR	Peru	626
PC 217	Isonami	PC/PB	Japan	458
218	Afif	YFU	Saudi Arabia	741
218	Al Burullus	AOTL/AWTL	Egypt	226
218	Aleksin	FFLM	Russian Federation	698
218	Chung Chi	LST	Taiwan	818
F 218	Mecklenburg-Vorpommern	FFGHM	Germany	286
G 218	Piccinni Leopardi	PB	Italy	423
PC 218	Chorrillos	PBR	Peru	626
PC 218	Nagozuki	PC/PB	Japan	458
219	RT 233	MHC	Russian Federation	708
219	RT 278	MSB	Russian Federation	708
F 219	Sachsen	FFGHM	Germany	289
G 219	Bianco	PB	Italy	423
PC 219	Chancay	PBR	Peru	626
PC 219	Yaezuki	PC/PB	Japan	458
220	Dheba	YFU	Saudi Arabia	742
F 220	Hamburg	FFGHM	Germany	289

IHS Jane's Fighting Ships 2014-2015

© 2014 IHS

Pennant list [83]

Number	Ship's name	Type	Country	Page
G 220	Starace	PB	Italy	423
PC 220	Camana	PBR	Peru	626
PC 220	Guillermo Endara	PB	Panama	613
PC 220	Hamayuki	PC/PB	Japan	458
PC 220	Jose Natividad Macias	PB	Mexico	536
221	Chung Chuan	LST	Taiwan	818
221	Georgi Rakovski	YH	Bulgaria	92
221	Priyadarshini	WPBO	India	355
221	Sobat	AFL	Sudan	793
F 221	Hessen	FFGHM	Germany	289
F 221	Regele Ferdinand	FFHM	Romania	660
G 221	Cultrona	PB	Italy	423
P 221	Kaman	PGGF	Iran	384
PC 221	Chala	PBR	Peru	626
PC 221	Ernesto Perez Balladares	PB	Panama	613
222	Dinder	AFL	Sudan	793
222	Razia Sultana	WPBO	India	355
222	Umlus	YFU	Saudi Arabia	742
222	Vasil Levski	YH	Bulgaria	92
F 222	Baden-Württemberg	FFGHM	Germany	290
F 222	Regina Maria	FFHM	Romania	660
G 222	Benvenuti	PB	Italy	423
P 222	Zoubin	PGGF	Iran	384
PC 222	Cancas	PBR	Peru	626
PC 222	Mireya Moscoso	PB	Panama	613
PC 222	Umigiri	PC/PB	Japan	458
223	Kapitan Vtori Rang Nikola Furnadjiev	YDT	Bulgaria	92
F 223	Nordrhein-Westfalen	FFGHM	Germany	290
G 223	Sanges	PB	Italy	423
P 223	Araz	PB	Azerbaijan	40
P 223	Khadang	PGGF	Iran	384
PC 223	Asagiri	PC/PB	Japan	458
PC 223	Martin Torrijos	PB	Panama	613
PM 223	Rio Chira	PB	Peru	626
TM 223	Libertador	YFL	Bolivia	69
224	Al Leeth	YFU	Saudi Arabia	742
224	Kamla Devi	WPBO	India	355
224	Proteo	ARS	Bulgaria	93
F 224	Sachsen-Anhalt	FFGHM	Germany	290
P 224	Peykan	PGGF	Iran	384
PC 224	Punta Arenas	PBR	Peru	626
PC 224	Yucatan	PB	Mexico	536
TM 224	Trinidad	YFL	Bolivia	69
225	Amrit Kaur	WPBO	India	355
F 225	Rheinland-Pfalz	FFGHM	Germany	290
P 225	Joshan	PGGF	Iran	384
PC 225	Tabasco	PB	Mexico	536
PL 225	Santa Rosa	PBR	Peru	626
TM 225	J Chavez Suarez	YFL	Bolivia	69
226	Al Quonfetha	YFU	Saudi Arabia	742
226	Chung Chih	LST	Taiwan	818
226	Kanak Lata Barua	WPBO	India	355
P 226	Falakhon	PGGF	Iran	384
PC 226	Cochimie	PB	Mexico	536
PC 226	Pacasmayo	PBR	Peru	626
227	Bhikaji Cama	WPBO	India	355
227	Chung Ming	LST	Taiwan	818
P 227	Shamshir	PGGF	Iran	384
PC 227	Barranca	PBR	Peru	626
228	Suchetra Kripalani	WPBO	India	355
P 228	Gorz	PGGF	Iran	384
P 228	Toxotis	PB	Greece	307
PC 228	Coishco	PBR	Peru	626
229	Sarojini Naidu	WPBO	India	356
DE 229	Abukuma	FFGM/DE	Japan	441
F 229	Lancaster	FFGHM	UK	904
P 229	Gardouneh	PGGF	Iran	384
P 229	Tolmi	PG	Greece	308
PC 229	Independencia	PBR	Peru	626
230	Chung Pang	LST	Taiwan	818
230	Durgabai Deshmukh	WPBO	India	356
230	Shaladein	ARL	Egypt	226
DE 230	Jintsu	FFGM/DE	Japan	441
P 230	Khanjar	PGGF	Iran	384
P 230	Ormi	PG	Greece	308
PL 230	San Nicolas	PBR	Peru	626
231	Chung Yeh	LST	Taiwan	818
231	Halaib	AEL	Egypt	226
231	Kapitan Parvi Rang Boris Rogev	AGSC	Bulgaria	92
231	Kasturba Gandhi	WPBO	India	356
DE 231	Ooyodo	FFGM/DE	Japan	441
F 231	Argyll	FFGHM	UK	904
GC 231	El Mounkid I	SAR	Algeria	8
NL 231	Bayraktar	LST	Turkey	864
P 231	Neyzeh	PGGF	Iran	384
SSV 231	Vassily Tatischev	AGIM	Russian Federation	712
232	Aruna Asaf Ali	WPBO	India	356
232	Chung Ho	LSTH	Taiwan	818
232	Kalmykia	FFLM	Russian Federation	698
DE 232	Sendai	FFGM/DE	Japan	441
GC 232	El Mounkid II	SAR	Algeria	8
NL 232	Sancaktar	LST	Turkey	864
P 232	Tabarzin	PGGF	Iran	384
PP 232	Río Santa	PBR	Peru	626
233	Chung Ping	LSTH	Taiwan	818
233	Subhadra Kumari Chauhan	WPBO	India	356
A 233	Maistros	AXS	Greece	311
DE 233	Chikuma	FFGM/DE	Japan	441
GC 233	El Mounkid III	SAR	Algeria	8
P 233	Derafsh	PGGF	Iran	384
PP 233	Río Majes	PBR	Peru	626
234	Meera Behn	WPBO	India	356
A 234	Sorokos	AXS	Greece	311
DE 234	Tone	FFGM/DE	Japan	441
F 234	Iron Duke	FFGHM	UK	904
GC 234	El Mounkid IV	SAR	Algeria	8
PC 234	Matarani	PBR	Peru	626
235	Hirsala	AKSL	Finland	242
235	Savitri Bai Phule	WPBO	India	356
F 235	Monmouth	FFGHM	UK	904
PC 235	Río Viru	PBR	Peru	627
236	Aadesh	PBO	India	357
F 236	Montrose	FFGHM	UK	904
237	Abheek	PBO	India	357
237	Hila	AKSL	Finland	243
F 237	Westminster	FFGHM	UK	904
PP 237	Rio Surco	PBR	Peru	626
238	Abhinav	PBO	India	357
238	Haruna	AKSL	Finland	243
A 238	Zefiros	AXS	Greece	311
F 238	Northumberland	FFGHM	UK	904
PC 238	Sama	PBR	Peru	626
239	Abhiraj	PBO	India	357
239	RT 252	MHC	Russian Federation	708
F 239	Richmond	FFGHM	UK	904
PC 239	Atico	PBR	Peru	627
240	Achook	PBO	India	357
240	Kaszub	FSM	Poland	640
F 240	Yavuz	FFGHM	Turkey	859
PC 240	Malabrigo	PBR	Peru	627
U 240	Feodosiya	YDT/YFL/YPT	Ukraine	881
241	Agrim	PBO	India	357
241	Askeri	YFB	Finland	243
241	Slazak	PSO	Poland	641
F 241	Turgutreis	FFGHM	Turkey	859
PC 241	Catarindo	PBR	Peru	627
PC 241	Démocrata	PBO	Mexico	536
F 242	Fatih	FFGHM	Turkey	859
PC 242	Francisco I Madero	PBO	Mexico	536
PC 242	Punta Pariñas	PBR	Peru	627
243	MPK 227	FFLM	Russian Federation	698
F 243	Yildirim	FFGHM	Turkey	859
PM 243	Rio Nepeña	WPB	Peru	626
F 244	Barbaros	FFGHM	Turkey	858
PM 244	Rio Tambo	WPB	Peru	626
245	MPK 105	FFLM	Russian Federation	698
A 245	Leeuwin	AGS	Australia	34
F 245	Orucreis	FFGHM	Turkey	858
PM 245	Rio Ocoña	WPB	Peru	626
A 246	Melville	AGS	Australia	34
F 246	Salihreis	FFGHM	Turkey	858
LD 246	Morrosquillo	LCU	Colombia	176
PM 246	Rio Huarmey	WPB	Peru	626
A 247	Pelikanen	YPT	Sweden	804
F 247	Kemalreis	FFGHM	Turkey	858
P 247	Onitsha	PBF	Nigeria	581
PM 247	Rio Zaña	WPB	Peru	626
LD 248	Bahía Honda	LCU	Colombia	176
PC 248	Casma	PBR	Peru	627
LD 249	Bahía Portete	LCU	Colombia	176
PC 249	Tortugas	PBR	Peru	627
250	Kazakhstan	PBO	Kazakhstan	463
P 250	Alpha Yaya Diallo	PB	Guinea	318
251	Oral	PBO	Kazakhstan	463
251	Wodnik	AXTH	Poland	644
251	Zulfiquar	FFGH	Pakistan	602
GC 251	El Mouderrib I	AXL	Algeria	8
LD 251	Bahía Solano	LCU	Colombia	176
P 251	Ghantut	PGG	UAE	887
252	Shamsheer	FFGH	Pakistan	602
GC 252	El Mouderrib II	AXL	Algeria	8
LD 252	Bahía Cupica	LCU	Colombia	176
P 252	Salahlah	PGG	UAE	887
253	Iskra	AXS	Poland	644
253	Saif	FFGH	Pakistan	602
GC 253	El Mouderrib III	AXL	Algeria	8
LD 253	Bahía Utria	LCU	Colombia	176
254	Aslat	FFGH	Pakistan	602
GC 254	El Mouderrib IV	AXL	Algeria	8
LD 254	Bahía Malaga	LCU	Colombia	176
255	Mutiara	AGSH	Malaysia	514
GC 255	El Mouderrib V	AXL	Algeria	8
P 255	Ikot-Abasi	PBF	Nigeria	581
GC 256	El Mouderrib VI	AXL	Algeria	8
P 256	Benin	PBF	Nigeria	581
GC 257	El Mouderrib VII	AXL	Algeria	8
P 257	Clyde	PSOH	UK	908
P 258	Torie	PB	Nigeria	582
P 259	Egede	PB	Nigeria	582
260	Admiral Petre Barbuneanu	FS	Romania	663
260	Alamgir	–	Pakistan	605
F 260	Braunschweig	FSGHM	Germany	290
M 260	Edincik	MHC	Turkey	865
PF 260	Río Huallaga	PBR	Peru	627
F 261	Magdeburg	FSGHM	Germany	290
GC 261	El Mourafek	WARL	Algeria	8
M 261	Edremit	MHC	Turkey	865
P 261	Aparajeya	PBO	Bangladesh	57
PF 261	Río Santiago	PBR	Peru	627
262	Nawigator	AGI	Poland	644
F 262	Erfurt	FSGHM	Germany	290
M 262	Enez	MHC	Turkey	865
P 262	Adomya	PBO	Bangladesh	57
P 262	Tainha	PB	Cape Verde	114
PF 262	Río Putumayo	PBR	Peru	627
263	Hydrograf	AGI	Poland	644
263	Vice Admiral Eugeniu Rosca	FS	Romania	663
F 263	Oldenburg	FSGHM	Germany	290
M 263	Erdek	MHC	Turkey	865
P 263	Atondro	PBO	Bangladesh	57
PF 263	Río Nanay	PBR	Peru	627
264	Contre Admiral Eustatiu Sebastian	FSH	Romania	662
A 264	Trossö	AGP	Sweden	803
F 264	Ludwigshafen	FSGHM	Germany	290
M 264	Erdemli	MHC	Turkey	865

© 2014 IHS

[84] Pennant list

Number	Ship's name	Type	Country	Page
P 264	Archer	PB/AXL	UK	907
PF 264	Rio Napo	PBR	Peru	627
265	Admiral Horia Macelariu	FSH	Romania	662
265	Heweliusz	AGS	Poland	643
M 265	Alanya	MHSC	Turkey	865
PF 265	Rio Yavari	PBR	Peru	627
266	Arctowski	AGS	Poland	643
A 266	Sirius	AORH	Australia	35
M 266	Amasra	MHSC	Turkey	865
P 266	Machitis	PG	Greece	308
PF 266	Rio Matador	PBR	Peru	627
M 267	Ayvalik	MHSC	Turkey	865
P 267	Nikiforos	PG	Greece	308
M 268	Akçakoca	MHSC	Turkey	865
P 268	Aittitos	PG	Greece	308
M 269	Anamur	MHSC	Turkey	865
P 269	Krateos	PG	Greece	308
M 270	Akçay	MHSC	Turkey	865
P 270	Biter	PB/AXL	UK	907
PF 270	Rio Itaya	PBR	Peru	627
271	Gagah Samudera	AX	Malaysia	515
A 271	Gold Rover	AORLH	UK	916
PF 271	Rio Patayacu	PBR	Peru	627
272	General Kazimierz Pułaski	FFGHM	Poland	638
272	Teguh Samudera	AX	Malaysia	515
P 272	Smiter	PB/AXL	UK	907
PC 272	Cabo Corzo	PB	Mexico	536
PF 272	Rio Zapote	PBR	Peru	627
273	General Tadeusz Kościuszko	FFGHM	Poland	638
A 273	Black Rover	AORLH	UK	916
P 273	Pursuer	PB/AXL	UK	907
PC 273	Cabo Catoche	PB	Mexico	536
PF 273	Rio Chambira	PBR	Peru	627
274	Vice Admiral Constantin Balescu	ML/MCS	Romania	665
P 274	Tracker	PB/AXL	UK	907
PF 274	Rio Tambopata	PBR	Peru	627
P 275	CC Karamoko Cheik Conde	PB	Guinea	318
P 275	Raider	PB/AXL	UK	907
278	Bukhansan	PBO	Korea, South	488
279	Chulmasan	PBO	Korea, South	488
P 279	Blazer	PB/AXL	UK	907
280	Iroquois	DDGH	Canada	102
P 280	Dasher	PB/AXL	UK	907
281	Constanta	AETLMH	Romania	667
281	Piast	ARS	Poland	645
P 281	Tyne	PSO	UK	907
PC 281	Punta Morro	PB	Mexico	536
282	Athabaskan	DDGH	Canada	102
282	Lech	ARS	Poland	645
P 282	Severn	PSO	UK	907
PC 282	Punta Mastun	PB	Mexico	536
283	Algonquin	DDGH	Canada	102
283	Midia	AETLMH	Romania	667
P 283	Mersey	PSO	UK	907
P 284	Scimitar	PB	UK	907
P 285	Sabre	PB	UK	907
286	PSKA 286	PBO	Russian Federation	729
P 286	Diopos Antoniou	PB	Greece	306
P 287	Kelefstis Stamou	PB	Greece	306
T-AKR 287	Algol	AKRH	US	983
288	Mircea	AXS	Romania	666
T-AKR 288	Bellatrix	AKRH	US	983
T-AKR 289	Denebola	AKRH	US	983
T-AKR 290	Pollux	AKRH	US	983
291	Orzeł	SSK	Poland	638
AT 291	Subanon	LCU	Philippines	632
P 291	Puncher	PB/AXL	UK	907
T-AKR 291	Altair	AKRH	US	983
P 292	Charger	PB/AXL	UK	907
T-AKR 292	Regulus	AKRH	US	983
AT 293	Bagobo	LCU	Philippines	632
P 293	Ranger	PB/AXL	UK	907
PL 293	Juli	PBR	Peru	626
T-AKR 293	Capella	AKRH	US	983
294	Sokół	SSK	Poland	637
P 294	Trumpeter	PB/AXL	UK	907
PL 294	Moho	PBR	Peru	626
T-AKR 294	Antares	AKRH	US	983
295	Sep	SSK	Poland	637
AT 295	Tausug	LCU	Philippines	632
T-AKR 295	Shughart	AKR	US	980
296	Bielik	SSK	Poland	637
AT 296	Tagbanua	LCU	Philippines	633
T-AKR 296	Gordon	AKR	US	981
297	Kondor	SSK	Poland	637
297	PSKA 297	PBO	Russian Federation	729
T-AKR 297	Yano	AKR	US	980
298	Magnetica	ADG/AGI	Romania	666
T-AKR 298	Gilliland	AKR	US	981
P 300	Rayyan	PB	Kuwait	493
S 300	Ula	SSK	Norway	585
T-AKR 300	Bob Hope	AKR	US	980
301	Astana	PB	Kazakhstan	462
301	Denden	LST	Eritrea	233
301	Huracan	PTG	Mexico	532
301	Jebel Antar	PB	Algeria	8
301	Rio Segovia	PB	Nicaragua	578
301	Teanoai	PB	Kiribati	465
301	Tripoli	PB	Lebanon	497
A 301	Drakensberg	AORH	South Africa	764
MSO 301	Yaeyama	MHS	Japan	445
P 301	Bizerte	PBOM	Tunisia	851
P 301	Butinah	PB	UAE	888
P 301	Inttisar	PB	Kuwait	492
P 301	Panquiaco	PB	Panama	612
PA 301	Almirante Didiez Burgos	PBO/WMEC	Dominican Republic	205

Number	Ship's name	Type	Country	Page
S 301	Utsira	SSK	Norway	585
T-AKR 301	Fisher	AKR	US	980
302	Jebel Hando	PB	Algeria	8
302	Jounieh	PB	Lebanon	497
302	Okba	PG	Morocco	547
302	Oral	PB	Kazakhstan	462
302	RT 231	MHC	Russian Federation	708
302	Sirius	YL	Bulgaria	93
302	Tormenta	PTG	Mexico	532
ABH 302	Morona	ABH	Peru	624
MSO 302	Tsushima	MHS	Japan	445
P 302	Al Bazam	PB	UAE	888
P 302	Aman	PB	Kuwait	492
P 302	Horria	PBOM	Tunisia	851
P 302	Ligia Elena	PB	Panama	612
S 302	Utstein	SSK	Norway	585
T-AKR 302	Seay	AKR	US	980
303	Akin	AOTL	Bulgaria	92
303	Almaty	PB	Kazakhstan	462
303	Batroun	PB	Lebanon	497
303	Triki	PG	Morocco	547
ABH 303	Corrientes	AH	Peru	625
MSO 303	Hachijyo	MHS	Japan	445
P 303	Maimon	PB	Kuwait	492
S 303	Utvaer	SSK	Norway	585
T-AKR 303	Mendonca	AKR	US	980
W 303	Svalbard	WPSOH	Norway	592
304	Al Riyadh	WPBF	Saudi Arabia	742
304	Antares	YL	Bulgaria	93
304	Byblos	PB	Lebanon	497
304	Commandant El Khattabi	PGG	Morocco	547
304	Ras Djenad	PB	Algeria	8
304	Urengoy	FFLM	Russian Federation	698
ABH 304	Curaray	AH	Peru	625
OR 304	Success	AORH	Australia	35
P 304	Al Muroom	PB	UAE	888
P 304	Mobark	PB	Kuwait	492
P 304	Monastir	PBOM	Tunisia	851
S 304	Uthaug	SSK	Norway	585
T-AKR 304	Pililaau	AKR	US	980
305	Atyrau	PB	Kazakhstan	462
305	Beirut	PB	Lebanon	497
305	Commandant Boutouba	PGG	Morocco	547
305	Ras Tenes	PB	Algeria	8
305	Zulurab	WPBF	Saudi Arabia	742
ABH 305	Pastaza	AH	Peru	625
P 305	AG 5	ABU	Turkey	868
P 305	Al Hamryah	PB	UAE	888
P 305	Al Shaheed	PB	Kuwait	492
S 305	Uredd	SSK	Norway	585
T-AKR 305	Brittin	AKR	US	980
306	Commandant El Harty	PGG	Morocco	547
306	Ras Tekkouch	PB	Algeria	8
306	Sidon	PB	Lebanon	497
ABH 306	Puno	AH	Peru	627
P 306	AG 6	ABU	Turkey	868
P 306	Al Khan	PB	UAE	888
P 306	Bayan	PB	Kuwait	492
P 306	Taboga	PB	Panama	612
T-AKR 306	Benavidez	AKR	US	980
307	Commandant Azouggarh	PGG	Morocco	547
307	Oskemen	PB	Kazakhstan	462
307	Ras Sisli	PB	Algeria	8
307	Sarafand	PB	Lebanon	497
A 307	Thetis	YNT	Greece	312
P 307	Al Zawraa	PB	UAE	888
P 307	Dasman	PB	Kuwait	492
308	El Hahiq	PBO	Morocco	547
308	Nakoura	PB	Lebanon	498
308	Ras Nouh	PB	Algeria	8
308	Zelenodolsk	FFLM	Russian Federation	698
P 308	Subahi	PB	Kuwait	493
309	El Tawfiq	PBO	Morocco	547
309	Ras Bougaroni	PB	Algeria	8
P 309	Jaberi	PB	Kuwait	493
310	L V Rabhi	PBO	Morocco	548
310	Ras Tamentfoust	PB	Algeria	8
F 310	Fridtjof Nansen	FFGHM	Norway	586
P 310	Saad	PB	Kuwait	493
T-AKR 310	Watson	AKR	US	977
U 310	Chernigiv	MSO	Ukraine	879
311	Errachiq	PBO	Morocco	548
311	Kazanets	FFLM	Russian Federation	698
311	Prabparapak	PTFG	Thailand	838
311	Ras Oullis	PB	Algeria	8
F 311	Roald Amundsen	FFGHM	Norway	586
P 311	Ahmadi	PB	Kuwait	493
P 311	Bishkhali	PB	Bangladesh	57
P 311	Weeraya	PB	Sri Lanka	787
T-AKR 311	Sisler	AKR	US	977
U 311	Cherkasy	MSO	Ukraine	879
312	El Akid	PBO	Morocco	548
312	Hanhak Sattru	PTFG	Thailand	838
F 312	Otto Sverdrup	FFGHM	Norway	586
P 312	Naif	PB	Kuwait	493
P 312	Padma	PBO	Bangladesh	57
T-AKR 312	Dahl	AKR	US	977
W 312	Alesund	WPSO	Norway	591
313	El Maher	PBO	Morocco	548
313	Suphairin	PTFG	Thailand	838
ALY 313	Marte	AXS	Peru	624
CP 313	Dante Novaro	PB	Italy	425
F 313	Helge Ingstad	FFGHM	Norway	586
M 313	Admiral Cowan	MHC	Estonia	234
P 313	Surma	PBO	Bangladesh	57
P 313	Thafir	PB	Kuwait	493
P 313-1	Shahid Mehdavi	PTFG	Iran	384
P 313-2	Shahid Kord	PTFG	Iran	384
P 313-3	Shahid Shafihi	PTFG	Iran	384
P 313-4	Shahid Towsali	PTFG	Iran	384

Pennant list [85]

Number	Ship's name	Type	Country	Page
P 313-5	Shahid Hejat Zadeh	PTFG	Iran	384
P 313-6	Shahid Dara	PTFG	Iran	384
P 313-7	Shahid Absalan	PTFG	Iran	384
P 313-8	Shahid Rahisi Raisi	PTFG	Iran	384
P 313-9	Shahid Golzam	PTFG	Iran	384
P 313-10	Shahid Sahrabi	PTFG	Iran	384
T-AKR 313	Red Cloud	AKR	US	977
WLI 313	Bluebell	WLI/ABU	US	990
314	El Majid	PBO	Morocco	548
F 314	Thor Heyerdahl	FFGHM	Norway	586
M 314	Sakala	MHC	Estonia	234
P 314	Karnaphuli	PBO	Bangladesh	56
P 314	Marzoug	PB	Kuwait	493
T-AKR 314	Charlton	AKR	US	977
315	Al Khyber	SS	Libya	499
315	El Bachir	PBO	Morocco	548
M 315	Ugandi	MHC	Estonia	234
P 315	Jagatha	PB	Sri Lanka	787
P 315	Mash'noor	PB	Kuwait	493
P 315	Tista	PBO	Bangladesh	56
T-AKR 315	Watkins	AKR	US	977
TM 315	Ingeniero Palacios	YFL	Bolivia	69
WLIC 315	Smilax	WLIC	US	991
316	Al Hunain	SS	Libya	499
316	El Hamiss	PBO	Morocco	547
316	RT 57	MHC	Russian Federation	708
P 316	Abeetha II	PB	Sri Lanka	787
P 316	Wadah	PB	Kuwait	493
T-AKR 316	Pomeroy	AKR	US	977
317	Assir	WPB	Saudi Arabia	742
317	El Karib	PBO	Morocco	547
317	Ikrimah	PB	Libya	501
P 317	Edithara II	PB	Sri Lanka	787
T-AKR 317	Soderman	AKR	US	977
318	Aldhahran	WPB	Saudi Arabia	742
318	Raïs Bargach	PSO	Morocco	548
P 318	Wickrama II	PB	Sri Lanka	787
W 318	Harstad	ARS	Norway	592
319	Alkahrj	WPB	Saudi Arabia	742
319	MPK 178	FFLM	Russian Federation	697
319	Raïs Britel	PSO	Morocco	548
320	Arar	WPB	Saudi Arabia	742
320	Raïs Charkaoui	PSO	Morocco	548
A 320	Furusund	ARS	Sweden	803
W 320	Nordkapp	WPSOH	Norway	592
321	Aheloi	ATS	Bulgaria	93
321	Raïs Maaninou	PSO	Morocco	548
321	Ratcharit	PGGF	Thailand	836
GC 321	Nedjmet El Kotb	PBF	Algeria	8
P 321	Denizkuşu	PTFG	Turkey	861
W 321	Senja	WPSOH	Norway	592
322	Raïs Al Mounastiri	PSO	Morocco	548
322	Witthayakhom	PGGF	Thailand	836
A 322	Heros	YTM	Sweden	804
ART 322	San Lorenzo	YPT	Peru	624
P 322	Ranarisi	PB	Sri Lanka	787
P 322	Shaheed Ali	PB	Maldives	522
W 322	Andenes	WPSOH	Norway	592
323	El Ayouk	PB	Algeria	8
323	Foros	YDT	Bulgaria	92
323	Metel	FFLM	Russian Federation	697
323	Udomdet	PGGF	Thailand	836
P 323	Şahin	PTFG	Turkey	861
T-AK 323	TSGT John A Chapman	AK	US	978
324	Atija	YH	Bulgaria	92
324	Zouhel	PB	Algeria	8
A 324	Hera	YTM	Sweden	804
A 324	Protea	AGSH	South Africa	764
325	Kolokita	YH	Bulgaria	92
GC 325	El Hamil	PBF	Algeria	8
GC 326	El Assad	PBF	Algeria	8
P 326	Pelikan	PTFG	Turkey	861
327	Janzur	PB	Libya	501
GC 327	Markhad	PBF	Algeria	8
P 327	Albatros	PTFG	Turkey	861
WIX 327	Eagle	WIX/AXS	US	991
GC 328	Etair	PBF	Algeria	8
P 328	Intrépide	PB	Guinea	318
P 328	Şimşek	PTFG	Turkey	861
PRF 328	Rio Arica	PB	Colombia	174
GC 329	Akhir Nahr	PBF	Algeria	8
P 329	Kasirga	PTFG	Turkey	861
PRF 329	Rio Pinillos	PB	Colombia	174
330	Halifax	FFGHM	Canada	100
F 330	Vasco Da Gama	FFGH	Portugal	650
P 330	Kiliç	PGFG	Turkey	861
P 330	Ranajaya	PB	Sri Lanka	787
PRF 330	Rio Angosturas	PB	Colombia	174
U 330	Melitopol	MHSC	Ukraine	879
W 330	Nornen	PBO	Norway	592
331	Chon Buri	PG	Thailand	836
331	Kontraadmiral Sava Ivanov	AGSC	Bulgaria	92
331	Requin	PB	Algeria	8
331	Sri Gaya	AP	Malaysia	514
331	Vancouver	FFGHM	Canada	100
331	Wallaby	WFL/AOTL	Australia	36
F 331	Álvares Cabral	FFGH	Portugal	650
P 331	Ranadeera	PB	Sri Lanka	787
P 331	Kalkan	PGFG	Turkey	861
PC 331	Tenochtitlan	PB	Mexico	537
TM 331	TF R Rios V	YFL	Bolivia	69
U 331	Mariupol	MHSC	Ukraine	879
W 331	Farm	PBO	Norway	592
Y 331	Søløven	MHCD/AX	Denmark	202
332	Erkutskiy Komsomolets	FFLM	Russian Federation	697
332	Songkhla	PG	Thailand	836
332	Sri Tiga	AP	Malaysia	514
332	Ville De Québec	FFGHM	Canada	100
332	Wombat	WFL/AOTL	Australia	36
F 332	Corte Real	FFGH	Portugal	650
P 332	Mizrak	PGFG	Turkey	861
P 332	Ranawickrama	PB	Sri Lanka	787
PC 332	Teotihuacan	PB	Mexico	537
W 332	Heimdal	PBO	Norway	592
333	Marsouin	PB	Algeria	8
333	Phuket	PG	Thailand	836
333	Toronto	FFGHM	Canada	100
333	Warrigal	WFL/AOTL	Australia	36
F 333	Bartolomeu Dias	FFGHM	Portugal	650
P 333	Tufan	PGFG	Turkey	861
PC 333	Palenque	PB	Mexico	537
W 333	Njord	PBO	Norway	592
334	Deneb	PB	Algeria	9
334	Murene	PB	Algeria	8
334	Regina	FFGHM	Canada	100
334	Wyulda	WFL/AOTL	Australia	36
F 334	D. Francisco De Almeida	FFGHM	Portugal	650
P 334	Meltem	PGFG	Turkey	861
PC 334	Mitla	PB	Mexico	537
W 334	Tor	PBO	Norway	592
335	Calgary	FFGHM	Canada	100
335	Mizar	PB	Algeria	9
P 335	Imbat	PGFG	Turkey	861
PC 335	Uxmal	PB	Mexico	537
W 335	Magnus Lagabøte	PBO	Norway	589
336	Alkaid	PB	Algeria	9
336	Montreal	FFGHM	Canada	100
P 336	Zipkin	PGFG	Turkey	861
337	Altair	PB	Algeria	9
337	Fredericton	FFGHM	Canada	100
P 337	Atak	PGFG	Turkey	861
338	Mirfak	PB	Algeria	9
338	Winnipeg	FFGHM	Canada	100
P 338	Bora	PGFG	Turkey	861
339	Alnair	PB	Algeria	9
339	Charlottetown	FFGHM	Canada	100
340	St John's	FFGHM	Canada	100
P 340	Doğan	PGFG	Turkey	862
P 340	Prathpa	PB	Sri Lanka	787
W 340	Barentshav	ARS	Norway	592
341	Bin An Zaran	PSO	Morocco	548
341	El Yadekh	PG	Algeria	6
341	Ottawa	FFGHM	Canada	100
M 341	Karmøy	MHCM/MSCM	Norway	590
P 341	Marti	PGFG	Turkey	862
P 341	Udara	PB	Sri Lanka	787
TM 341	Ingeniero Gumucio	YFL	Bolivia	69
W 341	Bergen	ARS	Norway	592
342	El Mourakeb	PG	Algeria	6
M 342	Måløy	MHCM/MSCM	Norway	590
P 342	Tayfun	PGFG	Turkey	862
W 342	Sortland	ARS	Norway	592
343	El Kechef	PG	Algeria	6
M 343	Hinnøy	MHCM/MSCM	Norway	590
P 343	Volkan	PGFG	Turkey	862
Y 343	Lunden	YAG	Denmark	202
344	El Moutarid	PG	Algeria	6
A 344	Loke	AKL	Sweden	804
P 344	Rüzgar	PGFG	Turkey	862
Y 344	Arvak	YTL	Denmark	201
345	El Rassed	PG	Algeria	6
P 345	Poyraz	PGFG	Turkey	862
Y 345	Alsin	YTL	Denmark	201
346	El Djari	PG	Algeria	6
P 346	Gurbet	PGFG	Turkey	862
347	El Saher	PG	Algeria	6
P 347	Firtina	PGFG	Turkey	862
S 347	Atilay	SSK	Turkey	855
348	El Moukadem	PG	Algeria	6
348	RT 248	MHC	Russian Federation	708
P 348	Yildiz	PGFG	Turkey	862
S 348	Saldiray	SSK	Turkey	855
349	El Tinai	PG	Algeria	6
P 349	Karayel	PGFG	Turkey	862
S 349	Batiray	SSK	Turkey	855
350	El Kanass	PG	Algeria	6
350	Sovetskaya Gavani	FFLM	Russian Federation	697
DCB 350	La Cruz	PBR	Peru	627
M 350	Alta	MHCM/MSCM	Norway	590
S 350	Yildiray	SSK	Turkey	855
351	Ahmad Yani	FFGHM	Indonesia	360
351	Al Jouf	WPBF	Saudi Arabia	743
351	Djebel Chenoua	FSG	Algeria	5
DCB 351	Cabo Blanco	PBR	Peru	627
M 351	Otra	MHCM/MSCM	Norway	590
S 351	Doğanay	SSK	Turkey	855
352	El Chihab	FSG	Algeria	5
352	Slamet Riyadi	FFGHM	Indonesia	360
352	Turaif	WPBF	Saudi Arabia	743
DCB 352	Colán	PBR	Peru	627
M 352	Rauma	MHCM/MSCM	Norway	590
S 352	Dolunay	SSK	Turkey	855
353	Al Kirch	FSG	Algeria	5
353	Hail	WPBF	Saudi Arabia	743
353	Yos Sudarso	FFGHM	Indonesia	360
DCB 353	Samanco	PBR	Peru	627
S 353	Preveze	SSK	Turkey	856
354	El Mahir	PG	Algeria	6
354	MPK 221	FFLM	Russian Federation	697
354	Najran	WPBF	Saudi Arabia	743
354	Oswald Siahaan	FFGHM	Indonesia	360
DCB 354	Besique	PBR	Peru	627
S 354	Sakarya	SSK	Turkey	856
355	Abdul Halim Perdanakusuma	FFGHM	Indonesia	360
DCB 355	Salinas	PBR	Peru	627
S 355	18 Mart	SSK	Turkey	856
356	El Azoum	PG	Algeria	6
356	Karel Satsuitubun	FFGHM	Indonesia	360

© 2014 IHS

IHS Jane's Fighting Ships 2014-2015

[86] Pennant list

Number	Ship's name	Type	Country	Page
DCB 356	Ancón	PBR	Peru	627
S 356	Anafartalar	SSK	Turkey	856
357	El Djasur	PG	Algeria	6
DCB 357	Paracas	PBR	Peru	627
F 357	Thetis	FFHM	Denmark	196
S 357	Gür	SSK	Turkey	856
358	El Hamis	PG	Algeria	6
F 358	Triton	FFHM	Denmark	196
S 358	Çanakkale	SSK	Turkey	856
A 359	Ostria	AXS	Greece	311
F 359	Vaedderen	FFHM	Denmark	196
S 359	Burakreis	SSK	Turkey	856
360	El Mountassir	PG	Algeria	6
F 360	Hvidbjørnen	FFHM	Denmark	196
P 360	Viana Do Castelo	PSOH	Portugal	652
S 360	Inönü	SSK	Turkey	856
U 360	Genichesk	MHC	Ukraine	879
361	Fatahillah	FFG/FFGH	Indonesia	361
F 361	Iver Huitfeldt	FFGHM	Denmark	194
P 361	Kvarnen	AXL	Norway	590
P 361	Figueira Da Foz	PSOH	Portugal	652
362	Malahayati	FFG/FFGH	Indonesia	361
362	Ust-Ilimsk	FFLM	Russian Federation	697
F 362	Peter Willemoes	FFGHM	Denmark	194
P 362	Nordnes	AXL	Norway	590
363	Nala	FFG/FFGH	Indonesia	361
F 363	Niels Juel	FFGHM	Denmark	194
364	Ki Hajar Dewantara	FFGH/FFT	Indonesia	372
365	Agalina	YH	Bulgaria	92
365	Diponegoro	FS	Indonesia	363
366	Sultan Hasanuddin	FS	Indonesia	363
367	Sultan Iskandar Muda	FS	Indonesia	363
368	Frans Kaisiepo	FS	Indonesia	363
369	Kholmsk	FFLM	Russian Federation	697
370	Achernar	PB	Algeria	9
P 370	Rio Minho	PBR	Portugal	653
PG 370	José Andrada	PB	Philippines	631
371	Kaus Australe	PB	Algeria	9
M 371	Ohue	MHSC	Nigeria	583
PG 371	Enrique Jurado	PB	Philippines	631
372	Formalhaut	PB	Algeria	9
M 372	Barama	MHSC	Nigeria	583
PG 372	Alfredo Peckson	PB	Philippines	631
373	Markab	PB	Algeria	9
A 373	Gregos	AXS	Greece	311
374	Betelgeuse	PB	Algeria	9
374	Lambung Mangkurat	FS	Indonesia	362
A 374	Prometheus	AORH/MCCS	Greece	311
PG 374	Simeon Castro	PB	Philippines	631
375	Cut Nyak Dien	FS	Indonesia	362
375	Kiffa Borealis	PB	Algeria	9
375	MPK 82	FFLM	Russian Federation	697
A 375	Zeus	AOTL	Greece	311
PG 375	Carlos Albert	PB	Philippines	631
376	Hamal	PB	Algeria	9
A 376	Orion	AOTL	Greece	311
PG 376	Heracleo Alano	PB	Philippines	631
377	Deneb Algedi	PB	Algeria	9
PG 377	Liberato Picar	PB	Philippines	631
378	Arnab	PB	Algeria	9
PG 378	Hilario Ruiz	PB	Philippines	631
379	Mirzam	PB	Algeria	9
PG 379	Rafael Pargas	PB	Philippines	631
380	Peacock	PB	Algeria	9
DLS 380	Punta Capones	DLS/PBF	Peru	623
P 380	Olav Tryggvason	PBO	Norway	589
PG 380	Nestor Reinoso	PB	Philippines	631
381	Nairal Zaurac	PB	Algeria	9
DLS 381	Punta Malpelo	DLS/PBF	Peru	623
PG 381	Dioscoro Papa	PB	Philippines	631
382	Pleidas	PB	Algeria	9
DLS 382	Punta Mero	DLS/PBF	Peru	623
383	Iman Bonjol	FS	Indonesia	362
383	Sirius	PB	Algeria	9
DLS 383	Punta Sal	DLS/PBF	Peru	623
PG 383	Ismael Lomibao	PB	Philippines	631
384	L'Étoile Polaire	PB	Algeria	9
384	Pati Unus	FS	Indonesia	362
PG 384	Leovigildo Gantioqui	PB	Philippines	631
385	Teuku Umar	FS	Indonesia	362
A 385	Fort Rosalie	AFSH	UK	917
PG 385	Federico Martir	PB	Philippines	631
386	Silas Papare	FS	Indonesia	362
A 386	Fort Austin	AFSH	UK	917
PG 386	Filipino Flojo	PB	Philippines	631
A 387	Fort Victoria	AORH	UK	916
PG 387	Anastacio Cacayorin	PB	Philippines	631
PG 388	Manuel Gomez	PB	Philippines	631
Y 388	Tulugaq	PB	Denmark	198
A 389	Wave Knight	AORH	UK	915
PG 389	Teotimo Figuracion	PB	Philippines	631
390	Korets	FFLM	Russian Federation	697
390	La Habana	AG	Cuba	191
A 390	Wave Ruler	AORH	UK	915
PG 390	José Loor Sr	PB	Philippines	631
PG 392	Juan Magluyan	PB	Philippines	631
PG 393	Florencio Inigo	PB	Philippines	631
PG 394	Alberto Navarette	PB	Philippines	631
PG 395	Felix Apolinario	PB	Philippines	631
PG 396	Abraham Campo	PB	Philippines	631
P 400	Cabo San Juan	PBO	Equatorial Guinea	232
401	Al Basra	PSO	Iraq	391
401	Cakra	–	Indonesia	359
401	General Jose Dolores Estrada	PB	Nicaragua	577
CP 401	Oreste Cavallari	PB	Italy	423
L 401	Al Soumood	LCU	Kuwait	494
L 401	Ertuğrul	LSTH/ML	Turkey	863
P 401	Cassiopea	PSOH	Italy	412
PG 401	Gavion	WPB	Venezuela	1007
TNBH 401	Julian Apaza	–	Bolivia	69
U 401	Kirovograd	LSM	Ukraine	878
402	Al Fayhaa	PSO	Iraq	391
402	Cacique Agateyte	PB	Nicaragua	577
402	Daoud Ben Aicha	LSMH	Morocco	549
402	Nanggala	–	Indonesia	359
402	Polyarny	MHSC/MHSCM	Russian Federation	708
A 402	General Esteban Huertas	LCU	Panama	614
CP 402	Renato Pennetti	PB	Italy	423
L 402	Al Tahaddy	LCU	Kuwait	494
P 402	Libra	PSOH	Italy	412
PG 402	Alca	WPB	Venezuela	1007
U 402	Konstantin Olshansky	LST	Ukraine	879
403	Ahmed Es Sakali	LSMH	Morocco	549
ASR 403	Chihaya	ASRH	Japan	449
CP 403	Walter Fachin	PB	Italy	423
L 403	Saffar	LCU	Kuwait	494
P 403	Spica	PSOH	Italy	412
PG 403	Bernacla	WPB	Venezuela	1007
Y 403W	RP 101	YTM	Italy	421
404	Abou Abdallah El Ayachi	LSMH	Morocco	549
404	Cacique Diriangen	PB	Nicaragua	577
CP 404	Gaetano Magliano	PB	Italy	423
P 404	Vega	PSOH	Italy	412
PG 404	Chaman	WPB	Venezuela	1007
Y 404	RP 102	YTM	Italy	421
405	El Aigh	AKS	Morocco	550
405	Vologda	SSK	Russian Federation	682
AS 405	Chiyoda	AS/ASRH	Japan	448
CP 405	Francisco Mazzinghi	PB	Italy	424
P 405	Esploratore	PB	Italy	413
PG 405	Cormoran	WPB	Venezuela	1007
406	Xia	SSBN	China	128
CP 406	Antonio Scialoja	PB	Italy	424
P 406	Sentinella	PB	Italy	413
PG 406	Colimbo	WPB	Venezuela	1007
SB 406	Vikr	ATS	Russian Federation	724
Y 406	RP 103	YTM	Italy	421
407	Sidi Mohammed Ben Abdallah	LSTH	Morocco	549
CP 407	Michele Lolini	PB	Italy	424
P 407	Vedetta	PB	Italy	413
PG 407	Fardela	WPB	Venezuela	1007
Y 407	RP 104	YTM	Italy	421
408	Dakhla	AKS	Morocco	550
CP 408	Mario Grabar	PB	Italy	424
P 408	Staffetta	PB	Italy	413
PG 408	Fumarel	WPB	Venezuela	1007
Y 408	RP 105	YTM	Italy	421
409	Magnitogorsk	SSK	Russian Federation	682
409	Moroz	FSG	Russian Federation	701
CP 409	Giulio Ingianni	PB	Italy	424
P 409	Sirio	PSOH	Italy	412
PG 409	Negron	WPB	Venezuela	1007
410	Stefan Karadja	ATS	Bulgaria	93
LP 410	Capitán Bretel	PBR	Bolivia	68
P 410	Orione	PSOH	Italy	412
PG 410	Pigargo	WPB	Venezuela	1007
Y 410	RP 106	YTM	Italy	421
411	Kangan	AWT	Iran	390
A 411	Rio Papaloapan	LSTH	Mexico	538
LP 411	Teniente Soliz	PBR	Bolivia	68
P 411	Shaheed Daulat	PC	Bangladesh	56
PG 411	Pagaza	WPB	Venezuela	1007
412	Taheri	AWT	Iran	390
A 412	Aias	YTM/YTL	Greece	312
A 412	Usumacinta	LSTH	Mexico	538
MSC 412	Addriyah	MHSC	Saudi Arabia	741
P 412	Shaheed Farid	PC	Bangladesh	56
PG 412	Serreta	WPB	Venezuela	1007
413	Pin Klao	FFT	Thailand	843
LP 413	San Pedro	PBR	Bolivia	68
P 413	Shaheed Mohibullah	PC	Bangladesh	56
Y 413	Porto Fossone	YTB	Italy	420
MSC 414	Al Quysumah	MHSC	Saudi Arabia	741
P 414	Shaheed Aktheruddin	PC	Bangladesh	56
416	Tariq Ibn Ziyad	FSGM	Libya	500
A 416	Ouranos	AOTL	Greece	311
MSC 416	Al Wadeeah	MHSC	Saudi Arabia	741
Y 416	Porto Torres	YTB	Italy	420
417	Admiral Gorshkov	FFGH	Russian Federation	698
A 417	Hyperion	AOTL	Greece	311
Y 417	Porto Corsini	YTB	Italy	420
418	Inej	FSG	Russian Federation	701
418	Kotelnich	MHSC/MHSCM	Russian Federation	708
MSC 418	Safwa	MHSC	Saudi Arabia	741
SSV 418	Ekvator	AGI/AGIM	Russian Federation	712
A 419	Pandora	AG	Greece	312
420	Al Jawf	MHC	Saudi Arabia	741
A 420	Pandrosos	AG	Greece	312
U 420	Donetsk	ACV/LCUJM	Ukraine	878
421	Bandar Abbas	AORLH	Iran	391
421	Cornwall	PB	Jamaica	427
421	Naresuan	FFGHM	Thailand	829
421	Orkan	FSGM	Poland	639
L 421	Canterbury	AKRH/AX	New Zealand	576
Y 421	Porto Empedocle	YTB	Italy	420
422	Bushehr	AORLH	Iran	391
422	Middlesex	PB	Jamaica	427
422	Piorun	FSGM	Poland	639
422	Shaqra	MHC	Saudi Arabia	741
422	Taksin	FFGHM	Thailand	829
A 422	Kadmos	YTM/YTL	Greece	312
AOE 422	Towada	AOE/AORH	Japan	449
P 422	Ristna	PC	Estonia	234
Y 422	Porto Pisano	YTB	Italy	420
423	Grom	FSGM	Poland	639
423	Smerch	FSG	Russian Federation	701
423	Surrey	PB	Jamaica	427
AOE 423	Tokiwa	AOE/AORH	Japan	449

Pennant list [87]

Number	Ship's name	Type	Country	Page
Y 423	Porto Conte	YTB	Italy	420
424	Al Kharj	MHC	Saudi Arabia	741
424	Daylam	AEL/AKL/AWT	Iran	390
A 424	Iason	YTM/YTL	Greece	312
AOE 424	Hamana	AOE/AORH	Japan	449
425	Jaroslavl	SSK	Russian Federation	682
A 425	Odisseus	YTM/YTL	Greece	312
AOE 425	Mashuu	AOE/AORH	Japan	448
Y 425	Porto Ferraio	YTB	Italy	420
426	Kolomna	MHSC/MHSCM	Russian Federation	708
426	Mineralny Vodi	MHSC/MHSCM	Russian Federation	708
AOE 426	Oumi	AOE/AORH	Japan	448
Y 426	Porto Venere	YTB	Italy	420
Y 428	Porto Salvo	YTB	Italy	420
429	Lipetsk	SSK	Russian Federation	682
430	Al Nour	PC	Egypt	222
A 430	Atlas	YTM/YTL	Greece	312
431	Kharg	AORH	Iran	390
431	Tapi	FS	Thailand	834
431	Vladikavkaz	SSK	Russian Federation	682
A 431	Achilleus	YTM/YTL	Greece	312
432	Khirirat	FS	Thailand	834
A 432	Gigas	YTM/YTL	Greece	312
A 432	Tasuja	MLC	Estonia	234
433	Al Hadi	PC	Egypt	222
433	Makut Rajakumarn	FFH	Thailand	831
A 433	Kerkini	YW	Greece	312
A 433	Wambola	MLC	Estonia	234
434	Admiral Ushakov	DDGHM	Russian Federation	693
A 434	Prespa	YW	Greece	311
A 435	Kekrops	YTM/YTL	Greece	312
436	Al Hakim	PC	Egypt	222
A 437	Pelias	YTM/YTL	Greece	312
438	Leytenant Ilin	MHSC/MHSCM	Russian Federation	708
A 438	Antaios	YTM/YTL	Greece	312
439	Al Wakil	PC	Egypt	222
A 439	Atrefs	YTM/YTL	Greece	312
440	Novosibirsk	SSK	Russian Federation	682
A 440	Diomidis	YTM/YTL	Greece	312
441	Rattanakosin	FSGM	Thailand	833
A 441	Theseus	YTM/YTL	Greece	312
442	Al Qatar	PC	Egypt	222
442	Sukhothai	FSGM	Thailand	833
442	Yan Myat Aung	PC	Myanmar	556
A 442	Romaleos	YTM/YTL	Greece	312
443	Yan Nyein Aung	PC	Myanmar	556
A 443	Titan	YTM/YTL	Greece	312
444	Yan Khwin Aung	PC	Myanmar	556
445	Al Gabbar	PC	Egypt	222
445	Yan Min Aung	PC	Myanmar	556
446	Yan Ye Aung	PC	Myanmar	556
447	Yan Paing Aung	PC	Myanmar	556
448	Al Salam	PC	Egypt	222
448	Yan Win Aung	PC	Myanmar	556
449	Yan Aye Aung	PC	Myanmar	556
A 449	Istros	YTR	Greece	312
450	Razliv	FSG	Russian Federation	701
450	Yan Zwe Aung	PC	Myanmar	556
A 450	Pineios	YTR	Greece	312
F 450	Elli	FFGH	Greece	305
451	Al Rafa	PC	Egypt	222
A 451	Acheloos	YTR	Greece	312
F 451	Limnos	FFGH	Greece	305
F 452	Hydra	FFGH	Greece	304
Y 452	RP 108	YTM	Italy	421
F 453	Spetsai	FFGH	Greece	304
454	Yelnya	MHSC/MHSCM	Russian Federation	708
F 454	Psara	FFGH	Greece	304
455	Chao Phraya	FFG/FFGH	Thailand	830
F 455	Salamis	FFGH	Greece	304
456	Bangpakong	FFG/FFGH	Thailand	830
Y 456	RP 109	YTM	Italy	421
457	Kraburi	FFG/FFGH	Thailand	830
458	Saiburi	FFG/FFGH	Thailand	830
Y 458	RP 110	YTM	Italy	421
F 459	Adrias	FFGH	Greece	305
A 460	Evrotas	YPT	Greece	312
F 460	Aegeon	FFGH	Greece	305
Y 460	RP 111	YTM	Italy	421
461	Phuttha Yotfa Chulalok	FFGHM	Thailand	831
A 461	Arachthos	YPT	Greece	312
F 461	Navarinon	FFGH	Greece	305
462	Phuttha Loetla Naphalai	FFGHM	Thailand	831
F 462	Kountouriotis	FFGH	Greece	305
Y 462	RP 112	YTM	Italy	421
A 463	Nestos	YPT	Greece	312
MST 463	Uraga	MSTH/ML	Japan	444
Y 463	RP 113	YTM	Italy	421
A 464	Axios	ARL/AOR/MCCS	Greece	311
F 464	Kanaris	FFGH	Greece	305
MST 464	Bungo	MSTH/ML	Japan	444
Y 464	RP 114	YTM	Italy	421
F 465	Themistocles	FFGH	Greece	305
Y 465	RP 115	YTM	Italy	421
466	Avangard	MHSC/MHSCM	Russian Federation	708
A 466	Trichonis	YW	Greece	312
F 466	Nikiforos Fokas	FFGH	Greece	305
Y 466	RP 116	YTM	Italy	421
A 467	Doirani	YW	Greece	311
Y 467	RP 123	YTM	Italy	421
468	Kaluga	SSK	Russian Federation	682
A 468	Kalliroe	YW	Greece	311
Y 468	RP 118	YTM	Italy	421
469	Vyborg	SSK	Russian Federation	682
469	Yadryn	MHSC/MHSCM	Russian Federation	708
A 469	Stymfalia	AOTL	Greece	311
A 470	Aliakmon	ARL/AOR/MCCS	Greece	311
Y 470	RP 119	YTM	Italy	421
471	Delvar	AEL/AKL/AWT	Iran	390
471	Maga	PTG	Myanmar	555
F 471	António Enes	FSH	Portugal	651
Y 471	RP 120	YTM	Italy	421
472	Kalaat Beni Hammad	LSTH	Algeria	6
472	Saittra	PTG	Myanmar	555
472	Sirjan	AEL/AKL/AWT	Iran	390
Y 472	RP 121	YTM	Italy	421
473	Kalaat Beni Rached	LSTH	Algeria	6
473	Duwa	PTG	Myanmar	555
Y 473	RP 122	YTM	Italy	421
474	Kalaat Beni-Abbes	LHD	Algeria	7
474	Zeyda	PTG	Myanmar	555
A 474	Pytheas	AGOR	Greece	310
475	Hantha	PTG	Myanmar	555
F 475	João Coutinho	FSH	Portugal	651
476	Banda	PTG	Myanmar	555
A 476	Stravon	AGSC	Greece	311
F 476	Jacinto Cândido	FSH	Portugal	651
477	Sankt Peterburg	SSK	Russian Federation	681
F 477	General Pereira D'Eça	FSH	Portugal	651
Y 477	RP 124	YTM	Italy	421
A 478	Naftilos	AGS	Greece	310
Y 478	RP 125	YTM	Italy	421
A 479	I Theophilopoulos-Karavogiannos	ABUH	Greece	312
Y 479	RP 126	YTM	Italy	421
Y 480	RP 127	YTM	Italy	421
481	Charak	AEL/AKL/AWT	Iran	390
A 481	St Lykoudis	ABUH	Greece	312
Y 481	RP 128	YTM	Italy	421
482	Chiroo	AEL/AKL/AWT	Iran	390
Y 482	RP 129	YTM	Italy	421
483	Soroo	AEL/AKL/AWT	Iran	390
ARC 483	Muroto	ARC	Japan	450
Y 483	RP 130	YTM	Italy	421
Y 484	RP 131	YTM	Italy	421
Y 485	RP 132	YTM	Italy	421
F 486	Baptista De Andrade	FSH	Portugal	651
Y 486	RP 133	YTM	Italy	421
487	Dimitrov	SSK	Russian Federation	682
F 487	João Roby	FSH	Portugal	651
Y 487	RP 134	YTM	Italy	421
488	Ho Shan	LCU	Taiwan	817
F 488	Afonso Cerqueira	FSH	Portugal	651
489	Ho Chuan	LCU	Taiwan	817
490	Ho Seng	LCU	Taiwan	817
F 490	Gaziantep	FFGHM	Turkey	857
P 490	Comandante Cigala Fulgosi	PSOH	Italy	412
491	Ho Meng	LCU	Taiwan	817
F 491	Giresun	FFGHM	Turkey	857
P 491	Comandante Borsini	PSOH	Italy	412
492	Ho Mou	LCU	Taiwan	817
F 492	Gemlik	FFGHM	Turkey	857
P 492	Comandante Bettica	PSOH	Italy	412
493	Ho Shou	LCU	Taiwan	817
F 493	Gelibolu	FFGHM	Turkey	857
P 493	Comandante Foscari	PSOH	Italy	412
F 494	Gökçeada	FFGHM	Turkey	857
F 495	Gediz	FFGHM	Turkey	857
F 496	Gokova	FFGHM	Turkey	857
F 497	Göksu	FFGHM	Turkey	857
LCU 497	Ho Fong	LCU	Taiwan	818
A 498	Lana	AGS	Nigeria	583
LCU 498	Ho Hu	LCU	Taiwan	818
Y 498	Mario Marino	YDT	Italy	420
A 499	Commander Apayi Joe	YTB/YTL	Nigeria	583
Y 499	Alcide Pedretti	YDT	Italy	420
500	Grozavu	ATA	Romania	667
A 500	Commander Rudolph	YTB/YTL	Nigeria	583
F 500	Bozcaada	FFLG	Turkey	860
U 500	Donbas	AGF/AR	Ukraine	880
501	BT 212	MHSC/MHSCM	Russian Federation	708
501	Eilat	FSGHM	Israel	396
501	German Ugryumov	MHSC/MHSCM	Russian Federation	708
501	La Galité	PGGF	Tunisia	851
501	Lieutenant Colonel Errhamani	FFGM	Morocco	545
A 501	Altair	YXT	Sweden	802
A 501	Kyanwa	PBO	Nigeria	580
ETG 501	Bocachica	AGP	Colombia	178
F 501	Bodrum	FFLG	Turkey	860
HQ 501	Tran Khanh Du	LST	Vietnam	1014
LP 501	Santa Cruz De La Sierra	PBR	Bolivia	68
LT 501	Laguna	LST	Philippines	632
M 501	Fethiye	MSI	Turkey	865
SS 501	Souryu	SSK	Japan	429
502	Kurmuk	PBR	Sudan	793
502	Lahav	FSGHM	Israel	396
502	Tunis	PGGF	Tunisia	851
A 502	Antares	YXT	Sweden	802
A 502	Ologbo	PBO	Nigeria	580
ETG 502	Arturus	–	Colombia	178
F 502	Bandirma	FFLG	Turkey	860
HQ 502	Vung Tau	LST	Vietnam	1014
M 502	Fatsa	MSI	Turkey	865
SS 502	Unryu	SSK	Japan	429
503	Carthage	PGGF	Tunisia	851
503	Hanit	FSGHM	Israel	396
503	Qaysan	PBR	Sudan	793
503	Teluk Amboina	–	Indonesia	369
A 503	Arcturus	YXT	Sweden	802
A 503	Nwamba	PBO	Nigeria	580
ETG 503	Pedro David Salas	–	Colombia	178
F 503	Beykoz	FFLG	Turkey	860
HQ 503	Qui Nonh	LST	Vietnam	1014
M 503	Finike	MSI	Turkey	865
SS 503	Hakuryu	SSK	Japan	429
504	Chita	SSK	Russian Federation	682
504	Qeshm	LSL	Iran	389
504	Rumbek	PBR	Sudan	793

[88] Pennant list

Number	Ship's name	Type	Country	Page
A 504	Argo	YXT	Sweden	802
A 504	Obula	PBO	Nigeria	580
ETG 504	Sirius	–	Colombia	178
F 504	Bartin	FFLG	Turkey	860
SS 504	Kenryu	SSK	Japan	429
505	Aleksey Lebedev	MHSC/MHSCM	Russian Federation	708
505	Hamilcar	PG	Tunisia	852
505	Hormuz	LSL	Iran	389
505	Mayom	PBR	Sudan	793
505	Naiza	PB	Kazakhstan	463
A 505	Astrea	YXT	Sweden	802
F 505	Bafra	FFLG	Turkey	860
SS 505	Zuiryu	SSK	Japan	429
506	Dauriya	AKH/AGF	Russian Federation	714
506	Forur	LSL	Iran	389
506	Hannon	PG	Tunisia	852
506	Yesil	PB	Kazakhstan	463
SS 506	Kokuryu	SSK	Japan	429
507	Daqhiliya	MSO	Egypt	224
507	Egemen	PB	Kazakhstan	463
507	Himilcon	PG	Tunisia	852
507	Mogochey	SSK	Russian Federation	682
ETG 507	Calima	–	Colombia	178
LT 507	Benguet	LST	Philippines	632
508	Hannibal	PG	Tunisia	852
ETG 508	Bahía Santa Catalina	–	Colombia	178
509	Hasdrubal	PG	Tunisia	852
509	Teluk Ratai	LST	Indonesia	369
AOR 509	Protecteur	AORH	Canada	105
ETG 509	Móvil I	–	Colombia	178
510	BT 230	MHSC/MHSCM	Russian Federation	708
510	Giscon	PG	Tunisia	852
AOR 510	Preserver	AORH	Canada	105
ETG 510	Sula	–	Colombia	178
U 510	Slavutich	AGFHM	Ukraine	881
511	Al Siddiq	PGGF	Saudi Arabia	740
511	Hengam	LSLH	Iran	389
511	Jymy	YFB	Finland	243
511	Kontradmiral Xawery Czernicki	MCCS	Poland	644
511	Pattani	PBOH	Thailand	832
511	Teluk Bone	LST	Indonesia	369
A 511	Elbe	ARLHM	Germany	296
A 511	Shaheed Ruhul Amin	PBO/AX	Bangladesh	55
F 511	Heybeliada	FFLG	Turkey	860
P 511	Guardião	PBO	Cape Verde	114
U 511	Simferopol	AGS	Ukraine	879
512	Larak	LSLH	Iran	389
512	Narathiwat	PBOH	Thailand	832
512	Raju	YFB	Finland	243
512	Teluk Semangka	LSTH	Indonesia	369
A 512	Mosel	ARLHM	Germany	296
A 512	Shahayak	YR	Bangladesh	60
F 512	Büyükada	FFLG	Turkey	860
513	Al Farouq	PGGF	Saudi Arabia	740
513	Huayin	FFG	China	147
513	Krajmorje	PB	Bulgaria	93
513	Sinai	MSO	Egypt	224
513	Teluk Penyu	LSTH	Indonesia	369
513	Tonb	LSLH	Iran	389
A 513	Rhein	ARLHM	Germany	296
A 513	Shahjalal	AG	Bangladesh	60
F 513	Burgazada	FFLG	Turkey	860
514	Burgas	PB	Bulgaria	93
514	Lavan	LSLH	Iran	389
514	PSKR 639	PTF	Russian Federation	728
514	Teluk Mandar	LSTH	Indonesia	369
A 514	Werra	ARLHM	Germany	296
F 514	Kinaliada	FFLG	Turkey	860
M 514	Silifke	MSC	Turkey	865
515	Abdul Aziz	PGGF	Saudi Arabia	740
515	Emine	PB	Bulgaria	93
515	Teluk Sampit	LSTH	Indonesia	369
A 515	Khan Jahan Ali	AOTL	Bangladesh	60
A 515	Main	ARLHM	Germany	296
M 515	Saros	MSC	Turkey	865
516	Assiyut	MSO	Egypt	224
516	Jiujiang	FFG	China	147
516	Teluk Banten	LSTH	Indonesia	369
A 516	Donau	ARLHM	Germany	296
A 516	Imam Gazzali	AOTL	Bangladesh	59
M 516	Sigacik	MSC	Turkey	865
517	Faisal	PGGF	Saudi Arabia	740
517	Teluk Ende	LSTH	Indonesia	369
M 517	Sapanca	MSC	Turkey	865
518	Jian	FFG	China	147
M 518	Sariyer	MSC	Turkey	865
519	Changzhi	FFG	China	147
519	Khalid	PGGF	Saudi Arabia	740
520	Bisma	PBO	Indonesia	377
520	Rassvet	FSG	Russian Federation	701
A 520	Sagres	AXS	Portugal	655
P 520	Diana	PB	Denmark	198
SSV 520	Feodor Golovin	AGIM	Russian Federation	712
521	Al Siddiq	MHC	Egypt	225
521	Amyr	PGGF	Saudi Arabia	740
521	Baladewa	PBO	Indonesia	377
521	Jiaxin	FFGHM	China	143
521	Krasnokamensk	SSK	Russian Federation	682
521	Sattahip	PG	Thailand	837
A 521	Schultz Xavier	ABU	Portugal	656
P 521	Freja	PB	Denmark	198
P 521	Vigilante	PBO	Cape Verde	114
522	B 394	SSK	Russian Federation	682
522	Klongyai	PG	Thailand	837
522	Lianyungang	FFGHM	China	143
522	Sergey Kolbassev	MHSC/MHSCM	Russian Federation	708
A 522	D. Carlos I	AGS	Portugal	654
P 522	Havfruen	PB	Denmark	198
S 522	Salvatore Pelosi	SSK	Italy	402
523	Kiiski 3	MSI	Finland	241
523	Putian	FFGHM	China	143
523	Takbai	PG	Thailand	837
523	Tariq	PGGF	Saudi Arabia	740
A 523	Almirante Gago Coutinho	AGS	Portugal	654
P 523	Najaden	PB	Denmark	198
S 523	Giuliano Prini	SSK	Italy	402
524	Al Farouk	MHC	Egypt	225
524	Balchik	PB	Bulgaria	94
524	Kantang	PG	Thailand	837
524	Kiiski 4	MSI	Finland	241
524	Sanming	FFGHM	China	143
P 524	Nymfen	PB	Denmark	198
S 524	Primo Longobardo	SSK	Italy	402
525	BT 232	MHSC/MHSCM	Russian Federation	708
525	Kiiski 5	MSI	Finland	241
525	Maanshan	FFGHM	China	142
525	Obzor	PB	Bulgaria	93
525	Oqbah	PGGF	Saudi Arabia	740
525	Sozopol	PB	Bulgaria	94
525	Thepha	PG	Thailand	837
525	Wu Kang	AKM	Taiwan	820
P 525	Rota	PB	Denmark	198
S 525	Gianfranco Gazzana Priaroggia	SSK	Italy	402
526	Kiiski 6	MSI	Finland	241
526	Nakat	FSG	Russian Federation	701
526	Nesebar	PB	Bulgaria	94
526	Taimuang	PG	Thailand	837
526	Wenzhou	FFGHM	China	142
S 526	Salvatore Todaro	SSK	Italy	401
527	Abu Obaidah	PGGF	Saudi Arabia	740
527	Kiiski 7	MSI	Finland	241
527	Luoyang	FFGHM	China	143
S 527	Sciré	SSK	Italy	401
528	Mianyang	FFGHM	China	143
S 528	Pietro Venuti	SSK	Italy	401
529	B 187	SSK	Russian Federation	682
529	Zhoushan	FFGHM	China	144
S 529	Romeo Romei	SSK	Italy	401
530	Giza	MSO	Egypt	225
530	Steregushchiy	FFGHM	Russian Federation	696
530	Wu Yi	AOEHM	Taiwan	820
530	Xuzhou	FFGHM	China	144
A 530	Lood	YTD	Estonia	234
531	Kavarna	PB	Bulgaria	93
531	Khamronsin	FS	Thailand	833
531	Soobrazitelny	FFGHM	Russian Federation	696
531	Syöksy	YFB	Finland	243
531	Syvatitel Nikolay Chudotvorets	SSK	Russian Federation	682
531	Teluk Gilimanuk	LSM	Indonesia	370
532	Boiky	FFGHM	Russian Federation	696
532	Panshih	AOEH	Taiwan	820
532	Teluk Celukan Bawang	LSM	Indonesia	370
532	Thayanchon	FS	Thailand	833
532	Tulcea	AOT	Romania	666
533	Aswan	MSO	Egypt	225
533	DKA 182	LCMs	Russian Federation	706
533	Longlom	FS	Thailand	833
533	Taizhou	FFG	China	147
533	Teluk Cendrawasih	LSM	Indonesia	370
A 533	Norge	YAC	Norway	591
534	Jinhua	FFG	China	147
534	Shafak	PGGF	Libya	500
534	Varna	PB	Bulgaria	93
535	Aysberg	FSG	Russian Federation	701
535	Kaliakra	PB	Bulgaria	93
535	Teluk Peleng	LSM	Indonesia	370
A 535	Valkyrien	–	Norway	591
536	Qina	MSO	Egypt	225
536	Teluk Sibolga	LSM	Indonesia	370
537	Cangzhou	FFG	China	146
537	Teluk Manado	LSM	Indonesia	370
538	Yantai	FFGHM	China	144
538	Teluk Hading	LSM	Indonesia	370
539	Anqing	FFGHM	China	146
539	Sohag	MSO	Egypt	225
539	Teluk Parigi	LSM	Indonesia	370
540	Huainan	FFGHM	China	146
540	Teluk Lampung	LSM	Indonesia	370
A 540	Dannebrog	YAC	Denmark	201
A 540	Hansaya	LCP	Sri Lanka	791
U 540	Chigirin	AXL	Ukraine	880
541	Hua Hin	PSO	Thailand	837
541	Huaibei	FFGHM	China	146
541	Teluk Jakarta	LSM	Indonesia	370
541	Vinha	YFB	Finland	243
A 541	Birkholm	MSD/AXL/AGSC	Denmark	198
P 541	Aboubekr Ben Amer	PBO	Mauritania	525
U 541	Smila	AXL	Ukraine	880
542	BT 114	MHSC/MHSCM	Russian Federation	708
542	Dat Assawari	MHC	Egypt	225
542	Klaeng	PSO	Thailand	837
542	Teluk Sangkulirang	LSM	Indonesia	370
542	Tongling	FFGHM	China	146
542-051	Ercsi	PBR	Hungary	325
542-054	Baja	PBR	Hungary	325
A 542	Fyrholm	MSD/AXL/AGSC	Denmark	198
ETG 542	Playa Blanca	–	Colombia	178
U 542	Nova Kahovka	AXL	Ukraine	880
543	Dandong	FFG	China	147
543	Marshal Shaposhnikov	DDGHM	Russian Federation	692
543	Si Racha	PSO	Thailand	837
543	Teluk Cirebon	AKL/ARL	Indonesia	374
A 543	Ertholm	MSD/AXL/AGSC	Denmark	198
ETG 543	Tierra Bomba	–	Colombia	178
544	Lushun	FFGH	China	160
544	Teluk Sabang	AKL/ARL	Indonesia	374

Pennant list [89]

Number	Ship's name	Type	Country	Page
A 544	Alholm	MSD/AXL/AGSC	Denmark	198
ETG 544	Bell Salter	–	Colombia	178
545	Linfen	FFG	China	147
545	Navarin	MHC	Egypt	225
545	Stoiky	FFGHM	Russian Federation	696
546	Yancheng	FFGHM	China	144
ETG 546	Orion	–	Colombia	178
547	Linyi	FFGHM	China	144
547	Ust-Kamshats	SSK	Russian Federation	682
548	Admiral Panteleyev	DDGHM	Russian Federation	692
548	Burullus	MHC	Egypt	225
548	Yiyang	FFGHM	China	144
549	Changzhou	FFGHM	China	144
549	Ust-Bolsheretsk	SSK	Russian Federation	682
550	Weifang	FFGHM	China	144
C 550	Cavour	CV	Italy	404
LC 550	Bacolod City	LSVH	Philippines	632
551	Krabi	PSO	Thailand	834
551	Liven	FSG	Russian Federation	701
ATF 551	Ta Wan	ATF/ARS	Taiwan	821
B 551	Voum-Legleita	PBO	Mauritania	525
C 551	Giuseppe Garibaldi	LPH	Italy	413
LC 551	Dagupan City	LSVH	Philippines	632
P 551	Sadd	PBF	Pakistan	611
WLM 551	Ida Lewis	WLM/ABU	US	989
552	Ta Hu	ARS	Taiwan	820
F 552	Urania	FSM	Italy	410
P 552	Shabhaz	PBF	Pakistan	611
WLM 552	Katherine Walker	WLM/ABU	US	989
553	Shaoguan	FFG	China	147
ATF 553	Ta Han	ATF/ARS	Taiwan	821
D 553	Andrea Doria	DDGHM	Italy	406
P 553	Vaqar	PBF	Pakistan	611
WLM 553	Abigail Burgess	WLM/ABU	US	989
554	Alrosa	SSK	Russian Federation	682
ATF 554	Ta Kang	ATF/ARS	Taiwan	821
D 554	Caio Duilio	DDGHM	Italy	406
F 554	Sfinge	FSM	Italy	410
P 554	Burq	PBF	Pakistan	611
WLM 554	Marcus Hanna	WLM/ABU	US	989
555	Geyzer	FSG	Russian Federation	701
555	Nikolay Rubtsov	LCMs	Russian Federation	706
555	Zhaotong	FFG	China	147
ATF 555	Ta Fung	ATF/ARS	Taiwan	821
F 555	Driade	FSM	Italy	410
WLM 555	James Rankin	WLM/ABU	US	989
YTB 555	Tillicum	YTB/YTL/YTM	Canada	105
F 556	Chimera	FSM	Italy	410
WLM 556	Joshua Appleby	WLM/ABU	US	989
F 557	Fenice	FSM	Italy	410
WLM 557	Frank Drew	WLM/ABU	US	989
558	Zigong	FFG	China	147
ETG 558	Juanchaco	–	Colombia	178
WLM 558	Anthony Petit	WLM/ABU	US	989
559	Beihai	FFG	China	147
A 559	Sleipner	AKS	Denmark	201
ETG 559	Libertad	–	Colombia	178
WLM 559	Barbara Mabrity	WLM/ABU	US	989
560	BT 256	MHSC/MHSCM	Russian Federation	708
560	Dongguan	FFG	China	147
560	Won San	MLH	Korea, South	485
560	Zyb	FSG	Russian Federation	701
A 560	Gunnar Thorson	YPC/ABU	Denmark	202
D 560	Luigi Durand De La Penne	DDGHM	Italy	407
ETG 560	Isla Tesoro	–	Colombia	178
WLM 560	William Tate	WLM/ABU	US	989
561	BT 115	MHSC/MHSCM	Russian Federation	708
561	Kang Kyeong	MHSC	Korea, South	485
561	Multatuli	AGFH	Indonesia	373
561	Shantou	FFG	China	147
A 561	Gunnar Seidenfaden	YPC/ABU	Denmark	202
D 561	Francesco Mimbelli	DDGHM	Italy	407
WLM 561	Harry Claiborne	WLM/ABU	US	989
YTR 561	Firebird	YTR	Canada	106
562	Jiangmen	FFG	China	147
562	Kang Jin	MHSC	Korea, South	485
A 562	Mette Miljø	AKL	Denmark	201
WLM 562	Maria Bray	WLM/ABU	US	989
YTR 562	Firebrand	YTR	Canada	106
563	Foshan	FFG	China	147
563	Ko Ryeong	MHSC	Korea, South	485
A 563	Marie Miljø	AKL	Denmark	201
ATF 563	Ta Tai	ATF/ARS	Taiwan	821
WLM 563	Henry Blake	WLM/ABU	US	989
564	Admiral Tributs	DDGHM	Russian Federation	692
564	Magomed Gadgiev	MHSC/MHSCM	Russian Federation	708
564	Yichang	FFGHM	China	143
WLM 564	George Cobb	WLM/ABU	US	989
565	BT 100	MHSC/MHSCM	Russian Federation	708
565	Huludao	FFGHM	China	143
565	Kim Po	MHSC	Korea, South	485
566	Huaihua	FFGHM	China	143
566	Ko Chang	MHSC	Korea, South	485
567	Kum Wha	MHSC	Korea, South	485
567	Xiangfan	FFGHM	China	143
568	Hengyang	FFGHM	China	144
569	Yulin	FFGHM	China	144
570	DKA 106	LCMs	Russian Federation	706
570	Huangshan	FFGHM	China	144
570	Passat	FSG	Russian Federation	701
A 570	Taşkizak	AORL	Turkey	867
P 570	Knud Rasmussen	PGBH	Denmark	197
571	Yang Yang	MSC/MHC	Korea, South	485
571	Yuncheng	FFGHM	China	144
A 571	Albay Hakki Burak	AOTL	Turkey	867
F 571	Grecale	FFGHM	Italy	409
HQ 571	Truong Sa	AP	Vietnam	1015
P 571	Ejnar Mikkelsen	PGBH	Denmark	197
SSV 571	Belomore	AGIM	Russian Federation	712
572	Admiral Vinogradov	DDGHM	Russian Federation	692
572	Hengshui	FFGHM	China	144
572	Ongjin	MSC/MHC	Korea, South	485
A 572	Yuzbasi Ihsan Tulunay	AOTL	Turkey	867
F 572	Libeccio	FFGHM	Italy	409
573	Hae Nam	MSC/MHC	Korea, South	485
573	Liuzhou	FFGHM	China	144
A 573	Binbasi Sadettin Gürcan	AOTL	Turkey	867
F 573	Scirocco	FFGHM	Italy	409
574	Sanya	FFGHM	China	144
F 574	Aliseo	FFGHM	Italy	409
575	DKA 144	LCU	Russian Federation	705
575	Yueyang	FFGHM	China	144
F 575	Euro	FFGHM	Italy	409
576	Huangshi	FFGHM	China	144
A 576	Değirmendere	ATA	Turkey	870
F 576	Espero	FFGHM	Italy	409
577	BT 262	MHSC/MHSCM	Russian Federation	708
A 577	Sokullu Mehmet Paşa	AX	Turkey	866
F 577	Zeffiro	FFGHM	Italy	409
578	DKA 148	LCMs	Russian Federation	706
A 578	Darica	ATR	Turkey	869
A 579	Cezayirli Gazi Hasan Paşa	AX	Turkey	866
580	Datong	FSG	China	148
580	Dore	LCU	Indonesia	370
A 580	Akar	AORH	Turkey	866
581	Yingkou	FSG	China	148
A 581	Darshak	LCU/LCP	Bangladesh	57
A 581	Çinar	AWT	Turkey	867
582	Bengbu	FSG	China	148
582	Kupang	LCU	Indonesia	369
A 582	Tallashi	LCU/LCP	Bangladesh	57
583	Dili	LCU	Indonesia	369
583	Shangrao	FSG	China	148
A 583	Agradoot	AGS	Bangladesh	59
F 583	Aviere	FFGHM	Italy	410
584	Meizhou	FSG	China	148
584	Nusa Utara	LCU	Indonesia	369
A 584	LCT 101	LCU/LCP	Bangladesh	57
F 584	Bersagliere	FFGHM	Italy	410
H 584	Anushandhan	AGS	Bangladesh	59
585	Baise	FSG	China	148
A 585	Akin	ASR	Turkey	868
A 585	LCT 102	LCU/LCP	Bangladesh	57
586	Jian	FSG	China	148
A 586	Akbaş	ATS	Turkey	870
SFP 586	Akademik Isanin	AGS	Russian Federation	710
587	Jieyang	FSG	China	148
A 587	Gazal	ATS	Turkey	870
A 587	LCT 104	LCU/LCP	Bangladesh	57
588	PSKA 588	PBO	Russian Federation	729
588	Quanzhou	FSG	China	148
A 588	Çandarli	AGS	Turkey	866
589	Qingyuan	FSG	China	148
A 589	Işin	ARS	Turkey	868
590	DKA 464	LCMs	Russian Federation	706
590	Makassar	LPD/APCR	Indonesia	368
590	Weihai	FSG	China	148
A 590	Inebolu	ATF	Turkey	869
F 590	Carlo Bergamini	FFGH	Italy	408
SS 590	Oyashio	SSK	Japan	430
YTL 590	Lawrenceville	YTB/YTL/YTM	Canada	105
591	Surabaya	LPD/APCR	Indonesia	368
F 591	Virginio Fasan	FFGH	Italy	408
SS 591	Michishio	SSK	Japan	430
YTL 591	Parksville	YTB/YTL/YTM	Canada	105
592	Banjarmasin	LPD/APCR	Indonesia	368
F 592	Carlo Margottini	FFGH	Italy	408
SS 592	Uzushio	SSK	Japan	430
YTL 592	Listerville	YTB/YTL/YTM	Canada	105
593	Banda Aceh	LPD/APCR	Indonesia	368
593	Suqian	FSG	China	148
F 593	Carabiniere	FFGH	Italy	408
SS 593	Makishio	SSK	Japan	430
YTL 593	Merrickville	YTB/YTL/YTM	Canada	105
A 594	Çubuklu	AGS	Turkey	866
F 594	Alpino	FFGH	Italy	408
SS 594	Isoshio	SSK	Japan	430
YTL 594	Granville	YTB/YTL/YTM	Canada	105
A 595	Yarbay Kudret Güngör	AORH	Turkey	866
F 595	Luigi Rizzo	FFGH	Italy	408
SS 595	Narushio	SSK	Japan	430
596	Huizhou	FSG	China	148
SS 596	Kuroshio	SSK	Japan	430
597	Qinzhou	FSG	China	148
SS 597	Takashio	SSK	Japan	430
A 598	Söğüt	AWT	Turkey	867
SS 598	Yaeshio	SSK	Japan	430
A 599	Çesme	AGS	Turkey	866
SS 599	Setoshio	SSK	Japan	430
600	Zvezdochka	AGE/ASR	Russian Federation	711
A 600	Kavak	AWT	Turkey	867
SS 600	Mochishio	SSK	Japan	430
601	23 Of July	PGGF	Egypt	221
601	PSKA 585	PBO	Russian Federation	729
601	Pyi Daw Aye	AK	Myanmar	558
601	Ras El Blais	PBO	Tunisia	853
A 601	Monge	AGMH	France	270
LG-601	Rio Jubones	WPBR	Ecuador	215
MSC 601	Hirashima	MHSC	Japan	445
P 601	Élorn	PB	France	279
P 601	Jayesagara	PB	Sri Lanka	786
P 601	Limam El Hadrami	PB	Mauritania	525
PAF 601	Filigonio Hichamón	PBR	Colombia	173
S 601	Rubis	SSN/SNA	France	248
U 601	Alchevsk	AGS	Ukraine	880
602	6 Of October	PGGF	Egypt	221
602	Junon	PB	Seychelles	748
602	Nizhny Novgorod	SSN	Russian Federation	675
602	Ras Ajdir	PBO	Tunisia	853

© 2014 IHS

IHS Jane's Fighting Ships 2014-2015

Pennant list

Number	Ship's name	Type	Country	Page
LG 602	Rio Chongon	WPBR	Ecuador	216
MSC 602	Yakushima	MHSC	Japan	445
P 602	Verdon	PB	France	279
PAF 602	SSIM Manuel A Moyar	PBR	Colombia	173
S 602	Saphir	SSN/SNA	France	248
603	21 Of October	PGGF	Egypt	221
603	Aiyar Lunin	–	Myanmar	557
603	Jin Chiang	PCG	Taiwan	816
603	PSKA 584	PBO	Russian Federation	729
603	Ras El Edrak	PBO	Tunisia	853
LG 603	Rio Valdivia	WPBR	Ecuador	216
MSC 603	Takashima	MHSC	Japan	445
P 603	Adour	PB	France	279
PAF 603	Igaraparaná	PBR	Colombia	173
S 603	Casabianca	SSN/SNA	France	248
Y 603	Nymphea	YFL	France	275
604	18 Of June	PGGF	Egypt	221
604	Aiyar Mai	LCU	Myanmar	557
604	PSKA 290	PBO	Russian Federation	729
604	PSKA 583	PBO	Russian Federation	729
604	Ras El Manoura	PBO	Tunisia	853
LG 604	Rio Yacuambi	WPBR	Ecuador	216
MSC 604	Enoshima	MSC	Japan	446
P 604	Scarpe	PB	France	279
PAF 604	SSIM Julio Correa Hernández	PBR	Colombia	173
S 604	Émeraude	SSN/SNA	France	248
605	25 Of April	PGGF	Egypt	221
605	Admiral Levchenko	DDGHM	Russian Federation	692
605	Aiyar Maung	LCU	Myanmar	557
605	Andromache	PB	Seychelles	748
605	Ras Enghela	PBO	Tunisia	853
605	Tan Chiang	PCG	Taiwan	816
MSC 605	Chichijima	MSC	Japan	446
P 605	Vertonne	PB	France	279
PAF 605	Manacacías	PBR	Colombia	173
S 605	Améthyste	SSN/SNA	France	248
Y 605	Gendarme Perez	YFL	France	275
606	Aiyar Minthamee	LCU	Myanmar	557
606	Hsin Chiang	PCG	Taiwan	816
606	Ras Ifrikia	PBO	Tunisia	853
606	Topaz	PBO	Seychelles	748
MSC 606	Hatushima	MSC	Japan	446
P 606	Dumbea	PB	France	279
PAF 606	Cotuhe	PBR	Colombia	173
S 606	Perle	SSN/SNA	France	248
Y 606	Lavande	YFL	France	275
607	Aiyar Minthar	LCU	Myanmar	557
607	Feng Chiang	PCG	Taiwan	816
A 607	Meuse	AORHM	France	272
P 607	Yser	PB	France	279
PAF 607	SSCIM Senen Alberto Araujo	PBR	Colombia	174
608	MDK 18	ACV/LCUJ	Russian Federation	707
608	Tseng Chiang	PCG	Taiwan	816
A 608	Var	AORHM	France	272
P 608	Argens	PB	France	279
PAF 608	CPCIM Guillermo Londoño Vargas	PBR	Colombia	174
609	Kao Chiang	PCG	Taiwan	816
609	MDK 88	ACV/LCUJ	Russian Federation	707
P 609	Hérault	PB	France	279
PAF 609	Ariarí	PBR	Colombia	173
610	Jing Chiang	PCG	Taiwan	816
610	Nastoychivy	DDGHM	Russian Federation	693
P 610	Gravona	PB	France	279
PAF 610	Mario Villegas	PBR	Colombia	174
YDT 610	Sechelt	YTT/YPT/YDT	Canada	104
611	Hsian Chiang	PCG	Taiwan	816
611	Mohammed V	FFGHM	Morocco	544
LG 611	Rio Verde	WPBF	Ecuador	215
M 611	Vulcain	MCD	France	268
P 611	Odet	PB	France	279
P 611	Tawheed	PC	Bangladesh	56
PAF 611	Tony Pastrana Contreras	PBR	Colombia	174
Y 611	MDLC Richard	YFL	France	275
YPT 611	Sikanni	YTT/YPT/YDT	Canada	104
612	Badr	–	Saudi Arabia	739
612	Hassan II	FFGHM	Morocco	544
612	Tsi Chiang	PCG	Taiwan	816
LG 612	Rio Bulu Bulu	WPBF	Ecuador	215
P 612	Maury	PB	France	279
P 612	Tawfiq	PC	Bangladesh	56
PAF 612	CTCIM Jorge Moreno Salazar	PBR	Colombia	174
YDT 612	Sooke	YTT/YPT/YDT	Canada	104
613	Tarik Ben Ziyad	FSG	Morocco	546
A 613	Achéron	MCD	France	268
LG 613	Rio Macara	WPBF	Ecuador	215
P 613	Charente	PB	France	279
P 613	Tamjeed	PC	Bangladesh	56
PAF 613	Juan Ricardo Oyola Vera	PBR	Colombia	174
YPT 613	Stikine	YTT/YPT/YDT	Canada	104
614	Al Yarmook	–	Saudi Arabia	739
614	Po Chiang	PCG	Taiwan	816
614	PSKA 595	PBO	Russian Federation	729
614	Sultan Moulay Ismael	FSG	Morocco	546
D 614	Cassard	DDGHM	France	256
LG 614	Rio Yaguachi	WPBF	Ecuador	215
M 614	Styx	MCD	France	268
P 614	Tanveer	PC	Bangladesh	56
P 614	Tech	PB	France	279
PAF 614	Alexander Pérez Rodriguez	PBR	Colombia	174
615	Allal Ben Abdallah	FSG	Morocco	546
615	Bora	PGGJM	Russian Federation	699
615	Chan Chiang	PCG	Taiwan	816
615	PSKA 581	PBO	Russian Federation	729
D 615	Jean Bart	DDGHM	France	256
P 615	Penfeld	PB	France	279
PAF 615	Cristian Reyes Holguín	PBR	Colombia	174
WMEC 615	Reliance	PSOH/WMEC	US	985
616	Hitteen	–	Saudi Arabia	739
616	Samum	PGGJM	Russian Federation	699
LG 616	Rio San Miguel	WPBF	Ecuador	215
P 616	Trieux	PB	France	279
PAF 616	Alejandro Ledesma Ortiz	PB	Colombia	176
S 616	Le Triomphant	SSBN/SNLE-NG	France	250
WMEC 616	Diligence	PSOH/WMEC	US	985
617	Chu Chiang	PCG	Taiwan	816
617	Mirazh	FSG	Russian Federation	701
AD 617	Yakal	ARL	Philippines	633
LG 617	Rio Quinindé	WPBF	Ecuador	215
P 617	Vésubie	PB	France	279
PAF 617	Harrys Tous Cataño	PB	Colombia	176
S 617	Le Téméraire	SSBN/SNLE-NG	France	250
WMEC 617	Vigilant	PSOH/WMEC	US	985
618	Obninsk	SSN	Russian Federation	680
618	Tabuk	–	Saudi Arabia	739
618	Tuo Jiang	PGGF	Taiwan	817
LG 618	Rio Catamayo	WPBF	Ecuador	215
P 618	Escaut	PB	France	279
S 618	Le Vigilant	SSBN/SNLE-NG	France	250
WMEC 618	Active	PSOH/WMEC	US	985
619	Severomorsk	DDGHM	Russian Federation	692
P 619	Huveaune	PB	France	279
S 619	Le Terrible	SSBN/SNLE-NG	France	250
WMEC 619	Confidence	PSOH/WMEC	US	985
620	Bespokoiny	DDGHM	Russian Federation	693
620	Shtyl	FSG	Russian Federation	701
D 620	Forbin	DDGHM	France	254
P 620	Sayura	PSOH	Sri Lanka	786
P 620	Sévre	PB	France	279
WMEC 620	Resolute	PSOH/WMEC	US	985
621	Flaming	MHCM	Poland	642
621	Mandau	PTFG	Indonesia	365
621	Thalang	MCS	Thailand	842
D 621	Chevalier Paul	DDGHM	France	254
P 621	Aber-Wrach	PB	France	279
P 621	Samudura	PSOH	Sri Lanka	787
WMEC 621	Valiant	PSOH/WMEC	US	985
622	Karp	SSN	Russian Federation	674
622	Rencong	PTFG	Indonesia	365
M 622	Pluton	MCD	France	268
P 622	Estéron	PB	France	279
P 622	Sagara	PSOH	Sri Lanka	789
623	Badik	PTFG	Indonesia	365
623	Mewa	MHCM	Poland	642
P 623	Mahury	PB	France	279
WMEC 623	Steadfast	PSOH/WMEC	US	985
624	Czajka	MHCM	Poland	642
624	Keris	PTFG	Indonesia	365
P 624	Organabo	PB	France	279
WMEC 624	Dauntless	PSOH/WMEC	US	985
WMEC 625	Venturous	PSOH/WMEC	US	985
626	Vitse Admiral Kulakov	DDGHM	Russian Federation	692
WMEC 626	Dependable	PSOH/WMEC	US	985
WMEC 627	Vigorous	PSOH/WMEC	US	985
WMEC 629	Decisive	PSOH/WMEC	US	985
630	DKA 156	LCU	Russian Federation	705
630	Goplo	MHC	Poland	643
A 630	Marne	AORHM	France	272
WMEC 630	Alert	PSOH/WMEC	US	985
631	Bang Rachan	MHSC	Thailand	842
631	DKA 131	LCU	Russian Federation	705
631	Gardno	MHC	Poland	643
631	Todak	PGG	Indonesia	365
A 631	Somme	AORHM	France	272
632	Bukowo	MHC	Poland	643
632	Nongsarai	MHSC	Thailand	842
633	Dąbie	MHC	Poland	643
633	Lat Ya	MHSC	Thailand	842
A 633	Taape	AG/ATS/YDT/YPC/YPT	France	273
634	DKA 56	LCU	Russian Federation	705
634	Jamno	MHC	Poland	643
634	Tha Din Daeng	MHSC	Thailand	842
635	Mielno	MHC	Poland	643
A 635	Revi	AFL	France	273
U 635	Skvyra	YDT/YFL/YPT	Ukraine	881
636	Wicko	MHC	Poland	643
A 636	Maïto	YTM	France	276
637	Resko	MHC	Poland	643
A 637	Maroa	YTM	France	276
638	Kutilang	PBO	Indonesia	377
638	Sarbsko	MHC	Poland	643
A 638	Manini	YTM	France	276
Y 638	Lardier	YTM	France	276
639	Bangau	PBO	Indonesia	377
639	Necko	MHC	Poland	643
Y 639	Giens	YTM	France	276
640	Balibis	PBO	Indonesia	377
640	D 145	ACV/LCUJ	Russian Federation	705
640	Nakło	MHC	Poland	643
Y 640	Mengam	YTM	France	276
YTB 640	Glendyne	YTB/YTL/YTM	Canada	105
641	Clurit	PBM	Indonesia	367
641	Drużno	MHC	Poland	643
641	Pelikan	PBO	Indonesia	377
A 641	Esterel	YTM	France	276
D 641	Dupleix	DDGHM	France	257
M 641	Éridan	MHC	France	269
Y 641	Balaguier	YTM	France	276
YTB 641	Glendale	YTB/YTL/YTM	Canada	105
642	DKA 185	LCMs	Russian Federation	706
642	Hańcza	MHC	Poland	643
642	Kujang	PBM	Indonesia	367
642	Punai	PBO	Indonesia	377
A 642	Lubéron	YTM	France	276

Pennant list [91]

Number	Ship's name	Type	Country	Page
D 642	Montcalm	DDGHM	France	257
M 642	Cassiopée	MHC	France	269
UAM 642	Calmaria	YP	Portugal	656
WLI 642	Buckthorn	WLI/ABU	US	990
Y 642	Taillat	YTM	France	276
YTB 642	Glenevis	YTB/YTL/YTM	Canada	105
643	Beladau	PBM	Indonesia	367
643	Mamry	MHSCM	Poland	642
D 643	Jean De Vienne	DDGHM	France	257
M 643	Andromède	MHC	France	269
UAM 643	Cirro	YP	Portugal	656
Y 643	Nividic	YTM	France	276
YTB 643	Glenbrook	YTB/YTL/YTM	Canada	105
644	Alamang	PBM	Indonesia	367
644	Wigry	MHSCM	Poland	642
D 644	Primauguet	DDGHM	France	258
M 644	Pégase	MHC	France	269
UAM 644	Vendaval	YP	Portugal	656
YTB 644	Glenside	YTB/YTL/YTM	Canada	105
645	DKA 172	LCU	Russian Federation	705
645	PSKA 291	PBO	Russian Federation	729
645	Sniardwy	MHSCM	Poland	642
A 645	Alizé	YDT	France	273
D 645	La Motte-Picquet	DDGHM	France	258
M 645	Orion	MHC	France	269
UAM 645	Monção	YP	Portugal	656
646	Wdzydze	MHSCM	Poland	642
D 646	Latouche-Tréville	DDGHM	France	258
M 646	Croix Du Sud	MHC	France	269
UAM 646	Suão	YP	Portugal	656
M 647	Aigle	MHC	France	269
UAM 647	Macareu	YP	Portugal	656
Y 647	Le Four	YTM	France	276
648	Hayabusa	PB	Indonesia	377
648	Kostroma	SSN	Russian Federation	674
M 648	Lyre	MHC	France	269
UAM 648	Preia-Mar	YP	Portugal	656
649	Anis Madu	PB	Indonesia	377
A 649	L'Étoile	AXS	France	271
UAM 649	Baixa-Mar	YP	Portugal	656
Y 649	Port Cros	YTM	France	276
650	Admiral Chabanenko	DDGHM	Russian Federation	691
650	DKA 107	LCU	Russian Federation	705
650	Taka	PB	Indonesia	377
A 650	La Belle Poule	AXS	France	271
D 650	Aquitaine	DDGHM	France	255
M 650	Sagittaire	MHC	France	269
651	Singa	PBO	Indonesia	366
D 651	Normandie	DDGHM	France	255
GC 651	Tecun Uman	PB	Guatemala	317
A 652	Mutin	AXS	France	271
D 652	Provence	DDGHM	France	255
GC 652	Kaibil Balan	PB	Guatemala	317
M 652	Céphée	MHC	France	269
653	Ajak	PBO	Indonesia	366
A 653	La Grande Hermine	AXS	France	271
D 653	Languedoc	DDGHM	France	255
GC 653	Azumanche	PB	Guatemala	317
M 653	Capricorne	MHC	France	269
D 654	Auvergne	DDGHM	France	255
GC 654	Tzacol	PB	Guatemala	317
D 655	Alsace	DDGHM	France	255
GC 655	Bitol	PB	Guatemala	317
656	Orienburg	SSAN	Russian Federation	684
BH 656	Gucumaz	PB	Guatemala	317
D 656	Bretagne	DDGHM	France	255
Y 656	Phaéton	YAG	France	274
657	PSKA 276	PBO	Russian Federation	729
D 657	Lorraine	DDGHM	France	255
Y 657	Machaon	YAG	France	274
659	DK-143	ACV	Russian Federation	733
661	Tambov	SSN	Russian Federation	680
663	Pskov	SSN	Russian Federation	675
663	PSKA 589	PBO	Russian Federation	729
A 664	Malabar	ATA	France	276
665	PSKR 274	PBO	Russian Federation	729
A 669	Tenace	ATA	France	276
670	Ramadan	PGGF	Egypt	222
Y 670	Las	YTR	France	275
671	Petrozavodsk	SSN	Russian Federation	680
P 671	Glaive	PB	France	278
Y 671	Douffine	YTR	France	275
672	Khyber	PGGF	Egypt	222
673	PSKA 282	PBO	Russian Federation	729
674	El Kadessaya	PGGF	Egypt	222
675	DKA 57	LCMs	Russian Federation	706
A 675	Fréhel	YTM	France	276
P 675	Arago	PBO	France	265
676	El Yarmouk	PGGF	Egypt	222
A 676	Saire	YTM	France	276
MSC 676	Kumejima	MHSC	Japan	445
P 676	Flamant	PBO	France	265
UAM 676	Guia	ABU	Portugal	655
677	DKA 70	LCMs	Russian Federation	706
A 677	Armen	YTM	France	276
MSC 677	Makishima	MHSC	Japan	445
P 677	Cormoran	PBO	France	265
678	Badr	PGGF	Egypt	222
A 678	La Houssaye	YTM	France	276
MSC 678	Tobishima	MHSC	Japan	445
P 678	Pluvier	PBO	France	265
A 679	Kéréon	YTM	France	276
MSC 679	Yugeshima	MHSC	Japan	445
680	Hettein	PGGF	Egypt	222
A 680	Sicié	YTM	France	276
MSC 680	Nagashima	MHSC	Japan	445
Y 680	La Mitre	YFL	France	275
681	Kojoon Bong	LSTH	Korea, South	484
A 681	Taunoa	YTM	France	276
MSC 681	Sugashima	MHC	Japan	445
P 681	Albatros	PSO	France	264
Y 681	Tour Royale	YFL	France	275
682	Biro Bong	LSTH	Korea, South	484
682	S Ezzat	PCFG	Egypt	222
A 682	Rascas	YTM	France	276
MSC 682	Notojima	MHC	Japan	445
Y 682	Léon	YFL	France	275
683	F Zekry	PCFG	Egypt	222
683	Hyangro Bong	LSTH	Korea, South	484
MSC 683	Tsunoshima	MHC	Japan	445
Y 683	Cornouailles	YFL	France	275
684	Danil Moskovskiy	SSN	Russian Federation	680
684	M Fahmy	PCFG	Egypt	222
MSC 684	Naoshima	MHC	Japan	445
P 684	La Capricieuse	PBO	France	265
Y 684	Contentin	YFL	France	275
685	A Gad	PCFG	Egypt	222
685	Seongin Bong	LSTH	Korea, South	484
MSC 685	Toyoshima	MHC	Japan	445
MSC 686	Ukushima	MHC	Japan	445
P 686	La Glorieuse	PBO	France	265
MSC 687	Izushima	MHC	Japan	445
P 687	La Gracieuse	PBO	France	265
688	Cheonwangbong	LST	Korea, South	484
MSC 688	Aishima	MHC	Japan	445
P 688	La Moqueuse	PBO	France	265
MSC 689	Aoshima	MHC	Japan	445
690	Fabian Wrede	AX	Finland	242
MSC 690	Miyajima	MHC	Japan	445
691	Tatarstan	FFGM	Russian Federation	696
691	Wilhelm Carpelan	AX	Finland	242
MSC 691	Shishijima	MHC	Japan	445
692	Axel Von Fersen	AX	Finland	242
MSC 692	Kuroshima	MHC	Japan	445
Y 692	Telenn Mor	ABU	France	273
693	Dagestan	FFGM	Russian Federation	696
A 695	Bélier	YTB	France	277
A 696	Buffle	YTB	France	277
A 697	Bison	YTB	France	277
698	PSKA 294	PBO	Russian Federation	729
SSN 698	Bremerton	SSN	US	938
699	DK-447	ACV	Russian Federation	733
SSN 699	Jacksonville	SSN	US	938
700	Kingston	MM	Canada	103
700	R 32	FSGM	Russian Federation	700
A 700	Khaireddine	AGS	Tunisia	852
SSN 700	Dallas	SSN	US	938
SSV 700	Temryuk	AGS/AGI/AGE	Russian Federation	709
U 700	Netisin	YDT	Ukraine	881
701	El Moundjid	ARS	Algeria	9
701	Glace Bay	MM	Canada	103
701	Karachejevo-Cherkessia	PGGK	Russian Federation	703
701	Kilat 1	PB	Malaysia	519
701	Mohammad VI	FFGHM	Morocco	546
701	Pelindung 1	PB	Malaysia	519
701	Thar	WPB	Egypt	228
A 701	N N O Salammbo	AGOR/AX	Tunisia	852
P 701	Le Malin	PBO	France	265
P 701	Nandimithra	PGG	Sri Lanka	786
SG 701	Dost	PSOH	Turkey	871
SSN 701	La Jolla	SSN	US	938
T 701	Nooradheen	PB	Maldives	521
702	Budenovsk	PGGK	Russian Federation	703
702	El Mous'if	ARS	Algeria	9
702	Fateh	PBO	Iraq	392
702	Kilat 2	PB	Malaysia	519
702	Madina	FFGHM	Saudi Arabia	737
702	Nanaimo	MM	Canada	103
702	Pelindung 2	PB	Malaysia	519
L 702	Chikoko I	LCU	Malawi	505
P 702	Suranimila	PGG	Sri Lanka	786
SG 702	Güven	PSOH	Turkey	871
T 702	Iskandhar	PB	Maldives	521
703	Edmonton	MM	Canada	103
703	El Moussanid	ARS	Algeria	9
703	Kilat 3	PB	Malaysia	519
703	Nasir	PBO	Iraq	392
703	Nur	WPB	Egypt	228
703	Pelindung 3	PB	Malaysia	519
P 703	Kasungu	PB	Malawi	505
SG 703	Umut	PSOH	Turkey	871
Y 703	Lilas	YFL	France	275
704	Hofouf	FFGHM	Saudi Arabia	737
704	Kilat 4	PB	Malaysia	519
704	Majed	PBO	Iraq	392
704	Pelindung 4	PB	Malaysia	519
704	Shawinigan	MM	Canada	103
P 704	Kaning'a	PB	Malawi	505
SG 704	Yaşam	PSOH	Turkey	871
705	Kilat 5	PB	Malaysia	519
705	Pelindung 5	PB	Malaysia	519
705	Shimookh	PBO	Iraq	392
705	Stupinets	FSGM	Russian Federation	700
705	Whitehorse	MM	Canada	103
SSN 705	City Of Corpus Christi	SSN	US	[938]
U 705	Kremenets	ATA/YTM	Ukraine	882
706	Abha	FFGHM	Saudi Arabia	737
706	Borovsk	PGGK	Russian Federation	703
706	Kilat 6	PB	Malaysia	519
706	Yellowknife	MM	Canada	103
SSN 706	Albuquerque	SSN	US	938
U 706	Izyaslav	ATA/YTM	Ukraine	882
707	Goose Bay	MM	Canada	103
707	Kilat 7	PB	Malaysia	519
708	Kilat 8	PB	Malaysia	519
708	Moncton	MM	Canada	103
708	Taif	FFGHM	Saudi Arabia	737
709	Saskatoon	MM	Canada	103
709	Kilat 9	PB	Malaysia	519
A 709	Ain Zaghouan	AWT	Tunisia	853

[92] Pennant list

Number	Ship's name	Type	Country	Page
710	Brandon	MM	Canada	103
710	Kilat 10	PB	Malaysia	519
F 710	La Fayette	FFGHM	France	259
Y 710	General Delfosse	YFL	France	275
711	Kilat 11	PB	Malaysia	519
711	Pulau Rengat	MHSC	Indonesia	370
711	Summerside	MM	Canada	103
711	Yoon Young-Ha	PGGF	Korea, South	482
F 711	Surcouf	FFGHM	France	259
P 711	Barkat	PC	Bangladesh	56
SSN 711	San Francisco	SSN	US	938
712	Han Sang Guk	PGGF	Korea, South	482
712	Kilat 12	PB	Malaysia	519
712	Neustrashimy	FFHM	Russian Federation	695
712	Pulau Rupat	MHSC	Indonesia	370
A 712	Athos	YFRT	France	273
F 712	Courbet	FFGHM	France	259
P 712	Salam	PB	Bangladesh	56
713	Nisr	WPB	Egypt	228
713	Jo Cheon Hyeong	PGGF	Korea, South	482
713	Kilat 13	PB	Malaysia	519
713	Kerch	CGHM	Russian Federation	690
A 713	Aramis	YFRT	France	273
F 713	Aconit	FFGHM	France	259
P 713	Sangu	PBO/AX	Bangladesh	55
SSN 713	Houston	SSN	US	938
Y 713	Capitaine Moulié	YFL	France	275
714	Kilat 14	PB	Malaysia	519
F 714	Guépratte	FFGHM	France	259
P 714	Turag	PBO/AX	Bangladesh	55
SSN 714	Norfolk	SSN	US	938
715	Bystry	DDGHM	Russian Federation	693
715	Hwang Dohyun	PGGF	Korea, South	482
715	Kilat 15	PB	Malaysia	519
SSN 715	Buffalo	SSN	US	938
716	Kilat 16	PB	Malaysia	519
716	Suh Hoowon	PGGF	Korea, South	482
P 716	MDLC Jacques	YFL	France	275
717	Kilat 17	PB	Malaysia	519
717	Park Donghyuk	PGGF	Korea, South	482
SSN 717	Olympia	SSN	US	938
WHEC 717	Mellon	PSOH/WHEC	US	984
718	Hyun Sihak	PGGF	Korea, South	482
718	Kilat 18	PB	Malaysia	519
718	MT 265	MSOM	Russian Federation	707
719	Jung Geungmo	PGGF	Korea, South	482
719	Kilat 19	PB	Malaysia	519
719	Nimr	WPB	Egypt	228
SSN 719	Providence	SSN	US	938
WHEC 719	Boutwell	PSOH/WHEC	US	984
720	Kilat 20	PB	Malaysia	519
P 720	Géranium	PB	France	279
SSN 720	Pittsburgh	SSN	US	938
WHEC 720	Sherman	PSOH/WHEC	US	984
721	Ji Deokchil	PGGF	Korea, South	482
721	Kilat 21	PB	Malaysia	519
721	Pulau Rote	MSC	Indonesia	371
721	Sichang	LSTH	Thailand	840
A 721	Khadem	ATA	Bangladesh	60
P 721	Jonquille	PB	France	279
SSN 721	Chicago	SSN	US	938
WHEC 721	Gallatin	PSOH/WHEC	US	984
722	Kilat 22	PB	Malaysia	519
722	Lim Byeongrae	PGGF	Korea, South	482
722	Pulau Raas	MSC	Indonesia	371
722	Surin	LSTH	Thailand	840
722	Vaarlahti	AKSL	Finland	242
A 722	Sebak	YTM	Bangladesh	60
P 722	Violette	PB	France	279
SSN 722	Key West	SSN	US	938
U 722	Borsziv	YTR	Ukraine	881
WHEC 722	Morgenthau	PSOH/WHEC	US	984
723	Hong Siuk	PGGF	Korea, South	482
723	Kilat 23	PB	Malaysia	519
723	Pulau Romang	MSC	Indonesia	371
723	Vänö	AKSL	Finland	242
A 723	Rupsha	YTM	Bangladesh	60
P 723	Jasmin	PB	France	279
SSN 723	Oklahoma City	SSN	US	938
WHEC 723	Rush	PSOH/WHEC	US	984
724	Kilat 24	PB	Malaysia	519
724	Pulau Rimau	MSC	Indonesia	371
A 724	Shibsha	YTM	Bangladesh	60
SSN 724	Louisville	SSN	US	938
WHEC 724	Munro	PSOH/WHEC	US	984
725	Hong Daeson	PGGF	Korea, South	482
725	Kilat 25	PB	Malaysia	519
P 725	L'Adroit	FS	France	264
SSN 725	Helena	SSN	US	938
726	Han Munsik	PGGF	Korea, South	482
726	Kilat 26	PB	Malaysia	519
726	Pulau Rusa	MSC	Indonesia	371
SSGN 726	Ohio	SSGN	US	936
WHEC 726	Midgett	PSOH/WHEC	US	984
727	Kilat 27	PB	Malaysia	519
727	Kim Changhak	PGGF	Korea, South	482
727	Pulau Rangsang	MSC	Indonesia	371
727	Yaroslav Mudryy	FFHM	Russian Federation	695
SSGN 727	Michigan	SSGN	US	936
728	Kilat 28	PB	Malaysia	519
728	Park Dongjin	PGGF	Korea, South	482
MCL 728	Ieshima	MCSD	Japan	444
SSGN 728	Florida	SSGN	US	936
U 728	Evpatoriya	YTR	Ukraine	881
729	Kilat 29	PB	Malaysia	519
729	Pulau Rempang	MSC	Indonesia	371
MCL 729	Maejima	MCSD	Japan	444
SSGN 729	Georgia	SSGN	US	936
730	Haukipää	YTM	Finland	244
730	Kilat 30	PB	Malaysia	519
F 730	Floréal	FFGHM	France	261
SSBN 730	Henry M Jackson	SSBN	US	935
731	Kilat 31	PB	Malaysia	519
F 731	Prairial	FFGHM	France	261
SSBN 731	Alabama	SSBN	US	935
732	Kilat 32	PB	Malaysia	519
F 732	Nivôse	FFGHM	France	261
SSBN 732	Alaska	SSBN	US	935
733	Kilat 33	PB	Malaysia	519
F 733	Ventôse	FFGHM	France	261
SSBN 733	Nevada	SSBN	US	935
734	Kilat 34	PB	Malaysia	519
F 734	Vendémiaire	FFGHM	France	261
SSBN 734	Tennessee	SSBN	US	935
735	Kilat 35	PB	Malaysia	519
F 735	Germinal	FFGHM	France	261
SSBN 735	Pennsylvania	SSBN	US	935
736	Kilat 36	PB	Malaysia	519
SSBN 736	West Virginia	SSBN	US	935
737	Kilat 37	PB	Malaysia	519
SSBN 737	Kentucky	SSBN	US	935
738	Kilat 38	PB	Malaysia	519
738	MT 264	MSOM	Russian Federation	707
SSBN 738	Maryland	SSBN	US	935
739	Hästö	AKSL	Finland	243
739	Kilat 39	PB	Malaysia	519
SSBN 739	Nebraska	SSBN	US	935
740	Kilat 40	PB	Malaysia	519
P 740	Fulmar	PB	France	265
SSBN 740	Rhode Island	SSBN	US	935
741	Kilat 41	PB	Malaysia	519
SSBN 741	Maine	SSBN	US	935
742	Kilat 42	PB	Malaysia	519
SSBN 742	Wyoming	SSBN	US	935
743	Kilat 43	PB	Malaysia	519
A 743	Denti	AETL	France	270
SSBN 743	Louisiana	SSBN	US	935
744	Kilat 44	PB	Malaysia	519
745	Kilat 45	PB	Malaysia	519
746	DKA 465	LCMs	Russian Federation	706
746	Kilat 46	PB	Malaysia	519
747	DKA 67	LCU	Russian Federation	705
747	Kilat 47	PB	Malaysia	519
748	Kilat 48	PB	Malaysia	519
A 748	Léopard	AXL	France	271
749	Kilat 49	PB	Malaysia	519
A 749	Panthère	AXL	France	271
750	Kilat 50	PB	Malaysia	519
A 750	Jaguar	AXL	France	271
SSN 750	Newport News	SSN	US	938
WMSL 750	Bertholf	PSOH/WMSL	US	984
751	Kilat 51	PB	Malaysia	519
A 751	Lynx	AXL	France	271
SSN 751	San Juan	SSN	US	938
WMSL 751	Waesche	PSOH/WMSL	US	984
Y 751	Engageante	AXL	France	271
752	Kilat 52	PB	Malaysia	519
A 752	Guépard	AXL	France	271
SSN 752	Pasadena	SSN	US	938
WMSL 752	Stratton	PSOH/WMSL	US	984
Y 752	Vigilante	AXL	France	271
753	Kilat 53	PB	Malaysia	519
A 753	Chacal	AXL	France	271
SSN 753	Albany	SSN	US	938
WMSL 753	Hamilton	PSOH/WMSL	US	984
754	Bezboyaznennyy	DDGHM	Russian Federation	693
A 754	Tigre	AXL	France	271
SSN 754	Topeka	SSN	US	938
WMSL 754	James	PSOH/WMSL	US	984
Y 754	Taina	YFL	France	274
A 755	Lion	AXL	France	271
WMSL 755	Munro	PSOH/WMSL	US	984
SSN 756	Scranton	SSN	US	938
U 756	Sudak	AWT	Ukraine	880
WMSL 756	Kimball	PSOH/WMSL	US	984
SSN 757	Alexandria	SSN	US	938
758	Kyong Ju	FS/FSG	Korea, South	481
A 758	Beautemps-Beaupré	AGOR	France	269
SSN 758	Asheville	SSN	US	938
Y 758	Kermeur	YFB	France	274
759	Mok Po	FS/FSG	Korea, South	481
A 759	Dupuy De Lôme	AGIH	France	269
SSN 759	Jefferson City	SSN	US	938
Y 759	Kernaleguen	YFB	France	274
SSN 760	Annapolis	SSN	US	938
761	Kim Chon	FS/FSG	Korea, South	481
P 761	Kara	PB	Togo	848
SSN 761	Springfield	SSN	US	938
762	Chung Ju	FS/FSG	Korea, South	481
L 762	Lachs	LCU	Germany	292
P 762	Mono	PB	Togo	848
SSN 762	Columbus	SSN	US	938
Y 762	L'Etoile De Mer	YFL	France	274
763	Jin Ju	FS/FSG	Korea, South	481
SSN 763	Santa Fe	SSN	US	938
SSN 764	Boise	SSN	US	938
765	Yo Su	FS/FSG	Korea, South	481
L 765	Schlei	LCU	Germany	292
SSN 765	Montpelier	SSN	US	938
Y 765	Avel Mor	YFL	France	274
766	Jin Hae	FS/FSG	Korea, South	481
SSN 766	Charlotte	SSN	US	938
767	Sun Chon	FS/FSG	Korea, South	481
SSN 767	Hampton	SSN	US	938
768	Yee Ree	FS/FSG	Korea, South	481
A 768	Élan	AG/ATS/YDT/YPC/YPT	France	273
SSN 768	Hartford	SSN	US	938
769	Won Ju	FS/FSG	Korea, South	481
SSN 769	Toledo	SSN	US	938

IHS Jane's Fighting Ships 2014-2015

© 2014 IHS

Pennant list [93]

Number	Ship's name	Type	Country	Page
770	Valentin Pikul	MSOM	Russian Federation	707
770	Yangjiang	PTG	China	151
770	Yevgeniy Kocheshkov	ACVM/LCUJM	Russian Federation	706
A 770	Glycine	AXL	France	271
M 770	Antarès	MHI	France	268
SSN 770	Tucson	SSN	US	938
Y 770	Morse	YT	France	276
771	An Dong	FS/FSG	Korea, South	481
771	Anawrahta	FSG	Myanmar	554
771	Shunde	PTG	China	151
771	Thong Kaeo	LCU	Thailand	840
A 771	Eglantine	AXL	France	271
M 771	Altaïr	MHI	France	268
SSN 771	Columbia	SSN	US	938
Y 771	Otarie	YT	France	276
YTB 771	Keokuk	YTB	US	974
772	Bayintnaung	FSG	Myanmar	554
772	Nanhai	PTG	China	151
772	Thong Lang	LCU	Thailand	840
M 772	Aldébaran	MHI	France	268
SSN 772	Greeneville	SSN	US	938
Y 772	Loutre	YT	France	276
773	Panyu	PTG	China	151
773	Song Nam	FS/FSG	Korea, South	481
773	Wang Nok	LCU	Thailand	840
P 773	Njambuur	PBO	Senegal	744
SSN 773	Cheyenne	SSN	US	938
Y 773	Phoque	YT	France	276
774	Lianjiang	PTG	China	151
774	Wang Nai	LCU	Thailand	840
SSN 774	Virginia	SSN	US	937
775	Bu Chon	FS/FSG	Korea, South	481
775	Xinhui	PTG	China	151
A 775	Gazelle	AG/ATS/YDT/YPC/YPT	France	273
SSN 775	Texas	SSN	US	937
776	Jae Chon	FS/FSG	Korea, South	481
SSN 776	Hawaii	SSN	US	937
777	Dae Chon	FS/FSG	Korea, South	481
777	Porkkala	MLI	Finland	241
SSN 777	North Carolina	SSN	US	937
Y 777	Palangrin	YFL	France	274
778	Sok Cho	FS/FSG	Korea, South	481
SSN 778	New Hampshire	SSN	US	937
779	Yong Ju	FS/FSG	Korea, South	481
SSN 779	New Mexico	SSN	US	937
SSN 780	Missouri	SSN	US	937
781	Man Nok	LCU	Thailand	840
781	Nam Won	FS/FSG	Korea, South	481
SSN 781	California	SSN	US	937
782	Kwan Myong	FS/FSG	Korea, South	481
782	Man Klang	LCU	Thailand	840
782	Mordoviya	ACVM/LCUJM	Russian Federation	706
SSN 782	Mississippi	SSN	US	937
U 782	Sokal	YH/TFL	Ukraine	881
783	Man Nai	LCU	Thailand	840
783	Sin Hung	FS/FSG	Korea, South	481
SSN 783	Minnesota	SSN	US	937
U 783	Illichivsk	YDT/YFL/YPT	Ukraine	881
Y 783	Avel Aber	YTR	France	274
784	Mattapon	LCU	Thailand	842
SSN 784	North Dakota	SSN	US	937
Y 784	La Loude	YTR	France	274
785	Kong Ju	FS/FSG	Korea, South	481
785	Rawi	LCU	Thailand	842
A 785	Thétis	MCD/BEGM	France	270
SSN 785	John Warner	SSN	US	937
Y 785	La Divette	YTR	France	274
SSN 786	Illinois	SSN	US	937
Y 786	Auté	YFL	France	274
SSN 787	Washington	SSN	US	937
Y 787	Tiaré	YFL	France	274
SSN 788	Colorado	SSN	US	937
F 789	Lieutenant De Vaisseau Le Hénaff	FS	France	260
SSN 789	Indiana	SSN	US	937
A 790	Coralline	AGE	France	274
F 790	Lieutenant De Vaisseau Lavallée	FS	France	260
SSN 790	South Dakota	SSN	US	937
Y 790	Dionée	YDT	France	274
791	Angthong	LPD	Thailand	841
791	Hai Shih	SS	Taiwan	812
A 791	Lapérouse	AGS	France	270
F 791	Commandant L'Herminier	FS	France	260
SSN 791	Delaware	SSN	US	937
Y 791	Myosotis	YDT	France	274
792	Hai Bao	SS	Taiwan	812
A 792	Borda	AGS	France	270
F 792	Premier Maître L'Her	FS	France	260
P 792	Pavois	PB	France	278
Y 792	Gardénia	YDT	France	274
793	Hai Lung	SSK	Taiwan	811
A 793	Laplace	AGS	France	270
F 793	Commandant Blaison	FS	France	260
P 793	Écu	PB	France	278
Y 793	Liseron	YDT	France	274
794	Hai Hu	SSK	Taiwan	811
F 794	Enseigne De Vaisseau Jacoubet	FS	France	260
P 794	Rondache	PB	France	278
Y 794	Magnolia	YDT	France	274
F 795	Commandant Ducuing	FS	France	260
P 795	Harnois	PB	France	278
Y 795	Ajonc	YDT	France	274
F 796	Commandant Birot	FS	France	260
P 796	Haubert	PB	France	278
Y 796	Genêt	YDT	France	274
F 797	Commandant Bouan	FS	France	260
P 797	Heaume	PB	France	278

Number	Ship's name	Type	Country	Page
Y 797	Giroflée	YDT	France	274
798	Matelot Brice Kpomasse	PB	Benin	67
P 798	Brigantine	PB	France	278
Y 798	Acanthe	YDT	France	274
YTB 798	Opelika	YTB	US	974
799	DKA 325	LCMs	Russian Federation	706
799	Hylje	YPC	Finland	244
799	La Sota	PB	Benin	67
P 799	Gantelet	PB	France	278
L 800	Rotterdam	LPD	Netherlands	567
WLIC 800	Pamlico	WLIC	US	991
YTB 800	Manhattan	YT	US	974
801	Ladny	FFM	Russian Federation	694
801	Pandrong	PBO	Indonesia	366
801	Pusan	AGOR	Korea, South	485
801	Rais Hamidou	PTGM	Algeria	5
801	Te Mataili	PB	Tuvalu	874
L 801	Johan De Witt	LPD	Netherlands	568
MHV 801	Aldebaran	PB	Denmark	198
P 801	Huravee	PBO	Maldives	521
UAM 801	Coral	YGS	Portugal	654
WLIC 801	Hudson	WLIC	US	991
YTB 801	Washtucna	YT	US	974
802	Hamzah	PBO	Iran	388
802	Salah Rais	PTGM	Algeria	5
802	Sura	PBO	Indonesia	366
A 802	Sidi Bou Said	ABU	Tunisia	852
A 802	Snellius	AGSH	Netherlands	570
F 802	De Zeven Provincien	FFGHM	Netherlands	564
MHV 802	Carina	PB	Denmark	198
P 802	Ghazee	PB	Maldives	522
S 802	Walrus	SSK	Netherlands	562
UAM 802	Atlanta	YGS	Portugal	654
WLIC 802	Kennebec	WLIC	US	991
YT 802	Valiant	YTB	US	974
803	Rais Ali	PTGM	Algeria	5
A 803	Luymes	AGSH	Netherlands	570
F 803	Tromp	FFGHM	Netherlands	564
MHV 803	Aries	PB	Denmark	198
S 803	Zeeleeuw	SSK	Netherlands	562
TRV 803	Tailor	YPT	Australia	35
WLIC 803	Saginaw	WLIC	US	991
YT 803	Reliant	YTB	US	974
804	Hiu	PGG	Indonesia	365
804	Huoqiu	MCMV	China	157
A 804	Pelikaan	AP	Netherlands	572
A 804	Tabarka	ABU	Tunisia	853
F 804	De Ruyter	FFGHM	Netherlands	564
MHV 804	Andromeda	PB	Denmark	198
YT 804	Defiant	YTB	US	974
805	Layang	PGG	Indonesia	365
805	Tula	SSBN	Russian Federation	672
805	Zhangjiagang	MCMV	China	157
A 805	Taguermess	ABU	Tunisia	853
F 805	Evertsen	FFGHM	Netherlands	564
MHV 805	Gemini	PB	Denmark	198
UAM 805	Fisália	YGS	Portugal	654
YT 805	Seminole	YTB	US	974
806	Lemadang	PGG	Indonesia	365
MHV 806	Dubhe	PB	Denmark	198
YT 806	Puyallup	YTB	US	974
807	Boa	PB	Indonesia	366
807	Ekaterinburg	SSBN	Russian Federation	672
807	Phraongkamrop	PB	Thailand	846
807	Yay Bo	AGSC	Myanmar	558
BG 807	Matros Mikola Mushnirov	PBR	Ukraine	883
MHV 807	Jupiter	PB	Denmark	198
YT 807	Menominee	YTB	US	974
808	Komendor	MSOM	Russian Federation	707
808	Picharnpholakit	PB	Thailand	846
808	Pytlivy	FFM	Russian Federation	694
808	Welang	PB	Indonesia	366
MHV 808	Lyra	PB	Denmark	198
S 808	Dolfijn	SSK	Netherlands	562
809	Raminthra	PB	Thailand	846
809	Suluh Pari	PB	Indonesia	366
MHV 809	Antares	PB	Denmark	198
810	Jingjiang	MCMV	China	157
810	Katon	PB	Indonesia	366
810	Smetlivy	DDGM	Russian Federation	690
MHV 810	Luna	PB	Denmark	198
P 810	Jaguar	PB	Netherlands	573
S 810	Bruinvis	SSK	Netherlands	562
811	Chanthara	AGS	Thailand	843
811	Incheon	FFGHM	Korea, South	480
811	Kakap	PBOH	Indonesia	366
811	V Gumanenko	MHSO	Russian Federation	708
MHV 811	Apollo	PB	Denmark	198
P 811	Durjoy	PTG	Bangladesh	57
P 811	Panter	PB	Netherlands	573
U 811	Balta	ADG	Ukraine	880
Y 811	Knurrhahn	APB	Germany	296
812	Al Riyadh	FFGHM	Saudi Arabia	738
812	Gyeonggi	FFGHM	Korea, South	480
812	Kerapu	PBOH	Indonesia	366
812	Suk	AGOR	Thailand	843
812	Voronezh	SSGN	Russian Federation	676
MHV 812	Hercules	PB	Denmark	198
P 812	Nirbhoy	PC	Bangladesh	55
P 812	Poema	PB	Netherlands	573
Y 812	Lütje Hörn	YTM	Germany	297
813	Jeonbuk	FFGHM	Korea, South	480
813	Pharuehatsabodi	AGSH	Thailand	843
813	Tongkol	PBOH	Indonesia	366
MHV 813	Baunen	PB	Denmark	198
P 813	Nirmul	PTG	Bangladesh	57
UAM 813	Bellatrix	AXS	Portugal	655
814	Barakuda	PBOH	Indonesia	366
814	Liaoyang	ML/MST	China	157
814	Makkah	FFGHM	Saudi Arabia	738

© 2014 IHS

IHS Jane's Fighting Ships 2014-2015

[94] Pennant list

Number	Ship's name	Type	Country	Page
MHV 814	Budstikken	PB	Denmark	198
P 814	Cocle	PBR	Panama	613
UAM 814	Canopus	AXS	Portugal	655
Y 814	Knechtsand	YTM	Germany	297
815	Sanca	PB	Indonesia	366
MHV 815	Kureren	PB	Denmark	198
Y 815	Scharhörn	YTM	Germany	297
816	Al Dammam	FFGHM	Saudi Arabia	738
816	Smolensk	SSGN	Russian Federation	676
816	Warakas	PB	Indonesia	366
MHV 816	Patrioten	PB	Denmark	198
Y 816	Vogelsand	YTM	Germany	297
817	Panana	PB	Indonesia	366
MHV 817	Partisan	PB	Denmark	198
Y 817	Nordstrand	YTM	Germany	297
818	Kalakae	PB	Indonesia	366
818	Kunshan	MCMV	China	157
819	Alexander Nevsky	SSBN	Russian Federation	671
819	R 47	FSGM	Russian Federation	700
819	Tedong Naga	PB	Indonesia	366
Y 819	Langeness	YTM	Germany	297
820	Briansk	SSBN	Russian Federation	672
820	Viper	PB	Indonesia	367
YTB 820	Wanamassa	YTB	US	974
821	Ch'ungnam	AGOR	Korea, South	485
821	Lublin	LST/ML	Poland	642
821	Piton	PB	Indonesia	367
821	Suriya	ABU	Thailand	843
822	Gniezno	LST/ML	Poland	642
822	Weling	PB	Indonesia	367
823	Krakow	LST/ML	Poland	642
823	Matacora	PB	Indonesia	367
823	Sprut	PSO	Russian Federation	726
YTB 823	Canonchet	YTB	US	974
824	Dmitriy Donskoy	SSBN	Russian Federation	674
824	Hayabusa	PGGF	Japan	443
824	Poznan	LST/ML	Poland	642
824	Tedung Selar	PB	Indonesia	367
PS 824	Purga	PSO	Russian Federation	731
SSV 824	Liman	AGI/AGIM	Russian Federation	712
YTB 824	Santaquin	YTB	US	974
825	Boiga	PB	Indonesia	367
825	Dimitrovgrad	FSGM	Russian Federation	700
825	Toruń	LST/ML	Poland	642
825	Wakataka	PGGF	Japan	443
826	Isku	MLI	Finland	243
826	Kelabang	MSC	Indonesia	371
826	Ootaka	PGGF	Japan	443
827	Krait	PB	Indonesia	367
827	Kumataka	PGGF	Japan	443
827	Verchoture	SSBN	Russian Federation	672
828	Edermen	FSGM	Turkmenistan	873
828	Kala Hitam	MSC	Indonesia	371
828	Umitaka	PGGF	Japan	443
F 828	Van Speijk	FFGHM	Netherlands	563
829	Gayratly	FSGM	Turkmenistan	873
829	Shirataka	PGGF	Japan	443
829	Tarihu	PB	Indonesia	367
830	Alkura	PB	Indonesia	367
830	Högsåra	AKSL	Finland	243
U 830	Korets	ATA/YTM	Ukraine	882
831	Birang	PB	Indonesia	367
831	Chula	AORL	Thailand	844
831	Kallanpää	YTM	Finland	244
831	Kangwon	AGOR	Korea, South	485
F 831	Van Amstel	FFGHM	Netherlands	563
U 831	Kovel	ATA/YTM	Ukraine	882
YTB 831	Dekanawida	YTB	US	974
832	Mulga	PB	Indonesia	367
832	Samui	YO	Thailand	844
833	Prong	YO	Thailand	844
833	R 257	FSGM	Russian Federation	700
A 833	Karel Doorman	AFSH	Netherlands	571
834	Proet	YO	Thailand	844
835	Gepard	SSN	Russian Federation	678
835	Samed	YO	Thailand	844
Y 835	Todendorf	YFRT	Germany	296
YTB 835	Skenandoa	YTB	US	974
836	Matra	AOL	Thailand	843
A 836	Amsterdam	AORH	Netherlands	570
L 836	Ranavijaya	LCM	Sri Lanka	790
Y 836	Putlos	YFRT	Germany	296
YTB 836	Pokagon	YTB	US	974
Y 837	Baumholder	YFRT	Germany	296
838	Yuri Dolgoruky	SSBN	Russian Federation	671
839	Karelia	SSBN	Russian Federation	672
839	Liuyang	MCMV	China	157
L 839	Ranagaja	LCM	Sri Lanka	790
Y 839	Munster	YFRT	Germany	296
840	Luxi	MCMV	China	157
840	P 191	PB/YTD	Russian Federation	704
P 840	Holland	PSO	Netherlands	566
841	Badau	PB	Indonesia	367
841	Chuang	YW	Thailand	844
841	P 349	PB/YTD	Russian Federation	704
841	Xiaoyi	MCMV	China	157
P 841	Zeeland	PSO	Netherlands	566
842	Chik	YO	Thailand	844
842	P 350	PB/YTD	Russian Federation	704
842	Salawaku	PB	Indonesia	367
842	Taishan	MCMV	China	157
P 842	Friesland	PSO	Netherlands	566
Y 842	Dock B	–	Germany	297
843	Changshou	MCMV	China	157
P 843	Bocas Del Toro	PB	Panama	613
P 843	Groningen	PSO	Netherlands	566
844	Heshan	MCMV	China	157
845	Qingzhou	MCMV	China	157
847	Orel	SSGN	Russian Federation	676
847	Sibarau	PB	Indonesia	367
PG 847	Leopoldo Regis	PBF	Philippines	632
848	Siliman	PB	Indonesia	367
849	Novomoskovsk	SSBN	Russian Federation	672
849	Pari	PB	Indonesia	368
850	Sembilang	PB	Indonesia	368
A 850	Soemba	YDT	Netherlands	572
851	Beijixing	AGM/AGI	China	158
851	Klueng Badaan	YTL	Thailand	845
A 851	Cerberus	YDT	Netherlands	571
GC 851	Utatlan	PB	Guatemala	316
MHV 851	Sabotøren	PB	Denmark	198
PG 851	Apollo Tiano	PBF	Philippines	632
852	Kuznetsk	FSGM	Russian Federation	700
852	Marn Vichai	YTL	Thailand	845
A 852	Argus	YDT	Netherlands	571
GC 852	Subteniente Osorio Saravia	PB	Guatemala	316
U 852	Shostka	ABU	Ukraine	881
853	Rin	YTB	Thailand	844
853	Tianwangxing	AGM/AGI	China	158
853	Tigr	SSN	Russian Federation	678
A 853	Nautilus	YDT	Netherlands	571
PG 853	Sulpicio Fernandez	PBF	Philippines	632
U 853	Shulyavka	YDT/YFL/YPT	Ukraine	881
854	Rang	YTB	Thailand	844
A 854	Hydra	YDT	Netherlands	571
855	Kontradmiral Vlasov	MSOM	Russian Federation	707
855	Samaesan	YTR	Thailand	845
855	Zarechny	FSGM	Russian Federation	700
856	Raet	YTR	Thailand	845
857	Sigalu	PB	Indonesia	367
M 857	Makkum	MHC	Netherlands	569
858	Silea	PB	Indonesia	367
859	Siribua	PB	Indonesia	367
M 860	Schiedam	MHC	Netherlands	569
861	Changxingdao	ASRH	China	162
M 861	Urk	MHC	Netherlands	569
Y 861	Kronsort	AG	Germany	294
862	Chongmingdao	ASRH	China	162
862	Ryazan	SSBN	Russian Federation	673
862	Siada	PB	Indonesia	367
M 862	Zierikzee	MHC	Netherlands	569
Y 862	Helmsand	AG	Germany	294
863	Sikuda	PB	Indonesia	367
863	Yongxingdao	ASRH	China	162
M 863	Vlaardingen	MHC	Netherlands	569
Y 863	Stollergrund	AG	Germany	294
864	Haiyangdao	ASR	China	164
864	Sigurot	PB	Indonesia	367
M 864	Willemstad	MHC	Netherlands	569
Y 864	Mittelgrund	AG	Germany	294
865	Liugongdao	ASR	China	164
866	Cucut	PB	Indonesia	365
866	Daishandao	AHH	China	163
Y 866	Breitgrund	AG	Germany	294
867	Changdao	ASR	China	164
867	Kobra	PB	Indonesia	366
867	Volk	SSN	Russian Federation	678
868	Anakonda	PB	Indonesia	366
869	Patola	PB	Indonesia	366
870	R 2	FSGM	Russian Federation	700
870	Taliwangsa	PB	Indonesia	366
871	Kapitan Pattimura	FS	Indonesia	362
871	Similan	AORH	Thailand	844
872	Untung Suropati	FS	Indonesia	362
872	Zhu Kezhen	AGS	China	159
872	Leopard	SSN	Russian Federation	678
873	Qian Sanqiang	AGS	China	159
873	Sultan Nuku	FS	Indonesia	362
874	Morshansk	FSGM	Russian Federation	700
A 874	Linge	YTM	Netherlands	572
875	Pyhäranta	MLI	Finland	241
A 875	Regge	YTM	Netherlands	572
Y 875	Hiev	–	Germany	297
876	Pansio	MLI	Finland	241
876	Sultan Thaha Syaifuddin	FS	Indonesia	362
876	Victoria	SSK	Canada	98
A 876	Hunze	YTM	Netherlands	572
Y 876	Griep	–	Germany	297
877	Kampela 3	LCU/AKSL	Finland	242
877	Sutanto	FS	Indonesia	362
877	Windsor	SSK	Canada	98
A 877	Rotte	YTM	Netherlands	572
878	Corner Brook	SSK	Canada	98
878	Pantera	SSN	Russian Federation	678
878	Sutedi Senoputra	FS	Indonesia	362
A 878	Gouwe	YTM	Netherlands	572
879	Chicoutimi	SSK	Canada	98
879	Wiratno	FS	Indonesia	362
L 880	Shakthi	LSM	Sri Lanka	790
881	Hongzhu	AORH	China	161
881	Tjiptadi	FS	Indonesia	362
882	Hasan Basri	FS	Indonesia	362
882	Poyang Hu	AORH	China	161
885	Qinghai Hu	AORH	China	161
886	Qiandao Hu	AORH	China	161
887	Weishan Hu	AORH	China	161
889	P 104	PB/YTD	Russian Federation	704
889	Tai Hu	AORH	China	161
890	Chao Hu	AORH	China	161
890	Vepr	SSN	Russian Federation	678
891	Bi Sheng	AGOR/AGE	China	158
AG 891	Corregidor	ABU	Philippines	634
U 891	Kherson	YDT/YFL/YPT	Ukraine	881
Y 891	Altmark	APB	Germany	297
892	Hua Luogeng	AGOR/AGE	China	158
893	Qianlao	AGOR/AGE	China	158
Y 895	Wische	APB	Germany	297
899	Halli	YPC	Finland	244
900	Beidiao	AGI	China	158
A 900	Mercuur	ASL/YTT	Netherlands	571

IHS Jane's Fighting Ships 2014-2015

© 2014 IHS

Pennant list [95]

Number	Ship's name	Type	Country	Page
L 900	Shah Amanat	LSL	Bangladesh	58
901	A Zheleznyakov	MHSO	Russian Federation	708
901	Balikpapan	AOTL	Indonesia	373
901	Mourad Rais	FFLM	Algeria	4
901	Pengaman 1	PB	Malaysia	518
901	Sharm El Sheikh	FFGHM	Egypt	218
901	Sriyanont	PB	Thailand	846
901	Tareq	SSK	Iran	379
CP 901	Saettia	SAR	Italy	424
L 901	Shah Poran	LCU	Bangladesh	58
MHV 901	Enø	PB	Denmark	199
P 901	Castor	PBO	Belgium	64
WMEC 901	Bear	PSOH/WMEC	US	985
902	Boraida	AORH	Saudi Arabia	741
902	Noor	SSK	Iran	379
902	Rais Kellich	FFLM	Algeria	4
902	Sambu	AOTL	Indonesia	373
902	Tomsk	SSGN	Russian Federation	676
A 902	Van Kinsbergen	AXL	Netherlands	570
L 902	Shah Makhdum	LCU	Bangladesh	58
MHV 902	Manø	PB	Denmark	199
P 902	Pollux	PBO	Belgium	64
WMEC 902	Tampa	PSOH/WMEC	US	985
903	Arun	AORLH	Indonesia	373
903	Rais Korfou	FFLM	Algeria	4
903	Yunes	SSK	Iran	379
MHV 903	Hjortø	PB	Denmark	199
WMEC 903	Harriet Lane	PSOH/WMEC	US	985
904	Yunbou	AORH	Saudi Arabia	741
CP 904	Michele Fiorillo	SAR	Italy	424
MHV 904	Lyø	PB	Denmark	199
WMEC 904	Northland	PSOH/WMEC	US	985
CP 905	Antonio Peluso	SAR	Italy	424
MHV 905	Askø	PB	Denmark	199
WMEC 905	Spencer	PSOH/WMEC	US	985
906	Toushka	FFGHM	Egypt	218
CP 906	Orazio Corsi	SAR	Italy	424
MHV 906	Faenø	PB	Denmark	199
WMEC 906	Seneca	PSOH/WMEC	US	985
MHV 907	Hvidsten	PB	Denmark	199
WMEC 907	Escanaba	PSOH/WMEC	US	985
908	Vitse-Admiral Zakharin	MSOM	Russian Federation	707
908	Yandanshang	LSTH	China	154
MHV 908	Brigaden	PB	Denmark	199
WMEC 908	Tahoma	PSOH/WMEC	US	985
909	Jiuhuashan	LSTH	China	154
909	Vitseadmiral Zhukov	MSOM	Russian Federation	707
MHV 909	Speditøren	PB	Denmark	199
WMEC 909	Campbell	PSOH/WMEC	US	985
910	Huanggangshan	LSTH	China	154
MHV 910	Ringen	PB	Denmark	199
WMEC 910	Thetis	PSOH/WMEC	US	985
911	Alexandria	FFGHM	Egypt	218
911	Chakri Naruebet	CVM	Thailand	828
911	Dongxiu	ASL	China	165
911	Ivan Golubets	MSOM	Russian Federation	707
911	Sorong	AOTL	Indonesia	373
911	Tianzhushan	LSTH	China	154
MHV 911	Bopa	PB	Denmark	199
P 911	Madhumati	PSO	Bangladesh	55
WMEC 911	Forward	PSOH/WMEC	US	985
912	Daqingshan	LSTH	China	154
912	Turbinist	MSOM	Russian Federation	707
MHV 912	Holger Danske	PB	Denmark	199
P 912	Kapatakhaya	PBO/AX	Bangladesh	55
WMEC 912	Legare	PSOH/WMEC	US	985
913	Baxianshan	LSTH	China	154
913	Kovrovets	MSOM	Russian Federation	707
P 913	Karatoa	PBO/AX	Bangladesh	55
WMEC 913	Mohawk	PSOH/WMEC	US	985
P 914	Gomati	PBO/AX	Bangladesh	55
915	Podolsk	SSBN	Russian Federation	673
M 915	Aster	MHC/AEL	Belgium	64
916	R 29	FSGM	Russian Federation	700
916	Taba	FFGHM	Egypt	218
M 916	Bellis	MHC/AEL	Belgium	64
M 917	Crocus	MHC/AEL	Belgium	64
920	Tver	SSGN	Russian Federation	676
CP 920	Gregoretti	PSO	Italy	424
921	El Fateh	AXT	Egypt	225
921	R 20	FSGM	Russian Federation	700
M 921	Lobelia	MHC/AEL	Belgium	64
SB 921	Paradoks	ATS	Russian Federation	724
SB 922	Shakhter	ATS	Russian Federation	724
923	Soputan	ATF	Indonesia	374
M 923	Narcis	MHC/AEL	Belgium	64
924	Leuser	ATF	Indonesia	374
924	R 14	FSGM	Russian Federation	700
M 924	Primula	MHC/AEL	Belgium	64
927	Yuntaishan	LST	China	155
928	Wufengshan	LST	China	155
929	Zijinshan	LST	China	155
930	Lingyanshan	LST	China	155
930	P 351	PB/YTD	Russian Federation	704
F 930	Leopold I	FFGHM	Belgium	63
931	Burujulasad	AGORH	Indonesia	372
931	Dongtingshan	LST	China	155
F 931	Louise-Marie	FFGHM	Belgium	63
932	Chin Yang	FFGH	Taiwan	815
932	Dewa Kembar	AGSH	Indonesia	372
932	Helanshan	LST	China	155
933	Fong Yang	FFGH	Taiwan	815
933	Jalanidhi	AGOR	Indonesia	372
933	Liupanshan	LST	China	155
934	Danxiashan	LSTH	China	154
934	Feng Yang	FFGH	Taiwan	815
934	Lampo Batang	YTM	Indonesia	374
935	Lan Yang	FFGH	Taiwan	815
935	Tambora	YTM	Indonesia	374
935	Xuefengshan	LSTH	China	154
936	Bromo	YTM	Indonesia	374
936	Hae Yang	FFGH	Taiwan	815
936	Haiyangshan	LSTH	China	154
937	Hwai Yang	FFGH	Taiwan	815
937	Qingchengshan	LSTH	China	154
937	R 18	FSGM	Russian Federation	700
937	Soummam	AXH	Algeria	7
938	Ning Yang	FFGH	Taiwan	815
939	Putuoshan	LSTH	China	154
939	Yi Yang	FFGH	Taiwan	815
940	R 11	FSGM	Russian Federation	700
940	Tiantaishan	LSTH	China	154
CP 940	Luigi Dattilo	PSO	Italy	424
941	Shengshan	LSM	China	155
CP 941	Ubaldo Diciotti	PSO	Italy	424
F 941	Abu Qir	FFGM	Egypt	219
942	Lushan	LSM	China	155
943	Yushan	LSM	China	155
944	Mengshan	LSM	China	155
945	Huashan	LSM	China	155
946	R 24	FSGM	Russian Federation	700
946	Songshan	LSM	China	155
F 946	El Suez	FFGM	Egypt	219
947	Blåtunga	LCPFM	Sweden	801
947	Omsk	SSGN	Russian Federation	676
U 947	Krasnoperekopsk	ATA/YTM	Ukraine	882
948	Xueshan	LSM	China	155
949	Hengshan	LSM	China	155
950	Taishan	LSM	China	155
A 950	Valcke	YTM	Belgium	66
951	Kuzbass	SSN	Russian Federation	678
951	Najim Al Zaffer	FFG	Egypt	220
951	R 297	FSGM	Russian Federation	700
951	Ulsan	FFG	Korea, South	478
952	R 109	FSGM	Russian Federation	700
952	Seoul	FFG	Korea, South	478
A 952	Wesp	YTL	Belgium	65
953	Chung Nam	FFG	Korea, South	478
953	Groza	FSGM	Russian Federation	700
U 953	Dubno	ATA/YTM	Ukraine	882
954	Ivanovets	FSGM	Russian Federation	700
A 954	Zeemeeuw	YTL	Belgium	65
955	Burya	FSGM	Russian Federation	700
955	Masan	FFG	Korea, South	478
A 955	Mier	YTL	Belgium	65
956	El Nasser	FFG	Egypt	220
956	Kyong Buk	FFG	Korea, South	478
957	Chon Nam	FFG	Korea, South	478
958	Che Ju	FFG	Korea, South	478
A 958	Zenobe Gramme	AXS	Belgium	65
959	Pusan	FFG	Korea, South	478
960	Karimata	AKL	Indonesia	374
A 960	Godetia	AGFH	Belgium	65
P 960	Skjold	PTGMF	Norway	588
961	Chung Ju	FFG	Korea, South	478
961	Damyat	FFGH	Egypt	219
961	Wagio	AKL	Indonesia	374
P 961	Storm	PTGMF	Norway	588
962	R 71	FSGM	Russian Federation	700
A 962	Belgica	AGOR/PBO	Belgium	65
P 962	Skudd	PTGMF	Norway	588
A 963	Stern	AGFH	Belgium	65
P 963	Steil	PTGMF	Norway	588
P 964	Glimt	PTGMF	Norway	588
P 965	Gnist	PTGMF	Norway	588
966	Rasheed	FFGH	Egypt	219
HQ 966	Truong	AKL	Vietnam	1015
970	Samara	SSN	Russian Federation	678
971	Kwanggaeto Daewang	DDGHM	Korea, South	477
971	R 298	FSGM	Russian Federation	700
971	Tanjung Kambani	AP	Indonesia	370
972	Euljimundok	DDGHM	Korea, South	477
973	Tanjung Nusanive	AP	Indonesia	368
973	Yangmanchun	DDGHM	Korea, South	477
974	Tanjung Fatagar	AP	Indonesia	368
975	Chungmugong Yi Sun-Shin	DDGHM	Korea, South	476
976	Moonmu Daewang	DDGHM	Korea, South	476
977	Daejoyoung	DDGHM	Korea, South	476
978	R 19	FSGM	Russian Federation	700
978	Wang Geon	DDGHM	Korea, South	476
979	Gang Gam Chan	DDGHM	Korea, South	476
981	Choi Young	DDGHM	Korea, South	476
981	Karang Pilang	AP	Indonesia	369
982	Karang Tekok	AP	Indonesia	369
983	Karang Banteng	AP	Indonesia	369
985	Karang Unarang	AP	Indonesia	369
985	Kashalot	SSN	Russian Federation	678
989	Changbaishan	LPD	China	153
990	Bratsk	SSN	Russian Federation	678
990	Dr Soeharso	LPD/APCR	Indonesia	368
990	Wudangshan	LSM	China	154
991	Emeishan	LSTH	China	154
991	Sejong Daewang	DDGHM	Korea, South	475
992	Huadingshan	LSTH	China	154
992	Träskö	YFB	Finland	243
992	Yulgok Yi I	DDGHM	Korea, South	475
993	Luoxiaoshan	LSTH	China	154
993	Syvatoy Giorgiy Pobedonosets	SSBN	Russian Federation	673
993	Yu Seong-Ryong	DDGHM	Korea, South	475
994	Daiyunshan	LSTH	China	154
995	Wanyang-Shan	LSTH	China	154
996	Laotieshan	LSTH	China	154
996	P 377	PB/YTD	Russian Federation	704
996	R 79	FSGM	Russian Federation	700
A 996	Albatros	YTM	Belgium	66
997	Magadan	SSN	Russian Federation	678
997	Yunwashan	LSTH	China	154
998	Kunlunshan	LPD	China	153

© 2014 IHS

IHS Jane's Fighting Ships 2014-2015

[96] Pennant list

Number	Ship's name	Type	Country	Page
999	Jinggangshan	LPD	China	153
999	Louhi		Finland	242
DDG 1000	Zumwalt	DDGH	US	950
DDG 1001	Michael Monsoor	DDGH	US	950
T-AKR 1001	Adm W M H Callaghan	AKR	US	982
DDG 1002	Lyndon B Johnson	DDGH	US	950
PC 1005	Han Kang	PG	Korea, South	487
1006	Fantome	YGS	Australia	34
PC 1006	Sumjinkang	PSO	Korea, South	489
1007	Meda	YGS	Australia	34
1008	Duyfken	YGS	Australia	34
1009	Tom Thumb	YGS	Australia	34
1010	John Gowlland	YGS	Australia	34
1011	Casuarina	YGS	Australia	34
P 1011	Titas	PBF	Bangladesh	56
1012	Geographe	YGS	Australia	34
P 1012	Kusiyara	PBF	Bangladesh	56
1013	Azmat	PTG	Pakistan	608
P 1013	Chitra	PBF	Bangladesh	56
P 1014	Dhansiri	PBF	Bangladesh	56
1021	Conder	YGS	Australia	34
1023	Jurrat	PGG	Pakistan	607
1026	Essequibo	PBO	Guyana	318
1028	Quwwat	PGG	Pakistan	607
1029	Jalalat	PTG	Pakistan	606
1030	Shujaat	PTG	Pakistan	606
AB 1050	Coconut Queen	–	Australia	37
AB 1051	Seahorse 2	–	Australia	37
D 1051	Al Gaffa	YDT	UAE	890
FNH 1051	Guaymuras	PB	Honduras	319
GC 1051	Kukulkán	PB	Guatemala	316
FNH 1052	Hibueras	PB	Honduras	319
AB 1053	Sea Widow	–	Australia	37
FNH 1053	Honduras	PB	Honduras	319
AB 1056	Charlie Brown	–	Australia	37
AB 1055	Midnight Sun	–	Australia	37
AB 1058	Anzac Cove	–	Australia	37
M 1058	Fulda	MHC	Germany	293
AB 1059	Snoopy	–	Australia	37
M 1059	Weilheim	MHC	Germany	293
1060	Barkat	PBO	Pakistan	611
AB 1060	The Wanderer	–	Australia	37
1061	Rehmat	PBO	Pakistan	611
AB 1061	Northern Warrior	–	Australia	37
M 1061	Rottweil	MCD	Germany	293
1062	Nusrat	PBO	Pakistan	611
AB 1062	Ocean Marie	–	Australia	37
M 1062	Sulzbach-Rosenberg	MHC	Germany	293
1063	Vehdat	PBO	Pakistan	611
AB 1063	Ice Fire	–	Australia	37
M 1063	Bad Bevensen	MHC	Germany	293
AB 1064	Sea Trojan	–	Australia	37
M 1064	Grömitz	MHC	Germany	293
AB 1065	Viking Sun	–	Australia	37
M 1065	Dillingen	MHC	Germany	293
AB 1066	Hastings	–	Australia	37
P 1066	Sabqat	PB	Pakistan	611
AB 1067	Charisma	–	Australia	37
M 1067	Bad Rappenau	MCD	Germany	293
M 1068	Datteln	MHC	Germany	293
P 1068	Rafaqat	PB	Pakistan	611
M 1069	Homburg	MHC	Germany	293
P 1069	Sadaqat	PB	Pakistan	611
FNH 1071	Tegucigalpa	PB	Honduras	320
M 1090	Pegnitz	MHCD	Germany	294
M 1092	Hameln	MHCD	Germany	294
M 1093	Auerbach	MHCD	Germany	294
M 1094	Ensdorf	MHCD	Germany	294
M 1095	Überherrn	MHC	Germany	293
M 1098	Siegburg	MHCD	Germany	294
M 1099	Herten	MHC	Germany	293
1101	Chasanyabadee	PB	Thailand	845
1101	Cheng Kung	FFGHM	Taiwan	813
PI 1101	Polaris	PBF	Mexico	535
WPC 1101	Bernard C Webber	PBO/WPC	US	987
1102	Chawengsak Songkram	PB	Thailand	845
PI 1102	Sirius	PBF	Mexico	535
WPC 1102	Richard Etheridge	PBO/WPC	US	987
1103	Cheng Ho	FFGHM	Taiwan	813
1103	Phromyothee	PB	Thailand	845
PI 1103	Capella	PBF	Mexico	535
WPC 1103	William Flores	PBO/WPC	US	987
PI 1104	Canopus	PBF	Mexico	535
WPC 1104	Robert Yered	PBO/WPC	US	987
1105	Chi Kuang	FFGHM	Taiwan	813
1105	Kaoh Chhlam	PBR	Cambodia	94
MAI 1105	Stefan Cel Mare	PSO	Romania	668
PI 1105	Vega	PBF	Mexico	535
WPC 1105	Margaret Norvell	PBO/WPC	US	987
1106	Kaoh Rong	PBR	Cambodia	94
1106	Yueh Fei	FFGHM	Taiwan	813
PI 1106	Achernar	PBF	Mexico	535
WPC 1106	Paul Clark	PBO/WPC	US	987
1107	Tzu-i	FFGHM	Taiwan	813
PI 1107	Rigel	PBF	Mexico	535
WPC 1107	Charles David	PBO/WPC	US	987
1108	Pan Chao	FFGHM	Taiwan	813
PI 1108	Arcturus	PBF	Mexico	535
WPC 1108	Charles Sexton	PBO/WPC	US	987
1109	Chang Chien	FFGHM	Taiwan	813
PI 1109	Alpheratz	PBF	Mexico	535
WPC 1109	Kathleen Moore	PBO/WPC	US	987
1110	Tien Tan	FFGHM	Taiwan	813
PI 1110	Procyón	PBF	Mexico	535
WPC 1110	Raymond Evans	PBO/WPC	US	987
PI 1111	Avior	PBF	Mexico	535
WPC 1111	William Trump	PBO/WPC	US	987
PI 1112	Deneb	PBF	Mexico	535
WPC 1112	Isaac Mayo	PBO/WPC	US	987
PI 1113	Fomalhaut	PBF	Mexico	535

Number	Ship's name	Type	Country	Page
WPC 1113	Richard Dixon	PBO/WPC	US	987
PI 1114	Pollux	PBF	Mexico	535
WPC 1114	Heriberto Hernandez	PBO/WPC	US	987
PI 1115	Régulus	PBF	Mexico	535
PI 1116	Acrux	PBF	Mexico	535
PI 1117	Spica	PBF	Mexico	535
PI 1118	Hadar	PBF	Mexico	535
PI 1119	Shaula	PBF	Mexico	535
PI 1120	Mirfak	PBF	Mexico	535
PI 1121	Ankaa	PBF	Mexico	535
PI 1122	Bellatrix	PBF	Mexico	535
PI 1123	Elnath	PBF	Mexico	535
PI 1124	Alnilán	PBF	Mexico	535
PI 1125	Peacock	PBF	Mexico	535
T-AOT 1125	Lawrence H Gianella	AOT	US	981
PI 1126	Betelgeuse	PBF	Mexico	535
PI 1127	Adhara	PBF	Mexico	535
PI 1128	Alioth	PBF	Mexico	535
PI 1129	Rasalhague	PBF	Mexico	535
PI 1130	Nunki	PBF	Mexico	535
1131	Mondolkiri	PBF	Cambodia	94
PI 1131	Hamal	PBF	Mexico	535
PI 1132	Suhail	PBF	Mexico	535
PI 1133	Dubhe	PBF	Mexico	535
1134	Ratanakiri	PBF	Cambodia	94
PI 1134	Denebola	PBF	Mexico	535
PI 1135	Alkaid	PBF	Mexico	535
PI 1136	Alphecca	PBF	Mexico	535
PI 1137	Eltanin	PBF	Mexico	535
PI 1138	Kochab	PBF	Mexico	535
PI 1139	Enif	PBF	Mexico	535
P 1140	Cacine	PBO	Portugal	652
PI 1140	Schedar	PBF	Mexico	535
PI 1141	Markab	PBF	Mexico	535
M 1142	Umzimkulu	MHC	South Africa	764
PI 1142	Megrez	PBF	Mexico	535
PI 1143	Mizar	PBF	Mexico	535
P 1144	Cuanza	PBO	Portugal	652
PI 1144	Phekda	PBF	Mexico	535
PI 1145	Acamar	PBF	Mexico	535
P 1146	Zaire	PBO	Portugal	652
PI 1146	Diphda	PBF	Mexico	535
PI 1147	Menkar	PBF	Mexico	535
PI 1148	Sabik	PBF	Mexico	535
P 1150	Argos	PBR	Portugal	653
P 1151	Dragão	PBR	Portugal	653
P 1152	Escorpião	PBR	Portugal	653
P 1153	Cassiopeia	PBR	Portugal	653
P 1154	Hidra	PBR	Portugal	653
P 1155	Centauro	PBR	Portugal	653
P 1156	Orion	PBR	Portugal	653
P 1157	Pégaso	PBR	Portugal	653
P 1158	Sagitario	PBR	Portugal	653
1161	Maroub	PBR	Sudan	793
1162	Fijab	PBR	Sudan	793
1163	Salak	PBR	Sudan	793
1164	Halote	PBR	Sudan	793
P 1165	Águia	PBR	Portugal	653
P 1167	Cisne	PBR	Portugal	653
P 1200	Tuzla	PC	Turkey	862
1201	Baklan	HSIC	Yemen	1019
P 1201	Karaburun	PC	Turkey	862
1202	Kang Ding	FFGHM	Taiwan	814
1202	Siyan	HSIC	Yemen	1019
P 1202	Köyceğiz	PC	Turkey	862
1203	Si Ning	FFGHM	Taiwan	814
1203	Zuhrab	HSIC	Yemen	1019
P 1203	Kumkale	PC	Turkey	862
1204	Akissan	HSIC	Yemen	1019
P 1204	Tarsus	PC	Turkey	862
1205	Kun Ming	FFGHM	Taiwan	814
1205	Hunaish	HSIC	Yemen	1019
P 1205	Karabiga	PC	Turkey	862
1206	Di Hua	FFGHM	Taiwan	814
1206	Zakr	HSIC	Yemen	1019
P 1206	Karşiyaka	PC	Turkey	862
1207	Wu Chang	FFGHM	Taiwan	814
P 1207	Tekirdağ	PC	Turkey	862
1208	Chen Te	FFGHM	Taiwan	814
P 1208	Kaş	PC	Turkey	862
P 1209	Kilimli	PC	Turkey	862
P 1210	Türkeli	PC	Turkey	862
P 1211	Taşucu	PC	Turkey	862
M 1212	Umhloti	MHC	South Africa	764
P 1212	Karataş	PC	Turkey	862
P 1213	Karpaz	PC	Turkey	862
P 1214	Kozlu	PC	Turkey	862
P 1215	Kuşadasi	PC	Turkey	862
1280	Petir 80	PB	Malaysia	518
1281	Petir 81	PB	Malaysia	518
1282	Petir 82	PB	Malaysia	518
1283	Petir 83	PB	Malaysia	518
1285	Petir 85	PB	Malaysia	518
1301	Yung Feng	MHC	Taiwan	819
PI 1301	Acuario	PBF	Mexico	537
WPB 1301	Farallon	WPB	US	988
1302	Yung Chia	MHC	Taiwan	819
PI 1302	Aguila	PBF	Mexico	537
1303	Yung Ting	MHC	Taiwan	819
PI 1303	Aries	PBF	Mexico	537
PI 1304	Auriga	PBF	Mexico	537
WPB 1304	Maui	WPB	US	988
1305	Yung Shun	MHC	Taiwan	819
PI 1305	Cancer	PBF	Mexico	537
1306	Yung Yang	MSO	Taiwan	819
PI 1306	Capricorno	PBF	Mexico	537
1307	Yung Tzu	MSO	Taiwan	819
PI 1307	Centauro	PBF	Mexico	537
WPB 1307	Ocracoke	WPB	US	988
1308	Yung Ku	MSO	Taiwan	819

Pennant list [97]

Number	Ship's name	Type	Country	Page	Number	Ship's name	Type	Country	Page
PI 1308	Geminis	PBF	Mexico	537	1433	Pengawal 36	PB	Malaysia	520
1309	Yung Teh	MSO	Taiwan	819	1434	Pengawal 37	PB	Malaysia	520
WPB 1309	Aquidneck	WPB	US	988	1435	Pengawal 38	PB	Malaysia	520
1310	Yung Jin	MHC	Taiwan	819	1436	Pengawal 39	PB	Malaysia	520
WPB 1310	Mustang	WPB	US	988	1437	Pengawal 40	PB	Malaysia	520
1311	Pengawal 11	PB	Malaysia	518	A 1437	Planet	AGE	Germany	294
1311	Yung An	MHC	Taiwan	819	1438	Pengawal 41	PB	Malaysia	520
WPB 1311	Naushon	WPB	US	988	1439	Pengawal 42	PB	Malaysia	520
1312	Pengawal 12	PB	Malaysia	518	A 1439	Baltrum	ATS/YDT	Germany	298
LST 1312	Ambe	LST	Nigeria	582	1440	Pengawal 43	PB	Malaysia	520
WPB 1312	Sanibel	WPB	US	988	A 1440	Juist	ATS/YDT	Germany	298
1313	Pengawal 13	PB	Malaysia	520	1441	Pengawal 44	PB	Malaysia	520
WPB 1313	Edisto	WPB	US	988	1442	Pengawal 45	PB	Malaysia	520
1314	Pengawal 14	PB	Malaysia	520	A 1442	Spessart	AOL	Germany	295
WPB 1314	Sapelo	WPB	US	988	A 1443	Rhön	AOL	Germany	295
1315	Pengawal 15	PB	Malaysia	520	A 1451	Wangerooge	ATS/YDT	Germany	298
WPB 1315	Matinicus	WPB	US	988	A 1452	Spiekeroog	ATS/YDT	Germany	298
1316	Pengawal 16	PB	Malaysia	520	A 1456	Alliance	AGOR	Nato	561
WPB 1316	Nantucket	WPB	US	988	A 1458	Fehmarn	ATR	Germany	297
1317	Pengawal 17	PB	Malaysia	520	FNH 1491	Punta Caxinas	LCU	Honduras	321
1318	Pengawal 18	PB	Malaysia	520	M 1499	Umkomaas	MHC	South Africa	764
WPB 1318	Baranof	WPB	US	988	1501	Jaemin I	ARSH	Korea, South	488
1319	Pengawal 19	PB	Malaysia	520	1501	Pengawal 46	PB	Malaysia	520
WPB 1319	Chandeleur	WPB	US	988	1502	Jaemin II	ARS	Korea, South	488
1320	Pengawal 20	PB	Malaysia	520	1502	Pengawal 47	PB	Malaysia	520
WPB 1320	Chincoteague	WPB	US	988	1503	Jaemin III	ARSH	Korea, South	489
1321	Pengawal 21	PB	Malaysia	520	1503	Pengawal 48	PB	Malaysia	520
WPB 1321	Cushing	WPB	US	988	1503	Sri Indera Sakti	AOR/AE/AXH	Malaysia	513
1322	Pengawal 22	PB	Malaysia	520	1504	Mahawangsa	AOR/AE/AXH	Malaysia	513
WPB 1322	Cuttyhunk	WPB	US	988	1504	Pengawal 49	PB	Malaysia	520
WPB 1323	Drummond	WPB	US	988	1505	Jaemin V	ARSH	Korea, South	489
WPB 1324	Key Largo	WPB	US	988	1506	Jaemin VI	ARSH	Korea, South	489
WPB 1326	Monomoy	WPB	US	988	1507	Jaemin VII	ARSH	Korea, South	489
WPB 1327	Orcas	WPB	US	988	1508	Jaemin VIII	ARSH	Korea, South	489
WPB 1329	Sitkinak	WPB	US	988	1509	Jaemin IX	ARSH	Korea, South	489
WPB 1330	Tybee	WPB	US	988	1510	Jaemin X	ARSH	Korea, South	489
WPB 1331	Washington	WPB	US	988	1511	Jaemin XI	ARSH	Korea, South	489
WPB 1332	Wrangell	WPB	US	988	1512	Jaemin XII	ARSH	Korea, South	489
WPB 1333	Adak	WPB	US	988	1513	Jaemin XIII	ARSH	Korea, South	489
WPB 1334	Liberty	WPB	US	988	A 1531	E 1	AXL	Turkey	866
WPB 1335	Anacapa	WPB	US	988	A 1532	E 2	AXL	Turkey	866
WPB 1336	Kiska	WPB	US	988	A 1533	E 3	AXL	Turkey	866
WPB 1337	Assateague	WPB	US	988	A 1534	E 4	AXL	Turkey	866
WPB 1338	Grand Isle	WPB	US	988	A 1535	E 5	AXL	Turkey	866
WPB 1339	Key Biscayne	WPB	US	988	A 1536	E 6	AXL	Turkey	866
WPB 1340	Jefferson Island	WPB	US	988	A 1537	E 7	AXL	Turkey	866
WPB 1341	Kodiak Island	WPB	US	988	A 1538	E 8	AXL	Turkey	866
WPB 1342	Long Island	WPB	US	988	A 1542	Sönderen 2	YTB/YTM/YTL	Turkey	870
WPB 1343	Bainbridge Island	WPB	US	988	A 1543	Sönderen 3	YTB/YTM/YTL	Turkey	870
WPB 1344	Block Island	WPB	US	988	A 1544	Sönderen 4	YTB/YTM/YTL	Turkey	870
WPB 1345	Staten Island	WPB	US	988	P 1552	Tobie	PB	South Africa	763
WPB 1346	Roanoke Island	WPB	US	988	P 1553	Tern	PB	South Africa	763
WPB 1347	Pea Island	WPB	US	988	P 1554	Tekwane	PB	South Africa	763
WPB 1348	Knight Island	WPB	US	988	P 1563	Adam Kok	PG	South Africa	763
WPB 1349	Galveston Island	WPB	US	988	P 1565	Isaac Dyobha	PG	South Africa	763
AM 1353	Coral Snake	–	Australia	38	P 1567	Galeshewe	PG	South Africa	763
1401	Hendijan	PBO	Iran	390	P 1569	Makhanda	PG	South Africa	763
FNH 1401	Lempira	PB	Honduras	321	1571	Penyelamat 1	PB	Malaysia	518
PI 1401	Miaplacidus	PBF	Mexico	536	1572	Penyelamat 2	PB	Malaysia	518
1402	Sirik	PBO	Iran	390	1573	Penyelamat 3	PB	Malaysia	518
FNH 1402	Morazan	PB	Honduras	321	1574	Penyelamat 4	PB	Malaysia	518
PI 1402	Algol	PBF	Mexico	536	A 1600	Iskenderun	AP	Turkey	866
1403	Konarak	PBO	Iran	390	1601	Ta Kuan	AGOR	Taiwan	820
PI 1403	Beaver	PBF	Mexico	536	LSG 1603	Alacalufe	WPB	Chile	126
1404	Gavatar	PBO	Iran	390	LSG 1604	Hallef	WPB	Chile	126
PI 1404	Merak	PBF	Mexico	536	LSG 1609	Aysén	WPB	Chile	126
1405	Mooam	PBO	Iran	390	LSG 1610	Corral	WPB	Chile	126
PI 1405	Caph	PBF	Mexico	536	LSG 1611	Conceptión	WPB	Chile	126
1406	Bahregan	PBO	Iran	390	LSG 1612	Caldera	WPB	Chile	126
PI 1406	Mirach	PBF	Mexico	536	LSG 1613	San Antonio	WPB	Chile	126
1407	Kalat	PBO	Iran	390	LSG 1614	Antofagasta	WPB	Chile	126
PI 1407	Alhena	PBF	Mexico	536	LSG 1615	Arica	WPB	Chile	126
1408	Genaveh	PBO	Iran	390	LSG 1616	Coquimbo	WPB	Chile	126
PI 1408	Saiph	PBF	Mexico	536	LSG 1617	Puerto Natales	WPB	Chile	126
1409	Rostani	PBO	Iran	390	LSG 1618	Valparaíso	WPB	Chile	126
A 1409	Wilhelm Pullwer	YAG	Germany	294	LSG 1619	Punta Arenas	WPB	Chile	126
PI 1409	Algorab	PBF	Mexico	536	LSG 1620	Talcahuano	WPB	Chile	126
1410	Nayband	PBO	Iran	390	LSG 1621	Quintero	WPB	Chile	126
PI 1410	Albireo	PBF	Mexico	536	LSG 1622	Chiloé	WPB	Chile	126
1411	Pengawal 1	PB	Malaysia	518	LSG 1623	Puerto Montt	WPB	Chile	126
A 1411	Berlin	AFSH	Germany	295	LSG 1624	Iquique	WPB	Chile	126
PI 1411	Alnitak	PBF	Mexico	536	LSG 1625	Ona	PB	Chile	126
1412	Pengawal 2	PB	Malaysia	518	1630	Penggalang 30	PB	Malaysia	520
A 1412	Frankfurt Am Main	AFSH	Germany	295	1631	Penggalang 31	PB	Malaysia	520
PI 1412	Mintaka	PBF	Mexico	536	1632	Penggalang 32	PB	Malaysia	520
1413	Pengawal 3	PB	Malaysia	518	1633	Penggalang 33	PB	Malaysia	520
A 1413	Bonn	AFSH	Germany	295	1634	Penggalang 34	PB	Malaysia	520
PI 1413	Alfirk	PBF	Mexico	536	1635	Penggalang 35	PB	Malaysia	520
1414	Pengawal 4	PB	Malaysia	518	1636	Penggalang 36	PB	Malaysia	520
PI 1414	Alderamin	PBF	Mexico	536	1637	Penggalang 37	PB	Malaysia	520
1415	Pengawal 5	PB	Malaysia	518	1638	Penggalang 38	PB	Malaysia	520
PI 1415	Menkalinan	PBF	Mexico	536	1639	Penggalang 39	PB	Malaysia	520
1416	Pengawal 6	PB	Malaysia	518	1640	Penggalang 40	PB	Malaysia	520
1417	Pengawal 7	PB	Malaysia	518	1641	Penggalang 41	PB	Malaysia	520
1418	Pengawal 8	PB	Malaysia	518	1642	Penggalang 42	PB	Malaysia	520
1420	Pengawal 23	PB	Malaysia	520	1643	Penggalang 43	PB	Malaysia	520
1421	Pengawal 24	PB	Malaysia	520	Y 1643	Bottsand	YPC	Germany	297
1422	Pengawal 25	PB	Malaysia	520	1644	Penggalang 44	PB	Malaysia	520
1423	Pengawal 26	PB	Malaysia	520	Y 1644	Eversand	YPC	Germany	297
1424	Pengawal 27	PB	Malaysia	520	1645	Penggalang 46	PB	Malaysia	520
1425	Pengawal 28	PB	Malaysia	520	Y 1659	Warnow	YFL	Germany	296
A 1425	Ammersee	AOL	Germany	296	Y 1671	AK 1	YFL	Germany	296
1426	Pengawal 29	PB	Malaysia	520	Y 1675	AM 8	YFL	Germany	296
A 1426	Tegernsee	AOL	Germany	296	Y 1676	MA 2	YFL	Germany	296
1427	Pengawal 30	PB	Malaysia	520	Y 1677	MA 3	YFL	Germany	296
1428	Pengawal 31	PB	Malaysia	520	Y 1678	MA 1	YFL	Germany	296
1429	Pengawal 32	PB	Malaysia	520	Y 1679	AM 7	YFL	Germany	296
1430	Pengawal 33	PB	Nigeria	520	Y 1683	AK 6	YFL	Germany	296
1431	Pengawal 34	PB	Malaysia	520	Y 1685	Aschau	YFL	Germany	296
1432	Pengawal 35	PB	Malaysia	520	Y 1686	AK 2	YFL	Germany	296

© 2014 IHS

IHS Jane's Fighting Ships 2014-2015

[98] Pennant list

Number	Ship's name	Type	Country	Page
Y 1687	Borby	YFL	Germany	296
Y 1689	Bums	YAG	Germany	295
1701	Peninjau 1	PB	Malaysia	518
LSR 1703	Pelluhue	WPB	Chile	125
LSR 1704	Arauco	WPB	Chile	125
LSR 1705	Chacao	WPB	Chile	125
LSR 1706	Queitao	WPB	Chile	125
LSR 1707	Guaiteca	WPB	Chile	125
LSR 1708	Curaumila	WPB	Chile	125
1801	Keelung	DDGHM	Taiwan	812
1801	Penggalang 1	PB	Malaysia	518
DT 1801	Quokka	YTM	Australia	37
1802	Damrong Rachanuphap	PBO	Thailand	845
1802	Penggalang 2	PB	Malaysia	518
1802	Suao	DDGHM	Taiwan	812
1803	Lopburi Rames	PBO	Thailand	845
1803	Tsoying	DDGHM	Taiwan	812
1804	Srinakarin	PSO	Thailand	845
1805	Makung	DDGHM	Taiwan	812
1810	Penggalang 10	PBF	Malaysia	519
1811	Penggalang 11	PBF	Malaysia	519
1812	Penggalang 12	PBF	Malaysia	519
1813	Penggalang 13	PBF	Malaysia	519
1814	Penggalang 14	PBF	Malaysia	519
LPC 1814	Diaz	PB	Chile	125
1815	Penggalang 15	PBF	Malaysia	519
1816	Penggalang 16	PBF	Malaysia	519
LPC 1816	Salinas	PB	Chile	125
1817	Penggalang 17	PBF	Malaysia	519
1818	Penggalang 18	PBF	Malaysia	519
1819	Penggalang 19	PBF	Malaysia	519
1820	Penggalang 20	PBF	Malaysia	519
LPC 1820	Machado	PB	Chile	125
1821	Penggalang 21	PBF	Malaysia	519
1822	Penggalang 22	PBF	Malaysia	519
1823	Penggalang 23	PBF	Malaysia	519
LPC 1823	Hudson	PB	Chile	125
1824	Penggalang 24	PBF	Malaysia	519
1825	Penggalang 25	PBF	Malaysia	519
1826	Penggalang 26	PBF	Malaysia	519
1827	Penggalang 27	PBF	Malaysia	519
1958	Geofjord	AGS	Norway	590
1978	Oljevern 01	AGS	Norway	590
1978	Oljevern 02	AGS	Norway	590
1978	Oljevern 03	AGS	Norway	590
1978	Oljevern 04	AGS	Norway	590
1992	Shun Hu 1	WPSO	Taiwan	822
1992	Shun Hu 5	WPB	Taiwan	823
1992	Shun Hu 6	WPB	Taiwan	823
2001	Seal	YDT/PB	Australia	36
LCU 2001	Runnymede	LCU-ARMY	US	968
LCU 2001	Yusoutei-Ichi-Gou	LCU	Japan	444
LCU 2002	Kennesaw Mountain	LCU-ARMY	US	968
LCU 2002	Yusoutei-Ni-Gou	LCU	Japan	444
2003	Malu Baizam	YDT/PB	Australia	36
LCU 2003	Macon	LCU-ARMY	US	968
2004	Shark	YDT/PB	Australia	36
LCU 2004	Aldie	LCU-ARMY	US	968
2005	Penyelamat 5	PB	Malaysia	519
LCU 2005	Brandy Station	LCU-ARMY	US	968
2006	Penyelamat 6	PB	Malaysia	519
LCU 2006	Bristoe Station	LCU-ARMY	US	968
2007	Penyelamat 7	PB	Malaysia	519
LCU 2007	Broad Run	LCU-ARMY	US	968
2008	Penyelamat 8	PB	Malaysia	519
LCU 2008	Buena Vista	LCU-ARMY	US	968
2009	Penyelamat 9	PB	Malaysia	519
LCU 2009	Calaboza	LCU-ARMY	US	968
2010	Penyelamat 10	PB	Malaysia	519
LCU 2010	Cedar Run	LCU-ARMY	US	968
LCU 2011	Chickahominy	LCU-ARMY	US	968
LCU 2012	Chickasaw Bayou	LCU-ARMY	US	968
LCU 2013	Churubusco	LCU-ARMY	US	968
LCU 2014	Coamo	LCU-ARMY	US	968
LCU 2015	Contreras	LCU-ARMY	US	968
LCU 2016	Corinth	LCU-ARMY	US	968
LCU 2017	El Caney	LCU-ARMY	US	968
LCU 2018	Five Forks	LCU-ARMY	US	968
LCU 2019	Fort Donelson	LCU-ARMY	US	968
LCU 2020	Fort McHenry	LCU-ARMY	US	968
LCU 2021	Great Bridge	LCU-ARMY	US	968
LCU 2022	Harpers Ferry	LCU-ARMY	US	968
LCU 2023	Hobkirk	LCU-ARMY	US	968
LCU 2024	Homigueros	LCU-ARMY	US	968
LCU 2025	Malvern Hill	LCU-ARMY	US	968
LCU 2026	Matamoros	LCU-ARMY	US	968
LCU 2027	Mechanicsville	LCU-ARMY	US	968
LCU 2028	Missionary Bridge	LCU-ARMY	US	968
LCU 2029	Molino Del Ray	LCU-ARMY	US	968
LCU 2030	Monterrey	LCU-ARMY	US	968
LCU 2031	New Orleans	LCU-ARMY	US	968
LCU 2032	Palo Alto	LCU-ARMY	US	968
LCU 2033	Paulus Hook	LCU-ARMY	US	968
LCU 2034	Perryville	LCU-ARMY	US	968
LCU 2035	Port Hudson	LCU-ARMY	US	968
T-AKR 2044	Cape Orlando	AKR	US	982
LCAC 2101	Air Cushion-Tei - 1 - Gou	LCAC	Japan	444
LCAC 2102	Air Cushion-Tei - 2 - Gou	LCAC	Japan	444
LCAC 2103	Air Cushion-Tei - 3 - Gou	LCAC	Japan	444
LCAC 2104	Air Cushion-Tei - 4 - Gou	LCAC	Japan	444
LCAC 2105	Air Cushion-Tei - 5 - Gou	LCAC	Japan	444
LCAC 2106	Air Cushion-Tei - 6 - Gou	LCAC	Japan	444
2161	Semilang	PB	Malaysia	518
2162	Alu-Alu	PB	Malaysia	518
2163	Mersuji	PB	Malaysia	518
2164	Siakap	PB	Malaysia	518
2201	Nusa	PB	Malaysia	517
2202	Rentap	PB	Malaysia	517
2203	Renggis	PB	Malaysia	520
2204	Sugut	PB	Malaysia	520
2205	Balung	PB	Malaysia	520
2210	Tugau	PB	Malaysia	520
2211	Mukah	PB	Malaysia	520
2212	Tatau	PB	Malaysia	520
2213	Nyalau	PB	Malaysia	520
2214	Niah	PB	Malaysia	520
2215	Kindurong	PB	Malaysia	520
2216	Jepak	PB	Malaysia	520
2217	Sikuati	PB	Malaysia	520
2218	Tambisan	PB	Malaysia	520
2219	Bagahak	PB	Malaysia	520
2220	Siagut	PB	Malaysia	520
2221	Mangalum	PB	Malaysia	520
2222	Medang	PB	Malaysia	520
2223	Memmon	PB	Malaysia	520
2224	Subuan	PB	Malaysia	520
GN 2301	Utique	PB	Tunisia	854
2344	Al Amane	SAR	Morocco	550
2345	Ait Baâmrane	SAR	Morocco	550
2551	Malawali	PB	Malaysia	516
2552	Serasan	PB	Malaysia	516
2553	Manjung	PB	Malaysia	516
2554	Tebrau	PB	Malaysia	516
2601	Rhu	PB	Malaysia	517
2601	Sana'a	WPB	Yemen	1018
DT 2601	Tammar	YTM	Australia	37
2602	Aden	WPB	Yemen	1018
2602	Stapa	PB	Malaysia	517
P 2701	Al Noketha	PBF	Kuwait	491
P 2703	Bubiyan	PBF	Kuwait	491
P 2705	Kubbar	PBF	Kuwait	491
P 2707	Bayan	PBF	Kuwait	491
P 2709	Al Shoaie	PBF	Kuwait	491
P 2711	Al Saffar	PBF	Kuwait	491
P 2713	Al Seep	PBF	Kuwait	491
P 2715	Al Bateel	PBF	Kuwait	491
P 2717	Al Tawash	PBF	Kuwait	491
P 2719	Al Boom	PBF	Kuwait	491
2860	Tuba	PB	Malaysia	519
2861	Rimau	PB	Malaysia	519
2862	Kendi	PB	Malaysia	519
2863	Serimbum	PB	Malaysia	519
2864	Aur	PB	Malaysia	519
2865	Matanani	PB	Malaysia	519
2866	Layang 2	PB	Malaysia	519
2867	Selingan	PB	Malaysia	519
2868	Gador	PB	Malaysia	519
2950	Gemia	PB	Malaysia	519
2951	Rawa	PB	Malaysia	519
2952	Peringgi	PB	Malaysia	519
2953	Redang	PB	Malaysia	519
2954	Kapas	PB	Malaysia	519
2955	Libaran	PB	Malaysia	519
2956	Mabul	PB	Malaysia	519
2957	Tenggol	PB	Malaysia	519
2958	Sebatik	PB	Malaysia	519
3001	Tae Pung Yang I	ARSH	Korea, South	488
3002	Tae Pung Yang II	ARSH	Korea, South	489
3003	Tae Pung Yang III	ARSH	Korea, South	488
3005	Tae Pung Yang V	ARSH	Korea, South	488
T-AK 3005	SGT Matej Kocak	AKH	US	981
3006	Tae Pung Yang VI	ARSH	Korea, South	488
T-AK 3006	PFC Eugene A Obregon	AKH	US	981
3007	Tae Pung Yang VII	ARSH	Korea, South	488
L 3007	Lyme Bay	LSD	UK	919
T-AK 3007	Maj Stephen W Pless	AKH	US	981
3008	Tae Pung Yang VIII	ARSH	Korea, South	488
L 3008	Mounts Bay	LSD	UK	919
T-AK 3008	2nd Lt John P Bobo	AKRH	US	977
3009	Tae Pung Yang IX	ARSH	Korea, South	490
L 3009	Cardigan Bay	LSD	UK	919
T-AK 3009	PFC Dewayne T Williams	AKRH	US	977
3010	Tae Pung Yang X	ARSH	Korea, South	490
T-AK 3010	1st Lt Baldomero Lopez	AKRH	US	977
3011	Tae Pung Yang XI	ARSH	Korea, South	490
T-AK 3011	1st Lt Jack Lummus	AKRH	US	977
3012	Tae Pung Yang XII	ARSH	Korea, South	490
T-AK 3012	SGT William R Button	AKRH	US	977
L 3015	Leconi II	LCU	Gabon	280
T-AK 3015	1st Lt Harry L Martin	AK	US	981
T-AK 3016	L/CPL Roy M Wheat	AK	US	981
T-AK 3017	GYSGT Fred W Stockham	AKR	US	977
P 3123	Harambee	PBO	Kenya	465
P 3124	Jasiri	PSO	Kenya	464
P 3126	Nyayo	PGGF	Kenya	464
P 3127	Umoja	PGGF	Kenya	464
P 3130	Shujaa	PBO	Kenya	464
P 3131	Shupavu	PBO	Kenya	464
3132	Lang	PB	Malaysia	516
3133	Segantang	PB	Malaysia	516
3134	Jarak	PB	Malaysia	516
3135	Kukup	PB	Malaysia	516
3136	Sempadi	PB	Malaysia	516
3137	Labas	PB	Malaysia	516
3138	Nyireh	PB	Malaysia	516
3139	Kuraman	PB	Malaysia	516
3140	Siamil	PB	Malaysia	516
3141	Pemanggil	PB	Malaysia	516
3142	Bidong	PB	Malaysia	516
3143	Satang	PB	Malaysia	516
3144	Rumbia	PB	Malaysia	516
3145	Ligitan	PB	Malaysia	516
3221	Ramunia	PB	Malaysia	517
3222	Marudu	PB	Malaysia	517
3223	Danga	PB	Malaysia	517
3224	Siangin	PB	Malaysia	517
3225	Kimanis	PB	Malaysia	517
3226	Burau	PB	Malaysia	517
3227	Nipah	PB	Malaysia	517
P 3301	Ardhana	PB	UAE	887

IHS Jane's Fighting Ships 2014-2015

© 2014 IHS

Pennant list [99]

Number	Ship's name	Type	Country	Page
P 3302	Zurara	PB	UAE	887
P 3303	Murban	PB	UAE	887
P 3304	Al Ghullan	PB	UAE	887
P 3305	Radoom	PB	UAE	887
P 3306	Ghanadhah	PB	UAE	887
3501	Ilocos Norte	PB	Philippines	634
3501	Perdana	PTFG	Malaysia	513
A 3501	Annad	YTB	UAE	890
3502	Nueva Vizcaya	PB	Philippines	634
3502	Serang	PTFG	Malaysia	513
3503	Ganas	PTFG	Malaysia	513
3503	Romblon	PB	Philippines	634
3504	Davao Del Norte	PB	Philippines	634
3504	Ganyang	PTFG	Malaysia	513
3505	Jerong	PB	Malaysia	513
3506	Todak	PB	Malaysia	513
3507	Paus	PB	Malaysia	513
3508	Yu	PB	Malaysia	513
TV 3508	Kashima	AXH/TV	Japan	447
3509	Baung	PB	Malaysia	513
3510	Pari	PB	Malaysia	513
3511	Handalan	PTFG	Malaysia	513
3512	Perkasa	PTFG	Malaysia	513
3513	Pendekar	PTFG	Malaysia	513
TV 3513	Shimayuki	AXGHM/TV	Japan	447
3514	Gempita	PTFG	Malaysia	513
TV 3517	Shirayuki	AXGHM/TV	Japan	447
TV 3518	Setoyuki	AXGHM/TV	Japan	447
P 3568	Pukaki	PBI	New Zealand	575
P 3569	Rotoiti	PBI	New Zealand	575
P 3570	Taupo	PBI	New Zealand	575
P 3571	Hawea	PBI	New Zealand	575
TSS 3601	Asashio	SSK	Japan	430
TSS 3607	Fuyushio	SSK	Japan	430
P 3711	Um Almaradim	PBM	Kuwait	490
P 3713	Ouha	PBM	Kuwait	490
P 3715	Failaka	PBM	Kuwait	490
P 3717	Maskan	PBM	Kuwait	490
P 3719	Al-Ahmadi	PBM	Kuwait	490
P 3721	Alfahaheel	PBM	Kuwait	490
P 3723	Al-Yarmouk	PBM	Kuwait	490
P 3725	Garoh	PBM	Kuwait	490
3901	Gagah	PBF	Malaysia	516
3902	Tabah	PBF	Malaysia	516
3903	Cekal	PBF	Malaysia	516
3904	Berani	PBF	Malaysia	516
3905	Setia	PBF	Malaysia	516
3906	Amanah	PBF	Malaysia	516
3907	Jujur	PBF	Malaysia	516
3908	Ikhlas	PBF	Malaysia	516
3909	Budiman	PBF	Malaysia	516
3910	Tegas	PBF	Malaysia	516
3911	Mulia	PBF	Malaysia	516
3912	Bijak	PBF	Malaysia	516
3913	Adil	PBF	Malaysia	516
3914	Pintar	PBF	Malaysia	516
3915	Bistari	PBF	Malaysia	516
4001	Marlin	AX	Malaysia	520
LST 4001	Oosumi	LPD/LSTH	Japan	443
LST 4002	Shimokita	LPD/LSTH	Japan	443
LST 4003	Kunisaki	LPD/LSTH	Japan	443
ATS 4202	Kurobe	AVM/TV	Japan	448
ATS 4203	Tenryu	AVHM/TV	Japan	448
AMS 4301	Hiuchi	YTT	Japan	450
AMS 4302	Suou	YTT	Japan	450
AMS 4303	Amakusa	YTT	Japan	450
AMS 4304	Genkai	YTT	Japan	450
AMS 4305	Enshuu	YTT	Japan	450
T-AK 4396	Maj Bernard F Fisher	AK	US	978
FNH 4401	Cisne	WPB	Honduras	321
FNH 4402	Utila	WPB	Honduras	321
FNH 4403	Roatan	WPB	Honduras	321
FNH 4404	Guanaja	WPB	Honduras	321
P 4505	Al Sanbouk	PGGF	Kuwait	491
L 4512	Hellen	LCP	Norway	589
L 4513	Torås	LCP	Norway	589
L 4514	Møvik	LCP	Norway	589
L 4520	Skrolsvik	LCP	Norway	589
L 4521	Kråkenes	LCP	Norway	589
L 4522	Stangnes	LCP	Norway	589
L 4525	Kopås	LCP	Norway	589
L 4526	Tangen	LCP	Norway	589
L 4527	Oddane	LCP	Norway	589
L 4531	Søviknes	LCP	Norway	589
L 4532	Osternes	LCP	Norway	589
L 4541	Kamøy	PB	Norway	589
L 4542	Bondøy	PB	Norway	589
L 4543	Rypøy	PB	Norway	589
T-AK 4543	LTC John U D Page	AK	US	977
T-AK 4544	SSGT Edward A Carter	AK	US	977
L 4545	Kjelmøy	PB	Norway	589
L 4546	Reinøy	PB	Norway	589
HSV 4676	Westpac Express	HSV	US	981
5001	Sambongho	PSO	Korea, South	489
T-AG 5001	VADM K R Wheeler	AG	US	977
AGB 5003	Shirase	AGBH	Japan	450
T-AKR 5051	Cape Ducato	AKR	US	982
T-AKR 5052	Cape Douglas	AKR	US	982
T-AKR 5053	Cape Domingo	AKR	US	982
T-AKR 5054	Cape Decision	AKR	US	982
T-AKR 5055	Cape Diamond	AKR	US	982
T-AKR 5062	Cape Isabel	AKR	US	982
T-AKR 5063	Cape May	AK/AKR	US	982
T-AKR 5065	Cape Mohican	AK/AKR	US	982
T-AKR 5066	Cape Hudson	AKR	US	982
T-AKR 5067	Cape Henry	AKR	US	982
T-AKR 5068	Cape Horn	AKR	US	982
T-AKR 5069	Cape Edmont	AKR	US	982
T-AKR 5076	Cape Inscription	AKR	US	982
T-AKR 5082	Cape Knox	AKR	US	982
T-AKR 5083	Cape Kennedy	AKR	US	982
AGS 5103	Suma	AGS	Japan	446
AGS 5104	Wakasa	AGS	Japan	446
AGS 5105	Nichinan	AGS	Japan	446
AGS 5106	Syounan	AGS	Japan	446
AOS 5201	Hibiki	AGOSH	Japan	446
AOS 5202	Harima	AGOSH	Japan	446
A 5203	Andrómeda	AGSC	Portugal	654
A 5204	Polar	AXS	Portugal	654
A 5205	Auriga	AGSC	Portugal	654
A 5210	Bérrio	AORLH	Portugal	656
A 5302	Caroly	AXS	Italy	417
A 5303	Ammiraglio Magnaghi	AGSH	Italy	416
A 5304	Aretusa	AGS	Italy	416
A 5305	Murena	YFT	Italy	420
A 5308	Galatea	AGS	Italy	416
A 5309	Anteo	ARSH	Italy	419
A 5311	Palinuro	AXS	Italy	417
A 5312	Amerigo Vespucci	AXS	Italy	417
A 5313	Stella Polare	AXS	Italy	417
A 5315	Raffaele Rossetti	AG/AGOR	Italy	416
A 5316	Corsaro II	AXS	Italy	417
A 5320	Vincenzo Martellotta	AG/AGE	Italy	416
A 5322	Capricia	AXS	Italy	417
A 5323	Orsa Maggiore	AXS	Italy	417
A 5324	Titano	ATR	Italy	421
A 5325	Polifemo	ATR	Italy	421
A 5326	Etna	AORH	Italy	418
A 5327	Stromboli	AORH	Italy	419
A 5329	Vesuvio	AORH	Italy	419
A 5330	Saturno	ATR	Italy	421
A 5340	Elettra	AGORH/AGE/AGI	Italy	416
A 5347	Gorgona	AKL	Italy	419
A 5348	Tremiti	AKL	Italy	419
A 5349	Caprera	AKL	Italy	419
A 5351	Pantelleria	AKL	Italy	419
A 5352	Lipari	AKL	Italy	419
A 5353	Capri	AKL	Italy	419
A 5359	Bormida	AWT	Italy	418
A 5364	Ponza	ABU	Italy	420
A 5366	Levanzo	ABU	Italy	420
A 5367	Tavolara	ABU	Italy	420
A 5368	Palmaria	ABU	Italy	420
A 5370	Panarea	AWT	Italy	418
A 5371	Linosa	AWT	Italy	418
A 5372	Favignana	AWT	Italy	418
A 5373	Salina	AWT	Italy	418
A 5376	Ticino	AWT	Italy	418
A 5377	Tirso	AWT	Italy	418
A 5379	Astice	AXL	Italy	417
A 5380	Mitilo	AXL	Italy	417
A 5382	Porpora	AXL	Italy	417
A 5383	Procida	ABU	Italy	420
A 5390	Leonardo	(AGOR(C))	Italy	416
S 5509	Al Dorrar	–	Kuwait	492
M 5552	Milazzo	MHSC	Italy	415
M 5553	Vieste	MHSC	Italy	415
M 5554	Gaeta	MHSC	Italy	415
M 5555	Termoli	MHSC	Italy	415
M 5556	Alghero	MHSC	Italy	415
M 5557	Numana	MHSC	Italy	415
M 5558	Crotone	MHSC	Italy	415
M 5559	Viareggio	MHSC	Italy	415
M 5560	Chioggia	MHSC	Italy	415
M 5561	Rimini	MHSC	Italy	415
P 5702	Istiqlal	PGGF	Kuwait	491
ASE 6102	Asuka	AGEH	Japan	447
6111	Dokdo	LPD	Korea, South	483
P 6121	Gepard	PGGFM	Germany	292
P 6122	Puma	PGGFM	Germany	292
P 6123	Hermelin	PGGFM	Germany	292
P 6125	Zobel	PGGFM	Germany	292
P 6126	Frettchen	PGGFM	Germany	292
P 6128	Ozelot	PGGFM	Germany	292
P 6129	Wiesel	PGGFM	Germany	292
P 6130	Hyäne	PGGFM	Germany	292
FNH 6501	Nacaome	PB	Honduras	320
FNH 6502	Goascoran	PB	Honduras	320
FNH 6503	Patuca	PB	Honduras	320
FNH 6504	Ulua	PB	Honduras	320
FNH 6505	Choluteca	PB	Honduras	320
FNH 6506	Rio Coco	PBI	Honduras	321
7501	Langkawi	PSOH	Malaysia	517
7502	Banggi	PSOH	Malaysia	517
Y 8018	Breezand	YTL	Netherlands	572
Y 8019	Balgzand	YTL	Netherlands	572
Y 8050	Urania	AXS	Netherlands	570
Y 8055	Schelde	YTL	Netherlands	572
Y 8056	Wierbalg	YTL	Netherlands	572
Y 8057	Malzwin	YTL	Netherlands	572
Y 8058	Zuidwal	YTL	Netherlands	572
Y 8059	Westwal	YTL	Netherlands	572
P 8111	Durbar	PTFG	Bangladesh	55
P 8112	Duranta	PTFG	Bangladesh	55
P 8113	Durvedya	PTFG	Bangladesh	55
P 8114	Durdam	PTFG	Bangladesh	55
P 8125	Durdharsha	PTFG	Bangladesh	54
P 8126	Durdanta	PTFG	Bangladesh	54
T-AKR 5128	Dordania	PTFG	Bangladesh	54
P 8131	Anirban	PTFG	Bangladesh	54
P 8141	Uttal	PTFG	Bangladesh	55
A 8201	Punta Barima	PB	Venezuela	1008
A 8202	Punta Mosquito	PB	Venezuela	1008
A 8203	Punta Mulatos	PB	Venezuela	1008
A 8204	Punta Perret	PB	Venezuela	1008
A 8205	Punta Cardon	PB	Venezuela	1008
A 8206	Punta Playa	PB	Venezuela	1008
P 8221	TB 1	PTL	Bangladesh	56
P 8222	TB 2	PTL	Bangladesh	56
P 8223	TB 3	PTL	Bangladesh	56

© 2014 IHS
IHS Jane's Fighting Ships 2014-2015

[100] Pennant list

Number	Ship's name	Type	Country	Page
P 8224	TB 4	PTL	Bangladesh	56
P 8235	TB 35	PTK	Bangladesh	54
P 8236	TB 36	PTK	Bangladesh	54
P 8237	TB 37	PTK	Bangladesh	54
P 8238	TB 38	PTK	Bangladesh	54
A 8307	Punta Macoya	PB	Venezuela	1008
A 8308	Punta Moron	PB	Venezuela	1008
A 8309	Punta Unare	PB	Venezuela	1008
A 8310	Punta Ballena	PB	Venezuela	1008
A 8311	Punta Macuro	PB	Venezuela	1008
A 8312	Punta Mariusa	PB	Venezuela	1008
B 8421	Rio Arauca II	PB	Venezuela	1008
B 8422	Rio Catatumbo II	PB	Venezuela	1008
B 8423	Rio Apure II	PB	Venezuela	1008
B 8424	Rio Negro II	PB	Venezuela	1008
B 8425	Rio Meta II	PB	Venezuela	1008
B 8426	Rio Portuguesa II	PB	Venezuela	1008
B 8427	Rio Sarare	PB	Venezuela	1008
B 8428	Rio Uribante	PB	Venezuela	1008
B 8429	Rio Sinaruco	PB	Venezuela	1008
B 8430	Rio Icabaru	PB	Venezuela	1008
B 8431	Rio Guarico II	PB	Venezuela	1008
B 8432	Rio Yaracuy	PB	Venezuela	1008
FNH 8501	Chamelecon	PB	Honduras	320
Y 8760	Patria	AOTL	Netherlands	571
WM1-9001	Jonge Prins	YFL	Netherlands	572
WM1-9002	Jonge Jan	YFL	Netherlands	572
L 9012	Siroco	LSDH/TCD 90	France	267
L 9013	Mistral	LHDM/BPC	France	266
L 9014	Tonnerre	LHDM/BPC	France	266
L 9015	Dixmude	LHDM/BPC	France	266
L 9032	Dumont d'Urville	LSTH	France	267
L 9034	La Grandière	LSTH	France	267
L 9062	Hallebarde	LCT	France	268
L 9090	Gapeau	LSL	France	273
9101	Soldado Canave	LCT	Chile	121
9102	Cabo Reyes	LCM	Chile	122
9103	Soldado Fuentes	LCM	Chile	122
T-AOT 9109	Petersburg	AOT	US	982
9423	Nesbitt	YGS	UK	913
9424	Pat Barton	YGS	UK	913
9425	Cook	YGS	UK	913
9426	Owen	YGS	UK	913
T-AKR 9666	Cape Vincent	AKR	US	982
T-AKR 9678	Cape Rise	AKR	US	982
T-AKR 9679	Cape Ray	AKR	US	982
T-AKR 9701	Cape Victory	AKR	US	982
T-AKR 9711	Cape Trinity	AKR	US	982
L 9892	San Giorgio	LPD	Italy	414
L 9893	San Marco	LPD	Italy	414
L 9894	San Giusto	LPD	Italy	414
T-AKR 9960	Cape Race	AKR	US	982
T-AKR 9961	Cape Washington	AKR	US	982
T-AKR 9962	Cape Wrath	AKR	US	982
21689	Dugong	YDT/PB	Australia	36
WLI 65400	Bayberry	WLI/ABU	US	990
WLI 65401	Elderberry	WLI/ABU	US	990
WLR 65501	Ouachita	WLR	US	991
WLR 65502	Cimarron	WLR	US	991
WLR 65503	Obion	WLR	US	991
WLR 65504	Scioto	WLR	US	991
WLR 65505	Osage	WLR	US	991
WLR 65506	Sangamon	WLR	US	991
WYTL 65601	Capstan	WYTL	US	992
WYTL 65602	Chock	WYTL	US	992
WYTL 65604	Tackle	WYTL	US	992
WYTL 65607	Bridle	WYTL	US	992
WYTL 65608	Pendant	WYTL	US	992
WYTL 65609	Shackle	WYTL	US	992
WYTL 65610	Hawser	WYTL	US	992
WYTL 65611	Line	WYTL	US	992
WYTL 65612	Wire	WYTL	US	992
WYTL 65614	Bollard	WYTL	US	992
WYTL 65615	Cleat	WYTL	US	992
WLIC 75301	Anvil	WLIC	US	992
WLIC 75302	Hammer	WLIC	US	992
WLIC 75303	Sledge	WLIC	US	992
WLIC 75304	Mallet	WLIC	US	992
WLIC 75305	Vise	WLIC	US	992
WLIC 75306	Clamp	WLIC	US	992
WLR 75307	Wedge	WLR	US	991
WLIC 75309	Hatchet	WLIC	US	992
WLIC 75310	Axe	WLIC	US	992
WLR 75401	Gasconade	WLR	US	991
WLR 75402	Muskingum	WLR	US	991
WLR 75403	Wyaconda	WLR	US	991
WLR 75404	Chippewa	WLR	US	991
WLR 75405	Cheyenne	WLR	US	991
WLR 75406	Kickapoo	WLR	US	991
WLR 75407	Kanawha	WLR	US	991
WLR 75408	Patoka	WLR	US	991
WLR 75409	Chena	WLR	US	991
WLR 75500	Kankakee	WLR	US	991
WLR 75501	Greenbrier	WLR	US	991
87301	Barracuda	WPB	US	987
87301	Barracuda	WPB	US	987
87302	Hammerhead	WPB	US	987
87303	Mako	WPB	US	987
87304	Marlin	WPB	US	987
87305	Stingray	WPB	US	987
87306	Dorado	WPB	US	987
87307	Osprey	WPB	US	987
87308	Chinook	WPB	US	987
87309	Albacore	WPB	US	987
87310	Tarpon	WPB	US	987
87311	Cobia	WPB	US	987
87312	Hawksbill	WPB	US	987
87313	Cormorant	WPB	US	987
87314	Finback	WPB	US	987
87315	Amberjack	WPB	US	987
87316	Kittiwake	WPB	US	987
87317	Blackfin	WPB	US	987
87318	Bluefin	WPB	US	987
87319	Yellowfin	WPB	US	987
87320	Manta	WPB	US	987
87321	Coho	WPB	US	987
87322	Kingfisher	WPB	US	987
87323	Seahawk	WPB	US	987
87324	Steelhead	WPB	US	987
87325	Beluga	WPB	US	987
87326	Blacktip	WPB	US	987
87327	Pelican	WPB	US	987
87328	Ridley	WPB	US	987
87329	Cochito	WPB	US	987
87330	Manowar	WPB	US	987
87331	Moray	WPB	US	987
87332	Razorbill	WPB	US	987
87333	Adelie	WPB	US	987
87334	Gannet	WPB	US	987
87335	Narwhal	WPB	US	987
87336	Sturgeon	WPB	US	987
87337	Sockeye	WPB	US	987
87338	Ibis	WPB	US	987
87339	Pompano	WPB	US	987
87340	Halibut	WPB	US	987
87341	Bonito	WPB	US	987
87342	Shrike	WPB	US	987
87343	Tern	WPB	US	987
87344	Heron	WPB	US	987
87345	Wahoo	WPB	US	987
87346	Flyingfish	WPB	US	987
87347	Haddock	WPB	US	987
87348	Brant	WPB	US	987
87349	Shearwater	WPB	US	987
87350	Petrel	WPB	US	987
87352	Sea Lion	WPB	US	987
87353	Skipjack	WPB	US	987
87354	Dolphin	WPB	US	987
87355	Hawk	WPB	US	987
87356	Sailfish	WPB	US	987
87357	Sawfish	WPB	US	987
87358	Swordfish	WPB	US	987
87359	Tiger Shark	WPB	US	987
87360	Blue Shark	WPB	US	987
87361	Sea Horse	WPB	US	987
87362	Sea Otter	WPB	US	987
87363	Manatee	WPB	US	987
87364	Ahi	WPB	US	987
87365	Pike	WPB	US	987
87366	Terrapin	WPB	US	987
87367	Sea Dragon	WPB	US	987
87368	Sea Devil	WPB	US	987
87369	Crocodile	WPB	US	987
87370	Diamondback	WPB	US	987
87371	Reef Shark	WPB	US	987
87372	Alligator	WPB	US	987
87373	Sea Dog	WPB	US	987
87374	Sea Fox	WPB	US	987

IHS Jane's Fighting Ships 2014-2015 © 2014 IHS

FREE ENTRY/CONTENT IN THIS PUBLICATION

Having your products and services represented in our titles means that they are being seen by the professionals who matter - both by those involved in the procurement and by those working for the companies that are likely to affect your business. We therefore feel that it is very much in the interest of your organisation, as well as IHS Jane's, to ensure your data is current and accurate.

- **Don't forget** - You may be missing out on business if your entry in an IHS Jane's product is incorrect because you have not supplied the latest information to us.

- **Ask yourself** - Can you afford not to be represented in IHS Jane's printed and electronic products? And if you are listed, can you afford for your information to be out of date?

- **And most importantly** - The best part of all is that your entries in IHS Jane's products are TOTALLY FREE OF CHARGE.

Please provide (using a photocopy of this form or by email to the below address) the information on the following categories where appropriate:

1. Organisation name:

2. Division name:

3. Location address:

4. Mailing address if different:

5. Telephone (please include switchboard and main departmental contact numbers, for example Public Relations, Sales, and so on):

6. Facsimile:

7. E-mail:

8. Web sites:

9. Contact name and job title:

10. A brief description of your organisation's activities, products and services:

11. IHS Jane's publications in which you would like to be included:

Please send this information to:
Research and Information Services, IHS Jane's,
IHS Global Limited, Sentinel House, 163 Brighton Road, Coulsdon, Surrey CR5 2YH, UK
Tel: (+44 0) 20 32 53 22 62
Fax: (+44 0) 20 87 00 39 59
e-mail: yearbook@ihs.com

Copyright enquiries:
e-mail: copyright@ihs.com

Please tick this box if you do not wish your organisation's staff to be included in IHS Jane's mailing lists ☐

JFS

How to use

(see also Glossary and Type abbreviations)

(1) Details of major warships are grouped under six separate non-printable headings. These are:-
  (a) **Number and Class name**. Totals of vessels per class are listed as 'in service + building (proposed)' or 'in service + transfer (proposed)'.
  (b) **Building programme**. This includes builders' names and key dates. In general the 'laid down' column reflects keel laying but modern shipbuilding techniques make it difficult to be specific about the start date of actual construction. Launching and christening can be similarly confusing, now that many ships are lowered into the water and formally christened some time later. Some nations commission their ships on completion of building, others after the ships have completed trials. In this hardcopy edition any date after April 2014 is projected or estimated and therefore liable to change.
  (c) **Hull**. This section tends to have only specification and performance parameters and contains little free text. Hull related details such as **Military lift** and **Cargo capacity** may be included when appropriate. **Displacement** and **Measurement** tonnages, **Dimensions**, **Horsepower** and so on, are defined in the Glossary. Throughout the life of a ship its displacement tends to creep upwards as additional equipment is added and redundant fixtures and fittings are left in place. For the same reasons, ships of the same class, active in different navies, frequently have different displacements and other dissimilar characteristics. Unless otherwise stated the lengths and widths given are overall and the draught is at full load. Sustained maximum horsepower is given where the information is available and may not be the same for similar engines operating in different hulls under different conditions. **Speed** is the maximum obtainable under trials conditions.
  (d) **Weapon systems**. This section contains operational details and some free text on weapons and sensors which are laid out in a consistent order using the same subheadings throughout the book. The titles are:- **Missiles** (subdivided into SLBM, SSM, SAM, A/S); **Guns** (numbers of barrels are given and the rate of fire is 'per barrel' unless stated otherwise); **Torpedoes**; **A/S mortars**; **Depth charges**; **Mines**; **Physical countermeasures**; **Electronic countermeasures**; **Combat data systems**; **Weapons control**; **Electro-optic systems**; **Radars**; **Sonars**. The Weapons control heading is used for weapons' direction equipment. In most cases the performance specifications are those of the manufacturer and may therefore be considered to be at the top end of the spectrum of effective performance. So-called 'operational effectiveness' is difficult to define, depends upon many variables and in the context of range may be considerably less than the theoretical maximum. Numbers inserted in the text refer to corresponding numbers included on line drawings.
  (e) **Aircraft**. Only the types and numbers are included here. Where appropriate each country has a separate section listing overall numbers and operational parameters of front-line shipborne and land-based maritime aircraft, normally included after the Frigate section if there is one.
  (f) **General comments**. A maximum of six sub-headings are used to sweep up the variety of additional information which is available but has no logical place in the other sections. These headings are: **Programmes**; **Modernisation**; **Structure**; **Operational**; **Sales**, and **Opinion**. The last of these allows space for informed comment. Some ships remain theoretically in the order of battle in some navies even though they never go to sea and could be more accurately described as in reserve. Where this is known comment is made under **Operational**.

(2) Minor or less important ship entries follow the same format except that there is often much less detail in the first four headings and all additional remarks are put together under the single heading of **Comment**. The distinction between major and minor depends upon editorial judgement and is primarily a function of firepower. The age of the ship or class and its relative importance within the navy concerned is also taken into account.

(3) The space devoted to front-line maritime aircraft reflects the importance of air power as an addition to the naval weapon systems armoury, but the format used is necessarily brief and covers only numbers, roles, and operational characteristics. Greater detail can be found in *IHS Jane's All the World's Aircraft: Development & Production* and the appropriate volume of the *IHS Jane's Weapons* series.

(4) For most larger navies, tables are included at the front of each country section with details such as strength of the fleet, senior appointments, personnel numbers, bases, and so on. There is also a list of pennant numbers and a deletions column covering the previous three years. If you cannot find your favourite ship, always look in the **Deletions** list first.

(5) No addenda is included because modern typesetting technology allows changes to the main text to be made up to a few weeks before publication.

(6) Shipbuilding companies and weapons manufacturers frequently change their names by merger or takeover. As far as possible the published name shows the title when the ship was built or weapon system installed. It is therefore historically accurate.

(7) Like many descriptive terms in international naval nomenclature, differences between Coast Guards, Maritime Police, Customs, and other paramilitary maritime forces are often indistinct and particular to an individual nation. Such vessels are usually included if they have a paramilitary function and are armed.

(8) When selecting photographs for inclusion, priority is given to those that have been taken most recently. A glossy picture five years old may look nice but often does not show the ship as it is now.

(9) The navies by country section is geared to the professional user who needs to be able to make an assessment of the fighting characteristics of a navy or class of ship without having to cross refer to other navies and sections of the book. Much effort has also been made to prevent entries spilling across from one page to another.

(10) Regular updates can be found online.

(11) Photographs are dated and where * appears a new or re-scanned photograph has been substituted or added. Many are followed by a seven digit number to ease identification.

WORLD NAVIES

A–Z

Albania
FORCE DETAR

Country Overview

After being governed by a communist regime since 1946, democratic elections in the Republic of Albania took place in 1991 although since then there have been periods of instability. Situated in western part of the Balkan Peninsula, the country has an area of 11,100 square miles and is bordered to the north by Montenegro and Serbia, to the east by FYRO Macedonia and to the south by Greece. There is a coastline of 195 n miles with the Adriatic Sea on which Durrës and Vlorë are the principal ports. The capital and largest city is Tirana. Territorial waters (12 n miles) are claimed but an EEZ has not been claimed. Italy provides strong operational, training and administrative support. Joint Coast Guard and Customs patrols are mounted within territorial waters while other personnel training is conducted in Italy.

Headquarters Appointments

Commander of the Navy:
Brigadier General Qemal Shkurti

Personnel

2014: 1,900 approximately

Bases

HQ: Durrës
Districts: Durrës (1st), Vlorë (2nd).
Bases: Shengyin, Himarë, Saranda, Sazan Island, Porto Palermo, Vlorë (Pasha Liman).

PATROL FORCES

Notes: (1) Pennant numbers beginning with '1' indicate units from the Durrës district. Those beginning with '2' are from the Vlorë district.
(2) There are plans to procure two 60–80 m offshore patrol vessels by 2020.

2 SHANGHAI II CLASS (FAST ATTACK CRAFT—GUN) (PC)

P 115   P 208

Displacement, tonnes: 115 standard; 136 full load
Dimensions, metres (feet): 38.8 × 5.4 × 1.7 *(127.3 × 17.7 × 5.6)*
Speed, knots: 30
Range, n miles: 700 at 16.5 kt
Complement: 34

Machinery: 2 Type L-12V-180 diesels; 2,400 hp(m) *(1.76 MW)* (forward) 2 Type 12-D-6 diesels; 1,820 hp(m) *(1.34 MW)* (aft); 4 shafts
Guns: 4 China 37 mm/63 (2 twin); 180 rds/min to 8.5 km *(4.6 n miles)*; weight of shell 1.42 kg. 4 USSR 25 mm/60 (2 twin); 270 rds/min to 3 km *(1.6 n miles)*; weight of shell 0.34 kg.
Torpedoes: 2—21 in *(533 mm)* tubes; Yu-1; 9.2 km *(5 n miles)* at 39 kt; warhead 400 kg.
Mines: Rails can be fitted; probably only 10 mines.
Depth charges: 2 projectors; 8 depth charges in lieu of torpedo tubes.
Radars: Surface search/fire control: Pot Head; I-band.
Sonars: Hull-mounted set probably fitted.

Comment: Four transferred from China in mid-1974 and two in 1975. One ship escaped to Italy in early 1997, returned in early 1998 and was reported repaired in 2000. Have torpedo tubes on the stern taken from deleted Huchuan class. Both reported active in 2010.

SHANGHAI II (China colours)   6/1992 / 0081445

1 KRONSHTADT (PROJECT 122) CLASS
(LARGE PATROL CRAFT) (PG)

P 207

Displacement, tonnes: 303 standard; 335 full load
Dimensions, metres (feet): 52.1 × 6.5 × 2.1 *(170.9 × 21.3 × 6.9)*
Speed, knots: 18
Range, n miles: 1,400 at 12 kt
Complement: 51 (4 officers)

Machinery: 3 Kolomna Type 9-D-8 diesels; 3,000 hp(m) *(2.2 MW)* sustained; 3 shafts
Guns: 1—3.5 in *(85 mm)*/52; 18 rds/min to 15.5 km *(8.5 n miles)*; weight of shell 9.5 kg. 1—37 mm/63; 160 rds/min to 4 km *(2.2 n miles)*; weight of shell 0.7 kg. 6—12.7 mm (3 vertical twin) MGs.
A/S Mortars: 2 RBU 1200 five-tubed rocket launchers; range 1,200 m; warhead 34 kg.
Mines: 2 rails; approx 8 mines.
Depth charges: 2 projectors; 2 racks.
Radars: Surface search: Ball Gun; E/F-band.
Navigation: Neptun; I-band.
IFF: High Pole.

Programmes: Four were transferred from the USSR in 1958. This sole survivor was returned by Italy in late 1998. A second of class was beyond repair and was towed back at the same time. Reported active in 2010.

KRONSHTADT   1989 / 0505952

2 NYRYAT I (PROJECT 1896) CLASS (PB)

P 218 (ex-R 218)   P 219 (ex-R 219)

Displacement, tonnes: 118 full load
Dimensions, metres (feet): 28.6 × 5.2 × 1.7 *(93.8 × 17.1 × 5.6)*
Speed, knots: 12.5
Range, n miles: 1,500 at 10 kt
Complement: 15
Machinery: 1 diesel; 450 hp(m) *(331 kW)* sustained; 1 shaft
Radars: Surface search: Spin Trough; I-band.

Comment: Both entered service in about 1960 and, having transferred to the Coast Guard in 2003, returned to the Navy in 2010. Both employed as patrol craft.

AUXILIARIES

Notes: In addition there are a Project 368 Poluchat survey and torpedo recovery craft of 70 tons (R 110 (ex-A 110)), an old ex-USSR Shalanda class tender *Marinza* (A 210), a water-barge (A 211), two tugs and a floating dock *(Vlorë)*.

COAST GUARD (ROJA BREGDETARE)

Notes: An Italian Coast Guard craft CP 224 transferred in 2008.

2 COASTAL PATROL CRAFT (PB)

R 117   R 217

Displacement, tonnes: 18 full load
Dimensions, metres (feet): 13.9 × 4 × 0.9 *(45.6 × 13.1 × 3.0)*
Speed, knots: 34
Range, n miles: 200 at 30 kt
Complement: 4
Machinery: 2 diesels; 1,300 hp *(942 kW)*; 2 waterjets
Guns: 1—12.7 mm MG.
Radars: Surface search: Raytheon; I-band.

Comment: Transferred from the US on 27 February 1999. Reported operational.

R 217   6/2007, Massimo Annati / 1166505

7 TYPE 2010 INSHORE PATROL CRAFT (PBR)

| R 125 (ex-CP 2008) | R 127 (ex-CP 2021) | R 224 (ex-CP 2010) | R 228 (ex-CP 2023) |
| R 126 (ex-CP 2020) | R 128 (ex-CP 2034) | R 227 (ex-CP 2007) | |

Displacement, tonnes: 15 full load
Dimensions, metres (feet): 12.5 × 3.6 × 1.1 *(41 × 11.8 × 3.6)*
Speed, knots: 24
Range, n miles: 533 at 20 kt
Complement: 5
Machinery: 2 AIFO diesels; 1,072 hp *(800 kW)*; 2 shafts
Radars: Surface search: I-band.

Comment: Former harbour launches built in Italy in the 1970s. GRP construction. One transferred from Italian Coast Guard to Albanian Coast Guard in 2002 and a further six in 2004.

Coast guard < **Albania** — Introduction < **Algeria**

3 + 1 DAMEN STAN PATROL 4207 (PB)

| ILIRIA P 131 | ORIKU P 132 | LISUS P 133 | BUTRINDI P 134 |

Displacement, tonnes: 208 standard
Dimensions, metres (feet): 42.8 × 7.11 × 2.52 *(140.4 × 23.3 × 8.3)*
Speed, knots: 26
Machinery: 2 Caterpillar 3516B DI-TA; 5,600 hp *(4.17 MW)*; 2 cp props
Guns: To be announced.

Comment: Contract signed with Damen Shipyards, Gorinchem on 13 November 2007 for the acquisition of four Stan Patrol 4207 offshore patrol vessels. The first vessel was built in Holland whilst the remaining three are being built at Pashaliman Shipyard near Vlorë. *Oriku* commissioned on 17 July 2012 and *Lisus* in February 2013. *Butrindi* is to be delivered in 2014. The contract also includes refurbishment of the shipyard, training and maintenance services. Details are based on those in UK Customs service and in Jamaica.

ILIRIA 7/2007, A A de Kruijf / 1335320

8 V 4000 (FAST PATROL CRAFT) (PBF)

Displacement, tonnes: 28 full load
Dimensions, metres (feet): 16.5 × 4.5 × 0.8 *(54.1 × 14.8 × 2.6)*
Speed, knots: 48
Range, n miles: 420 at 35 kt
Complement: 5
Machinery: 2 Isotta Fraschini ID 36 SS 16V diesels; 2,450 hp(m) *(1.8 MW)* sustained
Radars: Surface search: GEM BX 132; I-band.

Comment: Eight Drago craft transferred from the Italian Guardia di Finanza in 2006.

V 4000 CRAFT 6/2006, Guardia di Finanza / 1164418

4 TYPE 227 INSHORE PATROL CRAFT (PBR)

| R 123 (ex-CP 229) | R 124 (ex-CP 235) | R 225 (ex-CP 234) | R 226 (ex-CP 236) |

Displacement, tonnes: 16 full load
Dimensions, metres (feet): 13.4 × 4.8 × 1.3 *(44 × 15.7 × 4.3)*
Speed, knots: 24
Range, n miles: 400 at 24 kt
Complement: 5
Machinery: 2 AIFO 8281-SRM diesels; 1,770 hp *(1.32 MW)*; 2 shafts
Radars: Surface search: I-band.

Comment: Wooden construction. Built in Italy 1966–69. Transferred from Italian Coast Guard to Albanian Coast Guard in 2002.

3 SEA SPECTRE MK III (PB)

| R 118 | R 215 | R 216 |

Displacement, tonnes: 42 full load
Dimensions, metres (feet): 19.8 × 5.5 × 1.8 *(65 × 18.0 × 5.9)*
Speed, knots: 28
Range, n miles: 450 at 25 kt
Complement: 9
Machinery: 3 Detroit 8V-71 diesels; 690 hp *(515 kW)* sustained; 3 shafts
Guns: 2 — 12.7 mm MGs.
Radars: Surface search: Raytheon; I-band.

Comment: Transferred from the US on 27 February 1999.

R 215 6/2008 / 1335319

1 TYPE 303 COASTAL PATROL CRAFT (PB)

R 122 (ex-CP 303)

Displacement, tonnes: 20 full load
Dimensions, metres (feet): 13.4 × 3.8 × 1.1 *(44 × 12.5 × 3.6)*
Speed, knots: 13
Range, n miles: 350 at 13 kt
Complement: 5
Machinery: 2 GM6V53 diesels; 730 hp *(544 kW)*; 2 shafts
Radars: Surface search: I-band.

Comment: Built in US in 1965. Transferred from Italian Coast Guard to Albanian Coast Guard in 2002.

1 TYPE 246 CLASS (INSHORE PATROL CRAFT) (PBR)

– (ex-CP 249)

Displacement, tonnes: 22 full load
Dimensions, metres (feet): 15 × 4.85 × 1.65 *(49.2 × 15.9 × 5.4)*
Speed, knots: 27
Complement: 7
Machinery: 2 Isotta Fraschini ID 35 SS6V diesels; 1,350 hp(m) *(1.0 MW)*; 2 shafts
Radars: Surface search: I-band.

Comment: Built in Italy in 1980. Transferred from the Italian Coast Guard in 2008.

5 ARCHANGEL CLASS (RESPONSE BOATS) (PBF)

R 01–05

Displacement, tonnes: 9 standard
Dimensions, metres (feet): 13.4 × 4.3 × 0.76 *(44 × 14.1 × 2.5)*
Speed, knots: 36
Range, n miles: 300 at 25 kt
Complement: 6
Machinery: 2 Caterpillar C9 diesels; 2 Hamilton 322 waterjets
Guns: 2 — 7.62 mm MGs.
Radars: Navigation: I-band.

Comment: High-speed inshore patrol craft of aluminium construction and foam collar built by SAFE Boats International, Port Orchard, Washington. Three donated by the United States government in August 2010 and a further two in March 2013.

Algeria

MARINE DE LA REPUBLIQUE ALGERIENNE

Country Overview

Formerly a French colony, the People's Democratic Republic of Algeria gained independence in 1962. Situated in north Africa, it has an area of 919,595 square miles and is bordered to the east by Tunisia and Libya, to the south by Niger, Mali, and Mauritania and to the west by Morocco. It has a 540 n mile coastline with the Mediterranean. The capital, largest city and principal port is Algiers. Territorial seas (12 n miles) and Fishery zones (32/52 n miles) have been claimed but an EEZ has not been claimed.

Headquarters Appointments

Commander of the Navy: Rear Admiral Malek Necib
Inspector General of the Navy:
  Rear Admiral Abdelmadjid Taright

Personnel

(a) 2014: 7,500 (500 officers) (Navy) (includes at least 600 naval infantry); 500 (Coast Guard)
(b) Voluntary service

Bases

Algiers (1st Region), Mers-el-Kebir (2nd Region), Jijel (3rd Region), Annaba (CG HQ)

Coast Defence

Four batteries of truck-mounted SS-C-3 Styx twin launchers. Permanent sites at Algiers, Mers-el-Kebir and Jijel linked by radar.

Algeria > Submarines — Frigates

SUBMARINES

Notes: One decommissioned Romeo class is used for training.

4 + (2) KILO CLASS (PROJECT 877EKM/636) (SSK)

Name	No	Builders	Laid down	Launched	Commissioned
RAIS HADJ MUBAREK (ex-*B 861*)	012	Nizhny Novgorod	1985	1986	Oct 1987
EL HADJ SLIMANE (ex-*B 386*)	013	Nizhny Novgorod	1985	1987	Jan 1988
AKRAM PACHA	021	Admiralty Yard, St Petersburg	2006	20 Nov 2008	2010
SIDI AHMED RAIS	022	Admiralty Yard, St Petersburg	2007	9 Apr 2009	2010

Displacement, tonnes: 2,362 surfaced; 3,125 dived
Dimensions, metres (feet): 72.6, 73.8 (021–022) × 9.9 × 6.6 *(238.2, 242.1 × 32.5 × 21.7)*
Speed, knots: 10 surfaced; 9 snorting; 17 dived
Range, n miles: 6,000 at 7 kt snorting; 400 at 3 kt dived
Complement: 52 (13 officers)

Machinery: Diesel-electric; 2 diesels; 3,650 hp(m) *(2.68 MW)*; 2 generators; 1 motor; 5,900 hp(m) *(4.34 MW)*; 1 shaft; 2 auxiliary MT-168 motors; 204 hp(m) *(150 kW)*; 1 economic speed motor; 130 hp(m) *(95 kW)*
Missiles: SLCM: Novator Alfa Klub SS-N-27 (022, 023); active radar homing to 220 km *(120 n miles)* at 0.8 Mach (cruise) and 2.9 Mach (attack); warhead 400 kg.
Torpedoes: 6—21 in *(533 mm)* tubes. Combination of Russian TEST-71ME; anti-submarine active/passive homing to 15 km *(8.2 n miles)* at 40 kt; warhead 205 kg and 53–65; anti-surface ship passive wake homing to 19 km *(10.3 n miles)* at 45 kt; warhead 300 kg. Total of 18 weapons.
Mines: 24 in lieu of torpedoes.
Electronic countermeasures: ESM: Brick Pulp; radar warning.
Radars: Surface search: Snoop Tray; I-band.
Sonars: MGK 400 Shark Teeth/Shark Fin; hull-mounted; passive/active search and attack; medium frequency. MG 519 Mouse Roar; active attack; high frequency.
Weapon control systems: MVU 110 TFCS.

Programmes: The Project 877EKM were new construction hulls which replaced the decommissioned Romeo class. A contract for the construction of two missile-armed Project 636 boats was signed with Admiralty Shipyards in mid-2006. The first was handed over in March 2010 and the second in August 2010. The first was delivered to Mers-el-Kebir in late 2010 and the second in late 2011. Procurement of a further two Type 636 class is reportedly under consideration.
Modernisation: Following refits in 1993–96, both Project 877 submarines underwent further two-year refits at Admiralty Yard, St Petersburg. Work on the first boat, which is reported to have included upgrade of the sonar system, began in November 2005 and completed in 2008. Refit of the second boat completed in 2012.
Structure: Diving depth, 790 ft *(240 m)*. 9,700 kWh batteries. Pressure hull 169.9 ft *(51.8 m)*. May be fitted with SA-N-5/8 portable SAM launcher.
Operational: Based at Mers El Kebir.

TYPE 636 022   8/2010, Lemachko Collection / 1366592

FRIGATES

3 MOURAD RAIS (KONI) CLASS (PROJECT 1159.2) (FFLM)

Name	No	Builders	Commissioned
MOURAD RAIS	901	Zelenodolsk Shipyard, Kazan	20 Dec 1980
RAIS KELLICH	902	Zelenodolsk Shipyard, Kazan	24 Mar 1982
RAIS KORFOU	903	Zelenodolsk Shipyard, Kazan	3 Jan 1985

Displacement, tonnes: 1,463 standard; 1,930 full load
Dimensions, metres (feet): 96.4 × 12.6 × 3.5 *(316.3 × 41.3 × 11.5)*
Speed, knots: 27; 22 diesel
Range, n miles: 1,800 at 14 kt
Complement: 130

Machinery: CODAG; 1 SGW, Nikolayev, M8B gas turbine (centre shaft); 18,000 hp(m) *(13.25 MW)* sustained; 2 Russki B-68 diesels; 15,820 hp(m) *(11.63 MW)* sustained; 3 shafts
Guns: 4—3 in *(76 mm)*/59 AK 726 (2 twin) ❶; 90 rds/min to 16 km *(8.5 n miles)*; weight of shell 5.9 kg. 4—30 mm/65 (2 twin) ❷; 500 rds/min to 5 km *(2.7 n miles)*; weight of shell 0.54 kg.
Torpedoes: 4—533 mm (2 twin) ❹.
A/S Mortars: 2—12-barrelled RBU 6000 ❸; range 6,000 m; warhead 31 kg.
Mines: Rails; capacity 22.
Depth charges: 2 racks.
Physical countermeasures: Decoys: 2 PJ 46 decoy launchers.
Electronic countermeasures: ESM: NRJ-6A.
Radars: Air/surface search: Pozitiv-ME1.2 ❺; I-band.
Navigation: Don 2; I-band.
Fire Control: Drum tilt ❻; H/I-band (for search/acquisition/FC). Pop Group ❼; F/H/I-band (for missile control).
IFF: High Pole B. 2 Square Head.
Sonars: Hercules (MG 322) hull-mounted; active search and attack; medium frequency.
Electro-optic systems: 2 optronic directors (901).

Programmes: New construction ships built in USSR with hull numbers 5, 7 and 10 in sequence. Others of the class built for Cuba, Yugoslavia, East Germany and Libya. Interest was shown in ex-GDR ships in 1991 but sale was rejected by the German government.
Modernisation: New generators fitted 1992–94. *Rais Korfou* in refit at Kronstadt from 1997 to November 2000. The refit included replacement of Strut Curve radar, removal of SA-N-4 SAM, removal of Hawk screech fire-control radar, fitting of torpedo tubes and a new electronic suite. Refit of *Mourad Rais* began in late 2007 and completed in 2011. *Rais Kellich* began refit in late 2008 and completed in 2012. Drum Tilt radar has been replaced by optronic director in these ships. A contract for the refit of *Rais Korfou* was let in 2012 and upgrade work is expected to be completed in 2014.
Structure: The deckhouse aft in Type II Konis houses air conditioning machinery.
Operational: All have been used for Training cruises. All based at Mers El Kebir.

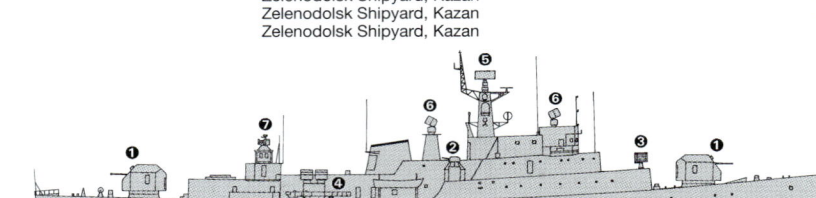

RAIS KORFOU   (Scale 1 : 900), Ian Sturton / 1366955

MOURAD RAIS   7/2011, M Declerck / 1406696

RAIS KORFOU   4/2005, Rafael Cabrera / 1167851

Frigates — Shipborne aircraft < **Algeria** 5

0 + 2 (2) MEKO A-200 CLASS (FFGHM)

Displacement, tonnes: 3,648 full load
Dimensions, metres (feet): 121 × 16.4 × 6.2 *(397 × 53.8 × 20.3)*
Speed, knots: 28
Range, n miles: 7,700 at 15 kt

Machinery: CODAG; 1 gas turbine; 2 diesels; 2 shafts; 1 centreline waterjet
Missiles: SSM: 8 Saab RBS 15 Mk 3.
SAM: Denel Umkhonto 16-cell VLS.
Guns: 1 Oto Melara 5 in *(127 mm)*. 2 Rheinmetall MLG 27 mm.
Torpedoes: 4 — 324 mm (2 twin) tubes; Eurotorp MU 90.
Physical countermeasures: To be announced.
Electronic countermeasures: To be announced.
Radars: Air/surface search: Saab Sea Giraffe AMB; G-band.
Fire control: Saab Ceros 200; J-band.
Navigation: To be announced.
Sonars: To be announced.
Combat data systems: Atlas Electronik.
Electro-optic systems: Saab Ceros 200 optronic director.
Helicopters: Super Lynx.

Comment: A contract for the construction of two MEKO A-200 frigates was placed with ThyssenKrupp Marine Systems on 26 March 2012. There is reported to be an option for two further ships. Details of the ships have not been released and, although reported to be broadly based on the South African Valour-class design, are illustrative. Delivery of the first of class is expected in 2016.

CORVETTES

3 NANUCHKA II (BURYA) CLASS (PROJECT 1234)
(MISSILE CORVETTES) (PTGM)

Name	No	Builders	Commissioned
RAIS HAMIDOU	801	Petrovsky Shipyard, Leningrad	4 Jul 1980
SALAH RAIS	802	Petrovsky Shipyard, Leningrad	9 Feb 1981
RAIS ALI	803	Petrovsky Shipyard, Leningrad	8 May 1982

Displacement, tonnes: 671 full load
Dimensions, metres (feet): 59.3 × 11.8 × 2.6 *(194.6 × 38.7 × 8.5)*
Speed, knots: 33
Range, n miles: 2,500 at 12 kt, 900 at 31 kt
Complement: 42 (7 officers)

Machinery: 6 M 504 diesels; 26,112 hp(m) *(19.2 MW)*; 3 shafts
Missiles: SSM: 8 Zvezda SS-N-25 Switchblade (2 quad) (Kh 35E Uran); active radar homing to 130 km *(70.2 n miles)* at 0.9 Mach; warhead 145 kg; sea skimmer.
SAM: SA-N-4 Gecko twin launcher; semi-active radar homing to 15 km *(8 n miles)* at 2.5 Mach; height envelope 9 – 3,048 m *(29.5 – 10,000 ft)*; warhead 50 kg; 20 missiles. Some anti-surface capability.
Guns: 2 — 57 mm/75 AK 725 (twin); 120 rds/min to 12.7 km *(6.8 n miles)*; weight of shell 2.8 kg.
1 — 30 mm/65 AK 630; 6 barrels per mounting; 3,000 rds/min combined to 2 km.
Physical countermeasures: Decoys: 2 PJ 46 decoy launchers.
Electronic countermeasures: ESM: NRJ-6A.
Radars: Surface search: Pozitiv-ME1.2; I-band.
Navigation: Don 2; I-band.
Fire control: Pop Group; F/H/I-band (SA-N-4). Drum Tilt; G/H-band. Plank Shave; E-band (SS-N-25).
IFF: Two Square Head. High Pole.

Programmes: Delivered as new construction.
Modernisation: *Salah Rais* refitted at Kronstadt 1997 to November 2000 with refurbished diesels, a replacement SSM system and electronic suite. Refit of *Rais Hamidou* began in late 2007 and completed in 2011. *Rais Ali* began refit in late 2008 and completed in 2012. A contract for the refit of *Salah Rais* was let in 2012 and work is expected to be completed in 2014.

SALAH RAIS (SS-N-25 not fitted)   12/2007, Diego Quevedo / 1335235

RAIS HAMIDOU   10/2010, Lemachko Collection / 1366590

0 + 3 CORVETTES (FS)

Name	No	Builders	Laid down	Launched	Commissioned
–	–	Hudong-Zhonghua, Shanghai	2013	2014	May 2015
–	–	Hudong-Zhonghua, Shanghai	2013	2014	Sep 2015
–	–	Hudong-Zhonghua, Shanghai	2013	2014	Feb 2016

Measurement, tonnes: 2,880 dwt
Dimensions, metres (feet): 120 × 14.4 × 3.87 *(393.7 × 47.2 × 12.7)*
Speed, knots: 28

Machinery: 4 MTU 16V 1163 TB93 diesels; 32,196 hp *(23.7 MW)*; 2 shafts
Guns: To be announced.
Radars: Surface search: To be announced.
Navigation: To be announced.
Fire control: To be announced.

Comment: A contract for the construction of three corvettes or light frigates was signed with China Shipbuilding Trading Company in May 2012. Construction at Hudong-Zhonghua Shipyard was expected to start in 2013 and all three ships are to be delivered in 2015. The design of the ships is expected to include a helicopter deck but details of weapons and sensors have yet to be announced. Dimensions suggest similarity to the Pakistan Navy's Sword class.

3 DJEBEL CHENOUA (C 58) CLASS (PROJECT 802) (FSG)

Name	No	Builders	Launched	Commissioned
DJEBEL CHENOUA	351	ECRN, Mers-el-Kebir	3 Feb 1985	Nov 1988
EL CHIHAB	352	ECRN, Mers-el-Kebir	Feb 1990	Jun 1995
AL KIRCH	353	ECRN, Mers-el-Kebir	Jul 2000	2002

Displacement, tonnes: 504 standard; 549 full load
Dimensions, metres (feet): 58.4 × 8.5 × 2.6 *(191.6 × 27.9 × 8.5)*
Speed, knots: 31
Complement: 52 (6 officers)

Machinery: 3 MTU 20V 538 TB92 diesels; 12,800 hp(m) *(9.4 MW)*; 3 shafts
Missiles: SSM: 4 China C 802 (CSS-N-8 Saccade) (2 twin); active radar homing to 120 km *(65 n miles)* at 0.9 Mach; warhead 165 kg.
Guns: 1 Russian 3 in *(76 mm)*/59 AK 176; 120 rds/min to 15 km *(8 n miles)*; weight of shell 5.9 kg. 1 — 30 mm/65 AK 630; 6 barrels per mounting; 3,000 rds/min combined to 2 km.
Physical countermeasures: Decoys: 2 chaff launchers.
Radars: Surface search: E/F-band.
Navigation: Racal Decca 1226; I-band.
Fire control: I-band.
Electro-optic systems: Optronic director.

Programmes: Ordered July 1983. Project 802 built with Bulgarian assistance. First one completed trials in 1988. Work on the second of class was suspended in 1992 due to shipyard debt problems but the ship completed in 1995. Main guns were fitted at a later date.
Structure: Hull size suggests association with Bazán Cormoran class.

AL KIRCH   6/2005, Marian Ferrette / 1127285

EL CHIHAB   7/2005, B Prézelin / 1129990

0 + 1 C 62 CLASS (CORVETTE) (FSG)

Name	No	Builders	Launched	Commissioned
GHARABA	–	ERCN, Mers-el-Kebir	Mar 2012	2014

Displacement, tonnes: 600 full load
Dimensions, metres (feet): 61.7 × 8.45 × 2.5 *(202.4 × 27.7 × 8.2)*
Speed, knots: 28

Machinery: 3 MTU 16V595 TE90 diesels; 3 shafts
Missiles: 4 C 802.
Guns: 1 — 76 mm AK-176. 1 — 30 mm AK-630.
Radars: Surface search: To be announced.
Fire control: To be announced.
Navigation: To be announced.

Comment: The new ship appears to be a stretched version of the C 58 Djebel Chenoua-class corvettes and to have similar armament. The missiles are situated forward of the AK-630 mounting, which is close to the stern. Entry into service is expected in 2014.

SHIPBORNE AIRCRAFT

Notes: It was reported in mid-2012 that six helicopters had been ordered from AgustaWestland for future operation from its new MEKO A200-class frigates. The helicopters could be further Super Lynx 300, of which four entered service in 2007, or Lynx Wildcat AW159.

8 Algeria > Tugs — Coast guard

TUGS

Notes: There are a number of harbour tugs of about 265 tons. These include *Kader* A 210, *El Chadid* A 211 and *Mazafran 1–4* Y 206–209.

MAZAFRAN 4 6/1994 / 0056454

COAST GUARD

Notes: (1) Six Kebir class were transferred from the Navy for Coast Guard duties but may have naval crews.
(2) There are also up to 12 small fishery protection vessels in the GC 301 series.

1 SUPPORT SHIP (WARL)

EL MOURAFEK GC 261

Displacement, tonnes: 610 full load
Dimensions, metres (feet): 59 × 8.4 × 2.1 *(193.6 × 27.6 × 6.9)*
Speed, knots: 14
Complement: 54
Machinery: 2 diesels; 2,200 hp(m) *(1.6 MW)*; 2 shafts
Guns: 2—12.7 mm MGs.
Radars: Surface search: I-band.

Comment: Delivered by transporter ship from China in April 1990. The design appears to be a derivative of the T43 minesweeper but with a stern gantry. May have a minelaying capability. Based at Algiers.

EL MOURAFEK 6/2009, Diego Quevedo / 1305756

7 EL MOUDERRIB (CHUI-E) CLASS (AXL)

EL MOUDERRIB I GC 251	**EL MOUDERRIB V** GC 255
EL MOUDERRIB II GC 252	**EL MOUDERRIB VI** GC 256
EL MOUDERRIB III GC 253	**EL MOUDERRIB VII** GC 257
EL MOUDERRIB IV GC 254	

Displacement, tonnes: 394 full load
Dimensions, metres (feet): 58.8 × 7.2 × 2.2 *(192.9 × 23.6 × 7.2)*
Speed, knots: 24
Range, n miles: 1,400 at 15 kt
Complement: 42 (25 trainees)
Machinery: 3 PCR/Kolomna diesels; 6,600 hp(m) *(4.92 MW)*; 3 shafts
Guns: 4 China 14.5 mm (2 twin).
Radars: Surface search: Type 756; I-band.

Comment: Two delivered by transporter ship from China in April 1990 and described as training vessels. Two more acquired in January 1991, the last three in July 1991. Hainan class hull with modified propulsion and superstructure similar to some Chinese paramilitary vessels. Used for training when boats are carried aft in place of the second 14.5 mm gun. GC 255 and 257 are reported non-operational.

EL MOUDERRIB IV 3/2006, M Declerck / 1164475

6 BAGLIETTO TYPE 20 (PBF)

EL HAMIL GC 325	**MARKHAD** GC 327	**AKHIR NAHR** GC 329
EL ASSAD GC 326	**ETAIR** GC 328	**NEDJMET EL KOTB** GC 321

Displacement, tonnes: 45 full load
Dimensions, metres (feet): 20.4 × 5.2 × 1.7 *(66.9 × 17.1 × 5.6)*
Speed, knots: 36
Range, n miles: 445 at 20 kt
Complement: 11 (3 officers)
Machinery: 2 CRM 18DS diesels; 2,660 hp(m) *(2 MW)*; 2 shafts
Guns: 1 Oerlikon 20 mm.

Comment: The first pair delivered by Baglietto, Varazze in August 1976 and six further in pairs at two monthly intervals. Fitted with radar and optical fire control. Four others of the class cannibalised for spares.

BAGLIETTO 20 3/2006, M Declerck / 1164477

6 BAGLIETTO MANGUSTA CLASS (PB)

EL AYOUK 323 **ZOUHEL** 324 **REQUIN** 331 – 332 **MARSOUIN** 333 **MURENE** 334

Displacement, tonnes: 92 full load
Dimensions, metres (feet): 30 × 5.8 × 2.2 *(98.4 × 19.0 × 7.2)*
Speed, knots: 32.5
Range, n miles: 800 at 24 kt
Complement: 14 (3 officers)
Machinery: 3 MTU diesels; 4,000 hp *(3.0 MW)*; 3 shafts
Guns: 2—25 mm (1 twin). 1—12.7 mm MG.
Radars: Navigation: I-band.

Comment: Six patrol craft first of which delivered to Algeria in early 1977. All six craft reported to have been decommissioned between 1998 and 2001 but brought back to service in 2006–08.

REQUIN 3/2006 / 1164476

4 EL MOUNKID CLASS (SAR)

EL MOUNKID I GC 231	**EL MOUNKID III** GC 233
EL MOUNKID II GC 232	**EL MOUNKID IV** GC 234

Comment: First three delivered by transporter ship from China which arrived in Algiers in April 1990, a fourth followed a year later. Used for SAR.

GC 231–233 1991 / 0056457

12 JEBEL ANTAR CLASS (PB)

JEBEL ANTAR 301	**RAS TENES** 305	**RAS BOUGARONI** 309
JEBEL HANDO 302	**RAS TEKKOUCH** 306	**RAS TAMENTFOUST** 310
– 303	**RAS SISLI** 307	**RAS OULLIS** 311
RAS DJENAD 304	**RAS NOUH** 308	– 312

Dimensions, metres (feet): 17 × ? × ? *(55.8 × ? × ?)*
Speed, knots: 15
Machinery: 2 diesels; 2 shafts
Guns: To be announced.
Radars: Navigation: I-band.

Comment: Patrol craft reported constructed at Mers-el-Kebir 1982–83.

IHS Jane's Fighting Ships 2014-2015 © 2014 IHS

Coast guard < **Algeria** — Patrol forces < **Angola** 9

21 OCEA FPB 98 CLASS (PATROL CRAFT) (PB)

DENEB 334	KAUS AUSTRALE 371	ARNAB 378
MIZAR 335	FORMALHAUT 372	MIRZAM 379
ALKAID 336	MARKAB 373	PEACOCK 380
ALTAIR 337	BETELGEUSE 374	NAIRAL ZAURAC 381
MIRFAK 338	KIFFA BOREALIS 375	PLEIDAS 382
ALNAIR 339	HAMAL 376	SIRIUS 383
ACHERNAR 370	DENEB ALGEDI 377	L'ÉTOILE POLAIRE 384

Displacement, tonnes: 118 full load
Dimensions, metres (feet): 31.8 × 6.3 × 1.2 *(104.3 × 20.7 × 3.9)*
Speed, knots: 30. **Range, n miles:** 900 at 14 kt
Complement: 11 (3 officers)
Machinery: 2 Caterpillar C 32 diesels; 3,660 hp *(2.7 MW)*; 2 waterjets
Guns: 1 — 30 mm.
Radars: Navigation: 2 Furuno; I-band.

Comment: The contract with OCEA, reported to have been signed in 2007, for the construction of 21 patrol craft was announced in October 2008. Delivery of the first vessel was made in late 2008 and the programme was completed in 2011. *L'Etoile Polaire* has been modified for VIP duties. The vessels were built at St Nazaire.

MIRZAM 11/2010, B Prézelin / 1406807

3 SALVAGE TUGS (ARS)

EL MOUNDJID 701	EL MOUS'IF 702	EL MOUSSANID 703

Measurement, tonnes: 3,398 gt
Dimensions, metres (feet): 86 × 17.5 × 6.0 *(282.2 × 57.4 × 19.7)*
Speed, knots: 19
Machinery: 4 Bergen B32 diesels; 21,752 hp *(16.0 MW)*; 2 shafts; cp props; 4 bow thrusters
Radars: Navigation: I-band.

Comment: The ships are similar to the French Bourbon (Ulstein 515)-class salvage ships and were constructed at STX Tulcea in Romania and STX Brattvag Shipyard in Norway. Steel construction. The first ship was delivered in December 2011, the second in May 2012, and the third in September 2012.

EL MOUNDJID 2/2012, Andrew Knox / 1455754

CUSTOMS

Notes: The Customs service is a paramilitary organisation employing a number of patrol craft armed with small MGs. These include *Bouzagza*, *Djurdjura*, *Hodna*, *Aures* and *Hoggar*. The first three are P 1200 class 39 ton craft capable of 33 kt. The next pair are P 802 class. They were built by Watercraft, Shoreham and delivered in November 1985.

Angola
MARINHA DE GUERRA

Country Overview

Formerly known as Portuguese West Africa, the Republic of Angola became independent in 1975 but was ravaged by civil war until 2002. Elections were held in September 2008. With an area of 481,354 square miles it has borders to the south with Namibia, to the east with Zambia and to the north and east with the Democratic Republic of the Congo which separates a small exclave, Cabinda, from the rest of the country. Angola has a coastline with the south Atlantic Ocean of some 864 n miles. The capital, largest city and principal port is Luanda. Territorial seas (12 n miles) and a fisheries zone (200 n miles) are claimed. A 200 n mile Exclusive Economic Zone (EEZ) has been claimed but the limits have not been published.

Headquarters Appointments

Commander of the Navy: Admiral Augusto da Silva Cunha

Personnel

(a) 2014: 890
(b) Voluntary service

Bases

Luanda, Lobito, Namibe. (There are other good harbours available on the 1,000 mile coastline.) Naval HQ at Luanda on Ila de Luanda is in an old fort, as is Namibe.

Naval Aviation

Seven EADS-CASA C 212-300MP and one Fokker F27 maritime patrol aircraft are operated by the Air Force.

PATROL FORCES

Notes: Negotiations to procure both a 54 m Macaé class and a 46.5 m Grajaú-class patrol ship from Brazil were in progress in late 2012. A contract was expected in 2013 but has not been confirmed.

5 OFFSHORE PATROL VESSELS (PB)

COMANDANTE IMPERIAL SANTANA	COMANDANTE VALODIA
COMANDANTE KASSANJE	LUCRETIA PAIM
COMANDANTE PAIVA DOMINGOS DA SILVA	

Measurement, tonnes: 107 gt
Dimensions, metres (feet): 30.07 × 5.7 × 1.55 *(98.7 × 18.7 × 5.1)*
Speed, knots: 20
Machinery: 2 MTU 12V 2000M70 diesels; 2,142 hp *(1.58 MW)*; 2 shafts

Comment: Constructed by Zhuhai Jianglong Shipbuilding, Zuhai, and delivered in September 2009. GRP construction. Owned and operated by the Ministry of Fisheries.

DA SILVA 6/2012 / 1455755

5 OFFSHORE PATROL VESSELS (PBO)

REI BULA MATADI	REI KINGURY KA BANGUELA	REI MPANZU A NIMI
REI EKUIKI II	REI MANDUME NDEMUFAYO	

Measurement, tonnes: 322 gt
Dimensions, metres (feet): 47.4 × 7.6 × 2.55 *(155.5 × 24.9 × 8.4)*
Speed, knots: 20. **Range, n miles:** 420 at 20 kt
Machinery: 2 MTU 12V 4000M70 diesels; 4,732 hp *(3.48 MW)*; 2 shafts

Comment: Constructed by Xijiang Shipyard, Luzhou, and delivered in September 2009. Steel construction. Owned and operated by the Ministry of Fisheries.

REI MANDUME NDEMUFAYO 12/2009, Annati Collection / 1406001

5 ARESA PVC-170 CLASS (PB)

P 618–622

Displacement, tonnes: 18 full load
Dimensions, metres (feet): 16 × 4.48 × 2.3 *(52.5 × 14.7 × 7.5)*
Speed, knots: 35
Range, n miles: 420 at 20 kt
Complement: 5
Machinery: 2 diesels; 1,650 hp *(1.2 MW)*; 2 shafts

Comment: Constructed by ARESA Shipbuilders, Barcelona, in 2009 for Angolan fishery protection duties under Spanish Fisheries Department aid programme.

P 620 12/2008, Adolfo Ortigueira Gil / 1367007

© 2014 IHS IHS Jane's Fighting Ships 2014-2015

10 Angola > Patrol forces — **Antigua and Barbuda** > Introduction

4 MANDUME CLASS (COASTAL PATROL CRAFT) (PC)

Name	No	Builders	Laid down	Launched	Commissioned
MANDUME	P 100	Bazán Shipyard, San Fernando	Nov 1991	11 Sep 1992	28 Jan 1993
POLAR	P 102	Bazán Shipyard, San Fernando	Nov 1991	11 Sep 1992	28 Jan 1993
ATLANTICO	P 104	Bazán Shipyard, San Fernando	–	Feb 1993	Apr 1993
GOLFINHO	P 106	Bazán Shipyard, San Fernando	–	Feb 1993	Apr 1993

Displacement, tonnes: 112 full load
Dimensions, metres (feet): 31.6 × 5.9 × 1.5 *(103.7 × 19.4 × 4.9)*
Speed, knots: 27. **Range, n miles:** 8,000 at 15 kt
Complement: 11 (1 officer)
Machinery: 2 Paxman Vega 16CM diesels; 3,840 hp *(2.86 MW)* sustained; 2 shafts
Guns: 1 Oerlikon GAM-BO1 20 mm. 2 – 12.7 mm MGs.
Radars: Surface search: Racal Decca; I-band.

Comment: Ordered 27 March 1991. First two laid down November 1991 at Bazán Shipyard, San Fernando; launched 11 September 1992. First one handed over in April 1993, the others at three month intervals. This is an Alcotan 30 design with steel hulls and an aluminium superstructure. These craft have a controlled clutch hydraulic drive system for slow speed operations. All four craft reported refitted in 2008 with Korean assistance.

MANDUME — 11/2012*, M Globke / 1486559

2 NAMACURRA CLASS (INSHORE PATROL CRAFT) (PB)

Displacement, tonnes: 5 full load
Dimensions, metres (feet): 9 × 2.7 × 0.8 *(29.5 × 8.9 × 2.6)*
Speed, knots: 32. **Range, n miles:** 180 at 20 kt
Complement: 4

Machinery: 2 Yamaha outboards; 380 hp(m) *(2.79 kW)*
Guns: 1 – 12.7 mm MG. 2 – 7.62 mm MGs.
Depth charges: 1 rack.
Radars: Surface search: Furuno; I-band.

Comment: Built in South Africa in 1980–81. Can be transported by road. Donated by South Africa in 2006.

NAMACURRA (South Africa colours) — 8/2001, van Ginderen Collection / 0132783

2 DAMEN 6210 CLASS (PATROL SHIP) (PSO)

NGOLA KILUANGE   NZINGA MBANDI

Displacement, tonnes: 1,022 full load
Dimensions, metres (feet): 61.94 × 9.7 × 3.25 *(203.2 × 31.8 × 10.7)*
Speed, knots: 18. **Range, n miles:** 420 at 20 kt
Complement: 45 (13 officers)
Machinery: 2 MTU 16V 4000 M70 diesels; 6,222 hp *(4.64 MW)*; 2 shafts; cp props; bow thruster
Radars: Surface search/navigation: E/F/I-bands.
Helicopters: Platform for 1 medium.

Comment: Ordered from Damen by the Ministry of Fisheries. Equipped to conduct inspection, surveillance, SAR and firefighting, the ships are fitted with a 12-berth hospital. *Ngola Kiluange* was launched at Galati, Romania, in December 2012 and *Nzinga Mbandi* in May 2012. They were delivered in August and November 2012 respectively.

NZINGA MBANDI — 11/2012*, Damen / 1525511

Anguilla

Country Overview

British dependency since 1971 following secession from associated state of St Kitts-Nevis-Anguilla. With an area of 35 square miles, the island is situated at the northern end of the Leeward Islands in the Lesser Antilles and bordered by the Caribbean to the west and Atlantic to the east. Territorial seas (3 n miles) and a fishery zone (200 n miles) are claimed.

Headquarters Appointments
Inspector of Marine:
  Superintendent Allan Coppin

Personnel
2014: 99

POLICE

1 HALMATIC M160 CLASS (INSHORE PATROL CRAFT) (PB)

DOLPHIN

Displacement, tonnes: 18 standard
Dimensions, metres (feet): 16 × 4.7 × 1.4 *(52.5 × 15.4 × 4.6)*
Speed, knots: 32. **Range, n miles:** 575 at 23 kt
Complement: 8
Machinery: 2 MAN V10 diesels; 820 hp *(610 kW)* sustained; 2 shafts
Guns: 1 – 12.7 mm MG.
Radars: Surface search: JRC 2254; I-band.

Comment: Built by Halmatic and delivered 22 December 1989. GRP hull. Rigid inflatable boat launched by electrical davit. Returned to service on 30 August 2004 after refit.

1 BOSTON WHALER (INSHORE PATROL CRAFT) (PB)

LAPWING

Displacement, tonnes: 2 full load
Dimensions, metres (feet): 8.2 × 3 × 0.5 *(26.9 × 9.8 × 1.6)*
Speed, knots: 35
Complement: 4
Machinery: 2 Yamaha outboards; 300 hp *(225 kW)*

Comment: Delivered in 1990 and re-engined in 2008.

DOLPHIN — 6/2010, Anguilla Police / 1366497

LAPWING — 6/2010, Anguilla Police / 1366496

Antigua and Barbuda

Country Overview

Independent since 1981, the British monarch, represented by a governor-general, is head of state. Situated at the southern end of the Leeward Antilles chain, the country comprises Antigua (108 square miles), Barbuda to the north and uninhabited Redonda to the southwest. The capital, largest town, and main port is St John's. An archipelagic state, territorial seas (12 n miles) and a fishery zone (200 n miles) are claimed. A 200 n mile Exclusive Economic Zone (EEZ) has also been claimed but the limits are not defined. The Antigua Barbuda Defence Force (ABDF) took over the Coast Guard on 1 May 1995.

Headquarters Appointments
Commanding Officer, Coast Guard:
  Lieutenant Commander Auden Nicholas

Personnel
2014: 50 (3 officers)

Bases
HQ: Deepwater Harbour, St Johns
Maintenance: Camp Blizzard

IHS Jane's Fighting Ships 2014-2015 © 2014 IHS

COAST GUARD

Notes: (1) In addition there is a Hurricane RIB, *CG 081* with a speed of 35 kt and two Boston Whalers, *CG 071-2*, with speeds of 30 kt. All were acquired in 1988/90.
(2) A 920 Zodiac RHIB, *CG 091*, was donated by the US government in 2003. It is capable of over 40 kt.

CG 091   9/2004, ABDFCG / 0587690

CG 081   9/2004, ABDFCG / 0587691

2 SPECIAL PURPOSE CRAFT (PBF)

D 7   D 8

Displacement, tonnes: 7 full load
Dimensions, metres (feet): 10.1 × 3.0 × 0.7 *(33.1 × 9.8 × 2.3)*
Speed, knots: 50
Complement: 4
Machinery: 3 Mercury outboard motors; 825 hp *(615 kW)*
Guns: 2—7.62 mm MGs.
Radars: Navigation: Raymarine; I-band.

Comment: Built by SAFE Boats International and very similar to the craft in service with the US Coast Guard. Both craft donated by US Southern Command under the US foreign Military Sales Program in November 2012.

1 SWIFT 65 ft CLASS (PB)

Name	No	Builders	Commissioned
LIBERTA	P 01	Swiftships, Morgan City	30 Apr 1984

Displacement, tonnes: 37 full load
Dimensions, metres (feet): 20 × 5.6 × 1.5 *(65.6 × 18.4 × 4.9)*
Speed, knots: 22. **Range, n miles:** 250 at 18 kt
Complement: 9
Machinery: 2 Detroit Diesel 12V-71TA diesels; 840 hp *(616 kW)* sustained; 2 shafts
Guns: 1—12.7 mm MG. 2—7.62 mm MGs.
Radars: Surface search: Furuno; I-band.

Comment: Ordered in November 1983. Aluminium construction. Funded by US. Refitted in 2001.

LIBERTA   5/2003 / 0568341

1 DAUNTLESS CLASS (PB)

Name	No	Builders	Commissioned
PALMETTO	P 02	SeaArk Marine, Monticello	7 Jul 1995

Displacement, tonnes: 11 full load
Dimensions, metres (feet): 12.2 × 4.3 × 1.3 *(40 × 14.1 × 4.3)*
Speed, knots: 27. **Range, n miles:** 600 at 18 kt
Complement: 4
Machinery: 2 Caterpillar 3208TA diesels; 870 hp *(650 kW)* sustained; 2 shafts
Guns: 1—7.62 mm MG.
Radars: Surface search: Raytheon R40; I-band.

Comment: Funded by USA. Similar craft delivered to several Caribbean countries in 1994–98.

PALMETTO   9/2004, ABDFCG / 0587689

Argentina

ARMADA ARGENTINA

Country Overview

The Argentine Republic is in southern South America. With an area of 1,068,302 square miles it has borders to the north with Bolivia and Paraguay, to the east with Brazil and Uruguay and to the south and west with Chile. The country includes the Tierra del Fuego territory which comprises the eastern half of the Isla Grande de Tierra del Fuego and a number of adjacent islands to the east, including Isla de los Estados. It also claims sovereignty of the Falkland Islands. The capital, largest city and principal port is Buenos Aires. There are further ports at La Plata, Bahia Blanca, Comodoro Rivadavia and a river port at Rosario. There are some 5,940 n miles of navigable internal waterways. Territorial Seas (12 n miles) are claimed. An EEZ (200 n miles) is claimed but its limits are only partly defined by boundary agreements.

Headquarters Appointments

Chief of Naval General Staff:
  Vice Admiral Fernando Gastón Erice
Deputy Chief of Naval Staff:
  Rear Admiral Álvaro Manuel González Lonzieme
Director General Plans and Programmes:
  Rear Admiral Jorge Alberto Martino

Senior Appointments

Commander Fleet: Rear Admiral Juan Carlos Temperoni
Commander, Marine Infantry: Rear Admiral Jorge Luís García
Commander Naval Aviation:
  Rear Admiral Guillermo Rolando Bellido
Commander, Naval Area Austral:
  Rear Admiral Carlos Enrique Aguilera
Commander, Submarines: Captain Gabriel Eduardo Attis
Commander, Naval Area Atlantic:
  Commodore Diego Orlando Dalmiro
Commander, Naval Area Fluvial:
  Captain Miguel Adrian Gonzalez

Personnel

2014: 18,249 (2,531 officers)

Organisation

Naval Area Austral covers coastal area from latitude 46° to 60° south.
Naval Area Atlantic covers coastal area from latitude 36° 18′ to 46° south.
Naval Area Fluvial includes the rivers Paraná, Uruguay and Plate.
Naval Area Antarctica is activated when *Almirante Irizar* deploys.

Special Forces

Consists of tactical divers who operate from submarines and other naval units, and amphibious commandos who are trained in parachuting and behind the lines operations. Both groups consist of about 150. Based at Mar del Plata naval base.

Bases

Buenos Aires (Dársena Norte): Some naval training.
Rio Santiago (La Plata): Schools.
Mar del Plata: Submarine base plus Maritime Patrol Division and Hydrographic ships.
Puerto Belgrano: Main naval base, schools. Fleet Marine Force.
Ushuaia (new naval base to be built), Deseado, Dársena Sur, Zárate, Caleta Paula; Small naval bases.
Rio Gallegos (Santa Cruz). A small base opened in 2012.

Prefix to Ships' Names

ARA (Armada Republica Argentina)

Naval Aviation

Personnel: 2,500
The Naval Air Command is at Puerto Belgrano.
1st Naval Air Wing (Punta Indio Naval Air Base): Naval Reconnaissance Group with Beech 200s. Naval Aviation School with Beech T-34 Turbo Mentor.
2nd Naval Air Wing (Comandante Espora Naval Air Base): ASW Squadron with Grumman S-2T Trackers; 2nd Naval Helicopter Squadron with Agusta/Sikorsky SH-3H and AS-61D Sea Kings; 2nd Naval Attack Squadron with Super Etendards; 1st Naval Helicopter Squadron with Alouette III and Fennecs.
3rd Naval Air Wing (Almirante Zar Naval Air Base, Trelew): 6th Naval Reconnaissance and Surveillance Squadron with Lockheed P-3C Orions, Beechcraft B 200 and Pilatus PC-6B.
52 Logistic Support Flight (Almirante Izar Naval Air Base): Fokker F-28s.

Marine Corps

Personnel: 2,800
2nd Marine Infantry Battalion (Puerto Belgrano)
3rd Marine Infantry Battalion (Zarate)
4th Marine Infantry Battalion (Ushuaia)
5th Marine Infantry Battalion (Training) (Rio Grande)
Marine Field Artillery Battalion (Puerto Belgrano)
Command and Logistics Support Battalion (Puerto Belgrano)
Amphibious Vehicles Battalion (Puerto Belgrano)
Communications Battalion (Puerto Belgrano)
Marine A/A Battalion (Puerto Belgrano)
Amphibious Engineers Company (Puerto Belgrano)
Amphibious Commandos Group (Puerto Belgrano)
There are Marine Security Battalions at Naval Bases in Buenos Aires and Puerto Belgrano.
There are Marine Security Companies at Naval Bases in Mar del Plata, Trelew, Ushuaia, Punta Indio and Zarate.

Strength of the Fleet

Type	Active (Reserve)	Building
Patrol Submarines	3	—
Destroyers	4	—
Frigates	9	—
Patrol Ships	4	(4)
Fast Attack Craft (Gun/Missile)	2	—
Coastal Patrol Craft	6	—
Survey/Oceanographic Ships	4	—
Survey Launches	1	—
Transports/Tankers	8	—
Training Ships	8	—

Argentina > Introduction — Submarines

PENNANT LIST

Submarines

S 31	Salta
S 41	Santa Cruz
S 42	San Juan

Destroyers

D 10	Almirante Brown
D 11	La Argentina
D 12	Heroina
D 13	Sarandi

Frigates

31	Drummond
32	Guerrico
33	Granville

41	Espora
42	Rosales
43	Spiro
44	Parker
45	Robinson
46	Gomez Roca

Patrol Forces

A 2	Teniente Olivieri
A 3	Francisco de Gurruchaga
A 6	Suboficial Castillo
A 9	Alferez Sobral
P 20	Murature
P 21	King
P 61	Baradero
P 62	Barranqueras
P 63	Clorinda

P 64	Concepción del Uruguay
P 65	Punta Mogotes
P 66	Rio Santiago
P 85	Intrepida
P 86	Indomita

Auxiliaries

B 1	Patagonia
B 3	Canal Beagle
B 4	Bahia San Blas
B 5	Cabo de Hornos
B 13	Ingeniero Julio Krause
B 52	Hercules
Q 2	Libertad
Q 5	Almirante Irizar
Q 11	Comodoro Rivadavia
Q 15	Cormoran

Q 20	Puerto Deseado
Q 61	Ciudad de Zarate
Q 62	Ciudad de Rosario
Q 63	Punta Alta
Q 73	Itati
Q 74	Fortuna I
Q 75	Fortuna II
Q 76	Fortuna III
R 2	Querandi
R 3	Tehuelche
R 5	Mocovi
R 6	Calchaqui
R 7	Ona
R 8	Toba
R 10	Chulupi
R 12	Mataco
R 16	Capayán
R 18	Chiquillán
R 19	Morcoyán

SUBMARINES

Notes: (1) Cosmos and Havas underwater chariots in service. Cosmos types are capable of carrying limpet or ground mines.
(2) There are no known plans to replace the current submarine force.

2 SANTA CRUZ (TR 1700) CLASS (SSK)

Name	No	Builders	Laid down	Launched	Commissioned
SANTA CRUZ	S 41	Thyssen Nordseewerke, Emden	6 Dec 1980	28 Sep 1982	18 Oct 1984
SAN JUAN	S 42	Thyssen Nordseewerke, Emden	18 Mar 1982	20 Jun 1983	19 Nov 1985

Displacement, tonnes: 2,150 surfaced; 2,300 dived
Dimensions, metres (feet): 66 × 7.3 × 6.5 *(216.5 × 24.0 × 21.3)*
Speed, knots: 15 surfaced; 12 snorting; 26 dived
Range, n miles: 12,000 at 8 kt surfaced, 20 at 25 kt dived, 460 at 6 kt dived
Complement: 29 (5 officers)

Machinery: Diesel-electric; 4 MTU 16V 6,720 hp diesels; 6,720 hp(m) *(4.94 MW)* sustained; 4 alternators; 4.4 MW; 1 Siemens Type 1HR4525 + 1HR 4525 4-circuit DC motor; 6.6 MW; 1 shaft
Torpedoes: 6—21 in *(533 mm)* bow tubes. 22 AEG SST 4; wire-guided; active/passive homing to 12/28 km *(6.5/15 n miles)* at 35/23 kt; warhead 260 kg; automatic reload in 50 seconds.
Mines: Capable of carrying 34 ground mines.
Electronic countermeasures: ESM: Kollmorgen Sea Sentry III; radar warning.
Radars: Navigation: Thomson-CSF Calypso IV; I-band.
Sonars: Atlas Elektronik CSU 3/4; active/passive search and attack; medium frequency. Thomson Sintra DUUX 5; passive ranging.
Weapon control systems: Signaal Simbads; can handle 5 targets and 3 torpedoes simultaneously.

Programmes: Contract signed 30 November 1977 with Thyssen Nordseewerke for two submarines to be built at Emden. Parts and technical oversight were also to be provided for the construction of four further boats in Argentina by Astilleros Domecq Garcia, Buenos Aires. Work on units three and four was initiated and S 43 *(Santa Fe)* was reported as 70% complete by 2004. Reports in 2013 suggested that work on the boat may be resumed and that the boat will be completed as an SSK. Work on S 44 *(Santiago del Estero)* was reported as 30% complete in 1996 but further work since then has not been reported. Equipment for numbers five and six has been used for spares.
Modernisation: *Santa Cruz* underwent a mid-life modernisation at Arsenal de Marinha, Rio de Janeiro between September 1999 and 2001. The work included replacement of the engines, batteries, and sonar. *San Juan* is being similarly modernised at Domecq Garcia shipyard. The refit, which began on 17 August 2007, is expected to be completed in 2014. On completion, *Santa Cruz* is to undertake a full refit, including battery change.
Structure: Diving depth, 270 m *(890 ft)*.
Operational: Maximum endurance is 70 days. Both can be used for Commando insertion operations. They are based at Mar del Plata.

SAN JUAN — 5/2004, A E Galarce / 1044065

SANTA CRUZ — 7/2004, A E Galarce / 1044064

Submarines — Destroyers ◂ **Argentina** 13

1 SALTA (TYPE 209/1200) CLASS (SSK)

Name	No	Builders	Laid down	Launched	Commissioned
SALTA	S 31	Howaldtswerke, Kiel	30 Apr 1970	9 Nov 1972	7 Mar 1974

Displacement, tonnes: 1,158 surfaced; 1,268 dived
Dimensions, metres (feet): 55.9 × 6.3 × 5.5
 (183.4 × 20.7 × 18.0)
Speed, knots: 10 surfaced; 11 snorting; 22 dived
Range, n miles: 6,000 at 8 kt surfaced; 400 at 4 kt dived
Complement: 31 (5 officers)

Machinery: Diesel-electric; 4 MTU 12V 493 AZ80 diesels; 2,400 hp(m) *(1.76 MW)* sustained; 4 alternators; 1.7 MW; 1 motor; 4,600 hp(m) *(3.36 MW)*; 1 shaft
Torpedoes: 8—21 in *(533 mm)* bow tubes. 14 AEG SST 4 Mod 1; wire-guided; active/passive homing to 12/28 km *(6.5/15 n miles)* at 35/23 kt; warhead 260 kg.
Mines: Capable of carrying ground mines.
Electronic countermeasures: ESM: Thomson CSF DR 2000; radar warning.
Radars: Navigation: Thomson-CSF Calypso II.
Sonars: Atlas Elektronik CSU 3 (AN 526/AN 5039/41); active/passive search and attack; medium frequency. Thomson Sintra DUUX 2C and DUUG 1D; passive ranging.
Weapon control systems: Signaal M8 digital; computer-based; up to 3 targets engaged simultaneously.

Programmes: Ordered in 1968. Built in sections by Howaldtswerke Deutsche Werft AG, Kiel from the IK 68 design of Ingenieurkontor, Lübeck. Sections were shipped to Argentina for assembly at Tandanor, Buenos Aires. Second of class *(San Luis)* has been used for spares since 1997. However, re-activation remains a possibility; a full refit, including insertion of a new section, is reportedly under consideration.
Modernisation: *Salta* completed a mid-life modernisation at the Domecq Garcia Shipyard in May 1995. New engines, weapons and electrical systems fitted. Installation of new batteries began at Domecq Garcia in 2004 and completed in August 2005. A further refit 2009–10 included upgrade of the sonar and combat system. The boat underwent an engineering overhaul in 2013 and is to be completed in 2014. Replacement of the CSU 3 sonar with L-3 ELAC NAUTIK LOPAS 8300 passive sonar is under consideration.
Structure: Diving depth, 250 m *(820 ft)*.
Operational: Operational and based at Mar del Plata.

SALTA   *12/2002, A E Galarce* / 0529819

DESTROYERS

4 ALMIRANTE BROWN (MEKO 360 H2) CLASS (DDGHM)

Name	No	Builders	Laid down	Launched	Commissioned
ALMIRANTE BROWN	D 10	Blohm + Voss, Hamburg	8 Sep 1980	28 Mar 1981	26 Jan 1983
LA ARGENTINA	D 11	Blohm + Voss, Hamburg	30 Mar 1981	25 Sep 1981	4 May 1983
HEROINA	D 12	Blohm + Voss, Hamburg	24 Aug 1981	17 Feb 1982	31 Oct 1983
SARANDI	D 13	Blohm + Voss, Hamburg	9 Mar 1982	31 Aug 1982	16 Apr 1984

Displacement, tonnes: 2,947 standard; 3,667 full load
Dimensions, metres (feet): 125.9 × 14 × 5.8 screws
 (413.1 × 45.9 × 19.0)
Speed, knots: 30. **Range, n miles:** 4,500 at 18 kt
Complement: 200 (26 officers)

Machinery: COGOG; 2 RR Olympus TM3B gas turbines; 50,000 hp *(37.4 MW)* sustained 2 RR Tyne RM1C gas turbines; 9,900 hp *(7.4 MW)* sustained; 2 shafts; cp props
Missiles: SSM: 8 Aerospatiale MM 40 Exocet (2 quad) launchers ❶; inertial cruise; active radar homing to 70 km *(40 n miles)*; warhead 165 kg; sea-skimmer.
 SAM: Selenia/Elsag Albatros octuple launcher ❷; 24 Aspide; semi-active homing to 13 km *(7 n miles)* at 2.5 Mach; height envelope 15–5,000 m *(49.2–16,405 ft)*; warhead 30 kg.
Guns: 1 Oto Melara 5 in *(127 mm)*/54 automatic ❸; 45 rds/min to 23 km *(12.42 n miles)* anti-surface; 7 km *(3.6 n miles)* anti-aircraft; weight of shell 32 kg; also fires chaff and illuminants.
 8 Breda/Bofors 40 mm/70 (4 twin) ❹; 300 rds/min to 12.6 km *(6.8 n miles)* anti-surface; 4 km *(2.2 n miles)* anti-aircraft; weight of shell 0.96 kg; 2 Oerlikon 20 mm.
Torpedoes: 6—324 mm ILAS 3 (2 triple) tubes ❺. Whitehead A 244; anti-submarine; active/passive homing to 7 km *(3.8 n miles)* at 33 kt; warhead 34 kg (shaped charge); 18 reloads.
Depth charges: 10 Mk 9 depth charges.
Physical countermeasures: Decoys: CSEE Dagaie double mounting; Graseby G1738 towed torpedo decoy system. 2 Breda 105 mm SCLAR chaff rocket launchers; 20 tubes per launcher; can be trained and elevated; chaff to 5 km *(2.7 n miles)*; illuminants to 12 km *(6.6 n miles)*.
Electronic countermeasures: ESM/ECM: Sphinx/Scimitar.
Radars: Air/surface search: Signaal DA08A ❻; F-band; range 204 km *(110 n miles)* for 2 m² target.
 Surface search: Signaal ZW06 ❼; I-band.
 Navigation: Decca 1226; I-band.
 Fire control: Signaal STIR ❿; I/J/K-band; range 140 km *(76 n miles)*; Signaal WM25 ❼; I/J-band.
Sonars: Atlas Elektronik 80 (DSQS-21BZ); hull-mounted; active search and attack; medium frequency.
Combat data systems: Signaal SEWACO; Link 10/11. SATCOMs can be fitted.
Electro-optic systems: 2 Signaal LIROD radar/optronic systems ❾ each controlling 2 twin 40 mm mounts.
Helicopters: AS 555 Fennec or SH-3D Sea King (D 10, D 11, D 13) ⓫.

Programmes: Six were originally ordered in 1978, but later restricted to four when Meko 140 frigates were ordered in 1979. Similar to Nigerian frigate *Aradu*.
Modernisation: *Almirante Brown* completed a two-year refit in 2005 and she was followed by *La Argentina* in 2006. Upgrades included reconfiguration of the flight deck to facilitate Sea King operations. *Sarandi* completed a similar refit in 2009. *Heroina* completed refit in 2008 but did not receive the flight deck modification. Upgrade of the command and control system is under consideration. MM 40 Exocet missiles are being modified by CITADEF with new Gradicom motors, which are reported to extend missile range.
Operational: *Almirante Brown* took part in allied Gulf operations in late 1990. Fennec helicopters delivered in 1996 provide over the horizon targeting for SSMs and have the potential to improve ASW capability. All are active and form 2nd Destroyer Squadron based at Puerto Belgrano. All can be used as Flagships.

ALMIRANTE BROWN   *(Scale 1 : 1,200), Ian Sturton* / 0569252

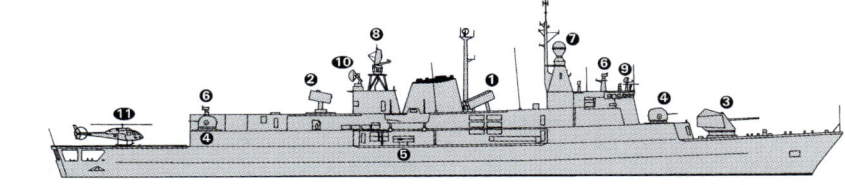

SARANDI   *5/2010, A E Galarce* / 1366820

SARANDI   *5/2010, A E Galarce* / 1366821

© 2014 IHS   *IHS Jane's Fighting Ships 2014-2015*

FRIGATES

3 DRUMMOND (TYPE A 69) CLASS (FFG)

Name	No	Builders	Laid down	Launched	Commissioned
DRUMMOND (ex-*Good Hope*; ex-*Lieutenant de Vaisseau le Hénaff*)	31 (ex-F 789)	Lorient Naval Dockyard, Lorient	12 Mar 1976	5 Mar 1977	Mar 1978
GUERRICO (ex-*Transvaal*; ex-*Commandant l'Herminier*)	32 (ex-F 791)	Lorient Naval Dockyard, Lorient	1 Oct 1976	13 Sep 1977	Oct 1978
GRANVILLE	33	Lorient Naval Dockyard, Lorient	1 Dec 1978	28 Jun 1980	22 Jun 1981

Displacement, tonnes: 1,170 standard; 1,230 full load
Dimensions, metres (feet): 80 × 10.3 × 3 hull; 5.5 sonar *(262.5 × 33.8 × 9.8; 18.0)*
Speed, knots: 23. **Range, n miles:** 4,500 at 15 kt, 3,000 at 18 kt
Complement: 93 (10 officers)

Machinery: 2 SEMT-Pielstick 12 PC2.2 V 400 diesels; 12,000 hp(m) *(8.82 MW)* sustained; 2 shafts; LIPS cp props
Missiles: SSM: 4 Aerospatiale MM 38 Exocet (2 twin) launchers ❶; inertial cruise; active radar homing to 42 km *(23 n miles)*; warhead 165 kg; sea-skimmer.
Guns: 1 Creusot-Loire 3.9 in *(100 mm)*/55 Mod 1953 ❷; 80° elevation; 60 rds/min to 17 km *(9 n miles)* anti-surface; 8 km *(4.4 n miles)* anti-aircraft; weight of shell 13.5 kg. 2 Breda 40 mm/70 (twin) ❸; 300 rds/min to 12.5 km *(6.8 n miles)*; weight of shell 0.96 kg; ready ammunition 736 (or 444) using AP tracer, impact or proximity fuzing. 2 Nexter F2A 20 mm ❹. 2 – 12.7 mm MGs.
Torpedoes: 6 – 324 mm Mk 32 (2 triple) tubes ❺. Whitehead A 244; anti-submarine; active/passive homing to 7 km *(3.8 n miles)* at 33 kt; warhead 34 kg.
Physical countermeasures: Decoys: CSEE Dagaie double mounting; 10 or 6 replaceable containers; trainable; chaff to 12 km *(6.5 n miles)*; illuminants to 4 km *(2.2 n miles)*; decoys in H- to J-bands or Corvus sextuple launchers for chaff.
Electronic countermeasures: ESM: DR 2000/DALIA 500; radar warning.
ECM: Thomson-CSF Alligator; jammer.

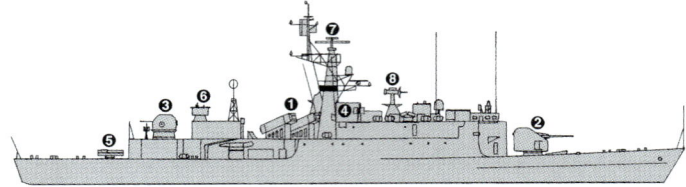

GRANVILLE *(Scale 1 : 900)*, Ian Sturton / 0506262

Radars: Air/surface search: Thomson-CSF DRBV 51A ❼ with UPX12 IFF; G-band.
Navigation: Concilium Selesmar; I-band.
Fire control: Thomson-CSF DRBC 32E ❽; I/J-band (for 100 mm gun).
Sonars: Thomson Sintra Diodon; hull-mounted; active search and attack.
Combat data systems: MINIACO C 31.
Weapon control systems: Thomson-CSF Vega system. CSEE Panda Mk 2 optical director ❻. Naja optronic director (for 40 mm guns).

Programmes: The first pair was originally built for the French Navy and sold to the South African Navy in 1976 while under construction. As a result of a UN embargo on arms sales to South Africa this sale was cancelled. Purchased by Argentina in Autumn 1978. Both arrived in Argentina 2 November 1978 (third ship being ordered shortly afterwards) and all have proved very popular ships in the Argentine Navy. The transfer of a further three of the class from the French Navy is very unlikely.
Modernisation: *Drummond* has had her armament updated to the same standard as the other two, replacing the Bofors 40/60. Fitted with MINIACO C 31 combat data system by 2008. All three ships reported equipped with refurbished MM-38 missiles during 2010.
Operational: Endurance, 15 days. Very economical in fuel consumption. Employed on EEZ patrol operations. All based at Mar del Plata.

DRUMMOND 5/2004, A E Galarce / 1044066

GRANVILLE 12/2002, A E Galarce / 0529818

6 ESPORA (MEKO 140 A16) CLASS (FFGH)

Name	No	Builders	Laid down	Launched	Commissioned
ESPORA	41	AFNE, Rio Santiago	3 Oct 1980	23 Jan 1982	5 Jul 1985
ROSALES	42	AFNE, Rio Santiago	1 Jul 1981	4 Mar 1983	14 Nov 1986
SPIRO	43	AFNE, Rio Santiago	4 Jan 1982	24 Jun 1983	24 Nov 1987
PARKER	44	AFNE, Rio Santiago	2 Aug 1982	31 Mar 1984	17 Apr 1990
ROBINSON	45	AFNE, Rio Santiago	8 Jun 1983	15 Feb 1985	28 Aug 2000
GOMEZ ROCA	46	AFNE, Rio Santiago	1 Dec 1983	14 Nov 1986	20 May 2004

Displacement, tonnes: 1,494 standard; 1,880 full load
Dimensions, metres (feet): 91.2 × 11.1 × 3.4 *(299.2 × 36.4 × 11.2)*
Speed, knots: 28
Range, n miles: 4,000 at 18 kt
Complement: 93 (11 officers)

Machinery: 2 SEMT-Pielstick 16 PC2-5 V 400 diesels; 20,400 hp(m) *(15 MW)* sustained; 2 shafts
Missiles: SSM: 4 Aerospatiale MM 38 Exocet ❶ inertial cruise; active radar homing to 42 km *(23 n miles)*; warhead 165 kg; sea-skimmer.
Guns: 1 Oto Melara 3 in *(76 mm)*/62 compact ❷; 85 rds/min to 16 km *(8.7 n miles)* anti-surface; 12 km *(6.5 n miles)* anti-aircraft; weight of shell 6 kg; also fires chaff and illuminants. 4 Breda 40 mm/70 (2 twin) ❸; 300 rds/min to 12.5 km *(6.8 n miles)*; weight of shell 0.96 kg; ready ammunition 736 (or 444) using AP tracer, impact or proximity fuzing. 2—12.7 mm MGs.
Torpedoes: 6—324 mm ILAS 3 (2 triple) tubes ❹. Whitehead A 244/S; anti-submarine; active/passive homing to 7 km *(3.8 n miles)* at 33 kt; warhead 34 kg (shaped charge).
Physical countermeasures: Decoys: CSEE Dagaie double mounting; 10 or 6 replaceable containers; trainable; chaff to 12 km *(6.5 n miles)*; illuminants to 4 km *(2.2 n miles)*; decoys in H- to J-bands.
Electronic countermeasures: ESM: Elettronica RQN-3B; radar warning.
ECM: Elettronica TQN-2X; jammer.
Radars: Air/surface search: Signaal DA05 ❻; E/F-band; range 137 km *(75 n miles)* for 2 m² target.
Navigation: Decca TM 1226 (41–45); I-band. Concilium Celestar (46); I-band.
Fire control: Signaal WM28 ❼; I/J-band; range 46 km *(25 n miles)*. Signaal WM 22/41; I/J-band.
IFF: Mk 10.
Sonars: Atlas Elektronik ASO 4; hull-mounted; active search and attack; medium frequency.
Combat data systems: Signaal SEWACO.
Electro-optic systems: 1 LIROD 8 optronic director ❺.

Helicopters: 1 SA 319B Alouette III or AS 555 Fennec ❽ (in 44–46).

Programmes: A contract was signed with Blohm + Voss on 1 August 1979 for this group of ships which are scaled down Meko 360s. All have been fabricated in AFNE, Rio Santiago. The last pair were to have been scrapped, but on 8 May 1997 a decision was taken to complete them some 14 years after each was first launched. A formal restart ceremony was held on 18 July 1997 and *Robinson* became operational in 2001. *Gomez Roca* became operational in late 2005.
Modernisation: Plans to fit MM 40 Exocet from Meko 360. Flight deck extensions for AS 555 helicopters. *Robinson* and *Gomez Roca* equipped with different EW suite. Upgrade of the command and control system is under consideration. *Gomez Roca* was refitted 2012–13 and was followed by *Rosales*.
Structure: The last three ships were fitted on build with a telescopic hangar. The first three ships may be retro-fitted at a later date.
Operational: Mostly used for offshore patrol and fishery protection duties. Form 2nd Frigate Squadron based at Puerto Belgrano. *Espora* underwent generator repairs at Simonstown in late 2012 following withdrawal from Exercise Altasur IX. She returned to Argentina in early 2013.

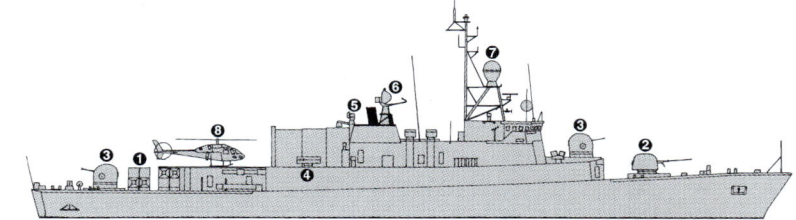

PARKER (Scale 1 : 900), Ian Sturton / 0012007

ROSALES 5/2008, M Declerck / 1335607

ROBINSON 5/2008, Guy Toremans / 1335603

ROBINSON 6/2008, Robert Pabst / 1335602

SHIPBORNE AIRCRAFT

Notes: There are plans to acquire four Eurocopter AS-565 helicopters to operate from the Almirante Brown-class frigates.

Numbers/Type: 3 Aerospatiale AS 555 SN Fennec.
Operational speed: 121 kt *(225 km/h)*.
Service ceiling: 13,125 ft *(4,000 m)*.
Range: 389 n miles *(722 km)*.
Role/Weapon systems: Principal role OTHT with potential ASW capability. Delivered in 1996. Four further aircraft, not fitted with radar, are to be acquired. Sensors: Bendix RDR 1500 radar; Mk 3 MAD. Weapons: ASW; 2 × A 244 torpedoes or 4 depth bombs may be fitted.

FENNEC   7/2004, A E Galarce / 1044071

Numbers/Type: 2/4 Agusta-Sikorsky ASH-3H/UH-3H Sea King.
Operational speed: 120 kt *(222 km/h)*.
Service ceiling: 12,205 ft *(3,720 m)*.
Range: 630 n miles *(1,165 km)*.
Role/Weapon systems: Six aircraft are operational: Two ASH-3H (possibly armed with AM 39) and four UH-3H acquired from US Navy in 2006. There are two further UH-3H used for spares and two SH-3D in reserve. Sensors (ASH variants): APS-705 search radar, Bendix AQS 18 sonar. Weapons: ASW; up to 4 × A 244 torpedoes or 4 × depth bombs. ASV: 1 AM 39 Exocet ASM.

SEA KING   8/2002, A E Galarce / 0529816

LAND-BASED MARITIME AIRCRAFT

Notes: (1) In addition there are two (plus one in reserve) Fokker F28 (possibly to be replaced by EADS-CASA C-295 transport aircraft) for Logistic Support; one Pilatus PC-6B for reconnaissance and 10 Beech T-34 Turbo Mentor training aircraft. The four Lockheed Electra L-188 are no longer in service.
(2) Thirty-six (32 A-4M and 4 TA-4F) ex-USMC Fighting Hawks with radar APG-66 acquired by the Air Force by July 1998. First 18 delivered in crates in 1995–96 and remainder modernised before delivery in 1997–98. Three have been lost and about 12 are reported to be operational. A reactivation programme may be in progress.
(3) Acquisition of 20 second-hand Mirage F1M aircraft from Spain was reported in August 2013 to be under consideration.
(4) There are plans to acquire up to eight LMAASA AT-63 Pampa III training/light attack aircraft to replace the MB-326 fleet. Two are expected in 2014, two in 2015, and one in 2016. Three more are to follow at a later date.

Numbers/Type: 5 Dassault-Breguet Super Etendard.
Operational speed: Mach 1.
Service ceiling: 44,950 ft *(13,700 m)*.
Range: 920 n miles *(1,700 km)*.
Role/Weapon systems: Strike Fighter with anti-shipping ability. In the past have flown from US or Brazilian aircraft carriers. Five aircraft are reported operational out of a total of 11. An upgrade programme to modify 10 aircraft to SEM 4/5 standard is planned to start in 2014. The SEM package includes Anemone radar, inertial navigation, modernised cockpit screens and controls. Strike, air defence and ASV roles. Hi-lo-hi combat radius 460 n miles *(850 km)*. Sensors: Thomson-CSF Agave multimode radar, ECM. Weapons: Strike; 2.1 tons of 'iron' bombs. ASVW; 1 AM 39 Exocet or 1 × Martin Pescador missiles. Self-defence; 2 × Magic AAMs. Standard; 2 × 30 mm cannon.

SUPER ETENDARD   10/2007, Argentine Navy / 1335605

Numbers/Type: 3 Grumman S-2ET Tracker.
Operational speed: 130 kt *(241 km/h)*.
Service ceiling: 25,000 ft *(7,620 m)*.
Range: 1,350 n miles *(2,500 km)*.
Role/Weapon systems: Three aircraft reported serviceable with a fourth being refurbished. A fifth aircraft crashed on 20 April 2011. Used for MR and EEZ patrol. One shipped to Israel in 1989 for Garrett turboprop installation. Prototype for fleet conversion in Argentina when completed in 2000. Sensors: EL/M-2022 search radar up to 32 sonobuoys, ALD-2B or AES 210/E ESM, echo-ranging depth charges. Weapons: ASW; A 244 torpedoes, bombs and depth charges.

S-2 TRACKER (landing on São Paulo)   5/2002, Walter Lastra/Fuerzas Navales / 0528430

Numbers/Type: 4 Beechcraft B 200M Cormoran.
Operational speed: 260 kt *(482 km/h)*.
Service ceiling: 31,000 ft *(9,448 m)*.
Range: 2,000 n miles *(3,705 km)*.
Role/Weapon systems: Multipurpose converted to Cormoran version for maritime patrol. There are three other unconverted B 200F aircraft. Sensors: Search radar. Weapons: Unarmed.

BEECH CORMORAN   5/2004 / 0570789

Numbers/Type: 4 Lockheed P-3B Orion.
Operational speed: 410 kt *(760 km/h)*.
Service ceiling: 28,300 ft *(8,625 m)*.
Range: 4,000 m *(7,410 km)*.
Role/Weapon systems: Two acquired in 1997 from US; four more in 1998, and two for spares in 1999. Four are serviceable and two in reserve. Sensors: APS-115 radar; ESM. Weapons: Three aircraft modified by 2007 under Gran Explorador programme. Upgrades included restoration of ASW capabilities, addition of AM-39 Exocet, new ECM pods and radar modifications. FLIR has been fitted to one aircraft. A new ventral radar is to be installed in all four aircraft.

ORION   6/2002, Argentine Navy / 0528429

PATROL FORCES

Notes: Following the reported abandonment of plans to acquire four 80 m Fassmer-designed offshore patrol vessels to fulfil the requirements of Project POM, a contract was signed in September 2013 by the Argentine defence minister with Tandanor and Rio Santiago Shipyards for the construction of four 35 m patrol craft.

2 CHEROKEE CLASS (PATROL SHIPS) (PSO)

Name	No	Builders	Commissioned
FRANCISCO DE GURRUCHAGA (ex-*Luiseno*)	A 3 (ex-ATF 156)	Charleston SB and DD Co, Charleston	16 Jun 1945
SUBOFICIAL CASTILLO (ex-*Takelma*)	A 6 (ex-ATF 113)	United Engineering Co, Alameda	3 Aug 1944

Displacement, tonnes: 1,255 standard; 1,759 full load
Dimensions, metres (feet): 62.5 × 11.7 × 5.2 *(205.1 × 38.4 × 17.1)*
Speed, knots: 16
Range, n miles: 6,500 at 15 kt, 15,000 at 8 kt
Complement: 85
Machinery: Diesel-electric; 4 GM 12—278 diesels; 4,400 hp *(3.28 MW)*; 4 generators; 1 motor; 3,000 hp *(2.24 MW)*; 1 shaft
Guns: 2 Bofors 40 mm/60 (A 3); 1 Bofors 40 mm/60 (A 6); 4 Oerlikon 20 mm (A 3, A 6); 2—12.7 mm MGs (A 6).
Radars: Surface search: Racal Decca 626; I-band.
Navigation: Racal Decca 1230; I-band.

Comment: Fitted with powerful pumps and other salvage equipment. *Francisco De Gurruchaga* transferred on 24 July 1975 by sale, *Suboficial Castillo* on 30 September 1993 by grant aid. *Gurruchaga* fitted with two new diesel engines in 2008. The ships appear to be fitted for but not with armament. *Gurruchaga* base at Ushuaia and *Castillo* at Mar del Plata. A third ship, *Irigoyen*, decommissioned in 2009.

SUBOFICIAL CASTILLO   11/2007, A E Galarce / 1335601

1 OLIVIERI CLASS (PATROL SHIP) (PBO)

Name	No	Builders	Commissioned
TENIENTE OLIVIERI (ex-*Marsea 10*)	A 2	Quality SB, Louisiana	1981

Displacement, tonnes: 1,666 full load
Dimensions, metres (feet): 56.3 × 12.2 × 4.3 *(184.7 × 40.0 × 14.1)*
Speed, knots: 14
Range, n miles: 2,800 at 10 kt
Complement: 15 (4 officers)
Machinery: 2 GM/EMD 16-645 E6; 3,230 hp *(2.4 MW)* sustained; 2 shafts; bow thruster
Guns: 2—12.7 mm MGs.

Comment: Built by Quality Shipyards, New Orleans, as an oilfield support ship but rated as an Aviso. Acquired from US Maritime Administration 15 November 1987. Capable of carrying 600 tons of stores and 800 tons of liquids. Based at Puerto Belgrano.

TENIENTE OLIVIERI   3/2000 / 0104168

1 SOTOYOMO CLASS (PATROL SHIP) (PBO)

Name	No	Builders	Commissioned
ALFEREZ SOBRAL (ex-*Salish*)	A 9 (ex-ATA 187)	Levingstone Shipbuilding Co, Orange	9 Sep 1944

Displacement, tonnes: 813 full load
Dimensions, metres (feet): 43.6 × 10.3 × 4 *(143 × 33.8 × 13.1)*
Speed, knots: 12.5
Range, n miles: 16,500 at 8 kt
Complement: 49
Machinery: Diesel-electric; 2 GM 12-278A diesels; 2,200 hp *(1.64 MW)*; 2 generators; 1 motor; 1,500 hp *(1.12 MW)*; 1 shaft
Guns: 1 Bofors 40 mm/60. 2 Oerlikon 20 mm.
Radars: Surface search: Decca 1226; I-band.

Comment: Former US ocean tug transferred on 10 February 1972. Paid off in 1987 but back in service by 1996. Armament has been reduced. Refitted in 2011. Based at Mar del Plata.

ALFEREZ SOBRAL   2/2001, Eric Grove / 1127024

4 BARADERO (DABUR) CLASS (COASTAL PATROL CRAFT) (PB)

Name	No	Builders	Commissioned
BARADERO	P 61	Israel Aircraft Industries	1978
BARRANQUERAS	P 62	Israel Aircraft Industries	1978
CLORINDA	P 63	Israel Aircraft Industries	1978
CONCEPCIÓN DEL URUGUAY	P 64	Israel Aircraft Industries	1978

Displacement, tonnes: 34 standard; 40 full load
Dimensions, metres (feet): 19.8 × 5.5 × 1.8 *(65 × 18.0 × 5.9)*
Speed, knots: 19
Range, n miles: 450 at 13 kt
Complement: 9
Machinery: 2 GM 12V-71TA diesels; 840 hp *(627 kW)* sustained; 2 shafts
Guns: 2 Oerlikon 20 mm. 4—12.7 mm MGs.
Depth charges: 2 portable rails.
Radars: Navigation: I-band.

Comment: Of all-aluminium construction. Employed in 1991 and 1992 as part of the UN Central American peacekeeping force. Based at Ushuaia.

BARADERO CLASS   12/2000, Eric Grove / 1044073

18 Argentina > Patrol forces — Amphibious forces

2 INTRÉPIDA CLASS (TYPE TNC 45)
(FAST ATTACK CRAFT — GUN/MISSILE) (PGGF)

Name	No	Builders	Launched	Commissioned
INTRÉPIDA	P 85	Lürssen, Bremen-Vegesack	2 Dec 1973	20 Jul 1974
INDOMITA	P 86	Lürssen, Bremen-Vegesack	8 Apr 1974	12 Dec 1974

Displacement, tonnes: 272 full load
Dimensions, metres (feet): 44.9 × 7.4 × 2.4 (147.3 × 24.3 × 7.9)
Speed, knots: 25
Range, n miles: 1,450 at 20 kt
Complement: 39 (5 officers)

Machinery: 4 MTU MD 16V 538 TB90 (Intrépida) diesels; 12,000 hp(m) (8.82 MW); 4 shafts. 2 Volvo Penta D34MT diesels (Indomita); 2 shafts
Missiles: SSM: 2 Aerospatiale Exocet MM 38 (Intrepida); active radar homing to 42 km (23 n miles); warhead 165 kg.
Guns: 1 Oto Melara 3 in (76 mm)/62 compact; 85 rds/min to 16 km (9 n miles) anti-surface; 12 km (6.5 n miles) anti-aircraft; weight of shell 6 kg. 1 Bofors 40 mm/70; 330 rds/min to 12 km (6.5 n miles) anti-surface; 4 km (2.2 n miles) anti-aircraft; weight of shell 0.89 kg. 2 — 12.7 mm MGs.
Torpedoes: 2 — 21 in (533 mm) launchers (Intrepida). AEG SST-4; wire-guided; active/passive homing to 28 km (15 n miles) at 23 kt; warhead 250 kg.
Rockets: 2 Oerlikon 81 mm rocket launchers for illuminants.
Electronic countermeasures: ESM: Racal RDL 1; radar warning.
Radars: Surface search: Decca 626; I-band.
Weapon control systems: Signaal WM22 optronic for guns/missiles. Signaal M11 for torpedo guidance and control.

Comment: These two vessels were ordered in 1970. Intrépida is painted in a brown/green camouflage paint scheme and can also be fitted with camouflage netting. The vessel was equipped with Exocet SSMs in place of the forward of the two Bofors guns in 1995. Recent refits to Intrépida have been conducted at Puerto Belgrano from 2012 and again in Ushuaia in mid-2013. Work is reported to have included overhaul/replacement of MTU diesels, maintenance of weapons and sensors (including Exocet and torpedo tubes) and removal of the after Bofors 40/70. Indomita started conversion to a patrol craft role in January 2008 when she began refit at Domecq Garcia (now Cinar) shipyard. Modifications, which are expected to be completed in late 2014, include replacement of main machinery with two Volvo Penta D34MT diesels (giving speed of 22 kt), replacement of WM 22 and Decca 626 radars with I-band surface search radar, installation of Miniaco combat data system, and removal of torpedo tubes. The armament of one 76 mm gun and one 40 mm Bofors is to be controlled by an electro-optic gun control system. The ship, which is to be painted grey, is to be equipped with a telescopic crane for launch and recovery of a RHIB.

INTREPIDA 6/2001, Argentine Navy / 0130735

INTREPIDA (with camouflage netting) 3/2001 / 0126381

2 POINT CLASS (PB)

Name	No	Builders	Commissioned
PUNTA MOGOTES (ex-Point Hobart)	P 65 (ex-82377)	J M Martinac, Tacoma	13 Jul 1970
RIO SANTIAGO (ex-Point Carrew)	P 66 (ex-82374)	US Coast Guard Yard, Curtis Bay	18 May 1970

Displacement, tonnes: 68 full load
Dimensions, metres (feet): 25.3 × 5.2 × 1.8 (83 × 17.1 × 5.9)
Speed, knots: 22
Range, n miles: 1,200 at 8 kt
Complement: 10
Machinery: 2 Caterpillar diesels; 1,600 hp (1.19 MW); 2 shafts.
Guns: 2 — 12.7 mm MGs.
Radars: Surface search: Raytheon SPS 64; I-band.

Comment: Punta Mogotes transferred from US Coast Guard on 8 July 1999 and is based at Mar del Plata. Rio Santiago transferred 22 August 2000.

IHS Jane's Fighting Ships 2014-2015

RIO SANTIAGO 5/2011, Hartmut Ehlers / 1406512

AMPHIBIOUS FORCES

Notes: (1) Marine Corps acquired two Guardian craft in October 1999 and two more in February 2000. Powered by twin 150 hp Johnson outboards. Carry 1 — 12.7 mm MG and 4 — 7.62 mm MGs, Raytheon radar.
(2) The first two of a new class of eight indigenously built LCVPs entered service in December 2006. Their names are Corbeta Uruguay and Rompehielos General San Martín.
(3) Procurement of a multirole ship is under consideration.

GUARDIAN 35 5/2004, A E Galarce / 1044075

1 HERCULES (TYPE 42) CLASS (LCC)

Name	No	Builders	Laid down	Launched	Commissioned
HERCULES	B 52 (ex-D 1, ex-28)	Vickers Shipbuilding & Engineering, Barrow-in-Furness	16 Jun 1971	24 Oct 1972	12 Jul 1976

Displacement, tonnes: 3,201 standard; 4,166 full load
Dimensions, metres (feet): 125.6 × 14.3 × 5.8 (412.1 × 46.9 × 19.0)
Flight deck, metres: 26 × 13 (85.3 × 42.7)
Speed, knots: 29; 18 Tynes. **Range, n miles:** 4,000 at 18 kt
Complement: 418

Machinery: COGOG; 2 RR Olympus TM3B gas turbines; 50,000 hp (37.3 MW) sustained 2 RR Tyne RM1A gas-turbines; 9,900 hp (7.4 MW) sustained; 2 shafts; cp props
Guns: 1 Vickers 4.5 in (115 mm)/55 Mk 8 automatic; 25 rds/min to 22 km (12 n miles); weight of shell 21 kg; also fires chaff and illuminants. 2 Oerlikon 20 mm Mk 7. 2 — 12.7 mm MGs.
Depth charges: 1 rack for Mk 9.
Physical countermeasures: Decoys: Graseby towed torpedo decoy. Knebworth Corvus 8-tubed trainable launchers for chaff.
Electronic countermeasures: ESM: Racal RDL 257; radar intercept.
ECM: Racal RCM 2; jammer.
Radars: Air search: Marconi Type 965P with double AKE2 array and 1010/1011 IFF; A-band.
Surface search: Marconi Type 992Q; E/F-band.
Navigation, HDWS and helicopter control: Kelvin Hughes Type 1006; I-band.
Sonars: Graseby Type 184M; hull-mounted; active search and attack; medium frequency 6–9 kHz. Kelvin Hughes Type 162M classification set; sideways looking; active; high frequency.
Combat data systems: Plessey-Ferranti ADAWS-4; Link 10.
Helicopters: 2 Sea King.

Programmes: Contract signed 18 May 1970 between the Argentine government and Vickers Ltd.
Modernisation: Combat Data System has been improved with local modifications. Refitted in Chile from November 1999 to July 2000 to make flight deck and hanger Sea King capable. Further modifications have included removal of MM38 launchers to be replaced by assault boats, removal of the Sea Dart launcher in 2009, modification of the Sea Dart magazine to accommodate marines/special forces and removal of the Type 909 fire-control radar. One depth charge rack for Mk 9 depth charges has been added. The Type 965 radar is not now to be replaced by LW-08 (ex-25 de Mayo) and, instead, is to be refurbished by CITEFA. The second of class, Santisima Trinidad, has been decommissioned and sank at her mooring on 21 January 2013.
Operational: Based at Puerto Belgrano. Officially described as an Amphibious command and control ship.

HERCULES 6/2001, Argentine Navy / 0130745

© 2014 IHS

Amphibious forces — Training ships < **Argentina** 19

16 LCVPS

EDVP 30–37   +8

Displacement, tonnes: 13 full load
Dimensions, metres (feet): 10.9 × 3.2 × 1.1 (35.8 × 10.5 × 3.6)
Speed, knots: 9
Range, n miles: 110 at 9 kt
Military lift: 3.5 tons
Machinery: 1 Gray 64 HN9 diesel; 165 hp (123 kW) sustained; 1 shaft
Guns: 2 – 12.7 mm MGs.

Comment: Details are for the eight LCVPs acquired from the US in 1970. There is a smaller variant built locally since 1971.

LCVP 1 and 4   9/2007, A E Galarce / 1167921

SURVEY AND RESEARCH SHIPS

Notes: (1) There are also two Fisheries Research Ships employed by the government. These are *Oca Balda* and *Eduardo Holmberg*.
(2) Two 10 m hydrographic launches, *Monte Blanco* and *Kualchink* entered service in 2004.
(3) An agreement was signed in September 2010 with Chernomorsky Shipyard, Ukraine, for a project to build Polar-class ships at Astillero Rio Santiago Shipyard. There have also been reports of discussions with a Finnish shipyard.

1 SURVEY SHIP (AGOB)

Name	No	Builders	Commissioned
PUERTO DESEADO	Q 20 (ex-Q 8)	Astillero Astarsa, San Fernando	26 Feb 1979

Displacement, tonnes: 2,167 standard; 2,439 full load
Dimensions, metres (feet): 76.8 × 15.8 × 6.5 (252 × 51.8 × 21.3)
Speed, knots: 14
Range, n miles: 12,000 at 12 kt
Complement: 61 (12 officers) + 20 scientists
Machinery: 2 MAN 9L20/27 diesels; 2,450 hp (1.8 MW); 2 shafts
Radars: Navigation: Decca 1629; I-band.

Comment: Laid down on 17 March 1976 for Consejo Nacional de Investigaciones Tecnicas y Scientificas. Launched on 4 December 1976. For survey work fitted with: four Hewlett-Packard 2108-A, gravimeter, magnetometer, seismic systems, high-frequency sonar, geological laboratory. Omega and NAVSAT equipped.

PUERTO DESEADO   11/2004, A E Galarce / 1151098

1 RESEARCH SHIP (AGOR)

Name	No	Builders	Commissioned
COMODORO RIVADAVIA	Q 11	Mestrina, Tigre	6 Dec 1974

Displacement, tonnes: 833 full load
Dimensions, metres (feet): 52.2 × 8.8 × 2.9 (171.3 × 28.9 × 9.5)
Speed, knots: 12
Range, n miles: 6,000 at 12 kt
Complement: 34 (8 officers)
Machinery: 2 Stork Werkspoor RHO-218K diesels; 1,160 hp(m) (853 kW); 2 shafts; cp props

Comment: Laid down on 17 July 1971 and launched on 2 December 1972. Used for research. Reportedly re-engined in 2009.

COMODORO RIVADAVIA   3/2001 / 0126380

1 SURVEY CRAFT (AGSC)

Name	No	Builders	Commissioned
CORMORAN	Q 15	AFNE, Rio Santiago	20 Feb 1964

Displacement, tonnes: 104 full load
Dimensions, metres (feet): 25.3 × 5 × 1.8 (83 × 16.4 × 5.9)
Speed, knots: 11
Complement: 19 (3 officers)
Machinery: 2 GM 6-71 diesels; 440 hp(m) (323 kW); 2 shafts
Radars: Navigation: Decca TM1226; I-band.

Comment: Launched 10 August 1963. Classified as a coastal launch.

CORMORAN   5/2003, A E Galarce / 0572406

TRAINING SHIPS

Notes: There are also three small yachts: *Itati* (Q 73), *Fortuna I* (Q 74) and *Fortuna II* (Q 75) plus a 25 ton yawl *Tijuca* acquired in 1993. *Fortuna III* was commissioned in 2004. A further yacht, *Irene* was acquired in 2005.

1 SAIL TRAINING SHIP (AXS)

Name	No	Builders	Commissioned
LIBERTAD	Q 2	AFNE, Rio Santiago	28 May 1963

Displacement, tonnes: 3,074 standard; 3,825 full load
Dimensions, metres (feet): 91.7 oa; 79.9 wl × 13.8 × 6.6 (300.9; 262.1 × 45.3 × 21.7)
Speed, knots: 13.5 diesel
Range, n miles: 12,000 at 8 kt
Complement: 350 (150 cadets)
Machinery: 2 Sulzer diesels; 2,400 hp(m) (1.76 MW); 2 shafts
Guns: 4 Hotchkiss 47 mm saluting guns.
Radars: Navigation: Decca; I-band.

Comment: Launched 30 May 1956. She set record for crossing the North Atlantic under sail in 1966. Sail area, 26,835 m^2. Based at Puerto Belgrano. Mid-life refit at Rio Santiago Shipyard completed in 2006. The refit is reported to have included new engines. The ship was impounded in Ghana in October 2012 following a court order by holders of Argentine bonds. She was later released on 19 December 2012.

LIBERTAD   9/2008, Chris Sattler / 1335604

20 Argentina > Training ships — Icebreakers

2 KING CLASS (AX)

Name	No	Builders	Launched	Commissioned
MURATURE	P 20	Base Nav Rio Santiago, Rio Santiago	5 Jul 1943	12 Apr 1945
KING	P 21	Base Nav Rio Santiago, Rio Santiago	2 Nov 1943	28 Jul 1946

Displacement, tonnes: 928 standard; 1,049 full load
Dimensions, metres (feet): 77 × 9 × 4 *(252.6 × 29.5 × 13.1)*
Speed, knots: 18
Range, n miles: 9,000 at 12 kt
Complement: 130
Machinery: 2 Werkspoor diesels; 2,500 hp(m) *(1.8 MW)*; 2 shafts
Guns: 3 Vickers 4 in *(105 mm)*/45; 16 rds/min to 19 km *(10 n miles)*; weight of shell 16 kg. 4 Bofors 40 mm/60 (1 twin, 2 single); 120 rds/min/barrel to 10 km *(5.5 n miles)*; weight of shell 0.89 kg. 5—12.7 mm MGs.
Radars: Surface search: Racal Decca 1226; I-band.

Comment: Named after Captain John King, an Irish follower of Admiral Brown, who distinguished himself in the war with Brazil, 1826–28; and Captain Jose Murature, who performed conspicuous service against the Paraguayans at the Battle of Cuevas in 1865. *King* laid down June 1938. *Murature* March 1940. Used both for cadet training and as river gunships on the Paraná, Uruguay and Río de la Plata rivers.

MURATURE 5/2011, Hartmut Ehlers / 1406513

AUXILIARIES

Notes: There is a fishery protection vessel *Luisito* Q 51. Painted yellow, it is based at Mar del Plata.

1 DURANCE CLASS (AORH)

Name	No	Builders	Launched	Commissioned
PATAGONIA (ex-*Durance*)	B 1 (ex-A 629)	Brest Naval Dockyard	6 Sep 1975	1 Dec 1976

Displacement, tonnes: 18,086 full load
Dimensions, metres (feet): 157.3 × 21.2 × 10.8 *(516.1 × 69.6 × 35.4)*
Speed, knots: 15
Range, n miles: 9,000 at 15 kt
Complement: 164 (10 officers) + 29 spare berths

Cargo capacity: 9,000 tons fuel; 500 tons Avcat; 140 distilled water; 170 victuals; 150 munitions; 50 naval stores
Machinery: 2 SEMT-Pielstick 16 PC2.5 V 400 diesels; 20,800 hp(m) *(15.3 MW)* sustained; 2 shafts; LIPS cp props
Guns: 2 Bofors 40 mm/60. 4—12.7 mm MGs.
Radars: Navigation: 2 Racal Decca 1226; I-band.
Helicopters: 1 Sea King (platform only).

Comment: Acquired from France on 12 July 1999 having been in reserve for two years. Entered Argentine Navy service in July 2000 after short refit.

PATAGONIA 5/2000, A E Galarce / 0104175

1 CHARTERED SHIP (AKS/AOTL)

Name	No	Builders	Commissioned
INGENIERO JULIO KRAUSE	B 13	Astillero Astarsa, Tigre	1981

Displacement, tonnes: 8,480 full load
Dimensions, metres (feet): 111.8 × 17.2 × 6.7 *(366.8 × 56.4 × 22.0)*
Speed, knots: 14
Complement: 32
Cargo capacity: 7,500 tons fuel
Machinery: 1 Sulzer diesel; 5,800 hp *(4.3 MW)*; 1 shaft

Comment: Chartered by the navy on 5 March 1993. Capable of stern replenishment at sea. Refitted by Cinar Shipyard in 2009 and operational in 2010.

IHS Jane's Fighting Ships 2014-2015

---

3 COSTA SUR CLASS (TRANSPORT) (AKS)

Name	No	Builders	Laid down	Launched	Commissioned
CANAL BEAGLE	B 3	Astillero Principe y Menghi SA	10 Jan 1977	19 Oct 1977	29 Apr 1978
BAHIA SAN BLAS	B 4	Astillero Principe y Menghi SA	11 Apr 1977	29 Apr 1978	27 Nov 1978
CABO DE HORNOS (ex-*Bahia Camarones*)	B 5	Astillero Principe y Menghi SA	29 Apr 1978	4 Nov 1978	28 Jun 1979

Displacement, tonnes: 11,069 full load
Dimensions, metres (feet): 119 × 17.5 × 7.5 *(390.4 × 57.4 × 24.6)*
Speed, knots: 16.5
Complement: 40
Machinery: 2 AFNE-Sulzer diesels; 6,400 hp(m) *(4.7 MW)*; 2 shafts

Comment: Three ships ordered December 1975. Laid down 10 January 1977 (B 3), 11 April 1977 (B 4) and 29 April 1978 (B 5). Launched 19 October 1977 (B 3), 29 April 1978 (B 4) and 4 November 1978 (B 5). Used to supply offshore research installations in Naval Area South. *Bahia San Blas* painted grey in 1998 indicating an active naval role in amphibious support operations. Capable of carrying up to eight LCVPs on deck. 132 troops can be accommodated in containers. The ship has been fitted with a helicopter (light) landing deck near the bow. *Cabo de Hornos* underwent refit at Astillero Rio Santiago in late 2007 and has operated in the amphibious role (with LCVPs) in 2009 and 2010. *Canal Beagle* started refit at Cinar in 2010 and was observed operating with two LCVPs in 2012.

BAHIA SAN BLAS 7/2007, A E Galarce / 1167922

CANAL BEAGLE 8/1999, P Marsan / 0081446

3 RED CLASS (BUOY TENDERS) (ABU)

Name	No	Builders	Launched
PUNTA ALTA (ex-*Red Birch*)	Q 63 (ex-WLM 687)	CG Yard, Maryland	19 Feb 1965
CIUDAD DE ZARATE (ex-*Red Cedar*)	Q 61 (ex-WLM 688)	CG Yard, Maryland	1 Aug 1970
CIUDAD DE ROSARIO (ex-*Red Wood*)	Q 62 (ex-WLM 685)	CG Yard, Maryland	4 Apr 1964

Displacement, tonnes: 533 full load
Dimensions, metres (feet): 49.1 × 10.1 × 1.8 *(161.1 × 33.1 × 5.9)*
Speed, knots: 12. **Range, n miles:** 2,248 at 11 kt
Complement: 31 (6 officers)
Machinery: 2 Caterpillar D398 diesels; 1,800 hp *(1.34 MW)*; 2 shafts; cp props; bow thruster
Guns: 2—12.7 mm MGs.

Comment: Ex-USCG buoy tenders. First one transferred on 10 June 1998 and recommissioned on 17 November 1998. Two more transferred 30 July 1999. Strengthened hull for light ice breaking. Equipped with a 10 ton boom. *Punta Alta* used as supply ship in the southern archipelago. The other pair are used as river supply ships and are based at Buenos Aires and Zárate River naval base.

CIUDAD DE ZARATE 5/2011, Hartmut Ehlers / 1406514

3 FLOATING DOCK

Dimensions, metres (feet): 215.8 × 46 × 45.5 *(708 × 150.9 × 149.3)*

Comment: Based at Puerto Belgrano, capacity 750 tons. All other docks have been sold.

ICEBREAKERS

Notes: During the repair of *Almirante Irizar*, ships are being leased as required to support Antarctic operations. These include the Russian icebreakers *Kapitan Dravitsyn*, *Vasily Golovnin* (which may be purchased) and the Chinese icebreaker, *Xue Long*.

© 2014 IHS

1 SUPPORT SHIP (AGB/AGOB)

Name	No	Builders	Launched	Commissioned
ALMIRANTE IRIZAR	Q 5	Wärtsilä, Helsinki	3 Feb 1978	15 Dec 1978

Displacement, tonnes: 15,139 full load
Dimensions, metres (feet): 121.3 × 25 × 9.5 *(398 × 82.0 × 31.2)*
Speed, knots: 17
Complement: 135 + 45 spare berths

Machinery: Diesel-electric; 4 Wärtsilä-SEMT-Pielstick 8 PC2.5 L diesels; 18,720 hp(m) *(13.77 MW)* sustained; 4 generators; 2 Stromberg motors; 16,200 hp(m) *(11.9 MW)*; 2 shafts
Radars: Air/surface search: Plessey AWS 2; E/F-band.
Navigation: 2 Decca; I-band.
Helicopters: 2 ASH-3H Sea King.

Comment: Fitted for landing craft with two 16 ton cranes, fin stabilisers, Wärtsilä bubbling system and a 60 ton towing winch. RAST helicopter securing system. Designed for Antarctic support operations and able to remain in polar regions throughout the Winter with 210 people aboard. Used as a transport to South Georgia in December 1981 and as a hospital ship during the Falklands war April to June 1982. Has been used as a Patagonian supply ship, and for other activities associated with the Navy in the region. The ship completed a refit including the installation of Satcom, by early 2005. 40 mm guns have been removed. Following a generator-room fire on 11 April 2007, the ship requires substantial repair work. The contract for this work, which includes new MAN diesel main engines and new generators, was signed with CINAR in September 2009. The ship is expected to become operational again by 2015.

ALMIRANTE IRIZAR
3/2006, A E Galarce
1040738

TUGS

Notes: (1) Procurement of a salvage tug, to be based at Ushuaia, is under consideration.
(2) An order for two 30 m and four 20 m tugs was made in September 2013.

11 TUGS (YTB/YTL)

QUERANDI R 2	**ONA** R 7	**CAPAYÁN** R 16
TEHUELCHE R 3	**TOBA** R 8	**CHIQUILLÁN** R 18
MOCOVI R 5	**CHULUPI** R 10	**MORCOYÁN** R 19
CALCHAQUI R 6	**MATACO** R 12	

Comment: R 2–3 and R 7–8 and R 12 are coastal tugs of about 250 tons. The remainder are harbour tugs transferred from the USA.

MATACO
5/2000, A E Galarce / 0104177

PREFECTURA NAVAL ARGENTINA – COAST GUARD

Headquarters Appointments

Commander:
Prefecto General Luis Alberto Heiler
Vice Commander:
Prefecto General Gerardo Horacio Crespo

Personnel

2014: 11,900 (1,600 officers)

Tasks

Under the General Organisation Act the PNA is charged with:
1. Enforcement of Federal Laws on the high seas and waters subject to the Argentine Republic.
2. Enforcement of environmental protection laws in Federal waters.
3. Safety of ships in EEZ. Search and Rescue.
4. Security of waterfront facilities and vessels in port.
5. Operation of certain Navaids.
6. Operation of some Pilot Services.
7. Management and operation of Aviation Service; Coastguard Vessels; Salvage, Fire and Anti-Pollution Service; Yachtmaster School; National Diving School; several Fire Brigades and Anti-Narcotics Department.
8. Operation of some Customs activities.

Organisation

Formed in 10 districts; High Parana River, Upper Parana and Paraguay Rivers, Lower Parana River, Upper Uruguay River, Lower Uruguay River, Delta, River Plate, Northern Argentine Sea, Southern Argentine Sea, Lakes and Comahue.

History

The Spanish authorities in South America established similar organisations to those in Spain. In 1756 the Captainship of the Port came into being in Buenos Aires—in 1810 the Ship Registry office was added to this title. On 29 October 1896 the title of Capitania General de Puertos was established by Act of Congress, the beginning of the PNA. Today, as a security and safety force, it has responsibilities throughout the rivers of Argentina, the ports and harbours as well as within territorial waters out to the 200 mile EEZ. An attempt was made in January 1992 to restrict operations to a 12 mile limit but the legislation was cancelled. The Coast Guard was placed under the Interior Ministry in 1996.

Identity markings

Two unequal blue stripes with, superimposed, crossed white anchors followed by the title Prefectura Naval.

Strength of Prefectura

Patrol Ships	6
Large Patrol Craft	3
Coastal Patrol Craft	20
Inshore Patrol Craft	77
Training Ships	4
Pilot Stations	1
Pilot and Patrol Craft	5

PATROL FORCES

Notes: (1) In addition to the ships and craft listed below the PNA operates 400 craft, including floating cranes, runabouts and inflatables of all types.
(2) There are plans to acquire up to 20 new 30 m patrol craft with aluminium hulls.
(3) There are 25 Zodiac Hurricane 929 (without cabin) and 920 (with cabin). Pennant numbers SR 9201–9225. They were acquired 2005–09.
(4) There are approximately 60 RHIBs of 5 to 6 m commissioned in 2011.

5 HALCON (TYPE B 119) CLASS (WPSO)

Name	No	Builders	Commissioned
MANTILLA	GC 24	Bazán, El Ferrol	20 Dec 1982
AZOPARDO	GC 25	Bazán, El Ferrol	28 Apr 1983
THOMPSON	GC 26	Bazán, El Ferrol	20 Jun 1983
PREFECTO FIQUE	GC 27	Bazán, El Ferrol	29 Jul 1983
PREFECTO DERBES	GC 28	Bazán, El Ferrol	20 Nov 1983

Displacement, tonnes: 925 standard; 1,101 full load
Dimensions, metres (feet): 67 × 10.5 × 4.2 *(219.8 × 34.4 × 13.8)*
Speed, knots: 20
Range, n miles: 5,000 at 18 kt
Complement: 33 (10 officers)
Machinery: 2 Bazán-MTU 16V 956 TB91 diesels; 7,500 hp(m) *(5.52 MW)* sustained; 2 shafts
Guns: 3—12.7 mm MGs.
Radars: Navigation: Decca 1226 ARPA; I-band.

Comment: Ordered in 1979 from Bazán, El Ferrol, Spain. Hospital with four beds. Carry one rigid rescue craft *(6 m)* with a 90 hp MWM diesel powering a Hamilton water-jet and a capacity for 12 and two inflatable craft *(4.1 m)* with Evinrude outboard. All ships, starting with Prefecto Derbes in August 2012, are being refitted at Cinar Shipyard. The 40 mm gun is being replaced with a third 12.7 mm MG and the engines are being overhauled. New navigation radars and communications equipment are being installed. A crane is being fitted on the helicopter deck.

MANTILLA
5/2010, A E Galarce / 1366823

45 + 7 TORO CLASS (FAST INTERVENTION CRAFT) (WPB)

LF 8701–8736   +16

Dimensions, metres (feet): 8.7 × 2.1 × 0.6 *(28.5 × 6.9 × 2.0)*
Speed, knots: 33
Complement: 10
Machinery: 1 diesel; waterjet propulsion
Guns: 1—7.62 mm MG.

Comment: Built to a local design. Began entering service in 2007. Fitted with armoured protection in 2011. A further seven were being acquired in 2013.

22 Argentina > Prefectura Naval Argentina — Patrol forces

1 PATROL SHIP (WPSO)

Name	No	Builders	Commissioned
DELFIN	GC 13	Ijsselwerf	14 May 1957

Displacement, tonnes: 711 standard; 1,016 full load
Dimensions, metres (feet): 59 × 9.1 × 4.2 (193.6 × 29.9 × 13.8)
Speed, knots: 15
Range, n miles: 6,720 at 10 kt
Complement: 27
Machinery: 2 MAN diesels; 2,300 hp(m) (1.69 MW); 2 shafts
Guns: 1 Oerlikon 20 mm (fitted for). 2—12.7 mm Browning MGs.
Radars: Navigation: Decca; I-band.

Comment: Whaler acquired for PNA in 1969. Commissioned 23 January 1970.

DELFIN 7/2003, A E Galarce / 0572409

2 LYNCH CLASS (LARGE PATROL CRAFT) (WPB)

Name	No	Builders	Commissioned
LYNCH	GC 21	AFNE, Rio Santiago	20 May 1964
TOLL	GC 22	AFNE, Rio Santiago	7 Jul 1966

Displacement, tonnes: 102 standard; 119 full load
Dimensions, metres (feet): 30 × 6.4 × 2.1 (98.4 × 21.0 × 6.9)
Speed, knots: 22
Range, n miles: 2,000
Complement: 14 (3 officers)
Machinery: 2 MTU Maybach diesels; 2,700 hp(m) (1.98 MW); 2 shafts
Guns: 1 Oerlikon 20 mm (can be carried). 1—7.62 mm MG.
Radars: Surface search: Decca; I-band.

LYNCH 1/1997, Prefectura Naval / 0012018

1 LARGE PATROL CRAFT (WAX)

Name	No	Builders	Commissioned
MANDUBI	GC 43	Base Nav Rio Santiago, Rio Santiago	1940

Displacement, tonnes: 274 full load
Dimensions, metres (feet): 33.2 × 6.3 × 1.9 (108.9 × 20.7 × 6.2)
Speed, knots: 14
Range, n miles: 800 at 14 kt, 3,400 at 10 kt
Complement: 12
Machinery: 2 MAN G6V-23.5/33 diesels; 500 hp(m) (367 kW); 1 shaft
Guns: 2—12.7 mm Browning MGs.
Radars: Surface search: Decca; I-band.

Comment: Since 1986 has acted as training craft for PNA Cadets School carrying 20 cadets.

MANDUBI 8/1994, Mario Diaz / 0056488

1 RIVER PATROL SHIP (WARS)

Name	No	Builders	Commissioned
TONINA	GC 47	SANYM SA San Fernando	30 Jun 1978

Displacement, tonnes: 105 standard; 155 full load
Dimensions, metres (feet): 25.5 × 6.5 × 3.3 (83.7 × 21.3 × 10.8)
Speed, knots: 10
Range, n miles: 2,800 at 10 kt
Complement: 11 (3 officers)
Machinery: 2 GM 16V-71TA diesels; 1,000 hp (746 kW) sustained; 2 shafts
Guns: 1 Oerlikon 20 mm.
Radars: Navigation: Decca 1226; I-band.

Comment: Served as training ship for PNA Cadets School until 1986. Now acts as salvage ship with salvage pumps and recompression chamber. Capable of operating divers and underwater swimmers. Also used as a patrol ship. Refit of the ship began in 2013.

TONINA 1/1998, Hartmut Ehlers / 0017541

18 MAR DEL PLATA (Z-28) CLASS (COASTAL PATROL CRAFT) (WPB)

MAR DEL PLATA GC 64	RIO DE LA PLATA GC 70	INGENIERO WHITE GC 76
MARTIN GARCIA GC 65	LA PLATA GC 71	GOLFO SAN MATIAS GC 77
RIO LUJAN GC 66	BUENOS AIRES GC 72	MADRYN GC 78
RIO URUGUAY GC 67	CABO CORRIENTES GC 73	RIO DESEADO GC 79
RIO PARAGUAY GC 68	RIO QUEQUEN GC 74	USHUAIA GC 80
RIO PARANA GC 69	BAHIA BLANCA GC 75	CANAL DE BEAGLE GC 81

Displacement, tonnes: 82 full load
Dimensions, metres (feet): 28 × 5.3 × 1.6 (91.9 × 17.4 × 5.2)
Speed, knots: 22
Range, n miles: 1,200 at 12 kt, 780 at 18 kt
Complement: 14 (3 officers)
Machinery: 2 MTU 8V-331-TC92 diesels; 1,770 hp(m) (1.3 MW) sustained; 2 shafts
Guns: 1 Oerlikon 20 mm. 2—12.7 mm Browning MGs.
Radars: Navigation: Decca 1226; I-band.

Comment: Ordered 24 November 1978 from Blohm + Voss to a Z-28 design. First delivered in June 1979 and then at monthly intervals. Steel hulls. GC 82 and 83 were captured by the British Forces in 1982.

CABO CORRIENTES 10/2011, A E Galarce / 1482859

1 COASTAL PATROL CRAFT (WPB)

Name	No	Builders	Commissioned
DORADO	GC 101	Base Nav Rio Santiago, Rio Santiago	17 Dec 1939

Displacement, tonnes: 44 full load
Dimensions, metres (feet): 21.2 × 4.3 × 1.5 (69.6 × 14.1 × 4.9)
Speed, knots: 12
Range, n miles: 1,550
Complement: 7 (1 officer)
Machinery: 2 GM 6071-6A diesels; 360 hp (268 kW); 1 shaft
Radars: Navigation: Furuno; I-band.

DORADO 12/1999, R O Rivero / 0056490

IHS Jane's Fighting Ships 2014-2015 © 2014 IHS

Prefectura Naval Argentina — Patrol forces < **Argentina** 23

25 SMALL PATROL CRAFT (WPB)

ESTRELLEMAR GC 48	**TIBURON** GC 57	**ROBALDO** GC 92
REMORA GC 49	**MELVA** GC 58	**CAMARON** GC 93
CONGRIO GC 50	**LENGUADO** GC 59	**GAVIOTA** GC 94
MERO GC 51	**ORCA** GC 60	**ABADEJO** GC 95
MARSOPA GC 52	**PINGUINO** GC 61	**ANCHOA** GC 102
PETREL GC 53	**MEDUSA** GC 88	**NARVAL** GC 106
SALMON GC 54	**PERCA** GC 89	**DON LUCAS** GC 114
BIGUA GC 55	**CALAMAR** GC 90	
FOCA GC 56	**HIPOCAMPO** GC 91	

Displacement, tonnes: 15 full load
Dimensions, metres (feet): 12.5 × 3.6 × 1.1 *(41 × 11.8 × 3.6)*
Speed, knots: 20
Range, n miles: 400 at 18 kt
Complement: 3
Machinery: 2 GM diesels; 514 hp *(383 kW)*; 2 shafts
Guns: 12.7 mm Browning MG.
Radars: Navigation: I-band.

Comment: First delivered September 1978. First 14 built by Cadenazzi, Tigre 1977–79, most of the remainder by Ast Belen de Escobar 1984–86. *GC 102* built in 1941 and *GC 106* in 1944.

PERCA *11/2004, A E Galarce* / 1151093

1 BAZAN TYPE (WPBF)

SUREL GC 142

Displacement, tonnes: 15 full load
Dimensions, metres (feet): 11.9 × 3.8 × 0.7 *(39 × 12.5 × 2.3)*
Speed, knots: 38
Range, n miles: 300 at 25 kt
Complement: 4
Machinery: 2 MAN D2848 LXE diesels; 1,360 hp(m) *(1 MW)* sustained; 2 Hamilton 362 waterjets
Guns: 1 — 12.7 mm MG.
Radars: Navigation: Furuno; I-band.

Comment: Acquired in 1997 from Bazán, San Fernando. Similar to Spanish Bazán 39 class for Spanish Maritime Police. Plans to acquire further craft were not fulfilled.

SUREL *12/2001, A E Galarce* / 0529809

1 PATROL SHIP (PBO)

Name	No	Builders	Commissioned
MARIANO MORENO (ex-*Martes Santo*)	GC 186	Construcciones Santodomingo, Vigo	1974

Measurement, tonnes: 251 gt
Dimensions, metres (feet): 29.1 × 7.5 × ? *(95.5 × 24.6 × ?)*
Speed, knots: 10.5
Machinery: 1 diesel; 750 hp *(552 kW)*; 1 shaft
Guns: 1 — 12.7 mm MG.

Comment: Former fishing vessel converted for Coast Guard patrol duties by CINAR in 2013.

4 TRAINING SHIPS (WAXL/WAXS)

ESPERANZA YR 7	**TALITA II** GC 115
ADHARA II GC 117	**DR BERNARDO HOUSSAY** (ex-*El Austral*) MOV 1

Displacement, tonnes: 34 standard
Dimensions, metres (feet): 19 × 4.3 × 2.7 *(62.3 × 14.1 × 8.9)*
Complement: 12 (6 cadets)
Machinery: 1 VM diesel; 90 hp(m) *(66 kW)*; 1 shaft

Comment: Details given are for *Esperanza* built by Ast Central de la PNA. Launched and commissioned 20 December 1968 as a sail training ship. The 30 ton training craft *Adhara II* and *Talita II* are of similar dimensions. *Dr Bernardo Houssay* is a Danish-built ketch built in 1930. Displacement 460 tons and has a crew of 25 (five officers). Acquired by the PNA in 1996 and completed refit (including a new steel hull) at Tandanor Shipyard in 2011.

TALITA II *6/1998, Prefectura Naval* / 0017545

DR BERNARDO HOUSSAY *3/2010, A E Galarce* / 1366827

10 ALUCAT 1050 CLASS (WPB)

CORMORAN GC 137	**SURUBI** GC 143	**HUALA** GC 146	**MANDURUYU** GC 148
CISNE GC 138	**BOGA** GC 144	**PACU** GC 147	**CORVINA** GC 149
PEJERREY GC 139	**SABALO** GC 145		

Displacement, tonnes: 9 full load
Dimensions, metres (feet): 11.5 × 3.8 × 0.6 *(37.7 × 12.5 × 2.0)*
Speed, knots: 18
Complement: 4
Machinery: 2 Volvo 61 ALD; 577 hp(m) *(424 kW)*; 2 Hamilton 273 waterjets
Radars: Navigation: Furuno 12/24; I-band.

Comment: First three delivered in September 1994. Seven more ordered in 1999.

HUALA *4/2000, Hartmut Ehlers* / 0104180

33 ALUCAT 850 CLASS (WPB)

GC 152–184 (ex-LS 9201–9233)

Displacement, tonnes: 7 full load
Dimensions, metres (feet): 9.2 × 3.3 × 0.6 *(30.2 × 10.8 × 2.0)*
Speed, knots: 26
Complement: 4
Machinery: 2 Volvo TAMD 41B; 400 hp(m) *(294 kW)*; 2 waterjets
Radars: Navigation: Furuno; I-band.

Comment: Alucat 850 class built by Damen. First six delivered in 1995, six more in February 1996, five more in December 1996 and five in December 1997. Five more ordered in 1999.

GC 181 *10/2005, A E Galarce* / 1151092

© 2014 IHS IHS Jane's Fighting Ships 2014-2015

SUBMARINES

Notes: Project Sea 1000 is for the replacement of the current Collins class fleet with a new class of 12 submarines. The role of the new submarines is to be anti-submarine and anti-surface warfare, land-attack, intelligence collection and special forces operations. Nuclear propulsion has been ruled out and the 2013 Defence White Paper narrowed the future design to two options: an evolved design based on the Collins class; and a 'clean sheet' concept. The boats are to be built in Australia. Based on the need for an eight-year design phase and a seven-year construction phase, entry into service is expected in about 2030.

6 COLLINS CLASS (SSK)

Name	No	Builders	Laid down	Launched	Commissioned
COLLINS	73	Australian Submarine Corp, Adelaide	14 Feb 1990	28 Aug 1993	27 Jul 1996
FARNCOMB	74	Australian Submarine Corp, Adelaide	1 Mar 1991	15 Dec 1995	31 Jan 1998
WALLER	75	Australian Submarine Corp, Adelaide	19 Mar 1992	14 Mar 1997	10 Jul 1999
DECHAINEUX	76	Australian Submarine Corp, Adelaide	4 Mar 1993	12 Mar 1998	23 Feb 2001
SHEEAN	77	Australian Submarine Corp, Adelaide	17 Feb 1994	1 May 1999	23 Feb 2001
RANKIN	78	Australian Submarine Corp, Adelaide	12 May 1995	7 Nov 2001	29 Mar 2003

Displacement, tonnes: 3,100 surfaced; 3,407 dived
Dimensions, metres (feet): 77.8 × 7.8 × 7 (255.2 × 25.6 × 23.0)
Speed, knots: 10 surfaced; 10 snorting; 20 dived
Range, n miles: 11,500 at 10 kt surfaced; 9,000 at 10 kt snorting; 400 at 4 kt dived
Complement: 58 (8 officers)

Machinery: Diesel-electric; 3 Hedemora/Garden Island Type V18B/14 diesels; 6,020 hp *(4.42 MW)*; 3 Jeumont Schneider generators; 4.2 MW; 1 Jeumont Schneider motor; 7,344 hp(m) *(5.4 MW)*; 1 shaft; 1 MacTaggart Scott DM 43006 hydraulic motor for emergency propulsion
Missiles: SSM: McDonnell Douglas Sub Harpoon Block 1B (UGM 84C); active radar homing to 92 km *(50 n miles)* at 0.9 Mach; warhead 227 kg.
Torpedoes: 6—21 in *(533 mm)* fwd tubes. Gould Mk 48 Mod 4/6/7; dual purpose; wire-guided; active/passive homing to 38 km *(21 n miles)* at 55 kt or 50 km *(27 n miles)* at 40 kt; warhead 295 kg. Air turbine pump discharge. Total of 22 weapons including Mk 48 and Sub Harpoon.
Mines: 44 in lieu of torpedoes.
Physical countermeasures: Decoys: 2 SSE.
Electronic countermeasures: ESM: Condor CS-5600; intercept and warning.
Radars: Navigation: Kelvin Hughes Type 1007; I-band.
Sonars: Thomson Sintra Scylla active/passive bow array and passive flank, intercept and ranging arrays. Thales SHORTAS retractable, passive.
Weapon control systems: AN-BYG 1. Link 11.

Programmes: Contract signed on 3 June 1987 for construction of six Swedish-designed Kockums Type 471. Fabrication work started in June 1989; bow and midships (escape tower) sections of the *Collins* built in Sweden.
Structure: Scylla is an updated Eledone sonar suite. Diving depth, 250 m *(820 ft)*. Anechoic tiles are fitted. Pilkington Optronics CK 43 search and CH 93 attack periscopes fitted.
Modernisation: The Replacement Combat System AN-BYG 1 evolved from Raytheon's CCS Mk 2. The first seagoing system was installed in *Waller* in 2006 and in *Farncomb*, *Dechaineux*, and *Sheean* by late 2013. *Rankin* is to be upgraded in 2014 and *Collins* in 2016. Collaborative development of the US Mk 48 Mod 7 ADCAP torpedo is being progressed. Two boats have received modifications to increase their special forces capabilities. Further upgrades under the Collins Continuous Improvement Programme are to include improvements to communications and EW capabilities, a periscope system upgrade and sonar upgrades.

Operational: All submarines are based at Fleet Base West with one or two deploying regularly to the east coast. The first firing of a M48 Mod 7 torpedo was conducted by *Waller* in July 2008. A further firing was conducted by *Waller* during RIMPAC 2012.

FARNCOMB — 10/2013*, Chris Sattler / 1525692

SHEEAN — 2/2013*, Chris Sattler / 1525693

FARNCOMB — 11/2013*, Chris Sattler / 1525694

WALLER — 7/2013*, Chris Sattler / 1525691

Destroyers — Frigates < **Australia** 27

DESTROYERS

0 + 3 HOBART CLASS (DESTROYERS) (DDGHM)

Name	No	Builders	Laid down	Launched	Commissioned
HOBART	39	ASC, Osborne	6 Sep 2012	2014	2016
BRISBANE	41	ASC, Osborne	3 Feb 2014	2015	2017
SYDNEY	42	ASC, Osborne	2014	2017	2019

Displacement, tonnes: 6,350 full load
Dimensions, metres (feet): 146.7 oa; 133.2 pp × 18.6 × 4.9 *(481.3; 437 × 61.0 × 16.1)*
Flight deck, metres: 26.4 × 17 *(86.6 × 55.8)*
Speed, knots: 28. **Range, n miles:** 5,000 at 18 kt
Complement: 205 (19 air crew) + 29 spare berths

Machinery: CODOG; 2 GE LM 2500 gas turbines; 47,328 hp(m) *(34.8 MW)* sustained; 2 Bazan/Caterpillar diesels; 12,240 hp(m) *(9 MW)* sustained; 2 shafts; LIPS cp props
Missiles: SSM: 8 Boeing Harpoon Block 2 ❶; active radar homing to 124 km *(67 n miles)* at 0.9 Mach; warhead 227 kg.
SAM: Mk 41 VLS (48 cells) ❷; 32 Raytheon SM-2 Block III; command/inertial guidance; semi-active radar homing to 167 km *(90 n miles)* at 3 Mach. 64 Evolved Sea Sparrow RIM 162B (in quadpacks); semi-active radar homing to 18 km *(9.7 n miles)* at 3.6 Mach; warhead 38 kg.
Guns: 1 FMC 5 in *(127 mm)*/54 Mk 45 Mod 4 ❸; 20 rds/min to 24.7 km *(13.3 n miles)*; weight of shell 32 kg. 1 Raytheon 20 mm Vulcan Phalanx Block 1B ❹; 6 barrels per launcher; 4,500 rds/min combined to 1.5 km. 2 Rafael Typhoon 25 mm.
Torpedoes: 4—323 mm (2 twin) Mk 32 Mod 9 fixed launchers ❺. Eurotorp MU 90; anti-submarine; active/passive homing to 25 km *(13.5 n miles)* at 29/50 kt; warhead 32 kg.
Physical countermeasures: Decoys: 4 Mk 37 Mod 1 SKWS for SRBOC/NATO Sea Gnat ❻. LESCUT torpedo decoys. 4 Mk 169 Nulka expendable decoy launchers. SLQ-25 Nixie TTCM.
Electronic countermeasures: ESM: EDO ES 3701-02S. ECM: To be announced.
Radars: Air/surface search: Aegis SPY-1D(V) ❼; 3D; E/F-band.
Surface search: Sperry Marine AN/SPQ-9B ❽; I-band.
Fire control: 2 Raytheon SPG-62 Mk 99 (for SAM) ❾. I/J-band.
Navigation: 2 L3 Radarpilot 1100; I-band.
Sonars: Ultra integrated sonar suite comprising Type 2150 hull mounted sonar and VDS with torpedo detection.
Combat data systems: Lockheed Aegis Baseline 7.1; Link 11/16. USG-2A CEC.
Electro-optic systems: Mk 20 EOSS. 2 Sagem VAMPIR IRST. 2 Ultra 2500 EO tracking systems.
Helicopters: 1 Sikorsky MH-60R Seahawk ❿.

Programmes: The Navantia F-100 was selected by the Australian government as the platform for the Hobart class Air Warfare Destroyers on 20 June 2007. The contract to build the ships was signed on 4 October 2007. The Combat System is to be an Australian version of Aegis; subsystems yet to be selected include communications and electronic warfare. The project is being executed under an alliance arrangement between the Australian government, ASC AWD Shipbuilder Pty Ltd and Raytheon Australia Pty Ltd. The headquarters of the Alliance is the AWD Systems Centre in Adelaide. Hull blocks are being manufactured around Australia. Contractors include: ASC (27 forward blocks); BAE Systems (36 hull); Forgacs (30 aft). Completed blocks are being consolidated at ASC Shipyard, Adelaide. The overall programme has been delayed by over a year due to problems with hull block fabrication. Current dates reflect a major re-baselining schedule announced in September 2012. The keel-to-keel interval has been extended to 18 months.

HOBART CLASS *(Scale 1 : 1,200), Ian Sturton / 1366678*

FRIGATES

Notes: Project Sea 5000 is for a class of eight new frigates to replace the current ANZAC class. The ships, whose primary role is to be anti-submarine warfare, are to be of the order of 6,000 tons and are to be capable of operating helicopters and unmanned aerial vehicles. First Pass Approval for the project is planned in 2018–21 timeframe with a view to achieving an initial operational capability by 2030.

4 ADELAIDE (OLIVER HAZARD PERRY) CLASS (FFGHM)

Name	No	Builders	Laid down	Launched	Commissioned
SYDNEY	03	Todd Pacific Shipyard Corporation, Seattle	16 Jan 1980	26 Sep 1980	29 Jan 1983
DARWIN	04	Todd Pacific Shipyard Corporation, Seattle	3 Jul 1981	26 Mar 1982	21 Jul 1984
MELBOURNE	05	Australian Marine Eng (Consolidated), Williamstown	12 Jul 1985	5 May 1989	15 Feb 1992
NEWCASTLE	06	Australian Marine Eng (Consolidated), Williamstown	21 Jul 1989	21 Feb 1992	11 Dec 1993

Displacement, tonnes: 4,267 full load
Dimensions, metres (feet): 138.1 × 13.7 × 4.5 hull; 7.5 sonar *(453.1 × 44.9 × 14.8; 24.6)*
Speed, knots: 29. **Range, n miles:** 4,500 at 20 kt
Complement: 184 (15 officers)

Machinery: 2 GE LM 2500 gas turbines; 41,000 hp *(30.6 MW)* sustained; 1 shaft; cp prop; 2 auxiliary electric retractable propulsors fwd; 650 hp *(484 kW)*
Missiles: SSM: 8 McDonnell Douglas Harpoon Block 2; active radar homing to 124 km *(67 n miles)* at 0.9 Mach; warhead 227 kg.
SAM: GDC Pomona Standard SM-2 Block IIIA; Mk 13 Mod 4 launcher for both SAM and SSM systems ❶; command guidance; semi-active radar homing to 165 km *(90 n miles)* at 2 Mach; 40 missiles (combined SSM and SAM). 32 Raytheon RIM-162 ESSM; Mk 41 8-cell VLS launcher ❷; semi-active radar homing to 18.5 km *(10 n miles)* at 3.6 Mach; warhead 227 kg.
Guns: 1 Oto Melara 3 in *(76 mm)*/62 US Mk 75 compact ❸; 85 rds/min to 16 km *(9 n miles)* anti-surface; 12 km *(6.5 n miles)* anti-aircraft; weight of shell 6 kg. 1 Raytheon 20 mm Mk 15 Vulcan Phalanx Block 1B ❹; anti-missile system with 6 barrels; 4,500 rds/min combined to 1.5 km. Up to 6—12.7 mm MGs. 2 Rafael Mini-Typhoon 12.7 mm remote-controlled guns (for selected deployments).
Torpedoes: 6—324 mm Mk 32 (2 triple) tubes ❺. Eurotorp MU 90; active/passive homing to 25 km *(13.5 n miles)* at 29/50 kt.
Physical countermeasures: Decoys: 4 six-barrelled Loral Hycor SRBOC Mk 36 chaff and IR decoy launchers or TERMA SKWS; range 1–4 km. 3 (4 in *Sydney*) BAe Nulka quad expendable decoy launchers. 2 Rafael long-range chaff rocket launchers (fixed 2-barrel system). LESCUT torpedo countermeasures.
Electronic countermeasures: ESM/ECM: Elbit EA-2118 jammer. Rafael C-Pearl ❻; intercept.
Radars: Air search: Raytheon SPS-49 A(V)1 ❼; C-band.
Surface search/navigation: ISC Cardion SPS-55 ❽; I-band.
Fire control: Lockheed SPG-60 ❾; I/J-band; range 110 km *(60 n miles)*; Doppler search and tracking. Sperry Mk 92 Mod 12 ❿; I/J-band.
IFF: AIMS Mk XII.
Sonars: Thales Spherion (TMS 4131); active search and attack; medium frequency; hull mounted. Petrel (TMS 5424) high frequency mine-avoidance. Albatros (TMS 4350) towed-array torpedo-warning system.
Combat data systems: ADACS. OE-2 SATCOM; Link 11. Link 16.
Weapon control systems: Sperry Mk 92 Mod 12 gun and missile control (Signaal derivative). Radamec 2500 optronic director with TV, laser and IR imager.
Helicopters: 2 Sikorsky S-70B-2 Seahawks ⓫ or 1 Seahawk and 1 Squirrel.

Programmes: US numbers: *Sydney* FFG 35; *Darwin* FFG 44.
Modernisation: The original ship design was modified to provide improved helicopter facilities. The improvements resulted in angling the transom, increasing the ship's overall length by 8 ft and fitting the RAST helo recovery system. The modifications also included longitudinal strengthening and buoyancy upgrades. The FFG Upgrade Program (FFG-UP) was delivered by Project Sea 1390. The lead ship *Sydney* returned to service in April 2006. Work on *Melbourne* completed in 2007 and *Darwin* and *Newcastle* completed in 2008. All four ships returned to full operational service in 2009. The modification included major upgrades to the combat system and sensors including installation of the Mk 41 VLS and integration of ESSM. The first firing of ESSM from an FFG was conducted by *Sydney* on 20 August 2007. All ships upgraded to SM-2 Block IIIA by 2011. A successful test firing was conducted by *Melbourne* on 8 December 2009 and further firings by *Sydney* in June 2011.
Operational: *Canberra* decommissioned on 12 November 2005 and *Adelaide* on 19 January 2008. The four remaining upgraded ships of the class are based at Fleet Base East. For operational tasking the ships are fitted with enhanced communications, TopLite Electro-Optical sights and the Mini Typhoon weapon system. All ships are fighter and air control capable.

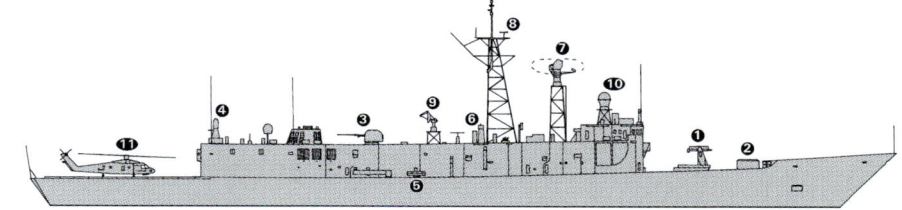

SYDNEY *(Scale 1 : 1,200), Ian Sturton / 1153837*

SYDNEY *11/2013*, Michael Nitz / 1525695*

MELBOURNE *4/2013*, Chris Sattler / 1529101*

© 2014 IHS — IHS Jane's Fighting Ships 2014-2015

28 Australia > Frigates

8 ANZAC (MEKO 200) CLASS (FFGHM)

Name	No
ANZAC	150
ARUNTA (ex-*Arrernte*)	151
WARRAMUNGA (ex-*Warumungu*)	152
STUART	153
PARRAMATTA	154
BALLARAT	155
TOOWOOMBA	156
PERTH	157

Builders	Laid down	Launched	Commissioned
Transfield, Williamstown	5 Nov 1993	16 Sep 1994	18 May 1996
Transfield, Williamstown	22 Jul 1995	28 Jun 1996	12 Dec 1998
Tenix Defence Systems, Williamstown	26 Jul 1997	23 May 1998	31 Mar 2001
Tenix Defence Systems, Williamstown	25 Jul 1998	17 Apr 1999	17 Aug 2002
Tenix Defence Systems, Williamstown	5 Jun 1999	17 Jun 2000	4 Oct 2003
Tenix Defence Systems, Williamstown	4 Aug 2000	25 May 2002	26 Jun 2004
Tenix Defence Systems, Williamstown	26 Jul 2002	16 May 2003	8 Oct 2005
Tenix Defence Systems, Williamstown	24 Jul 2003	20 Mar 2004	26 Aug 2006

Displacement, tonnes: 3,759 (150–156), 3,810 (157) full load
Dimensions, metres (feet): 118 oa; 109 wl × 14.8 × 4.35 *(387.1; 357.6 × 48.6 × 14.3)*
Speed, knots: 27
Range, n miles: 6,000 at 18 kt
Complement: 191 (24 officers)

Machinery: CODOG: 1 GE LM 2500 gas turbine; 30,172 hp *(22.5 MW)* sustained; 2 MTU 12V 1163 TB83 diesels; 8,840 hp(m) *(6.5 MW)* sustained; 2 shafts; cp props
Missiles: SSM: 8 McDonnell Douglas Harpoon Block 2 ❶; active radar homing to 124 km *(67 n miles)* at 0.9 Mach; warhead 227 kg.
SAM: Lockheed Martin Mk 41 Mod 5 octuple vertical launcher ❷. Quadpack Evolved Sea Sparrow RIM-162 for 32 missiles; semi-active homing to 18.0 km *(9.7 n miles)* at 3.6 Mach; warhead 38 kg.
Guns: 1 United Defense 5 in *(127 mm)*/54/62 Mk 45 Mod 2 ❸; 20 rds/min to 23 km *(12.6 n miles)*; weight of shell 32 kg. 4—12.7 mm MGs. 2 Rafael Mini Typhoon 12.7 mm remote-controlled guns (for selected deployments).
Torpedoes: 6—324 mm (2 triple) Mk 32 Mod 5 tubes ❹. Eurotorp MU 90; active/passive homing to 25 km *(13.5 n miles)* at 29/50 kt.
Physical countermeasures: Decoys: Loral Hycor SRBOC Mk 36 Mod 1 decoy launchers ❺ for SRBOC/NATO Sea Gnat. 4 BAE Nulka quad expendable decoy launchers. FEL SLQ-25C towed torpedo decoy.
Electronic countermeasures: RESM: Thales Centaur; radar intercept.
CESM: Telefunken PST-1720 Telegon 10; comms intercept.
Radars: Air search: Raytheon SPS-49(V)8 ANZ ❻; C-band.
Air/surface search: Ericsson Sea Giraffe ❼; G/H-band. CEAFAR active phased array (157) ❽; 3D; E/F-band.
Navigation: Atlas Elektronik 9600 ARPA; I-band. Kelvin Hughes Sharp Eye (157).
Fire control: CEAMOUNT illuminators (157) ❾; I/J-band. CelsiusTech Ceros 200 ❿; J-band.
IFF: Cossor AIMS Mk XII.
Sonars: Thomson Sintra Spherion B Mod 5; hull-mounted; active search and attack; medium frequency. Thales UMS 5424 Petrel; active mine avoidance; very high frequency.
Combat data systems: Saab Systems 9LV 453 Mk 3 (Mk 3E in 157). Link 11. Link 16 (157).
Electro-optic systems: Saab Systems Ceros 200 optronic director with CEA SSCWI (for RIM-162). Sagem VAMPIR NG IRST (157 only).

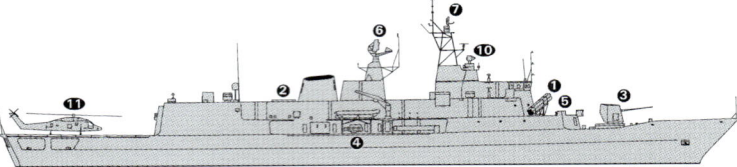

ARUNTA *(Scale 1 : 1,200), Ian Sturton / 1153838*

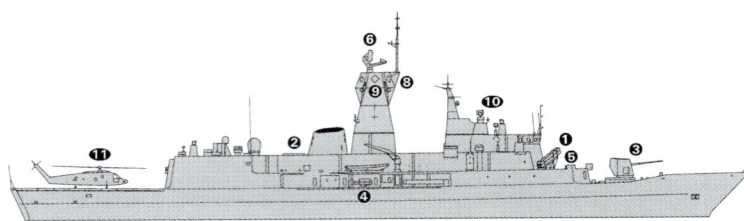

PERTH *(Scale 1 : 1,200), Ian Sturton / 1406511*

Helicopters: 1 S-70B-2 Seahawk ⓫.

Programmes: Contract signed with Australian Marine Engineering Consolidated (now BAE Systems) on 10 November 1989 to build eight Blohm + Voss designed MEKO 200 ANZAC frigates for Australia and two for New Zealand. First ship started construction 27 March 1992. Modules were constructed at Whangarei and shipped to Williamstown for assembly. The second and fourth ships of the class were delivered to New Zealand.
Modernisation: Evolved Seasparrow missile (ESSM) was integrated in *Warramunga*, the world's first warship to be so fitted (first missile launched 21 January 2003). All remaining ships have since been similarly equipped. Petrel MOAS (Mine Obstacle Avoidance Sonar) was introduced in 2005, the MU 90 torpedo in 2008 and Harpoon has now been progressively installed across the entire class. *Perth* is the first of class to be fitted with the 9LV Mk 3E Combat Management System, which forms the foundation of the ASMD Upgrade programme in progress 2010–17. Other key elements of the upgrade include replacement of Sea Giraffe radar with CEAFAR active phased array radar system, addition of four CEAMOUNT illuminators (Ceros 200 fire control is retained), installation of the Sagem Vampir NG IRST (Infra-Red Search and Track) system, replacement of the Atlas Elektronik navigation radar with a dual Kelvin Hughes Sharp Eye system and a significant modernisation and upgrade to the Operations Room. A major communications upgrade (including installation of Link 16) is also in progress. *Perth* completed ASMD upgrade in late 2010, conducted a successful ESSM firing against supersonic sea-skimming targets in 2013 and a successful firing of Harpoon Block 2 in 2012. *Arunta* and *Anzac* are the next ships to complete the ASMD upgrade. A successful firing of a war shot MU 90 torpedo was conducted by *Stuart* on 16 August 2013.
Structure: Space and weight have been reserved for the installation of Mini Typhoon, an additional octuple VLS, towed array sonar, offboard active ECM, extended ESM frequency coverage, Helo datalink and SATCOM. The ASMD upgrade involves significant structural changes including new forward and aft masts and complete enclosure of the quarter deck.
Operational: Two RHIBs are carried on all ships. 150, 154 and 155 are based at Sydney; the remainder at Perth.

PERTH *11/2013*, Michael Nitz / 1525700*

BALLARAT *8/2013*, John Mortimer / 1525699*

IHS Jane's Fighting Ships 2014-2015 © 2014 IHS

Frigates < **Australia** 29

PERTH 7/2013*, Chris Sattler / 1525698

PARRAMATTA 9/2013*, Chris Sattler / 1525696

STUART 7/2013*, Chris Sattler / 1525697

SHIPBORNE AIRCRAFT

Notes: Project Air 9000 is for the procurement of 24 new Seahawk MH-60R to replace the existing S-70B Seahawks. The primary role of the aircraft is to be ASW but there is also to be an anti-surface warfare capability. The aircraft are to be armed with Mk 54 torpedoes and AGM-114 Hellfire ASM. The first two aircraft were delivered in late 2013 for testing and evaluation; Initial Operating Capability is planned to be achieved in mid-2015.

Numbers/Type: 16 Sikorsky S-70B-2 Seahawk.
Operational speed: 135 kt (250 km/h).
Service ceiling: 10,000 ft (3,050 m).
Range: 600 n miles (1,110 km).
Role/Weapon systems: Seahawk SH-60F derivative aircraft designed by Sikorsky to meet RAN specifications for ASW and ASuW operations. Eight assembled by ASTA in Victoria. Helicopters embarked in FFG-7 and in ANZAC frigates. A two-phase Seahawk Capability Assurance Programme is in progress; obsolescent parts are to be replaced in the first phase and systems capability to be upgraded in the second. To be replaced by MH-60R from 2015. Sensors: Raytheon AAQ 27 FLIR, Tracor ALE 47 countermeasures, Northrop Grumman AN/AAR-54 MAWS and Elisra AES 210 ESM. Thales Super Searcher Surface surveillance radar, CDC Sonobuoy Processor and Barra Side Processor, and CAE Magnetic Anomaly Detector Set controlled by a Rockwell Collins Tactical Data System. Weapons: ASW; two Mk 46 Mod 5 torpedoes. ASV; one Mag 58 MG.

SEAHAWK   11/2013*, Michael Nitz / 1528967

Numbers/Type: 6 Eurocopter MRH-90.
Operational speed: 165 kt (305 km/h).
Service ceiling: 10,000 ft (3,050 m).
Range: 648 n miles (1,200 km).
Role/Weapon systems: Contract let with Australian Aerospace to provide a total of 46 MRH-90 to the Australian Defence Force (ADF). Six MSH (Maritime Support Helicopters) entered service in 2011 to replace the Sea King fleet and there are to be 40 in service with the army. They are to be capable of operating from future amphibious ships. Primary missions are to be afloat logistics support, SAR and MEDIVAC and boarding party operations. Sensors: Honeywell PRIMUS 701A weather radar, piloting FLIR, EW Self Protection System (Thales RWR, EADS Laser Warner System, LFK AN/AAR-60 Missile Launch Detection System (MILDS), MBDA Saphir-M chaff/flare dispenser system), Thales 'Top Owl' Helmet Mounted Sight and Display (HMSD) with integrated night vision device. Weapons: 2 – 7.62 mm MGs.

MSH   10/2013*, Chris Sattler / 1528966

LAND-BASED MARITIME AIRCRAFT

Notes: (1) Eight Boeing P-8A Poseidon aircraft are planned to replace the AP-3C Orion by 2017–18 (the Orion will reach its 'life of type' around 2019, by when it will have been in service with the Air Force for over 50 years). Mission systems are to include an advanced multimode radar, a high definition electro-optic camera, an acoustic system and an advanced electronic support system that is a derivative of the system fitted to the EA-18G Growler. Initial Operational Capability (IOC) for the P-8A is scheduled for the period FY17–18 through FY19–20.
(2) It was announced on 25 November 2009 that Australia is to procure an initial batch of 14 F-35A Joint Strike Fighters. The first two aircraft are to be delivered for test and evaluation in 2014 with the remaining 12 to follow by 2017. IOC is to be achieved in 2018. Acquisition of a further tranche of about 58 aircraft is to be considered in 2013 and a further tranche, to bring the total to 100 aircraft, is to be considered at a later date. However, if the JSF programme were to be delayed, acquisition of further F/A-18F Super Hornets may be considered as an alternative.
(3) It was announced on 3 May 2013 that 12 new-build EF-18G Growlers are to be procured to supplement the existing 24 F/A-18F Super Hornet fleet. Initial operating capability is to be achieved in 2018.

Numbers/Type: 6 Boeing 737 AEW&C 'Wedgetail'.
Operational speed: 470 kt (870 km/h).
Service ceiling: 41,000 ft (12,500 m).
Range: 3,780 n miles (7,000 km).
Role/Weapon systems: Contract for four aircraft (adaptation of Boeing Business Jet) signed on 20 December 2000. Two additional aircraft, under option, were added in 2004. Delivery of first two aircraft was originally scheduled for November 2006 but was delayed until November 2009. Initial Operating Capability (IOC) was achieved in 2012. AAR capable. Sensors: Northrop Grumman ESSD L-band multirole electronically scanned array (MESA) radar (fuselage mounted); electronic warfare self-protection (EWSP) system (including IR countermeasures, chaff and flares); Links 11 and 16; Satcom.

BOEING WEDGETAIL   11/2013*, Michael Nitz / 1528970

Numbers/Type: 18 Lockheed AP-3C Orion.
Operational speed: 360 kt (667 km/h).
Service ceiling: 28,300 ft (8,625 m).
Range: 4,000 n miles (7,410 km).
Role/Weapon systems: Operated by Air Force for long-range maritime patrol, ASW, maritime strike and ISR. All aircraft upgraded to AP-3C standard by late 2004. Sensors: Elta EL/M-2022A(V)3 radar, GDC UYS-503 acoustic system, Star Safire HD electro-optics, ELTA ALR-2001 ESM, up to 84 sonobuoys. Weapons: eight Mk 46(V) torpedoes, up to six Harpoon missiles.

ORION AP-3C   3/2013*, Paul Jackson / 1528964

Numbers/Type: 55/16 McDonnell Douglas F/A-18A Hornet/F/A-18B Hornet.
Operational speed: 1,032 kt (1,910 km/h).
Service ceiling: 50,000 ft (15,240 m).
Range: 1,000 n miles (1,829 km).
Role/Weapon systems: Air defence and strike aircraft operated by Air Force but with fleet defence and anti-shipping secondary roles. An upgrade programme is being conducted in three phases. Phase 1 modifications, completed in 2002, included new radios, upgraded mission computers, EW upgrade and GPS. Phase 2-1, completed in 2003, included installation of the AN/APG-73 radar and upgraded aircraft software. In Phase 2-2, completed by late 2007, the aircraft were equipped with Link 16, improved avionics and helmet mounted sight. In Phase 2-3, the EW suite (RWR and jammer) is to be upgraded and in Phase 2-4, a new target designation system (HDTS) is to be installed. Phase 3, structural modifications, is in progress. Upgraded F/A-18A/Bs are to complement the 24 F/A-18F Block 2 Super Hornets acquired as a bridging aircraft capability on withdrawal of the F-111 in 2010. Sensors: APG-73 attack radar, Litening Pod radar warning receiver. Weapons: ASV; 4 × Harpoon missiles. Strike; 1 × 20 mm cannon, up to 7.7 tons of 'iron' bombs. Fleet defence; 4 × AMRAAM and 4 × ASRAAM.

F/A-18   10/2013*, John Mortimer / 1528965

Numbers/Type: 6 Aerospatiale AS 350B Squirrel.
Operational speed: 125 kt (232 km/h).
Service ceiling: 10,000 ft (3,050 m).
Range: 275 n miles (510 km).
Role/Weapon systems: Support helicopter for utility tasks and training duties. Forms part of 723 Squadron. Sensors: None. Weapons: None.

SQUIRREL   11/2013*, Michael Nitz / 1528968

Numbers/Type: 24 Boeing F/A-18F Super Hornet.
Operational speed: 930 kt *(1,721 km/h).*
Service ceiling: 50,000 ft *(15,240 m).*
Range: 1,320 n miles *(2,376 km).*
Role/Weapon systems: Acquisition of 24 aircraft confirmed on 17 March 2008. Entered service from 2010 with all aircraft delivered by September 2011. They are to act as an interim replacement for the F-111. Details are for those in US Navy service. There are 12 aircraft wired for conversion to EF/-18G configuration but plans to convert these aircraft have been superseded by a decision to acquire 12 new E/F-18G. Sensors: APG-79 AESA radar, ALR-67(V)3 RWR. ECM: ALQ-165 ASPJ, ALQ-214 RFCM, towed decoys. Weapons: 11 wing stations for 8,680 kg of weapons (same armament as C/D) plus 20 mm gun.

Numbers/Type: 3 Bell 429.
Operational speed: 147 kt *(273 km/h).*
Service ceiling: 10,000 ft *(3,050 m).*
Range: 390 n miles *(722 km).*
Role/Weapon systems: Three Agusta A 109E were leased to 723 Squadron in 2007 for pilots' skill-retention training. These were replaced by Bell 429 aircraft in 2012.

F/A-18F 9/2005, *US Navy* / 1154040

BELL 429 8/2013*, *Chris Sattler* / 1528969

AMPHIBIOUS FORCES

Notes: (1) Replacements for the current amphibious capability are being procured under Joint Project (JP) 2048. *Tobruk* and the decommissioned LPA amphibious transports (*Kanimbla* and *Manoora*) are to be replaced by the Canberra-class LHDs. *Choules*, acquired in 2011, is to be retained as a third amphibious vessel.
(2) There are four 13 m LCVPs, capable of transporting 36 troops or a vehicle.

1 BAY CLASS (LANDING SHIPS DOCK) (LSD)

Name	No	Builders	Laid down	Launched	Commissioned
CHOULES (ex-*Largs Bay*)	L 100 (ex-L 3007)	Swan Hunter Shipbuilders, Wallsend-on-Tyne	28 Jan 2002	18 Jul 2005	28 Nov 2006

Displacement, tonnes: 16,419 full load
Dimensions, metres (feet): 176.6 × 26.4 × 5.8 *(579.4 × 86.6 × 19.0)*
Speed, knots: 18. **Range, n miles:** 10,000 at 15 kt
Complement: 158 + 350 embarked forces

Military lift: 1,130 linear metres of space for vehicles equating to 24 MBTs or 150 light trucks plus 200 tons ammunition or 24—24 ft containers, mexeflote and landing craft
Machinery: Diesel-electric; 2 Wärtsilä 8L26 generators; 6,000 hp *(4.5 MW)*; 2 Wärtsilä 12V26 generators; 9,000 hp *(6.7 MW)*; 2 steerable propulsors; bow thruster
Guns: 2—30 mm. 2 Mk 44 7.63 Miniguns. 6—7.62 mm MGs.
Radars: Navigation: E/F/I-bands.

Helicopters: Capable of operating Black Hawk and Seahawk.

Programmes: Ordered from Swan Hunter on 18 December 2000 and served in UK Royal Fleet Auxiliary 2006–2011. The decision to acquire the ship was announced by the Australian government in April 2011 and the ship arrived in Australia in mid-December.
Structure: Based on the Royal Schelde Enforcer design, the ship is designed to transport troops, vehicles, ammunition and stores in support of amphibious operations. Offload is enabled by a flight deck capable of operating heavy helicopters and an amphibious dock capable of operating landing craft and two mexeflotes which can be stowed on the ship's side. The ship is fitted with a Rubb fabric-covered hangar. There is no beaching capability. Davit-launched infantry landing craft (LCVPs) are not fitted but two can be carried in the dock or on deck. There are two 30 t cranes.
Operational: Failure of one of the main propulsion transformers in June 2012 necessitated temporary withdrawal of the ship from service. Following replacement of all four transformers, the ship became operational again in 2013.

CHOULES 4/2013*, *Chris Sattler* / 1528971

0 + 2 CANBERRA CLASS (AMPHIBIOUS ASSAULT SHIPS) (LHD)

Name	No	Builders	Laid down	Launched	Commissioned
CANBERRA	L 02	Navantia, Ferrol; Tenix	23 Sep 2009	18 Feb 2011	2014
ADELAIDE	L 01	Navantia, Ferrol; Tenix	2 Feb 2010	4 Jul 2012	2015

Displacement, tonnes: 26,800 full load
Dimensions, metres (feet): 231.4 × 32.0 × 7.08 *(759.2 × 105.0 × 23.2)*
Flight deck, metres: 202.3 × 32.0 *(663.7 × 105.0)*
Speed, knots: 19
Range, n miles: 9,000 at 15 kt
Complement: 1,221

Machinery: 1 GE LM 2500 gas turbine generator; 25,693 hp *(19.16 MW)*; 2 MAN 16V32/40 diesel generators; 19,975 hp *(14.9 MW)*; 2 Siemens podded propulsors; 29,500 hp *(22 MW)*; 2 bow thrusters
Guns: 4 Rafael Typhoon 25 mm. 6—12.7 mm MGs.
Physical countermeasures: Decoys: SLQ-25C towed torpedo decoy. 4 BAE Nulka decoy launchers.
Radars: Air/surface search: Saab Sea Giraffe; G/H-band.
Navigation: To be announced.
Combat data systems: Saab 9LV Mk 3E. Link 16.

Helicopters: Landing spots for 6 helicopters.

Programmes: Tenix/Navantia announced on 20 June 2007 as the preferred tenderer for Project 2048, the procurement of two helicopter-capable assault ships. A contract for the design and build of the ships was signed on 9 October 2007. The design of the ships is based on the Strategic Projection Ship constructed for the Spanish Navy. The ships' hulls from keel to flight deck are being built at Ferrol, Spain. Once built, they are to be transported to Tenix's Williamstown shipyard in Melbourne where the locally built superstructure will be joined to the hull. *Canberra* arrived in Australia in 2012 and *Adelaide* in early 2014. The majority of combat system design and integration work is taking place at Adelaide.
Structure: The hangar (990 m^2) is to be capable of accommodating 11 NH90s. Below the hangar, there is to be a 1,880 m^2 'garage' to accommodate 150 vehicles (including main battle tanks), provisions or containers. The landing dock (69.3 × 16 m) is to be capable of operating four LCM-1E or at least one landing craft air cushion. Medical facilities will include operating rooms, intensive care unit, and sick bay (22 beds). The 'ski jump' deck is also suitable for launching fixed-wing UAVs and will also enable cross-decking of STOVL aircraft operated by allies.
Operational: The principal roles are amphibious, strategic projection of land forces and disaster relief. The ships are to replace the capabilities of *Tobruk*, *Kanimbla*, and *Manoora*.

CANBERRA 3/2014*, *Chris Sattler* / 1529270

Australia > Amphibious forces

1 HEAVY LIFT SHIP (LSLH)

Name	No	Builders	Laid down	Launched	Commissioned
TOBRUK	L 50	Carrington Slipways Pty Ltd, Carrington	7 Feb 1978	1 Mar 1980	23 Apr 1981

Displacement, tonnes: 3,353 standard; 5,791 full load
Dimensions, metres (feet): 127 × 18.3 × 4.9 *(416.7 × 60.0 × 16.1)*
Speed, knots: 18. **Range, n miles:** 8,000 at 15 kt
Complement: 148 (13 officers)
Military lift: 314 troops (prolonged embarkation); 1,300 tons cargo or 330 lane-metres of vehicles; 70 tons capacity derrick; 2 – 4.25 ton cranes; 2 LCVP; 2 LCM 8
Machinery: 2 Mirrlees Blackstone KDMR8 diesels; 9,600 hp *(7.2 MW)*; 2 shafts

Guns: 2 – 12.7 mm MGs. 2 Mini Typhoon 12.7 mm guns. 2 Mini Typhoon 25 mm guns.
Radars: Surface search: Kelvin Hughes Type 1006; I-band. Navigation: Kelvin Hughes 1007; I-band.
Helicopters: Platform for one Sea King. Second Chinook capable spot on forward flight deck (clear of cargo).
Structure: The design is an update of the British Sir Bedivere class and provides facilities for the operation of helicopters, landing craft, amphibians for ship-to-shore movement.

A special feature is the ship's heavy lift derrick system for handling heavy loads. Able to embark a squadron of M1A1 tanks plus a number of wheeled vehicles and artillery in addition to its troop lift. Bow and stern ramps are fitted. Two LCM 8 carried on deck and two LCVPs at davits.

Operational: A basic communications fit enables participation in amphibious operations but not in command role. Based at Sydney. To be replaced by the Canberra class.

TOBRUK   2/2013*, Chris Sattler / 1528975

TOBRUK   6/2013*, Chris Sattler / 1528974

3 LANDING CRAFT (HEAVY) (LCH/LSM)

Name	No	Builders	Commissioned
BRUNEI	L 127	Walkers Ltd, Queensland	5 Jan 1973
LABUAN	L 128	Walkers Ltd, Queensland	9 Mar 1973
TARAKAN	L 129	Walkers Ltd, Queensland	15 Jun 1973

Displacement, tonnes: 364 standard; 517 full load
Dimensions, metres (feet): 44.5 × 10.1 × 2 *(146 × 33.1 × 6.6)*
Speed, knots: 10
Range, n miles: 3,000 at 10 kt
Complement: 16 (2 officers)
Military lift: 2 M1A1
Machinery: 2 Caterpillar 3406E diesels; 442 hp *(330 kW)* sustained; 2 shafts
Guns: 2 – 12.7 mm MGs.
Radars: Navigation: Racal Decca Bridgemaster; I-band.

Comment: Originally this class was ordered for the Army but only *Balikpapan* (decommissioned) saw Army service until being commissioned into the Navy on 27 September 1974. The remainder were built for the Navy. All were given a life extension refit, 2000–02 and also installation of Caterpillar diesels 2005–07. *Balikpapan*, *Betano* and *Wewak* were decommissioned in December 2012. The last three remaining vessels are based at Cairns.

TARAKAN   10/2013*, John Mortimer / 1528973

Amphibious forces — Survey ships (hydrographic survey) < **Australia** 33

4 LANDING CRAFT (LIGHT) (LCVP)

T 4–7

Displacement, tonnes: 7 full load
Dimensions, metres (feet): 13.2 × 3.5 × 0.7 (43.3 × 11.5 × 2.3)
Speed, knots: 22; 15 full load
Complement: 3
Military lift: 4.5 tons cargo or 1 Land Rover or 36 troops
Machinery: 2 Volvo Penta Sterndrives; 400 hp(m) (294 kW)

Comment: Prototype built by Geraldton, Western Australia. Trials conducted in late 1992. Three more delivered in July 1993. Two for *Tobruk*, one for *Success* and one spare attached to Defence Maritime Services at Garden Island, Sydney.

T6   5/2010, Chris Sattler / 1366694

0 + (12) LANDING CRAFT (LCM)

Displacement, tonnes: 108 full load
Dimensions, metres (feet): 23.3 × 6.4 × 1.1 (76.4 × 21.0 × 3.6)
Speed, knots: 14. **Range, n miles:** 160 at 12 kt
Complement: 3
Military lift: 100 tons or one main battle tank
Machinery: 2 MAN 2842-LE 402X diesels; 2,200 hp(m) (1.62 MW); 2 waterjets

Comment: Following selection of the design of the Canberra class amphibious assault ships in June 2007, Navantia selected in May 2009 to tender for the build 12 LCM-1E landing craft (four for each LPD, two for training and two for trials). Similar craft are in service with the Spanish Navy. The decision was confirmed in October 2011, construction began in October 2012 and delivery of the first four is to be made in 2014.

PATROL FORCES

Notes: Project Sea 1180 is for a class of about 20 offshore combatant vessels to replace the current capabilities of patrol craft, mine countermeasures and survey vessels. Using a modular approach, they are to be capable of operating unmanned underwater systems for both mine warfare and survey tasks. It is envisaged that these capabilities will be containerised. The ships are also to be capable of undertaking offshore and littoral warfighting roles, border protection, counter-terrorism and counter-piracy operations, special forces operations and regional stabilisation tasks. The ships are also to be capable of operating helicopters and/or unmanned aerial vehicles. First Pass Approval is planned by 2018.

14 ARMIDALE CLASS (PATROL CRAFT) (PB)

Name	No	Builders	Commissioned
ARMIDALE	83	Austal Ships, Fremantle	24 Jun 2005
LARRAKIA	84	Austal Ships, Fremantle	10 Feb 2006
BATHURST	85	Austal Ships, Fremantle	10 Feb 2006
ALBANY	86	Austal Ships, Fremantle	15 Jul 2006
PIRIE	87	Austal Ships, Fremantle	29 Jul 2006
MAITLAND	88	Austal Ships, Fremantle	29 Sep 2006
ARARAT	89	Austal Ships, Fremantle	10 Nov 2006
BROOME	90	Austal Ships, Fremantle	10 Feb 2007
BUNDABERG	91	Austal Ships, Fremantle	3 Mar 2007
WOLLONGONG	92	Austal Ships, Fremantle	23 Jun 2007
CHILDERS	93	Austal Ships, Fremantle	7 Jul 2007
LAUNCESTON	94	Austal Ships, Fremantle	22 Sep 2007
MARYBOROUGH	95	Austal Ships, Fremantle	8 Dec 2007
GLENELG	96	Austal Ships, Fremantle	22 Feb 2008

Displacement, tonnes: 300 standard
Dimensions, metres (feet): 56.8 × 9.7 × 2.7 (186.4 × 31.8 × 8.9)
Speed, knots: 25. **Range, n miles:** 3,000 at 12 kt
Complement: 21

Machinery: 2 MTU 4000 16V diesels; 6,225 hp (4.64 MW); 2 shafts
Guns: 1 – 25 mm Rafael M242 Bushmaster. 2 – 12.7 mm MGs.
Electronic countermeasures: RESM: BAE Systems Prism III; intercept.
Radars: Surface search/navigation: Bridgemaster E; E/F/I-band.
Electro-optic systems: Rafael Toplite optronic director.

Comment: Austal Ships in conjunction with Defence Maritime Services (DMS) contracted on 17 December 2003 to supply patrol boats to replace the Fremantle class under Project Sea 1444. The craft are of monohull design and carry two RHIBs. DMS is contracted to provide through-life logistics and maintenance support over 15 years. The craft are named after Australian cities and towns. Ten of the craft are based at Darwin, Northern Territory and four at Cairns, Queensland. The ships are operated by 21 crews under a multicrewing regime to maximise operational availability of the hulls.

BUNDABERG   9/2013*, Chris Sattler / 1528978

MINE WARFARE FORCES

6 HUON (GAETA) CLASS (MINEHUNTERS—COASTAL) (MHC)

Name	No	Builders	Launched	Commissioned
HUON	82	Intermarine, Sarzanaâ; ADI	25 Jul 1997	15 May 1999
HAWKESBURY	83	ADI, Newcastle	24 Apr 1998	12 Feb 2000
NORMAN	84	ADI, Newcastle	3 May 1999	26 Aug 2000
GASCOYNE	85	ADI, Newcastle	11 Mar 2000	2 Jun 2001
DIAMANTINA	86	ADI, Newcastle	2 Dec 2000	4 May 2002
YARRA	87	ADI, Newcastle	19 Jan 2002	1 Mar 2003

Displacement, tonnes: 732 full load
Dimensions, metres (feet): 52.5 × 9.9 × 3.0 (172.2 × 32.5 × 9.8)
Speed, knots: 14; 6 electrohydraulic
Range, n miles: 1,600 at 12 kt
Complement: 40 (6 officers) + 9 spare berths

Machinery: 1 Fincantieri GMT diesel; 1,986 hp(m) (1.46 MW); 1 shaft; LIPS cp prop; 3 Isotta Fraschini 1300 diesels, 1,440 hp(m) (1,058 kW); 3 electrohydraulic motors; 506 hp(m) (372 kW); Riva Calzoni retractable/rotatable APUs
Guns: 1 MSI DS 30B 30 mm/75. 650 rds/min to 10 km (5.4 n miles) anti-surface; 3 km (1.6 n miles) anti-aircraft; weight of shell 0.36 kg.
Physical countermeasures: MCM: 2 Bofors SUTEC Double-Eagle Mk 2 mine disposal vehicles with DAMDIC charges; ADI double Oropesa mechanical sweep and capable of towing the Australian developed Mini-Dyad influence sweep.
Decoys: 2 MEL Aviation Super Barricade; chaff launchers.
Electronic countermeasures: ESM: AWADI Prism.
Radars: Navigation: Kelvin Hughes 1007; I-band.
Sonars: GEC-Marconi Type 2093; VDS; VLF-VHF multifunction with five arrays; mine search and classification.
Combat data systems: GEC-Marconi NAUTIS 2M with Link 11 receive only.
Weapon control systems: Radamec 1400N optronic surveillance system.

Programmes: The Force Structure Review of May 1991 recommended the acquisition of coastal minehunters of proven design. A contract was signed with Australian Defence Industries (ADI) on 12 August 1994 to build six Intermarine-designed Gaeta-class derivatives. The hull of the first ship was constructed at Intermarine's Sarzana Shipyard in Italy and arrived in Australia as deck cargo on 31 August 1995 for fitting out in Newcastle, where the remaining five ships were built at ADI's Throsby Basin. Local content for this project was about 69%.
Structure: Monocoque GRP construction. A recompression chamber, one RIB and an inflatable diving boat are carried to support a six-man diving team.
Operational: This class which is named after Australian rivers, is based at HMAS *Waterhen* in Sydney. Two ships, *Hawkesbury* and *Norman* are at extended readiness.

GASCOYNE   7/2013*, Chris Sattler / 1528977

YARRA   2/2013*, Chris Sattler / 1528976

SURVEY SHIPS (HYDROGRAPHIC SURVEY)

Notes: In addition to the ships listed below, there are three civilian survey capable vessels; *Southern Surveyor*, *Solander* and *Cape Fergusson*. The Australian Antarctic Division also lease-operates the Antarctic supply ship *Aurora Australis*. This ship commenced operations in the Antarctic in 1990, is capable of carrying 70 scientists and is fitted with a helicopter hangar.

AURORA AUSTRALIS   6/2011, Chris Sattler / 1406525

IHS Jane's Fighting Ships 2014-2015

4 PALUMA CLASS (AGSC)

Name	No	Builders	Commissioned
PALUMA	A 01	Eglo, Adelaide	27 Feb 1989
MERMAID	A 02	Eglo, Adelaide	4 Dec 1989
SHEPPARTON	A 03	Eglo, Adelaide	24 Jan 1990
BENALLA	A 04	Eglo, Adelaide	20 Mar 1990

Displacement, tonnes: 325 full load
Dimensions, metres (feet): 36.6 × 12.8 × 2.65 *(120.1 × 42.0 × 8.7)*
Speed, knots: 11
Range, n miles: 3,600 at 11 kt
Complement: 14 (3 officers)
Machinery: 2 Detroit 12V-92TA diesels; 1,100 hp *(820 kW)* sustained; 2 shafts
Radars: Navigation: Kelvin Hughes 1007; I-band.
Sonars: Thales Petrel three-dimensional forward-looking active; high frequency.

Comment: Catamaran design based on Prince class ro-ro passenger ferries. Steel hulls and aluminium superstructure. Contract signed in November 1987. Fitted with Klein 5000 V2 towed side-scan sonar (455 kHz), Reson Seabat 7125 multibeam echosounder (200 and 400 kHz, 256 or 512 beams), Kongsberg EA 400/600 hull side-scan sonar (210 kHz), Kongsberg single beam echosounder (15 and 120 kHz), Thales Petrel three dimensional forward looking sonar (100 kHz) and MVP200 moving vessel profiler with salinity, temperature, depth probe, grab and free fall penetrometer. Positioning sensors: Fugro Seastar – WAdGPS, Leica dGPS, Trimble RTK dGPS and Microfix terrestrial based positioning system. All ships based at Cairns and normally operate in pairs when undertaking survey operations.

2 LEEUWIN CLASS (AGS)

Name	No	Builders	Launched	Commissioned
LEEUWIN	A 245	NQEA, Cairns	19 Jul 1997	27 May 2000
MELVILLE	A 246	NQEA, Cairns	23 Jun 1998	27 May 2000

Displacement, tonnes: 2,205 full load
Dimensions, metres (feet): 71.2 × 15.2 × 4.3 *(233.6 × 49.9 × 14.1)*
Speed, knots: 14
Range, n miles: 18,000 at 9 kt
Complement: 61 (10 officers; 5 trainees)

Machinery: Diesel-electric; 4 GEC Alsthom 6RK 215 diesel generators; 4,290 hp *(3.2 MW)* sustained; 2 Alsthom motors; 1.94 MW; 2 shafts; 1 Schottel bow thruster
Radars: Navigation: STN Atlas 9600 ARPA; I-band.
Sonars: C-Tech CMAS 36/39; hull mounted; high frequency active.
Helicopters: 1 AS 350B (not permanently embarked).

Comment: Contract awarded 2 April 1996 to North Queensland Engineers & Agents (NQEA). Fitted with Atlas Fansweep-20 multibeam echo sounder and one Atlas Hydrographic Deso single beam echo sounder. Also fitted with Edgetex MPX 4300 high speed towed sidescan sonar. The ships are capable of various small boat configurations utilising the three SMB davits. The ships are also fitted with an additional RHIB and two light utility boats. Based at Cairns.

LEEUWIN *9/2013*, Chris Sattler* / 1528980

BENALLA *10/2013*, Chris Sattler* / 1528981

9 SURVEY MOTOR BOATS (YGS)

FANTOME 1006	TOM THUMB 1009	CASUARINA 1011
MEDA 1007	JOHN GOWLLAND 1010	CONDER 1021
DUYFKEN 1008	GEOGRAPHE 1012	WYATT EARP 0329

Dimensions, metres (feet): 10.7 × 2.9 × 1.7 *(35.1 × 9.5 × 5.6)*
Speed, knots: 24
Range, n miles: 300 at 12 kt
Complement: 4 (1 officer)
Machinery: 2 Volvo Penta AQAD-41A diesel stern drives; 400 hp(m) *(294 kW)*; 2 props
Radars: Navigation: JRC; I-band.

Comment: Six Survey Motor Boats (SMB) built by Pro Marine, Victoria 1992–1993. Two additional SMBs (*Casuarina* and *Geographe*) were built in 1997 to supplement the Leeuwin-class AGS. One SMB has been taken out of service. The remaining seven SMBs are fitted with an Atlas Hydrographic Fansweep 20 multibeam echo sounder and Atlas Hydrographic Deso 15 single beam echo sounder. Three SMBs are fitted for the Klein 2000 towed lightweight side scan sonar. SMB *Conder* built by North Queensland Engineers and Agents in 2003 as a prototype replacement SMB is fitted with an Atlas Hydrographic Fansweep 20 multibeam echo sounder and Atlas Hydrographic Deso 15 single beam echo sounder. Six SMBs are allocated to the Leeuwin-class AGS in Cairns and two to the hydrographic school at HMAS *Penguin*. The Antarctic Survey Vessel (ASV) *Wyatt Earp*, a 9 m craft purpose built by Pro Marine, Victoria in 1992 for operations in the Antarctic. *Wyatt Earp* is allocated to the Deployable Geospatial Support Team (DGST 1) in Sydney and is fitted with L3 tactical hydrographic survey system consisting of Reson 7125 multibeam echo sounder, Kongsberg EA400SP single-beam echo sounder and Klein 3200 towed lightweight side-scan sonar.

DEEP SUBMERGENCE VEHICLES

1 RESCUE SUBMERSIBLE (DSRV)

LR 5

Displacement, tonnes: 21.5 surfaced
Dimensions, metres (feet): 9.2 × 3.0 × 3.5 *(30.2 × 9.8 × 11.5)*
Speed, knots: 2.5 dived
Complement: 2 + 1 personnel
Machinery: 2 electric motors; 16 hp *(12 kW)*; 2 hydraulic transverse thrusters; 9 hp *(6.7 kW)*; 2 hydraulic tiltable side thrusters; 9 hp *(6.7 kW)*

Comment: LR 5 replaced *Remora* as the Australian submarine rescue vehicle in 2012 and is operated by James Fisher Submarine Rescue Systems (JFSRS). Elements of JFSRS are based at Henderson, West Australia. LR 5 was part of the UK Submarine Rescue System (UKSRS) until replaced by the NATO Submarine Rescue System in 2008. In 2000 James Fisher Rumic Ltd rebuilt LR 5, replacing the original GRP hull by incorporating a new steel rescue chamber, thereby increasing rescue payload capability from three to 16 personnel and enabling rescue at greater pressure and depth. In 2005, a replacement steel command module and battery pod assemblies were fitted as part of a routine major survey. Capable of operating down to 400 m depth, it is to be operated from the deck of the new Australian Submarine Rescue Ship, to be completed in 2016, or any suitable mother ship. Its role is to rescue up to 16 survivors at a time from a disabled submarine on the seabed and bring them back to the surface. This can be done at normal atmospheric pressure and at increased pressure up to 5 bar. Mating with the disabled submarine can be achieved at up to 10° heel on LR 5 or at up to 60° bow up. LR 5 is complemented by a Universal Deck Reception Chamber allowing casualties to be transferred at pressure to recompression chambers. It is also supported by an ROV, Scorpio 45, which is attached to a 1,000 m umbilical. This is used to locate the disabled submarine, clear obstructions from the escape hatches and replenish life support stores.

MEDA *8/2013*, Chris Sattler* / 1528979

LR 5 *1/2009*, James Fisher* / 1375481

TRAINING SHIPS

Notes: In addition to *Young Endeavour* (navy operated), there are five Fleet class yachts. Of 36.1 ft *(11 m)*. GRP yachts named *Charlotte of Cerberus*, *Friendship of Leeuwin*, *Scarborough of Cerberus*, *Lady Penrhyn of Nirimba* and *Alexander of Creswell*. The names are a combination of Australia's first colonising fleet and the training base to which each yacht is allocated.

1 SAIL TRAINING SHIP (AXS)

Name	Builders	Launched	Commissioned
YOUNG ENDEAVOUR	Brooke Yachts, Lowestoft	2 Jun 1987	25 Jan 1988

Displacement, tonnes: 243 full load
Dimensions, metres (feet): 44 × 7.8 × 4 *(144.4 × 25.6 × 13.1)*
Speed, knots: 14.5; 10 diesel
Range, n miles: 2,880 at 8 kt
Complement: 33 (24 cadets)
Machinery: 2 Perkins V8 diesels; 334 hp *(294 kW)*; 2 shafts

Comment: Built to Lloyds 100 Al LMC yacht classification by Brooke Yachts, Lowestoft. Sail area 707.1 m². Presented to Australia by UK government as a bicentennial gift. Operated by RAN on behalf of the Young Endeavour Youth Scheme.

YOUNG ENDEAVOUR 8/2013*, Chris Sattler / 1528984

1 TRAINING SHIP (AXL)

Name	Builders	Launched
SEAHORSE MERCATOR	Tenix Shipbuilding, Henderson	15 Oct 1998

Displacement, tonnes: 212 full load
Dimensions, metres (feet): 32.5 × 4.18 × 2.6 *(106.6 × 13.7 × 8.5)*
Speed, knots: 14.5
Range, n miles: 2,700 at 10 kt
Complement: 26 (18 trainees)
Machinery: 2 Caterpillar 3412 diesels; 2,104 hp *(1.56 MW)*; 2 shafts

Comment: Operated by Defence Maritime Services as a Navigation training ship based at Sydney. Similar to Pacific class patrol craft.

SEAHORSE MERCATOR 5/2012, Chris Sattler / 1482942

AUXILIARIES

Notes: (1) Only *Sirius* and *Success* are navy operated. The rest have been contracted to the Defence Maritime Services. These craft have blue hulls and buff superstructures, and are chartered as required.
(2) In addition to the vessels listed there are some 24 workboats (AWB and NWB numbers), a VIP launch *Tresco II* and an admiral's barge *Admiral Hudson*.
(3) An ex-landing craft, *Seahorse Kultarr*, is based at Darwin. The 16.7 m vessel is used for ammunitioning and general utility duties.
(4) The Spanish Navy replenishment tanker *Cantabria* deployed to Australia for most of 2013 to support RAN operations while *Success* was in refit.

1 SIRIUS CLASS (REPLENISHMENT TANKER) (AORH)

Name	No	Builders	Launched	Commissioned
SIRIUS (ex-*Delos*)	A 266	Hyundai Mipo Dockyard	12 Apr 2004	16 Sep 2006

Displacement, tonnes: 46,755 full load
Dimensions, metres (feet): 189.5 × 31.0 × 10.5 *(621.7 × 101.7 × 34.4)*
Speed, knots: 16.5
Range, n miles: 16,000 at 14 kt
Complement: 56 (8 officers)

Cargo capacity: Total volume in excess of 36,000 m³. Dry cargo capacity 240 tonnes
Machinery: 1 Hyundai B&W 6S 50MC diesel; 1 shaft; bow thruster
Guns: 5 — 12.7 mm MGs.
Radars: 2 Sperry Marine Bridgemaster-E; E/F/I-bands.

Comment: Acquired as the replacement for the single-hulled *Westralia*, *Sirius* is a double-hulled ship built to Lloyd's standard. Bought new in June 2004 as *MT Delos* and subsequently leased for use as an oil tanker until September 2005. Contract for the conversion of the ship to military use awarded to Tenix Defence on 15 March 2005. The conversion included the addition of a flight deck and RAS equipment. The first RAN ship to carry the name *Sirius*, she is named after the flagship of the First Fleet which arrived in Australia in 1788. To remain in service until 2020.

SIRIUS 7/2013*, Chris Sattler / 1528983

1 DURANCE CLASS
(UNDERWAY REPLENISHMENT TANKER) (AORH)

Name	No	Builders	Laid down	Launched	Commissioned
SUCCESS	OR 304	Cockatoo Dockyard, Sydney	9 Aug 1980	3 Mar 1984	19 Feb 1986

Displacement, tonnes: 18,221 full load
Dimensions, metres (feet): 157.2 × 21.2 × 8.6 *(515.7 × 69.6 × 28.2)*
Speed, knots: 20
Range, n miles: 8,616 at 15 kt
Complement: 237 (25 officers)

Cargo capacity: 10,200 tons: 8,707 dieso; 975 Avcat; 116 distilled water; 57 victuals; 250 munitions including SM1 missiles and Mk 46 torpedoes; 95 naval stores and spares
Machinery: 2 SEMT-Pielstick 16 PC2.5 V 400 diesels; 20,800 hp(m) *(15.3 MW)* sustained; 2 shafts; LIPS cp props
Guns: 1 Raytheon Vulcan Phalanx Mk 15 CIWS (fitted for but not with). 7 — 12.7 mm MGs.
Radars: Navigation: 2 Kelvin Hughes Type 100G; E/F/I-bands.
Helicopters: 1 AS 350B Squirrel, Sea King or Seahawk.

Comment: Based on French Durance class design. Replenishment at sea from four beam positions (two having heavy transfer capability) and vertrep. One LCVP is carried on the starboard side aft. The ship has been modified into a double-hull vessel, to meet IMO standards for oil tankers. The work was carried out by ST Marine, Singapore from December 2010 to April 2011. *Success* underwent a major refit during 2013 during which time her duties were undertaken by Spanish replenishment tanker, *Cantabria*. In late 2013, *Success* conducted first-of-class flight trials for MRH 90 helicopter.

SUCCESS 11/2013*, Chris Sattler / 1528982

1 FISH CLASS (TORPEDO RECOVERY VESSELS) (YPT)

TAILOR TRV 803

Displacement, tonnes: 94 full load
Dimensions, metres (feet): 26.8 × 6.1 × 1.1 *(87.9 × 20.0 × 3.6)*
Speed, knots: 9.5
Complement: 9
Machinery: 3 GM diesels; 890 hp *(664 kW)*; 3 shafts
Radars: Navigation: I-band.

Comment: Built at Williamstown completed between January 1970 and April 1971. Can transport eight torpedoes. Based at Fleet Base West.

FISH CLASS 5/2006, Bob Fildes / 1159952

Australia > Auxiliaries

2 TRIALS AND SAFETY VESSELS (ASR)

Name	Builders	Commissioned
SEAHORSE STANDARD (ex-*British Viking*)	Marystown Shipyard, Newfoundland	1980
SEAHORSE SPIRIT (ex-*British Magnus*)	Marystown Shipyard, Newfoundland	1980

Measurement, tonnes: 2,124 gt; 1,661 dwt
Dimensions, metres (feet): 72 × 16 × 6.5 *(236.2 × 52.5 × 21.3)*
Speed, knots: 12
Complement: 20 + 44 spare berths
Machinery: 2 MLW-ALCO Model 251 V-12 diesels; 5,480 hp(m) *(4.03 MW)*; 1 shaft; cp prop; 2 stern and 2 bow thrusters
Radars: Surface search/navigation: JMA 7000; I-band. JMA 5320; I-band.

Comment: Acquired 2 December 1998 by Defence Maritime Services to support RAN trials in Western and Southern Australian waters. Dynamic Positioning system. These ships are also used for weapon recovery and can embark LR5 submarine rescue suites. Likely to be replaced by new submarine rescue ships in 2015.

SEAHORSE SPIRIT   *8/2008, Chris Sattler / 1335632*

1 TRIALS AND SAFETY VESSEL (ASR)

Name	No	Builders	Commissioned
SEAHORSE HORIZON (ex-*Protector*; ex-*Blue*, ex-*Nabilla*, ex-*Osprey*)	– (ex-ASR 241)	Stirling Marine Services, WA	1984

Displacement, tonnes: 681 full load
Dimensions, metres (feet): 42.7 × 9.5 × 3 *(140.1 × 31.2 × 9.8)*
Speed, knots: 9.5. **Range, n miles:** 3,192 at 9.5 kt
Complement: 9 (est.) + 6 personnel
Machinery: 2 Detroit 12V-92TA diesels; 2,440 hp *(1.82 MW)* sustained; 2 Heimdal cp props
Radars: Navigation: JRC 310; I-band. Decca RM 970BT; I-band.
Sonars: Klein; side scan; high frequency.
Helicopters: Platform for 1 light.

Comment: A former National Safety Council of Australia vessel commissioned into the Navy in November 1990. Used to support contractor's sea trials of the Collins class submarines, and for mine warfare trials and diving operations. LIPS dynamic positioning, two ROVs and a recompression chamber. Helicopter deck and a submersible were removed in 1992. Based at Jervis Bay. Decommissioned in early 1998 and run as part of the commercial support programme. Also used for junior officer training.

SEAHORSE HORIZON   *7/2013*, Chris Sattler / 1528989*

4 SELF-PROPELLED LIGHTERS (WFL/AOTL)

WARRIGAL 333 (ex-WFL 8001)	**WOMBAT** 332 (ex-WFL 8003)
WALLABY 331 (ex-WFL 8002)	**WYULDA** 334 (ex-WFL 8004)

Displacement, tonnes: 269 standard; 1,210 full load
Dimensions, metres (feet): 39.25 × 11.0 × 4.8 *(128.8 × 36.1 × 15.7)*
Speed, knots: 7
Cargo capacity: 560 tons dieso and 200 tons water
Machinery: 2 Dorman diesels; 2 azimuth propulsion steering units

Comment: First three were laid down at Williamstown in 1978. The fourth, for HMAS *Stirling*, was ordered in 1981 from Williamstown Dockyard. Used for water/fuel transport. Steel hulls with twin, swivelling, outboard propellers. *Warrigal* at Darwin; *Wombat* and *Wallaby* at Fleet Base East; *Wyulda* at Fleet Base West.

WOMBAT   *8/2013*, Chris Sattler / 1528988*

3 WATTLE CLASS STORES LIGHTERS (YE)

WATTLE CSL 01   **BORONIA** CSL 02   **TELOPEA** CSL 03

Displacement, tonnes: 149 full load
Dimensions, metres (feet): 24.2 × 10.0 × 1.66 *(79.4 × 32.8 × 5.4)*
Speed, knots: 8. **Range, n miles:** 320 at 8 kt
Machinery: 2 Caterpillar D333C diesels; 600 hp *(447 kW)*
Radars: Navigation: 1 JRC JMA-2253; I-band.

Comment: Built by Cockatoo DY, Sydney and delivered in 1972. Employed to transport ammunition and stores. Equipped with 3-ton electric crane. CSL 02 and 03 based at Sydney and CSL 01 at Darwin.

TELOPEA   *9/2013*, Chris Sattler / 1528987*

4 DIVING TENDERS (YDT/PB)

SEAL 2001   **MALU BAIZAM** 2003   **SHARK** 2004   **DUGONG** 21689

Displacement, tonnes: 44 full load
Dimensions, metres (feet): 21.18 × 5.35 × 1.64 *(69.5 × 17.6 × 5.4)*
Speed, knots: 26. **Range, n miles:** 1,200 at 15 kt
Complement: 22 (16 divers)
Machinery: 2 Caterpillar 3406E diesels; 901 hp *(672 kW)*; 2 shafts

Comment: Built by Geraldton Boat Builders, Western Australia and completed in August 1993. Carry 2 tons of diving equipment to support 24 hour diving operations in depths of 54 m. *Shark* based at *Stirling*, *Seal* at *Waterhen* and *Dugong* at Sydney, *Malu Baizam* is based at Thursday Island in the Torres Strait and is navy manned. *Porpoise* grounded in 1995 and was assessed as being beyond economical repair. Replacement built in 1996. Run as part of the commercial support operation from 1997. Sister craft *Coral Snake* and *Red Viper* are operated by the Army.

SEAL   *6/2013*, Chris Sattler / 1528986*

1 DISASTER RELIEF VESSEL

Name	Builders	Laid down	Launched	Commissioned
OCEAN SHIELD (ex-*Skandi Bergen II*)	Aker Tulcea	11 Apr 2011	22 Oct 2011	22 May 2012

Displacement, tonnes: 8,500 full load
Measurement, tonnes: 8,368 gt
Dimensions, metres (feet): 110.9 × 22.05 × 6.6 *(363.8 × 72.3 × 21.7)*
Speed, knots: 16
Complement: 22 + 50 spare berths
Machinery: Diesel-electric; 4 Wärtsilä 6L32 diesels; 16,316 hp *(12.0 MW)*; 2 motors; 6,438 hp *(4.8 MW)*; 2 directional props
Guns: 2 – 12.7 mm MGs.
Radars: Surface search/navigation: E/F/I-bands.
Helicopters: Platform for 1 medium.

Comment: The acquisition of the former *Skandi Bergen II* was announced on 8 June 2012. The ship's primary role is to transport troops and supplies in support of humanitarian and disaster relief operations domestically and in the region. Originally built in 2012 as a multipurpose offshore support vessel, the ship is of steel construction, is equipped with emergency medical facilities and can provide basic accommodation for up to 120 people in addition to the crew. The ship has a 1,100 m² deck area, is equipped with a 140 t crane and carries two 8.5 m response craft. The ship is operated as an Australian Defence Vessel (government owned and civilian operated) until the Canberra-class LHDs enter service. She is then to replace her sister ship *Ocean Protector*, which is chartered by the Customs and Border Protection Service.

OCEAN SHIELD   *11/2013*, Michael Nitz / 1528985*

1 WORKBOAT (YFRT)

SEAHORSE PLATYPUS

Displacement, tonnes: 24 full load
Dimensions, metres (feet): 15.84 × 4.4 × 1.3 (52 × 14.4 × 4.3)
Speed, knots: 15
Machinery: 1 diesel; 643 hp (480 kW); 1 shaft

Comment: Utility craft used for target towing, submarine and trials support.

2 SUBMARINE RESCUE SHIPS (ASR)

Name	Builders	Laid down	Launched	Commissioned
STOKER	Damen, Haiphong	2013	2014	2015
BESSANT	Damen, Haiphong	2013	2014	2015

Displacement, tonnes: 2,540 full load
Measurement, tonnes: 3,600 gt
Dimensions, metres (feet): 83 × 16.0 × 4.25 (272.3 × 52.5 × 13.9)
Speed, knots: 14
Complement: 16
Machinery: 2 diesels; 4,000 hp (3.0 MW); 2 shafts; 1 bow thruster
Radars: Surface search/navigation: E/F/I-bands.
Helicopters: Platform for 1 medium.

Comment: The order for two new submarine rescue ships was made in 2012. The design is based on the Damen Support Ship 8316 design (on which details are based) and is similar to *SD Victoria*, operated by Serco Denholm for UK military training and diving support operations. The ships are to be delivered in 2015 and are to be based in West Australia. They are to be the primary platform for the Australian submarine rescue submersible LR 5, operated by James Fisher Submarine Rescue Services. The ships are named after the commanding officers of Australia's first two submarines AE1 and AE2.

TUGS

Notes: In addition the two MSCD are used as tugs. Details under Mine Warfare Forces.

6 HARBOUR TUGS (YTM)

TAMMAR DT 2601	SEAHORSE QUENDA –	ELWING –
QUOKKA DT 1801	SEAHORSE CHUDITCH –	WAREE –

Comment: *Tammar* is 26 m long, has a bollard pull of 35 tons and is based at Stirling; *Quokka* is 19 m long, has a bollard pull 8 tons, is based at Darwin. *Seahorse Chuditch* and *Seahorse Quenda* were built in Malaysia and delivered in 2003. 23 m long they have a bollard pull of 16 tons. *Elwing* and *Waree* are 25 m Damen ASD 2411 built in Vietnam. They arrived in Sydney on 8 November 2012.

WAREE   9/2013*, Chris Sattler / 1528991

QUOKKA   9/2012, Chris Sattler / 1482936

4 SMALL TUGS (YTL)

SEAHORSE BETONG	SEAHORSE KOWARI
SEAHORSE QUOLL	SEAHORSE PARMA

Displacement, tonnes: 44 full load
Dimensions, metres (feet): 14.98 × 5.75 × 2.0 (49.1 × 18.9 × 6.6)
Speed, knots: 11
Machinery: 2 Caterpillar 3406TA diesels; 800 hp (596 kW); 2 shafts

Comment: Small tugs also used for general harbour services, diving platforms and target towing. Details are for *Betong* and *Quoll*. *Kowari* and *Parma* are to a slightly smaller modified design.

© 2014 IHS

SEAHORSE QUOLL   9/2013*, Chris Sattler / 1528990

ARMY

Notes: (1) Operated by Royal Australian Army Corps of Transport. Personnel: About 300 as required.
(2) In addition to the craft listed below there are 159 assault boats 16.4 ft (5 m) in length and capable of 30 kt. Can carry 12 troops or 1,200 kg of equipment. Also there are 12 ex-US Army LARC-V amphibious wheeled lighters, which have limited capability to operate with *Tobruk* and LCHs. They will be able to operate with LHDs.
(3) All LCM are to be replaced in about 2016 by new amphibious watercraft (JP 2048).

6 LCM 2000 CLASS (AMPHIBIOUS WATERCRAFT) (LCM)

AB 2000–2005

Displacement, tonnes: 137 full load
Dimensions, metres (feet): 25.4 × 7.6 × 1.0 (83.3 × 24.9 × 3.3)
Speed, knots: 11. **Range, n miles:** 720 at 10 kt
Complement: 5
Machinery: 2 Detroit 6062 diesels; 2 Doen waterjets
Guns: 2—12.7 mm MGs.

Comment: Contract signed with ADI in June 2002 to provide Army watercraft. The sixth craft was delivered in mid-2005. Designed to be operated from the now decommissioned Kanimbla class and to be capable of tranferring vehicles to and from the ships, the craft are of aluminium construction and they have through-deck, roll-on/roll-off design and bow and stern ramps. With 65 tonne cargo capacity, the craft can carry five armoured vehicles. Following sea trials, the craft were found to have structural issues and have yet to reach operational status. Meanwhile, they are used for training and are based at Townsville.

AB 2000   12/2004, Bob Fildes / 1153864

15 LCM 8 CLASS

COCONUT QUEEN AB 1050	ANZAC COVE AB 1058	ICE FIRE AB 1063
SEAHORSE 2 AB 1051	SNOOPY AB 1059	SEA TROJAN AB 1064
SEA WIDOW AB 1053	THE WANDERER AB 1060	VIKING SUN AB 1065
MIDNIGHT SUN AB 1055	NORTHERN WARRIOR AB 1061	HASTINGS AB 1066
CHARLIE BROWN AB 1056	OCEAN MARIE AB 1062	CHARISMA AB 1067

Displacement, tonnes: 120 full load
Dimensions, metres (feet): 22.4 × 6.4 × 1.6 (73.5 × 21.0 × 5.2)
Speed, knots: 12
Range, n miles: 500 at 10 kt
Complement: 4
Military lift: 55 tons
Machinery: 2 8V92TA diesels; 543 hp (405 kW); 2 shafts
Guns: 2—12.7 mm MGs.

Comment: Built by North Queensland Engineers, Cairns and Dillinghams, Fremantle to US design. Based at Townsville and Darwin. *AB 1057* transferred to Tonga 1982, *AB 1052* and *AB 1054* sold to civilian use in 1992. All upgraded to Mod 2 standard by late 1999 with new engines and with endurance increased.

CHARLIE BROWN   10/2002, John Mortimer / 0528383

38 **Australia** > Army — Customs and border protection

2 SAFCOL CRAFT

CORAL SNAKE AM 1353   **RED VIPER** –

Displacement, tonnes: 41 full load
Dimensions, metres (feet): 20 × 6.1 × 1.4 *(65.6 × 20.0 × 4.6)*
Speed, knots: 28
Range, n miles: 350 at 25 kt
Complement: 3
Machinery: 2 General Motors Detroit 8V92 diesels; 1,800 hp *(1.34 MW)*

Comment: Sister to Seal class built at Geraldton Boat Builders. *Coral Snake* delivered in 1994 and *Red Viper* in 1996. Used as Special Action Forces Craft Offshore Large (SAFCOL) to support dives and transport of stores and personnel.

RED VIPER   3/2010, Chris Sattler / 1966711

9 EXPRESS SHARK CAT CLASS (PB)

AM 237–244   AM 428

Displacement, tonnes: 7 full load
Dimensions, metres (feet): 9.8 × 2.8 × 1.0 *(32.2 × 9.2 × 3.3)*
Speed, knots: 40
Range, n miles: 180 at 40 kt
Complement: 2
Machinery: 2 Verado Mercury outboards; 550 hp *(410 kW)*

Comment: Built by NoosaCat, Queensland and delivered by 1995. Trailer transportable.

AM 243   11/1997, van Ginderen Collection / 0012946

NON-NAVAL PATROL CRAFT

4 SHARK CAT 800 CLASS (WORKBOATS) (YFL)

0801–0803   0805

Displacement, tonnes: 8 full load
Dimensions, metres (feet): 8.35 × 2.8 × 1.0 *(27.4 × 9.2 × 3.3)*
Speed, knots: 30
Complement: 1 + 11 spare berths
Machinery: 2 Mercury outboard engines; 450 hp *(335 kW)*

Comment: Built by Shark Cat, Noosaville, Queensland and delivered in 1980s. GRP construction. Used for target-towing, naval police and range clearance duties. *0801* and *0802* based at Fleet Base East; *0803* and *0805* at HMAS *Creswell*.

SHARK CAT 800   8/2013*, Chris Sattler / 1528994

IHS Jane's Fighting Ships 2014-2015

4 NOOSACAT 930 WORKBOATS (YFL)

0901–0904

Dimensions, metres (feet): 9.3 × 3.5 × 0.7 *(30.5 × 11.5 × 2.3)*
Speed, knots: 30
Range, n miles: 240 at 20 kt
Machinery: 2 Volvo Penta ADQ41DP diesels; 400 hp *(298 kW)*; 2 props

Comment: Built by Noosacat, Queensland and delivered in 1994. GRP hulled craft. *0903* and *0904* based at Sydney, *0902* at HMAS *Creswell* and *0901* at HMAS *Cerberus*. Used for force protection duties.

NOOSACAT 930   9/2013*, Chris Sattler / 1528992

10 STEBER CLASS WORKBOATS (YFL/YDT)

BUNDEENA NGPWB 01   **RELIANCE** NGPWB 06
ELOUERA NGPWB 02   **PATONGA** NGPWB 07
SHOALHAVEN NGPWB 03   **BILGOLA** NGPWB 08
SEA DRAGON NGPWB 04   **SEA WITCH** NGPWB 09
ETHEL JOY NGPWB 05   **BRUTUS** NGPWB 10

Displacement, tonnes: 17 full load
Dimensions, metres (feet): 13.2 × 4.7 × 1.3 *(43.3 × 15.4 × 4.3)*
Speed, knots: 25 (NGPWB 1–6), 20 (NGPWB 7–10)
Machinery: 2 Caterpillar 3208TA diesels (01–06). 1 Caterpillar 3406E diesel (07–10)

Comment: Built by Steber craft and delivered in 1997. GRP hulled craft for general purpose stores and personnel transport and for use as diving tenders. Most have radars 01, 02, 07 and 08 based at Sydney, 03 at HMAS *Cresswell*, 04 and 09 at Fleet Base West and 06 at HMAS *Cerberus*.

BILGOLA   9/2013*, Chris Sattler / 1528993

CUSTOMS AND BORDER PROTECTION

Notes: Aircraft: The Customs and Border Protection Service manages its civil aerial surveillance programme through commercial contracts with Surveillance Australia (fixed wing), Australian Helicopters (Torres Strait) and Helicopters Australia (Gove). The new fixed-wing aircraft fleet consists of six De Havilland Dash 8-202 and four Dash 8-315 equipped with radar, IR and EO sensors. These aircraft are either new or have been upgraded under Project Sentinel which provided the Dash 8 with new electro-optics and the Raytheon 2022 SAR/ISAR radar, as well as an integrated information management system and a range of other electronic sensors. The aircraft are based in Broome (WA), Darwin (NT), Horn Island (QLD – Torres Strait), Weipa (QLD) and Cairns (QLD). The ACS helicopter surveillance fleet consists of a Bell 412 and an AS350 Squirrel (Australian Helicopters) which are both based in the Torres Strait. A Eurocopter-145 helicopter is based in Gove (NT) in a rapid response and surveillance role.

DASH 8-200   6/2005, Massimo Annati / 1153871

© 2014 IHS

Customs and border protection < **Australia** 39

1 + 7 CAPE CLASS (OFFSHORE PATROL VESSELS) (PBO)

CAPE ST GEORGE	CAPE SORELL	CAPE WESSEL
CAPE BYRON	CAPE JERVIS	CAPE YORK
CAPE NELSON	CAPE LEVEQUE	

Dimensions, metres (feet): 57.8 × 10.3 × 3.0 *(189.6 × 33.8 × 9.8)*
Speed, knots: 25
Range, n miles: 4,000 at 12 kt
Complement: 18
Machinery: 2 Caterpillar 3516C diesels; 6,770 hp *(5.05 MW)*; 2 shafts; 1 bow thruster *(160 kW)*
Radars: Surface search: To be announced.
Navigation: To be announced.

Comment: Following initiation in mid-2010 of a programme to replace the Bay-class patrol craft, a contract for the design, construction and through-life support of eight new Cape-class patrol boats was signed with Austal Shipbuilding on 12 August 2011. The launch of the first and keel-laying of the second of class took place at Henderson, Western Australia, in January 2013 and delivery of all eight ships is to be by August 2015. The roles of the ships are to include EEZ protection, counter-piracy and interception of suspect ships. Two RHIB interception craft are to be carried at the stern and one smaller craft midships.

CAPE ST GEORGE *4/2013*, Austal Shipbuilding* / 1405991

8 BAY CLASS (PB)

ROEBUCK BAY ACV 10	HERVEY BAY ACV 40	DAME ROMA MITCHELL ACV 70
HOLDFAST BAY ACV 20	CORIO BAY ACV 50	STORM BAY ACV 80
BOTANY BAY ACV 30	ARNHEM BAY ACV 60	

Displacement, tonnes: 136 standard
Dimensions, metres (feet): 38.2 × 7.2 × 2.4 *(125.3 × 23.6 × 7.9)*
Speed, knots: 24
Range, n miles: 1,000 at 20 kt
Complement: 12
Machinery: 2 MTU 16V 2000M 70 diesels; 2,856 hp(m) *(2.1 MW)* sustained; 2 shafts. 1 Vosper Thornycroft bow thruster
Radars: Surface search: Racal Decca; E/F- and I-band.
Sonars: Wesmar SS 390E dipping sonar.

Comment: Built by Austal Ships and delivered from February 1999 to August 2000. The craft carry two RIBs capable of 35 kt.

HOLDFAST BAY *8/2010, John Mortimer* / 1366712

1 OFFSHORE PATROL VESSEL (PSOH)

| Name | Builders | Commissioned |
| TRITON | Vosper Thornycroft, Woolston | Sep 2000 |

Displacement, tonnes: 1,118 full load
Measurement, tonnes: 2,272 gt
Dimensions, metres (feet): 98.7 × 22.5 × 3.2 *(323.8 × 73.8 × 10.5)*
Speed, knots: 20
Range, n miles: 17,000 at 10 kt
Complement: 14 + 30 personnel

Machinery: Diesel-electric; 2 Paxman 12V 185 diesel generators; 5,364 hp *(4 MW)*; 1 HMA motor; 4,700 hp *(3.5 MW)*; 1 shaft (centreline); 2 HMA motors; 938 hp *(700 kW)*; 2 Schottel propulsors (outer hulls)
Guns: 2 – 12.7 mm MGs.
Radars: Surface search/navigation: Grumman Sperry Marine Bridgemaster E; E/F/I-bands.
Helicopters: Platform for 1 medium.

Comment: Originally built as trimaran hull demonstrator vessel for the UK MoD research agency. Following five years of trials, sold to Gardline Shipping in 2005 and thereafter acted as a hydrographic survey vessel for the UK Maritime and Coast Guard Agency. Contracted in early 2007 by the Australian Customs and Border Protection Service to act as an offshore patrol vessel in northern waters from Broome, West Australia, to Cairns, Queensland. It carries two 7 m high-speed interception craft.

TRITON *10/2009, Bob Fildes* / 1305757

1 OFFSHORE PATROL VESSEL (PBO)

ASHMORE GUARDIAN ACV 110

Measurement, tonnes: 344 gt
Dimensions, metres (feet): 34.9 × 8 × ? *(114.5 × 26.2 × ?)*
Speed, knots: 10
Complement: 6 + 10 personnel
Machinery: 2 diesels; 2 shafts
Radars: Surface search/navigation: To be announced.

Comment: Modified commercial fleet support ship chartered and operated by Customs and Border Protection to protect offshore maritime areas off north-western Australia. Priority tasks are environmental protection and the prevention of illegal fishing and people smuggling. The vessel is stationed at the Ashmore Reef National Nature Reserve and Cartier Island Marine Reserves. The ship is equipped with two 7 m RHIBs.

ASHMORE GUARDIAN *12/2008, Australian Customs Service* / 1335635

1 OFFSHORE PATROL VESSEL (PSO)

Name	Builders	Laid down	Launched	Commissioned
OCEAN PROTECTOR	Aker Tulcea	25 Aug 2006	28 Jan 2007	3 Aug 2007
(ex-*Skandi Bergen*)				

Displacement, tonnes: 8,500 full load
Measurement, tonnes: 6,596 gt
Dimensions, metres (feet): 105.9 × 21.02 × 6.6 *(347.4 × 69.0 × 21.7)*
Speed, knots: 16
Complement: 22 + 50 personnel

Machinery: Diesel-electric; 4 Wärtsilä 6L32 diesels; 14,804 hp *(11.04 MW)*; 2 motors; 6,438 hp *(4.8 MW)*; 2 directional props
Guns: 2 – 12.7 mm MGs.
Radars: Surface search/navigation: E/F/I-bands.
Helicopters: Platform for 1 medium.

Comment: Originally built in 2007 as a multipurpose offshore support vessel, the ship was chartered by the Customs and Border Protection Service for a period of four to six years in mid-2010. The principal role of the vessel is to undertake patrols in the Southern Ocean for which it is to be available for 300 days per year. The ship, which is of steel construction is equipped with emergency medical facilities and can provide basic accommodation for up to 120 passengers in addition to the crew. The ship has a 1,100 m² deck area, is equipped with a 140 t crane and carries two 8.5 m response craft.

OCEAN PROTECTOR *8/2011, Chris Sattler* / 1406531

IHS Jane's Fighting Ships 2014-2015

MINE WARFARE FORCES

2 YEVGENYA CLASS (PROJECT 1258) (MINEHUNTERS) (MHC)

M 328 (ex-RT 136)   **M 327** (ex-RT 473)

Displacement, tonnes: 78 standard; 91 full load
Dimensions, metres (feet): 24.6 × 5.5 × 1.5 *(80.7 × 18.0 × 4.9)*
Speed, knots: 11
Range, n miles: 300 at 10 kt
Complement: 10

Machinery: 2 Type 3-D-12 diesels; 600 hp(m) *(440 kW)* sustained; 2 shafts
Guns: 2 — 25 mm (twin).
Physical countermeasures: Minehunting gear is lowered on a crane at the stern.
Radars: Navigation: Don 2; I-band.
Sonars: MG 7 lifted over the stern.

Comment: Ex-Russian craft built in the 1970s.

M 328   6/2008 / 1335323

2 SONYA (YAKHONT) (PROJECT 1265) CLASS
(COASTAL MINEHUNTER) (MHC)

M 325 (ex-BT 103)   **M 326** (ex-BT 16)

Displacement, tonnes: 457 full load
Dimensions, metres (feet): 48 × 8.8 × 2 *(157.5 × 28.9 × 6.6)*
Speed, knots: 15
Range, n miles: 3,000 at 10 kt
Complement: 43 (5 officers)

Machinery: 2 Kolomna Type 9-D-8 diesels; 2,000 hp(m) *(1.47 MW)* sustained; 2 shafts
Missiles: 2 quad SA-N-5 launchers.
Guns: 2 — 30 mm/65 AK 630 or 2 — 30 mm/65 (twin) and 2 — 25 mm/80 (twin).
Mines: 8.
Radars: Don 2 or Kivach or Nayada; I-band.
IFF: 2 Square Head. High Pole B.
Sonars: MG 69/79; hull-mounted; active minehunting; high frequency.

Comment: Wooden hull with GRP sheath. Transferred from Russia in 1992. One further vessel is reported non-operational.

M 326   6/2012, Hartmut Ehlers / 1455698

AUXILIARIES

Notes: A variety of auxiliary craft is reported to be in Azerbaijan service although operational status has not been confirmed. Vessels include an Emba class cable ship (T 750) and four survey ships (one Kamenka, one Finik, one Vadim Popov and one Valeryan Uryvayev). There is also a Neftegaz class salvage tug S 003, a B-92 class salvage tug (S 701), three Toplivo class coastal tankers, two Pozharny class firefighting craft, an SK-620 class A 343 and two Tamyr-class icebreakers *Kapitan Izmaylov* and *Kapitan A Radzhabov*.

1 RESEARCH SHIP (PROJECT 10470) (AGS)

A 671 (ex-Svyaga)

Dimensions, metres (feet): 126 × 16.6 × 4.2 *(413.4 × 54.5 × 13.8)*
Machinery: 2 diesels; 1,315 hp *(17.65 MW)*; 2 shafts

Comment: Former civilian Project 1677 Oleg Koshevoy class river/sea tanker converted by the Soviet Union in 1985 to undertake underwater research. Taken over by the Azerbaijan Navy in 1992. Possibly used as a platform for the operation of submersibles.

A 671   5/2008, M Globke / 1335204

1 FLAMINGO (TANYA) (PROJECT 1415) CLASS
(TENDERS) (YTD)

A 648

Displacement, tonnes: 43 standard
Dimensions, metres (feet): 22.2 × 3.9 × 1.4 *(72.8 × 12.8 × 4.6)*
Speed, knots: 12
Complement: 8
Machinery: 1 Type 3-D-12 diesel; 300 hp(m) *(220 kW)* sustained; 1 shaft

Comment: Successor to Nyryat II. Likely to be employed as a diving tender and workboat.

A 648   6/2012, Hartmut Ehlers / 1455699

TUGS

1 PROJECT V 820 CLASS (YTB)

T 758 (ex-RB 33)

Displacement, tonnes: 140 full load
Dimensions, metres (feet): 22 × 6.0 × 1.95 *(72.2 × 19.7 × 6.4)*
Speed, knots: 9
Complement: 6
Machinery: 2 E 680/193/1 diesels; 330 hp(m) *(246 kW)* sustained; 2 shafts

Comment: Former Soviet tug built by Ustka, Gdansk.

T 758   6/2012, Hartmut Ehlers / 1455700

1 PROJECT 73 CLASS (YTB)

T 759

Displacement, tonnes: 34 full load
Dimensions, metres (feet): 15.8 × 4.05 × 1.41 *(51.8 × 13.3 × 4.6)*
Speed, knots: 9
Complement: 6
Machinery: 1 3D6 diesel; 150 hp(m) *(112 kW)* sustained; 1 shaft

Comment: Former Soviet tug built by Rizhskiy Shipyard, Odessa.

T 759   6/2012, Hartmut Ehlers / 1455701

BORDER GUARD

3 STENKA (PROJECT 205P) CLASS (PBF)

S 005   S 006 (ex-AK 374)   S 007 (ex-AK 234)

Displacement, tonnes: 257 full load
Dimensions, metres (feet): 39.4 × 7.9 × 2.5 *(129.3 × 25.9 × 8.2)*
Speed, knots: 37
Range, n miles: 2,300 at 14 kt
Complement: 25

Machinery: 3 diesels; 14,100 hp(m) *(10.36 MW)*; 3 shafts
Guns: 4 – 30 mm/65 (2 twin) AK 230.
Radars: Surface search: Pot Drum; H/I-band.
Fire control: Drum Tilt; H/I-band.
Navigation: Palm Frond; I-band.

Comment: Ex-Russian craft built in the 1970s. Sonar and torpedo tubes removed. Operated by the Border Guard.

STENKA S 006   *6/2008* / 1335327

2 SILVER SHIPS 48 ft CLASS (PB)

S 11   S 12

Displacement, tonnes: 13 standard
Dimensions, metres (feet): 14.6 × 3.7 × 1.1 *(47.9 × 12.1 × 3.6)*
Speed, knots: 40
Range, n miles: 385 at 36 kt
Complement: 6
Machinery: 2 Caterpillar 3196D diesels; 1,140 hp *(850 kW)*; 2 surface piercing props
Radars: Surface search: I-band.

Comment: Constructed by Silver Ships of Theodore, Alabama. Acquired in 2001, although the details of the purchase are unclear.

S 11 and S 12   *6/2008* / 1335321

2 BALTIC 150 PATROL CRAFT (WPB)

Displacement, tonnes: 16 full load
Dimensions, metres (feet): 16.15 × 4.4 × 0.75 *(53 × 14.4 × 2.5)*
Speed, knots: 35
Range, n miles: 350 at 15 kt
Complement: 5
Machinery: 2 MAN R6-730 diesels; 1,440 hp *(1.07 MW)*; 2 shafts
Radars: Navigation: I-band.

Comment: Built by Baltic Workboats A/S, Estonia and delivered in 2010. Aluminium construction. Based at Baku.

1 OSA II (PROJECT 205) CLASS (PB)

S 008

Displacement, tonnes: 249 full load
Dimensions, metres (feet): 38.6 × 7.6 × 2.7 *(126.6 × 24.9 × 8.9)*
Speed, knots: 37. **Range, n miles:** 500 at 35 kt
Complement: 30

Machinery: 3 Type M 504 diesels; 10,800 hp(m) *(7.94 MW)* sustained; 3 shafts
Guns: 4 USSR 30 mm/65 AK 230 (2 twin); 500 rds/min to 5 km *(2.7 n miles)*; weight of shell 0.54 kg.
Radars: Surface search: I-band.
Fire control: Drum Tilt; H/I-band.

Comment: Probably transferred from the Russian Caspian Flotilla in 1992. SS-N-2B missiles have been removed.

OSA S 008   *6/2008* / 1335328

1 POINT CLASS (PB)

Name	No	Builders	Commissioned
– (ex-*Point Brower*)	S 14 (ex-201, ex-82372)	US Coast Guard Yard, Curtis Bay	21 Apr 1970

Displacement, tonnes: 68 full load
Dimensions, metres (feet): 25.3 × 5.3 × 1.8 *(83 × 17.4 × 5.9)*
Speed, knots: 22. **Range, n miles:** 1,200 at 8 kt
Complement: 10
Machinery: 2 Caterpillar diesels; 1,600 hp *(1.19 MW)*; 2 shafts
Guns: 2 – 12.7 mm MGs.
Radars: Surface search: Hughes/Furuno SPS-73; I-band.

Comment: Transferred from US Coast Guard on 28 February 2003.

POINT CLASS   *6/2008* / 1335329

1 VIKHR (IVA) (PROJECT B-99) CLASS
(FIREFIGHTING TUG) (ARS)

S 703

Displacement, tonnes: 2,337 full load
Dimensions, metres (feet): 72.3 × 14.3 × 4.6 *(237.2 × 46.9 × 15.1)*
Speed, knots: 16. **Range, n miles:** 2,500 at 12 kt
Complement: 25
Machinery: 2 diesels; 5,900 hp(m) *(4.4 MW)*; 2 shafts; cp props; 2 bow thrusters

Comment: Built in Gdansk, Poland, in mid-1980s.

S 703   *6/2008* / 1335322

4 DEFENDER CLASS (PATROL CRAFT) (PBF)

S 18   +3

Displacement, tonnes: 3.85 full load
Dimensions, metres (feet): 12.9 × 4.05 × 0.76 *(42.3 × 13.3 × 2.5)*
Speed, knots: 40. **Range, n miles:** 250 at 30 kt
Complement: 4
Machinery: 2 Yanmar diesels; 1,000 hp *(746 kW)*; 2 Hamilton waterjets

Comment: Four craft manufactured by SAFE boats and donated by the US government in 2009.

Bahamas

Country Overview

The Commonwealth of the The Bahamas gained independence in 1973; the British monarch, represented by a governor-general, is head of state. Situated in the west Atlantic Ocean, it comprises about 700 islands and islets, and nearly 2,400 cays and rocks which stretch between Florida and Hispaniola. About 30 of the islands are inhabited. The capital, Nassau, is on New Providence Island which contains more than half of the total population. Grand Bahama, the most northerly of the group, is the second major island. An archipelagic regime, territorial seas (12 n miles) and a fishery zone (200 n miles) are claimed. A 200 n mile Exclusive Economic Zone (EEZ) has been claimed but the limits are not defined.

Headquarters Appointments

Commander Royal Bahamas Defence Force:
 Commodore Roderick Bowe
Captain, HMBS Coral Harbour:
 Captain Clyde Sawyer
Commander, Operations:
 Commander Nedley Martinborough

Maritime Patrol

Air assets include a Vulcan Air P 68 Observer, a King Air and Cessna Caravan 208B.

Bases

HMBS *Coral Harbour* (New Providence Island)
HMBS *Matthew Town* (Great Inagua Island)
Northern Command (Grand Bahama)
A six-year programme to upgrade/expand bases was signed with Damen Shipyards in April 2013.

Personnel

2014: 1,200

Prefix to Ships' Names

HMBS (Her Majesty's Bahamian Ship)

PATROL FORCES

Notes: There are three interception craft P 121–123 of unknown type. Six Nor-Tech 14 m interception craft (EF 11, 12, 17–20) were acquired in 2009.

2 BAHAMAS CLASS

Name	No	Builders	Commissioned
BAHAMAS	P 60	Moss Point Marine, Escatawpa	27 Jan 2000
NASSAU	P 61	Moss Point Marine, Escatawpa	27 Jan 2000

Displacement, tonnes: 381 full load
Dimensions, metres (feet): 60.6 × 8.9 × 2.6 *(198.8 × 29.2 × 8.5)*
Speed, knots: 24. **Range, n miles:** 3,000 at 10 kt
Complement: 35 + 28 spare berths
Machinery: 3 Caterpillar 3516B diesels; 6,600 hp(m) *(4.85 MW)*; 3 shafts
Guns: 1 Bushmaster 25 mm. 3—12.7 mm MGs.
Radars: Surface search/Navigation: Decca Bridgemaster Type 656-14/CAB; I-band.

Comment: Order placed 14 March 1997 with Halter Marine Group. Aluminium superstructures fabricated at Equitable Shipyards while hulls built at Moss Point. The design is an adapted Vosper International Europatrol 250 with a RIB and launching crane at the stern. Based at Nassau. As part of the contract with Damen Shipyards in 2013 for nine new vessels, both ships are to receive a mid-life upgrade.

BAHAMAS   *10/2008, Marco Ghiglino* / 1305791

1 PROTECTOR CLASS (PB)

Name	No	Builders	Commissioned
YELLOW ELDER	P 03	Fairey Marine, Cowes	20 Nov 1986

Displacement, tonnes: 112 standard; 183 full load
Dimensions, metres (feet): 33 × 6.7 × 2.1 *(108.3 × 22.0 × 6.9)*
Speed, knots: 30
Range, n miles: 300 at 24 kt, 600 at 14 kt
Complement: 20 (3 officers) + 5 spare berths
Machinery: 3 Detroit 16V-149TI diesels; 3,483 hp *(2.6 MW)* sustained; 3 shafts
Guns: 1 Rheinmetall 20 mm. 3—7.62 mm MGs.
Radars: Surface search: Furuno; I-band.

Comment: Ordered December 1984. Steel hull. One RIB is carried and can be launched by a trainable crane. Based at Coral Harbour. *Port Nelson* and *Samana* decommissioned in 2007.

PROTECTOR CLASS   *4/1996, RBDF* / 0056527

3 CHALLENGER CLASS (PB)

P 38–40

Displacement, tonnes: 8 full load
Dimensions, metres (feet): 8.2 × 1.7 × 0.3 *(26.9 × 5.6 × 1.0)*
Speed, knots: 45
Complement: 4
Machinery: 2 Yamaha outboards; 450 hp *(330 kW)*
Guns: 1—7.62 mm MG.

Comment: Built by Boston Whaler Edgewater, Florida and delivered in September 1995. GRP hull.

P 40   *10/2008, Marco Ghiglino* / 1305792

1 ELEUTHERA (KEITH NELSON) CLASS (PB)

Name	No	Builders	Commissioned
INAGUA	P 27	Vosper Thornycroft	10 Dec 1979

Displacement, tonnes: 30 standard; 38 full load
Dimensions, metres (feet): 18.3 × 4.8 × 1.4 *(60 × 15.7 × 4.6)*
Speed, knots: 20
Range, n miles: 650 at 16 kt
Complement: 11
Machinery: 2 Caterpillar 3408BTA diesels; 1,070 hp *(800 kW)* sustained; 2 shafts
Guns: 3—7.62 mm MGs.
Radars: Surface search: Furuno; I-band.

Comment: The survivor of a class of five. Light machine guns mounted in sockets either side of the bridge. One more is used as a museum. Main engine replaced in 1990.

INAGUA   *6/1998, RBDF* / 0017574

Patrol forces — Auxiliaries < **Bahamas** 45

2 DAUNTLESS CLASS
(INSHORE PATROL CRAFT) (PB)

P 42   P 43

Displacement, tonnes: 11 full load
Dimensions, metres (feet): 12.3 × 4.3 × 1.3 (40.4 × 14.1 × 4.3)
Speed, knots: 25
Range, n miles: 600 at 18 kt
Complement: 5
Machinery: 2 Caterpillar 3208TA diesels; 870 hp (650 kW) sustained; 2 shafts
Guns: 2 — 7.62 mm MGs.
Radars: Furuno 1761; I-band.

Comment: Built by SeaArk Marine, Monticello, Arkansas and delivered in January 1996. Aluminium construction. Used primarily for medium-range search and rescue missions. Based at Coral Harbour.

P 43   6/1999, RBDF / 0081453

0 + 4 DAMEN SPA 3007 (PATROL CRAFT) (PB)

Dimensions, metres (feet): 30.8 × 7.0 × ? (101 × 23.0 × ?)
Speed, knots: 20
Range, n miles: 2,000 at 10 kt
Complement: 13
Machinery: 2 diesels; 2 shafts
Radars: Navigation: To be announced.

Comment: It was announced on 19 April 2013 that four Damen SPa 3007 patrol craft were to be acquired as part of an order for nine vessels. SPa 3007 is a Sea Axe design and is optimised for coastal and inshore operations. There is a stern ramp for the launch and recovery of an interception craft.

4 BOSTON WHALERS (PBF)

P 110–113

Displacement, tonnes: 2 full load
Dimensions, metres (feet): 6.1 × 2.2 × 0.4 (20 × 7.2 × 1.3)
Speed, knots: 45 (P 110–111), 38 (P 112–113)
Complement: 3
Machinery: 2 Evinrude outboards; 180 hp (134 kW) (P 110–111); 2 Mariner outboards; 150 hp (120 kW) (P 112–113)

Comment: P 110 and 111 are Impact designs commissioned 25 September 1995. P 112 and 113 are Wahoo types commissioned 23 October 1995.

P 113   6/1999, RBDF / 0081454

P 110 and P 111   9/1997, RBDF / 0012053

2 SEA ARK 49 ft CUTTERS (PB)

P 48   P 49

Displacement, tonnes: 17 full load
Dimensions, metres (feet): 14.9 × 5.25 × 1.4 (48.9 × 17.2 × 4.6)
Speed, knots: 28
Range, n miles: 300 at 12 kt
Complement: 6
Machinery: 2 Caterpillar C-12 diesels; 1,320 hp (984 kW); 2 shafts
Radars: Navigation: Furuno; I-band.

Comment: Sea Ark Dauntless RAM design. Aluminium construction. Donated by the US on 26 May 2006. Delivered on 18 July 2008.

P 48   6/2008, SeaArk Marine / 1298814

0 + 4 DAMEN STAN PATROL 4207 (PB)

Displacement, tonnes: 208 standard
Dimensions, metres (feet): 42.8 × 7.11 × 2.52 (140.4 × 23.3 × 8.3)
Speed, knots: 26
Complement: 18 (4 officers)
Machinery: 2 Caterpillar 3516B DI-TA; 5,600 hp (4.17 MW); 2 cp props
Guns: 2 — 12.7 mm MGs.

Comment: A Letter of Intent signed with Damen Shipyards on 19 April 2013 for the acquisition of four Damen Stan Patrol 4207. Details are based on those in Jamaican service.

2 SEA ARK 40 ft CUTTERS (PB)

P 44   P 45

Displacement, tonnes: 12 full load
Dimensions, metres (feet): 12.2 × 4.1 × 1.32 (40 × 13.5 × 4.3)
Speed, knots: 33
Complement: 4
Machinery: 2 diesels; 2 shafts
Radars: Navigation: Raymarine; I-band.

Comment: SeaArk Marine Dauntless RAM patrol craft acquired in 2008.

AUXILIARIES

0 + 1 DAMEN STAN LANDER 5612
(TRANSPORT SHIPS) (AKL)

Measurement, tonnes: 722 gt
Dimensions, metres (feet): 58.8 × 12.0 × 2.7 (192.9 × 39.4 × 8.9)
Speed, knots: 11
Complement: 11
Machinery: 2 Caterpillar C32 diesels; 2,230 hp (1.64 MW); 2 shafts

Comment: It was announced on 19 April 2013 that one Damen Stan Lander 5612 was to be acquired from Damen Shipyards. The ship is to be modified to embark disaster relief equipment containers when required.

Bahrain

Country Overview
Formerly under British control from 1861, Bahrain gained its independence in 1971. Situated in the southern Gulf, with which it has a coastline of 87 n miles, the country comprises a group of 33 islands between the Qatar Peninsula to the east and Saudi Arabia to the west. The principal islands include Bahrain (217 square miles), Al Muharraq; Umm an Na'san; Sitrah; Jiddah and the Hawar group. The capital, largest city and principal port is Manama. Territorial seas (12 n miles) are claimed. An EEZ has not been claimed.

Headquarters Appointments
Chief of Staff: Major General Sheikh Duaij Bin Salman Al Khalifa
Commander of Navy: Brigadier Abdulla al Mansoori
Director of Coast Guard: Brigadier Ala Abdulla Seyadi

Personnel
(a) 2014: 1,000 (Navy), 770 (Coast Guard 260 seagoing)
(b) Voluntary service

Bases
Mina Sulman (Navy)
Bandar-Dar (CG base)
Muharraq (CG HQ)

Coast Guard
This unit is under the direction of the Ministry of the Interior.

Prefix to Ships' Names
BRNS (Bahrain Royal Navy Ship)

FRIGATES

1 OLIVER HAZARD PERRY CLASS (FFGHM)

Name	No	Builders	Laid down	Launched	Commissioned	Recommissioned
SABHA (ex-*Jack Williams*)	90 (ex-FFG 24)	Bath Iron Works, Maine	25 Feb 1980	30 Aug 1980	19 Sep 1981	25 Feb 1997

Displacement, tonnes: 2,794 standard; 3,696 full load
Dimensions, metres (feet): 135.6 × 13.7 × 4.5 hull; 7.5 sonar *(444.9 × 44.9 × 14.8; 24.6)*
Speed, knots: 29
Range, n miles: 4,500 at 20 kt
Complement: 206 (13 officers; 19 air crew)

Machinery: 2 GE LM 2500 gas turbines; 41,000 hp *(30.59 MW)* sustained; 1 shaft; cp prop 2 auxiliary retractable props; 650 hp *(484 kW)*
Missiles: SSM: 4 McDonnell Douglas Harpoon; active radar homing to 90 km *(52 n miles)* at 0.9 Mach; warhead 227 kg.
SAM: 36 GDC Standard SM-1MR Block VI; command guidance; semi-active radar homing to 38 km *(20.5 n miles)* at 2 Mach. 1 Mk 13 Mod 4 launcher for both SSM and SAM missiles ❶.
Guns: 1 Oto Melara 3 in *(76 mm)*/62 Mk 75 ❷; 85 rds/min to 16 km *(8.7 n miles)* anti-surface; 12 km *(6.6 n miles)* anti-aircraft; weight of shell 6 kg. 1 Raytheon 20 mm/76 6-barrelled Mk 15 Vulcan Phalanx Block 1B ❸; 4,500 rds/min (4,500 in Block 1) combined to 1.5 km. 4—12.7 mm MGs.
Torpedoes: 6—324 mm Mk 32 Mod 7 (2 triple) tubes ❹. 24 Honeywell Mk 46; anti-submarine; active/passive homing to 11 km *(5.9 n miles)* at 40 kt; warhead 44 kg.
Physical countermeasures: Decoys: 2 Loral Hycor SRBOC 6-barrelled fixed Mk 36 ❺; IR flares and chaff to 4 km *(2.2 n miles)*. SLQ-25 Nixie; torpedo decoy.
Electronic countermeasures: ESM/ECM: SLQ-32(V)2 ❻; radar warning. Sidekick modification adds jammer and deception system.
Radars: Air search: Raytheon SPS-49(V)4 ❼; C-band; range 457 km *(250 n miles)*.
Surface search: ISC Cardion SPS-55 ❽; I-band.
Fire control: Lockheed STIR (modified SPG-60) ❾; I/J-band; range 110 km *(60 n miles)*. Sperry Mk 92 (Signaal WM28) ❿; I/J-band.
Tacan: URN 25.
Sonars: Raytheon SQS-56; hull-mounted; active search and attack; medium frequency.
Combat data systems: NTDS with Link 14. INMARSAT.
Weapon control systems: SWG-1 Harpoon LCS. Mk 92 (Mod 4). The Mk 92 is the US version of the Signaal WM28 system. Mk 13 weapon direction system. 2 Mk 24 optical directors.

Helicopters: 1 Eurocopter BO 105 ⓫.

Programmes: *Sabha* transferred from the US by grant 18 September 1996. Arrived in the Gulf in June 1997 for a work-up and training period.
Structure: Apart from the removal of the US SATCOM aerials there are no visible changes from US service.

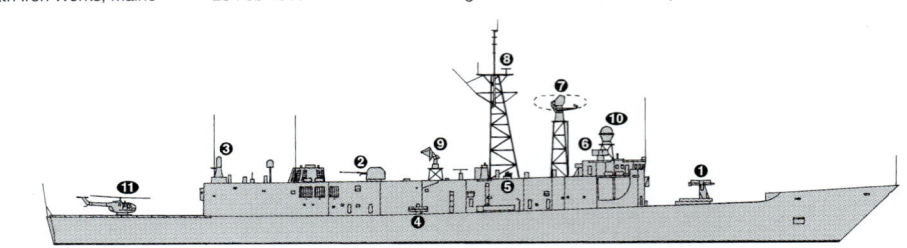

SABHA *(Scale 1 : 1,200), Ian Sturton / 0056532*

SABHA *4/2000, Guy Toremans / 0104200*

SABHA *6/2003, A Sharma / 0568881*

IHS Jane's Fighting Ships 2014-2015

CORVETTES

2 AL MANAMA (MGB 62) CLASS (FSGH)

Name	No	Builders	Commissioned
AL MANAMA	50	Lürssen	14 Dec 1987
AL MUHARRAQ	51	Lürssen	3 Feb 1988

Displacement, tonnes: 642 full load
Dimensions, metres (feet): 63 × 9.3 × 2.9 *(206.7 × 30.5 × 9.5)*
Speed, knots: 32. **Range, n miles:** 4,000 at 16 kt
Complement: 43 (7 officers)

Machinery: 4 MTU 20V 538 TB92 diesels; 12,820 hp(m) *(9.42 MW)* sustained; 4 shafts
Missiles: SSM: 4 Aerospatiale MM 40 Exocet launchers (2 twin) ❶; inertial cruise; active radar homing to 70 km *(40 n miles)* at 0.9 Mach; warhead 165 kg; sea-skimmer.
Guns: 1 Oto Melara 3 in *(76 mm)*/62 compact ❷; 85 rds/min to 16 km *(8.7 n miles)* anti-surface; 12 km *(6.5 n miles)* anti-aircraft; weight of shell 6 kg. 2 Breda 40 mm/70 (twin) ❸; 300 rds/min to 12.5 km *(6.8 n miles)*; weight of shell 0.96 kg. 2—7.62 mm MGs.
Physical countermeasures: Decoys: CSEE Dagaie ❹; chaff and IR flares.
Electronic countermeasures: ESM/ECM: Racal Decca Cutlass/Cygnus ❺; intercept and jammer.
Radars: Air/surface search: Philips Sea Giraffe 50 HC ❼; G-band.

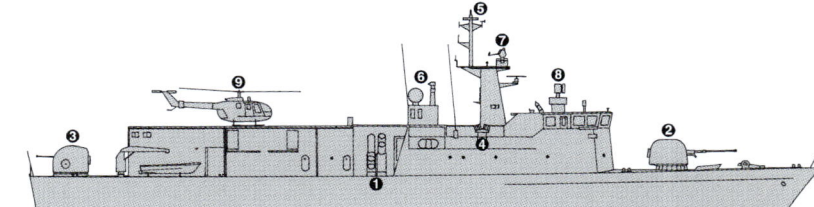

AL MANAMA *(Scale 1 : 600), Ian Sturton / 0104201*

Navigation: Racal Decca 1226; I-band.
Fire control: Philips 9LV 331 ❽; J-band.
Weapon control systems: CSEE Panda Mk 2 optical director. Philips TV/IR optronic director ❻.
Helicopters: 1 Eurocopter BO 105 ❾.

Programmes: Ordered February 1984.
Structure: Similar to Singapore and UAE designs. Steel hull, aluminium superstructure. Fitted with a helicopter platform which incorporates a lift to lower the aircraft into the hangar.
Operational: Planned SA 365F helicopters were not acquired.

AL MUHARRAQ *9/2008, Shaun Jones / 1335640*

AL MANAMA *11/2001, Royal Australian Navy / 0526836*

SHIPBORNE AIRCRAFT

Numbers/Type: 2 Eurocopter BO 105.
Operational speed: 113 kt *(210 km/h)*.
Service ceiling: 9,845 ft *(3,000 m)*.
Range: 407 n miles *(754 km)*.
Role/Weapon systems: Acquired in August 1994 as the first aircraft of a Naval Air Arm. Sensors: Bendix RDR 1500B radar. Weapons: Unarmed.

BO 105
6/1995
0056541

48 Bahrain > Patrol forces — Amphibious forces

PATROL FORCES

4 AHMAD EL FATEH (TNC 45) CLASS
(FAST ATTACK CRAFT—MISSILE) (PGGF)

Name	No	Builders	Commissioned
AHMAD EL FATEH	20	Lürssen	5 Feb 1984
AL JABIRI	21	Lürssen	3 May 1984
ABDUL RAHMAN AL FADEL	22	Lürssen	10 Sep 1986
AL TAWEELAH	23	Lürssen	25 Mar 1989

Displacement, tonnes: 263 full load
Dimensions, metres (feet): 44.9 × 7 × 2.5 (147.3 × 23.0 × 8.2)
Speed, knots: 40
Range, n miles: 1,600 at 16 kt
Complement: 36 (6 officers)

Machinery: 4 MTU 16V 538 TB92 diesels; 13,640 hp(m) (10 MW) sustained; 4 shafts
Missiles: SSM: 4 Aerospatiale MM 40 Exocet (2 twin); inertial cruise; active radar homing to 70 km (40 n miles) at 0.9 Mach; warhead 165 kg; sea-skimmer.
Guns: 1 Oto Melara 3 in (76 mm)/62; compact; 85 rds/min to 16 km (8.7 n miles) anti-surface; 12 km (6.5 n miles) anti-aircraft; weight of shell 6 kg. 2 Breda 40 mm/70 (twin); 300 rds/min to 12.5 km (6.8 n miles); weight of shell 0.96 kg. 3—7.62 mm MGs.
Physical countermeasures: Decoys: CSEE Dagaie launcher; trainable mounting; 10 containers firing chaff decoys and IR flares.
Electronic countermeasures: ESM: Thales Sealion.
ECM: Racal Cygnus (not in 20 and 21); jammer.
Radars: Air/surface search: Philips Sea Giraffe 50 HC; G-band.
Fire control: Philips 9LV 226/231; J-band.
Navigation: Racal Decca 1226; I-band.
Weapon control systems: 1 Panda optical director for 40 mm guns.

Programmes: First pair ordered in 1979, second pair in 1985. Similar craft in service with Ecuador, Kuwait and UAE navies.
Structure: Only the second pair have the communication radome on the after superstructure.
Operational: Refits from 2000 by Lürssen at Abu Dhabi.

AHMAD EL FATEH   4/2003, A Sharma / 0568844

ABDUL RAHMAN AL FADEL   1/2009, Guy Toremans / 1367008

2 AL JARIM (FPB 20) CLASS (FAST ATTACK CRAFT—GUN) (PB)

Name	No	Builders	Commissioned
AL JARIM	30	Swiftships, Morgan City	9 Feb 1982
AL JASRAH	31	Swiftships, Morgan City	26 Feb 1982

Displacement, tonnes: 34 full load
Dimensions, metres (feet): 19.2 × 5.6 × 2 (63 × 18.4 × 6.6)
Speed, knots: 30
Range, n miles: 1,200 at 18 kt
Machinery: 2 Detroit 12V-71TA diesels; 840 hp(m) (627 kW) sustained; 2 shafts
Guns: 1 Oerlikon GAM-BO1 20 mm.
Radars: Surface search: Decca 110; I-band.

Comment: Aluminium hulls.

AL JARIM   5/2003, A Sharma / 0568879

2 AL RIFFA (FPB 38) CLASS (FAST ATTACK CRAFT—GUN) (PB)

Name	No	Builders	Commissioned
AL RIFFA	10	Lürssen	3 Mar 1982
HAWAR	11	Lürssen	3 Mar 1982

Displacement, tonnes: 208 full load
Dimensions, metres (feet): 38.5 × 7 × 2.2 (126.3 × 23.0 × 7.2)
Speed, knots: 32
Range, n miles: 1,100 at 16 kt
Complement: 27 (3 officers)

Machinery: 2 MTU 16V 538 TB92 diesels; 6,810 hp(m) (5 MW) sustained; 2 shafts
Guns: 2 Breda 40 mm/70 (twin); dual purpose; 300 rds/min to 12 km (6.5 n miles) anti-surface; 4 km (2.2 n miles); weight of shell 0.96 kg.
Rockets: 1—57 mm Starshell rocket launcher.
Mines: Mine rails fitted.
Physical countermeasures: Decoys: 1 Wallop Barricade chaff launcher.
Electronic countermeasures: ESM: Racal RDL-2 ABC; radar warning.
Radars: Surface search: Philips 9GR 600; I-band.
Navigation: Racal Decca 1226; I-band.
Weapon control systems: CSEE Lynx optical director with Philips 9LV 126 optronic system.

Comment: Ordered in 1979. Al Riffa launched April 1981. Hawar launched July 1981.

HAWAR   6/2003, A Sharma / 0568880

AMPHIBIOUS FORCES

1 AJEERA CLASS (SUPPLY SHIP) (YFU)

Name	No	Builders	Commissioned
AJEERA	41	Swiftships, Morgan City	21 Oct 1982

Displacement, tonnes: 427 full load
Dimensions, metres (feet): 39.6 × 11 × 1.8 (129.9 × 36.1 × 5.9)
Speed, knots: 13
Range, n miles: 1,500 at 10 kt
Complement: 21
Machinery: 2 General Motors 16V-71 diesels; 1,800 hp (1.34 MW) sustained; 2 shafts
Guns: 2—12.7 mm MGs.
Radars: Navigation: Racal Decca; I-band.

Comment: Used as general purpose cargo ships and can carry up to 200 tons of fuel and water. Built to an LCU design with a bow ramp and 15 ton crane.

AJEERA   4/2003, A Sharma / 0568843

4 LCU 1466 CLASS (LCU)

MASHTAN 42   RUBODH 43   SUWAD 44   JARADAH 45

Displacement, tonnes: 366 full load
Dimensions, metres (feet): 36.3 × 10.4 × 1.8 (119.1 × 34.1 × 5.9)
Speed, knots: 8
Range, n miles: 800 at 8 kt
Complement: 15

Cargo capacity: 150 tons
Machinery: 3 Gray Marine 64 YTL diesels; 675 hp (504 kW); 3 shafts
Guns: 2—12.7 mm MGs.
Radars: Navigation: Racal Decca; I-band.

Comment: Transferred from US in 1991.

RUBODH   4/2003, A Sharma / 0568842

IHS Jane's Fighting Ships 2014-2015   © 2014 IHS

Amphibious forces — Coast guard < **Bahrain** 49

2 LANDING CRAFT (LCU)

DINAR 47   **AL HAMRA** 48

Displacement, tonnes: 386 standard
Dimensions, metres (feet): 42 × 10.0 × 3.5 *(137.8 × 32.8 × 11.5)*
Speed, knots: 10
Range, n miles: 1,000 at 8.5 kt
Complement: 51 (3 officers)
Military lift: military vehicles
Machinery: 2 Caterpillar CAT 3406 TA diesels; 730 hp *(544 kW)*; 2 shafts

Comment: Contract with Abu Dhabi Shipbuilding for the construction of two landing craft announced on 11 November 2008. The vessels, designed in the UAE, are to be of steel construction and based on those in service in the UAE Navy. Delivery of the first vessel was made on 14 July 2010 and the second on 28 December 2010.

LCU (UAE colours)   *6/2006, ADSB* / 1159231

2 HALMATIC WORK BOATS (PB)

Displacement, tonnes: 14 standard
Dimensions, metres (feet): 16 × 3.9 × 0.7 *(52.5 × 12.8 × 2.3)*
Speed, knots: 24
Complement: 5
Machinery: 2 Caterpillar diesels; 2 Rolls Royce waterjets
Radars: Navigation: I-band.

Comment: Contract with Abu Dhabi Shipbuilding for the construction of two work boats announced on 11 November 2008. Based on the VT Halmatic Sea Keeper design with an asymmetric catamaran hull, the craft are highly manoeuvrable and are capable of carrying an 8.5 tonne payload. The vessels were delivered on 21 October 2009.

WORK BOAT   *2/2007, IHS/Patrick Allen* / 1321982

1 LANDING CRAFT (LCU)

AL ZUBARA (ex-*Sabha*) 40

Displacement, tonnes: 152 full load
Dimensions, metres (feet): 22.5 × 7.5 × 1.2 *(73.8 × 24.6 × 3.9)*
Speed, knots: 6
Complement: 8
Machinery: 2 General Motors 8V92N diesels; 780 hp *(575 kW)*; 2 shafts
Radars: Navigation: I-band.

Comment: Fairey Marine Cowes, UK Loadmaster II class which entered service in 1981.

AL ZUBARA   *4/2003, A Sharma* / 0568839

© 2014 IHS

AUXILIARIES

Notes: There are also two RTK Medevac boats and one Diving Boat (512).

1 PERSONNEL TRANSPORT CRAFT (YFL)

TIGHATLIB 46

Comment: Catamaran hulled transport craft. Details not known.

TIGHATLIB   *6/2003, A Sharma* / 0568877

COAST GUARD

Notes: (1) There are 10 coastal patrol craft, Haris 5–8, (8 m, 25 kt) and Fajr 1–6 (8.3 m, 35 kt).
(2) There are 16 interceptor craft. Jarada 1–2 (11 m, 35 kt); Jarada 4–5 (11 m, 36 kt), Haris 11–16 (9.7 m, 42 kt) and Saham 1–6 (11.6 m, 55 kt). There are also two 7.4 m patrol craft Jida 1–2 and two 8 m inshore patrol craft Jida 4–5.
(3) The procurement of six 16 m patrol craft from ARES Shipyard, Turkey, is in progress. All are to be delivered in early 2014.
(4) There are four 36 × 12 m barges used as inspection and security checkpoints on the seaborne approaches to Bahrain. Built by Arab Ship Repair Yard, Bahrain, they became operational in 2011. Other roles include forward command post and base and stores pre-positioning.
(5) There are nine jet skis.
(6) Five 7.6 m Boston Whalers (Fajr 11–15) were donated by the United States in 2013.

JIDA 4   *6/2011, Bahrain Coast Guard* / 1406698

CHECKPOINT BARGE   *10/2011, Bahrain Coast Guard* / 1406697

1 WASP 30 METRE CLASS (WPB)

AL MUHARRAQ

Displacement, tonnes: 91 standard; 105 full load
Dimensions, metres (feet): 30 × 6.4 × 1.6 *(98.4 × 21.0 × 5.2)*
Speed, knots: 25
Range, n miles: 500 at 22 kt
Complement: 9
Machinery: 2 Detroit 16V-149TI diesels; 2,322 hp *(1.73 MW)* sustained; 2 shafts
Guns: 2 – 7.62 mm MGs. 1 Hughes chain 7.62 mm.
Radars: Surface search: Racal Decca; I-band.

Comment: Ordered from Souters, Cowes, Isle of Wight in 1984. Laid down November 1984, launched 12 August 1985, shipped 21 October 1985. GRP hull. A mid-life refit is planned.

AL MUHARRAQ   *4/2003, A Sharma* / 0568841

IHS Jane's Fighting Ships 2014-2015

4 HALMATIC 20 METRE CLASS (WPB)

DERA'A 2 **DERA'A 6–8**

Displacement, tonnes: 32 full load
Dimensions, metres (feet): 20.1 × 5.3 × 1.5 *(65.9 × 17.4 × 4.9)*
Speed, knots: 29
Range, n miles: 500 at 20 kt
Complement: 6
Machinery: 2 MTU 8V 2000 M92 diesels; 2,170 hp *(1.6 MW)* sustained; 2 shafts
Guns: 1 – 12.7 mm MG. 2 – 7.62 mm MGs.

Comment: Three delivered in late 1991, the last in early 1992. GRP hulls. All four craft underwent a mid-life refit at Abu Dhabi shipbuilding 2008–09.

DERA'A 8   11/2008, John Fidler / 1335639

2 SOUTER 20 METRE CLASS (WPB)

DERA'A 4 **DERA'A 5**

Displacement, tonnes: 37 full load
Dimensions, metres (feet): 20 × 5 × 1.5 *(65.6 × 16.4 × 4.9)*
Speed, knots: 24.5
Range, n miles: 500 at 20 kt
Complement: 8
Machinery: 2 Detroit 12V-71TA diesels; 840 hp *(626 kW)* sustained; 2 shafts
Guns: 2 – 7.62 mm MGs.
Radars: Surface search: Racal Decca; I-band.

Comment: Built by Souters, Cowes, Isle of Wight. Delivered 1983. GRP hulls.

DERA'A 4   10/2009, Bahrain Coast Guard / 1367010

6 HALMATIC 160 CLASS (WPB)

SAIF 5–10

Displacement, tonnes: 17 full load
Dimensions, metres (feet): 14.4 × 3.9 × 1.2 *(47.2 × 12.8 × 3.9)*
Speed, knots: 27
Range, n miles: 500 at 20 kt
Complement: 4
Machinery: 2 MTU S6062 06N04M diesels; 950 hp *(708 kW)* sustained; 2 shafts
Guns: 1 – 7.62 mm MG.
Radars: Surface search: Furuno; I-band.

Comment: Built by Halmatic, UK, and delivered in 1990–91. GRP hulls. All six craft underwent a mid-life refit at Abu Dhabi Shipbuilding 2008–09.

SAIF 10   10/2008, John Fidler / 1335637

2 AL DHAEN 12 M CLASS (PB)

HAWAR 1 **HAWAR 2**

Displacement, tonnes: 11 full load
Dimensions, metres (feet): 12.4 × 4.0 × 0.7 *(40.7 × 13.1 × 2.3)*
Speed, knots: 30
Machinery: 2 Cummins 6CTA8.3 diesels
Guns: 1 – 7.62 mm MG.

Comment: Entered service in 2003.

HAWAR 1   6/2003, John Fidler / 0567903

1 SUPPORT CRAFT (YAG)

SAFRA 3

Displacement, tonnes: 168 full load
Dimensions, metres (feet): 25.9 × 7.9 × 1.6 *(85 × 25.9 × 5.2)*
Speed, knots: 13
Range, n miles: 700 at 12 kt
Complement: 6
Machinery: 2 Detroit 16V-92TA diesels; 1,380 hp *(1.03 MW)*; 2 shafts
Radars: Navigation: Racal Decca; I-band.

Comment: Built by Halmatic, Havant and delivered in early 1992. Logistic support work boat equipped for towing and firefighting. Can carry 15 tons.

SAFRA 3   4/2003, A Sharma / 0568840

1 LANDING CRAFT (LCM)

SAFRA 2

Displacement, tonnes: 152 full load
Dimensions, metres (feet): 22.5 × 7.5 × 1.2 *(73.8 × 24.6 × 3.9)*
Speed, knots: 6
Complement: 8
Machinery: 2 General Motors 8V92N diesels; 780 hp *(575 kW)*; 2 shafts
Radars: Navigation: Furuno; I-band.

Comment: Fairey Marine Loadmaster II class which was delivered in 1981. Based at Bandar-Dar. Similar to craft in naval service.

SAFRA 2   10/2008, John Fidler / 1305328

Coast guard < **Bahrain** — Submarines < **Bangladesh**   51

4 RODMAN 20 M CLASS (PB)

DERA'A 11–14

Displacement, tonnes: 34 full load
Dimensions, metres (feet): 20.5 × 4.93 × 2.25 (67.3 × 16.2 × 7.4)
Speed, knots: 30
Complement: 11
Machinery: 2 MTU 8V 2000 M92 diesels; 2,170 hp (1.6 MW); 2 shafts
Guns: 1 — 12.7 mm MG. 2 — 7.62 mm MGs.
Radars: Navigation: I-band.

Comment: Built by Rodman, Spain, and delivered 2008–09.

SAIF 3   11/1999, Bahrain Coast Guard / 0056543

4 MAHAR 31 CLASS (PB)

HARIS 1–4

Dimensions, metres (feet): 9.4 × 2.4 × 0.3 (30.8 × 7.9 × 1.0)
Speed, knots: 40
Complement: 2
Machinery: 2 Yamaha outboard motors.
Guns: 1 — 7.62 mm MG.

Comment: Built by Pearl craft, Bahrain and delivered in 2012.

DERA'A 14   6/2009, Bahrain Coast Guard / 1367011

4 FAIREY SWORD CLASS (WPB)

SAIF 1–4

Displacement, tonnes: 15 standard
Dimensions, metres (feet): 13.7 × 4.1 × 1.3 (44.9 × 13.5 × 4.3)
Speed, knots: 22
Complement: 6
Machinery: 2 GM 8V-71 diesels; 590 hp (440 kW) sustained; 2 shafts
Radars: Navigation: Furuno; I-band.

Comment: Purchased in 1980. Built by Fairey Marine Ltd. To be replaced by new craft.

HARIS 1   6/2012, Bahrain Coast Guard / 1482959

Bangladesh

Country Overview

The People's Republic of Bangladesh gained independence from Pakistan in 1971. Situated in south Asia and with an area of 55,598 square miles, most of its land border is with India (almost cutting off north-east India from the rest). There is a short border with Myanmar to the south-east. Its 386 n mile coastline is with the Bay of Bengal on which the principal ports of Chittagong and Mongla are situated. The capital and largest city is Dhaka. Territorial waters (12 n miles) are claimed. An EEZ (200 n miles) has been claimed: maritime delimitation with Myanmar was resolved through the International Tribunal for the Law of the Sea in 2012 and arbitral proceedings with India are in progress.

Headquarters Appointments

Chief of Naval Staff:
   Vice Admiral Mohammad Farid Habib
Assistant Chief of Naval Staff (Operations):
   Rear Admiral A M M M Auranezeb Chowdhury
Assistant Chief of Naval Staff (Personnel):
   Rear Admiral Makbul Hossain
Assistant Chief of Naval Staff (Materials):
   Commodore S A M A Abedin
Assistant Chief of Naval Staff (Logistics):
   Rear Admiral M Saiful Kabir

Senior Appointments

Naval Administrative Authority, Dhaka:
   Commodore Shah Aslam Parvez
Commodore Commanding BN Flotilla:
   Commodore Shawkat Imran
Commodore Commanding Chittagong:
   Commodore Akhtar Habib
Commodore Commanding Khulna:
   Commodore Muhammad Shaheen Iqbal
Commodore Naval Aviation:
   Commodore Shawkat Imran
Commodore Superintendent, Dockyard:
   Commodore Kazi Kamrul Hassan

Bases

Chittagong (BNS *Issa Khan*, BN Dockyard, Naval Stores Depot, Chittagong, BNS *Ulka*, Bangladesh Naval Academy, BNS *Patenga*, BNS *Bhatiary*, BNS *Nirvik*, Naval Units *Cox's Bazar*, *Chanua* and *St Martins*), Kaptai (BNS *Shaheed Moazzam*).
   Dhaka (NHQ, BNS *Haji Mohsin* and Naval Units *Pagla* and *Khilkhet*).
   Khulna (BNS *Titumir*, BNS *Mongla*, BNS *Upasham*, Forward Bases *Khepupara* and *Hiron Point*.

Personnel

(a)   2014: 15,000 (1,500 officers)
(b)   Voluntary service

Strength of the Fleet

Type	Active	Building
Frigates	5	—
Corvettes	2	2
Fast Attack Craft (Missile)	11	—
Fast Attack Craft (Torpedo)	8	—
Fast Attack Craft (Gun)	12	—
Coastal Patrol Craft	16	3
Riverine Patrol Craft	22	—
LCT/LCU/LSL	9	4
Minesweepers	5	—
Training Ships	1	—
Repair Ship	1	—
Tankers	2	1
Survey Ships	2	—

Coast Guard

Formed on 19 December 1995 with two ships on loan from the Navy. Bases at Chittagong (East Zone) and Khulna (West Zone). Personnel 721 (54 officers). Colours thick red and thin blue diagonal stripes on hull with COAST GUARD on ships side.

Prefix to Ships' Names

Navy: BNS
Coast Guard: CGS

PENNANT LIST

Frigates

F 15	Abu Bakr
F 16	Umar Farooq
F 17	Ali Haider
F 18	Osman
F 25	Bangabandhu
F 28	Somudra Joy

Corvettes

F 35	Bijoy
F 36	Dhaleswari

Patrol Forces

P 211	Meghna
P 212	Jamuna
P 261	Aparajeya
P 262	Adomya
P 263	Atondo
P 311	Bishkhali
P 312	Padma
P 313	Surma
P 314	Karnaphuli
P 315	Tista
P 411	Shaheed Daulat
P 412	Shaheed Farid
P 413	Shaheed Mohibullah
P 414	Shaheed Aktheruddin
P 711	Barkat
P 712	Salam
P 713	Sangu
P 714	Turag
P 811	Durjoy
P 812	Nirbhoy
P 813	Nirmul
P 911	Madhumati
P 912	Kapatakhaya
P 913	Karatoa
P 914	Gomati
P 1011	Titas
P 1012	Kusiyara
P 1013	Chitra
P 1014	Dhansiri
P 8111	Durbar
P 8112	Duranta
P 8113	Durvedya
P 8114	Durdam
P 8125	Durdharsha
P 8126	Durdanta
P 8128	Dordanda
P 8131	Anirban
P 8141	Uttal
P 8221	TB 1
P 8222	TB 2
P 8223	TB 3
P 8224	TB 4
P 8235	TB 35
P 8236	TB 36
P 8237	TB 37
P 8238	TB 38

Mine Warfare Forces

M 91	Sagar
M 95	Shapla
M 96	Saikat
M 97	Surovi
M 98	Shaibal

Survey Ships

A 583	Agradoot
H 584	Anushandhan

Auxiliaries

A 511	Shaheed Ruhul Amin
A 512	Shahayak
A 513	Shahjalal
A 515	Khan Jahan Ali
A 516	Imam Gazzali
A 581	Darshak
A 582	Tallashi
A 584	LCT-101
A 585	LCT-102
A 587	LCT-104
A 711	Sundarban
A 721	Khadem
A 722	Sebak
A 723	Rupsha
A 724	Shibsha
A 731	Balaban
L 900	Shah Amanat
L 901	Shah Paran
L 902	Shah Makhdum

SUBMARINES

Notes: Negotiations with China to procure a submarine capability were announced in January 2013 and it was confirmed in December 2013 that an agreement to procure two Ming-class submarines had been concluded. Delivery of the first of class is expected in about 2019. There is also a requirement for construction of the necessary base facilities.

FRIGATES

1 MODIFIED ULSAN CLASS

Name	No	Builders	Laid down	Launched	Commissioned	Recommissioned
BANGABANDHU (ex-*Khalid Bin Walid*; ex-*Bangabandhu*)	F 25	Daewoo	12 May 1999	29 Aug 2000	20 Jun 2001	12 Jul 2007

Displacement, tonnes: 2,205 standard; 2,408 full load
Dimensions, metres (feet): 103.7 × 12.5 × 3.8 *(340.2 × 41.0 × 12.5)*
Speed, knots: 25. **Range, n miles:** 4,000 at 18 kt
Complement: 186 (16 officers)

Machinery: CODAD: 4 SEMT-Pielstick 12V PA6V280 STC diesels; 22,501 hp *(16.78 MW)* sustained; 2 shafts
Missiles: 4 Otomat Mk 2 ❶; command guidance; active radar homing to 180 km *(97.2 n miles)*, at 0.9 Mach; warhead 210 kg; sea-skimmer.
SAM: 1 HQ-7 (FM-90N) ❷; line of sight guidance to 13 km *(7 n miles)* at 2.4 Mach; warhead 14 kg.
Guns: 1 Otobreda 3 in *(76 mm)*/62 Super Rapid ❸; 120 rds/min to 16 km *(8.7 n miles)*; weight of shell 6 kg. 4 Otobreda 40 mm/70 (2 twin) compact ❹; 300 rds/min to 12.5 km *(6.8 n miles)*; weight of shell 0.96 kg. 2—12.7 mm MGs.
Torpedoes: 6—324 mm B-515 (2 triple) tubes ❺; Whitehead A244S; anti-submarine; active/passive homing to 7 km *(3.8 n miles)*; warhead 34 kg (shaped charge).
Physical countermeasures: Decoys: 2 Super Barricade launchers ❻.
Electronic countermeasures: ESM: Racal Cutlass 242; intercept. ECM: Racal Scorpion; jammer.
Radars: Air search: Signaal DA08 ❼; F-band.

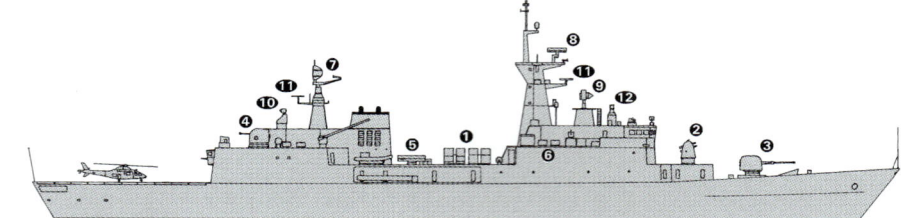

BANGABANDHU *(Scale 1 : 900), Ian Sturton / 1455668*

Surface search: Thales Variant ❽; G-band.
Fire control: Type 345 ❾; I/J-band (for FM-90N). Thales Lirod Mk 2 ❿; K-band.
Navigation: 2 KH-1007 ⓫; I-band.
Sonars: STN Atlas ASO 90; hull-mounted; active search; medium frequency.
Combat data systems: Thales TACTICOS.
Electro-optic systems: Thales Mirador optronic director ⓬.

Helicopters: 1 AgustaWestland AW 109.

Programmes: Modified Ulsan class ordered from Daewoo in March 1998. Arrived at Chittagong on 16 June 2001.
Operational: The ship was decommissioned on 13 February 2002 for design modification, warranty repairs and capability upgrades. This included installation of FM-90N, the export version of the Chinese HQ-7 SAM system. A period of uncertainty, during which the ship's future was kept under review, followed. This ended on 12 July 2007 when the ship was recommissioned. The original name was restored in 2009.

BANGABANDHU *12/2008, US Navy / 1482961*

BANGABANDHU *6/2012 / 1482962*

IHS Jane's Fighting Ships 2014-2015

Frigates ‹ **Bangladesh** 53

1 OSMAN (JIANGHU I) CLASS (TYPE 053 H1) (FFG)

Name	No	Builders	Laid down	Launched	Commissioned
OSMAN (ex-*Xiangtan*)	F 18 (ex-556)	Hudong Shipyard, Shanghai	1986	Dec 1988	4 Nov 1989

Displacement, tonnes: 1,448 standard; 1,729 full load
Dimensions, metres (feet): 103.2 × 10.7 × 3.1 *(338.6 × 35.1 × 10.2)*
Speed, knots: 26. **Range, n miles:** 2,700 at 18 kt
Complement: 300 (27 officers)

Machinery: 2 Type 12 E 390V diesels; 16,000 hp(m) *(11.9 MW)* sustained; 2 shafts
Missiles: SSM: 8 C-802 (YJ-83 (CSS-N-8 Saccade)) ❶ mid-course guidance and active radar homing to 120 km *(65 n miles)* at 0.9 Mach; warhead 165 kg.
Guns: 4 China 3.9 in *(100 mm)*/56 (2 twin) ❷; 25 rds/min to 22 km *(12 n miles)*; weight of shell 15.6 kg. 8 China 37 mm/76 (4 twin) ❸; 180 rds/min to 8.5 km *(4.6 n miles)* anti-aircraft; weight of shell 1.42 kg.
A/S Mortars: 2 RBU 1200 5-tubed fixed launchers ❹; range 1,200 m; warhead 34 kg.
Mines: Can carry up to 60.
Depth charges: 2 BMB-2 projectors; 2 racks.
Physical countermeasures: Decoys: 2 Loral Hycor SRBOC Mk 36 6-barrelled chaff launchers.
Electronic countermeasures: ESM: Watchdog; radar warning.
Radars: Air/surface search: MX 902 Eye Shield (922-1) ❻; G-band.

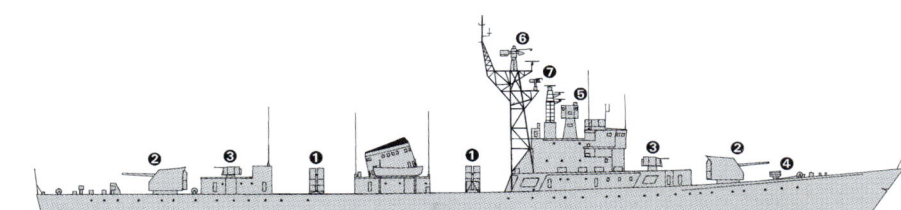

OSMAN *(Scale 1 : 900), Ian Sturton / 1305605*

Surface search/fire control: Square Tie (254) ❼; I-band.
Navigation: Fin Curve (352); I-band.
IFF: High Pole A.
Sonars: Echo Type 5; hull-mounted; active search and attack; medium frequency.
Weapon control systems: Wok Won director (752A) ❺.

Programmes: Transferred 26 September 1989 from China, arrived Bangladesh 8 October 1989. Second order expected in 1991 was cancelled.

Modernisation: C-802 missiles replaced HY-2 (C-201) missiles in 2007.
Structure: This is a Jianghu Type I (version 4) hull with twin 100 mm guns (vice the 57 mm in the ships sold to Egypt), Wok Won fire-control system and a rounded funnel.
Operational: A test-firing of C-802 was carried out on 12 May 2008.

OSMAN *6/2011, Bangladesh Navy / 1406699*

1 HAMILTON AND HERO CLASS (FFH)

Name	No	Builders	Laid down	Launched	Commissioned
SOMUDRA JOY (ex-*Jarvis*)	F 28 (ex-WHEC 725)	Avondale Shipyards, New Orleans	9 Sep 1970	24 Apr 1971	30 Dec 1971

Displacement, tonnes: 3,353 full load
Dimensions, metres (feet): 115.2 × 13.1 × 6.1 *(378 × 43.0 × 20.0)*
Flight deck, metres: 26.8 × 12.2 *(87.9 × 40.0)*
Speed, knots: 29. **Range, n miles:** 9,600 at 15 kt
Complement: 162 (19 officers)

Machinery: CODOG: 2 Pratt & Whitney FT4A-6 gas turbines; 36,000 hp *(26.86 MW)*; 2 Fairbanks-Morse 38TD8-1/8-12 diesels; 7,000 hp *(5.22 MW)* sustained; 2 shafts; cp props; retractable bow propulsor; 350 hp *(261 kW)*
Guns: 1 Oto Melara 3 in *(76 mm)*/62 Mk 75 Compact; 85 rds/min to 16 km *(8.7 n miles)* anti-surface; 12 km *(6.6 n miles)* anti-aircraft; weight of shell 6 kg. 4—12.7 mm MGs.

Radars: Surface search: Hughes/Furuno SPS-73; E/F- and I-bands.
Fire control: Sperry Mk 92; I/J-band.
Tacan: URN 25.
Combat data systems: SCCS 378.
Weapon control systems: Mk 92 Mod 1 GFCS.

Helicopters: Platform for 1 medium.

Programmes: Twelve built of a total of 36 originally planned.
Modernisation: FRAM programme from October 1985 to October 1992. Work included standardising the engineering plants, improving the clutching systems, and replacing the Mk 56 fire-control system and 5 in/38 gun mount with the Mk 92 system and a single 76 mm Oto Melara Compact gun. The flight deck and other aircraft facilities upgraded including a telescopic hangar.
Structure: These ships have clipper bows, twin funnels enclosing a helicopter hangar, helicopter platform aft. All are fitted with elaborate communications equipment. Superstructure is largely of aluminium construction. Bridge control of manoeuvring is by aircraft-type joystick rather than wheel.
Operational: Transferred from the US Coast Guard on 23 May 2013.

SOMUDRA JOY *10/2013*, M Mazumdar / 1525557*

Bangladesh > Frigates — Patrol forces

2 JIANGHU III (TYPE 053 H2) CLASS (FFG)

Name	No	Builders	Launched	Commissioned
ABU BAKR (ex-*Wuhu*)	F 15 (ex-536)	Hudong Shipyard, Shanghai	9 Aug 1986	29 Dec 1987
ALI HAIDER (ex-*Huangshi*)	F 17 (ex-535)	Hudong Shipyard, Shanghai	28 Dec 1985	15 Dec 1986

Displacement, tonnes: 1,955 standard
Dimensions, metres (feet): 103.2 × 10.8 × 3.1 *(338.6 × 35.4 × 10.2)*
Speed, knots: 26
Range, n miles: 4,000 at 15 kt, 2,700 at 18 kt
Complement: 200 (30 officers)

Machinery: 2 Type 18E 390V diesels; 14,400 hp(m) *(10.6 MW)* sustained; 2 shafts
Missiles: SSM: 4 C-802 (YJ-83/CSS-N-8 Saccade); mid-course guidance and active radar homing to 120 km *(65 n miles)* at 0.9 Mach; warhead 165 kg; sea skimmer.
Guns: 4 China 3.9 in *(100 mm)*/56 (2 twin); 25 rds/min to 22 km *(12 n miles)*; weight of shell 15.6 kg. 8 China 37 mm/63 (4 twin); 180 rds/min to 8.5 km *(4.6 n miles)* anti-aircraft; weight of shell 1.42 kg.
A/S Mortars: 2 RBU 1200 5-tubed fixed launchers; range 1,200 m; warhead 34 kg.
Mines: Can carry up to 60.
Depth charges: 2 BMB-2 projectors; 2 racks.
Physical countermeasures: Decoys: 2 China 26-barrelled chaff launchers.
Electronic countermeasures: ESM: Elettronica Newton; radar warning.
ECM: Elettronica 929 (Type 981); jammer.
Radars: Air search: Type 517 Knife Rest; A-band.
Air/surface search: Type 354 Eye Shield (MX 902); G-band.
Surface search/fire control: Type 352 Square Tie; I-band.
Navigation: Fin Curve; I-band.
Fire control: Type 347G Rice Bowl; I/J-band. Type 343G (Wok Won) (Wasp Head); I-band.
IFF: High Pole A. Square Head.
Sonars: Echo Type 5; hull-mounted; active search and attack; medium frequency.
Combat data systems: ZKJ-3.

Comment: The agreement to transfer two former Chinese Jianghu-class frigates was signed by the Bangladesh government in September 2012. Details are based on the ships in Chinese service. The transfer is to take place in 2014. The ships replace the Leopard-class frigates of the same names.

CORVETTES

Notes: A steel cutting ceremony for two corvettes was conducted at Wuchang Shipyard in China on 8 January 2013. The ships are reported to be of the order of 90 m long and 1,300-tonnes displacement which suggests they are to be Type 056 corvettes. Entry into service is expected in 2015.

2 CASTLE CLASS (FSGH)

Name	No	Builders	Launched	Commissioned
BIJOY (ex-*Dumbarton Castle*)	F 35 (ex-P 265)	Hall Russell, Aberdeen	3 Jun 1981	26 Mar 1982
DHALESHWARI (ex-*Leeds Castle*)	F 36 (ex-P 258)	Hall Russell, Aberdeen	29 Oct 1980	27 Oct 1981

Displacement, tonnes: 1,630 full load
Dimensions, metres (feet): 81 × 11.5 × 3.6 *(265.7 × 37.7 × 11.8)*
Speed, knots: 19.5. **Range, n miles:** 10,000 at 12 kt
Complement: 49 (6 officers)

Machinery: 2 Ruston 12RKC diesels; 5,640 hp *(4.21 MW)* sustained; 2 shafts; cp props
Missiles: 2 C-704; inertial guidance and active terminal homing to 38 km *(20.5 n miles)* at 0.8 Mach; warhead 130 kg.
Guns: 1—76 mm NG-16-1.
Physical countermeasures: Decoys: Outfit DLE; 2 or 4 Plessey Shield 102 mm 6-tubed launchers.
Electronic countermeasures: ESM: 'Orange Crop'; intercept.
Radars: Surface search: Type 360 Seagull S; E/F-band.
Fire control: Type 347G(1); I-band.
Navigation: Kelvin Hughes Type 1007; I-band. Furuno FR-2117; I-band.
Combat data systems: Racal CANE DEA-3 action data automation.
Weapon control systems: Radamec 2000 series optronic director.
Helicopters: Platform for operating Sea King or Lynx.

Comment: Agreement to acquire two ex-Royal Navy Castle-class patrol ships was reached in April 2010. Both ships transferred to Bangladesh following refits at A & P Tyne which started in May 2010. The original design included the capability to lay mines, while in *Bijoy* the 76 mm (never fitted) gunbay has been converted to an operations room. She is also fitted with two cranes to facilitate launch/recovery of RIBs. *Dhaleswari* was equipped with C-704 missiles in 2012 and a test firing was conducted in January 2013. *Bijoy* is to be similarly refitted.

DHALESHWARI   3/2013*, Chris Sattler / 1486509

SHIPBORNE AIRCRAFT

Numbers/Type: 2 AgustaWestland AW 109E Power.
Operational speed: 144 kt *(267 km/h)*.
Service ceiling: 15,000 ft *(4,572 m)*.
Range: 418 n miles *(774 km)*.
Role/Weapon systems: Contract signed with Agusta, Italy for two aircraft to be operated from the frigate *Bangabandhu*. They were delivered on 17 July 2011. Weapons: 1 — 7.62 mm MG.

AW 109E   4/2011*, AgustaWestland / 1486510

LAND-BASED MARITIME AIRCRAFT

Notes: Two Dornier 228 Maritime Patrol aircraft were ordered on 23 June 2011. Delivery was made on 29 August 2013.

PATROL FORCES

4 DURDHARSHA (HUANGFEN) CLASS (TYPE 021)
(FAST ATTACK CRAFT — MISSILE) (PTFG)

DURDHARSHA P 8125	DURDANTA P 8126	DORDANDA P 8128	ANIRBAN P 8131

Displacement, tonnes: 174 standard; 208 full load
Dimensions, metres (feet): 38.6 × 7.6 × 2.7 *(126.6 × 24.9 × 8.9)*
Speed, knots: 35
Range, n miles: 800 at 30 kt
Complement: 35 (5 officers)

Machinery: 3 diesels; 12,000 hp(m) *(8.8 MW)*; 3 shafts
Missiles: SSM: 4 HY-2 (C 201 Seersucker); active radar or IR homing to 95 km *(51 n miles)* at 0.9 Mach; warhead 513 kg.
Guns: 4 USSR 30 mm/65 (2 twin).
Radars: Surface search: Square Tie; I-band.
IFF: High Pole A.

Comment: Built in China. First four commissioned in Bangladesh Navy on 10 November 1988. Chinese equivalent of the Soviet Osa class which started building in 1985. All damaged in April 1991 typhoon but recovered and repaired (*Durnibar* was converted to a patrol craft). A fifth vessel *Anirban* was delivered in June 1992. Original main machinery replaced.

DORDANDA   6/2003, Bangladesh Navy / 0572413

4 HUCHUAN CLASS (TYPE 026)
(FAST ATTACK CRAFT — TORPEDO) (PTK)

TB 35 P 8235	TB 36 P 8236	TB 37 P 8237	TB 38 P 8238

Displacement, tonnes: 47 full load
Dimensions, metres (feet): 22.5 × 5 × 2.1 over foils *(73.8 × 16.4 × 6.9)*
Speed, knots: 50
Complement: 23 (3 officers)

Machinery: 3 Type L-12V-180 diesels; 3,600 hp(m) *(2.64 MW)*; 3 shafts
Guns: 4 China 14.5 mm (2 twin); 600 rds/min to 7 km *(3.8 km)*.
Torpedoes: 2 — 21 in *(533 mm)* China YU-1; anti-ship; to 9.2 km *(5 n miles)* at 39 kt or 3.7 km *(2.1 n miles)* at 51 kt; warhead 400 kg.
Radars: Surface search: China Type 753; I-band.

Comment: Chinese Huchuan class. Two damaged in April 1991 typhoon but were repaired. All reported operational.

TB 38   6/2003, Bangladesh Navy / 0572415

6 ISLAND CLASS
(COASTAL PATROL VESSEL/TRAINING CRAFT) (PBO/AX)

Name	No	Builders	Commissioned	Recommissioned
SHAHEED RUHUL AMIN (ex-*Jersey*)	A 511 (ex-P 295)	Hall Russell, Aberdeen	15 Oct 1976	1994
KAPATAKHAYA (ex-*Shetland*)	P 912 (ex-P 298)	Hall Russell, Aberdeen	14 Jul 1977	4 May 2003
KARATOA (ex-*Alderney*)	P 913 (ex-P 278)	Hall Russell, Aberdeen	6 Oct 1979	4 May 2003
GOMATI (ex-*Anglesey*)	P 914 (ex-P 277)	Hall Russell, Aberdeen	1 Jun 1979	3 Oct 2004
SANGU (ex-*Guernsey*)	P 713 (ex-P 297)	Hall Russell, Aberdeen	28 Oct 1977	3 Oct 2004
TURAG (ex-*Lindisfarne*)	P 714 (ex-P 300)	Hall Russell, Aberdeen	3 Mar 1978	3 Oct 2004

Displacement, tonnes: 940 standard; 1,280 full load
Dimensions, metres (feet): 59.5 oa; 53.7 wl × 11 × 4.5 *(195.2; 176.2 × 36.1 × 14.8)*
Speed, knots: 16.5. **Range, n miles:** 7,000 at 12 kt
Complement: 39

Machinery: 2 Ruston 12RKC diesels; 5,640 hp *(4.21 MW)* sustained; 1 shaft; cp prop
Guns: 1 Bofors 40 mm/60 Mk 3. 2 FN 7.62 mm MGs.
Electronic countermeasures: ESM: Orange Crop; intercept.
Radars: Navigation: Kelvin Hughes Type 1006; I-band.
Combat data systems: Racal CANE DEA-1 action data automation.

Comment: *Shaheed Ruhul Amin* transferred as a training craft in 1993. Five further former UK Island class acquired as patrol craft. *Kapatakhaya* transferred 31 July 2002, *Karatoa* 31 October 2002, *Gomati* on 12 September 2003 and *Sangu* and *Turag* on 29 January 2004.

KARATOA   *4/2009, Shaun Jones* / 1305481

1 MADHUMATI (SEA DRAGON) CLASS
(OFFSHORE PATROL CRAFT) (PSO)

Name	No	Builders	Commissioned
MADHUMATI	P 911	Hyundai, Ulsan	18 Feb 1998

Displacement, tonnes: 645 full load
Dimensions, metres (feet): 60.8 × 8 × 2.7 *(199.5 × 26.2 × 8.9)*
Speed, knots: 24
Range, n miles: 6,000 at 15 kt
Complement: 43 (7 officers)

Machinery: 2 SEMT-Pielstick 12 PA6 diesels; 9,600 hp(m) *(7.08 MW)* sustained; 2 shafts
Guns: 1 Bofors 57 mm/70 Mk 1; 220 rds/min to 17 km *(9.3 n miles)*; weight of shell 2.4 kg.
1 Bofors 40 mm/70. 2 Oerlikon 20 mm.
Radars: Surface search: Kelvin Hughes KH 1007; I-band.
Navigation: GEM Electronics SPN 753B; I-band.
Weapon control systems: Optronic director.

Comment: Ordered in 1995 and delivered in October 1997. Very similar to the South Korean Coast Guard vessels, but with improved fire-control equipment. Vosper stabilisers.

MADHUMATI   *1/2012*, Felipe Salles* / 1525558

16 DEFENDER CLASS (RESPONSE BOATS) (PBF)

Displacement, tonnes: 3.85 full load
Dimensions, metres (feet): 8.23 × 2.6 × 1.0 *(27 × 8.5 × 3.3)*
Speed, knots: 46
Range, n miles: 175 at 35 kt
Complement: 4
Machinery: 2 Mercury outboard motors; 450 hp *(335 kW)*
Guns: 1 – 12.7 mm MG.
Radars: To be announced.

Comment: Ex-US Coast Guard high-speed inshore patrol craft donated to the Bangladesh Navy on 13 April 2010. Of aluminium construction and foam collar, the craft were originally built by SAFE Boats International, Port Orchard, Washington. The craft are likely to be used by the Special Operations Force of the Bangladesh Navy for counter-terrorist and -drug operations in coastal areas. Five additional craft are operated by the Coast Guard.

© 2014 IHS

5 DURBAR (HEGU) CLASS (TYPE 024)
(FAST ATTACK CRAFT – MISSILE) (PTFG)

DURBAR P 8111	DURVEDYA P 8113	UTTAL P 8141
DURANTA P 8112	DURDAM P 8114	

Displacement, tonnes: 69 standard; 80 full load
Dimensions, metres (feet): 27 × 6.3 × 1.3 *(88.6 × 20.7 × 4.3)*
Speed, knots: 37.5
Range, n miles: 400 at 30 kt
Complement: 17 (4 officers)

Machinery: 4 Type L-12V-180B diesels; 4,800 hp(m) *(3.57 MW)*; 4 shafts
Missiles: SSM: 2 SY-1; active radar or IR homing to 45 km *(24.3 n miles)* at 0.9 Mach; warhead 513 kg.
Guns: 2 – 25 mm/80 (twin); 270 rds/min to 3 km *(1.6 n miles)*; weight of shell 0.34 kg.
Radars: Surface search: Square Tie; I-band.

Comment: Built in China. First pair commissioned in Bangladesh Navy on 6 April 1983, second pair on 10 November 1983. Two badly damaged in April 1991 typhoon but were repaired. *Uttal* was delivered in June 1992. Missiles are seldom embarked. All have been refitted with new versions of original engines.

UTTAL   *3/1998* / 0017590

1 DURJOY (HAINAN) CLASS (TYPE 037)
(OFFSHORE PATROL CRAFT) (PC)

NIRBHOY P 812

Displacement, tonnes: 381 standard; 398 full load
Dimensions, metres (feet): 58.8 × 7.2 × 2.2 *(192.9 × 23.6 × 7.2)*
Speed, knots: 30.5
Range, n miles: 1,300 at 15 kt
Complement: 70

Machinery: 4 diesels; 4,000 hp(m) *(2.94 MW)* sustained; 4 shafts
Guns: 4 China 57 mm/70 (2 twin); 120 rds/min to 12 km *(6.5 n miles)*; weight of shell 6.31 kg.
4 – 25 mm/60 (2 twin); 270 rds/min to 3 km *(1.6 n miles)* anti-aircraft.
A/S Mortars: 4 RBU 1200 fixed 5-barrelled launchers; range 1,200 m; warhead 34 kg.
Mines: Fitted with rails for 12 mines.
Depth charges: 2 racks; 2 throwers. 18 DCs.
Radars: Surface search: I-band.
IFF: High Pole.
Sonars: Tamir II; hull-mounted; short-range attack; high frequency.

Comment: Transferred from China and commissioned 1 December 1985. Forms part of Escort Squadron 81 at Chittagong. *Durjoy* damaged beyond repair by cyclone in 1991. *Nirbhoy* refitted with new main machinery.

NIRBHOY   *6/2011, Bangladesh Navy* / 1406700

6 HIGH-SPEED INTERCEPTION CRAFT (HSIC)

Displacement, tonnes: 3 full load
Dimensions, metres (feet): 10.4 × 2.8 × 0.6 *(34.1 × 9.2 × 2.0)*
Speed, knots: 54
Range, n miles: 200 at 35 kt
Complement: 10
Machinery: 2 VM diesels; 640 hp *(480 kW)*; 2 shafts
Guns: 1 – 7.62 mm MG.

Comment: RIB33SC design by FB Design, Italy. Funded by UN for riverine and coastal patrol in southern Sudan. Craft in UN livery but commissioned in Bangladesh Navy in 2005.

INTERCEPTION CRAFT   *6/2007, Massimo Annati* / 1166507

58 Bangladesh > Amphibious forces — Mine warfare forces

1 LANDING SHIP (LCT)

Name	No	Builders	Commissioned
SHAKTI SANCHAR	–	Dockyard and Engineering Works, Narayanganj	2012

Displacement, tonnes: 2,200 full load
Dimensions, metres (feet): 65.7 × 12.0 × 3.5 *(215.6 × 39.4 × 11.5)*
Complement: 45
Military lift: 9 tanks and 150 troops
Machinery: 2 diesels; 2 shafts
Helicopters: Platform for 1 medium.

Comment: Designed by GB Marine Pte Limited, Singapore, and built in Bangladesh. Owned and operated by the Bangladesh Army.

SHAKTI SANCHAR 6/2013*, Bangladesh Navy / 1529104

1 LANDING CRAFT LOGISTIC (LSL)

SHAH AMANAT L 900

Displacement, tonnes: 372 full load
Dimensions, metres (feet): 47 × 10.4 × 2.4 *(154.2 × 34.1 × 7.9)*
Speed, knots: 9.5
Complement: 31 (3 officers)
Military lift: 150 tons
Machinery: 2 Caterpillar D 343 diesels; 730 hp *(544 kW)* sustained; 2 shafts
Guns: 2 – 12.7 mm MGs.

Comment: Australian civil vessel confiscated by the Navy while engaged in smuggling in 1988. Transferred to the Navy and commissioned in 1990.

SHAH AMANAT 6/1996, Bangladesh Navy / 0056562

2 LCU 1512 CLASS (LCU)

SHAH PORAN (ex-*Cerro Gordo*) L 901   **SHAH MAKHDUM** (ex-*Cadgel*) L 902

Displacement, tonnes: 381 full load
Dimensions, metres (feet): 41.1 × 8.8 × 1.9 *(134.8 × 28.9 × 6.2)*
Speed, knots: 11. **Range, n miles:** 1,200 at 8 kt
Complement: 14 (2 officers)

Military lift: 170 tons
Machinery: 4 Detroit 6-71 diesels; 696 hp *(508 kW)* sustained; 2 shafts
Guns: 2 – 12.7 mm MGs.
Radars: Navigation: LN 66; I-band.

Comment: Ex-US Army landing craft transferred in April 1991 and commissioned 30 January 1993 after refit.

SHAH MAKHDUM 6/1996, Bangladesh Navy / 0056563

3 LCVP

L 011–013

Displacement, tonnes: 84 full load
Dimensions, metres (feet): 21.3 × 5.2 × 1.5 *(69.9 × 17.1 × 4.9)*
Speed, knots: 12
Complement: 10 (1 officer)
Machinery: 2 Cummins diesels; 730 hp *(544 kW)*; 2 shafts

Comment: First two built at Khulna Shipyard and *013* at DEW Narayangong; all completed in 1984.

L 011 6/1996, Bangladesh Navy / 0056564

MINE WARFARE FORCES

4 SHAPLA (RIVER) CLASS
(MINESWEEPERS/PATROL CRAFT/SURVEY SHIPS) (MHSC/PBO/AGS)

Name	No	Builders	Commissioned
SHAPLA (ex-*Waveney*)	M 95	Richards Ltd, Lowestoft	12 Jul 1984
SAIKAT (ex-*Carron*)	M 96	Richards Ltd, Great Yarmouth	30 Sep 1984
SUROVI (ex-*Dovey*)	M 97	Richards Ltd, Great Yarmouth	30 Mar 1985
SHAIBAL (ex-*Helford*)	M 98	Richards Ltd, Great Yarmouth	7 Jun 1985

Displacement, tonnes: 904 full load
Dimensions, metres (feet): 47.5 × 10.5 × 2.9 *(155.8 × 34.4 × 9.5)*
Speed, knots: 14. **Range, n miles:** 4,500 at 10 kt
Complement: 30 (7 officers)
Machinery: 2 Ruston 6RKC diesels; 3,100 hp *(2.3 MW)* sustained; 2 shafts; cp props
Guns: 1 Bofors 40 mm/60 Mk 3.
Radars: Navigation: 2 Racal Decca TM 1226C; I-band.

Comment: These ships are four of a class of 12 of which seven are in service with Brazil. Transferred from the UK on 3 October 1994 and recommissioned on 27 April 1995. Steel hulled for deep-armed team sweeping with wire sweeps, and intended for use both as minesweepers and as patrol craft. Fitted with Racal Integrated Minehunting System. *Shaibal* converted for hydrographic survey duties but retains minesweeping gear. Fitted with echo sounders, side-scan sonar and a laboratory.

SUROVI 3/1998 / 0017593

1 SAGAR (T 43) CLASS (MINESWEEPER) (MSO)

Name	No	Builders	Commissioned
SAGAR	M 91	Wuhan Shipyard	27 Apr 1995

Displacement, tonnes: 528 standard; 599 full load
Dimensions, metres (feet): 60 × 8.8 × 2.3 *(196.9 × 28.9 × 7.5)*
Speed, knots: 14. **Range, n miles:** 3,000 at 10 kt
Complement: 70 (10 officers)

Machinery: 2 CXZ MAN B&W Type 9L 20-27 diesels; 2,400 hp *(1.8 MW)* sustained; 2 shafts; cp props
Guns: 4 China 37 mm/63 (2 twin); 180 rds/min to 8.5 km *(4.6 n miles)*; weight of shell 1.42 kg. 4 – 25 mm/60 (2 twin); 270 rds/min to 3 km *(1.6 n miles)*. 4 China 14.5 mm/93 (2 twin); 600 rds/min to 7 km *(3.8 n miles)*.
Mines: Can carry 12–16.
Depth charges: 2 BMB-2 projectors; 20 depth charges.
Physical countermeasures: MCMV MPT-1 paravanes; MPT-3 mechanical sweep; acoustic and magnetic gear.
Radars: Surface search: Fin Curve; I-band.
Sonars: Celcius Tech CMAS 36/39; active high frequency mine detection.

Comment: Ordered from China in 1993. Based on Type 010G minesweeper design. Used mostly as a patrol ship. New sonar fitted in 1998.

SAGAR 3/1998 / 0017594

IHS Jane's Fighting Ships 2014-2015

SURVEY AND RESEARCH SHIPS

1 SURVEY SHIP (AGS)

Name	No	Builders	Commissioned
AGRADOOT (ex-Kodan)	A 583	Khulna Shipyard	19 Mar 2002

Displacement, tonnes: 698 full load
Dimensions, metres (feet): 47.8 × 7.8 × 3.5 (156.8 × 25.6 × 11.5)
Speed, knots: 12.5
Complement: 70 (8 officers)
Machinery: 2 Baudouin diesels
Guns: 1 Oerlikon 20 mm.
Radars: Furuno HR 2110. Kelvin Hughes HR-3000A.

Comment: Former Thai trawler converted into a Survey vessel by Khulna shipyard. Fitted with two dual frequency digital hydrographic echo sounders, side-scan sonar and laboratories. Carries a survey launch.

1 ROEBUCK CLASS (AGS)

Name	No	Builders	Launched	Commissioned
ANUSHANDHAN (ex-Roebuck)	H 584 (ex-H 130)	Brooke Marine, Lowestoft	14 Nov 1985	3 Oct 1986

Displacement, tonnes: 1,059 standard; 1,430 full load
Dimensions, metres (feet): 63.9 × 13 × 4 (209.6 × 42.7 × 13.1)
Speed, knots: 14
Range, n miles: 4,000 at 10 kt
Complement: 46 (6 officers)
Machinery: 4 Mirrlees Blackstone ESL8 Mk 1 diesels; 3,040 hp (2.27 MW); 2 shafts; cp props
Guns: 1—20 mm. 2 M323 Mk 44 7.62 mm Miniguns (fitted for).
Radars: Navigation: Kelvin Hughes Nucleus 2–6000; I-band.

Comment: Designed for hydrographic surveys to full modern standards. Air conditioned. Carries Survey Motor Launch. A Ship Life Extension Programme started in September 2004 and was completed in mid-2005. The upgrade included refurbishment and renewal of engineering systems and habitability improvements. A 20 mm gun system has been installed. Roles include Rapid Environmental Assessment and Amphibious Warfare survey. The new survey suite consists of EM 1002 hull-mounted multibeam sonar, EA 600 SBES and 2094 towed side-scan sonar and adaptive planning system, Moving Vessel Profiler (MVP) 200 and WECDIS. Decommissioned from the Royal Navy in 2010, she departed for Bangladesh on 4 June 2010.

AGRADOOT   6/2003, Bangladesh Navy / 0572419

ANUSHANDHAN   6/2010, Maritime Photographic / 1366446

TRAINING SHIPS

1 SALISBURY CLASS (TYPE 61) (FF)

Name	No
UMAR FAROOQ (ex-Llandaff)	F 16

Builders	Laid down	Launched	Commissioned
Hawthorn Leslie Ltd	27 Aug 1953	30 Nov 1955	11 Apr 1958

Displacement, tonnes: 2,205 standard; 2,447 full load
Dimensions, metres (feet): 103.6 × 12.2 × 4.7 screws (339.9 × 40.0 × 15.4)
Speed, knots: 24
Range, n miles: 2,300 at 24 kt, 7,500 at 16 kt
Complement: 237 (14 officers)

Machinery: 8 16 VTS ASR 1 diesels; 14,400 hp (10.7 MW) sustained; 2 shafts
Guns: 2 Vickers 4.5 in (115 mm)/45 (twin) Mk 6 ❶; dual purpose; 20 rds/min to 19 km (10 n miles) anti-surface; 6 km (3.3 n miles) anti-aircraft; weight of shell 25 kg. 2 Bofors 40 mm/60 Mk 9 ❷; 120 rds/min to 3 km (1.6 n miles) anti-aircraft; 10 km (5.5 n miles) maximum.
Physical countermeasures: Decoys: Corvus chaff launchers.
Radars: Air search: Marconi Type 965 with double AKE 2 array ❸; A-band.
Air/surface search: Plessey Type 993 ❹; E/F-band.
Heightfinder: Type 278M ❺; E-band.
Surface search: Decca Type 978 ❻; I-band.
Navigation: Decca Type 978; I-band.
Fire control: Type 275 ❼; F-band.
Sonars: Type 174; hull-mounted; active search; medium frequency. Graseby Type 170B; hull-mounted; active attack; 15 kHz.
Weapon control systems: 1 Mk 6M gun director.

Programmes: Transferred from UK at Royal Albert Dock, London 10 December 1976.
Modernisation: The ship is to be re-engined and accommodation is to be upgraded to prolong service life.
Operational: The radar Type 982 aerial is still retained on the after mast but the set is non-operational. The ship has been modified as a training ship and is expected to remain in service until about 2025.

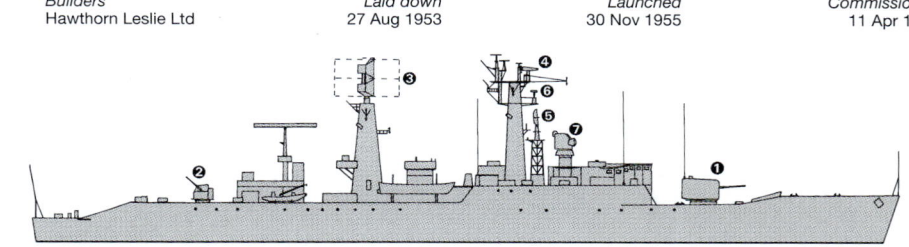

UMAR FAROOQ   (Scale 1 : 900), Ian Sturton / 1366676

UMAR FAROOQ   3/2007, Paul Daly / 1166506

AUXILIARIES

Notes: Floating Dock A 711 (Sundarban) acquired from Brodogradiliste Joso Lozovina-Mosor, Trogir, Yugoslavia in 1980; capacity 3,500 tons. Has a complement of 85 (5 officers). Floating crane A 731 (Balaban) is self-propelled at 9 kt and has a lift of 70 tons; built at Khulna Shipyard and commissioned 18 May 1988, she has a complement of 29 (two officers).

1 TANKER (AOTL)

IMAM GAZZALI A 516

Displacement, tonnes: 216 full load
Dimensions, metres (feet): 44.8 × 7 × 3.4 (147 × 23.0 × 11.2)
Speed, knots: 8
Complement: 30 (2 officers)
Machinery: 1 Cummins diesel; 1 shaft

Comment: An oil tanker of some 600,000 litres capacity acquired in 1994 and commissioned 6 May 1997.

IMAM GAZZALI   6/1999, Bangladesh Navy / 0056556

60 Bangladesh > Auxiliaries — Tugs

1 TANKER (AOTL)

Name	No	Commissioned
KHAN JAHAN ALI	A 515	14 Jul 1987

Displacement, tonnes: 2,947 full load
Measurement, tonnes: 1,365 gt
Dimensions, metres (feet): 76.4 × 11.4 × 5.6 *(250.7 × 37.4 × 18.4)*
Speed, knots: 12
Complement: 26 (3 officers)
Cargo capacity: 1,500 tons
Machinery: 1 diesel; 1,350 hp(m) *(992 kW)*; 1 shaft
Guns: 2 Oerlikon 20 mm.

Comment: Completed in Japan in 1983. Can carry out stern replenishment at sea but is seldom used in this role. To be replaced by a new ship of the same name in 2014.

KHAN JAHAN ALI   3/1998 / 0017595

1 REPAIR SHIP (YR)

SHAHAYAK A 512

Displacement, tonnes: 485 full load
Dimensions, metres (feet): 44.7 × 8 × 2 *(146.7 × 26.2 × 6.6)*
Speed, knots: 11.5. **Range, n miles:** 3,800 at 11.5 kt
Complement: 45 (1 officer)
Machinery: 1 Cummins 12 VTS 6 diesel; 425 hp *(317 kW)*; 1 shaft
Guns: 1 Oerlikon 20 mm.

Comment: Re-engined and modernised at Khulna Shipyard and commissioned on 23 November 1978 to act as repair vessel.

SHAHAYAK   6/1996, Bangladesh Navy / 0056557

1 TENDER (AG)

SHAHJALAL A 513

Displacement, tonnes: 610 full load
Dimensions, metres (feet): 40.2 × 9.1 × 3.8 *(131.9 × 29.9 × 12.5)*
Speed, knots: 12
Range, n miles: 7,000 at 12 kt
Complement: 55 (3 officers)
Machinery: 1 V 16-cyl type diesel; 1 shaft
Guns: 1 Oerlikon 20 mm.

Comment: Ex-Thai fishing vessel SMS *Gold 4*. Probably built in Tokyo. Commissioned on 15 January 1987 and used as a diving/salvage tender.

SHAHJALAL   6/1996, Bangladesh Navy / 0056558

1 HARBOUR TENDER (YAG)

SANKET

Displacement, tonnes: 81 full load
Dimensions, metres (feet): 29.4 × 6.1 × 1.8 *(96.5 × 20.0 × 5.9)*
Speed, knots: 16. **Range, n miles:** 1,000 at 16 kt
Complement: 16 (1 officer)
Machinery: 2 Deutz diesels; 2,400 hp(m) *(1.76 MW)*; 2 shafts
Guns: 1 Oerlikon 20 mm.

Comment: Former harbour craft of the Chittagong Port Authority taken over by the navy in 1984. It is used as a utility harbour craft. No pennant number has been allocated.

SANKET   3/1996 / 0056559

0 + 1 TANKER (AOTL)

Name	No	Builders	Laid down	Commissioned
KHAN JAHAN ALI	A 515	Ananda Shipyard and Slipways Ltd	11 Mar 2013	2014

Displacement, tonnes: 4,089 full load
Dimensions, metres (feet): 79.8 × 12.5 × 5.0 *(261.8 × 41.0 × 16.4)*
Speed, knots: 13
Complement: 34 (4 officers)

Cargo capacity: 2,750 m³ dieso; 150 m³ aviation fuel
Machinery: 2 Cummins QSK60-M diesels; 4,000 hp *(2.98 MW)* sustained; 2 shafts
Guns: 2 Oerlikon 20 mm.
Radars: Navigation: To be announced.

Comment: Contract with Ananda Shipyard Ltd for the construction of one tanker. Delivery is expected in 2014.

TUGS

1 HUJIU CLASS (OCEAN TUG) (ATA)

KHADEM A 721

Displacement, tonnes: 1,496 full load
Dimensions, metres (feet): 60.2 × 11.6 × 4.9 *(197.5 × 38.1 × 16.1)*
Speed, knots: 14. **Range, n miles:** 7,200 at 14 kt
Complement: 56 (7 officers)
Machinery: 2 LVP 24 diesels; 1,800 hp(m) *(1.32 MW)*; 2 shafts
Guns: 2—12.7 mm MGs.
Radars: Navigation: China Type 756; I-band.

Comment: Commissioned 6 May 1984 after transfer from China.

KHADEM   6/1996, Bangladesh Navy / 0056565

3 COASTAL TUGS (YTM)

SEBAK A 722	RUPSHA A 723	SHIBSHA A 724

Displacement, tonnes: 335 full load
Dimensions, metres (feet): 30 × 8.4 × 3.5 *(98.4 × 27.6 × 11.5)*
Speed, knots: 12. **Range, n miles:** 1,800 at 12 kt
Complement: 23 (3 officers)
Machinery: 2 Caterpillar 12V 3512B diesels; 2,700 hp *(2.0 MW)*; 2 shafts
Guns: 2—7.62 mm MGs (fitted for).

Comment: Details are for *Rupsha* and *Shibsha* built to a Damen Stan Tug 3008 design by Khulna Shipyard. Construction started in 2001, completed in 2003 and commissioned on 3 October 2004. *Sebak* built in Narayangang Dockyard in 1993 and commissioned on 23 December 1993.

SHIBSHA   6/2003, Bangladesh Navy / 0572422

IHS Jane's Fighting Ships 2014-2015   © 2014 IHS

COAST GUARD

Senior Appointments

Director General:
 Rear Admiral Kazi Sarwar Hossain
Deputy Director General:
 Commodore Mohammad Mohiuddin Razib

Personnel

2014: 2,255 (22 officers)

General

The Bangladesh Coast Guard was established on 14 February 1995. The service is organised into four operational commands, each of which has a small number of sub-stations: East, West and South Zones, and Dhaka Sub Zone.

Bases

Coast Guard Base Chittagong (East Zone)
Coast Guard Base Mongla (West Zone)
Coast Guard Base Bhola (South Zone)

PATROL FORCES

Notes: In addition to the ships listed, four Shaheed (Shanghai II)-class patrol vessels are listed under naval patrol forces: *Tawheed* (P 611), *Tawfiq* (P 612), *Tamjeed* (P 613), and *Tanveer* (P 614).

1 RUPOSHI BANGLA CLASS (COASTAL PATROL CRAFT) (PB)

Name	No	Builders	Launched	Commissioned
RUPOSHI BANGLA	P 201	Hong Leong-Lürssen	28 Jun 1999	23 Jan 2000

Displacement, tonnes: 198 full load
Dimensions, metres (feet): 38.5 × 7 × 1.8 *(126.3 × 23.0 × 5.9)*
Speed, knots: 30
Complement: 45
Machinery: 2 Paxman 12VP 185 diesels; 6,729 hp(m) *(4.95 MW)* sustained; 2 shafts
Guns: 1 Oto Melara 25 mm KBA. 2 – 7.62 mm MGs.
Radars: Surface search: Furuno; I-band.

Comment: Ordered in June 1998 and laid down 11 August 1998. Based on the PZ design for the Malaysian Police. Operated by the Coast Guard.

RUPOSHI BANGLA *10/1999, Hong Leong-Lürssen /* 0064625

5 PABNA CLASS (RIVERINE PATROL CRAFT) (PBR)

Name	No	Builders	Commissioned
PABNA	P 111	DEW Narayangonj, Dhaka	12 Jun 1972
NOAKHALI	P 112	DEW Narayangonj, Dhaka	10 Jul 1972
PATUAKHALI	P 113	DEW Narayangonj, Dhaka	27 Mar 1975
RANGAMATI	P 114	DEW Narayangonj, Dhaka	12 Feb 1977
BOGRA	P 115	DEW Narayangonj, Dhaka	15 Jul 1977

Displacement, tonnes: 71 full load
Dimensions, metres (feet): 22.9 × 6.1 × 1.1 *(75.1 × 20.0 × 3.6)*
Speed, knots: 10.8. **Range, n miles:** 700 at 8 kt
Complement: 33 (3 officers)
Machinery: 2 Cummins diesels; 2 shafts
Guns: 1 Oerlikon 20 mm.
Radars: Navigation: Furuno; I-band.

Comment: The first indigenous naval craft built in Bangladesh. Form River Patrol Squadron 11 at Mongla. All operated by the Coast Guard from 2003.

PABNA *12/2010*, Mazumdar Collection /* 1486512

2 PATROL CRAFT (PB)

SHET GANG P 101   PORTE GRANDE P 102

Displacement, tonnes: 81 full load
Dimensions, metres (feet): 31.2 × 5.4 × 1.4 *(102.4 × 17.7 × 4.6)*
Speed, knots: 25
Complement: 30
Machinery: 2 MTU diesels; 3,000 hp *(2.2 MW)*; 2 shafts
Guns: 1 – 12.7 mm MG.
Radars: Navigation: Furuno; I-band.

Comment: Both ships constructed by Ananda Shipyard and commissioned on 28 May 2006. Operated by the Coast Guard.

PORTE GRANDE *6/2012*, Mazumdar Collection /* 1486511

20 DOLPHIN CLASS (PATROL CRAFT) (PBF)

Displacement, tonnes: 1.5 full load
Dimensions, metres (feet): 8.69 × 2.59 × 0.45 *(28.5 × 8.5 × 1.5)*
Speed, knots: 40
Complement: 4
Machinery: 2 diesels; 230 hp *(171 kW)*; 2 shafts

Comment: Built by BF International Bangladesh in 2008. Principal role of anti-smuggling operations.

5 DEFENDER CLASS (PATROL CRAFT) (PBF)

Displacement, tonnes: 3.85 full load
Dimensions, metres (feet): 8.23 × 2.6 × 1.0 *(27 × 8.5 × 3.3)*
Speed, knots: 40
Complement: 4
Machinery: 2 Mercury outboard motors; 450 hp *(335 kW)*

Comment: Five craft manufactured by the United States and received as a grant in 2011. There are 16 similar craft in service with the Bangladesh Navy.

8 TORNADO CLASS (PATROL CRAFT) (PBF)

Displacement, tonnes: 7.42 full load
Dimensions, metres (feet): 14.8 × 2.68 × 0.8 *(48.6 × 8.8 × 2.6)*
Speed, knots: 38
Range, n miles: 300 at 30 kt
Complement: 6
Machinery: 2 diesels; 2 shafts

Comment: Built in 2009 by Brunswick Asia Pacific Group, Mercury Marine Singapore Ptd, Ltd. These boats are used for anti-smuggling operations.

3 METAL SHARK DEFIANT 38 CLASS (PATROL CRAFT) (PBF)

Displacement, tonnes: 7.24 full load
Dimensions, metres (feet): 11.6 × 3.53 × 0.76 *(38.1 × 11.6 × 2.5)*
Speed, knots: 45
Range, n miles: 200 at 30 kt
Complement: 6
Machinery: 3 outboard motors; 750 hp *(559 kW)*

Comment: Built by Metal Shark, Jeanerette, Louisiana, and donated by the US government in 2012. Employed on anti-smuggling SAR duties.

0 + 6 TYPHOON CLASS (PATROL CRAFT) (PBF)

Displacement, tonnes: 5.6 full load
Dimensions, metres (feet): 10.8 × 3.56 × 0.47 *(35.4 × 11.7 × 1.5)*
Speed, knots: 35. **Range, n miles:** 200 at 15 kt
Complement: 4
Machinery: 2 Volvo Penta D6 diesels; 750 hp *(559 kW)*; 2 shafts

Comment: Under construction in late 2013 at Adriatic Workboats, Croatia.

0 + 3 HARBOUR PATROL CRAFT (PBR)

Displacement, tonnes: 25 full load
Dimensions, metres (feet): 17.47 × 5.5 × 2.56 *(57.3 × 18.0 × 8.4)*
Speed, knots: 25. **Range, n miles:** 350 at 15 kt
Complement: 12
Machinery: 2 diesels; 2 shafts

Comment: Under construction in late 2013 at Dockyard & Engineering Works Ltd, Narayangonj.

62 Barbados > Introduction — Belgium > Introduction

Barbados

Country Overview

Barbados gained independence in 1966; the British monarch, represented by a governor-general, is head of state. The easternmost island of the Windward Islands of the Lesser Antilles chain, it consists of a single island of 166 square miles. The capital, largest town and principal port is Bridgetown, located on the southwestern coast. Territorial seas (12 n miles) are claimed. A 200 n mile Exclusive Economic Zone (EEZ) has also been claimed but the limits are not defined. A Coast Guard was formed in 1973 and became the naval arm of the Barbados Defence Force in 1979.

Headquarters Appointments

Chief of Staff, Barbados Defence Force:
  Colonel Alvin Quintyne
Commanding Officer Coast Guard:
  Lieutenant Commander Neville Springer

Personnel

2014:
(a) 215 (18 officers)
(b) Voluntary service

Bases

Spring Garden, Bridgetown (HMBS *Pelican*).

Prefix to Ships' Names

HMBS

COAST GUARD

Notes: (1) Three 10 m Damen RIB 1000 capable of 35 kt were delivered in June 2007.
(2) A Zodiac 920 RHIB was donated by the US in 2004. A Zodiac 921 RHIB was purchased in 2005.
(3) The former *Trident*, decommissioned in 2009, is used as a training platform.

3 DAMEN STAN PATROL 4207 (PB)

Name	No	Builders	Commissioned
TRIDENT	P 01	Damen Shipyard, Gorinchem	25 Apr 2009
LEONARD C BANFIELD	P 02	Damen Shipyard, Gorinchem	14 Sep 2007
RUDYARD LEWIS	P 03	Damen Shipyard, Gorinchem	13 Sep 2008

Displacement, tonnes: 208 standard
Dimensions, metres (feet): 42.8 × 7.11 × 2.52 *(140.4 × 23.3 × 8.3)*
Speed, knots: 26
Complement: 14
Machinery: 2 Caterpillar 3516B DI-TA; 5,600 hp *(4.17 MW)*; 2 cp props
Guns: 2 — 12.7 mm MGs. 2 — 7.62 mm MGs.

Comment: Contract signed with Damen Shipyards, Gorinchem for construction of three Damen Stan Patrol 4207 offshore patrol craft. *Trident* arrived Barbados on 2 March 2009. Steel hull with aluminium superstructure. Capable of carrying a 7 m RIB. Similar craft in service in Jamaica Coast Guard.

RUDYARD LEWIS   *11/2010, Shaun Jones* / 1406002

2 DAMEN STAN PATROL 1204 (PB)

ENTERPRISE P 05   EXCELLENCE P 06

Dimensions, metres (feet): 11.98 × 3.7 × 0.66 *(39.3 × 12.1 × 2.2)*
Speed, knots: 24
Complement: 4
Machinery: 2 Caterpillar C7 diesels; 740 hp *(550 kW)*; 2 Hamilton waterjets

Comment: Contract signed with Damen Shipyards, Gorinchem for construction of three Damen Stan Patrol 1204 patrol craft. Both commissioned on 13 September 2008. Aluminium hull with GRP superstructure.

EXCELLENCE   *2/2008, Damen Shipyards* / 1305301

1 DAUNTLESS CLASS (INSHORE PATROL CRAFT) (PB)

Name	No	Builders	Commissioned
ENDEAVOUR	P 04	SeaArk Marine, Monticello	5 Jun 1999

Displacement, tonnes: 11 full load
Dimensions, metres (feet): 12.2 × 4.3 × 1.3 *(40 × 14.1 × 4.3)*
Speed, knots: 27
Range, n miles: 600 at 18 kt
Complement: 4
Machinery: 2 Caterpillar 3208TA diesels; 870 hp *(650 kW)*; 2 shafts
Guns: 1 — 12.7 mm MG. 1 — 7.62 mm MG.
Radars: Surface search: Raytheon R40; I-band.

Comment: Aluminium construction. Underwent refit in 2009. Sister ship *Excellence* has been decommissioned.

ENDEAVOUR   *11/2006, M Declerck* / 1167124

Belgium

Country Overview

The Kingdom of Belgium is situated in north-western Europe. With an area of 11,787 square miles, it is bordered to the north by the Netherlands and to the south by France. It has a 35 n mile coastline with the North Sea. The capital and largest city is Brussels while the principal port is Antwerp which is accessible via the Schelde and Meuse estuaries, which lie within the Netherlands. Antwerp is also connected to an extensive canal system. Territorial seas (12 n miles) are claimed and an EEZ has also been claimed.

Headquarters Appointments

Commander, Maritime Command:
  Rear Admiral Michel Hofman
Deputy Commander, Maritime Command:
  Captain Edwin Van Den Haute

Personnel

(a) 2014: 2,127
(b) Voluntary service

Bases

Zeebrugge: Frigates, MCMV, Reserve Units, Training Ships, Logistics, Diving Centre. Mine Warfare Operational Sea Test centre (MOST).

Oostende: Belgium-Netherlands Mine-warfare school (EGUERMIN).
Koksijde: Naval aviation.
Brugge: Naval training centre.

Fleet Disposition

Operational control of Belgian and Netherlands surface forces is under Admiral Benelux Command at Den Helder.

Frigates < **Belgium** 63

FRIGATES

2 M CLASS (FFGHM)

Name	No	Builders	Laid down	Launched	Commissioned	Recommissioned
LEOPOLD I (ex-*Karel Doorman*)	F 930 (ex-F 827)	Koninklijke Maatschappij De Schelde, Flushing	26 Feb 1985	20 Apr 1988	31 May 1991	26 Mar 2007
LOUISE-MARIE (ex-*Willem Van Der Zaan*)	F 931 (ex-F 829)	Koninklijke Maatschappij De Schelde, Flushing	6 Nov 1985	21 Jan 1989	28 Nov 1991	4 Apr 2008

Displacement, tonnes: 3,373 full load
Dimensions, metres (feet): 123.8 oa; 114.2 wl × 14.4 × 4.3 *(406.2; 374.7 × 47.2 × 14.1)*
Flight deck, metres: 24.5 × 14.4 *(80.4 × 47.2)*
Speed, knots: 30; 21 diesel
Range, n miles: 5,000 at 18 kt
Complement: 149 (15 officers) + 31 spare berths

Machinery: CODOG; 2 RR Spey SM1C; 33,800 hp *(25.2 MW)* sustained; 2 Stork-Wärtsilä 12SW280 diesels; 9,790 hp(m) *(7.2 MW)* sustained; 2 shafts; LIPS cp props
Missiles: SSM: 8 McDonnell Douglas Harpoon Block 1C (2 quad) launchers ❶; active radar homing to 124 km *(67 n miles)* at 0.9 Mach; warhead 227 kg.
SAM: Raytheon Sea Sparrow RIM 7P Mk 48 vertical launchers ❷; semi-active radar homing to 16 km *(8.5 n miles)* at 2.5 Mach; warhead 38 kg; 16 missiles. Canisters mounted on port side of hangar.
Guns: 1—3 in *(76 mm)*/62 Oto Melara compact Mk 100 ❸; 100 rds/min to 16 km *(8.6 n miles)* anti-surface; 12 km *(6.5 n miles)* anti-aircraft; weight of shell 6 kg. 1 Signaal SGE-30 Goalkeeper with General Electric 30 mm 7-barrelled ❹; 4,200 rds/min combined to 2 km. 2 Oerlikon 20 mm; 800 rds/min to 2 km.
Torpedoes: 4—324 mm US Mk 32 Mod 9 (2 twin) tubes (mounted inside the after superstructure) ❺. Honeywell Mk 46 Mod 5; anti-submarine; active/passive homing to 11 km *(5.9 n miles)* at 40 kt; warhead 44 kg.
Physical countermeasures: Decoys: 2 Loral Hycor SRBOC 6-tubed fixed Mk 36 quad launchers; IR flares and chaff to 4 km *(2.2 n miles)*. SLQ-25 Nixie towed torpedo decoy.
Electronic countermeasures: ESM/ECM: Argo APECS II (includes AR 700 ESM); intercept and jammers.
Radars: Air/surface search: Thales SMART-S ❻; 3D; F-band.
Air search: Thales LW08 ❼; D-band.
Surface search: Thales Scout ❽; I-band. Thales Seastar ❾; I/J-band.
Navigation: Racal Decca 1226; I-band.
Fire control: 2 Thales STIR ❿; I/J/K-band; range 140 km *(76 n miles)* for 1 m² target.
Sonars: Signaal PHS-36; hull-mounted; active search and attack; medium frequency. TNO IRLFAS (fitted for); towed array; active/passive low frequency.
Combat data systems: SEWACO XI; Link 16/22. SATCOM ⓫. WSC-6 twin aerials.
Electro-optic systems: Thales Gatekeeper ⓬.

Helicopters: 1 NH90/Alouette III ⓭.

Programmes: The purchase of two ex-Netherlands frigates was approved by Belgium's Council of Ministers on 20 July 2005 and a contract for their supply, a support package, weapons transfer, joint upgrades and crew training was signed on 21 December 2005.

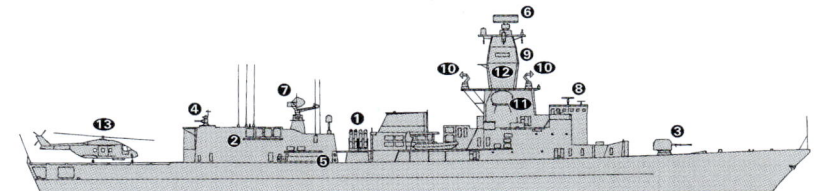

LEOPOLD I
(Scale 1 : 1,200), Ian Sturton / 1455742

LEOPOLD I
9/2013*, M Declerck / 1529107

Modernisation: Modification of flight decks and hangars to operate the NH90 helicopter has been completed in both ships. In conjunction with the Netherlands programme, both ships are undergoing a two-year mid-life modernisation period. Upgrades include replacement of the combat data system by SEWACO XI, addition of Thales Seastar radar, provision for installation of a low-frequency active sonar, a redesigned mast including Gatekeeper electro-optical surveillance system, installation of Link 16/22 and upgrade of navigation systems. Platform management systems are also being upgraded. Refit of *Leopold I* was completed in September 2013 and is to be completed in *Louise-Marie* in 2015.
Structure: The VLS SAM is similar to Canadian Halifax and Greek MEKO classes. The ship is designed to reduce radar and IR signatures and has extensive CBRN arrangements. Full automation and roll stabilisation fitted.

LEOPOLD I
9/2013*, M Declerck / 1529106

LEOPOLD I
9/2013*, M Declerck / 1529108

© 2014 IHS
IHS Jane's Fighting Ships 2014-2015

Belgium > Shipborne aircraft — Mine warfare forces

SHIPBORNE AIRCRAFT

Numbers/Type: 3 Aerospatiale SA 316B Alouette III.
Operational speed: 113 kt *(210 km/h)*.
Service ceiling: 10,500 ft *(3,200 m)*.
Range: 290 n miles *(540 km)*.
Role/Weapon systems: CG helicopter; used for close-range search and rescue and support for commando forces. Weapons: 1 — 7.62 mm MG.

ALOUETTE III   7/2008, Maritime Photographic / 1335242

Numbers/Type: 4 NHIndustries NH90.
Operational speed: 157 kt *(291 km/h)*.
Service ceiling: 13,940 ft *(4,250 m)*.
Range: 621 n miles *(1,150 km)*.
Role/Weapon systems: Four NFH shipborne aircraft, also to be used as SAR helicopters, the first of which entered service in August 2013. The aircraft are to be equipped with ISAR, FLIR, ESM and Link 11.

NH90   6/2001, NHIndustries / 0094462

LAND-BASED MARITIME AIRCRAFT

Numbers/Type: 4 Westland Sea King Mk 48.
Operational speed: 140 kt *(260 km/h)*.
Service ceiling: 10,500 ft *(3,200 m)*.
Range: 630 n miles *(1,165 km)*.
Role/Weapon systems: SAR helicopter; operated by air force; used for SAR tasks. Upgraded in 1995 with new radar, FLIR and GPS. One decommissioned in 2005. Sensors: Bendix RDR 1500B search radar. FLIR 2000F. Weapons: Unarmed.

SEA KING   6/2008, Michael Nitz / 1335241

PATROL FORCES

Notes: (1) Three 7 m RIC were acquired in May 1994 from RIBTEC, Swanwick.
(2) Two range safety craft, *Werl* A 998 and *Spich* A 995, have replaced the hovercraft *Barbara* A 999.

A 998   6/2005, M Declerck / 1151239

0 + 2 OFFSHORE PATROL SHIPS (PBO)

Name	No	Builders	Laid down	Launched	Commissioned
CASTOR	P 901	Socarenam, Boulogne	2013	2014	Jun 2014
POLLUX	P 902	Socarenam, Boulogne	2013	2014	Jan 2015

Displacement, tonnes: 448 full load
Dimensions, metres (feet): 52 × 9.3 × 3.38 *(170.6 × 30.5 × 11.1)*
Speed, knots: 21
Range, n miles: 2,000 at 10 kt
Complement: 12 + 18 spare berths

Machinery: 2 MTU 16V 4000 diesels; 7,724 hp *(5.76 MW)*; 2 shafts
Guns: 1 — 12.7 mm remotely controlled MG.
Radars: Surface search: To be announced.
Navigation: To be announced.
Combat data systems: Thales SEWACO.

Comment: A contract with Socarenam, Boulogne, for the construction of two Ready Duty Ships was placed in December 2012. The design, which includes space for two RHIBs, is based on *Kermovan* built for the French Customs service. The new ships are to replace *Sterne*, *Valcke* and *Albatros*.

MINE WARFARE FORCES

6 FLOWER CLASS (TRIPARTITE)
(MINEHUNTERS — COASTAL) (MHC/AEL)

Name	No	Builders	Launched	Commissioned
ASTER	M 915	Beliard, Ostend	6 Jun 1985	17 Dec 1985
BELLIS	M 916	Beliard, Ostend	14 Feb 1986	14 Aug 1986
CROCUS	M 917	Beliard, Ostend	6 Aug 1986	5 Feb 1987
LOBELIA	M 921	Beliard, Ostend	6 Jan 1988	9 May 1989
NARCIS	M 923	Beliard, Ostend	30 Mar 1990	27 Sep 1990
PRIMULA	M 924	Beliard, Ostend	17 Dec 1990	29 May 1991

Displacement, tonnes: 630 standard; 660 full load
Dimensions, metres (feet): 51.5 × 8.9 × 2.5 *(169 × 29.2 × 8.2)*
Speed, knots: 15. **Range, n miles:** 3,000 at 12 kt
Complement: 38 (4 officers; 4 divers; 1 medical personnel)

Machinery: 1 Stork Wärtsilä A-RUB 215W-12 diesel; 1,860 hp(m) *(1.37 MW)* sustained; 1 shaft; LIPS cp prop; 2 motors; 240 hp(m) *(176 kW)*; 2 active rudders; 2 bow thrusters (port and starboard)
Guns: 1 DCN 20 mm/20; 720 rds/min to 10 km *(5.5 n miles)*. 1 — 12.7 mm MG.
Physical countermeasures: MCM: Atlas Seafox MIDS.
Radars: Navigation: Consilium Selesmar T250; I-band.
Sonars: Thales TSM 2022 Mk III; hull-mounted; active minehunting; 100, 200, and 400 kHz and Double Eagle Mk III Mod 1 variable depth sonar.
Combat data systems: Atlas Elektronic IMCMS.

Programmes: Developed in co-operation with France and the Netherlands. A 'ship factory' for the hulls was built at Ostend and the hulls were towed to Rupelmonde for fitting out. Each country built its own hulls but France provided all MCM gear and electronics, Belgium electrical installation and the Netherlands the engine room equipment.
Modernisation: Propulsion system upgrade completed in 1999 for all of the class. Capability upgrade to extend service life of six ships to 2020 is in progress at Zeebrugge. Modifications include an MCM command and control system, an Integrated Mine Countermeasures System (comprising hull-mounted and self-propelled variable-depth sonar (installed in Double Eagle Mk III Mod 1 ROV)) and a Mine-Identification and Disposal System (MIDS) based on the STN Atlas Seafox. Linked to the ship by a 3,000 m fibre optic tether, one variant (Seafox-C) is used for mine disposal and another (Seafox-I) is used for identification. The equipment was first installed in HrMS *Hellevoetsluis*. Completion dates for capability upgrades were: *Primula* (February 2006), *Aster* (October 2006), *Lobelia* (October 2007), *Bellis* (February 2008), *Narcis* (October 2008), and *Crocus* (February 2009).
Structure: GRP hull fitted with active tank stabilisation, full CBRN protection and air conditioning. Automatic pilot integrated with IMCMS.
Operational: A 5 ton container can be carried, stored for varying tasks-HQ support, research, patrol, extended diving, drone control. The ship's company includes four divers and one medical assistant. All of the class are based at Zeebrugge.
Sales: Three of the class paid off for sale in July 1993 and were bought by France in 1997. *Myosotis* sold to Bulgaria in 2009.

PRIMULA   7/2013*, Michael Nitz / 1528998

BELLIS   7/2013*, A A de Kruijf / 1528997

IHS Jane's Fighting Ships 2014-2015   © 2014 IHS

SURVEY SHIPS

Notes: In addition to *Belgica* there are five small civilian manned survey craft: *Ter Streep, Scheldewacht II, De Parel II, Veremans* and *Prosper*.

1 SURVEY SHIP (AGOR/PBO)

Name	No	Builders	Laid down	Launched	Commissioned
BELGICA	A 962	Boelwerf, Temse	17 Oct 1983	6 Jan 1984	5 Jul 1984

Displacement, tonnes: 1,102 full load
Dimensions, metres (feet): 51.1 × 10 × 4.4 *(167.7 × 32.8 × 14.4)*
Speed, knots: 13.5. **Range, n miles:** 5,000 at 12 kt
Complement: 26 + 11 personnel
Machinery: 1 ABC 6M DZC diesel; 1,600 hp(m) *(1.18 MW)* sustained; 1 Kort nozzle prop
Radars: Navigation: Racal Decca 1229; I-band.

Comment: Ordered 1 December 1982. Laid down 17 October 1983. Used for hydrography, oceanography, meteorology and fishery control. Marisat fitted. Equipped with Kongsberg EM1002 and EM3002 multibeam sonars. Based at Zeebrugge.

BELGICA 4/2010, M Declerck / 1406004

TRAINING SHIPS

1 SAIL TRAINING VESSEL (AXS)

Name	No	Builders	Laid down	Launched	Commissioned
ZENOBE GRAMME	A 958	Boel and Zonen, Temse	7 Oct 1960	23 Oct 1961	27 Dec 1961

Displacement, tonnes: 151 full load
Dimensions, metres (feet): 28 × 6.8 × 2.1 *(91.9 × 22.3 × 6.9)*
Speed, knots: 8
Complement: 17 (1 officer)
Machinery: 1 MWM diesel; 200 hp(m) *(147 kW)*; 1 shaft
Radars: Navigation: Racal Decca; I-band.

Comment: Auxiliary sail ketch. Laid down 7 October 1960 and launched 23 October 1961. Designed for scientific research but now only used as a training ship.

ZENOBE GRAMME 7/2007, Adolfo Ortigueira Gil / 1167854

AUXILIARIES

Notes: Plans to acquire a Command and Support Ship (MCS) to replace BNS *Godetia* in about 2018 are under review.

1 SUPPORT SHIP (AGFH)

Name	No	Builders	Commissioned
STERN (ex-*KBV 171*)	A 963	Karlskronavarvet	3 Sep 1980

Displacement, tonnes: 381 full load
Dimensions, metres (feet): 50 × 8.5 × 2.4 *(164 × 27.9 × 7.9)*
Speed, knots: 18. **Range, n miles:** 3,000 at 12 kt
Complement: 13 (1 officer)
Machinery: 2 Hedemora V16A diesels; 4,480 hp(m) *(3.28 MW)* sustained; 2 shafts; cp props
Radars: Navigation: 2 Kelvin Hughes; E/F- and I-band.
Helicopters: Platform for 1 light.

Comment: Transferred from Swedish Coast Guard on 6 October 1998. GRP hull indentical to Landsort class. The vessel serves as a Ready Duty Ship, fishery protection ship and performs SAR duties.

STERN 5/2013*, Derek Fox / 1529086

1 COMMAND AND SUPPORT SHIP (AGFH)

Name	No	Builders	Laid down	Launched	Commissioned
GODETIA	A 960	Boelwerf, Temse	15 Feb 1965	7 Dec 1965	23 May 1966

Displacement, tonnes: 2,032 standard; 2,296 full load
Dimensions, metres (feet): 91.8 × 14 × 3.5 *(301.2 × 45.9 × 11.5)*
Speed, knots: 19. **Range, n miles:** 8,700 at 12.5 kt
Complement: 92 (8 officers)

Machinery: 4 ACEC-MAN diesels; 5,400 hp(m) *(3.97 MW)*; 2 shafts; cp props
Guns: 4 — 12.7 mm MGs.
Radars: Surface search: Racal Decca 1229; I-band.
Helicopters: 1 Alouette III.

Comment: Laid down 15 February 1965. Rated as Command and Logistic Support Ship. Refit (1979–80) and mid-life conversion (1981–82) included helicopter deck and replacement cranes. Refitted in 1992 and again in 2006. Plans to instal a mine-avoidance sonar have been cancelled. Minesweeping cables fitted either side of helo deck have been removed. Replacement plans are under review.

GODETIA 7/2013*, Michael Nitz / 1528999

TUGS

3 HARBOUR TUGS (YTL)

WESP A 952	ZEEMEEUW A 954	MIER A 955

Displacement, tonnes: 198 full load
Dimensions, metres (feet): 26.23 × 7.5 × 3.25 *(86.1 × 24.6 × 10.7)*
Speed, knots: 11
Complement: 4
Machinery: 2 ABC 6 MDUS diesels; 1,000 hp *(746 kW)*

Comment: Details given are for A 952 and A 955. A 954 is 146 tons.

WESP 7/2008, Maritime Photographic / 1335236

ZEEMEEUW 10/2006, M Declerck / 1164726

66 Belgium > Tugs — Benin > Introduction

2 COASTAL TUGS (YTM)

Name	No	Launched
VALCKE (ex-Steenbank; ex-Astroloog)	A 950	1960
ALBATROS (ex-Westgat)	A 996	1967

Displacement, tonnes: 186 full load
Dimensions, metres (feet): 30.4 × 7.6 × 3.6 (99.7 × 24.9 × 11.8)
Speed, knots: 11
Complement: 8
Machinery: Diesel-electric; 2 Deutz diesel generators; 1,240 hp(m) (911 kW); 1 shaft; 1 bow thruster

Comment: Known as Ready Duty Ships. Details given are for A 950 which was launched in 1960. A 996 is 206 tons and was launched in 1967.

VALCKE 10/2006, M Declerck / 1164728

ALBATROS 5/2008, A A de Kruijf / 1335237

Belize

Country Overview

Formerly known as British Honduras, Belize became an independent state in 1981. The British monarch, represented by a governor-general, is head of state. With an area of 8,867 square miles, it has borders with Mexico to the north and Guatemala to the west; its 208 n mile coastline is on the Caribbean Sea and fringed by numerous coral barrier reefs and cays. The capital city is Belmopan while the largest city and major port is Belize City. Territorial seas (12 n miles) are claimed. A 200 n mile Exclusive Economic Zone (EEZ) has been claimed but the limits are not defined. The Belize National Coast Guard Service was formed on 29 November 2005.

Headquarters Appointments

Commander of the Coast Guard:
Rear Admiral John Borland

Personnel

(a) 2014: 152 (2 officers)
(b) The Maritime Wing of the Belize Defence Force comprises volunteers from the Army.

Bases

Belizean Beach (HQ), Hunting Cay, Calabash Cay (Turneffe Atoll)

Maritime Patrol

Two Air Force operated Pilatus Britten-Norman Defenders are used for maritime surveillance.

PATROL FORCES

Notes: Current assets include:
1. Two Halmatic 22 ft RIBs with twin Yamaha 115 hp outboards. Names *Stingray Commando* and *Blue Marlin Ranger*.
2. Two Pelikan 35 ft craft with twin Yamaha 200 hp outboards. Built at Bradleys Boatyard in 1996 and called *Ocean Sentinel* and *Reef Sniper*.
3. Six Colombian 32 ft skiffs with twin Yamaha 200 hp outboards, confiscated and commissioned in service 1995–97.
4. One 36 ft skiff.
5. Two US donated craft Stinger I and Stinger II and two 11 m Boston Whalers (2010). Two 11 m Boston 370 Outrage and two 7.5 m SAFE Boats Defender were donated in 2011.
6. An 8 m radar-equipped RHIB was donated in January 2009.

DEFENDER CLASS 11/2011*, US Southcom 1525512

Benin

FORCES NAVALES

Country Overview

Formerly part of French West Africa, the republic gained full independence in 1960 as the Republic of Dahomey; it was renamed The Republic of Benin in 1975. With an area of 43,484 square miles it has borders to the east with Nigeria and to the west with Togo. Benin has a short coastline of 65 n miles with the Gulf of Guinea. The capital is Porto-Novo while Cotonou is the largest city and principal port. Benin has not claimed an Exclusive Economic Zone (EEZ) but is one of a few coastal states which claims a 200 n mile territorial sea. The naval force was established in 1978.

Headquarters Appointments

Commander of the Navy: Captain Maxime Ahoyo

Aircraft

1 Boeing 727, 2 BAE 748, 1 DHC-6 Twin Otter, 4 Agusta 109 helicopters, 1 Eurocopter AS 350B helicopter.

Bases

Cotonou
Grand-popo

Personnel

2014: 600 (50 officers)

IHS Jane's Fighting Ships 2014-2015
© 2014 IHS

PATROL FORCES

Notes: There are four river patrol craft of South Korean origin and two US-built Boston Whalers (P 14, P 15).

2 CHINESE 27 METRE CLASS (PATROL CRAFT) (PB)

| MATELOT BRICE KPOMASSE 798 | LA SOTA 799 |

Displacement, tonnes: 81 full load
Dimensions, metres (feet): 27 × 4.1 × 1.4 (88.6 × 13.5 × 4.6)
Complement: 13
Machinery: 2 diesels; 1,000 hp (746 kW)
Guns: 4—14.5 mm (2 twin) MGs.
Radars: Navigation: I-band.

Comment: Understood to have been transferred from China in 2000. A similar craft is in service in Cape Verde.

3 OCEA FPB 98 CLASS (PATROL CRAFT) (PB)

| ALIBORI P 109 | OUEME P 110 | ZOU P 111 |

Displacement, tonnes: 116 full load
Dimensions, metres (feet): 35.2 × 6.8 × 1.2 (115.5 × 22.3 × 3.9)
Speed, knots: 30. **Range, n miles:** 1,000 at 12 kt
Complement: 12 (3 officers)
Machinery: 2 Caterpillar diesels; 3,586 hp (2.6 MW); 2 Kamewa waterjets
Guns: 1—20 mm. 2—12.7 mm MGs.
Radars: Navigation: Furuno FAR 2117; I-band.

Comment: The contract with OCEA for the construction of three patrol craft was announced in March 2011. Delivery of all three vessels was made in 2012. Details similar to craft of the same class in Algerian service.

ALIBORI — 4/2012, B Prézelin / 1455682

2 DEFENDER 27 CLASS (PB)

018   019

Displacement, tonnes: 2.7 standard
Dimensions, metres (feet): 7.62 × 2.6 × 1.0 (25 × 8.5 × 3.3)
Speed, knots: 35
Complement: 2
Machinery: 2 Mercury outboard motors; 500 hp
Guns: 2—12.7 mm MGs.
Radars: Navigation: Furuno; I-band.

Comment: Donated by the US government in 2010.

KPOMASSE and SOTA — 2001, Benin Navy / 0114348

Bermuda

Country Overview

A British self-governing dependency, a Governor, appointed by the British Crown, is responsible for external affairs, internal security, defence, and the police. Situated in the north Atlantic Ocean some 650 n miles southeast of Cape Hatteras, the country consists of six principal islands, of which the largest is 14 miles long, linked by bridges and a causeway; there are some 150 other small islands, islets, and rocks, of which about 20 are inhabited. Hamilton is the capital, chief port and largest town. Territorial seas (12 n miles) and an Exclusive Economic Zone (EEZ) (200 n miles) are claimed.

Headquarters Appointments

Commanding Officer: Inspector Philip Lewis

Bases

Hamilton

POLICE

Notes: In addition to patrol craft, three tugs, *Powerful*, *Faithful* and *Refit* are operated by the Department of Marine and Port Services.

1 AUSTAL PATROL CRAFT (PB)

GUARDIAN

Dimensions, metres (feet): 16.3 × 4.9 × 1.2 (53.5 × 16.1 × 3.9)
Speed, knots: 28. **Range, n miles:** 400 at 25 kt
Complement: 3 + 8 spare berths
Machinery: 2 Caterpillar C12 diesels; 1,300 hp (970 kW); 2 shafts

Comment: Contract with Austal Ships in August 2005 to build aluminium hull craft for operations up to 200 n miles from shore. Similar to craft operated by the New South Wales police. Delivery was made on 25 September 2006.

HERON II — 6/1997, Bermuda Police / 0012079

2 SAR CRAFT (SAR)

RESCUE I   RESCUE II

Comment: *Rescue I* replaced the craft of the same name in November 1998 and *Rescue II* replaced the craft of the same name in 2001. Both are Halmatic 24 ft Arctic RIBs with twin 200 hp Yamaha outboards and a complement of three.

GUARDIAN — 9/2006, Austal / 1335333

4 PATROL CRAFT (PBI)

HERON I–IV

Comment: *Heron I*, delivered in July 1997 to replace the previous craft of the same name, and *Heron III* delivered in June 1992 are 22 ft Boston Whalers fitted with twin Yamaha 225 hp and twin Yamaha 115 hp outboards, respectively. *Heron II* delivered in August 1996 to replace the previous craft of the same name, is a 27 ft Boston Whaler with twin Yamaha 250 hp(m) outboard engines. *Heron IV*, delivered in 2001, is a further 22 ft Boston Whaler with twin 115 hp outboards.

HALMATIC ARCTIC RIB — 2001, Bermuda Police / 0109933

Brazil > Introduction — Submarines

PENNANT LIST

Submarines		**Corvettes**		P 14	Penedo	P 74	Miramar (bldg)	H 41	Almirante Maximiano
				P 15	Poti	P 75	Magé (bldg)	H 44	Ary Rongel
S 30	Tupi	V 30	Inhaúma	P 20	Pedro Teixeira	P 76	Marogogipe (bldg)	U 14	Aspirante Moura
S 31	Tamoio	V 31	Jaceguai	P 21	Raposo Tavares				
S 32	Timbira	V 32	Julio de Noronha	P 30	Roraima	**Mine Warfare Forces**		**Auxiliaries**	
S 33	Tapajó	V 33	Frontin	P 31	Rondônia				
S 34	Tikuna	V 34	Barroso	P 32	Amapá	M 15	Aratú	G 15	Paraguassú
				P 40	Grajaú	M 16	Anhatomirim	G 17	Potengi
		Amphibious Forces		P 41	Guaiba	M 17	Atalaia	G 23	Almirante Gastao Motta
Aircraft Carriers				P 42	Graúna	M 18	Araçatuba	G 27	Marajo
		G 25	Almirante Saboía	P 43	Goiana	M 19	Abrolhos	K 11	Felinto Perry
A 12	São Paulo	G 28	Mattoso Maia	P 44	Guajará	M 20	Albardão	R 21	Tritão
		G 29	Garcia d'Avila	P 45	Guaporé			R 22	Tridente
		G 30	Ceará	P 46	Gurupá	**Survey Ships and Tenders**		R 23	Triunfo
		G 31	Rio de Janeiro	P 47	Gurupi			R 24	Almirante Guilhem
		GED 10	Guarapari	P 48	Guanabara	H 18	Comandante Varella	R 25	Almirante Guillobel
Destroyers/Frigates		GED 11	Tambaú	P 49	Guarujá	H 19	Tenente Castelo	U 10	Aspirante Nascimento
		GED 12	Camboriú	P 50	Guaratuba	H 20	Comandante Manhães	U 11	Guarda Marinha Jansen
F 40	Niteroi			P 51	Gravataí	H 21	Sirius	U 12	Guarda Marinha Brito
F 41	Defensora	**Patrol Forces**		P 60	Bracui	H 25	Tenente Boanerges	U 15	Pará
F 42	Constituição			P 61	Benevente	H 26	Faroleiro Mário Seixas	U 16	Doutor Montenegro
F 43	Liberal	V 15	Imperial Marinheiro	P 62	Bocaina	H 34	Almirante Graça Aranha	U 17	Parnaiba
F 44	Independência	V 19	Caboclo	P 63	Babitonga	H 35	Amorim do Valle	U 18	Oswaldo Cruz
F 45	União	P 10	Piratini	P 70	Macaé	H 36	Taurus	U 19	Carlos Chagas
F 46	Greenhalgh	P 11	Pirajá	P 71	Macau	H 37	Garnier Sampaio	U 20	Cisne Branco
F 48	Bosisio	P 12	Pampeiro	P 72	Maracaná	H 38	Cruzeiro do Sul	U 27	Brasil
F 49	Rademaker	P 13	Parati	P 73	Mangaratiba (bldg)	H 40	Antares	U 28	Teniente Maximiano
								U 29	Piraim

SUBMARINES

Notes: Following the revival by President Lula in June 2007 of plans to acquire a nuclear-powered submarine, the programme was formally re-launched on 26 September 2008 by the Commander of the Brazilian Navy. This was followed on 3 September 2009 by an agreement with the French government for DCNS to provide assistance on the non-nuclear aspects of the submarine's design. Construction of the first boat, which is expected to be a derivative of the Barracuda class design (approximately 100 m long and 6,000 tons), is to begin in 2016 at a new shipyard at Itaguai, Rio de Janeiro State. The main construction facility at this site was opened in November 2012. The submarine is to enter service in 2025 and there are plans for two further boats. The submarine, possibly to be named *Alvaro Alberto*, is to be equipped with DCNS SUBTICS combat system and is to be conventionally armed with missiles (probably Exocet SM 39) and torpedoes. Alongside the nuclear submarine programme, a nuclear reactor and two industrial plants are to be built at Aramar and Resende.

0 + 4 MODIFIED SCORPENE CLASS (SSK)

Name	No	Builders	Laid down	Launched	Commissioned
RIACHUELO	S 40	Itaguai Construçoes Navais, Sepetiba	16 Jul 2011	2015	2017
HUMAITA	S 41	Itaguai Construçoes Navais, Sepetiba	2013	2016	2018
TONELERO	S 42	Itaguai Construçoes Navais, Sepetiba	2015	2018	2020
ANGOSTURA	S 43	Itaguai Construçoes Navais, Sepetiba	2017	2019	2021

Displacement, tonnes: 1,709 surfaced; 1,870 dived
Dimensions, metres (feet): 75 × 6.2 × 5.8 *(246.1 × 20.3 × 19.0)*
Speed, knots: 11 surfaced; 20 dived
Complement: 31 (6 officers)

Machinery: Diesel-electric; 4 MTU 12V 396 diesels; 2,992 hp(m) *(2.4 MW)*; 1 Jeumont Schneider motor; 3,808 hp *(2.8 MW)*; 1 shaft
Missiles: SSM: MBDA Exocet SM-39.
Torpedoes: 6—21 in *(533 mm)* bow tubes. 18 WASS Black Shark torpedoes; wire (fibre-optic cable) guided; active/passive homing to 50 km *(27 n miles)* at 50 kt; warhead 250 kg.
Physical countermeasures: DCNS Contralto-S torpedo defence system.
Electronic countermeasures: ESM: To be announced.
Radars: Navigation: To be announced.
Sonars: Thales S-Cube sonar suite incorporating bow, flank and towed arrays.
Weapon control systems: UDS International SUBTICS.

Programmes: Following the Franco-Brazilian arms package agreement signed between the Presidents of France and Brazil on 12 February 2008, a contract for the construction of four new submarines was signed on 3 September 2009. The design is a stretched version of the Scorpene class. First steel was cut at Cherbourg on 27 May 2010 and at Itaguaí on 16 July 2011. The forward section (including sail) of *Riachuelo* were shipped from Cherbourg to Brazil in 2013. All sections of the other three boats are to be built in Brazil.
Structure: Diving depth 350 m.

1 TIKUNA (TYPE 209/1450) CLASS (SSK)

Name	No	Builders	Laid down	Launched	Commissioned
TIKUNA	S 34	Arsenal de Marinha, Rio de Janeiro	11 Jun 1996	9 Mar 2005	16 Dec 2005

Displacement, tonnes: 1,490 surfaced; 1,620 dived
Dimensions, metres (feet): 61 × 6.2 × 5.5 *(200.1 × 20.3 × 18.0)*
Speed, knots: 11 surfaced; 11 snorting; 22 dived
Range, n miles: 11,000 at 8 kt surfaced; 400 at 4 kt dived
Complement: 41 (8 officers)

Machinery: Diesel-electric; 4 MTU 12V 396SE84 diesels; 5,364 hp(m) *(4.0 MW)*; 4 Siemens alternators; 1 Siemens motor; 1 shaft
Torpedoes: 8—21 in *(533 mm)* bow tubes. Gould Mk 48 ADCAP Mod 6AT; active/passive homing to 50 km *(27 n miles)*/38 km *(21 n miles)* at 40/55 kt; warhead 267 kg; depth 900 m *(2,950 ft)*; 16 weapons.
Mines: 32 IPqM/Consub MCF-01/100 carried in lieu of torpedoes.
Electronic countermeasures: ESM: Argos AR-900; radar warning.
Radars: Navigation: Terma Scanter; I-band.
Sonars: Atlas Elektronik CSU-83/1; hull-mounted; passive/active search and attack; medium frequency. STN Atlas Elektronik FAS-3 flank array.
Weapon control systems: Lockheed Martin ICS.

Programmes: Planned intermediate stage between Tupi class and the first SSN. Designed by the Naval Engineering Directorate. Contract effective with HDW in October 1995. Plans for a second of class have been cancelled.
Modernisation: Tigerfish torpedoes are being replaced by Mk 48 Mod 6 and a Lockheed Martin integrated combat system AN/BYG-501 Mod 1D is to be installed. The upgrade is also likely to include new periscopes, radar and a new flank array. Work was to be completed by 2011.
Structure: Improved Tupi design similar to Turkish Gur class. Diving depth, 300 m *(985 ft)*. Very high-capacity batteries with GRP lead-acid cells by Microlite. More powerful engines than *Tupi*. Fitted with two Kollmorgen Mod 76 periscopes.
Operational: Endurance, 60 days. Sea trials began on 10 November 2005.

TIKUNA
10/2006, Brazilian Navy / 1170093

Submarines < **Brazil** 71

TIKUNA   6/2012, M Declerck / 1482967

4 TUPI (TYPE 209/1400) CLASS (SSK)

Name	No	Builders	Laid down	Launched	Commissioned
TUPI	S 30	Howaldtswerke-Deutsche Werft, Kiel	8 Mar 1985	28 Apr 1987	6 May 1989
TAMOIO	S 31	Arsenal de Marinha, Rio de Janeiro	15 Jul 1986	18 Nov 1993	17 Jul 1995
TIMBIRA	S 32	Arsenal de Marinha, Rio de Janeiro	15 Sep 1987	5 Jan 1996	16 Dec 1996
TAPAJÓ	S 33	Arsenal de Marinha, Rio de Janeiro	6 Aug 1992	5 Jun 1998	21 Dec 1999

Displacement, tonnes: 1,476 surfaced; 1,616 dived
Dimensions, metres (feet): 61.2 × 6.2 × 5.5 *(200.8 × 20.3 × 18.0)*
Speed, knots: 11 surfaced; 11 snorting; 21.5 dived
Range, n miles: 8,200 at 8 kt surfaced; 400 at 4 kt dived
Complement: 41 (8 officers)

Machinery: Diesel-electric; 4 MTU 12V 493 TY60 GA31L diesels; 3,200 hp(m) *(2.39 MW)*; 4 Siemens alternators; 1.7 MW; 1 Siemens motor; 4,600 hp(m) *(3.36 MW)* sustained; 1 shaft
Torpedoes: 8 – 21 in *(533 mm)* bow tubes. Gould Mk 48 ADCAP Mod 6AT; active/passive homing to 50 km *(27 n miles)*/38 km *(21 n miles)* at 40/55 kt; warhead 267 kg; depth 900 m *(2,950 ft)*; 16 weapons.

Electronic countermeasures: ESM: IPqM/Elebra Defensor ET/SLR-1X; radar intercept.
Radars: Navigation: Terma Scanter; I-band.
Sonars: Atlas Elektronik CSU-83/1; hull-mounted; passive/active search and attack; medium frequency. STN Atlas Elektronik FAS-3 flank array.
Weapon control systems: Lockheed Martin ICS.

Programmes: Contract signed with Howaldtswerke in February 1984. Financial negotiations were completed with the West German Government in October 1984. Original plans included building four in Brazil followed by two improved Tupis for a total of six. In the end only three were constructed in Brazil.

Modernisation: A programme (Mod Sub) to upgrade auxiliary machinery, sonars, weapon control, countermeasures and navigation systems was announced in 2003. Refit work on *Tamoio* was completed in June 2005, while work on *Timbira* was completed in January 2007. Tigerfish torpedoes have been replaced by Mk 48 Mod 6 and a Lockheed Martin integrated combat system AN/BYG-501 Mod 1D has been installed. The upgrade is also likely to have included new periscopes, radar and a new flank array. Work on *Tapajó* began in 2009 and was completed in January 2010. Work on *Tupi* began in mid-2012.
Structure: Hull constructed of HY 80 steel. Single hull. Diving depth, 250 m *(820 ft)*. Equipped with Sperry Mk 29 Mod 3 SINS and two Kollmorgen Mod 76 periscopes.
Operational: Based at Niteroi, Rio de Janeiro.

TAPAJÓ   10/2005, Mario R V Carneiro / 1153025

TIMBIRA   11/2013*, Marco Ghiglino / 1529110

© 2014 IHS   IHS Jane's Fighting Ships 2014-2015

AIRCRAFT CARRIERS

1 CLEMENCEAU CLASS (CVM)

Name	No	Builders	Laid down	Launched	Commissioned
SÃO PAULO (ex-*Foch*)	A 12 (ex-R 99)	Chantiers de l'Atlantique, St. Nazaire	15 Feb 1957	23 Jul 1960	15 Jul 1963

Displacement, tonnes: 27,745 standard; 34,213 full load
Dimensions, metres (feet): 265 oa; 238 pp × 51.2 oa; 31.7 hull × 8.6 *(869.4; 780.8 × 168.0; 104.0 × 28.2)*
Flight deck, metres: 259 × 47 *(849.7 × 154.2)*
Aircraft lift: 2; 16 × 10.97 m
Speed, knots: 30
Range, n miles: 7,000 at 18 kt, 4,800 at 24 kt, 3,500 at 30 kt
Complement: 1,578 (160 officers; 358 air crew)

Machinery: 6 La Valle boilers; 640 psi *(45 kg/cm²)*; 840°F *(450°C)*; 2 GEC Alsthom turbines; 126,000 hp(m) *(93 MW)*; 2 shafts
Missiles: SAM: 2 Matra Sadral; Mistral missiles; IR homing to 4 km *(2.2 n miles)* at 2.5 Mach; warhead 3 kg.
Guns: 5 — 12.7 mm MGs.
Physical countermeasures: Decoys: 2 CSEE AMBL 2A Sagai (10 barrelled trainable launchers); chaff and IR flares.

Radars: Air search: Thomson-CSF DRBV 23B ❶; D-band.
Air/surface search: Thomson-CSF DRBV 15A ❷; E/F-band.
Heightfinder: 2 DRBI 10 ❸; E/F-band.
Navigation: Racal Decca 1226; I-band.
Fire control: 2 Thomson-CSF DRBC 32C.
Tacan: NRBP-2B.
Landing approach control: NRBA 51 ❹; I-band.
Combat data systems: IPqM/Elebra SICONTA Mk 4 tactical system; Links YB and 14. Inmarsat.
Weapon control systems: 2 Sagem DMa optical directors.

Fixed-wing aircraft: 10–15 A-4 Skyhawks. 2 Tracker/Trader.
Helicopters: 3 Agusta SH-3A/D Sea Kings; 3 Aerospatiale UH-12/13; 2 UH-14 Cougar.

Programmes: Acquired from France on 15 November 2000 and following modifications in Brest, arrived in Brazil in February 2001.

Modernisation: The jet deflectors are enlarged (this implies reducing the area of the forward lift). Crotale and Sadral systems disembarked before transfer. Refit in 2003 included re-tubing of boilers and refurbishment of catapults. A further refit 2005–10 included a full machinery overhaul, flight deck modifications, and the installation of two twin Matra SAM. The combat data system is to be upgraded to SICONTA Mk 5.
Structure: Flight deck, island superstructure and bridges, hull (over machinery spaces and magazines) are all armour plated. There are three bridges: Flag, Command and Aviation. Two Mitchell-Brown steam catapults; Mk BS 5; able to launch 20 ton aircraft at 110 kt. The flight deck is angled at 8°. Two lifts 52.5 × 36 ft *(16 × 10.97 m)* one of which is on the starboard deck edge. Dimensions of the hangar are 590.6 × 78.7 × 23 ft *(180 × 24 × 7 m)*.
Operational: Oil fuel capacity is 3,720 tons. Service life 2025.

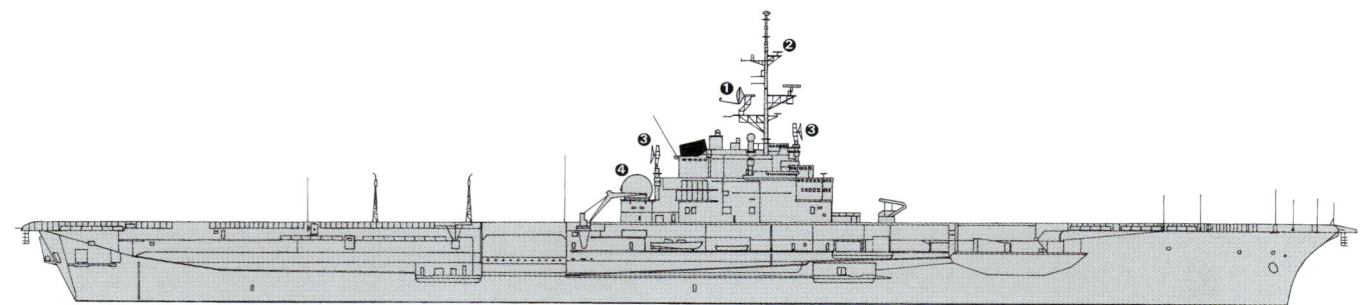

SÃO PAULO *(Scale 1 : 1,500), Ian Sturton / 0130381*

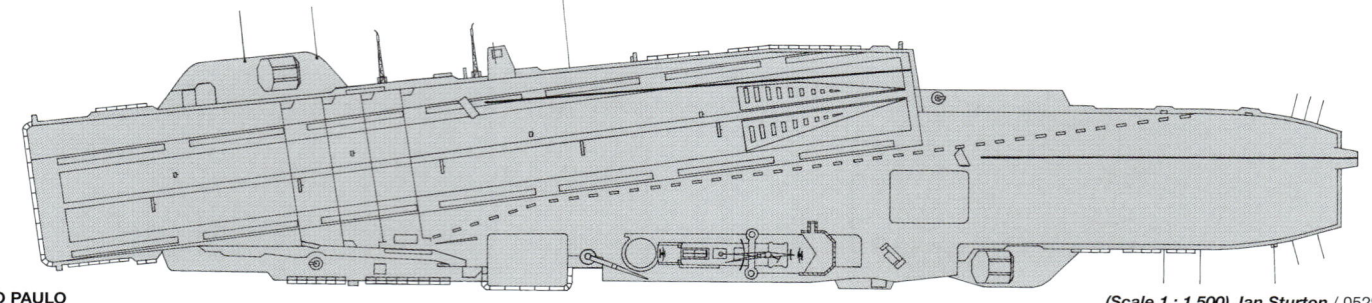

SÃO PAULO *(Scale 1 : 1,500), Ian Sturton / 0529159*

SÃO PAULO *2/2001, Mario R V Carneiro / 0059752*

Aircraft carriers < **Brazil** 73

SÃO PAULO 9/2003, S C Neto/Mario R V Carneiro / 0569158

SÃO PAULO 9/2011, Mario R V Carneiro / 1406703

SÃO PAULO 9/2007, Mario R V Carneiro / 1335447

74 Brazil > Frigates

FRIGATES

Notes: Acquisition of up to five new frigates under the PROSUPER programme is under consideration. The French and Italian variants of FREMM are potential contenders and proposals from UK and Germany have also been made. If funding proves to be an obstacle, procurement of second-hand vessels may be taken forward as a stop-gap solution.

3 BROADSWORD CLASS (TYPE 22) (FFGHM)

Name	No	Builders	Laid down	Launched	Commissioned	Recommissioned
GREENHALGH (ex-*Broadsword*)	F 46 (ex-F 88)	Yarrow Shipbuilders, Glasgow	7 Feb 1975	12 May 1976	3 May 1979	30 Jun 1995
BOSISIO (ex-*Brazen*)	F 48 (ex-F 91)	Yarrow Shipbuilders, Glasgow	18 Aug 1978	4 Mar 1980	2 Jul 1982	31 Aug 1996
RADEMAKER (ex-*Battleaxe*)	F 49 (ex-F 89)	Yarrow Shipbuilders, Glasgow	4 Feb 1976	18 May 1977	28 Mar 1980	30 Apr 1997

Displacement, tonnes: 3,556 standard; 4,807 full load
Dimensions, metres (feet): 131.2 oa; 125 wl × 14.8 × 6 (430.4; 410.1 × 48.6 × 19.7)
Speed, knots: 30; 18 Tynes
Range, n miles: 4,500 at 18 kt
Complement: 273 (17 officers)

Machinery: COGOG; 2 RR Olympus TM3B gas turbines; 50,000 hp (37.3 MW) sustained; 2 RR Tyne RM1C gas turbines; 9,900 hp (7.4 MW) sustained; 2 shafts; cp props
Missiles: SSM: 4 Aerospatiale MM 40 Block II Exocet ❶; inertial cruise; active radar homing to 70 km (42 n miles) at 0.9 Mach; warhead 165 kg; sea-skimmer.
SAM: 2 British Aerospace 6-barrelled Seawolf GWS 25 Mod 4 ❷; command line of sight (CLOS) TV/radar tracking to 5 km (2.7 n miles) at 2+ Mach; warhead 14 kg; 32 rounds.
Guns: 2 Bofors SAK 40 mm/L 70–350 A-3 ❸; 300 rds/min to 12 km (6.5 n miles) 2 Oerlikon BMARC 20 mm GAM-BO1; 1,000 rds/min to 2 km.
Torpedoes: 6—324 mm Plessey STWS Mk 2 (2 triple) tubes ❹. Honeywell Mk-46 Mod 5; active/passive homing to 11 km (5.9 n miles) at 40 kt; warhead 44 kg.
Physical countermeasures: Decoys: 4 Loral Hycor SRBOC Mk 36; 6-barrelled fixed launchers ❺; for chaff. Graseby Type 182; towed torpedo decoy.
Electronic countermeasures: ESM: MEL UAA-2; intercept. ECM: Type 670 Guardian.
Radars: Air/surface search: Marconi Type 967/968 ❻; D/E-band.
Navigation: Kelvin Hughes Type 1006; I-band.
Fire control: Two Marconi Type 910 ❼; I/Ku-band (for Seawolf).
Sonars: Plessey Type 2050; hull-mounted; search and attack; medium frequency.
Combat data systems: Siconta IV; Link YB, Link 11, Link 14, Inmarsat.
Weapon control systems: GWS 25 Mod 4 (for SAM); GWS 50 (Exocet).
Helicopters: 2 Westland Super Lynx AH-11A ❽.

Programmes: Contract signed on 18 November 1994 to transfer four Batch I Type 22 frigates from the UK, one in 1995, two in 1996 and one in 1997.
Modernisation: Plans to fit a single 57 mm gun on the bow were shelved in favour of a 40 mm gun on each beam. These guns were taken from the Niteroi class. A modernisation programme for all three ships started in mid-2009. F 49 completed in 2012. Upgrades include replacement of Exocet MM 38 with MM 40, modernisation of the Seawolf SAM system, overhaul of main machinery, replacement of navigation radar, communications upgrades, and overhaul of the ESM/ECM systems. The command system is being replaced by Siconta IV. The Seawolf missiles are also undergoing a refurbishment programme at Avibras-Brazil.
Structure: Accommodation modified in UK service to take 65 officers under training.
Operational: Primary role is ASW. Form part of Second Escort Squadron at Niteroi, Rio de Janeiro. F 47 decommissioned in 2005.

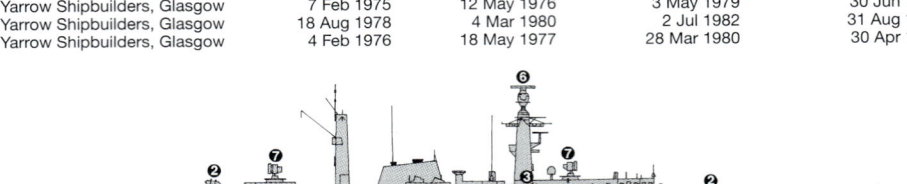

GREENHALGH (Scale 1 : 1,200), Ian Sturton / 0012084

BOSISIO 9/2010, Mario R V Carneiro / 1406704

GREENHALGH 9/2007, Mario R V Carneiro / 1335449

IHS Jane's Fighting Ships 2014-2015 © 2014 IHS

6 NITERÓI CLASS (FFGHM)

Name	No	Builders	Laid down	Launched	Commissioned
NITERÓI	F 40	Vosper Thornycroft Ltd	8 Jun 1972	8 Feb 1974	20 Nov 1976
DEFENSORA	F 41	Vosper Thornycroft Ltd	14 Dec 1972	27 Mar 1975	5 Mar 1977
CONSTITUIÇÃO	F 42	Vosper Thornycroft Ltd	13 Mar 1974	15 Apr 1976	31 Mar 1978
LIBERAL	F 43	Vosper Thornycroft Ltd	2 May 1975	7 Feb 1977	18 Nov 1978
INDEPENDÊNCIA	F 44	Arsenal de Marinha, Rio de Janeiro	11 Jun 1972	2 Sep 1974	3 Sep 1979
UNIÃO	F 45	Arsenal de Marinha, Rio de Janeiro	11 Jun 1972	14 Mar 1975	12 Sep 1980

Displacement, tonnes: 3,251 standard; 3,766 full load
Dimensions, metres (feet): 129.2 × 13.5 × 5.5 sonar *(423.9 × 44.3 × 18.0)*
Speed, knots: 30; 22 diesel.
Range, n miles: 5,300 at 17 kt, 4,200 at 19 kt, 1,300 at 28 kt
Complement: 209 (22 officers)

Machinery: CODOG; 2 RR Olympus TM3B gas turbines; 50,880 hp *(37.9 MW)* sustained; 4 MTU 16V 956 TB 91 diesels; 15,000 hp(m) *(11.0 MW)* sustained; 2 shafts; cp props
Missiles: SSM: 4 Aerospatiale MM 40 Block II Exocet (2 twin) launchers ❶; inertial cruise; active radar homing to 70 km *(40 n miles)* at 0.9 Mach; warhead 165 kg; sea-skimmer.
SAM: AESN Albatros (8 cell, 2 reloads) ❷; Aspide 2000; semi-active radar homing to 21 km *(11 n miles)* at 2.5 Mach.
Guns: 1 Vickers 4.5 in *(115 mm)*/55 Mk 8 ❸; 25 rds/min to 22 km *(12 n miles)* anti-surface; 6 km *(3.2 n miles)* anti-aircraft; weight of shell 21 kg. 2 Bofors SAK 40 mm/L 70-600 Mk 3 Sea Trinity ❹; 330 rds/min to 4 km *(2.2 n miles)*. 4—12.7 mm MGs.
Torpedoes: 6—324 mm Mk 32 (2 triple) tubes ❺. Honeywell Mk 46 Mod 5; anti-submarine; active/passive homing to 11 km *(5.9 n miles)* at 40 kt; warhead 44 kg.
A/S Mortars: 1 Bofors 375 mm trainable rocket launcher (twin-tube) ❻; automatic loading; range 1,600 m.
Physical countermeasures: Decoys: 4 IPqM/Elebra MDLS 12-barrel chaff launchers ❼.
Electronic countermeasures: ESM: Racal-Thorn Cygnus; radar intercept.
ECM: IPqm/Omnisys ET/SLQ-1A (F 41, 43), IPqm/Omnisys ET/SLQ-2X (F 44)
Radars: Air/surface search: AESN RAN 20 S (3L) ❽; D-band.
Surface search: Terma Scanter 4100 ❾; I-band.
Fire control: 2 AESN RTN 30X ❿; I/J-band.
Navigation: Furuno FR-1942 Mk 2; I-band.
Sonars: EDO 997F; hull-mounted; active search and attack; medium frequency. EDO 700E VDS (F 40, F 41); active search and attack.
Combat data systems: IPqM/Elebra Siconta II. Link YB.
Weapon control systems: Saab/Combitech EOS-400/10B optronic director. WSA 401. FCS.

Helicopters: 1 Westland Super Lynx AH-11A ⓫.

Programmes: A contract announced on 29 September 1970 was signed between the Brazilian government and Vosper Thornycroft for the design and building of six Vosper Thornycroft Mark 10 frigates. Seventh ship with differing armament was ordered from Navy Yard, Rio de Janeiro in June 1981 and is used as a training ship.
Modernisation: The modernisation plan (Mod Frag) first signed in March 1995 included replacing Seacat by Aspide, Plessey AWS 2 radar by Alenia RAN 20S, RTN 10X by RTN 30X, ZW06 radar by Terma Scanter, new 40 mm mountings, new EW equipment, combat data system and hull-mounted sonar. Ikara removed. Work was undertaken by Elebra. *Liberal* completed 2001. *Defensora* (2002), *Independência* (2004), and *Niterói* (2004). *Constituição* and *União* were completed in 2005. *Uniao* completed a further refit in mid-2011. She was followed by *Defensora*.
Structure: Originally F 40, 41, 44, and 45 were of the A/S configuration. F 42 and 43 general purpose design. Fitted with retractable stabilisers.
Operational: Endurance, 45 days' stores, 60 days' provisions. The helicopter has Sea Skua ASM. All are based at Niterói and form the First Escort Squadron.

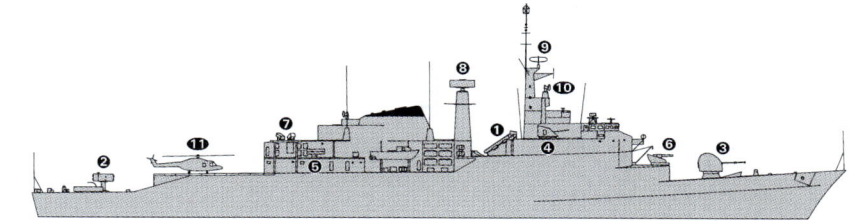

LIBERAL (Scale 1 : 1,200), Ian Sturton / 1170228

NITEROI 9/2013*, Mario R V Carneiro / 1529002

INDEPENDÊNCIA 6/2012, M Declerck / 1482966

UNIÃO 5/2013*, Mario R V Carneiro / 1529001

CORVETTES

1 + (5) BARROSO CLASS (FSGH)

Name	No	Builders	Laid down	Launched	Commissioned
BARROSO	V 34	Arsenal de Marinha, Rio de Janeiro	21 Dec 1994	20 Dec 2002	23 Nov 2009

Displacement, tonnes: 1,814 standard; 2,388 full load
Dimensions, metres (feet): 103.4 × 11.4 × 3.95 hull; 5.3 sonar *(339.2 × 37.4 × 13.0; 17.4)*
Speed, knots: 29. **Range, n miles:** 4,000 at 12 kt
Complement: 145 (15 officers)

Machinery: CODOG; 1 GE LM 2500 gas turbine; 27,500 hp *(20.52 MW)* sustained; 2 MTU 20V 1163 TB83 diesels; 11,780 hp(m) *(8.67 MW)* sustained; 2 shafts; Kamewa cp props
Missiles: SSM: 4 Aerospatiale MM 40 Exocet Block II ❶; inertial cruise; active radar homing to 70 km *(40 n miles)* at 0.90 Mach; warhead 165 kg; sea-skimmer.
Guns: 1 Vickers 4.5 in *(115 mm)* Mk 8 ❷; 55° elevation; 25 rds/min to 22 km *(12 n miles)*; weight of shell 21 kg. 1 Bofors SAK Sea Trinity CIWS 40 mm/70 Mk 3 ❸; 330 rds/min to 4 km *(2.2 n miles)*; anti-aircraft; 2.5 km *(1.4 n miles)* anti-missile; weight of shell 0.96 kg; with '3P' improved ammunition. 2—12.7 mm MGs.
Torpedoes: 6 ARES/DSAM SLT Mod 400 324 mm (2 triple) tubes ❹; Honeywell Mk 46 Mod 5; anti-submarine; active/passive homing to 11 km *(5.9 n miles)* at 40 kt; warhead 44 kg.
Physical countermeasures: Decoys: 2 IPqM/Elebra MDLS 101 12-tubed decoy launchers ❺.
Electronic countermeasures: ESM: IPqM/Elebra ET/SLR-1X ❻; radar warning.
ECM: IPqM/Elebra ET/SLQ-2 ❼; jammer.
Radars: Surface search: AESN RAN-20S ❿; F-band.
Navigation: Terma Scanter 4100; E/F/I-band.
Fire control: AESN RTN-30-X ⓫; I/J-band (for Albatross and guns).
Sonars: EDO 997(F); hull-mounted; active; medium frequency.
Combat data systems: IPqM/Esca Siconta Mk III with Link YB.
Weapon control systems: Saab/Combitech EOS-400 FCS with optronic director ❽; two OFDLSE optical directors ❾.

Helicopters: 1 AH-11A Westland Super Lynx ⓬.

Programmes: Ordered in 1994 as a follow-on to the Inhauma programme. The building programme of *Barroso* was beset by funding difficulties. However, it was confirmed in October 2012 that a further four or five ships, to a modified design, were to be built. Construction of the first of these ships is expected to start in late 2014.
Structure: The hull is some 4.2 m longer than the Inhauma class to improve sea-keeping qualities and allow extra space in the engine room. The design allows the use of containerised equipment to aid modernisation. Efforts have been made to incorporate stealth technology. Vosper stabilisers.
Operational: A successful Exocet firing was conducted on 18 May 2012.

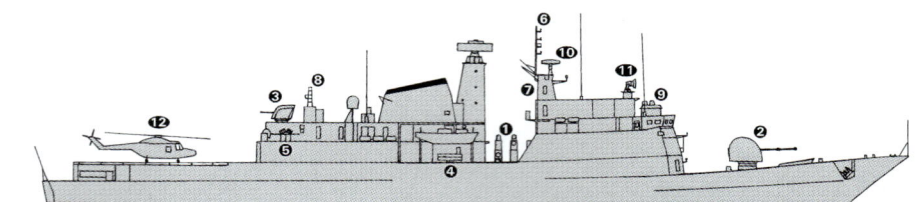

BARROSO *(Scale 1 : 900), Ian Sturton* / 1366675

BARROSO *9/2011, Mario R V Carneiro* / 1406706

4 INHAÚMA CLASS (FSGH)

Name	No	Builders	Laid down	Launched	Commissioned
INHAÚMA	V 30	Arsenal de Marinha, Rio de Janeiro	23 Sep 1983	13 Dec 1986	12 Dec 1989
JACEGUAI	V 31	Arsenal de Marinha, Rio de Janeiro	15 Oct 1984	8 Jun 1987	2 Apr 1991
JULIO DE NORONHA	V 32	Verolme, Angra dos Reis	8 Dec 1986	15 Dec 1989	27 Oct 1992
FRONTIN	V 33	Verolme, Angra dos Reis	14 May 1987	6 Feb 1992	11 Mar 1994

Displacement, tonnes: 1,626 standard; 2,174 full load
Dimensions, metres (feet): 95.8 × 11.4 × 3.7 hull; 5.3 sonar *(314.3 × 37.4 × 12.1; 17.4)*
Speed, knots: 27. **Range, n miles:** 4,000 at 15 kt
Complement: 145 (20 officers)

Machinery: CODOG; 1 GE LM 2500 gas turbine; 27,500 hp *(20.52 MW)* sustained; 2 MTU 16V 396 TB 91 diesels; 7,500 hp(m) *(5.5 MW)* sustained; 2 shafts; Kamewa cp props
Missiles: SSM: 4 Aerospatiale MM 40 Exocet Block II ❶; inertial cruise; active radar homing to 70 km *(40 n miles)* at 0.9 Mach; warhead 165 kg; sea-skimmer.
Guns: 1 Vickers 4.5 in *(115 mm)* Mk 8 ❷; 55° elevation; 25 rds/min to 22 km *(12 n miles)*, weight of shell 21 kg. 2 Bofors 40 mm/70 ❸; 300 rds/min to 12 km *(6.5 n miles)* anti-surface; 4 km *(2.2 n miles)* anti-aircraft; weight of shell 0.96 kg. 2—12.7 mm MGs.
Torpedoes: 6—324 mm Mk 32 (2 triple) tubes ❹. Honeywell Mk 46 Mod 5; anti-submarine; active/passive homing to 11 km *(5.9 n miles)* at 40 kt; warhead 44 kg.
Physical countermeasures: Decoys: 2 Plessey Shield chaff launchers ❺; fires chaff and IR flares in distraction, decoy or centroid patterns.
Electronic countermeasures: ESM: IPqM/Elebra Defensor ET/SLR-1X ❻; radar intercept.
ECM: IPqM/Elebra ET SLQ-1; jammer ❼.
Radars: Surface search: Plessey AWS 4 ❿; E/F-band.
Navigation: Kelvin Hughes Type 1007; I/J-band.
Fire control: Selenia Orion RTN 10X ⓫; I/J-band.
Sonars: Atlas Elektronik DSQS-21C; hull-mounted; active; medium frequency.
Combat data systems: Ferranti CAAIS 450/WSA 421; Link YB.
Weapon control systems: Saab EOS-400 FCS with optronic director ❽ and two OFDLSE optical ❾ directors.

Helicopters: 1 Westland Super Lynx ⓬ or UH-12/13 Ecureuil.

Programmes: Designed by Brazilian Naval Design Office with advice from West German private Marine Technik design company. Signature of final contract on 1 October 1981. First pair ordered on 15 February 1982 and second pair 9 January 1986. In mid-1986 the government approved, in principle, construction of a total of 16 ships but this was reduced to four.
Modernisation: A modernisation programme began (with V 32) in October 2008 and was still in progress in 2013. Refit of V 30 began in 2009 but the refits of V 31 and V 33 appear to have been delayed. The upgrade includes replacement of AWS 4 with SELEX RAN 20S search radar, RTN 10X with SELEX RTN 30X fire-control radar and KH 1007 with Furuno 8252 navigation radar. The Siconta IV combat data system is to replace CAAIS. The main machinery is also to be overhauled.
Operational: Form part of First Frigate Squadron based at Niterói, Rio de Janeiro.

INHAÚMA *(Scale 1 : 900), Ian Sturton* / 0017617

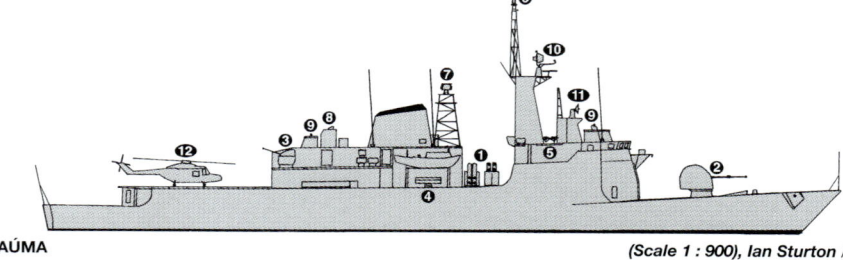

JULIO DE NORONHA *10/2005, Mario R V Carneiro* / 1153018

INHAÚMA
9/2007, Mario R V Carneiro
1335446

SHIPBORNE AIRCRAFT (FRONT LINE)

Numbers/Type: 15/3 McDonnell Douglas AF-1/AF-1A Skyhawk.
Operational speed: 560 kt *(1,040 km/h)*.
Service ceiling: 45,000 ft *(13,780 m)*.
Range: 1,060 n miles *(1,965 km)*.
Role/Weapon systems: Acquired from Kuwait Air Force in September 1998 to restore carrier fixed wing flying. A further five aircraft are kept as spares. There are 12 aircraft being upgraded to AF-1B/C standard by Embraer/Elbit and the first is to be completed in 2013. Modifications include a new radar and avionics. Sensors: APQ 145B radar; ESM/ECM. Weapons: AAM; 4 AIM 9H; 2 Colt 20 mm cannon; ASVW; bombs and rocket pods.

AF-1   10/2001, S C Neto/Mario R V Carneiro / 0569157

Numbers/Type: 1/2 Sikorsky SH-3A/SH-3B.
Operational speed: 125 kt *(230 km/h)*.
Service ceiling: 12,200 ft *(3,720 m)*.
Range: 400 n miles *(740 km)*.
Role/Weapon systems: ASW helicopter; carrierborne and shore-based for medium-range ASW, ASVW and SAR. Sixteen delivered between 1970 and 1997. Four have been lost and a further nine had been withdrawn from service by late 2012. Sensors: SMA APS-705(V)II or Northrop Grumman LN-66 HP search radar; Bendix AQS 13B or AQS 18(V) dipping sonar. Weapons: ASW; up to 2 × Mk 46 torpedoes, or 4 Mk II depth bombs. ASVW; 2 × AM 39 Exocet missiles.

Numbers/Type: 4/2 Aerospatiale UH-14 (AS 332F1 Super Puma)/UH-14 (AS 532 SC Cougar).
Operational speed: 120 kt *(222 km/h)*.
Service ceiling: 12,000 ft *(3,657 m)*.
Range: 445 n miles *(825 km)*.
Role/Weapon systems: SAR, troop transport and ASVW. Sensors: Bendix RDR-1400C search radar. Weapons: None.

UH-14   6/2003, S C Neto/Mario R V Carneiro / 0569155

Numbers/Type: 6 MH-16 (Sikorsky SH-60B) Seahawk.
Operational speed: 135 kt *(250 km/h)*.
Service ceiling: 10,000 ft *(3,050 m)*.
Range: 600 n miles *(1,110 km)*.
Role/Weapon systems: Four ex-US Navy aircraft ordered, under FMS funding arrangements, in June 2008. A further two ordered in 2011. The first four entered service in August 2012 and a further two are to be delivered by 2014. A further two aircraft were ordered on 31 October 2013 for delivery in 2017. The aircraft are to have an ASW/ASUW role. Sensors: APS-143 search radar; Helras dipping sonar; AAS-44 FLIR; ASQ-81 MAD; 25 sonobuoys; LR-100 ESM suite. Weapons: ASW: 2 Mk 46 Mod 5 torpedoes; Penguin Mk 2 Mod 7 ASM.

Numbers/Type: 12 AH-IIA Westland Super Lynx.
Operational speed: 125 kt *(232 km/h)*.
Service ceiling: 12,000 ft *(3,650 m)*.
Range: 130 n miles *(240 km)*.
Role/Weapon systems: ASW/ASV roles. First batch upgraded in 1994–97 to Super Lynx standard with Mk 3 radar and Racal Kestrel EW suite. Further modifications (2010–15) include installation of Star Safire FLIR. Sensors: Sea Spray Mk 1/Mk 3 radar; Racal MIR 2 ESM; Sea Star III FLIR. Weapons: ASW; 2 × Mk 46 torpedoes, or Mk II depth bombs. ASV; 4 × BAe/Ferranti Sea Skua missiles.

AH-IIA   9/2011, Mario R V Carneiro / 1406707

Numbers/Type: 4 Grumman/Embraer C-1T Trader.
Operational speed: 265 kt *(426 km/h)*.
Service ceiling: 21,000 ft *(6,400 m)*.
Range: 1,130 n miles *(2,900 km)*.
Role/Weapon systems: Ex-US Navy aircraft being modified by Marsh Aviation to KC-2 Turbo configuration. This includes two years logistic support. To be delivered 2014–15. To be used in COD and REVO role from *São Paulo*. Sensors: IMI/Elta EL/M-2032 search radar; Link 11; Link 16 and Link 22; Elettronica SpA ELT-553 Mk-1 ECM; Elettronica ALR-733 ESM; RWR. Weapons: unarmed.

Numbers/Type: 15 Aerospatiale UH-12 Esquilo (AS-350BA Ecureuil).
Operational speed: 147 kt *(272 km/h)*.
Service ceiling: 10,000 ft *(3,050 m)*.
Range: 240 n miles *(445 km)*.
Role/Weapon systems: Support helicopters for Fleet liaison and Marine Corps transportation. Sensors: None. Weapons: 2 × axial 7.62 mm MGs or 1 × lateral MG or 1 × rocket pod.

UH-12   12/2002, Mario R V Carneiro / 0569156

Numbers/Type: 7 Aerospatiale UH-13 Esquilo (AS 355F2 Ecureuil 2).
Operational speed: 121 kt *(224 km/h)*.
Service ceiling: 11,150 ft *(3,400 m)*.
Range: 240 n miles *(445 km)*.
Role/Weapon systems: SAR, liaison and utility in support of Marine Corps. One transferred to Uruguay in 2006. Sensors: Search radar. Weapons: 2 × axial 7.62 mm MGs or 1 × lateral MG or 1 × rocket pod.

UH-13   3/2008, M Declerck / 1335336

Numbers/Type: 17 IH-6B (Bell JetRanger III).
Operational speed: 115 kt *(213 km/h)*.
Service ceiling: 20,000 ft *(6,100 m)*.
Range: 368 n miles *(682 km)*.
Role/Weapon systems: Utility and training helicopters. One lost in June 2005. Two transferred from the Air Force in 2013. Sensors: None. Weapons: 2 × 7.62 mm MGs or 1 lateral 12.7 mm MG or 2 × rocket pods.

IH-6B   12/2002, Mario R V Carneiro / 0569154

Numbers/Type: 4 Grumman/Embraer S-2T Tracker.
Operational speed: 265 kt *(426 km/h)*.
Service ceiling: 21,000 ft *(6,400 m)*.
Range: 1,130 n miles *(2,900 km)*.
Role/Weapon systems: Ex-US Navy aircraft modified with new engines, avionics and radar. Some of these are to come from Australia and at least one from Uruguay. Air defence early warning aircraft, COMINT and ELINT with secondary role to provide long-range surveillance patrol. Sensors: Thales-Racal Searchwater 2000; Raytheon Sea Vue or FIAR Eliradar HEW-784; Sea Star Safire III FLIR; Elettronica SpA ELT-553 Mk-1 (ECM); Elettronica ALR-733 (ESM); Links 11, 16 and 22. Weapons: none.

Numbers/Type: 16 Helibras/Eurocopter EC-725 UH-15 Super Cougar Mk II.
Operational speed: 175 kt *(324 km/h)*.
Service ceiling: 20,000 ft *(6,095 m)*.
Range: 463 n miles *(857 km)*.
Role/Weapon systems: Contract awarded on 7 September 2009 for 16 aircraft being built by Helibras. Eight aircraft for ASUW and eight for combat SAR, troop transport and utility roles. Sensors: AN/APS-143(V)3 (search radar); Sea Star Safire III (FLIR); Dassault Electronique EWR-99 FRUIT (RWR); Alkan ELIPS (multipurpose chaffs/flares dispenser). Weapons: 2 AM-39 Block 2 Mod 2 Exocet; 2 Avibras SBAT-70/7 7-tube rocket launchers; 2 GIAT 20 mm cannons; 2 — 7.62 mm MGs.

EC-725   9/2011, Mario R V Carneiro / 1406708

LAND-BASED MARITIME AIRCRAFT (FRONT LINE)

Notes: Two Beech King Air B-350 maritime patrol aircraft are to be acquired.

Numbers/Type: 9 Lockheed P-3AM Orion.
Operational speed: 411 kt (761 km/h).
Service ceiling: 28,300 ft (8,625 m).
Range: 4,000 m (7,410 km).
Role/Weapon systems: Twelve P-3 A/B acquired by the Air Force from the US Navy in 2002. Nine being upgraded to P-3AM standard by EADS/CASA. Contract awarded in April 2005 and the seventh was delivered in November 2013. The remaining four aircraft are to be used for spare parts. Sensors: IMI/Elta EL/M-2022 radar, Star Safire II FLIR, ASQ-81 MAD, EADS/CASA FITS, AAR-47 warning receiver, AN/ALR-66(V)3 ESM, AN/ALQ-78A countermeasures suite. Weapons: ASW; four Mk 46 torpedoes, four Mk 14 depth charges. ASuW: 4 Boeing AGM-84C Harpoon.

P-3AM   8/2013*, Adolfo Ortigueira Gil / 1529003

Numbers/Type: 10/9 Bandeirante P-95A (EMB-111(A))/P-95B (EMB-111(B)).
Operational speed: 194 kt (360 km/h).
Service ceiling: 25,500 ft (7,770 m).
Range: 1,590 n miles (2,945 km).
Role/Weapon systems: Air Force operated for coastal surveillance role by four squadrons. Nine P-95B and four P-95A aircraft are to be upgraded with Elbit avionics. Sensors: MEL Super Searcher (P-95B) or Eaton/AIL APS-28 Sea Searcher (P-95A) search radar, searchlight pod on starboard wing, EFIS-74 (electronic flight instrumentation) and Collins APS-65 (autopilot); ESM Thomson-CSF DR2000A/Dalia 1000A Mk II, GPS (Trimble). Weapons: 4 or 6 × 127 mm rockets, or up to 28 × 70 mm rockets.

EMB-111   6/1995 / 0503428

Numbers/Type: 52 A-1 (Embraer/Alenia/Aermacchi) AMX.
Operational speed: 493 kt (914 km/h).
Service ceiling: 42,650 ft (13,000 m).
Range: 1,800 n miles (3,336 km).
Role/Weapon systems: Air Force operated for strike, reconnaissance and anti-shipping attack; shore-based for fleet air defence and ASV primary roles; operated by 1st/16th Group at Santa Cruz and by 1st/10th and 3rd/10th Groups at Santa Maria Air Base (KS). Upgrade of 43 A-1AM aircraft is to include new Scipio SPC-01 radar. Sensors: Tecnasa/SMA SCP-01 Scipio radar. ECM suite/ESM flares and chaffs; GPS and IFF. Weapons: Strike; up to 3,800 kg of 'IRON' bombs; Self-defence; AAM; 2 × MAA-1 Piranha or 2 × AIM-9 Sidewinder missiles; 2 DEFA 30 mm cannon.

AMX   6/1998 / 0013614

Numbers/Type: 51/6 Northrop/Embraer F-5EM/F-5FM Tiger II.
Operational speed: 939 kt (1,763 km/h).
Service ceiling: 51,800 ft (15,790 m).
Range: 190 n miles (306 km).
Role/Weapon systems: Air Force operated for strike, reconnaissance and maritime attack, fleet air defence and ASUW primary roles. Upgrade of 43 F-5EM and three F-5FM is planned and 11 (8 + 3) for Jordanian Air Force are also to be modified bringing the total to 57. Sensors: Alenia Fiar Grifo Doppler radar; Rafael RecceLite; Rafael Litening III; Rafael Sky Shield; TDMA datalink; Links 11, 16 and 22; Elettronica SpA ELT-553 Mk-1 (ECM); Elisra SP 1002 RWR; automatic flare release. Weapons: a variety of external stores including 3,175 kg of bombs, cluster bombs or missiles; 1 Pontiac M-39A3 20 mm cannon.

PATROL FORCES

Notes: (1) The NPO 1800 project is for up to five Navio Patrulha Oceânico (NPO). The broad requirement is for a 1,800 ton ship armed with a medium calibre gun and capable of operating a helicopter. Likely contenders include Fassmer (probably offering its Chilean 80 m design), Navantia, DCNS and Thyssen-Krupp. Following a contract, construction is expected to be completed in three years and subsequent deliveries are to follow at six-month intervals.
(2) There are 131 LAEP series Instruction and Support craft. 26 LAEP-10 are 11 m long and 105 (plus 31 on order) LAEP-7 are 7 m long.
(3) There are 181 LPN series River patrol craft of 3 to 15 m length.
(4) There are 18 8 m aluminium hulled LAR (fast insertion craft) in service with the Brazilian Marines. The first of a new class of 7.6 m LAR-E craft was delivered in January 2012. Large numbers are expected. There are two other variants of the class: LIN are operated by port authorities and LAM are ambulance craft.
(5) Four Tracker II (LPPN-21) 21 m patrol craft are employed as police patrol craft.
(6) Bidding for the construction of four NAPA 200 coastal patrol vessels was in progress in October 2012. The 210 tonne vessels are reported to have a length of 37.8 m, a breadth of 7.2 m and to be capable of 17 kt.
(7) Approximately 28 small patrol craft were transferred from the Ministry of Agriculture and Fisheries in 2012. Names include: *Jandia, Gurijuba, Dourada, Curimata,* and *Cavala.*

6 + (4) MARLIN (MEATINI) CLASS (PB)

MARLIN SE-01	**DOURADO** L-01	**ANEQUIM** NE-02	+4
BARRACUDA NE-01	**ALBACORA** SE-02	**TUCUNARÉ** (ex-*Pargo*) N-01	

Displacement, tonnes: 46 full load
Dimensions, metres (feet): 20.8 × 5.5 × 1.06 (68.2 × 18.0 × 3.5)
Speed, knots: 27. **Range, n miles:** 850 at 15 kt
Complement: 9 (1 officer)
Machinery: 2 MTU 8V 2000 M92 diesels; 2,000 hp (1.49 MW); 2 shafts
Guns: 3 — 7.62 mm MGs.
Radars: Surface search: Furuno; I-band.

Comment: The first of a new class of instruction and patrol craft entered service in 2005. Slightly longer version of Italian Meatini class design in service with Guardia di Finanzia. Built by Inace Shipyard, Brazil. GRP hull with aluminium superstructure. Five further craft were delivered by mid-2010 and a class of 10 is expected.

MARLIN (old number)   6/2007, L Frangetto / 1170090

3 + 4 (6) MACAÉ (NAPA 500) CLASS (PBO)

Name	No	Builders	Laid down	Launched	Commissioned
MACAÉ	P 70	INACE, Fortalesa	26 Nov 2006	2008	19 Dec 2009
MACAU	P 71 (ex-P 75)	INACE, Fortalesa	17 Jul 2007	2008	30 Nov 2010
MARACANÁ	P 72	Estaleiro Shipyard	26 Nov 2009	2011	2014
MANGARATIBA	P 73	Estaleiro Shipyard	15 Sep 2010	2012	2014
MIRAMAR	P 74	Estaleiro Shipyard	15 Feb 2012	2013	2014
MAGÉ	P 75	Estaleiro Shipyard	2013	2014	2014
MARAGOGIPE	P 76	Estaleiro Shipyard	2013	2014	2015

Displacement, tonnes: 413 standard; 485 full load
Dimensions, metres (feet): 54.2 × 8.0 × 2.48 (177.8 × 26.2 × 8.1)
Speed, knots: 21
Range, n miles: 2,500 at 15 kt
Complement: 43 (8 officers)

Machinery: 2 MTU 16V 4000 M 90 diesels; 8,000 hp(m) (5.9 MW); 2 shafts
Guns: 1 Bofors SAK-40 mm/L70 NADM 330; 330 rds/min to 4 km (2.2 n miles); weight of shell 0.96 kg. 2 Oerlikon/Royal Ordnance 20 mm GAM-BO1; 1,000 rds/min to 2 km.
Physical countermeasures: Decoys: 2 IPqM/Elebra MDLS 101 (12-tubed) launchers.
Electronic countermeasures: ESM: To be announced.
Radars: Surface search/navigation: Sperry Vision Master FT 250; E/F/I-bands.
Combat data systems: Sinconta V.
Weapon control systems: ARES/DSAM AO optical sight.

Comment: Following an invitation to tender in June 2006, contract awarded on 28 September 2006 to Indústria Naval do Ceará (INACE), Fortalesa, for the construction of two patrol ships in partnership with the French company CMN. The ships, designated NAPA 500, are to CMN's 54 m Vigilante 400CL 54 design and are to be similar in configuration to the three Al Bushra class in service in the Royal Navy of Oman. CMN is to provide technical assistance and integrated logistic support. Steel was first cut on 1 November 2006. The ships are to be employed on EEZ patrol duties. The contract for a further four (increased to five in December 2011) vessels was let with Estaleiro Shipyard on 25 September 2009. A further six ships are planned in Phase 1 of the programme. A further 15 ships are projected in Phase 2 and then 19 in Phase 3.

MACAÉ   9/2013*, Marco Ghiglino / 1529109

Patrol forces < **Brazil** 79

12 GRAJAÚ CLASS (LARGE PATROL CRAFT) (PBO)

Name	No	Builders	Launched	Commissioned
GRAJAÚ	P 40	Arsenal de Marinha, Rio de Janeiro	21 May 1993	1 Dec 1993
GUAIBA	P 41	Arsenal de Marinha, Rio de Janeiro	10 Dec 1993	12 Sep 1994
GRAÚNA	P 42	Estaleiro Mauá, Niteroi	10 Nov 1993	15 Aug 1994
GOIANA	P 43	Estaleiro Mauá, Niteroi	26 Jan 1994	26 Feb 1997
GUAJARÁ	P 44	Peenewerft	24 Oct 1994	28 Apr 1995
GUAPORÉ	P 45	Peenewerft	23 Jan 1995	29 Aug 1995
GURUPÁ	P 46	Peenewerft	11 May 1995	8 Dec 1995
GURUPI	P 47	Peenewerft	6 Sep 1995	23 Apr 1996
GUANABARA	P 48	INACE, Fortalesa	5 Nov 1997	9 Jul 1999
GUARUJÁ	P 49	INACE, Fortalesa	24 Apr 1998	25 Nov 1999
GUARATUBA	P 50	Peenewerft	16 Jun 1999	1 Dec 1999
GRAVATAÍ	P 51	Peenewerft	26 Aug 1999	17 Feb 2000

Displacement, tonnes: 200 standard; 220 full load
Dimensions, metres (feet): 46.5 × 7.5 × 2.3 (152.6 × 24.6 × 7.5)
Speed, knots: 26. **Range, n miles:** 2,200 at 12 kt
Complement: 29 (4 officers)

Machinery: 2 MTU 16V 396 TB94 diesels; 5,800 hp(m) (4.26 MW) sustained; 2 shafts
Guns: 1 Bofors 40 mm/L70-350. 2 Oerlikon 20 mm.
Radars: Surface search: Furuno ARPA M-1954C; I-band.
Navigation: Racal Decca 1290A; I-band.
Weapon control systems: ARES/DSAM AO optronic director may be fitted in due course.

Comment: Two ordered in late 1987 to a Vosper QAF design similar to Bangladesh Meghna class. Technology transfer in February 1988 and construction started in July 1988 for the first pair; second pair started construction in September 1990. Class name changed in 1993 when the first four were renumbered to reflect revised delivery dates. Building problems are also reflected in the replacing of the order for the third pair with Peenewerft in November 1993 and the fourth pair in August 1994. Two more ordered from Inace in September 1996 and from Peenewerft in 1998. Used for EEZ patrol duties and diver support. Carry one RIB and telescopic launching crane. A similar vessel has been built for Namibia.

GUAJARÁ 9/2010, Mario R V Carneiro / 1406710

GURUPÁ 9/2011, Mario R V Carneiro / 1406709

3 OFFSHORE PATROL VESSELS (PSO)

Name	No	Builders	Launched	Commissioned
AMAZONAS (ex-Port of Spain)	P 120 (ex-CG 50)	BAE Systems Surface Ships	7 Aug 2009	29 Jun 2012
APA (ex-Scarborough)	P 121 (ex-CG 51)	BAE Systems Surface Ships	19 Nov 2009	30 Nov 2012
ARAGUARI (ex-San Fernando)	P 122 (ex-CG 52)	BAE Systems Surface Ships	16 Jul 2010	19 Jun 2013

Displacement, tonnes: 2,286 full load
Dimensions, metres (feet): 90.5 × 13.5 × 3.5 (296.9 × 44.3 × 11.5)
Speed, knots: 25. **Range, n miles:** 5,500 at 12 kt
Complement: 89 (5 trainees)

Machinery: 2 MAN 16RK 280 diesels; 2 shafts; cp props; bow thruster
Guns: 1 MSI DS 30M 30 mm. 2 MSI DS 25M 25 mm. 2 – 12.7 mm MGs.
Radars: Air/surface search: Terma 4100; E/F-band.
Navigation: I-band.
Weapon control systems: Ultra Osiris optronic director.
Helicopters: Platform for one medium.

Comment: It was announced on 2 January 2012 that three offshore patrol vessels, originally built for Trinidad and Tobago, were to be sold to Brazil; the contract is reported to have included support services and the rights to build additional (perhaps nine) ships under licence in Brazil. These may be modified to include a hangar and a heavier gun. The original contract for the construction of the three ships was signed between the government of Trinidad and Tobago and VT Shipbuilding (now BAE Systems Surface Ships) on 5 April 2007. However, the contract was formally terminated in September 2010, following a change of government in Trinidad. The ships are an improved River-class design and are equipped with a 16-tonne crane, space for ISO containers (for disaster relief or military stores) and a 39 kt Pacific 24 RIB. A similar ship has been built for the Royal Thai Navy.

ARAGUARI 7/2013*, Maritime Photographic / 1529004

2 IMPERIAL MARINHEIRO CLASS (COASTAL PATROL SHIPS) (PG/ATR)

Name	No	Builders	Commissioned
IMPERIAL MARINHEIRO	V 15	Smit, Kinderdijk	8 Jun 1955
CABOCLO	V 19	Smit, Kinderdijk	5 Apr 1955

Displacement, tonnes: 926 standard; 1,041 full load
Dimensions, metres (feet): 56 × 9.3 × 3.6 (183.7 × 30.5 × 11.8)
Speed, knots: 16
Complement: 64 (6 officers)

Machinery: 2 Sulzer 6TD36 diesels (V 15); 2,160 hp(m) (1.59 MW); 2 Volvo Penta D49A-MS diesels (V 19), 2,394 hp (1.79 MW); 2 shafts
Guns: 1 – 3 in (76 mm)/50 Mk 33; 50 rds/min to 12.8 km (6.9 n miles); weight of shell 6 kg. 2 or 4 Oerlikon 20 mm.
Radars: Surface search: Furuno M-1831; I-band.
Navigation: Furuno FR-1505; I-band.

Comment: Fleet tugs classed as corvettes. Equipped for firefighting. Imperial Marinheiro has acted as a submarine support ship but gave up the role in 1990. V 21 and V 23 withdrawn from service in 2002, V 24 in 2003 and V 20 in 2004. V 19 has been re-engined and returned to service.

IMPERIAL MARINHEIRO 5/2010, Jurg Kürsener / 1406712

4 + (2) LPR-40 CLASS (RIVER PATROL VESSELS) (PBR)

Displacement, tonnes: 14 full load
Dimensions, metres (feet): 12.72 × 2.8 × 0.7 (41.7 × 9.2 × 2.3)
Speed, knots: 29. **Range, n miles:** 513 at 25 kt
Complement: 4
Machinery: 2 Caterpillar C9 diesels; 503 hp (375 kW); 2 waterjets
Guns: 3 – 12.7 mm MGs.
Helicopters: Platform for 1 small.

Comment: The announcement that Brazil was to acquire four river patrol vessels from Colombia was made on 3 May 2012. Following a contract in December 2012, the vessels were delivered in late 2013. The vessels are to be used for security patrols along the Amazon River. Two are operated by the navy. An order for two further vessels is expected.

LPR-40 CLASS 6/2012*, Cotecmar / 1455768

2 PEDRO TEIXEIRA CLASS (RIVER PATROL SHIPS) (PBR)

Name	No	Builders	Launched	Commissioned
PEDRO TEIXEIRA	P 20	Arsenal de Marinha, Rio de Janeiro	14 Oct 1970	17 Dec 1973
RAPOSO TAVARES	P 21	Arsenal de Marinha, Rio de Janeiro	11 Jun 1972	17 Dec 1973

Displacement, tonnes: 582 standard; 962 full load
Dimensions, metres (feet): 63.6 × 9.7 × 1.7 (208.7 × 31.8 × 5.6)
Speed, knots: 16. **Range, n miles:** 5,000 at 13 kt
Complement: 80 (7 officers)

Machinery: 2 Volvo Penta D49A-MS diesels; 4,788 hp (3.57 MW); 2 shafts; Kort nozzles
Guns: 1 Bofors 40 mm/60; 120 rds/min to 12 km (6.5 n miles). 6 – 12.7 mm MGs.
A/S Mortars: 2 – 81 mm Mk 2 mortars.
Radars: Surface search: Furuno FR-1505; I-band.
Navigation: Racal-Decca TM 1226C; I-band.
Helicopters: 1 Bell JetRanger or UH-12 Esquilo.

Comment: Built in Rio de Janeiro. Belong to Amazon Flotilla. Can carry two armed LCVPs and 85 marines in deck accommodation. All re-engined with Volvo engines.

PEDRO TEIXEIRA 6/1997, Brazilian Navy / 0012091

80 Brazil > Patrol forces — Amphibious forces

3 RORAIMA CLASS (RIVER PATROL SHIPS) (PBR)

Name	No	Builders	Launched	Commissioned
RORAIMA	P 30	Maclaren, Niteroi	2 Nov 1972	21 Feb 1975
RONDÔNIA	P 31	Maclaren, Niteroi	10 Jan 1973	3 Dec 1975
AMAPÁ	P 32	Maclaren, Niteroi	9 Mar 1973	12 Jan 1976

Displacement, tonnes: 345 standard; 371 full load
Dimensions, metres (feet): 46.3 × 8.5 × 1.4 (151.9 × 27.9 × 4.6)
Speed, knots: 17. **Range, n miles:** 6,000 at 15 kt
Complement: 48 (5 officers)

Machinery: 2 Volvo-Penta D49A-MS diesels; 2,394 hp(m) (1.79 MW); 2 shafts; Kort nozzles
Guns: 1 Bofors 40 mm/60; 120 rds/min to 12 km (6.5 n miles). 2 Oerlikon 20 mm. 6—12.7 mm MGs.
A/S Mortars: 2—81 mm mortars.
Radars: Surface search: Furuno FR-1525; I-band.
Navigation: Racal-Decca TM 1226C; I-band.

Comment: Carry two LAR fast insertion craft. Can carry 10 marines in deck accommodation. Belong to Amazon Flotilla. All re-engined with Volvo engines.

RORAIMA 6/1998, Brazilian Navy / 0017623

4 BRACUI (RIVER) CLASS (COASTAL PATROL CRAFT) (PBO)

Name	No	Builders	Commissioned
BRACUI (ex-Itchen)	P 60 (ex-M 2009)	Richards Ltd, Lowestoft	12 Oct 1985
BENEVENTE (ex-Blackwater)	P 61 (ex-M 2008)	Richards Ltd, Great Yarmouth	5 Jul 1985
BOCAINA (ex-Spey)	P 62 (ex-(ex-M 2013), ex-)	Richards Ltd, Lowestoft	4 Apr 1986
BABITONGA (ex-Arun)	P 63 (ex-(ex-M 2014), ex-)	Richards Ltd, Lowestoft	29 Aug 1986

Displacement, tonnes: 782 standard; 904 full load
Dimensions, metres (feet): 47.5 × 10.5 × 2.9 (155.8 × 34.4 × 9.5)
Speed, knots: 14. **Range, n miles:** 4,500 at 10 kt
Complement: 32 (4 officers)

Machinery: 2 Ruston 6 RKC diesels; 3,100 hp(m) (2.3 MW) sustained; 2 shafts
Guns: 1 Bofors 40 mm/60. 2 Oerlikon 20 mm.
Mines: Rails for up to 20.
Radars: Surface search: 2 Racal Decca TM 1226C; I-band.

Comment: Second batch of ex-UK River class minesweepers transferred in 1998. These four were converted as patrol craft in UK service. Recommissioned 6 April, 10 July, 10 July and 9 September respectively. Three others transferred in 1995 are listed as Survey Ships.

BOCAINA 7/1998, Maritime Photographic / 0056608

1 PARNAIBA CLASS (RIVER MONITOR) (PGRH)

Name	No	Builders	Commissioned
PARNAIBA	U 17 (ex-P 2)	Arsenal de Marinha, Rio de Janeiro	6 Nov 1938

Displacement, tonnes: 630 standard; 732 full load
Dimensions, metres (feet): 55 × 10.1 × 1.6 (180.4 × 33.1 × 5.2)
Speed, knots: 12. **Range, n miles:** 1,350 at 10 kt
Complement: 74 (6 officers)

Machinery: 2 diesels; 2 shafts
Guns: 1 US 76 mm. 2 Bofors 40 mm/70. 6 Oerlikon 20 mm.
Radars: Surface search: Racal Decca; I-band.
Navigation: Furuno 3600; I-band.
Helicopters: Platform for one IH-6B Jet Ranger.

Comment: Laid down 11 June 1936. Launched 2 September 1937. In Mato Grosso Flotilla. Re-armed with new guns in 1960. 3 in (76 mm) side armour and partial deck protection. Refitted in 1995/96 with improved armament, and with diesel engines replacing the steam reciprocating propulsion plant. Converted again in 1998 with Bofors 40 mm/70 guns taken from Niterói-class frigates and a helo deck at the stern. Facilities to refuel and re-arm a UH-12 helicopter. Recommissioned 6 May 1999.

PARNAIBA 6/2011, Brazilian Navy / 1406711

6 PIRATINI CLASS (COASTAL PATROL CRAFT) (PB)

Name	No	Builders	Commissioned
PIRATINI (ex-PGM 109)	P 10	Arsenal de Marinha, Rio de Janeiro	30 Nov 1970
PIRAJÁ (ex-PGM 110)	P 11	Arsenal de Marinha, Rio de Janeiro	8 Mar 1971
PAMPEIRO (ex-PGM 118)	P 12	Arsenal de Marinha, Rio de Janeiro	16 Jun 1971
PARATI (ex-PGM 119)	P 13	Arsenal de Marinha, Rio de Janeiro	29 Jul 1971
PENEDO (ex-PGM 120)	P 14	Arsenal de Marinha, Rio de Janeiro	30 Sep 1971
POTI (ex-PGM 121)	P 15	Arsenal de Marinha, Rio de Janeiro	29 Oct 1971

Displacement, tonnes: 107 standard; 148 full load
Dimensions, metres (feet): 29 × 5.8 × 2 (95.1 × 19.0 × 6.6)
Speed, knots: 17. **Range, n miles:** 1,700 at 12 kt
Complement: 16 (2 officers)

Machinery: 4 Cummins VT-12M diesels; 1,100 hp (820 kW); 2 shafts
Guns: 1 Oerlikon 20 mm. 2—12.7 mm MGs.
Radars: Surface search: Racal Decca 1070; I-band.
Navigation: Furuno 3600; I-band.

Comment: Built under offshore agreement with the USA and similar to the US Cape class. 81 mm mortar removed in 1988. Carries an inflatable launch. P 10, P 11, P 14 and P 15 are based at Ladário Fluvial Base, Mato Grosso, the other two at Belém.

PIRAJÁ 5/2011, Hartmut Ehlers / 1406713

AMPHIBIOUS FORCES

Notes: (1) One 25 m river support craft, Leverger, is based at Cadario.
(2) There are 32 Zodiac RIBs for special operations.
(3) There are 146 Flexboat general purpose RIBs. A further 41 are under construction.
(4) There are eight EDVP-11 LCVPs based at Ladario.
(5) There are 11 EDVM-17 LCM (LCM 6 type).
(6) There are plans to acquire two second hand LSD/LPD to replace Río de Janeiro. The most likely contenders are ex-US Navy Whidbey Island class as they decommission.

3 + 3 LCU 1610 CLASS (EDCG/LCU)

Name	No	Builders	Commissioned
GUARAPARI	GED 10 (ex-L 10)	Arsenal de Marinha, Rio de Janeiro	27 Mar 1978
TAMBAÚ	GED 11 (ex-L 11)	Arsenal de Marinha, Rio de Janeiro	27 Mar 1978
CAMBORIÚ	GED 12 (ex-L 12)	Arsenal de Marinha, Rio de Janeiro	6 Jan 1981

Displacement, tonnes: 208 standard; 396 full load
Dimensions, metres (feet): 41 × 8.4 × 2.0 (134.5 × 27.6 × 6.6)
Speed, knots: 11. **Range, n miles:** 1,200 at 8 kt
Complement: 14 (2 officers)

Military lift: 172 tons
Machinery: 2 GM 12V-71 diesels; 874 hp (650 kW) sustained; 2 shafts; cp props
Guns: 3—12.7 mm MGs.
Radars: Navigation: Furuno 3600; I-band.

Comment: Based at Niteroi. Three similar EDCG-41 craft are under construction at AMRJ, Rio de Janeiro.

CAMBORIÚ 6/2001, Brazilian Navy / 0130473

Amphibious forces < **Brazil** 81

1 NEWPORT CLASS (LSTH)

Name	No	Builders	Laid down	Launched	Commissioned	Recommissioned
MATTOSO MAIA (ex-*Cayuga*)	G 28 (ex-LST 1186)	National Steel & Shipbuilding Co, San Diego	28 Sep 1968	12 Jul 1969	8 Aug 1970	30 Aug 1994

Displacement, tonnes: 5,242 standard; 8,898 full load
Dimensions, metres (feet): ? × 21.2 × 5.3
  (? × 69.6 × 17.4)
Speed, knots: 20. **Range, n miles:** 14,250 at 14 kt
Complement: 267 (17 officers)
Military lift: 351 (33 officers); 2,600 tons vehicles; 2 LCPL on davits
Machinery: 6 ALCO 16-251 diesels; 16,500 hp *(12.3 MW)* sustained; 2 shafts; cp props; bow thruster; 800 hp *(596 kW)*

Guns: 1 Raytheon 20 mm Vulcan Phalanx Mk 15 Block 0; 3,000 rds/min combined to 1.5 km. 8 – 12.7 mm MGs.
Radars: Surface search: Raytheon SPS-10F; G-band.
Navigation: Raytheon SPS-64(V)6 and Furuno FR 2120; I-band.
Helicopters: Platform only.

Programmes: Transferred from the US Navy by lease 26 August 1994, arriving in Brazil in late October. Purchased outright on 19 September 2000.

Modernisation: A major refit, to extend life by 10 years, began in early 2012 and is to be completed in 2014. Service life extended to 2029.
Structure: The ramp is supported by twin derrick arms. A stern gate to the tank deck permits unloading of amphibious tractors into the water, or unloading of other vehicles into an LCU or onto a pier. Vehicle stowage covers 19,000 sq ft. Length over derrick arms is 562 ft *(171.3 m)*; full load draught is 11.5 ft forward and 17.5 ft aft.

MATTOSO MAIA 6/2003, S C Neto/Mario R V Carneiro / 0569153

1 CEARÁ (THOMASTON) CLASS (LSDH)

Name	No	Builders	Laid down	Launched	Commissioned	Recommissioned
CEARÁ (ex-*Hermitage*)	G 30 (ex-LSD 34)	Ingalls Shipbuilding, Pascagoula	11 Apr 1955	12 Jun 1956	14 Dec 1956	28 Nov 1989

Displacement, tonnes: 6,990 standard; 12,345 full load
Dimensions, metres (feet): 155.5 × 25.6 × 5.8
  (510.2 × 84.0 × 19.0)
Speed, knots: 22.5. **Range, n miles:** 14,800 at 12 kt
Complement: 223 (21 officers)
Military lift: 340 troops; 3 EDCG-41 (LCU) or 6 EDVM-25 (LCM 8) or 50 amphibious tractors in tank-deck and 30 amphibious tractors on upper deck
Machinery: 2 Babcock & Wilcox boilers; 580 psi *(40.8 kg/cm²)*; 2 GE turbines; 24,000 hp *(17.9 MW)*; 2 shafts
Guns: 6 USN 3 in *(76 mm)*/50 (3 twin) Mk 33; 50 rds/min to 12.8 km *(7 n miles)*; weight of shell 6 kg. 4 – 12.7 mm MGs.
Radars: Air/surface search: Plessey AWS-2; E/F-band.
Surface search: Raytheon SPS-10F; G-band.
Navigation: Raytheon CRP 3100; I-band. Furuno ARPA M-1942; E/F/I-band.
Helicopters: Platform for Super Puma.

Programmes: The original plan to build a 4,500 ton LST was overtaken by the acquisition of two LSDs from the US initially on a lease and finally by purchase on 24 January 2001.

CEARÁ 9/2007, Mario R V Carneiro / 1335451

Modernisation: Ceará started a major refit programme in 2008 and was completed in 2014. Service life extended to 2024.
Structure: Has two 50 ton capacity cranes and a docking well of 391 × 48 ft *(119.2 × 14.6 m)*. Two LCVPs and two LCP(L)s on davits. Ice-strengthened bow. SATCOM fitted. Phalanx guns and SRBOC chaff launchers removed before transfer.
Operational: *Rio de Janeiro* was decommissioned on 15 June 2012.

1 SIR GALAHAD CLASS (LSLH)

Name	No	Builders	Laid down	Launched	Commissioned
GARCIA D'ÁVILA (ex-*Sir Galahad*)	G 29 (ex-L 3005)	Swan Hunter Shipbuilders, Wallsend-on-Tyne	12 May 1985	13 Dec 1986	25 Nov 1987

Displacement, tonnes: 6,200 standard; 8,723 full load
Dimensions, metres (feet): 140.5 × 19.5 × 4.3
  (461 × 64.0 × 14.1)
Speed, knots: 18. **Range, n miles:** 13,000 at 15 kt
Complement: 49 (15 officers)
Military lift: 343 troops (537 overload); 16 MBT, 34 mixed vehicles
Machinery: 2 Mirrlees-Blackstone diesels; 13,320 hp *(9.94 MW)*; 2 shafts; cp props; 1 bow thruster; 400 hp *(298 kW)*
Guns: 2 Oerlikon/Royal Ordnance 20 mm GAM-BO3 (twin); 650 rds/min to 10 km *(5.4 n miles)*; weight of shell 0.36 kg. 2 – 7.62 mm MGs.
Physical countermeasures: Decoys: 2 Plessey Shield 200 (6-tubed launchers).
Radars: Surface search: Racal-Decca ARPA 2690; I-band.
Navigation: Kelvin-Hughes Type 1007; I-band.
Combat data systems: Racal CANE data automation.
Helicopters: Platform for 1 medium.

GARCIA D'ÁVILA 2/2008, Maritime Photographic / 1335335

Comment: Former UK Royal Fleet Auxiliary decommissioned in July 2006 and recommissioned into the Brazilian Navy on 4 December 2007 following a refit at Portsmouth. The work included overhaul of the engines and controllable-pitch propellers and upgrade of communications equipment. The ship is equipped with bow and stern ramps, a 25-tonne crane and three 8-tonne cranes. Up to four mexeflote pontoons can be attached to the hull.

10 EDVM 25 CLASS (LCM)

GED 801–805   CAGARRAS –   COMANDATUBA –
CAIEIRAS –   CATAGUAZES –   COTUNDUBA –

Displacement, tonnes: 62 standard; 132 full load
Dimensions, metres (feet): 21.7 × 6.4 × 1.5 *(71.2 × 21.0 × 4.9)*
Speed, knots: 9. **Range, n miles:** 95 at 9 kt
Complement: 5
Military lift: 150 troops plus 72 tons equipment
Machinery: 2 Detroit diesels; 400 hp *(294 kW)* sustained; 2 shafts

Comment: Five vessels constructed by Inace and delivered 1993–94. LCM 8 type. Construction of a further five craft started at AMRJ in 2008. *Caieiras* was delivered in March 2012, *Cagarras* in December 2012, *Cataguazes* in May 2013, *Comandatuba* in August 2013, and *Cotunduba* in late 2013. Based at Niteroi.

GED 801 6/2001, Brazilian Navy / 0130472

© 2014 IHS IHS Jane's Fighting Ships 2014-2015

Brazil > Amphibious forces — Survey and research ships

1 SIR BEDIVERE CLASS (LANDING SHIP LOGISTIC) (LSLH)

Name	No	Builders	Laid down	Launched	Commissioned
ALMIRANTE SABOÍA (ex-Sir Bedivere)	G 25 (ex-L 3004)	Hawthorn Leslie, Hebburn-on-Tyne	Oct 1965	20 Jul 1966	18 May 1967

Displacement, tonnes: 4,983 standard; 6,748 full load
Dimensions, metres (feet): 134.4 × 18.2 × 4 (440.9 × 59.7 × 13.1)
Speed, knots: 17
Range, n miles: 9,200 at 15 kt
Complement: 65 (21 officers)

Military lift: 340 troops (534 hard lying); 18 MBTs; 34 mixed vehicles; 120 tons POL; 30 tons ammunition; 1–25 ton crane; 2–4.5 ton cranes. Capacity for 20 helicopters (11 tank deck and 9 vehicle deck)

Machinery: 2 Wärtsilä 280 V12 diesels; 9,928 hp(m) *(7.3 MW)* sustained; 2 shafts; bow thruster; 980 hp(m) *(720 kW)*
Guns: 4 Oerlikon 20 mm. 4–7.62 mm MGs. 2 Mk 44 7.62 mm Miniguns.
Physical countermeasures: Decoys: 2 Plessey Shield chaff launchers.
Radars: Surface search: Racal-Decca ARPA 2690; I-band.
Navigation: Furuno ARPA M-1942; I-band. Kelvin Hughes Type 1007; I-band.
Helicopters: Platform to operate Lynx, Chinook or Sea King.

Comment: Former UK Royal Fleet Auxiliary decommissioned on 18 February 2008 and recommissioned into the Brazilian Navy on 21 May 2009. Fitted for bow and stern loading with drive-through facilities and deck-to-deck ramps. Facilities provided for onboard maintenance of vehicles and for laying out pontoon equipment. Mexeflote self-propelled floating platforms can be strapped one on each side. SLEP in Rosyth from December 1994 to January 1998 included lengthening by 29 ft an enlarged flight deck, new main engines and a new bridge. The helicopter platform was lowered by one deck, which reduced the size of the stern ramp.

ALMIRANTE SABOÍA (UK colours) *4/2007, Shaun Jones / 1170256*

MINE WARFARE FORCES

Notes: The tug *Triunfo* R 23 is reported to have a mine-laying capability.

6 ARATU (SCHÜTZE) CLASS (MINESWEEPERS—COASTAL) (MSC)

Name	No	Builders	Commissioned
ARATU	M 15	Abeking & Rasmussen, Lemwerder	5 May 1971
ANHATOMIRIM	M 16	Abeking & Rasmussen, Lemwerder	30 Nov 1971
ATALAIA	M 17	Abeking & Rasmussen, Lemwerder	13 Dec 1972
ARAÇATUBA	M 18	Abeking & Rasmussen, Lemwerder	13 Dec 1972
ABROLHOS	M 19	Abeking & Rasmussen, Lemwerder	25 Feb 1976
ALBARDÃO	M 20	Abeking & Rasmussen, Lemwerder	25 Feb 1976

Displacement, tonnes: 245 standard; 284 full load
Dimensions, metres (feet): 47.2 × 7.2 × 2.1 (154.9 × 23.6 × 6.9)
Speed, knots: 24. **Range, n miles:** 710 at 20 kt
Complement: 32 (4 officers)

Machinery: 2 MTU 16V 652 SB80 diesels; 4,500 hp(m) *(3.3 MW)*; 2 shafts; 2 Escher-Weiss cp props
Guns: 1 Bofors TY 39 40 mm/70; 240 rds/min to 12 km *(6.5 n miles)*; weight of shell 0.96 kg.
Radars: Surface search: Racal-Decca 1070A; I-band.
Navigation: Furuno M-1831; I-band.

Comment: Wooden hulled. First four ordered in April 1969 and last pair in November 1973. Same design as the now deleted German Schütze class. Can carry out wire, magnetic and acoustic sweeping. A life-extension refit programme started in 2001. M 15 completed in 2002 and M17, 18 and 19 by 2005. M 16 completed in 2006 and M 20 completed in 2007. Modifications include replacement of the surface search radar, communications upgrade and hull preservation measures. Based at Aratu, Bahia.

ABROLHOS *3/1998, Brazilian Navy / 0017625*

SURVEY AND RESEARCH SHIPS

Notes: (1) Survey ships are painted white except for those operating in the Antarctic which have red hulls.
(2) There are also 31 buoy tenders of between 15 and 26 m: 10 LB 15, two LB 17 (*Lufada* and *Piracema*), two LB 19 (*Santana* and *Braz de Aguiar*), 10 LB 20 (*Achernar, Aldebaran, Betelgeuse, Capella, Denébola, Formalhaut, Regulus, Rigel, Vega,* and *Pollux*), two LB 23 (*Suboficial Oliveira* and *Marco Zero*) and five LB 26.
(3) There are four 16 m inshore survey craft: *Camocin* (ex-H 16); *Paraibano* (ex-H 11); *Rio Branco* (ex-H 12); *Caravelas* (ex-H 17).
(4) Four 24.5 m river survey vessels were built at Inace. The first, *Rio Tocantins* (H 12) was delivered in July 2012. Three further vessels, *Rio Xingu* (H 13), *Rio Solimões* (H 14), and *Rio Negro* (H 15) followed in 2013.
(5) There is one 10 m hydrographic launch *Pinguim*.

1 LIGHTHOUSE TENDER (ABUH)

Name	No	Builders	Launched	Commissioned
ALMIRANTE GRAÇA ARANHA	H 34	Ebin, Niteroi	23 May 1974	9 Sep 1976

Displacement, tonnes: 1,087 standard; 2,479 full load
Dimensions, metres (feet): 74.8 × 13 × 4.2 *(245.4 × 42.7 × 13.8)*
Speed, knots: 14
Complement: 81 (8 officers)
Machinery: 1 MWM TDB-441-V16 diesel; 2,440 hp(m) *(1.8 MW)*; 1 shaft; bow thruster
Radars: Navigation: 2 Racal Decca TM 1226C; I-band. Furuno M-1832; I-band.
Helicopters: 1 Bell JetRanger.

Comment: Laid down in 1971. Fitted with telescopic hangar, 10 ton crane, two landing craft, one RHIB and two Land Rovers. Omega navigation system.

ALMIRANTE GRAÇA ARANHA *4/2000, Hartmut Ehlers / 0104234*

1 BUOY TENDER (ABU)

Name	No	Commissioned
FAROLEIRO MÁRIO SEIXAS (ex-*Mestre Jerânimo*)	H 26	31 Jan 1984

Displacement, tonnes: 238 standard; 299 full load
Dimensions, metres (feet): 35.5 × 6.6 × 3.6 *(116.5 × 21.7 × 11.8)*
Speed, knots: 10
Complement: 19 (2 officers)
Machinery: 2 Scania DSI 14 MO3 diesels; 900 hp *(671 kW)*; 2 shafts
Radars: Navigation: Furuno M-1832; I-band.

Comment: Former fishing vessel built in Vigo, Spain in 1964. Acquired by Brazilian Navy in 1979 and rebuilt as a buoy tender. Commissioned 31 January 1984.

FAROLEIRO MÁRIO SEIXAS *6/2002, Brazilian Navy / 0529148*

Survey and research ships — **Brazil** 83

1 POLAR RESEARCH SHIP (AGOBH)

Name	No	Builders	Commissioned
ARY RONGEL (ex-*Polar Queen*)	H 44	Eides	22 Jan 1981

Displacement, tonnes: 1,959 standard; 3,686 full load
Dimensions, metres (feet): 75.3 × 13 × 5.3 *(247 × 42.7 × 17.4)*
Speed, knots: 14.5
Range, n miles: 17,000 at 12 kt
Complement: 70 (19 officers) + 22 scientists

Cargo capacity: 2,400 m³
Machinery: 2 MAK 6M-453 diesels; 4,500 hp(m) *(3.3 MW)*; 1 shaft; cp prop; 2 Brunvoll bow thrusters; 1 Brunvoll stern thruster
Radars: Surface search: Furuno FR-2115; I-band.
Navigation: Raytheon ARPA M 34; I-band. Furuno M-1832; I-band. Furuno M-1623; I-band.
Helicopters: Platform for UH-13 Esquilo.

Comment: Acquired by sale 19 April 1994. Ice-strengthened hull fitted with Simrad Albatross dynamic positioning system. Equipped with three Zodiac RHIBs. Based at Rio de Janeiro.

ARY RONGEL *6/2002, Carlos Veras, Brazilian Navy* / 0572424

3 AMORIM DO VALLE (RIVER) CLASS (SURVEY SHIPS) (AGS)

Name	No	Builders	Commissioned
AMORIM DO VALLE (ex-*Humber*)	H 35 (ex-M 2007)	Richards Ltd, Lowestoft	7 Jun 1985
TAURUS (ex-*Helmsdale*; ex-*Jorge Leite*)	H 36 (ex-M 2010)	Richards Ltd, Lowestoft	1 Mar 1986
GARNIER SAMPAIO (ex-*Ribble*)	H 37 (ex-M-2012)	Richards Ltd, Great Yarmouth	19 Feb 1986

Displacement, tonnes: 782 standard; 904 full load
Dimensions, metres (feet): 47.5 × 10.5 × 2.9 *(155.8 × 34.4 × 9.5)*
Speed, knots: 14
Range, n miles: 4,500 at 10 ktH 37H 35H 36
Complement: 36 (4 officers)
Machinery: 2 Ruston 6RKC diesels; 3,100 hp *(2.3 MW)* sustained; 2 shafts; cp props
Radars: Navigation: 2 Racal Decca TM 1226C; I-band. Furuno M-1832; I-band.

Comment: Three ships transferred from the UK on 31 January 1995. The contract was signed on 18 November 1994. Steel hulled. All minesweeping gear and the 40 mm gun removed on transfer. Used as hydrographic ships. H 35 and H 36 fitted with a stern gantry and second crane amidships for oceanographic research. Equipment includes Simrad EM-1000 multibeam echo-sounders. Four others of the class transferred in 1998 are listed under Patrol Forces. The class is also in service with the Bangladesh Navy. H 35 and H 36 based at Rio de Janeiro and H 37 at Belém.

AMORIM DO VALLE *6/1995, David Cullen* / 1153035

GARNIER SAMPAIO *6/2002, Brazilian Navy* / 0529149

1 RESEARCH SHIP (AGS)

Name	No	Builders	Commissioned
ANTARES (ex-*M/V Lady Harrison*)	H 40	Mjellem & Karlsen AS, Bergen	Aug 1984

Displacement, tonnes: 869 standard; 1,268 full load
Dimensions, metres (feet): 55 × 10.3 × 4.3 *(180.4 × 33.8 × 14.1)*
Speed, knots: 13.5
Range, n miles: 10,000 at 12 kt
Complement: 58 (12 officers) + 12 personnel
Machinery: 1 Burmeister & Wain Alpha diesel; 1,860 hp(m) *(1.37 MW)*; 1 shaft; cp prop; bow thruster
Radars: Surface search: Racal Decca RMS 1230C; E/F-band.
Navigation: Racal Decca RM 914C; I-band.

Comment: Research vessel acquired from Racal Energy Resources. Equipped with side-scan sonar for route survey, Atlas Krupp deep echo sounder and Kongsberg/Simrad EA-500 deep echo sounder. Used for seismographic survey. Recommissioned 6 June 1988.

ANTARES *9/2010, Mario R V Carneiro* / 1406714

4 BUOY TENDERS (ABU)

Name	No	Builders	Commissioned
COMANDANTE VARELLA	H 18	Arsenal de Marinha, Rio de Janeiro	20 May 1982
TENENTE CASTELO	H 19	Estanave, Manaus	15 Aug 1984
COMANDANTE MANHÃES	H 20	Estanave, Manaus	15 Dec 1983
TENENTE BOANERGES	H 25	Estanave, Manaus	29 Mar 1985

Displacement, tonnes: 305 standard; 440 full load
Dimensions, metres (feet): 37.5 × 8.6 × 2.6 *(123 × 28.2 × 8.5)*
Speed, knots: 12
Range, n miles: 2,880 at 9 kt
Complement: 22 (2 officers)
Machinery: 2 MAN R8V16-18TL 8 cylinder diesels; 1,300 hp(m) *(955 kW)*; 2 shafts
Radars: Navigation: Racal Decca TM 1226C; I-band. Furuno M-1832; I-band.

Comment: Dual-purpose minelayers. *Comandante Varella* is based at Rio Grande, *Tenente Castelo* at Belém, *Comandante Manhães* at Natal and *Tenente Boanerges* at Salvador.

COMANDANTE VARELLA *5/2008, A E Galarce* / 1305677

1 RESEARCH VESSEL (AGS)

Name	No	Builders	Commissioned
ASPIRANTE MOURA (ex-*Finder*)	U 14	Brodogradiliste Split	13 May 1987

Displacement, tonnes: 653 full load
Dimensions, metres (feet): 36.1 × 9.2 × 2.6 *(118.4 × 30.2 × 8.5)*
Speed, knots: 11
Machinery: 1 MAN 9L20/27 diesel; 1,213 hp *(892 kW)*; 1 azimuth thruster
Radars: Navigation: Furuno; I-band.

Comment: Originally constructed as a buoy tender, the ship was converted into a survey vessel in 2005. Acquired by the Brazilian Navy on 25 January 2010.

ASPIRANTE MOURA *4/2010, I J Plokker* / 1366447

84 Brazil > Survey and research ships — Training ships

1 POLAR RESEARCH SHIP (AGOBH)

Name	No	Builders	Laid down	Launched	Commissioned
ALMIRANTE MAXIMIANO (ex-*Ocean Express*)	H 41	Todd Pacific Shipyard Corporation, Seattle	20 Aug 1973	13 Feb 1974	19 Dec 1978

Displacement, tonnes: 5,537 full load
Dimensions, metres (feet): 93.4 × 13.4 × 6.2 *(306.4 × 44.0 × 20.3)*
Speed, knots: 13
Range, n miles: 10,000 at 11 kt
Complement: 54

Machinery: 2 Caterpillar 3616 diesels; 8,670 hp *(6.5 MW)*; 2 shafts; 3 Brunvoll bow thrusters; 1 Aquamaster azimuth thruster
Radars: Surface search: Raytheon 3425; I-band.
Navigation: Raytheon 3410; I-band.
Helicopters: 2 UH-13 Esquilo.

Comment: Recommissioned on 3 February 2009 after undergoing conversion to an Antarctic support ship role at BREDO Shipyard in Germany. The refit included provision of a hangar and flight deck to operate two helicopters and the installation of five laboratories. Sonar equipment includes Teledyne Benthos side-scan sonar, bathymetry for botttom mapping and Simrad FS 3000 surveillance sonar.

ALMIRANTE MAXIMIANO   2/2009, Michael Nitz / 1305329

1 SIRIUS CLASS (SURVEY SHIP) (AGSH)

Name	No	Builders	Launched	Commissioned
SIRIUS	H 21	Ishikawajima Co Ltd, Tokyo	30 Jul 1957	17 Jan 1958

Displacement, tonnes: 1,471 standard; 1,915 full load
Dimensions, metres (feet): 78 × 12.1 × 3.7 *(255.9 × 39.7 × 12.1)*
Speed, knots: 14. **Range, n miles:** 4,000 at 11 kt
Complement: 129 (16 officers) + 14 scientists

Machinery: 2 Vilares-Burmeister & Wain diesels; 1,550 hp *(1.15 MW)*; 2 shafts; cp props
Radars: Surface search: Racal Decca RMS 1230C; E/F-band.
Navigation: Furuno M 1942; E/F/I-band. Furuno; I-band.
Helicopters: 1 Bell JetRanger or UH-12.

Comment: Laid down 1955–56. Equipped with side-scan sonar for route survey, Simrad EM-302 multibeam echo-sounder, Simrad EA-600 deep echo-sounder, temperature and depth profilers. There are three Flexboat RHIBs and there is space for two 20 ft containers.

SIRIUS   9/2007, Mario R V Carneiro / 1335445

1 RESEARCH SUPPORT VESSEL (AGS)

Name	No	Builders	Laid down	Launched	Commissioned
CRUZEIRO DO SUL (ex-*DSND Surveyor*)	H 38	Løngva Mek, Verksted	1 Mar 1986	1 Jul 1986	31 Jul 1986

Measurement, tonnes: 1,744 gt
Dimensions, metres (feet): 65.7 × 11.0 × 4.5 *(215.6 × 36.1 × 14.8)*
Speed, knots: 13.5
Range, n miles: 10,000 at 12 kt
Complement: 43 (8 officers)
Machinery: 1 Bergen KRMB-9 diesel; 1 shaft; Ulstein cp prop; 1 Ulstein forward thruster *(368 kW)*; 1 Brunvoll forward thruster *(600 kW)*; 1 Ulstein retractable azimuth thruster *(880 kW)*; 2 Ulstein bow thrusters *(552 kW and 368 kW)*
Radars: Surface search: Raytheon R 84; I-band.
Navigation: Raytheon R 81; I-band.

Comment: Originally built as a multirole inspection/survey vessel and converted in 1991 into a ROV support vessel. Acquired by the Brazilian Navy and commissioned on 28 February 2008. The ship is capable of performing a range of tasks including pipeline inspection, structural inspection, geophysical and geotechnical operations and other support services. The principal features of the ship include a 6-ton Hydralift crane, large survey/inspection and data processing offices, a wet and dry lab space and a photo lab. The ship has high station-keeping performance. There is a large work-deck area and a moontube for deploying survey transducers.

CRUZEIRO DO SUL   9/2011, Mario R V Carneiro / 1406716

0 + 1 SURVEY VESSEL (AGS)

Name	No	Builders	Laid down	Launched	Commissioned
RIO BRANCO	–	INACE, Fortalesa	23 Apr 2013	2014	2015

Displacement, tonnes: 530 full load
Dimensions, metres (feet): 55 × 9.0 × 2.0 *(180.4 × 29.5 × 6.6)*
Range, n miles: 3,000 at 12 kt
Complement: 36 (6 officers)
Machinery: To be announced
Radars: To be announced.

Comment: The contract for a new survey ship was signed with INACE on 5 December 2012. The ship has been designed to undertake hydrographic survey and environmental data collection in the Amazon River in concert with the Brazilian Army and Air Force and with the Geological Survey of Brazil. The ship is to be delivered in mid-2015 and is to be based at Manaus (Ninth Naval District).

0 + 1 SURVEY VESSEL (AGS)

Name	No	Builders	Laid down	Launched	Commissioned
VITAL DE OLIVEIRA	–	Hantong Shipbuilding, Guangzhou	2013	2014	2014

Displacement, tonnes: 3,500 full load
Dimensions, metres (feet): 78 × 20.0 × ? *(255.9 × 65.6 × ?)*
Speed, knots: 12
Range, n miles: 17,280 at 12 kt
Complement: 90 + 50 scientists
Machinery: To be announced
Radars: To be announced.

Comment: The announcement that a new survey vessel was to be built in China was made in September 2013. The purchase is as a result of a public-private partnership between the navy, Petrobas, the Ministry of Science, Technology and Innovation, and mining company VALE. Designed by ASK Subsea, the ship is to be equipped with five laboratories and the facilities to operate a remotely controlled vehicle with the capability to dive to 4,000 m.

TRAINING SHIPS

Notes: (1) There are 21 small sail training ships.
(2) There are three 18 m training craft (*Rosca Fina, Voga Picada, Leva Arriba*).

1 MODIFIED NITERÓI CLASS (AXH)

Name	No	Builders	Commissioned
BRASIL	U 27	Arsenal de Marinha, Rio de Janeiro	21 Aug 1986

Displacement, tonnes: 2,589 standard; 3,789 full load
Dimensions, metres (feet): 131.3 × 13.5 × 4.2 *(430.8 × 44.3 × 13.8)*
Speed, knots: 18
Range, n miles: 7,000 at 15 kt
Complement: 422 (27 officers; 204 midshipmen)

Machinery: 2 Pielstick/Ishikawajima (Brazil) 6 PC2.5 L 400 diesels; 7,020 hp(m) *(5.17 MW)* sustained; 2 shafts
Guns: 2 Bofors 40 mm/70. 4 saluting guns.
Physical countermeasures: Decoys: 2 CBV 50.8 mm flare launchers.
Electronic countermeasures: ESM: Racal RDL-2 ABC; radar intercept.
Radars: Surface search: Racal Decca RMS 1230C; E/F-band.
Navigation: Racal Decca TM 1226C and TMS 1230; I-band.
Weapon control systems: Saab Scania TVT 300 optronic director.
Helicopters: Platform for 1 Sea King.

Comment: A modification of the Vosper Thornycroft Mk 10 Frigate design ordered in June 1981. Laid down 18 September 1981, launched 23 September 1983. Designed to carry midshipmen and other trainees from the Naval and Merchant Marine Academies. Minimum electronics as required for training. There are two 51 mm launchers for flares and other illuminants.

BRASIL   8/2011, Michael Nitz / 1406715

IHS Jane's Fighting Ships 2014-2015   © 2014 IHS

Training ships — Auxiliaries < Brazil 85

3 NASCIMENTO CLASS (AXL)

Name	No	Builders	Commissioned
ASPIRANTE NASCIMENTO	U 10	Ebrasa, Santa Catarina	13 Dec 1980
GUARDA MARINHA JANSEN	U 11	Ebrasa, Santa Catarina	22 Jul 1981
GUARDA MARINHA BRITO	U 12	Ebrasa, Santa Catarina	22 Jul 1981

Displacement, tonnes: 110 standard; 138 full load
Dimensions, metres (feet): 28 × 6.5 × 1.8 (91.9 × 21.3 × 5.9)
Speed, knots: 10
Range, n miles: 700 at 10 kt
Complement: 16 (2 officers; 10 midshipmen)
Machinery: 2 MWM D232V12 diesels; 805 hp(m) (600 kW); 2 shafts
Guns: 2—12.7 mm MGs.
Radars: Navigation: Furuno FR 8252; I-band.

Comment: Can carry 10 trainees overnight. All of the class are attached to the Naval Academy at Rio de Janeiro. One RHIB is carried.

GUARDA MARINHA JANSEN 5/2012, Mario R V Carneiro / 1482965

1 SAIL TRAINING SHIP (AXS)

Name	No	Builders	Launched	Commissioned
CISNE BRANCO	U 20	Damen Shipyards, Gorinchem	4 Aug 1999	28 Feb 2000

Displacement, tonnes: 733 standard; 1,055 full load
Dimensions, metres (feet): 76 × 10.5 × 4.8 (249.3 × 34.4 × 15.7)
Speed, knots: 17; 11 diesel
Complement: 81 (9 officers; 31 trainees)
Machinery: 1 Caterpillar 3508B DI-TA diesel; 1,015 hp(m) (746 kW) sustained; 1 shaft; Berg cp prop; bow thruster; 408 hp(m) (300 kW)
Radars: Navigation: Furuno FR 1510 Mk 3; I-band.

Comment: Ordered in 1998. Maximum sail area 2,195 m².

CISNE BRANCO 5/2013*, Michael Nitz / 1529005

AUXILIARIES

Notes: (1) There are four 36 m Rio Pardo class transport vessels (*Rio Pardo*, *Rio Negro*, *Rio Chuí* and *Rio Oiapoque*). Capable of carrying 600 passengers, they are all based at Rio de Janeiro.
(2) One 23 m Torpedo Recovery Craft, *Almirante Hess*, is based at Niteroí.
(3) There is a requirement for a transport and logistic ship of approximately 9,000 tons. The ship is to be helicopter-capable with accommodation for 500 marines.
(4) There is a requirement for a replenishment tanker, to replace *Marajo*. The requirement is for a helicopter-capable ship with four (two each side) transfer stations.

1 SUBMARINE RESCUE SHIP (ASRH)

Name	No	Builders	Commissioned
FELINTO PERRY (ex-*Holger Dane*; ex-*Wildrake*)	K 11	Stord Verft	Dec 1979

Displacement, tonnes: 2,940 standard; 4,173 full load
Dimensions, metres (feet): 78.2 × 17.5 × 4.6 (256.6 × 57.4 × 15.1)
Speed, knots: 14.5
Complement: 65 (9 officers)
Machinery: Diesel-electric; 2 BMK KVG B12 and 2 KVG B16 diesels; 11,400 hp(m) (8.4 MW); 2 Daimler-Benz motors; 7,000 hp(m) (5.15 MW); 2 shafts; cp props; 2 bow thrusters; 2 stern thrusters
Radars: Navigation: 2 Raytheon; I-band.
Helicopters: Platform only.

Comment: Former oilfield support ship acquired 28 December 1988. Has an octagonal heliport (62.5 ft diameter) above the bridge. Equipped with a moonpool for saturation diving, and rescue and recompression chambers as the submarine rescue ship. A Deep Ocean Phantom DS4 ROV, capable of operating to 610 m, is also carried. Dynamic positioning system. Based at Niteroi, Rio de Janeiro.

© 2014 IHS

FELINTO PERRY 9/2010, Mario R V Carneiro / 1406719

1 RIVER TRANSPORT SHIP (AP)

Name	No	Builders	Commissioned
PARAGUASSÚ (ex-*Garapuava*)	G 15	Amsterdam Drydock, Amsterdam	1951

Displacement, tonnes: 203 standard; 235 full load
Dimensions, metres (feet): 40 × 7 × 1.5 (131.2 × 23.0 × 4.9)
Speed, knots: 13. **Range, n miles:** 2,500 at 10 kt
Complement: 35 (4 officers)

Military lift: 178 troops
Machinery: 3 diesels; 2,505 hp(m) (1.84 MW); 1 shaft
Guns: 4—12.7 mm MGs.
Radars: Furuno 3600; I-band.

Comment: Passenger ship converted into a troop carrier in 1957 and acquired on 20 June 1972.

PARAGUASSÚ 5/2000, Hartmut Ehlers / 0104240

1 RIVER TRANSPORT (YFBH)

Name	No	Builders	Commissioned
PIRAIM (ex-*Guaicuru*)	U 29	Estaleiro SNBP, Mato Grosso	10 Mar 1982

Displacement, tonnes: 74 standard; 93 full load
Dimensions, metres (feet): 25 × 5.5 × 0.97 (82 × 18.0 × 3.2)
Speed, knots: 7
Range, n miles: 700 at 7 kt
Complement: 17 (2 officers)

Machinery: 2 MWM diesels; 400 hp(m) (294 kW); 2 shafts
Guns: 4—7.62 mm MG.
Radars: Navigation: Furuno 3600; I-band.
Helicopters: Platform for UH-12.

Comment: Used as a logistics support ship for the Mato Grosso Flotilla. Can carry two platoons of marines and two rigid inflatable boats.

PIRAIM 5/2011, Hartmut Ehlers / 1406718

1 PARÁ CLASS (RIVER TRANSPORT SHIP) (YFB)

Name	No	Builders	Commissioned
PARÁ	U 15	Inconav; MacLaren	19 Jan 2005

Displacement, tonnes: 1,081 standard; 1,348 full load
Dimensions, metres (feet): 56.1 × 21.4 × 3.97 (184.1 × 70.2 × 13.0)
Speed, knots: 11
Range, n miles: 2,380 at 10 kt
Complement: 78 (7 officers; 12 medical personnel)

Machinery: 2 Ishibras-Daihatsu diesels; 1,050 hp (783 kW); 2 shafts
Guns: 4 Oerlikon 20 mm.
Radars: Navigation: Furuno 1830 and 1942; I-band.
Helicopters: Platform for 1 UH-12 or UH-13 Esquilo.

Comment: Ex-civilian catamaran hull vessel capable of carrying 175 marines and 350 tons cargo.

IHS Jane's Fighting Ships 2014-2015

86 Brazil > Auxiliaries

2 HOSPITAL SHIPS (AHH)

Name	No	Builders	Commissioned
OSWALDO CRUZ	U 18	Arsenal de Marinha, Rio de Janeiro	29 May 1984
CARLOS CHAGAS	U 19	Arsenal de Marinha, Rio de Janeiro	7 Dec 1984

Displacement, tonnes: 366 standard; 498 full load
Dimensions, metres (feet): 47.2 × 8.5 × 1.8 (154.9 × 27.9 × 5.9)
Speed, knots: 17. **Range, n miles:** 4,000 at 12 kt
Complement: 46 (4 officers; 21 medical personnel)
Machinery: 2 Volvo diesels; 714 hp(m) (525 kW); 2 shafts
Radars: Navigation: Racal Decca; I-band.
Helicopters: Platform for 1 UH-12/13 Esquilo.

Comment: Oswaldo Cruz launched 11 July 1983, and Carlos Chagas 16 April 1984. Has two sick bays, dental surgery, a laboratory, two clinics and X-ray centre. The design is a development of the Roraima class with which they operate in the Amazon Flotilla. Since 1992 both ships painted grey with dark green crosses on the hull.

OSWALDO CRUZ   6/2004, Brazilian Navy / 1044086

1 HOSPITAL SHIP (AH)

Name	No	Builders	Commissioned
DOUTOR MONTENEGRO	U 16	CONAVE Shipyard, Manaus	17 May 2000

Displacement, tonnes: 305 standard; 353 full load
Dimensions, metres (feet): 41 × 11 × 2.4 (134.5 × 36.1 × 7.9)
Speed, knots: 5
Complement: 58 (8 officers; 11 medical personnel)
Machinery: 2 Cummins NT 855M diesels; 560 hp (418 kW); 2 shafts
Radars: Navigation: Furuno 1942 Mk 2.

Comment: U 16 was built in January 1997 and belonged to the government of the Acre state before transfer to the Brazilian Navy. The ship has two wards, a pediatric ICU, an operating theatre, an X-ray room, a dentist office, a lab for clinical analysis, a trauma room and a pharmacy.

DOUTOR MONTENEGRO   6/2007, Brazilian Navy / 1170088

1 REPLENISHMENT TANKER (AOR)

Name	No	Builders	Commissioned
ALMIRANTE GASTÃO MOTTA	G 23	Ishibras, Rio de Janeiro	26 Nov 1991

Displacement, tonnes: 4,543 standard; 10,486 full load
Dimensions, metres (feet): 135 × 19 × 7.5 (442.9 × 62.3 × 24.6)
Speed, knots: 20
Range, n miles: 9,000 at 15 kt
Complement: 121 (13 officers) + 12 spare berths

Cargo capacity: 5,920 tons dieso; 950 tons JP-5; 200 tons dry
Machinery: Diesel-electric; 2 Wärtsilä 12V32 diesel generators; 11,700 hp(m) (8.57 MW) sustained; 1 motor; 1 shaft; Kamewa cp prop
Guns: 2—12.7 mm MGs.
Radars: Surface search: Furuno FR 1760; I-band.
Navigation: Furuno M 1832; I-band.

Comment: Ordered March 1987. Laid down 11 December 1989 and launched 1 June 1990. Fitted for abeam and stern refuelling. Carries two whalers and two RHIBs.

ALMIRANTE GASTÃO MOTTA   9/2010, Mario R V Carneiro / 1406720

1 REPLENISHMENT TANKER (AOR)

Name	No	Builders	Launched	Commissioned
MARAJO	G 27	Ishikawajima do Brasil	31 Jan 1968	8 Jan 1969

Displacement, tonnes: 7,620 standard; 15,352 full load
Dimensions, metres (feet): 134.4 × 19.3 × 7.3 (440.9 × 63.3 × 24.0)
Speed, knots: 13. **Range, n miles:** 9,200 at 13 kt
Complement: 80 (13 officers)

Cargo capacity: 7,470 tons fuel
Machinery: 1 Sulzer GRD 68 diesel; 8,000 hp(m) (5.88 MW); 1 shaft
Radars: Surface search: Racal Decca TM 1226C; I-band.
Navigation: Racal Decca BT 503; I-band.

Comment: Fitted for abeam replenishment with two stations on each side. Was to have been replaced by Gastão Motta but is to be retained in service. A refit at AMRJ began in 2010 and completed in 2011.

MARAJO   9/2013*, Mario R V Carneiro / 1529006

1 RIVER TENDER (AG)

Name	No	Builders	Commissioned	Recommissioned
POTENGI	G 17	Papendrecht	28 Jun 1938	6 May 1999

Displacement, tonnes: 152 standard; 604 full load
Dimensions, metres (feet): 54.5 × 7.5 × 1.8 (178.8 × 24.6 × 5.9)
Speed, knots: 10. **Range, n miles:** 600 at 8 kt
Complement: 19 (2 officers)

Cargo capacity: 460 tons of general cargo including fuel, frozen and dry stores
Machinery: 2 Krohout diesels; 550 hp(m) (404 kW); 2 shafts
Guns: 4—7.62 mm MGs.
Radars: Furuno FR-1525; I-band.

Comment: Launched 16 March 1938. Employed in the Mato Grosso Flotilla on river service. Converted to logistic support ship and recommissioned 6 May 1999.

POTENGI   5/2000, Hartmut Ehlers / 0104241

4 FLOATING DOCKS

CIDADE DE NATAL —   ALMIRANTE MARIO CARNEIRO —
ALMIRANTE SCHIECK —   ALMIRANTE JERONIMO GONÇALVES —

Comment: The first two are floating docks loaned to Brazil by US Navy in the mid-1960s and purchased 11 February 1980. Ship lifts of 2,800 tons and 1,000 tons respectively. Cidade de Natal based at Natal and Almirante Jeronimo Gonçalves at Manaus. Almirante Schieck of 3,600 tons displacement was built by Arsenal de Marinha, Rio de Janeiro and commissioned 12 October 1989. Almirante Mario Carneiro acquired from US and based at Val-de-Caes (Para).

1 HOSPITAL SHIP (AH)

Name	No	Builders	Commissioned
TENIENTE MAXIMIANO (ex-Scorpion)	U 28	Decenzo e Hipolito, Ladario	17 Mar 2009

Displacement, tonnes: 160 standard
Dimensions, metres (feet): 30 × 6.4 × 1.2 (98.4 × 21.0 × 3.9)
Speed, knots: 25
Complement: 22 (3 officers)
Machinery: 2 MWM TD229-EC-6 diesels; 360 hp (268 kW); 2 shafts
Radars: Navigation: Furuno FR-8252; I-band.
Helicopters: Platform for 1 Jet Ranger or 1 UH-12/13.

Comment: Ex-commercial vessel converted into hospital ship at Ladario. Equipped with two sick-bays, an operating theatre and dental surgery. There are two boats and two RIBs.

1 HOSPITAL SHIP (AH)

Name	No	Builders	Commissioned
SOARES DE MEIRELLES (ex-Ludovico Celani)	U 21	Comércio e Transporte e Navegação, Ltda	Jan 2011

Displacement, tonnes: 1,388 full load
Dimensions, metres (feet): 63 × 12.0 × 3.2 (206.7 × 39.4 × 10.5)
Speed, knots: 12
Complement: 44 (4 officers)
Machinery: 2 Yanmar 6AYM-ETE diesels; 1,640 hp (1.22 MW); 2 shafts
Radars: Navigation: To be announced.

Comment: Former river cruise ship, built at Manaus in 2008, purchased by the Brazilian Navy in August 2010. The ship was subsequently converted into a hospital ship and commissioned in January 2011. In 2012, the ship began a second phase of modification that is to include a helicopter deck and expansion of hospital facilities.

Tugs < Brazil — Introduction < Brunei 87

TUGS

Notes: (1) In addition to the vessels listed below there are eight harbour tugs: *Comandante Marroig* (BNRJ 03), *Comandante Didier* (BNRJ 04), *Tenente Magalhães* (BNA 06), *Cabo Schram* (BNVC 01), *Intrépido* (BNRJ 16), *Arrojado* (BNRJ 17), *Valente* (BNRJ 18) and *Impávido* (BNRJ 19).
(2) There are plans to procure six ocean tugs. These are also to serve as offshore patrol ships.

2 ALMIRANTE GUILHEM CLASS (FLEET OCEAN TUGS) (ATF)

Name	No	Builders	Commissioned
ALMIRANTE GUILHEM (ex-*Superpesa 4*)	R 24	Sumitomo, Uraga	1976
ALMIRANTE GUILLOBEL (ex-*Superpesa 5*)	R 25	Sumitomo, Uraga	1976

Displacement, tonnes: 2,431 standard; 2,779 full load
Dimensions, metres (feet): 63.2 × 13.4 × 4.5 *(207.3 × 44.0 × 14.8)*
Speed, knots: 14
Range, n miles: 10,000 at 13 kt
Complement: 40 (4 officers)
Machinery: 2 GM EMD 20-645F7B diesels; 7,120 hp *(5.31 MW)* sustained; 2 shafts; cp props; bow thruster
Guns: 2 Oerlikon 20 mm (not always carried)
Radars: Racal Decca; I-band. Furuno; I-band.

Comment: Originally built as civilian tugs. Bollard pull, 84 tons. Commissioned into the Navy 22 January 1981.

3 TRITÃO CLASS (FLEET OCEAN TUGS) (ATA)

Name	No	Builders	Commissioned
TRITÃO (ex-*Sarandi*)	R 21	Estanave, Manaus	19 Feb 1987
TRIDENTE (ex-*Sambaiba*)	R 22	Estanave, Manaus	8 Oct 1987
TRIUNFO (ex-*Sorocaba*)	R 23	Estanave, Manaus	5 Jul 1986

Displacement, tonnes: 832 standard; 1,707 full load
Dimensions, metres (feet): 55.4 × 11.6 × 3.4 *(181.8 × 38.1 × 11.2)*
Speed, knots: 13
Complement: 44 (6 officers)
Machinery: 2 Vilares-Burmeister and Wain Alpha diesels; 2,480 hp(m) *(1.82 MW)*; 2 shafts; bow thruster
Guns: 2 Oerlikon 20 mm.
Radars: Navigation: 1 Racal Decca TM 1226C; I-band. 2 Furuno; I-band.

Comment: Offshore supply vessels acquired from National Oil Company of Brazil and converted for naval use. Assumed names of previous three ships of Sotoyomo class. Fitted to act both as tugs and patrol vessels. *Triunfo* has a mine-laying capability. Bollard pull, 23.5 tons. Firefighting capability. Endurance, 45 days.

ALMIRANTE GUILLOBEL   9/2010, Mario R V Carneiro / 1406721

TRIDENTE   5/2012, Mario R V Carneiro / 1482964

British Indian Ocean Territory

Country Overview

The British Indian Ocean Territory was established as a British dependency in 1965 and is administered by a Commissioner and Administrator who reside in the UK. Situated in the Indian Ocean, halfway between Africa and Indonesia, the territory comprises six atolls of the Chagos Archipelago which consist of the order of 1,000 uninhabited islands. The largest island is Diego Garcia (17 square miles) which was leased to the United States in 1971 in order to build an air and naval base. Adjacent to the small military port, the lagoon provides a protected anchorage for US pre-positioned forces while the island is also home to a number of communications and space-related facilities. Exclusively occupied by military (largely US) forces and contractors, the base includes a small British garrison, whose commanding officer represents the Commissioner. Territorial waters (3 n miles) are claimed as is a 200 n mile fishery zone.

PATROL FORCES

1 FISHERY PATROL SHIP (PSO)

PACIFIC MARLIN (ex-*Bigorange XI*)

Measurement, tonnes: 1,219 gt
Dimensions, metres (feet): 57.7 × 12.2 × 3.8 *(189.3 × 40.0 × 12.5)*
Speed, knots: 12.5
Complement: 20 + 13 spare berths
Machinery: 2 Yanmar G250-E diesels; 2,600 hp *(1.9 MW)*; 2 shafts; 1 Kamome TF30DLN bow thruster; 300 hp *(225 kW)*
Radars: Surface search/navigation: JRC JMA-3210; I-band.

Comment: Former Production Testing Vessel built by Teraoka Zosen, Japan in 1978. Converted for fishery protection duties and chartered from Swire Pacific Offshore until 1 January 2015. Equipped with 32 ton deck crane and two fast rescue craft. Steel construction.

PACIFIC MARLIN
6/2008, Swire Pacific Offshore
1336058

Brunei

ANGKATAN TENTERA LAUT DIRAJA BRUNEI

Country Overview

Formerly a British dependency, the Nation of Brunei is a sultanate that gained full independence in 1984. Situated on the northern coast of the island of Borneo, the country has a total area of 2,226 square miles and is bordered and divided into two halves by the Malaysian state of Sarawak. It has an 87 n mile coastline with the South China Sea. The capital and largest town is Bandar Seri Begawan which also has port facilities. There are further ports at Kuala Belait and Muara.

Territorial seas (3 n miles) and an EEZ (200 n mile) are claimed.

Headquarters Appointments

Commander of the Navy: Brigadier Abd Halim bin Haji Mohd Hanifah

Personnel

(a) 2014: 747 (58 officers)
(b) Voluntary service

Bases

Muara

Prefix to Ships' Names

KDB (Kapal Di-Raja Brunei)

© 2014 IHS   IHS Jane's Fighting Ships 2014-2015

88 Brunei > Patrol forces — Auxiliaries

PATROL FORCES

Notes: There are also up to 15 Rigid Raider assault boats operated by the River Division for infantry battalions. These boats are armed with 1 — 7.62 mm MG.

3 PERWIRA CLASS (COASTAL PATROL CRAFT) (PB)

Name	No	Builders	Launched	Commissioned
PERWIRA	P 14	Vosper Ltd	5 May 1974	9 Sep 1974
PEMBURU	P 15	Vosper Ltd	30 Jan 1975	17 Jun 1975
PENYERANG	P 16	Vosper Ltd	20 Mar 1975	24 Jun 1975

Displacement, tonnes: 39 full load
Dimensions, metres (feet): 21.7 × 6.1 × 1.2 *(71.2 × 20.0 × 3.9)*
Speed, knots: 32
Range, n miles: 600 at 22 kt, 1,000 at 16 kt
Complement: 14 (2 officers)
Machinery: 2 MTU MB 12V 331 TC81 diesels; 2,450 hp(m) *(1.8 MW)* sustained; 2 shafts
Guns: 2 Oerlikon/BMARC 20 mm GAM-BO1; 800 rds/min to 2 km; weight of shell 0.24 kg. 2 — 7.62 mm MGs.
Radars: Surface search: Racal Decca RM 1290; I-band.

Comment: Of all-wooden construction on laminated frames. Fitted with enclosed bridges- modified July 1976. A high speed RIB is launched from a stern ramp. New guns fitted in mid-1980s. All three ships operational in 2013.

PEMBURU   6/2005 / 1167117

4 IJHTIHAD (FPB 41) CLASS (PATROL CRAFT) (PB)

IJHTIHAD 17   **BERKAT** 18   **SYAFAAT** 19   **AFIAT** 20

Displacement, tonnes: 262 full load
Dimensions, metres (feet): 41.3 × 7.7 × 1.7 *(135.5 × 25.3 × 5.6)*
Speed, knots: 33
Range, n miles: 1,500 at 12 kt
Complement: 16

Machinery: 2 MTU 16V 4000 M93L diesels; 9,225 hp *(6.9 MW)*; 2 shafts
Guns: 1 Rheinmetall 27 mm. 2 — 7.62 mm MGs.
Electronic countermeasures: ESM: Thales Cutlass.
Radars: Furuno 2217; I-band.
Electro-optic systems: Zeiss MEOS II.

Comment: Constructed by Lürssen Werft. First completed in May 2009 when she started sea trials. The first two ships became operational in March 2010 and the second two in August 2010. One 4 m RHIB is carried at a davit.

AFIAT   5/2010, Michael Nitz / 1406008

4 DARUSSALAM (PV 80) CLASS (PSO)

DARUSSALAM 06   **DARULEHSAN** 07   **DARULAMAN** 08   **DARUTTAQWA** 09

Displacement, tonnes: 1,625 standard
Dimensions, metres (feet): 80 × 13.0 × 3.0 *(262.5 × 42.7 × 9.8)*
Speed, knots: 22
Range, n miles: 7,500 at 12 kt
Complement: 54

Machinery: 2 MTU 12V 1163 TB 93 diesels; 11,400 hp *(8.5 MW)*; 2 shafts; cp props
Missiles: 4 Exocet MM40 Block 3.
Guns: 1 BAE Systems 57 mm Mk 3. 2 GAM BO1 20 mm.
Physical countermeasures: Decoys: Terma SKWS.
Electronic countermeasures: ESM: EDO ES-3601.
Radars: Surface search/navigation: Terma Scanter 4100; I-band.
Fire control: Thales STING; I/J-bands.
Electro-optic systems: Thales STING Mk 2 optronic and radar tracker. Zeiss MEOS II EO surveillance.
Helicopters: Platform for 1 medium.

Comment: The first of class constructed by Lürssen Werft and launched in late March 2010. The first two ships were accepted on 7 January 2011 and the third ship was delivered in late 2011. A 10 m RHIB is carried in a stern ramp and a 6 m RHIB at a davit. A fourth ship was launched in August 2013 for delivery in May 2014.

DARUSSALAM   3/2013*, Chris Sattler / 1525513

1 + (2) MUSTAED CLASS (PATROL CRAFT) (PB)

Name	No	Builders	Launched	Commissioned
MUSTAED	21	Marinteknik Shipyard, Tuas	29 Sep 2011	25 Nov 2011

Dimensions, metres (feet): 27.2 × 6.2 × 1.2 *(89.2 × 20.3 × 3.9)*
Speed, knots: 40
Complement: 15 (3 officers)

Machinery: 2 MTU 16V 2000 M93 diesels; 4,800 hp *(3.58 MW)*; 2 MJP waterjets
Guns: To be announced.
Radars: Navigation: Furuno; I-band.
Electro-optic systems: To be announced.

Comment: Contract signed on 26 March 2010 between the Brunei government and Lürssen Asia Pte. Ltd. for the construction of one patrol vessel. The design is known as FIB25-012 and is of aluminium construction. Two further vessels are expected.

MUSTAED   11/2011, Royal Brunei Navy / 1406722

LAND-BASED MARITIME AIRCRAFT

Notes: (1) There are also six BO-105, four S-70A and ten Bell 212 utility helicopters.
(2) Plans to acquire a maritime patrol capability were announced in the 2011 Defence White Paper. Options include ATR-42, CN235-300, C 295, Beech 1900D, and a Dash 8 derivative.

AUXILIARIES

2 SERASA CLASS (LCU)

Name	No	Builders	Commissioned
SERASA	L 33	Transfield, Perth	8 Nov 1996
TERABAN	L 44	Transfield, Perth	8 Nov 1996

Displacement, tonnes: 224 full load
Dimensions, metres (feet): 36.5 × 8 × 1.5 *(119.8 × 26.2 × 4.9)*
Speed, knots: 12
Complement: 12
Military lift: 100 tons
Machinery: 2 diesels; 2 shafts
Radars: Navigation: Racal; I-band.

Comment: Ordered in November 1995 and delivered in December 1996. Used as utility transports. Bow and side ramps are fitted. Reported active.

SERASA   5/2010, Mick Prendergast / 1406006

IHS Jane's Fighting Ships 2014-2015   © 2014 IHS

Auxiliaries < **Brunei** — Frigates < **Bulgaria** 89

2 CHEVERTON LOADMASTERS (YFU)

Name	No	Builders	Commissioned
DAMUAN	L 31	Cheverton Ltd, Isle of Wight	May 1976
PUNI	L 32	Cheverton Ltd, Isle of Wight	Feb 1977

Displacement, tonnes: 61 (L 31), 65 (L 32) standard
Dimensions, metres (feet): 19.8 (L 31), 22.8 (L 32) × 6.1 × 1.1 *(65, 74.8 × 20.0 × 3.6)*
Speed, knots: 9
Range, n miles: 1,000 at 9 kt
Complement: 8
Military lift: 32 tons
Machinery: 2 Detroit 6-71 diesels; 442 hp *(305 kW)* sustained; 2 shafts
Radars: Navigation: Racal Decca RM 1216; I-band

PUNI 5/2010, Mick Prendergast / 1406007

POLICE

Notes: In addition to the vessels listed below there are two 12 m Rotork type *Behagia* 07 and *Selamat* 10.

7 INSHORE PATROL CRAFT

PDB 11–15   PDB 63   PDB 68

Displacement, tonnes: 20 full load
Dimensions, metres (feet): 14.5 × 4.2 × 1.2 *(47.6 × 13.8 × 3.9)*
Speed, knots: 30. **Range, n miles:** 310 at 22 kt
Complement: 7
Machinery: 2 MAN D 2840 LE diesels; 1,040 hp(m) *(764 kW)* sustained; 2 shafts
Guns: 1 – 7.62 mm MG.
Radars: Surface search: Furuno; I-band.

Comment: Built by Singapore SBEC. First three handed over in October 1987, second pair in 1988, last two in 1996. Aluminium hulls.

PDB 15 3/1999, John Webber / 0056631

3 BENDEHARU CLASS (PB)

BENDEHARU P 21   MAHARAJALELA P 22   KEMAINDERA P 23

Displacement, tonnes: 69 full load
Dimensions, metres (feet): 28.5 × 5.4 × 1.7 *(93.5 × 17.7 × 5.6)*
Speed, knots: 29
Machinery: 2 MTU diesels; 2,260 hp *(1.7 MW)*; 2 shafts
Guns: 1 – 12.7 mm MG.
Radars: Navigation: I-band.

Comment: Constructed by PT Pal, Surabaya, and entered service in 1991.

Bulgaria
VOENNOMORSKI SILI

Country Overview

Situated in the Balkan Peninsula, the Republic of Bulgaria has an area of 42,823 square miles and is bordered to the north by Romania and to the south by Turkey and Greece. The River Danube forms much of the northern border. Bulgaria has a coastline of 191 n miles with the Black Sea on which Varna and Burgas are the principal ports. The capital is Sofia.

Territorial waters (12 n miles) are claimed. An Exclusive Economic Zone (EEZ) was declared in 1987; limits have been established with Turkey but have yet to be fully agreed and defined with Romania.

Headquarters Appointments

Commander of the Navy:
Rear Admiral Rumen Nikolov

Deputy Commander of the Navy:
Commodore Dimitar Denev

Personnel

(a) 2014: 4,002 (610 officers)
(b) Reserves 10,000

Bases

Varna (HQ), Burgas

Coast Defence

One battalion with six truck-mounted SS-C-3 Styx twin launchers. One unit of coastal artillery with 130 mm guns. A coastal surveillance system, EKRAN, became operational in 2012.

FRIGATES
1 KONI CLASS (PROJECT 1159) (FFLM)

SMELI (ex-*Delfin*) 11

Displacement, tonnes: 1,463 standard; 1,930 full load
Dimensions, metres (feet): 96.4 × 12.6 × 3.5 *(316.3 × 41.3 × 11.5)*
Speed, knots: 27; 22 diesel. **Range, n miles:** 1,800 at 14 kt
Complement: 110

Machinery: CODAG; 1 SGW, Nikolayev M8B gas turbine (centre shaft); 18,000 hp(m) *(13.25 MW)* sustained; 2 Russki B-68 diesels; 15,820 hp(m) *(11.63 MW)* sustained; 3 shafts
Missiles: SAM: SA-N-4 Gecko twin launcher ❶; semi-active radar homing to 15 km *(8 n miles)* at 2.5 Mach; warhead 14.5 kg; altitude 9.1–3,048 m *(30–10,000 ft)*; 20 missiles.
Guns: 4 – 3 in *(76 mm)*/59 AK 726 (2 twin) ❷; 90 rds/min to 16 km *(8.6 n miles)*; weight of shell 5.9 kg. 4 – 30 mm/65 (2 twin) ❸; 500 rds/min to 5 km *(2.7 n miles)*; weight of shell 0.54 kg.
A/S Mortars: 2 RBU 6000 12-tubed trainable ❹; range 6,000 m; warhead 31 kg.
Mines: Capacity for 22.
Depth charges: 2 racks.
Physical countermeasures: Decoys: 2 PK 16 chaff launchers.
Electronic countermeasures: ESM: 2 Watch Dog; radar warning.
Radars: Air search: Strut Curve ❺; F-band; range 110 km *(60 n miles)* for 2 m² target.
Surface search: Don 2; I-band.
Fire control: Hawk Screech ❻; I-band (for 76 mm). Drum Tilt ❼; H/I-band (for 30 mm). Pop Group ❽; F/H/I-band (for SA-N-4).
IFF: High Pole B.
Sonars: Hercules (MG 322); hull-mounted; active search and attack; medium frequency.

Programmes: First reported in the Black Sea in 1976. Type I retained by the USSR for training foreign crews but transferred in February 1990 when the Koni programme terminated. Others of the class acquired by the former East German Navy (now deleted), Serbia (deleted but for sale), Algeria, Cuba (deleted) and Libya.
Modernisation: Marisat fitted in 1996. Reported to be RAS capable. Communications upgrade planned to achieve NATO interoperability.
Operational: Based at Varna.

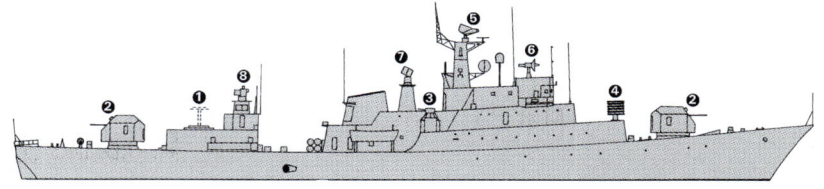

SMELI *(Scale 1 : 900), Ian Sturton / 0114505*

SMELI 6/2009, Bulgarian Navy / 1366168

© 2014 IHS IHS Jane's Fighting Ships 2014-2015

Bulgaria > Frigates — Corvettes

3 WIELINGEN CLASS (TYPE E-71) (FFGM)

Name	No	Builders	Laid down	Launched	Commissioned
DRAZKI (ex-*Wandelaar*)	41 (ex-F 912)	Boelwerf, Temse	28 Mar 1975	21 Jun 1977	27 Oct 1978
VERNI (ex-*Wielingen*)	42 (ex-F 910)	Boelwerf, Temse	5 Mar 1974	30 Mar 1976	20 Jan 1978
GORDI (ex-*Westdiep*)	43 (ex-F 911)	Cockerill, Hoboken	2 Sep 1974	8 Dec 1975	20 Jan 1978

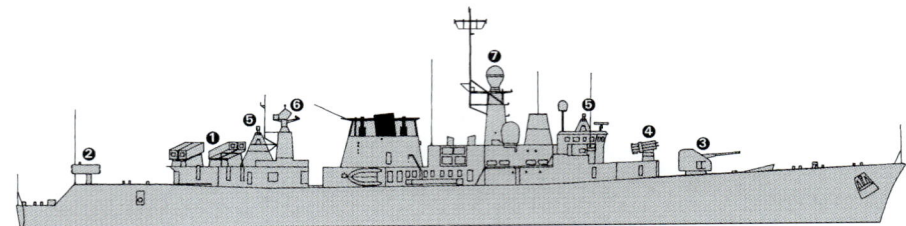

DRAZKI *(Scale 1 : 900), Ian Sturton / 1164332*

Displacement, tonnes: 1,971 standard; 2,469 full load
Dimensions, metres (feet): 106.4 × 12.3 × 5.6 *(349.1 × 40.4 × 18.4)*
Speed, knots: 26; 20 diesel
Range, n miles: 4,500 at 18 kt, 6,000 at 15 kt
Complement: 159 (13 officers)

Machinery: CODOG; 1 RR Olympus TM3B gas-turbine; 25,440 hp *(19 MW)* sustained; 2 Cockerill 240 CO V 12 diesels; 6,000 hp(m) *(4.4 MW)*; 2 shafts; LIPS cp props
Missiles: SSM: 4 Aerospatiale MM 38 (2 twin) launchers ❶; inertial cruise; active radar homing to 42 km *(23 n miles)* at 0.9 Mach; warhead 165 kg; sea-skimmer.
SAM: Raytheon Sea Sparrow RIM-7P; Mk 29 octuple launcher ❷; semi-active radar homing to 16 km *(8.5 n miles)* at 2.5 Mach; warhead 38 kg.
Guns: 1 Creusot-Loire 3.9 in *(100 mm)*/55 Mod 68 ❸; 80 rds/min to 17 km *(9 n miles)* anti-surface; 8 km *(4.4 n miles)* anti-aircraft; weight of shell 13.5 kg.
Torpedoes: 2—21 in *(533 mm)* launchers. ECAN L5 Mod 4; anti-submarine; active/passive homing to 9.5 km *(5 n miles)* at 35 kt; warhead 150 kg; depth to 550 m *(1,800 ft)*.
A/S Mortars: 1 Creusot-Loire 375 mm 6-barrelled trainable launcher ❹; Bofors rockets to 1,600 m; warhead 107 kg.
Physical countermeasures: Decoys: 2 Tracor MBA SRBOC 6-barrelled Mk 36 launchers; chaff (Mk 214 Seagnat) and IR flares to 4 km *(2.2 n miles)*. Nixie SLQ-25; towed anti-torpedo decoy.
Electronic countermeasures: ESM; Argos AR 900; intercept.
Radars: Air/surface search: Signaal DA05 ❻; E/F-band.
Surface search/fire control: Signaal WM25 ❼; I/J-band.
Navigation: Signaal Scout; I/J band.
IFF: Mk XII.
Sonars: Computing Devices Canada SQS 510; hull-mounted; active search and attack; medium frequency.
Combat data systems: Signaal SEWACO IV action data automation; Link 11. SATCOM.
Weapon control systems: Sagem Vigy 105 optronic director ❺.

Programmes: A compact, well-armed class of frigate originally designed by and for the Belgian Navy. Following the signature of a letter of intent on 4 December 2004, the Bulgarian government gave final approval on 17 March 2005 for transfer of ex-*Wandelaar* to Bulgarian service in October 2005. The procurement of ex-*Westdiep* and ex-*Wielingen* was confirmed on 7 December 2007. Ex-*Westdiep* transferred on 22 August 2008 and ex-*Wielingen* in February 2009.
Modernisation: The ships completed a major upgrade programme before leaving Belgian service. This included update of Sea Sparrow to 7P, modification of WM25 radar to include improved ECCM and MTI capabilities and a new navigation radar and sonar. A new optronic director, IFF and communications facilities were also installed. Platform improvements included new diesel engines and alternators. Further modernisation is under consideration.
Structure: Fully air conditioned. Fin stabilisers fitted.
Operational: The ships are used for surveillance missions in the Black Sea, maritime interdiction and contributions to international peace-support operations, both under the NATO flag and as part of the Black Sea Naval Co-operation Task Group (BLACKSEAFOR).

VERNI *11/2012*, Selim San / 1525515*

CORVETTES

1 TARANTUL II CLASS (PROJECT 1241.1M) (FSGM)

MULNAYA 43 (ex-101)

Displacement, tonnes: 391 standard; 462 full load
Dimensions, metres (feet): 56.1 × 11.5 × 2.5 *(184.1 × 37.7 × 8.2)*
Speed, knots: 36
Range, n miles: 400 at 36 kt, 2,000 at 20 kt
Complement: 34 (5 officers)

Machinery: COGAG; 2 Nikolayev Type DR 77 gas turbines; 16,016 hp(m) *(11.77 MW)* sustained; 2 Nikolayev Type DR 76 gas turbines with reversible gearboxes; 4,993 hp(m) *(3.67 MW)* sustained; 2 shafts
Missiles: SSM: 4 Raduga SS-N-2C Styx (2 twin) launchers; active radar or IR homing to 83 km *(45 n miles)* at 0.9 Mach; warhead 513 kg; sea-skimmer.
SAM: SA-N-5 Grail quad launcher; manual aiming; IR homing to 6 km *(3.2 n miles)* at 1.5 Mach; altitude to 2,500 m *(8,000 ft)*; warhead 1.5 kg.
Guns: 1—3 in *(76 mm)*/59 AK 176; 120 rds/min to 15 km *(8.1 n miles)*; weight of shell 5.9 kg. 2—30 mm/65; 6 barrels per mounting; 3,000 rds/min to 2 km.
Physical countermeasures: Decoys: 2 PK 16 chaff launchers.
Electronic countermeasures: ESM: 2 Half Hat; intercept.
Radars: Air/surface search: Plank Shave; E-band.
Navigation: Kivach; I-band.
Fire control: Bass Tilt; H/I-band. Band Stand (Mineral ME); D-band (for SSN 2C).
IFF: Square Head. High Pole.
Weapon control systems: Hood Wink optronic director. Band Stand datalink for SSM.

Comment: Built at Volodarski, Rybinsk. Transferred from USSR in December 1989. Name means Thunderbolt. Based at Atiya.

MULNAYA (old number) *7/2000, van Ginderen Collection / 0104245*

IHS Jane's Fighting Ships 2014-2015 © 2014 IHS

Corvettes — Mine warfare forces < **Bulgaria** 91

2 RESHITELNI (PAUK I) (PROJECT 1241P) CLASS (FSM)

RESHITELNI 13   **BODRI** 14

Displacement, tonnes: 447 full load
Dimensions, metres (feet): 59.5 × 10.2 × 3.3 *(195.2 × 33.5 × 10.8)*
Speed, knots: 32
Range, n miles: 2,200 at 14 kt
Complement: 38

Machinery: 2 Type 521 diesels; 16,180 hp(m) *(11.9 MW)* sustained; 2 shafts
Missiles: SAM: SA-N-5 Grail quad launcher; manual aiming; IR homing to 6 km *(3.2 n miles)* at 1.5 Mach; altitude to 2,500 m *(8,000 ft)*; warhead 1.5 kg; 8 missiles.
Guns: 1 – 3 in *(76 mm)*/59 AK 176; 120 rds/min to 15 km *(8 n miles)*; weight of shell 5.9 kg.
1 – 30 mm/65; 6 barrels; 3,000 rds/min combined to 2 km.
Torpedoes: 4 – 16 in *(406 mm)* tubes. Type 40; anti-submarine; active/passive homing up to 15 km *(8 n miles)* at up to 40 kt; warhead 100–150 kg.
A/S Mortars: 2 RBU 1200 5-tubed fixed; range 1,200 m; warhead 34 kg.
Depth charges: 2 racks (12).
Physical countermeasures: Decoys: 2 PK 16 chaff launchers.
Electronic countermeasures: ESM: 3 Brick Plug; intercept.
Radars: Air/surface search: Peel Cone; E-band.
Surface search: Spin Trough; I-band.
Fire control: Bass Tilt; H/I-band.
Sonars: Foal Tail VDS (mounted on transom); active attack; high frequency.

Comment: *Reshitelni* transferred from USSR in September 1989, *Bodri* in December 1990. Based at Varna.

RESHITELNI   4/2012, C D Yaylali / 1455756

LAND-BASED MARITIME AIRCRAFT (FRONT LINE)

Numbers/Type: 3 Mil Mi-14PL 'Haze A'.
Operational speed: 120 kt *(222 km/h)*.
Service ceiling: 15,000 ft *(4,570 m)*.
Range: 240 n miles *(445 km)*.
Role/Weapon systems: Primary role as inshore/coastal ASW and Fleet support helicopter; one converted as transport. All three aircraft in reserve. Based at Asparukhovo airport. Sensors: Search radar, MAD, sonobuoys, dipping sonar. Weapons: ASW; up to 2 × torpedoes, or mines, or depth bombs.

HAZE   6/2006, Bulgarian Navy / 1164494

Numbers/Type: 3 Eurocopter AS 565MB Panther.
Operational speed: 150 kt *(278 km/h)*.
Service ceiling: 15,420 ft *(4,700 m)*.
Range: 464 n miles *(859 km)*.
Role/Weapon systems: Six aircraft ordered on 28 January 2005. Three had been delivered by late 2011 but the other three were cancelled. The aircraft are shore-based and are used for maritime surveillance, ASW, anti-surface and SAR roles. Sensors and weapons to be announced.

AS 565 PANTHER   6/2009, Bulgarian Navy / 1366167

© 2014 IHS

MINE WARFARE FORCES

Notes: Mine warfare is a priority area for the Bulgarian Navy. There are plans to acquire autonomous underwater vehicles and to establish a mine warfare data centre. Side-scan sonar is also to be procured.

3 BRIZ (SONYA) (PROJECT 12650) CLASS (MINESWEEPERS — COASTAL) (MSC)

BRIZ 61   **SHKVAL** 62   **PRIBOY** 63

Displacement, tonnes: 457 full load
Dimensions, metres (feet): 48 × 8.8 × 2 *(157.5 × 28.9 × 6.6)*
Speed, knots: 15
Range, n miles: 1,500 at 14 kt
Complement: 43 (5 officers)

Machinery: 2 Kolomna Type 9-D-8 diesels; 2,000 hp(m) *(1.47 MW)* sustained; 2 shafts
Guns: 2 USSR 30 mm/65 (twin); 500 rds/min to 5 km *(2.7 n miles)*; weight of shell 0.54 kg.
2 USSR 25 mm/80 (twin); 270 rds/min to 3 km *(1.6 n miles)*; weight of shell 0.34 kg.
Mines: 5.
Radars: Surface search/navigation: Kivach; I-band.
IFF: Two Square Head. High Pole B.
Sonars: MG 69/79; hull-mounted; active minehunting; high frequency.

Comment: Wooden hulled ships transferred from USSR in 1981–84. Two ships (*Briz* and *Shtorm*) decommissioned by 2011 but *Briz* re-activated in 2013. Based at Atiya.

SHKVAL   8/2012, C D Yaylali / 1455757

PRIBOY   3/2002 / 0114506

1 FLOWER (TRIPARTITE) CLASS (MINEHUNTER) (MHC)

Name	No	Builders	Laid down	Launched	Commissioned
TSIBAR (ex-*Myosotis*)	32 (ex-M 922)	Beliard, Ostend	6 Jul 1987	4 Aug 1988	14 Dec 1989

Displacement, tonnes: 571 standard; 605 full load
Dimensions, metres (feet): 51.5 × 8.9 × 2.6 *(169 × 29.2 × 8.5)*
Speed, knots: 15; 7 electric
Range, n miles: 3,000 at 12 kt
Complement: 46

Machinery: 1 Stork Wärtsilä A-RUB 215X-12 diesel; 1,860 hp(m) *(1.35 MW)* sustained; 1 shaft; LIPS cp prop; 2 active rudders; 2 motors; 240 hp(m) *(179 kW)*; 2 bow thrusters
Guns: 1 DCN 20 mm/20; 720 rds/min to 10 km. 2 – 12.7 mm MGs.
Physical countermeasures: MCM: 2 PAP 104 remote-controlled mine locators; 39 charges. Mechanical minesweeping gear.
Radars: Navigation: Racal Decca TM 1229C; I-band.
Sonars: Thomson Sintra DUBM 21A; hull-mounted; minehunting; 100 kHz (±10 kHz).

Programmes: Originally procured for the Belgian Navy in co-operation with France and the Netherlands. The hull was built at Ostend and the ship was fitted out at Rupelmonde. It was subsequently modified to act as an ammunition transport. The ship was decommissioned from Belgian service in 2004 and, following an announcement on 7 December 2007 was re-activated and transferred to Bulgaria in early 2009. Details are based on the ships in Belgian service before modernisation.
Modernisation: The ship is to be overhauled before entering Bulgarian service.
Structure: GRP hull fitted with active tank stabilisation, full NBC protection and air conditioning. Has automatic pilot and buoy tracking. A 5-ton container can be carried for varying tasks.

TSIBAR   5/2009, Frank Findler / 1305679

IHS Jane's Fighting Ships 2014-2015

92　Bulgaria > Mine warfare forces — Auxiliaries

6 OLYA (PROJECT 1259) CLASS
(MINESWEEPERS—INSHORE) (MSB)

ISKAR 51	BALIK 54
DOBROTICH 52	KAPITAN-LEYTENANT EVSTATI VINAROV 55
KAPITAN-LEYTENANT KIRIL MINKOV 53	KAPITAN PARVI RANG DIMITAR PASKALEV 56

Displacement, tonnes: 65 full load
Dimensions, metres (feet): 25.8 × 4.5 × 1 *(84.6 × 14.8 × 3.3)*
Speed, knots: 12
Range, n miles: 300 at 10 kt
Complement: 15
Machinery: 2 Type 3D 6S11/235 diesels; 471 hp(m) *(346 kW)* sustained; 2 shafts
Guns: 2 — 12.7 mm MGs (twin).
Radars: Navigation: Pechora; I-band.

Comment: First five built between 1988 and 1992 in Bulgaria to the Russian Olya design. *56* completed in 1996. Minesweeping equipment includes AT-6, SZMT-1 and 3 PKT-2 systems. Based at Varna.

OLYA 55　　　　　　　　　　　　　　　　6/2009, Bulgarian Navy / 1366165

SURVEY SHIPS

2 COASTAL SURVEY VESSELS (PROJECT 612) (AGSC)

KAPITAN PARVI RANG BORIS ROGEV 231
KONTRAADMIRAL SAVA IVANOV 331

Displacement, tonnes: 116 full load
Dimensions, metres (feet): 26.7 × 5.8 × 1.5 *(87.6 × 19.0 × 4.9)*
Speed, knots: 12
Range, n miles: 600 at 10 kt
Complement: 9 (2 officers)
Machinery: 2 Type 3-D-12 diesels; 600 hp(m) *(440 kW)* sustained; 2 shafts
Radars: Navigation: I-band.

Comment: Built in Bulgaria in 1986 and 1988 respectively. Can carry 2 tons of equipment. *231* is based at Varna and *331* at Atiya.

AGSC 331　　　　　　　　　　　　　　　6/1996, Bulgarian Navy / 0506274

AUXILIARIES

2 SUPPORT TANKER (AOTL)

AKIN 303 (ex-203)　　BALCHIK 203

Displacement, tonnes: 1,270 full load
Dimensions, metres (feet): 55.4 × 11 × 3.5 *(181.8 × 36.1 × 11.5)*
Speed, knots: 12
Range, n miles: 1,000 at 8 kt
Complement: 23

Cargo capacity: 650 tons fuel
Machinery: 2 Sulzer 6AL-20-24 diesels; 1,500 hp(m) *(1.1 MW)*; 2 shafts
Guns: 2 ZU-23-2F Wrobel 23 mm (twin).
Radars: Navigation: I-band.

Comment: *Balchik* laid down 1989, launched 1993 and completed in 1994 at Burgas Shipyards, Burgas. *Akin* entered service in 2000. Based at Varna.

IHS Jane's Fighting Ships 2014-2015

BALCHIK　　　　　　　　　　　　　　　　5/2009, Frank Findler / 1305680

2 DIVING TENDERS (PROJECT 245) (YDT)

KAPITAN VTORI RANG NIKOLA FURNADJIEV 223　　FOROS 323

Displacement, tonnes: 114 full load
Dimensions, metres (feet): 27.9 × 5.2 × 2.2 *(91.5 × 17.1 × 7.2)*
Speed, knots: 10. **Range, n miles:** 400 at 10 kt
Complement: 13 (7 divers)
Machinery: Diesel-electric; 2 MCK 83-4 diesel generators; 1 motor; 300 hp(m) *(220 kW)*; 1 shaft
Radars: Navigation: Don 2; I-band.

Comment: Built in Bulgaria in mid-1980s. A twin 12.7 mm MG can be fitted. Capable of bell diving to 60 m. *223* based at Varna. *323* based at Atiya.

YDT 323　　　　　　　　　　　　　　　　6/2006, Bulgarian Navy / 1164485

1 BEREZA (PROJECT 130) CLASS (ADG/AX)

KAPITAN PARVI RANG DIMITRI DOBREV 206

Displacement, tonnes: 2,084 full load
Dimensions, metres (feet): 69.5 × 13.8 × 4 *(228 × 45.3 × 13.1)*
Speed, knots: 13. **Range, n miles:** 1,000 at 13 kt
Complement: 48
Machinery: 2 Zgoda-Sulzer 8 AL 25/30 diesels; 2,925 hp(m) *(2.16 MW)* sustained; 2 shafts; cp props
Radars: Navigation: Kivach; I-band.

Comment: New construction built in Poland and transferred July 1988. Used as a degaussing ship. Fitted with an NBC citadel and upper deck wash-down system. The ship has three laboratories. Has also been used as a training ship. Based at Varna.

KAPITAN PARVI RANG DIMITRI DOBREV　　　　　7/2007, Bob Fildes / 1166829

5 PROJECT 160 AMBULANCE CRAFT (YH)

GEORGI RAKOVSKI 221	ATIJA 324	AGALINA 365
VASIL LEVSKI 222	KOLOKITA 325	

Displacement, tonnes: 112 full load
Dimensions, metres (feet): 26.6 × 5.8 × 1.5 *(87.3 × 19.0 × 4.9)*
Speed, knots: 12
Machinery: 1 diesel; 290 hp(m) *(216 kW)*; 1 shaft

Comment: Capable of carrying five torpedoes.

222　　　　　　　　　　　　　　　　6/2007, Maritime Photographic / 1167807

5 AUXILIARIES (ATS)

| HADJI DIMITAR 211 | AHELOI 321 | – 421 |
| HRISTO BOTEV 212 | STEFAN KARADJA 410 | |

Comment: *421* is a merchant vessel converted to a training ship. *321* is a firefighting vessel. *211*, *212*, and *410* are tugs.

421 6/2003, Schaeffer/Marsan / 0567877

AHELOI 6/2006, Bulgarian Navy / 1164483

1 SALVAGE SHIP (ARS)

Name	No	Builders	Commissioned
PROTEO (ex-*Perseo*)	224 (ex-A 5310)	Cantieri Navali Riuniti, Ancona	24 Aug 1951

Displacement, tonnes: 1,895 standard; 2,181 full load
Dimensions, metres (feet): 75.6 × 11.6 × 6.4 *(248 × 38.1 × 21.0)*
Speed, knots: 16. **Range, n miles:** 7,500 at 13 kt
Complement: 122 (8 officers)
Machinery: 2 Fiat diesels; 4,800 hp(m) *(3.53 MW)*; 1 shaft
Radars: Navigation: SMA-748; I-band.

Comment: Transferred to Bulgaria on 3 June 2004 having been decommissioned from the Italian Navy in 2002. Originally laid down in 1943, construction was suspended until restarted in 1949. Details are those of the ship when in Italian service.

PROTEO 5/2009, Frank Findler / 1305681

2 VYDRA (PROJECT 106K) CLASS (YL)

| SIRIUS 302 | ANTARES 304 |

Displacement, tonnes: 432 standard; 559 full load
Dimensions, metres (feet): 54.8 × 7.7 × 2 *(179.8 × 25.3 × 6.6)*
Speed, knots: 12. **Range, n miles:** 2,500 at 10 kt
Complement: 20

Military lift: 200 tons or 100 troops or 3 MBTs
Machinery: 2 Type 3-D-12 diesels; 600 hp(m) *(440 kW)* sustained; 2 shafts
Guns: 1 – 14.5 mm.
Radars: Navigation: Don 2; I-band.
IFF: High Pole.

Comment: Originally a class of 19, the last two remaining vessels are former LCUs converted to auxiliary role as lighters.

VYDRA 6/2006, Bulgarian Navy / 1164492

© 2014 IHS

BORDER GUARD

1 DAMEN STAN PATROL 4207 (PB)

OBZOR 525

Displacement, tonnes: 205 full load
Dimensions, metres (feet): 42.8 × 7.11 × 2.52 *(140.4 × 23.3 × 8.3)*
Speed, knots: 26
Machinery: 2 Caterpillar 3516B DI-TA; 5,600 hp *(4.17 MW)*; 2 cp props
Guns: To be announced.

Comment: Constructed by Damen Shipyards and delivered in September 2010.

OBZOR 7/2012, C D Yaylali / 1455758

5 LÜRSSEN 21 METRE PATROL CRAFT (PB)

| BURGAS 514 | VARNA 534 | SHABLA |
| KAVARNA 531 | KRAJMORJE 513 | |

Displacement, tonnes: 51 standard
Dimensions, metres (feet): 21 × 5.8 × 1.4 *(68.9 × 19.0 × 4.6)*
Speed, knots: 30
Machinery: 2 Deutz MWM TBD 616 diesels; 2,970 hp(m) *(2.2 MW)*; 2 shafts

Comment: Contract awarded in November 2002 to Lürssen, Berne-Bardenfleth. Delivery of the first two craft made in 2003 and of the third in October 2005. Two further craft were delivered by 2011.

KAVARNA 6/2008 / 1455759

8 BALTIC 130 PATROL CRAFT (PB)

| EMINE 515 | VIDIN | NIKOPOL | RUSE |
| KALIAKRA 535 | LOM | SVISHTOV | SILISTRA |

Displacement, tonnes: 12.5 full load
Dimensions, metres (feet): 13.75 × 4.0 × 0.75 *(45.1 × 13.1 × 2.5)*
Speed, knots: 40
Complement: 4
Machinery: 2 Volvo Penta D11 diesels; 1,320 hp *(986 kW)*; 2 waterjets

Comment: Ordered in 2009, constructed by Baltic Workboats, Estonia, and delivered in 2010.

BALTIC 130 10/2012*, Hartmut Ehlers / 1525514

94 Bulgaria > Border guard — **Cambodia** > Patrol forces

3 NEUSTADT CLASS (PB)

SOZOPOL (ex-*Rosenheim*) 525 (ex-BG 18) **BALCHIK** (ex-*Duderstadt*) 524 (ex-BG 14)
NESEBAR (ex-*Neustadt*) 526 (ex-BG 11)

Dimensions, metres (feet): 38.5 × 7 × 2.2 *(126.3 × 23.0 × 7.2)*
Speed, knots: 30
Range, n miles: 450 at 27 kt
Complement: 17

Machinery: 2 MTU MD diesels; 6,000 hp(m) *(4.41 MW)*; 1 MWM diesel; 685 hp(m) *(500 kW)*; 3 shafts
Guns: 2 — 7.62 mm MGs.
Radars: Surface search: Selenia ARP 1645; I-band.
Navigation: Racal Decca Bridgemaster MA 180/4; I-band.

Comment: Built in 1970 by Lürssen, Vegesack. 525 transferred from German Border Guard in June 2002, 526 on 16 April 2004 and 524 in December 2004. Operational status doubtful.

7 ZHUK (PROJECT 1400M) CLASS
(COASTAL PATROL CRAFT) (PB)

511 512 521–523 532 533

Displacement, tonnes: 40 full load
Dimensions, metres (feet): 24 × 5 × 1.2 *(78.7 × 16.4 × 3.9)*
Speed, knots: 30
Range, n miles: 1,100 at 15 kt
Complement: 11 (3 officers)
Machinery: 2 Type M 401B diesels; 2,200 hp(m) *(1.6 MW)* sustained; 2 shafts
Guns: 4 USSR 14.5 mm (2 twin) MGs.
Radars: Surface search: Spin Trough; I-band.

Comment: Transferred from USSR 1980–81.

NESEBAR 5/2004, *Martin Mokrus* / 0587692

ZHUK 532 6/2008 / 1455761

Cambodia

Country Overview

Formerly a French protectorate, the south-east Asian Kingdom of Cambodia was ravaged by the Vietnam War and then by the Khmer Rouge regime before relative stability followed the nation's first multiparty elections in 1993. With an overall land area of 69,898 square miles, the country is bordered to the north by Thailand and Laos and to the east by Vietnam. There is a 239 n mile coastline with the Gulf of Thailand. The capital and largest city is Phnom Penh while the principal port is Sihanoukville. There are extensive inland waterways. Territorial seas (12 n miles) are claimed. An EEZ (200 n miles) is claimed but the limits have not been fully defined.

Headquarters Appointments

Commander of Navy: Vice Admiral Tea Vinh
Chief of Naval Staff: Rear Admiral Sao Sarin

Personnel

2014: 2,800 (780 officers) including marines

Bases

Ream (Sihanoukville) (ocean), Phnom Penh (river), Sihanoukville (civil)

Organisation

Ocean Division has nine battalions and the River Division seven battalions. Command HQ is at Phnom Penh.

PATROL FORCES

Notes: (1) There are also about 170 motorised and manual canoes.
(2) Six patrol craft of unknown type were donated by China on 9 January 2005. They may be operated by the Marine Police.
(3) Three 20 m patrol craft (1107–1109) and an approximately 25 m landing craft (1401) were delivered by China on 7 November 2007.

2 MODIFIED STENKA CLASS (PROJECT 205P)
(FAST ATTACK CRAFT — PATROL) (PBF)

MONDOLKIRI 1131 **RATANAKIRI** 1134

Displacement, tonnes: 214 standard; 257 full load
Dimensions, metres (feet): 39.4 × 7.9 × 2.5 *(129.3 × 25.9 × 8.2)*
Speed, knots: 37
Range, n miles: 800 at 24 kt, 500 at 35 kt
Complement: 25 (5 officers)

Machinery: 3 Caterpillar diesels; 14,000 hp(m) *(10.29 MW)*; 3 shafts
Guns: 2 — 23 mm/87 (twin). 1 Bofors 40 mm/70.
Radars: Surface search: Racal Decca Bridgemaster; I-band.
Fire control: Muff Cob; G/H-band.
Navigation: Racal Decca; I-band.
IFF: High Pole. 2 Square Head.

Comment: Four transferred from USSR in November 1987. Export model without torpedo tubes and sonar. One pair were modernised in Hong Leong Shipyard, Butterworth, from early 1995 to April 1996. New engines, guns and radars were fitted. The second pair similarly refitted by August 1997. By late 1998 only two were operational although it was reported in 2000 that a third may have undergone a further refit. Pennant numbers were changed for UN operations but changed back again in November 1993.

2 KAOH CLASS (RIVER PATROL CRAFT) (PBR)

KAOH CHHLAM 1105 **KAOH RONG** 1106

Displacement, tonnes: 45 full load
Dimensions, metres (feet): 23.3 × 6.1 × 1.2 *(76.4 × 20.0 × 3.9)*
Speed, knots: 34
Range, n miles: 400 at 30 kt
Complement: 13 (3 officers)
Machinery: 2 Deutz/MWM TBD 616 V16 diesels; 2,992 hp(m) *(2.2 MW)*; 2 shafts
Guns: 2 — 14.5 mm MG (twin). 2 — 12.7 mm MGs.
Radars: Surface search: Racal Decca Bridgemaster; I-band.

Comment: Ordered from Hong Leong Shipyard, Butterworth to a German design in 1995 and delivered 20 January 1997. Aluminium construction.

KAOH CHHLAM 1/1997, *Hong Leong Shipyard* / 0056667

2 SHERSHEN (PROJECT 206T) CLASS (PATROL CRAFT) (PBF)

1123 1124

Displacement, tonnes: 147 standard; 173 full load
Dimensions, metres (feet): 34.7 × 6.7 × 1.5 *(113.8 × 22.0 × 4.9)*
Speed, knots: 45
Range, n miles: 850 at 30 kt, 460 at 42 kt
Complement: 23

Machinery: 3 Type 503A diesels; 8,025 hp(m) *(5.9 MW)* sustained; 3 shafts
Guns: 4 — 25 mm (2 twin).
Radars: Surface search: Pot Drum; H/I-band.
IFF: High Pole A. Square Head.

Comment: Former Vietnamese patrol craft transferred in about 2006.

MONDOLKIRI 8/1997, *Hong Leong Shipyard* / 0056666

IHS Jane's Fighting Ships 2014-2015 © 2014 IHS

4 PATROL CRAFT (PBO)

1141–1144

Displacement, tonnes: 238 full load
Dimensions, metres (feet): 46.8 × 7.2 × 1.82 (153.5 × 23.6 × 5.97)
Speed, knots: 27
Range, n miles: 1,200 at 15 kt
Complement: 36 (6 officers)
Machinery: 4 MTU 16V2000 M90 diesels; 7,200 hp (5.37 MW) sustained; 4 shafts
Guns: 2—37 mm (twin). 2—14.5 mm (twin).
Radars: Surface search/navigation: JMA 5312 and JMA 5106; I-band.

Comment: Four patrol craft built in China and delivered in 2007. Similar craft are in service in Ghana and Congo-Brazzaville.

PATROL CRAFT
6/2012, US Navy
1455710

Cameroon
MARINE NATIONALE RÉPUBLIQUE

Country Overview

The Republic of Cameroon became a unitary republic in 1972 and replaced the federation of East Cameroon (formerly French Cameroons) and West Cameroon (formerly part of British Cameroons). With an area of 183,569 square miles, the country has borders to the west with Nigeria and to the south with Gabon and Equatorial Guinea. It has a 217 n mile coastline with Atlantic Ocean on the Bight of Bonny. The capital is Yaoundé while Douala is the principal port which also serves adjacent landlocked states. Kribi is the country's second port. Cameroon is the only coastal state to claim territorial seas of 50 n miles. It has not been declared an Exclusive Economic Zone (EEZ) and claims to jurisdiction would be complicated by the offshore islands of Bioko (Equatorial Guinea), São Tomé and Principe.

Headquarters Appointments

Chief of Naval Staff:
    Rear Admiral Jean Mendoua

Personnel

2014: 1,250

Bases

Douala (HQ), Limbe, Kribi

PATROL FORCES

Notes: (1) There were 10 Rodman 6.5 m craft delivered in 2000. All have speeds in excess of 25 kt.
(2) There are some eight Simmoneau 9 m and 10 m craft in service.
(3) Three Boston Whaler Justice Craft were purchased from the United States in early 2012.
(4) Two patrol craft, of approximately 60 m, were reported under construction in China in early 2014. One is called *La Sanaga* P 109.

1 COASTAL PATROL CRAFT (PB)

QUARTIER MAÎTRE ALFRED MOTTO

Displacement, tonnes: 98 full load
Dimensions, metres (feet): 29.1 × 6.2 × 1.9 (95.5 × 20.3 × 6.2)
Speed, knots: 14
Complement: 17 (2 officers)
Machinery: 2 Baudouin diesels; 1,290 hp(m) (948 kW); 2 shafts
Guns: 2—7.62 mm MGs.
Radars: Surface search: I-band.

Comment: Built at Libreville, Gabon in 1974. Discarded as a derelict hulk in 1990 but refurbished and brought back into service with assistance from the French Navy in 1995–96.

QUARTIER MAÎTRE ALFRED MOTTO
2/1996, French Navy / 0056670

1 BAKASSI (TYPE P 48S) CLASS
(OFFSHORE PATROL CRAFT) (PBO)

Name	No	Builders	Launched	Commissioned
BAKASSI	P 104	SFCN, Villeneuve la Garenne	22 Oct 1982	9 Jan 1984

Displacement, tonnes: 313 full load
Dimensions, metres (feet): 52.6 × 7.2 × 2.4 (172.6 × 23.6 × 7.8)
Speed, knots: 25
Range, n miles: 2,000 at 16 kt
Complement: 39 (6 officers)

Machinery: 2 SACM 195 V16 CZSHR diesels; 8,000 hp(m) (5.88 MW) sustained; 2 shafts
Guns: 2 Bofors 40 mm/70; 300 rds/min to 12.8 km (7 n miles); weight of shell 0.96 kg.
Radars: 2 Furuno; I-band.
Weapon control systems: 2 Naja optronic systems. Racal Decca Cane 100 command system.

Comment: Ordered January 1981. Laid down 16 December 1981. Six month major refit by Raidco Marine (Lorient) in 1999. This included removing the Exocet missile system and EW equipment, and fitting new propellers and a funnel aft of the mainmast to replace the waterline exhausts. New radars were also installed. Two RIBs are carried.

BAKASSI
7/1999, H M Steele / 0121304

2 SWIFT PBR CLASS (RIVER PATROL CRAFT) (PBR)

PR 001   PR 005

Displacement, tonnes: 12 full load
Dimensions, metres (feet): 11.6 × 3.8 × 1.0 (38.1 × 12.5 × 3.2)
Speed, knots: 32
Range, n miles: 210 at 20 kt
Complement: 4
Machinery: 2 Stewart and Stevenson 6V-92TA diesels; 520 hp (388 kW) sustained; 2 shafts
Guns: 2—12.7 mm MGs. 2—7.62 mm MGs.

Comment: Last two survivors of 30 built by Swiftships and supplied under the US Military Assistance Programme. First 10 delivered in March 1987, second 10 in September 1987 and the remainder in March 1988. Several others have been cannibalised for spares.

PBR class
4/1992 / 0056671

98 Canada > Introduction — Submarines

Canada

Country Overview

Canada is the world's second-largest country. The British monarch, represented by a governor-general, is head of state. With an area of 3,849,652 square miles, it occupies most of northern North America and is bordered to the south by the United States and to the west by the US state of Alaska. It has a coastline of 131,647 n miles with the Pacific, Arctic and Atlantic Oceans and with Baffin Bay and the Davis Strait. Numerous coastal islands include the Arctic Archipelago to the north, Newfoundland, Cape Breton, Prince Edward, and Anticosti to the east and Vancouver Island and the Queen Charlotte Islands to the west. Hudson Bay contains Southampton Island and many smaller islands. The 2,035 n mile St Lawrence-Great Lakes navigation system enables ocean-going vessels to sail between the Atlantic Ocean and the Great Lakes via the St Lawrence Seaway (opened 1959). Ottawa is the capital while Toronto is the largest city. Major ports include Vancouver, Montreal, Halifax, Sept-Îles, Port-Cartier, Quebec City, Saint John (New Brunswick), Thunder Bay, Prince Rupert, and Hamilton. Territorial seas (12 n miles) are claimed. A 200 n mile EEZ has been claimed but the limits have only been partly defined by boundary agreements.

Headquarters Appointments

Commander, Royal Canadian Navy:
Vice Admiral M A G Norman, CMM, CD
Deputy Commander, Royal Canadian Navy:
Rear Admiral M F R Lloyd, CMM, CD
Director General Naval Personnel:
Commodore B W N Santarpia, CD

Flag Officers

Commander, Maritime Forces, Atlantic:
Rear Admiral J F Newton, OMM, MSM, CD
Commander, Maritime Forces, Pacific:
Rear Admiral W S Truelove, OMM, CD
Commander, Naval Reserve: Commodore D W Craig, CD
Commander, Canadian Fleet Atlantic:
Commodore S E G Bishop, OMM, CD
Commander, Canadian Fleet Pacific:
Commodore J R Auchterlonie, CD

Diplomatic Representation

Commander, Canadian Defence Liaison Staff, Washington:
Major General E N Matern
Naval Attaché, Washington:
Captain J R P Gravel, CD
Naval Adviser, London: Captain T C Tulloch

Establishment

The Royal Canadian Navy (RCN) was officially established on 4 May 1910, when Royal Assent was given to the Naval Service Act. On 1 February 1968 the Canadian Forces Reorganisation Act unified the three branches of the Canadian Forces and the title 'Royal Canadian Navy' was dropped. On 16 August 2011, in order to recognise the distinct heritage of each command within the Unified Canadian Forces structure, the historical titles of each branch, including the Royal Canadian Navy, were reinstated.

Personnel

2014: 9,086 (Regular), 5,130 (Reserves)

Prefix to Ships' Names

HMCS

Bases

Halifax and Esquimalt

Fleet Deployment

Atlantic
Canadian Fleet Atlantic (destroyer, frigates, AOR)
Maritime Operations Group Five (maritime warfare forces, submarines, training ships)

Pacific
Canadian Fleet Pacific (destroyer, frigates, submarines, AOR)
Maritime Operations Group Four (maritime warfare forces, training ships)

Maritime Air Components (MAC)

1 Canadian Air Division HQ Detachment Air Component Co-ordination Element (ACCE) Atlantic (Halifax)
1 Canadian Air Division HQ Detachment ACCE Pacific (Esquimault)

Squadron/Unit	Base	Aircraft	Function
MP 404 (MP&T)	Greenwood, NS	Aurora/Arcturus	LRMP/Training
MP 405 (MP)	Greenwood, NS	Aurora	LRMP
HT 406 (M) OTS	Shearwater, NS	Sea King	Training
MP 407 (MP)	Comox, BC	Aurora	LRMP
MH 423 (MH)	Shearwater, NS	Sea King	General
MH 443 (MH)	Victoria, BC	Sea King	General
HOTEF	Shearwater, NS	Sea King	Test
MP & EU	Greenwood, NS	Aurora	Test

Notes

1. Detachments from 423 and 443 meet ships' requirements in Atlantic and Pacific Fleets respectively. Sea King helicopters are now classified as General Purpose vice the former ASW designation.
2. 413 Squadron based in Greenwood, NS, and 442 Squadron based in Comox, BC, are two maritime search and rescue squadrons under the command of 1 Canadian Air Division (CAD).
3. Combat training support provided by commercial contract from March 2002.

Strength of the Fleet

Type	Active	Building
Submarines	4	—
Destroyers	3	—
Frigates	12	—
Mine Warfare Forces	12	—
Survey Ships	1	—
Support Ships	2	(2)

PENNANT LIST

Submarines
876	Victoria
877	Windsor
878	Corner Brook
879	Chicoutimi

Destroyers
280	Iroquois
282	Athabaskan
283	Algonquin

Frigates
330	Halifax
331	Vancouver
332	Ville de Québec
333	Toronto
334	Regina
335	Calgary
336	Montreal
337	Fredericton
338	Winnipeg
339	Charlottetown
340	St John's
341	Ottawa

Mine Warfare Forces
700	Kingston
701	Glace Bay
702	Nanaimo
703	Edmonton
704	Shawinigan
705	Whitehorse
706	Yellowknife
707	Goose Bay
708	Moncton
709	Saskatoon
710	Brandon
711	Summerside

Training Ships
55	Orca
56	Raven
57	Caribou
58	Renard
59	Wolf
60	Grizzly
61	Cougar
62	Moose

Auxiliaries
172	Quest
509	Protecteur
510	Preserver
610	Sechelt
611	Sikanni
612	Sooke
613	Stikine

SUBMARINES

4 VICTORIA (UPHOLDER) CLASS (TYPE 2400) (SSK)

Name	No	Builders	Launched	Commissioned	Recommissioned
VICTORIA	876	Cammell Laird, Birkenhead	14 Nov 1989	7 Jun 1991	2 Dec 2000
WINDSOR	877	Cammell Laird, Birkenhead (VSEL)	16 Apr 1992	25 Jun 1993	4 Oct 2003
CORNER BROOK	878	Cammell Laird, Birkenhead (VSEL)	28 Feb 1991	8 May 1992	29 Jun 2003
CHICOUTIMI	879	Vickers Shipbuilding & Engineering, Barrow-in-Furness	2 Dec 1986	9 Jun 1990	2 Oct 2004

Displacement, tonnes: 2,203 surfaced; 2,494 dived
Dimensions, metres (feet): 70.3 × 7.6 × 5.5 *(230.6 × 24.9 × 18.0)*
Speed, knots: 12 surfaced; 12 snorting; 20 dived
Range, n miles: 8,000 at 8 kt snorting
Complement: 48 (7 officers) + 11 spare berths

Machinery: Diesel-electric; 2 Paxman Valenta 16SZ diesels; 3,620 hp *(2.7 MW)* sustained; 2 GEC alternators; 2.8 MW; 1 GEC motor; 5,400 hp *(4 MW)*; 1 shaft
Torpedoes: 6—21 in *(533 mm)* bow tubes. Up to 18 Raytheon Mk 48 Mod 4M; dual purpose; active/passive homing to 50 km *(27 n miles)*/38 km *(21 n miles)* at 40/55 kt; warhead 267 kg. Air turbine pump discharge.
Physical countermeasures: Decoys: 2 SSE launchers.
Electronic countermeasures: ESM: Sea Search II; intercept.
Radars: Navigation: Kelvin Hughes Type 1007; I-band. Furuno (portable); I-band.
Sonars: Thomson Sintra Type 2040; hull-mounted; passive search and intercept; medium frequency. BAE Type 2007; flank array; passive; low frequency. Thales Type 2046; towed array; passive very low frequency. Lockheed Martin AN/BQQ 10(V7); passive ranging.
Weapon control systems: Lockheed Martin SFCS.

Programmes: First ordered 2 November 1983. Further three ordered on 2 January 1986. Laid up after post Cold War defence cuts in 1994 and acquired from the UK on 6 April 1998. Refitted at Vickers, Barrow, for delivery from June 2000.
Modernisation: A modernisation programme is underway. This is to update/replace components of the combat system (torpedo, sonar suite, communications) and platform sub-systems. A more extensive Submarine Equipment Life Extension (SELEX) programme is in abeyance pending a feasibility study of platform life extension beyond the mid-2020s.
Structure: Single-skinned NQ1 high tensile steel hull, tear dropped shape 9:1 ratio, five man lock-out chamber in fin. Fitted with elastomeric acoustic tiles. Diving depth, greater than 200 m *(650 ft)*. Fitted with Pilkington Optronics CK 35 search and CH 85 attack optronic periscopes.

CORNER BROOK *8/2007, Blake Rodgers, RCN* / 1166835

Operational: *Victoria* returned to the Pacific Fleet on completion of an Extended Docking Work Period (EDWP) in 2011. *Windsor* completed EDWP at Halifax in 2012 and subsequently returned to Atlantic Fleet service in 2012. *Chicoutimi* suffered a serious fire in October 2004. She commenced repairs and EDWP at Esquimault in July 2010 and returned to full service in 2014. *Corner Brook* commenced EDWP at Esquimault in 2014.

CORNER BROOK 6/2010, Michael Nitz / 1366716

CORNER BROOK 4/2009, M Declerck / 1367015

WINDSOR 11/2009, Roxanne Clowe, RCN / 1155887

FRIGATES

12 HALIFAX CLASS (FFGHM)

Name	No	Builders	Laid down	Launched	Commissioned
HALIFAX	330	Saint John SB Ltd, New Brunswick	19 Mar 1987	30 Apr 1988	29 Jun 1992
VANCOUVER	331	Saint John SB Ltd, New Brunswick	19 May 1988	8 Jul 1989	23 Aug 1993
VILLE DE QUÉBEC	332	Marine Industries Ltd, Sorel	17 Jan 1989	16 May 1991	14 Jul 1994
TORONTO	333	Saint John SB Ltd, New Brunswick	24 Apr 1989	18 Dec 1990	29 Jul 1993
REGINA	334	Marine Industries Ltd, Sorel	6 Oct 1989	25 Oct 1991	30 Sep 1994
CALGARY	335	Marine Industries Ltd, Sorel	15 Jun 1991	28 Aug 1992	12 May 1995
MONTREAL	336	Saint John SB Ltd, New Brunswick	8 Feb 1991	28 Feb 1992	21 Jul 1994
FREDERICTON	337	Saint John SB Ltd, New Brunswick	25 Apr 1992	13 Mar 1993	10 Sep 1994
WINNIPEG	338	Saint John SB Ltd, New Brunswick	19 Mar 1993	5 Dec 1993	23 Jun 1995
CHARLOTTETOWN	339	Saint John SB Ltd, New Brunswick	5 Dec 1993	10 Jul 1994	9 Sep 1995
ST JOHN'S	340	Saint John SB Ltd, New Brunswick	24 Aug 1994	12 Feb 1995	26 Jun 1996
OTTAWA	341	Saint John SB Ltd, New Brunswick	29 Apr 1995	22 Nov 1995	28 Sep 1996

Displacement, tonnes: 4,847 full load
Dimensions, metres (feet): 134.7 oa; 124.5 pp × 16.4 × 5 hull; 7.1 screws (441.9; 408.5 × 53.8 × 16.4; 23.3)
Speed, knots: 29
Range, n miles: 9,500 at 13 kt, 3,930 at 18 kt
Complement: 215 (25 officers; 17 air crew)

Machinery: CODOG; 2 GE LM 2500 gas turbines; 47,494 hp (35.43 MW) sustained 1 SEMT-Pielstick 20 PA6 V 280 diesel; 8,800 hp(m) (6.48 MW) sustained; 2 shafts; cp props
Missiles: SSM: 8 McDonnell Douglas Harpoon Block 2 (2 quad) launchers ❶; active radar homing to 124 km (67 n miles) at 0.9 Mach; warhead 227 kg.
SAM: Raytheon RIM-162 Evolved Sea Sparrow; 2 Mk 48 octuple vertical launchers ❷; semi-active homing to 18 km (9.7 n miles) at 3.6 Mach; warhead 38 kg; 16 missiles.
Guns: 1 Bofors 57 mm/70 Mk 3 ❸; 220 rds/min to 17 km (9 n miles); weight of shell 2.4 kg. 1 Raytheon 20 mm Vulcan Phalanx Block 1B ❹; anti-missile; 4,500 rds/min (6 barrels combined) to 1.5 km. 6—12.7 mm MGs.
Torpedoes: 4—324 mm Mk 32 Mod 9 (2 twin) tubes ❺. 24 Honeywell Mk 46 Mod 5; anti-submarine; active/passive homing to 11 km (5.9 n miles) at 40 kt; warhead 44 kg.
Physical countermeasures: Decoys: 4 Rheinmetall MASS-4L decoy launchers ❻. Nixie SLQ-25A; towed acoustic decoy.
Electronic countermeasures: ESM: ELISRA ❼; intercept.
ECM: MEL/Lockheed Ramses SLQ-503 ❽; jammer.
Radars: Air search: Thales SMART-S Mk 2 ❾; 3D; E/F-band.
Air/surface search: Saab Sea Giraffe HC 180 ❿; G/H-band.
Fire control: 2 Saab Ceros 200 ⓫; J-band.
Navigation: Sperry Mk 340 being replaced by Kelvin Hughes 1007; I-band.
Tacan: URN 25. IFF Mk XII.
Sonars: Westinghouse SQS-510; hull-mounted; active search and attack; medium frequency. General Dynamics SQR-501 CANTASS towed array (uses part of Martin Marietta SQR-19 TACTASS).
Combat data systems: Lockheed Martin CMS 330. Links 11, 16, and 22.
Electro-optic systems: Rheinmetall SEOSS. Thales Sirius IRST.

Helicopters: 1 CH-124A ASW ⓬.

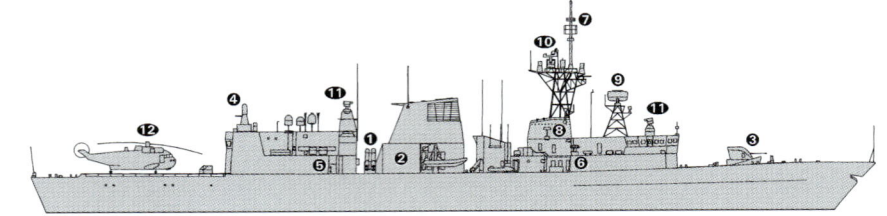

HALIFAX (Scale 1 : 1,200), Ian Sturton / 1525659

Programmes: On 29 June 1983 Saint John Shipbuilding Ltd won the competition for the first six of a new class of patrol frigates. Combat system design and integration was subcontracted to Loral Canada (formerly Paramax, a subsidiary of Unisys). Three ships were subcontracted to Marine Industries Ltd in Lauzon and Sorel. On 18 December 1987 six additional ships of the same design were ordered from Saint John SB Ltd.
Modernisation: The Halifax Class Modernisation (HCM)/ Frigate Life Extension (FELEX) programme subsumes all maintenance, sustainment and stand-alone projects planned to ensure the continued operation of the class for the duration of its life. In general, combat system enhancements reflect increasing emphasis on littoral operations in a joint force and a coalition. A Combat Systems Integrator contract was signed with Lockheed Martin for the design and construction of a new Command Management System (CMS 330) and the upgrade of associated radars and sensors. Major equipment acquisitions through HCM/FELEX include CMS 330 based on the SAAB 9LV, IBM Multi-Link (including Link 11, 16 and 22), Telephonics IFF Mode S/5, replacement of SPS 49 with Thales SMART S Mk II, replacement of SG-150 with SG-180 hybrid, replacement of SPG 503 with CEROS 200, replacement of SLQ-501 with Elisra ESM system, upgrades to internal communications, upgrade of Harpoon to Block II and Bofors 57 mm gun to Mk 3 standard, replacement of Plessey decoy system with Rheinmetall MASS, addition of SIRIUS Infra-Red Search and Track (IRST) and improvement of the degaussing system. In addition, a CTG capability is to be added to selected ships. ASW projects include improvement of torpedo defence (SLQ-25A). Integration of the Cyclone helicopter will also make a significant contribution to ASW improvements. The HCM/FELEX Project will also be responsible for platform changes arising from the planned modifications. These include upper deck and operations room reconfiguration, power supplies, heating/ventilation/air-conditioning, chilled water supplies, hull strength, and stability. All ships are being modified to achieve a common equipment and systems baseline before beginning the HCM/FELEX upgrade which began with *Halifax* in 2010. She was followed in 2011 by *Calgary* at Victoria Shipyard and *Fredericton* at Halifax and in 2012 by *Winnipeg* (Victoria) and *Montreal* (Halifax). In 2013, *Charlottetown* (Halifax) and *Vancouver* (Victoria) began their upgrades.
Structure: Much effort has gone into stealth technology. Gas turbine engines are raft mounted. Dresball IR suppression is fitted. Indal RAST helicopter handling system.
Operational: Problems on first of class trials included higher than designed radiated noise levels which were reported as speed associated. These have been rectified and the ships are stable and quiet in all sea conditions. *Vancouver*, *Regina*, *Calgary*, *Winnipeg* and *Ottawa* are Pacific based.

WINNIPEG 6/2010, Guy Toremans / 1366721

OTTAWA 7/2011, Chris Sattler / 1406534

VILLE DE QUÉBEC
4/2012, Adolfo Ortigueira Gil / 1482974

ST JOHN'S
6/2010, Michael Nitz / 1366719

CHARLOTTETOWN
6/2010, Michael Nitz / 1366718

DESTROYERS

Notes: The Canadian Surface Combatants (CSC) project is to recapitalise the surface combatant fleet by replacing the capabilities of the Iroquois and Halifax classes, and providing the necessary integrated logistic support and infrastructure. An Options Analysis Phase was completed in 2012 when the CSC project entered Definition Phase. The first CSC is planned to be delivered in 2025.

3 IROQUOIS CLASS (DDGH)

Name	No	Builders	Laid down	Launched	Commissioned
IROQUOIS	280	Marine Industries Ltd, Sorel	15 Jan 1969	28 Nov 1970	29 Jul 1972
ATHABASKAN	282	Davie Shipbuilding, Lauzon	1 Jun 1969	27 Nov 1970	30 Sep 1972
ALGONQUIN	283	Davie Shipbuilding, Lauzon	1 Sep 1969	23 Apr 1971	3 Nov 1973

Displacement, tonnes: 5,385 full load
Dimensions, metres (feet): 129.8 oa; 121.4 wl × 15.2 × 4.7 hull; 6.6 screws *(425.9; 398.3 × 49.9 × 15.4; 21.7)*
Speed, knots: 27
Range, n miles: 4,500 at 15 kt
Complement: 285 (32 officers; 30 air crew)

Machinery: COGOG; 2 Pratt & Whitney FT4A2 gas turbines; 50,000 hp *(37 MW)*; 2 GM Allison 570-KF gas turbines; 12,700 hp *(9.5 MW)* sustained; 2 shafts; LIPS cp props
Missiles: SAM: 1 Martin Marietta Mk 41 VLS ❶ for 29 GDC Standard SM-2MR Block III/IIIA; command/inertial guidance; semi-active radar homing to 167 km *(90 n miles)* at Mach 2.5.
Guns: 1 Oto Melara 3 in *(76 mm)*/62 Super Rapid ❷; 120 rds/min to 16 km *(8.7 n miles)*; weight of shell 6 kg. 6—12.7 mm MGs. 1 Raytheon 20 mm/76 6-barrelled Vulcan Phalanx Block 1B ❸; 4,500 rds/min combined to 1.5 km.
Torpedoes: 6—324 mm Mk 32 (2 triple) tubes ❹. Honeywell Mk 46 Mod 5; anti-submarine; active/passive homing to 11 km *(5.9 n miles)* at 40 kt; warhead 44 kg.
Physical countermeasures: Decoys: 4 Plessey Shield Mk 2 6-tubed fixed launchers ❺. P 8 chaff or P 6 IR flares. BAe Nulka offboard decoys in quad pack launchers. SLQ-25 Nixie; torpedo decoy.
Electronic countermeasures: ESM: MEL SLQ-501 Canews ❻; radar warning.
ECM: BAe Nulka.
Radars: Air search: Signaal SPQ-502 (LW08) ❾; D-band.
Surface search: Signaal SPQ-501 (DA08) ❿; E/F-band.
Fire control: 2 Signaal SPG-501 (STIR 1.8) ⓫; I/J-band.
Navigation: 2 Raytheon Pathfinder; I-band.
Tacan: URN 26.

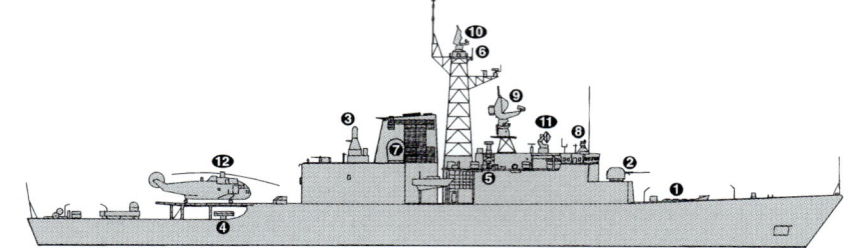

IROQUOIS *(Scale 1 : 1,200), Ian Sturton / 0056677*

Sonars: General Dynamics SQS-510; combined VDS and hull-mounted; active search and attack; medium frequency.
Combat data systems: SHINPADS, automated data handling with UYQ-504 and UYK-507 processors. Links 11, 14 and 16. JMCIS and Marconi Matra SHF SATCOM ❼.
Weapon control systems: Signaal LIROD 8 ❽ optronic director. UYS-503(V) sonobuoy processor.

Helicopters: 2 CH-124A Sea King ASW ⓬.

Modernisation: A contract for the Tribal Class Update and Modernisation Project (TRUMP) was awarded to Litton Systems Canada Limited in June 1986. The equipment reflected the changing role of the ship and replaced systems that did not meet the air defence requirement. *Algonquin* completed modernisation in October 1991, followed by *Iroquois* in May 1992 and *Athabaskan* in August 1994.

Sonar upgraded from 1998. Nulka system replaced ULQ-6 in 1999. JMCIS has been fitted vice JOTS II, with SHF SATCOM in all three ships. Shipboard Electro-Optic Surveillance System (SEOSS) is being fitted in all ships. Vulcan Phalanx upgraded to Block 1B from 2003. A programme to upgrade/overhaul the SPQ-501, SPQ-502 and SPG-501 radars and the LIROD 8 optronic director began in 2008.
Structure: These ships are also fitted with a landing deck equipped with double hauldown and Beartrap, pre-wetting system to counter NBC conditions, enclosed citadel and bridge control of machinery. The flume type anti-roll tanks have been replaced during modernisation with a water displaced fuel system. Design weight limit has been reached.
Operational: Helicopters can carry 12.7 mm MGs and ESM/FLIR instead of ASW gear. *Huron* was decommissioned in 2005. The remaining three are to remain in service until 2015.

IROQUOIS *5/2013*, Michael Nitz / 1525559*

ALGONQUIN *10/2012, M Mazumdar / 1482972*

SHIPBORNE AIRCRAFT

Notes: The five HELTAS Sea Kings have been converted to a battlefield/utility transport role. All acoustic systems have been removed.

Numbers/Type: 28 Sikorsky CH-148 Cyclone.
Operational speed: 165 kt *(305 km/h)*.
Service ceiling: 11,320 ft *(3,450 m)*.
Range: 444 n miles *(821 km)*.
Role/Weapon systems: Contract for the acquisition of 28 helicopters to replace the Sea King (by 2012) made on 23 November 2004. First flight in November 2008 and first ship-helo trials undertaken in December 2009. Trials and tests continue. The schedule for delivery of operational aircraft has not been announced. Multimission maritime helicopter for ASW and ASuW and secondary missions of SAR, special forces operations, and Medevac. Sensors: L3 HELRAS sonar, Telephonics APS-143B(V)3 ISAR radar, GDC acoustic signal processor, FLIR Systems Star Safire III electro-optics system, Rockwell-Collins ARC-210 communications suite, ATK Alliant AN/ARR-47 MAWS, Lockheed Martin AN/ALQ-210 ESM/radar warning, Lockheed Martin AN/ALR-47 laser warning, BAE AN/ALQ-144 IR jammer, BAE AN/ALE-47 countermeasures dispenser system; Links 11 and 22. Weapons: Two Mk 46 torpedoes and C6 7.62 mm MG.

CH-148   *6/2010, Royal Canadian Navy* / 1406537

Numbers/Type: 22/5 Sikorsky CH-124A ASW/CH-124B SCF Utility Sea King.
Operational speed: 110 kt *(203 km/h)*.
Service ceiling: 10,000 ft *(3,030 m)*.
Range: 380 n miles *(705 km)*.
Role/Weapon systems: ASW, surface surveillance and support (A variant), convertible for carriage of 13 troops (B variant only); deployed from shore or from three classes of ships (Halifax class FFG (1 aircraft), Iroquois class DDG (2 aircraft) and 'Protecteur' AOR (3 aircraft)). Sensors: CH-124A/B: APS-503 radar, ASN-123 mission computer, GPS, ARA-5 direction finder, APX-77A IFF, HF/VHF/UHF comms (with secure voice capability), ALQ-144 IR countermeasures (fitted for but not with). CH-124A: AQS-502 dipping sonar, ARR-52A sono receiver and ARR-1047 OTPI. CH-124B: AN/ARC-210 communications, AN/AAR-47 MAWS, AN/ALQ-144 IR jammer and AN/ALE-47 CDS. Weapons: Two Mk 46 torpedoes and C6 7.62 mm MGs for both aircraft types.

CH-124A   *10/2012, A Sheldon-Duplaix* / 1482971

LAND-BASED MARITIME AIRCRAFT (FRONT LINE)

Notes: The Canadian Multimission Aircraft (CMA) programme is for the acquisition of a new maritime patrol capability to replace the Aurora. Options include a single or mixed aircraft fleet.

Numbers/Type: 18 Lockheed CP-140 Aurora.
Operational speed: 405 kt *(750 km/h)*.
Service ceiling: 34,000 ft *(9,930 m)*.
Range: 4,400 n miles *(8,148 km)*.
Role/Weapon systems: Operated for long-range C4ISR missions in both maritime and overland roles. Maritime roles include ASW, ASuW, OTHT, MCT and constabulary duties. Overland roles include surveillance, targeting, intelligence and photo-mapping. SAR can be carried out in either environment. The Aurora Incremental Modernisation Programme (AIMP), currently underway, is to be completed by 2020. The programme includes upgrades to avionics, communications and sensors. In parallel, an ASLEP programme addresses airframe structural issues. Sensors: APS-506 (to be replaced by APS-508), ALR-502 (to be replaced by ALQ-217), L-3 Wescam MX 20, AN/UYS 504 MVASP acoustic processor, ASQ 502 MAD. Weapons: 8 Mk 46 Mod 5 torpedoes.

AURORA   *10/2007, Michael Nitz* / 1335647

© 2014 IHS

PATROL FORCES

0 + 8 ARCTIC OFFSHORE PATROL SHIPS (AGB)

Displacement, tonnes: 6,400 full load
Dimensions, metres (feet): 103.6 × 29.0 × 5.75 *(339.9 × 95.1 × 18.9)*
Speed, knots: 17. **Range, n miles:** 2,700 at 10 kt
Complement: 47 (6 officers) + 38 personnel

Machinery: Diesel-electric; 2 shafts; bow thruster
Guns: 1—25 mm.
Radars: To be announced.
Helicopters: 1 medium.

Comment: It was announced on 9 July 2007 that up to eight Arctic/Offshore Patrol Ships (AOPS) are to be acquired. The ships' primary role is to conduct surveillance and sovereignty enforcement operations in Canada's maritime estate. Their secondary role is to provide support to other Canadian Forces and government department operations in Canada's Exclusive Economic Zone, including the Arctic in the navigable season. The ships are to be constructed to commercial Polar class 4/5 standards and are to be equipped with a 20-tonne crane for autonomous loading or unloading. Other design features are likely to include two boat bays for 8.4 m multirole rescue boats and two additional bays for enclosed motor-propelled lifeboats or large rigid-hulled inflatable boats. There is to be space on the quarterdeck for a 10-man diving boat, an 11 m LCVP, and up to six TEUs of modular payloads. An additional two TEUs can be carried on the flight deck in lieu of aircraft. Modular payloads include a route survey system, bottom objection inspection vehicle, two-compartment containerised diving system, six-person accommodation modules or general cargo. There is also to be garage space for one 1-ton truck, assorted ATVs, or snow-mobiles. Construction at Irving Shipbuilding Industries, Halifax, is scheduled to begin in 2015 with first ship delivery due in 2018. Associated infrastructure projects include modifications to port facilities in Esquimault and Halifax, as well as establishment of a naval facility in Nanisivik, Nunavut. The Nanisivik component will see refurbishment of an existing jetty as well as the installation of basic refuelling and logistic support facilities. A contract for this work should be let in 2014, with work expected to be completed in time for delivery of the first ship.

AOPS   *6/2012*, Royal Canadian Navy* / 1366725

MINE WARFARE FORCES

Notes: The Remote Minehunting and Disposal Project is considering options to meet future capability requirements. Analysis includes review of current Unmanned Underwater Vehicles (UUV) and expendable mine disposal systems in service with other navies and/or available through industry. Delivery of both UUVs and improved Route Survey assets is expected 2014–15. The Interim Minehunting and Disposal System, also known as Seakeeper, has been placed at extended readiness.

12 KINGSTON CLASS (MM)

Name	No	Builders	Laid down	Launched	Commissioned
KINGSTON	700	Halifax Shipyards	15 Dec 1994	12 Aug 1995	21 Sep 1996
GLACE BAY	701	Halifax Shipyards	28 Apr 1995	22 Jan 1996	26 Oct 1996
NANAIMO	702	Halifax Shipyards	11 Aug 1995	17 May 1996	10 May 1997
EDMONTON	703	Halifax Shipyards	8 Dec 1995	16 Aug 1996	21 Jun 1997
SHAWINIGAN	704	Halifax Shipyards	26 Apr 1996	15 Nov 1996	14 Jun 1997
WHITEHORSE	705	Halifax Shipyards	26 Jul 1996	24 Feb 1997	17 Apr 1998
YELLOWKNIFE	706	Halifax Shipyards	7 Nov 1996	5 Jun 1997	18 Apr 1998
GOOSE BAY	707	Halifax Shipyards	22 Feb 1997	4 Sep 1997	26 Jul 1998
MONCTON	708	Halifax Shipyards	31 May 1997	5 Dec 1997	12 Jul 1998
SASKATOON	709	Halifax Shipyards	5 Sep 1997	30 Mar 1998	21 Nov 1998
BRANDON	710	Halifax Shipyards	6 Dec 1997	3 Sep 1998	5 Jun 1999
SUMMERSIDE	711	Halifax Shipyards	28 Mar 1998	4 Oct 1998	18 Jul 1999

Displacement, tonnes: 977 full load
Dimensions, metres (feet): 55.3 × 11.3 × 3.4 *(181.4 × 37.1 × 11.2)*
Speed, knots: 15; 10 sweeping. **Range, n miles:** 5,000 at 8 kt
Complement: 37

Machinery: Diesel-electric; 4 Wärtsilä UD 23V12 diesels; 4 Jeumont ANR-53-50 alternators; 7.2 MW; 2 Jeumont CI 560L motors; 3,000 hp(m) *(2.2 MW)*; 2 LIPS Z drive azimuth thrusters
Guns: 1 Bofors 40 mm/60 Mk 5C. 2—12.7 mm MGs.
Physical countermeasures: Three positions on the sweep deck can receive a variety of mission payloads on a 20 ft ISO footprint including: ISE Ltd Trailblazer 25 Bottom Object Inspection Vehicle (1 system); Fullerton and Crane Ltd 6-man, 2-compartment Containerised Diving System (2 systems), naval engineered 6-person accommodation modules (6 systems), L3/Klein 5500 high-definition side-scan sonar (2 systems), L3/Klein K3000 dual frequency side-scan sonar (4 systems), Seabotix vLBV 950 remotely operated vehicle (ROV) (8 systems) and Deep Ocean Engineering Phantom 4 ROVs.
Radars: Surface search/navigation: Sperry Bridgemaster E; E/F/I-bands.

Programmes: Contract awarded to Fenco MacLaren on 15 May 1992. Halifax Shipyards is owned by Saint John Shipbuilding. Known as Maritime Coastal Defence Vessels (MCDV) combining MCM with general patrol duties.
Modernisation: Modernisation of the degaussing system started in 2013. Modernisation of the AN-SQQ 511 route survey payload is to start in 2014. Acquisition of a remote minehunting and disposal system is planned in 2017.
Operational: Predominantly manned by reservists. Six on each coast (700, 701, 704, 707, 708 and 711 Atlantic, remainder Pacific). One ship per coast is kept at extended readiness on a rotational basis.

KINGSTON   *6/2013*, Marc Piché* / 1525560

Canada > Survey and research ships — Auxiliaries

SURVEY AND RESEARCH SHIPS

1 RESEARCH SHIP (AGORH)

Name	No	Builders	Launched	Commissioned
QUEST	AGOR 172	Burrard, Vancouver	9 Jul 1969	21 Aug 1969

Displacement, tonnes: 2,164 full load
Dimensions, metres (feet): 76.8 × 12.8 × 5.6 (252 × 42.0 × 18.4)
Speed, knots: 14.5
Range, n miles: 10,000 at 12 kt
Complement: 24 + 21 scientists
Machinery: Diesel-electric; 2 Fairbanks-Morse diesel generators; 2 GE motors; 2 shafts

Comment: Used by Defence Research and Development Canada (DRDC) for acoustic, hydrographic and general oceanographic research activities. Designed with special acoustic quieting (anechoic tiles, rotating machinery on resilient mounts, propulsion and service diesels resiliently mounted and acoustically enclosed, various operational quiet states). Capable of operating in summer ice conditions (Ice Class I). Based in Halifax, NS, operates mainly in North and Mid Atlantic. Mid-life update in 1997–99 included new communications and navigation equipment, improved noise insulation, updated deck cranes and hardware, and modernised laboratories.

QUEST　　　　　　　　　　　　　　　　　　　9/2011, Michael Nitz / 1406539

TRAINING SHIPS

8 ORCA CLASS (TRAINING SHIPS) (AXL)

ORCA 55	CARIBOU 57	WOLF 59	COUGAR 61
RAVEN 56	RENARD 58	GRIZZLY 60	MOOSE 62

Displacement, tonnes: 213 full load
Dimensions, metres (feet): 33 × 8.4 × 2.5 (108.3 × 27.6 × 8.2)
Speed, knots: 21
Range, n miles: 750 at 15 kt
Complement: 20 (16 trainees)
Machinery: 2 Caterpillar 3516 diesels; 5,000 hp (3.7 MW); 2 shafts
Guns: 1 – 12.7 mm MG (fitted for).
Radars: 2 Raytheon NSC 1810; I-band.

Comment: Contract awarded to Victoria Shipyards, BC, on 8 November 2004 for the construction of six training vessels. The option to build a further two has been exercised. Based on the Australian *Seahorse Mercator* design. Construction of the first vessel began on 8 September 2005 with formal acceptance on 17 November 2006. The eighth and final vessel was delivered in late 2008. All vessels based at Esquimalt.

WOLF　　　　　　　　　　　　　　　　　　　6/2010, Guy Toremans / 1366726

1 SAIL TRAINING SHIP (AXS)

Name	No	Builders	Launched
ORIOLE	YAC 3	Owens	4 Jun 1921

Displacement, tonnes: 93 full load
Dimensions, metres (feet): 31.1 × 5.8 × 2.7 (102 × 19.0 × 8.9)
Speed, knots: 8
Complement: 24 (1 officer; 18 trainees)
Machinery: 1 Cummins diesel; 165 hp (123 kW); 1 shaft

Comment: Commissioned in the Navy in 1948 and based at Esquimalt. Sail area (with spinnaker) 11,000 sq ft. Height of mainmast 94 ft (28.7 m), mizzen 55.2 ft (16.8 m).

ORIOLE　　　　　　　　　　　　　　　　　　　6/2008, RCN / 1335644

AUXILIARIES

2 GRANBY CLASS
(GENERAL PURPOSE DIVING TENDERS) (YDT)

– YDT 11　　　　GRANBY YDT 12

Displacement, tonnes: 112 standard
Dimensions, metres (feet): 27.3 × 6.2 × 2.6 (89.6 × 20.3 × 8.5)
Speed, knots: 11
Complement: 13
Machinery: Diesel; 228 hp (170 kW); 1 shaft
Radars: Navigation: Racal Decca; I-band.
Sonars: Fitted for L3/Klein K 5500.

Comment: Built to provide platform for underwater engineering and 100 m surface supplied diving operations. Secondary role is support of MCM operations and maritime explosive ordnance disposal operations. Both ships are equipped to deploy the deep ocean remote vehicle SEABOTIX.

YDT 11　　　　　　　　　　　　　　　　　　　6/2010, Guy Toremans / 1366730

4 SECHELT CLASS (YTT/YPT/YDT)

Name	No	Builders	Commissioned	Homeport
SECHELT	YDT 610	West Coast Manly	10 Nov 1990	Halifax
SIKANNI	YPT 611	West Coast Manly	10 Nov 1990	–
SOOKE	YDT 612	West Coast Manly	10 Nov 1990	Esquimault
STIKINE	YPT 613	West Coast Manly	10 Nov 1990	–

Displacement, tonnes: 295 full load
Dimensions, metres (feet): 33.1 × 8.5 × 2.4 (108.6 × 27.9 × 7.9)
Speed, knots: 12.5
Complement: 12 (YDT 610, 612), 4 (YPT 611, 613)
Machinery: 2 Caterpillar 3412T diesels; 1,080 hp (806 kW) sustained; 2 shafts
Sonars: Fitted for (YDT 610 and 612) L3/Klein K 3000 or K 5500 side scan sonar.

Comment: *Sikanni* and *Stikine* based at the Nanoose Bay Maritime Experimental and Test Range. *Sechelt* and *Sooke* converted to diving tenders in 1997 with a transportable 6-man recompression chamber. Diving operations supported to 80 m. Capable of limited support to MCM operations, all ships are fitted with SEABOTIX ROVs. *Sechelt* based at Halifax, Novia Scotia, *Sooke* at Esquimault, British Columbia.

SOOKE (with containerised diving system)　　　　　　　　6/2002, CDF / 0528415

IHS Jane's Fighting Ships 2014-2015　　　　　　　　　　　　　　　　© 2014 IHS

Auxiliaries — Tugs and tenders < **Canada** 105

2 PROTECTEUR CLASS (AORH)

Name	No	Builders	Laid down	Launched	Commissioned
PROTECTEUR	AOR 509	St John Dry Dock Co, NB	17 Oct 1967	18 Jul 1968	30 Aug 1969
PRESERVER	AOR 510	St John Dry Dock Co, NB	17 Oct 1967	29 May 1969	30 Jul 1970

Displacement, tonnes: 9,408 standard; 26,088 full load
Dimensions, metres (feet): 171.9 × 23.2 × 10.46 (564 × 76.1 × 34.3)
Speed, knots: 21
Range, n miles: 4,100 at 20 kt, 7,500 at 11.5 kt
Complement: 335 (38 officers; 45 air crew)

Cargo capacity: 13,036 tons fuel; 506 tons aviation fuel; 352 tons dry cargo; 300 tons ammunition; 2 cranes (15 ton lift)
Machinery: 2 Babcock & Wilcox boilers; 1 GE Canada turbine; 21,000 hp (15.7 MW); 1 shaft; bow thruster

Guns: 2 GE/GDC 20 mm/76 6-barrelled Vulcan Phalanx Mk 15. 6—12.7 mm MGs.
Physical countermeasures: Decoys: 6 Loral Hycor SRBOC chaff launchers.
Electronic countermeasures: ESM: Racal Kestrel SLQ-504; radar warning.
Radars: Surface search: Norden SPS-502 with Mk XII IFF.
Navigation: Racal Decca 1630 and 1629; I-band.
Tacan: URN 20.
Combat data systems: EDO Link 11; SATCOM WSC-3(V).

Helicopters: 3 CH-124A or CH-124B Sea King.

Comment: Four replenishment positions. Both have been used as Flagships and troop carriers. They can carry military vehicles and bulk equipment for sealift purposes; also two LCVPs. For the Gulf deployment in 1991, the 76 mm gun was remounted, two Vulcan Phalanx and two Bofors 40/60 guns were fitted, four Plessey Shield chaff launchers and ESM equipment were provided for *Protecteur*. Bofors and 76 mm guns are unlikely to be fitted again. *Protecteur* transferred to the Pacific Fleet November 1992. Both ships to remain in service until replaced by Joint Support Ships from 2019.

PRESERVER   6/2012, M Declerck / 1482969

0 + 2 (1) JOINT SUPPORT SHIPS (AFSH)

Name	No	Builders	Laid down	Launched	Commissioned
QUEENSTON	—	Vancouver Shipyards Co Ltd, Vancouver	2016	2018	2019
CHÂTEAUGAY	—	Vancouver Shipyards Co Ltd, Vancouver	2017	2019	2020

Displacement, tonnes: 20,675 full load
Dimensions, metres (feet): 174 × 24.3 × 7.4 (570.9 × 79.7 × 24.3)
Speed, knots: 20. **Range, n miles:** 168 at 15 kt

Cargo capacity: 7,912 tonnes fuel; 400 tonnes water; 280 tonnes cargo; 220 tonnes ammunition
Machinery: 2 diesels; 2 shafts; cp props, bow thruster

Guns: To be announced.
Radars: Air/surface search: To be announced.
Navigation: To be announced.

Helicopters: 2 medium.

Comment: The Joint Support Ship (JSS) is to replace the Protecteur-class (AORs) with a globally deployable naval support capability. Following termination of an earlier process in 2008, the JSS Project was re-launched in 2010. In June 2013, a ThyssenKrupp Marine Systems Canada Inc Berlin class Batch II design (based on FGS *Bonn*) was selected. The ships are to be capable of carrying up to 30 days of fuel and supplies to sustain a Canadian task group at sea. A production contract is expected by 2016 and the first ship is expected to enter service in 2019. There is an option for a third ship.

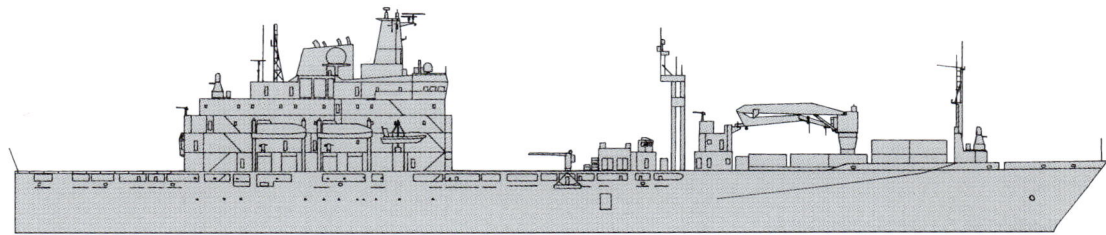

QUEENSTON   (Scale 1 : 1,200), Ian Sturton / 1525660

TUGS AND TENDERS

Notes: There are six 10 m landing craft YFU 121–126 used to transport personnel and equipment.

11 COASTAL TUGS (YTB/YTL/YTM)

GLENDYNE YTB 640	**GLENSIDE** YTB 644	**MERRICKVILLE** YTL 593
GLENDALE YTB 641	**LAWRENCEVILLE** YTL 590	**GRANVILLE** (ex-*Marysville*) YTL 594
GLENEVIS YTB 642	**PARKSVILLE** YTL 591	**TILLICUM** YTM 555
GLENBROOK YTB 643	**LISTERVILLE** YTL 592	

Comment: Glen class are 255 ton tugs built in the mid-1970s. Ville class are 70 ton tugs built in mid-1970s. The YTM is a 160 ton tug.

YFU 128   7/2011, Harald Carstens / 1406540

TILLICUM   6/2010, Guy Toremans / 1366729

106 **Canada** > Tugs and tenders — (COAST GUARD) Heavy icebreakers

6 DIVING SUPPORT CRAFT (YDT)

| FORTUNE | RESOLUTE | TONNERRE |
| ABALONE | DUNGENESS | SCULPIN |

Displacement, tonnes: 2 full load
Dimensions, metres (feet): 11.9 × 3.8 × 0.7 *(39 × 12.5 × 2.3)*
Speed, knots: 36
Range, n miles: 600 at 29 kt
Complement: 17 (14 divers)
Machinery: 2 Caterpillar 3126TA diesels; 740 hp(m) *(548 kW)*; 2 WMC 357 waterjets
Sonars: Fitted for L3/Klein K 3000 and K 5500 side scan sonars.

Comment: Built by Celtic Shipyards and delivered in early 1997. Landing craft bows for launching unmanned submersibles (fitted for SEABOTIX ROV). Bollard pull 6,560 lb. 1,000 kg hydraulic crane. *Fortune, Resolute* and *Tonnerre* based at Halifax, Nova Scotia, and the remainder at Esquimault, British Columbia.

2 FIRE CLASS (YTR)

FIREBIRD YTR 561   **FIREBRAND** YTR 562

Displacement, tonnes: 140 full load
Dimensions, metres (feet): 23.1 × 6.4 × 2.6 *(75.8 × 21.0 × 8.5)*
Speed, knots: 11
Complement: 8
Machinery: 1 diesel; 1 Voith-Schneider propulsor; 2,400 hp *(1.76 MW)*

Comment: Equipped with three water cannons. Both vessels entered service in 1978. *Firebird* based at Halifax and *Firebrand* at Esquimault.

TONNERRE   6/2010, Michael Nitz / 1366728

FIREBRAND   6/2010, Guy Toremans / 1366727

COAST GUARD

Administration

Commissioner Canadian Coast Guard: Marc Grégoire
Deputy Commissioner, Operations: Jody Thomas
Deputy Commissioner, Procurement: Michel Vermette
Director General, Operations: Wade Spurrell
Assistant Commissioner, Western Region: Roger Girouard
Assistant Commissioner, Central and Arctic Region: Mario Pelletier
Assistant Commissioner, Atlantic Region: John Butler

Establishment

In January 1962, the ships owned and operated by the Department of Transport along with vessels operated by some other government agencies were amalgamated into a new organisation to be known as the Canadian Coast Guard. This reflected the increase in duties that had occurred since 1945, especially in the Arctic. Further expansion and diversification followed: notably of the dedicated search and rescue facilities, vessel traffic management and pollution prevention and response.

On 1 April 1995, the fleet of the Department of Fisheries and Oceans was merged with the Coast Guard under the direction of the Minister of Fisheries and Oceans. Its headquarters are in Ottawa while operations are administered from regional offices in Victoria, British Colombia (Western Region), Montreal, Quebec (Central and Arctic Region) and St John's, Newfoundland (Atlantic Region).

Personnel

2014: 4,749

Missions

The Canadian Coast Guard's mandate is stated in the Oceans Act and the Canada Shipping Act (2001). Principal responsibilities include:
1. Aids to navigation
2. Vessel traffic services
3. Marine communications and traffic management services
4. Icebreaking and ice-management services
5. Sable Island management
6. Channel maintenance
7. Marine search and rescue
8. Marine pollution response
9. Support of other government departments, boards, and agencies by providing ships, aircraft, and other services.

Bases

St John's, Newfoundland; Darmouth, Nova Scotia; Québec City, Québec; Trois-Rivières, Québec (hovercrafts); Prescott, Ontario; Burlington, Ontario (in conjuction with the Canadian Centre for Inland Waters); Hay River, Northwest Territories; Inuvik, Nunavut; Victoria, British Columbia; Patricia Bay, British Columbia; Prince Rupert, British Columbia; Sea Island (Vancouver) (hovercrafts).

Shipborne Aircraft

A total of 21 helicopters can be operated from vessels equipped with flight decks. There are 12 MBB BO 105, six Bell 212 and three Bell 206L. All helicopters are operated and maintained by Transport Canada Aircraft Services. A programme to replace the helicopter fleet was announced in March 2012. Contract awards are expected in 2014.

DELETIONS

2011 *Wilfrid Templeman, Point Henry, Pointe Race, Skua, Nahidik,* CG 117, CG 118
2012 *Provo Wallis, E P Le Québécois, Shamook, Isle Rouge, Gull Isle, Tembah, Sterne, Tuebor,* CG 119
2013 *A H Chevarie, Opilio, G & D Mallard, Spilsbury* (ex-*Osprey), Louisbourg*

HEAVY ICEBREAKERS

Notes: The programme for a new polar icebreaker, to replace *Louis St Laurent* was announced in February 2008. Construction of the new ship, to be named *John G Diefenbaker*, is expected to start in 2019–20 and entry into service is planned for 2022.

1 GULF CLASS (TYPE 1300)

| Name | Builders | Launched | Commissioned |
| LOUIS S ST LAURENT | Canadian Vickers Ltd, Montreal | 3 Dec 1966 | Oct 1969 |

Displacement, tonnes: 14,733 full load
Measurement, tonnes: 11,345 gt; 5,456 net
Dimensions, metres (feet): 119.7 × 24.4 × 9.8 *(392.7 × 80.1 × 32.2)*
Speed, knots: 20
Range, n miles: 23,000 at 16 kt
Complement: 42 (13 officers) + 57 personnel
Machinery: Diesel-electric; 5 Krupp MaK 16 M 453C diesels; 39,400 hp(m) *(28.96 MW)*; 5 Siemens alternators; 3 GE motors; 27,000 hp(m) *(19.85 MW)*; 3 shafts; bow thruster
Radars: Navigation: 3 Sperry Bridgemaster; E/F/I-bands.
Helicopters: 2 BO 105 CBS.

Comment: Larger than any of the former Coast Guard icebreakers. Two 49.2 ft *(15 m)* landing craft embarked. Mid-life modernisation July 1988 to early 1993 included replacing main engines with a diesel-electric system, adding a more efficient *Henry Larsen* type icebreaking bow (adds 8 m to length) with an air bubbler system and improving helicopter facilities with a fixed hangar. In addition the complement was reduced. Based in the Atlantic Region at Argentia, Newfoundland. On 22 August 1994 became the first Canadian ship to reach the North Pole, in company with USCG *Polar Sea*. Operational life is to be extended to 2022.

LOUIS S ST LAURENT   6/2010, Globke Collection / 1366832

LOUIS ST LAURENT
7/2006*, Canadian Coast Guard
1525561

IHS Jane's Fighting Ships 2014-2015

© 2014 IHS

1 TERRY FOX CLASS (TYPE 1200)

Name	Builders	Launched	Commissioned
TERRY FOX	Burrard Yarrow, Vancouver	23 Apr 1982	16 Sep 1983

Displacement, tonnes: 7,214 full load
Measurement, tonnes: 4,301 gt; 1,986 net
Dimensions, metres (feet): 88 × 17.9 × 8.3 (288.7 × 58.7 × 27.2)
Speed, knots: 16. **Range, n miles:** 20,000 at 13.5 kt
Complement: 23 (10 officers) + 11 personnel
Machinery: 4 Stork-Werkspoor 8TM 410 diesels; 23,200 hp(m) (17 MW); 2 shafts; cp props; bow and stern thrusters
Radars: Navigation: 2 Sperry Bridgemaster; E/FI-bands.

Comment: Initially leased for two years from Gulf Canada Resources during the completion of *Louis S St Laurent* conversion but subsequently been retained. Commissioned in Coast Guard colours 1 November 1991 and purchased 1 November 1993. Based in the Atlantic region.

TERRY FOX 6/2010, Globke Collection / 1366830

MEDIUM ICEBREAKERS

3 R CLASS

Name	Builders	Launched	Commissioned
PIERRE RADISSON	Burrard, Vancouver	3 Jun 1977	26 May 1978
AMUNDSEN	Burrard, Vancouver	10 Mar 1978	29 Mar 1979
(ex-*Sir John Franklin*)			
DES GROSEILLIERS	Port Weller, Ontario	7 Aug 1982	14 Oct 1982

Displacement, tonnes: 6,503 standard; 8,311 full load
Measurement, tonnes: 5,911 (Amundsen), 6,097 (Des Groseilliers), 5,775 (Pierre Radisson) gt; 1,678 (Amundsen), 1,799 (Des Groseilliers), 1,732 (Pierre Radisson) net
Dimensions, metres (feet): 98.2 × 19.5 (Amundsen), 19.8 (Des Groseilliers), 19.2 (Pierre Radisson) × 7.3 (322.2 × 64.0, 65.0, 63.0 × 24.0)
Speed, knots: 16.5
Range, n miles: 35,000 at 14 kt (Amundsen), 30,600 at 14 kt (Des Groseilliers), 15,000 at 14 kt (Pierre Radisson)
Complement: 31 (11 officers)
Machinery: Diesel-electric; 6 Alco M251F diesels; 13,600 hp (10.14 MW); 2 shafts; bow and stern thrusters
Radars: Navigation: 3 Sperry Marine Bridgemaster; E/F/I-bands.
Helicopters: 1 MBB BO-105S.

Comment: The conversion of *Sir John Franklin* at Les Mechins, Québec, funded by the Canada Foundation for Innovation, included transformation of some storage holds into laboratory space and addition of a moon pool. The ship was re-commissioned in 2003 as *Amundsen*. All three vessels are based at Québec City, Québec, in the Central and Arctic region.

DES GROSEILLIERS 10/2013*, Marc Piché / 1525562

1 MODIFIED R CLASS (TYPE 1200)

Name	Builders	Launched	Commissioned
HENRY LARSEN	Versatile Pacific, Vancouver	3 Jan 1987	12 Jul 1988

Displacement, tonnes: 5,891 standard; 8,423 full load
Measurement, tonnes: 6,167 gt; 1,757 net
Dimensions, metres (feet): 99.8 × 19.7 × 7.3 (327.4 × 64.6 × 24.0)
Speed, knots: 16
Range, n miles: 20,000 at 13.5 kt
Complement: 31 (11 officers) + 40 spare berths
Machinery: Diesel-electric; 3 Wärtsilä Vasa 16V32 diesel generators; 22,970 hp (17.13 MW); 3 motors; 16,320 hp(m) (12 MW); 3 shafts
Radars: Navigation: Racal Decca Bridgemaster; I-band.
Helicopters: 1 BO-105.

Comment: Contract date 25 May 1984, laid down 23 August 1985. Although similar in many ways to *Pierre Radisson*, *Amundsen* and *Des Groseilliers* she has a different hull form particularly at the bow and a very different propulsion system. Fitted with Wärtsilä air bubbling system. Based at St John's in the Atlantic Region.

© 2014 IHS

HENRY LARSEN 6/2013*, Canadian Coast Guard / 1525564

LIGHT ICEBREAKERS

6 MARTHA L BLACK CLASS (TYPE 1100)

Name	Builders	Commissioned
MARTHA L BLACK	Versatile Pacific, Vancouver	30 Apr 1986
GEORGE R PEARKES	Versatile Pacific, Vancouver	17 Apr 1986
EDWARD CORNWALLIS	Marine Industries Ltd, Tracy	14 Aug 1986
SIR WILLIAM ALEXANDER	Marine Industries Ltd, Tracy	13 Feb 1987
SIR WILFRID LAURIER	Canadian Shipbuilding Ltd, Ontario	15 Nov 1986
ANN HARVEY	Halifax Industries Ltd, Halifax	29 Jun 1987

Displacement, tonnes: 4,737 full load
Measurement, tonnes: 3,818 (Martha L Black), 3,809 (George R Pearkes), 3,812 (Sir Wilfrid Laurier), 3,727 (Edward Cornwallis, Sir William Alexander), 3,854 (Ann Harvey) gt
Dimensions, metres (feet): 83 × 16.2 × 5.8 (272.3 × 53.1 × 19.0)
Speed, knots: 15.5. **Range, n miles:** 14,000 at 14 kt (Martha L Black)
Complement: 25 (10 officers)
Machinery: Diesel-electric; 3 Bombardier/Alco 12V-251 diesels; 8,019 hp (6 MW) sustained; 3 Canadian GE generators; 6 MW; 2 Canadian GE motors; 7,040 hp (5.25 MW); 2 shafts; bow thrusters
Radars: Navigation: 2 Sperry Bridgemaster; E/F/I-bands.
Helicopters: 1 light type, such as Bell 206L.

Comment: *Black* based at Quebec in the Central and Arctic Region, *Cornwallis* and *Alexander* in the Atlantic Region at Dartmouth, *Ann Harvey* and *Pearkes* in the Atlantic Region at St Johns and *Laurier* in the Western Region at Victoria.

SIR WILLIAM ALEXANDER 7/2010, Michael Nitz / 1366732

SIR WILFRID LAURIER 6/2010, Guy Toremans / 1366731

1 GRIFFON CLASS (TYPE 1100)

Name	Builders	Commissioned
GRIFFON	Davie Shipbuilding, Lauzon	Dec 1970

Displacement, tonnes: 3,146 full load
Measurement, tonnes: 2,212 gt; 752 net
Dimensions, metres (feet): 71.3 × 14.9 × 4.7 (233.9 × 48.9 × 15.4)
Speed, knots: 13. **Range, n miles:** 5,500 at 11 kt
Complement: 25 (9 officers) + 26 spare berths
Machinery: Diesel-electric; 4 Fairbanks-Morse diesel generators; 7,920 hp (5.9 MW); 2 motors; 3,982 hp(m) (2.97 MW); 2 shafts; bow thruster
Radars: Navigation: 2 Sperry Bridgemaster; E/F/I-bands.
Helicopters: Platform for 1 MBB BO-105.

Comment: Based in the Central and Arctic Region at Prescott, Ontario.

GRIFFON 11/2013*, Marc Piché / 1525563

IHS Jane's Fighting Ships 2014-2015

108 Canada (COAST GUARD) > Multirole vessels — Offshore patrol vessels

MULTIROLE VESSELS

2 SAMUEL RISLEY CLASS (TYPE 1050)

Name	Builders	Commissioned
SAMUEL RISLEY	Vito Construction Ltd, Delta	4 Jul 1985
EARL GREY	Pictou Shipyards Ltd, Pictou	30 May 1986

Displacement, tonnes: 2,982 full load
Measurement, tonnes: 1,967 *(Samuel Risley)*, 1,972 *(Earl Grey)* gt; 660 *(Samuel Risley)*, 652 *(Earl Grey)* net
Dimensions, metres (feet): 69.7 × 13.7 × 5.2 *(228.7 × 44.9 × 17.0)*
Speed, knots: 13
Range, n miles: 17,000 at 11 kt
Complement: 22 (9 officers)
Machinery: Diesel-electric; 4 Wärtsilä 4SA 12-cyl diesels; 8,644 hp(m) *(6.4 MW)* *(Samuel Risley)*; 4 Deutz 4SA 9-cyl diesels; 8,836 hp(m) *(6.5 MW)* *(Earl Grey)*; 2 shafts; cp props; bow and stern thrusters
Radars: Navigation: 3 Sperry Bridgemaster; E/F/I-bands.

Comment: *Risley* based in the Central and Arctic Region at Pary Sound, Ontario, *Grey* in the Atlantic Region at Charlottetown, Prince Edward Island.

EARL GREY   7/2011, Harald Carstens / 1406724

1 PROVO WALLIS CLASS (TYPE 1000)

Name	Builders	Commissioned
BARTLETT	Marine Industries, Sorel	Dec 1969

Displacement, tonnes: 1,646 full load
Measurement, tonnes: 1,317 gt; 491 net
Dimensions, metres (feet): 57.7 × 13 × 4.1 *(189.3 × 42.7 × 13.5)*
Speed, knots: 15
Range, n miles: 3,300 at 11 kt
Complement: 24 (9 officers) + 11 spare berths
Machinery: 2 National Gas 6-cyl diesels; 2,100 hp *(1.55 MW)*; 2 shafts; LIPS cp props; bow thruster
Radars: Navigation: 2 Sperry Bridgemaster; E/F/I-bands.

Comment: Based in Western Region at Victoria. Modernised in 1988.

PROVO WALLIS CLASS   6/2008, M Mazumdar / 1335247

1 TRACY CLASS (TYPE 1000)

Name	Builders	Commissioned
TRACY	Port Weller Drydocks, Ontario	17 Apr 1968

Displacement, tonnes: 1,321 full load
Measurement, tonnes: 837 gt; 251 net
Dimensions, metres (feet): 55.2 × 11.6 × 3.7 *(181.1 × 38.1 × 12.1)*
Speed, knots: 13
Range, n miles: 5,000 at 12 kt
Complement: 22 (8 officers) + 12 spare berths
Machinery: Diesel-electric; 2 Fairbanks-Morse 38D8-1/8-8 diesel generators; 2,700 hp *(1.94 MW)*; 2 motors; 2,000 hp *(1.49 MW)*; 2 shafts; bow thruster
Radars: Navigation: 2 Sperry Bridgemaster; E/F/I-bands.

Comment: Based in Central and Arctic Region at Sorel.

TRACY   6/2010, Globke Collection / 1366833

IHS Jane's Fighting Ships 2014-2015

OFFSHORE PATROL VESSELS

2 CAPE ROGER CLASS (TYPE 600)

Name	Builders	Commissioned
CYGNUS	Marystown SY, Newfoundland	7 May 1981
CAPE ROGER	Ferguson Industries, Pictou NS	25 Aug 1977

Displacement, tonnes: 1,489 full load
Measurement, tonnes: 1,255 gt; 357 net
Dimensions, metres (feet): 62.5 × 12.2 × 4.1 *(205.1 × 40.0 × 13.5)*
Speed, knots: 16
Range, n miles: 10,000 at 12 kt
Complement: 19 (8 officers) + 10 spare berths
Machinery: 2 Wärtsilä Nohab F 212V diesels, 4,461 hp(m) *(3.28 MW)*; 1 shaft; bow thrusters
Radars: Navigation: 2 Sperry Bridgemaster; E/F/I-bands.

Comment: *Cygnus* based at Dartmouth and *Cape Roger* at St John's in the Atlantic Region. Half-life refits completed in 1995–97. The flight decks in both ships have been inactivated and the hangars have been converted to SAR triage areas and workshops.

CAPE ROGER   4/2011*, Marc Piché / 1525566

1 LEONARD J COWLEY CLASS (TYPE 600)

Name	Builders	Commissioned
LEONARD J COWLEY	West Coast Manly Shipyards Ltd	17 Jun 1985

Displacement, tonnes: 2,113 full load
Measurement, tonnes: 2,188 gt; 655 net
Dimensions, metres (feet): 72 × 14.2 × 4.5 *(236.2 × 46.6 × 14.8)*
Speed, knots: 15
Range, n miles: 12,600 at 12 kt
Complement: 19 (8 officers)

Machinery: 2 Wärtsilä Nohab F 312A diesels; 2,325 hp(m) *(1.71 MW)*; 1 shaft; bow thrusters
Guns: 2 — 12.7 mm MGs.
Radars: Navigation: 2 Sperry Bridgemaster; E/F/I-bands.
Helicopters: 1 MBB BO-105.

Comment: Based in the Atlantic Region at St John's.

LEONARD J COWLEY   4/2012, Adolfo Ortigueira Gil / 1482973

1 TANU CLASS (TYPE 500)

Name	Builders	Commissioned
TANU	Yarrows Ltd, Victoria BC	Sep 1968

Displacement, tonnes: 940 full load
Measurement, tonnes: 758 gt; 206 net
Dimensions, metres (feet): 52.1 × 9.9 × 3.5 *(170.9 × 32.5 × 11.5)*
Speed, knots: 13.5. **Range, n miles:** 5,000 at 11 kt
Complement: 15 (6 officers)
Machinery: 2 Fairbanks-Morse 38D8 diesels; 2,624 hp *(1.96 MW)*; 1 shaft; bow and stern thrusters
Radars: Navigation: 2 Sperry Bridgemaster; E/F/I-bands.

Comment: Based in Western Region at Patricia Bay.

TANU   7/2004, M K Mitchell / 1042125

© 2014 IHS

Offshore patrol vessels — Midshore patrol vessels < **Canada** (COAST GUARD) 109

1 SIR WILFRED GRENFELL (TYPE 600)

Name	Builders	Launched	Commissioned
SIR WILFRED GRENFELL	Marystown SY, Newfoundland	24 Aug 1985	11 Apr 1986

Displacement, tonnes: 3,813 full load
Measurement, tonnes: 2,404 gt; 665 net
Dimensions, metres (feet): 68.5 × 15 × 5 *(224.7 × 49.2 × 16.4)*
Speed, knots: 16
Range, n miles: 11,000 at 11.5 kt
Complement: 20 (9 officers)
Machinery: 4 Deutz 4SA (2—16-cyl, 2—9-cyl) diesels; 12,862 hp(m) *(9.46 MW)*; 2 shafts; cp props; bow and stern thruster
Radars: Navigation: 2 Sperry Bridgemaster; E/F/I-bands.
Helicopters: Platform for 1 MBB BO-105

Comment: Originally laid down as an offshore supply vessel, but subsequently purchased by the Canadian government for Coast Guard service. Modified to include an 85 tonne towing winch and additional SAR accommodation and equipment. Ice strengthened hull. Based in the Atlantic Region at St John's.

SIR WILFRED GRENFELL 6/2010, *Canadian Coast Guard* / 1366741

MIDSHORE PATROL VESSELS

1 LOUISBOURG CLASS (TYPE 500)

Name	Builders	Commissioned
LOUIS M LAUZIER (ex-*Cape Harrison*)	Breton Industries, Port Hawkesbury	Feb 1977

Displacement, tonnes: 467 full load
Measurement, tonnes: 300 gt; 66 net
Dimensions, metres (feet): 37.1 × 8.2 × 2.1 *(121.7 × 26.9 × 6.9)*
Speed, knots: 13.0
Range, n miles: 1,800 at 12 kt
Complement: 14 (5 officers)
Machinery: 2 MTU 12V 538 TB91 diesels; 4,600 hp(m) *(3.38 MW)*; 2 shafts
Radars: Navigation: 2 Sperry Bridgemaster; E/F/I-bands.

Comment: Based at Quebec in the Central and Arctic Region. *Louis M Lauzier* returned to service from charter (to Memorial University) in 2005.

LOUIS M LAUZIER 10/2012*, *Marc Piché* / 1525565

3 POST CLASS

Name	Builders	Commissioned
ATLIN POST	Philbrooks Shipyard Ltd, Sidney	1975
KITIMAT II	Philbrooks Shipyard Ltd, Sidney	1974
SOOKE POST	Philbrooks Shipyard Ltd, Sidney	1973

Measurement, tonnes: 58 gt; 15 net
Dimensions, metres (feet): 19.8 × 5.2 × 2.0 *(65 × 17.1 × 6.5)*
Speed, knots: 15
Range, n miles: 400 at 12 kt
Complement: 4 (3 officers)
Machinery: 2 General Motors V12-71 diesels; 800 hp *(596 kW)*; 2 shafts
Radars: Navigation: 2 Sperry Bridgemaster; E/F/I-bands.

Comment: *Atlin Post* based at Patricia Bay, British Columbia, *Kitimat II* at Prince Rupert, British Columbia and *Sooke Post* at Port Hardy, BC. To be replaced by Hero-class patrol vessels.

© 2014 IHS

ATLIN POST 6/2001, *Canadian Coast Guard* / 0126355

1 ARROW POST CLASS

Name	Builders	Commissioned
ARROW POST	Hike Metal Products, Wheatley	1991

Measurement, tonnes: 232 gt; 95 net
Dimensions, metres (feet): 28.9 × 8.8 × 3.35 *(94.8 × 28.9 × 11.0)*
Speed, knots: 13. **Range, n miles:** 2,800 at 11 kt
Complement: 6 (3 officers)
Machinery: 2 Caterpillar 3512 diesels; 1,279 hp *(953 kW)*; 1 shaft; bow thruster
Radars: Navigation: 2 Sperry Bridgemaster; E/F/I-bands.

Comment: Based in Western Region at Prince Rupert, British Columbia. To be replaced by a Hero-class vessel 2014–15.

ARROW POST 6/2010, *Canadian Coast Guard* / 1366739

6 + 3 (3) HERO (DAMEN STAN PATROL 4207) CLASS (PB)

Name	Builders	Launched	Commissioned
PRIVATE ROBERTSON V C	Irving Shipbuilding, Halifax	12 May 2012	31 Jul 2012
CAPORAL KAEBLE V C	Irving Shipbuilding, Halifax	22 Sep 2012	7 Nov 2012
CORPORAL TEATHER C V	Irving Shipbuilding, Halifax	15 Dec 2012	5 Feb 2013
CONSTABLE CARRIÈRE	Irving Shipbuilding, Halifax	24 Mar 2013	30 Apr 2013
G PEDDLE	Irving Shipbuilding, Halifax	15 Jun 2013	1 Aug 2013
CORPORAL MCLAREN MMV	Irving Shipbuilding, Halifax	14 Sep 2013	31 Oct 2013
A LEBLANC	Irving Shipbuilding, Halifax	2014	2014
M CHARLES	Irving Shipbuilding, Halifax	2014	2014
CAPTAIN GODDARD MSM	Irving Shipbuilding, Halifax	2014	2014

Displacement, tonnes: 208 standard
Dimensions, metres (feet): 42.95 × 7.31 × 2.85 *(140.9 × 24.0 × 9.4)*
Speed, knots: 26. **Range, n miles:** 2,000 at 12 kt
Complement: 18
Machinery: 2 MTU 4000M diesels; 6,788 hp *(5.0 MW)*; 2 shafts; bow thruster
Radars: Navigation: 2 Sperry Marine Visionmaster; E/F/I-bands.

Comment: Contract for the construction of nine (with an option for three further vessels) Damen Stan Patrol 4207, modified for Canadian use, awarded to Irving Shipbuilding Inc on 2 September 2009. Steel hull with aluminium superstructure. Capable of deploying a 7 m RHIB from a deck-mounted cradle. The first vessel was delivered in 2012 and all nine vessels by mid-2014. Five of the vessels are to be used in the Atlantic, Western and Central and Arctic regions while the other four are to be used for security operations on the Great Lakes and St Lawrence Seaway in a joint programme with the RCMP. Similar vessels are in service with Barbados, Jamaica, Albania and UK Border Agency.

CONSTABLE CARRIÈRE 6/2013*, *Marc Piché* / 1525568

IHS *Jane's Fighting Ships* 2014-2015

SPECIAL NAVAIDS VESSELS

1 ECKALOO CLASS (TYPE 700)

Name	Builders	Commissioned
ECKALOO	Vancouver SY Ltd	31 Aug 1988

Displacement, tonnes: 543 full load
Measurement, tonnes: 661 gt; 213 net
Dimensions, metres (feet): 49 × 13.4 × 1.4 (160.8 × 44.0 × 4.6)
Speed, knots: 13. **Range, n miles:** 2,000 at 11 kt
Complement: 10 (4 officers)
Machinery: 2 Caterpillar 3512TA; 2,420 hp (1.8 MW) sustained; 2 shafts
Radars: Navigation: 2 Sperry Bridgemaster; I-band.
Helicopters: Platform for 1 Bell 206L/L-1.

Comment: Replaced vessel of the same name. Similar design to Dumit. Based in Central and Arctic Region at Hay River, Northwest Territories.

ECKALOO　　　　　9/1994, van Ginderen Collection / 0056697

1 DUMIT CLASS (TYPE 700)

Name	Builders	Commissioned
DUMIT	Allied Shipbuilders Ltd, N Vancouver	5 Sep 1979

Displacement, tonnes: 639 full load
Measurement, tonnes: 569 gt; 176 net
Dimensions, metres (feet): 48.8 × 12.2 × 1.6 (160.1 × 40.0 × 5.2)
Speed, knots: 13.5. **Range, n miles:** 7,700 at 11 kt
Complement: 10 (4 officers)
Machinery: 2 Caterpillar 3512TA; 2,420 hp (1.8 MW) sustained; 2 shafts
Radars: Navigation: 2 Sperry Bridgemaster; I-band.

Comment: Similar to Eckaloo. Based in Central and Arctic Region at Hay River, Northwest Territories.

DUMIT　　　　　6/2010, Canadian Coast Guard / 1366736

SPECIALTY VESSELS

1 VAKTA CLASS

Name	Builders	Commissioned
VAKTA	Hike Metal Products Ltd, Wheatley	2004

Measurement, tonnes: 35 gt; 26 net
Dimensions, metres (feet): 16.3 × 4.5 × 1.4 (53.5 × 14.8 × 4.6)
Speed, knots: 21. **Range, n miles:** 320 at 15 kt
Complement: 3
Machinery: 2 Caterpillar C12 diesels; 980 hp (731 kW); 2 shafts

Comment: Replaced Namao in 2005. Provides navigational aids and SAR services on Lake Winnipeg. Based in the Central and Arctic Region at Gimli, Manitoba.

VAKTA　　　　　6/2005, Canadian Coast Guard / 1151242

3 COVE ISLAND CLASS (TYPE 800)

Name	Builders	Commissioned
COVE ISLE	Canadian D and D, Kingston	1980
ILE SAINT-OURS	Breton Industries, Port Hawkesbury	15 May 1986
CARIBOU ISLE	Breton Industries, Port Hawkesbury	16 Jun 1986

Displacement, tonnes: 140 full load
Measurement, tonnes: 93, 79.7 (Cove Isle) gt; 37, 33.3 (Cove Isle) net
Dimensions, metres (feet): 23 (Caribou Isle, Ile Saint-Ours), 20 (Cove Isle) × 6 × 1.4 (75.5, 65.6 × 19.7 × 4.6)
Speed, knots: 10. **Range, n miles:** 1,800 at 11 kt
Complement: 5 (2 officers)
Machinery: 2 Detroit 8V-92 diesels; 845 hp (630 kW); 2 shafts
Radars: Navigation: Sperry 1270; I-band.

Comment: Cove Isle and Caribou Isle are based in the Central and Arctic Region at Parry Sound, Amherstburg and Prescott respectively. Ile Saint-Ours is based in the Central and Arctic Region at Sorel. Ile des Barques was decommissioned in 2008, Tsekoa II in 2009, and Gull Isle in 2012.

COVE ISLE　　　　　10/2011*, Canadian Coast Guard / 1525569

1 TRAVERSE CLASS

Name	Builders	Commissioned
TRAVERSE	Metal Craft Marine Ltd, Kingston	1998

Measurement, tonnes: 55.2 gt; 14.1 net
Dimensions, metres (feet): 19.8 × 7.3 × 0.6 (65 × 24.0 × 2.0)
Speed, knots: 9. **Range, n miles:** 800 at 8 kt
Complement: 4
Machinery: 2 Volvo Penta inboard/outboard diesels; 181 hp(m) (135 kW); 2 shafts

Comment: Provides navigation aids on Lake of the Woods. Based in Central and Arctic Region at Kenora, Ontario.

TRAVERSE　　　　　6/2010, Canadian Coast Guard / 1366734

1 CAPE HURD CLASS (TYPE 400)

Name	Builders	Commissioned
CAPE HURD	Breton Industrial and Machinery, Pt Hawkesbury	1982

Displacement, tonnes: 99 full load
Measurement, tonnes: 55.2 gt; 14 net
Dimensions, metres (feet): 21.6 × 5.5 × 1.3 (70.9 × 18.0 × 4.3)
Speed, knots: 18. **Range, n miles:** 1,400 at 16 kt
Complement: 4
Machinery: 2 MTU 8V 396 TC82 diesels; 1,740 hp(m) (1.28 MW) sustained; 2 shafts
Radars: Navigation: 2 Sperry Bridgemaster; I-band.

Comment: Aluminium alloy hull. Based in Central and Arctic Region at Parry Sound.

CAPE HURD　　　　　6/2010, Canadian Coast Guard / 1366738

Specialty vessels < **Canada** (COAST GUARD) 111

2 KELSO CLASS

Name	Builders	Commissioned
KELSO	ABCO Industries Ltd, Lunenburg	2009
VIOLA M DAVIDSON	ABCO Industries Ltd, Lunenburg	2010

Measurement, tonnes: 59.7 gt; 45 net
Dimensions, metres (feet): 17.7 × 5.5 × 1.5 *(58.1 × 18.0 × 4.9)*
Speed, knots: 16
Range, n miles: 400 at 14 kt
Complement: 2
Machinery: 2 Volvo Penta diesels; 998 hp(m) *(744 kW)*; 2 shafts
Radars: Navigation: 1 Sperry Bridgemaster. 1 Raymarine; I-band.

Comment: Both craft provide support to federal science. *Kelso* based in Central and Arctic Region at Burlington, Ontario, and *Viola M Davidson* in Atlantic Region at St Andrews, New Brunswick.

KELSO 5/2009*, Canadian Coast Guard / 1525570

1 HOOD CLASS (TYPE 200)

Name	Builders	Commissioned
HARP	Georgetown SY, PEI	12 Dec 1986

Displacement, tonnes: 229 full load
Measurement, tonnes: 182 gt; 70 net
Dimensions, metres (feet): 24.5 × 7.6 × 2.5 *(80.4 × 24.9 × 8.2)*
Speed, knots: 11
Range, n miles: 500 at 10 kt
Complement: 7 (3 officers)
Machinery: 2 Caterpillar 3408 diesels; 850 hp *(634 kW)*; 2 Kort nozzle props
Radars: Navigation: 2 Sperry Bridgemaster; E/F/I-bands.

Comment: Ordered 26 April 1985. Ice strengthened hull. Based in the Atlantic Region at St Anthony.

HARP 6/2010, Canadian Coast Guard / 1366751

4 CAPE LIGHT CLASS

Name	Builders	Commissioned
CAPE LIGHT	Metal Craft Marine, Kingston	2001
GELIGET	ABCO Industries Ltd, Lunenburg	2002
POINT CAVEAU	ABCO Industries Ltd, Lunenburg	2003
S DUDKA	ABCO Industries Ltd, Lunenburg	2013

Measurement, tonnes: 31.9 gt; 23.9 net
Dimensions, metres (feet): 14.6 × 4.9 *(Cape Light, Geliget, Point Caveau)*, 5.1 *(S Dudka)* × 0.8 *(47.9 × 16.1, 16.7 × 2.6)*
Speed, knots: 30
Range, n miles: 250 at 20 kt
Machinery: 2 Volvo Penta D12-650 (D13-700 *S Dudka*) diesels; 1,300 (1,400 *S Dudka*) hp *(969 (1,043) kW)*; 2 waterjets

Comment: Provide support to federal conservation and protection. Based in Atlantic Region: *Cape Light* at Westhead, *Geliget* at Yarmouth, *Point Caveau* at Cheticamp, and *S Dudka* at Alberton, Prince Edward Island.

© 2014 IHS

POINT CAVEAU 7/2011, Harald Carstens / 1406725

1 SIGMA T CLASS

Name	Builders	Commissioned
SIGMA T	Samson Shipyard	2008

Measurement, tonnes: 13.3 gt
Dimensions, metres (feet): 12.1 × 4.66 × 1.1 *(39.7 × 15.3 × 3.6)*
Speed, knots: 15. **Range, n miles:** 200 at 14 kt
Complement: 2
Machinery: 1 Volvo Penta D9-575 diesel; 575 hp *(429 kW)*; 1 shaft
Radars: Navigation: 1 Raymarine; I-band.

Comment: Provides support to federal science. Based in Atlantic Region at Dartmouth.

SIGMA T 6/2010, Canadian Coast Guard / 1366756

1 GORDON REID CLASS (TYPE 500)

Name	Builders	Commissioned
GORDON REID	Versatile Pacific, Vancouver	13 Dec 1991

Measurement, tonnes: 880 gt; 257 net
Dimensions, metres (feet): 49.9 × 11 × 4.0 *(163.7 × 36.1 × 13.1)*
Speed, knots: 16
Range, n miles: 2,500 at 15 kt
Complement: 14 (6 officers)
Machinery: 4 Deutz SBV-6M-628 diesels; 2,475 hp(m) *(1.82 MW)* sustained; 2 shafts; bow thruster; 400 hp *(294 kW)*
Radars: Navigation: 2 Sperry Bridgemaster; I-band.

Comment: Designed for long-range patrols along the British Columbian coast out to 200 mile limit. Has a stern ramp for launching Zodiac Hurricane 733 rigid inflatables in up to Sea State 6. The Zodiac has a speed of 50 kt and is radar equipped. Based in the Western Region at Victoria.

GORDON REID 6/2010, Canadian Coast Guard / 1366752

IHS Jane's Fighting Ships 2014-2015

SAR LIFEBOATS

Notes: There are also at least 15 Inshore Rescue boats with CG numbers.

10 LIFEBOATS (TYPE 300A)

Name	Builders	Commissioned
BICKERTON	Halmatic, Havant	Aug 1989
SPINDRIFT	Georgetown SY, PEI	Oct 1993
SPRAY	Industrie Raymond, Quebec	Sep 1994
COURTENAY BAY (ex-*Spume*)	Industrie Raymond, Quebec	Oct 1994
W JACKMAN (ex-*Cap Aux Meules*)	Industrie Raymond, Quebec	Sep 1995
W G GEORGE	Industrie Raymond, Quebec	Sep 1995
CAP AUX MEULES	Hike Metal Products Ltd, Ontario	Oct 1996
CLARK'S HARBOUR	Hike Metal Products Ltd, Ontario	Sep 1996
SAMBRO	Hike Metal Products Ltd, Ontario	Jan 1997
WESTPORT	Hike Metal Products Ltd, Ontario	May 1997

Measurement, tonnes: 35 gt
Dimensions, metres (feet): 15.9 × 5.3 × 1.5 *(52.2 × 17.4 × 4.9)*
Speed, knots: 20. **Range, n miles:** 200 at 18 kt
Complement: 4 (2 officers)
Machinery: 2 Caterpillar 3408BTA diesels; 1,070 hp *(786 kW)* sustained; 2 shafts
Radars: Navigation: Sperry Bridgemaster. Furuno; I-band.

Comment: High-endurance self-righting lifeboats. Nine based in Atlantic Region, one in Central and Arctic Region. *Bickerton* was a steel-hulled prototype vessel based on the Arun class and built in the UK. The remainder are aluminium hulled and were built in Canada.

CAPE FAREWELL   6/2010, *Canadian Coast Guard* / 1366749

SPINDRIFT   2/2007*, *Canadian Coast Guard* / 1525571

1 CAPE GOÉLAND (TYPE 300) CLASS

Name	Builders	Launched	Commissioned
GOÉLAND (ex-*Cape Goéland*)	Hike Metal Products Ltd	1985	1985

Measurement, tonnes: 17.4 gt; 13 net
Dimensions, metres (feet): 13.5 × 3.9 *(44.3 × 12.8)*
Speed, knots: 14.5. **Range, n miles:** 150 at 12.0 kt
Complement: 4 (1 officer)
Machinery: Diesel; 2 Caterpillar 3208 diesels; 456 hp(m) *(340 kW)* sustained; 2 shafts
Radars: Navigation: 1 Furuno FR-1050; I-band.

Comment: High endurance self-righting SAR lifeboat. Based at the Canadian Coast Guard College as a training vessel. Can be utilised as a hot standby SAR vessel for Atlantic region.

36 CAPE CLASS (TYPE 300B)

Name	Builders	Commissioned
CAPE SUTIL	Metalcraft Marine, Kingston	Dec 1998
CAPE CALVERT	Metalcraft Marine, Kingston	Aug 1999
CAPE ST JAMES	Metalcraft Marine, Kingston	Nov 1999
CAPE MERCY	Metalcraft Marine, Kingston	Dec 2000
CAPE LAMBTON	Metalcraft Marine, Kingston	Jul 2001
CAPE STORM	Metalcraft Marine, Kingston	Nov 2002
THUNDER CAPE	Metalcraft Marine, Kingston	Aug 2000
CAPE FOX	Victoria Shipyard Co Ltd, Victoria	May 2003
CAPE NORMAN	Victoria Shipyard Co Ltd, Victoria	May 2003
CAP DE RABAST	Victoria Shipyard Co Ltd, Victoria	Aug 2003
CAP ROZIER	Victoria Shipyard Co Ltd, Victoria	Aug 2003
CAPE MUDGE	Victoria Shipyard Co Ltd, Victoria	Nov 2003
CAPE FAREWELL	Victoria Shipyard Co Ltd, Victoria	Nov 2003
CAPE COCKBURN	Victoria Shipyard Co Ltd, Victoria	Jan 2004
CAPE SPRY	Victoria Shipyard Co Ltd, Victoria	Apr 2004
CAP NORD	Victoria Shipyard Co Ltd, Victoria	Apr 2004
CAP BRETON	Victoria Shipyard Co Ltd, Victoria	Apr 2004
CAPE MCKAY	Victoria Shipyard Co Ltd, Victoria	Jun 2004
CAPE CHAILLON	Victoria Shipyard Co Ltd, Victoria	Oct 2004
CAPE PROVIDENCE	Victoria Shipyard Co Ltd, Victoria	Oct 2004
CAPE COMMODORE	Victoria Shipyard Co Ltd, Victoria	Oct 2004
CAPE ANN	Victoria Shipyard Co Ltd, Victoria	Nov 2004
CAPE CAUTION	Victoria Shipyard Co Ltd, Victoria	Dec 2004
CAPE DISCOVERY	Victoria Shipyard Co Ltd, Victoria	Jan 2005
CAPE HEARNE	Victoria Shipyard Co Ltd, Victoria	Feb 2005
CAPE DUNDAS	Victoria Shipyard Co Ltd, Victoria	Mar 2005
CAPE TOURMENTE	Victoria Shipyard Co Ltd, Victoria	Apr 2005
CAP D'ESPOIR	Victoria Shipyard Co Ltd, Victoria	Jun 2005
CAP PERCÉ	Victoria Shipyard Co Ltd, Victoria	Aug 2005
CAPE EDENSAW	Victoria Shipyard Co Ltd, Victoria	Sep 2005
CAPE KUPER	Victoria Shipyard Co Ltd, Victoria	Oct 2005
CAPE NADEN	Victoria Shipyard Co Ltd, Victoria	2011
CAPE RESCUE	Victoria Shipyard Co Ltd, Victoria	2011
CAP AUPALUK	Victoria Shipyard Co Ltd, Victoria	2011
CAPE DAUPHIN	Victoria Shipyard Co Ltd, Victoria	2011
CAPE PALMERSTON	Victoria Shipyard Co Ltd, Victoria	2011

Measurement, tonnes: 34 gt
Dimensions, metres (feet): 14.6 × 4.27 × 1.37 *(47.9 × 14.0 × 4.5)*
Speed, knots: 25. **Range, n miles:** 200 at 22 kt
Complement: 4
Machinery: 2 Caterpillar C12 diesels; 905 hp *(675 kW)* sustained; 2 shafts
Radars: Navigation: Raymarine; I-band.

Comment: High-speed, self-righting lifeboats. Based in all three regions.

AIR CUSHION VEHICLES

3 AP1-88/400 TYPE

SIPU MUIN   SIYAY   MAMILOSSA

Displacement, tonnes: 70 full load
Dimensions, metres (feet): 28.5 on cushion × 12 *(93.5 × 39.4)*
Speed, knots: 50
Complement: 6 (2 officers)
Cargo capacity: 22.6 tons
Machinery: 4 Caterpillar 3412 TTA diesels; 3,650 hp(m) *(2.68 MW)* sustained

Comment: Contract awarded for two hovercraft to GKN Westland in May 1996. Built at Hike Metal Products, Wheatley, Ontario and completed in August and December 1998 respectively. A third vessel, *Mamilossa*, was manufactured by Griffon Hoverwork, Isle of Wight, and completed in 2009. Well-deck size 8.2 × 4.6 m. There is a 5,000 kg load crane. *Sipu Muin* and *Mamilossa* are based at Trois Rivières and *Siyay* at Sea Island, BC.

MAMILOSSA   2/2009, *Maritime Photographic* / 1305806

1 AP1-88/100 TYPE (TRAINING SHIP) (AXL)

PENAC (ex-*Liv Viking*)

Displacement, tonnes: 46 full load
Dimensions, metres (feet): 24.5 on cushion × 11.9 *(80.4 × 39.0)*
Speed, knots: 50
Complement: 7 (2 officers)
Cargo capacity: 5.3 tons
Machinery: 2 Deutz BF 12L513 diesels; 1,050 hp(m) *(785 kW)*. 2 MTU 12V 183TB32 diesels; 1,640 hp(m) *(1.25 MW)* sustained

Comment: Built by Hoverworks Ltd, Isle of Wight, UK in 1984. Procured by Canadian Coast Guard in 2004. Based in the Western Region at Sea Island, BC. To be replaced by a new Griffon hovercraft in 2014.

PENAC   6/2004, *Canadian Coast Guard* / 1042123

FISHERY RESEARCH SHIPS

8 + 3 FISHERY RESEARCH SHIPS

ALFRED NEEDLER	W E RICKER	TELEOST
CALANUS II	NEOCALIGUS	VLADYKOV
M PERLEY	LEIM	

Comment: *Alfred Needler* has a measurement of 959 gt and was commissioned in 1982. Based at Dartmouth, NS. *W E Ricker* has a measurement of 1,104 gt and was commissioned in 1978. Based at Nanaimo, BC. *Teleost* has a measurement of 2,405 gt and was commissioned in 1988. Based at St John's, NL. *Calanus II* has a measurement of 138 gt and was commissioned in 1991. Based at Rimouski, QC. *Neocaligus* has a measurement of 98 gt and was commissioned in 1989. Based at Patricia Bay, BC.
Alfred Needler, W E Ricker, and *Teleost* are classified as Offshore Fishery Science vessels. The other five are near-shore Fishery Research Vessels. Three new 67 m offshore fishery science vessels were funded in the 2006 and 2007 budgets. They are to be built by Vancouver Shipyards and are to replace *Alfred Needler, Teleost,* and *W E Ricker*. Three new 25 m vessels, *Vladykov, M Perley,* and *Leim* entered service in 2012. They have a measurement of 254 gt.

TELEOST   6/2010, Canadian Coast Guard / 1366746

SURVEY AND RESEARCH SHIPS

9 + 1 RESEARCH SHIPS

| MATTHEW | HUDSON | VECTOR | FREDERICK G CREED | GC-03 |
| F C G SMITH | JOHN P TULLY | LIMNOS | OTTER BAY | |

Comment: *Matthew* has a measurement of 856 gt and was commissioned in 1990. Based at Dartmouth, NS. *FCG Smith* has a measurement of 438 gt and was commissioned in 1985. Based at Quebec, QC. *Hudson* has a measurement of 3,740 gt and was commissioned in 1963. Based at Dartmouth, NS. *John P Tully* has a measurement of 2,195 gt and was commissioned in 1985. Based at Patricia Bay, BC. *Vector* has a measurement of 515 gt and was commissioned in 1967. Based at Patricia Bay, BC. *Limnos* has a measurement of 489 gt and was commissioned in 1968. Based at Burlington, ON. *Frederick G Creed* has a measurement of 151 gt and was commissioned in 1988. Based at Mont Joli, QC. *Otter Bay* has a measurement of 21 gt and was commissioned in 1992. Based at Mont Joli, QC. *GC-03* has a measurement of 57 gt and was commissioned in 1973. Based at Quebec Region. *Hudson* and *Tully* are classified as Offshore Oceanographic Science vessels. *Hudson* is to be replaced by a new 90 m vessel, to be built by Vancouver Shipyards, by 2017. *Matthew, Frederick G Creed, Limnos, Vector* and *Otter Bay* are classified as Midshore Science Vessels. *F C G Smith* and *GC-03* are classified as a Channel Survey and Sounding vessels.

VECTOR   6/2012*, Marc Piché / 1525572

ROYAL CANADIAN MOUNTED POLICE

Notes: The Marine Branch of the Royal Canadian Mounted Police is responsible for enforcement of Customs, Immigration, Shipping and Drug regulations as well as for standard policing duties in areas that are difficult to access by land. In addition there are some 377 smaller craft for use on inland waterways.

2 PATROL CRAFT (PB)

| INKSTER | MURRAY |

Measurement, tonnes: 65 gt; 49 net
Dimensions, metres (feet): 21.62 × ? × ? (70.9 × ? × ?)
Speed, knots: 25
Complement: 4
Machinery: 2 Mann diesels 1,640 hp *(1.2 MW)(Inkster)*. 2 Caterpillar diesels 2,100 hp *(1.6 MW) (Murray)*; 2 Arneson surface drives

Comment: Catamaran design patrol craft. *Inkster* based on the Pacific Coast and *Murray* on the Atlantic coast at Burin, Newfoundland.

MURRAY   6/2010, Globke Collection / 1366834

4 PATROL CRAFT (PB)

| NADON | HIGGIT | LINDSAY | SIMMONDS |

Measurement, tonnes: 62 gt; 47 net
Dimensions, metres (feet): 17.7 × 6.7 × 0.67 (58.1 × 22.0 × 2.2)
Speed, knots: 29
Complement: 4
Machinery: 2 MAN diesels 1,640 hp *(1.2 MW)*; 2 Arneson surface drives

Comment: Catamaran design patrol craft. *Simmonds* based in Newfoundland and the other three based on the Pacific coast.

LINDSAY   6/2006, RCMP / 1159228

Cape Verde

Country Overview

A former Portuguese colony, the Republic of Cape Verde became independent in 1975. Situated in the Atlantic Ocean some 335 n miles due west of the western point of Africa, it has a land area of 1,557 square miles and consists of ten islands and a number of islets. These are divided into the northerly windward (Barlavento) and southerly leeward (Sotavento) groups. The windward group includes the islands of Santo Antão, São Vicente, Santa Luzia, São Nicolau, Sal and Boa Vista and the islets of Branco and Raso; the leeward group includes the islands of Santiago, Brava, Fogo and Maio and the islets of the Secos group. Mindelo, on São Vicente, is the principal port and economic centre while Praia on Santiago is the capital and largest town. An archipelagic state, territorial seas (12 n miles) are claimed. A 200 n mile Exclusive Economic Zone (EEZ) has been claimed but the limits are not fully defined.

Headquarters Appointments

Commander, Coast Guard:
Lieutenant Colonel Antonio Duarte Monteiro

Personnel

2014: 50

Bases

Mindelo (Isle de São Vicente), naval base and repair yard.

Maritime Aircraft

One Dornier 228-212 and one Embraer EMB 110 Bandeirante are used for maritime surveillance.

PATROL FORCES

1 SEA RAY PATROL CRAFT (PB)

GUAIADO

Displacement, tonnes: 7.4 full load
Dimensions, metres (feet): 11 × 4.24 × 0.71 (36.1 × 13.9 × 2.3)
Speed, knots: 23
Range, n miles: 400 at 18 kt
Complement: 4
Machinery: 2 Yamaha diesels; 2 shafts
Radars: Surface search/navigation: JMA-2343; I-band.

Comment: Sea Ray cruiser design vessel adapted for Coast Guard use in June 2008.

GUAIADO   6/2010, Cape Verde Coast Guard / 1366594

1 ESPADARTE CLASS (PETERSON MK 4 TYPE)
(COASTAL PATROL CRAFT) (PB)

Name	No	Builders	Commissioned
ESPADARTE	P 151	Peterson Builders Inc, Sturgeon Bay	19 Aug 1993

Displacement, tonnes: 22 full load
Dimensions, metres (feet): 15.6 × 4.5 × 1.3 (51.2 × 14.8 × 4.3)
Speed, knots: 24. **Range, n miles:** 500 at 20 kt
Complement: 6 (1 officer)
Machinery: 2 Detroit 6V-92TA diesels; 520 hp (388 kW) sustained; 2 shafts
Guns: 2—12.7 mm MGs (twin). 2—7.62 mm MGs.
Radars: Surface search: Raytheon; I-band.

Comment: Ordered from Peterson Builders Inc, under FMS programme on 25 September 1992. Option on three more not taken up. Aluminium hulls. The 12.7 mm mounting is aft with the smaller guns on the bridge roof.

ESPADARTE   6/2010, Cape Verde Coast Guard / 1366595

1 CHINESE 27 METRE CLASS (PATROL CRAFT) (PB)

Name	No	Commissioned
TAINHA	P 262	2000

Displacement, tonnes: 52 full load
Dimensions, metres (feet): 26.8 × 4.4 × 1.2 (87.9 × 14.4 × 3.9)
Speed, knots: 18. **Range, n miles:** 1,200 at 14 kt
Complement: 9 (1 officer)
Machinery: 2 MWM TBD 234V8 diesels; 984 hp (734 kW)
Guns: 2—12.7 mm MGs. 2—7.62 mm MGs.
Radars: Surface search/navigation: JMA-2343; I-band.

Comment: Built in China in 1998 and acquired by Cape Verde in April 2000. Similar craft in service in Benin.

TAINHA   6/2010, Cape Verde Coast Guard / 1366596

1 ARCHANGEL CLASS (RESPONSE BOATS) (PBF)

REI

Displacement, tonnes: 14.5 full load
Dimensions, metres (feet): 13.4 × 4.3 × 0.58 (44 × 14.1 × 1.9)
Speed, knots: 35. **Range, n miles:** 400 at 21 kt
Complement: 4
Machinery: 2 Caterpillar C9 diesels; 2 Hamilton 322 waterjets
Guns: 1—12.7 mm MG.
Radars: Surface search/navigation: Furuno 1834C; I-band.

Comment: High-speed inshore patrol craft of aluminium construction and foam collar built by SAFE Boats International, Port Orchard, Washington. Acquired in June 2010.

REI   6/2010, Cape Verde Coast Guard / 1366598

1 KONDOR I CLASS (COASTAL PATROL CRAFT) (PBO)

Name	No	Builders	Commissioned
VIGILANTE (ex-Kühlungsborn)	P 521 (ex-BG 32, ex-GS 07)	Peenewerft, Wolgast	1970

Displacement, tonnes: 367 full load
Dimensions, metres (feet): 52 × 6.2 × 2.4 (170.6 × 20.3 × 7.9)
Speed, knots: 18. **Range, n miles:** 1,800 at 15 kt
Complement: 17 (3 officers)
Machinery: 2 Russki/Kolomna Type 40DM diesels; 4,408 hp(m) (3.24 MW) sustained; 2 shafts; cp props
Guns: 1 ZU 23 mm. 1—7.62 mm MG.
Radars: Surface search: Kelvin Hughes Nucleus 2 5000A; I-band.
Helicopters: Platform for 1 medium.

Comment: Former GDR minesweeper taken over by the German Coast Guard, and then acquired by Cape Verde in September 1998. Armament refitted in Cape Verde in 1999. Refitted in 2007.

VIGILANTE   6/2010, Cape Verde Coast Guard / 1366597

1 OFFSHORE PATROL VESSEL (PBO)

Name	No	Builders	Commissioned
GUARDIÃO	P 511	Damen Shipyards	7 Jan 2012

Displacement, tonnes: 290 full load
Dimensions, metres (feet): 50 × 9.4 × 3.5 (164 × 30.8 × 11.5)
Speed, knots: 23.
Complement: 18 (4 officers)
Machinery: 4 Caterpillar C32 diesels; 5,800 hp (4.32 kW); 4 shafts; 2 bow thrusters; 200 kW
Radars: Surface search/navigation: To be announced.

Comment: Based on the Damen Sea Axe 5009 design, the ship has open deck space aft from which a 7.5 m RHIB can be launched and recovered. Steel hull and aluminium superstructure. The ship also has a firefighting capability. Launched on 1 June 2011. Based at Mindelo.

GUARDIÃO   1/2012, Bram Plokker / 1406888

Cayman Islands

Country Overview

A British dependency since 1962, the island group is situated south of Cuba in the Caribbean Sea. It comprises three islands: Grand Cayman, containing the capital George Town, Little Cayman and Cayman Brac, located about 80 miles northeast of Grand Cayman. Territorial seas (12 n miles) and a Fishery Zone (200 n miles) are claimed. A governor, appointed by the British Crown, is responsible for external affairs, internal security, defence and the police. The Marine unit consists of officers from the Royal Cayman Islands Police Service (RCIPS), Immigration and HM Customs. The unit is commanded by RCIPS.

Headquarters Appointments

Commander Joint Marine Unit: Andre Tahal

Personnel

2014: 19 (mixture of police and customs)

Bases

Grand Cayman (North Sound), Little Cayman, Cayman Brac.

POLICE

Notes: (1) There are five Yamaha VX cruiser wave runners, *Lima 2-6*, used for coastal and shallow water rescue and patrols. Capable of 55 kt, three are based at Grand Cayman and two at Cayman Brac.
(2) *Lima 7* is a 5 m Starcraft Marine aluminium skiff, capable of 30 kt. She is based at Grand Cayman.
(3) *Typhoon* is a 7 m Zodiac Hurricane capable of 50 kt. She is based at Grand Cayman.
(4) Three Customs craft are based at Cayman Brac and Little Cayman. *Interceptor* is an 11 m Eduardono Panga capable of 50 kt; *Defender* is a 7 m Boston Whaler capable of 30 kt; *E-5* is a 6 m Dusky capable of 30 kt.

2 SAFE BOATS DEFENDER 38 CLASS (PB)

NIVEN D	TORNADO

Displacement, tonnes: 6 full load
Dimensions, metres (feet): 12.3 × 3.0 × 0.75 (40.4 × 9.8 × 2.5)
Speed, knots: 65
Complement: 4
Machinery: 4 Mercury outboard motors; 1,200 hp (895 kW)
Radars: Navigation: I-band.

Comment: Both acquired from SafeBoats International in 2009. *Niven D* is a full cabin interceptor and *Tornado* is a T Top interceptor. Both based at Grand Cayman.

NIVEN D 6/2010, RCIP / 1366599

1 SEA ARK 65 ft CUTTER (PB)

CAYMAN GUARDIAN

Displacement, tonnes: 27 standard
Dimensions, metres (feet): 19.8 × 5.5 × 1.7 (65 × 18.0 × 5.6)
Speed, knots: 30
Complement: 6
Machinery: 2 MAN diesels; 2 shafts
Radars: Navigation: Raymarine; I-band.

Comment: SeaArk Marine Dauntless RAM patrol craft acquired in December 2008. To be employed on border protection tasks. Aluminium construction.

CAYMAN GUARDIAN 12/2008, SeaArk Marine / 1335342

1 SEA ARK 38 ft CUTTER (PB)

CAYMAN DEFENDER

Displacement, tonnes: 10 full load
Dimensions, metres (feet): 11.6 × 4.0 × 1.1 (38.1 × 13.1 × 3.6)
Speed, knots: 33
Complement: 4
Machinery: 2 MAN diesels; 1,100 hp (820 kW)
Radars: Navigation: Raymarine; I-band.

Comment: SeaArk Marine Dauntless RAM patrol craft acquired on 7 October 2008. To be employed on border protection tasks. Aluminium construction. Based at Grand Cayman.

CAYMAN DEFENDER 10/2008, SeaArk Marine / 1335341

1 DAUNTLESS CLASS (PB)

CAYMAN PROTECTOR

Displacement, tonnes: 17 full load
Dimensions, metres (feet): 14.6 × 4.3 × 1 (47.9 × 14.1 × 3.3)
Speed, knots: 26
Range, n miles: 400 at 20 kt
Complement: 4
Machinery: 2 MAN diesels; 720 hp(m) (529 kW) sustained; 2 shafts
Guns: 2 — 7.62 mm MGs.
Radars: Raytheon R40; I-band.

Comment: Built by SeaArk Marine, Monticello and acquired in July 1994. Aluminium construction. Based at Grand Cayman.

CAYMAN PROTECTOR 6/2010, RCIP / 1366600

Chile
ARMADA DE CHILE

Country Overview

The Republic of Chile is situated in western South America. With an area of 292,135 square miles it has borders to the north with Peru and to the east with Bolivia and Argentina. Off the 2,305 n mile coastline with the Pacific Ocean lie the Chiloé Island, Wellington Island and the western portion of Tierra del Fuego. Chilean islands in the south Pacific include the Juan Fernández Islands, Easter Island, and Salas y Gómez. The capital and largest city is Santiago. Principal ports include Valparaíso, Talcahuano, Mejillones, Quintero, Antofagasta, San Antonio, Arica, Iquique, Coquimbo, San Vicente, Puerto Montt, and Punta Arenas. Territorial seas (12 n miles) and an EEZ (200 n miles) are claimed.

Headquarters Appointments

Commander-in-Chief:
 Admiral Enrique Larrañaga Martin
Chief of Naval Staff: Vice Admiral Francisco
 García-Huidobro Campos
Director General, Logistics: Vice Admiral
 Cristián de la Maza Riquelme
*Director General, Maritime Territory and
 Merchant Marine:* Rear Admiral Humberto
 Ramírez Navarro
Commander, Naval Operations Command:
 Vice Admiral José Miguel Romero Aguirre
Director General, Naval Personnel:
 Vice Admiral Kenneth Pugh Olavarría
Commander, Surface Squadron:
 Rear Admiral José Miguel Rivera Sariego
Commander, Marine Corps:
 Rear Admiral Tuilio Rojas

Headquarters Appointments — continued

Flag Officer, 1st Naval Zone:
 Rear Admiral Julio Leíva
Flag Officer, 2nd Naval Zone: Rear Admiral
 Osvaldo Scharzenberg Ashton
Flag Officer, 3rd Naval Zone:
 Rear Admiral Kurt Hartung
Flag Officer, 4th Naval Zone:
 Rear Admiral Jorge Rodriguez
Commander, 5th Naval Zone:
 Rear Admiral Luis Felipe Bertolotto
Commander, Aviation: Rear Admiral
 Arturo Undurraga Díaz
Commander, Submarines: Rear Admiral
 Julio Silva

Personnel

(a) 2014: 26,162 (2,194 officers)
(b) 2,053 Marines (178 officers)
(c) 2 years' national service (794)

Command Organisation

1st Naval Zone. HQ at Valparaiso. From 26° S to 35° S.
2nd Naval Zone. HQ at Talcahuano. From 35° S to 40° S.
3rd Naval Zone. HQ at Punta Arenas. From 49° S to Antarctica.
4th Naval Zone. HQ at Iquique. From 18° S to 26° S.
5th Naval Zone. HQ at Puerto Montt. From 40° S to 49° S.

Naval Air Stations and Organisation

The main naval air station is at Viña del Mar. There are detachments at Iquique, Puerto Montt, Puerto Arenas and Puerto Williams.
 Five Squadrons: VP1: EMBP-111, P-3A and C-295
 HA1: NAS 332C Cougar
 VC1: EMBC-111, CASA-212 and O-2A
 HU1: BO 105C, Bell 206B, AS-365
 VT1: PC 7

Infanteria de Marina

Organisation: 4 detachments each comprising Amphibious Warfare, Coast Defence and Local Security. Also embarked are detachments of commandos, engineering units and a logistic battalion.
1st Marine Infantry Detachment 'Patricio Lynch'. At Iquique.
2nd Marine Infantry Detachment 'Miller'. At Viña del Mar.
3rd Marine Infantry Detachment 'Sargento Aldea'. At Talcahuano.
4th Marine Infantry Detachment 'Cochrane'. At Punta Arenas.
Special forces (Navy SEALs units and Commando Group). At Viña del Mar.

Bases

Valparaiso. Main naval base, schools, repair yard. HQ 1st Naval Zone. Air station.
Talcahuano. Naval base, schools, major repair yard (two dry docks, three floating docks), two floating cranes. HQ 2nd Naval Zone. Submarine HQ and base.
Punta Arenas. Naval base. Dockyard with slipway having building and repair facilities. HQ 3rd Naval Zone. Air station.
Iquique. Small naval base. HQ 4th Naval Zone.
Puerto Montt. Small naval base. HQ 5th Naval Zone.
Puerto Williams (Beagle Channel). Small naval base. Air Detachment.
Dawson Island (Magellan Straits). Small naval base.

Strength of the Fleet (including Coast Guard)

Type	Active	Building
Patrol Submarines	4	—
Frigates	8	—
Landing Ship Dock	1	—
Landing Ships (Tank)	4	—
Landing Craft	2	—
Fast Attack Craft (Missile)	6	—
Large Patrol Craft	9	1
Coastal Patrol Craft	84	—
Survey Ships	2	—
Training Ships	1	—
Transports	2	—
Tankers	2	—
Tenders	4	—

DELETIONS

Patrol Forces

2012 *Riquelme, Maullín* (CG), *Troncoso* (CG)

Amphibious Forces

2011 *Valdivia*
2012 *Orompello, Elicura*

PENNANT LIST

Notes: From 1997 pennant numbers have been painted on major warship hulls.

Submarines

20	Thomson						
21	Simpson						
22	Carrera						
23	O'Higgins						

Frigates

05	Almirante Cochrane
06	Almirante Condell
07	Almirante Lynch
11	Capitán Prat
14	Almirante Latorre
15	Almirante Blanco Encalada
18	Almirante Riveros
19	Almirante Williams

Patrol Forces

30	Casma
31	Chipana
34	Angamos
37	Orella
38	Serrano
39	Uribe
71	Micalvi
72	Ortiz
73	Isaza
78	Sibbald
81	Piloto Pardo
82	Comandante Toro
83	Marinero Fuentealba (bldg)
1603	Alacalufe (CG)
1604	Hallef (CG)
1609	Aysen (CG)
1610	Corral (CG)
1611	Concepcion (CG)
1612	Caldera (CG)
1613	San Antonio (CG)
1614	Antofagasta (CG)
1615	Arica (CG)
1616	Coquimbo [1616] (CG)
1617	Natales (CG)
1618	Valparaiso (CG)
1619	Punta Arenas (CG)
1620	Talcahuano (CG)
1621	Quintero (CG)
1622	Chiloe (CG)
1623	Puerto Montt (CG)
1624	Iquique (CG)
1625	Ona (CG)
1814	Diaz (CG)
1816	Salinas (CG)
1820	Machado (CG)
1823	Hudson (CG)

Survey Ships

46	Contre-almirante Oscar Viel Toro
61	Cabo de Hornos (bldg)
63	George Slight Marshall
77	Cabrales

Training Ships

| 43 | Esmeralda |

Amphibious Forces

| 91 | Sargento Aldea |
| 92 | Rancagua |

94	Orompello
95	Chacabuco
9101	Soldado Canave
9102	Cabo Reyes
9103	Soldado Fuentes

Auxiliaries

41	Aquiles
42	Almirante Merino
52	Almirante Montt
53	Araucano
74	Cirujano Videla
110	Meteoro
113	Buzo Sobenes
114	Grumete Perez
118	Guardián Brito

Tugs/Supply Ships

| 66 | Galvarino |
| 67 | Lautaro |

SUBMARINES

Notes: (1) There are some Swimmer Delivery Vehicles French Havas Mk 8 in service. This is the two-man version.
(2) Acquisition of a Crocodile 250 submarine is reported to be under consideration. The 33 m boat is capable of diving to 300 m and has rescue and research potential in addition to military applications.

2 SCORPENE CLASS (SSK)

Name	No	Builders	Laid down	Launched	Commissioned
O'HIGGINS	23	DCN, Cherbourg; IZAR	18 Nov 1999	1 Nov 2003	9 Sep 2005
CARRERA	22	IZAR, Cartagena; DCN	Nov 2000	24 Nov 2004	20 Jul 2006

Displacement, tonnes: 1,577 surfaced; 1,711 dived
Dimensions, metres (feet): 66.4 × 6.2 × 5.8 *(217.8 × 20.3 × 19.0)*
Speed, knots: 12 surfaced; 20 dived
Range, n miles: 6,500 at 8 kt surfaced; 550 at 4 kt dived
Complement: 31 (6 officers)

Machinery: Diesel electric; 4 MTU 12V 396 SE84 diesels; 5,364 hp(m) *(4.0 MW)*; 1 Jeumont Schneider motor; 3,808 hp(m) *(2.8 MW)*; 1 shaft
Missiles: MBDA Exocet SM39 Block 2; launched from 21 in *(533 mm)* tubes; inertial cruise; active terminal homing to 50 km *(27 n miles)* at 0.9 Mach; warhead 165 kg.
Torpedoes: 6—21 in *(533 mm)* tubes. 18 WASS Black Shark torpedoes; wire (fibre-optic cable) guided; active/passive homing to 50 km *(27 n miles)* at 50 kt; warhead 250 kg, or AEG SUT Mod 1; wire-guided; active homing to 12 km *(6.5 n miles)* at 35 kt; passive homing to 28 km *(15 n miles)* at 25 kt; warhead 250 kg.
Electronic countermeasures: ESM: Argos AR 900; intercept.
Radars: Navigation: Sagem; I-band.
Sonars: Hull mounted; active/passive search and attack, medium frequency.
Weapon control systems: UDS International SUBTICS.

CARRERA *10/2011, Michael Nitz / 1406541*

Programmes: Project Neptune. Contract awarded to DCN and Bazán on 18 December 1997 and became effective in April 1998. The bows of both boats were built at Cherbourg and the sterns at Cartagena. *O'Higgins* arrived at Valparaiso on 10 December 2005 and *Carrera* at Talcahuano on 13 December 2006.

Modernisation: Procurement of Exocet SM 39 is under consideration.
Structure: Equipped with Sagem APS attack periscope, an SMS optronic search periscope and SISDEF datalink terminal. Diving depth more than 300 m *(984 ft)*. AIP is not fitted.
Operational: Based at Talcahuano.

O'HIGGINS *3/2007, Ships of the World / 1305003*

Submarines — Frigates < **Chile** 117

2 THOMSON (TYPE 209/1400) CLASS (SSK)

Name	No	Builders	Laid down	Launched	Commissioned
THOMSON	20	Howaldtswerke	1 Nov 1980	28 Oct 1982	31 Aug 1984
SIMPSON	21	Howaldtswerke	15 Feb 1982	29 Jul 1983	18 Sep 1984

Displacement, tonnes: 1,520 surfaced; 1,614 dived
Dimensions, metres (feet): 61.2 × 6.2 × 5.5 *(200.8 × 20.3 × 18.0)*
Speed, knots: 11 surfaced; 21.5 dived
Range, n miles: 8,200 at 8 kt snorting, 400 at 4 kt dived, 16 at 21.5 kt dived
Complement: 32 (5 officers)

Machinery: Diesel-electric; 4 MTU 12V 493 AZ80 GA31L diesels; 2,400 hp(m) *(1.76 MW)* sustained; 4 Piller alternators; 1.7 MW; 1 Siemens motor; 4,600 hp(m) *(3.38 MW)* sustained; 1 shaft
Missiles: MBDA Exocet SM 39 Block 2; launched from 21 in *(533 mm)* torpedo tubes; inertial cruise; active terminal homing to 50 km *(27 n miles)* at 0.9 Mach; warhead 165 kg.

Torpedoes: 8—21 in *(533 mm)* bow tubes. 14 WASS Black Shark torpedoes; wire (fibre-optic cable) guided; active/passive homing to 50 km *(27 n miles)* at 50 kt; warhead 250 kg, or AEG SUT Mod 1; wire-guided; active homing to 12 km *(6.5 n miles)* at 35 kt; passive homing to 28 km *(15 n miles)* at 23 kt; warhead 250 kg.
Electronic countermeasures: ESM: Thomson-CSF DR 2000U; radar warning.
Radars: Surface search: SAGEM; I-band.
Sonars: Atlas Elektronik CSU 90-36; hull-mounted; active/passive search and attack; low/medium frequency.
Weapon control systems: UDS International SUBTICS.

Programmes: Ordered from Howaldtswerke, Kiel in 1980.

Modernisation: *Thomson* refit completed at Talcahuano in late 1990, *Simpson* in 1991. Refit duration about 10 months each. A major programme to upgrade and extend the service life of both boats to 2027 has been completed. The work included the fitting of a UDS Subtics combat management system and a new fire-control system. Torpedo tubes were upgraded to enable Whitehead Black Shark torpedoes and anti-ship missiles to be fired while platform improvements include a new engine-control system and battery set. Work on *Thomson* started in 2006 and was completed by early 2008. Modernisation of *Simpson* was undertaken 2008–12 (work was interrupted by a tsunami in February 2010).
Structure: Fin and associated masts lengthened by 50 cm to cope with wave size off Chilean coast.
Operational: Based at Talcahuano.

THOMSON   11/2012*, Albert Campanera i Rovira / 1525580

FRIGATES

2 BLANCO ENCALADA (M) CLASS (FFGHM)

Name	No	Builders	Laid down	Launched	Commissioned
ALMIRANTE BLANCO ENCALADA (ex-*Abraham van der Hulst*)	15 (ex-F 832)	Koninklijke Maatschappij De Schelde, Flushing	8 Feb 1989	7 Sep 1991	15 Dec 1993
ALMIRANTE RIVEROS (ex-*Tjerk Hiddes*)	18 (ex-F 830)	Koninklijke Maatschappij De Schelde, Flushing	28 Oct 1986	9 Dec 1989	3 Dec 1992

Displacement, tonnes: 3,373 full load
Dimensions, metres (feet): 122.3 oa; 114.2 wl × 14.4 × 4.3 *(401.2; 374.7 × 47.2 × 14.1)*
Flight deck, metres: 22 × 14.4 *(72.2 × 47.2)*
Speed, knots: 30; 21 diesel
Range, n miles: 5,000 at 18 kt
Complement: 156 (16 officers) + 7 spare berths

Machinery: CODOG; 2 RR Spey SM1C; 33,800 hp *(25.2 MW)* sustained; 2 Stork-Wärtsilä 12SW280 diesels; 9,790 hp(m) *(7.2 MW)* sustained; 2 shafts; LIPS cp props
Missiles: SSM: 4 McDonnell Douglas Harpoon Block II launchers ❶; active radar homing to 130 km *(70 n miles)* at 0.9 Mach; warhead 227 kg.
SAM: Raytheon RIM-7P Sea Sparrow Mk 48 vertical launchers ❷; semi-active radar homing to 16 km *(8.5 n miles)* at 2.5 Mach; warhead 38 kg; 16 missiles. Canisters mounted on port side of hangar.
Guns: 1—3 in *(76 mm)*/62 Oto Melara compact Mk 100 ❸; 100 rds/min to 16 km *(8.6 n miles)* anti-surface; 12 km *(6.5 n miles)* anti-aircraft; weight of shell 6 kg.
Torpedoes: 4—324 mm US Mk 32 Mod 9 (2 twin) tubes (mounted inside the after superstructure) ❹. Honeywell Mk 46 Mod 5; anti-submarine; active/passive homing to 11 km *(5.9 n miles)* at 40 kt; warhead 44 kg.
Physical countermeasures: Decoys: 2 Loral Hycor SRBOC 6-tubed fixed Mk 36 quad launchers; IR flares and chaff to 4 km *(2.2 n miles)*. SLQ-25 Nixie towed torpedo decoy.

Electronic countermeasures: ESM/ECM: Argo APECS II (includes AR 700 ESM) ❺; intercept and jammers.
Radars: Air/surface search: Signaal SMART-S ❻; 3D; F-band.
Air search: Signaal LW08 ❼; D-band.
Surface search: Signaal Scout ❽; I-band.
Navigation: Racal Decca 1226; I-band.
Fire control: 2 Signaal STIR 180 ❾; I/J/K-band.
Sonars: Signaal PHS-36; hull-mounted; active search and attack; medium frequency. Thomson Sintra Anaconda DSBV 61; towed array; passive low frequency. LFAS may be fitted in due course.
Combat data systems: Signaal SEWACO VIIB action data automation; Link 11. WSC-6 twin aerials.

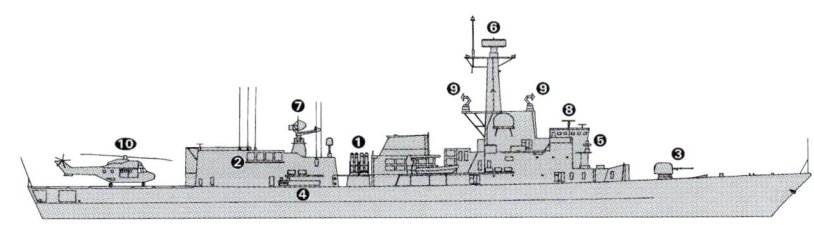

ALMIRANTE BLANCO ENCALADA   (Scale 1 : 1,200), Ian Sturton / 1455811

Helicopters: 1 NAS 332SC Cougar ❿.

Programmes: Contract signed on 26 March 2004 for the acquisition of two frigates. *Blanco Encalada* transferred on 16 December 2005 and arrived in Chile on 3 March 2006. *Riveros* was handed over on 18 April 2007 and arrived in Chile on 1 August 2007.
Structure: The VLS SAM is similar to Canadian Halifax and Greek MEKO classes. Both ships modified to operate Cougar helicopters. This includes lengthening and partly raising the helicopter hangar and replacement of the flight-deck grid with the ASIST system which includes 35 m traverse rails. A new horizon bar has also been installed.
Operational: Based at Valparaiso.

ALMIRANTE BLANCO ENCALADA   5/2009, Michael Nitz / 1366758

© 2014 IHS   IHS Jane's Fighting Ships 2014-2015

122 **Chile** > Amphibious forces

2 MAIPO (BATRAL) CLASS (LSTH)

Name	No	Builders	Launched	Commissioned
RANCAGUA	92	ASMAR, Talcahuano	6 Mar 1982	8 Aug 1983
CHACABUCO	95 (ex-93)	ASMAR, Talcahuano	16 Jul 1985	15 Apr 1986

Displacement, tonnes: 887 standard; 1,432 full load
Dimensions, metres (feet): 79.4 × 13 × 2.5
  (260.5 × 42.7 × 8.2)
Speed, knots: 14
Range, n miles: 3,500 at 13 kt
Complement: 57 (8 officers)

Military lift: 180 troops; 12 vehicles; 350 tons
Machinery: 2 Caterpillar diesels; 4,012 hp(m) *(2.95 MW)* sustained; 2 shafts; cp props
Guns: 2 Bofors 40 mm/60. 1 Oerlikon 20 mm.
A/S Mortars: 2 — 81 mm mortars.
Radars: Navigation: Decca 1229; I/J-band.

Helicopters: Platform for 1 Bell 206B or BO 105C.

Comment: First laid down in 1980 to standard French design with French equipment. Have 40 ton bow ramps and vehicle stowage above and below deck. Both ships underwent life-extension refits in 2002–03. Both based at Talcahuano.

CHACABUCO   *12/2004, Globke Collection* / 1047869

1 FOUDRE CLASS (LANDING SHIPS DOCK) (LSDH)

Name	No	Builders	Laid down	Launched	Commissioned
SARGENTO ALDEA (ex-*Foudre*)	91 (ex-L 9011)	DCN, Brest	26 Mar 1986	19 Nov 1988	7 Dec 1990

Displacement, tonnes: 8,321 standard; 12,599 full load
Dimensions, metres (feet): 168 × 23.5 × 5.2
  (551.2 × 77.1 × 17.1)
Aircraft lift: 1
Speed, knots: 21
Range, n miles: 11,000 at 15 kt
Complement: 243 (21 officers)

Military lift: 470 (up to 2,000 for 3 days) troops plus 1,880 tons load; 1 LCU plus 4 LCMs (typical) or 2 LCUs, or 10 LCMs, or 20 LARC XV amphibious vehicles; 150 vehicles
Machinery: 2 SEMT-Pielstick 16 PC2.5 V 400 diesels; 20,800 hp(m) *(15.3 MW)* sustained; 2 shafts; LIPS cp props; bow thruster; 1,000 hp(m) *(735 kW)*

Missiles: SAM: 3 MBDA Matra Simbad twin launchers; Mistral; IR homing to 4 km *(2.2 n miles)*; warhead 3 kg.
Guns: 3 Breda/Mauser 30 mm/70. 800 rds/min to 3 km *(1.6 n miles)*; weight of shell 0.36 kg. 4 — 12.7 mm MGs.
Physical countermeasures: Decoys: SLQ-25 Nixie towed torpedo decoy.
Electronic countermeasures: ECM: 2 Thales ARBB 36A jammers.
Radars: Air/surface search: Thomson-CSF DRBV 21A Mars; D-band.
Surface search: DRBN 34; I-band.
Navigation: DRBN 34 APP; I-band (for helo control). Sperry Bridgemaster; I-band.
Combat data systems: STIDAV/SENIT 8-01 for close range air defence. Link 11 (receive only). INMARSAT.

Weapon control systems: 2 Sagem DIBC-2A VIGY-105 optronic systems (for 30 mm guns).

Helicopters: 4 AS 532UL Cougar or SA 330B Puma.

Programmes: Former French LSD which entered French naval service in 1990. Formally transferred to the Chilean Navy on 22 December 2011 and recommissioned on 10 March 2012.
Structure: Hospital facilities (500 m²) include two operating suites and 50 beds. Well dock of 122 × 14.2 m *(1,732 m²)* which can be used to dock a 400 tonne ship. Crane of 37 tons and lift of 52 tons. Flight deck of 1,450 m² with two landing spots. Additional landing spot on the (removable) well rolling cover. Flume stabilisation fitted in 1993.

SARGENTO ALDEA   *6/2013*, Chilean Navy* / 1525578

2 LCMS (LCM)

SOLDADO FUENTES 9103 (ex-CTM 19)   **CABO REYES** 9102 (ex-CTM 24)

Displacement, tonnes: 60 standard; 152 full load
Dimensions, metres (feet): 23.8 × 6.4 × 1.3 *(78.1 × 21.0 × 4.3)*
Speed, knots: 9.5. **Range, n miles:** 600 at 5 kt
Complement: 4 + 100 embarked forces

Military lift: 90 tons (maximum); 48 tons (normal)
Machinery: 2 Poyaud V8520NS diesels; 450 hp(m) *(331 kW)*; 2 shafts
Guns: 2 — 12.7 mm MGs.
Radars: Navigation: I-band.

Comment: Former French LCMs built at CMN, Cherbourg 1982–92. Equipped with a bow ramp. Transferred to Chile in January 2012 with *Sargento Aldea* and one CDIC-class LCT.

SOLDADO FUENTES
6/2013*, Chilean Navy
1525576

IHS Jane's Fighting Ships 2014-2015   © 2014 IHS

SURVEY SHIPS

1 BUOY TENDER (ABU)

Name	No	Builders	Commissioned
GEORGE SLIGHT MARSHALL (ex-*M V Vigilant*)	BRS 63	Pattje, Waterhuizen	Jul 1978

Displacement, tonnes: 1,200 full load
Dimensions, metres (feet): 53 × 11.2 × 3.5 *(173.9 × 36.7 × 11.5)*
Speed, knots: 10. **Range, n miles:** 4,600 at 5 kt
Complement: 32
Machinery: 2 Ruston 6AP230 diesels; 1,360 hp *(1 MW)*; 2 shafts; bow thruster
Guns: 2 Oerlikon 20 mm.
Radars: Navigation: Decca 252/6; I-band.

Comment: Acquired from the UK Mersey Harbour Board and recommissioned 5 February 1997. Carries a 15 ton derrick. Based at Puerto Montt.

GEORGE SLIGHT MARSHALL   6/2008, *Chilean Navy* / 1335349

1 TYPE 1200 CLASS (AGS/AGOBH)

Name	No	Builders	Commissioned
CONTRE-ALMIRANTE OSCAR VIEL TORO (ex-*Norman McLeod Rogers*)	AP 46	Canadian Vickers, Montreal	Oct 1960

Displacement, tonnes: 6,421 full load
Measurement, tonnes: 4,246 gt; 1,877 net
Dimensions, metres (feet): 89.9 × 19.1 × 6.1 *(294.9 × 62.7 × 20.0)*
Speed, knots: 15
Range, n miles: 12,000 at 12 kt
Complement: 33
Machinery: 4 Fairbanks-Morse 38D8-1/8-12 diesels; 8,496 hp *(6.34 MW)* sustained; 4 GE generators; 4.8 MW; 2 Ruston RK3CZ diesels; 7,250 hp *(5.6 MW)* sustained; 2 GE generators; 2.76 MW; 2 GE motors; 12,000 hp *(8.95 MW)*; 2 shafts
Guns: 2 Oerlikon 20 mm.
Helicopters: 1 BO 105C.

Comment: Acquired from the Canadian Coast Guard on 16 February 1995. The ship was formerly based on the west coast at Victoria, BC, and was laid up in 1993. The ship completed an extensive refit in 2010 which included installation of an automated propulsion and platform control system developed by SISDEF. There are no plans to replace the ship. Based at Punta Arenas.

CONTRE-ALMIRANTE OSCAR VIEL TORO   6/2004, *Chilean Navy* / 1044093

1 OCEANOGRAPHIC RESEARCH SHIP (AGOR)

Name	No	Builders	Laid down	Launched	Commissioned
CABO DE HORNOS	AGS 61	ASMAR, Talcahuano	20 Feb 2009	30 Jan 2011	27 Feb 2013

Measurement, tonnes: 3,068 gt
Dimensions, metres (feet): 74.1 × 15.6 × 5.8 *(243.1 × 51.2 × 19.0)*
Speed, knots: 14.5
Range, n miles: 10,000 at 12 kt
Complement: 48 (9 officers) + 25 scientists
Machinery: Diesel electric; 3 Wärtsilä 8L20 diesel generators; 6,435 hp *(4.8 MW)*; 2 Ansaldo motors; 4,023 hp *(3.0 MW)*; 1 shaft; 1 bow thruster *(450 kW)*; 1 stern thruster *(450 kW)*
Radars: Surface search: Kongsberg; E/F-band.
Navigation: Kongsberg; I-band.

Comment: Project Medusa: Contract signed with ASMAR Talcahuano on 28 December 2007 for the construction of an oceanographic and fisheries research vessel to replace *Vidal Gormaz*. The ST-367 design was developed by Skipsteknisk of Norway. The ship is equipped with four laboratories while the hydroacoustic research and positioning equipment is to be provided by Kongsberg Simrad. This includes: multibeam echosounders for deep and medium depth water, singlebeam echosounder for deep water, sub-bottom profiler, omni-directional sonar for biomass, surface sound velocity profiler and an acoustic Doppler current profiler. Damage sustained in an earthquake and subsequent tsunami in February 2010 was slight and the ship was delivered in 2013.

CABO DE HORNOS   6/2013*, *ASMAR* / 1525575

TRAINING SHIPS

1 SAIL TRAINING SHIP (AXS)

Name	No	Builders	Commissioned
ESMERALDA (ex-*Don Juan de Austria*)	BE 43	Bazán, Cadiz	15 Jun 1954

Displacement, tonnes: 3,439 standard; 3,673 full load
Dimensions, metres (feet): 113.1 × 13.1 × 7 *(371.1 × 43.0 × 23.0)*
Speed, knots: 13
Range, n miles: 11,600 at 10 kt
Complement: 310 (21 officers; 97 cadets)
Machinery: 1 Burmeister & Wain diesel; 1,400 hp(m) *(1.03 MW)*; 1 shaft
Guns: 4 Hotchkiss saluting guns.

Comment: Four-masted schooner originally intended for the Spanish Navy. Near sister ship of *Juan Sebastian de Elcano* in the Spanish Navy. Refitted Saldanha Bay, South Africa, 1977. Sail area, 26,910 sq ft. Based at Valparaiso.

ESMERALDA   6/2012, *Chris Sattler* / 1455765

AUXILIARIES

Notes: A small landing craft *Guardián Brito* (BSG 118) was commissioned on 9 November 2010. She replaced *Pisagua*, which was destroyed in the tsunami of February 2010. The craft is used to provide logistic support to Quiriquina Island, off Talcahuano.

3 FLOATING DOCKS (YFD)

Name	No	Commissioned	Recommissioned
INGENIERO MERY	131	1944	1973
MUTILLA	132	1944	1960
TALCAUANO	133	1944	1999

Comment: There is also a Floating Dock *Marinero Gutierrez* with a 1,200 ton lift. Built in 1991.

1 PERSONNEL TENDER (YFB)

BUZO SOBENES (ex-*Patagonia Express*) YFB 113

Displacement, tonnes: 110 full load
Dimensions, metres (feet): 28 × 8.6 × 1.4 *(91.9 × 28.2 × 4.6)*
Speed, knots: 29
Machinery: 2 GM 12V-149-TI diesels; 1,055 hp *(776 kW)*; 2 shafts
Radars: Navigation: I-band.

Comment: Built in 1991 by Astillero Asenav (Valdivia). Based at Talcahuano naval base and Quiriquina island and entered service in November 2013 to replace former craft of the same name destroyed in 2010 tsunami. Aluminium catamaran hull.

124   **Chile** > Auxiliaries

1 TRANSPORT SHIP (APH)

Name	No	Builders	Launched	Commissioned
AQUILES	AP 41	ASMAR, Talcahuano	4 Dec 1987	15 Jul 1988

Displacement, tonnes: 2,769 standard; 4,760 full load
Dimensions, metres (feet): 103 × 17 × 5.5 (337.9 × 55.8 × 18.0)
Speed, knots: 18
Complement: 80
Military lift: 250 troops
Machinery: 2 Krupp MaK 8 M 453B diesels; 7,080 hp(m) (5.10 MW) sustained; 1 shaft; bow thruster
Helicopters: Platform for up to Cougar size.

Comment: Ordered 4 October 1985. Can be converted rapidly to act as hospital ship. Based at Valparaiso.

AQUILES   11/2009, Chris Sattler / 1366163

1 ÄLVSBORG CLASS (SUPPORT SHIP) (AGP/ASH)

Name	No	Builders	Launched	Commissioned
ALMIRANTE MERINO (ex-Älvsborg)	42 (ex-A 234, ex-M 02)	Karlskronavarvet	11 Nov 1969	6 Apr 1971

Displacement, tonnes: 2,703 full load
Dimensions, metres (feet): 92.5 × 14.7 × 4 (303.5 × 48.2 × 13.1)
Speed, knots: 16
Complement: 52 + 153 spare berths

Machinery: 2 Nohab-Polar 112 VS diesels; 4,200 hp(m) (3.1 MW); 1 shaft; cp prop; bow thruster; 350 hp(m) (257 kW)
Guns: 3 Bofors 40 mm/70 SAK 48.
Physical countermeasures: Decoys: 2 Philax chaff/IR launchers.
Radars: Raytheon; E/F-band.
Surface search: Philips 9GR 600; I-band.
Fire control: Philips 9LV 200 Mk 2; I/J-band.
Navigation: Terma Scanter 009; I-band.
Helicopters: Platform for 1 medium.

Comment: Ordered in 1968 as a minelayer. Transferred from the Swedish Navy in November 1996, having been paid off in 1995. Recommissioned 7 February 1997. Originally designed as a minelayer with a capacity of 300 mines. Converted to act as a general support ship with improved accommodation and workshops. Acts as a depot ship for submarines and attack craft. The full name is *Almirante José Toribio Merino Castro*.

MERINO   7/2001, Maritime Photographic / 0121328

1 REPLENISHMENT SHIP (AOR)

Name	No	Builders	Commissioned
ARAUCANO (ex-Alpaca)	AO 53	Uddevallavarvet AB, Uddevalla	1985

Displacement, tonnes: 19,700 full load
Dimensions, metres (feet): 156.4 × 23.1 × 10.2 (513.1 × 75.8 × 33.5)
Speed, knots: 13
Complement: 66 (14 officers)
Cargo capacity: 23,637 m³ fuel.
Machinery: 1 B&W 4L67GBE diesel; 7,200 hp(m) (5.3 MW) sustained; 1 shaft; cp prop
Radars: Navigation: Liton BME ATA 252/6; I-band. Furuno FR-2115; I-band.

Comment: Former chemical/products tanker built in 1984 acquired by the Chilean Navy on 20 December 2010. Converted to fleet tanker role. The double-bottomed ship is of steel construction. Based at Valparaiso.

IHS Jane's Fighting Ships 2014-2015

ARAUCANO   3/2013*, L-G Nilsson / 1525574

2 HARBOUR TRANSPORTS (YFB)

Name	No	Builders	Commissioned
METEORO	YFB 110	ASMAR, Talcahuano	3 Jun 1968
GRUMETE PEREZ	YFB 114	ASMAR, Talcahuano	12 Dec 1975

Displacement, tonnes: 168 full load
Dimensions, metres (feet): 24.4 × 6.7 × 2.6 (80.1 × 22.0 × 8.5)
Speed, knots: 10
Complement: 6
Machinery: 1 diesel; 370 hp(m) (272 kW); 1 shaft
Guns: 1 Oerlikon 20 mm can be carried.
Radars: Navigation: Furuno; I-band.

Comment: Transferred to Seaman's School as harbour transports. Modified fishing boat design.

GRUMETE PEREZ   8/1997, Chilean Navy / 0012168

1 REPLENISHMENT SHIP (AOR)

Name	No	Builders	Laid down	Launched	Commissioned
ALMIRANTE MONTT (ex-Andrew J Higgins)	52 (ex-T-AO 190)	Avondale Shipyards, New Orleans	21 Nov 1985	17 Jan 1987	22 Oct 1987

Displacement, tonnes: 40,900 full load
Dimensions, metres (feet): 206.5 × 29.7 × 10.9 (677.5 × 97.4 × 35.8)
Speed, knots: 20
Range, n miles: 6,000 at 18 kt
Complement: 74

Cargo capacity: 180,000 barrels of fuel oil or aviation fuel
Machinery: 2 Colt-Pielstick 10 PC4.2 V 570 diesels; 34,422 hp(m) (24.3 MW) sustained; 2 shafts; cp props
Physical countermeasures: Decoys: SLQ-25 Nixie, towed torpedo decoy.
Radars: Navigation: 2 Raytheon; I-band.
Helicopters: Platform only.

Comment: Ex-US Navy Henry J Kaiser-class tanker. Single hull construction. After being withdrawn from US service in May 1996 the ship was laid up until being sold to Chile on 19 May 2009. A five-month refit of the ship was completed at Atlantic Marine Alabama Shipyard, Mobile, Alabama in February 2010. There are stations on both sides for underway replenishment of fuel and solids. The ship has replaced the former *Araucano*.

ALMIRANTE MONTT   6/2011, Chilean Navy / 1406547

© 2014 IHS

TUGS

Notes: Small harbour tugs *Reyes*, *Cortés* (both 100 tons and built in 1960) and *Galvez* (built in 1975) are also in commission.

2 VERITAS CLASS (TUG/SUPPLY VESSELS) (ATF)

Name	No	Builders	Commissioned
GALVARINO (ex-*Maersk Traveller*)	ATF 66	Aukra Bruk, Aukra	1974
LAUTARO (ex-*Maersk Tender*)	ATF 67	Aukra Bruk, Aukra	1973

Displacement, tonnes: 956 standard; 2,418 full load
Dimensions, metres (feet): 58.3 × 12.6 × 3.9 *(191.3 × 41.3 × 12.8)*
Speed, knots: 14
Complement: 11 + 12 spare berths
Cargo capacity: 1,400 tons

Machinery: 2 Krupp MaK 8 M 453AK diesels; 6,400 hp(m) *(4.7 MW)*; 2 shafts; cp props; bow thruster
Guns: 1 Bofors 40 mm/70 can be carried.
Radars: Navigation: Terma Pilot 7T-48; Furuno FR 240; I-band.

Comment: *Janequero* and *Galvarino* delivered from Maersk and commissioned into Navy 26 January 1988. *Lautaro* delivered in 1991. *Janequero* since deleted. Bollard pull, 70 tonnes; towing winch, 100 tons. Fully air conditioned. Designed for towing large semi-submersible platform in extreme weather conditions. Ice strengthened. *Lautaro* underwent refit at ASMAR October 2006 to January 2007. *Galvarino* based at Valparaiso and *Lautaro* at Punta Arenas.

GALVARINO — 7/2001, *Maritime Photographic* / 0121330

COAST GUARD

Notes: There are also large numbers of harbour and SAR craft.

26 ARCHANGEL CLASS (RESPONSE BOAT) (PBF)

LPM 4400–4413   LSR 4420–4431

Displacement, tonnes: 13 standard
Dimensions, metres (feet): 13.5 × 4.3 × 2.3 *(44.3 × 14.1 × 7.5)*
Speed, knots: 36
Range, n miles: 300 at 25 kt
Complement: 3
Machinery: 2 Caterpillar C9 diesels; 550 hp *(409 kW)*; 2 Hamilton 322 waterjets
Guns: 1 — 7.62 mm MG.
Radars: Navigation: Furuno; I-band.

Comment: High-speed inshore patrol craft of aluminium construction and foam collar built by SAFE Boats International, Port Orchard, Washington. There had been 23 delivered by late 2013 and a further three are expected by mid-2014.

ARCHANGEL CLASS — 6/2008, *Chilean Navy* / 1335343

6 TYPE 44 CLASS (WPB)

PELLUHUE LSR 1703	CHACAO LSR 1705	GUAITECA LSR 1707
ARAUCO LSR 1704	QUEITAO LSR 1706	CURAUMILA LSR 1708

Displacement, tonnes: 18 full load
Dimensions, metres (feet): 13.5 × 3.9 × 1.1 *(44.3 × 12.8 × 3.6)*
Speed, knots: 10
Range, n miles: 215 at 10 kt
Complement: 3
Machinery: 2 Detroit 6V-38 diesels; 185 hp *(136 kW)*; 2 shafts

Comment: Acquired from the US and recommissioned on 31 May 2001.

QUEITAO — 6/2008, *Chilean Navy* / 1335346

4 GRUMETE DIAZ (DABUR) CLASS
(COASTAL PATROL CRAFT) (PB)

DIAZ LPC 1814	SALINAS LPC 1816	MACHADO LPC 1820	HUDSON LPC 1823

Displacement, tonnes: 40 full load
Dimensions, metres (feet): 19.8 × 5.5 × 1.8 *(65 × 18.0 × 5.9)*
Speed, knots: 15. **Range, n miles:** 450 at 13 kt
Complement: 8 (2 officers)
Machinery: 2 Detroit 12V 71TA diesels; 840 hp *(627 kW)* sustained; 2 shafts
Guns: 2 Oerlikon 20 mm or 2 — 12.7 mm MGs.
Radars: Surface search: Racal Decca Super 101 Mk 3; I-band.

Comment: All have LPC numbers and Grumete precedes the ships' names. First six transferred from Israel and commissioned 3 January 1991. Second batch of four more transferred and commissioned 17 March 1995. A RIB inspection boat is carried on the stern. All underwent life extension refits in 2001–02 at Valparaiso and Puerto Montt. Two craft deleted in 2006, one in 2012, and three in 2013. Four craft refitted 2011–12 to extend life to 2020.

HUDSON — 7/2001, *Maritime Photographic* / 0121321

126 Chile > Coast guard

18 PROTECTOR CLASS (WPB)

ALACALUFE LSG 1603	SAN ANTONIO LSG 1613	PUNTA ARENAS LSG 1619
HALLEF LSG 1604	ANTOFAGASTA LSG 1614	TALCAHUANO LSG 1620
AYSÉN LSG 1609	ARICA LSG 1615	QUINTERO LSG 1621
CORRAL LSG 1610	COQUIMBO LSG 1616	CHILOÉ LSG 1622
CONCEPTIÓN LSG 1611	PUERTO NATALES LSG 1617	PUERTO MONTT LSG 1623
CALDERA LSG 1612	VALPARAÍSO LSG 1618	IQUIQUE LSG 1624

Displacement, tonnes: 122 full load
Dimensions, metres (feet): 33.1 × 6.6 × 2.0 (108.6 × 21.7 × 6.6)
Speed, knots: 23. **Range, n miles:** 3,600 at 12 kt
Complement: 10 (2 officers)
Machinery: 2 MTU MDEC 2,000 diesels; 5,200 hp(m) (3.82 MW); 2 shafts
Guns: 1—12.7 mm MG.
Radars: Navigation: Litton BME 252/6; I-band.

Comment: All built under licence from FBM at ASMAR, Talcahuano, in conjunction with FBM Marine. There are minor differences between LEP 1603-4 and the rest. First commissioned 24 June 1989 and last on 10 March 2004. All conduct coastal patrols between Arica and Puerto Williams.

ARICA — 12/2004, Globke Collection / 1047868

SAN ANTONIO — 6/2013*, Chilean Navy / 1525579

1 ONA CLASS (PATROL CRAFT) (PB)

ONA (ex-*Pudú*) LSG 1625

Displacement, tonnes: 110 full load
Dimensions, metres (feet): 26.6 × 6.0 × 2.0 (87.3 × 19.7 × 6.6)
Speed, knots: 12. **Range, n miles:** 1,430 at 10 kt
Complement: 8 (2 officers)
Machinery: 2 Caterpillar 3412 12V diesels; 1,070 hp (798 kW); 2 shafts
Radars: To be announced.

Comment: Built by ASENAV, Valdivia, in 1992 and operated as a pilot vessel by Ultratug until transferred to the Chilean Navy on 5 July 2011. The roles of the ship include surveillance, environmental monitoring and SAR. Based at Punta Arenas.

ONA — 7/2011, Chilean Navy / 1406548

12 RODMAN 800 CLASS (WPB)

PM 2031–2033   PM 2035–2038   PM 2040   PM 2041   PM 2043–2045

Displacement, tonnes: 4 full load
Dimensions, metres (feet): 8.9 × 3 × 1.1 (29.2 × 9.8 × 3.6)
Speed, knots: 28
Range, n miles: 150 at 25 kt
Complement: 3
Machinery: 2 Volvo diesels; 300 hp(m) (220 kW); 2 shafts
Radars: Navigation: Raytheon; I-band.

Comment: Built by Rodman Polyships, Vigo and all delivered by 17 May 1996.

RODMAN 800 CLASS — 7/2001, Maritime Photographic / 0121331

15 DEFENDER CLASS (RESPONSE BOATS) (PBF)

PM 2501–2515

Displacement, tonnes: 4.2 full load
Dimensions, metres (feet): 7.6 × 2.6 × 1.1 (24.9 × 8.5 × 3.6)
Speed, knots: 45
Range, n miles: 175 at 35 kt
Complement: 2
Machinery: 2 Honda outboard motors; 450 hp (335 kW)
Guns: 1—7.62 mm MG.
Radars: Navigation: Furuno 1834; I-band.

Comment: High-speed inshore patrol craft of aluminium construction and foam collar built by SAFE Boats International, Port Orchard, Washington. Four delivered 2007–08 and a further six by 2011. A further five were delivered in 2012.

DEFENDER CLASS (old number) — 6/2007, Chilean Navy / 1292774

IHS Jane's Fighting Ships 2014-2015

© 2014 IHS

China
PEOPLE'S LIBERATION ARMY NAVY (PLAN)

Country Overview

The People's Republic of China, proclaimed on 1 October 1949, is the world's third-largest country by area (3,695,000 square miles) and the largest by population. It is bordered to the north by Kyrgyzstan, Kazakhstan, Mongolia and Russia, to the south by Vietnam, Laos, Myanmar, India, Bhutan, Nepal and North Korea and to the west by Pakistan, Afghanistan and Tajikistan. It has a 7,830 n mile coastline with the Yellow, East China and South China seas. There are more than 3,400 offshore islands of which Hainan is the largest. Sovereignty over Taiwan, still formally a province of China, is also claimed. Ownership of some or all of the Spratly Islands is disputed between China, Brunei, Taiwan, Vietnam, Malaysia and the Philippines although a code of conduct was mutually brokered in 2002. The principal ports are Shanghai (largest city), Qingdao, Shantou, Xiamen, Tianjin, Guangzhou, Hong Kong, Dalian, and Ningbo-Zhoushan. Overall there are 54,000 n miles of navigable inland waterways including the Yangtze River on which the port of Wuhan is situated. Territorial seas (12 n miles) are claimed. A 200 n mile EEZ has also been claimed but the limits have not been defined.

Headquarters Appointments

Commander-in-Chief of the Navy:
  Admiral Wu Shengli
Political Commissar of the Navy:
  Admiral Liu Xiaojiang
Deputy Commanders-in-Chief of the Navy:
  Vice Admiral Zhang Zhannan
  Vice Admiral Ding Yiping
  Vice Admiral Gu Wengen
  Vice Admiral Zhang Yongyi
Chief of Naval Staff:
  Vice Admiral Su Shiliang

Fleet Commanders

North Sea Fleet:
  Vice Admiral Tian Zhong
East Sea Fleet:
  Vice Admiral Su Zhiqian
South Sea Fleet:
  Vice Admiral Jiang Weilie

Personnel

(a) 2014: 255,000 officers and men, including 25,000 naval air force, 8–10,000 marines (28,000 in time of war) and 28,000 for coastal defence
(b) 2 years' national service for sailors afloat; 3 years for those in shore service. Some stay on for up to 15 years. 41,000 conscripts

Operational Numbers

Because numbers of vessels are kept in operational reserve, the Chinese version of the order of battle tends to show fewer ships than are counted by Western observers.

Organisation

Each of the North, East and South Sea Fleets has two submarine divisions, three DD/FF divisions and one MCMV division. The North also has one Amphibious Division, and the other Fleets have two each. The South has two Marine Infantry Brigades.

Bases

North Sea Fleet. Major bases: Qingdao (HQ), Huludao, Jianggezhuang, Guzhen Bay, Lushun, Xiaopingdao. Minor bases: Weihai Wei, Qingshan, Luda, Lianyungang, Ling Shan, Ta Ku Shan, Changshandao, Liuzhuang, Dayuanjiadun, Dalian
East Sea Fleet. Major bases: Ningbo (HQ), Zhoushan, Shanghai, Daxie, Fujan. Minor bases: Zhenjiangguan, Wusong, Xinxiang, Wenzhou, Sanduao, Xiamen, Xingxiang, Quandou, Wen Zhou SE, Wuhan, Dinghai, Jiaotou
South Sea Fleet. Major bases: Zhanjiang (HQ), Yulin (Hainan Island), Huangfu, Hong Kong, Yalong (Hainan Island), Guangzhou (Canton). Minor bases: Haikou, Shantou, Humen, Kuanchuang, Tsun, Kuan Chung, Mawai, Beihai, Ping Tan, San Chou Shih, Tang-Chiah Huan, Longmen, Bailong, Dongcun, Baimajing, Xiachuandao, Yuchi

Coast Defence

A large number of HY-2 (CSSC-3) and HY-3 (CSSC-301) SSMs in 20 semi-fixed armoured sites. 35 Coastal Artillery regiments.

Equipment Procurement

Although often listed under the name of the designer, equipment has not necessarily been supplied direct from the parent company. It may have been acquired from a third party or by reverse engineering.

Training

The main training centres are:
Dalian: Naval Academy
Guangzhou (Canton): Naval Arms Command College
Qingdao: Submarine Academy
Wuhan: Naval Engineering University
Nanjing: Naval Command College
Yantai: Naval Aeronautical Engineering College
Bengbu: Naval School for Non-Commissioned Officers.

Marines

There are two brigades based at Zhanjiang and subordinate to the Navy. Each has three Infantry regiments and one Artillery regiment.

Naval Air Force

With 25,000 officers and men and over 800 aircraft, this is a considerable naval air force primarily land-based. There is a total of eight Divisions with 27 Regiments split between the three Fleets. Some aircraft are laid up unrepaired.

Air bases include:

North Sea Fleet: Dalian, Qingdao, Jinxi, Jiyuan, Laiyang, Jiaoxian, Xingtai, Laishan, Anyang, Changzhi, Liangxiang and Shan Hai Guan
East Sea Fleet: Danyang, Daishan, Shanghai (Dachang), Ningbo, Luqiao, Feidong and Shitangqiao
South Sea Fleet: Foluo, Haikou, Lingshui, Sanya, Guiping, Jialaishi and Lingling

Strength of the Fleet

Type	Active (Reserve)	Building (Planned)
SSBN	4	2
SSN	5	4
SSK	53	—
SSA	2	—
Aircraft carriers	1	(1)
Destroyers	19	6 (4)
Frigates	49	4
Corvettes	8	12 (10)
Fast Attack Craft (Missile)	86	—
Patrol Craft	119	—
Minesweepers (Ocean)	40	2
Mine Warfare Drones	6	—
Minelayer	1	—
Hovercraft	15	2
LPD	3	—
LSTs	27	—
LSMs	30	—
LCMs-LCUs	165	—
Survey/Research	18	—
Training Ships	3	—
Troop Transports (AP/AH)	4	—
Submarine Support Ships	13	—
Salvage and Repair Ships	1	—
Supply Ships	22	—
Fleet Replenishment Ships	7	2
Support Tankers	73	—
Hospital Ship	8	—
Icebreakers	2	—

DELETIONS

Submarines

2012   Ming 352–354

Destroyers

2012   *Nanjing, Yinchuan, Nanning* (to Coast Guard)
2013   *Xining, Hefei, Chongqing, Nanchang, Quilin, Zunyi*

Frigates

2012   *Anshun, Jishou* (both to Myanmar), *Nantong, Wuxi, Maoming, Yibin*
2013   *Wuhu, Huangshi* (both to Bangladesh)

Patrol Forces

2012   35 Shanghai II, 11 Huangfen

Amphibious Forces

2012   1 Yudao class

Research Ships

2012   *Yuan Wang 2, Shuguang 203*

PENNANT LIST

Submarines
406	Xia

Aircraft Carriers
16	Liaoning

Destroyers
109	Kaifeng
110	Dalian
112	Harbin
113	Qingdao
115	Shenyang
116	Shijiazhuang
117	Yinchuan (bldg)
118	Taiyuan (bldg)
136	Hangzhou
137	Fuzhou
138	Taizhou
139	Ningbo
150	Changchun
151	Zhengzhou
152	Jinan (bldg)
153	Xian (bldg)
154	Hefei (bldg)
155	Nanjing (bldg)
165	Zhanjiang
166	Zhuhai
167	Shenzhen
168	Guangzhou
169	Wuhan
170	Lanzhou
171	Haikou
172	Kunming (bldg)
173	Changsha (bldg)
175	Guiyang (bldg)
176	Chengdu (bldg)

Frigates
513	Huayin
514	Zhenjiang
516	Jiujiang
518	Jian
519	Changzhi
521	Jiaxing
522	Lianyungang
523	Putian
524	Sanming
525	Maanshan
526	Wenzhou
527	Luoyang
528	Mianyang
529	Zhoushan
530	Xuzhou
533	Taizhou
534	Jinhua
537	Cangzhou
538	Yantai
539	Anqing
540	Huainan
541	Huaibei
542	Tongling
543	Dandong
545	Linfen
546	Yancheng
547	Linyi
548	Yiyang
549	Changzhou
550	Weifang
553	Shaoguan
555	Zhaotong
558	Zigong
559	Beihai
560	Dongguan
561	Shantou
562	Jiangmen
563	Foshan
564	Yichang
565	Huludao
566	Huaihua
567	Xiangfan
568	Hengyang
569	Yulin
570	Huangshan
571	Yuncheng
572	Hengshui
573	Liuzhou
574	Sanya
575	Yueyang
576	Huangshi (bldg)

Corvettes
580	Datong
581	Yingkou
582	Bengbu
583	Shangrao
584	Meizhou
586	Jian
596	Huizhou
597	Qinzhou

Patrol Forces
770	Yangjiang
771	Shunde
772	Nanhai
773	Panyu
774	Lianjiang
775	Xinhui

Amphibious Forces
908	Yandanshan
909	Jiuhuashan
910	Huangganshan
911	Tianzhushan
912	Daqingshan
913	Baxianshan
918	—
927	Yuntaishan
928	Wufengshan
929	Zijinshan
930	Lingyanshan
931	Dongtingshan
932	Helanshan
933	Liupanshan
934	Danxiashan
935	Xuefengshan
936	Haiyangshan
937	Qingchengshan
939	Putuoshan
940	Tiantaishan
941	Shengshan
942	Lushan
943	—
944	Yushan
945	Huashan
946	Songshan
947	—
948	Xueshan
949	Hengshan
950	Taishan
989	Changbaishan
990	Wudangshan

128　China > Introduction — Submarines

PENNANT LIST – continued

991	Emeishan	814	Liaoyang	892	Hua Luogeng	—	Haiyangdao
992	Huadingshan	818	Kunshan	893	Qianlao	866	Daishandao
993	Luoxiaoshan	839	Liuyang	900	Beidiao	867	Changdao
994	Daiyunshan	840	Luxi			881	Hongzhu
995	Wangyangshan	841	Xiaoyi	**Training Ships**		882	Poyang Hu
996	Laotieshan	842	—	81	Zhenghe	885	Qinghai Hu
997	Yunwashan	843	—	82	Shichang	886	Qiandao Hu
998	Kunlunshan			544	Lushun	887	Weishan Hu
999	Jinggangshan	**Survey and Research Ships**				889	Tai Hu (bldg)
		851	Beijixing	**Principal Auxiliaries**		890	Chao Hu (bldg)
Mine Warfare Forces		853	Tianwangxing			911	Dongxiu
804	Huoqiu	871	Li Siguang	861	Changxingdao		
805	Zhangjiagang	872	Zhu Zheken	862	Chongmingdao		
810	Jingjiang	891	Bi Sheng	863	Yongxingdao		

SUBMARINES

Strategic Missile Submarines

Notes: The fourth test flight of a JL-2 missile was successfully accomplished on about 12 June 2005. The firing was made from a submarine, probably the Golf class SSB, off Qingdao and impacted in the western desert. Further tests are reported to have been conducted in 2008, 2009, 2010 and 2011. A further test was conducted on 16 August 2012 from a Jin-class SSBN. The missile is expected to become operational in 2014.

3 + 2 JIN CLASS (TYPE 094) (SSBN)

Name	No	Builders	Laid down	Launched	Commissioned
—	411	Bohai Shipyard, Huludao	2001	28 Jul 2004	Mar 2007
—	412	Bohai Shipyard, Huludao	2003	2006	2010
—	413	Bohai Shipyard, Huludao	2004	Dec 2009	2012
—	414	Bohai Shipyard, Huludao	2006	2011	2013

Dimensions, metres (feet): 137 × 11.8 × 2.3 *(449.5 × 38.7 × 7.5)*
Complement: 140

Machinery: Nuclear: 2 PWR; 150 MW; 2 turbines; 1 shaft
Missiles: SLBM; 12 JL-2 (CSS-NX-5); 3-stage solid-fuel rocket; stellar inertial guidance to over 8,000 km *(4,320 n miles)*; single nuclear warhead of 1 MT or 3–8 MIRV of smaller yield. CEP 300 m approximately.
Torpedoes: 6—21 in (533 mm tubes).
Electronic countermeasures: Decoys: ESM.
Radars: Surface search/navigation: Type 359; I-band.
Sonars: Hull mounted passive/active; flank arrays. Passive intercept array.

Programmes: The first of class became operational as a submarine in mid-2007 but will not become fully operational until the JL-2 missile enters service. Although a class of six boats was at first expected, the building of further Shang-class SSN at Huludao suggests that construction may have slowed. Up to five boats are now expected before the next-generation Type 096 SSBN begins to enter service in the Jin class construction programme.
Structure: Likely to be based on the Type 093 SSN design which in turn is believed to be derived from the Russian Victor III design. The dimensions of the hull assume the incorporation of a 30 m 'missile plug' of 12 tubes for the 42 ton JL-2 missiles.
Operational: The first two boats are to be based at Yalong, Hainan Island. Sea trials of the third of class are reported to have started in 2011 and completed in 2012 when she became operational as a submarine. Trials of the fourth boat began in 2013. While the performance of the missile is speculative, its range may prompt a change in operating concept to a 'bastion' patrol approach.

JIN CLASS　　　　　　　　　　　　　　　　　　　　　　　　　　12/2006 / 1167755

JIN CLASS　　　　　　　　　　　　　　　　　　　　　　　　　　10/2007 / 1169381

1 XIA CLASS (TYPE 092) (SSBN)

Name	No	Builders	Laid down	Launched	Commissioned
XIA	406	Bohai Shipyard, Huludao	1978	30 Apr 1981	1987

Displacement, tonnes: 6,604 dived
Dimensions, metres (feet): 120 × 10 × 8 *(393.7 × 32.8 × 26.2)*
Speed, knots: 22 dived
Complement: 140

Machinery: Nuclear; turbo-electric; 1 PWR; 90 MW; 1 shaft
Missiles: SLBM: 12 JL-1 (CSS-N-3); inertial guidance to 2,150 km *(1,160 n miles)*; warhead single nuclear 250 kT.
Torpedoes: 6—21 in *(533 mm)* bow tubes. Yu-3 (SET-65E); active/passive homing to 15 km *(8.1 n miles)* at 40 kt; warhead 205 kg.
Electronic countermeasures: ESM: Type 921-A; radar warning.
Radars: Surface search: Snoop Tray; I-band.
Sonars: Trout Cheek; hull-mounted; active/passive search and attack; medium frequency.

Programmes: A second of class was reported launched in 1982 and an unconfirmed report suggests that one of the two was lost in an accident in 1985.
Modernisation: Started major update in late 1995 at Huludao, thought to include fitting improved JL-1A missile with increased range but this has not been confirmed.
Structure: Diving depth 300 m *(985 ft)*.
Operational: First test launch of the JL-1 missile took place on 30 April 1982 from a submerged pontoon near Huludao (Yellow Sea). Second launched on 12 October 1982, from the Golf class trials submarine. The first firing from *Xia* was in 1985 and was unsuccessful (delaying final acceptance into service of the submarine) and it was not until 27 September 1988 that a satisfactory launch took place. Based in the North Sea Fleet at Jianggezhuang. Following a refit which completed in late 1998, was reported to be operational as a submarine in 2003 although firing of a JL-1 missile has not been reported and its status as a ballistic-missile submarine is uncertain.

XIA　　　　　　　　　　　　　　　　　　　　　　　　　　2002, Ships of the World / 0529138

IHS Jane's Fighting Ships 2014-2015　　　　　　　　　　　　　　　© 2014 IHS

Attack Submarines

2 + 4 SHANG CLASS (TYPE 093/093A) (SSN)

Name	No	Builders	Laid down	Launched	Commissioned
–	407	Bohai Shipyard, Huludao	1994	24 Dec 2002	Dec 2006
–	408	Bohai Shipyard, Huludao	2000	Dec 2003	Jun 2007
–	409	Bohai Shipyard, Huludao	2009	2012	2014

Displacement, tonnes: 6,096 dived
Dimensions, metres (feet): 107 × 11 × 7.5 *(351 × 36.1 × 24.6)*
Speed, knots: 30 dived
Complement: 100

Machinery: Nuclear: 1 PWR; 150 MW; 2 turbines; 1 shaft
Missiles: SSM: C-801A (YJ-82); radar active homing to 40 km *(22 n miles)* at 0.9 Mach; warhead 165 kg.
Torpedoes: 6–21 in *(533 mm)* bow tubes; combination of Yu-3 (SET-65E); active/passive homing to 15 km *(8.1 n miles)* at 40 kt; warhead 205 kg and Yu-4; active/passive homing to 15 km *(8.1 n miles)* at 30 kt; warhead 309 kg. Yu-6 wake-homing torpedo may also be carried.
Physical countermeasures: Decoys.
Electronic countermeasures: ESM.
Radars: Surface search/navigation: Type 359; I-band.
Sonars: Hull mounted passive/active; flank array. Passive intercept array.

Programmes: Designed in conjunction with Russian experts. The first two boats (Type 093) entered service in 2006 and 2007, and reports suggest that up to four further boats (Type 093A), the first of which was reported launched in 2012, are under construction. These are reported to be a modified design (commercial imagery suggests that Type 093A may be slightly longer than Type 093).
Structure: Performance is likely to be similar to the double-hulled Russian Victor III design.
Operational: Sea trials of the first of class began in 2005 and of the second boat in 2006. Both based at Yalong, Hainan Island.

SHANG CLASS 6/2007 / 1166716

SHANG CLASS 6/2007 / 1166715

130 China > Submarines

3 HAN CLASS (TYPE 091/091G) (SSN)

No	Builders	Laid down	Launched	Commissioned
403	Bohai Shipyard, Huludao	1980	1983	21 Sep 1984
404	Bohai Shipyard, Huludao	1984	1987	Nov 1988
405	Bohai Shipyard, Huludao	1987	8 Apr 1990	Dec 1990

Displacement, tonnes: 4,572 surfaced; 5,639 dived
Dimensions, metres (feet): 96 (403), 101 (404–405) × 10 × 7.4 *(315, 331.4 × 32.8 × 24.3)*
Speed, knots: 12 surfaced; 25 dived
Complement: 75

Machinery: Nuclear; turbo-electric; 1 PWR; 90 MW; 1 shaft
Missiles: SSM: C-801A (YJ-82); inertial cruise; active radar homing to 40 km *(22 n miles)* at 0.9 Mach; warhead 165 kg.
Torpedoes: 6—21 in *(533 mm)* bow tubes; combination of Yu-3 (SET-65E); active/passive homing to 15 km *(8.1 n miles)* at 40 kt; warhead 205 kg and Yu-4; active/passive homing to 15 km *(8.1 n miles)* at 36 kt; warhead 309 kg.
Mines: 36 in lieu of torpedoes.
Electronic countermeasures: ESM: Type 921-A; radar warning.
Radars: Surface search: Snoop Tray; I-band.
Sonars: Trout Cheek; hull-mounted; active/passive search and attack; medium frequency. DUUX-5; passive ranging and intercept; low frequency.

Programmes: First of this class delayed by problems with the power plant. Although completed in 1974 she was not fully operational until the 1980s.
Modernisation: The basic Russian ESM equipment was replaced by a French design. A French intercept sonar set has been fitted.
Structure: From *404* onwards the hull has been extended by some 5 m although this was not to accommodate missile tubes as previously reported. SSMs may be fired from the torpedo tubes. Diving depth 300 m *(985 ft)*.
Operational: All based in North Sea Fleet at Jianggezhuang. *403* and *404* started mid-life refits in 1998 which completed in early 2000. *405* started mid-life refit in 2000 and was reported completed in 2002. Torpedoes are a combination of older straight running and more modern Russian homing types. The first of class *401* was reported to have been decommissioned in 2003 and *402* in 2007.

HAN 404   *5/1996, Ships of the World* / 0506277

HAN CLASS   1990 / 0506276

Patrol Submarines

Notes: (1) An unknown number of midget submarines are reported in service.
(2) It was reported in March 2013 that agreement had been reached to procure four Lada-class submarines from Russia. The building yard(s) and construction timetable have not been confirmed.

16 MING CLASS (TYPE 035) (SS)

305–308   310–313   356–363

Displacement, tonnes: 1,609 surfaced; 2,147 dived
Dimensions, metres (feet): 76 × 7.6 × 5.1 *(249.3 × 24.9 × 16.7)*
Speed, knots: 15 surfaced; 10 snorting; 18 dived
Range, n miles: 8,000 at 8 kt snorting; 330 at 4 kt dived
Complement: 57 (10 officers)

Machinery: Diesel-electric; 2 diesels; 5,200 hp(m) *(3.82 MW)*; 2 shafts
Torpedoes: 8—21 in *(533 mm)* (6 fwd, 2 aft) tubes. Combination of Yu-4 (SAET-50); passive homing to 15 km *(8.1 n miles)* at 30 kt; warhead 309 kg, and Yu-1 (53–51) to 9.2 km *(5 n miles)* at 39 kt or 3.7 km *(2.1 n miles)* at 51 kt; warhead 400 kg; 16 weapons.
Mines: 32 in lieu of torpedoes.
Radars: Surface search: Snoop Tray; I-band.
Sonars: Pike Jaw; hull-mounted; active/passive search and attack; medium frequency DUUX 5; passive ranging and intercept; low frequency

Programmes: First three completed between 1971 and 1979 one of which was scrapped after a fire and another *(232)* has been decommissioned. These were Type ES5C/D. Building resumed at Wuhan Shipyard in 1987 at the rate of one per year to a modified design ES5E. The programme was thought to have ended with hull number 14 *(363)* launched in May 1996, but *305* was launched in June 1997 followed by *306* in September 1997, *307* in May 1998, *308* in October 1998, *310* in June 2000, *311* in September 2000, *312* in May 2001 and *313* in April 2002.
Structure: Diving depth, 300 m *(985 ft)*. Only the later models have the DUUX 5 sonar. Hull 20 is reported to have a 2 m extension to its machinery space. Early units have a small 'step' in the forward fin.
Operational: Basing: South *(305–308, 310–313)*; North *(356–363)*. Fitted with Magnavox SATNAV. All on board *361* (70 officers and men) killed in an accident in April 2003. The cause of the accident is believed to have been carbon monoxide poisoning. After repairs at Dalian, the submarine became operational again in 2004. Older units are being decommissioned as the Yuan class become operational. Two are to be sold to Bangladesh.

MING CLASS   6/2012* / 1525581

IHS Jane's Fighting Ships 2014-2015   © 2014 IHS

13 SONG CLASS (TYPE 039/039G) (SSG)

No	Builders	Laid down	Launched	Commissioned
320	Wuhan Shipyard	1991	25 May 1994	Jun 1999
321	Wuhan Shipyard	1995	11 Nov 1999	Apr 2001
322	Wuhan Shipyard	1996	28 Jun 2000	Dec 2001
323	Wuhan Shipyard	1998	May 2002	Nov 2003
324	Wuhan Shipyard	1999	28 Nov 2002	Dec 2003
325	Wuhan Shipyard	2001	3 Dec 2002	2004
314	Wuhan Shipyard	2001	19 May 2003	2004
315	Wuhan Shipyard	2002	29 Sep 2003	2004
316	Wuhan Shipyard	2002	28 Aug 2004	2005
326	Wuhan Shipyard	2002	Jul 2004	2005
328	Jiangnan Shipyard, Shanghai	2002	Aug 2004	2005
327	Wuhan Shipyard	2003	Sep 2004	2006
329	Jiangnan Shipyard, Shanghai	2003	Nov 2004	2006

Displacement, tonnes: 1,727 surfaced; 2,286 dived
Dimensions, metres (feet): 74.9 × 7.5 × 5.3 *(245.7 × 24.6 × 17.4)*
Speed, knots: 15 surfaced; 22 dived
Complement: 60 (10 officers)

Machinery: Diesel-electric; 4 MTU 16V 396 SE; 6,092 hp(m) *(4.48 MW)* diesels; 4 alternators; 1 motor; 1 shaft
Missiles: SSM: C-801A (YJ-82); radar active homing to 40 km *(22 n miles)* at 0.9 Mach; warhead 165 kg.
Torpedoes: 6—21 in *(533 mm)* tubes. Combination of Yu-4 (SAET-50); passive homing to 15 km *(8.1 n miles)* at 30 kt; warhead 309 kg and Yu-3 (SET-65E); active/passive homing to 15 km *(8.1 n miles)* at 40 kt; warhead 205 kg. Yu-6 wake-homing torpedoes may also be fitted.
Mines: In lieu of torpedoes.
Electronic countermeasures: ESM: Type 921-A; radar warning.
Radars: Surface search: I-band.
Sonars: Bow-mounted; passive/active search and attack; medium frequency. Flank array; passive search; low frequency.

Programmes: First of class (Type 039) started sea trials in August 1995, as a result of which substantial modifications were made. Second of class (Type 039G) trials started in early 2000 and third in early 2001. Fourth commissioned in 2003 while fifth and sixth conducted trials in late 2003. Construction of the seventh hull is understood to have started in 2001 and of the eighth, ninth and tenth hulls in 2002. The twelfth hull is reported to have started construction at Wuhan in 2003. The building programme appears to have been switched to Jiangnan Shipyard, Shanghai, where the eleventh and thirteenth boats were built. Further units of the class are not expected.
Structure: Comparable in size to Ming class but with a single skew propeller and an integrated spherical bow sonar. The forward hydroplanes are mounted below the bridge, which is on a step lower than the part of the fin that contains the masts in earlier boats. The fin is of a different shape (no cutaway) in later boats. Some of the details are speculative and the latest hulls of the class may have benefited from experience gained with the Kilos. The diesel engines are likely to be reverse engineered. Sonars are reported to be of French design.
Operational: Basing: North *(320, 321, 322, 323, 327, 328)*; East *(314, 324, 325)*; South *(315, 316, 326, 329)*.

SONG CLASS   4/2004, Ships of the World / 1042142

SONG CLASS   1/2007, Ships of the World / 1166772

SONG CLASS 315 and 316   6/2005, Hachiro Nakai / 1153050

SONG CLASS   6/2012* / 1525582

Auxiliary Submarines

1 TYPE 032 (QING) CLASS (SSA)

No	Builders	Laid down	Launched	Commissioned
–	Wuhan Shipyard	2008	9 Sep 2010	2011

Displacement, tonnes: 5,000 surfaced (est.)
Dimensions, metres (feet): 92 × 10.6 × 7.0 *(301.8 × 34.8 × 23.0)*
Machinery: Diesel-electric.
Missiles: To be announced.
Radars: To be announced.
Sonars: To be announced.

Comment: The first and possibly only unit of this new class of submarine was launched at Wuhan in September 2010. Approximately one-third larger than the Yuan class, its design includes an unusually large 22 m long sail. It is believed to be conventionally powered but it is unclear whether the boat is equipped with Air-Independent Propulsion (AIP). A mid-May 2011 image of the submarine in a dry dock indicates that there is an extension at the base of the hull under the sail similar to that seen in early Soviet ballistic missile-carrying submarines. Such an extension is required in order to maintain the stability of the boat if a ballistic missile is fitted. This suggests that the principal role of the boat may be to replace the long-serving Golf-class SSB, built in the 1960s, which has been serving as a trials platform for the JL-2 and/or projected JL-3 missile. There is speculation that the boat may also be used to test the DF-21D anti-ship ballistic missile. In addition, the submarine could be used for other research roles. Based at Xiapingdao.

QING CLASS   5/2011 / 1406377

QING CLASS   5/2011 / 1406378   QING CLASS   5/2011 / 1406379

1 GOLF CLASS (TYPE 031) (SSB)

200

Displacement, tonnes: 2,388 surfaced; 2,997 dived
Dimensions, metres (feet): 97.5 × 8.6 × 6.6 *(319.9 × 28.2 × 21.7)*
Speed, knots: 17 surfaced; 13 dived
Range, n miles: 6,000 at 15 kt surfaced
Complement: 86 (12 officers)
Machinery: Diesel-electric; 3 Type 37-D diesels; 6,000 hp(m) *(4.41 MW)*; 3 motors; 5,500 hp(m) *(4 MW)*; 3 shafts
Missiles: SLBM: 1 JL-2 (CSS-NX-5); 3-stage solid fuel; stellar inertial guidance to 8,000 km *(4,320 n miles)*; single nuclear warhead of 1 MT or 3-8 MIRV of smaller yield. CEP 300 m approx.
Torpedoes: 10–21 in *(533 mm)* tubes (6 bow, 4 stern). 12 Type Yu-4 (SAET-50); passive homing to 15 km *(8.1 n miles)* at 30 kt; warhead 309 kg.
Radars: Navigation: Snoop Plate; I-band.
Sonars: Pike Jaw; hull-mounted; active/passive search; medium frequency.

Programmes: Ballistic missile submarine similar but not identical to the deleted USSR Golf class. Built at Dalian and launched in September 1966.
Modernisation: Refitted in 1995 to take the JL-2 missile.
Operational: This was the trials submarine for the JL-1 ballistic missile which was successfully launched to 1,800 km in October 1982. Continues to be available as a trials platform for the successor missile JL-2 and probably conducted a test firing on 12 June 2005. Further testings were reported in 2011 but it is possible that the boat is to be replaced by the Qing class. Based in the North Sea Fleet.

GOLF 200   2002, Ships of the World / 0529137

Aircraft carriers < China

AIRCRAFT CARRIERS

Notes: (1) The former Russian aircraft carrier *Minsk* is a tourist attraction at Shenzhen.
(2) Building of an indigenous aircraft carrier (possibly Type 089) is expected to start by 2018 with a view to entering service in about 2025.

1 KUZNETSOV (OREL) (PROJECT 1143.5/6) CLASS (CVGM)

Name	No	Builders	Laid down	Launched	Commissioned
LIAONING (ex-*Varyag*; ex-*Riga*)	16	Nikolayev South	6 Dec 1985	6 Dec 1988	25 Sep 2012

Displacement, tonnes: 46,637 standard; 59,439 full load
Dimensions, metres (feet): 304.5 oa; 280 wl × 70 oa; 37 wl × 10.5 *(999; 918.6 × 229.7; 121.4 × 34.4)*
Flight deck, metres: 304.5 × 70 *(999 × 229.7)*
Aircraft lift: 2
Speed, knots: 30
Range, n miles: 3,850 at 29 kt, 8,500 at 18 kt
Complement: 1,960 (200 officers; 626 air crew) + 40 flag staff

Machinery: 8 boilers; 4 turbines; 200,000 hp(m) *(147 MW)*; 4 shafts
Missiles: SAM: 4 FL-3000N 24-cell launchers ❶; passive IR/anti-radiation homing to 9.0 km *(4.8 n miles)*.
Guns: 2—30 mm Type 1030 ❷; 10 barrels per mounting; 4,200 rds/min combined to 1.5 km.
A/S Mortars: 2 RBU 6000 ❸.
Physical countermeasures: Decoys: 4 Type 726 chaff launchers ❹.
Electronic countermeasures: ESM/ECM: To be announced.
Radars: Air search: Type 381B Top Plate (Fregate MAE-3) ❺; 3D; E/F-band.
Air search/fire control: Type 346 (Dragon Eye) ❻; 3D; G-band.
Surface search: To be announced.
Fire control: 2 Band Stand (Mineral ME) ❼; I-band.
Navigation: To be announced.
CCA: To be announced.
Tacan: To be announced.
IFF: To be announced.
Sonars: To be announced.
Weapon control systems: To be announced.

Fixed-wing aircraft: 18 J-15.
Helicopters: To be announced.

Programmes: Procurement of an aircraft carrier capability has been a high priority for the Chinese Navy since the 1990s. Ex-*Varyag*, the second of the Kuznetsov class (the first of class, *Admiral Kuznetsov*, remains in service in the Russian Navy) was between 70 and 80% complete by early 1993 when building was terminated after an unsuccessful attempt by the Russian Navy to fund completion. Subsequently the ship was bought by China and, having been towed through the Bosporus on 2 November 2001, arrived at Dalian in March 2002. Once the ship emerged from dock in mid-2005 painted in military colours, it became clear that it was intended to bring the ship into operational service. Work in 2006 included the application of a non-skid surface to the flight deck and, by mid-2008, the exterior of the ship was looking relatively shipshape. The ship was moved to a dry-dock on 27 April 2009 for a refit during which shafts and/or propellers are likely to have been fitted. Other work included substantial rebuild of the upper island structure which has involved installation of the Dragon Eye radar. The work was completed in mid-March 2010 when the ship was moved out of dock to a degaussing berth. Meanwhile a full-scale replica of the flight deck at Wuhan became available for flight-deck training.

Structure: The hangar is 183 × 29.4 × 7.5 m and can hold up to 18 Flanker aircraft. There are two starboard side lifts, a ski jump of 14° and angled deck of 7°. There are four arrester wires. The ship has some 16.5 m of freeboard. The FL-3000N launchers are sited on sponsons fore and aft and the Type 730 CIWS are on the port and starboard quarters.

Operational: Initial sea trials started on 10 August 2011 and the ship returned to Dalian after four days. A further docking period followed during which the flight-deck was stripped and repainted. It is not known whether there were other problems that required a docking. Between November 2011 and August 2012, nine sets of sea trials, culminating in a 25-day period in July 2012, were carried out. Subsequently, the ship was handed over to the PLAN on 23 September 2012 and formally commissioned on 25 September 2012. While the ship is now capable of limited operations, its initial role is likely to be as an aviation training ship with a view to working the ship up to a full aircraft carrier capability in due course. This is expected to take several years. A significant step in this process was the successful recovery of a J-15 aircraft on 20 November 2012. The ship arrived at her base at Dazhu Shan (50 km south of Qingdao) on 27 February 2013.

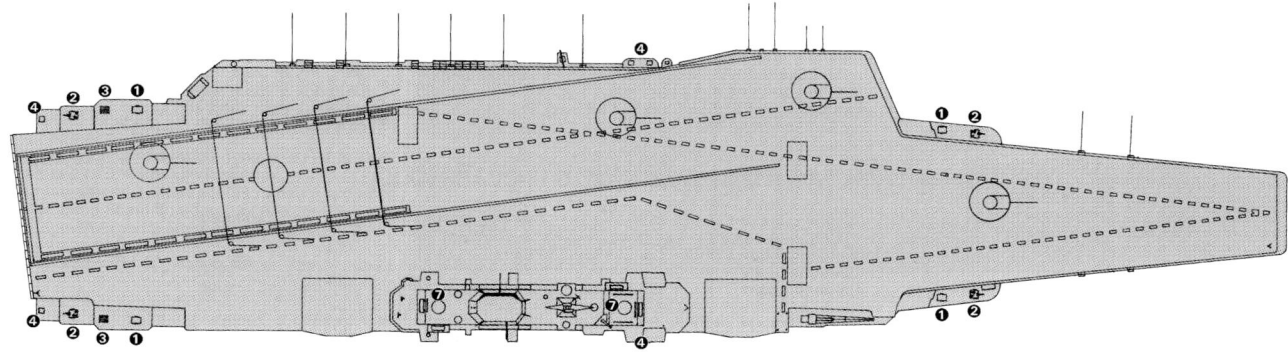

LIAONING (Scale 1 : 1,800), Ian Sturton / 1483595

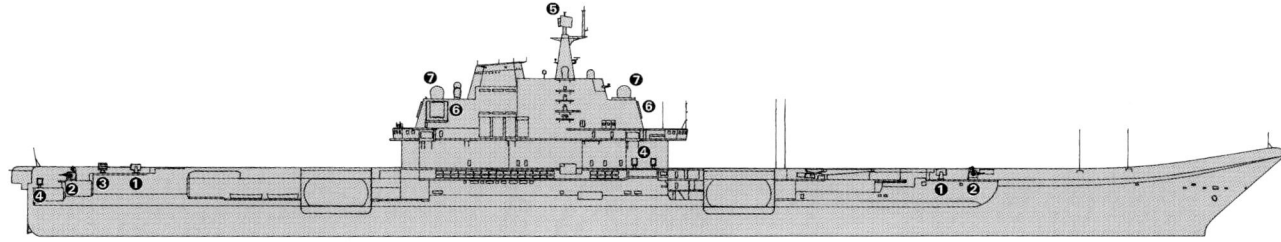

LIAONING (Scale 1 : 1,800), Ian Sturton / 1483593

LIAONING 3/2013*, *Ships of the World* / 1525662

2 LUYANG I (TYPE 052B) CLASS (DDGHM)

Name	No	Builders	Laid down	Launched	Commissioned
GUANGZHOU	168	Jiangnan Shipyard, Shanghai	2001	25 May 2002	18 Jul 2004
WUHAN	169	Jiangnan Shipyard, Shanghai	2001	9 Sep 2002	18 Jul 2004

Displacement, tonnes: 7,112 full load
Dimensions, metres (feet): 155 × 17 × 6 (508.5 × 55.8 × 19.7)
Speed, knots: 29
Range, n miles: 4,500 at 15 kt
Complement: 280 (40 officers)

Machinery: 2 Ukraine DA80 gas turbines; 48,600 hp(m) *(35.7 MW)*; 2 MTU 1163 TB 83 diesels; 13,400 hp(m) *(10.0 MW)*; 2 shafts; cp props
Missiles: SSM: 16 C-802 (YJ-83/C SS-N-8 Saccade) 4 quad ❶; mid-course guidance and active radar homing to 120 km *(65 n miles)* at 0.9 Mach; warhead 165 kg; sea skimmer.
SAM: SA-N-12 Grizzly (Shtil-1) 9M38M2 ❷; command/semi-active radar and IR homing to 35 km *(18.9 n miles)* at 3 Mach; warhead 70 kg; 2 magazines (forward and aft). 48 missiles.
Guns: 1—3.9 in *(100 mm)*/56 ❸; 25 rds/min to 22 km *(12 n miles)*; weight of shell 15.6 kg. 2—30 mm Type 730A ❹; 7 barrels per mounting; 4,200 rds/min combined to 1.5 km.
Torpedoes: 6—324 mm B 515 (2 triple) tubes ❺; Yu-7; active/passive homing to 14 km *(7.6 n miles)* at 42 kt; warhead 45 kg.
A/S Mortars: 4 multiple rocket launchers (possibly multirole) ❻.
Physical countermeasures: Decoys: 4—18 tube 100 mm launchers ❼.
Electronic countermeasures: ESM: SRW 210A.
ECM: Type 984 (I-band jammer). Type 985 (E/F-band jammer).
Radars: Air search: Top Plate (Fregat MAE-3); 3D ❾; E/F-band.
Air/surface search: Type 364 Seagull C ❿; G-band.
Fire control: 4 Front Dome (Orekh) ⓫; H/I-band (for SA-N-12). Band Stand (Mineral ME) ❽; I-band (for C-802). Type 344 (MR 34) ⓬; I-band (for 100 mm). 2 Type 347G(2) (LR 66); I-band (for Type 730).
Navigation: To be announced.
Sonars: Bow mounted. To be announced.
Combat data systems: To be announced. SATCOM.
Weapon control systems: Band Stand (Mineral ME) ❽; I-band; datalink (for C-803).
Helicopters: 1 Harbin Zhi-9A Haitun or Kamov KA-28 Helix ⓭.

Programmes: Construction of new multirole destroyers with medium-range air defence capability started in 2001.
Structure: Based on 'Luhai' design but with more advanced stealth features. The aft superstructure contains the hangar on the port side and aft missile magazine to starboard.
Operational: Based in the South Sea Fleet.

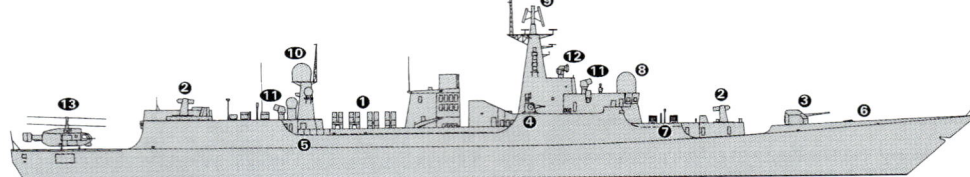
GUANGZHOU (Scale 1 : 1,200), Ian Sturton / 1170050

WUHAN 6/2012, Ships of the World / 1483601

GUANGZHOU 9/2007, R G Sharpe / 1166778

GUANGZHOU 8/2009, Michael Nitz / 1305808

Destroyers < **China** 139

4 + 2 LUYANG II (TYPE 052C) CLASS (DDGHM)

Name	No	Builders	Laid down	Launched	Commissioned
LANZHOU	170	Jiangnan Shipyard, Shanghai	Jun 2002	29 Apr 2003	18 Jul 2004
HAIKOU	171	Jiangnan Shipyard, Shanghai	Nov 2002	29 Oct 2003	20 Jul 2005
CHANGCHUN	150	Jiangnan Shipyard, Shanghai	2009	28 Nov 2010	31 Jan 2013
ZHENGZHOU	151	Jiangnan Shipyard, Shanghai	2009	20 Jul 2011	26 Dec 2013
JINAN	152	Jiangnan Shipyard, Shanghai	2011	16 Oct 2011	2014
XIAN	153	Jiangnan Shipyard, Shanghai	2011	28 May 2012	2014

Displacement, tonnes: 7,112 full load
Dimensions, metres (feet): 155 × 17 × 6 (508.5 × 55.8 × 19.7)
Speed, knots: 29
Range, n miles: 4,500 at 15 kt
Complement: 280 (40 officers)

Machinery: 2 Ukraine DA80 (QC 280 in 150–153) gas turbines; 48,600 hp(m) (35.7 MW); 2 MTU 1163 TB 83 diesels; 13,400 hp(m) (10.0 MW); 2 shafts; cp props
Missiles: SSM: 8 C-602 (YJ-62) ❶ 2 quad; inertial-GPS guidance and terminal active radar homing to 280 km (151 n miles) at 0.8 Mach; warhead 300 kg.
SAM: HHQ-9 ❷; 8 vertical fixed sextuple launchers (6 forward, 2 aft); command guidance; semi-active radar homing to 100 km (54 n miles) at 3 Mach; warhead 90 kg; 48 missiles.
Guns: 1–3.9 in (100 mm)/56 ❸; 25 rds/min to 22 km (12 n miles); weight of shell 15.6 kg. 2–30 mm Type 730 ❹; 7 barrels per mounting; 4,200 rds/min combined to 1.5 km.
Torpedoes: 6–324 mm B 515 (2 triple) tubes ❺; Yu-7; active/passive homing to 14 km (7.6 n miles) at 42 kt; warhead 45 kg.
A/S Mortars: 4 multiple rocket launchers (possibly multirole) ❻.
Electronic countermeasures: ESM/ECM; NRJ-6A.
Radars: Air search: Type 517B Knife Rest ❽; A-band.
Air search/fire control: Type 346 Dragon Eye phased arrays ❾; 3D; G-band.
Air/surface search: Type 364 Seagull C ❿; G-band.

Fire control: Type 344 (MR 34) ⓫; I-band (for 100 mm). Band Stand ❼; I-band (for YJ-62). 2 Type 347G(2) (LR 66); I-band (for Type 730).
Navigation: To be announced.
Sonars: Bow mounted. To be announced.
Combat data systems: ZBJ-1.
Weapon control systems: Band Stand (Mineral ME) ❼; I-band; datalink for YJ-62.
Electro-optic systems: OFC-3.

Helicopters: 2 Harbin Zhi-9A Haitun or Kamov KA-28 Helix ⓬.

Programmes: The first two ships comprised second phase of the destroyer construction programme which introduced the long-range HHQ-9 missile system into service. A third ship, reported to be to a modified design, was launched on 28 November 2010. She started sea trials on 16 October 2011 and three further ships are under construction.
Structure: Appears to share the same basic hull design as the Type 052B destroyers which in turn are based on the Luhai class. As well as incorporating stealth features, the design includes a taller forward superstructure in which the four phased array antennas are installed. The helicopter hangar is on the port side of the aft superstructure. The CIWS systems are on raised platforms forward and on top of the hangar. The main machinery in the third and subsequent ships includes a Chinese-licensed variant of DA80 gas turbines (QC 280).
Operational: Basing: 170 and 171 in South Sea Fleet. 150 and 151 in East Sea Fleet.

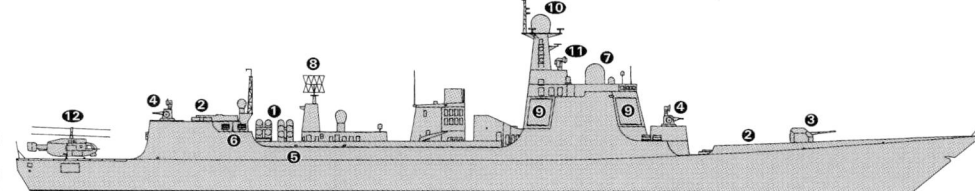

LANZHOU (Scale 1 : 1,200), Ian Sturton / 1170051

LANZHOU 6/2013*, Ships of the World / 1525665

CHANGCHUN 6/2013*, Ships of the World / 1525663

© 2014 IHS IHS Jane's Fighting Ships 2014-2015

140 China > Destroyers

0 + 8 (4) LUYANG III (TYPE 052D) CLASS (DDGHM)

Name	No	Builders	Laid down	Launched	Commissioned
KUNMING	172	Jiangnan Shipyard, Shanghai	2010	29 Aug 2012	2014
CHANGSA	173	Jiangnan Shipyard, Shanghai	2010	28 Dec 2012	2014
GUIYANG	175	Jiangnan Shipyard, Shanghai	2011	1 Jul 2013	2014
CHENGDU	176	Jiangnan Shipyard, Shanghai	2012	21 Nov 2013	2015
HEFEI	154	Jiangnan Shipyard, Shanghai	2013	2014	2015
NANJING	155	Jiangnan Shipyard, Shanghai	2013	2014	2016
YINCHUAN	117	Dalian Shipyard	2013	2014	2016
TAIYUAN	118	Jiangnan Shipyard, Shanghai	2013	2015	2017

Displacement, tonnes: 7,500 full load
Dimensions, metres (feet): 157 × 17.0 × 6.0 *(515.1 × 55.8 × 19.7)*
Speed, knots: 30. **Range, n miles:** 4,500 at 15 kt
Complement: 280 (40 officers)

Machinery: CODOG: 2 China QC 280 gas turbines; 56,000 hp(m) *(41.8 MW)*; 2 diesels; 8,840 hp(m) *(6.5 MW)*; 2 shafts; cp props
Missiles: SSM: To be announced (VLS launched) 18-cell launcher.
SAM: HHQ-9B; 64 VLS cells (32 forward, 32 aft) ❶; command guidance; semi-active radar homing to 100 km *(54 n miles)*; warhead 90 kg. 1 FL-3000N 8-cell launcher ❸; passive IR/anti-radiation homing to 9 km *(4.8 n miles)*.
Guns: 1 H/PJ38 130 mm ❷; 60–80 rds/min to 17 km *(9.3 n miles)*; weight of shell 13.5 kg. 1 Type 730 30 mm ❹; 7 barrels per mounting; 4,200 rds/min combined to 1.5 km.
Torpedoes: 6–324 mm (2 triple); Yu-7; active/passive homing to 14 km *(7.6 n miles)* at 42 kt; warhead 45 kg.
Physical countermeasures: To be announced.
Electronic countermeasures: ESM/ECM: NRJ-6A.
Radars: Air search: Type 517B Knife Rest ❺; A-band.
Air search/fire control: Type 346 Dragon Eye (modified) phased arrays ❻; 3D; E/F-band.
Surface search: Type 364 Seagull C ❼; G-band.
Fire control: Type 344 (MR 34) ❽; I-band (for 130 mm). Band Stand (Mineral ME) ❾; I-band (for SSM).
Navigation: 2 Type 760 series; I-band.
Sonars: Bow mounted: to be announced. Towed array (to be confirmed).
Combat data systems: To be announced, SATCOM.
Weapon control systems: Datalink (Band Stand?) for YJ-83.

Helicopters: 2 Harbin Zhi-9A Haitun or Kamov KA-28 Helix ❿.

Programmes: The first two ships of a new class of destroyers were observed under construction at Jiangnan Shipyard, Changxingdao, in 2012. While further ships are expected, it is likely that both ships will undergo exhaustive testing and trials before construction of follow-on ships is initiated. A class of 12 is expected in due course.
Structure: The ships appear to be a development of the Type 052C design destroyers. The principal features of the new class include a development of the Dragon Eye phased array radar incorporated in the forward superstructure and a new vertical launching system (VLS) with rectangular cells

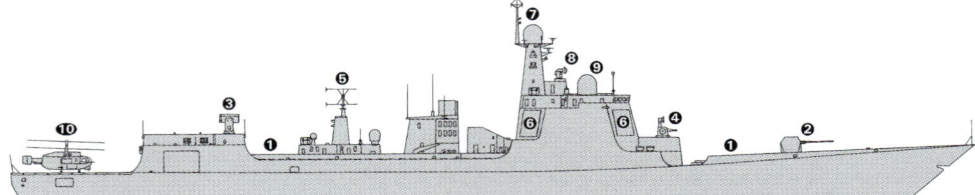

LUYANG III *(Scale : 1, 200), Ian Sturton / 1525661*

LUYANG III *12/2013*, Ships of the World / 1529102*

capable of housing air-to-surface, surface-to-surface, and anti-submarine missiles. Reports suggest that there is a total of 64 rectangular cells (32 forward, 32 aft). Armament forward includes a single-barrelled 130 mm gun and a Type 730 CIWS. Aft there is a flight deck and centre-line hangar on which there is a short-range missile system (FL-3000). There is an open-deck area starboard side aft (below the flight deck) which suggests there may be a towed array.
Operational: The first four are likely to be based in the South Sea Fleet.

1 LUHAI CLASS (TYPE 051B) (DDGHM)

Name	No	Builders	Laid down	Launched	Commissioned
SHENZHEN	167	Dalian Shipyard	Jul 1996	16 Oct 1997	4 Jan 1999

Displacement, tonnes: 6,096 full load
Dimensions, metres (feet): 154 × 16 × 6 *(505.2 × 52.5 × 19.7)*
Speed, knots: 29
Range, n miles: 4,500 at 14 kt
Complement: 250 (42 officers)

Machinery: 2 Ukraine gas turbines; 48,600 hp(m) *(35.7 MW)*; 2 MTU 12V 1163 TB 83 diesels; 8,840 hp(m) *(6.5 MW)* sustained; 2 shafts; cp props
Missiles: SSM: 16 C-802 (YJ-83/CSS-N-8 Saccade) ❶; mid-course guidance and active radar homing to 120 km *(65 n miles)* at 0.9 Mach; warhead 165 kg; sea skimmer.
SAM: 1 HQ-7 (Crotale) CSA-N-4 octuple launcher ❷; line of sight guidance to 13 km *(7 n miles)* at 2.4 Mach; warhead 14 kg. Possible reloading hatch aft of the HQ-7 launcher.
Guns: 2–3.9 in *(100 mm)*/56 (twin) ❸; 25 rds/min to 22 km *(12 n miles)*; weight of shell 15.6 kg. 8–37 mm/63 Type 76A (4 twin) ❹; 180 rds/min to 8.5 km *(4.6 n miles)* anti-aircraft; weight of shell 1.42 kg.
Torpedoes: 6–324 mm B515 (2 triple) tubes ❺ Yu-7; active/passive homing to 14 km *(7.6 n miles)* at 42 kt; warhead 45 kg.
Physical countermeasures: Decoys: 2 Type 946 15-tube 100 mm chaff launchers ❻. 2 Type 947 10-tube 130 mm chaff launchers.
Electronic countermeasures: ESM: Type 826.
ECM: Type 984; I-band jammer; Type 985; E/F-band jammer.
Radars: Air search: Type 517A Knife Rest ❼; A-band.
Air search: Type 381A Rice Shield ❽; E/F-band.
Air/surface search: Type 360 Seagull S ❾; E/F-band.
Fire control: Type 344 (MR 34) ❿; I-band (for SSM and 100 mm). 2 Type 347G(1) Rice Bowl ⓫; I-band (for 37 mm). Type 345 ⓬; I/J-band (for HQ-7).
Navigation: Racal/Decca 1290; I-band.
Sonars: DUBV-23; hull mounted; active search and attack; medium frequency.
Combat data systems: Thomson-CSF Tavitac; SATCOM.
Weapon control systems: 2 GDG 776 optronic directors.

Helicopters: 2 Harbin Zhi-9C Haitun ⓭ or Kamov Ka-28 Helix.

Programmes: Follow-on from the Luhu class. Although the only ship of its class, it would appear to be the baseline design for the Type 051C destroyers.
Structure: Apart from the second funnel and octuple SSM launchers, there are broad similarities with the smaller Luhu. Anti-aircraft guns are all mounted aft allowing more space in front of the bridge which seems to show a reloading hatch for HQ-7.
Operational: Based at Zhanjiang in South Sea Fleet.

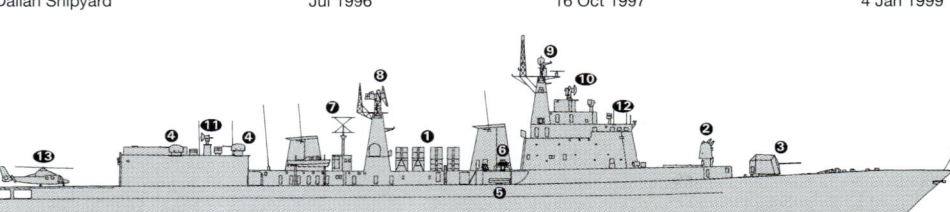

SHENZHEN *(Scale 1 : 1,200), Ian Sturton / 0569249*

SHENZHEN *12/2007, Hachiro Nakai / 1166774*

IHS Jane's Fighting Ships 2014-2015

Destroyers ‹ China   141

2 LUHU (TYPE 052A) CLASS (DDGHM)

Name	No	Builders	Laid down	Launched	Commissioned
HARBIN	112	Jiangnan Shipyard, Shanghai	Nov 1990	Oct 1991	Jul 1994
QINGDAO	113	Jiangnan Shipyard, Shanghai	Jan 1993	Oct 1993	Mar 1996

Displacement, tonnes: 4,674 full load
Dimensions, metres (feet): 144 × 16 × 5.1 *(472.4 × 52.5 × 16.7)*
Speed, knots: 31. **Range, n miles:** 5,000 at 15 kt
Complement: 266 (38 officers)

Machinery: 2 GE LM 2500 gas turbines (112); 55,000 hp *(41 MW)* sustained or 2 Ukraine gas turbines (113) 48,600 hp(m) *(35.7 MW)*; 2 MTU 12V 1163 TB83 diesels; 8,840 hp(m) *(6.5 MW)* sustained; 2 shafts; cp props
Missiles: SSM: 16 C-802 (YJ-83/CSS-N-8 Saccade) ❶; mid-course guidance and active radar homing to 120 km *(65 n miles)* at 0.9 Mach; warhead 165 kg; sea-skimmer.
SAM: 1 HQ-7 (Crotale) CSA-N-4 octuple launcher ❷; line of sight guidance to 13 km *(7 n miles)* at 2.4 Mach; warhead 14 kg. 32 missiles.
Guns: 2 – 3.9 in *(100 mm)*/56 (twin) ❸; 25 rds/min to 22 km *(12 n miles)*; weight of shell 15.6 kg. 2 – 30 mm Type 730 ❹; 7 barrels per mounting; 4,200 rds/min combined to 1.5 km; 180 rds/min to 8.5 km *(4.6 n miles)* anti-aircraft; weight of shell 1.42 kg.
Torpedoes: 6 – 324 mm Whitehead B515 (2 triple) tubes ❺. Yu-7; active/passive homing to 14 km *(7.6 n miles)* at 42 kt; warhead 45 kg.
A/S Mortars: 2 FQF 2500 ❻ 12-tubed fixed launchers; range 1,200 m; warhead 34 kg. 120 rockets.
Physical countermeasures: Decoys: 2 Type 726 ⓫; 24 barrelled 100 mm chaff launchers.
Electronic countermeasures: ESM: Rapids.
ECM: Scimitar.
Radars: Air search: Type 517B Knife Rest; A-band ❼.
Air/surface search: Type 363S Sea Tiger ❽; E/F-band.
Surface search: Type 364 Seagull C ❾; G-band.
Fire control: Type 344 (MR 34) ❿; I-band (for SSM and 100 mm). Type 345 ⓬; I/J-band (for HQ-7).
Navigation: Racal Decca 1290; I-band.
Sonars: SJD-8/9 (DUBV 23); hull-mounted; active search and attack; medium frequency. Towed array.
Combat data systems: Thomson-CSF Tavitac action data automation. SATCOM. Link W.
Electro-optic systems: 2 GDG-775 optronic directors ⓭. OFC-3.
Helicopters: 2 Harbin Zhi-9C Haitun ⓮.

Programmes: Class of two ordered in 1985 but delayed by priority being given to export orders for Thailand.
Modernisation: Harbin completed refit in early 2003. Qingdao completed similar refit in 2005. Both fitted with a new low radar profile 100 mm gun turret. Harbin began a further major refit programme in 2010 and Qingdao in 2011. The scope of work has included replacement of the gas turbines, replacement of Type 518 search radar with Type 517 replacement of the Type 362 radar with Type 364, replacement of the four twin 37 mm guns with two Type 730 CIWS and replacement of the DUBV-43 VDS with a linear towed array of unknown type.
Structure: The most notable features are the SAM launcher, improved radar and fire-control systems and a modern 100 mm gun. The HQ-7 launcher is a Chinese copy of Crotale. DCN Samahe 110N helo handling system. The quarterdeck was completely enclosed during the 2010–12 refits.
Operational: Both based in North Sea Fleet at Dazhu Shan. Qingdao made port visits in the Black Sea during 2012.

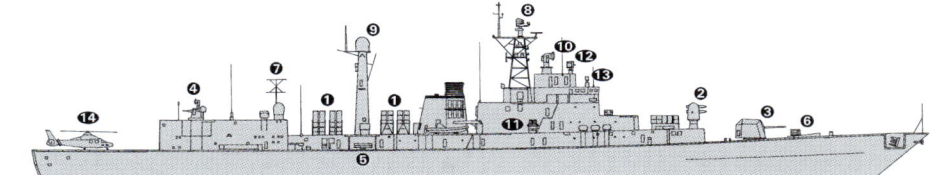

QINGAO (Scale 1 : 1,200), Ian Sturton / 1406510

QINGDAO 10/2013*, John Mortimer / 1525584

QINGDAO 7/2012, Frank Bottema / 1483006

HARBIN 11/2011 / 1406889

© 2014 IHS   IHS Jane's Fighting Ships 2014-2015

FRIGATES

2 JIANGKAI I (TYPE 054) CLASS (FFGHM)

Name	No	Builders	Laid down	Launched	Commissioned
MAANSHAN	525	Hudong-Zhonghua Shipyard, Shanghai	Dec 2001	11 Sep 2003	18 Feb 2005
WENZHOU	526	Huangpu Shipyard, Guangzhou	Feb 2002	30 Nov 2003	26 Sep 2006

Displacement, tonnes: 3,556 standard; 3,963 full load
Dimensions, metres (feet): 132 × 15 × 5.0 *(433.1 × 49.2 × 16.4)*
Speed, knots: 27. **Range, n miles:** 3,800 at 18 kt
Complement: 190

Machinery: CODAD; 4 SEMT-Pielstick diesels; 2 shafts
Missiles: SSM: 8 C-802 (YJ-83/CSS-N-8 Saccade) ❶; mid-course guidance and active radar homing to 120 km *(65 n miles)* at 0.9 Mach; warhead 165 kg; sea skimmer.
SAM: 1 HQ-7 (Crotale) CSA-N-4 ❷; line-of-sight guidance to 13 km *(7 n miles)* at 2.4 Mach; warhead 14 kg.
Guns: 1 – 3.9 in *(100 mm)*/56 ❸; 25 rds/min to 22 km *(12 n miles)*; weight of shell 15.6 kg. 4 – 30 mm/65 AK 630 ❹; 6 barrels per mounting; 3,000 rds/min combined to 2 km.
Torpedoes: 6 – 324 mm B515 (2 triple) tubes; Yu-7; active/passive homing to 14 km *(7.6 n miles)* at 42 kt; warhead 45 kg.
A/S Mortars: 1 Type 87.
Physical countermeasures: To be announced.
Radars: Air/surface search: Type 360 Seagull S ❺; E/F-band.
Surface search: Type 364 Seagull C ❻; G-band.
Fire control: Type 344 (MR 34) ❼; I-band (for SSM and 100 mm). Type 345 ❽; I/J-band (for HQ-7). Type 347G(1) Rice Bowl ❾; I-band (for AK 630).
Navigation: RM-1290; I-band.
Sonars: to be announced.
Combat data systems: to be announced.
Helicopters: 1 Harbin Zhi-9C Haitun ❿.

Programmes: Two vessels of a new general-purpose frigate class which followed the Jiangwei II class. Further ships are unlikely.
Structure: A new design incorporating stealth features.
Operational: Based in the East Sea Fleet.

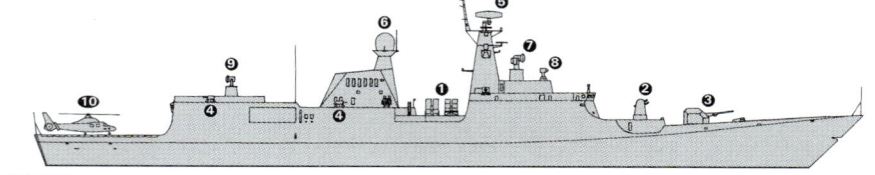

MAANSHAN *(Scale 1 : 1,200), Ian Sturton* / 1164338

WENZHOU *1/2007, Ships of the World* / 1167734

WENZHOU *1/2007, Ships of the World* / 1167733

10 JIANGWEI II (TYPE 053H3) CLASS (FFGHM)

Name	No	Builders	Laid down	Launched	Commissioned
JIAXIN	521 (ex-597)	Hudong Shipyard, Shanghai	Oct 1996	10 Aug 1997	Nov 1998
LIANYUNGANG	522	Hudong Shipyard, Shanghai	Dec 1996	8 Aug 1997	Feb 1999
PUTIAN	523	Hudong Shipyard, Shanghai	Jun 1997	10 Aug 1998	Oct 1999
SANMING	524	Hudong Shipyard, Shanghai	Dec 1997	Dec 1998	Nov 1999
YICHANG	564	Huangpu Shipyard, Guangzhou	Dec 1997	Oct 1998	Dec 1999
HULUDAO (ex-*Sanya*; ex-*Yulin*)	565	Huangpu Shipyard, Guangzhou	May 1998	Apr 1999	Jul 2000
HUAIHUA (ex-*Yuxi*)	566	Hudong Shipyard, Shanghai	May 2000	Jan 2001	Mar 2002
XIANGFAN	567	Huangpu Shipyard, Guangzhou	Mar 2001	Aug 2001	Sep 2002
LUOYANG	527	Hudong Shipyard, Shanghai	2003	1 Oct 2004	Jan 2005
MIANYANG	528	Huangpu Shipyard, Guangzhou	2003	30 May 2004	Oct 2004

Displacement, tonnes: 2,286 full load
Dimensions, metres (feet): 111.7 × 12.4 × 4.8 *(366.5 × 40.7 × 15.7)*
Speed, knots: 27
Range, n miles: 4,000 at 18 kt
Complement: 170

Machinery: 2 Type 18E 390 diesels; 24,000 hp(m) *(17.65 MW)* sustained; 2 shafts
Missiles: SSM: 8 C-802 (YJ-83/CSS-N-8 Saccade) (2 quad) launchers ❶; mid-course guidance and active radar homing to 120 km *(65 n miles)* at 0.9 Mach; warhead 165 kg; sea-skimmer.
SAM: 1 HQ-7 (Crotale) octuple launcher ❷; CSA-N-4 line of sight guidance to 13 km *(7 n miles)* at 2.4 Mach; warhead 14 kg.
Guns: 2 China 3.9 in *(100 mm)*/56 (twin) ❸; 25 rds/min to 22 km *(12 n miles)*; weight of shell 15.6 kg. 8 China 37 mm/63 Type 76A (4 twin) ❹; 180 rds/min to 8.5 km *(4.6 n miles)* anti-aircraft; weight of shell 1.42 kg.
A/S Mortars: 2 RBU 1200 ❺; 5-tubed fixed launchers; range 1,200 m; warhead 34 kg.
Physical countermeasures: Decoys: 2 SRBOC Mk 36 6-barrelled chaff launchers ❻; 2 China 26-barrelled chaff launchers ❼.
Electronic countermeasures: ESM: SR-210; intercept.
ECM: 981-3 noise jammer. RWD-8 deception jammer.
Radars: Air search: Type 517B Knife Rest ❽; A-band.
Air/surface search: Type 360 Seagull S ❾; E/F-band.
Fire control: Type 343G (Wok Won). Type 344 (MR 34) (527, 528) ❿; I-band (for SSM and 100 mm). Type 345 ⓫; I/J-band (for HQ-7). Type 347G(1) Rice Bowl ⓬; I/J-band (for 37 mm).
Navigation: 2 RM-1290; I-band.
Sonars: Echo Type 5; hull-mounted; active search and attack; medium frequency.
Combat data systems: ZKJ 3C. SATCOM.
Weapon control systems: JM-83H optronic director.

Helicopters: 2 Harbin Z-9C (Dauphin) ⓭.

Programmes: Follow-on to the Jiangwei class, building some four years later. The building programme appeared to have been terminated after eight ships but two further ships were subsequently built.
Structure: An improved SAM system, updated fire-control radars and a redistribution of the after anti-aircraft guns are the obvious differences from the original Jiangwei. New Type 99 turret fitted in 522 and to be retro-fitted to the remainder of the class.
Operational: Basing: 527–528 and 564–565 in North Sea Fleet; 521–524 in East Sea Fleet; 566–567 in South Sea Fleet.
Sales: Four modified designs built for Pakistan (Sword class).

JIANGWEI II *(Scale 1 : 900), Ian Sturton* / 1335658

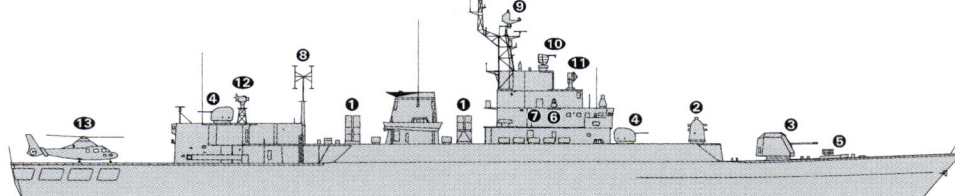

MIANYANG *9/2010, Chris Sattler* / 1366772

LUOYANG *6/2010, Ships of the World* / 1366835

LUOYANG *6/2012, Ships of the World* / 1483598

144 China > Frigates

16 + 4 JIANGKAI II (TYPE 054A) CLASS (FFGHM)

Name	No	Builders	Laid down	Launched	Commissioned
XUZHOU	530	Huangpu Shipyard, Guangzhou	2005	30 Sep 2006	27 Jan 2008
ZHOUSHAN	529	Hudong-Zhonghua Shipyard, Shanghai	2006	21 Dec 2006	3 Jan 2008
HUANGSHAN	570	Huangpu Shipyard, Guangzhou	2006	18 Mar 2007	13 May 2008
HENGYANG	568	Hudong-Zhonghua Shipyard, Shanghai	2006	23 May 2007	30 Jun 2008
YUNCHENG	571	Huangpu Shipyard, Guangzhou	2007	8 Feb 2009	17 Jan 2010
YULIN	569	Hudong-Zhonghua Shipyard, Shanghai	2007	28 Apr 2009	1 Feb 2010
YIYANG	548	Huangpu Shipyard, Guangzhou	2008	17 Nov 2009	26 Oct 2010
CHANGZHOU	549	Hudong-Zhonghua Shipyard, Shanghai	Dec 2009	21 May 2010	30 May 2011
YANTAI	538	Huangpu Shipyard, Guangzhou	2009	24 Aug 2010	9 Jun 2011
YANCHENG	546	Hudong-Zhonghua Shipyard, Shanghai	2010	27 Apr 2011	5 Jun 2012
HENGSHUI	572	Huangpu Shipyard, Guangzhou	2010	21 May 2011	9 Jul 2012
LIUZHOU	573	Hudong-Zhonghua Shipyard, Shanghai	2011	10 Dec 2011	28 Dec 2012
LINYI	547	Huangpu Shipyard, Guangzhou	2011	13 Dec 2011	22 Dec 2012
YUEYANG	575	Huangpu Shipyard, Guangzhou	2011	9 May 2012	3 May 2013
WEIFANG	550	Hudong-Zhonghua Shipyard, Shanghai	2011	9 Jul 2012	22 Jun 2013
SANYA	574	Huangpu Shipyard, Guangzhou	2011	30 Nov 2012	13 Dec 2013
HUANGSHI	576	Hudong-Zhonghua Shipyard, Shanghai	2012	28 Sep 2013	2014
–	–	Hudong-Zhonghua Shipyard, Shanghai	2012	28 Apr 2013	2014
SANMENXIA	–	Hudong-Zhonghua Shipyard, Shanghai	2012	30 Sep 2013	2014
HUANGGANG	–	Huangpu Shipyard, Guangzhou	2013	8 Oct 2013	2015

Displacement, tonnes: 3,556 standard; 3,963 full load
Dimensions, metres (feet): 134 × 16.0 × 5.0 *(439.6 × 52.5 × 16.4)*
Speed, knots: 27
Range, n miles: 3,800 at 18 kt
Machinery: CODAD; 4 SEMT Pielstick 16PA 6V 280 STC; 28,200 hp *(20.7 MW)*; 2 shafts
Missiles: SSM: 8 C-802 (YJ-83/CSS-N-8 Saccade) ❶; mid-course guidance and active radar homing to 120 km *(65 n miles)* at 0.9 Mach; warhead 165 kg; sea skimmer.
SAM: HHQ-16 ❷. 1 (forward) 32 cell vertical launch system (possible cold launch).
Guns: 1 – 3 in *(76 mm)* ❸. 2 – 30 mm Type 730A ❹; 7 barrels per mounting; 4,200 rds/min combined to 1.5 km.
Torpedoes: 6 – 324 mm B515 (2 triple) tubes; Yu-7; active/passive homing to 14 km *(7.6 n miles)* at 42 kt; warhead 45 kg.
A/S Mortars: 2 RBU 1200; 5-tubed fixed launchers; range 1,200 m; warhead 34 kg.
Physical countermeasures: Decoys 2 Type 726 24-barrelled launchers.
Radars: Air search: Top Plate (Fregat MAE-3) ❻; 3D; E/F-band.
Air/surface search: Type 364 Seagull C ❼; G-band.

Fire control: 4 Front Dome (Orekh) ❽; H/I-band for HHQ-16. Band Stand (Mineral ME) ❺; I-band (for YJ-83). Typec 344 (MR 34) ❾; I-band (for 76 mm gun). 2 Type 347G(2) (LR 66) ❿; I-band for Type 730.
Navigation: RM-1290; I-band.
Sonars: To be announced.
Combat data systems: To be announced.
Weapon control systems: Band Stand (Mineral ME) ❺; I-band; datalink for C-803.
Helicopters: 1 Ka 28 PL Helix ⓫.

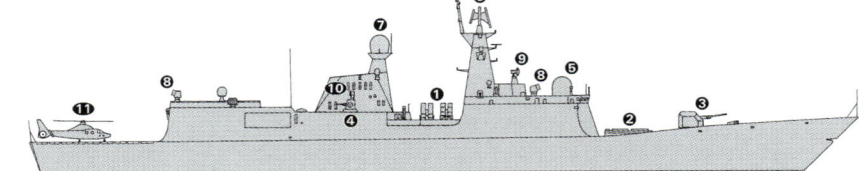

JIANGKAI II *(Scale 1 : 1,200), Ian Sturton / 1335659*

Programmes: Follow-on ships to the two ships of the Jiangkai I class. The modified design includes a VLS launcher for the SAM system. The construction of hull 6 at Hudong Shipyard was briefly delayed by the collapse of a crane in 2008. Twelve units comprise Batch 1 and a class of at least 20 is expected.
Operational: Basing: *Yantai, Yancheng, Linyi,* and *Weifang,* in the North Sea Fleet; *Zhoushan, Xuzhou, Yiyang, Changzhou* in the East; *Huangshan, Hengyang, Yuncheng, Yulin, Hengshui, Liuzhou, Yueyang,* and *Sanya* in the South.

YIYANG *12/2012, Chris Sattler / 1483008*

YULIN *2/2013*, Ships of the World / 1525671*

IHS Jane's Fighting Ships 2014-2015

Frigates < **China** 145

HENGSHUI
6/2013*, Ships of the World / 1525676

YULIN
6/2012, Ships of the World / 1483602

YULIN
6/2013*, Ships of the World / 1525675

CHANGZHOU
12/2012, Chris Sattler / 1483007

© 2014 IHS
IHS Jane's Fighting Ships 2014-2015

146 China > Frigates

4 JIANGWEI I (TYPE 053 H2G) CLASS (FFGHM)

Name	No	Builders	Laid down	Launched	Commissioned
ANQING	539	Hudong Shipyard, Shanghai	Nov 1990	Jul 1991	Dec 1991
HUAINAN	540	Hudong Shipyard, Shanghai	Jan 1991	Oct 1991	Jul 1992
HUAIBEI	541	Hudong Shipyard, Shanghai	Jul 1992	Apr 1993	Aug 1993
TONGLING	542	Hudong Shipyard, Shanghai	Dec 1992	Sep 1993	Apr 1994

Displacement, tonnes: 2,286 full load
Dimensions, metres (feet): 111.7 × 12.4 × 4.8 (366.5 × 40.7 × 15.7)
Speed, knots: 27. **Range, n miles:** 4,000 at 18 kt
Complement: 170

Machinery: 2 Type 18E 390 diesels; 24,000 hp(m) *(17.65 MW)* sustained; 2 shafts
Missiles: SSM: 6 C-802 (YJ-83/CSS-N-8 Saccade) (2 triple) launchers ❶; mid-course guidance and active radar homing to 120 km *(65 n miles)* at 0.9 Mach; warhead 165 kg; sea-skimmer.
SAM: 1 HQ-61 CSA-N-2 sextuple launcher ❷; semi-active radar homing to 10 km *(5.5 n miles)* at 2 Mach. Similar to Sea Sparrow. May be replaced in due course.
Guns: 2 China 3.9 in *(100 mm)*/56 (twin) ❸; 25 rds/min to 22 km *(12 n miles)*; weight of shell 15.6 kg. 8 China 37 mm/63 Type 76A (4 twin) ❹; 180 rds/min to 8.5 km *(4.6 n miles)* anti-aircraft; weight of shell 1.42 kg.
A/S Mortars: 2 Type 87 ❺ 6-tubed launchers.
Physical countermeasures: Decoys: 2 China Type 945 26-barrelled chaff launchers ❻.
Electronic countermeasures: ESM: RWD8; intercept. ECM: NJ81-3; jammer. Similar to Scimitar.
Radars: Air search: Type 517 Knife Rest ❼; A-band.

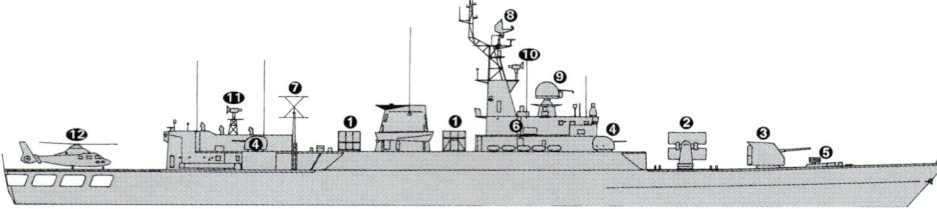

HUAIBEI *(Scale 1 : 900), Ian Sturton / 0130723*

Air/surface search: Type 360 Seagull S ❽; E/F-band.
Fire control: Type 343 (Wok Won) (Wasp Head) ❾; I-band (for 100 mm). Type 342 (Fog Lamp) ❿; I/J-band (for SAM). Type 347G(1) Rice Bowl ⓫; I/J-band (for 37 mm).
Navigation: Racal Decca 1290; I-band.
Sonars: Echo Type 5; hull-mounted; active search and attack; medium frequency.

Helicopters: 2 Harbin Z-9C (Dauphin) ⓬.

Programmes: Programme started in 1988. First one conducted sea trials in late 1991. Four of the class built before the design moved on to the Jiangwei II.
Modernisation: SAM system has been unsatisfactory and may be replaced in due course.
Structure: The sextuple launcher is a multiple launch SAM system using the CSA-N-2 missile.
Operational: All based in the East Sea Fleet at Dinghai.

ANQING *9/2012, A Sheldon-Duplaix / 1483009*

1 JIANGHU III (TYPE 053 H2) CLASS (FFG)

CANGZHOU (ex-*Zhoushan*) 537

Displacement, tonnes: 1,955 full load
Dimensions, metres (feet): 103.2 × 10.8 × 3.1 *(338.6 × 35.4 × 10.2)*
Speed, knots: 26
Range, n miles: 4,000 at 15 kt, 2,700 at 18 kt
Complement: 200 (30 officers)

Machinery: 2 Type 18E 390V diesels; 14,400 hp(m) *(10.6 MW)* sustained; 2 shafts
Missiles: SSM: 4 C-802 (YJ-83/CSS-N-8 Saccade) ❶; mid-course guidance and active radar homing to 120 km *(65 n miles)* at 0.9 Mach; warhead 165 kg; sea skimmer.
Guns: 4 China 3.9 in *(100 mm)*/56 (2 twin) ❷; 25 rds/min to 22 km *(12 n miles)*; weight of shell 15.6 kg. 8 China 37 mm/63 (4 twin) ❸; 180 rds/min to 8.5 km *(4.6 n miles)* anti-aircraft; weight of shell 1.42 kg.
A/S Mortars: 2 RBU 1200 5-tubed fixed launchers ❹; range 1,200 m; warhead 34 kg.
Mines: Can carry up to 60.
Depth charges: 2 BMB-2 projectors; 2 racks.
Physical countermeasures: Decoys: 2 China 26-barrelled chaff launchers.
Electronic countermeasures: ESM: Elettronica Newton; radar warning.
ECM: Elettronica 929 (Type 981); jammer.
Radars: Air search: Type 517 Knife Rest ❺; A-band.
Air/surface search: Type 354 Eye Shield (MX 902) ❻; G-band.
Surface search/fire control: Type 352 Square Tie ❼; I-band.
Navigation: Fin Curve; I-band.
Fire control: Type 347G Rice Bowl ❽; I/J-band. Type 343G (Wok Won) (Wasp Head) ❾; I-band.
IFF: High Pole A. Square Head.
Sonars: Echo Type 5; hull-mounted; active search and attack; medium frequency.
Combat data systems: ZKJ-3.

Programmes: Built at Hudong, Shanghai. *Cangzhou* was completed in 1989 one of the first three Chinese warships to be equipped with a computerised combat system.
Structure: The main deck is higher in the midships section and the lower part of the mast is solid. The arrangement of the launchers is side by side, as opposed to the staggered

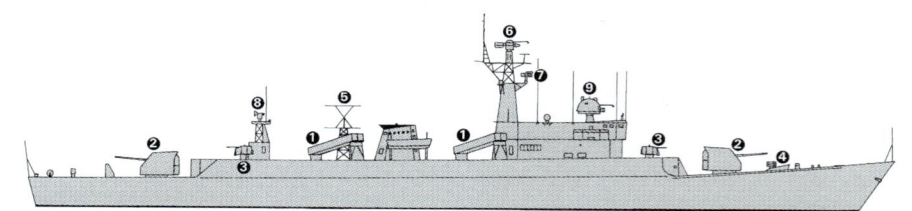

CANGZHOU *(Scale 1 : 900), Ian Sturton / 0130726*

pairings in the first two ships. These were the first all-enclosed, air conditioned ships built in China.
Operational: Based in East Sea Fleet at Dinghai.

Sales: Four modified Type III to Thailand in 1991–92. *Wuhu* and *Huangshi* were decommissioned on 27 April 2013 and are likely to be transferred to Bangladesh in 2014.

CANGZHOU *10/1992, Ships of the World / 0056766*

IHS Jane's Fighting Ships 2014-2015 © 2014 IHS

16 JIANGHU I/II/V (TYPE 053H/053H1/053H1G) CLASS (FFG)

Name	No	Name	No	Name	No
HUAYIN	513	DANDONG	543	DONGGUAN	560
JIUJIANG	516	LINFEN	545	SHANTOU	561
JIAN	518	SHAOGUAN	553	JIANGMEN	562
CHANGZHI	519	ZHAOTONG	555	FOSHAN	563
TAIZHOU	533	ZIGONG	558		
JINHUA	534	BEIHAI	559		

Displacement, tonnes: 1,448 standard; 1,729 full load
Dimensions, metres (feet): 103.2 × 10.8 × 3.1 (338.6 × 35.4 × 10.2)
Speed, knots: 26
Range, n miles: 4,000 at 15 kt, 2,700 at 18 kt
Complement: 200 (30 officers)

Machinery: 2 Type 12E 390V diesels; 14,400 hp(m) (10.6 MW) sustained; 2 shafts
Missiles: SSM: 4 HY-2 (CSSC-3 Seersucker) (2 twin) launchers ❶; active radar or IR homing to 95 km (51 n miles) at 0.9 Mach; warhead 513 kg.
Guns: 2 or 4 China 3.9 in (100 mm)/56 (2 single ❷ or 2 twin ❸); 25 rds/min to 22 km (12 n miles); weight of shell 15.6 kg. 12 China 37 mm/63 (6 twin) ❹ (8 (4 twin), in some); 180 rds/min to 8.5 km (4.6 n miles) anti-aircraft; weight of shell 1.42 kg.
A/S Mortars: 2 RBU 1200 5-tubed fixed launchers (4 in some) ❺; range 1,200 m; warhead 34 kg.
Mines: Can carry up to 60.
Depth charges: 2 BMB-2 projectors; 2 racks (in some).
Physical countermeasures: Decoys: 2 RBOC Mk 33 6-barrelled chaff launchers or 2 China 26-barrelled launchers.
Electronic countermeasures: ESM: Jug Pair or Watchdog; radar warning.
Radars: Air search: Type 517 Knife Rest ❼; A-band.
Air/surface search: Type 354 Eye Shield (MX 902) ❽; G-band. Type (unknown) ❾; I-band.
Surface search/fire control: Type 352 Square Tie ❿; I-band.
Navigation: Don 2 or Fin Curve or Racal Decca; I-band.
Fire control: Type 347G Rice Bowl (in some) ⓫; I/J-band. Type 343 (Wok Won) (Wasp Head) (in some) ⓬; I-band.
IFF: High Pole A. Yard Rake or Square Head.
Sonars: Echo Type 5; hull-mounted; active search and attack; medium frequency.
Weapon control systems: Wok Won director (in some) ❻.

Programmes: Pennant numbers changed in 1979. All built in Shanghai starting in the mid-1970s at the Hudong, Jiangnan and Huangpu shipyards. Ships were completed in the following order: 516, 513, 518, 519, 533, 543, 553, 555, 545, 544, 558, 560, 561, 559, 562 and 563. The last of class 563 completed in February 1996.
Modernisation: Equipment varies considerably from ship to ship. The Type 053H ships are 513, 515, 516, 518, 519 and 545. These are equipped with SY-1 or SY-2 SSM, single 100 mm guns and SJD-3 sonar. Type 053H1 ships are 533, 534, 543, 553 and 555. These are similar to Type 053H but are equipped with twin 100 mm guns and SJD-5 (Echo 5) sonar. Type 053H1G ships are 558–563. These are similar to Type 053H1 but are equipped with 37 mm enclosed gun mounts. A larger bridge structure suggests a possible CIC compartment. The designation of the Air/Surface search radar in Type 053H1G is not yet known but it bears similarities to the I-band MR-36A which has been promoted as a replacement for Type 352 'Square Tie'. 516 appears to have been modified for a shore bombardment role having been fitted with a new twin 100 mm mounting and seven 122 mm Multiple Launch Rocket Systems (MLRs). 559 may also be similarly converted. 560 reportedly completed a refit in 2010; modifications included replacement of the 100 mm twin mounting with a modern mounting and C-201 missiles with C-802 (YJ-83).
Structure: All of the class have the same hull dimensions. Previously reported Type numbers have been superseded by the following designations:
Type I has at least five versions: Version 1 has an oval funnel and square bridge wings; version 2 has a square funnel with bevelled bridge face; version 3 an octagonal funnel; version 4 reverts back to the oval funnel and version 5 has a distinctive fluting arrangement with cowls on the funnel, as well as gunhouses on the 37 mm guns. Some have bow bulwarks.
Types III and IV. See separate entries.
Operational: 520 was scrapped in 1985; 509 and 510 transferred to the Coast Guard in 2007; 515 and 517 were decommissioned in 2010; 511, 512, 551 and 552 were decommissioned in 2012. 514 was decommissioned in 2013. Basing: 519 in North Sea Fleet; 513, 516, 518, 558, and 559 in East Sea Fleet; 533, 534, 543, 545, 553, 555, 560–563 in South Sea Fleet.
Sales: Two have been transferred to Egypt, one in September 1984, the other in March 1985, and one, *Xiangtan* 556, to Bangladesh in November 1989. *Anshun* 554 and *Jishou* 557 were transferred to Myanmar in 2012.

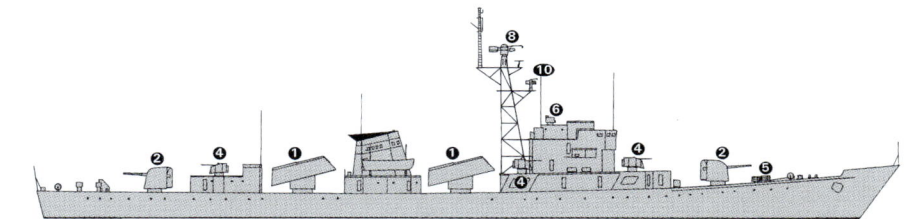
JIUJIANG (TYPE 053H) *(Scale 1 : 900), Ian Sturton / 0529151*

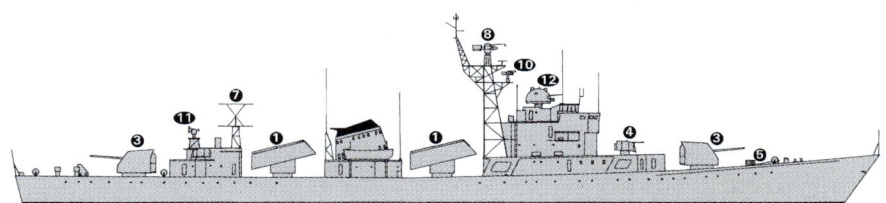
TAIZHOU (TYPE 053H1) *(Scale 1 : 900), Ian Sturton / 0130728*

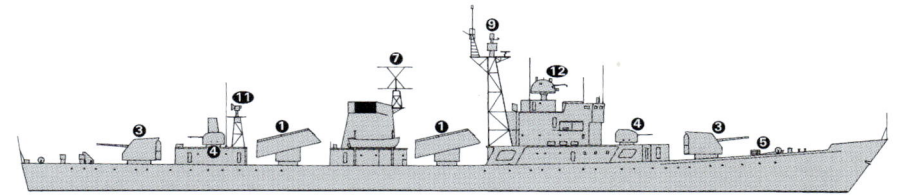
DONGGUAN (TYPE 053H1G) *(Scale 1 : 900), Ian Sturton / 0130727*

HUAYIN (TYPE 053H) *12/2007, Chris Sattler / 1170059*

HUAYIN *3/2011, Jurg Kürsener / 1406550*

CORVETTES

11 + 9 (10) JIANGDAO (TYPE 056) CLASS (FSG)

Name	No	Builders	Laid down	Launched	Commissioned
BENGBU	582	Hudong-Zhonghua Shipyard, Shanghai	2010	23 May 2012	25 Feb 2013
HUIZHOU	596	Huangpu Shipyard, Guangzhou	2010	3 Jun 2012	1 Jul 2013
MEIZHOU	584	Wuhan Shipyard	2010	31 Jul 2012	30 Jul 2013
DATONG	580	Liao Nan Shipyard, Lushun	2011	10 Aug 2012	18 May 2013
SHANGRAO	583	Hudong-Zhonghua Shipyard, Shanghai	2011	19 Aug 2012	10 Jun 2013
QINZHOU	597	Huangpu Shipyard, Guangzhou	2011	30 Aug 2012	1 Jul 2013
BAISE	585	Wuhan Shipyard	2011	25 Oct 2012	12 Oct 2013
YINGKOU	581	Liao Nan Shipyard, Lushun	2012	18 Nov 2012	1 Aug 2013
JIEYANG	587	Huangpu Shipyard, Guangzhou	2012	28 Jan 2013	26 Jan 2014
JIAN	586	Hudong-Zhonghua Shipyard, Shanghai	2012	25 Feb 2013	8 Jan 2014
QINGYUAN	589	Huangpu Shipyard, Guangzhou	2012	31 May 2013	2014
QUANZHOU	588	Hudong-Zhonghua Shipyard, Shanghai	2012	26 Jun 2013	2014
–	592	Wuhan Shipyard	2012	16 Jul 2013	2014
WEIHAI	590	Liao Nan Shipyard, Lushun	2012	1 Aug 2013	15 Mar 2014
–	591	Liao Nan Shipyard, Lushun	2012	1 Aug 2013	2014
–	595	Wuhan Shipyard	2013	14 Nov 2013	2014
SUQIAN	593	Hudong-Zhonghua Shipyard, Shanghai	2013	20 Nov 2013	2014
–	594	Huangpu Shipyard, Guangzhou	2013	30 Nov 2013	2014
–	–	Huangpu Shipyard, Guangzhou	2013	2014	2015
–	–	Hudong-Zhonghua Shipyard, Shanghai	2013	2014	2015

Displacement, tonnes: 1,500 full load
Dimensions, metres (feet): 90 × 11.0 × 4.4 *(295.3 × 36.1 × 14.4)*
Speed, knots: 25. **Range, n miles:** 3,500 at 16 kt
Complement: 60 (18 officers)
Machinery: CODAD: 2 diesels; 2 shafts; cp props
Missiles: SSM: 4 C-802 (YJ-83); mid-course guidance and active radar homing to 120 km *(65 n miles)* at 0.9 Mach; warhead 165 kg ❶.
SAM: 1 FL-3000N 8-cell launcher ❷; passive IR/anti-radiation homing to 9.0 km *(4.8 n miles)*.
Guns: 1 — 76 mm AK-176 ❸. 2 Norinco HP/J17 30 mm ❹.
Torpedoes: 6 — 324 mm (2 triple) tubes; Yu-7; active-passive homing to 14 km *(7.6 n miles)* at 42 kt; warhead 45 kg.
Physical countermeasures: Decoys: 2 9-barrelled launchers
Electronic countermeasures: ESM: To be announced.
Radars: Air/surface search: Type 360 Seagull 5 ❺; E/F-band.
Navigation: Type 760; I-band ❻.
Fire control: Type 347G (LR 66) ❼; I-band.
Sonars: Bow mounted.
Combat data systems: To be announced, SATCOM.
Electro-optic systems: IR-17 optronic sensor.
Helicopters: Platform for 1 medium.

Comment: Indications of a new corvette design emerged in November 2010. Details of the ship, known as the Type 056, appear to be broadly based on those of the Pattini-class corvette, built for Thailand in 2005–06. Series production is underway at four shipyards and a class of at least 30 is expected if the class is to consolidate replacement of older classes such as the Jianghu-class frigates and Houxin-class attack craft. 596 and 597 are based at Hong Kong.

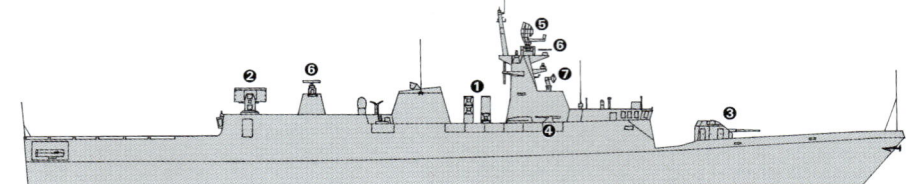

TYPE 056 *(Scale 1 : 900), Ian Sturton / 1483594*

TYPE 056 *8/2012 / 1483003*

TYPE 056 *6/2013*, Ships of the World / 1525668*

SHIPBORNE AIRCRAFT

Numbers/Type: 17 Changhe Z-8 Super Frelon.
Operational speed: 134 kt (248 km/h).
Service ceiling: 10,000 ft (3,100 m).
Range: 440 n miles (815 km).
Role/Weapon systems: ASW helicopter; Eight SA 321G delivered from France in 1977 but supplemented by 12 locally built Zhi-8, of which the first operational aircraft was delivered in late 1991. Thomson Sintra HS-12 in four SA 321Gs for SSBN escort role. Sensors: HS-12 dipping sonar and processor, some have French-built search radar. Weapons: ASW; Yu-7 torpedo. ASV; C-802K ASM, rockets and MGs.

Z-8   9/2002, Paul Jackson / 0525833

Numbers/Type: 11/3 Hafei Z-9C Haitun/Z-9D Haitun (Panther).
Operational speed: 140 kt (260 km/h).
Service ceiling: 15,000 ft (4,575 m).
Range: 410 n miles (758 km).
Role/Weapon systems: Eurocopter AS 365 Panther 2 aircraft built under licence. All delivered by about 2000. An anti-ship missile variant, Z-9D, was rolled out in mid-2008. The missile is believed to be the 4–15 km range TL-10, similar to the Iranian Kosar. Sensors: Thomson-CSF Agrion; HS-12 dipping sonar; Crouzet MAD. Weapons: ASW; Yu-7 torpedoes.

Z-9C   12/2007, Hachiro Nakai / 1166773

Numbers/Type: 6/4 Kamov Ka 28PL Helix A/Ka 28PS Helix D.
Operational speed: 135 kt (250 km/h).
Service ceiling: 19,685 ft (6,000 m).
Range: 432 n miles (800 km).
Role/Weapon systems: First pair are (Ka 28PL) ASW helicopters acquired in 1997 for evaluation. Four more ASW versions and four (Ka 28PS) for SAR delivered in late 1999. Sensors: Splash Drop radar; VGS-3 dipping sonar; MAD; ESM. Weapons: three torpedoes or depth bombs or mines.

Ka-28   6/2004 / 1042165

Numbers/Type: 9 Shenyang J-15 Flying Shark.
Operational speed: 1,555 kt (2,878 km/h).
Service ceiling: 65,700 ft (20,000 m).
Range: 2,050 n miles (3,500 km).
Role/Weapon systems: Carrier-based aircraft in development by the Shenyang Aircraft Corporation. The aircraft is reported to be a development of and bears a similarity to the Russian Sukhoi Su-33. Performance details have not been confirmed. Production aircraft are likely to be fitted with Chinese sensors and weapons. The date of the maiden flight of the prototype is not known but first images were released in July 2010. An arrested landing on Liaoning was successfully achieved on 20 November 2012.

J-15   6/2012* / 1525585

Numbers/Type: 8 Kamov Ka-31 Helix B.
Operational speed: 119 kt (220 km/h).
Service ceiling: 11,480 ft (3,500 m).
Range: 325 n miles (600 km).
Role/Weapon systems: AEW helicopter. Eight aircraft reportedly delivered by 2011.

Ka-31   6/2005, IHS/Patrick Allen / 1136991

LAND-BASED MARITIME AIRCRAFT (FRONT LINE)

Notes: In addition to those listed there are about 170 training and transport aircraft.

Numbers/Type: 5 KJ-2000 AWACS.
Operational speed: 425 kt (785 km/h).
Service ceiling: 34,440 ft (10,500 m).
Range: 2,753 n miles (5,100 km).
Role/Weapon systems: Airborne Warning And Control System (AWACS) aircraft based on the Russian-made A-50 (Mainstay) airframe which itself is based on the Ilyushin Il-76 transport aircraft. The non-rotating radome houses three Chinese-made (AESA) phased array antennas in a triangular configuration. A SATCOM antenna may be installed inside a fairing on top of forward cabin.

KJ-2000   6/2010 / 1366761

Numbers/Type: 24 Sukhoi Su-30 MK 2 Flanker.
Operational speed: 1,345 kt (2,500 km/h).
Service ceiling: 59,000 ft (18,000 m).
Range: 2,160 n miles (4,000 km).
Role/Weapon systems: 24 delivered in 2004. The air force operates at least 150 of the similar Su-27 which also might be used for fleet air-defence. Sensors: Doppler radar. Weapons: One 30 mm cannon; 10 AAMs. AS-17A/B anti-ship missiles may be fitted to some aircraft.

Su-30   6/2010 / 1366764

150 China > Land-based maritime aircraft

Numbers/Type: 83 XAC JH-7.
Operational speed: 653 kt *(1,210 km/h)*.
Service ceiling: 51,180 ft *(15,600 m)*.
Range: 891 n miles *(1,650 km)*.
Role/Weapon systems: All-weather dual seat 'Flounder' type attack fighter first delivered in 1998. A second batch of 18 JH-7A was delivered in 2004. Sensors: Letri JL-10A Shen-Ying pulse Doppler fire-control radar capable of tracking four targets to 29 n miles *(54 km)* in look-down mode simultaneously. Weapons: AAM; PL-5b, PL-7 and 23 mm gun. ASM: Two C-801 or C-802 anti-ship missiles; C-701 anti-ship missile and 500 kg LGBs. AS-17 may be fitted in due course.

JH-7   6/2010 / 1366763

Numbers/Type: 4 Harbin SH-5.
Operational speed: 243 kt *(450 km/h)*.
Service ceiling: 23,000 ft *(7,000 m)*.
Range: 2,563 n miles *(4,750 km)*.
Role/Weapon systems: Multipurpose amphibian introduced into service in 1986. Sensors: Doppler radar; MAD; sonobuoys. Weapons: ASV; four C 101, two gun turret, bombs. ASW; Yu-2 (Mk 46 Mod 1) torpedoes, mines, depth bombs.

SH-5   9/2007 / 1335686

Numbers/Type: 4/5/1 SAC Y-8X (Cub)/Y-8JB/Y-8Q.
Operational speed: 351 kt *(650 km/h)*.
Service ceiling: 34,120 ft *(10,400 m)*.
Range: 3,020 n miles *(5,600 km)*.
Role/Weapon systems: Maritime patrol version of An-12 Cub transport; first flown 1985. There are reported to be four Y-8J variants equipped with Searchwater radar in a dropped nose radome. In addition there are five Y-8JB Elint variants in service. There have also been reports of a Y-8Q ASW variant featuring a MAD tailboom and a large chin radome. At least one prototype aircraft is thought to exist. Sensors: Litton APSO-504(V)3 search radar in undernose radome. Two Litton LTN 72R INS and Omega/Loran. Weapons: No weapons carried.

Y-8X   6/2010 / 1366766

Y-8JB   6/2011, Ships of the World / 1406685

Numbers/Type: 90/80/50/80/40 SAC J-8 Finback B/J-8 Finback D/J-8 Finback E/J-8 Finback F/J-8 Finback H.
Operational speed: 701 kt *(1,300 km/h)*.
Service ceiling: 65,620 ft *(20,000 m)*.
Range: 1,187 n miles *(2,200 km)*.
Role/Weapon systems: Dual role, all-weather fighter introduced into service in 1990. Some Finback A/B have been upgraded to Finback F/H. There are at least 170 more in service with the Air Force. Weapons: 23 mm twin-barrel cannon; PL-2/7 AAM; ASM. PL-2 has some ASM capability.

J-8   6/2010 / 1366765

Numbers/Type: 35/3 XAC H-6G/H-6U (Tu-16 Badger).
Operational speed: 535 kt *(992 km/h)*.
Service ceiling: 40,350 ft *(12,300 m)*.
Range: 2,605 n miles *(4,800 km)*.
Role/Weapon systems: Three regiments of H-6G bomber and maritime reconnaissance aircraft. Some (H-6U) converted as tankers. Sensors: Search/attack radar; ECM. Weapons: ASV; two underwing anti-shipping (possibly C-802) missiles. Up to five 23 mm cannon; bombs.

H-6G   6/2010 / 1366767

Numbers/Type: 29 CAC J-7.
Operational speed: 1,175 kt *(2,175 km/h)*.
Service ceiling: 61,680 ft *(18,800 m)*.
Range: 804 n miles *(1,490 km)*.
Role/Weapon systems: Land-based Fleet air defence fighter with limited strike role against enemy shipping or beachhead. Sensors: Search attack radar, some ECM. Weapons: ASV; 500 kg bombs or 36 rockets. Standard; two 30 mm cannon. AD; two 'Atoll' AAMs.

J-7E   6/2002, Ships of the World / 0554725

Numbers/Type: 23/2 CAC J-10AH/SH J-10AH/J-10SH.
Operational speed: 793 kt *(1,468 km/h)*.
Service ceiling: 55,780 ft *(17,000 m)*.
Range: 891 n miles *(1,650 km)*.
Role/Weapon systems: Single-seat fighter developed for the Air Force. One regiment operated by the PLAN. Sensors and Weapons: to be announced.

J-10   5/2013* / 1488275

Numbers/Type: 24/5 Shenyang J-11BH/BSH J-11BH/J-11BSH.
Operational speed: 1,555 kt *(2,878 km/h)*.
Service ceiling: 63,523 ft *(19,000 m)*.
Range: 2,070 n miles *(3,530 km)*.
Role/Weapon systems: Land-based aircraft, reportedly an indigenously developed version of the Sukhoi SU-27 Flanker multirole aircraft. Performance details not confirmed. Sensors and Weapons: to be announced.

J-11   9/2013* / 1488277

IHS Jane's Fighting Ships 2014-2015   © 2014 IHS

PATROL FORCES

Notes: Many patrol craft carry the HN-5 shoulder-launched Chinese version of the SA-N-5 SAM.

60 HOUBEI (TYPE 022) CLASS
(FAST ATTACK CRAFT—MISSILE) (PGGF)

2101–2112   2208–2231   2308–2331

Displacement, tonnes: 224 full load
Dimensions, metres (feet): 42.6 × 12.2 × 1.5 *(139.8 × 40.0 × 4.9)*
Speed, knots: 40
Complement: 12

Machinery: 2 diesels; 6,865 hp *(5.1 MW)*; 4 Kamewa waterjet propulsors
Missiles: 8 C-802 (YJ-83) (CSS-N-8 Saccade); mid-course guidance and active radar homing to 120 km *(65 n miles)* at 0.9 Mach; warhead 165 kg; sea skimmer.
Guns: 1—30 mm/65 AK 630; 6 barrels; 3,000 rds/min combined to 2 km; 12 missiles.
Radars: Surface search: Type 348 (LR 66); I-band.
Navigation: I-band.
Weapon control systems: Optronic director.

Comment: A new fast attack craft, the first of which was launched at Qiuxin Shipyard, Shanghai in April 2004. The design is believed to be based on a 42 m hull developed by AMD Marine Consulting, Sydney. This was further progressed by its joint venture company in Guangzhou, Sea Bus International (SBI), into a patrol boat configuration which was selected by the Chinese Navy after a five-year investigation into various platform contenders. The craft has a wave-piercing catamaran hull form and a centre bow. Likely to be of aluminium alloy construction, the design clearly incorporates RCS reduction measures. Following extensive first of class trials, full production was reported to have taken place in at least six shipyards. Production appeared to stop in 2009 and it is not known whether further craft are to be built. Dimensions are based on the original AMD design. The installation of C-802 (or possibly C-705) missiles suggests that there may be a datalink to facilitate over-the-horizon targeting. The craft are based in all three fleets and are likely to use a sector-defence concept. Basing: 2101–2112 (North); 2208–2231 (East); 2308–2331 (South).

HOUBEI CLASS   1/2008 / 1335685

HOUBEI CLASS   12/2007, Chris Sattler / 1335684

HOUBEI CLASS   6/2007 / 1166866

20 HOUXIN (TYPE 037/1G) CLASS
(FAST ATTACK CRAFT—MISSILE) (PTG)

651–656   751–760   764–767

Displacement, tonnes: 486 full load
Dimensions, metres (feet): 62.8 × 7.2 × 2.4 *(206 × 23.6 × 7.9)*
Speed, knots: 28
Range, n miles: 750 at 18 kt
Complement: 71

Machinery: 4 China PR 230ZC diesels; 4,000 hp(m) *(2.94 MW)*; 4 shafts
Missiles: SSM: 4 YJ-1 (Eagle Strike) (C-801) (CSS-N-4 Sardine) (2 twin); active radar homing to 40 km *(22 n miles)* at 0.9 Mach; warhead 165 kg; sea-skimmer.
Guns: 4—37 mm/63 (Type 76A) (2 twin); 180 rds/min to 8.5 km *(4.6 n miles)* anti-aircraft; weight of shell 1.42 kg. 4—14.5 mm (Type 69) (2 twin); 600 rds/min to 7 km *(3.8 n miles)*.
Electronic countermeasures: ESM/ECM: Intercept and jammer.
Radars: Surface search: Type 352C (Square Tie); I-band.
Fire control: Type 341 (Rice Lamp); I-band.
Navigation: Anritsu Type 723; I-band.

Programmes: First seen in 1991 and built at the rate of up to three per year at Qiuxin and Huangpu Shipyards to replace the Houku class and for export. Building may have stopped in mid-1999.
Structure: This is a missile armed version of the Hainan class. There are some variations in the bridge superstructure in later ships of the class.
Operational: Basing: 651–656 (North); 751–756, 766, 767 (South); 757–760, 764, 765 (East).
Sales: Two to Burma in December 1995, two in July 1996 and two in late 1997.

HOUXIN 758   3/2003, Bob Fildes / 0569184

HOUXIN 765   5/2004 / 1042140

6 HOUJIAN (OR HUANG) (TYPE 037/2) CLASS
(FAST ATTACK CRAFT—MISSILE) (PTG)

Name	No	Builders	Launched	Commissioned
YANGJIANG	770	Huangpu Shipyard	Jan 1991	May 1991
SHUNDE	771	Huangpu Shipyard	Jul 1994	Feb 1995
NANHAI	772	Huangpu Shipyard	Feb 1995	Apr 1995
PANYU	773	Huangpu Shipyard	May 1995	Jul 1995
LIANJIANG	774	Huangpu Shipyard	Sep 1998	Feb 1999
XINHUI	775	Huangpu Shipyard	Apr 1999	Nov 1999

Displacement, tonnes: 528 standard
Dimensions, metres (feet): 65.4 × 8.4 × 2.4 *(214.6 × 27.6 × 7.9)*
Speed, knots: 32
Range, n miles: 1,800 at 18 kt
Complement: 75

Machinery: 3 SEMT-Pielstick 12 PA6 280 diesels; 15,840 hp(m) *(11.7 MW)* sustained; 3 shafts
Missiles: SSM: 6 C-802 (YJ-83/CSS-N-8 Saccade); mid-course guidance and active radar homing to 120 km *(65 n miles)* at 0.9 Mach; warhead 165 kg; sea skimmer.
Guns: 2—37 mm/63 (twin) Type 76A; 180 rds/min to 8.5 km *(4.6 n miles)* anti-aircraft; weight of shell 1.42 kg. 4—30 mm/65 (2 twin) Type 69; 500 rds/min to 5 km *(2.7 n miles)*; weight of shell 0.54 kg.
Physical countermeasures: Decoys: 2 Type 945G 26-barrelled launcher.
Electronic countermeasures: ESM: Type 928; intercept.
Radars: Surface search: Type 363; E/F-band.
Fire control: Type 344 (MR 34); I-band.
Navigation: Type 765; I-band.
Weapon control systems: Type JM-83 optronic director.

Programmes: First of class laid down in 1989 and built in a very short time. Sometimes called the Huang class.
Modernisation: Some may be fitted with Type 363 search radar and Type 344 (MR 34) fire-control radar rather than Type 347G.
Operational: Based on South Sea Fleet at Hong Kong from mid-1997. One possibly sunk in late 1997. *Lianjiang* severely damaged in a collision with a freighter on 26 June 2006 but was later repaired at Guangdong Shipyard in 2008.

SHUNDE   6/2007 / 1166848

2 HAIJIU (TYPE 037/1) CLASS (LARGE PATROL CRAFT) (PC)

688   693

Displacement, tonnes: 498 full load
Dimensions, metres (feet): 64 × 7.2 × 2.2 (210 × 23.6 × 7.2)
Speed, knots: 28
Range, n miles: 750 at 18 kt
Complement: 72

Machinery: 4 diesels; 8,800 hp(m) (6.47 MW); 4 shafts
Guns: 2 China 57 mm/70 (1 twin); 120 rds/min to 12 km (6.5 n miles); weight of shell 6.31 kg.
2 USSR 30 mm/65 (1 twin); 500 rds/min to 5 km (2.7 n miles) anti-aircraft; weight of shell 0.54 kg.
A/S Mortars: 4 RBU 1200 5-tubed fixed launchers; range 1,200 m; warhead 34 kg.
Depth charges: 2 rails.
Radars: Surface search: Type 351; Pot Head; I-band.
Fire control: Round Ball; I-band.
Sonars: Stag Ear or Thomson Sintra SS 12 (688, 693).

Comment: A lengthened version of the Hainan class probably used as a prototype for the Houxin class. Based in East Sea Fleet. Two have been scrapped.

HAIJIU 688   6/2008 / 1335706

39 HAINAN (TYPE 037) CLASS
(FAST ATTACK CRAFT—PATROL) (PC)

621–622	678	695–696	720–724	740
624–630	680–683	699–701	726–729	
650	685–689	706–708	732	

Displacement, tonnes: 381 standard; 398 full load
Dimensions, metres (feet): 58.8 × 7.2 × 2.2 (192.9 × 23.6 × 7.2)
Speed, knots: 30.5
Range, n miles: 1,300 at 15 kt
Complement: 78

Machinery: 4 PCR/Kolomna Type 9-D-8 diesels; 4,000 hp(m) (2.94 MW) sustained; 4 shafts
Missiles: Can be fitted with 4 YJ-1 launchers in lieu of the after 57 mm gun.
Guns: 4 China 57 mm/70 (2 twin); 120 rds/min to 12 km (6.5 n miles); weight of shell 6.31 kg.
4 USSR 25 mm/60 (2 twin); 270 rds/min to 3 km (1.6 n miles) anti-aircraft; weight of shell 0.34 kg.
A/S Mortars: 4 RBU 1200 5-tubed fixed launchers; range 1,200 m; warhead 34 kg.
Mines: Rails fitted for 12.
Depth charges: 2 BMB-2 projectors; 2 racks. 18 DCs.
Radars: Surface search: Type 351; Pot Head or Skin Head; E/F-band.
IFF: High Pole.
Sonars: Stag Ear; hull-mounted; active search and attack; high frequency. Thomson Sintra SS 12 (in some); VDS.

Programmes: A larger Chinese-built version of the former Soviet SO 1. Low freeboard. Programme started 1963–64 and continued with new hulls replacing the first ships of the class. There are at least six variants with minor differences.
Structure: Later ships have a tripod or solid foremast in place of a pole and a short stub mainmast. Two trials SS 12 sonars fitted in 1987.
Operational: Basing: 621–622, 624–630, 700–701, 706–708 in the North Sea Fleet; 650, 680–683, 685–689, 695–696, 699 in the East Sea Fleet; 678, 720–729, 732, 740 in the South Sea Fleet.
Sales: Two to Bangladesh, one in 1982 and one in 1985; eight to Egypt in 1983–84; six to North Korea 1975–78; four to Pakistan, two in 1976 and two in 1980; six to Burma in 1991 and four in 1993.

HAINAN 689   9/2011, A Sheldon-Duplaix / 1483011

HAINAN 686   10/2006, E & M Laursen / 1164869

22 HAIQING (TYPE 037/1S) CLASS
(FAST ATTACK CRAFT—PATROL) (PC)

| 611–614 | 631–635 | 710–713 | 743–744 | 761–763 | 786–789 |

Displacement, tonnes: 486 full load
Dimensions, metres (feet): 62.8 × 7.2 × 2.4 (206 × 23.6 × 7.9)
Speed, knots: 28
Range, n miles: 1,300 at 15 kt
Complement: 71

Machinery: 4 Chinese PR 230ZC diesels; 4,000 hp(m) (2.94 MW) sustained; 4 shafts
Guns: 4 China 37 mm/63 (2 twin) Type 76. 4 China 14.5 mm (2 twin) Type 69.
A/S Mortars: 2 Type 87 6-tubed launchers.
Radars: Surface search: Anritsu RA 723; I-band.
Sonars: Hull mounted; active search and attack; medium frequency Thomson Sintra SS 12; VDS.

Programmes: Starting building at Qiuxin Shipyard in 1992 and replaced the Hainan class programme. First one completed in November 1993. Production continued at Qingdao, Chongqing and Huangpu as well as Qiuxin.
Structure: Based on the Hainan class, but the large A/S mortars suggest a predominantly ASW role, and this may explain the rapid building rate.
Operational: Basing: 611–614, 631–635, 710–711 in North Sea Fleet; 712–713, 743–744, 761–763 in East Sea Fleet; 786–789 in South Sea Fleet.
Sales: One to Sri Lanka in December 1995.

HAIQING 743   6/2008 / 1335705

56 HAIZHUI/SHANGHAI III (TYPE 062/1) CLASS
(COASTAL PATROL CRAFT) (PC)

| 1101–1108 | 1201–1208 | 1231–1245 | 1271–1282 | 1301–1313 |

Displacement, tonnes: 173 full load
Dimensions, metres (feet): 41 × 5.3 × 1.8 (134.5 × 17.4 × 5.9)
Speed, knots: 25
Range, n miles: 750 at 17 kt
Complement: 43

Machinery: 4 Chinese L12-180A diesels; 4,400 hp(m) (3.22 MW) sustained; 4 shafts
Guns: 4 China 37 mm/63 (2 twin); 180 rds/min to 8.5 km (4.6 n miles); weight of shell 1.42 kg.
4 China 14.5 mm (2 twin) Type 69 or 4 China 25 mm (2 twin).
Radars: Surface search: Type 351 Pot Head or Anritsu 726; I-band.
Sonars: Stag Ear; hull-mounted; active search; high frequency (in some).

Programmes: First seen in 1992 and built for Chinese use and for export. Sometimes referred to as Shanghai III class when not fitted with ASW equipment.
Structure: Lengthened Shanghai II hull. Inclined pole mast and a pronounced step at the back of the bridge superstructure are recognition features. Much reduced top speed. Some may be equipped with RBU 1200 launchers in place of other armament.
Operational: Basing: 1101–1108 in North Sea Fleet; 1201–1208, 1231–1245, 1271–1282 in East Sea Fleet; 1301–1313 in South Sea Fleet.
Sales: Three of a variant to Tunisia in 1994, three to Sri Lanka in August 1995, three more in May 1996 and three more in August 1998. One to Bangladesh in mid-1996. One to Sierra Leone in 1997.

HAIZHUI 1206   10/2013*, A Sheldon-Duplaix / 1525586

HAIZHUI 1202   3/2007 / 1166862

Patrol forces — Amphibious forces < **China** 153

4 TYPE 026H (HARBOUR PATROL CRAFT) (PBI)

7358–7361

Displacement, tonnes: 81 full load
Dimensions, metres (feet): 25 × 4.1 × 1.4 (82 × 13.5 × 4.6)
Speed, knots: 28
Machinery: 2 diesels; 2 shafts
Guns: 2 — 14.5 mm (twin).
Radars: Surface search: I-band.

Comment: Four new patrol craft arrived at Hong Kong on 1 July 1997. There may be more of the class, which are similar to some of the paramilitary patrol craft, but much faster.

TYPE 026H
12/2009, Annati Collection
1366768

AMPHIBIOUS FORCES

Notes: (1) In addition to the ships listed below there are up to 500 minor LCM/LCVP types used to transport stores and personnel.
(2) Eight Yuchai class (USSR T 4 design) and ten T4 LCMs are still in reserve in the South Sea Fleet.
(3) A 20 m WIG (Wing-In-Ground effect) craft assembled at Shanghai and completed in late 1997. Resembles Russian Volga II passenger ferry and may enter naval service if it proves to be reliable.
(4) There are reports that construction of a Type 081 LHD is under consideration. The ship is believed to be of the order of 20,000 tonnes and may be based on the Type 071 hull. Armament is likely to include Type 730 Close-In Weapon System (CIWS) and FL-3000N short-range Surface-to-Air Missiles (SAM).

3 YUZHAO (TYPE 071) CLASS (ASSAULT SHIP) (LPD)

Name	No	Builders	Laid down	Launched	Commissioned
KUNLUNSHAN	998	Hudong-Zhonghua Shipyard, Shanghai	Jun 2006	21 Dec 2006	13 Nov 2007
JINGGANGSHAN	999	Hudong-Zhonghua Shipyard, Shanghai	2009	18 Nov 2010	30 Oct 2011
CHANGBAISHAN	989	Hudong-Zhonghua Shipyard, Shanghai	2010	26 Sep 2011	23 Sep 2012

Displacement, tonnes: 18,500 standard
Dimensions, metres (feet): 210 × 28.0 × 7.0 (689 × 91.9 × 23.0)
Speed, knots: 20
Complement: 120

Military lift: Four Yuyi air-cushion vehicles plus 35–45 vehicles and troops
Machinery: CODAD; 4 SEMT Pielstick 16 PC2.6 V 400 diesels; 47,000 hp (35.2 MW); 2 shafts
Guns: 1 — 76 mm ❶. 4 — 30 mm/65 AK 630 ❷.
Physical countermeasures: Decoys: 2 launchers ❸.

Radars: Air search: Type 363 (Sea Tiger) ❹; E/F-band.
Air/surface search: Type 364 Seagull C ❺; G-band.
Fire control: Type 347G(2) (LR 66) ❻; J-band for 76 mm. Type 347G(1) (Rice Bowl) ❼; J-band for AK 630.
Navigation: Type NR 2000 ❽; I-band.

Helicopters: 4 Z-8 Super Frelon.

Programmes: After several years' speculation, the existence of the programme was confirmed when construction of the first of class was initiated in mid-2006. The programme constitutes a key component of the PLA(N)'s plan to improve its sealift and power projection capabilities. Three ships have been completed.
Structure: The principal features of the design include a large well deck area to accommodate four air cushion vehicles (ACV) in the aft two-thirds of the ship. The ACVs access the ship through a stern gate. The ship may have to ballast down for operation. There is a large stern helicopter flight deck and a hangar. An internal garage deck for vehicles may be accessed via side ramps (port and starboard). There is space for the HQ7 launcher which may be fitted at a later date. Two LCVPs are carried.
Operational: All three based at Zhanjiang (South Sea Fleet).

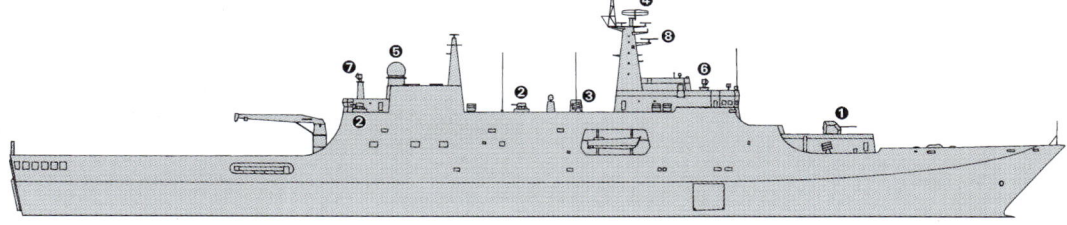

KUNLUNSHAN

(Scale 1 : 1,500), Ian Sturton / 1166825

KUNLUNSHAN *9/2007 /* 1166864

KUNLUNSHAN *9/2007 /* 1166863

JINGGANGSHAN *6/2013*, Ships of the World /* 1525664

© 2014 IHS IHS Jane's Fighting Ships 2014-2015

154 China > Amphibious forces

1 YUDENG (TYPE 073 III) CLASS (LSM)

Name	No	Builders	Launched	Commissioned
WUDANGSHAN	990	Zhonghua Shipyard	Mar 1991	Aug 1994

Displacement, tonnes: 1,880 full load
Dimensions, metres (feet): 87 × 13 × 3.8 (285.4 × 42.7 × 12.5)
Speed, knots: 14
Complement: 35

Military lift: 500 troops; 9 tanks
Machinery: 2 diesels; 2 shafts
Guns: 2 China 57 mm/50 (twin). 4—25 mm (2 twin).
Radars: Navigation: China Type 753; I-band.

Comment: The only one of the class. Based in the South Sea Fleet. Production may have been for export or the design was overtaken by the smaller Wuhu-A class.

WUDANGSHAN 4/2008 / 1335681

10 YUTING I (TYPE 072 II) CLASS (LSTH)

Name	No	Builders	Launched	Commissioned
EMEISHAN	991	Zhonghua Shipyard, Shanghai	Sep 1991	Sep 1992
DANXIASHAN	934	Zhonghua Shipyard, Shanghai	Apr 1995	Sep 1995
XUEFENGSHAN	935	Zhonghua Shipyard, Shanghai	Jul 1995	Dec 1995
HAIYANGSHAN	936	Zhonghua Shipyard, Shanghai	Dec 1995	May 1996
QINGCHENGSHAN	937	Zhonghua Shipyard, Shanghai	Apr 1996	Aug 1996
YANDANSHANG	908 (ex-938)	Zhonghua Shipyard, Shanghai	Aug 1996	Jan 1997
JIUHUASHAN	909 (ex-939)	Zhonghua Shipyard, Shanghai	Nov 1999	Apr 2000
HUANGGANGSHAN	910	Zhonghua Shipyard, Shanghai	May 2000	Dec 2001
PUTUOSHAN	939	Zhonghua Shipyard, Shanghai	Apr 2001	Aug 2001
TIANTAISHAN	940	Zhonghua Shipyard, Shanghai	Dec 2001	Apr 2002

Displacement, tonnes: 3,830 standard; 4,877 full load
Dimensions, metres (feet): 120 × 16 × 3.2 (393.7 × 52.5 × 10.5)
Speed, knots: 17
Range, n miles: 3,000 at 14 kt
Complement: 120

Military lift: 250 troops; 10 tanks; 4 LCVP
Machinery: 2 SEMT Pielstick 12PA6V-280MPC diesels; 19,200 hp (14.31 MW); 2 shafts
Guns: 6 China 37 mm/63 (3 twin); 180 rds/min to 8.5 km (4.6 n miles); weight of shell 1.42 kg.
Radars: Navigation: 2 China Type 753; I-band.
Helicopters: Platform for 2 medium.

Comment: To augment amphibious lift capabilities and provide helicopter lift. Bow and bridge structures are very similar to the Yukan class but there is a large helicopter deck. 934–937 and 991 based in South Sea Fleet. 908–910 and 939–940 based in East Sea Fleet.

HUANGGANGSHAN 1/2008, A Sheldon-Duplaix / 1335683

PUTUOSHAN 6/2007 / 1166861

7 YULIANG (TYPE 079) CLASS (LSM)

973–975 977 979–980 986

Displacement, tonnes: 1,118 full load
Dimensions, metres (feet): 63 × 10 × 2.4 (206.7 × 32.8 × 7.9)
Speed, knots: 14
Complement: 60

Military lift: 3 tanks
Machinery: 2 diesels; 2 shafts
Guns: 4—25 mm/60 (2 twin); 270 rds/min to 3 km (1.6 n miles).
Rockets: 2 BM 21 MRL rocket launchers; range about 9 km (5 n miles).
Radars: Navigation: Fin Curve; I-band.

Comment: Production started in 1980 in three or four smaller shipyards but later stopped in favour of Yuhai class. The last remaining seven are based in the South Sea Fleet.

IHS Jane's Fighting Ships 2014-2015

YULIANG 986 6/2008 / 1335702

10 YUTING II (TYPE 072 III) CLASS (LSTH)

Name	No	Builders	Launched	Commissioned
BAXIANSHAN	913	Zhonghua Shipyard, Shanghai	23 Apr 2003	Oct 2003
TIANZHUSHAN	911	Dalian Shipyard	1 Jul 2003	2004
HUADINGSHAN	992	Wuhan Shipyard	Jun 2003	2004
—	918	Wuhan Shipyard	Apr 2004	2004
LUOXIAOSHAN	993	Zhonghua Shipyard, Shanghai	18 Jul 2003	Jan 2004
DAQINGSHAN	912	Dalian Shipyard	Sep 2003	2004
DAIYUNSHAN	994	Wuhan Shipyard	16 Dec 2003	2004
WANYANG-SHAN	995	Zhonghua Shipyard, Shanghai	26 Nov 2003	2004
LAOTIESHAN	996	Dalian Shipyard	1 Jan 2004	2004
YUNWASHAN	997	Wuhan Shipyard	2004	2005

Displacement, tonnes: 3,830 standard; 4,877 full load
Dimensions, metres (feet): 120 × 16.4 × 3.2 (393.7 × 53.8 × 10.5)
Speed, knots: 17
Range, n miles: 3,000 at 14 kt
Complement: 120

Military lift: 250 troops; 10 tanks; 4 LCVP
Machinery: 2 diesels; 2 shafts
Guns: 2—37 mm (twin).
Radars: Navigation: 2 China Type 753; I-band.
Helicopters: Platform for 2 medium.

Comment: Details are speculative but reported to be an improved version of the Yuting I class with similar dimensions. Design differences include modifications to the stern, including the ramp and a taller funnel. A tunnel in the centre of the superstructure connects the main and after decks. With construction undertaken at three shipyards, a pause in the programme after 10 ships may be temporary. 992–997 based in the South Sea Fleet; 911 and 912 in the North and 913 and 918 in the East.

BAXIANSHAN 5/2011 / 1406551

DAQINGSHAN 6/2007 / 1166854

12 YUHAI (TYPE 074) (WUHU-A) CLASS (LSM)

3111–3113 3115–3117 3229 3231 3244 7593–7595

Displacement, tonnes: 812 full load
Dimensions, metres (feet): 58.4 × 10.4 × 2.7 (191.6 × 34.1 × 8.9)
Speed, knots: 14
Complement: 56

Military lift: 2 tanks; 250 troops
Machinery: 2 MAN-8L 20/27 diesels; 4,900 hp(m) (3.6 MW); 2 shafts
Guns: 2—25 mm/80 (1 twin). 4—14.5 mm (2 twin).
Radars: Navigation: I-band.

Comment: First one completed in Wuhu Shipyard in 1995. One sold to Sri Lanka in December 1995. Basing: 3111–3113, 3115 – 3117 in the North Sea Fleet; 3229, 3231 and 3244 in the East Sea Fleet and 7593 – 7595 in the South Sea Fleet.

YUHAI 7595 10/2012, Bill Clements / 1483014

© 2014 IHS

Amphibious forces < **China** 155

7 YUKAN (TYPE 072) CLASS (LST)

YUNTAISHAN 927	LINGYANSHAN 930	HELANSHAN 932
WUFENGSHAN 928	DONGTINGSHAN 931	LIUPANSHAN 933
ZIJINSHAN 929		

Displacement, tonnes: 3,160 standard; 4,237 full load
Dimensions, metres (feet): 120 × 15.3 × 2.9 *(393.7 × 50.2 × 9.5)*
Speed, knots: 18
Range, n miles: 3,000 at 14 kt
Complement: 109

Military lift: 200 troops; 10 tanks; 2 LCVP; total of 500 tons
Machinery: 2 Type 12E 390 diesels; 14,400 hp(m) *(10.6 MW)* sustained; 2 shafts
Guns: 2 China 57 mm/50 (1 twin); 120 rds/min to 12 km *(6.5 n miles)*; weight of shell 6.31 kg. 4, 6 or 8—37 mm (2, 3 or 4 twin); 180 rds/min to 8.5 km *(4.6 n miles)*; weight of shell 1.42 kg. 4—25 mm/60 (2 twin) (some also have 4—25 mm (2 twin) mountings amidships above the tank deck); 270 rds/min to 3 km *(1.6 n miles)*.
Radars: Navigation: 2 China Type 753; I-band.

Comment: First completed in 1980 at Wuhan Shipyard. Building appeared to terminate in November 1995. Bow and stern ramps fitted. Carry two LCVPs. Bow ramp maximum load 50 tons, stern ramp 20 tons. All based in the East Sea Fleet.

DONGTINGSHAN 6/2012* / 1525588

DONGTINGSHAN 11/2011, A Sheldon-Duplaix / 1483015

10 YUNSHU (TYPE 073 IV) CLASS (LSM)

Name	No	Builders	Launched	Commissioned
SONGSHAN	946	Hudong Zhonghua Shipyard, Shanghai	Jun 2003	2004
–	947	Qingdao Naval Dockyard	1 Aug 2003	2004
XUESHAN	948	Lushun Shipyard	Sep 2003	2004
MENGSHAN	944	Lushun Shipyard	20 Mar 2004	2004
HUASHAN	945	Wuhu Shipyard	1 Jul 2003	2004
SHENGSHAN	941	Hudong Zhonghua Shipyard, Shanghai	Dec 2003	2004
HENGSHAN	949	Lushun Shipyard	Feb 2004	2004
LUSHAN	942	Wuhu Shipyard	2004	2004
YUSHAN	943	Qingdao Naval Dockyard	2004	2004
TAISHAN	950	Hudong Zhonghua Shipyard, Shanghai	Mar 2004	2004

Displacement, tonnes: 1,483 standard; 1,880 full load
Dimensions, metres (feet): 87 × 12.6 × 2.25 *(285.4 × 41.3 × 7.38)*
Speed, knots: 17
Range, n miles: 1,500 at 14 kt
Complement: 70

Military lift: 6 tanks or 12 trucks or 250 tons dry stores
Machinery: 2 diesels; 2 shafts
Guns: 2—37 mm (twin).
Radars: Navigation: I-band.

Comment: A new class of LSM, based on the Yudeng class, built at Zhonghua, Wuhu, Qingdao and Lushun. 941–944 based in the East Sea Fleet and 945–950 in the South Sea Fleet.

MENGSHAN 6/2008 / 1335704

10 YUBEI (TYPE 074A) CLASS (LCU)

No	Builders	Launched	Commissioned
3128	Qingdao Naval Dockyard	Sep 2003	2004
3315	Zhanjiang Shipyard North	2003	2004
3232	Shanghai Shipyard International	Sep 2003	2004
3129	Qingdao Naval Dockyard	Dec 2003	2004
3316	Dinghai Naval Dockyard	Sep 2003	2004
3317	Dinghai Naval Dockyard	Nov 2003	2004
3318	Dinghai Naval Dockyard	Jan 2004	2004
3233	Qingdao Naval Dockyard	2004	2004
3234	–	2004	2005
3235	–	2004	2005

Displacement, tonnes: 914 standard; 1,219 full load
Dimensions, metres (feet): 65 × 11.0 × 2.7 *(213.3 × 36.1 × 8.9)*

Military lift: 10 tanks; 150 troops
Machinery: 2 diesels; 2 shafts
Guns: 4—14.5 mm (2 twin).
Radars: To be announced.

Comment: Built at Qingdao, Zhanjiang, Shanghai and Dinghai. Catamaran hull with superstructure on the starboard side. Basing: *3128* and *3129* in the North Sea Fleet; *3232–35* in the South Sea Fleet; *3315–3318* in the East Sea Fleet.

YUBEI 3316 4/2013*, E & M Laursen / 1525587

120 YUNNAN CLASS (TYPE 067) (LCU)

Displacement, tonnes: 86 standard; 137 full load
Dimensions, metres (feet): 28.6 × 5.4 × 1.5 *(93.8 × 17.7 × 4.9)*
Speed, knots: 12
Range, n miles: 500 at 10 kt
Complement: 12

Military lift: 46 tons
Machinery: 2 12V150 diesels; 600 hp(m) *(441 kW)*; 2 shafts
Guns: 4—14.5 mm (2 twin) MGs.
Radars: Navigation: Fuji; I-band.

Comment: Built in China 1968–72 although a continuing programme was reported in 1982. Pennant numbers in 3000 series (3313, 3321, 3344 seen). 5000 series (5526 seen) and 7000 series (7566 and 7568 seen). One to Sri Lanka in 1991 and a second in 1995. Estimation of numbers is difficult but most are believed to be in reserve or in non-naval service. Some may have 12.7 mm MGs. There are 12 in the East Sea Fleet, the remainder in the South.

YUNNAN 3221 6/2008 / 1335701

10 JINGSHA II CLASS (HOVERCRAFT) (UCAC)

452 +9

Displacement, tonnes: 71 standard; 79 full load
Dimensions, metres (feet): 22 on cushion × ? *(72.2 × ?)*
Speed, knots: 55
Military lift: 15 tons
Machinery: 2 propulsion motors; 2 lift motors
Guns: 4—14.5 mm (2 twin) MGs.

Comment: The prototype was built at Dagu in 1979. This may now have been scrapped and been superseded by this improved version which has a bow door for disembarkation. Numbers are uncertain.

JINGSHA II 1993, Ships of the World / 0056783

20 YUCH'IN (TYPE 068/069) CLASS (LCM)

Displacement, tonnes: 59 standard; 86 full load
Dimensions, metres (feet): 24.8 × 5.2 × 1.3 (81.4 × 17.1 × 4.3)
Speed, knots: 11.5
Range, n miles: 450 at 11.5 kt
Complement: 12
Military lift: Up to 150 troops
Machinery: 2 Type 12V 150C diesels; 600 hp(m) (441 kW); 2 shafts
Guns: 4 — 14.5 mm (2 twin) MGs.

Comment: Built in Shanghai 1962–72. Smaller version of Yunnan class with a shorter tank deck and longer poop deck. Primarily intended for personnel transport. Based in South Sea Fleet. Six sold to Bangladesh and two to Tanzania in 1995.

YUCH'IN 3201   10/2008, Chris Sattler / 1335679

25 TYPE 271 (LANDING CRAFT) (LCU)

Displacement, tonnes: 620 standard; 813 full load
Dimensions, metres (feet): 56.5 × 10.4 × 2.3 (185.4 × 34.1 × 7.5)
Speed, knots: 13
Complement: 25

Military lift: 150 tons
Machinery: 2 diesels; 2,250 hp (1.7 MW); 2 shafts
Guns: 4 — 14.5 mm (2 twin).
Radars: Navigation: I-band.

Comment: Utility landing craft widely used for the transport of troops, vehicles and stores. The first variant (Type 271-I) entered service in about 1970 and this was followed in the late 1970s by Type 270-II and in the late 1980s by Type 271-III. Building continued in the 1990s to replace decommissioned craft but current numbers are approximate. Details are based on the latest generation of craft.

TYPE 271   10/2008, Chris Sattler / 1335678

3 YUYI (TYPE 726) CLASS
(LANDING CRAFT — AIR CUSHION) (LCAC)

3320

Displacement, tonnes: 170 full load
Dimensions, metres (feet): 33 on cushion × ? (108.3 × ?)
Speed, knots: 40
Range, n miles: 200 at 40 kt
Military lift: Armoured Fighting Vehicle plus troops or 60 tons approx
Machinery: 2 QC-70 gas turbines for propulsion and lift; 18,775 hp (14 MW); 2 shrouded reversible-pitch airscrews (propulsion)
Radars: To be announced.

Comment: A new class of air cushion landing craft design, intended for operation from the Yuzhao (Type 071) class LHD. The craft appears to be similar to but larger than the US Navy LCAC. The vehicle is capable of transporting two ZBD-05 AAVs and troops. In contrast to the US Navy LCAC, the driving/command module is located on the port side instead of the starboard side. The main cargo deck is about 6 m wide and there are bow and stern ramps. Propulsion is provided by two 4 m shrouded reversible-pitch propellers. Built at Qiuxin Shipyard, the first vessel was launched in January 2008 and two further craft were reported in 2013.

LCAC   6/2008 / 1335703

2 + 2 POMORNIK (ZUBR) (PROJECT 1232.2) CLASS
(ACVM/LCUJM)

3325   +3

Displacement, tonnes: 559 full load
Dimensions, metres (feet): 57.6 × 25.6 × ? (189 × 84.0 × ?)
Speed, knots: 63
Range, n miles: 300 at 55 kt
Complement: 31 (4 officers)

Military lift: 3 MBT or 10 APC plus 230 troops (total 130 tons)
Machinery: 5 Type NK-12MV gas-turbines; 2 for lift, 23,672 hp(m) (17.4 MW) nominal; 3 for drive, 35,508 hp(m) (26.1 MW) nominal
Missiles: SAM: 2 SA-N-5 Grail quad launchers; manual aiming; IR homing to 6 km (3.2 n miles) at 1.5 Mach; altitude to 2,500 m (8,000 ft); warhead 1.5 kg.
Guns: 2 — 30 mm/65 AK 630; 6 barrels per mounting; 3,000 rds/min combined to 2 km.
Rockets: 2 — 140 mm A-22 Ogon 22-barrelled rocket launchers.
Mines: 2 rails can be carried for 80.
Physical countermeasures: Decoys: MS227 chaff launcher.
Electronic countermeasures: ESM: Tool Box; intercept.
Radars: Surface search: Curl Stone; I-band.
Fire control: Bass Tilt; H/I-band.
IFF: Salt Pot A/B. Square Head.
Weapon control systems: Quad Look (DWU-3) (modified Squeeze Box) optronic director.

Comment: Negotiations with Almaz, St Petersburg, for the procurement of air cushion landing craft were first reported during 2005 and further discussions with Ukraine were opened in 2008. It was announced on 7 August 2009 that four vessels were to be procured. Construction of the first two began at Morye Shipyard, Ukraine, in September 2010 and a further two are to be built in China. Further vessels are likely to be built. The craft, which first entered service in 1986, are operated by the Russian and Ukrainian navies and four had also been exported to Greece by 2005. The world's largest air cushion craft have bow and stern ramps for Ro-Ro working. Details of the craft are as for those in Russian service and Chinese requirements may be different. The first of class was delivered in November 2012 and the second began sea trials in Ukraine in 2013. Two further craft are under construction in China.

MINE WARFARE FORCES

Notes: (1) There are also some 50 auxiliary minesweepers of various types including trawlers and motor-driven junks. Up to 20 Shanghai II class, known as the Fushun class, may be used.
(2) There are six Type 529 drone minesweepers (8041, 8042, 8043, 8181, 8183) based in the East Sea Fleet. They entered service in 2010 and are controlled by Wozang-class MCMVs.

DRONE 8181   9/2012, A Sheldon-Duplaix / 1483017

14 WOSAO (TYPE 082) CLASS
(MINESWEEPERS — COASTAL) (MSC)

800–803   806–807   816–817   820–821   824–827

Displacement, tonnes: 295 standard; 325 full load
Dimensions, metres (feet): 44.8 × 6.8 × 2.3 (147 × 22.3 × 7.5)
Speed, knots: 15
Range, n miles: 500 at 8 kt
Complement: 28

Machinery: 2 diesels; 2,000 hp (1.5 MW); 2 shafts
Guns: 2 China 25 mm/60 (twin); 270 rds/min to 3 km (1.6 n miles).
Mines: 6.
Physical countermeasures: Acoustic, magnetic and mechanical sweeps.
Radars: Navigation: China Type 753; I-band.
Sonars: Hull-mounted; active minehunting.

Comment: Building started in 1986. First of class commissioned in 1988 but second, with modified bridge structure, not seen until 1997. Steel hull with low magnetic properties. Equipped with mechanical (Type 316), magnetic (Type 317), acoustic (Type 318) and infrasonic (Type 319) sweeps. Basing: 820–821 in the North, 824–827 in the South and 800–803, 806–807 and 816–817 in the East Sea Fleet.

WOSAO 821   12/2009, Annati Collection / 1366770

Mine warfare forces < **China** 157

2 + 2 WOZANG CLASS (MCMV)

Name	No	Builders	Launched	Commissioned
HUOQIU	804	Qiuxin Shipyard, Shanghai	Apr 2004	Jul 2005
KUNSHAN	818	Jiangnan Shipyard, Shanghai	2 Nov 2010	2012

Displacement, tonnes: 584 full load
Dimensions, metres (feet): 55 × 9.3 × 2.6 (180.4 × 30.5 × 8.5)

Machinery: 2 diesels; 2 shafts
Guns: 2—25 mm (twin).
Physical countermeasures: To be announced.
Radars: To be announced.
Sonars: To be announced.
Combat data systems: To be announced.

Comment: A new class of mine-countermeasures vessel which was thought to be a successor to the T43 class before the appearance of the Wochi class. Little is known about the capabilities of the vessel. A second ship, to a modified design and possibly of GRP construction was launched in 2010. Two further ships are reported to be under construction. Based in the East Sea Fleet.

HUOQIU   4/2009, Ships of the World / 1305361

KUNSHAN   9/2012, A Sheldon-Duplaix / 1483016

15 (+22 RESERVE) T 43 CLASS (TYPE 6610) (MINESWEEPERS—OCEAN) (MSO)

808–809 811–813 830–838 850

Displacement, tonnes: 528 standard; 599 full load
Dimensions, metres (feet): 60 × 8.8 × 2.3 (196.9 × 28.9 × 7.5)
Speed, knots: 14
Range, n miles: 3,000 at 10 kt
Complement: 70 (10 officers)

Machinery: 2 PCR/Kolomna Type 9-D-8 diesels; 2,000 hp(m) (1.47 MW); 2 shafts
Guns: 2 or 4 China 37 mm/63 (1 or 2 twin) (3 of the class have a 65 mm/52 forward instead of one twin 37 mm/63); dual purpose; 180 rds/min to 8.5 km (4.6 n miles); weight of shell 1.42 kg. 4 USSR 25 mm/60 (2 twin); 270 rds/min to 3 km (1.6 n miles). 4 China 14.5 mm/93 (2 twin); 600 rds/min to 7 km (3.8 n miles). Some also carry 1—85 mm/52 Mk 90K; 18 rds/min to 15 km (8 n miles); weight of shell 9.6 kg.
Mines: Can carry 12–16.
Depth charges: 2 BMB-2 projectors; 20 depth charges.
Physical countermeasures: MCMV MPT-1 paravanes; MPT-3 mechanical sweep; acoustic and magnetic gear.
Radars: Surface search: Fin Curve or Type 756; F-band.
IFF: High Pole or Yard Rake.
Sonars: Tamir II; hull-mounted; active search and attack; high frequency.

Programmes: Started building in 1956 and continued intermittently until about 1987 at Wuhan and at Guangzhou.
Structure: Based on the USSR T 43s, some of which transferred in the mid-1950s but have all now been deleted.
Operational: Some are used as patrol ships with sweep gear removed. Three units reported as having a 65 mm/52 gun forward. Basing: 811 – 813 in the North Sea Fleet; 808, 830 – 834 in the East Sea Fleet; 809, 835 – 838 and 850 in the South Sea Fleet. There are approximately 22 of the class in reserve.
Sales: One to Bangladesh in 1995.

T 43 833   12/2005, Massimo Annati / 1153106

T 43 832   10/2008, Chris Sattler / 1335676

1 WOLEI (TYPE 918) CLASS (MINELAYER) (ML/MST)

LIAOYANG 814

Displacement, tonnes: 2,337 standard; 3,150 full load
Dimensions, metres (feet): 94.9 × 14.4 × 4 (311.4 × 47.2 × 13.1)
Speed, knots: 18
Range, n miles: 7,000 at 14 kt
Complement: 180

Machinery: 2 diesels; 4,300 hp (3.2 MW); 2 shafts
Guns: 2 China 57 mm/50 (twin). 6 China 37 mm/63 (3 twin); 180 rds/min to 8.5 km (4.6 n miles); weight of shell 1.42 kg.
Mines: 300.
Radars: Surface search. Fire control. Navigation.

Comment: Built at Dalian Shipyard and completed successful sea trials in 1988. Resembles the deleted Japanese Souya class and may be used as a support ship as well as a minelayer. Based in the North Sea Fleet.

LIAOYANG   6/2011 / 1406552

9 WOCHI (TYPE 081A) CLASS (MCMV)

ZHANGJIAGANG 805 (ex-328)	LUXI 840	CHANGSHOU 843
JINGJIANG 810 (ex-438)	XIAOYI 841	HESHAN 844
LIUYANG 839 (ex-329)	TAISHAN 842	QINGZHOU 845

Dimensions, metres (feet): 68 × 10 × ? (223.1 × 32.8 × ?)
Machinery: To be announced
Guns: 2—37 mm.
Physical countermeasures: To be announced.
Radars: To be announced.
Sonars: To be announced.
Combat data systems: To be announced.

Comment: A new class of mine-countermeasures vessel (805) which, although outwardly similar to the T43 class is approximately 5 m longer. Construction has taken place at Qiuxin Shipyard, Shanghai, and at Wuhan. Little is known about the details or capabilities of the vessel. A new ship (841), to a modified design, was launched in 2010 and 842 and 843 followed in 2011 and 2012 respectively. 844 and 845 entered service in 2013. Basing: 805 and 810 in the East Sea Fleet; 839–843 in the South Sea Fleet.

WOCHI 805   5/2011 / 1406555

HESHAN   10/2013*, A Sheldon-Duplaix / 1525589

SURVEY AND RESEARCH SHIPS

Notes: (1) The existence of a possible trimaran trials ship (length 60 m approximately) was reported in September 2011. The ship was built at Wuzhou and has pennant number 143.
(2) A SWATH (Small Waterplane Area Twin Hull) survey vessel 429 was completed at Huangpu Shipyard, Guangzhou, in 2011. It is reported to be a prototype oceanographic survey ship.

SWATH 429   6/2012, Ships of the World / 1455684

3 DAHUA (TYPE 909) CLASS (AGOR/AGE)

BI SHENG 891 (ex-970, ex-909)   HUA LUOGENG 892   QIANLAO 893

Displacement, tonnes: 6,096 full load
Dimensions, metres (feet): 130 × 17.5 × 7 (426.5 × 57.4 × 23.0)
Speed, knots: 20
Complement: 80
Machinery: 2 diesels; 2 shafts
Helicopters: Platform for one medium.

Comment: First ship launched on 9 March 1997 at Zhonghua, and completed in August 1997. There is a helicopter deck aft. It is currently fitted with Top Plate air search radar and Front Dome missile fire-control radars. A second unit, also constructed by Hudong-Zhonghua Shipyard, was launched on 30 March 2006 and a third to a modified possibly Type 636B design was launched on 12 November 2011. These key units, all based in the North Sea Fleet, are utilised for radar and missile trials.

BI SHENG   6/2012 / 1483019

HUA LUOGENG   6/2012 / 1483018

1 DADIE (TYPE 814A) CLASS (AGI)

BEIDIAO 900 (ex-841)

Displacement, tonnes: 2,591 full load
Dimensions, metres (feet): 94 × 11.3 × 4 (308.4 × 37.1 × 13.1)
Speed, knots: 17
Complement: 170 (18 officers)
Machinery: 2 diesels; 2 shafts
Guns: 4 – 14.5 mm (2 twin)
Radars: Navigation: 2 Type 753; I-band.

Comment: Built at Wuhan shipyard, Wuchang and commissioned in 1986. North Sea Fleet and seen regularly in Sea of Japan and East China Sea.

BEIDAO 900   4/2008 / 1335671

2 DONGDIAO (TYPE 815) CLASS (AGM/AGI)

BEIJIXING 851 (ex-232)   TIANWANGXING 853

Displacement, tonnes: 6,096 full load
Dimensions, metres (feet): 130 × 16.4 × 6.5 (426.5 × 53.8 × 21.3)
Speed, knots: 20
Complement: 250
Machinery: 2 SEMT-Pielstick diesels; 2 shafts
Guns: 1 – 37 mm. 2 – 14.5 mm.
Helicopters: Platform for one medium.

Comment: The first ship (851) was seen fitting out in 1999. The design appears to be based on the Daxin class training ship. Two radar (possibly missile tracking) arrays have been replaced by three radomes. In service in March 2000. A second ship (853) was launched on 10 December 2009 at Shanghai. 851 is based in the East Sea Fleet and 853 in the South.

DONGDIAO 853   3/2013*, Ships of the World / 1525666

2 KAN CLASS (AGOR)

101   102

Displacement, tonnes: 1,118 full load
Dimensions, metres (feet): 68.6 × 6.9 × 2.7 (225.1 × 22.6 × 8.9)
Speed, knots: 18
Complement: 150
Machinery: 2 diesels; 2 shafts
Radars: Navigation: Fin Curve; I-band.

Comment: Details given are for 102 which is believed built in 1985–87, possibly at Shanghai. Large open stern area. Aft main deck area covered and may have cable reel system. 101 is similar but slightly larger and may have been built in 1965 as an ASR. Operate in East China Sea and Sea of Japan.

KAN 101   5/2000, van Ginderen Collection / 0103684

1 GANZHU CLASS (AGS)

420

Displacement, tonnes: 1,016 full load
Dimensions, metres (feet): 65 × 9 × 3 (213.3 × 29.5 × 9.8)
Speed, knots: 20
Complement: 125
Machinery: 4 diesels; 4,400 hp(m) (3.23 MW); 2 shafts
Guns: 4 – 37 mm/63 (2 twin); 8 – 14.5 mm (4 twin).

Comment: Built at Zhujiang in 1973–75. Long refit in 1996 for up to two years.

GANZHU 420   8/1998 / 0056802

Survey and research ships — Training ships < **China** 159

4 YENLAI (TYPE 635) CLASS (AGS)

226–227   427   943

Displacement, tonnes: 1,057 full load
Dimensions, metres (feet): 73.7 × 9.8 × 3 (241.8 × 32.2 × 9.8)
Speed, knots: 16. **Range, n miles:** 4,000 at 14 kt
Complement: 25
Machinery: 2 PRC/Kolomna Type 9-D-8 diesels; 2,000 hp(m) (1.47 MW) sustained; 2 shafts
Guns: 4 China 37 mm/63 (2 twin). 4—25 mm/80 (2 twin).
Radars: Navigation: Fin Curve; I-band.

Comment: Built at Zhonghua Shipyard, Shanghai in early 1970s. Carries four survey motor boats.

YENLAI 226   6/2005, Hachiro Nakai / 1153052

1 SPACE EVENT SHIP (AGMH)

YUAN WANG 3

Displacement, tonnes: 17,059 full load
Dimensions, metres (feet): 180 × 22.2 × 8.0 (590.6 × 72.8 × 26.2)
Speed, knots: 19. **Range, n miles:** 18,000 at 12 kt
Complement: 470
Machinery: 2 Mitsubishi 8L42MC diesels; 18,554 hp (13.65 MW); 2 shafts
Helicopters: Platform for 1 medium.

Comment: A second-generation space tracking ship launched in 1994 and commissioned in April 1995. Equipped with E/F-band tracking radar. The ship is normally positioned in the South Atlantic off the West African coast for ShenZhou flight missions.

YUAN WANG 3   6/2008 / 1335673

2 SPACE EVENT SHIPS (AGMH)

YUAN WANG 5   YUAN WANG 6

Measurement, tonnes: 22,686 gt
Dimensions, metres (feet): 222.2 × 25.2 × 8.2 (729 × 82.7 × 26.9)
Speed, knots: 20. **Range, n miles:** 20,000 at 12 kt
Machinery: 2 Wärtsilä 8L 46C diesels; 31,078 hp (22.86 MW); 2 shafts
Helicopters: 1 medium.

Comment: Two new third-generation space tracking ships built at Jiangnan Shipyard in Shanghai. Yuan Wang 5 was launched on 15 September 2006 and started undergoing sea trials in September 2007. The ship entered service in September 2007. Yuan Wang 6 was launched on 16 March 2007 and entered service on 12 April 2008. Yuan Wang 6 reportedly differs from Yuan Wang 5 in that it includes a large mission control hall occupying two decks. The ships are equipped with E/F-band and G-band monopulse tracking radars. They are also fitted with a range of meteorological instruments including weather radar, radiosonde and weather balloons.

YUAN WANG 6   8/2009, Ships of the World / 1305687

2 SHUPANG (TYPE 636A) CLASS (AGS)

ZHU KEZHEN (ex-Haiyang 20) 872
QIAN SANQIANG (ex-Haiyang 22) 873

Measurement, tonnes: 4,335 gt
Dimensions, metres (feet): 106 × 17.4 × 5.5 (347.8 × 57.1 × 18.0)
Speed, knots: 13
Machinery: Diesel-electric: 2 diesels; 2 motors; 5,710 hp (4.2 MW); 2 shafts
Radars: To be announced.
Helicopters: 1 medium.

Comment: Naval manned research ships first reported in 2005.

© 2014 IHS

ZHU KEZHEN (ex-HAIYANG 20)   12/2005, Massimo Annati / 1153098

DEEP SUBMERGENCE VEHICLES

1 RESCUE SUBMERSIBLE

LR 7

Dimensions, metres (feet): 9.6 × 3.2 × 3.4 (31.5 × 10.5 × 11.2)
Speed, knots: 3
Complement: 3 (2 pilots)
Machinery: 2 electric motors; 26.8 hp (20 kW); 4 tiltable side thrusters; 16 hp (12 kW)

Comment: Powered by two external lead-acid battery pods, the Perry Slingsby LR 7 is a development of the LR 5 rescue submersible, originally built for North Sea commercial operations and subsequently purchased by the Royal Navy for submarine rescue operations. Capable of operating down to 500 m depth, it can be deployed anywhere in the world and operated from the deck of any suitable mother ship. Its role is to rescue up to 18 survivors at a time from a disabled submarine on the seabed and bring them back to the surface. This can be done at normal atmospheric pressure and at increased pressue up to 5 bar. Mating with the disabled submarine can be achieved at up to 60° bow up. LR 7 is complemented by an ROV, Scorpio 45, which is attached to a 1,000 m umbilical. This is used to locate the disabled submarine, clear obstructions from the escape hatches and replenish life support stores. Following tests in Scotland, LR 7 entered service in the Chinese Navy in 2009.

2 DSRV (SALVAGE SUBMARINES) (DSRV)

Dimensions, metres (feet): 14.9 × 2.6 × 2.6 (48.9 × 8.5 × 8.5)
Speed, knots: 4
Complement: 3
Machinery: 2 silver-zinc batteries; 1 mortar; 1 shaft

Comment: First tested in 1986 and can be carried on large salvage ships. Capable of 'wet' rescue at 200 m and of diving to 600 m. Capacity for six survivors. Underwater TV, high-frequency active sonar and a manipulator arm are all fitted. Life support duration is 1,728 man-hours.

DSRV   1991, CSSC / 0056786

TRAINING SHIPS

1 DAXIN (TYPE 795) CLASS (AXH)

Name	No	Builders	Launched	Commissioned
ZHENGHE	81	Qiuxin, Shanghai	12 Jul 1986	27 Apr 1987

Displacement, tonnes: 5,558 full load
Dimensions, metres (feet): 130 × 16.0 × 4.8 (426.5 × 52.5 × 15.7)
Speed, knots: 15
Range, n miles: 5,000 at 15 kt
Complement: 400 (30 instructors; 200 midshipmen)

Machinery: 2 SEMT Pielstick 6PC2-5L diesels; 7,800 hp(m) (5.73 MW); 2 shafts
Guns: 2 China 57 mm/70 (twin). 4—30 mm AK 230 (2 twin). 4—12.7 mm MGs.
A/S Mortars: 2 FQF 2500 fixed 12-tubed launchers; range 1,200 m; warhead 34 kg.
Radars: Air/surface search: Eye Shield; E-band.
Surface search: China Type 756; I-band.
Navigation: Racal Decca 1290; I-band.
Fire control: Round Ball; I-band.
Sonars: Echo Type 5; hull-mounted; active; high frequency.
Helicopters: Platform only.

Comment: Resembles a small cruise liner. Subordinate to the Naval Academy and replaced Huian. Based in the North Sea Fleet. A similar ship sold to Algeria in 2006.

ZHENGHE   9/2010, John Mortimer / 1366771

IHS Jane's Fighting Ships 2014-2015

160 China > Training ships — Auxiliaries

1 JIANGHU IV (TYPE 053HTH) CLASS (FFGH)

Name	No	Builders	Laid down	Launched	Commissioned
LUSHUN (ex-Siping)	544	Hudong Shipyard, Shanghai	1984	Sep 1985	Nov 1986

Displacement, tonnes: 1,575 standard; 1,895 full load
Dimensions, metres (feet): 103.2 × 10.8 × 3.1 (338.6 × 35.4 × 10.2)
Speed, knots: 26
Range, n miles: 4,000 at 15 kt, 2,700 at 18 kt
Complement: 185 (30 officers)

Machinery: 2 Type 12E 390V diesels; 14,400 hp(m) (10.6 MW) sustained; 2 shafts
Missiles: SSM: 2 HY-2 (C-201) (CSSC-3 Seersucker) (twin) launchers ❶; active radar or IR homing to 95 km (51 n miles) at 0.9 Mach; warhead 513 kg.
Guns: 1 Creusot-Loire 3.9 in (100 mm)/55 ❷; 60–80 rds/min to 17 km (9.3 n miles); weight of shell 13.5 kg. 8 China 37 mm/63 (4 twin) ❸; 180 rds/min to 8.5 km (4.6 n miles) anti-aircraft; weight of shell 1.42 kg.
Torpedoes: 6—324 mm ILAS (2 triple) tubes ❹. Yu-2 (Mk 46 Mod 1) active/passive homing to 11 km (5.9 n miles) at 40 kt; warhead 44 kg.
A/S Mortars: 2 RBU 1200 5-tubed fixed launchers ❺; range 1,200 m; warhead 34 kg.
Physical countermeasures: Decoys: 2 SRBOC Mk 33 6-barrelled chaff launchers or 2 China 26-barrelled launchers.
Electronic countermeasures: ESM: Jug Pair or Watchdog; radar warning.

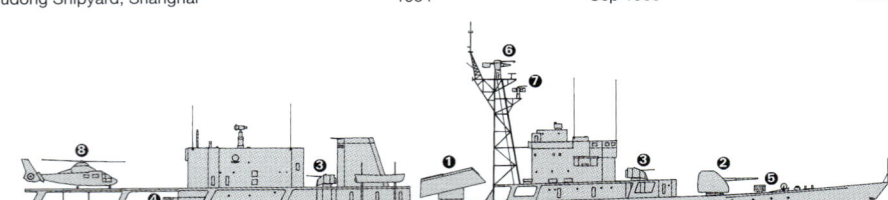

LUSHUN   (Scale 1 : 900), Ian Sturton / 0572397

Radars: Air/surface search: Type 354 Eye Shield (MX 902) ❻; G-band.
Surface search/fire control: Type 352 Square Tie ❼; I-band.
Navigation: Don 2 or Fin Curve; I-band.
IFF: High Pole A. Yard Rake or Square Head.
Sonars: Echo Type 5; hull-mounted; active search and attack; medium frequency.
Weapon control systems: CSEE Naja optronic director for 100 mm gun.
Helicopters: Harbin Z-9C (Dauphin) ❽.

Programmes: Built as a standard Jianghu I and then converted, probably as a helicopter trials ship for the Luhu and Jiangwei classes, before being commissioned.
Structure: The after part of the ship has been rebuilt to take a hangar and flight deck for a single helicopter. Alcatel 'Safecopter' landing aid. This ship also has a French 100 mm gun and optronic director, and Italian triple torpedo tubes mounted on the quarterdeck.
Operational: Based in North Sea Fleet at Guzhen Bay. Formally changed to a new role as training ship for Dalian Naval Academy on 28 July 2010.

LUSHUN   6/2003 / 0569166

1 SHICHANG CLASS (HSS/AHH)

Name	No	Builders	Launched	Commissioned
SHICHANG	82	Qiuxin, Shanghai	Apr 1996	27 Jan 1997

Displacement, tonnes: 10,160 full load
Dimensions, metres (feet): 120 × 18 × 7 (393.7 × 59.1 × 23.0)
Speed, knots: 17.5. **Range, n miles:** 8,000 at 17 kt
Complement: 370 (200 trainees)
Military lift: 300 containers
Machinery: 2 diesels; 2 shafts
Helicopters: 2 Zhi-9A Haitun.

Comment: China's first air training ship described officially as a defence mobilisation vessel which can be used for civilian freight, for helicopter or navigation training, or as a hospital ship. The vessel looks like a scaled down version of the UK Argus with the bridge superstructure forward and an after funnel on the starboard side of the flight deck. There are two landing spots. Based in the North Sea Fleet.

SHICHANG   5/1998, RAN / 0017739

AUXILIARIES

Notes: (1) There are four 104 m supply tankers 640, 968, 971, and 972.
(2) There are three tankers 637 (94 m), 625 and 960 (100 m). All of unknown type.
(3) There is an auxiliary of unknown type 870.
(4) There are numerous army tankers whose pennant numbers are prefixed by N or NU.

6 QIONGSHA CLASS (4 AP + 2 AH)

| 830–831 | NANYI 832 – (ex-832) | NANYI 09 – (ex-833) | Y 834 | Y 835 |

Displacement, tonnes: 2,185 full load
Dimensions, metres (feet): 86 × 13.5 × 4 (282.2 × 44.3 × 13.1)
Speed, knots: 16
Complement: 59

Military lift: 400 troops; 350 tons cargo
Machinery: 3 SKL 8 NVD 48 A-2U diesels; 3,960 hp(m) (2.91 MW) sustained; 3 shafts
Guns: 8 China 14.5 mm/93 (4 twin); 600 rds/min to 7 km (3.8 n miles).
Radars: Navigation: Fin Curve; I-band.

Comment: Personnel attack transports begun about 1980. Previous numbers of this class were overestimated. All South Sea Fleet. Has four sets of davits, light cargo booms serving forward and aft. No helicopter pad. Twin funnels. Carries a number of LCAs. Nanyi 832 (ex-832) and Nanyi 09 (ex-833) converted to Hospital Ships (AH) and painted white.

SHICHANG   5/1998, Sattler/Steele / 0017738

NANYI 832   6/2008 / 1335698

IHS Jane's Fighting Ships 2014-2015   © 2014 IHS

Auxiliaries < **China** 161

2 FUQING (TYPE 905) CLASS
(REPLENISHMENT SHIPS) (AORH)

HONGZHU (ex-*Taicang*) 881 (ex-575)   **POYANG HU** (ex-*Fencang*) 882 (ex-615)

Displacement, tonnes: 7,620 standard; 22,099 full load
Dimensions, metres (feet): 168.2 × 21.8 × 9.4 (551.8 × 71.5 × 30.8)
Speed, knots: 18. **Range, n miles:** 18,000 at 14 kt
Complement: 130 (24 officers)

Cargo capacity: 10,550 tons fuel; 1,000 tons dieso; 200 tons feed water; 200 tons drinking water; 4 small cranes
Machinery: 1 Sulzer 8RL B66 diesel; 15,000 hp(m) (11 MW) sustained; 1 shaft
Guns: 8—37 mm (4 twin) (fitted for but not with).
Radars: Navigation: Fin Curve or Racal Decca 1290; I-band.
Helicopters: Platform for 1 medium.

Comment: Operational in late 1979. This is the first class of ships built for underway replenishment in the Chinese Navy. Helicopter platform but no hangar. Both built at Dalian. Two liquid replenishment positions each side with one solid replenishment position each side by the funnel. A third of the class *Hongcang* (X 950) was converted to merchant use in 1989 and renamed *Hai Lang*, registered at Dalian. A fourth (X 350) was sold to Pakistan in 1987. 882 based in the North and 881 in the East. *Fengcang* appears to have a command role.

HONGZHU   6/2012, Ships of the World / 1483599

POYANG HU   6/2011, Ships of the World / 1406478

1 NANYUN CLASS (REPLENISHMENT SHIP) (AORH)

Name	No	Builders	Launched	Commissioned
QINGHAI HU (ex-*Nancang*; ex-*Vladimir Peregudov*)	885 (ex-953)	Kherson Shipyard	Apr 1992	2 Jun 1996

Displacement, tonnes: 37,594 full load
Measurement, tonnes: 29,211 dwt
Dimensions, metres (feet): 178.9 × 25.3 × 11 (586.9 × 83.0 × 36.1)
Speed, knots: 16
Complement: 125
Cargo capacity: 9,630 tons fuel
Machinery: 1 B&W diesel; 11,600 hp(m) (8.53 MW); 1 shaft
Helicopters: 1 Super Frelon.

Comment: Sometimes referred to as Fusu class. One of a class of 11 built at Kherson Shipyard, Crimea. Laid down in January 1989. Sailed from Ukraine to Dalian Shipyard in 1993. Completed fitting out in China and joined the South Sea Fleet. RAS rigs on both sides and stern refuelling. Similar to Indian *Jyoti* but with better helicopter facilities.

QINGHAI HU (old number)   8/2000, Robert Pabst / 0103677

QINGHAI HU   6/2005, A Sheldon-Duplaix / 1153101

1 DADONG (TYPE 946A) CLASS (SALVAGE SHIPS) (ARS)

304

Displacement, tonnes: 1,524 full load
Dimensions, metres (feet): 84.4 × 11 × 2.7 (276.9 × 36.1 × 8.9)
Speed, knots: 18
Complement: 150
Machinery: 2 diesels; 7,400 hp(m) (5.44 MW); 2 shafts
Guns: 4—25 mm/80 (2 twin).
Radars: Navigation: Type 756; F-band.

Comment: Built at Hudong Shipyard, Shanghai and entered service in December 1982. Has a large and conspicuous crane aft. Principal role is wreck location and salvage. Based in the East Sea Fleet.

DADONG 304   6/2011 / 1483020

4 FUCHI (TYPE 903) CLASS (AORH)

Name	No	Builders	Laid down	Launched	Commissioned
QIANDAO HU (ex-*Fuchi*)	886	Hudong Shipyard, Shanghai	2002	29 Mar 2003	30 Apr 2004
WEISHAN HU	887	Guangzhou Shipyard International	–	Jun 2003	2004
TAI HU	889	Donglang Shipyard, Guangzhou	2010	22 Mar 2012	18 Jun 2013
CHAO HU	890	Hudong Shipyard, Shanghai	2010	6 May 2012	11 Sep 2013

Displacement, tonnes: 23,369 full load
Dimensions, metres (feet): 178.5 × 24.8 × 8.7 (585.6 × 81.4 × 28.5)
Speed, knots: 19
Range, n miles: 10,000 at 14 kt
Complement: 130

Cargo capacity: 10,500 tons fuel, 250 tons of water, 680 tons of ammunition and stores
Machinery: 2 SEMT-Pielstick diesels; 24,000 hp (17.9 MW); 2 shafts
Guns: 8—37 mm (4 twin).
Radars: To be announced.
Helicopters: Platform for 1 medium.

Comment: Ships which bear a marked resemblance to Type R22T Similan class tanker built for Thailand in 1996. Fitted with two RAS stations (one liquids, one solids) on each side. Basing: 886 and 890 in the East Sea Fleet, 887 in the South and 889 in the North.

QIANDAO HU   12/2012, Chris Sattler / 1483023

WEISHAN HU   8/2012, C D Yaylali / 1483022

4 DALANG (TYPE 922 II/III) CLASS
(SUBMARINE SUPPORT SHIPS) (ASL)

122   138   332   510

Displacement, tonnes: 3,759 standard; 4,267 full load
Dimensions, metres (feet): 111.9 × 14.6 × 4.3 (367.1 × 47.9 × 14.1)
Speed, knots: 16. **Range, n miles:** 8,000 at 14 kt
Complement: 180
Machinery: 2 diesels; 4,000 hp(m) (2.94 MW); 2 shafts
Guns: 2—25 mm/80 (1 twin) or 2—14.5 mm/93 (1 twin).
Radars: Navigation: Fin Curve; I-band.

Comment: Construction of the first Type 922-II class (305 (ex-503) began at Guangzhou Shipyard in September 1971. It was commissioned in November 1975 and later transferred to the fisheries agency in 2006 and subsequently to the Coast Guard in 2013. As a result of experience gained, development of an improved Type 922-III version began in 1978. Construction of the first of these ships (122) began in December 1982 at Wuchang Shipyard, Wuhan, and the ship later commissioned in 1986. Subsequently, three further modified ships were built: 332 (1989), 138 (1992) and 510 (1995). These modifications include changes to upper deck design and the possible incorporation of a decompression chamber. Basing: 122 and 138 in North Sea Fleet; 332 in East Sea Fleet and 510 in South Sea Fleet.

DALANG 332   6/2010, Ships of the World / 1366773

© 2014 IHS   IHS Jane's Fighting Ships 2014-2015

162 China > Auxiliaries

3 DAJIANG (TYPE 925) CLASS
(SUBMARINE SUPPORT SHIPS) (ASRH)

CHANGXINGDAO 861 (ex-J 121) **YONGXINGDAO** 863 (ex-J 506)
CHONGMINGDAO 862 (ex-J 302)

Displacement, tonnes: 12,167 full load
Dimensions, metres (feet): 156.2 × 20.6 × 6.8 *(512.5 × 67.6 × 22.3)*
Speed, knots: 20
Complement: 308

Machinery: 2 MAN K9Z60/105E diesels; 9,000 hp(m) *(6.6 MW)*; 2 shafts
Guns: Light MGs. Can carry 6 — 37 mm (3 twin).
Radars: Surface search: Eye Shield; E-band.
Navigation: 2 Fin Curve; I-band.
Helicopters: 2 Aerospatiale SA 321G Super Frelon.

Comment: Submarine support and salvage ships built at Shanghai. First launched in mid-1973, operational in 1976. *Yongxingdao* has a smoke deflector on funnel. Provision for DSRV on forward well-deck aft of launching crane. Foremast on *Yongxingdao* suggests long-range communications capability, possibly for submarine command. Basing: 861 in the North Sea Fleet; 862 in the East Sea Fleet; 863 in the South Sea Fleet.

CHONGMINGDAO 6/2011, Ships of the World / 1406479

2 DAZHOU (TYPE 946) CLASS (SUBMARINE TENDERS) (ASL)

502 137 (ex-504)

Displacement, tonnes: 1,118 full load
Dimensions, metres (feet): 79 × 9.5 × 2.6 *(259.2 × 31.2 × 8.5)*
Speed, knots: 18
Complement: 130
Machinery: 2 diesels; 2 shafts
Guns: 2 China 37 mm/63 (twin). 4 — 14.5 mm/93 (2 twin).
Radars: Navigation: Fin Curve; I-band.

Comment: The first, *502*, commissioned in 1977; the second in 1978. Both built at Guangzhou Shipyard. *502* based in the South Sea Fleet and *137* in the North Sea Fleet. Both have been used as AGIs.

DAZHOU 502 6/2008, / 1335697

3 YANTAI (TYPE 073) CLASS (SUPPLY SHIPS) (AK)

745 (ex-801) 757 938 (ex-800)

Displacement, tonnes: 3,383 full load
Dimensions, metres (feet): 78 × 11.5 × 3.0 *(255.9 × 37.7 × 9.8)*
Speed, knots: 17
Range, n miles: 3,000 at 16 kt
Complement: 100
Machinery: 2 diesels; 9,600 hp(m) *(7.06 MW)*; 2 shafts
Guns: 2 China 37 mm/63 (twin).
Radars: Navigation: Type 756; I-band.

Comment: First seen in 1992. Appears to be based on Yukan-class landing ship design but without a bow door. Fitted with cargo-handling cranes fore and aft. A ship with pennant number 938 has also been reported unloading missile containers. It is not known whether this is an additional ship or a change of pennant number. 745 and 938 based in South Sea Fleet and 757 in East Sea Fleet.

YANTAI 938 6/2007 / 1166855

2 DAYUN (TYPE 904) CLASS SUPPLY SHIPS (AKH)

883 (ex-951) 884 (ex-952)

Displacement, tonnes: 8,636 standard; 11,151 full load
Dimensions, metres (feet): 124.2 × 12.8 × 3.8 *(407.5 × 42.0 × 12.5)*
Speed, knots: 22
Complement: 240

Machinery: 2 diesels; 9,000 hp(m) *(6.6 MW)*; 2 shafts
Guns: 4 — 37 mm/63 (2 twin). 4 — 25 mm/80 (2 twin).
Radars: Navigation: 2 Type 756; I-band.
Helicopters: 2 SA 321 Super Frelon.

Comment: First of class completed at Hudong Shipyard in March 1992, second in August 1992. Four landing craft are embarked. Both based in South Sea Fleet. A reported third of class was in fact the first of the larger Nanyun class. Dayun 883 has also served as Yuzheng 21 (fishery protection).

DAYUN CLASS 6/2005, A Sheldon-Duplaix / 1153100

9 DANLIN CLASS (SUPPLY SHIPS) (AK/AOT)

531 591–592 594 794 827 834–835 975

Displacement, tonnes: 1,311 full load
Dimensions, metres (feet): 60.5 × 9 × 4 *(198.5 × 29.5 × 13.1)*
Speed, knots: 15
Complement: 35

Cargo capacity: 750–800 tons
Machinery: 1 USSR/PRC Type 6DRN 30/50 diesel; 750 hp(m) *(551 kW)*; 1 shaft
Guns: 4 — 25 mm/80 (2 twin). 4 — 14.5 mm (2 twin).
Radars: Navigation: Fin Curve or Skin Head; I-band.

Comment: Built in China in early 1960–62. The AKs have refrigerated stores capability and serve in the South Sea Fleet. The AOTs are split between the Fleets. Not all are armed.

DANLIN 794 5/1992, Henry Dodds / 0056790

11 DANDAO (TYPE 917) CLASS (AK/AOT)

455 484–485 529 758–759 802–803 841 844–845

Displacement, tonnes: 1,626 full load
Dimensions, metres (feet): 65.7 × 12.5 × 4 *(215.6 × 41.0 × 13.1)*
Speed, knots: 12
Complement: 40
Machinery: 1 diesel; 1 shaft
Guns: 4 China 37 mm/63 (2 twin). 4 China 14.5 mm/93 (2 twin).
Radars: Navigation: Fin Curve; I-band.

Comment: Built in the late 1970s. Similar to the Danlin class. Basing: 455, 484, 485, and 529 in the North Sea Fleet; 758, 759, 802, and 803 in the East Sea Fleet; 841, 844 and 845 in the South Sea Fleet.

DANDAO 529 9/2007 / 1335667

Auxiliaries < **China** 163

7 HONGQI CLASS (AK)

| 443 | 528 | 755–756 | 771 | 835–836 |

Displacement, tonnes: 1,981 full load
Dimensions, metres (feet): 62 × 12 × 4.4 (203.4 × 39.4 × 14.4)
Speed, knots: 14. **Range, n miles:** 2,500 at 11 kt
Complement: 35
Machinery: 1 diesel; 1 shaft
Guns: 4 China 25/80 (2 twin).

Comment: Used to support offshore military garrisons. A further ship, L 202, appears to be similar but carries no armament. Others of this type in civilian use. Basing: 443 and 528 in the North Sea Fleet; 755, 756 and 771 in the East Sea Fleet; 835 and 836 in the South Sea Fleet.

HONGQI 755   3/2003, Bob Fildes / 0569175

8 YOUDIAN (TYPE 991) CLASS (CABLE SHIPS) (ARC)

| 233–234 | 764–765 | 868 | 873–874 | 882 |

Displacement, tonnes: 1,350 full load
Dimensions, metres (feet): 71.4 × 10.5 × 3.4 (234.3 × 34.4 × 11.2)
Speed, knots: 14
Complement: 95
Machinery: 2 diesels; 2 shafts
Guns: 4 — 37 mm. 4 — 12.7 mm MGs.
Radars: Navigation: Fin Curve; I-band.

Comment: Built in 1970s for both military and civil use. Basing: 233 and 234 (South); 764 and 765 (North); 868, 873, 874 and 882 (East).

4 YULIN CLASS (TANKERS) (AOL)

| 648–649 | 964–965 |

Dimensions, metres (feet): 71.3 × 10.4 × 3.4 (233.9 × 34.1 × 11.2)
Machinery: To be announced

Comment: Basing: 964–965 in the South Sea Fleet; 648–649 in the East Sea Fleet.

YULIN 649   9/2012, A Sheldon-Duplaix / 1483021

6 SUPPLY TANKERS (AOL)

| 572 | 633 | 636 | 958 | 963 | 976 |

Dimensions, metres (feet): 54 × ? × ? (177.2 × ? × ?)
Machinery: To be announced
Radars: To be announced.

Comment: Six supply tankers of an unknown type. Basing: 572 in the North Sea Fleet; 633 and 636 in the East Sea Fleet; 958, 963 and 976 in the South Sea Fleet.

6 SUPPLY TANKERS (AOL)

| 565 | 631 | 641 | 957 | 959 | 973 |

Dimensions, metres (feet): 53 × ? × ? (173.9 × ? × ?)
Machinery: To be announced
Radars: To be announced.

Comment: Supply tankers of an unknown type. Basing: 565 in the North Sea Fleet; 631 and 641 in the East Sea Fleet; 957, 959 and 973 in the South Sea Fleet.

AOL 631   5/2011 / 1406556

19 FULIN CLASS (REPLENISHMENT SHIPS) (AOT)

| 400 | 566 | 581 | 634–635 | 642 | 941 | 966 |
| 561–562 | 572 | 632 | 638–639 | 650 | 961–962 | 969–970 |

Displacement, tonnes: 2,337 standard
Dimensions, metres (feet): 66 × 13 × 4 (216.5 × 42.7 × 13.1)
Speed, knots: 10
Range, n miles: 1,500 at 8 kt
Complement: 30
Machinery: 1 diesel; 600 hp(m) (441 kW); 1 shaft
Guns: 4 — 14.5 mm/93 (2 twin).
Radars: Navigation: Fin Curve; I-band.

Comment: A total of 20 of these ships built at Hudong, Shanghai, beginning in 1972. There are 14 oil tankers and five water tankers. Basing: 400, 561, 562, 566, 572, and 581 in the North Sea Fleet; 630, 632, 634, 635, 638, 639, 642, and 650 in the East Sea Fleet; 961, 962, 966, 969, and 970 in the South Sea Fleet.

FULIN 630   5/2011 / 1406557

2 SHENGLI CLASS (AOT)

| 620 | 621 |

Displacement, tonnes: 3,353 standard; 5,029 full load
Dimensions, metres (feet): 101 × 13.8 × 5.5 (331.4 × 45.3 × 18.0)
Speed, knots: 14
Range, n miles: 2,400 at 11 kt
Complement: 48

Cargo capacity: 3,400 tons dieso
Machinery: 1 6 ESDZ 43/82B diesel; 2,600 hp(m) (1.91 MW); 1 shaft
Guns: 2 — 37 mm/63 (twin). 4 — 25 mm/80 (2 twin).
Radars: Navigation: Fin Curve; I-band.

Comment: Built at Hudong SY, Shanghai in late 1970s. Others of the class in commercial service. Both based in the East Sea Fleet.

1 ANWEI (TYPE 920) CLASS (HOSPITAL SHIP) (AHH)

Name	No	Builders	Laid down	Launched	Commissioned
DAISHANDAO	866	Guangzhou Shipyard International	2006	29 Aug 2007	2008

Displacement, tonnes: 23,369 full load
Dimensions, metres (feet): 180 × 24.6 × 9 (590.6 × 80.7 × 29.5)
Speed, knots: 19
Range, n miles: 10,000 at 14 kt
Complement: 130
Machinery: 2 diesels; 2 shafts
Radars: To be announced.
Helicopters: 1 medium.

Comment: The first purpose-built hospital ship for the Chinese Navy was launched in August 2007 and commissioned in 2008. The design seems to be based on the Fuchi class replenishment ships. The ship is equipped with eight operating theatres and is fitted with a flight deck and hangar capable of operating a medium size helicopter. Based in the East Sea Fleet.

DAISHANDAO   12/2009, Annati Collection / 1366774

2 JINYOU CLASS (AOT)

| 625 | 960 |

Displacement, tonnes: 4,877 full load
Dimensions, metres (feet): 99 × 31.8 × 5.7 (324.8 × 104.3 × 18.7)
Speed, knots: 15
Range, n miles: 4,000 at 10 kt
Complement: 40
Machinery: 1 SEMT-Pielstick 8PC2.2L diesel; 3,000 hp (2.24 MW); 1 shaft
Radars: Navigation: I-band.

Comment: Built by Kanashashi Shipyard, Japan and entered service 1989–90. Basing: 625 in the East Sea Fleet; 960 in the South Sea Fleet.

164 China > Auxiliaries

18 FUZHOU CLASS (AOT/AWT)

| 555 | 560 | 570 | 589 | 626 | 643–644 | 940–941 |
| 557 | 563 | 573 | 606–607 | 628–629 | 937–938 | |

Displacement, tonnes: 2,134 full load
Dimensions, metres (feet): 63.5 × 12.6 × 3.8 *(208.3 × 41.3 × 12.5)*
Speed, knots: 11
Complement: 35

Cargo capacity: 600 tons
Machinery: 1 diesel; 600 hp(m) *(441 kW)*; 1 shaft
Guns: 4—25 mm/80 (2 twin). 4—14.5 mm/93 (2 twin).
Radars: Navigation: Fin Curve; I-band.

Comment: Built 1964–70. Transport ships for liquids, 11 for oil and seven for water. Basing: 555, 557, 560, 563, 570, 573, and 589 in the North Sea Fleet; 606, 607, 626, 628, 629, 643, and 644 in the East Sea Fleet; 937, 938, 940, and 941 in the South Sea Fleet.

FUZHOU 629   6/2007 / 1166852

5 GUANGZHOU CLASS (AOTL/AWTL)

| 571 | 590 | 593 | 645 | 954 |

Displacement, tonnes: 539 full load
Dimensions, metres (feet): 49 × 7.5 × 3.0 *(160.8 × 24.6 × 9.8)*
Speed, knots: 10
Complement: 19
Machinery: 1 diesel; 1 shaft
Guns: 4—14.5 mm/93 (2 twin).

Comment: Coastal tankers built in the 1970s and 1980s. 571 is an oil tanker and the remainder water tankers. Basing: 571, 590 and 593 in the North Sea Fleet; 645 in the East Sea Fleet; 954 in the South Sea Fleet.

GUANGZHOU 645   10/2008, Chris Sattler / 1335669

5 ANKANG CLASS (HOSPITAL SHIPS) (AHH)

| 01 | 10–13 |

Dimensions, metres (feet): 60 × 9 × ? *(196.9 × 29.5 × ?)*
Machinery: 2 diesels; 2 shafts
Radars: To be announced.

Comment: Small hospital ships that work in conjunction with the larger Anwei class. Basing: 01 based in the North Sea Fleet; 12 and 13 based in the East Sea Fleet; 10 and 11 in the South Sea Fleet.

ANKANG 13   10/2009 / 1305807

4 YANNAN CLASS (BUOY TENDERS) (ABU)

| 263 | 463 | 982–983 |

Displacement, tonnes: 1,778 standard
Dimensions, metres (feet): 72.3 × 11.8 × 4 *(237.2 × 38.7 × 13.1)*
Speed, knots: 12
Complement: 95
Machinery: 2 diesels; 2,640 hp(m) *(1.94 MW)*; 2 shafts
Radars: Navigation: Fin Curve; I-band.

Comment: Built 1978–79; commissioned 1980. Basing: 982 and 983 in the North Sea Fleet; 263 in the East Sea Fleet; 463 in the South Sea Fleet.

YANNAN CLASS   3/2004, L-G Nilsson / 1042153

7 YEN PAI (YANCI) CLASS (ADG)

| 203 | 205 | 735–736 | 860 | 863–864 |

Displacement, tonnes: 758 standard
Dimensions, metres (feet): 65 × 9 × 2.6 *(213.3 × 29.5 × 8.5)*
Speed, knots: 16
Range, n miles: 800 at 15 kt
Complement: 55
Machinery: Diesel-electric; 2 12VE 230ZC diesels; 2,200 hp(m) *(1.62 MW)*; 2 ZDH-99/57 motors; 2 shafts
Guns: 4—37 mm/63 (2 twin). 4—25 mm/80 (2 twin).
Radars: Navigation: Type 756; I-band.

Comment: Enlarged version of T 43 MSF with larger bridge and funnel amidships. Reels on quarterdeck for degaussing function. Not all the guns are embarked. Basing: 735 and 736 in the North Sea Fleet; 860, 863 and 864 in the East Sea Fleet; 203 and 205 in the South Sea Fleet.

YEN PAI 864   10/2008, Chris Sattler / 1335666

3 DALAO (TYPE 926) CLASS (SUBMARINE RESCUE SHIP) (ASR)

Name	No	Builders	Laid down	Launched	Commissioned
HAIYANGDAO	864	Guangzhou Shipyard International	2007	25 Nov 2008	2010
LIUGONGDAO	865	Guangzhou Shipyard International	2010	Oct 2011	2013
CHANGDAO	867	Guangzhou Shipyard International	2010	9 Aug 2011	2013

Displacement, tonnes: 9,652 full load
Dimensions, metres (feet): 136 × 18.5 × ? *(446.2 × 60.7 × ?)*
Machinery: To be announced
Radars: To be announced.

Comment: A new class of purpose-built submarine rescue ships to carry and deploy the new LR7 rescue submersible. The first ship was commissioned in 2010 and two further ships were commissioned in 2013. One ship is likely to be allocated to each fleet.

HAIYANGDAO   6/2010, Ships of the World / 1366438

1 TYPE 648 SUBMARINE TENDER (ASL)

DONGXIU 911

Displacement, tonnes: 3,556 standard; 4,064 full load
Dimensions, metres (feet): 86 × 14.0 × 4 *(282.2 × 45.9 × 13.1)*
Speed, knots: 16

Machinery: 2 diesels; 2 shafts
Guns: 8—25 mm (4 twin).
Radars: Navigation: I-band.
Helicopters: Platform for one medium.

Comment: The first and only hull of its class was built at Wuhu and commissioned in 1985. The role of the ship is to provide conventional submarines with repair and maintenance facilities in addition to fuel and water. Based in the East Sea Fleet.

DONGXIU　　　　　　　　　　　　　　　　　　　　　5/2011 / 1406558

2 DACHOU (TYPE 917A) CLASS (YPT)

530　　　846

Dimensions, metres (feet): 67 × 10 × ? *(219.8 × 32.8 × ?)*
Machinery: To be announced
Guns: To be announced.
Physical countermeasures: To be announced.
Radars: To be announced.
Sonars: To be announced.
Combat data systems: To be announced.

Comment: A new class of torpedo recovery vessel which appears very similar in design to that of the Wochi class mine countermeasures vessels on which outline details are based. The first was built at Wuzhou Shipyard and entered service in 2006. Basing: 530 in the North Sea Fleet; 846 in the South Sea Fleet.

DACHOU 846　　　　　　　　　　　　　　　　6/2008, Ships of the World / 1335653

1 DANYAO (TYPE 904A) CLASS (SUPPORT SHIP) (AF)

888

Displacement, tonnes: 15,241 full load
Dimensions, metres (feet): 152 × 19 × ? *(498.7 × 62.3 × ?)*
Machinery: 2 SEMT Pielstick 16PC V 400 diesels; 24,000 hp *(17.9 MW)*; 2 shafts
Guns: 2—37 mm (twin).
Helicopters: Platform for one medium.

Comment: Support ship constructed at Guangzhou, and launched on 28 December 2006. The ship is equipped with two pairs of davits, capable of handling small landing craft, and a flight deck for medium helicopters. Following sea trials in 2007, the ship is reported to have been commissioned in late 2007. In 2009, the ship transferred to the Fishery Law Enforcement Command (FLEC) and changed its pennant number to 88 before returning to combat support duties in the South Sea Fleet.

DANYAO 888　　　　　　　　　　　　　　　　12/2009, Annati Collection / 1366160

ICEBREAKERS

2 YANHA CLASS (AGB/AGI)

721　　　722

Displacement, tonnes: 3,251 full load
Dimensions, metres (feet): 88.4 × 16.2 × 5.2 *(290 × 53.1 × 17.1)*
Speed, knots: 17.5
Complement: 90
Machinery: Diesel-electric; 2 diesel generators; 1 motor; 1 shaft
Guns: 8—37 mm/63 Type 61/74 (4 twin). 4—25 mm/80 Type 61.
Radars: Navigation: Fin Curve; I-band.

Comment: Icebreakers used as AGIs in the North Sea Fleet.

© 2014 IHS

YANHA CLASS　　　　　　　　　　　　　　　　　10/1991, G Jacobs / 0505974

TUGS

Notes: The vessels below represent a cross-section of the craft available.

3 TUQIANG CLASS (TUGS) (ATA)

181　　　189　　　721

Dimensions, metres (feet): 80 × ? × ? *(262.5 × ? × ?)*
Machinery: To be announced
Radars: To be announced.

Comment: Ocean-going tugs built 2004–08. Basing: 721 in the North Sea Fleet; 181 and 189 in the South Sea Fleet.

TUQIANG 721　　　　　　　　　　　　　　　　6/2012, Ships of the World / 1366842

1 DAOZHA CLASS (ATF)

Displacement, tonnes: 4,064 full load
Dimensions, metres (feet): 84 × 12.6 × 5.4 *(275.6 × 41.3 × 17.7)*
Speed, knots: 18
Complement: 125
Machinery: 2 diesels; 8,600 hp(m) *(6.32 MW)*; 2 shafts

Comment: Built in 1993–94 probably as a follow-on to the Tuzhong class. Based in South Sea Fleet.

DAOZHA　　　　　　　　　　　　　　　　　　9/1993, Hachiro Nakai / 0506142

19 HUJIU CLASS (ATF)

147	164	185	635	715	836–837	875
155–156	174–175	622	711–712	717	842–843	877

Displacement, tonnes: 1,494 full load
Dimensions, metres (feet): 60.2 × 11.6 × 4.4 *(197.5 × 38.1 × 14.4)*
Speed, knots: 15. **Range, n miles:** 7,200 at 14 kt
Complement: 56
Machinery: 2 LVP 24 diesels; 1,800 hp(m) *(1.32 MW)*; 2 shafts
Radars: Navigation: Fin Curve or Type 756; I-band.

Comment: Built at Wuhu in 1980s. One sold to Bangladesh in 1984 and a second in 1995. Basing: 622, 635, 711, 712, 715, and 717 in the North Sea Fleet; 836, 837, 842, 843, 875, and 877 in the East Sea Fleet; 147, 155, 156, 164, 174, 175, and 185 in the South Sea Fleet.

HUJIU 877　　　　　　　　　　　　　　　　　10/2006, E & M Laursen / 1164866

1 OFFSHORE PATROL SHIP (PSO)

Name	Builders	Laid down	Launched	Commissioned
HAIJING 2151 (ex-*Haijian 51*)	Wuchang Shipyard, Wuhan	30 Dec 2004	5 Jun 2005	16 Nov 2005

Measurement, tonnes: 1,937 gt
Dimensions, metres (feet): 88 × 12.0 × 3.5 *(288.7 × 39.4 × 11.5)*
Speed, knots: 18
Machinery: 2 Caterpillar 9M20 diesels; 4,650 hp *(3.42 MW)*; 2 shafts
Radars: Surface search: E/F-band.
Navigation: I-band.

Comment: Steel construction. Based in the East China Sea Branch.

HAIJING 2151   10/2013*, A Sheldon-Duplaix / 1525591

4 OFFSHORE PATROL SHIPS (PSO)

Name	Builders	Laid down	Launched	Commissioned
HAIJING 1123 (ex-*Haijian 23*)	Guangzhou Shipyard	23 May 2010	2010	24 Dec 2010
HAIJING 1126 (ex-*Haijian 26*)	Guangzhou Shipyard	28 May 2010	2010	29 Mar 2011
HAIJING 2166 (ex-*Haijian 66*)	Guangzhou Shipyard	16 May 2010	2010	24 Jan 2011
HAIJING 3175 (ex-*Haijian 75*)	Guangzhou Shipyard	20 Apr 2010	2010	26 Oct 2010

Measurement, tonnes: 1,149 gt
Dimensions, metres (feet): 77.7 × 10.2 × 4.0 *(254.9 × 33.5 × 13.1)*
Speed, knots: 20
Machinery: 2 MAN 7L27/38 diesels; 6,472 hp *(4.76 MW)*; 2 shafts
Radars: Surface search: E/F-band.
Navigation: I-band.

Comment: Steel construction. *Haijing 1123* and *1126* based in the North, *Haijing 2166* in the East and *Haijing 3175* in the South China Sea Branch.

HAIJING 2166   10/2013*, Ships of the World / 1525672

3 OFFSHORE PATROL SHIPS (PSO)

Name	Builders	Laid down	Launched	Commissioned
HAIJING 1127 (ex-*Haijian 27*)	Guangzhou Shipyard	3 Nov 2003	2004	28 Dec 2004

Measurement, tonnes: 1,124 gt
Dimensions, metres (feet): 75.8 × 10.2 × 4.0 *(248.7 × 33.5 × 13.1)*
Speed, knots: 16
Machinery: 2 MAN 7L27/38 diesels; 6,472 hp *(4.76 MW)*; 2 shafts
Radars: Surface search: E/F-band.
Navigation: I-band.

Comment: Steel construction. Based in the North China Sea Branch.

HAIJING 1127   1/2013*, Ships of the World / 1486607

1 OFFSHORE PATROL SHIP (PSO)

HAIJING 3210 (ex-*Yuzheng 310*)

Displacement, tonnes: 2,580 full load
Dimensions, metres (feet): 110 × 14.0 × ? *(360.9 × 45.9 × ?)*
Speed, knots: 22

Machinery: To be announced
Radars: Surface search: E/F-band.
Navigation: I-band.
Helicopters: 1 Z-9A.

Comment: Built at Guangzhou and entered service with the Fisheries Law Enforcement Service in 2010. Transferred to the China Coast Guard in 2013.

HAIJING 3210 (FLEC colours)   6/2012* / 1525590

3 OFFSHORE PATROL SHIPS (PSO)

Name	Builders	Laid down	Launched	Commissioned
HAIJING 1117 (ex-*Haijian 17*)	Wuchang Shipyard, Wuhan	17 Sep 2004	5 Jan 2005	9 May 2005
HAIJING 2146 (ex-*Haijian 46*)	Wuchang Shipyard, Wuhan	20 Aug 2004	2005	8 May 2005
HAIJING 3171 (ex-*Haijian 71*)	Wuchang Shipyard, Wuhan	28 Sep 2004	1 Feb 2005	17 Aug 2005

Measurement, tonnes: 1,111 gt
Dimensions, metres (feet): 73.9 × 10.2 × 2.6 *(242.5 × 33.5 × 8.5)*
Speed, knots: 16
Machinery: 1 Caterpillar 9M20 diesel; 2,325 hp *(1.71 MW)*; 1 shaft
Radars: Surface search: E/F-band.
Navigation: I-band.

Comment: Steel construction. *Haijing 1117* based in the North, *Haijing 2146* in the East and *Haijing 3171* in the South China Sea Branch.

6 OFFSHORE PATROL SHIPS (PSO)

HAIJING 1112 (ex-*Yuzheng 118*)   **HAIJING 3101** (ex-*Yuzheng 301*)
HAIJING 2101 (ex-*Yuzheng 201*)   **HAIJING 3102** (ex-*Yuzheng 302*)
HAIJING 2102 (ex-*Yuzheng 202*)   **HAIJING 3103** (ex-*Yuzheng 303*)

Displacement, tonnes: 968 standard
Dimensions, metres (feet): 69.38 × 9.6 × 3.29 *(227.6 × 31.5 × 10.8)*
Complement: 35
Machinery: 2 MAN 6L28/32 diesels; 3,551 hp *(2.65 MW)*; 2 shafts; bow thruster
Radars: Surface search: E/F-band.
Navigation: I-band.

Comment: Built by Wuchang Shipyard, Wuhan for the Fisheries Law Enforcement Command and transferred to the Coast Guard in 2013.

HAIJING 2101   10/2013*, Ships of the World / 1525673

3 OFFSHORE PATROL SHIPS (PSO)

Name	No	Builders	Laid down	Launched	Commissioned
– (ex-*Haijian 18*)	–	Wuchang Shipyard, Wuhan	19 Nov 1994	15 Aug 1995	25 Mar 1996
HAIJING 2149 (ex-*Haijian 49*)	–	Wuchang Shipyard, Wuhan	20 Dec 1993	30 Oct 1994	30 Jun 1995
HAIJING 3174 (ex-*Haijian 74*)	–	Wuchang Shipyard, Wuhan	12 Nov 1994	2 Apr 1995	18 May 1996

Measurement, tonnes: 912 gt
Dimensions, metres (feet): 71.4 × 10.2 × 3.37 *(234.3 × 33.5 × 11.05)*
Speed, knots: 15.5
Machinery: 1 Zhenjiang 6L28/32 diesel; 1,800 hp *(1.32 MW)*; 1 shaft
Radars: Surface search: E/F-band.
Navigation: I-band.

Comment: Steel construction. Ex-*Haijian 18* based in the North, *Haijian 2149* in the East and *Haijian 3174* in the South China Sea Branch.

1 SHUPANG (TYPE 636A) CLASS (AGE)

HAIJING 2506 (ex-*Hai Yang 18*)

Measurement, tonnes: 4,335 gt
Dimensions, metres (feet): 106 × 17.4 × 5.5 *(347.8 × 57.1 × 18.0)*
Speed, knots: 13
Machinery: Diesel-electric: 2 diesels; 2 motors; 5,710 hp *(4.2 MW)*; 2 shafts
Radars: To be announced.
Helicopters: 1 medium.

Comment: Former research ship transferred from the Navy to the FLEC in 2012. Subsequently transferred to the Coast Guard in 2013.

HAIJING 2506   10/2013*, Ships of the World / 1525669

2 OFFSHORE PATROL SHIPS (PSO)

Name	Builders	Laid down	Launched	Commissioned
HAIJING 1115 (ex-*Haijian 15*)	Wuchang Shipyard, Wuhan	21 Oct 2009	28 Apr 2010	28 Nov 2010
HAIJING 3184 (ex-*Haijian 84*)	Wuchang Shipyard, Wuhan	10 Dec 2009	Jul 2010	28 Apr 2011

Measurement, tonnes: 1,819 gt
Dimensions, metres (feet): 88 × 12.0 × 3.5 *(288.7 × 39.4 × 11.5)*
Speed, knots: 14
Machinery: 2 Caterpillar 9M20 diesels; 4,650 hp *(3.42 MW)*; 2 shafts
Radars: Surface search: E/F-band.
Navigation: I-band.

Comment: Steel construction. *Haijing 1115* based in the North and *Haijing 3184* in the South China Sea Branch.

HAIJING 1115   10/2013*, Ships of the World / 1525674

14 OFFSHORE PATROL SHIPS (PSO)

HAIJING 2112 (ex-*Haijian 8002*) –   – (ex-*Haijian 1001*) –
HAIJING 2113 (ex-*Haijian 5001*) –   – (ex-*Haijian 2169*) –
HAIJING 3111 (ex-*Haijian 9020*) –   – (ex-*Haijian 1118*) –
HAIJING 2112 (ex-*Haijian 9030*) –   – (ex-*Haijian 1013*) –
– (ex-*Haijian 4072*) –   – (ex-*Haijian 1002*) –
– (ex-*Haijian 2032*) –   – (ex-*Haijian 4001*) –
– (ex-*Haijian 2168*) –   – (ex-*Haijian 4002*) –

Measurement, tonnes: 1,337 gt
Dimensions, metres (feet): 79.9 × ? × ? *(262.1 × ? × ?)*
Speed, knots: 20
Machinery: To be announced
Radars: Surface search: E/F-band.
Navigation: I-band.

Comment: Steel construction. All built for the China Marine Surveillance in 2012–13 and subsequently transferred to the newly formed China Coast Guard.

HAIJING 2112   10/2013*, Ships of the World / 1525670

1 YANBING (MOD YANHA) CLASS (AGB/PSO)

HAIJING 1411 (ex-*Hai Bing 723*)

Displacement, tonnes: 4,491 full load
Dimensions, metres (feet): 102 × 17.1 × 5.9 *(334.6 × 56.1 × 19.3)*
Speed, knots: 17
Complement: 95
Machinery: Diesel-electric; 2 diesel generators; 2 motors; 2 shafts
Guns: 8 — 37 mm/63 Type 61/74 (4 twin).
Radars: Navigation: 2 Fin Curve; I-band.

Comment: Enlarged version of Yanha class icebreaker, built in 1982, with greater displacement, longer and wider hull, added deck level and curved upper funnel. Transferred to China Coast Guard in 2013. Based in the north.

© 2014 IHS

Government maritime forces < China   169

YANBING 723   12/2001, Ships of the World / 0529115

6 OFFSHORE PATROL SHIPS (PSO)

HAIJING 1112 (ex-*Yuzheng 118*)   HAIJING 3101 (ex-*Yuzheng 301*)
HAIJING 2101 (ex-*Yuzheng 201*)   HAIJING 3102 (ex-*Yuzheng 302*)
HAIJING 2102 (ex-*Yuzheng 202*)   HAIJING 3103 (ex-*Yuzheng 303*)

Displacement, tonnes: 968 standard
Dimensions, metres (feet): 69.38 × 9.6 × 3.29 *(227.6 × 31.5 × 10.8)*
Speed, knots: 20
Complement: 35
Machinery: 2 MAN diesels; 3,551 hp *(2.65 MW)*; 2 shafts; bow thruster
Radars: Surface search: E/F-band.
Navigation: I-band.

Comment: All vessels built by Wuchang Shipyard for the FLEC and subsequently transferred to the Coast Guard in 2013.

18 TYPE 618 PATROL CRAFT (PB)

HAIJING 13101–13102 (ex-*Haijing 13001–13002*)   HAIJING 35103 (ex-*Haijing 35003*)
HAIJING 21101–21102 (ex-*Haijing 21001–21002*)   HAIJING 37101–37102 (ex-*Haijing 37001–37002*)
HAIJING 31102 (ex-*Haijing 31001*)   HAIJING 44103 (ex-*Haijing 44068*)
HAIJING 32101–32102 (ex-*Haijing 32001–32002*)   HAIJING 45101–45102 (ex-*Haijing 45001–45002*)
HAIJING 33101–33102 (ex-*Haijing 33001–33002*)   HAIJING 46101 (ex-*Haijing 46003*)
HAIJING 35101 (ex-*Haijing 35002*)   HAIJING 46102 (ex-*Haijing 46006*)

Speed, knots: 20
Machinery: 2 diesels; 2 shafts
Guns: 4 — 37 mm (2 twin).
Radars: Surface search: E/F-band.
Navigation: I-band.

Comment: Medium-sized patrol vessels that entered service from 2009.

1 OFFSHORE PATROL SHIP (PSO)

Name	No	Builders	Laid down	Launched	Commissioned
HAIJING 3383 (ex-*Haijian 83*)	–	Jiangnan Shiyard, Shanghai	24 Jun 2004	15 Oct 2004	8 Aug 2005

Measurement, tonnes: 3,276 gt
Dimensions, metres (feet): 98 × 15.2 × 5.5 *(321.5 × 49.9 × 18.0)*
Speed, knots: 18
Machinery: 3 diesels; 5,400 hp *(3.97 MW)*; 3 shafts
Radars: Surface search: E/F-band.
Navigation: I-band.

Comment: Steel construction. Based in the South China Sea Branch.

1 OFFSHORE PATROL SHIP (PSO)

Name	No	Builders	Commissioned
HAIJING 3172 (ex-*Haijian 72*)	–	Wuchang Shipyard, Wuhan	1989

Measurement, tonnes: 881 gt
Dimensions, metres (feet): 70 × 9.4 × 3.0 *(229.7 × 30.8 × 9.8)*
Speed, knots: 19
Machinery: 2 Zhenjiang 6L28/32 diesels; 3,600 hp *(2.65 MW)*; 1 shaft
Radars: Surface search: E/F-band.
Navigation: I-band.

Comment: Steel construction. Based in the South China Sea Branch.

1 TYPE 825C OFFSHORE PATROL SHIPS (PSO)

HAIJING 3368 (ex-*Haijian 168*; ex-*Haiyang 11*, ex-*Nan Diao 11*)

Dimensions, metres (feet): 105 × 14.0 × 5.2 *(344.5 × 45.9 × 17.06)*
Machinery: To be announced
Radars: Surface search: E/F-band.
Navigation: I-band.

Comment: Built as oceanographic survey ship by Hudong Shipyard (now Hudong-Zhonghua Shipyard). Launched in June 1981 and commissioned by the South Sea Fleet of the PLA Navy in November 1981. Transferred to the China Coast Guard in 2013.

1 OFFSHORE PATROL SHIP (PSO)

Name	No	Builders	Commissioned
HAIJING 2153 (ex-*Haijian 53*)	–	Zhonghua Shipyard, Shanghai	1976

Measurement, tonnes: 949 gt
Dimensions, metres (feet): 71.4 × 10.5 × 3.6 *(234.3 × 34.4 × 11.8)*
Speed, knots: 14
Machinery: 2 Hongwei 8300ZC diesels; 1,496 hp *(1.1 MW)*; 2 shafts
Radars: Surface search: E/F-band.
Navigation: I-band.

Comment: Steel construction. Based in the East China Sea Branch.

IHS Jane's Fighting Ships 2014-2015

Colombia
ARMADA DE LA REPUBLICA

Country Overview

The Republic of Colombia is the only South American country that fronts both the Caribbean Sea and the Pacific Ocean with coastlines of 950 n miles and 782 n miles respectively. With an area of 440,831 square miles, it is bordered to the north by Panama, to the east by Venezuela and Brazil and to the south by Peru and Ecuador. The capital and largest city is Bogotá. Buenaventura and Tumaco are the main Pacific ports while Cartagena, Santa Marta and Barranquilla, which is near the mouth of the principal river and transport artery, the Magdalena, are on the Caribbean side. Territorial seas (12 n miles) are claimed but while it has claimed a 200 n mile EEZ, its limits have not been fully defined.

Headquarters Appointments

Commander of the Navy:
   Vice Admiral Hernándo Wills Velez
Deputy Commander and Chief of Staff of the Navy:
   Vice Admiral Cesar Augusto Narváez
Inspector General:
   Vice Admiral Henry John Blain Garzón
Chief of Naval Operations:
   Vice Admiral Rodolfo Amaya Kerguelen
Commander Caribbean Force:
   Rear Admiral Leonardo Santamaria Gaitán
Commander Marine Corps:
   Rear Admiral Luís Gómez Vásquez
Commander Pacific Force:
   Rear Admiral Pablo Emilio Romero Rojas
Commander Eastern Naval Force:
   Rear Admiral Hector Alfonso Medina Torres
Commander, Marine Force:
   Brigadier General Héctor Julio Pachón Cañon

Personnel

(a) 2014: 12,000 (Navy); 9,000 (Marines); 200 (Coast Guard); 100 (Aircrew)
(b) 2 years' national service (few conscripts in the Navy)

Organisation

Caribbean Force Command: HQ at Cartagena.
Pacific Force Command: HQ at Bahia Malaga.
Eastern Force Command: HQ at Puerto Carreño
Naval Force South: HQ at Puerto Leguízamo.
Riverine Brigade: HQ at Bogotá, DC.
Coast Guard: HQ at Bogotá.

Bases

ARC Bolivar, Cartagena, Main naval base (floating dock, 1 slipway), schools.
ARC Bahía Málaga: Major Pacific base.
ARC Barranquilla: Naval training base.
ARC Puerto Leguízamo: Putumayo River base.
Turbo: Minor River base.
Puerto López: Minor River base.
Puerto Carreño: Minor River base.
Irida: Minor River base.
San Andrés y Providencia: Specific Command

Aviation

Baranquilla: new base opened in September 2009.

Marine Corps

Organisation: First Brigade (Corozal):
BAFIM 1 (San Andrés)
BAFIM 2 (Cartagena)
BAFIM 3 (Malagana)
BAFIM 4 (Corozal)
CFENIM Training Battalion (Coveñas)
Second Riverine Brigade (Bogotá)
BASFLIM 3 (Bahía Solano)
BASFLIM 4 (Bahía Málaga)
No. 70 Battalion (Tumaco)
No. 80 Battalion (Buenaventura)
No. 10 Battalion (Guapí)

Strength of the Fleet

Type	Active	Building (planned)
Patrol Submarines	2	—
Midget Submarines	2	—
Frigates	4	—
Large Patrol Ships and Fast Attack Craft (Gun)	16	1
Coast Patrol Craft	2	—
Amphibious Forces	10	—
Inshore/River Patrol Craft	98	47
River Patrol Craft Support	7	—
River Assault Boats	169	—
Survey Vessels	4	—
Auxiliaries	18	—
Training Ships	6	—

Prefix to Ships' Names

ARC (Armada Republica de Colombia)

Dimar

Maritime authority in charge of hydrography and navigational aids.

Coast Guard and Customs (DIAN)

The Coast Guard was established in 1979 but then gave way to the Customs Service before being re-established in January 1992 under the control of the Navy. Headquarters at Bogotá. Main bases are Cartagena, Buenaventura y Turbo and Valle. Ships have a red and yellow diagonal stripe on the hull and patrol craft have a PM number. Customs craft were absorbed into the Coast Guard but by 1995 were again independent as part of the DIAN (Direccion de Impuestos y Aduanas Nacionales). Customs craft have Aduana written on the ship's side, a thick and two thin diagonal stripes and have AN numbers.

PENNANT LIST

Submarines
SO 28 Pijao
SO 29 Tayrona
— Intrépido
— Indomable

Frigates
FL 51 Almirante Padilla
FL 52 Caldas
FL 53 Antioquia
FL 54 Independiente
— Nariño

Patrol Forces
PAF 601 Filigonio Hichamón
PAF 602 SSIM Manuel Antonio Moyar
PAF 603 Igaraparana
PAF 604 SSIM Julio Correa Hernández
PAF 605 Manacacias
PAF 606 Cotuhe
PAF 607 SSCIM Senen Alberto Araujo
PAF 608 CPCIM Guillermo Londoño Vargas
PAF 609 Ariari
PAF 610 Mario Villegas
PAF 611 Tony Pastrana Contreras
PAF 612 CTCIM Jorge Moreno Salazar
PAF 613 Juan Ricarda Oyola Vera
PAF 614 Alexander Pérez Rodriguez
PAF 615 Cristian Reyes Holguín
PAF 616 Alejandro Ledesma Ortiz
PAF 617 Harrys Tous Cataño
PC 141 Cabo Corrientes
PC 142 Cabo Manglares
PC 143 Cabo Tiburon
PC 144 Cabo de la Vella
PC 145 11 De Noviembre
PO 44 Valle del Cauca
PO 45 San Andrés
PO 46 20 de Julio
PO 47 7 de Agosto (bldg)
PM 102 Rafael del Castillo y Rada
PM 103 TN José María Palas
PM 105 S2 Jaime Gómez Castro
PM 106 S2 Juan Nepomuceno Peña
PM 112 Quitasueño
PM 113 José María García y Toledo
PM 114 Juan Nepomuceno Eslava
PM 115 TECIM Jaime E Cárdenas Gomez
PB 446 Capella
PF 121 Diligente
PF 122 Juan Lucio
PF 123 Alfonso Vargas
PF 124 Fritz Hagale
PF 125 Vengadora
PF 128 Carlos Galindo
PF 136 Leticia
PF 137 Arauca

Amphibious Forces
LD 246 Morrosquillo
LD 248 Bahía Honda
LD 249 Bahía Portete
LD 251 Bahía Solano
LD 252 Bahía Cupica
LD 253 Bahía Utría
LD 254 Bahía Málaga

Auxiliaries
BL 161 Cartagena de Indias
BL 162 Buenaventura
ETG 501 Bocachica
ETG 502 Arturus
ETG 503 Pedro David Salas
ETG 504 Sirius
ETG 507 Calima
ETG 508 Bahí Santa Catalina
ETG 509 Móvil I
ETG 510 Sula
ETG 542 Playa Blanca
ETG 543 Tierra Bouba
ETG 544 Bell Salter
ETG 546 Orion
ETG 558 Juanchaco
ETG 559 Libertád
ETG 560 Isla Tesoro
DF 170 Mayor Jaime Arias Arango

Survey Vessels
BO 154 Gorgona
BO 155 Providencia
BO 156 Malpelo
BH 153 Quindio
BB 33 Abadía Médez
BB 34 Ciénaga de Mayorquin
BB 35 Isla Palma

Training Ships
BE 160 Gloria
YT 230 Comodoro
YT 231 Tridente
YT 232 Cristina
YT 233 Albatros
YT 234 Poseidon

Tugs
RB 76 Josué Alvarez
RB 77 Don Vizo
RB 78 Portete
RB 80 Cienaga de San Juan
RF 83 Joves Fiallo
RF 85 Miguel Silva
RF 86 Capitán Rigoberto Giraldo
RF 87 Vladimir Valek
RF 88 Teniente Luis Bernal
RF 93 Sejeri
RF 96 Inirida
RM 75 Andagoya

SUBMARINES

Notes: Three 7 m Swimmer Delivery Vehicles were acquired in 1970: *Defensora* (SL 15), *Poderosa* (SL 16) and *Protectora* (SL 17). Capable of 6 kt on the surface, they are capable of carrying two 250 kg weapons.

PIJAO

6/2000, Colombian Navy / 0103689

Submarines < **Colombia** 171

2 TYPE 206A (SSK)

Name	No	Builders	Laid down	Launched	Commissioned
INTREPIDO (ex-*U 23*)	– (ex-S 172)	Thyssen Nordseewerke	5 Mar 1973	25 May 1974	2 May 1975
INDOMABLE (ex-*U 24*)	– (ex-S 173)	Thyssen Nordseewerke	20 Mar 1972	26 Jun 1973	16 Oct 1974

Displacement, tonnes: 450 surfaced; 498 dived
Dimensions, metres (feet): 48.6 × 4.6 × 4.5 *(159.4 × 15.1 × 14.8)*
Speed, knots: 10 surfaced; 17 dived
Range, n miles: 4,500 at 5 kt surfaced
Complement: 22 (4 officers)

Machinery: Diesel-electric; 2 MTU 12V 493 AZ80 GA31L diesels; 1,200 hp *(882 kW)*; 1 Siemens motor; 1,800 hp *(1.32 MW)*; 1 shaft
Torpedoes: 8 – 21 in *(533 mm)* bow tubes. AEG SUT; dual purpose; wire-guided active/passive homing to 12 km *(6.5 n miles)* at 35 kt; 28 km *(15 n miles)* at 23 kt; warhead 250 kg; or STN Atlas DM2A3 Seehecht; wire-guided active homing to 13 km *(7 n miles)* at 35 kt; passive homing to 28 km *(15 n miles)* at 18 kt; warhead 260 kg.
Mines: GRP containers, each containing 12 mines, can be secured outside hull. 16 mines may be carried in place of torpedoes.
Electronic countermeasures: ESM: Thomson-CSF DR 2000U with THORN EMI Sarie 2; intercept.
Radars: Surface search: Thomson-CSF Calypso II; I-band.
Sonars: Atlas Elektronik DBQS-21D; passive/active search and attack; medium frequency.
Weapon control systems: SLW 83 (TFCS).

Programmes: Originally built for the Federal German Navy in the 1970s and decommissioned in late 2010. Both boats transferred to Colombia on 28 August 2012. Two further boats, U 16 (S 195) and U 18 (S 197) were transferred as spares. Procurement of torpedoes is reported to be part of the package but this has not been confirmed.
Modernisation: The whole class was extensively modernised 1987–92 and the Colombian boats are expected to be 'tropicalised' for operations in Caribbean waters.
Structure: Built of high-tensile non-magnetic steel. Diving depth 160 m.
Operational: When both boats become operational in 2014, they will replace submarines of the same name, which were decommissioned in 2013.

INTREPIDO 8/2012, Michael Nitz / 1455766

2 PIJAO (TYPE 209/1200) CLASS (SS)

Name	No	Builders	Laid down	Launched	Commissioned
PIJAO	SO 28	Howaldtswerke, Kiel	1 Apr 1972	10 Apr 1974	18 Apr 1975
TAYRONA	SO 29	Howaldtswerke, Kiel	1 May 1972	16 Jul 1974	16 Jul 1975

Displacement, tonnes: 1,199 surfaced; 1,306 dived
Dimensions, metres (feet): 55.9 × 6.3 × 5.4 *(183.4 × 20.7 × 17.7)*
Speed, knots: 11 surfaced; 22 dived
Range, n miles: 8,000 at 8 kt surfaced; 4,000 at 4 kt dived
Complement: 34 (7 officers)

Machinery: Diesel-electric; 4 MTU 12V 493 AZ80 diesels; 2,400 hp(m) *(1.76 MW)* sustained; 4 AEG alternators; 1.7 MW; 1 Siemens motor; 4,600 hp(m) *(3.38 MW)* sustained; 1 shaft
Torpedoes: 8 – 21 in *(533 mm)* bow tubes. 14 AEG SUT; dual purpose; wire-guided; active/passive homing to 12 km *(6.5 n miles)* at 35 kt; 28 km *(15 n miles)* at 23 kt; warhead 250 kg. Swim-out discharge or STN Atlas DM2A3 Seehecht; wire-guided active homing to 13 km *(7 n miles)* at 35 kt; passive homing to 28 km *(15 n miles)* at 18 kt; warhead 260 kg.
Electronic countermeasures: ESM: Saab UME-100; intercept.
Radars: Surface search: Thomson-CSF Calypso II; I-band.
Sonars: Krupp Atlas PSU 83-55; hull-mounted; active/passive search and attack; medium frequency. Atlas Elektronik PRS 3-4; passive ranging; integral with CSU 3.
Weapon control systems: Atlas Elektronik ISUS 90-III.

Programmes: Ordered in 1971.
Modernisation: Both boats were refitted by HDW at Kiel 1990–91; main batteries were replaced. Further refits were carried out at Cotecmar; *Pijao* 1999–2002 and *Tayrona* 2003–06. A contract for a major modernisation programme was signed with HDW on 14 January 2009. *Tayrona* started refit at Cotecmar in mid-2009 and *Pijao* followed in mid-2010. The scope of work includes replacement of the weapons control system with Atlas Elektronik ISUS 90-III, new search (Cassidian Optronics SERO-250) and attack periscopes, new sonars, main machinery and batteries. The boats are also to receive new torpedoes. *Tayrona* was completed in mid-2013 and *Pijao* in August 2013.
Structure: Single-hulled. Diving depth, 820 ft *(250 m)*.
Operational: Both boats employed on counter-drug operations.

PIJAO 4/2008, Marco Ghiglino / 1335712

© 2014 IHS IHS Jane's Fighting Ships 2014-2015

FRIGATES

Notes: Replacement of the Almirante Padilla class from about 2025 is under consideration. Meanwhile, Plan Puente is to procure second-hand ships until new ships can be built.

4 ALMIRANTE PADILLA CLASS (TYPE FS 1500) (FLGHM)

Name	No	Builders	Laid down	Launched	Commissioned
ALMIRANTE PADILLA	FL 51	Howaldtswerke, Kiel	17 Mar 1981	6 Jan 1982	31 Oct 1983
CALDAS	FL 52	Howaldtswerke, Kiel	14 Jun 1981	23 Apr 1982	14 Feb 1984
ANTIOQUIA	FL 53	Howaldtswerke, Kiel	22 Jun 1981	28 Aug 1982	30 Apr 1984
INDEPENDIENTE	FL 54	Howaldtswerke, Kiel	22 Jun 1981	21 Jan 1983	24 Jul 1984

Displacement, tonnes: 1,524 standard; 2,134 full load
Dimensions, metres (feet): 99.1 × 11.3 × 3.7 (325.1 × 37.1 × 12.1)
Speed, knots: 27; 18 diesel
Range, n miles: 7,000 at 14 kt, 5,000 at 18 kt
Complement: 94

Machinery: 4 MTU 16V 1163 TB 73L diesels; 27,892 hp(m) (20.8 MW) sustained; 2 shafts; cp props
Missiles: SSM: 4 (fitted for 8) Aerospatiale MM 40 Exocet ❶; inertial cruise; active radar homing to 70 km (40 n miles) at 0.9 Mach; warhead 165 kg; sea-skimmer.
SAM: 2 Matra Simbad twin launchers ❷; Mistral; IR homing to 4 km (2.2 n miles); warhead 3 kg; anti-sea-skimmer.
Guns: 1 Oto Melara Strales 3 in (76 mm)/62 compact ❸; 85 rds/min to 16 km (8.7 n miles); weight of shell 6 kg. 2 Breda 40 mm/70 (twin) ❹; 300 rds/min to 12.5 km (6.8 n miles) anti-surface; weight of shell 0.96 kg. 2–12.7 mm MGs.
Torpedoes: 6–324 mm ILAS 3 (2 triple) tubes ❺; Whitehead A244S; anti-submarine; active/passive homing to 7 km (3.8 n miles); warhead 38 kg (shaped charge).
Physical countermeasures: Decoys: 1 Terma SKWS.
Electronic countermeasures: ESM: Thales Vigile; intercept. ECM: Racal Scimitar; jammer.
Radars: Air/surface search: Thales SMART-S ❻; 3D; E/F-band.
Surface search/navigation: Terma Scanter 2001 ❼; E/F/I-bands.
Fire control: Thales Sting ❽; I/K-band.
IFF: Mk 10.
Sonars: Atlas Elektronik ASO 4-2; hull-mounted; active attack; medium frequency.
Combat data systems: Thomson-CSF TAVITAC action data automation. Possibly Link Y fitted.
Electro-optic systems: Thales Mirador optronics director.

Helicopters: 1 MBB BO 105 CB ❾ or 1 Bell 412.

Programmes: Order for four Type FS 1500 placed late 1980. Reclassified as light frigates in 1999. Similar to Malaysian Kasturi-class frigates.
Modernisation: Mistral SAM system fitted. Helicopter deck lengthened by 2 m to take Bell 412 aircraft. There have also been minor modifications to ship systems and superstructure. A contract for a major modernisation programme was signed with Cotecmar in February 2009. The scope of work includes replacement of the air search radar with Thales SMART S, the fire-control radar with Thales Sting, the navigation radar with Terma Scanter 2001, the ESM system with Thales Vigile, the installation of Thales Mirador TEOOS and Terma SKWS chaff. The 76 mm guns are to be replaced/upgraded with the Oto Melara Strales system which includes DART precision guided ammunition. The Breda 40 mm/70 guns are being refurbished. The main machinery is also being replaced with new MTU engines. Work on the first two ships, *Antioquia* and *Caldas* took place 2009–11 and is in progress for *Almirante Padilla* and *Independiente* in 2012–14. In a separate contract, there have been reports that Exocet SSMs are to be replaced by Hae Sung 1 (SSM-700K) Sea Star missiles in 2014.

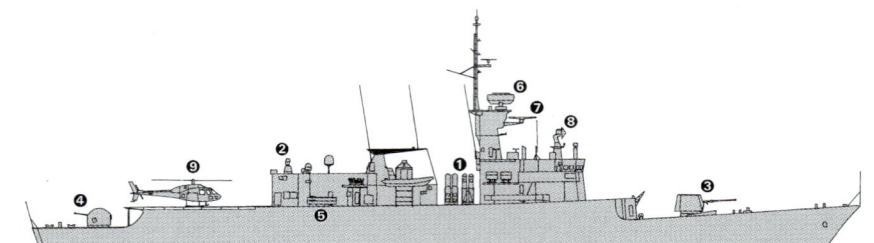

ANTIOQUIA *(Scale 1 : 1,200), Ian Sturton / 1529100*

ANTIOQUIA *9/2012, US Navy / 1455767*

ANTIOQUIA *9/2012*, US Navy / 1529112*

Frigates — Patrol forces < **Colombia** 173

1 DONG HAE CLASS (FS/FSG)

Name	No	Builders	Commissioned
NARIÑO (ex-*An Yang*)	– (ex-755)	Daewoo, Okpo	Dec 1983

Displacement, tonnes: 1,093 full load
Dimensions, metres (feet): 78.1 × 9.6 × 2.6 *(256.2 × 31.5 × 8.5)*
Speed, knots: 31
Range, n miles: 4,000 at 15 kt
Complement: 95 (10 officers)

Machinery: CODOG; 1 GE LM 2500 gas turbine; 26,820 hp *(20 MW)* sustained; 2 MTU 12V 956 TB82 diesels; 6,260 hp(m) *(4.6 MW)* sustained; 2 shafts; Kamewa cp props
Guns: 1 Oto Melara 3 in *(76 mm)*/62 compact; 85 rds/min to 16 km *(8.6 n miles)* anti-surface; 12 km *(6.5 n miles)* anti-aircraft; weight of shell 6 kg. 4 Emerlec 30 mm (2 twin). 2 Bofors 40 mm/60 (twin).
Torpedoes: 6—324 mm Mk 32 (2 triple) tubes. Honeywell Mk 46; anti-submarine; active/passive homing to 11 km *(5.9 n miles)* at 40 kt; warhead 44 kg.
Depth charges: 12.
Physical countermeasures: Decoys: 4 MEL Protean fixed launchers; 36 grenades.
Electronic countermeasures: ESM/ECM: THORN EMI or NobelTech; intercept/jammer.
Radars: Surface search: Raytheon SPS-64; I-band.
Fire control: Signaal WM28; I/J-band.
Sonars: Signaal PHS-32; hull-mounted; active search and attack; medium frequency.
Combat data systems: Signaal Sewaco ZK.
Weapon control systems: Signaal Liod optronic director.

Comment: Following decommissioning from the South Korean Navy it was announced in May 2013 that the ship is to be transferred to Colombia following a refit. This is expected to be completed in 2014. The details of weapons and sensors are based on the ship in South Korean service and have not been confirmed.

SHIPBORNE AIRCRAFT

Numbers/Type: 2 Bolkow BO 105CB.
Operational speed: 113 kt *(210 km/h)*.
Service ceiling: 9,854 ft *(3,000 m)*.
Range: 407 n miles *(754 km)*.
Role/Weapon systems: Surface search and limited ASW helicopter. Sensors: Search/weather radar. Weapons: ASW; provision to carry depth bombs. ASV; light attack role with machine gun pods.

BO 105　　　　　　　　　　　　　　　*6/2011, Colombian Navy* / 1406566

Numbers/Type: 3 Bell 412.
Operational speed: 122 kt *(226 km/h)*.
Service ceiling: 10,000 ft *(3,300 m)*.
Range: 500 n miles *(744 km)*.
Role/Weapon systems: Multipurpose used mostly for surveillance, troop transport and logistic support. One aircraft lost in January 2013. Four ASW aircraft reportedly ordered in December 2012. Sensors: Weather radar. Weapons: ASV 7.62 mm MG can be carried.

BELL 412　　　　　　　　　　　　　　*6/2011, Colombian Navy* / 1406564

Numbers/Type: 6 Bell UH-1N Twin Huey.
Operational speed: 110 kt *(204 km/h)*.
Service ceiling: 10,000 ft *(3,048 m)*.
Range: 230 n miles *(426 km)*.
Role/Weapon systems: Light Utility platform for all-weather assault, transport, airborne command and control, armed reconnaissance and SAR. Can carry eight marines. Entered service, after being refurbished, from 2010. Sensors: BRITE Star FLIR. Weapons: Can be armed with 12.7 mm or 7.62 mm machine guns and 2.75 in rockets.

UH-1N (US Navy colours)　　　　　　　*5/1999, A Sharma* / 0084120

Numbers/Type: 2 Eurocopter AS 555 Fennec.
Operational speed: 121 kt *(225 km/h)*.
Service ceiling: 13,125 ft *(4,000 m)*.
Range: 389 n miles *(722 km)*.
Role/Weapon systems: OTHT capability for surface-to-surface role. Also used for logistic support. More are being acquired. Sensors: Bendix RDR 1500B radar. Weapons: Torpedoes may be fitted in due course.

AS 555　　　　　　　　　　　　　　　*6/2011, Colombian Navy* / 1406565

LAND-BASED MARITIME AIRCRAFT

Notes: The navy operates the following fixed-wing aircraft for maritime surveillance and transport: one CASA 212-100, two Gavilan G358M, one Cessna 152, one Piper PA-31 Navajo, one Piper PA-34, four Piper PA-28 Cherokee, four Cessna 206 Stationair, two Cessna Grand Caravan C-208, two Beech B-350 (equipped with Elint equipment). There are also one Bell 212 and one BK-117 helicopters for training and transport.

Numbers/Type: 4 Casa CN-235 200/300.
Operational speed: 210 kt *(384 km/h)*.
Service ceiling: 24,000 ft *(7,315 m)*.
Range: 2,000 n miles *(3,218 km)*.
Role/Weapon systems: EEZ surveillance. First two CN-235 200 delivered in 2003. One of these remains operational. One further new aircraft ordered in 2007. Both operational CN-235 200 to be modified and fitted with FITS system. One CN-235 300, equipped with FITS, ventral radar (Telephonics APS-143 C(V)3 Ocean Eye) and FLIR delivered in 2010. A further aircraft ordered in 2013. Sensors: Search radar Bendix APS 504(V)5; FLIR. Weapons: Unarmed.

CN-235　　　　　　　　　　　　　　　*6/2003, CASA* / 0587695

PATROL FORCES

Notes: (1) There are five river patrol craft of unknown type, PRF 323–327.
(2) There are two 8 m Mako Marine 261 craft named *Escorpion* and *Libra*.

9 RIVER PATROL CRAFT (PBR)

FILIGONIO HICHAMÓN PAF 601 (ex-NF 141)
SSIM MANUEL A MOYAR PAF 602 (ex-NF 144)
IGARAPARANÁ PAF 603
SSIM JULIO CORREA HERNÁNDEZ PAF 604 (ex-NF 143)
MANACACÍAS PAF 605
COTUHE PAF 606 (ex-RR 98)
ARIARÍ PAF 609
CAPITÁN RIGOBERTO GIRALDO – (ex-RF 86)
INÍRIDA – (ex-RF 96)

Comment: Miscellaneous patrol/support craft of unknown characteristics. *Capitán Rigoberto Giraldo* and *Inírida* are both former tugs converted for patrol duties.

FILIGONIO HICHAMÓN　　　　　　　　*6/2011, Colombian Navy* / 1406577

IHS Jane's Fighting Ships 2014-2015

2 OFFSHORE PATROL VESSELS (PSO)

Name	No	Builders	Laid down	Launched	Commissioned
20 DE JULIO	PO 46	Cotecmar, Cartagena	2009	23 Jul 2010	3 Feb 2012
7 DE AGOSTO	PO 47	Cotecmar, Cartagena	2011	4 Sep 2013	2013

Displacement, tonnes: 1,723 full load
Dimensions, metres (feet): 80.6 × 13.0 × 3.8 (264.4 × 42.7 × 12.5)
Speed, knots: 20
Range, n miles: 8,600 at 12 kt
Complement: 40

Machinery: 2 Wärtsilä 12V26 diesels; 10,940 hp (8.2 MW); 2 shafts; LIPS cp props; 2 bow thrusters
Guns: 1 Bofors 40/70 mm. 4—12.7 mm MGs.
Radars: Surface search: Terma Scanter; I-band.
Navigation: Furuno; I-band.
Helicopters: Platform for one medium.

Comment: Construction of first of class began at Cotecmar in 2009. A second ship was ordered on 20 October 2011. The Fassmer OPV 80 design, similar to the Piloto Pardo class of the Chilean Coast Guard, includes a stern ramp for the launch/recovery of interception craft. Manual 12.7 mm MGs are to be replaced by THOR/DARET remotely controlled weapons with optronic directors.

20 DE JULIO   6/2012, FASSMER / 1455769

2 ARAUCA CLASS (RIVER GUNBOATS) (PBR)

Name	No	Builders	Commissioned
LETICIA	PF 136 (ex-36)	Union Industrial de Barranquilla	6 Sep 1956
ARAUCA	PF 137 (ex-37)	Union Industrial de Barranquilla	6 Sep 1956

Displacement, tonnes: 279 full load
Dimensions, metres (feet): 49.9 × 8.3 × 2.7 (163.7 × 27.2 × 8.9)
Speed, knots: 14
Range, n miles: 1,890 at 14 kt
Complement: 43
Machinery: 2 Caterpillar diesels; 916 hp (683 kW); 2 shafts
Guns: 2 USN 3 in (76 mm)/50 Mk 26. 4 Oerlikon 20 mm (Arauca). 1—40 mm; 4—20 mm (Leticia).

Comment: Launched in 1955. Based in Naval Force South.

LETICIA   6/2011, Colombian Navy / 1406567

8 NODRIZA CLASS (RIVER PATROL VESSELS) (PBR)

SSCIM SENEN ALBERTO ARAUJO PAF 607 (ex-NF 147)
CPCIM GUILLERMO LONDOÑO VARGAS PAF 608 (ex-NF 146)
MARIO VILLEGAS PAF 610
TONY PASTRANA CONTRERAS PAF 611 (ex-NF 149)
CTCIM JORGE MORENO SALAZAR PAF 612
JUAN RICARDO OYOLA VERA PAF 613
ALEXANDER PÉREZ RODRIGUEZ PAF 614
CRISTIAN REYES HOLGUÍN PAF 615

Displacement, tonnes: 376 standard
Dimensions, metres (feet): 40.3 × 9.5 × 1.26 (132.2 × 31.2 × 4.1)
Speed, knots: 9
Complement: 33 (4 officers)

Military lift: 39 troops
Machinery: 2 diesels; 900 hp (670 kW); 2 Schottel propulsors
Guns: 8—12.7 mm MGs (4 twin). 1 Mk 19 grenade launcher.
Helicopters: Platform (ARC 612–615) for 1 small.

Comment: Powerfully armed river patrol vessels. Built to an innovative design by Cotecmar, Cartagena, in three batches: Batch I (PAF 607, 608) delivered in 2000; Batch II (PAF 610) delivered in 2003 and Batch III (PAF 611–613) delivered by 2007. Batch III ships have a helicopter deck. Two Batch IV (PAF 614, 615) were delivered in March 2009. The two Batch I ships were upgraded in late 2010.

OYOLA VERA   6/2008 / 1335708

4 + 3 (43) LPR-40 CLASS (RIVER PATROL CRAFT) (PB)

RIO ARICA PRF 328
RIO PINILLOS PRF 329
RIO ANGOSTURAS PRF 330
+47

Displacement, tonnes: 14 full load
Dimensions, metres (feet): 12.72 × 2.8 × 0.7 (41.7 × 9.2 × 2.3)
Speed, knots: 29
Range, n miles: 513 at 25 kt
Complement: 4
Machinery: 2 Caterpillar C9 diesels; 503 hp (375 kW); 2 waterjets
Guns: 3—12.7 mm MGs.
Radars: Surface search: Raytheon R70; I-band.

Comment: New class of inshore patrol craft designed by Cotecmar. Aluminium construction. Transportable on a C-130 aircraft. The construction programme at Cotecmar began in 2007 and the first three entered service in 2009. Four further craft were ordered in November 2012, the first of which was delivered on 4 September 2013. Some 50 of the class are expected.

RIO PINILLOS   6/2012, Cotecmar / 1455768

2 RAFAEL DEL CASTILLO Y RADA (SWIFT 105) CLASS (LARGE PATROL CRAFT) (PB)

Name	No	Builders	Commissioned
RAFAEL DEL CASTILLO Y RADA	PM 102 (ex-GC 102, ex-AN 202)	Swiftships Inc, Berwick	28 Feb 1983
TECIM JAIME E CÁRDENAS GOMEZ (ex-Olaya Herrera)	PM 115 (ex-AN 21, ex-AN 201)	Swiftships Inc, Berwick	16 Oct 1981

Displacement, tonnes: 117 full load
Dimensions, metres (feet): 31.5 × 6.7 × 2.1 (103.3 × 22.0 × 6.9)
Speed, knots: 25
Range, n miles: 1,200 at 18 kt
Complement: 19 (3 officers)

Machinery: 4 MTU 12V 331 TC92 diesels; 5,320 hp(m) (3.97 MW) sustained; 4 shafts
Guns: 1 Bofors 40 mm/60 Mk 3 (PM 102). 2—12.7 mm MGs.
Radars: Surface search: Raytheon; I-band.
Weapon control systems: 1 COAR optronic director.

Comment: Delivered for the Customs service. PM 102 is part of the Coast Guard. PM 115 was paid off, but returned unarmed as part of the resurrected Customs service until being transferred back to the Coast Guard in 1997.

RAFAEL DEL CASTILLO Y RADA   6/1999, Colombian Navy / 0056826

Patrol forces < **Colombia** 175

1 JOSÉ MARIA PALAS (SWIFT 110) CLASS
(LARGE PATROL CRAFT) (PB)

Name	No	Builders	Commissioned
JOSÉ MARIA PALAS	PM 103 (ex-GC 103)	Swiftships Inc, Berwick	Sep 1989

Displacement, tonnes: 101 full load
Dimensions, metres (feet): 33.5 × 7.5 × 2 *(109.9 × 24.6 × 6.6)*
Speed, knots: 25
Range, n miles: 2,250 at 15 kt
Complement: 19 (3 officers)
Machinery: 4 Detroit 12V-71TI diesels; 2,400 hp *(1.79 MW)*; 4 shafts
Guns: 1 Bofors 40 mm/70. 1—12.7 mm MG. 2—7.62 mm MGs.
Radars: Surface search: Furuno FR 8100D; I-band.

Comment: Acquired under US FMS programme. This ship belongs to the Coast Guard.

JOSÉ MARIA PALAS *1/1996, van Ginderen Collection* / 0056824

3 SWIFTSHIPS CLASS (RIVER PATROL CRAFT) (PBR)

PRF 320–322

Displacement, tonnes: 17 full load
Dimensions, metres (feet): 13.9 × 3.6 × 0.6 *(45.6 × 11.8 × 2.0)*
Speed, knots: 22. **Range, n miles:** 600 at 22 kt
Complement: 4
Machinery: 2 Detroit 6V-92TA diesels; 900 hp *(671 kW)*; 2 Hamilton water-jets
Guns: 2 M2HB 12.7 mm MGs; 2 M60D 7.62 mm MGs.
Radars: Surface search: Raytheon 40; I-band.

Comment: Acquired in 2000. Hard chine modified V hull form. Can carry up to eight troops.

SWIFTSHIPS CLASS *6/2001, Ecuador Coast Guard* / 0114516

0 + 3 OFFSHORE PATROL VESSELS (PBO)

Name	No	Builders	Laid down	Launched	Commissioned
–	–	STX Shipbuilding and Marine	5 Feb 2014	Jun 2014	2014
–	–	STX Shipbuilding and Marine	5 Feb 2014	Jun 2014	2014
–	–	Cotecmar, Cartagena	2015	2016	2017

Displacement, tonnes: 284 full load
Dimensions, metres (feet): 46.25 × 7.09 × 1.8 *(151.7 × 23.3 × 5.9)*
Speed, knots: 23
Complement: 18
Machinery: 1 MTU 12V4000 M70 diesel; 2,366 hp *(1.74 MW)*; 1 shaft
Guns: 1—25 mm.
Radars: Surface search: I-band.

Comment: The contract for the construction of two steel patrol ships was signed with STX O&S on 7 February 2013. Both ships are to be built at Jinhae Shipyard and to be delivered in late 2014. A third ship, to be built at Cotecmar, is also to be ordered.

11 RIO CLASS (RIVER PATROL CRAFT) (PBR)

PRF 300–304   PRF 314–319

Displacement, tonnes: 7 full load
Dimensions, metres (feet): 9.8 × 3.5 × 0.6 *(32.2 × 11.5 × 2.0)*
Speed, knots: 24. **Range, n miles:** 150 at 22 kt
Complement: 4
Machinery: 2 Detroit 6V-53 diesels; 296 hp *(221 kW)* sustained; 2 waterjets
Guns: 2—12.7 mm (twin) MGs. 1—7.62 mm MG.
A/S Mortars: 1—60 mm mortar.
Radars: Surface search: Raytheon 1900; I-band.

Comment: Acquired in 1989–90. Ex-US PBR Mk II built by Uniflite in 1970. All recommissioned in September 1990. GRP hulls. Names were dropped and new pennant numbers assigned in 2006.

PRF 319 *6/2011, Colombian Navy* / 1406569

9 TENERIFE CLASS (RIVER PATROL CRAFT) (PBR)

PRF 305–313

Displacement, tonnes: 12 full load
Dimensions, metres (feet): 12.4 × 2.9 × 0.6 *(40.7 × 9.5 × 2.0)*
Speed, knots: 29
Range, n miles: 530 at 15 kt
Complement: 17
Machinery: 2 Caterpillar 3208 TA diesels; 850 hp *(634 kW)* sustained; 2 shafts
Guns: 3—12.7 mm MGs (1 twin, 1 single). 1 Mk 19 grenade launcher. 1—7.62 mm MGs.
Radars: Surface search: Raytheon 1900; I-band.

Comment: Built by Bender Marine, Mobile, Alabama. Acquired in October 1993 for anti-narcotics patrols. Aluminium hulls. Can be transported by aircraft. Names were dropped and new pennants numbers assigned in 2006. All craft were undergoing life-extension refits in 2013.

PRF 313 *2000, Colombian Navy* / 0103698

1 BALSAM CLASS (PSO)

Name	No	Builders	Commissioned
SAN ANDRES (ex-*Gentian*)	PO 45 (ex-WIX 290)	Zenith Dredge Co, Duluth	3 Nov 1942

Displacement, tonnes: 1,051 full load
Dimensions, metres (feet): 54.9 × 11.3 × 3.8 *(180.1 × 37.1 × 12.5)*
Speed, knots: 13
Range, n miles: 8,000 at 12 kt
Complement: 53
Machinery: Diesel electric; 2 diesels; 1,402 hp *(1.06 MW)*; 1 motor; 1,200 hp *(895 kW)*; 1 shaft; bow thruster
Guns: 1 Boeing 25 mm Bushmaster. 2—12.7 mm MGs.
Radars: Navigation: SPS-73; I-band.

Comment: Following overhaul at Boston, transferred from the US Coast Guard on 12 October 2007. Based in the Caribbean.

SAN ANDRES *1/2010, Marco Ghiglino* / 1406010

© 2014 IHS IHS Jane's Fighting Ships 2014-2015

176 Colombia > Patrol forces — Amphibious forces

1 + (3) FASSMER 40 METRE PATROL VESSEL (PBO)

Name	No	Builders	Laid down	Launched	Commissioned
11 DE NOVIEMBRE	PC 145	Fassmer Werft, Bremen	2010	16 Dec 2010	22 Apr 2011

Displacement, tonnes: 245 full load
Dimensions, metres (feet): 40 × 7.56 × 2.35 (131.2 × 24.8 × 7.7)
Speed, knots: 22
Range, n miles: 2,000 at 12 kt
Complement: 24
Machinery: 2 MTU 12V 4000 M73 diesels; 5,150 hp (3.8 MW); 2 shafts
Guns: 1 — 25 mm Typhoon. 2 — 12.7 mm MGs.
Radars: To be announced.

Comment: Contract with Fassmer for the construction of one coastal patrol vessel signed in December 2009. The ship is being built in Bremen and is of steel construction and aluminium superstructure. The design includes a stern ramp for launch/recovery of an Avon Sea Raider interception craft. Three further ships are expected and are likely to be built in Colombia.

11 DE NOVIEMBRE   6/2011, Michael Nitz / 1406380

6 RIVER PATROL CRAFT (PBR)

DILIGENTE PF 121 (ex-LR 121)   **FRITZ HAGALE** PF 124
JUAN LUCIO PF 122   **VENGADORA** PF 125 (ex-LR 125)
ALFONSO VARGAS PF 123   **CARLOS GALINDO** PF 128

Displacement, tonnes: 85 full load
Dimensions, metres (feet): 32 × 5.7 × 1.47 (105 × 18.7 × 4.8)
Speed, knots: 15
Range, n miles: 1,880 at 15 kt
Complement: 14
Machinery: 2 Caterpillar diesels; 166 hp (124 kW); 2 shafts
Guns: 2 — 12.7 mm MGs. 2 — 7.6 mm MGs.
Radars: Surface search: Raytheon SPS-73; I-band.

Comment: Various designs and ages, but all are armed with two 12.7 mm MGs and most have 7.62 mm MGs as well.

DILIGENTE   6/2011, Colombian Navy / 1406573

2 + 2 PAF-L CLASS (RIVER PATROL CRAFT) (PB)

ALEJANDRO LEDESMA ORTIZ PAF 616   **HARRYS TOUS CATAÑO** PAF 617

Displacement, tonnes: 116 full load
Dimensions, metres (feet): 30 × 7.0 × 0.75 (98.4 × 23.0 × 2.5)
Speed, knots: 9
Range, n miles: 840 at 8 kt
Complement: 33 (3 officers)

Military lift: 28 troops
Machinery: 2 Caterpillar C12 diesels; 640 hp (477 kW); 2 waterjets
Guns: 2 — 12.7 mm MGs. 2 — 7.62 mm MGs. 2 Mk 19 grenade launchers.
Radars: Surface search/navigation: I-band.

Comment: New class of inshore patrol craft designed and built by Cotecmar. The first was commissioned on 25 August 2010 and the second on 21 November 2010. There are two armoured turrets (fore and aft). Two further craft were ordered in September 2012.

ALEJANDRO LEDESMA ORTIZ   6/2010, Cotecmar / 1455770

AMPHIBIOUS FORCES

Notes: (1) Two new Cotecmar-designed LCUs are expected to enter service in 2014.
(2) Two hovercraft, possibly Griffon 2000 TD, were ordered in October 2012.
(3) There are 14 armoured troop transport craft TBT 381–394. They are 6.8 m craft construction by Boston Whaler.

348 RIVER ASSAULT BOATS (RAB) (PBR)

Comment: There are some 100 command and 248 assault boats. Most are 6.8 m Boston Whaler craft armed with one 12.7 mm and two 7.62 mm MGs. They operate in conjunction with the river patrol vessels (PAF classes).

ASSAULT BOAT   6/2011, Colombian Navy / 1406571

7 MORROSQUILLO (LCU 1466A) CLASS (LCU)

MORROSQUILLO LD 246   **BAHÍA SOLANO** LD 251   **BAHÍA UTRIA** LD 253
BAHÍA HONDA LD 248   **BAHÍA CUPICA** LD 252   **BAHÍA MALAGA** LD 254
BAHÍA PORTETE LD 249

Displacement, tonnes: 353 full load
Dimensions, metres (feet): 36.3 × 10.4 × 1.8 (119.1 × 34.1 × 5.9)
Speed, knots: 7
Range, n miles: 700 at 7 kt
Complement: 14

Cargo capacity: 167 tons or 300 troops
Machinery: 3 Detroit 6-71 diesels; 522 hp (389 kW) sustained; 3 shafts
Guns: 2 — 12.7 mm MGs.
Radars: Navigation: Raytheon; I-band.

Comment: Former US Army craft built in 1954 and transferred in 1991 and 1992 with new engines. Used as inshore transports. Speed quoted is fully laden. Numbers split between each coast.

MORROSQUILLO   4/2009, Marco Ghiglino / 1305813

IHS Jane's Fighting Ships 2014-2015

Amphibious forces — Training ships — **Colombia** 177

3 + 5 GRIFFON 2000 TD HOVERCRAFT (UCAC)

Displacement, tonnes: 5 full load
Dimensions, metres (feet): 12.7 on cushion × 6.1 *(41.7 × 20.0)*
Speed, knots: 35
Complement: 3
Machinery: 1 Deutz BF8L diesel; 355 hp *(265 kW)*
Radars: Furuno 1000C; I-band.

Comment: The announcement that Griffon Hoverwork Ltd has been awarded the contract for the construction of eight craft was made in January 2013. Similar craft are in service in Estonia, Finland, Lithuania, Pakistan, Peru, and Sweden. The first batch of three entered service in October 2013.

GRIFFON 2000TD *6/2013*, Griffon Hoverwork / 1529007*

0 + 2 (4) LANDING CRAFT TANK (LCT)

Displacement, tonnes: 626 full load
Dimensions, metres (feet): 45.8 × 11.0 × 1.5 *(150.3 × 36.1 × 4.9)*
Speed, knots: 12
Military lift: 200 tons
Machinery: 2 Caterpillar C18 diesels; 1,104 hp *(824 kW)*
Guns: To be announced.

Comment: A contract to build two locally designed landing ships was signed with shipbuilder Cotecmar on 21 March 2013. The new ships are to begin replacing the Morrosquillo class. A class of six is expected.

SURVEY SHIPS

Notes: (1) There are also three small buoy tenders: *Abadía Médez* BB 33, *Ciénaga de Mayorquin* BB 34, and *Isla Palma* BB 35.
(2) There are plans to acquire 10 survey craft. Based on the SeaArk Dauntless RAM design, six of the 17-m craft are to be built in the US and four by Cotecmar.

2 PROVIDENCIA CLASS (AGOR)

Name	No	Builders	Commissioned
PROVIDENCIA	BO 155	Martin Jansen SY, Leer	24 Jul 1981
MALPELO	BO 156	Martin Jansen SY, Leer	24 Jul 1981

Displacement, tonnes: 1,176 full load
Dimensions, metres (feet): 50.3 × 10 × 4 *(165 × 32.8 × 13.1)*
Speed, knots: 13. **Range, n miles:** 15,000 at 12 kt
Complement: 48 (5 officers) + 6 scientists
Machinery: 2 MAN-Augsburg diesels; 1,570 hp(m) *(1.15 MW)*; 1 Kort nozzle prop; bow thruster
Radars: Navigation: Raytheon; I-band.

Comment: Both launched in January 1981. *Malpelo* employed on fishery research and *Providencia* on geophysical research. Both are operated by DIMAR, the naval authority in charge of hydrographic, pilotage, navigational and ports services.

MALPELO *2000, Colombian Navy / 0103704*

1 BUOY TENDER

Name	No	Builders	Commissioned
QUINDIO (ex-*YFR 443*)	BH 153	Niagara Shipbuilding, Buffalo	11 Nov 1943

Displacement, tonnes: 610 full load
Dimensions, metres (feet): 40 × 9.1 × 2.7 *(131.2 × 29.9 × 8.85)*
Speed, knots: 10
Complement: 17 (2 officers)
Machinery: 2 Union diesels; 600 hp *(448 kW)*; 2 shafts

Comment: Transport ship transferred by lease from the US in July 1964 and by sale on 31 March 1979. Used as a buoy tender.

QUINDIO *2000, Colombian Navy / 0103705*

1 SURVEY SHIP (AGSC)

Name	No	Builders	Commissioned
GORGONA	BO 154 (ex-BB 31, ex-BO 161, ex-FB 161)	Lidingoverken	28 May 1954

Displacement, tonnes: 583 full load
Dimensions, metres (feet): 41.2 × 9 × 2.8 *(135.2 × 29.5 × 9.19)*
Speed, knots: 13
Complement: 45 (2 officers)
Machinery: 2 Wärtsilä Nohab diesels; 910 hp(m) *(669 kW)*; 2 shafts

Comment: Paid off in 1982 but after a complete overhaul at Cartagena naval base was back in service in late 1992. A further major refit took place 2005–06. This included work on the hull and possible changes to the superstructure.

GORGONA (old number) *1993, Colombian Navy / 0056835*

TRAINING SHIPS

Notes: There are also nine sail training yachts *Comodoro* YT 230, *Tridente* YT 231, *Cristina* YT 232, *Albatros* YT 233, *Poseidon* YT 234, *Atenea* YT 235, *Cronos* YT 236, *Apolo* YT 237 and *Brigadier* YT 238.

1 SAIL TRAINING SHIP (AXS)

Name	No	Builders	Launched	Commissioned
GLORIA	BE 160	AT Celaya, Bilbao	6 Sep 1966	16 May 1969

Displacement, tonnes: 1,270 full load
Dimensions, metres (feet): 76 oa × 64.6 wl × 10.6 × 6.6 *(249.3; 211.9 × 34.8 × 21.7)*
Speed, knots: 10.5
Complement: 139 (10 officers; 88 trainees)
Machinery: 1 auxiliary diesel; 530 hp(m) *(389 kW)*; 1 shaft

Comment: Sail training ship. Barque rigged. Hull is entirely welded. Sail area, 1,675 sq yds *(1,400 sq m)*. Endurance, 60 days. Similar to Ecuador, Mexico, and Venezuelan vessels.

GLORIA *9/2013*, Hachiro Nakai / 1529087*

AUXILIARIES

Notes: There are 15 river stations EMAF 1531–37 and EMAF 1541–48. These are mobile bases capable of supporting river operations.

EMAF 1532   6/2011, Colombian Navy / 1406576

8 TRANSPORTS

BOCACHICA ETG 501   **SIRIUS** ETG 504 (ex-TM 62)   **MÓVIL I** ETG 509
ARTURUS ETG 502   **CALIMA** ETG 507 (ex-TM 49)   **SULA** (ex-*Móvil II*) ETG 510
PEDRO DAVID SALAS ETG 503 (ex-TM 101)   **BAHÍA SANTA CATALINA** ETG 508

Comment: Small supply ships of various characteristics from 3 tons to 30 tons. Some have transferred to an inshore patrol craft role.

BOCACHICA   6/2011, Colombian Navy / 1406575

2 LUNEBURG CLASS (TYPE 701) (SUPPORT SHIPS) (AGP)

Name	No	Builders	Commissioned
CARTAGENA DE INDIAS (ex-*Luneburg*)	BL 161 (ex-A 1411)	Flensburger Schiffbau	31 Jan 1966
BUENAVENTURA (ex-*Nienburg*)	BL 162 (ex-A 1416)	Bremer Vulcan, Vegesack	1 Aug 1968

Displacement, tonnes: 3,539 full load
Dimensions, metres (feet): 104 × 13.2 × 4.2 *(341.2 × 43.3 × 13.8)*
Speed, knots: 16
Range, n miles: 3,200 at 14 kt
Complement: 70 (9 officers)
Cargo capacity: 1,100 tons
Machinery: 2 MTU MD 16V 538 TB90 diesels; 6,000 hp(m) *(4.1 MW)* sustained; 2 shafts; cp props; bow thruster
Guns: 4 Bofors 40 mm/70 (2 twin).
Radars: Navigation: I-band.
Helicopters: Platform for 1 medium (*Cartagena de Indias*).

Comment: BL 161 paid off from the German Navy in 1994. Taken in hand for refit by HDW, Kiel in August 1997. Recommissioned on 2 November 1997. Guns were cocooned in German service. The ship acts as a depot ship for patrol craft. BL 162 paid off and was transferred the same day on 27 March 1998. She is now based at Málaga. *Cartagena de Indias* has been refitted with a helicopter deck in order to operate Bell 412 helicopters.

BUENAVENTURA   8/2013*, Martin Mokrus / 1529088

1 FLOATING DOCK (ASL)

MAYOR JAIME ARIAS ARANGO DF 170 (ex-DF 41, ex-170)

Comment: Capacity of 165 tons, length 140 ft *(42.7 m)*, displacement 700 tons. Used as a non-self-propelled depot ship for the midget submarines.

MAYOR JAIME ARIAS ARANGO   6/2001, Maritime Photographic / 0114513

7 BAY SUPPORT CRAFT

PLAYA BLANCA ETG 542   **ORION** ETG 546   **LIBERTAD** ETG 559
TIERRA BOMBA ETG 543   **JUANCHACO** ETG 558   **ISLA TESORO** ETG 560
BELL SALTER ETG 544

Comment: Mostly small craft of less than 10 tons. The largest is ETG 544 which is 87 tons and has previously been listed as an Admiral's Yacht.

BELL SALTER   6/1999, Colombian Navy / 0056840

TUGS

10 TUGS (YTL)

ANDAGOYA RM 75   **JOVES FIALLO** RF 83
JOSUÉ ALVAREZ RB 76   **MIGUEL SILVA** RF 85
DON VIZO RB 77   **CAPITÁN VLADIMIR VALEK MOURE** RF 87
PORTETE RB 78   **TENIENTE LUÍS BERNAL BAQUERO** RF 88
CIENAGA DE SAN JUAN RB 80   **SEJERI** RF 93

Comment: River craft of various types described as 'Remolcador Bahia (RB), Fluvial (RF) or Mar (RM)'. Used for transport and ferry duties in harbours and rivers. RF 86 modified as a support vessel and armed with 12.7 mm MGs. RF 86 and 96 modernised in 2012 to undertake river patrol support roles.

JOSUÉ ALVAREZ   6/1999, Colombian Navy / 0056841

COAST GUARD

Notes: There are plans to acquire up to 44 Eduardoño Type 11.7 m interception craft powered by three Yamaha 250 hp engines. There were 14 ordered in late 2012, of which eight had been delivered by mid-2013.

4 POINT CLASS (PB)

Name	No	Builders	Commissioned
CABO CORRIENTES POINT WARDE (ex-*Point Warde*)	PC 141 (ex-82368)	J M Martinac, Tacoma	14 Aug 1967
CABO MANGLARES POINT WELLS (ex-*Point Welis*)	PC 142 (ex-82343)	US Coast Guard Yard, Curtis Bay	20 Nov 1963
CABO TIBURON (ex-*Point Estero*)	PC 143 (ex-82344)	US Coast Guard Yard, Curtis Bay	11 Dec 1963
CABO DE LA VELLA (ex-*Point Sal*)	PC 144 (ex-82352)	J M Martinac, Tacoma	5 Dec 1966

Displacement, tonnes: 67 standard; 70 full load
Dimensions, metres (feet): 25.3 × 5.2 × 1.8 *(83 × 17.1 × 5.9)*
Speed, knots: 23.5. **Range, n miles:** 1,500 at 8 kt
Complement: 10 (1 officer)
Machinery: 2 Caterpillar 3412 diesels; 1,600 hp *(1.19 MW)*; 2 shafts
Guns: 2—12.7 mm MGs.
Radars: Surface search: Hughes/Furuno SPS-73; I-band.

Comment: Steel hulled craft with aluminium superstructure built in United States 1960–70. *Cabo Corrientes* transferred on 29 June 2000 followed by *Cabo Manglares* on 13 October 2000. *Cabo Tiburon* and *Cabo de la Vella* transferred on 8 February 2001 and 29 May 2001 respectively.

CABO TIBURON *4/2009, Marco Ghiglino* / 1406009

1 RELIANCE CLASS

Name	No	Builders	Commissioned
VALLE DEL CAUCA (ex-*Durable*)	PO 44 (ex-WMEC 628)	Coast Guard Yard, Baltimore	8 Dec 1967

Displacement, tonnes: 1,147 full load
Dimensions, metres (feet): 64.2 × 10.4 × 3.2 *(210.6 × 34.1 × 10.5)*
Speed, knots: 18. **Range, n miles:** 6,100 at 14 kt, 2,700 at 18 kt
Complement: 75 (12 officers)
Machinery: 2 Alco 16V-251 diesels; 6,480 hp *(4.83 MW)* sustained; 2 shafts; LIPS cp props
Guns: 1 Bofors 40 mm/70 Mk 38 Bushmaster. 2—12.7 mm MGs.
Radars: Surface search: Hughes/Furuno SPS-73; I-band.
Helicopters: Platform for one medium.

Comment: Transferred to Colombia on 4 September 2003. During 34 years in USCG service, underwent Major Maintenance Availability (MMA) in 1989. The exhausts for main engines, ship service generators and boilers were run in a vertical funnel which reduced flight deck size. Capable of towing ships up to 10,000 tons. Based in the Pacific.

VALLE DE CAUCA *6/2011, Colombian Navy* / 1406563

2 TOLEDO CLASS (LARGE PATROL CRAFT) (PB)

Name	No	Builders	Commissioned
JOSÉ MARIA GARCIA Y TOLEDO	PM 113	Bender Marine, Mobile	15 Jul 1994
JUAN NEPOMUCENO ESLAVA	PM 114	Bender Marine, Mobile	25 May 1994

Displacement, tonnes: 144 full load
Dimensions, metres (feet): 35.4 × 7.6 × 2.1 *(116.1 × 24.9 × 6.9)*
Speed, knots: 25
Range, n miles: 1,200 at 15 kt
Complement: 25 (5 officers)
Machinery: 2 MTU 12V 396 TE94 diesels; 8,240 hp(m) *(6.1 MW)*; 2 shafts
Guns: 1 Bushmaster 25 mm/87 Mk 96. 2—12.7 mm MGs.
Radars: Surface search: Furuno FR 1510D; I-band.

Comment: Acquired under US FMS programme.

JUAN NEPOMUCENO ESLAVA *6/2001, Maritime Photographic* / 0114511

1 ASHEVILLE CLASS (FAST ATTACK CRAFT—GUN) (PGF)

Name	No	Builders	Commissioned
QUITASUEÑO (ex-*Tacoma*)	PM 112	Tacoma Boatbuilding Co, Tacoma	14 Jul 1969

Displacement, tonnes: 229 standard; 249 full load
Dimensions, metres (feet): 50.1 × 7.3 × 2.9 *(164.4 × 24.0 × 9.5)*
Speed, knots: 40. **Range, n miles:** 1,700 at 16 kt, 325 at 37 kt
Complement: 24
Machinery: CODOG; 2 Cummins VT12-875M diesels; 1,450 hp *(1.08 MW)*; 1 GE LM 1500 gas turbine; 13,300 hp *(9.92 MW)*; 2 shafts; cp props
Guns: 1 US 3 in *(76 mm)*/50 Mk 34; 50 rds/min to 12.8 km *(7 n miles)*; weight of shell 6 kg. 1 Bofors 40 mm/56; 160 rds/min to 11 km *(5.9 n miles)* anti-aircraft; weight of shell 0.96 kg. 2—12.7 mm (twin) MGs.
Radars: Surface search: Raytheon 3100; I-band.

Comment: Transferred from US by lease 16 May 1983 and recommissioned 6 September 1983 and by sale August 1989. Fire-control system removed. Unreliable propulsion system prevented further transfers of this class and it is unlikely the gas turbine is operational, which reduces the top speed to 16 kt. The ship was refitted in 2008–09.

QUITASUEÑO *2000, Colombian Navy* / 0103696

2 JAIME GÓMEZ (MK III PB) CLASS
(COASTAL PATROL CRAFT) (PB)

Name	No	Builders	Commissioned
JAIME GÓMEZ CASTRO	PM 105 (ex-GC 105)	Peterson Builders Inc, Sturgeon Bay	1975
JUAN NEPOMUCENO PEÑA	PM 106 (ex-GC 106)	Peterson Builders Inc, Sturgeon Bay	1977

Displacement, tonnes: 35 full load
Dimensions, metres (feet): 19.8 × 5.5 × 1.6 *(65 × 18.0 × 5.2)*
Speed, knots: 28
Range, n miles: 450 at 26 kt
Complement: 7 (1 officer)
Machinery: 3 Detroit 8V-71 diesels; 690 hp *(515 kW)* sustained; 3 shafts
Guns: 2—12.7 mm MGs. 2—7.62 mm MGs. 1 Mk 19 grenade launcher.
Radars: Surface search: 2 Furuno FR 1510D; I-band.

Comment: Acquired from the US. Recommissioned in December 1989 and February 1990 respectively. Original 40 mm and 20 mm guns replaced by lighter armament. Both based at Leticia, Rio Amazonas.

JUAN NEPOMUCENO PEÑA *6/2011, Colombian Navy* / 1406572

184　Cote d'Ivoire > Introduction — **Croatia** > Introduction

Cote d'Ivoire
MARINE CÔTE D'IVOIRE

Country Overview

Formerly a French colony, The Republic of Côte d'Ivoire gained full independence in 1960. Located in west Africa, the country has an area of 133,425 square miles and a 281 n mile coastline with the Gulf of Guinea. It is bordered to the east by Ghana and to the west by Liberia and Guinea. The capital is Yamoussoukro while the former capital, Abidjan, is the largest city, principal port and commercial centre. A further port at San Pedro is linked to Mali by rail. Territorial seas (12 n miles) are claimed. A 200 n mile EEZ has been claimed but the limits have not been defined by boundary agreements.

Following the rebellion of September 2002, a Government of National Conciliation has restored a level of stability although internal tensions continue. While the navy remains unchanged, operational effectiveness is likely to have suffered.

Headquarters Appointments

Chief of Naval Staff:
　Captain Djakaridja Konaté

Bases

Use made of ports at Locodjo (Abidjan), Sassandra, Tabouand San-Pédro

Personnel

2014: 950 (75 officers)

PATROL FORCES

Notes: (1) There is a Raidco 14 m patrol craft *Bian* operated by the police.
(2) A 33 m Japanese-built stern trawler *Golfe de Guinée* is used for fishery protection.

1 PATRA CLASS (LARGE PATROL CRAFT) (PBO)

Name	No	Builders	Laid down	Launched	Commissioned
L'INTRÉPIDE	–	Auroux, Arcachon	7 Jul 1977	21 Jul 1978	6 Oct 1978

Displacement, tonnes: 150 full load
Dimensions, metres (feet): 40.4 × 5.9 × 1.6 *(132.5 × 19.4 × 5.2)*
Speed, knots: 26. **Range, n miles:** 1,750 at 10 kt, 750 at 20 kt
Complement: 19 (2 officers)
Machinery: 2 SACM AGO 195 V12 CZSHR diesels; 4,340 hp(m) *(3.19 MW)* sustained; 2 shafts; cp props
Guns: 1 Breda 40 mm/70. 1 Oerlikon 20 mm. 2 — 7.62 mm MGs.
Radars: Surface search: Racal Decca 1226; I-band.

Comment: Of similar design to French Patra class. Laid down 7 July 1977. Patrol endurance of five days. SS-12M missiles are no longer carried. Sister ship *L'Ardent* decommissioned in 2003 to provide spares. Operational status doubtful.

PATRA CLASS　　　　　　　　　　　　　　　　　　　　　3/*1994* / 0080123

AUXILIARIES

Notes: (1) There are also some Rotork 412 craft supplied in 1980. Some are naval, some civilian.
(2) Two French harbour tugs *Merisier* and *Meronnior* were acquired in September 1999.
(3) A Yunnan class LCM *Atchan* may still be in limited service.

2 CTM (LCM)

ABY (ex-*CTM 15*)　　　**TIAGHA** (ex-*CTM 16*)

Displacement, tonnes: 152 full load
Dimensions, metres (feet): 23.8 × 6.4 × 1.3 *(78.1 × 21.0 × 4.3)*
Speed, knots: 9.5
Range, n miles: 350 at 8 kt
Complement: 6
Military lift: 48 tons
Machinery: 2 Poyaud 520 V8 diesels; 225 hp(m) *(165 kW)*; 2 shafts

Comment: Transferred from France in March 1999. Built in about 1968. Bow ramps are fitted. Probably not operational.

AFFAIRES MARITIMES

2 RODMAN 890 (PBR)

AMOUGNA AF 003　　　**MONSEKELA** AF 004

Dimensions, metres (feet): 8.9 × 3 × 0.8 *(29.2 × 9.8 × 2.6)*
Speed, knots: 28
Range, n miles: 150 at 25 kt
Complement: 3
Machinery: 2 Volvo diesels; 300 hp(m) *(220 kW)*; 2 shafts
Guns: 1 — 7.62 mm MG.
Radars: Surface search: I-band.

Comment: Two craft delivered by Rodman in 1997. Employed on Fishery Protection duties.

CTM (French colours)　　　　　　　　　　　　6/*1995* / 0012960

AMOUGNA　　　　　　　　　　　　　6/*1997, Rodman* / 0583296

Croatia
HRVATSKA RATNA MORNARICA

Country Overview

Formerly a constituent republic of the Federal Republic of Yugoslavia, Croatia declared its independence in 1991. With an area of 21,829 square miles, it is situated in south-east Europe in the Balkan Peninsula and bordered to the north by Slovenia and Hungary, to the east and south by Bosnia and Herzegovina and to the east by Montenegro. There are some 1,100 offshore islands and there is an overall coastline of 3,127 n miles with the Adriatic Sea on which Dubrovnik, Split, Ploče and Rijeka are the principal ports. The capital and largest city is Zagreb. Territorial waters (12 n miles) are claimed and an Ecological and Fishery Zone was declared in 2004.

Headquarters Appointments

Commander of the Navy:
　Commodore Robert Hranj
Commander, Coast Guard:
　Captain Marin Stošić
Commander of the Flotilla:
　Captain Ivo Raffanelli

Personnel

2014: 1,710 (388 officers)

General

The Navy was established on 12 September 1991. The law to establish a Coast Guard, as a component of the navy, was passed on 3 October 2007. Its roles include fishery protection, counter-drugs and smuggling operations and environmental protection.

Organisation

The Navy is organised into Surface Ships, Support and Mine Warfare Divisions. There are also a Coastal Surveillance Battalion and Training Centre. The Coast Guard is organised into two divisions.

Bases

Navy headquarters and main base: Split.
Minor base: Ploče.
Coast guard headquarters and main base: Split.
Minor base: Pula.

Coast Defence

Three mobile RBS 15 batteries on trucks are likely to be decommissioned. Jadran command system for coastal defence using Italian (Gem) built and US (More) radars installed in 2003. Sites include the islands of Vis, Lastovo, Dugi Otok and Mljet.

IHS Jane's Fighting Ships 2014-2015　　　　　　　　　© 2014 IHS

SUBMARINES

2 R-2 MALA CLASS
(TWO-MAN SWIMMER DELIVERY VEHICLES) (LDW)

Dimensions, metres (feet): 4.9 × 1.4 × 1.3 *(16.1 × 4.6 × 4.3)*
Speed, knots: 4.4
Complement: 2
Machinery: 1 motor; 4.7 hp(m) *(3.5 kW)*; 1 shaft
Mines: 250 kg of limpet mines.

Comment: Free-flood craft with the main motor, battery, navigation pod and electronic equipment housed in separate watertight cylinders. Instrumentation includes aircraft type gyrocompass, magnetic compass, depth gauge (with 0 to 100 m scale), echo-sounder, sonar and two searchlights. Constructed of light aluminium and plexiglass, it is fitted with fore and after-hydroplanes, the tail being a conventional cruciform with a single rudder abaft the screw. Large perspex windows give a good all-round view. Operating depth, 60 m *(196.9 ft)*, maximum. Two reported sold to Syria and one to Sweden.

Notes: There is also an R-1 craft which is 3.7 m long and capable of 2.8 kt down to 50 m. It has a range of 4 n miles. There may also be some locally built SDVs.

R-1   2/2002, RH-Alan / 0528427

R-2   2/2002, RH-Alan / 0528428

LAND-BASED MARITIME AIRCRAFT

Notes: Six Pilatus PC9 aircraft, four Mi-8 helicopters and one unmanned aircraft are used for fishery protection and counter-pollution tasks. These are Air Force assets assigned primarily to the Coast Guard.

PATROL FORCES

Notes: (1) There are plans to acquire two corvettes of up to 120 m length. These may be new ships or second-hand. Acquisition is unlikely to begin before 2020.
(2) There are plans to acquire five new coastal patrol craft by 2018.
(3) Two RHIBs were acquired in 2008 for special forces. Their names are *Bljesak* and *Oluja*.

2 HELSINKI CLASS (FAST ATTACK CRAFT—MISSILE) (PTGM)

Name	No	Builders	Commissioned
VUKOVAR (ex-*Oulu*)	RTOP 41 (ex-62)	Wärtsilä, Helsinki	1 Oct 1985
DUBROVNIK (ex-*Kotka*)	RTOP 42 (ex-63)	Wärtsilä, Helsinki	16 Jun 1986

Displacement, tonnes: 284 standard; 305 full load
Dimensions, metres (feet): 45 × 8.9 × 3 *(147.6 × 29.2 × 9.8)*
Speed, knots: 30
Complement: 30

Machinery: 3 MTU 16V 538 TB92 diesels; 10,230 hp(m) *(7.52 MW)* sustained; 3 shafts
Missiles: SSM: 8 Saab RBS 15 ❶; inertial guidance; active radar homing to 70 km *(37.8 n miles)* at 0.8 Mach; warhead 150 kg; sea-skimmer.
Guns: 1 Bofors 57 mm/70 ❷; 200 rds/min to 17 km *(9.3 n miles)*; weight of shell 2.4 kg. 2 Sako 23 mm/87 (twin) ❸.
Rockets: 6—103 mm rails for rocket illuminants.
Depth charges: 2 rails.
Physical countermeasures: Decoys: Philax chaff and IR flare launcher.
Electronic countermeasures: ESM: Argo; radar intercept.
Radars: Surface search: 9GR 600 ❹; I-band.
Fire control: Philips 9LV 225 ❺; J-band.
Navigation: Raytheon ARPA; I-band.
Sonars: Simrad Marine SS 304; high-resolution active scanning. Finnyards Sonac/PTA towed array; low frequency.
Weapon control systems: Saab EOS 400 optronic director.

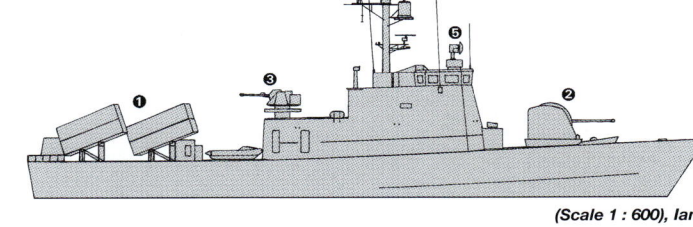

VUKOVAR   (Scale 1 : 600), Ian Sturton / 1335453

Programmes: Both ordered for the Finnish Navy on 13 January 1983. Decommissioned in 2007 and sold to Croatia in 2008. Details are as for ships in Finnish service and may be different.
Modernisation: A Sako barbette can take either twin 23 mm guns or a Sadral SAM launcher. The Sako mounting has replaced the original ZU version. Upgrade of RBS 15 missiles is under consideration.
Structure: The light armament can be altered to suit the planned role. Missile racks can also be replaced by mine rails. Hull and superstructure of light alloy.

DUBROVNIK   6/2009, Croatian Navy / 1406014

186 Croatia > Patrol forces

2 KRALJ (TYPE R-03) CLASS (FSG)

Name	No	Builders	Launched	Commissioned
KRALJ PETAR KRESIMIR IV	RTOP 11	Kraljevica Shipyard	21 Mar 1992	7 Jul 1992
KRALJ DMITAR ZVONIMIR	RTOP 12	Kraljevica Shipyard	30 Mar 2001	16 Sep 2001

Displacement, tonnes: 388 (RTOP 11), 396 (RTOP 12) full load
Dimensions, metres (feet): 54.2 × 8.6 × 3.6 *(177.8 × 28.2 × 11.8)*
Speed, knots: 32
Range, n miles: 1,700 at 18 kt
Complement: 32 (5 officers)

Machinery: 3 M 504B-2 diesels; 12,500 hp(m) *(9.2 MW)* sustained; 3 shafts
Missiles: SSM: 4 or 8 Saab RBS15 (2 or 4 twin) ❶; active radar homing to 70 km *(37.8 n miles)* at 0.8 Mach; warhead 83 kg.
Guns: 1 Bofors 57 mm/70 ❷; 200 rds/min to 17 km *(9.3 n miles)*; weight of shell 2.4 kg. Launchers for illuminants on side of mounting. 1—30 mm/65 AK 630M ❸; 6 barrels; 3,000 rds/min combined to 4 km.
Mines: 4 AIM-70 magnetic or 6 SAG-1 acoustic in lieu of SSMs.
Physical countermeasures: Decoys: 2 Wallop Barricade chaff/IR launchers.
Radars: Surface search: Racal BT 502 ❹; E/F-band.
Fire control: PEAB 9LV 249 Mk 2 ❺; I/J-band.
Navigation: Racal 1290A; I-band.
Sonars: RIZ PP10M; hull-mounted; active search; high frequency.
Weapon control systems: Philips PEAB 9LV 249 Mk 2 director. Kolonka for AK 630M.

Programmes: The building of this class (formerly called Kobra by NATO) was officially announced as 'suspended' in 1989 but was restarted in 1991. Designated as a missile Gunboat.
Modernisation: Both ships are to be modernised with new diesel engines, probably of German origin. Upgrade of the RBS 15 missiles is under consideration.
Structure: Derived from the Koncar class with a stretched hull and a new superstructure. Either missiles or mines may be carried. The second of class is 0.6 m longer than the first ship and incorporates modifications to the bridge structure.
Operational: Based at Split.

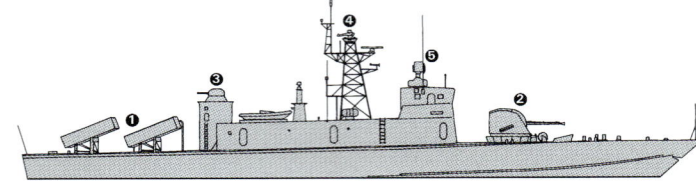
KRALJ DMITAR ZVONIMIR (Scale 1 : 600), Ian Sturton / 1044094

KRALJ PETAR KRESIMIR IV 2/2002, Hrvatski Vojnik / 0528426

KRALJ DMITAR ZVONIMIR 10/2007, Croatian Navy / 1170105

KRALJ PETAR KRESIMIR IV 10/2007, Croatian Navy / 1170104

1 RIVER PATROL CRAFT (PBR)

ŠOKADIJA OB 93

Displacement, tonnes: 49 full load
Dimensions, metres (feet): 19.4 × 4.4 × 1.0 *(63.6 × 14.4 × 3.3)*
Speed, knots: 12
Complement: 11
Machinery: 2 Torpedo B 536RM diesels; 300 hp *(224 kW)*; 2 shafts
Guns: 1—20 mm.
Radars: Surface search/navigation: Furuno M 1942 Mk 2; I-band.

Comment: Former minesweeper launched in 1971 at Mačvanska Mitrovica. Used as a river patrol vessel. Operated by the army and based in Osijek on River Drava.

OB 93
10/2007, Croatian Navy
1170102

IHS Jane's Fighting Ships 2014-2015

Patrol forces — Amphibious forces < **Croatia** 187

1 KONČAR (TYPE R-02) CLASS
(FAST ATTACK CRAFT—MISSILE) (PTGF)

Name	No	Builders	Launched	Commissioned	Recommissioned
SIBENIK (ex-*Vlado Četković*)	RTOP 21 (ex-402)	Tito SY, Kraljevica	20 Aug 1977	Mar 1978	28 Sep 1991

Displacement, tonnes: 268 full load
Dimensions, metres (feet): 45.8 × 8.4 × 3.0 *(150.3 × 27.6 × 9.8)*
Speed, knots: 38; 23 diesel
Range, n miles: 500 at 35 kt, 880 at 23 kt
Complement: 31 (5 officers)

Machinery: CODAG; 2 RR Proteus 52-M558 gas turbines; 7,200 hp *(5.37 MW)* sustained; 2 MTU 16V 538 TB91 diesels; 7,200 hp(m) *(5.29 MW)* sustained; 4 shafts; cp props
Missiles: SSM: 4 Saab RBS15; active radar homing to 70 km *(37.8 n miles)* at 0.8 Mach; warhead 83 kg.
Guns: 1 Bofors 57 mm/70; 200 rds/min to 17 km *(9.3 n miles)*; weight of shell 2.4 kg. 1—30 mm/65 AK 630M; 6 barrels; 3,000 rds/min to 4 km.
Rockets: 128 mm rocket launcher for illuminants.
Physical countermeasures: Decoys: 2 Wallop Barricade double layer chaff launchers.
Radars: Surface search: Decca 1226; I-band.
Fire control: Philips TAB; I/J-band.
Weapon control systems: PEAB 9LV 202 GFCS.

Programmes: Type name, Raketna Topovnjaca. Recommissioned into the Croatian Navy on 28 September 1991. Others of the class serve with the Yugoslav Navy.
Modernisation: The original Styx missiles have been replaced by RBS 15 and the after 57 mm gun by a 30 mm AK 630. Fire-control radar was updated in 1994 and a new surface search radar was acquired in 2008.
Structure: Aluminium superstructure. Designed by the Naval Shipping Institute in Zagreb based on Swedish Spica class with bridge amidships like Malaysian boats.
Operational: Based at Split. To remain in service until 2018.

ŠIBENIK   *10/2004, Croatian Navy* / 1170103

AMPHIBIOUS FORCES

2 CETINA (SILBA) CLASS (LCT/ML)

Name	No	Builders	Launched	Commissioned
CETINA	DBM 81	Brodosplit, Split	18 Jul 1992	19 Feb 1993
KRKA	DBV 82	Brodosplit, Split	17 Sep 1994	9 Mar 1995

Displacement, tonnes: 894 full load
Dimensions, metres (feet): 49.7 oa; 43.9 wl × 10.2 × 3.2 *(163.1; 144 × 33.5 × 10.5)*
Speed, knots: 12
Range, n miles: 1,200 at 12 kt
Complement: 27 (5 officers)

Military lift: 460 tons or 6 medium tanks or 7 APCs or 4—130 mm guns plus towing vehicles or 300 troops with equipment
Machinery: 2 Alpha 10V23L-VO diesels; 3,100 hp(m) *(2.28 MW)* sustained; 2 shafts; cp props
Missiles: SAM: 1 SA-N-5 Grail quad mounting *(Cetina)*.
Guns: 4—30 mm/65 (2 twin) AK 230 *(Cetina)*. 2 *(Krka)* Hispano 20 mm M71.
Mines: SAG-2 (152 DBM 81, 114 DBV 82); MNS 90 (124 DBM 81, 92 DBV 82); AIM M70 (72 DBM 81, 52 DBV 82).
Radars: Surface search: Racal Decca 1290A; I-band.

Comment: Ro-ro design with bow and stern ramps. *Cetina*'s two 30 mm guns are either side of the bridge. *Krka* appears to be fitted with a tank gun forward. Can be used for minelaying, transporting weapons or equipment and personnel. *Krka* is being used as a water carrier. Both are operational and based at Split.

CETINA   *9/2005, Croatian Navy* / 1170106

KRKA   *6/2007, Freivogel Collection* / 1167948

3 TYPE 21 (LCVP)

DJB 103   **DJB 104**   **DJB 107**

Displacement, tonnes: 39 full load
Dimensions, metres (feet): 21.3 × 4.3 × 1.1 *(69.9 × 14.1 × 3.6)*
Speed, knots: 21
Range, n miles: 320 at 18 kt
Complement: 6

Military lift: 6 tons or 40 troops
Machinery: 1 (2 in 103) MTU 12V 331 TC81 diesel; 1,450 hp(m) *(1.07 MW)*; 1 shaft (2 waterjets in 103)
Guns: 1—20 mm M71. 1—30 mm grenade launcher.
Radars: Navigation: Decca 1213; I-band.

Comment: Built at Greben Shipyard 1987-88. *DJB 103* upgraded with new main machinery in 1991.

DJB 103   *5/1997, Dario Vuljanić* / 0012246

1 TYPE 22 (LCVPF)

DJC 106 (ex-624)

Displacement, tonnes: 43 full load
Dimensions, metres (feet): 22.3 × 4.8 × 1 *(73.2 × 15.7 × 3.3)*
Speed, knots: 35
Range, n miles: 320 at 22 kt
Complement: 8

Military lift: 40 troops or 15 tons cargo
Machinery: 2 MTU MWM 604 TDV8 diesels; 1,740 hp(m) *(1.28 MW)*; 2 waterjets
Guns: 2 Hispano 20 mm. 1—30 mm grenade launcher.
Radars: Navigation: Decca 150; I-band.

Comment: Built at Greben Shipyard in 1987 of polyester and glass fibre.

DJC 106   *8/1998, N A Sifferlinger* / 0038489

188 Croatia > Mine warfare forces — Coast guard

MINE WARFARE FORCES

Notes: There are plans to acquire two second-hand minehunters.

1 MPMB CLASS (MINEHUNTER—INSHORE) (MHI)

Name	No	Builders	Launched	Commissioned
KORCULA	LM 51	Greben, Vela Luka	22 Apr 2006	20 Apr 2007

Displacement, tonnes: 176 full load
Dimensions, metres (feet): 25.7 × 6.8 × 2.6 (84.3 × 22.3 × 8.5)
Speed, knots: 11
Range, n miles: 1,000 at 9 kt
Complement: 14 (3 officers)

Machinery: 2 MTU 8V 183TE62 diesels; 993 hp(m) (730 kW); 2 Holland Roerpropeler stern azimuth thrusters; bow thruster; 190 hp(m) (140 kW)
Missiles: SAM: SA-N-10 (Igla).
Guns: 1—20 mm M71.
Physical countermeasures: Minehunting: 1 Super Sea Rover (Benthos); Minesweeping: MDL3 mechanical sweep.
Radars: Navigation: Kelvin Hughes 5000 ARPA, NINAS Mod.
Sonars: Reson mine avoidance; active; high frequency. Klein 2000 side scan; active for route survey; high frequency.

Comment: Ordered in 1995. The ship has a trawler appearance with a gun on the forecastle and a hydraulic crane on the sweep deck. GRP hull. Due to a shortage of funds, building had stopped by late 1999 but was later revived. Became fully operational in mid-2008.

KORCULA   9/2007, Croatian Navy / 1170101

AUXILIARIES

Notes: In addition there are two diving tenders *BRM-81* and *BRM-83*, auxiliary transport ship *PDS-713*, five harbour transport boats *BMT-1/5*, and two yachts *Učka* (ex-*Podgorka*) and *Jadranka* (ex-civilian *Smile*). *Jadranka* was involved in a grounding incident in April 2006.

BRM 83   2/2007, Croatian Navy / 1335354

1 PT 71 TYPE (TRANSPORT) (AKL)

PT 71 (ex-*Meduza*)

Displacement, tonnes: 721 full load
Dimensions, metres (feet): 46.4 × 7.2 × 5.2 (152.2 × 23.6 × 17.1)
Speed, knots: 10
Complement: 16 (2 officers)
Machinery: 1 Burmeister & Wain diesel; 930 hp(m) (684 kW); 1 shaft
Guns: 1 Bofors 40 mm/60. 2 Hispano 20 mm M71 can be carried.
Radars: Navigation: Racal Decca 1216A; I-band.

Comment: Built in 1953. Underwent refit at Marina Punat in 2007. Water capacity 320 tons.

IHS Jane's Fighting Ships 2014-2015

PT 71   2/2007, Croatian Navy / 1170097

COAST GUARD

Notes: There are plans to acquire a new patrol craft.

4 MIRNA (TYPE 140) CLASS
(FAST ATTACK CRAFT—PATROL) (PCM)

Name	No	Builders	Launched
NOVIGRAD (ex-*Biokovo*)	OB 61 (ex-171)	Kraljevica Shipyard	18 Dec 1980
SOLTA (ex-*Mukos*)	OB 62 (ex-176)	Kraljevica Shipyard	11 Nov 1982
CAVTAT (ex-*Vrlika*; ex-*Cer*)	OB 63 (ex-180)	Kraljevica Shipyard	27 Sep 1984
HRVATSKA KOSTAJNICA (ex-*Durmitor*)	OB 64 (ex-181)	Kraljevica Shipyard	10 Jan 1985

Displacement, tonnes: 144 full load
Dimensions, metres (feet): 32.6 × 6.7 × 2.3 (107 × 22.0 × 7.5)
Speed, knots: 25. **Range, n miles:** 600 at 24 kt
Complement: 19 (3 officers)

Machinery: 2 SEMT-Pielstick 12 PA4 200 VGDS diesels; 5,292 hp(m) (3.89 MW) sustained; 2 shafts
Missiles: SAM: 1 SA-N-5 Grail quad mounting; manual aiming; IR homing to 6 km (3.2 n miles) at 1.5 Mach; altitude to 2,500 m (8,000 ft); warhead 1.5 kg.
Guns: 1 Bofors 40 mm/70. 2—12.7 mm MGs.
Depth charges: 8 DCs.
Physical countermeasures: Decoys: chaff launcher (OB 62).
Radars: Surface search: Racal Decca 1216C; I-band.
Sonars: Simrad SQS-3D/SF; active high frequency.

Comment: An electric outboard motor has been removed. Two were captured after sustaining heavy damage, one by a missile and the other by a torpedo fired from the island of Brač. Both fully repaired and all four are operational and transferred to the Coast Guard in 2009. The 20 mm guns and illuminant launchers have been removed in order to accommodate a RHIB for boarding parties.

NOVIGRAD   6/2009, Massimo Annati / 1305815

1 MOMA (PROJECT 861) CLASS (AX)

Name	No	Builders	Commissioned
ANDRIJA MOHOROVIČIĆ	BS 72 (ex-PH 33)	Northern Shipyard, Gdansk	1972

Displacement, tonnes: 1,538 full load
Dimensions, metres (feet): 73.3 × 11.2 × 3.9 (240.5 × 36.7 × 12.8)
Speed, knots: 17. **Range, n miles:** 9,000 at 11 kt
Complement: 27 (4 officers)
Machinery: 2 Zgoda-Sulzer 6TD48 diesels; 3,300 hp(m) (2.4 MW) sustained; 2 shafts; cp props
Radars: Navigation: Racal Decca BT 502; I-band.

Comment: Built in 1971 for the Yugoslav Navy as a survey vessel. Based at Split. Has a 5 ton crane and carries a launch. Transferred to the Coast Guard in 2009 but continues to be used as the Naval Academy training ship.

ANDRIJA MOHOROVIČIĆ   1/2007, Croatian Navy / 1170100

© 2014 IHS

1 SPASILAC CLASS (ASR)

Name	No	Builders	Commissioned
FAUST VRANČIČ (ex-*Spasilac*)	BS 73 (ex-PS 12)	Tito Shipyard, Belgrade	10 Sep 1976

Displacement, tonnes: 1,616 full load
Dimensions, metres (feet): 55.5 × 12 × 3.8 *(182.1 × 39.4 × 12.5)*
Speed, knots: 13
Range, n miles: 4,000 at 12 kt
Complement: 28 (4 officers)

Cargo capacity: 350 tons fuel; 300 tons deck cargo
Machinery: 2 diesels; 4,340 hp(m) *(3.19 MW)*; 2 shafts; Kort nozzle props; bow thruster
Guns: 2 — 20 mm M 71.
Radars: Navigation: Kelvin Hughes Nucleus 5000R; I-band.

Comment: Former salvage ship now employed as a training and command unit. All salvage equipment has been removed. Underwent refit during 2005. Transferred to the Coast Guard in 2009.

FAUST VRANČIČ   12/2006, Croatian Navy / 1170098

2 PATROL CRAFT (PB)

LR 71   LR 73

Displacement, tonnes: 86 standard; 130 full load
Dimensions, metres (feet): 21.39 × 4.92 × 2.1 *(70.2 × 16.1 × 6.9)*
Speed, knots: 13
Machinery: 1 diesel; 245 hp *(180 kW)*; 1 shaft
Guns: To be announced.
Radars: Surface search/navigation: To be announced.

Comment: Former tugs originally built in 1960. Transferred to the Coast Guard for patrol duties in 2009.

LR 73   6/2009, Croatian Navy / 1406015

MINISTRY OF INTERIOR

Notes: (1) A Ministry of Interior maritime force polices inshore waters. These vessels are in five types:
Type 1: 3 — 24 m craft capable of 30 kt; P-1 *(Srd)*, P-2 *(Marino)*, P-101 *(Sveti Mihovic)*
Type 2: 6 — 13 m craft capable of 23 kt; P-11 to P-16
Type 3: 6 — 11 m craft capable of 23 kt; P-111 to P-116
Type 4: 4 — 14 m craft capable of 30 kt; P-201, P 202, P 203 and P 207
Type 5: Numerous small craft under 10 m; RIB or inflatable construction
(2) In addition there are civilian registered base port craft with PU (Pula), SB (Sibenic), ST (Split) and so on markings.

SVETI MIHOVIC   9/2008, Per Körnefeldt / 1335353

P 201   6/2009, Massimo Annati / 1305818

P 114   9/2008, Per Körnefeldt / 1335351

Cuba

MARINA DE GUERRA REVOLUCIONARIA

Country Overview

The Republic of Cuba is an independent republic located in the Caribbean Sea with which it has a 2,020 n mile coastline. The most westerly of the Greater Antilles group, the country comprises two main islands, Cuba (40,519 square miles) and Isla de la Juventud (849 square miles), and more than 1,600 small coral cays and islets. To the west, Cuba commands the approaches to the Gulf of Mexico; the Straits of Florida and the Yucatán Channel separate the country from Florida and Mexico respectively. To the east, the Windward Passage separates the island from Hispaniola (Haiti and the Dominican Republic). Jamaica lies to the south and the Bahamas to the north-east. Havana is the capital, largest city and principal port. Territorial seas (6 n miles) are claimed. A 200 n mile EEZ has been claimed but the limits have not been defined.

The Navy is in a parlous state and has no capability to sustain operations beyond territorial waters.

Headquarters Appointments

Commander of the Navy:
Vice Admiral Pedro Perez Miguel Betancourt
Chief of Naval Staff:
Rear Admiral Luis González Navarro

Personnel

2014: 2,800 (approximately) (including 550 marines)

Command Organisation

Western Naval District (HQ Cabanas).
Eastern Naval District (HQ Holguin).

Naval Aviation

Four Kamov Ka-28 and 14 Mi-14PL Haze A have been reported but operational status is not known.

Coast Defence

Truck mounted SS-N-2B and SS-N-2D Styx. One battery near Havana and another at Nicaro.

Bases

Cabanas, Nicaro, Cienfuegos, Havana, Santiago de Cuba, Banes.
The Naval Academy is at Punta Santa Ana.

DELETIONS

Notes: Some vessels have been disposed of. Others are decaying alongside in harbour.

Denmark
DEN KONGELIGE DANSKE MARINE

Country Overview

The Kingdom of Denmark is a constitutional monarchy. The southernmost of the Scandinavian countries, it comprises most of the Jutland peninsula and more than 400 islands, the principal of which are Sjaelland (the largest), Fyn, Lolland, Falster, Langeland and Møn. The island of Bornholm lies in the Baltic about 70 n miles east of Sjaelland. With an area of 16,639 square miles, the country is bordered to the south by Germany. Its 1,825 n mile coastline is with the North Sea to the west, the Skagerrak to the north and the Kattegatt, which is linked to the Baltic Sea by the Øresund, to the east. The capital, largest city and principal port is Copenhagen. There are further ports at Århus, Odense and Ålborg. There are also two external territories in the north Atlantic ocean: Greenland and the Faroe Islands are both internally self-governing. Territorial seas (12 n miles) are claimed. It has claimed a 200 n mile EEZ for the mainland and 200 n mile Fishery Zones for the Faroes and Greenland.

Headquarters Appointments

Admiral Fleet:
  Rear Admiral Frank Trojahn
Commander, Naval Home Guard:
  Captain J Walter

Diplomatic Representation

Defence Attaché, Washington and Ottawa:
  Brigadier J Jacobsen
Defence Attaché, London, Dublin and The Hague:
  Captain M la Cour-Andersen
Defence Attaché, Paris:
  Colonel J O Rossen-Jørgensen
Defence Attaché, Berlin and Prague:
  Colonel C Wessel-Tolvig
Defence Attaché, Moscow and Minsk:
  Brigadier L Fredskov Hansen
Defence Attaché, Warsaw:
  Colonel C Rasmussen
Defence Attaché, Vilnius:
  Colonel S Frausig
Defence Attaché, Addis Ababa:
  Colonel F Johansen
Defence Attaché, Kabul:
  Colonel C P Brixensen
Defence Attaché, Islamabad:
  Colonel S Bornholdt-Andersen

Diplomatic Representation – continued

Defence Attaché, Abu Dhabi:
  Lieutenant Colonel J G Thygesen

Personnel

(a) 2014: 3,179 (577 officers) including 101 national service
     Reserves: 70.
     Naval Home Guard: 4,985.
(b) 4 months' national service

Bases

Korsør (Frigates, Patrol Craft), Frederikshavn (OPV, Combat Support Ships), Nuuk (Greenland)

Naval Air Arm

Naval helicopters are formed in 723rd Squadron which is part of the Air Force Command helicopter wing.

Naval Home Guard

Established in 1952 as a separate service under the operational control of the navy. Duties include surveillance, harbour control, search and rescue and the guarding of naval installations ashore. Following the Defence Agreement 2004, the service is to play a greater role in home defence and further tasks include environmental survey, pollution control and support of the police and customs services.

Coast Defence

The coastal radar system, known as KYRA is based on the Terma Scanter 2001 and 4000 radars in conjunction with electro-optical camera equipped lookout stations. The system is operated from the maritime headquarters at Aarhus and from two maritime surveillance centres at Fredrikshavn and Bornhölm Island.

Command and Control

The Royal Danish Navy, on behalf of the Ministry of Defence, runs and maintains two environmental protection divisions based in Frederikshavn and Korsør respectively. Responsibility for environmental survey, protection and pollution fighting in maritime areas around Denmark is executed by the Royal Danish Navy. Survey ships are run by the Danish Maritime Authority under the Ministry of Defence, and the Directorate of Fisheries has four rescue vessels.

Appearance

Ships are painted in six different colours as follows:
  Grey: frigates, corvettes and patrol frigates.
  Orange: survey vessels and pollution control vessels.
  White: Royal Yacht and the sail training yawls.
  Black/yellow: service vessels, tugs and ferryboats.

Strength of the Fleet

Type	Active	Building (Projected)
Frigates	7	—
Large Patrol Craft	3	1
Patrol Craft	36	—
Minehunters and Drones	6	—
Support Ships	2	—
Training Ships	4	—
Research Ships	1	—
Transport Ship	1	—
Royal Yacht	1	—
Diving Ship	1	—

Prefix to Ships' Names

HDMS

DELETIONS

Patrol Forces

2011   VTS 3, VTS 4

Auxiliaries

2011   Havkatten

Icebreakers

2013   Thorbjørn, Danbjørn, Isbjørn

PENNANT LIST

Frigates		Patrol Forces		P 571	Ejnar Mikkelsen	A 543		Ertholm	MSD 5	Hirsholm
F 357	Thetis	P 520	Diana	Y 388	Tulugaq	A 544		Alholm	MSD 6	Saltholm
F 358	Triton	P 521	Freja			A 559		Sleipner	Y 331	Søloven
F 359	Vaedderen	P 522	Havfruen	**Auxiliaries**		A 560		Gunnar Thorson	Y 101	Svanen
F 360	Hvidbjørnen	P 523	Najaden			A 561		Gunnar Seidenfaden	Y 102	Thyra
F 361	Iver Huitfeldt	P 524	Nymfen	A 540	Dannebrog	A 562		Mette Miljø	Y 343	Lunden
F 362	Peter Willemoes	P 525	Rota	A 541	Birkholm	A 563		Marie Miljø	Y 344	Arvak
F 363	Niels Juel	P 570	Knud Rasmussen	A 542	Fyrholm	L 16		Absalon	Y 345	Alsin
						L 17		Esbern Snare		

FRIGATES

3 IVER HUITFELDT CLASS (FFGHM)

Name	No	Builders	Laid down	Launched	Commissioned
IVER HUITFELDT	F 361	Odense Shipyard, Lindø	2 Jun 2008	11 Mar 2010	6 Feb 2012
PETER WILLEMOES	F 362	Odense Shipyard, Lindø	12 Mar 2009	22 Dec 2010	Jan 2013
NIELS JUEL	F 363	Odense Shipyard, Lindø	22 Dec 2009	22 Dec 2010	Jun 2014

Displacement, tonnes: 5,944 standard; 6,645 full load
Dimensions, metres (feet): 138.7 × 19.8 × 6.3
  (455.1 × 65.0 × 20.7)
Speed, knots: 28
Complement: 100 + 65 spare berths

Machinery: CODAD; 4 MTU 20V M70 diesels; 44,000 hp (32.8 MW); 2 shafts; cp props; bow thruster
Missiles: SSM: 16 Boeing Harpoon Block 2 (2 octuple AHWCS VLS launchers) ❶; active radar homing to 124 km (67 n miles) at 0.9 Mach; warhead 227 kg.
  SAM: 32 GDC Standard SM-2 MR Block IIIA ❷; command/inertial guidance; semi-active radar homing to 167 km (90 n miles) at 2.5 Mach. Lockheed Martin Mk 41 VLS (32 cells). 24 Evolved Sea Sparrow RIM 162B ❸; semi-active radar homing to 18 km (9.7 n miles) at 3.6 Mach; warhead 38 kg. 2 Raytheon Mk 56 VLS (2 — 12 cells).
Guns: 2 Oto Melara 76 mm Super Rapid; 120 rds/min to 16 km (8.7 n miles); weight of shell 6 kg ❹. 1 Oerlikon Contraves 35 mm ❺. 4—12.7 mm MGs.
Torpedoes: 4—324 mm (2 twin) launchers ❻; Eurotorp MU 90 Impact; active/passive homing to 15 km (8 n miles) at 29/50 kt.
Physical countermeasures: Decoys: Terma 130 mm Decoy Launching System; 8 Mk 137 decoy launchers (48 barrels).
Electronic countermeasures: ESM: EDO ES 3701.
Radars: Air search: Thales Smart-L ❼; 3D; D-band.
  Air/surface search/fire control: Thales APAR ❽; E/F/I-bands.
  Surface search: Terma Scanter 6000; I-band.
  Fire control (guns): Saab Ceros 200 ❾; J/K-band.

Helo-control: Terma Scanter 6002; I-band.
Navigation: Furuno FAR-2117; I-band.
Sonars: Atlas ASO 94 hull mounted.
Combat data systems: Terma C-Flex Combat Management System.
Electro-optic systems: FLIR systems SeaStar Safire III.

Helicopters: 1 medium ❿.

Programmes: Construction of three frigates was approved in the 2004 Defence Agreement. The contract for construction was signed with Odense Shipyard on 20 December 2006. The blocks of the ships are under construction at Klaipeda, Lithuania, and at Loksa, Estonia. The first four blocks were delivered to Odense on 20 May 2008.
Structure: Built to DNV standards. The design is based on the Absalon class Flexible Support Ships and utilises the same hull (with one fewer deck) and the majority of equipment. There are dedicated staff facilities for national or NATO task group commanders. Four Stanflex container positions are located on the weapons deck and one at B-position. There is cargo space for four 20 ft TEU containers. The flight deck is capable of operating 20 ton helicopters and prepared to operate UAVs. 'A' gun position is suitable for upgrade to a 127 mm gun if and when required.
Operational: The ships are to have a global, expeditionary role and to be capable of providing area air-defence and support of land forces. Sea trials of *Niels Juel* began in April 2013. *Iver Huitfeldt* completed her maiden deployment to the Indian Ocean in June 2013. The ships will not become fully operational until Standard SM-2 missiles are installed. In the meantime, the ships are to achieve an interim capability with ESSM.

IVER HUITFELDT   *(Scale 1 : 1,200), Ian Sturton* / 1525658

Frigates < **Denmark** 195

IVER HUITFELDT
8/2012, Royal Danish Navy / 1455711

IVER HUITFELDT
1/2012, Erik Laursen / 1455744

PETER WILLEMOES
9/2012, Guy Toremans / 1455712

Denmark > Frigates

4 THETIS CLASS (FFHM)

Name	No
THETIS	F 357
TRITON	F 358
VAEDDEREN	F 359
HVIDBJØRNEN	F 360

Builders	Laid down	Launched	Commissioned
Svendborg Vaerft	10 Oct 1988	14 Jul 1989	1 Jul 1991
Svendborg Vaerft	27 Jun 1989	16 Mar 1990	2 Dec 1991
Svendborg Vaerft	19 Mar 1990	21 Dec 1990	9 Jun 1992
Svendborg Vaerft	2 Jan 1991	11 Oct 1991	30 Nov 1992

Displacement, tonnes: 2,642 standard; 3,556 full load
Dimensions, metres (feet): 112.5 oa; 99.8 wl × 14.4 × 6 (369.1; 327.4 × 47.2 × 19.7)
Speed, knots: 20; 8 thruster
Range, n miles: 8,500 at 15.5 kt
Complement: 60 (12 officers) + 30 spare berths

Machinery: 3 MAN/Burmeister & Wain Alpha 12V 28/32A diesels; 10,800 hp(m) (7.94 MW) sustained; 1 shaft; Kamewa cp prop; bow and azimuth thrusters; 880 hp(m) (647 kW), 1,100 hp(m) (800 kW)
Guns: 1 Oto Melara 3 in (76 mm)/62; Super Rapid ❶; dual purpose; 120 rds/min to 16 km (8.7 n miles); SAPOMER round weight 12.7 kg. 2—12.7 mm MGs.
Depth charges: 1 rail (door in stern).
Electronic countermeasures: ESM: Racal Sabre; intercept.
Radars: Air/surface search: Plessey AWS 6 ❹; G-band.
Surface search: Furuno 2135; E/F-band.
Navigation: Furuno FR-2130; I-band.
Fire control: CelsiusTech 9LV Mk 3 ❺; I/J-band.
Sonars: Thomson Sintra TSM 2640 Salmon; VDS; active search and attack; medium frequency. Celsius Tech; hull-mounted; active search; medium frequency.
Combat data systems: Terma C3; SATCOM ❷.
Weapon control systems: Bofors 9LV 200 Mk 3 director. FSI Safire surveillance director ❸.
Helicopters: 1 Westland Lynx Mk 90B ❻.

Programmes: Preliminary study by YARD in 1986 led to Dwinger Marine Consultants being awarded a contract for a detailed design completed in mid-1987. All four ordered in October 1987.
Modernisation: A modernisation plan is under consideration. AWS 6 radar is to be replaced by Terma Scanter 4100; Celsius Tech 9LV Mk 3 is to be replaced by Ceros 200 and the combat data system is also to be replaced by Terma C-Flex. The Lynx helicopter is to be replaced by MH-60R Seahawk from 2017.
Structure: The hull is some 30 m longer than the decommissioned Hvidbjørnen class to improve sea-keeping qualities and allow considerable extra space for additional armament. The design allows the use of containerised equipment to be shipped depending on role. The hull is ice strengthened to enable penetration of 1 m thick ice and efforts have been made to incorporate stealth technology, for instance by putting anchor equipment, bollards and winches below the upper deck. There is a double skin up to 2 m below the waterline. A rigid inflatable boarding craft plumbed by a hydraulic crane is fitted alongside the fixed hangar. The bridge and ops room are combined.
Operational: Primary role is sovereignty patrol and fishery protection in the North Atlantic.

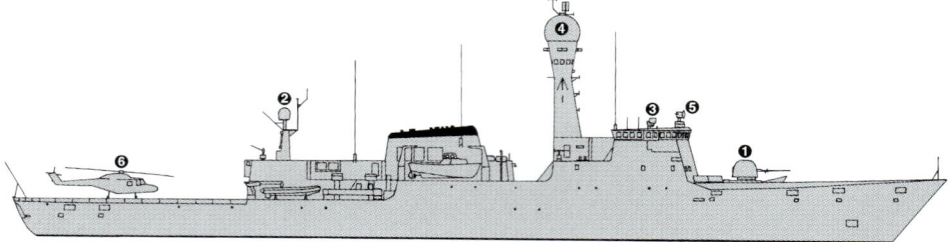

THETIS (Scale 1 : 900), Ian Sturton / 0012258

HVIDBJØRNEN 9/2012, Tony Roper / 1455743

VAEDDEREN 4/2013*, Tony Roper / 1525594

THETIS 6/2012*, Michael Nitz / 1525593

IHS Jane's Fighting Ships 2014-2015

SHIPBORNE AIRCRAFT

Notes: The Defence Agreement (2010–14) provides for the procurement of a new maritime helicopter to replace the current Lynx inventory. The new aircraft is to be capable of operating from the Iver Huitfeldt, Absalon and Thetis classes and also to utilise the Knud Rasmussen class for short periods. The decision to procure nine Sikorsky MH-60R was announced in November 2012. Delivery is to be made 2016–18.

Numbers/Type: 8 Westland Lynx Mk 90B.
Operational speed: 125 kt *(232 km/h)*.
Service ceiling: 12,500 ft *(3,810 m)*.
Range: 320 n miles *(593 km)*.
Role/Weapon systems: Shipborne helicopter for EEZ and surface search tasks. All upgraded to Super Lynx standard with first delivered November 2000. Sensors: Ferranti Seaspray; Racal Kestrel ESM; FLIR 2000. Weapons: 1 — 12.7 mm MG.

LYNX   *8/2012, Freddy Philips* / 1455714

LAND-BASED MARITIME AIRCRAFT

Numbers/Type: 3 Challenger 604.
Operational speed: 470 kt *(870 km/h)*.
Service ceiling: 41,000 ft *(12,497 m)*.
Range: 3,769 n miles *(6,980 km)*.
Role/Weapon systems: Maritime reconnaissance for EEZ patrol in the Baltic and off Greenland. Sensors: Terma SLAR radar; IR/UV scanner. Weapons: unarmed.

CHALLENGER 604   *6/2005, Massimo Annati* / 1153495

Numbers/Type: 8 AgustaWestland EH 101 Mk 512.
Operational speed: 160 kt *(296 km/h)*.
Service ceiling: 15,000 ft *(4,572 m)*.
Range: 550 n miles *(1,019 km)*.
Role/Weapon systems: Contract on 7 December 2001 for a total of 14 utility variants of the EH 101. Eight are configured for SAR duties and six for troop-carrying although the aircraft are designed for rapid role-change. By agreement with the UK, the delivery of six aircraft was delayed in order to meet a high-priority UK operational requirement. They were delivered to Denmark in late 2009. Military lift is 28 troops and up to four tonnes underslung. Sensors: Telephonics RDR-1600 SAR Weather Avoidance Radar.

EH 101   *7/2006, IHS/Patrick Allen* / 1184189

PATROL FORCES

6 LCP CLASS (COASTAL PATROL CRAFT) (PB)

LCP 1–4	DAGMA SAR 1	NAJA SAR 2

Displacement, tonnes: 7 full load
Dimensions, metres (feet): 11.9 × 2.9 × 0.7 *(39 × 9.5 × 2.3)*
Speed, knots: 38
Complement: 3
Machinery: 1 Scania DSI 14 V8 diesel; 625 hp *(465 kW)*; 1 Kamewa water-jet
Guns: 1 — 12.7 mm or 7.62 mm MG.

Comment: Based on the Swedish Combatboat 90E, these craft were developed as a joint venture between Forsvarets Materielverk and Storebro by whom LCP 1–4 were constructed and completed in 2004. Used as fast landing craft from the Absalon-class support ships, they can carry 10 fully equipped soldiers or four stretchers. Two ice-strengthened variants, *Dagma* and *Naja* are operated by the Knud Rasmussen class. Painted orange, they were delivered in 2006.

LCP 3   *6/2010, Michael Nitz* / 1406022

2 + 1 KNUD RASMUSSEN CLASS
(ARCTIC PATROL SHIPS) (PGBH)

Name	No	Builders	Laid down	Launched	Commissioned
KNUD RASMUSSEN	P 570	Karstensens Skibsvaerft, Skagen	21 Nov 2005	19 Oct 2006	18 Feb 2008
EJNAR MIKKELSEN	P 571	Karstensens Skibsvaerft, Skagen	2005	1 Jul 2007	16 Jan 2009
–	–	–	2014	2015	2016

Displacement, tonnes: 1,748 standard
Dimensions, metres (feet): 71.8 × 14.6 × 4.95 *(235.6 × 47.9 × 16.2)*
Speed, knots: 17
Complement: 18 + 25 spare berths

Machinery: 2 MAN B&W ALPHA 8L 27/38 diesels; 7,300 hp *(5.4 MW)*; 1 shaft; cp prop
Guns: 1 — 76 mm. 2 — 12.7 mm MGs.
Electronic countermeasures: To be announced.
Radars: Surface/air search: Terma Scanter 4100; I-band.
Navigation: Furuno FAR-2117; I-band.
Sonars: Reson; hull mounted (retractable).
Combat data systems: Terma C-Flex.
Electro-optic systems: FLIR Systems SeaFlir II.

Helicopters: Platform for 1 medium.

Programmes: Contract for the construction of the first two ships let in December 2004. Production started in September 2005. The hulls and propulsion were manufactured/installed by the Stocznia Pólnocna (Northern) Shipyard in Gdansk and the ships subsequently completed at Skagen. Installation of military equipment was undertaken by Naval Material Command. Approval for a third ship was included in Defence Agreement 2013–17.
Structure: Built to DNV Navy ICE 1A standards. A high-speed long-range rescue craft, an ice-strengthened version of the Combat Boat 90E, can be launched from a bay in the stern. Fitted with four Flex container positions for equipment and weapons, the design has the flexibility to operate in its (lightly armed) primary role or in a more heavily armed secondary role. The ships are to be equipped with SeaFLIR infrared imaging system.
Operational: Have replaced Agdlek class. The principal role is sovereignty patrol in the arctic waters off Greenland while secondary roles, such as command and control of a small force, might be exercised globally. Containerised weapons including a 76 mm gun, a Mk 56 launcher with evolved Sea Sparrow missiles and MU 90 torpedoes may be fitted.

EJNAR MIKKELSEN   *7/2010, Michael Nitz* / 1406017

6 DIANA (SF MK II) CLASS (LARGE PATROL CRAFT) (PB)

Name	No	Builders	Commissioned
DIANA	P 520	Faaborg Vaerft; Kockums	12 Dec 2007
FREJA	P 521	Faaborg Vaerft; Kockums	30 May 2008
HAVFRUEN	P 522	Faaborg Vaerft; Kockums	25 Sep 2008
NAJADEN	P 523	Faaborg Vaerft; Kockums	11 Dec 2008
NYMFEN	P 524	Faaborg Vaerft; Kockums	4 May 2009
ROTA	P 525	Faaborg Vaerft; Kockums	12 Dec 2009

Displacement, tonnes: 280 full load
Dimensions, metres (feet): 43 × 8.2 × 2.2 (141.1 × 26.9 × 7.2)
Speed, knots: 25
Complement: 12 + 3 spare berths
Machinery: 2 MTU 396 16V TB94 diesels; 2,700 hp (2 MW); 2 shafts; cp props
Guns: 2 — 12.7 mm MGs.
Radars: Navigation: Furuno FAR-2117; I-band.
Electro-optic systems: FLIR Systems Safire III.

Comment: GRP vessels that have replaced the Ø class. Ordered on 3 December 2004 from Faaborg Vaerft, Denmark, the hull, superstructure and machinery were built by Kockums, Karlskrona. Fitted with one Stanflex container position.

ROTA 3/2012*, E & M Laursen / 1525596

6 HOLM CLASS (MULTIROLE CRAFT) (MSD/AXL/AGSC)

BIRKHOLM A 541	ERTHOLM A 543	HIRSHOLM MSD 5
FYRHOLM A 542	ALHOLM A 544	SALTHOLM MSD 6

Displacement, tonnes: 140 full load
Dimensions, metres (feet): 28.9 × 6.4 × 2.0 (94.8 × 21.0 × 6.6)
Speed, knots: 12
Complement: 3 + 6 spare berths
Machinery: 2 Scania DC 16 diesels; 1,005 hp(m) (750 kW); 2 azimuth thrusters
Radars: Navigation: Furuno FR-2117; I-band.

Comment: Multirole GRP vessels constructed by Danish Yacht A/S, Skagen. One Stanflex container position. Two vessels (A 541 and A 542) are inshore survey craft to replace SKA 11 and SKA 15, two (A 543 and A 544) are training vessels. Two MCM drones MSD 5 and MSD 6 were delivered by late 2007. Both the latter vessels are to be upgraded to control the MSF-class drones.

BIRKHOLM 6/2008, Frank Findler / 1335458

HIRSHOLM 6/2012, L-G Nilsson / 1455717

1 AGDLEK CLASS (LARGE PATROL CRAFT) (PB)

Name	No	Builders	Commissioned
TULUGAQ	Y 388	Svendborg Vaerft	26 Jun 1979

Displacement, tonnes: 400 full load
Dimensions, metres (feet): 31.4 × 7.7 × 3.4 (103 × 25.3 × 11.2)
Speed, knots: 12
Complement: 14 (3 officers)
Machinery: 1 Burmeister & Wain Alpha A08-26 VO diesel; 800 hp(m) (588 kW); 1 shaft
Guns: 2 — 12.7 mm MGs.
Radars: Surface search: Furuno FR-2135; E/F-band.

Comment: Ice strengthened. SATCOM fitted. Agdlek decommissioned in 2008 and replaced by Knud Rasmussen. Agpa decommissioned in 2009. Last remaining craft, Tulugaq, stationed in Greenland.

TULUGAQ 9/2009, Marco Ghiglino / 1406023

17 MHV 800 CLASS (COASTAL PATROL CRAFT) (PB)

Name	No	Builders	Commissioned
ALDEBARAN	MHV 801	Søby Shipyard	9 Jul 1992
CARINA	MHV 802	Søby Shipyard	30 Sep 1992
ARIES	MHV 803	Søby Shipyard	30 Mar 1993
ANDROMEDA	MHV 804	Søby Shipyard	30 Sep 1993
GEMINI	MHV 805	Søby Shipyard	28 Feb 1994
DUBHE	MHV 806	Søby Shipyard	1 Jul 1994
JUPITER	MHV 807	Søby Shipyard	30 Nov 1994
LYRA	MHV 808	Søby Shipyard	30 May 1995
ANTARES	MHV 809	Søby Shipyard	30 Nov 1995
LUNA	MHV 810	Søby Shipyard	30 May 1996
APOLLO	MHV 811	Søby Shipyard	30 Nov 1996
HERCULES	MHV 812	Søby Shipyard	28 May 1997
BAUNEN	MHV 813	Søby Shipyard	17 Dec 1997
BUDSTIKKEN	MHV 814	Søby Shipyard	30 Aug 1998
KUREREN	MHV 815	Søby Shipyard	30 May 1999
PATRIOTEN	MHV 816	Søby Shipyard	25 Feb 2000
PARTISAN	MHV 817	Søby Shipyard	29 Nov 2000

Displacement, tonnes: 84 full load
Dimensions, metres (feet): 23.7 × 5.6 × 2.0 (77.8 × 18.4 × 7.2)
Speed, knots: 13. **Range, n miles:** 990 at 11 kt
Complement: 8 + 4 spare berths
Machinery: 2 Saab Scania DSI-14 diesels; 900 hp(m) (661 kW); 2 shafts
Guns: 2 — 7.62 mm MGs. 2 — 12.7 mm MGs (can be fitted).
Radars: Navigation: Furuno FR-2117; I-band.

Comment: First six ordered in April 1991, second six in July 1992, six more in 1997. Steel hulls with a moderate ice capability. Operated by the Naval Home Guard. MHV 818 subsequently re-classified as MHV 850 class.

DUBHE 6/2013*, Frank Findler / 1525677

1 MHV 850 CLASS (COASTAL PATROL CRAFT) (PB)

Name	No	Builders	Commissioned
SABOTØREN	MHV 851 (ex-818)	Søby Shipyard	13 Oct 2001

Displacement, tonnes: 97 full load
Dimensions, metres (feet): 27.2 × 5.6 × 2.0 (89.2 × 18.4 × 7.2)
Speed, knots: 13
Complement: 10
Machinery: 2 Saab Scania DSI-14; 900 hp(m) (661 kW); 2 shafts
Guns: 2 — 7.62 mm MGs.

Comment: Former MHV 800 class lengthened by 3 m to act as a test platform for environmental test equipment and subsequently reverted to patrol craft role. Operated by the Home Guard.

12 MHV 900 CLASS (COASTAL PATROL CRAFT) (PB)

Name	No	Builders	Commissioned
ENØ	MHV 901	Søby Shipyard	18 Oct 2003
MANØ	MHV 902	Søby Shipyard	8 May 2004
HJORTØ	MHV 903	Søby Shipyard	29 Jan 2005
LYØ	MHV 904	Søby Shipyard	30 Sep 2005
ASKØ	MHV 905	Søby Shipyard	5 Jul 2006
FAENØ	MHV 906	Søby Shipyard	14 Apr 2007
HVIDSTEN	MHV 907	Søby Shipyard	8 Mar 2008
BRIGADEN	MHV 908	Søby Shipyard	15 Jun 2008
SPEDITØREN	MHV 909	Søby Shipyard	18 Jan 2009
RINGEN	MHV 910	Søby Shipyard	16 May 2009
BOPA	MHV 911	Søby Shipyard	28 Nov 2009
HOLGER DANSKE	MHV 912	Søby Shipyard	14 Jun 2011

Displacement, tonnes: 97 full load
Dimensions, metres (feet): 27.2 × 5.7 × 2.5 (89.2 × 18.7 × 8.2)
Speed, knots: 13
Complement: 10
Machinery: 2 Saab Scania DI 16V8 diesels; 980 hp(m) (730 kW); 2 shafts
Guns: 2 – 7.62 mm MGs.
Radars: Navigation: Furuno FR-2117; I-band.

Comment: Similar to but 3.5 m longer than the MHV 800 class. The added length is to allow space for environmental protection equipment. Steel construction. Operated by the Naval Home Guard.

SPEDITØREN — 6/2013*, Michael Nitz / 1525595

MINE WARFARE FORCES

Notes: As a part of the Defence Agreement 2010–14, a new modular MCM concept (MCM Denmark) has been implemented. MCM Denmark is designed to fulfil various tailor-made mission packages, of which one will include the use of a Logistic Support Platform carrying a containerised command and diving module together with upgraded remotely controlled drones of the MSF and the Holm class.

4 MSF CLASS (MRD)

MSF 1-4

Displacement, tonnes: 127 full load
Dimensions, metres (feet): 26.5 × 7 × 2.1 (86.9 × 23.0 × 6.9)
Speed, knots: 12
Complement: 4
Machinery: 2 Scania DSI 14 diesels; 1,000 hp(m) (736 kW); 2 Schottel waterjets or 2 Schottel azimuth thrusters
Radars: Navigation: Furuno FAR-2117; I-band.
Sonars: Thomson Marconi STS 2054 side scan active; high frequency.
Combat data systems: IN-SNEC/INFOCOM.

Comment: MSF (Minor Standard Vessel). Ordered in January 1997 from Danyard, Aalborg, and five delivered June 1998 to January 1999. Used primarily as MCM drones although built as multipurpose platform (with one Stanflex container position). Fitted with containerised MCM gear. GRP hulls. Units will continue as MCM drones as part of the future modular MCM concept, controlled by one Holm-class MSD unit or a containerised control unit.

MSF 4 — 6/2012, Michael Winter / 1455720

SURVEY SHIPS

1 RESEARCH SHIP (AGE)

DANA

Displacement, tonnes: 3,759 full load
Dimensions, metres (feet): 78.5 × 14.8 × 6 (257.5 × 48.6 × 19.7)
Speed, knots: 15. **Range, n miles:** 8,000 at 14 kt
Complement: 27 + 12 scientists
Machinery: 2 Burmeister and Wain Alpha 16V23-LU diesels; 4,960 hp(m) (3.65 MW); 1 shaft cp prop; bow and stern thrusters

Comment: Built by Dannebrog, Aarhus in 1982. Owned and operated by the Danish National Institute of Aquatic Resources. Has an ice-strengthened hull and three 6 ton cranes.

DANA — 6/2002, Royal Danish Navy / 0533223

3 SURVEY LAUNCHES (YGS)

SKA 12-14

Displacement, tonnes: 53 full load
Dimensions, metres (feet): 20 × 5.2 × 2.1 (65.6 × 17.1 × 6.9)
Speed, knots: 12
Complement: 6 (1 officer)
Machinery: 1 GM diesel; 540 hp (403 kW); 1 shaft
Radars: Navigation: Furuno FR-2115; I-band.

Comment: GRP hulls. Built 1981–84 by Rantsausminde. SKA 12 has strengthened hull and is permanently deployed to Naval Station Grønnedal (Greenland) for surveying of Greenland waters. SKA 11 was lost off Greenland on 3 May 2006. Multibeam echo sounders are fitted. SKA 13 and 14 have been modified for other tasks at the Naval Bases. SKA 11 and 15 were replaced by two Holm class. The survey launches can work alone or in pairs.

SKA 14 — 9/2012, Frank Findler / 1455705

TRAINING SHIPS

Notes: There are two small Sail Training Ships, *Svanen* Y 101 and *Thyra* Y 102. Of 32 tons they have a sail area of 480 m² and an auxiliary diesel of 72 hp(m) (53 kW). Built in 1960 by Molich yacht builders, Hundested. Used to train midshipmen before attending the naval academy.

THYRA — 6/2013*, Michael Nitz / 1525597

AUXILIARIES

Notes: (1) The OPLOG organisation consists of the former Mobile Logistic Unit and parts of the maintenance and supply facilities of the naval bases at Frederickshavn and Korsör. The mobile capability includes containerised workshops, stores, accommodation and helicopter refuelling facilities carried on approximately 40 trucks and trailers.
(2) Sealift: Denmark launched the ARK project in 2003 with the aim of securing availability of strategic sealift to NATO. Germany joined the project in 2006. Three ships are on call for strategic sealift: 183 m *Ark Futura* (2,308 lane-metres + 644 TEU) 198 m *Britannia Seaways* and *Suecia Seaways* (3,000 lane-metres + 647 TEU). Two new German-built RoRo vessels were to be on call from November 2013 and April 2014 respectively.

2 ABSALON CLASS (COMBAT SUPPORT SHIPS) (AGF/AKR/AH)

Name	No	Builders	Laid down	Launched	Commissioned
ABSALON	L 16	Odense Shipyard, Lindø	28 Nov 2003	25 Feb 2004	19 Oct 2004
ESBERN SNARE	L 17	Odense Shipyard, Lindø	2004	21 Jun 2004	18 Apr 2005

Displacement, tonnes: 6,401 full load
Dimensions, metres (feet): 137 × 19.5 × 6.3 *(449.5 × 64.0 × 20.7)*
Speed, knots: 23. **Range, n miles:** 11,500 at 14 kt
Complement: 99 + 70 personnel

Machinery: CODAD. 2 MTU 8000 M 70 diesels; 22,300 hp *(16.63 MW)*; 2 shafts; CP propellers; bow thruster
Missiles: SSM: 16 Boeing Harpoon Block II ❶ (2 octuple AHWCS VLS launchers); active radar homing to 124 km *(67 n miles)* at 0.9 Mach; warhead 227 kg.
SAM: 36 Evolved Sea Sparrow RIM 162B ❷; semi-active radar homing to 18 km *(9.7 n miles)* at 3.6 Mach; warhead 38 kg. 3 Raytheon Mk 56 VLS (3 × 12 cells).
Guns: United Defense 5 in *(127 mm)*/62 Mk 45 Mod 4 ❸; 20 rds/min to 23 km *(12.6 n miles)*; weight of shell 32 kg. Prepared for extended range capable munitions. 2 Oerlikon Contraves 35 mm GDM08 Millenium guns ❹. 4 – 12.7 mm MGs.
Torpedoes: 4 – 324 mm (2 twin) launchers ❺; Eurotorp Mu 90 Impact; active/passive homing to 15 km *(8 n miles)* at 29/50 kt.
Physical countermeasures: Decoys: Terma 130 mm Decoy Launching System; 8 Mk 137 decoy launchers (48 barrels).
Electronic countermeasures: ESM: EDO ES 3701.
Radars: Air/surface search: Thales SMART-S 3D ❼; E/F-band.
Surface search/navigation: Terma Scanter 2001 ❽; I-band.
Fire control: 4 (up to) SaabTech Ceros 200 Mk 3 ❾; J/K-band.
Navigation: Furuno FAR-2117 ❿; I-band.
Sonars: Atlas ASO 94 hull mounted. VDS/DTAS/ATAS to be decided.
Combat data systems: Terma C-Flex Link 11.
Electro-optic systems: FLIR Systems Sea Star Safire III ❻.
Helicopters: 1–2 Westland Lynx Mk 90B ⓫.

Programmes: Contract on 16 October 2001 for detailed design and construction of two multirole support ships. Construction of first of class started on 30 April 2003.
Structure: Built to DNV Navy standards with five Stanflex container positions. Ro-Ro ramp aft gives access to 900 m² of multipurpose deck (vehicles (including 62 ton MBT), logistics, ammunition, up to 34 TEU containers). 2 Combat Boat 90E high-speed insertion craft carried on cargo deck. Flight deck capable of operating 20 ton helicopters.
Operational: Capable of acting as a command platform, transporting up to 200 personnel and equipment, provision of joint logistic support, and as a hospital ship. *Absalon* achieved full operational capability in 2007 and *Esbern Snare* in mid-2008.

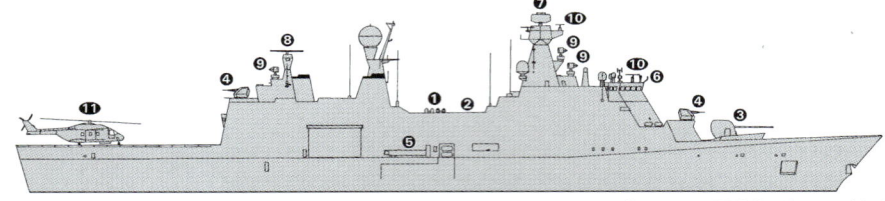

ABSALON   *(Scale 1 : 1,200), Ian Sturton / 1525657*

ESBERN SNARE   *4/2013*, Tony Roper / 1525599*

ABSALON   *7/2010, Michael Nitz / 1406024*

ABSALON   *6/2012*, Michael Nitz / 1525598*

Auxiliaries < **Denmark** 201

1 TRANSPORT SHIP (AKS)

Name	No	Builders	Commissioned
SLEIPNER	A 559	Åbenrå Vaerft og A/S, Aabenraa	18 Jul 1986

Displacement, tonnes: 465 full load
Dimensions, metres (feet): 36.5 × 7.6 × 3.1 (119.8 × 24.9 × 10.2)
Speed, knots: 10
Range, n miles: 10,000 at 10 kt
Complement: 7 (1 officer)
Cargo capacity: 150 tons
Machinery: 1 Callesen diesel; 540 hp(m) (403 kW); 1 shaft
Radars: Navigation: Furuno FAR-21x7; E/F/I-band.

SLEIPNER 9/2012, Guy Toremans / 1455719

1 ROYAL YACHT (YAC)

Name	No	Builders	Laid down	Launched	Commissioned
DANNEBROG	A 540	R Dockyard, Copenhagen	2 Jan 1931	10 Oct 1931	20 May 1932

Displacement, tonnes: 1,133 full load
Dimensions, metres (feet): 75 × 10.4 × 3.7 (246.1 × 34.1 × 12.1)
Speed, knots: 14
Complement: 55 (9 officers)
Machinery: 2 Burmeister & Wain Alpha T23L-KVO diesels; 1,800 hp(m) (1.32 MW); 2 shafts; cp props
Radars: Navigation: Furuno FR-2115; I-band.

Comment: Major refit 1980 included new engines and electrical gear. A major refit in 2012 included new auxiliary machinery and control room.

DANNEBROG 8/2013*, Selim San / 1525600

2 POLLUTION CONTROL CRAFT (YPC)

MILJØ 101 MILJØ 102

Displacement, tonnes: 16 full load
Dimensions, metres (feet): 16.2 × 4.2 × 2.2 (53.1 × 13.8 × 7.2)
Speed, knots: 15
Range, n miles: 350 at 8 kt
Complement: 3 (1 officer)
Machinery: 1 MWM TBD232V12 diesel; 454 hp(m) (334 kW) sustained; 1 shaft

Comment: Built by Ejvinds Plastikbodevaerft, Svendborg. Carry derricks and booms for framing oil slicks and dispersant fluids. Naval manned. Delivered 1 November and 1 December 1977.

MILJØ 102 9/2011, Frank Findler / 1406929

© 2014 IHS

4 RESCUE VESSELS (PBO)

NORDSØEN VESTKYSTEN HAVØRNEN VIBEN

Measurement, tonnes: 594 (Nordsøen), 657 (Vestkysten), 188 (Havørnen), 23 (Viben) gt
Dimensions, metres (feet): 53.2 (Nordsøen), 49.9 (Vestkysten), 30.9 (Havørnen), 17.2 (Viben) × 10.3 (Nordsøen), 10 (Vestkysten), 6.6 (Havørnen), 3.6 (Viben) × 3.3 (Nordsøen), 4.2 (Vestkysten), 1.6 (Viben) (174.5, 163.7, 101.4, 56.4 × 33.8, 32.8, 21.7, 11.8 × 10.8, 13.8, 5.2)

Comment: Non-naval ships operated by the Ministry of Food and Fisheries. Nordsøen and Vestkysten operate primarily in the North Sea and Kattegat area, Havørnen in the Baltic Sea around Bornholm and Viben in shallow waters. Capable of 14–18 kt.

NORDSØEN 1/1999, Harald Carstens / 0056889

2 SEA TRUCKS (AKL)

METTE MILJØ A 562   MARIE MILJØ A 563

Displacement, tonnes: 247 full load
Dimensions, metres (feet): 29.8 × 8 × 1.6 (97.8 × 26.2 × 5.2)
Speed, knots: 10
Complement: 9 (1 officer)
Machinery: 2 Grenaa diesels; 660 hp(m) (485 kW); 2 shafts

Comment: Built by Carl B Hoffmann A/S, Esbjerg and Søren Larsen & Sønners Skibsvaerft A/S, Nykøbing Mors. Delivered 22 February 1980. Have orange and yellow superstructure.

METTE MILJØ 9/2011, Guy Toremans / 1406957

2 ARVAK CLASS (HARBOUR TUGS) (YTL)

ARVAK Y 344   ALSIN Y 345

Displacement, tonnes: 80 full load
Dimensions, metres (feet): 16 × 6.6 × 2.5 (52.5 × 21.7 × 8.2)
Speed, knots: 10
Machinery: 1 MTU 12V 183TE62 diesel; 737 hp(m) (550 kW)

Comment: Built by Hvide Sande Skibs & Baadebyggeri and delivered on 18 November 2002. In service at Korsør and Frederikshavn. Fitten with Stanflex container position aft to facilitate transport of containerised stores and equipment between naval bases.

ALSIN 9/2012, M Declerck / 1455723

IHS Jane's Fighting Ships 2014-2015

202 Denmark > Auxiliaries

2 OIL POLLUTION CRAFT (YPC/ABU)

GUNNAR THORSON A 560 **GUNNAR SEIDENFADEN** A 561

Displacement, tonnes: 762 full load
Dimensions, metres (feet): 56 × 12.3 × 3.9 (183.7 × 40.4 × 12.8)
Speed, knots: 12.5
Complement: 16 (7 officers)
Machinery: 2 Burmeister and Wain Alpha 8V23L-VO diesels; 2,320 hp(m) (1.7 MW); 2 shafts; cp props; bow thruster

Comment: Built by Ørnskov Stålskibsvaerft, Frederikshavn. Delivered 8 May and 2 July 1981 respectively. G Thorson at Copenhagen, G Seidenfaden at Korsør. Carry firefighting equipment. Large hydraulic crane fitted in 1988 for the secondary task of buoy tending.

GUNNAR THORSON   6/2013*, Frank Findler / 1525678

GUNNAR SEIDENFADEN   8/2010, Frank Findler / 1366969

1 FLYVEFISKEN CLASS
(MINEHUNTER/DIVING VESSEL) (MHCD/AX)

Name	No	Builders	Commissioned
SØLØVEN	Y 331 (ex-P 563)	Danyard A/S, Aalborg	28 May 1996

Displacement, tonnes: 488 full load
Dimensions, metres (feet): 54 × 9 × 2.5 (177.2 × 29.5 × 8.2)
Speed, knots: 18
Range, n miles: 2,400 at 18 kt
Complement: 10 (2 officers)
Machinery: 2 MTU 16V 396 TB94 diesels; 5,800 hp(m) (4.26 MW) sustained; 2 shafts; cp props; bow thruster. Auxiliary propulsion by hydraulic motors on outer gearboxes; hydraulic pumps driven by 1 GM 12V-71 diesel; 500 hp (375 kW)
Radars: Surface search: Terma Scanter Mil; I-band.
Navigation: Furuno; I-band.

Programmes: A total of 14 of the class were built of which this last one remains following the decommissioning of the class.
Structure: GRP sandwich hull.
Operational: The original concept of a class of 14 ships was to be able to re-role by interchanging mission-specific containers for different taskings (ASUW, ASW, MCM and Patrol). This was abandoned in 2004 when role-specialisation was re-introduced and Svaerdfisken, Flyvefisken, Hajen and Lommen were subsequently decommissioned. Of the remaining ten ships, a further eight ships were decommissioned in 2010 under the terms of the Defence Agreement 2010–14. Havkatten was subsequently decommissioned in 2011 after completing MCM Concept trials. Søløven has been converted to a diving training ship at Faaberg Vaerft. The work included removal of all weapons systems, extension of the existing superstructure aft to house a decompression chamber and installation of a crane at the stern to facilitate operation of a diving bell. The ship became operational in its new role in 2012.

SØLØVEN   6/2012, Royal Danish Navy / 1455722

1 POLLUTION CONTROL CRAFT (YPC)

MILJØ 103

Displacement, tonnes: 28 full load
Dimensions, metres (feet): 16.3 × 4.6 × 0.85 (53.5 × 15.1 × 2.8)
Speed, knots: 20
Complement: 3
Machinery: 2 Scania DI12 59M diesels; 1,080 hp(m) (808 kW); 2 Rolls Royce FF500 waterjets

Comment: Designed for pollution control in shallow water, built by Hvide Sande Skibs og Bådebyggeri. Aluminium hull and glass-fibre superstructure. Entered service in February 2008.

MILJØ 103   10/2009, Royal Danish Navy / 1367020

2 SAV CLASS (YFU)

MRD 3   **MRD 4**

Displacement, tonnes: 33 full load
Dimensions, metres (feet): 18.2 × 4.8 × 1.2 (59.7 × 15.7 × 3.9)
Speed, knots: 12
Complement: 4

Machinery: 2 Detroit diesels; 350 hp(m) (257 kW); 2 Schottel waterjet propulsors
Radars: Navigation: Furuno; I-band.
Sonars: Thomson Sintra TSM 2054 side scan; active minehunting; high frequency.
Combat data systems: Terma link to Flyvefisken class (in MCMV configuration).

Comment: Former minehunter drones of which six were originally built by Danyard 1991–96. MRD 4 is used as a station vessel at Korsør and MRD 3 is stored ashore at Frederikshavn. MRD 1, 2, 5 and 6 decommissioned in 2009.

MRD 4   5/2008, E & M Laursen / 1335367

1 SERVICE CRAFT (YAG)

LUNDEN (ex-Bopa) Y 343 (ex-MHV 90)

Displacement, tonnes: 86 full load
Dimensions, metres (feet): 19.8 × 5.7 × 2.5 (65 × 18.7 × 8.2)
Speed, knots: 11
Complement: 12
Machinery: 1 Burmeister & Wain diesel; 400 hp(m) (294 kW); 1 shaft
Guns: 2 – 7.62 mm MGs.
Radars: Navigation: Furuno 1505; I-band.

Comment: Built in 1973. Ex-Naval Home Guard patrol vessel decommissioned in 2009. Recommissioned in 2011 as a target-towing craft and general tender.

LUNDEN   8/2012, Freddy Philips / 1455721

Djibouti
MARINE NATIONALE DJIBOUTIENNE

Country Overview

Formerly the French territory of French Somaliland and later the Afars and the Issas, Djibouti became independent in 1977. With an area of 8,957 square miles and a coastline of 170 n miles, the country is situated in a strategic position on the Bab el Mandeb, the strait that links the Red Sea with the Gulf of Aden. It is bordered to the north by Eritrea, to the west by Ethiopia and to the south by Somalia. The largest town and capital is also called Djibouti whose port serves as an international transhipment and refuelling centre. It also provides Ethiopia with its only rail link to the sea. Territorial seas (12 n miles) are claimed. A 200 n mile Exclusive Economic Zone (EEZ) has been claimed but the limits are not fully defined.

Headquarters Appointments

Commander of Navy:
Colonel Abdourahman Aden Cher

Personnel

2014: 380

Bases

Djibouti

French Navy

The permanent French naval contingent usually includes up to three frigates and a repair ship.

PATROL FORCES

Notes: Up to six RIBs are in use. Zodiac and Avon types.

2 BATTALION 17 (PBF)

P 16   P 17

Displacement, tonnes: 36 full load
Dimensions, metres (feet): 17.05 × 5.2 × 1.6 *(55.9 × 17.1 × 5.2)*
Speed, knots: 35.2
Range, n miles: 680 at 30 kt
Complement: 9
Machinery: 2 MTU 12V 183 TE 92 diesels; 1,970 hp *(1.47 MW)*; 2 shafts
Guns: 2 — 14.5 mm MGs (1 twin).
Radars: Surface search: Raytheon; I-band.

Comment: Australian design craft built by Harena Boat Yard at Assab, Eritrea and delivered in 2001. Five similar craft in service in Eritrea.

P 16   10/2009, Djibouti Navy / 1367032

4 PATROL CRAFT (PC)

P 02–05

Displacement, tonnes: 18 full load
Dimensions, metres (feet): 13.4 × 3.8 × 1.2 *(44 × 12.5 × 3.9)*
Speed, knots: 13
Range, n miles: 200 at 11 kt
Complement: 4
Machinery: 2 General Motors Detroit 6V53 diesels; 500 hp *(370 kW)*; 2 shafts

Comment: Former US Coast Guard lifeboats constructed in the 1960s. Four were delivered in June 2006 under FMS funding arrangements. A further craft was transferred as spares.

P 02   7/2013*, A Sheldon-Duplaix / 1525518

1 PATROL CRAFT (PB)

P 14

Displacement, tonnes: 14 full load
Dimensions, metres (feet): 13.4 × 4.8 × 1.3 *(44 × 15.7 × 4.3)*
Speed, knots: 26
Complement: 7
Machinery: 2 Iveco Aifo diesels; 2 shafts

Comment: Acquired in about 2006.

P 14   10/2009, Djibouti Navy / 1367099

1 PATROL CRAFT (PB)

P 15

Displacement, tonnes: 25 full load
Dimensions, metres (feet): 16.8 × 5.0 × 1.7 *(55.1 × 16.4 × 5.6)*
Speed, knots: 30
Complement: 4
Machinery: 2 Iveco Aifo 1380 diesels; 2,760 hp *(2.1 MW)*; 2 shafts

Comment: Acquired in about 2006.

P 15   10/2009, Djibouti Navy / 1367036

204 Djibouti > Patrol forces — Dominica > Coast guard

2 CANTIERI DEL GOLFO CLASS (PB)

P 6   P 7

Displacement, tonnes: 9 full load
Dimensions, metres (feet): 10.8 × 3.66 × 1.17 (35.4 × 12.0 × 3.8)
Speed, knots: 32
Complement: 4
Machinery: 2 Isotta Fraschini 1306T2MLLCCL diesels; 860 hp (640 kW); 2 shafts

Comment: Acquired in 2006.

AUXILIARIES

1 EDIC 700 CLASS (LCT)

Name	No	Builders	Commissioned
– (ex-Dague)	– (ex-L 9052)	SFCN, Villeneuve la Garenne	19 Dec 1987

Displacement, tonnes: 330 standard; 748 full load
Dimensions, metres (feet): 59 × 11.6 × 1.7 (193.6 × 38.1 × 5.6)
Speed, knots: 12
Range, n miles: 1,800 at 12 kt
Complement: 10 + 180 spare berths

Military lift: 200 tons, 11 trucks or 5 AMX 30.
Machinery: 2 SACM Uni Diesel UD 30 V12 M3 diesels; 1,400 hp(m) (1 MW) sustained; 2 shafts
Guns: 2 Giat 20F2 20 mm. 2 — 12.7 mm MGs.
Radars: Navigation: Racal Decca 1229; I-band.

Comment: Former French Navy vessel based at Djibouti and transferred to Djibouti Navy in 2012. Originally ordered 10 March 1986. Rated as Engins de Débarquement d'Infanterie et Chars (EDIC III).

P 6   10/2009, Djibouti Navy / 1367035

2 SEA ARK 1739 (PATROL CRAFT) (PB)

P 18   P 19

Displacement, tonnes: 28 full load
Dimensions, metres (feet): 17 × 4.2 × 1.2 (55.8 × 13.8 × 3.9)
Speed, knots: 28.5
Complement: 8
Machinery: 2 MTU diesels; 2,760 hp (2.1 MW); 2 shafts
Guns: 2 — 14.5 mm.

Comment: Acquired in about 2006.

Ex-DAGUE   6/2013*, A Sheldon-Duplaix / 1525519

P 19   10/2009, Djibouti Navy / 1367034

L 9051   5/2003, Per Körnefeldt / 0569951

 # Dominica

Country Overview

Formerly a British colony, the Commonwealth of Dominica became an independent republic in 1978. With an area of 290 sq miles and coastline of 80 n miles, it is the largest and most northerly of the Windward Islands in the Lesser Antilles chain and is situated in the Caribbean Sea between the French possessions of Guadeloupe to the north and Martinique to the south. The capital, major town, and port is Roseau. Territorial seas (12 n miles) are claimed. A 200 n mile Exclusive Economic Zone (EEZ) has been claimed but the limits are not fully defined.

Headquarters Appointments

Head of Police Coast Guard:
   Inspector Eric Elizee

Personnel
2014: 35

Bases
Roseau

COAST GUARD

1 SWIFT 65 ft CLASS (PB)

Name	No	Builders	Commissioned
MELVILLE	D 4	Swiftships, Morgan City	1 May 1984

Displacement, tonnes: 34 full load
Dimensions, metres (feet): 19.8 × 5.6 × 2 (65 × 18.4 × 6.6)
Speed, knots: 23
Range, n miles: 250 at 18 kt
Complement: 10
Machinery: 2 Detroit 12V-71TA diesels; 840 hp (616 kW) sustained; 2 shafts
Guns: 1 — 7.62 mm MG.
Radars: Furuno; I/J-band.

Comment: Donated by US government. Similar craft supplied to Antigua and St Lucia. Aluminium construction.

MELVILLE
11/1993, Maritime Photographic
0506143

Coast guard < **Dominica** — Patrol forces < **Dominican Republic** 205

2 DAUNTLESS CLASS (PB)

Name	No	Builders	Commissioned
UKALE	D 05	SeaArk Marine, Monticello	8 Nov 1995
DOMINIQUE	D 06	SeaArk Marine, Monticello	4 Mar 2009

Displacement, tonnes: 11 full load
Dimensions, metres (feet): 12.2 × 4.3 × 1.3 *(40 × 14.1 × 4.3)*
Speed, knots: 27
Range, n miles: 600 at 18 kt
Complement: 6
Machinery: 2 Caterpillar 3208TA diesels; 870 hp *(650 kW)* sustained; 2 shafts
Guns: 1 — 7.62 mm MG (can be carried).
Radars: Raytheon; I-band.

Comment: First craft delivered in 1995 and the second in 2009. Aluminium construction.

OBSERVER　　　　　　　　　　　　　　　　　　*11/1993, Maritime Photographic / 0506227*

2 SPECIAL PURPOSE CRAFT (PBF)

D 07　　　　D 08

Displacement, tonnes: 7 full load
Dimensions, metres (feet): 10.1 × 3.0 × 0.7 *(33.1 × 9.8 × 2.3)*
Speed, knots: 50
Complement: 4
Machinery: 3 Mercury outboard motors; 825 hp *(615 kW)*
Guns: 2 — 7.62 mm MGs.
Radars: Navigation: Raymarine; I-band.

Comment: Built by SAFE Boats International and very similar to the craft in service with the US Coast Guard. Both craft donated by US Southern Command under the US foreign Military Sales Program in 2012.

UKALE　　　　　　　　　　　　　　　　　　*11/1995, SeaArk / 0056897*

3 PATROL CRAFT (PBR)

VIGILANCE　　　　OBSERVER　　　　RESCUER

Displacement, tonnes: 2 full load
Dimensions, metres (feet): 8.2 × 2.6 × 0.3 *(26.9 × 8.5 × 1.0)*
Speed, knots: 28 *(Observer, Vigilance)*, 45 *(Rescuer)*
Complement: 3
Machinery: 1 Evinrude outboard; 225 hp *(168 kW)* sustained or 2 Johnson outboards *(Rescuer)*; 280 hp *(205 kW)*

Comment: First two are Boston Whalers acquired in 1988. *Rescuer* is of similar size but is an RHIB acquired in 1994.

D 07　　　　　　　　　　　　　　　　　　*6/2012, US SOUTHCOM / 1455771*

Dominican Republic
MARINA DE GUERRA

Country Overview

The Dominican Republic is an independent state whose constitution was promulgated in 1966. With an area of 18,816 square miles, it occupies the eastern two thirds of the island of Hispaniola, which it shares with Haiti to the west. There are also a number of adjacent islands, notably Beata and Saona. It has a 697 n mile coastline and is bordered to the north by the Atlantic Ocean, to the east by the Mona Passage, which separates it from Puerto Rico, and to the south by the Caribbean Sea. Santo Domingo is the capital, largest city and principal port. Territorial seas (6 n miles) are claimed. A 200 n mile EEZ has been claimed but the limits have not been defined by boundary agreements.

Headquarters Appointments

Chief of Naval Staff:
　Vice Admiral Edwin Rafael Dominici Rosario

Personnel

(a)　2014: 9,900 officers and men (including naval infantry)
(b)　Selective military service

Bases

27 de Febrero, Santo Domingo: HQ of CNS, Naval School. Supply base.
Las Calderas, Las Calderas, Bani: Naval dockyard, 700 ton synchrolift. Training centre. Supply base.
Haina: Dockyard facility. Supply base.
Puerto Plata. Small naval base.

Organisation

There are three naval zones:
North: Haitian border east to the Mona passage.
South: Mona passage west to the Haitian border.
Santo Domingo: Naval establishments in the capital and its environs.

DELETIONS

Notes: *Melia* still flies an ensign as a museum ship.

PATROL FORCES

Notes: There are two 8.5 m fast interception craft *Enif* LR 159 and *Rigel* LR 160.

1 BALSAM CLASS (PBO/WMEC)

Name	No	Builders	Commissioned
ALMIRANTE DIDIEZ BURGOS (ex-*Buttonwood*)	PA 301	Duluth Shipyard, Minnesota	24 Sep 1943

Displacement, tonnes: 1,051 full load
Dimensions, metres (feet): 54.9 × 11.3 × 3.8 *(180.1 × 37.1 × 12.5)*
Speed, knots: 13
Complement: 54 (4 officers)
Machinery: Diesel-electric; 2 Cooper Bessemer diesels; 1,402 hp *(1.06 MW)*; 2 motors; 1,200 hp *(895 kW)*; 1 shaft; bow thruster
Guns: 1 — 4 in; 2 — 20 mm.
Radars: Surface search: Raytheon SPS-64(V)1; I-band.

Comment: PA 301 transferred from US Coast Guard on 30 June 2001 and refitted in 2010. PA 302 decommissioned in 2012.

ALMIRANTE DIDIEZ BURGOS　　　　　　　*5/2013*, A Sheldon-Duplaix / 1525520*

206 Dominican Republic > Patrol forces

2 WHITE SUMAC CLASS (ABU)

Name	No	Builders	Commissioned
TORTUGUERO (ex-White Pine)	PM 203 (ex-BA 1, ex-WLM 547)	Erie Concrete and Steel, Erie	11 Jul 1944
CAPOTILLO (ex-White Sumac)	PM 204 (ex-BA 2, ex-WLM 504)	Niagara Shipbuilding, Buffalo	1943

Displacement, tonnes: 493 full load
Dimensions, metres (feet): 40.5 × 9.5 × 2.7 *(132.9 × 31.2 × 8.9)*
Speed, knots: 9
Complement: 24
Machinery: 2 Caterpillar 353 diesels; 600 hp *(448 kW)*; 2 shafts

Comment: PM 203 transferred from US Coast Guard in 1999 and PM 204 on 20 September 2002. Fitted with a 10 ton capacity boom. Reclassified as patrol ships in 2006.

TORTUGUERO   5/2013*, A Sheldon-Duplaix / 1525521

2 CANOPUS (SWIFTSHIPS 110 ft) CLASS
(LARGE PATROL CRAFT) (PB)

Name	No	Builders	Commissioned
CANOPUS (ex-Cristobal Colon)	GC 107	Swiftships, Morgan City	Jun 1984
ORION	GC 109	Swiftships, Morgan City	Aug 1984

Displacement, tonnes: 95 full load
Dimensions, metres (feet): 33.5 × 7.3 × 1.8 *(109.9 × 24.0 × 5.9)*
Speed, knots: 23
Range, n miles: 1,500 at 12 kt
Complement: 19 (3 officers)
Machinery: 3 Caterpillar 3412E diesels; 1,700 hp *(1.3 MW)* sustained; 3 shafts
Guns: 1 – 20 mm or 2 – 12.7 mm MGs.
Radars: Surface search: Raytheon; I-band.

Comment: Built of aluminium. GC 107 completely rebuilt and reconditioned by Swiftships in 2003. GC 109 was similarly refitted in 2004.

ORION   5/2013*, A Sheldon-Duplaix / 1525522

3 POINT CLASS (PB)

Name	No	Builders	Commissioned
ARIES (ex-Point Martin)	GC 101 (ex-82379)	US Coast Guard Yard, Curtis Bay	20 Aug 1970
ANTARES (ex-Point Baton)	GC 105 (ex-82340)	J M Martinac, Tacoma	20 Aug 1970
SIRIUS (ex-Point Spencer)	GC 110 (ex-82349)	J M Martinac, Tacoma	25 Oct 1966

Displacement, tonnes: 68 full load
Dimensions, metres (feet): 25.3 × 5.2 × 1.8 *(83 × 17.1 × 5.9)*
Speed, knots: 22. **Range, n miles:** 1,200 at 8 kt
Complement: 10
Machinery: 2 Caterpillar D3412 diesels; 1,600 hp *(1.19 MW)*; 2 shafts
Guns: 2 – 12.7 mm MGs.
Radars: Surface search: Hughes/Furuno SPS-73; I-band.

Comment: *Antares* transferred from US Coast Guard 1 October 1999 and *Sirius* transferred 12 December 2000. *Aries* reported decommissioned in 2005 but returned to service after refit in 2007.

SIRIUS   6/2004, A Sheldon-Duplaix / 0587698

IHS Jane's Fighting Ships 2014-2015

4 BELLATRIX CLASS (COASTAL PATROL CRAFT) (PB)

Name	No	Builders	Commissioned
PROCION	GC 103	Sewart Seacraft Inc, Berwick	1967
ALDEBARÁN	GC 104	Sewart Seacraft Inc, Berwick	1972
BELLATRIX	GC 106	Sewart Seacraft Inc, Berwick	1967
CAPELLA	GC 108	Sewart Seacraft Inc, Berwick	1968

Displacement, tonnes: 61 full load
Dimensions, metres (feet): 25.9 × 5.5 × 1.5 *(85 × 18.0 × 4.9)*
Speed, knots: 18.7. **Range, n miles:** 800 at 15 kt
Complement: 12
Machinery: 2 Caterpillar 3412E diesels; 1,700 hp *(1.3 MW)* sustained; 2 shafts
Guns: 3 – 12.7 mm MGs.
Radars: Surface search: Raytheon SPS-64; I-band.

Comment: Transferred to the Dominican Navy by the US. *Procion* was taken out of service in 1995 but returned in 1997 after a long refit. GC 103 and GC 106 completely rebuilt and reconditioned by Swiftships, Morgan City, in 2003. GC 104 and GC 108 were similarly refitted in 2004.

PROCION   6/2012, A Sheldon-Duplaix / 1455774

2 SWIFTSHIPS 35M CLASS (LARGE PATROL CRAFT) (PB)

Name	No	Builders	Commissioned
ALTAIR	GC 112	Swiftships, Morgan City	Oct 2003
ARCTURUS	GC 114	Swiftships, Morgan City	Mar 2004

Displacement, tonnes: 97 standard
Dimensions, metres (feet): 35.1 × 7.3 × 1.5 *(115.2 × 24.0 × 4.9)*
Speed, knots: 25
Machinery: 3 CAT 3412 diesels; 3,600 hp *(2.7 MW)*; 3 Hamilton HM 651 waterjets
Guns: 1 – 25 mm. 2 – 12.7 mm MGs.

Comment: Two craft ordered from Swiftships, Morgan City, LA as part of wider programme to increase capability to conduct counter-smuggling and drug-trafficking operations. Fitted with launching ramp for 4.7 m RIB.

ALTAIR   3/2013*, A Sheldon-Duplaix / 1525523

4 INTERCEPTOR CRAFT (PBF)

POLLUX LR 155   **CASTOR** LR 156   **SHAULA** LR 157   **ATRIA** LR 158

Dimensions, metres (feet): 13.4 × 2.75 × 0.9 *(44 × 9.0 × 3.0)*
Speed, knots: 60
Range, n miles: 600 at 25 kt
Complement: 6
Machinery: 3 Yanmar DE 315 diesels; 945 hp *(704 kW)*; Bravo X drives

Comment: Manufactured by Nor-Tech, Fort Myers, Florida. Composite and glass fibre V-bottomed hull. The first two donated by the US Southern Command on 13 July 2007 and the second two on 6 September 2007. Employed on counter drugs, arms trafficking and illegal immigration duties.

INTERCEPTOR CRAFT   6/2007, US Southern Command / 1167968

© 2014 IHS

Patrol forces — Tugs < **Dominican Republic** 207

3 DAMEN 1505 PATROL CRAFT (LARGE PATROL CRAFT) (PB)

Name	No	Builders	Commissioned
HAMAL	LR 151	Astilleros Navales de la Bahia de las Calderas	Dec 2004
DENEB	LR 153	Astilleros Navales de la Bahia de las Calderas	14 Apr 2005
ACAMAR	LR 154	Astilleros Navales de la Bahia de las Calderas	14 Apr 2005

Displacement, tonnes: 16 standard
Dimensions, metres (feet): 15.1 × 4.5 × 1.0 (49.5 × 14.8 × 3.3)
Speed, knots: 34
Complement: 6
Machinery: 2 Caterpillar 2406 diesels; 1,800 hp (1.3 MW)
Guns: 1 — 7.62 mm MG (fitted for).

Comment: Damen Stan Patrol 1505 design craft constructed in the Dominican Republic. Aluminium construction. Employed as patrol craft on counter-drugs and illegal immigration duties. *Vega* sank during a storm in February 2011.

DAMEN 1505   6/2006, Damen Shipyards / 1164479

7 INTERCEPTOR CRAFT (PBF)

ELNATH LI 161	NUNKI LI 163	REGULUS LI 165	ACRUX LI 167
POLARIS LI 162	DUBHE LI 164	DENEBOLA LI 166	

Displacement, tonnes: 2 full load
Dimensions, metres (feet): 9.8 × 3.1 × 0.55 (32.2 × 10.2 × 1.8)
Speed, knots: 55
Complement: 3
Machinery: 2 Mercury outboards; 600 hp (477 kW)

Comment: Boston Whaler 32 Justice craft delivered in 2008–11. Glass-fibre laminate construction. Employed on counter drugs, arms trafficking and illegal immigration duties.

NUNKI   6/2009, A Sheldon-Duplaix / 1305482

LAND-BASED MARITIME AIRCRAFT

Numbers/Type: 2 Bell OH-58 Kiowa.
Operational speed: 102 kt (188 km/h).
Service ceiling: 14,000 ft (4,267 m).
Range: 260 n miles (481 km).
Role/Weapon systems: Light observation helicopters. The first was acquired in November 2003 and two further were received from the US in 2006. Two remain in service. Fitted for 7.62 mm MG.

OH-58 (Australian colours)   6/1996, Rockwell Australia / 1164548

© 2014 IHS

AUXILIARIES

Notes: (1) There are also two dredgers manned by the Navy. *Puerto Plata* BD 11, *San Pedro* BD 12.
(2) There are eight auxiliary craft: *Nizao* LA 1, *Cayo Vigia* LA 2, *Chavon* LA 3, *Yuma* LA 4, *Cayo Arena* LA 5, *Rio Ozama* LA 7, *Beata* LA 8, *Cayo Levantado* LA 9.

CAYO VIGIA   6/2012 / 1455775

1 FLOATING DOCK (YFD)

ENDEAVOR DF 1 (ex-AFDL 1)

Comment: Lift, 1,000 tons. Commissioned in 1943. Transferred from US on loan 8 March 1986 and approved for transfer 10 June 1997. DF 2 (ex-AFDM 2), previously reported, was not acquired.

1 LCU 1600 CLASS (UTILITY LANDING CRAFT) (LCU)

NEYBA (ex-*Commando*) LD 31 (ex-LDM 4, ex-LCU 1675)

Displacement, tonnes: 203 standard; 381 full load
Dimensions, metres (feet): 41.1 × 8.8 × 1.9 (134.8 × 28.9 × 6.2)
Speed, knots: 11. **Range, n miles:** 1,200 at 8 kt
Complement: 14 (2 officers)

Military lift: 134 tons or 400 troops
Machinery: 4 Detroit 6-71 diesels; 696 hp (519 kW) sustained; 2 shafts; Kort nozzles; 2 Detroit 12V-71 diesels (LCU 1680–1681); 680 hp (508 kW) sustained; 2 shafts; Kort nozzles
Guns: 2 — 12.7 mm MGs.
Radars: Navigation: Furuno; I-band.

Comment: Steel hulled construction. Built by General Ship and Engineering Works in 1978. Formerly operated by the US Army and transferred in 2004.

NEYBA   6/2012, A Sheldon-Duplaix / 1455777

TUGS

4 COASTAL/HARBOUR TUGS (YTM/YTL)

GUAROA RM 1	GUARIONEX RM 2	GUAROCUYA RM 3	MAGUA RM 4

Displacement, tonnes: 269 full load
Dimensions, metres (feet): 26.1 × 7.9 × 4.05 (85.6 × 25.9 × 13.3)
Speed, knots: 12.7
Complement: 6
Machinery: 2 Caterpillar 3512B diesels; 3,500 hp (2.6 MW); 2 shafts

Comment: Details given are for RM 1 and RM 2, Damen Stantug 2608, built at Astilleros Navales de la Bahia de las Calderas and commissioned in April 2004 and June 2005 respectively. RM 3 and RM 4 are former US Navy tugs.

GUAROCUYA   6/2011 / 1455776

IHS Jane's Fighting Ships 2014-2015

Ecuador > Frigates — Corvettes

FRIGATES

2 LEANDER CLASS (FFGHM)

Name	No	Builders	Laid down	Launched	Commissioned
PRESIDENTE ELOY ALFARO (ex-*Almirante Condell*)	FM 01 (ex-06)	Yarrow Shipbuilders, Scotstoun	5 Jun 1971	12 Jun 1972	21 Dec 1973
MORAN VALVERDE (ex-*Almirante Lynch*)	FM 02 (ex-07)	Yarrow Shipbuilders, Scotstoun	6 Dec 1971	6 Dec 1972	25 May 1974

Displacement, tonnes: 2,540 standard; 3,187 full load
Dimensions, metres (feet): 113.4 oa; 109.7 wl × 13.1 × 5.5 screws *(372; 359.9 × 43.0 × 18.0)*
Speed, knots: 27. **Range, n miles:** 4,500 at 12 kt
Complement: 248 (20 officers)

Machinery: 2 Babcock & Wilcox boilers; 550 psi *(38.7 kg/cm²)*; 850°F *(450°C)*; 2 White/English Electric turbines; 30,000 hp *(22.4 MW)*; 2 shafts
Missiles: SSM: 4 Aerospatiale MM 40 Exocet ❶; inertial cruise; active radar homing to 70 km *(40 n miles)* at 0.9 Mach; warhead 165 kg.
SAM: 3 twin Matra Simbad launchers for Mistral (may be fitted); IR homing to 4 km *(2.2 n miles)*; warhead 3 kg.
Guns: 2 Vickers 4.5 in *(115 mm)*/45 Mk 6 (twin) ❷ semi-automatic; 20 rds/min to 19 km *(10 n miles)* anti-surface; 6 km *(3.2 n miles)* anti-aircraft; weight of shell 25 kg. 4 Oerlikon 20 mm Mk 9 (2 twin) ❸; 800 rds/min to 2 km. 1 Raytheon 20 mm/76 Vulcan Phalanx ❹; 3,000 rds/min combined to 1.5 km.
Torpedoes: 6—324 mm Mk 32 (2 triple) tubes ❺; Whitehead A 244; anti-submarine; pattern running to 7 km *(3.8 n miles)* at 33 kt; warhead 34 kg.
Physical countermeasures: Decoys: 2 Corvus 8-barrelled trainable chaff rocket launchers ❻; distraction or centroid patterns to 1 km. Wallop Barricade double layer chaff launchers.
Electronic countermeasures: ESM/ECM: Elta EW system; intercept and jammer.

Radars: Air search: Marconi Type 965/966 ❼; A-band.
Surface search: Marconi Type 992 Q ❽; E/F-band.
Navigation: Liton Type 1006 ❾; I-band.
Fire control: Selenia ❿; I-band (for guns).
Sonars: Graseby Type 184 M/P; hull-mounted; active search and attack; medium frequency (6/9 kHz).
Combat data systems: Sisdef Imagen SP 100 includes datalink. Link 11 receive.
Weapon control systems: Maiten-1/CH for gunnery.

Helicopters: 1 Bell 230 ⓫.

Programmes: Following service in the Chilean Navy, both ships were decommissioned in 2007 and subsequently acquired by the Ecuador Navy. Following overhaul and modification in Chilean yards, they were transferred on 18 April 2008 (FM 01)

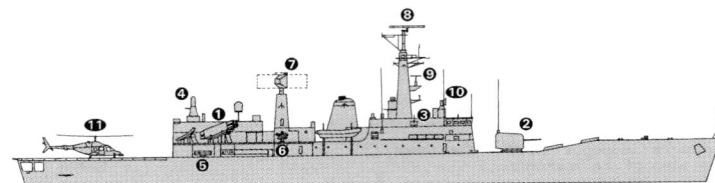

PRESIDENTE ELOY ALFARO *(Scale 1 : 1,200), Ian Sturton / 1445597*

and 15 October 2008 (FM 02). They replaced two ex-British Leanders (*Penelope* and *Danae*) originally built in the 1960s and acquired in 1991.
Modernisation: While in Chilean service, *Lynch* (1989) and *Condell* (1993) were both modernised by ASMAR, Talcahuano. Upgrades included enlargement of the hangar and flight deck, the fitting of the Indal Assist helicopter recovery system, mounting of two twin MM 40 Exocet launchers on each side of the hangar and moving the torpedo tubes down one deck. Other modifications included a new combat data system, improvements to the fire-control radars and the installation of Israeli EW systems. *Lynch* was further modernised in 2002 and *Condell* in 2004. Upgrades included complete overhaul of propulsion and machinery systems. Vulcan Phalanx has been installed on the hangar roof.

PRESIDENTE ELOY ALFARO *4/2008, Ecuador Navy / 1335375*

CORVETTES

6 ESMERALDAS CLASS (FSGHM)

Name	No	Builders	Laid down	Launched	Commissioned
ESMERALDAS	CM 11	Fincantieri, Muggiano	27 Sep 1979	1 Oct 1980	7 Aug 1982
MANABI	CM 12	Fincantieri, Ancona	19 Feb 1980	9 Feb 1981	9 Apr 1983
LOS RIOS	CM 13	Fincantieri, Muggiano	5 Dec 1979	27 Feb 1981	1 Oct 1983
EL ORO	CM 14	Fincantieri, Ancona	20 Mar 1980	9 Feb 1981	10 Dec 1984
GALAPÁGOS	CM 15	Fincantieri, Muggiano	4 Dec 1980	4 Jul 1981	27 May 1984
LOJA	CM 16	Fincantieri, Ancona	25 Mar 1981	27 Feb 1982	28 May 1984

Displacement, tonnes: 650 standard; 696 full load
Dimensions, metres (feet): 62.3 × 9.3 × 2.5 *(204.4 × 30.5 × 8.2)*
Speed, knots: 37
Range, n miles: 4,400 at 14 kt
Complement: 67

Machinery: 4 MTU 20V 956 TB92 diesels; 22,140 hp(m) *(16.27 MW)* sustained; 4 shafts
Missiles: SSM: 6 Aerospatiale MM 40 Exocet (2 triple) launchers ❶; inertial cruise; active radar homing to 70 km *(40 n miles)* at 0.9 Mach; warhead 165 kg; sea-skimmer.
SAM: Selenia Elsag Albatros quad launcher ❷; Aspide; semi-active radar homing to 13 km *(7 n miles)* at 2.5 Mach; warhead 30 kg.
Guns: 1 Oto Melara 3 in *(76 mm)*/62 compact ❸; 85 rds/min to 16 km *(8.7 n miles)*; weight of shell 6 kg. 2 Breda 40 mm/70 (twin) ❹; 300 rds/min to 12.5 km *(6.8 n miles)* anti-surface; weight of shell 0.96 kg.
Torpedoes: 6—324 mm ILAS-3 (2 triple) tubes ❺; Whitehead Motofides A244; anti-submarine; self-adaptive patterns to 7 km *(3.8 n miles)* at 33 kt; warhead 34 kg shaped charge. Not fitted in all.
Physical countermeasures: Decoys: 1 Breda 105 mm SCLAR launcher ❻; chaff to 5 km *(2.7 n miles)*; illuminants to 12 km *(6.6 n miles)*.
Electronic countermeasures: ESM/ECM: Elettronika Gamma ED; radar intercept and jammer.

Radars: Air/surface search: Selenia RAN 10S ❻; E/F-band; range 155 km *(85 n miles)*.
Navigation: Furuno 2115; I-band.
Fire control: 2 Selenia Orion 10X ❼; I/J-band; range 40 km *(22 n miles)*.
Sonars: Thomson Sintra Diodon; hull-mounted; active search and attack; 11, 12 or 13 kHz.
Combat data systems: Selenia IPN 10 action data automation. Link Y.
Weapon control systems: 2 Selenia NA21 with C03 directors.

ESMERALDAS *(Scale 1 : 600), Ian Sturton / 0505980*

Helicopters: Platform for 1 Bell 206B.

Programmes: Ordered in 1979.
Modernisation: A modernisation programme began in 2006. *Esmeraldas* completed a refit in November 2008 in which she was reportedly fitted with new Israeli systems, including the combat data system; the engines were also refurbished. *El Oro* has *Los Rios* had completed refits by late 2013 and *Manabi* is next to be modified.
Operational: Torpedo tubes removed from two of the class to refit in frigates. CM 16 took part in Exercise Unitas during 2008.

Corvettes — Shipborne aircraft < **Ecuador** 211

EL ORO 2/2000 / 0103731

MANABI 6/2002, Ecuador Navy / 0533898

SHIPBORNE AIRCRAFT

Numbers/Type: 1 Bell 230T.
Operational speed: 145 kt (269 km/h).
Service ceiling: 18,000 ft (5,500 m).
Range: 307 n miles (568 km).
Role/Weapon systems: Support helicopter for afloat reconnaissance and SAR. Navalised Bell 230s acquired in 1995. One lost in October 2009. Sensors: Surveillance radar. Weapons: None.

BELL 230 6/2003, Ecuador Navy / 0568886

Numbers/Type: 2 Kaman Seasprite SH-2G.
Operational speed: 130 kt (241 km/h).
Service ceiling: 10,000 ft (3,048 m).
Range: 350 n miles (650 km).
Role/Weapon systems: The acquisition of two SH-2G by Foreign Military Sale was notified to US Congress on 29 September 2011 and confirmed in 2013. They are to be delivered in 2016. The aircraft are to be refurbished by Kaman and the package would include associated equipment, parts, training and logistical support. Sensors: HELRAS dipping sonar, AAQ-22 Forward-Looking Infrared Radar (FLIR), APS-143C(V)3 radar. Weapons: to be announced.

Numbers/Type: 3/3 Bell 206 Jet Ranger/206TH 57 Sea Ranger.
Operational speed: 115 kt (213 km/h).
Service ceiling: 13,500 ft (4,115 m).
Range: 368 n miles (682 km).
Role/Weapon systems: Support helicopter for afloat reconnaissance and SAR. Sensors: None. Weapons: Depth bombs, 7.62 mm MG.

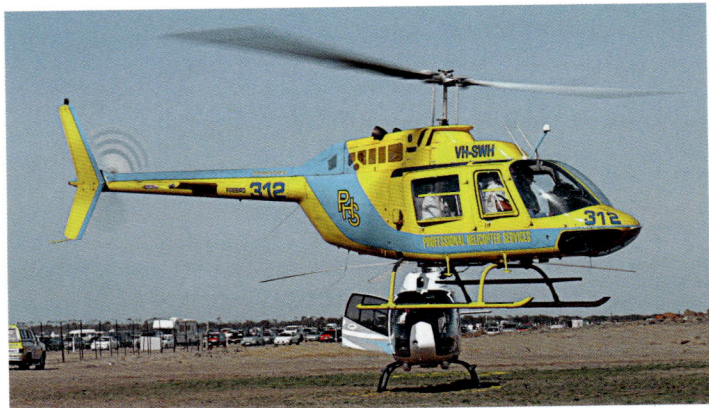

BELL 206 5/2004, Paul Jackson / 0569619

Numbers/Type: 2 Bell 430.
Operational speed: 145 kt (269 km/h).
Service ceiling: 14,600 ft (4,450 m).
Range: 324 n miles (600 km).
Role/Weapon systems: Acquired in 2010 to replace lost Bell 230. Reported not radar-fitted.

IHS Jane's Fighting Ships 2014-2015

212 Ecuador > Land-based maritime aircraft — Training ships

LAND-BASED MARITIME AIRCRAFT (FRONT LINE)

Notes: (1) The Navy operates four ENAER T-35 Pillan training aircraft, one Beech King Air B-350 and two Beech T-34C.
(2) Two IAI Heron and four IAI Searcher UAVs were acquired in November 2008. They became operational at Manta air base in June 2009. They are likely to be equipped with synthetic aperture radars and EO sensors for surveillance of maritime areas.

Numbers/Type: 2 Casa CN-235-300MP Persuader.
Operational speed: 210 kt *(384 km/h).*
Service ceiling: 24,000 ft *(7,315 m).*
Range: 2,000 n miles *(3,218 km).*
Role/Weapon systems: EEZ surveillance. Delivered in 2005/06. A CASA CN-235M-100 transport aircraft converted to maritime patrol role in 2010. Sensors: Seaspray radar. Weapons: unarmed.

CN 235-300MP   6/2008, Ecuador Navy / 1335374

Numbers/Type: 5 Beech King Air B-200.
Operational speed: 239 kt *(443 km/h).*
Service ceiling: 9,144 m *(30,000 ft).*
Range: 2,000 n miles *(3,218 km).*
Role/Weapon systems: There are two B-200, one CATPASS and one B-300 maritime patrol aircraft equipped with Seaspray radar. A B-350 was reported to have been acquired in 2013.

B-200   6/1999, Ecuador Navy / 0054061

PATROL FORCES

Notes: There are plans to acquire an offshore patrol ship of 1,500–1,800 tons.

3 QUITO (LÜRSSEN 45) CLASS

Name	No	Builders	Launched	Commissioned
QUITO	LM 21	Lürssen, Bremen-Vegesack	20 Nov 1975	13 Jul 1976
GUAYAQUIL	LM 23	Lürssen, Bremen-Vegesack	5 Apr 1976	22 Dec 1977
CUENCA	LM 24	Lürssen, Bremen-Vegesack	6 Dec 1976	17 Jul 1977

Displacement, tonnes: 259 standard
Dimensions, metres (feet): 45 × 7 × 2.5 *(147.6 × 23.0 × 8.2)*
Speed, knots: 40
Range, n miles: 700 at 40 kt, 1,800 at 16 kt
Complement: 35

Machinery: 4 MTU 16V 396 diesels; 13,600 hp(m) *(10 MW)* sustained; 4 shafts
Missiles: SSM: 4 Aerospatiale MM 38 Exocet; inertial cruise; active radar homing to 42 km *(23 n miles)* at 0.9 Mach; warhead 165 kg; sea-skimmer.
Guns: 1 Oto Melara 3 in *(76 mm)/62* compact; 85 rds/min to 16 km *(8.7 n miles);* weight of shell 6 kg. 2 Oerlikon 35 mm/90; 550 rds/min to 6 km *(3.3 n miles);* weight of shell 1.55 kg.
Electronic countermeasures: ESM: ELISRA NS-9010; intercept.
Radars: Air/surface search: Thomson-CSF Triton; G-band; range 33 km *(18 n miles)* for 2 m² target.
Fire control: Thomson-CSF Pollux; I/J-band; range 31 km *(17 n miles)* for 2 m² target.
Navigation: Furuno 2115; I-band.
Weapon control systems: Thomson-CSF Vega system.

Modernisation: New engines fitted during refits in 1994–95 at Guayaquil. The 35 mm guns were not replaced by 40 mm as planned but were modified for AHEAD ammunition. All three vessels are being updated. Quito completed refit in 2012 and the other two ships are to follow. MM 38 missiles had been refurbished by October 2013.

IHS Jane's Fighting Ships 2014-2015

CUENCA   2/2000 / 0103732

SURVEY AND RESEARCH SHIPS

1 SURVEY SHIP (YGS)

Name	No	Builders	Commissioned
ORION (ex-*Dometer*)	BI 91 (ex-HI 91, ex-HI 92)	Ishikawajima, Tokyo	10 Nov 1982

Measurement, tonnes: 1,123 gt
Dimensions, metres (feet): 64.2 pp × 10.7 × 3.6 *(210.6 × 35.1 × 11.8)*
Speed, knots: 12.6. **Range, n miles:** 6,000 at 12 kt
Complement: 45 (6 officers) + 14 personnel

Machinery: Diesel-electric; 3 Caterpillar 3412 diesel generators; 2,380 hp *(1.77 MW)* sustained; 2 motors; 1,900 hp *(1.42 MW);* 1 shaft
Radars: Surface search/Navigation: Furuno 2837; E/F-band.
Navigation: Sperry Marine Bridgemaster; I-band.
Helicopters: Platform for one medium.

Comment: Research vessel for oceanographic, hydrographic and meteorological work. A refit 2007–08 included installation of a flight deck and new engines.

ORION   6/2008, Ecuador Navy / 1335373

1 HYDROGRAPHIC VESSEL (AGSC)

SIRIUS LH 96

Displacement, tonnes: 245 standard
Dimensions, metres (feet): 32.5 × 8.0 × 2.0 *(106.6 × 26.2 × 6.6)*
Speed, knots: 12
Machinery: 2 Caterpillar C18670 diesels; 2 shafts

Comment: A new craft, built at Astinave, Guayaquil, entered service in 2011. It is based in the Galapagos Islands.

TRAINING SHIPS

1 SAIL TRAINING SHIP (AXS)

Name	No	Builders	Commissioned
GUAYAS	BE 91 (ex-BE 01)	Ast Celaya	23 Jul 1977

Measurement, tonnes: 949 gt; 238 dwt
Dimensions, metres (feet): 80 × 10.2 × 4.2 *(262.5 × 33.5 × 13.8)*
Speed, knots: 11.3
Complement: 130 (80 trainees)
Machinery: 1 GM 12V-149T diesel; 875 hp *(652 kW)* sustained; 1 shaft

Comment: Three masted. Launched 23 September 1976. Has accommodation for 180. Similar to ships in service with Colombia, Mexico and Venezuela. Modernised 2006–07.

GUAYAS   6/2012, M Declerck / 1483025

© 2014 IHS

Auxiliaries — Coast guard < **Ecuador** 213

AUXILIARIES

Notes: (1) A new dredger, *Francisco de Orellana* DR 90, was commissioned in July 2008.
(2) There are plans to acquire a logistic ship to replace *Hualcopo* decommissioned in 2006.

1 ARMAMENT STORES CARRIER (AETL)

Name	No	Builders	Commissioned
CALICUCHIMA (ex-*Throsk*)	TR 62 (ex-A 379)	Cleland SB Co, Wallsend	20 Sep 1977

Displacement, tonnes: 2,219 full load
Dimensions, metres (feet): 70.5 × 11.9 × 4.6 *(231.3 × 39.0 × 15.1)*
Speed, knots: 11
Range, n miles: 4,000 at 11 kt
Complement: 29 (5 officers)
Cargo capacity: 785 tons
Machinery: 2 General Motors diesels; 2,100 hp *(1.56 MW)*; 1 shaft
Radars: Navigation: Decca 926; I-band.

Comment: Acquired from the UK in November 1991. Recommissioned 24 March 1992. Completed refit in December 2009.

CALICUCHIMA *11/2004, Globke Collection* / 1129995

1 WATER CLASS (WATER TANKER) (AWT)

Name	No	Builders	Commissioned
QUISQUIS (ex-*Waterside*)	TR 64 (ex-Y 20)	Drypool Engineering, Hull	1968

Measurement, tonnes: 527 gt
Dimensions, metres (feet): 40.1 × 7.7 × 3.5 *(131.6 × 25.3 × 11.5)*
Speed, knots: 10
Range, n miles: 1,585 at 9 kt
Complement: 20 (4 officers)
Cargo capacity: 150 tons
Machinery: 1 Lister-Blackstone ERS-8-MCR diesel; 660 hp *(492 kW)*; 1 shaft
Radars: Navigation: Furuno; I-band.

Comment: Acquired from the UK in November 1991.

QUISQUIS *2/1992, A J Moorey* / 0056909

2 ARD 12 CLASS (FLOATING DOCKS) (YFD)

Name	No	Commissioned
RIO NAPO	DF 82 (ex-ARD 24)	1944
RIO ORELLANA (ex-*Cenepa*)	DF 83 (ex-ARD 26)	1944

Dimensions, metres (feet): 150 × 24.7 × 5.4 *(492.1 × 81.0 × 17.7)*

Comment: *Napo* bought from the US in 1988. Suitable for docking ships up to 3,200 tons. *Orellana* is 48 ft longer and was transferred from US service in 2000.

1 SUPPLY SHIP (AKL)

Name	Builders	Commissioned
GALAPAGOS (ex-*Arca Foz*; ex-*Riveira*)	Astilleros de Huelva	1982

Measurement, tonnes: 2,659 gt
Dimensions, metres (feet): 74.7 × 14.2 × 4.6 *(245.1 × 46.6 × 15.1)*
Speed, knots: 12
Machinery: 1 diesel; 2,100 hp *(1.56 MW)*; 1 cp prop

Comment: Spanish-built refrigerated cargo ship acquired in July 2008. The ship is used to supply the Galapagos Islands. Operated by navy owned company TRANSNAVE.

© 2014 IHS

ISLA SAN CRISTÓBAL *11/2004, Globke Collection* / 1129994

© 2014 IHS

1 BASENTO CLASS (WATER TANKER) (AWT)

Name	No	Builders	Commissioned
ATAHUALPA (ex-*Basento*)	TR 63 (ex-A 5336)	Inma di La Spezia	19 Jul 1971

Displacement, tonnes: 1,945 full load
Dimensions, metres (feet): 68.7 × 10.1 × 3.9 *(225.4 × 33.1 × 12.8)*
Speed, knots: 13
Range, n miles: 1,650 at 12 kt
Complement: 24 (3 officers)
Machinery: 2 Fiat LA 230 diesels; 1,730 hp *(1.27 MW)*; 2 shafts
Radars: Navigation: SPN-703; I-band.

Comment: Ex-Italian water-tanker decommissioned in 2004. Donated to Ecuador in April 2009 and replaced 1945 vintage ship of the same name and pennant number. Based in the Galapagos Islands.

ATAHUALPA *10/2009, Diego Quevedo* / 1305819

TUGS

5 HARBOUR TUGS (YTM/YTL)

MARANON (ex-*Sangay*) RB 72	ILINIZA RB 75	QUILOTOA RB 78
COTOPAXI RB 73	ALTAR RB 76	

Comment: Mostly built in the 1950s and 1960s. *Maranon* (380 tons built in the US in 1938 and is also used as a training vessel. *Cotopaxi* (180 tons) built in the US in 1941, *Iliniza* and *Altar* built in Ecuador in 1977.

1 CHEROKEE CLASS (ATF)

Name	No	Builders	Commissioned
CHIMBORAZO (ex-*Chowanoc* ATF 100)	RA 70 (ex-R 71, ex-R 105)	Charleston SB and DD Co, Charleston	21 Feb 1945

Displacement, tonnes: 1,255 standard; 1,666 full load
Dimensions, metres (feet): 62.5 × 11.7 × 5.2 *(205.1 × 38.4 × 17.1)*
Speed, knots: 12
Range, n miles: 7,000 at 15 kt
Complement: 67 (5 officers)
Machinery: Diesel-electric; 4 Caterpillar D 399 diesels; 4 generators; 1 motor; 3,000 hp *(2.24 MW)*; 1 shaft
Guns: 1—40 mm. 2—12.7 mm MGs.
Radars: Navigation: Simrad; I-band.

Comment: Launched 20 August 1943 and transferred 1 October 1977.

CHIMBORAZO *6/2001, Maritime Photographic* / 0114524

COAST GUARD

Notes: (1) In addition to the vessels listed below, there are up to 40 river patrol launches operated by both the Coast Guard and the Army.
(2) There are 18 8 m Amazonas class and smaller Marañon-class patrol craft.
(3) There are three floating river stations (15.5 × 13.5 m): *Lago Yaguarcocha* EGM 601; *Lago Cuyabeno* EGM 602; *Lago San Pablo* EGM 603. All entered service in 2009.

IHS Jane's Fighting Ships 2014-2015

RIO DAULE *6/2012, Ecuador Coast Guard* / 1486479

IHS Jane's Fighting Ships 2014-2015

ROMEO 852 and 855

3/2007, Marco Ghiglino / 1166536

IHS Jane's Fighting Ships 2014-2015

214 Ecuador ▸ Coast guard

3 ISLA FERNANDINA (VIGILANTE) CLASS
(OFFSHORE PATROL CRAFT) (PBO)

Name	No	Builders	Commissioned
ISLA FERNANDINA	LG 39	Astilleros de Murueta	

216 Ecuador ▸ Coast guard

3 ALBATROS 830 CLASS (WPBR)

Name	No	Builders	Commissioned
RIO COANGOS	LG-161	Astilleros SITECNA	2008
RIO MUISNE	LG-162	Astilleros SITECNA	

4 RIO COCA CLASS (WPB)

RIO COCA LG117 RIO CONONACO LG119
RIO CONAMBO LG118 RIO CURARAY LG120

218 Egypt ▸ Frigates

FRIGATES

4 OLIVER HAZARD PERRY CLASS (FFGHM)

Name	No	Builders	Laid down	Launched	Commissioned
ALEXANDRIA (ex-*Mubarak*; ex-*Copeland*)	911 (ex-FFG 25)	Todd Pacific Shipyard Corporation, San Pedro	24 Oct 1979	26 Jul 1980	7 Aug 1982
TABA (ex-*Gallery*)	916 (ex-FFG 26)	Bath Iron Works, Maine	17 May 1980	20 Dec 1980	5 Dec 1981
SHARM EL SHEIKH (ex-*Fahrion*)	901 (ex-FFG 22)	Todd Pacific Shipyard Corporation, Seattle	1 Dec 1978	24 Aug 1979	16 Jan 1982
TOUSHKA (ex-*Lewis B Puller*)	906 (ex-FFG 23)	Todd Pacific Shipyard Corporation, San Pedro	23 May 1979	15 Mar 1980	17 Apr 1982

Displacement, tonnes: 2,794 standard; 3,696 full load
Dimensions, metres (feet): 135.6 × 13.7 × 4.5 hull; 7.5 sonar *(444.9 × 44.9 × 14.8; 24.6)*
Speed, knots: 29. **Range, n miles:** 4,500 at 20 kt
Complement: 206 (13 officers; 19 air crew)

Machinery: 2 GE LM 2500 gas turbines; 41,000 hp *(30.59 MW)* sustained; 1 shaft; cp prop 2 auxiliary retractable props; 650 hp *(484 kW)*
Missiles: SSM: 4 McDonnell Douglas Harpoon Block 1B; active radar homing to 92 km *(50 n miles)* at 0.9 Mach; warhead 227 kg.
SAM: 36 GDC Standard SM-1MR Block VI; command guidance; semi-active radar homing to 38 km *(20.5 n miles)* at 2 Mach. 1 Mk 13 Mod 4 launcher for both SSM and SAM missiles ❶.
Guns: 1 Oto Melara 3 in *(76 mm)*/62 Mk 75 ❷; 85 rds/min to 16 km *(8.7 n miles)* anti-surface; 12 km *(6.6 n miles)* anti-aircraft; weight of shell 6 kg. 1 Raytheon 20 mm/76 6-barrelled Mk 15 Vulcan Phalanx ❸; 3,000 rds/min combined to 1.5 km. 4 – 12.7 mm MGs.
Torpedoes: 6 – 324 mm Mk 32 (2 triple) tubes ❹. 24 Alliant Mk 46 Mod 5; anti-submarine; active/passive homing to 11 km *(5.9 n miles)* at 40 kt; warhead 44 kg.
Physical countermeasures: Decoys: 2 Loral Hycor SRBOC 6-barrelled fixed Mk 36 ❺; IR flares and chaff to 4 km *(2.2 n miles)*. T-Mk-6 Fanfare/SLQ-25 Nixie; torpedo decoy. ESM/ECM: Raytheon SLQ-32 ❻ radar warning.
Electronic countermeasures: ESM/ECM: Raytheon SLQ-32.
Radars: Air search: Raytheon SPS-49(V)4 ❼; C/D-band.
Surface search: ISC Cardion SPS-55 ❽; I-band.
Fire control: Lockheed STIR (modified SPG-60) ❾; I/J-band; range 110 km *(60 n miles)*. Sperry Mk 92 (Signaal WM28) ❿; I/J-band. Navigation: Furuno; I-band ⓫. JRC; I-band. Tacan: URN 25. IFF Mk XII AIMS UPX-29.
Sonars: Raytheon SQS-56; hull-mounted; active search and attack; medium frequency.
Combat data systems: NTDS with Link Y.
Weapon control systems: SWG-1 Harpoon LCS. Mk 92 (Mod 4). Mk 13 weapon direction system. 2 Mk 24 optical directors.

Helicopters: 2 Kaman SH-2G Seasprite ⓬.

Programmes: First one acquired from US on 18 September 1996, second on 28 September 1996, third on 31 March 1998, and fourth on 30 September 1998.
Modernisation: JRC radar fitted on hangar roof.
Operational: First pair arrived in Egypt in mid-1997 after working up, third in late 1998 and fourth in 1999. All reported active, at least one in the Red Sea.

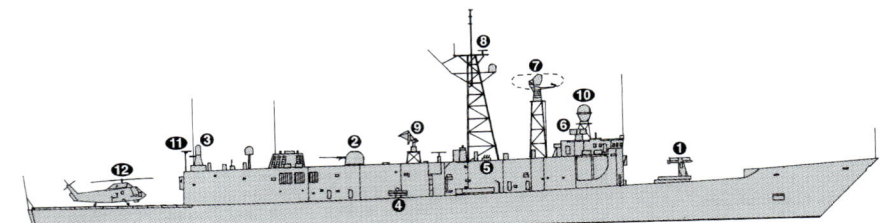

ALEXANDRIA

(Scale 1 : 1,200), Ian Sturton / 0103734

TABA 6/2006 / 1167111

TABA 6/2006 / 1167112

TABA 6/2006 / 1167113

© 2014 IHS

IHS Jane's Fighting Ships 2014-2015

Frigates — Egypt

2 KNOX CLASS (FFGH)

Name	No	Builders	Laid down	Launched	Commissioned	Recommissioned
DAMYAT (ex-*Jesse L Brown*)	961 (ex-FF 1089)	Avondale Shipyards, New Orleans	8 Apr 1971	18 Mar 1972	17 Feb 1973	1 Oct 1994
RASHEED (ex-*Moinester*)	966 (ex-FF 1097)	Avondale Shipyards, New Orleans	25 Aug 1972	12 May 1973	2 Nov 1974	1 Oct 1994

Displacement, tonnes: 3,059 standard; 4,328 full load
Dimensions, metres (feet): 134 × 14.3 × 4.6 hull; 7.8 sonar (439.6 × 46.9 × 15.1; 25.6)
Speed, knots: 27
Range, n miles: 4,000 at 22 kt
Complement: 288 (17 officers)

Machinery: 2 Combustion Engineering/Babcock & Wilcox boilers; 1,200 psi (84.4 kg/cm^2); 950°F (510°C); 1 turbine; 35,000 hp (26 MW); 1 shaft
Missiles: A/S: Honeywell ASROC Mk 16 octuple launcher with reload system (has 2 cells modified to fire Harpoon) ❶; inertial guidance to 1.6–10 km (1–5.4 n miles); payload Mk 46.
Guns: 1 FMC 5 in (127 mm)/54 Mk 42 Mod 9 ❷; 20–40 rds/min to 24 km (13 n miles) anti-surface; 14 km (7.7 n miles) anti-aircraft; weight of shell 32 kg. 1 Raytheon 20 mm/76 6-barrelled Mk 15 Vulcan Phalanx ❸; 3,000 rds/min combined to 1.5 km.
Torpedoes: 4—324 mm Mk 32 (2 twin) fixed tubes ❹. 22 Alliant Mk 46 Mod 5; anti-submarine; active/passive homing to 11 km (5.9 n miles) at 40 kt; warhead 44 kg.
Physical countermeasures: Decoys: 2 Loral Hycor SRBOC 6-barrelled fixed Mk 36 ❺; IR flares and chaff to 4 km (2.2 n miles). T Mk 6 Fanfare/SLQ-25 Nixie; torpedo decoy. Prairie Masker hull and blade rate noise suppression.
Electronic countermeasures: ESM/ECM: Elettronica ❻ intercept and jammer.
Radars: Air search: Lockheed SPS-40B ❼; B-band; range 320 km (175 n miles).
Surface search: Raytheon SPS-10 or Norden SPS-67 ❽; G-band.
Navigation: Marconi LN66; I-band.
Fire control: Western Electric SPG-53A/D/F ❾; I/J-band.
Tacan: SRN 15.
Sonars: EDO/General Electric SQS-26 CX; bow-mounted; active search and attack; medium frequency.
Combat data systems: FFISTS mini NTDS with Link Y.
Weapon control systems: SWG-1A Harpoon LCS. Mk 68 GFCS. Mk 114 ASW FCS. Mk 1 target designation system.

Helicopters: 1 Kaman SH-2G Seasprite ❿.

Programmes: Lease agreed from USA in mid-1993 and signed 27 July 1994 when both ships sailed for Egypt. Two others were transferred for spares in 1996. Ships of this class have been transferred to Greece, Taiwan, Turkey and Thailand.
Modernisation: Vulcan Phalanx fitted in the mid-1980s. EW suite replaced.
Structure: Four torpedo tubes are fixed in the midship superstructure, two to a side, angled out at 45°. A lightweight anchor is fitted on the port side and an 8,000 lb anchor fits in to the after section of the sonar dome.

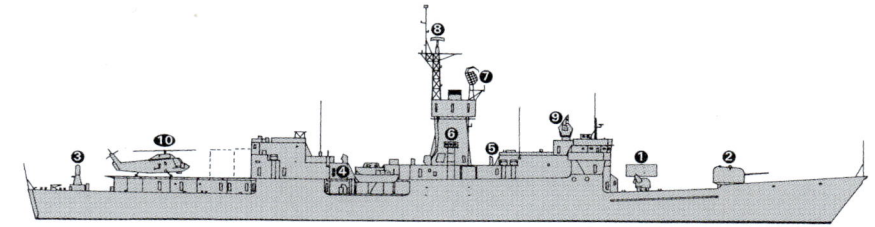
DAMYAT (Scale 1 : 1,200), Ian Sturton / 0506185

RASHEED 3/2007, Marco Ghiglino / 1166535

Operational: These ships have had boiler problems in Egyptian service. Refits may be undertaken with US assistance if and when funds become available. An ASROC firing was conducted by one of the class on 30 October 2011.

2 DESCUBIERTA CLASS (FFGM)

Name	No	Builders	Laid down	Launched	Commissioned
EL SUEZ (ex-*Serviola*)	F 946	Bazán, Ferrol	28 Feb 1979	20 Dec 1979	27 Oct 1984
ABU QIR (ex-*Centinela*)	F 941	Bazán, Ferrol	31 Oct 1978	6 Oct 1979	21 May 1984

Displacement, tonnes: 1,253 standard; 1,503 full load
Dimensions, metres (feet): 88.8 × 10.4 × 3.8 (291.3 × 34.1 × 12.5)
Speed, knots: 28
Range, n miles: 4,000 at 18 kt
Complement: 116 (10 officers)

Machinery: 4 MTU-Bazán 16V 956 TB91 diesels; 15,000 hp(m) (11 MW) sustained; 2 shafts; cp props
Missiles: SSM: 8 McDonnell Douglas Harpoon (2 quad) launchers ❶; active radar homing to 130 km (70 n miles) at 0.9 Mach; warhead 227 kg.
SAM: Selenia Elsag Albatros octuple launcher ❷; 24 Aspide; semi-active radar homing to 13 km (7 n miles) at 2.5 Mach; height envelope 15–5,000 m (49.2–16,405 ft); warhead 30 kg.
Guns: 1 Oto Melara 3 in (76 mm)/62 compact ❸; 85 rds/min to 16 km (8.7 n miles); weight of shell 6 kg. 2 Bofors 40 mm/70 ❹; 300 rds/min to 12.5 km (6.8 n miles); weight of shell 0.96 kg.
Torpedoes: 6—324 mm Mk 32 (2 triple) tubes ❺. MUSL Stingray; anti-submarine; active/passive homing to 11 km (5.9 n miles) at 45 kt; warhead 35 kg (shaped charge); depth to 750 m (2,460 ft).
A/S Mortars: 1 Bofors 375 mm twin-barrelled trainable launcher ❻; automatic loading; range 1,600 or 3,600 m depending on type of rocket.
Physical countermeasures: Prairie Masker; acoustic signature suppression.
Electronic countermeasures: ESM/ECM: Elettronica SpA Beta; intercept and jammer.
Radars: Air/surface search: Signaal DA05 ❼; E/F-band; range 137 km (75 n miles) for 2 m^2 target.
Navigation: Signaal ZW06; I-band.
Fire control: Signaal WM25 ❽; I/J-band.
Sonars: Raytheon 1160B; hull-mounted; active search and attack; medium frequency. Raytheon 1167 ❾; VDS; active search; 12–7.5 kHz.
Combat data systems: Signaal SEWACO action data automation. Link Y.

Programmes: Ordered September 1982 from Bazán, Spain. The two Spanish ships *Centinela* and *Serviola* were sold to Egypt prior to completion and transferred after completion at Ferrol and modification at Cartagena. *El Suez* completed 28 February 1984 and *Abu Qir* on 31 July 1984.
Modernisation: The combat data system, air search and fire-control radars were updated in 1995–96.
Operational: Stabilisers fitted. Modern noise insulation of main and auxiliary machinery. *Abu Qir* reported operational.

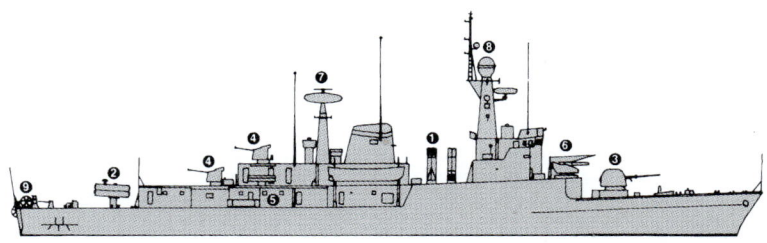
EL SUEZ (Scale 1 : 900), Ian Sturton / 0505984

EL SUEZ 4/2012, US Navy / 1486483

220 Egypt > Frigates

2 JIANGHU I CLASS (FFG)

Name	No	Builders	Commissioned
NAJIM AL ZAFFER	951	Hudong, Shanghai	27 Oct 1984
EL NASSER	956	Hudong, Shanghai	16 Apr 1985

Displacement, tonnes: 1,448 standard; 1,729 full load
Dimensions, metres (feet): 103.2 × 10.8 × 3.1
(338.6 × 35.4 × 10.2)
Speed, knots: 26
Range, n miles: 4,000 at 15 kt
Complement: 195

Machinery: 2 Type 12 E 390V diesels; 14,400 hp(m) (10.6 MW) sustained; 2 shafts
Missiles: SSM: 4 HY-2 (C-201) (2 twin) ❶; active radar or passive IR homing to 80 km (43.2 n miles) at 0.9 Mach; warhead 513 kg.
Guns: 4 China 57 mm/70 (2 twin) ❷; 120 rds/min to 12 km (6.5 n miles); weight of shell 6.31 kg. 12 China 37 mm/63 (6 twin) ❸; 180 rds/min to 8.5 km (4.6 n miles); weight of shell 1.42 kg.
A/S Mortars: 4 RBU 1200 5-tubed fixed launchers ❹; range 1,200 m; warhead 34 kg.
Mines: Up to 60.
Depth charges: 4 projectors.
Electronic countermeasures: ESM/ECM: Elettronica SpA Beta or Litton Triton; intercept and jammer.
Radars: Air search: Type 765 ❺; A-band.
Surface search: Eye Shield ❻; G-band.
Surface search/gun direction: Square Tie; I-band.
Fire control: Fog Lamp.
Navigation: Decca RM 1290A; I-band.
Sonars: China Type E5; hull-mounted; active search and attack; high frequency.

Programmes: Ordered from China in 1982. This is a Jianghu I class modified with 57 mm guns vice the standard 100 mm. These were the 17th and 18th hulls of the class.

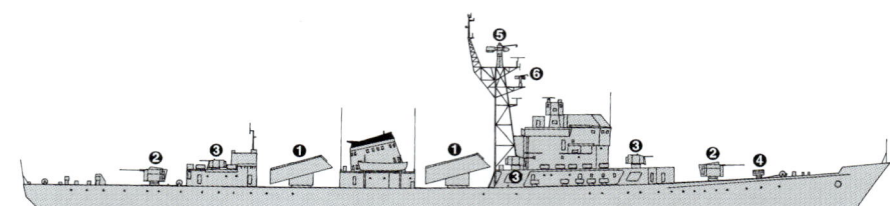

NAJIM AL ZAFFER   (Scale 1 : 900), Ian Sturton / 0056914

Modernisation: Combat data system to be fitted together with CSEE Naja optronic fire-control directors. There are also plans, confirmed in October 1994, to remove the after superstructure and guns and build a flight deck for an SH-2G Seasprite helicopter. Although a refit programme is reported to have been proposed by China, there is still no sign yet of work being done.
Structure: The funnel is the rounded version of the Jianghu class.
Operational: Both ships are active.

EL NASSER

5/2006, B Prézelin / 1040733

NAJIM AL ZAFFER

5/2007, Camil Busquets i Vilanova / 1166549

IHS Jane's Fighting Ships 2014-2015

© 2014 IHS

SHIPBORNE AIRCRAFT

Numbers/Type: 10 Kaman SH-2G(E) Seasprite.
Operational speed: 130 kt *(241 km/h)*.
Service ceiling: 22,500 ft *(6,860 m)*.
Range: 367 n miles *(679 km)*.
Role/Weapon systems: Total of 10 upgraded SH-2F aircraft transferred under FMS by September 1998. New engines and avionics. A further avionics upgrade was reportedly under consideration in 2004. Sensors: LN66/HP radar; ALR-66 ESM; ALE-39 ECM; ARN-118 Tacan; Ocean Systems AQS-18A dipping sonar. Possible mine detection optronic sensor. Weapons: 2 × Mk 46 torpedoes or a depth bomb.

SEASPRITE   1/2004, Kaman / 0566188

LAND-BASED MARITIME AIRCRAFT (FRONT LINE)

Notes: There are also 24 Westland Commando WS 61 helicopters. Some refitted in 1997/98.

Numbers/Type: 9 Aerospatiale SA 342L Gazelle.
Operational speed: 142 kt *(264 km/h)*.
Service ceiling: 14,105 ft *(4,300 m)*.
Range: 407 n miles *(755 km)*.
Role/Weapon systems: Air Force helicopter for coastal anti-shipping strike, particularly against FAC and insurgents. Sensors: SFIM sight. Weapons: ASV; 2 × AS-12 wire-guided missiles.

Numbers/Type: 7 Grumman E-2C Hawkeye 2000.
Operational speed: 323 kt *(598 km/h)*.
Service ceiling: 37,000 ft *(11,278 m)*.
Range: 1,540 n miles *(2,852 km)*.
Role/Weapon systems: Air Force airborne early warning and control tasks; capable of handling up to 30 tracks over water or land. In late 1999, Northrop Grumman contracted to upgrade five E-2Cs to Hawkeye 2000E export configuration. In June 2001, a contract for the supply of an additional (sixth) aircraft to Hawkeye 2000E standard was signed. In 2008, a contract for the upgrade on an additional (seventh) aircraft was concluded. Sensors: APS-138 search/warning radar being replaced by APS-145 from October 2002 as part of major upgrade programme. The first upgraded aircraft delivered in February 2003, second in early 2004, third in August 2004, fourth in May 2005 and fifth in December 2006. Various ESM/ECM systems. Weapons: Unarmed.

HAWKEYE 2000   3/2003, Northrop Grumman / 0530203

Numbers/Type: 2 Westland Sea King Mk 47.
Operational speed: 112 kt *(208 km/h)*.
Service ceiling: 14,700 ft *(4,480 m)*.
Range: 664 n miles *(1,230 km)*.
Role/Weapon systems: Air Force helicopter for ASW and surface search; secondary role as SAR helicopter. Airframe and engine refurbishment in 1990. Seven more are in reserve and out of service. Sensors: MEL search radar. Weapons: ASW; 4 × Mk 46 or Stingray torpedoes or depth bombs. ASV; Otomat.

Numbers/Type: 2 Beechcraft 1900C.
Operational speed: 267 kt *(495 km/h)*.
Service ceiling: 25,000 ft *(7,620 m)*.
Range: 1,569 n miles *(2,907 km)*.
Role/Weapon systems: Two (of six) Air Force aircraft acquired in 1988 and used for maritime surveillance. Sensors: Litton search radar; Motorola multimode SLAMMR radar; Singer S-3075 ESM; Datalink Y. Weapons: Unarmed.

PATROL FORCES

5 TIGER CLASS (TYPE 148)
(FAST ATTACK CRAFT—MISSILE) (PGGF)

Name	No	Builders	Commissioned
23 OF JULY (ex-*Alk*)	601 (ex-P 6155)	CMN, Cherbourg	7 Jan 1975
6 OF OCTOBER (ex-*Fuchs*)	602 (ex-P 6146)	CMN, Cherbourg	17 Oct 1973
21 OF OCTOBER (ex-*Löwe*)	603 (ex-P 6148)	CMN, Cherbourg	9 Jan 1974
18 OF JUNE (ex-*Dommel*)	604 (ex-P 6156)	CMN, Cherbourg	12 Feb 1975
25 OF APRIL (ex-*Weihe*)	605 (ex-P 6157)	CMN, Cherbourg	3 Apr 1975

Displacement, tonnes: 238 standard; 269 full load
Dimensions, metres (feet): 47 × 7 × 2.7 *(154.2 × 23.0 × 8.9)*
Speed, knots: 36
Range, n miles: 570 at 30 kt, 1,600 at 15 kt
Complement: 30 (4 officers)

Machinery: 4 MTU MD 16V 538 TB90 diesels; 12,000 hp(m) *(8.82 MW)* sustained; 4 shafts
Missiles: SSM: 4 Aerospatiale MM 38 Exocet (2 twin) launchers; inertial cruise; active radar homing to 42 km *(23 n miles)* at 0.9 Mach; warhead 165 kg; sea-skimmer.
Guns: 1 Oto Melara 3 in *(76 mm)*/62 compact; 85 rds/min to 16 km *(8.6 n miles)* anti-surface; 12 km *(6.5 n miles)* anti-aircraft; weight of shell 6 kg. 1 Bofors 40 mm/70; 330 rds/min to 12 km *(6.5 n miles)* anti-surface; 4 km *(2.2 n miles)* anti-aircraft; weight of shell 0.96 kg; fitted with GRP dome (1984) (see *Modernisation*).
Mines: Laying capability.
Physical countermeasures: Decoys: Wolke chaff launcher. Hot Dog IR launcher.
Radars: Air/surface search: Thomson-CSF Triton; G-band; range 33 km *(18 n miles)* for 2 m² target.
Navigation: SMA 3 RM 20; I-band; range 73 km *(40 n miles)*.
Fire control: Thomson-CSF Castor; I/J-band.
Combat data systems: PALIS and Link 11.
Weapon control systems: CSEE Panda optical director. Thomson-CSF Vega PCET system, controlling missiles and guns.

Programmes: 601 transferred from Germany in July 2002 and the remainder in March 2003. Weapons and sensors have also been transferred with the possible exception of EW equipment.
Modernisation: Triton search and Castor fire-control radars fitted to the whole class.
Structure: Steel-hulled craft. Similar to Combattante II craft.

21 OF OCTOBER   4/2003, Michael Nitz / 0552773

25 OF APRIL   4/2003, Michael Nitz / 0552771

25 OF APRIL   2/2011, Ships of the World / 1406808

222 Egypt > Patrol forces

2 + 2 AMBASSADOR IV CLASS
(FAST ATTACK CRAFT — MISSILE) (PCFG)

Name	No	Builders	Laid down	Launched	Commissioned
S EZZAT	682	VT Halter Marine, Pascagoula	7 Apr 2010	25 Oct 2011	19 Nov 2013
F ZEKRY	683	VT Halter Marine, Pascagoula	2011	4 Mar 2012	Dec 2013
M FAHMY	684	VT Halter Marine, Pascagoula	2012	2013	2014
A GAD	685	VT Halter Marine, Pascagoula	2012	2014	2014

Displacement, tonnes: 779 full load
Dimensions, metres (feet): 62.61 × 9.98 × 2.74 (205.4 × 32.7 × 9.0)
Speed, knots: 35
Range, n miles: 2,000 at 15 kt
Complement: 38 (10 officers)

Machinery: 4 MTU 16V 595 TE90 diesels; 21,690 hp(m) (16.2 MW); 4 shafts
Missiles: SSM: 8 (2 quad) McDonnell Douglas Harpoon Block II; active radar homing to 130 km (76 n miles) at 0.9 Mach; warhead 227 kg.
SAM: 1 Raytheon Mk 31 RAM (RIM-116) launcher; 21 RAM block 1A missile; passive IR/anti-radiation homing to 9.6 km (5.2 n miles) at 2 Mach; warhead 9.1 kg.
Guns: 1 Oto Melara 3 in (76 mm)/62 Super Rapid; 120 rds/min to 16 km (8.6 n miles) anti-surface; 12 km (6.6 n miles) anti-aircraft; weight of shell 6 kg. 1 Raytheon Mk 15 Mod 21 (block 1B) Phalanx; 4,500 rds/min combined to 1.5 km. 2 — 7.62 mm MGs.
Physical countermeasures: Decoys: 4 Super Barricade decoy launchers.
Electronic countermeasures: ESM: Argon ST WRR-2000.
ECM: To be announced.
Radars: Air/surface search: Thales MRR; 3D; G-band.
Fire control: Thales STING; I-band.
Navigation: Thales Scout; I-band.
Combat data systems: TACTICOS ICMS.
Weapon control systems: Thales STING optronic director.
Electro-optic systems: To be announced.

Comment: Following responses to an ITT issued in 1999, the Egyptian Navy placed an order in January 2001 for four fast attack craft (missile). These craft were to have been built by Halter Marine. However, following suspension of the project in 2002 and the subsequent purchase of the shipbuilder by Singapore Technologies, the project was revived in 2004 and a contract for the design of a new craft was then let to VT Halter Marine in late 2005. This was followed on 22 November 2006 by a contract (modified in September 2008) for the construction of three craft. A fourth unit was contracted in March 2010. The design includes a steel hull and aluminium superstructure. A 5.5 m RHIB is carried forward of the RAM mounting. Sea trials of the first of class began in January 2013.

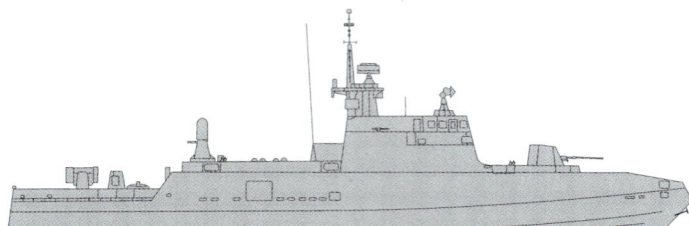

AMBASSADOR IV (Scale 1 : 600), Ian Sturton / 1366998

S EZZAT 6/2012, VT Halter Marine / 1486484

8 HAINAN CLASS (FAST ATTACK CRAFT — PATROL) (PC)

AL NOUR 430	AL HADI 433	AL HAKIM 436	AL WAKIL 439
AL QATAR 442	AL GABBAR 445	AL SALAM 448	AL RAFA 451

Displacement, tonnes: 381 standard; 398 full load
Dimensions, metres (feet): 58.8 × 7.2 × 2.2 (192.9 × 23.6 × 7.2)
Speed, knots: 30.5
Range, n miles: 1,300 at 15 kt
Complement: 69

Machinery: 4 PRC/Kolomna Type 9-D-8 diesels; 4,000 hp (2.94 MW) sustained; 4 shafts
Guns: 4 China 57 mm/70 (2 twin); 120 rds/min to 12 km (6.5 n miles); weight of shell 6.31 kg. 4 — 23 mm (2 twin); locally constructed to fit the 25 mm mountings.
Torpedoes: 6 — 324 mm (2 triple) tubes (in two of the class). Mk 44 or MUSL Stingray.
A/S Mortars: 4 RBU 1200 fixed 5-tubed launchers; range 1,200 m; warhead 34 kg.
Mines: Rails fitted. 12 mines.
Depth charges: 2 projectors; 2 racks. 18 DCs.
Radars: Surface search: Pot Head or Skin Head; I-band.
Navigation: Decca; I-band.
IFF: High Pole.
Sonars: Stag Ear; hull-mounted; active search and attack; high frequency.

Programmes: First pair transferred from China in October 1983, next three in February 1984 (commissioned 21 May 1984) and last three late 1984.
Modernisation: Two fitted with torpedo tubes and with Singer Librascope fire control. No sign of the remainder being similarly equipped. New sonar reported being fitted.
Operational: Based at Alexandria.

AL WAKIL 4/2007, Marco Ghiglino / 1166523

6 RAMADAN CLASS
(FAST ATTACK CRAFT — MISSILE) (PGGF)

Name	No	Builders	Launched	Commissioned
RAMADAN	670	Vosper Thornycroft	6 Sep 1979	20 Jul 1981
KHYBER	672	Vosper Thornycroft	31 Jan 1980	15 Sep 1981
EL KADESSAYA	674	Vosper Thornycroft	19 Feb 1980	6 Apr 1982
EL YARMOUK	676	Vosper Thornycroft	12 Jun 1980	18 May 1982
BADR	678	Vosper Thornycroft	17 Jun 1981	17 Jun 1982
HETTEIN	680	Vosper Thornycroft	25 Nov 1980	28 Oct 1982

Displacement, tonnes: 312 full load
Dimensions, metres (feet): 52 × 7.6 × 2.3 (170.6 × 24.9 × 7.5)
Speed, knots: 40
Range, n miles: 1,600 at 18 kt
Complement: 30 (4 officers)

Machinery: 4 MTU 20V 538 TB91 diesels; 15,360 hp(m) (11.29 MW) sustained; 4 shafts
Missiles: SSM: 4 Oto Melara/Matra Otomat Mk 2; active radar homing to 160 km (86.4 n miles) at 0.9 Mach; warhead 210 kg.
Guns: 1 Oto Melara 3 in (76 mm) compact; 85 rds/min to 16 km (8.7 n miles); weight of shell 6 kg. 2 Breda 40 mm/70 (twin); 300 rds/min to 12.5 km (6.8 n miles) anti-surface; weight of shell 0.96 kg.
Physical countermeasures: Decoys: 4 Protean fixed launchers each with 4 magazines containing 36 chaff decoy and IR flare grenades.
Electronic countermeasures: ESM: Racal Cutlass; radar intercept.
ECM: Racal Cygnus; jammer.
Radars: Air/surface search: Marconi S 820; E/F-band; range 73 km (40 n miles).
Navigation: Marconi S 810; I-band.
Fire control: 2 Marconi ST 802; I-band.
Combat data systems: AMS NAUTIS 3.
Weapon control systems: Marconi Sapphire System with 2 radar/TV and 2 optical directors.

Programmes: The contract was carried out at the Porchester yard of Vosper Thornycroft Ltd with some hulls built at Portsmouth Old Yard, being towed to Porchester for fitting out.
Modernisation: Contracts for the modernisation of these craft was let in 2001. Alenia Marconi Systems upgraded the Otomat missiles to Mk 2, renovated the S 820 and ST 802 radars and replaced the CAAIS combat system by NAUTIS 3. Work carried out 2002–2007.
Operational: Portable SAM SA-N-5 sometimes carried.

EL YARMOUK 3/2006, M Declerck / 1167109

6 KAAN 20 CLASS (FAST INTERVENTION CRAFT) (PBF)

61 +5

Displacement, tonnes: 38 full load
Dimensions, metres (feet): 22.55 × 4.76 × 1.3 (74 × 15.6 × 4.3)
Speed, knots: 60
Range, n miles: 350 at 35 kt
Complement: 5

Machinery: 2 MTU 12V 2000 M92 diesels; 3,600 hp(m) (4.8 MW); 2 Arneson ASD 12 B1L surface drives
Guns: 1 General Dynamics GAU-19 3-barrelled 12.7 mm Gatling gun.

Comment: Yonuk MRTP 20 (enlarged MRTP 15) design built at Yonca Shipyard. The craft are of advanced composites construction and feature a Deep V hull. Three to be built at Istanbul and three at Alexandria Shipyard. The first was delivered in December 2011 and delivery of all six craft was completed in 2013.

KAAN 20 12/2011, Turkships / 1486432

IHS Jane's Fighting Ships 2014-2015 © 2014 IHS

Patrol forces < **Egypt** 223

12 OSA I (PROJECT 205) CLASS
(FAST ATTACK CRAFT—MISSILE) (PTFG)

| 631 | 643 | 651 (ex-306) | – (ex-304) | 767 (ex-11) | – (ex-14) |
| 633 | 649 (ex-305) | 653 (ex-307) | – (ex-308) | 768 (ex-12) | – (ex-15) |

Displacement, tonnes: 174 standard; 213 full load
Dimensions, metres (feet): 38.6 × 7.6 × 2.7 *(126.6 × 24.9 × 8.9)*
Speed, knots: 35
Range, n miles: 400 at 34 kt
Complement: 30

Machinery: 3 MTU diesels; 12,000 hp(m) *(8.82 MW)*; 3 shafts
Missiles: SSM: 4 SS-N-2A Styx; active radar or IR homing to 46 km *(25 n miles)* at 0.9 Mach; altitude preset up to 300 m *(984.3 ft)*; warhead 513 kg.
SAM: SA-N-5 Grail; manual aiming; IR homing to 6 km *(3.2 n miles)* at 1.5 Mach; altitude to 2,500 m *(8,000 ft)*; warhead 1.5 kg.
Guns: 4 USSR 30 mm/65 (2 twin); 500 rds/min to 5 km *(2.7 n miles)* anti-aircraft; weight of shell 0.54 kg. 2—12.7 mm MGs.
Electronic countermeasures: ESM: Thomson-CSF DR 875; radar warning.
ECM: Racal; jammer.
Radars: Air/surface search: Kelvin Hughes; I-band.
Navigation: Racal Decca 916; I-band.
Fire control: Drum Tilt; H/I-band.
IFF: High Pole. Square Head.

Programmes: Of the 13 reported to have been delivered to Egypt by the Soviet Navy in 1966–68, three remained in service in 2003. Acquisition of additional craft has been made from two sources. Five Osa I class, originally acquired by the Yugoslav Navy in the 1960s, were delivered by May 2007. *651* and *653* were reported commissioned on 28 October 2007. Also, four Osa II were transferred from Finland in late 2006. These were originally acquired from the then Soviet Union by Finland in 1974; subsequently converted in 1993 to a minelaying role and decommissioned in 2000. These vessels are not equipped with missiles.
Modernisation: Refitted with MTU diesels, two machine guns, improved radars and EW equipment.
Operational: Three more *637* and *639* and *641* are laid up. The operational status of *631*, *633* and *643* is doubtful. All the ex-Yugoslav and Finnish craft are reported to have an operational role.

OSA 768   6/2010, Egyptian Navy / 1406034

OSA 649   5/2007, Freivogel Collection / 1166529

OSA 653   5/2007, Freivogel Collection / 1166530

4 OCTOBER CLASS (FAST ATTACK CRAFT—MISSILE) (PTFG)

781   783   787   789

Displacement, tonnes: 83 full load
Dimensions, metres (feet): 25.5 × 6.1 × 1.3 *(83.7 × 20.0 × 4.3)*
Speed, knots: 38
Range, n miles: 400 at 30 kt
Complement: 20

Machinery: 4 CRM 12 D/SS diesels; 5,000 hp(m) *(3.67 MW)* sustained; 4 shafts
Missiles: SSM: 2 Oto Melara/Matra Otomat Mk 2; active radar homing to 160 km *(86.4 n miles)* at 0.9 Mach; warhead 210 kg; can be carried.
Guns: 4 BMARC/Oerlikon 30 mm/75 (2 twin); 650 rds/min to 10 km *(5.5 n miles)* anti-surface; 3 km *(1.6 n miles)* anti-aircraft; weight of shell 1 kg and 0.36 kg mixed.
Physical countermeasures: Decoys: 2 Protean fixed launchers each with 4 magazines containing 36 chaff decoy and IR flare grenades.
Electronic countermeasures: ESM: Racal Cutlass; radar warning.
Radars: Air/surface search: Marconi S 810; range 48 km *(25 n miles)*.
Fire control: Marconi/ST 802; I-band.
Weapon control systems: Marconi Sapphire radar/TV system.

Programmes: Built in Alexandria 1975–76. Hull of same design as USSR Komar class. Refitted by Vosper Thornycroft, completed 1979–81. *791* was washed overboard on return trip, recovered and returned to Portsmouth for refit. Left UK after repairs on 12 August 1982. Probably Link fitted.
Modernisation: Alenia Marconi systems to upgrade Otomat missiles to Mk 2 between 2002–2007.
Operational: *791* reported non-operational and *785* is laid up.

OCTOBER 783   2/2004 / 1044123

4 HEGU CLASS (FAST ATTACK CRAFT—MISSILE) (PTFG)

609   611   613   615

Displacement, tonnes: 69 standard; 80 full load
Dimensions, metres (feet): 27 × 6.3 × 1.3 *(88.6 × 20.7 × 4.3)*
Speed, knots: 37.5
Range, n miles: 400 at 30 kt
Complement: 17 (2 officers)

Machinery: 4 Type L-12V-180 diesels; 4,800 hp(m) *(3.53 MW)*; 4 shafts
Missiles: SSM: 2 SY-1; active radar or passive IR homing to 40 km *(22 n miles)* at 0.9 Mach; warhead 513 kg.
Guns: 2—23 mm (twin); locally constructed to fit 25 mm mounting.
Electronic countermeasures: ESM: Litton Triton; radar intercept.
Radars: Surface search/fire control: Square Tie; I-band or Decca; I-band.
IFF: High Pole A.

Programmes: Acquired from China and commissioned in Egypt on 27 October 1984. The Hegu is the Chinese version of the deleted Komar.
Modernisation: ESM fitted in 1995–96.
Operational: *619* and *617* are reported laid up.

HEGU 609   3/2000 / 0103740

HEGU 609 and 611   3/2007, Marco Ghiglino / 1166521

224 Egypt > Patrol forces — Mine warfare forces

5 SHERSHEN CLASS (FAST ATTACK CRAFT—GUN) (PTFM)

753   755   757   759   761

Displacement, tonnes: 147 standard; 173 full load
Dimensions, metres (feet): 34.7 × 6.7 × 1.5 (113.8 × 22.0 × 4.9)
Speed, knots: 45. **Range, n miles:** 850 at 30 kt
Complement: 23

Missiles: SAM: SA-N-5 Grail (755–761); manual aiming; IR homing to 6 km (3.2 n miles) at 1.5 Mach; warhead 1.5 kg.
Guns: 4 USSR 30 mm/65 (2 twin); 500 rds/min to 5 km (2.7 n miles); weight of shell 0.54 kg.
Rockets: 2 USSR 122 mm rocket launchers (755–761 in lieu of torpedo tubes); 20 barrels per launcher; range 9 km (5 n miles).
Depth charges: 12.
Electronic countermeasures: ESM: Thomson-CSF DR 875; radar warning.
Radars: Surface search: Pot Drum; H/I-band.
Fire control: Drum Tilt, H/I-band (in some).
IFF: High Pole.

Programmes: Five delivered from USSR in 1967 and two more in 1968. One deleted. 753 completed an extensive refit at Ismailia in 1987; 751 in 1988.
Structure: The last four have had their torpedo tubes removed to make way for multiple BM21 rocket launchers and one SA-N-5 Grail, which are not always carried. Some have Drum Tilt radars removed. 753 has also had its torpedo tubes removed but these may be replaced.
Operational: Based at Alexandria, Port Said and Mersa Matru. 751 reported non-operational.

SHERSHEN 757   3/2007, Marco Ghiglino / 1166522

4 SHANGHAI II CLASS (FAST ATTACK CRAFT—GUN) (PB)

793   795   797   799

Displacement, tonnes: 115 standard; 133 full load
Dimensions, metres (feet): 38.8 × 5.4 × 1.7 (127.3 × 17.7 × 5.6)
Speed, knots: 30. **Range, n miles:** 700 at 16.5 kt
Complement: 34

Machinery: 2 Type L12-180 diesels; 2,400 hp(m) (1.76 MW) (forward); 2 Type L12-180Z diesels; 1,820 hp(m) (1.34 MW) (aft); 4 shafts
Guns: 4 China 37 mm/63 (2 twin); 180 rds/min to 8.5 km (4.6 n miles); weight of shell 1.42 kg. 2—23 mm (1 twin); locally constructed to fit the 25 mm mountings.
Mines: Rails can be fitted for 10 mines.
Electronic countermeasures: ESM: Thomson-CSF; radar warning.
Radars: Surface search: Decca; I-band.
IFF: High Pole.

Programmes: Transferred from China in 1984.
Operational: Three based at Suez and one (799) at Mersa Matru. 795 refitted in 1998.

SHANGHAI 797   6/1997, J W Currie / 0012295

AMPHIBIOUS FORCES

Notes: (1) Acquisition of LSTs is a high priority.
(2) Ro-Ro ferries are chartered for amphibious exercises.
(3) Rigid Raiders with Johnson outboards are also in service.
(4) Three small hovercraft similar to Slingsby SAH 2200 reported to be in service.

3 POLNOCHNY A (PROJECT 770) CLASS (LSM)

301   303   305

Displacement, tonnes: 813 full load
Dimensions, metres (feet): 73 × 8.5 × 1.8 (239.5 × 27.9 × 5.9)
Speed, knots: 19. **Range, n miles:** 1,000 at 18 kt
Complement: 40

Military lift: 6 tanks; 350 tons.
Machinery: 2 Kolomna Type 40-D diesels; 4,400 hp(m) (3.2 MW) sustained; 2 shafts
Guns: 2 USSR 30 mm/65 (twin); 500 rds/min to 5 km (2.7 n miles); weight of shell 0.54 kg. 2—140 mm rocket launchers; 18 barrels to 9 km (4.9 n miles).
Radars: Surface search: Decca; I-band.
Fire control: Drum Tilt; H/I-band.

Comment: Built at Northern Shipyard, Gdansk and transferred from USSR 1973–74. All used for Gulf logistic support in 1990–91. SA-N-5 may be carried. A radar has replaced the twin 30 mm mounting in at least one ship. All are active.

POLNOCHNY 303   10/2000, F Sadek / 0103742

5 SEAFOX TYPE (SWIMMER DELIVERY CRAFT) (LDW)

21   23   26–28

Displacement, tonnes: 11 full load
Dimensions, metres (feet): 11 × 3 × 0.8 (36.1 × 9.8 × 2.6)
Speed, knots: 30. **Range, n miles:** 200 at 20 kt
Complement: 3
Machinery: 2 GM 6V-92TA diesels; 520 hp (388 kW) sustained; 2 shafts
Guns: 2—12.7 mm MGs. 2—7.62 mm MGs.
Radars: Surface search: LN66; I-band.

Comment: Ordered from Uniflite, Washington in 1982. GRP construction painted black. There is a strong underwater team in the Egyptian Navy which is also known to use commercial two-man underwater chariots. Based at Abu Qir. 26 and 28 are not fully operational and others of the class are in various states of repair. RIBs are also in service.

SEAFOX   1999 / 0056917

9 VYDRA CLASS (LCU)

330   332   334   336   338   340   342   344   346

Displacement, tonnes: 432 standard; 610 full load
Dimensions, metres (feet): 54.8 × 7.7 × 2 (179.8 × 25.3 × 6.6)
Speed, knots: 11. **Range, n miles:** 2,500 at 10 kt
Complement: 20

Military lift: 200 troops; 250 tons.
Machinery: 2 Type 3-D-12 diesels; 600 hp(m) (440 kW) sustained; 2 shafts
Guns: 2 or 4—37 mm/63 (1 or 2 twin) (may be fitted).
Radars: Navigation: Decca; I-band.

Comment: Built in late 1960s, transferred from USSR 1968–69. For a period after the Israeli war of October 1973 several were fitted with rocket launchers and two 37 or 40 mm guns, some of which have now been removed. All still in service.

VYDRA 346   10/2009*, US Navy / 1529008

MINE WARFARE FORCES

3 T 43 CLASS (MINESWEEPERS—OCEAN) (MSO)

DAQHILIYA 507   SINAI 513   ASSIYUT 516

Displacement, tonnes: 589 full load
Dimensions, metres (feet): 58 × 8.4 × 2.1 (190.3 × 27.6 × 6.9)
Speed, knots: 15
Range, n miles: 3,000 at 10 kt
Complement: 65

Machinery: 2 Kolomna Type 9-D-8 diesels; 2,000 hp(m) (1.47 MW) sustained; 2 shafts
Guns: 4—37 mm/63 (2 twin); 160 rds/min to 9 km (5 n miles); weight of shell 0.7 kg. 8—12.7 mm (4 twin) MGs.
Mines: Can carry 20.
Radars: Navigation: Don 2; I-band.
Sonars: Stag Ear; hull-mounted; active search; high frequency.

Comment: Delivered in the early 1970s from the USSR. Others of the class have been sunk or used as targets or cannibalised for spares. The plan to fit them with VDS sonars and ROVs has been shelved.

DAQAHLIYA   3/2000 / 0103745

IHS Jane's Fighting Ships 2014-2015   © 2014 IHS

Mine warfare forces — Training ships < **Egypt** 225

2 SWIFTSHIPS TYPE (ROUTE SURVEY VESSELS) (MSI)

Name	No	Builders	Commissioned
SAFAGA	RSV 1 (ex-610)	Swiftships	1 Oct 1994
ABU EL GHOSON	RSV 2 (ex-613)	Swiftships	1 Oct 1994

Displacement, tonnes: 168 full load
Dimensions, metres (feet): 27.4 × 7.6 × 2.4 (89.9 × 24.9 × 7.9)
Speed, knots: 12. **Range, n miles:** 1,500 at 10 kt
Complement: 16 (2 officers)

Machinery: 2 MTU 12V 183 TA61 diesels; 928 hp(m) (682 kW); 2 shafts; bow thruster; 60 hp(m) (44 kW)
Guns: 1—12.7 mm MG.
Radars: Navigation: Furuno 2020; I-band.
Sonars: EG & G side scan; active; high frequency.

Comment: Route survey vessels ordered from Swiftships in November 1990 and delivered in September 1993. Unisys improved SYQ-12 command system. Provision for both shallow and deep towed bodies. The names have been taken from the obsolete K 8 class.

ABU EL GHOSON   3/2006, M Declerck / 1167106

2 OSPREY CLASS (MINEHUNTERS—COASTAL) (MHC)

Name	No	Builders	Launched	Commissioned
AL SIDDIQ (ex-Cardinal)	521 (ex-MHC 60)	Intermarine, Savannah	9 Mar 1996	18 Oct 1997
AL FAROUK (ex-Raven)	524 (ex-MHC 61)	Intermarine, Savannah	28 Sep 1996	5 Sep 1998

Displacement, tonnes: 945 full load
Dimensions, metres (feet): 57.2 × 11 × 2.9 (187.7 × 36.1 × 9.5)
Speed, knots: 13
Range, n miles: 1,500 at 10 kt
Complement: 51 (5 officers)

Machinery: 2 Isotta Fraschini ID 36 SS 8V AM diesels; 1,600 hp(m) (1.18 MW) sustained; 2 Voith-Schneider props; 3 Isotta Fraschini ID 36 diesel generators; 984 kW
Guns: 2—12.7 mm MGs.
Physical countermeasures: MCM: Alliant SLQ-48 mine neutralisation system ROV (with 1,070 m cable). Degaussing DGM-4.
Radars: Surface search: Raytheon SPS-64(V)9; I-band.
Navigation: R41XX; I-band.
Sonars: Raytheon/Thomson Sintra SQQ-32(V)3; VDS; active minehunting; high frequency.
Combat data systems: Unisys SYQ 13 and SYQ 109; integrated combat and machinery control system. USQ-119E(V), UHF Dama, and OTCIXS provide GCCS connectivity.

Programmes: Original design contract for Lerici-class minehunters was awarded in August 1986 to Intermarine USA which built eight of the 12 ships of the class for the US Navy. Transferred to Egypt on 7 January 2007 and recommissioned on 28 October 2007.
Structure: Construction is of monocoque GRP throughout hull, with frames eliminated. Main machinery is mounted on GRP cradles and provided with acoustic enclosures. SQQ-32 is deployed from a central well forward. Fitted with Voith cycloidal propellers which eliminate need for forward thrusters during station keeping.

AL SIDDIQ and AL FAROUK   3/2007, Paul Daly / 1167736

4 YURKA CLASS (MINESWEEPERS—OCEAN) (MSO)

GIZA 530   ASWAN 533   QINA 536   SOHAG 539

Displacement, tonnes: 549 full load
Dimensions, metres (feet): 52.4 × 9.4 × 2.6 (171.9 × 30.8 × 8.5)
Speed, knots: 17
Range, n miles: 1,500 at 12 kt
Complement: 45

Machinery: 2 Type M 503 diesels; 5,350 hp(m) (3.91 MW) sustained; 2 shafts
Guns: 4 USSR 30 mm/65 (2 twin); 500 rds/min to 5 km (2.7 n miles); weight of shell 0.54 kg.
Mines: Can lay 10.
Radars: Navigation: Don; I-band.
Sonars: Stag Ear; hull-mounted; active search; high frequency.

Comment: Steel-hulled minesweepers transferred from the USSR in 1969. Built 1963–69. Egyptian Yurka class do not carry Drum Tilt radar and have a number of ship's-side scuttles. The plan to equip them with VDS sonar has been shelved. At least one operates an ROV.

SOHAG   3/2007, Marco Ghiglino / 1166525

3 SWIFTSHIPS TYPE (COASTAL MINEHUNTERS) (MHC)

Name	No	Builders	Launched	Commissioned
DAT ASSAWARI	542 (ex-CMH 1)	Swiftships, Morgan City	4 Oct 1993	13 Jul 1997
NAVARIN	545 (ex-CMH 2)	Swiftships, Morgan City	13 Nov 1993	13 Jul 1997
BURULLUS	548 (ex-CMH 3)	Swiftships, Morgan City	4 Dec 1993	13 Jul 1997

Displacement, tonnes: 206 full load
Dimensions, metres (feet): 33.8 × 8.2 × 2.3 (110.9 × 26.9 × 7.5)
Speed, knots: 12.4. **Range, n miles:** 2,000 at 10 kt
Complement: 25 (5 officers)

Machinery: 2 MTU 12V 183 TE61 diesels; 1,068 hp(m) (786 kW); 2 Schottel steerable props; 1 White Gill thruster; 300 hp (224 kW)
Guns: 2—12.7 mm MGs.
Radars: Navigation: Sperry; I-band.
Sonars: Thoray/Thomson Sintra TSM 2022; hull-mounted; active minehunting; high frequency.

Comment: MCM vessels with GRP hulls ordered from Swiftships in December 1990 with FMS funding. First one acceptance trials in June 1994 and completion in August. Fitted with a Unisys command data handling system which is an improved version of SYQ-12. GPS and line of sight navigation system. Dynamic positioning. A side scan sonar body and Gaymarine Pluto ROV can be streamed from a deck crane. Portable decompression chamber carried. Two delivered 29 November 1995 and the third in April 1996. All were finally commissioned after delays caused by problems with the minehunting equipment.

NAVARIN   9/2010, Bob Fildes / 1406035

TRAINING SHIPS

Notes: (1) *Al Kousser* P 91 is a 1,000 ton vessel belonging to the Naval Academy. *Intishat* is a 500 ton training ship. Pennant number 160 is a 41 m USSR Sekstan class used as a cadet training ship. Two YSB training craft acquired from the US in 1989. A 3,300 ton training ship *Aida IV* presented by Japan in 1988 for delivery in March 1992 belongs to the Arab Maritime Transport Academy.
(2) A campaign to transfer the Black Swan-class sloop *Tariq*, formerly HMS *Whimbrel*, to Liverpool, to become a floating memorial to the Battle of the Atlantic, was unsuccessful.

1 Z CLASS (AXT)

Name	No	Builders	Laid down	Launched	Commissioned
EL FATEH (ex-Zenith; ex-Wessex)	921	Wm Denny & Bros Ltd, Dumbarton	19 May 1942	5 Jun 1944	22 Dec 1944

Displacement, tonnes: 1,758 standard; 2,616 full load
Dimensions, metres (feet): 110.6 × 10.9 × 4.9 (362.9 × 35.8 × 16.1)
Speed, knots: 24. **Range, n miles:** 2,800 at 20 kt
Complement: 186

Machinery: 2 Admiralty boilers; 2 Parsons turbines; 40,000 hp (30 MW); 2 shafts
Radars: Air/surface search: Marconi SNW 10; D-band.
Navigation: Racal Decca 916; I-band.
Fire control: Marconi Type 275; F-band.

Programmes: Purchased from the UK in 1955.
Operational: Used primarily for harbour training, and the intention is to keep the ship in service. Last seen at sea in 1994. The last survivor of its class, the ship may be preserved as a museum. The ship has been disarmed.

EL FATEH   3/2006, M Declerck / 1167102

IHS Jane's Fighting Ships 2014-2015

230 El Salvador > Patrol forces

1 SWIFTSHIPS 65 ft CLASS (COASTAL PATROL CRAFT) (PB)

PM 10 (ex-GC 10)

Displacement, tonnes: 37 full load
Dimensions, metres (feet): 20 × 6 × 1.5 (65.6 × 19.7 × 4.9)
Speed, knots: 23
Range, n miles: 600 at 18 kt
Complement: 6
Machinery: 2 Detroit 12V-71TA diesels; 840 hp (626 kW) sustained; 2 shafts
Guns: 1 Oerlikon 20 mm. 1 or 2—12.7 mm MGs. 1—81 mm mortar.
Radars: Surface search: Furuno; I-band.

Comment: Aluminium hull. Delivered by Swiftships, Morgan City 14 June 1984. Was laid up for a time in 1989–90 but became operational again in 1991. Refitted in 1996.

PM 10   6/2011, El Salvador Navy / 1406581

4 TYPE 44 CLASS (PBI)

PRM 01–04

Displacement, tonnes: 18 full load
Dimensions, metres (feet): 13.5 × 3.9 × 1.1 (44.3 × 12.8 × 3.6)
Speed, knots: 14
Range, n miles: 215 at 10 kt
Complement: 3
Machinery: 2 Detroit 6V-38 diesels; 185 hp (136 kW); 2 shafts

Comment: Ex-USCG craft similar to those transferred to Uruguay. Operational status doubtful.

PRM 04   11/2001, Julio Montes / 0130482

8 AIR PATROL BOATS (PBI)

PFR 1–8

Comment: Purchased in Miami for SAR on inland waters.

PFR 04   4/2005, Julio Montes / 1166720

9 PROTECTOR CLASS (RIVER PATROL CRAFT) (PBR)

PC 01–09

Displacement, tonnes: 9 full load
Dimensions, metres (feet): 12.3 × 4 × 0.4 (40.4 × 13.1 × 1.3)
Speed, knots: 28
Range, n miles: 350 at 20 kt
Complement: 4
Machinery: 2 Caterpillar 3208TA diesels; 680 hp (507 kW) sustained; 2 shafts
Guns: 2—12.7 mm MGs. 2—7.62 mm MGs.
Radars: Surface search: Furuno 3600; I-band.

Comment: Ordered in December 1987 from SeaArk Marine (ex-MonArk). Four delivered in December 1988 and four in February and March 1989. Seven reported operational, one in maintenance and one non-operational.

PC 03   3/2006, Julio Montes / 1166721

2 MERCOUGAR INTERCEPT CRAFT (PBR)

PA 01   **PA 02**

Comment: Two remaining of five 40 ft craft delivered by Mercougar in 1988. Powered by two Ford Merlin diesels; 600 hp (448 kW) giving speeds of up to 40 kt and range of 556 km (300 n miles). Radar fitted.

PA 02   5/2001, Julio Montes / 0109938

8 RODMAN 890 (PBR)

Displacement, tonnes: 3 full load
Dimensions, metres (feet): 8.9 × 3 × 0.8 (29.2 × 9.8 × 2.6)
Speed, knots: 28
Range, n miles: 150 at 25 kt
Complement: 3
Machinery: 2 Volvo diesels; 300 hp(m) (220 kW); 2 shafts
Guns: 1—7.62 mm MG.
Radars: Surface search: I-band.

Comment: Eleven craft delivered by Rodman in 1998. Eight remaining operational craft were transferred to the navy in 2010.

RODMAN 890 (Police colours)   6/1998, Rodman / 0576109

IHS Jane's Fighting Ships 2014-2015   © 2014 IHS

4 DEFENDER CLASS (RESPONSE BOATS) (PBF)

PA 06–09

Displacement, tonnes: 3 full load
Dimensions, metres (feet): 10.5 × 3.05 × 0.8 (34.4 × 10.0 × 2.6)
Speed, knots: 60
Complement: 4
Machinery: 3 outboard motors; 825 hp (615 kW)
Guns: 2 — 12.7 mm MGs.
Radars: To be announced.

Comment: High-speed inshore patrol craft of aluminium construction and foam collar built by SAFE Boats International, Port Orchard, Washington. Four were delivered in 2011.

JUSTICE CLASS — 10/2011, El Salvador Navy / 1406580

AUXILIARIES

4 LCM 8 CLASS

BD 02 (ex-LD 02)   BD 04 (ex-LD 04)   BD 05 (ex-LD 05)   +1

Displacement, tonnes: 46 full load
Dimensions, metres (feet): 21.5 × 4.6 × 1.6 (70.5 × 15.1 × 5.2)
Speed, knots: 15
Complement: 6
Machinery: 2 Detroit 12V 71TA diesels; 840 hp (626 kW) sustained; 2 shafts
Guns: 2 — 12.7 mm MGs. 2 — 7.62 mm MGs.
Radars: Navigation: Furuno; I-band.

Comment: First one delivered by SeaArk Marine in January 1987, second pair in May 1996. A fourth was donated by the United States in 2010.

PA 07 — 6/2011, El Salvador Navy / 1406579

6 BOSTON WHALER JUSTICE CLASS (PBF)

Displacement, tonnes: 3.8 full load
Dimensions, metres (feet): 9.8 × 3.09 × 0.55 (32.2 × 10.1 × 1.8)
Speed, knots: 50
Complement: 4
Machinery: 2 Mercury outboard motors; 600 hp (447 kW)
Guns: 1 — 7.62 mm MG.
Radars: To be announced.

Comment: High-speed inshore patrol craft of glass fibre construction built by Boston Whaler, Florida, US. Four delivered in 2007 and two further in 2011.

BD 04 — 6/2003, El Salvador Navy / 0568336

Equatorial Guinea

Country Overview

The Republic of Equatorial Guinea became independent in 1968 as a federation of the two former Spanish provinces of Fernando Po and Río Muni. It became a unitary state in 1973. Located in west Africa, the country has an overall area of 10,831 square miles and includes a mainland section which is bordered to the north by Cameroon and to the east and south by Gabon. It has a 160 n mile coastline with the Gulf of Guinea in which lie the islands of Bioko (formerly Fernando Po), Annobón, Corisco, Elobey Grande and Elobey Chico. The administrative capital on the mainland is Bata while Malabo, on the north coast of Bioko, is capital of the republic, largest city and prinicpal port. Territorial waters (12 n miles) are claimed. A 200 n mile Exclusive Economic Zone (EEZ) has been claimed but the boundaries have not been agreed.

Personnel

2014: 400 (approx) officers and men

Bases

Malabo, Bata.

PATROL FORCES

Notes: (1) The Lantana 68 class *Isla de Bioko* and 20 m patrol craft *Riowele* are believed to be non-operational.
(2) The sale of a 103 m Barroso-class corvette to Equatorial Guinea was announced by Brazilian President Lula in July 2010. If the sale goes ahead, it would represent a major upgrade in capability.

1 DAPHNE CLASS (PB)

Name	No	Builders	Commissioned
URECA (ex-*Nymfen*)	P 31 (ex-P 535)	Royal Dockyard, Copenhagen	4 Oct 1963

Displacement, tonnes: 173 full load
Dimensions, metres (feet): 36.9 × 6.1 × 2.0 (121.1 × 20.0 × 6.6)
Speed, knots: 20
Complement: 23
Machinery: 3 diesels; 3 shafts
Guns: 2 — 14.5 mm.
Radars: Navigation: Furuno; I-band.

Comment: Acquired in 1999.

2 ZHUK (GRIF) CLASS (PROJECT 1400M) (PB)

MIGUEL ELA EDJODJOMO LP 039   HIPOLITO MICHA LP 041

Displacement, tonnes: 40 full load
Dimensions, metres (feet): 24 × 5 × 1.2 (78.7 × 16.4 × 3.9)
Speed, knots: 30
Range, n miles: 1,100 at 15 kt
Complement: 13 (1 officer)
Machinery: 2 diesels; 2 shafts
Guns: 2 — 14.5 mm (twin, fwd) MGs. 1 — 12.7 mm (aft) MG.
Radars: Surface search: Furuno; I-band.

Comment: Reported to have been transferred from Ukraine in 2000.

2 KALKAN (PROJECT 50030) M CLASS
(INSHORE PATROL CRAFT) (PBR)

GASPAR OBIANG ESONO 43   FERNANDO NUARA ENGONDA 45

Displacement, tonnes: 9 full load
Dimensions, metres (feet): 11.6 × 3.3 × 0.6 (38.1 × 10.8 × 2.0)
Speed, knots: 34
Complement: 2
Machinery: 1 Type 475K diesel; 496 hp (370 kW); 1 waterjet

Comment: Built by Morye Feodosiya and reportedly acquired in 2001.

KALKAN CLASS (Ukraine colours) — 6/2003, Morye / 0572655

2 OPV 62 CLASS (PATROL SHIPS) (PBO)

KIE-NTEM   LITORAL

Displacement, tonnes: 470 full load
Dimensions, metres (feet): 61.7 × 7.62 × 2.77 (202.4 × 25.0 × 9.1)
Speed, knots: 32
Range, n miles: 3,200 at 12 kt
Complement: 35

Machinery: 4 MTU diesels; 4 shafts
Guns: 1 Typhoon 23 mm.
Radars: Surface search: To be announced.
Navigation: To be announced.
Electro-optic systems: Rafael Toplite.
Helicopters: Platform for 1 small.

Comment: Construction of both vessels, which appear to be based on the SAAR 4 design, began at Israel Shipyards, Haifa, in late 2009. The ships are equipped with a small crane for handling a RHIB and are capable of operating a small helicopter. Sea trials of the first vessel began in October 2010 and delivery of both ships was made on 22 February 2011.

KIE-NTEM   4/2011, Israel Shipyards / 1394740

2 PV50M CLASS (PATROL SHIP) (PBO)

ESTUARIO DE MUNI (ex-*Dolphin V1*) –   CABO SAN JUAN (ex-*Dolphin V2*) P 400

Dimensions, metres (feet): 49.1 × 9.0 × ? (161.1 × 29.5 × ?)
Speed, knots: 25
Range, n miles: 3,000 at 14 kt
Complement: 30

Machinery: 2 Caterpillar 3516C diesels; 7,302 hp (5.37 MW); 2 shafts
Guns: 2 – 30 mm (twin) AK 230. 4 – 25 mm (2 twin).
Radars: Surface search: To be announced.
Navigation: To be announced.
Helicopters: Platform for 1 small.

Comment: Both vessels built at MTG-Dolphin, Varna, Bulgaria and possibly fitted out in Ukraine. Both were completed in 2007 and entered service in 2008.

ESTUARIO DE MUNI   6/2008, USCG / 1455782

2 SHALDAG MK II CLASS (PBF)

ISLA DE CORISCO   ISLA DE ANNOBON

Displacement, tonnes: 58 full load
Dimensions, metres (feet): 24.8 × 6.0 × 1.2 (81.4 × 19.7 × 3.9)
Speed, knots: 45
Range, n miles: 700 at 32 kt
Complement: 10 (2 officers)
Machinery: 2 MTU diesels; 4,570 hp(m) (3.36 MW); 2 waterjets
Guns: 2 – 12.7 mm MGs.
Radars: Navigation: Furuno; I-band.

Comment: Two craft reportedly delivered from Israel in 2005.

1 PV88 CLASS (PSO)

Name	No	Builders	Laid down	Launched	Commissioned
BATA (ex-*Kasatka*)	047	MTG Dolphin, Varna	Jun 2010	Jan 2011	3 Feb 2012

Displacement, tonnes: 1,360 full load
Dimensions, metres (feet): 87.3 × 11.8 × 4.05 (286.4 × 38.7 × 13.3)
Speed, knots: 25
Range, n miles: 3,500 at 12 kt
Complement: 35 + 20 spare berths

Machinery: 2 Caterpillar C280-16 diesels; 13,760 hp (8.6 MW); 2 shafts
Guns: 1 – 76 mm AK 176. 2 – 30 mm AK 630. 2 Katran-M.
Radars: Surface search: Kvant Delta; D-band.
Navigation: Cascade; I-band.
Helicopters: Platform for 1 medium.

Comment: The ship is reported to be to an OPV88 design marketed by Fast Craft Naval Supplies. The ship was built in Bulgaria and possibly fitted out in Ukraine. The ship is of steel construction and is equipped with a flight deck (but no hangar). A payload and stern ramp for two RHIBs aft is incorporated in the stern.

BATA   6/2012, Mazumar Collection / 1455780

AUXILIARIES

1 TRANSPORT SHIP (AK)

Name	No	Builders	Laid down	Commissioned
CAPITAN DAVID EYAMA ANGUE OSA (ex-*Salamandra*)	106	Linhai Hangchang Shipbuilding	Aug 2008	Jun 2009

Measurement, tonnes: 2,800 gt
Dimensions, metres (feet): 91.45 × 14.7 × 3.7 (300 × 48.2 × 12.1)

Machinery: 2 Caterpillar 3516B-HD diesels; 4,584 hp (3.37 MW); 2 shafts
Guns: 1 – 76 mm AK 176. 2 – 25 mm. 2 Katran-M.
Radars: Surface search: To be announced.
Navigation: To be announced.

Comment: The ship is of steel construction and was built in China as a Ro-Ro cargo ship. Subsequently she was reportedly fitted out in Ukraine.

CAPITAN DAVID EYAMA ANGUE OSA   6/2012, Mazumdar Collection / 1455781

Eritrea

Country Overview

A British protectorate from 1941, The State of Eritrea was federated with Ethiopia in 1952 and incorporated as a province in 1962. The following war of liberation culminated in independence in 1993. The country is situated on the southwest shore of the Red Sea with which it has a 621 n mile coastline with an area of 46,842 square miles, it is bordered to the north by Sudan, to the west by Ethiopia and to the south by Djibouti. The largest town and capital is Asmara and the principal port is Massawa. There are no claims to maritime jurisdiction over territorial seas or Exclusive Economic Zone (EEZ).

All vessels of the former Ethiopian Navy were put up for sale at Djibouti from 16 September 1996. All were either taken over by Eritrea, sold to civilian firms or scrapped.

Headquarters Appointments

Commander Eritrean Navy:
    Major General Hummed Mohammed Karikare
Chief of Staff:
    Brigadier General Fitsum Gebrehiwet

Personnel

2014: 1,100 including 500 conscripts

Bases

Assab, Massawa, Dahlak.

PATROL FORCES

Notes: (1) There are also about 50 rigid raiding craft.
(2) The Osa II class FMB 161 is reported non-operational.

4 SUPER DVORA CLASS (FAST ATTACK CRAFT—GUN) (PTF)

P 101–104

Displacement, tonnes: 59 full load
Dimensions, metres (feet): 25 × 5.7 × 0.9 (82 × 18.7 × 3.0)
Speed, knots: 40. **Range, n miles:** 1,200 at 17 kt
Complement: 10 (1 officer)

Machinery: 2 MTU 8V 396 TE 94 diesels; 3,046 hp(m) (2.24 MW); 2 shafts; ASD 14 surface drives
Guns: 2—23 mm (twin). 2—12 mm MGs.
Depth charges: 1 rail.
Radars: Surface search: Raytheon; I-band.
Weapon control systems: Optronic sight.

Comment: Built by Israel Aircraft Industries and delivered from July 1993 to a modified Super Dvora design. The original order may have been for six of the class. All are based at Massawa and all are active.

SUPER DVORA P 104 — 6/2000, Eritrean Navy / 0103787

3 SWIFTSHIPS 105 ft CLASS (LARGE PATROL CRAFT) (PB)

P 151–153

Displacement, tonnes: 120 full load
Dimensions, metres (feet): 32 × 7.2 × 2 (105 × 23.6 × 6.6)
Speed, knots: 30. **Range, n miles:** 1,200 at 18 kt
Complement: 21
Machinery: 2 MTU MD 16V 538 TB90 diesels; 6,000 hp(m) (4.41 MW) sustained; 2 shafts
Guns: 4 Emerlec 30 mm (2 twin) (P 151); 600 rds/min to 6 km (3.3 n miles); weight of shell 0.35 kg. 4—23 mm/60 (2 twin) (P 152/153). 2—12.7 mm (twin).
Radars: Surface search: Decca RM 916; I-band.

Comment: Six ordered in 1976 of which four were delivered in April 1977 before the cessation of US arms sales to Ethiopia. Built by Swiftships, Louisiana. One deserted to Somalia and served in that Navy for a time. Based at Massawa and in reasonable condition. All are active.

P 153 — 1/1998 / 0017825

5 BATTALION 17 (PBF)

P 084–088

Displacement, tonnes: 36 full load
Dimensions, metres (feet): 17.05 × 5.2 × 1.6 (55.9 × 17.1 × 5.2)
Speed, knots: 35.2. **Range, n miles:** 680 at 30 kt
Complement: 9
Machinery: 2 MTU 12V 183 TE 92 diesels; 1,970 hp (1.47 MW); 2 shafts
Guns: 2—14.5 mm MGs (1 twin).
Radars: Surface search: Raytheon; I-band.

Comment: Australian design craft built by Harena Boat Yard at Assab, Eritrea. Five craft delivered in 2000 with possible further orders since then.

P 086 — 6/2000, Eritrean Navy / 0103788

AMPHIBIOUS FORCES

Notes: (1) Two obsolete ex-USSR T4 LCUs (LST-63 and 64) are in harbour service at Massawa.
(2) The passenger vessel *Harat* arrived at Massawa on 10 February 2006. The 118 m vessel has a helicopter landing deck and accommodation for 2,800. Inspected by the commander of the navy on arrival, the ship may have a military role.

1 ASHDOD CLASS (LST)

P 63 (ex-302)

Displacement, tonnes: 406 standard; 742 full load
Dimensions, metres (feet): 62.7 × 10 × 1.8 (205.7 × 32.8 × 5.9)
Speed, knots: 10.5
Complement: 20
Machinery: 3 MWM diesels; 1,900 hp(m) (1.4 MW); 3 shafts
Guns: 2—23 mm (1 twin). 2—12.7 mm MGs.

Comment: Former Ethiopian commercial LST acquired from Israel in 1993, taken over by Eritrea in 1997 and subsequently transferred to the Navy. Reported operational.

P 63 (Israeli pennant number) — 1995, Eritrean Navy / 0103789

1 CHAMO CLASS (LST)

DENDEN 301

Displacement, tonnes: 898 full load
Dimensions, metres (feet): 60.2 × 12 × 1.44 (197.5 × 39.4 × 4.7)
Speed, knots: 10
Complement: 23
Machinery: 2 MTU 6V 396 TB 63; 1,350 hp(m) (1 MW); 2 shafts
Guns: 2—23 mm (1 twin); 2—12.7 mm MGs.

Comment: German built former Ethiopian commercial LST taken over by Eritrea in 1997 and subsequently transferred to the Navy. Reported operational.

Estonia

EESTI MEREVÄGI

Country Overview

The Republic of Estonia regained independence in 1991 after 51 years as a Soviet republic. Situated in northeastern Europe, the country includes more than 1,500 islands, the largest of which are Saaremaa and Hiiumaa. With an area of 17,462 square miles it has borders to the east with Russia and to the south with Latvia. It has a 750 n mile coastline with the Baltic Sea and Gulf of Finland. Tallinn is the capital, largest city and principal port. Territorial seas (12 n miles) are claimed but while it has claimed a 200 n mile Exclusive Economic Zone (EEZ), its limits have not been fully defined by boundary agreements.

The Navy was founded in 1918 and re-established on 22 April 1994. The Border Guard comes under the Ministry of Internal Affairs and is responsible for SAR and Pollution Prevention.

Headquarters Appointments

Chief of Staff:
Commander Jüri Saska

Personnel

(a) 2014: 404 (64 officers) including 178 conscripts
(b) 8–11 months' national service

Bases

Major: Miinisadam (Tallinn)
Minor: Kopli (Tallinn) (Border Guard)

PATROL FORCES

Notes: Following the decommissioning of the frigate *Admiral Pitka* in mid-2013, procurement of a corvette-sized replacement vessel is reported to be under consideration.

1 DIVING TENDER (YTD)

LOOD A 530 (ex-EVA 321)

Displacement, tonnes: 31 full load
Dimensions, metres (feet): 16.09 × 4.62 × 0.7 *(52.8 × 15.2 × 2.3)*
Speed, knots: 20
Complement: 5
Machinery: 2 MTU 12V 183TE72 diesels; 818 hp *(610 kW)*; 2 shafts
Radars: Navigation: I-band.

Comment: Used for harbour operations, navigational training, and diving support.

1 RIHTNIEMI CLASS (PC)

Name	No	Builders	Commissioned
RISTNA (ex-*Rihtniemi*)	P 422 (ex-51)	Rauma-Repola	21 Feb 1957

Displacement, tonnes: 91 standard; 112 full load
Dimensions, metres (feet): 31 × 5.6 × 1.8 *(101.7 × 18.4 × 5.9)*
Speed, knots: 15
Complement: 24 (4 officers)

Machinery: 2 MTU MB diesels; 2,500 hp(m) *(1.84 MW)*; 2 shafts; cp props
Guns: 2 ZU-23-2 (twin). 3—12.7 mm MGs.
A/S Mortars: 2 RBU 1200 fixed 5-tubed launchers.
Mines: Can lay mines.
Radars: Navigation: Decca 1226; I-band.
Sonars: Simrad; hull-mounted; active search and attack; high frequency.

Comment: *Suurop* (ex-*Rymättylä*) and *Ristna* transferred from Finland on 8 July 1999. Both ships subsequently decommissioned in 2005. *Ristna* re-activated in 2009.

LOOD 6/2013*, Estonian Navy / 1525601

RISTNA 8/2011*, Estonian Navy / 1486577

MINE WARFARE FORCES

2 LINDORMEN CLASS
(COASTAL MINELAYER) (MLC)

Name	No	Builders	Launched	Commissioned
TASUJA (ex-*Lindormen*)	A 432 (ex-N 43)	Svendborg Vaerft	7 Jun 1977	16 Feb 1978
WAMBOLA (ex-*Lossen*)	A 433 (ex-N 44)	Svendborg Vaerft	11 Oct 1977	14 Jun 1978

Displacement, tonnes: 579 full load
Dimensions, metres (feet): 44.5 × 9.0 × 2.6 *(146 × 29.5 × 8.5)*
Speed, knots: 14
Complement: 27 (4 officers)
Machinery: 2 Frichs diesels; 1,600 hp *(1.2 MW)*; 2 shafts
Guns: 2—12.7 mm MGs.
Radars: Navigation: I-band.

Comment: Both former Danish minelayers. *Tasuja* was handed over on 12 April 2006. *Wambola* (ex-*Lossen*) was also procured as a civilian training ship for the Estonian Maritime Academy. She was handed over to the Estonian Navy in 2010 and a decision on her future is awaited.

TASUJA 9/2011, Per Körnefeldt / 1406446

3 SANDOWN CLASS (MINEHUNTERS) (MHC)

Name	No	Builders	Launched	Commissioned
ADMIRAL COWAN (ex-*Sandown*)	M 313 (ex-M 101)	Vosper Thornycroft, Woolston	16 Apr 1988	9 Jun 1989
SAKALA (ex-*Inverness*)	M 314 (ex-M 102)	Vosper Thornycroft, Woolston	27 Feb 1990	24 Jan 1991
UGANDI (ex-*Bridport*)	M 315 (ex-M 105)	Vosper Thornycroft, Woolston	20 Jul 1992	6 Nov 1993

Displacement, tonnes: 457 standard; 492 full load
Dimensions, metres (feet): 52.5 × 10.5 × 2.3 *(172.2 × 34.4 × 7.5)*
Speed, knots: 13; 6.5 electric drive. **Range, n miles:** 2,500 at 12 kt
Complement: 34 (5 officers) + 6 casualty berths

Machinery: 2 Paxman-Valenta 6RP200E/M diesels; 1,523 hp *(1.14 MW)* sustained; Voith-Schneider propulsion; 2 Schottel bow thrusters
Guns: 2 Sako 23 mm/87 (twin). 3—12.7 mm MGs.
Physical countermeasures: MCM Seafox C
Radars: Navigation: Kelvin Hughes Type 1007; I-band.
Sonars: Marconi Type 2093; VDS; VLF-VHF multifunction with 5 arrays; mine search and classification. Klein 5000 towed side-scan sonar.
Combat data systems: BAE Insyte NAUTIS M.

Programmes: Single-role minehunter originally designed for deep water operations and built by Vosper Thornycroft for the UK Royal Navy. All three ships withdrawn from RN service following force-level reductions announced in 2004. Preliminary agreement for the regeneration and transfer of the three ships made between the UK and Estonian governments in late 2005. Following a letter of intent on 11 April 2006, a final agreement was signed on 14 September 2006. *Admiral Cowan* handed over on 26 April 2007 and *Sakala* on 28 January 2008. *Ugandi* was handed over in January 2009 and undertakes navigation and training roles.
Modernisation: The modernisation package upgraded the two operational ships to the similar equipment standards as those in service in the RN. Principal components include Sonar Type 2093, Seafox C submersibles, Drumgrange Precise Fixing System and NAUTIS combat data system.
Structure: GRP hull. Combines vectored thrust units with bow thrusters and remote-control submersibles. The sonar is deployed from a well in the hull.

ADMIRAL COWAN 6/2013*, Michael Nitz / 1525524

POLICE AND BORDER GUARD BOARD

Notes: (1) *Director General:* Colonel Raivo Küüt
(2) The Aviation Group was formed in February 1993 and includes two L-410 maritime patrol aircraft and one Cessna 172 aircraft; three Agusta-Westland 139 and one Enström 480 helicopters.

L-410   3/2012, Hartmut Ehlers / 1483028

AW 139   3/2012, Hartmut Ehlers / 1483027

1 SILMÄ CLASS (LARGE PATROL CRAFT) (PBO)

Name	No	Builders	Commissioned
KOU (ex-*Silmä*)	PVL 107	Laivateollisuus, Turku	19 Aug 1963

Displacement, tonnes: 539 full load
Dimensions, metres (feet): 48.3 × 8.3 × 4.3 *(158.5 × 27.2 × 14.1)*
Speed, knots: 11
Complement: 10
Machinery: 1 Werkspoor diesel; 1,800 hp(m) *(1.32 MW)*; 1 shaft
Guns: 2 – 14.5 mm (twin). 1 – 7.62 mm MG.
Radars: Surface search/navigation: Transas Navi-radar 4000; E/F/I-bands.

Comment: Transferred from Finland Frontier Guard in January 1995.

KOU   4/2007, E & M Laursen / 1305004

1 PIKKER CLASS (COASTAL PATROL CRAFT) (PB)

Name	No	Builders	Launched	Commissioned
PIKKER	PVL 103	AS Tallinna Meretehas	23 Dec 1995	Apr 1996

Displacement, tonnes: 91 full load
Dimensions, metres (feet): 30 × 5.8 × 1.5 *(98.4 × 19.0 × 4.9)*
Speed, knots: 20
Complement: 7
Machinery: 2 MTU 10V 2000 CRM 72 diesels; 2,415 hp(m) *(1.8 MW)*; 2 shafts
Guns: 1 – 14.5 mm MG.
Radars: Surface search: Transas Navi-radar 4000; E/F/I-bands.

Comment: Steel hull and superstructure. Carries a RIB with a hydraulic launch crane aft.

PIKKER   6/2003, Hartmut Ehlers / 0561494

1 BALTIC 24 (PATROL CRAFT) (WPB)

Name	No	Builders	Commissioned
VALVE	PVL 112	Baltic Workboats, Talinn	2010

Displacement, tonnes: 51 full load
Dimensions, metres (feet): 23.7 × 5.65 × 1.35 *(77.8 × 18.5 × 4.4)*
Speed, knots: 21
Range, n miles: 500 at 15 kt
Complement: 6
Machinery: 2 MTU 10V2000 M 92 diesels; 2,722 hp *(2.03 MW)*; 2 shafts
Guns: 1 – 12.7 mm MG (fitted for).
Radars: Surface search/navigation: JMA 5300; I-band.

Comment: Contract with Baltic Workboats A/S, Estonia in 2009 for the construction of one 24 m workboat. Of aluminium construction, the craft is equipped with an aft launch/recovery ramp for a 5 m RHIB. Delivered in 2010.

VALVE   6/2010, Estonian Police and Border Guard / 1366606

1 POLLUTION CONTROL CRAFT

KATI PVL 202 (ex-KBV 003)

Displacement, tonnes: 471 full load
Dimensions, metres (feet): 40 × 6.6 × 3.3 *(131.2 × 21.7 × 10.8)*
Speed, knots: 12
Complement: 8
Machinery: 1 Wärtsilä SF16RS diesel; 750 hp *(660 kW)*; 1 shaft
Radars: Surface search/navigation: Transas Navi-radar 4000; E/F/I-bands.

Comment: Built by Rosslauer Shipyard, Germany, as a trawler in 1966. Taken over by the Swedish Coast Guard and rebuilt as a pollution control craft 1972–73. Transferred to Estonia in 2002.

KATI   6/2010, Estonian Police and Border Guard / 1366605

1 KBV 236 CLASS (PB)

PVK 003 (ex-KBV 246)

Displacement, tonnes: 17 full load
Dimensions, metres (feet): 19.2 × 4 × 1.3 *(63 × 13.1 × 4.3)*
Speed, knots: 22
Complement: 5
Machinery: 2 Volvo Penta TAMD120A diesels; 700 hp(m) *(515 kW)*; 2 shafts
Guns: 1 – 7.62 mm MG.

Comment: Transferred on 6 December 1993. Former Swedish Coast Guard vessel built in 1970. Similar craft to Latvia and Lithuania. KBV 001 decommissioned in 2013.

PVK 001   6/2011, Estonian Police and Border Guard / 1406443

236 Estonia › Police and border guard board

13 INSHORE PATROL CRAFT (PBI)

| PVK 010 | PVK 020–021 | MP 15 | MP 37 |
| PVK 012 | PVK 025 | MP 29 | MP 39–43 |

Comment: PVK 010 is a 15 m patrol craft built in 1997. PVK 025 is a 19 m ex-Swedish craft (KBV 275) acquired in 1997. Further craft under 12 m have numbers PVK 012, 020, and 021. There are also eight Boomeranger 10 m RIBs: MP 15, 29, 37, and 39–43.

PVK 010   6/2002, Baltic Ship Repairers / 0526817

MP 39   6/2011, Estonian Police and Border Guard / 1406442

1 GRIFFON 2000 TDX MK II (HOVERCRAFT) (UCAC)

PVH 1

Displacement, tonnes: 7 full load
Dimensions, metres (feet): 11.7 on cushion × ? (38.4 × ?)
Speed, knots: 33
Range, n miles: 300 at 25 kt
Complement: 2
Military lift: 16 troops or 2 tons
Machinery: 1 Deutz BF8L 513 diesel; 320 hp (293 kW) sustained
Guns: 1—7.62 mm MG.

Comment: Similar to craft supplied to Finland. Acquired in 1999.

PVH 1   6/2010, Estonian Police and Border Guard / 1366601

1 GRIFFON 1000 (HOVERCRAFT) (UCAC)

PVH 2

Displacement, tonnes: 3.5 full load
Dimensions, metres (feet): 9 on cushion × ? (29.5 × ?)
Speed, knots: 36
Complement: 2
Machinery: 1 Deutz BF6L 142 diesel; 190 hp (142 kW)

Comment: Acquired in 2007.

PVH 2   6/2010, Estonian Police and Border Guard Board / 1366604

1 MULTIPURPOSE PATROL SHIP (PSO)

KINDRAL KURVITS PVL 101

Displacement, tonnes: 1,200 full load
Dimensions, metres (feet): 63.9 × 10.2 × 4.2 (209.6 × 33.5 × 13.8)
Speed, knots: 15
Complement: 16
Machinery: 2 Wärtsilä 8L20 diesels; 4,290 hp (3.2 MW); 2 Scania DI 16 429 diesels; 1,150 hp (858 kW); 2 Rolls Royce azimuth thrusters; 1 Rolls Royce bow thruster

Comment: Based on the design of the Finnish Tursas class, the ship was ordered in late 2010 and was constructed by Finnish Shipyard Työvene Oy. Delivery was made on 3 August 2012. The ship operates as a border patrol vessel in the Baltic Sea with the capability to act as an oil-recovery and/or firefighting ship if required. The ship is equipped with a 7.5 m workboat and a RIB.

KINDRAL KURVITS   8/2012, Työvene / 1486485

5 BALTIC 101 INSHORE PATROL CRAFT (PBI)

MP 44–48

Displacement, tonnes: 5.6 full load
Dimensions, metres (feet): 10.8 × 3.56 × 0.47 (35.4 × 11.7 × 1.5)
Speed, knots: 45. **Range, n miles:** 300 at 15 kt
Complement: 6
Machinery: 2 Volvo Penta D6 diesels; 2 Rolls Royce waterjets

Comment: Built by Workboats A/S, Estonia. Aluminium construction. Delivered in 2010.

MP 46   6/2011, Police and Border Guard / 1406441

IHS Jane's Fighting Ships 2014-2015

MARITIME ADMINISTRATION
(EESTI VEETEEDE AMET (EVA))

Notes: The Maritime Administration (EVA) was re-established in 1990 and is responsible for hydrographic work, aids to navigation, ice-breaking and control of shipping. The main base is at Tallinn. Ships are painted with a blue hull and white superstructure and are as follows:
Tarmo, icebreaker built in 1963 and acquired from Finland in 1992. Fleet flagship.
Sektori, buoy tender and hydrographic vessel built in Finland in 1985
EVA 019, port control launch built in Estonia in 1997
EVA 021–026, control launches built 2004–08
EVA 300 (ex-*Tormilind*), hydrographic ship built in Russia in 1983
EVA 301, hydrographic ship built in Finland in 2007
EVA 302, buoy ship built in Finland in 2007
EVA 303 (ex-*Kaater*), buoy ship built in Poland in 1988
EVA 316 (ex-*Lonna*), buoy ship built in Finland in 1980
EVA 317–318, buoy ships built in Finland in 1994
EVA 320, hydrographic ship built in Finland in 1997
EVA 322, service launch built in Finland in 1997
EVA 324, service boat built in Finland in 2009
EVA 325, hydrographic ship built in Finland in 2002
EVA 326–328, service boats built in Finland in 2006–07
Jakob Prei, SWATH-hulled hydrographic ship built in Germany in 2012.

EVA 316   *6/2011, EVA* / 1406445

TARMO   *3/2012, Hartmut Ehlers* / 1483032

JAKOB PREI   *2/2012, Frank Findler* / 1483603

Falkland Islands

Country Overview

The Falkland Islands are a self-governing British Overseas Territory administered by a Governor and an elected Legislative Authority. Situated in the south Atlantic Ocean 323 n miles northeast of Cape Horn, approximately 200 islands are divided into two main groups on the east and west by the narrow Falkland Sound. The two largest islands are West Falkland Island (2,090 square miles) and East Falkland Island (2,610 square miles) on which the capital, largest town and principal port, Stanley, is situated. Territorial waters (12 n miles) are claimed as is a 200 n mile fishery zone.

Maritime Aircraft

There is one Pilatus Britten-Norman Defender unarmed maritime surveillance aircraft.

PATROL FORCES

Notes: The ex-Northern Lighthouse Board vessel *Pharos*, renamed *Pharos SG* has been on charter since November 2006, to the Government of South Georgia and the South Sandwich Islands as a fishery patrol and logistics support vessel. The Falkland Islands government continues to provide the fishery officer and support facilities.

1 FISHERY PATROL SHIP (PSO)

PROTEGAT (ex-*Sumiyoshi Maru 35*; ex-*Chokyo Moru 35*)

Measurement, tonnes: 1,207 gt
Dimensions, metres (feet): 70.17 × 10.6 × 5.6 (230.2 × 34.8 × 18.4)
Speed, knots: 13.5. **Range, n miles:** 19,000 at 10 kt
Complement: 16 + 5 spare berths

Machinery: 1 Niigata NHP30AH diesel; 1,800 hp *(1.3 MW)*; 1 shaft; 1 Kamome bow thruster; 200 hp *(150 kW)*
Guns: 1 — 12.7 mm MG.
Radars: Surface search/navigation: 1 JRC JMA 527; 1 Furuno FR 2135; 1 Furuno FR 1525; E/F/I-bands.

Comment: Former fishing vessel built by Miho Zoshenko KK-Shimizu in 1987. Converted for fishery protection duties and first chartered in May 2008. The current charter will expire in January 2017. Steel construction with bulbous bow.

PROTEGAT   *6/2010, Falkland Island Fisheries* / 1366491

Faroe Islands

Country Overview

The Faroe Islands are a self-governing island group that is an integral part of Denmark which retains control of foreign relations. Located in the North Atlantic Ocean, about midway between the Shetland Islands and Iceland, there are 18 islands, of which the most important are Østerø, Suderø, Sandø, Vagø, Bordø and Strømø, on which the capital and principal port, Tórshavn, is situated. Territorial waters (12 n miles) are claimed. A 200 n mile fishery zone has also been claimed although the limits have only been partly defined by boundary agreements.

The Faroe Islands Fishery Inspection Service comes under the Landsstyri which is the islands' local government. Vessels work closely with the Danish Navy.

Headquarters Appointments

Head of Coast Guard:
Captain Elmar Hojgaard

Personnel

2014: 60

Bases

Tórshavn (Isle of Streymoy)

COAST GUARD

Notes: There is also a 6 m inshore patrol vessel *Spogsvin* with a maximum speed of 9 kt.

1 PATROL SHIP (PSO)

BRIMIL

Displacement, tonnes: 2,032 full load
Dimensions, metres (feet): 59.96 × 12.6 × 4.3 *(196.7 × 41.3 × 14.1)*
Speed, knots: 17
Complement: 12 (3 divers) + 30 spare berths
Machinery: 2 Bergen diesels; 5,452 hp *(4.06 MW)*
Radars: Surface search: 2 Furuno.

Comment: Built for Faroese government as a patrol vessel by Myclebust Mek. Verksted, Norway. Entered service in April 2001.

1 PATROL SHIP (PBO)

TJALDRID

Displacement, tonnes: 660 full load
Dimensions, metres (feet): 42.1 × 10.02 × 3.2 *(138.1 × 32.9 × 10.5)*
Speed, knots: 14.5
Complement: 22 (4 divers)
Machinery: 2 MWM diesels; 2,400 hp(m) *(1.76 MW)*; 2 shafts
Guns: 1 Oerlikon 20 mm can be carried.
Radars: Surface search: Raytheon TM/TCPA; I-band.

Comment: Originally a commercial tug built in 1976 by Svolvaer, Verksted and acquired by the local government in 1987. The old 57 mm gun has been replaced. A decompression chamber can be carried.

BRIMIL   7/2008, Marco Ghiglino / 1353022

TJALDRID   12/1999, Faroes Coast Guard / 0080652

Fiji

Country Overview

A former British colony, the Republic of Fiji gained independence in 1970. Part of Melanesia, it is situated in the south Pacific Ocean some 972 n miles north of New Zealand and comprises more than 300 islands and islets, 100 of which are inhabited. The largest and most important of these are Viti Levu and Vanua Levu, which together contain more than 85 per cent of the total land area. To the southeast lie Taveuni, Kandavu, Koro and the Lau group while to the northwest lie Rotuma and the Yasawa group. The capital, largest town and principal port is Suva. An archipelagic state, territorial seas (12 n miles) are claimed. An Exclusive Economic Zone (EEZ) (200 n miles) is also claimed but limits have yet to be fully defined by boundary agreements.

Headquarters Appointments

Commander, Navy:
Commander John Fox

Personnel

2014: 300

Bases

RFNS *Viti*, at Togalevu (Training).
RFNS *Stanley Brown*.
Operation base at Walu Bay, Suva.
Forward base at Lautoka.

Prefix to Ships' Names

RFNS (Republic of Fiji naval ship)

PATROL FORCES

3 PACIFIC CLASS (LARGE PATROL CRAFT) (PB)

Name	No	Builders	Commissioned
KULA	201	Transfield Shipbuilding	28 May 1994
KIKAU	202	Transfield Shipbuilding	27 May 1995
KIRO	203	Transfield Shipbuilding	14 Oct 1995

Displacement, tonnes: 165 full load
Dimensions, metres (feet): 31.5 × 8.1 × 2.1 *(103.3 × 26.6 × 6.9)*
Speed, knots: 20
Range, n miles: 2,500 at 12 kt
Complement: 17 (4 officers)
Machinery: 2 Caterpillar 3516TA diesels; 2,820 hp *(2.09 MW)* sustained; 2 shafts
Guns: 1—20 mm Oerlikon. 2—12.7 mm MGs.
Radars: Surface search: Furuno; I-band.

Comment: Ordered in December 1992. These are hulls 17, 19 and 20 of the class offered by the Australian government under Defence Co-operation Programme. *Kikau* underwent a half-life refit at Gladstone in 2001 followed by *Kula* and *Kiro* in 2002. Following the decision by the Australian government to extend the Pacific Patrol Boat project until 2025, life extension refits was due for *Kikau* in 2011 and for *Kula* and *Kiro* in 2012. *Kula* completed refit at Walu Bay in October 2013.

KIRO
9/1998, van Ginderen Collection
0017831

2 COASTAL PATROL CRAFT (PB)

Name	No	Builders	Recommissioned
LEVUKA	101	Beaux's Bay Craft, Louisiana	22 Oct 1987
LAUTOKA	102	Beaux's Bay Craft, Louisiana	28 Oct 1987

Displacement, tonnes: 99 full load
Dimensions, metres (feet): 33.8 × 7.4 × 1.5 *(110.9 × 24.3 × 4.9)*
Speed, knots: 12
Complement: 12 (2 officers)
Machinery: 4 GM 12V-71TA diesels; 1,680 hp *(1.25 MW)* sustained; 4 shafts
Guns: 1 – 12.7 mm MG.
Radars: Surface search: Racal Decca; I-band.

Comment: Built in 1979–80 as oil rig support craft. Purchased in September 1987. All aluminium construction. Both assigned to hydrographic operations.

LEVUKA   *1/2012, Chris Sattler* / 1455671

Finland

SUOMEN MERIVOIMAT

Country Overview

The Republic of Finland is situated in northern Europe. Nearly one third of the country lies north of the Arctic Circle. With an area of 130,559 square miles, which includes some 60,000 lakes, it has borders to the west with Sweden, to the north with Norway and to the east with Russia. It has a 675 n mile coastline with the Baltic Sea and Gulf of Finland. The Ahvenanmaa archipelago (Åland Islands), consisting of some 6,500 islands, lies southwest of the mainland. Helsinki is the capital, largest city and principal port. Territorial Seas and a Fishing Zone, both of 12 n miles, have been claimed but not an EEZ.

Headquarters Appointments

Commander-in-Chief Finnish Navy:
Rear Admiral Kari Takanen

Chief of Staff FNHQ:
Commodore Juha Vauhkonen

Diplomatic Representation

Defence Attaché in London:
Colonel Simo Hautala

Personnel

(a) 2014: 1,810 regulars
(b) 3,000 conscripts (6–12 months' national service)

Fleet Organisation

Naval Headquarters: Turku
Gulf of Finland Naval Command; main base Upinniemi, Helsinki.
Archipelago Sea Naval Command; main base at Pansio, near Turku.
Uusimaa Brigade at Tammisaari.
The Navy is undergoing re-structuring. From 1 January 2015, the coastal fleet is to be based at Pansio, the coastal brigade at Upinniemi, and Uusimaa Brigade at Tammisaari.

Coast Defence

Coastal Artillery and naval infantry troops. RBS 15 truck-mounted quadruple SSM launchers. 130 mm fixed and mobile guns.

Frontier Guard

All Frontier Guard vessels come under the Ministry of the Interior. From 2014, a new paint scheme is to reflect the colours of the national flag. Hulls and superstructures are to be white, with some blue details, and the hull is to have a blue diagonal stripe with a thin white stripe superimposed. Personnel numbers: 600.

DELETIONS

Patrol Forces
2011 *Kiisla, Kurki*

Mine Warfare Forces
2013 *Pohjanmaa, Kuha 21-22, Kiiski 1-2*

Training Ships
2012 *Lokki*

Auxiliaries
2011 *Kampela 2*
2012 *Vahakari, Valas, Kala 6, Hauki, Hankoniemi*
2013 *Putsaari, Mursu, Havouri, Torsö, Alskär, Houtskär*

PENNANT LIST

Patrol Forces		Mine Warfare Forces				Auxiliaries		241	Askeri	730	Haukipää
70	Rauma	02	Hämeenmaa	875	Pyhäranta			511	Jymy	739	Hästö
71	Raahe	05	Uusimaa	876	Pansio	56	Kajava	512	Raju	799	Hylje
72	Porvoo	23–26	Kuha 23-26			92	Putsaari	531	Syöksy	826	Isku
73	Naantali	40	Katanpää			235	Hirsala	541	Vinha	830	Högsåra
80	Hamina	41	Purunpää			237	Hila	690	Fabian Wrede	831	Kallanpää
81	Tornio	42	Vahterpää			238	Haruna	691	Wilhelm Carpelan	877	Kampela 3
82	Hanko	523–527	Kiiski 3-7					692	Axel Von Fersen	899	Halli
83	Pori	777	Porkkala					722	Vaarlahti	992	Träskö
								723	Vänö	999	Louhi

PATROL FORCES

4 HAMINA CLASS (FAST ATTACK CRAFT – MISSILE) (PTGM)

Name	No	Builders	Commissioned
HAMINA	80 (ex-74)	Aker Finnyards, Rauma	24 Aug 1998
TORNIO	81	Aker Finnyards, Rauma	12 May 2003
HANKO	82	Aker Finnyards, Rauma	22 Jun 2005
PORI	83	Aker Finnyards, Rauma	19 Jun 2006

Displacement, tonnes: 274 full load
Dimensions, metres (feet): 50.8 × 8.3 × 2 *(166.7 × 27.2 × 6.6)*
Speed, knots: 32
Range, n miles: 500 at 30 kt
Complement: 29 (5 officers)

Machinery: 2 MTU 16V 538 TB93 diesels; 7,510 hp(m) *(5.52 MW)* sustained; 2 Kamewa 90SII waterjets
Missiles: SSM: 4 Saab RBS15F; active radar homing to 100 km *(54 n miles)* at 0.8 Mach; warhead 200 kg.
SAM: Denel Umkhonto 8 cell VLS; inertial guidance with mid-course guidance and IR homing to 12 km *(6.5 n miles)* at 2.4 Mach; warhead 23 kg.
Guns: 1 Bofors 57 mm/L 70 Mk 3; 220 rds/min to 17 km *(9.2 n miles)*; weight of shell 2.6 kg.
2 – 12.7 mm MGs.
Mines: 1 rail for 10 mines.
Depth charges: 1 rail.
Physical countermeasures: Decoys: 2 Rheinmetall MASS-2L; decoy launchers.
Electronic countermeasures: ESM/ECM: Thales SIEWS; radar intercept.
Radars: Air/surface search: EADS TRS-3D; G-band.
Fire control: SAAB Ceros 200; J-band.
Navigation: Furuno; I-band.
Sonars: Simrad Subsea Toadfish sonar; search and attack; active high frequency. Finnyards Sonac/PTA towed array; low frequency.
Combat data systems: EADS Advanced Naval Combat System (ANCS SQ 2000).
Weapon control systems: Saab Ceros electro-optic director. Sagem EOMS IR scanner.

Programmes: First ordered on 31 December 1996, second in February 2000, third on 3 December 2003 and a fourth on 15 February 2005 for delivery in 2006.
Structure: A continuation of the Rauma design with aluminium hull, composite superstructure and RAM coating. Signature reduction is aided by RAM coatings on the superstructure, submerged engine exhausts, upper deck pre-wetting, resilient mountings for all machinery, waterjet propulsion and conductive sealings on doors and hatches to prevent electromagnetic leakage.
Operational: Umkhonto missile fired from *Hanko* on 26 May 2006. The squadron is based at Upinniemi and became operational in 2008.

HANKO   *9/2013*, E & M Laursen* / 1529018

TORNIO   *7/2012, Freddy Philips* / 1483033

4 RAUMA CLASS (FAST ATTACK CRAFT—MISSILE) (PTGM)

Name	No	Builders	Commissioned
RAUMA	70	Hollming, Rauma	18 Oct 1990
RAAHE	71	Hollming, Rauma	20 Aug 1991
PORVOO	72	Finnyards, Rauma	27 Apr 1992
NAANTALI	73	Finnyards, Rauma	23 Jun 1992

Displacement, tonnes: 218 standard; 240 full load
Dimensions, metres (feet): 48 × 8 × 1.5 *(157.5 × 26.2 × 4.9)*
Speed, knots: 30
Complement: 24 (5 officers)

Machinery: 2 MTU 16V 538 TB93 diesels; 8,850 hp(m) *(6.6 MW)* sustained; 2 Riva Calzoni IRC 115 water-jets
Missiles: SSM: 6 Saab RBS15F (could embark 8); active radar homing to 150 km *(80 n miles)* at 0.8 Mach; warhead 200 kg.
Guns: 1 Bofors 40 mm/70; 300 rds/min to 12 km *(6.6 n miles)*; weight of shell 0.96 kg. 2—12.7 mm MGs.
A/S Mortars: 4 Saab Elma LLS-920 9-tubed launchers; range 300 m; warhead 4.2 kg shaped charge.
Depth charges: 1 rail.
Physical countermeasures: Decoys: Rheinmetall MASS.
Electronic countermeasures: ESM/ECM: Thales SIEWS.
Radars: Surface search: Ericsson Sea Giraffe; G/H-band.
Fire control: Bofors Electronic 9LV 225; J-band.
Navigation: Furuno; I-band.
Sonars: Simrad ST2400 VDS.
Combat data systems: Saab 9LV Mk 4.
Electro-optic systems: Bofors Electronic 9LV Mk 3 optronic director with TV camera; infra-red and laser telemetry.

Programmes: Ordered 27 August 1987.
Modernisation: A mid-life upgrade was conducted 2010–13. The principal features included installation of a new combat management system, a new variable-depth sonar, and decoy system.
Structure: Developed from Helsinki class. Hull and superstructure of light alloy. SAM and 23 mm guns are interchangeable within the same Sako barbette which has replaced the ZU mounting.
Operational: Primary function is the anti-ship role but there is some ASW capability. Mine rails can be fitted in place of the missile launchers. Towed array cable is 78 m with 24 hydrophones and can be used at speeds between 3 and 12 kt.

RAUMA 9/2010, Guy Toremans / 1406038

MINE WARFARE FORCES

2 + 1 KATANPÄÄ CLASS (MINEHUNTERS) (MHSC)

Name	No	Builders	Laid down	Launched	Commissioned
KATANPÄÄ	40	Intermarine, Sarzana	Jul 2007	16 Jun 2009	4 May 2012
PURUNPÄÄ	41	Intermarine, Sarzana	Mar 2008	26 Aug 2010	20 Aug 2013
VAHTERPÄÄ	42	Aker Finnyards, Rauma	Feb 2009	3 Nov 2011	2014

Displacement, tonnes: 708 full load
Dimensions, metres (feet): 52.5 × 9.9 × 3.1 *(172.2 × 32.5 × 10.2)*
Speed, knots: 13
Range, n miles: 1,500 at 12 kt
Complement: 36 (6 officers)

Machinery: 2 MTU 8V 396 TE74K diesels (for transit); 2,680 hp *(2 MW)*; 2 motors (for minehunting); 2 Voith Schneider cycloidal propellers
Guns: 1 Bofors 40 mm/70.
Physical countermeasures: MCM Atlas Sea Fox C MIDS.
Radars: Navigation: I-band.
Sonars: Atlas Elektronik SQS-12M hull-mounted; LF/HF/VHF minehunting sonar; Double Eagle Mk 3 dual-frequency UDS.
Combat data systems: Atlas Integrated mine countermeasure system (IMCMS).

Programmes: Contract signed with Intermarine SPA on 23 November 2006 for the construction of three MCMVs. The first two ships are being built at Sarzana; the third ship was built by Intermarine and is to be completed at Aker Shipyards, Rauma. The contract includes training and logistic support. The principal components of the minehunting combat system are: a command system, a hull-mounted sonar and self-propelled variable depth sonar (installed in Saab Double Eagle Mk III ROV), a Mine Identification and Disposal System (MIDS) based on the Atlas Sea Fox; both Hydroid Remus and Kongsberg Hugin 1000 AUVs are used for seabed survey and reconnaissance.
Structure: Monocoque GRP construction. The design is similar to the Gaeta class built for the Italian and Australian navies. A new superstructure accommodates the command-and-control suite in the forward-central superstructure and mine-detection and hunting housing and recovery equipment in the central and stern sections.
Operational: The three vessels are expected to become operational in 2014.

PURUNPÄÄ 1/2013*, Giorgio Ghiglione / 1529089

2 HÄMEENMAA CLASS (MINELAYERS) (ML)

Name	No	Builders	Laid down	Launched	Commissioned
HÄMEENMAA	02	Finnyards, Rauma	2 Apr 1991	11 Nov 1991	15 Apr 1992
UUSIMAA	05	Finnyards, Rauma	12 Nov 1991	Jun 1992	2 Dec 1992

Displacement, tonnes: 1,473 full load
Dimensions, metres (feet): 77.8 oa; 69.6 wl × 11.6 × 3.2 *(255.2; 228.3 × 38.1 × 10.5)*
Speed, knots: 20
Complement: 66

Machinery: 2 Wärtsilä 16V22 diesels; 6,300 hp(m) *(4.64 MW)* sustained; 2 Kamewa cp props; bow thruster; 247 hp(m) *(184 kW)*
Missiles: SAM: Denel Umkhonto 8 cell VLS; inertial guidance with mid-course guidance and IR homing to 12 km *(6.5 n miles)* at 2.4 Mach; warhead 23 kg.
Guns: 1 Bofors 57 mm/70 Mk 1; 220 rds/min to 17 km *(9.2 n miles)*; weight of shell 2.6 kg. 2—40 mm grenade launchers. 2—12.7 mm MGs.
A/S Mortars: 2 RBU 1200 fixed 5-tubed launchers; range 1,200 m; warhead 34 kg.
Mines: 4 rails for 200 contact (S 43–55, 541, 558) or influence (Seamine 2004 and PM 90) mines.
Depth charges: 2 racks for 8 DCs.
Physical countermeasures: Decoys: 2 Rheinmetall MASS-2L; decoy launchers.
Electronic countermeasures: ESM/ECM: Thales SIEW.
Radars: Air/surface search: EADS TRS-3D; G-band.
Surface search and Navigation: 3 Furuno 2827/2837S; E/F/I-bands.
Sonars: Simrad; hull-mounted; active mine detection; high frequency.
Combat data systems: EADS ANCS 2000.
Weapon control systems: Saab Ceros electro-optic director. Sagem EOMS IR scanner.
Helicopters: Platform for 1 light.

Programmes: First one ordered 29 December 1989 after the original order in July from Wärtsilä had been cancelled. Second ordered 13 February 1991.
Modernisation: A contract for the mid-life upgrade of both vessels was awarded to EADS Defence and Security Division in April 2006. Modernisation, undertaken by Aker Shipyards, included EADS ANCS 2000 combat data system, EADS TRS 3D radar, Sagem EOMS and Umkhonto point defence missile system. *Hämeenmaa* completed upgrade on 13 April 2007 and *Uusimaa* in September 2007. Both vessels became fully operational in 2008.
Structure: Steel hull and alloy superstructure. Ice strengthened (Ice class 1A) and capable of breaking up to 40 mm ice. During the modernisation period (2006–07), a new fixed bow, stabilisers, two new masts and a new combat information system were added. The flight deck can operate light helicopters.
Operational: Dual role as a transport and support ship.

UUSIMAA 9/2013*, E & M Laursen / 1529019

4 KUHA CLASS (MINESWEEPERS—INSHORE) (MSI)

Name	No	Builders
KUHA 23	23	Laivateollisuus, Turku
KUHA 24	24	Laivateollisuus, Turku
KUHA 25	25	Laivateollisuus, Turku
KUHA 26	26	Laivateollisuus, Turku

Displacement, tonnes: 91 full load
Dimensions, metres (feet): 31.7 oa; 26.6 wl × 6.9 × 2 *(104; 87.3 × 22.6 × 6.6)*
Speed, knots: 12
Complement: 15 (3 officers)

Machinery: 2 Cummins MT-380M diesels; 600 hp(m) *(448 kW)*; 1 shaft; cp prop; active rudder
Guns: 2 ZU 23 mm/60 (twin). 1—12.7 mm MG.
Radars: Navigation: Decca; I-band.
Sonars: Reson Seabat 6012 mine avoidance; active high frequency.

Comment: All ordered 1972. First one completed 28 June 1974, and last on 13 November 1975. Fitted for magnetic, acoustic and pressure-mine clearance. Hulls are of GRP. May carry a Pluto ROV. Four of the class were lengthened in 1997/98 and remaining two by 2000 to take a new minesweeping control system, and new magnetic and acoustic sweeps. New sonars installed. Armament not fitted in all of the class. Kuha 21–22 decommissioned in 2013.

KUHA CLASS 6/2001, Finnish Navy / 0114724

Mine warfare forces — Training ships < **Finland** 241

3 PANSIO CLASS (MINELAYERS—LCU TYPE) (MLI)

Name	No	Builders	Commissioned
PANSIO	876 (ex-576)	Olkiluoto Shipyard	25 Sep 1991
PYHÄRANTA	875 (ex-575, ex-475)	Olkiluoto Shipyard	26 May 1992
PORKKALA	777	Olkiluoto Shipyard	29 Oct 1992

Displacement, tonnes: 457 standard
Dimensions, metres (feet): 44 oa; 39.2 wl × 10 × 2 (144.4; 128.6 × 32.8 × 6.6)
Speed, knots: 10
Complement: 12

Machinery: 2 MTU 12V 183 TE62 diesels; 1,500 hp(m) (1.1 MW); 2 shafts; bow thruster
Guns: 2 ZU 23 mm/87 (twin). 2—12.7 mm MGs.
Mines: 50.
Radars: Navigation: Raytheon ARPA; I-band.

Comment: Ordered in May 1990. Used for inshore minelaying and transport with a capacity of 100 tons. Ice strengthened with ramps in bow and stern. Has a 15 ton crane fitted aft.

PYHÄRANTA   7/2012, Guy Toremans / 1483037

5 KIISKI CLASS (MINESWEEPERS—INSHORE) (MSI)

Name	No	Builders
KIISKI 3	523	Fiskars, Turku
KIISKI 4	524	Fiskars, Turku
KIISKI 5	525	Fiskars, Turku
KIISKI 6	526	Fiskars, Turku
KIISKI 7	527	Fiskars, Turku

Displacement, tonnes: 20 full load
Dimensions, metres (feet): 15.2 × 4.1 × 1.2 (49.9 × 13.5 × 3.9)
Speed, knots: 11
Range, n miles: 260 at 11 kt
Complement: 4
Machinery: 2 Valmet 611 CSMP diesels; 340 hp(m) (250 kW); 2 Hamilton water-jets

Comment: Ordered January 1983. All completed by 24 May 1984. GRP hull. Built to be used with Kuha class for unmanned teleguided sweeping, but this was not successful and they are now used for manned magnetic and acoustic sweeping operations with crew of four. *Kiiski 1–2* decommissioned in 2013.

KIISKI 5   6/2001, Finnish Navy / 0114725

LAND-BASED MARITIME AIRCRAFT

Numbers/Type: 2 Dornier Do 228-212.
Operational speed: 223 kt (413 km/h).
Service ceiling: 29,600 ft (9,020 m).
Range: 939 n miles (1,740 km).
Role/Weapon systems: Maritime surveillance, SAR and pollution control. Acquired in 1995. Sensors: Selex Galileo Seaspray radar; Star SAFIRE IR/EO; FLIR; TV camera; Cobham 935-11 direction finder; MSS 6000 data system. Weapons: Unarmed.

DORNIER 228   6/2013*, Finnish Border Guard / 1529017

© 2014 IHS

Numbers/Type: 5 Agusta AB 412 Griffon.
Operational speed: 122 kt (226 km/h).
Service ceiling: 17,000 ft (5,180 m).
Range: 354 n miles (656 km).
Role/Weapon systems: Operated by Coast Guard/Frontier force for patrol and SAR. Sensors: Radar and FLIR. Weapons: Unarmed at present but mountings for machine guns.

AB 412   6/2005, Finnish Navy / 1133422

Numbers/Type: 3 Eurocopter AS 332L1 Super Puma.
Operational speed: 130 kt (240 km/h).
Service ceiling: 15,090 ft (4,600 m).
Range: 672 n miles (1,245 km).
Role/Weapon systems: Coastal patrol, surveillance and SAR helicopters. Sensors: Surveillance radar, FLIR, tactical navigation systems and SAR equipment. Weapons: Unarmed.

AS 332   6/2005, Finnish Navy / 1133423

Numbers/Type: 4 AgustaWestland AW119 Koala.
Operational speed: 150 kt (278 km/h).
Service ceiling: 15,000 ft (4,572 m).
Range: 535 n miles (990 km).
Role/Weapon systems: Border surveillance helicopters. Weapons: unarmed.

KOALA   6/2013*, Finnish Border Guard / 1529016

TRAINING SHIPS

1 LOKKI CLASS (AX)

Name	No	Builders	Commissioned
KAJAVA	56	Valmet, Lavateollisuus	3 Oct 1981

Displacement, tonnes: 64 standard
Dimensions, metres (feet): 26.8 × 5.5 × 1.9 (87.9 × 18.0 × 6.2)
Speed, knots: 25
Complement: 6
Machinery: 2 MTU 8V 396 TB84 diesels; 2,100 hp(m) (1.54 MW) sustained; 2 shafts
Guns: 2 ZU 23 mm/60 can be carried.
Sonars: Simrad SS 242; hull-mounted; active search; high frequency.

Comment: Transferred from the Frontier Guard to the Navy in 1999 and used as a training vessel. Built in light metal alloy. A third of class to Lithuania in 1997 and a fourth to Latvia in 2001. *Lokki* decommissioned in 2012.

LOKKI CLASS   6/2001, Finnish Navy / 0114723

IHS Jane's Fighting Ships 2014-2015

242   Finland > Training ships — Auxiliaries

3 FABIAN WREDE CLASS (AX)

Name	No	Builders	Commissioned

---

244   Finland > Auxiliaries — Border guard

23 RAIDING CRAFT (LCVP)

G 100 series

Displacement, tonnes: 3 full load
Dimensions, metres (feet): 8 × 2.1 × 0.3 *(26.2 × 6.9 × 1.0)*
Speed, knots: 30
Complement: 1
Military lift: 9 troops with equipment
Machinery: 1 Yanmar 4LHA-STE diesel; 240 hp *(179 kW)*; 1 RR FF-jet 240 waterjet

Comment: First batch of 23 units ordered in February 2001. Based on Swedish Gruppbåt and built by Alutech Ltd. Delivered late 2001.

RAIDING CRAFT   6/2001, Finnish Navy / 0114721

2 POLLUTION CONTROL VESSELS (YPC)

HYLJE 799   HALLI 899

Displacement, tonnes: 1,524 (799), 1,626 (899) full load
Dimensions, metres (feet): 54 (799), 61.5 (899) × 12.5 × 3 *(177.2, 201.8 × 41.0 × 9.8)*
Speed, knots: 7 (799), 13 (899)
Machinery: 2 Saab diesels; 680 hp(m) *(500 kW)*; 2 shafts; active rudders; bow thruster *(Hylje)*; 2 Wärtsilä diesels; 2,650 hp(m) *(19.47 MW)*; 2 shafts; active rudders *(Halli)*

Comment: Painted grey. Strengthened for ice. Owned by Ministry of Environment, civilian-manned but operated by Navy from Turku. *Hylje* commissioned 3 June 1981, *Halli* in January 1987. Capacity is about 550 m³ *(Hylje)* and 1,400 m³ *(Halli)* of contaminated seawater. The ships have slightly different superstructure lines aft.

HALLI   8/2013*, Selim San / 1529021

0 + 12 U 700 CLASS (LCP)

Displacement, tonnes: 29 full load
Dimensions, metres (feet): 19.3 × 4.4 × 0.96 *(63.3 × 14.4 × 3.1)*
Speed, knots: 40
Complement: 6 + 25 embarked forces
Machinery: 2 Scania diesels; 1,800 hp *(1.34 MW)*; 2 waterjets
Guns: 1 — 12.7 mm remotely-controlled MG. 1 — 7.62 mm remotely controlled MG.
Radars: Navigation: I-band.

Comment: The order for 12 new Watercat M18 armoured modular craft high-speed landing vessels was placed with Marine Alutech in October 2012. The vessels are fitted with nuclear, biological and chemical filtration systems and the design includes ballistic protection. Delivery is to start in 2014 and to be completed in 2016.

U 700 CLASS   10/2012, Marine Alutech / 1486486

IHS Jane's Fighting Ships 2014-2015

TUGS

2 HARBOUR TUGS (YTM)

HAUKIPÄÄ 730   KALLANPÄÄ 831

Displacement, tonnes: 39 full load
Dimensions, metres (feet): 14 × 5 × 2.3 *(45.9 × 16.4 × 7.5)*
Speed, knots: 9
Complement: 2
Machinery: 2 diesels; 360 hp(m) *(265 kW)*; 2 shafts

Comment: Delivered by Teijon Telakka Oy in December 1985. Similar to Hauki class. Also used as utility craft.

HAUKIPÄÄ (old number)   6/2000, Finnish Navy / 0103813

BORDER GUARD

2 TURSAS CLASS (OFFSHORE PATROL VESSELS) (WPBO)

TURSAS   UISKO

Displacement, tonnes: 1,270 full load
Dimensions, metres (feet): 61.45 × 10.2 × 4.85 *(201.6 × 33.5 × 15.9)*
Speed, knots: 14
Complement: 32
Machinery: Diesel electric; 2 Rolls-Royce azimuth thrusters; 4,360 hp *(3.2 MW)*
Guns: 2 Sako 23 mm/60 (twin).
Sonars: Simrad SS105; active scanning; 14 kHz.

Comment: First ordered from Rauma-Repola on 21 December 1984, launched 31 January 1986 and delivered 6 June 1986. Second ordered 20 March 1986, launched 19 June 1986 and delivered 27 January 1987. Both ships underwent conversion at Uusikaupunki Workboat Ltd 2004–06. The ships were lengthened by 12 m and modified to conduct anti-pollution operations.

TURSAS   6/2013*, Finnish Border Guard / 1529015

1 UVL 10 CLASS (OFFSHORE PATROL SHIP) (PSO)

Name	Builders	Laid down	Launched	Commissioned
TURVA	STX, Rauma	25 Feb 2013	2 Aug 2013	Nov 2013

Measurement, tonnes: 4,600 gt
Dimensions, metres (feet): 95.8 × 17.7 × 5.5 *(314.3 × 58.1 × 18.0)*
Speed, knots: 18
Machinery: 2 Wärtsilä 6R34DF diesels, 1 Wärtsilä 12V34DF diesel; 15,600 hp *(12 MW)*; 1 shaft; cp prop; 2 Rolls Royce azimuth thrusters; 2 bow thrusters
Radars: Surface search/navigation: TRS-3D; E/F-band.
Navigation: Consillium; I-band.
Helicopters: Platform for 1 medium.

Comment: Multipurpose patrol ship of steel construction equipped with liquid natural gas (LNG)-fuelled engines. Design features include a 1,200 m³ tank for oilspill recovery and chemical accidents. Design work was completed in 2012 and construction at STX Rauma Shipyard was completed in November 2013.

TURVA (artist's impression)   11/2013*, Finnish Border Guard / 1529014

© 2014 IHS

Border guard < **Finland** 245

1 IMPROVED TURSAS CLASS
(OFFSHORE PATROL VESSEL) (WPBO)

MERIKARHU

Displacement, tonnes: 1,118 full load
Dimensions, metres (feet): 57.8 × 11 × 4.6 (189.6 × 36.1 × 15.1)
Speed, knots: 15. **Range, n miles:** 2,000 at 15 kt
Complement: 30
Machinery: 2 Wärtsilä Vasa 8R26 diesels; 3,808 hp(m) (2.8 MW) sustained; 1 shaft; cp prop; bow and stern thrusters
Guns: 2—23 mm/87 (twin) can be carried.
Radars: Surface search. Navigation.

Comment: Ordered 17 June 1993 from Finnyards, and completed 28 October 1994. Capable of 5 kt in 50 cm of ice. Used as an all-weather patrol ship in the Baltic, capable of Command, SAR, tug work with 30 ton bollard pull, and environmental pollution cleaning up. Carries an RIB launched from a hydraulic crane.

MERIKARHU   8/2013*, Selim San / 1529022

4 SLINGSBY SAH 2200 (HOVERCRAFT) (UCAC)

Displacement, tonnes: 6 full load
Dimensions, metres (feet): 10.6 on cushion × ? (34.8 × ?)
Speed, knots: 40. **Range, n miles:** 400 at 30 kt
Complement: 2
Military lift: 2.2 tons or 12 troops
Machinery: 1 Cummins 6CTA-8-3M-1 diesel; 300 hp (224 kW)
Radars: Navigation: Raytheon R41; I-band.

Comment: First one acquired from Slingsby Amphibious Hovercraft Company in March 1993. Three more ordered in February 1998 and delivered in late 1999.

SLINGSBY 2200   6/2013*, Finnish Border Guard / 1529012

3 GRIFFON 2000 TDX(M) (HOVERCRAFT) (UCAC)

Displacement, tonnes: 7 full load
Dimensions, metres (feet): 11 on cushion × ? (36.1 × ?)
Speed, knots: 33. **Range, n miles:** 300 at 25 kt
Complement: 2
Military lift: 16 troops or 2 tons
Machinery: 1 Deutz BF8L513 diesel; 320 hp (239 kW) sustained
Radars: Navigation: I-band.

Comment: First two acquired from Griffon, UK and commissioned 1 December 1994; third one bought in June 1995. Can be embarked in an LCU. Speed indicated is at Sea State 3 with a full load.

GRIFFON 2000   6/2013*, Finnish Border Guard / 1529013

© 2014 IHS

24 PV08 CLASS (PATROL CRAFT) (PB)

Measurement, tonnes: 17 gt
Dimensions, metres (feet): 15.2 × 3.8 × 0.85 (49.9 × 12.5 × 2.8)
Complement: 3
Machinery: 2 diesels; 1,282 hp (956 kW); 2 shafts
Radars: Surface search/navigation: To be announced.

Comment: Entered service 2006–13.

PV 265   6/2013*, Finnish Border Guard / 1529011

6 PATROL CRAFT (PB)

Measurement, tonnes: 25 gt
Dimensions, metres (feet): 15 × 4.0 × 2.0 (49.2 × 13.1 × 6.6)
Speed, knots: 10
Complement: 2
Machinery: 1 diesel; 469 hp (350 kW); 1 shaft
Radars: Surface search/navigation: To be announced.

Comment: There are four RV 90 class and two RV 06 class. Entered service 1990–2006.

RV 150 (RV 90)   6/2013*, Finnish Border Guard / 1529010

RV 155 (RV 06)   6/2013*, Finnish Border Guard / 1529009

IHS Jane's Fighting Ships 2014-2015

France

MARINE NATIONALE

Country Overview

The French Republic, which includes the island of Corsica, is situated in western Europe. With an area of 210,026 square miles, the mainland is bordered to the north by Belgium, Luxembourg and Germany, to the southeast by Switzerland and Italy and to the south-west by Spain. It has a 1,852 n mile coastline with the Atlantic Ocean, Mediterranean Sea, North Sea and English Channel. Overseas departments are French Guiana, Martinique, Guadeloupe and Réunion. Dependencies include St Pierre and Miquelon, Mayotte, New Caledonia, French Polynesia, the French Southern and Antarctic Territories, and Wallis and Futuna Islands. The capital and largest city is Paris while the principal ports are Marseille, Le Havre, Dunkirk, St Nazaire and Rouen. Strasbourg is a port on the Rhine. Territorial seas (12 n miles) are claimed. An EEZ (200 n miles) has also been claimed but not all the large number of boundaries have been defined by agreements.

Headquarters Appointments

Chief of the Naval Staff: Amiral Bernard Rogel
Inspector General of the Armies: Amiral Xavier Magne
Major General of the Navy:
 Vice-Amiral d'Escadre Stéphane Verwaerde
Director of Personnel: Vice-Amiral d'Escadre Christophe Prazuck
Inspector of the Navy: Vice-Amiral d'Escadre Loïc Raffaelli

Senior Appointments

C-in-C Atlantic Theatre (CECLANT):
 Vice-Amiral d'Escadre Jean Pierre Labonne
C-in-C Mediterranean Theatre (CECMED):
 Vice-Amiral d'Escadre Yves Joly
Flag Officer, French Forces Polynesia (ALPACI):
 Contre-Amiral Anne Cullerre
Flag Officer, Naval Forces Indian Ocean (ALINDIEN):
 Contre-Amiral Antoine Beaussant
Flag Officer, Antilles: Contre-Amiral George Bosselut
Flag Officer, Cherbourg: Vice-Amiral d'Escadre Emmanuel Carlier
Flag Officer, Submarines (ALFOST):
 Vice-Amiral d'Escadre Charles-Edouard de Coriolos
Flag Officer, Naval Action Force (ALFAN):
 Vice-Amiral d'Escadre Philippe Coindreau
Deputy Flag Officer, Naval Action Force:
 Contre-Amiral Eric Chaperon
Chief of Staff, Naval Action Force: Vice-Amiral Denis Beraud
Flag Officer Naval Aviation (ALAVIA):
 Contre-Amiral Hervé Denys de Bonnaventure

Senior Appointments — *continued*

Flag Officer Lorient and Commandant Marines (Alfusco):
 Contre-Amiral Olivier Coupry

Diplomatic Representation

Naval Attaché in London:
 Capitaine de Vaisseau Yves le Corre
Head of French Military Mission to EU:
 Général Yves Rouby
Naval Attaché in Washington:
 Capitaine de Vaisseau Philippe Alquier

Personnel

(a) 2014: 32,032 (3,836 officers)
(b) 2014: civilians in direct support: 3,098

Bases

Brest: Main Atlantic base. SSBN base
Toulon: Mediterranean Command base
Cherbourg: Channel base
Bayonne: Landes firing range
Small bases at Papeete (Tahiti), Fort-de-France (Martinique), Nouméa (New Caledonia), Degrad-des-Cannes (French Guiana), Port-des-Galets (La Réunion), Abu Dhabi and Dzaoudi (Comoros).

Shipyards (Naval)

All former naval shipbuilding facilities are privatised and are operated by DCNS. Main facilities are at:
Cherbourg: Submarines and Fast Attack Craft (private shipyard)
Brest: Major warships and refitting
Lorient: Destroyers and Frigates, MCMVs, Patrol Craft
Toulon: Major refits.

Reserve

A ship in 'Complément' has a reduced complement and is able to return to service although in most cases it is the first step in the withdrawal of a ship. 'Reserve Normale' has no complement but is available at short notice. 'Reserve Speciale' means that a major refit will be required before the ship can go to sea again. This situation precedes 'Condamnation'. 'Condamnation' is the state before being broken up or sold; at this stage a Q number is allocated.

Prefix to Ships' Names

FS is used in NATO communications but is not official.

Strength of the Fleet

Type	Active (Reserve)	Building (Projected)
Submarines (SSBN)	4	—
Submarines (SSN)	6	6
Aircraft Carriers	1	—
Destroyers	11	10
Frigates	20	—
Patrol Craft	12	—
LPH	3	—
LPD	1	—
LST/LCT	9	—
LCMs	17	—
Route Survey Vessels	3	—
Minehunters	11	—
Diving Tenders	16	—
Survey/Research Ships	9	—
Tankers (AOR)	4	(3)
Ocean tugs/salvage	10	(8)
Supply Tenders	4	—
Transports	1	—
Training Ships	15	2

DELETIONS

Destroyers

2011	*Tourville*
2013	*De Grasse*, *Georges Leygues*
2014	*Dupleix*

Patrol Forces

2011	*La Rieuse* (to Kenya), *La Railleuse*
2012	*La Tapageuse*

Amphibious Forces

2011	*Foudre* (to Chile), *Rapière* (to Chile)
2012	*Dague* (to Djibouti)
2013	*Jacques Cartier*

Fleet Air Arm Bases

Notes: (1) In addition to the following squadrons, there are three other squadrons operating with mixed Air Force and Navy crews on behalf of both services:
- Helicopter Squadron EH-1/67 "Pyrénées", based at Cazaux AFB, for the combat SAR (CSAR) role, operating eight specialised Aerospatiale SA-330 Puma helicopters and four Eurocopter EC 725 R2 Cougar Mk 2 Plus Resco delivered in 2005–06. These helicopters regularly embark on *Charles de Gaulle*.
- Army special operations helicopter flight EOS 3 based at Pau, equipped with Aerospatiale AS 532 Cougar helicopters to be replaced from 2006 by eight new Eurocopter EC 725 R2 Cougar Mk 2 Plus HUS. Roles include counter-terrorism and they can embark on *Charles de Gaulle*, LSDs and eventually on La Fayette and Floréal-class frigates.
- Training Squadron EAT-319, based at Avord AFB, with Embraer 121 Xingu light transport (some coming from the Navy) for pilot basic training.

(2) There are also naval sections within various Air Force training units where trainees fly CAP 231 light aircraft, Aerospatiale TB-30 Epsilon or Embraer Tucano basic and Dassault/Dornier Alphajet advanced trainers. Basic helicopter training is performed within the Army Aviation at Dax with Aerospatiale Gazelles and Ecureuils.
(3) Fighter pilots are trained to carrier operations in the US at NAS Meridian, flying BAE Systems/MDD (Boeing) T-45C Goshawk.

Embarked Squadrons

Base/Squadron No	Aircraft	Task
Lann Bihoué/4F	E-2C Hawkeye	AEW
Landivisiau/11F	Rafale F3	Assault, Recce
Landivisiau/12F	Rafale F3	Air Defence
Landivisiau/17F	Super Étendard	Assault, Recce
Hyères/31F	NH90	ASW
Lanvéoc-Poulmic/34F	Lynx	ASW
Hyères/36F	Panther	Surveillance

Support Squadrons

Base/Squadron No	Aircraft	Task
Hyères/CEPA/10S	NH90	Research, trials
Lanvéoc-Poulmic/22S (detachments on ships)	Alouette III	Training, Support Atlantic Region
Lanvéoc-Poulmic/32F (detachment at Hyères)	EC 225	Transport, SAR
Lanvéoc-Poulmic/33F	NH90	Transport, SAR
Hyères/35F (detachments at various locations and ships)	Dauphin 2, Alouette III, Dauphin N3	Surveillance, SAR, Carrier-borne SAR
Landivisiau/57S	Falcon 10 MER	Support, Training

Maritime Patrol Squadrons

Base/Squadron No	Aircraft	Task
Lann Bihoué/21F	Atlantique Mk 2	MP
Lann Bihoué/23F	Atlantique Mk 2	MP
Lann Bihoué/24F	Falcon 50M	Surveillance, SAR
Lann Bihoué/28F	Xingu	Logistic support
Faaa (Papeete)/25F (detachment at Tontouta, New Caledonia)	Gardian	Surveillance, SAR

Training Squadrons

Base/Squadron No	Aircraft	Task
Lanvéoc-Poulmic/EIP/50S	MS 880 Rallye/CAP 10	Initial Flying School, Recreational

Approximate Fleet Dispositions 1 May 2014

		Channel	Atlantic	Mediterranean	Indian Ocean*	Pacific	Antilles F. Guiana
Carriers	FAN	—	—	1	—	—	—
SSBN	FOST	—	4	—	—	—	—
SSN	FOST	—	—	6	—	—	—
DDG/DDH	FAN	—	4	7	—	—	—
FFG	FAN	—	0	5	2	2	2
MCMV (incl tenders)	FAN	1	12	5	—	—	—
Patrol Forces**	FAN/GM	4	5	6	3	3	2
LHD/LSD	FAN	—	—	4	—	—	—
LST/LCT	FAN	—	—	5	1	1	1
AOR	FAN	—	1	3	—	—	—

FAN = Force d'Action Navale (HQ at Toulon). All surface ships based at Toulon, Brest or overseas.
FOST = Force Océanique Stratégique (HQ at Brest). SSBNs based at l'Île Longue near Brest. All SSNs based at Toulon.
GM = Gendarmerie Maritime
*Plus one or two DDG/DDH/FFG regularly deployed from Toulon.
**Patrol forces include vessels manned by the Navy and major craft from the Gendarmerie Maritime.

PENNANT LIST

Submarines		F 735	Germinal	P 605	Vertonne (GM)	**Amphibious Forces**		A 696	Buffle
		F 789	Lieutenant de	P 606	Dumbéa (GM)	L 9012	Siroco	A 697	Bison
S 601	Rubis		Vaisseau le Hénaff	P 607	Yser (GM)	L 9013	Mistral	A 712	Athos
S 602	Saphir	F 790	Lieutenant de	P 608	Argens (GM)	L 9014	Tonnerre	A 713	Aramis
S 603	Casabianca		Vaisseau Lavallée	P 609	Hérault (GM)	L 9015	Dixmude	A 743	Denti
S 604	Émeraude	F 791	Commandant l'Herminier	P 610	Gravona (GM)	L 9032	Dumont D'Urville	A 748	Léopard
S 605	Améthyste	F 792	Premier Maître l'Her	P 611	Odet (GM)	L 9034	La Grandière	A 749	Panthère
S 606	Perle	F 793	Commandant Blaison	P 612	Maury (GM)	L 9062	Hallebarde	A 750	Jaguar
S 616	Le Triomphant	F 794	Enseigne de	P 613	Charente (GM)	L 9090	Gapeau	A 751	Lynx
S 617	Le Téméraire		Vaisseau Jacoubet	P 614	Tech (GM)			A 752	Guépard
S 618	Le Vigilant	F 795	Commandant Ducuing	P 615	Penfeld (GM)	**Major Auxiliaries Survey and Support Ships**		A 753	Chacal
S 619	Le Terrible	F 796	Commandant Birot	P 616	Trieux (GM)			A 754	Tigre
		F 797	Commandant Bouan	P 617	Vésubie (GM)			A 755	Lion
Aircraft and Helicopter Carriers				P 618	Escaut (GM)	A 601	Monge	A 758	Beautemps-Beaupré
				P 619	Huveaune (GM)	A 607	Meuse	A 759	Dupuy de Lôme
R 91	Charles de Gaulle	**Mine Warfare Forces**		P 620	Sèvre (GM)	A 608	Var	A 768	Élan
Destroyers				P 621	Aber Wrach (GM)	A 613	Achéron	A 770	Glycine
		M 611	Vulcain	P 622	Estéron (GM)	A 630	Marne	A 771	Églantine
		M 614	Styx	P 623	Mahury (GM)	A 631	Somme	A 775	Gazelle
D 614	Cassard	M 622	Pluton	P 624	Organabo (GM)	A 633	Taape	A 785	Thétis
D 615	Jean Bart	M 641	Éridan	P 671	Glaive	A 635	Revi	A 790	Coralline
D 620	Forbin	M 642	Cassiopée	P 675	Arago	A 636	Maito	A 791	Lapérouse
D 621	Chevalier Paul	M 643	Andromède	P 676	Flamant	A 637	Maroa	A 792	Borda
D 642	Montcalm	M 644	Pégase	P 677	Cormoran	A 638	Manini	A 793	Laplace
D 643	Jean de Vienne	M 645	Orion	P 678	Pluvier	A 641	Esterel	Y 638	Lardier
D 644	Primauguet	M 646	Croix du Sud	P 681	Albatros	A 642	Lubéron	Y 639	Giens
D 645	La Motte-Picquet	M 647	Aigle	P 684	La Capricieuse	A 645	Alizé	Y 640	Mengam
D 646	Latouche-Tréville	M 648	Lyre	P 686	La Glorieuse	A 649	L'Étoile	Y 641	Balaguier
D 650	Aquitaine	M 650	Sagittaire	P 687	La Gracieuse	A 650	La Belle Poule	Y 642	Taillat
D 651	Normandie	M 652	Céphée	P 688	La Moqueuse	A 652	Mutin	Y 643	Nividic
Frigates		M 653	Capricorne	P 701	Le Malin	A 653	La Grand Hermine	Y 647	Le Four
		M 770	Antarès	P 716	MDLC Jaques (GM)	A 664	Malabar	Y 649	Port Cros
		M 771	Altaïr	P 720	Géranium (GM)	A 669	Tenace	Y 670	Las
F 710	La Fayette	M 772	Aldébaran	P 721	Jonquille (GM)	A 675	Fréhel	Y 671	Douffine
F 711	Surcouf			P 722	Violette (GM)	A 676	Saire	Y 692	Telenn Mor
F 712	Courbet	**Patrol Forces**		P 723	Jasmin (GM)	A 677	Armen	Y 751	Engageante
F 713	Aconit			P 725	L'Adroit	A 678	La Houssaye	Y 752	Vigilante
F 714	Guépratte	A 789	Melia (GM)	P 740	Fulmar (GM)	A 679	Kéréon	Y 758	Kermeur
F 730	Floréal	P 601	Elorn (GM)	P 778	Réséda (GM)	A 680	Sicié	Y 759	Kernaleguen
F 731	Prairial	P 602	Verdon (GM)	P 791	Hortensia	A 681	Taunoa	Y 770	Morse
F 732	Nivôse	P 603	Adour (GM)			A 682	Rascas	Y 771	Otarie
F 733	Ventôse	P 604	Scarpe (GM)	GM = Gendarmerie Maritime		A 695	Bélier	Y 772	Loutre
F 734	Vendémiaire							Y 773	Phoque

SUBMARINES

Attack Submarines

Notes: (1) The Agosta class submarine *Ouessant* was re-introduced into service on 5 August 2005 following a refit. It was used as a training vessel by DCNS to support submarine sales to Malaysia and was de-activated in July 2009. She was transferred to Malaysia in 2011.

(2) France signed an MoU with Norway and UK on 5 August 2003 for the procurement of the NATO Submarine Rescue System (NSRS). Based in UK, the system entered service in November 2008.

(3) An Unmanned Underwater Vehicle (UUV), ASMX, is under development by DCNS. Based on an F17 torpedo, it can be launched from a 533 mm tube. ASMX is undergoing trials in the DCNS ship *Topaze*.

0 + 6 SUFFREN (BARRACUDA) CLASS (SSN)

Name	Builders	Laid down	Launched	Commissioned
SUFFREN	DCN, Cherbourg	19 Dec 2007	2015	2017
DUGUAY-TROUIN	DCN, Cherbourg	26 Jun 2009	2017	2020
TOURVILLE	DCN, Cherbourg	28 Jun 2011	2019	2022
DUQUESNE	DCN, Cherbourg	2014	2021	2025
DE GRASSE	DCN, Cherbourg	2016	2023	2027
DUPETIT-THOUARS	DCN, Cherbourg	2018	2025	2029

Displacement, tonnes: 4,725 surfaced; 5,200 dived
Dimensions, metres (feet): 99.5 × 8.8 × 7.3 *(326.4 × 28.9 × 24.0)*
Speed, knots: 25 dived
Complement: 60 (12 officers) + 15 embarked forces

Machinery: Nuclear; 1 DCNS/AREVA-TA PWR (derivative of K-15); 50 MW; turbo-electric; 2 motors; 2 emergency DA SEMT-Pielstick diesels; 1 shaft; pump jet propulsor
Missiles: SLCM: Up to 12 MBDA Scalp-Naval land-attack missile launched in capsule from torpedo tubes; inertial cruise and tercom, electro-optic homing to 1,000 km *(540 n miles)* at 0.9 Mach; warhead 300 kg.
SSM: Aerospatiale SM 39 Exocet Block 2 Mod 2 launched from 21 in *(533 mm)* torpedo tubes; inertial cruise; active radar homing to 50 km *(27 n miles)* at 0.9 Mach; warhead 165 kg.
Torpedoes: 4–21 in *(533 mm)* bow tubes. Future heavyweight torpedo (F 21 Artemis). Total of 20 torpedoes/missiles in mixed load.
Mines: Type FG 29. In lieu of torpedoes.
Physical countermeasures: DCNS Contralto-S torpedo decoys.
Electronic countermeasures: ESM: DCNS Canto acoustic decoy.
Radars: Surface search: I-band.
Sonars: Thales UMS 3000 comprising bow sonar, wide aperture flank array and a reelable thin-line towed array.
Combat data systems: SYCOBS, Link 22, Syracuse satcom, ELF comms.

Programmes: Studies for a new generation SSN (Project Barracuda) funded under the 1997–2002 budget. Programme launched on 14 October 1998 and development phase in November 2002. *Suffren* was ordered on 22 December 2006. First steel of *Duguay-Trouin* cut on 26 June 2009 and of *Tourville* on 28 June 2011 *Duquesne* is to be ordered in 2014. The contract includes a six-year integrated support package.
Structure: Much of the technology emanates from the Le Triomphant design as well as new features developed for the Scorpene design. A high level of automation is planned to reduce complement to 60. A hybrid propulsion system uses electric propulsion at cruise speeds and turbo-mechanical propulsion for higher speeds. The boats are to have dry dock hangar capability. A SAGEM-SAFRAN optronic mast is to be installed instead of a periscope. There is likely to be a mixture of 20 weapons.
Operational: Sea trials for *Suffren* are scheduled for 2016 and entry into service in 2017. The submarines are to be available for 240 days per year and refits with core refuelling are planned at 10-year intervals. Diving depth 350–400 m.

SUFFREN CLASS (artist's impression) 11/2004, DCN / 0590253

6 RUBIS AMÉTHYSTE CLASS (SSN/SNA)

Name	No	Builders	Laid down	Launched	Commissioned
RUBIS	S 601	Cherbourg Naval Dockyard	11 Dec 1976	7 Jul 1979	23 Feb 1983
SAPHIR	S 602	Cherbourg Naval Dockyard	1 Sep 1979	1 Sep 1981	6 Jul 1984
CASABIANCA	S 603	Cherbourg Naval Dockyard	19 Sep 1981	22 Dec 1984	21 Apr 1987
ÉMERAUDE	S 604	Cherbourg Naval Dockyard	4 Mar 1983	12 Apr 1986	15 Sep 1988
AMÉTHYSTE	S 605	Cherbourg Naval Dockyard	31 Oct 1984	14 May 1988	3 Mar 1992
PERLE	S 606	Cherbourg Naval Dockyard	27 Mar 1987	22 Sep 1990	7 Jul 1993

Displacement, tonnes: 2,449 surfaced; 2,713 dived
Dimensions, metres (feet): 73.6 × 7.6 × 6.4 *(241.5 × 24.9 × 21.0)*
Speed, knots: 25
Complement: 68 (8 officers)

Machinery: Nuclear; turbo-electric; 1 PWR CAS 48; 48 MW; 2 turbo-alternators; 1 motor; 9,500 hp(m) *(7 MW)*; SEMT-Pielstick/Jeumont Schneider 8 PA4 V 185 SM diesel-electric auxiliary propulsion; 450 kW; 1 emergency motor (500 kW); 1 pump jet propulsor
Missiles: SSM: Aerospatiale SM 39 Exocet; launched from 21 in *(533 mm)* torpedo tubes; inertial cruise; active radar homing to 50 km *(27 n miles)* at 0.9 Mach; warhead 165 kg.
Torpedoes: 4—21 in *(533 mm)* tubes. ECAN F17 Mod 2 (to be replaced by F-21); wire-guided; active/passive homing to 20 km *(10.8 n miles)* at 40 kt; warhead 250 kg; depth 600 m *(1,970 ft)*. Total of 14 torpedoes and missiles carried in a mixed load.
Mines: Up to 32 FG 29 in lieu of torpedoes.
Electronic countermeasures: ESM: Thomson-CSF ARUR 13/DR 3000U; intercept.
Radars: Navigation: 1 Thomson-CSF DRUA-33A; I-band; 1 Kelvin Hughes 1007; I-band.
Sonars: Thomson Sintra DMUX 20 multifunction; passive search; low frequency. DSUV 62C; towed passive array; very low frequency. DSUV 22 (Saphir); listening suite. DUUG 7A sonar intercept.
Combat data systems: TIT (Traitement des Informations Tactiques) data system (to be replaced by TITLAT); OPSMER command support system; Syracuse 2 SATCOM. Link 11 (receive only). 2 Sigma 40XP inertial guidance systems (S 602, 606 and other boats by 2012).
Weapon control systems: LAT (Lancement des Armes Tactiques) system (to be combined with TIT as TITLAT).

Programmes: The programme was terminated early by defence economies with the seventh of class *Turquoise* and eighth of class *Diamant* being cancelled.
Modernisation: Between 1989 and 1995 the first four of this class converted under operation Améthyste (AMÉlioration Tactique HYdrodynamique Silence Transmission Ecoute) to bring them to the same standard of ASW (included new

SAPHIR *7/2009, B Prézelin / 1305694*

sonars) efficiency as *Améthyste* and *Perle* rather than that required for the original anti-surface ship role. Two F17 torpedoes can be guided simultaneously against separate targets. *Saphir* recommissioned 1 July 1991, *Rubis* in February 1993; *Casabianca* in June 1994 and *Émeraude* in March 1996. A new radar added on a telescopic mast. A modernisation programme began in 2004. Upgrades include improvements to the tactical system (TITLAT programme) installation of a pump jet propulsor and a new ESM suite. Refits completed for *Casabianca* (October 2006), *Emeraude* and *Saphir* (2008) and *Perle* (2010). A nuclear refuelling programme for *Casabianca* (2010–2012), *Rubis* and *Emeraude* is to be completed in 2014. S 603–606 to be refitted with ECAN F 21 torpedoes 2014–19.

Structure: Diving depth, greater than 300 m *(984 ft)*. There has been a marked reduction in the size of the reactor compared with the L'Inflexible class. On completion of the modernisation programme, all six of the class are virtually identical.
Operational: All operational SSNs are assigned to Escadrille des Sous-Marins nucléaires d'attaque (ESNA) based at Toulon but frequently deploy to the Atlantic or overseas. Endurance rated at 45 days, limited by amount of food carried. *Rubis* had an underwater collision on 30 March 2007. Repairs at Brest were completed in July 2008. *Émeraude* had a bad steam leak on 30 March 1994 which caused casualties amongst the crew. Service life 35 years. To be replaced by the Suffren class from 2017.

PERLE *9/2010, B Prézelin / 1366844*

RUBIS *5/2009, B Prézelin / 1305695*

CASABIANCA 9/2007, B Prézelin / 1305053

ÉMERAUDE 2/2009, B Prézelin / 1305693

PERLE 8/2009, Adolfo Ortigueira Gil / 1305836

250 France > Submarines

Strategic Missile Submarines

Notes: Studies for a new SSBN to replace the Le Triomphant class from about 2030 are in progress.

4 LE TRIOMPHANT CLASS (SSBN/SNLE-NG)

Name	No	Builders	Laid down	Launched	Commissioned
LE TRIOMPHANT	S 616	DCN, Cherbourg	9 Jun 1989	26 Mar 1994	21 Mar 1997
LE TÉMÉRAIRE	S 617	DCN, Cherbourg	18 Dec 1993	21 Jan 1998	23 Dec 1999
LE VIGILANT	S 618	DCN, Cherbourg	Jan 1996	19 Sep 2003	26 Nov 2004
LE TERRIBLE	S 619	DCN, Cherbourg	Nov 2002	10 Sep 2008	27 Sep 2010

Displacement, tonnes: 12,843 surfaced; 14,565 dived
Dimensions, metres (feet): 138 × 17 × 12.5 (452.8 × 55.8 × 41.0)
Speed, knots: 25 dived
Complement: 111 (15 officers)

Machinery: Nuclear; turbo-electric; 1 PWR Type K15 (enlarged CAS 48); 150 MW; 2 turbo-alternators; 1 motor; 41,500 hp(m) (30.5 MW); diesel-electric auxiliary propulsion; 2 SEMT-Pielstick 8 PA4 V 200 SM diesels; 900 kW; 1 emergency motor; 1 shaft; pump jet propulsor
Missiles: SLBM: 16 Aerospatiale M45/TN 75 (S 617); 3-stage solid-fuel rockets; inertial guidance to 6,000 km (3,240 n miles); thermonuclear warhead with 6 MRV each of 100 kT. 16 EADS Astrium M51/TN 75 (S 616, 618–619); inertial guidance to 9,000 km (4,860 n miles); thermonuclear warhead with 6 MRV each of 100 kT.
SSM: Aerospatiale SM 39 Exocet; launched from 21 in (533 mm) torpedo tubes; inertial cruise; active radar homing to 50 km (27 n miles) at 0.9 Mach; warhead 165 kg.
Torpedoes: 4—21 in (533 mm) Q72 iD70 AT tubes. ECAN F 17 Mod 2 (to be replaced by F-21); dual purpose; active/passive homing to 9.5 km (5.1 n miles) at 35 kt; warhead 150 kg; depth to 550 m (1,800 ft); total of 18 torpedoes and SSM carried in a mixed load.
Electronic countermeasures: ESM: Thomson-CSF ARUR 13/DR 3000U; intercept.
Radars: Search: Dassault; I-band.
Sonars: Thomson Sintra DMUX 80 'multifunction' passive bow and flank arrays (S 616–618). Thales UMS 300 (S 619) comprising bow, flank and towed arrays. DUUX 5; passive ranging and intercept; low frequency. DSUV 61 (S 616–618); towed array; very low frequency.
Weapon control systems: SAD (Système d'Armes de Dissuasion) strategic data system (for SLBMs) SAD M5I will be fitted in S 619; SAT (Système d'Armes Tactique) tactical data system and DLA 4A weapon control system (for SSM and torpedoes). SYCOBS in S 619, to be installed in other three boats.

Programmes: *Le Triomphant* ordered 10 March 1986. *Le Téméraire* ordered 18 October 1989. *Le Vigilant* ordered 27 May 1993. *Le Terrible* ordered 28 July 2000. Class of six originally planned, but reduced to four after the end of the Cold War. Sous-marins Nucléaires Lanceurs d'Engins-Nouvelle Génération (SNLE-NG).
Modernisation: All four submarines are being modified to fire the M51 missile which is to replace the M45. The M51.1 missile was installed in *Le Terrible* on build and in *Le Vigilant* during her first refit 2010–13. *Le Triomphant* and *Le Téméraire* are to be modified during their second refits in 2014–16 and 2016–19 respectively. *Le Téméraire* is to be the first boat to be equipped with the uprated M51.2 missile which thereafter will be phased into the other three boats. A yet further development of the missile, to improve the performance of the third stage, is under consideration and may be introduced from 2020. A new warhead TNO (Tête Nucléaire Océanique), delivered by the M51.2 missile, is to be first introduced in *Le Téméraire* in 2016 and in the other boats as they are equipped with the M51.2 missile.
Structure: Built of HLES 100 steel capable of withstanding pressures of more than 100 kg/mm². Diving depth 500 m (1,640 ft). Height from keel to top of fin is 21.3 m (69.9 ft). Plans to lengthen the hull in later ships of the class have been shelved.
Operational: First sea cruise of *Le Triomphant* 16 July to 22 August 1995. First submerged M45 launch on 14 February 1995, second on 19 September 1996. *Le Triomphant* completed 30 month first refit in April 2005 and conducted test launch of M45 missile on 2 February 2005. *Le Triomphant* involved in a collision with HMS *Vanguard* in February 2009 and returned to service in October 2009. Her second refit,

LE TERRIBLE 3/2009, B Prézelin / 1305692

LE TÉMÉRAIRE 6/2002, French Navy / 0529140

LE TÉMÉRAIRE 1/2014*, B Prézelin / 1531193

including conversion to fire the M 51 missile, is to start in 2014. *Le Téméraire* official trials started April 1998, first submerged M 45 launch 4 May 1999. *Le Téméraire* completed 22-month refit in October 2007. *Le Vigilant* underwent her first 30-month refit 2010–13 and re-entered service in late 2013.

Sea trials of *Le Terrible* conducted in 2009. The first underwater test launch of the M 51.1 missile was conducted from *Le Terrible* on 27 January 2010. The test-firing of an M51.1 missile from *Le Vigilant* in May 2013 was unsuccessful. All submarines based at Ile Longue, Brest.

LE VIGILANT 1/2014*, B Prézelin / 1531194

LE TÉMÉRAIRE

1/2014*, B Prézelin / 1531192

LE TERRIBLE

1/2014*, B Prézelin / 1531195

LE TRIOMPHANT

5/2008, B Prézelin / 1335767

AIRCRAFT CARRIERS

Notes: A second aircraft carrier (PA 2) was planned under the 2003–08 Defence Programming Law and it later was announced on 13 February 2004 that the ship was to be built in co-operation with the UK carrier programme. This was updated by the 2009–14 Defence Programming Law which deferred the decision to proceed with construction for a few years. However, as no formal announcements were made in 2012, as expected, and in light of budget constraints, it is unlikely that a second carrier is to be built. If, as is now more likely, a future carrier will be required to replace *Charles de Gaulle* at the end of her service life in the 2040s, it is unlikely that firm plans for a future ship will emerge until about 2020. Meanwhile, studies continue and a DCNS/STX proposed design for a 280 m, nuclear-powered RXX was unveiled in late 2012.

1 CHARLES DE GAULLE CLASS (CVNM/PAN)

Name	No	Builders	Laid down	Launched	Commissioned
CHARLES DE GAULLE	R 91	DCN, Brest	14 Apr 1989	7 May 1994	18 May 2001

Displacement, tonnes: 37,680 standard; 43,182 full load
Dimensions, metres (feet): 261.5 oa; 238 wl × 64.4 oa; 31.5 wl × 9.4 *(857.9; 780.8 × 211.3; 103.3 × 30.8)*
Flight deck, metres: 261.5 × 64.4 *(857.9 × 211.3)*
Aircraft lift: 2; 19 × 12.5 m, 36 tonnes
Speed, knots: 27
Complement: 1,862 (est.) (107 officers; 542 air crew) + 60 flag staff + 88 spare berths

Machinery: Nuclear; 2 PWR Type K15; 300 MW; 2 GEC Alsthom 61 SW turbines; 83,000 hp(m) *(61 MW)* sustained; 2 shafts
Missiles: SAM: EUROSAAM SAAM/F system with 4 (2 port, 2 starboard) DCN Sylver A43 octuple VLS launchers ❶; MBDA ASTER 15; inertial guidance and mid-course update; active radar homing at 3 Mach to 30 km *(16.2 n miles)*; warhead 13 kg. 32 weapons. 2 Matra Sadral PDMS sextuple launchers ❷; Mistral; IR homing to 4 km *(2.2 n miles)*; warhead 3 kg; anti-sea-skimmer; able to engage targets down to 10 ft above sea level.
Guns: 4 Giat 20F2 20 mm; 720 rds/min to 8 km *(4.3 n miles)*; weight of shell: 0.25 kg.
Physical countermeasures: Decoys: 4 CSEE Sagaie AMBL-2A 10-barrelled trainable launchers ❸; medium range; chaff to 8 km *(4.3 n miles)*; IR flares to 3 km *(1.6 n miles)*. Dassault LAD offboard decoys. SLAT torpedo decoys from 2006.
Electronic countermeasures: ESM: Thomson-CSF ARBR 21; intercept. 1 DIBV 2A Vampir MB; (IRST) ❹.
EDM: 2 ARBB 33B ❺; jammers.
Radars: Air search: Thomson-CSF DRBJ 11B ❼; 3D; E/F-band; range 366 km *(200 n miles)* for aircraft. Thales DRBV 26D Jupiter ❽; D-band; range 183 km *(100 n miles)* for 2 m² target.
Air/surface search: Thomson-CSF DRBV 15C Sea Tiger Mk 2 ❾; E/F-band; range 110 km *(60 n miles)* for 2 m² target.
Navigation: 2 Racal 1229 (DRBN 34A) ❿; I-band.
Fire control: Thomson-CSF Arabel 3D ⓫; I/J-band (for SAAM); range 70 km *(38 n miles)* for 2 m² target.
Tacan: NRBP 20A ⓬.
Sonars: SLAT torpedo attack warning.

Combat data systems: SENIT 8-05; Links 11, 14 and 16. Syracuse 3 and FLEETSATCOM ❻. AIDCOMER and MCCIS command support systems.
Electro-optic systems: 2 DIBC 2A (Sagem VIGY-105) optronic directors.

Fixed-wing aircraft: 12–16 Super Étendard, 2 E-2C Hawkeye. 10–14 Rafale F2 and F3.
Helicopters: 2 AS 565 Panther or 2 AS 322 Cougar (AF) or 2 Puma/Super Puma plus 2 Dauphin SAR.

Programmes: On 23 September 1980 the Defence Council decided to build two nuclear-propelled carriers to replace *Clemenceau* in 1996 and *Foch* some years later. First of class ordered 4 February 1986, first metal cut 24 November 1987. Hull floated for technical trials on 19 December 1992, and back in dock on 8 January 1993. A 19.8 m *(65 ft)* long one-twelfth scale model was used for hydrodynamic trials. Building programme delayed three years due to defence budget cuts.
Modernisation: From October 1999 to March 2000 modifications included additional radiation shielding, and lengthening of angled flight deck by 4.4 m. A 43 launchers to be replaced by A 50 (for ASTER 15 and ASTER 30) in due course. During her IPER 2007–08, she was fitted with new propellers, Syracuse 3 Satcom, and modifications to operate Rafale F2 and F3. The next IPER is planned 2016–17.
Structure: Two lifts 62.3 × 41 ft *(19 × 12.5 m)* of 36 tons capacity. Hangar for 20–25 aircraft; dimensions 454.4 × 96.5 × 20 ft *(138.5 × 29.4 × 6.1 m)*. Angled deck 8.5° and 655.7 ft *(200 m)* overall length. Catapults: 2 USN Type C13-3; length 246 ft *(75 m)* for Super Étendards and up to 23 tonne aircraft. Enhanced weight capability of flight deck to allow operation of AEW aircraft. Island placed well forward so that both lifts can be protected from the weather. CSEE Dallas (Deck Approach and Landing Laser System) fitted, later to be replaced by MLS system. Active fin stabilisers. Bunkerage of 3,000 cum of avgas and 1,500 cum dieso.
Operational: Seven years continuous steaming at 25 kt available before refuelling (same reactors as *Le Triomphant*). Both reactors self-sustaining by 10 June 1998. Sea trials started 26 January 1999 and continued until 9 November 2000 when a large section of the port propeller was lost while steaming at high speed. Trials resumed on 26 March 2001 with spare propellers from decommissioned *Clemenceau*. During post-refit trials in January 2009, propulsion problems led to a further six-month docking period which was completed in August 2009. Based at Toulon.

CHARLES DE GAULLE 3/2010, B Prézelin / 1366845

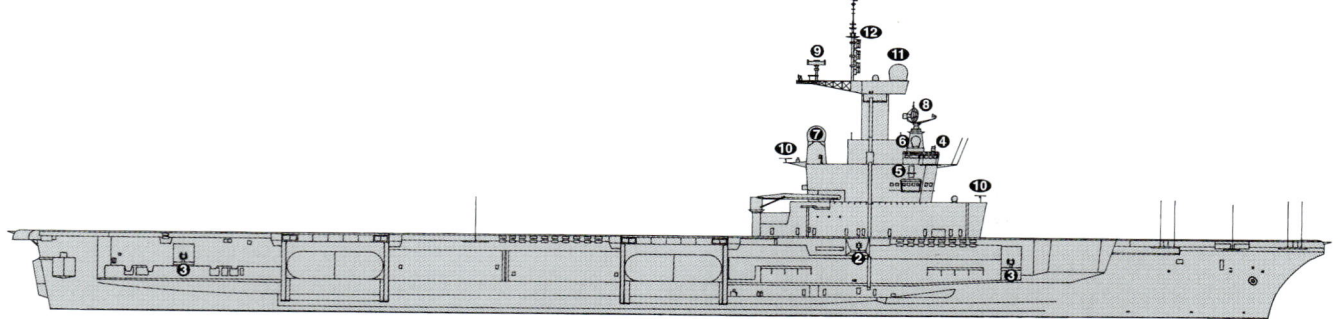
CHARLES DE GAULLE (Scale 1 : 1,500), Ian Sturton / 0069903

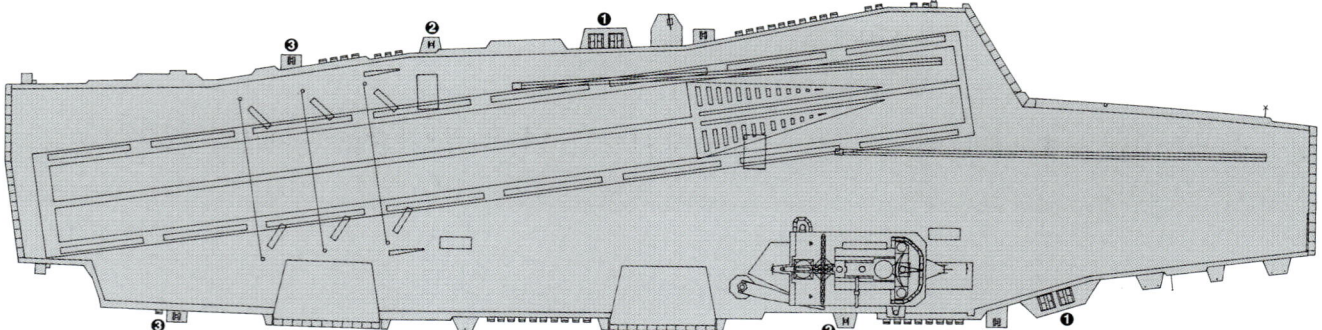
CHARLES DE GAULLE (Scale 1 : 1,500), Ian Sturton / 0104438

CHARLES DE GAULLE 2/2009, B Prézelin / 1305712

Aircraft carriers < **France** 253

CHARLES DE GAULLE 3/2006 / 1167143

CHARLES DE GAULLE 4/2006, Guy Toremans / 1167142

DESTROYERS

2 FORBIN (HORIZON) CLASS (DDGHM)

Name	No	Builders	Laid down	Launched	Commissioned
FORBIN	D 620	DCN, Lorient	16 Jan 2004	10 Mar 2005	14 Oct 2010
CHEVALIER PAUL	D 621	DCN, Lorient	13 Jan 2005	12 Jul 2006	10 Jun 2011

Displacement, tonnes: 5,791 standard; 7,163 full load
Dimensions, metres (feet): 152.9 oa; 141.7 wl × 20.3 × 8 *(501.6; 464.9 × 66.6 × 26.2)*
Speed, knots: 31; 18 diesel. **Range, n miles:** 7,000 at 18 kt
Complement: 195 (27 officers)

Machinery: CODOG: 2 Fiat Avio/GE LM 2500 gas turbines; 63,030 hp *(47 MW)*; 2 SEMT-Pielstick 12PA 6STC; 11,700 hp(m) *(9.4 MW)*; 2 shafts; cp props; bow thruster *(550 kW)*
Missiles: SSM: 8 MBDA Exocet MM 40 Block 3 ❶; inertial cruise; active radar homing to 180 km *(97 n miles)* at 0.9 Mach; warhead 165 kg; sea-skimmer.
SAM: EUROPAAMS PAAMS with DCN Sylver A50 VLS ❷ for Aerospatiale Matra Aster 15 (16 missiles) and Aster 30 (32 missiles); 48 cells (six octuple launcher modules); inertial guidance, mid-course update and active homing; range (Aster 15) 30 km *(16.2 n miles)* at 3 Mach; (Aster 30) 120 km *(65 n miles)* at 4.5 Mach. 1 MBDA Tetral quadruple launcher (fitted for but not with) ❸ for Mistral SR SAMs; IR homing to 6 km; warhead 3 kg; anti-sea-skimmer; able to engage targets down to 10 ft above sea level.
Guns: 2 Oto Melara 76 mm/62 Super Rapid ❹; 120 rds/min to 16 km *(8.7 n miles)*; weight of shell 6 kg. 2 Giat 20F2 20 mm ❺; 720 rds/min to 2 km.
Torpedoes: 2 Eurotorp TLS 324 mm fixed launchers ❻. Up to 24 Eurotorp Mu 90 Impact torpedoes; active/passive homing to 25 km *(13.5 n miles)* at 29 kt or 12 km *(6.5 n miles)* at 50 kt; warhead 32 kg.
Physical countermeasures: Decoys: 2 EADS NGDS multifunction decoy launchers ❼. SLAT torpedo defence system.
Electronic countermeasures: ESM/ECM: SIGEN EW suite comprising, radar warning equipment, a high-power jammer ❽ and an ESM/ECM support aid.
Radars: Air/surface search: Thales/Marconi DRBV 27 (S 1850M) Astral ⓫; D-band.
Surveillance/fire control: Alenia Marconi SPY-790 EMPAR ⓬; G-band; multifunction.
Surface search: 2 SPN 753 ⓭; I-band.
Fire control: Alenia Marconi NA 25 XP (RTN-30X) ⓮; J-band.
Sonars: Thales TUS-WASS 4110CL; hull-mounted; active search and attack; medium frequency. Thales passive towed array.
Combat data systems: EUROSYSNAV; 2 Link 11 (Link 22 in the future) and Link 16; OPSMER or SIC 21 follow-on command support system; Syracuse 3 SATCOM ❾.
Electro-optic systems: Sagem Vampir optronic director ❿.
Helicopters: 1 NH90 ⓯.

Programmes: Classified as 'Frégates de défense aérienne' (FDA). Initially a three-nation project with Italy and UK. Joint project office established in 1993. After UK withdrew in April 1999, an agreement was signed on 7 September 1999 between France and Italy to continue. Following a French/Italian MoU on 22 September 2000 to build four destroyers, the French government ordered two ships to be built by DCN Lorient and delivered in December 2006 and April 2008. Plans to build a second pair of ships were shelved in favour of two AAW variants of the FREMM design.

Modernisation: Plans to equip both ships with Scalp Naval land-attack missiles are under consideration.
Structure: Details given are subject to change. Space available for two additional missile launcher modules, possibly with Sylver A70 VLS.
Operational: Sea trials for *Forbin* started in July 2006 and for *Chevalier Paul* on 15 October 2007. Commissioning was delayed until integration of the combat management system and PAAMS was completed. Both based at Toulon.

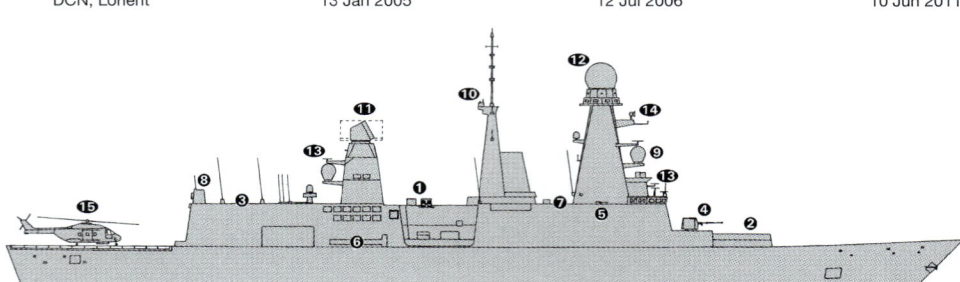

FORBIN (Scale 1 : 1,200), Ian Sturton / 1461757

FORBIN 3/2010, B Prézelin / 1366846

FORBIN 4/2010, Michael Nitz / 1366776

CHEVALIER PAUL 3/2012, B Prézelin / 1455783

IHS Jane's Fighting Ships 2014-2015

1 + 7 (3) AQUITAINE CLASS (DDGHM)

Name	No	Builders	Laid down	Launched	Commissioned
AQUITAINE	D 650	DCN, Lorient	Dec 2007	29 Apr 2010	2014
NORMANDIE	D 651	DCN, Lorient	Oct 2009	18 Oct 2012	2014
PROVENCE	D 652	DCN, Lorient	Dec 2010	18 Sep 2013	2015
LANGUEDOC	D 653	DCN, Lorient	Sep 2011	2014	2016
AUVERGNE	D 654	DCN, Lorient	Aug 2012	2014	2016
ALSACE	D 655	DCN, Lorient	2014	2015	2019
BRETAGNE	D 656	DCN, Lorient	2014	2016	2020
LORRAINE	D 657	DCN, Lorient	2015	2017	2021

Displacement, tonnes: 5,217 standard; 6,096 full load (est.)
Dimensions, metres (feet): 142.2 oa; 137.1 wl × 19.7 × 5.4 *(466.5; 449.8 × 64.6 × 17.7)*
Speed, knots: 27.5; 16 motors. **Range, n miles:** 6,000 at 15 kt
Complement: 94 (22 officers, 14 air crew) + 37 spare berths

Machinery: CODLOG; 1 Fiat/GE LM 2500 T4 gas turbine; 43,520 hp(m) *(32 MW)*; 4 MTU 16V 4000 M63L diesel generators; 11,264 hp *(8.4 MW)*; 2 Jeumont motors; 5,766 hp *(4.3 MW)*; 2 shafts. 1 Brunvoll azimuth retractable propulsor (for emergency)
Missiles: SLCM: 16 (2 octuple) cell Sylver A70 VLS (F-ASM variants) ❶ for MBDA Scalp-Naval; inertial/terrain following navigation with GPS and high precision IIR terminal guidance to 1,000 km *(540 n miles)*; warhead 300 kg.
SAM: 16 (2 octuple) cell Sylver A43 (A50 (4 octuple) in FREDA) VLS for MBDA Aster 15 (and Aster 30 in FREDA) ❶; inertial guidance, mid-course update and final active homing to 30 km *(16.2 n miles)* at 3 Mach
SSM: 8 MBDA Exocet MM 40 Block 3 ❷; inertial cruise; active radar homing to 180 km *(100 n miles)* at 0.9 Mach; warhead 165 kg.
Guns: 1 Oto Melara 76 mm/62 Super Rapid ❸. 2—20 mm. 2—12.7 mm MGs.
Torpedoes: 2 twin 324 mm Eurotorp B 515/2H/F fixed launchers for Eurotorp MU 90; active/passive homing to 25 km *(13.5 n miles)* at 29 kt or 12 km *(6.5 n miles)* at 50 kt; warhead 32 kg. 19 weapons (F-ASM).
Physical countermeasures: Decoys: 2 EADS NGDS 12-barrelled chaff, IR and anti-torpedo decoy launchers.
Torpedo defence: SLAT (Thales TUS WASSB 525/12) and Alto torpedo warning system.
Electronic countermeasures: ESM/ECM: Thales/Electronica Sigen CESM and RESM suite.
Radars: Air/surface search: Thales Herakles 3-D multifunction ❹; E/F-band.
Fire control: Alenia Marconi NA-25XP (RTN-30X); J-band.
Navigation: 2 Scanter 2001; I-band ❺.
Sonars: Thales TUS/MFS 4110CL; hull mounted (bow dome); active search and attack. Thales CAPTAS UMS-4249 VDS active/passive towed array (F-ASM).
Combat data systems: DCN/Thales SETIS CMS. Links 11 and 16, 22 and JSAT.
Electro-optic systems: 1 optronic FCS. Thales Artémis IRST. 1 Sagem Nagic optronic director.

Helicopters: 1 NH-90 ❻. ASW aircraft in ASW variant.

Programmes: Agreement reached on 7 November 2002 for a 27-ship collaborative programme with Italy. The original French requirement was for 17 FREMM of which there would be eight ASW (F-ASM) variants and nine land-attack (F-AVT). This plan was later modified by the 2009–14 Defence Programming Law in which the total number of ships was reduced to 11, the F-AVT variants were dropped and two AAW variants (FREDA) were included. Contract for the first phase awarded on 16 November 2005 to Armaris (DCN/Thales joint venture) for the construction of a first batch of eight F-ASM ships. The contract for a second batch of three ships (comprising one F-ASM and two FREDA (hulls 10 and 11)) was announced on 8 October 2009. The programme was again modified by the 2014–19 Defence Programming Law: *Bretagne* and *Lorraine* are to be FREDA variants and the building of hulls 9–11 was suspended until 2016 when a decision on the way-ahead will be made.
Structure: FREMM has a conventional hull design. The main engine room contains the gas turbine and two diesel generators while the aft machinery space contains the motors. Particular attention has been paid to signature reduction. The radar signature is expected to be comparable to that of the La Fayette class while exhaust cooling measures are expected to achieve a comparatively low IR signature. Acoustic quietening is to be achieved by the rafting of engines and motors and the use of electric propulsion. The original design for a single integrated mast has been abandoned in favour of a two mast configuration. The Herakles radar is housed in the foremast and communications and IFF in the after mast. The FREMM-ER design, which is likely to be chosen to meet the FREDA requirement, features a 32-cell Sylver A 50 launcher and a single integrated mast housing a four-panel Thales Seafire 500 phased array radar. Seafire 500 is a variant of the Herakles radar.
Operational: Sea trials for *Aquitaine* began on 18 April 2011. She was delivered to the French Navy on 23 November 2012 and will become operational in 2014. Sea trials of *Normandie* began 25 October 2013. Basing: Brest: *Aquitaine, Normandie, Bretagne, Lorraine*. Toulon: *Provence, Auvergne, Languedoc, Alsace*.

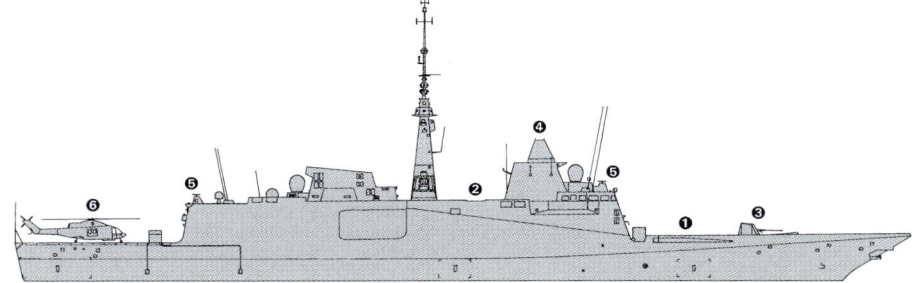

AQUITAINE *(Scale 1 : 1,200), Ian Sturton* / 1406944

AQUITAINE *5/2011, B Prézelin* / 1406383

AQUITAINE *3/2012, B Prézelin* / 1455786

AQUITAINE *4/2011, B Prézelin* / 1406381

256 France > Destroyers

2 CASSARD CLASS (TYPE F 70 (A/A)) (DDGHM)

Name	No	Builders	Laid down	Launched	Commissioned
CASSARD	D 614	Lorient Naval Dockyard, Lorient	3 Sep 1982	6 Feb 1985	28 Jul 1988
JEAN BART	D 615	Lorient Naval Dockyard, Lorient	12 Mar 1986	19 Mar 1988	21 Sep 1991

Displacement, tonnes: 4,298 standard; 5,080 full load
Dimensions, metres (feet): 139 × 15.0 × 6.5 sonar *(456 × 49.2 × 21.3)*
Speed, knots: 29.5. **Range, n miles:** 8,000 at 17 kt
Complement: 250 (25 officers) + 3 spare berths

Machinery: 4 SEMT-Pielstick 18 PA6 V 280 BTC diesels; 43,200 hp(m) *(31.75 MW)* sustained; 2 shafts
Missiles: SSM: 8 (4 carried in peacetime) Aerospatiale MM 40 Exocet Block 2 ❶; inertial cruise; active radar homing to 70 km *(40 n miles)* at 0.9 Mach; warhead 165 kg; sea-skimmer.
SAM: 40 GDC Pomona Standard SM-1MR Block VI; semi-active radar homing to 38 km *(20.5 n miles)* at 2 Mach; height envelope 45–18,288 m *(150–60,000 ft)*. Mk 13 Mod 5 launchers ❷ taken from T 47 (DDG) ships. 2 Matra Sadral PDMS sextuple launchers ❸; 39 Mistral; IR homing to 4 km *(2.2 n miles)*; warhead 3 kg; anti-sea-skimmer; able to engage targets down to 10 ft above sea level.
Guns: 1 DCN/Creusot-Loire 3.9 in *(100 mm)*/55 Mod 68 CADAM automatic ❹; 78 rds/min to 17 km *(9 n miles)* anti-surface; 8 km *(4.4 n miles)* anti-aircraft; weight of shell 13.5 kg. 2 Giat 20F2 20 mm ❺; 720 rds/min to 2 km *(1.1 n miles)*. 4—12.7 mm MGs.
Torpedoes: 2 fixed launchers model KD 59E ❻. 10 ECAN L5 Mod 4; anti-submarine; active/passive homing to 9.5 km *(5.1 n miles)* at 35 kt; warhead 150 kg; depth to 550 m *(1,800 ft)*.
Physical countermeasures: Decoys: 2 CSEE AMBL 1B Dagaie ❼ and 2 AMBL 2A (D 614) or 2B (D 615) Sagaie 10-barrelled trainable launchers ❽; fires a combination of chaff and IR flares. Dassault LAD offboard decoys.
Torpedo defence: SLQ-25A Nixie.
Electronic countermeasures: ESM: Thomson-CSF ARBR 17B (DR 4000) ❾; radar intercept. DIBV 1A Vampir ❿; IR detector (integrated with search radar for active/passive tracking in all weathers). ARBG-1A (Saigon) comms intercept at masthead.
ECM: 2 Dassault Electronique ARBB 33; jammers; H-, I- and J-bands.

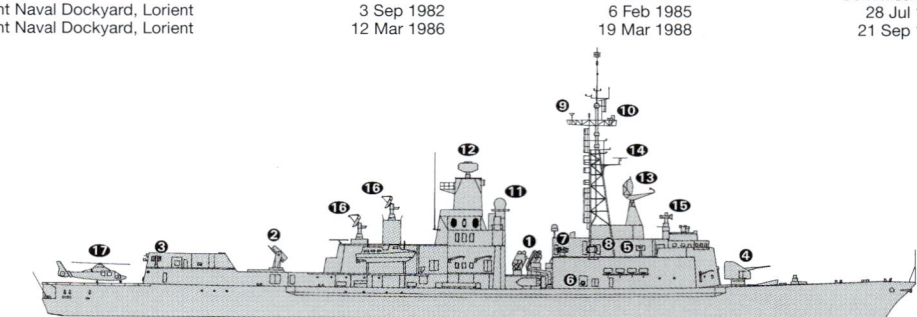

JEAN BART *(Scale 1 : 1,200), Ian Sturton / 1455812*

Radars: Air search: Thales SMART-S Mk 2 ⓬; 3D; E/F-band.
Air/surface search: Thomson-CSF DRBV 26C ⓭; D-band.
Navigation: 2 Racal DRBN 34A; I-band (1 for close-range helicopter control ⓮).
Fire control: Thomson-CSF DRBC 33A ⓯; I-band (for guns). 2 Raytheon SPG-51C ⓰; G/I-band (for missiles).
Sonars: Thomson Sintra DUBA 25A (13 kHz) (D 614); DUBA 24C (5 kHz) (D 615); hull-mounted; active search and attack; medium frequency.
Combat data systems: SENIT 6/8; Links 11, 14 and 16. Syracuse 2 SATCOM ⓫. OPSMER command support system. SIC-21 command assistance tool.
Electro-optic systems: DCN CTMS optronic/radar system with DIBC 1A Piranha II IR/TV tracker; CSEE Najir optronic secondary director.

Helicopters: 1 AS 565SA Panther ⓱.

Programmes: The building programme was considerably slowed down by finance problems and doubts about the increasingly obsolescent Standard SM 1 missile system and was curtailed at two units. Re-rated F 70 (ex-C 70) on 6 June 1988, officially 'frégates anti-aériennes (FAA)'.
Modernisation: DRBJ 15 radar initially fitted in *Cassard* but this was replaced in 1992 by DRBJ 11. Panther has replaced Lynx helicopter. *Cassard* refitted 2000–2001. Upgrade included hull strengthening, fitting of new propellers and SENIT 68 combat direction system (SENIT 6 core augmented by SENIT 8 data-link processing component (for Link 16 and data forwarding). *Jean Bart* similarly refitted October 2002 to September 2003. *Cassard* underwent a six-month refit which completed in April 2009. *Jean Bart* was refitted in 2012. This included the replacement of DRBJ 11 radar with SMART-S. *Cassard* is to be similarly modernised in 2014. Plans to fit ASTER 30 have been abandoned.
Structure: Samahe 210 helicopter handling system.
Operational: Both ships are based at Toulon. Service lives: *Cassard*, 2022; *Jean Bart*, 2023. To be replaced by FREDA (AAW variants of FREMM).

CASSARD
4/2010, Michael Nitz / 1366777

JEAN BART
4/2013*, C D Yaylali / 1529023

IHS Jane's Fighting Ships 2014-2015

© 2014 IHS

3 GEORGES LEYGUES CLASS (TYPE F 70 (ASW)) (DDGHM)

Name	No	Builders	Laid down	Launched	Commissioned
DUPLEIX	D 641	Brest Naval Dockyard	17 Oct 1975	2 Dec 1978	13 Jun 1981
MONTCALM	D 642	Brest Naval Dockyard	5 Dec 1975	31 May 1980	28 May 1982
JEAN DE VIENNE	D 643	Brest Naval Dockyard	26 Oct 1979	17 Nov 1981	25 May 1984

Displacement, tonnes: 3,942 standard; 4,908 full load
Dimensions, metres (feet): 139 × 14 × 5.9 *(456 × 45.9 × 19.4)*
Speed, knots: 30; 20 diesel
Range, n miles: 8,000 at 15 kt, 2,500 at 28 kt
Complement: 235 (22 officers)

Machinery: CODOG; 2 RR Olympus TM3B gas turbines; 52,000 hp *(38.2 MW)* sustained; 2 SEMT-Pielstick 16 PA6 V280 diesels; 11,200 hp(m) *(8.3 MW)* sustained; 2 shafts; LIPS cp props
Missiles: SSM: 4 MBDA Exocet MM 38 (D 641) or 8 Exocet MM 40 (D 642 and D 643) ❶; inertial cruise; active radar homing to 42 km *(23 n miles)* (MM 38) or 70 km *(40 n miles)* (MM 40) at 0.9 Mach; warhead 165 kg; sea-skimmer. 4 additional Exocet MM 40 missiles can be carried as a warload (D 642 and D 643).
SAM: Thomson-CSF Crotale Naval EDIR octuple launcher ❷; command line of sight guidance; radar/IR homing to 13 km *(7 n miles)* at 2.4 Mach; warhead 14 kg; 26 missiles. 2 Matra Sadral sextuple launchers for Mistral SR SAMs; IR homing to 6 km *(3.2 n miles)*; warhead 3 kg.
Guns: 1 DCN/Creusot-Loire 3.9 in *(100 mm)*/55 Mod 68 CADAM automatic ❸; dual purpose; 78 rds/min to 17 km *(9 n miles)* anti-surface; 8 km *(4.4 n miles)* anti-aircraft; weight of shell 13.5 kg. 2 Breda/Mauser 30 mm ❹. 800 rds/min to 3 km; weight of shell 0.37 kg. 2—12.7 mm MGs.
Torpedoes: 2 DCN KD-59E fixed tubes for 533 mm (21 in) DCN L5 Mod 4 torpedoes; active/passive homing to 7 km *(3.8 n miles)*; 8 to 10 torpedoes. Honeywell Mk 46 mod 2 or EuroTorp MU 90 Impact lightweight torpedoes for helicopters.
Physical countermeasures: Decoys: EADS AMBL-1C (Dagaie Mk 2) ❺; 2 10-barrel trainable launchers; chaff and IR flares. 4 AMBL-3A (Replica); offboard decoys.
Torpedo defence: AN/SLQ-25A Nixie (2 towed decoys); Prairie-Masker noise suppression system.
Electronic countermeasures: ESM: Thomson-CSF ARBR-10X and ARBR-16B (DR 2000) or (D 643) ARBR-17 (DR 4000) ❻ radar intercept; Sagem DIBV-2A (Vampir MB) IRST.
ECM: Thales ARBB-36A (D 641-643) jammer.
Radars: Air search: Thomson-CSF DRBV 26A (Jupiter) ❽; D-band.
Air/surface search: Thales DRBV-15A or -15B (Sea Tiger) ❾; E/F-band.
Navigation: 1 DRBN-34A (RM 1290) (D 641) or Kelvin Hughes DRBN 37 (KH 1007 Nucleus) (D 642, D 643); I-band; one for helo control.
Fire control: Thomson-CSF Castor 2J ❿ for Crotale Naval SAM; J-band. Thomson-CSF DRBC-32E ⓫ (Castor 2B) for gun FCS; I-band.
Sonars: Thomson-Sintra DUBV-23D (D 641) bow mounted; active search and attack; 5 kHz. 1 UMS 4110 CL (D 642, D 643). Thomson-Sintra DUBV-43B (D 641-642) or -43C (D 643) VDS ⓬; active search and attack; 5 kHz; paired with DUBV-23D; tows at up to 24 kt down to 300 m *(985 ft)* for DUBV-43B or 700 m *(2,300 ft)* for -43C. TUS DSBV-62C (D 641) (Lamproie) passive linear towed array with URDT-1A torpedo warning equipment; very low frequency.
Combat data systems: DCN SENIT 4 CDS and STIDAV/SENIT 8-01 added for anti-air/anti-missile defence; Link 11. ACOM/Opsmer command support system. Syracuse ❼ and Inmarsat satcomms.
Electro-optic systems: Thomson-CSF CTH (Vega) radar/optronic FCS and CSEE DM-Ab (Panda) optical director for 100 mm guns; 2 Sagem DIBC-2A (VIGY 105) optronic FCSs for 30 mm guns.
Helicopters: 2 Westland WG 13 Lynx Mk 4 (FN) ⓭ (one normally carried in peacetime).

Programmes: Design of a new ASW escort vessel approved in December 1971 under the designation of 'Corvette anti-sous-marine type 1970 (C 70)'. Re-rated 'Frégate anti-sous-marine type 1970 (FASM 70 or F 70)' on 6 June 1988. First four ships on the 1970–75 Defence Programming Law.
Modernisation: Ships of this class have received regular upgrades. Most important was the Opération programmée amélioration autodéfense antimissiles (OP3A, air defence upgrade programme) completed in March 1996 for Jean de Vienne, April 1999 for Dupleix and April 2000 for Montcalm; large command structure fitted above the bridge, SENIT 8-01 CDS package added to current CDS to command and control air-defence weapons and systems, 2 MBDA Sadral SAM launchers and 2 Oto Melara/Mauser 30 mm gun mounts (controlled by Sagem VIGY 105 optronic directors) added; new ESM suite, new ECM equipment and Replica offboard decoys. Plans to fit MBDA Milas ASW missiles have been shelved. Due to her new role (see below). All ships modified to receive female crew.
Structure: Hull and main deck have been strengthened to cope with fatigue problems; to restore seaworthiness, 235 tonnes of ballast have been embarked and two fuel tanks turned into water-ballasts; completed 2002–03 on all four ships. DCN SPHEX helicopter handling system.
Operational: Based at Toulon. Endurance 45 days. Service lives: Dupleix 2014; Montcalm 2015, and Jean de Vienne 2018. Camcopter S-100 UAV recovered to Montcalm on 10 October 2008.

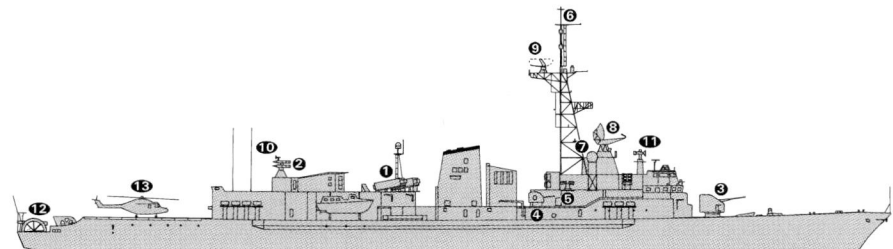

DUPLEIX *(Scale 1 : 1,200), Ian Sturton / 0581795*

JEAN DE VIENNE *3/2010, B Prézelin / 1366847*

MONTCALM *7/2011, B Prézelin / 1406961*

France > Destroyers

3 MODIFIED GEORGES LEYGUES CLASS (TYPE F 70 (ASW)) (DDGHM)

Name	No	Builders	Laid down	Launched	Commissioned
PRIMAUGUET	D 644	Brest Naval Dockyard	17 Nov 1981	17 Mar 1984	5 Nov 1986
LA MOTTE-PICQUET	D 645	Brest Naval Dockyard, Lorient	12 Feb 1982	6 Feb 1985	18 Feb 1988
LATOUCHE-TRÉVILLE	D 646	Brest Naval Dockyard, Lorient	15 Feb 1984	19 Mar 1988	16 Jul 1990

Displacement, tonnes: 4,074 standard; 4,989 full load
Dimensions, metres (feet): 139 × 15.0 × 5.7 *(456 × 49.2 × 18.7)*
Speed, knots: 30; 21 diesel
Range, n miles: 8,000 at 15 kt, 2,500 at 28 kt
Complement: 233 (21 officers)

Machinery: CODOG; 2 RR Olympus TM3B gas turbines; 52,000 hp *(38.2 MW)* sustained; 2 SEMT-Pielstick 16 PA6 V280 diesels; 11,200 hp(m) *(8.3 MW)* sustained; 2 shafts; LIPS cp props
Missiles: SSM: 8 MBDA Exocet MM 40 (only 4 in peacetime) ❶; inertial cruise and active radar homing to 72 km *(39 n miles)* at 0.93 Mach; warhead 165 kg.
SAM: Thomson-CSF Crotale Naval EDIR system ❷; octuple launcher; radar/IR command to line-of-sight to 13 km *(7 n miles)* at 2.4 Mach; warhead 14 kg; total of 26 V5S missiles carried. 2 MBDA Simbad twin launchers for Mistral SR SAMs; IR homing to 6 km *(3.2 n miles)*; warhead 3 kg.
Guns: 1 DCN 100 mm/55 (3.9 in/55) Modèle 68 CADAM automatic ❸; dual purpose; 78 rds/min to 17 km *(9 n miles)* anti-surface; 6 km *(3.2 n miles)* anti-aircraft; weight of shell 13.5 kg. 2 Giat 20F2 20 mm ❹; 720 rds/min to 2 km; 4—12.7 mm MGs.
Torpedoes: Two 324 mm EuroTorp B515/1H/F fixed torpedo tubes for EuroTorp MU 90 Impact lightweight ASW torpedoes; active/passive homing to 25 km *(13.5 n miles)* at 29 kt or 12 km *(6.5 n miles)* at 50 kt; warhead 32 kg of TATB explosive (shaped charge); depth to 1,000 m; same torpedoes for the helicopters.
Physical countermeasures: Decoys: EADS AMBL-1C (Dagaie Mk 2); two 10-barrel trainable launchers ❺; chaff and IR flares. Four AMBL-3A (Replica) (D 645) offboard decoys.
Torpedo defence: AN/SLQ-25A Nixie (two torpedo decoys); Prairie-Masker noise suppression system.
Electronic countermeasures: ESM: Thales ARBR-17 ❻ (DR 4000) radar intercept; ARBG-1A (Saigon) comms intercept; Sagem DIBV-2A (Vampir MB) IRST.
ECM: Thales ARBV-36A; jammer.
Radars: Air/surface search: Thomson-CSF DRBV-15A (D 645) or -15B (Sea Tiger) (D 644, D 646) ❽; E/F-band.
Navigation: 2 DRBN 34A (D 646); 2 DRBN 37 (D 644, D 645); I-band.

LA MOTTE-PICQUET *(Scale 1 : 1,200), Ian Sturton / 0581796*

Fire control: Thomson-CSF Castor 2J ❾ for Crotale Naval SAM; J-band. Thomson-CSF DRBC-33A (Castor 2C) ❿ for gun FCS; I-band.
Sonars: Thomson-Sintra DUBV-24C bow-mounted; active search and attack; 5 kHz Thomson-Sintra DUBV-43C VDS ⓫; active search and attack; 5 kHz; paired with DUBV-24C; tows at up to 24 kt down to 700 m *(2,300 ft)*. TUS DSBV-61B passive linear towed array with URDT-1A torpedo warning equipment; very low frequency. PAF sonobuoy data processing system.
Combat data systems: DCN SENIT 4 CDS; Link 11 (Link 22 in due course). ACOM/Opsmer command support system; Syva ASW decision aid. Syracuse ❼ and Inmarsat satcomms.
Electro-optic systems: DCN CTMS radar/optronic FCS (with DRBC-33A radar, DIBC-1A Pirana IR tracker, TV tracker) and CSEE DM-Ab (Panda) optical director for 100 mm gun. Alcatel DLT-L5 for torpedoes.
Helicopters: 2 Westland WG 13 Lynx Mk 4 (FN) ⓬ (one normally carried in peacetime).
Programmes: 'Frégates anti-sous-marines type 1970' (FASM 70 or F 70). Authorised on the 1975–80 Defence Programming Law. Fourth unit cancelled before construction had started. *La Motte-Picquet* and *Latouche-Tréville* started building at Brest and towed to Lorient for outfitting. To be replaced by FREMM/ASM in the late 2010s.
Modernisation: The ships have been upgraded by the OP3A (Opération programmée amélioration autodéfense antimissiles) air defence upgrade programme, limited to the upgrade of sensors and ESM equipment and the installation of two MBDA Simbad twin launchers for Mistral SR SAMs; completed 1997–99. In 2004–06, further modernisation include the replacement of the two KD-59E launchers for 533 mm (21 in) L 5 torpedoes by two 324 mm B 515 fixed tubes for EuroTorp MU 90 Impact lightweight torpedoes. Plans to fit MBDA Milas ASW missiles have been shelved. All vessels modified to receive female crew. Plans to install LFTASS (ATBF 2) very low frequency towed active sonar have been postponed indefinitely.
Structure: Bridge raised one deck as compared to first four ships of the class. Hull and main deck have been strengthened to cope with fatigue problems; to restore seaworthiness, 235 tonnes of ballast have been embarked and two fuel tanks turned into water-ballasts; completed 2002–03 on all ships. DCN SPHEX helicopter handling system.
Operational: All based at Brest. Service lives: 2021–23.

LATOUCHE-TRÉVILLE

10/2012, Michael Nitz / 1455784

PRIMAUGET

2/2013, Michael Nitz / 1529024*

IHS Jane's Fighting Ships 2014-2015

© 2014 IHS

FRIGATES

5 LA FAYETTE CLASS (FFGHM)

Name	No	Builders	Laid down	Launched	Commissioned
LA FAYETTE	F 710	DCN, Lorient	15 Dec 1990	13 Jun 1992	23 Mar 1996
SURCOUF	F 711	DCN, Lorient	6 Jul 1992	3 Jul 1993	7 Feb 1997
COURBET	F 712	DCN, Lorient	15 Sep 1993	12 Mar 1994	1 Apr 1997
ACONIT (ex-*Jauréguiberry*)	F 713	DCN, Lorient	1 Aug 1996	8 Jun 1997	3 Jun 1999
GUÉPRATTE	F 714	DCN, Lorient	1 Oct 1998	3 Mar 1999	27 Oct 2001

Displacement, tonnes: 3,353 standard; 3,810 full load
Dimensions, metres (feet): 124.2 oa; 115 pp × 15.4 × 5.8 screws *(407.5; 377.3 × 50.5 × 19.0)*
Speed, knots: 25
Range, n miles: 7,000 at 15 kt, 9,000 at 12 kt
Complement: 178 (15 officers)

Machinery: CODAD; 4 SEMT-Pielstick 12 PA6 V 280 STC diesels; 21,107 hp(m) *(15.52 MW)* sustained; 2 shafts; LIPS cp props; bow thruster
Missiles: SSM: 8 Aerospatiale MM 40 Block 3 Exocet ❶; inertial cruise; active radar homing to 180 km *(97 n miles)* at 0.9 Mach; warhead 165 kg; sea-skimmer.
SAM: Thomson-CSF Crotale Naval CN 2 octuple launcher ❷; command line of sight guidance; radar/IR homing to 13 km *(7 n miles)* at 3.5 Mach; warhead 14 kg. 26 missiles. Space for 2 – 8 cell VLS ❸.
Guns: 1 DCN 3.9 in *(100 mm)*/55 TR ❹; 78 rds/min to 17 km *(9 n miles)*; weight of shell 13.5 kg. 2 Giat 20F2 20 mm ❺; 720 rds/min to 10 km *(5.5 n miles)*.
Physical countermeasures: Decoys: 2 CSEE AMBL-1C (Dagaie Mk 2) ❻ 10-barrelled trainable launchers; chaff and IR flares.
Torpedo defence: SLQ-25A Nixie.
Electronic countermeasures: ESM: Thomson-CSF ARBR 21A (DR 3000-S) ❼; radar intercept. ARBG-1 (Saigon) (F 710–712) or ARBG 2A (F 713–714) (Maigret); comms intercept. DIBV 10 Vampir ❽; IR detector (can be fitted).
ECM: Dassault ARBB 33; jammer (can be fitted).
Radars: Air/surface search: Thales DRBV-15C (Sea Tiger 2) ❿; E/F-band; range 110 km *(60 n miles)* for 2 m^2 target.
Navigation: 2 Racal Decca 1229 (DRBN 34B) ⓫; I-band. One set for helicopter control.
Fire control: Thomson-CSF Castor 2J/C ⓬; J-band; range 17 km *(9.2 n miles)* for 1 m^2 target. Crotale ⓭; J-band (for SAM).
Combat data systems: Thomson-CSF TAVITAC 2000. Link 11. Syracuse 2 SATCOM ❾. OPSMER command support system. INMARSAT.
Electro-optic systems: Sagem TDS 90 VIGY optronic system.
Helicopters: 1 Aerospatiale AS 565MA Panther ⓮ or platform for 1 Super Frelon. NH90 in due course.

Programmes: Originally described as 'Frégates Légères' but this was changed in 1992 to 'Frégates type La Fayette'. First three ordered 25 July 1988; three more 24 September 1992 but the last of these was cancelled in May 1996. The construction timetable was delayed by several months because of funding problems.
Modernisation: Fitted with VIGISCAN optronic system from October 2009. Dillon M134 mini-guns may be fitted (two per ship). Modernisation of the class to mitigate the effects of the delay and possible curtailment of the FREMM programme is under consideration. Possibilities include installation of Aster 15 or VL-MICA SAM, a sonar, torpedo tubes, and upgrade of the command system. A decision is expected in 2016.
Structure: Constructed from high-tensile steel with a double skin from waterline to upperdeck. 10 mm plating protects vital spaces. External equipment and upper deck fittings are concealed or placed in low positions. Superstructure inclined at 10° to vertical to reduce REA. Extensive use of radar absorbent paint. DCN Samahe helicopter handling system. RHIB assault craft fitted – these are launched and recovered from a stern access. The design includes potential to install new and/or replace old weapon systems in the future. This includes the SAAM/F system to replace Crotale (space is available forward of the bridge to install Sylver A43 octuple VLS launchers for Aster 15 missiles).
Operational: *La Fayette* started sea trials 27 September 1993, *Surcouf* 4 July 1994, *Courbet* 14 September 1995, *Aconit* 14 April 1998 and *Guépratte* on 16 January 2001. All based at Toulon. Service life 30 years.
Sales: Three of an improved design to Saudi Arabia, six for Taiwan, and six for Singapore.

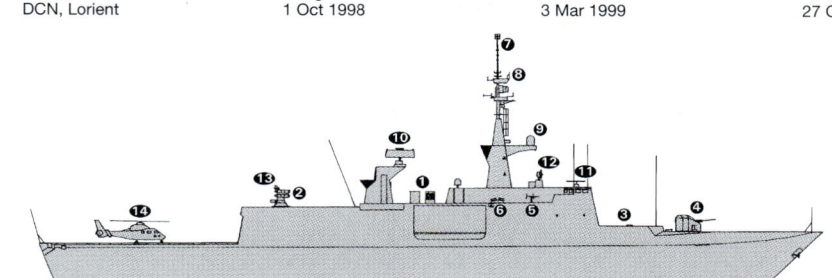

LA FAYETTE *(Scale 1 : 1,200), Ian Sturton / 0581797*

GUÉPRATTE *8/2013*, Shaun Jones / 1529025*

COURBET *2/2009, B Prézelin / 1305707*

ACONIT *5/2013*, Giorgio Ghiglione / 1529090*

9 D'ESTIENNE D'ORVES (TYPE A 69) CLASS (FS)

Name	No	Builders	Laid down	Launched	Commissioned
LIEUTENANT DE VAISSEAU LE HÉNAFF	F 789	Lorient Naval Dockyard, Lorient	21 Mar 1977	16 Sep 1978	13 Feb 1980
LIEUTENANT DE VAISSEAU LAVALLÉE	F 790	Lorient Naval Dockyard, Lorient	30 Nov 1977	12 May 1979	9 Oct 1980
COMMANDANT L'HERMINIER	F 791	Lorient Naval Dockyard, Lorient	29 May 1979	7 Mar 1981	19 Jan 1986
PREMIER MAÎTRE L'HER	F 792	Lorient Naval Dockyard, Lorient	15 Dec 1978	28 Jun 1980	5 Dec 1981
COMMANDANT BLAISON	F 793	Lorient Naval Dockyard, Lorient	15 Nov 1979	7 Mar 1981	28 Apr 1982
ENSEIGNE DE VAISSEAU JACOUBET	F 794	Lorient Naval Dockyard, Lorient	8 Jul 1980	26 Sep 1981	23 Oct 1982
COMMANDANT DUCUING	F 795	Lorient Naval Dockyard, Lorient	1 Oct 1980	26 Sep 1981	17 Mar 1983
COMMANDANT BIROT	F 796	Lorient Naval Dockyard, Lorient	23 Mar 1981	22 May 1982	14 Mar 1984
COMMANDANT BOUAN	F 797	Lorient Naval Dockyard, Lorient	12 Oct 1981	23 Apr 1983	31 Oct 1984

Displacement, tonnes: 1,194 standard; 1,270 (F 789–791), 1,311 (F 792–793), 1,351 (F 794–797) full load
Dimensions, metres (feet): 80.5 × 10.3 × 5.5 sonar *(264.1 × 33.8 × 18.0)*
Speed, knots: 24 (F 789–790, 792–797), 25 (F 791)
Range, n miles: 4,500 at 15 kt
Complement: 108 (est.) (7 officers)

Machinery: 2 SEMT-Pielstick 12 PC2 V 400 diesels; 12,000 hp(m) *(8.82 MW)* (F 792–797); 13,200 hp(m) *(9.8 MW)* (F 789–791); 2 shafts; LIPS cp props 2 SEMT-Pielstick 12 PA6 V 280 BTC diesels; 14,400 hp(m) *(10.6 MW)* sustained; 2 shafts; LIPS cp props (F 791)
Missiles: SAM: Matra Simbad twin launcher for Mistral ❶; IR homing to 4 km *(2.2 n miles)*; warhead 3 kg.
Guns: 1 DCN/Creusot-Loire 3.9 in *(100 mm)*/55 Mod 68 CADAM automatic ❷; 80 rds/min to 17 km *(9 n miles)* anti-surface; 8 km *(4.4 n miles)* anti-aircraft; weight of shell 13.5 kg. 2 Giat 20 mm ❸; 720 rds/min to 10 km *(5.5 n miles)*. 4 – 12.7 mm MGs.
Torpedoes: 4 fixed tubes ❹ not fitted with torpedoes.
Physical countermeasures: Decoys: SLQ-25 Nixie; torpedo decoy.
Electronic countermeasures: ESM: ARBR 16; radar warning.
Radars: Air/surface search: Thomson-CSF DRBV 51A ❺; G-band.
Navigation: Kelvin Hughes 1007; I-band.
Fire control: Thomson-CSF DRBC 32E ❼; I-band.
Sonars: Thomson Sintra DUBA 25 (F 789–793); hull-mounted; search and attack; medium frequency.
Combat data systems: Syracuse 2 SATCOM (F 792, F 793, F 794, F 795, F 796, F 797) ❻. OPSMER command support system. INMARSAT.
Weapon control systems: Thomson-CSF Vega system; CSEE DM-Ab (Panda) optical secondary director.

Programmes: Classified as 'Avisos'.
Modernisation: In 1985 Commandant L'Herminier, F 791, fitted with 12PA6 BTC Diesels Rapides as trial for Type F 70. Most have dual MM 38/MM 40 ITL (Installation de Tir Légère) capability. Weapon fit depends on deployment and operational requirement. Those without ITL are fitted with ITS (Installation de Tir Standard). Syracuse 2 SATCOM fitted in F 792–797, and accommodation provided for commandos. Matra Simbad launchers have been fitted aft of the Syracuse SATCOM for operations. Fast raiding craft fitted to Commandant Birot and to others in due course.
Operational: F 794–797 based at Toulon and F 789–793 at Brest. Following the removal of missiles (but not launchers), all nine ships have been reduced to a patrol ship role, although the Brest-based ships retain sonars and are fitted for but not with torpedoes. Ships' lives have been extended to 35 years.

Sales: The original Lieutenant de Vaisseau Le Hénaff and Commandant l'Herminier sold to South Africa in 1976 while under construction. As a result of the UN embargo on arms sales to South Africa, they were sold to Argentina in September 1978 followed by a third, specially built. Six ships were sold to Turkey in October 2000. All delivered by July 2002 after refit at Brest. The last one, Second Maître Le Bihan, decommissioned from the French Navy on 26 June 2002. No further sales are planned.

PREMIER MAÎTRE L'HER *(Scale 1 : 900), Ian Sturton / 1305882*

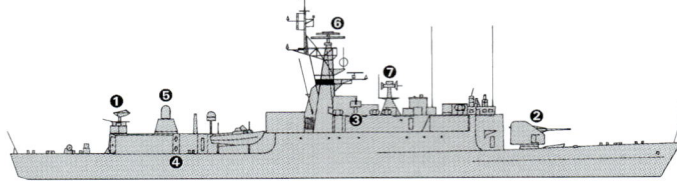

COMMANDANT L'HERMINIER *12/2011, Derek Fox / 1406943*

COMMANDANT DUCUING *4/2010, C D Yaylali / 1366780*

COMMANDANT BLAISON *6/2012, Michael Winter / 1455785*

6 FLORÉAL CLASS (FFGHM)

Name	No	Builders	Laid down	Launched	Commissioned
FLORÉAL	F 730	Chantiers de l'Atlantique, St Nazaire	2 Apr 1990	6 Oct 1990	27 May 1992
PRAIRIAL	F 731	Chantiers de l'Atlantique, St Nazaire	11 Sep 1990	16 Mar 1991	20 May 1992
NIVÔSE	F 732	Chantiers de l'Atlantique, St Nazaire	16 Jan 1991	10 Aug 1991	16 Oct 1992
VENTÔSE	F 733	Chantiers de l'Atlantique, St Nazaire	28 Jun 1991	14 Mar 1992	5 May 1993
VENDÉMIAIRE	F 734	Chantiers de l'Atlantique, St Nazaire	17 Jan 1992	23 Aug 1992	21 Oct 1993
GERMINAL	F 735	Chantiers de l'Atlantique, St Nazaire	17 Aug 1992	14 Mar 1993	17 May 1994

Displacement, tonnes: 2,642 standard; 2,997 full load
Dimensions, metres (feet): 93.5 × 14 × 4.3 (306.8 × 45.9 × 14.1)
Speed, knots: 20
Range, n miles: 10,000 at 15 kt
Complement: 83 (11 officers) + 24 embarked forces + 13 spare berths

Machinery: CODAD; 4 SEMT-Pielstick 6 PA6 L 280 BTC diesels; 8,820 hp(m) (6.5 MW) sustained; 2 shafts; LIPS cp props; 272 hp (200 kW) bow thruster; 340 hp(m) (250 kW)
Missiles: SSM: 2 Aerospatiale MM 38 Exocet ❶; inertial cruise; active radar homing to 42 km (23 n miles) at 0.9 Mach; warhead 165 kg; sea-skimmer.
SAM: 1 or 2 Matra Simbad twin launchers can replace 20 mm guns or Dagaie launcher.
Guns: 1 DCN 3.9 in (100 mm)/55 Mod 68 CADAM ❷; 78 rds/min to 17 km (9 n miles); weight of shell 13.5 kg. 2 Giat 20 F2 20 mm ❸; 720 rds/min to 10 km (5.5 n miles).
Physical countermeasures: Decoys (fitted for but not with): 2 CSEE AMBL-1C (Dagaie Mk II); 10-barrelled trainable launchers ❹; chaff and IR flares.
Electronic countermeasures: ESM: Thomson-CSF ARBR 16A (F 735) ❺; radar intercept. ARBG 1A (Saigon); comms intercept (F 730 and F 733).
Radars: Air/surface search: Thomson-CSF Mars DRBV 21C ❻; D-band.
Navigation: 2 Racal Decca 1229 (DRBN 34A); I-band (1 for helicopter control ❾).
Combat data systems: ACOM/OPSMER command support system (F 735). Syracuse (F 730 and F 733) and INMARSAT ❼ SATCOM.

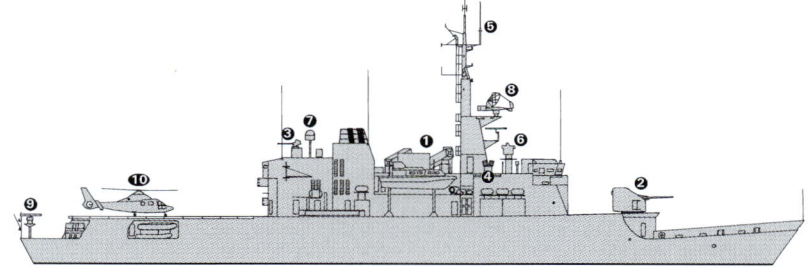

PRAIRIAL (Scale 1 : 900), Ian Sturton / 0529161

Electro-optic systems: CSEE Najir optronic director ❺.
Helicopters: 1 AS 565MA Panther or platform for 1 AS 332F Super Puma ❿.
Programmes: Officially described as 'Frégates de Surveillance' or 'Ocean capable patrol vessel' and designed to operate in the offshore zone in low-intensity operations. First two ordered on 20 January 1989; built at Chantiers de l'Atlantique, St Nazaire, with weapon systems fitted by DCAN Lorient. Second pair ordered 9 January 1990; third pair in January 1991. Named after the months of the Revolutionary calendar.
Modernisation: Najir optronic director is being replaced by Sagem EOMS-NG optronic system from 2012.
Structure: Built to merchant passenger marine standards with stabilisers and air conditioning. New funnel design improves air flow over the flight deck. Has one freight bunker aft for about 100 tons cargo. Second-hand Exocet MM 38 has been fitted instead of planned MM 40.
Operational: Endurance, 50 days. Able to operate a helicopter up to Sea State 5. Stations as follows: *Ventose* and *Germinal* in Antilles, *Prairial* in Tahiti. *Floréal* and *Nivôse* at La Réunion and *Vendémiaire* at Noumea (New Caledonia). *Floréal* refitted in floating dry-dock at Papeete in 2003 and again in 2013. Service life 30 years.
Sales: Two delivered to Morocco in 2002 and 2003.

VENDÉMIAIRE 2/2012, Hachiro Nakai / 1455813

PRAIRIAL 8/2012*, A A de Kruijf / 1529026

SHIPBORNE AIRCRAFT

Notes: The naval drone concepts for the French Navy have moved to a common navy/army project for a VTOL drone that could perform both tactical and long-range missions. It should be able to operate from the flight deck of a frigate. Contenders contracted on 10 November 2006 are Thales, Boeing Little Bird, Sagem/Bell/Rheinmetall with an Eagle Eye derivative and EADS/Vertivision with Orka 2000. Following a feasibility study, a demonstration phase will test a UAV operationally. Up to 50 UAV may be required in due course. On 1 December 2005, a contract was awarded to DCN to study and develop the integration of UAVs on board naval ships. In August 2010, DGA completed successful trials of Thales/DCNS/Schiebel S-100 Camcopter on Levant Island and, in September 2011, a three-years trials period began on the patrol ship *L'Adroit*. This programme is called Système de Drone Aérien de la Marine (SDAM). Delivery of the first SDAM is planned for 2019.

Numbers/Type: 35 Dassault Aviation ACM Rafale M.
Operational speed: Mach 2.
Service ceiling: 50,000 ft *(15,240 m)*.
Range: 2,000 n miles *(3,700 km)*.
Role/Weapon systems: Procurement of 60 Rafale M single-seaters (air superiority and ground/surface attack) was reduced to a total of 40 by Defence Programming Law 2014–19. First of two Rafale M naval prototypes (single seaters) flown 12 December 1992. First deck trials in *Foch* in 1993. First production Rafale M flown 7 July 1999. Of 10 (M 1–10) aircraft at standard F1 (air superiority role; crash programme carried out to enable tanker role), one aircraft used for trials and nine others in preservation pending upgrade to standard F3 (with full air superiority, air-to-ground, air-to-surface, nuclear strike and reconnaissance capabilities). This is to be completed in 2014–17. Of 16 (M 11–26) standard F2 (with limited air-to-ground capabilities) delivered, 14 subsequently upgraded to standard F3 and two (M 22 and 25) lost at sea on 24 September 2009. Two further aircraft (M 18 and M 24) were lost on 28 November 2010 and 2 July 2012 respectively. The first five of 12 (M 27–38) standard F3 were delivered by November 2010 and a further seven in 2012. Nine (M 39–47) were ordered on 12 November 2009. Two were delivered in 2013 and two in 2014. Sensors: Thales/Dassault Electronique RBE2 (RBE-AESA from M34) multirole radar; Thales/Dassault Electronique/MBDA SPECTRA integrated EW/IR countermeasure suite; Thales/Sagem OSF optronic surveillance and target acquisition equipment (from standard F2); MIDSCO MIDS-LVT terminal for Link 16 (from standard F2); Thales Reco NG optronic reconnaissance pod (for eight specially wired standard F3 Rafales). Weapons: Giat M791 30 mm cannon; up to eight AAMs (air defence role), including MBDA Magic 2 short range and MBDA Mica EM medium range AAMs (standard F1); MBDA Mica IR replacing Magic 2 from standard F2 (later, MBDA Meteor to replace Mica EM by 2017); MBDA SCALP-EG stand-off precision guided ASM (from standard F2); Sagem AASM general purpose precision ammunition (from standard F2); MBDA Exocet AM 39 Block 2 Mod 2 ASM (one carried) and ASMP-A 350 kT nuclear strike missile (standard F3); nacelle for air-to-air refuelling. Up to 8 tons of military load on 13 hardpoints.

RAFALE M *4/2013*, Tony Roper / 1529032*

Numbers/Type: 24 Dassault-Bréguet Super Étendard.
Operational speed: Mach 1.
Service ceiling: 45,000 ft *(13,700 m)*.
Range: 1,460 n miles *(2,700 km)*.
Role/Weapon systems: Carrierborne all-weather strike fighter with nuclear strike capabilities and limited air defence role; tactical recce role to be added. All aircraft still in inventory modernised 1994–1999 to Standard F3. Standard F4 for all the fleet from mid-2000 to early 2005; tactical recce role added; standard 5 upgrade for a total of 34 aircraft by 2008. Of these, 11 have subsequently been put in storage. Service life extended to 2018. Sensors: Dassault Electronique Anémone radar, DRAX (standard 3) or Thales-Detexis Sherloc-F ESM (standard 4), SAGEM UAT 90 computer, Thomson-CSF Barracuda jammer, Phimat chaff dispenser, Alkan IR decoy dispenser; Thales Optrosys photo/optronic chassis (with Omera 40 panoramic camera and SDS-250 digital camera) in a ventral bay (Standard F4); Thales Improved Damocles day/night FLIR/designator (Standard F4 and F5). Weapons: air defence and self protection: two Matra BAe Dynamic Magic 2 short range AAMs and two DEFA 30 mm cannon; nuclear strike: one Aerospatiale ASMP nuclear ASM; air-to-surface: one Aerospatiale AM 39 Exocet anti-ship missile; air-to-ground: bombs and MBDA CEMB/BANG 125 or 250 bombs with Raytheon Enhanced Paveway 2 precision guidance (Standard 4 and 5) or one Aerospatiale AS 30L laser guided missile. 7 hardpoints (from Standard 4).

SUPER ÉTENDARD *8/2013*, Adolfo Ortigueira Gil / 1529030*

Numbers/Type: 4/10/2 Eurocopter EC 725 R2 Cougar Mk 2 Plus Resco/EC 725 HUS/EC 225.
Operational speed: 154 kt *(285 km/h)*.
Service ceiling: 13,120 ft *(4,000 m)*.
Range: 421 n miles *(780 km)*.
Role/Weapon systems: Cougar Resco perform the Combat SAR (C-SAR) mission with AF Squadron EH 1/67 'Pyrénées' (with mixed Air Force/Navy crews). One or two to embark in *Charles de Gaulle* for every deployment. A total of 14 such aircraft is expected. Cougar HUS (Hélicoptère Unite Spéciale) are operated by Army Aviation Flight no 3 for special operations, including maritime counter-terrorism, and could be embarked in *Charles de Gaulle*, LHDs, LSDs, FREMM/AVT, La Fayette and Floréal class frigates. Two EC 225 ordered in December 2009 to replace aging Super Frelon. Sensors: Bendix 1400C radar; Thales Chlio FLIR; Thales Sherloc radar warning; Thales MWS 20 Damien missile warning; Marconi laser detector; Alkan Elips chaff dispenser; Link 16 in due course (Resco helicopters). Weapons: 2 FN 7.62 mm MGs (possibly 12.7 mm MGs or 20 mm cannons on HUS variant). Capable of carrying 29 passengers or 11 stretchers.

EC 225 *1/2013*, B Prézelin / 1529029*

EC 725 RESCO *2/2009, B Prézelin / 1305729*

Numbers/Type: 3 Grumman E-2C Hawkeye Group 2.
Operational speed: 320 kt *(593 km/h)*.
Service ceiling: 37,000 ft *(11,278 m)*.
Range: 1,540 n miles *(2,852 km)*.
Role/Weapon systems: Used for AEW, and direction of AD and strike operations. First pair ordered in May 1995 and delivered in April and December 1998 respectively. Third delivered in December 2003. First two aircraft completed upgrade programme (including eight-bladed propellers) in 2006. Procurement of a fourth aircraft was discontinued in December 2007. A sensor modernisation programme is under consideration. Sensors: APS-145 radar, ESM, ALR-73 PDS, ALQ-108 airborne tactical data system with Links 11 and 16. Weapons: Unarmed.

E-2C *4/2013*, Tony Roper / 1529031*

Numbers/Type: 9 NHIndustries NH90 CAIMAN MARINE.
Operational speed: 162 kt *(300 km/h)*.
Service ceiling: 13,940 ft *(4,250 m)*.
Range: 621 n miles *(1,150 km)*.
Role/Weapon systems: Total of 27 NH90 ordered 30 June 2000 for the French Navy in two variants: 13 NHS support helicopters with secondary ASuW role; 14 NHC combat helicopters for ASW and ASuW. First production aircraft flown on 12 May 2006. Delivery programme: Two NHS in 2010, four in 2011, three in 2012, and three in 2014. First embarked in *Aquitaine* in 2013. All NHC are to be delivered by 2013. Sensors: both variants: Thales ENR surveillance radar; Sagem OLOSP tactical FLIR; MBDA Saphir decoy dispenser; Link 11; NHC: TUS FLASH dipping sonar, and UMS 2000-TSM 8203 sonobuoy processing system. Weapons: ASM (NHC and NHS); 2 MU 90 Impact torpedoes (NHC).

NH90 *2/2009, B Prézelin / 1305724*

Shipborne aircraft — Land-based maritime aircraft ‹ **France** 263

Numbers/Type: 3/8/16 Eurocopter (Aerospatiale) SA 365F Dauphin 2/SA 365N Dauphin 2/AS 565MA Panther.
Operational speed: 165 kt *(305 km/h)*.
Service ceiling: 16,700 ft *(5,100 m)*.
Range: 486 n miles *(900 km)*.
Role/Weapon systems: New-built SA 365F Dauphin 2s acquired to replace Alouette IIIs for carrierborne SAR. They feature the same ORB-32 radars as Panthers. Six SA 365Ns are second-hand helicopters purchased for SAR, general surveillance and public service roles from various locations in metropolitan France. They do not have any radar. A further two radar fitted aircraft delivered by 2012 for service in Tahiti. Fifteen AS 565 Panthers purchased in several batches to operate from Cassard class DDGs, La Fayette and Floréal class frigates. 16th aircraft acquired from the Armée de l'Air (French Air Force). All Panthers to be modernised (first delivered in May 2011) to Standard 2 2008–2012 by 2016 with new avionics, comprehensive countermeasures suite (laser, radar and missile warning systems, decoy dispenser), FLIR and datalink. Follow-on Standard 3 are to feature a new surveillance radar and lightweight anti-ship missiles. Service life to 2025 (AS 565MA). Sensors: (AS 565MA and SA 365F) Thales ORB-32 radar and (AS 565MA) Thales Chlio FLIR on some helicopters (all fitted for); Titus tactical situation management aid (with encrypted data link). Weapons: (AS 565MA) provision for internally mounted 7.62 mm MG.

SA 365F DAUPHIN 2   *1/2011, Marco Ghiglino* / 1406962

AS 565 PANTHER   *3/2010, B Prézelin* / 1366853

Numbers/Type: 90 Aerospatiale SA 330Ba Puma.
Operational speed: 139 kt *(257 km/h)*.
Service ceiling: 15,750 ft *(4,800 m)*.
Range: 297 n miles *(550 km)*.
Role/Weapon systems: Troop carrying helicopter owned by French Army and operable from amphibious ships.

SA 330BA   *7/2013*, A Sheldon-Duplaix* / 1529028

Numbers/Type: 16 Westland Lynx Mk 4 (FN).
Operational speed: 125 kt *(232 km/h)*.
Service ceiling: 12,500 ft *(3,810 m)*.
Range: 320 n miles *(593 km)*.
Role/Weapon systems: Sole French ASW helicopter, all now of the Mk 4 variant; embarked in destroyers and deployed on training tasks. Of 21 aircraft, 16 are in service and the remainder in storage. All to be replaced by NH90. A limited modernisation (Link 11 and Thales Chlio FLIR) programme will be applied to a small number of aircraft. Sensors: Omera 31 search radar, Alcatel (DUAV 4) dipping sonar, sonobuoys, Sextant Avionique MAD. Weapons: ASW; two Mk 46 Mod 1 (all aircraft being modified to launch EuroTorp Mu 90 Impact) torpedoes, or depth charges. ASV: 1–7.62 mm MG.

LYNX   *1/2013*, B Prézelin* / 1529027

Numbers/Type: 23 Aerospatiale SA 319B Alouette III.
Operational speed: 113 kt *(210 km/h)*.
Service ceiling: 10,500 ft *(3,200 m)*.
Range: 327 n miles *(605 km)*.
Role/Weapon systems: Twelve general purpose helicopters, SA 316B with Turboméca Artouste engine; 13 SA 319B with Astazou engine; one VSV aircraft replaced by Lynx for ASW; now used for trials, surveillance and training tasks. Sensors: Some radar. Weapons: Unarmed.

ALOUETTE III   *6/2013*, A Sheldon-Duplaix* / 1529033

LAND-BASED MARITIME AIRCRAFT (FRONT LINE)

Notes: In addition to frontline aircraft, the naval inventory includes 11 Embraer EMB-121 Xingu executive aircraft used for communications (Flotilles 24F and 28F) and seven CAP 10 and six Morane-Saulnier (SOCATA) Rallye for initial in-flight training with EIP/Escadrille 50S.

Numbers/Type: 4/4 Dassault Falcon 50MI/Dassault Falcon 50MS.
Operational speed: 475 kt *(880 km/h)*.
Service ceiling: 49,000 ft *(14,930 m)*.
Range: 3,500 n miles *(6,480 km)*.
Role/Weapon systems: Maritime reconnaissance and SAR roles in the Atlantic and overseas stations (replaced deleted Atlantic Mk 1). First aircraft delivered in December 1999 (for Opeval), second in March 2000, third in March 2001; fourth in late 2002. Four further aircraft from the VIP transport fleet are being converted to the maritime surveillance role; the first entered service in 2013 and the remainder by 2015. Being fitted with a Spationav VI terminal to share common picture with maritime surveillance assets. A Standard 2 modernisation programme was implemented from 2007. Allocated to Flotille 24F (Lann-Bihoué). Sensors: Thales/DASA Ocean Master 100(V) search radar, Thales Chlio FLIR, Inmarsat C. Weapons: Unarmed (two SAR chains). Endurance: six hours 30 minutes at 100 n miles *(185 km)* from base, four hours at 500 n miles *(926 km)* or one hour at 1,200 n miles *(2,222 km)*.

FALCON 50M   *1/2007, B Prézelin* / 1305025

Numbers/Type: 22 Dassault Aviation Atlantique Mk 2.
Operational speed: 355 kt *(658 km/h)*.
Service ceiling: 32,800 ft *(10,000 m)*.
Range: 11 hours patrol at 600 n miles from base; 8 hours patrol at 1,000 n miles from base; 4 hours patrol at 1,500 n miles from base.
Role/Weapon systems: Maritime reconnaissance. ASW, ASV, COMINT/ELINT roles. Last one of 27 built delivered in January 1998. Assigned to Flottilles 21F and 23F. Four aircraft are in long-term storage. Sensors: Thomson-CSF DRAA-10B Iguane radar, ARAR 13 ESM, ECM, FLIR, MAD, sonobuoys (with DSAX-1 Thomson-CSF Sadang processing equipment). Link 11 (being fitted in all). COMINT/ELINT equipment optional. Integrated sensor/weapon system built around a CIMSA 15/125X computer. Weapons: Two AM 39 Exocet ASMs in ventral bay, or up to eight lightweight torpedoes (Mu 90), or depth charges, mines or laser-guided bombs (GBU-12). Aircraft modified to drop Mu 90 torpedoes in 2010. More extensive modernisation for 15 aircraft by 2023 (first to be delivered in 2018). This includes upgrade of avionics and acoustics and replacement of Iguane radar by a development of Thales Ocean Master. Four other aircraft are to be limited to a reconnaissance role. Aircraft deployed to Dakar, Djibouti, Mombasa and (occasionally) Chad.

ATLANTIQUE III   *3/2010, B Prézelin* / 1366855

© 2014 IHS   IHS Jane's Fighting Ships 2014-2015

264 France > Land-based maritime aircraft — Patrol forces

Numbers/Type: 4 Boeing E-3F Sentry AWACS.
Operational speed: 460 kt *(853 km/h)*.
Service ceiling: 30,000 ft *(9,145 m)*.
Range: 870 n miles *(1,610 km)*.
Role/Weapon systems: Air defence early warning aircraft with secondary role to provide coastal AEW for the Fleet; 6 hours endurance at the range given above. Modernised 2003–06 under the Radar System Improvement Programme (RSIP). A further upgrade programme is started in 2013. This will modernise the mission computing system of all four aircraft to the Block 40/45 standard. Sensors: Westinghouse APY-2 surveillance radar, Bendix weather radar, Mk XII IFF, Yellow Gate, ESM, ECM. Weapons: Unarmed. Operated by the Air Force.

E-3F   6/2002, Armée de l'Air / 0118289

Numbers/Type: 6 Dassault-Aviation Falcon 10MER.
Operational speed: 492 kt *(912 km/h)*.
Service ceiling: 35,500 ft *(10,670 m)*.
Range: 1,920 n miles *(3,560 km)*.
Role/Weapon systems: Primary aircrew/ECM training role but also has overwater surveillance role. Avionics upgrade (Standard 2) programme started in 2006. Sensors: Search radar. Weapons: Unarmed. Allocated to Flottille 57S (Landivisiau).

FALCON 10MER   4/2013*, Tony Roper / 1529035

Numbers/Type: 5 Dassault-Aviation Falcon 200/Gardian.
Operational speed: 470 kt *(870 km/h)*.
Service ceiling: 45,000 ft *(13,715 m)*.
Range: 2,425 n miles *(4,490 km)*.
Role/Weapon systems: Assigned to Flotilla 25F based at Tahiti with permanent detachments at Tontouta (New Caledonia) and Martinique. Maritime reconnaissance role. Service life 2015; modernisation/replacement (AVISMAR programme) is under consideration. Sensors: Thomson-CSF Varan radar, Omega navigation, ECM/ESM pods. Weapons: Unarmed.

GARDIAN   9/2009, Chris Sattler / 1305837

PATROL FORCES

Notes: (1) 'Sauvegarde Maritime' is the organisation that encompasses the surveillance and traffic control of all maritime approaches around continental France and overseas territories. It also includes pollution control. Although all naval ships could participate in surveillance tasks, specialised vessels include the OPVs manned by the navy, patrol vessels and patrol craft of the 'Gendarmerie Maritime', French Customs and 'Affaires Maritimes'. In addition there are merchant support vessels on long-term charter (see *Government Maritime Forces*). All these ships, including specialised naval ships, display blue/white/red stripes on hull sides. In December 2009, it was announced that a Coast Guard (Fonction Gardes-Côtes FGC) was to be formed under naval leadership.
(2) Naval patrol ships (OPVs) are referred to as 'Patrouilleurs de Service Public' (PSP, Public Service Special Patrol Vessel). All PSPs and other government service craft are to be fitted with Spationav VI terminals to share a common maritime picture.
(3) The potential use of Unmanned Surface Vehicles (USV) is under investigation. One prototype 'Inspector', built by Couach, is undergoing trials.
(4) There are some 60 RHIBs in service for harbour and ship protection.
(5) The BATSIMAR (Batiments de Surveillance et d'Intervention Maritime) programme is for up to 20 patrol ships to replace the A 69-class frigates and the current inventory of offshore patrol ships. The Gowind corvette, currently under trial, is a contender. A decision on the way-ahead has been deferred until at least 2019.
(6) Two new patrol craft are required for patrol duties off French Guiana and the approaches to the Guiana Space Centre at Kourou. An order for both vessels is expected in 2014.

IHS Jane's Fighting Ships 2014-2015   © 2014 IHS

1 GOWIND CORVETTE (FS)

Name	No	Builders	Laid down	Launched	Commissioned
L'ADROIT	P 725	DCNS	7 May 2010	18 May 2011	19 Mar 2012

Displacement, tonnes: 1,100 standard; 1,500 full load
Dimensions, metres (feet): 87 × 13.0 × 3.0 *(285.4 × 42.7 × 9.8)*
Speed, knots: 21
Range, n miles: 8,000 at 12 kt
Complement: 32 + 27 spare berths

Machinery: 2 ABC 12VDZC diesels; 8,160 hp *(6.0 MW)*; 2 shafts; cp props
Guns: 1 Nexter Narwhal remotely operated 20 mm. 2—12.7 mm MGs. 2—6.72 mm MGs.
Physical countermeasures: Decoys: Lacroix Sylena (RF, IR and optronic decoys).
Electronic countermeasures: RESM: Thales Vigile.
CESM: Thales Altesse.
Radars: Air/surface search: Terma Scanter 4100; I-band.
Surface search/navigation: Terma Scanter 6002; I-band.
Combat data systems: Polaris.
Electro-optic systems: Thales AGILE. Sagem EOMS NG optronic director.

Comment: Project Hermes is a DCNS private venture for an offshore patrol vessel demonstrator that is being made available to the French Navy for three years (2012–15). The aim is to prove the design of the Gowind family of corvettes. The design is of steel construction and includes an integrated mast, a 360° bridge, a stern-ramp for rapid launch and recovery of a 9 m RHIB and/or unmanned surface vessels, a helicopter platform and full provision for the stowage and operation of unmanned aerial vehicles (UAV). Sea trials began on 27 July 2011 and trials of a Camcopter S-100 UAV were conducted in November 2011 and in 2012–13. The ship was deployed to the Indian Ocean and Mediterranean in 2013–14. Following postponement of the BATSIMAR programme, the ship is to be returned to DCNS in late 2014 and may then be sold.

L'ADROIT   10/2013*, Diego Quevedo / 1529034

1 TRAWLER TYPE (PSO)

Name	No	Builders	Commissioned
ALBATROS (ex-*Nevé*)	P 681	Ch de la Seine Maritime	1967

Displacement, tonnes: 1,971 standard; 2,845 full load
Dimensions, metres (feet): 85 × 13.5 × 6.0 *(278.9 × 44.3 × 19.7)*
Speed, knots: 15
Range, n miles: 14,700 at 14 kt
Complement: 50 (8 officers) + 15 spare berths

Machinery: Diesel-electric; 2 SACM UD 33 V12 S4 diesel generators; 3,050 hp(m) *(2.24 MW)* sustained; 2 MEP MCC motors, 2,200 hp(m) *(1.62 MW)*; 1 shaft
Guns: 1 Bofors 40 mm/60. 2—12.7 mm MGs.
Electronic countermeasures: ESM: ARBR 16A; radar detector.
Radars: Surface search: 2 DRBN 38A; I-band.

Comment: Former trawler bought in April 1983 from Compagnie Nav. Caennaise for conversion into a patrol ship. Commissioned 19 May 1984. Conducts patrols from Réunion to Kerguelen, Crozet, St Paul and Amsterdam Islands with occasional deployments to South Pacific. Vertrep facilities. Can carry 200 tons cargo, and has 4 tonne telescopic crane. Hospital with six berths and operating room. Major refit in Lorient from June 1990 to March 1991 included new diesel-electric propulsion. A further major overhaul was undertaken in France August 2001–April 2002. Maintenance now carried out in Indian Ocean Shipyards. Service life: 2017.

ALBATROS   4/2002, B Prézelin / 0528841

0 + 3 (1) PATROL SHIPS (PSO)

Displacement, tonnes: 1,500 standard; 2,300 full load
Dimensions, metres (feet): 65 × 14.0 × 4.2 *(213.3 × 45.9 × 13.8)*
Speed, knots: 15
Complement: 20
Machinery: 2 diesels; 2 shafts
Guns: 2—12.7 mm MGs.
Radars: Surface search/navigation: I-band.

Comment: The award of a contract to Piriou and DCNS for the construction of three B2M-class offshore patrol ships (with an option for a fourth) was announced on 9 January 2014. The ships are to replace the Batral-class landing ships deployed in French overseas territories. The contract covers design, construction, and in-service support. Based on an offshore supply ship design, the ships are to be optimised for EEZ patrol operations (30 days autonomy) and to have a multipurpose role. The ships are to feature a large work deck aft, are to be equipped with fire-fighting equipment, a 12-tonne crane for loading and unloading containers and are to be capable of operating an 8-m RHIB. They will also be capable of deploying divers and/or security forces. The ships are to enter service from 2015.

1 LAPÉROUSE CLASS (PBO)

Name	No	Builders	Launched	Commissioned
ARAGO	P 675 (ex-A 795)	Lorient Naval Dockyard, Lorient	9 Sep 1990	9 Jul 1991

Displacement, tonnes: 843 standard; 996 full load
Dimensions, metres (feet): 59 × 10.9 × 3.6 *(193.6 × 35.8 × 11.8)*
Speed, knots: 15
Range, n miles: 5,200 at 12 kt
Complement: 30 (3 officers)
Machinery: 2 Unidiesel UD 30 V12 M6D diesels; 2,500 hp(m) *(1.84 MW)*; 2 cp props; bow thruster; 160 hp(m) *(120 kW)*
Guns: 2—12.7 mm MGs.
Radars: Navigation: 1 Decca E 250 (DRBN 38A); 1 Furuno; I-band.

Comment: Ex-survey ship converted in 2002 for patrol duties. Based at Papeete, Tahiti, since 1 September 2011. Equipped with raiding craft, Inmarsat, TVSAT and Spationav. Service life expires in 2016.

ARAGO 1/2011, Marco Ghiglino / 1406963

4 P 400 CLASS (LARGE PATROL CRAFT) (PBO)

Name	No	Builders	Commissioned
LA CAPRICIEUSE	P 684	CMN, Cherbourg	13 Mar 1987
LA GLORIEUSE	P 686	CMN, Cherbourg	18 Apr 1987
LA GRACIEUSE	P 687	CMN, Cherbourg	17 Jul 1987
LA MOQUEUSE	P 688	CMN, Cherbourg	18 Apr 1987

Displacement, tonnes: 413 standard; 488 full load
Dimensions, metres (feet): 54.8 × 8 × 2.5 *(179.8 × 26.2 × 8.2)*
Speed, knots: 23
Range, n miles: 4,200 at 15 kt
Complement: 26 (3 officers) + 20 spare berths
Machinery: 2 SEMT-Pielstick 16 PA4 200 VGDS diesels; 8,000 hp(m) *(5.88 MW)* sustained; 2 shafts
Guns: 1 Bofors 40 mm/60; 1 Giat 20F2 20 mm; 2—7.62 mm MGs.
Radars: Surface search: 1 Racal Decca DRBN-38A (Bridgemaster E 250); I-band.

Programmes: First six ordered in May 1982, with further four in March 1984. The original propulsion system was unsatisfactory. Modifications were ordered and construction slowed. This class relieved the Patra fast patrol craft which transferred to the Gendarmerie.
Structure: Steel hull and superstructure protected by an upper deck bulwark. Design modified from original missile craft configuration. Now capable of transporting personnel with appropriate store rooms. Of more robust construction than previously planned and used as overseas patrol craft. Twin funnels replaced the unsatisfactory submerged diesel exhausts in 1990–91.
Operational: Deployments: Antilles; P 687, French Guiana; P 684. Nouméa; P 686, 688. P 687 completed refit at Lorient in 2006 and subsequently based at Fort de France. Endurance, 15 days with 45 people aboard. P 685 deleted in 2009, P 682 and P 683 in 2010 and P 689 in 2011. P 690 transferred to Kenya in 2011 and P 691 decommissioned in 2012. The remaining four are to be refitted to extend service lives to 2020.
Sales: To Gabon, Kenya and Oman.

LA MOQUEUSE 10/2010, Chris Sattler / 1366783

1 LE MALIN CLASS (PBO)

LE MALIN (ex-*Apache*) P 701 (ex-A 616)

Displacement, tonnes: 1,146 full load
Dimensions, metres (feet): 49.99 × 10.7 × 5.0 *(164 × 35.1 × 16.4)*
Speed, knots: 14
Range, n miles: 1,700 at 12 kt
Complement: 16 (2 officers)
Machinery: 1 Caterpillar 3806 diesel; 2,550 hp *(1.9 MW)*; 1 shaft; bow thruster
Guns: 2—12.7 mm MGs.
Radars: Navigation: Furuno FR 2110 and FR 2155; I-band.

Comment: Ex-fishing vessel built in Gdansk in 1997, seized on 23 June 2004 and acquired by the French Navy on 7 September 2005 at Port des Galets (La Réunion). Has a reinforced hull for ice. Refitted at Toulon in 2006 and entered French naval service as a diving tender replacing *Poséidon* in that role in April 2006. Further modified by Piriou, Concarneau, to conduct a patrol ship role. Based at La Réunion from October 2011.

LE MALIN (old number) 10/2006, B Prézelin / 1040705

3 FLAMANT (OPV 54) CLASS (PBO)

Name	No	Builders	Launched	Commissioned
FLAMANT	P 676	CMN, Cherbourg	24 Apr 1995	18 Dec 1997
CORMORAN	P 677	Leroux & Lotz, Lorient	15 May 1995	29 Oct 1997
PLUVIER	P 678	CMN, Cherbourg	2 Dec 1996	18 Dec 1997

Displacement, tonnes: 319 standard; 396 full load
Dimensions, metres (feet): 54.8 × 10 × 2.8 *(179.8 × 32.8 × 9.2)*
Speed, knots: 22
Range, n miles: 4,500 at 14 kt
Complement: 20 (3 officers)
Machinery: CODAD; 2 Deutz/MWM 16V TBD 620 diesels and 2 MWM 12V TBD 234 diesels; 7,230 hp(m) *(5.32 MW)* sustained; 2 shafts; LIPS cp props
Guns: 2—12.7 mm MGs.
Radars: Surface search: 1 Racal Decca Bridgemaster 250 (DRBN 38A); I-band.
Navigation: Racal Decca 20V90 (DRBN 34B); I-band. Racal Decca DRBN 34A (Bridgemaster E 250); I-band.

Comment: Authorised in July 1992 and ordered in August 1993 to a Serter Deep V design. Has a stern door for a 7 m EDL 700 fast assault craft or a Zodiac Hurricane RIB, capable of 30 kt. Two passive stabilisation tanks are fitted, and a remotely operated water-jet gun for firefighting. Deck area of 12 × 9 m for Vertrep. Similar to craft built for Mauritania in 1994. Hulls of all three ships strengthened by DCN Brest by late 2004. Service life: 2022. All based at Cherbourg.

PLUVIER 4/2013*, Frank Findler / 1529091

1 FULMAR CLASS (COASTAL PATROL CRAFT) (PB)

FULMAR (ex-*Jonathan*) P 740

Displacement, tonnes: 559 standard; 691 full load
Dimensions, metres (feet): 36.8 × 8.5 × 4.7 *(120.7 × 27.9 × 15.4)*
Speed, knots: 13
Range, n miles: 3,500 at 12 kt
Complement: 9 (1 officer)
Machinery: 1 Stork Wärtsilä 8 FDH 240G diesel; 1,200 hp(m) *(882 kW)*; 1 shaft. Bow thruster
Guns: 1—12.7 mm MG.
Radars: Surface search: 2 Furuno; I-band.

Comment: Former trawler built in 1990, acquired in October 1996 and converted for patrol duties by April 1997. Recommissioned 28 October 1997 and is based at St Pierre and Miquelon for western Atlantic Fishery Protection duties.

FULMAR 6/2010, Globke Collection / 1366874

266 France > Amphibious forces

AMPHIBIOUS FORCES

Notes: (1) About 25 LCVPs are still in service (from 59 built). Most are used on board *Siroco*, Batral LCTs and AORs.
(2) Replacement of the Batral class LSTs is under consideration. Vessels are required for service in New Caledonia, Tahiti, Antilles and (possibly) La Réunion. The requirement is unlikely to be met by amphibious vessels and, in light of suspension of the BATSIMAR programme, the B2M programme may provide a stop-gap.

3 MISTRAL CLASS (AMPHIBIOUS ASSAULT SHIPS) (LHDM/BPC)

Name	No	Builders	Laid down	Launched	Commissioned
MISTRAL	L 9013	DCN, Brest	10 Jul 2003	6 Oct 2004	15 Dec 2006
TONNERRE	L 9014	DCN, Brest	26 Aug 2003	26 Jul 2005	1 Aug 2007
DIXMUDE	L 9015	STX, Saint-Nazaire	29 Jan 2010	18 Dec 2010	27 Jul 2012

Displacement, tonnes: 16,794 standard; 21,947 full load
Dimensions, metres (feet): 199 × 32 × 6.2 *(652.9 × 105.0 × 20.3)*
Flight deck, metres: 199 × 32 *(652.9 × 105.0)*
Aircraft lift: 2
Speed, knots: 19
Range, n miles: 11,000 at 15 kt, 6,000 at 18 kt
Complement: 160 (18 officers)

Military lift: 450 (up to 900 in austerity conditions) troops and 60 armoured vehicles/(13 MBTs) (approx 1,200 tons of cargo). 4 CTM (LCU) or 2 EDA-Rs or 2 LCACs.
Machinery: Electric propulsion: 4 (3 Wärtsilä 16V32 and 1 Wärtsilä 18V200) diesel generators provide total of 20.8 MW for propulsion and services. 2 Alstom Mermaid podded propulsors trainable through 360°; 19,040 hp(m) *(14 MW)* sustained; 1 (2 in L 9015) bow thruster; 2,040 hp(m) *(1.5 MW)*.
Missiles: SAM: 2 MBDA Simbad twin PDMS launchers for Matra BAE Dynamics Mistral; IR homing to 6 km *(3.2 n miles)*; warhead 3 kg; anti-sea-skimmer.
Guns: 2 Breda Mauser 30 mm/70; 800 rds/min to 3 km; weight of shell 0.36 kg. 4 – 12.7 mm MGs.
Physical countermeasures: Torpedo defence: SLAT system.
Electronic countermeasures: ESM: ARBR 21; intercept.
Radars: Air/surface search: Thales MRR; 3-D; G-band.
Navigation: 2 Racal-Decca Bridgemaster E 250 (DRBN 38A); I-band.
Combat data systems: SENIT 9 combat data system, SIC 21 command support system for joint operations; space available for afloat CJTF command; Syracuse III, Fleetsatcom and Inmarsat. Link 11, Link 16.
Electro-optic systems: 2 Sagem VIGY-20 optronic systems. Sagem EOMS-NG.

Helicopters: Up to 16 NH90 or SA 330 Puma or AS 532U2 Cougar or AS 665 Tigre attack helicopters.

Programmes: Designated BPC (Bâtiment de Projection et de Commandement, support and command ship for force projection), ex-NTCD (new LHDs); which replaced *Ouragan* and *Orage*. Design and definition phase launched 12 November 1999; building contract notified 22 December 2000; ordered from DCN (prime contractor) and Alstom Marine-Chantiers de l'Atlantique. Forward sections built at St Nazaire, and middle and aft blocks at Brest where final construction and outfitting took place. Sixty per cent of the aft section subcontracted to Stocznia Remontowa, Gdansk, and shipped to Brest by barge. The contract for construction of a third ship was placed with STX France on 10 April 2009. The ship was fitted out by DCNS at Toulon and has replaced *Foudre*.
Modernisation: Measures to improve self-defence capabilities are under consideration.
Structure: Built to merchant marine standards. Flight deck has 6 spots, one of which calibrated for CH-53 or MV-22 operations. One 1,800 m² hangar for helicopters or vehicles (2 lifts), one 2,650 m² hangar for vehicles only (1 lift); up to 1,200 tons load on vehicle deck. Well dock 885 m². Hospital: 69 beds; additional modular field hospital may be embarked for humanitarian missions. Other modular facilities could also be embarked according to missions.
Operational: Roles: forward presence, force projection, logistic support for deployed force (ashore or at sea), humanitarian aid, disaster relief, command ship for combined operations. Endurance: 45 days.
Sales: Two ships ordered by the Russian Navy in 2011 for delivery in 2014 and 2015.

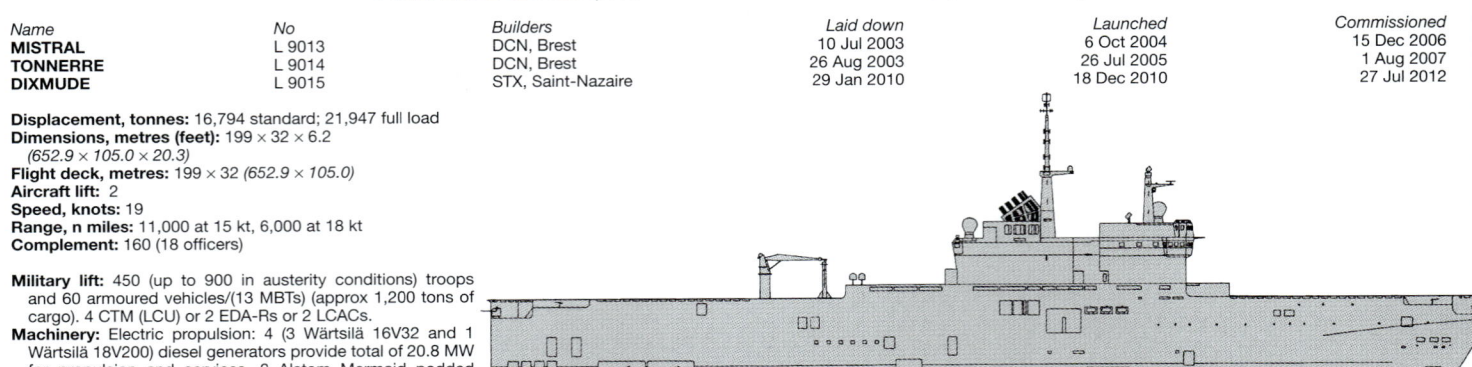

MISTRAL (Scale 1 : 1,500), Ian Sturton / 1042093

MISTRAL 9/2012*, M Declerck / 1529037

TONNERRE 3/2013*, Maritime Photographic / 1529036

DIXMUDE 4/2011, B Prézelin / 1406966

IHS Jane's Fighting Ships 2014-2015 © 2014 IHS

1 FOUDRE CLASS (LANDING SHIPS DOCK) (LSDH/TCD 90)

Name	No	Builders	Laid down	Launched	Commissioned
SIROCO	L 9012	DCN, Brest	9 Oct 1994	14 Dec 1996	21 Dec 1998

Displacement, tonnes: 8,362 standard; 12,599 full load
Dimensions, metres (feet): 168 × 23.5 × 5.2
  (551.2 × 77.1 × 17.1)
Aircraft lift: 1
Speed, knots: 21
Range, n miles: 11,000 at 15 kt
Complement: 218 (18 officers)

Military lift: 470 (up to 2,000 for 3 days) troops plus 1,880 tons load; 1 EDIC/CDIC plus 4 CTMs (typical) or 2 CDIC or 10 CTMs or 20 LARC XV amphibious vehicles; 150 vehicles
Machinery: 2 SEMT-Pielstick 16 PC2.5 V 400 diesels; 20,800 hp(m) *(15.3 MW)* sustained; 2 shafts; LIPS cp props; bow thruster; 1,000 hp(m) *(735 kW)*
Missiles: SAM: 2 MBDA Matra Simbad twin launchers ❶; Mistral; IR homing to 4 km *(2.2 n miles)*; warhead 3 kg.
Guns: 3 Breda/Mauser 30 mm/70 ❷. 800 rds/min to 3 km *(1.6 n miles)*; weight of shell 0.36 kg. 4—12.7 mm MGs.
Physical countermeasures: Decoys: SLQ-25 Nixie towed torpedo decoy
Electronic countermeasures: ECM: 2 Thales ARBB 36A jammers.
Radars: Air/surface search: Thomson-CSF DRBV 21A Mars ❹; D-band.
Navigation: 2 Racal-Decca DRBN 34A; I-band (1 for helo control) ❺.
Combat data systems: STIDAV/SENIT 8-01 for close range air defence; Syracuse SATCOM ❸. OPSMER command support system. Link 11 (receive only). INMARSAT.
Weapon control systems: 2 Sagem DIBC-2A VIGY-105 optronic systems (for 30 mm guns).
Helicopters: 4 AS 532UL Cougar or SA 330B Puma ❻.

Programmes: Ordered 11 April 1994. Transports de Chalands de Débarquement (TCD).
Modernisation: Sadral SAM replaced by two lightweight Simbad SAMs either side of bridge. New air search radar. Sagem optronic fire control fitted in 1997.
Structure: Designed to take a mechanised regiment of the Rapid Action Force and act as a logistic support ship. Extensive command (OPSMER and other systems) and hospital facilities (500 m²) include two operating suites and 47 beds. Modular field hospital may be embarked on *Siroco*. Well dock of 122 × 14.2 m *(1,732 m²)* which can be used to dock a 400 tons ship. Crane of 37 tons and lift 38 tons. Landing deck extended aft up to the lift to give a 1,740 m² area.
Operational: Two landing spots on flight deck plus one on deck well rolling cover. Can operate Super Frelons or Super Pumas. Could carry up to 1,600 troops in emergency. Endurance, 30 days (with 700 persons aboard). Assigned to FAN and based at Toulon. Typical loads: one CDIC, four CTM, 10 AMX 10RC armoured cars and 50 vehicles or total of 180 to 200 vehicles (without landing craft). *Foudre* sold and transferred to Chile in early 2012. *Siroco* is to be decommissioned in 2015 and is likely to be sold, possibly to Chile.

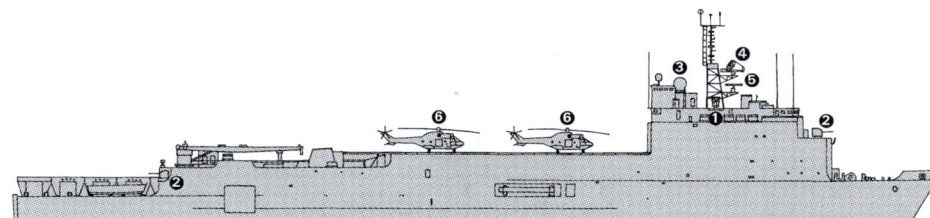

SIROCO (Scale 1 : 1,500), Ian Sturton / 0529157

SIROCO 5/2013*, Derek Fox / 1529092

2 BATRAL TYPE (LIGHT TRANSPORTS AND LANDING SHIPS) (LSTH)

Name	No	Builders	Commissioned
DUMONT D'URVILLE	L 9032	Français de l'Ouest	5 Feb 1983
LA GRANDIÈRE	L 9034	Français de l'Ouest	20 Jan 1987

Displacement, tonnes: 762 standard; 1,605 full load
Dimensions, metres (feet): 80 × 13 × 2.4
  (262.5 × 42.7 × 7.9)
Speed, knots: 14.5
Range, n miles: 4,500 at 13 kt
Complement: 52 (5 officers)

Military lift: 180 troops; 12 vehicles; 350 tons load; 10 ton crane
Machinery: 2 SACM AGO 195 V12 diesels; 3,600 hp(m) *(2.65 MW)* sustained; 2 shafts; cp props
Missiles: SAM: 2 Matra Simbad twin launchers (may be fitted).
Guns: 2 Giat 20F2 20 mm. 2—12.7 mm MGs.
Radars: Navigation: DRBN 32; I-band.
Helicopters: Platform for Lynx or Panther.

Programmes: Classified as Batral 3F. Bâtiments d'Assaut et de TRansport Légers (BATRAL). *Dumont D'Urville* floated out 27 November 1981, *La Grandière* 15 December 1985.
Structure: 40 ton bow ramp; stowage for vehicles above and below decks. One LCVP and one LCPS carried. Helicopter landing platform.
Operational: Deployment: *Dumont D'Urville*, Martinique; *La Grandière*, Indian Ocean. Service lives of *Dumont D'Urville* (2014), and *La Grandière* (2014) extended. *Champlain* placed in reserve in Martinique 2004 and later sunk as a target. *Francis Garnier* decommissioned in 2011 and *Jacques Cartier* in 2013.
Sales: Ships of this class built for Chile, Gabon, Ivory Coast, and Morocco. *La Grandière* was also built for Gabon under Clause 29 arrangements but funds were not available.

BATRAL CLASS 5/2006, M Declerck / 1167137

268 France > Amphibious forces — Mine warfare forces

1 CDIC CLASS (LCT)

Name	No	Builders	Commissioned
HALLEBARDE	L 9062	SFCN, Villeneuve la Garenne	17 Feb 1989

Displacement, tonnes: 386 standard; 762 full load
Dimensions, metres (feet): 59.4 × 11.9 × 1.8 *(194.9 × 39.0 × 5.9)*
Speed, knots: 10.5
Range, n miles: 1,000 at 10 kt
Complement: 18 (1 officer) + 230 embarked forces

Military lift: 340 tons
Machinery: 2 SACM Uni Diesel UD 30 V12 M1 diesels; 1,200 hp(m) *(882 kW)* sustained; 2 shafts
Guns: 2 Giat 20F2 20 mm. 2—12.7 mm MGs.
Radars: Navigation: Racal Decca 1229; I-band.

Comment: CDIC (Chaland de Débarquement d'Infanterie et de Chars) built to work with Foudre class. The well-dock is 40 × 10.4 m. The wheelhouse can be lowered to facilitate docking manoeuvres in the LPDs. Assigned to FAN at Toulon. *Rapière* transferred to Chile in early 2012 and *Hallebarde* is expected to decommission in 2014.

CDIC CLASS   1/2009, B Prézelin / 1305720

4 + (4) EDA-R CLASS (LCT)

L 9092–9095

Displacement, tonnes: 305 full load
Dimensions, metres (feet): 30 × 12.6 × 2.4 *(98.4 × 41.3 × 7.9)*
Speed, knots: 25; 18 full load
Range, n miles: 600 at 12 kt

Military lift: 2 main battle tanks or 100 tons
Machinery: 4 MTU 12V2000 M93 diesels; 7,187 hp *(5.36 MW)*; 4 MJP 650 waterjets
Guns: 2—12.7 mm MGs. 2—7.62 mm MGs.
Radars: Navigation: I-band.

Comment: Following evaluation of the CNIM L-Cat design in 2008–09, four craft ordered from CNIM, on 15 June 2009. They are designed for operation from the Mistral class. The craft are of aluminium construction with a catamaran design and bow and stern ramps. The hydraulically adjustable pontoon deck operates in three modes: in transit, the craft operates as a catamaran; in beaching, the pontoon is lowered; and for docking the platform is partially or totally flooded. Construction of the first craft began in June 2010 and all four entered service on 18 April 2013. Four further craft are likely to be ordered. These may be larger 44 m craft.

L 9094   10/2012, A Sheldon-Duplaix / 1455787

19 CTMs (LCM)

CTM 12–18   CTM 20–23   CTM 25–31

Displacement, tonnes: 60 standard; 152 full load
Dimensions, metres (feet): 23.8 × 6.4 × 1.3 *(78.1 × 21.0 × 4.3)*
Speed, knots: 9.5. **Range, n miles:** 380 at 8 kt
Complement: 4 + 200 embarked forces

Military lift: 90 tons (maximum); 48 tons (normal)
Machinery: 2 Poyaud V8520NS diesels; 450 hp(m) *(331 kW)*; 2 shafts
Guns: 2—12.7 mm MGs.
Radars: Navigation: I-band.

Comment: First series of 16 built 1966–70 and all have been deleted. Second series CTM 17–18 built by Auroux, Arachon; CTM 19–31 built at CMN, Cherbourg 1982–92. All have a bow ramp. Chalands de Transport de Matériel (CTM). *CTM 17* based at Lorient, *CTM 17* (army) at Dakar, *CTM 18* at Mayotte, *CTM 25* at Djibouti, and the remaining 14 (including five ex-army) at Toulon. Of the army craft *(CTM 12–17)*, all but *CTM 17* transferred to the navy in July 2010. CTMs 19 and 24 transferred to Chile in late 2012. *CTM 26* returned to Toulon in October 2012. Ten craft are to be retained in service until 2020.

IHS Jane's Fighting Ships 2014-2015

CTM   3/2010, B Prézelin / 1366862

MINE WARFARE FORCES

Notes: Replacement of the capabilities of the current mine-countermeasures force is under consideration. The new system, SLAM-F (Système de Lutte Anti-Mines Futur), to be introduced from about 2018, is likely to be based on unmanned surface (Project 'Espadon') and underwater (Project 'Alister') vehicles rather than special purpose ships. Under Project 'Espadon', a 17 m prototype unmanned surface vehicle, *Sterenn Du*, is under evaluation. Two such vehicles, capable of deploying autonomous underwater vehicles for mine-detection and destruction, might be operated from a mother ship. Under Project 'Alister', there are a number of experimental autonomous underwater vehicles capable of deploying a variety of sonars, cameras, and counter-mine charges.

3 ANTARÈS (BRS) CLASS (ROUTE SURVEY VESSELS) (MHI)

Name	No	Builders	Commissioned
ANTARÈS	M 770	Socarenam, Boulogne	15 Dec 1993
ALTAÏR	M 771	Socarenam, Boulogne	9 Jul 1994
ALDÉBARAN	M 772	Socarenam, Boulogne	10 Mar 1995

Displacement, tonnes: 254 standard; 345 full load
Dimensions, metres (feet): 28.3 × 7.7 × 4 *(92.8 × 25.3 × 13.1)*
Speed, knots: 12. **Range, n miles:** 3,600 at 10 kt
Complement: 23 (1 officer)

Machinery: 1 Baudouin 12P15-2SR diesel; 800 hp(m) *(590 kW)*; 1 shaft; cp prop; bow thruster
Guns: 1—12.7 mm MG.
Radars: Navigation: 1 Racal-Decca Bridgemaster E 180; I-band.
Sonars: 1 TUS DUBM-44 towed sidescan.

Comment: The vessels' role is to conduct surveillance operations in the approaches to Brest in support of the SSBN fleet. BRS Bâtiments Remorqueurs de Sonars. Trawler type similar to Glycine class (see *Training Ships* section). The DUBM 41B towed bodies have been replaced by TUS DUBM-44 Synthetic Aperture Sonar. A mechanical sweep is also carried. There are two 4.5 ton hydraulic cranes. Original dual navigation training role has been lost. Service life 2020.

ALTAIR   1/2010, B Prézelin / 1367102

4 MCM DIVING TENDERS (MCD)

Name	No	Builders	Launched	Commissioned
VULCAIN	M 611	La Perrière, Lorient	17 Jan 1986	11 Oct 1986
PLUTON	M 622	La Perrière, Lorient	13 May 1986	12 Dec 1986
ACHÉRON	A 613	CMN, Cherbourg	19 Nov 1986	21 Apr 1987
STYX	M 614	CMN, Cherbourg	3 Mar 1987	22 Jul 1987

Displacement, tonnes: 416 standard; 513 full load
Dimensions, metres (feet): 41.6 × 7.5 × 3.8 *(136.5 × 24.6 × 12.5)*
Speed, knots: 13.7. **Range, n miles:** 2,800 at 13 kt, 7,400 at 9 kt
Complement: 26 (1 officer; 12 divers)
Machinery: 2 SACM-Wärtsilä UD 30V 16M3 diesels; 2,200 hp(m) *(1.62 MW)*; 2 shafts; bow thruster; 70 hp(m) *(51 kW)*
Guns: 1—12.7 mm MG. 2—7.62 mm MGs.
Radars: Navigation: Decca DRBN-38; I-band.

Comment: First pair ordered in December 1984. Second pair ordered July 1985. Designed to act as support ships for clearance divers. (Bâtiments Bases pour Plongeurs Démineurs – BBPD). *Vulcain* based at Cherbourg, *Pluton* at Toulon, *Achéron* at Toulon as a diving school tender and *Styx* at Brest. Modified Chamois (BSR) class design. 5 ton hydraulic crane.

STYX   6/2012, Derek Fox / 1455815

© 2014 IHS

Mine warfare forces — Survey and research ships < **France** 269

11 ÉRIDAN (TRIPARTITE) CLASS (MINEHUNTERS) (MHC)

Name	No	Laid down	Launched	Commissioned
ÉRIDAN	M 641	20 Dec 1977	2 Feb 1979	16 Apr 1984
CASSIOPÉE	M 642	26 Mar 1979	26 Sep 1981	5 May 1984
ANDROMÈDE	M 643	6 Mar 1980	22 May 1982	18 Oct 1984
PÉGASE	M 644	22 Dec 1980	23 Apr 1983	30 May 1985
ORION	M 645	17 Aug 1981	6 Feb 1985	14 Jan 1986
CROIX DU SUD	M 646	22 Apr 1982	6 Feb 1985	14 Nov 1986
AIGLE	M 647	2 Dec 1982	8 Mar 1986	1 Jul 1987
LYRE	M 648	13 Oct 1983	14 Nov 1986	16 Dec 1987
SAGITTAIRE	M 650	1 Feb 1993	14 Jan 1995	2 Apr 1996
CÉPHÉE (ex-*Fuchsia*)	M 652	28 Oct 1985	23 Oct 1987	18 Feb 1988
CAPRICORNE (ex-*Dianthus*)	M 653	17 Apr 1985	26 Feb 1987	14 Aug 1987

Displacement, tonnes: 571 standard; 625 full load
Dimensions, metres (feet): 51.5 × 8.9 × 3.8 *(169 × 29.2 × 12.5)*
Speed, knots: 18; 7 electric motors
Range, n miles: 3,000 at 12 kt
Complement: 49 (5 officers)

Machinery: 1 Stork Wärtsilä A-RUB 215V-12 diesel; 1,860 hp(m) *(1.37 MW)* sustained; 1 shaft; LIPS cp prop. Auxiliary propulsion; 2 motors; 240 hp(m) *(179 kW)*; 2 active rudders; 120 hp *(90 kW)*; 1 bow thruster
Guns: 1 Giat 20F2 20 mm; 720 rds/min to 2 km; 1 – 12.7 mm MG. 2 – 7.62 mm MGs.
Physical countermeasures: MCM: 2 ECA PAP 104 Mod 4 ROVs; Bofors Double Eagle Mk 2 ROV.
Radars: Navigation: Racal Decca DRBN-38A (Bridgemaster E 250); I-band.
Sonars: 1 DUBM 21E (TUS 2022 Mk III) sonar (hull-mounted) and one SPIV PVDS on Bofors Double Eagle Mk 2 ROV; dual frequency.
Combat data systems: TSM 2061.

Programmes: All French ships built in Lorient. Belgium, France and the Netherlands each agreed to build 15 (10 in Belgium with option on five more). Subsequently the French programme was cut to 10. Belgium provided all the electrical installations, France all the minehunting gear and some electronics and the Netherlands the propulsion systems. Replacement for the last of class (sold to Pakistan) was ordered in January 1992. Three Belgian ships of the class acquired between March and August 1997 after being in reserve since 1990.
Modernisation: A modernisation programme started in 2001 and was completed in December 2005. Modernisation included replacement of sonar by TUS 2022 Mk III, fitting of a Bofors Double Eagle Mk 2 ROV, a new tactical data system and upgrade of radar and comms.
Structure: GRP hull. Equipment includes: autopilot and hovering; automatic radar navigation; navigation aids by Loran and Syledis.
Operational: Minehunting, minesweeping, patrol, training, directing ship for unmanned minesweeping, HQ ship for diving operations and pollution control. Prepacked 5 ton modules of equipment embarked for separate tasks. M 645 and 653 based at Toulon, remainder at Brest. M 649 and M 651 placed in reserve in 2009 and subsequently deleted in 2010. To be retained in service until at least 2020 when progressive replacement by autonomous systems is planned to begin.
Sales: The original tenth ship of the class, completed in 1989, was transferred to Pakistan 24 September 1992 as part of an order for three; the second built in Lorient, the third in Karachi.

SURVEY AND RESEARCH SHIPS

Notes: (1) These ships are painted white. A total of about 100 officers and technicians with oceanographic and hydrographic training is employed in addition to the ships' companies listed here. They occupy the extra billets marked as 'scientists'.
(2) In addition to the ships listed below there is a civilian-manned 25 m trawler *L'Aventurière II* (launched July 1986) operated by GESMA, Brest for underwater research which comes under DCN.
(3) Two 9 m survey launches, *Matthew* and *Hunter* were built in 1980.
(4) In New Caledonia, there is a 30 m buoy-tender *Louis Hénin* and a 7 m survey launch *Chambeyron*.
(5) There are three ROVs used for research and salvage: *Erato* can operate to a depth of 1,200 m; *Achille* to 400 m; *Ulisse* to 1,000 m.

1 BEAUTEMPS-BEAUPRÉ CLASS (BHO HYDROGRAPHIC AND OCEANOGRAPHIC SURVEY SHIP) (AGOR)

Name	No	Builders	Laid down	Launched	Commissioned
BEAUTEMPS-BEAUPRÉ	A 758	Alstom Marine, Lorient	17 Jul 2001	26 Apr 2002	13 Dec 2003

Displacement, tonnes: 2,159 standard; 3,383 full load
Dimensions, metres (feet): 80.6 × 14.9 × 7 *(264.4 × 48.9 × 23.0)*
Speed, knots: 14
Range, n miles: 8,300 at 12 kt
Complement: 26 (5 officers) + 30 scientists

Machinery: Diesel-electric; four 1,500 hp(m) *(1.1 MW)* Mitsubishi diesels; 2 Alstom electric motors; 2,950 hp *(2.2 MW)*; 1 shaft; 3,000 hp(m) *(2.2 MW)*. 2 active rudders 300 hp(m) *(220 kW)* each; bow thruster 600 hp(m) *(440 kW)*.
Guns: 2 – 7.62 mm MGs.
Radars: Navigation: 2 Kongsberg ARPA; I-band.
Sonars: EG & G side looking towed sonar; Kongsberg/Simrad EM 120 deep multipath echo sounder (12 kHz); Kongsberg/Simrad EA 600 deep echo sounder (12 kHz); Kongsberg/Simrad EM 1002S shallow waters multipath echo-sounder (95 kHz); Kongsberg/Simrad EA 400-210 shallow waters echo sounder (33 kHz); Kongsberg/Simrad SBP 120 (3 to 7 kHz) narrow beam and SHOM 9 TR 109 (3.5 kHz) wide beam sediment echo sounders. Bodenseewerk KSS31 gravimeter; Thales SMM II magnetometer; acoustic current profiler. Most sensor transducers mounted on a removable chassis fixed underneath the hull. Oceanographic buoys; Sippican Mk 21. SDIV system.

Comment: Contracted to Alstom-Leroux Naval 13 March 2001. Derived from the civilian research ship *Thalassa* built in 1995 by Leroux & Lotz (now part of Alstom Marine) for the French government civilian agency IFREMER. 95% funded by the MoD and 5% by the Ministry of civilian research on behalf of IFREMER that will use the ship 10 days per year. First steel cut 17 July 2001. Started builder sea trials 17 October 2002 and official acceptance trials late December. Two VH 8 survey launches. 10 tonne stern gantry and 10 tonne crane; up to 5 shelters can be shipped and bolted on the deck to increase lab surfaces; up to 4 vehicles can be stored in the hold. TVSTA and Inmarsat communications. Endurance 45 days. Operated by two crews. Bâtiment hydrographique et océanographique (BHO, hydrographic and oceanographic survey ship).

CASSIOPÉE 6/2010, Michael Winter / 1366860

BEAUTEMPS-BEAUPRÉ 8/2008, B Prézelin / 1335741

1 DUPUY DE LÔME INTELLIGENCE COLLECTION SHIP (AGIH)

Name	No	Builders	Laid down	Launched	Commissioned
DUPUY DE LÔME	A 759	Royal Niestern Sander, Delfzijl	1 Dec 2002	27 Mar 2004	23 Jun 2006

Displacement, tonnes: 3,150 standard; 3,600 full load
Dimensions, metres (feet): 101.7 × 15.8 × 4.9 *(333.7 × 51.8 × 16.1)*
Speed, knots: 16. **Range, n miles:** 3,400 at 16 kt
Complement: 30 + 80 personnel
Machinery: 2 MaK 9M25 diesels; 8,030 hp *(5.98 MW)*; 2 shafts; 2 bow thrusters
Guns: 2 – 12.7 mm MGs.
Radars: Navigation: 2 Racal-Decca DRBN-38A; I-band.

Programmes: Programme initiated 29 October 2001. Contract awarded 14 January 2002 to Thales Naval France (for the mission system) and Compagnie Nationale de Navigation to procure and maintain the vessel for initial five year period. Installation of the MINREM mission system started at Toulon in January 2005. After trials, the ship was delivered to the navy on 15 December 2005 and replaced *Bougainville* in April 2006.
Structure: The ship has a design life of 30 years and is fitted with a flight deck and underway replenishment facilities.
Operational: Fitted with both COMINT and ELINT equipment. By 2012, some sensors may be deployed in unmanned air vehicles. The ship is to be available for 350 days a year and active for 240 days. There are two complements. Based at Brest.

ORION 3/2010, B Prézelin / 1366863

ANDROMÈDE 9/2010, Derek Fox / 1366859

DUPUY DE LÔME 8/2008, Ships of the World / 1335740

© 2014 IHS

IHS Jane's Fighting Ships 2014-2015

AUXILIARIES

Notes: (1) The programme to procure 'Bâtiments de Soutien et d'assistance hauturiers' (BSAH) is for eight ships to replace the three Chamois class, *Rari, Tenace, Malabar*, the two UT 711 class *Alcyon* and *Ailette*, and (probably) *Argonaute* and *Jason*. However, by late 2013 the programme appeared to have been suspended and alternative solutions may be sought.

(2) Inshore transport duties at Brest and Toulon have been chartered to civilian companies. At Brest, Société Morbihannaise de Navigation (SMN) awarded a five-year contract from 1 July 2004 to transport 2,300 daily passengers from Brest to and from Lanvéoc-Poulmic and I'lle Longue. The company has progressively introduced five purposely-built light transports; *Bindy, Tibidy, Trébéron, Arun* and *Térénez* built 2004-05 by Gamelin, La Rochelle (aluminium hull and superstructure, 35.4 × 9 × 1.7 m, 20 kt, 400 passengers).

(3) There are plans to acquire three new AOR to replace the Durance class from about 2018. The 180–205 m BRAVE (Bâtiment Ravitailleur d'Escadre) multipurpose logistics support vessel was proposed by DCNS in 2012. The concept design is a versatile replenishment ship capable of carrying two helicopters.

4 DURANCE CLASS (UNDERWAY REPLENISHMENT TANKERS) (AORHM)

Name	No	Builders	Laid down	Launched	Commissioned
MEUSE	A 607	Brest Naval Dockyard	2 Jun 1977	2 Dec 1978	21 Nov 1980
VAR	A 608	Brest Naval Dockyard	8 May 1979	1 Jun 1981	29 Jan 1983
MARNE	A 630	Brest Naval Dockyard	4 Aug 1982	2 Feb 1985	16 Jan 1987
SOMME	A 631	Normed, la Seyne	3 May 1985	3 Oct 1987	7 Mar 1990

Displacement, tonnes: 7,722 (A 607), 7,925 (A 608, 630–631) standard; 18,187 (A 607), 18,797 (A 608, 630–631) full load
Dimensions, metres (feet): 157.3 × 21.2 × 10.8 *(516.1 × 69.6 × 35.4)*
Speed, knots: 19. **Range, n miles:** 9,000 at 15 kt
Complement: 162 (11 officers) + 29 spare berths

Cargo capacity: 1,300 tons FFO; 5,200 diesel; 3,000 TR5 Avcat; 130 distilled water; 170 victuals; 150 munitions; 50 naval stores *(Meuse)*. 1,300 tons FFO; 8,400 diesel; 1,090 TR5 Avcat; 260 distilled water; 170 munitions; 250 tons spare parts *(Var, Somme* and *Marne)*
Machinery: 2 SEMT-Pielstick 16 PC2.5 V 400 diesels; 20,800 hp(m) *(15.3 MW)* sustained; 2 shafts; LIPS cp props
Missiles: SAM: 3 (1 in A 607) Matra Simbad twin launchers; Mistral; IR homing to 4 km *(2.2 n miles)*; warhead 3 kg.
Guns: 1 Bofors 40 mm/L 60. 6–12.7 mm MGs.
Electronic countermeasures: ESM/ECM.
Radars: Navigation: 2 Racal Decca Bridgemaster (DRBN 38A; DRBN 34A on A 607); I-band.
Combat data systems: AIDCOMER command support system (fitted for BCR ships). Syracuse and INMARSAT SATCOM.

Helicopters: 1 SA 319B Alouette III.

Programmes: One classed as Pétrolier Ravitailleur d'Escadres (PRE). Other three classed as Bâtiments de Commandement et de Ravitaillement (BCR; Command and Replenishment Ships).
Modernisation: EW equipment fitted to improve air defences under the 3A programme in 1996–99. Oerlikon 20 mm replaced by 12.7 mm MGs.
Structure: Single-hull ships. Four beam transfer positions and two astern, two of the beam positions having heavy transfer capability. *Var, Marne* and *Somme* differ from *Meuse* in several respects. The bridge extends further aft, boats are located either side of the funnel and a crane is located between the gantries. Also fitted with Syracuse 3 SATCOM.
Operational: *Var, Marne* and *Somme* are designed to carry a Maritime Zone staff or Commander of a Logistic Formation and a commando unit of up to 45 men. Capable of accommodating 250 men. Assigned to FAN with one of the three BCR ships deployed to the Indian Ocean as a Flagship. *Somme* replaced *Var* in that role for two years from August 2009. *Var* has conducted trials to act as an MCM support ship to replace the decommissioned *Loire*. To be replaced after 2018 by new ships, which are likely to be multipurpose command and replenishment ships. *Meuse* to be placed in reserve 2014–15. *Somme* will then move from Brest to Toulon where the other two ships are based.
Sales: One to Australia built locally; two of similar but smaller design to Saudi Arabia. One to Argentina in July 1999.

VAR
5/2009, Guy Toremans / 1305839

MARNE
4/2010, Michael Nitz / 1366785

Auxiliaries < **France** 273

4 CHAMOIS CLASS
(SUPPLY TENDERS) (AG/ATS/YDT/YPC/YPT)

Name	No	Builders	Commissioned
TAAPE	A 633	La Perrière, Lorient	2 Nov 1983
ÉLAN	A 768	La Perrière, Lorient	7 Apr 1978
– (ex-*Chevreuil*)	Y 693 (ex-A 774)	La Perrière, Lorient	7 Oct 1977
GAZELLE	A 775	La Perrière, Lorient	13 Jan 1978

Displacement, tonnes: 383 (A 633), 320 (A 768, 775, Y 693) standard; 505 (A 633), 500 (A 768, 775, Y 693) full load
Dimensions, metres (feet): 41.5 × 7.5 × 3.2 *(136.2 × 24.6 × 10.5)*
Speed, knots: 14.5
Range, n miles: 6,000 at 12 kt (A 633), 7,200 at 12 kt (A 768, 775)
Complement: 20 + 12 spare berths
Machinery: 2 SACM AGO 175 V16 diesels; 2,850 hp(m) *(2.06 MW)*; 2 shafts; cp props; bow thruster
Radars: Navigation: Racal Decca 1226; I-band.

Comment: Similar to the standard fish oil rig support ships. Can act as tugs, oil pollution vessels, salvage craft (two 30 ton and two 5 ton winches), coastal and harbour controlled minelaying, torpedo recovery, diving tenders and a variety of other tasks. Bollard pull 25 tons. Can carry 100 tons of stores on deck or 125 tons of fuel and 40 tons of water or 65 tons of fuel and 120 tons of water. *Taape* ordered in March 1982 from La Perrière-of improved design but basically similar with bridge one deck higher. *Elan* based at Cherbourg, remainder at Toulon. Three paid off so far, one of which (ex-*Chamois*) transferred to Madagascar in May 1996. *Chevreuil* was converted to act as a mooring vessel in 2010. All three vessels are likely to be decommissioned in 2014.

Ex-CHEVREUIL (old number)   9/2009, M Declerck / 1305829

1 MOORING VESSEL (ABU)

TELENN MOR Y 692

Displacement, tonnes: 398 standard; 528 full load
Dimensions, metres (feet): 41.4 × 9.1 × 1.9 *(135.8 × 29.9 × 6.2)*
Speed, knots: 8
Machinery: 2 Baudouin diesels; 900 hp(m) *(670 kW)*
Radars: 1 Furuno M 1832 (DRBN 39); I-band.

Comment: Commissioned on 16 January 1986 and based at Brest. Equipped with 18 ton hydraulic crane. To be decommissioned in 2017.

TELENN MOR   8/2006, B Prézelin / 1040706

1 TRANSPORT LANDING SHIP (LSL)

Name	No	Builders	Commissioned
GAPEAU	L 9090	Chantier Serra, la Seyne	2 Oct 1987

Displacement, tonnes: 572 standard; 1,107 full load
Dimensions, metres (feet): 66 × 12.5 × 3.4 *(216.5 × 41.0 × 11.2)*
Speed, knots: 11. **Range, n miles:** 1,900 at 10 kt
Complement: 6 + 30 scientists
Cargo capacity: 460 tons
Machinery: 2 diesels; 1,600 hp(m) *(1.17 kW)*; 2 shafts
Radars: Navigation: Racal Decca 1226 and Furuno FRS 1000; I-band.

Comment: Supply ship with bow doors. Based at Toulon. Operates for Centre d'Essais de la Mediterranée. Conducts transfers between Toulon or Port Pothuau and Levant Island (missile range). There are plans to replace the main machinery.

GAPEAU   5/2003, Per Körnefeldt / 0569982

1 ALIZE CLASS (DIVING TENDER) (YDT)

Name	No	Builders	Commissioned
ALIZÉ	A 645	Socarénam, Boulogne	8 Nov 2005

Displacement, tonnes: 1,118 standard; 1,727 full load
Dimensions, metres (feet): 60 × 13.8 × 5.0 *(196.9 × 45.3 × 16.4)*
Speed, knots: 14
Range, n miles: 8,000 at 12 kt
Complement: 17 (3 officers) + 30 spare berths

Machinery: 2 ABC 6DZC diesels; 3,800 hp *(2.8 MW)*; 2 shafts; bow thruster
Guns: 2 — 12.7 mm MGs.
Radars: Navigation: 2 Racal Decca Bridgemaster (DRBN 38A); I-band.
Helicopters: Platform for one medium.

Comment: Ordered in November 2003. Replaced *Isard* in diving support role in 2006. Equipped with recompression chamber and medical facilities. Based at Toulon.

ALIZÉ   7/2013*, E & M Laursen / 1529039

1 RR 4000 TYPE (SUPPLY TENDERS) (AFL)

Name	No	Builders	Commissioned
REVI	A 635	Breheret, Couéron	9 Mar 1985

Displacement, tonnes: 1,052 standard; 1,602 full load
Dimensions, metres (feet): 51 × 12.6 × 4 *(167.3 × 41.3 × 13.1)*
Speed, knots: 14.5
Range, n miles: 5,800 at 12 kt
Complement: 26 (2 officers) + 8 spare berths
Machinery: 2 SACM-Wärtsilä AGO 195 V12 M6 diesels; 4,410 hp(m) *(3.24 MW)*; 2 shafts; cp props; 2 bow thrusters; 400 hp *(300 kW)*
Radars: Navigation: Racal Decca Bridgemaster (DRBN 38A); I-band.

Comment: 'Remorqueurs ravitailleurs' built for le Centre d'Expérimentation du Pacifique. Can carry 400 tons of cargo or six 20 ft containers. 50 ton gantry and 18 ton crane. Two water cannons on deck. Bollard pull 47 tons. Based at Papeete.

REVI   7/2010, Chris Sattler / 1366784

2 RANGE SUPPORT VESSELS (YFRT)

ATHOS A 712   **ARAMIS** A 713

Displacement, tonnes: 90 standard; 110 full load
Dimensions, metres (feet): 32.1 × 6.5 × 1.9 *(105.3 × 21.3 × 6.2)*
Speed, knots: 28
Range, n miles: 1,200 at 15 kt
Complement: 13 + 6 spare berths
Machinery: 2 SACM UD 33V12 M5 diesels; 4,500 hp(m) *(3.31 MW)*; 2 shafts
Guns: 1 — 12.7 mm MG.
Radars: Navigation: Racal Decca 1226 (A 712); Furuno (A 713); I-band.

Comment: Built by Chantiers Navals de l'Esterel for Missile Trials Centre of des Landes (CELM). Based at Bayonne, forming Groupe des Vedettes de l'Adour. A 712 commissioned 20 November 1979 and A 713 on 9 September 1980. Classified as Range Safety Craft from July 1995. *Aramis* completed refit at Cherbourg in 2002 followed by *Athos* in 2003. Service life extended to 2015. Replacement is under consideration.

ARAMIS   6/2010, B Prézelin / 1366866

© 2014 IHS   IHS Jane's Fighting Ships 2014-2015

274 France > Auxiliaries

9 VIP 21 DIVING TENDERS (YDT)

DIONÉE Y 790	LISERON Y 793	GENÊT Y 796
MYOSOTIS Y 791	MAGNOLIA Y 794	GIROFLÉE Y 797
GARDÉNIA Y 792	AJONC Y 795	ACANTHE Y 798

Displacement, tonnes: 36 standard; 50 full load
Dimensions, metres (feet): 21.7 × 4.9 × 1.6 *(71.2 × 16.1 × 5.2)*
Speed, knots: 13
Range, n miles: 500 at 12 kt
Complement: 18 (14 divers)
Machinery: 2 Baudouin 12F11M or V6 TI 330 diesels; 530 hp(m) *(390 kW)*; 3 shafts (1 for loitering)
Radars: Navigation: Racal Decca RD 170.

Comment: Diving tenders built at Lorient. Entered service 1990–96. Y 794 and Y 798 based at Cherbourg. Y 790, Y 791, Y 792, Y 795 and Y 797 based at Toulon. Y 793 and Y 796 based at Brest. Rated as 'Vedettes d'Instruction Plongée de 21 m (VIP 21)', divers training craft, and 'Vedettes d'Intervention Plongeurs-Démineurs (VIPD 21)', clearance diving team support craft.

GIROFLÉE   *3/2010, B Prézelin / 1366868*

9 TYPE V14 (HARBOUR CRAFT) (YFL)

TAINA Y 754	PALANGRIN Y 777	AUTÉ Y 786
L'ETOILE DE MER Y 762	– Y 780	TIARÉ Y 787
AVEL MOR Y 765	– Y 781	HORTENSIA –

Displacement, tonnes: 15 standard; 20 full load
Dimensions, metres (feet): 14.6 × 4.6 × 1.9 *(47.9 × 15.1 × 6.2)*
Speed, knots: 25
Range, n miles: 400 at 11 kt
Complement: 4
Machinery: 2 Baudouin diesels; 1,000-750 hp(m) *(735-551 kW)*; 2 shafts

Comment: Y 754, Y 786 (both at Brest) and Y 787 (Nouméa) are small personnel transport craft. Y 762 (Toulon) and Y 765 (Brest) VIP transport craft. Y 777 (Brest) is a radiological monitoring craft. Y 780 (Brest) and Y 781 (Toulon) are pilot craft. Design by DCN Cherbourg. Same hull and similar arrangement as for PBs manned by the Gendarmerie Maritime (see *Government forces*). Built under control of DCN Lorient by Stento Shipyard, Balaruc-les-Bains in 1987–88, or by Chantiers Alan Sibiril, Carantec, in 1990–93.

AUTÉ   *4/2010, B Prézelin / 1366867*

PALANGRIN   *3/2009, B Prézelin / 1305696*

2 VTP CLASS (TRANSPORTS) (YFB)

KERMEUR Y 758   KERNALEGUEN Y 759

Displacement, tonnes: 15 standard; 21 full load
Dimensions, metres (feet): 14 × ? × ? *(45.9 × ? × ?)*
Speed, knots: 8
Complement: 3 + 45 spare berths
Machinery: 2 MAN diesels; 800 hp *(600 kW)*; 2 shafts
Radars: Navigation: Furuno; I-band.

Comment: Built by Raidco Marine and delivered 19 September 2006. GRP hull and superstructure. Used as transport craft at I'lle Longue.

KERMEUR   *8/2007, B Prézelin / 1305044*

2 PHAÉTON CLASS (TOWED ARRAY TENDERS) (YAG)

PHAÉTON Y 656   MACHAON Y 657

Displacement, tonnes: 70 standard; 76 full load
Dimensions, metres (feet): 19.2 × 6.8 × 1.2 *(63 × 22.3 × 3.9)*
Speed, knots: 8
Range, n miles: 300 at 8 kt
Complement: 4
Machinery: 2 SACM diesel; 720 hp(m) *(530 kW)*; waterjet
Radars: Navigation: Decca C181; I-band.

Comment: 18.6 m catamarans built in 1993–94 at Brest. Water-jet propulsion, speed 8 kt. Hydraulic crane and winch to handle submarine towed arrays. *Phaéton* based at Toulon, *Machaon* at Brest.

PHAETON   *6/2009, A A de Kruijf / 1305835*

3 VIR FIREFIGHTING CRAFT (YTR)

AVEL ABER (ex-*Elorn*) Y 783   LA LOUDE Y 784   LA DIVETTE Y 785

Displacement, tonnes: 14 standard; 23 full load
Dimensions, metres (feet): 14.6 × 4.6 × 1.9 *(47.9 × 15.1 × 6.2)*
Speed, knots: 17
Range, n miles: 110 at 12 kt
Complement: 4
Machinery: 2 Baudouin V6 TI450 diesels; 750 hp *(550 kW)*; 2 shafts
Radars: Navigation: Furuno 1832; I-band.

Comment: Firefighting craft built by Alan Sibril, Carantec, and delivered in 1993–94. Similar to V 14 craft. Equipped with two water cannons. Y 783 at Brest; Y 784 at Toulon and Y 785 at Cherbourg.

AVEL ABER   *8/2008, B Prézelin / 1335731*

1 RADIOLOGICAL MONITORING CRAFT (AGE)

CORALLINE A 790

Displacement, tonnes: 42 standard; 50 full load
Dimensions, metres (feet): 21.7 × 4.9 × 1.6 *(71.2 × 16.1 × 5.2)*
Speed, knots: 13
Range, n miles: 500 at 12 kt
Complement: 7
Machinery: 2 Baudouin 12F11M diesels; 530 hp *(390 kW)*; 2 shafts
Radars: Navigation: Furuno 1832; I-band.

Comment: Built by DCN Lorient and delivered 1 December 1990. Similar to VIP 21 diving tenders but with different superstructure. Employed on radiation monitoring tasks at Cherbourg. Service life 2015.

Auxiliaries < France 275

33 HARBOUR SUPPORT CRAFT

Comment: There are 12 oil barges (CICGH), one of which is of 1,200 tonnes and the rest between 100 and 800 tonnes, eight 400 tonne oily bilge barges (CIEM), three anti-pollution barges (800 tonne BAPM, and two 400 tonne CIEP), and seven water barges (CIE, 120 to 400 tonnes). Some self-propelled and one 16 m Sea Truck craft (*Anthias*).

4 COUACH-PLASCOA 980 (SERVICE CRAFT) (YFL)

| NYMPHEA Y 603 (ex-P 706) | MDLC RICHARD Y 611 (ex-P 709) |
| GENDARME PEREZ Y 605 (ex-P 708) | GENERAL DELFOSSE Y 710 (ex-P 710) |

Displacement, tonnes: 6 standard; 7 full load
Dimensions, metres (feet): 9.9 × 3.73 × 1.0 *(32.5 × 12.2 × 3.3)*
Speed, knots: 28
Range, n miles: 200 at 15 kt
Complement: 4
Machinery: 2 Volvo Penta TAMD 61 diesels; 500 hp *(370 kW)*; 2 shafts
Radars: Navigation: Furuno 2400; I-band.

Comment: Former Gendarmerie Maritime craft built in 1985 and transferred in 2004–05. Based at Brest.

MDLC RICHARD (old number)   6/2007, B Prézelin / 1305011

7 ARCOR 34 (SERVICE CRAFT) (YFL)

LAVANDE Y 606 (ex-P 717)	STEREDEN –	MDLC JACQUES P 716
LILAS Y 703 (ex-P 703)	AN HEOL –	CAPITAINE MOULIÉ Y 713
GENTIANE –		

Displacement, tonnes: 7 standard; 8 full load
Dimensions, metres (feet): 10.3 × 3.7 × 1.0 *(33.8 × 12.1 × 3.3)*
Speed, knots: 26
Range, n miles: 200 at 20 kt
Complement: 3
Machinery: 2 Volvo Penta TAMD 61 diesels; 500 hp *(370 kW)*; 2 shafts
Radars: Navigation: Furuno 1830; I-band.

Comment: Built in 1989–90 by CN d'Aquitaine. *Stereden* and *An Heol* are used as transport craft in the submarine base at I'Ile Longue. The others are former Gendarmerie Maritime craft transferred in 2004–05. Based at l'Ecole Navale (*Lilas*), Saint Mandrier (*Lavande*), Hyères (*Gentiane*), Toulon (*MDLC Jacques*), and COMAR Paris (*Capitaine Moulié*).

5 SELF-PROPELLED FLOATING CRANES (YD)

Y 675–679

Dimensions, metres (feet): 21.4 × 9.9 × 1.7 *(70.2 × 32.5 × 5.6)*
Speed, knots: 6
Complement: 3
Machinery: 2 diesels; 300 hp *(220 kW)*; 2 shafts

Comment: Ordered from Socarenam, Boulogne-sur-Mer on 16 January 2007 and delivered in March 2008. Equipped with a crane with a capacity of 8.3 tonnes to 8.5 metres and a winch with a capacity of 12 tonnes. Capable of carrying 24 tonnes of cargo. One based at Cherbourg (*Y 679*) and two each at Brest (*Y 677, 678*) and Toulon (*Y 675, 676*).

Y 677   4/2010, B Prézelin / 1366872

2 FIREFIGHTING CRAFT (YTR)

| LAS Y 670 | DOUFFINE Y 671 |

Displacement, tonnes: 68 full load
Dimensions, metres (feet): 20 × 5.6 × 1.6 *(65.6 × 18.4 × 5.2)*
Speed, knots: 14
Complement: 2 + 6 personnel
Machinery: 2 diesels; 2 shafts
Radars: Navigation: I-band.

Comment: Two new VIR craft ordered from Socarenam, Boulogne, on 10 September 2008. Both entered service in April 2010. *Las* based at Toulon and *Douffine* at Brest.

LAS   11/2010, B Prézelin / 1366871

5 HARBOUR AND PILOT CRAFT (YFL)

| LA MITRE Y 680 | LÉON Y 682 | CONTENTIN Y 684 |
| TOUR ROYALE Y 681 | CORNOUAILLES Y 683 | |

Displacement, tonnes: 9.5 full load
Dimensions, metres (feet): 12.2 × 3.55 × 0.8 *(40 × 11.6 × 2.6)*
Speed, knots: 30
Complement: 2 + 6 spare berths
Machinery: 2 Yanmar 6PLA-ST2P, 510 hp *(375 kW)*; 2 shafts
Radars: Navigation: I-band.

Comment: Built by Delta Power Group, Stockport, UK. All entered service February-March 2010. Y 680 and Y 681 at Toulon; Y 682 and Y 683 at Brest; Y 684 at Cherbourg.

LA MITRE   11/2011, B Prézelin / 1455790

8 LANDING CRAFT (LIGHTER) (YFU)

| CHA 28–30 | CHA 34–38 |

Displacement, tonnes: 20 standard; 30 full load
Dimensions, metres (feet): 16.4 × 5.13 × 1.6 *(53.8 × 16.8 × 5.2)*
Speed, knots: 9.5
Complement: 5
Machinery: 1 Poyaud-Wärtsilä diesel; 250 hp *(185 kW)*; 1 shaft

Comment: Cargo lighters constructed by CIB Brest 1987–89. Equipped with a bow ramp and crane. Used for ammunition and personnel transport. To be replaced by 2015.

CHA 35   5/2011, Jordi Montoro / 1406968

5 FLOATING REPAIR FACILITIES

Comment: There is one 150 × 33 m floating dock of 3,800 tons capacity, built at Brest in 1975. Based at Papeete. There are two floating cranes: one 15 ton crane at Toulon (GFA 3) and one at Brest (GFA 6 *Alpaga*).

TUGS

Notes: There is a 21 m RP 10-class tug *Papayer* Y 740 that is to remain in service until 2015. There are also two 8 m PSS 6 tugs P1 and P2 based at Cherbourg.

2 ESTEREL (TYPE RPC 50) CLASS
(COASTAL/HARBOUR TUGS) (YTM)

ESTEREL A 641 (ex-Y 601)	LUBÉRON A 642 (ex-Y 602)

Displacement, tonnes: 518 standard; 681 full load
Dimensions, metres (feet): 36.3 oa; 35.5 wl × 11.6 × 5 *(119.1; 116.5 × 38.1 × 16.4)*
Speed, knots: 14
Range, n miles: 1,500 at 12 kt
Complement: 8
Machinery: 2 ABC 8 DZ 1000. 179 diesels; 2 Voith-Schneider 28 GII propulsors; 5,120 hp(m) *(3,812 kW)*
Radars: Navigation: Furuno DRBN 39; I-band.

Comment: Ordered 15 December 2000; built by SOCARENAM, Boulogne. *Esterel* delivered 27 March 2002 and *Lubéron* 4 July 2002. Based at Toulon to assist *Charles de Gaulle* in harbour. Bollard pull 52 tonnes; 1,350 kN towing winch; fire fighting equipment; 20 cubic metre tank for pollution control dispersal agent. Classified as 'Remorqueurs portuaires et côtiers de 50 tonnes de traction' (RPC 50, 50 tonne bollard pull harbour tugs).

LUBÉRON   *11/2010, B Prézelin / 1366870*

2 OCEAN TUGS (ATA)

MALABAR A 664	TENACE A 669

Displacement, tonnes: 1,097 standard; 1,477 full load
Dimensions, metres (feet): 51 × 11.5 × 5.7 *(167.3 × 37.7 × 18.7)*
Speed, knots: 15
Range, n miles: 9,500 at 13 kt
Complement: 56 (2 officers)
Machinery: 2 Krupp MaK 9 M 452 AK diesels; 4,600 hp(m) *(3.38 MW)*; 1 shaft; Kort nozzle
Radars: Navigation: Racal Decca RM 1226 (A 669); Racal Decca 060 (A 664); I-band. Racal Decca 060; I-band.

Comment: *Malabar* and *Tenace* built by J. Oelkers, Hamburg. *Tenace* commissioned 15 November 1973, and *Malabar* on 3 February 1976. Based at Brest. Can carry firefighting and oil-pollution control equipment. Bollard pull, 60 tons. One of the class to Turkey in 1999. Likely to be decommissioned by 2017.

TENACE   *5/2012, Michael Nitz / 1455791*

16 FRÉHEL CLASS (COASTAL TUGS) (YTM)

LARDIER Y 638	LE FOUR Y 647	KÉRÉON (ex-*Sicie*) A 679
GIENS Y 639	PORT CROS Y 649	SICIÉ A 680
MENGAM Y 640	FRÉHEL A 675	TAUNOA A 681
BALAGUIER Y 641	SAIRE A 676	RASCAS A 682
TAILLAT Y 642	ARMEN A 677	
NIVIDIC Y 643	LA HOUSSAYE A 678	

Displacement, tonnes: 224 standard; 263 full load
Dimensions, metres (feet): 25 × 8.4 × 3.4 *(82 × 27.6 × 11.2)*
Speed, knots: 11
Range, n miles: 800 at 10 kt
Machinery: 2 SACM-Wärtsilä UD 30 V12 M3 diesels (A 675 and 676); 2 Baudouin P 15 2S (others) 1,320 hp *(984 kW)*; 2 Voith-Schneider propulsors; 1,280 hp(m) *(941 kW)*; 1,360 hp(m) *(1 MW)* in later vessels
Radars: 1 Racal Decca RM 170 or Bridgemaster C 181; I-band.

Comment: Built at Lorient Naval et Industries shipyard (formerly Chantiers et Ateliers de la Perrière, now part of Leroux et Lotz) and at Boulogne by SOCARENAM. *Fréhel* in service 23 May 1989, based at Cherbourg, *Saire* 6 October 1989 at Cherbourg, *Armen* 6 December 1991 at Brest, *La Houssaye* 30 October 1992 at Brest, *Kéréon* 5 December 1992 at Brest. *Mengam* 6 October 1994 at Brest and *Sicié* 6 October 1994 at Toulon, *Giens* 2 December 1994 at Toulon, *Lardier* 12 March 1995 at Toulon, *Balaguier* 8 July 1995 at Toulon, *Taillat* 18 October 1995 at Toulon. *Taunoa* completed 9 March 1996 at Brest, *Nividic* on 13 February 1996 at Brest, *Port Cros* on 21 June 1997 at Toulon, *Le Four* on 13 March 1998 at Brest and *Rascas* on 22 November 2003 at Toulon. Bollard pull 12 tons. Type RPC 12 coastal tugs, with 'A' pennant numbers and a crew of eight. Type RP12 harbour tugs with 'Y' pennant numbers and a crew of five. A further order for six craft has been abandoned.

RASCAS   *9/2009, M Declerck / 1305824*

4 TYPE RP 10 HARBOUR TUGS/PUSHERS (YT)

MORSE Y 770	OTARIE Y 771	LOUTRE Y 772	PHOQUE Y 773

Displacement, tonnes: 84 standard; 99 full load
Dimensions, metres (feet): 15.5 × 6.4 × 2.1 *(50.9 × 21.0 × 6.9)*
Speed, knots: 8. **Range, n miles:** 160 at 7 kt
Complement: 4
Machinery: 2 Baudouin 6R123S diesels; 800 hp *(588 kW)*; 2 shafts
Radars: Navigation: Furuno M 1832; I-band.

Comment: Ordered on 21 November 2003. All entered service on 5 October 2005. Built by SOCARENAM, Boulogne. Bollard pull 10 tons. *Morse* based at Mayotte and the remainder at Toulon.

LOUTRE   *9/2009, M Declerck / 1305823*

3 MAÏTO CLASS (YTM)

MAÏTO A 636	MAROA A 637	MANINI A 638

Displacement, tonnes: 232 standard; 284 full load
Dimensions, metres (feet): 27.6 × 8.3 × 3.5 *(90.6 × 27.2 × 11.5)*
Speed, knots: 11
Range, n miles: 1,200 at 11 kt
Complement: 6 + 4 spare berths
Machinery: 2 SACM-Wärtsilä UD 30 L6 M6 diesels; 1,280 hp(m) *(940 kW)*; 2 Voith-Schneider propulsors
Radars: Navigation: Racal-Decca 1226; I-band.

Comment: Built by SFCN, Villeneuve-La-Garenne, and formerly used at the CEP Nuclear Test Range. *Maïto* commissioned 25 July 1984 and is based at Martinique. *Maroa* (commissioned 28 July 1984) and *Manini* (commissioned 12 September 1985) are both based at Papeete, Tahiti. Bollard pull, 12 tons. Fire-fighting water cannon.

MAÏTO   *11/2006, M Declerck / 1167128*

3 BÉLIER CLASS (YTB)

| BÉLIER A 695 | BUFFLE A 696 | BISON A 697 |

Displacement, tonnes: 362 standard; 508 full load
Dimensions, metres (feet): 31.8 × 9.2 × 4.2 *(104.3 × 30.2 × 13.8)*
Speed, knots: 11
Complement: 12
Machinery: 2 SACM-Wärtsilä UD 33V12 M4 diesels; 2,600 hp(m) *(1.91 MW)*; 2 Voith-Schneider props
Radars: Navigation: Racal-Decca C 810; I-band.

Comment: Built by DCN at Cherbourg. *Bélier* commissioned 10 July 1980, *Buffle* on 19 July 1980, *Bison* on 16 April 1981. A 695 and 697 based at Toulon and A 696 at Brest. Bollard pull, 25 tons. Service life: 2015.

BISON *11/2010, B Prézelin* / 1366869

4 TYPE PSS 10 PUSHER TUGS (YTL)

P 101–104

Displacement, tonnes: 45 standard; 70 full load
Dimensions, metres (feet): 17.5 × 6.4 × 2.0 *(57.4 × 21.0 × 6.6)*
Speed, knots: 10
Range, n miles: 480 at 6 kt
Complement: 3
Machinery: 2 Poyaud UD 25 L06 M4D diesels; 800 hp *(588 kW)*; 2 shafts in Kort nozzles

Comment: Built by Leroux Naval Industrie, Lorient, and commissioned in 1993. Designed to handle Le Triomphant class SSBNs. All based at Brest.

P 101 *8/2008, B Prézelin* / 1335727

26 TYPE PS4 PUSHER TUGS (YTL)

| P 6 | P 13–24 | P 26–38 |

Displacement, tonnes: 28 standard; 30 full load
Dimensions, metres (feet): 11.9 × 4.4 × 2.1 *(39 × 14.4 × 6.9)*
Speed, knots: 9
Range, n miles: 540 at 8 kt
Complement: 3
Machinery: 2 Poyaud UD 6 (P 6); 430 hp *(316 kW)* or Poyaud UD 18 (P 13–30); 440 hp *(324 kW)* or Baudouin V6 TI 330 (P 31–38); 480 hp *(353 kW)* diesels; 2 shafts

Comment: Built by shipyards at Brest and Lorient. *P 6* (ex-*P 12*)entered service in 1973, *P 13–18* 1982–83 and the remainder 1989–97. Based at naval bases in France and overseas. Bollard pull 5 tons.

P 33 *1/2009, B Prézelin* / 1305726

GOVERNMENT MARITIME FORCES

Notes: (1) 'Action de l'Etat en Mer (AEM)' encompasses all activities regarding maritime surveillance and sea traffic control, fishery protection and policing, SAR, safety of navigation, pollution control and so on. It involves the Marine Nationale (navy), the Gendarmerie Maritime, the Affaires Maritimes, the Douanes françaises (customs), the Administration des Phares et Balises (lighthouses and navigation aids management organisation) and some local police forces. The organisation is for French mainland as well as for overseas territories. In homeland waters, it is under the direct control of the flag officers (C-in-Cs) at Cherbourg, Brest and Toulon. All ships and craft involved in AEM tasks display 'AEM markings' (inclined blue/white/red stripes on their hull sides). This also applies to naval manned patrol vessels (patrouilleurs de service public, PSP) and to the Eurocopter SA 365N Dauphin 2 helicopters acquired by the naval air arm for SAR duties.
(2) For the 'Sauvegarde maritime' (maritime approaches surveillance organisation), most AEM tasked vessels (including naval OPVs) and coastal VTS are connected, through a dedicated datalink, to the Spationav common maritime picture network. Spationav display terminals are also being fitted to the Dassault Falcon 50M land-based maritime surveillance aircraft.
(3) It was announced in December 2009 that a Coast Guard, Fonction Guardes-Côtes (FGC), was to be formed. A Coast Guard Functions Management Centre was established on 20 September 2010. Its purpose is to share and analyse maritime information rather than command operations which remains the responsibility of the appropriate Préfet Maritime in their regions (Mediterranean, English Channel and North Sea, Atlantic).

AUXILIARIES

Notes: (1) Permanently chartered vessels for AEM tasks include four salvage and rescue tugs, and four support and pollution fighting vessels. They perform civilian tasks such as safety of navigation, SAR, pollution control, and military missions in support of the fleet: torpedo recovery, diving operation support, submarine crew rescue, experiments, and so on.
(2) A contract was renewed with Bourbon/Abeilles International in July 2002 for emergency use of a large fleet of harbour tugs. Similar contracts are concluded with local fishing associations.
(3) The DGA (Directorate General for Armament) charters the 67 m *Langevin* for submarine associated trials.
(4) On 29 November 2005, the European Maritime Safety Agency (EMSA) contracted Louis Dreyfus Armateurs (LDA) for the standby charter of the cable repair vessel *Ile de Bréhat* (built 2001, 14,960 UMS, 140 m, 15 kt) for oil recovery during emergencies. Currently chartered for the maintenance and repairs of transatlantic submarine cables, this ship is based at Brest. It will receive some modifications to be classified as '(standby) oil recovery vessel'; it will be capable of recovering and storing up to 4,000 m^3 of polluted water and of deploying a ROV.
(5) The existing salvage apparatus can be boosted by civilian coastal tugs which are at three hours readiness. The tugs are operated by the following ports: Dunkirk, Calais, Boulogne, Dieppe, Le Havre, St Malo, Lorient, Nantes-St Nazaire, Bayonne, Port-la-Nouvelle, Sète, Marseilles-Fos, Nice, Ajaccio and Bastia.

2 ULSTEIN UT 515 (SALVAGE AND RESCUE TUGS) (ARS)

| ABEILLE BOURBON | ABEILLE LIBERTÉ |

Displacement, tonnes: 3,251 standard; 4,064 full load
Dimensions, metres (feet): 80 × 16.5 × 6.5 *(262.5 × 54.1 × 21.3)*
Speed, knots: 19.5
Complement: 12
Machinery: 4 MaK 8M32C diesels; 21,700 hp(m) *(16 MW)*; 2 shafts; 2 cp props; 2 bow and 2 stern thrusters
Radars: Navigation: 2 Furuno; I-band.

Comment: Contract awarded in November 2003 to Abeilles International for the procurement and operation (over eight years) of two Ulstein 515 salvage tugs, classified as Remorques d'Intervention, d'Assistance et de Sauvetage (RIAS). Equipped with a 500 ton towing winch and with extensive fire-fighting and pollution control equipment. Vessels built by Maritim, Gdansk and outfitted by Myklebust, Norway. The first ship, *Abeille Bourbon*, entered service on 21 May 2005 and is based at Brest. She was fitted with a reinforced bow in November 2007. *Abeille Liberté* entered service on 25 October 2005 and is based at Cherbourg. There are two crews of 12 per ship and the ships are maintained at 40 minutes readiness.

ABEILLE BOURBON *8/2008, B Prézelin* / 1335725

1 ULSTEIN UT 710 (SALVAGE AND RESCUE TUG) (ARS/BSAD)

ARGONAUTE (ex-*Island Patriot*)

Displacement, tonnes: 2,409 standard; 4,491 full load
Dimensions, metres (feet): 68.9 × 15.5 × 7 *(226 × 50.9 × 23.0)*
Speed, knots: 16. **Range, n miles:** 19,000 at 10 kt
Complement: 9 + 22 spare berths
Machinery: 2 Rolls Royce Bergen BRM-9 diesels; 10,800 hp(m) *(8.1 MW)*; 2 shafts; 2 cp props with Kort nozzles; 2 bow and 1 stern thrusters
Radars: Navigation: 2 Raytheon; I-band.

Comment: Built by Aker-Brevik Construction AS, Norway, the ship was launched on 7 July 2003 and entered service with Island Offshore on 12 December 2003. Chartered by the French government from 1 January 2004 and modified in June 2004 to meet naval requirements. Based at Brest. Fitted for pollution control and with fire-fighting equipment. Capable of operating an ROV.

ARGONAUTE *5/2012, Frank Findler* / 1455816

278 France > Government maritime forces

2 ULSTEIN UT 507 CLASS (SALVAGE TUGS) (ARS)

ABEILLE FLANDRE (ex-*Neptun Suecia*) **ABEILLE LANGUEDOC** (ex-*Neptun Gothia*)

Displacement, tonnes: 3,048 standard; 3,556 full load
Dimensions, metres (feet): 63.4 × 14.4 × 7.3 *(208 × 47.2 × 24.0)*
Speed, knots: 17
Range, n miles: 36,000 at 10 kt
Complement: 12
Machinery: 4 MaK 8M453AK diesels; 23,000 hp *(16.9 MW)*; 2 cp props; 2 Ulstein bow thrusters
Radars: Navigation: 1 Racal Decca Bridgemaster; 1 Racal Decca Bright Track 90; E/F-band.

Comment: Built by Ulstein Hatho A/S, Norway and entered service in 1978 and 1979. On long-term charter from Abeilles International since 14 December 1979. Bollard pull 160 tons. *Abeille Flandre* based at Toulon since 30 May 2005, *Abeille Languedoc* was based at La Pallice, near La Rochelle, since 25 October 2005 until September 2011 when she transferred to Boulogne. Both ships refitted in 2005.

ABEILLE FLANDRE   5/2009, B Prézelin / 1305727

1 JASON CLASS (BUOY TENDER) (ABU)

JASON (ex-*Bourbon Aspara*)

Displacement, tonnes: 2,100 standard
Dimensions, metres (feet): 67 × 15.4 × 5.95 *(219.8 × 50.5 × 19.5)*
Speed, knots: 14
Complement: 30
Machinery: 2 Caterpillar 3612 DITA diesels; 10,890 hp *(8.1 MW)*; 2 cp props; 2 bow thrusters; 1,600 hp *(1.2 MW)*; 1 stern thruster; 800 hp *(600 kW)*
Radars: Navigation: I-band.

Comment: Anchor handling vessel constructed by Keppel Singmarine, Singapore, and completed in 2005. Equipped with towing winch (bollard pull 125 t), firefighting gear, standby rescue equipment for 150 survivors and a dynamic positioning system. Replaced *Carangue* on 1 July 2009 and based at Toulon.

JASON   6/2010, Marco Ghiglino / 1406969

1 JIF XPLORER CLASS (AGMS)

JIF XPLORER (ex-*Jean-Marie Christian V*)

Measurement, tonnes: 489 gt
Dimensions, metres (feet): 45.6 × 11.1 × 3.6 *(149.6 × 36.4 × 11.8)*
Speed, knots: 14
Machinery: 2 MWM TBD604BL6 diesels; 1,716 hp *(1.26 MW)*; 2 azimuth props
Radars: Navigation: I-band.
Sonars: Klein 3000.

Comment: Former tunny fishing vessel built by Construction Navales Stento, Balarucles-Bains in 1991. Chartered from Jifmar Offshore Services by DGA (Direction Générale de l'Armement) for the missile trial center CEL (Centre d'Essai des Landes). Modified with Klein 3000 sonar and two cranes for its new role of missile recovery. Based at Bayonne.

2 ULSTEIN UT 711 CLASS (BUOY TENDERS) (ABU/BSAD)

ALCYON (ex-*Bahram*)   **AILETTE** (ex-*Cyrus*)

Displacement, tonnes: 1,229 standard; 1,930 full load
Dimensions, metres (feet): 53 × 13.3 × 6.8 *(173.9 × 43.6 × 22.3)*
Speed, knots: 14.5
Range, n miles: 5,400 at 14 kt
Complement: 7 + 15 spare berths
Machinery: 2 Bergens-Normo KVMB-12 diesels; 5,200 hp *(3.9 MW)*; 2 cp props; 1 Ulstein 90 bow thruster; 500 hp *(370 kW)*; 1 Rolls Royce bow thruster (Ailette only); 400 hp *(300 kW)*
Radars: Navigation: Racal-Decca Bridgemaster 252C and Furuno FR 2120; I-band.

Comment: Built by A & C de la Manche, Dieppe and entered service in 1981 and 1982. On long-term charter from SURF (Groupe Bourbon). Former oil-field supply vessels both modernised in 2002–03 by Chantiers Piriou, Concarneau, for limited oil-pollution control activities: TRANSREC 250 sea skimming system and polluted water storage capacity of 500 m³. New 23 ton hydraulic deck crane fitted; dynamic positioning system fitted to *Ailette*. Deck capacity 480 tons and bollard pull of 64 tons. *Alcyon* based at Brest and *Ailette* at Toulon. To be replaced by BSAH.

IHS Jane's Fighting Ships 2014-2015

AILETTE   5/2009, B Prézelin / 1305728

RESEARCH SHIPS

Note: Several government agencies use research vessels for various purposes. Most of them are operated by GENAVIR on their behalf. Main agency is IFREMER (Institut Français de Recherche pour l'Exploitation de la Mer) that operates four large ocean-going vessels; *Pourquoi Pas ?* (2005, 6,600 tons), *Thalassa* (1996, 3,022 tons), *L'Atalante* (1989, 3,550 tons), *Le Suroît* (1975, modernised 1999, 1,132 tons); and three coastal operations vessels: *L'Europe* (1993, 264 tons catamaran), *Thalia* (1978, 135 grt, trawler type) and *Gwen Drez* (1976, 249 tons, trawler type). IRD (Institut de Recherche pour le Développement, ex-ORSTOM) operates in the Pacific two research vessels: *Antéa* (1995, 421 grt, catamaran) and *Alis* (1987, 198 grt UMS, trawler type), and two smaller craft. INSU (Institut National des Sciences de l'Univers) operates five coastal vessels (12.5 to 24.9 m) along the French coasts. TAAF (Administration des Terres Australes et Antarctiques Françaises) uses *Marion Dufresne* (1995, 9,403 GRT UMS, 120 m long), a large support ship for Antarctic operations also fitted for scientific research work, *L'Astrolabe*, (ex-*Austral Fish*, 1986, 1,370 grt) and *La Curieuse* (1989, 150 grt UMS, trawler type).

POLICE (GENDARMERIE MARITIME)

Note: (1) The Gendarmerie Maritime is a force of 1,104 officers and men belonging to the Gendarmerie Nationale but acting under the operational control of the French Navy. The Force is tasked to safeguard, supervise and control shipping traffic.
(2) The Gendarmerie Nationale also operates about 50 small patrol craft whose pennant numbers are prefixed 'G'.

1 PATRA CLASS (COASTAL PATROL CRAFT) (PB)

Name	No	Builders	Commissioned
GLAIVE	P 671	Auroux, Arcachon	2 Apr 1977

Displacement, tonnes: 117 standard; 150 full load
Dimensions, metres (feet): 40.4 × 5.9 × 1.6 *(132.5 × 19.4 × 5.2)*
Speed, knots: 26
Range, n miles: 1,750 at 10 kt, 750 at 20 kt
Complement: 18 (1 officer)
Machinery: 2 SACM-Wärtsilä UD 33 V12 M5 diesels; 4,000 hp(m) *(2.94 MW)*; 2 shafts; cp props
Guns: 1 Bofors 40 mm/60. 2—7.5 mm MGs.
Radars: Surface search: Racal Decca 1226; I-band.

Comment: Based at Cherbourg. Service life extended following replacement of main machinery in 2012.

PATRA CLASS   6/2006, B Prézelin / 1040756

8 VSMP PATROL CRAFT (PB)

PAVOIS P 792	RONDACHE P 794	HAUBERT P 796	BRIGANTINE P 798
ÉCU P 793	HARNOIS P 795	HEAUME P 797	GANTELET P 799

Displacement, tonnes: 11 full load
Dimensions, metres (feet): 12.64 × 3.88 × 1.5 *(41.5 × 12.7 × 4.9)*
Speed, knots: 35
Complement: 8
Machinery: 2 Cummins diesels; 760 hp *(560 kW)*; 2 waterjets
Guns: 1—7.62 mm MG.
Radars: Navigation: Furuno; I-band.

Comment: Coastal surveillance craft designated Vedette de Sûreté Maritime et Portuaire (VSMP). Ordered on 30 September 2009 and built by Raidco Marine, Lorient, and UFAST at Quimper. Delivered 2010–11. Employed on patrol duties at Marseille-Fos, Nantes-St Nazaire, Le Havre and Dunkirk.

4 GERANIUM CLASS (PB)

Name	No	Builders	Commissioned
GÉRANIUM	P 720	DCN, Lorient	19 Feb 1997
JONQUILLE	P 721	Chantiers Guy Couach Plascoa	15 Nov 1997
VIOLETTE	P 722	DCN, Lorient	4 Dec 1997
JASMIN	P 723	Chantiers Guy Couach Plascoa	15 Nov 1997

Displacement, tonnes: 81 (P720, 722), 83 (P721, 723) standard; 102 full load
Dimensions, metres (feet): 32.2 × 6.1 × 1.9 (105.6 × 20.0 × 6.2)
Speed, knots: 30
Range, n miles: 1,500 at 15 kt
Complement: 15 (2 officers)
Machinery: 2 Deutz/MWM TBD 516 V16; 1 Deutz/MWM TBD 516 V12; 3,960 hp *(2.95 MW)*; 2 shafts; 1 Hamilton 422 water-jet
Guns: 1 – 12.7 mm MG. 1 – 7.62 mm MG.
Radars: Navigation: Racal-Decca CH 180/6; E/F-band.

Comment: There are some minor differences between the DCN (details shown) and the Plascoa craft. *Géranium* based at Cherbourg; *Jonquille* at Toulon; *Violette* at Pointe-à-Pitre, Guadeloupe; *Jasmin* at Papeete, Tahiti. Two similar craft built for Affaires Maritimes.

GERANIUM 5/2009, Marco Ghiglino / 1305833

24 TYPE VCSM (PATROL CRAFT) (PB)

ÉLORN P 601	YSER P 607	CHARENTE P 613	HUVEAUNE P 619
VERDON P 602	ARGENS P 608	TECH P 614	SÉVRE P 620
ADOUR P 603	HÉRAULT P 609	PENFELD P 615	ABER-WRACH P 621
SCARPE P 604	GRAVONA P 610	TRIEUX P 616	ESTÉRON P 622
VERTONNE P 605	ODET P 611	VÉSUBIE P 617	MAHURY P 623
DUMBEA P 606	MAURY P 612	ESCAUT P 618	ORGANABO P 624

Displacement, tonnes: 43 standard
Dimensions, metres (feet): 20 × 5.2 × 1.5 (65.6 × 17.1 × 4.9)
Speed, knots: 28
Range, n miles: 530 at 15 kt
Complement: 5
Machinery: 2 MAN V12 diesels; 2 shafts; 2,000 hp(m) *(1,470 kW)*
Guns: 2 – 7.62 mm MG.
Radars: Navigation: Furuno; I-band.

Comment: Designated 'Vedette Côtière de Surveillance Maritime' (VCSM), coastal surveillance craft. Raidco RPB 20. Ordered in two batches of 11 on 6 Dec 2001 and 6 June 2002. Built at l'Herbaudière by Raidco Marine with the co-operation of Chantiers Beneteau. Bear names of rivers. First of class (P 601) entered service on 20 June 2003 followed by P 602–604 in 2003, P 605–610 in 2004 and P 611–615 in 2005. The remainder entered service by March 2007. Replace VSC 14 and VSC 10 craft. GRP hull and superstructure. One 4.9 m RIB fitted aft on an inclined ramp. Also fitted with water-cannon. P 606 based at Noumea, New Caledonia, P 623 and P 624 in French Guiana and P 602 at Mayotte. Two similar craft in service in Morocco and two in Senegal.

TECH 9/2011, Adolfo Ortigueira Gil / 1406970

CUSTOMS (DOUANES FRANÇAISES)

Notes: The French customs service, a division of the Ministry of Finance, has a number of tasks not normally associated with such an organisation. In addition to the usual duties of dealing with ships entering either its coastal area or ports it also has certain responsibilities for rescue at sea, control of navigation, fishery protection and pollution protection. Operated by about 650 personnel and about 200 airmen, the fleet comprises 41 (a further four 12 m vessels are to enter service in 2014) vessels of which 21 are large patrol vessels (21 to 43 m). These include DFP 1 *Jacques Oudart Fourmentin* (400 tons), DFP 2 *Kermovan* (400 tons), DF 48 *Arafenua* (105 tons), DF 42 *Suroît* (67 tons), DF 31 *Pitera* (64 tons), DF 37 *Nordet* (64 tons), DF 43 *Haize Hegoa* (64 tons), DF 44 *Rakawa* (64 tons), DF 45 *Sirocco* (64 tons), DF 46 *Avel Sterenn* (64 tons), DF 47 *Lissero* (64 tons), and DF 40 *Vent d'Amont* (61 tons), DF 30 *Marinada* (56 tons), DF 16 *Cers* (56 tons), DF 49 *Orsuro* (56 tons), DF 25 *Libecciu* (39 tons), DF 14 *Muntese* (39 tons), DF 12 *U'Marinu* (39 tons), DF 24 *Sua Luiga* (56 tons), DF 21 *Alizé* (39 tons). All vessels have DF numbers painted on the bow and 'AEM markings' (blue/white/red inclined stripes). There are also 13 Reims-Cessna F406 lightweight patrol aircraft, including three equipped for pollution control, two Eurocopter AS 350B1 Ecureuil helicopters and seven Eurocopter EC 135 helicopters equipped for maritime surveillance.

KERMOVAN 8/2008, B Prézelin / 1335720

AFFAIRES MARITIMES

Notes: The Affaires Maritimes (to be subsumed within the Coast Guard organisation FGC) is a force administered and funded by the Ministry of Transport to enforce safety of navigation, SAR, fishery protection and pollution control. The force also contributes to surveillance against terrorist activities. Operational control is vested locally in Préfet Maritimes who are naval flag officers. SAR is coordinated through a network of Maritime Rescue Coordination Centres (MRCC) at Gris Nez (Dover Strait), Jobourg (Western Channel), Corsen (Brittany), Etel (Bay of Biscay), La Garde (Mediterranean, Gulf of Lion), Aspretto (Corsica), Port des Galets (La Réunion), Fort-de-France (Caribbean). CROSS Etel is responsible for monitoring all fishing activity in French waters. Vessels operated by Affaires Maritimes are usually unarmed and manned by civilians. They are painted with grey/blue hulls, grey superstructure and display the AEM blue/white/red stripes, PM pennant numbers and 'Affaires Maritimes' written on the superstructure. The fleet comprises:

- Six large patrol vessels (30–52 m): PM 41 *Themis* (400 tons), PM 40 *Iris* (230 tons), PM 32 *Armoise* (91 tons), PM 30 *Gabian* (76 tons), PM 29 *Mauve* (65 tons); *Osiris*, a seized fishing trawler, is based at La Réunion.
- 21 patrol launches (8–17 m): These include four Callisto class PM 100–103 which are 16 m FPB 50 Mk II patrol boats built 2000–01 by OCEA, Les Sables d'Olonne.
- Service craft which may be identified by 'Phares & Balises' written on the superstructure. Larger vessels include: *Armorique* (500 tons), *Hauts de France* (450 tons), *Gascogne, Provence* (326 tons), *Chef de Caux* (128 tons), *Louis Henin* (73 tons) and *Le Kahouanne* (73 tons).

THEMIS 7/2007, B Prézelin / 1305048

HAUTS DE FRANCE 8/2009, Bob Fildes / 1305832

PM 103 6/2010, B Prézelin / 1366875

Gabon
MARINE GABONAISE

Country Overview
A former French colony, the Gabonese Republic achieved independence in 1960. Located astride the Equator, the country has an area of 103,347 square miles and has borders to the north with Cameroon and Equatorial Guinea and to the east and south with Congo. It has a 480 n mile coastline with the Atlantic Ocean. The capital, largest city and principal port is Libreville and there is a further port at Port-Gentil. Territorial seas (12 n miles) are claimed. A 200 n mile Exclusive Economic Zone (EEZ) has been claimed but the limits are not defined; jurisdiction is complicated by the offshore islands of Isla de Annobon (Equatorial Guinea) and São Tomé and Principe.

Headquarters Appointments
Chief of Naval Staff:
  Rear Admiral Hervé Nambo Ndouany

Bases
Port Gentil, Mayumba

Personnel
2014: 600 (65 officers)

Maritime Aircraft
One Embraer EMB 110 Bandeirante is used for Maritime Surveillance.

PATROL FORCES

Notes: Two Rodman 14 m craft were delivered in January 2006. Their names are: *Mbanie* and *Kouango*.

2 P 400 CLASS (LARGE PATROL CRAFT) (PBO)

Name	No	Builders	Commissioned
GÉNÉRAL D'ARMÉE BA-OUMAR	P 07	CMN, Cherbourg	27 Jun 1988
COLONEL DJOUE-DABANY	P 08	CMN, Cherbourg	14 Sep 1990

Displacement, tonnes: 453 full load
Dimensions, metres (feet): 54.6 × 8 × 2.5 *(179.1 × 26.2 × 8.2)*
Speed, knots: 24
Range, n miles: 4,200 at 15 kt
Complement: 32 (4 officers)

Military lift: 20 troops
Machinery: 2 Wärtsilä UD 33 V16 diesels; 8,000 hp(m) *(5.88 MW)* sustained; 2 shafts; cp props
Guns: 1 Bofors 57 mm/70 SAK 57 Mk 2 (P 07); 220 rds/min to 17 km *(9 n miles)*; weight of shell 2.4 kg. Not in P 08 which has a second Oerlikon 20 mm. 2 Giat F2 20 mm (P 08).
Radars: Racal Decca 1226C; I-band.
Weapon control systems: CSEE Naja optronic director (P 07).

Programmes: Contract signed May 1985 with CMN Cherbourg. First laid down 2 July 1986, launched 18 December 1987 and arrived in Gabon 6 August 1988 for a local christening ceremony. Second ordered in February 1989 and launched 29 March 1990.
Structure: There is space on the quarterdeck for two MM 40 Exocet surface-to-surface missiles. These craft are similar to the French vessels but with different engines. *Djoue-Dabany* had twin funnels fitted in 1992, similar to French P 400 class conversions.

COLONEL DJOUE-DABANY *6/2000, Gabon Navy* / 0104491

1 PATRA CLASS (FAST ATTACK CRAFT – MISSILE) (PTM)

Name	No	Builders	Commissioned
GÉNÉRAL NAZAIRE BOULINGUI	P 10	Chantiers Naval de l'Estérel	7 Aug 1978
(ex-*Président Omar Bongo*)			

Displacement, tonnes: 163 full load
Dimensions, metres (feet): 42 × 7.7 × 1.9 *(137.8 × 25.3 × 6.2)*
Speed, knots: 32
Range, n miles: 1,500 at 15 kt
Complement: 20 (3 officers)

Machinery: 3 SACM 195 V12 CSHR diesels; 5,400 hp(m) *(3.97 MW)* sustained; 3 shafts
Missiles: SSM: 4 Aerospatiale SS 12M; wire-guided to 5.5 km *(3 n miles)* subsonic; warhead 30 kg.
Guns: 1 Bofors 40 mm/60. 1 DCN 20 mm.
Radars: Surface search: Racal Decca RM1226; I-band.

Comment: Re-activated in 2000.

GÉNÉRAL NAZAIRE BOULINGUI *6/2000, Gabon Navy* / 0104492

4 VCSM CLASS (PB)

COCO-BEACH V 7	PORT GENTIL V 8	OMBOUE V 9	MAYUMBA V 10

Displacement, tonnes: 43 standard
Dimensions, metres (feet): 20 × 5.2 × 1.5 *(65.6 × 17.1 × 4.9)*
Speed, knots: 25
Range, n miles: 530 at 15 kt
Complement: 5
Machinery: 2 MAN V12 diesels; 2,000 hp *(1.47 MW)*; 2 shafts
Guns: 1 – 7.62 mm MG.
Radars: Navigation: Furuno; I-band.

Comment: Four Raidco Marine RPB 20 class were procured in 2010. Similar to French VCSM class.

3 DEFENDER CLASS (RESPONSE BOATS) (PBF)

Displacement, tonnes: 2.7 full load
Dimensions, metres (feet): 8.2 × 2.6 × 0.55 *(26.9 × 8.5 × 1.8)*
Speed, knots: 46
Range, n miles: 175 at 35 kt
Complement: 4
Machinery: 2 outboard motors; 450 hp *(335 kW)*
Guns: 1 – 12.7 mm MG.
Radars: To be announced.

Comment: High-speed inshore patrol craft of aluminium construction and foam collar built by SAFE Boats International, Port Orchard, Washington. Donated by the US government in 2010.

4 RODMAN 66 CLASS (PB)

CAPITAINE DE FRÉGATE AWORE PAUL P 01
LIEUTENANT DE VAISSEAU MANGOYE JEAN-BAPTISTE P 02
ENSEIGNE DE VAISSEAU BETSENG ADRIEN P 03
ENSEIGNE DE VAISSEAU MENDENE P 04

Displacement, tonnes: 36 full load
Dimensions, metres (feet): 20.5 × 4.9 × 0.96 *(67.3 × 16.1 × 3.1)*
Speed, knots: 30
Range, n miles: 450 at 22 kt
Complement: 6
Machinery: 2 Caterpillar diesels; 1,200 hp *(880 kW)*; 2 shafts
Radars: Navigation: Furuno; I-band.

Comment: Built by Rodman and delivered in January 2006.

LV MANGOYE JEAN-BAPTISTE *4/2009, US Navy* / 1455792

AMPHIBIOUS FORCES

1 LCU MK 9 (LCU)

LECONI II L 3015

Displacement, tonnes: 75 standard; 176 full load
Dimensions, metres (feet): 25.7 × 6.5 × 1.7 *(84.3 × 21.3 × 5.6)*
Speed, knots: 10
Machinery: 2 diesels; 440 hp *(325 kW)*; 2 waterjets
Radars: To be announced.

Comment: Former British landing craft built in the 1960s. Entered service in 2011 following refit at Brest which included replacement of engines.

Amphibious forces < **Gabon** — Patrol forces < **Gambia** 281

1 BATRAL TYPE (LSTH)

Name	No	Builders	Launched	Commissioned
PRESIDENT EL HADJ OMAR BONGO	L 05	Français de l'Ouest, Rouen	16 Apr 1984	26 Nov 1984

Displacement, tonnes: 782 standard; 1,357 full load
Dimensions, metres (feet): 80 × 13 × 2.4 *(262.5 × 42.7 × 7.9)*
Speed, knots: 16
Range, n miles: 4,500 at 13 kt
Complement: 39

Military lift: 188 troops; 12 vehicles; 350 tons cargo
Machinery: 2 SACM Type 195 V12 CSHR diesels; 3,600 hp(m) *(2.65 MW)*; 2 shafts; cp props
Guns: 1 Bofors 40 mm/60; 300 rds/min to 12 km *(6.5 n miles)*; weight of shell 0.89 kg. 2 Browning 12.7 mm MGs. 1—7.62 mm MG.
A/S Mortars: 2—81 mm mortars
Radars: Surface search: Racal Decca 1226; I-band.
Helicopters: Capable of operating up to SA 330 Puma size.

Comment: Sister to French *La Grandière*. Carries one LCVP and one LCP. Started refit by Denel, Cape Town in April 1996, and returned to service in 1997 with bow doors welded shut. Completed repair and cleaning at Abidjan during 2000.

PRESIDENT EL HADJ OMAR BONGO *6/1993, Gabon Navy /* 0069977

12 LANDING CRAFT PERSONNEL (LCVP)

Comment: Two 12 m craft built by Tanguy Marine, Le Havre in 1985. Equipped with two Volvo Penta 165 hp(m) *(121 kW)* engines. There are also 10 Simonneau craft: one of 12 m, two of 8 m and seven of 7 m.

Gambia

Country Overview

The Republic of Gambia was a British protectorate until 1965 when it gained independence. With an area of 4,361 square miles, it has a short 43 n mile coastline with the Atlantic Ocean but is otherwise completely surrounded by Senegal. The two countries united in 1981 to form the confederation of Senegambia but this collapsed in 1989 when the countries reverted to being separate states. The capital, largest city and principal port is Banjul (formerly Bathurst). Territorial seas (12 n miles) and a 200 n mile fishing zone are claimed. The patrol craft came under 3 Marine Company of the National Army until 1996 when a navy was established.

Headquarters Appointments

Commander, Navy:
   Commodore Madani Senghore

Personnel

(a) 2014: 233
(b) Voluntary service

Bases

Banjul

PATROL FORCES

Notes: Three high-speed interception craft were donated by Taiwan in August 2013. The 10-m craft are equipped with 115 hp outboard motors, giving a top speed of 45 kt, and are armed with 12.7 mm machine guns. Their names are *Berre Kuntu*, *Kennyeh Kennyeh Jamang-Oh*, and *Sanimentereng*.

1 PETERSON MK 4 CLASS (PB)

Name	No	Builders	Commissioned
BOLONG KANTA	P 14	Peterson Builders Inc, Sturgeon Bay	15 Oct 1993

Displacement, tonnes: 24 full load
Dimensions, metres (feet): 15.5 × 4.5 × 1.3 *(50.9 × 14.8 × 4.3)*
Speed, knots: 24
Range, n miles: 500 at 20 kt
Complement: 6
Machinery: 2 Detroit 6V-92A diesels; 520 hp *(388 kW)* sustained; 2 shafts
Guns: 2—12.7 mm MGs.
Radars: Raytheon R41X; I-band.

Comment: Reported non-operational and awaiting repairs. Similar craft in service in Egypt, Cape Verde and Senegal.

BOLONG KANTA *8/2009, Gambia Navy /* 1305463

2 PATROL CRAFT (PB)

FATIMAH I PT 01   SULAYMAN JUN-KUNG PT 02

Displacement, tonnes: 25 standard
Dimensions, metres (feet): 16.1 × 4.5 × 1.6 *(52.8 × 14.8 × 5.2)*
Speed, knots: 40
Machinery: 2 Caterpillar diesels; 1,600 hp *(1.2 MW)*
Guns: 3—7.62 mm MGs.
Radars: Surface search: Furuno; I-band.

Comment: Procured from Taiwan in 1999.

FATIMAH *6/2000, Gambian Navy /* 0104493

4 HAI OU CLASS (PB)

TAIPEI FPB 1	KINDY CAMARA FPB 3
KUNTA KINTEH FPB 2	MUSA MALIYA JAMMEH FPB 4

Displacement, tonnes: 48 full load
Dimensions, metres (feet): 21.6 × 5.5 × 2.02 *(70.9 × 18.0 × 6.6)*
Speed, knots: 30
Range, n miles: 700 at 22 kt
Complement: 10 (2 officers)
Machinery: 2 MTU 12V 331 TC82 diesels; 2,605 hp(m) *(1.92 MW)* sustained; 2 shafts
Guns: 1—12.7 mm MG.
Radars: Surface search: Furuno; I-band.

Comment: Four former Taiwan Navy Hai Ou class donated by the Taiwan government on 26 June 2009. One vessel is reported non-operational.

MUSA MALIYA JAMMEH *8/2009, Gambia Navy /* 1305462

© 2014 IHS   IHS Jane's Fighting Ships 2014-2015

Georgia

Country Overview

Formerly part of the USSR, the Republic of Georgia declared independence in 1991. Situated in the Transcaucasia region of western Asia, the country has an area of 26,900 square miles and is bordered to the north by Russia and to the south by Turkey, Armenia and Azerbaijan. It has a coastline of 167 n miles with the Black Sea on which Poti and Batumi are the principal ports. T'bilisi is the capital and largest city. The country includes two autonomous republics, Abkhazia and Ajaria, and one autonomous region, South Ossetia. USSR legislation appears still to apply to maritime claims. Territorial waters (12 n miles) are claimed, as is an EEZ (200 n miles) although the limits of the latter are not defined.

Much of the Georgian Navy and Coast Guard was destroyed during the Georgia-Russia conflict in August 2008. All remaining units were merged into a single Coast Guard force in 2009. The new order of battle remains unconfirmed and it is possible that some vessels, previously thought destroyed, may be repaired.

Headquarters Appointments

Commander of the Coast Guard:
Captain Lasha Kharabadze

Personnel

2014: 685 (276 officers)

Bases

Poti (HQ), Batumi.

COAST GUARD

PATROL FORCES

2 DILOS CLASS (PB)

Name	No	Builders	Commissioned
IVERIA (ex-*Lindos*)	P 105 (ex-201, ex-P 269)	Hellenic Shipyards, Skaramanga	1978
MESTIA (ex-*Dilos*)	P 106 (ex-203, ex-P 267)	Hellenic Shipyards, Skaramanga	1978

Displacement, tonnes: 76 standard; 87 full load
Dimensions, metres (feet): 29 × 5.0 × 1.7 *(95.1 × 16.4 × 5.6)*
Speed, knots: 27
Range, n miles: 1,600 at 24 kt
Complement: 15
Machinery: 2 MTU 12V 331 TC92 diesels; 2,660 hp(m) *(1.96 MW)* sustained; 2 shafts
Guns: 4—23 mm ZSU (2 twin). 2—12.7 mm MGs.
Radars: Surface search: Racal Decca 1226C; I-band.

Comment: First one transferred from the Greek Navy in February 1998, second in September 1999. Reported to have been refitted in Greece in 2004.

IVERIA and MESTIA   10/2002, Hartmut Ehlers / 0552757

1 KAAN 33 (FAST ATTACK CRAFT) (PBF)

SOKHUMI P 24

Displacement, tonnes: 115 full load
Dimensions, metres (feet): 35.6 × 6.7 × 1.4 *(116.8 × 22.0 × 4.6)*
Speed, knots: 45
Range, n miles: 450 at 14 kt
Complement: 14 (5 officers)
Machinery: 2 MTU 12V 4000 M90 diesels; 7,396 hp(m) *(5.44 MW)*; 2MJP 753 DD waterjets
Guns: 1—12.7 mm MG (stabilised).
Radars: Navigation: Furuno FR-2127; I-band.

Comment: With advanced composites structure, the craft are modified versions of those in service in the Turkish Coast Guard. The craft are suitable for use as the patrol of littoral waters, maritime interdiction and special forces operations. The vessel was delivered in mid-2008.

SOKHUMI   4/2011, C D Yaylali / 1406729

3 ZHUK (PROJECT 1400) CLASS (WPB)

P 103–104   PAZISI P 109

Displacement, tonnes: 25 full load
Dimensions, metres (feet): 23 × 4.5 × 0.7 *(75.5 × 14.8 × 2.3)*
Speed, knots: 28
Range, n miles: 250 at 12 kt
Complement: 9 (3 officers)
Machinery: 2 diesels; 2,300 hp *(1.7 MW)*; 2 shafts
Guns: 1—23 mm (P 103). 2—23 mm (P 104). 1—12.7 mm MG (P 103, P 109).
Radars: Navigation: Furuno; I-band.

Comment: P 103 and P 104 constructed at Batumi 1997–99. P 109 modernised under the 'orbi' programme and returned to service in 2012.

P 104   6/2012* / 1525526

2 POINT CLASS (WPB)

Name	No	Builders	Commissioned
TSOTNE DADIANI (ex-*Point Countess*)	P 101 (ex-P 210, ex-82335)	US Coast Guard Yard, Curtis Bay	8 Aug 1962
GENERAL MAZNIASHVILI (ex-*Point Baker*)	P 102 (ex-P 211, ex-82342)	US Coast Guard Yard, Curtis Bay	30 Oct 1963

Displacement, tonnes: 67 standard; 70 full load
Dimensions, metres (feet): 25.3 × 5.3 × 1.8 *(83 × 17.4 × 5.9)*
Speed, knots: 23.5
Range, n miles: 1,500 at 8 kt
Complement: 10 (1 officer)
Machinery: 2 Caterpillar 3412 diesels; 1,600 hp *(1.19 MW)*; 2 shafts
Guns: 2—12.7 mm MGs.
Radars: Surface search: Hughes/Furuno SPS-73; I-band.

Comment: Steel hulled craft with aluminium superstructure. First transferred from United States in June 2000 and second on 12 February 2002.

GENERAL MAZNIASHVILI   10/2002, Hartmut Ehlers / 0589740

Coast guard < **Georgia** 283

2 DAUNTLESS CLASS (WPB)

P 001 (ex-P 209)   P 106 (ex-P 208)

Displacement, tonnes: 11 full load
Dimensions, metres (feet): 12.2 × 4.3 × 1.3 *(40 × 14.1 × 4.3)*
Speed, knots: 27. **Range, n miles:** 600 at 18 kt
Complement: 3
Machinery: 2 Caterpillar 3208TA diesels; 870 hp *(650 kW)*; 2 shafts
Guns: 1—12.7 mm MG.
Radars: Surface search: Raytheon; I-band.

Comment: Aluminium construction. Acquired in July 1999 from SeaArk Marine.

P 209 (old number)   10/2002, Hartmut Ehlers / 0552751

P 209 (old number)   7/1999 / 0088751

1 TORNADO CLASS (PATROL CRAFT) (PB)

P 107 (ex-P 201)

Displacement, tonnes: 30 full load
Dimensions, metres (feet): 19.6 × 3.9 × 0.9 *(64.3 × 12.8 × 2.95)*
Speed, knots: 40. **Range, n miles:** 350 at 18 kt
Complement: 5 (2 officers)
Machinery: 2 Deutz 16V diesels; 3,286 hp *(2.45 MW)*; 2 shafts
Guns: 2—12.7 mm MGs.
Radars: Navigation: Furuno; I-band.

Comment: Patrol craft of unknown origin.

4 ARES CLASS (PATROL CRAFT) (PB)

P 005   P 007–009

Displacement, tonnes: 11 full load
Dimensions, metres (feet): 13.3 × 4.1 × 0.75 *(43.6 × 13.5 × 2.46)*
Speed, knots: 40
Range, n miles: 350 at 18 kt
Complement: 3
Machinery: 2 Volvo Penta D12-650 diesels; 1,300 hp *(969 kW)*; 2 shafts
Radars: Navigation: Furuno; I-band.

Comment: Built at Batumi 2006–07.

1 KAAN 20 CLASS
(FAST INTERVENTION CRAFT) (PBF)

POTI P 108

Displacement, tonnes: 38 full load
Dimensions, metres (feet): 22.55 × 4.76 × 1.3 *(74 × 15.6 × 4.3)*
Speed, knots: 60
Range, n miles: 350 at 35 kt
Complement: 5
Machinery: 2 MTU 12V 2000 M92 diesels; 3,600 hp(m) *(4.8 MW)*; 2 Arneson ASD 12 B1L surface drives
Guns: 2—12.7 mm MGs.

Comment: Yonuk MRTP 20 (enlarged MRTP 15) design built at Yonca Shipyard. The craft are of advanced composites construction and feature a Deep V hull. Delivered in 2009.

POTI   6/2012* / 1525525

1 TURK (AB 25) CLASS (PB)

Name	No	Builders	Commissioned
KUTAISI	P 22 (ex-P 202, ex-AB 30)	Haliç Shipyard	21 Feb 1969

Displacement, tonnes: 165 full load
Dimensions, metres (feet): 40.2 × 6.4 × 1.7 *(131.9 × 21.0 × 5.6)*
Speed, knots: 16
Range, n miles: 600 at 12 kt
Complement: 18 (6 officers)
Machinery: 2 AGO V12 diesels; 4,000 hp(m) *(2.98 MW)*; 2 shafts
Guns: 2 Bofors 40 mm/60. 2 ZSU 23 mm. 2—12.7 mm MGs.
Radars: Surface search: JRC; I-band.

Comment: Transferred from Turkish Navy on 5 December 1998 to the Navy but returned to service in 2012.

KUTAISI (old number)   10/2002, Hartmut Ehlers / 0552754

Germany
DEUTSCHE MARINE

Country Overview

The Federal Republic of Germany (FRG) is situated in central Europe. The country was re-unified in 1990 when the German Democratic Republic became part of the FRG. With an area of 137,823 square miles, it is bordered to the north by Denmark, to the east by Poland and the Czech Republic, to the south by Austria and Switzerland and to the west by France, Luxembourg, Belgium and the Netherlands. It has a 1,290 n mile coastline with the North and Baltic Seas which are linked by the Kiel Canal. The capital and largest city is Berlin. North Sea ports include Hamburg, Wilhelmshaven, Bremen, Nordenham and Emden, while the main Baltic ports are Lübeck, Wismar, Rostock and Stralsund. The Rhine is the principal inland waterway on which Duisburg is the largest port. Territorial seas (12 n miles) are claimed. An EEZ (200 n miles) has also been claimed.

Headquarters Appointments

Chief of German Navy:
  Vice Admiral Axel Schimpf
Deputy Chief of German Navy:
  Vice Admiral Andreas Krause
Director, Operations:
  Rear Admiral Michael Mollenhauer

Diplomatic Representation

Naval Attaché in Paris:
  Captain Ralf Schmitt-Raiser
Naval Attaché in London:
  Captain Jan Hackstein
Naval Attaché in Washington:
  Captain Michael Setzer
Naval Attaché in Moscow:
  Captain Jochen Ungethüm
Naval Attaché in Pretoria:
  Captain Rainer Kümpel
Naval Attaché in Brasilia:
  Captain Bernd Kuhbier
Defence Attaché in Islamabad:
  Commander Carsten Klenke
Defence Attaché in Abuja:
  Commander Wofl Kinzel
Naval Attaché in Ankara:
  Commander Björn Kohlhass
Naval Attaché in Madrid:
  Commander Jens Müller
Naval Attaché in Oslo:
  Commander Michael Sauerborn
Defence Attaché in Bogota:
  Commander Martin Piechot

Diplomatic Representation – continued

Naval Attaché in Beijing:
  Commander Helmut Greve
Defence Attaché in Tel Aviv:
  Commander Stephan Jütten
Defence Attaché in Lisbon:
  Commander F-J Birkel

Personnel

(a) 2014: 23,000 (5,375 officers) (including naval air arm) plus 1,100 conscripts
(b) 9 months' national service (suspended)

Fleet Disposition

Naval Command (Rostock)
1st Flotilla (Kiel)
1st Corvette Squadron (Warnemünde); Type 130
7th FPB Squadron (Warnemünde); Type 143A
3rd and 5th Mine Warfare Squadron (Kiel); Type 332, 333 and 352
1st Submarine Squadron (Eckernförde); Type 212
Fleet Service Ships (Eckernförde); Type 423
2nd Flotilla (Wilhelmshaven)
2nd Frigate Squadron; Type 123 and 124
4th Frigate Squadron; Type 122
Auxiliary Squadron; Type 702 (AORH), 703 (AOL), 704 (AOL), 720 (ATR), 722 (ATS), 760 (AEL)

Bases

Baltic: Kiel, Warnemünde, Eckernförde.
North Sea: Wilhelmshaven.
Naval Arsenal: Wilhelmshaven, Kiel.
Training (other than in bases above): Bremerhaven, Parow, Flensburg, Plön.

Naval Air Arm

51 (DEU) Tactical Reconnaissance Wing (Jagel): GAF PA 200 (Tactical Air Support of Maritime Operations)
  DEU Naval Air Command (Nordholz)
  3 (DEU) Naval Air Wing: P-3C Orion; Sea Lynx; Dornier 228
  5 (DEU) Naval Air Wing: Sea King Mk 41.

Strength of the Fleet

Type	Active	Building (Projected)
Submarines—Patrol	4	2
Frigates	12	4
Corvettes	5	(6)

Strength of the Fleet – continued

Type	Active	Building (Projected)
Fast Attack Craft—Missile	8	—
LCM/LCU	2	—
Minehunters	10	—
Minesweepers	5	—
Minesweepers—Drones	18	—
Tenders	6	—
Replenishment Ships	7	—
Tugs—Icebreaking	3	—
AGIs	3	—
Research ships	8	—
Sail Training Ships	1	—
Diver Support Vessel	2	—

Prefix to Ships' Names

Prefix FGS is used in communications.

Hydrographic Service

This service, under the direction of the Ministry of Transport, is civilian-manned with HQ at Hamburg. Survey ships are listed at the end of the section.

DELETIONS

Submarines

2011   U 16, U 18, U 23, U 24

Frigates

2012   Köln
2013   Rheinland-Pfalz, Emden

Patrol Forces

2012   Dachs, Nerz

Mine Warfare Forces

2012   Laboe, Kulmbach
2013   Passau

Survey and Research Vessels

2011   Schwedeneck

Auxiliaries

2011   Wurstrow, Dranske
2012   Langeoog

PENNANT LIST

Submarines

S 181	U 31
S 182	U 32
S 183	U 33
S 184	U 34
S 185	U 35 (bldg)
S 186	U 36 (bldg)

Frigates

F 207	Bremen
F 208	Niedersachsen
F 212	Karlsruhe
F 213	Augsburg
F 214	Lübeck
F 215	Brandenburg
F 216	Schleswig-Holstein
F 217	Bayern
F 218	Mecklenburg-Vorpommern
F 219	Sachsen
F 220	Hamburg
F 221	Hessen

Corvettes

F 260	Braunschweig
F 261	Magdeburg
F 262	Erfurt
F 263	Oldenburg
F 264	Ludwigshafen

Patrol Forces

P 6121	S 71 Gepard
P 6122	S 72 Puma
P 6123	S 73 Hermelin
P 6125	S 75 Zobel
P 6126	S 76 Frettchen
P 6128	S 78 Ozelot
P 6129	S 79 Wiesel
P 6130	S 80 Hyäne

Mine Warfare Forces

M 1058	Fulda
M 1059	Weilheim
M 1061	Rottweil
M 1062	Sulzbach-Rosenberg
M 1063	Bad Bevensen
M 1064	Grömitz
M 1065	Dillingen
M 1067	Bad Rappenau
M 1068	Datteln
M 1069	Homburg
M 1090	Pegnitz
M 1092	Hameln
M 1093	Auerbach
M 1094	Ensdorf
M 1095	Überherrn
M 1098	Siegburg
M 1099	Herten

Amphibious Forces

L 762	Lachs
L 765	Schlei

Auxiliaries

A 50	Alster
A 52	Oste
A 53	Oker
A 60	Gorch Fock
A 511	Elbe
A 512	Mosel
A 513	Rhein
A 514	Werra
A 515	Main
A 516	Donau
A 1409	Wilhelm Pullwer
A 1411	Berlin
A 1412	Frankfurt Am Main
A 1413	Bonn (bldg)
A 1425	Ammersee
A 1426	Tegernsee
A 1437	Planet
A 1439	Baltrum
A 1440	Juist
A 1442	Spessart
A 1443	Rhön
A 1451	Wangerooge
A 1452	Spiekeroog
A 1458	Fehmarn
Y 811	Knurrhahn
Y 812	Lütje Hörn
Y 814	Knechtsand
Y 815	Scharhörn
Y 816	Vogelsand
Y 817	Nordstrand
Y 819	Langeness
Y 835	Todendorf
Y 836	Putlos
Y 837	Baumholder
Y 839	Munster
Y 842	Schwimmdock A
Y 861	Kronsort
Y 862	Helmsand
Y 863	Stollergrund
Y 864	Mittelgrund
Y 866	Breitgrund
Y 875	Hiev
Y 876	Griep
Y 891	Altmark
Y 895	Wische
Y 1643	Bottsand
Y 1644	Eversand
Y 1659	Warnow
Y 1671	AK 1
Y 1675	AM 8
Y 1676	MA 2
Y 1677	MA 3
Y 1678	MA 1
Y 1679	AM 7
Y 1683	AK 6
Y 1685	Aschau
Y 1686	AK 2
Y 1687	Borby
Y 1689	Bums

SUBMARINES

4 + 2 TYPE 212A (SSK)

Name	No	Builders	Laid down	Launched	Commissioned
U 31	S 181	HDW, Kiel	Feb 2000	20 Mar 2002	19 Oct 2005
U 32	S 182	TNSW, Emden	Jan 2002	4 Dec 2003	19 Oct 2005
U 33	S 183	HDW, Kiel	Oct 2002	13 Sep 2004	13 Jun 2006
U 34	S 184	TNSW, Emden	Jun 2003	1 Jul 2005	3 May 2007
U 35	S 185	HDW, Kiel	Aug 2007	15 Nov 2011	2014
U 36	S 186	HDW, Kiel	Feb 2008	2013	2014

Displacement, tonnes: 1,473 surfaced; 1,859 dived
Dimensions, metres (feet): 55.9 (S 181–184), 57.1 (S 185–186) × 7 × 6 *(183.4, 187.3 × 23.0 × 19.7)*
Speed, knots: 12 surfaced; 20 dived
Range, n miles: 8,000 at 8 kt surfaced
Complement: 27 (8 officers)

Machinery: Diesel-electric; 1 MTU 16V 396 SE 84 diesel; 1,810 hp(m) *(1.35 MW)*; 1 alternator; 1 Siemens Permasyn motor; 3,875 hp(m) *(2.85 MW)*; 1 shaft; 9 Siemens/HDW PEM fuel cell (AIP) modules; 306 kW; sodium sulphide high-energy batteries
Torpedoes: 6 – 21 in *(533 mm)* bow tubes; water ram discharge; Atlas Elektronik DM 2 A4 torpedoes; wire guided active/passive homing to 50 km *(27 n miles)* at 50 kt; warhead 250 kg. Total 12 weapons.
Electronic countermeasures: DASA FL 1800U or EADS MRBR 800 (Batch 2); radar warning.
Radars: Navigation: Kelvin Hughes 1007; I-band.
Sonars: STN Atlas Elektronik DBQS-40; passive ranging and intercept; FAS-3 flank and passive towed array. STN Atlas Elektronik MOA 3070 or Allied Signal ELAK; mine detection; active; high frequency.
Weapon control systems: Kongsberg MSI-90U (Batch 1). Atlas Elektronik ISUS (Batch 2).

Programmes: Design phase first completed in 1992 by ARGE 212 (HDW/TNSW) in conjunction with IKL. Authorisation for the first four of the class was given on 6 July 1994, but the first steel cut was delayed to 1 July 1998 because of modifications needed to achieve commonality with the Italian Navy. The order for Batch 2 of two modified boats was made on 22 September 2006 and steel for *U 35* was cut on 21 August 2007. The submarines are to enter service by 2014.
Modernisation: Improvements to the fifth and sixth boats include EFA flank array sonar, Carl Zeiss SERO 400 periscope and OMS 100 non-penetrating optronic mast, a four-man lock-out chamber to improve safety and efficiency for the deployment of special forces, a towed communications buoy 'Callisto' to provide communications from deep and a new command and weapons control system.
Structure: The design combines a conventional diesel generator set and lead-acid batteries with an Air Independent Propulsion (AIP) system based on Siemens fuel-cell technology. The submarine has partial double hull which has a larger diameter forward. This is joined to the after end by a conical section which houses the fuel cell plant. Two LOX tanks and hydrogen stored in metal cylinders are carried around the circumference of the smaller hull section. Zeiss search and attack periscopes.
Operational: Maximum speed on AIP is 8 kt without use of main battery. *U 32* conducted a submerged transit from the German Bight to the Bay of Cadiz 11–25 April 2006. The entire passage was conducted using air-independent propulsion and without snorkelling, a speed of advance of 4–6 kt. Based at Eckernförde as part of the First Submarine Squadron.
Sales: Two identical submarines have been built in Italy and two further are under contract.

U 31   *9/2013*, L-G Nilsson* / 1525602

U 34   *5/2012, Michael Nitz* / 1486261

U 34   *5/2012, Harald Carstens* / 1486265

FRIGATES

4 BRANDENBURG CLASS (TYPE 123) (FFGHM)

Name	No	Builders	Laid down	Launched	Commissioned
BRANDENBURG	F 215	Blohm + Voss, Hamburg	11 Feb 1992	28 Aug 1992	14 Oct 1994
SCHLESWIG-HOLSTEIN	F 216	Howaldtswerke, Kiel	1 Jul 1993	8 Jun 1994	2 Nov 1995
BAYERN	F 217	Thyssen Nordseewerke, Emden	16 Dec 1993	30 Jun 1994	15 Jun 1996
MECKLENBURG-VORPOMMERN	F 218	Bremer Vulkan, Vegesack; Thyssen Nordseewerke	23 Nov 1993	8 Jul 1995	6 Dec 1996

Displacement, tonnes: 5,487 full load
Dimensions, metres (feet): 138.9 oa; 126.9 wl × 16.7 × 6.8 (455.7; 416.3 × 54.8 × 22.3)
Speed, knots: 29; 21 diesel
Range, n miles: 4,000 at 18 kt
Complement: 243 (31 officers) + 14 scientists

Machinery: 2 GE 7LM2500SA-ML gas turbines; 51,000 hp (38 MW) sustained; 2 MTU 20V 956 TB92 diesels; 11,070 hp(m) (8.14 MW) sustained; 2 shafts; Escher Wyss; cp props
Missiles: SSM: 4 Aerospatiale MM 38 Exocet (2 twin) ❶ (from Type 101A); inertial cruise; active radar homing to 42 km (23 n miles) at 0.9 Mach; warhead 165 kg; sea-skimmer.
SAM: Martin Marietta VLS Mk 41 Mod 3 ❷ for 16 NATO Sea Sparrow RIM-7P; semi-active radar homing to 16 km (8.5 n miles) at 2.5 Mach; warhead 38 kg. 2 Raytheon RAM RIM-116 21 cell Mk 49 launchers ❸; passive IR/anti-radiation homing to 9.6 km (5.2 n miles) at 2.5 Mach; warhead 9.1 kg; 42 missiles.
Guns: 1 Oto Melara 3 in (76 mm)/62 Mk 75 ❹; 105 rds/min to 16 km (8.6 n miles) anti-surface; 12 km (6.5 n miles) anti-aircraft; weight of shell 6 kg. 2 Rheinmetall 20 mm Rh 202 to be replaced by Mauser 27 mm.
Torpedoes: 4—324 mm Mk 32 Mod 9 (2 twin) tubes ❺; anti-submarine. Honeywell Mk 46 Mod 2; anti-submarine; active/passive homing to 11 km (5.9 n miles) at 40 kt; warhead 44 kg.
Physical countermeasures: Decoys: 4 Rheinmetall MASS-4L decoy launchers ❻.
Electronic countermeasures: ESM/ECM: EADS FL 1800S Stage II; intercept and jammers.
Radars: Air search: Thales LW08 ❽; D-band.
Air/surface search: Thales SMART ❾; 3D; F-band.
Fire control: 2 Thales STIR 180 trackers ❿.
Navigation: 2 Sperry Bridgemaster E; I-band.
Sonars: Atlas Elektronik DSQS-21B Mod 2; hull-mounted; active search and attack; medium frequency.
Combat data systems: Thales SABRINA 21; Link 11. Link 16. SATCOM ❼.
Weapon control systems: Thales MWCS. 2 optical sights. RDE MSP 500 optronic sensor/tracker.
Helicopters: 2 Westland Super Lynx Mk 88A ⓫.

Programmes: Four ordered 28 June 1989. Developed by Blohm + Voss whose design was selected in October 1988. Replaced deleted Hamburg class.
Modernisation: SCOT 3 SATCOM and STN optronic sensor fitted from 1998. All four ships are to undergo a major modernisation programme to extend service life to at least 2025. A contract was signed on 21 September 2005 for Phase 1, the replacement of the combat data system by the Thales SABRINA 21 system, which incorporates Tacticos-NC and Sewaco-DDS technology. In Phase 2, the Link and IFF systems are to be upgraded. In a separate contract the DSQS-23BZ was upgraded to DSQS-21B Mod 2 2005–09. There are plans to replace MM 38 Exoceet with Harpoon and to complete installation of MASS decoys. RIM-78 Sea Sparrow is also to be replaced by RIM-162 ESSM.
Structure: The design is a mixture of MEKO and improved serviceability Type 122 having the same propulsion as the Type 122. Contemporary stealth features. All steel. Fin stabilisers. Space allocated for a Task Group Commander and Staff.
Operational: 2nd Frigate Squadron based at Wilhelmshaven. One RIB is carried for boarding operations.

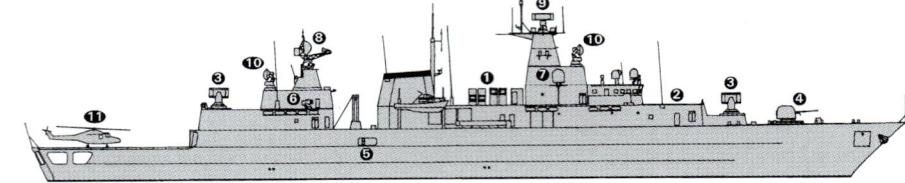

BRANDENBURG (Scale 1 : 1,200), Ian Sturton / 1153487

BRANDENBURG 7/2013*, Michael Nitz / 1525604

MECKLENBURG-VORPOMMERN 5/2012, Michael Nitz / 1486263

MECKLENBURG-VORPOMMERN

BRANDENBURG

BAYERN

SCHLESWIG-HOLSTEIN

5 BREMEN CLASS (TYPE 122) (FFGHM)

Name	No	Builders	Laid down	Launched	Commissioned
BREMEN	F 207	Bremer Vulkan, Vegesack	9 Jul 1979	27 Sep 1979	7 May 1982
NIEDERSACHSEN	F 208	AG Weser, Bremen; Bremer Vulkan	9 Nov 1979	9 Jun 1980	15 Oct 1982
KARLSRUHE	F 212	Howaldtswerke, Kiel; Bremer Vulkan	10 Mar 1981	8 Jan 1982	19 Apr 1984
AUGSBURG	F 213	Bremer Vulkan, Vegesack	4 Apr 1987	17 Sep 1987	3 Oct 1989
LÜBECK	F 214	Thyssen Nordseewerke, Emden; Bremer Vulkan	1 Jun 1987	15 Oct 1987	19 Mar 1990

Displacement, tonnes: 3,739 full load
Dimensions, metres (feet): 130 × 14.5 × 6.5 (426.5 × 47.6 × 21.3)
Speed, knots: 30; 20 diesel. **Range, n miles:** 4,000 at 18 kt
Complement: 219 (26 officers)

Machinery: CODOG; 2 GE LM 2500 gas turbines; 51,000 hp (38 MW) sustained; 2 MTU 20V 956 TB92 diesels; 11,070 hp(m) (8.14 MW) sustained; 2 shafts; cp props
Missiles: SSM: 8 McDonnell Douglas Harpoon (2 quad) launchers ❶; active radar homing to 130 km (70 n miles) at 0.9 Mach; warhead 227 kg.
SAM: 8 Raytheon NATO Sea Sparrow RIM-7P; Mk 29 octuple launcher ❷; semi-active radar homing to 16 km (8.5 n miles) at 2.5 Mach; warhead 38 kg. 2 Raytheon RAM RIM-116 21 cell Mk 49 launchers ❸; passive IR/anti-radiation homing to 9.6 km (5.2 n miles) at 2.5 Mach; warhead 9.1 kg.
Guns: 1 Oto Melara 3 in (76 mm)/62 Mk 75 ❹; 105 rds/min to 16 km (8.6 n miles) anti-surface; 12 km (6.5 n miles) anti-aircraft; weight of shell 6 kg. 2 Mauser 27 mm. 4—12.7 mm MGs.
Torpedoes: 4—324 mm Mk 32 (2 twin) tubes ❺. 8 Honeywell Mk 46 Mod 2; anti-submarine; active/passive homing to 11 km (5.9 n miles) at 40 kt; warhead 44 kg.
Physical countermeasures: Decoys: 4 Loral Hycor SRBOC ❻ 6-barrelled fixed Mk 36; chaff and IR flares to 4 km (2.2 n miles). SLQ-25 Nixie; towed torpedo decoy. Prairie bubble noise reduction.
Electronic countermeasures: ESM/ECM: EADS FL 1800 Stage II ❼; intercept and jammer.
Radars: Air/surface search: EADS TRS-3D/32 ❾; C-band.
Navigation: Kelvin Hughes Nucleus 2 5000A; I-band.
Fire control: Thales WM25 ❿; I/J-band. Thales STIR ⓫; I/J/K-band; range 140 km (76 n miles) for 1 m² target.
Sonars: Atlas Elektronik DSQS-21BZ (BO); hull-mounted; active search and attack; medium frequency.
Combat data systems: SATIR action data automation; Link 11; Link 16; Matra Marconi SCOT 1A SATCOM ❽ (3 sets for the class).
Weapon control systems: Thales WM25/STIR. RDE MSP 500 optronic sensor/tracker.
Helicopters: 2 Westland Super Lynx Mk 88A ⓬.

Programmes: Approval given in early 1976 for first six of this class, a modification of the Netherlands Kortenaer class.

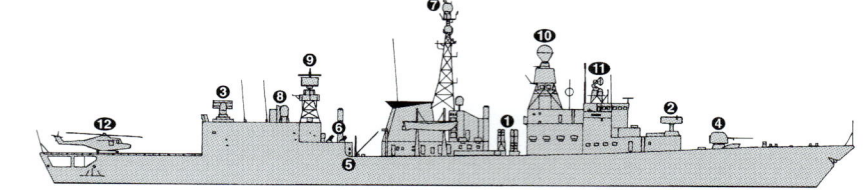

BREMEN CLASS (Scale 1 : 1,200), Ian Sturton / 0012400

AUGSBURG 9/2011, Michael Nitz / 1406894

Replaced the deleted Fletcher and Köln classes. Equipment ordered February 1986 after order placed 6 December 1985 for last pair. Hulls and some engines provided in the five building yards. Ships were then towed to the prime contractor Bremer Vulkan where weapon systems and electronics were fitted and trials conducted. The three names for F 210–212 were changed from the names of Länder to take the well known town names of the Köln class as they were paid off.
Modernisation: RAM fitted from 1993–1996; Updated EW fit from 1994. 20 mm guns, taken from Type 520 LCUs, fitted aft of the bridge on each side. TRS-3D/32 radar has replaced DA 08 in all ships. STN optronic sensor fitted from 1998. 27 mm guns to replace 20 mm in due course.
Operational: Form 4th Frigate Squadron based at Wilhelmshaven. Three containerised SCOT 1A terminals acquired in 1988 and when fitted are mounted on the hangar roof. *Köln* decommissioned on 31 July 2012 and *Rheinland-Pfalz* and *Emden* in 2013. The remaining ships are to be decommissioned as follows: *Bremen* 2014; *Niedersachsen* 2015; *Karlsruhe* 2017; *Lübeck* 2018; *Augsburg* 2019.

BREMEN CLASS 3/2012, Frank Findler / 1483621

KARLSRUHE 9/2013*, Michael Nitz / 1525605

IHS Jane's Fighting Ships 2014-2015

Frigates ‹ Germany

3 SACHSEN CLASS (TYPE 124) (FFGHM)

Name	No	Builders	Laid down	Launched	Commissioned
SACHSEN	F 219	Blohm + Voss, Hamburg	1 Feb 1999	1 Dec 1999	4 Nov 2004
HAMBURG	F 220	Howaldtswerke, Kiel	1 Sep 2000	16 Aug 2002	13 Dec 2004
HESSEN	F 221	Thyssen Nordseewerke, Emden	14 Sep 2002	27 Jun 2003	15 Dec 2005

Displacement, tonnes: 5,690 full load
Dimensions, metres (feet): 143 oa; 132.2 wl × 17.4 × 6.9 *(469.2; 433.7 × 57.1 × 22.6)*
Speed, knots: 29
Range, n miles: 4,000 at 18 kt
Complement: 255 (39 officers)

Machinery: CODAG; 1 GE LM 2500 gas turbine; 31,514 hp *(23.5 MW)*; 2 MTU 20V 1163 TB 93 diesels; 20,128 hp(m) *(14.8 MW)*; 2 shafts; cp props
Missiles: SSM: 8 McDonnell Douglas Harpoon Block 1D ❶ 2 (twin); active radar homing to 95 km *(51 n miles)* at 0.9 Mach; warhead 227 kg.
SAM: Mk 41 VLS (32 cells) ❷ 24 Raytheon Standard SM-2 Block IIIA; command/inertial guidance; semi-active radar homing to 167 km *(90 n miles)* at 2.5 Mach. 32 Evolved Sea Sparrow RIM 162B; semi-active radar homing to 18 km *(9.7 n miles)* at 3.6 Mach; warhead 39 kg. 2 RAM RIM-116 launchers ❸. 21 cell Mk 49 launchers; passive IR/anti-radiation homing to 9.6 km *(5.2 n miles)* at 2.5 Mach; warhead 9.1 kg. 42 missiles.
Guns: 1 Otobreda 76 mm/62 Compact ❹; 108 rds/min to 16 km *(8.6 n miles)* anti-surface; 12 km *(6.5 n miles)* anti-aircraft; weight of shell 6 kg. 2 Mauser 27 mm ❺. 4—12.7 mm MGs.
Torpedoes: 6—324 mm (2 triple) Mk 32 Mod 7 tubes ❻. Eurotorp Mu 90 Impact.
Physical countermeasures: Decoys: 4 Rheinmetall MASS-4L decoy launchers (F 220, 221) ❼.
Electronic countermeasures: ESM/ECM: EADS Fl 1800S-II; intercept ❽ and jammer.

Radars: Air search: SMART L ❿ 3D; D-band.
Air/surface search/fire control: Thales APAR ⓫; E/F/I-bands.
Navigation: 2 SAM 9600M ⓬; E/I-band.
IFF: Mk XII.
Sonars: Atlas DSQS-21B (Mod); bow-mounted; active search; medium frequency.
Combat data systems: CDS F 124; Link 11/16.
Electro-optic systems: MSP optronic director ❾.
Helicopters: 2 NH90 NFH ⓭ or 2 Westland Super Lynx Mk 88A.

Programmes: Type 124 air defence ships built to replace the Lütjens class. A collaborative design with the Netherlands. A Memorandum of Understanding (MoU) was signed in October 1993 between Blohm + Voss, Royal Schelde and Bazán shipyards. A contract to build three ships was authorised on 12 June 1996. An option for a fourth was not exercised.
Modernisation: SRBOC chaff launchers replaced by MASS. Upgrade of the class with a BMD capability is under consideration. Options include modification of SMART-L radar with an ELR mode and the installation of SM3 missiles. At the same time, renewal of obsolete segments of the command system and AAW sensor suite is likely to be undertaken.
Structure: Based on the Type 123 hull with improved stealth features. MBB-FHS helo handling system.
Operational: Successful sea-firings of Standard SM-2 and ESSM conducted at USN range off southern California in July/August 2004. Part of 2nd Frigate Squadron based at Wilhelmshaven.

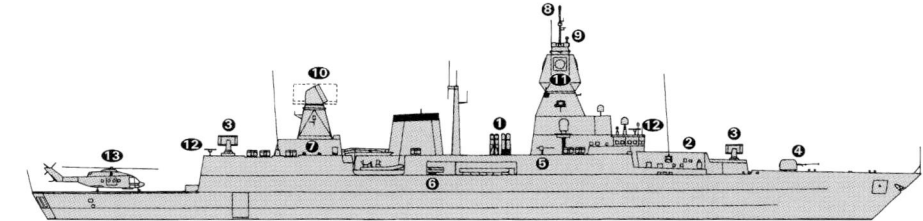

SACHSEN *(Scale 1 : 1,200), Ian Sturton* / 1353058

HESSEN *6/2012, M Declerck* / 1486268

SACHSEN *5/2013*, Michael Nitz* / 1525606

HAMBURG *7/2011, Michael Nitz* / 1486266

290 | Germany > Frigates — Corvettes

0 + 4 BADEN-WÜRTTEMBERG (TYPE 125) CLASS (FFGHM)

Name	No	Laid down	Launched	Commissioned
BADEN-WÜRTTEMBERG	F 222	2 Nov 2011	28 Mar 2014	2016
NORDRHEIN-WESTFALEN	F 223	Oct 2012	2015	2017
SACHSEN-ANHALT	F 224	2014	2016	2018
RHEINLAND-PFALZ	F 225	2015	2017	2018

Displacement, tonnes: 7,316 full load
Dimensions, metres (feet): 149.5 × 18.8 × 5.0
 (490.5 × 61.7 × 16.4)
Speed, knots: 26. **Range, n miles:** 4,000 at 18 kt
Complement: 140 (20 air crew) + 50 embarked forces

Machinery: CODLAG: 1 GE LM 2500 gas turbine; 26,820 hp *(20 MW)*; 4 MTU 20V 4000 diesels; 16,100 hp *(12 MW)*; 2 motors; 12,100 hp *(9.0 MW)*; 2 shafts; cp props; bow thruster
Missiles: SSM: 8 McDonnell Douglas Harpoon Block II ❶.
 SAM: 2 Raytheon RAM 21-cell Mk 49 launchers ❷.
Guns: 1 Oto Melara 5 in *(127 mm)/64 LW* ❸.
 2 Rheinmetall/Mauser MLG 27 mm. 5—12.7 mm remote-controlled MGs; 7—12.7 mm MGs.
Physical countermeasures: Decoys: 4 Rheinmetall MASS decoy launchers.
Radars: Air search: Cassidian TRS-4D ❺; G/H-band.
 Navigation: To be announced ❻.
Sonars: Atlas Elektronik Cerberus diver detection (HF).
Combat data systems: Atlas Elektronik (FuWES). Links 11, 16 and 22.
Electro-optic systems: 2 Rheinmetall MSP 500 optronic sensor/tracker. Diehl BGT Simone EO surveillance system ❹.

Helicopters: 2 MH 90 ❼.

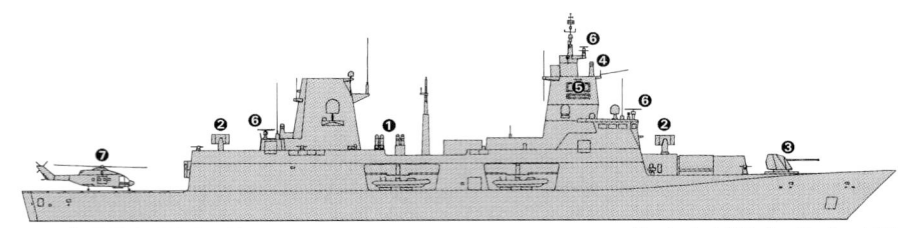

BADEN-WÜRTTEMBERG CLASS *(Scale 1 : 1,200), Ian Sturton* / 1166826

Programmes: The contract for the design and construction of four F 125 frigates was signed on 26 June 2007. The building consortium includes ThyssenKrupp Marine Systems and Lürssen Werft. The principal role of the ship is to conduct long-endurance crisis-management operations, particularly tactical naval gunfire support and support of special forces. Fabrication of the first of class began on 9 May 2011.
Structure: The ships are to be equipped with an 'innovative damage control concept'. There is to be accommodation for up to 50 special forces for whom there will be a dedicated operations room. The ship is to carry four high speed 11 m craft.
Operational: The ships will be designed to be able to deploy for up to two years without return to home-base. This will include a 50% reduced manning concept featuring two crews of about 100 each (plus 20 for the aviation detachment). These would relieve each other on a regular, four-month, rotating schedule.

CORVETTES

Notes: The Multi-Role Combat Ship (MKS 180) (ex-K 131) programme is for four to six medium-sized, surface combatants to replace the capabilities of the Type 122 frigates and Type 143A fast attack craft. After completion of the analysis phase in late March 2013, the programme entered a two-year Development and Proposals Phase. The principal requirements are: 2-D/3-D multisensor detection and tracking; medium calibre gun, surface-to-surface missiles (optional); point-defence missile system; unmanned air and underwater systems; towed array sonar; up to two helicopters; two special forces craft; complement of 100 with additional accommodation for 80; speed of 26 kt (18 kt sustained); ice class 1c. Potential solutions include commercial, government or military off-the-shelf products, improvement of ships in service or new ships. Decisions on the way ahead are expected in 2015–16 and the vessels are planned to enter service 2020–24.

5 BRAUNSCHWEIG (K130) CLASS (FSGHM)

Name	No	Builders	Laid down	Launched	Commissioned
BRAUNSCHWEIG	F 260	Blohm + Voss, Hamburg	2005	19 Apr 2006	16 Apr 2008
MAGDEBURG	F 261	Lürssen, Bremen-Vegesack	2005	6 Sep 2006	22 Sep 2008
ERFURT	F 262	Thyssen Nordseewerke, Emden	2006	29 Mar 2007	28 Feb 2013
OLDENBURG	F 263	Blohm + Voss, Hamburg	2006	28 Jun 2007	21 Jan 2013
LUDWIGSHAFEN	F 264	Lürssen, Bremen-Vegesack	2006	26 Sep 2007	21 Mar 2013

Displacement, tonnes: 1,870 full load
Dimensions, metres (feet): 88.8 × 13.2 × 4.8
 (291.3 × 43.3 × 15.7)
Speed, knots: 26. **Range, n miles:** 2,500 at 15 kt
Complement: 58 (8 officers)

Machinery: 2 MTU diesels; total of 19,850 hp(m) *(14.8 MW)*; 2 shafts
Missiles: SSM: 4 Saab RBS15 Mk 3 ❶; active radar homing to 200 km *(108 n miles)* at 0.9 Mach; warhead 200 kg.
 SAM: 2 Raytheon RAM RIM-116 21 cell Mk 49 launchers ❷; passive IR/anti-radiation homing to 9.6 km *(5.2 n miles)* at 2.5 Mach; warhead 9.1 kg; 42 missiles.
Guns: 1 Oto Melara 76 mm/62 Compact ❸; 108 rds/min to 16 km *(8.6 n miles)* anti-surface; weight of shell 6 kg; 2 Mauser 27 mm ❹.
Physical countermeasures: Decoys: 2 Rheinmetall MASS ❺; decoy launchers.
Electronic countermeasures: ESM/ECM: EADS UL 5000K; intercept and jammer.
Radars: Air/surface search: EADS TRS-3D ❼; C-band.
 Navigation: 2 Raymarine Pathfinder/ST 34 ❽; E/F/I-bands.
 Fire control: EADS TRS-3D; C-band.
Combat data systems: SEWACO; Link 11/16.
Electro-optic systems: 2 Thales Mirador Trainable Electro-Optical Observation System (TEOOS) ❻.

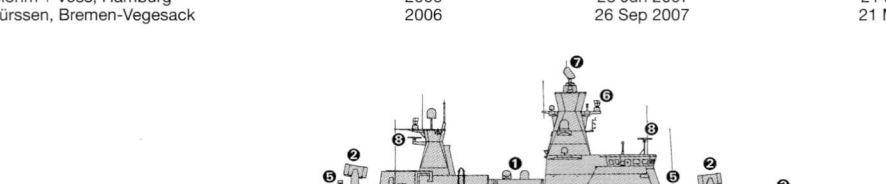

BRAUNSCHWEIG *(Scale 1 : 900), Ian Sturton* / 1166827

Helicopters: Platform for 1 medium and for UAV (possibly Schiebel Camcopter).

Programmes: Invitations to tender accepted at the end of 1998. Blohm + Voss selected as consortium leader 18 July 2000. Consortium includes Thyssen Nordseewerke and Lürssen. Batch of five ships ordered on 14 December 2001 and first steel cut for the first of class on 19 July 2004. The bow section of the first ship was launched on 6 September 2005. All bow sections were constructed at Emden, the aft sections at Lürssen and the superstructure at Blohm+Voss. There are to be no further ships of this class.
Modernisation: All five ships are to be fitted with bow-thrusters.
Structure: Measures to reduce radar, IR (water-cooled surface exhaust system) and noise signatures have been included in the design.
Operational: The ships form the 1st Corvette Squadron based at Rostock-Warnemünde. Due to defects on the gearing systems the whole class was temporarily de-activated until remedial work had been completed. Trials in *Braunschweig* began in August 2010 and the whole class returned to operational service in 2013. Sea trials of the RBS15 Mk 3 missile were expected to start in 2013.

MAGDEBURG *9/2013*, Michael Nitz* / 1525607

IHS Jane's Fighting Ships 2014-2015 © 2014 IHS

Corvettes — Land-based maritime aircraft < Germany

BRAUNSCHWEIG 2/2013*, Frank Findler / 1525679

OLDENBURG 2/2012*, A Campanera i Rovira / 1525608

SHIPBORNE AIRCRAFT

Notes: An MoU to acquire 18 MH 90 NTH Sea Lion has been signed by MoD Germany. The aircraft are to replace the Sea King Mk 41 fleet and are to begin entering service in 2017. The replacement of Sea Lynx Mk 88 for deployment in the F 123, F 124, and F 125 classes is under consideration, but it is not expected before 2025.

Numbers/Type: 21 Westland Sea King Mk 41 KWS.
Operational speed: 140 kt *(260 km/h)*.
Service ceiling: 10,500 ft *(3,200 m)*.
Range: 630 n miles *(1,165 km)*.
Role/Weapon systems: Used in shipborne role for Berlin class AFSH. Land-based roles include SAR, area surveillance and transport. To be replaced from 2017. Sensors: Ferranti Sea Spray Mk 3 radar, FLIR, RWR, chaff and flare dispenser. Weapons: 1 — 12.7 mm MG.

LYNX MK 88A 4/2013*, Michael Winter / 1525609

LAND-BASED MARITIME AIRCRAFT (FRONT LINE)

Numbers/Type: 8 Lockheed P-3C Orion CUP.
Operational speed: 405 kt *(750 km/h)*.
Service ceiling: 30,000 ft *(9,145 m)*.
Range: 4,875 n miles *(9,030 km)*.
Role/Weapon systems: Long-range maritime reconnaissance aircraft procured from the Netherlands 2005–06 and became fully operational in 2008. Aircraft updated under CUP programme. Sensors: AN/APS-137B(V)5 radar, AAQ 22 Safire FLIR, AN/ALR 95 ESM, AN/ALE 47 chaff dispenser, AN/AAR 47 missile warning system, AN/ASQ 227 central processor, AN/ASQ-78B acoustic processor, AQS 81 MAD. Weapons: 8 Mk 46 torpedoes (or Eurotorp MU 90 Impact in due course).

SEA KING 5/2012, Michael Nitz / 1486270

Numbers/Type: 22 Westland Super Lynx Mk 88A.
Operational speed: 125 kt *(232 km/h)*.
Service ceiling: 12,000 ft *(3,658 m)*.
Range: 260 n miles *(480 km)*.
Role/Weapon systems: ASW/ASuW role. Sensors: GEC Marine Sea Spray 3000 FLIR and Bendix AQS-18 dipping sonar. Weapons: ASW; up to two Mk 46 Mod 2 (or Eurotorp MU 90 Impact in due course) torpedoes. ASV; 2 light ASM, 1 — 12.7 mm MG.

P-3C 7/2006, Michael Winter / 1159960

Germany > Land-based maritime aircraft — Amphibious forces

Numbers/Type: 2 Dornier DO 228–212.
Operational speed: 156 kt *(290 km/h)*.
Service ceiling: 20,700 ft *(6,300 m)*.
Range: 667 n miles *(1,235 km)*.
Role/Weapon systems: Pollution control. Sensors: Weather radar; SLAR, IR/UR scanner, microwave radiometer, LLL TV camera and data downlink. Weapons: Unarmed.

DORNIER 228   *8/2006, Frank Findler / 1159886*

Numbers/Type: 13 Panavia Tornado ECR.
Operational speed: Mach 2.2.
Service ceiling: 80,000 ft *(24,385 m)*.
Range: 750 n miles *(1,389 km)*.
Role/Weapon systems: All former naval aircraft transferred to the Air Force in 2005. Electronic and Reconnaissance (ECR) aircraft continue to train for anti-ship operations. Sensors: Texas Instruments nav/attack system, Zeiss FLIR, MBB/Alenia multisensor recce pod. Weapons: 2 AGM-88 HARM, 2 AIM-9L Sidewinders, ECM pod, chaff dispensers.

TORNADO ECR   *4/2013*, Tony Roper / 1529103*

PATROL FORCES

Notes: Vessels in this section have an 'S' number as part of their name as well as a 'P' pennant number. The 'S' number is shown in the Pennant List at the front of this country.

8 GEPARD CLASS (TYPE 143 A)
(FAST ATTACK CRAFT—MISSILE) (PGGFM)

Name	No	Builders	Launched	Commissioned
GEPARD	P 6121	AEG, Lürssen	25 Sep 1981	13 Dec 1982
PUMA	P 6122	AEG, Lürssen	8 Feb 1982	24 Feb 1983
HERMELIN	P 6123	AEG, Kröger	8 Dec 1981	5 May 1983
ZOBEL	P 6125	AEG, Kröger	30 Jun 1982	25 Sep 1983
FRETTCHEN	P 6126	AEG, Lürssen	26 Jan 1983	15 Dec 1983
OZELOT	P 6128	AEG, Lürssen	7 Jun 1983	3 May 1984
WIESEL	P 6129	AEG, Lürssen	8 Aug 1983	12 Jul 1984
HYÄNE	P 6130	AEG, Lürssen	5 Oct 1983	13 Nov 1984

Displacement, tonnes: 397 full load
Dimensions, metres (feet): 57.6 × 7.8 × 2.6 *(189 × 25.6 × 8.5)*
Speed, knots: 40. **Range, n miles:** 2,600 at 16 kt, 600 at 33 kt
Complement: 34 (4 officers)

Machinery: 4 MTU MA 16V 956 SB80 diesels; 13,200 hp(m) *(9.7 MW)* sustained; 4 shafts
Missiles: SSM: 4 Aerospatiale MM 38 Exocet; inertial cruise; active radar homing to 42 km *(23 n miles)* at 0.9 Mach; warhead 165 kg; sea-skimmer.
SAM: 1 Raytheon RAM RIM-116 21 cell Mk 49 launcher; passive IR/anti-radiation homing to 9.6 km *(5.2 n miles)* at 2.5 Mach; warhead 9.1 kg.
Guns: 1 Otobreda 3 in *(76 mm)*/62 compact; 85 rds/min to 16 km *(8.6 n miles)* anti-surface; 12 km *(6.5 n miles)* anti-aircraft; weight of shell 6 kg. 2—12.7 mm MGs.
Mines: Can lay mines.
Physical countermeasures: Decoys: Buck-Wegmann Hot Dog/Silver Dog; IR/chaff dispenser.
Electronic countermeasures: ESM/ECM: Dasa FL 1800 Mk 2; radar intercept and jammer.
Radars: Surface search/fire control: Signaal WM27; I/J-band; range 46 km *(25 n miles)*.
Navigation: Sperry Bridgemaster; I-band.
Combat data systems: AGIS with Signaal update; Link 11.
Electro-optic systems: STN Atlas WBA optronic sensor.

Programmes: Ordered mid-1978 from AEG-Telefunken with subcontracting to Lürssen (P 6121, 6122, 6124–6128) and Kröger (P 6123, 6129, 6130).
Modernisation: Updated EW fit in 1994–95. RAM fitted in *Puma* in 1992, and to the rest from 1993–98. Combat data system update completed in 1999. Improved EW aerials fitted from 1999.
Structure: Wooden hulls on aluminium frames.
Operational: Form 7th Squadron based on the tender *Elbe* at Warnemünde. *Nerz* and *Dachs* decommissioned in 2012 and the remainder to be withdrawn from service by 2016.

HYÄNE   *5/2010, Frank Findler / 1366567*

FRETTCHEN   *4/2013*, Derek Fox / 1525680*

HERMELIN   *9/2013*, Per Körnefeldt / 1525612*

GEPARD   *8/2013*, Harald Carstens / 1525611*

AMPHIBIOUS FORCES

Notes: Plans to procure up to three Joint Support Ships, possibly using an LHD/LPD design have been postponed until about 2023.

2 TYPE 520 (LCU)

LACHS L 762	SCHLEI L 765

Displacement, tonnes: 437 full load
Dimensions, metres (feet): 40 × 8.8 × 2.2 *(131.2 × 28.9 × 7.2)*
Speed, knots: 11
Complement: 17
Military lift: 150 tons
Machinery: 2 MWM 12-cyl diesels; 1,020 hp(m) *(750 kW)*; 2 shafts
Radars: Navigation: Kelvin-Hughes; I-band.

Comment: Similar to the US LCU (Landing Craft Utility) type. Provided with bow and stern ramp. Built by Howaldtswerke, Hamburg, 1965–66. Two sold to Greece in November 1989 and six more in 1992. Based at Eckernförde. Guns have been removed.

LACHS   *6/2013*, Michael Nitz / 1525610*

MINE WARFARE FORCES

8 FRANKENTHAL CLASS (TYPE 332)
(MINEHUNTERS—COASTAL) (MHC)

Name	No	Builders	Launched	Commissioned
BAD BEVENSEN	M 1063	Lürssen	21 Jan 1993	9 Dec 1993
GRÖMITZ	M 1064	Krögerwerft, Rendsburg	29 Apr 1993	23 Aug 1994
DATTELN	M 1068	Lürssen	27 Jan 1994	8 Dec 1994
DILLINGEN	M 1065	Abeking & Rasmussen, Lemwerder	26 May 1994	25 Apr 1995
HOMBURG	M 1069	Krögerwerft, Rendsburg	21 Apr 1994	26 Sep 1995
SULZBACH-ROSENBERG	M 1062	Lürssen	27 Apr 1995	23 Jan 1996
FULDA	M 1058	Abeking & Rasmussen, Lemwerder	29 Sep 1997	16 Jun 1998
WEILHEIM	M 1059	Lürssen	26 Feb 1998	3 Dec 1998

Displacement, tonnes: 660 full load
Dimensions, metres (feet): 54.5 × 9.2 × 2.6 *(178.8 × 30.2 × 8.5)*
Speed, knots: 18
Complement: 37 (5 officers)

Machinery: 2 MTU 16V 396 TB84 diesels; 5,550 hp(m) *(4.08 MW)* sustained; 2 shafts; cp props; 1 motor (minehunting)
Missiles: SAM: 2 Stinger quad launchers.
Guns: 1 Mauser 27 mm. 3—12.7 mm MGs.
Radars: Navigation: Raytheon SPS-64 or Sperry Bridgemaster; I-band.
Sonars: Atlas Elektronik DSQS-11M; hull-mounted; high frequency.
Combat data systems: STN MWS 80-4.

Programmes: First 10 ordered in September 1988 with STN Systemtechnik Nord as main contractor. M 1066 laid down at Lürssen 6 December 1989. Two ordered 16 October 1995.
Modernisation: M 1058, 1059, 1062, 1065 and 1069 modified 2009–11 to carry six disposable ROV Sea Fox I for inspection and up to 27 ROV Sea Fox C for mine disposal.
Structure: Same hull, similar superstructure and high standardisation as Type 332 and 352. Built of amagnetic steel.
Sales: Six of the class built for Turkey from late 1999. M 1060 and M 1066 decommissioned in 2006 and have been sold to the UAE. M 1061 converted to diving support role in 2007.
Operational: M 1067 has replaced *Langeoog* in diving and naval special forces role. The remainder assigned to 3rd Minesweeping Squadron based at Kiel.

HOMBURG   6/2013*, Michael Nitz / 1525614

18 SEEHUND (MINESWEEPERS—DRONES) (MSD)

SEEHUND 1–18

Displacement, tonnes: 97 full load
Dimensions, metres (feet): 23.8 × 4.6 × 2.1 *(78.1 × 15.1 × 6.9)*
Speed, knots: 9
Range, n miles: 520 at 8 kt
Complement: 3
Machinery: 1 Deutz MWM D602 diesel; 446 hp(m) *(328 kW)*; 1 shaft

Comment: Built by MaK, Kiel and Blohm + Voss, Hamburg between August 1980 and May 1982. Modernised in conjunction with the Type 352 conversion programme 2000–2001. Assigned to 5th Minesweeping Squadron based at Kiel.

SEEHUND 10   8/2013*, Frank Findler / 1525681

2 KULMBACH CLASS (TYPE 333)
(MINEHUNTERS—COASTAL) (MHC)

Name	No	Builders	Launched	Commissioned
ÜBERHERRN	M 1095	Abeking & Rasmussen, Lemwerder	30 Aug 1988	19 Sep 1989
HERTEN	M 1099	Krögerwerft, Rendsburg	22 Dec 1989	26 Feb 1991

Displacement, tonnes: 645 full load
Dimensions, metres (feet): 54.4 × 9.2 × 2.5 *(178.5 × 30.2 × 8.2)*
Speed, knots: 18
Complement: 37 (4 officers)

Machinery: 2 MTU 16V 538 TB91 diesels; 6,140 hp(m) *(4.5 MW)* sustained; 2 shafts; cp props
Missiles: SAM: 2 Stinger quad launchers.
Guns: 2 Mauser 27 mm. 3—12.7 mm MGs.
Mines: 60.
Physical countermeasures: Decoys: 2 Silver Dog chaff rocket launchers (to be replaced by Rheinmetall MASS).
Electronic countermeasures: ESM: Thomson-CSF DR 2000 (to be replaced by SAAB Avitronic SME 100); radar warning.
Radars: Surface search/fire control: Signaal WM20/2; I/J-band.
Navigation: Raytheon SPS-64 or Sperry Bridgemaster; I-band.
Sonars: Atlas Elektronik DSQS-11M; hull-mounted; high frequency.
Combat data systems: PALIS with Link 11.

Programmes: On 3 January 1985 an STN Systemtechnik Nord-headed consortium was awarded the order. The German designation of 'Schnelles Minenkampfboot' was changed in 1989 to 'Schnelles Minensuchboot'. After modernisation redesignated 'Minenjagdboote'.
Modernisation: Five (two since decommissioned) ships of Hameln class converted to minehunters 1999–2001 and redesignated Kulmbach class (Type 333). Eight to ten disposable ROV Sea Fox I are carried for inspection and up to 30 Sea Fox C for mine disposal. It has a range of 500 m at 6 kt and uses a shaped charge.
Structure: Ships built of amagnetic steel adapted from submarine construction. Signaal M 20 System removed from the deleted Zobel class fast attack craft. PALIS active link.
Operational: Assigned to 3rd Minesweeping Squadron based at Kiel. *Laboe* and *Kulmbach* decommissioned in 2012 and *Passau* in 2013.

ÜBERHERRN   6/2013*, Michael Nitz / 1525613

2 DIVER SUPPORT SHIP (TYPE 332B) (MCD)

Name	No	Builders	Launched	Commissioned
ROTTWEIL	M 1061	Krögerwerft, Rendsburg	12 Mar 1992	7 Jul 1993
BAD RAPPENAU	M 1067	Abeking & Rasmussen, Lemwerder	3 Jun 1993	19 Apr 1994

Displacement, tonnes: 660 full load
Dimensions, metres (feet): 54.5 × 9.2 × 2.6 *(178.8 × 30.2 × 8.5)*
Speed, knots: 18
Complement: 27 (5 officers)

Machinery: 2 MTU 16V 396 TB84 diesels; 5,550 hp(m) *(4.08 MW)* sustained; 2 shafts; cp props; 1 motor (minehunting)
Missiles: SAM: 2 Stinger quad launchers.
Guns: 1 Mauser 27 mm.
Radars: Navigation: Sperry Bridgemaster SPS-64; I-band.
Sonars: Atlas Elektronik DSQS-11M; hull-mounted; high frequency.
Combat data systems: STN MWS 80-4.

Comment: Built and operated as minehunters until 2007 (M 1061) and 2012 (M 1067) when they were converted to a diving support role. Carries up to three diving teams. Capable of laying 24 mines. Amagnetic steel construction. Based at Kiel.

ROTTWEIL   6/2011, Per Körnefeldt / 1406595

5 ENSDORF CLASS (TYPE 352)
(MINESWEEPERS—COASTAL) (MHCD)

Name	No	Builders	Launched	Commissioned
HAMELN	M 1092	Lürssen	15 Mar 1988	29 Jun 1989
PEGNITZ	M 1090	Lürssen	13 Mar 1989	9 Mar 1990
SIEGBURG	M 1098	Krögerwerft, Rendsburg	14 Apr 1989	17 Jul 1990
ENSDORF	M 1094	Lürssen	8 Dec 1989	25 Sep 1990
AUERBACH	M 1093	Lürssen	18 Jun 1990	7 May 1991

Displacement, tonnes: 645 full load
Dimensions, metres (feet): 54.4 × 9.2 × 2.5 (178.5 × 30.2 × 8.2)
Speed, knots: 18
Complement: 38 (4 officers)

Machinery: 2 MTU 16V 538 TB91 diesels; 6,140 hp(m) (4.5 MW) sustained; 2 shafts; cp props
Missiles: SAM: 2 Stinger quad launchers.
Guns: 2 Mauser 27 mm. 3—12.7 mm MGs.
Mines: 60.
Physical countermeasures: Decoys: 2 Silver Dog chaff rocket launchers (to be replaced by Rheinmetall MASS).
Electronic countermeasures: ESM: Thomson-CSF DR 2000 (to be replaced by SAAB Avitronic SME 100); radar warning.
Radars: Surface search/fire control: Signaal WM20/2; I/J-band.
Navigation: Raytheon SPS-64 or Sperry Bridgemaster; I-band.
Sonars: STN ADS DSQS 15A mine-avoidance; active high frequency.
Combat data systems: PALIS with Link 11. STN C2 remote-control system for minesweeping drone Seehund.

Programmes: On 3 January 1985 an STN Systemtechnik Nord-headed consortium was awarded the order. The German designation of 'Schnelles Minenkampfboot' was changed in 1989 to 'Schnelles Minensuchboot'. After modernisation redesignated 'Hohlstablenkboote'.
Modernisation: Five minesweepers of Hameln class converted 2000–2001 to control up to four remotely controlled minesweeping drones (Seehund). ROV Sea Fox I carried for inspection. ROV Sea Fox C for mine disposal.
Structure: Ships built of amagnetic steel adapted from submarine construction. Signaal M 20 System removed from the deleted Zobel class fast attack craft. PALIS active link.
Operational: Assigned to 5th Minesweeping Squadron based at Kiel.

PEGNITZ 8/2013*, Frank Findler / 1525682

SURVEY AND RESEARCH SHIPS

Notes: A 12 ton midget submarine *Narwal* was recommissioned in April 1996 for research. Originally built by Krupp Atlas as an SDV.

1 TYPE 751 (AGE)

Name	No	Builders	Commissioned
PLANET	A 1437	Thyssen Nordseewerke, Emden	31 May 2005

Displacement, tonnes: 3,556 full load
Dimensions, metres (feet): 73 × 27.2 × 6.8 (239.5 × 89.2 × 22.3)
Speed, knots: 15
Range, n miles: 5,000 at 15 kt
Complement: 25 + 20 personnel
Machinery: Diesel electric; 2 permanent magnet motors; 6,034 hp(m) (4.5 MW); 2 shafts

Comment: Ex-Type 752 SWATH design which replaced the old *Planet*. The roles of the ship include both research and trials. It is run by Wehrtechnische Dienstelle (WTD 71) in Eckenförde. It supports both WTD 71 and Forschungsanstalt für Wasserschall und Geophysik (FWG) in Kiel. First authorised in April 1998 and contract placed with TNSW, Emden. After a delay of over two years, firm order finally made in December 2000. Launched on 12 August 2003, the ship has a sonar well, torpedo tubes and can carry five 20 ft containers.

PLANET 6/2011, A A de Kruijf / 1406594

---

2 SCHWEDENECK CLASS (TYPE 748) (MULTIPURPOSE) (AG)

Name	No	Builders	Commissioned
KRONSORT	Y 861	Elsflether Werft	2 Dec 1987
HELMSAND	Y 862	Krögerwerft, Rendsburg	4 Mar 1988

Displacement, tonnes: 1,034 full load
Dimensions, metres (feet): 56.5 × 10.8 × 5.2 (185.4 × 35.4 × 17.1)
Speed, knots: 13. **Range, n miles:** 2,400 at 13 kt
Complement: 13 + 10 personnel
Machinery: Diesel-electric; 3 MTU 6V 396 TB53 diesel generators; 1,485 kW 60 Hz sustained; 1 motor; 1 shaft
Radars: Navigation: 2 Raytheon; I-band.

Comment: Order for three placed in mid-1985. One more was planned after 1995 but was not funded. Y 860 decommissioned in 2011. Based at Eckernförde.

KRONSORT 4/2013*, Frank Findler / 1525683

3 STOLLERGRUND CLASS (TYPE 745) (MULTIPURPOSE) (AG)

Name	No	Builders	Commissioned
STOLLERGRUND	Y 863	Krögerwerft, Rendsburg	31 May 1989
MITTELGRUND	Y 864	Elsflether Werft	21 Sep 1989
BREITGRUND	Y 866	Elsflether Werft	23 Feb 1990

Displacement, tonnes: 457 full load
Dimensions, metres (feet): 33.5 × 9.2 × 3.2 (109.9 × 30.2 × 10.5)
Speed, knots: 12. **Range, n miles:** 1,000 at 12 kt
Complement: 7 + 6 personnel
Machinery: 1 Deutz-MWM SBV6M628 diesel; 1,690 hp(m) (1.24 MW) sustained; 1 shaft; bow thruster

Comment: Five ordered from Lürssen in November 1987; two subcontracted to Elsflether. Equipment includes two I-band radars and an intercept sonar. Based at the Armed Forces Technical Centre, Eckernförde. *Bant* decommissioned in 2003 and *Kalkgrund* in 2004. Both ships transferred to Israel.

STOLLERGRUND 9/2013*, Frank Findler / 1525684

1 TRIALS SHIP (TYPE 741) (YAG)

Name	No	Builders	Commissioned
WILHELM PULLWER	A 1409 (ex-Y 838)	Schürenstadt, Bardenfleth	18 Jul 1967

Displacement, tonnes: 163 full load
Dimensions, metres (feet): 31.5 × 7.5 × 2.2 (103.3 × 24.6 × 7.2)
Speed, knots: 12.5
Complement: 17
Machinery: 2 MTU MB diesels; 700 hp(m) (514 kW); 2 Voith-Schneider props

Comment: Wooden hulled trials ship for barrage systems. To be decommissioned in 2015.

WILHELM PULLWER 9/2004, Hartmut Ehlers / 1044260

1 TRIALS BOAT (TYPE 740) (YAG)

Name	No	Builders	Commissioned
BUMS	Y 1689	Howaldtswerke, Kiel	16 Feb 1970

Dimensions, metres (feet): 26.4 × 6.8 × 1.5 *(86.6 × 22.3 × 4.9)*

Comment: Single diesel engine. Has a 3 ton crane. Based at Eckernförde. To be decommissioned in 2015.

BUMS   *4/2012, Michael Nitz* / 1486282

INTELLIGENCE VESSELS

3 OSTE CLASS (TYPE 423) (AGI)

Name	No	Builders	Commissioned
ALSTER	A 50	Schiffsbaugesellschaft, Flensburg	5 Oct 1989
OSTE	A 52	Schiffsbaugesellschaft, Flensburg	30 Jun 1988
OKER	A 53	Schiffsbaugesellschaft, Flensburg	10 Nov 1988

Displacement, tonnes: 3,251 full load
Dimensions, metres (feet): 83.5 × 14.6 × 4.2 *(274 × 47.9 × 13.8)*
Speed, knots: 21; 8 motors
Complement: 51 (est.)
Machinery: 2 Deutz-MWM BV16M628 diesels; 8,980 hp(m) *(6.6 MW)* sustained; 2 shafts; 2 motors (for slow speed)
Missiles: SAM: 2 Stinger launchers.
Guns: 2 — 12.7 mm Mauser MGs.

Comment: Ordered in March 1985 and December 1986 and replaced the Radar Trials Ships of the same name (old *Oker* and *Alster* transferred to Greece and Turkey respectively). *Oste* launched 15 May 1987, *Oker* 24 September 1987, *Alster* 4 November 1988. Carry Atlas Elektronik passive sonar and optical ELAM and electronic surveillance equipment. Particular attention given to accommodation standards. Fitted for but not with light armaments. Expected to remain in service until 2020.

OKER   *6/2013*, Michael Nitz* / 1525618

TRAINING SHIPS

Notes: In addition to the one listed below there are 54 other sail training vessels (Types 910–915).

1 SAIL TRAINING SHIP (AXS)

Name	No	Builders	Launched	Commissioned
GORCH FOCK	A 60	Blohm + Voss, Hamburg	23 Aug 1958	17 Dec 1958

Displacement, tonnes: 2,038 full load
Dimensions, metres (feet): 89.3 × 12 × 4.9 *(293 × 39.4 × 16.1)*
Speed, knots: 15; 11 diesel
Range, n miles: 1,990 at 10 kt
Complement: 206 (10 officers; 140 cadets)
Machinery: Auxiliary 1 Deutz MWM BV6M628 diesel; 1,690 hp(m) *(1.24 MW)* sustained; 1 shaft; Kamewa cp prop

Comment: Sail training ship of the improved Horst Wessel type. Barque rig. Launched on 23 August 1958. Sail area, 21,141 sq ft. Major modernisation in 1985 at Howaldtswerke. Second major refit in 1991 at Motorenwerke, Bremerhaven included a new propulsion engine and three diesel generators, which increased displacement. Third major refit at Elsfleth-Werft in 2000–2001 included modernisation of electrical distribution system. Based at Naval Academy, Muerwik.

© 2014 IHS

GORCH FOCK   *5/2013*, Michael Nitz* / 1525617

AUXILIARIES

3 BERLIN (TYPE 702) CLASS (AFSH)

Name	No	Builders	Launched	Commissioned
BERLIN	A 1411	Flensburger Schiffbau	30 Apr 1999	11 Apr 2001
FRANKFURT AM MAIN	A 1412	Flensburger Schiffbau	5 Jan 2001	27 May 2002
BONN	A 1413	Peenewerft	27 Apr 2011	13 Sep 2013

Displacement, tonnes: 20,565 full load
Dimensions, metres (feet): 174 oa; 160.8 wl × 24.3 × 7.4 *(570.9; 527.6 × 79.7 × 24.3)*
Speed, knots: 20
Complement: 139 (12 officers) + 94 casualty berths

Cargo capacity: 7,850 tons fuel; 1,330 tons water; 280 tons cargo; 220 tons ammunition; 115 tons lubricants
Machinery: 2 MAN 12V 32/40 (A 1411, A 1412) diesels; 14,322 hp(m) *(10.68 MW)* sustained; 2 MTU 20V 8000 M71R (A 1413) diesels; 19,584 hp *(14.4 MW)*; 2 shafts; cp props; bow thruster; 1,000 hp(m) *(735 kW)*
Missiles: SAM: 2 RAM launchers fitted for but not with.
Guns: 4 Mauser 27 mm. 4 — 12.7 mm MGs.
Radars: Air/surface search: Thales Variant; G/I-bands.
Navigation and aircraft control: Sperry Bridgemaster; E/F/I-bands.
Helicopters: 2 Sea King Mk 41.

Comment: First ship ordered 15 October 1997, and second 3 July 1998. Hulls built by FSG, superstructure by Kröger and electronics by Lürssen. MBB-FHS helo handling system. Two RAS beam stations and stern refuelling. Two portable SAM launchers are carried. EW equipment may be fitted. These ships are designed to support UN type operations abroad. Trials with the Finnish 14 m Jurmo-class landing craft were conducted in A 1412 during 2007. There can be 26 containers mounted in two layers on the upper deck. This could include a containerised hospital unit for 50. The third ship incorporates improvements based on experience of the first two ships. These include increased power and accommodation. All three ships based at Wilhelmshaven.

BONN   *9/2013*, Michael Nitz* / 1525615

2 REPLENISHMENT TANKERS (TYPE 704) (AOL)

Name	No	Builders	Commissioned
SPESSART (ex-*Okapi*)	A 1442	Krögerwerft, Rendsburg	1974
RHÖN (ex-*Okene*)	A 1443	Krögerwerft, Rendsburg	1974

Displacement, tonnes: 14,396 full load
Measurement, tonnes: 6,201 gt; 10,973 dwt
Dimensions, metres (feet): 130.2 × 19.3 × 8.7 *(427.2 × 63.3 × 28.5)*
Speed, knots: 16. **Range, n miles:** 3,250 at 12 kt
Complement: 42
Cargo capacity: 11,000 m³ fuel; 400 m³ water
Machinery: 1 MaK 12-cyl diesel; 8,000 hp(m) *(5.88 MW)*; 1 shaft; cp prop
Radars: Navigation: Sperry Bridgemaster; E/F/I-bands.

Comment: Completed for Terkol Group as tankers. Acquired in 1976 for conversion (*Spessart* at Bremerhaven, *Rhön* at Kröger). The former commissioned for naval service on 5 September 1977 and the latter on 23 September 1977. Has two portable SAM positions. Civilian manned. Based at Kiel (*Spessart*) and Wilhelmshaven (*Rhön*).

RHÖN   *4/2013*, Tony Roper* / 1525616

298 Germany > Tugs — Coast guard

4 WANGEROOGE CLASS
(2 TYPE 722 AND 2 TYPE 754) (ATS/YDT)

Name	No	Builders	Commissioned
WANGEROOGE	A 1451	Schichau, Bremerhaven	9 Apr 1968
SPIEKEROOG	A 1452	Schichau, Bremerhaven	14 Aug 1968
BALTRUM	A 1439	Schichau, Bremerhaven	8 Oct 1968
JUIST	A 1440	Schichau, Bremerhaven	1 Oct 1971

Displacement, tonnes: 868 standard; 1,040 full load
Dimensions, metres (feet): 52 × 12.1 × 3.9 (170.6 × 39.7 × 12.8)
Speed, knots: 14
Range, n miles: 5,000 at 10 kt
Complement: 57 (A 1439–1440), 33 (A 1451–1452)
Machinery: Diesel-electric; 4 MWM 16-cyl diesel generators; 2 motors; 2,400 hp(m) (1.76 MW); 2 shafts
Guns: 1 Bofors 40 mm/70 (cocooned in some, not fitted in all).

Comment: First two are salvage tugs with firefighting equipment and ice-strengthened hulls. *Wangerooge* based at Wilhelmshaven and sometimes used for pilot training and *Spiekeroog* based at Kiel as submarine safety ship. The other two were converted 1974–78 to training ships with *Baltrum* and *Juist* being used as diving training vessels at Neustadt, with recompression chambers and civilian crews. *Langeoog* was decommissioned in 2012. A 1455 sold to Uruguay in 2002.

SPIEKEROOG   6/2013*, Frank Findler / 1525688

COAST GUARD (KÜSTENWACHE)

Notes: The Coast Guard was formed on 1 July 1974 and is a loose affiliation of the forces of several organisations including: seagoing units of the Border Guard (Bundespolizei); Fishery Protection (Fischereischutz); Maritime Police (Wasserschutzpolizei); Water and Navigation Board (Schiffahrtspolizei); Customs (Zoll). These organisations have responsibility for the operation and maintenance of their own craft but all have the inscription *Küstenwache* on the side.

BORDER GUARD (BUNDESPOLIZEI)

Notes: (1) The force consists of about 600 men. Headquarters at Neustadt and bases at Warnemunde and Cuxhaven. There are three Flotillas; one each at Neustadt, Cuxhaven and Warnemunde. The name of the force was changed from Bundesgrenzschutz-See to Bundespolizei on 1 July 2005.
(2) The force is augmented by a maritime section of the anti-terrorist force GSG 9.
(3) Craft have dark blue hulls and white superstructures with a black, red and yellow diagonal stripe and the inscription Küstenwache painted on the ship's side and Bundespolizei insignia.
(4) There is a total of some 60 helicopters including 13 Eurocopter EC 155, 9 EC 135, 13 Bell UH-1D, 8 Bell 212, 17 BO-105 and a number of AS 330 Puma.
(5) All 40 mm guns removed in 1997.

3 BAD BRAMSTEDT CLASS (WPSO)

Name	No	Builders	Commissioned
BAD BRAMSTEDT	BP 24 (ex-BG 24)	Abeking & Rasmussen, Lemwerder	8 Nov 2002
BAYREUTH	BP 25 (ex-BG 25)	Abeking & Rasmussen, Lemwerder	2 May 2003
ESCHWEGE	BP 26 (ex-BG 26)	Abeking & Rasmussen, Lemwerder	18 Dec 2003

Displacement, tonnes: 813 standard
Dimensions, metres (feet): 65.9 × 10.6 × 3.2 (216.2 × 34.8 × 10.5)
Speed, knots: 21.5
Complement: 14 + 10 casualty berths
Machinery: 1 MTU 16V 1163 diesel; 7,000 hp(m) (5.2 MW); 1 shaft; fixed propeller
Radars: Surface search: I-band.
Navigation: I-band.

Comment: Contract awarded in 2000 to Prime Contractor Abeking and Rasmussen for three craft to replace six ships of Neustadt class. Hulls constructed by Yantar, Kaliningrad and completed at Lemwerder. Steel hull with aluminium superstructure. The Russian Federal Border Guard Sprut-class offshore patrol vessel is based on this design.

BAYREUTH   5/2013*, Michael Nitz / 1525622

---

2 SASSNITZ CLASS (TYPE PB 50 EX-TYPE 153) (WPBO)

Name	No	Builders	Commissioned
NEUSTRELITZ (ex-*Sassnitz*)	BP 22 (ex-P 6165, ex-591)	Peenewerft, Wolgast	31 Jul 1990
BAD DÜBEN (ex-*Binz*)	BP 23 (ex-BG 23, ex-593)	Peenewerft, Wolgast	23 Dec 1990

Displacement, tonnes: 375 full load
Dimensions, metres (feet): 48.9 oa; 45 wl × 8.7 × 2.2 (160.4; 147.6 × 28.5 × 7.2)
Speed, knots: 25
Range, n miles: 2,400 at 20 kt
Complement: 33 (7 officers)
Machinery: 2 MTU 12V 595 TE90 diesels; 8,800 hp(m) (6.48 MW) sustained; 2 shafts
Guns: 2 — 7.62 mm MGs.
Radars: Surface search: Racal AC 2690 BT; I-band (BG 22 and 23).
Navigation: Racal ARPA; I-band (BG 22 and 23).

Comment: Ex-GDR designated Balcom 10 and seen for the first time in the Baltic in August 1988. The original intention was to build up to 50 for the USSR, Poland and the GDR. In 1991 the first three were transferred to the Border Guard, based at Neustadt. *Neustrelitz* fitted with German engines and electronics in 1992–93 and accommodation improved. *Bad Düben* similarly modified at Peenewerft in 1995–96. The original design had the SS-N-25 SSM and three engines. The third of class, *Sellin*, had been on loan to WTD 71 (weapons trials) at Eckernförde but was sold in 1999.

BAD DÜBEN   8/2011, Michael Nitz / 1486285

1 BREDSTEDT CLASS (TYPE PB 60) (WPSO)

Name	No	Builders	Laid down	Launched	Commissioned
BREDSTEDT (ex-BG 21)	BP 21	Elsflether Werft	3 Mar 1988	18 Dec 1988	24 May 1989

Displacement, tonnes: 684 full load
Dimensions, metres (feet): 65.4 × 9.2 × 3.2 (214.6 × 30.2 × 10.5)
Speed, knots: 21; 12 motor
Range, n miles: 2,000 at 25 kt, 7,000 at 10 kt
Complement: 17 + 4 casualty berths

Machinery: 1 MTU 20V 1163 TB93 diesel; 8,325 hp(m) (6.12 MW) sustained; 1 shaft; bow thruster; 1 auxiliary diesel generator; 1 motor
Guns: 1 — 40 mm (fitted for but not with).
Radars: Surface search: Racal AC 2690 BT; I-band.
Navigation: 2 Racal ARPA; I-band.
Helicopters: Platform for 1 light.

Comment: Ordered 27 November 1987, laid down 3 March 1988 and launched 18 December 1988. An Avon Searider rigid inflatable craft can be lowered by a stern ramp. A second RIB on the port side is launched by crane. Based at Cuxhaven.

BREDSTEDT   6/2013*, Michael Nitz / 1525621

1 TYPE P22 CLASS (WPBR)

Name	No	Builders	Commissioned
RETTIN	BP 5	Mützelfield Shipyard	20 Dec 1979

Measurement, tonnes: 109 gt
Dimensions, metres (feet): 21.8 × 6.8 × 2.9 (71.5 × 22.3 × 9.5)
Speed, knots: 9
Machinery: 2 MWM TRHS518V12 diesels; 582 hp(m) (434 kW); 2 Voith-Schneider propulsors
Radars: Kelvin Hughes; I-band.

Comment: Constructed at Mützelfield Shipyard and commissioned on 20 December 1979. Used for river patrols.

Coast guard < **Germany** 299

5 PRIGNITZ CLASS (WPB)

PRIGNITZ BP 61	ALTMARK BP 63	RHOEN BP 65
UCKERMARK BP 62	BÖRDE BP 64	

Displacement, tonnes: 39 full load
Dimensions, metres (feet): 21 × 5.2 × 2.6 *(68.9 × 17.1 × 8.5)*
Speed, knots: 23
Machinery: 2 diesels; 1,580 hp(m) *(1.2 MW)*; 1 shaft; fixed propeller

Comment: Built by Schiffs-und-Entwicklungsgesellschaft 2006–08.

ALTMARK   6/2013*, Frank Findler / 1525690

FISHERY PROTECTION SHIPS (Fischereischutz)

Notes: Operated by Ministry of Food and Agriculture.

3 PATROL SHIPS

MEERKATZE	SEEFALKE	SEEADLER

Comment: Fishery Protection Ships. Black hulls with grey superstructure and black, red and yellow diagonal stripes. An order for two 72 m vessels was made in December 2006. They were both delivered in 2009. *Seefalke* and *Meerkatze* are 73 m vessels of 1,981 tons and 20 kt. Completed November 2008 and April 2009 respectively. *Seeadler* is a 72 m vessel of 2,000 tons and 19 kt. Completed 2000. All three vessels capable of 19.5 kt.

SEEFALKE   5/2012, Michael Nitz / 1486289

MARITIME POLICE (Wasserschutzpolizei)

Notes: (1) Under the control of regional governments. Most have Küstenwache markings but colours vary from region to region.
(2) There are 17 seaward patrol craft: *WSP 1, 2, 3, 4,* and *5, Helgoland, Sylt, Fehmarn, Falshöft, Bürgermeister Brauer, Bürgermeister Weichmann, Warnow, Hoben, Granitz, Damerow, Lesmona,* and *Visura*.

FALSHÖFT   6/2012, Frank Findler / 1483634

SYLT   5/2012, Harald Carstens / 1486288

CUSTOMS (Zoll)

Notes: (1) Operated by Ministry of Finance with a total of over 100 craft. Green hulls with grey superstructure and sometimes carry machine guns. Some have Küstenwache markings.
(2) Seaward patrol craft include *Usedom, Kniepsand, Jade, Schleswig-Holstein, Emden, Glückstadt, Hiddensee, Rügen, Priwall, Borkum* and *Helgoland*. The latter two ships are of SWATH design and entered service 2009–10.

BORKUM   6/2013*, Michael Nitz / 1525623

WATER AND NAVIGATION BOARD (Schiffahrtspolizei)

Notes: (1) Comes under the Ministry of Transport. Most ships have black hulls with black/red/yellow stripes. Some have Küstenwache markings.
(2) Seven buoy tenders: *Gustav Meyer, Norden, Baumrönne, Vilm, Knechtsand, Strelasund, Triton*.
(3) Four oil recovery ships: *Scharhörn, Arkona, Mellum, Neuwerk*.
(4) Two SKB 64 and 601 types (ex-GDR). *Vogelsand, Ranzow*.
(5) One launch: *Friedrich Voss*.

VOGELSAND   5/2012, Frank Findler / 1483633

NEUWERK   5/2013*, Frank Findler / 1525689

CIVILIAN SURVEY AND RESEARCH SHIPS

Notes: The following ships operate for the Bundesamt für Seeschiffahrt und Hydrographie (BSH), either under the Ministry of Transport or the Ministry of Research and Technology (*Polarstern, Meteor, Poseidon, Sonne* and *Alkor*).
KOMET (survey and research) 1,590 tons completed by Krögerwerft in October 1998.
ATAIR (survey), **DENEB** (survey), **WEGA** (survey) 1,050 tons, diesel-electric, 11.5 kt. Complement 16 plus 6 scientists. Built by Krögerwerft and Peenewerft (*Deneb*), completed 3 August 1987, 24 November 1994 and 26 October 1990 respectively.
METEOR (research) 3,500 tons, diesel-electric, 14 kt, range 10,000 n miles. Complement 33 plus 29 research staff. Completed by Schlichting, Travemünde 15 March 1986.
WALTHER HERWIG III 2,400 tons. Completed 1993
CAPELLA 455 tons. Completed by Fassmerwerft in 2003.
POLARSTERN (polar research) 10,878 grt. Completed 1982.
SONNE (research) 1,200 grt. Completed by Rickmerswerft 1990.
ALKOR and **HEINKE** 1,200 tons. Completed 1990.
SOLEA 770 tons. Completed by Fassmer 2004.
MARIA S MERIAN 6,050 tons. Completed by Kröger in 2005.
POSEIDON 1,700 tons. Completed by Schichau in 1976.

WALTER HERWIG III   12/2006, Frank Findler / 1166799

POSEIDON   6/2012, Michael Nitz / 1486292

KOMET   5/2011, Frank Findler / 1406496

ALTAIR
9/2012, Hartmut Ehlers
1486291

Ghana

Country Overview

Formerly a British colony known as the Gold Coast, Ghana gained independence in 1957. Located in west Africa, the country has an area of 92,100 square miles and a 320 n mile coastline with the Gulf of Guinea. It is bordered to the east by Togo and to the west by Ivory Coast. The capital and largest city is Accra which has links to a deep-water port at Tema. There is a second port at Sekondi-Takoradi. Territorial seas (12 n miles) are claimed. A 200 n mile Exclusive Economic Zone (EEZ) has been claimed but the limits are not defined.

Headquarters Appointments

Commander, Navy: Rear Admiral Geoffrey Mawuli Biekro
Eastern Naval Command: Commodore Stephen Kwaku Darbo
Western Naval Command:
    Commodore Godson Kwabla Zowonoo

Personnel

(a) 2014: 2,618 (234 officers)
(b) Voluntary service

Bases

Burma Camp, Accra (Headquarters)
Sekondi (Western Naval Command)
Tema (near Accra) (Eastern Naval Command)

Maritime Aircraft

Three Diamond DA42, four Mi-171 SH and two Mi-17V5 are operated for Coastal Surveillance, SAR and shipping control.

PATROL FORCES

2 ALBATROS CLASS (TYPE 143B) (PB)

Name	No	Builders	Commissioned
YAA ASANTEWA (ex-*Bussard*)	P 38 (ex-P 6114)	Lürssen, Vegesack	14 Aug 1976
NAA GBEWAA (ex-*Albatros*)	P 39 (ex-P 6111)	Lürssen, Vegesack	1 Nov 1976

Displacement, tonnes: 398 full load
Dimensions, metres (feet): 57.6 × 7.8 × 2.6 *(189 × 25.6 × 8.5)*
Speed, knots: 40. **Range, n miles:** 1,300 at 30 kt
Complement: 42 (6 officers)
Machinery: 4 MTU 16V 956 TB 91 diesels; 17,700 hp(m) *(13 MW)* sustained; 4 shafts
Guns: 1—20 mm GAM BO1. 2—12.7 mm MGs.
Radars: Surface search/navigation: Furuno; I-band.

Comment: Withdrawn from German service in 2005 and approved for procurement by the Ghana government in mid-2010. Both ships were refitted before transfer in July 2012. Six sister ships were transferred to Tunisia in 2005.

2 LÜRSSEN PB 57 CLASS (FAST ATTACK CRAFT—GUN) (PG)

Name	No	Builders	Commissioned
ACHIMOTA	P 28	Lürssen, Bremen-Vegesack	27 Mar 1981
YOGAGA	P 29	Lürssen, Bremen-Vegesack	27 Mar 1981

Displacement, tonnes: 395 full load
Dimensions, metres (feet): 58.1 × 7.6 × 2.8 *(190.6 × 24.9 × 9.2)*
Speed, knots: 32
Complement: 55 (5 officers)

Machinery: 3 MTU 16V 538 TB91 diesels; 9,210 hp(m) *(6.78 MW)* sustained; 3 shafts
Guns: 1 Oto Melara 3 in *(76 mm)* compact; 85 rds/min to 16 km *(8.6 n miles)* anti-surface; 12 km *(6.5 n miles)* anti-air; weight of shell 6 kg; 250 rounds. 1 Breda 40 mm/70; 300 rds/min to 12.5 km *(6.8 n miles)* anti-surface; weight of shell 0.96 kg.
Radars: Surface search/fire control: Thomson-CSF Canopus A; I/J-band.
Navigation: Furuno FR-2125; I-band.
Weapon control systems: LIOD optronic director.

Comment: Ordered in 1977. *Yogaga* completed a major overhaul at Swan Hunter's Wallsend, Tyneside yard 8 May 1989. *Achimota* started a similar refit at CMN Cherbourg in May 1991 and was joined by *Yogaga* for repairs in late 1991. Both completed by August 1992. Employed on Fishery Protection duties. Refits for both ships are planned to start in 2014.

YAA ASANTEWA   7/2012, Frank Findler / 1455817

ACHIMOTA   6/2007, Ghana Navy / 1167861

Patrol forces < **Ghana** 301

2 BALSAM CLASS (PBO)

Name	No	Builders	Commissioned
ANZONE (ex-*Woodrush*)	P 30 (ex-WLB 407)	Duluth Shipyard, Minnesota	22 Sep 1944
BONSU (ex-*Sweetbrier*)	P 31 (ex-WLB 405)	Duluth Shipyard, Minnesota	26 Jul 1944

Displacement, tonnes: 950 standard; 1,041 full load
Dimensions, metres (feet): 54.9 × 11.3 × 3.8 *(180.1 × 37.1 × 12.5)*
Speed, knots: 13
Range, n miles: 8,000 at 12 kt
Complement: 60 (5 officers)
Machinery: Diesel electric; 2 diesels; 1,710 hp *(1.28 MW)*; 1 motor; 1,200 hp *(895 kW)*; 1 shaft; bow thruster
Guns: 4 — 14.5 mm (2 twin).
Radars: Navigation: Raytheon SPS-64(V)1; Furuno FR-2125; Furuno FA-2117; I-band.

Comment: Formed USCG buoy tenders. *Anzone* transferred from the US Coast Guard on 4 May 2001 and *Bonsu* on 27 August 2001. Both received new engines 1988–91. Employed on EEZ patrol, fishery protection and troop support duties.

ANZONE *10/2005*, US Navy / 1305464

2 LÜRSSEN FPB 45 CLASS
(FAST ATTACK CRAFT—GUN) (PBO)

Name	No	Builders	Commissioned
DZATA	P 26	Lürssen, Bremen-Vegesack	4 Dec 1979
SEBO	P 27	Lürssen, Bremen-Vegesack	2 May 1980

Displacement, tonnes: 273 full load
Dimensions, metres (feet): 44.9 × 7 × 2.7 *(147.3 × 23.0 × 8.9)*
Speed, knots: 27
Range, n miles: 1,800 at 16 kt, 700 at 25 kt
Complement: 45 (5 officers)
Machinery: 2 MTU 16V 538 TB91 diesels; 6,140 hp(m) *(4.5 MW)* sustained; 2 shafts
Guns: 2 Breda 40 mm/70; 300 rds/min to 12.5 km *(6.8 n miles)*; weight of shell 0.96 kg.
Radars: Surface search/navigation: Furuno FAR-2117; Furuno FR-1500; I-band.

Comment: Ordered in 1976. *Dzata* completed a major overhaul at Swan Hunter's Wallsend, Tyneside yard on 8 May 1989. *Sebo* started a similar refit at CMN Cherbourg in May 1991 which completed in August 1992. Employed in Fishery Protection role.

DZATA *5/2002* / 0533318

7 DEFENDER CLASS (RESPONSE BOATS) (PBF)

STEVE OBIMPEH PB 1 DZANG PB 4 OHENE-KWAPONG PB 6
PHILEMON QUAYE PB 2 OWUSU-ANSAH PB 5 TOM ANNAN PB 7
JOY AMEDUME PB 3

Displacement, tonnes: 3 full load
Dimensions, metres (feet): 7.6 × 2.6 × 2.7 *(24.9 × 8.5 × 8.9)*
Speed, knots: 46
Range, n miles: 175 at 35 kt
Complement: 4
Machinery: 2 Honda outboard motors; 450 hp *(335 kW)*
Guns: 1 — 12.7 mm MG.
Radars: Furuno 1824C; I-band.

Comment: High-speed inshore patrol craft of aluminium construction and foam collar built by SAFE Boats International, Port Orchard, Washington. Three craft handed over on 24 October 2008 and a further four in March 2010. All donated by US AFRICOM command to contribute to safety and security in Ghana's territorial waters. The vessels are named after former Chiefs of Naval Staff.

JOY AMEDUME *6/2010*, Ghana Navy / 1406044

1 INSHORE PATROL CRAFT (PBI)

DAVID HANSEN P 32

Displacement, tonnes: 32 standard; 42 full load
Dimensions, metres (feet): 19.8 × 5.5 × 1.8 *(65 × 18.0 × 5.9)*
Speed, knots: 28
Range, n miles: 450 at 26 kt
Complement: 10 (1 officer)
Machinery: 3 Detroit 8V 71 diesel; 690 hp *(515 kW)* sustained; 3 shafts
Guns: 2 — 14.5 mm MGs.
Radars: Navigation: JRC 3000; I-band.

Comment: Ex-US Navy PB Mk III Series built by Peterson Builders, Wisconsin in 1975–76. Aluminium construction. The design includes a pilot house offset to starboard to provide space to port for the installation of additional weapons. Transferred to the Ghana Navy in 2001 and employed on harbour and anchorage surveillance and security patrols.

DAVID HANSEN *6/2007*, Ghana Navy / 1167860

1 SEA DOLPHIN/WILDCAT CLASS
(FAST ATTACK CRAFT—PATROL) (PBF/PTF)

STEPHEN OTU P 33 (ex-PKM 237)

Displacement, tonnes: 150 full load
Dimensions, metres (feet): 33.1 × 6.92 × 2.38 *(108.6 × 22.7 × 7.8)*
Speed, knots: 37
Range, n miles: 600 at 20 kt
Complement: 31 (5 officers)
Machinery: 2 MTU MD 16V 538 TB90 diesels; 6,000 hp(m) *(4.41 MW)* sustained; 2 shafts
Guns: 1 Bofors 40 mm/60. 1 GE/GD 20 mm Sea Vulcan Gatlings. 2 — 20 mm.
Radars: Surface search: Raytheon 1645; I-band.
Weapon control systems: Optical director.

Comment: A refurbished Sea Dolphin-class fast attack craft was donated by South Korea and recommissioned on 21 January 2011. The ship was built by Korea Tacoma Marine and served in the RoK Navy 1982–2008.

STEPHEN OTU *6/2011*, Ann Till / 1406396

Ghana > Patrol forces — Greece > Introduction

4 SNAKE CLASS (PBO)

Name	No	Commissioned
BLIKA	P 34	20 Feb 2012
GARINGA	P 35	20 Feb 2012
CHEMLE	P 36	20 Feb 2012
EHWOR	P 37	20 Feb 2012

Displacement, tonnes: 247 full load
Dimensions, metres (feet): 46.8 × 7.2 × 1.82 (153.5 × 23.6 × 6.0)
Speed, knots: 27
Range, n miles: 1,200 at 15 kt
Machinery: 4 MTU 16V2000 M90 diesels; 7,200 hp (5.37 MW) sustained; 4 shafts
Guns: 1—37 mm (P 36, P 37). 2—14.5 mm (twin) (P 34, P 35). 2—14.5 mm (P 34, P 35, P 36, P 37).
Radars: Surface search/navigation: JMA 5310 and JMA 5106; I-band.

Comment: Four patrol craft built at Quindao Shipyard, China. The design includes a stern-ramp for launch and recovery of a 6.5-m RHIB. They were commissioned on 20 February 2012.

GARINGA   2/2012, Ghana Navy / 1455670

Greece
HELLENIC NAVY

Country Overview

The Hellenic Republic is situated in south-eastern Europe and occupies the southernmost part of the Balkan Peninsula. It includes more than 3,000 islands, most of which are in the Aegean Sea. With an area of 50,949 square miles, it has borders to north-west with Albania, to the north with the Former Yugoslav Republic of Macedonia and with Bulgaria and to the north-east with Turkey. It has a 7,387 n mile coastline with the Aegean, Mediterranean and Ionian Seas. The capital and largest city is Athens whose seaport, Piraeus, is also the largest. Other major ports include Thessaloníki, Patras and Iráklion. Territorial seas (6 n miles) are claimed but an EEZ is not claimed.

Headquarters Appointments

Chief of the Hellenic Navy:
 Vice Admiral Kosmas Christidis
Deputy Chief of Staff:
 Rear Admiral Panagiotis Pastouseas
Commander, Navy Training Command:
 Rear Admiral Spyridon Dimoulas
Commander, Navy Logistics Command:
 Rear Admiral Ilias Dimopoulos
Inspector General:
 Rear Admiral Georgios Bougioukos

Fleet Command

Commander of the Fleet:
 Vice Admiral Panagiotis Litsas
Deputy Commander of the Fleet:
 Rear Admiral Spiridon Konidaris

Personnel

(a) 2014: 19,000 (4,000 officers) including 2,900 conscripts
(b) 12 months' national service

Bases

Salamis and Souda Bay

Naval Commands

Commander of the Fleet has under his flag all combatant ships. Navy Logistic Command is responsible for the bases at Salamis and Souda Bay, the Supply Centre and all auxiliary ships. Navy Training Command is in charge of the Petty Officers' School, the naval staff and commanding officers course and two training centres.

Naval Districts

Aegean, Ionian and Northern Greece

Naval Aviation

Alouette III helicopters (Training).
AB 212ASW helicopters (No 1 Squadron).
S-70B-6 Seahawk (No 2 Squadron).

Prefix to Ships' Names

HS (Hellenic Ship)

Strength of the Fleet

Type	Active	Building (Planned)
Patrol Submarines	8	3 (2)
Frigates	13	(6)
Fast Attack Craft–Missile	16	3
Offshore Patrol Craft	10	—
Coastal Patrol Craft	6	—
LST/LSD/LSM	5	—
LCU/LCM	16	—
Hovercraft	3	—
Minehunters/Sweepers	4	2
Survey and Research Ships	3	—

Strength of the Fleet – continued

Type	Active	Building (Planned)
Command/Support ships	3	—
Training Ships	5	—
Tankers	4	—

DELETIONS

Notes: Some of the deleted ships are in unmaintained reserve in anchorages.

Submarines

2011   *Glavkos*

Frigates

2012   *Bouboulina*

Patrol Forces

2011   *Plotarchis Vlahavas, Plotarchis Sakipis, Ypoploiarchos Tournas*

Amphibious Forces

2011   *Ios*
2013   *Naxos, Serifos*

Mine Warfare Forces

2012   *Alkyon, Avra, Aidon*

Tugs

2011   *Adamastos*
2013   *Nestor, Perseus*

PENNANT LIST

Submarines

S 111	Nereus
S 112	Triton
S 113	Proteus
S 116	Poseidon
S 117	Amphitrite
S 118	Okeanos
S 119	Pontos
S 120	Papanikolis
S 121	Pipinos (bldg)
S 122	Matrozos (bldg)
S 123	Katsonis (bldg)

Frigates

F 450	Elli
F 451	Limnos
F 452	Hydra
F 453	Spetsai
F 454	Psara
F 455	Salamis
F 459	Adrias
F 460	Aegeon
F 461	Navarinon
F 462	Kountouriotis
F 464	Kanaris
F 465	Themistocles
F 466	Nikiforos Fokas

Patrol Forces

P 18	Armatolos
P 19	Navmachos
P 20	Anthypoploiarchos Laskos
P 21	Plotarchis Blessas
P 22	Ypoploiarchos Mikonios
P 23	Ypoploiarchos Troupakis
P 24	Simeoforos Kavaloudis
P 26	Ypoploiarchos Degiannis
P 27	Simeoforos Xenos
P 28	Simeoforos Simitzopoulos
P 29	Simeoforos Starakis
P 57	Kasos
P 61	Polemistis
P 67	Ypoploiarchos Roussen
P 68	Ypoploiarchos Daniolos
P 69	Ypoploiarchos Kristallidis
P 70	Ypoploiarchos Grigoropoulos (bldg)
P 71	Anthypoploiarchos Ritsos (bldg)
P 72	Ypoploiarchos Votsis
P 73	Anthyploiarchos Pezopoulos
P 75	Plotarchis Maridakis
P 196	Andromeda
P 198	Kyknos
P 199	Pigasos
P 228	Toxotis
P 229	Tolmi
P 230	Ormi
P 266	Machitis
P 267	Nikiforos
P 268	Aittitos
P 269	Krateos
P 286	Diopos Antoniou
P 287	Kelefstis Stamou

Amphibious Forces

L 169	Irakleia
L 170	Folegandros
L 173	Chios
L 174	Samos
L 175	Ikaria
L 176	Lesbos
L 177	Rodos
L 179	Paros
L 180	Kefallinia
L 181	Ithaki
L 182	Kerkira

Minesweepers/Hunters

M 61	Evniki
M 62	Evropi
M 63	Kallisto
M 64	Kalypso

Auxiliaries, Training and Survey Ships

A 233	Maistros
A 234	Sorokos
A 238	Zefiros
A 307	Thetis
A 359	Ostria
A 373	Gregos
A 374	Prometheus
A 375	Zeus
A 376	Orion
A 412	Aias
A 416	Ouranos
A 417	Hyperion
A 419	Pandora
A 420	Pandrosos
A 422	Kadmos
A 424	Iason
A 425	Odisseus
A 430	Atlas
A 431	Achilefs
A 432	Gigas
A 433	Kerkini
A 434	Prespa
A 435	Kekrops
A 437	Pelias
A 438	Antaios
A 439	Atrefs
A 440	Diomidis
A 441	Theseus
A 442	Romaleos
A 443	Titan
A 449	Istros
A 450	Pineios
A 451	Acheloos
A 460	Evrotas
A 461	Arachthos
A 463	Nestos
A 464	Axios
A 466	Trichonis
A 467	Doirani
A 468	Kalliroe
A 469	Stimfalia
A 470	Aliakmon
A 474	Pytheas
A 476	Stravon
A 478	Naftilos
A 479	I Theophilopoulos-Karavogiannos
A 481	St Lykoudis

IHS Jane's Fighting Ships 2014-2015   © 2014 IHS

SUBMARINES

1 + 3 PAPANIKOLIS (TYPE 214) CLASS (SSK)

Name	No	Builders	Laid down	Launched	Commissioned
PAPANIKOLIS	S 120	Howaldtswerke, Kiel	27 Feb 2001	22 Apr 2004	2 Nov 2010
PIPINOS	S 121	Hellenic Shipyards, Skaramanga	15 Oct 2002	15 Feb 2007	2015
MATROZOS	S 122	Hellenic Shipyards, Skaramanga	1 Apr 2003	27 Mar 2008	2015
KATSONIS	S 123	Hellenic Shipyards, Skaramanga	1 Apr 2004	29 Apr 2009	2015

Displacement, tonnes: 1,727 surfaced; 1,829 dived
Dimensions, metres (feet): 65 × 6.3 × 6.6 *(213.3 × 20.7 × 21.7)*
Speed, knots: 11 surfaced; 20 dived
Complement: 40 (6 officers)

Machinery: 2 MTU 16V 396 SE 84 diesels; 3,620 hp(m) *(2.7 MW)*; 1 Siemens Permasyn motor; 1 shaft; 2 HDW PEM fuel cells; 240 kW
Missiles: SSM: Boeing Sub Harpoon.
Torpedoes: 8—21 in *(533 mm)* bow tubes (4 fitted for Sub Harpoon discharge). Total of 16 weapons.
Physical countermeasures: Decoys: CIRCE torpedo countermeasures.
Electronic countermeasures: ESM Elbit TIMNEX II.
Radars: Surface search: Thales Sphynx; I-band.
Sonars: Bow and flank arrays. To be fitted for but not with towed array.
Weapon control systems: STN Atlas ISUS-90.

Programmes: Decision taken on 24 July 1998 and announced on 9 October to order three HDW-designed submarines with an option for a fourth. The first of class was built at Kiel and subsequent hulls at Hellenic Shipyards. Contracts to build signed 15 February 2000 and the fourth was ordered in 2002. The first of class was offered for acceptance in 2006 but was initially refused by the Greek Ministry of Defence, which claimed the boat did not perform to requirements. Following contractual negotiations between the Greek government and Thyssenkrupp Marine Systems, it was agreed in April 2010 that all four boats are to be accepted and that two further boats are to be built at Hellenic Shipyards. However, HDW is reported to have requested in May 2011 to withdraw from the agreement to build two further boats. It was subsequently announced in March 2014 that the three Hellenic Shipyard boats are to be completed in 2015.
Structure: Diving depth 400 m *(1,300 ft)*. Equipped with SERO-14 search periscope, SERO-15 attack periscope, and SATCOM.
Operational: *Papanikolis* started initial sea trials on 2 February 2005 and further trials were completed in September 2008.

PAPANIKOLIS
11/2010, Michael Nitz / 1421681

PAPANIKOLIS
11/2010, Michael Nitz / 1421680

7 GLAVKOS CLASS (TYPE 209/1100/1200) (SSK)

Name	No	Builders	Laid down	Launched	Commissioned
NEREUS	S 111	Howaldtswerke, Kiel	15 Jan 1969	7 Jun 1971	10 Feb 1972
TRITON	S 112	Howaldtswerke, Kiel	1 Jun 1969	14 Oct 1971	8 Aug 1972
PROTEUS	S 113	Howaldtswerke, Kiel	1 Oct 1969	1 Feb 1972	8 Aug 1972
POSEIDON	S 116	Howaldtswerke, Kiel	15 Jan 1976	21 Mar 1978	22 Mar 1979
AMPHITRITE	S 117	Howaldtswerke, Kiel	26 Apr 1976	14 Jun 1978	14 Sep 1979
OKEANOS	S 118	Howaldtswerke, Kiel	1 Oct 1976	16 Nov 1978	15 Nov 1979
PONTOS	S 119	Howaldtswerke, Kiel	25 Jan 1977	21 Mar 1979	29 Apr 1980

Displacement, tonnes: 1,143 (S 111–113), 1,219 (S 116–117, 119), 1,580 (S 118) surfaced; 1,255 (S 111–113), 1,306 (S 116–117, 119), 1,678 (S 118) dived
Dimensions, metres (feet): 54.4 (S 111–113), 55.9 (S 116–117, 119), 62.4 (S 118) × 6.2 × 5.6 (S 111–113), 5.7 (S 116–119) *(178.5, 183.4, 204.7 × 20.3 × 18.4, 18.7)*
Speed, knots: 11 surfaced; 21.5 dived
Complement: 38 (6 officers)

Machinery: Diesel-electric; 4 MTU 12V 493 AZ80 diesels; 2,400 hp(m) *(1.76 MW)* sustained; 4 Siemens alternators; 1.7 MW; 1 Siemens motor; 4,600 hp(m) *(3.38 MW)* sustained; 1 shaft; 2 HDW PEM fuel cells (S 118); 240 kW
Missiles: McDonnell Douglas Sub Harpoon; active radar homing to 130 km *(70 n miles)* at 0.9 Mach; warhead 258 kg. Can be discharged from 4 tubes only (S 111–113).
Torpedoes: 8—21 in *(533 mm)* bow tubes. 14 AEG SUT Mod 0; wire-guided; active/passive homing to 12 km *(6.5 n miles)* at 35 kt; warhead 250 kg. Swim-out discharge.
Electronic countermeasures: ESM: Argos AR-700-S5; radar warning (S 111–113). Thomson Arial DR 2000; radar warning (S 116–119).
Radars: Surface search: Thomson-CSF Calypso II (S 116–119). Thomson MILNAV (S 111–113). I-band.
Sonars: Atlas Elektronik CSU 83-90 (DBQS-21) (S 111–113); Atlas Elektronik CSU 3-4 (S 116–119); hull-mounted; active/passive search and attack; medium frequency. Atlas Elektronik PRS-3-4; passive ranging. STN Atlas flank array (S 118); passive low frequency.
Weapon control systems: Signaal Sinbads (S 116, 117, 119). Atlas Elektronik ISUS-90 (S 118). Unisys/Kanaris with UYK-44 computers (S 111–113).

Programmes: Designed by Ingenieurkontor, Lübeck for construction by Howaldtswerke, Kiel and sale by Ferrostaal, Essen all acting as a consortium.
Modernisation: Contract signed 5 May 1989 with HDW and Ferrostaal to implement a Neptune I update programme to bring first four up to an improved standard and along the same lines as the German S 206A class. Included Sub Harpoon, flank array sonar, Unisys FCS, Sperry Mk 29 Mod 3 inertial navigation system, GPS and Argos ESM. *Triton* completed refit at Kiel in May 1993, *Proteus* at Salamis in December 1995, *Glavkos* in November 1997, and *Nereus* in March 2000. A contract signed 31 May 2002 with Hellenic Shipyards (main sub-contractor HDW) for a Neptune II modernisation programme for S 117–119. S 116 is not to be modernised. *Okeanos* started refit in December 2004 and is expected to re-enter service in 2015. A 'plug-in' extension of 6.5 m was required to incorporate AIP (Siemens PEM fuel cell system). In addition an STN Atlas ISUS-90 combat management system, flank array sonar, electro-optic mast, SATCOM, Link II and Sub Harpoon were fitted. The planned refits of S 117 and S 119 were aborted in March 2009 in favour initially of the procurement of two new Type 209/1400 submarines. This was superseded in 2010 by an agreement to build two further Type 214 boats although this agreement was reportedly terminated by HDW in May 2011.
Structure: A single-hull design with four ballast tanks and forward and after trim tanks. Fitted with snort and remote machinery control. The single screw is slow revving. Very high-capacity batteries with GRP lead-acid cells and battery cooling by Wilh Hagen and VARTA. Diving depth, 250 m *(820 ft)*. Fitted with two periscopes.
Operational: Endurance, 50 days. A mining capability is reported but not confirmed. *Glavkos* was decommissioned in 2011 and the remaining three Type 209/1100 boats are likely to be decommissioned when the new Type 214 boats enter service.

TRITON
3/2010, Edward McDonnell / 1366629

FRIGATES

Notes: It was announced on 22 January 2009 that the Greek government would enter negotiations with France for the procurement of six frigates. The French shipbuilder DCNS is teamed with Greek shipyard Elefsis. The FREMM frigates are to be multirole and are to be capable of conducting both blue-water and littoral operations.

4 HYDRA CLASS (MEKO 200 HN) (FFGH)

Name	No	Builders	Laid down	Launched	Commissioned
HYDRA	F 452	Blohm + Voss, Hamburg	17 Dec 1990	25 Jun 1991	15 Oct 1992
SPETSAI	F 453	Hellenic Shipyards, Skaramanga	11 Aug 1992	9 Dec 1993	24 Oct 1996
PSARA	F 454	Hellenic Shipyards, Skaramanga	12 Dec 1993	20 Dec 1994	30 Apr 1998
SALAMIS	F 455	Hellenic Shipyards, Skaramanga	20 Dec 1994	15 May 1997	16 Dec 1998

Displacement, tonnes: 2,753 standard; 3,404 full load
Dimensions, metres (feet): 117 oa; 109 wl × 14.8 × 6 *(383.9; 357.6 × 48.6 × 19.7)*
Speed, knots: 31; 20 diesel
Range, n miles: 4,100 at 16 kt
Complement: 198 (27 officers) + 16 flag staff

Machinery: CODOG; 2 GE LM 2500 gas turbines; 60,000 hp *(44.76 MW)* sustained; 2 MTU 20V 956 TB82 diesels; 10,420 hp(m) *(7.66 MW)* sustained; 2 shafts; cp props
Missiles: SSM: 8 McDonnell Douglas Harpoon Block 1C; 2 quad launchers ❶; active radar homing to 130 km *(70 n miles)* at 0.9 Mach; warhead 227 kg.
SAM: Raytheon Evolved Sea Sparrow RIM-162 (F 452, 454) Mk 48 Mod 5 vertical launcher ❷; 16 missiles; semi-active radar homing to 18 km *(9.7 n miles)* at 3.6 Mach; warhead 38 kg.
Guns: 1 FMC 5 in *(127 mm)*/54 Mk 45 Mod 2A ❸ 20 rds/min to 24 km *(13 n miles)* anti-surface; 14 km *(7.7 n miles)* anti-aircraft; weight of shell 32 kg. 2 Raytheon Vulcan Phalanx Block 1 20 mm Mk 15 Mod 12 ❹; 6 barrels per mounting; 3,000 rds/min combined to 1.5 km.
Torpedoes: 6—324 mm Mk 32 Mod 5 (2 triple) tubes ❺. Honeywell Mk 46 Mod 5; anti-submarine; active/passive homing to 11 km *(5.9 n miles)* at 40 kt; warhead 44 kg.
Physical countermeasures: Decoys: 4 Mk 36 Mod 2 SRBOC chaff launchers ❻. SLQ-25 Nixie; torpedo decoy.
Electronic countermeasures: ESM: Argo AR 700; Telegon 10; intercept.
ECM: Argo APECS II; jammer.
Radars: Air search: Signaal MW08 ❼; 3D; F/G-band.
Air/surface search: Signaal/Magnavox; DA08 ❽; G-band.
Navigation: Racal Decca 2690 BT; ARPA; I-band.
Fire Control: 2 Signaal STIR ❾; I/J/K-band.
IFF: Mk XII Mod 4.
Sonars: Raytheon SQS-56/DE 1160; hull-mounted and VDS.
Combat data systems: Signaal STACOS Mod 2; Links 11 and 14.
Weapon control systems: 2 Signaal Mk 73 Mod 1 (for SAM). Vesta Helo transponder with datalink for OTHT. SAR-8 IR search. SWG 1 A(V) Harpoon LCS.
Helicopters: 1 Sikorsky S-70B-6 Aegean Hawk ❿.

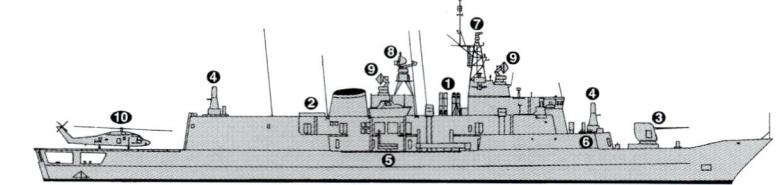

HYDRA *(Scale 1 : 1,200), Ian Sturton / 0052282*

Programmes: Decision to buy four Meko 200 HN announced on 18 April 1988. The first ship ordered 10 February 1989 built by Blohm + Voss, Hamburg and the remainder ordered 10 May 1989 at Hellenic Shipyards, Skaramanga. Programme was delayed by financial problems at Hellenic Shipyards in 1992 and some of the prefabrication of *Spetsai* was done in Hamburg.
Modernisation: A mid-life upgrade programme is a high priority programme. Enhancements are to include upgrades and replacement of sensors and trackers in addition to platform improvements. The upgrade is likely to be similar to that of six ships of the Elli class. As a separate programme Mk 48 launcher systems have been upgraded to Mod 5 to accommodate ESSM.
Structure: The design follows the Portuguese Vasco da Gama class. All steel fin stabilisers.
Operational: Aegean Hawk carried from 1995. *Hydra* and *Salamis* are part of the 1st Frigate Squadron and *Spetsai* and *Psara* part of the 2nd Frigate Squadron.

PSARA *10/2008, M Declerck / 1335381*

HYDRA *6/2005, Michael Winter / 1133492*

SALAMIS *6/2006, Marco Ghiglino / 1164520*

IHS Jane's Fighting Ships 2014-2015

Frigates < **Greece** 305

9 ELLI (KORTENAER) CLASS (FFGH)

Name	No	Builders	Laid down	Launched	Commissioned
ELLI (ex-*Pieter Florisz*)	F 450 (ex-F 812)	Koninklijke Maatschappij de Schelde, Flushing	1 Jul 1977	15 Dec 1979	10 Oct 1981
LIMNOS (ex-*Witte de With*)	F 451 (ex-F 813)	Koninklijke Maatschappij de Schelde, Flushing	13 Jun 1978	27 Oct 1979	18 Sep 1982
AEGEON (ex-*Banckert*)	F 460 (ex-F 810)	Koninklijke Maatschappij de Schelde, Flushing	25 Feb 1976	13 Jul 1978	29 Oct 1980
ADRIAS (ex-*Callenburgh*)	F 459 (ex-F 808)	Koninklijke Maatschappij de Schelde, Flushing	30 Jun 1975	12 Mar 1977	26 Jul 1979
NAVARINON (ex-*Van Kinsbergen*)	F 461 (ex-F 809)	Koninklijke Maatschappij de Schelde, Flushing	2 Sep 1975	16 Apr 1977	24 Apr 1980
KOUNTOURIOTIS (ex-*Kortenaer*)	F 462 (ex-F 807)	Koninklijke Maatschappij de Schelde, Flushing	8 Apr 1975	18 Dec 1976	26 Oct 1978
KANARIS (ex-*Jan van Brakel*)	F 464 (ex-F-825)	Koninklijke Maatschappij de Schelde, Flushing	16 Nov 1979	16 May 1981	14 Apr 1983
THEMISTOCLES (ex-*Philips Van Almonde*)	F 465 (ex-F-823)	Dok en Werfmaatschappij-Fijenoord	3 Oct 1977	11 Aug 1979	2 Dec 1981
NIKIFOROS FOKAS (ex-*Bloys van Treslong*)	F 466 (ex-F 824)	Dok en Werfmaatschappij-Fijenoord	27 Apr 1978	15 Nov 1980	25 Nov 1982

Displacement, tonnes: 3,099 standard; 3,688 full load
Dimensions, metres (feet): 130.5 × 14.6 × 6.2 screws *(428.1 × 47.9 × 20.3)*
Speed, knots: 30
Range, n miles: 4,700 at 16 kt
Complement: 186 (26 officers)

Machinery: COGOG; 2 RR Olympus TM3B gas turbines; 50,880 hp *(39.7 MW)* sustained; 2 RR Tyne RM1C gas turbines; 9,900 hp *(7.4 MW)* sustained; 2 shafts; LIPS cp props
Missiles: SSM: 8 McDonnell Douglas Harpoon (2 quad) launchers ❶; active radar homing to 130 km *(70 n miles)* at 0.9 Mach; warhead 227 kg.
SAM: Raytheon NATO Sea Sparrow RIM-7P ❷; Mk 29 octuple launcher; 8 missiles; semi-active radar homing to 16 km *(8.5 n miles)* at 2.5 Mach; warhead 38 kg. Portable Redeye; shoulder-launched; short range.
Guns: 1 (F 459–462, 464–466) or 2 (F 450, 451) Oto Melara 3 in *(76 mm)*/62 compact ❸; 85 rds/min to 16 km *(8.6 n miles)* anti-surface; 12 km *(6.5 n miles)* anti-aircraft; weight of shell 6 kg. 1 (F 459, 460, 461, 462, 465) or 2 (F 450, 451) Raytheon Vulcan Phalanx Block 1 20 mm Mk 15 6-barrelled ❹; 3,000 rds/min combined to 1.5 km.
Torpedoes: 4—324 mm Mk 32 (2 twin) tubes ❺. 16 Honeywell Mk 46 Mod 5; anti-submarine; active/passive homing to 11 km *(5.9 n miles)* at 40 kt; warhead 44 kg. Can be fitted.
Physical countermeasures: Decoys: 2 Loral Hycor Mk 36 SRBOC chaff launchers (Sippican ALEX in F 450, 451, 459, 460, 461, 462).
Electronic countermeasures: ESM: Elettronika Sphinx and MEL Scimitar; intercept. EDO CS-3701 (F 450, 451, 459, 460, 461, 462).
ECM: ELT 715; jammer.
Radars: Air search: Signaal LW08 ❻; D-band; range 264 km *(145 n miles)* for 2 m² target.
Surface search: Signaal ZW06 ❼; Thales Scout Mk 2 (F 450, 451, 459, 460, 461, 462); I-band.
Fire control: Signaal WM25 ❽; I/J-band; range 46 km *(25 n miles)*. Signaal STIR ❾; I/J/K-band; range 39 km *(21 n miles)* for 1 m² target.
Sonars: Canadian Westinghouse SQS-505; hull-mounted; active search and attack; 7 kHz.
Combat data systems: Signaal SEWACO II action data automation; Thales Tacticos (F 450, 451, 459, 460, 461, 462); Links 10, 11 and 14, SHF Satcom.
Electro-optic systems: Thales Mirador Trainable Electrical-Optical Observation System (TEOOS) (F 450, 451, 459, 460, 461, 462).
Helicopters: 2 AB 212ASW ❿ or 1 S-70B (F 451).

Programmes: A contract was signed with the Netherlands on 15 September 1980 for the purchase of *Elli*, a Kortenaer class, building for the Netherlands' Navy. An option for a second ship *Limnos* was exercised on 7 June 1981. On 9 November 1992, agreement was reached to transfer further ships of the class from the Netherlands Navy: *Aegeon* recommissioned on 14 May 1993, *Adrias* on 30 March 1994, *Navarinon* on 1 March 1995 and *Kountouriotis* on 15 December 1997. The first four ships are known as Batch I and the next two as Batch II. Four Batch III ships were later acquired on decommissioning from the Netherlands Navy: *Bouboulina* recommissioned on 14 December 2001, *Kanaris*, on 29 November 2002, *Themistocles* on 24 October 2003 and *Nikiforos Fokas* on 17 December 2003. *Bouboulina* was subsequently decommissioned in 2012.
Modernisation: Mid-Life Modernisation programme (MLM) extended life of the six Batch I and II ships to 2020. The upgrade was undertaken by Hellenic Shipyards with Thales Nederland acting as main sub-contractor. The MLM included replacement of the combat data system with Tacticos, replacement of ZW06 surface search radar with Scout, improvements to the tracking performance of LW08 and WM25/STIR and installation of the Mirador optronic director. Upgrades to the EW capability included EDO CS-3701 ESM receiver and upgrade of SRBOC. Upgrade of the Sea Sparrow system to RIM 162 ESSM has been postponed indefinitely. *Kountouriotis*, the first modernised frigate, was handed back on 12 September 2006 the second, *Adrias*, in February 2007, the third *Navarinon* in October 2007, the fourth, *Limnos*, in January 2009, the fifth *Elli* in July 2009 and sixth *Aegeon* in September 2010.
Structure: Hangar is 2 m longer than in the original Netherlands-designed ships to accommodate AB 212ASW helicopters. The helicopter deck and hangar of *Limnos*, *Adrias* and *Navarinon* have been modified to accommodate an S-70B helicopter.
Operational: Assignments; 1st FS (*Elli, Adrias, Kountouriotis, Themistocles*). 2nd FS (*Limnos, Aegeon, Navarinon, Kanaris, Nikiforos Fokas*).

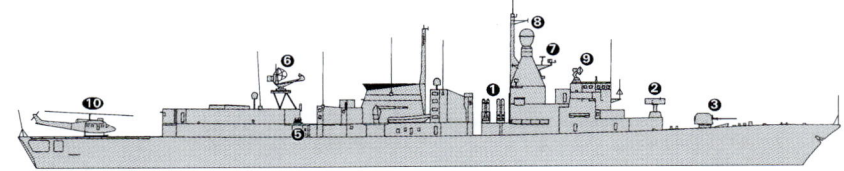

KANARIS *(Scale 1 : 1,200), Ian Sturton* / 1044255

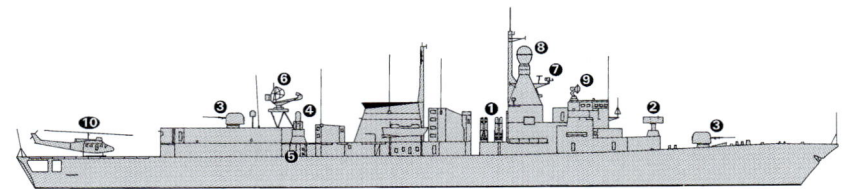

ELLI *(Scale 1 : 1,200), Ian Sturton* / 0126346

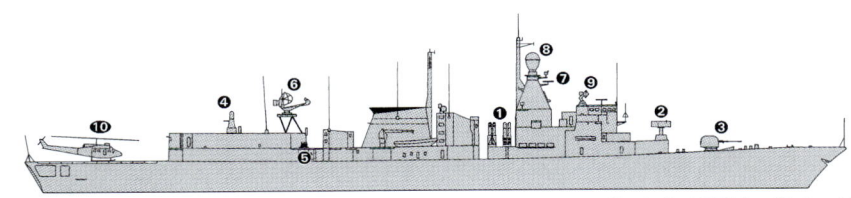

KOUNTOURIOTIS *(Scale 1 : 1,200), Ian Sturton* / 1335461

ELLI 2/2007, *Camil Busquets i Vilanova* / 1170148

KOUNTOURIOTIS 4/2009, *B Prézelin* / 1367005

THEMISTOCLES 7/2011, *A Campanera i Rovira* / 1406730

© 2014 IHS

IHS Jane's Fighting Ships 2014-2015

SHIPBORNE AIRCRAFT

Notes: There are also two Alouette IIIs used for SAR and training.

Numbers/Type: 8 Agusta AB 212ASW.
Operational speed: 106 kt *(196 km/h)*.
Service ceiling: 14,200 ft *(4,330 m)*.
Range: 230 n miles *(425 km)*.
Role/Weapon systems: Shipborne ASW and surface search role from escorts. Sensors: Selenia APS-705 radar, AlliedSignal AQS-18 dipping sonar (ASW version). Weapons: ASW; two Mk 46 or two A244/S homing torpedoes.

Numbers/Type: 11 Sikorsky S-70B-6 Aegean Hawk.
Operational speed: 135 kt *(250 km/h)*.
Service ceiling: 10,000 ft *(3,050 m)*.
Range: 600 n miles *(1,110 km)*.
Role/Weapon systems: Five ordered 17 August 1991. First one delivered 14 October 1994, remainder in July 1995. The option was taken up on three more of which one was delivered in 1997, and two more in 1998. Three further more modern aircraft ordered June 2000, all of which have been delivered (differences are indicated in brackets). All of the original eight aircraft are to be similarly upgraded. Sensors: Telephonica APS 143(V)3 search radar and AAQ-22 (or AAS 44) FLIR, AlliedSignal AQS 18(V)3 (or Ocean Systems HELRAS) dipping sonar, MAD, Litton ALR 606(V)2 (or LR 100) ESM, Litton ASN 150(V) tactical data system with CD22 or Link 11. Weapons: ASV; Kongsberg Penguin Mk 2 Mod 7, four AGM-114K Hellfire. ASW; two (or three) Mk 46 torpedoes.

AB 212 — *9/2011, Erik Laursen /* 1406731

AEGEAN HAWK — *10/2001, Diego Quevedo /* 0126292

LAND-BASED MARITIME AIRCRAFT

Notes: (1) A squadron of Air Force Mirage 2000 EG fighters is assigned to the naval strike role using Exocet AM 39 ASMs.
(2) Replacement of the decommissioned P-3B Orions with up to five maritime patrol aircraft is under consideration. Replacement is not expected before 2018. In the meantime, maritime patrol duties are occasionally undertaken by camera-equipped C-130H and C-27J transport aircraft.

PATROL FORCES

4 + 3 ROUSSEN (SUPER VITA) CLASS
(FAST ATTACK CRAFT — MISSILE) (PGG)

Name	No	Builders	Launched	Commissioned
YPOPLOIARCHOS ROUSSEN	P 67	Elefsis Shipyard	13 Nov 2002	20 Dec 2005
YPOPLOIARCHOS DANIOLOS	P 68	Elefsis Shipyard	8 Jul 2003	22 Feb 2006
YPOPLOIARCHOS KRISTALLIDIS	P 69	Elefsis Shipyard	5 Apr 2004	8 May 2006
YPOPLOIARCHOS GRIGOROPOULOS	P 70	Elefsis Shipyard	20 Dec 2005	1 Oct 2010
ANTHYPOPLOIARCHOS RITSOS	P 71	Elefsis Shipyard	9 Oct 2006	Mar 2014
KARATHANASIS	–	Elefsis Shipyard	–	Jun 2015
VLAHAKOS	–	Elefsis Shipyard	–	Apr 2016

Displacement, tonnes: 671 full load
Dimensions, metres (feet): 61.9 × 9.5 × 2.6 *(203.1 × 31.2 × 8.5)*
Speed, knots: 34
Range, n miles: 1,800 at 12 kt
Complement: 45 (8 officers)

Machinery: 4 MTU 16V 595 TE 90 diesels; 23,170 hp *(17.3 MW)*; 4 shafts
Missiles: SSM: 8 MBDA Exocet MM 40 Block 2 (Block 3 in P 70 and P 71) ❶; inertial cruise; active radar homing to 70 km *(40 n miles)* at 0.9 Mach; warhead 165 kg; sea skimmer.
SAM: 1 RAM RIM-116 ❷; Mk 49 launcher; passive IR/anti-radiation homing to 9.6 km *(5.2 n miles)* at 2.5 Mach; warhead 9.1 kg.
Guns: 1 Oto Melara 76 mm/62 Super Rapid ❸; 120 rds/min to 16 km *(8.7 n miles)*; weight of shell 6 kg. 2 Otobreda 30 mm ❹.
Physical countermeasures: Decoys: 2 Loral Hycor Mk 36 SRBOC chaff launchers ❺.
Electronic countermeasures: ESM: Thales DR 3000 ❻; intercept.
Radars: Air/surface search: Thales MW-08 ❽; G-band.
Surface search: Thales Scout Mk 2 LPI; I-band.
Navigation: Litton Marine Bridgemaster; I-band.
Fire control: Thales Sting ❾; I/J-band.
IFF: Mk XII.
Combat data systems: Thales Tacticos. Link 11.
Electro-optic systems: Thales Mirador Trainable Electro-Optical Observation System (TEOOS) ❼.

Programmes: Design selected 21 September 1999 based on Vosper Thornycroft Vita corvettes in service in Qatar. Contract signed 7 January 2000 for the building of first three vessels which started in March 2000. A contract for the construction of two further ships was signed on 23 August 2003. The contract for the sixth and seventh vessels was signed on 25 September 2008. Following the termination of the agreement by BAE Systems in 2011, the fifth vessel remains alongside, and the sixth and seventh vessels remain in the assembly hall.
Structure: A rigid inflatable boat is carried amidships.

DANIOLOS — *7/2006, Richard Scott /* 1159225

KRISTALLIDIS — *10/2013*, Harald Carstens /* 1529040

2 ANTONIOU CLASS (PB)

Name	No	Builders	Commissioned
DIOPOS ANTONIOU	P 286	Ch N de l'Esterel	4 Dec 1975
KELEFSTIS STAMOU	P 287	Ch N de l'Esterel	28 Jul 1975

Displacement, tonnes: 117 full load
Dimensions, metres (feet): 32 × 5.8 × 1.6 *(105 × 19.0 × 5.2)*
Speed, knots: 30
Range, n miles: 1,500 at 15 kt
Complement: 20 (2 officers)
Machinery: 2 MTU 12V 331 TC81 diesels; 2,610 hp(m) *(1.92 MW)* sustained; 2 shafts
Guns: 1 Rheinmetall 20 mm. 1—12.7 mm MG.
Radars: Surface search: Decca 1226; I-band.

Comment: Originally ordered for Cyprus, later transferred to Greece. Wooden hulls. Fast RIB carried on the stern. Surface-to-surface missiles no longer carried.

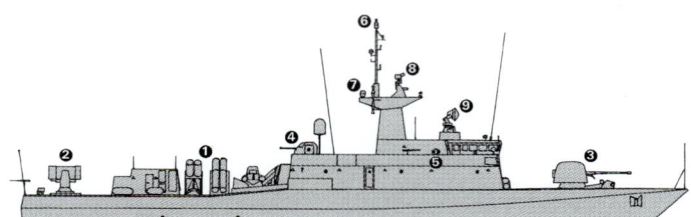

YPOPLOIARCHOS ROUSSEN — *(Scale 1 : 900), Ian Sturton /* 0126344

DIOPOS ANTONIOU — *7/2013*, A Campanera i Rovira /* 1529041

9 LASKOS (LA COMBATTANTE III) CLASS
(FAST ATTACK CRAFT — MISSILE) (PGGF/PGG)

Name	No	Builders	Commissioned
ANTHYPOPLOIARCHOS LASKOS	P 20	CMN, Cherbourg	20 Apr 1977
PLOTARCHIS BLESSAS	P 21	CMN, Cherbourg	7 Jul 1977
YPOPLOIARCHOS MIKONIOS	P 22	CMN, Cherbourg	10 Feb 1978
YPOPLOIARCHOS TROUPAKIS	P 23	CMN, Cherbourg	8 Nov 1977
SIMEOFOROS KAVALOUDIS	P 24	Hellenic Shipyards, Skaramanga	14 Jul 1980
YPOPLOIARCHOS DEGIANNIS	P 26	Hellenic Shipyards, Skaramanga	11 Dec 1980
SIMEOFOROS XENOS	P 27	Hellenic Shipyards, Skaramanga	31 Mar 1981
SIMEOFOROS SIMITZOPOULOS	P 28	Hellenic Shipyards, Skaramanga	30 Jun 1981
SIMEOFOROS STARAKIS	P 29	Hellenic Shipyards, Skaramanga	12 Oct 1981

Displacement, tonnes: 365 (P 20–23), 334 (P 24, 26–29) standard; 432 (P 20–23), 436 (P 24, 26–29) full load
Dimensions, metres (feet): 56.2 × 8 × 2.1 (184.4 × 26.2 × 6.9)
Speed, knots: 36 (P 20–23), 32.5 (P 24, 26–29)
Range, n miles: 700 at 32 kt, 2,700 at 15 kt
Complement: 43 (6 officers)

Machinery: 4 MTU 20V 538 TB92 diesels; 17,060 hp(m) (12.54 MW) sustained; 4 shafts (P 20–23). 4 MTU 20V 538 TB91 diesels; 15,360 hp(m) (11.29 MW) sustained; 4 shafts (P 24–29).
Missiles: SSM: 4 Aerospatiale MM 38 Exocet (P 20–P 23); inertial cruise; active radar homing to 42 km (23 n miles) at 0.9 Mach; warhead 165 kg. 6 Kongsberg Penguin Mk 2 Mod 3 (P 24–P 29); inertial/IR homing to 27 km (15 n miles) at 0.8 Mach; warhead 120 kg.
Guns: 2 Oto Melara 3 in (76 mm)/62 compact; 85 rds/min to 16 km (8.6 n miles) anti-surface; 12 km (6.5 n miles) anti-aircraft; weight of shell 6 kg. 4 Emerlec 30 mm (2 twin); multipurpose; 1,200 rds/min combined to 6 km (3.2 n miles); weight of shell 0.35 kg.
Torpedoes: 2—21 in (533 mm) aft tubes. AEG SST-4; anti-surface; wire-guided; active homing to 12 km (6.5 n miles) at 35 kt; passive homing to 28 km (15 n miles) at 23 kt; warhead 250 kg.
Physical countermeasures: Decoys: Wegmann chaff launchers.
Electronic countermeasures: ESM: Thomson CSF DR 2000S (P 20–23); intercept.
Radars: Air/surface search: Thomson-CSF Triton (P 24–29); Thales Variant (P 20–23); G-band. Thales Scout 2 (P 20–23); I-band.
Navigation: Decca 1226C (P 24–29); I-band. Sperry Bridgemaster (P 20–23); I-band.
Fire control: Thomson-CSF Pollux; I/J-band. Thales LIROD Mk 2; K-band.
Combat data systems: Tacticos (P 20–23). Link 11 (P 20–23).
Weapon control systems: 2 CSEE Panda optical directors for 30 mm guns. Mirador optronic director (P 20–P 23). NFT PFCS-2 (P 24–P 29).

Programmes: First four ordered in September 1974. Second group of six ordered 1978.
Modernisation: P 24–29 upgraded to fire Penguin Mk 2 Mod 3 missiles. A contract for the upgrade of P 20–23 was signed on 31 October 2003. Modernisation began in 2005 the first ship, *Laskos* was delivered in April 2008, *Blessas* and *Mikonios* in October 2009 and *Troupakis* in February 2011. The programme included installation of the Tacticos Combat Management System, the MIRADOR optronic director, SRBOC launchers, Thales DR 3000 ESM, Link 11 and Variant, Scout Mk 2 and Bridgemaster radars.
Structure: First four fitted with SSM Exocet; remainder have Penguin.
Operational: P 25 sunk after collision with a ferry in November 1996.

YPOPLOIARCHOS MIKONIOS (without Exocet) 3/2010, Edward McDonnell / 1366628

PLOTARCHIS BLESSAS (with Exocet) 7/2006, Marco Ghiglino / 1164515

SIMEOFOROS STARAKIS 11/2008, M Declerck / 1335380

3 VOTSIS (LA COMBATTANTE IIA) (TYPE 148) CLASS
(FAST ATTACK CRAFT — MISSILE) (PGFG)

Name	No	Builders	Commissioned	Recommissioned
YPOPLOIARCHOS VOTSIS (ex-*Iltis*)	P 72 (ex-P 51)	CMN, Cherbourg	8 Jan 1973	17 Feb 1994
ANTHYPOPLOIARCHOS PEZOPOULOS (ex-*Storch*)	P 73 (ex-P 30)	CMN, Cherbourg	17 Jul 1974	17 Feb 1994
PLOTARCHIS MARIDAKIS (ex-*Häher*)	P 75	CMN, Cherbourg	12 Jun 1974	30 Jun 1995

Displacement, tonnes: 269 full load
Dimensions, metres (feet): 47 × 7 × 2.7 (154.2 × 23.0 × 8.9)
Speed, knots: 36
Range, n miles: 570 at 30 kt, 1,600 at 15 kt
Complement: 41 (6 officers)

Machinery: 4 MTU MD 16V 538 TB90 diesels; 12,000 hp(m) (8.82 MW) sustained; 4 shafts
Missiles: SSM: 4 Aerospatiale MM 38 Exocet (2 twin) launchers (P 72–73); inertial cruise; active radar homing to 42 km (23 n miles) at 0.9 Mach; warhead 165 kg; sea-skimmer. 4 McDonnell Douglas Harpoon (2 twin) launchers (P 75); active radar homing to 130 km (70 n miles) at 0.9 Mach; warhead 227 kg.
Guns: 1 Oto Melara 3 in (76 mm)/62 compact; 85 rds/min to 16 km (8.6 n miles) anti-surface; 12 km (6.5 n miles) anti-aircraft; weight of shell 6 kg. 1 Bofors 40 mm/70; 330 rds/min to 12 km (6.5 n miles) anti-surface; 4 km (2.2 n miles) anti-aircraft; weight of shell 0.96 kg; fitted with GRP dome (1984).
Mines: Laying capability.
Physical countermeasures: Decoys: Wolke chaff launcher.
Electronic countermeasures: ESM: Thomson-CSF DR 2000S; intercept.
Radars: Air/surface search: Thomson-CSF Triton; G-band.
Navigation: Decca 1226 (P 71, 73); Decca 1690 (P 75); I-band.
Fire control: Thomson-CSF Pollux; I/J-band.
Combat data systems: PALIS and Link 11.
Weapon control systems: CSEE Panda optical director. Thomson-CSF Vega PCET system, controlling missiles and guns.

Programmes: Two transferred from Germany in September 1993 and recommissioned 17 February 1994. Two more transferred 16 March 1995 and recommissioned 30 June 1995. A third pair transferred from Germany in 2000. Three ships were subsequently decommissioned in 2011.
Modernisation: Mid-life updates in 1980s. P 75 fitted with Harpoon. New ESM fitted after transfer.
Structure: Steel hulls. Similar to Combattante II class. Spray rails have been fitted to improve hydrodynamic performance.

PLOTARCHIS MARIDAKIS 12/2008, M Declerck / 1367041

4 ANDROMEDA (NASTY) CLASS
(PATROL CRAFT) (PB)

Name	No	Builders	Commissioned
ANDROMEDA	P 196	Mandal, Norway	21 Nov 1966
KYKNOS	P 198	Mandal, Norway	25 Feb 1967
PIGASOS	P 199	Mandal, Norway	12 Apr 1967
TOXOTIS	P 228	Mandal, Norway	31 May 1966

Displacement, tonnes: 73 full load
Dimensions, metres (feet): 24.5 × 7.5 × 2.1 (80.4 × 24.6 × 6.9)
Speed, knots: 25
Range, n miles: 676 at 17 kt
Complement: 20 (2 officers)
Machinery: 2 MTU 12V 331 TC92 diesels; 2,660 hp(m) (1.96 MW) sustained; 2 shafts
Guns: 1 Bofors 40 mm/70. 1 Rheinmetall 20 mm.
Radars: Surface search: Decca 1226; I-band.
Navigation: Sperry Bridgemaster E251; I-band.

Comment: Six of the class acquired from Norway in 1967 and paid off into reserve in the early 1980s. Four re-engined and brought back into service in 1988. These craft continue to be active although top speed has been markedly reduced. Torpedo tubes have been removed.

PIGASOS 8/2013*, C D Yaylali / 1529042

308   Greece > Patrol forces

2 ARMATOLOS (OSPREY 55) CLASS
(LARGE PATROL CRAFT) (PG)

4 MACHITIS CLASS (LARGE PATROL CRAFT) (PG)

Name	No	Builders	Launched	Commissioned

314   Greece > Coast guard

13 GUARDIAN 53 CRAFT (WPB)

LS 114–119	LS 121–123	LS 125–128

Displacement, tonnes: 24 full load
Dimensions, metres (feet): 16.5 × 4.7 × 1.4 (54.1 × 15.4 × 4.6)
Speed, knots: 34
Range, n miles: 500 at 25 kt
Complement: 5 (1 officer)
Machinery: 2 MAN D2840 LE 401 diesels; 1,644 hp(m) (1.21 MW) sustained; 2 shafts
Guns: 1 — 12.7 mm MG. 1 — 7.62 mm MG.
Radars: Surface search: Raytheon; I-band.

Comment: Ordered from Colvic Craft, Colchester in 1993. Shipped to Motomarine, Glifada for engine and electronics installation. Completed in mid-1994. GRP hulls with a stern platform for recovery of divers.

LS 101   5/2000, van Ginderen Collection / 0104571

16 MOTOMARINE PANTHER 57 MK II CRAFT (WPB)

LS 601–616

Displacement, tonnes: 27 standard
Dimensions, metres (feet): 19.2 × 4.7 × 0.9 (63 × 15.4 × 3.0)
Speed, knots: 50
Machinery: 2 MTU 12V2000 M 91 diesels; 2 surface piercing propellers

Comment: Constructed by Motomarine, Koropi, Greece. LS 601 delivered November 2003. LS 609–615 delivered between February 2005 and March 2006.

LS 119   11/2004, M Declerck / 1133490

16 LS 51 CLASS (WPB)

LS 51–52	LS 55–56	+12

Displacement, tonnes: 13 full load
Dimensions, metres (feet): 13.4 × 3.5 × 1 (44 × 11.5 × 3.3)
Speed, knots: 25
Range, n miles: 400 at 18 kt
Complement: 4
Machinery: 2 diesels; 630 hp(m) (463 kW); 2 shafts
Guns: 1 — 7.62 mm MG.
Radars: Surface search: Racal Decca; I-band.

Comment: Built by Olympic Marine. GRP hulls.

3 COMBATBOAT 90HEX (WPBF)

LS 134–136

Displacement, tonnes: 19 full load
Dimensions, metres (feet): 15.9 × 3.8 × 0.8 (52.2 × 12.5 × 2.6)
Speed, knots: 45
Range, n miles: 240 at 30 kt
Complement: 3
Machinery: 2 Volvo Penta TAMD 163P diesels; 1,500 hp(m) (1.1 MW); 2 waterjets
Guns: 3 — 12.7 mm MGs.
Radars: Surface search: I-band.

Comment: Built by Dockstavarvet in Sweden and delivered 6 July 1998. Same design as Swedish naval craft but with more powerful engines. GRP construction with armoured protection for cockpit.

LS 616   7/2013*, A Campanera i Rovira / 1529046

94 COASTAL CRAFT

Comment: Included in the total are 38 of 8.2 m, 26 of 5.8 m. In addition there are at least 24 inflatable craft, and four FB Design MIL-40 (LS 130–133).

LS 132   4/2010, Edward McDonnell / 1366627

LS 136   7/2004, A Campanera i Rovira / 0587761

16 OL 44 CLASS (WPB)

LS 55	LS 84–88	LS 97	LS 103	LS 109–110
LS 65	LS 95	LS 101	LS 106–107	LS 112

Displacement, tonnes: 14 full load
Dimensions, metres (feet): 13.7 × 4.4 × 0.6 (44.9 × 14.4 × 2.0)
Speed, knots: 23
Complement: 4
Machinery: 2 diesels; 630 hp(m) (463 kW); 2 shafts
Guns: 1 — 7.62 mm MG.
Radars: Surface search: JRC; I-band.

Comment: Built by Olympic Marine. GRP hulls.

LS 214   7/2004, C D Yaylali / 0587760

IHS Jane's Fighting Ships 2014-2015   © 2014 IHS

Coast guard < **Greece** — Coast guard < **Grenada** 315

4 POLLUTION CONTROL SHIPS (YPC)

| LS 401 | LS 413–415 |

Displacement, tonnes: 234 full load
Dimensions, metres (feet): 29 × 6.2 × 2.5 (95.1 × 20.3 × 8.2)
Speed, knots: 15
Range, n miles: 500 at 13 kt
Complement: 12
Machinery: 2 CAT 3512 DITA diesels; 2,560 hp(m) (1.88 MW) sustained; 2 shafts
Radars: Navigation: Furuno; I-band.

Comment: Details given are for *LS 413–415*. Built by Astilleros Gondan, Spain in collaboration with Motomarine. Delivered in 1993–94. *LS 401* is a smaller, older pollution control ship.

LS 414   11/2005, M Declerck / 1164499

11 ARUN 60 CLASS (LIFEBOATS) (SAR)

| SAR 12–14 | SAR 510–511 | SAR 520 |
| SAR 17–19 | SAR 515–516 | |

Displacement, tonnes: 35 full load
Dimensions, metres (feet): 18 × 5.3 × 1.5 (59.1 × 17.4 × 4.9)
Speed, knots: 18
Complement: 5
Machinery: 2 Caterpillar 3408 diesels; 2 shafts

Comment: Built by Motormarine, Koropi, Greece. GRP hull moulded by Halmatic, UK. A stretched version of the lifeboat used in the UK and Canada. Entered service 1997–98.

SAR 516   5/2006, Marco Ghiglino / 1164498

36 MOTOMARINE PANTHER 57 MK I CRAFT (WPB)

LS 137–172

Displacement, tonnes: 28 full load
Dimensions, metres (feet): 18.2 × 4.68 × 0.92 (59.7 × 15.4 × 3.0)
Speed, knots: 44
Machinery: 2 MAN diesels; 2 shafts
Guns: 1 — 12.7 mm MG.

Comment: A development of the Guardian class. Constructed by Motomarine and delivered between about 1997 and 2006.

LS 172   8/2013*, C D Yaylali / 1529047

CUSTOMS

Notes: The Customs service also operates large numbers of coastal and inshore patrol. The craft have a distinctive Alpha Lambda (Α/Λ) on the hull and are sometimes armed with 7.62 mm MGs.

AL 20   6/2002, C D Yaylali / 0525874

Grenada

Country Overview

Grenada gained independence in 1974; the British monarch, represented by a governor-general, is the head of state. The southernmost of the Windward Islands in the Lesser Antilles chain, the country comprises the island of Grenada (311 square miles) and some of the southern Grenadines including Carriacou and Petit Martinique. The capital, largest town, and main port is St George's. Territorial seas (12 n miles) are claimed. A 200 n mile Exclusive Economic Zone (EEZ) has been claimed but the limits are not defined. The Coast Guard craft are operated under the direction of the Commissioner of Police.

Personnel
2014: 30

Bases
Prickly Bay

COAST GUARD

Notes: A 9 m RHIB, donated by the US government, entered service in 2004. Powered by two Mercury 300 hp engines, it is capable of 40 kt. Another 7 m RHIB, powered by two Mercury 200 hp engines, is capable of 35 kt.

1 DAUNTLESS CLASS (PB)

Name	No	Builders	Commissioned
LEVERA	PB 02	SeaArk Marine, Monticello	8 Sep 1995

Displacement, tonnes: 11 full load
Dimensions, metres (feet): 12.2 × 4.3 × 1.3 (40 × 14.1 × 4.3)
Speed, knots: 27
Range, n miles: 600 at 18 kt
Complement: 5
Machinery: 2 Caterpillar 3208TA diesels; 870 hp (650 kW) sustained; 2 shafts
Guns: 1 — 7.62 mm MG.
Radars: Raytheon R40X; I-band.

Comment: One of many of this type, provided by the US, throughout the Caribbean navies. Aluminium construction.

LEVERA   4/2008, Marco Ghiglino / 1305863

© 2014 IHS   IHS Jane's Fighting Ships 2014-2015

Grenada > Coast guard — **Guatemala** > Patrol forces

2 SPECIAL PURPOSE CRAFT (PBF)

PB 07–08

Displacement, tonnes: 7 full load
Dimensions, metres (feet): 10.1 × 3.0 × 0.7 (33.1 × 9.8 × 2.3)
Speed, knots: 50
Complement: 4
Machinery: 3 Mercury Verado outboard motors; 825 hp (615 kW)
Guns: 2 — 7.62 mm MGs.
Radars: Navigation: Raymarine; I-band.

Comment: Built by SAFE Boats International and very similar to the craft in service with the US Coast Guard. Both craft donated by US Southern Command under the US foreign Military Sales Program in May 2012.

PB 07   6/2012*, US Navy / 1525527

Guatemala

Country Overview

The Republic of Guatemala is situated in Central America between Mexico to the north, Belize to the east and Honduras and El Salvador to the south-east. With an area of 42,042 square miles, it has an 83 n mile coastline with the Caribbean and a 133 n mile coastline with the Pacific Ocean. The capital city is Guatemala City while the principal Caribbean ports are Puerto Barrios and Santo Tomás de Castilla and Pacific ports are Puerto Quetzal, San José and Champerico. Territorial seas (12 n miles) are claimed. A 200 n mile EEZ has been claimed but the limits are not defined.

Headquarters Appointments

Commander of the Navy:
  Vice Admiral Carlos Antonio Lainfiesta Soto
Commander Pacific Naval Region:
  Captain Carlo Adolfo Thimas Ramirez
Commander Caribbean Naval Region:
  Captain Danilo Mayen Garcia

Personnel

(a) 2014: 1,250 (130 officers) including 500 Marines (2 battalions) (mostly volunteers)
(b) 2¼ years' national service

Bases

Pacific: Puerto Quetzal (HQ), Puerto San Jose, Champerico
Atlantic: Santo Tomás de Castilla (HQ), Puerto Barrios, Livingston

Special Forces

A special forces unit, Fuerza Especial Naval (FEN), was established in 2012.

PATROL FORCES

Notes: (1) There is also a naval manned Ferry *15 de Enero* (T 691) and a 69 ft launch *Orca* which was built locally in 1996/97.
(2) Three sail training craft, *Mendieta, Margarita* and *Ostuncalco* are based at Santo Thomás de Castilla.
(3) Two launches were reported donated by the Guatemalan government and the US Embassy in 2005.
(4) There are two 11 m personnel landing craft *Picuda* D 361 and *Barracuda* D 362.
(5) The acquisition of 10 small patrol craft from Brazil was announced in April 2008.
(6) There are four former commercial craft that have been confiscated: *Mero, Pampana, Sardina, Mavro I*.

1 BROADSWORD CLASS (COASTAL PATROL CRAFT) (PB)

Name	No	Builders	Commissioned
KUKULKÁN	GC 1051 (ex-P 1051)	Halter Marine	4 Aug 1976

Displacement, tonnes: 92 standard; 112 full load
Dimensions, metres (feet): 32 × 6.2 × 1.9 (105 × 20.3 × 6.2)
Speed, knots: 22
Range, n miles: 1,150 at 20 kt
Complement: 20 (5 officers)
Machinery: 2 Detroit 8V 92TA Model 91; 1,300 hp (970 kW); 2 shafts
Guns: 2 Oerlikon GAM/204 GK 20 mm. 2 — 7.62 mm MGs.
Radars: Surface search: Furuno; I-band.

Comment: Rearmed with 20 mm guns in 1989. These were replaced by GAM guns in 1990–91 when the ship received a new radar. Refitted again in 1996 with new engines. Based in the Atlantic.

SUBTENIENTE OSORIO SARAVIA   12/2004, Julio Montes / 1129556

1 DAUNTLESS CLASS (PB)

IXIMCHE

Displacement, tonnes: 11 full load
Dimensions, metres (feet): 12.19 × 3.86 × 0.69 (40 × 12.7 × 2.3)
Speed, knots: 28
Range, n miles: 400 at 22 kt
Complement: 5
Machinery: 2 Caterpillar 3208TA diesels; 850 hp (635 kW); 2 shafts
Guns: 1 — 7.62 mm MG.
Radars: Surface search: Raytheon R40X; I-band.

Comment: Built by SeaArk, Monticello, of aluminium construction. Donated by US government as foreign aid in 1997.

KUKULKÁN   12/2004, Julio Montes / 1129555

2 SEWART CLASS (COASTAL PATROL CRAFT) (PB)

Name	No	Builders	Commissioned
UTATLAN	GC 851 (ex-P 851)	Sewart Seacraft Inc, Berwick	May 1967
SUBTENIENTE OSORIO SARAVIA	GC 852 (ex-P 852)	Sewart Seacraft Inc, Berwick	Nov 1972

Displacement, tonnes: 55 full load
Dimensions, metres (feet): 25.9 × 5.7 × 2.2 (85 × 18.7 × 7.2)
Speed, knots: 22
Range, n miles: 400 at 12 kt
Complement: 17 (4 officers)
Machinery: 2 Detroit 8V 92TA Model 91; 1,300 hp (970 kW); 2 shafts
Guns: 1 Oerlikon GAM/204 GK 20 mm. 2 — 7.62 mm MGs.
Radars: Surface search: Furuno; I-band.

Comment: Aluminium superstructure. Both rearmed with 20 mm guns, and 75 mm recoilless removed in 1990. P 851 is based in the Atlantic; P 852 in the Pacific. Refitted in 1995–96 with new engines.

DAUNTLESS CLASS (Cayman Islands colours)   6/2001, RCIS / 0121305

Patrol forces < **Guatemala** — Introduction < **Guinea** 317

16 RIVER PATROL CRAFT (PBR)

Group A	Group B	Group C
DENEB	LAGO DE ATITLAN	CHOCHAB
SIRIUS	MAZATENANGO	ALIOTH
PROCYON	RETALHULEU	MIRFA
VEGA	ESCUINTLA	SCHEDAR
POLLUX		COMAMEFA
SPICA		
STELLA MARIS		

Comment: Group A are wooden hull craft with a speed of 19 kt. Group B have aluminium hulls and a speed of 28 kt. Group C are probably of Israeli design. All can be armed with 7.62 mm MGs and are used by Marine battalions as well as the Navy.

CHOCHAB AND COMAMEFA　　　　　　　2/1996, Julio Montes / 0064686

6 CUTLASS CLASS
(5 COASTAL PATROL CRAFT AND 1 SURVEY CRAFT) (PB)

Name	No	Builders	Commissioned
TECUN UMAN	GC 651 (ex-P 651)	Halter Marine	26 Nov 1971
KAIBIL BALAN	GC 652 (ex-P 652)	Halter Marine	8 Feb 1972
AZUMANCHE	GC 653 (ex-P 653)	Halter Marine	8 Feb 1972
TZACOL	GC 654 (ex-P 654)	Halter Marine	10 Mar 1976
BITOL	GC 655 (ex-P 655)	Halter Marine	4 Aug 1976
GUCUMAZ	BH 656 (ex-GC 656)	Halter Marine	15 May 1981

Displacement, tonnes: 46 full load
Dimensions, metres (feet): 19.7 × 5.2 × 0.9 *(64.6 × 17.1 × 3.0)*
Speed, knots: 25
Range, n miles: 400 at 15 kt
Complement: 10 (2 officers)
Machinery: 2 Detroit 8V 92TA Model 91 diesels; 1,300 hp *(970 kW)*; 2 shafts
Guns: 2 Oerlikon GAM/204 GK 20 mm. 2 or 3 — 12.7 mm MGs.
Radars: Surface search: Furuno; I-band.

Comment: First five rearmed with 20 mm guns in 1991. P 651, 654 and 655 are in the Atlantic, remainder in the Pacific. Aluminium hulls. *Gucumaz* was used as a survey craft but by 1996 was again serving as a patrol craft with three MGs. Reverted to survey craft in 2004. 654 and 656 refitted in 1994–95, remainder in 1995–97. New engines fitted.

GUCUMAZ　　　　　　　　　　　　　　12/2004, Julio Montes / 1129558

AZUMANCHE　　　　　　　　　　　　　12/2004, Julio Montes / 1129557

6 VIGILANTE CLASS (PBI)

GC 271–276

Displacement, tonnes: 4 full load
Dimensions, metres (feet): 8.1 × 3 × 0.5 *(26.6 × 9.8 × 1.6)*
Speed, knots: 40 (est.)
Complement: 4
Machinery: 2 Evinrude outboards; 600 hp *(448 kW)*
Guns: 1 — 12.7 mm MG.
Radars: Surface search: Furuno; I-band.

Comment: Ordered in 1993 from Boston Whaler. Delivered in 1994 and divided three to each coast.

GC 275　　　　　　　　　　　　　　　　12/1999 / 0104574

8 JUSTICE INTERCEPTOR CRAFT (PBF)

FEN 03–05　　FEN 08　　+4

Displacement, tonnes: 2 full load
Dimensions, metres (feet): 9.8 × 3.1 × 0.55 *(32.2 × 10.2 × 1.8)*
Speed, knots: 55
Complement: 3
Machinery: 2 Mercury outboards; 600 hp *(477 kW)*
Guns: 1 — 7.62 mm MG.

Comment: Boston Whaler 32 Justice craft donated by the US 2009–12. Glass-fibre laminate construction. Operated by the Fuerza Especial Naval. Five based in the Pacific and three in the Caribbean.

JUSTICE CRAFT　　　　　　　　　　　　6/2012*, US Navy / 1529175

Guinea

Country Overview

A former French colony, The Republic of Guinea became independent in 1958. Located in west Africa, the country has an area of 94,926 square miles, a 173 n mile coastline with the Atlantic Ocean and includes the Iles de Los. It is bordered by the north by Guinea-Bissau and Senegal and to the south by Liberia and Sierra Leone. The capital, largest city and principal port is Conakry. Following the seizure of power by a military faction in 2008, a new constitution is to be introduced and presidential elections were held in November 2010. Territorial seas (12 n miles) are claimed. A 200 n mile Exclusive Economic Zone (EEZ) has been claimed but the limits have not been formally agreed. Fishery Protection may be provided by civilian contractors.

Headquarters Appointments

Chief of Naval Staff:
　Lieutenant Colonel Lamine Toure

Personnel

(a) 2014: 400 officers and men
(b) 2 years' conscript service

Bases

Conakry, Kakanda

Notes: (1) A number of craft, including two Zhuk, two Bogomol, two Stinger and one Swiftship *(Vigilante* P 300) are laid up alongside. Some of these might be resurrected to combat piracy problems in the region. A Damen 13 m patrol boat, *Matakang*, is reported to have been delivered in 1999 and there are two MonArk 8 m Stinger craft, P 30 and P 35, which were delivered in 1985. There is a 7 m skiff *Kassa* P 17.
(2) Development of the port of Conakry is under consideration.

© 2014 IHS　　　　　　　　　　　　　　　　　　　　　　　　IHS Jane's Fighting Ships 2014-2015

320 Honduras > Patrol forces

1 GUARDIAN CLASS (COASTAL PATROL CRAFT) (PB)

TEGUCIGALPA FNH 1071 (ex-FNH 107)

Displacement, tonnes: 96 full load
Dimensions, metres (feet): 32.3 × 6.3 × 2.1 *(106 × 20.7 × 6.9)*
Speed, knots: 30. **Range, n miles:** 1,500 at 18 kt
Complement: 17 (3 officers)

Machinery: 3 Detroit 16V-92TA diesels; 2,070 hp *(1.54 MW)* sustained; 3 shafts
Guns: 2 — 12.7 mm MGs.
Radars: Surface search: Furuno; I-band.
Weapon control systems: Kollmorgen 350 optronic director.

Comment: Delivered by Lantana Boatyard, Florida August 1986. Second of class, *Copan*, no longer in service. A third of the class, completed in May 1984, became the Jamaican *Paul Bogle*. Aluminium hulls.

TEGUCIGALPA   6/2011, Honduras Navy / 1406605

4 INTERCEPTOR CRAFT (PBF)

EF 15–16   EF 21–22

Displacement, tonnes: 4 full load
Dimensions, metres (feet): 13.4 × 2.75 × 0.9 *(44 × 9.0 × 3.0)*
Speed, knots: 60. **Range, n miles:** 170 at 40 kt
Complement: 6
Machinery: 3 Yanmar diesels; 945 hp *(704 kW)*; Bravo X drives

Comment: Manufactured by Nor-Tech, Fort Myers, Florida. V-bottomed hull of composite and glass-fibre construction. Donated by the US Southern Command in 2007.

INTERCEPTOR CRAFT   6/2011, Honduras Navy / 1406609

5 SWIFT 65 ft CLASS (COASTAL PATROL CRAFT) (PB)

NACAOME (ex-*Aguan*; ex-*Gral*) FNH 6501 (ex-FNH 651)
GOASCORAN (ex-*General J T Cabanas*) FNH 6502 (ex-FNH 652)
PATUCA FNH 6503 (ex-FNH 653)
ULUA FNH 6504 (ex-FNH 654)
CHOLUTECA FNH 6505 (ex-FNH 655)

Displacement, tonnes: 34 full load
Dimensions, metres (feet): 21.3 × 5.2 × 1.6 *(69.9 × 17.1 × 5.2)*
Speed, knots: 25 (FNH 6501–6502), 36 (FNH 6503–6505)
Range, n miles: 2,000 at 22 kt (FNH 6501–6502)
Complement: 9 (2 officers)
Machinery: 2 GM 12V-71TA diesels; 840 hp *(627 kW)* sustained; 2 shafts (FNH 6501–6502)
2 MTU 8V 396 TB93 diesels; 2,180 hp(m) *(1.6 MW)* sustained; 2 shafts (FNH 6503–6505)
Guns: 2 — 12.7 mm MGs. 3 — 7.62 mm MGs.
Radars: Surface search: Racal Decca; I-band.

Comment: First pair (Peterson Mk III) built by Swiftships, Morgan City originally for Haiti. Contract cancelled and Honduras bought the two that had been completed in 1973-74. Delivered in 1977. Three more ordered in 1979 and delivered 1980.

CHOLUTECA   6/2011, Honduras Navy / 1406603

1 SWIFT 85 ft CLASS (COASTAL PATROL CRAFT) (PB)

CHAMELECON (ex-*Rio Kuringwas*) FNH 8501

Displacement, tonnes: 61 full load
Dimensions, metres (feet): 25.9 × 6.1 × 1.8 *(85 × 20.0 × 5.9)*
Speed, knots: 25
Complement: 10 (2 officers)
Machinery: 2 Detroit diesels; 2 shafts
Radars: Surface search: Racal/Decca; I-band.

Comment: Built by Swiftships, Morgan City in about 1967 for Nicaragua from where it was transferred in 1979.

IHS Jane's Fighting Ships 2014-2015

CHAMELECON   6/2011, Honduras Navy / 1406604

5 OUTRAGE 25 ft CLASS (RIVER PATROL CRAFT) (PBR)

LP 01–05

Displacement, tonnes: 2 full load
Dimensions, metres (feet): 7.6 × 2.4 × 0.4 *(24.9 × 7.9 × 1.3)*
Speed, knots: 30. **Range, n miles:** 200 at 30 kt
Complement: 4
Machinery: 2 Evinrude outboards; 300 hp *(224 kW)*
Guns: 1 — 12.7 mm MG. 2 — 7.62 mm MGs.
Radars: Navigation: Furuno 3600; I-band.

Comment: Built by Boston Whaler in 1982. Seven deleted so far. Radar is sometimes embarked.

OUTRAGE 25 ft CLASS   6/2011, Honduras Navy / 1406611

14 EDUARDONO CLASS (PATROL CRAFT) (PB)

FNH 3205–3206   FNH 3208   +11

Displacement, tonnes: 2 full load
Dimensions, metres (feet): 32 × 7.87 × 1.31 *(105 × 25.8 × 4.3)*
Speed, knots: 30
Range, n miles: 200 at 30 kt
Complement: 4
Machinery: 2 outboard motors; 200 hp *(149 kW)*
Guns: 1 — 12.7 mm MGs. 2 — 7.62 mm MGs.

Comment: Donated by the US government from 2011.

EDUARDONO CLASS   6/2011, Honduras Navy / 1406608

8 PIRANHA CLASS (PATROL CRAFT) (PB)

No	Builders
LR 3601–3608	Lantana Boatyard, Florida

Displacement, tonnes: 8 full load
Dimensions, metres (feet): 11 × 3.05 × 0.53 *(36.1 × 10.0 × 1.7)*
Speed, knots: 30. **Range, n miles:** 200 at 30 kt
Complement: 5
Machinery: 2 Caterpillar 3208 diesels; 630 hp *(470 kW)*
Guns: 2 — 12.7 mm MGs. 2 — 7.62 mm MGs.
Radars: Navigation: Furuno 3600; I-band.

Comment: Built by Lantana Boatyard, Florida, and entered service in 1986. Aluminium construction with Kevlar armour.

PIRANHA 36 ft CLASS   2/2005, Honduras Navy / 1406607

© 2014 IHS

Patrol forces < **Honduras** — Introduction < **Hong Kong** 321

4 TYPE 44 CLASS (WPB)

| GUANAJA (ex-*44344*) FNH 4404 | UTILA (ex-*44351*) FNH 4402 |
| ROATAN (ex-*44390*) FNH 4403 | CISNE (ex-*44365*) FNH 4401 |

Displacement, tonnes: 18 full load
Dimensions, metres (feet): 13.5 × 3.9 × 1.1 *(44.3 × 12.8 × 3.6)*
Speed, knots: 14
Range, n miles: 215 at 10 kt
Complement: 3
Machinery: 2 Detroit 6V-38 diesels; 185 hp *(136 kW)*; 2 shafts

Comment: Acquired from the US in 1999.

TYPE 44 CLASS 9/2004, Honduras Navy / 1406612

1 INSHORE PATROL CRAFT (PBI)

RIO COCO FNH 6506

Displacement, tonnes: 32 standard; 41.9 full load
Dimensions, metres (feet): 19.8 × 5.5 × 1.8 *(65 × 18.0 × 5.9)*
Speed, knots: 28
Range, n miles: 450 at 26 kt
Complement: 10 (1 officer)
Machinery: 3 Detroit 8V 71 diesel; 690 hp *(515 kW)* sustained; 3 shafts
Guns: 1 — 14.5 mm MG.

Comment: Ex-US Navy PB Mk III Series built by Peterson Builders, Wisconsin in 1975–76. Aluminium construction. The design includes a pilot house offset to starboard to provide space to port for the installation of additional weapons. Acquired by the Honduran Navy in 2009.

RIO COCO 6/2011, Honduras Navy / 1406613

2 DAMEN STAN PATROL 4207 (PB)

LEMPIRA FNH 1401   MORAZAN FNH 1402

Displacement, tonnes: 208 standard
Dimensions, metres (feet): 42.8 × 7.11 × 2.52 *(140.4 × 23.3 × 8.3)*
Speed, knots: 26
Machinery: 2 Caterpillar 3516B DI-TA; 5,600 hp *(4.17 MW)*; 2 cp props
Guns: 1 — 30 mm. 1 — 12.7 mm MG.

Comment: Contract signed with Damen Shipyards, Gorinchem in October 2012 for the lease of two Stan Patrol 4207 offshore patrol vessels. The lease period is reported to be 13 years, with the option to extend to 20 years, and there is an option to buy. Both craft were delivered in 2013.

LEMPIRA 7/2013*, A A de Kruijf / 1516979

0 + 6 DAMEN DI-1102 INTERCEPTOR CRAFT (PBF)

Dimensions, metres (feet): 10.97 × 2.49 × 0.71 *(36 × 8.2 × 2.3)*
Speed, knots: 55
Range, n miles: 230 at 40 kt
Complement: 2 + 4 personnel
Machinery: 2 Volvo Penta D6 diesels; 740 hp *(552 kW)*; Volvo DP-R stern drives
Radars: Navigation: Raymarine; I-band.

Comment: The contract for the acquisition of six interception craft was announced in October 2012.

AUXILIARIES

Notes: In addition there are two ex-US LCM 8 (*Warunta* FNH 7302, *Tansin* FNH 7303) transferred in 1987. Both are used as transport vessels.

LCM 8 2000, Honduran Navy / 0105812

1 LANDING CRAFT (LCU)

PUNTA CAXINAS FNH 1491

Displacement, tonnes: 635 full load
Dimensions, metres (feet): 45.4 × 10 × 2 *(149 × 32.8 × 6.6)*
Speed, knots: 14
Range, n miles: 3,500 at 12 kt
Complement: 18 (3 officers)
Cargo capacity: 100 tons equipment or 50,000 gallons dieso plus 4 standard containers
Machinery: 3 Caterpillar 3412 diesels; 1,821 hp *(1.4 MW)* sustained; 3 shafts
Radars: Navigation: Furuno 3600; I-band.

Comment: Ordered in 1986 from Lantana, Florida, and commissioned in May 1988. The ship was modified in 2012 to act as a mother-ship for patrol craft.

PUNTA CAXINAS 12/2004, Julio Montes / 1129560

Hong Kong
POLICE MARINE REGION

Country Overview

Formerly a British colony, the Hong Kong Special Administrative Region of China reverted to Chinese sovereignty on 1 July 1997. While China has assumed responsibility for foreign affairs and defence, the territory is to maintain its own legal, social, and economic systems until at least 2047. Hong Kong comprises three main regions, Hong Kong Island (29 sq miles), Kowloon Peninsula and Stonecutters Island (6 sq miles) and the New Territories (380 sq miles). As with the remainder of China, territorial seas (12 n miles) are claimed. An EEZ (200 n mile) is also claimed but the limits have not been defined by boundary agreements. The role of the Marine Police is to maintain the integrity of the sea boundary and territorial waters of Hong Kong, enforce the laws of Hong Kong in territorial waters, prevent illegal immigration by sea, SAR in territorial and adjacent waters, and casualty evacuation.

Headquarters Appointments

Regional Commander (Marine):
   Ms Chiu Wai-yin

Organisation

Marine Police Regional HQ, Sai Wan Ho
Bases at Ma Liu Shui, Tui Min Hoi, Tai Lam Chung, Aberdeen, Sai Wan Ho

Personnel

(a) 2014: 2,600
(b) Voluntary service

© 2014 IHS   IHS Jane's Fighting Ships 2014-2015

POLICE

4 BARGE OPERATING PLATFORMS (YAG)

PB 1–4

Displacement, tonnes: 248 standard
Dimensions, metres (feet): 30 × 13 × 0.7 (98.4 × 42.7 × 2.3)
Complement: 4
Machinery: 2 Cummins 108 MXDG diesels (PB 3-4)
Radars: Surface search: Sperry Bridgemaster; I-band.

Comment: Steel-hulled barges. PB 1–2 (built by Guangzhou Waterway Bureau Shipyards) delivered in June 2002 and PB 3–4 (built by Leung Wan Kee Shipyards) in October 2007. PB 1–2 are moored in Deep Bay, PB 3 in Rocky Harbour and PB 4 at entrance to Tolo Channel.

PB 1   6/2004, Hong Kong Police / 0589752

1 TRAINING VESSEL (WAX)

PL 3

Displacement, tonnes: 427 full load
Dimensions, metres (feet): 40 × 8.6 × 3.2 (131.2 × 28.2 × 10.5)
Speed, knots: 14
Range, n miles: 1,500 at 14 kt
Complement: 7
Machinery: 2 Caterpillar 3512TA diesels; 2,350 hp (1.75 MW) sustained; 2 shafts
Radars: Surface search: 2 Racal Decca ARPA C342/8; I-band.

Comment: Built by Hong Kong SY in 27 July 1987 and commissioned 1 February 1988. Steel hull. Racal ARPA and GPS Electronic Chart system. 12.7 mm MGs removed in mid-1996. Can carry up to 30 armed police for short periods. Former command vessel converted to a training role.

PL 3   6/2004, Hong Kong Police / 0589753

6 KEKA CLASS (PATROL CRAFT) (WPB)

PL 60–65

Displacement, tonnes: 107 standard
Dimensions, metres (feet): 30 × 6.3 × 2.2 (98.4 × 20.7 × 7.2)
Speed, knots: 25
Range, n miles: 360 at 15 kt
Complement: 14
Machinery: 2 MTU 12V 396 TE84 diesels; 4,020 hp (3.0 MW) sustained; 1 MTU 8V 2000 M60 diesel; 536 hp (400 kW) sustained; 1 Hamilton waterjet
Radars: Surface search: Decca; I-band.

Comment: Aluminium-hulled craft built by Cheoy Lee Shipyards Ltd to replace Damen Mk 1-class patrol craft. Delivered in 2002, 2004 and 2005.

PL 63   11/2013*, Chris Sattler / 1529177

6 PROTECTOR (ASI 315) CLASS
(COMMAND/PATROL CRAFT) (WPB)

PL 51–56

Displacement, tonnes: 152 full load
Dimensions, metres (feet): 32.6 × 8.2 × 2.5 (107 × 26.9 × 8.2)
Speed, knots: 24
Range, n miles: 600 at 18 kt
Complement: 19
Machinery: 2 Caterpillar 3516TA diesels; 5,600 hp (4.17 MW) sustained; 2 shafts; 1 Caterpillar 3412TA; 764 hp (570 kW) sustained; Hamilton HM 521 waterjet (centreline)
Radars: Surface search: Racal Decca; I-band.
Weapon control systems: GEC V3901 optronic director.

Comment: Built by Transfield Australian Shipbuilding Industries and completed in 1993. As well as patrol work, the craft provide command platforms for Divisional commanders. 12.7 mm guns removed in 1996 and the optronic director is used for surveillance only.

PL 54   12/2007, Chris Sattler / 1170156

5 DAMEN COUGARTEK CLASS (INTERCEPTOR CRAFT) (HSIC)

PL 85–89

Displacement, tonnes: 9 full load
Dimensions, metres (feet): 14.8 × 2.7 × 1.2 (48.6 × 8.9 × 3.9)
Speed, knots: 60
Complement: 5
Machinery: 3 Innovation Marine Sledge Hammers; 1,590 hp(m) (1.2 MW); 3 shafts
Radars: Surface search: Raytheon; I-band.

Comment: Built by Damen, Gorinchem in 1999. Used by the Small Boat Division.

PL 85   10/2013*, Chris Sattler / 1529176

4 SEASPRAY CLASS (LOGISTIC CRAFT) (YFB)

PL 46–49

Displacement, tonnes: 11 full load
Dimensions, metres (feet): 11.4 × 4.2 × 1.2 (37.4 × 13.8 × 3.9)
Speed, knots: 25
Complement: 4
Machinery: 2 Caterpillar 3208TA diesels; 750 hp (560 kW) sustained; 2 shafts (PL 47-49)
Radars: Navigation: Koden; I-band.

Comment: Built by Seaspray Boats, Fremantle in 1992. Catamaran hulls capable of carrying 16 police officers.

PL 47   12/2007, Chris Sattler / 1170154

Police < Hong Kong 323

6 SEASPRAY CLASS (INSHORE PATROL CRAFT) (WPB)

| PL 22 | PL 24 | PL 26–27 | PL 29 | PL 31 |

Displacement, tonnes: 9 full load
Dimensions, metres (feet): 9.9 × 4.2 × 1.2 (32.5 × 13.8 × 3.9)
Speed, knots: 30
Complement: 4
Machinery: 2 Caterpillar C7 diesels (PL 26, 29); 910 hp (680 kW); 2 Caterpillar 3208 TAH diesels (PL 22, 24, 27, 31) 750 hp (560 kW); 2 shafts
Radars: Surface search: Koden; I-band.

Comment: Built by Seaspray Boats, Fremantle in 1992–93.

PL 28   12/2007, Chris Sattler / 1170152

7 INSHORE PATROL CRAFT (WPB)

PL 90–96

Displacement, tonnes: 2 standard
Dimensions, metres (feet): 7.5 (PL 90–92), 8.1 (PL 93–96) × 2.5 (PL 90–92), 3.1 (PL 93–96) × 0.9 (PL 90–92), 1 (PL 93–96) (24.6, 26.6 × 8.2, 10.2 × 3.0, 3.3)
Speed, knots: 33 (PL 90–92), 42 (PL 93–96)
Complement: 3
Machinery: 2 outboards; 230 hp (170 kW) (PL 90–92); 2 outboards; 500 hp (370 kW) (PL 93–96)
Radars: Surface search: Koden; I-band.

Comment: PL 90–92 are Boston Whaler Guardians and PL 93–96 are Boston Whaler Vigilants. All delivered in 1997.

PL 93   12/2007, Chris Sattler / 1335252

6 CHEOY LEE CLASS (INSHORE PATROL CRAFT) (WPB)

PL 40–45

Displacement, tonnes: 20 standard
Dimensions, metres (feet): 13.07 × 4.0 × 0.8 (42.9 × 13.1 × 2.6)
Speed, knots: 35
Complement: 4
Machinery: 2 MAN D2842LE403 diesels; 1,420 hp (1.06 MW) sustained; 2 Hamilton waterjets
Radars: Surface search: Bridgemaster E 180; I-band.

Comment: Based upon a design from Peterson Shipbuilders, these shallow draft vessels were constructed by Cheoy Lee Shipyards Ltd and delivered in 2000.

PL 42   12/2007, Chris Sattler / 1335251

© 2014 IHS

8 HIGH SPEED INTERCEPTORS (HSIC)

PV 30–37

Displacement, tonnes: 3 full load
Dimensions, metres (feet): 8.5 × 2.6 × 0.7 (27.9 × 8.5 × 2.3)
Speed, knots: 49
Complement: 3
Machinery: 2 Mercury outboards; 500 hp (373 kW)

Comment: Built by Queensland Ships in 1997. Used by the Small Boat Division.

PV 35   10/2013*, Chris Sattler / 1529181

17 MEDIUM PATROL LAUNCH (PATROL CRAFT) (PBF)

PL 5–21

Displacement, tonnes: 40 standard
Dimensions, metres (feet): 19 × 5.0 × 1.0 (62.3 × 16.4 × 3.3)
Speed, knots: 45
Complement: 8
Machinery: 3 Caterpillar C32 diesels; 3,300 hp (2.46 MW); 3 Hamilton waterjets

Comment: Contract for the construction of 17 patrol craft signed with Lung-Teh Shipyard, Taiwan on 19 September 2006. Aluminium construction. The first six entered service in November 2007 and the remainder by 2010.

PL 15   10/2013*, Chris Sattler / 1529180

12 DIVISIONAL FAST PATROL CRAFT (PBF)

PV 5–16

Displacement, tonnes: 5 standard
Dimensions, metres (feet): 9.88 × 2.55 × 0.6 (32.4 × 8.4 × 2.0)
Speed, knots: 50
Complement: 3
Machinery: 2 Mercury Verado outboards; 550 hp (410 kW)

Comment: Constructed by Brisbane Ship Constructions Ltd in 2007.

PV 9   10/2013*, Chris Sattler / 1529179

3 FB 55SC CLASS (INSHORE PATROL CRAFT) (PBF)

Displacement, tonnes: 10 standard
Dimensions, metres (feet): 16.43 × 2.85 × 0.84 (53.9 × 9.4 × 2.8)
Speed, knots: 65 (est.)
Complement: 8
Machinery: 3 Seatek diesels; 2,250 hp (1.7 MW); surface-piercing propeller

Comment: Designed by FB design of Italy. Delivered in 2003. Kevlar monohull fast inshore patrol craft for maritime law enforcement tasks. Eight seats in forward compartment.

FB 55SC   10/2013*, Chris Sattler / 1529178

IHS Jane's Fighting Ships 2014-2015

1 FB RIB 42SC CLASS (INSHORE PATROL CRAFT) (PB)

Displacement, tonnes: 7 standard (est.)
Dimensions, metres (feet): 13.2 × 3.55 × 0.7 *(43.3 × 11.6 × 2.3)*
Speed, knots: 63
Complement: 3
Machinery: 2 Caterpillar diesels; 1,400 hp *(1.04 MW)*; surface piercing propeller

Comment: Designed by FB design, Italy. Rigid inflatable Kevlar monohull for maritime law enforcement tasks. Delivered 2004.

RIB 42SC   *10/2013*, Chris Sattler* / 1529184

CUSTOMS

Headquarters Appointments

Senior Superintendent Ports and Marine Command: Wong, Sui-hang, Vincent

Notes: The Hong Kong Customs Fleet is mainly based at the marine base Stonecutters Island; with support of a subsidiary operational base at Tuen Mun. The Customs fleet comprises five Sector Command launches of which there are three 30 m Special Service Crafts (*Sea Glory II* (CE 6), *Sea Guardian II* (CE 5) and *Sea Leader II* (CE 2)) with gross tonnages of 171 tonnes and maximum speed of 26 kt. These were built by Lung Teh Shipbuilding Company, Taiwan and entered service in 2009. There are also two 32 m Challenger launches (*Sea Reliance* (CE 8) and *Sea Fidelity* (CE 9)) with gross tonnages of 180 tonnes and a maximum speed of 25 kt. These were built by Wang Tak Eng and SB Co. Ltd., Hong Kong, and entered service in 2000. All Sector Command launches are equipped with a 'sea-rider' and fitted with night vision aids as well as narcotics and explosives scanning devices. There are also three 17 m FB design high-speed pursuit craft (CE 15, 17 and 18), capable of 49 kt, and two Boston Whaler 10 m shallow-water launches (CE 12–13) capable of 39 kt.

SEA FIDELITY   *11/2013*, Chris Sattler* / 1529183

CE 12   *6/2010, Ports and Maritime Command* / 1366631

CE 15   *10/2013*, Chris Sattler* / 1529182

LAND-BASED MARITIME AIRCRAFT

Notes: (1) All aircraft belong to the Government Flying Service based at Hong Kong International Airport. In addition to the aircraft listed, there is a ZLIN Z242L fixed-wing training aircraft.
(2) Two Bombardier Challenger 605s were ordered in October 2011. They are likely to replace the Jetstreams.

Numbers/Type: 2 BAE Jetstream J 41.
Operational speed: 260 kt *(482 km/h)*.
Service ceiling: 26,000 ft *(7,925 m)*.
Range: 774 n miles *(1,433 km)*.
Role/Weapon systems: SAR (command and control), airborne surveillance, survey and photography. Sensors: Radar, FLIR, survey camera, VHF/UHF/DF.

JETSTREAM   *6/2011, Government Flying Service* / 1406451

Numbers/Type: 3 Eurocopter AS 332 L2 Super Puma.
Operational speed: 130 kt *(240 km/h)*.
Service ceiling: 15,090 ft *(4,600 m)*.
Range: 672 n miles *(1,245 km)*.
Role/Weapon systems: SAR/coastal surveillance, Medevac and transport. Sensors: radar, Spectrolab searchlight. Weapons: Unarmed. Medical equipment and up to six stretchers. Ordered on 17 September 1999. The aircraft entered service in April 2002.

AS 332   *6/2011, Government Flying Service* / 1406450

Numbers/Type: 4 Eurocopter EC 155B1.
Operational speed: 140 kt *(260 km/h)*.
Service ceiling: 16,760 ft *(5,110 m)*.
Range: 432 n miles *(800 km)*.
Role/Weapon systems: SAR, Medevac, VIP transport; enlarged variant of 'Dauphin'. Sensors: Radar, FLIR, searchlight, siren, loudspeaker. Weapons: Unarmed. Two stretchers. Ordered on 17 September 1999; aircraft delivered in late 2002.

EC 155   *6/2005, Government Flying Service* / 1127923

Hungary

Country Overview

A landlocked central European country, the Republic of Hungary has an area of 35,919 square miles and is bordered by Slovakia, Ukraine, Romania, Serbia, Croatia, Slovenia, and Austria. Budapest is the country's capital and largest city. The country is divided into two general regions by the principal river, the Danube, which flows for 145 n miles north-south through the centre of the country and serves as a major artery of the transport system.

Diplomatic Representation

Defence Attaché in London:
Colonel Jozsef Gulyas

Personnel

(a) 2014: 100
(b) National service replaced by a professional army on 3 November 2004.

Bases

Budapest.

MINE WARFARE FORCES

3 NESTIN CLASS (RIVER MINESWEEPERS) (MSR)

| ÓBUDA AM 22 | DUNAÚJVÁROS AM 31 | DUNAFOLDVAR AM 32 |

Displacement, tonnes: 73 full load
Dimensions, metres (feet): 26.5 × 6.5 × 1.2 *(86.9 × 21.3 × 3.9)*
Speed, knots: 15. **Range, n miles:** 810 at 11 kt
Complement: 17 (1 officer)

Machinery: 2 Torpedo 12-cyl diesels; 520 hp(m) *(382 kW)*; 2 shafts
Guns: 6 Hispano 20 mm (1 quad M75 fwd, 2 single M70 aft).
Mines: 24 ground mines.
Radars: Navigation: Furuno FR 8062; I-band.

Comment: Built by Brodotehnika, Belgrade in 1980–82. Full magnetic/acoustic and wire sweeping capabilities. Kram minesweeping system employs a towed sweep at 200 m. The ships form the first 'Honved' Ordnance Disposal and Warship Regiment.

DUNAFOLDVAR 8/2010, IHS/C Hollosi / 1406734

2 AN-2 CLASS (RIVER PATROL VESSELS) (PBR)

ERCSI 542-051   BAJA 542-054

Displacement, tonnes: 11 full load
Dimensions, metres (feet): 13.4 × 3.8 × 0.6 *(44 × 12.5 × 2.0)*
Speed, knots: 19
Complement: 7 (1 officer)
Machinery: 2 Volvo Penta diesels; 380 hp(m) *(283 kW)* sustained; 2 shafts

Comment: Last survivors of an original 45 units built at Duna Shipyard in 1953. Refitted 2005. Employed on river patrol, diving support and disaster relief duties.

AN-2 CLASS 9/2009, Hungary Maritime Wing / 1305474

Iceland
LANDHELGISGAESLAN

Country Overview

An island republic, the Republic of Iceland lies just south of the Arctic Circle in the North Atlantic Ocean about 162 n miles southeast of Greenland and 432 n miles northwest of Scotland. With an area of 39,769 square miles, the country has a 2,695 n mile coastline. Reykjavík is the capital, largest city and principal port. Territorial waters (12 n miles) are claimed. A 200 n mile Exclusive Economic Zone (EEZ) has also been claimed although the limits are not fully defined by boundary agreements. The Coast Guard Service deals with fishery protection, salvage, rescue, security, pollution control, hydrographic research, lighthouse duties and bomb disposal.

Headquarters Appointments

Director General of Coast Guard:
  Rear Admiral Georg K Lárusson

Personnel

2014:   250 officers and men

Colours

Since 1990 vessels have been marked with red, white and blue diagonal stripes on the ships' side and the Coast Guard name (Landhelgisgaeslan).

Bases

Reykjavík

Research Ships

A number of government Research Ships bearing RE pennant numbers operate off Iceland.

Maritime Patrol Aircraft

Maritime aircraft include a Bombardier Dash-8 maritime surveillance aircraft plus three AS 332 Super Puma.

COAST GUARD

1 THOR (ULSTEIN UT 512L) CLASS (PSO)

Name	Builders	Launched	Commissioned
THOR	ASMAR, Talcahuano	29 Apr 2009	23 Sep 2011

Displacement, tonnes: 4,064 full load
Dimensions, metres (feet): 93.8 × 15.5 × 4.9 *(307.7 × 50.9 × 16.1)*
Speed, knots: 19.5
Complement: 48 (18 officers)
Machinery: 2 Bergen B 32:40L diesels; 10,730 hp *(8 MW)*; 2 Kamewa Ulstein cp props; two bow thrusters; 1 Kamewa Ulstein 736 kW tunnel thruster; 1 Ulstein Aquamaster swing-up 883 kW azimuth thruster

Comment: Contract awarded on 1 December 2006 for the construction of a replacement vessel for *Odinn*. The ship, designated UT 512L, is an enlarged design of the Norwegian Coast Guard ship *Harstad*. The Rolls Royce design is for a variety of coastguard and EEZ management roles including offshore standby and rescue, firefighting, salvage, pollution prevention, general law enforcement operations and fishery control. One engine of the vessel was damaged by seawater ingress when a tidal wave hit the ship after an earthquake in southern Chile in February 2010. Subsequently, the ship was repaired and arrived in Reykjavik on 27 October 2011.

THOR 10/2011, Iceland Coast Guard / 1406626

1 BALDUR CLASS (AGS/PB)

Name	Builders	Commissioned
BALDUR	Velsmidja Seydisfjardar	8 May 1991

Displacement, tonnes: 55 full load
Dimensions, metres (feet): 20.7 × 5.2 × 1.7 *(67.9 × 17.1 × 5.6)*
Speed, knots: 12
Complement: 5
Machinery: 2 Caterpillar 3406TA diesels; 640 hp *(480 kW)*; 2 shafts
Radars: Navigation: Furuno; I-band.

Comment: Built in an Icelandic Shipyard. Used for survey work and patrol duties.

BALDUR 6/2005, Iceland Coast Guard / 1153888

2 AEGIR CLASS (PSOH)

Name	Builders	Commissioned
AEGIR	Aalborg Vaerft, Aalborg	1968
TYR	Dannebrog Vaerft	15 Mar 1975

Displacement, tonnes: 1,146 *(Aegir)*, 1,233 *(Tyr)* standard; 1,524 full load
Dimensions, metres (feet): 69.8 *(Aegir)*, 71.1 *(Tyr)* × 10 × 4.6 *(229, 233.3 × 32.8 × 15.1)*
Speed, knots: 19 *(Aegir)*, 20 *(Tyr)*. **Range, n miles:** 9,000 at 18 kt
Complement: 36 (18 officers)
Machinery: 2 MAN/Burmeister & Wain 8L 40/54 diesels; 13,200 hp(m) *(9.68 MW)* sustained; 2 shafts; cp props
Guns: 1 Bofors 40 mm/60 Mk 3.
Radars: Surface search/navigation: Sperry; E/F/I-band.

Comment: Similar ships but *Tyr* has a slightly improved design and *Aegir* has no sonar. The hangar is between the funnels. In 1994 a large crane was fitted on the starboard side at the forward end of the flight deck. In 1997 the helicopter deck was extended and a radome fitted on the top of the tower. *Aegir* refitted in Poland in 2005 and *Tyr* in 2006. Work included extension and modernisation of the bridge, upgrade of accommodation and the installation of helicopter-in-flight refuelling equipment. *Tyr* was chartered in 2011–12 by the European Fisheries Control Agency to conduct fishery inspections.

TYR 9/2008, Adolfo Ortigueira Gil / 1335788

India

Country Overview

The Republic of India is a federal democracy which gained independence in 1947. It consists of the entire Indian peninsula and parts of the Asian mainland. With an area of 1,269,219 square miles, it is bordered to the north by Pakistan, Tibet, Nepal, China, and Bhutan and to the east by Burma and Bangladesh, which almost separates north-east India from the rest of the country. The status of Jammu and Kashmir is disputed with Pakistan. It has a 4,104 n mile coastline with the Arabian Sea, the Gulf of Mannar (which separates it from Sri Lanka) and the Bay of Bengal. The capital is New Delhi while the largest city is Mumbai. The principal ports include Mumbai, Calcutta, Madras and Vishakapatnam. Territorial waters (12 n miles) are claimed. A 200 n mile EEZ has been claimed although the limits have only been partly defined by boundary agreements.

Headquarters Appointments

Chief of Naval Staff:
  Admiral R K Dhowan, PVSM, AVSM, YSM
Vice Chief of Naval Staff: To be announced
Deputy Chief of Naval Staff:
  Vice Admiral Pradeep K Chatterjee, AVSM, NM
Chief of Personnel:
  Vice Admiral P Murugesan, AVSM, VSM
Chief of Material: Vice Admiral N N Kumar, AVSM, VSM

Senior Appointments

Flag Officer Commanding Western Naval Command:
  To be announced
Flag Officer Commanding Eastern Naval Command:
  Vice Admiral Anil Chopra, PVSM, AVSM
Flag Officer Commanding Southern Naval Command:
  Vice Admiral Satish Soni, PVSM, AVSM, NM
Commander-in-Chief, Andaman and Nicobar:
  Air Marshal P K Roy, AVSM, VM, VSM
Flag Officer Commanding Western Fleet:
  Rear Admiral A R Karve
Flag Officer Commanding Eastern Fleet:
  Rear Admiral Ajit Kumar
Flag Officer, Naval Aviation and Goa Area (at Goa):
  Rear Admiral Balvinder Singh Prahar
Flag Officer, Submarines (Vishakapatnam):
  Rear Admiral S V Bokhare
Flag Officer, Sea Training:
  Rear Admiral M S Pawar

Personnel

(a) 2014: 58,700 (8,700 officers) (including 5,000 Naval Air Arm and 2,000 Marines)
(b) Voluntary service
(c) The Marine Commando Force was formed in 1986.

Naval Air Arm

Squadron	Aircraft	Role
300 (Goa)	Sea Harrier FRS. Mk 51 LUSH	Fighter/Strike
303 (Goa)	MiG-29 KUB	Training
310 (Goa)	Dornier 228	MRMP/IW
311 (Vizag)	Dornier 228	MRMP, Transport
312 (Arakonnam)	Tu-142M 'Bear F', P8I	LRMP/ASW
315 (Goa)	Il-38 SD May	LRMP/ASW
318 (Port Blair)	Dornier 228, HAL	MRMP
321 (Mumbai)	HAL Chetak	Utility/SAR
330 (Mumbai)	Sea King Mk 42B	ASW
333 (ships) (Vizag)	Kamov Ka-28 'Helix'	ASW
336 (Kochi)	Sea King Mk 42A/42B/42C	Training
339 (Goa)	Ka-31	AEW
342 (Kochi)	Heron, Searcher II	UAV
343 (Porbandar)	Heron, Searcher II	UAV
344 (Uchipuli)	Heron, Searcher II	UAV
350 (Vizag)	UH 3H	Shipborne Flights
550 (Kochi)	Dornier 228	Training
Deepak		
551 (Goa)	Kiran Mk I/II	Training, FRU
552 (Goa)	Sea Harrier T Mk 60	Training
561 (Arakonnam)	HAL Chetak	Training

Air Stations

Name	Location	Role
INS *Shikra*	Mumbai	Helicopters
INS *Garuda*	Willingdon Island, Kochi	Helicopters
INS *Hansa* Naval Air Stations, LRMP, Strike/Fighter	Goa	HQ Flag Officer
INS *Utkrosh*	Port Blair, Andaman Isles	Maritime Patrol
Maritime Patrol		
Maritime Patrol		
INS *Dega* maritime patrol	Vishakapatnam	Fleet support and
INS *Rajali* Training	Arakonnam	LRMP, Helo
NAS *Ramnad* Naval Air Technical School	Rameshwaram	LRMP
INS *Parundu*	Tamil Nadu	UAV

Prefix to Ships' Names

INS

Colour Scheme

Surface ship colour scheme was changed from dark grey to light grey in 2004.

Bases and Establishments

New Delhi, Integrated HQ of Ministry of Defence (Navy).
Mumbai, C-in-C **Western Command**, barracks and main Dockyard; with one 'Carrier' dock. Submarine base (INS *Vajrabahu*). Supply school (INS *Hamla*). The region includes Mazagon and Goa shipyards.
Vishakapatnam, C-in-C **Eastern Command**, submarine base (INS *Virbahu*), submarine school (INS *Satyavahana*) and major dockyard built with Soviet support and being extended. Naval Air Station (INS *Dega*). Marine Gas Turbine maintenance facility (INS *Eksila*). New entry training (INS *Chilka*). At Thirunelveli is the submarine VLF W/T station completed in September 1986. The region includes Hindustan and Garden Reach shipyards.
Kochi, C-in-C **Southern Command**, Naval Air Station, and professional schools (INS *Venduruthy*) (all naval training comes under Southern Command). Ship repair yard. Gunnery Training establishment (INS *Dronacharya*).
There are also limited support facilities including a floating dock at Port Blair in the Andaman Islands.
Goa is HQ Flag Officer Naval Aviation.
Karwar (near Goa) is the site for a new naval base; Phase 1 (which can base 11 major warships and 10 smaller ships) was opened on 31 May 2005 and operations began on 15 December 2005. Phase 2A (2012–19) is to include expansion of the berthing facilities to accommodate 32 major warships (including aircraft carriers) and the construction of a naval air station with a 6,000 ft runway. Phase 2B will increase capacity to 50 major warships. There is a small base at Minicoy Island, one of the Lakshadweep archipelago. Plans to build a new base on the east coast, 50 km south of Vishakapatnam, were announced in 2006.
Shipbuilding: Mumbai (submarines, destroyers, frigates, corvettes); Calcutta (frigates, corvettes, LSTs, auxiliaries); Goa (patrol craft, LCU, MCMV facility planned). Vishakapatnam (corvettes, patrol craft).

Marine Commando Force (MCF)

The MCF was formed in 1987. Known as MARCOS, elements are based in the three regional commands. The force is trained in counter-terrorism operations.

Coast Defence

Truck-mounted SS-3-C Styx missiles. Several fixed sites.

Strength of the Fleet

Type	Active	Building (Projected)
Strategic Submarine (SSBN)	—	2 (1)
Attack Submarine (SSN)	1	—
Patrol Submarines	13	6 (6)
Aircraft Carriers	2	1 (1)
Destroyers	8	3 (4)
Frigates	15	3 (7)
Corvettes	24	4 (8)
Patrol Ships	24	2 (16)
LPD	1	(4)
LST	5	—
LSM/LCM/LCU	14	8
Minesweepers—Ocean	8	—
Minehunters	—	8
Research and Survey Ships	10	5
Training Ships	4	2
Diving Support/Rescue Ship	1	—
Fleet Support Ships	—	(5)
Replenishment Tankers	4	—
Transport Ships	3	—
Support Tankers	6	—
Water Carriers	2	—
Ocean Tugs	1	—

DELETIONS

Submarines

2013 *Sindhurakshak*

Frigates

2011 *Vindhyagiri*
2013 *Taragiri*

Mine Warfare Forces

2012 *Ratnagiri*

Amphibious Forces

2011 *Sharabh*, L 32, L 34

Training Ships

2012 *Krishna*

PENNANT LIST

Strategic Missile Submarines

S 73	Arihant (bldg)
—	Aridaman (bldg)

Attack Submarines

S 71	Chakra

Patrol Submarines

S 44	Shishumar
S 45	Shankush
S 46	Shalki
S 47	Shankul
S 55	Sindhughosh
S 56	Sindhudhvaj
S 57	Sindhuraj
S 58	Sindhuvir
S 59	Sindhuratna
S 60	Sindhukesari
S 61	Sindhukirti
S 62	Sindhuvijay
S 65	Sindhushastra

Aircraft Carriers

R 22	Viraat
R 33	Vikramaditya

Destroyers

D 51	Rajput
D 52	Rana
D 53	Ranjit
D 54	Ranvir
D 55	Ranvijay
D 60	Mysore
D 61	Delhi
D 62	Mumbai
D 63	Kolkata (bldg)
D 64	Kochi (bldg)
D 65	Chennai (bldg)

Frigates

F 20	Godavari
F 21	Gomati
F 22	Ganga
F 31	Brahmaputra
F 37	Beas
F 39	Betwa
F 40	Talwar
F 41	Taragiri
F 43	Trishul
F 44	Tabar
F 45	Teg
F 46	Tarkash
F 47	Shivalik
F 48	Satpura
F 49	Sahyadri
F 51	Trikand

Corvettes

P 33	Abhay
P 34	Ajay
P 35	Akshay
P 36	Agray
P 44	Kirpan
P 46	Kuthar
P 47	Khanjar
P 49	Khukri
P 61	Kora
P 62	Kirch
P 63	Kulish
P 64	Karmukh
P 77	Kamorta
—	Kadmatt (bldg)
—	Kiltan (bldg)
—	Kavaratti (bldg)
K 40	Veer
K 41	Nirbhik
K 42	Nipat
K 43	Nishank
K 44	Nirghat
K 45	Vibhuti
K 46	Vipul
K 47	Vinash
K 48	Vidyut
K 83	Nashak
K 91	Pralaya
K 92	Prabal

Patrol Forces

P 50	Sukanya
P 51	Subhadra
P 52	Suvarna
P 53	Savitri
P 54	Saryu (bldg)
P 55	Sharada
P 56	Sujata
P 57	Sunayna (bldg)
P 58	Sumedha (bldg)
P 59	Sumitra (bldg)
T 61	Trinkat
T 63	Tarasa
T 65	Bangaram
T 66	Bitra
T 67	Batti Malv
T 68	Baratang
T 69	Car Nicobar
T 70	Chetlat
T 71	Cora Divh
T 72	Cheriyam
T 73	Cankarso
T 74	Kondul
T 75	Kalpeni
T 76	Kabra
T 77	Koswari
T 78	Karuva

Mine Warfare Forces

M 65	Alleppey
M 67	Karwar
M 68	Cannanore
M 69	Cuddalore
M 70	Kakinada
M 71	Kozhikode
M 72	Konkan

Amphibious Forces

L 15	Kesari
L 16	Shardul
L 18	Cheetah
L 19	Mahish
L 20	Magar
L 21	Guldar
L 22	Kumbhir
L 23	Gharial
L 24	Airavat
L 33	—
L 35	—
L 36	—
L 37	—
L 38	—
L 39	—
L 41	Jalashwa

Auxiliaries and Survey Ships

A 15	Nireekshak
A 50	Deepak
A 53	Matanga
A 57	Shakti
A 58	Jyoti
A 59	Aditya
A 72	Torpedo Recovery Vessel
A 74	Sagardhwani
A 75	Tarangini
A 86	Tir
J 14	Nirupak
J 15	Investigator
J 16	Jamuna
J 17	Sutlej
J 18	Sandhayak
J 19	Nirdeshak
J 21	Darshak
J 22	Sarvekshak
—	Meen (bldg)
J 31	Makar (bldg)

SUBMARINES

Notes: (1) A request for proposals for a class of six submarines (Project 75I), to follow the Scorpene programme, was made on 7 September 2010. Contenders include further Scorpene class, Spanish S 80, German Type 214 and Russian Lada class. A global tender is expected to be issued although a timetable has not been announced.

(2) India operates up to 11 Cosmos CE2F/FX100 swimmer delivery vehicles, delivered in 1991.
(3) Plans to procure a submarine rescue bell (SRB) system were announced in October 2013. The broad requirement is for an SRB with space for two crew and 10 escapees, capable of working to depths of 300 m. In addition approval for plans to acquire two DSRVs was given on 23 December 2013.

(4) There are plans to acquire six 150-tonne midget submarines. The requirement is for boats capable of special forces' operations in depths up to 60 m. Hindustan Shipyard was selected in November 2012 to build the first two boats and is expected to be contracted to build the additional four vessels.

Strategic Missile Submarines

0 + 2 (1) ARIHANT CLASS (SSBN/SSGN)

Name	No	Builders	Laid down	Launched	Commissioned
ARIHANT	S 73	Vishakhapatnam Dockyard	2004	26 Jul 2009	2013
ARIDAMAN	–	Vishakhapatnam Dockyard	2009	2014	2015

Dimensions, metres (feet): 120 × 14.0 × 10.4 *(393.7 × 45.9 × 34.1)*
Speed, knots: 10 surfaced; 24 dived
Complement: 95 (23 officers)

Machinery: Nuclear; 1 PWR; 82.5 MW; 1 turbine; 20,000 hp(m) *(14.9 MW)*; 1 shaft; 7-bladed prop; 1 auxiliary retractable prop
Missiles: SLBM: 4 VLS launch tubes for 12 K 15 Sagarika (ballistic) missiles; inertial guidance to 750 km *(405 n miles)*; nuclear warhead.
SLCM/SSM: Novator Alfa Klub SS-N-27 (3M-54E-1) and SS-N-30 (3M 14) for anti-ship and land attack.
Torpedoes: 6—12 in *(533 mm)* tubes.
Electronic countermeasures: ESM: Radar warning.
Radars: Surface search: I-band.
Sonars: USHUS; active/passive sonar suite including cylindrical bow, flank and towed arrays.
Weapon control systems: To be announced.

Programmes: All details are speculative (dimensions are based on the Victor/Akula class with a 10 m 'plug' of four missile tubes). Some aspects of the design also appear to be similar to the Delta I class. The so-called Advanced Technology Vessel (ATV) project was initiated in the 1980s. While the first of class is likely, at least initially, to be used as a technology demonstrator, the operational use of the boat is believed to be in a strategic role. Weapon options include nuclear-tipped cruise or ballistic missiles in addition to torpedo-tube launched conventional anti-ship and land-attack missiles. Companies in support of the project are reported to include Larsen and Toubro at Hazira, Mazagon Dock Ltd and Bharat Electronics. Official acknowledgement at the launch ceremony of Russian co-operation suggests that it is likely that the submarine is based on a Russian design with the possible incorporation of some technology and/or features of the Project 885 Yasen-class SSGN. The boat is reported to have some 40 per cent indigenous content. The nuclear propulsion system was to have been an Indo-Russian PWR but there have also been reports of a Russian supplied VM-5 PWR. A second boat was reported under construction in mid-2011 and further submarines are expected.
Structure: Double-hulled. Seven W/T compartments. Four 'universal' launcher tubes in after casing are reported to be able to house either three Sagarika missiles or one larger missile once it has been developed. This may be the K-4 missile. Optronic periscope. Diving depth 450 m approximately.

ARIHANT 7/2009 / 1366461

Operational: The submarine is located at the so-called 'Site Bravo', a covered test area. It is reported that the reactor achieved 'critical' status on 10 August 2013 and sea trials are expected to start in 2014.

Attack Submarines

1 AKULA (SCHUKA-B) CLASS (PROJECT 971) (SSN)

Name	No	Builders	Laid down	Launched	Commissioned
CHAKRA (ex-Nerpa)	S 71	Komsomolsk Shipyard	1986	24 Jun 2006	4 Apr 2012

Displacement, tonnes: 7,620 surfaced; 9,246 dived
Dimensions, metres (feet): 110 oa; 103 wl × 14 × 10.4 *(360.9; 337.9 × 45.9 × 34.1)*
Speed, knots: 10 surfaced; 28 dived
Complement: 62 (31 officers)

Machinery: Nuclear; 1 VM-5 PWR; 190 MW; 2 GT3A turbines; 47,600 hp(m) *(35 MW)*; 2 emergency propulsion motors; 750 hp(m) *(552 kW)*; 1 shaft; 2 spinners; 1,006 hp(m) *(740 kW)*
Missiles: SLCM/SSM: Novator Alfa Klub SS-N-27 (3m-54E-1 anti-ship); active radar homing to 180 km *(97.2 n miles)* at 0.7 Mach (cruise) and 2.5 Mach (attack); warhead 450 kg.
Torpedoes: 4—21 in *(533 mm)* and 4—25.6 in *(650 mm)* tubes.
Electronic countermeasures: ESM: Rim Hat; intercept.
Radars: Surface search: Snoop Pair or Snoop Half with back to back aerials on ESM mast; I-band.
Sonars: Shark Gill (Skat MGK 503); hull-mounted; passive/active search and attack; low/medium frequency. Mouse Roar; hull-mounted; active attack; high frequency. Skat 3 towed array; passive; very low frequency.

Programmes: The construction of Nerpa (K 152) began at Komsomolsk in 1986 but, following the collapse of the Soviet Union in 1991, work was suspended. Negotiations for the 10-year lease of the boat by the Indian Navy started in about 1996 and terms were subsequently agreed in September 2001 when construction, likely to have been at least partly financed by India, was restarted. The boat was subsequently launched in 2006 and, following sea trials and certification by the Russian Navy, was handed over on 24 January 2012 and arrived at Vishakapatnam on 30 March 2012. The contract included a training package and three crews are reported to have been trained at Sosnovy Bar near St Petersburg. The weapons and sensors of the submarine in Indian service are speculative and have not been confirmed. Lease of a second Akula class is reported to be under consideration. This would probably involve completion of Irbis, said to have been 60% completed at Komsomolsk Shipyard when construction was abandoned in the 1990s.
Structure: The very long fin is particularly notable. Diving depth 450 m approximately.
Operational: Chakra bears the same name as the Charlie-class SSN leased from the Soviet Union 1988–91. Initially, the principal role of the submarine is to be training of both sea-going and shore-based personnel in nuclear submarine operations and support. The boat is likely to carry a number of Russian crew which may place some restrictions on the boat's operational use. As experience is gained, the submarine is likely to be deployed on a broader range of SSN operations. Based at Vishakapatnam. Reports in late 2012 suggested that operational availability is being constrained by a shortage of critical spare components.

CHAKRA 4/2012, Indian Navy / 1455657

Patrol Submarines

0 + 6 SCORPENE CLASS (SSK)

Name	No	Builders	Laid down	Launched	Commissioned
—	—	Mazagon Dock Ltd, Mumbai	2010	2014	2017

Displacement, tonnes: 1,732 dived
Dimensions, metres (feet): 66.4 × 6.2 × 5.8 *(217.8 × 20.3 × 19.0)*
Speed, knots: 11 surfaced; 20 dived
Range, n miles: 6,500 at 8 kt surfaced; 550 at 4 kt dived
Complement: 31 (6 officers)

Machinery: Diesel electric; 4 MTU 12V 396 SE84 diesels; 2,992 hp(m) *(2.2 MW)*; 1 Jeumont Schneider motor; 3,808 hp(m) *(2.8 MW)*; 1 shaft
Missiles: MBDA Exocet SM 39. Block 2 launched from 21 in *(533 mm)* tubes; inertial cruise, active terminal homing to 50 km *(27 n miles)* at 0.9 Mach; warhead 165 kg.
Torpedoes: 6 — 21 in *(533 mm)* tubes.
Electronic countermeasures: ESM.
Radars: Navigation: Sagem; I-band.
Sonars: Hull mounted; active/passive search and attack, medium frequency.
Weapon control systems: UDS International SUBTICS.

Programmes: Project 75. Following protracted negotiations which began in 2002, a contract for the licensed production of six submarines at Mazagon Dock Ltd, Mumbai, was signed on 6 October 2005. The agreement is reported to include an option for a further nine boats. DCNS is to supply technical advisers and provide prefabricated hull elements and the combat systems, including the command system, underwater sensors, optronics, and communications. MBDA is to supply Exocet SM39 missiles as part of the package. Details are based on the boats built for Chile. AIP is not to be installed in the first two boats but a reassessment for the remaining submarines will be made at a later date. Delivery of all six boats was to have begun in 2012 and to have been completed in about 2018 but progress is slow. The programme is likely to have been delayed by at least four years.

Structure: Diving depth more than 300 m *(984 ft)*. AIP would require the addition of an 8 m 'plug' to incorporate the MESMA system.

SCORPENE (computer graphic)   *1998, DCN* / 0017689

4 SHISHUMAR (TYPE 209/1500) CLASS (SSK)

Name	No	Builders	Laid down	Launched	Commissioned
SHISHUMAR	S 44	Howaldtswerke, Kiel	1 May 1982	13 Dec 1984	22 Sep 1986
SHANKUSH	S 45	Howaldtswerke, Kiel	1 Sep 1982	11 May 1984	20 Nov 1986
SHALKI	S 46	Mazagon Dock Ltd, Mumbai	5 Jun 1984	30 Sep 1989	7 Feb 1992
SHANKUL	S 47	Mazagon Dock Ltd, Mumbai	3 Sep 1989	21 Mar 1992	28 May 1994

Displacement, tonnes: 1,687 surfaced; 1,880 dived
Dimensions, metres (feet): 64.4 × 6.5 × 6 *(211.3 × 21.3 × 19.7)*
Speed, knots: 11 surfaced; 22 dived
Range, n miles: 13,000 at 10 kt surfaced; 8,000 at 8 kt snorting
Complement: 40 (8 officers)

Machinery: Diesel-electric; 4 MTU 12V 493 AZ80 GA31L diesels; 2,400 hp(m) *(1.76 MW)* sustained; 4 Siemens alternators; 1.8 MW; 1 Siemens motor; 4,600 hp(m) *(3.38 MW)* sustained; 1 shaft
Torpedoes: 8 — 21 in *(533 mm)* tubes. 14 AEG SUT 266 (Mod 1); wire-guided; active/passive homing to 28 km *(15.3 n miles)* at 23 kt; 12 km *(6.6 n miles)* at 35 kt; warhead 250 kg.
Mines: External 'strap-on' type for 24 mines.
Physical countermeasures: Decoys: C 303 acoustic decoys.
Electronic countermeasures: ESM: Argo Phoenix II AR 700 or Kollmorgen Sea Sentry; radar warning.
Radars: Surface search: Thomson-CSF Calypso; I-band.
Sonars: Thales 2272; hull-mounted; passive/active search and attack; medium frequency. Thomson Sintra DUUX-5; passive ranging and intercept.
Weapon control systems: CSS-90-1/ISUS 90.

Programmes: Howaldtswerke concluded an agreement with the Indian Navy on 11 December 1981. This was in four basic parts: the building in West Germany of two Type 1500 submarines; the supply of 'packages' for the building of two more boats at Mazagon, Mumbai; training of various groups of specialists for the design and construction of the Mazagon pair; logistic services during the trials and early part of the commissions as well as consultation services in Mumbai. In 1984 it was announced that a further two submarines would be built at Mazagon for a total of six but this was overtaken by events in 1987–88 and the agreement with HDW terminated at four. This was reconsidered in 1992 and again in 1997.
Modernisation: Thales 2272 sonar replaced CSU 83 2000–09. Upgrade of at least two boats (probably *Shalki* and *Shankul* with anti-ship missiles is under consideration. Life-extension refits for these boats are also likely.
Structure: The Type 1500 has a central bulkhead and an IKL designed integrated escape sphere which can carry the full crew of up to 40 men, has an oxygen supply for 8 hours, and can withstand pressures at least as great as those that can be withstood by the submarine's pressure hull. Diving depth 260 m *(853 ft)*.
Operational: Form 10th Submarine Squadron based at Mumbai. *Shishumar* mid-life refit started in 1999 and had been completed by 2001. She undertook a further repair period in 2004 following a collision. *Shankul* underwent refit 2001-2005 and *Shankush* is reported to have started refit in 2001. *Shalki* underwent refit 2007–09. All four boats reported operational.

SHALKI   *12/2011, Hartmut Ehlers* / 1406895

SHANKUSH   *12/2009, K Chandni* / 1406048

IHS Jane's Fighting Ships 2014-2015

9 SINDHUGHOSH (KILO) (PROJECT 877EM/8773) CLASS (SSK)

Name	No	Builders	Laid down	Launched	Commissioned
SINDHUGHOSH	S 55	Sudomekh, Leningrad	29 May 1983	29 Jul 1985	30 Apr 1986
SINDHUDHVAJ	S 56	Sudomekh, Leningrad	1 Apr 1986	27 Jul 1986	12 Jun 1987
SINDHURAJ	S 57	Sudomekh, Leningrad	—	—	20 Oct 1987
SINDHUVIR	S 58	Sudomekh, Leningrad	15 May 1987	13 Sep 1987	26 Aug 1988
SINDHURATNA	S 59	Sudomekh, Leningrad	—	—	22 Dec 1988
SINDHUKESARI	S 60	Sudomekh, Leningrad	20 Apr 1988	16 Aug 1988	16 Feb 1989
SINDHUKIRTI	S 61	Sudomekh, Leningrad	5 Apr 1989	26 Aug 1989	4 Jan 1990
SINDHUVIJAY	S 62	Sudomekh, Leningrad	6 Apr 1990	27 Jul 1990	8 Mar 1991
SINDHUSHASTRA	S 65	Sudomekh, St Petersburg	12 Dec 1998	14 Oct 1999	19 Jul 2000

Displacement, tonnes: 2,362 surfaced; 3,125 dived
Dimensions, metres (feet): 72.6 × 9.9 × 6.6 *(238.2 × 32.5 × 21.7)*
Speed, knots: 10 surfaced; 9 snorting; 17 dived
Range, n miles: 6,000 at 7 kt snorting; 400 at 3 kt dived
Complement: 52 (13 officers)

Machinery: Diesel-electric; 2 Model 4-2DL-42M diesels; 3,650 hp(m) *(2.68 MW)*; 2 generators; 1 motor; 5,900 hp(m) *(4.34 MW)*; 1 shaft; 2 MT-168 auxiliary motors; 204 hp(m) *(150 kW)*; 1 economic speed motor; 130 hp(m) *(95 kW)*
Missiles: SLCM: Novator Alfa Klub SS-N-27 (3M-54 anti-ship missiles) (S 55, 57, 59, 60, 62, and 65); active radar homing to 220 km *(120 n miles)* at 0.8 Mach (cruise) and 2.9 Mach (attack); warhead 400 kg. Novator Klub SS-N-30 (3M 14) land-attack missiles (S 55, S 62); terrain following/SATNAV guidance to 300 km *(162 n miles)* at 0.7 Mach; warhead 450 kg.
SAM: SA-N-8 portable launcher; IR homing to 3.2 n miles *(6 km)*.
Torpedoes: 6 − 21 in *(533 mm)* tubes. Combination of Type 53-65; passive wake homing to 19 km *(10.3 n miles)* at 45 kt; warhead 305 kg and TEST 71/96; anti-submarine; active/passive homing to 15 km *(8.1 n miles)* at 40 kt or 20 km *(10.8 n miles)* at 25 kt; warhead 220 kg. Total of 18 weapons. Wire-guided on 2 tubes.
Mines: 24 DM-1 in lieu of torpedoes.
Electronic countermeasures: ESM: Squid Head; radar warning.
Radars: Navigation: Snoop Tray; I-band.
Sonars: Shark Teeth/Shark Fin; MGK-400; or Bel Ushus (S 55, 62); hull-mounted; active/passive search and attack; medium frequency. Mouse Roar; MG-519; hull-mounted; active search; high frequency.
Weapon control systems: Uzel MVU-119EM TFCS.

Programmes: The Kilo class was launched in the former Soviet Navy in 1979 and although India was the first country to acquire one they have since been transferred to Algeria, Poland, Romania, Iran and China. Because of the slowness of the S 209 programme, the original order in 1983 for six Kilo class expanded to 10 but was then cut back again to eight. Two further orders were confirmed in May 1997. S 63 was a spare Type 877 hull built for the Russian Navy, but never purchased. S 65 is a Type 8773 and was fitted for SLCM on build.
Modernisation: *Sindhuvir* completed major refit at Zvezdochka Shipyard, Severodvinsk from May 1997 to July 1999. *Sindhuraj* and *Sindhukesari* completed similar refits at Admiralty Yard, St Petersburg from May 1999 to November 2001. *Sindhuratna* completed a two-year refit at Severodvinsk in 2002. *Sindhughosh*, following refit work at Vishakapatnam from 1999, started modernisation at Severodvinsk in September 2002 which completed on 22 April 2005. *Sindhuvijay* started a two-year refit at Severodvinsk in May 2005 which was completed on 8 May 2007 although acceptance of the boat was delayed until August 2008 due to reported defects in the missile system. She became the sixth boat to be fitted with SS-N-27. Both *Sindhughosh* and *Sindhuvijay* are also equipped with the SS-N-30 (3M 14) land-attack missiles. *Sinhuraj, Sindhuvir, Sindhuratna* and *Sindhushastra* are to be similarly fitted. *Sindhukirti* began refit at Hindustan Shipyard, Vishakapatnam, in January 2006 and is expected to complete in 2014.
Structure: Diving depth, 300 m *(985 ft)*. Reported that from *Sindhuvir* onwards these submarines have an SA-N-8 SAM capability. The launcher is shoulder held and stowed in the fin for use when the submarine is surfaced. Two torpedo tubes can fire wire-guided torpedoes and four tubes have automatic reloading. Anechoic tiles are fitted on casings and fins.
Operational: *Sindhuvir* and *Sinhukirti* form the 11th Submarine Squadron based at Vishakapatnam. The remainder form the 12th Squadron based at Mumbai. *Sindhurakshak* suffered a major fire, possibly triggered by a weapons explosion, on 14 August 2013. She subsequently sank alongside at Mumbai. Damage was extensive and return to service is not expected.

SINDHUVIJAY   9/2008, Diego Quevedo / 1353085

SINDHUKESARI   12/2011, Guy Toremans / 1406896

SINDHURATNA   12/2011, Michael Nitz / 1455659

AIRCRAFT CARRIERS

Notes: The Maritime Capability Perspective Plan includes proposals to achieve a three-carrier force by 2022. Construction of a second indigenously built carrier which may be a conventional take-off and landing (rather than Short Take Off and Vertical Landing (STOVL)) ship is expected to start in about 2017. Details of the ship have not been finalised but she is expected to be of the order of 65,000 tonnes. The name of the ship is reported to be *Vishal*.

1 MODIFIED KIEV CLASS (PROJECT 11430) (CVGM)

Name	No	Builders	Laid down	Launched	Commissioned
VIKRAMADITYA (ex-*Admiral Gorshkov*; ex-*Baku*)	R 33	Nikolayev South	17 Feb 1978	1 Apr 1982	16 Nov 2013

Displacement, tonnes: 46,129 full load
Dimensions, metres (feet): 283 oa; 249.5 wl × 51 oa; 32.7 hull × 10 (928.5; 818.6 × 167.3; 107.3 × 32.8)
Aircraft lift: 1; 19.2 × 10.3 m, 30 tonnes
1; 18.5 × 4.7 m, 20 tonnes
Speed, knots: 29
Range, n miles: 13,800 at 18 kt
Complement: 1,326

Machinery: 8 KWG4 boilers; 4 GTZA 674 turbines; 200,000 hp(m) *(147 MW)*; 4 shafts
Missiles: SAM: Barak 1 (to be confirmed).
Guns: 4—30 mm AK 630 (to be confirmed).
Physical countermeasures: Decoys: 2 PK2 chaff launchers; 2 towed torpedo decoys.
Electronic countermeasures: ESM/ECM: Bharat intercept and jammers.
Radars: Air search: Flat Screen (Podberyozovik-ET2); 3D; E/F-band.
Air/surface search: Top Plate (Fregat MAE-3); 3D; E/F-band.

Surface search: 2 Strut Pair; F-band.
Navigation: Sperry Bridgemaster E; E/F/I-bands.
Sonars: Horse Jaw (MG 355); hull-mounted; active search; medium frequency.
Combat data systems: Lesorub E.

Fixed-wing aircraft: 12 MiG 29K.
Helicopters: 6 Helix 27/28/31.

Programmes: Last of the four Project 1143.4 aircraft carriers built for the Soviet Navy. First offered for sale to India by Russia in 1994. By 1999 the proposal was to gift the ship as long as India pays for the refit. Following a Government to Government agreement on 4 October 2000 and protracted negotiations, contract signed on 20 January 2004 for a five-year refit at a cost estimated to be USD625 million. However, it was announced in August 2007 that the refit had been delayed by three years and that the ship would not enter service until 2012. Following problems with the main power plant during sea trials in 2012, the refit was further extended until late 2013 in order to complete remedial work. Agreement to fund cost overruns was authorised by the Indian Cabinet Committee on Security (CCS) in December 2008 and approved by CCS on 10 March 2010.
Modernisation: New propulsion, power, and air conditioning systems have been fitted. All the original Russian weapons systems removed. Weapon systems and sensors have not been confirmed. The flight deck has been converted to a STOBAR configuration with a 14.3° ski-jump.
Structure: The ship has a 198 m angled deck with three arrestor wires. Flight deck lifts are 19.2 × 10.3 m and 18.5 × 4.7 m, and can lift 30 tons (aft) and 20 tons (midships) respectively. The hangar is 130 × 22.5 m.
Operational: The ship was re-launched on 4 December 2008. Sea trials started on 8 June 2012 and the first touch-and-go landing by a MiG-29K was conducted on 23 July 2012. Following remedial boiler work, sea trials recommenced in July 2013 and the ship was formally handed over on 16 November 2013 and arrived at her home base Karwar on 8 January 2014. She is expected to become operational in 2015.

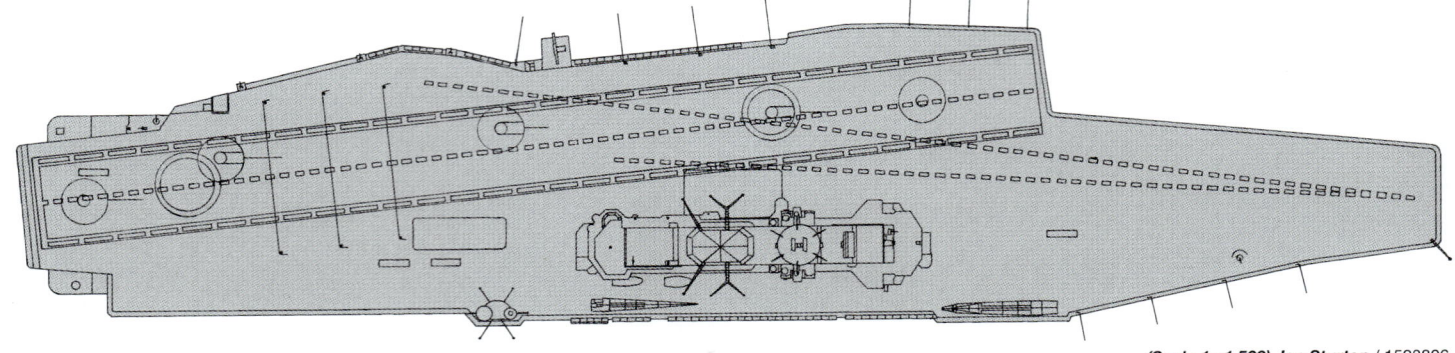

VIKRAMADITYA   *(Scale 1 : 1,500), Ian Sturton / 1528806*

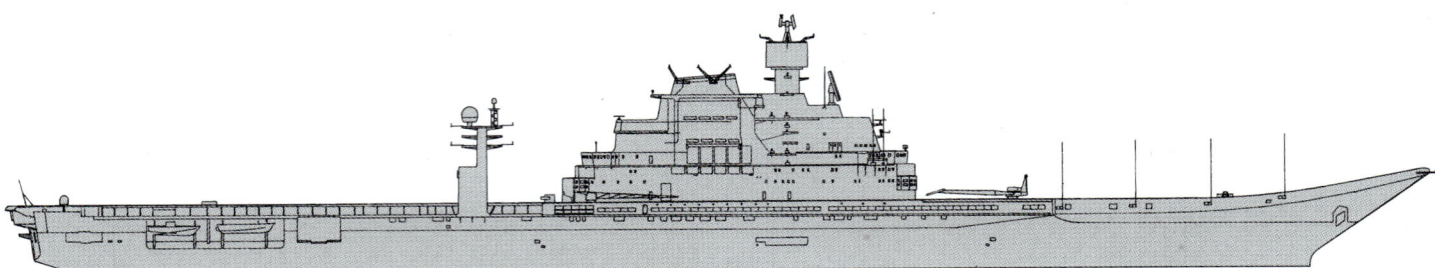

VIKRAMADITYA   *(Scale 1 : 1,500), Ian Sturton / 1528807*

VIKRAMADITYA   *1/2014*, Indian Navy / 1529187*

Aircraft carriers < **India** 331

VIKRAMADITYA
1/2014*, Indian Navy / 1529186

VIKRAMADITYA
1/2014*, Indian Navy / 1529185

VIKRAMADITYA
7/2013*, Ships of the World / 1529125

332 India > Aircraft carriers

1 HERMES CLASS (CVM)

Name	No	Builders	Laid down	Launched	Commissioned
VIRAAT (ex-*Hermes*)	R 22	Vickers Shipbuilding & Engineering, Barrow-in-Furness	21 Jun 1944	16 Feb 1953	18 Nov 1959

Displacement, tonnes: 24,284 standard; 29,161 full load
Dimensions, metres (feet): 226.9 oa; 208.8 wl × 48.8 oa; 27.4 hull × 8.7 *(744.4; 685 × 160.1; 89.9 × 28.5)*
Aircraft lift: 2
Speed, knots: 28
Complement: 1,350 (143 officers)

Machinery: 4 Admiralty boilers; 400 psi *(28 kg/cm²)*; 700°F *(370°C)*; 2 Parsons geared turbines; 76,000 hp *(57 MW)*; 2 shafts
Missiles: 2 Octuple IAI/Rafael Barak 1 VLS ❶, command line of sight radar or optical guidance to 10 km *(5.5 n miles)* at 2 Mach; warhead 22 kg.
Guns: 1—40 mm. 4—30 mm/65 (2 twin) AK 230 ❷; 500 rds/min to 5 km *(2.7 n miles)*; weight of shell 0.54 kg.
Physical countermeasures: Decoys: 2 Knebworth Corvus chaff launchers ❸.
Electronic countermeasures: ESM/ECM: SEWS V5 with C Pearl ❹.
Radars: Air search: Bharat RAWL-02 Mk 3 (LW08) ❺; D-band.
Air/surface search: Bharat RAWS (PFN 513) ❻; E/F-band.
Fire control: IAI/Elta EL/M-2221 ❼; Ka-band.
Navigation: 2 Bharat Rashmi ❽; I-band.
Tacan: FT 13-S/M.
Sonars: Graseby Type 184M; hull-mounted; active search and attack; 6–9 kHz.
Combat data systems: CAAIS action data automation. SATCOM.

Fixed-wing aircraft: 8 Sea Harriers FRS Mk 51 ❾ (capacity for 30).
Helicopters: 7 Sea King Mk 42B/C ❿ ASW/ASV/Vertrep and Ka-27 Helix. Ka-31 Helix.

Programmes: Purchased in May 1986 from the UK, thence to an extensive refit in Devonport Dockyard. Commissioned in Indian Navy 20 May 1987.
Modernisation: UK refit included new fire-control equipment, navigation radars and deck landing aids. Boilers were converted to take distillate fuel and the ship was given improved NBC protection. New search radar in 1995. Further modernisation in 1999–2001 refit, improved indigenous RAWL 02 (Mk II) and Rashmi radars for CCA/navigation, EW equipment and new communications systems. A further refit, completed in December 2004, included installation of Barak CIWS. This has replaced the previously fitted 40 mm guns. Another (fourth) refit, to extend ship-life to 2015, was completed in 2009.
Structure: Fitted with 12° ski jump. Reinforced flight deck (0.75 in); 1 to 2 in of armour over magazines and machinery spaces. Four LCVP on after davits. Magazine capacity includes 80 lightweight torpedoes. Barak launchers are recessed in the starboard side of the flight deck, aft of the island.
Operational: The Sea Harrier complement is likely to be of the order of six aircraft leaving room for a greater mix of Sea King and Helix helicopters. Based at Mumbai.

VIRAAT 10/2005, *Ships of the World* / 1153836

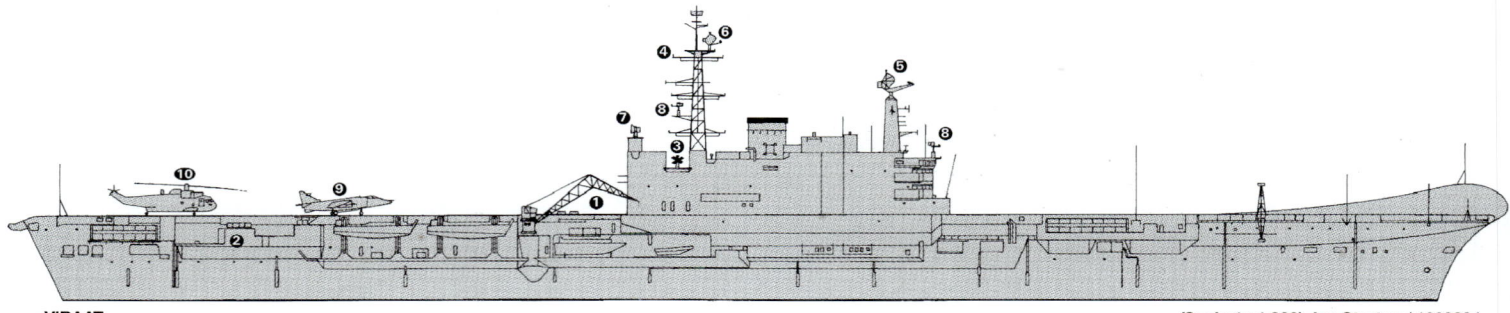

VIRAAT (Scale 1 : 1,200), Ian Sturton / 1303034

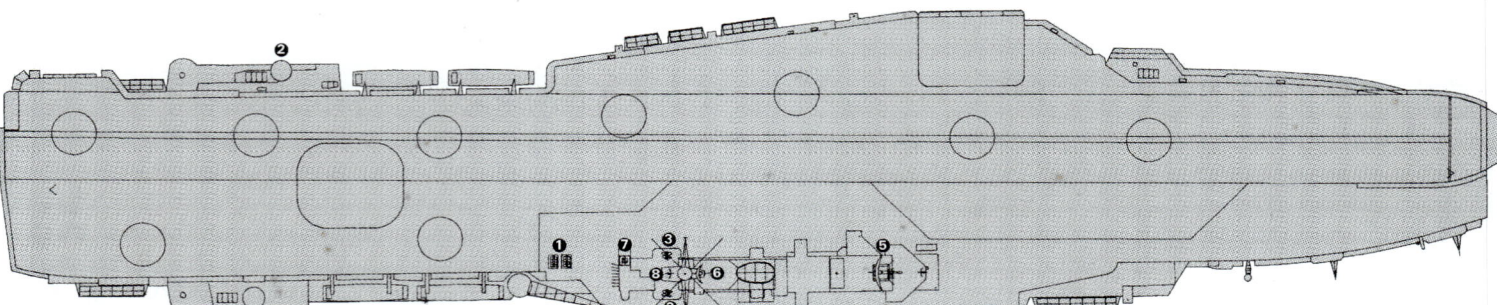

VIRAAT (Scale 1 : 1,200), Ian Sturton / 1166562

VIRAAT 12/2011, Guy Toremans / 1406897

IHS Jane's Fighting Ships 2014-2015

Aircraft carriers — Destroyers < **India** 333

0 + 1 INDIGENOUS AIRCRAFT CARRIER CLASS (PROJECT 71) (CVM)

Name	No	Builders	Laid down	Launched	Commissioned
VIKRANT	—	Kochi Shipyard Ltd	28 Feb 2009	12 Aug 2013	2018

Displacement, tonnes: 40,642 standard
Dimensions, metres (feet): 262.5 × 60.84 × 8.4 *(861.2 × 199.6 × 27.6)*
Aircraft lift: 2
Speed, knots: 28. **Range, n miles:** 7,500 at 18 kt
Complement: 1,400 (160 officers)

Machinery: COGAG: 4 General Electric LM 2500 gas turbines; 120,000 hp *(89.5 MW)*; 2 shafts; cp props
Missiles: SAM: IAI/Rafael Barak 2/8.
Guns: CIWS; 1 — 76 mm.
Radars: Air search: Selex RAN-40L; 3D; D-band.
Air/surface search: EL/M-2248 STAR.
Surface search: To be announced.
Navigation: To be announced.
Sonars: Hull mounted.

Fixed-wing aircraft: 20 MiG 29K.
Helicopters: 10 Ka-31 and HAL Dhruv.

VIKRANT *8/2013*, Indian Navy / 1529188*

Programmes: The plan announced in 1989 was to build two new aircraft carriers. The Indigenous Aircraft Carrier (IAC), formerly the Air Defence Ship (ADS), is to replace the former *Vikrant* (and will probably receive the same name) while *Vikramaditya* (ex-*Admiral Gorshkov*) is to replace *Viraat* in late 2012. A number of international companies including DCN, IZAR and Fincantieri are believed to have been involved in conceptual and design work of the ADS and it is understood that the shipbuilder, Cochin Shipyard Ltd (CSL), has sub-contracted specialist 'task forces' to collaborate in building the ship. Two contracts signed in mid-2004 with Fincantieri to finalise the ADS design and its ancillary propulsion systems and main power plants. Fincantieri is likely to be providing further assistance during the vessel's construction, tests and sea trials. First steel cut on 11 April 2005 and construction of building blocks started thereafter. The order for a second ship is expected after the launch of the first of class. It was announced in July 2012 that the overall programme had slipped three years.
Structure: The design includes a short take off (with 14° ski jump) and arrested recovery (STOBAR) system. There are two aircraft lifts.
Operational: Sea trials are to begin in 2016. The ship is to be based on the east coast possibly at a new base.

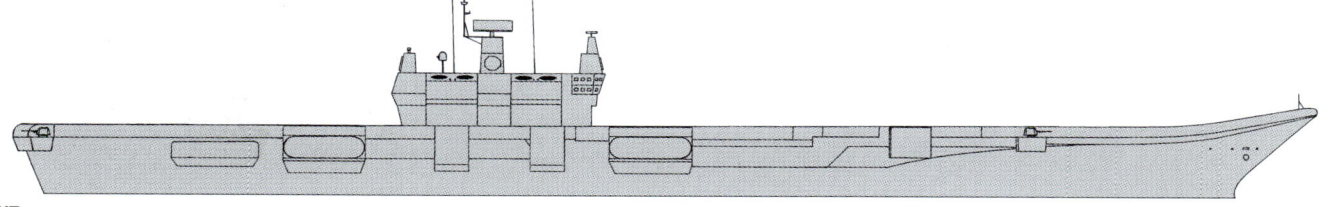

VIKRANT *(Scale 1 : 1,500), Ian Sturton / 1482997*

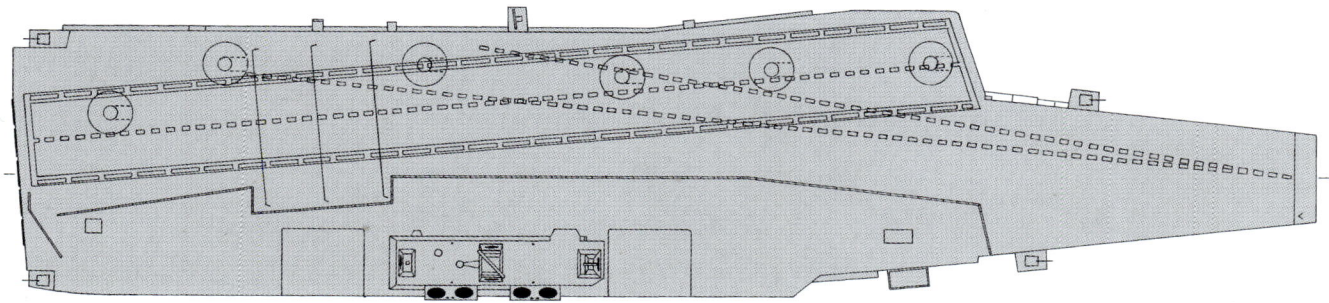

VIKRANT *(Scale 1 : 1,500), Ian Sturton / 1366991*

DESTROYERS

0 + 3 (4) KOLKATA (PROJECT 15A) CLASS (DDGHM)

Name	No	Builders	Laid down	Launched	Commissioned
KOLKATA	D 63	Mazagon Dock Ltd, Mumbai	26 Sep 2003	30 Mar 2006	2014
KOCHI	D 64	Mazagon Dock Ltd, Mumbai	25 Oct 2005	18 Sep 2009	2015
CHENNAI	D 65	Mazagon Dock Ltd, Mumbai	21 Feb 2006	1 Apr 2010	2016

Displacement, tonnes: 6,800 standard; 7,292 full load
Dimensions, metres (feet): 163.95 × 17.71 × 5.4 *(537.9 × 58.1 × 17.7)*
Speed, knots: 32. **Range, n miles:** 4,500 at 18 kt
Complement: 360 (40 officers)

Machinery: 4 Zorya/Mashprockt DT-59 gas turbines; 82,820 hp(m) *(61.7 MW)*; 2 shafts; cp props
Missiles: SSM: 16 Brahmos PJ-10 (2 octuple VLS) ❶; active/passive radar homing to 290 km *(157 n miles)* at 2.6 Mach; warhead 200 kg; sea skimmer in terminal phase.
SAM: IAI/Rafael Barak 8 ❷; inertial mid-course guidance and active radar homing to 80 km *(43.2 n miles)*; 1 — 16 cell VLS launcher (forward), 1 — 32 cell VLS launcher (aft); total of 48 missiles.
Guns: 1 Oto Melara 3 in *(76 mm)* Super Rapid ❸; 120 rds/min to 16 km *(8.7 n miles)*; weight of shell 6 kg. 4 — 30 mm/AK 630 ❹; 6 barrels per mounting; 3,000 rds/min combined to 2 km.
Torpedoes: 4 ITTL 21 in *(533 mm)* (2 twin) tubes ❺. Combination of SET 65E; anti-submarine; active/passive homing to 15 km *(8.1 n miles)* at 40 kt; warhead 205 kg and Type 53–65; passive wake homing to 19 km *(10.3 n miles)* at 45 kt; warhead 305 kg.
A/S Mortars: 2 RBU 6000 ❻; 12 tubed trainable; range 6,000 m; warhead 31 kg.
Physical countermeasures: Decoys: 4 Kavach chaff launchers ❼; Mareech torpedo decoy.
Electronic countermeasures: ESM/ECM: Ellora EW suite.
Radars: Air search: Bharat RAWL-02 Mk 3 (LW08) ❽; D-band.
Air/surface search: EL/M-2248 STAR; 3D; E/F-band ❾.
Fire control: ELTA EL-M 2221 STGR; I/J/K-band (for SAM); BEL Lynx; I-band (for 100 mm); Plank Shave (Granit Garpun B) (for SSM) ❿; I/J-band.
Navigation: Consilium Selesmar; E/F/I-bands.
Sonars: Bharat HUMSA; bow-mounted; medium frequency. Towed array (to be confirmed).
Combat data systems: CAIO-15A.

Helicopters: 2 Westland Sea Kings Mk 42B ⓫ or 2 HAL Dhruv.

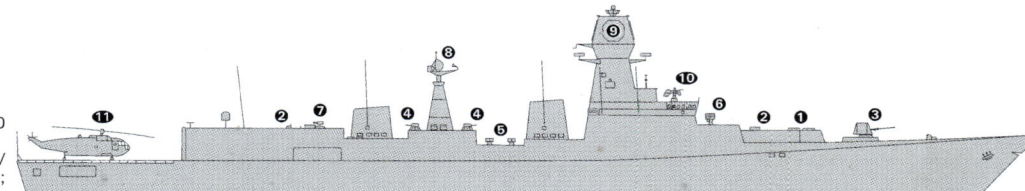

KOLKATA *(Scale 1 : 1,200), Ian Sturton / 1366999*

KOLKATA *9/2013*, Indian Navy / 1529173*

Programmes: The first of three modified Delhi class was laid down in 2003 but progress since her launch in 2006 has been very slow. Approval for construction of four Project 15B ships was given in May 2009. The ships are to be to a modified design of Project 15A ships. A contract is expected in 2014.
Structure: Designed by the Indian Naval Design Bureau, the design appears to be a development of the Delhi class incorporating some features of both the Talwar and Project 17 frigates.
Operational: Sea trials of *Kolkata* began in 2013.

334 India > Destroyers

3 DELHI CLASS (PROJECT 15) (DDGHM)

Name	No	Builders	Laid down	Launched	Commissioned
DELHI	D 61	Mazagon Dock Ltd, Mumbai	14 Nov 1987	1 Feb 1991	15 Nov 1997
MYSORE	D 60	Mazagon Dock Ltd, Mumbai	2 Feb 1991	4 Jun 1993	2 Jun 1999
MUMBAI	D 62	Mazagon Dock Ltd, Mumbai	14 Dec 1992	20 Mar 1995	22 Jan 2001

Displacement, tonnes: 6,808 full load
Dimensions, metres (feet): 163 × 17.4 × 6.5 (534.8 × 57.1 × 21.3)
Speed, knots: 32
Range, n miles: 4,500 at 18 kt
Complement: 360 (40 officers)
Machinery: 4 Zorya/Mashprockt DT-59 gas turbines; 82,820 hp(m) (61.7 MW); 2 shafts; cp props
Missiles: SSM 16 Zvezda SS-N-25 Switchblade (4 quad) (KH 35E Uran) ❶; active radar homing to 130 km (70.2 n miles) at 0.9 Mach; warhead 145 kg; sea skimmer.
SAM 2 SA-N-7 Gadfly (Kashmir/Uragan) ❷ command, semi-active radar and IR homing to 25 km (13.5 n miles) at 3 Mach; warhead 70 kg. Total of 48 missiles. 4 Octuple IAI/Rafael Barak 1 VLS (D 60, D 61) ❸; command line of sight radar or optical guidance to 10 km (5.5 n miles) at 2 Mach; warhead 22 kg.
Guns: 1 USSR 3.9 in (100 mm)/59 ❹. AK 100; 60 rds/min to 21.5 km (11.5 n miles); weight of shell 15.6 kg. 2 (4 in D 62) USSR 30 mm/65 ❺ AK 630; 6 barrels per mounting; 3,000 rds/min combined to 2 km. 4 – 12.7 mm MGs.
Torpedoes: 5 PTA 21 in (533 mm) (quin) tubes ❻. Combination of SET 65E; anti-submarine; active/passive homing to 15 km (8.1 n miles) at 40 kt; warhead 205 kg and Type 53-65; passive wake homing to 19 km (10.3 n miles) at 45 kt; warhead 305 kg.
A/S Mortars: 2 RBU 6000 ❼; 12 tubed trainable; range 6,000 m; warhead 31 kg.
Depth charges: 2 rails.

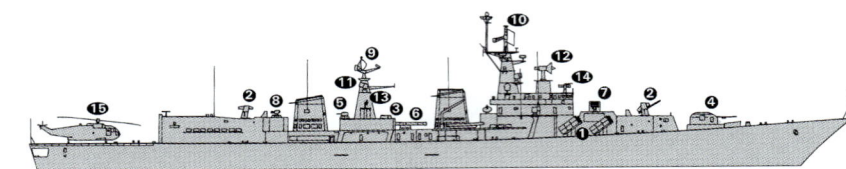

DELHI (Scale 1 : 1,500), Ian Sturton / 0572398

Physical countermeasures: Decoys: 2 PK2 chaff launchers ❽. Towed torpedo decoy.
Electronic countermeasures: ESM/ECM: Ellora EW suite.
Radars: Air search: Bharat RAWL-02 Mk 3 (LW08) (D 60); Bharat RAWL-02 Mk 2 (LW04) (D 61, D 62) ❾; D-band.
Air/surface search: Half Plate ❿; E-band.
Fire control: 6 Front Dome ⓫; H/I-band (for SAM); Kite Screech ⓬; I/J-band (for 100 mm); 2 Bass Tilt (MR-123) (D 62); I/J-band (for AK 630); EL/M-2221 STGR (D 60, D 61) ⓭ (for Barak); I/J/K-band; Plank Shave (Granit Garpun B) ⓮ (for SSM); I/J-band.
Navigation: 3 Nyada MR-212/201; I-band.
Sonars: Bharat HUMVAD; hull-mounted; active search; medium frequency. Bharat HUMSA; hull-mounted; medium frequency (D 62). Indal/Garden Reach Model 15-750 VDS. Thales ATAS; active towed array (D 62).
Combat data systems: Bharat IPN Shikari (IPN 10).
Helicopters: 2 Westland Sea Kings Mk 42B ⓯ or 2 HAL Dhruv.

Programmes: Built with Russian Severnoye Design Bureau assistance. *Delhi* ordered in March 1986. Programme was called Project 15. Much delay was caused by the breakdown in the central control of Russian export equipment.
Structure: The design is described as a 'stretched *Rajput*' with some *Godavari* features. A combination of Russian and Indian weapon systems fitted. Missile blast deflectors indicate an original intention to fit SS-N-22 Sunburn. Samahé helo handling system. Forward funnel offset to port and after funnel to starboard.
Modernisation: Barak has replaced the forward AK-630 mountings in D 60 and D 61. The two Bass Tilt radars have also been replaced by EL/M-2221 STGR. D 62 is to be similarly refitted. SS-N-25 may be replaced by Brahmos.
Operational: Based at Mumbai. Have Flag facilities.

DELHI 6/2009, B Prézelin / 1305609

MYSORE 12/2011, Michael Nitz / 1406900

IHS Jane's Fighting Ships 2014-2015 © 2014 IHS

MUMBAI   *12/2011, Hartmut Ehlers* / 1406899

MUMBAI   *4/2009, Hachiro Nakai* / 1305610

DELHI   *5/2011, Guy Toremans* / 1406615

336 India > Destroyers

5 RAJPUT (KASHIN II) CLASS (PROJECT 61ME) (DDGHM)

Name	No	Builders	Laid down	Launched	Commissioned
RAJPUT (ex-*Nadezhniy*)	D 51	Nikolayev North (61 Kommuna)	11 Sep 1976	17 Sep 1977	30 Sep 1980
RANA (ex-*Gubitelyniyy*)	D 52	Nikolayev North (61 Kommuna)	29 Nov 1976	27 Sep 1978	28 Jun 1982
RANJIT (ex-*Lovkiyy*)	D 53	Nikolayev North (61 Kommuna)	29 Jun 1977	16 Jun 1979	24 Nov 1983
RANVIR (ex-*Tverdyy*)	D 54	Nikolayev North (61 Kommuna)	24 Oct 1981	12 Mar 1983	28 Oct 1986
RANVIJAY (ex-*Tolkoviyy*)	D 55	Nikolayev North (61 Kommuna)	19 Mar 1982	1 Feb 1986	15 Jan 1988

Displacement, tonnes: 4,013 standard; 5,054 full load
Dimensions, metres (feet): 146.5 × 15.8 × 4.8 *(480.6 × 51.8 × 15.7)*
Speed, knots: 35
Range, n miles: 4,500 at 18 kt, 2,600 at 30 kt
Complement: 320 (35 officers)

Machinery: COGAG; 4 Ukraine gas turbines; 72,000 hp(m) *(53 MW)*; 2 shafts
Missiles: SSM: 2 (D 51) or 4 SS-N-2D Mod 2 Styx (D 52–55) ❶; IR homing to 83 km *(45 n miles)* at 0.9 Mach; warhead 513 kg; sea-skimmer. 4 (D 51) or 8 (D 54, 55) Brahmos PJ-10 ❷; active/passive radar terminal homing to 290 km *(157 n miles)* at 2.6 Mach; warhead 200 kg.
SAM: 2 (D 51, 52, 53) or 1 (D 54, 55) SA-N-1 Goa twin launchers ❸; command guidance to 31.5 km *(17 n miles)* at 2 Mach; height 91–22,860 m *(300–75,000 ft)*; warhead 60 kg; 44 missiles. Some SSM capability. 2 octuple IAI/Rafael Barak 1 VLS (D 54, D 55) ❹; command line of sight radar or optical guidance to 10 km *(5.5 n miles)* at 2 Mach; warhead 22 kg.
Guns: 2–3 in *(76 mm)*/59 AK 726 (twin, fwd) ❺; 90 rds/min to 16 km *(8.5 n miles)*; weight of shell 5.9 kg. 8–30 mm/65 (4 twin) AK 230 (D 51, 52, 53) ❻; 500 rds/min to 5 km *(2.7 n miles)*; weight of shell 0.54 kg. 2–30 mm/65 AK 630 (6 barrels per mounting) D 54, 55) ❼; 3,000 rds/min combined to 2 km.
Torpedoes: 5–21 in *(533 mm)* (quin) tubes ❽. Combination of SET-65E; anti-submarine; active/passive homing to 15 km *(8.1 n miles)* at 40 kt; warhead 205 kg and Type 53–65; passive wake homing to 19 km *(10.3 n miles)* at 45 kt; warhead 305 kg.
A/S Mortars: 2 RBU 6000 12-tubed trainable ❾; range 6,000 m; warhead 31 kg.
Physical countermeasures: Decoys: 4 PK 16 chaff launchers for radar decoy and distraction.
Electronic countermeasures: ESM: Bharat Ajanta Mk 2 (SEWS in D 54, D 55); intercept.
ECM: Elettronica TQN-2; jammer.
Radars: Air search: Big Net A (D 51) ❿; C-band; range 183 km *(100 n miles)* for 2 m² target. Bharat/Signaal RAWL-02 Mk 2 (LW04) (D 52–55) ⓫; D-band.
Air/surface search: Head Net C (D 51-53) ⓬; 3D; E-band. EL/M-2238 STAR (D 54, 55) ⓭; 3D; E/F-band.
Navigation: 2 Bharat Rashmi; I-band.
Fire control: 2 or 1 (D 54, D 55) Peel Group ⓮; H/I-band; range 73 km *(40 n miles)* for 2 m² target. Owl Screech ⓯; G-band. 2 Drum Tilt ⓰ or 2 Bass Tilt; H/I-band or 2 EL/M-2221 STGR ⓱; I/J/K-band.
IFF: 2 High Pole B.
Sonars: Vycheda MG 311 (D 51, 52, 53); hull-mounted; active search and attack; medium frequency. Mare Tail VDS; active search; medium frequency. Bharat Humsa (D 54, 55); hull-mounted; medium frequency.

Helicopters: 1 Ka-28 Helix ⓲.

Programmes: First batch of three ordered in the mid-1970s. *Ranvir* was the first of the second batch ordered on 20 December 1982.
Modernisation: New EW equipment installed on all ships refitted since 1993. It is possible that an Italian combat data system compatible with Selenia IPN-10 has been installed. D 51, 54 and 55 have undergone extensive modernisation since 2003. This includes installation of Brahmos SSM and Barak SAM. EL/M-2238 STAR search radar has replaced Head Net C in D 54 and D 55. All ships are Inmarsat fitted.
Structure: Originally built to a modified Kashin-class design, the ships are equipped with a helicopter hangar, reached by a lift from the flight deck, to replace the after 76 mm twin mount. Recent modifications have led to some differences in structure. Four Brahmos missile launchers have replaced the forward SS-N-2D mountings in D 51 while, in D 54 and D 55, an 8-cell Brahmos VLS system has been installed in lieu of the aft SA-N-1 launcher. The aft Peel Group radar has also been removed. D 54 and D 55 have also been fitted (port and starboard) with two octuple VLS silos for Barak SAM and associated EL/M-2221 STGR fire-control radars. These silos are in place of the forward AK-630 mountings. D 52 and D 53 may be similarly fitted in due course.

Operational: All based at Vishakapatnam. Dhanush (Prithvi) ballistic missile test launched from *Rajput* on 28 December 2005. Vertical launch of Brahmos was conducted from *Ranvir* on 18 December 2008.

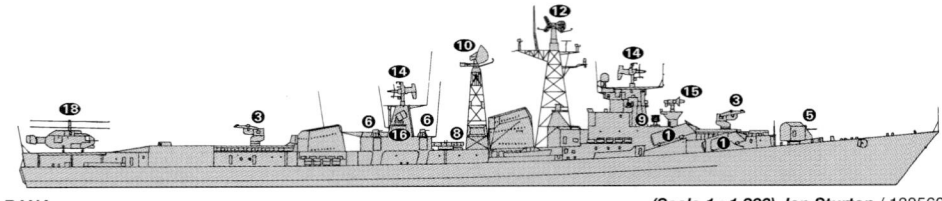

RANA *(Scale 1 : 1,200), Ian Sturton* / 1305601

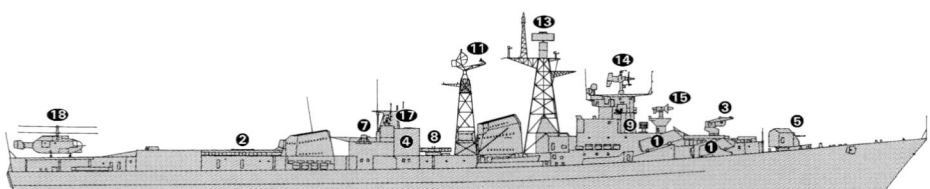

RANVIR *(Scale 1 : 1,200), Ian Sturton* / 1305602

RANVIR 12/2011, Michael Nitz / 1406898

RAJPUT 12/2009, M Mazumdar / 1406050

RANVIJAY 12/2010, M Mazumdar / 1406614

IHS Jane's Fighting Ships 2014-2015

FRIGATES

6 + (3) TALWAR (PROJECT 1135.6) CLASS (FFGHM)

Name	No	Builders	Laid down	Launched	Commissioned
TALWAR	F 40	Baltic Shipyard, St Petersburg	10 Mar 1999	12 May 2000	18 Jun 2003
TRISHUL	F 43	Baltic Shipyard, St Petersburg	24 Sep 1999	24 Nov 2000	25 Jun 2003
TABAR	F 44	Baltic Shipyard, St Petersburg	26 May 2000	25 May 2001	19 Apr 2004
TEG	F 45	Yantar Shipyard, Kaliningrad	27 Jul 2007	27 Nov 2009	27 Apr 2012
TARKASH	F 46	Yantar Shipyard, Kaliningrad	28 Nov 2007	23 Jun 2010	9 Nov 2012
TRIKAND	F 51	Yantar Shipyard, Kaliningrad	11 Jun 2008	29 May 2011	29 Jun 2013

Displacement, tonnes: 3,678 standard; 4,100 full load
Dimensions, metres (feet): 124.8 × 15.2 × 4.7 *(409.4 × 49.9 × 15.4)*
Speed, knots: 32
Range, n miles: 4,850 at 14 kt, 1,600 at 30 kt
Complement: 180 (18 officers)

Machinery: COGAG; 2 Zorya DN-59 gas turbines; 43,448 hp/m *(34.2 MW)*; 2 Zorya UGT 6000 gas turbines; 16,628 hp(m) *(12.4 MW)*; 2 shafts; fixed propellers
Missiles: SSM: 8 SS-N-27 Novator Alfa Klub-N (Batch 1) (3M-54-TE) ❶ active radar homing to 220 km *(120 n miles)* at 0.8 Mach (cruise) and 2.9 Mach (attack); warhead 400 kg. 8 Brahmos PJ-10 (Batch 2); active/passive radar terminal homing to 290 km *(157 n miles)* at 0.9 Mach; warhead 513 kg. VLS silo.
SAM: SA-N-7 Gadfly (Kashmir/Uragan) single launcher ❷ command, semi-active radar and IR homing to 25 km *(13.5 n miles)* at 3 Mach; warhead 70 kg. 24 9M 317 missiles.
Guns: SAM/Guns: 2 CADS-N-1 (Kashtan) (F 40, 43, 44) ❸ each has twin 30 mm Gatling combined with 8 SA-N-11 (Grisson) and Hot Flash/Hot Spot radar/optronic director. Laser beam guidance for missiles to 8 km *(4.4 n miles)* warhead 9 kg; 9,000 rds/min (combined) to 1.5 km for guns. 2−30 mm AK 630 (Batch 2); 6 barrels per mounting; 3,000 rds/min combined to 2 km. 4−12.7 mm MGs.
1−3.9 in *(100 mm)*/70 A 190E ❹; 80 rds/min to 21.5 km *(11.6 n miles)*; weight of shell 15.6 kg.

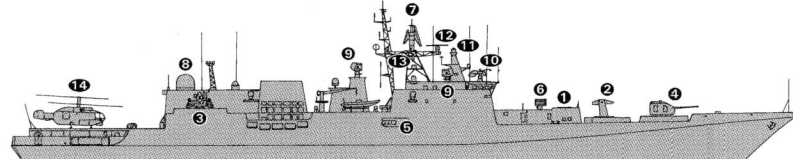

TALWAR (Scale 1 : 1,200), Ian Sturton / 1166541

Torpedoes: 4 DTA-53 21 in *(533 mm)* (2 twin) fixed launchers ❺.
A/S Mortars: 1 RBU 6000 12-barrelled launcher ❻ range 6 km; warhead 31 kg.
Physical countermeasures: Decoys: 2 PK 2 chaff launchers (to be fitted).
Electronic countermeasures: ESM: ASOR (TK-25E-5); jammer.
Radars: Air search: Top Plate (Fregat-M2EM) ❼ 3D; E/F-band.
Air/surface search: Cross Dome (Positiv-E) ❽; E/F-band.
Fire control: 4 Front Dome (MR-90) ❾; H/I-band (for SA-N-7). Plank Shave (Garpun-B) ❿; I/J-band (for SSM); Ratep 5P-10E Puma ⓫; I-band (for 100 mm gun).
Navigation: Kelvin Hughes Nucleus 6000 ⓬; E/F-band. 2 Nyada MR 212/201 (Palm Frond) ⓭; I-band.
Sonars: HUMSA; hull mounted; active/passive medium frequency. VDS (may be fitted in future).

Combat data systems: Trebovaniye-M.

Helicopters: 1 Ka-28/Ka-31 Helix ⓮ or HAL Dhruv.

Programmes: Contract placed in 1997 and confirmed 21 July 1998 for the first batch of three modified Krivak IIIs. Mutual interference difficulties reportedly delayed entry into service of first of class by one year. An option for a second batch of three ships was exercised on 14 July 2006. Construction at Yantar Shipyard, Kaliningrad, started in July 2007. Talks about the construction of a third batch of three ships were reportedly in progress in late 2012.
Structure: Batch 1 are the first surface units to be fitted with the SS-N-27 missile. This missile is replaced by Brahmos in Batch 2.
Operational: *Trikand* arrived at Mumbai in 2013.

TABAR 12/2011, Michael Nitz / 1406901

TRIKAND 7/2013*, Michael Nitz / 1529189

338 India > Frigates

3 + (7) SHIVALIK (PROJECT 17) CLASS (FFGHM)

Name	No	Builders	Laid down	Launched	Commissioned
SHIVALIK	F 47	Mazagon Dock Ltd, Mumbai	11 Jul 2001	18 Apr 2003	29 Apr 2010
SATPURA	F 48	Mazagon Dock Ltd, Mumbai	Oct 2002	4 Jun 2004	20 Aug 2011
SAHYADRI	F 49	Mazagon Dock Ltd, Mumbai	30 Sep 2003	27 May 2005	21 Jul 2012

Displacement, tonnes: 4,674 standard; 6,299 full load
Dimensions, metres (feet): 143 × 16.9 × 5.3 (469.2 × 55.4 × 17.4)
Speed, knots: 30
Range, n miles: 4,500 at 18 kt, 1,600 at 30 kt
Complement: 257 (35 officers)

Machinery: CODOG; 2 GE LM 2,500 gas turbines; 44,000 hp (32.8 MW); 2 SEMT-Pielstick 16PA 6V280 STC diesels; 15,200 hp (11.3 MW); 2 cp propellers.
Missiles: SSM: 8 SS-N-27 Novator Alfa Klub-N (3K-54-TE) ❶; active radar homing to 220 km (120 n miles) at 0.8 Mach (cruise) and 2.9 Mach (attack); warhead 400 kg; VLS silo.
SAM: SA-N-7 Shtil (9M317E) single launcher ❷ command, semi-active radar and IR homing to 25 km (13.5 n miles) at 3 Mach; warhead 70 kg. 24 9M38M1 missiles. 4 octuple Barak 1 VLS ❸; command line-of-sight radar or optical guidance to 10 km (5.5 n miles) at 2 Mach; warhead 22 kg.
Guns: 1 Oto Melara 3 in (76 mm)/62 Super Rapid ❹; 120 rds/min to 16 km (8.7 n miles); weight of shell 6 kg. 2—30 mm AK 630 ❺; 6 barrels per mounting; 3,000 rds/min combined to 2 km.
Torpedoes: 6—324 mm ILAS 3 (2 triple) ❻.
A/S Mortars: 2 RBU 6000 12-barrelled launcher ❼ range 6 km; warhead 31 kg.
Physical countermeasures: Decoys: 4 Kavach chaff launchers.
Electronic countermeasures: ESM/ECM: Ellora EW suite.
Radars: Air search: Elta 2238 AMDR-ER; 3D; E/F-band ❽.
Air/surface search: Top Plate (Fregat-M2EM) ❾ 3D; D/E-band.
Fire control: 2 BEL Shikari (based on Contraves Seaguard) ❿ (for 76 mm); I/K-bands. 1 BEL Aparna (modified Plank Shave/Garpun B) ⓫ (for SSMs); I/J-bands. 4 Front Dome (MR 90) ⓬ (for SA-N-7); H/I-band. 2 EL/M 2221 STGR; I/J/K-band (for Barak).
Navigation: 2 BEL Rashmi; I-band.
Sonars: Bharat HUMSA; hull-mounted; active search and attack; medium frequency. Thales Sintra ATAS.
Combat data systems: BEL CMS-17.
Electro-optic systems: EON-51 optronic director.

Helicopters: 1 Sea King Mk 42B or HAL Dhruv ⓭.

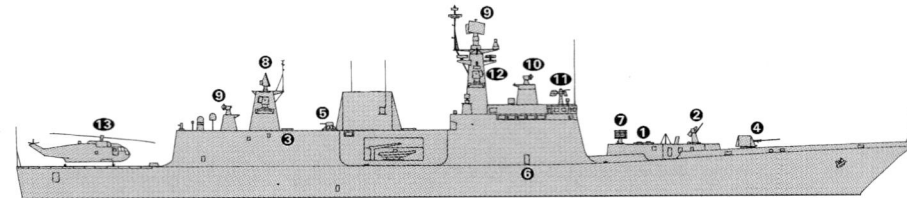

SHIVALIK (Scale 1 : 1,200), Ian Sturton / 1305906

Programmes: Three Project 17 ships approved in June 1999 and construction of the first of class began in 2001. Initially, building was rapid but the production process has taken much longer than first estimates. Project 17A, for seven follow-on ships, was approved in July 2009. The ships are to be built at Mazagon Dock (4) and Garden Reach (3) to a modified design. A construction contract is expected in 2017–18.
Structure: An enlarged and modified version of the Talwar class, the aft section resembles the Delhi class. Signature reduction (IR and RCS) features are believed to be incorporated. Details are speculative.
Operational: *Shivalik* based at Vishakapatnam.

SATPURA — 12/2011, Michael Nitz / 1406902

SHIVALIK — 1/2011, Kapil Chandni / 1406617

SAHYADRI — 10/2013*, Chris Sattler / 1529190

IHS Jane's Fighting Ships 2014-2015 © 2014 IHS

3 GODAVARI CLASS (PROJECT 16) (FFGHM)

Name	No	Builders	Laid down	Launched	Commissioned
GODAVARI	F 20	Mazagon Dock Ltd, Mumbai	2 Jun 1978	15 May 1980	10 Dec 1983
GOMATI	F 21	Mazagon Dock Ltd, Mumbai	1981	19 Mar 1984	16 Apr 1988
GANGA	F 22	Mazagon Dock Ltd, Mumbai	1980	21 Oct 1981	30 Dec 1985

Displacement, tonnes: 4,277 full load
Dimensions, metres (feet): 126.5 × 14.5 × 4.5 hull; 9 sonar *(415 × 47.6 × 14.8; 29.5)*
Speed, knots: 28
Range, n miles: 4,500 at 12 kt
Complement: 313 (40 officers; 13 air crew)

Machinery: 2 Babcock & Wilcox boilers; 550 psi *(38.7 kg/cm²)*; 850°F *(450°C)*; 2 turbines; 30,000 hp *(22.4 MW)*; 2 shafts
Missiles: SSM: 4 SS-N-2D Styx ❶; active radar (Mod 1) or IR (Mod 2) homing to 83 km *(45 n miles)* at 0.9 Mach; warhead 513 kg; sea-skimmer at end of run. Indian designation.
SAM: 3 octuple IAI/Rafael Barak VLS ❷; command line of sight radar or optical guidance to 10 km *(5.5 n miles)* at 2 Mach; warhead 22 kg.
Guns: 1 Oto Melara 76 mm/62 Super Rapid ❸; 120 rds/min to 16 km *(8.7 n miles)*; weight of shell 6 kg. 4—30 mm/65 AK 630 (6 barrels per mounting) ❹; 3,000 rds/min combined to 2 km *(1.08 n miles)*. 4—12.7 mm MGs. 2—7.62 mm MGs.
Torpedoes: 6—324 mm ILAS 3 (2 triple) tubes ❺. Whitehead A244S; anti-submarine; active/passive homing to 7 km *(3.8 n miles)* at 33 kt; warhead 34 kg (shaped charge). *Godavari* has tube modifications for the Indian NST 58 version of A244S.
Physical countermeasures: Decoys: 2 Rafael Deseaver decoy launchers. Graseby G738 towed torpedo decoy.
Electronic countermeasures: ESM/ECM: Rafael SEWS.
Radars: Air search: Thales LW-08 ❻; D-band.
Air/surface search: EL/M-2238 STAR ❼; 3D; E/F-band.
Navigation/helo control: 2 Signaal ZW06 ❽; or Don Kay; I-band.
Fire control: EL/M-2221 STGR ❾; I/J/K-band. Bel Lynx ❿; I-band (for 76 mm).
Sonars: Bharat APSOH; hull-mounted; active panoramic search and attack; medium frequency. Fathoms Oceanic VDS. Thomson Sintra DSBV 62 (in *Ganga*); passive towed array; very low frequency. Type 162M; bottom classification; high frequency.
Combat data systems: Selenia IPN-10 action data automation. Inmarsat communications (JRC) ⓫.
Weapon control systems: MR 301 MFCS. MR 103 GFCS.

Helicopters: 2 Sea King or 1 Sea King and 1 Chetak ⓬.

Modernisation: A mid-life update programme has been conducted for all three ships 2001–2011. Barak launchers have replaced SA-N-4. 57 mm gun has been replaced by Oto Melara 76 mm. The sonar may have been replaced by HUMSA.
Structure: A further modification of the original Leander design with an indigenous content of 72% and a larger hull. Poor welding is noticeable in *Godavari*. *Gomati* is the first Indian ship to have digital electronics in her combat data system.
Operational: French Samahé helicopter handling equipment is fitted. Usually only one helo is carried with more than one crew. These ships have a unique mixture of Russian, Western and Indian weapon systems which has inevitably led to some equipment compatibility problems.

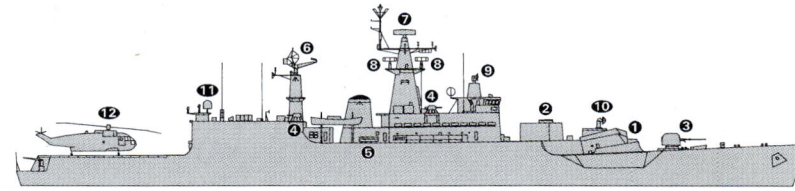

GANGA (Scale 1 : 1,200), *Ian Sturton* / 1482998

GANGA 3/2013*, *Bob Fildes* / 1529191

GOMATI 12/2011, *Michael Nitz* / 1406903

GODAVARI 1/2011, *Kapil Chandni* / 1406618

3 BRAHMAPUTRA CLASS (PROJECT 16A) (FFGHM)

Name	No	Builders	Laid down	Launched	Commissioned
BRAHMAPUTRA	F 31	Garden Reach Shipbuilding & Engineering, Kolkata	1989	29 Jan 1994	14 Apr 2000
BETWA	F 39	Garden Reach Shipbuilding & Engineering, Kolkata	22 Aug 1994	26 Feb 1998	7 Jul 2004
BEAS	F 37	Garden Reach Shipbuilding & Engineering, Kolkata	26 Feb 1998	2002	11 Jul 2005

Displacement, tonnes: 4,521 full load
Dimensions, metres (feet): 126.5 × 14.5 × 4.5 hull; 9 sonar (415 × 47.6 × 14.8; 29.5)
Speed, knots: 27
Range, n miles: 4,500 at 12 kt
Complement: 351 (31 officers; 13 air crew)

Machinery: 2 boilers; 550 psi (38.7 kg/cm²); 850°F (450°C); 2 Bhopal turbines; 30,000 hp (22.4 MW); 2 shafts
Missiles: SSM: 16 SS-N-25 Switchblade (4 quad) (KH-35E Uran) ❶; active radar homing to 130 km (70.2 n miles) at 0.9 Mach; warhead 145 kg; sea skimmer.
SAM: 3 octuple IAI/Rafael Barak VLS ❷; command line of sight radar or optical guidance to 10 km (5.5 n miles) at 2 Mach; warhead 22 kg.
Guns: Oto Melara 76 mm/62 Super Rapid ❸; 120 rds/min to 16 km (8.6 n miles) weight of shell 6 kg. 4—30 mm/65 AK 630 ❹; 6 barrels per mounting; 3,000 rds/min combined to 2 km.
Torpedoes: 6—324 mm ILAS 3 (2 triple) tubes ❺. Whitehead A244S; anti-submarine; active/passive homing to 7 km (3.8 n miles) at 33 kt; warhead 34 kg (shaped charge).
Physical countermeasures: Decoys: 2 Rafael Deseaver decoy launchers. Graseby G738 towed torpedo decoy.
Electronic countermeasures: ESM/ECM: SEWS (F 31, 39); Ellora (F 37) ❻.
Radars: Air search: Bharat RAWL-02 Mk 3 (LW08) ❼; D-band.
Air/surface search: EL/M-2238 STAR; 3D; E/F-band ❽.
Navigation/helo control: Decca Bridgemaster; I-band. BEL Rashmi (PIN 524) (using ZW 06 antenna); I-band.
Fire control: 2 BEL Shikari (based on Contraves Seaguard) ❾ (for 76 mm and AK 630); I/K-bands. EL/M-2221 STGR (for Barak); I/J/K-bands. Bharat Aparna (modified Plank Shave/Garpun B) ❿ (for SSM); I/J-band.
Sonars: Bharat HUMSA (APSOH); hull-mounted; active panoramic search and attack; medium frequency. Thales towed array.
Combat data systems: BEL EMCCA. Inmarsat communications (JRC).
Helicopters: 2 Sea King or 1 Sea King and 1 Chetak ⓫.

Programmes: Project 16A.
Structure: The principal change to the Godavari class on which the design is based, is the replacement of the Godavari SS-N-2 by SS-N-25. Following the cancellation of the Trishul SAM programme, Barak was fitted in its place. Gun armament has also improved.

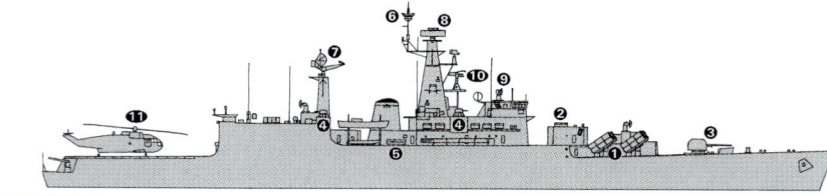

BRAHMAPUTRA (Scale 1 : 1,200), Ian Sturton / 1366993

BRAHMAPUTRA 3/2013*, Bob Fildes / 1529192

BEAS 12/2011, Michael Nitz / 1406906

BETWA 3/2012, Michael Nitz / 1482982

Corvettes < **India** 341

CORVETTES

0 + 4 (8) KAMORTA CLASS (PROJECT 28) (CORVETTES) (FFG)

Name	No	Builders	Laid down	Launched	Commissioned
KAMORTA	P 77	Garden Reach Shipbuilding & Engineering, Kolkata	18 Nov 2006	19 Apr 2010	2014
KADMATT	P 78	Garden Reach Shipbuilding & Engineering, Kolkata	27 Sep 2007	25 Oct 2011	2015
KILTAN	P 79	Garden Reach Shipbuilding & Engineering, Kolkata	10 Aug 2010	26 Mar 2013	2016
KAVARATTI	P 80	Garden Reach Shipbuilding & Engineering, Kolkata	20 Jan 2012	2014	2016

Displacement, tonnes: 3,150 full load
Dimensions, metres (feet): 109.2 × 12.8 × 3.72 *(358.3 × 42.0 × 12.2)*
Speed, knots: 25. **Range, n miles:** 4,000 at 12 kt
Complement: 123

Machinery: CODAD: 4 Pielstick 12PA 6 STC diesels; 22,030 hp *(16.2 MW)*; 2 shafts; cp props
Missiles: SSM: To be announced.
SAM: To be announced.
Guns: 1 Otobreda 3 in *(76 mm)*/62 Super Rapid; 120 rds/min to 16 km *(8.7 n miles)*; weight of shell 6 kg. 2 – 30 mm/65 AK 630; 6 barrels per mounting; 3,000 rds/min combined to 2 km.
Torpedoes: 4 ITTL (2 twin) 533 mm; Eurotorp MU-90.
A/S Mortars: 2 RBU 6000 12-barrelled launchers; range 6 km; warhead 31 kg.
Physical countermeasures: Decoys: 2 Kavach chaff/flare decoy launchers. Towed torpedo decoy.
Electronic countermeasures: ESM: Sanket Mk 3.
ECM: To be announced.
Radars: Surveillance: Bharat Revathi; 3D; E/F-band.
Fire control: 2 Bel Lynx (for 76 mm); I-band.
Navigation: Decca Bridgemaster; I-band.
Sonars: Bharat HUMSA; hull-mounted sonar. Active/passive towed array.
Combat data systems: BEL CMS-28. Datalinks. Satcom.
Weapon control systems: EON 51 optronic director.

Helicopters: 1 Ka-28PL or HAL Dhruv.

Programmes: Multipurpose corvette designed to operate in Indian offshore waters. First four units ordered in 2003 and first steel cut for first of class on 12 August 2005 but construction progress has been slow. A class of 12 is planned.
Structure: The design is understood to be the result of a joint venture by the Indian Navy's DGND SSG (Directorate General Naval Design Surface Ship Group) and Garden Reach Shipbuilder's in-house design team. Details have not been formally released and are speculative. Measures to reduce acoustic, magnetic, IR and radar cross-section signatures are reported to have been incorporated. The hull may use amagnetic steel. The third and subsequent ships may have composite superstructures.
Operational: *Kamorta* ran aground during sea trials off Kolkata on 25 October 2013.

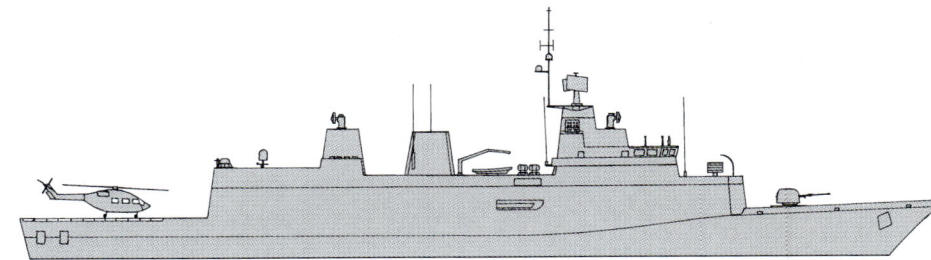

PROJECT 28 (Scale 1 : 900), Ian Sturton / 1353092

KAMORTA 1/2012, GRSE / 1450968

4 ABHAY (PROJECT 1241 PE) (PAUK II) CLASS (FSM)

Name	No	Builders	Commissioned
ABHAY	P 33	Volodarski, Rybinsk	10 Mar 1989
AJAY	P 34	Volodarski, Rybinsk	24 Jan 1990
AKSHAY	P 35	Volodarski, Rybinsk	10 Dec 1990
AGRAY	P 36	Volodarski, Rybinsk	30 Jan 1991

Displacement, tonnes: 493 full load
Dimensions, metres (feet): 58.5 × 10.2 × 3.4 *(191.9 × 33.5 × 11.2)*
Speed, knots: 28
Range, n miles: 2,400 at 14 kt
Complement: 32 (6 officers)

Machinery: 2 MTU 16V 1163 TB93 diesels (P 34–35); 12,900 hp(m) *(9.6 MW)* sustained; 2 Radial M 521-TM5 diesels (P 33, P 36); 2 shafts
Missiles: SAM: SA-N-5/8 Grail quad launcher; manual aiming, IR homing to 6 km *(3.2 n miles)* at 1.5 Mach; warhead 1.5 kg.
Guns: 1 USSR 3 in *(76 mm)*/59 AK 176; 120 rds/min to 15 km *(8 n miles)*; weight of shell 5.9 kg. 1 – 30 mm/65 AK 630; 6 barrels; 3,000 rds/min combined to 2 km.
Torpedoes: 4 – 21 in *(533 mm)* (2 twin) tubes. SET-65E; active/passive homing to 15 km *(8.1 n miles)* at 40 kt; warhead 205 kg.
A/S Mortars: 2 RBU 1200 5-tubed fixed; range 1,200 m; warhead 34 kg.
Physical countermeasures: Decoys: 2 PK 16 chaff launchers.
Radars: Air/surface search: Cross Dome; E/F-band.
Navigation: Pechora; I-band.
Fire control: Bass Tilt; H/I-band.
Sonars: Rat Tail VDS (on transom); attack; high frequency.

Programmes: Modified Pauk II class built in the USSR at Volodarski, Rybinsk for export. Original order in late 1983 but completion of the first delayed by lack of funds and the order for the others was not reinstated until 1987. Names associated with former coastal patrol craft.
Modernisation: Type M 521 diesels are being replaced by MTU diesels. The VDS system may be replaced by a western system.
Structure: Has a longer superstructure than the Pauk I, larger torpedo tubes and improved electronics.
Operational: Classified as ASW ships. Comprise 23rd Patrol Boat Squadron based at Mumbai. *Agray* was damaged by an onboard explosion on 6 February 2004 but was subsequently repaired at Mumbai. The torpedo tubes and VDS have been removed and she acts as a trials platform and patrol ship.

AJAY 12/2011, Michael Nitz / 1406908

© 2014 IHS

IHS Jane's Fighting Ships 2014-2015

342 India > Corvettes

4 KORA CLASS (PROJECT 25A) (FSGHM)

Name	No	Builders	Laid down	Launched	Commissioned
KORA	P 61	Garden Reach Shipbuilding & Engineering, Kolkata	10 Jan 1990	23 Sep 1992	10 Aug 1998
KIRCH	P 62	Garden Reach Shipbuilding & Engineering, Kolkata; Mazagon Dock	31 Jan 1990	28 Sep 1995	22 Jan 2001
KULISH	P 63	Garden Reach Shipbuilding & Engineering, Kolkata	4 Oct 1995	18 Aug 1997	20 Aug 2001
KARMUKH	P 64	Garden Reach Shipbuilding & Engineering, Kolkata; Mazagon Dock	27 Aug 1997	6 Apr 2000	4 Feb 2004

Displacement, tonnes: 1,483 full load
Dimensions, metres (feet): 91.1 × 10.5 × 4.5 *(298.9 × 34.4 × 14.8)*
Speed, knots: 25
Range, n miles: 4,000 at 16 kt
Complement: 134 (14 officers)

Machinery: 2 SEMT-Pielstick/Kirloskar 18 PA6 V 280 diesels; 14,400 hp(m) *(10.58 MW)* sustained; 2 shafts; LIPS cp props

Missiles: SSM: 8 (or 16) Zvezda SS-N-25 Switchblade (2 (or 4) quad) (Kh 35E Uran) ❶; active radar homing to 130 km *(70.2 n miles)* at 0.9 Mach; warhead 145 kg; sea skimmer.
SAM: 2 SA-N-5 Grail ❷; manual aiming; IR homing to 6 km *(3.2 n miles)* at 1.5 Mach; altitude to 2,500 m *(8,000 ft)*; warhead 1.5 kg.
Guns: 1 USSR 3 in *(76 mm)*/59 AK 176 (P 61) ❸; 120 rds/min to 15 km *(8.0 n miles)*; weight of shell 5.9 kg. 1 Otobreda 76 mm/62 Super Rapid (P 62, P 63 and P 64). 2—30 mm/65 AK 630 ❹; 6 barrels per mounting; 3,000 rds/min to 2 km.
Physical countermeasures: Decoys: 2 PK 10 chaff launchers ❺. 2 BEL TOTED; towed torpedo decoys.
Electronic countermeasures: ESM: Bharat Ajanta P Mk II intercept ❻.
Radars: Air search: Cross Dome ❼; E/F-band; range 130 km *(70 n miles)*.
Air/surface search: Plank Shave (Granit Harpun B) ❽; I/J-band.
Fire control: Bass Tilt (P 61) ❾; H/I-band; BEL Lynx (P62-64); I-band.
Navigation: Bharat 1245; I-band.
IFF: Square Head.
Combat data systems: Bharat Vympal IPN-10.
Helicopters: Platform only ❿ for Chetak or HAL Dhruv.

Programmes: First pair ordered in April 1990 and second pair in October 1994. Programme was slowed by delays in provision of Russian equipment.
Structure: Very similar to the original Khukri class except that SS-N-25 has replaced SS-N-2. Stabilisers fitted.

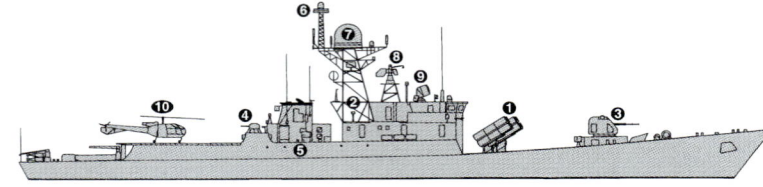
KORA *(Scale 1 : 900), Ian Sturton / 0064715*

KIRCH *5/2013*, Guy Toremans / 1529193*

KULISH *12/2010, M Mazumdar / 1406619*

KORA *12/2011, Hartmut Ehlers / 1482981*

IHS Jane's Fighting Ships 2014-2015

Corvettes < **India** 343

4 KHUKRI CLASS (PROJECT 25) (FSGHM)

Name	No	Builders	Laid down	Launched	Commissioned
KHUKRI	P 49	Mazagon Dock Ltd, Mumbai	27 Sep 1985	3 Dec 1986	23 Aug 1989
KUTHAR	P 46	Mazagon Dock Ltd, Mumbai	13 Sep 1986	15 Apr 1989	7 Jun 1990
KIRPAN	P 44	Garden Reach Shipbuilding & Engineering, Kolkata	15 Nov 1985	16 Aug 1988	12 Jan 1991
KHANJAR	P 47	Garden Reach Shipbuilding & Engineering, Kolkata	15 Nov 1985	16 Aug 1988	22 Oct 1991

Displacement, tonnes: 1,446 full load
Dimensions, metres (feet): 91.1 × 10.5 × 4 *(298.9 × 34.4 × 13.1)*
Speed, knots: 24
Range, n miles: 4,000 at 16 kt
Complement: 112 (12 officers)

Machinery: 2 SEMT-Pielstick/Kirloskar 18 PA6 V 280 diesels; 14,400 hp(m) *(10.58 MW)* sustained; 2 shafts; LIPS cp props
Missiles: SSM: 4 SS-N-2D Mod 1 Styx (2 twin) launchers ❶; IR homing to 83 km *(45 n miles)* at 0.9 Mach; warhead 513 kg.
SAM: SA-N-5 Grail ❷; manual aiming; IR homing to 6 km *(3.2 n miles)* at 1.5 Mach; altitude to 2,500 m *(8,000 ft)*; warhead 1.5 kg.
Guns: 1 USSR 3 in *(76 mm)*/59 AK 176 ❸; 120 rds/min to 15 km *(8.0 n miles)*; weight of shell 5.9 kg. 2 – 30 mm/65 AK 630 ❹; 6 barrels per mounting; 3,000 rds/min to 2 km.
Physical countermeasures: Decoys: 2 PK 16 chaff launchers ❺.
Electronic countermeasures: ESM: Bharat Ajanta P; intercept.
Radars: Air search: Cross Dome ❻; E/F-band; range 130 km *(70 n miles)*.
Air/surface search: Plank Shave ❼; I-band.
Fire control: Bass Tilt ❽; H/I-band.
Navigation: Bharat 1245; I-band.
Combat data systems: Selenia IPN-10 (Khukri); Bharat Vympal IPN-10 (remainder).
Helicopters: Platform only ❾ for Chetak (to be replaced by HAL Dhruv in due course).

Programmes: First two ordered December 1983, two in 1985. The diesels were assembled in India under licence by Kirloskar. Indigenous content of the whole ship is about 65%.
Operational: All based at Vishakapatnam.

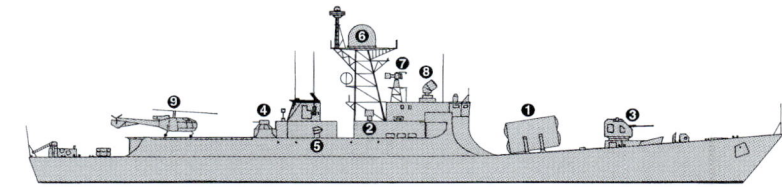

KHUKRI (Scale 1 : 900), Ian Sturton / 0064713

KHANJAR 12/2011, Hartmut Ehlers / 1406909

KHUKRI 8/2009, John Mortimer / 1305492

KIRPAN 12/2011, Hartmut Ehlers / 1482980

© 2014 IHS IHS Jane's Fighting Ships 2014-2015

India > Corvettes

12 VEER (TARANTUL I) CLASS (PROJECT 1241RE) (FSGM)

Name	No	Builders	Laid down	Launched	Commissioned
VEER	K 40	Volodarski, Rybinsk	1984	Oct 1986	26 Mar 1987
NIRBHIK	K 41	Volodarski, Rybinsk	1985	Oct 1987	21 Dec 1987
NIPAT	K 42	Volodarski, Rybinsk	1986	Nov 1988	5 Dec 1988
NISHANK	K 43	Volodarski, Rybinsk	1987	Jun 1989	12 Sep 1989
NIRGHAT	K 44	Volodarski, Rybinsk	1988	Mar 1990	15 Dec 1989
VIBHUTI	K 45	Mazagon Dock Ltd, Mumbai	28 Sep 1987	26 Apr 1990	3 Jun 1991
VIPUL	K 46	Mazagon Dock Ltd, Mumbai	29 Feb 1988	3 Jan 1991	16 Mar 1992
VINASH	K 47	Goa Shipyard Ltd, Goa	30 Jan 1989	24 Jan 1992	20 Nov 1993
VIDYUT	K 48	Goa Shipyard Ltd, Goa	27 May 1990	12 Dec 1992	16 Jan 1995
NASHAK	K 83	Mazagon Dock Ltd, Mumbai	21 Jan 1991	12 Nov 1993	29 Dec 1994
PRABAL	K 92	Mazagon Dock Ltd, Mumbai	31 Aug 1998	28 Sep 2000	11 Apr 2002
PRALAYA	K 91	Goa Shipyard Ltd, Goa	14 Nov 1998	14 Dec 2000	18 Dec 2002

Displacement, tonnes: 462 (K 40–48,83), 485 (K 91–92) full load
Dimensions, metres (feet): 56.1 × 11.5 × 2.5 (184.1 × 37.7 × 8.2)
Speed, knots: 36. **Range, n miles:** 2,000 at 20 kt, 400 at 36 kt
Complement: 41 (5 officers)
Machinery: COGAG (M15E); 2 Nikolayev Type DR 77 (DS 71 in K 92) gas turbines; 16,016 hp(m) *(11.77 MW)* sustained; 2 Nikolayev Type DR 76 gas turbines with reversible gearboxes; 4,993 hp(m) *(3.67 MW)* sustained; 2 shafts
Missiles: SSM: 4 SS-N-2D Mod 1 Styx; IR homing to 83 km *(45 n miles)* at 0.9 Mach; warhead 513 kg; sea-skimmer at end of run. 16 (4 quad) SS-N-25 Switchblade (Kh 35 Uran) in K 91 and K 92; active radar homing to 130 km *(70.2 n miles)* at 0.9 Mach; warhead 145 kg; sea skimmer.
SAM: SA-N-5 Grail quad launcher; manual aiming; IR homing to 6 km *(3.2 n miles)* at 1.5 Mach; warhead 1.5 kg.
Guns: 1 USSR 3 in *(76 mm)*/59 AK 176; 120 rds/min to 15 km *(8 n miles)*; weight of shell 5.9 kg. 1 Oto Melara 3 in *(76 mm)*/62 Super Rapid (K 91 and K 92); 120 rds/min to 16 km *(8.7 n miles)*; weight of shell 6 kg. 2—30 mm/65 AK 630; 6 barrels per mounting; 3,000 rds/min combined to 2 km. 2—7.62 mm MGs.
Physical countermeasures: Decoys: PK 16 chaff launcher.
Electronic countermeasures: ESM: Bharat Ajanta P Mk II; intercept.
Radars: Air/surface search: Plank Shave; E-band. Cross Dome (K 91 and K 92); E/F-band.
Navigation: Mius; I-band.
Fire control: Bass tilt; H/I-band. BEL Lynx (K 91 and K 92) (for guns); I-band; Bharat Aparna (modified Plank Shave/Harpun B) (for SSM); I/J-band.
IFF: Salt Pot, Square Head A.

Weapon control systems: Hood Wink optronic director.
Programmes: First five are USSR Tarantul I class built for export. Six further of the same type built in India. Two further craft, armed with the SS-N-25 missile were delivered in 2002.
Structure: K 92 and K 91 are to a modified design to accommodate the SS-N-25 missile. Principal differences are the bridge and mast configurations.
Operational: All form the 22nd Missile Vessel Squadron at Mumbai although K 41 and 47 are based at Vishakapatnam. *Prahar* sunk after a collision on 21 April 2006.

PRALAYA *3/2008, Guy Toremans* / 1305305

PRABAL *12/2011, Michael Nitz* / 1455660

NASHAK *12/2011, Michael Nitz* / 1482979

SHIPBORNE AIRCRAFT

Notes: (1) A Request for Information for carrier-based aircraft was issued to four companies in December 2009: Saab (Gripen JAS-39), EADS (Typhoon), Dassault (Rafale) and Boeing (F/A-18 Super Hornet). These aircraft are to be in addition to the MiG 29 already ordered.
(2) Replacement of the Sea King fleet was initiated in January 2006 when Requests for Proposals were issued to eight overseas suppliers.
(3) The maiden flight of the two-seat naval version of the HAL Tejas light combat aircraft was achieved on 27 April 2012. There are two Mk 1 demonstrators (NP1 and NP2) and two additional Mk 2 are to follow in 2015. Ski-jump trials of Mk 1 are to be conducted at Goa with a view to achieving an Interim Operating Capability in 2018.
(4) Clearance for the US to export E-2D Hawkeye AEW aircraft was granted in September 2009.

Numbers/Type: 41/4 MIG 29K Fulcrum/29 KUB.
Operational speed: 750 kt *(1,400 km/h)*.
Service ceiling: 57,000 ft *(17,400 m)*.
Range: 1,400 n miles *(2,600 km)*.
Role/Weapon systems: All-weather single-seat fighter with attack capability, optimised for ski-jump take off, is to be main weapon of *Admiral Gorshkov* aircraft carrier. Initial order for 12 aircraft and four trainers delivered following the maiden flight of a twin-seat MiG 29KUB on 20 January 2007. Four aircraft were successfully tested on board the Russian carrier *Kuznetsov* in September 2009. Four aircraft were delivered on 19 February 2010 and a further five in May 2011. Agreement in principle to acquire a further 29 aircraft was reached in September 2008 and the order was confirmed in March 2010. Sensors: Phazotron-NIIR Zhuk-ME radar; Elta EL/M-8222 jammer; OLS IR search and track. Weapons: AAM; RVV-AE and R-73. ASM: Kh-35 (possibly Club). Conventional bombs: KAB-500 Kr. 30 mm cannon.

MIG 29K *12/2011, Michael Nitz* / 1406915

Numbers/Type: 8/2 British Aerospace Sea Harrier FRS. Mk 51/Mk 60.
Operational speed: 640 kt *(1,186 km/h)*.
Service ceiling: 51,200 ft *(15,600 m)*.
Range: 800 n miles *(1,480 km)*.
Role/Weapon systems: Fleet air defence, strike and reconnaissance STOVL fighter. Three more acquired from UK in 1999 to make good losses. Following a fatal crash in August 2009, the fleet was reduced to eight. Sensors: Elta 2032 air interception radar, limited ECM/RWR (Elta 8420 in due course). Weapons: Air defence; two or Derby Magic AAMs, two 30 mm Aden cannon. Plans for a mid-life upgrade have been abandoned. Avionics are to be improved to extend life of aircraft to 2020.

SEA HARRIER *12/2011, Michael Nitz* / 1406916

Numbers/Type: 2/17/11 Westland Sea King Mk 42A/Mk 42B/Mk 42C.
Operational speed: 112 kt *(208 km/h)*.
Service ceiling: 11,500 ft *(3,500 m)*.
Range: 664 n miles *(1,230 km)*.
Role/Weapon systems: Mk 42A has primary ASW and 42B primary ASV capability; Mk 42C for commando assault/vertrep. Mk 42C numbers include six UH-3H aircraft acquired from the US in 2010. Not all aircraft are operational. Sensors: MEL Super Searcher radar, Thomson Sintra H/S-12 dipping sonar (Mk 42A and B), AQS 902B acoustic processor (Mk 42B), Marconi Hermes ESM (Mk 42B); Bendix weather radar (Mk 42C). Weapons: ASW; 2 Whitehead A244S or USSR APR-2 torpedoes; Mk 11 depth bombs, mines (Mk 42B only). ASV; two Sea Eagle (Mk 42B only). Unarmed (Mk 42C).

SEA KING MK 42B *12/2011, Hartmut Ehlers* / 1406914

Numbers/Type: 14 Kamov Ka-28 Helix A.
Operational speed: 110 kt *(204 km/h)*.
Service ceiling: 12,000 ft *(3,660 m)*.
Range: 270 n miles *(500 km)*.
Role/Weapon systems: ASW helicopter embarked in large escorts and *Vikramaditya*. Has replaced Ka-25. Sensors: Splash Drop search radar; VGS-3 dipping sonar, sonobuoys. Weapons: ASW; two Whitehead A244S or USSR APR-2 torpedoes or four depth bombs.

Ka-28 *12/2011, Hartmut Ehlers* / 1406913

Numbers/Type: 14 Kamov Ka-31 Helix B.
Operational speed: 119 kt *(220 km/h)*.
Service ceiling: 11,480 ft *(3,500 m)*.
Range: 325 n miles *(600 km)*.
Role/Weapon systems: AEW helicopter. First two delivered late 2002 and seven further in 2003. Five further aircraft delivered 2012–13. Radar antenna folds beneath fuselage. Sensors: OKO E-80/M radar.

Ka-31 *6/2005, IHS/Patrick Allen* / 1136991

Numbers/Type: 18 Aerospatiale (HAL) SA 319B Chetak (Alouette III).
Operational speed: 113 kt *(210 km/h)*.
Service ceiling: 10,500 ft *(3,200 m)*.
Range: 290 n miles *(540 km)*.
Role/Weapon systems: Several helicopter roles performed including embarked ASW and carrier-based SAR, utility and support to commando forces. 15 aircraft are operated by Coast Guard. Weapons: ASW; two Whitehead A244S torpedoes.

CHETAK *12/2011, Michael Nitz* / 1406912

Numbers/Type: 9 HAL Dhruv.
Operational speed: 156 kt *(290 km/h)*.
Service ceiling: 9,850 ft *(3,000 m)*.
Range: 216 n miles *(400 km)*.
Role/Weapon systems: Formerly known as Advanced Light Helicopter (ALH), full production was delayed by thrust and vibration problems which have now been overcome. The naval variant started trials in March 1995 and the first two were delivered in 2003. A total of eight utility variants were delivered by October 2008 and a further aircraft was ordered in 2013 but options for the ASW variant have been deferred.

DHRUV *12/2011, Michael Nitz* / 1406911

LAND-BASED MARITIME AIRCRAFT (FRONT LINE)

Notes: (1) There are 19 HAL Kiran jet trainers also used for FRU duties.
(2) There is a requirement for up to six Medium Range Maritime Patrol Aircraft.
(3) The first of six A-50 Mainstay AEW aircraft, equipped with Elta radar, was delivered in May 2009.

Numbers/Type: 14 Dornier 228.
Operational speed: 200 kt *(370 km/h)*.
Service ceiling: 28,000 ft *(8,535 m)*.
Range: 940 n miles *(1,740 km)*.
Role/Weapon systems: Coastal surveillance and EEZ protection duties for Navy and Coast Guard. A total of 30 aircraft delivered for the Navy and Coast Guard. Sensors: MEL Marec or THORN EMI Super Marec search radar with FLIR, cameras and searchlight. Weapons: Unarmed, but may carry anti-ship missiles in due course.

DORNIER 228   *12/2011, Michael Nitz* / 1406919

Numbers/Type: 5 Ilyushin Il-38SD (May).
Operational speed: 347 kt *(645 km/h)*.
Service ceiling: 32,800 ft *(10,000 m)*.
Range: 3,887 n miles *(7,200 km)*.
Role/Weapon systems: Shore-based long-range ASW reconnaissance into Indian Ocean. Following the loss of two aircraft in a mid-air collision in 2002, two replacement aircraft were donated by Russia. All five aircraft upgraded to Il-38SD standard with improved avionics, radar, ASM (probably Brahmos) and ASW capabilities. The first three aircraft had been delivered by 2008. Delivery of the final two was made by 2010. Sensors: Leninets Sea Dragon/Novella radar, MAD, sonobuoys, ESM. Weapons: ASW; various torpedoes, mines and depth bombs.

IL-38SD   *12/2011, Michael Nitz* / 1406918

Numbers/Type: 7 Tupolev Tu-142M (Bear F).
Operational speed: 500 kt *(925 km/h)*.
Service ceiling: 45,000 ft *(13,720 m)*.
Range: 6,775 n miles *(12,550 km)*.
Role/Weapon systems: First entered service in April 1988 for long-range surface surveillance and ASW. Air Force manned. The aircraft were upgraded in 2001 and a further upgrade was conducted by Beriev from about 2009. The first aircraft was completed in December 2011. The scope of the upgrade included new engines, a new mission suite and a structural refit. Sensors: Wet Eye search and attack radars, MAD, cameras. 75 active and passive sonobuoys. Weapons: ASW; 12 torpedoes, depth bombs. ASV; two 23 mm cannon. Avionics, ASM (possibly SS-N-25).

BEAR F   *12/2011, Michael Nitz* / 1406917

Numbers/Type: 8 SEPECAT/HAL Jaguar International.
Operational speed: 917 kt *(1,699 km/h)* (max).
Service ceiling: 36,000 ft *(11,000 m)*.
Range: 760 n miles *(1,408 km)*.
Role/Weapon systems: A maritime strike squadron. Air Force operated. Sensors: Thomson-CSF Agave radar. Weapons: ASV; 2 BAe Sea Eagle; 2 DEFA 30 mm cannon or up to 8–1,000 lb bombs. Can carry 2 Magic AAM overwing.

JAGUAR   *2/2001, Wingman Aviation* / 0121340

Numbers/Type: 12 Boeing P-8I Neptune.
Operational speed: 490 kt *(907 km/h)*.
Service ceiling: 41,000 ft *(12,500 m)*.
Range: 1,380 n miles *(2,555 km)*.
Role/Weapon systems: Contract for eight export variants of the P-8A aircraft signed on 31 December 2008. The first arrived in India on 15 May 2013 and the remainder are to be completed by 2016. A further four aircraft were ordered on 5 October 2010. To replace Tu-142 (Bear F). Design based on Boeing 737–800ERX. Crew of nine. Sensors: To be equipped with modern ASW, ASUW and intelligence, surveillance and reconnaissance (ISR) sensors. Weapons: To be announced.

P-8I NEPTUNE   *6/2012, Boeing* / 1482978

UNMANNED AIR VEHICLES

Numbers/Type: 12 Israel Aircraft Industries Heron.
Operational speed: 125 kt *(231 km/h)*.
Service ceiling: 26,500 ft *(8,075 m)*.
Range: 108 n miles *(200 km)*.
Role/Weapon systems: Capable of performing a variety of missions but primarily a real-time system for intelligence collection, surveillance, target acquisition/tracking, and communications/data relay. Several payloads can be carried simultaneously including real-time TV/FLIR, synthetic aperture radar or camera. Can be controlled from ground station via direct LOS data/command link. Part of UAV squadron commissioned on 6 January 2006. Based at Kochi but operated from other bases. Conducted sea trials with INS *Vindhyagiri* (now decommissioned). Endurance 50 hours.

HERON UAV   *12/2005, IAI* / 1116200

Numbers/Type: 18 Israel Aircraft Industries Searcher II.
Operational speed: 105 kt *(194 km/h)*.
Service ceiling: 20,000 ft *(6,100 m)*.
Range: 92 n miles *(170 km)*.
Role/Weapon systems: Can be configured for tactical surveillance or as communications relay aircraft. Several payloads can be carried simultaneously including real-time TV/FLIR, synthetic aperture radar or camera. Can be controlled from ground station via direct LOS data/command link. Part of UAV squadron commissioned on 6 January 2006. Based at Kochi but operated from other bases. Endurance 18 hours.

SEARCHER II   *6/2003, IHS/C Hoyle* / 0531011

Patrol forces < **India** 347

PATROL FORCES

Notes: (1) A contract to build five new 95 m offshore patrol vessels was announced by Pipavav Shipyard, Gujarat on 30 May 2011. Construction of the first two vessels had started by early 2014. The design, developed by Alion Science and Technology, includes a medium-calibre gun and a helicopter deck.
(2) Approval for indigenous development of a 700-tonne ASW patrol craft was given by the Defence Acquisition Council on 23 December 2013. Acquisition of 16 vessels, for operations in coastal waters, is under consideration. Garden Reach Shipbuilders and Engineers is to work in co-ordination with the navy on the design.

3 + 1 SARYU CLASS (PSOH)

Name	No	Builders	Laid down	Launched	Commissioned
SARYU	P 54	Goa Shipyard Ltd, Goa	15 Dec 2006	30 Mar 2009	21 Jan 2013
SUNAYNA	P 57	Goa Shipyard Ltd, Goa	25 Sep 2007	14 Nov 2009	15 Oct 2013
SUMEDHA	P 58	Goa Shipyard Ltd, Goa	7 May 2008	21 May 2011	2014
SUMITRA	P 59	Goa Shipyard Ltd, Goa	28 Apr 2010	6 Dec 2010	2014

Displacement, tonnes: 2,230 full load
Dimensions, metres (feet): 105.3 × 12.9 × 3.6 (345.5 × 42.3 × 11.8)
Speed, knots: 24. **Range, n miles:** 6,500 at 12 kt
Complement: 113 (18 officers)

Machinery: 2 Pielstick PA 6B STC diesels; 21,725 hp(m) (16.2 MW); 2 shafts; cp props
Guns: 1 Oto Melara 3 in (76 mm)/62 Super Rapid; 120 rds/min to 16 km (8.7 n miles); weight of shell 6 kg. 2—30 mm/65 AK 630 (6 barrels per mounting); 3,000 rds/min combined to 2 km.
Physical countermeasures: 4 Kavach decoy launchers.
Electronic countermeasures: ESM.
Radars: Surface search: To be announced.
Navigation: Sperry Bridgemaster; I-band.
Electro-optic systems: EON 51.

Helicopters: 1 HAL Dhruv.

Comment: The order for three offshore patrol vessels was announced in 2006 and the order for a fourth was confirmed on 20 April 2007. The ships are naval variants of the Coast Guard's Sankalp class under construction at Goa Shipyard.

SARYU   11/2013*, Shaun Jones / 1529194

Sumedha was constructed at Shoft Shipyard and was delivered to Goa for completion on 19 May 2010. Saryu was completed on 30 November 2012 and commissioned in 2013.

6 SUKANYA CLASS (PSOH)

Name	No	Builders	Launched	Commissioned
SUKANYA	P 50	Korea Tacoma, Masan	1989	31 Aug 1989
SUBHADRA	P 51	Korea Tacoma, Masan	1989	25 Jan 1990
SUVARNA	P 52	Korea Tacoma, Masan	22 Aug 1990	4 Apr 1991
SAVITRI	P 53	Hindustan Shipyard, Vishakapatnam	23 May 1989	27 Nov 1990
SHARADA	P 55	Hindustan Shipyard, Vishakapatnam	22 Aug 1990	27 Oct 1991
SUJATA	P 56	Hindustan Shipyard, Vishakapatnam	25 Oct 1991	3 Nov 1993

Displacement, tonnes: 1,920 full load
Dimensions, metres (feet): 101.1 oa; 96 wl × 11.5 × 4.4 (331.7; 315 × 37.7 × 14.4)
Speed, knots: 21
Range, n miles: 5,800 at 15 kt
Complement: 140 (15 officers)

Machinery: 2 SEMT-Pielstick 16 PA6 V 280 diesels; 12,800 hp(m) (9.41 MW) sustained; 2 shafts
Guns: 1 Bofors 40 mm/60. 4—12.7 mm MGs.
A/S Mortars: 4 RBU 2500 16-tubed trainable launchers; range 2,500 m; warhead 21 kg. Two launchers fitted in forward section.
Radars: Surface search: Racal Decca 2459; I-band.
Navigation: Bharat 1245; I-band.

Helicopters: 1 Chetak.

Programmes: First three ordered in March 1987 from Korea Tacoma to an Ulsan class design. Second four ordered in August 1987. The Korean-built ships commissioned at Masan and then sailed for India where the armament was fitted. Three others of a modified design have been built for the Coast Guard. P 54 transferred to Sri Lanka December 2000.
Structure: Lightly armed and able to 'stage' helicopters, they are fitted out for offshore patrol work only but have the capacity to be much more heavily armed. Fin stabilisers fitted. Firefighting pump on hangar roof aft.
Operational: These ships are used for harbour defence, protection of offshore installations and patrol of the EEZ. Potential for role change is considerable. Some ships modified to test fire Dhanush (naval version of Prithvi) SRBM from the flight deck. The most recent test was conducted on 11 March 2011 from Suvarna. On 7 November 2004, a 350 km range Prithvi 3 solid propellant missile was reportedly fired in the Bay of Bengal. P 50 and P 51 based at Karwar, P 52 at Mumbai, P 53 at Vishakapatnam and the other two at Kochi.
Sales: Saryu transferred to Sri Lanka in 2000.

SAVITRI   9/2009, Bob Fildes / 1305500

8 + 72 SOLAS MARINE INTERCEPTOR CRAFT (PBF)

T 301–304   T 401–404   +72

Displacement, tonnes: 11.5 full load
Dimensions, metres (feet): 16.33 × 3.8 × 0.8 (53.6 × 12.5 × 2.6)
Speed, knots: 45
Complement: 4 + 12 personnel
Machinery: 2 Caterpillar C18 diesels; 1,746 hp (1.3 MW); 2 Castoldi waterjets
Guns: 1—7.62 mm MG.
Radars: Navigation: I-band.

Comment: An order for 80 fast interceptor craft was placed with Solas Marine, Sri Lanka, in September 2011. The craft are to be manned by the Sagar Prahari naval security force. The first four were delivered in March 2013 and the second on 28 June 2013.

T 403   3/2013*, Indian Navy / 1461753

5 SUPER DVORA MK II CLASS (PBF)

No	Builders	Commissioned
T 80	IAI, Ramta	24 Jun 1998
T 81	Goa Shipyard Ltd, Goa	6 Jun 1999
T 82	IAI, Ramta	9 Oct 2003
T 83	Goa Shipyard Ltd, Goa	27 Nov 2003
T 84	Goa Shipyard Ltd, Goa	19 Apr 2004

Displacement, tonnes: 61 full load
Dimensions, metres (feet): 25.4 × 5.6 × 1.5 (83.3 × 18.4 × 4.9)
Speed, knots: 50. **Range, n miles:** 700 at 42 kt
Complement: 10 (1 officer)

Machinery: 2 MTU 12V 396 TE94 diesels; 4,570 hp(m) (3.36 MW); 2 Arneson ASD 16 surface drives
Guns: 1 Bushmaster 25 mm. 1—12.7 mm MG.
Radars: Surface search: Koden; I-band.
Weapon control systems: Elop MSIS optronic director.

Comment: Collaborative programme involving IAI, Ramta, Israel and Goa Shipyard Ltd. T 80 was built at Ramta and T 82 was procured by the Indian Navy from Israel. The other three (not six as previously reported) were assembled at Goa.

T 82   12/2011, Michael Nitz / 1482977

© 2014 IHS   IHS Jane's Fighting Ships 2014-2015

348 India > Patrol forces

10 CAR NICOBAR CLASS (PBO)

Name	No	Builders	Launched	Commissioned
CAR NICOBAR	T 69	Garden Reach Shipbuilding & Engineering, Kolkatta	23 Nov 2007	16 Feb 2009
CHETLAT	T 70	Garden Reach Shipbuilding & Engineering, Kolkatta	27 Nov 2007	16 Feb 2009
CORA DIVH (ex-*Cinque*)	T 71	Garden Reach Shipbuilding & Engineering, Kolkatta	16 Jul 2008	10 Sep 2009
CHERIYAM	T 72	Garden Reach Shipbuilding & Engineering, Kolkatta	16 Jul 2008	10 Sep 2009
CANKARSO	T 73	Rajabagan Shipyard, Kolkatta	27 Mar 2009	29 Jun 2010
KONDUL	T 74	Rajabagan Shipyard, Kolkatta	27 Mar 2009	29 Jun 2010
KALPENI	T 75	Rajabagan Shipyard, Kolkatta	27 Mar 2009	14 Oct 2010
KABRA	T 76	Garden Reach Shipbuilding & Engineering, Kolkatta	29 Mar 2010	8 Jun 2011
KOSWARI	T 77	Garden Reach Shipbuilding & Engineering, Kolkatta	29 Mar 2010	12 Jul 2011
KARUVA	T 78	Garden Reach Shipbuilding & Engineering, Kolkatta	29 Mar 2010	25 Aug 2011

Displacement, tonnes: 293 full load
Dimensions, metres (feet): 48.9 × 7.5 × 2.1 *(160.4 × 24.6 × 6.9)*
Speed, knots: 35
Range, n miles: 2,000 at 13 kt
Complement: 29 (6 officers)
Machinery: 3 MTU 16V 4000 M90 diesels; 11,238 hp(m) *(8.16 MW)* sustained; 3 Hamilton HM811 waterjets
Guns: 1—30 mm CRN-91. 2—12.7 mm MGs.
Radars: Surface search: To be announced.

Comment: The design is an improved version of the Bangaram class and other patrol craft built by Garden Reach and may also have drawn on the waterjet propelled Sarojini Naidu class built by Goa Shipyard for the coast guard. Steel hull and aluminium superstructure. Eight similar craft have been ordered for the Coast Guard. Two vessels are to be loaned to Mauritius in 2014 and 2015.

KARUVA
12/2011, Guy Toremans
1482975

6 SDB MK 5 (BANGARAM) CLASS
(LARGE PATROL CRAFT) (PBO)

Name	No	Launched	Commissioned
TRINKAT	T 61	28 Sep 2000	28 Sep 2000
TARASA	T 63	2000	24 Aug 2001
BANGARAM	T 65	11 Dec 2004	10 Feb 2006
BITRA	T 66	14 Dec 2004	28 Mar 2006
BATTI MALV	T 67	28 Jun 2005	31 Jul 2006
BARATANG	T 68	6 Aug 2005	12 Sep 2006

Displacement, tonnes: 264 full load
Dimensions, metres (feet): 46 × 7.5 × 2.5 *(150.9 × 24.6 × 8.2)*
Speed, knots: 30. **Range, n miles:** 2,000 at 12 kt
Complement: 34 (4 officers)
Machinery: 2 MTU 16V 538 TB92 diesels (T 61, 63); 6,820 hp(m) *(5 MW)* sustained; 2 MTU 16V 4000 M90 diesels (T 65–68); 7,492 hp *(5.6 MW)*; 2 shafts
Guns: 1 Medak 30 mm 2A42. 2—7.62 mm MGs.
Radars: Surface search: Bharat 1245; I-band.

Comment: T 61 and T 63 are the two survivors of four commissioned between September 2000 and March 2002. T 62 transferred to the Maldives in 2006 and T 64 to the Seychelles in June 2005. Four (T 65–68) of a modified design have since entered service.

ISV
6/2013*, Craftway Engineers / 1455838

15 + (7) COUACH PLASCOA 13 M PATROL CRAFT (PB)

Name	No	Builders	Commissioned
–	T 201	Chantier Naval Couach	2011
–	T 202	Chantier Naval Couach	2011
–	T 203	Chantier Naval Couach	2011
–	T 204	Chantier Naval Couach	2012
–	T 205	Chantier Naval Couach	2012
–	T 206	Chantier Naval Couach	2012
–	T 207	Chantier Naval Couach	2012
–	T 208	Chantier Naval Couach	2012
–	T 209	Chantier Naval Couach	2012
–	T 210	Chantier Naval Couach	2012
–	T 211	Chantier Naval Couach	2012
–	T 212	Chantier Naval Couach	2012
–	T 213	Chantier Naval Couach	2012
–	T 214	Chantier Naval Couach	2012
–	T 215	Chantier Naval Couach	2012

Displacement, tonnes: 12 full load
Dimensions, metres (feet): 13.65 × 3.8 × 0.76 *(44.8 × 12.5 × 2.5)*
Speed, knots: 50
Range, n miles: 200 at 20 kt
Complement: 5
Machinery: 2 MAN R6-800 diesels; 1,600 hp *(1.19 MW)*; 2 Arneson surface drives
Guns: 1—12.7 mm MG. 1—7.62 mm MG.

Comment: An order for 15 interceptor craft, with an option for a further seven, was made from Chantier Naval Couach in 2010. The first batch of three was delivered in June 2011 and a further 12 craft were delivered by the end of 2012. The craft are constructed of Aramat composite material (glass fibre, Kevlar and carbon fibre) and are to provide protection to naval vessels and naval installations in coastal waters.

BANGARAM
8/2009, Indian Navy / 1406058

TARASA
12/2011, Guy Toremans / 1482976

5 + 18 INTERMEDIATE SUPPORT VESSELS (PB)

T 11–15   +18

Dimensions, metres (feet): 23.1 × 5.6 × 0.93 *(75.8 × 18.4 × 3.05)*
Speed, knots: 40. **Range, n miles:** 200 at 20 kt
Complement: 5
Machinery: 2 Caterpillar diesels; 2 DOEN waterjets
Guns: To be announced.

Comment: An order for 23 Intermediate Support Vessels was made in 2013. Nine are to be built by Abu Dhabi Shipbuilding and the remaining 14 by Craftway Engineers, Mumbai. The craft are of GRP construction. While the vessels are to be operated by the Marine Commando Force, they were funded by the Oil and Natural Gas Corporation. The role of the vessels is to carry out day and night surveillance of offshore installations and general security operations. Five were delivered in 2013 and the remainder is to follow in 2014.

T 213
3/2013*, Bob Fildes / 1529195

IHS Jane's Fighting Ships 2014-2015

© 2014 IHS

AMPHIBIOUS FORCES

Notes: There are plans to acquire four amphibious ships. The broad requirement is for vessels capable of transporting up to 800 troops, six main battle tanks and up to 85 support vehicles. The dock is to be capable of accommodating two LCAC and four LCVP. Aviation facilities are to include a hangar for two helicopters and a flight deck with six spots. The ships are also to act as a command and control platform. Tenders were issued in November 2013 to Larsen and Toubro (teamed with Navantia), Pipavav (teamed with DCNS), and ABG (teamed with Avion). The winning design is to be built by Hindustan Shipyard.

1 AUSTIN CLASS (AMPHIBIOUS TRANSPORT DOCK) (LPD)

Name	No	Builders	Laid down	Launched	Commissioned
JALASHWA (ex-*Trenton*)	L 41 (ex-LPD 14)	Lockheed SB & Construction Co	8 Aug 1966	3 Aug 1968	6 Mar 1971

Displacement, tonnes: 9,277 standard; 17,521 full load (est.)
Dimensions, metres (feet): 178.8 × 30.5 × 7 *(586.6 × 100.1 × 23.0)*
Speed, knots: 21
Range, n miles: 7,700 at 20 kt
Complement: 329 (27 officers)

Military lift: 930 troops; 9 LCM 6s or 4 LCM 8s or 2 LCAC or 20 LVTs. 4 LCPL/LCVP
Machinery: 2 Foster-Wheeler boilers; 600 psi *(42.3 kg/cm²)*; 870°F *(467°C)*; 2 De Laval turbines; 24,000 hp *(18 MW)*; 2 shafts
Guns: 1 General Electric/General Dynamics 20 mm/76 6-barrelled Vulcan Phalanx Mk 15; 3,000 rds/min (4,500 in Block 1) combined to 1.5 km. 2—25 mm Mk 38. 8—12.7 mm MGs.
Physical countermeasures: Decoys: 4 Loral Hycor SRBOC 6-barrelled Mk 36; IR flares and chaff to 4 km *(2.2 n miles)*.
Electronic countermeasures: ESM: SLQ-32(V)1; intercept.
Radars: Air search: Lockheed SPS-40E; B-band.
Surface search: Norden SPS-67; G-band.
Navigation: Raytheon SPS-73(V)12; I-band.
Tacan: URN 25.
IFF: Mk XII UPX-36.
Combat data systems: SATCOM, WSC-3 (UHF), WSC-6 (SHF).

Helicopters: Up to 6 Sea King UH-3H can be carried. Hangar for only 1 light.

Programmes: Ex-LPD 14 authorised in the US Navy's FY65 new construction programme. Transferred to India on 17 January 2007 and formally recommissioned on 22 June 2007. The transfer of a second ship (ex-*Nashville*) is no longer considered likely.
Structure: One small telescopic hangar. Flight deck is 168 ft *(51.2 m)* in length. Well-deck 394 × 50 ft *(120.1 × 15.2 m)*.
Operational: Based at Vishakapatnam.

JALASHWA   12/2009, M Mazumdar / 1406062

5 MAGAR CLASS (LSTH)

Name	No	Builders	Launched	Commissioned
MAGAR	L 20	Garden Reach Shipbuilding & Engineering, Calcutta	7 Nov 1984	15 Jul 1987
GHARIAL	L 23	Hindustan Shipyard, Vishakapatnam; Garden Reach	1 Apr 1991	14 Jan 1997
SHARDUL	L 16	Hindustan Shipyard, Vishakapatnam; Garden Reach	3 Apr 2004	4 Jan 2007
KESARI	L 15	Garden Reach Shipbuilding & Engineering	8 Jun 2005	5 Apr 2008
AIRAVAT	L 24	Garden Reach Shipbuilding & Engineering	27 Mar 2006	19 May 2009

Displacement, tonnes: 5,746 full load
Dimensions, metres (feet): 124.8 oa; 120 wl × 17.5 × 4 *(409.4; 393.7 × 57.4 × 13.1)*
Speed, knots: 15
Range, n miles: 3,000 at 14 kt
Complement: 136 (16 officers)

Military lift: 10 tanks plus 11 APC plus 500 troops
Machinery: 2 SEMT-Pielstick 12 PA6 V280 STC diesels; 8,560 hp(m) *(6.29 MW)* sustained; 2 shafts
Guns: 2 CRN-91 30 mm.
Rockets: 2—122 mm multibarrel rocket launchers at the bow.
Electronic countermeasures: ESM: Bharat Ajanta; intercept.
Radars: Navigation: Bharat; I-band.
Electro-optic systems: Bharat EON-51.
Helicopters: 1 Sea King 42C; platform for 2.

Comment: Based on the *Sir Lancelot* design. *Magar* was built entirely at Garden Reach. *Gharial* ordered in 1985. Built at Hindustan Shipyard but fitted out at Garden Reach. Internal design differs from *Magar*. Carries four LCVPs on davits. Bow door. Can beach on gradients 1 in 40 or more. *Magar* refitted in 1995. *Shardul* and subsequent ships include major design changes. *Magar*, *Gharial* and *Airavat* are based at Vishakapatnam and *Shardul* at Karwar.

GHARIAL   12/2011, Michael Nitz / 1406923

KESARI   12/2011, Hartmut Ehlers / 1406922

AIRAVAT   12/2011, Hartmut Ehlers / 1406921

6 MK 2/3 LANDING CRAFT (LSM)

L 33   L 35–39

Displacement, tonnes: 508 full load
Dimensions, metres (feet): 57.5 oa; 53.2 pp × 8.2 × 1.6 *(188.6; 174.5 × 26.9 × 5.2)*
Speed, knots: 11
Range, n miles: 1,000 at 8 kt
Complement: 167

Military lift: 250 tons; 2 PT 76 or 2 APC. 120 troops
Machinery: 3 Kirloskar-MAN R8V 16/18TL diesels; 1,686 hp(m) *(1.24 MW)*; 3 shafts
Guns: 2 Bofors 40 mm/60 (aft).
Mines: Can be embarked.
Radars: Navigation: Decca 1229; I-band.

Comment: L 33 and 35 are Mk 2 craft built 1980–83. L 36–39 are Mk 3 craft built 1986–87. All built by Goa Shipyard. L 36–39 have a considerably modified superstructure and a higher bulwark on the cargo deck. L 33 and 35 commissioned 1980–83 and L 36–39 1986–87.

L 36   2/1999, 92 Wing RAAF / 0064719

4 LCM 8 TYPE

LCM 41   +3

Displacement, tonnes: 65.6 standard; 127 full load
Dimensions, metres (feet): 22.5 × 6.4 × 1.6 *(73.8 × 21.0 × 5.2)*
Speed, knots: 122
Range, n miles: 190 at 9 kt
Complement: 4
Military lift: 110 troops plus equipment or 67.5 tonnes of vehicles and stores or 1 MBT.
Machinery: 2 Detroit 12V-71 diesels; 400 hp *(298 kW)*; 2 shafts; Kort nozzles

Comment: Transferred by the US Navy in 2007 as part of the transfer of *Jalashwa*.

LCM 41   12/2010, M Mazumdar / 1406384

Indonesia

TENTARA NASIONAL

Country Overview

The Republic of Indonesia gained full independence from the Netherlands in 1949. Straddling the equator, the country comprises more than 13,670 islands, of which some 6,000 are inhabited. The major islands include Sumatra, Java, Sulawesi (Celebes), southern Borneo (Kalimantan) and western New Guinea (Papua). Smaller islands include Madura, western Timor, Lombok, Sumbawa, Flores, and Bali. The Moluccas and Lesser Sunda Islands are the largest island groups. The coastline of 29,550 n miles is with the South China Sea, the Celebes Sea, the Pacific Ocean and the Indian Ocean. The total land area is 741,903 square miles. The capital, largest city and principal port is Jakarta (Java). Further main ports are at Surabaya (Java), Medan (Sumatra) and Ujung Pandang (Sulawesi). An archipelagic state, territorial seas (12 n miles) are claimed. A 200 n mile EEZ has also been claimed but the limits are only partly defined by boundary agreements.

Headquarters Appointments

Chief of the Naval Staff: Admiral Marsetio
Vice Chief of the Naval Staff: Vice Admiral Hari Bowo
Inspector General of the Navy: Major General Djunaidi Djahri

Fleet Command

Commander-in-Chief Western Fleet (Jakarta):
 Rear Admiral Amarulla Octavian
Commander-in-Chief Eastern Fleet (Surabaya):
 To be announced
Commandant of Navy Marine Corps:
 Major General Muhammad Alfan Baharudin

Personnel

(a) 2014: 65,000 (including 20,000 Marine Commando Corps and 1,000 Naval Air Arm)
(b) Selective national service

Bases

Tanjung Priok (North Jakarta), Ujung (Surabaya), Sabang, Belawan (North Sumatera), Ujung Pandang (South Sulawesi), Balikpapan (East Kalimantan), Jayapura (Irian Jaya), Tanjung Pinang, Bitung (North Sulawesi), Teluk Ratai (South Sumatera), Banjarmasin (South Kalmantan). Naval Air Base at Juanda (Surabaya), Biak (Irian Jaya), Pekan Baru, Sam Ratulangi (North Sulawesi), Sabang, Natuna, P Aru.

Command Structure

Eastern Command (Surabaya)
Western Command (Jakarta)
Training Command
Military Sea Communications Command (Maritime Security Agency)
Military Sealift Command (Logistic Support)
Plans announced in July 2005 include creation of a third naval command based at Sorong, west Irian Jaya and for the Eastern and Western commands to move to Makassar and Tanjung Pinang, Sumatra, respectively. Dates for implementing the changes have not been announced.

Marine Corps

Reorganisation in March 2001 created the 1st Marine Corps Group (1st, 3rd and 5th battalions) based at Surabaya and the Independent Marine Corps Brigade (2nd, 4th and 6th battalions) based in Jakarta. A new formation (7th, 8th and 9th battalions) is to be based at Teluk Ratai, Sumatra. Equipment includes amphibious tanks, field artillery and anti-aircraft missiles and guns. There are plans to expand the Corps to 22,800 by 2009. Further reorganisation is expected to include relocation of the eastern command from Surabaya to Makassar and the central command from Jakarta to Surabaya.

Strength of the Fleet

Type	Active	Building (Projected)
Patrol Submarines	2	3 (9)
Frigates	6	(3)
Corvettes	22	2 (2)
Attack Craft—Missile	12	3
Large Patrol Craft	10	3
Patrol craft	37	3
LPD	5	1
LST/LSM	19	—
MCMV	9	—
Survey and Research Ships	8	—
Command Ship	1	—
Replenishment Tankers	2	1
Coastal Tankers	2	—
Support Ships	4	—
Transports	8	—
Training Ships	3	—

Prefix to Ships' Names

KRI (Kapal di Republik Indonesia)

PENNANT LIST

Submarines

401	Cakra
402	Nanggala

Frigates

351	Ahmad Yani
352	Slamet Riyadi
353	Yos Sudarso
354	Oswald Siahann
355	Abdul Halim Perdanakusuma
356	Karel Satsuitubun
364	Ki Hajar Dewantara

Corvettes

361	Fatahillah
362	Malahayati
363	Nala
365	Diponegoro
366	Sultan Hasanuddin
367	Sultan Iskandar Muda
368	Frans Kaisiepo
374	Lambung Mangkurat
375	Cut Nyak Dien
383	Iman Bonjol
384	Pati Unus
385	Teuku Umar
386	Silas Papare
871	Kapitan Pattimura
872	Untung Suropati
873	Sutan Nuku
876	Sultan Thaha Syaifuddin
877	Sutanto
878	Sutedi Senoputra
879	Wiratno
881	Tjiptadi
882	Hasan Basri

Patrol Forces

621	Mandau
622	Rencong
623	Badik
624	Keris
631	Todak
641	Clurit
642	Kujang
643	Beladau
651	Singa
653	Ajak
801	Pandrong
802	Sura
804	Hiu
805	Layang
806	Lemadang
807	Boa
808	Welang
809	Suluh Pari
810	Katon
811	Kakap
812	Kerapu
813	Tongkol
814	Barakuda
815	Sanca
816	Warakas
817	Panana
818	Kalakae
819	Tedong Naga
820	Viper
821	Piton
822	Weling
823	Matacora
824	Tedung Selar
825	Boiga
826	Kelabang
827	Krait
828	Kala Hitam
829	Tarihu
830	Alkura
831	Birang
832	Mulga
841	Badau
842	Salawaku
847	Sibarau
848	Siliman
857	Sigalu
858	Silea
859	Siribua
862	Siada
863	Sikuda
864	Sigurot
866	Cucut
867	Kobra
868	Anakonda
869	Patola
870	Taliwangsa

Amphibious Forces

503	Teluk Amboina
509	Teluk Ratai
511	Teluk Bone
512	Teluk Semangka
513	Teluk Penyu
514	Teluk Mandar
515	Teluk Sampit
516	Teluk Banten
517	Teluk Ende
531	Teluk Gilimanuk
532	Teluk Celukan Bawang
533	Teluk Cendrawasih
535	Teluk Peleng
536	Teluk Sibolga
537	Teluk Manado
538	Teluk Hading
539	Teluk Parigi
540	Teluk Lampung
541	Teluk Jakarta
542	Teluk Sangkuring
580	Dore
582	Kupang
583	Dili
584	Nusa Utara
590	Makassar
591	Surabaya
592	Banjarmasin
593	Banda Aceh
971	Tanjung Kambani
973	Tanjung Nusanive
974	Tanjung Fataga
981	Karang Pilang
982	Karang Tekok
983	Karang Banteng
985	Karang Unarang
990	Dr Soeharso

Survey Ships

KAL-IV-02	Baruna Jaya I
KAL-IV-03	Baruna Jaya II
KAL-IV-04	Baruna Jaya III
KAL-IV-05	Baruna Jaya IV
KAL-IV-06	Baruna Jaya VIII
931	Burujulasad
932	Dewa Kembar
933	Jalanidhi

Mine Warfare Forces

711	Pulau Rengat
712	Pulau Rupat
721	Pulau Rote
722	Pulau Raas
723	Pulau Romang
724	Pulau Rimau
726	Pulau Rusa
727	Pulau Rangsang
729	Pulau Rempang

Auxiliaries

543	Teluk Cirebon
544	Teluk Sabang
561	Multatuli
901	Balikpapan
902	Sambu
903	Arun
906	Sungai Gerong
911	Sorong
923	Soputan
924	Leuser
934	Lampo Batang
935	Tambora
936	Bromo
960	Karimata
961	Wagio

SUBMARINES

0 + 3 (9) TYPE 209/1400 CLASS (SSK)

Displacement, tonnes: 1,280 surfaced; 1,412 dived
Dimensions, metres (feet): 61.2 × 6.25 × 5.5 (200.8 × 20.5 × 18.0)
Speed, knots: 11 surfaced; 21.5 dived
Complement: 32 (5 officers)

Machinery: Diesel-electric; 4 diesels; 1 motor; 1 shaft
Missiles: To be announced.
Torpedoes: 8—21 in (533 mm) bow tubes.
Electronic countermeasures: To be announced.
Radars: To be announced.

Sonars: To be announced.
Weapon control systems: To be announced.

Comment: It was announced on 20 December 2011 that, following a competition, Daewoo Shipbuilding had been awarded a contract to build three new submarines for the Indonesian Navy. It is reported that the boats are to be built with technical assistance from Germany's HDW. The first two boats, to be delivered by 2018, are to be constructed in South Korea and the third, to be delivered by 2020, under licence at PT Pal, Surabaya. The industrial infrastructure to undertake this work is to be completed in 2016. According to plans announced in September 2012, a further nine submarines are to be procured by 2024. While details of the boats have not yet been released, it is expected that the submarines are to be stretched and more advanced versions of the Chang Bogo (Type 209/1200) class that are in service with the South Korean Navy and are likely to equate broadly to Type 209/1400-class submarines in service in Brazil, Chile, Greece, South Africa, and Turkey. However, this has not been confirmed and details remain speculative.

IHS Jane's Fighting Ships 2014-2015

© 2014 IHS

Submarines — **Indonesia** 359

2 CAKRA TYPE 209/1300 CLASS

Name	No	Builders	Laid down	Launched	Commissioned
CAKRA	401	Howaldtswerke, Kiel	25 Nov 1977	10 Sep 1980	19 Mar 1981
NANGGALA	402	Howaldtswerke, Kiel	14 Mar 1978	10 Sep 1980	6 Jul 1981

Displacement, tonnes: 1,306 surfaced; 1,412 dived
Dimensions, metres (feet): 59.5 × 6.2 × 5.4 (195.2 × 20.3 × 17.7)
Speed, knots: 11 surfaced; 21.5 dived
Complement: 34 (6 officers)

Machinery: Diesel-electric; 4 MTU 12V 493 AZ80 GA31L diesels; 2,400 hp(m) (1.76 MW) sustained; 4 Siemens alternators; 1.7 MW; 1 Siemens motor; 4,600 hp(m) (3.38 MW) sustained; 1 shaft
Torpedoes: 8—21 in (533 mm) bow tubes. 14 AEG SUT Mod 0; dual purpose; wire-guided; active/passive homing to 12 km (6.5 n miles) at 35 kt; 28 km (15 n miles) at 23 kt; warhead 250 kg.

Electronic countermeasures: ESM: Thomson-CSF DR 2000U; radar warning.
Radars: Surface search: Thomson-CSF Calypso; I-band.
Sonars: L-3 ELAC Nautik LOPAS 8300; passive search.
Weapon control systems: Signaal Sinbad system.

Programmes: Ordered on 2 April 1977. Designed by Ingenieurkontor, Lübeck for construction by Howaldtswerke, Kiel and sale by Ferrostaal, Essen—all acting as a consortium.
Modernisation: Major refits at HDW spanning three years from 1986 to 1989. These refits were expensive and lengthy and may have discouraged further orders at that time. *Cakra* refitted again at Surabaya from 1993 completing in April 1997, including replacement batteries and updated Sinbad TFCS. *Nanggala* received a similar refit from October 1997 to mid-1999. *Cakra* began a refit at Daewoo Shipyard, South Korea in 2004 which was completed in 2005. Work is reported to have included new batteries, overhaul of engines and modernisation of the combat system. A similar refit of *Nanggala* was completed in April 2006. Both boats have been fitted with L-3 LOPAS 8300 sonar to replace the Atlas Elektronik system.
Structure: Have high-capacity batteries with GRP lead-acid cells and battery cooling supplied by Wilhelm Hagen AG. Diving depth, 240 m (790 ft).
Operational: Endurance, 50 days. Operational status of both boats is doubtful.

NANGGALA
2/2013*, Bob Fildes / 1486523

NANGGALA
6/2012, Indonesian Navy / 1483041

FRIGATES

Notes: Three Nakhoda Ragam corvettes originally completed for Brunei in 2004, but not subsequently accepted, have been acquired by the Indonesian Navy. The ships are armed with a 76 mm gun and it is reported that the original Seawolf system has been replaced by VL-MICA. The new names of the three ships are: *Bung Tomo* (ex-*Jerambak*) 357 (ex-30); *John Lie* (ex-*Nakhoda Ragam*) 358 (ex-28); *Usman-Harun* (ex-*Bendahara Sakam*) 359 (ex-29). Delivery is expected in 2014.

6 AHMAD YANI (VAN SPEIJK) CLASS (FFGHM)

Name	No	Builders	Laid down	Launched	Commissioned
AHMAD YANI (ex-*Tjerk Hiddes*)	351	Nederlandse Dok en Scheepsbouw Mij, Amsterdam	1 Jun 1964	17 Dec 1965	16 Aug 1967
SLAMET RIYADI (ex-*Van Speijk*)	352	Nederlandse Dok en Scheepsbouw Mij, Amsterdam	1 Oct 1963	5 Mar 1965	14 Feb 1967
YOS SUDARSO (ex-*Van Galen*)	353	Koninklijke Maatschappij de Schelde, Flushing	25 Jul 1963	19 Jun 1965	1 Mar 1967
OSWALD SIAHAAN (ex-*Van Nes*)	354	Koninklijke Maatschappij de Schelde, Flushing	25 Jul 1963	26 Mar 1966	9 Aug 1967
ABDUL HALIM PERDANAKUSUMA (ex-*Evertsen*)	355	Koninklijke Maatschappij de Schelde, Flushing	6 Jul 1965	18 Jun 1966	21 Dec 1967
KAREL SATSUITUBUN (ex-*Isaac Sweers*)	356	Nederlandse Dok en Scheepsbouw Mij, Amsterdam	5 May 1965	10 Mar 1967	15 May 1968

Displacement, tonnes: 2,261 standard; 2,880 full load
Dimensions, metres (feet): 113.4 × 12.5 × 4.2 *(372 × 41.0 × 13.8)*
Speed, knots: 20
Range, n miles: 4,500 at 12 kt
Complement: 180

Machinery: 2 Caterpillar 3612 diesels (356); 12,512 hp *(9.2 MW)*; 2 Caterpillar 3616 diesels (351, 352, 353, 355); 14,617 hp *(10.9 MW)*; 2 SEMT Pielstick 12 PA6B (354); 14,000 hp *(10.6 MW)*; 2 shafts

Missiles: SSM: 8 McDonnell Douglas Harpoon (F 351, 352, 356); active radar homing to 130 km *(70 n miles)* at 0.9 Mach; warhead 227 kg. 4 (2 twin VLS) SS-N-26 (Yakhont) (F 354) ❶; inertial guidance and active/passive radar homing to 300 km *(162 n miles)* at 2.6 Mach; sea skimmer in terminal phase; warhead 250 kg. 4 C-802 (YJ-82) (F 353, 355); mid-course guidance and active radar homing to 120 km *(65 n miles)* at 0.9 Mach; warhead 165 kg.
SAM: 2 twin Matra Simbad launchers for Mistral; IR homing to 4 km *(2.2 n miles)*; warhead 3 kg.
Guns: 1 Oto Melara 3 in *(76 mm)*/62 compact ❷; 85 rds/min to 16 km *(8.7 n miles)* anti-surface; 12 km *(6.6 n miles)* anti-aircraft; weight of shell 6 kg. 4—12.7 mm MGs.
Torpedoes: 6—324 mm Mk 32 (2 triple) tubes ❸. Honeywell Mk 46; anti-submarine; active/passive homing to 11 km *(5.9 n miles)* at 40 kt; warhead 44 kg.
Physical countermeasures: Decoys: 2 Knebworth Corvus 8-tubed trainable; radar distraction or centroid chaff to 1 km.
Electronic countermeasures: ESM: UA 8/9; UA 13 (355 and 356); radar warning. FH5 D/F.
Radars: Air search: Signaal LW03 ❹; D-band; range 219 km *(120 n miles)* for 2 m² target.
Air/surface search: Signaal DA05 ❺; E/F-band; range 137 km *(75 n miles)* for 2 m² target.
Navigation: Racal Decca 1229; I-band.
Fire control: Signaal M 45 ❻; I/J-band (for 76 mm gun and SSM).
Sonars: Signaal CWE 610; hull-mounted; active search and attack; medium frequency. VDS; medium frequency.
Combat data systems: SEWACO V action data automation and Daisy data processing.
Weapon control systems: Signaal LIOD optronic director. Mk 2 fitted in 353, 354 and 356. SWG-1A Harpoon LCS.

Helicopters: 1 NBO-105C ❼.

Programmes: On 11 February 1986 agreement signed with the Netherlands for transfer of two of this class with an option on two more. Transfer dates: *Tjerk Hiddes*, 31 October 1986; *Van Speijk*, 1 November 1986; *Van Galen*, 2 November 1987; *Van Nes*, 31 October 1988. Contract of sale for the last two of the class signed 13 May 1989. *Evertsen* transferred 1 November 1989 and *Isaac Sweers* 1 November 1990. Ships provided with all spare parts but not towed arrays or helicopters.
Modernisation: This class underwent mid-life modernisation at Rykswerf Den Helder from 1976. This included replacement of 4.5 in turret by 76 mm, A/S mortar by torpedo tubes, new electronics and electrics, updating combat data system, improved communications, extensive automation with reduction in complement, enlarged hangar for Lynx and improved habitability. Harpoon for first two only initially because there was no FMS funding for the others. However the USN then provided sufficient SWG 1A panels for all of the class to be retrofitted with Harpoon missiles. LIOD optronic directors Mk 2 fitted in 353, 354 and 356 in 1996–97. Seacat replaced by Simbad twin launchers. *Ahmad Yani* and *Karel Satsuitubun* appear to have some additional superstructure in place of the Seacat launcher on the hangar roof. All six ships have been re-engined with diesel propulsion. 356 was refitted by Tesco Corp in 2003 and 354 by PT Mulia/PT Pal in 2006. 351 and 355 had been refitted by Tesco in mid-2007 and 352 and 353 were completed by Tesco by 2008. In March 2011, 354 was equipped with four SS-N-26 missiles. These are fitted in two VLS silos situated on either side of the hangar. A test firing was successfully conducted on 15 October 2012. Two other ships (353, 355) are reported to have been fitted with C-802 while replacement of Harpoon by C-705 is under consideration for the other three ships.
Operational: Harpoon missiles are reported to be time-expired.

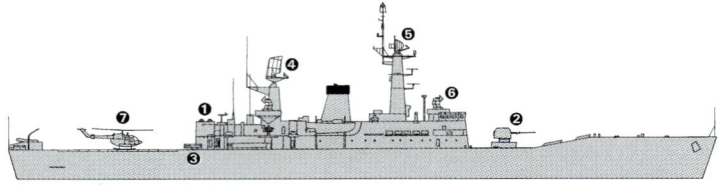

OSWALD SIAHAAN (Scale 1 : 1,200), Ian Sturton / 1406945

YOS SUDARSO 8/2009, John Mortimer / 1367071

AHMAD YANI 8/2009, Michael Nitz / 1390021

CORVETTES

3 FATAHILLAH CLASS (FFG/FFGH)

Name	No	Builders	Laid down	Launched	Commissioned
FATAHILLAH	361	Wilton Fijenoord, Schiedam	31 Jan 1977	22 Dec 1977	16 Jul 1979
MALAHAYATI	362	Wilton Fijenoord, Schiedam	28 Jul 1977	19 Jun 1978	21 Mar 1980
NALA	363	Wilton Fijenoord, Schiedam	27 Jan 1978	11 Jan 1979	4 Aug 1980

Displacement, tonnes: 1,219 standard; 1,473 full load
Dimensions, metres (feet): 84 × 11.1 × 3.3 *(275.6 × 36.4 × 10.8)*
Speed, knots: 30
Range, n miles: 4,250 at 16 kt
Complement: 89 (11 officers)

Machinery: CODOG; 1 RR Olympus TM3B gas turbine; 25,440 hp *(19 MW)* sustained; 2 MTU 20V 956 TB92 diesels; 11,070 hp(m) *(8.14 MW)* sustained; 2 shafts; LIPS cp props
Missiles: SSM: 4 Aerospatiale MM 38 Exocet ❶; inertial cruise; active radar homing to 42 km *(23 n miles)* at 0.9 Mach; warhead 165 kg; sea-skimmer.
Guns: 1 Bofors 4.7 in *(120 mm)*/46 ❷; 80 rds/min to 18.5 km *(10 n miles)*; weight of shell 21 kg. 1 or 2 Bofors 40 mm/70 (2 in *Nala*) ❸; 300 rds/min to 12 km *(6.6 n miles)*; weight of shell 0.96 kg. 2 Rheinmetall 20 mm; 1,000 rds/min to 2 km anti-aircraft; weight of shell 0.24 kg.
Torpedoes: 6—324 mm Mk 32 or ILAS 3 (2 triple) tubes (none in *Nala*) ❹. 12 Mk 46 (or A244S); anti-submarine; active/passive homing to 11 km *(5.9 n miles)* at 40 kt; warhead 44 kg.
A/S Mortars: 1 Bofors 375 mm twin-barrelled trainable ❺; 54 Erika; range 1,600 m and Nelli; range 3,600 m.
Physical countermeasures: Decoys: 2 Knebworth Corvus 8-tubed trainable chaff launchers ❻; radar distraction or centroid modes to 1 km. 1 T-Mk 6; torpedo decoy.
Electronic countermeasures: ESM: MEL Susie 1 (UAA-1); radar intercept.
Radars: Air/surface search: Signaal DA05 ❼; E/F-band; range 137 km *(75 n miles)* for 2 m² target.
Surface search: Racal Decca AC 1229 ❽; I-band.
Fire control: Signaal WM28 ❾; I/J-band; range 46 km *(25 n miles)*.
Sonars: Signaal PHS-32; hull-mounted; active search and attack; medium frequency.
Combat data systems: Signaal SEWACO-RI action data automation.
Weapon control systems: Signaal LIROD optronic director.

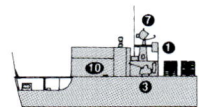

NALA

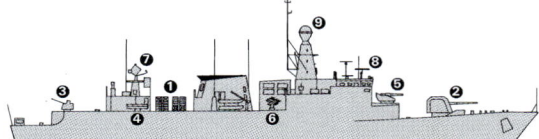

FATAHILLAH (Scale 1 : 1,200), Ian Sturton / 0121374 / 0126692

MALAHAYATI 8/2009, John Mortimer / 1367070

Helicopters: 1 small (*Nala* only) ❿.

Programmes: Ordered August 1975. Officially rated as Corvettes.
Structure: NEVESBU design. *Nala* is fitted with a folding hangar/landing deck.
Modernisation: A mid-life upgrade of these ships is expected. The combat data system is to be replaced by an Ultra system and the air/surface radar is to be replaced by Terma Scanter 4100.
Operational: These ships are the busiest of the larger warships. Three successful Exocet (locally modified after life-expiry) firings conducted on 25 August 2002.

FATAHILLAH 11/2004, Chris Gee / 1047876

NALA 6/2000, van Ginderen Collection / 0104593

Indonesia > Corvettes

15 KAPITAN PATTIMURA (PARCHIM I) CLASS (PROJECT 1331) (FS)

Name	No	Builders	Commissioned	Recommissioned
KAPITAN PATTIMURA (ex-*Prenzlau*)	871 (ex-371, ex-231)	Peenewerft, Wolgast	11 May 1983	23 Sep 1993
UNTUNG SUROPATI (ex-*Ribnitz*)	872 (ex-372, ex-233)	Peenewerft, Wolgast	29 Oct 1983	23 Sep 1993
SULTAN NUKU (ex-*Waren*)	873 (ex-224)	Peenewerft, Wolgast	23 Nov 1982	15 Dec 1993
LAMBUNG MANGKURAT (ex-*Angermünde*)	374 (ex-214)	Peenewerft, Wolgast	26 Jul 1985	12 Jul 1994
CUT NYAK DIEN (ex-*Lübz*)	375 (ex-P 6169, ex-221)	Peenewerft, Wolgast	12 Feb 1982	25 Feb 1994
SULTAN THAHA SYAIFUDDIN (ex-*Bad Doberan*)	876 (ex-376, ex-222)	Peenewerft, Wolgast	30 Jun 1982	25 Feb 1995
SUTANTO (ex-*Wismar*)	877 (ex-377, ex-P 6170, ex-241)	Peenewerft, Wolgast	9 Jul 1981	10 Mar 1995
SUTEDI SENOPUTRA (ex-*Parchim*)	878 (ex-242)	Peenewerft, Wolgast	9 Apr 1981	19 Sep 1994
WIRATNO (ex-*Perleberg*)	879 (ex-243)	Peenewerft, Wolgast	19 Sep 1981	19 Sep 1994
TJIPTADI (ex-*Bergen*)	881 (ex-381, ex-213)	Peenewerft, Wolgast	1 Feb 1985	10 May 1996
HASAN BASRI (ex-*Güstrow*)	882 (ex-223)	Peenewerft, Wolgast	10 Nov 1982	10 May 1996
IMAN BONJOL (ex-*Teterow*)	383 (ex-P 6168, ex-234)	Peenewerft, Wolgast	27 Jan 1984	26 Apr 1994
PATI UNUS (ex-*Ludwiglust*)	384 (ex-232)	Peenewerft, Wolgast	4 Jul 1983	21 Jul 1995
TEUKU UMAR (ex-*Grevesmühlen*)	385 (ex-212)	Peenewerft, Wolgast	21 Sep 1984	27 Oct 1996
SILAS PAPARE (ex-*Gadebusch*)	386 (ex-P 6167, ex-211)	Peenewerft, Wolgast	31 Aug 1984	27 Oct 1996

Displacement, tonnes: 781 standard
Dimensions, metres (feet): 75.2 × 9.8 × 3.5 *(246.7 × 32.2 × 11.5)*
Speed, knots: 20
Range, n miles: 1,750 at 18 kt
Complement: 64 (9 officers)

Machinery: 1 Zvezda M 504A diesel; 4,700 hp *(3.5 MW)* for centreline cp prop 2 Deutz TBD 620 V16 diesels (374, 872, 873, 877, 878, 881); 6,000 hp *(4.5 MW)* or 2 MTU 16V 4000 M 90 diesels (383, 386, 871, 879 and 882); 7,300 hp *(5.4 MW)* or 2 CAT 3516B diesels (375, 384, 385, 876); 5,200 hp *(3.9 MW)*; 2 outboard shafts
Missiles: SAM: SA-N-5/8 launchers fitted in some. May be replaced by twin Simbad launchers.
Guns: 2 USSR 57 mm/75 AK 725 (twin) ❶ automatic; 120 rds/min to 12.7 km *(6.8 n miles)*; weight of shell 2.8 kg. 2—30 mm (twin) ❷; 500 rds/min to 5 km *(2.7 n miles)* anti-aircraft; weight of shell 0.54 kg. 1—20 mm.
Torpedoes: 4—400 mm tubes ❸.
A/S Mortars: 2 RBU 6000 12-barrelled trainable launchers ❹; automatic loading; range 6,000 m; warhead 31 kg.
Mines: Mine rails fitted.
Physical countermeasures: Decoys: 2 PK 16 chaff rocket launchers.
Electronic countermeasures: ESM: 2 Watch Dog; radar warning.
Radars: Air/surface search: Strut Curve ❺; F-band; range 110 km *(60 n miles)* for 2 m² target.
Navigation: TSR 333; I-band.
Fire control: Muff Cob ❻; G/H-band.
IFF: High Pole B.
Sonars: MG 332T; hull-mounted; active search and attack; high frequency. Elk Tail; VDS system on starboard side (in some hulls).

Programmes: Ex-GDR ships mostly paid off in 1991. Formally transferred on 4 January 1993 and became Indonesian ships on 25 August 1993. First three arrived Indonesia in November 1993.
Modernisation: All refitted prior to sailing for Indonesia. Range increased and air conditioning added to accommodation. SAM launchers can be carried. A re-engining programme was completed in 2005.
Operational: *Memet Sastrawiria* was damaged by fire in 2008 and subsequently decommissioned.

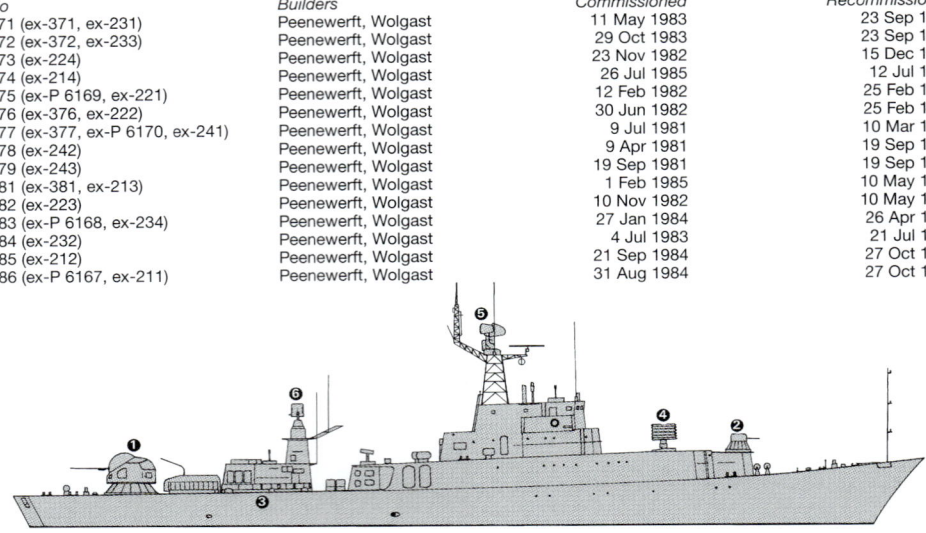

KAPITAN PATTIMURA *(Scale 1 : 600), Ian Sturton / 0506007*

PATI UNUS *2/2012, Indian Navy / 1483042*

IMAN BONJOL *2/2010, M Mazumdar / 1406064*

Corvettes — **Indonesia** 363

4 DIPONEGORO (SIGMA) CLASS (CORVETTES) (FS)

Name	No	Builders	Laid down	Launched	Commissioned
DIPONEGORO	365	Royal Schelde, Vlissengen	24 Mar 2005	16 Sep 2006	2 Jul 2007
SULTAN HASANUDDIN	366	Royal Schelde, Vlissengen	24 Mar 2005	16 Sep 2006	24 Nov 2007
SULTAN ISKANDAR MUDA	367	Royal Schelde, Vlissengen	8 May 2006	24 Nov 2007	18 Oct 2008
FRANS KAISIEPO	368	Royal Schelde, Vlissengen	8 May 2006	Jun 2008	7 Mar 2009

Displacement, tonnes: 1,719 full load
Dimensions, metres (feet): 90.7 × 13.0 × 3.6 (297.6 × 42.7 × 11.8)
Speed, knots: 28
Range, n miles: 4,000 at 18 kt
Complement: 80

Machinery: 2 SEMT Pielstick 20 PA6B diesels; 21,725 hp (16.2 MW); 2 shafts; cp props
Missiles: SAM: 2 quadruple Tetral launchers ❶; MBDA Mistral; IR homing to 4 km (2.2 n miles); warhead 3 kg.
SSM: 4 MBDA MM 40 Exocet Block II ❷; inertial cruise; active radar homing to 70 km (40 n miles) at 0.9 Mach; warhead 165 kg; sea-skimmer.
Guns: 1 Oto Melara 3 in (76 mm)/62 Super Rapid ❸; 120 rds/min to 16 km (8.7 n miles); weight of shell 6 kg. 2 Giat 20 mm ❹.
Torpedoes: 6—324 mm (2 B 515 triple) tubes ❺. Eurotorp Mu-90; active/passive homing to 25 km (13.5 n miles) at 29/50 kt.
Physical countermeasures: Decoys: 2 Terma SKWS 130 mm launchers.
Electronic countermeasures: ESM: Thales DR 3000; intercept.
ECM: Racal Scorpion; jammer.
Radars: Surface search: Thales MW 08 ❼; G-band.
Navigation: Sperry Marine Bridgemaster E ❽; E/F/I-band.
Sonars: Thales Kingclip; hull-mounted.
Combat data systems: Tacticos including Link Y.
Electro-optic systems: LIROD Mk 2 optronic tracker ❻.

Helicopters: Platform only.

Programmes: Contract for the construction of two Sigma 9113 corvettes, both to be built in the Netherlands, signed on 7 January 2004. The role of the ships is to conduct coastal security operations. Sea trials of *Diponegoro* started in April 2007 and the ship arrived in Indonesia in September 2007. *Hasanuddin* began sea-trials in November 2007. The option to build two further craft was exercised on 18 May 2005. These were also built in the Netherlands and were delivered on 2 December 2008 and April 2009 respectively.

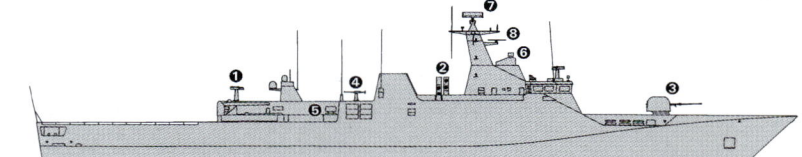

DIPONEGORO (Scale 1 : 900), Ian Sturton / 1353094

FRANS KAISIEPO 5/2013*, Guy Toremans / 1486522

SULTAN ISKANDAR MUDA 11/2013*, Michael Nitz / 1530904

SULTAN ISKANDAR MUDA 10/2013*, Chris Sattler / 1530905

© 2014 IHS IHS Jane's Fighting Ships 2014-2015

364 Indonesia > Corvettes — Land-based maritime aircraft

0 + 2 (2) SIGMA 10514 CLASS (CORVETTE) (FSG)

Name	No	Builders	Laid down	Launched	Commissioned
–	–	PT PAL, Surabaya	2014	2016	2017
–	–	PT PAL, Surabaya	2014	2017	2018

Displacement, tonnes: 2,365 full load
Dimensions, metres (feet): 105.1 × 14.0 × 3.7 (344.8 × 45.9 × 12.1)
Speed, knots: 30. **Range, n miles:** 4,000 at 18 kt
Complement: 100 + 20 spare berths

Machinery: CODAD: 4 diesels; 36,960 hp (27.5 MW); 2 shafts; cp props
Missiles: SAM: 12 (2 sextuple) MBDA VL-MICA ❶; command/inertial guidance; radar/IR homing to 20 km (10.8 n miles); warhead 12 kg.
SSM: 4 MBDA MM 40 Exocet Block II ❷; inertial cruise; active radar homing to 70 km (40 n miles) at 0.9 Mach; warhead 165 kg; sea-skimmer.
Guns: 1—76 mm ❸. 1 Rheinmetall Millenium 35 mm ❹.
Torpedoes: 6—324 mm.
A/S Mortars: 1 Bofors 375.
Physical countermeasures: Decoys: To be announced.
Electronic countermeasures: ESM/ECM: To be announced.
Torpedo defence: To be announced.
Radars: Air/surface search: Thales SMART-S ❺; 3D; E/F-band.
Fire control: Thales Stir ❻; I/J/K-bands.
Navigation: To be announced.

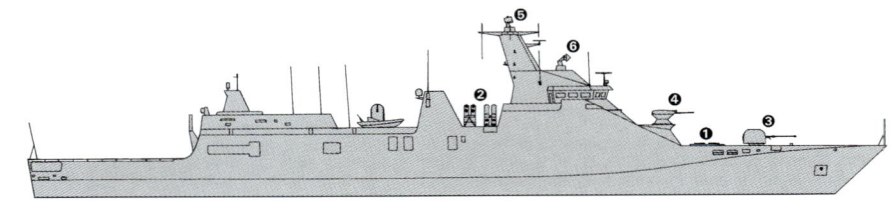

SIGMA 10514 (Scale 1 : 900), Ian Sturton / 1528808

Sonars: Thales UMS 4132 Kingclip; hull mounted; medium frequency.
Combat data systems: Thales Tacticos.
Electro-optic systems: Thales Stir.

Helicopters: To be announced.

Comment: The Perusak Kawai Rudal (PKR) programme is for a new class of 105 m Sigma 10514 corvettes. A contract for the construction of the first ship was signed with Damen Schelde Shipbuilding on 5 June 2012. An option for a second ship was subsequently exercised. Modules of the first ship are to be built in Europe and at PT PAL, Surabaya, where final assembly is to take place. The modules (except for the mast/bridge/operation block) of the second ship are to be built at PT PAL. First steel was cut for both vessels on 15 January 2014.

SHIPBORNE AIRCRAFT

Notes: Six Mi-17 medium lift helicopters for the Indonesian Marine Corps were acquired from Russia in 2008.

Numbers/Type: 6 Dirgantara (MBB) NBO 105CB.
Operational speed: 113 kt (210 km/h).
Service ceiling: 9,845 ft (3,000 m).
Range: 407 n miles (754 km).
Role/Weapon systems: Surveillance/support aircraft. A further three for SAR. Sensors: Thomson-CSF AMASCOS surveillance system; Chlio FLIR. Weapons: Unarmed.

BOEING 737 9/2003, Boeing / 0560018

Numbers/Type: 7 PZL Mielec M-28 Bryza.
Operational speed: 181 kt (335 km/h).
Service ceiling: 13,770 ft (4,200 m).
Range: 736 n miles (1,365 km).
Role/Weapon systems: Polish built aircraft based on the USSR Cash light transport. Contract on 18 August 2005 for seven maritime patrol aircraft delivered in late 2006. Sensors: PIT ARS-400M radar (SAR/ISAR modes).

Numbers/Type: 24/6 GAF Searchmaster Nomad B/Nomad L.
Operational speed: 168 kt (311 km/h).
Service ceiling: 21,000 ft (6,400 m).
Range: 730 n miles (1,352 km).
Role/Weapon systems: Nomad type built in Australia. Short-range maritime patrol, EEZ protection and anti-smuggler duties. 20 more acquired from Australian Army in August 1997 for use in maritime role. Not all are operational and NC-212 replacements are planned. Sensors: Nose-mounted search radar. Weapons: Unarmed.

NBO 105 8/2009, Michael Nitz / 1367066

Numbers/Type: 3 Dirgantara (Aerospatiale) NAS-332 Super Puma.
Operational speed: 151 kt (279 km/h).
Service ceiling: 15,090 ft (4,600 m).
Range: 335 n miles (620 km).
Role/Weapon systems: ASW and assault operations with secondary role in utility and SAR; ASVW development possible with Exocet or similar. Sensors: Thomson-CSF Omera radar and Alcatel dipping sonar in some. Weapons: ASW; two Mk 46 torpedoes or depth bombs.

NOMAD 8/2009, Michael Nitz / 1367065

Numbers/Type: 3 AirTech CN-235.
Operational speed: 236 kt (437 km/h).
Service ceiling: 25,000 ft (7,620 m).
Range: 1,565 n miles (2,519 km).
Role/Weapon systems: Maritime Patrol aircraft, first of which delivered in 2008. Operated by the Air Force. Sensors: Thales Ocean Master radar, Elettronica ALR 733 RWR, Chlio thermal imager, CAE AN/ASQ-508 MAD.

SUPER PUMA (French colours) 6/1994 / 0080008

LAND-BASED MARITIME AIRCRAFT (FRONT LINE)

Numbers/Type: 3 Boeing 737-2X9 Surveiller.
Operational speed: 462 kt (856 km/h).
Service ceiling: 50,000 ft (15,240 m).
Range: 2,530 n miles (4,688 km).
Role/Weapon systems: Land based for long-range maritime surveillance roles. Air Force manned. Sensors upgraded in 1993–94 to include IFF. Sensors: Motorola APS-135(v) SLAM MR radar, Thomson-CSF Oceanmaster radar. Weapons: Unarmed.

CN-235 2/2008, IHS/Patrick Allen / 1342647

Numbers/Type: 7 Northrop F-5E Tiger II.
Operational speed: 940 kt *(1,740 km/h)*.
Service ceiling: 51,800 ft *(15,790 m)*.
Range: 300 n miles *(556 km)*.
Role/Weapon systems: Fleet air defence and strike fighter, formed 'naval co-operation unit'. Planned to be replaced by BAe Hawk 200 in due course. Sensors: AI radar. Weapons: AD; two AIM-9 Sidewinder, two 20 mm cannon. Strike; 3,175 tons of underwing stores.

Numbers/Type: 6 Bell 412.
Operational speed: 122 kt *(226 km/h)*.
Service ceiling: 10,000 ft *(3,300 m)*.
Range: 500 n miles *(744 km)*.
Role/Weapon systems: Multipurpose aircraft built under license by Dirgantara and delivered from 1989 onwards. Six further aircraft were ordered in 2009.

BELL 412   8/2009, Michael Nitz / 1367064

PATROL FORCES

4 DAGGER CLASS (FAST ATTACK CRAFT – MISSILE) (PTFG)

Name	No	Builders	Commissioned
MANDAU	621	Korea Tacoma, Masan	20 Jul 1979
RENCONG	622	Korea Tacoma, Masan	20 Jul 1979
BADIK	623	Korea Tacoma, Masan	Feb 1980
KERIS	624	Korea Tacoma, Masan	Feb 1980

Displacement, tonnes: 274 full load
Dimensions, metres (feet): 50.2 × 7.3 × 2.3 *(164.7 × 24.0 × 7.5)*
Speed, knots: 41; 17 diesel
Range, n miles: 2,000 at 17 kt
Complement: 43 (7 officers)

Machinery: CODOG; 1 GE LM 2500 gas turbine; 23,000 hp *(17.16 MW)* sustained; 2 MTU 12V 331 TC81 diesels; 2,240 hp(m) *(1.65 MW)* sustained; 2 shafts; cp props
Missiles: SSM: 4 Aerospatiale MM 38 Exocet; inertial cruise; active radar homing to 42 km *(23 n miles)* at 0.9 Mach; warhead 165 kg; sea-skimmer.
Guns: 1 Bofors 57 mm/70 Mk 1; 200 rds/min to 17 km *(9.3 n miles)*; weight of shell 2.4 kg. Launchers for illuminants on each side. 1 Bofors 40 mm/70; 300 rds/min to 12 km *(6.6 n miles)*; weight of shell 0.96 kg. 2 Rheinmetall 20 mm.
Electronic countermeasures: ESM: Thomson-CSF DR 2000S (in 623 and 624); radar intercept.
Radars: Surface search: Racal Decca 1226; I-band.
Fire control: Signaal WM28; I/J-band.
Weapon control systems: Selenia NA-18 optronic director.

Programmes: PSMM Mk 5 type craft ordered in 1975.
Structure: Shorter in length and smaller displacement than decommissioned South Korean units. Mandau has a different shaped mast with a tripod base.

RENCONG   10/1998 / 0052358

1 PATROL CRAFT (PB)

Name	No	Builders	Launched	Commissioned
CUCUT (ex-Jupiter)	866 (ex-A 102)	Singapore SB and Marine	3 Apr 1990	19 Aug 1991

Displacement, tonnes: 173 full load
Dimensions, metres (feet): 35.8 × 7.1 × 2.3 *(117.5 × 23.3 × 7.5)*
Speed, knots: 14
Range, n miles: 200 at 14 kt
Complement: 33 (5 officers)
Machinery: 2 Deutz MWM TBD234V12 diesels; 1,360 hp(m) *(1 MW)* sustained; 2 shafts; bow thruster
Guns: 1 Oerlikon 20 mm GAM-BO1. 4—12.7 mm MGs.
Radars: Navigation: Racal Decca; I-band.

Comment: Designed as an underwater search and salvage craft, decommissioned from the Singapore Navy and transferred on 21 March 2002. Deployed as a patrol craft.

CUCUT (Singapore colours)   6/1994, van Ginderen Collection / 0084281

4 TODAK (PB 57) CLASS (NAV V)
(ATTACK CRAFT – MISSILE) (PGG)

Name	No	Builders	Commissioned
TODAK	631 (ex-803)	PT PAL, Surabaya	4 May 2000
HIU	804	PT PAL, Surabaya	Sep 2000
LAYANG	805	PT PAL, Surabaya	10 Jul 2002
LEMADANG (ex-Dorang)	806	PT PAL, Surabaya	Aug 2004

Displacement, tonnes: 454 full load
Dimensions, metres (feet): 58.1 × 7.6 × 2.8 *(190.6 × 24.9 × 9.2)*
Speed, knots: 27
Range, n miles: 6,100 at 15 kt, 2,200 at 27 kt
Complement: 53

Machinery: 2 MTU 16V 956 TB92 diesels; 8,850 hp(m) *(6.5 MW)* sustained; 2 shafts
Missiles: SSM: 2 C-802 (YJ-82) (fitted for but not with); mid-course guidance and active radar homing to 120 km *(65 n miles)* at 0.9 Mach; warhead 165 kg.
Guns: 1 Bofors SAK 57 mm/70 Mk 2; 220 rds/min to 14 km *(7.6 n miles)*; weight of shell 2.4 kg. 1 Bofors SAK 40 mm/70; 300 rds/min to 12 km *(6.6 n miles)*; weight of shell 0.96 kg. 2 Rheinmetall 20 mm.
Physical countermeasures: Decoys: CSEE Dagaie chaff launchers.
Electronic countermeasures: ESM: Thomson-CSF DR 3000 S1; intercept.
Radars: Air/surface search: Thales Variant; G-band.
Surface search: Thales Scout; I-band.
Fire control: Signaal LIROD Mk 2; K-band.
Navigation: Kelvin Hughes KH 1007; I-band.
Combat data systems: TACTICOS type.
Weapon control systems: Signaal LIOD 73 Ri Mk 2 optronic director.

Comment: Ordered in mid-1993 from PT PAL Surabaya. Weapon systems ordered in November 1994. Much improved combat data system is fitted. The after gun was intended to be a second 57 mm but this was changed to a 40 mm. C-802/YJ-82 missiles were temporarily installed and evaluated in Layang and Hiu in 2008.

HIU   2/2013*, Bob Fildes / 1486519

TODAK   12/2011, Chris Sattler / 1406972

TODAK   3/2013*, Chris Sattler / 1486520

4 KAKAP (PB 57) CLASS (NAV III AND IV)
(LARGE PATROL CRAFT) (PBOH)

Name	No	Builders	Commissioned
KAKAP	811	Lürssen; PT Pal	29 Jun 1988
KERAPU	812	Lürssen; PT Pal	5 Apr 1989
TONGKOL	813	PT Pal, Surabaya	Dec 1993
BARAKUDA (ex-Bervang)	814	PT Pal, Surabaya	Aug 1995

Displacement, tonnes: 430 full load
Dimensions, metres (feet): 58.1 × 7.6 × 2.8 *(190.6 × 24.9 × 9.2)*
Speed, knots: 28
Range, n miles: 6,100 at 15 kt, 2,200 at 27 kt
Complement: 49 + 8 spare berths

Machinery: 2 MTU 16V 956 TB92 diesels; 8,850 hp(m) *(6.5 MW)* sustained; 2 shafts
Guns: 1 Bofors 40 mm/70; 240 rds/min to 12.6 km *(6.8 n miles)*; weight of shell 0.96 kg. 2—12.7 mm MGs.
Electronic countermeasures: ESM: Thomson-CSF DR 3000 S1; intercept.
Radars: Surface search: Racal Decca 2459; I-band.
Navigation: KH 1007; I-band.
Helicopters: Platform for 1 NBO-105.

Comment: Ordered in 1982. First pair shipped from West Germany and completed at PT Pal Surabaya. Second pair assembled at Surabaya taking longer than expected to complete. The first three are NAV III SAR and Customs versions and by comparison with NAV I are very lightly armed and have a 13 × 7.1 m helicopter deck in place of the after guns and torpedo tubes. Vosper Thornycroft fin stabilisers are fitted. Can be used for Patrol purposes as well as SAR, and can transport two rifle platoons. There is also a fast seaboat with launching crane at the stern and two water guns for firefighting. The single NAV IV version has some minor variations and is used as Presidential Yacht manned by a special unit.

TONGKOL 2/2001, Sattler/Steele / 0121380

TONGKOL 6/2012 / 1483044

13 KAL-36 PATROL CRAFT (PB)

Name	No	Builders	Commissioned
KOBRA	867	Fasharkan, Mentigi	31 Mar 2003
ANAKONDA	868	Fasharkan, Jakarta	31 Mar 2003
PATOLA	869	PT Pelindo, Tanjung Pinang	Oct 2003
BOA	807	Fasharkan, Mentigi	6 Aug 2004
WELANG	808	Fasharkan, Mentigi	6 Aug 2004
TALIWANGSA	870	Fasharkan, Manokwari	6 Aug 2004
SULUH PARI	809	Fasharkan, Mentigi	20 Jan 2005
KATON	810	Fasharkan, Mentigi	20 Jan 2005
SANCA	815	Fasharkan, Manokwari	20 Jan 2005
WARAKAS	816	Fasharkan, Jakarta	20 Jan 2005
PANANA	817	Fasharkan, Makassar	20 Jan 2005
KALAKAE	818	Fasharkan, Makassar	20 Jan 2005
TEDONG NAGA	819	Fasharkan, Jakarta	20 Jan 2005

Displacement, tonnes: 91 full load
Dimensions, metres (feet): 36 × 7.0 × 1.35 *(118.1 × 23.0 × 4.4)*
Speed, knots: 38
Complement: 18
Machinery: 3 MAN D2842 LE 410 diesels; 3,300 hp *(2.46 MW)*; or 3 Caterpillar 3412E diesels; 3,600 hp *(2.7 MW)*
Guns: 1—20 mm. 2—12.7 mm MGs.
Radars: Navigation: I-band.

Comment: Kobra was the prototype vessel first demonstrated in late 2002. Glass fibre hull. There are some differences in armament and superstructure, some being fitted with a stern ramp for RIB. Patola funded by Bali province and others may have been similarly procured. Constructed by variety of shipbuilders and operated by the Indonesian Navy.

SULUH PARI 8/2009, Michael Nitz / 1367061

4 SINGA (PB 57) CLASS (NAV I AND II)
(LARGE PATROL CRAFT) (PBO)

Name	No	Builders	Commissioned
SINGA	651	Lürssen; PT Pal	Apr 1988
AJAK	653	Lürssen; PT Pal	5 Apr 1989
PANDRONG	801	PT Pal, Surabaya	1992
SURA	802	PT Pal, Surabaya	1993

Displacement, tonnes: 454 (651,653), 435 (801–802) full load
Dimensions, metres (feet): 58.1 × 7.6 × 2.8 *(190.6 × 24.9 × 9.2)*
Speed, knots: 27
Range, n miles: 6,100 at 15 kt, 2,200 at 27 kt
Complement: 42 (6 officers)

Machinery: 2 MTU 16V 956 TB92 diesels; 8,850 hp(m) *(6.5 MW)* sustained; 2 shafts
Guns: 1 Bofors SAK 57 mm/70 Mk 2; 220 rds/min to 14 km *(7.6 n miles)*; weight of shell 2.4 kg. 1 Bofors SAK 40 mm/70; 300 rds/min to 12 km *(6.6 n miles)*; weight of shell 0.96 kg. 2 Rheinmetall 20 mm.
Torpedoes: 2—21 in *(533 mm)* Toro tubes (651 and 653). AEG SUT; anti-submarine; wire-guided; active/passive homing to 12 km *(6.6 n miles)* at 35 kt; 28 km *(15 n miles)* at 23 kt warhead 250 kg.
Physical countermeasures: Decoys: CSEE Dagaie single trainable launcher; automatic dispenser for IR flares and chaff; H/J-band.
Electronic countermeasures: ESM: Thomson-CSF DR 2000 S3 with Dalia analyser; intercept. DASA Telegon VIII D/F.
Radars: Surface search: Racal Decca 2459; I-band; Signaal Scout; H/I-band (801 and 802).
Fire control: Signaal WM22; I/J-band (651 and 653).
Sonars: Signaal PMS 32 (NAV I); active search and attack; medium frequency.
Weapon control systems: Thales LIROD 2 (801, 802) optronic director. Signaal WM22 72 Ri WCS (651 and 653).

Comment: Class ordered from Lürssen in 1982. First launched and shipped incomplete to PT Pal Surabaya for fitting out in January 1984. Second shipped July 1984. The first two are NAV I ASW versions with torpedo tubes and sonars. The second pair are NAV II AAW versions with an augmented gun armament, an improved surveillance and fire-control radar, but without torpedo tubes and sonars and completed later than expected in 1992-93. Vosper Thornycroft fin stabilisers are fitted.

SINGA (NAV I) 5/1999, G Toremans / 0080009

AJAK (NAV I) 5/1998, John Mortimer / 0052359

SURA 5/2000, M Declerck / 0104597

Patrol forces < **Indonesia** 367

8 SIBARAU (ATTACK) CLASS (LARGE PATROL CRAFT) (PB)

Name	No	Builders	Commissioned
SIBARAU (ex-*Bandolier*)	847	Walkers Ltd	14 Dec 1968
SILIMAN (ex-*Archer*)	848	Walkers Ltd	15 May 1968
SIGALU (ex-*Barricade*)	857	Walkers Ltd	26 Oct 1968
SILEA (ex-*Acute*)	858	Evans Deakin	24 Apr 1968
SIRIBUA (ex-*Bombard*)	859	Walkers Ltd	5 Nov 1968
SIADA (ex-*Barbette*)	862	Walkers Ltd	16 Aug 1968
SIKUDA (ex-*Attack*)	863	Evans Deakin	17 Nov 1967
SIGUROT (ex-*Assail*)	864	Evans Deakin	12 Jul 1968

Displacement, tonnes: 148 full load
Dimensions, metres (feet): 32.8 × 6.1 × 2.2 *(107.6 × 20.0 × 7.2)*
Speed, knots: 21
Range, n miles: 1,220 at 13 kt
Complement: 19 (3 officers)

Machinery: 2 Paxman 16YJCM diesels; 4,000 hp *(2.98 MW)* sustained; 2 shafts
Guns: 1 Bofors 40 mm/60. 1 — 12.5 mm MG.
Electronic countermeasures: ESM: DASA Telegon VIII; intercept.
Radars: Surface search: Decca 916; I-band.

Comment: Transferred from Australia after refit— *Bandolier* 16 November 1973, *Archer* in 1974, *Barricade* March 1982, *Acute* 6 May 1983, *Bombard* September 1983, *Attack* 22 February 1985 (recommissioned 24 May 1985), *Barbette* February 1985, *Assail* February 1986. All carry rocket/flare launchers. Two similar craft with pennant numbers 860 and 861 were built locally in 1982/83 but have not been reported for some years.

SIGALU *4/1999* / 0080013

1 KRAIT CLASS (PATROL CRAFT) (PB)

Name	No	Builders	Commissioned
KRAIT	827	Fasharkan, Mentigi (Riau)	7 Jan 2009

Dimensions, metres (feet): 40 × 7.2 × 2.0 *(131.2 × 23.6 × 6.6)*
Speed, knots: 20
Machinery: 2 diesels; 2,500 hp *(1.86 MW)*; 2 shafts
Guns: 2 — 25 mm. 2 — 12.7 mm MGs.

Comment: The first of a new class of what is reported to be a national design. The design includes space for two RHIBs to be carried at the stern. Aluminium construction. Construction of further units is expected but has not been confirmed.

KRAIT *11/2008* / 1353098

10 KAL-40 CLASS (PATROL CRAFT) (PB)

Name	No	Builders	Commissioned
VIPER	820	Fasharkan, Jakarta	19 Oct 2006
PITON	821	Fasharkan, Mentigi (Riau)	19 Oct 2006
WELING	822	Fasharkan, Mentigi (Riau)	19 Oct 2006
MATACORA	823	Fasharkan, Mentigi (Riau)	14 Mar 2008
TEDUNG SELAR	824	Fasharkan, Jakarta	14 Mar 2008
BOIGA	825	Fasharkan, Manokwari	1 Aug 2007
TARIHU	829	Fasharkan, Mentigi (Riau)	7 Jan 2009
ALKURA	830	Fasharkan, Manokwari	20 Mar 2009
BIRANG	831	Fasharkan, Mentigi (Riau)	26 Jan 2010
MULGA	832	Fasharkan, Manokwari	26 Jan 2010

Displacement, tonnes: 102 full load
Dimensions, metres (feet): 40 × 7.3 × ? *(131.2 × 24.0 × ?)*
Speed, knots: 29
Complement: 25
Machinery: To be announced
Guns: 2 — 25 mm. 2 — 12.7 mm MGs.

Comment: A successor to the PC-36-class patrol craft and a building programme is in progress.

WELING *2/2013*, Bob Fildes* / 1486518

4 KCR-40 CLASS (ATTACK CRAFT—MISSILE) (PBM)

Name	No	Builders	Laid down	Launched	Commissioned
CLURIT	641	PT Palindo Marine Industry, Batam	2009	26 Apr 2011	Apr 2011
KUJANG	642	PT Palindo Marine Industry, Batam	2009	2011	Feb 2012
BELADAU	643	PT Palindo Marine Industry, Batam	2010	2012	25 Jan 2013
ALAMANG	644	PT Palindo Marine Industry, Batam	2011	2013	23 Dec 2013

Displacement, tonnes: 248 full load
Dimensions, metres (feet): 44 × 7.4 × 2.4 *(144.4 × 24.3 × 7.9)*
Speed, knots: 30
Complement: 35

Machinery: 3 MAN D2862 LE433 V12 diesels; 5,400 hp *(4.03 MW)*; 3 shafts
Missiles: SSM: 4 C-705; inertial guidance and radar homing to 75 km *(40.4 n miles)* at 0.9 Mach; warhead 110 kg.
Guns: 1 — 20 mm Vektor G12. 2 — 12.7 mm MGs.
Radars: Surface search/navigation: Furuno 1832; Furuno RPU 013; I-band.
Combat data systems: SEWACO.

Comment: An initial order for two missile-armed craft was made in 2009 and plans for a further 22 vessels were confirmed on 5 January 2012. With a steel hull and aluminium superstructure, the ship was designed in collaboration with the Institute of Technology in Surabaya. It is possible a development of the *Krait* design. There is a davit for a RHIB. Possibly fitted with a Chinese Type 360 series radar.

BELADAU *4/2013*, Chris Sattler* / 1486521

2 BADAU (WASPADA) CLASS (PB)

Name	No	Builders	Launched	Commissioned
BADAU (ex-*Pejuang*)	841 (ex-P 03)	Vosper Ltd	15 Mar 1978	25 Mar 1979
SALAWAKU (ex-*Waspada*)	842 (ex-P 02)	Vosper Ltd	3 Aug 1977	2 Aug 1978

Displacement, tonnes: 209 full load
Dimensions, metres (feet): 36.9 × 7.2 × 1.8 *(121.1 × 23.6 × 5.9)*
Speed, knots: 32
Range, n miles: 1,200 at 14 kt
Complement: 24 (4 officers)

Machinery: 2 MTU 20V 538 TB91 diesels; 7,680 hp(m) *(5.63 MW)* sustained; 2 shafts
Guns: 2 Oerlikon 30 mm GCM-B01 (twin); 650 rds/min to 10 km *(5.5 n miles)*; weight of shell 1 kg. 2 — 7.62 mm MGs. 2 MOD(N) 2 in launchers for illuminants.
Electronic countermeasures: ESM: Decca RDL; radar warning.
Radars: Surface search: Kelvin Hughes Type 1007; I-band.
Weapon control systems: Sea Archer system with Sperry Co-ordinate Calculator and 1412A digital computer. Radamec 2500 optronic director.

Modernisation: Started in 1988 and included improved gun fire control and ESM equipment. Further improvements in 1998-2000 included Type 1007 radar and a Radamec 2500 optronic director.
Structure: Welded steel hull with aluminium alloy superstructure. *Salawaku* has an enclosed upper bridge for training purposes.
Operational: Both ships transferred from Brunei in 2011.

BADAU (Brunei colours) *7/2000* / 0104244

© 2014 IHS   IHS Jane's Fighting Ships 2014-2015

368 Indonesia > Patrol forces — Amphibious forces

0 + 3 KCR-60M CLASS (ATTACK CRAFT—MISSILE) (PGG)

Name	No	Builders	Commissioned
—	628	PT PAL, Surabaya	2014
—	—	PT PAL, Surabaya	2014
—	—	PT PAL, Surabaya	2014

Displacement, tonnes: 460 full load
Dimensions, metres (feet): 59.8 × 8.1 × 2.6 (196.2 × 26.6 × 8.5)
Speed, knots: 28
Range, n miles: 2,400 at 20 kt
Complement: 43

Machinery: 2 MTU 20V 4000M73L diesels; 9,656 hp (7.2 MW); 2 shafts
Missiles: SSM: 4 (2 twin).
Guns: 1—57 mm. 2—20 mm.
Radars: Surface search/navigation: To be announced.

Comment: First steel was cut on 24 May 2012 for the construction of the first of three missile-armed patrol craft. The first ship was launched on 23 December 2013.

0 + 3 KLEWANG CLASS (PATROL SHIPS)

Dimensions, metres (feet): 63 × 16.0 × 1.2 (206.7 × 52.5 × 3.9)
Speed, knots: 35
Range, n miles: 2,000 at 16 kt
Complement: 29

Machinery: 4 MAN V12 diesel engines; 7,200 hp (5.37 MW); 4 MJP 550 water jets
Missiles: To be announced.
Guns: To be announced.
Radars: Surface search/navigation: To be announced.

Comment: The contract for the construction of four trimaran patrol craft was signed with PT Lundin (North Sea Boats, Sweden) in 2009. It is understood that the contract followed a 24-month research, design and development collaboration with New Zealand naval architects, LOMOcean Design Ltd. The boats are under construction at the Lundin Shipyard, Banyuwangi, East Java. The design includes a wave-piercing trimaran hull constructed of infused vinylester carbon fibre cored sandwich materials with external stealth features. The design has been optimised to facilitate high-speed operations in the short, steep seas of Indonesian littoral waters. Two water jets are located in the centre hull and one each in the side hulls. The ship is capable of operating an 11 m RHIB capable of 50 kt. Weapons and sensors have not yet been announced. The first of class Klewang was launched on 31 August 2012 but was subsequently destroyed by fire on 29 September 2012. The programme has been suspended pending the results of the enquiry into the causes of the fire.

KLEWANG 8/2012, North Sea Boats / 1459392

2 PC-43 CLASS (PATROL CRAFT) (PB)

Name	No	Builders	Launched	Commissioned
PARI	849	PT Palindo, Tanjung Pinang	24 Apr 2013	2013
SEMBILANG	850	PT Palindo, Tanjung Pinang	24 Apr 2013	2013

Displacement, tonnes: 250 full load
Dimensions, metres (feet): 43 × 7.4 × ? (141.1 × 24.3 × ?)
Speed, knots: 24
Machinery: 3 MAN diesels; 4,200 hp (3.13 MW); 3 shafts
Guns: 1—30 mm. 2—12.7 mm MG.
Radars: Navigation: I-band.

Comment: Similar to the KCR-40 class, the first two of what is expected to be a large class of gun-armed patrol craft were launched in 2013. Pari is to be based in the eastern Region, Sembilang KRI-850 is to be based in the western region.

AMPHIBIOUS FORCES

Notes: (1) This section includes some vessels of the Military Sealift Command-Kolinlamil.
(2) There are also some 20 LCM 6 type and 30 LCVPs in service.
(3) There are at least two 23 m LCUs which operate from the LPDs.
(4) Plans to construct two new LSTs were reported in May 2012.

LCU 990-2 8/2009, Michael Nitz / 1367050

5 MULTIROLE VESSELS (LPD/APCR)

Name	No	Builders	Laid down	Launched	Commissioned
DR SOEHARSO (ex-Tanjung Dalpele)	990 (ex-972)	Dae Sun Shipbuilders, Busan	2002	17 May 2003	Sep 2003
MAKASSAR	590	Dae Sun Shipbuilders, Busan	2005	7 Dec 2006	29 Apr 2007
SURABAYA	591	Dae Sun Shipbuilders, Busan	7 Dec 2006	23 Mar 2007	1 Aug 2007
BANJARMASIN	592	PT Pal, Surabaya	19 Oct 2006	28 Aug 2008	28 Nov 2009
BANDA ACEH	593	PT Pal, Surabaya	7 Dec 2007	19 Mar 2010	21 Mar 2011

Displacement, tonnes: 7,417 standard; 11,583 full load
Dimensions, metres (feet): 122 (590–591,990), 125 (592–593) × 22 × 4.9 (400.3, 410.1 × 72.2 × 16.1)
Speed, knots: 15
Range, n miles: 8,600 at 12 kt
Complement: 126

Military lift: 13 tanks; 507 troops; 2 LCVPs
Machinery: CODAD; 2 B&W 8L28/32A diesels; 5,250 hp (3.9 MW); 2 shafts
Guns: 2—20 mm.
Radars: Navigation: 2-I-band.
Helicopters: 2 Super Puma.

Programmes: Officially designated a Multipurpose Hospital Ship. Following delivery of the first vessel in mid-2003, a contract for a further four vessels, to a modified design, was finalised on 21 December 2004. The first two of these are being built in South Korea and the second two in Indonesia. First steel was cut for the first Indonesian vessel on 19 October 2006.
Structure: Has a docking well, capable of accommodating two LCU-23M, stern and side ramps and hospital facilities.

SURABAYA 8/2009, Michael Nitz / 1367059

DR SOEHARSO 8/2009, Michael Nitz / 1367058

DR SOEHARSO 8/2009, Michael Nitz / 1367057

2 TROOP TRANSPORT SHIPS (AP)

Name	No	Builders	Commissioned
TANJUNG NUSANIVE (ex-Kambuna)	973	Meyer Werft, Papenburg	1984
TANJUNG FATAGAR (ex-Rinjani)	974	Meyer Werft, Papenburg	1984

Measurement, tonnes: 14,178 gt
Dimensions, metres (feet): 144 × 23.4 × 5.9 (472.4 × 76.8 × 19.4)
Speed, knots: 20
Range, n miles: 5,500 at 12 kt
Complement: 119
Machinery: 2 MaK diesels; 16,760 hp (12.5 MW); 2 shafts; bow thruster
Guns: To be announced.
Radars: Navigation: I-band.

Comment: Converted passenger ships originally delivered to the Directorate of Sea Communications, Jakarta, in 1984. Capable of transporting 1,600 passengers and used to serve the Indonesian islands in their civilian configuration. Acquired by the Indonesian Navy in early 2005, converted into troop transports and commissioned on 1 September 2005.

Amphibious forces ◁ **Indonesia** 369

4 TROOP TRANSPORT SHIPS (AP)

Name	No	Builders	Commissioned
KARANG PILANG (ex-*Ambulu*)	981	Lürssen, Lemwerder	1998
KARANG TEKOK (ex-*Mahakam*)	982	Lürssen, Lemwerder	1998
KARANG BANTENG (ex-*Serayu*)	983	Lürssen, Lemwerder	1998
KARANG UNARANG (ex-*Barito*)	985	Lürssen, Lemwerder	1998

Displacement, tonnes: 501 standard
Dimensions, metres (feet): 69.8 × 10.4 × 2.0 *(229 × 34.1 × 6.6)*
Speed, knots: 38
Range, n miles: 550 at 35 kt
Machinery: 4 MTU 16V 595 TE 70L diesels; 20,400 hp *(15.2 MW)*; 2 Kamewa waterjets
Guns: 2—20 mm.
Radars: Navigation: I-band.

Comment: Converted passenger ferries of aluminium construction transferred from PT ASDP ferry company. Capable of transporting 600 troops and their equipment. Used to serve the Indonesian islands in their civilian configuration and acquired by the Indonesian Navy between September 2005 and April 2006. *Karang Galang* has been decommissioned.

TROOP TRANSPORT SHIP   12/2006 / 1164967

2 LST 1–511 AND 512–1152 CLASSES (LST)

Name	No	Builders	Commissioned
TELUK RATAI	509	American Bridge, PA	30 Jun 1944
TELUK BONE (ex-*Iredell County*; ex-*LST 839*)	511	American Bridge, PA	6 Dec 1944

Displacement, tonnes: 1,680 standard; 4,145 full load
Dimensions, metres (feet): 100 × 15.2 × 4.3 *(328.1 × 49.9 × 14.1)*
Speed, knots: 11.6
Range, n miles: 11,000 at 10 kt
Complement: 119 + 147 spare berths
Military lift: 2,100 tons
Machinery: 2 GM 12-567A diesels; 1,800 hp *(1.34 MW)*; 2 shafts
Guns: 7—40 mm. 8—37 mm (remainder).
Radars: Surface search: SPS-21 *(Teluk Ratai)*; SPS-53 *(Teluk Bone)*.

Comment: *Teluk Bone* transferred from US in June 1961 (and purchased 22 February 1979). These ships are used as transports and stores carriers.

TELUK RATAI   1/2005, David Boey / 1154407

6 TACOMA TYPE (LSTH)

Name	No	Builders	Commissioned
TELUK SEMANGKA	512	Korea Tacoma, Masan	20 Jan 1981
TELUK PENYU	513	Korea Tacoma, Masan	20 Jan 1981
TELUK MANDAR	514	Korea Tacoma, Masan	Jul 1981
TELUK SAMPIT	515	Korea Tacoma, Masan	Jun 1981
TELUK BANTEN	516	Korea Tacoma, Masan	May 1982
TELUK ENDE	517	Korea Tacoma, Masan	2 Sep 1982

Displacement, tonnes: 3,810 full load
Dimensions, metres (feet): 100 × 14.4 × 4.2 *(328.1 × 47.2 × 13.8)*
Speed, knots: 15
Range, n miles: 7,500 at 13 kt
Complement: 90 (13 officers)
Military lift: 1,800 tons (including 17 MBTs); 2 LCVPs; 200 troops
Machinery: 2 diesels; 12,800 hp(m) *(9.41 MW)* sustained; 2 shafts
Guns: 2 or 3 Bofors 40 mm/70. 2 Rheinmetall 20 mm.
Radars: Surface search: Raytheon; E/F-band *(Teluk Banten* and *Teluk Ende)*.
Navigation: Racal Decca; I-band.
Helicopters: 1 Westland Wasp; 3 NAS-332 Super Pumas can be carried in last pair.

Comment: First four ordered in June 1979, last pair June 1981. No hangar in *Teluk Semangka* and *Teluk Mandar*. Two hangars in *Teluk Ende*. The last pair differ in silhouette having drowned exhausts in place of funnels and having their LCVPs carried forward of the bridge. They also have only two 40 mm guns and an additional radar fitted above the bridge. Battalion of marines can be embarked if no tanks are carried. *Teluk Ende* and *Teluk Banten* act as Command ships, the former also able to serve as a hospital ship.

TELUK MANDAR   8/2009, Michael Nitz / 1367055

TELUK SAMPIT   1/2005, David Boey / 1154405

TELUK ENDE   8/2009, John Mortimer / 1367054

1 LST

Name	No	Builders	Commissioned
TELUK AMBOINA	503	Sasebo Heavy Industries	Jun 1961

Displacement, tonnes: 2,416 standard; 4,267 full load
Dimensions, metres (feet): 99.7 × 15.3 × 4.6 *(327.1 × 50.2 × 15.1)*
Speed, knots: 13.1
Range, n miles: 4,000 at 13.1 kt
Complement: 88
Military lift: 212 troops; 2,100 tons; 4 LCVP on davits
Machinery: 2 MAN V6V 22/30 diesels; 3,425 hp(m) *(2.52 MW)*; 2 shafts
Guns: 6—37 mm; anti-aircraft.

Comment: Launched on 17 March 1961 and transferred from Japan in June 1961. A faster copy of US LST 511 class with 30 ton crane forward of bridge. Serves with the Military Sealift Command.

TELUK AMBOINA   8/1995, van Ginderen Collection / 0080018

3 KUPANG CLASS (LCU)

KUPANG 582   DILI 583   NUSA UTARA 584

Displacement, tonnes: 406 full load
Dimensions, metres (feet): 42.9 × 9.1 × 1.4 *(140.7 × 29.9 × 4.6)*
Speed, knots: 12
Range, n miles: 700 at 11 kt
Complement: 17
Military lift: 200 tons
Machinery: 4 diesels; 2 shafts

Comment: Built at Naval Training Centre, Surabaya in 1978–80.

NUSA UTARA   8/2009, Michael Nitz / 1367051

IHS Jane's Fighting Ships 2014-2015

Indonesia > Amphibious forces — Mine warfare forces

11 FROSCH I CLASS (TYPE 108) (LSM)

Name	No	Commissioned	Recommissioned
TELUK GILIMANUK (ex-*Hoyerswerda*)	531 (ex-611)	12 Nov 1976	12 Jul 1994
TELUK CELUKAN BAWANG (ex-*Hagenow*)	532 (ex-632)	1 Dec 1976	25 Feb 1994
TELUK CENDRAWASIH (ex-*Frankfurt/Oder*)	533 (ex-613)	2 Feb 1977	9 Dec 1994
TELUK PELENG (ex-*Lübben*)	535 (ex-631)	15 Mar 1978	23 Sep 1993
TELUK SIBOLGA (ex-*Schwerin*)	536 (ex-612)	19 Oct 1977	15 Dec 1993
TELUK MANADO (ex-*Neubrandenburg*)	537 (ex-633)	28 Dec 1977	2 Jun 1995
TELUK HADING (ex-*Cottbus*)	538 (ex-614)	26 May 1978	12 Jul 1994
TELUK PARIGI (ex-*Anklam*)	539 (ex-635)	14 Jul 1978	21 Jul 1995
TELUK LAMPUNG (ex-*Schwedt*)	540 (ex-636)	7 Sep 1979	26 Apr 1994
TELUK JAKARTA (ex-*Eisenhüttenstadt*)	541 (ex-615)	4 Jan 1979	19 Sep 1994
TELUK SANGKULIRANG (ex-*Grimmen*)	542 (ex-616)	4 Jan 1979	9 Dec 1994

Displacement, tonnes: 1,981 full load
Dimensions, metres (feet): 98 × 11.1 × 2.8 *(321.5 × 36.4 × 9.2)*
Speed, knots: 18
Complement: 46

Military lift: 600 tons
Machinery: 2 diesels; 5,000 hp(m) *(3.68 MW)*; 2 shafts
Guns: 1 — 40 mm/60. 2 — 37 mm/63 (1 twin). 4 — 25 mm (2 twin).
Mines: Can lay 40 mines through stern doors.
Physical countermeasures: Decoys: 2 PK 16 chaff launchers.
Radars: Air/surface search: Strut Curve; F-band.
Navigation: TSR 333; I-band.

Comment: All built by Peenewerft, Wolgast. Former GDR ships transferred from Germany on 25 August 1993. Demilitarised with all guns removed, but 37 mm guns have replaced the original 57 mm and 30 mm twin guns. All refitted in Germany prior to sailing. First two arrived Indonesia in late 1993, remainder throughout 1994 and 1995. *Teluk Lampung* damaged by heavy seas during transit in June 1994 but was repaired. *Teluk Berau* decommissioned on 29 September 2012.

TELUK SANGKULIRANG 8/2009, Michael Nitz / 1367052

FROSCH I CLASS 8/2009, John Mortimer / 1367053

TELUK PELENG 1/2005, David Boey / 1164966

1 TRANSPORT SHIP (AP)

Name	No	Builders	Commissioned
TANJUNG KAMBANI (ex-*Dong Yang 6*)	971	Sanuki Shipbuilding	1982

Displacement, tonnes: 7,253 standard
Dimensions, metres (feet): 114.5 × 19.8 × 6.0 *(375.7 × 65.0 × 19.7)*
Speed, knots: 18

Military lift: To be announced
Machinery: 2 Makita diesels; 8,200 hp *(6.1 MW)*; 2 shafts; cp props
Radars: Navigation: 2 I-band.
Helicopters: Platform for 2 medium.

Comment: Former Ro-Ro ferry converted for military use by Daesun Shipbuilders, Pusan, and delivered to the Indonesian Navy on 9 November 2000. Reported to be capable of carrying one battalion which may be disembarked by four LCVPs and/or helicopter.

TANJUNG KAMBANI 8/2009, John Mortimer / 1367056

1 TRANSPORT SHIP (LCU)

Name	No	Builders	Commissioned
DORE	580	Korneuberg	1968

Displacement, tonnes: 185 standard; 279 full load
Dimensions, metres (feet): 38.3 × 10.0 × 1.8 *(125.7 × 32.8 × 5.9)*
Speed, knots: 8
Range, n miles: 600 at 8 kt
Machinery: 2 diesels; 420 hp *(309 kW)*; 2 shafts
Radars: Navigation: I-band.

Comment: Sister ship *Amurang* was lost at sea in 1992.

MINE WARFARE FORCES

2 PULAU RENGAT (TRIPARTITE) CLASS (MHSC)

Name	No	Builders	Launched	Commissioned
PULAU RENGAT	711	van der Giessen-de Noord	23 Jul 1987	26 Mar 1988
PULAU RUPAT	712	van der Giessen-de Noord	27 Aug 1987	26 Mar 1988

Displacement, tonnes: 510 standard; 577 full load
Dimensions, metres (feet): 51.5 × 8.9 × 2.5 *(169 × 29.2 × 8.2)*
Speed, knots: 15; 7 gas-turbine
Range, n miles: 3,000 at 12 kt
Complement: 46 + 4 spare berths

Machinery: 2 MTU 12V 396 TC82 diesels; 2,610 hp(m) *(1.92 MW)* sustained; 1 shaft; LIPS cp prop; auxiliary propulsion; 3 Turbomeca gas-turbine generators; 2 motors; 2,400 hp(m) *(1.76 MW)*; 2 retractable Schottel propulsors; 2 bow thrusters; 150 hp(m) *(110 kW)*
Guns: 2 Rheinmetall 20 mm. Matra Simbad SAM launcher may be added for patrol duties or a third 20 mm gun.
Physical countermeasures: MCM: OD3 Oropesa mechanical sweep gear; Fiskars F-82 magnetic and SA Marine AS 203 acoustic sweeps; Ibis V minehunting system; 2 PAP 104 Mk 4 mine disposal systems.
Radars: Navigation: Racal Decca AC 1229C; I-band.
Sonars: Thomson Sintra TSM 2022; active minehunting; high frequency.
Combat data systems: Signaal SEWACO-RI action data automation.

Programmes: First ordered on 29 March 1985, laid down 22 July 1985, second ordered 30 August 1985 and laid down 15 December 1985. More were to have been built in Indonesia up to a total of 12 but this programme was cancelled due to lack of funds.
Structure: There are differences in design between these ships and the European Tripartites, apart from their propulsion. Deckhouses and general layout are different as they are required to act as minehunters, minesweepers and patrol ships. Hull construction is GRP shock-proven.
Operational: Endurance, 15 days. Automatic operations, navigation and recording systems, Thomson-CSF Naviplot TSM 2060 tactical display. A 5 ton container can be shipped, stored for varying tasks — research; patrol; extended diving; drone control.

PULAU RENGAT 8/2009, John Mortimer / 1367049

Mine warfare forces — Survey and research ships < **Indonesia** 371

9 KONDOR II (TYPE 89) CLASS
(MINESWEEPERS—COASTAL) (MSC)

Name	No	Builders	Commissioned
PULAU ROTE (ex-*Wolgast*)	721 (ex-V 811)	Peenewerft, Wolgast	1 Jun 1971
PULAU RAAS (ex-*Hettstedt*)	722 (ex-353)	Peenewerft, Wolgast	22 Dec 1971
PULAU ROMANG (ex-*Pritzwalk*)	723 (ex-325)	Peenewerft, Wolgast	26 Jun 1972
PULAU RIMAU (ex-*Bitterfeld*)	724 (ex-332, ex-M 2672)	Peenewerft, Wolgast	7 Aug 1972
KELABANG (ex-*Pulau Rondo*; ex-*Zerbst*)	826 (ex-725, ex-335)	Peenewerft, Wolgast	30 Sep 1972
PULAU RUSA (ex-*Oranienburg*)	726 (ex-341)	Peenewerft, Wolgast	1 Nov 1972
PULAU RANGSANG (ex-*Jüterbog*)	727 (ex-342)	Peenewerft, Wolgast	7 Apr 1973
KALA HITAM (ex-*Pulau Raibu*; ex-*Sömmerda*)	828 (ex-728, ex-311)	Peenewerft, Wolgast	9 Aug 1973
PULAU REMPANG (ex-*Grimma*)	729 (ex-336)	Peenewerft, Wolgast	10 Nov 1973

Displacement, tonnes: 315 full load
Dimensions, metres (feet): 56.7 × 7.5 × 2.4 *(186 × 24.6 × 7.9)*
Speed, knots: 17
Range, n miles: 2,000 at 14 kt
Complement: 31 (6 officers)

Machinery: 2 Russki Kolomna Type 40-DM diesels; 4,408 hp(m) *(3.24 MW)* sustained; 2 shafts; cp props
Guns: 6—25 mm/80 (3 twin). 1—12.7 mm MG.
Mines: 2 rails.
Radars: Navigation: TSR 333; I-band.
Sonars: Bendix AQS 17 VDS; minehunting; active; high frequency (in some).

Comment: Former GDR minesweepers transferred from Germany in Russian dockship *Trans-Shelf* arriving 22 October 1993. Patrol duties take precedence over MCM and ex-*Pulau Rondo* and ex-*Pulau Raibu* formally converted in 2008 when new names and pennant numbers were allocated. There are some variations in armament. *Pulau Rempang*, *Pulau Rote* and *Pulau Romang* are also used for survey duties. ADI Dyads can be embarked for MCM.

PULAU RIMAU 4/2004, *Chris Sattler* / 1044128

PALAU RUSA 8/1995, *van Ginderen Collection* / 0080021

PULAU RONDO (old number) 4/2004, *John Mortimer* / 1153200

SURVEY AND RESEARCH SHIPS

4 RESEARCH SHIPS (AGS/AGOR)

Name	No	Builders	Commissioned
BARUNA JAYA I	KAL-IV-02	CMN, Cherbourg	10 Aug 1989
BARUNA JAYA II	KAL-IV-03	CMN, Cherbourg	25 Sep 1989
BARUNA JAYA III	KAL-IV-04	CMN, Cherbourg	3 Jan 1990
BARUNA JAYA IV	KAL-IV-05	CMN, Cherbourg	2 Nov 1995

Displacement, tonnes: 1,199 (KAL-IV-2–4), 1,448 (KAL-IV-5) full load
Dimensions, metres (feet): 60.4 × 12.1 × 4.2 *(198.2 × 39.7 × 13.8)*
Speed, knots: 14
Range, n miles: 7,500 at 12 kt
Complement: 37 (8 officers) + 26 scientists
Machinery: 2 Niigata/SEMT-Pielstick 5 PA5 L 255 diesels; 2,990 hp(m) *(2.2 MW)* sustained; 1 shaft; cp prop; bow thruster

Comment: First three ordered from La Manche, Dieppe in February 1985 by the office of Technology, Ministry of Industry and Research. Badly delayed by the closing down of the original shipbuilders (ACM, Dieppe) and construction taken over by CMN at Cherbourg. Fourth of class ordered in 1993 to a slightly enlarged design and with a more enclosed superstructure. *Baruna Jaya 1* is employed on hydrography, the second on oceanography and the third combines both tasks. *Baruna Jaya IV* is operated by the Agency responsible for developing new technology. All are part of the Naval Auxiliary Service.

BARUNA JAYA II 4/1998, *John Mortimer* / 0052362

BARUNA JAYA IV 11/1995, *van Ginderen Collection* / 0080023

1 RESEARCH SHIP (AGOR)

Name	No	Builders	Commissioned
BARUNA JAYA VIII	KAL-IV-06	Mjellem & Karlsen AS, Bergen	1998

Displacement, tonnes: 1,500 full load
Dimensions, metres (feet): 53.2 × 12.5 × 4.3 *(174.5 × 41.0 × 14.1)*
Speed, knots: 13
Range, n miles: 7,500 at 12 kt
Complement: 42 (11 officers) + 23 scientists
Machinery: 1 Caterpillar 3516BTA diesel; 2,026 bhp *(1.5 MW)*; 1 shaft; cp prop; 1 Schottel SPJ-82TL bow thruster
Radars: Navigation: Furuno FAR-2835S; E/F-band. Furuno FR-2110; I-band.

Comment: Multipurpose survey vessel equipped to conduct fisheries research, geophysics and seabed mapping. Delivered to Indonesia on 28 September 1998. Sensors include Simrad SD570 sonar, EM 1000 multibeam echo sounder and EA 500 single beam echo sounder.

BARUNA JAYA VIII 9/1998, *Maritime Photographic* / 0044067

© 2014 IHS

IHS Jane's Fighting Ships 2014-2015

374 Indonesia > Auxiliaries — Tugs

2 FROSCH II CLASS (TYPE 109) (SUPPORT SHIPS) (AKL/ARL)

Name	No	Builders	Commissioned
TELUK CIREBON (ex-Nordperd)	543 (ex-E 171)	Peenewerft, Wolgast	3 Oct 1979
TELUK SABANG (ex-Südperd)	544 (ex-E 172)	Peenewerft, Wolgast	26 Feb 1980

Displacement, tonnes: 1,727 full load
Dimensions, metres (feet): 90.7 × 11.1 × 2.8 *(297.6 × 36.4 × 9.2)*
Speed, knots: 18

Cargo capacity: 650 tons
Machinery: 2 diesels; 4,408 hp(m) *(3.24 MW)* sustained; 2 shafts
Guns: 4 — 37 mm/63 (2 twin). 4 — 25 mm (2 twin).
Physical countermeasures: Decoys: 2 PK 16 chaff launchers.
Radars: Air/surface search: Strut Curve; F-band.
Navigation: I-band.

Comment: Ex-GDR ships disarmed and transferred from Germany 25 August 1993. 5 ton crane amidships. In GDR service these ships had two twin 57 mm and two twin 25 mm guns plus Muff Cob fire-control radar. Both refitted at Rostock and recommissioned 25 April 1995. 37 mm guns fitted after transfer. Rocket launchers are mounted forward of the bridge.

TELUK SABANG *12/2009, John Mortimer* / 1406067

TELUK CIBERON *5/2011, Guy Toremans* / 1406973

2 TISZA CLASS (SUPPORT SHIPS) (AKL)

KARIMATA 960   **WAGIO** 961

Displacement, tonnes: 2,439 full load
Dimensions, metres (feet): 78.8 × 10.8 × 4.6 *(258.5 × 35.4 × 15.1)*
Speed, knots: 12
Range, n miles: 3,000 at 11 kt
Complement: 26

Cargo capacity: 875 tons dry; 11 tons liquid
Machinery: 1 MAN diesel; 1,000 hp(m) *(735 kW)*; 1 shaft
Guns: 4 — 14.5 mm (2 twin) MGs.
Radars: Navigation: Spin Trough; I-band.

Comment: Built in Hungary. Transferred in 1963–64. Military Sealift Command since 1978.

KARIMATA *12/2009, John Mortimer* / 1406066

0 + 1 REPLENISHMENT TANKER (AOR)

Name	No	Builders	Commissioned
–	–	PT Anugrah Buana Marine	2014

Dimensions, metres (feet): 122.4 × 16.5 × ? *(401.6 × 54.1 × ?)*
Speed, knots: 18

Cargo capacity: 5,500 m³ oil
Machinery: To be announced.
Guns: To be announced.
Radars: Navigation: To be announced.

Comment: First steel for a new replenishment tanker was cut on 8 May 2012. There are replenishment positions on both sides.

0 + 1 LOGISTICS SUPPORT SHIP (AKR)

Name	No	Builders	Commissioned
–	–	PT Daya Radar Utama, Jakarta	2015

Measurement, tonnes: 6,000 gt
Dimensions, metres (feet): 120 × 18.0 × 3.0 *(393.7 × 59.1 × 9.8)*
Military lift: To be announced.
Machinery: 2 diesels; 7,940 hp *(5.84 MW)*; 2 shafts
Radars: Navigation: 2 — I-band.

Comment: An order for a new logistics support ship was placed in December 2012. The ship is reported to have a Ro-Ro design and is expected to enter service in 2015.

TUGS

Notes: Two BIMA VIII class of 423 tons completed in 1991 are not naval. Names *Merapi* and *Merbabu*.

3 HARBOUR TUGS (YTM)

Name	No	Builders	Commissioned
LAMPO BATANG	934	Ishikawajima-Harima	Sep 1961
TAMBORA	935	Ishikawajima-Harima	Jun 1961
BROMO	936	Ishikawajima-Harima	Aug 1961

Comment: All of 250 tons displacement. There are a number of other naval tugs in the major ports.

1 NFI CLASS (ATF)

Name	No	Builders	Commissioned
SOPUTAN	923	Dae Sun Shipbuilders, Busan	11 Aug 1995

Measurement, tonnes: 1,300 gt
Dimensions, metres (feet): 66.2 × 11.9 × 5.2 *(217.2 × 39.0 × 17.1)*
Speed, knots: 13.5
Complement: 42
Machinery: Diesel-electric; 4 SEMT-Pielstick diesel generators; 1 motor; 12,240 hp(m) *(9 MW)*; 1 shaft; bow thruster
Guns: 1 — 40 mm. 2 — 20 mm.
Radars: Navigation: Racal Decca; I-band.

Comment: Ocean Cruiser class NFI. Bollard pull 120 tons.

SOPUTAN *8/2009, Michael Nitz* / 1367045

1 FLEET TUG (ATF)

Name	No	Builders	Commissioned
LEUSER	924	PT Dok & Perkapalan Kodja Bahari, Jakarta	2002

Measurement, tonnes: 1,604 gt
Dimensions, metres (feet): 71.5 × 13 × ? *(234.6 × 42.7 × ?)*
Speed, knots: 15
Machinery: 2 Pielstick 16PA5V diesels; 7,700 hp *(5.7 MW)*; 2 shafts
Radars: Navigation: I-band.

Comment: Fleet tug also employed on hydrographic duties.

LEUSER *12/2009, John Mortimer* / 1406068

CUSTOMS

Notes: Identified by BC (Tax and Customs) preceding the pennant number.

14 COASTAL PATROL CRAFT (WPB)

BC 2001–2007   BC 3001–3007

Displacement, tonnes: 71 full load
Dimensions, metres (feet): 28.5 × 5.4 × 1.7 *(93.5 × 17.7 × 5.6)*
Speed, knots: 34 (est.)
Complement: 19
Machinery: 2 MTU 12V 331 TC92 diesels; 2,660 hp(m) *(1.96 MW)* sustained; 2 shafts
Guns: 1—20 mm or 1—12.7 mm MG.

Comment: Built CMN Cherbourg. Delivered in 1980 and 1981.

BC 2007   *1/1990, 92 Wing RAAF* / 0506011

10 LÜRSSEN VSV 15 CLASS (WHSIC)

BC 1601–1610

Displacement, tonnes: 11 full load
Dimensions, metres (feet): 16 × 2.8 × 1 *(52.5 × 9.2 × 3.3)*
Speed, knots: 50
Range, n miles: 750 at 30 kt
Complement: 5 (1 officer)
Machinery: 2 MTU diesels; 600 hp(m) *(441 kW)*; 2 shafts
Guns: 1—7.62 mm MG.

Comment: Built in Germany and delivered between November 1998 and June 1999.

BC 1608   *5/1999, Lürssen* / 0080032

36 LÜRSSEN 28 METRE TYPE (WPB)

BC 4001–4006   BC 6001–6006   BC 8001–8006
BC 5001–5006   BC 7001–7006   BC 9001–9006

Displacement, tonnes: 69 full load
Dimensions, metres (feet): 28 × 5.4 × 1.8 *(91.9 × 17.7 × 5.9)*
Speed, knots: 30
Range, n miles: 1,100 at 15 kt, 860 at 28 kt
Complement: 19 (6 officers)
Machinery: 2 Deutz diesels; 2,720 hp(m) *(2 MW)*; or 2 MTU diesels; 2,260 hp(m) *(1.66 MW)*; 2 shafts
Guns: 1—12.7 mm MG.

Comment: Lürssen design, some built by Fulton Marine and Scheepswerven van Langebrugge of Belgium, some by Lürssen Vegesack and some by PT PAL Surabaya (which also assembled most of them). Programme started in 1980. Some of these craft are operated by the Navy, the Police, and the Coast Guard.

BC 9006   *2/2013*, Bob Fildes* / 1486516

5 LÜRSSEN NEW 28 METRE TYPE (WHSIC)

BC 10001–10002   BC 20001–20003

Displacement, tonnes: 86 full load
Dimensions, metres (feet): 28.2 × 6.6 × 1.4 *(92.5 × 21.7 × 4.6)*
Speed, knots: 40. **Range, n miles:** 1,100 at 30 kt
Complement: 11 (3 officers)
Machinery: 2 MTU 16V 396 TE94 diesels; 2,955 hp(m) *(2.14 MW)* sustained; 2 shafts
Guns: 2—7.62 mm MGs.
Radars: Surface search: Furuno FR 8731; I-band.

Comment: First pair built in Germany and delivered between May 1999 and November 1999. Last three built by PT Pal Surabaya and delivered between September 1999 and November 1999. Aluminium construction.

BC 10001   *5/1999, Lürssen* / 0080034

BC 20001   *9/1999, PT Pal* / 0075857

COAST GUARD

Notes: (1) Established in 1978 as the Maritime Security Agency to control the 200 mile EEZ and to maintain navigational aids. The official name is the Indonesian Sea and Coast Guard.
(2) There are numerous small patrol craft, most of which have white hulls with a diagonal red and blue stripe on the side. Pennant numbers include: KNP 323, 325, 326, 329, 330, 331, 337, 345. Some craft have KPLP on the superstructure.

KNP 345   *2/2013*, Bob Fildes* / 1486515

2 DAMEN 6210 (OFFSHORE PATROL SHIPS) (WPSO)

Name	Builders	Commissioned
TRISULA	PT Dumas, Surabaya	28 Jun 2004
SAROTAMA	PT Dumas, Surabaya	6 Jul 2004

Measurement, tonnes: 878 gt
Dimensions, metres (feet): 61.8 × 9.7 × 3.2 *(202.8 × 31.8 × 10.5)*
Speed, knots: 18
Machinery: 2 MTU 16V4000 M70 diesels; 6,308 hp *(4.64 MW)*; 2 shafts
Radars: Surface search/navigation: To be announced.

Comment: Principal role as patrol ships but also equipped to conduct counter-pollution tasks.

1 MANDALIKA CLASS (BUOY TENDER) (WABU)

Name	Builders	Commissioned
MANDALIKA	Usuki Iron Works	30 May 1975

Measurement, tonnes: 607 gt
Dimensions, metres (feet): 44.07 × 9.8 × 3.15 *(144.6 × 32.2 × 10.3)*
Speed, knots: 11
Machinery: 1 Niigata 6L28X diesel; 1,201 hp *(883 kW)*; 1 shaft
Radars: Surface search/navigation: To be announced.

376 Indonesia — Coast guard

2 DISASTER RESPONSE SHIPS (WPSO)

ARDA DEDALI   ALUGARA

Measurement, tonnes: 539 gt
Dimensions, metres (feet): 60 × 8.0 × 3.2 (196.9 × 26.2 × 10.5)
Speed, knots: 19.3
Range, n miles: 3,000 at 17 kt
Machinery: 2 MTU 16V4000 M60 diesels; 2 shafts; cp props
Radars: Surface search/navigation: To be announced.

Comment: Built by Niigata Shipbuilding & Repair Inc., a wholly owned subsidiary of Mitsui Engineering & Shipbuilding Co., *Arda Dedali* delivered to the Directorate General of Sea Communication (DGSC) on 27 January 2005. *Alugara* delivered in mid-2005. The ships are designed to undertake disaster relief operations and are equipped to deal with accidents at sea, including rescue and firefighting, and counter-pollution tasks. The ships are likely to be deployed in the Malacca/Singapore Strait region.

ALUGARA   6/2005, Ships of the World / 1153202

1 BUOY TENDER (WABU)

Name	Builders	Commissioned
JADAYAT	Niigata Shipbuilding and Repair	10 Oct 2003

Measurement, tonnes: 872 gt
Dimensions, metres (feet): 56.9 × 11.0 × 3.5 (186.7 × 36.1 × 11.5)
Speed, knots: 10.5
Complement: 45
Machinery: 1 diesel; 985 hp (735 MW); 1 shaft
Radars: Navigation: I-band.

Comment: Funded by the Nippon Foundation through the Malacca Strait Council.

JADAYAT   7/2004, Ian Edwards / 1040686

3 DAMEN 4810 (BUOY TENDERS) (WABU)

Name	Builders	Commissioned
ANDROMEDA	PT Dumas, Surabaya	Dec 2008
ALPHARD	PT Dumas, Surabaya	May 2009
ALNILAM	PT Dumas, Surabaya	May 2009

Measurement, tonnes: 851 gt
Dimensions, metres (feet): 51.4 × 10.37 × 3.0 (168.6 × 34.0 × 9.8)
Speed, knots: 12
Machinery: 1 MaK 6M20 diesel; 1,387 hp (1.02 MW); 1 shaft
Radars: Surface search/navigation: To be announced.

1 DAMEN 5811 (BUOY TENDER) (WABU)

Name	Builders	Commissioned
BIMASAKTI UTARA	PT Dumas, Surabaya	15 Nov 2008

Measurement, tonnes: 1,271 gt
Dimensions, metres (feet): 59.85 × 11.57 × 3.49 (196.4 × 38.0 × 11.5)
Speed, knots: 12
Machinery: 1 MaK 6M20C diesel; 1,387 hp (1.02 MW); 1 shaft
Radars: Surface search/navigation: To be announced.

Comment: Entered service in 2008.

2 BUOY TENDERS (WABU)

Name	Builders	Commissioned
KUMBA	Niigata Engineering	1 Mar 1973
MITHUNA	Niigata Engineering	Mar 1975

Measurement, tonnes: 569 gt
Dimensions, metres (feet): 50.5 × 10.0 × 3.71 (165.7 × 32.8 × 12.2)
Speed, knots: 11
Machinery: 1 Niigata 6L28X diesel; 850 hp (625 kW); 1 shaft
Radars: Surface search/navigation: To be announced.

5 KUJANG CLASS (WPB)

KUJANG 201   **CELURIT** 203   **BELATI** 205
PARANG 202   **CUNDRIK** 204

Displacement, tonnes: 165 full load
Dimensions, metres (feet): 38.3 × 6 × 2.1 (125.7 × 19.7 × 6.9)
Speed, knots: 28
Range, n miles: 1,500 at 18 kt
Complement: 18
Machinery: 2 AGO SACM 195 V12 CZSHR diesels; 4,410 hp(m) (3.24 MW); 2 shafts
Guns: 1—20 mm.

Comment: Built by SFCN, Villeneuve la Garenne. Completed April 1981 (*Kujang* and *Parang*), August 1981 (*Celurit*), October 1981 (*Cundrik*), December 1981 (*Belati*). Pennant numbers are preceded by PAT.

CUNDRIK   11/1998, van Ginderen Collection / 0052366

4 GOLOK CLASS (WSAR)

GOLOK 206   **PANAN** 207   **PEDANG** 208   **KAPAK** 209

Displacement, tonnes: 193 full load
Dimensions, metres (feet): 37.5 × 7.2 × 2 (123 × 23.6 × 6.6)
Speed, knots: 25. **Range, n miles:** 1,500 at 18 kt
Complement: 18
Machinery: 2 MTU 16V 652 TB91 diesels; 4,610 hp(m) (3.39 MW) sustained; 2 shafts
Guns: 1 Rheinmetall 20 mm. 1—7.62 mm MG.

Comment: All launched 5 November 1981. First pair completed 12 March 1982. Last pair completed 12 May 1982. Built by Deutsche Industrie Werke, Berlin. Fitted out by Schlichting, Travemünde. Used for SAR and have medical facilities. Pennant numbers preceded by PAT.

KAPAK   11/1998, van Ginderen Collection / 0052367

15 HARBOUR PATROL CRAFT (WPB)

PAT 01–15

Displacement, tonnes: 12 full load
Dimensions, metres (feet): 12.2 × 4.3 × 1 (40 × 14.1 × 3.3)
Speed, knots: 14
Complement: 4
Machinery: 1 Renault diesel; 260 hp(m) (191 kW); 1 shaft
Guns: 1—7.62 mm MG.

Comment: First six built at Tanjung Priok Shipyard 1978–79. Four more of a similar design built in 1993–94 by Mahalaya Utama Shipyard and delivered from 1995.

HARBOUR PATROL CRAFT TYPE   11/1998, van Ginderen Collection / 0052368

Army — Police ⊲ **Indonesia** 377

ARMY

Notes: The Army (ADRI) craft have mostly been transferred to the Military Sealift Command (Logistic Support).

27 LANDING CRAFT LOGISTICS (LCL)

ADRI XXXII–LVIII

Displacement, tonnes: 589 full load
Dimensions, metres (feet): 42 × 10.7 × 1.8 *(137.8 × 35.1 × 5.9)*
Speed, knots: 10. **Range, n miles:** 1,500 at 10 kt
Complement: 15
Military lift: 122 tons equipment
Machinery: 2 Detroit 6-71 diesels; 348 hp(m) *(260 kW)* sustained; 2 shafts

Comment: A variety of LCL built in Tanjung Priok Shipyard 1979-82. Details are for *Adri XL*. XXXI sank in February 1993.

ADRI XXXIII *10/1999, David Boey* / 0080037

POLICE

Notes: The police operate numerous craft of varying sizes. There is a patrol ship *Tekukur* 643 and three patrol ships 5015, 5016 and 5017. There are four 14 m Simonneau craft, approximately 32 7.6 m patrol craft, propelled by waterjet, and numerous other small craft. Five former Singapore Police Shark class were transferred on 9 February 2012.

PERANJAK 5017 *2/2013*, Bob Fildes* / 1486517

7 FPB 28 CLASS (PATROL CRAFT) (PB)

619–625

Displacement, tonnes: 68.5 full load
Dimensions, metres (feet): 28 × 5.4 × 1.6 *(91.9 × 17.7 × 5.2)*
Speed, knots: 30
Machinery: 2 MTU diesels; 2,720 hp *(2.0 MW)*; 2 shafts
Guns: 1 — 12.7 mm MGs.
Radars: Surface search/navigation: To be announced.

Comment: Constructed by PT PAL, Surabaya, and delivered in the 1980s.

2 OFFSHORE PATROL CRAFT (PBO)

Name	No	Builders	Commissioned
BISMA	520	Astilleros Gondan, Castropol	May 2003
BALADEWA	521	Astilleros Gondan, Castropol	Jun 2003

Dimensions, metres (feet): 61 × 9.9 × 2.6 *(200.1 × 32.5 × 8.5)*
Speed, knots: 22. **Range, n miles:** 3,500 at 12 kt
Machinery: 2 MTU 12V 595TE 90 diesels; 8,700 hp *(6.5 MW)*
Helicopters: Platform for one medium.

Comment: Primary role Search and Rescue.

BISMA *5/2003, Astilleros Gondan* / 0569201

5 NS-395 PATROL CRAFT (PBO)

KUTILANG 638   **BANGAU** 639   **BALIBIS** 640   **PELIKAN** 641   **PUNAI** 642

Displacement, tonnes: 200 full load
Dimensions, metres (feet): 36.4 × 6.8 × 1.9 *(119.4 × 22.3 × 6.2)*
Speed, knots: 26
Machinery: 2 MTU 12V4000 M90 diesels; 5,550 hp *(4.08 MW)*; 2 shafts
Guns: 3 — 12.7 mm MGs.
Radars: Surface search/navigation: To be announced.

Comment: Constructed by Stocznia Marynarki Wojennej, Poland, and delivered in 2007.

PUNAI *4/2007, J Cislak* / 1170185

3 NATSUGIRI CLASS (PB)

HAYABUSA 648   **ANIS MADU** 649   **TAKA** 650

Displacement, tonnes: 69 standard
Dimensions, metres (feet): 27 × 5.6 × 1.2 *(88.6 × 18.4 × 3.9)*
Speed, knots: 27
Machinery: 2 diesels; 3,000 hp *(2.21 MW)*; 2 shafts
Guns: 1 — 12.7 mm MG.
Radars: Surface search/navigation: To be announced.

Comment: Based on the Natsugiri-class patrol craft operated by the Japanese Coast Guard. Delivered in 2007.

POLICE 622 *8/1995* / 0080038

© 2014 IHS

IHS Jane's Fighting Ships 2014-2015

Iran

Country Overview

Formerly a constitutional monarchy ruled by a shah, The Islamic Republic of Iran was established in 1979. With an area of 636,296 square miles, it is situated in the Middle East and is bordered to the north by Armenia, Azerbaijan and Turkmenistan, to the west by Iraq and Turkey and to the east by Afghanistan and Pakistan. It has a 1,318 n mile coastline with the Gulf, the Gulf of Oman and the Caspian Sea. The capital and largest city is Tehran. The principal Caspian ports are Bandar-e Anzali and Bandar-e Torkeman while those in the Gulf include the oil-shipping facilities on Kharg Island, Khorramshahr, Bandar-Khomeini and Bandar-Abbas on the strategic Strait of Hormuz. Territorial Seas (12 n miles) are claimed. An EEZ (200 n miles) has been claimed but the limits have not been defined.

Headquarters Appointments

Commander of Navy:
Rear Admiral Habibollah Sayyari
Head of IRCG(N) (Sepah):
Rear Admiral Ali Fadavi

Personnel

2014: 18,000 Navy (including 2,000 Naval Air and Marines), 20,000 IRGCN

Bases

Persian Gulf: Bandar Abbas (MHQ and 1st Naval District), Boushehr (2nd Naval District and also a Dockyard), Kharg Island, Qeshm Island, Bandar Lengeh
Indian Ocean: Chah Bahar (Bandar Beheshti) (3rd Naval District and forward base), Jask
Caspian Sea: Bandar Anzali (4th Naval District)
Pasdaran: Al Farsiyah, Halileh, Sirri, Abu Musa, Larak

Organisation

The regular navy is responsible for naval operations outside the Strait of Hormuz and into the Indian Ocean; the IRGC is responsible for naval operations within the Gulf.

Coast Defence

Three Navy and one IRGCN brigades with many fixed installations and command posts. Approximately 100 truck-mounted C 802 and 80 CSSC-3 (Seersucker) Chinese SSMs in at least four sites. The indigenously developed Ra'ad cruise missile and C-701 Kosak may be based at launching bases under construction at Bandar Abbas, Bandar Lengeh, Boushehr and Bandar Khomeini.

Mines

Stocks of up to 3,000 mines are reported including Chinese EM 52 rising mines.

Strength of the Fleet

Type	Active	Building
Submarines	3	—
Mini Submarines	17	2 (8)
Frigates	4	3 (2)
Corvettes	2	—
Fast Attack Craft—Missile	23	4
Large Patrol Craft	7	—
Coastal Patrol Craft	138+	—
Inshore Patrol Craft	300+	—
Landing Ships (Logistic)	10	—
Landing Ships (Tank)	5	—
Landing Craft (Tank)	3	—
Hovercraft	8	—
Replenishment Ship	1	—
Supply Ships	2	—
Support Ships	9	—
Water Tankers	4	—
Tenders	12	—

Prefix to Ships' Names

IS

PENNANT LIST

Submarines		Patrol Forces		P 233	Derafsh	24	Farsi
901	Tareq	201	Kayvan	P 313-1	Fath	25	Sardasht
902	Noor	202	Tiran	P 313-2	Nasr	26	Sab Sahel
903	Yunes	204	Mahan	P 313-3	Saf	101	Fouque
		211	Parvin	P 313-4	Ra'd	411	Kangan
Frigates		212	Bahram	P 313-5	Fajr	412	Taheri
		213	Nahid	P 313-6	Shams	421	Bandar Abbas
71	Alvand	P 221	Kaman	P 313-7	Me'raj	422	Bushehr
72	Alborz	P 222	Zoubin	P 313-8	Falaq	424	Daylam
73	Sabalan	P 223	Khadang	P 313-9	Hadid	431	Kharg
76	Jamaran	P 224	Peykan	P 313-10	Qadr	471	Delvar
77	Damavand	P 225	Joshan			472	Sirjan
—	Sahand	P 226	Falakhon	**Mine Warfare Forces**		481	Charak
		P 227	Shamshir	301	Hamzeh	482	Chiroo
Corvettes		P 228	Gorz			483	Soroo
		P 229	Gardouneh	**Amphibious Warfare Forces**		511	Hengam
		P 230	Khanjar	**and Auxiliaries**		512	Larak
81	Bayandor	P 231	Neyzeh	21	Hejaz	513	Tonb
82	Naghdi	P 232	Tabarzin	22	Karabala	514	Lavan
						802	Hamzah

SUBMARINES

Notes: A programme to build a large submarine, known as the Besat class, was first announced in 2008. This is likely to be in development with a view to replacing the Kilo class in due course.

1 NAHANG CLASS (MIDGET SUBMARINES) (SSM)

Displacement, tonnes: 100 surfaced; 115 dived
Dimensions, metres (feet): 25 × 3.0 × 2.5 *(82 × 9.8 × 8.2)*
Machinery: To be announced

Comment: Little is known about this submarine whose existence was noted April 2006. Dimensions are approximate. Whereas it was reported that perhaps two further boats were to be constructed, this is now considered unlikely. It is claimed that the submarine has been indigenously designed and built. The submarine is designed for shallow water operations and potential roles include acting as mothership to swimmer delivery vehicles. Sonars and torpedoes are not fitted.

AL SABEHAT 15   4/2006 / 1164700

NAHANG CLASS   6/2011 / 1406456

8 SWIMMER DELIVERY VEHICLES (LDW)

Comment: On 29 August 2000, the first Iranian-built Swimmer Delivery Vehicle (SDV) *Al Sabehat 15* was launched at Bandar Abbas. The 8 m craft can accommodate a two-man crew and has the capability to carry three additional divers. It is well suited to coastal reconnaissance, Special Forces insertion/extraction and mining (it can carry 14 limpet mines) of ports and anchorages but not to open water operations. Four further craft have been reported and four of a different design have also been observed. The Hengam-class LSLs act as motherships.

SDV (new type)   4/2006 / 1164699

3 KILO CLASS (PROJECT 877 EKM) (SSK)

Name	No	Builders	Laid down	Launched	Commissioned
TAREQ (ex-*B 175*)	901	Admiralty Yard, St Petersburg	1988	25 Sep 1991	21 Nov 1992
NOOR (ex-*B 224*)	902	Admiralty Yard, St Petersburg	1989	16 Oct 1992	6 Jun 1993
YUNES (ex-*B 220*)	903	Admiralty Yard, St Petersburg	1990	12 Jul 1994	25 Nov 1996

Displacement, tonnes: 2,394 surfaced; 3,125 dived
Dimensions, metres (feet): 72.6 × 9.9 × 6.6 *(238.2 × 32.5 × 21.7)*
Speed, knots: 10 surfaced; 9 snorting; 17 dived
Range, n miles: 6,000 at 7 kt snorting; 400 at 3 kt dived
Complement: 53 (12 officers)

Machinery: Diesel-electric; 2 diesels; 3,650 hp(m) *(2.68 MW)*; 2 generators; 1 motor; 5,500 hp(m) *(4.05 MW)*; 1 economic speed motor; 130 hp(m) *(95 kW)*; 1 shaft; 2 auxiliary propulsion motors; 204 hp(m) *(150 kW)*
Torpedoes: 6 – 21 in *(533 mm)* tubes; combination of TEST-71/96; wire-guided active/passive homing to 15 km *(8.1 n miles)* at 40 kt; warhead 220 kg and 53-65; passive wake homing to 19 km *(10.3 n miles)* at 45 kt; warhead 350 kg. Total of 18 weapons.
Mines: 24 in lieu of torpedoes.
Electronic countermeasures: ESM: Squid Head; radar warning. Quad Loop D/F.
Radars: Surface search: Snoop Tray MRP-25; I-band.
Sonars: Sharks Teeth MGK-400; hull-mounted; passive/active search and attack; medium frequency. Mouse Roar MG-519; active attack; high frequency.
Weapon control systems: MVU-119EM Murena TFCS.

Programmes: Contract signed in 1988 for three of the class. The first submarine to be transferred sailed from the Baltic in October 1992 flying the Russian flag and with a predominantly Russian crew. The second sailed in June 1993. The third completed in 1994 but delivery delayed by funding problems. She arrived in Iran in mid-January 1997.
Modernisation: Chinese YJ-1 or Russian Novator Alfa SSMs may be fitted in due course.
Structure: Diving depth, 240 m *(787 ft)* normal. Has a 9,700 kW/h battery. SA-N-10 SAM system may be fitted, but this is not confirmed.
Operational Based at Bandar Abbas. Plans to move to Chah Bahar (Bandar Beheshti) on the northern shore of the Gulf of Oman appear to have been abandoned, at least temporarily, in favour of using the port as a forward operating base. A jetty has been extended to meet this requirement. Operational effectiveness has been adversely affected by technical difficulties, although previously reported problems with battery cooling and air conditioning were understood to have been overcome using Indian batteries. Following negotiations with Rosoboronexport (Russian arms agency) to upgrade the boats, *Tareq* began refit at Bandar Abbas in mid-2005 and was re-launched in September 2012. She became operational in 2013. *Noor* is expected to be the next boat to receive a major refit. A maintenance period for *Yunes* began in 2009 and completed in 2010. *Yunes* conducted a deployment to the Gulf of Aden and Red Sea April to June 2011.

KILO CLASS   4/2006 / 1164704

KILO CLASS   4/2006 / 1164703

0 + 2 (8) FATEH CLASS (COASTAL SUBMARINES) (SSC)

Displacement, tonnes: 500 dived
Dimensions, metres (feet): 48 × 6.0 × ? *(157.5 × 19.7 × ?)*
Machinery: Diesel-electric
Torpedoes: 4 – 21 in *(533 mm)* tubes (to be confirmed).
Sonars: To be announced.

Programmes: The announcement of a new class of coastal submarines was first made on 25 August 2008. Known initially as the Qaaem class, the boats were originally thought likely to be a development of the North Korean Sang-O class but the larger size of the boat and official announcements suggest that the submarines may have been both indigenously designed and built. If so, this would represent a significant achievement by Iran's naval industry. The existence of the new building programme was confirmed in October 2013 by commercial imagery, which revealed the construction of two boats. The first was in the water at Bostanu Shipyard, some 15 n miles west of Bandar Abbas, and is probably being fitted out before being officially unveiled, in early 2014. A second boat is under construction at Bandar Anzali, a major development in the Caspian region. The size of the programme is not known but about 10 boats are expected.
Structure: Initial impressions are of a distinctive casing design. The foreplanes are mounted on the hull. The propulsion was concealed at launch but is believed to include a conventional propller.

FATEH CLASS   2/2014*, IRIN / 1525864

16 YONO (IS 120) CLASS (MIDGET SUBMARINES) (SSM)

941–954   +2

Displacement, tonnes: 117 surfaced; 125 dived
Dimensions, metres (feet): 29 × 2.75 × 2.5 (95.1 × 9.0 × 8.2)
Speed, knots: 10 surfaced; 8 dived
Complement: 7
Machinery: Diesel-electric
Torpedoes: 2—21 in (533 mm) tubes.
Sonars: To be announced.
Programmes: Also known as the Ghadir class, the existence of these submarines was first noted in February 2004. If indigenously built, as has been claimed, this would represent a significant technological development. It is more likely that another country, possibly North Korea, has been involved in the project. Based at Bandar Abbas, they are employed in shallow areas of the Gulf such as the Strait of Hormuz. The fourth was launched on 28 November 2007, the fifth on 27 November 2008 and the sixth and seventh in 2009. Four further boats were commissioned on 8 August 2010 and three further on 26 November 2011. Two further craft were commissioned in February 2012. The design includes a retractable secondary propeller.

YONO 947
9/2009, FARS News Agency
1367081

FRIGATES

Notes: The Loghman project is for a new class of frigates to follow the Mowj (modified Alvand)-class programme. The ships are expected to be larger (approximate length of 120 m) and may be a stretched version of the Alvand class and is to feature a hangar and helicopter deck in addition to a medium calibre gun (probably 76 mm) and surface-to-surface missiles. The timetable for production has not been announced but it is unlikely that the first of class will be launched until about 2017.

4 + 3 (2) ALVAND (VOSPER MK 5) CLASS (FFG)

Name	No	Builders	Laid down	Launched	Commissioned
ALVAND (ex-Saam)	71	Vosper Thornycroft, Woolston	22 May 1967	25 Jul 1968	20 May 1971
ALBORZ (ex-Zaal)	72	Vickers Shipbuilding & Engineering, Barrow-in-Furness	3 Mar 1968	25 Jul 1969	1 Mar 1971
SABALAN (ex-Rostam)	73	Vickers Shipbuilding & Engineering, Newcastle & Barrow	10 Dec 1967	4 Mar 1969	28 Feb 1972
JAMARAN	76	Bandar Abbas	2004	28 Nov 2007	19 Feb 2010
DAMAVAND	77	Bandar Anzali	2009	15 Mar 2013	2014
SAHAND	–	Bandar Abbas	2010	2014	2015
–	–	Bostanu Shipyard	2012	2014	2016

Displacement, tonnes: 1,372 full load
Dimensions, metres (feet): 94.5 × 11.1 × 4.3 (310 × 36.4 × 14.1)
Speed, knots: 39 (71–73), 28 (76); 18 diesel
Range, n miles: 3,650 at 18 kt, 550 at 36 kt
Complement: 125 + 21 spare berths

Machinery: CODOG (71, 72, 73); 2 RR Olympus TM2A gas turbines; 40,000 hp (29.8 MW) sustained; 2 Paxman 16YJCM diesels; 3,800 hp (2.83 MW) sustained; 2 shafts; cp props; 2 diesels (Jamaran); 20,000 hp (14.9 MW); 2 shafts
Missiles: SSM: 4 Noor C-802 (2 twin) ❶; active radar homing to 120 km (65 n miles) at 0.9 Mach; warhead 165 kg; sea-skimmer.
SAM: 4 Standard SM-1 Block 5 (76, 77 in boxes) ❷; semi-active radar homing to 38 km (20.5 n miles); warhead 115 kg.
Guns: 1 Vickers 4.5 in (114 mm)/55 Mk 8 (71–73) ❸; 25 rds/min to 22 km (12 n miles) anti-surface; 6 km (3.3 n miles) anti-aircraft; weight of shell 21 kg. 1—76 mm (76) ❹; 50 rds/min to 16 km (8.6 n miles); weight of shell 6 kg. 1—40 mm (76) ⓫. 2 Oerlikon 35 mm/90 (twin) (71–73) ❺; 550 rds/min to 6 km (3.3 n miles); weight of shell 1.55 kg. 1 Oerlikon GAM-BO1 20 mm (71–73) ❻. 2—12.7 mm MGs.
Torpedoes: 6—324 mm Mk 32 (2 triple) tubes ❼.
Physical countermeasures: Decoys: 2 UK Mk 5 rocket flare launchers.
Electronic countermeasures: ESM: Decca RDL 2AC; radar warning. Racal FH 5-HF/DF.
Radars: Air/surface search: Plessey AWS 1 (71, 72, 73, 76) ❽; E/F-band. Asr (77) ⓬; 3D; F-band.
Surface search: Racal Decca 1226 ❾; I-band.
Navigation: Decca 629; I-band.
Fire control: Contraves Sea Hunter ❿; I/J-band. Signaal WM 28 (77) ⓭; I/J-band.
IFF: UK Mk 10.
Sonars: Graseby 174; hull-mounted; active search; medium/high frequency. Graseby 170; hull-mounted; active attack; high frequency.

Helicopters: Platform only (Jamaran).

Programmes: The first three units were ordered from UK Shipyards on 25 August 1966. Construction of a fourth unit (Jamaran) at Bandar Abbas was completed in 2010. Known as the Mowj project, the design is very similar to the original Vosper Mk 5 design. The ship is reported to have all diesel propulsion. Construction of a second Mowj unit (Damavand) at Bandar Anzali on the Caspian Sea and of a third unit (Sahand) at Bandar Abbas is in progress. Construction of a fourth at Bostanu Shipyard has also been reported. A further two ships are expected.
Modernisation: Major refits of 71–73 including replacement of 4.5 in Mk 5 gun by Mk 8 completed 1977. Modifications in 1988 included replacing Seacat with a 23 mm gun and boat davits with minor armaments. By mid-1991 the 23 mm and both boats had been replaced by GAM-BO1 20 mm guns and the SSM launcher had effectively become a twin launcher. In 1996/97 two of the class had the Sea Killer SSM replaced by C-802 launchers and a new communications mast fitted between the two fire-control radars. The third has been similarly modified. Sabalan appears to be fitted with Rice Screen air/surface search radar. Torpedo tubes which replaced the mortars in Alvand were probably taken from decommissioned Babr class. A new indigenous frequency-scanned 3D radar, Asr, has been observed in Damavand, rather than an AWS 1 radar. It is likely that this will be fitted to the other Mowj ships. Damavand is also equipped with a WM 28 fire-control radar.
Structure: Air conditioned throughout. Fitted with Vosper stabilisers. Jamaran is fitted with a flight-deck.
Operational: Sahand sunk by USN on 18 April 1988. Sabalan had her back broken by a laser-guided bomb in the same skirmish but was out of dock by the end of 1990 and was operational again in late 1991. ASW mortars probably unserviceable. Alborz deployed to the Gulf of Aden to conduct anti-piracy operations in mid-2009. Test firing of Noor missile conducted by Jamaran in March 2010.

ALVAND (Scale 1 : 900), Ian Sturton / 1335484

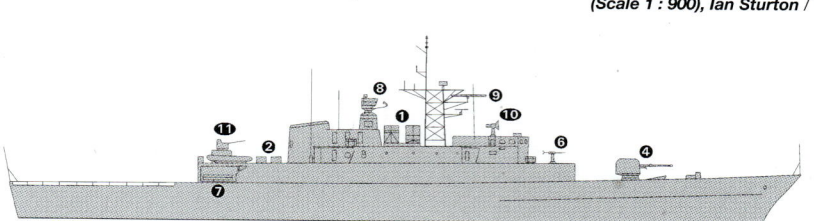

JAMARAN (Scale 1 : 900), Ian Sturton / 1483592

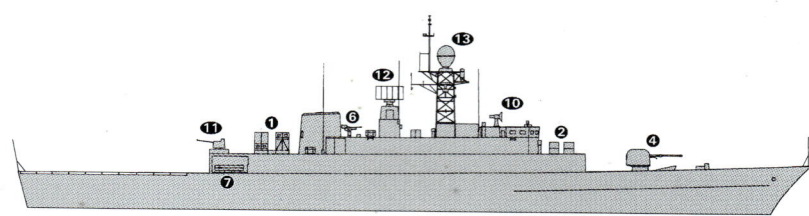

DAMAVAND (Scale 1 : 900), Ian Sturton / 1531002

JAMARAN 3/2010 / 1366304

SABALAN 2/1998 / 0052371

ALVAND 1/2002 / 0569203

CORVETTES

2 BAYANDOR (PF 103) CLASS (FS)

Name	No	Builders	Laid down	Launched	Commissioned
BAYANDOR (ex-US PF 103)	81	Levingstone Shipbuilding Co, Orange	20 Aug 1962	7 Jul 1963	18 May 1964
NAGHDI (ex-US PF 104)	82	Levingstone Shipbuilding Co, Orange	12 Sep 1962	10 Oct 1963	22 Jul 1964

Displacement, tonnes: 914 standard; 1,153 full load
Dimensions, metres (feet): 84 × 10.1 × 3.1 *(275.6 × 33.1 × 10.2)*
Speed, knots: 20
Range, n miles: 2,400 at 18 kt, 4,800 at 12 kt
Complement: 140

Machinery: 2 Fairbanks-Morse 38TD8-1/8-9 diesels; 5,250 hp *(3.92 MW)* sustained; 2 shafts
Missiles: 4 C-802 Noor ❶; active radar homing to 120 km *(65 n miles)* at 0.9 Mach; warhead 165 kg.
Guns: 1 Oto Melara 76 mm/62 compact ❷; 85 rds/min to 16 km *(8.7 n miles)*; weight of shell 6 kg. 2 Bofors 40 mm/60 (twin) ❸; 120 rds/min to 10 km *(5.5 n miles)*; weight of shell 0.89 kg. 2 Oerlikon GAM-BO1 20 mm ❹. 2−12.7 mm MGs.
Torpedoes: 6−324 mm Mk 32 (2 triple) tubes.
Radars: Air/surface search: Westinghouse SPS-6C ❺; D-band; range 146 km *(80 n miles)* (for fighter).
Surface search: Racal Decca ❻; I-band.
Navigation: Raytheon 1650 ❼; I/J-band.
Fire control: Western Electric Mk 36; I/J-band.
IFF: UPX-12B.
Sonars: EDO SQS-17A; hull-mounted; active attack; high frequency.
Weapon control systems: Mk 63 for 76 mm gun. Mk 51 Mod 2 for 40 mm guns.

Programmes: Transferred from the US to Iran under the Mutual Assistance programme in 1964.
Modernisation: *Naghdi* change of engines and reconstruction of accommodation completed in mid-1988. *Naghdi* completed a substantial modernisation programme in 2009. Upgrades included the replacement of the 76 mm gun with a modern weapon, with an electro-optic sight, installation of four C-802 missile tubes aft of the funnel and of Mk 32 torpedo tubes aft. *Bayandor* completed a similar refit in April 2013.

Operational: *Milanian* and *Khanamuie* sunk in 1982 during war with Iraq. Both remaining ships are very active. Sonars may have been removed.

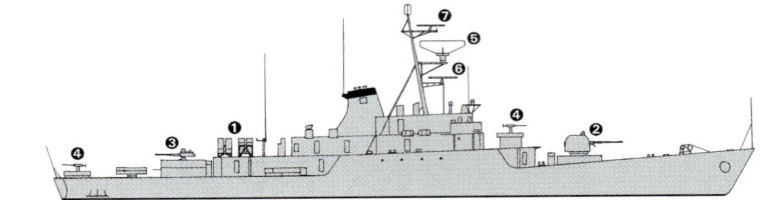

NAGHDI *(Scale 1 : 900), Ian Sturton / 1305473*

NAGHDI *7/2011, FARS News Agency / 1406455*

NAGHDI *9/2009 / 1367075*

SHIPBORNE AIRCRAFT

Numbers/Type: 9 Agusta AB 204ASW/212.
Operational speed: 104 kt (193 km/h).
Service ceiling: 11,500 ft (3,505 m).
Range: 332 n miles (615 km).
Role/Weapon systems: Mainly engaged in ASV operations in defence of oil installations. Numbers are uncertain. Sensors: APS 705 search radar, dipping sonar (if carried). Weapons: ASW; two China YU-2 torpedoes. ASV; two AS 12 missiles.

AB 212   6/2009, Annati Collection / 1366303

Numbers/Type: 9 Agusta-Sikorsky ASH-3D Sea King.
Operational speed: 120 kt (222 km/h).
Service ceiling: 12,200 ft (3,720 m).
Range: 630 n miles (1,165 km).
Role/Weapon systems: Shore-based ASW helicopter to defend major port and oil installations. Six are reported serviceable. Can be embarked in *Kharg*. Sensors: Selenia search radar, dipping sonar. Weapons: ASW; four A244/S torpedoes or depth bombs. ASV; trials of an anti-ship missile 'Fajr-e-Darya' are reported to have taken place. Capabilities not known but could be a development of Sea Killer.

SEA KING   6/2009, Annati Collection / 1366636

LAND-BASED MARITIME AIRCRAFT (FRONT LINE)

Notes: (1) The Air Force also has up to six F-4 Phantoms equipped with C 80IK ASMs for the maritime role.
(2) Two F-27 Fokker Friendship aircraft are used in a utility MPA role.
(3) Two Dornier 228 are also in service but are reported not to be very active.
(4) The Iranian Air Force operates some 14 (plus 18 ex-Iraqi) Su-24 Fencer ground attack, some of which may be 'marinised' for an anti-ship role.
(5) An-140 transport aircraft are under licensed production at Esfahan. The first aircraft flew in January 2001. These are used as a multipurpose platform including maritime patrol. These are known as Oghab.
(6) At least seven Harbin Y-12 utility aircraft are used for maritime patrol.
(7) There is an unknown number of UAVs; details of their sensors are not known.
(8) There were 24 ex-Iraqi Dassault Mirage aircraft impounded in 1991. Of these at least six F1BQ two-seaters and 15 F1EQ single-seaters are believed to be operational.
(9) There are approximately 45 Bell AH-1J Sea Cobra (Toufan). Refurbishment of these aircraft begain in 1998 and 10 upgraded aircraft were reportedly delivered in 2010.

OGHAB   6/2012 / 1483505

Numbers/Type: 6 Sikorsky RH/MH-53D Sea Stallion.
Operational speed: 125 kt (232 km/h).
Service ceiling: 11,100 ft (3,385 m).
Range: 405 n miles (750 km).
Role/Weapon systems: Surface search helicopter which could be used for mine clearance but so far has only been used for Logistic purposes. Can be carried on Hengam class flight deck. Sensors: Weather radar. Weapons: Unarmed.

SEA STALLION   6/2009, Annati Collection / 1366635

Numbers/Type: 2 Lockheed P-3F Orion.
Operational speed: 410 kt (760 km/h).
Service ceiling: 28,300 ft (8,625 m).
Range: 4,000 n miles (7,410 km).
Role/Weapon systems: Air Force manned. One of the remaining two aircraft can be used for early warning and control duties for strikes. Replacements are being sought. Sensors: Search radar, sonobuoys. Weapons: ASW; various weapons can be carried. ASV; C-802 SSM.

P-3F   11/2008, Guy Toremans / 1367079

Numbers/Type: 1 HESA Shahed 285.
Operational speed: 129 kt (240 km/h).
Service ceiling: 21,000 ft (6,400 m).
Range: 184 n miles (340 km).
Role/Weapon systems: Unveiled on 24 May 2009, the Shahed 285 is probably based on the Shahed 278 which in turn has been developed from the Bell 206 Jet Ranger design. There are reported to be two variants: a light attack/reconnaissance version and a maritime patrol/anti-ship version. The prototype maritime version appears to be fitted with a small chin-mounted search radar (possibly commercial) and is armed with a Kosar 3 missile, which is based on the C-701 anti-ship missile. Performance details are based on those of the Bell Jet Ranger and are speculative. Further aircraft, to be operated by the IRGC (Pasdaran), are expected.

SHAHED 285   5/2009, FARS News Agency / 1363241

PATROL FORCES

Notes: (1) There are at least one 13 m (RIB 42SC), one 16.5 m (FB 55), two 12 m (FB 38), two 11 m (RIB 36) and two 16.7 m (RIB 55SC) craft used for maritime enforcement tasks. Designed by FB design of Italy, they are capable of 60–70 kt. Additional units to similar designs have been built in Iran.
(2) The test of a high-speed rocket-torpedo was announced on 2 April 2006. The weapon appears to resemble the Russian Shkval which has a speed of 195 kt and a range of 3.75 n miles.
(3) A number of Båtservice 17 m and 20 m craft are in service with the Coast Guard.

RIB 42   6/2007, FB design / 1166564

10 TIR (IPS 18) CLASS (COASTAL PATROL CRAFT) (PTF)

Displacement, tonnes: 29 standard
Dimensions, metres (feet): 21.1 × 5.8 × 0.9 (69.2 × 19.0 × 3.0)
Speed, knots: 52
Complement: 6
Machinery: 3 diesels; 3,600 hp (2.7 MW)
Guns: 1—12.7 mm MG.
Torpedoes: 2—533 mm (unknown type).

Comment: Up to ten of this class in service with the Pasdaran. Built in North Korea, two craft were reported to have been delivered on 8 December 2002 on the Iranian freighter *Iran Meead*. Anti-surface ship role.

TIR and BOGHAMMAR   1/2006, RAAF / 1167756

10 THONDOR (HOUDONG) CLASS
(FAST ATTACK CRAFT—MISSILE) (PTFG)

SHAHID MEHDAVI (ex-*Fath*) P 313-1		SHAHID DARA (ex-*Shams*) P 313-6	
SHAHID KORD (ex-*Nasr*) P 313-2		SHAHID ABSALAN (ex-*Me'raj*) P 313-7	
SHAHID SHAFIHI (ex-*Saf*) P 313-3		SHAHID RAHISI RAISI (ex-*Falaq*) P 313-8	
SHAHID TOWSALI (ex-*Ra'd*) P 313-4		SHAHID GOLZAM (ex-*Hadid*) P 313-9	
SHAHID HEJAT ZADEH (ex-*Fajr*) P 313-5		SHAHID SAHRABI (ex-*Qadr*) P 313-10	

Displacement, tonnes: 174 standard; 208 full load
Dimensions, metres (feet): 38.6 × 6.8 × 2.7 *(126.6 × 22.3 × 8.9)*
Speed, knots: 35. **Range, n miles:** 800 at 30 kt
Complement: 28 (3 officers)

Machinery: 3 diesels; 8,025 hp(m) *(7.94 MW)* sustained; 3 shafts
Missiles: SSM: 4 Noor C-802; active radar homing to 120 km *(65 n miles)* at 0.9 Mach; warhead 165 kg; sea-skimmer.
Guns: 2—30 mm/65 (twin) AK 230. 2—23 mm/87 (twin).
Radars: Surface search: China SR-47A; I-band.
Navigation: China RM 1070A; I-band.
Fire control: Rice Lamp Type 341; I/J-band.

Programmes: Negotiations for sale started in 1991 but were held up by arguments over choice of missile. Built at Zhanjiang Shipyard. First five delivered in September 1994 by transporter vessel, second batch in March 1996. Original pennant numbers 301–310. More may be built in Iran under licence.
Structure: The hull is a shortened version of the Chinese Huangfen (Osa 1) class but the superstructure has a lattice mast to support two I-band radars and there is a separate director plinth for the fire-control system. A twin 23 mm gun is fitted aft of the mast.
Operational: Manned by the Pasdaran. Renamed in approximately 2006.

SHAHID GOLZAM   4/2006 / 1164698

SHAHID ABSALAN   4/2006 / 1164697

13 US MK III CLASS (COASTAL PATROL CRAFT) (PB)

Displacement, tonnes: 42 full load
Dimensions, metres (feet): 19.8 × 5.5 × 1.8 *(65 × 18.0 × 5.9)*
Speed, knots: 30
Range, n miles: 500 at 28 kt
Complement: 8
Machinery: 3 GM 8V-71TI diesels; 690 hp *(515 kW)* sustained; 3 shafts
Guns: 1—20 mm GAM-BO1. 1—12.7 mm MG.
Radars: Surface search: RCA LN66; I-band.

Comment: Twenty ordered from Marinette Marine Corporation, Wisconsin, USA; the first delivered in December 1975 and the last in December 1976. A further 50 were ordered in 1976 to be shipped out and completed in Iran. It is not known how many were finally assembled. Six lost in the Gulf War, others have been scrapped. These last 13 are based at Boushehr and Bandar Abbas. Continue to be active.

US Mk III   5/1999 / 0080041

13 + 4 KAMAN (COMBATTANTE II) CLASS
(FAST ATTACK CRAFT—MISSILE) (PGGF)

Name	No	Builders	Commissioned
KAMAN	P 221	CMN, Cherbourg	12 Aug 1977
ZOUBIN	P 222	CMN, Cherbourg	12 Sep 1977
KHADANG	P 223	CMN, Cherbourg	15 Mar 1978
PEYKAN	P 224	Bandar Anzali	2004
JOSHAN	P 225	Bandar Anzali	2006
FALAKHON	P 226	CMN, Cherbourg	31 Mar 1978
SHAMSHIR	P 227	CMN, Cherbourg	31 Mar 1978
GORZ	P 228	CMN, Cherbourg	22 Aug 1978
GARDOUNEH	P 229	CMN, Cherbourg	11 Sep 1978
KHANJAR	P 230	CMN, Cherbourg	1 Aug 1981
NEYZEH	P 231	CMN, Cherbourg	1 Aug 1981
TABARZIN	P 232	CMN, Cherbourg	1 Aug 1981
DERAFSH	P 233	Bandar Anzali	2009

Displacement, tonnes: 253 standard; 279 full load
Dimensions, metres (feet): 47 × 7.1 × 1.9 *(154.2 × 23.3 × 6.2)*
Speed, knots: 37.5
Range, n miles: 2,000 at 15 kt, 700 at 33.7 kt
Complement: 31

Machinery: 4 MTU 16V 538 TB91 diesels; 12,280 hp(m) *(9.03 MW)* sustained; 4 shafts
Missiles: SSM: 2 or 4 Noor C-802 (1 or 2 twin); active radar homing to 120 km *(66 n miles)* at 0.9 Mach; warhead 165 kg; sea-skimmer or 4 McDonnell Douglas Harpoon (2 twin); active radar homing to 40 km *(22 n miles)* at 0.9 Mach; warhead 165 kg; sea-skimmer or Standard SM1-MR box launchers *(Gorz)*.
Guns: 1 Oto Melara 3 in *(76 mm)*/62 compact; 85 rds/min to 16 km *(8.7 n miles)* anti-surface; 12 km *(6.6 n miles)* anti-aircraft; weight of shell 6 kg; 320 rounds. 1 Breda Bofors 40 mm/70; 300 rds/min to 12 km *(6.6 n miles)*; weight of shell 0.96 kg; 900 rounds. Some have a 23 mm or 20 mm gun in place of the 40 mm. 2—12.7 mm MGs.
Electronic countermeasures: ESM: Thomson-CSF TMV 433 Dalia; radar intercept.
ECM: Thomson-CSF Alligator; jammer.
Radars: Surface search/fire control: Signaal WM28; I/J-band.
Navigation: Racal Decca 1226; I-band.
IFF: UPZ-27N/APX-72.

Programmes: Twelve ordered in February 1974. The transfer of the last three craft was delayed by the French government after the Iranian revolution. On 12 July 1981 France decided to hand them over. This took place on 1 August, on 2 August they sailed and soon after *Tabarzin* was seized by a pro-Royalist group off Cadiz. After the latter surrendered to the French in Toulon further problems were prevented by sending all three to Iran in a merchant ship. Further indigenously built craft have been developed for operations in the Caspian Sea. Known as the SINA programme, SINA 1 *(Peykan)* was launched on 29 September 2003, SINA 2 *(Joshan)* commissioned in 2006 and SINA 3 *(Derafsh)* in 2009. Four further craft (SINA 4–7) are under construction, one reportedly in the Caspian and the remainder at Bandar Abbas. Entry into service is expected in 2016.
Modernisation: Most of the class fitted with C-802 SSM in 1996-98. *Gorz* has been used for trials, first with Harpoon, and now with SM 1 launchers taken from the deleted Sumner class destroyers.
Structure: Portable SA-7 launchers may be embarked in some.
Operational: The original *Peykan* P 224 was sunk in 1980 by Iraq; *Joshan* P 225 in April 1988 by the US Navy. The new *Peykan* P 224, *Joshan* P 225 and *Derafsh* P 233 are based in the Caspian Sea.

GARDOUNEH (with Harpoon)   11/2001, Royal Australian Navy / 0528433

SHAMSHIR   4/2006 / 1164696

PEYKAN   6/2008, IRNA / 1367078

3 PARVIN (PGM-71) CLASS (LARGE PATROL CRAFT) (PC)

Name	No	Builders	Commissioned
PARVIN (ex-PGM 103)	211	Peterson Builders Inc, Sturgeon Bay	1967
BAHRAM (ex-PGM 112)	212	Peterson Builders Inc, Sturgeon Bay	1969
NAHID (ex-PGM 122)	213	Peterson Builders Inc, Sturgeon Bay	1970

Displacement, tonnes: 100 standard; 150 full load
Dimensions, metres (feet): 30.8 × 6.5 × 2.5 (101 × 21.3 × 8.2)
Speed, knots: 22. **Range, n miles:** 1,140 at 17 kt
Complement: 20

Machinery: 8 GM 6-71 diesels; 2,040 hp (1.52 MW) sustained; 2 shafts
Missiles: SSM: 2 launchers.
Guns: 1 Bofors 40 mm/60 (old variant). 2 GAM-BO1 20 mm. 2—12.7 mm MGs.
Depth charges: 4 racks (8 US Mk 6).
Radars: Surface search: I-band.
Sonars: SQS-17B; hull-mounted active attack; high frequency.

Comment: All three ships have been modified with two missile launchers (of unknown type). The sonar is unlikely to be serviceable.

NAHID 6/2012, Annati Collection / 1486491

3 KAJAMI CLASS (SEMI-SUBMERSIBLE CRAFT) (PTF)

Displacement, tonnes: 30.4 standard (est.)
Dimensions, metres (feet): 21 × ? × ? (68.9 × ? × ?)
Speed, knots: 50 (est.)
Torpedoes: 2 lightweight.

Comment: Originally reported as the Taedong-B high-speed infiltration craft, two of these craft were reported delivered from North Korea on 8 December 2002 on the Iranian freighter *Iran Meead*. Little is known about the design of the craft except that its concept of operations is likely to include a high speed surface approach to a target before submerging to a depth of about 3 m to conduct the attack phase using a snort mast.

KAJAMI (submerged approach) (artist's impression) 10/2005 / 1151268

KAJAMI 6/2006 / 1164691

10 GHAEM (MIG-S-1800) CLASS
(COASTAL PATROL CRAFT) (PB)

Displacement, tonnes: 61 full load
Dimensions, metres (feet): 18.7 × 5.8 × 1.1 (61.4 × 19.0 × 3.6)
Speed, knots: 18
Complement: 10
Machinery: 2 MWM TBD 234 V12 diesels; 1,646 hp(m) (1.21 MW); 2 shafts
Guns: 1 Oerlikon 20 mm. 2—7.62 mm MGs.
Radars: Surface search: I-band.

Comment: Assembled in Iran as general purpose patrol craft. Numbers uncertain. Pasdaran craft.

GHAEM CLASS 1996, Joolaee Marine Industries / 0506299

3 KAYVAN (CAPE) CLASS (LARGE PATROL CRAFT) (PB)

KAYVAN 201 (ex-61)   TIRAN 202 (ex-63)   MAHAN 204 (ex-64)

Displacement, tonnes: 100 standard; 150 full load
Dimensions, metres (feet): 28.9 × 6.2 × 2 (94.8 × 20.3 × 6.6)
Speed, knots: 21. **Range, n miles:** 2,324 at 8 kt
Complement: 15
Machinery: 2 Cummins NYHMS-1200 diesels; 2,120 hp (1.58 MW); 2 shafts
Missiles: SSM: To be announced.
Guns: 1 Bofors 40 mm/60. 2 USSR 23 mm/80 (twin). 2—12.7 mm MGs.

Comment: Three patrol craft originally built by the US Coast Guard, Curtis Bay, Maryland in the 1950s were withdrawn from Iranian service in approximately 1995. It is reported that they have been refitted and recommissioned. They may have been armed with missiles. Details are as for the craft in 1994 but it is likely that machinery and armament may now be different.

MAHAN (old number) 6/1975, Iranian Navy / 1293457

10 PASHE (MIG-G-1900) CLASS
(COASTAL PATROL CRAFT) (PBF)

Displacement, tonnes: 30 full load
Dimensions, metres (feet): 19.5 × 4.2 × 0.9 (64 × 13.8 × 3.0)
Speed, knots: 36
Complement: 8
Machinery: 2 MWM TBD 234 V12 diesels; 1,646 hp(m) (1.21 MW); 2 shafts
Guns: 2—23 mm/80 (twin).
Radars: Surface search: I-band.

Comment: Building in Iran to a modified US Mk II design. Numbers uncertain. Pasdaran craft.

MIG-G-1900 1992, Iranian Marine Industries / 0080042

17 PEYKAAP I (IPS 16) CLASS
(COASTAL PATROL CRAFT) (PTF)

Displacement, tonnes: 14 standard
Dimensions, metres (feet): 17 × 3.75 × 0.7 (55.8 × 12.3 × 2.3)
Speed, knots: 52
Complement: 3
Machinery: 2 diesels; 2,400 hp (1.79 MW); surface piercing propeller
Guns: 1—12.7 mm MG.
Torpedoes: 2 lightweight

Comment: Approximately 17 of this class in service with the Pasdaran. Built in North Korea, six craft were reported to have been delivered on 8 December 2002 on the Iranian freighter *Iran Meead*. An apparently stealthy craft whose unusual armament of 324 mm lightweight torpedoes suggest a ship-disabling role.

PEYKAAP I CLASS 4/2006 / 1164695

30 PEYKAAP II (IPS 16 MOD) CLASS
(COASTAL PATROL CRAFT) (PTG)

Displacement, tonnes: 14 standard (est.)
Dimensions, metres (feet): 17.3 × 3.75 × 0.7 (56.8 × 12.3 × 2.3)
Speed, knots: 52 (est.)
Machinery: 2 diesels; 2,400 hp (1.79 MW); surface piercing propeller
Missiles: 2 FL-10 or C-701 or C-704 (Nasr); inertial guidance and active terminal homing to 38 km (20.5 n miles) at 0.8 Mach, warhead 130 kg.
Torpedoes: 2 lightweight.

Comment: Slightly larger versions of the Peykaap I class armed with FL-10 or C-701 (Kosar) or C-704 (Nasr) missiles in addition to torpedoes. Approximately 29 of this class in service with the Pasdaran. Probably built in Iran as a development of the original North Korean design.

PEYKAAP II   9/2009, FARS News Agency / 1367074

15 PEYKAAP III (IPS 16 MOD) CLASS
(COASTAL PATROL CRAFT) (PTF)

Displacement, tonnes: 13.75 standard (est.)
Dimensions, metres (feet): 17.3 × 3.75 × 0.7 (56.8 × 12.3 × 2.3)
Speed, knots: 52 (est.)
Machinery: 2 diesels; 2,400 hp (1.79 MW); surface piercing propeller
Missiles: 2 FL-10 or C-701 or C-704 (Nasr); inertial guidance and terminal homing to 38 km (56.7 n miles) at 0.8 Mach; warhead 130 kg.
Guns: 2 – 12.7 mm MGs.

Comment: Similar to the Peykaap II class armed with FL-10 or C-701 (Kosar) or C-704 (Nasr) missiles and with 12.7 mm MGs rather than torpedoes. Approximately six of this class are in service with the Pasdaran. Further craft are expected. Numbers are uncertain but the building programme is reported to be continuing.

PEYKAAP III   6/2010 / 1366588

15 TARLAN CLASS (INSHORE PATROL CRAFT) (PTF)

Displacement, tonnes: 9 standard
Dimensions, metres (feet): 11.9 × 3.1 × 0.65 (39 × 10.2 × 2.1)
Speed, knots: 50
Complement: 2
Machinery: 2 diesels; 1,320 hp (985 kW); 2 surface piercing propellers

Comment: A new class of indigenously built inshore attack craft first reported in 2005. Design features include an aluminium, catamaran hull, probably adapted from a commercial craft, and a 1.5 m high pedestal in the after part of the vessel. This might support a wire/laser guided weapon similar to an Anti-Tank Guided Missile (ATGM). Numbers of craft are uncertain. Two craft of a modified version, known as the Dalaam class, have also been reported. These craft may be capable of firing a high-speed rocket torpedo resembling the Russian Shkval.

TARLAN CLASS (under construction)   6/2007 / 1166572

IHS Jane's Fighting Ships 2014-2015

3 GAHJAE CLASS (SEMI-SUBMERSIBLE CRAFT) (PTF)

Displacement, tonnes: 7 standard (est.)
Dimensions, metres (feet): 15 × 3.0 × 0.7 (49.2 × 9.8 × 2.3)
Speed, knots: 50 (est.)
Torpedoes: 2 lightweight.

Comment: Originally reported as the Taedong-C semi-submersible torpedo boat, three of these craft were reported delivered from North Korea on 8 December 2002 on the Iranian freighter *Iran Meead*. The stealthy design appears to be based on the Peykaap class inshore patrol craft on which the dimensions, which are speculative, are based. The concept of operations is likely to include a high speed surface approach to a target before submerging to a depth of about 3 m to conduct the attack phase using a snort mast.

GAHJAE CLASS (artist's impression)   10/2005 / 1151265

12 C 14 CLASS (PTGF)

611–614   +8

Displacement, tonnes: 17 standard
Dimensions, metres (feet): 13.65, 13.75 (611–614) × 4.8 × 0.7 (44.8, 45.1 × 15.7 × 2.3)
Speed, knots: 50
Complement: 5

Machinery: 2 diesels; 2,300 hp (1.7 MW); 2 surface piercing propellers
Missiles: SSM: 4 C-704 (Nasr) launchers; inertial guidance and active terminal homing to 38 km (20.5 n miles) at 0.8 Mach; warhead 130 kg.
Guns: 1 – 20 mm. 1 – 12.7 mm MG. 1 – 12-barrelled 122 mm MRL (611–614).
Radars: Surface search: I-band.
Weapon control systems: Optronic director.

Comment: There are two known variants of this catamaran-hulled class, the prototype of which was reported delivered in late 2000 and commissioned in 2001. Five missile-armed craft are operated by the Pasdaran. The type of missile has not been confirmed but is probably C-704 (Nasr). Four (611–614) slightly longer (13.75 m) craft are fitted with BM-21 122 mm MRL on the bridge roof and are operated by the navy.

C 14 (missile variant)   5/2010, IAIO / 1363821

C 14 (MRL variant)   9/2009, IRNA / 1367073

20 BOGHAMMAR CRAFT (PBF)

Displacement, tonnes: 7 full load
Dimensions, metres (feet): 13 × 2.7 × 0.7 (42.7 × 8.9 × 2.3)
Speed, knots: 46
Range, n miles: 500 at 40 kt
Complement: 6 (est.)

Machinery: 2 Seatek 6-4V-9 diesels; 1,160 hp (853 kW); 2 shafts
Guns: 3 – 12.7 mm MGs.
Rockets: 1 RPG-7 rocket launcher or 106 mm recoilless rifle. 1 – 12-barrelled 107 mm rocket launcher (MRL).
Radars: Surface search: I-band.

Comment: Ordered in 1983 and completed in 1984–85 for Customs Service. Total of 51 originally delivered. Used extensively by the Pasdaran. Maximum payload 450 kg. Speed is dependent on load carried. They can be transported by Amphibious Lift Ships and can operate from bases at Farsi, Sirri and Abu Musa Islands with a main base at Bandar Abbas. Re-engined with Seatek diesels from 1991. There are also 10 further 11 m craft with similar characteristics. Known as TORAGH boats and manned by the Pasdaran and the Navy. Numbers approximate.

Patrol forces < **Iran** 387

10 US MK II CLASS (COASTAL PATROL CRAFT) (PB)

Displacement, tonnes: 23 full load
Dimensions, metres (feet): 15.2 × 4.6 × 1.3 (49.9 × 15.1 × 4.3)
Speed, knots: 28
Range, n miles: 750 at 26 kt
Complement: 8
Machinery: 2 GM 8V-71TI diesels; 460 hp (343 kW) sustained; 2 shafts
Guns: 2—12.7 mm MGs.
Radars: Surface search: Furuno; I-band.

Comment: Twenty-six ordered from Peterson, USA in 1976-77. Six were for the Navy and the remainder for the Imperial Gendarmerie. All were built in association with Arvandan Maritime Corporation, Abadan. The six naval units operate in the Caspian Sea. Of the remaining 20, six were delivered complete and the others were only 65 per cent assembled on arrival in Iran. Some were lost when the Iraqi Army captured Koramshahr. Others have been lost at sea. Numbers uncertain.

US Mk II   3/1996 / 0080043

100 ASHOORA I (MIG-G-0800) CLASS
(INSHORE PATROL CRAFT) (PBF)

Displacement, tonnes: 1 full load
Dimensions, metres (feet): 6.7 × 2.3 × 0.4 (22 × 7.5 × 1.3)
Speed, knots: 40 (est.)
Complement: 4
Machinery: 2 outboards; 240 hp (179 kW)
Guns: Various, but can include 1—12-barrelled 107 mm MRL or 1—12.7 mm MG.

Comment: Boston Whaler type craft based on a Watercraft (UK) design. Numerous indigenously constructed GRP hulls. Numbers uncertain but revised estimates suggest that there are at least 100 craft although these are probably in course of replacement by Bladerunners and other craft. Some are used as minelayers. Manned by the Pasdaran and the Navy.

ASHOORA MINELAYER   6/2013* / 1529051

RIVER ROADSTEAD PATROL AND HOVERCRAFT (PBR)

Comment: Numerous craft used by the Revolutionary Guard include:
Type 2: Dimensions, feet (metres): 22.0 × 7.2 (6.7 × 2.2); single outboard engine; 1–12.7 mm MG.
Type 3: Dimensions, feet (metres): 16.4 × 5.2 (5.0 × 1.6); single outboard engine; small arms.
Type 4: Dimensions, feet (metres): 13.1–26.2 × 7.9 (4–8 × 1.6); two outboard engines; multipurpose craft including MRL and mine laying.
Type 5: Dimensions, feet (metres): 24.6 × 9.2 (7.5 × 2.8); assault craft.
Type 6: Dimensions, feet (metres): 30.9 × 11.8 (9.4 × 3.6); single outboard engine; 1–12.7 mm MG.
Dhows: Dimensions, feet (metres): 77.1 × 20 (23.5 × 6.1); single diesel engine; mine rails.
Yunus: Dimensions, feet (metres): 27.6 × 9.8 (8.4 × 3); speed 32 kt.
Ashoora II: Dimensions, feet (metres): 26.6 × 7.9 (8.1 × 2.4); two outboards; speed 42 kt; 1–7.62 mm MG.
Kuch: Dimensions, feet (metres): 29.5 × 9.8 (9.0 × 3.0); two outboards. Some have been modified with a stern dock for jet ski and are reported to be known as the Daryai class. These craft may be used as explosive craft from which the crew can escape by jet ski.
Bladerunner: There are approximately 50 10 m and 15 m Bladerunner craft. Armed with MRLS and a 12.7 mm MG, the craft are capable of 75 kt. Iranian versions of the craft are known as Seraj.
Jet Skis: RPGs.
There is an unknown number of 13 m catamaran-hulled patrol craft armed with two twin ZU-23-2 guns.

SERAJ   6/2012 / 1483507

CATAMARAN CRAFT   6/2013* / 1529050

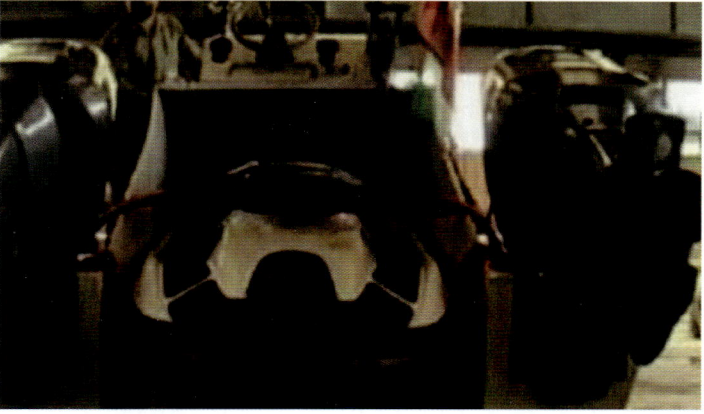

DARYAI (with stern dock)   6/2007 / 1166576

30 MURCE (MIG-G-0900) CLASS
(INSHORE PATROL CRAFT) (PBI)

Displacement, tonnes: 4 full load
Dimensions, metres (feet): 9.2 × 2.8 × 0.45 (30.2 × 9.2 × 1.5)
Speed, knots: 30
Complement: 3

Machinery: 2 Volvo Penta diesels; 1,260 hp (940 kW)
Guns: 3—12.7 mm MGs. 1 RPG-7 rocket launcher or 106 mm recoilless rifle.
Rockets: 1—12-barrelled 107 mm rocket launcher (MRL).
Radars: Surface search: I-band.

Comment: Built by MiG, the unarmed variant has been produced in relatively large numbers since the mid-1990s. Numbers are declining and this approximate number of armed variant is believed to be in Pasdaran or naval service.

MURCE CLASS   6/2000, MiG / 0126375

15 KASHDOM II CLASS (INSHORE PATROL CRAFT) (PBF)

Displacement, tonnes: 17.8 standard (est.)
Dimensions, metres (feet): 16 × 3.0 × 1.1 (52.5 × 9.8 × 3.6)
Speed, knots: 50 (est.)
Complement: 5
Machinery: 2 diesels; 2,400 hp (1.8 MW); surface piercing propeller
Guns: 1—23 mm. 1—12.7 mm MG.

Comment: Probably a development of the C 14 class design, the catamaran-hulled inshore patrol craft. A MRL launcher may also be mounted on the cabin roof and some may be missile-armed. Later missile-armed vessels of the class are known as Kashdom III and IV although precise numbers are uncertain.

KASHDOM IV   6/2013* / 1529049

© 2014 IHS   IHS Jane's Fighting Ships 2014-2015

AUXILIARIES

Notes: (1) There is also an inshore survey vessel *Abnegar*.
(2) Two 65 ton training vessels of Kialas-C-Qasem class are reported to have commissioned mid-2000. There may be further craft. No other details are known.

4 KANGAN CLASS (WATER TANKERS) (AWT)

| KANGAN 411 | TAHERI 412 | SHAHID MARJANI | AMIR |

Displacement, tonnes: 12,193 full load
Measurement, tonnes: 9,581 dwt
Dimensions, metres (feet): 148 × 21.5 × 5 *(485.6 × 70.5 × 16.4)*
Speed, knots: 15
Complement: 14

Cargo capacity: 9,000 m³ of water
Machinery: 1 MAN 7L52/55A diesel; 7,385 hp(m) *(5.43 MW)* sustained; 1 shaft
Guns: 2 USSR 23 mm/80 (twin). 2—12.7 mm MGs.
Radars: Navigation: Decca 1229; I-band.

Comment: The first two were built in Mazagon Dock, Bombay in 1978 and 1979. The second pair to a slightly modified design was acquired in 1991–92 but may be civilian manned. Some of the largest water tankers afloat and can be used to supply remote coastal towns and islands. Accommodation is air conditioned. All have a 10 ton boom crane.

TAHERI 5/1989 / 0506013

6 DELVAR CLASS (SUPPORT SHIPS) (AEL/AKL/AWT)

| DAYLAM 424 | SIRJAN 472 | CHIROO 482 |
| DELVAR 471 | CHARAK 481 | SOROO 483 |

Measurement, tonnes: 904 gt; 777 dwt
Dimensions, metres (feet): 64 × 10.5 × 3.3 *(210 × 34.4 × 10.8)*
Speed, knots: 11
Complement: 20
Machinery: 2 MAN G6V 23.5/33ATL diesels; 1,560 hp(m) *(1.15 MW)*; 2 shafts
Guns: 1 GAM-BO1 20 mm. 2—12.7 mm MGs.
Radars: Navigation: Decca 1226; I-band.

Comment: All built by Karachi SY in 1980–82. *Delvar* and *Sirjan* are ammunition ships (AEL), *Daylam* is a water carrier (AWT) and *Charak*, *Chiroo* and *Soroo* are general cargo ships (AKL). The water carriers have only one crane (against two on the other types), and have rounded sterns (as opposed to transoms). Re-armed.

DAYLAM 5/2003, A Sharma / 0569202

CHIROO 8/2010, US Navy / 1483508

2 FLOATING DOCKS

400 (ex-*US ARD 29*; ex-*FD 4*)   DOLPHIN

Dimensions, metres (feet): 149.9 *(400)*, 240 *(Dolphin)* × 24.7 *(400)*, 52.5 *(Dolphin)* × 10 *(400)*, 17.8 *(Dolphin)* *(491.8, 787.4 × 81.0, 172.2 × 32.8, 58.4)*

Comment: *400* is an ex-US ARD 12 class built by Pacific Bridge, California and transferred in 1977; lift 3,556 tons. *Dolphin* built by MAN-GHH Nordenham, West Germany and completed in November 1985; lift 28,000 tons.

12 HENDIJAN CLASS (TENDERS) (PBO)

HENDIJAN 1401	MOOAM 1405	ROSTANI 1409
SIRIK 1402	BAHREGAN (ex-*Geno*) 1406	NAYBAND 1410
KONARAK 1403	KALAT 1407	MACHAM —
GAVATAR 1404	GENAVEH 1408	KORAMSHAHR —

Displacement, tonnes: 420 full load
Dimensions, metres (feet): 47 × 8.55 × 2.86 *(154.2 × 28.1 × 9.4)*
Speed, knots: 25
Complement: 15 + 90 scientists

Cargo capacity: 40 tons on deck; 95 m³ of liquid/solid cargo space
Machinery: 2 Mitsubishi S16MPTK diesels; 6,600 hp(m) *(4.9 MW)*; 2 shafts
Missiles: 4 Noor C-802 *(Kalat)*; active radar homing to 120 km *(65 n miles)* at 0.9 Mach; warhead 165 kg.
Guns: 1—20 mm (sometimes fitted in patrol craft). 2—12.7 mm MGs.
Radars: Navigation: Racal Decca or China RM 1070A; I-band.

Comment: First eight built by Damen, Netherlands 1988–91. Remainder built at Bandar Abbas under the MIG-S-4700 programme. Last pair launched on 25 November 1995. Reports of three more being built may be caused by confusion with new corvettes. Variously described in the Iranian press as 'frigates' or 'patrol ships', they are regularly used for coastal surveillance. One is used as a training ship and at least one *(Kalat)* has been converted to carry C-802 missiles.

KALAT 4/2009, RAAF / 1367076

1 REPLENISHMENT SHIP (AORH)

Name	No	Builders	Commissioned
KHARG	431	Swan Hunter Shipbuilders, Wallsend-on-Tyne	5 Oct 1984

Displacement, tonnes: 11,242 standard; 33,544 full load
Measurement, tonnes: 18,880 gt; 9,517 dwt
Dimensions, metres (feet): 207.2 × 26.5 × 9.2 *(679.8 × 86.9 × 30.2)*
Speed, knots: 21.5
Complement: 248

Machinery: 2 Babcock & Wilcox boilers; 2 Westinghouse turbines; 26,870 hp *(19.75 MW)*; 1 shaft
Guns: 1 Oto Melara 76 mm/62 compact. 4 USSR 23 mm/80 (2 twin). 2—12.7 mm MGs.
Radars: Navigation: Decca 1229; I-band
Tacan: URN 20.
Helicopters: 3 Sea Kings (twin hangar).

Comment: Ordered October 1974. Laid down 27 January 1976. Launched 3 February 1977. Ship handed over to Iranian crew on 25 April 1980 but remained in UK. In 1983 Iranian Government requested this ship's transfer. The UK Government delayed approval until January 1984. On 10 July 1984 began refit at Tyne Ship Repairers. Trials began 4 September 1984 and ship was then delivered without guns which were subsequently fitted. A design incorporating some of the features of the British Ol class but carrying ammunition and dry stores in addition to fuel. Inmarsat fitted.

KHARG 6/1998 / 0052380

KHARG 5/1997 / 0052379

IHS Jane's Fighting Ships 2014-2015   © 2014 IHS

Auxiliaries < Iran — Patrol forces < Iraq 391

2 FLEET SUPPLY SHIPS (AORLH)

Name	No	Builders	Commissioned
BANDAR ABBAS	421	C Lühring Yard, Brake	Apr 1974
BUSHEHR	422	C Lühring Yard, Brake	Nov 1974

Displacement, tonnes: 4,748 full load
Measurement, tonnes: 3,237 gt; 3,302 dwt
Dimensions, metres (feet): 108 × 16.6 × 4.5 (354.3 × 54.5 × 14.8)
Speed, knots: 20
Range, n miles: 3,500 at 16 kt
Complement: 59

Machinery: 2 MAN 6L 52/55 diesels; 12,060 hp(m) (8.86 MW) sustained; 2 shafts
Guns: 3 GAM-BO1 20 mm can be carried. 2—12.7 mm MGs.
Radars: Navigation: 2 Decca 1226; I-band.
Helicopters: 1 AB 212.

Comment: Bandar Abbas launched 11 August 1973, Boushehr launched 23 March 1974. Combined tankers and store-ships carrying victualling, armament and general stores. There are no RAS facilities. Telescopic hangar. Both carry 2 SA-7 portable SAM and 20 mm guns have replaced the former armament. Bandar Abbas damaged by an explosion in early 1999 but has been repaired.

BANDAR ABBAS   10/2004 / 1151276

3 NASER CLASS (SUPPORT VESSELS) (YAG)

Dimensions, metres (feet): 33 × 8.0 × 1.5 (108.3 × 26.2 × 4.9)
Speed, knots: 27
Machinery: 2 diesels; 2 shafts
Guns: 1—12.7 mm MG.
Electro-optic systems: To be announced.

Comment: Built by Arvandan Shipyard. Aluminium construction. The third unit was delivered in March 2011 and further units are expected. The vessels are capable of carrying 86 passengers or 35 tonnes of supplies. Operated by the Republican Guard.

NASER CLASS   6/2013* / 1486524

TUGS

17 HARBOUR TUGS (YTB/YTM)

HAAMOON	SEFID-RUD	ILAM	HANGAM
ALBAN	DEHLORAN	ARVAND	KARKHEH
ASLAM	KHANDAG	HARI-RUD	ARAS
DARYAVAND II	MENAB	ABAD	
HIRMAND	ATRAK		

Comment: All between 70 and 90 ft in length, built since 1984.

Iraq

Country Overview

The Republic of Iraq was proclaimed in 1958 following a coup d'état. With an area of 168,754 square miles, it is situated in the Middle East and is bordered to the north by Turkey, to the east by Iran (with which it was at war 1980-88), to the west by Jordan and Syria and to the south by Saudi Arabia and Kuwait (which it invaded and occupied 1990-91 until expelled in the Gulf War 1991). It has a 31 n mile coastline with the Gulf. Baghdad is the capital and largest city. There are two ports on the Khawr Abd Allah Channel at Umm Qasr and Khawr al Zubayr. Territorial Seas (12 n miles) are claimed. An EEZ has not been claimed.

In the wake of the US-led occupation in March-April 2003, Iraq remained under coalition control until 30 June 2004 when full authority was handed over to an Iraqi Interim Government. Following elections on 30 January 2005, a new constitution was ratified by public referendum on 15 October 2005. This was followed by a general election on 15 December 2005 to elect a permanent Iraqi National Assembly. All naval coastal defence units, surface ships and aircraft were destroyed or disabled during the war and are unlikely to be resurrected. An oiler (Agnadeen) at Alexandria could be reclaimed but this is unlikely to be a high priority. The Iraqi Coastal Defence Force (ICDF), now known as the Iraqi Navy, was formally established at Umm Qasr on 30 September 2004. Key tasks include defence of the Khawr al Amaya (KAAOT) and Al Basra (ABOT) offshore oil terminals.

Headquarters Appointments

Commander Iraqi Navy:
Rear Admiral Ali Hussein Al-Rubaye

Bases

Baghdad (Navy HQ), Khor Az Zubayr, Umm Qasr.

Personnel

2014: 3,650 (including 1,500 marines)

Marines

The Iraqi Marines Corps comprises two battalions. The 1st Battalion is responsible for point-defence of Iraq's oil terminals while the 2nd Battalion provides port security at Umm Qasr and Az Zubayr. A third battalion may be transferred from the army.

PATROL FORCES

2 OFFSHORE PATROL VESSELS (PSO)

AL BASRA 401   **AL FAYHAA** 402

Displacement, tonnes: 1,400 full load
Dimensions, metres (feet): 60 × 11.2 × 3.8 (196.9 × 36.7 × 12.5)
Speed, knots: 16
Range, n miles: 4,000 at 10 kt
Complement: 42
Machinery: 2 Caterpillar 3516C diesels; 6,300 hp (4.7 MW); 2 shafts
Guns: 1 MS1-DS30MA2 30 mm. 4—12.7 mm MGs. 6—7.62 mm MGs.

Comment: It was announced on 5 March 2010 that RiverHawk Fast Sea Frames, LLC, Tampa, Florida, had been awarded a contract for the detail, design and construction of two offshore support vessels and associated equipment and services. The role of the ships is to contribute to the support and defence of Iraq's offshore oil terminals. The work is being undertaken at Houma, Louisiana, by Gulf Island Marine Fabricators, LLC (a subsidiary of Gulf Island Fabrication, Inc.). The ships are a development of an AMP-137 design, the basis of a joint venture agreement between RiverHawk and Jordan's King Abdullah II Design and Development Bureau (KADDB). The ships are of steel construction and are capable of launching, recovering, refuelling and maintaining three 9 m interception craft. Al Basra was launched on 4 November 2011 and Al Fayhaa on 1 February 2012. The first ship was completed on 23 July 2012 and both ships were handed over on 20 December 2012.

AL BASRA
1/2012, RHFSF
1455704

© 2014 IHS   IHS Jane's Fighting Ships 2014-2015

392 Iraq > Patrol forces

4 FATEH (SAETTIA MK IV) CLASS (PBO)

FATEH 702   **NASIR** 703   **MAJED** 704   **SHIMOOKH** 705

Displacement, tonnes: 397 standard
Dimensions, metres (feet): 53.4 × 8.1 × 2.0 (175.2 × 26.6 × 6.6)
Speed, knots: 23. **Range, n miles:** 2,100 at 16 kt
Complement: 38 (4 officers)

Machinery: 2 Isotto Fraschini V1716 T2 MSD diesels; 6,335 hp (4.7 MW); 2 shafts
Guns: 1 Oto Melara 30 mm. 2 – 12.7 mm MGs. 4 – 7.62 mm MGs.
Radars: Surface search: E/F-band.
Navigation: I-band.

Comment: Contract signed with Fincantieri, Muggiano in September 2006 for the construction of four vessels. Two built at Riva Trigoso and two at Muggiano. The steel-hulled design is based on Diciotti (mod Saettia)-class vessels in service with the Italian Coast Guard and is generally similar to the ship supplied to Malta in 2005 but with a stern ramp for launching an 11 m RIB. The contract included a training and logistic support package and some additional ex-Italian Coast Guard craft. *Fateh* delivered in May 2009, *Nasir* in October 2009 and *Majed* and *Shimookh* on 16 December 2009.

MAJED   11/2009, Giorgio Ghiglione / 1366579

5 PREDATOR CLASS (INSHORE PATROL CRAFT) (PB)

P 101–105

Displacement, tonnes: 68 full load
Dimensions, metres (feet): 27.1 × 2.8 × 1.8 (88.9 × 9.2 × 5.9)
Speed, knots: 32
Complement: 6
Machinery: 2 MTU 12V 396 TE742; 4,025 hp (3 MW)
Guns: 1 – 7.62 mm MG.

Comment: Built at Wuhan by Nanhua High-Speed Engineering Company and originally acquired in 2002. Maintained in dry-dock at Jebel Ali, UAE, until the first two were commissioned on 4 April 2004. Three further craft followed in May 2004. Acquisition and refit costs funded by the US. The craft were used initially for training of Iraqi personnel but are used increasingly for patrol duties as the navy develops. Based at Umm Qasr.

P 102   5/2004, US Navy / 0580527

P 104   8/2008, Shaun Jones / 1335385

24 SEASPRAY ASSAULT BOATS (PB)

F 1–24

Dimensions, metres (feet): 9.5 × 3.4 × 0.5 (31.2 × 11.2 × 1.6)
Speed, knots: 50
Range, n miles: 450 at 17 kt
Complement: 5
Machinery: 2 Mercury outboards; 500 hp (375 kW)
Radars: Navigation: I-band.

Comment: 24 craft donated by the UAE government in 2005. Designed by Sea Spray Aluminium Boats.

IHS Jane's Fighting Ships 2014-2015

ASSAULT BOAT   9/2008, IHS/Tim Fish / 1294748

12 + (3) SWIFTSHIPS 35 M CLASS
(LARGE PATROL CRAFT) (PB)

P 301–312

Displacement, tonnes: 97 standard
Dimensions, metres (feet): 35 × 7.25 × 2.6 (114.8 × 23.8 × 8.5)
Speed, knots: 30
Range, n miles: 1,500 at 12 kt
Complement: 25 (4 officers)
Machinery: 3 MTU 16V 2000 diesels; 7,200 hp (5.37 MW); 3 shafts
Guns: 1 MSI 30 mm. 1 – 12.7 mm MG. 2 – 7.62 mm MGs.

Comment: Contract awarded on 25 September 2009 to Swiftships, Morgan City, LA, for the design and construction of nine patrol boats under the US government's Foreign Military Sales programme. The contract was modified in March 2011 to allow for the construction of three further craft with an option for a further three. Aluminium alloy hull and superstructure. Stern ramp for launch/recovery of 7 m RHIB. The 12th vessel was delivered on 5 July 2013.

P 304   2/2011, Swiftships / 1455669

26 DEFENDER 2710 CLASS (RESPONSE BOATS) (PBF)

Displacement, tonnes: 3 full load
Dimensions, metres (feet): 8.9 × 3.07 × 0.82 (29.2 × 10.1 × 2.7)
Speed, knots: 46. **Range, n miles:** 175 at 35 kt
Complement: 4
Machinery: 2 Yamaha outboard motors; 450 hp (335 kW)
Guns: 1 – 12.7 mm MG. 1 – 7.62 mm MG.
Radars: To be announced.

Comment: High-speed inshore patrol craft of aluminium construction and foam collar built by SAFE Boats International, Port Orchard, Washington. Acquired from the US government and delivered by early 2009. The new patrol craft are used for patrol of the Khor Abd Allah waterway.

DEFENDER CLASS   6/2009*, US Navy / 1394595

© 2014 IHS

2 TYPE 200 (INSHORE PATROL CRAFT) (PBR)

P 201 (ex-CP 250)   P 202 (ex-CP 247)

Displacement, tonnes: 22 standard
Dimensions, metres (feet): 15 × 4.4 × 1.6 *(49.2 × 14.4 × 5.2)*
Speed, knots: 31
Range, n miles: 350 at 31 kt
Complement: 7
Machinery: 2 Isotta Fraschini ID36-SS-6V diesels; 1,380 hp *(1.0 MW)*; 2 shafts
Radars: Navigation: I-band.

Comment: Former Italian Coast Guard craft built by Navaltechnica Anzio, 1977–81. Transferred to Iraq in 2007 as part of the contract to procure four Diciotti-class offshore patrol craft.

4 TYPE 2010 (INSHORE CRAFT) (PBR)

P 203 (ex-CP 2036)   P 205 (ex-CP 2067)
P 204 (ex-CP 2037)   P 206 (ex-CP 2069)

Displacement, tonnes: 11 standard
Dimensions, metres (feet): 12.5 × 3.6 × 1.3 *(41 × 11.8 × 4.3)*
Speed, knots: 25. **Range, n miles:** 600 at 12 kt
Complement: 4
Machinery: 2 AIFO 8362SRM27 diesels; 550 hp *(398 kW)*; 2 shafts
Radars: Navigation: I-band.

Comment: Former Italian Coast Guard craft built 1973-85. GRP construction. Transferred to Iraq in 2007 as part of the contract to procure four Diciotti-class offshore patrol craft. Others of the class have been transferred to Albania.

Ireland
AN SEIRBHIS CHABHLAIGH

Country Overview

The Republic of Ireland comprises about five sixths of the island of Ireland. Situated west of Great Britain, the country consists of the provinces of Leinster, Munster, Connaught and three counties of the province of Ulster. The remaining six counties of Ulster form Northern Ireland, a constituent part of the United Kingdom. With an area of 27,136 square miles, the country has a 783 n mile coastline with the Atlantic Ocean and Irish Sea. Dublin is the capital, largest city and principal port. There is another major port at Cork. Territorial waters (12 n miles) are claimed. A 200 n mile Fishery zone has also been claimed.

Headquarters Appointments
Flag Officer Commanding Naval Service:
  Commodore Mark Mellett
Bases
Haulbowline Island, Cork Harbour–Naval HQ, Base and Dockyard
Personnel
(a) 2014: 1,094 (183 officers)
(b) Voluntary service
(c) Reserves: 400 (one unit in each of the following cities: Dublin, Waterford, Cork and Limerick)

Operational Role

The duties of the Irish Naval Service include maritime security operations, peace support operations, fishery protection, European Union Common Security and Defence Policy (CSDP) missions, and aid to the civil power/authority missions.

Prefix to Ships' Names

LÉ (Long Éirennach = Irish Ship)

PATROL FORCES

1 EITHNE CLASS (PSOH)

Name	No	Builders	Laid down	Launched	Commissioned
EITHNE	P 31	Verolme, Cork	15 Dec 1982	19 Dec 1983	7 Dec 1984

Displacement, tonnes: 1,788 standard; 1,941 full load
Dimensions, metres (feet): 80.8 × 12 × 4.3 *(265.1 × 39.4 × 14.1)*
Speed, knots: 20 (est.). **Range, n miles:** 7,000 at 15 kt
Complement: 73 (10 officers)

Machinery: 2 Ruston 12RKC diesels; 6,800 hp *(5.07 MW)* sustained; 2 shafts; cp props
Guns: 1 Bofors 57 mm/70 Mk 1; 200 rds/min to 17 km *(9.3 n miles)*; weight of shell 2.4 kg. 2 Rheinmetall Mk 20 DM 5 20 mm/20. 2–7.62 mm MGs. 2 Wallop 57 mm launchers for illuminants.
Radars: Air/surface search: Signaal DA05 Mk 4; E/F-band
Surface search: Kelvin Hughes; E/F-band
Navigation: 2 Kelvin Hughes; 6000A; I-band
Tacan: MEL RRB transponder
Weapon control systems: Signaal LIOD director. 2 Signaal optical sights.

Helicopters: Not routinely carried.

Programmes: Ordered 23 April 1982 from Verolme, Cork, this was the last ship to be built at this yard.
Structure: Fitted with retractable stabilisers. Closed circuit TV for flight deck operations. Satellite navigation and communications. CTD tactical displays.
Operational: Helicopter no longer operational. Two Delta 7.5 m inboard diesel RIBs fitted in addition to two 5.4 m RIBs in 2003. Long refit (SLEP) in 1998/99.

EITHNE   *10/2013*, Derek Fox* / 1530907

2 ROÍSÍN CLASS (PSO)

Name	No	Builders	Laid down	Launched	Commissioned
ROÍSÍN	P 51	Appledore Shipbuilders, Bideford	Dec 1998	12 Aug 1999	15 Dec 1999
NIAMH	P 52	Appledore Shipbuilders, Bideford	Jun 2000	10 Feb 2001	18 Sep 2001

Displacement, tonnes: 1,727 full load
Dimensions, metres (feet): 78.9 × 14 × 3.9 *(258.9 × 45.9 × 12.8)*
Speed, knots: 23
Range, n miles: 6,000 at 15 kt
Complement: 44 (6 officers)

Machinery: 2 Wärtsilä 16V26 diesels; 6,800 hp(m) *(5 MW)* sustained; 2 shafts; LIPS cp props; bow thruster; 462 hp(m) *(340 kW)*
Guns: 1 Oto Melara 3 in *(76 mm)*/62 Compact; 85 rds/min to 16 km *(8.6 n miles)*; weight of shell 6 kg. 2 Rheinmetall Mk 20 DM 3 20 mm/20. 2–12.7 mm MGs. 4–7.62 mm MGs.
Radars: Surface search: Kelvin Hughes; E/F-band
Navigation: Kelvin Hughes; I-band.
Weapon control systems: Radamec 1500 optronic director.

Programmes: Contract for first ship signed on 16 December 1997 with 65 per cent of EU funding. Option on a second of class taken up on 6 April 2000.
Operational: Designated Large Patrol Vessel, the design is a modification of the Mauritius ship *Vigilant* but without the hangar or flight deck. Two Delta 6.5 m and one Avon 5.4 m RIBs are carried. CTD tactical displays.

ROÍSÍN   *9/2011, Frank Findler* / 1406931

2 P 41 PEACOCK CLASS (COASTAL PATROL VESSELS) (PSO)

Name	No	Builders	Commissioned
ORLA (ex-*Swift*)	P 41	Hall Russell, Aberdeen	3 May 1985
CIARA (ex-*Swallow*)	P 42	Hall Russell, Aberdeen	17 Oct 1984

Displacement, tonnes: 723 full load
Dimensions, metres (feet): 62.6 × 10 × 2.7 *(205.4 × 32.8 × 8.9)*
Speed, knots: 25. **Range, n miles:** 2,500 at 17 kt
Complement: 39 (5 officers)

Machinery: 2 Crossley SEMT-Pielstick 18 PA6 V 280 diesels; 14,400 hp(m) *(10.58 MW)* sustained; 2 shafts; auxiliary drive; Schottel prop; 181 hp(m) *(133 kW)*
Guns: 1—3 in *(76 mm)*/62 Oto Melara compact; 85 rds/min to 16 km *(8.6 n miles)*; weight of shell 6 kg. 2 Rheinmetall Mk 20 DM 3 20 mm/20. 2—12.7 mm MGs. 4—7.62 mm MGs.
Radars: Surface search: Kelvin Hughes; I-band.
Navigation: Kelvin Hughes Nucleus 5000A/6000A; I-band.
Weapon control systems: Radamec 1500 optronic director (for 76 mm).

Programmes: *Orla* launched 11 September 1984 and *Ciara* 31 March 1984. Both served in Hong Kong from mid-1985 until early 1988. Acquired from UK and commissioned 21 November 1988. Others of the class acquired by the Philippines in 1997.
Modernisation: New radars fitted in 1993. CTD tactical display fitted.
Structure: Have loiter drive. Displacement increased after building by the addition of more electronic equipment.

ORLA   10/2013*, Derek Fox / 1530906

2 P 21 CLASS (OFFSHORE PATROL VESSELS) (PSO)

Name	No	Builders	Launched	Commissioned
AOIFE	P 22	Verolme, Cork	12 Apr 1979	29 Nov 1979
AISLING	P 23	Verolme, Cork	3 Oct 1979	21 May 1980

Displacement, tonnes: 1,036 standard
Dimensions, metres (feet): 65.2 × 10.5 × 4.4 *(213.9 × 34.4 × 14.4)*
Speed, knots: 17. **Range, n miles:** 4,000 at 17 kt, 6,750 at 12 kt
Complement: 47 (6 officers)

Machinery: 2 SEMT-Pielstick 6 PA6 L 280 diesels; 4,800 hp *(3.53 MW)*; 1 shaft; bow thruster (*Aoife* and *Aisling*)
Guns: 1 Bofors 40 mm/L 70; 300 rds/min to 12 km *(6.5 n miles)*; weight of shell 0.88 kg. 2 Rheinmetall Mk 20 DM 3 20 mm/20. 2—12.7 mm MGs. 2—7.62 mm MGs.
Radars: Surface search: Kelvin Hughes I-band
Navigation: Kelvin Hughes Nucleus 6000A; I-band.

Modernisation: New search radars were fitted in 1994–95. CTD tactical display fitted.
Structure: Stabilisers fitted. Both are equipped with a bow thruster.
Operational: *Aoife* refitted in 1996/97 and *Aisling* in 1997/98. Sonars have been removed. *Emer* decommissioned in September 2013 and *Aoife* is to follow in September 2014.

AISLING   6/2008, Frank Findler / 1335790

AISLING   6/2008, Martin Mokrus / 1335789

AUXILIARIES

Notes: (1) In addition there are a number of mostly civilian manned auxiliaries including: *Tailte* a Dufour 35 ft sail training yacht bought in 1979; *Creidne*, a 15 m Bermudan ketch (built in 1967) was refitted in 2009 and has a permanent naval crew of three and eight training berths.
(2) *Granuaile* is an 80 m lighthouse tender with a helicopter flight deck forward operated by the Commissioners of Irish Lights. Launched on 14 August 1999 this ship replaced a previous vessel of the same name on 23 March 2000.

GRANUAILE   3/2000, Commissioners of Irish Lights / 0093593

0 + 2 (1) MOD ROÍSÍN (PV 90) CLASS
(OFFSHORE PATROL VESSELS) (PSO)

Name	No	Builders	Laid down	Launched	Commissioned
SAMUEL BECKETT	P 61	Babcock; Appledore	18 May 2012	3 Nov 2013	2014
JAMES JOYCE	P 62	Babcock; Appledore	4 Nov 2013	2014	2015

Dimensions, metres (feet): 89.5 × 14.0 × 3.8 *(293.6 × 45.9 × 12.5)*
Speed, knots: 23. **Range, n miles:** 6,000 at 15 kt
Complement: 54

Machinery: 2 Wärtsilä 16V26 diesels; 6,800 hp(m) *(5.0 MW)*; 2 shafts; cp props
Guns: 1 Oto Melara 3 in *(76 mm)*/62; 85 rds/min to 16 km *(8.6 n miles)*; weight of shell 6 kg. 2 Rheinmetall Mk 20 DM3 20 mm/20. 2—12.7 mm MGs. 2—7.62 mm MGs.
Radars: Surface search/navigation: Kelvin Hughes SharpEye; E/F/I-bands.
Weapon control systems: Optronic director.

Comment: As the first step in the Naval Service Replacement Programme, a Request for Proposals for two offshore patrol vessels was issued on 24 August 2007. Following the announcement in April 2010 that Babcock International was the preferred bidder, a contract was awarded in October 2010. The ships are to an STX Marine PV 90 design, a lengthened version of the PV 80 design used for the Roísín class. Delivery of the two ships is expected in 2014 and 2015. There is likely to be an option for a third vessel to be ordered at a later date.

MOD ROÍSÍN CLASS   5/2011, STX MARINE / 1406373

LAND-BASED MARITIME AIRCRAFT

Notes: Four civilian operated Sikorsky S-92A helicopters provide long-range SAR services. They are based at Shannon, Sligo, Dublin, and Waterford.

Numbers/Type: 2 Casa CN-235 MP Persuader.
Operational speed: 210 kt *(384 km/h)*.
Service ceiling: 24,000 ft *(7,315 m)*.
Range: 2,000 n miles *(3,218 km)*.
Role/Weapon systems: EEZ surveillance. First one delivered in June 1992 but returned to Spain in 1995. Two more delivered in December 1994. Sensors: Search radar Bendix APS 504(V)5; FLIR. Weapons: Unarmed.

CN-235 MP   7/2003, Paul Jackson / 0568896

Israel
HEYL HAYAM

Country Overview

Established in 1948, The State of Israel is situated on the eastern shore of the Mediterranean Sea and has borders to the north with Lebanon, to the north-east with Syria, to the east with Jordan and to the south-west with Egypt. It has coastlines with the Mediterranean (142 n miles) and with the Gulf of Aqaba (5 n miles) in the northern Red Sea. A land area of 8,463 square miles includes East Jerusalem and other territory (including Gaza Strip, the West Bank region of Jordan, the Golan Heights area of south-western Syria) annexed in 1967. Jerusalem is the largest city but, although claimed as the capital, is not so recognised by the United Nations. Many nations maintain embassies at Tel Aviv. Haifa is the principal port. Territorial seas (12 n miles) are claimed but an EEZ is not claimed.

Headquarters Appointments

Commander-in-Chief: Vice Admiral Ram Rothberg
Chief of Naval Staff: Rear Admiral Ben Yehuda

General

Less than 5% of Israeli defence budget is allocated to the Navy.

Personnel

2014:
(a) 9,500 (880 officers) of whom 2,500 are conscripts. Includes a Naval Commando of 300
(b) 3 years' national service for Jews and Druzes
Notes: An additional 5,000 Reserves available on mobilisation.

Bases

Haifa, Ashdod, Eilat
(The repair base at Eilat has a synchrolift)

Coast Defence

There are ten integrated coastal radar and electro-optical surveillance stations. These are to be converted to an unmanned, remote-controlled system employing a wideband communications network.

Prefix to Ships' Names

INS (Israeli Naval Ship)

SUBMARINES

5 + 1 DOLPHIN (TYPE 800) CLASS (SSK)

Name	No	Builders	Laid down	Launched	Commissioned
DOLPHIN	–	Howaldtswerke; Thyssen Nordseewerke	7 Oct 1994	12 Apr 1996	27 Jul 1999
LEVIATHAN	–	Howaldtswerke; Thyssen Nordseewerke	13 Apr 1995	25 Apr 1997	15 Nov 1999
TEKUMA	–	Howaldtswerke; Thyssen Nordseewerke	12 Dec 1996	26 Jun 1998	25 Jul 2000
TANNIN	–	Thyssenkrupp Marine Systems	2007	19 Feb 2012	4 May 2012
RAHAV	–	Thyssenkrupp Marine Systems	2008	11 Mar 2013	29 Apr 2013
–	–	Thyssenkrupp Marine Systems	2012	2016	2017

Displacement, tonnes: 1,666 surfaced; 1,930 dived
Dimensions, metres (feet): 57.3, 68.6 (*Rahav, Tannin*) × 6.8 × 6.2 (*188, 225.1 × 22.3 × 20.3*)
Speed, knots: 11 snorting; 20 dived
Range, n miles: 8,000 at 8 kt surfaced; 420 at 8 kt dived
Complement: 30 (6 officers)

Machinery: 3 MTU 16V 396 SE 84 diesels; 5,430 hp(m) *(4.05 MW)* sustained; 3 alternators; 2.91 MW; 1 Siemens motor; 3,875 hp(m) *(2.85 MW)* sustained; 2 HDW PEM fuel cells (AIP) modules (*Tannin, Rahav*); 240 kW; 1 shaft
Missiles: SSM: Sub Harpoon; UGM-84C; active radar or GPS homing to 130 km *(70 n miles)* at 0.9 Mach; warhead 227 kg.
SAM: Fitted for Triten anti-helicopter system.
Torpedoes: 4—25.6 in *(650 mm)* and 6—21 in *(533 mm)* bow tubes. STN Atlas DM2A4 Seehecht; wire-guided active homing to 13 km *(7 n miles)* at 35 kt; passive homing to 28 km *(15 n miles)* at 23 kt; warhead 260 kg. Total of 16 torpedoes and 5 SSMs. The four 650 mm tubes may be for SDVs, but could carry torpedoes if liners are fitted.
Mines: In lieu of torpedoes.
Electronic countermeasures: ESM: Elbit Timnex 4CH(V)2; intercept.
Radars: Surface search: Elta; I-band.
Sonars: Atlas Elektronik CSU 90; hull-mounted; passive/active search and attack. Atlas Elektronik PRS-3; passive ranging. FAS-3; flank array; passive search.
Weapon control systems: STN/Atlas Elektronik ISUS 90-1 TCS.

Programmes: In mid-1988 Ingalls Shipbuilding Division of Litton Corporation was chosen as the prime contractor for two IKL-designed Dolphin-class submarines to be built in West Germany with FMS funds by HDW in conjunction with Thyssen Nordseewerke. Funds approved in July 1989 with an effective contract date of January 1990 but the project was cancelled in November 1990 due to pressures on defence funds. After the Gulf War in April 1991 the contract was resurrected, this time with German funding for two submarines with an option on a third taken up in July 1994. A contract for the construction of two further modified Dolphin II-class submarines was signed on 6 July 2006. The new submarines are about 11 m longer in order to incorporate air-independent propulsion. The boats are under construction at HDW and TNSW. Israel is funding two thirds of the budget while the German government is funding the remaining third. Construction of the first boat is reported to have started in 2007 and she was launched in February 2012. A contract to take up the option on a sixth boat was signed on 21 March 2012.

DOLPHIN   6/1999, Michael Nitz / 0080058

Modernisation: Installation of air-independent propulsion in the first three boats is under consideration. An anti-torpedo defence system, Rafael Torbuster, is to be installed from 2012. The system includes externally mounted tubes capable of launching up to 10 decoys.
Structure: Diving depth, 350 m *(1,150 ft)*. Similar to German Type 212 in design but with a 'wet and dry' compartment for underwater swimmers. Two Kollmorgen periscopes. Probably fitted for Triten anti-helicopter SAM system.
Operational: Endurance, 30 days. Used for interdiction, surveillance and special boat operations. Development of a submarine-launched cruise missile would complete the final part of a triad of nuclear deterrents. However, while Israel probably has the expertise and technology to deploy SLCM, little information exists to confirm or deny such a programme. Adaptation of the indigenous Delilah and Popeye groups of missiles is a possible option although encapsulation of the missile would pose a significant challenge. Painted blue/green to aid concealment in the eastern Mediterranean. Some other NT 37E torpedoes are embarked until full Seehecht outfits are available. The boats form Flotilla 7 based at Haifa. *Leviathan* transited the Suez Canal in June/July 2009 in order to conduct operations in the Red Sea. *Tannin* began sea trials on 12 May 2013 and is expected to become operational in 2014.

TANNIN   5/2013*, Michael Nitz / 1486525

CORVETTES

Notes: Plans to procure up to four corvettes were first initiated in 2003 and have included consideration of the US Navy's Littoral Combat Ship and the German MEKO A-100 designs. In 2012 South Korea emerged as another potential supplier.

3 EILAT (SAAR 5) CLASS (FSGHM)

Name	No	Builders	Laid down	Launched	Commissioned
EILAT	501	Ingalls Shipbuilding, Pascagoula	24 Feb 1992	9 Feb 1993	24 May 1994
LAHAV	502	Ingalls Shipbuilding, Pascagoula	25 Sep 1992	20 Aug 1993	23 Sep 1994
HANIT	503	Ingalls Shipbuilding, Pascagoula	5 Apr 1993	4 Mar 1994	7 Feb 1995

Displacement, tonnes: 1,092 standard; 1,316 full load
Dimensions, metres (feet): 85 × 11.9 × 3.2 *(278.9 × 39.0 × 10.5)*
Speed, knots: 33; 20 diesel
Range, n miles: 3,500 at 17 kt
Complement: 74 (20 officers; 10 air crew)
Machinery: CODOG; 1 GE LM 2500 gas turbine; 30,000 hp *(22.38 MW)* sustained; 2 MTU 12V 1163 TB82 diesels; 6,600 hp(m) *(4.86 MW)* sustained; 2 shafts; Kamewa cp props
Missiles: SSM: 8 McDonnell Douglas Harpoon (2 quad) launchers ❶; active radar homing to 130 km *(70 n miles)* at 0.9 Mach; warhead 227 kg.
SAM: 2 Israeli Industries Barak I (vertical launch) ❷; 2—32 cells; command line of sight radar or optical guidance to 10 km *(5.5 n miles)* at 2 Mach; warhead 22 kg (see *Operational*).
Guns: Oto Melara 3 in *(76 mm)*/62 compact ❸; 85 rds/min to 16 km *(8.7 n miles)*; weight of shell 6 kg. The main gun is interchangeable with a Bofors 57 mm gun or Raytheon Vulcan Phalanx ❹. 2 Sea Vulcan 20 mm ❺; range 1 km.
Torpedoes: 6—324 mm Mk 32 (2 triple) tubes ❻. Honeywell Mk 46; anti-submarine; active/passive homing to 11 km *(5.9 n miles)* at 40 kt; warhead 44 kg. Mounted in the superstructure.
Physical countermeasures: Decoys: 3 Elbit/Deseaver 72-barrelled chaff and IR launchers ❼; Rafael ATC-1 towed torpedo decoy.
Electronic countermeasures: ESM: Elisra NS 9003; intercept. Tadiran NATACS.
ECM: 2 Rafael 1010; Elisra NS 9005; jammers.
Radars: Air search: Cardion Type 5544 ❾; D-band.
Air/surface search: Elta EL/M-2228S ❿; E/F-band.
Navigation: ❽ I-band.
Fire control: 2 Elta EL/M-2221 GM STGR ⓫; I/J/K-bands.
Sonars: EDO Type 796 Mod 1; hull-mounted; search and attack; medium frequency. Rafael towed array (fitted for).

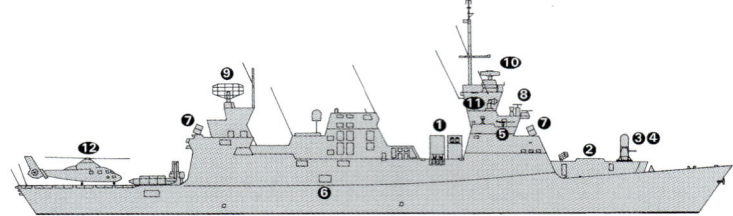

EILAT

(Scale 1 : 900), Ian Sturton / 1151070

Combat data systems: Elbit NTCCS using Elta EL/S-9000 computers. Reshet datalink.
Weapon control systems: 2 Elop MSIS optronic directors ❽.
Electro-optic systems: 2 optronic directors.
Helicopters: 1 Dauphin SA 366G ⓬ or Sea Panther can be carried.

Programmes: A design by John J McMullen Associates Inc for Israeli Shipyards, Haifa in conjunction with Ingalls Shipbuilding Division of Litton Corporation which was authorised to act as main contractor using FMS funding. Contract awarded 8 February 1989. All delivered to Israel for combat system installation, first two completed in 1996 and last one in mid-1997. Major refits of these ships are reported to be under consideration. The option for a fourth SAAR 5 was not taken up and plans to procure a further five new ships (SAAR 5+) under similar FMS funding are now unlikely to be taken forward.
Modernisation: Trials of a compact EL/M-2248 phased array radar were reportedly conducted in *Eilat* in 2011 with a view to being fitted in all three ships. Plans to equip all three ships with Barak-8 were reported in 2013 but have not been confirmed.
Structure: Steel hull and aluminium superstructure. Stealth features including resilient mounts for main machinery, funnel exhaust cooling, Radar Absorbent Material (RAM), NBC washdown and Prairie Masker Bubbler system. A secondary operations room is fitted aft. There are some Flag capabilities. Plans to carry Gabriel SSMs have been scrapped because of topweight problems. The planned third MSIS director has not yet been seen on the platform aft of the air search radar.
Operational: Endurance, 20 days. The main role is to counter threats in shipping routes. ICS-2 integrated communications system. The position of the satellite aerial suggests that the SAM after VLS launchers are not used. Barak has still to be installed, because of lack of funds. For the same reason the normal Harpoon load may be reduced to four. *Hanit* damaged by missile attack off Lebanon on 14 July 2006. Repairs were completed on 6 August 2006. All three ships allocated to Flotilla 3.

EILAT

7/2009, Shaun Jones / 1305864

HANIT

7/2009, Shaun Jones / 1305865

SHIPBORNE AIRCRAFT

Numbers/Type: 7 Eurocopter AS 565SA Sea Panther.
Operational speed: 165 kt *(305 km/h)*.
Service ceiling: 16,700 ft *(5,100 m)*.
Range: 483 n miles *(895 km)*.
Role/Weapon systems: Built by American Eurocopter in Texas. Three delivered by October 1998 with one more in 1999. Roles include reconnaissance, targetting and SAR. Sensors: Telephonics search radar; Elop MSIS for OTHT. Weapons: Unarmed.

AS 565SB   *6/2002, Adolfo Ortigueira Gil / 0567461*

LAND-BASED MARITIME AIRCRAFT

Notes: (1) Air Force helicopters can be used including some 116 AH-1F Vipers and 60 AH-64A Apaches.
(2) There are three Gulfstream G-1159D Elint aircraft.

Numbers/Type: 3 IAI 1124N Sea Scan.
Operational speed: 471 kt *(873 km/h)*.
Service ceiling: 45,000 ft *(13,725 m)*.
Range: 2,500 n miles *(4,633 km)*.
Role/Weapon systems: Acquired in 1977. Air Force manned. Coastal surveillance tasks with long endurance; used for intelligence gathering. Sensors: Elta EL/M-2022 radar, IFF, MAD, Sonobuoys, and various EW systems of IAI manufacture.

SEA SCAN   *4/2002, IHS/Paul Tompkins / 0000518*

PATROL FORCES

Notes: (1) There are about 12 'Firefish' type fast attack boats in service with Special Forces.
(2) A 50 ft *(15.2 m)* shallow draft Stealth craft was built in a Vancouver Shipyard and delivered in late 1998. In addition, approximately 20 Alligator semi-submersible craft, have reportedly been completed by Oregon Iron Works, Portland in 1999. They are powered by two diesels giving 35 kt, are 20 m long and are equipped with a Rafael optronic surveillance system. Crew of five.

5 SHALDAG CLASS (FAST ATTACK CRAFT—GUN) (PBF)

840–844

Displacement, tonnes: 59 full load
Dimensions, metres (feet): 24.8 × 6 × 1.2 *(81.4 × 19.7 × 3.9)*
Speed, knots: 50. **Range, n miles:** 700 at 32 kt
Complement: 10

Machinery: 2 Deutz 620 TB 16V or MTU 396 TE diesels; 5,000 hp(m) *(3.68 MW)*; 2 LIPS or MJP water-jets
Guns: 1 Rafael Typhoon 25 mm. 1—20 mm.
Radars: Surface search: Sperry Marine Bridgemaster; E/F/I-bands.
Electro-optic systems: Taman POP 300.

Comment: Order in January 2002 for two craft, with option for two further hulls, made from Israel Shipyards, Haifa. Details reflect those in Sri Lankan service and are thus speculative. Both delivered in late 2003. An option for a further three craft was exercised in 2005. The first of these was delivered in early 2008. Installation of a missile-defence capability is a high priority. EL/M-2222S Nav-Guard radar, under development, is one possibility.

SHALDAG 840   *5/2010, Richard Scott / 1406070*

8 HETZ (SAAR 4.5) CLASS
(FAST ATTACK CRAFT—MISSILE) (PGGM)

Name	Builders	Launched	Commissioned
ROMAT	Israel Shipyards, Haifa	30 Oct 1981	Oct 1981
KESHET	Israel Shipyards, Haifa	Oct 1982	Nov 1982
HETZ (ex-*Nirit*)	Israel Shipyards, Haifa	Oct 1990	Feb 1991
KIDON	Israel Shipyards, Haifa	1993	7 Feb 1994
TARSHISH	Israel Shipyards, Haifa	1995	Jun 1995
YAFFO	Israel Shipyards, Haifa	1998	1 Jul 1998
HEREV	Israel Shipyards, Haifa	2002	Jun 2002
SUFA	Israel Shipyards, Haifa	2002	Aug 2002

Displacement, tonnes: 496 full load
Dimensions, metres (feet): 58 (*Herev, Kidon, Sufa, Tarshish, Yaffo*), 61.7 (*Hetz, Keshet, Romat*) × 7.6 × 2.5 *(190.3, 202.4 × 24.9 × 8.2)*
Speed, knots: 31
Range, n miles: 3,000 at 17 kt, 1,500 at 30 kt
Complement: 53

Machinery: 4 MTU 16V 538 TB93 or 4 MTU 16V 396 TE diesels; 16,600 hp(m) *(12.2 MW)*; 4 shafts
Missiles: SSM: 4 or 8 McDonnell Douglas Harpoon ❶; active radar homing to 130 km *(70 n miles)* at 0.9 Mach; warhead 227 kg. 6 IAI Gabriel II (removed from some ships) ❷; radar or optical guidance; semi-active radar plus anti-radiation homing to 36 km *(19.4 n miles)* at 0.7 Mach; warhead 75 kg.
SAM: Israeli Industries Barak I (vertical launch) ❸; 32 or 16 cells in 2- or 4–8 pack launchers; command line of sight radar or optical guidance to 10 km *(5.5 n miles)* at 2 Mach; warhead 22 kg. Most fitted for but not with.
Guns: 1 Oto Melara 3 in *(76 mm)*/62 Compact ❹; 85 rds/min to 16 km *(8.7 n miles)*; weight of shell 6 kg. 2 Oerlikon 20 mm; 800 rds/min to 2 km. 1 Rafael Typhoon 25 mm (Herev). 1 Raytheon Vulcan Phalanx 6-barrelled 20 mm Mk 15 ❺; 3,000 rds/min combined to 1.5 km anti-missile. 2 or 4—12.7 mm (twin or quad) MGs.
Physical countermeasures: Decoys: Elbit/Deseaver 72-barrelled launchers for chaff and IR flares ❻.
Electronic countermeasures: ESM/ECM: Elisra NS 9003/5; intercept and jammer.
Radars: Air/surface search: Thomson-CSF TH-D 1040 Neptune ❽; G-band.
Fire control: 2 Elta EL/M-2221 GM STGR ❾; I/K/J-band.
Combat data systems: IAI Reshet datalink.
Weapon control systems: Galileo OG 20 optical director; Elop MSIS optronic director ❼.

Programmes: *Hetz* started construction in 1984 as the fifth of the SAAR 4 class but was not completed, as an economy measure. Taken in hand again in 1989 and fitted out as the trials ship for some of the systems installed in the Eilat class.
Modernisation: *Romat* and *Keshet* were modernised to same standard as *Hetz* in what was called the Nirit programme. The remaining craft were new build and some of these have been given names previously allocated to decommissioned/transferred SAAR 4s.
Structure: The CIWS is mounted in the eyes of the ship replacing the 40 mm gun. The eight pack Barak launchers are fully containerised and require no deck penetration or onboard maintenance. They are fitted aft in place of two of the Gabriel launchers where these are still fitted. The fire-control system for Barak is fitted on the platform aft of the bridge on the port side. Davits can be installed aft of the Gabriel missiles for special forces boats.

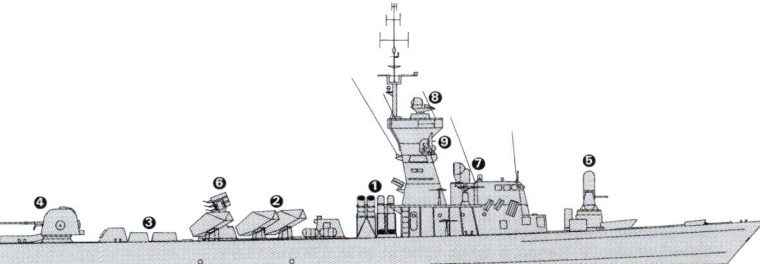

HETZ   *(Scale 1 : 600), Ian Sturton / 0126347*

KESHET   *4/2006, M Declerck / 1158546*

HEREV   *4/2006, M Declerck / 1158547*

Italy
MARINA MILITARE

Country Overview

Italy is situated in southern Europe and comprises, in addition to the Italian mainland, the islands of Sardinia, Sicily, Elba and many smaller islands. Enclaves within mainland Italy are the independent countries of San Marino and Vatican City. With an area of 116,341 square miles, it is bordered to the north by France, Switzerland, Austria and Slovenia. It has a 2,700 n mile coastline with the Mediterranean, Ionian, Adriatic, Tyrrhenian Sea and Ligurian Seas. The capital and largest city is Rome while the principal ports are Genoa, Naples, Trieste, Taranto, Palermo and Venice. Territorial waters (12 n miles) are claimed but an EEZ has not been claimed.

Headquarters Appointments

Chief of Naval Staff: Admiral Giuseppe de Giorgi
Vice Chief of Naval Staff: Vice Admiral Claudio Gaudiosi
Chief of Procurement: Engineer Vice Admiral Ernesto Nencioni
Chief of Naval Personnel: Vice Admiral Raffaele Caruso

Flag Officers

Commander, Logistic Support Command:
  Engineer Vice Admiral Stefano Tortora
Commander-in-Chief of Fleet (Rome):
  Vice Admiral Filippo Foffi
Commander, Training and Schools Command (COMSCUOLE) (Ancona): Vice Admiral Gerard Talarico
Commander, High Seas Fleet (COMFORAL):
  Rear Admiral Paolo Treu
Commander, Submarine Force (COMFORSUB) (Rome):
  Commodore Michele La Fortezza
Commander (1st Frigate Squadron) (Taranto):
  Captain Alessandro Cimino
Commander, Naval Group (2nd Frigate Squadron) (La Spezia):
  Captain Rosario La Pira
Commander, Submarine Flotilla (COMFLOTSOM) (Taranto):
  Captain Mario Berardocco
Commander, MCM Forces (COMFORDRAG) and Commander, Auxiliary Force (COMFORAUS) (La Spezia):
  Captain Massimo Vianello
Commander, Amphibious Force (COMFORSBARC) (Brindisi):
  Commodore Pasquale Guerra
Commander, Coastal and Patrol Forces (COMFORPAT) (Augusta): Commodore Mario Culcasi
Commander, Fleet Air Arm (COMFORAER) (Rome):
  Commodore Giorgio Gomma
Commander, Naval Special Forces (COMSUBIN) (La Spezia):
  Commodore Francesco Chionna

Diplomatic Representation

Naval Attaché in Peking: Captain Massimo Arigoni
Defence Attaché in London: Rear Admiral Dario Giacomin
Naval Attaché in Paris: Captain Andrea Romani
Naval Attaché in Washington: Captain Walter Zappellini
Defence Attaché in Madrid: Commodore Roberto Ciammaichella
Naval Attaché in Cairo: Captain Gianluca Vuccilli
Naval Attaché in Santiago: Captain David Volpe
Defence Attaché in Jakarta: Captain Maurizio Bonora
Naval Attaché in Kuala Lumpur: Captain Luca Bercini
Naval Attaché in Muscat: Captain Captain Stefano Clementieri
Naval Attaché in Rabat: Captain Cosimo Russo
Naval Attaché in Tunis: Captain Danilo Murciano
Defence Attaché in Ankara: Rear Admiral Emanuele Bottazzi
Naval Attaché in Tripoli: Captain Stefano Bovara
Naval Attaché in Lima: Captain Maurizio Mele
Defence Attaché in New Delhi: Rear Admiral Franco Favre

Bases

Area Logistic Commands: La Spezia (Tyrrhenian Sea), Taranto (Ionian Sea), Augusta (Sicily).
Main bases (Major Arsenals/Navy Shipyards): Taranto, La Spezia.
Secondary base (Minor Arsenal/Navy Shipyard): Augusta.
Minor bases: Brindisi.

Organisation

CINCNAV is responsible for all operational activities. There are six subordinate commands:
High-Sea Forces Command (COMFORAL) including all Major and Amphibious Ships. Based in Taranto with subordinated command COMGRUPNAVIT.
Patrol Forces Command (COMFORPAT) Corvettes and OPVs. Based in Augusta.
Fleet Air Command. Based at Santa Rosa, Rome.
Submarine Force Command. Based at Taranto.
Mine Countermeasures Command. Based at La Spezia.
COMFORSBARC with San Marco Regiment, Carlotto (logistic) regiment and one assault boat group. Based at Brindisi.
Special Forces Command (COMSUBIN) Commandos and support craft. Based near La Spezia. Controlled directly by Chief of Naval Staff.

Prefix to Ships' Names

ITS (Italian Ship)

Strength of the Fleet

Type	Active	Building (Planned)
Submarines	6	2
Aircraft Carriers	1	—
Destroyers	4	—
Frigates	12	5 (2)
Corvettes	5	—
Offshore Patrol Vessels	10	(12)
Coastal Patrol Craft	4	—
LPH	1	—
LPD/LHD	3	(2)
Minehunters/sweepers	10	—
Survey/Research Ships	7	—

Strength of the Fleet – continued

Type	Active	Building (Planned)
Replenishment Tankers	3	(2)
Coastal Tankers	12	—
Coastal Transports	6	—
Sail Training Ships	8	—
Training Ships	5	—
Lighthouse Tenders	5	—
Salvage Ships	1	(1)

Personnel

2014: 34,000 (4,150 officers) including 1,573 fleet air arm and 2,100 naval infantry (amphib). This is to be reduced to 30,421 by 2016 and 26,800 by 2024.

Fleet Air Arm Command

Naval Air Stations
Catania (Fontanarossa): AB-212 (2nd), EH-101 (3rd)
La Spezia (Luni): EH-101 (1st), AB-212/SH-90A (5th), Fleet Air Arm Experimental Centre (CSA)
Taranto (Grottaglie): AV-8B (Embarked Aircraft Group), AB-212 (4th Group); AB-212 and EH-101 (Air Assault Group)
Air Force Base hosting maritime aircraft
Catania (Sigonella): Atlantic (41st)
Pratica di Mare (Roma): Detachment 9-01 (P 180)

Naval Infantry and Army Amphibious Units

The San Marco Infantry Brigade comprises the 1st San Marco Regiment (amphibious assault, combat support and combat service support), 2nd San Marco Regiment (afloat Force Protection, maritime interdiction/boarding, harbour protection), 3rd San Marco Regiment (all base protection detachments).
  The 1st San Marco Regiment and the Army Amphibious Regiment 'serenissima', with their support units, form the core of the national seaborne force projection capability.

DELETIONS

Frigates

| 2012 | Artigliere |
| 2013 | Granatiere, Maestrale |

Corvettes

| 2012 | Minerva, Sibilla |
| 2013 | Danaide |

Mine Warfare Forces

| 2012 | Lerici, Sapri |

Tugs

| 2012 | Prometeo |
| 2013 | Ciclope, Tenace, Gigante |

PENNANT LIST

Submarines

S 522	Salvatore Pelosi
S 523	Giuliano Prini
S 524	Primo Longobardo
S 525	Gianfranco Gazzana Priaroggia
S 526	Salvatore Todaro
S 527	Scire
S 528	Pietro Venuti (bldg)
S 529	Romeo Romei (bldg)

Light Aircraft Carriers

| C 550 | Cavour |
| C 551 | Giuseppe Garibaldi |

Destroyers

D 553	Andrea Doria
D 554	Caio Duilio
D 560	Luigi Durand de la Penne
D 561	Francesco Mimbelli

Frigates

F 571	Grecale
F 572	Libeccio
F 573	Scirocco
F 574	Aliseo
F 575	Euro
F 576	Espero
F 577	Zeffiro
F 583	Aviere
F 584	Bersagliere
F 590	Carlo Bergamini
F 591	Virginio Fasan
F 592	Carlo Margottini
F 593	Carabiniere (bldg)
F 594	Alpino (bldg)
F 595	Luigi Rizzo (bldg)

Corvettes

F 552	Urania
F 554	Sfinge
F 555	Driade
F 556	Chimera
F 557	Fenice

Patrol Forces

P 401	Cassiopea
P 402	Libra
P 403	Spica
P 404	Vega
P 405	Esploratore
P 406	Sentinella
P 407	Vedetta
P 408	Staffetta
P 409	Sirio
P 410	Orione
P 490	Comandante Cigala Fulgosi
P 491	Comandante Borsini
P 492	Comandante Bettica
P 493	Comandante Foscari

Minehunters

M 5552	Milazzo
M 5553	Vieste
M 5554	Gaeta
M 5555	Termoli
M 5556	Alghero
M 5557	Numana
M 5558	Crotone
M 5559	Viareggio
M 5560	Chioggia
M 5561	Rimini

Amphibious Forces

L 9892	San Giorgio
L 9893	San Marco
L 9894	San Giusto

Survey and Research Ships

A 5303	Ammiraglio Magnaghi
A 5304	Aretusa
A 5308	Galatea
A 5315	Raffaele Rossetti
A 5320	Vincenzo Martellotta
A 5340	Elettra

Auxiliaries

A 5301	Leonardo
A 5302	Caroly
A 5305	Murena
A 5309	Anteo
A 5311	Palinuro
A 5312	Amerigo Vespucci
A 5313	Stella Polare
A 5316	Corsaro II
A 5322	Capricia
A 5323	Orsa Maggiore
A 5324	Titano
A 5325	Polifemo
A 5326	Etna
A 5327	Stromboli
A 5329	Vesuvio
A 5330	Saturno
A 5347	Gorgona
A 5348	Tremiti
A 5349	Caprera
A 5351	Pantelleria
A 5352	Lipari
A 5353	Capri
A 5359	Bormida
A 5364	Ponza
A 5366	Levanzo
A 5367	Tavolara
A 5368	Palmaria
A 5370	Panarea
A 5371	Linosa
A 5372	Favignana
A 5373	Salina
A 5376	Ticino
A 5377	Tirso
A 5379	Astice
A 5380	Mitilo
A 5382	Porpora
A 5383	Procida
Y 413	Porto Fossone
Y 416	Porto Torres
Y 417	Porto Corsini
Y 421	Porto Empedocle
Y 422	Porto Pisano
Y 423	Porto Conte
Y 425	Porto Ferraio
Y 426	Porto Venere
Y 428	Porto Salvo
Y 498	Mario Marino
Y 499	Alcide Pedretti

SUBMARINES

2 + 2 TODARO (TYPE 212A) CLASS (SSK)

Name	No	Builders	Laid down	Launched	Commissioned
SALVATORE TODARO	S 526	Fincantieri, Muggiano	3 Jul 1999	6 Nov 2003	29 Mar 2006
SCIRÈ	S 527	Fincantieri, Muggiano	10 May 2000	18 Dec 2004	19 Feb 2007
PIETRO VENUTI	S 528	Fincantieri, Muggiano	7 Aug 2009	2014	2015
ROMEO ROMEI	S 529	Fincantieri, Muggiano	1 Jul 2010	2015	2016

Displacement, tonnes: 1,450 surfaced; 1,830 dived
Dimensions, metres (feet): 57 × 7 × 6 *(187 × 23.0 × 19.7)*
Speed, knots: 12 surfaced; 20 dived
Range, n miles: 8,000 at 8 kt surfaced
Complement: 27 (6 officers)

Machinery: Diesel-electric; 1 MTU 16V 396 SE 84 diesel; 1,810 hp (m) *(1.35 MW)*; 1 alternator; 1 Siemens PEM motor; 3,875 hp(m) *(2.85 MW)*; 1 shaft; Siemens/HDW PEM 9 fuel cell (AIP) modules; 306 kW
Torpedoes: 6 − 21 in *(533 mm)* bow tubes; water ram discharge; 12 Whitehead A 184 Mod 3; dual purpose; wire guided; active passive homing to 25 km *(13.7 n miles)* at 24 kt; 17 km *(9.2 n miles)* at 38 kt; warhead 250 kg. To be replaced by NSP (national version of WASS Black Shark); wire (fibre-optic cable) guided; active/passive homing to 50 km *(27 n miles)* at 50 kt; warhead 250 kg.
Mines: In lieu of torpedoes.

Physical countermeasures: Decoys: Fitted for CIRCE Torpedo countermeasures.
Electronic countermeasures: ESM: EADS FL 1800U; intercept.
Radars: Navigation: KH 1007; I-band.
Sonars: STN Atlas Elektronik DBQS-40; cylindrical array, flank array, towed array. Mine avoidance sonar, passive ranging sonar and cylindrical intercept array.
Weapon control systems: Kongsberg MSI-90U.

Programmes: German design phase first completed in 1992 by ARGE 212 (HDW/TNSW) in conjunction with IKL. MoU signed with Germany 22 April 1996 for a common design. First pair ordered from Fincantieri in August 1997. Government approval to procure second pair was given in March 2008 and a contract with Fincantieri was signed on 21 April 2008. These two boats are to be similar to but upgraded versions of the first two. Plans for a fifth boat were cancelled in 2012.
Modernisation: Weapon integration of the new NSP torpedo began in S 526 in late 2010. The third and fourth boats are to be equipped with the Kongsberg MSI-90U Mk 2 weapon control system, STN Atlas CSU 90-138 integrated sonar system, WASS NSP torpedoes, optronic search periscope and Indra MRBR-800 ESM.
Structure: Equipped with a hybrid fuel cell/battery propulsion based on the Siemens PEM fuel cell technology. The submarine is designed with a partial double hull which has a larger diameter forward. This is joined to the after end by a short conical section which houses the fuel cell plant. Two LOX tanks and hydrogen stored in metal cylinders are carried around the circumference of the smaller hull section. Italian requirements included a greater diving depth, improved external communications, and better submerged escape facilities. The final design is identical to the German submarines. Fitted with Zeiss search and attack periscopes.
Operational: Maximum speed on AIP is 8 kt without use of main battery. A transit speed of 4–6 kt was achieved by a German Type 212A in 2006.

SCIRE 9/2012, M Declerck / 1483509

SCIRE 7/2012, Giorgio Ghiglione / 1483611

SCIRE 6/2011, Guy Toremans / 1406735

4 IMPROVED SAURO CLASS (SSK)

Name	No	Builders	Laid down	Launched	Commissioned
SALVATORE PELOSI	S 522	Fincantieri, Monfalcone	23 Jul 1986	29 Nov 1986	14 Jul 1988
GIULIANO PRINI	S 523	Fincantieri, Monfalcone	30 Jul 1987	12 Dec 1987	17 May 1989
PRIMO LONGOBARDO	S 524	Fincantieri, Monfalcone	19 Dec 1991	20 Jun 1992	14 Dec 1993
GIANFRANCO GAZZANA PRIAROGGIA	S 525	Fincantieri, Monfalcone	12 Nov 1992	26 Jun 1993	12 Apr 1995

Displacement, tonnes: 1,500 (S 522–523), 1,680 (S 524–525) surfaced; 1,689 (S 522–523), 1,892 (S 524–525) dived
Dimensions, metres (feet): 64.4 (S 522–523), 66.4 (S 524–525) × 6.8 × 5.6 *(211.3, 217.8 × 22.3 × 18.4)*
Speed, knots: 11 surfaced; 12 snorting; 19 dived
Range, n miles: 11,000 at 11 kt surfaced; 250 at 4 kt dived
Complement: 51 (7 officers)

Machinery: Diesel-electric; 3 Fincantieri GMT 210.16 SM diesels; 3,672 hp(m) *(2.7 MW)* sustained; 3 generators; 2.16 MW; 1 motor; 3,128 hp(m) *(2.3 MW)*; 1 shaft
Torpedoes: 6—21 in *(533 mm)* bow tubes. 12 Whitehead A184 Mod 3; dual purpose; wire-guided; active/passive homing to 25 km *(13.7 n miles)* at 24 kt; 17 km *(9.2 n miles)* at 38 kt; warhead 250 kg. Swim-out discharge.

Electronic countermeasures: ESM: Elettronica BLD-727; radar warning; 2 aerials-1 on a mast, second in search periscope.
Radars: Search/navigation: SMA BPS 704(V)2; I-band; also periscope radar for attack ranging.
Sonars: STN Atlas Elektronic ISUS 90-20; conformal array, passive intercept sonar and cylindrical transducer array.
Weapon control systems: STN Atlas ISUS 90-20.

Programmes: The first two were ordered in March 1983 and the second pair in July 1988.
Modernisation: An upgrade programme included replacement of acoustic sensors, weapons control system (STN Atlas ISUS 90-20) and communications. Work on all four boats was completed in late 2004.

Structure: Pressure hull of HY 80 steel with a central bulkhead for escape purposes. Diving depth, 300 m *(985 ft)* (test) and 600 m *(1,970 ft)* (crushing). The second pair has a slightly longer hull to give space for SSMs.
Periscopes: Kollmorgen; S 76 Mod 322 with laser rangefinder attack; S 76 Mod 323 with ESM-search. Wave contour snort head has a very low radar profile. All boats have anechoic tiles.
Operational: Two Lital Mk 39 inertial navigation; Sepa autopilot. Endurance, 45 days. Service lives: 2015–16 (S 522 and S 523); 2018–20 (S 524 and S 525).

PRIMO LONGOBARDO 2/2011, A Campanera i Rovira / 1406736

GIULIANO PRINI 6/2005, John Mortimer / 1153233

Submarines < **Italy** 403

GIULIANO PRINI   4/2002, H M Steele / 0076765

GIULIANO PRINI   6/2005, John Brodie / 1153247

GIANFRANCO GAZZANA PRIAROGGIA   6/2001, Giorgio Ghiglione / 0130332

IHS Jane's Fighting Ships 2014-2015

404 Italy > Aircraft carriers

AIRCRAFT CARRIERS

1 CAVOUR CLASS (CV)

Name	No	Builders	Laid down	Launched	Commissioned
CAVOUR (ex-*Andrea Doria*)	C 550	Fincantieri, Muggiano	17 Jul 2001	20 Jul 2004	10 Jun 2009

Displacement, tonnes: 27,535 full load
Dimensions, metres (feet): 235.6 oa; 215.6 wl × 39 oa; 29.5 wl × 7.5 (773; 707.3 × 128.0; 96.8 × 24.6)
Flight deck, metres: 220 × 34 (721.8 × 111.5)
Aircraft lift: 2
Speed, knots: 28. **Range, n miles:** 7,000 at 16 kt
Complement: 696 (168 air group) + 145 flag staff + 325 embarked forces + 39 spare berths

Military lift: (garage only): 100 wheeled vehicles or 60 armoured vehicles or 24 MBTs (Ariete) or mixture
Machinery: COGAG: 4 GE/Fiat LM 2500 gas turbines; 118,000 hp(m) *(88 MW)* sustained; 2 shafts; cp props; bow and stern thrusters; 6 Wärtsilä 2.2 MW diesel generators and 2 Ansaldo Sistemi Industriali shaft generators
Missiles: SAM: 32 (4−8 cell Sylver VLS) Aster 15 ❶; inertial mid-course guidance; active radar homing to 30 km *(16 n miles)* at 3 Mach; warhead 15 kg.
Guns: 2 Oto Melara 3 in *(76 mm)*/62 Super Rapid Strales ❷. 120 rds/min to 16 km *(8.7 n miles)*; weight of shell 6 kg. 3 Oto Melara KBA 25/80 mm.
Physical countermeasures: Decoys: 2 Breda SCLAR-H 20-barrel trainable chaff/decoy launchers.
TCM: 2 SLAT TCM launchers.
Electronic countermeasures: ESM/ECM: MM/SLQ-750 ❸.
Radars: Long-range air search: SPS 798 (RAN-40L); D-band ❺.
Air search and missile guidance: Alenia Marconi EMPAR; G-band ❻.
CCA: SPN-41; J-band.
CCA: Finmeccanica SPN 720(V)5; I-band.
Surface search: SPS-791 (RAN-30X/I) ❼; I-band.
Fire control: 2 Alenia Marconi NA 25XP ❽; I-band.
Navigation: SPN-753G(V) ❾; I-band.
Tacan: SRN-15A.
Sonars: WASS SNA-2000 mine avoidance sonar (bow dome).
Combat data systems: 'Horizon' derivative flag and command support system. Links 11 and 16; provision for Link 22. Satcom ❹.
Weapon control systems: Galileo Avionica SASS optronic director. 2 Alenia NA25XP.

Fixed-wing aircraft: 8 AV-8B Harrier II or JSF.
Helicopters: 12 EH 101 (fitted also for AB 212, NH90 and SH-3D).

Programmes: Following a study phase which included significant changes to the initial configuration of the Nuova Unita Maggiore (NUM) design, the Italian government placed a contract with Fincantieri for the construction of a ship to replace *Vittorio Veneto* in 2007. Capabilities include afloat command, air and amphibious operations. The bow section of the ship was constructed at Muggiano and the centre and stern sections at Riva Trigoso. The ship was joined, outfitted and tested at Muggiano. A second contract, for the development and supply of the combat system was signed with an AMS-led industrial group in October 2002.
Modernisation: Modifications to operate the F-35B are to be completed by 2017.
Structure: The flight deck features six helicopter take-off spots, one spot for SAR, eight parking spots and a 12° ski jump. A notional air group includes 12 EH-101 helicopters and 8 AV-8B Harrier IIs. There is provision in the design to operate JSF and UAVs. The hangar/garage can accommodate various combinations of aircraft and vehicles (including MBT and trucks). There are two 30 ton lifts, one forward of the island and the other starboard side aft. Two Ro-Ro ramps are positioned aft and starboard side. Two 15 ton and one 7 ton lifts are fitted for ordnance and logistic needs respectively. The VLS silos for Aster are located on the port quarter and starboard bow. There is a 430 m³ hospital facility. Two 76 mm guns with Strales guidance were installed by January 2013.
Operational: Sea trials started in 2006. Deployed to Haiti in early 2010 for earthquake relief operations. Home port Taranto. Flight deck qualified for STOVL operations in September 2011. Following the decision to convert *Garibaldi* to an amphibious role, *Cavour* is now the only fixed-wing carrier.

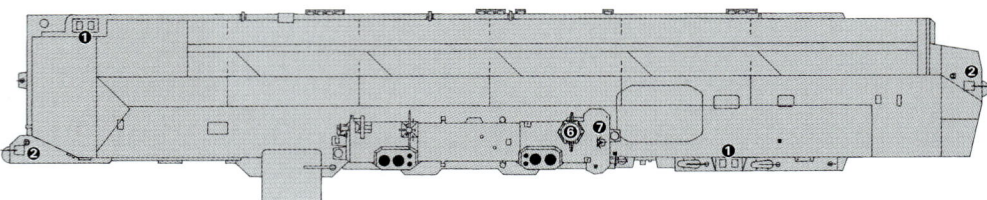

CAVOUR (Scale 1 : 1,800), Ian Sturton / 1366155

CAVOUR (Scale 1 : 1,800), Ian Sturton / 1366156

CAVOUR 6/2011, Guy Toremans / 1406737

CAVOUR 6/2011, Guy Toremans / 1406738

CAVOUR 6/2011, M Declerck / 1406739

IHS Jane's Fighting Ships 2014-2015 © 2014 IHS

CAVOUR *6/2012, Annati Collection* / 1483511

CAVOUR *6/2011, Annati Collection* / 1483510

DESTROYERS

2 ANDREA DORIA (HORIZON) CLASS (DDGHM)

Name	No	Builders	Laid down	Launched	Commissioned
ANDREA DORIA (ex-*Carlo Bergamini*)	D 553	Fincantieri, Riva Trigoso; Muggiano	19 Jul 2002	14 Oct 2005	22 Dec 2008
CAIO DUILIO	D 554	Fincantieri, Riva Trigoso; Muggiano	19 Sep 2003	23 Oct 2007	3 Apr 2009

Displacement, tonnes: 6,741 full load
Dimensions, metres (feet): 152.9 oa; 141.7 wl × 20.3 × 8 *(501.6; 464.9 × 66.6 × 26.2)*
Speed, knots: 29
Range, n miles: 7,000 at 18 kt
Complement: 200 (35 officers)

Machinery: CODOG: 2 GE LM 2500 gas turbines; 55,750 hp(m) *(41 MW)*; 2 SEMT Pielstick 12 PA6B STC diesels; 11,700 hp(m) *(8.6 MW)*; 2 shafts; cp props
Missiles: SSM: 8 (2 quad) Oto Melara Teseo Mk 2A ❶; mid-course guidance; active radar homing to 160 km *(86 n miles)* at 0.9 Mach; warhead 210 kg.
SAM: DCN Sylver VLS ❷ PAAMS (principal anti-air missile system); 48 cells for Aster 15 and Aster 30 weapons; range (Aster 30) 120 km *(65 n miles)*.
Guns: 3 Oto Melara 76 mm/62 Super Rapid Strales ❸. 2 Breda Oerlikon 25 mm/80 ❹.
Torpedoes: 2 fixed launchers ❺. Eurotorp Mu 90 Impact torpedoes.
Physical countermeasures: Decoys: 2 Otobreda SCLAR-H chaff/IR flare launchers ❻. SLAT torpedo defence system.
Electronic countermeasures: ESM/ECM: MM/SLQ-750 ❼.
Radars: Air/surface search: S 1850M ❽; D-band.
Surveillance/fire control: Alenia EMPAR MM/SPY 790 ❾; G-band; multifunction.
Surface search: SPS 791 (RAN-30X) ❿; I-band.
Fire control: 2 Alenia Marconi NA 25XP ⓫; J-band.
Navigation: Alenia SPN 753(V)4 ⓬; I-band.
Sonars: Thomson Marconi 4110CL; hull-mounted; active search and attack; medium frequency.
Combat data systems: DCN/Alenia CMS; Link 16. Link 14 SATCOM ⓭.
Weapon control systems: Sagem Vampir optronic director ⓮.
Helicopters: 1 Augusta/Westland EH 101 Merlin ⓯ or NH-90.

Programmes: Three-nation project for a new air defence ship with Italy, France and UK. Joint project office established in 1993. Memorandum of Understanding for joint development signed 11 July 1994. After UK withdrew in April 1999, an agreement was signed on 7 September 1999 between France and Italy to continue. Following a preliminary agreement on 2 August 2000, a Memorandum of Understanding was signed by the French and Italian Defence Ministries on 22 September 2000 for the joint development of the 'Horizon' destroyer. A Horizon Joint Venture Company was created by DCN/Thomson-CSF and Fincantieri/Finmeccanica on 16 October 2000. The first batch of two vessels for each country was ordered on 27 October 2000. Plans for a second batch of two ships have been cancelled.
Modernisation: Both ships are to be equipped with Strales guidance system for 76 mm guns: D 553 in 2014, D 554 in 2016.
Operational: *Andrea Doria* started sea trials in October 2006 and combat system qualification trials were completed in February 2011.

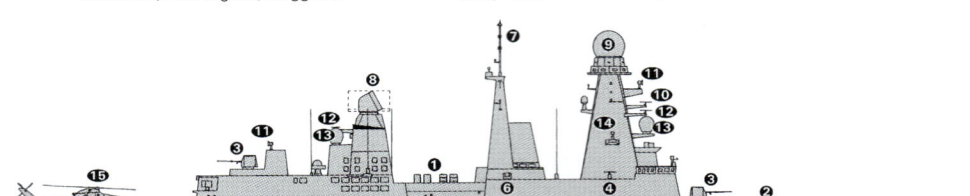

ANDREA DORIA *(Scale 1 : 1,200), Ian Sturton* / 1483596

CAIO DUILIO *5/2012, Giorgio Ghiglione* / 1483604

ANDREA DORIA *2/2013*, B Prézelin* / 1529054

CAIO DUILIO *9/2013*, Guy Toremans* / 1529055

2 DE LA PENNE (EX-ANIMOSO) CLASS (DDGHM)

Name	No	Builders	Laid down	Launched	Commissioned
LUIGI DURAND DE LA PENNE (ex-*Animoso*)	D 560	Fincantieri, Riva Trigoso; Muggiano	20 Jan 1988	29 Oct 1989	18 Mar 1993
FRANCESCO MIMBELLI (ex-*Ardimentoso*)	D 561	Fincantieri, Riva Trigoso; Muggiano	15 Nov 1989	13 Apr 1991	19 Oct 1993

Displacement, tonnes: 4,399 standard; 5,487 full load
Dimensions, metres (feet): 147.7 × 16.1 × 8.6 sonar *(484.6 × 52.8 × 28.2)*
Flight deck, metres: 24 × 13 *(78.7 × 42.7)*
Speed, knots: 31; 21 diesel
Range, n miles: 7,000 at 18 kt
Complement: 331 (25 officers)

Machinery: CODOG; 2 Fiat/GE LM 2500 gas turbines; 54,000 hp *(40.3 MW)* sustained; 2 GMT BL 230.20 DVM diesels; 12,600 hp(m) *(9.3 MW)* sustained; 2 shafts; cp props
Missiles: SSM: 4 or 8 Oto Melara/Matra Teseo Mk 2/A (TG 2) (2 or 4 twin) ❶; mid-course guidance; active radar homing to 160 km *(86.4 n miles)* at 0.9 Mach; warhead 210 kg; sea-skimmer. Mk 3 with radar/IR homing to 300 km *(162 n miles)*; warhead 160 kg in due course.
A/S: Oto Melara/Matra Milas launcher; inertial guidance with command update to 55 km *(29.8 n miles)* at 0.9 Mach; payload Mk 46 Mod 5 or Mu 90 torpedo; 4 weapons (see *Modernisation*).
SAM: 40 Raytheon Standard SM-1MR Block VI; Mk 13 Mod 4 launcher ❷; command guidance; semi-active radar homing to 38 km *(20.5 n miles)* at 2 Mach. Selenia Albatros Mk 2 octuple launcher for Aspide ❸; semi-active radar homing to 13 km *(7 n miles)* at 2.5 Mach; 16 missiles. Automatic reloading.
Guns: 1 Oto Melara 5 in *(127 mm)*/54 ❹; 45 rds/min to 23 km *(12.42 n miles)*; weight of shell 32 kg. 3 Oto Melara 3 in *(76 mm)*/62 Super Rapid ❺; 120 rds/min to 16 km *(8.7 n miles)*; weight of shell 6 kg. 2—12.7 mm MGs.
Torpedoes: 6—324 mm B-515 (2 triple) tubes ❻. Honeywell Mk 46; anti-submarine; active/passive homing to 11 km *(5.9 n miles)* at 40 kt; warhead 44 kg. May be replaced by Whitehead Mu 90 in due course.
Physical countermeasures: Decoys: 2 SCLAR-H chaff launchers ❼. 1 SLQ-25 Nixie anti-torpedo system.
Electronic countermeasures: ESM/ECM: Elettronica SLQ-732 Nettuno ❽; integrated intercept and jamming system.
Radars: Air search: AMS RAN-40L (SPS 798); 3D ❾; D-band.
Air/surface search: AMS RAN-21S (SPS 794) ❿; E/F-band.
Surface search: SMA SPS-702 LE ⓫; I-band.
Fire control: 3 Selenia SPG-76 (RTN 30X) ⓬; I/J-band (for Dardo). 2 Raytheon SPG-51D ⓭; G/I-band (for SAM).
Navigation: SMA SPN-753D; I-band.
IFF: Mk X/XII. Tacan: SRN-15A.
Sonars: Raytheon DE 1164 LF-VDS; integrated bow and VDS; active search and attack; medium frequency (3.75 kHz (hull); 7.5 kHz (VDS)).
Combat data systems: Selenia Elsag IPN 20 (SADOC 2); Links 11, 16 and 22. SHF SATCOM.
Weapon control systems: 4 Dardo-F systems (3 channels for Aspide). Milas TFCS.
Helicopters: 2 AB 212ASW ⓮; SH-3D Sea King and EH 101 Merlin capable.

Programmes: Order placed 9 March 1986 with Riva Trigoso. All ships built at Riva Trigoso are completed at Muggiano after launching. Names changed on 10 June 1992 to honour former naval heroes. Acceptance dates were delayed by reduction gear radiated noise problems which have been resolved.
Modernisation: Milas ASW launchers fitted by late 2004. New sonar dome fitted in D 560 in 2000 increased draft by 1.5 m. A major 2-year upgrade of D 561 was completed in March 2009. D 560 was completed in March 2011. The 127 mm

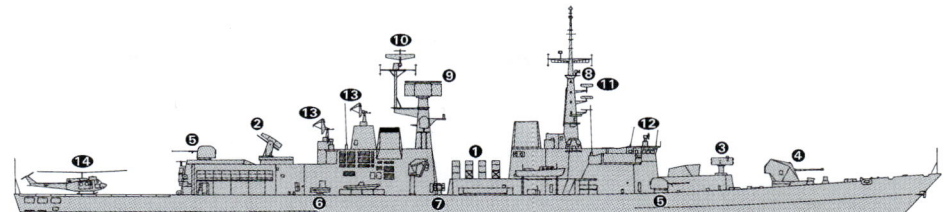

FRANCESCO MIMBELLI (Scale 1 : 1,200), Ian Sturton / 1366157

FRANCESCO MIMBELLI 9/2013*, Guy Toremans / 1529056

LUIGI DURAND DE LA PENNE 6/2012, Giorgio Ghiglione / 1483605

gun is to be upgraded to fire Vulcano guided munitions in 2014 (D 560) and 2016 (D 561). The SM-1 missiles have been refurbished but will cease to be supported from 2016. Removal of some or all of the SM-1 system to make space for other capabilities, including a CATF planning room, is under consideration. Both ships are to be adapted to a fire-support role.

Structure: Kevlar armour fitted. Steel alloys used in superstructure. Prairie Masker noise suppression system. The 127 mm guns are ex-Audace class B turrets. Fully stabilised. Hangar is 18.5 m in length. The aft mast was rebuilt to accommodate SPS-798 and SPS-794 radars.
Operational: The ships are expected to remain in service until about 2025.

LUIGI DURAND DE LA PENNE 6/2012, Italian Navy / 1486494

FRIGATES

3 + 5 (2) BERGAMINI CLASS (MULTIMISSION FRIGATES) (FFGH)

Name	No	Builders	Laid down	Launched	Commissioned
CARLO BERGAMINI	F 590	Fincantieri, Riva Trigoso	4 Feb 2008	16 Jul 2011	29 May 2013
VIRGINIO FASAN	F 591	Fincantieri, Riva Trigoso	12 May 2009	31 Mar 2012	19 Dec 2013
CARLO MARGOTTINI	F 592	Fincantieri, Riva Trigoso	21 Apr 2010	29 Jun 2013	27 Feb 2014
CARABINIERE	F 593	Fincantieri, Riva Trigoso	6 Apr 2011	29 Mar 2014	2015
ALPINO	F 594	Fincantieri, Riva Trigoso	23 Feb 2012	2015	2016
LUIGI RIZZO	F 595	Fincantieri, Riva Trigoso	5 Mar 2013	2016	2017
–	F 596	Fincantieri, Riva Trigoso	2014	2017	2018
–	F 597	Fincantieri, Riva Trigoso	2015	2018	2019

Displacement, tonnes: 6,700 full load
Dimensions, metres (feet): 143.9 × 19.4 × 5.4 (472.1 × 63.6 × 17.7)
Speed, knots: 27
Range, n miles: 6,000 at 15 kt
Complement: 131 (23 air crew) + 25 spare berths

Machinery: CODLAG; 1 General Electric LM 2500+ gas turbine; 42,912 hp (32 MW); 4 Isotta Fraschini diesels; 11,270 hp (8.4 MW); 2 motors; 5,900 hp (4.4 MW); 2 shafts; cp props; 1 azimuth thruster (1 MW).
Missiles: SLCM: to be decided.
SAM: 16 Sylver A50 cell VLS ❶. Mix of Aster 30; active pulse Doppler radar homing to 120 km (65 n miles) at 4.5 Mach; warhead 15 kg and Aster 15; active pulse Doppler radar homing to 30 km (16 n miles) at 3.0 Mach.
SSM: 4 (8 in GP variant) Teseo Mk 2/A ❷; mid-course guidance and active radar homing to 160 km (86.4 n miles); warhead 210 kg.
A/S: MBDA Milas launcher; inertial guidance with command update to 55 km (29.8 n miles) at 0.9 Mach; payload Mk 46 or MU-90 torpedo.
Guns: 1 Oto Melara 127 mm/64LW Vulcano (F 590, 597) ❸; 35 rds/min to 100 km (54 n miles); weight of shell 32 kg. 2 (F 591–594) (1 (F 590, 597)) Oto Melara 76 mm Super Rapid Strales ❹. 2 – 25 mm ❺.
Torpedoes: 6 (2 triple) tubes; MU-90 ❻.
Physical countermeasures: Decoys: 2 Breda SCLAR-H 20-barrel trainable chaff/decoy launchers.
TCM: SLAT launchers.
Electronic countermeasures: ESM: Radar and Comms intercept.
ECM: jammer.
Radars: Air search/fire control: Alenia EMPAR; G-band ❼.
Surface search: SPS 791 (RAN-30X/I) ❽; I-band.
Navigation: 1 SPN-753; I-band. SPN-741; I-band.
Fire control: 2 Alenia Marconi NA-25XP ❾; J-band.
Sonars: Thales TUS 4110CL; hull-mounted (bow dome). Thales TUS 4249 (ASW) active/passive towed array. Mine avoidance sonar.
Combat data systems: Cavour derivative system. Links 11, 16 and 22.
Weapon control systems: Galileo Avionica SASS IRST optronic director ❿.
Helicopters: 2 NH 90 or 1 NH 90 ⓫ plus 1 EH 101.

Programmes: Agreement reached on 7 November 2002 for a 27 ship collaborative programme with France. The original Italian requirement was for 10 frigates with common hull and machinery in two variants. Four ASW and six GP (general purpose/land-attack) ships were to replace the Lupo and Maestrale classes. Contract for the first phase awarded on 16 November 2005 to Orizzonte Sistemi Navali (Fincantieri/Finmeccanica joint venture) for the construction of a first batch of two (1 GP (F 590), 1 ASW) ships. Plans to procure a second batch of four ships (3 ASW, 1 GP (F 595)) were confirmed by the Italian government in 2008. The first two of a third batch of four GP ships were ordered in September 2013. These ships may be upgraded with an active fixed-array radar, subject to funding. All ships are AAW capable as Aster 15 and 30 can be fired from the A50 launcher.
Structure: The class has a conventional hull design. The main engine room contains the gas turbine and two diesel generators while the aft machinery space contains the motors. The Italian variants have a higher foredeck (an extra deck) than their French counterparts. Particular attention has been paid to signature reduction. The radar signature is expected to be comparable to that of the French La Fayette class while exhaust cooling measures are expected to achieve a comparatively low IR signature. Acoustic quietening is to be achieved by the rafting of engines and motors and the use of electric propulsion. The Italian variants are fitted with controllable pitch propellers. All ships of the class are being lengthened to 143.9 m by the insertion of a 3.6 m 'mainframe module' at the extreme stern of the hull. The Batch 2 units are being stretched on build, while work on the Batch 1 ships was completed in 2013. Stern-located equipment includes a RHIB launch/recovery ramp (F 590, F 595), VDS (F 591–594) and towed array (all ships). There are davits for an 11 m RHIB (port) and 7 m RHIB (starboard). GP units are fitted with a fully automatic ammunition magazine with a capacity of 350 rounds and a further 56 in the feeding drums. F 596 and F 597 are to be equipped with more powerful diesels, to improve cruise speed, and F 597 is to be equipped with an integrated mast (UNIMAST).
Operational: *Virginio Fasan* started sea trials in June 2012.

CARLO BERGAMINI (Scale 1 : 1,200), Ian Sturton / 1406805

VIRGINIO FASAN 6/2012, Giorgio Ghiglione / 1483607

VIRGINIO FASAN 6/2012, Giorgio Ghiglione / 1483606

CARLO BERGAMINI 7/2013*, Giorgio Ghiglione / 1529095

Frigates — Italy

7 MAESTRALE CLASS (FFGHM)

Name	No	Builders	Laid down	Launched	Commissioned
GRECALE	F 571	Fincantieri, Muggiano	21 Mar 1979	12 Sep 1981	5 Feb 1983
LIBECCIO	F 572	Fincantieri, Riva Trigoso	1 Aug 1979	7 Sep 1981	5 Feb 1983
SCIROCCO	F 573	Fincantieri, Riva Trigoso	26 Feb 1980	17 Apr 1982	20 Sep 1983
ALISEO	F 574	Fincantieri, Riva Trigoso	10 Aug 1980	29 Oct 1982	7 Sep 1983
EURO	F 575	Fincantieri, Riva Trigoso	15 Apr 1981	25 Apr 1983	24 Jan 1984
ESPERO	F 576	Fincantieri, Riva Trigoso	29 Jul 1982	19 Nov 1983	4 May 1984
ZEFFIRO	F 577	Fincantieri, Riva Trigoso	15 Mar 1983	19 May 1984	4 May 1985

Displacement, tonnes: 2,540 standard; 3,251 full load
Dimensions, metres (feet): 122.7 × 12.9 × 4.6 (402.6 × 42.3 × 15.1)
Flight deck, metres: 27 × 12 (88.6 × 39.4)
Speed, knots: 32; 21 diesel. **Range, n miles:** 6,000 at 16 kt
Complement: 205 (16 officers)

Machinery: CODOG; 2 Fiat/GE LM 2500 gas turbines; 50,000 hp (37.3 MW) sustained; 2 GMT B 230.20 MVM diesels; 11,000 hp(m) (8.1 MW) sustained; 2 shafts; LIPS cp props
Missiles: SSM: 4 Oto Melara Teseo Mk 2 (TG 2) ❶; mid-course guidance; active radar homing to 160 km (86.4 n miles); warhead 210 kg; sea-skimmer. Mk 3 with radar/IR homing to 300 km (162 n miles); warhead 160 kg in due course.
SAM: Selenia Albatros octuple launcher; 16 Aspide ❷; semi-active homing to 13 km (7 n miles) at 2.5 Mach; height envelope 15–5,000 m (49.2–16,405 ft); warhead 30 kg.
Guns: 1 Oto Melara 5 in (127 mm)/54 automatic ❸; 45 rds/min to 23 km (12.42 n miles) anti-surface; 7 km (3.8 n miles) anti-aircraft; weight of shell 32 kg; fires chaff and illuminants. 4 Breda 40 mm/70 (2 twin) compact ❹; 300 rds/min to 12.5 km (6.8 n miles) anti-surface; 4 km (2.2 n miles) anti-aircraft; weight of shell 0.96 kg. 2—12.7 mm MGs.
Torpedoes: 6—324 mm US Mk 32 (2 triple) tubes ❺. Honeywell Mk 46; anti-submarine; active/passive homing to 11 km (5.9 n miles) at 40 kt; warhead 44 kg. To be replaced by MU 90.
Physical countermeasures: Decoys: 2 Breda 105 mm SCLAR 20-tubed trainable chaff rocket launchers ❻; chaff to 5 km (2.7 n miles); illuminants to 12 km (6.6 n miles). 2 Dagaie chaff launchers. SLQ-25; towed torpedo decoy. Prairie Masker; noise suppression system.
Electronic countermeasures: ESM/ECM: SLQ-746.
Radars: Air/surface search: Selenia SPS-774 (RAN 10S) (F 574–576); SPS-794 (RAN 21S) (F 571–573, 577) ❽; E/F-band.

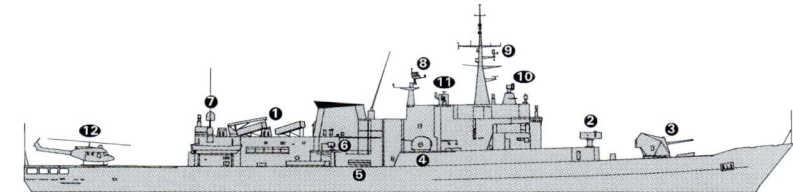
MAESTRALE (Scale 1 : 1,200), Ian Sturton / 0569915

Surface search: SMA SPS-702 ❾; I-band.
Navigation: GEM SPN 753; I-band.
Fire control: Selenia SPG-75 (RTN 30X) ❿; I/J-band (for Albatros and 12.7 mm gun). 2 Selenia SPG-74 (RTN 20X) ⓫; I/J-band; range 15 km (8 n miles) (for Dardo).
IFF: Mk XII.
Sonars: Raytheon DE 1164; hull-mounted; VDS; active/passive attack; medium frequency. VDS can be towed at up to 28 kt. Maximum depth 300 m. Modified to include mine detection active high frequency.
Combat data systems: IPN 20 (SADOC 2) action data automation; Link 11. SATCOM ❼.
Weapon control systems: NA 30 for Albatros and 5 in guns. 2 Dardo (Dardo-F in F 572) for 40 mm guns. Galileo Avionica SASS IRST (F 575).
Helicopters: 2 AB 212ASW ⓬.

Programmes: First six ordered December 1976 and last pair in October 1980. All Riva Trigoso ships completed at Muggiano after launch.

Modernisation: Hull and VDS sonars modified from 1994 to give better shallow water performance and a mine detection capability. A major upgrade of four of the class has been undertaken. F 573 and F 577 were completed in early 2007. F 572 was completed in late 2008. F 571 was completed in October 2010. SPS-774 replaced by AMS RAN-21S (SPS-794). Dardo replaced by two new fire-control systems (Dardo-F with RTN-30X), New combat data system installed.
Structure: There has been a notable increase of 34 ft in length and 5 ft in beam over the Lupo class to provide for the fixed hangar and VDS, the result providing more comfortable accommodation but a small loss of top speed. Fitted with stabilisers.
Operational: A towed passive LF array may be attached to the VDS body. Aft A 184 torpedo tubes have been removed. F 571, F 572, F 573 and F 577 to remain in service until 2018–2021. These ships are to be used to operated ScanEagle UAV system. F 570 was decommissioned in 2013, F 576 is to follow in 2014, F 574 in 2015, and F 575 in 2016. All are likely to be offered for sale. Ecuador, Peru, and Philippines are potential buyers.

ALISEO 9/2013*, Guy Toremans / 1529057

GRECALE 6/2011, Guy Toremans / 1406743

LIBECCIO 7/2012, Erik Laursen / 1483512

Italy > Frigates — Corvettes

2 ARTIGLIERE (LUPO) CLASS (FFGHM)

Name	No
AVIERE (ex-*Thi Qar*)	F 583 (ex-F 15)
BERSAGLIERE (ex-*Al Yarmouk*)	F 584 (ex-F 17)

Builders	Laid down	Launched	Commissioned
Fincantieri, Ancona	3 Sep 1982	19 Dec 1984	4 Jan 1995
Fincantieri, Riva Trigoso	12 Mar 1984	18 Apr 1985	8 Nov 1995

Displacement, tonnes: 2,243 standard; 2,566 full load
Dimensions, metres (feet): 113.2 × 11.3 × 3.7 *(371.4 × 37.1 × 12.1)*
Speed, knots: 35; 21 diesel. **Range, n miles:** 5,000 at 15 kt
Complement: 177 (13 officers)

Machinery: CODOG; 2 Fiat/GE LM 2500 gas turbines; 50,000 hp *(37.3 MW)* sustained; 2 GMT BL 230.20 M diesels; 7,800 hp(m) *(5.7 MW)* sustained; 2 shafts; LIPS cp props
Missiles: SSM: 8 Oto Melara Teseo Mk 2 (TG 2) ❶; mid-course guidance; active radar homing to 160 km *(86.4 n miles)* at 0.9 Mach; warhead 210 kg; sea-skimmer.
SAM: Selenia Elsag Aspide octuple launcher ❷; semi-active radar homing to 14.6 km *(8 n miles)* at 2.5 Mach; warhead 39 kg. 8 reloads.
Guns: 1 Oto Melara 5 in *(127 mm)*/54 Compact Vulcano ❸; 35 rds/min to 100 km *(54 n miles)* anti-surface; weight of shell 32 kg. 4 Breda 40 mm/70 (2 twin) compact ❹; 300 rds/min to 12.5 km *(6.8 n miles)* anti-surface; 4 km *(2.2 n miles)* anti-aircraft; weight of shell 0.96 kg. 2–12.7 mm MGs.
Physical countermeasures: Decoys: 2 Breda SCLAR-H 20-barrel chaff/decoy trainable launcher ❺.
Electronic countermeasures: ESM/ECM: Selenia SLQ-747 (INS-3M); intercept and jammer.
Radars: Air search: Selenia SPS-774 (RAN 10S) ❻; E/F-band.
Surface search: Selenia SPQ-712 (RAN 12 L/X) ❼; I-band.
Navigation: SMA SPN-703; I-band.
Fire control: 2 Selenia SPG-70 (RTN 10X) ❽; I/J-band; range 40 km *(22 n miles)* (for Argo). 2 Selenia SPG-74 (RTN 20X) ❾; I/J-band; range 15 km *(8 n miles)* (for Dardo).
IFF: Mk XII.
Combat data systems: IPN 10 mini SADOC action data automation; Link 11. SATCOM.
Weapon control systems: 2 Elsag Mk 10 Argo with NA 21 directors for missiles and 5 in gun. 2 Dardo for 40 mm guns.
Helicopters: 1 AB 212 ❿.

Programmes: On 20 January 1992 it was decided to transfer the four ships built for Iraq to the Italian Navy. The original sale to Iraq was first delayed by payment problems and then cancelled in 1990 when UN embargoes were placed on military sales to Iraq. After several attempts by the Italian Defence Committee to cancel the project, finance was finally authorised in July 1993.
Modernisation: F 584 modified in 2013 with 127 mm Compact gun and Vulcano ammunition and with SCLAR-H decoy launchers.
Operational: Trials of the Schiebel Camcopter S-100 VTUAV were conducted in *Bersagliere* in 2012. F 582 and F 585 decommissioned 2012–13. F 583 and 584 are to be decommissioned in 2015 and 2017 respectively.

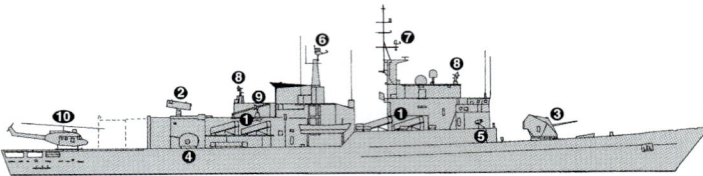

ARTIGLIERE   (Scale 1 : 1,200), Ian Sturton / 0506300

ARTIGLIERI CLASS   9/2009, M Declerck / 1305912

BERSAGLIERE   12/2011, Giorgio Ghiglione / 1406810

CORVETTES

Notes: The future corvette programme has been superseded by the requirement for eight (approved by Financial Stability Law December 2013) dual-use offshore patrol ships (Unità Pattagliamento Altura Duale: UPAD). A lightly-armed variant would be optimised for constabulary roles and humanitarian operations while a more heavily armed variant would be armed with anti-air and anti-ship missiles, two medium-calibre guns (127 mm and 76 mm) and a payload area for containerised special-mission packages. To meet these requirements, a 5,000-tonne ship of 129 m length, is under consideration. The design is to include an integrated mast (UNIMAST) and a mission bay below the flight deck.

5 MINERVA CLASS (FSM)

Name	No	Builders	Laid down	Launched	Commissioned
URANIA	F 552	Fincantieri, Riva Trigoso	4 Apr 1985	21 Jun 1986	1 Jun 1987
SFINGE	F 554	Fincantieri, Muggiano	2 Sep 1986	16 May 1987	13 Feb 1988
DRIADE	F 555	Fincantieri, Riva Trigoso	18 Mar 1988	11 Mar 1989	19 Apr 1990
CHIMERA	F 556	Fincantieri, Riva Trigoso	21 Dec 1988	7 Apr 1990	15 Jan 1991
FENICE	F 557	Fincantieri, Riva Trigoso	6 Sep 1988	9 Sep 1989	11 Sep 1990

Displacement, tonnes: 1,046 standard; 1,306 full load
Dimensions, metres (feet): 86.6 × 10.5 × 3.2 *(284.1 × 34.4 × 10.5)*
Speed, knots: 24. **Range, n miles:** 3,500 at 18 kt
Complement: 106 (8 officers)

Machinery: 2 Fincantieri GMT BM 230.20 DVM diesels; 11,000 hp(m) *(8.1 MW)* sustained; 2 shafts; cp props
Missiles: SAM: Selenia Elsag Albatros octuple launcher (F 555–557) ❶; 8 Aspide; semi-active radar homing to 13 km *(7 n miles)* at 2.5 Mach; height envelope 15–5,000 m *(49.2–16,405 ft)*; warhead 30 kg. Capacity for larger magazine.
Guns: 1 Oto Melara 3 in *(76 mm)*/62 Compact ❷; 85 rds/min to 16 km *(8.7 n miles)* anti-surface; 12 km *(6.6 n miles)* anti-aircraft; weight of shell 6 kg. 1 Oto Breda 25 mm/80 (F 552, 554, 556, 557). 2–12.7 mm MGs (F 555).
Torpedoes: 6—324 mm Whitehead B 515 (2 triple) tubes (F 555–557) ❸. Honeywell Mk 46; active/passive homing to 11 km *(5.9 n miles)* at 40 kt; warhead 44 kg. Being replaced by Whitehead Mu 90.
Physical countermeasures: Decoys: 2 Wallop Barricade double layer launchers for chaff and IR flares. SLQ-25 Nixie; towed torpedo decoy.
Electronic countermeasures: ESM/ECM: Selenia SLQ-747 intercept and jammer.
Radars: Air/surface search: Selenia SPS-774 (RAN 10S) ❺; E/F-band.
Navigation: SMA SPN-728(V)2 ❻; I-band.
Fire control: Selenia SPG-76 (RTN 30X) ❼; I/J-band (for Albatros and gun).
Sonars: Raytheon/Elsag DE 1167; hull-mounted; active search and attack; 7.5–12 kHz.
Combat data systems: Selenia IPN 10 Mini SADOC action data automation; Link 11. SATCOM.
Weapon control systems: 1 Elsag Dardo E system. Selenia/Elsag NA 18L Pegaso optronic director ❹. Elmer TLC system.

Programmes: First four ordered in November 1982, second four in January 1987. A third four were planned, but this plan was overtaken by the acquisition of the Artigliere class.
Structure: The funnels remodelled to reduce turbulence and IR signature. Two fin stabilisers. The ships are not fitted for or with SSM.
Operational: Omega transit fitted. Intended for a number of roles including EEZ patrol, fishery protection and Commanding Officers' training. SAM launchers and torpedo tubes removed from some. All based at Augusta, Sicily. F 551 and 558 were decommissioned in 2012 and F 553 in 2013. The remainder are to be decommissioned as follows: F 552 (2014), F 554 (2015), F 555 (2016), F 556 and F 557 (2018).

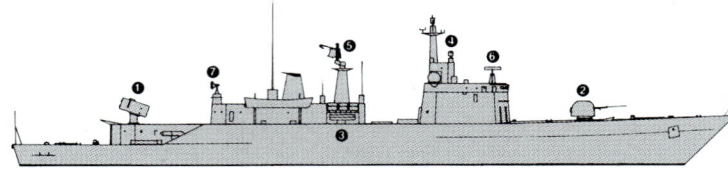

DRIADE   (Scale 1 : 900), Ian Sturton / 0506019

DRIADE   4/2002, Schaeffer/Marsan / 0528348

SHIPBORNE AIRCRAFT

Notes: (1) It is planned to procure 15 STOVL (12 based in Italy and three in the US for joint training) variants (F-35B) of the Joint Strike Fighter in naval livery to enter service from about 2018. In addition, the Italian Air Force is to procure 15 F-35B and 60 land-based F-35A. All aircraft are to be based at Grottaglie NAS (Taranto).
(2) A training programme to integrate AH-129D attack helicopters on board carriers and amphibious vessels is in progress. The aircraft will continue to be operated and maintained by army personnel.

Numbers/Type: 14/2 McDonnell Douglas AV-8B/TAV-8B Harrier II Plus.
Operational speed: 585 kt *(1,083 km/h)*.
Service ceiling: 43,000 ft *(12,192 m)*.
Range: 800 n miles *(1,480 km)*.
Role/Weapon systems: Two trainers delivered in July 1991 plus 15 front-line aircraft from 1994 to December 1997. One aircraft lost in 2010. Expected to remain in service until 2024. Sensors: APG-65 radar, FLIR ALR-67 ESM, ALQ-164 DECM. Weapons: Maverick ASM; AIM-9L1 Sidewinder; AMRAAM AIM-120B AAM; LGB (Paveway) and JDAM bombs and 25 mm cannon.

HARRIER PLUS 9/2013*, Guy Toremans / 1529059

Numbers/Type: 22 Agusta/Westland AW 101 Merlin.
Operational speed: 150 kt *(277.5 km/h)*.
Service ceiling: 15,000 ft *(4,572 m)*.
Range: 550 n miles *(1,019 km)*.
Role/Weapon systems: Primary anti-submarine role with secondary anti-surface and troop carrying capabilities. 16 ordered in October 1995 and approved in July 1997. Six delivered by mid-2002 and further 10 by June 2004. Total of 10 for ASW/ASuW, four for AEW and eight amphibious support (ASH). Of the latter, four are equipped for special operations with an additional fuel tank, a DASS integrated protection suite and personal locator system for combat-SAR. Sensors: APS-784 (ASW/ASV version), APS-717 (ASH version); Eliradar HEW-784 (AEW version) radar, L-3 HELRAS dipping sonar, Star Safire FLIR, ALR 735 ESM, Marconi RALM 1 decoys, Link 11, sonobuoy acoustic processor. Weapons: ASW; four Mu 90 torpedoes. ASuW; four Marte Mk 2/S ASM capability for guidance of ship-launched SSM.

EH 101 11/2008, B Prézelin / 1366053

Numbers/Type: 6 NHIndustries SH90 NFH.
Operational speed: 157 kt *(291 km/h)*.
Service ceiling: 13,940 ft *(4,250 m)*.
Range: 621 n miles *(1,150 km)*.
Role/Weapon systems: Two variants of NH90 (known as SH90 in Italy) to replace the AB-212: 46 SH-90A combat helicopters for ASW/ASV; 10 SH-90B TTH utility/assault helicopters. Sensors (SH-90A): Galileo Avionica ENR radar, Sagem OLOSP FLIR, L3 HELRAS dipping sonar, OTS-90 acoustic system processor. Weapons: Mu-90 torpedoes, Marte Mk 2/S ASM. SH-90B variant has no radar but is fitted with Galileo FLIR weather radar.

SH90 6/2012, Annati Collection / 1483513

Numbers/Type: 25 Agusta-Bell 212.
Operational speed: 120 kt *(222 km/h)*.
Service ceiling: 15,000 ft *(4,373 m)*.
Range: 360 n miles *(667 km)*.
Role/Weapon systems: Assault (seven aircraft) and ASW/ASUW (18 aircraft); mainly deployed to escorts, but also shore-based. Sensors (ASW/ASUW): APS-705 radar; ASQ 13-B sonar; STAR SAFIRE II FLIR; datalink for Teseo SSM. Weapons (ASW/ASUW): 2 Mk 46 torpedoes; (assault): 2 – 7.62 mm MGs; 2 – 12.7 mm MGs; 2 – 70 mm rocket pods.

AB-212 11/2008, M Declerck / 1305922

Numbers/Type: 6 Agusta-Sikorsky SH-3D/H Sea King.
Operational speed: 120 kt *(222 km/h)*.
Service ceiling: 14,000 ft *(4,267 m)*.
Range: 540 n miles *(1,000 km)*.
Role/Weapon systems: Six remaining aircraft are for amphibious assault. Have armoured cabins, no sensors, and are armed with two 7.62 mm MGs.

SEA KING 6/2003, Adolfo Ortigueira Gil / 0570676

LAND-BASED MARITIME AIRCRAFT

Notes: (1) Plans to acquire up to eight Boeing P-8A Poseidon Maritime Patrol Aircraft have been abandoned.
(2) It is planned to procure three Boeing 737 AEW Wedgetail aircraft (with option for one further) to be operated by a joint Navy/Air Force Squadron.
(3) One Agusta A 109 transport helicopter procured in 2002 for liaison duties.
(4) Five RQ-1B Predator are owned and maintained by the Italian Air Force. These can be controlled from *Giuseppe Garibaldi*.
(5) A contract for delivery of two ScanEagle UAV systems (10 aircraft total) was signed with Boeing-Insitu in September 2013. Deliveries are to be made in late 2014. The systems are to be operated from Maestrale-class frigates. Two VTUAV Schiebel Camcopter S-100 are on lease and have been operated from *Bersagliere*.

Numbers/Type: 15 Panavia Tornado IDS.
Operational speed: 2.2 Mach.
Service ceiling: 80,000 ft *(24,385 m)*.
Range: 1,500 n miles *(2,780 km)*.
Role/Weapon systems: Air Force swing wing strike and recce; part of a force of a total of 100 aircraft of which 15 are used for maritime operations based at Gioia de Colle. Sensors: Texas Instruments nav/attack systems. Weapons: ASV; four Kormoran missiles; two 27 mm cannon. AD; four AIM-9L Sidewinder.

TORNADO IDS 8/2001, IHS/C Hoyle / 0034970

Numbers/Type: 3 Piaggio P-180 Avanti Maritime.
Operational speed: 260 kt *(482 km/h)*.
Service ceiling: 39,000 ft *(11,885 m)*.
Range: 1,195 n miles *(2,213 km)*.
Role/Weapon systems: Maritime version of business aircraft. Two aircraft procured in 2002 for liaison duties since retrofitted with FLIR to conduct surveillance. Third aircraft ordered in 2005. Sensors: FLIR.

P-180 MARITIME 7/2005, Massimo Annati / 1127625

Numbers/Type: 4 EADS P 72A (ATR-72-600).
Operational speed: 255 kt *(472 km/h)*.
Service ceiling: 22,000 ft *(6,705 m)*.
Range: 1,200 n miles *(2,200 km)*.
Role/Weapon systems: Four aircraft ordered in 2005 for operation by the Navy in surveillance and SAR roles. The first two are to be delivered in 2013. Sensors: Airborne Tactical Observation and Surveillance System (ATOS), with two tactical consoles and one communication console; Selex Galileo Sea Spray 7300E radar, Galileo EOST-23 FLIR, Elettronica ALR-733 ESM, Link-11 and 16 datalink, defensive suite with chaffs/flare launchers. Weapons: pod-mounted MG.

P 72A 4/2009, Alenia Aeronautica / 1294940

Italy > Amphibious forces

3 SAN GIORGIO CLASS (LPD)

Name	No	Builders	Laid down	Launched	Commissioned
SAN GIORGIO	L 9892	Fincantieri, Riva Trigoso	27 Jun 1985	25 Feb 1987	9 Oct 1987
SAN MARCO	L 9893	Fincantieri, Riva Trigoso	28 Jun 1986	21 Oct 1987	18 Mar 1988
SAN GIUSTO	L 9894	Fincantieri, Riva Trigoso	30 Nov 1992	2 Dec 1993	9 Apr 1994

Displacement, tonnes: 6,794 standard; 8,088 (L 9892–9893), 8,128 (L 9894) full load
Dimensions, metres (feet): 133.3 (L 9892–9893), 137 (L 9894) × 20.5 × 5.3 (437.3, 449.5 × 67.3 × 17.4)
Flight deck, metres: 100 × 20.5 (328.1 × 67.3)
Speed, knots: 21
Range, n miles: 7,500 at 16 kt, 4,500 at 20 kt
Complement: 168 (L 9892–9893), 167 (L 9894) (12 officers (L 9892–9893); 15 officers (L 9894))

Military lift: Battalion of 400 plus 30–36 APCs or 30 medium tanks. 2 LCMs in stern docking well. 3 (San Giusto) or 2 (San Giorgio and San Marco) LCVPs on sponsons. 1 LCPL
Machinery: 2 Fincantieri GMT A 420.12 diesels; 16,800 hp(m) (12.35 MW) sustained; 2 shafts; LIPS cp props; bow thruster
Guns: 1 Oto Melara 3 in (76 mm)/62 (Compact in San Giusto); 60 rds/min to 16 km (8.7 n miles); weight of shell 6 kg. 2 Breda Oerlikon 25 mm/90. 2—12.7 mm MGs.
Electronic countermeasures: ESM: SLR 730; intercept. ESM/ECM: SLQ-747 (San Giusto).
Radars: Surface search: SMA SPS-702; I-band
Navigation: SMA SPN-748; I-band
Fire control: Selenia SPG-70 (RTN 10X); I/J-band
Combat data systems: Selenia IPN 20 (San Giusto). Marisat. SATCOM.
Weapon control systems: Elsag NA 10.

Helicopters: 3 SH-3D Sea King or EH 101 Merlin or 5 AB 212.

Programmes: San Giorgio ordered 26 November 1983, San Marco on 5 March 1984 and San Giusto 1 March 1991. Launching dates of the first two are slightly later than the 'official' launching ceremony because of poor weather and for the third because of industrial problems.
Modernisation: 25 mm guns replaced 20 mm from 1999. Modifications to San Giorgio include removal of the 76 mm gun, movement of LCVPs from davits to a new sponson, and lengthening and enlargement of the flight deck to allow two Merlin and two AB 212 to operate simultaneously on deck. Work completed in early 2003. Similar work on San Marco completed in March 2004. San Giusto has been fitted with an MCC data system to enable her to act as CJTF. San Giusto is to undergo a mid-life modernisation programme although the dates are to be decided.
Structure: Aircraft carrier type flight deck with island to starboard. Following modernisation, San Giorgio and San Marco have four landing spots, a stern docking well (20.5 × 7 m), 2 LCVPs on a port side sponson, a 30 ton lift and two 40 ton travelling cranes for LCMs. San Giusto is of similar design, but was 300 tons heavier on build to include extra accommodation, 3 LCVP sponsons and a slightly longer island. Bow doors and beaching capability removed from San Marco and San Giorgio in refit.
Operational: San Marco was paid for by the Ministry of Civil Protection, is specially fitted for disaster relief but is run by the Navy. All are based at Brindisi and assigned to COMFORAL. One of the three ships carries out the annual Summer cruise for officer and petty officer cadets.

SAN GIUSTO 6/2011, Annati Collection / 1406744

SAN MARCO 9/2013*, Guy Toremans / 1529061

SAN GIUSTO
9/2004, John Brodie
1044369

9 MTM 217 CLASS (LCM)

MEN 217–222 MEN 227–228 MEN 551

Displacement, tonnes: 66 full load
Dimensions, metres (feet): 18.5 × 5.1 (60.7 × 16.7)
Speed, knots: 9
Range, n miles: 300 at 9 kt
Complement: 3
Military lift: 30 tons
Machinery: 2 Fiat diesels; 560 hp(m) (412 kW); 2 shafts

Comment: First six built at Muggiano, La Spezia by Fincantieri. Three completed 9 October 1987 for San Giorgio, three completed 8 March 1988 for San Marco. Three more ordered in March 1991 from Balzamo Shipyard and completed in 1993 for San Giusto. Others of this class are also in service with the Army.

MEN 219 and 220 2000, M Annati / 0104881

17 MTP 96 CLASS (LCVP)

MDN 94–104 MDN 108–109 MDN 114–117

Displacement, tonnes: 15 full load
Dimensions, metres (feet): 13.7 × 3.8 (44.9 × 12.5)
Speed, knots: 29 (est.)
Range, n miles: 100 at 12 kt
Complement: 3
Machinery: 2 diesels; 700 hp(m) (515 kW); 2 shafts or 2 water-jets

Comment: Built by Technomatic Ancona in 1985 (two), Technomatic Bari in 1987–88 (six) and Technoplast Venezia 1991–94 (nine). Can carry 45 men or 4.5 tons of cargo. These craft have Kevlar armour. The most recent versions have water-jet propulsion which gives a top speed of 29 kt (22 kt fully laden). This is being backfitted to all GRP LCVPs.

MDN 101 10/2001, Chris Sattler / 0130326

Amphibious forces — Mine warfare forces < **Italy** 415

0 + (1) AMPHIBIOUS ASSAULT SHIPS (LHD)

Displacement, tonnes: 20,000 full load
Dimensions, metres (feet): 190 oa; 167 wl × 28.0 × 6.3 (623.4; 547.9 × 91.9 × 20.7)
Speed, knots: 21. **Range, n miles:** 7,000 at 16 kt
Complement: 210

Military lift: (garage only): 1,200 lane-metres or 360 tons; 750 troops
Machinery: CODAD; 4 diesels; 2 shafts; bow thruster (to be confirmed)
Guns: 2—76 mm. 3—25 mm.
Physical countermeasures: Decoys: To be announced.
TCM: To be announced.
Electronic countermeasures: ESM/ECM: To be announced.
Radars: Air/Surface search: To be announced.
Navigation: To be announced.
Sonars: Mine avoidance.
Combat data systems: To be announced.
Electro-optic systems: To be announced.

Helicopters: 12–15 aircraft (depending on mix).

Programmes: Funding of one amphibious assault ship was approved by the Financial Stability Law of December 2013. The timescale for the programme has not been announced but a one-year design contract is to be the first step in the acquisition process. This is likely to be followed by a project definition phase leading to a construction contract.
Structure: The flight deck is to provide six helicopter landing spots and is to be served by two aircraft lifts, one at the stern, the other forward of the island. There is a well-dock for four LCACs, an aircraft maintenance hangar and a garage deck accessed from the well-dock or through a door on the starboard side. Four LCVPs are to be carried on the port side of the flight deck. There are to be command and medical facilities. The ship is to be designed with civil-protection tasks in mind; a large generating capacity and water-purification equipment are likely to be included.

4 MODIFIED MTM 217 CLASS (LCM)

MEN 562–565

Displacement, tonnes: 35 standard; 66 full load
Dimensions, metres (feet): 19.5 × 5.1 × 0.9 (64 × 16.7 × 2.9)
Speed, knots: 9
Range, n miles: 300 at 9 kt
Complement: 3
Machinery: 2 diesels; 2 shafts

Comment: Built at Vittoria Shipyard, Adria. Slightly larger versions than MTM 217 class to accommodate latest army armoured vehicles. Ballistic protection fitted. Commissioned in April 2010.

LCM   6/2012*, Italian Navy / 1529060

LHD (artist's impression)
10/2010, Annati Collection
1366960

MINE WARFARE FORCES

Notes: Acquisition of a future MCM capability is under consideration. The way ahead is likely to involve the use of modular elements on board frigates and or/OPVs and the use of airborne MCM rather than specialist ships. Unmanned underwater vehicles are liktely to include: Remus 100, Pluto Gigas, Hugin 1000, Mi-Ki (Plutino) one-shot mine-destructors and Miniranger USV with sidescan sonar.

10 LERICI/GAETA CLASS (MINEHUNTERS/SWEEPERS) (MHSC)

Name	No	Builders	Launched	Commissioned
MILAZZO	M 5552	Intermarine, Sarzana	4 Jan 1985	6 Aug 1985
VIESTE	M 5553	Intermarine, Sarzana	18 Apr 1985	2 Dec 1985
GAETA	M 5554	Intermarine, Sarzana	28 Jul 1990	3 Jul 1992
TERMOLI	M 5555	Intermarine, Sarzana	15 Dec 1990	13 Nov 1992
ALGHERO	M 5556	Intermarine, Sarzana	11 May 1991	31 Mar 1993
NUMANA	M 5557	Intermarine, Sarzana	26 Oct 1991	30 Jul 1993
CROTONE	M 5558	Intermarine, Sarzana	11 Apr 1992	19 Jan 1994
VIAREGGIO	M 5559	Intermarine, Sarzana	3 Oct 1992	1 Jul 1994
CHIOGGIA	M 5560	Intermarine, Sarzana	9 May 1994	19 May 1996
RIMINI	M 5561	Intermarine, Sarzana	17 Sep 1994	26 Nov 1996

Displacement, tonnes: 630 (M 5552–5553), 708 (M 5554–5561) full load
Dimensions, metres (feet): 50 (M 5552–5553), 52.5 (M 5554–5561) × 9.9 × 2.6 (164, 172.2 × 32.5 × 8.5)
Speed, knots: 146 hunting. **Range, n miles:** 1,500 at 14 kt
Complement: 44 (4 officers; 7 divers)

Machinery: 1 Fincantieri GMT BL 230.8 M diesel (passage); 1,985 hp(m) (1.46 MW) sustained; 1 shaft; LIPS cp prop; 3 Isotta Fraschini ID 36 SS 6V diesels (hunting); 1,481 hp(m) (1.1 MW) sustained; 3 hydraulic 360° rotating thrust props; 506 hp(m) (372 kW) (1 fwd, 2 aft)
Guns: 1—20 mm.
Physical countermeasures: MCM: 1 Plutogigas and 1 Pluto standard RoV; 1 Pluto Plus (Gaeta onwards); diving equipment and Galeazzi recompression chamber; Galeazzi Z1 two-man recompression chamber (Gaeta onwards).
Minesweeping: Oropesa Mk 4 wire sweep.
Radars: Navigation: SMA SPN 753V(1), SPN 754; I-band.
Sonars: FIAR SQQ-14(IT) VDS (lowered from keel forward of bridge); classification and route survey; high frequency.
Combat data systems: Motorola MRS III/GPS, NAVM code P(Y) precision navigation system with Datamat SMA SSN-714V(3) automatic plotting. Datamat SMA SSN-714 V(2).

Programmes: First four (Lerici class) ordered 7 January 1978 under Legge Navale. Next six (Gaeta class) ordered from Intermarine 30 April 1988 and two more in 1991. The Gaeta-class ships are 2 m longer and are of an improved design. Construction of Gaetas started in 1988. The last pair delayed by budget cuts but re-ordered on 17 September 1992.
Modernisation: Improvements to Gaeta class include a better minehunting sonar system which was backfitted to the Lerici class in 1991. Other Gaeta upgrades include a third hydraulic system, improved electrical generators, Pluto Gigas ROV, a new type of recompression chamber, and a reduced magnetic signature. Modernisation of the eight Gaeta class was authorised in August 2009 and is being implemented 2010–2018. Upgrades include: replacement of SQQ-14 sonar with Thales 2093; a new combat data system Datamat 712(V)4; replacement of the SMIN Mk 2 ROVs with Pluto Gigas ROV and provision of Plutino - MIKI expendable mine neutralisation system. Hydroid Remus 100 and Hugin 1000 (Vieste test platform) unmanned vehicles are also carried for mine reconnaissance and survey respectively. Some alterations to the superstructure will be required.
Structure: Of heavy GRP throughout hull, decks and bulkheads, with frames eliminated. All machinery is mounted on vibration dampers and main engines made of a magnetic material. Fitted with crane for launching RoVs and for diving operations.
Operational: Endurance, 12 days. For long passages passive roll-stabilising tanks can be used for extra fuel increasing range to 4,000 miles at 12 kt. The Lerici class are being decommissioned. Lerici and Sapri were paid off in 2012 and Milazzo (2015) and Vieste (2016) are to follow.
Sales: Four to Malaysia, two to Nigeria, two to Thailand and three to Finland. 12 of a modified design built by the US and six by Australia.

MILAZZO   3/2010, M Declerck / 1366640

NUMANA   10/2011, Selim San / 1406745

© 2014 IHS   IHS Jane's Fighting Ships 2014-2015

SURVEY AND RESEARCH SHIPS

1 SURVEY SHIP (AGORH/AGE/AGI)

Name	No	Builders	Commissioned
ELETTRA	A 5340	Fincantieri, Muggiano	2 Apr 2003

Displacement, tonnes: 3,231 full load
Dimensions, metres (feet): 93 × 15.2 × 5.2 *(305.1 × 49.9 × 17.1)*
Speed, knots: 17. **Range, n miles:** 8,000 at 12 kt
Complement: 94 (12 officers)
Machinery: Diesel electric; 2 Wartsila CW 12V 200 diesel generators; 5,750 kVA. 2 ABB motors; 4,023 hp *(3 MW)*; 2 shafts; bow thruster
Radars: Navigation: I-band.
Helicopters: Platform for one medium.

Comment: Ordered on 1 December 1999; construction started in March 2000 and launch on 24 July 2002. The design is derived from that of the NATO *Alliance* but is equipped as an intelligence collector. The propulsion system, based on two multi permanent magnet electric motors, is the first of its type to be fitted in a surface vessel.

ELETTRA   9/2008, Giorgio Ghiglione / 1335470

1 SURVEY SHIP (AGSH)

Name	No	Builders	Commissioned
AMMIRAGLIO MAGNAGHI	A 5303	Fincantieri, Riva Trigoso	2 May 1975

Displacement, tonnes: 1,727 full load
Dimensions, metres (feet): 82.7 × 13.7 × 3.5 *(271.3 × 44.9 × 11.5)*
Speed, knots: 16. **Range, n miles:** 6,000 at 12 kt, 4,200 at 16 kt
Complement: 148 (14 officers) + 15 police

Machinery: 2 GMT B 306 SS diesels; 3,000 hp(m) *(2.2 MW)*; 1 shaft; cp prop; auxiliary motor; 240 hp(m) *(176 kW)*; bow thruster
Guns: 1 Oto Breda 25 mm/80 (not fitted).
Radars: Navigation: SMA 3 RM 20; I-band.
Helicopters: Platform only.

Comment: Ordered under 1972 programme. Laid down 13 June 1973. Launched 11 October 1974. Full air conditioning, bridge engine controls, flume-type stabilisers. Equipped for oceanographic studies including laboratories and underwater TV. Two Qubit Trac V integrated navigation and logging systems and a Chart V data processing system installed in 1992 to augment the existing Trac 100-based HODAPS. Carries four surveying motor boats and up to two RHIBs. To be decommissioned when replaced by a new 110 m ship in 2016.

AMMIRAGLIO MAGNAGHI   3/2009, M Declerck / 1305920

1 RESEARCH SHIP (AG/AGOR)

Name	No	Builders	Launched	Commissioned
RAFFAELE ROSSETTI	A 5315	Picchiotti, Viareggio	12 Jul 1986	20 Dec 1986

Displacement, tonnes: 325 full load
Dimensions, metres (feet): 44.6 × 7.9 × 2.1 *(146.3 × 25.9 × 6.9)*
Speed, knots: 17.5
Range, n miles: 700 at 15 kt
Complement: 17 (2 officers)
Machinery: 2 Fincantieri Isotta Fraschini ID 36 N 6V diesels; 3,520 hp(m) *(2.55 kW)* sustained; 2 shafts; cp props; bow thruster

Comment: Five different design torpedo tubes fitted for above and underwater testing and trials. Other equipment for research into communications, surface and air search as well as underwater weapons. There is a stern doorway which is partially submerged and the ship has a set of 96 batteries to allow 'silent' propulsion. Operated by the Permanent Commission for Experiments of War Materials at La Spezia.

RAFFAELE ROSSETTI   3/2009, Guy Toremans / 1305919

2 SURVEY SHIPS (AGS)

Name	No	Builders	Commissioned
ARETUSA	A 5304	Intermarine	10 Jan 2002
GALATEA	A 5308	Intermarine	10 Jan 2002

Displacement, tonnes: 422 full load
Dimensions, metres (feet): 39.2 × 12.6 × 2.5 *(128.6 × 41.3 × 8.2)*
Speed, knots: 13. **Range, n miles:** 1,700 at 13 kt
Complement: 29 (4 officers)
Machinery: Diesel electric; 2 Isotta Fraschini V170812 ME diesels; 2 ABB generators 1,904 hp(m) *(1.4 MW)*; 2 shafts; Schottel props; 2 bow thrusters
Guns: 2—7.62 mm MGs.
Radars: 2 Navigation; I-band.

Comment: GRP catamaran design. Ordered in January 1998. *Aretusa* launched 8 May 2000 and *Galatea* 7 June 2000. Fitted with Kongsberg EA 500 single-beam echo sounder, towed sidescan sonar and dynamic positioning system.

GALATEA   1/2004, Giorgio Ghiglione / 1044371

1 RESEARCH SHIP (AG/AGE)

Name	No	Builders	Commissioned
VINCENZO MARTELLOTTA	A 5320	Picchiotti, Viareggio	22 Dec 1990

Displacement, tonnes: 345 full load
Dimensions, metres (feet): 44.6 × 7.9 × 2.3 *(146.3 × 25.9 × 7.5)*
Speed, knots: 17. **Range, n miles:** 700 at 15 kt
Complement: 19 (2 officers)
Machinery: 2 Fincantieri Isotta Fraschini ID 36 SS 16V diesels; 3,520 hp(m) *(2.59 MW)* sustained; 2 shafts; cp props; bow thruster

Comment: Launched on 28 May 1988. Has one 21 in *(533 mm)* and three 12.75 in *(324 mm)* torpedo tubes and acoustic equipment to operate a 3-D tracking range for torpedoes or underwater vehicles. Like *Rossetti* she is operated by the Commission for Experiments at La Spezia.

VINCENZO MARTELLOTTA   9/2008, Giorgio Ghiglione / 1335471

1 COASTAL RESEARCH VESSEL (AGOR(C))

Name	No	Builders	Commissioned
LEONARDO	A 5390	McTay Marine Ltd, Bromborough	6 Sep 2002

Displacement, tonnes: 399 full load
Dimensions, metres (feet): 28.6 × 9.0 × 2.5 *(93.8 × 29.5 × 8.2)*
Speed, knots: 11
Range, n miles: 1,500 at 11 kt
Complement: 5 + 7 police
Machinery: Diesel-electric; 1,570 hp *(1,170 kW)*; 2 azimuth thrusters; 1—360° bow thruster
Radars: Navigation: 2 sets; I-band.
Sonars: Kongsberg Simrad multibeam echo-sounders.

Comment: The order for a coastal underwater research vessel was placed by NATO Underwater Research Centre in December 2000. Designed by Corlett and Partners, construction of the hull was undertaken by Remontowa in Poland while the superstructure and final assembly was undertaken by the prime contractor, McTay Marine Ltd. The ship is equipped with a moon pool, oceanographic winches, two cranes and Kongsberg navigation/research suite. A 20 ft container can be embarked to augment the main scientific laboratory. Based at La Spezia, the vessel is operated on behalf of NATO by the Italian Navy.

LEONARDO   5/2012, Giorgio Ghiglione / 1455685

TRAINING SHIPS

Notes: In addition to the ships listed the LPDs are used in a training role.

1 SAIL TRAINING SHIP (AXS)

Name	No	Builders	Commissioned
AMERIGO VESPUCCI	A 5312	Castellammare	15 May 1931

Displacement, tonnes: 3,600 standard; 4,213 full load
Dimensions, metres (feet): 82.4 oa; 70 pp × 15.5 × 7 *(270.3; 229.7 × 50.9 × 23.0)*
Speed, knots: 10. **Range, n miles:** 5,450 at 6.5 kt
Complement: 243 (13 officers)
Machinery: Diesel-electric; 2 Fiat B 306 ESS diesel generators; 2 Marelli motors; 2,000 hp(m) *(1.47 MW)*; 1 shaft
Radars: Navigation: 2 SMA SPN-748; I-band.

Comment: Launched on 22 March 1930. Hull, masts and yards are of steel. Sail area, 22,604 sq ft, length including bowsprit 100 m (330 ft). Extensively refitted at La Spezia Naval Dockyard in 1973 and again in 1984. Used for Naval Academy Summer cruise with up to 150 trainees.

AMERIGO VESPUCCI *8/2013*, Michael Nitz* / 1529062

3 ARAGOSTA (HAM) CLASS (AXL)

ASTICE A 5379	MITILO A 5380	PORPORA A 5382

Displacement, tonnes: 191 full load
Dimensions, metres (feet): 32.5 × 6.4 × 1.8 *(106.6 × 21.0 × 5.9)*
Speed, knots: 14. **Range, n miles:** 2,000 at 9 kt
Complement: 13 (2 officers)
Machinery: 2 Fiat-MTU 12V 493 TY7 diesels; 2,200 hp(m) *(1.62 MW)* sustained; 2 shafts
Radars: Navigation: BX 732; I-band.

Comment: Builders: CRDA, Monfalcone: *Astice*. Picchiotti, Viareggio: *Mitilo*. Costaguta, Voltri: *Porpora*. Similar to the late UK Ham class. All constructed to the order of NATO in 1955–57. Designed armament of one 20 mm gun not mounted. Originally class of 20. Remaining three converted for training 1986. *Porpora* used by the Naval Academy. *Astice* has a modified bridge structure. To be decommissioned in 2015 without replacement.

ASTICE *1/2009, Giorgio Ghiglione* / 1366052

PORPORA *10/2007, Marco Ghiglino* / 1170075

1 SAIL TRAINING SHIP (AXS)

ITALIA

Displacement, tonnes: 33 full load
Dimensions, metres (feet): 61 × 9.2 × ? *(200.1 × 30.2 × ?)*
Complement: 10 + 100 spare berths
Machinery: 1 diesel; 480 hp *(358 kW)*

Comment: The world's largest brigantine donated by the Italian Yacht Club to the Italian Navy in 2008. The vessel has 1,300 m² of sails.

ITALIA *6/2008, Annati Collection* / 1335388

1 SAIL TRAINING SHIP (AXS)

Name	No	Builders	Commissioned
PALINURO (ex-*Commandant Louis Richard*)	A 5311	Ch Dubigeon, Nantes	1934

Displacement, tonnes: 1,059 standard; 1,473 full load
Measurement, tonnes: 872 gt
Dimensions, metres (feet): 59 × 10 × 4.8 *(193.6 × 32.8 × 15.7)*
Speed, knots: 7.5
Range, n miles: 5,390 at 7.5 kt
Complement: 69 (6 officers)
Machinery: 1 GMT A 230.6N diesel; 600 hp *(447 kW)*; 1 shaft
Radars: Navigation: SPN-748; I-band.

Comment: Barquentine launched in 1934. Purchased in 1951. Rebuilt in 1954–55 and commissioned in Italian Navy on 1 July 1955. Sail area, 1,152 sq ft. She was one of the last two French Grand Bank cod-fishing barquentines. Owned by the Armement Glâtre she was based at St Malo until bought by Italy. Used for seamanship basic training.

PALINURO *10/2008, Camil Busquets i Vilanova* / 1305918

5 SAIL TRAINING YACHTS (AXS)

Name	No	Builders	Commissioned
CAROLY	A 5302	Baglietto, Varazze	1948
STELLA POLARE	A 5313	Sangermani, Chiavari	8 Oct 1965
CORSARO II	A 5316	Costaguta, Voltri	5 Jan 1961
CAPRICIA	A 5322	Bengt-Plym	1963
ORSA MAGGIORE	A 5323	Tencara, Venezia	1994

Comment: The first three are sail training yachts between 40 and 60 tons with a crew including trainees of about 16. *Capricia* is a yawl of 55 tons and was donated by the Agnelli foundation as replacement for *Cristoforo Colombo II* which was not completed when the shipyard building her went bankrupt. *Capricia* commissioned in the Navy 23 May 1993. *Orsa Maggiore* is a ketch of 70 tons.

STELLA POLARE *7/2007, Giorgio Ghiglione* / 1166587

422 Italy > Government maritime forces

1 ANTONIO ZARA CLASS (PB)

Name	No	Builders	Commissioned
GIOVANNI DENARO	P 03	Fincantieri, Muggiano	20 Mar 1998

Displacement, tonnes: 345 full load
Dimensions, metres (feet): 51 × 7.5 × 1.9 *(167.3 × 24.6 × 6.2)*
Speed, knots: 36
Range, n miles: 3,800 at 15 kt
Complement: 33 (3 officers)

Machinery: 4 MTU 16V 396 TB94 diesels; 13,924 hp(m) *(10.38 MW)* sustained; 2 shafts
Guns: 1 Breda 30 mm/70 (single or twin). 2—7.62 mm MGs.
Radars: Surface search: Gemant 2 ARPA and SPN 749; I-band.
Weapon control systems: AESN Medusa optronic director.

Comment: Similar to the Ratcharit class built for Thailand in 1976–79. First pair ordered in August 1987. Third ordered in October 1995. Fitted with an infrared search and surveillance sensor (AMS SVIR). The first pair were decommissioned in 2007.

GIOVANNI DENARO 6/2006, Guardia di Finanza / 1158732

9 MAZZEI CLASS (PB/YXT)

Name	No	Builders	Commissioned
MAZZEI	G 01	Intermarine, Sarzana	1 Apr 1998
VACCARO	G 02	Intermarine, Sarzana	29 Apr 1998
DI BARTOLO	G 03	Intermarine, Sarzana	3 Oct 2003
AVALLONE	G 04	Intermarine, Sarzana	29 Jan 2004
OLTRAMONTI	G 05	Intermarine, Sarzana	24 Jun 2004
BARBARISO	G 06	Intermarine, Sarzana	12 Jul 2007
PAOLINI	G 07	Intermarine, Sarzana	20 Dec 2007
GRECO	G 08	Intermarine, Sarzana	15 May 2008
CINUS	G 09	Intermarine, Sarzana	14 Oct 2008

Displacement, tonnes: 137 full load
Dimensions, metres (feet): 35.5 (G 01–07), 36.5 (G 08–09) × 7.6 × 1.1 *(116.5, 119.8 × 24.9 × 3.6)*
Speed, knots: 38. **Range, n miles:** 920 at 22 kt
Complement: 19 (G 03–09), 37 (G 01–02) (18 trainees (G 01–02))

Machinery: 2 MTU 16V 396 TB94 (MTU 16V 4000 M 90 G 08, G 09) diesels; 5,800 hp(m) *(4.26 MW)* sustained; 2 shafts
Guns: 1 Breda Mauser 30 mm/70. 2—7.62 mm MGs.
Radars: GEM 3072A ARPA; I-band.
Navigation: GEM 1410; I-band.
Weapon control systems: AESN Medusa optronic director.

Comment: Based on the Bigliani class but with an extended hull. G 01 and G 02 used as training ships. All are being fitted with an infra-red search and surveillance sensor (AMS SVIR). G 03–09 are used as command units for air-naval task group.

DI BARTOLO 8/2007, Marco Ghiglino / 1170074

19 BIGLIANI CLASS (PB)

OTTONELLI G 78	SANNA G 117	LA SPINA G 122	ZACCOLA G 127
BARLETTA G 79	INZUCCHI G 118	SALONE G 123	STANISCI G 128
LA MALFA G 88	VITALI G 119	CAVATORTO G 124	SOTTILE G 129
ROSATI G 89	CALABRESE G 120	FUSCO G 125	DE FALCO G 130
LAGANÀ G 116	URSO G 121	DE ROSA G 126	

Displacement, tonnes: 88 (G 116–125, G 78–79, 88–89), 100 (G 126–130) full load
Dimensions, metres (feet): 26.4 (G 116–125, G 78–79, 88–89), 28.8 (G 126–130) × 7 × 1.1 *(86.6, 94.5 × 23.0 × 3.6)*
Speed, knots: 42. **Range, n miles:** 770 at 18 kt
Complement: 12

Machinery: 2 MTU 16V 396 TB94 diesels; 6,850 hp(m) *(5.12 MW)* sustained; 2 shafts
Guns: 1 Breda Mauser 30 mm/80. 2—7.62 mm MGs. 1 Breda 12.7 mm.
Radars: Surface search: GEM 3072A ARPA; I-band.
Navigation: GEM 1410; I-band.
Combat data systems: AMS IPNS.
Weapon control systems: Elsag Medusa Mk 4 optronic director.

Comment: First eight built by Crestitalia and delivered from October 1987 to September 1992. Three more were ordered from Crestitalia/Intermarine in October 1994 and were delivered from December 1996 to April 1997. A fourth was delivered in late 1999. There are minor structural differences between Series II (G 82–87) and Series III (G 78–79, G 88–89). Ten series IV (G 116–125) craft ordered from Intermarine, Sarzana, for delivery in 2004–06. These include Kevlar armour and are fitted with a remote-control Breda 12.7 mm gun and 40 mm grenade launcher. All are fitted with an infrared search and surveillance sensor (AMS SVIR). A further five craft (G 126-130) are stretched versions delivered in 2009. G 82, 83, 84, 85, 86 and 87 donated to the Libyan Coast Guard in 2009.

OTTONELLI 6/2011, M Declerck / 1406753

24 CORRUBIA CLASS (PBF)

ALBERTI G 92	LIPPI G 100	CONVERSANO G 108
ANGELINI G 93	LOMBARDI G 101	INZERILLI G 109
CAPPELLETTI G 94	MICCOLI G 102	LETIZIA G 110
CIORLIERI G 95	TREZZA G 103	MAZZARELLA G 111
D'AMATO G 96	APRUZZI G 104	NIOI G 112
FAIS G 97	BALLALI G 105	PARTIPILO G 113
FELICIANI G 98	BOVIENZO G 106	PULEO G 114
GARZONI G 99	CARRECA G 107	ZANNOTTI G 115

Displacement, tonnes: 93 full load
Dimensions, metres (feet): 26.8 × 7.6 × 1.2 *(87.9 × 24.9 × 3.9)*
Speed, knots: 43. **Range, n miles:** 700 at 20 kt
Complement: 12 (1 officer)

Machinery: 2 Isotta Fraschini ID 36 SS 16V diesels; 6,400 hp(m) *(4.7 MW)*; 2 shafts (G 90–91) 2 MTU 16V 396 TB94; 5,800 hp(m) *(4.26 MW)* sustained; 2 shafts (G 92-103)
Guns: 1 Breda Mauser 30 mm/70 (G 90–103). 1 Astra 20 mm (G 104–115). 2—7.62 mm MGs.
Radars: Surface search: GEM 3072A ARPA; I-band.
Navigation: GEM 1210; I-band.
Weapon control systems: Elsag Medusa optronic director.

Comment: First two built by Cantieri del Golfo, Gaeta, delivered in 1990 and decommissioned in 2007. Others built by Cantieri del Golfo (G 92–100), and Crestitalia (G 101–103), and Intermarine from 1995 onwards. G 115 completed in 1999. There are minor structural differences between the second series (G 92–103) and the third batch (G 104–115). All are being fitted with an infra-red search and surveillance sensor (AMS SVIR).

CONVERSANO 6/2005, Marco Ghiglino / 1153210

8 MEATINI CLASS

MAZZEO G 44	IGNESTI G 47	TRIDENTI G 56	FIDONE G 60
SILANOS G 46	FIORE G 51	ATZEI G 58	DARIDA G 64

Displacement, tonnes: 41 full load
Dimensions, metres (feet): 20.1 × 5.2 × 1 *(65.9 × 17.1 × 3.3)*
Speed, knots: 34. **Range, n miles:** 550 at 20 kt
Complement: 11 (1 officer)
Machinery: 2 CRM 18D/52 diesels; 2,500 hp(m) *(1.84 MW)*; 2 shafts
Guns: 1—12.7 mm MG.
Radars: Surface search: 1 GEM 1210; I-band.

Comment: Fifty-six of the class built from 1970 to 1978. *Darida* is to be retained in service for historical purposes, the remaining seven are to be decommissioned.

DARIDA 4/2004, Giorgio Ghiglione / 1044378

IHS Jane's Fighting Ships 2014-2015 © 2014 IHS

Government maritime forces — Italy

24 BURATTI (N 23) CLASS (PB)

BURATTI G 200	CASOTTI G 208	VERDECCHIA G 216
DE IANNI G 201	PRATA G 209	DE SANTIS G 217
SALERNO G 202	MARRA G 210	PICCINNI LEOPARDI G 218
ROSSI G 203	GOTTARDI G 211	BIANCO G 219
GARULLI G 204	LA PICCIRELLA G 212	STARACE G 220
SANGES G 205	PERISSINOTTO G 213	CULTRONA G 221
CORRIAS G 206	ROCCA G 214	BENVENUTI G 222
CORTILE G 207	BERTOLDI G 215	SANGES G 223

Displacement, tonnes: 56 full load
Dimensions, metres (feet): 22 × 5.4 × 2.0 (72.2 × 17.7 × 6.6)
Speed, knots: 33
Range, n miles: 850 at 25 kt
Complement: 8
Machinery: 2 MTU 12V2000-M93 diesels; 3,595 hp(m) (2.68 MW) sustained; 2 shafts
Radars: Surface search: To be announced.

Comment: A new class of craft to replace the Meatini class ordered from Intermarine in September 2006. The first two craft commissioned in March 2008.

CORRIAS   3/2013*, Bob Fildes / 1529066

22 V 5000 CLASS (FAST PATROL CRAFT) (HSIC)

V 5000–5020   V 5100

Displacement, tonnes: 27 full load
Dimensions, metres (feet): 16.46 × 4.55 × 0.8 (54 × 14.9 × 2.6)
Speed, knots: 52
Complement: 5
Machinery: 2 MTU 8V 396 TE94 diesels; 2,500 hp (1.84 MW); 2 waterjets
Radars: Surface search: I-band.

Comment: Delivered 1992–2002.

V 5006   6/2001, Guardia di Finanzia / 0130143

62 V 2000 (N 61) CLASS (FAST PATROL CRAFT) (PBF)

V 2000–2051   V 2056–2057   V 2071–2078

Displacement, tonnes: 11 full load
Dimensions, metres (feet): 13.2 × 3.4 × 0.9 (43.3 × 11.2 × 3.0)
Speed, knots: 45
Range, n miles: 380 at 33 kt
Complement: 4
Machinery: 2 Seatek 600 diesels; 1,240 hp(m) (925 kW) sustained; 2 Kamewa waterjets
Radars: Surface search: GEM SC412; I-band.

Comment: Constructed by Intermarine. After an initial batch of 14 craft, a further 48 craft were ordered for a total of 62 craft.

V 2072   3/2013*, Bob Fildes / 1529065

0 + 2 DAMEN STAN PATROL 5509 (OFFSHORE PATROL SHIPS) (PSO)

MONTE SPERONE P 01   MONTE CIMONE P 02

Displacement, tonnes: 460 full load
Dimensions, metres (feet): 55 × 9.0 × ? (180.4 × 29.5 × ?)
Speed, knots: 28
Complement: 27
Machinery: 2 Caterpillar diesels; 2 shafts; 2 bow thrusters
Guns: 1 — 12.7 mm MG.
Radars: Surface search: To be announced.
Navigation: To be announced.

Comment: The contract for the construction of two Damen 5509 offshore patrol vessels was announced on 22 May 2013. The ships are under construction at Adria Shipyards, Rovigo. The design includes a 'sea-axe' bow and a work-deck aft from which two long-range interception RHIBs can be launched and recovered.

10 V 6000 CLASS (INTERCEPTION CRAFT) (PBF)

V 6003–6012

Displacement, tonnes: 16 full load
Dimensions, metres (feet): 16.43 × 2.84 × 0.8 (53.9 × 9.3 × 2.6)
Speed, knots: 70
Complement: 4
Machinery: 4 Seatek 6 4V 10D diesels; 2,856 hp (2.13 MW); 4 surface-piercing propellers
Radars: Surface search: To be announced.

Comment: Designed by FB Design, built by Intermarine, Sarzana, in 2002.

V 6006   6/2005, Marco Ghiglino / 1153209

8 V 3000 CLASS (INTERCEPTION CRAFT) (PBF)

V 3000–3007

Displacement, tonnes: 6 full load
Dimensions, metres (feet): 11.85 × 2.81 × 0.8 (38.9 × 9.2 × 2.6)
Speed, knots: 50
Complement: 4
Machinery: 2 Mercury Verado outboard motors; 700 hp (521 kW)
Radars: Surface search: To be announced.
Navigation: To be announced.

Comment: Designed and built by FB Design in 2011. Operated on the north Italian lakes.

COAST GUARD
(GUARDIA COSTIERA — CAPITANERIE DI PORTO)

Notes: This is a force which is affiliated with the Marina Militare under whose command it would be placed in an emergency. The Coast Guard denomination was given after the Sea Protection Law in 1988. The force is responsible for the Italian Maritime Rescue Co-ordination Centre (MRCC) in Rome and 13 sub-centres (MRSC). The SAR network consists of 109 stations, three air stations and one helicopter station. All vessels have a red diagonal stripe painted on the white hull and many are armed with 7.62 mm MGs. There are some 10,500 naval personnel including 1,200 officers of which about half are doing national service. Ranks are the same as the Navy.
(1) There are numerous high-speed 8–9 m RHIBs.
(2) Aircraft include seven Piaggio P 166 DL3-SEM and three ATR 42MP maritime patrol, four AW 139 helicopters and nine Agusta AB-412-CP helicopters.
(3) CP 210 and CP 211 are airboats used for SAR in the Venice Lagoon area.

AW 139   6/2011, Annati Collection / 1406754

4 CP 401 CLASS (PATROL VESSELS) (PB)

| ORESTE CAVALLARI CP 401 | WALTER FACHIN CP 403 |
| RENATO PENNETTI CP 402 | GAETANO MAGLIANO CP 404 |

Displacement, tonnes: 100 full load
Dimensions, metres (feet): 28.6 × 6.2 × 2.0 (93.8 × 20.3 × 6.6)
Speed, knots: 14
Complement: 13

Machinery: 4 Isotta-Fraschini ID 36 SS 8V 200 diesels; 3,520 hp (2.59 MW); 2 shafts
Guns: 2 — 7.62 mm MGs.
Radars: Surface search: To be announced.
Navigation: To be announced.

Comment: Constructed by CNR, Ancona, 1987–91.

IHS Jane's Fighting Ships 2014-2015

Italy > Government maritime forces — Jamaica > Introduction

5 CP 2201 CLASS (PATROL VESSELS) (PB)

CP 2201–2205

Displacement, tonnes: 15 full load
Dimensions, metres (feet): 12.57 × 3.64 (41.2 × 11.9)
Speed, knots: 26
Complement: 4
Machinery: 2 Iveco AIFO C78ENTM diesels; 791 hp (590 kW); 2 shafts
Radars: Surface search: To be announced.
Navigation: To be announced.

Comment: Constructed at Motormar, Palermo.

11 CP 814 CLASS (PATROL VESSELS) (PB)

CP 814–824

Displacement, tonnes: 12.5 full load
Dimensions, metres (feet): 11.9 × 4.1 (39 × 13.5)
Speed, knots: 17
Complement: 3
Machinery: 2 Cummins C 420 diesels; 885 hp (660 kW); 2 waterjets
Radars: Surface search: To be announced.
Navigation: To be announced.

Comment: Constructed by Cantieri Navale Vittoria, Adria, 1996–97.

12 CP 601 CLASS (PATROL VESSELS) (PB)

CP 601–612

Displacement, tonnes: 6.5 full load
Dimensions, metres (feet): 11.43 × 3.1 (37.5 × 10.2)
Speed, knots: 43
Complement: 4
Machinery: 2 Iveco AIFO FTP N 67 diesels; 938 hp (700 kW); 2 shafts
Radars: Surface search: To be announced.
Navigation: To be announced.

Comment: Constructed at FB Design 2006–08.

27 CP 801 CLASS (PATROL VESSELS) (PB)

CP 801–813   CP 863–871   CP 882–883   CP 890–892

Displacement, tonnes: 10 full load
Dimensions, metres (feet): 10.6 × 4.1 (34.8 × 13.5)
Speed, knots: 30
Complement: 3
Machinery: 2 Volvo Penta TAMD 74C diesels; 885 hp (660 kW); 2 waterjets
Radars: Surface search: To be announced.
Navigation: To be announced.

Comment: Constructed by Cantieri Navale Codecasa, Viareggio, 1992–2001.

12 CP 701 CLASS (PATROL VESSELS) (PB)

CP 701–712

Displacement, tonnes: 6 full load
Dimensions, metres (feet): 9.6 × 3.15 (31.5 × 10.3)
Speed, knots: 43
Complement: 4
Machinery: 2 Iveco AIFO 8061 SRM diesels; 590 hp (440 kW); 2 waterjets
Radars: Surface search: To be announced.
Navigation: To be announced.

Comment: Constructed at Novomarine Due, Olbia, 1992–95.

CP 701   8/2013*, Italian Coast Guard / 1517291

72 CP 506 CLASS (PATROL VESSELS) (PB)

CP 506   CP 512   CP 514–583

Displacement, tonnes: 6.5 full load
Dimensions, metres (feet): 9.73 × 3.5 (31.9 × 11.5)
Speed, knots: 34
Complement: 3
Machinery: 2 Isotta Fraschini ID 32 SS-6 LM diesels; 560 hp (410 kW); 2 waterjets
Radars: Surface search: To be announced.
Navigation: To be announced.

Comment: Constructed at Cantieri Navali del Golfo, Gaeta, Cantieri Tencara, Venezia and Cantieri Stanisci, Taranto, 1998–2002.

14 CP 713 CLASS (PATROL VESSELS) (PB)

CP 713–726

Displacement, tonnes: 6.7 full load
Dimensions, metres (feet): 9.85 × 3.3 (32.3 × 10.8)
Speed, knots: 43
Complement: 2
Machinery: 2 Volvo Penta D6 330 diesels; 651 hp (486 kW); 2 waterjets
Radars: Surface search: To be announced.
Navigation: To be announced.

Comment: Constructed at Arimar Solemar.

POLICE (SERVIZIO NAVALE CARABINIERI)

Notes: (1) The Carabinieri established its maritime force in 1969 and has some 600 personnel. There are 172 craft in service or building which operate in coastal waters within the 3 mile limit and in inshore waters. Craft currently in service include: 27—800 class of 28 tons; 6—700 class of 15 tons; 30—600 class of 12 tons; 30 N 500 class of 6 tons; 3 S 500 class of 18 tons; 74—200 class of 2 tons, 28 minor craft and 30 RHIBs.
Most are capable of 20 to 25 kt except the 800 class at 35 kt.
(2) There is also a Sea Police Force of the State. All craft have POLIZIA written on the side. Vessels include 37 Squalo class of 14 tons, 4 Nelson class of 11 tons, 7 Intermarine class of 8.4 tons, 37 Crestitalia class of 6 tons and 25 Aquamaster/Drago classes of 3 tons. Speeds vary between 23 and 45 kt.

809   11/2007, Marco Ghiglino / 1170072

820   9/2005, P Marsan / 1153235

Jamaica

Country Overview

Jamaica gained independence in 1962; the British monarch, represented by a governor-general, is head of state. The island country (area 4,244 square miles), third-largest of the Greater Antilles, is situated south of Cuba and has a 552 n mile coastline with the Caribbean Sea. Kingston is the capital, largest town and principal port. An archipelagic state, territorial seas (12 n miles) are claimed. A 200 n mile Exclusive Economic Zone (EEZ) has been claimed but the limits are not fully defined.

Headquarters Appointments

Commanding Officer Coast Guard:
Commander D P Chin-Fong

Personnel

(a) 2014: 290 (22 officers) Regulars
(b) 52 (8 officers) Reserve Forces

Aviation

Seven helicopters (three Bell 412 and four Bell 407) are used for SAR and land operations. One Cessna 210M is used for liaison duties. Two Diamond DA 40 fixed-wing aircraft and one Bell 406B are used for training at the Flight Training School.

Bases

Headquarters: HMJS *Cagway*, Port Royal.
Bases: Discovery Bay, Pedro Cays, Port Antonio, Port Morant, Montego Bay, and Black River.

COAST GUARD

3 COUNTY (DAMEN STAN PATROL 4207) CLASS (PB)

Name	No	Builders	Commissioned
CORNWALL	421	Damen Shipyard, Gorinchem	27 Oct 2005
MIDDLESEX	422	Damen Shipyard, Gorinchem	7 Apr 2006
SURREY	423	Damen Shipyard, Gorinchem	26 Jun 2007

Displacement, tonnes: 208 standard
Dimensions, metres (feet): 42.8 × 7.11 × 2.52 (140.4 × 23.3 × 8.3)
Speed, knots: 26
Complement: 18 (4 officers)
Machinery: 2 Caterpillar 3516B DI-TA; 5,600 hp (4.17 MW); 2 cp props
Guns: 2—12.7 mm MGs.

Comment: Contract signed on 21 April 2004 with Damen Shipyard Gorinchem for construction of three Damen 4207 offshore patrol craft. Details are based on those in UK Customs service.

SURREY (on trials) *11/2006, Martyn Westers* / 1164414

1 HERO CLASS (PB)

Name	No	Builders	Commissioned
PAUL BOGLE	P 8	Lantana Boatyard Inc, Lake Worth	17 Sep 1985

Displacement, tonnes: 94 full load
Dimensions, metres (feet): 32 × 6.3 × 2.1 (105 × 20.7 × 6.9)
Speed, knots: 32
Complement: 20 (4 officers)

Machinery: 3 MTU 8V 396 TB93 diesels; 3,270 hp(m) (2.4 MW) sustained; 3 shafts
Guns: 1 Oerlikon 20 mm. 2—12.7 mm MGs.
Radars: Surface search: Furuno 2400; I-band.
Navigation: Sperry 4016; I-band.

Comment: Of all-aluminium construction, launched in 1984. *Paul Bogle* was originally intended for Honduras as the third of the Guardian class. Similar to patrol craft in Honduras and Grenada navies. Refitted in March 1998 at Network Marine, Louisiana and further refitted in 2004–05 by Damen Shipyards, Gorinchem.

PAUL BOGLE *6/1999, JDFCG* / 0080126

4 FAST COASTAL INTERCEPTORS (PBF)

CG 134–137

Displacement, tonnes: 8 full load
Dimensions, metres (feet): 13.4 × 2.75 × 0.9 (44 × 9.0 × 3.0)
Speed, knots: 45. **Range, n miles:** 600 at 25 kt
Complement: 4
Machinery: 3 Yanmar diesels; 945 hp (704 kW); Bravo X drives

Comment: Manufactured by Nor-Tech, Fort Myers, Florida. Composite and glass-fibre hull with V-bottomed hull. The first two donated by the US Southern Command in February 2008 and the second two in October 2008. Employed on counter drugs duties.

CG 134 *6/2008, Jamaica Coast Guard* / 1335391

3 FAST COASTAL INTERCEPTORS (PBF)

CG 131–133

Displacement, tonnes: 11 full load
Dimensions, metres (feet): 13.4 × 3.2 × 0.92 (44 × 10.5 × 3.0)
Speed, knots: 37
Range, n miles: 400 at 20 kt
Complement: 6
Machinery: 2 Caterpillar 3196 diesels; 1,140 hp (850 kW); two twin disc waterjets
Guns: 1—7.62 mm M60 MG.
Radars: Surface search: Raytheon Pathfinder; I-band.

Comment: Aluminium construction. Built by Silver Ships, Mobile, Alabama. Funded by the US State Department, Narcotics Affairs Section. Delivered in March 2003.

CG 131 *6/2003, JDFCG* / 0568335

4 DAUNTLESS CLASS (INSHORE PATROL CRAFT) (PB)

CG 121–124

Displacement, tonnes: 11 full load
Dimensions, metres (feet): 12.2 × 4.3 × 1.3 (40 × 14.1 × 4.3)
Speed, knots: 27
Range, n miles: 600 at 18 kt
Complement: 5
Machinery: 2 Caterpillar 3208TA diesels; 870 hp (650 kW); 2 shafts
Guns: 1—7.62 mm MG (can be carried).
Radars: Surface search: Raytheon 40X; I-band.

Comment: Delivered in September and November 1992, January 1993 and May 1994. Built by SeaArk Marine, Monticello. Aluminium construction. Craft of this class have been distributed throughout the Caribbean under FMS funding. Two craft were refitted during 2006.

CG 123 *5/2010*, US Navy* / 1525528

Japan
MARITIME SELF-DEFENCE FORCE (MSDF)
KAIJOU JIEI-TAI

Country Overview
Japan is a constitutional monarchy in East Asia that comprises four main islands: Hokkaido, Honshu, Shikoku and Kyushu. It also includes the Ryukyu Islands to the southwest and more than 1,000 lesser islands. The sovereignty of the South Kuril Islands (Etorofu, Kunashiri, Shikotan and the Habomai Group) is disputed with Russia. With an overall area of 145,850 square miles it has a coastline of 16,065 n miles, with the Pacific Ocean, Sea of Japan, the La Perouse Strait (which separates it from Sakhalin Island), Sea of Okhotsk, East China Sea and the Korea Strait (which separates it from South Korea). The capital and largest city is Tokyo while the principal ports are Yokohama, Osaka and Kobe. Territorial seas of 12 n miles (3 n miles in Korea Strait) are claimed. A 200 n mile EEZ has also been claimed but the limits have not been defined.

Headquarters Appointments
Chief of Staff, Maritime Self-Defence Force:
  Admiral Katsutoshi Kawano
Commander-in-Chief, Self-Defence Fleet:
  Vice Admiral Yasushi Matsushita

Senior Appointments
Commander Fleet Escort Force:
  Vice Admiral Toshihiro Ikeda
Commander Fleet Air Force:
  Vice Admiral Yasuhiro Shigeoka
Commander Fleet Submarine Force:
  Vice Admiral Masakazu Kaji

Diplomatic Representation
Defence (Naval) Attaché in London:
  Captain Atsushi Minami

Personnel
2014: 45,517 (including Naval Air) plus 3,269 civilians

Organisation of the Major Surface Units of Japan (MSDF)
The Fleet Escort and Air Forces were reorganised on 26 March 2008. There are also two Submarine Flotillas (Kure and Yokosuka), one Minesweeper Flotilla (Yokosuka), one Transport Command (Kure), one Sea Supply Command (Yokosuka) and five District Flotillas (Yokosuka, Kure, Sasebo, Maizuru and Oominato). The District Flotillas are composed of an AMS and a number of MSC and patrol craft. The Fleet Training Group is based at Kure.

Fleet Escort Force (Yokosuka)
Escort Flotilla 1 (Yokosuka)
Escort Division 1 (Y)
Hyuga (DDH 181) (Y)
Shimakaze (DDG 172) (S)
Murasame (DD 101) (Y)
Ikazuchi (DD 107) (Y)
Escort Division 5 (S)
Kongou (DDG 173) (S)
Akebono (DD 108) (S)
Akizuki (DD 115) (S)
Sawagiri (DD 157) (S)

Escort Flotilla 2 (Sasebo)
Escort Division 2 (S)
Kurama (DDH 144) (S)
Ashigara (DDG 178) (S)
Harusame (DD 102) (Y)
Amagiri (DD 154) (S)
Escort Division 6 (S)
Choukai (DDH 176) (S)
Takanami (DD 110) (Y)
Oonami (DD 111) (Y)
Teruzuki (DD 116) (Y)

Escort Flotilla 3 (Maizuru)
Escort Division 3 (O)
Shirane (DDH 143) (M)
Atago (DDG 177) (M)
Makinami (DD 112) (O)
Suzunami (DD 114) (O)
Escort Division 7 (M)
Myoukou (DDG 175) (M)
Yuudachi (DD 103) (S)
Ariake (DD 109) (S)
Setogiri (DD 156) (O)

Escort Flotilla 4 (Kure)
Escort Division 4 (K)
Ise (DDH 182) (K)
Hatakaze (DDG 171) (Y)
Umigiri (DD 158) (K)
Samidare (DD 106) (K)
Escort Division 8 (K)
Kirishima (DDG 174) (Y)
Inazuma (DD 105) (K)
Sazanami (DD 113) (K)
Kirisame (DD 104) (S)
Escort Division 11 (Y)
Yamagiri (DD 129)
Yamagiri (DD 152)
Yuugiri (DD 153) (O)
Escort Division 12 (K)
Abukuma (DE 229)
Sendai (DE 232)
Tone (DE 234)
Escort Division 13 (S)
Isoyuki (DD 127)
Harayuki (DD 128)
Asayuki (DD 132)
Jintsu (DE 230)
Escort Division 14 (M)
Asagiri (DD 151)
Matsuyuki (DD 130)
Escort Division 15 (O)
Ooyodo (DE 231)
Chikuma (DE 233)
Hamagiri (DE 230)

Bases
Naval-Yokosuka, Kure, Sasebo, Maizuru, Ohminato
Naval Air-Atsugi, Hachinohe, Iwakuni, Kanoya, Komatsujima, Naha, Ozuki, Oominato, Ohmura, Shimofusa, Minami-Torishima, Tokushima, Ioujima, Maizuru

Coast Defence
The Army controls approximately 100 SSM-1 truck-mounted sextuple launchers.

Strength of the Fleet (31 March 2014)

Type	Active (Auxiliary)	Building (Projected)
Submarines	16 (2)	4 (1)
Helicopter carriers	2	2
Destroyers	41	1 (3)
Frigates	6	—
Patrol Forces	6	—
LSTs	3	—
LCUs	2	—
LCACs	6	—
Landing Craft (LCM)	12	—
MCM Tenders/Controllers	4	—
MCMV—Ocean	3	(1)
MCMV—Coastal	22	1
Training ships	4	—
Survey/Research ships	6	—
Major Auxiliaries	16	—

New Construction Programme (Warships)
2012 1 — 19,500 ton DDH, 1 — 2,900 ton SS
2013 1 — 5,000 ton DD, 1 — 2,900 ton SS, 1 — 690 ton MSO
2014 2 — 5,000 ton DD, 1 — 5,000 ton DD, 1 — 2,900 ton SS,
      1 — 5,600 ton ASR, 1 — 690 ton MSO

Naval Air Force
10 Air Patrol Sqns: P-3C, EP-3, OP-3C, SH-60J/K
Four Air Training Sqns: P-3C, YS-11, TC-90, T-5, OH-6D, SH-60J, TH-135
One Air Training Support Squadron: U-36A, UP-3D, LC-90
One Transport Sqn: YS-11, LC-90
One MCM Sqn: MH-53E, MCH-101
Fleet Air Force (Atsugi)
Air Training Command (Shimofusa)
Air Wings at Kanoya (Wing 1), Hachinohe (Wing 2), Atsugi (Wing 4), Naha (Wing 5), Tateyama (Wing 21), Ohmura (Wing 22), Iwakuni (Wing 31)

DELETIONS AND CONVERSIONS

Submarines
2011 *Hayashio, Fuyushio* (converted)
2012 *Arashio*
2013 *Wakashio*

Destroyers
2011 *Hiei, Shirayuki* (converted)
2012 *Hamayuki, Setoyuki* (converted)
2013 *Sawayuki, Mineyuki*

Amphibious Forces
2012 *Noto*
2013 *Yura*

Mine Warfare Forces
2012 *Tsukishima*
2013 *Sakushima, Maejima* (converted)

Survey and Research Vessels
2012 *Kurihama*

Training Ships
2011 *Yamagiri* (re-converted to destroyer)
2012 *Asagiri* (re-converted to destroyer)

Auxiliaries
2012 *Muroto*
2013 *Kurihama*

PENNANT LIST

Submarines—Patrol
SS 501	Souryu
SS 502	Unryu
SS 503	Hakuryu
SS 504	Kenryu
SS 505	Zuiryu
SS 506	Kokuryu (bldg)
SS 507	— (bldg)
SS 508	— (bldg)
SS 590	Oyashio
SS 591	Michishio
SS 592	Uzushio
SS 593	Makishio
SS 594	Isoshio
SS 595	Narushio
SS 596	Kuroshio
SS 597	Takashio
SS 598	Yaeshio
SS 599	Setoshio
SS 600	Mochishio

Submarines—Auxiliary
TSS 3601	Asashio
TSS 3607	Fuyushio

Helicopter Carriers
DDH 181	Hyuga
DDH 182	Ise
DDH 183	Izumo (bldg)
DDH 184	— (bldg)

Destroyers
DD 101	Murasame
DD 102	Harusame
DD 103	Yuudachi
DD 104	Kirisame
DD 105	Inazuma
DD 106	Samidare
DD 107	Ikazuchi
DD 108	Akebono
DD 109	Ariake
DD 110	Takanami
DD 111	Oonami
DD 112	Makinami
DD 113	Sazanami
DD 114	Suzunami
DD 115	Akizuki
DD 116	Teruzuki
DD 117	Suzutsuki (bldg)
DD 118	Fuyuzuki (bldg)
DD 119	— (bldg)
DD 127	Isoyuki
DD 128	Haruyuki
DD 129	Yamayuki
DD 130	Matsuyuki
DD 132	Asayuki
DDH 143	Shirane
DDH 144	Kurama
DD 151	Asagiri
DD 152	Yamagiri
DD 153	Yuugiri
DD 154	Amagiri
DD 155	Hamagiri
DD 156	Setogiri
DD 157	Sawagiri
DD 158	Umigiri
DDG 171	Hatakaze
DDG 172	Shimakaze
DDG 173	Kongou
DDG 174	Kirishima
DDG 175	Myoukou
DDG 176	Choukai
DDG 177	Atago
DDG 178	Ashigara

Frigates
DE 229	Abukuma
DE 230	Jintsu
DE 231	Ooyodo
DE 232	Sendai
DE 233	Chikuma
DE 234	Tone

Patrol Forces
PG 824	Hayabusa
PG 825	Wakataka
PG 826	Ootaka
PG 827	Kumataka
PG 828	Umitaka
PG 829	Shirataka

Minehunters/Sweepers—Ocean
MSO 301	Yaeyama
MSO 302	Tsushima
MSO 303	Hachijyo
MSO 304	— (bldg)

Minesweepers—Coastal
MSC 601	Hirashima
MSC 602	Yakushima
MSC 603	Takashima
MSC 604	Enoshima
MSC 605	Chichijima
MSC 606	Hatsushima (bldg)
MSC 676	Kumejima
MSC 677	Makishima
MSC 678	Tobishima
MSC 679	Yugeshima
MSC 680	Nagashima
MSC 681	Sugashima
MSC 682	Notojima
MSC 683	Tsunoshima
MSC 684	Naoshima
MSC 685	Toyoshima
MSC 686	Ukushima
MSC 687	Izushima
MSC 688	Aishima
MSC 689	Aoshima
MSC 690	Miyajima
MSC 691	Shishijima
MSC 692	Kuroshima

MCM Tenders/Control Ships
MCL 728	Ieshima
MCL 729	Maejima
MST 463	Uraga
MST 464	Bungo

Amphibious Forces
LCU 2001	Yusotei-Ichi-Go
LCU 2002	Yusotei-Ni-Go
LST 4001	Oosumi
LST 4002	Shimokita
LST 4003	Kunisaki

Submarine Depot/Rescue Ships
AS 405	Chiyoda
ASR 403	Chihaya

Fleet Support Ships
AOE 422	Towada
AOE 423	Tokiwa
AOE 424	Hamana
AOE 425	Mashuu
AOE 426	Oumi

Training Ships
TV 3508	Kashima
TV 3513	Shimayuki
TV 3517	Shirayuki
TV 3518	Setoyuki

Training Support Ships
ATS 4202	Kurobe
ATS 4203	Tenryu
AMS 4301	Hiuchi
AMS 4302	Suou
AMS 4303	Amakusa
AMS 4304	Genkai
AMS 4305	Enshuu

Cable Repair Ship
ARC 483	Muroto

Icebreakers
AGB 5003	Shirase

Survey and Research Ships
AGS 5103	Suma
AGS 5104	Wakasa
AGS 5105	Nichinan
AGS 5106	Syounan
ASE 6102	Asuka

Ocean Surveillance Ships
AOS 5201	Hibiki
AOS 5202	Harima

Tenders
ASY 91	Hashidate
YDT 01–06	—

Submarines < **Japan** (MSDF) 429

SUBMARINES

5 + 4 (1) SOURYU CLASS (SSK)

Name	No	Builders	Laid down	Launched	Commissioned
SOURYU	SS 501	Mitsubishi, Kobe	31 Mar 2005	5 Dec 2007	30 Mar 2009
UNRYU	SS 502	Kawasaki	31 Mar 2006	15 Oct 2008	25 Mar 2010
HAKURYU	SS 503	Mitsubishi, Kobe	6 Feb 2007	16 Oct 2009	14 Mar 2011
KENRYU	SS 504	Kawasaki, Kobe	31 Mar 2008	15 Nov 2010	16 Mar 2012
ZUIRYU	SS 505	Mitsubishi, Kobe	16 Mar 2009	20 Oct 2011	6 Mar 2013
KOKURYU	SS 506	Kawasaki, Kobe	21 Jan 2011	31 Oct 2013	Mar 2015
–	SS 507	Mitsubishi, Kobe	14 Feb 2012	Oct 2014	Mar 2016
–	SS 508	Kawasaki, Kobe	15 Mar 2013	Oct 2015	Mar 2017
–	SS 509	Mitsubishi, Kobe	2014	2016	2018

Displacement, tonnes: 2,947 surfaced; 4,100 dived
Dimensions, metres (feet): 84 × 9.1 × 10.3 *(275.6 × 29.9 × 33.8)*
Speed, knots: 12 surfaced; 20 dived
Complement: 70

Machinery: Diesel-stirling-electric; 2 diesels; 4 Kockums Stirling AIP; 408 hp *(300 kW)*; 1 motor; 8,000 hp *(5.96 MW)*; 1 shaft
Missiles: SSM: McDonnell Douglas Sub-Harpoon; active radar homing to 130 km *(70 n miles)* at 0.9 Mach; warhead 227 kg.
Torpedoes: 6—21 in *(533 mm)* bow tubes. Japanese Type 89; wire-guided (option); active/passive homing to 50 km *(27 n miles)* at 40/55 kt; warhead 267 kg. Type 80 ASW. SSM and torpedoes (total unknown).
Physical countermeasures: To be announced.
Radars: Surface search: JRC ZPS-6F; I-band.
Sonars: Hughes/OKI ZQQ 7; hull and flank arrays; active/passive search and attack; medium/low frequency. Towed array.
Weapon control systems: To be announced.

Programmes: First of new class authorised in fiscal year 2004 budget, second in FY 2005, third in FY 2006, fourth in FY 2007, fifth in FY 2008, sixth in FY 2010, seventh in FY 2011, eighth in FY 2012, and ninth in FY 2013 budget. A tenth has been proposed in FY 2014 budget.
Structure: The hull design is based on the Oyashio class and incorporates the Swedish Stirling air-independent propulsion system. Components of this system are provided by Kockums for assembly by KHI.

ZUIRYU 8/2012, Hachiro Nakai / 1482874

UNRYU 7/2011, Hachiro Nakai / 1406815

KENRYU 7/2013*, Hachiro Nakai / 1529127

430　Japan (MSDF) > Submarines

11 OYASHIO CLASS (SSK)

Name	No
OYASHIO	SS 590
MICHISHIO	SS 591
UZUSHIO	SS 592
MAKISHIO	SS 593
ISOSHIO	SS 594
NARUSHIO	SS 595
KUROSHIO	SS 596
TAKASHIO	SS 597
YAESHIO	SS 598
SETOSHIO	SS 599
MOCHISHIO	SS 600

Builders	Laid down	Launched	Commissioned
Kawasaki, Kobe	26 Jan 1994	15 Oct 1996	16 Mar 1998
Mitsubishi, Kobe	16 Feb 1995	18 Sep 1997	10 Mar 1999
Kawasaki, Kobe	6 Mar 1996	26 Nov 1998	9 Mar 2000
Mitsubishi, Kobe	26 Mar 1997	22 Sep 1999	29 Mar 2001
Kawasaki, Kobe	9 Mar 1998	27 Nov 2000	14 Mar 2002
Mitsubishi, Kobe	2 Apr 1999	4 Oct 2001	3 Mar 2003
Kawasaki, Kobe	27 Mar 2000	23 Oct 2002	8 Mar 2004
Mitsubishi, Kobe	30 Jan 2001	1 Oct 2003	9 Mar 2005
Kawasaki, Kobe	15 Jan 2002	4 Nov 2004	9 Mar 2006
Mitsubishi, Kobe	23 Jan 2003	5 Oct 2005	28 Feb 2007
Kawasaki, Kobe	23 Feb 2004	6 Nov 2006	6 Mar 2008

Displacement, tonnes: 2,794 surfaced; 3,556 dived
Dimensions, metres (feet): 81.7 × 8.9 × 7.4 *(268 × 29.2 × 24.3)*
Speed, knots: 12 surfaced; 20 dived
Complement: 70 (10 officers)

Machinery: Diesel-electric; 2 Kawasaki 12V25S diesels; 5,520 hp(m) *(4.1 MW)*; 2 Kawasaki alternators; 3.7 MW; 2 Toshiba motors; 7,750 hp(m) *(5.7 MW)*; 1 shaft
Missiles: SSM: McDonnell Douglas Sub-Harpoon; active radar homing to 130 km *(70 n miles)* at 0.9 Mach; warhead 227 kg.
Torpedoes: 6—21 in *(533 mm)* tubes; Type 89; wire-guided; active/passive homing to 50 km *(27 n miles)*/38 km *(21 n miles)* at 40/55 kt; warhead 267 kg and Type 80 ASW. Total of 20 SSM and torpedoes.
Electronic countermeasures: ESM: NZLR-1B; radar warning.
Radars: Surface search: JRC ZPS 6D; I-band.
Sonars: Hughes/Oki ZQQ 6; hull and flank arrays; active/passive search and attack; medium/low frequency. Towed array; passive search; very low frequency.
Weapon control systems: SMCS type TFCS.

Programmes: First of a new class approved in the 1993 budget and then one a year up to FY03.
Structure: Fitted with large flank sonar arrays which are reported as the reason for the increase in displacement over the Harushio class. Double hull sections forward and aft and anechoic tiles on the fin. A new type of deck casing and faired fin are other distinguishing features. Diving depth 650 m *(2,130 ft)*.

UZUSHIO　　　　　　　　　　　　　　　　　　3/2012, Hachiro Nakai / 1482871

ISOSHIO
7/2010, Jin Tetsuya
1366880

ISOSHIO　　　　　　　　　　　　　　　　　　10/2012, Hachiro Nakai / 1482870

2 HARUSHIO CLASS (SSK)

Name	No
FUYUSHIO	TSS 3607 (ex-SS 588)
ASASHIO	TSS 3601 (ex-SS 589)

Builders	Laid down	Launched	Commissioned
Kawasaki, Kobe	12 Dec 1991	16 Feb 1994	7 Mar 1995
Mitsubishi, Kobe	24 Dec 1992	12 Jul 1995	12 Mar 1997

Displacement, tonnes: 2,489 (TSS 3607), 2,947 (TSS 3601) surfaced; 3,251 (TSS 3607), 3,759 (TSS 3601) dived
Dimensions, metres (feet): 77, 87 (TSS 3601) × 10 × 7.7 *(252.6, 285.4 × 32.8 × 25.3)*
Speed, knots: 12 surfaced; 20 dived
Complement: 75, 70 (TSS 3601) (10 officers)

Machinery: Diesel-electric; 2 Kawasaki 12V25/25S diesels; 5,520 hp(m) *(4.1 MW)*; 2 Kawasaki alternators; 3.7 MW; 2 Fuji motors; 7,200 hp(m) *(5.3 MW)*; 1 shaft; 4 Stirling engines (TSS 3601) Kockums V4-275R Mk 2; 348 hp *(260 kW)*
Missiles: SSM: McDonnell Douglas Sub-Harpoon; active radar homing to 130 km *(70 n miles)* at 0.9 Mach; warhead 227 kg.
Torpedoes: 6—21 in *(533 mm)* tubes. Japanese Type 89; wire-guided (option); active/passive homing to 50 km *(27 n miles)*/38 km *(21 n miles)* at 40/55 kt; warhead 267 kg; depth to 900 m, and Type 80 ASW. Total of 20 SSM and torpedoes.
Electronic countermeasures: ESM: NZLR-1; radar warning.
Radars: Surface search: JRC ZPS 6; I-band.
Sonars: Hughes/Oki ZQQ 5B; hull-mounted; active/passive search and attack; medium/low frequency. ZQR 1 towed array similar to BQR 15; passive search; very low frequency.

Programmes: First approved in 1986 estimates and then one per year until 1992.

FUYUSHIO　　　　　　　　　　　　　　　　11/2013*, Hachiro Nakai / 1529128

Structure: The slight growth in all dimensions is a natural evolution from the Yuushio class and includes more noise reduction, towed sonar and wireless aerials, as well as anechoic coating. Double hull construction. *Asashio* had a slightly larger displacement on build and a small cutback in the crew as a result of greater systems automation for machinery and snorting control. The hull was extended in 2001 to accommodate an AIP module (Stirling engine) which was fitted by Mitsubishi, Kobe. Diving depth 550 m *(1,800 ft)*.
Operational: *Asashio* is an experimental submarine which has been used for testing of AIP propulsion. *Fuyushio* converted to a training role on 15 March 2011. *Hayashio* decommissioned in 2011, *Arashio* in 2012 and *Wakashio* in 2013.

HELICOPTER CARRIERS

2 HYUGA CLASS (CVHG)

Name	No	Builders	Laid down	Launched	Commissioned
HYUGA	DDH 181	IHI Marine United, Yokohama	11 May 2006	23 Aug 2007	18 Mar 2009
ISE	DDH 182	IHI Marine United, Yokohama	30 May 2008	21 Aug 2009	16 Mar 2011

Displacement, tonnes: 13,950 standard; 18,289 full load
Dimensions, metres (feet): 197 × 33.0 × 9.7 (646.3 × 108.3 × 31.8)
Aircraft lift: 2
Speed, knots: 30. **Range, n miles:** 6,000 at 20 kt
Complement: 347 + 25 flag staff

Machinery: COGAG; 4 LM 2500 gas turbines; 2 shafts
Missiles: SAM: Lockheed Martin Marietta Mk 41 Mod 5 sixteen cell vertical launcher ❶; Raytheon Sea Sparrow RIM-162 ESSM; semi-active radar homing to 18.0 km (9.7 n miles) at 3.6 Mach; warhead 38 kg. 16 missiles.
A/S: Vertical launch ASROC; inertial guidance to 1.6–10 km (1–5.4 n miles); payload Mk 46 Mod 5 Neartip. 12 rounds.
Guns: 2 Raytheon 20 mm/76 Vulcan Phalanx Block 1B ❷; 4,500 rds/min combined to 1.5 km. 7 – 12.7 mm MGs.
Torpedoes: 6 – 324 mm (2 triple) HOS-303 tubes ❸.
Physical countermeasures: Decoys: 4 Hycor Mk 137 sextuple RBOC chaff launchers ❹.
Electronic countermeasures: ESM/ECM: NOLQ-3C ❺.
Radars: Air search/Fire control: Melco FCS-3 ❻; G/H/I-band. Navigation: JRC OPS-20C ❼; I-band.
Sonars: Bow-mounted sonar. OQQ 21.
Combat data systems: Link 16.

Helicopters: 3 SH-60K plus 7 SH-60K or 7 MCH-101.

Programmes: Two new aviation capable ships to replace the Haruna class authorised in the FY01–05 and FY05–09 programmes. The first authorised in the FY04 budget and the second in the FY06 budget.
Structure: Broadly similar to the Spanish light carrier *Principe de Asturias* although not fitted with a ski jump and VSTOL capability. The flight deck has two lifts and four helicopter spots. The Mk 41 VLS launcher is situated on the starboard quarter.
Operational: Capable of acting as Command Vessels.

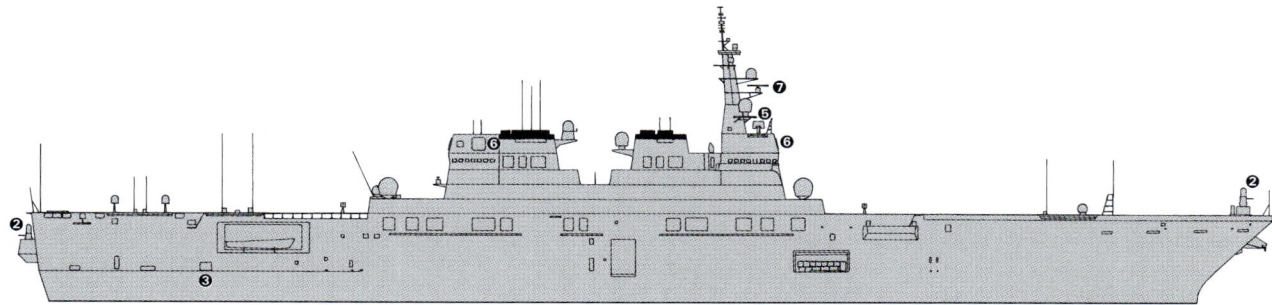

HYUGA (Scale 1 : 1,200), Ian Sturton / 1366152

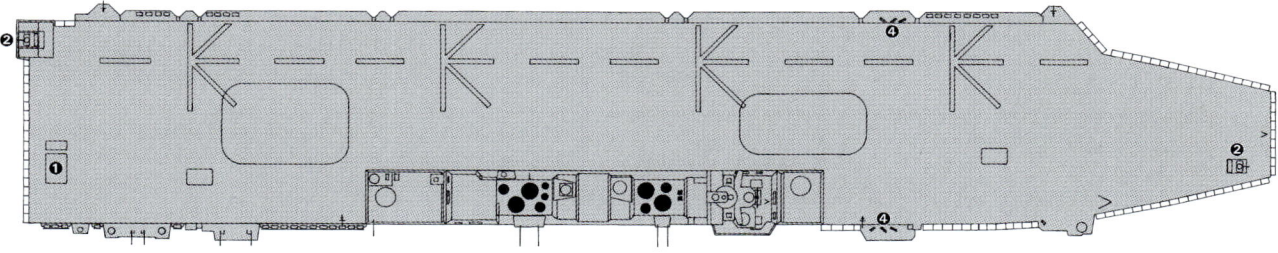

HYUGA (Scale 1 : 1,200), Ian Sturton / 1366153

HYUGA 11/2012, Hachiro Nakai / 1482869

ISE 9/2013*, Hirotoshi Yamamoto / 1529129

432 Japan (MSDF) > Helicopter carriers — Destroyers

0 + 2 IZUMO CLASS (HELICOPTER CARRIERS) (CVHG)

Name	No	Builders	Laid down	Launched	Commissioned
IZUMO	DDH 183	IHI Marine United, Yokohama	27 Jan 2012	6 Aug 2013	Mar 2015
—	DDH 184	IHI Marine United, Yokohama	Jan 2014	Aug 2015	Mar 2017

Displacement, tonnes: 19,813 standard; 24,000 full load
Dimensions, metres (feet): 248 × 38 × 7.3
(813.6 × 124.7 × 24.0)
Speed, knots: 30
Complement: 470

Machinery: COGAG: 4 LM 2500 gas turbines; 112,000 hp *(83.5 MW)*; 2 shafts
Missiles: SAM: 2 Raytheon Sea RAM RIM-116.
Guns: 2 GE 20 mm/76 Mk 15 Vulcan Phalanx.
Torpedoes: To be announced.
Physical countermeasures: To be announced.
Electronic countermeasures: ESM/ECM: NOLQ-3C.
Radars: Air/surface search: OPS-50.
Surface search: OPS-28.
Sonars: OQQ-22. Bow mounted.
Combat data systems: To be announced.
Electro-optic systems: To be announced.

Helicopters: 7 SH-60K. 7 MCH-101.

Programmes: In September 2009, the Ministry of Defence proposed the construction of two 'helicopter-capable destroyers'. Significantly larger than the Hyuga class, they will become the largest ships in the Japanese Navy. A request for the first of the new helicopter carriers, known as 22DDH, was authorised in the fiscal year 2010 budget and a second in FY 2012. The ships are to replace the Shirane-class destroyers, which were built in the 1970s.

IZUMO 8/2013*, Tetsuya Kakitani / 1517294

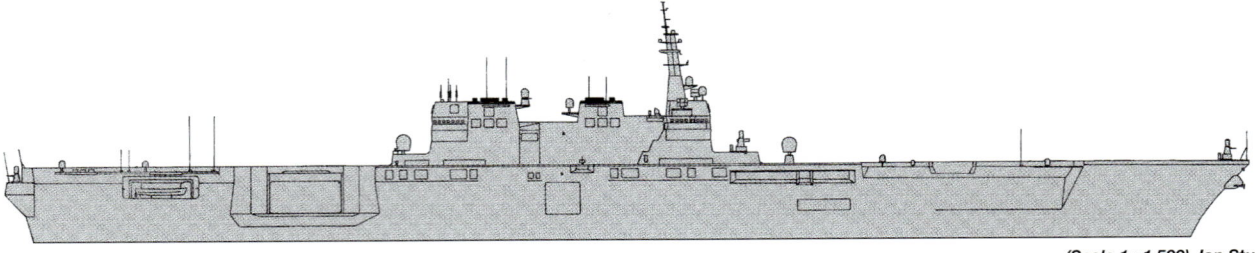

IZUMO (Scale 1 : 1,500), Ian Sturton / 1528809

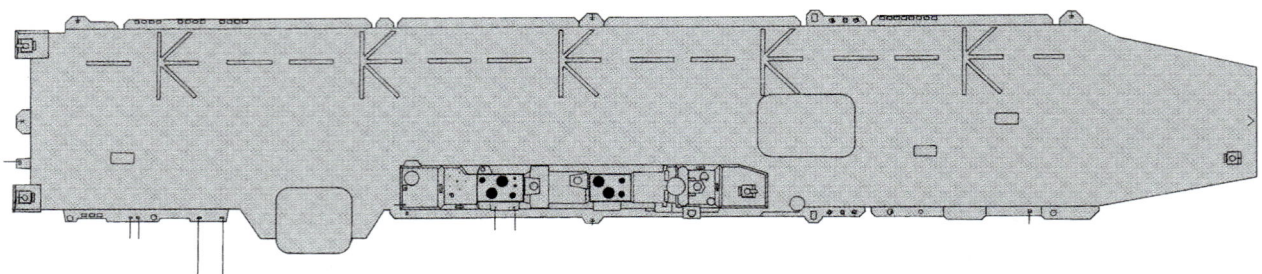

IZUMO (Scale 1 : 1,500), Ian Sturton / 1528810

DESTROYERS

2 + (2) ATAGO CLASS (DDGHM)

Name	No	Builders	Laid down	Launched	Commissioned
ATAGO	DDG 177	Mitsubishi, Nagasaki	5 Apr 2004	24 Aug 2005	15 Mar 2007
ASHIGARA	DDG 178	Mitsubishi, Nagasaki	6 Apr 2005	30 Aug 2006	13 Mar 2008

Displacement, tonnes: 7,824 standard; 10,160 full load
Dimensions, metres (feet): 164.9 × 21.0 × 6.2
(541 × 68.9 × 20.3)
Speed, knots: 30. **Range, n miles:** 4,500 at 20 kt
Complement: 309 (27 officers)

Machinery: COGAG; 4 GE LM 2500 gas turbines; 102,160 hp *(76.21 MW)* sustained; 2 shafts; cp props
Missiles: SSM: 8 Mitsubishi Type 90 SSM-1B (2 quad) ❶; active radar homing to 200 km *(108 n miles)* at 0.9 Mach; warhead 270 kg.
SAM: Raytheon Standard SM-2MR Block IIIB. FMC Mk 41 VLS; 64 cells forward ❷ 32 cells aft ❸; command/inertial guidance; semi-active radar homing to 167 km *(90 n miles)* at 2.5 Mach.
A/S: Vertical launch ASROC; inertial guidance to 1.6–10 km *(1–5.4 n miles)*; payload Mk 46 Mod 5 Neartip.
Guns: 1 United States Mk 45 Mod 4 5 in *(127 mm)*/62 ❹; 20 rds/min to 23 km *(12.6 n miles)*; weight of shell 32 kg.
2 Raytheon 20 mm/76 Mk 15 Vulcan Phalanx Block IB ❺; 4,500 rds/min combined to 1.5 km.
Torpedoes: 6—324 mm (2 triple) HOS 302 tubes ❻. Honeywell Mk 46 Mod 5 Neartip; anti-submarine; active/passive homing to 11 km *(5.9 n miles)* at 40 kt; warhead 44 kg.
Physical countermeasures: Decoys: 4 Mk 36 SRBOC ❼ 6-barrelled Mk 36 chaff launchers; Type 4 towed torpedo decoy.
Electronic countermeasures: ESM/ECM: NOLQ-2B ❽.
Radars: Air search: RCA SPY 1D(V) ❾; 3D; F-band.
Surface search: JRC OPS-28E ❿; G-band.
Navigation: JRC OPS-20; I-band.
Fire control: 3 SPG-62 ⓫; 1 FCS 2/21 ⓬; I/J-band. 2 AN/UPS-2; J-band.
Sonars: SQS-53C bow sonar.
Combat data systems: Aegis NTDS with Link 11. AN/USC-42 SATCOM.

Helicopters: 1 Mitsubishi/Sikorsky SH-60J/K ⓭.

Programmes: Two ships authorised in the fiscal year 2001–2005 programme. The first authorised in the FY 2002 budget and the second in FY 2003 budget. Two further ships, possibly to a modified design, are likely to be ordered. Both are likely to be equipped with a ballistic missile defence (BMD) capability.
Modernisation: Both ships are being upgraded to undertake BMD missions.
Structure: The upgrade from the Kongou class includes one hangar for embarked helicopters. The arrangement of vertical launchers is different to that of the Kongou class: there are 64 cells forward (rather than 29) and 32 aft (61).
Operational: Atago sailed for sea trials on 17 May 2006.

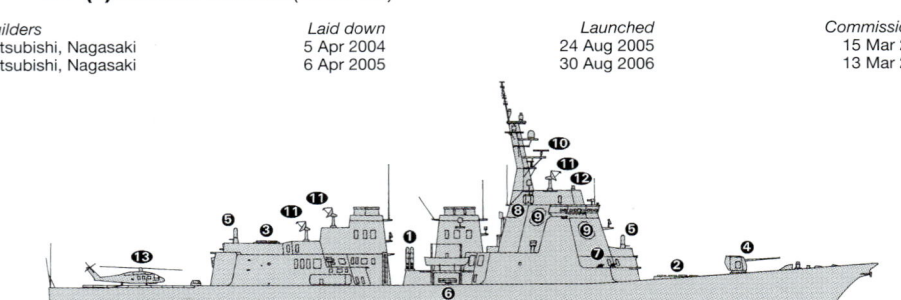

ATAGO (Scale 1 : 1,500), Ian Sturton / 1167684

ASHIGARA 7/2013*, Ryoji Okimoto / 1529130

IHS Jane's Fighting Ships 2014-2015 © 2014 IHS

ASHIGARA   4/2011, Hachiro Nakai / 1406819

2 SHIRANE CLASS (DDHM)

Name	No	Builders	Laid down	Launched	Commissioned
SHIRANE	DDH 143	Ishikawajima Harima, Tokyo	25 Feb 1977	18 Sep 1978	17 Mar 1980
KURAMA	DDH 144	Ishikawajima Harima, Tokyo	17 Feb 1978	20 Sep 1979	27 Mar 1981

Displacement, tonnes: 5,283 standard; 7,316 full load
Dimensions, metres (feet): 159 × 17.5 × 5.3 *(521.7 × 57.4 × 17.4)*
Speed, knots: 32 (DDH 143), 31 (DDH 144)
Complement: 370 (DDH 143), 380 (DDH 144) + flag staff

Machinery: 2 IHI boilers; 850 psi *(60 kg/cm²)*; 900°F *(480°C)*; 2 IHI turbines; 70,000 hp(m) *(51.5 MW)*; 2 shafts
Missiles: SAM: Raytheon Sea Sparrow RIM-7M; Type 3 launcher ❶; semi-active radar homing to 16 km *(8.5 n miles)* at 2.5 Mach; warhead 38 kg; 24 missiles.
A/S: Honeywell ASROC Mk 112 octuple launcher ❷; inertial guidance to 10 km *(5.4 n miles)* at 0.9 Mach; payload Mk 46 Mod 5 Neartip.
Guns: 2 FMC 5 in *(127 mm)*/54 Mk 42 automatic ❸; 20–40 rds/min to 24 km *(13 n miles)* anti-surface; 14 km *(7.6 n miles)* anti-aircraft; weight of shell 32 kg. 2 Raytheon 20 mm Phalanx Mk 15 CIWS ❹; 6 barrels per mounting; 4,500 rds/min combined to 1.5 km.
Torpedoes: 6 – 324 mm HOS 301 (2 triple) tubes ❺. Honeywell Mk 46 Mod 5 Neartip; anti-submarine; active/passive homing to 11 km *(5.9 n miles)* at 40 kt; warhead 44 kg.
Physical countermeasures: Decoys: 4 Mk 36 SRBOC chaff launchers. Prairie Masker; blade rate suppression system.
Electronic countermeasures: ESM/ECM: Melco NOLQ 1; intercept/jammer. Fujitsu OLR 9B; intercept.

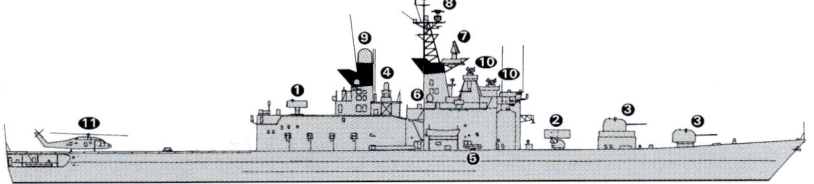

SHIRANE   (Scale 1 : 1,500), Ian Sturton / 1153010

Radars: Air search: Nec OPS-12 ❼; 3D; D-band.
Surface search: JRC OPS-28 ❽; G-band.
Navigation: JRC OPS-20; I-band.
Fire control: FCS 2-12 ❾; I/J-band. 2 FCS 72-1A ❿; I/J-band.
Tacan: ORN-6C/6C-Y.
Sonars: EDO/Nec SQS-35(J); VDS; active/passive search; medium frequency. Nec OQS 101; bow-mounted; low frequency. EDO/Nec SQR-18A; towed array; passive; very low frequency.
Combat data systems: OYQ-3B; Links 11 and 14. SATCOM ❻.
Weapon control systems: Singer Mk 114 for ASROC and TFCS; FCS 72-1A.

Helicopters: 3 SH-60J Seahawk ⓫.

Programmes: One each in 1975 and 1976 programmes.
Modernisation: DDH 143 refit in 1989–90. Both fitted with CIWS and towed array sonars by mid-1990. DDH 144 upgraded with Type 3 launcher to fire RIM-7M during 2003–04 refit at Mitsubishi, Nagasaki. DDH 143 similarly upgraded at IHI Yokohama in 2004.
Structure: Fitted with Vosper Thornycroft fin stabilisers. The after funnel is set to starboard and the forward one to port. The crane is on the starboard after corner of the hangar. Bear Trap helicopter hauldown gear.
Operational: *Shirane* badly damaged by a fire in December 2007.

KURAMA   4/2011, Hachiro Nakai / 1406820

4 KONGOU CLASS (DDGHM)

Name	No	Builders	Laid down	Launched	Commissioned
KONGOU	DDG 173	Mitsubishi, Nagasaki	8 May 1990	26 Sep 1991	25 Mar 1993
KIRISHIMA	DDG 174	Mitsubishi, Nagasaki	7 Apr 1992	19 Aug 1993	16 Mar 1995
MYOUKOU	DDG 175	Mitsubishi, Nagasaki	8 Apr 1993	5 Oct 1994	14 Mar 1996
CHOUKAI	DDG 176	Ishikawajima Harima, Tokyo	29 May 1995	27 Aug 1996	20 Mar 1998

Displacement, tonnes: 7,366 standard; 9,637 full load
Dimensions, metres (feet): 161 × 21 × 6.2 hull; 10 sonar (528.2 × 68.9 × 20.3; 32.8)
Speed, knots: 30. **Range, n miles:** 4,500 at 20 kt
Complement: 300 (27 officers)

Machinery: COGAG; 4 GE LM 2500 gas turbines; 102,160 hp (76.21 MW) sustained; 2 shafts; cp props
Missiles: SSM: 8 McDonnell Douglas Harpoon Block 1B (2 quad) ❶ launchers; active radar homing to 92 km (50 n miles) at 0.9 Mach; warhead 227 kg.
SAM: Raytheon Standard SM-2MR Block IIIA. FMC Mk 41 VLS (29 cells) forward ❷. Martin Marietta Mk 41 VLS (61 cells) aft ❸; command/inertial guidance; semi-active radar homing to 167 km (90 n miles) at 2.5 Mach. Standard SM-3 Block 1A (DDG 173, 175, 176); command/inertial/GPS guidance and IR homing to 650 n miles (1,200 km) at 3 Mach. Total of 90 Standard and ASROC weapons.
A/S: Vertical launch ASROC; inertial guidance to 1.6–10 km (1–5.4 n miles); payload Mk 46 Mod 5 Neartip.
Guns: 1 Oto Melara 5 in (127 mm)/54 Compatto ❹; 45 rds/min to 23 km (12.42 n miles); weight of shell 32 kg. 2 Raytheon 20 mm/76 Mk 15 Vulcan Phalanx Block IB ❺. 6 barrels per mounting; 4,500 rds/min combined to 1.5 km.
Torpedoes: 6–324 mm (2 triple) HOS 302 tubes ❻. Honeywell Mk 46 Mod 5 Neartip; anti-submarine; active/passive homing to 11 km (5.9 n miles) at 40 kt; warhead 44 kg.
Physical countermeasures: Decoys: 4 Mk 36 SRBOC ❼ 6-barrelled Mk 36 chaff launchers; Type 4 towed torpedo decoy.
Electronic countermeasures: ESM/ECM: Melco NOLQ 2; intercept/jammer.

Radars: Air search: RCA SPY 1D ❿; 3D; F-band.
Surface search: JRC OPS-28D ⓫; G-band.
Navigation: JRC OPS-20; I-band.
Fire control: 3 SPG-62 ⓬; 1 FCS 2-21 ⓭; I/J-band.
IFF: UPX 29.
Sonars: Nec OQS 102 (SQS-53B/C) bow-mounted; active search and attack. Oki OQR 2 (SQR-19A (V)) TACTASS; towed array; passive; very low frequency.
Combat data systems: Aegis NTDS with Link 11. SATCOM WSC-3/OE-82C ❽. ORQ-1 helicopter datalink ❾.
Weapon control systems: 3 Mk 99 Mod 1 MFCS. FCS 2–21 GFCS. Mk 116 Hitachi OYQ 102 (Mod 7 for ASW).

Helicopters: Platform ⓮ and fuelling facilities for SH-60J.

Programmes: Proposed in the FY87 programme; first one accepted in FY88 estimates, second in FY90, third in FY91, fourth in FY93. Designated as destroyers but these ships are of cruiser size. The combination of cost and US Congressional reluctance to release Aegis technology slowed the programme down. The ships' names were last used by battleships and cruisers of the Second World War era.
Modernisation: All four ships have completed an upgrade programme that included Aegis Baseline 3.6.1 and Standard SM-3 Block 1A missiles. Kongou was completed in 2007 and a successful SM-3 test-firing was conducted on 17 December 2007. Choukai completed upgrade in 2008 and Myoukou in 2009 and Kirishima in 2010.
Structure: This is an enlarged and improved version of the USN Arleigh Burke with a lightweight version of the Aegis system. There are two missile magazines. OQS 102 plus OQR 2 towed array is the equivalent of SQQ-89. Prairie-Masker acoustic suppression system.
Operational: As well as air defence of the Fleet, these ships contribute to the air defences of mainland Japan.

KONGOU (Scale 1 : 1,500), Ian Sturton / 0130387

KONGOU 11/2012, Hachiro Nakai / 1482867

MYOUKOU 11/2011, Hachiro Nakai / 1406821

CHOUKAI 4/2011, Hachiro Nakai / 1406822

Destroyers < **Japan** (MSDF) 435

4 AKIZUKI CLASS (DESTROYER) (DDHM)

Name	No	Builders	Laid down	Launched	Commissioned
AKIZUKI	DD 115	Mitsubishi, Nagasaki	17 Jul 2009	13 Oct 2010	14 Mar 2012
TERUZUKI	DD 116	Mitsubishi, Nagasaki	3 Jun 2010	15 Sep 2011	7 Mar 2013
SUZUTSUKI	DD 117	Mitsubishi, Nagasaki	18 May 2011	17 Oct 2012	12 Mar 2014
FUYUZUKI	DD 118	Mitsui, Tamano	14 Jun 2011	22 Aug 2012	13 Mar 2014

Displacement, tonnes: 5,050 standard
Dimensions, metres (feet): 151 × 18.3 × 5.4 *(495.4 × 60.0 × 17.7)*
Speed, knots: 30
Complement: 200

Machinery: COGAG: 4 RR Spey SM1C gas turbines; 64,000 hp *(47.72 MW)*; 2 shafts
Missiles: SSM: 8 Mitsubishi Type 90 (2 quad) ❶; active radar homing to 130 km *(70 n miles)*; at 0.9 Mach; warhead 227 kg.
SAM: Lockheed Martin Mk 41 Mod 29 vertical launcher (32 cells forward) ❷; Raytheon RIM-162 ESSM; semi-active homing to 18 km *(9.7 n miles)*; at 3.6 Mach; warhead 38 kg.
Guns: 1—5 in *(127 mm)*/62 Mk 45 Mod 4; 20 rds/min to 23 km *(12.6 n miles)*; weight of shell 32 kg ❸. 2 Raytheon 20 mm/76 Vulcan Phalanx Block 1B ❹. Mk 46 Mod 5; anti-submarine; active/passive homing to 11 km *(5.9 n miles)* at 40 kt; warhead 44 kg.
Torpedoes: 6—324 mm (2 triple) HOS-303 tubes ❺. Mk 46 Mod 5; anti-submarine; active/passive homing to 11 km *(5.9 n miles)* at 40 kt; warhead 44 kg.
Physical countermeasures: Decoys: 2 Mk 36 SRBOC chaff launchers.

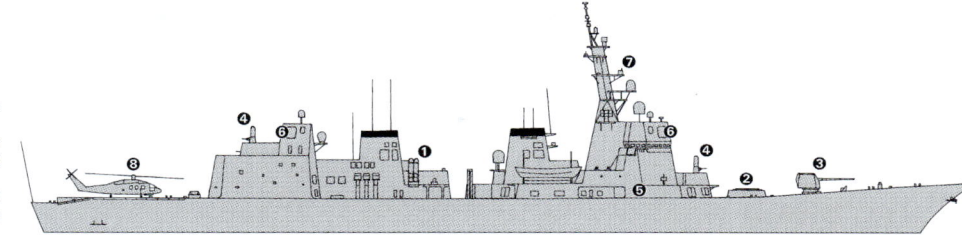

AKIZUKI (Scale 1 : 1,200), Ian Sturton / 1406803

Electronic countermeasures: ESM/ECM: NOLQ-3D EW suite.
Radars: Air search: Melco FCS-3A ❻; G/H/I-band.
Fire control: Melco FCS-3; G/H/I-band.
Navigation: JRC OPS-20C ❼; I-band.
Sonars: OQQ-22 sonar suite comprising hull-mounted sonar and OQR-3 TACTAS.
Combat data systems: OYQ-11, Link 16.

Helicopters: 2 Mitsubishi/Sikorsky SH-60K ❽ or 1 MCH-101.

Programmes: First authorised in FY07 and second in FY08 budget. Two more authorised in FY09 budget. The ships are to replace the Hatsuyuki class in the current inventory.
Structure: Measures to reduce the radar cross-section include a new design mast.

FUYUZUKI 9/2013*, Hachiro Nakai / 1529131

TERUZUKI 3/2013*, Hachiro Nakai / 1529132

© 2014 IHS — IHS Jane's Fighting Ships 2014-2015

436 Japan (MSDF) > Destroyers

2 HATAKAZE CLASS (DDGHM)

Name	No	Builders	Laid down	Launched	Commissioned
HATAKAZE	DDG 171	Mitsubishi, Nagasaki	20 May 1983	9 Nov 1984	27 Mar 1986
SHIMAKAZE	DDG 172	Mitsubishi, Nagasaki	30 Jan 1985	30 Jan 1987	23 Mar 1988

Displacement, tonnes: 4,674 (DDG 171), 4,725 (DDG 172) standard; 5,995 full load
Dimensions, metres (feet): 150 × 16.4 × 4.8 (492.1 × 53.8 × 15.7)
Speed, knots: 30
Complement: 260 (23 officers)

Machinery: COGAG; 2 RR Olympus TM3B gas turbines; 49,400 hp (36.8 MW) sustained; 2 RR Spey SM1A (DDG 171), SM1C (DDG 172) gas turbines; 26,650 hp (19.9 MW) sustained; 2 shafts; Kamewa cp props
Missiles: SSM: 8 McDonnell Douglas Harpoon Block 1B ❶; active radar homing to 92 km (50 n miles) at 0.9 Mach; warhead 227 kg.
SAM: 40 Raytheon Standard SM-1MR Block VIA; Mk 13 Mod 4 launcher ❷; command guidance; semi-active radar homing to 38 km (20.5 n miles) at 2 Mach; height envelope 45–18,288 m (150–60,000 ft).
A/S: Honeywell ASROC Mk 112 octuple launcher ❸; inertial guidance to 1.6–10 km (1–5.4 n miles) at 0.9 Mach; payload Mk 46 Mod 5 Neartip. Reload capability.
Guns: 2 FMC 5 in (127 mm)/54 Mk 42 automatic ❹; 20–40 rds/min to 24 km (13 n miles) anti-surface; 14 km (7.6 n miles) anti-aircraft; weight of shell 32 kg. 2 Raytheon 20 mm Phalanx Mk 15 CIWS ❺; 6 barrels per mounting, 4,500 rds/min combined to 1.5 km.
Torpedoes: 6–324 mm Type 68 or HOS 301 (2 triple) tubes ❻. Honeywell Mk 46 Mod 5 Neartip; anti-submarine; active/passive homing to 11 km (5.9 n miles) at 40 kt; warhead 44 kg.
Physical countermeasures: Decoys: 2 Loral Hycor SRBOC 6-barrelled Mk 36 chaff launchers; range 4 km (2.2 n miles).
Electronic countermeasures: ESM/ECM: Melco NOLQ-1; intercept/jammer. Fujitsu OLR 9B; intercept.
Radars: Air search: Hughes SPS-52C ❽; 3D; E/F-band. Melco OPS-11C ❾; B-band.
Surface search: JRC OPS-28B ❿; G/H-band.
Navigation: JRC OPS-20; I-band.
Fire control: 2 Raytheon SPG-51C ⓫; G-band. Melco 2-21 ⓬; I/J-band. FCS 2-12 ⓭; I-band.
Sonars: Nec OQS 4 Mod 1; bow-mounted; active search and attack; medium frequency.
Combat data systems: OYQ-4 Mod 1 action data automation; Link 11. SATCOM ❼.
Weapon control systems: FCS 2-21C for 127 mm guns. General Electric Mk 74 Mod 13 for Standard.

Helicopters: Platform for 1 SH-60J Seahawk ⓮.

Programmes: DDG 171 provided for in 1981 programme. DDG 172 provided for in 1983 programme, ordered 29 March 1984.

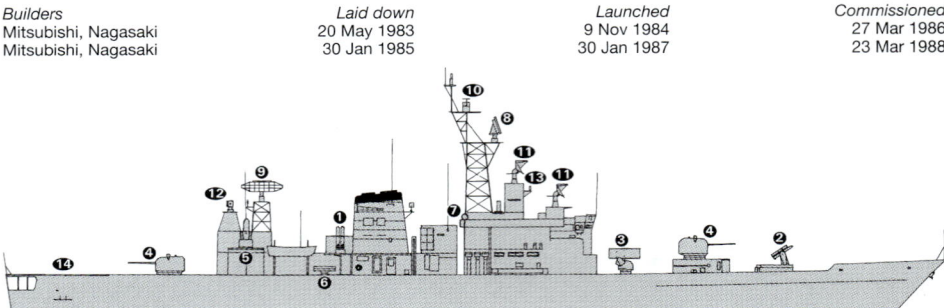

HATAKAZE   (Scale 1 : 1,200), Ian Sturton / 0506023

HATAKAZE   4/2013*, Hachiro Nakai / 1529133

9 MURASAME CLASS (DDGHM)

Name	No	Builders	Laid down	Launched	Commissioned
MURASAME	DD 101	Ishikawajima Harima, Tokyo	18 Aug 1993	23 Aug 1994	12 Mar 1996
HARUSAME	DD 102	Mitsui, Tamano	11 Aug 1994	16 Oct 1995	24 Mar 1997
YUUDACHI	DD 103	Marine United (Sumitomo), Uraga	18 Mar 1996	19 Aug 1997	4 Mar 1999
KIRISAME	DD 104	Mitsubishi, Nagasaki	3 Apr 1996	21 Aug 1997	18 Mar 1999
INAZUMA	DD 105	Mitsubishi, Nagasaki	8 May 1997	9 Sep 1998	15 Mar 2000
SAMIDARE	DD 106	Marine United (Ishikawajima Harima), Tokyo	11 Sep 1997	24 Sep 1998	21 Mar 2000
IKAZUCHI	DD 107	Hitachi, Maizuru	25 Feb 1998	24 Jun 1999	14 Mar 2001
AKEBONO	DD 108	Marine United (Ishikawajima Harima), Tokyo	29 Oct 1999	25 Sep 2000	19 Mar 2002
ARIAKE	DD 109	Mitsubishi, Nagasaki	18 May 1999	16 Oct 2000	6 Mar 2002

Displacement, tonnes: 4,623 standard; 6,299 full load
Dimensions, metres (feet): 151 × 17.4 × 5.2 (495.4 × 57.1 × 17.1)
Speed, knots: 30
Complement: 165

Machinery: COGAG; 2 RR Spey SM1C gas turbines; 26,600 hp (19.9 MW) sustained; 2 GE LM 2500 gas turbines; 32,500 hp (24.3 MW) sustained; 2 shafts
Missiles: SSM: 8 Type 90 SSM-1B ❶ (Harpoon); active radar homing to 130 km (70 n miles) at 0.9 Mach; warhead 227 kg.
SAM: Raytheon Mk 48 VLS 16 cells ❷ Sea Sparrow RIM-7P; semi-active radar homing to 16 km (8.5 n miles) at 2.5 Mach; warhead 38 kg.
A/S: Mk 41 VL ASROC 16 cells ❸. Total of 29 missiles can be carried.
Guns: 1 Otobreda 3 in (76 mm)/62 compact ❹; 85 rds/min to 16 km (8.6 n miles) anti-surface; 12 km (6.5 n miles) anti-aircraft; weight of shell 6 kg. 2 Raytheon 20 mm Phalanx Mk 15 CIWS ❺; 6 barrels per mounting, 3,000 rds/min combined to 1.5 km.
Torpedoes: 6–324 mm HOS 302 (2 triple) tubes ❻ Mk 46 Mod 5; anti-submarine; active/passive homing to 11 km (5.9 n miles) at 40 kt; warhead 44 kg.
Physical countermeasures: Decoys: 4 Mk 36 SRBOC chaff launchers ❼. Type 4 towed torpedo decoy.
Electronic countermeasures: ESM/ECM: Nec NOLQ 3; intercept and jammer.
Radars: Air search: Melco OPS-24B ❾; 3D; D-band.
Surface search: JRC OPS-28D ❿; G-band.
Fire control: 2 FCS 2-31 ⓫.
Navigation: OPS-20; I-band.
Sonars: Mitsubishi OQS-5; hull-mounted; active search and attack; low frequency. OQR-1 towed array; passive search; very low frequency.
Combat data systems: OYQ-9B with Link 11. ORQ-1 helicopter datalink ❽.
Weapon control systems: Hitachi OYQ-103 ASW control system.

Helicopters: 1 SH-60J Seahawk ⓬.

Programmes: First one approved in fiscal year 1991 as an addition to the third Aegis-type destroyer. Second approved in FY 1992. Two more approved in FY 1994, two in FY 1995, one in FY 1996 and two in FY 1997. The programme was given added priority as the Kongou class was reduced to four ships because of the cost of Aegis.
Modernisation: DD 103, 108 and 109 converted to fire ESSM 2007–08. One further ship reported converted in 2009.
Structure: More like a mini-Kongou than an enlarged Asagiri class, with VLS and a much reduced complement. Stealth features are evident in sloping sides and rounded superstructure. Indal RAST helicopter hauldown.
Operational: ASROC missiles are not carried. *Kirisame* deployed to Indian Ocean in November 2001 to provide non-combatant support to US forces.

MURASAME   (Scale 1 : 1,200), Ian Sturton / 0506235

IHS Jane's Fighting Ships 2014-2015   © 2014 IHS

Destroyers < **Japan** (MSDF) 437

AKEBONO   *11/2012, Hachiro Nakai* / 1482863

IKAZUCHI   *4/2013*, Hachiro Nakai* / 1529134

YUUDACHI   *3/2012, Hachiro Nakai* / 1482875

© 2014 IHS   *IHS Jane's Fighting Ships 2014-2015*

Japan (MSDF) > Destroyers

5 TAKANAMI CLASS (DDGHM)

Name	No	Builders	Laid down	Launched	Commissioned
TAKANAMI	DD 110	IHI Marine United, Yokosuka (Uraga)	25 Apr 2000	26 Jul 2001	12 Mar 2003
OONAMI	DD 111	Mitsubishi, Nagasaki	17 May 2000	20 Sep 2001	13 Mar 2003
MAKINAMI	DD 112	IHI Marine United, Yokohama	7 Jul 2001	8 Aug 2002	18 Mar 2004
SAZANAMI	DD 113	Mitsubishi, Nagasaki	4 Apr 2002	29 Aug 2003	16 Feb 2005
SUZUNAMI	DD 114	IHI Marine United, Yokohama	24 Sep 2003	26 Aug 2004	16 Feb 2006

Displacement, tonnes: 4,725 standard; 6,401 full load
Dimensions, metres (feet): 151 × 17.4 × 5.3 *(495.4 × 57.1 × 17.4)*
Speed, knots: 30
Complement: 176

Machinery: COGAG; 2 RR Spey SM1C gas turbines; 26,600 hp *(19.9 MW)* sustained; 2 GE LM 2500 gas turbines; 32,500 hp *(24.3 MW)* sustained; 2 shafts
Missiles: SSM: 8 Mitsubishi Type 90 SSM-1B (2 quad) ❶; active radar homing to 150 km *(81 n miles)* at 0.9 Mach; warhead 225 kg.
SAM: Mk 41 VLS 32 cells ❷ Sea Sparrow RIM 162 ESSM (PIP); semi-active radar homing to 18 km *(9.7 n miles)* at 3.6 Mach; warhead 38 kg and VL ASROC; internal guidance to 1.6–10 km *(1–5.4 n miles)*; payload Mk 46 Mod 5 Neartip.
Guns: 1 Otobreda 5 in *(127 mm)*/54 ❸; 45 rds/min to 24 km *(12.42 n miles)*; weight of shell 32 kg. 2 Raytheon 20 mm Phalanx Mk 15 CIWS ❹; 6 barrels per mounting; 4,500 rds/min combined to 1.5 km.
Torpedoes: 6—324 mm HOS-302 (2 triple) tubes ❺ Mk 46 Mod 5; anti-submarine; active/passive homing to 11 km *(5.9 n miles)* at 40 kt; warhead 44 kg.
Physical countermeasures: Decoys: 4 Mk 36 SRBOC chaff launchers ❻. SLQ-25 Nixie towed torpedo decoy.
Electronic countermeasures: ESM/ECM: Nec NOLQ 3; intercept and jammer.

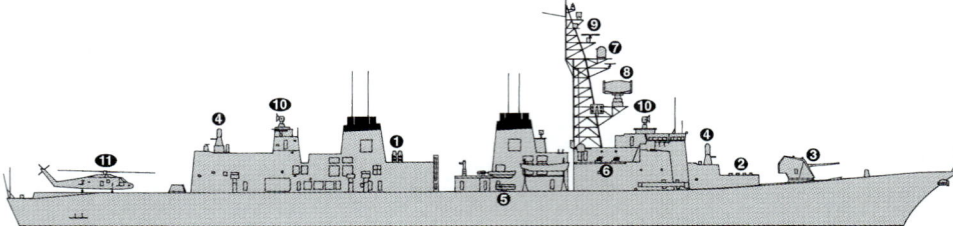

Radars: Air search: Melco OPS-24B ❽; 3D; D-band.
Surface search: JRC OPS-28D ❾; G-band.
Fire control: 2 FCS 2-31B ❿.
Navigation: OPS-20; I-band.
Sonars: OQS-5; Bow-mounted; active search and attack; low frequency. OQR-2; towed array; passive search; very low frequency.
Combat data systems: OYQ-9 with Link 11. ORQ-1B helicopter datalink ❼.

Weapon control systems: Hitachi OYQ-103 ASW control system.

Helicopters: 1 Mitsubishi/Sikorsky SH-60J/K ⓫.

Programmes: First two approved in fiscal year 1998, then one a year up to FY 2001.
Structure: Murasame class modified to fit a Mk 41 VLS, improved missile fire control and new sonar.

TAKANAMI CLASS *(Scale 1 : 1,200), Ian Sturton / 0080138*

SUZUNAMI *7/2013*, Hachiro Nakai / 1529135*

MAKINAMI *10/2013*, John Mortimer / 1530908*

8 ASAGIRI CLASS (DDGHM)

Name	No	Builders	Laid down	Launched	Commissioned
ASAGIRI	DD 151	Ishikawajima Harima, Tokyo	13 Feb 1985	19 Sep 1986	17 Mar 1988
YAMAGIRI	DD 152	Matsui, Tamano	5 Feb 1986	8 Oct 1987	25 Jan 1989
YUUGIRI	DD 153	Sumitomo, Uraga	25 Feb 1986	21 Sep 1987	28 Feb 1989
AMAGIRI	DD 154	Ishikawajima Harima, Tokyo	3 Mar 1986	9 Sep 1987	17 Mar 1989
HAMAGIRI	DD 155	Hitachi, Maizuru	20 Jan 1987	4 Jun 1988	31 Jan 1990
SETOGIRI	DD 156	Sumitomo, Uraga	9 Mar 1987	12 Sep 1988	14 Feb 1990
SAWAGIRI	DD 157	Mitsubishi, Nagasaki	14 Jan 1987	25 Nov 1988	6 Mar 1990
UMIGIRI	DD 158	Ishikawajima Harima, Tokyo	31 Oct 1988	9 Nov 1989	12 Mar 1991

Displacement, tonnes: 3,556 (DD 151–154), 3,607 (DD 155–158) standard; 4,979 (DD 151–154), 5,029 (DD 155–158) full load
Dimensions, metres (feet): 137 × 14.6 × 4.5 (449.5 × 47.9 × 14.8)
Speed, knots: 30 (est.)
Complement: 220

Machinery: COGAG; 4 RR Spey SM1C gas turbines; 53,300 hp (39.8 MW) sustained; 2 shafts; cp props
Missiles: SSM: 8 McDonnell Douglas Harpoon (2 quad) launchers ❶; active radar homing to 130 km (70 n miles) at 0.9 Mach; warhead 227 kg.
SAM: Raytheon Sea Sparrow RIM-7M Mk 29 (Type 3/3A) octuple launcher ❷; semi-active radar homing to 16 km (8.5 n miles) at 2.5 Mach; warhead 38 kg; 20 missiles.
A/S: Honeywell ASROC Mk 112 octuple launcher ❸; inertial guidance to 1.6–10 km (1–5.4 n miles) at 0.9 Mach; payload Mk 46 Mod 5 Neartip. Reload capability.
Guns: 1 Otobreda 3 in (76 mm)/62 compact ❹; 85 rds/min to 16 km (8.6 n miles) anti-surface; 12 km (6.5 n miles) anti-aircraft; weight of shell 6 kg. 2 Raytheon 20 mm Phalanx Mk 15 CIWS ❺; 6 barrels per mounting; 3,000 rds/min combined to 1.5 km.
Torpedoes: 6—324 mm Type 68 (2 triple) HOS 301 tubes ❻. Honeywell Mk 46 Mod 5 Neartip; anti-submarine; active/passive homing to 11 km (5.9 n miles) at 40 kt; warhead 44 kg.
Physical countermeasures: Decoys: 2 Loral Hycor SRBOC 6-barrelled Mk 36 chaff launchers ❼; range 4 km (2.2 n miles). 1 SLQ-25 Nixie or Type 4; towed torpedo decoy.
Electronic countermeasures: ESM: Nec NOLR 6C or NOLR 8 (DD 152) ❽; intercept.
ECM: Fujitsu OLT-3; jammer.
Radars: Air search: Melco OPS-14C (DD 151–154). Melco OPS-24 (DD 155–158) ❿; 3D; D-band.
Surface search: JRC OPS-28C ⓫; G-band (DD 151–152, 155–158). JRC OPS-28C-Y; G-band (DD 153–154).
Navigation: JRC OPS-20; I-band.
Fire control: FCS 2–22 (for guns) ⓬. FCS 2-12E (for SAM) (DD 151–154); FCS 2-12G (for SAM) ⓭ (DD 155–158).
Tacan: ORN-6D (URN 25).
Sonars: Mitsubishi OQS 4A (II); hull-mounted; active search and attack; low frequency. OQR-1; towed array; passive search; very low frequency.
Combat data systems: OYQ-7B data automation; Link 11/14. SATCOM. ORQ-1 helicopter datalink ❾ for SH-60J.
Helicopters: 1 SH-60J Seahawk ⓮.

Programmes: DD 153–154 in 1984 estimates, DD 155–157 in 1985 and DD 158 in 1986.
Modernisation: The last four were fitted on build with improved air search radar, updated fire-control radars and a helicopter datalink. Plans to fit the first four may have been postponed. Umigiri also commissioned with a sonar towed array which has been fitted to the rest of the class. All eight ships to be fitted with SM1C gas turbines (to replace SM1A) 2014–15.
Structure: Because of the enhanced IR signature and damage to electronic systems on the mainmast caused by after funnel gases there were modifications to the original design to help contain the problem. The mainmast is now slightly higher and has been offset to port. The forward funnel is also offset slightly to port and the after funnel to the starboard side of the superstructure. The hangar structure is asymmetrical extending to the after funnel on the starboard side but only to the mainmast to port. SATCOM is fitted at the after end of the hangar roof.
Operational: Beartrap helicopter hauldown system. Yamagiri (D 152) converted to training ship on 18 March 2004 and Asagiri (D 151) on 16 February 2005. Yamagiri re-converted to destroyer role on 16 March 2011 and Asagiri on 14 March 2012.

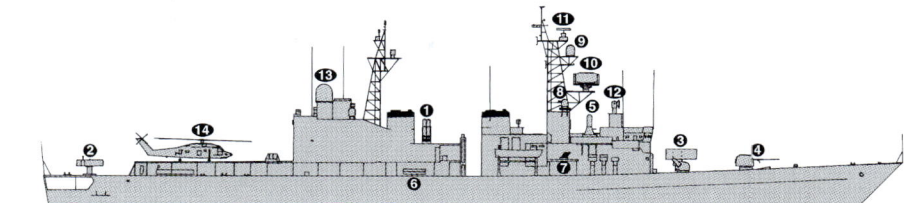

UMIGIRI (Scale 1 : 1,200), Ian Sturton / 0012635

YAMAGIRI 6/2012, Hachiro Nakai / 1482878

UMIGIRI 4/2013*, Hachiro Nakai / 1529136

5 HATSUYUKI CLASS (DDGHM)

Name	No	Builders	Laid down	Launched	Commissioned
ISOYUKI	DD 127	Ishikawajima Harima, Tokyo	20 Apr 1982	19 Sep 1983	23 Jan 1985
HARUYUKI	DD 128	Sumitomo, Uraga	11 Mar 1982	6 Sep 1983	14 Mar 1985
YAMAYUKI	DD 129	Hitachi, Maizuru	25 Feb 1983	10 Jul 1984	3 Dec 1985
MATSUYUKI	DD 130	Ishikawajima Harima, Tokyo	7 Apr 1983	25 Oct 1984	19 Mar 1986
ASAYUKI	DD 132	Sumitomo, Uraga	22 Dec 1983	16 Oct 1985	20 Feb 1987

Displacement, tonnes: 2,997 (DD 127–128), 3,099 (DD 129–130, 132) standard; 4,064 (DD 127–128), 4,267 (DD 129–130,132) full load
Dimensions, metres (feet): 130 × 13.6 × 4.2 (DD 127–128), 4.4 (DD 129–130,132) (426.5 × 44.6 × 13.8, 14.4)
Speed, knots: 30; 19 Tynes
Complement: 200

Machinery: COGOG; 2 Kawasaki-RR Olympus TM3B gas turbines; 49,400 hp *(36.8 MW)* sustained; 2 RR Type RM1C gas turbines; 9,900 hp *(7.4 MW)* sustained; 2 shafts; cp props
Missiles: SSM: 8 McDonnell Douglas Harpoon (2 quad) launchers ❶; active radar homing to 130 km *(70 n miles)* at 0.9 Mach; warhead 227 kg.
SAM: Raytheon Sea Sparrow RIM-7M Mk 29 Type 3A launcher ❷; semi-active radar homing to 16 km *(8.5 n miles)* at 2.5 Mach; warhead 38 kg; 12 missiles.
A/S: Honeywell ASROC Mk 112 octuple launcher ❸; inertial guidance to 1.6–10 km *(1–5.4 n miles)* at 0.9 Mach; payload Mk 46 Mod 5 Neartip.
Guns: 1 Otobreda 3 in *(76 mm)*/62 compact ❹; 85 rds/min to 16 km *(8.6 n miles)* anti-surface; 12 km *(6.5 n miles)* anti-aircraft; weight of shell 6 kg. 2 Raytheon 20 mm Phalanx Mk 15 CIWS ❺; 6 barrels per mounting; 3,000 rds/min combined to 1.5 km.
Torpedoes: 6—324 mm Type 68 or HOS 301 (2 triple) tubes ❻. Honeywell Mk 46 Mod 5 Neartip; anti-submarine; active/passive homing to 11 km *(5.9 n miles)* at 40 kt; warhead 44 kg.

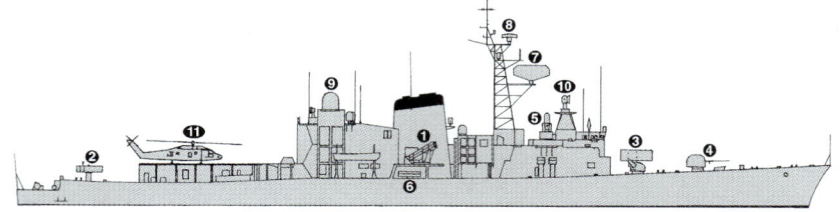

HATSUYUKI   (Scale 1 : 1,200), Ian Sturton / 0506301

Physical countermeasures: Decoys: 2 Loral Hycor SRBOC 6-barrelled Mk 36 chaff launchers; range 4 km *(2.2 n miles)*.
Electronic countermeasures: ESM: Nec NOLR 6C; intercept.
ECM: Fujitsu OLT 3; jammer.
Radars: Air search: Melco OPS-14B ❼; D-band.
Surface search: JRC OPS-18C ❽; G-band.
Navigation: JRC OPS-20; I-band.
Fire control: FCS 2-12 A ❾; I/J-band (for SAM). 2 FCS 2-21/21A ❿; I/J-band (for guns).
Tacan: ORN-6C-Y (DD 132); ORN-6C (remainder).
Sonars: Nec OQS 4A (II) (SQS-23 type); bow-mounted; active search and attack; low frequency. OQR 1 TACTASS (in some); passive; low frequency.
Combat data systems: OYQ-5B action data automation. SATCOM.

Helicopters: 1 SH-60J Seahawk ⓫.

Modernisation: All ships fitted with Phalanx by 1996. *Matsuyuki* first to get sonar towed array in 1990. All of the class converted to carry Seahawk helicopters.
Structure: Fitted with fin stabilisers. Steel in place of aluminium alloy for bridge and so on after DD 129 which increased displacement.
Operational: Canadian Beartrap helicopter landing aid. Improved Electronic Countermeasures (ECM) equipment in the last two of the class. Last of class *Shimayuki* converted to a training ship 18 March 1999, *Shirayuki* on 16 March 2011 and *Setoyuki* in 2012. *Hatsuyuki* decommissioned on 25 June 2010, *Hamayuki* on 14 March 2012 and *Mineyuki* and *Sawayuki* in 2013.

ISOYUKI   8/2013*, Michael Nitz / 1530909

HARUYUKI   2/2012, Hachiro Nakai / 1482879

0 + 1 (1) IMPROVED AKIZUKI CLASS (DESTROYER) (DDHM)

Name	No	Laid down	Launched	Commissioned
–	DD 119	2014	2015	2017
–	DD 120	2015	2016	2018

Displacement, tonnes: 5,000 standard
Dimensions, metres (feet): 150 × ? × ? *(492.1 × ? × ?)*
Speed, knots: 30

Machinery: COGAG; 4 gas turbines; 2 shafts
Missiles: SSM: 8 Mitsubishi Type 90 (2 quad).
SAM: Lockheed Martin Mk 41 vertical launcher. Raytheon RIM 162 ESSM.
A/S: Vertical launch ASROC.

Guns: 1–5 in *(127 mm)*/62. 2 Raytheon 20 mm Vulcan Phalanx.
Torpedoes: 6—324 mm (2 triple) tubes.
Electronic countermeasures: ESM/ECM: NOLQ-3D.
Radars: Air search: FCS-3.
Surface search: To be announced.
Fire control: To be announced.
Navigation: To be announced.
Sonars: OQQ-24 bow mounted. OQR-4 TACTAS.

Helicopters: 1 Mitsubishi/Sikorsky SH-60J/K.

Programmes: The first of a new class of improved Akizuki-class destroyers was authorised in the fiscal year 2013 budget.

FRIGATES

Notes: The MSDF classifies these ships as Destroyer Escorts.

6 ABUKUMA CLASS (FFGM/DE)

Name	No	Builders	Laid down	Launched	Commissioned
ABUKUMA	DE 229	Mitsui, Tamano	17 Mar 1988	21 Dec 1988	12 Dec 1989
JINTSU	DE 230	Hitachi, Maizuru	14 Apr 1988	31 Jan 1989	28 Feb 1990
OOYODO	DE 231	Mitsui, Tamano	8 Mar 1989	19 Dec 1989	23 Jan 1991
SENDAI	DE 232	Sumitomo, Uraga	14 Apr 1989	26 Jan 1990	15 Mar 1991
CHIKUMA	DE 233	Hitachi, Maizuru	14 Feb 1991	22 Jan 1992	24 Feb 1993
TONE	DE 234	Sumitomo, Uraga	8 Feb 1991	6 Dec 1991	8 Feb 1993

Displacement, tonnes: 2,032 standard; 2,591 full load
Dimensions, metres (feet): 109 × 13.4 × 3.8 *(357.6 × 44.0 × 12.5)*
Speed, knots: 27
Complement: 120
Machinery: CODOG; 2 RR Spey SM1A gas turbines; 26,650 hp *(19.9 MW)* sustained; 2 Mitsubishi S12U-MTK diesels; 6,000 hp(m) *(4.4 MW)*; 2 shafts
Missiles: SSM: 8 McDonnell Douglas Harpoon (2 quad) launchers ❶; active radar homing to 130 km *(70 n miles)* at 0.9 Mach; warhead 227 kg.
A/S: Honeywell ASROC Mk 112 octuple launcher ❷; inertial guidance to 1.6–10 km *(1–5.4 n miles)* at 0.9 Mach; payload Mk 46 Mod 5 Neartip.
Guns: 1 Otobreda 3 in *(76 mm)*/62 compact ❸; 85 rds/min to 16 km *(8.6 n miles)* anti-surface; 12 km *(6.5 n miles)* anti-aircraft; weight of shell 6 kg. 1 Raytheon 20 mm Phalanx CIWS Mk 15 ❹; 6 barrels per mounting; 3,000 rds/min combined to 1.5 km.
Torpedoes: 6—324 mm HOS 301 (2 triple) tubes ❺. Honeywell Mk 46 Mod 5 Neartip; anti-submarine; active/passive homing to 11 km *(5.9 n miles)* at 40 kt; warhead 44 kg.
Physical countermeasures: Decoys: 2 Loral Hycor SRBOC 6-barrelled Mk 36 chaff launchers.
Electronic countermeasures: ESM: Nec NOLR-8; intercept.
Radars: Air search: Melco OPS-14C ❻; D-band.
Surface search: JRC OPS-28D (DE 233–234); JRS OPC-28C (remainder) ❼; G-band.
Fire control: FCS 2–21 ❽.
Sonars: Hitachi OQS-8; hull-mounted; active search and attack; medium frequency. SQR-19A towed passive array in due course.
Combat data systems: OYQ-6. SATCOM.
Weapon control systems: FCS 2–21.

Programmes: First pair of this class approved in 1986 estimates, ordered March 1987; second pair in 1987 estimates, ordered February 1988; last two in 1989 estimates, ordered 24 January 1989. The name of the first of class commemorates that of a light cruiser which was sunk in the battle of Leyte Gulf in October 1944.
Structure: Stealth features include non-vertical and rounded surfaces. German RAM PDMS may be fitted later, although this now seems unlikely, and space has been left for a towed sonar array. SATCOM fitted aft of the after funnel.

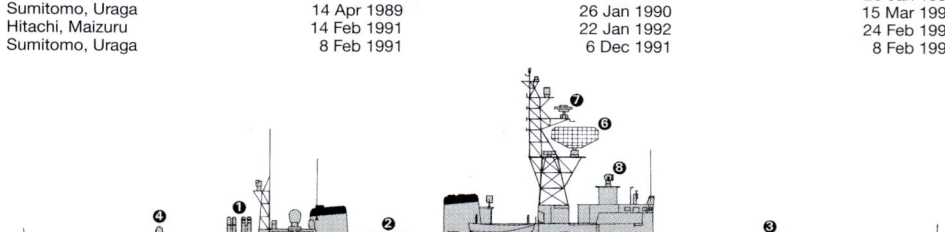

ABUKUMA (Scale 1 : 900), Ian Sturton / 0506197

CHIKUMA 7/2011, Hachiro Nakai / 1406833

SHIPBORNE AIRCRAFT

Numbers/Type: 49/37 Sikorsky/Mitsubishi SH-60J/SH-60K (Seahawk).
Operational speed: 139 kt *(257 km/h)*.
Service ceiling: 13,500 ft *(4,090 m)*.
Range: 600 n miles *(1,110 km)*.
Role/Weapon systems: ASW helicopter; started replacing HSS-2B in July 1991; built in Japan; prototypes fitted by Mitsubishi with Japanese avionics and mission equipment. Overall requirement for 103 aircraft. SH-60K are upgraded aircraft with an improved tactical data processing system. Sensors: Texas Instruments APS 124 search radar; sonobuoys plus datalink; Bendix AQS 18/Nippon HQS 103 dipping sonar, ECM, HLR 108 ESM. Weapons: ASW; two Mk 46 torpedoes or depth bombs. 2 Hellfire ASM (SH-60K).

SH-60K 7/2013*, Hachiro Nakai / 1529137

LAND-BASED MARITIME AIRCRAFT

Notes: Aircraft type names are not used by the MSDF.

Numbers/Type: 3 NAMC YS-11.
Operational speed: 248 kt *(459 km/h)*.
Service ceiling: 21,500 ft *(6,580 m)*.
Range: 1,960 n miles *(3,629 km)*.
Role/Weapon systems: First flew in 1967. Of 182 aircraft constructed, three YS-11M remain in service.

YS-11M 4/2013*, Hachiro Nakai / 1529138

442 Japan (MSDF) > Land-based maritime aircraft

Numbers/Type: 5/2 Agusta Westland/Kawasaki MCH-101/CH-101.
Operational speed: 150 kt *(278 km/h)*.
Service ceiling: 15,000 ft *(4,575 m)*.
Range: 610 n miles *(1,129 km)*.
Role/Weapon systems: Based on the Agusta Westland EH-101. There are to be 10 MCH-101 AMCM and cargo carrying aircraft to replace the MH-53E. Two CH-101 are to replace the S-61A support aircraft deployed in the ice-patrol ship.

MCH-101   9/2013*, Hachiro Nakai / 1529139

CH-101   9/2010, Hachiro Nakai / 1366892

Numbers/Type: 3 Shinmeiwa US-1A Rescue.
Operational speed: 265 kt *(491 km/h)*.
Service ceiling: 28,400 ft *(8,655 m)*.
Range: 2,300 n miles *(4,260 km)*.
Role/Weapon systems: Turboprop amphibian designed for maritime patrol and SAR missions. Crew of 12. Accommodation for 16 survivors or 12 stretchers.

US-1A   10/2012, Hachiro Nakai / 1482881

Numbers/Type: 6 Sikorsky/Mitsubishi S-80M-1 (Sea Dragon) (MH53E).
Operational speed: 170 kt *(315 km/h)*.
Service ceiling: 18,500 ft *(5,640 m)*.
Range: 1,120 n miles *(2,000 km)*.
Role/Weapon systems: Three-engined AMCM helicopter tows Mk 103, 104, 105, and 106 MCM sweep equipment; self-deployed. Weapons: Two 12.7 mm guns for mine disposal.

MH-53E   9/2013*, Hachiro Nakai / 1529140

Numbers/Type: 77/5/1/3/5 Lockheed/Kawasaki P-3C/EP-3/UP-3C/UP-3D/OP-3C.
Operational speed: 395 kt *(732 km/h)*.
Service ceiling: 28,300 ft *(8,625 m)*.
Range: 3,300 n miles *(6,100 km)*.
Role/Weapon systems: Long-range MR/ASW and surface surveillance and attack. Most maritime surveillance is done by these aircraft. Sensors: APS-115 radar, ASQ-81 MAD, AQA 7 processor, Unisys CP 2044 computer, IFF, ECM, ALQ 78, ESM, ALR 66, sonobuoys. Weapons: ASW; eight Mk 46 torpedoes, depth bombs or mines, four underwing stations for Harpoon and ASM-1.

P-3C   4/2013*, Hachiro Nakai / 1529141

EP-3   9/2011, Hachiro Nakai / 1406842

Numbers/Type: 2 Kawasaki P-1.
Operational speed: 448 kt *(830 km/h)*.
Service ceiling: 36,100 ft *(11,000 m)*.
Range: 4,320 n miles *(8,000 km)*.
Role/Weapon systems: The first test version of the XP-1 future maritime patrol aircraft was rolled out on 4 July 2007. The aircraft has been under full development since 2001 and is to replace the P-3C inventory in due course. The aircraft incorporates the world's first Fly-By-Light (FBL) system. The aircraft is to be equipped with new acoustic and radar systems. Following flight testing, the aircraft was delivered to the Ministry of Defence by the end of 2008. The first two operational aircraft were delivered to the JMSDF in March 2013. Up to 70 aircraft may be acquired.

P-1   4/2013*, Hachiro Nakai / 1529143

Numbers/Type: 5 Shinmeiwa US-2.
Operational speed: 300 kt *(556 km/h)*.
Service ceiling: 28,400 ft *(8,655 m)*.
Range: 2,500 n miles *(4,630 km)*.
Role/Weapon systems: Following trials, two former experimental aircraft entered service in 2007. A further aircraft was authorised in the FY07 budget. The US-2 is an upgraded version of the US-1A and is designed for maritime patrol and SAR missions. Sensors: Thales Ocean Master radar.

US-2   4/2013*, Hachiro Nakai / 1529144

PATROL FORCES

6 HAYABUSA CLASS (PGGF)

Name	No	Builders	Launched	Commissioned
HAYABUSA	824	Mitsubishi, Shimonoseki	13 Jun 2001	25 Mar 2002
WAKATAKA	825	Mitsubishi, Shimonoseki	13 Sep 2001	25 Mar 2002
OOTAKA	826	Mitsubishi, Shimonoseki	13 May 2002	24 Mar 2003
KUMATAKA	827	Mitsubishi, Shimonoseki	2 Aug 2002	24 Mar 2003
UMITAKA	828	Mitsubishi, Shimonoseki	21 May 2003	24 Mar 2004
SHIRATAKA	829	Mitsubishi, Shimonoseki	8 Aug 2003	24 Mar 2004

Displacement, tonnes: 203 standard; 244 full load
Dimensions, metres (feet): 50.1 × 8.4 × 4.2 *(164.4 × 27.6 × 13.8)*
Speed, knots: 44
Complement: 18 (3 officers)

Machinery: 3 LM 500-G07 gas turbines 16,200 hp *(12.08 MW)*; 3 water jets
Missiles: 4 Mitsubishi Type 90 SSM-1B; active radar homing to 130 km *(70 n miles)* at 0.9 Mach; warhead 227 kg.
Guns: 1 Oto Melara 3 in *(76 mm)*/62 compact; 85 rds/min to 16 km *(8.7 n miles)* anti-surface; 12 km *(6.6 n miles)* anti-aircraft; weight of shell 6 kg. 2—12.7 mm MGs.
Physical countermeasures: Decoys: Chaff launchers.
Electronic countermeasures: ESM/ECM: NOLR-9B.
Radars: Surface search: OPS-18-3; G-band.
Fire control: FCS 2-31C.
Navigation: OPS-20; I-band.

Comment: First pair authorised in fiscal year 1999 budget, second pair in FY 2000 and third pair in FY 2001. Single hull.

HAYABUSA   7/2013*, Hachiro Nakai / 1529145

AMPHIBIOUS FORCES

3 OOSUMI CLASS (LPD/LSTH)

Name	No	Builders	Laid down	Launched	Commissioned
OOSUMI	LST 4001	Mitsui, Tamano	6 Dec 1995	18 Nov 1996	11 Mar 1998
SHIMOKITA	LST 4002	Mitsui, Tamano	30 Nov 1999	29 Nov 2000	12 Mar 2002
KUNISAKI	LST 4003	Universal, Maizuru	7 Sep 2000	13 Dec 2001	26 Feb 2003

Displacement, tonnes: 9,043 standard; 14,225 full load
Dimensions, metres (feet): 178 × 25.8 × 6 *(584 × 84.6 × 19.7)*
Flight deck, metres: 130 × 23 *(426.5 × 75.5)*
Speed, knots: 22
Complement: 135

Military lift: 330 troops; 2 LCAC; 10 Type 90 tanks or 1,400 tons cargo
Machinery: 2 Mitsui 16V42MA diesels; 26,000 hp(m) *(19.4 MW)*; 2 shafts; 2 bow thrusters
Guns: 2 Raytheon 20 mm Vulcan Phalanx Mk 15 ❶. 6 barrels per mounting; 4,500 rds/min combined to 1.5 km.
Electronic countermeasures: ESM/ECM.
Radars: Air search: Mitsubishi OPS-14C ❷; C-band.
Surface search: JRC OPS-28D ❸; G-band.
Navigation: JRC OPS-20; I-band.

Helicopters: Platform for CH-47J.

Programmes: A 5,500 ton LST was requested and not approved in the 1989 or 1990 estimates. The published design resembled the Italian San Giorgio with a large flight deck and a stern dock. No further action was taken for two years but the FY93 request included a larger ship showing the design of a USN LPH, although smaller in size. This vessel, with some modifications, was authorised in the 1993 estimates. A second of class approved in FY98 and third in FY99.
Structure: Through deck, flight deck and stern docking well make this more like a mini LHA than an LST, except that the ship is described as providing only 'platform and refuelling facilities for helicopters'.

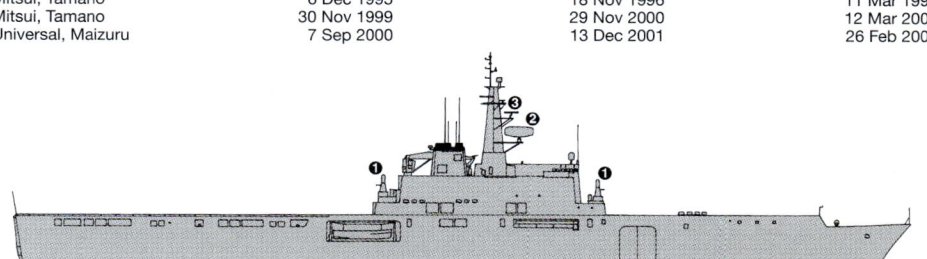

OOSUMI   (Scale 1 : 1,500), Ian Sturton / 0012652

KUNISAKI   10/2011, Hachiro Nakai / 1406843

SHIMOKITA   11/2010, Hachiro Nakai / 1406844

2 YUSOUTEI CLASS (LCU)

Name	No	Builders	Commissioned
YUSOUTEI-ICHI-GOU	LCU 2001	Sasebo Heavy Industries	17 Mar 1988
YUSOUTEI-NI-GOU	LCU 2002	Sasebo Heavy Industries	11 Mar 1992

Displacement, tonnes: 427 standard; 549 full load
Dimensions, metres (feet): 52 × 8.7 × 1.6 (170.6 × 28.5 × 5.2)
Speed, knots: 12
Complement: 28
Machinery: 2 Mitsubishi S6U-MTK diesels; 3,000 hp(m) (2.23 MW); 2 shafts
Guns: 1 GE 20 mm/76 Sea Vulcan; 3 barrels per mounting; 1,500 rds/min combined to 4 km (2.2 n miles).
Radars: Navigation: OPS-9B/26; I-band.

Comment: First approved in 1986 estimates, laid down 11 May 1987, launched 9 October 1987. Second approved in FY90 estimates, laid down 15 May 1991, launched 7 October 1991; plans for a third have been scrapped. Official names are LCU 01 and LCU 02.

YUSOTEI-ICHI-GOU 10/2010, Hachiro Nakai / 1366906

6 LANDING CRAFT AIR CUSHION (LCAC)

AIR CUSHION-TEI - 1 - GOU	LCAC 2101	AIR CUSHION-TEI - 4 - GOU	LCAC 2104
AIR CUSHION-TEI - 2 - GOU	LCAC 2102	AIR CUSHION-TEI - 5 - GOU	LCAC 2105
AIR CUSHION-TEI - 3 - GOU	LCAC 2103	AIR CUSHION-TEI - 6 - GOU	LCAC 2106

Displacement, tonnes: 102 standard; 183 full load
Dimensions, metres (feet): 26.8 on cushion; 24.7 between hard structures × 14.3; 13.1 (87.9; 81 × 46.9; 43)
Speed, knots: 40 (est.). **Range, n miles:** 300 at 35 kt, 200 at 40 kt
Complement: 5
Military lift: 24 troops; 1 MBT or 60–75 tons
Machinery: 4 Avco-Lycoming TF-40B gas turbines; 2 for propulsion and 2 for lift; 16,000 hp (12 MW) sustained; 2 shrouded reversible-pitch airscrews (propulsion); 4 double entry fans, centrifugal or mixed flow (lift)
Radars: Navigation: LN-66; I-band.

Comment: Built by Textron Marine, New Orleans for embarkation in LPDs. Approval for sale given by US on 8 April 1994. First two commissioned in March 1998, second two in March 2002 and third in February 2003. Cargo space capacity is 1,809 sq ft.

LCAC 2102 2/2013*, Hachiro Nakai / 1529146

10 LCM TYPE (LCM)

YF 2121 YF 2124–2125 YF 2127–2129 YF 2132 YF 2135 YF 2138 YF 2141

Displacement, tonnes: 25 standard
Dimensions, metres (feet): 17 × 4.3 × 0.7 (55.8 × 14.1 × 2.3)
Speed, knots: 10
Range, n miles: 130 at 9 kt
Complement: 3
Military lift: 34 tons or 80 troops
Machinery: 2 Isuzu E120-MF6R diesels; 480 hp(m) (353 kW); 2 shafts

Comment: Built in Japan. YF 2127–29 commissioned in March 1992, 2132 in March 1993, 2135 in March 1995, 2138 in March 1996 and 2141 in March 1997. YF 2150–51 are 50 ton vessels built by Yokohama Yacht and completed in March 2003. With a military lift of 100 tons they are capable of 16 kt.

YF 2127 4/2011, Hachiro Nakai / 1406846

2 YF 2150 CLASS LCM (LCM)

YF 2150–2151

Displacement, tonnes: 51 standard
Dimensions, metres (feet): 19.8 × 5.4 × 2.3 (65 × 17.7 × 7.5)
Speed, knots: 16
Complement: 4
Military lift: 100 troops or 1 vehicle
Machinery: 2 Mitsubishi S12R-MTK diesels; 3,000 hp (2.24 MW); 2 waterjets

Comment: Built in Japan by Universal, Keihin and commissioned on 19 March 2003.

YF 2150 9/2009, Hachiro Nakai / 1366085

MINE WARFARE FORCES

2 URAGA CLASS (MINESWEEPER TENDERS) (MSTH/ML)

Name	No	Builders	Launched	Commissioned
URAGA	MST 463	Hitachi, Maizuru	22 May 1996	19 Mar 1997
BUNGO	MST 464	Mitsui, Tamano	24 Apr 1997	23 Mar 1998

Displacement, tonnes: 5,741 standard; 6,960 full load
Dimensions, metres (feet): 141 × 22 × 5.4 (462.6 × 72.2 × 17.7)
Speed, knots: 22
Complement: 160
Machinery: 2 Mitsui 16V42MA diesels; 19,500 hp(m) (14.33 MW); 2 shafts
Guns: 1 Oto Melara 3 in (76 mm)/62 compact (MST 464); 85 rds/min to 16 km (8.6 n miles); weight of shell 6 kg.
Mines: Laying capability; 4 rails (Type 3). 200 mines.
Radars: Air search: OPS-14C; C-band.
Fire control: FCS 2–23; I/J-band.
Navigation: JRC OPS-39C; I-band.
Helicopters: Platform for 1 MH-53E.

Comment: First one authorised 15 February 1994 and laid down 19 May 1995; second authorised in FY95 and laid down 4 July 1996. Capable of laying mines, from four internal rails. Phalanx is planned to be fitted forward of the bridge and on the superstructure aft of the funnel.

BUNGO 5/2012, Hachiro Nakai / 1482889

2 NIIJIMA CLASS (DRONE CONTROL SHIPS) (MCSD)

Name	No	Builders	Commissioned
IESHIMA	MCL 728 (ex-MSC 673)	Hitachi, Kanagawa	19 Dec 1990
MAEJIMA	MCL 729 (ex-MSC 675)	Hitachi, Kanagawa	15 Dec 1993

Displacement, tonnes: 447 standard; 518 full load
Dimensions, metres (feet): 55 × 9.4 × 2.5 (180.4 × 30.8 × 8.2)
Speed, knots: 14
Complement: 28
Machinery: 2 Mitsubishi 12ZC diesels; 1,440 hp(m) (1.06 MW); 2 shafts
Guns: 1 GE 20 mm/76 Sea Vulcan 20; 3 barrels per mounting; 1,500 rds/min combined to 4 km (2.2 n miles).
Radars: Surface search: Fujitsu OPS-9B; I-band.

Comment: Both converted to act as minesweeper control ship (MCLs) and equipped to operate SAM remote controlled drones. All minesweeping gear removed. Ieshima converted as MCL on 26 February 2010 and Maejima in 2013.

MAEJIMA 9/2013*, Hirotoshi Yamamoto / 1529147

Mine warfare forces < **Japan** (MSDF) 445

3 YAEYAMA CLASS (MINEHUNTERS—OCEAN) (MHS)

Name	No	Builders	Launched	Commissioned
YAEYAMA	MSO 301	Hitachi Zosen, Kanagawa	29 Aug 1991	16 Mar 1993
TSUSHIMA	MSO 302	Nippon Kokan, Tsurumi	20 Sep 1991	23 Mar 1993
HACHIJYO	MSO 303	Nippon Kokan, Tsurumi	15 Dec 1992	24 Mar 1994

Displacement, tonnes: 1,016 standard; 1,219 full load
Dimensions, metres (feet): 67 × 11.8 × 3.1 (219.8 × 38.7 × 10.2)
Speed, knots: 14
Complement: 60

Machinery: 2 Mitsubishi 6NMU-TA1 diesels; 2,400 hp(m) (1.76 MW); 2 shafts; 1 hydrojet bow thruster; 350 hp(m) (257 kW)
Guns: 1 JM-61 20 mm/76 Sea Vulcan; 3 barrels per mounting; 1,500 rds/min combined to 4 km (2.2 n miles).
Radars: Surface search: Fujitsu OPS-39B; I-band.
Sonars: Raytheon SQQ-32 VDS; high frequency; active.

Comment: First two approved in 1989 estimates, third in 1990. First laid down 30 August 1990, second 20 July 1990 and third 17 May 1991. Wooden hulls. Fitted with S 7 deep sea minehunting system, S 8 (SLQ-48) deep sea moored minesweeping equipment and ADI Dyad sweeps. Appears to be a derivative of the USN Avenger class. An integrated tactical system is fitted. Termination of the programme at three of the class suggests similar problems to US ships of the same class.

HACHIJYO 9/2013*, Hirotoshi Yamamoto / 1529148

12 SUGASHIMA CLASS (MINEHUNTER (COASTAL)) (MHC)

Name	No	Builders	Launched	Commissioned
SUGASHIMA	MSC 681	Nippon Kokan, Tsurumi	25 Aug 1997	16 Mar 1999
NOTOJIMA	MSC 682	Hitachi, Kanagawa	3 Sep 1997	16 Mar 1999
TSUNOSHIMA	MSC 683	Hitachi, Kanagawa	22 Oct 1998	13 Mar 2000
NAOSHIMA	MSC 684	Nippon Kokan, Tsurumi	7 Oct 1999	16 Mar 2001
TOYOSHIMA	MSC 685	Hitachi, Kanagawa	13 Sep 2000	4 Mar 2002
UKUSHIMA	MSC 686	Universal, Keihin (Tsurumi)	17 Sep 2001	18 Mar 2003
IZUSHIMA	MSC 687	Universal, Keihin (Kanawaga)	31 Oct 2001	18 Mar 2003
AISHIMA	MSC 688	Universal, Keihin (Tsurumi)	8 Oct 2002	16 Feb 2004
AOSHIMA	MSC 689	Universal, Keihin (Kanawaga)	16 Sep 2003	9 Feb 2005
MIYAJIMA	MSC 690	Universal, Keihin (Tsurumi)	10 Oct 2003	9 Feb 2005
SHISHIJIMA	MSC 691	Universal, Keihin (Tsurumi)	29 Sep 2004	8 Feb 2006
KUROSHIMA	MSC 692	Universal, Keihin (Tsurumi)	31 Aug 2005	23 Feb 2007

Displacement, tonnes: 518 standard; 599 full load
Dimensions, metres (feet): 54 × 9.4 × 3.0 (177.2 × 30.8 × 9.8)
Speed, knots: 14
Range, n miles: 2,500 at 10 kt
Complement: 45

Machinery: 2 Mitsubishi 6 NMU-TAI diesels; 1,800 hp(m) (1.33 MW); 2 shafts; bow thrusters
Guns: 1 JM-61 20 mm/76 Sea Vulcan; 3 barrels for mounting; 1,500 rds/min combined to 4 km (2 n miles).
Radars: Surface search: Fujitsu OPS-39B; I-band.
Sonars: THALES Hitachi GEC Type 2093 VDS; high frequency; active.
Combat data systems: AMS/NEC NAUTIS-M type MCM control system.

Comment: First pair authorised in fiscal year 1995, third in FY 1996, fourth in FY 1997, fifth in FY 1998, sixth, and seventh in FY 1999, eighth in FY 2000, ninth and tenth in FY 2001, eleventh in FY 2002, and twelfth in FY 2003. Hull is similar to *Uwajima* but the upper deck is extended aft to provide more stowage for mine disposal gear, and there are twin funnels. PAP 104 Mk 5 ROVs are carried and ADI Dyad minesweeping gear fitted.

NOTOJIMA 7/2013*, Hachiro Nakai / 1529149

6 SAM CLASS (MSD)

SAM 01–06

Displacement, tonnes: 20 full load
Dimensions, metres (feet): 18 × 6.1 × 1.6 (59.1 × 20.0 × 5.2)
Speed, knots: 8
Range, n miles: 330 at 8 kt
Machinery: 1 Volvo Penta TAMD 70D diesel; 210 hp(m) (154 kW); 1 Schottel prop

Comment: First pair acquired from Karlskronavarvet, Sweden in February 1998 followed by two more in December 1998 and two more in 2000. Remote controlled magnetic and acoustic catamaran sweepers operated by *Kamishima* and *Ogishima*.

SAM 02 9/2011, Hachiro Nakai / 1406853

3 HIRASHIMA CLASS (MINESWEEPERS/MINEHUNTERS—COASTAL) (MHSC)

Name	No	Builders	Launched	Commissioned
HIRASHIMA	MSC 601	Universal, Keihin (Tsurumi)	27 Sep 2006	11 Mar 2008
YAKUSHIMA	MSC 602	Universal, Keihin (Tsurumi)	26 Sep 2007	6 Mar 2009
TAKASHIMA	MSC 603	Universal, Keihin (Tsurumi)	25 Sep 2008	26 Feb 2010

Displacement, tonnes: 579 standard; 660 full load
Dimensions, metres (feet): 57 × 9.8 × 4.4 (187 × 32.2 × 14.4)
Speed, knots: 14
Complement: 45
Machinery: 2 Mitsubishi 6 NMU diesels; 2,200 hp (1.64 MW); 2 shafts; bow thrusters
Guns: 1—20 mm Sea Vulcan.
Sonars: Hitachi ZQS 4; hull-mounted; high frequency.

Comment: First authorised in FY04 budget, second in FY05 budget and third in FY06 budget. Wooden hull. Equipped with S-10 minesweeping and disposal system.

TAKASHIMA 11/2012, Hachiro Nakai / 1482892

5 UWAJIMA CLASS (MINEHUNTERS/SWEEPERS—COASTAL) (MHSC)

Name	No	Builders	Commissioned
KUMEJIMA	MSC 676	Nippon Kokan, Tsurumi	12 Dec 1994
MAKISHIMA	MSC 677	Hitachi, Kanagawa	12 Dec 1994
TOBISHIMA	MSC 678	Nippon Kokan, Tsurumi	10 Mar 1995
YUGESHIMA	MSC 679	Hitachi, Kanagawa	11 Mar 1996
NAGASHIMA	MSC 680	Nippon Kokan, Tsurumi	25 Dec 1996

Displacement, tonnes: 498 standard; 579 full load
Dimensions, metres (feet): 58 × 9.4 × 2.5 (190.3 × 30.8 × 8.2)
Speed, knots: 14
Range, n miles: 2,500 at 10 kt
Complement: 45

Machinery: 2 Mitsubishi 6NMU-TAI diesels; 1,800 hp(m) (1.3 MW); 2 shafts
Guns: 1 JM-61 20 mm/76 Sea Vulcan 20; 3 barrels per mounting; 1,500 rds/min combined to 4 km (2.2 n miles).
Radars: Surface search: Fujitsu OPS-39; I-band.
Sonars: Nec/Hitachi ZQS 3; hull-mounted; minehunting; high frequency.

Programmes: First ordered in 1976. Last two authorised in FY94.
Structure: Hulls are made of wood.
Operational: Fitted with S 7 mine detonating equipment, a remote-controlled counter-mine charge. Four clearance divers are carried. Earlier vessels of the class converted to drone control or paid off at a rate of one or two a year. *Tsukishima* decommissioned on 21 March 2012 and *Maejima* converted to MCL role in 2013.

MAEJIMA 7/2011, Hachiro Nakai / 1406854

446　Japan (MSDF) > Mine warfare forces — Survey and research ships

2 + 1 ENOSHIMA CLASS (MINESWEEPERS—COASTAL) (MSC)

Name	No	Builders	Laid down	Launched	Commissioned
ENOSHIMA	MSC 604	Universal, Keihin (Tsurumi)	14 May 2009	25 Oct 2010	21 Mar 2012
CHICHIJIMA	MSC 605	Universal, Keihin (Tsurumi)	24 May 2010	24 Nov 2011	21 Mar 2013
HATUSHIMA	MSC 606	Universal, Keihin (Tsurumi)	26 Apr 2012	6 Dec 2013	Mar 2015

Displacement, tonnes: 570 standard
Dimensions, metres (feet): 60 × 10.1 × 4.5 *(196.9 × 33.1 × 14.8)*
Speed, knots: 14
Complement: 45
Machinery: 2 6NMU-TK(B)E diesels; 2,200 hp *(1.64 MW)*; 2 shafts
Guns: 1 JM-61 *20 mm*/76 Sea Vulcan.

Comment: A larger, improved version of the Hirashima class. FRP construction. First authorised in FY08 budget, second in FY09 budget and third in FY11 budget.

ENOSHIMA　　　　　　　　　　　　　　　　6/2012, *Ships of the World* / 1455667

0 + 1 690 TON CLASS (MINEHUNTERS—COASTAL) (MHSC)

Name	No	Builders	Launched	Commissioned
–	MSO 304	Japan Marine United, Tsurumi	2015	Mar 2017

Displacement, tonnes: 690 standard
Dimensions, metres (feet): 67 × ? × ? *(219.8 × ? × ?)*
Speed, knots: 14
Machinery: 2 diesels; 2 shafts
Guns: 1—20 mm Sea Vulcan.
Sonars: Hitachi ZQS-4 VDS.

Comment: First authorised in fiscal year 2013 budget. FRP construction.

SURVEY AND RESEARCH SHIPS

Notes: Survey ships are also included in the Coast Guard section.

2 HIBIKI CLASS (OCEAN SURVEILLANCE SHIPS) (AGOSH)

Name	No	Builders	Launched	Commissioned
HIBIKI	AOS 5201	Mitsui, Tamano	27 Jul 1990	30 Jan 1991
HARIMA	AOS 5202	Mitsui, Tamano	11 Sep 1991	10 Mar 1992

Displacement, tonnes: 2,896 standard; 3,048 full load
Dimensions, metres (feet): 67 × 29.9 × 7.5 *(219.8 × 98.1 × 24.6)*
Speed, knots: 11; 3 towing. **Range, n miles:** 3,800 at 10 kt
Complement: 40

Machinery: Diesel-electric; 4 Mitsubishi Stu diesels; 3,000 hp(m) *(2.2 MW)*; 4 generators; 2 motors; 3,000 hp(m) *(2.2 MW)*; 2 shafts
Radars: Surface search: JRC OPS-16; G-band.
Navigation: Koden OPS-9; I-band.
Sonars: UQQ 2 SURTASS; passive surveillance.
Helicopters: Platform only.

Comment: First authorised 24 January 1989, laid down 28 November, second approved in fiscal year 1990, laid down 26 December 1990. Auxiliary Ocean Surveillance (AOS) ships to a SWATH design similar to USN TAGOS-19 class. A data collection station is based at Yokosuka Bay using WSC-6 satellite data relay to the AOS.

HIBIKI　　　　　　　　　　　　　　　　　9/2013*, *Hachiro Nakai* / 1529150

1 SYOUNAN CLASS (SURVEY SHIP) (AGS)

Name	No	Builders	Laid down	Launched	Commissioned
SYOUNAN	AGS 5106	Mitsui, Tamano	16 Dec 2008	29 Jun 2009	17 Mar 2010

Displacement, tonnes: 3,251 standard
Dimensions, metres (feet): 103 × 16.4 × 4.5 *(337.9 × 53.8 × 14.8)*
Speed, knots: 16
Complement: 80
Machinery: Diesel-electric; 4,825 hp *(3.6 MW)*; 2 shafts
Radars: Navigation: I-band.

Comment: New survey ship authorised in FY07 budget.

SYOUNAN　　　　　　　　　　　　　　　11/2010, *Hachiro Nakai* / 1406855

1 NICHINAN CLASS (SURVEY SHIP) (AGS)

Name	No	Builders	Launched	Commissioned
NICHINAN	AGS 5105	Mitsubishi, Shimonoseki	11 Jun 1998	24 Mar 1999

Displacement, tonnes: 3,353 standard; 4,572 full load
Dimensions, metres (feet): 111 × 17 × 4.5 *(364.2 × 55.8 × 14.8)*
Speed, knots: 18
Complement: 90
Machinery: Diesel-electric; 2 Mitsubishi S16U diesel generators; 3 motors; 3,600 hp(m) *(2.7 MW)*; 2 shafts; bow and stern thrusters

Comment: Authorisation approved in FY96. Combination cable repair and hydrographic survey ship. Equipped with one ROV.

NICHINAN　　　　　　　　　　　　　　　8/2007, *Hachiro Nakai* / 1305104

1 SUMA CLASS (AGS)

Name	No	Builders	Launched	Commissioned
SUMA	AGS 5103	Hitachi, Maizuru	1 Sep 1981	30 Mar 1982

Displacement, tonnes: 1,199 standard; 1,727 full load
Dimensions, metres (feet): 72 × 12.8 × 3.4 *(236.2 × 42.0 × 11.2)*
Speed, knots: 15
Complement: 64 + 5 scientists
Machinery: 2 Fuji 6L27.5XF diesels; 3,000 hp(m) *(2.24 MW)*; 2 shafts; cp props; bow thruster
Electronic countermeasures: ESM: NOLR-6.
Radars: Navigation: OPS-20; I-band.

Comment: Laid down 24 September 1980. Carries an 11 m launch for surveying work.

SUMA　　　　　　　　　　　　　　　　　3/2011, *Hachiro Nakai* / 1406856

1 FUTAMI CLASS (AGS)

Name	No	Builders	Laid down	Launched	Commissioned
WAKASA	AGS 5104	Hitachi, Maizuru	21 Aug 1984	21 May 1985	25 Feb 1986

Displacement, tonnes: 2,083 standard; 3,226 full load
Dimensions, metres (feet): 97 × 15 × 4.2 *(318.2 × 49.2 × 13.8)*
Speed, knots: 16
Complement: 95
Machinery: 2 Fuji 8L27.5XF diesels; 3,250 hp(m) *(2.39 MW)*; 2 shafts; cp props; bow thruster
Radars: Navigation: JRC OPS-18-3; G-band.

Comment: Built to merchant marine design. Carries an RCV-225 remote-controlled rescue/underwater survey submarine.

WAKASA　　　　　　　　　　　　　　　11/2009, *Hachiro Nakai* / 1366092

IHS Jane's Fighting Ships 2014-2015　　　　　　　　　　　　　　　　　　　　　　© 2014 IHS

1 ASUKA CLASS (AGEH)

Name	No	Builders	Launched	Commissioned
ASUKA	ASE 6102	Sumitomo, Uraga	21 Jun 1994	22 Mar 1995

Displacement, tonnes: 4,318 standard; 6,299 full load
Dimensions, metres (feet): 151 × 17.3 × 5 (495.4 × 56.8 × 16.4)
Speed, knots: 27
Complement: 70 + 100 scientists

Machinery: COGLAG; 2 IHI/GE LM 2500 gas turbines; 43,000 hp (31.6 MW); 2 shafts; cp props
Missiles: SAM: 8 cell VLS.
Radars: Air search: SPY-1D type; E/F-band.
Air/surface search: Melco OPS-14C; D-band.
Surface search: JRC OPS-18-1; G-band.
Fire control: Type 3; I/J-band.
Sonars: Bow-mounted; active search; medium frequency. Towed passive/active array in due course.
Weapon control systems: Type 3 FCS.
Helicopters: 1 SH-60J Seahawk.

Comment: Included in the FY92 programme and laid down 21 April 1993. For experimental and weapon systems testing which started with the FCS 3 in 1996. The bow sonar dome extends aft to the bridge. The VLS system is on the forecastle. Surveillance and countermeasures systems are also evaluated.

ASUKA   10/2012, Hachiro Nakai / 1482893

RESCUE VEHICLES

2 RESCUE SUBMARINES (DSRV)

Displacement, tonnes: 41 standard
Dimensions, metres (feet): 12.4 × 3.2 × 4.6 (40.7 × 10.5 × 15.1)
Speed, knots: 4
Complement: 2
Machinery: Electric; 30 hp (22 kW); single shaft

Comment: Rescue submersibles built by Kawasaki Heavy Industries, Kobe and delivered on 27 August 1999. Space for 12 people. Sonars are fitted on the bow, upper and lower casings for depth sounding and obstacle avoidance. Can be deployed in the submarine rescue ships Chiyoda (AS 405) and Chihaya (ASR 403).

DSRV   10/2012, Hachiro Nakai / 1482894

TRAINING SHIPS

1 KASHIMA CLASS (TRAINING SHIP) (AXH/TV)

Name	No	Builders	Launched	Commissioned
KASHIMA	TV 3508	Hitachi, Maizuru	23 Feb 1994	26 Jan 1995

Displacement, tonnes: 4,115 standard; 5,487 full load
Dimensions, metres (feet): 143 × 18 × 4.6 (469.2 × 59.1 × 15.1)
Speed, knots: 25
Range, n miles: 7,000 at 18 kt
Complement: 360 (125 midshipmen)

Machinery: CODOG; 2 RR Spey SM1C gas-turbines; 27,000 hp (20.1 MW) sustained; 2 Mitsubishi S16U-MTK diesels; 8,000 hp(m) (5.88 MW); 2 shafts
Guns: 1 Oto Melara 76 mm/62. 2—40 mm saluting guns.
Torpedoes: 6—324 mm (2 triple) tubes.
Radars: Air/surface search: Melco OPS-14C; D-band.
Surface search: JRC OPS-18-1; D-band.
Navigation: Fujitsu OPS-20; I-band.
Fire control: Type 2-23; I/J-band.
Sonars: Hull-mounted; active search and attack; medium frequency. OQS-4.
Helicopters: Platform for 1 medium.

Comment: Approved in fiscal year 1991 as a dedicated training ship but the project postponed to FY 1992 as a budget saving measure. Laid down 20 April 1993.

KASHIMA   8/2013*, Michael Nitz / 1530911

3 SHIMAYUKI CLASS (TRAINING SHIP) (AXGHM/TV)

Name	No	Builders	Laid down	Launched	Commissioned
SHIMAYUKI	TV 3513 (ex-DD 133)	Mitsubishi, Nagasaki	8 May 1984	29 Jan 1986	17 Feb 1987
SHIRAYUKI	TV 3517 (ex-DD 123)	Hitachi, Maizuru	3 Dec 1979	4 Aug 1981	8 Feb 1983
SETOYUKI	TV 3518 (ex-DD 131)	Mitsui, Tamano	26 Jan 1984	3 Jul 1985	17 Dec 1986

Displacement, tonnes: 3,099 standard; 4,267 full load
Dimensions, metres (feet): 130 × 13.6 × 4.4 (426.5 × 44.6 × 14.4)
Speed, knots: 30; 19 Tynes
Complement: 200

Machinery: COGOG; 2 Kawasaki-RR Olympus TM3B gas turbines; 45,000 hp (33.5 MW) sustained; 2 RR Type RM1C gas turbines; 9,900 hp (7.4 MW) sustained; 2 shafts; cp props
Missiles: SSM: 8 McDonnell Douglas Harpoon (2 quad) launchers; active radar homing to 130 km (70 n miles) at 0.9 Mach; warhead 227 kg.
SAM: Raytheon Sea Sparrow RIM-7M Mk 29 Type 3A launcher; semi-active radar homing to 16 km (8.5 n miles) at 2.5 Mach; warhead 38 kg; 12 missiles.
A/S: Honeywell ASROC Mk 112 octuple launcher; inertial guidance to 1.6–10 km (1–5.4 n miles) at 0.9 Mach; payload Mk 46 Mod 5 Neartip.
Guns: 1 Oto Melara 3 in (76 mm)/62 compact; 85 rds/min to 16 km (8.6 n miles) anti-surface; 12 km (6.5 n miles) anti-aircraft; weight of shell 6 kg. 2 Raytheon 20 mm Phalanx Mk 15 CIWS; 6 barrels per mounting; 3,000 rds/min combined to 1.5 km.
Torpedoes: 6—324 mm Type 68 (2 triple) tubes. Honeywell Mk 46 Mod 5 Neartip; anti-submarine; active/passive homing to 11 km (5.9 n miles) at 40 kt; warhead 44 kg.
Physical countermeasures: Decoys: 2 Loral Hycor SRBOC 6-barrelled Mk 36 chaff launchers; range 4 km (2.2 n miles).
Electronic countermeasures: ESM: NOLR 6C; intercept.
ECM: Fujitsu OLT 3; jammer.
Radars: Air search: Melco OPS-14B; D-band.
Surface search: JRC OPS-18-1; G-band.
Fire control: Type 2–12 A; I/J-band (for SAM). 2 Type 2-21/21A; I/J-band (for guns).
Tacan: ORN-6C.
Sonars: Nec OQS 4A (II) (SQS-23 type); bow-mounted; active search and attack; low frequency.
Combat data systems: OYQ-5 action data automation; Link 14 (receive only). SATCOM.
Helicopters: Platform for 1 SH-60J Seahawk.

Comment: Shimayuki converted to training ship in March 1999, Shirayuki on 16 March 2011 and Setoyuki in March 2012. Helicopter hangar converted to lecture rooms.

SHIRAYUKI   7/2013*, Maritime Photographic / 1530910

1 TRAINING TENDER (YXT)

YTE 12

Displacement, tonnes: 112 standard; 183 full load
Dimensions, metres (feet): 35 × 7.5 × 1.5 (114.8 × 24.6 × 4.9)
Speed, knots: 14
Machinery: 2 diesels; 2 shafts
Radars: To be announced.

Comment: Entered service in 1983.

YTE 12   6/2013*, Hirotoshi Yamamoto / 1529151

448 Japan (MSDF) > Training ships — Auxiliaries

1 TRAINING TENDER (YXT)

YTE 13

Displacement, tonnes: 182 standard
Dimensions, metres (feet): 35.3 × 7.4 × 1.72 (115.8 × 24.3 × 5.6)
Speed, knots: 16
Machinery: 2 Yanmar 12 LAK ST2 diesels; 2,200 hp(m) (1.16 MW); 2 shafts

Comment: Approved in FY00 budget and commissioned in 2002. Assigned to 1st Maritime Service School for cadet training.

YTE 13   10/2011, Hachiro Nakai / 1406859

1 TENRYU CLASS (TRAINING SUPPORT SHIP) (AVHM/TV)

Name	No	Builders	Launched	Commissioned
TENRYU	ATS 4203	Sumitomo, Uraga	14 Apr 1999	17 Mar 2000

Displacement, tonnes: 2,489 standard; 2,794 full load
Dimensions, metres (feet): 106 × 16.5 × 4.1 (347.8 × 54.1 × 13.5)
Speed, knots: 22
Complement: 140

Machinery: 4 Niigata 8MG28H diesels; 12,800 hp(m) (9.5 MW) sustained; 2 shafts
Guns: 1 Oto Melara 3 in (76 mm)/62 compact; 85 rds/min to 16 km (8.6 n miles); weight of shell 6 kg.
Radars: Air/surface search: Melco OPS-14; D-band.
Surface search: OPS-28D; G/H-band.
Fire control: FCS 2–22; I/J-band.
Helicopters: 1 medium.

Comment: Authorised in 1997 budget as a replacement for *Azuma* and laid down 19 June 1998. Carries four BQM-34J drones and four Northrop Chukar III drones used for evaluating performance of ships SAM systems. Improved 'Kurobe' design.

TENRYU   10/2012, Hachiro Nakai / 1482897

1 KUROBE CLASS (TRAINING SUPPORT SHIP) (AVM/TV)

Name	No	Builders	Laid down	Launched	Commissioned
KUROBE	ATS 4202	Nippon Kokan, Tsurumi	31 Jul 1987	23 May 1988	23 Mar 1989

Displacement, tonnes: 2,235 standard; 2,794 full load
Dimensions, metres (feet): 101 × 16.5 × 4 (331.4 × 54.1 × 13.1)
Speed, knots: 20
Complement: 155 (17 officers)

Machinery: 4 Fuji 8L27.5XF diesels; 9,160 hp(m) (6.8 MW); 2 shafts; cp props
Guns: 1 FMC/Oto Melara 3 in (76 mm)/62 Mk 75; 85 rds/min to 16 km (8.6 n miles) anti-surface; 12 km (6.5 n miles) anti-aircraft; weight of shell 6 kg.
Radars: Air search: Melco OPS-14C; D-band.
Surface search: JRC OPS-18-1; G-band.
Fire control: Type 2–22; I/J-band.

Comment: Approved under 1986 estimates, laid down 31 July 1987, launched 23 May 1988. Carries four BQM-34AJ high-speed drones and four Northrop Chukar II drones with two stern launchers. Used for training crews in anti-aircraft operations and evaluating the effectiveness and capability of ships' anti-aircraft missile systems.

KUROBE   10/2011, Hachiro Nakai / 1406861

IHS Jane's Fighting Ships 2014-2015

AUXILIARIES

2 MASHUU CLASS
(FAST COMBAT SUPPORT SHIPS) (AOE/AORH)

Name	No	Builders	Laid down	Launched	Commissioned
MASHUU	AOE 425	Mitsui, Tamano	21 Jan 2002	5 Feb 2003	15 Mar 2004
OUMI	AOE 426	Universal, Maizuru	7 Feb 2003	19 Feb 2004	3 Mar 2005

Displacement, tonnes: 13,717 standard; 25,401 full load
Dimensions, metres (feet): 221 × 27 × 8.3 (725.1 × 88.6 × 27.2)
Speed, knots: 24
Complement: 145

Machinery: 2 Kawasaki RR Spey SM1C gas turbines; 40,000 hp (29.8 MW); 2 shafts
Guns: 2—20 mm CIWS (to be fitted).
Physical countermeasures: Decoys: 4 SRBOC Mk 36 chaff and IR launchers.
Radars: Navigation: I-band.
Helicopters: 2 medium.

Comment: First ship approved in FY00 and second in FY01. Capacity for 30 containers. Cranes capable of lifting 15 tons. Three replenishment at sea positions on each side.

OUMI   11/2012, Hachiro Nakai / 1482899

6 300 TON CLASS (EOD TENDERS) (YDT)

YDT 01–06

Displacement, tonnes: 305 standard
Dimensions, metres (feet): 46 × 8.6 × 2.2 (150.9 × 28.2 × 7.2)
Speed, knots: 15
Complement: 30 (15 divers)
Machinery: 2 Niigata 6NSDL diesels; 1,500 hp(m) (1.1 MW); 2 shafts
Radars: Navigation: I-band.

Comment: Built by Maehata Zousen. First pair approved in fiscal year 1998, third in FY 1999, fourth in FY 2000 and fifth and sixth in FY 2001. First two commissioned 24 March 2000, third on 21 March 2001, fourth in December 2001, and last two on 14 March 2003. Used as diving tenders.

YDT 05   3/2013*, Hachiro Nakai / 1529152

1 CHIYODA CLASS (SUBMARINE TENDER DEPOT AND RESCUE SHIP) (AS/ASRH)

Name	No	Builders	Launched	Commissioned
CHIYODA	AS 405	Mitsui, Tamano	7 Dec 1983	27 Mar 1985

Displacement, tonnes: 3,709 standard; 5,487 full load
Dimensions, metres (feet): 113 × 17.6 × 4.6 (370.7 × 57.7 × 15.1)
Speed, knots: 17
Complement: 120 + 80 spare berths

Machinery: 2 Mitsui 8L42M diesels; 11,500 hp(m) (8.6 MW); 2 shafts; cp props; bow and stern thrusters
Radars: Navigation: JRC OPS-16; G-band.
Sonars: SQS-36D.
Helicopters: Platform for up to MH-53 size.

Comment: Laid down 19 January 1983. Carries a 40 ton Deep Submergence Rescue Vehicle (DSRV), which is lowered and recovered through a centreline moonpool. The DSRV can mate to a decompression chamber. A personnel transfer capsule can also be deployed. Flagship Second Submarine Flotilla based at Yokosuka.

CHIYODA   8/2011, Hachiro Nakai / 1406870

© 2014 IHS

Auxiliaries < **Japan** (MSDF) 449

1 CHIHAYA CLASS (SUBMARINE RESCUE SHIP) (ASRH)

Name	No	Builders	Launched	Commissioned
CHIHAYA	ASR 403	Mitsui, Tamano	8 Oct 1998	23 Mar 2000

Displacement, tonnes: 5,537 standard; 7,011 full load
Dimensions, metres (feet): 128 × 20 × 5.1 (419.9 × 65.6 × 16.7)
Speed, knots: 21
Complement: 125
Machinery: 2 Mitsui 12V 42M-A diesels; 19,500 hp(m) (14.33 MW); 2 shafts; 2 bow and 2 stern thrusters
Radars: Navigation: OPS-20; I-band.
Helicopters: Platform for up to MH-53 size.

Comment: Authorisation approved in the 1996 budget as a replacement for *Fushimi*. Laid down 13 October 1997. Fitted with a search sonar and carries a 40 ton DSRV. Also used as a hospital ship.

CHIHAYA   9/2009, Hachiro Nakai / 1366101

3 TOWADA CLASS
(FAST COMBAT SUPPORT SHIPS) (AOE/AORH)

Name	No	Builders	Laid down	Launched	Commissioned
TOWADA	AOE 422	Hitachi, Maizuru	17 Apr 1985	25 Mar 1986	24 Mar 1987
TOKIWA	AOE 423	Ishikawajima Harima, Tokyo	12 May 1988	23 Mar 1989	12 Mar 1990
HAMANA	AOE 424	Hitachi, Maizuru	8 Jul 1988	18 May 1989	29 Mar 1990

Displacement, tonnes: 8,281 standard; 16,104 full load
Dimensions, metres (feet): 167 × 22 × 8.2 (547.9 × 72.2 × 26.9)
Speed, knots: 22
Range, n miles: 10,500 at 20 kt
Complement: 140

Cargo capacity: 5,700 tons
Machinery: 2 Mitsui 16V42MA diesels; 26,000 hp(m) (19.4 MW); 2 shafts
Physical countermeasures: Decoys: 2 chaff launchers can be fitted.
Radars: Surface search: JRC OPS-18-1/28C; G-band.
Helicopters: Platform for MH-53 size.

Comment: First approved under 1984 estimates, laid down 17 April 1985. Second and third of class in 1987 estimates. AOE 423 laid down 12 May 1988, and AOE 424 8 July 1988. Three replenishment at sea positions on each side (two fuel only, one stores).

HAMANA   4/2013*, Hachiro Nakai / 1529153

29 HARBOUR TANKERS (YO/YW/YG)

YO 25–27   YO 29–31   YO 33–40   YW 17–24   YO 28   YO 32   YG 203–207

Comment: There are: 14 of 490 tonnes (YO 25–27, 29–31, 33–40); eight of 310 tonnes (YW 17–24); seven of 270 tonnes (YO 28, YO 32, YG 203–207).

YO 36   5/2013*, Hachiro Nakai / 1529154

2 FIREFIGHTING TENDERS (YTR)

YR 01–02

Displacement, tonnes: 61 standard
Dimensions, metres (feet): 25 × 5.5 × 1.1 (82 × 18.0 × 3.6)
Speed, knots: 19
Complement: 10
Machinery: 1 Isuzu Marine UM6WGITCG diesels; 750 hp (560 kW); 2 Isuzu Marine UM6RB diesels; 1,040 hp (775 kW); 3 shafts

Comment: Built in Japan by Ishikawajima-Harima Heavy Industries. *YR 01* approved in FY99 budget and commissioned in 2001. *YR 02* approved in FY00 budget and commissioned in 2002. Fitted with three waterjets forward and a crane aft.

YR 01   9/2002, Takatoshi Okano / 0570888

1 HASHIDATE CLASS (ASY/YAC)

Name	No	Builders	Launched	Commissioned
HASHIDATE	ASY 91	Hitachi, Kanagawa	26 Jul 1999	30 Nov 1999

Displacement, tonnes: 406 standard; 498 full load
Dimensions, metres (feet): 62 × 9.4 × 2.0 (203.4 × 30.8 × 6.6)
Speed, knots: 20
Range, n miles: 1,000 at 12 kt
Complement: 29 + 130 spare berths
Machinery: 2 Niigata 16V 16FX diesels; 5,500 hp(m) (4.04 MW); 2 shafts

Comment: Authorised in FY97 budget. Laid down 28 October 1998. Replaced *Hiyodori* as a ceremonial yacht. Has facilities for disaster relief. Based at Yokosuka.

HASHIDATE   8/2012, Kazumasa Watanabe / 1482903

7 LANDING CRAFT (LIGHTER) (YL)

YL 9–15

Displacement, tonnes: 122 full load
Dimensions, metres (feet): 27 × 7.0 × 1.04 (88.6 × 23.0 × 3.4)
Speed, knots: 10
Complement: 5
Machinery: 2 Isuzu diesels; 560 hp (410 kW); 2 shafts

Comment: Cargo lighters constructed by Ishihara, Takasogo. First entered service in 1980 and latest in 1998. Equipped with a bow ramp and two 2 ton cranes.

YL 10   10/2011, Hachiro Nakai / 1406869

5 HIUCHI CLASS (MULTIPURPOSE SUPPORT SHIPS) (YTT)

Name	No	Builders	Launched	Commissioned
HIUCHI	AMS 4301	Nippon Kokan, Tsurumi	4 Sep 2001	27 Mar 2002
SUOU	AMS 4302	Universal, Keihin (Tsurumi)	25 Apr 2003	16 Mar 2004
AMAKUSA	AMS 4303	Universal, Keihin (Tsurumi)	6 Aug 2003	16 Mar 2004
GENKAI	AMS 4304	Universal, Keihin (Tsurumi)	24 May 2007	20 Feb 2008
ENSHUU	AMS 4305	Universal, Keihin (Tsurumi)	9 Aug 2007	20 Feb 2008

Displacement, tonnes: 996 standard
Dimensions, metres (feet): 65 × 12 × 3.5 *(213.3 × 39.4 × 11.5)*
Speed, knots: 15
Complement: 40
Machinery: 2 Daihatsu 6 DKM-28 (L) diesels; 5,000 hp(m) *(3.67 MW)*; 2 shafts
Radars: Navigation: OPS-26B; I-band.

Comment: First authorised in fiscal year 1999, two more in FY 2001 and two further in FY 2005 budget. Equipped for torpedo launch and recovery. Replaced ASU 81 class. Used as an ocean tug.

HIUCHI *7/2013*, Hachiro Nakai* / 1529155

1 CABLE REPAIR SHIP (ARC)

Name	No	Builders	Laid down	Launched	Commissioned
MUROTO	ARC 483	Mitsubishi, Shimonoseki	Sep 2011	5 Jul 2012	15 Mar 2013

Displacement, tonnes: 4,950 standard; 6,400 full load
Dimensions, metres (feet): 131 × 19.0 × 5.7 *(429.8 × 62.3 × 18.7)*
Speed, knots: 16
Machinery: Diesel-electric; 3 diesels; 7,300 hp *(5.44 MW)*; 2 shafts
Radars: To be announced.

Comment: Authorised in fiscal year 2009 budget. Has replaced old *Muroto*.

MUROTO *3/2013*, Kenji Konishi* / 1529156

ICEBREAKERS

1 ICEBREAKER (AGBH)

Name	No	Builders	Laid down	Launched	Commissioned
SHIRASE	AGB 5003	Universal, Maizuru	15 Mar 2007	16 Apr 2008	20 May 2009

Displacement, tonnes: 12,701 standard; 20,321 full load
Dimensions, metres (feet): 138 × 28.0 × 9.2 *(452.8 × 91.9 × 30.2)*
Speed, knots: 19.5
Range, n miles: 30,000 at 15 kt
Complement: 179 (34 officers) + 80 scientists

Cargo capacity: 1,100 tons
Machinery: Diesel-electric; 4 Mitsui 16V42M-B diesels; 4 generators; 4 motors; 30,000 hp *(22 MW)*; 2 shafts
Radars: Surface search: JRC OPS-18-3; G/H-band.
Navigation: Fujitsu OPS-39D; I-band.
Tacan: ORN-6E.
Helicopters: 2 CH-101.

Comment: New Antarctic expedition ship which has replaced the decommissioned ship of the same name.

SHIRASE *3/2013*, Chris Sattler* / 1530912

TUGS

28 OCEAN TUGS (ATA/YT)

| YT 58 | YT 78–79 | YT 84 | YT 89–90 | YT 94–97 | YT 01–02 |
| YT 63–74 | YT 81 | YT 86 | YT 92 | YT 99 | |

Displacement, tonnes: 264 standard
Dimensions, metres (feet): 28.4 × 8.6 × 2.5 *(93.2 × 28.2 × 8.2)*
Speed, knots: 11
Complement: 10
Machinery: 2 Niigata 6L25B diesels; 1,800 hp(m) *(1.32 MW)*; 2 shafts

Comment: YT 58 entered service on 31 October 1978, YT 63 on 27 September 1982, YT 64 on 20 September 1983, YT 65 on 20 September 1984, YT 66 on 20 September 1985, YT 67 on 30 September 1986, YT 68 on 9 September 1987, YT 69 on 16 September 1987, YT 70 on 2 September 1988, YT 71 on 28 July 1989, YT 72 on 27 July 1990, YT 73 on 31 July 1991, YT 74 on 30 September 1991, YT 78 on 28 July 1994, YT 79 on 29 September 1994, YT 81 on 30 July 1996, YT 84 on 20 October 1998, YT 86 on 21 March 2000, YT 89 and 90 on 16 March 2001, YT 92 on 17 March 2006 and YT 94 on 16 March 2007. YT 95 and 96 entered service on 15 March 2010, YT 97 on 30 May 2011, YT 99 on 19 March 2012, YT 01 on 21 January 2013, and YT 02 on 22 January 2013.

YT 02 *7/2013*, Hachiro Nakai* / 1529157

16 COASTAL AND HARBOUR TUGS (YTM/YTB)

| YT 03 | YT 75–77 | YT 82–83 | YT 87–88 | YT 93 |
| YT 60–62 | YT 80 | YT 85 | YT 91 | YT 98 |

Displacement, tonnes: 54 standard
Dimensions, metres (feet): 17 × 4.8 × 2.4 *(55.8 × 15.7 × 7.9)*
Speed, knots: 8
Complement: 4
Machinery: 2 Isuzu UM6SD1TCB diesels; 500 hp *(373 kW)*; 2 shafts

Comment: Details given are for 50 ton class (YT 75–77, YT 80, YT 82–83, YT 85, YT 87–88, YT 91, YT 93, YT 98, and YT 03). There are also one of 30 ton class (YT 62) and three of 35 ton class (YT 60–62).

YT 75 *3/2011, Hachiro Nakai* / 1406872

COAST GUARD
KAIJYOU HOANCHOU

Headquarters Appointments

Commandant of the Coast Guard: Sato Yuji

Establishment

The Japan Coast Guard (Maritime Safety Agency before 1 April 2000) was established on 1 May 1948. Its five missions are Maintenance of Maritime Order, Maritime Search and Rescue, Maritime Environmental Protection and Enforcement, Maritime Traffic Safety and Co-operation with other national and international agencies. The HQ is at Tokyo, the Coast Guard Academy is at Kure and the Coast Guard School is at Maizuru.

The main operational branches are the Guard and Rescue, the Hydrographic and the Maritime Traffic Departments. The nation is divided into 11 regions to facilitate operations. Each region has a regional HQ under which there are various Coast Guard Offices, Coast Guard Stations, Air Stations, Hydrographic Observatory and Traffic Advisory Service Centres. The 11 Regional HQs are at: Otaru (1st); Shiogama (2nd); Yokohama (3rd); Nagoya (4th); Kobe (5th); Hiroshima (6th); Kitakyushu (7th); Maizuru (8th); Niigata (9th); Kagoshima (10th) and Naha, Okinawa (11th).

Personnel

2014: 12,258 (2,630 officers)

Strength of the Fleet

Type	Active	Building
GUARD AND RESCUE SERVICE		
Patrol Vessels:		
Large with helicopter (PLH)	14	—
Large (PL)	40	8 (6)
Medium (PM)	38	2
Small (PS)	27	—
Firefighting Vessels (FL)	1	—
Patrol Craft:		
Patrol Craft (PC)	69	9
Patrol Craft (CL)	170	—
Special Service Craft:		
Monitoring Craft (MS)	2	—
Guard Boats (GS)	2	—
Surveillance Craft (SS)	42	—

Strength of the Fleet – *continued*

Type	Active	Building
HYDROGRAPHIC SERVICE		
Surveying Vessels:		
Large (HL)	5	—
Small (HS)	7	—
AIDS TO NAVIGATION SERVICE		
Aids to Navigation Tenders:		
Medium (LM)	7	—
Small (LS)	5	—

DELETIONS

2011 *Okitsu, Misasa, Takachiho, Kitagumo, Katsuragi, Bizan, Komayuki, Yukigumo,* CL 233, CL 239, CL 256, CL 259, CL 261, CL 264, SS 32, LS 170, LS 195, LS 216, LS 219, LS 221

2012 *Shiretoko, Shimokita, Shirakami, Bizan, Shizuki, Kongou, Asoyuki, Hatagumo, Kawagiri, Tosagiri, Tsushima, Ginga, Zuiun, Katsura*

2013 *Shoryu, Suiryu, Shiraito, Minoo, Ryusei, Kiyotaki, Kairyu, Nanryu, Natsui, Bihoro*

LARGE PATROL VESSELS

2 SHIKISHIMA CLASS (PLH/PSOH)

Name	No	Builders	Laid down	Launched	Commissioned
SHIKISHIMA	PLH 31	Ishikawajima Harima, Tokyo	24 Aug 1990	27 Jun 1991	8 Apr 1992
AKITSUSHIMA	PLH 32	IHI Marine United, Yokohama	5 Oct 2011	4 Jul 2012	27 Nov 2013

Displacement, tonnes: 6,604 standard; 9,500 full load
Dimensions, metres (feet): 150 × 16.5 × 9.0 *(492.1 × 54.1 × 29.5)*
Speed, knots: 25
Range, n miles: 20,000 at 18 kt
Complement: 140 (30 air crew)

Machinery: 4 SEMT-Pielstick 12 PC2 6V400; 36,000 hp(m) *(26.48 MW)*; 2 shafts; bow thruster
Guns: 4 Oerlikon 35 mm/90 Type GDM-C (2 twin); 1,100 rds/min to 6 km *(3.2 n miles)*; weight of shell 1.55 kg. 2 JM-61 MB 20 mm Gatling.
Radars: Air/surface search: Melco Ops 14; D-band.
Surface search: JMA 1576; I-band.
Navigation: JMA 1596; I-band.
Helo control: JMA 3000; I-band.
Tacan: ORN-6 (URN 25).

Helicopters: 2 Aerospatiale A 332 L1.

Comment: *Shikishima* authorised in the fiscal year 1989 programme in place of the third Mizuho class. Used to escort the plutonium transport ship. SATCOM fitted. A second ship to a modified design was authorised in FY 2010 budget.

AKITSUSHIMA 12/2013*, Hiroshi Kobayashi / 1529126

2 MIZUHO CLASS (PLH/PSOH)

Name	No	Builders	Launched	Commissioned
MIZUHO	PLH 21	Mitsubishi, Nagasaki	5 Jun 1985	19 Mar 1986
YASHIMA	PLH 22	Nippon Kokan, Tsurumi	20 Jan 1988	1 Dec 1988

Displacement, tonnes: 4,979 standard; 5,288 full load
Dimensions, metres (feet): 130 × 15.5 × 5.4 *(426.5 × 50.9 × 17.7)*
Speed, knots: 23. **Range, n miles:** 8,500 at 22 kt
Complement: 100 (30 air crew)

Machinery: 2 SEMT-Pielstick 14 PC2.5 V 400 diesels; 18,200 hp(m) *(13.38 MW)* sustained; 2 shafts; cp props; bow thruster
Guns: 1 Oerlikon 35 mm/90; 550 rds/min to 6 km *(3.2 n miles)* anti-surface; 5 km *(2.7 n miles)* anti-aircraft; weight of shell 1.55 kg. 1 JM-61 MB 20 mm Gatling.
Radars: Surface search: JMA 8303; I-band.
Navigation/helo control: 2 JMA 3000; I-band.
Helicopters: 2 Fuji-Bell 212.

Comment: PLH 21 ordered under the FY83 programme laid down 27 August 1984. PLH 22 in 1986 estimates, laid down 3 October 1987. Two sets of fixed electric fin stabilisers that have a lift of 26 tons × 2 and reduce rolling by 90 per cent at 18 kt. Employed in search and rescue outside the 200 mile economic zone.

10 SOYA CLASS (PLH/PSOH)

Name	No	Builders	Commissioned
SOYA	PLH 01	Nippon Kokan, Tsurumi	22 Nov 1978
TSUGARU	PLH 02	IHI Marine United, Tokyo	17 Apr 1979
OOSUMI	PLH 03	Mitsui, Tamano	18 Oct 1979
HAYATO (ex-*Uraga*)	PLH 04	Hitachi, Maizuru	5 Mar 1980
ZAO	PLH 05	Mitsubishi, Nagasaki	19 Mar 1982
OKINAWA (ex-*Chikuzen*)	PLH 06	Kawasaki, Kobe	28 Sep 1983
SETTSU	PLH 07	Sumitomo, Oppama	27 Sep 1984
ECHIGO	PLH 08	Mitsui, Tamano	28 Feb 1990
RYUKYU	PLH 09	Mitsubishi, Nagasaki	31 Mar 2000
DAISEN	PLH 10	Nippon Kokan, Tsurumi	1 Oct 2001

Displacement, tonnes: 3,251 standard; 4,102 full load
Dimensions, metres (feet): 105.4 (PLH 02–10), 98.6 (PLH 01) × 14.6 (PLH 02–10), 15.6 (PLH 01) × 4.8 (PLH 02–10), 5.2 (PLH 01) *(345.8, 323.5 × 47.9, 51.2 × 15.7, 17.1)*
Speed, knots: 21 (PLH 01), 22 (PLH 02–10). **Range, n miles:** 5,700 at 18 kt
Complement: 69 (PLH 05–10), 71 (PLH 01–04)

Machinery: 2 SEMT-Pielstick 12 PC2.5 V 400 diesels; 15,604 hp(m) *(11.47 MW)* sustained; 2 shafts; cp props; bow thruster
Guns: 1 Bofors 40 mm or Oerlikon 35 mm. 1 Oerlikon 20 mm (PLH 01, 02, 05–07) or 1—20 mm JM61MB Gatling gun.
Radars: Surface search: JMA 1576; I-band.
Navigation: JMA 1596; I-band.
Helo control: JMA 1596; I-band.
Helicopters: 1 Fuji-Bell 212.

Comment: PLH 01 has an icebreaking capability while the other ships are only ice strengthened. Fitted with both fin stabilisers and anti-rolling tanks of 70 tons capacity. The fixed electric hydraulic fins have a lift of 26 tons × 2 at 18 kt which reduces rolling by 90% at that speed. At slow speed the reduction is 50%, using the tanks. PLH 04 name changed on 27 March 1997 and PLH 06 on 2 October 2013.

YASHIMA 6/2012, Hachiro Nakai / 1482907

ECHIGO 6/2012, Hachiro Nakai / 1482908

Numbers/Type: 3 Aerospatiale AS 332L1 Super Puma.
Operational speed: 125 kt *(231 km/h).*
Service ceiling: 15,090 ft *(4,600 m).*
Range: 500 n miles *(926 km).*
Role/Weapon systems: Medium lift, support and SAR. One aircraft damaged by tsunami in March 2011. Sensors: Search radar. Weapons: Unarmed.

AS 332L1   6/2012, *Hachiro Nakai* / 1482913

Numbers/Type: 4 Sikorsky S-76C.
Operational speed: 135 kt *(250 km/h).*
Service ceiling: 11,800 ft *(3,505 m).*
Range: 607 n miles *(1,125 km).*
Role/Weapon systems: Utility aircraft acquired in 1994–98. One aircraft lost on 10 January 2005. Up to 20 required to replace Bell 212s. Sensors: Search radar. Weapons: Unarmed.

S-76C   6/2012, *Hachiro Nakai* / 1482916

Numbers/Type: 4 Bell 206B Jet Ranger.
Operational speed: 115 kt *(213 km/h).*
Service ceiling: 13,500 ft *(4,115 m).*
Range: 368 n miles *(682 km).*
Role/Weapon systems: Support helicopter for reconnaissance and SAR.

Bell 206B   6/2005, *Japan Coast Guard* / 1154399

Numbers/Type: 2 Eurocopter EC 225.
Operational speed: 149 kt *(276 km/h).*
Service ceiling: 13,120 ft *(4,000 m).*
Range: 500 n miles *(926 km).*
Role/Weapon systems: SAR and coastal surveillance helicopter ordered on 5 December 2005 and delivered in September 2007. To replace the AS 332 Super Puma in due course.

EC 225   7/2012, *Hachiro Nakai* / 1482914

Numbers/Type: 5 AgustaWestland AW 139.
Operational speed: 167 kt *(309 km/h).*
Service ceiling: 19,460 ft *(5,931 m).*
Range: 307 n miles *(568 km).*
Role/Weapon systems: Medium-range support and SAR helicopter selected in late 2006 as the replacement for the Bell fleet. A total of 24 aircraft is expected.

AW 139   4/2009, *Hachiro Nakai* / 1366113

LAND-BASED MARITIME AIRCRAFT (FRONT LINE)

Notes: There is also a Cessna U 206G.

Numbers/Type: 2/9 Beech Super King Air 200T/Super King Air 350.
Operational speed: 200 kt *(370 km/h).*
Service ceiling: 35,000 ft *(10,670 m).*
Range: 1,460 n miles *(2,703 km).*
Role/Weapon systems: Visual reconnaissance in support of EEZ. Two are trainers. One King Air 350 damaged by tsunami in March 2011. Sensors: Weather/search radar. Weapons: Unarmed.

BEECH 350   5/2005, *Mitsuhiro Kadota* / 1153337

Numbers/Type: 2 Gulfstream Aerospace G-V.
Operational speed: 510 kt *(945 km/h).*
Service ceiling: 41,000 ft *(12,500 m).*
Range: 6,500 n miles *(12,040 km).*
Role/Weapon systems: Reconnaissance version of long-range business jet ordered on 14 November 2001 and delivered in 2004. Sensors: Ocean Master radar, FLIR, AMASCOS mission system. Can also drop liferafts.

GULFSTREAM G-V   6/2012, *Hachiro Nakai* / 1482917

Numbers/Type: 2 Dassault Falcon 900.
Operational speed: 428 kt *(792 km/h).*
Service ceiling: 51,000 ft *(15,544 m).*
Range: 4,170 n miles *(7,722 km).*
Role/Weapon systems: Maritime surveillance. Sensors: Weather/search radar. Weapons: Unarmed.

FALCON 900   6/2012, *Hachiro Nakai* / 1482918

Numbers/Type: 4 SAAB 340B.
Operational speed: 250 kt *(463 km/h).*
Service ceiling: 25,000 ft *(7,620 m).*
Range: 570 n miles *(1,056 km).*
Role/Weapon systems: Patrol aircraft procured in 1997. Two SAR variants were delivered in 2007.

SAAB 340B   9/2011, *Hachiro Nakai* / 1406932

Numbers/Type: 8 Bombardier DHC-8-315.
Operational speed: 265 kt *(491 km/h).*
Service ceiling: 14,775 ft *(4,503 m).*
Range: 1,630 n miles *3,020 km).*
Role/Weapon systems: Maritime surveillance variant of the Dash-8 Q300 regional airliner selected by the Japanese Coast Guard in December 2006 and delivered in 2008. Sensors: not confirmed but likely to include surveillance radar and FLIR.

DHC-8-315 *5/2010, Hachiro Nakai* / 1366932

MEDIUM PATROL VESSELS

11 TESHIO CLASS (PM/PSO)

Name	No	Builders	Commissioned
KITAKAMI (ex-*Oirose*)	PM 02	Naikai	29 Aug 1980
TOKACHI	PM 04	Narazaki	24 Mar 1981
HITACHI	PM 05	Tohoku	19 Mar 1981
ISAZU	PM 07	Naikai	18 Feb 1982
CHITOSE	PM 08	Shikoku	15 Mar 1983
MABECHI (ex-*Kuwano*)	PM 09	Naikai	10 Mar 1983
SORACHI	PM 10	Tohoku	30 Aug 1984
YUBARI	PM 11	Usuki	28 Nov 1985
MOTOURA	PM 12	Shikoku	21 Nov 1986
MUROMI (ex-*Ishikari*; ex-*Kano*)	PM 13	Naikai	13 Nov 1986
TSURUMI (ex-*Sendai*)	PM 14	Shikoku	1 Jun 1988

Displacement, tonnes: 640 standard; 681 full load
Dimensions, metres (feet): 67.8 × 7.9 × 2.7 *(222.4 × 25.9 × 8.9)*
Speed, knots: 18
Range, n miles: 3,200 at 16 kt
Complement: 33
Machinery: 2 Fuji 6S32F or Arakata 6M31E diesels; 3,650 hp(m) *(2.69 MW);* 2 shafts
Guns: 1 JN-61B 20 mm Gatling.
Radars: Navigation: 2 JMA 159B; I-band.

Comment: First three built under fiscal year 1979 programme and second three under FY 1980, seventh under FY 1981, PM 08–09 under FY 1982, PM 10 under FY 1983, PM 11 under FY 1984, PM 12–13 under FY 1985, PM 14 under FY 1987. *Isazu* has an additional structure aft of the mainmast which is used as a classroom. PM 03 changed name on 30 March 2008, PM 09 on 1 March 2009, PM 13 on 19 November 2009 and again on 15 October 2011. PM 14 changed name on 20 December 2013. PM 06 decommissioned on 13 June 2011. PM 03 decommissioned on 5 March 2013 and PM 01 on 3 June 2013.

TOKACHI *10/2009, Tsukasa Sasaki* / 1482919

2 TAKATORI CLASS (PM/PBO)

Name	No	Builders	Commissioned
TAKATORI	PM 89	Naikai	24 Mar 1978
KUMANO	PM 94	Namura	23 Feb 1979

Displacement, tonnes: 644 standard
Dimensions, metres (feet): 46.5 × 9.2 × 2.9 *(152.6 × 30.2 × 9.5)*
Speed, knots: 15
Range, n miles: 700 at 14 kt
Complement: 34
Machinery: 2 Niigata 6M31EX diesels; 3,000 hp(m) *(2.21 MW);* 2 shafts; cp props
Radars: Navigation: JMA 1596 and JMA 1576; I-band.

Comment: SAR vessels equipped for salvage and firefighting.

TAKATORI *5/2010, Hachiro Nakai* / 1366933

4 AMAMI CLASS (PM/PBO)

Name	No	Builders	Commissioned
AMAMI	PM 95	Hitachi, Kanagawa	28 Sep 1992
KUROKAMI (ex-*Matsuura*)	PM 96	Hitachi, Kanagawa	24 Nov 1995
ISHIKARI (ex-*Kunashiri*)	PM 97	Mitsubishi, Shimonoseki	26 Aug 1998
HOROBETSU (ex-*Minabe*)	PM 98	Mitsubishi, Shimonoseki	26 Aug 1998

Displacement, tonnes: 234 standard
Dimensions, metres (feet): 56 × 7.5 × 2 *(183.7 × 24.6 × 6.6)*
Speed, knots: 25
Machinery: 2 Fuji 8S40B diesels; 8,120 hp(m) *(5.97 MW);* 2 shafts; cp props
Guns: 1—20 mm JM-61B Gatling.
Radars: Navigation: I-band.

Comment: First one authorised in the FY91 programme; laid down 22 October 1991. Second authorised in FY93 programme; laid down 7 October 1994. Last pair authorised in FY96 programme and both laid down 30 September 1997. Stern ramp for launching RIB. PM 96 changed name 3 April 2000. PM 95 damaged in incident with possible North Korean intelligence collection ship on 22 December 2001. PM 98 changed name on 9 February 2010 and PM 97 on 27 January 2012.

AMAMI *3/2012, Hachiro Nakai* / 1482920

1 TESHIO CLASS (ICEBREAKER) (PM/AGOB)

Name	No	Builders	Commissioned
TESHIO	PM 15	Nippon Kokan, Tsurumi	19 Oct 1995

Displacement, tonnes: 559 standard
Dimensions, metres (feet): 55 × 10.6 × 3.9 *(180.4 × 34.8 × 12.8)*
Speed, knots: 14.5
Complement: 35
Machinery: 2 diesels; 3,600 hp(m) *(2.65 MW);* 2 shafts; bow thruster
Guns: 1—20 mm JM-61B Gatling.
Radars: Navigation: 2 sets; I-band.

Comment: Authorised in FY93; laid down 7 October 1994, launched 20 April 1995. Has an icebreaker bow.

TESHIO *6/2002, Japan Coast Guard* / 0570891

20 + 2 TOKARA CLASS (PM/PBO)

Name	No	Builders	Commissioned
TOKARA	PM 21	Universal, Keihin (Tsurumi)	12 Mar 2003
FUKUE	PM 22	Mitsubishi, Shimonoseki	12 Mar 2003
OIRASE	PM 23	Universal, Keihin (Tsurumi)	18 Mar 2004
FUJI	PM 24	Universal, Keihin (Tsurumi)	30 Apr 2008
ECHIZEN	PM 25	Universal, Keihin (Tsurumi)	30 Apr 2008
KIKUCHI	PM 26	Universal, Keihin (Tsurumi)	17 Feb 2009
YOSHINO	PM 27	Universal, Keihin (Tsurumi)	26 Mar 2009
ISUZU	PM 28	Universal, Keihin (Tsurumi)	26 Mar 2009
YAMAKUNI	PM 29	Universal, Keihin (Tsurumi)	29 Jun 2009
KANO	PM 30	Universal, Keihin (Tsurumi)	16 Dec 2009
ABUKUMA	PM 31	Universal, Keihin (Tsurumi)	8 Mar 2010
MINABE	PM 32	Universal, Keihin (Tsurumi)	8 Mar 2010
MATSUURA	PM 33	Universal, Keihin (Tsurumi)	14 Sep 2010
CHIKUGO	PM 34	Universal, Keihin (Tsurumi)	1 Dec 2010
KUROSE	PM 35	Universal, Keihin (Tsurumi)	5 Apr 2011
OKITSU	PM 36	Universal, Keihin (Tsurumi)	28 Jun 2011
KUNASHIRI	PM 37	Universal, Keihin (Tsurumi)	20 Feb 2012
OUMI	PM 38	Universal, Keihin (Tsurumi)	31 May 2012
OKUSHIRI	PM 39	Universal, Keihin (Tsurumi)	29 Mar 2013
NATSUI	PM 40	Universal, Keihin (Tsurumi)	Jun 2013
–	PM 41	Universal, Keihin (Tsurumi)	Mar 2014
–	PM 42	Universal, Keihin (Tsurumi)	Mar 2014

Displacement, tonnes: 340 standard
Dimensions, metres (feet): 56 × 8.5 × 4.4 *(183.7 × 27.9 × 14.4)*
Speed, knots: 30 (est.)
Machinery: 3 diesels; 3 waterjets
Guns: 1 – 20 mm Gatling gun. 1 – 12.7 mm MG.

Comment: First two authorised in fiscal year 2001, third in FY 2002, six in FY 2006, three in FY 2007, four in FY 2008, and four in FY 2009.

CHIKUGO 3/2013*, Hachiro Nakai / 1529161

SMALL PATROL VESSELS

19 MIHASHI AND RAIZAN CLASS (PS/PBF)

Name	No	Builders	Commissioned
SHINZAN (ex-*Akiyoshi*; ex-*Mihashi*)	PS 01	Mitsubishi, Shimonoseki	9 Sep 1988
SAROMA	PS 02	Hitachi, Kanagawa	24 Nov 1989
INASA	PS 03	Mitsubishi, Shimonoseki	31 Jan 1990
KIRISHIMA	PS 04	Hitachi, Kanagawa	22 Mar 1991
KAMUI	PS 05	Mitsubishi, Shimonoseki	31 Jan 1994
RAIZAN (ex-*Banna*; ex-*Bizan*)	PS 06	Hitachi, Kanagawa	31 Jan 1994
ASHITAKI	PS 07	Mitsui, Tamano	30 Sep 1994
KARIBA (ex-*Kurama*)	PS 08	Mitsubishi, Shimonoseki	29 Aug 1995
ARASE	PS 09	Mitsubishi, Shimonoseki	29 Jan 1997
SANBE	PS 10	Hitachi, Kanagawa	29 Jan 1997
MIZUKI	PS 11	Mitsui, Tamano	9 Jun 2000
KOUYA	PS 12	Universal, Keihin	18 Mar 2004
TSUKUBA	PS 13	Mitsubishi, Shimonoseki	4 Mar 2009
AKAGI	PS 14	Mitsubishi, Shimonoseki	4 Mar 2009
BIZAN	PS 15	Mitsubishi, Shimonoseki	26 Apr 2011
NOBARU	PS 16	Mitsubishi, Shimonoseki	26 Apr 2011
TAKACHIHO	PS 17	Mitsubishi, Shimonoseki	24 Aug 2011
SANREI	PS 18	Mitsubishi, Shimonoseki	12 Mar 2012
ASAJI	PS 19	Mitsubishi, Shimonoseki	12 Mar 2012

Displacement, tonnes: 198 standard
Dimensions, metres (feet): 43 × 7.5 × 1.7 *(141.1 × 24.6 × 5.6)*
Speed, knots: 35
Range, n miles: 650 at 34 kt
Complement: 34
Machinery: 2 SEMT-Pielstick 16 PA4 V 200 VGA diesels; 7,072 hp(m) *(5.2 MW)*; 2 shafts. 1 SEMT-Pielstick 12 PA4 V 200 VGA diesel; 2,720 hp(m) *(2 MW)*; Kamewa 80 water-jet
Guns: 1 – 12.7 mm MG or 1 – 20 mm JM 61 Gatling (being progressively fitted).
Radars: Navigation: Furuno; I-band.

Comment: Capable of 15 kt on the water-jet alone. PS 01 name changed 28 January 1997 and again on 24 January 2001, PS 06 on 17 April 1999 and again on 1 August 2008. PS 08 on 29 March 2004. PS 11 authorised in FY98 programme, PS 12 in FY02 programme. PS 13–14 authorised in FY07 budget, PS 15–16 in FY09 and PS 17–19 in FY09 extra budget.

ASHITAKI 6/2012, Hachiro Nakai / 1482921

---

2 TAKATSUKI CLASS (PS/PBF)

Name	No	Builders	Commissioned
TAKATSUKI	PS 108	Mitsubishi, Shimonoseki	23 Mar 1992
KATSURAGI (ex-*Nobaru*)	PS 109	Hitachi, Kanagawa	22 Mar 1993

Displacement, tonnes: 117 standard; 183 full load
Dimensions, metres (feet): 35 × 6.7 × 1.3 *(114.8 × 22.0 × 4.3)*
Speed, knots: 35
Complement: 13
Machinery: 2 MTU 16V 396 TB94 diesels; 5,200 hp(m) *(3.82 MW)*; 2 Kamewa 71 water-jets
Guns: 1 – 12.7 mm MG.
Radars: Navigation: I-band.

Comment: First authorised in the fiscal year 1991 programme, second in FY 1992. Aluminium hulls. PS 109 changed name on 1 April 2011.

KATSURAGI 10/2013*, Hachiro Nakai / 1529162

6 TSURUUGI CLASS (PS/PBOF)

Name	No	Builders	Commissioned
TSURUUGI	PS 201	Hitachi, Kanagawa	15 Feb 2001
HOTAKA	PS 202	Mitsubishi, Shimonoseki	16 Mar 2001
NORIKURA	PS 203	Mitsui, Tamano	16 Mar 2001
KAIMON	PS 204	Mitsui, Tamano	21 Apr 2004
ASAMA	PS 205	Mitsui, Tamano	21 Apr 2004
HOUOU	PS 206	Mitsui, Tamano	27 Jan 2005

Displacement, tonnes: 224 standard
Dimensions, metres (feet): 50 × 8.0 × 4.0 *(164 × 26.2 × 13.1)*
Speed, knots: 35
Machinery: 3 diesels; 3 waterjets
Guns: 1 – 20 mm JM-61 RFS Gatling.

Comment: First three authorised in FY99 budget, fourth and fifth in FY02 budget and sixth in FY03 budget.

HOTAKA 7/2011, Hachiro Nakai / 1406935

COASTAL PATROL CRAFT

2 + 6 KOTONAMI CLASS (PC/PB)

Name	No	Builders	Commissioned
KOTONAMI	PC 31	Sumidagawa	2 Mar 2012
HATAGUMO	PC 32	Sumidagawa	2 Mar 2012
–	PC 33	–	Mar 2014
–	PC 34	–	Mar 2014
–	PC 35	–	Mar 2014
–	PC 36	–	Mar 2014
–	PC 37	–	Mar 2014
–	PC 38	–	Mar 2014

Displacement, tonnes: 64 standard
Dimensions, metres (feet): 27 × 5.6 × 2.8 *(88.6 × 18.4 × 9.2)*
Machinery: 2 diesels; 2 shafts
Guns: To be announced.

Comment: Two ships authorised in FY10 budget. Six further ships (PC 33–38) authorised in FY12 extra budget.

KOTONAMI 6/2012, Hachiro Nakai / 1482926

10 YODO CLASS (PC/YTR)

Name	No	Builders	Launched	Commissioned
YODO	PC 51	Sumidagawa	2 Oct 2001	29 Mar 2002
KOTOBIKI	PC 52	Sumidagawa	23 Oct 2002	27 Mar 2003
NACHI	PC 53	Ishihara	29 Jan 2003	27 Mar 2003
NUNOBIKI	PC 54	Sumidagawa	4 Dec 2002	27 Mar 2003
FUDOU	PC 55	Sumidagawa	29 Jun 2012	28 Jan 2013
RYUSEI	PC 56	Sumidagawa	17 Aug 2012	14 Feb 2013
TAKATAKI	PC 57	Niigata	8 Sep 2012	15 Feb 2013
AOTAKI	PC 58	Sumidagawa	–	21 Mar 2013
NACHI	PC 59	Niigata	–	29 Mar 2013
MINOO	PC 60	Nagasaki	–	29 Mar 2013

Displacement, tonnes: 127 standard
Dimensions, metres (feet): 37 × 6.7 × 3.4 (121.4 × 22.0 × 11.2)
Speed, knots: 25
Machinery: 2 MTU 16V 4000 M90 diesels; 6,524 hp (4.8 MW); 2 waterjets

Comment: The first authorised in FY00 budget, three more in FY01 budget and six more in FY11 extra budget. Also equipped for firefighting and replaced firefighting vessel of the same name.

YODO   6/2012, Hachiro Nakai / 1482922

20 HAYAGUMO CLASS (PC/PBF)

Name	No	Builders	Commissioned
HAYAGUMO (ex-Hamayuki; ex-Kagayuki)	PC 105	Mitsubishi, Shimonoseki	24 Dec 1999
MURAKUMO	PC 106	Hitachi, Kanagawa	19 Aug 2002
IZUNAMI	PC 107	Mitsui, Tamano	18 Mar 2003
YAEGUMO	PC 108	Sumidagawa	4 Mar 2008
NATSUGUMO	PC 109	Sumidagawa	4 Mar 2008
AKIGUMO	PC 110	Sumidagawa	10 Mar 2008
HAGINAMI (ex-Tatsugumo)	PC 111	Sumidagawa	10 Mar 2009
IKIGUMO	PC 112	Sumidagawa	10 Mar 2009
NATSUZUKI	PC 113	Sumidagawa	10 Mar 2009
OKIGUMO	PC 114	Sumidagawa	31 Jul 2009
AWAGUMO	PC 115	Sumidagawa	31 Jul 2009
SHIMAGUMO	PC 116	Sumidagawa	24 Aug 2009
YUKIGUMO	PC 117	Niigata	10 Mar 2011
KITAGUNI	PC 118	Niigata	10 Mar 2011
KOMAYUKI	PC 119	Sumidagawa	25 Feb 2011
KAWAGIRI	PC 120	Niigata	9 Mar 2012
WAKAZUKI	PC 121	Universal, Keihin	10 Jul 2012
ISONAMI	PC 122	Sumidagawa	30 Jan 2014
NAGOZUKI	PC 123	Sumidagawa	30 Jan 2014
YAEZUKI	PC 124	Niigata	28 Feb 2014

Displacement, tonnes: 102 standard
Dimensions, metres (feet): 32 × 6.5 × 3.3 (105 × 21.3 × 10.8)
Speed, knots: 36
Complement: 10
Machinery: 2 diesels; 5,200 hp(m) (3.82 MW); 2 waterjets
Guns: 1 – 12.7 mm MG.

Comment: Larger version of Asogiri class with waterjet propulsion and higher top speed. PC 105 changed name on 22 February 2001 and again on 18 March 2006. PC 111 changed name on 7 February 2012. PC 106 authorised in FY01 budget and PC 107 in FY01 extra budget. PC 108-110 authorised in FY06 budget, PC 111–116 in FY07 budget, PC 117–119 in FY09 extra budget and PC 120–121 in FY10 budget. PC 122-124 authorised in FY12 budget.

OKIGUMO   6/2012, Hachiro Nakai / 1482923

3 SHIMAGIRI CLASS (PC/PB)

Name	No	Builders	Commissioned
SHIMAGIRI	PC 83	Hitachi, Kanagawa	7 Feb 1985
OKINAMI (ex-Setogiri)	PC 84	Hitachi, Kanagawa	22 Mar 1985
HAYAGIRI	PC 85	Mitsubishi, Shimonoseki	22 Feb 1985

Displacement, tonnes: 52 standard
Dimensions, metres (feet): 23 × 5.3 × 1.9 (75.5 × 17.4 × 6.2)
Speed, knots: 30
Complement: 10
Machinery: 2 Ikegai 12V 175 RTC diesels; 3,000 hp(m) (2.21 MW); 2 shafts
Guns: 1 – 12.7 mm MG (not in all).
Radars: Navigation: FRA 10 Mk 2; I-band.

Comment: Aluminium hulls. PC 84 name changed 1 October 2000.

SHIMAGIRI   11/2010, Hachiro Nakai / 1366940

8 AKIZUKI CLASS (PC/SAR)

Name	No	Builders	Commissioned
URAYUKI	PC 72	Mitsubishi, Shimonoseki	31 May 1975
MAKIGUMO	PC 76	Mitsubishi, Shimonoseki	19 Mar 1976
HAMAZUKI	PC 77	Mitsubishi, Shimonoseki	29 Nov 1976
ISOZUKI	PC 78	Mitsubishi, Shimonoseki	18 Mar 1977
SHIMANAMI	PC 79	Mitsubishi, Shimonoseki	23 Dec 1977
YUZUKI	PC 80	Mitsubishi, Shimonoseki	22 Mar 1979
TAMANAMI (ex-Hanayuki)	PC 81	Mitsubishi, Shimonoseki	27 Mar 1981
AWAGIRI	PC 82	Mitsubishi, Shimonoseki	24 Mar 1983

Displacement, tonnes: 78 standard
Dimensions, metres (feet): 26 × 6.3 × 2.1 (85.3 × 20.7 × 6.9)
Speed, knots: 22
Range, n miles: 220 at 21.5 kt
Complement: 10
Machinery: 3 Mitsubishi 12DM20MTK diesels; 3,000 hp(m) (2.21 MW); 3 shafts
Radars: Navigation: FRA 10 Mk 2; I-band.

Comment: Aluminium hulls. Used mostly for SAR. PC 75 paid off on 14 February 2012.

AWAGIRI   9/2012, Hachiro Nakai / 1482924

2 NATSUGIRI CLASS (PC/PB)

Name	No	Builders	Commissioned
NATSUGIRI	PC 86	Sumidagawa	29 Jan 1990
SUGANAMI	PC 87	Sumidagawa	29 Jan 1990

Displacement, tonnes: 69 standard
Dimensions, metres (feet): 27 × 5.6 × 1.2 (88.6 × 18.4 × 3.9)
Speed, knots: 27
Complement: 10
Machinery: 2 diesels; 3,000 hp(m) (2.21 MW); 2 shafts
Radars: Navigation: I-band.

Comment: Built under FY88 programme. Steel hulls.

SUGANAMI   5/2010, Hachiro Nakai / 1366944

Japan (COAST GUARD) > Coastal patrol craft

7 MURAKUMO CLASS (PC/PB)

Name	No	Builders	Commissioned
ISEYUKI (ex-Hamayuki)	PC 216	Hitachi, Kanagawa	27 Feb 1981
ISONAMI	PC 217	Mitsubishi, Shimonoseki	19 Mar 1981
NAGOZUKI	PC 218	Hong Leong-Lurssen Shipyard (1992) Sdn Bhd, Kanagawa	29 Jan 1981
YAEZUKI	PC 219	Hitachi, Kanagawa	19 Mar 1981
HAMAYUKI (ex-Yamayuki)	PC 220	Hitachi, Kanagawa	16 Feb 1982
UMIGIRI	PC 222	Hitachi, Kanagawa	17 Feb 1983
ASAGIRI	PC 223	Mitsubishi, Shimonoseki	23 Feb 1983

Displacement, tonnes: 86 standard
Dimensions, metres (feet): 30 × 6.3 × 2.2 (98.4 × 20.7 × 7.2)
Speed, knots: 30
Range, n miles: 350 at 28 kt
Complement: 13
Machinery: 2 Ikegai MTU MB 16V 652 SB70 diesels; 4,400 hp(m) (3.23 MW) sustained; 2 shafts
Guns: 1 — 12.7 mm MG.
Radars: Navigation: I-band.

Comment: PC 211 name changed on 17 April 1999 and again on 1 October 2004. P 216 changed name on 22 February 2001 and PC 220 on 18 March 2006. PC 206 paid off on 21 February 2008, PC 207–208 on 15 February 2008, PC 212, PC 215 on 20 February 2009, PC 214 on 9 March 2009, PC 202 and 203 on 21 March 2011, PC 210 on 21 February 2012, PC 211 on 17 February 2012 and PC 221 on 24 February 2011.

MURAKUMO CLASS 5/2004, Mitsuhiro Kadota / 1044450

MURAKUMO CLASS 5/2008, Hachiro Nakai / 1353150

15 HAYANAMI CLASS (PC/PB/YTR)

Name	No	Builders	Commissioned
HAYANAMI	PC 11	Sumidagawa	25 Mar 1993
SETOGIRI (ex-Shikinami)	PC 12	Sumidagawa	24 Mar 1994
MIZUNAMI	PC 13	Ishihara	24 Mar 1994
IYONAMI	PC 14	Sumidagawa	30 Jun 1994
KURINAMI	PC 15	Sumidagawa	30 Jan 1995
HAMANAMI	PC 16	Sumidagawa	28 Mar 1996
SHINONOME	PC 17	Ishihara	29 Feb 1996
HARUNAMI	PC 18	Ishihara	28 Mar 1996
KIYOZUKI	PC 19	Sumidagawa	23 Feb 1996
AYANAMI	PC 20	Yokohama Yacht	28 Mar 1996
TOKINAMI	PC 21	Yokohama Yacht	28 Mar 1996
HAMAGUMO	PC 22	Sumidagawa	27 Aug 1999
AYUZUKI (ex-Awani)	PC 23	Sumidagawa	27 Aug 1999
YUFUGIRI (ex-Uranami)	PC 24	Sumidagawa	24 Jan 2000
TOMONAMI (ex-Shikinami)	PC 25	Ishihara	24 Oct 2000

Displacement, tonnes: 112 standard; 193 full load
Dimensions, metres (feet): 35 × 6.3 × 2.3 (114.8 × 20.7 × 7.5)
Speed, knots: 25
Complement: 13
Machinery: 2 diesels; 4,000 hp(m) (2.94 MW); 2 shafts
Guns: 1 — 12.7 mm MG.
Radars: Navigation: I-band.

Comment: One more authorised in FY99 budget. From PC 22 onwards these craft are equipped for firefighting. PC 12 changed name 1 October 2000. PC 23 changed name on 29 January 2013, PC 24 on 11 January 2013 and PC 25 on 12 Februaty 2013.

TOKINAMI 4/2011, Hachiro Nakai / 1406936

4 ASOGIRI CLASS (PC/PB)

Name	No	Builders	Commissioned
ASOGIRI	PC 101	Yokohama Yacht	19 Dec 1994
MUROZUKI	PC 102	Ishihara	27 Jul 1995
URAZUKI (ex-Wakagumo)	PC 103	Ishihara	17 Jul 1996
KAGAYUKI (ex-Naozuki)	PC 104	Sumidagawa	23 Jan 1997

Displacement, tonnes: 89 standard
Dimensions, metres (feet): 33 × 6.3 × 1.4 (108.3 × 20.7 × 4.6)
Speed, knots: 30
Complement: 10
Machinery: 2 diesels; 5,200 hp(m) (3.82 MW); 2 shafts
Guns: 1 — 12.7 mm MG.

Comment: First pair authorised in FY93 programme, third and fourth in FY95. PC 104 changed names on 1 April 2006 and PC 103 on 23 January 2012.

MUROZUKI 8/2001, Hachiro Nakai / 0130250

214 COASTAL PATROL AND RESCUE CRAFT (CL/PB)

CL 01–09   CL 11–171   GS 01–02   SS 34–46   SS 51–79

Comment: Some have firefighting capability. Built by Shigi, Ishihara, Sumidagawa, Yokohama Yacht Co and Yamaha. For coastal patrol and rescue duties. Built of high tensile steel. Fourteen CL 11 class authorised in FY01 budget, eight in FY05 extra budget, eight in FY06 extra budget and eight in FY07 extra budget. CL 05–09 (ex-LS 231–235) were converted in 2008.

CL 132 3/2013*, Hachiro Nakai / 1529163

SS 51 10/2009, Hachiro Nakai / 1366124

1 MATSUNAMI CLASS (PC/PB)

Name	No	Builders	Commissioned
MATSUNAMI	PC 01	Mitsubishi, Shimonoseki	22 Feb 1995

Displacement, tonnes: 168 standard
Dimensions, metres (feet): 35 × 8 × 3.3 (114.8 × 26.2 × 10.8)
Speed, knots: 25
Complement: 30
Machinery: 2 diesels; 5,200 hp(m) (3.82 MW); 2 water-jets
Radars: Navigation: I-band.

Comment: Has replaced old craft of the same name. Laid down 10 May 1994. Used for patrol and for VIPs.

MATSUNAMI 6/2012, Hachiro Nakai / 1482925

FIREFIGHTING VESSELS AND CRAFT

1 MODIFIED HIRYU CLASS (FL/YTR)

Name	No	Builders	Launched	Commissioned
HIRYU	FL 01	Nippon Kokan, Tsurumi	5 Sep 1997	24 Dec 1997

Displacement, tonnes: 284 standard
Dimensions, metres (feet): 35 × 12.2 × 2.7 (114.8 × 40.0 × 8.9)
Speed, knots: 14
Complement: 15
Machinery: 2 diesels; 4,000 hp(m) (2.94 MW); 2 shafts

Comment: Authorised in FY96 programme. Catamaran design. Replaced ship of the same name and pennant number.

HIRYU   3/2011, Hachiro Nakai / 1406938

SURVEY SHIPS

1 SHOYO CLASS (AGS)

Name	No	Builders	Launched	Commissioned
SHOYO	HL 01	Mitsui, Tamano	23 Jun 1997	20 Mar 1998

Displacement, tonnes: 3,048 standard
Dimensions, metres (feet): 98 × 15.2 × 3.6 (321.5 × 49.9 × 11.8)
Speed, knots: 17
Complement: 60
Machinery: Diesel-electric; 2 Mitsui 6ADD30V diesels; 8,100 hp(m) (5.95 MW); 2 motors; 5,712 hp(m) (4.2 MW); 2 shafts; cp props

Comment: Authorised in FY95 programme. Laid down 4 October 1996.

SHOYO   6/2012, Hachiro Nakai / 1482930

1 TENYO CLASS (AGS)

Name	No	Builders	Commissioned
TENYO	HL 04	Sumitomo, Oppama	27 Nov 1986

Displacement, tonnes: 782 standard
Dimensions, metres (feet): 56 × 9.8 × 2.9 (183.7 × 32.2 × 9.5)
Speed, knots: 13
Range, n miles: 5,400 at 12 kt
Complement: 43 (18 officers)
Machinery: 2 Akasaka diesels; 1,300 hp(m) (955 kW); 2 shafts
Radars: Navigation: 2 JMA 1596; I-band.

Comment: Laid down 11 April 1986, launched 5 August 1986. Based at Tokyo.

TENYO   6/2006, Okano Takatoshi / 1040613

1 TAKUYO CLASS (AGS)

Name	No	Builders	Commissioned
TAKUYO	HL 02	Nippon Kokan, Tsurumi	31 Aug 1983

Displacement, tonnes: 3,048 standard
Dimensions, metres (feet): 96 × 14.2 × 4.6 (315 × 46.6 × 15.1)
Speed, knots: 17
Range, n miles: 12,000 at 16 kt
Complement: 60 (24 officers)
Machinery: 2 Fuji 6S40B diesels; 6,090 hp(m) (4.47 MW); 2 shafts; cp props
Radars: Navigation: 2 sets; I-band.

Comment: Laid down on 14 April 1982, launched on 24 March 1983. Based at Tokyo. Side scan sonar fitted. Two survey launches.

TAKUYO   5/2008, Hachiro Nakai / 1353160

2 MEIYO CLASS (AGS)

Name	No	Builders	Commissioned
MEIYO	HL 03	Kawasaki, Kobe	24 Oct 1990
KAIYO	HL 05	Mitsubishi, Shimonoseki	7 Oct 1993

Displacement, tonnes: 559 standard
Dimensions, metres (feet): 60 × 10.5 × 3.1 (196.9 × 34.4 × 10.2)
Speed, knots: 15
Range, n miles: 5,280 at 11 kt
Complement: 25 + 13 scientists
Machinery: 2 Daihatsu 6 DLM-24 diesels; 3,000 hp(m) (2.2 MW); 2 shafts; bow thruster
Radars: Navigation: 2 sets; I-band.

Comment: Meiyo laid down 24 July 1989 and launched 29 June 1990; Kaiyo laid down 7 July 1992 and launched 26 April 1993. Have anti-roll tanks and resiliently mounted main machinery. Has a 12 kHz bottom contour sonar. A large survey launch is carried on the port side.

MEIYO   12/2012*, Hachiro Nakai / 1529164

7 HAMASHIO CLASS (YGS)

Name	No	Builders	Commissioned
HAMASHIO	HS 21	Yokohama Yacht	25 Mar 1991
ISOSHI	HS 22	Yokohama Yacht	25 Mar 1993
UZUSHIO	HS 23	Yokohama Yacht	22 Dec 1995
OKISHIO	HS 24	Ishihara	4 Mar 1999
ISESHIO	HS 25	Ishihara	10 Mar 1999
HAYASHIO	HS 26	Ishihara	10 Mar 1999
KURUSHIMA	HS 27	Nissui Marine	26 Mar 2003

Displacement, tonnes: 43 standard
Dimensions, metres (feet): 20.3 × 4.5 × 1.2 (66.6 × 14.8 × 3.9)
Speed, knots: 15
Complement: 10
Machinery: 3 diesels; 1,015 hp(m) (746 kW); 3 shafts
Radars: Navigation: I-band.

Comment: Survey launches. HS 27 authorised in FY01 extra budget.

UZUSHIO   10/2010, Hachiro Nakai / 1366951

AIDS TO NAVIGATION SERVICE

7 SUPPLY CRAFT (AKSL)

Name	No	Builders	Commissioned
SEIUN	LM 202	Sumidagawa	22 Feb 1989
SEKIUN	LM 203	Ishihara	12 Mar 1991
SOUUN (ex-Houun)	LM 204	Ishihara	22 Feb 1991
REIUN	LM 205	Ishihara	28 Feb 1992
GENUN	LM 206	Wakamatsu	19 Mar 1996
AYABANE	LM 207	Ishihara	9 Mar 2000
KOUN	LM 208	Sumidagawa	16 Mar 2001

Displacement, tonnes: 59 full load
Dimensions, metres (feet): 23 × 6 × 1 *(75.5 × 19.7 × 3.3)*
Speed, knots: 14
Range, n miles: 250 at 14 kt
Complement: 9
Machinery: 2 GM 12V-71TA diesels; 840 hp *(627 kW)* sustained; 2 shafts
Radars: Navigation: FRA 10 Mk III; I-band.

Comment: LM 114 decommissioned on 31 March 2006 and LM 201 on 31 March 2007. LM 204 changed name on 1 April 2011.

LS 223 — 5/2010, Hachiro Nakai / 1366950

KOUN — 10/2013*, Hachiro Nakai / 1529165

5 SMALL TENDERS (YAG)

LS 201   LS 217   LS 220   LS 222   LS 223

Displacement, tonnes: 27 full load
Dimensions, metres (feet): 20 × 4.5 × 2.3 *(65.6 × 14.8 × 7.5)*
Speed, knots: 25
Complement: 8
Machinery: 2 diesels; 1,820 hp(m) *(1.34 MW)*; 2 shafts

Comment: Details given are for LS 231–233. Others with varying characteristics. LS 161, LS 164–167 and LS 212 decommissioned on 31 March 2006, LS 189 on 8 December 2006, LS 188–193. LS 213–215 converted to coastal patrol craft on 31 March 2007. LS 169, 194 and 218 decommissioned in 2009.

ENVIRONMENT MONITORING CRAFT

2 SERVICE CRAFT (YPC)

Name	No	Builders	Commissioned
SAIKAI	MS 02	Ishihara, Takasago	4 Feb 1994
KATSUREN	MS 03	Sumidagawa	18 Dec 1997

Displacement, tonnes: 26 standard
Dimensions, metres (feet): 18.1 × 4.3 × ? *(59.4 × 14.1 × ?)*
Speed, knots: 25
Complement: 8
Machinery: 2 diesels; 1,800 hp(m) *(1.3 MW)*; 2 shafts

Comment: *Saikai* and *Katsuren* are both monohulls of 26 tons. Used for monitoring pollution and radiation.

SAIKAI — 4/2012*, Hachiro Nakai / 1529166

Jordan

Country Overview

The Hashemite Kingdom of Jordan is situated in the Middle East. With an area of 34,492 square miles, it has borders to the north with Syria, to the east with Iraq, to the west with Israel and the West Bank and to the east and south with Saudi Arabia. It has a 14 n mile coastline with the Gulf of Aqaba (in the northern Red Sea) on which Aqaba, the only seaport, is situated. Amman is the capital and largest city. Territorial seas (3 n miles) are claimed but an Exclusive Economic Zone (EEZ) is not claimed.

Headquarters Appointments

Commander Naval Forces:
  Major General Fadeel Nahar Tanashat

Headquarters Appointments – continued

Deputy Commander:
  Colonel Abdelkareem Fdoul

Organisation

The Royal Jordanian Naval Force comes under the Director of Operations at General Headquarters.

Bases

Aqaba

Personnel

(a) 2014: 500 officers and men
(b) Voluntary service

PATROL FORCES

3 AL HUSSEIN (HAWK) CLASS
(FAST ATTACK CRAFT – GUN) (PB)

AL HUSSEIN 101   AL HASSAN 102   KING ABDULLAH 103

Displacement, tonnes: 126 full load
Dimensions, metres (feet): 30.5 × 6.9 × 1.5 *(100.1 × 22.6 × 4.9)*
Speed, knots: 32
Range, n miles: 750 at 15 kt, 1,500 at 11 kt
Complement: 16 (3 officers)

Machinery: 2 MTU 16V 396 TB94 diesels; 5,800 hp(m) *(4.26 MW)* sustained; 2 shafts
Guns: 1 Oerlikon GCM-A03 30 mm. 1 Oerlikon GAM-BO1 20 mm. 2 – 12.7 mm MGs.
Physical countermeasures: Decoys: 2 Wallop Stockade chaff launchers.
Radars: Surface search: Kelvin Hughes 1007; I-band.
Combat data systems: Racal Cane 100.
Weapon control systems: Radamec Series 2000 optronic director for 30 mm gun.

Comment: Ordered from Vosper Thornycroft in December 1987. GRP structure. First one on trials in May 1989 and completed December 1989. Second completed in March 1990 and the third in early 1991. All transported to Aqaba in September 1991.

AL HUSSEIN — 4/2006, M Declerck / 1164802

Patrol forces < **Jordan** 461

4 FAYSAL CLASS (INSHORE PATROL CRAFT) (PB)

FAYSAL   HUSSEIN (ex-*Han*)   HASSAN (ex-*Hasayu*)   MUHAMMED

Displacement, tonnes: 8 full load
Dimensions, metres (feet): 11.6 × 4 × 0.5 *(38.1 × 13.1 × 1.6)*
Speed, knots: 22
Range, n miles: 240 at 20 kt
Complement: 8
Machinery: 2 6M 8V715 diesels; 600 hp *(441 kW)*; 2 shafts
Guns: 1—12.7 mm MG. 1—7.62 mm MG.
Radars: Surface search: Decca; I-band.

Comment: Acquired from Bertram, Miami in 1974. GRP construction. Still operational and no replacements are planned yet.

MUHAMMED   *3/2004, Bob Fildes* / 0587768

2 HASHIM (ROTORK) CLASS (PB)

HASHIM   FAISAL

Displacement, tonnes: 9 full load
Dimensions, metres (feet): 12.7 × 3.2 × 0.9 *(41.7 × 10.5 × 3.0)*
Speed, knots: 28
Complement: 5

Military lift: 30 troops
Machinery: 2 Deutz diesels; 240 hp *(179 kW)*; 2 shafts
Guns: 1—12.7 mm MG. 1—7.62 mm MG.
Radars: Surface search: Furuno; I-band.

Comment: Delivered in late 1990 for patrolling the Dead Sea. Due to the annual decrease of water depth, the original three craft were moved to Aqaba in 2000. *Hamza* scrapped in 2006.

HASHIM CLASS   *3/2004, Bob Fildes* / 0587769

1 INTERCEPTION CRAFT (PBF)

Displacement, tonnes: 15.7 full load
Dimensions, metres (feet): 13.5 × 4.2 × 0.55 *(44.3 × 13.8 × 1.8)*
Complement: 4
Machinery: 2 Caterpillar C12 diesels; 1,400 hp *(1.1 MW)*; 2 Arneson ASD11 drives
Guns: To be announced.

Comment: Built by Jordan Boats and delivered in 2006. Based at Aqaba.

INTERCEPTION CRAFT   *6/2006*, Jordan Boats* / 1486576

4 FAISAL (COMMANDER) CLASS (PATROL CRAFT) (PB)

FAISAL 1–4

Displacement, tonnes: 3 full load
Dimensions, metres (feet): 8.1 × 3.1 × 0.45 *(26.6 × 10.2 × 1.5)*
Speed, knots: 46
Complement: 3
Machinery: 2 Evinrude outboard motors; 500 hp *(375 kW)*
Guns: 2—12.7 mm MGs.
Radars: Navigation: Raymarine RL70C; I-band.

Comment: Sea Ark Commander design acquired in 2006.

FAISAL 1   *3/2012, Jürg Kürsener* / 1483514

1 ASSAULT CRAFT (PBF)

Displacement, tonnes: 2.6 full load
Dimensions, metres (feet): 10 × 3.0 × 0.45 *(32.8 × 9.8 × 1.5)*
Complement: 1 + 8 embarked forces
Machinery: 2 outboard motors; 500 hp *(372 kW)*
Guns: To be announced.

Comment: Built by Jordan Boats and delivered in 2006. Based at Aqaba.

ASSAULT CRAFT   *6/2006*, Jordan Boats* / 1486575

1 LANDING CRAFT (LCP)

Displacement, tonnes: 7 full load
Dimensions, metres (feet): 12.5 × 3.7 × 0.5 *(41 × 12.1 × 1.64)*
Complement: 2
Machinery: 2 outboard motors; 450 hp *(335 kW)*
Guns: To be announced.

Comment: Built by Jordan Boats and delivered in 2006. Based at Aqaba.

LANDING CRAFT   *6/2006*, Jordan Boats* / 1486574

4 ABDULLAH (DAUNTLESS) CLASS (PATROL CRAFT) (PB)

68171–68174

Displacement, tonnes: 15 full load
Dimensions, metres (feet): 13.2 × 4.2 × 1.35 (43.3 × 13.8 × 4.4)
Speed, knots: 35
Complement: 4
Machinery: 2 Cummins QSM-11 diesels; 1,160 hp (865 kW); 2 shafts
Guns: 2—12.7 mm MGs. 2—7.62 mm MGs.
Radars: Navigation: Raymarine RL70C; I-band.

Comment: Sea Ark Dauntless design acquired in 2006.

ABDULLAH 68174
4/2006, M Declerck
1164801

Kazakhstan

Country Overview

Formerly part of the USSR, the Republic of Kazakhstan declared its independence in 1991. Situated in Central Asia, it has an area of 1,049,155 square miles and is bordered to the north and west with Russia, to the east with China and to the south with Kyrgyzstan, Uzbekistan, and Turkmenistan. It has a 755 n mile coastline with the Caspian Sea on which Aktau, the principal port, is situated. Astana became the capital city in 1995 while Almaty, the former capital, is the largest city. Maritime claims in the Caspian Sea are not clear. The 2003 defence reform programme planned a three-service structure of the armed forces with the navy as an independent service. This was re-affirmed in 2007 with the announcement that a new naval strategy through to 2025 was under development.

Headquarters Appointments

Commander, Navy:
 Rear Admiral Zhandarbek Zhanzakov

Bases

Aktau (Caspian) (HQ)
Bautino (Caspian)

Personnel

2014: 3,000

PATROL FORCES

Notes: (1) Plans to acquire further patrol vessels were announced in early 2010. These include three corvettes of the order of 500–700 tons. The latter are to be built with a foreign partner and contenders are reported to include STX, with a version of the Gumdoksuri class, and Damen.
(2) Five Guardian class Boston Whalers delivered in November 1995 are reported operational.

1 TURK (AB 25) CLASS (PB)

Displacement, tonnes: 173 full load
Dimensions, metres (feet): 40.2 × 6.4 × 1.7 (131.9 × 21.0 × 5.6)
Speed, knots: 22
Complement: 31
Machinery: 4 SACM-AGO V16 CSHR diesels; 9,600 hp(m) (7.06 MW); 2 cruise diesels; 300 hp(m) (220 kW); 2 shafts
Guns: 1 Bofors 40 mm/70. 1 Oerlikon 20 mm.
Radars: Surface search: Racal Decca; I-band.

Comment: AB 32 and AB 26 presented by the Turkish Navy on 3 July 1999 and 25 July 2001 respectively. One has since been decommissioned. May have retained active sonar and ASW rocket launcher but this is unlikely.

TURK CLASS (Turkish colours) 10/2000, Selim San/ 0106636

7 ZHUK (PROJECT 1400) CLASS (PB)

ASTANA 301	– 304	TARAZ –
ORAL 302	ATYRAU 305	OSKEMEN 307
ALMATY 303		

Displacement, tonnes: 40 full load
Dimensions, metres (feet): 24 × 5 × 1.2 (78.7 × 16.4 × 3.9)
Speed, knots: 30
Range, n miles: 1,100 at 15 kt
Complement: 11
Machinery: 2 Type M401B diesels; 2,200 hp(m) (1.6 MW) sustained; 2 shafts
Guns: 1—12.7 mm MG.
Radars: Surface search: Spin Trough; I-band.

Comment: Built at the Zenit Shipyard, Uralsk, and commissioned 15 July 1998.

OSKEMEN 6/2013*, Kazakh Navy / 1486572

1 DAUNTLESS CLASS (PB)

ABAY 106

Displacement, tonnes: 11 full load
Dimensions, metres (feet): 12.8 × 4.3 × 1.3 (42 × 14.1 × 4.3)
Speed, knots: 35
Range, n miles: 600 at 18 kt
Complement: 5
Machinery: 2 Detroit 8V-92TA diesels; 1,270 hp (935 kW); 2 shafts
Guns: 1—12.7 mm MG. 2—7.62 mm MGs.
Radars: Surface search: Furuno; I-band.

Comment: Ordered under US funding in November 1995. Built by SeaArk, Monticello. Used to interdict the smuggling of nuclear materials across the Caspian Sea.

DAUNTLESS CLASS 7/1996, SeaArk Marine / 0080220

4 SUNKAR (PROJECT 14081) CLASS (PB)

501–504

Displacement, tonnes: 11.5 full load
Dimensions, metres (feet): 14.1 × 3.5 × 0.9 (46.3 × 11.5 × 3.0)
Speed, knots: 35
Range, n miles: 135 at 35 kt
Complement: 2
Machinery: 1 diesel; 980 hp(m) (720 kW); 1 water-jet
Guns: 2—7.62 mm MGs.
Radars: Surface search: I-band.

Comment: Built at Zenith Shipyard, Uralsk, and delivered in 2006.

PROJECT 14081 (Russian colours) 7/1996, Hartmut Ehlers / 0052520

3 SEA DOLPHIN CLASS (PBF)

SHAPSHAL 031 **BATYR** 032 **IZET** 033

Displacement, tonnes: 150 full load
Dimensions, metres (feet): 37 × 6.9 × 1.7 *(121.4 × 22.6 × 5.6)*
Speed, knots: 37
Range, n miles: 600 at 20 kt
Complement: 31 (5 officers)
Machinery: 2 MTU MD 16V 538 TB90 diesels; 6,000 hp(m) *(4.41 MW)*; 2 shafts
Guns: 2 Emerson Electric 30 mm (twin). 2—20 mm.
Radars: Surface search/navigation: Raytheon 1645; I-band.

Comment: Transferred from the South Korean Navy in 2006. Details are based on those in Korean service and may differ.

SEA DOLPHIN 031   *6/2012, Mazumdar Collection* / 1483515

0 + 2 KATRAN (PROJECT 20970) CLASS (PGG)

Displacement, tonnes: 320 full load
Dimensions, metres (feet): 46 × 8.9 × 2.0 *(150.9 × 29.2 × 6.6)*
Speed, knots: 30
Range, n miles: 1,200 at 12 kt
Complement: 26

Machinery: 2 MTU 20V4000M93L diesels; 11,500 hp *(8.6 kW)*; 2 Kamewa waterjets
Missiles: SSM: 8 (2 quad) SS-N-25 Switchblade (Uran); active radar homing to 130 km *(70.2 n miles)* at 0.9 Mach; warhead 145 kg.
SAM: 1 Ghibka launcher for SA-N-10 (Gimlet); IR homing to 5 km *(2.7 n miles)*; warhead 1.5 kg; 12 missiles.
Guns: 1—57 mm/75 A-220; 120 rds/min to 12.7 km *(6.8 n miles)*; weight of shell 6.6 kg. 1—30 mm/65 AK 630; 6 barrels; 3,000 rds/min combined to 2 km.
Physical countermeasures: Decoys: 2 PK-10 decoy launchers.
Radars: Air/Surface search: Cross Dome (Pozitiv ME); 3D; I-band.
Fire control: Bass Tilt (MR123); I-band.
Navigation: Furuno; I-band.

Comment: Designed by Almaz and under construction at Zenit Shipyard, Uralsk. Reports in mid-2012 suggested the project was delayed and the delivery date of the first of class is not known.

KATRAN CLASS   *(Scale 1 : 600), Ian Sturton* / 1366956

4 SARDAR (PROJECT 22180) CLASS (PB)

SARDAR 201 **SAKSHI** 202 **ZHENIS** 203 **SEMSER** 204

Displacement, tonnes: 270 full load
Dimensions, metres (feet): 40 × 7.5 × 1.65 *(131.2 × 24.6 × 5.4)*
Speed, knots: 28
Range, n miles: 2,300 at 12 kt
Complement: 14
Machinery: 2 Deutz TBD620V16 diesels; 6,000 hp *(4.5 MW)*; 2 shafts
Guns: To be announced.
Radars: To be announced.

Comment: Designed by Severnoye Design Bureau and constructed at Zenit Shipyard, Uralsk. *Zhenis* was launched on 5 May 2010. The vessels combine the roles of supply ship, patrol ship and SAR. A 7 m RHIB can be launched and recovered from a stern slipway. The ships are equipped with a 2-ton crane.

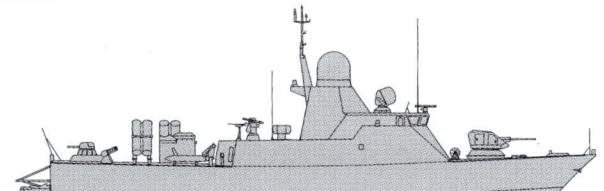

SARDAR   *6/2011* / 1406976

0 + 7 SOBOL (PROJECT 12200) CLASS (PBF)

Displacement, tonnes: 57.5 full load
Dimensions, metres (feet): 27.96 × 5.82 × 1.34 *(91.7 × 19.1 × 4.4)*
Speed, knots: 48. **Range, n miles:** 500 at 40 kt
Complement: 6
Machinery: 2 MTU 12V2000 M93 diesels; 3,644 hp *(2.7 MW)*; 2 Arneson drives
Guns: 2—12.7 mm MGs.

Comment: Seven craft are under construction at Zenith Shipyard, Uralsk, and are expected to enter service in 2014.

2 + 1 KAZAKHSTAN CLASS (PATROL SHIPS) (PBO)

KAZAKHSTAN 250 **ORAL** 251 **+1**

Displacement, tonnes: 230 full load
Dimensions, metres (feet): 46 × 9.0 × 2.5 *(150.9 × 29.5 × 8.2)*
Speed, knots: 28
Range, n miles: 1,200 at 12 kt
Complement: 31

Machinery: 2 diesels; 2 shafts
Missiles: SAM: 1 Arbalet-K mounting; 4 SA-N-10 Igla; electro-optic aiming; IR homing to 6 km *(3.2 n miles)* at 1.5 Mach; warhead 1.5 kg.
Guns: 2—23 mm. 1—12-tube MRLS.
Radars: Surface search/navigation: To be announced.

Comment: The first of three patrol ships was launched at Zenit Shipyard, Uralsk in April 2012 and the second on 30 April 2013. Although few details of the ship have been released, the design is believed to be a an enlarged variant of the Project 22180 Sardar class. One further ship is expected.

KAZAKHSTAN   *4/2012, ZENIT* / 1483518

3 ARCHANGEL CLASS (RESPONSE BOATS) (PBF)

620–622

Displacement, tonnes: 11 full load
Dimensions, metres (feet): 12.9 × 4.0 × 0.76 *(42.3 × 13.1 × 2.5)*
Speed, knots: 40
Range, n miles: 300 at 25 kt
Complement: 4
Machinery: 2 Caterpillar diesels; 2 Hamilton 322 waterjets
Guns: 1—12.7 mm MG.
Radars: To be announced.

Comment: High-speed inshore patrol craft of aluminium construction and foam collar built by SAFE Boats International, Port Orchard, Washington. Donated by the United States government in 2006.

ARCHANGEL 621   *4/2012, Mazumdar Collection* / 1483517

3 FC-19 CLASS (PATROL SHIPS) (PB)

NAIZA 505 **YESIL** 506 **EGEMEN** 507

Displacement, tonnes: 26 full load
Dimensions, metres (feet): 19.8 × 4.3 × 0.95 *(65 × 14.1 × 3.1)*
Speed, knots: 53
Range, n miles: 400 at 12 kt
Complement: 8
Machinery: 2 diesels; 2 shafts
Guns: 1—12.7 mm MG.
Radars: Surface search/navigation: To be announced.

Comment: Built at Zenit Shipyard, Uralsk and delivered in 2009–11.

NAIZA   *6/2013*, Kazakh Navy* / 1486573

SURVEY AND RESEARCH SHIPS

1 SURVEY SHIP (AGS)

Name	Builders	Commissioned
ZHAIK	Astrakhanskaya Sudoverf, Astrakhan	19 May 2008

Measurement, tonnes: 196 gt
Dimensions, metres (feet): 31.8 × 6.9 × 1.8 *(104.3 × 22.6 × 5.9)*
Speed, knots: 10
Machinery: 1 Deutz BF6M1015MC diesel; 337 hp *(248 kW)*; 1 shaft
Radars: To be announced.

Comment: Former fishing vessel laid down in 1994 and converted for use as a survey ship in 2008.

ZHAIK
6/2012, Mazumdar Collection
1483516

Kenya

Country Overview

A former British colony, The Republic of Kenya gained independence in 1963. Located astride the Equator, the country has an area of 224,082 square miles and has borders to the north with Somalia and Ethiopia and to the south with Tanzania. It has a 292 n mile coastline with the Indian Ocean. The country includes almost all of Lake Turkana (Lake Rudolf) and a small portion of Lake Victoria. The capital and largest city is Nairobi and the main seaport is Mombasa. Kisumu is a port on Lake Victoria. Perhaps the first proponent of the Exclusive Economic Zone (EEZ) concept, Kenya claims a 200 n mile EEZ whose limits have been partly defined. An extension to 350 n miles was claimed in May 2009. Territorial seas (12 n miles) are claimed.

Headquarters Appointments

Commander, Navy:
Major General Ngewa Mukala

Personnel

(a) 2014: 1,370 plus 120 marines
(b) Voluntary service

Bases

Mombasa (Mtongwe port), Manda, Malindi, Lamu, Kisumu (Lake Victoria)

Coast Defence

There are nine Masura coastal radar stations spread along the coast. Each station has 30 ft fast boats to investigate contacts.

Customs/Police

There are some 14 Customs and Police patrol craft of between 12 and 14 m. Mostly built by Cheverton, Performance Workboats and Fassmer in the 1980s. One Cheverton 18 m craft acquired in early 1997.

PATROL FORCES

Notes: There are also five Spanish built inshore patrol craft of 12 m armed with 12.7 mm MGs and driven by twin 538 hp diesels for a speed of 16 kt. Acquired in 1995, they have pennant numbers P 943–947.

2 NYAYO CLASS (FAST ATTACK CRAFT—MISSILE) (PGGF)

Name	No	Builders	Launched	Commissioned
NYAYO	P 3126	Vosper Thornycroft	20 Aug 1986	23 Jul 1987
UMOJA	P 3127	Vosper Thornycroft	5 Mar 1987	16 Sep 1987

Displacement, tonnes: 315 standard; 437 full load
Dimensions, metres (feet): 56.7 × 8.2 × 2.4 *(186 × 26.9 × 7.9)*
Speed, knots: 40. **Range, n miles:** 2,000 at 18 kt
Complement: 40

Machinery: 4 Paxman Valenta 18CM diesels; 15,000 hp *(11.19 MW)* sustained; 4 shafts; 2 motors (slow speed patrol); 100 hp *(74.6 kW)*
Electronic countermeasures: ESM: Racal Cutlass; radar warning.
ECM: Racal Cygnus; jammer.
Radars: Surface search: Plessey AWS 4; E/F-band; range 101 km *(55 n miles)*.
Navigation: Furuno; I-band.
Weapon control systems: CAAIS 450.

Programmes: Ordered in September 1984. Sailed in company from the UK, arriving at Mombasa 30 August 1988. Similar to Omani Province class.
Operational: Form Squadron 86. Both ships refitted by Fincantieri 2009-11. Work included removal of the Otomat SSM, the 76 mm gun and fire-control radar, refurbishment of the main machinery and replacement of the navigation and communication systems. They returned to Mombasa on 16 August 2011.

UMOJA
6/2011, Guy Toremans / 1406458

2 SHUPAVU CLASS (LARGE PATROL CRAFT) (PBO)

SHUJAA P 3130   SHUPAVU P 3131

Displacement, tonnes: 488 full load
Dimensions, metres (feet): 58 × 8.2 × 2.8 *(190.3 × 26.9 × 9.2)*
Speed, knots: 22
Complement: 24

Machinery: 2 diesels; 2 shafts
Guns: 1 Oto Melara 3 in *(76 mm)*/62; 85 rds/min to 16 km *(8.7 n miles)* anti-surface; 12 km *(6.5 n miles)* anti-aircraft; weight of shell 6 kg. 1 Mauser 30 mm.
Radars: Surface search: I-band.
Weapon control systems: Breda optronic director.

Comment: Built to civilian standards at Astilleros Gondan, Castropol and delivered in 1997 when they were taken over by the Navy. Armament fitted in Kenya.

SHUJAA
2/2001, Michael Nitz / 0137788

1 PATROL SHIP (PSO)

Name	No	Builders	Laid down	Launched
JASIRI	P 3124	Astilleros Gondan, Castropol	30 Jun 2004	26 Jan 2005

Displacement, tonnes: 1,020 full load
Dimensions, metres (feet): 85 × 12.2 × 3.0 *(278.9 × 40.0 × 9.8)*
Speed, knots: 28
Range, n miles: 5,500 at 12 kt
Complement: 41 + 20 spare berths

Machinery: 2 MTU 20V 1163 TB93 diesels; 10,000 hp *(7.4 MW)*; 2 shafts
Guns: 1—76 mm. 1—30 mm. 4—12.7 mm MGs.
Radars: Surface search: Furuno FAR-28355; E/F-band.
Navigation: Kelvin Hughes Type 2007; I-band.
Weapon control systems: Optronic director.

Comment: Contract for the procurement of a new patrol ship was signed between the government of Kenya and Euromarine on 15 July 2003. Subsequently, Astilleros Gondan was subcontracted to undertake construction of the vessel. However, following the return of the standby crew to Kenya in July 2005, the future of the ship became uncertain and the ship was laid up at Sada, near La Coruña for six years. Agreement for purchase of the ship was finally reached in April 2012 and the ship arrived in Mombasa on 29 August 2012.

JASIRI
5/2007 / 1335393

Patrol forces < **Kenya** — Patrol forces < **Kiribati** 465

1 ARCHANGEL CLASS (RESPONSE BOAT) (PBF)

Displacement, tonnes: 13 full load
Dimensions, metres (feet): 12.9 × 4.1 × 2.3 (42.3 × 13.5 × 7.5)
Speed, knots: 36. **Range, n miles:** 300 at 25 kt
Complement: 6
Machinery: 2 Caterpillar C9 diesels; 550 hp (409 kW); 2 Hamilton 322 waterjets
Guns: 2 — 7.62 mm MGs.
Radars: Navigation: Furuno; I-band.

Comment: High-speed inshore patrol craft of aluminium construction and foam collar built by SAFE Boats International, Port Orchard, Washington. Donated by the US government on 9 October 2006. The new patrol craft is to be used for monitoring the coastline and deterrence of criminal activity including illegal arms and drug running.

ARCHANGEL CLASS   6/2006, SAFE Boats / 1164947

9 DEFENDER CLASS (RESPONSE BOATS) (PBF)

PB 211   PB 212–219

Displacement, tonnes: 3 full load
Dimensions, metres (feet): 7.6 × 2.6 × 1.1 (24.9 × 8.5 × 3.6)
Speed, knots: 46. **Range, n miles:** 175 at 35 kt
Complement: 4
Machinery: 2 Honda outboard motors; 450 hp (335 kW)
Guns: 1 — 12.7 mm MG.
Radars: To be announced.

Comment: High-speed inshore patrol craft of aluminium construction and foam collar built by SAFE Boats International, Port Orchard, Washington. Five craft donated by the US government on 9 October 2006 and a further four craft in 2010. The patrol craft are used for monitoring the coastline and deterrence of criminal activity including illegal arms and drug running.

PB 212   6/2006, SAFE Boats / 1335394

1 P 400 CLASS (LARGE PATROL CRAFT) (PBO)

Name	No	Builders	Commissioned
HARAMBEE (ex-*La Rieuse*)	P 3123 (ex-P 690)	CMN, Cherbourg	13 Jun 1987

Displacement, tonnes: 413 standard; 488 full load
Dimensions, metres (feet): 54.8 × 8 × 2.5 (179.8 × 26.2 × 8.2)
Speed, knots: 23. **Range, n miles:** 4,200 at 15 kt
Complement: 26 (3 officers) + 20 spare berths
Machinery: 2 SEMT-Pielstick 16 PA4 200 VGDS diesels; 8,000 hp(m) (5.88 MW) sustained; 2 shafts
Guns: 1 Bofors 40 mm/60; 1 Giat 20F2 20 mm; 2 — 7.62 mm MGs.
Radars: Surface search: 1 Racal Decca DRBN-38A (Bridgemaster E 250); I-band.

Programmes: One of 12 patrol craft that entered service with the French Navy in the late 1980s. Transferred to Kenya at La Réunion on 7 June 2011.
Structure: Steel hull and superstructure protected by an upper deck bulwark. Design modified from original missile craft configuration.

AUXILIARIES

2 GALANA CLASS (LCM)

Name	No	Builders	Commissioned
GALANA	L 38	Astilleros Gondan, Castropol	Feb 1994
TANA	L 39	Astilleros Gondan, Castropol	Feb 1994

Displacement, tonnes: 1,422 full load
Dimensions, metres (feet): 63.5 × 13.3 × 2.4 (208.3 × 43.6 × 7.9)
Speed, knots: 12.5
Complement: 30
Machinery: 2 MTU/Bazán diesels; 2,700 hp(m) (1.98 MW) sustained; 2 shafts; bow thruster
Radars: Navigation: Racal Decca; I-band.

Comment: Acquired by Galway Ltd for civilian use and taken over by the Navy for logistic support. The 4 m wide ramp is capable of taking 70 ton loads.

TANA   2/1999 / 0052523

2 TENDER (LCM)

Dimensions, metres (feet): 18.3 × 4.8 × 1.5 (60 × 15.7 × 4.9)
Speed, knots: 10
Range, n miles: 200 at 10 kt
Complement: 2 + 136 personnel
Machinery: 2 Caterpillar 3306B-DIT diesels; 880 hp(m) (647 kW); 2 shafts

Comment: Built by Souters, Cowes and delivered in 1998. Personnel tenders.

Kiribati

Country Overview

The Republic of Kiribati, formerly the Gilbert Islands, is a south Pacific island group which gained independence in 1979 after the other part of the former British colony, the Ellice Islands, became independent as Tuvalu the previous year. Straddling the equator some 1,385 n miles southwest of Hawaii, it comprises from west to east Banaba (Ocean Island) and three detached island groups: the 16 Gilbert Islands, including Tarawa, on which the capital, Bairiki, is located, nine Phoenix Islands and eight of the 11 Line Islands. About 20 of the 34 islands are permanently inhabited. An archipelagic state, territorial seas (12 n miles) are claimed. An Exclusive Economic Zone (EEZ) (200 n miles) is also claimed but limits have not been fully defined by boundary agreements.

Headquarters Appointments

Head of Police Maritime Unit:
Assistant Superintendent John Mote

Bases

Tarawa

PATROL FORCES

1 PACIFIC CLASS (LARGE PATROL CRAFT) (PB)

Name	No	Builders	Commissioned
TEANOAI	301	Transfield Shipbuilding	22 Jan 1994

Displacement, tonnes: 168 full load
Dimensions, metres (feet): 31.5 × 8.1 × 2.1 (103.3 × 26.6 × 6.9)
Speed, knots: 18
Range, n miles: 2,500 at 12 kt
Complement: 18 (3 officers)
Machinery: 2 Caterpillar 3516TA diesels; 4,400 hp (3.28 MW) sustained; 2 shafts
Guns: Can carry 1 — 12.7 mm MG but is unarmed.
Radars: Navigation: Furuno 1011; I-band.

Comment: The Pacific Patrol Boat programme was started by Australia in 1987. *Teanoai*, the 16th of the class, was handed over to Kiribati in 1994. The Australian government has announced that the programme will be extended so that all 22 boats will be able to operate for 30 years. *Teanoai* completed a half-life refit at Gladstone in 2001 and a life extension refit in 2008.

TEANOAI
9/2008,
Kiribati Marine Police
1335214

© 2014 IHS   IHS Jane's Fighting Ships 2014-2015

Korea, North
PEOPLE'S DEMOCRATIC REPUBLIC

Country Overview

The Democratic People's Republic of Korea (DPRK) was proclaimed in 1948 and occupies the northern part of the Korean peninsula. Located in north-eastern Asia and with an area of 46,540 square miles, it is bordered to the north by China and Russia and to the south by South Korea. It has a 1,350 n mile coastline with the Sea of Japan and the Yellow Sea. The capital and largest city is Pyóngyang while the principal ports are Nampo and Haeju on the west coast and Chojin and Wónsan on the east coast. Territorial seas (12 n miles) are claimed. A 200 n mile EEZ has also been claimed but the limits have not been defined. A source of tension at sea is the dispute concerning the status of the *Northern Limit Line* and a number of South Korean islands off the south-west coast of DPRK.

The Korean People's Army Naval Force is principally a coastal force and is the lowest priority military service. Ships are allocated to East or West Fleet Command. The Navy is manpower intensive and most equipment is technologically outdated and incapable of bluewater operations. Nevertheless, considerable emphasis has been placed on high speed infiltration and assault craft and the ability to conduct special forces operations. Fishing vessels are likely to be converted and/or commandeered for military use while ocean-going merchant vessels are likely to have military roles including arms transfers and intelligence gathering.

Headquarters Appointments

Commander of the Navy:
Admiral Jyung Myung-do
Commander West Sea Fleet:
Rear Admiral Han Sang-soon
Commander East Sea Fleet:
Rear Admiral Park Won-shik

Bases

Naval Headquarters: Pyongyang.
East Fleet Command (HQ T'oejo-dong (Nagwon-up)).
East coast: T'oejo-dong, Ch'aho (submarines), Munch'on-up, Mayang-do (submarines), Najin.
Minor bases: Chakto-dong (Chakto-ri), Hwangt'o-do, Hodo-ri, Kosong-up (Changjon-ni), Mugye-ri, Ohang-ni, Puam-dong, Sinch'ang-nodongjagu, Chongjin, Songjin (Kimch'aek), Songjon-pardo, Wonsan, Yoho-ri, Yongam-ni, and Yukt'aedong-ni.
West Fleet Command: (HQ Namp'o).
West coast: Namp'o (Chinnamp'o), Pipa-got (submarines) and Sagon-ni (Sa-got).
Minor bases: Cho-do, Haeju, Kwangyang-ni, Sunwi-do, Yongdok, and Yongamp'o.

Personnel

(a) 2014: 60,000 officers and other ranks
(b) 5 years' national service

Maritime Security Battalions

In addition to the Navy there is a Coastal and Port Security Police Force which would be subordinate to the Navy in war. It is reported that the strength of this force is 10–15 Chong-Jin patrol craft and 130 patrol boats of various types.

Naval Aviation

There is believed to be a battalion-sized naval support/ASW air unit containing ASW, helicopter and transport elements. The ASW element consists of 10-20 Mi-14PL Haze-A ASW helicopters acquired during the late 1980s and early 1990s. The majority are thought to be subordinated to the East Sea Fleet although there are no details as to how they are organised and deployed. In addition, there are reported to be a small number of Ka-32S Helix although their role is unclear.

Coastal Defence

Considerable emphasis is given to coastal defence. There are believed to be two missile regiments (one in each fleet), a large number of surveillance radar companies and numerous artillery batteries. Missile sites are reported to be located at An-gol, Chakto-dong, Mayang-do, Sinsang-ni, and Unami-ni on the East Sea coast; and Chungsan, Hwajin-ni, Pip'a-got and Tungsan-got on the West Sea coast. Target acquisition is provided by organic target acquisition radar and ESM. There are numerous other soft sites available for redeployment and truck-mounted mobility is a key feature of the system. Major ports and naval bases are likely to be heavily defended.

Strength of the Fleet

Type	Active
Submarines—Patrol	20
Submarines—Coastal	44
Submarines—Midgets	23
Frigates	3
Corvettes	4
Patrol Forces	400+
Amphibious Craft	129
Hovercraft (LCPA)	135
Minesweepers	24
Depot Ships for Midget Submarines	8
Survey Vessels	4

DELETIONS

Notes: The order of battle and fleet dispositions represent the best estimates that can be made based on incomplete information.

SUBMARINES

Notes: (1) There are four obsolete ex-Soviet Whiskey class based at Mayang-do used for training. Probably restricted to periscope depth when dived.
(2) Reports of a sea-based ballistic missile capability have not been substantiated. A surface-ship based system is considered more likely than a submarine-launched missile which would present considerable technical challenges.
(3) There is an unknown number of Daedong-class midget submarines (length 17 m, breadth 4 m). It is reported that the boats are equipped with two exterior 4 m tubes for lightweight torpedoes.

12 YONO (P4) CLASS (MIDGET SUBMARINES) (SSM)

Displacement, tonnes: 117 surfaced; 125 dived
Dimensions, metres (feet): 29 × 2.75 × 2.5 *(95.1 × 9.0 × 8.2)*
Speed, knots: 10 surfaced; 7 dived
Complement: 7
Machinery: Diesel-electric

Torpedoes: 2 — 12 in *(533 mm)*.
Sonars: To be announced.

Comment: Probably a development of the Yugo class, this submarine has been in service since the 1990s. Numbers are approximate. The design has been exported to Iran where 16 similar boats (also known as the Ghadir class) are now in operation. The Iranian boats feature a retractable secondary propeller.

20 ROMEO (PROJECT 033) CLASS (SS)

Displacement, tonnes: 1,499 surfaced; 1,859 dived
Dimensions, metres (feet): 76.6 × 6.7 × 5.2 *(251.3 × 22.0 × 17.1)*
Speed, knots: 15 surfaced; 13 dived
Range, n miles: 9,000 at 9 kt surfaced
Complement: 54 (10 officers)

Machinery: Diesel-electric; 2 Type 37-D diesels; 4,000 hp(m) *(2.94 MW)*; 2 motors; 2,700 hp(m) *(1.98 MW)*; 2 creep motors; 2 shafts
Torpedoes: 8 — 21 in *(533 mm)* tubes (6 bow, 2 stern). 14 probably SAET-60; passive homing up to 15 km *(8.1 n miles)* at 40 kt; warhead 400 kg. Also some 53–56 may be carried.
Mines: 28 in lieu of torpedoes.
Electronic countermeasures: ESM: China Type 921A Golf Ball (Stop Light); radar warning.
Radars: Surface search: Snoop Plate/Tray; I-band.
Sonars: Pike Jaw; hull-mounted; active. Feniks; hull-mounted; passive.

Programmes: Two transferred from China 1973, two in 1974 and three in 1975. First three of class built at Sinpo and Mayang-do shipyards in 1976. Programme ran at about one every 14 months until 1995 when it stopped in favour of the Sang-O class. One reported sunk in February 1985.
Operational: Approximately 17 are stationed on east coast and have occasionally operated in Sea of Japan. The remainder, including ex-Chinese units, are based on the west coast. By modern standards these are basic attack submarines with virtually no anti-submarine performance or potential and their operational status is doubtful.

ROMEO CLASS

38 SANG-O CLASS (SSC)

Displacement, tonnes: 260 surfaced; 281 dived
Dimensions, metres (feet): 35.5 × 3.8 × 3.7 *(116.5 × 12.5 × 12.1)*
Speed, knots: 7.6 surfaced; 7.2 snorting; 8.8 dived
Complement: 25 (2 officers; 6 divers)

Machinery: 1 Russian diesel generator; 1 North Korean motor; 1 shaft; shrouded prop
Torpedoes: 2 or 4 — 21 in *(533 mm)* tubes (in some). Probably Russian Type 53–56.
Mines: 16 can be carried (in some).
Radars: Surface search: Furuno; I-band.
Sonars: Russian hull-mounted; passive/active search and attack.

Programmes: Started building in 1995 at Sinpo accelerating up to about four to six a year by 1996. Reported to have been building at about three a year from 1997. One reported delivered in 2002 and one in 2003 and overall numbers reflect an estimated building rate of almost two per year.
Structure: A variation of a reverse engineered Yugoslav design. There are at least two types, one with torpedo tubes and one capable of carrying up to 16 externally-fitted bottom mines. There is a single periscope and a VLF radio receiver in the fin. Rocket launchers and a 12.7 mm MG can be carried. Diving depth 180 m *(590 ft)*. A longer (39 m) variant submarine may replace older boats.
Operational: Used extensively for infiltration operations. The submarine can bottom, and swimmer disembarkation is reported as being normally exercised from periscope depth. One of the class grounded and was captured by South Korea on 18 September 1996. Some crew members may be replaced by special forces for short operations. 17 stationed on east coast.

SANG-O CLASS
9/1996
0080223

6 SANG-O II (K-300) CLASS (SSC)

Displacement, tonnes: 295 surfaced; 320 dived
Dimensions, metres (feet): 39 × 3.8 × 3.7 *(128 × 12.5 × 12.1)*
Speed, knots: 13
Complement: 25 (2 officers)

Machinery: 1 diesel generator; 1 motor; 1 shaft
Torpedoes: 4 — 21 in *(533 mm)* tubes.
Mines: 16 can be carried.
Sonars: Active/passive search and attack.

Comment: The existence of a new class of submarine, believed to be a stretched version of the Sang-O class, was confirmed by South Korean officials in March 2011. Work on the new design, which includes insertion of a 5 m plug aft of the sail, probably began during the late 1990s, with production initiated by the mid-2000s. The new design is reported to include an improved diesel-electric powerplant and additional space for special forces. However details remain speculative. The number of operational boats is also uncertain but it is likely that the submarines are progressively to replace the ageing Romeo class, most of which are based in the East Sea Fleet.

11 (+10 RESERVE) YUGO CLASS (MIDGET SUBMARINES) (SSW)

Displacement, tonnes: 91 surfaced; 112 dived
Dimensions, metres (feet): 20 × 3.1 × 4.6 *(65.6 × 10.2 × 15.1)*
Speed, knots: 12 surfaced; 8 dived
Range, n miles: 550 at 10 kt surfaced; 50 at 4 kt dived
Complement: 11 (est.) (7 divers (est.))
Machinery: 2 diesels; 320 hp(m) *(236 kW)*; 1 shaft
Torpedoes: 2 — 406 mm tubes.
Radars: Navigation: I-band.

Comment: Built at Yukdaeso-ri shipyard since early 1960s. More than one design. Details given are for the latest type, at least one of which has been exported to Iran, and have been building since 1987 to a Yugoslavian design. Some have two short external torpedo tubes and some have a snort mast. The conning tower acts as a wet and dry compartment for divers. There is a second and smaller propeller for slow speed manoeuvring while dived. Two were exported to Vietnam in 1997.

YUGO P-4
6/1998, Ships of the World / 0052525

FRIGATES

Notes: The hull of what is probably an ex-Russian Krivak III frigate is at Nampo naval shipyard. All weapons and sensors have been removed from the ship and the future of the vessel is unclear. If the ship were to be re-armed and activated, it would represent a significant increase in the capabilities of the surface fleet.

1 SOHO CLASS (FFGH)

Name	No	Builders	Laid down	Launched	Commissioned
—	823	Najin Shipyard	Jun 1980	Nov 1981	May 1982

Displacement, tonnes: 1,666 full load
Dimensions, metres (feet): 73.8 × 15.5 × 3.8 *(242.1 × 50.9 × 12.5)*
Speed, knots: 23
Complement: 189 (17 officers)

Machinery: 2 diesels; 15,000 hp(m) *(11.03 MW)*; 2 shafts
Missiles: SSM: 4 CSS-N-2 ❶; active radar or IR homing to 46 km *(25 n miles)* at 0.9 Mach; warhead 513 kg.
Guns: 1—3.9 in *(100 mm)*/56 ❷; 40° elevation; 15 rds/min to 16 km *(8.6 n miles)*; weight of shell 15.6 kg. 4—37 mm/63 (2 twin) ❸. 4—30 mm/65 (2 twin) ❹. 4—25 mm/60 (2 twin) ❺.
A/S Mortars: 2 RBU 1200 5-tubed fixed launchers ❻; range 1,200 m; warhead 34 kg.
Electronic countermeasures: ESM: China RW-23 Jug Pair (Watch Dog); intercept.
Radars: Surface search: Square Tie ❼; I-band.
Fire control: Drum Tilt ❽; H/I-band.
Navigation: I-band.
Sonars: Stag Horn; hull-mounted; active search and attack; high frequency.

Helicopters: Platform for 1 medium.

Programmes: Planned class of six but only one was ordered.

SOHO *(Scale 1 : 600), Ian Sturton / 0506237*

Structure: One of the largest warships built anywhere with a twin hull design and a helicopter deck aft. Has a large central superstructure to carry the heavy gun armament.

Operational: Probably very weather limited like many catamaran designs. Base and operational status not known.

2 NAJIN CLASS (FFG)

531	591

Displacement, tonnes: 1,524 full load
Dimensions, metres (feet): 102 × 10 × 2.7 *(334.6 × 32.8 × 8.9)*
Speed, knots: 24
Range, n miles: 4,000 at 13 kt
Complement: 180 (16 officers)

Machinery: 3 SEMT-Pielstick Type 16 PA6 280 diesels; 18,000 hp(m) *(13.2 MW)*; 3 shafts
Missiles: SSM: 2 SY-1 Scrubbrush CSS-N-1 ❶; active radar or IR homing to 40 km *(22 n miles)* at 0.9 Mach; warhead 513 kg HE. Replaced torpedo tubes on both ships.
Guns: 2—3.9 in *(100 mm)*/56 ❷; 40° elevation; 15 rds/min to 16 km *(8.6 n miles)*; weight of shell 15.6 kg. 4—57 mm/80 (2 twin) ❸; 120 rds/min to 6 km *(3.2 n miles)*; weight of shell 2.8 kg. 12 or 4—30 mm/60 (6 or 2 twin) ❹ (see *Structure*). 12—25 mm (6 twin) ❺.
A/S Mortars: 2 RBU 1200 5-tubed fixed launchers ❻; range 1,200 m; warhead 34 kg (not in 531).
Mines: 30 (estimated).
Depth charges: 2 projectors; 2 racks. 30 weapons.
Physical countermeasures: Decoys: 6 chaff launchers.

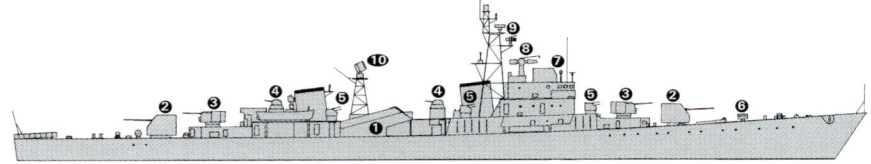

NAJIN *(Scale 1 : 900), Ian Sturton / 0506153*

Electronic countermeasures: ESM: China RW-23 Jug Pair (Watch Dog); intercept.
Radars: Air search: Square Tie ❽; I-band.
Surface search: Pot Head ❾; I-band.
Navigation: Pot Drum; H/I-band.
Fire control: Drum Tilt ❿; H/I-band.
IFF: High Pole. Square Head.
Sonars: Stag Horn; hull-mounted; active search; high frequency.
Weapon control systems: Optical director ❼.

Programmes: Built at Najin and Nampo shipyards. First completed 1973, second 1975.
Structure: There is some resemblance to the ex-Soviet Kola class, now deleted. The original torpedo tubes were replaced by CSS-N-1 missile launchers in the mid-1980s and the RBU 1200 mortars have been removed in at least one of the class. Gun armaments differ, one having six twin 30 mm while the other only has one twin 30 mm and six twin 25 mm.
Operational: One based on each coast but seldom seen at sea.

NAJIN 531 *5/1993, JMSDF / 0080224*

CORVETTES

4 SARIWON CLASS (FS)

611–614

Displacement, tonnes: 660 full load
Dimensions, metres (feet): 62.1 × 7.3 × 2.4 *(203.7 × 24.0 × 7.9)*
Speed, knots: 16. **Range, n miles:** 2,700 at 16 kt
Complement: 60 (7 officers)

Machinery: 2 diesels; 3,000 hp(m) *(2.21 MW)*; 2 shafts
Guns: 4—57 mm/80 (2 twin). 4—37 mm/6 (2 twin). 16—14.5 mm (4 quad).

A/S Mortars: 2 RBU 1200 5-tubed fixed launchers.
Mines: 30
Depth charges: 2 rails.
Radars: Surface search: Pot Head or Don 2; I-band.
Navigation: Model 351; I-band.
IFF: Ski Pole.
Sonars: Stag Horn; hull-mounted; active; high frequency.
Programmes: Four Sariwon class built in North Korea in the mid-1960s.

Structure: Sariwon design based on the original USSR fleet minelayer Tral or Fugas class which entered service in the 1930s. One Sariwon is reported as having sonar and ASW armament. Minelaying rails are visible along the upper deck aft of the superstructure.
Operational: Based on the east coast at Najin or Kosong-up.

1 TRAL CLASS (FS)

671

Displacement, tonnes: 589 full load
Dimensions, metres (feet): 62.1 × 7.3 × 2.4 *(203.7 × 24.0 × 7.9)*
Speed, knots: 16. **Range, n miles:** 2,700 at 16 kt
Complement: 60 (7 officers)

Machinery: 2 diesels; 3,000 hp(m) *(2.21 MW)*; 2 shafts
Guns: 1—85 mm/52 tank turret ❶. 2—37 mm/6 (single) ❷. 16—14.5 mm ❸; 4 quad.
Mines: 30.
Radars: Surface search: Pot Head or Don 2 ❹; I-band.
Navigation: Model 351; I-band.
IFF: Ski Pole.

Programmes: Two Tral class fleet minesweepers of 1930s vintage were transferred from the USSR in the mid-1950s, were paid off in the early 1980s but one returned to service in the early 1990s.
Structure: Minelaying rails are visible along the whole of upper deck aft of the bridge superstructure.

Operational: Based on the east coast (Najin or Kosong-up).

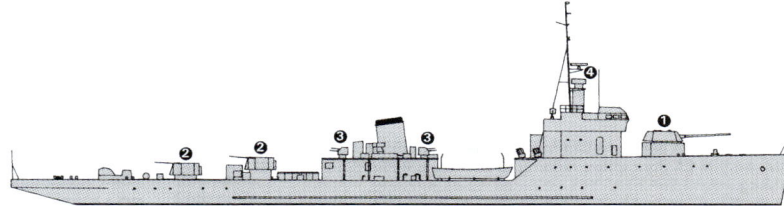

TRAL 671   *(Scale 1 : 600), Ian Sturton* / 0506198

TRAL 671   *5/1993, JMSDF* / 0080225

PATROL FORCES

Notes: There is reported to be a class of hovercraft or Surface Effect Ship (SES) designed for patrol duties. The 38 × 12 m craft have a displacement of 170 tons and are reported to have a speed of 48 kt. They are armed with a 57 mm gun forward and a 39 mm gun aft.

33 SINPO CLASS (FAST ATTACK CRAFT—TORPEDO) (PTF/PTK)

Displacement, tonnes: 65 standard; 74 full load
Dimensions, metres (feet): 26 × 6.1 × 1.5 *(85.3 × 20.0 × 4.9)*
Speed, knots: 45
Complement: 15

Machinery: 4 Type M 50 diesels; 4,400 hp(m) *(3.2 MW)* sustained; 4 shafts
Guns: 4—25 mm/80 (2 twin) (original). 2—37 mm (others). 6—14.5 mm MGs (Sinpo class).
Torpedoes: 2—21 in *(533 mm)* tubes (in some). Sinpo class has no tubes.
Depth charges: 8 in some.
Radars: Surface search: Skin Head; I-band (some have Furuno).
IFF: Dead Duck. High Pole.

Comment: Thirteen craft remain of the 27 P 6 class transferred from the USSR and 15 Shantou class transferred from China. Some of the P 6s have hydrofoils and one sank in June 1999. The Sinpo (or Sinnam) class are locally built versions of these craft of which 20 now remain. Based in both fleets.

SINPO 471   0506038

P 6   0506037

12 KOMAR CLASS (PROJECT 183) (FAST ATTACK CRAFT—MISSILE) (PTFG)

Displacement, tonnes: 76 standard; 86 full load
Dimensions, metres (feet): 25.6 × 7.3 × 1.8 *(84 × 24.0 × 5.9)*
Speed, knots: 40
Range, n miles: 400 at 30 kt
Complement: 19

Machinery: 4 Type M 50 diesels; 4,400 hp(m) *(3.3 MW)* sustained; 4 shafts
Missiles: SSM: 2 SS-N-2A Styx or CSS-N-1; active radar or IR homing to 46 km *(25 n miles)* at 0.9 Mach; warhead 513 kg.
Guns: 2—25 mm/80 (twin); 270 rds/min to 3 km *(1.6 n miles)*; weight of shell 0.34 kg. 2—14.5 mm (twin) MGs.
Radars: Surface search: Square Tie; I-band.
IFF: Square Head.

Programmes: There are six Komar class remaining of 10 transferred from the USSR. Wooden hulls have been replaced by steel. There are also six Sohung class, North Korean copies of the Komar class, first built in 1980–81 and no longer in production. The 'Komars' and four 'Sohung' are based on the east coast.

KOMAR   0506032

10 SOJU CLASS (FAST ATTACK CRAFT—MISSILE) (PTG)

Displacement, tonnes: 269 full load
Dimensions, metres (feet): 42.5 × 7.5 × 1.7 *(139.4 × 24.6 × 5.6)*
Speed, knots: 34
Range, n miles: 600 at 30 kt
Complement: 32 (4 officers)

Machinery: 3 Type M 503A diesels; 8,025 hp(m) *(5.9 MW)* sustained; 3 shafts
Missiles: SSM: 4 SS-N-2 Styx; active radar or IR homing to 46 km *(25 n miles)* at 0.9 Mach; warhead 513 kg.
Guns: 4—30 mm/65 (2 twin) AK 230; 500 rds/min to 5 km *(2.7 n miles)*; weight of shell 0.54 kg.
Electronic countermeasures: ESM: China BM/HZ 8610; intercept.
Radars: Surface search: Square Tie; I-band.
Fire Control: Drum Tilt; H/I-band.

Comment: North Korean built and enlarged version of Osa class. First completed in 1981; built at about one per year at Nampo, Najin and Yongampo shipyards, but the programme terminated in 1996. Six based on the east coast and four on the west.

474 Korea, South > Submarines

3 + 6 KSS-2 (TYPE 214) CLASS (SSK)

Name	No
SOHN WON-IL	072
JEONGJI	073
AHN JUNG-GEUN	075
KIM JWA-JIN	076
–	077
–	078
–	079
–	081

Builders	Laid down	Launched	Commissioned
Hyundai, Ulsan	2003	9 Jun 2006	26 Dec 2007
Hyundai, Ulsan	2004	13 Jun 2007	2 Dec 2008
Hyundai, Ulsan	2005	4 Jun 2008	1 Dec 2009
Daewoo, Okpo	2010	13 Aug 2013	2014
Hyundai, Ulsan	2010	2014	2015
Daewoo, Okpo	2011	2015	2016
Hyundai, Ulsan	2012	2016	2017
Daewoo, Okpo	2013	2017	2018

Displacement, tonnes: 1,727 surfaced; 1,890 dived
Dimensions, metres (feet): 65 × 6.3 × 6 *(213.3 × 20.7 × 19.7)*
Speed, knots: 12 surfaced; 20 dived
Complement: 27 (5 officers)

Machinery: 2 MTU 16V 396 SE 84 diesels; 4,243 hp (m) *(3.12 MW)*; 1 Siemens Permasyn motor; 3,875 hp(m) *(2.85 MW)*; 1 shaft; 2 HDW PEM fuel cells; 240 kW; sodium sulphide high-energy batteries
Torpedoes: 8 – 21 in *(533 mm)* bow tubes. LIG NEX1 White Shark; active/passive homing to 30 km *(16.3 n miles)*; warhead 250 kg; total of 16 weapons.
Physical countermeasures: Decoys.
Electronic countermeasures: ESM: SAAB UME 100 (UME 200 in Batch 2).
Radars: Surface search: I-band.
Sonars: Bow, flank and towed arrays.
Weapon control systems: STN Atlas. ISUS 90–61.

Programmes: Decision taken in November 2000 to order three (Batch 1) HDW designed Air Independent Propulsion (AIP) submarines. The boats were built by Hyundai Heavy Industries with the German Submarine Corporation, led by HDW, providing construction plans, materials and other equipment. First steel cut for the first of class in November 2002. A contract for the supply of six further material packages (Batch 2) was signed with HDW in December 2008. Construction of this second batch started in 2009. The first boat is being constructed by Daewoo and successor boats are likely to be built at the rate of one per year.
Structure: The Type 214 is a synthesis of the proven Type 209 design with AIP from the Type 212. South Korea is the second customer for the Type 214 after Greece. Details given are mainly for the Type 214 as advertised by HDW but changes may have been made. The boats are equipped with an attack periscope and an optronic mast, which includes the ESM system and electro-optic devices. Diving depth 400 m.

SOHN WON-IL *10/2008, Guy Toremans* / 1353187

SOHN WON-IL
10/2008, Michael Nitz
1353188

9 CHANG BOGO (TYPE 209/1200) CLASS (SSK)

Name	No
CHANG BOGO	061
YI CHON	062
CHOI MUSON	063
PARK WI	065
LEE JONGMU	066
JUNG WOON	067
LEE SUNSIN	068
NA DAEYONG	069
LEE EOKGI	071

Builders	Laid down	Launched	Commissioned
HDW, Kiel	1989	18 Jun 1992	2 Jun 1993
Daewoo, Okpo	1990	14 Oct 1992	30 Apr 1994
Daewoo, Okpo	1991	25 Aug 1993	27 Feb 1995
Daewoo, Okpo	1992	20 May 1994	3 Feb 1996
Daewoo, Okpo	1993	17 Apr 1995	29 Aug 1996
Daewoo, Okpo	1994	7 May 1996	29 Aug 1997
Daewoo, Okpo	1995	21 May 1998	15 Jun 1999
Daewoo, Okpo	1996	15 Jun 1999	Nov 2000
Daewoo, Okpo	1997	26 May 2000	30 Nov 2001

Displacement, tonnes: 1,118 surfaced; 1,306 dived
Dimensions, metres (feet): 56.4 × 6.2 × 5.5 *(185 × 20.3 × 18.0)*
Speed, knots: 11 surfaced; 11 snorting; 22 dived
Range, n miles: 7,500 at 8 kt surfaced
Complement: 33 (6 officers)

Machinery: Diesel-electric; 4 MTU 8V 396 SE diesels; 3,218 hp(m) *(2.4 MW)* sustained; 4 alternators; 1 motor; 4,600 hp(m) *(3.38 MW)* sustained; 1 shaft
Missiles: SSM: McDonnell Douglas UGM-84B Sub Harpoon; active radar homing to 130 km *(70 n miles)* at 0.9 Mach; warhead 227 kg; fitted to at least three boats).
Torpedoes: 8 – 21 in *(533 mm)* bow tubes. LIG NEX1 White Shark; active/passive homing to 30 km *(16.3 n miles)*; warhead 250 kg; total of 14 weapons.
Mines: 28 in lieu of torpedoes.
Electronic countermeasures: ESM: Argo; radar warning.
Radars: Navigation: I-band.
Sonars: Atlas Elektronik CSU 83; hull-mounted; passive search and attack; medium frequency.
Weapon control systems: Atlas Elektronik ISUS 83 TFCS.

Programmes: First three ordered in late 1987, one built at Kiel by HDW, and two assembled at Okpo by Daewoo from material packages transported from Germany. Second three ordered in October 1989 and a further batch of three in January 1994.
Modernisation: Mid-life upgrade of all nine boats is under consideration. It is envisaged that AIP propulsion and Sub-Harpoon SSM may be fitted in stretched hulls. Installation of a Sagem 40XP inertial navigation system throughout the class began in late 2011.
Structure: Type 1200 similar to those built for the Turkish Navy with a heavy dependence on Atlas Elektronik sensors and STN torpedoes. Diving depth 250 m *(820 ft)*. A passive towed array may be fitted in due course.
Operational: An indigenous torpedo based on the Honeywell NP 37 may be available in due course. The class is split between the three Fleets. Operations conducted off Hawaii from 1997 to improve operating standards.

PARK WI *10/2008, Guy Toremans* / 1353189

LEE SUNSIN *5/2008, US Navy* / 1353190

YI CHON
9/2013, Kenji Konishi*
1529169

IHS Jane's Fighting Ships 2014-2015 © 2014 IHS

DESTROYERS

Notes: Plans for a batch of six destroyers were announced on 13 October 2009. The ships, unofficially designated KDX-2X, are to be a development of the KDX-2 class design and are to incorporate the Aegis system. Following the announcement in December 2013 that a further three KDX-3 class are to be built, the ships are unlikely to start entering service before 2023.

3 + (3) SEJONG DAEWANG (KDX-3) CLASS (DDGHM)

Name	No	Builders	Laid down	Launched	Commissioned
SEJONG DAEWANG	991	Hyundai, Ulsan	12 Nov 2004	25 May 2007	22 Dec 2008
YULGOK YI I	992	Daewoo, Okpo	25 Jul 2007	14 Nov 2008	31 Aug 2010
YU SEONG-RYONG	993	Hyundai, Ulsan	2009	24 Mar 2011	3 Sep 2012

Displacement, tonnes: 7,773 standard; 10,455 full load
Dimensions, metres (feet): 165.9 × 21 × 10.5 (544.3 × 68.9 × 34.4)
Speed, knots: 30
Range, n miles: 5,000 at 14 kt

Machinery: COGAG; 4 GE LM 2500 gas turbines; 105,000 hp (78.33 MW) sustained; 2 shafts; cp props
Missiles: SLCM: 32 Cheon Ryong land-attack missiles ❶; inertial/GPS guidance to 500 km (270 n miles) at 0.7 Mach; warhead 500 kg.
SSM: 8 McDonnell Douglas Harpoon Block 1C ❷; active homing to 124 km (67 n miles) at 0.9 Mach; warhead 227 kg.
SAM: Mk 41 VLS; 80 cells for Standard SM-2 MR Block IIIB ❸; command/inertial guidance; semi-active radar homing to 167 km (90 n miles) at 2.5 Mach; 2 magazines; 48 missile tubes forward, 32 aft. 1 GMLS Mk 49 RAM RIM-116 ❹; 21 rounds; passive IR/anti-radiation homing to 9.6 km (5.2 n miles) at 2 Mach; warhead 9.1 kg.
A/S: 16 Loral ASROC (Red Shark) VLA ❶; inertial guidance 1.6–16.6 km (1–9 miles).
Guns: 1 United Defence 5 in (127 mm)/54 Mk 45 Mod 4 ❺; 20 rds/min to 23 km (12.6 miles); anti-surface; weight of shell 32 kg. 1 Signaal/General Electric 30 mm 7-barrelled Goalkeeper ❻; 4,200 rds/min to 1.5 km.
Torpedoes: 6—324 mm (2 triple) Mk 32 tubes ❼; K745 LW (Blue Shark); anti-submarine; active/passive homing to 11 km (5.9 n miles) at 40 kt; warhead 44 kg.
Electronic countermeasures: ESM/ECM: SLQ-200(V)5.
Radars: Air search/fire control: SPY 1D(V) phased arrays ❽; 3D; F-band.
Surface search: Norden/DRS SPS-67(V) ❾; G-band.
Fire control: 3 Raytheon SPG-62 ❿; I/J-band.
Navigation: To be announced.
Sonars: Lockheed Martin SQQ-89(V); underwater combat system with SQS-53C bow mounted; active search and attack.
Combat data systems: Aegis Baseline 7.1.
Weapon control systems: To be announced.
Helicopters: 2 Westland Super Lynx Mk 99 ⓫.

Programmes: The KDX-3 programme is the third phase of a surface ship modernisation programme that began with the KDX-1 programme in the early 1990s. Plans to build a class of six vessels were confirmed in December 2013 and are now subject to budget approval. The new ships are likely to have a ballistic missile defence capability and to enter service by 2023. Lockheed Martin was selected on 24 July 2002 to supply the combat data system and multifunction radar and South Korea is the fifth nation to operate the AEGIS system. The details of the Cheon Ryong cruise missile are speculative.
Modernisation: SM-2 missiles are to be replaced by SM-6 from 2012.
Structure: A development of the Arleigh Burke class, the South Korean variant also incorporates the AN/SPY-1D AEGIS system but the design has been enlarged to accommodate additional weapon systems. The ships have three magazines: the forward Mk 41 VLS launcher consists of 48 cells for SM-2 missiles which may also be launched from a 32-cell Mk 41 VLS aft. A separate, indigenous 48-cell VLS launcher aft contains 32 Cheon Ryong land attack cruise missiles and 16 ASROC anti-submarine missiles. There are hangar facilities for two helicopters.

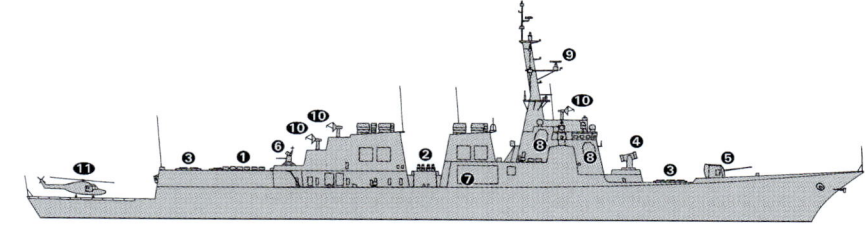

SEJONG DAEWANG (Scale 1 : 1,500), Ian Sturton / 1167965

SEJONG DAEWANG 10/2008, Michael Nitz / 1353198

SEJONG DAEWANG 10/2008, Michael Nitz / 1353199

476 Korea, South > Destroyers

6 KDX-2 CLASS (DDGHM)

Name	No	Builders	Laid down	Launched	Commissioned
CHUNGMUGONG YI SUN-SHIN	975	Daewoo, Okpo	2001	20 May 2002	2 Dec 2003
MOONMU DAEWANG	976	Hyundai, Ulsan	2002	11 Apr 2003	30 Sep 2004
DAEJOYOUNG	977	Daewoo, Okpo	2002	12 Nov 2003	30 Jun 2005
WANG GEON	978	Hyundai, Ulsan	2003	4 May 2005	2 Oct 2007
GANG GAM CHAN	979	Daewoo, Okpo	2004	16 Mar 2006	10 Nov 2006
CHOI YOUNG	981	Hyundai, Ulsan	2005	20 Oct 2006	5 Sep 2008

Displacement, tonnes: 4,572 standard; 5,588 full load
Dimensions, metres (feet): 154.4 × 16.9 × 4.3 (506.6 × 55.4 × 14.1)
Speed, knots: 29
Range, n miles: 4,000 at 18 kt
Complement: 200 (18 officers)

Machinery: CODOG; 2 GE LM 2500 gas turbines; 58,200 hp (43.42 MW) sustained; 2 MTU 20V 956 TB92 diesels; 8,000 hp(m) (5.88 MW); 2 shafts
Missiles: SSM: 8 Harpoon Block 1C (2 quad) ❶; active radar homing to 124 km (67 n miles) at 0.9 Mach; warhead 227 kg.
SAM: Mk 41 Mod 2 VLS ❷ 32 cells for Raytheon SM-2MR (Block IIIA); command/inertial guidance; semi-active radar homing to 167 km (90 n miles) at 2.5 Mach. 1 Raytheon RAM M 49 launcher RIM 116 ❸; 21 rounds per launcher; passive IR/anti-radiation homing to 9.6 km (5.2 n miles) at 2.5 Mach; warhead 9.1 kg.
A/S: ASROC VLS; inertial guidance 1.6–10 km (1–5.4 n miles) at 0.9 Mach; payload Mk 48.
Guns: 1 United Defense 5 in (127 mm)/62 Mk 45 Mod 4 ❹; 20 rds/min to 23 km (12.6 n miles); weight of shell 32 kg. 1 Signaal Goalkeeper 30 mm ❺; 7 barrels per mounting; 4,200 rds/min to 1.5 km.
Torpedoes: 6—324 mm Mk 32 (2 triple) tubes ❻; Alliant techsystems Mk 46 Mod 5; anti-submarine; active/passive homing to 11 km (5.9 n miles) at 40 kt; warhead 44 kg.
Physical countermeasures: Decoys: 4 chaff launchers.
Electronic countermeasures: ESM/ECM: SLQ-200(V)5.
Radars: Air search: Raytheon SPS-49(V)5 ❼; C/D-band.
Surface search: Signaal MW08 ❽; G-band.
Navigation: I-band ❾.
Fire control: 2 Signaal STIR 240 ❿; I/J/K-band.
Sonars: DSQS-23; hull-mounted; active search; medium frequency. Daewoo Telecom towed array; passive low frequency.
Combat data systems: BAeSema/Samsung KD COM-2; Link 11.
Weapon control systems: Marconi Mk 14 weapons direction system.

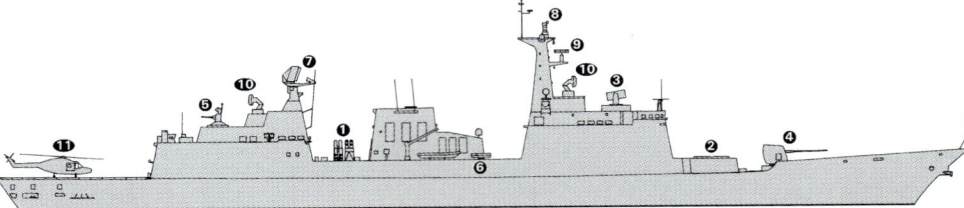
CHUNGMUGONG YI SUN-SHIN (Scale 1 : 1,200), Ian Sturton / 1153009

GANG GAM CHAN 12/2011, Chris Sattler / 1406980

Helicopters: 1 Westland Super Lynx Mk 99 ⓫.

Programmes: Approval for first three given in late 1996 but the final decision was not taken until 1998. Contract to design and build the first of class won by Daewoo in November 1999. The first of a second batch of three was launched at Hyundai in May 2005.
Operational: A successful test firing of a Cheon Ryong land-attack missile was conducted from *Wang Geon* on 19 February 2013.

DAEJOYOUNG 10/2013*, Marc Piché / 1530913

CHUNGMUGONG YI SUN-SHIN 9/2012, M Mazumdar / 1483519

Destroyers < **Korea, South** 477

3 KWANGGAETO DAEWANG (KDX-1) CLASS (DDGHM)

Name	No	Builders	Laid down	Launched	Commissioned
KWANGGAETO DAEWANG	971	Daewoo, Okpo	Jun 1995	28 Oct 1996	24 Jul 1998
EULJIMUNDOK	972	Daewoo, Okpo	Jan 1996	16 Oct 1997	20 Jun 1999
YANGMANCHUN	973	Daewoo, Okpo	Aug 1997	19 Oct 1998	29 Jun 2000

Displacement, tonnes: 3,917 full load
Dimensions, metres (feet): 135.4 × 14.2 × 4.2 (444.2 × 46.6 × 13.8)
Speed, knots: 30. **Range, n miles:** 4,000 at 18 kt
Complement: 170 (15 officers)

Machinery: CODOG; 2 GE LM 2500 gas turbines; 58,200 hp (43.42 MW) sustained; 2 MTU 20V 956 TB92 diesels; 8,000 hp(m) (5.88 MW); 2 shafts
Missiles: SSM: 8 McDonnell Douglas Harpoon Block 1C (2 quad) launchers ❶; active radar homing to 130 km (70 n miles) at 0.9 Mach; warhead 227 kg.
SAM: Raytheon Sea Sparrow; Mk 48 Mod 2 VLS launcher ❷ for 16 cells RIM-7P; semi-active radar homing to 16 km (8.5 n miles) at 2.5 Mach; warhead 38 kg.
Guns: 1 Otobreda 5 in (127 mm)/54 ❸; 45 rds/min to 23 km (12.4 n miles); weight of shell 32 kg. 2 Signaal 30 mm Goalkeeper ❹; 7 barrels per mounting; 4,200 rds/min combined to 2 km.
Torpedoes: 6—324 mm (2 triple) Mk 32 tubes ❺; Alliant Techsystems Mk 46 Mod 5; anti-submarine; active/passive homing to 11 km (5.9 n miles) at 40 kt; warhead 44 kg.
Physical countermeasures: Decoys: 4 CSEE Dagaie Mk 2 chaff launchers ❻. SLQ-25 Nixie towed torpedo decoy.
Electronic countermeasures: ESM/ECM: Argo AR 700/APECS II ❼; intercept and jammer.

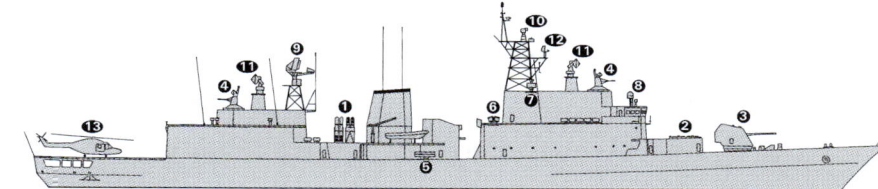

KWANGGAETO DAEWANG (Scale 1 : 1,200), Ian Sturton / 0572485

Radars: Air search: Raytheon SPS-49V5 ❾; C/D-band.
Surface search: Signaal MW08 ❿; G-band.
Fire control: 2 Signaal STIR 180 ⓫; I/J/K-band.
Navigation: Daewoo DTR 92 (SPS 55M) ⓬; I-band, IFF: UPX-27.
Sonars: Atlas Elektronik DSQS-21BZ; hull-mounted active search medium frequency. Daewoo Telecom towed array; passive low frequency.
Combat data systems: BAeSEMA/Samsung SSCS Mk 7; Litton NTDS (Link 11). SATCOM ❽.
Helicopters: 1 Westland Super Lynx ⓭.

Programmes: Project KDX-1. A much delayed programme. The first keel was to have been laid down at Daewoo in late 1992 for completion in 1996, but definition studies extended to late 1993, when contracts started to be signed for the weapon systems. First steel cut at Daewoo Okpo in April 1994.
Structure: Emphasis is on air defence but the design took so long to reach fulfilment that it was overtaken by the KDX-2. McTaggart Scott Trigon 5 helo handling system.
Operational: The Goalkeepers are also used against close-in surface threats using FAPDS (Frangible Armour Penetrating Discarding Sabot).

YANGMANCHUN 10/2008, Michael Nitz / 1353197

YANGMANCHUN 8/2010, Chris Sattler / 1406071

KWANGGAETO DAEWANG 10/2008, Michael Nitz / 1353196

© 2014 IHS IHS Jane's Fighting Ships 2014-2015

FRIGATES

9 ULSAN CLASS (FFG)

Name	No	Builders	Laid down	Launched	Commissioned
ULSAN	951	Hyundai, Ulsan	1979	8 Apr 1980	1 Jan 1981
SEOUL	952	Hyundai, Ulsan	1982	24 Apr 1984	30 Jun 1985
CHUNG NAM	953	Korean SEC, Pusan	1984	26 Oct 1984	1 Jun 1986
MASAN	955	Korea Tacoma	1983	26 Oct 1984	20 Jul 1985
KYONG BUK	956	Daewoo, Okpo	1984	15 Jan 1986	30 May 1986
CHON NAM	957	Hyundai, Ulsan	1986	19 Apr 1988	17 Jun 1989
CHE JU	958	Daewoo, Okpo	1986	3 May 1988	1 Jan 1990
PUSAN	959	Hyundai, Ulsan	1990	20 Feb 1992	1 Jan 1993
CHUNG JU	961	Daewoo, Okpo	1990	20 Mar 1992	1 Jun 1993

Displacement, tonnes: 1,520 standard; 2,215 (951–953, 955–956), 2,337 (957–959, 961) full load
Dimensions, metres (feet): 102 × 11.5 × 3.5 *(334.6 × 37.7 × 11.5)*
Speed, knots: 34; 18 diesel
Range, n miles: 4,000 at 15 kt
Complement: 150 (16 officers)
Machinery: CODOG; 2 GE LM 2500 gas turbines; 53,640 hp *(40 MW)* sustained; 2 MTU 16V 538 TB82 diesels; 5,940 hp(m) *(4.37 MW)* sustained; 2 shafts; cp props
Missiles: SSM: 8 McDonnell Douglas Harpoon (4 twin) launchers ❶; active radar homing to 130 km *(70 n miles)* at 0.9 Mach; warhead 227 kg.
Guns: 2—3 in *(76 mm)*/62 Oto Melara compact ❷; 85 rds/min to 16 km *(8.6 n miles)* anti-surface; 12 km *(6.5 n miles)* anti-aircraft; weight of shell 6 kg. 8 Emerlec 30 mm (4 twin) (FF 951–955) ❸; 6 Breda 40 mm/70 (3 twin) (FF 956–961) ❹.
Torpedoes: 6—324 mm Mk 32 (2 triple) tubes ❺. Honeywell Mk 46 Mod 1; anti-submarine; active/passive homing to 11 km *(5.9 n miles)* at 40 kt; warhead 44 kg.
Depth charges: 12.
Physical countermeasures: Decoys: 4 Loral Hycor SRBOC 6-barrelled Mk 36 launchers ❻; range 4 km *(2.2 n miles)*. SLQ-25 Nixie; towed torpedo decoy.
Electronic countermeasures: ESM: ULQ-11K; intercept.
Radars: Air/surface search: Signaal DA05 ❾; E/F-band.
Surface search: Signaal ZW06 (FF 951–956) ❿; Marconi S 1810 (FF 957–961) ⓫; I-band.
Fire control: Signaal WM28 (FF 951–956) ⓬; Marconi ST 1802 (FF 957–961) ⓭; I/J-band.
Navigation: Raytheon SPS-10C (FF 957–961) ⓮; I-band.
Tacan: SRN 15.
Sonars: Raytheon DE 1167; hull-mounted; active search and attack; medium frequency.
Combat data systems: Samsung/Ferranti WSA 423 action data automation (FF 957–961). Litton systems retrofitted to others. Link 11 in three of the class. WSC-3 SATCOM (F 957).

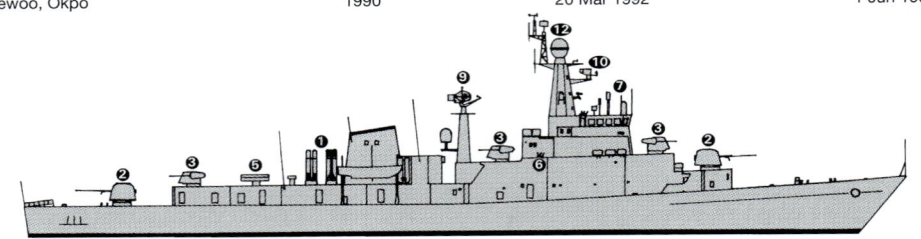

ULSAN *(Scale 1 : 900), Ian Sturton / 0506154*

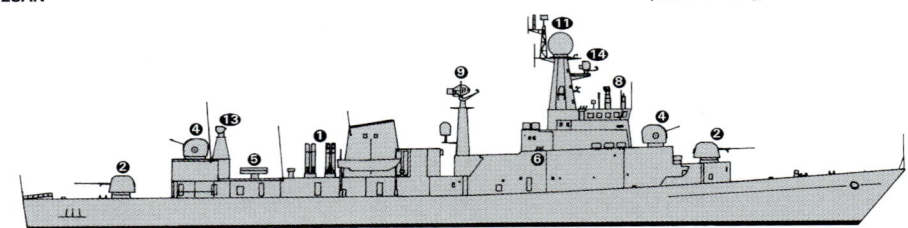

CHE JU *(Scale 1 : 900), Ian Sturton / 0506155*

Weapon control systems: 1 Signaal Liod optronic director (FF 951–956) ❼; 1 Radamec System 2400 optronic director (FF 957–961) ❽.

Modernisation: New sonars fitted. WSC-3 SATCOM fitted in *Chon Nam*.
Structure: Steel hull with aluminium alloy superstructure. There are three versions. The first five ships are the same but *Kyong Buk* has the four Emerson Electric twin 30 mm guns replaced by three Breda twin 40 mm, and the last four of the class have a built-up gun platform aft and a different combination of surface search, target indication and navigation radars. Weapon systems integration caused earlier concern and a Ferranti combat data system has been installed in the last five; Litton Systems Link 11 fitted in three of the class.
Operational: *Che Ju* and *Chung Nam* conducted the first ever deployment of South Korean warships to Europe during a four month tour from September 1991 to January 1992. Trainees were embarked. Three of the class have a shore datalink and act as local area commanders to control attack craft carrying out coastal protection patrols.

CHE JU *10/2002, Guy Toremans / 0528915*

ULSAN *8/2000, van Ginderen Collection / 0104996*

Frigates < **Korea, South** 479

CHON NAM   *10/2008, Michael Nitz* / 1353200

CHUNG NAM   *10/2008, Michael Nitz* / 1353201

CHUNG JU   *2/2001, Ships of the World* / 0130106

1 + 5 (9) INCHEON CLASS (FFGHM)

Name	No	Builders	Laid down	Launched	Commissioned
INCHEON	811	Hyundai, Ulsan	20 Jan 2009	29 Apr 2011	17 Jan 2013
GYEONGGI	812	Hyundai, Ulsan	2011	18 Jul 2013	2014
JEONBUK	813	Hyundai, Ulsan	2012	13 Nov 2013	2015
—	815	STX Shipbuilding and Marine	2012	2014	2015
—	816	STX Shipbuilding and Marine	2012	2014	2015
—	817	STX Shipbuilding and Marine	2013	2015	2016

Displacement, tonnes: 2,337 standard; 3,251 full load
Dimensions, metres (feet): 114 × 14 × ?
 (374 × 45.9 × ?)
Speed, knots: 30
Range, n miles: 4,500 at 13 kt
Complement: 140

Machinery: CODOG: 2 General Electric LM 2500 gas turbines (Batch 1); 58,200 hp *(43.42 MW)*; 1 Rolls Royce MT-30 gas turbine (Batch 2); 48,275 hp *(36 MW)*; 2 MTU 12V 1163 TB83 diesels; 8,840 hp *(6.5 MW)*; 2 shafts
Missiles: SLCM: 4 Cheon Ryong land-attack missiles ❶; inertial/GPS guidance to 500 km *(270 n miles)* at 0.7 Mach; warhead 500 kg.
 SSM: 4 Hae Sung SSM-700K ❷; active homing to 124 km *(67 n miles)* at 0.9 Mach; warhead 227 kg.
 SAM: 1 Raytheon Mk 49 RAM RIM-116 ❸; 21 rounds; passive IR/anti-radiation homing to 9.6 km *(5.2 n miles)* at 2 Mach, warhead 9.1 kg.
Guns: 1 — 127 mm/L 62 Mk 45 Mod 4 ❹. 1 Raytheon 20 mm/76 6-barrelled Mk 15 Block 1B Vulcan Phalanx; 4,500 rds/min combined to 1.5 km ❺.
Torpedoes: 6 — 324 mm (2 triple) tubes.
Physical countermeasures: Decoys: 4 KDAGAIE Mk 2.
Electronic countermeasures: LIG SLQ-200(V)K Sonata EW Suite.
Radars: Air/surface search: Thales SMART-S ❻; 3D; E/F-band.
 Surface search: To be announced.
 Fire control: Saab Ceros ❼; J/K-band.
 Navigation: To be announced.
Sonars: Hull-mounted and towed-array.
Combat data systems: To be announced.
Electro-optic systems: To be announced ❽.

Helicopters: 1 AgustaWestland AW159 Wildcat.

Programmes: Hyundai Heavy Industries awarded the contract in early 2009 for the construction of the lead ship of a new FFX class of frigates. Hyundai was selected to build the second and third ships in September 2010; the first three ships comprise Batch 1. The next six ships comprise Batch 2. STX Marine contracted for the fourth and fifth hulls on 30 November 2011 and for the sixth on 27 September 2012. A further nine ships are projected to enter service by 2020 to replace the Po Hang and Dong Hae classes.

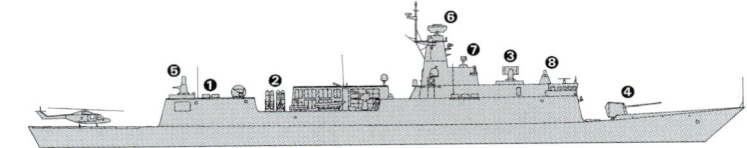

INCHEON *(Scale 1 : 1,200), Ian Sturton / 1528816*

INCHEON
10/2013*
Ships of the World
1529167

INCHEON *10/2013*, Ships of the World / 1529168*

CORVETTES

21 PO HANG CLASS (FS/FSG)

Name	No	Builders	Commissioned
KYONG JU	758	Hyundai, Ulsan	Nov 1986
MOK PO	759	Daewoo, Okpo	Aug 1986
KIM CHON	761	Korea SEC, Pusan	May 1985
CHUNG JU	762	Korea Tacoma	May 1985
JIN JU	763	Hyundai, Ulsan	Jun 1988
YO SU	765	Daewoo, Okpo	Nov 1988
JIN HAE	766	Korea SEC, Pusan	Feb 1989
SUN CHON	767	Korea Tacoma	Jun 1989
YEE REE	768	Hyundai, Ulsan	Jun 1989
WON JU	769	Daewoo, Okpo	Aug 1989
AN DONG	771	Korea SEC, Pusan	Nov 1989
SONG NAM	773	Daewoo, Okpo	May 1989
BU CHON	775	Hyundai, Ulsan	Apr 1989
JAE CHON	776	Korea SEC, Pusan	May 1989
DAE CHON	777	Korea Tacoma	Apr 1989
SOK CHO	778	Korea SEC, Pusan	Feb 1990
YONG JU	779	Hyundai, Ulsan	Mar 1990
NAM WON	781	Daewoo, Okpo	Apr 1990
KWAN MYONG	782	Korea Tacoma	Jul 1990
SIN HUNG	783	Korea SEC, Pusan	Mar 1993
KONG JU	785	Korea Tacoma	Jul 1993

Displacement, tonnes: 1,240 full load
Dimensions, metres (feet): 88.3 × 10 × 2.9 (289.7 × 32.8 × 9.5)
Speed, knots: 32
Range, n miles: 4,000 at 15 kt
Complement: 95 (10 officers)

Machinery: CODOG; 1 GE LM 2500 gas turbine; 26,820 hp (20 MW) sustained; 2 MTU 12V 956 TB82 diesels; 6,260 hp(m) (4.6 MW) sustained; 2 shafts; Kamewa cp props
Missiles: SSM: 2 Aerospatiale MM 38 Exocet (758–759) ❶; inertial cruise; active radar homing to 42 km (23 n miles) at 0.9 Mach; warhead 165 kg; sea-skimmer. 4 McDonnell Douglas Harpoon (762, 769, 777, 779) (2 twin) launchers ❷; active radar homing to 130 km (70 n miles) at 0.9 Mach; warhead 227 kg.
Guns: 1 or 2 Oto Melara 3 in (76 mm)/62 compact ❸; 85 rds/min to 16 km (8.6 n miles) anti-surface; 12 km (6.5 n miles) anti-aircraft; weight of shell 6 kg. 4 Emerlec 30 mm (2 twin) (758–759) ❹; 4 Breda 40 mm/70 (2 twin) (761 onwards) ❺.
Torpedoes: 6—324 mm Mk 32 (2 triple) tubes ❻. Honeywell Mk 46; anti-submarine; active/passive homing to 11 km (5.9 n miles) at 40 kt; warhead 44 kg.
Depth charges: 12 (761 onwards).
Physical countermeasures: Decoys: 4 MEL Protean fixed launchers; 36 grenades. 2 Loral Hycor SRBOC 6-barrelled Mk 36 launchers (in some); range 4 km (2.2 n miles).
Electronic countermeasures: ESM/ECM: THORN EMI or NobelTech; intercept/jammer.
Radars: Surface search: Marconi 1810 ❽ and/or Raytheon SPS-64 ❾; I-band.
Fire control: Signaal WM28 ❿; I/J-band; or Marconi 1802 ⓫; I/J-band.
Sonars: Signaal PHS-32; hull-mounted; active search and attack; medium frequency.
Combat data systems: Signaal Sewaco ZK (758–759); Ferranti WSA 423 (761 onwards).
Weapon control systems: Signaal Liod or Radamec 2400 (766 onwards) optronic director ❼.

Programmes: First laid down early 1983. The programme terminated in 1993.
Structure: The first three are Exocet fitted and have a different weapon systems arrangement. The remainder have an improved combat data system with Ferranti/Radamec/Marconi fire-control systems and radars as in the later versions of the Ulsan class.
Operational: *Chon An* 772 sank after an explosion on 26 March 2010 off Baengnyeong Island. The cause of the explosion is likely to have been a torpedo fired from a North Korean midget submarine. *Po Hang* was decommissioned in 2009. *Kun San* 757 was decommissioned on 29 September 2011.

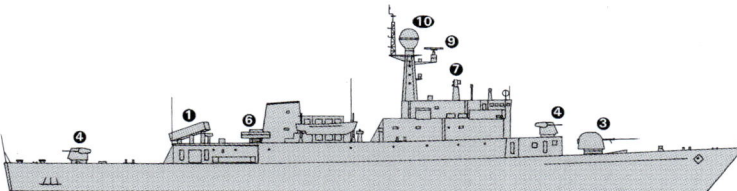

PO HANG *(Scale 1 : 900), Ian Sturton / 0572484*

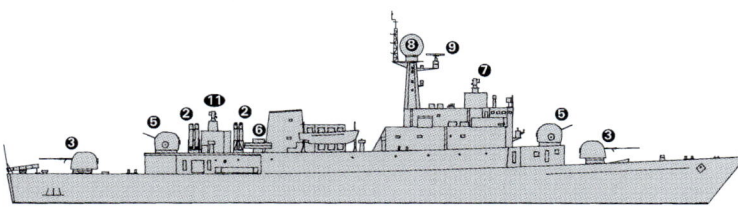

WON JU *(Scale 1 : 900), Ian Sturton / 0569920*

KUN SAN *10/2008, Guy Toremans / 1353202*

YO SU *4/2012, John Mortimer / 1483520*

WON JU *7/2002, Chris Sattler / 0528917*

SHIPBORNE AIRCRAFT

Numbers/Type: 10/13 Westland Lynx Mk 99/Mk 99A.
Operational speed: 125 kt (231 km/h).
Service ceiling: 12,000 ft (3,660 m).
Range: 320 n miles (593 km).
Role/Weapon systems: 12 Mk 99 helicopters delivered by 1991; 13 Mk 99A ordered in June 1997 and delivered in 1999/2000. One Mk 99 lost in May 2010. Sensors: Ferranti Sea Spray Mk 3 radar and Racal ESM. Bendix AQS 18(V) dipping sonar and ASQ 504(V) MAD in ASW versions. Weapons: 4 BAe Sea Skua missiles. Mk 46 (Mod 5) torpedo (in ASW version). Sea Skua may be replaced in due course.

LYNX MK 99A *10/2008, Michael Nitz / 1353206*

Numbers/Type: 5 Aerospatiale SA 316B/SA 319B Alouette III.
Operational speed: 113 kt (210 km/h).
Service ceiling: 10,500 ft (3,200 m).
Range: 290 n miles (540 km).
Role/Weapon systems: Marine support helicopter; operated by RoK Marine Corps. Sensors: None. Weapons: Unarmed.

ALOUETTE III *6/2008, Annati Collection / 1353207*

Numbers/Type: 19 Sikorsky UH-60P Blackhawk.
Operational speed: 145 kt (268 km/h).
Service ceiling: 18,700 ft (5,070 m).
Range: 315 n miles (583 km).
Role/Weapon systems: Korean built variant of Sikorsky UH-60L. Naval version used for SAR and operations from *Dokdo*.

UH-60P *10/2008, Michael Nitz / 1353208*

Numbers/Type: 8 AgustaWestland AW 159 Lynx Wildcat.
Operational speed: 165 kt (306 km/h).
Service ceiling: 20,000 ft (6,096 m).
Range: 675 n miles (1,250 km).
Role/Weapon systems: Eight aircraft are to be procured to augment 23 Lynx helicopters in service. Entry into service is to being in 2015. Primary role anti-surface warfare. Sensors: Selex Galileo Seaspray 7000E radar; ESM, ECM. Weapons: Future Air-to-Surface missiles; 7.62 or 12.7 mm machine guns.

482 Korea, South > Land-based maritime aircraft — Amphibious forces

LAND-BASED MARITIME AIRCRAFT (FRONT LINE)

Notes: (1) F-16 fighters are capable of firing Harpoon ASV missiles.
(2) There are also 13 UH-1 utility helicopters.
(3) Plans to acquire up to 18 S-3 Vikings, retired from US Navy servcie, were announced in 2013. The aircraft would be used in the reconnaissance and surveillance role.

Numbers/Type: 16 Lockheed P-3C Orion Update III.
Operational speed: 411 kt (761 km/h).
Service ceiling: 28,300 ft (8,625 m).
Range: 4,000 n miles (7,410 km).
Role/Weapon systems: Eight maritime patrol aircraft ordered in December 1990. First pair delivered April 1995, remainder April 1996. A further eight upgraded P-3CK were received in 2010. The Update III version is fitted with ASQ-212 tactical computer. Sensors: APS-134 or 137(V)6 search radar; AAS-36 IR. Weapons: four Harpoon ASM.

P-3C   8/2008, Michael Nitz / 1353210

Numbers/Type: 5 Rheims-Cessna F 406 Caravan II.
Operational speed: 229 kt (424 km/h).
Service ceiling: 30,000 ft (9,145 m).
Range: 1,153 m (2,135 km).
Role/Weapon systems: Maritime surveillance version ordered in 1997 with first one delivered in mid-1999. Sensors: APS 134 radar; Litton FLIR. Weapons: none.

F 406   6/2008, Annati Collection / 1353209

PATROL FORCES

72 CHAMSURI CLASS
(FAST ATTACK CRAFT – PATROL) (PBF/PTF)

281   296   362   +69

Displacement, tonnes: 150 full load
Dimensions, metres (feet): 37 × 6.9 × 1.7 (121.4 × 22.6 × 5.6)
Speed, knots: 37. **Range, n miles:** 600 at 20 kt
Complement: 31 (5 officers)

Machinery: 2 MTU MD 16V 538 TB90 diesels; 6,000 hp(m) (4.41 MW) sustained; 2 shafts
Guns: 2 Emerson Electric 30 mm (twin) or USN 3 in (76 mm)/50 or Bofors 40 mm/60. 2 GE/GD 20 mm Sea Vulcan Gatlings (in most). 2 – 12.7 mm MGs. Rocket launchers in lieu of after Gatling in some.
Radars: Surface search: Raytheon 1645; I-band.
Weapon control systems: Optical director.

Comment: Fifty-four Sea Dolphins built by Korea SEC, and 47 Wildcats by Korea Tacoma. Both are known collectively as the Chamsuri class. First laid down 1978. The class has some gun armament variations and some minor superstructure changes in later ships. These craft form the basis of the coastal patrol effort against incursions by North Korean amphibious units. Six sold to the Philippines in 1995 and two further in 2006. Four transferred to Bangladesh in 2000–04. One transferred to Ghana in 2011, three to East Timor in 2011 and three to Kazakhstan in 2006. Some deleted so far, others are in reserve. One decommissioned after collision on 10 November 2010. PKM 212–375 series.

CHAMSURI 281   10/2008, Guy Toremans / 1353212

CHAMSURI 296   10/2008, Guy Toremans / 1353211

CHAMSURI 362   10/2008, Michael Nitz / 1353213

15 + 3 (2) GUMDOKSURI CLASS
(FAST ATTACK CRAFT – MISSILE) (PGGF)

Name	No	Builders	Laid down	Launched	Commissioned
YOON YOUNG-HA	711	Hanjin Heavy Industries, Pusan	2005	28 Jun 2007	17 Dec 2008
HAN SANG GUK	712	STX Jinhae	2007	23 Sep 2009	14 Sep 2011
JO CHEON HYEONG	713	STX Jinhae	2007	23 Sep 2009	14 Sep 2011
HWANG DOHYUN	715	STX Jinhae	2008	11 Dec 2009	13 Jan 2012
SUH HOOWON	716	STX Jinhae	2008	11 Dec 2009	28 Nov 2011
PARK DONGHYUK	717	Hanjin Heavy Industries, Pusan	2009	28 Jul 2010	28 Nov 2011
HYUN SIHAK	718	Hanjin Heavy Industries, Pusan	2009	28 Jul 2010	2011
JUNG GEUNGMO	719	Hanjin Heavy Industries, Pusan	2009	2 Nov 2010	19 Dec 2011
JI DEOKCHIL	721	Hanjin Heavy Industries, Pusan	2010	2 Nov 2010	19 Dec 2011
LIM BYEONGRAE	722	STX Jinhae	2011	20 Nov 2012	3 Jul 2013
HONG SIUK	723	STX Jinhae	2011	20 Nov 2012	10 Oct 2013
HONG DAESON	725	STX Jinhae	2011	20 Nov 2012	Jul 2013
HAN MUNSIK	726	Hanjin Heavy Industries, Pusan	2012	24 Apr 2013	28 Jan 2014
KIM CHANGHAK	727	Hanjin Heavy Industries, Pusan	2012	24 Apr 2013	2014
PARK DONGJIN	728	Hanjin Heavy Industries, Pusan	2012	24 Apr 2013	2014
–	–	STX Jinhae	–	2014	2015
–	–	STX Jinhae	–	2014	2015
–	–	STX Jinhae	–	2014	2015

Displacement, tonnes: 447 standard; 579 full load
Dimensions, metres (feet): 63 × 9.0 × 5.0 (206.7 × 29.5 × 16.4)
Speed, knots: 41. **Range, n miles:** 2,000 at 15 kt
Complement: 40

Machinery: CODAG: 2 GE LM 500 gas turbines; 10,900 hp (8.1 MW); 2 MTU 16V 1163 diesels; 15,880 hp (11.8 MW)
Missiles: SSM: 4 Hae Sung SSM-700K; active radar homing to 124 km (67 n miles) at 0.9 Mach; warhead 227 kg.
Guns: 1 – 3 in (76 mm). 1 – 40 mm.
Electronic countermeasures: LIG SLQ-200(V)K Sonata EW suite.
Radars: Air/surface search: Thales MW 08; G-band.
Fire control: Saab Ceros 200; J-band.
Navigation: I-band.

Comment: A new class of patrol craft to replace Sea Dolphin class. Following construction of the first of class by Hanjin, hulls 2–5 were built by STX Shipbuilding, Jinhae, hulls 6–9 by Hanjin, hulls 10–12 by STX, 13–15 by Hanjin, and 16–18 by STX. A class of 20 is projected.

YOON YOUNG-HA   10/2008, Michael Nitz / 1353214

AMPHIBIOUS FORCES

Notes: Future plans for amphibious forces were outlined in 2013. Following the second Dokdo-class LPH, which is to be equipped with a ski-jump for the launch of VSTOL aircraft, construction of a larger amphibious assault ship is under consideration. This might be similar in size to the Spanish Navy's *Juan Carlos*.

MISCELLANEOUS LANDING CRAFT

Comment: A considerable number of US LCVP type built of GRP in South Korea. In addition there were plans to build up to 20 small hovercraft for special forces; first two reported building in 1994, and one seen on sea trials in May 1995. Also 56 combat support boats of 8 m were ordered from FBM Marine for assembly by Hanjin Heavy Industries.

HOVERCRAFT   5/1995, David Jordan / 0081167

IHS Jane's Fighting Ships 2014-2015   © 2014 IHS

1 + 1 AMPHIBIOUS TRANSPORT DOCK (LPD)

Name	No	Builders	Laid down	Launched	Commissioned
DOKDO	6111	Hanjin Heavy Industries, Pusan	2003	12 Jul 2005	3 Jul 2007

Displacement, tonnes: 13,209 standard; 19,305 full load
Dimensions, metres (feet): 200 × 32.0 × 6.5 (656.2 × 105.0 × 21.3)
Speed, knots: 22
Complement: 1,100

Military lift: 700 troops, 10 tanks and two air-cushion landing craft
Machinery: CODAD: 4 SEMT Pielstick 16PC 2.5 STC diesels; 41,615 hp(m) (30.6 MW) sustained; 2 shafts

Missiles: 1 Raytheon Mk 49 launcher RAM 116 ❶; 21 rds; passive IR/anti-radiation homing to 9.6 km (5.2 n miles) at 2.5 Mach; warhead 9.1 kg.
Guns: 2 TNNL Goalkeeper ❷ 30 mm; 4,200 rds/min to 1.5 km.
Radars: Air search: Thales SMART L ❸; 3D; D-band.
Surface search: Signaal MW 08 ❹; G-band.
Navigation: To be announced.
CCA: Galileo Avionica SPN-720; I-band.
Combat data systems: Based on Tacticos.

Helicopters: 10 UH-60.

Programmes: The contract for an amphibious assault ship was placed with Hanjin Heavy Industries on 28 October 2002. An order for a second ship, possibly to be called *Marado*, is expected in 2014 to meet a delivery date of 2018. This ship may be equipped with a ski-jump.
Structure: The design includes a well dock.

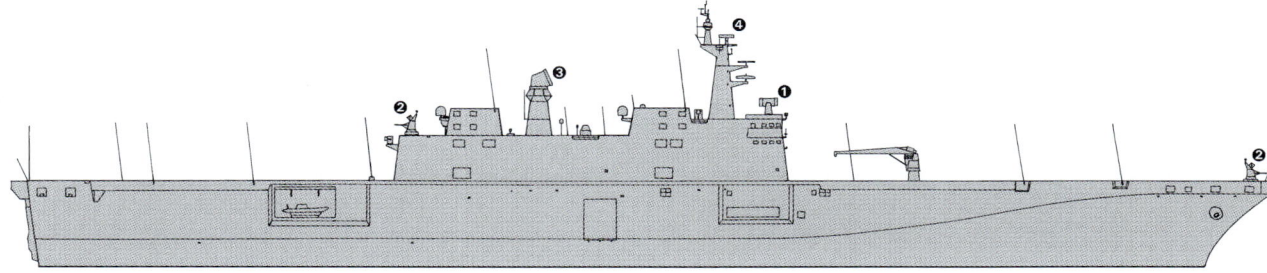

DOKDO   *(Scale 1 : 1,200), Ian Sturton* / 1366154

DOKDO   *10/2008, Michael Nitz* / 1353216

DOKDO   *12/2007, Michael Nitz* / 1170070

DOKDO   *12/2009, Michael Nitz* / 1406072

484 Korea, South > Amphibious forces

10 LCM 8 CLASS (LCM)

Displacement, tonnes: 117 full load
Dimensions, metres (feet): 22.7 × 6.4 × 1.4 *(74.5 × 21.0 × 4.6)*
Speed, knots: 11
Complement: 11
Military lift: 55 tons
Machinery: 4 GM 6-71 diesels; 696 hp *(519 kW)* sustained; 2 shafts

Comment: Previously US Army craft. Transferred in September 1978.

LCM 87   *10/2008, Michael Nitz* / 1353219

3 TSAPLYA (MURENA E) (PROJECT 12061) CLASS (ACV)

621–623

Displacement, tonnes: 110 standard; 152 full load
Dimensions, metres (feet): 31.3 on cushion × ? *(102.7 × ?)*
Speed, knots: 50
Range, n miles: 500 at 50 kt
Complement: 111 (3 officers)
Machinery: 2 PR-77 gas turbines for lift and propulsion; 8,000 hp *(5.88 MW)*
Guns: 2—30 mm AK 306M. 2—30 mm grenade launchers. 2—12.7 mm MGs.

Comment: Ordered on 5 August 2002. Designed by Almaz, all built at Khabarovsk. First laid down on 26 April 2004 and delivered to Inchon on 11 November 2005. The second and third delivered in November and December 2006 respectively. Capable of carrying one medium tank or 130 troops.

TSAPLYA 621   *6/2006* / 1164765

4 ALLIGATOR CLASS (LSTH)

Name	No	Builders	Launched	Commissioned
KOJOON BONG	681	Korea Tacoma, Masan	Sep 1992	Jun 1993
BIRO BONG	682	Korea Tacoma, Masan	Dec 1996	Nov 1997
HYANGRO BONG	683	Korea Tacoma, Masan	Oct 1998	Aug 1999
SEONGIN BONG	685	Korea Tacoma, Masan	Feb 1999	Nov 1999

Displacement, tonnes: 1,930 standard; 4,347 full load
Dimensions, metres (feet): 112.5 × 15.3 × 3 *(369.1 × 50.2 × 9.8)*
Speed, knots: 16
Range, n miles: 4,500 at 12 kt
Complement: 169

Military lift: 200 troops; 15 MBT; 6—3 ton vehicles; 4 LCVPs.
Machinery: 2 SEMT-Pielstick 16 PA6 V 280; 12,800 hp(m) *(9.41 MW)* sustained; 2 shafts; cp props
Guns: 2 Breda 40 mm/70 (LST 683, 685). 2—30 mm (1 twin) (LST 681). 2 Vulcan 20 mm Gatlings.
Physical countermeasures: Decoys: 1 RBOC chaff launcher.
Electronic countermeasures: ESM: radar intercept.
Radars: Surface search: Raytheon SPS 64; E/F-band.
Navigation: Raytheon SPS 64; I-band.
Tacan: SRN 15.
Weapon control systems: Selenia NA 18. Optronic director. Daeyoung WCS-86.
Helicopters: Platform for 1 UH-60A.

Comment: First one ordered in June 1990 from Korea Tacoma, Masan but delayed by financial problems. Korea Tacoma became Hanjin Heavy Industries. Design improvements include stern ramp for underway launching of LVTs, helicopter deck, and a lengthened bow ramp. There are unlikely to be further orders. *Hyangro Bong* was not loaned to Malaysia in 2010 as previously reported.

IHS Jane's Fighting Ships 2014-2015

KOJOON BONG   *10/2008, Michael Nitz* / 1353217

2 LSF-II LANDING CRAFT AIR CUSHION (LCAC)

631   632

Displacement, tonnes: 102 standard; 157 full load
Dimensions, metres (feet): 26.8 on cushion × ? *(87.9 × ?)*
Speed, knots: 40
Range, n miles: 300 at 35 kt
Complement: 5

Military lift: 23 troops; 1 main battle tank or 55 tons
Machinery: 4 Vericor Power Systems ETF40B gas turbines for propulsion and lift; 15,800 hp *(11.8 MW)*; 2 shrouded reversible-pitch airscrews (propulsion); 4 double-entry centrifugal fans (lift)
Guns: 1—20 mm.
Radars: To be announced.

Comment: LSF II is a high-speed air-cushion craft of aluminium construction developed and manufactured by Hanjin Heavy Industries for operations in conjunction with the LPD *Dokdo*. The design appears to be based on the US Navy's LCAC design. Both delivered in mid-2007.

LCAC 632   *10/2008, Michael Nitz* / 1353220

0 + 4 LANDING SHIPS (LST)

Name	No	Builders	Laid down	Launched	Commissioned
CHEONWANGBONG	688	Hanjin Heavy Industries, Pusan	2011	11 Sep 2013	2014
–	–	Hyundai, Ulsan	2013	2015	2016
–	–	–	2014	2015	2016
–	–	–	2015	2016	2017

Displacement, tonnes: 4,950 standard; 7,140 full load
Dimensions, metres (feet): 126.9 × 19.4 × 5.4 *(416.3 × 63.6 × 17.7)*
Speed, knots: 18. **Range, n miles:** 8,000 at 12 kt

Military lift: To be announced.
Machinery: CODAD: 4 MAN 12V28/33D diesels; 16,286 hp(m) *(21.8 MW)*; 2 shafts
Guns: 1—40 mm. 2—20 mm.
Physical countermeasures: Decoys: Rheinmetall MASS.
Radars: Surface search: To be announced.
Navigation: To be announced.
Helicopters: Platform for 1 UH-60A.

Comment: The LST-2 programme is for four new LSTs. Construction of the first was started at Hanjin Heavy Industries in 2011 with delivery planned for 2014. Three further ships are to follow at annual intervals. The ships are likely to replace the LST 512-1152 class.

CHEONWANGBONG   *9/2013*, Hanjin* / 1530914

MINE WARFARE FORCES

6 SWALLOW CLASS (MINEHUNTERS) (MHSC)

Name	No	Builders	Commissioned
KANG KYEONG	561	Kangnam Corporation	Dec 1986
KANG JIN	562	Kangnam Corporation	May 1991
KO RYEONG	563	Kangnam Corporation	Nov 1991
KIM PO	565	Kangnam Corporation	Apr 1993
KO CHANG	566	Kangnam Corporation	Oct 1993
KUM WHA	567	Kangnam Corporation	Apr 1994

Displacement, tonnes: 478 standard; 528 full load
Dimensions, metres (feet): 50 × 8.3 × 2.6 *(164 × 27.2 × 8.5)*
Speed, knots: 15. **Range, n miles:** 2,000 at 10 kt
Complement: 48 (5 officers; 4 divers)

Machinery: 2 MTU diesels; 2,040 hp(m) *(1.5 MW)* sustained; 2 Voith-Schneider props; bow thruster; 102 hp(m) *(75 kW)*
Guns: 1 Oerlikon 20 mm. 2—7.62 mm MGs.
Physical countermeasures: MCM: 2 Gaymarine Pluto remote-control submersibles (possibly to be replaced by Double Eagle).
Radars: Navigation: Raytheon SPS 64; I-band.
Sonars: GEC-Marconi 193M Mod 1 or Mod 3; minehunting; high frequency.
Combat data systems: Racal MAINS 500.

Comment: Built to a design developed independently by Kangnam Corporation but similar to the Italian Lerici class. GRP hull. Single sweep gear deployed at 8 kt. Decca/Racal plotting system. First delivered at the end of 1986 for trials. Two more with some modifications ordered in 1988, three more in 1990.

KANG KYEONG   10/2008, Michael Nitz / 1353221

3 YANG YANG CLASS (MSC/MHC)

Name	No	Builders	Commissioned
YANG YANG	571	Kangnam Corporation	Dec 1999
ONGJIN	572	Kangnam Corporation	Feb 2004
HAE NAM	573	Kangnam Corporation	Mar 2005

Displacement, tonnes: 894 full load
Dimensions, metres (feet): 59.4 × 10.5 × 3.0 *(194.9 × 34.4 × 9.8)*
Speed, knots: 15. **Range, n miles:** 3,000 at 12 kt
Complement: 61 (7 officers; 5 divers)

Machinery: 2 MTU diesels; 4,000 hp(m) *(2.98 MW)* sustained; 2 Voith-Schneider props; bow thruster; 134 hp(m) *(100 kW)*
Guns: 1—20 mm Sea Vulcan Gatling. 2—7.62 mm MGs.
Physical countermeasures: MCM: BAE Systems deep mechanical and combined influence sweep system. 2 Gayrobot Pluto GIGAS ROVs.
Radars: Navigation: Raytheon; I-band.
Sonars: Thomson Marconi Type 2093 VDS; minehunting; active multifrequency.
Combat data systems: Thomson Marconi TSM 2061 Mk 3.

Comment: The first one ordered in late 1995 and a second pair delivered by 2005. A large version of the Swallow class built to a design developed by Kangnam Corporation. GRP hull. The integrated navigation and dynamic positioning system developed by Kongsberg Simrad. The design is likely to be the basis for eight MCMVs to be built for the Indian Navy.

ONGJIN   10/2008, Michael Nitz / 1353222

1 WON SAN CLASS (MINELAYER) (MLH)

Name	No	Builders	Launched	Commissioned
WON SAN	560	Hyundai, Ulsan	Sep 1996	Sep 1997

Displacement, tonnes: 3,353 full load
Dimensions, metres (feet): 103.8 × 15 × 3.4 *(340.6 × 49.2 × 11.2)*
Speed, knots: 22. **Range, n miles:** 4,500 at 15 kt
Complement: 160

Machinery: CODAD; 4 SEMT-Pielstick 12 PA6 diesels; 17,200 hp(m) *(12.64 MW)*; 2 shafts
Guns: 1 Oto Melara 3 in *(76 mm)*/62 Compact; 85 rds/min to 16 km *(8.6 n miles)*; weight of shell 6 kg. 2 Breda 40 mm/70.
Torpedoes: 6—324 mm Mk 32 (2 triple) launchers.
Mines: 2 stern launchers. Up to 300.
Physical countermeasures: Decoys: 2 chaff launchers.
Electronic countermeasures: ESM/ECM.
Radars: Air/surface search: E/F-band.
Fire control: Marconi 1802; I/J-band.
Navigation: I-band.
Sonars: Bow-mounted; active search and attack; medium frequency.
Weapon control systems: Radamec optronic director.
Helicopters: Platform only.

Comment: Project design contract ordered October 1991 and completed July 1993 by Hyundai. Order to build given in October 1994.

WON SAN   11/2011, Hachiro Nakai / 1168716

SURVEY SHIPS

18 SURVEY SHIPS (AGOR)

| PUSAN 801 | – 805–806 | CH'UNGNAM 821 | – 201–204 | – 215–217 |
| – 802–803 | – 810 | KANGWON 831 | – 208–209 | – 220 |

Comment: All ships are painted white with a distinctive yellow coloured crest on the funnel. Most were commissioned in the 1980s. The Hydrographic Service is responsible to the Ministry of Transport.

201   10/2008, Guy Toremans / 1353224

202   4/2000, M Declerck / 0105000

217   10/2008, Guy Toremans / 1353223

RESCUE VEHICLES

1 RESCUE SUBMERSIBLE

DSRV II

Dimensions, metres (feet): 9.6 × 2.7 × 3.8 *(31.5 × 8.9 × 12.5)*
Speed, knots: 3
Complement: 3 (2 pilots)
Machinery: 2 electric motors; 26.8 hp *(20 kW)*; 4 tiltable side thrusters; 16 hp *(12 kW)*

Comment: James Fisher Defence Limited announced on 13 December 2006 that it had been awarded a contract to deliver a new Submarine Rescue Vehicle (SRV) to the Republic of Korea Navy. It was delivered in November 2008. The new submarine rescue vehicle, DSAR-5, is based on the LR5/DSAR-1, in-service with the Royal Navy until replaced in 2008. Lithium based battery technology enables the underwater endurance to be increased significantly over conventional lead-acid based systems. DSAR-5 has two compartments. The forward compartment houses the pilot and co-pilot while the aft compartment carries the RCO and up to 16 rescuees. Capable of operating at a depth of 500 m in currents of up to 3 kt, rescuees may be transferred under pressure to the medical and decompression facilities onboard the submarine rescue mothership *Cheong Hae Jin*.

AUXILIARIES

Notes: There are approximately 35 small service craft in addition to the YO-type tankers listed and the harbour tugs. These craft include open lighters, floating cranes, diving tenders, dredgers, ferries, non self-propelled fuel barges, pontoon barges, and sludge removal barges; most are former US Navy craft.

1 CHEONG HAE JIN CLASS (ARS)

Name	No	Builders	Launched	Commissioned
CHEONG HAE JIN	21	Daewoo, Okpo	Oct 1995	30 Nov 1996

Displacement, tonnes: 4,369 full load
Dimensions, metres (feet): 102.8 × 16.4 × 4.6 *(337.3 × 53.8 × 15.1)*
Speed, knots: 18
Range, n miles: 9,500 at 15 kt
Complement: 130

Machinery: Diesel-electric; 4 MAN Burmeister & Wain 16V 28/32 diesels; 11,800 hp(m) *(8.67 MW)*; 2 motors; 5,440 hp(m) *(4 MW)*; 2 shafts; cp props; 3 bow and 2 stern thrusters
Guns: 1 GE/GD 20 mm Vulcan Gatling (can be fitted). 6—12.7 mm MGs.
Radars: Navigation: I-band.
Sonars: Hull-mounted; active search; high frequency.
Helicopters: Platform for 1 light.

Comment: Ordered in 1992. Laid down December 1994. A multipurpose salvage and submarine rescue ship which can carry DSRV II as well as two LCVPs on davits plus a diving bell for nine men and a decompression chamber. Two large hydraulic cranes fore and aft and one towing winch. There are also two salvage ships which belong to the Coast Guard.

CHEONG HAE JIN *9/2003, Hartmut Ehlers / 0570936*

2 EDENTON CLASS (SALVAGE SHIPS) (ATS)

Name	No	Builders	Commissioned
PYONG TAEK (ex-*Beaufort*)	27	Brooke Marine, Lowestoft	22 Jan 1972
KWANG YANG (ex-*Brunswick*)	28	Brooke Marine, Lowestoft	19 Dec 1972

Displacement, tonnes: 2,976 full load
Dimensions, metres (feet): 86.1 × 15.2 × 4.6 *(282.5 × 49.9 × 15.1)*
Speed, knots: 16
Range, n miles: 10,000 at 13 kt
Complement: 129 (7 officers)
Machinery: 4 Paxman 12YJCM diesels; 6,000 hp *(4.48 MW)* sustained; 2 shafts; cp props; bow thruster
Guns: 2 Oerlikon 20 mm Mk 68.
Radars: Navigation: Sperry SPS-53; I/J-band.

Comment: Transferred from USA on 29 August 1996. Capable of (1) ocean towing, (2) supporting diver operations to depths of 850 ft, (3) lifting submerged objects weighing as much as 600,000 lb from a depth of 120 ft by static tidal lift or 30,000 lb by dynamic lift, (4) fighting ship fires. Fitted with 10 ton capacity crane forward and 20 ton capacity crane aft. Both recommissioned 28 February 1997.

PYONG TAEK (US colours) *12/1995, Giorgio Arra / 0506303*

3 CHUN JEE CLASS (LOGISTIC SUPPORT SHIPS) (AORH)

Name	No	Builders	Launched	Commissioned
CHUN JEE	57	Hyundai, Ulsan	May 1990	Dec 1990
DAE CHUNG	58	Hyundai, Ulsan	Jan 1997	1 Dec 1997
HWA CHUN	59	Hyundai, Ulsan	Jul 1997	Mar 1998

Displacement, tonnes: 4,247 standard; 9,327 full load
Dimensions, metres (feet): 130 × 17.8 × 6.5 *(426.5 × 58.4 × 21.3)*
Speed, knots: 20
Range, n miles: 4,500 at 15 kt

Cargo capacity: 4,200 tons liquids; 450 tons solids
Machinery: 2 SEMT-Pielstick 16 PA6 V 280 (57) or 12 PC2.5 diesels; 12,800 hp(m) *(9.4 MW)* sustained; 2 shafts
Guns: 4 Emerlec 30 mm (2 twin) or 2 Breda 40 mm/70. 2 GE/GD 20 mm Vulcan Gatlings.
Radars: Navigation: 2 Racal Decca; I-band.
Helicopters: Platform for 1 medium.

Comment: *Chun Jee* laid down September 1989. Underway replenishment stations on both sides. Helicopter for Vertrep but no hangar. There are three 6 ton lifts. Possibly based on Italian Stromboli class. Second of class was to have followed on but was eventually ordered together with the third in May 1995, to a slightly different design. More may be built when funds are available.

DAE CHUNG *12/2012, Chris Sattler / 1483522*

HWA CHUN *8/2010, Chris Sattler / 1406073*

1 TRIALS SUPPORT SHIP (AGE)

Name	No	Builders	Launched	Commissioned
SUNJIN	AGS 11	Hyundai, Ulsan	Nov 1992	Apr 1993

Displacement, tonnes: 325 full load
Dimensions, metres (feet): 34.5 × 15 × 3.7 *(113.2 × 49.2 × 12.1)*
Speed, knots: 21
Range, n miles: 600 at 16 kt
Complement: 5 + 20 scientists
Machinery: 1 MTU 16V 396 TE74L diesel; 2,680 hp(m) *(2 MW)*; 1 shaft; cp prop; 2 bow thrusters
Guns: 1—20 mm Gatling.
Radars: Navigation: I-band.

Comment: Experimental design built by Hyundai. Ordered June 1991, laid down June 1992. Aluminium SWATH hull with dynamic positioning system. Fitted with various trials equipment including an integrated navigation system and torpedo tracking pinger system. VDS and towed arrays. Used by the Defence Development Agency and civilian operated.

SUNJIN *1993, Hyundai / 0081169*

0 + 1 (1) SALVAGE SHIPS (ARSH)

Name	No	Builders	Laid down	Launched	Commissioned
TONGYEONG	31	Daewoo Shipbuilding	Nov 2010	4 Sep 2012	2014

Displacement, tonnes: 3,500 standard; 4,700 full load
Dimensions, metres (feet): 107.5 × 16.8 × 8.0 *(352.7 × 55.1 × 26.2)*
Speed, knots: 21

Machinery: Diesel-electric; 4 diesels; 2 shafts; cp props; three bow thrusters.
Guns: 1—20 mm.
Radars: Navigation: I-band.
Helicopters: Platform for one medium.

Comment: Equipped with a dynamic positioning system and four-point mooring system, the ship is also fitted with a side-scanning sonar and a remotely operated underwater vehicle capable of searching to a depth of 3,000 m. A decompression chamber can accommodate eight people. The design includes a flight deck capable of operating medium-sized helicopters. The ship is to replace *Pyong Taek* and *Kwang Yang* in the submarine rescue role and a second ship is expected.

TONGYEONG *9/2012, DSME / 1451222*

TUGS

Notes: There are approximately 16 tugs of which there is one coastal tug (A 73), eight coastal tugs (A 51–55, 62, 65–66 and 68) and seven harbour tugs (A 13, 22–23, 25–26, 30 and 38).

A 73   10/2008, Guy Toremans / 1353226

COAST GUARD

Notes: (1) The Republic of Korea Coast Guard was originally established as the Maritime Safety Division on 12 December 1953. After becoming the Maritime Police Unit in 1962, it separated from the national police in 1996 and changed its name to the Coast Guard in December 2000. It is responsible for Maritime Security (including maritime counter-terrorism), Search and Rescue, Marine Environmental Protection, Marine Pollution Response and Maritime Safety. With its Headquarters at Songo, Incheon, it has four regional headquarters at Incheon, Mokpo, Busan and Donghae. There are 16 coast guard stations including those at Sogcho, Donghae, Pohang, Ulsan, Busan, Tongyeong, Yeosu, Wando, Cheju, Mokpo, Gunsan, Taean, and Incheon. Most larger ships are painted with white hulls with a red, yellow, and blue diagonal stripe on the side of the hull. Some smaller craft have blue hulls and white superstructure with the word "Police" on the side of the superstructure. These are being progressively changed to the white paint scheme. Some specialist vessels (tugs, environmental vessels) have blue hulls with a red, yellow, and blue diagonal stripe.
(2) Aviation assets include a Bombardier Challenger 604 maritime surveillance aircraft, Kamov Ka-27 Helix and Bell 412 SAR helicopters, and AS 565 MB helicopters.
(3) There are reported to be new 50- and 100-ton classes of patrol vessel under construction at several yards including Samwong Shipbuilding, San Kwang Shipbuilding, and Il Heung Shipbuilding.

3 MAZINGER CLASS (PSO)

PC 1001–1003

Displacement, tonnes: 1,219 full load
Dimensions, metres (feet): 80.5 × 9.8 × 3.2 (264.1 × 32.2 × 10.5)
Speed, knots: 22
Range, n miles: 7,000 at 18 kt
Complement: 69 (11 officers)
Machinery: 2 SEMT-Pielstick 12 PA6 V 280 diesels; 9,600 hp(m) (7.08 MW) sustained; 2 shafts
Guns: 1 Bofors 40 mm/70. 4 Oerlikon 20 mm (2 twin).
Radars: Surface search: Raytheon; I-band.

Comment: Ordered 7 November 1980 from Korea Tacoma and Hyundai. PC 1001 delivered 29 November 1981. PC 1002 31 August 1982 and PC 1003 on 31 August 1983. All-welded mild steel construction. Used for offshore surveillance and general coast guard duties. PC 1001 is the Coast Guard Command ship. Only three of this class were completed.

MAZINGER (old colours)   1987, Korea Tacoma / 0506048

6 430 TON CLASS (PBO)

300–303   402–403

Displacement, tonnes: 437 full load
Dimensions, metres (feet): 53.7 × 7.4 × 2.4 (176.2 × 24.3 × 7.9)
Speed, knots: 19
Range, n miles: 2,100 at 17 kt
Complement: 14
Machinery: 2 MTU 16V 396 TB83 diesels; 1,990 hp(m) (1.49 MW); 2 shafts; cp props
Guns: 1 or 2 GD/GE 20 mm Vulcan Gatlings. 4—12.7 mm MGs.
Radars: Surface search: Raytheon; I-band.

Comment: All built between 1990 and 1995 by Hyundai except 301 which was built by Daewoo. Multipurpose patrol ships. Operational status of 300–303 doubtful.

300   3/1996, D Swetnam / 0081172

301   8/2000, van Ginderen Collection / 0097741

1 1,200 TON CLASS (PG)

HAN KANG PC 1005

Displacement, tonnes: 1,199 full load
Dimensions, metres (feet): 88.3 × 10 × 2.9 (289.7 × 32.8 × 9.5)
Speed, knots: 32
Range, n miles: 4,000 at 15 kt
Complement: 72 (11 officers)

Machinery: CODOG; 1 GE LM 2500 gas turbine; 26,820 hp (20 MW) sustained; 2 MTU 12V 956 TB82 diesels; 6,260 hp(m) (4.6 MW) sustained; 3 shafts
Guns: 1 Oto Melara 76/62 compact. 1 Bofors 40 mm/70. 2 GE/GD 20 mm Vulcan Gatlings.
Radars: Surface search: Raytheon SPS-64(V); I-band.
Fire control: Signaal WM28; I/J-band.
Weapon control systems: Signaal LIOD optronic director.

Comment: Built between May 1984 and December 1985 by Daewoo. Same hull as Po Hang class but much more lightly armed. Only one of the class was completed.

1,200 TON CLASS   9/2000 / 0097740

INSHORE PATROL CRAFT (PBR)

P 01   P 135

Displacement, tonnes: 48 full load
Dimensions, metres (feet): 21.3 × 5.4 × 1.4 (69.9 × 17.7 × 4.6)
Speed, knots: 22
Range, n miles: 400 at 12 kt
Complement: 11
Machinery: 2 diesels; 1,800 hp(m) (1.32 MW); 2 shafts
Guns: 1 Rheinmetall 20 mm. 3—12.7 mm MGs.
Radars: Surface search: Furuno; I-band.

Comment: Details are for the largest design of patrol craft. There are numbers of this type of vessel used for inshore patrol work. All Police craft have P pennant numbers. Armaments vary.

P 01   10/2008, Michael Nitz / 1353230

P 135   10/2008, Guy Toremans / 1353231

488　**Korea, South** > Coast guard

4 BUKHANSAN CLASS (PBO)

| BUKHANSAN 278 | CHULMASAN 279 | P 281 | P 282 |

Displacement, tonnes: 386 full load
Dimensions, metres (feet): 53.1 × 7.3 × 2.2 (174.2 × 24.0 × 7.2)
Speed, knots: 28. **Range, n miles:** 2,500 at 15 kt
Complement: 35 (3 officers)
Machinery: 2 MTU diesels; 8,300 hp(m) (6.1 MW) sustained; 2 shafts
Guns: 1 Breda 40 mm/70. 1 GE/GD 20 mm Vulcan Gatling. 2 — 12.7 mm MGs.
Radars: Surface search: I-band.
Weapon control systems: Radamec optronic director.

Comment: Follow on to Sea Wolf class developed by Hyundai in 1987. Ordered in 1988 from Hyundai and Daewoo respectively. First pair in service in 1989, and second pair in 1990.

CHULMASAN (old colours)　　　　　　　　　　　　　　　　　*1989, Daewoo / 0506049*

5 HYUNDAI TYPE (PB)

| 105 | 113 | 118 | 121 | 125 |

Displacement, tonnes: 112 full load
Dimensions, metres (feet): 32.2 × 6 × 1.4 (105.6 × 19.7 × 4.6)
Speed, knots: 25
Complement: 19
Machinery: 2 diesels; 2 shafts
Guns: 1 Rheinmetall 20 mm. 2 — 12.7 mm MGs.
Radars: Surface search: Furuno; I-band.

Comment: Ordered in 1996 and delivered from June 1997.

HYUNDAI 125　　　　　　　　　　　　　　　　　*3/2009, Bob Fildes / 1305924*

1 SALVAGE SHIP (ARSH)

Name	No	Builders	Launched	Commissioned
TAE PUNG YANG I	3001	Hyundai, Ulsan	Oct 1991	18 Feb 1993

Displacement, tonnes: 3,251 standard; 4,369 full load
Dimensions, metres (feet): 104.7 × 15 × 5.2 (343.5 × 49.2 × 17.1)
Speed, knots: 21. **Range, n miles:** 8,500 at 15 kt
Complement: 121
Machinery: 4 Ssangyong MAN Burmeister & Wain 16V 28/32 diesels; 4,800 hp(m) (3.53 MW); 2 shafts; cp props; bow and stern thrusters
Guns: 1 GD/GE 20 mm Vulcan Gatling. 6 — 12.7 mm MGs.
Radars: Navigation: I-band.
Helicopters: 1 light.

Comment: Laid down February 1991. Has a helicopter deck and hangar, an ROV capable of diving to 300 m and a firefighting capability. Dynamic positioning system. Can be used for cable laying. Operates for the Marine Police.

TAE PUNG YANG 1　　　　　　　　　　　　　　　　　*10/2008, Guy Toremans / 1353228*

1 SALVAGE SHIP (ARSH)

Name	No	Builders	Commissioned
JAEMIN I	1501	Daewoo, Okpo	28 Dec 1992

Displacement, tonnes: 2,105 full load
Dimensions, metres (feet): 77.6 × 13.5 × 4.2 (254.6 × 44.3 × 13.8)
Speed, knots: 18
Range, n miles: 4,500 at 12 kt
Complement: 92
Machinery: 2 MTU diesels; 8,000 hp(m) (5.88 MW); 2 shafts; cp props
Guns: 1 GD/GE 20 mm Vulcan Gatling.
Radars: Navigation: I-band.

Comment: Ordered in 1990. Fitted with diving equipment and has a four point mooring system. Carries two LCVPs.

JAEMIN I　　　　　　　　　　　　　　　　　*8/2000 / 0097745*

1 SALVAGE SHIP (ARS)

Name	No	Builders	Launched	Commissioned
JAEMIN II	1502	Hyundai, Ulsan	15 Jul 1995	Apr 1996

Displacement, tonnes: 2,540 full load
Dimensions, metres (feet): 88 × 14.5 × 4.6 (288.7 × 47.6 × 15.1)
Speed, knots: 20
Range, n miles: 4,500 at 15 kt
Complement: 81
Machinery: 2 MTU diesels; 12,662 hp(m) (9.31 MW); 2 shafts; Kamewa cp props; bow and stern thrusters
Guns: 1 GE/GD 20 mm Vulcan Gatling.
Radars: Navigation: I-band.

Comment: Ordered in December 1993 for Maritime Police. A general purpose salvage ship capable of towing, firefighting, supply or patrol duties.

JAEMIN II　　　　　　　　　　　　　　　　　*8/1999, Ships of the World / 0081176*

5 SALVAGE SHIPS (ARSH)

Name	No	Builders	Commissioned
TAE PUNG YANG III	3003	Hanjin Heavy Industries, Pusan	2003
TAE PUNG YANG V	3005	Hanjin Heavy Industries, Pusan	2005
TAE PUNG YANG VI	3006	Hanjin Heavy Industries, Pusan	20 Dec 2005
TAE PUNG YANG VII	3007	Hanjin Heavy Industries, Pusan	28 Aug 2006
TAE PUNG YANG VIII	3008	Hanjin Heavy Industries, Pusan	23 Mar 2008

Displacement, tonnes: 4,064 full load
Dimensions, metres (feet): 110.5 × 15.4 × 4.9 (362.5 × 50.5 × 16.1)
Speed, knots: 21
Range, n miles: 4,500 at 15 kt
Complement: 120
Machinery: 2 MTU 12V 1163 TB93 diesels; 11,800 hp (8.8 MW); 2 shafts
Guns: 1 — 20 mm. 2 — 12.7 mm MGs.
Radars: Navigation: I-band.
Helicopters: Platform for 1 large.

Comment: Multipurpose EEZ patrol and salvage ships believed to be of the same class.

TAE PUNG YANG VI　　　　　　　　　　　　　　　　　*10/2008, Guy Toremans / 1353232*

Coast guard < **Korea, South** 489

1 SALVAGE SHIP (ARSH)

Name	No	Builders	Commissioned
JAEMIN III	1503	Hyundai, Ulsan	Nov 1998

Displacement, tonnes: 4,267 full load
Dimensions, metres (feet): 110.5 × 15.4 × 4.9 (362.5 × 50.5 × 16.1)
Speed, knots: 18
Complement: 120
Machinery: 2 diesels; 2 shafts
Guns: 2 GE 20 mm Vulcan Gatlings. 6—12.7 mm MGs.
Radars: Navigation: I-band.

Comment: Ordered in 1996, from Hyundai, Ulsan. Large helicopter deck but no hangar.

JAEMIN III   10/2008, Guy Toremans / 1353229

1 DAEWOO TYPE (PSO)

SUMJINKANG PC 1006

Displacement, tonnes: 1,676 full load
Dimensions, metres (feet): 84 × 10.4 × 3.6 (275.6 × 34.1 × 11.8)
Speed, knots: 21. **Range, n miles:** 4,500 at 18 kt
Complement: 57 (7 officers)
Machinery: 2 Wärtsilä Nohab 16V25 diesels; 10,000 hp(m) (7.35 MW) sustained; 2 shafts
Guns: 1—20 mm Sea Vulcan Gatling. 4—12.7 mm MGs.
Radars: Surface search: I-band.

Comment: Ordered in 1997 from Daewoo. Described as a multipurpose patrol ship. Launched 22 January 1999, and delivered 20 June 1999.

SUMJINKANG   6/2006, Korea Coast Guard / 1159985

1 SAMBONGHO CLASS (PATROL SHIP) (PSO)

Name	No	Builders	Commissioned
SAMBONGHO	5001	Hyundai, Ulsan	23 Apr 2002

Displacement, tonnes: 7,000 full load (est.)
Dimensions, metres (feet): 145.5 × 16.5 × ? (477.4 × 54.1 × ?)
Speed, knots: 23. **Range, n miles:** 7,300 at 15 kt
Complement: 93
Machinery: 2 MTU diesels; 12,000 hp (8.84 MW); 2 shafts
Guns: 2—20 mm Sea Vulcan. 6—7.62 mm MGs.
Radars: Navigation: I-band.
Helicopters: 1 large.

Comment: The largest ship in the Coast Guard.

SAMBONGHO   6/2006, Korea Coast Guard / 1159986

1 SALVAGE SHIP (ARSH)

Name	No	Builders	Commissioned
TAE PUNG YANG II	3002	Hyundai, Ulsan	31 Oct 1998

Displacement, tonnes: 3,963 standard
Dimensions, metres (feet): 110.5 × 15.4 × 4.9 (362.5 × 50.5 × 16.1)
Speed, knots: 18
Complement: 120

Machinery: 2 diesels; 2 shafts
Guns: 2—20 mm Vulcan Gatlings. 6—12.7 mm MGs.
Radars: Surface search: I-band.
Helicopters: Platform for 1 large.

Comment: Ordered from Hyundai in mid-1996. Also used for SAR operations.

© 2014 IHS

TAE PUNG YANG II   6/2006, Korea Coast Guard / 1159984

9 SALVAGE SHIPS (ARSH)

Name	No	Builders	Commissioned
JAEMIN V	1505	–	–
JAEMIN VI	1506	Hyundai Heavy Industries	–
JAEMIN VII	1507	Hyundai Heavy Industries	–
JAEMIN VIII	1508	Hyundai Heavy Industries	20 Oct 2005
JAEMIN IX	1509	Hanjin Heavy Industries	2006
JAEMIN X	1510	Hanjin Heavy Industries	2007
JAEMIN XI	1511	Hyundai Heavy Industries	2008
JAEMIN XII	1512	STX, Jinhae	2010
JAEMIN XIII	1513	STX, Jinhae	2011

Displacement, tonnes: 2,772 full load
Dimensions, metres (feet): 98.1 × 14.0 × 4.3 (321.9 × 45.9 × 14.1)
Speed, knots: 21
Range, n miles: 4,500 at 15 kt
Machinery: 2 MTU 16V 1163 TB83 diesels; 12,875 hp (9.6 MW); 2 shafts
Guns: 1—20 mm. 2—12.7 mm MGs.
Radars: Navigation: I-band.

Comment: Salvage ships reported to be of the same class and which entered service 2000–09. Details are those published for Jaemin VII. The latest variants may have waterjet propulsion.

JAEMIN VII   6/2006, Korea Coast Guard / 1159983

2 P 127 CLASS (PB)

P 127   P 128

Machinery: To be announced
Guns: To be announced.
Radars: To be announced.

Comment: Patrol craft of unknown type.

P 128   3/2009, Bob Fildes / 1305923

14 HAE URI CLASS (PB)

308–313   315–322

Displacement, tonnes: 500 full load
Speed, knots: 30
Machinery: 2 MTU 16V 4000; 7,295 hp (5.4 MW); 2 waterjets
Guns: 1—20 mm Sea Vulcan. 2—12.7 mm MGs.
Radars: Navigation: I-band.

Comment: A new design of patrol vessel which had probably replaced the Sea Wolf/Shark class by 2010. Built at Hanjin and Hyundai shipyards. A class of 24 is expected. Armament varies.

P 311   3/2009, Bob Fildes / 1406074

IHS Jane's Fighting Ships 2014-2015

Korea, South > Coast guard — Kuwait > Patrol forces

12 GUNSAN CLASS (SALVAGE SHIPS) (ARSH)

HAN KANG 1007–1013 **HAN KANG 1015–1018**

Displacement, tonnes: 1,600 full load
Dimensions, metres (feet): 91 × 11.8 × 3.3 *(298.6 × 38.7 × 10.8)*
Speed, knots: 30
Complement: 57

Machinery: 3 MTU 20V 1163 TB94 diesels; 29,770 hp *(22.2 MW)*; 3 waterjets
Guns: 1 — 20 mm Sea Vulcan. 4 — 7.62 mm MGs.
Radars: Navigation: I-band.
Helicopters: Platform for 1 large.

Comment: A new design of EEZ patrol and salvage ship built at Hanjin and Hyundai shipyards.

TAE PUNG YANG IX   6/2010 / 1406979

HANG KANG 1008   6/2010 / 1406978

15 TAE GEUK CLASS (PB)

501–503   **505–513**   **515–517**

Displacement, tonnes: 630 full load
Dimensions, metres (feet): 56.4 × 9.1 × 2.6 *(185 × 29.9 × 8.5)*
Speed, knots: 30
Machinery: CODAD: 2 MTU 16V 4000; 2 MTU 12V 1163 TB93 diesels; 2 waterjets
Guns: 1 — 20 mm Sea Vulcan. 4 — 7.62 mm MGs.
Radars: Navigation: I-band.

Comment: A new design of patrol vessel which had probably replaced the Sea Dragon/Whale class by 2010. Built at Hanjin and Hyundai shipyards. Two further ships are under construction at Hanjin. Armament varies.

4 SALVAGE SHIPS (ARSH)

Name	No	Builders	Commissioned
TAE PUNG YANG IX	3009	Hyundai Heavy Industries	2010
TAE PUNG YANG X	3010	Hyundai Heavy Industries	2010
TAE PUNG YANG XI	3011	STX, Jinhae	2012
TAE PUNG YANG XII	3012	Hyundai Heavy Industries	2012

Displacement, tonnes: 4,000 full load (est.)
Dimensions, metres (feet): 112.7 × 14.2 × ? *(369.8 × 46.6 × ?)*
Speed, knots: 30
Complement: 120

Machinery: CODAD: 4 MTU 20V 1163 TB94 diesels; 39,690 hp *(29.6 MW)*; 2 shafts
Guns: 1 — 40 mm. 1 — 20 mm Sea Vulcan. 4 — 7.62 mm MGs.
Radars: Navigation: I-band.
Helicopters: Platform for 1 large.

Comment: A new design of EEZ patrol and heavy salvage ship to follow on from previous Tae Pung Yang series. A further two ships are reported under construction.

TAE GEUK 512   6/2010 / 1406977

Kuwait

Country Overview

Formerly a British protectorate, the Kingdom of Kuwait gained independence in 1961. Situated on the northwestern coast of the Gulf, it is bordered to the north by Iraq and to the south by Saudi Arabia. The country's total area, including the islands of Bubiyan, Warbah, and Faylakah, is 6,880 square miles. It has a 269 n mile coastline with the Gulf. The capital, largest city and principal port is Kuwait City. The country was annexed by Iraq from August 1990 to February 1991 when the country was liberated. Territorial seas (12 n miles) are claimed. An EEZ has not been claimed.

Headquarters Appointments

Commander of the Navy:
  Major General Jassim Mohammad Al Ansari

Deputy Commander of the Navy:
  Brigadier Monoor Al Massad

Personnel

2014: 2,700 (including 500 Coast Guard)

Aviation

The Air Force operates five Eurocopter AS 532C Cougar helicopters armed with Exocet AM 39 ASMs and 40 F/A-18C/D Hornets.

Bases

Navy: Ras Al Qalayah
Coast Guard: Shuwaikh, Umm Al-Hainan, Al-Bida

PATROL FORCES

8 UM ALMARADIM (COMBATTANTE I) CLASS (PBM)

Name	No	Builders	Launched	Commissioned
UM ALMARADIM	P 3711	CMN, Cherbourg	27 Feb 1997	31 Jul 1998
OUHA	P 3713	CMN, Cherbourg	29 May 1997	31 Jul 1998
FAILAKA	P 3715	CMN, Cherbourg	29 Aug 1997	19 Dec 1998
MASKAN	P 3717	CMN, Cherbourg	6 Jan 1998	19 Dec 1998
AL-AHMADI	P 3719	CMN, Cherbourg	2 Apr 1998	1 Jul 1999
ALFAHAHEEL	P 3721	CMN, Cherbourg	16 Jun 1998	1 Jul 1999
AL-YARMOUK	P 3723	CMN, Cherbourg	3 Mar 1999	7 Jun 2000
GAROH	P 3725	CMN, Cherbourg	Jun 1999	7 Jun 2000

Displacement, tonnes: 249 full load
Dimensions, metres (feet): 42 oa; 37 wl × 8.2 × 1.9 *(137.8; 121.4 × 26.9 × 6.2)*
Speed, knots: 30
Range, n miles: 1,350 at 14 kt
Complement: 29 (5 officers)

Machinery: 2 MTU 16V 538 TB93 diesels; 4,000 hp(m) *(2.94 MW)*; 2 Kamewa waterjets
Missiles: SSM: 4 BAe Sea Skua (2 twin). Semi-active radar homing to 15 km *(8.1 n miles)* at 0.9 Mach.
  SAM: Sadral sextuple launcher fitted for only.
Guns: 1 Otobreda 40 mm/70; 120 rds/min to 12.5 km *(6.8 n miles)*; weight of shell 0.96 kg.
  1 Giat 20 mm M 621. 2 — 12.7 mm MGs.
Physical countermeasures: Decoys: 2 Dagaie Mk 2 chaff launchers fitted for only.
Electronic countermeasures: ESM:Thomson-CSF DR 3000 S1; intercept.
Radars: Air/surface search: Thomson-CSF MRR; 3D; G-band.
  Fire control: BAe Seaspray Mk 3; I/J-band (for SSM).
  Navigation: Litton Marine 20V90; I-band.
Combat data systems: Thomson-CSF TAVITAC NT; Link Y.
Weapon control systems: CS Defence Najir Mk 2 optronic director.

Programmes: Contract signed with CMN Cherbourg on 27 March 1995. First steel cut 9 June 1995. Names are taken from former Kuwaiti patrol craft.
Structure: Late decisions were made on the missile system which has been fitted in the last pair on build and to the remainder from 2000. Provision is also made for Simbad SAM and Dagaie decoy launchers, which may be fitted later. Positions of smaller guns are uncertain.
Operational: Training done in France. The aim is to have 10 crews capable of manning the eight ships. First four arrived in the Gulf in mid-August 1999, second four arrived in mid-2000.

AL-AHMADI   3/2007, Edward McDonnell / 1170189

MAKSAN   2/2012* / 1530915

Patrol forces — Amphibious forces < **Kuwait** 491

1 TNC 45 TYPE (FAST ATTACK CRAFT — MISSILE) (PGGF)

Name	No	Builders	Commissioned
AL SANBOUK	P 4505	Lürssen, Bremen-Vegesack	26 Apr 1984

Displacement, tonnes: 259 full load
Dimensions, metres (feet): 44.9 × 7 × 2.3 *(147.3 × 23.0 × 7.5)*
Speed, knots: 41
Range, n miles: 1,800 at 16 kt
Complement: 35 (5 officers)

Machinery: 4 MTU 16V 538 TB92 diesels; 13,640 hp(m) *(10 MW)* sustained; 4 shafts
Missiles: SSM: 4 Aerospatiale MM 40 Exocet; inertial cruise; active radar homing to 70 km *(40 n miles)* at 0.9 Mach; warhead 165 kg; sea-skimmer.
Guns: 1 Oto Melara 3 in *(76 mm)*/62 compact; 85 rds/min to 16 km *(8.6 n miles)* anti-surface; 12 km *(6.5 n miles)* anti-aircraft; weight of shell 6 kg. 2 Breda 40 mm/70 (twin); 300 rds/min to 12.5 km *(6.6 n miles)*; weight of shell 0.96 kg.
Physical countermeasures: Decoys: CSEE Dagaie; IR flares and chaff; H/J-band.
Electronic countermeasures: ESM: Racal Cutlass; intercept.
Radars: Air/surface search: Ericsson Sea Giraffe 50HC; G/H-band.
Fire control: Philips 9LV 200; J-band.
Navigation: Decca TM 1226C; I-band.
Weapon control systems: PEAB 9LV 228 system; Link Y; CSEE Lynx optical sight.

Programmes: Six ordered from Lürssen in 1980 and delivered in 1983–84.
Operational: *Al Sanbouk* escaped to Bahrain when the Iraqis invaded in August 1990, but the rest of this class was taken over by the Iraqi Navy, and either sunk or severely damaged by Allied forces in February 1991. The ship was refitted by Lürssen in 1995 and again in 2004.

AL SANBOUK *3/2003, A Sharma* / 0568872

1 FPB 57 TYPE (FAST ATTACK CRAFT — MISSILE) (PGGF)

Name	No	Builders	Commissioned
ISTIQLAL	P 5702	Lürssen, Bremen-Vegesack	9 Aug 1983

Displacement, tonnes: 417 full load
Dimensions, metres (feet): 58.1 × 7.6 × 2.7 *(190.6 × 24.9 × 8.9)*
Speed, knots: 36
Range, n miles: 1,300 at 30 kt
Complement: 40 (5 officers)

Machinery: 4 MTU 16V 956 TB91 diesels; 15,000 hp(m) *(11 MW)* sustained; 4 shafts
Missiles: SSM: 4 Aerospatiale MM 40 Exocet; inertial cruise; active radar homing to 70 km *(40 n miles)* at 0.9 Mach; warhead 165 kg; sea-skimmer.
Guns: 1 Oto Melara 3 in *(76 mm)*/62 compact; 85 rds/min to 16 km *(8.6 n miles)* anti-surface; 12 km *(6.5 n miles)* anti-aircraft; weight of shell 6 kg. 2 Breda 40 mm/70 (twin); 300 rds/min to 12.5 km *(6.6 n miles)*; weight of shell 0.96 kg.
Mines: Fitted for minelaying.
Physical countermeasures: Decoys: CSEE Dagaie trainable mounting; automatic dispenser; IR flares and chaff; H/J-band.
Electronic countermeasures: ESM: Racal Cutlass; radar intercept.
ECM: Racal Cygnus; jammer.
Radars: Surface search: Marconi S 810 (after radome); I-band; range 43 km *(25 n miles)*.
Navigation: Decca TM 1226C; I-band.
Fire control: Philips 9LV 200; J-band.
Weapon control systems: PEAB 9LV 228 system; Link Y; CSEE Lynx optical sight.

Programmes: Two ordered from Lürssen in 1980.
Operational: *Istiqlal* escaped to Bahrain when the Iraqis invaded in August 1990. The second of this class was captured and sunk in February 1991. Having been laid up since 1997 *Istiqlal* was refitted at Lürssen 2003-2005. In addition to operational roles, it is also used as a training ship.

ISTIQLAL *4/2005, Michael Nitz* / 1121416

10 MK V CLASS (INTERCEPTION CRAFT) (PBF)

AL NOKETHA P 2701	BAYAN P 2707	AL SEEP P 2713	AL BOOM P 2719
BUBIYAN P 2703	AL SHOAIE P 2709	AL BATEEL P 2715	
KUBBAR P 2705	AL SAFFAR P 2711	AL TAWASH P 2717	

Displacement, tonnes: 55 full load
Dimensions, metres (feet): 27.4 × 5.5 × 1.5 *(89.9 × 18.0 × 4.9)*
Speed, knots: 45
Range, n miles: 600 at 35 kt
Complement: 12 (2 officers)

Machinery: 2 MTU 12V 396 TE94 diesels; 4,500 hp *(3.36 MW)*; 2 Kamewa waterjets
Guns: 1 Rheinmetall MLG 27 mm (remotely operated).
Electronic countermeasures: ESM: To be announced.
Radars: Navigation: I-band.

Comment: The US Congress was advised on 17 December 2005 of the possible sale of 12 (later reduced to 10) interception craft. With a higher superstructure, the craft are to be a modified version of the US Mk V Pegasus class. A contract for their construction was awarded to US Marine Inc, Gulfport, Mississippi on 22 May 2009. *Al Noketha* was delivered on 25 August 2011. The second, third and fourth boats were delivered on 24 January 2012, the fifth and sixth on 16 July 2012 and seventh and eighth on 22 January 2013. The final two were delivered on 31 July 2013.

AL NOKETHA *4/2011** / 1530916

AMPHIBIOUS FORCES

0 + 2 LANDING CRAFT (LCT)

Displacement, tonnes: 864 standard
Dimensions, metres (feet): 64 × 12.0 × 2.7 *(210 × 39.4 × 8.9)*
Speed, knots: 11
Complement: 19 + 56 embarked forces
Military lift: Military vehicles
Machinery: 2 Caterpillar 3508 diesels; 2 shafts

Comment: It was announced on 21 February 2013 that Abu Dhabi Shipbuilding had been awarded the contract for the Kuwaiti Navy's future landing craft programmes. Contract includes an order for two 64-m landing craft that are reported to be based on similar craft built for the UAE Navy in 2004. The UAE vessels (on which details are based) were of steel construction. Delivery is expected in 2015.

0 + 1 LANDING CRAFT (LCU)

Measurement, tonnes: 386 dwt
Dimensions, metres (feet): 44.4 × 10.0 × 2.2 *(145.7 × 32.8 × 7.2)*
Speed, knots: 10
Range, n miles: 1,000 at 8.5 kt
Complement: 11 (3 officers) + 40 embarked forces
Military lift: Military vehicles.
Machinery: 2 Caterpillar CAT 3406 TA diesels; 730 hp *(544 kW)*; rated 365 hp at 1,800 rpm each. To be announced.

Comment: It was announced on 21 February 2013 that Abu Dhabi Shipbuilding had been awarded the contract for the Kuwaiti Navy's future landing craft programmes. Contract includes an order for one 42-m landing craft that is reported to be based on similar craft built for the UAE Navy in 2006. The UAE vessels (on which details are based) were designed as multimission craft equipped with three hydraulic deck cranes. Delivery is expected in 2015.

0 + 5 FAST LANDING CRAFT (LCP)

Displacement, tonnes: 13.3 standard
Dimensions, metres (feet): 16 × 4.0 × 0.7 *(52.5 × 13.1 × 2.3)*
Speed, knots: 24
Complement: 5
Machinery: 2 diesels; 2 waterjets
Radars: Navigation: I-band.

Comment: It was announced on 21 February 2013 that Abu Dhabi Shipbuilding had been awarded the contract for the Kuwaiti Navy's future landing craft programmes. Contract includes an order for five fast landing craft that are reported to be based on the workboats procured for the UAE Coast Guard in 2007. These craft are based on the VT Halmatic Sea Keeper design with an asymmetric catamaran hull. Highly manoeuvrable, the vessels are capable of carrying a 10-tonne payload. Delivery is expected in 2015.

AUXILIARIES

1 SAWAHIL CLASS (SUPPORT SHIP)

AL DORRAR (ex-*Qaruh*; ex-*Sawahil 35*) S 5509

Measurement, tonnes: 554 dwt
Dimensions, metres (feet): 55.4 × 9.6 × 2 *(181.8 × 31.5 × 6.6)*
Speed, knots: 9
Complement: 40
Machinery: 2 diesels; 2,400 hp(m) *(1.76 MW)*; 2 shafts
Guns: 2—12.7 mm MGs.
Radars: Navigation: Racal Decca; I-band.

Comment: This is a Sawahil class oil rig replenishment and accommodation ship which was built in South Korea in 1986. She escaped to Bahrain during the Iraqi invasion, and is back in service. High-level helicopter platform aft. Used as a utility transport. Refitted in 1996/97.

AL DORRAR   3/2007, *Edward McDonnell* / 1170188

COAST GUARD

PATROL FORCES

Headquarters Appointments

Director of Coast Guard:
  Brigadier Jassem Al-Failakawi

16 VICTORY TEAM P 46 CLASS (PATROL CRAFT) (PBF)

Displacement, tonnes: 9 standard
Dimensions, metres (feet): 14 × 3.23 × 0.8 *(45.9 × 10.6 × 2.6)*
Speed, knots: 52
Range, n miles: 200 at 50 kt
Complement: 4
Machinery: 2 Yanmar 6CX diesels; 930 hp *(690 kW)*; 2 Arneson ASD 8 surface drives
Guns: 2—12.7 mm MGs.
Radars: Navigation.

Comment: Contract for 16 craft signed in April 2004 with delivery of the final vessel expected by mid-2006. The Victory Team of Dubai design is a twin-stepped deep-'vee' monohull developed from its offshore power boats. The hull, deck and internal assembly are built from a sandwich composite comprising a glass fibre, kevlar and carbon mix to provide structural integrity at a minimum weight. The cockpit is protected by 17 mm Dyneema Ballistic panelling.

P 46   3/2005, *Victory Team* / 1127034

4 INTTISAR (OPV 310) CLASS (PB)

Name	No	Builders	Commissioned
INTTISAR	P 301	Australian Shipbuilding Industries	20 Jan 1993
AMAN	P 302	Australian Shipbuilding Industries	20 Jan 1993
MAIMON	P 303	Australian Shipbuilding Industries	7 Aug 1993
MOBARK	P 304	Australian Shipbuilding Industries	7 Aug 1993

Displacement, tonnes: 152 full load
Dimensions, metres (feet): 31.5 oa; 27.1 wl × 6.5 × 2 *(103.3; 88.9 × 21.3 × 6.6)*
Speed, knots: 28
Range, n miles: 300 at 28 kt
Complement: 11 (3 officers)
Machinery: 2 MTU 16V 396 TB94 diesels; 5,800 hp(m) *(4.26 MW)* sustained; 2 shafts; 1 MTU 8V 183 TE62 diesel; 750 hp(m) *(550 kW)* maximum; 1 Hamilton 422 water-jet
Guns: 1 Oerlikon 20 mm. 1—12.7 mm MG.
Radars: Surface search: 2 Racal Decca; I-band.

Comment: First two ordered from Australian Shipbuilding Industries in 1991. Second pair ordered in July 1992. Steel hulls, aluminium superstructure. The third engine drives a small waterjet to provide a loiter capability. Carries an RIB. Used by the Coast Guard.

AMAN   1992, *Australian Shipbuilding Industries* / 0081178

3 INSHORE PATROL CRAFT (PBR)

KASSIR T 205   **DASTOOR** T 210   **MAHROOS** T 215

Dimensions, metres (feet): 21.6 × 5.96 × 1.5 *(70.9 × 19.6 × 4.9)*
Speed, knots: 25
Range, n miles: 325 at 25 kt
Complement: 3 + 41 spare berths
Machinery: 2 MTU 12V 183 TE92 diesels; 1,800 hp *(1.45 MW)*; 2 shafts
Radars: Navigation: to be announced.

Comment: Order for three craft for the Coast Guard announced on 7 January 2003. Based on the 22 m craft in service with the New South Wales Police, the vessels were constructed by Austal Ships subsidiary, Image Marine and delivered in June 2004. Aluminium hull.

DASTOOR   6/2004, *Austal Ships* / 0587772

3 AL SHAHEED CLASS (PB)

Name	No	Builders	Commissioned
AL SHAHEED	P 305	OCEA, Les Sables d'Olonne	Jul 1997
BAYAN	P 306	OCEA, Les Sables d'Olonne	Apr 1999
DASMAN	P 307	OCEA, Les Sables d'Olonne	2001

Displacement, tonnes: 106 full load
Dimensions, metres (feet): 33.3 × 7 × 1.2 *(109.3 × 23.0 × 3.9)*
Speed, knots: 30
Range, n miles: 360 at 25 kt
Complement: 11 (3 officers)

Machinery: 2 MTU 12V 396 TE94; 4,352 hp(m) *(3.2 MW)* sustained; 2 shafts.
Guns: 1 Oerlikon 20 mm. 2—12.7 mm MGs.
Radars: Surface search: Racal Decca 20V 90 TA; E/F-band.
Navigation: Racal Decca Bridgemaster ARPA; I-band.

Comment: Built by OCEA, France to FPB 100K design. Operated by the Coast Guard.

AL SHAHEED   10/1997, *Ships of the World* / 0012718

Patrol forces — Auxiliaries < **Kuwait** (COAST GUARD) 493

10 SUBAHI CLASS (PB)

Name	No	Builders	Commissioned
RAYYAN	P 300	OCEA, St Nazaire	23 Aug 2005
SUBAHI	P 308	OCEA, St Nazaire	6 Aug 2003
JABERI	P 309	OCEA, St Nazaire	Dec 2003
SAAD	P 310	OCEA, St Nazaire	Feb 2004
AHMADI	P 311	OCEA, St Nazaire	Mar 2004
NAIF	P 312	OCEA, St Nazaire	May 2004
THAFIR	P 313	OCEA, St Nazaire	Jul 2004
MARZOUG	P 314	OCEA, St Nazaire	Sep 2004
MASH'NOOR	P 315	OCEA, St Nazaire	Jan 2005
WADAH	P 316	OCEA, St Nazaire	May 2005

Displacement, tonnes: 118 full load
Dimensions, metres (feet): 35.2 × 6.8 × 1.2 *(115.5 × 22.3 × 3.9)*
Speed, knots: 32
Range, n miles: 300 at 28 kt
Complement: 11 (3 officers)
Machinery: 2 MTU 12V 4000 M70 diesels; 4,600 hp *(3.43 MW)*; 2 Kamewa waterjets
Guns: 1 Oerlikon 20 mm. 2—12.7 mm MGs.
Radars: Sperry Bridgemaster E; I-band.

Comment: Built by OCEA, France based on Al Shaheed class design. Aluminium construction. Operated by the Coast Guard. P 300 is a VIP variant equipped with three cabins.

MARZOUG   8/2004, B Prézelin / 1133080

RAYYAN   7/2006, B Prézelin / 1040681

33 AL-SHAALI TYPE (INSHORE PATROL CRAFT) (PBF)

Comment: Ten 10 m and 23 8.5 m patrol craft built by Al-Shaali Marine, Dubai, and delivered in June 1992.

12 MANTA CLASS (INSHORE PATROL CRAFT) (PBF)

Displacement, tonnes: 10 full load
Dimensions, metres (feet): 14 × 3.8 × 0.7 *(45.9 × 12.5 × 2.3)*
Speed, knots: 40
Range, n miles: 180 at 35 kt
Complement: 4
Machinery: 2 Caterpillar 3208 diesels; 810 hp(m) *(595 kW)* sustained; 2 shafts
Guns: 3 Herstal M2HB 12.7 mm MGs.
Radars: Surface search: Furuno; I-band.

Comment: Original craft ordered in September 1992 from Simonneau Marine and delivered in 1993. Aluminium construction. This version has two inboard engines. Pennant numbers are in odd number sequence. All the class reported to be inoperable due to technical problems. An underlying cause may be that the boats were fitted with inboard engines although designed for outboards.

MANTA 1509   5/2009, Edward McDonnell / 1305925

6 COUGAR ENFORCER 40 CLASS
(INSHORE PATROL CRAFT) (PBF)

Displacement, tonnes: 6 full load
Dimensions, metres (feet): 12.2 × 2.8 × 0.8 *(40 × 9.2 × 2.6)*
Speed, knots: 45
Range, n miles: 250 at 35 kt
Complement: 4
Machinery: 2 Sabre 380 S diesels; 760 hp(m) *(559 kW)*; 2 Arneson ASD 8 surface drives; 2 shafts
Guns: 1—12.7 mm MG.
Radars: Surface search: Koden; I-band.

Comment: First one completed in July 1996 for the Coastguard by Cougar Marine, Warsash. The craft has a V monohull design.

ENFORCER 40   7/1996, Cougar Marine / 0081179

17 COUGAR TYPE (INSHORE PATROL CRAFT) (PBF)

Comment: Three Cat 900 (32 ft) three Predator 1100 (35 ft) and three Predator 1000 (33 ft) all powered by two Yamaha outboards (400 hp(m) *(294 kW)*. Four Type 1200 (38 ft) and four Type 1300 (41 ft) all powered by two Sabre diesels (760 hp(m) *(559 kW)*). All based on the high-performance planing hull developed for racing, and acquired in 1991–92. Most have a 7.62 mm MG and a Kroden I-band radar. Used by the Coast Guard.

COUGAR 1200   1991, Cougar Marine / 0081180

AUXILIARIES

Notes: (1) There is a logistic craft P 140.
(2) There is an ex-oilrig supply vessel *Abdul Jaal* with pennant number B 45.

P 140   10/2002 / 0587770

1 LANDING SUPPLY CRAFT (LCU)

L 404

Measurement, tonnes: 305 dwt
Dimensions, metres (feet): 49 × ? × ? *(160.8 × ? × ?)*
Speed, knots: 12
Complement: 12
Machinery: 2 diesels; 2 shafts
Radars: Navigation: I-band.

Comment: Contract for the design and build of a landing craft signed with Singapore Technologies Marine Ltd (ST Marine) on 8 October 2004. The multipurpose vessel is to be used for transport and supply operations as well as law enforcement duties in the Arabian Gulf. In addition to carrying roll-on roll-off goods on the main deck, the vessel is also designed to transport liquid, refrigeration and general cargoes. Delivery of the ship was made in late 2005.

494   Kuwait (COAST GUARD) > Auxiliaries — Latvia > Patrol forces

1 LOADMASTER MK 2 (LOGISTIC SUPPORT CRAFT) (LCU)

SAFFAR (ex-*Jalbout*) L 403

Displacement, tonnes: 427 full load
Dimensions, metres (feet): 33 × 10.2 × 1.75 *(108.3 × 33.5 × 5.7)*
Speed, knots: 10
Complement: 7 (1 officer)
Machinery: 2 Caterpillar V12 diesels; 1,000 hp *(745 kW)*; 2 props
Radars: Navigation: I-band.

Comment: Built by Fairey Marine Cowes, UK and entered service in 1985. Captured by Iraqi forces in 1990 and subsequently recovered and reactivated in 1992.

AL TAHADDY   *1/1999, Maritime Photographic* / 0053294

1 SAWAHIL CLASS (AGH)

SAWAHIL (ex-*Sawahil 50*) B 50

Measurement, tonnes: 554 dwt
Dimensions, metres (feet): 55.4 × 9.6 × 2 *(181.8 × 31.5 × 6.6)*
Speed, knots: 9
Complement: 40
Machinery: 2 diesels; 2,400 hp(m) *(1.76 MW)*; 2 shafts
Guns: 2 — 12.7 mm MGs.
Radars: Navigation: Racal Decca; I-band.

Comment: This is a Sawahil class oil rig replenishment and accommodation ship which was built in South Korea in 1986 and taken on by the Coast Guard in 1990. She escaped to Bahrain during the Iraqi invasion, and is back in service. High-level helicopter platform aft. Used as a utility transport. Refitted in 1996/97. A similar vessel, *Al Dorrar*, is operated by the navy.

SAFFAR   *5/2001* / 0525907

2 AL TAHADDY CLASS (LCU)

Name	No	Builders	Commissioned
AL SOUMOOD	L 401	Singapore SB and Marine	Jul 1994
AL TAHADDY	L 402	Singapore SB and Marine	Jul 1994

Displacement, tonnes: 218 full load
Dimensions, metres (feet): 43 × 10 × 1.9 *(141.1 × 32.8 × 6.2)*
Speed, knots: 13
Complement: 12
Military lift: 80 tons
Machinery: 2 MTU diesels; 2 shafts
Radars: Navigation: Racal Decca; I-band.

Comment: Ordered in 1993 and launched on 15 April 1994. Multipurpose supply ships with cargo tanks for fuel, fresh water, refrigerated stores and containers on the main deck. Has 3 ton crane. Capable of beaching. Used by the Coast Guard.

SAWAHIL CLASS   *11/1997, Kuwait Navy* / 0012721

Latvia
LATVIJAS JURAS SPEKI

Country Overview

The Republic of Latvia regained independence in 1991 after 51 years as a Soviet republic. Situated in northeastern Europe, the country has an area of 24,938 square miles and borders to the north with Estonia, east with Russia and to the south with Belarus and Lithuania. It has a 286 n mile coastline with the Baltic Sea. Riga is the capital, largest city and principal port. Territorial seas (12 n miles) are claimed but while it has claimed a 200 n mile Exclusive Economic Zone (EEZ), its limits have not been fully defined by boundary agreements.

Headquarters Appointments

Commander of the Navy:
   Captain Rimant Strimaitis

Bases

Liepaja, Riga

Personnel

2014:   600 Navy (including Coast Guard)

Coastal Surveillance

Work began in 2002 on a maritime sea surveillance system which includes 13 fixed and two mobile surveillance stations. Each station is equipped with radar and day and infra-red cameras. The system is to become operational in June 2011. The Latvian AIS (Automatic Indentification System) was commissioned in 2005 and is part of the HELCOM network that links other Baltic and Scandinavian navies. The Maritime Search and Rescue Coordination Centre (MRCC) is based at Riga.

Coast Guard

These ships have a diagonal thick white and thin white line on the hull, and have KA numbers. They operate as part of the Navy.

PATROL FORCES

5 SKRUNDA CLASS (PB)

Name	No	Builders	Launched	Commissioned
SKRUNDA	P 05	Abeking and Rasmussen	20 Jan 2011	18 Apr 2011
CĒSIS	P 06	Abeking and Rasmussen	23 Nov 2011	2 Apr 2012
VIESĪTE	P 07	Riga Shipyard	11 Apr 2012	22 Aug 2012
JELGAVA	P 08	Abeking and Rasmussen	11 Apr 2013	10 Jul 2013
RĒZEKNE	P 09	Riga Shipyard	14 Oct 2013	20 Mar 2014

Displacement, tonnes: 127 full load
Dimensions, metres (feet): 25.71 × 13.0 × 2.7 *(84.4 × 42.7 × 8.9)*
Speed, knots: 20
Range, n miles: 1,000 at 12 kt
Complement: 10 (2 officers)
Machinery: Diesel-electric: 2 MAN D 2842 diesels; 2,170 hp *(1.62 MW)*; 2 motors; 2 shafts; cp props
Guns: 2 — 12.7 mm MGs.
Radars: Navigation: Raytheon; I-band.

Comment: Contract signed with Abeking & Rasmussen, Lemwerder, on 23 June 2008 for the construction of five SWATH patrol vessels. The ships are being constructed at Lemwerder and delivered in co-operation with Riga Shipyard. The vessels are derived from the design of SWATH pilot boats which have been in operation since 1999. The roles of the ships are to be patrol and surveillance of territorial waters and EEZ and feature a modular mission bay in the forward section. Options include a 35 mm gun and an SAR module. Delivery of the first-of-class was made in April 2011, the next two in 2012, the fourth in 2013 and fifth in 2014.

JELGAVA   *6/2013*, Michael Nitz* / 1525531

IHS Jane's Fighting Ships 2014-2015   © 2014 IHS

MINE WARFARE FORCES

5 ALKMAAR (TRIPARTITE) CLASS (MINEHUNTERS) (MHC)

Name	No	Laid down	Launched	Commissioned
IMANTA (ex-*Harlingen*)	M 04 (ex-*M 854*)	30 Nov 1981	9 Jul 1983	12 Apr 1984
VIESTURS (ex-*Scheveningen*)	M 05 (ex-*M 855*)	24 May 1982	2 Dec 1983	18 Jul 1984
TALIVALDIS (ex-*Dordrecht*)	M 06 (ex-*M 852*)	5 Jan 1981	26 Feb 1983	16 Nov 1983
VISVALDIS (ex-*Delfzijl*)	M 07 (ex-*M 851*)	30 Jan 1979	18 May 1982	28 May 1983
RŪSIŅŠ (ex-*Alkmaar*)	M 08 (ex-*M 850*)	30 Jan 1979	18 May 1982	28 May 1983

Displacement, tonnes: 571 standard; 605 full load
Dimensions, metres (feet): 51.5 × 8.9 × 3.74 *(169 × 29.2 × 12.3)*
Speed, knots: 12; 7 electric
Range, n miles: 3,000 at 12 kt
Complement: 42 (est.)

Machinery: 1 Stork Wärtsilä A-RUB 215X-12 diesel; 1,860 hp(m) *(1.35 MW)* sustained; 1 shaft; LIPS cp props; 2 active rudders; 2 motors; 240 hp(m) *(179 kW)*; 2 bow thrusters
Guns: 1 Giat 20 mm. 2—12.7 mm MGs.
Physical countermeasures: MCM: 2 PAP 104 remote-controlled submersibles.
Radars: Navigation: Furuno FR-2115, Consilium Selesmar MM 950; I-band.
Sonars: Thomson Sintra DUBM 21A; hull-mounted; minehunting; 100 kHz (±10 kHz).
Combat data systems: Signaal Sewaco IX. SATCOM.

Programmes: Originally procured for the Royal Netherlands Navy, these ships were part of the Netherlands commitment to a tripartite co-operative plan between Netherlands, Belgium and France for GRP hulled minehunters. All ships built by van der Giessen-de Noord. Ex-*Alkmaar*, *Delfzijl* and *Dordrecht* were withdrawn from RNLN service in 2000 and *Harlingen* and *Scheveningen* in 2003. *Imanta* was handed over on 6 March 2007, *Viesturs* on 5 September 2007, *Talivaldis* in January 2008 and *Visvaldis* in October 2008. The fifth vessel *Rūsiņš* (ex-*Alkmaar*) entered Latvian service in 2011.
Modernisation: The ships were overhauled before entering Latvian service and a mid-life upgrade may also be considered.
Structure: A 5 ton container can be shipped, stored for varying tasks-research; patrol; extended diving; drone control.
Operational: Endurance, 15 days. Automatic radar navigation system. Automatic data processing and display. EVEC 20. Decca Hi-fix positioning system. Alcatel dynamic positioning system.

VISVALDIS *6/2011, Frank Findler / 1406432*

VIESTURS *9/2013*, E & M Laursen / 1525530*

AUXILIARIES

1 VIDAR CLASS (MCCS/AG)

Name	No	Builders	Launched	Commissioned
VIRSAITIS (ex-*Vale*)	A 53 (ex-*N 53*)	Mjellem & Karlsen AS, Bergen	5 Aug 1977	10 Feb 1978

Displacement, tonnes: 1,524 standard; 1,700 full load
Dimensions, metres (feet): 64.8 × 12 × 4.6 *(212.6 × 39.4 × 15.1)*
Speed, knots: 14
Complement: 50

Machinery: 2 Wichmann 7AX diesels; 4,200 hp(m) *(3.1 MW)*; 2 shafts; auxiliary motor; 425 hp(m) *(312 kW)*; bow thruster
Guns: 1 Bofors 40 mm/70; 300 rds/min to 12 km *(6.6 n miles)*; weight of shell 0.96 kg.
Radars: Surface search: Kongsberg DataBridge 2000; E/F/I-bands.
Sonars: Simrad; hull-mounted; search and attack; medium/high frequency.
Weapon control systems: TVT optronic director.

Programmes: Decommissioned from Norwegian Navy in 2001 and transferred to Latvia on 27 January 2003.
Operational: Former minelayer modified to undertake mine countermeasures command and support roles. Additional tasks are likely to include training and support of diving operations.

VIRSAITIS *6/2013*, Michael Nitz / 1525532*

1 LOGISTICS VESSEL (AKS/AXL)

Name	No	Builders	Commissioned
VARONIS (ex-*Buyskes*)	A 90 (ex-*A 904*)	Boele's Scheepswerven	9 Mar 1973

Displacement, tonnes: 983 standard; 1,050 full load
Dimensions, metres (feet): 60 × 11.1 × 3.7 *(196.9 × 36.4 × 12.1)*
Speed, knots: 13.5
Range, n miles: 3,000 at 11.5 kt
Complement: 43 (6 officers)
Machinery: Diesel-electric; 3 Paxman 12 RPH diesel generators; 2,100 hp *(1.57 MW)*; 1 motor; 1,400 hp(m) *(1.03 MW)*; 1 shaft
Radars: Navigation: JRC 9922; E/F/I-bands.
Sonars: Side scanning and wreck search.

Comment: Originally designed and operated as a hydrographic vessel by the Royal Netherlands Navy from which she was decommissioned in 2003. Donated to Latvia on 8 November 2004 for use as a logistic and training vessel. Hydrographic launches were not transferred and the ship is fitted with an inflatable boat.

VARONIS *6/2013*, Frank Findler / 1525507*

COAST GUARD

5 KBV 236 CLASS (WPB)

KRISTAPS KA 01 (ex-*KBV 244*) **AUSMA** KA 07 (ex-*KBV 260*) **KLINTS** KA 09 (ex-*KBV 250*)
GAISMA KA 06 (ex-*KBV 249*) **SAULE** KA 08 (ex-*KBV 256*)

Displacement, tonnes: 17 full load
Dimensions, metres (feet): 19.2 × 4 × 1.3 *(63 × 13.1 × 4.3)*
Speed, knots: 20
Complement: 3 (1 officer)
Machinery: 2 Volvo Penta TMD 100C diesels; 526 hp(m) *(387 kW)*; 2 shafts
Radars: Navigation: Furuno; I-band.

Comment: Former Swedish Coast Guard vessel built in 1964. First one recommissioned 5 March 1993, second pair 9 November 1993 and last pair 27 April 1994.

SAULE *6/2006, Latvian Navy / 1164481*

496 Latvia > Coast guard — State border security service

1 PATROL CRAFT (WPB)

ASTRA KA 14

Displacement, tonnes: 22 full load
Dimensions, metres (feet): 22.8 × 5.6 × 1.1 *(74.8 × 18.4 × 3.6)*
Speed, knots: 25
Range, n miles: 575 at 25 kt
Complement: 4 (1 officer)
Machinery: 3 Scania D91 1467M diesels; 1,850 hp *(1.38 MW)*
Radars: Navigation: Furuno; I-band.

Comment: Built in Finland in 1996. Commissioned on 12 March 2001.

ASTRA   *4/2007, E & M Laursen / 1166813*

LAND-BASED MARITIME AIRCRAFT

Numbers/Type: 2 Mi-8 MTV1 Hip H.
Operational speed: 124 kt *(230 km/h)*.
Service ceiling: 16,400 ft *(5,000 m)*.
Range: 324 n miles *(600 km)*.
Role/Weapon systems: SAR aircraft acquired in 1999. Operated by the Air Force.

Mi-8   *4/2007, Freddy Philips / 1166812*

STATE BORDER SECURITY SERVICE

Notes: (1) There are five Finnish-built Boomeranger A-3500 10 m RIBs with pennant numbers RK 15–19.
(2) There is a small hovercraft RK 14.

RK 14   *6/2013*, Latvian State Security Service / 1525529*

RK 15   *8/2013*, C D Yaylali / 1525534*

---

1 LOKKI CLASS (PB)

TIIRA RK 03

Displacement, tonnes: 77 full load
Dimensions, metres (feet): 26.8 × 5.5 × 1.9 *(87.9 × 18.0 × 6.2)*
Speed, knots: 25
Complement: 6
Machinery: 2 MTU 8V 396 TB82 diesels; 1,740 hp(m) *(1.28 MW)* sustained 2 MTU 8V 396 TB84 diesels; 2,100 hp(m) *(1.54 MW)* sustained; 2 shafts

Comment: Donated by Finland in 2001.

TIIRA   *6/2012, Latvian Navy / 1483524*

1 BALTIC 24 PATROL CRAFT (WPB)

RĀNDA RK 20

Displacement, tonnes: 51 full load
Dimensions, metres (feet): 23.7 × 5.65 × 1.35 *(77.8 × 18.5 × 4.4)*
Speed, knots: 28
Range, n miles: 500 at 15 kt
Complement: 6
Machinery: 2 MTU 10V2000 M 92 diesels; 2,722 hp *(2.03 MW)*; 2 shafts
Guns: 1 — 12.7 mm MG (fitted for).
Radars: Surface search/Navigation: JMA 5300; I-band.

Comment: Contract with Baltic Workboats A/S, Estonia in 2007 for the construction of one 24 m workboat. Of aluminium construction, the craft is equipped with an aft launch/recovery ramp for a 5 m RHIB. Delivered in 2008.

RĀNDA   *6/2012, Martin Mokrus / 1483614*

4 BALTIC 101 INSHORE PATROL CRAFT (PBI)

RK 21–24

Displacement, tonnes: 5.6 full load
Dimensions, metres (feet): 10.8 × 3.56 × 0.47 *(35.4 × 11.7 × 1.5)*
Speed, knots: 45
Range, n miles: 300 at 15 kt
Complement: 6
Machinery: 2 Volvo Penta D6 diesels; 2 waterjets

Comment: Built by Baltic Workboats A/S, Estonia. Aluminium construction. Delivered in 2010.

RK 22   *6/2012, Baltic Workboats / 1483523*

IHS Jane's Fighting Ships 2014-2015   © 2014 IHS

1 VALPAS CLASS (OFFSHORE PATROL VESSEL) (WPBO)

VALPAS RK 12

Displacement, tonnes: 554 full load
Dimensions, metres (feet): 48.5 × 8.5 × 3.8 *(159.1 × 27.9 × 12.5)*
Speed, knots: 15
Complement: 18
Machinery: 1 Werkspoor diesel; 2,000 hp(m) *(1.47 MW)*; 1 shaft; cp prop
Guns: 1 Oerlikon 20 mm.
Sonars: Simrad SS105; active scanning; 14 kHz.

Comment: An improvement on the *Silmä* design. Built by Laivateollisuus, Turku, and commissioned 21 July 1971. Ice strengthened. Donated by Finland on 25 September 2002.

VALPAS 6/2013*, Michael Nitz / 1525533

Lebanon

Country Overview

The Lebanese Republic gained independence from France in 1946 but was devastated by civil war between 1975–1991. Situated on the eastern shore of the Mediterranean Sea, it has an area of 4,015 square miles and is bordered to the north and east by Syria and to the south by Israel. It has a 121 n mile coastline with the Mediterranean Sea. The capital, largest city and principal port is Beirut. Other important ports include Tripoli and Sidon. Territorial seas (12 n miles) are claimed but an EEZ is not claimed.

Headquarters Appointments

Navy Commander: Rear Admiral Nazih Baroudi

Personnel

2014: 1,100 (395 officers)

Bases

Beirut (HQ), Jounieh

PATROL FORCES

Notes: (1) There is a patrol craft, *Emanuela*, of unknown type with pennant number 501.
(2) Ten interceptor craft were donated by the UAE in 2008. There are four 14.5 m diesel-engined craft (1401–1404) and six 12.5 m petrol-engined craft.
(3) There is a patrol craft of unknown type with pennant number 1406.

P 501 11/2008, M Declerck / 1305866

INTERCEPT CRAFT 11/2008, M Declerck / 1305867

25 INSHORE PATROL CRAFT (PBR)

| 401 | 403–418 | 420–427 |

Displacement, tonnes: 6 full load
Dimensions, metres (feet): 8.2 × 2.5 × 0.6 *(26.9 × 8.2 × 2.0)*
Speed, knots: 22
Range, n miles: 154 at 22 kt
Complement: 4
Machinery: 2 Sabre 212 diesels; 212 hp(m) *(156 kW)*; 2 waterjets
Guns: 3—5.56 mm MGs.

Comment: M-boot type used by the US Army on German rivers and 27 were transferred in January 1994. Called Combat Support Boats, there are 20 operational and five laid up. Two were decommissioned in 2002.

404 11/2008, M Declerck / 1305869

7 TRACKER MK 2 CLASS (COASTAL PATROL CRAFT) (PB)

TRIPOLI (ex-*Attacker*) 301
JOUNIEH (ex-*Fencer*) 302
BATROUN (ex-*Safeguard*) 303
BYBLOS (ex-*Chaser*) 304
BEIRUT (ex-*HunterII*) 305
SIDON (ex-*Striker*) 306
SARAFAND (ex-*Swift*) 307

Displacement, tonnes: 39 full load
Dimensions, metres (feet): 20 × 5.2 × 1.5 *(65.6 × 17.1 × 4.9)*
Speed, knots: 21
Range, n miles: 650 at 14 kt
Complement: 13 (1 officer)
Machinery: 2 Detroit 12V-71TA diesels; 840 hp *(616 kW)* sustained; 2 shafts
Guns: 3—12.7 mm MGs.
Radars: Surface search: Racal Decca 1216; I-band.

Comment: All built at Cowes and Southampton. GRP construction. The ex-Royal Naval Units were originally commissioned in March 1983. Three transferred from UK on 17 July 1992 after serving as patrol craft for British bases in Cyprus. The other two were acquired in 1993. The two ex-UK Customs Craft (*Batroun* and *Sarafand*) were originally commissioned in 1979 and acquired in late 1993.

JOUNIEH 7/2006, Marco Ghiglino / 1164961

1 PATROL SHIP (PB)

AAMCHIT (ex-*Bremen 2*) 41

Dimensions, metres (feet): 34 × 5.2 × 1.8 *(111.5 × 17.1 × 5.9)*
Speed, knots: 28
Machinery: 2 diesels; 3,900 hp *(2.9 MW)*; 2 shafts

Comment: Former City of Bremen Maritime Police vessel built by Schiffswerft Ernst Menzer, Hamburg-Bergedorf in 1974 and transferred to Lebanon on 7 June 2007. The contract includes a training package.

AAMCHIT 11/2008, M Declerck / 1305870

498 Lebanon > Patrol forces — Libya > Introduction

1 PATROL SHIP (PB)

NAKOURA (ex-*Bremen 9*) 308

Displacement, tonnes: 33 full load
Dimensions, metres (feet): 20 × 5.6 × 1.45 *(65.6 × 18.4 × 4.8)*
Speed, knots: 32. Range, n miles: 300 at 25 kt
Complement: 6
Machinery: 2 MTU 12V 183TE92 diesels; 1,970 hp *(1.5 MW)*; 2 shafts

Comment: Former City of Bremen Maritime Police vessel built by Fassmer Werft in 1992 and transferred to Lebanon on 7 June 2007. Aluminium construction. The contract includes a training package. The design includes space for a 3 m interceptor craft.

NAKOURA 11/2008, M Declerck / 1305871

1 PATROL SHIP (PB)

Name	No	Builders	Commissioned
TABARJA (ex-*Bergen*)	42 (ex-Y 838)	Lürssen, Bremen-Vegesack	19 May 1994

Displacement, tonnes: 128 full load
Dimensions, metres (feet): 27.8 × 6.0 × 1.4 *(91.2 × 19.7 × 4.6)*
Speed, knots: 16
Complement: 15
Machinery: 2 KHD TBD 234 diesels; 2,054 hp *(1.51 MW)*; 2 shafts

Comment: Former German Range Safety Craft donated by the German Navy on 17 June 2008.

TABARJA 11/2008, M Declerck / 1305868

1 PATROL SHIP (PB)

Name	No	Builders	Commissioned
AL KALAMOUN (ex-*Avel Gwalarn*)	43 (ex-Y 838)	CN de l'Esterel, Cannes	1984

Displacement, tonnes: 68 full load
Dimensions, metres (feet): 30.35 × 5.8 × 1.83 *(99.6 × 19.0 × 6.0)*
Speed, knots: 28
Range, n miles: 1,200 at 15 kt
Machinery: 2 Poyaud UD 23 diesels; 2,640 hp *(1.97 MW)*; 2 shafts
Guns: 1 — 12.7 mm MG.

Comment: Former French Customs vessel transferred by the French government on 29 May 2009.

1 + (2) OFFSHORE PATROL VESSELS (PSO)

TRABLOUS +2

Displacement, tonnes: 265 full load
Dimensions, metres (feet): 43.5 × 8.5 × 2 *(142.7 × 27.9 × 6.6)*
Speed, knots: 25
Range, n miles: 2,600 at 11 kt
Complement: 6

Machinery: 2 MTU 4000M93 diesels; 2 Hamilton HT1000 waterjets
Guns: 1 Oto Melara 25 mm. 4 — 12.7 mm MGs.
Radars: To be announced.
Electro-optic systems: To be announced.

Comment: A contract was awarded to Maritime Security Strategies on 14 January 2011 for the construction of an offshore patrol vessel. The deal was executed under the US governments Foreign Military Sales program. The design is based on the RiverHawk Fast Sea Frames AMP 145 design. The construction involves a lightweight epoxy-resin hull, provided by Vectorworks Marine and an aluminium superstructure. The ship was completed at RiverHawk, Tampa. A 7 m RHIB can be deployed and recovered via a stern ramp. The ship was delivered on 26 November 2012. Two further ships may be ordered.

TRABLOUS 6/2012, RiverHawk / 1483525

AMPHIBIOUS FORCES

2 FRENCH EDIC 700 CLASS (LCT)

Name	No	Builders	Commissioned
SOUR	21	SFCN, Villeneuve la Garenne	28 Mar 1985
DAMOUR	22	SFCN, Villeneuve la Garenne	28 Mar 1985

Displacement, tonnes: 681 full load
Dimensions, metres (feet): 59 × 12 × 1.3 *(193.6 × 39.4 × 4.3)*
Speed, knots: 10
Range, n miles: 1,800 at 9 kt
Complement: 20 (2 officers)

Military lift: 96 troops; 11 trucks or 8 APCs
Machinery: 2 SACM MGO 175 V12 M1 diesels; 1,200 hp(m) *(882 kW)*; 2 shafts
Guns: 2 Oerlikon 20 mm. 2 — 12.7 mm MGs. 1 — 7.62 mm MG.
A/S Mortars: 1 — 81 mm mortar.
Radars: Navigation: Decca; I-band.

Comment: Both were damaged in early 1990 but repaired in 1991 and are fully operational. Used by the Marine Regiment formed in 1997.

DAMOUR 12/2008, M Declerck / 1305872

Libya

Country Overview

The Socialist People's Libyan Arab Jamahiriyah is situated in north Africa. With an area of 679,362 square miles, it has a 956 n mile coastline with the Mediterranean Sea and is bordered to the east by Egypt, to the south by Sudan, Chad and Niger and to the west by Algeria and Tunisia. The capital and largest city is Tripoli which, with Benghazi, is a principal port. Territorial seas (12 n miles) are claimed. An EEZ has not been claimed. The status of the Gulf of Sirte, which Libya claims as internal waters, is disputed by numerous states including USA, United Kingdom, France, Italy and Greece. Following the revolution, which culminated in the overthrow of the Ghadaffi regime, a new interim government was formed on 22 November 2011. The future size and shape of the armed forces in general and navy in particular are yet to be determined.

Headquarters Appointments

Chief of Staff Navy:
  Brigadier General Hassan Ali Bushnak

Personnel

(a) 2014: 6,000 officers and ratings, including Coast Guard
(b) Voluntary service

Bases

Naval HQ at Al Khums.
Operating Ports at Tripoli, Darnah (Derna) and Benghazi.
Naval bases at Al Khums and Tobruq.
Submarine base at Ras Hilal.
Naval air station at Al Girdabiyah.
Naval infantry battalion at Sidi Bilal.

Coast Defence

Batteries of truck-mounted SS-C-3 Styx missiles.

IHS Jane's Fighting Ships 2014-2015 © 2014 IHS

SUBMARINES

2 FOXTROT CLASS (PROJECT 641) (SS)

AL KHYBER 315   **AL HUNAIN** 316

Displacement, tonnes: 1,981 surfaced; 2,515 dived
Dimensions, metres (feet): 91.3 × 7.5 × 6 *(299.5 × 24.6 × 19.7)*
Speed, knots: 16 surfaced; 15 dived
Range, n miles: 20,000 at 8 kt surfaced; 380 at 2 kt dived
Complement: 75 (8 officers)

Machinery: Diesel-electric; 3 Type 37-D diesels (1 × 2,700 and 2 × 1,350); 6,000 hp(m) *(4.4 MW)*; 3 motors; 5,400 hp(m) *(3.97 MW)*; 3 shafts; 1 auxiliary motor; 140 hp(m) *(103 kW)*
Torpedoes: 10—21 in *(533 mm)* (6 bow, 4 stern) tubes. SAET-60; passive homing to 15 km *(8.1 n miles)* at 40 kt; warhead 400 kg, and SET-65E; active/passive homing to 15 km *(8.1 n miles)* at 40 kt; warhead 205 kg or Type 53–56. Total of 22 torpedoes.
Mines: 44 in place of torpedoes.
Electronic countermeasures: ESM: Stop Light; radar warning.
Radars: Surface search: Snoop Tray; I-band.
Sonars: Herkules; hull-mounted; active; medium frequency. Feniks; hull-mounted; passive.

Programmes: Six of the class originally transferred from USSR. Four (311–314) have since been decommissioned.

FOXTROT   *6/1992, van Ginderen Collection / 0081190*

Operational: Libyan crews trained in the USSR and much of the maintenance was done by Russian personnel. No routine patrols have been seen since 1984 although both boats have been reported to conduct surface patrols. One submarine was reported to be in dry dock at Tripoli during 2003 and *Al Khyber* reported to be sea-going. However a return to full operational capability remains highly unlikely, particularly in the wake of the revolutionary conflict.

FRIGATES

Notes: The *Dat Assawari* F 211 is a training hulk alongside in Tripoli.

1 KONI (PROJECT 1159) CLASS (FFGM)

AL HANI PF 212

Displacement, tonnes: 1,463 standard; 1,930 full load
Dimensions, metres (feet): 96.4 × 12.6 × 3.5 *(316.3 × 41.3 × 11.5)*
Speed, knots: 27; 22 diesel. **Range, n miles:** 1,800 at 14 kt
Complement: 120

Machinery: CODAG; 1 SGW, Nikolayev, M8B gas turbine (centre shaft); 18,000 hp(m) *(13.25 MW)* sustained; 2 Russki B-68 diesels; 15,820 hp(m) *(11.63 MW)* sustained; 3 shafts
Missiles: SSM: 4 Soviet SS-N-2C Styx (2 twin) launchers ❶; active radar/IR homing to 83 km *(45 n miles)* at 0.9 Mach; warhead 513 kg; sea-skimmer at end of run.
SAM: SA-N-4 Gecko twin launcher ❷; semi-active radar homing to 15 km *(8 n miles)* at 2.5 Mach; altitude 9.1–3,048 m *(29.5–10,000 ft)*; warhead 14.5 kg; 20 missiles.
Guns: 4 USSR 3 in *(76 mm)*/59 AK 726 (2 twin) ❸; 90 rds/min to 16 km *(8.5 n miles)* anti-surface; weight of shell 5.9 kg.
4 USSR 30 mm/65 (2 twin) automatic ❹; 500 rds/min to 2 km *(1.1 n miles)*; weight of shell 0.54 kg.
Torpedoes: 4—406 mm (2 twin) tubes amidships ❺. USET-95; active/passive homing to 10 km *(5.5 n miles)* at 30 kt; warhead 100 kg.
A/S Mortars: 1 RBU 6000 12-tubed trainable launcher ❻; automatic loading; range 6,000 m; warhead 31 kg.
Mines: Capacity for 20.
Depth charges: 2 racks.
Physical countermeasures: Decoys: 2—16-barrelled chaff launchers. Towed torpedo decoys.
Electronic countermeasures: ESM: 2 Watch Dog; radar warning.
Radars: Air search: Strut Curve ❼; F-band; range 110 km *(60 n miles)* for 2 m² target.
Surface search: Plank Shave ❽; E/F-band.
Navigation: Don 2; I-band.
Fire control: Drum Tilt ❾; H/I-band (for 30 mm). Hawk Screech ❿; I-band; range 27 km *(15 n miles)* (for 76 mm). Pop Group ⓫; F/H/I-band (for SAM).
IFF: High Pole B. Square Head.
Sonars: Hercules (MG 322); hull-mounted; active search and attack; medium frequency.

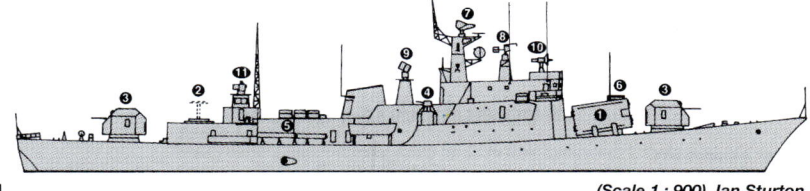

AL HANI   *(Scale 1 : 900), Ian Sturton / 0506050*

Programmes: Type III Konis built at Zelenodolsk and transferred from the Black Sea. 212 commissioned 28 June 1986 and 213 on 24 October 1987.
Structure: SSMs mounted either side of small deckhouse on forecastle behind gun. A deckhouse amidships contains air conditioning machinery. Changes to the standard Koni include SSM, four torpedo tubes, only one RBU 6000 and Plank Shave surface search and target indication radar. Camouflage paint applied in 1991.
Operational: One of the class fired an exercise Styx missile in September 1999. *Al Qirdabiyah* was badly damaged in the revolutionary war in 2011 and is reported to have sunk. *Al Hani* arrived at Cassar Ship Repair, Malta, in October 2013 for unspecified repairs.

AL HANI   *7/1999, van Ginderen Collection / 0081191*

504 **Lithuania** > State border security service — **Madagascar** > Introduction

1 KBV 101 CLASS (PB)

LILIAN 101 (ex-KBV 101)

Displacement, tonnes: 70 full load
Dimensions, metres (feet): 25 × 5 × 2 *(82 × 16.4 × 6.6)*
Speed, knots: 18
Range, n miles: 1,000 at 15 kt
Complement: 5
Machinery: 2 Cummins KTA38-M diesels; 2,120 hp(m) *(1.56 MW)*; 2 shafts
Radars: Navigation: Furuno FR 2010 and FCR 1411; I-band.

Comment: Built in Sweden in 1969. Transferred from Swedish Coast Guard on 24 June 1996. Used in Swedish service as a salvage diving vessel and had a high frequency active hull-mounted sonar.

2 CHRISTINA (GRIFFON 2000 TD) CLASS HOVERCRAFT (UCAC)

CHRISTINA   VIESULAS

Displacement, tonnes: 5 full load
Dimensions, metres (feet): 12.6 on cushion × ? *(41.3 × ?)*
Speed, knots: 35
Complement: 3
Machinery: 1 Deutz BF8L diesel; 355 hp *(265 kW)*
Radars: Furuno 1000C; I-band.

Comment: *Christina* built by Griffon UK and delivered in 2000. *Viesulas* built by Griffon in 1999 for Klaipeda Port Authority. Transferred to the Border Security Service in 2008 for SAR duties on Curonian Lagoon. Similar to crafts supplied to Estonia and Finland.

LILIAN   6/2003, Hartmut Ehlers / 0561503

CHRISTINA   6/2001, Lithuanian Navy / 0114364

Macedonia, Republic of

Country Overview

The Former Yugoslav Republic of Macedonia declared its independence in 1991. A land-locked country with an area of 9,928 square miles, it is situated in south-eastern Europe and is bordered to the north by Serbia, to the east by Bulgaria, to the south by Greece and to the west by Albania. Parts of the borders with Albania and Greece pass through the two principal lakes, Ohrid and Prespa. The capital and largest city is Skopje.

PATROL FORCES

Notes: The Macedonian Lake Service (Ezerska sluzba - EZ) consists of about 400 soldiers and is nominally an independent arm of the Army although in practice it is almost integrated with Land Forces. In addition up to five ex-Yugoslavian Army patrol boats on Lake Ohrid, there are two further small craft on Lake Prespa although their operational status is doubtful.

2 BOTICA CLASS (TYPE 16) (RIVER PATROL CRAFT) (PBR)

303   304

Displacement, tonnes: 23 full load
Dimensions, metres (feet): 17 × 3.6 × 0.8 *(55.8 × 11.8 × 2.6)*
Speed, knots: 15
Range, n miles: 340 at 14 kt
Complement: 7

Military lift: 3 tons or 30 troops
Machinery: 2 RHS 518A diesels; 464 hp *(340 kW)*; 2 shafts
Guns: 1 Oerlikon 20 mm. 2 — 7.62 mm MGs.
Radars: Decca 110; I-band.

Comment: Former Yugoslavian craft which entered service in the 1970s.

BOTICA 304
6/2007, Freivogel Collection
1167944

Madagascar

MALAGASY REPUBLIC MARINE

Country Overview

Formerly a French Protectorate, the Malagasy Republic became self-governing in 1958 and fully independent in 1960. It adopted the name Democratic Republic of Madagascar in 1975. Situated in the Indian Ocean and separated from the southeastern coast of Africa by the Mozambique Channel, it comprises Madagascar Island, the fourth largest island in the world, and several small islands. The country's total area is 226,658 square miles and it has a coastline of 2,608 n miles.

Antananarivo is the capital while Toamasina is the principal commercial port. There are further ports at Antsiranana, Mahajanga and Toliara. Territorial seas (12 n miles) are claimed. An Exclusive Economic Zone (EEZ) has been claimed but boundaries have not been agreed.

Headquarters Appointments

Head of Navy: Rear Admiral Ratsimitsetra

Personnel

2014: 430 officers and men (including Marine Company of 120 men)

Bases

Antsiranana (main), Toamasina, Mahajanga, Toliara, Nosy Bé, Tolanoro, Manakara.

IHS Jane's Fighting Ships 2014-2015   © 2014 IHS

PATROL FORCES

6 PATROL CRAFT (PB)

V 11–16

Displacement, tonnes: 18 full load
Dimensions, metres (feet): 13.4 × 3.8 × 1.2 *(44 × 12.5 × 3.9)*
Speed, knots: 13
Range, n miles: 200 at 11 kt
Complement: 3
Machinery: 2 General Motors Detroit 6V53 diesels; 2 shafts
Radars: Furuno; I-band.

Comment: Former US Coast Guard lifeboats (MLB) constructed in the 1960s. Formally donated on 12 February 2003 for use as coastal surveillance and SAR vessels. All six craft refitted at Galveston, Texas, before transfer and a further unit was transferred as spares.

AUXILIARIES

Notes: (1) There are two Aigrette class harbour tugs, *Engoulevent* and *Martin-Pêcheur* were acquired from France in May 1996.
(2) There is also a 400 ton coastal tug *Trozona*.
(3) A former trawler, *Daikannon Maru*, is employed on fishery protection duties.

MLBs (Seychelles colours)   9/2003, Seychelles Coast Guard / 0568334

Malawi

Country Overview

Formerly the British Protectorate of Nyasaland, the Republic of Malawi gained independence in 1964. A landlocked country situated in east Central Africa, it is bordered to the north by Tanzania, to the west by Zambia and to the south and east by Mozambique. The country's total area is 45,747 square miles, nearly a quarter of which is water. The principal lake is Lake Malawi (formerly Lake Nyasa), with which there is a shoreline of some 475 n miles. The largest city is Blantyre and the capital, since 1975, is Lilongwe. The naval base at Monkey Bay is situated on a peninsula at the south of the lake.

Headquarters Appointments

Commander of the Malawi Army Marine Unit:
To be announced

Bases
Monkey Bay, Lake Malawi

Personnel
2014: 225

PATROL FORCES

Notes: One survey craft built in France in 1988 is operated on Lake Malawi by Department of Surveys.

1 ANTARES CLASS (PB)

KASUNGU (ex-*Chikala*) P 703

Displacement, tonnes: 42 full load
Dimensions, metres (feet): 21 × 4.9 × 1.5 *(68.9 × 16.1 × 4.9)*
Speed, knots: 22
Range, n miles: 650 at 15 kt
Complement: 16
Machinery: 2 Poyaud 520 V12 M2 diesels; 1,300 hp(m) *(956 kW)*; 2 shafts
Guns: 1 MG 21 20 mm. 2—7.62 mm MGs.
Radars: Surface search: Decca; I-band.

Comment: Built in prefabricated sections by SFCN Villeneuve-la-Garenne and shipped to Malawi for assembly on 17 December 1984. Commissioned May 1985. Operational status doubtful.

KASUNGU   6/1996, Malawi Navy / 0012737

2 NAMACURRA CLASS (PB)

KANING'A (ex-*Y 1520*) P 704   +1

Displacement, tonnes: 5 full load
Dimensions, metres (feet): 9 × 2.7 × 0.8 *(29.5 × 8.9 × 2.6)*
Speed, knots: 32
Range, n miles: 180 at 20 kt
Complement: 4
Machinery: 2 BMW 3.3 outboards; 380 hp(m) *(279 kW)*
Guns: 1—12.7 mm MG. 2—7.62 mm MGs.
Radars: Surface search: Decca; I-band.

Comment: First craft donated by South Africa on 29 October 1988. A second was donated in February 2008.

KANING'A   6/1997, Malawi Navy / 0012736

1 ROTORK CLASS (LCU)

CHIKOKO I L 702

Displacement, tonnes: 9 full load
Dimensions, metres (feet): 12.7 × 3.2 × 0.5 *(41.7 × 10.5 × 1.6)*
Speed, knots: 24
Range, n miles: 3,000 at 15 kt
Complement: 8
Machinery: 2 Volvo diesels; 260 hp(m) *(191 kW)*; 2 shafts
Guns: 3—7.62 mm MGs.

Comment: Built by Rotork Marine. Needs a refit but no funds are available.

CHIKOKO I   6/1996, Malawi Navy / 0012738

7 NAUTIC GUARDIAN BR850 INTERCEPTOR CRAFT (PBF)

Displacement, tonnes: 3.8 full load
Dimensions, metres (feet): 8.5 × 2.8 × 0.6 *(27.9 × 9.2 × 2.0)*
Speed, knots: 42
Range, n miles: 378 at 2 kt
Complement: 2 + 6 embarked forces
Machinery: 1 diesel; 500 hp *(373 kW)*; 1 shaft
Guns: 1—12.7 mm MG.

Comment: It was announced on 12 November 2013 that seven Guardian interceptor craft had entered service. They are based at Monkey Bay and are to be used for patrol, SAR, and disaster relief operations on Lake Malawi.

Malaysia
TENTERA LAUT DIRAJA

Country Overview

The Federation of Malaysia was formed in 1963. Situated in south-east Asia, its two regions are separated by some 350 n miles of the South China Sea. Peninsular Malaysia (formerly West Malaysia) is bordered to the north by Thailand and to the south by Singapore (which left the federation in 1965) and includes 11 states occupying the southern half of the Malay Peninsula. To the east, the states (former British colonies) of Sabah and Sarawak (which surrounds the sultanate of Brunei) occupy the northern third of the island of Borneo, the remainder of which forms the Indonesian province of Kalimantan. With an overall land area of 127,320 square miles, Malaysia has a coastline of 2,425 n miles with the Strait of Malacca, the South China Sea, the Sulu and Celebes Seas. Kuala Lumpur is the capital and largest city while the principal ports are Penang, Port Klang, Tanjung Pelepas, Kuantan, Kota Kinabalu and Kuching. Territorial seas (12 n miles) are claimed. An EEZ (200 n miles) is claimed but the limits have not been fully defined.

Headquarters Appointments

Chief of Navy: Admiral Tan Sri Abdul Aziz Bin Haji Jaafar
Deputy Chief of Navy: Vice Admiral Ahmad Bin Haji Ahmad Kamarulzaman Badaruddin
Fleet Commander: Vice Admiral Dato' Abd Hadi A Rashid
Commander Naval Area I (Kuantan):
    Rear Admiral Dato' Azhari Bid Abdul Rashid
Commander Naval Area II (Kota Kinabalu):
    Rear Admiral Dato' Pahlawan Mohammad Rosland bin Omar
Commander Naval Area III (Langkawi): Commodore Muhamad Ridzwan Bin Abdul Rahman

Personnel

(a) 2014: 17,918 (2,436 officers)
(b) Voluntary service: Royal Malaysian Navy Voluntary Reserve (RMNVR): Total, 2,697 (1,211 officers)

Coastal Defence

Procurement of a coastal surveillance system is under consideration.

Bases

1. Lumut: HQ Fleet Operations, HQ Fleet System, HQ Support, HQ Air, RMN Mine Warfare and Diving Centre and HQ Special Forces.
2. Kuantan: HQ Commander Naval Area 1 (west of longitude 109E).
3. Kota Kinabalu: HQ Commander Naval Area 2 (east of longitude 109E) and HQ submarine force.
4. Langkawi: HQ Commander Naval Area 3 (Northern Malacca Strait).
5. Pulau Indah: National Hydrographic Centre and KD *Sultan Abdul Aziz Shah*.
6. Tanjung Pengelih: KD *Sultan Ismael* (Recruitment Training Centre).
7. Sandakan: KD *Sri Sandakan*.
8. Semporna: KD *Sri Semporna*.
9. Tawau: KD *Sri Tawau*.
10. Bintulu: A new base and HQ COMNAV IV is to be established.

Prefix to Ships' Names

The names of Malaysian warships are prefixed by KD (Kapal DiRaja meaning His Majesty's Ship).

Maritime Patrol Craft

There are large numbers of armed patrol craft belonging to the Police, Customs and Fisheries Departments. Details at the end of the section.

Strength of the Fleet

Type	Active	Building (Planned)
Submarines	2	—
Frigates	2	(2)
Corvettes	12	6
Fast Attack Craft—Missile	8	—
Fast Attack Craft—Gun	6	—
Patrol Craft	19	—
Minehunters	4	—
Survey Ships	2	1
Training Ships	4	—
Logistic Support	4	(3)
Salvage Vessels	1	—

PENNANT LIST

Submarines

| — | Tunku Abdul Rahman |
| — | Tun Razak |

Frigates

| 29 | Jebat |
| 30 | Lekiu |

Corvettes

25	Kasturi
26	Lekir
134	Laksamana Hang Nadim
135	Laksamana Tun Abdul Jamil
136	Laksamana Muhammad Amin
137	Laksamana Tan Pusmah
171	Kedah
172	Pahang
173	Perak
174	Terengganu
175	Kelantan
176	Selangor

Patrol Forces

47	Sri Perlis
49	Sri Johor
3501	Perdana
3502	Serang
3503	Ganas
3504	Ganyang
3505	Jerong
3506	Todak
3507	Paus
3508	Yu
3509	Baung
3510	Pari
3511	Handalan
3512	Perkasa
3513	Pendekar
3514	Gempita

Mine Warfare Forces

11	Mahamiru
12	Jerai
13	Ledang
14	Kinabalu

Amphibious Forces

331	Sri Gaya
322	Sri Tiga
1503	Sri Indera Sakti
1504	Mahawangsa

Training Ships

76	Hang Tuah
A 13	Tunas Samudera
271	Gagah Samudera
272	Teguh Samudera

Survey Ships

| 151 | Perantau |
| 255 | Mutiara |

Auxiliaries

4	Penyu
5	Bunga Mas Lima
6	Bunga Mas Enam
8	Kepah

SUBMARINES

Notes: The French Agosta-class submarine *Ouessant* was on loan to Malaysia to provide initial submarine training which started in early 2005. The boat was formally transferred to Malaysia on 23 September 2011 and is to become a museum at Klebang.

TUN RAZAK

12/2009, Diego Quevedo / 1365905

Submarines — **Malaysia** 507

2 + (3) SCORPENE CLASS (SSK)

Name	No	Builders	Laid down	Launched	Commissioned
TUNKU ABDUL RAHMAN	–	DCN, Cherbourg	25 Apr 2004	23 Oct 2007	27 Jan 2009
TUN RAZAK	–	Navantia, Cartagena	25 Apr 2005	8 Oct 2008	25 Oct 2009

Displacement, tonnes: 1,586 surfaced; 1,755 dived
Dimensions, metres (feet): 67.56 × 6.2 × 5.4
 (221.7 × 20.3 × 17.7)
Speed, knots: 12 surfaced; 20.5 dived
Range, n miles: 6,500 at 8 kt surfaced; 550 at 4 kt dived
Complement: 32 (7 officers)

Machinery: Diesel electric; 2 SEMT-Pielstick 12 PA4 200 SM DS diesels; 1 Jeumont Industrie motor; 4,290 hp *(3.2 MW)*; 1 shaft
Missiles: SSM: Aerospatiale SM39 Exocet; launched from 21 in *(533 mm)* torpedo tubes; inertial cruise; active radar homing to 50 km *(27 n miles)* at 0.9 Mach; warhead 165 kg.
Torpedoes: 6—21 in *(533 mm)* tubes. WASS Black Shark torpedoes; wire (fibre-optic cable) guided; active and passive homing to 50 km *(27 n miles)* at 50 kt, warhead 250 kg. Total of 18 weapons.
Electronic countermeasures: ESM: Thales DR 3000; intercept.
Radars: Navigation: I-band.
Sonars: Hull mounted; active/passive search and attack, medium frequency.
Weapon control systems: UDS International SUBTICS.

Programmes: Contract for the construction of two submarines awarded to Armaris and IZAR on 5 June 2002. A four-year training programme aboard an Agosta-70 (ex-*Ouessant*) was included in the package. The two forward modules were constructed by DCN and the two aft modules by Navantia. Up to three more submarines are reported to be under consideration.
Structure: Similar in design to the Chilean boats. Diving depth more than 300 m *(984 ft)*. Option to retrofit AIP at a later date.
Operational: Following sea trials, the first boat arrived in Malaysia on 3 September 2009 and the second on 2 July 2010. Based at Kota Kinabalu Naval Base, Sabah.

TUN RAZAK 12/2011, Chris Sattler / 1406984

TUN RAZAK 2/2009, Diego Quevedo / 1305931

TUNKU ABDUL RAHMAN 3/2013*, Chris Sattler / 1486526

© 2014 IHS IHS Jane's Fighting Ships 2014-2015

FRIGATES

Notes: Plans to build two new frigates was announced on 17 July 2006. However, the project was suspended in 2009 and is unlikely to be resurrected before 2016.

2 LEKIU CLASS

Name	No	Builders	Laid down	Launched	Commissioned
JEBAT	29	Yarrow Shipbuilders, Glasgow	Nov 1994	27 May 1995	20 Nov 1999
LEKIU	30	Yarrow Shipbuilders, Glasgow	Mar 1994	3 Dec 1994	9 Oct 1999

Displacement, tonnes: 1,875 standard; 2,390 full load
Dimensions, metres (feet): 105.5 oa; 97.5 wl × 12.8 × 3.6 *(346.1; 319.9 × 42.0 × 11.8)*
Speed, knots: 28
Range, n miles: 5,000 at 14 kt
Complement: 146 (18 officers)

Machinery: CODAD; 4 MTU 20V 1163 TB93 diesels; 33,300 hp(m) *(24.5 MW)* sustained; 2 shafts; Kamewa cp props
Missiles: SSM: 8 Aerospatiale MM 40 Exocet Block II ❶; inertial cruise; active radar homing to 70 km *(40 n miles)* at 0.9 Mach; warhead 165 kg; sea-skimmer.
SAM: British Aerospace VLS Seawolf; 16 launchers ❷; command line of sight (CLOS) radar/TV tracking to 6 km *(3.3 n miles)* at 2.5 Mach; warhead 14 kg.
Guns: 1 Bofors 57 mm/70 SAK Mk 2 ❸; 220 rds/min to 17 km *(9.3 n miles)*; weight of shell 2.4 kg. 2 MSI 30 mm/75 DS 30B ❹; 650 rds/min to 10 km *(5.4 n miles)*; weight of shell 0.36 kg.
Torpedoes: 6 Whitehead B 515 324 mm (2 triple) tubes ❺; anti-submarine; Marconi Stingray; active/passive homing to 11 km *(5.9 n miles)* at 45 kt; warhead 35 kg (shaped charge).
Physical countermeasures: Decoys: 2 Super Barricade 12-barrelled launchers for chaff ❻; Graseby Sea Siren torpedo decoy.
Electronic countermeasures: ESM: AEG Telefunken/Marconi Mentor; intercept.
Radars: Air search: Signaal DA08 ❽; E/F-band.
Surface search: Ericsson Sea Giraffe 150HC ❾; G/H-band.
Navigation: Racal Decca; I-band.
Fire control: 2 Marconi 1802 ❿; I/J-band.
Sonars: Thomson Sintra Spherion; hull-mounted active search and attack; medium frequency.
Combat data systems: GEC-Marconi NAUTIS-F; Signaal Link Y Mk 2.
Electro-optic systems: Radamec 2400 Optronic director ❼. Thomson-CSF ITL 70 (for Exocet); GEC-Marconi Type V 3901 thermal imager.
Helicopters: 1 Westland Super Lynx ⓫.

Programmes: GEC Naval Systems Frigate 2000 design with a modern combat data system and automated machinery control.
Operational: Delivery dates were delayed by weapon system integration problems but both arrived in Malaysia in early 2000. Form 23rd Frigate Squadron.

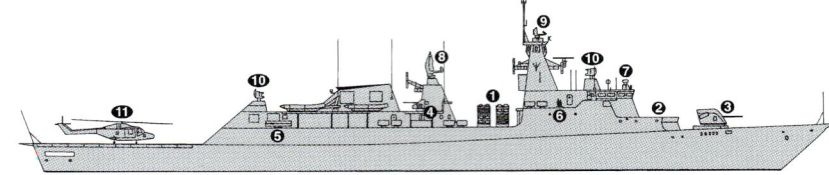

LEKIU *(Scale 1 : 900), Ian Sturton / 0081204*

LEKIU *12/2011, Chris Sattler / 1406986*

JEBAT *9/2013*, Chris Sattler / 1530917*

JEBAT *11/2013*, Michael Nitz / 1530918*

CORVETTES

2 KASTURI (TYPE FS 1500) CLASS (FSGH)

Name	No	Builders	Laid down	Launched	Commissioned
KASTURI	25	Howaldtswerke, Kiel	3 Jan 1983	14 May 1983	15 Aug 1984
LEKIR	26	Howaldtswerke, Kiel	3 Jan 1983	14 May 1983	15 Aug 1984

Displacement, tonnes: 1,524 standard; 1,850 full load
Dimensions, metres (feet): 97.3 × 11.3 × 3.5 *(319.2 × 37.1 × 11.5)*
Speed, knots: 28; 18 diesel
Range, n miles: 3,000 at 18 kt, 5,000 at 14 kt
Complement: 124 (13 officers)

Machinery: 4 MTU 20V 1163 TB92 diesels; 23,400 hp(m) *(17.2 MW)* sustained; 2 shafts
Missiles: SSM: 8 Aerospatiale MM 40 Exocet Block II ❶; inertial cruise; active radar homing to 70 km *(40 n miles)* at 0.9 Mach; warhead 165 kg; sea-skimmer.
Guns: 1 Creusot-Loire 3.9 in *(100 mm)*/55 Mk 2 compact ❷; 20/45/90 rds/min to 17 km *(9.2 n miles)* anti-surface; 6 km *(3.2 n miles)* anti-aircraft; weight of shell 13.5 kg. 1 Bofors 57 mm/70 ❸; 200 rds/min to 17 km *(9.2 n miles)*; weight of shell 2.4 kg. Launchers for illuminants. 4 Emerlec 30 mm (2 twin) ❹; 1,200 rds/min combined to 6 km *(3.2 n miles)*; weight of shell 0.35 kg.
A/S Mortars: 1 Bofors 375 mm twin trainable launcher ❺; automatic loading; range 3,600 m.
Physical countermeasures: Decoys: 2 CSEE Dagaie trainable systems; replaceable containers for IR or chaff.
Electronic countermeasures: ESM: Rapids.
ECM: MEL Scimitar; jammer.
Radars: Air/surface search: Signaal DA08 ❻; F-band.
Navigation: Kelvin Hughes 1007; I-band.
Fire control: Signaal WM22 ❼; I/J-band.
IFF: US Mk 10.
Sonars: Atlas Elektronik DSQS-24C; hull-mounted; active search and attack; medium frequency.
Combat data systems: Signaal Sewaco-MA. Link Y Mk 2.
Electro-optic systems: 2 Signaal LIOD optronic directors.

Helicopters: Platform for 1 medium ❽.

Programmes: First two ordered in February 1981. Fabrication began early 1982.
Modernisation: An extensive Ship Life Extension Programme (SLEP) for both ships started at Boustead Naval Shipyard in 2009. The upgrade includes a new Combat Data System (Tacticos), IFF, Target Designation Sight (TDS), Mirador optronic director, and underwater telephone. Bofors 375 A/S launchers are to be replaced by 324 mm torpedo tubes with A244S torpedoes while the 30 mm armament is to be replaced by new 30 mm MSI DS30B REMSIG guns. The 100 mm gun forward is to be replaced by a Bofors 57 mm gun while the aft 57 mm gun is to be removed to facilitate extension of the flight deck. Super Barricade is to replace the Dagaie chaff system while the MEL Scimitar jammer is to be removed. The refits were to have been completed by 2013 (*Kasturi*) and 2014 (*Lekir*) but progress has not been confirmed.
Structure: Near sisters to the Colombian ships with differing armament.
Operational: Form 22nd Corvette Squadron.

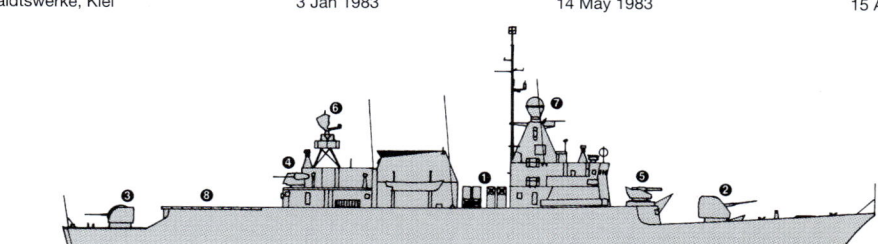

KASTURI *(Scale 1 : 900), Ian Sturton / 0506055*

KASTURI *12/2005, Chris Sattler / 1153383*

LEKIR *12/2007, Michael Nitz / 1170224*

LEKIR *12/2009, Chris Sattler / 1305929*

6 KEDAH (MEKO 100 RMN) CLASS (FSGHM)

Name	No	Builders	Laid down	Launched	Commissioned
KEDAH	171	Blohm + Voss, Hamburg; Penang Shipbuilding	13 Nov 2001	21 Mar 2003	5 Jun 2006
PAHANG	172	Blohm + Voss, Hamburg; Penang Shipbuilding	21 Dec 2001	2 Oct 2003	3 Aug 2006
PERAK	173	Boustead Naval Shipyard, Lumut	2 Jan 2003	12 Nov 2007	3 Jun 2009
TERENGGANU	174	Boustead Naval Shipyard, Lumut	Aug 2004	6 Dec 2007	8 Dec 2009
KELANTAN	175	Boustead Naval Shipyard, Lumut	Jul 2005	24 Nov 2008	6 May 2010
SELANGOR	176	Boustead Naval Shipyard, Lumut	Jul 2006	23 Jul 2009	28 Dec 2010

Displacement, tonnes: 1,676 full load
Dimensions, metres (feet): 91.1 × 12.85 × 3.4 (298.9 × 42.2 × 11.2)
Speed, knots: 22
Range, n miles: 6,050 at 12 kt
Complement: 68 (11 officers)

Machinery: 2 Caterpillar 3616 diesels; 14,617 hp(m) (10.9 MW) sustained; 2 shafts; cp propellors
Missiles: Fitted for SSM (MM40) ❶ and SAM (RAM CIWS) ❷.
Guns: 1 Otobreda 3 in (76 mm)/62 ❸; Super Rapid; 120 rds/min to 16 km (8.7 n miles); weight of shell 6 kg. 1 — 30 mm Otobreda/Mauser ❹. 2 — 12.7 mm MGs.
Physical countermeasures: Decoys: RBOC chaff launcher.
Radars: Air/surface search: EADS TRS-3D/16ES ❻; G-band.
Fire control: Contraves TMX; I-band.
Navigation: Atlas Elektronik 9600 ARPA; I-band.
Sonars: Fitted for.
Combat data systems: STN Atlas Cosys 110M1.
Electro-optic systems: Contraves TMEO optronic director ❺.

Helicopters: Platform for medium helicopter.

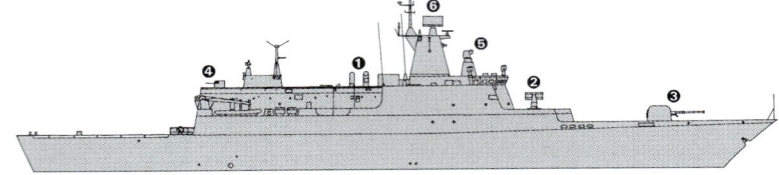

KEDAH

(Scale 1 : 900), Ian Sturton / 1366158

Programmes: An agreement between the Malaysian government, the Penang Shipbuilding Corporation (PSC) and German Naval Group consortium (led by Blohm + Voss) was reached in November 2000 for the supply of six vessels. The first two OPVs were built in Germany for shipment to Malaysia and assembly and fitting out at Lumut. The other four ships were constructed in Malaysia. Following technical problems and construction delays PSC reverted to state control as Boustead Naval Shipyard which took over the programme in September 2005.

Modernisation: There are plans to replace TRS-3D radar and to install SSM and (possibly) SAM systems.
Structure: Design based on Blohm + Voss MEKO 100 including measures to reduce the radar and IR signatures. Space has been included for future enhancements which may include SSM, SAM, sonar and an EW suite.
Operational: Principal tasks are maritime surveillance and patrol duties in the Malaysian EEZ. The first two are based at Kota Kinabalu, the second two at Lumut and the last two at Kuantan.

KELANTAN

3/2013*, Chris Sattler / 1486528

SELANGOR

12/2011, Chris Sattler / 1406988

IHS Jane's Fighting Ships 2014-2015

Corvettes < **Malaysia** 511

4 LAKSAMANA (ASSAD) CLASS (FSGM)

Name	No	Builders	Laid down	Launched	Commissioned
LAKSAMANA HANG NADIM (ex-*Khalid Ibn Al Walid*)	F 134 (ex-F 216)	Fincantieri, Breda	3 Jun 1982	5 Jul 1983	28 Jul 1997
LAKSAMANA TUN ABDUL JAMIL (ex-*Saad Ibn Abi Waccade*)	F 135 (ex-F 218)	Fincantieri, Breda	17 Sep 1982	2 Dec 1983	28 Jul 1997
LAKSAMANA MUHAMMAD AMIN (ex-*Abdulla Ben Abi Sarh*)	F 136 (ex-F 214)	Fincantieri, Breda	22 Mar 1982	5 Jul 1983	31 Jul 1999
LAKSAMANA TAN PUSMAH (ex-*Salahi Ad Deen Alayoori*)	F 137 (ex-F 220)	Fincantieri, Breda	17 Sep 1982	30 Mar 1984	31 Jul 1999

Displacement, tonnes: 716 full load
Dimensions, metres (feet): 62.3 × 9.3 × 2.5 *(204.4 × 30.5 × 8.2)*
Speed, knots: 36
Range, n miles: 2,300 at 18 kt
Complement: 47

Machinery: 4 MTU 20V 956 TB92 diesels; 20,120 hp(m) *(14.8 MW)* sustained; 4 shafts
Missiles: SSM: 6 Oto Melara/Matra Otomat Teseo Mk 2 (TG 2) (3 twin) ❶; command guidance; active radar homing to 180 km *(98.4 n miles)* at 0.9 Mach; warhead 210 kg; sea-skimmer.
SAM: 1 Selenia/Elsag Albatros launcher ❷ (4 cell-2 reloads); Aspide; semi-active radar homing to 13 km *(7 n miles)* at 2.5 Mach; warhead 30 kg.
Guns: 1 Oto Melara 3 in *(76 mm)*/62 Super Rapid ❸; 120 rds/min to 16 km *(8.7 n miles)* anti-surface; 12 km *(6.6 n miles)* anti-aircraft; weight of shell 6 kg. 2 Breda 40 mm/70 (twin) ❹; 300 rds/min to 12.5 km *(6.8 n miles)*; weight of shell 0.96 kg.
Physical countermeasures: Decoys: 2 Breda 105 mm 6-tubed multipurpose launchers; chaff to 5 km *(2.7 n miles)*; illuminants to 12 km *(6.6 n miles)*.
Electronic countermeasures: ESM: Selenia INS-3; intercept. ECM: Selenia TQN-2; jammer.
Radars: Air/surface search: Selenia RAN 12L/X ❺; D/I-band; range 82 km *(45 n miles)*.
Navigation: Kelvin Hughes 1007; I-band.
Fire control: 2 Selenia RTN 10X ❻; I/J-band.

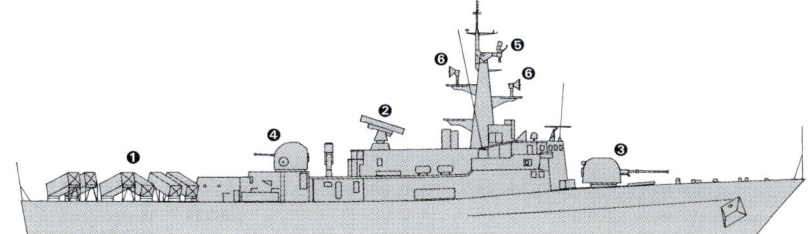

LAKSAMANA TAN PUSMAH *(Scale 1 : 600), Ian Sturton* / 1406946

Sonars: Atlas Elektronik ASO 84-41; hull-mounted; active search and attack.
Combat data systems: Selenia IPN 10 (136, 137); Alenia IPN-S (134, 135); Signaal/AESN Link Y Mk 2.
Weapon control systems: 2 Selenia NA 21; Dardo.

Programmes: Ordered in February 1981 for the Iraqi Navy and fell foul of UN sanctions before they could either be paid for or delivered. Subsequently completed in 1988 and maintained by Fincantieri. Two near sister ships were paid for by Iraq and remain laid up at La Spezia. Contract signed on 26 October 1995, and confirmed on 26 July 1996, to transfer two of the class to the Malaysian Navy after refit at Muggiano and three months training in Italy. Contract for two more signed on 20 February 1997 for conversion and delivery.
Modernisation: Super Rapid 76 mm gun, datalink, new navigation radar and GPS fitted in 1996. Bridge wings are extended to the after gun deck. Contract signed with Alenia Marconi on 11 April 2002 to upgrade command systems of F 134 and F 135 to IPN-S.
Structure: NBC citadel and full air conditioning fitted. Torpedo tubes have been removed.
Operational: First pair arrived in Malaysia in September 1997. Second pair in September 1999. Constitute 24th Corvette Squadron.

LAKSAMANA TUN ABDUL JAMIL *3/2013*, Chris Sattler* / 1530919

LAKSAMANA HANG NADIM *12/2009, Chris Sattler* / 1305956

© 2014 IHS IHS Jane's Fighting Ships 2014-2015

514 Malaysia > Amphibious forces — Training ships

130 DAMEN ASSAULT CRAFT 540

Dimensions, metres (feet): 5.4 × 1.8 × 0.6 *(17.7 × 5.9 × 2.0)*
Speed, knots: 12
Military lift: 10 troops
Machinery: 1 outboard; 40 hp(m) *(29.4 kW)*

Comment: First 65 built by Damen Gorinchem, Netherlands in 1986. Remainder built by Limbungan Timor SY. Army assault craft. Manportable and similar to Singapore craft. Used by the Army. Some have been deleted.

2 FAST TROOP VESSELS (AP)

SRI GAYA 331	SRI TIGA 332	

Displacement, tonnes: 118 full load
Dimensions, metres (feet): 37.5 × 7.0 × 1.1 *(123 × 23.0 × 3.6)*
Speed, knots: 25. **Range, n miles:** 540
Complement: 12
Military lift: 32 troops + stores
Machinery: 4 MAN D2842 LE 408 diesels; 2,080 hp *(1.55 MW)*; 4 water-jets
Radars: Navigation: Furuno; I-band.

Comment: Design based on Australian Wave Master fast-ferry monohull. Procured to transport troops and stores particularly in Sabah and Sarawak waters. Built by Naval Dockyard, Lumut and commissioned on 29 May 2001. Based at Kota Kinabalu.

SRI GAYA *12/2005, Chris Sattler* / 1153396

MINE WARFARE FORCES

4 MAHAMIRU (LERICI) CLASS (MINEHUNTERS) (MHC)

Name	No	Builders	Launched	Commissioned
MAHAMIRU	11	Intermarine	23 Feb 1984	11 Dec 1985
JERAI	12	Intermarine	5 Jan 1984	11 Dec 1985
LEDANG	13	Intermarine	14 Jul 1983	11 Dec 1985
KINABALU	14	Intermarine	19 Mar 1983	11 Dec 1985

Displacement, tonnes: 620 full load
Dimensions, metres (feet): 51 × 9.9 × 2.8 *(167.3 × 32.5 × 9.2)*
Speed, knots: 16; 7 thrust jet. **Range, n miles:** 2,000 at 12 kt
Complement: 42 (5 officers)

Machinery: 2 MTU 12V 396 TC82 diesels (passage); 2,605 hp(m) *(1.91 MW)* sustained; 2 shafts; Kamewa cp props; 3 Fincantieri Isotta Fraschini ID 36 SS 6V diesels; 1,481 hp(m) *(1.09 MW)* sustained; 2 Riva Calzoni hydraulic thrust jets
Guns: 1 Bofors 40 mm/70; 300 rds/min to 12.5 km *(6.8 n miles)*; weight of shell 0.96 kg.
Physical countermeasures: MCM: Thomson-CSF IBIS II minehunting system; 2 improved PAP 104 ROVs. Oropesa 'O' MIS-4 mechanical sweep.
Radars: Navigation: Kelvin Hughes 1007; Thomson-CSF Tripartite III; I-band.
Sonars: Thomson Sintra TSM 2022 Mk III with Display 2060; minehunting; high frequency.

Comment: Ordered on 20 February 1981. All arrived in Malaysia on 26 March 1986. Heavy GRP construction without frames. Snach active tank stabilisers. Draeger Duocom decompression chamber. Slightly longer than Italian sisters. Endurance, 14 days. Upgrade of tactical data system completed in 2001; Minehunting Tactical Display System (MTDS) installed by Altech Defence System, South Africa. A SLEP for *Mahamiru* and *Ledang* was completed in 2007. Upgrades included TSM Mk III sonar, MTDS adaptation and Kongsberg navigation echo-sounder. The other two ships are to be similarly modified. Form the 26th Mine Countermeasures Squadron.

KINABALU *12/2011, Chris Sattler* / 1406992

SURVEY SHIPS

1 SURVEY VESSEL (AGSH)

Name	No	Builders	Commissioned
MUTIARA	255 (ex-152)	Hong Leong-Lürssen, Butterworth	12 Jan 1978

Displacement, tonnes: 1,936 full load
Dimensions, metres (feet): 71 × 13 × 4 *(232.9 × 42.7 × 13.1)*
Speed, knots: 16. **Range, n miles:** 4,500 at 16 kt
Complement: 155 (14 officers)

Machinery: 2 Deutz SBA12M528 diesels; 4,000 hp(m) *(2.94 MW)*; 2 shafts
Guns: 4 Oerlikon 20 mm (2 twin) (fitted for but not with).
Radars: Navigation: 2 Racal Decca 1226/1229; I-band.
Helicopters: Platform only.

Comment: Ordered in early 1975. Carries satellite navigation, auto-data system and computerised fixing system. Davits for six survey launches. Forms part of 36 Squadron. The ship was badly damaged by a fire on 12 January 2012 while undergoing maintenance. The ship is to be repaired.

MUTIARA *12/2005, Chris Sattler* / 1153397

1 SURVEY VESSEL (AGS)

Name	No	Builders	Commissioned
PERANTAU	151	Hong Leong-Lürssen, Butterworth	12 Oct 1998

Displacement, tonnes: 2,028 full load
Dimensions, metres (feet): 67.8 × 13.3 × 4 *(222.4 × 43.6 × 13.1)*
Speed, knots: 16
Range, n miles: 6,000 at 10 kt
Complement: 94 (17 officers)
Machinery: 2 Deutz/MWM SBV8 M628 diesels; 4,787 hp(m) *(3.52 MW)*; 2 shafts; Berg cp props; Schottel bow thruster
Guns: 4 Oerlikon 20 mm (2 twin) (fitted for but not with).
Radars: Navigation: STN Atlas; I-band.

Comment: Ordered from Krogerwerft in 1996. The ship is equipped with two survey launches and four multipurpose boats and has three winches and two cranes, including a hoist for a STN Atlas side scan sonar. Full range of hydrographic and mapping equipment embarked. Forms part of 36 Squadron.

PERANTAU *3/2013*, Chris Sattler* / 1486533

TRAINING SHIPS

1 HANG TUAH (TYPE 41/61) CLASS (FFH/AX)

Name	No	Builders	Commissioned
HANG TUAH (ex-*Mermaid*)	76 (ex-F 76)	Yarrow Shipbuilders, Glasgow	16 May 1973

Displacement, tonnes: 2,337 standard; 2,560 full load
Dimensions, metres (feet): 103.5 × 12.2 × 4.9 sonar *(339.6 × 40.0 × 16.1)*
Speed, knots: 24
Range, n miles: 4,800 at 15 kt
Complement: 210

Machinery: 2 Stork Wärtsilä 12SW28 diesels; 9,928 hp(m) *(7.3 MW)* sustained; 2 shafts; cp props
Guns: 1 Bofors 57 mm/70 Mk 1; 200 rds/min to 17 km *(9.2 n miles)*; weight of shell 2.4 kg.
2 Bofors 40 mm/70; 300 rds/min to 12 km *(6.5 n miles)* anti-surface; 4 km *(2.2 n miles)* anti-aircraft; weight of shell 0.96 kg.
Radars: Navigation: Kelvin Hughes 1007; I-band.
Helicopters: Platform for 1 medium.

Comment: Originally built for Ghana as a display ship for ex-President Nkrumah but put up for sale after his departure. She was launched without ceremony on 29 December 1966 and completed in 1968. Commissioned in Royal Navy 16 May 1973 and transferred to Royal Malaysian Navy May 1977. Refitted in 1991–92 to become a training ship. Main gun and main engines replaced in 1995–96. Sonars and Limbo ASW mortars removed. There are no plans for further modifications. Forms 21st Frigate Squadron. Likely to be decommissioned when new training ships become operational.

HANG TUAH *12/2007, Chris Sattler* / 1168000

Training ships — Tugs < **Malaysia** 515

1 SAIL TRAINING SHIP (AXS)

Name	No	Builders	Commissioned
TUNAS SAMUDERA	A 13	Brooke Yacht, Lowestoft	16 Oct 1989

Displacement, tonnes: 243 full load
Dimensions, metres (feet): 35 × 7.8 × 4 *(114.8 × 25.6 × 13.1)*
Speed, knots: 9
Complement: 36 (26 trainees)
Machinery: 2 Perkins diesels; 370 hp *(272 kW)*; 2 shafts
Radars: Navigation: Racal Decca; I-band.

Comment: Laid down 1 December 1988 and launched 4 August 1989. Two-masted brig manned by the Navy but used for training all sea services.

TUNAS SAMUDERA   12/2009, Michael Nitz / 1305948

2 TRAINING SHIPS (AX)

Name	No	Builders	Laid down	Launched	Commissioned
GAGAH SAMUDERA	271	NGV Tech; Daewoo	2012	14 Dec 2012	15 May 2013
TEGUH SAMUDERA	272	NGV Tech; Daewoo	2012	27 Feb 2013	30 Jun 2013

Displacement, tonnes: 1,270 full load
Dimensions, metres (feet): 75.9 × 11.0 × 3.0 *(249 × 36.1 × 9.8)*
Speed, knots: 20. **Range, n miles:** 2,500 at 12 kt
Complement: 50 (9 officers; 60 trainees)

Machinery: 2 MAN 9L27/38 diesels; 9,051 hp *(6.57 MW)*; 2 shafts; cp prop
Guns: 1 Seahawk DS-30M 30 mm. 2—7.62 mm MGs.
Radars: Surface search/navigation: Kelvin Hughes Sharp Eye; E/F/I-bands.
Electro-optic systems: Samsung/Thales optronic director.
Helicopters: Platform for one medium.

Comment: A contract for the construction of two training vessels was signed between the Malaysian government and NGV Tech Sdn Bhd in February 2012. The ships are based on a design by Daewoo (DSME) and blocks of the first vessel were built at DSME and assembled and fitted out by NGV Tech Sdn Bhd. The second vessel is to be constructed in Malaysia. While the ships' primary role is to be training, they are also to be capable of conducting patrols and SAR. Both ships are expected to be based at Lumut.

GAGAH SAMUDERA   12/2012 / 1483529

AUXILIARIES

Notes: There are six miscellaneous personnel launches: *Kempong, Kuramah, Mangkasa, Patak, Selar* and *Tepuruk*.

2 SUPPORT SHIPS (AG)

Name	No	Builders	Commissioned
BUNGA MAS LIMA	5	Malaysia Shipyard and Engineering	1997
BUNGA MAS ENAM	6	Malaysia Shipyard and Engineering	8 Aug 2011

Measurement, tonnes: 9,101 gt
Dimensions, metres (feet): 132.8 × 22.7 × 7.5 *(435.7 × 74.5 × 24.6)*
Speed, knots: 17
Complement: 57

Machinery: 1 Sulzer 5RTA52U diesel; 10,455 hp *(7.7 MW)*; 1 shaft
Guns: 2—7.62 mm MGs.
Radars: Navigation: I-band.
Helicopters: Platform for Super Lynx.

Comment: Both ships are converted container ships constructed by Malaysia Shipyard and Engineering. *Bunga Mas Lima 5* was built in 1997 and converted in 2009 to act as an auxiliary in support of anti-piracy patrols in the Gulf of Aden. The conversion included installation of a helicopter flight deck, communications, medical facilities, weapons mounts and facilities to launch and recover small boats. *Bunga Mas 6* was built in 1996 and commissioned into the RMN on 8 August 2011. She carries out similar tasks.

BUNDA MAS LIMA   8/2013*, Shaun Jones / 1530920

1 SUBMARINE RESCUE SHIP (ASR)

Name	Builders	Laid down	Launched	Commissioned
MEGA BAKTI	Keppel Singmarine	21 Dec 2010	14 Dec 2011	11 Jun 2012

Displacement, tonnes: 3,549 full load
Dimensions, metres (feet): 79.5 × 17.0 × 7.8 *(260.8 × 55.8 × 25.6)*
Speed, knots: 17
Range, n miles: 5,000 at 14 kt
Complement: 40 (10 officers)
Machinery: 4 Caterpillar 8M25C diesels; 14,155 hp *(10.56 kW)*; 2 shafts; cp props
Radars: Navigation: Furuno FAR-2137S; E/F-band. Furuno FAR-2117; I-band.

Comment: The ship was designed and built in Singapore to provide the platform for a submarine rescue and diving capability. The ship is operated by Target Resources Sendirian Berhad, in technical collaboration with Perry Slingsby Systems Limited, on behalf of the Royal Malaysian Navy. Acting as a mother-ship, the ship is capable of embarking and operating GPS Intelligent Buoys portable tracking system (GIB), Intervention Remotely Operated Vehicle (IROV), the Distress Submarine Ventilation and Depressurization System (DSVDS) and Emergency Life Support Stores (ELSS) pods. The ship is also equipped with Divers Decompression Chamber (DDC), Diver Attendance Recompression Transportable (DART) chamber and Shallow Water Diving System (SWS). The ship's secondary role is to support disaster relief and humanitarian operations.

MEGA BAKTI   6/2012, Royal Malaysian Navy / 1483530

TUGS

1 HARBOUR TUG (YTM)

TUNDA SATU

Measurement, tonnes: 150 gt
Dimensions, metres (feet): 25.91 × 7.6 × 3.7 *(85 × 24.9 × 12.1)*
Speed, knots: 6
Complement: 15
Machinery: 2 Cummins VTA 1710M2; 620 hp *(456 kW)*; 2 props
Radars: Navigation: Furuno 1931; I-band.

Comment: Built by Ironwoods Shipyard, Kuching, in 1978.

TUNDA SATU   6/2007, Royal Malaysian Navy / 1170006

1 OCEAN TUG (AT)

KEPAH (ex-*Asiatic Success*) 8

Measurement, tonnes: 202 gt
Dimensions, metres (feet): 33.5 × 8.0 × 3.2 *(109.9 × 26.2 × 10.5)*
Speed, knots: 11
Complement: 21
Machinery: 2 Yanmar 6UA-UT diesels; 1,500 hp *(1.1 MW)*; 2 shafts
Radars: Navigation: Furuno 1505; I-band.

Comment: Built in 1976 by Asahi Zosen, Sumoto.

1 OCEAN TUG (AT)

PENYU (ex-*Salvigilant*) 4

Measurement, tonnes: 398 gt
Dimensions, metres (feet): 39 × 10.04 × 4.3 *(128 × 32.9 × 14.1)*
Speed, knots: 11
Complement: 21
Machinery: 2 MWM TBD440-6 diesels; 1,800 hp *(1.32 MW)*; 2 shafts
Radars: Navigation: Furuno 1505; I-band.

Comment: Built in 1975 by Selco Shipyard, Singapore.

PENYU
10/2003,
Hartmut Ehlers
0567883

COAST GUARD (MARITIME ENFORCEMENT AGENCY)

Headquarters Appointments

Director General:
Admiral Datuk Mohd Amdan bin Kurish

Establishment

The Malaysian Maritime Enforcement Agency (MMEA) or *Agensi Penguatkuasaan Maritim Malaysia* (APMM) commenced its operations on 30 November 2005 and is the principal government agency tasked with maintaining law and order in the Malaysia Maritime Zone (MMZ) and coordinate Search And Rescue (SAR) operations in the Malaysian Maritime Search and Research Region (SRR). The Agency was formally established in May 2004 with the enactment of the MMEA Act 2004 (Act 633). Under the provisions of Act 633 which came into force on 15 February 2005, the MMEA shall be placed under the command and control of the Malaysian Armed Forces in time of war, special crisis and emergency.

MMEA's mission is to protect and safeguard peace, security and national sovereignty in the MMZ and to save life and properties at sea. Towards meeting this mission, MMEA has embarked on a maritime security strategy based on the principles of: Maritime Domain Awareness, visible deterrence, swift response, ever present, forward reach and maritime community cooperation.

Principal Missions

- Enforcement of law and order under Malaysian Federal Law
- Prevent and suppress the commission of an offence
- Maritime search and rescue
- Air and coastal surveillance
- Maintenance of maritime safety and security
- Control and prevention of maritime pollution
- Prevention and suppression of human trafficking and smuggling, piracy and illicit traffic in narcotic drugs

Personnel

2014: 4,067. There are plans to build up to a force of about 5,170 personnel.

Organisation

The Malaysian Maritime Zone is divided into five maritime regions which consist of 18 maritime districts.

Bases

MMEA HQ: Putrajaya
Northern Peninsula: Regional HQ: Langkawi (HQ)
District: Langkawi, Pulau Pinang, Lumut
Southern Peninsula: Regional HQ: Johor Bahru
District: Johor Bahru, Port Klang, Kuala Linggi, Tanjung Sedili
Eastern Peninsula: Regional HQ: Kuantan
District: Kuantan, Kuala Terengganu, Tok Bali
Sarawak: Regional HQ: Kuching
District: Kuching, Bintulu, Miri
Sabah and Labuan: Regional HQ: Kota Kinabalu
District: Kota Kinabalu, Labuan, Kudat, Sandakan, Tawau

Prefix to Ships' Names

KM: Kapal Maritim

PATROL FORCES

14 SIPADAN CLASS (PB)

Name	No	Builders	Commissioned
LANG (ex-*Kris*)	3132 (ex-34)	Vosper Ltd, Portsmouth	1 Jan 1966
SEGANTANG (ex-*Sundang*)	3133 (ex-36)	Vosper Ltd, Portsmouth	29 Nov 1966
JARAK (ex-*Badek*)	3134 (ex-37)	Vosper Ltd, Portsmouth	15 Dec 1966
KUKUP (ex-*Panah*)	3135 (ex-42)	Vosper Ltd, Portsmouth	27 Jul 1967
SEMPADI (ex-*Kelewang*)	3136 (ex-45)	Vosper Ltd, Portsmouth	4 Oct 1967
LABAS (ex-*Sri Sabah*)	3137 (ex-3144)	Vosper Ltd, Portsmouth	2 Sep 1964
NYIREH (ex-*Sri Negri Sembilan*)	3138 (ex-3146)	Vosper Ltd, Portsmouth	28 Sep 1964
KURAMAN (ex-*Renchong*)	3139 (ex-38)	Vosper Ltd, Portsmouth	17 Jan 1967
SIAMIL (ex-*Tombak*)	3140 (ex-39)	Vosper Ltd, Portsmouth	2 Mar 1967
PEMANGGIL (ex-*Kerambit*)	3141 (ex-43)	Vosper Ltd, Portsmouth	28 Jul 1967
BIDONG (ex-*Beladau*)	3142 (ex-44)	Vosper Ltd, Portsmouth	12 Sep 1967
SATANG (ex-*Rentaka*)	3143 (ex-46)	Vosper Ltd, Portsmouth	22 Sep 1967
RUMBIA (ex-*Sri Melaka*)	3144 (ex-3147)	Vosper Ltd, Portsmouth	2 Nov 1964
LIGITAN (ex-*Lembing*)	3145 (ex-40)	Vosper Ltd, Portsmouth	12 Apr 1967

Displacement, tonnes: 98 standard; 111 full load
Dimensions, metres (feet): 31.4 × 6 × 1.7 *(103 × 19.7 × 5.6)*
Speed, knots: 27
Range, n miles: 1,400 at 14 kt (3132–3136, 3138–3145), 1,660 at 14 kt (3137)
Complement: 23 (4 officers)
Machinery: 2 MTU MD 655/18 diesels; 3,500 hp(m) *(2.57 MW)*; 2 shafts
Guns: 1 Bofors 40 mm/70. 2 — 7.62 mm MGs.
Radars: Surface search: Racal Decca Bridgemaster ARPA; I-band.

Comment: The four ex-Sabah class were ordered in 1963 for delivery in 1964. The boats of the ex-Kris class were ordered in 1965 for delivery between 1966 and 1968. All are of prefabricated steel construction and are fitted with air conditioning and Vosper roll damping equipment. The differences between the classes are minor, the later ones having improved radar, communications, evaporators and engines of MTU, as opposed to Bristol Siddeley construction. All have been refitted to extend their operational lives and transferred from the Malaysian Navy to MMEA by June 2006. Similar craft in service in Panama.

NYIREH 6/2012, MMEA / 1484146

4 MALAWALI CLASS (PB)

MALAWALI (ex-*Bintang Utara*) 2551 **MANJUNG** (ex-*Bintang Manjung*) 2553
SERASAN (ex-*Bintang Timur*) 2552 **TEBRAU** (ex-*Bintang Baru*) 2554

Displacement, tonnes: 65 full load
Dimensions, metres (feet): 25 × 6 × 1.1 *(82 × 19.7 × 3.6)*
Speed, knots: 25
Complement: 14 (4 officers)
Machinery: 2 Deutz 16M 816CR diesels; 2 shafts
Radars: Navigation: I-band.

Comment: Built in 1999 and transferred from the Marine Department to the MMEA in April 2006.

MALAWALI 12/2011, Chris Sattler / 1406993

15 GAGAH CLASS (PBF)

GAGAH (ex-*Lang Malam*) 3901 (ex-PZ 2)	BUDIMAN (ex-*Mersuji*) 3909 (ex-PZ 14)
TABAH (ex-*Lang Lebah*) 3902 (ex-PZ 3)	TEGAS (ex-*Lang Hitam*) 3910 (ex-PZ 1)
CEKAL (ex-*Lang Kuik*) 3903 (ex-PZ 4)	MULIA (ex-*Balong*) 3911 (ex-PZ 5)
BERANI (ex-*Kurita*) 3904 (ex-PZ 7)	BIJAK (ex-*Belian*) 3912 (ex-PZ 6)
SETIA (ex-*Serangan Batu*) 3905 (ex-PZ 8)	ADIL (ex-*Harimau Belang*) 3913 (ex-PZ 11)
AMANAH (ex-*Harimau Bintang*) 3906 (ex-PZ 9)	PINTAR (ex-*Perangan*) 3914 (ex-PZ 13)
JUJUR (ex-*Harimau Kimbang*) 3907 (ex-PZ 10)	BISTARI (ex-*Alu-Alu*) 3915 (ex-PZ 15)
IKHLAS (ex-*Harimau Akar*) 3908 (ex-PZ 12)	

Displacement, tonnes: 234 full load
Dimensions, metres (feet): 39.5 × 7 × 1.8 *(129.6 × 23.0 × 5.9)*
Speed, knots: 23
Range, n miles: 1,200 at 15 kt
Complement: 30 (5 officers)
Machinery: 2 MTU 20V 538 TB92 diesels; 8,360 hp(m) *(6.14 MW)* sustained; 2 shafts
Guns: 1 Bofors 40 mm/70. 1 Oerlikon 20 mm. 2 FN 7.62 mm MGs.
Radars: Navigation: Kelvin Hughes; I-band.

Comment: Ordered from Hong Leong-Lürssen, Butterworth, Malaysia in 1979. First delivered August 1980, last in April 1983. All transferred from the Marine Police to the Maritime Enforcement Agency and became operational by November 2005.

TEGAS 3/2013*, Chris Sattler / 1486536

IHS Jane's Fighting Ships 2014-2015

Patrol forces < **Malaysia** (COAST GUARD) 517

2 LANGKAWI CLASS (OFFSHORE PATROL VESSELS) (PSOH)

Name	No	Builders	Launched	Commissioned
LANGKAWI (ex-*Musytari*)	7501 (ex-160)	Korea Shipbuilders, Pusan	20 Jul 1984	19 Dec 1985
BANGGI (ex-*Marikh*)	7502 (ex-161)	Malaysia SB and E Co, Johore	21 Jan 1985	9 Apr 1987

Displacement, tonnes: 1,321 full load
Dimensions, metres (feet): 75 × 10.8 × 3.7 *(246.1 × 35.4 × 12.1)*
Speed, knots: 22. **Range, n miles:** 5,000 at 15 kt
Complement: 70 (10 officers)

Machinery: 2 SEMT-Pielstick diesels; 12,720 hp(m) *(9.35 MW)*; 2 shafts

Guns: 1—57 mm.
Radars: Air/surface search: Signaal DA05; E/F-band; range 137 km *(75 n miles)* for 2 m² target.
Navigation: Kelvin Hughes 1007; I-band.
Fire control: Philips 9LV; J-band.
Weapon control systems: PEAB 9LV 230 optronic system.

Helicopters: Platform for 1 medium.

Programmes: Ordered in June 1983.
Structure: Flight deck suitable for Sikorsky S-61A Nuri army support helicopter.
Operational: These ships were transferred from the Malaysian Navy on 23 June 2006 and became operational in 2007. *Langkawi* based in Eastern Maritime Region and *Banggi* in Sabah and Labuan Maritime Region.

LANGKAWI   12/2011, Chris Sattler / 1406996

7 RAMUNIA CLASS (PB)

RAMUNIA (ex-*Bahtera Kinabalu*) 3221 (ex-K 35)
MARUDU (ex-*Bahtera Bayu*) 3222 (ex-K 37)
DANGA (ex-*Bahtera Hijau*) 3223 (ex-K 38)
SIANGIN (ex-*Bahtera Jerai*) 3224 (ex-K 40)
KIMANIS (ex-*Bahtera Juang*) 3225 (ex-K 33)
BURAU (ex-*Perak*) 3226 (ex-K 36)
NIPAH 3227 (ex-K 39)

Displacement, tonnes: 145 full load
Dimensions, metres (feet): 32.4 × 7.2 × 1.8 *(106.3 × 23.6 × 5.9)*
Speed, knots: 20; 8 diesel
Range, n miles: 2,000 at 8 kt
Complement: 24 (4 officers)
Machinery: 2 Paxman Valenta 16CM diesels; 6,650 hp *(5 MW)* sustained; 2 shafts 1 Cummins diesel; 575 hp *(423 kW)*; 1 shaft
Guns: 1—40 mm. 2—7.62 mm MGs.
Radars: Surface search: Kelvin Hughes; I-band.

Comment: Vosper 32 m craft ordered February 1981 from Malaysia Shipyard and Engineering Company with technical support from Vosper Thornycroft (Private) Ltd, Singapore. Two completed 1982, the remainder in 1983–84. Five transferred from the Customs Service to the MMEA in June 2005 and two from the Customs service in 2012.

DANGA   3/2013*, Chris Sattler / 1486535

4 PETIR 10 CLASS (PB)

1210–1211   1250–1251

Displacement, tonnes: 0.5 standard
Dimensions, metres (feet): 12 × 3.3 × 1.1 *(39.4 × 10.8 × 3.6)*
Speed, knots: 40
Complement: 6
Machinery: 2 outboard motors; 600 hp *(447 kW)*

Comment: Built by DALAC Marine Engineering Shipyard, Johor. Entered service in MMEA in May 2010.

PETIR 1210   6/2010, MMEA / 1366647

2 RHU CLASS (PB)

RHU 2601 (ex-P 202)   STAPA 2602 (ex-P 204)

Displacement, tonnes: 101 full load
Dimensions, metres (feet): 26.9 × 5.8 × 1.9 *(88.3 × 19.0 × 6.2)*
Speed, knots: 20
Complement: 14 (4 officers)
Machinery: 2 Deutz 16M 816CR diesels; 2 shafts
Radars: Navigation: I-band.

Comment: Former Fisheries Department craft built in 1990 and transferred in 2006.

RHU   6/2010, MMEA / 1366653

2 NUSA 22 CLASS (PATROL CRAFT) (PB)

NUSA (ex-*Rajawali 11*) 2201   RENTAP (ex-*Rajawali 111*) 2202

Displacement, tonnes: 54 full load
Dimensions, metres (feet): 22 × 6 × 1.3 *(72.2 × 19.7 × 4.3)*
Speed, knots: 25
Complement: 14 (4 officers)
Machinery: 2 Deutz 16M 816CR diesels; 2 shafts
Radars: Navigation: I-band.

Comment: Built in 1993 and transferred from the Marine Department to the MMEA in April 2006.

RENTAP   6/2010, MMEA / 1366652

518 Malaysia (COAST GUARD) > Patrol forces

4 SEMILANG CLASS (PB)

SEMILANG 2161 (ex-P 101) **MERSUJI** 2163 (ex-P 103)
ALU-ALU 2162 (ex-P 102) **SIAKAP** 2164 (ex-P 104)

Displacement, tonnes: 78 full load
Dimensions, metres (feet): 21 × 5.5 × 1.5 *(68.9 × 18.0 × 4.9)*
Speed, knots: 25
Complement: 14 (4 officers)
Machinery: 2 Deutz SBA 12M 816SR diesels; 2 shafts
Radars: Navigation: I-band.

Comment: Former Fisheries Department craft. Built in 1986 and transferred in 2006.

ALU-ALU 6/2010, MMEA / 1366651

1 PENINJAU CLASS (PB)

PENINJAU 1 1701 (ex-P 301)

Displacement, tonnes: 65 standard
Dimensions, metres (feet): 17 × 5.8 × 1.5 *(55.8 × 19.0 × 4.9)*
Speed, knots: 20
Complement: 10
Machinery: 2 Deutz SBA 12M 816CR diesels; 2 shafts
Radars: Navigation: I-band.

Comment: Former Fisheries Department craft transferred in 2006.

PENINJAU 1 6/2010, MMEA / 1366650

2 PENGGALANG 18 CLASS (PB)

PENGGALANG 1 1801 **PENGGALANG 2** 1802

Displacement, tonnes: 11 standard
Dimensions, metres (feet): 18 × 4.5 × 0.6 *(59.1 × 14.8 × 2.0)*
Speed, knots: 40
Complement: 10
Machinery: 2 CAT C 18 diesels; 2 shafts
Guns: 2—7.62 mm MGs.
Radars: Navigation: I-band.

Comment: Built by Destination Marine Services Shipyard at Port Klang and entered service on 10 September 2007.

PENGGALANG 2 12/2009, Michael Nitz / 1305943

4 PENYELAMAT 15 CLASS (PB)

PENYELAMAT 1 (ex-*Chendering*) 1571 **PENYELAMAT 3** (ex-*Murau*) 1573
PENYELAMAT 2 (ex-*Rhu*) 1572 **PENYELAMAT 4** (ex-*Lanngun*) 1574

Displacement, tonnes: 15 standard
Dimensions, metres (feet): 13.6 × 4.0 × 1.1 *(44.6 × 13.1 × 3.6)*
Speed, knots: 25
Complement: 7
Machinery: 2 Deutz SBA 12M 800 diesels; 2 shafts
Radars: Navigation: I-band.

Comment: Transferred from the Marine Department to the MMEA in 2006.

IHS Jane's Fighting Ships 2014-2015

PENYELAMAT 1 6/2008, MMEA / 1305340

10 PENGAWAL CLASS (PB)

PENGAWAL 1 (ex-*Labian*) 1411 **PENGAWAL 6** (ex-*Subis*) 1416
PENGAWAL 2 (ex-*Bidadari*) 1412 **PENGAWAL 7** (ex-*Niah*) 1417
PENGAWAL 3 (ex-*Kubung*) 1413 **PENGAWAL 8** (ex-*Murud*) 1418
PENGAWAL 4 (ex-*Serapi*) 1414 **PENGAWAL 11** (ex-*Matang*) 1311
PENGAWAL 5 (ex-*Memerang Laut*) 1415 **PENGAWAL 12** (ex-*Nyabau*) 1312

Displacement, tonnes: 13 standard
Dimensions, metres (feet): 13.6 × 4.0 × 1.2 *(44.6 × 13.1 × 3.9)*
Speed, knots: 25
Complement: 7
Machinery: 2 Deutz SBA 12M 800 diesels; 2 shafts
Radars: Navigation: I-band.

Comment: Transferred from the Marine Department to the MMEA in 2006.

PENGAWAL 2 6/2008, MMEA / 1305333

1 PENGAMAN CLASS (PB)

PENGAMAN 1 (ex-*Mastura*) 901

Displacement, tonnes: 9 standard
Dimensions, metres (feet): 9.2 × 3.0 × 0.4 *(30.2 × 9.8 × 1.3)*
Speed, knots: 28
Complement: 6
Machinery: 2 CAT C15 diesels; 2 shafts
Radars: Navigation: I-band.

Comment: Ex-Marine Department vessel originally commissioned in 1994. Transferred to the MMEA in June 2006.

PENGAMAN 1 6/2010, MMEA / 1366649

6 PETIR OPEN CLASS (PB)

PETIR 80 1280 **PETIR 82** 1282 **PETIR 84**
PETIR 81 1281 **PETIR 83** 1283 **PETIR 85** 1285

Displacement, tonnes: 0.5 standard
Dimensions, metres (feet): 12 × 3.3 × 1.1 *(39.4 × 10.8 × 3.6)*
Speed, knots: 50
Complement: 6
Machinery: 2 MAN 12V D2842 LE 406 diesels; 2,400 hp *(1.79 MW)*; 2 shafts

Comment: Built by DALAC Marine Engineering Shipyard, Johor. Entered service in 2010.

PETIR 85 3/2013*, Chris Sattler / 1486540

© 2014 IHS

Patrol forces < **Malaysia** (COAST GUARD) 519

53 KILAT CLASS (PB)

KILAT 1 701	KILAT 12 712	KILAT 23 723	KILAT 34 734	KILAT 45 745
KILAT 2 702	KILAT 13 713	KILAT 24 724	KILAT 35 735	KILAT 46 746
KILAT 3 703	KILAT 14 714	KILAT 25 725	KILAT 36 736	KILAT 47 747
KILAT 4 704	KILAT 15 715	KILAT 26 726	KILAT 37 737	KILAT 48 748
KILAT 5 705	KILAT 16 716	KILAT 27 727	KILAT 38 738	KILAT 49 749
KILAT 6 706	KILAT 17 717	KILAT 28 728	KILAT 39 739	KILAT 50 750
KILAT 7 707	KILAT 18 718	KILAT 29 729	KILAT 40 740	KILAT 51 751
KILAT 8 708	KILAT 19 719	KILAT 30 730	KILAT 41 741	KILAT 52 752
KILAT 9 709	KILAT 20 720	KILAT 31 731	KILAT 42 742	KILAT 53 753
KILAT 10 710	KILAT 21 721	KILAT 32 732	KILAT 43 743	
KILAT 11 711	KILAT 22 722	KILAT 33 733	KILAT 44 744	

Displacement, tonnes: 2 standard
Dimensions, metres (feet): 7.62 × 2.1 × 1.4 (25 × 6.9 × 4.6)
Speed, knots: 45
Complement: 6
Machinery: 2 outboard motors; 450 hp (335 kW)
Radars: Navigation: I-band.

Comment: RHIB craft built by Malaysia Shipyard and Engineering (MSET), Kuala Trengganu. Open type and semi-cabin variants delivered in January 2006. There were 15 further craft delivered by Kay Marine in December 2008.

KILAT 16   3/2007, MMEA / 1455753

5 PELINDUNG CLASS (PB)

PELINDUNG 1 701	PELINDUNG 3 703	PELINDUNG 5 705
PELINDUNG 2 702	PELINDUNG 4 704	

Displacement, tonnes: 2 standard
Dimensions, metres (feet): 6.4 × 2.5 × 0.8 (21 × 8.2 × 2.6)
Speed, knots: 35
Complement: 6
Machinery: 2 Mercury outboard motors; 300 hp (223 kW)
Radars: Navigation: I-band.

Comment: Ex-Fisheries Department craft originally commissioned in 1982. Transferred to the MMEA in June 2006.

PELINDUNG 3   6/2012, MMEA / 1455752

6 PENYELAMAT 20 CLASS (PATROL CRAFT) (PB)

PENYELAMAT 5 2005	PENYELAMAT 7 2007	PENYELAMAT 9 2009
PENYELAMAT 6 2006	PENYELAMAT 8 2008	PENYELAMAT 10 2010

Dimensions, metres (feet): 20 × 4.5 × 0.95 (65.6 × 14.8 × 3.1)
Speed, knots: 30. **Range, n miles:** 350 at 30 kt
Complement: 6
Machinery: 2 MTU 8V2000M92 diesels; 1,086 hp (809 kW); 2 Hamilton HM641 waterjets

Comment: Search and rescue craft designed by Camarc and built by Perlis Marine Engineering. Aluminium construction. Delivered in 2011.

PENYELAMAT 5   3/2013*, Chris Sattler / 1486544

18 PENGGALANG 17 (MRTP 16) (FAST INTERVENTION CRAFT) (PBF)

PENGGALANG 10 1810	PENGGALANG 16 1816	PENGGALANG 22 1822
PENGGALANG 11 1811	PENGGALANG 17 1817	PENGGALANG 23 1823
PENGGALANG 12 1812	PENGGALANG 18 1818	PENGGALANG 24 1824
PENGGALANG 13 1813	PENGGALANG 19 1819	PENGGALANG 25 1825
PENGGALANG 14 1814	PENGGALANG 20 1820	PENGGALANG 26 1826
PENGGALANG 15 1815	PENGGALANG 21 1821	PENGGALANG 27 1827

Displacement, tonnes: 26 full load
Dimensions, metres (feet): 17.73 × 4.19 × 1.2 (58.2 × 13.7 × 3.9)
Speed, knots: 65. **Range, n miles:** 350 at 35 kt
Complement: 4
Machinery: 2 MTU 10V2000M93 diesels; 3,040 hp (2.26 MW); 2 twin disc Arneson ASD12B1L surface drives
Guns: 1 Oto Melara 12.7 mm Hitrole.

Comment: Following the formation of a joint venture (BYO Marine) between Yonca-Onuk and Boustead Heavy Industries, BYO Marine awarded a contract for 10 (Penggalang 10–19) fast intervention craft. The MRTP 16 design is a development of the Kaan 15 class (MRTP 15) in service with the Turkish Coast Guard. The craft are of advanced composites construction and feature a Deep V hull. The first four craft were delivered in December 2011. A further eight craft (Penggalang 20–27) were constructed by Delex Marine Service and delivered in 2012.

PENGGALANG 11   3/2013*, Chris Sattler / 1486541

9 SIPADAN STEEL CLASS (PB)

GEMIA 2950 (ex-PX 5)	LIBARAN (ex-Dungun) 2955 (ex-PX 30)
RAWA 2951 (ex-PX 9)	MABUL (ex-Tioman) 2956 (ex-PX 31)
PERINGGI 2952 (ex-PX 16)	TENGGOL (ex-Tumpat) 2957 (ex-PX 32)
REDANG (ex-Sangitan) 2953 (ex-PX 28)	SEBATIK (ex-Juang) 2958 (ex-PX 33)
KAPAS (ex-Sabahan) 2954 (ex-PX 29)	

Displacement, tonnes: 116 full load
Dimensions, metres (feet): 29 × 6.0 × 1.7 (95.1 × 19.7 × 5.6)
Speed, knots: 17
Complement: 16 (4 officers)
Machinery: 2 Paxman Valenta 6CM diesels; 2,250 hp (1.68 MW); 2 shafts
Guns: 1 Oerlikon 20 mm.
Radars: Navigation: I-band.

Comment: Former police craft built by Penang Shipyard to a Brooke Marine design and delivered 1981–82. Transferred to the MMEA in 2011.

GEMIA   3/2013*, Chris Sattler / 1486538

9 SIPADAN KAYU CLASS (PB)

TUBA 2860 (ex-PX 13)	SERIMBUM 2863 (ex-PX 8)	LAYANG 2 2866 (ex-PX 17)
RIMAU 2861 (ex-PX 2)	AUR 2864 (ex-PX 22)	SELINGAN 2867 (ex-PX 18)
KENDI 2862 (ex-PX 10)	MATANANI 2865 (ex-PX 23)	GADOR 2868 (ex-PX 15)

Dimensions, metres (feet): 28 × 7.0 × 1.57 (91.9 × 23.0 × 5.2)
Speed, knots: 30
Complement: 17
Machinery: 2 MTU 820 DB 493 diesels; 2 shafts
Guns: 1 – 20 mm.
Radars: Navigation: I-band.

Comment: Former Marine Department craft transferred to the MMEA in 2011.

SERIBUM (Police colours)   10/2012, MMEA / 1455747

520 Malaysia (COAST GUARD) > Patrol forces — Training ships

8 ICARUS INTERCEPTOR CRAFT (PB)

1720–1727

Displacement, tonnes: 21 full load
Dimensions, metres (feet): 16.8 × 4.7 × 1.2 *(55.1 × 15.4 × 3.9)*
Speed, knots: 60
Complement: 3
Machinery: 2 MTU 10V2000 M93 diesels; 3,000 hp *(2.24 MW)*; 2 Arneson ASD14 steerable surface drives
Radars: Navigation: I-band.

Comment: Designed and built by Icarus Marine. Delivered in 2012.

ICARUS 1726 *3/2013*, Chris Sattler / 1486542*

3 NUSA 28 CLASS (PB)

RENGGIS 2203 (ex-KA 43) **SUGUT** 2204 (ex-KA 45) **BALUNG** 2205 (ex-KA 50)

Displacement, tonnes: 60 full load
Dimensions, metres (feet): 28.8 × 5.9 × 2.0 *(94.5 × 19.4 × 6.6)*
Speed, knots: 20
Complement: 14 (4 officers)
Machinery: 2 MTU 12V 396 TE84 diesels; 2 waterjets

Comment: Former Customs craft built by Limbungan Timor, Terengganu, in 1993 and transferred to the MMEA in 2011.

SUGUT *3/2013*, Chris Sattler / 1486537*

15 TUGAU CLASS (PB)

TUGAU 2210 (ex-PA 3)	**KINDURONG** 2215 (ex-PA 25)	**SIAGUT** 2220 (ex-PA 35)
MUKAH 2211 (ex-PA 5)	**JEPAK** 2216 (ex-PA 29)	**MANGALUM** 2221 (ex-PA 36)
TATAU 2212 (ex-PA 19)	**SIKUATI** 2217 (ex-PA 30)	**MEDANG** 2222 (ex-PA 37)
NYALAU 2213 (ex-PA 20)	**TAMBISAN** 2218 (ex-PA 31)	**MEMMON** 2223 (ex-PA 38)
NIAH 2214 (ex-PA 22)	**BAGAHAK** 2219 (ex-PA 33)	**SUBUAN** 2224 (ex-PA 41)

Dimensions, metres (feet): 22.5 × 6.3 × 1.2 *(73.8 × 20.7 × 3.9)*
Speed, knots: 38
Complement: 6 (1 officer)
Machinery: 2 MTU 12V 396 TE84 diesels; 2 waterjets

Comment: Former Marine Department craft transferred to the MMEA in 2011.

SIKUATI *3/2013*, Chris Sattler / 1486539*

10 PENGAWAL 13 CLASS (PB)

PENGAWAL 13 1313 (ex-KE 60)	**PENGAWAL 18** 1318 (ex-KE 65)
PENGAWAL 14 1314 (ex-KE 61)	**PENGAWAL 19** 1319 (ex-KE 66)
PENGAWAL 15 1315 (ex-KE 62)	**PENGAWAL 20** 1320 (ex-KE 67)
PENGAWAL 16 1316 (ex-KE 63)	**PENGAWAL 21** 1321 (ex-KE 68)
PENGAWAL 17 1317 (ex-KE 64)	**PENGAWAL 22** 1322 (ex-KE 69)

Dimensions, metres (feet): 13.5 × 4.5 × 0.8 *(44.3 × 14.8 × 2.6)*
Speed, knots: 30
Complement: 6
Machinery: 1 MTU 8V 183 TE76 diesel; 1 waterjet

Comment: Former Customs craft transferred to the MMEA in 2011.

IHS Jane's Fighting Ships 2014-2015

PENGAWAL 19 *6/2012, MMEA / 1455750*

16 PENGGALANG 16 CLASS (PB)

PENGGALANG 30 1630 (ex-KB 52)	**PENGGALANG 38** 1638 (ex-KB 71)
PENGGALANG 31 1631 (ex-KB 53)	**PENGGALANG 39** 1639 (ex-KB 72)
PENGGALANG 32 1632 (ex-KB 54)	**PENGGALANG 40** 1640 (ex-KB 73)
PENGGALANG 33 1633 (ex-KB 55)	**PENGGALANG 41** 1641 (ex-KB 74)
PENGGALANG 34 1634 (ex-KB 56)	**PENGGALANG 42** 1642 (ex-KB 75)
PENGGALANG 35 1635 (ex-KB 58)	**PENGGALANG 43** 1643 (ex-KB 76)
PENGGALANG 36 1636 (ex-KB 59)	**PENGGALANG 44** 1644 (ex-KB 77)
PENGGALANG 37 1637 (ex-KB 70)	**PENGGALANG 46** 1645 (ex-KB 78)

Dimensions, metres (feet): 16 × 4.5 × 1.1 *(52.5 × 14.8 × 3.6)*
Speed, knots: 55
Complement: 7
Machinery: 2 MAN 12V D2842 LE 406 diesels; 2,400 hp *(1.79 MW)*; 2 shafts
Guns: 1—7.62 mm MG.

Comment: Former Customs craft built by Perlis Marine Engineering in 2002 and transferred to the MMEA in 2011.

PENGGALANG 32 *6/2012, MMEA / 1455749*

27 PENGAWAL 23 CLASS (PB)

PENGAWAL 23 1420 (ex-PC 1)	**PENGAWAL 37** 1434 (ex-PC 24)
PENGAWAL 24 1421 (ex-PC 6)	**PENGAWAL 38** 1435 (ex-PC 25)
PENGAWAL 25 1422 (ex-PC 7)	**PENGAWAL 39** 1436 (ex-PC 26)
PENGAWAL 26 1423 (ex-PC 8)	**PENGAWAL 40** 1437 (ex-PC 27)
PENGAWAL 27 1424 (ex-PC 10)	**PENGAWAL 41** 1438 (ex-PC 28)
PENGAWAL 28 1425 (ex-PC 11)	**PENGAWAL 42** 1439 (ex-PC 29)
PENGAWAL 29 1426 (ex-PC 12)	**PENGAWAL 43** 1440 (ex-PC 32)
PENGAWAL 30 1427 (ex-PC 13)	**PENGAWAL 44** 1441 (ex-PC 33)
PENGAWAL 31 1428 (ex-PC 15)	**PENGAWAL 45** 1442 (ex-PC 34)
PENGAWAL 32 1429 (ex-PC 17)	**PENGAWAL 46** 1501 (ex-PC 14)
PENGAWAL 33 1430 (ex-PC 20)	**PENGAWAL 47** 1502 (ex-PC 15)
PENGAWAL 34 1431 (ex-PC 21)	**PENGAWAL 48** 1503 (ex-PC 16)
PENGAWAL 35 1432 (ex-PC 22)	**PENGAWAL 49** 1504 (ex-PC 31)
PENGAWAL 36 1433 (ex-PC 23)	

Dimensions, metres (feet): 14.8 × 2.7 × 0.8 *(48.6 × 8.9 × 2.6)*
Speed, knots: 30
Complement: 7
Machinery: 1 MTU 8V 183 TE76 diesel; 1 waterjet

Comment: Former Marine Department craft transferred to the MMEA in 2011.

TRAINING SHIPS

1 MARLIN CLASS (TRAINING VESSEL) (AX)

MARLIN 4001

Displacement, tonnes: 274 standard
Dimensions, metres (feet): 40 × 7.2 × 1.2 *(131.2 × 23.6 × 3.9)*
Speed, knots: 17
Complement: 28 (6 officers)
Machinery: 2 diesels; 1,974 hp *(1.47 MW)*; 2 shafts
Radars: Navigation: I-band.

Comment: A new training ship built and donated by The Nippon Foundation. The ship was handed over at Port Klang on 1 June 2006. Based at Lumut.

MARLIN *3/2013*, Chris Sattler / 1486543*

LAND-BASED MARITIME AIRCRAFT

Numbers/Type: 3 Agusta Westland AW139.
Operational speed: 167 kt *(309 km/h).*
Service ceiling: 20,000 ft *(6,098 m).*
Range: 675 n miles *(1,250 km).*
Role/Weapon systems: Multipurpose aircraft equipped for maritime law enforcement interdiction, marine pollution patrol, general surveillance and special forces operations. Sensors: search radar, FLIR. Weapons: 1 – 7.62 mm MG.

Numbers/Type: 3 Eurocopter AS-365N3 Dauphin.
Operational speed: 150 kt *(278 km/h).*
Service ceiling: 19,200 ft *(5,852 m).*
Range: 498 n miles *(922 km).*
Role/Weapon systems: Multipurpose aircraft equipped for maritime law enforcement interdiction, marine pollution patrol, general surveillance and special forces operations. Sensors: search radar, FLIR.

Numbers/Type: 2 Bombardier CL 415MP.
Operational speed: 187 kt *(346 km/h).*
Service ceiling: 14,700 ft *(4,500 m).*
Range: 1,319 n miles *(2,443 km).*
Role/Weapon systems: Multipurpose amphibious aircraft equipped for maritime law enforcement and maritime patrol missions. Capable of launching jet boat on water for special forces and maritime rescue operations. Capable of water-scooping and water bombing 6,000 litres in fire-fighting missions. Sensors: search radar, SLAR, FLIR.

BOMBARDIER CL 415 *3/2013*, Chris Sattler* / 1486545

AS-365 *3/2013*, Chris Sattler* / 1486546

GOVERNMENT MARITIME FORCES

Notes: As of late 2012, MMEA had taken over the majority of vessels from the Marine Operations Force and Royal Malaysian Customs as the MMEA progresses towards becoming the single maritime law enforcement agency.

Maldives

Country Overview

Formerly a British Protectorate, The Maldives gained independence in 1965 and a republic was established in 1968. Situated in the northern Indian Ocean, southwest of the southern tip of India, the country comprises a 468 n mile long chain of nearly 2,000 small coral islands that are grouped together into clusters of atolls. The capital and principal commercial centre is Malé and other populous atolls include Suvadiva and Tiladummati. An archipelagic state, territorial waters (12 n miles) are claimed. A 200 n mile Exclusive Economic Zone (EEZ) has been claimed although the limits have only been partly defined by boundary agreements.

Headquarters Appointments

Director General of Coast Guard:
 Colonel Ismail Shareef

Bases

Malé, Kaadeddhoo

Personnel

2014: 400

COAST GUARD

Notes: (1) The ex-UK patrol craft *Kingfisher* was acquired by a civilian company in early 1997. It is painted white and is used as a survey ship.
(2) There are also four RIBs in service.

2 ISKANDHAR CLASS (PB)

NOORADHEEN (ex-*Ghazee*) T 701 (ex-214)   **ISKANDHAR** T 702 (ex-223)

Displacement, tonnes: 59 full load
Dimensions, metres (feet): 24.4 × 5.8 × 1.3 *(80.1 × 19.0 × 4.3)*
Speed, knots: 37. **Range, n miles:** 600 at 25 kt
Complement: 18
Machinery: 2 Paxman diesels; 8,506 hp(m) *(6.26 MW)*; 2 Kamewa waterjets
Guns: 1 – 20 mm MG. 2 – 7.62 mm MGs.
Radars: Surface search/navigation: JRC-JMA 3625; JRC-JMA 3204; I-band.

Comment: Ordered from Colombo Dockyard in 1997. *Nooradheen* commissioned (originally as *Ghazee*) on 20 January 1998 and *Iskandhar* on 7 December 1998. Employed on security, fishery protection and SAR tasks.

HURAVEE *6/2009, Shaun Jones* / 1305928

3 TRACKER II CLASS (PB)

KAANI 133 (ex-11)   **MIDHILI** 151 (ex-13)   **NIROLHU** 106 (ex-14)

Displacement, tonnes: 40 full load
Dimensions, metres (feet): 20.2 × 5.2 × 1.5 *(66.3 × 17.1 × 4.9)*
Speed, knots: 25. **Range, n miles:** 450 at 20 kt
Complement: 10
Machinery: 2 Detroit 12V-71TA diesels; 840 hp *(627 kW)* sustained; 2 shafts
Guns: 1 – 12.7 mm MG. 1 – 7.62 mm MG.
Radars: Surface search: JRC-JMA; I-band.

Comment: First one ordered June 1985 from Fairey Marine, UK and commissioned in April 1987. Three more acquired July 1987 ex-UK Customs craft. GRP hulls. Used for fishery protection and security patrols. *Kuredhi* decommissioned in 2002.

NOORADHEEN *6/2005, Maldives Coast Guard* / 1133514

1 SDB MK 5 CLASS (LARGE PATROL CRAFT) (PBO)

HURAVEE (ex-*Tillanchang*) P 801 (ex-T 62)

Displacement, tonnes: 264 full load
Dimensions, metres (feet): 46 × 7.5 × 2.5 *(150.9 × 24.6 × 8.2)*
Speed, knots: 30. **Range, n miles:** 2,000 at 12 kt
Complement: 34 (4 officers)
Machinery: 2 MTU 16V 538 TB92 diesels; 6,820 hp(m) *(5 MW)* sustained; 2 shafts
Guns: 1 Medak 30 mm 2A42.
Radars: Surface search: Bharat 1245; I-band.

Comment: Built at Garden Reach and first commissioned in 2002. Transferred from the Indian Navy and recommissioned on 16 April 2006.

NIROLHU *6/2005, Maldives Coast Guard* / 1133513

522 Maldives > Coast guard — Malta > Introduction

1 CHEVERTON CLASS (PB)

BUREVI 115 (ex-7)

Displacement, tonnes: 26 full load
Dimensions, metres (feet): 17.3 × 4.4 × 1.3 *(56.8 × 14.4 × 4.3)*
Speed, knots: 23
Range, n miles: 590 at 18 kt
Complement: 10
Machinery: 2 MAN B&W diesels; 850 hp *(634 kW)* sustained; 2 shafts
Guns: 1—12.7 mm MG. 1—7.62 mm MG.
Radars: Surface search: JRC; I-band.

Comment: GRP hull and aluminium superstructure. Originally built by Fairey Marine, UK, for Kiribati and subsequently sold to Maldives and commissioned on 11 September 1981. Used for security and SAR operations.

BUREVI   6/2005, Maldives Coast Guard / 1133512

3 HARBOUR PATROL CRAFT (PB)

HC 106   HC 110   HC 112

Displacement, tonnes: 6 full load
Dimensions, metres (feet): 11 × 2.3 × 0.5 *(36.1 × 7.5 × 1.6)*
Speed, knots: 30
Range, n miles: 90 at 25 kt
Complement: 8
Machinery: 2 Yamaha outboard engines; 500 hp *(375 kW)*
Guns: 1—7.62 mm MG.

Comment: Built by Gulf Craft Service based in the Maldives. GRP hull. First craft commissioned 12 December 1999. Used for harbour patrol and SAR duties.

GULF CRAFT   2/2009, US Navy / 1455793

1 LANDING CRAFT (LCM)

L 301 (ex-LC 1)

Displacement, tonnes: 39 standard
Dimensions, metres (feet): 20.9 × 5.0 × 0.7 *(68.6 × 16.4 × 2.3)*
Speed, knots: 20
Range, n miles: 500 at 18 kt
Complement: 7
Machinery: 2 MAN B&W D 2842 LE 401 diesels; 2 Hamilton HM 422 waterjets
Guns: 2—7.62 mm MGs.
Radars: Surface search/Navigation: JRC-JMA 3253; I-band.

Comment: Built by Colombo Dockyard and commissioned on 12 December 1999. Aluminium hull and superstructure. Used for carrying troops and supplies.

L 301   6/2012 / 1455794

1 40 M PATROL CRAFT (PB)

GHAZEE P 802

Measurement, tonnes: 400 gt
Dimensions, metres (feet): 42.2 × 7.21 × 1.58 *(138.5 × 23.7 × 5.2)*
Speed, knots: 30
Complement: 27 (9 officers)
Machinery: 3 Cummins KTA50M2 diesels; 5,487 hp *(4.03 MW)*; 2 shafts; 1 (centreline) Hamilton HM721 waterjet
Guns: 1—7.62 mm MG.
Radars: Navigation: JMA 2344 and JMA 5320-7; I-band.

Comment: Contract with Colombo Shipyard, Sri Lanka, in January 2005 for the construction of a 35 m patrol craft. Keel laid in August 2006, launched in June 2007 and delivered on 22 December 2007. Steel construction.

GHAZEE   6/2013*, Maldives Coast Guard / 1525538

1 35 M PATROL CRAFT (PB)

SHAHEED ALI P 322

Measurement, tonnes: 300 gt
Dimensions, metres (feet): 35.3 × 7.0 × 1.35 *(115.8 × 23.0 × 4.4)*
Speed, knots: 28
Complement: 15 (7 officers)
Machinery: 2 Cummins KTA50M2 diesels; 3,658 hp *(2.69 MW)*; 2 Hamilton HM721 waterjets
Guns: 1—7.62 mm MG.
Radars: Navigation: JMA 2344 and JMA 5320-7; I-band.

Comment: Contract with Colombo Shipyard, Sri Lanka, in January 2005 for the construction of a patrol craft. Keel laid in May 2005, launched in December 2006 and delivered on 19 April 2007. Steel construction.

SHAHEED ALI   6/2013*, Maldives Coast Guard / 1525537

Malta

Country Overview

Formerly a British colony, the Republic of Malta gained independence in 1964. Situated 45 n miles south of Sicily, the country comprises the islands of Malta (95 square miles), Gozo (26 square miles), Comino, Kemmunett, and Filfla. It has a 76 n mile coastline with the Mediterranean Sea. The capital, largest town and principal port is Valletta. Territorial seas (12 n miles) are claimed. A fishery management and conservation zone of 25 n miles is also claimed.

Headquarters Appointments

Officer Commanding Maritime Squadron:
  Lieutenant Colonel Andrew Mallia

General

The Maritime Squadron of the Armed Forces of Malta was established in November 1970. An independent unit of the Armed Forces of Malta, it is employed primarily as a Coast Guard.

Personnel

2014: 302 (12 officers)

Bases

Haywharf (Malta)
Mgarr (Gozo)

IHS Jane's Fighting Ships 2014-2015   © 2014 IHS

Patrol forces — **Malta** 523

PATROL FORCES

2 MARINE PROTECTOR CLASS (PB)

P 51   P 52

Displacement, tonnes: 92 full load
Dimensions, metres (feet): 26.5 × 5.8 × 1.6 *(86.9 × 19.0 × 5.2)*
Speed, knots: 25
Range, n miles: 900 at 8 kt
Complement: 12 (1 officer)
Machinery: 2 MTU 8V 396 TE94 diesels; 2,680 hp(m) *(1.97 MW)* sustained; 2 shafts
Radars: Navigation: I-band.
Electro-optic systems: Selex Janus-N.

Comment: Built by Bollinger Shipyards to US Coast Guard specifications. The vessels are based on the hull of the Damen Stan Patrol 2600 in service with the Hong Kong police. Steel hull with GRP superstructure. A stern ramp is used for launching a 5.5 m RHIB (Zodiac ZH 558). P 51 was commissioned 18 November 2002 and P 52 was commissioned 7 July 2004.

P 52   *4/2012, Frank Findler* / 1483615

1 SAETTIA MK III CLASS (OFFSHORE PATROL VESSEL) (PBO)

P 61

Displacement, tonnes: 399 full load
Dimensions, metres (feet): 53.4 × 8.1 × 5.4 *(175.2 × 26.6 × 17.7)*
Speed, knots: 23
Range, n miles: 2,100 at 16 kt
Complement: 25 (4 officers)

Machinery: 2 Isotto Fraschini V1716 T2 MSD diesels; 6,335 hp *(4.7 MW)*; 2 shafts
Guns: 1 Otobreda 25 mm. 2 — 12.7 mm MGs.
Radars: Surface search: E/F-band.
Navigation: I-band.
Helicopters: Platform for 1 medium.

Comment: Financed from the 5th Italo-Maltese Protocol, contract signed on 12 March 2004 with Fincantieri, Muggiano, Italy for the construction of one vessel. The ship was commissioned on 3 November 2005. The contract included a training and logistic support package. Design based on vessels in service with the Italian Coast Guard. Steel hull with helicopter deck and stern ramp for launching a 6.5 m RIB.

P 61   *5/2009, Frank Findler* / 1305619

2 SUPERVITTORIA 800 CLASS (SAR)

MELITA I   MELITA II

Displacement, tonnes: 13 full load
Dimensions, metres (feet): 11.5 × 4.9 × 0.8 *(37.7 × 16.1 × 2.6)*
Speed, knots: 34. **Range, n miles:** 160 at 34 kt
Complement: 4
Machinery: 2 Cummins 6CTA 8.3 DIAMONS; 840 hp(m) *(618 kW)*; 2 Kamewa FF310 waterjets
Radars: Surface search: Raytheon Pathfinder SL 70; I-band.

Comment: Built in 1998 by Vittoria Naval Shipyard, Italy, for the Civil Protection Department of Malta. Transferred to the Armed Forces of Malta (AFM) in May 1999 for search and rescue duties. Both craft were refitted with new engines in 2011.

MELITA 1   *4/2012, Frank Findler* / 1483617

1 HIGH-SPEED INTERCEPTION CRAFT (HSIC)

P 01

Displacement, tonnes: 3 full load
Dimensions, metres (feet): 10.37 × 2.67 × 0.8 *(34 × 8.8 × 2.6)*
Speed, knots: 50 (est.)
Range, n miles: 200 at 35 kt
Complement: 10
Machinery: 2 VM diesels; 600 hp *(450 kW)*; 2 shafts
Guns: 1 — 7.62 mm MG.
Radars: Navigation: Furuno; I-band.

Comment: Co-financed by the EU. RIB 33SC designed and built by FB Design, Italy, and commissioned in February 2006. The high-speed interception craft is to provide support to maritime law enforcement agencies and special forces.

P 01   *4/2012, Frank Findler* / 1483618

4 PATROL CRAFT (PB)

P 21–24

Dimensions, metres (feet): 21.2 × 5.5 × 1.8 *(69.6 × 18.0 × 5.9)*
Speed, knots: 28
Complement: 8
Machinery: 2 MAN D2842 LE 410 diesels; 2,170 hp *(1.62 MW)*; 2 shafts
Guns: 1 — 12.7 mm MG; 2 — 7.62 mm MGs.
Radars: Surface search/navigation: To be announced.

Comment: Austal awarded the contract for the construction of four vessels, to replace the Bremse and Swift classes, on 18 February 2009. The contract was co-financed by the EU's External Borders Fund. The craft have a planning aluminium hull and are equipped with a stern ramp to facilitate launch/recovery of a 3.5 m rigid hull inflatable boat. All four were delivered in November 2009 and commissioned on 29 March 2010.

P 23   *4/2012, Frank Findler* / 1483616

3 BOOMERANGER INTERCEPTION CRAFT (PBF)

P 02–04

Displacement, tonnes: 4.6 full load
Dimensions, metres (feet): 11.8 × 3.5 × ? *(38.7 × 11.5 × ?)*
Speed, knots: 35 (est.)
Range, n miles: 200 at 35 kt
Complement: 6 + 8 personnel
Machinery: 3 Mercury Verado outboard motors; 1,050 hp *(783 kW)*
Guns: 1 — 7.62 mm MG.
Radars: Navigation: Ray Marine; I-band.

Comment: Co-financed from the EU External Borders fund and delivered in February 2013. The craft were built by Boomeranger Boats, Finland. Equipped with shock-absorbing seating for an eight-man boarding team.

P 04   *6/2013*, AFM* / 1455832

© 2014 IHS   *IHS Jane's Fighting Ships 2014-2015*

524 Malta › Patrol forces — Mauritania › Introduction

2 SAFEBOATS DEFENDER 310X CLASS
(INTERCEPTION CRAFT) (PBF)

P 05   P 06

Displacement, tonnes: 4.5 full load
Dimensions, metres (feet): 10.7 × 3.0 × 0.64 (35.1 × 9.8 × 2.1)
Speed, knots: 30
Range, n miles: 150 at 30 kt
Complement: 4 + 6 personnel

Machinery: 2 Mercury Verado outboard motors; 450 hp (335 kW)
Guns: 1 — 7.62 mm MG.
Radars: Navigation: I-band.
Electro-optic systems: FLIR.

Comment: Financed from US government Export Control and Related Border Security (EXBS) funding and built by SAFE Boats International. Similar to craft in service with US Coast Guard but with extended cabin and onboard generator.

LAND-BASED MARITIME AIRCRAFT

Notes: The Armed Forces of Malta operate one Britten-Norman BN-2B maritime patrol aircraft, two King Air B200, equipped with Telephonics radar, three SA.316B/D Alouette III and two AW 139 helicopters.

ALOUETTE III   6/2013*, AFM / 1455830

KING AIR B 200   6/2013*, AFM / 1455829

P 05   6/2013*, AFM / 1455831

Marshall Islands

Country Overview

The Republic of the Marshall Islands was a US-administered UN Trust territory from 1947 before becoming a self-governing republic in 1979. In 1986, a Compact of Free Association, delegating to the US the responsibility for defence and foreign affairs, came into effect. The country consists of some 1,200 atolls and reefs in the central Pacific. There are two main island groups: the Ratak and Ralik chains. Majuro is the capital island. Kwajalein is the largest atoll and is leased as a US missile test range. Bikini and Enewetak are former US nuclear test sites. An archipelagic state, territorial seas (12 n miles) are claimed. An Exclusive Economic Zone (EEZ) (200 n miles) is also claimed but limits have not been fully defined.

Headquarters Appointments

Sea Patrol Advisor:
  Lieutenant Commander Peter Metcalfe, RAN

Personnel

2014: 30

Bases

Majuro

PATROL FORCES

1 PACIFIC CLASS (LARGE PATROL CRAFT) (PB)

Name	No	Builders	Commissioned
LOMOR	03	Australian Shipbuilding Industries	29 Jun 1991

Displacement, tonnes: 165 full load
Dimensions, metres (feet): 31.5 × 8.1 × 2.1 (103.3 × 26.6 × 6.9)
Speed, knots: 20
Range, n miles: 2,500 at 12 kt
Complement: 17 (3 officers)
Machinery: 2 Caterpillar 3516TA diesels; 4,400 hp (3.3 MW) sustained; 2 shafts
Guns: 1 — 12.7 mm MG.
Radars: Surface search: Furuno 8111; I-band.

Comment: The 14th craft to be built in this series for a number of Pacific Island coast guards. Ordered in 1989. Following the decision by the Australian government to extend the Pacific Patrol Boat project to a 30-year life for each boat, *Lomor* completed a half-life refit in 1999 and a life-extension refit in December 2008.

LOMOR
12/2008, Chris Sattler
1335795

Mauritania

MARINE MAURITANIENNE

Country Overview

A former French colony, The Islamic Republic of Mauritania gained full independence in 1960. With an area of 397,955 square miles, it is situated in northwestern Africa and has borders to the north with western Sahara and Algeria, to the east with Mali and to the south with Senegal. It has a 405 n mile coastline with the Atlantic Ocean. The capital and largest city is Nouakchott while Nouadhibou is the principal port. Territorial seas (12 n miles) are claimed but while it has claimed a 200 n mile Exclusive Economic Zone (EEZ), its limits have not been defined by boundary agreements.

Headquarters Appointments

Commander of Navy:
  Captain Isselkou Ould Cheikh El-Weli

Personnel

(a) 2014: 700 (40 officers) plus 200 marines
(b) Voluntary service

Bases

Port Etienne, Nouadhibou (new quay began construction in 2007)
Port Friendship, Nouakchott

IHS Jane's Fighting Ships 2014-2015   © 2014 IHS

PATROL FORCES

Notes: (1) Two 16 m Rodman 55M (ex-M 02 and M 05) and two 12 m Saeta 12 (ex-L 01 and L 03) were transferred from the Spanish Guardia Civil in 2006.
(2) Two Raidco RPB 20 vessels *Oum Tounsi* and *Megsem Bobakkar* are operated by the coastal rescue service.
(3) A Chinese-built 60 m patrol ship, *Awkar*, was completed in December 2012 and was delivered in early 2013. It is operated by the Coast Guard.

1 OPV 54 CLASS (PBO)

Name	No	Builders	Launched	Commissioned
ABOUBEKR BEN AMER	P 541	Leroux & Lotz, Lorient	17 Dec 1993	7 Apr 1994

Displacement, tonnes: 380 full load
Dimensions, metres (feet): 54 × 10 × 2.8 *(177.2 × 32.8 × 9.2)*
Speed, knots: 23; 8 motor. **Range, n miles:** 4,500 at 12 kt
Complement: 21 (3 officers)
Machinery: 2 MTU 16V 396 TE94 diesels; 5,712 hp(m) *(4.2 MW)* sustained; 2 auxiliary motors; 250 hp(m) *(184 kW)*; 2 shafts; cp props
Guns: 2 – 12.7 mm MGs.
Radars: Surface search: Racal Decca Bridgemaster 250; I-band.

Comment: Ordered in September 1992. This is the prototype to a Serter design of three similar craft built for the French Navy. Stern ramp for a 30 kt RIB. Option on a second of class not taken up. Refitted at Lorient 2001.

ABOUBEKR BEN AMER 7/2001, Peron/Marsan / 0137787

1 LARGE PATROL CRAFT (PBO)

Name	No	Builders	Commissioned
VOUM-LEGLEITA (ex-*Poseidon*)	B 551 (ex-A 12)	Bazán	8 Aug 1964

Displacement, tonnes: 1,086 full load
Dimensions, metres (feet): 55.9 × 10 × 4 *(183.4 × 32.8 × 13.1)*
Speed, knots: 15. **Range, n miles:** 4,640 at 14 kt
Complement: 60
Machinery: 2 Sulzer diesels; 3,200 hp *(2.53 MW)*; 1 shaft; cp prop
Guns: 2 Oerlikon 20 mm.
Radars: Navigation: 2 Decca TM 626; I-band.

Comment: Ocean going tug transferred from Spain in January 2000, about a year later than planned. Used primarily as an OPV and for fishery protection.

VOUM-LEGLEITA 1/2000, Diego Quevedo / 0081240

1 HUANGPU CLASS (PB)

LIMAM EL HADRAMI P 601

Displacement, tonnes: 437 full load
Dimensions, metres (feet): 60 × 8.2 × 4.5 *(196.9 × 26.9 × 14.8)*
Speed, knots: 20

Machinery: 3 MTU 12V 4000 diesels; 3 shafts
Guns: 2 – 37 mm (1 twin). 4 – 14.5 mm (2 twin).
Radars: Navigation: I-band.
Fire control: Type 347G; I-band.

Comment: Delivered from China on 20 April 2002.

LIMAN EL HADRAMI 12/2009, Annati Collection / 1406078

1 ARGUIN CLASS (PBO)

Name	Builders	Commissioned
ARGUIN	Fassmer Werft, Berne; Motzen	17 Jul 2000

Measurement, tonnes: 1,016 dwt
Dimensions, metres (feet): 54.5 × 10.9 × 4.5 *(178.8 × 35.8 × 14.8)*
Speed, knots: 16.5. **Range, n miles:** 15,000 at 12 kt
Complement: 13
Machinery: 2 MaK 6M20 diesels; 2,735 hp *(2.04 MW)*; 1 shaft; cp prop

Comment: Ordered in 1998. Hull construction at Yantar, Kaliningrad. Steel hull and superstructure. Equipped with interception craft on centreline ramp in mother-daughter configuration.

ARGUIN 7/2000, Fassmer Werft / 1044268

4 MANDOVI CLASS (INSHORE PATROL CRAFT) (PB)

Displacement, tonnes: 15 full load
Dimensions, metres (feet): 15 × 3.6 × 0.8 *(49.2 × 11.8 × 2.6)*
Speed, knots: 24
Range, n miles: 250 at 14 kt
Complement: 8
Machinery: 2 Deutz MWM TBD232V12 Marine diesels; 750 hp(m) *(551 kW)*; 2 Hamilton water-jets
Guns: 1 – 7.62 mm MG.
Radars: Navigation: Furuno FR 8030; I-band.

Comment: Built by Garden Reach, Calcutta and delivered from India in 1990. Some may not be operational.

2 CONEJERA CLASS (PB)

Name	No	Builders	Commissioned
– (ex-*Espalmador*)	– (ex-P 33)	Bazán, Ferrol	31 Dec 1981
– (ex-*Alcanada*)	– (ex-P 34)	Bazán, Ferrol	10 May 1982

Displacement, tonnes: 86 full load
Dimensions, metres (feet): 32.2 × 5.3 × 1.4 *(105.6 × 17.4 × 4.6)*
Speed, knots: 13
Range, n miles: 1,200 at 13 kt
Complement: 12
Machinery: 2 MTU-Bazán MA 16V 362 SB80 diesels; 2,450 hp *(1.8 MW)*; 2 shafts
Guns: 1 Oerlikon 20 mm/120 Mk 10. 1 – 12.7 mm MG.
Radars: Surface search: Furuno; I-band.

Comment: Former Spanish coastal patrol craft transferred in 2007. The details are as for the vessels in Spanish service and may differ.

Ex-ALCANADA 3/2012, A Campanera i Rovira / 1455795

1 RAIDCO RPB 18 CLASS (PATROL CRAFT) (PB)

YACOUB OULD RAJEL

Dimensions, metres (feet): 17.8 × 4.6 × 1.25 *(58.4 × 15.1 × 4.1)*
Speed, knots: 23
Machinery: 2 MAN V12 diesels; 1,500 hp *(1.1 MW)*; 2 shafts
Radars: Navigation: I-band.

Comment: Donated by the European Union, the vessel was built by Raidco Marine and delivered in 2000. Steel hull and aluminium superstructure. Used for fishery protection. A similar craft, *Dah Ould Bah*, is operated by the Customs Service.

LAND-BASED MARITIME AIRCRAFT

Numbers/Type: 2 Piper Cheyenne II.
Operational speed: 283 kt *(524 km/h)*.
Service ceiling: 31,600 ft *(9,630 m)*.
Range: 1,510 n miles *(2,796 km)*.
Role/Weapon systems: Coastal surveillance and EEZ protection acquired 1981. Sensors: Bendix 1400 weather radar; cameras. Weapons: Unarmed.

2 BRAVO (BRONSTEIN) CLASS (FFH)

Name	No	Builders	Laid down	Launched	Commissioned
NICOLAS BRAVO (ex-*McCloy*)	F 201 (ex-E 40, ex-FF 1038)	Avondale Shipyards, New Orleans	15 Sep 1961	9 Jun 1962	21 Oct 1963
HERMENEGILDO GALEANA (ex-*Bronstein*)	F 202 (ex-E 42, ex-FF 1037)	Avondale Shipyards, New Orleans	16 May 1961	31 Mar 1962	16 Jun 1963

Displacement, tonnes: 2,398 standard; 2,693 full load
Dimensions, metres (feet): 113.2 × 12.3 × 4.1 hull; 7 sonar *(371.4 × 40.4 × 13.5; 23.0)*
Speed, knots: 23.5.
Range, n miles: 3,924 at 15 kt
Complement: 207 (17 officers)

Machinery: 2 Foster-Wheeler boilers; 1 De Laval geared turbine; 20,000 hp *(14.92 MW)*; 1 shaft
Missiles: A/S: Honeywell ASROC Mk 112 octuple launcher ❶.
Guns: 2 USN 3 in *(76 mm)*/50 (twin) Mk 33 ❷; 50 rds/min to 12.8 km *(7 n miles)*; weight of shell 6 kg, or 1 Bofors 57 mm/70 Mk 2; 220 rds/min to 17 km *(9.3 n miles)*; weight of shell 2.4 kg.
Torpedoes: 6—324 mm US Mk 32 Mod 7 (2 triple) tubes ❸. 14 Honeywell Mk 46; anti-submarine; active/passive homing to 11 km *(5.9 n miles)* at 40 kt; warhead 44 kg.
Physical countermeasures: Decoys: 2 Loral Hycor 6-barrelled fixed Mk 33; IR flares and chaff to 4 km *(2.2 n miles)*. T-Mk 6 Fanfare; torpedo decoy system.
Radars: Air search: Lockheed SPS-40D ❹; B-band; range 320 km *(175 n miles)*.
Surface search: Raytheon SPS-10F ❺; G-band.
Navigation: Marconi LN66; I-band.
Fire control: General Electric Mk 35 ❻; I/J-band.
Sonars: EDO/General Electric SQS-26 AXR; bow-mounted; active search and attack; medium frequency.

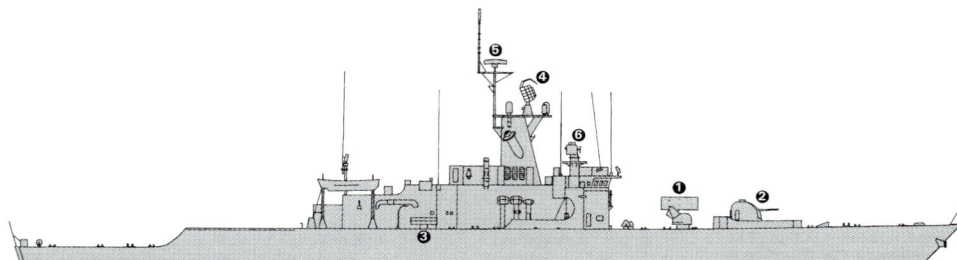

NICOLAS BRAVO *(Scale 1 : 900), Ian Sturton / 0506240*

Weapon control systems: Mk 56 GFCS. Mk 114 ASW FCS. Mk 1 target designation system. Elsag NA 18 optronic director may be fitted.
Helicopters: Platform and some facilities but no hangar.
Programmes: Transferred from the US to Mexico by sale 12 November 1993 having paid off in December 1990.
Modernisation: Bofors 57 mm SAK may be fitted to replace the Mk 33 gun, possibly with an Elsag NA 18 optronic director.
Structure: Position of stem anchor and portside anchor (just forward of gun mount) necessitated by large bow sonar dome. As built, a single 3 in (Mk 34) open mount was aft of the helicopter deck; removed for installation of towed sonar which has since been taken out.
Operational: ASROC is non-operational. Pennant numbers changed in 2001. Both based at Manzanillo.

NICOLAS BRAVO
6/2010, US Navy / 1406079

HERMENEGILDO GALEANA
4/2012, US Navy / 1528825*

SHIPBORNE AIRCRAFT

Numbers/Type: 6 MD 902 Explorer.
Operational speed: 113 kt *(210 km/h)*.
Service ceiling: 9,845 ft *(3,000 m)*.
Range: 407 n miles *(754 km)*.
Role/Weapon systems: Coastal patrol helicopter acquired 1999-2000 for patrol, fisheries protection and EEZ protection duties; SAR as secondary role. Sensors: Bendix search radar. Weapons: MGs or rocket pods.

BO 105CB   *9/1994, Mexican Navy* / 0052606

Numbers/Type: 11 Bolkow BO 105.
Operational speed: 100 kt *(185 km/h)*.
Service ceiling: 17,000 ft *(5,180 m)*.
Range: 160 n miles *(296 km)*.
Role/Weapon systems: Coastal patrol helicopter acquired 1982-86 for patrol, fisheries protection and EEZ protection duties; SAR as secondary role. A modernisation programme was announced in October 2003; the first upgraded aircraft was delivered in 2004 and the programme was completed by late 2006. Sensors: Bendix search radar. Weapons: MGs or rocket pods.

BO 105   *6/2010, US Navy* / 1406080

Numbers/Type: 2 Eurocopter AS 555 AF Fennec.
Operational speed: 121 kt *(225 km/h)*.
Service ceiling: 13,120 ft *(4,000 m)*.
Range: 389 n miles *(722 km)*.
Role/Weapon systems: Patrol helicopter for EEZ protection and SAR. Operated from Oaxaca class patrol ships. More may be acquired when funds are available. Sensors: Bendix 1500 search radar. Weapons: Can carry up to two torpedoes, rocket pods or an MG.

AS 555 AF   *6/2005, Mexican Navy* / 1133541

Numbers/Type: 4 Eurocopter AS 565ME Panther.
Operational speed: 165 kt *(305 km/h)*.
Service ceiling: 15,223 ft *(4,640 m)*.
Range: 200 n miles *(370 km)*.
Role/Weapon systems: Transport and reconnaissance helicopter. Two delivered in June 2005, three in 2010. One was lost in June 2012. Four further aircraft are to be acquired. Capable of carrying eight passengers or 1,000 kg load.

AS 565 ME   *6/2005, Mexican Navy* / 1133540

LAND-BASED MARITIME AIRCRAFT (FRONT LINE)

Notes: (1) Transport aircraft used include two Learjets, one Dash 8-200, 24 Mil Mi-17 and four MD 500E.
(2) Training aircraft include seven (of which four reported operational) Aeromacchi M-290TP Redigos, eight Maule MX-Z-180, seven Zlin Z242L, one Cessna 402C, three Lancair and five Schweizer 300C.
(3) Six UH-60M Black Hawk are to be transferred from the US under FMS funding arrangements. Three delivered in 2013.
(4) Four EADS-CASA C-295 transport aircraft delivered by late 2011.

Mi-2   *6/2005, Mexican Navy* / 1133539

Mi-17   *6/2009, Annati Collection* / 1367106

Numbers/Type: 7 CASA C-212 Aviocar.
Operational speed: 190 kt *(353 km/h)*.
Service ceiling: 24,000 ft *(7,315 m)*.
Range: 1,650 n miles *(3,055 km)*.
Role/Weapon systems: Acquired from 1987 and used for Maritime Surveillance. Two aircraft upgraded in Spain with EADS/CASA Integrated Tactical System (FITS) in 2003. The remainder upgraded in Mexico by late 2006. One aircraft lost in November 2006. Sensors: Search radar; APS 504, Sisdef SP 100 datalink. Weapons: Unarmed.

C-212   *6/2005, Mexican Navy* / 1133536

Numbers/Type: 4 Rockwell Turbo Commander.
Operational speed: 250 kt *(463 km/h)*.
Service ceiling: 31,000 ft *(9,450 m)*.
Range: 480 n miles *(890 km)*.
Role/Weapon systems: Acquired in 1992. Used for reconnaissance and transport.

TURBO COMMANDER   *6/2004, Mexican Navy* / 1133544

532 Mexico > Land-based maritime aircraft — Patrol forces

Numbers/Type: 1 Antonov AN-32B.
Operational speed: 250 kt *(463 km/h).*
Service ceiling: 26,000 ft *(7,925 m).*
Range: 750 n miles *(1,390 km).*
Role/Weapon systems: Acquired 1997-99. Used for transport and reconnaissance. Three aircraft refurbished in Ukraine by 2010. Two replaced by Ocean Sentry aircraft by 2013. Two aircraft fitted with FLIR.

Numbers/Type: 6 EADS/CASA HC-144A (CNB-235-300) Ocean Sentry.
Operational speed: 236 kt *(437 km/h).*
Service ceiling: 25,000 ft *(7,620 m).*
Range: 1,565 n miles *(2,519 km).*
Role/Weapon systems: Six aircraft acquired (the first in November 2011 and the remainder in 2012) from the US. Likely to be broadly similar to aircraft in USCG service. Sensors: Surface search/weather radar; electro-optical/infrared sensors.

AN-32B *12/2009, Annati Collection* / 1305965

CNB-235 *6/2012, Annati Collection* / 1486293

PATROL FORCES

Notes: The Caribe project was initiated in about 1998 and involves the construction of 31 m patrol craft of 110 tons. One was reported 30% completed at ASTIMAR 3 in September 2006. The project has been abandoned.

2 HURACAN (SAAR 4.5) CLASS (FAST ATTACK CRAFT — MISSILE) (PTG)

Name	No	Builders	Launched	Commissioned
HURACAN (ex-*Aliya*)	301	Israel Shipyards, Haifa	11 Jul 1980	11 Jul 1980
TORMENTA (ex-*Geoula*)	302	Israel Shipyards, Haifa	Oct 1980	31 Dec 1980

Displacement, tonnes: 506 full load
Dimensions, metres (feet): 61.7 × 7.6 × 2.8 *(202.4 × 24.9 × 9.2)*
Speed, knots: 31
Range, n miles: 3,000 at 17 kt, 1,500 at 30 kt
Complement: 53

Machinery: 4 MTU/Bazán 16V 956 TB91 diesels; 15,000 hp(m) *(11.03 MW)* sustained; 4 shafts
Missiles: SSM: 4 IAI Gabriel II; radar or optical guidance; semi-active radar plus anti-radiation homing to 36 km *(19.4 n miles)* at 0.7 Mach; warhead 75 kg.
Guns: 2 Oerlikon 20 mm; 800 rds/min to 2 km. 1 Raytheon Vulcan Phalanx 6-barrelled 20 mm Mk 15; 3,000 rds/min combined to 1.5 km anti-missile. 2 – 12.7 mm MGs.
Physical countermeasures: Decoys: 1 – 45-tube, 4 – 24-tube, 4 single-tube chaff launchers.
Electronic countermeasures: ESM/ECM: Elisra NS 9003/5; intercept and jammer.
Radars: Air/surface search: Thomson-CSF TH-D 1040 Neptune; G-band.
Fire control: Selenia Orion RTN-10X; I/J-band.
Combat data systems: IAI Reshet datalink.

Programmes: First two of the original class of five Saar 4.5s, before conversions from Saar 4s were started. Transferred to Mexico on 1 June 2004.
Structure: The CIWS mounted in the eyes of the ship replaced a 40 mm gun.
Operational: Test-firing of a Gabriel missile took place in June 2010. Both based at Coatzacoalcos.

TORMENTA *7/2004, Diego Quevedo* / 0584000

HURACAN *7/2004, Diego Quevedo* / 0583999

IHS Jane's Fighting Ships 2014-2015 © 2014 IHS

Patrol forces < Mexico

4 HOLZINGER (ÁGUILA) CLASS (GUNSHIPS) (PSOH)

Name	No	Builders	Laid down	Launched	Commissioned
CAPITÁN DE NAVIO SEBASTIAN JOSÉ HOLZINGER (ex-*Uxmal*)	P 131 (ex-C 01, ex-GA 01)	ASTIMAR 20, Salina Cruz	1 Jun 1985	1 Jun 1988	1 May 1991
CAPITÁN DE NAVIO BLAS GODINEZ (ex-*Mitla*)	P 132 (ex-C 02, ex-GA 02)	ASTIMAR 1, Tampico	1 Jul 1985	22 Mar 1988	1 Nov 1991
BRIGADIER JOSÉ MARIÁ DE LA VEGA (ex-*Peten*)	P 133 (ex-C 03, ex-GA 03)	ASTIMAR 20, Salina Cruz	22 Sep 1986	1 Jun 1988	16 Mar 1994
GENERAL FELIPE B BERRIOZÁBAL (ex-*Anahuac*)	P 134 (ex-C 04, ex-GA 04)	ASTIMAR 1, Tampico	9 Mar 1988	21 Apr 1991	18 Mar 1994

Displacement, tonnes: 1,311 full load
Dimensions, metres (feet): 74.4 × 10.5 × 3.4
(244.1 × 34.4 × 11.2)
Speed, knots: 22. **Range, n miles:** 3,820 at 16 kt
Complement: 75 (11 officers)

Machinery: 2 MTU 20V 956 TB92 diesels; 11,700 hp(m) *(8.6 MW)* sustained; 2 shafts
Guns: 2 Bofors 40 mm/60 (1 twin).
Radars: Surface search: Raytheon SPS-64(V)6A; I-band.
Navigation: Kelvin Hughes Nucleus; I-band.
Combat data systems: Elsag 2 CSDA-10.
Electro-optic systems: Elsag NA 18 optronic director (P 133, P 134). Garfio 2.0 (P 131).

Helicopters: 1 MBB BO 105 CB.

Programmes: Originally four were ordered from Tampico and Veracruz. First laid down November 1983, second in 1984 but then there were delays caused by financial problems. Named after military heroes.
Structure: An improved variant of the Bazán Halcon (Uribe) class with a flight deck extended to the stern.
Operational: Pennant numbers changed in 2001. All based at Lazaro.

CAPITÁN DE NAVIO SEBASTIAN JOSÉ HOLZINGER   6/2010, Mexican Navy / 1406082

3 SIERRA CLASS (GUNSHIPS) (PSOH)

Name	No	Builders	Laid down	Launched	Commissioned
JUSTO SIERRA MENDEZ	P 141 (ex-C 2001)	ASTIMAR 1, Tampico	19 Jan 1998	1 Jun 1998	1 Jun 1998
GUILLERMO PRIETO	P 143 (ex-C 2003)	ASTIMAR 1, Tampico	1 Jun 1998	18 Sep 1999	18 Sep 1999
MATIAS ROMERO	P 144 (ex-C 2004)	ASTIMAR 20, Salina Cruz	23 Jul 1998	17 Sep 1999	17 Sep 1999

Displacement, tonnes: 1,366 full load
Dimensions, metres (feet): 70.4 × 10.5 × 2.8
(231 × 34.4 × 9.2)
Speed, knots: 18
Complement: 76 (10 officers)

Machinery: 2 Caterpillar 3616 V16 diesels; 6,197 hp(m) *(4.55 MW)*; 2 shafts
Missiles: SA-N-10 (P 144); IR homing to 5 km *(2.7 n miles)* at 1.7 Mach; warhead 1.5 kg.
Guns: 1 Bofors 57 mm/70 Mk 2; 220 rds/min to 17 km *(9.3 n miles)*; weight of shell 2.4 kg.
Radars: Air/surface search: E/F-band.
Surface search: I-band.
Combat data systems: Alenia 2.
Weapon control systems: Saab EOS 450 optronic director.

Helicopters: 1 MD 902 Explorer.

Programmes: Follow on to the Holzinger class. Ordered in 1997.
Structure: Derived from the Holzinger class but with a markedly different superstructure. All ships carry 11 m interceptor craft capable of 50 kt.
Operational: All based at Acapulco. P 142 *Benito Juarez* badly damaged by fire in October 2003 and subsequently decommissioned. The ship was sunk as a target in July 2007.

JUSTO SIERRA MENDEZ   6/2005, Mexican Navy / 1133535

GUILLERMO PRIETO   6/2010, Mexican Navy / 1406081

4 DURANGO CLASS (GUNSHIPS) (PSOH)

Name	No	Builders	Laid down	Launched	Commissioned
DURANGO	P 151	ASTIMAR 1, Tampico	18 Dec 1999	11 Sep 2000	11 Sep 2000
SONORA	P 152	ASTIMAR 20, Salina Cruz	14 Dec 1999	4 Sep 2000	4 Sep 2000
GUANAJUATO	P 153	ASTIMAR 1, Tampico	2000	13 Dec 2001	13 Dec 2001
VERACRUZ	P 154	ASTIMAR 20, Salina Cruz	4 Sep 2000	17 Dec 2001	17 Dec 2003

Displacement, tonnes: 1,494 full load
Dimensions, metres (feet): 81.8 × 10.5 × 2.8 (268.4 × 34.4 × 9.2)
Speed, knots: 18
Complement: 76 (10 officers)

Machinery: 2 Caterpillar 3616 V16 diesels; 6,197 hp(m) (4.55 MW); 2 shafts

Guns: 1 Bofors 57 mm/70 Mk 3; 220 rds/min to 17 km (9.3 n miles); weight of shell 2.4 kg.
Radars: Air/surface search: E/F-band. Surface search: I-band.
Combat data systems: Alenia 2.
Weapon control systems: Saab EOS 450 optronic director.

Helicopters: 1 MD 902 Explorer.

Programmes: Follow on to the Sierra class. Ordered on 1 June 1998.
Structure: Derived from the Holzinger class but with a markedly different superstructure. Durango class slightly larger than the Sierra class. All ships carry 11 m interceptor craft capable of 50 kt.
Operational: P 151 and P 152 based at Guaymas and P 153 and P 154 at Coatzacoalcos.

SONORA   6/2010, Mexican Navy / 1406088

4 + (2) OAXACA CLASS (GUNSHIPS) (PSOH)

Name	No	Builders	Laid down	Launched	Commissioned
OAXACA	P 161	ASTIMAR 20, Salina Cruz	17 Dec 2001	11 Apr 2003	1 May 2003
BAJA CALIFORNIA	P 162	ASTIMAR 1, Tampico	13 Dec 2001	21 May 2003	1 Aug 2003
BICENTENARIO DE LA INDEPENDENCIA	P 163	ASTIMAR 20, Salina Cruz	11 Apr 2003	23 Jul 2009	Nov 2011
CENTENARIO DE LA REVOLUCIÓN	P 164	ASTIMAR 1, Tampico	21 May 2003	23 Nov 2009	Oct 2012

Displacement, tonnes: 1,707 standard
Dimensions, metres (feet): 86 × 10.5 × 3.6 (282.2 × 34.4 × 11.8)
Speed, knots: 20
Complement: 77

Machinery: 2 Caterpillar 3916 V16 diesels; 2 shafts
Guns: 1 Oto Melara 3 in (76 mm)/62 Compact; 85 rds/min to 16 km (8.7 n miles); weight of shell 6 kg. 1 Oto Melara 25 mm. 2 – 12.7 mm MGs.

Radars: Surface search/navigation: Terma Scanter 2001; I-band. Fire control: Alenia NA-25; I-band.
Combat data systems: Sisdef SP-21K.

Helicopters: Eurocopter AS 565 Panther.

Programmes: The programme was originally for six ships. Construction of P 163 and P 164 was suspended for about two years but was resumed in 2007. A further two ships are expected.

Structure: A further derivation of the basic Holzinger class and a slightly longer version of the Durango class. Capable of operating a helicopter and equipped with a fast 11 m interception boat capable of 50 kt.
Operational: P 161 and P 162 based at Coatzacoalcos. P 163 and P 164 based at Salina Cruz.

OAXACA   4/2009, Guy Toremans / 1305962

IHS Jane's Fighting Ships 2014-2015

Patrol forces < **Mexico** 535

5 URIBE CLASS (GUNSHIPS) (PSOH)

Name	No	Builders	Laid down	Launched	Commissioned
TENIENTE JOSÉ AZUETA	P 122 (ex-C 12, ex-GH 02)	Bazán, San Fernando	7 Sep 1981	12 Dec 1981	23 Sep 1982
CAPITÁN DE FRAGATA PEDRO SÁINZ DE BARANDA	P 123 (ex-C 13, ex-GH 03)	Bazán, San Fernando	22 Oct 1981	29 Jan 1982	1 May 1983
COMODORO CARLOS CASTILLO BRETÓN	P 124 (ex-C 14, ex-GH 04)	Bazán, San Fernando	11 Nov 1981	26 Feb 1982	24 Feb 1983
VICEALMIRANTE OTHÓN P BLANCO	P 125 (ex-C 15, ex-GH 05)	Bazán, San Fernando	18 Dec 1981	26 Mar 1982	24 Feb 1983
CONTRALMIRANTE ANGEL ORTIZ MONASTERIO	P 126 (ex-C 16, ex-GH 06)	Bazán, San Fernando	30 Dec 1981	4 May 1982	24 Feb 1983

Displacement, tonnes: 1,004 full load
Dimensions, metres (feet): 67 × 10.5 × 3.5 *(219.8 × 34.4 × 11.5)*
Speed, knots: 22. **Range, n miles:** 5,000 at 13 kt
Complement: 46 (7 officers)

Machinery: 2 MTU-Bazán 16V 958 TB92 diesels; 7,500 hp(m) *(5.52 MW)* sustained; 2 shafts
Guns: 1 Bofors 40 mm/70; 300 rds/min to 12.5 km *(6.7 n miles)*; weight of shell 0.96 kg.

Radars: Surface search: Decca AC 1226; I-band.
Navigation: I-band.
Tacan: SRN 15.
Weapon control systems: Naja optronic director.

Helicopters: 1 MBB BO 105 CB.

Programmes: Ordered in 1980 to a Halcon class design. Contracts for a further eight of the class have been shelved. Pennant numbers changed in 1992. Named after naval heroes.

Modernisation: All five ships are being modernised with a new machinery management system.
Structure: Flight deck extends to the stern. Similar ships built for Argentina.
Operational: Used for EEZ patrol. Pennant numbers changed in 2001. Basing: Lázaro (P 122); Manzanillo (P 123); Ensenada (P 124, 125, 126).

COMODORO CARLOS CASTILLO BRETÓN   9/2008, Julio Montes / 1353234

10 VALLE (AUK) CLASS (COAST GUARD) (PG/PGH)

JUAN DE LA BARRERA (ex-*Guillermo Prieto*; ex-*Symbol*, ex-*MSF 123*) P 102 (ex-C 71, ex-G-02)
MARIANO ESCOBEDO (ex-*Champion*; ex-*MSF 314*) P 103 (ex-C 72, ex-G-03)
MANUEL DOBLADO (ex-*Defense*; ex-*MSF 317*) P 104 (ex-C 73, ex-G-05)
SANTOS DEGOLLADO (ex-*Gladiator*; ex-*MSF 319*) P 106 (ex-C 75, ex-G-07)
JUAN N ALVARES (ex-*Ardent*; ex-*MSF 340*) P 108 (ex-C 77, ex-G-09)
MANUEL GUTIERREZ ZAMORA (ex-*Roselle*; ex-*MSF 379*) P 109 (ex-C 78, ex-G-10)
VALENTIN GOMEZ FARIAS (ex-*Starling*; ex-*MSF 64*) P 110 (ex-C 79, ex-G-11)
IGNACIO L VALLARTA (ex-*Velocity*; ex-*MSF 128*) P 113 (ex-C 82, ex-G-14)
JESUS GONZALEZ ORTEGA (ex-*Chief*; ex-*MSF 315*) P 114 (ex-C 83, ex-G-15)
MARIANO MATAMOROS (ex-*Hermenegildo Galeana*; ex-*Sage*, ex-*MSF 111*) P 117 (ex-C 86, ex-G-19)

Displacement, tonnes: 1,082 standard; 1,270 full load
Dimensions, metres (feet): 67.5 × 9.8 × 2.8 *(221.5 × 32.2 × 9.2)*
Speed, knots: 18
Range, n miles: 6,900 at 10 kt
Complement: 73 (9 officers)

Machinery: Diesel-electric; 2 Caterpillar diesels; 2 shafts
Guns: 1 USN 3 in *(76 mm)*/50. 4 Bofors 40 mm/60 (2 twin). 4—12.7 mm (2 twin) MGs (in some on quarterdeck).
Radars: Surface search: Kelvin Hughes 14/9 (in most); I-band.
Helicopters: Platform for 1 BO 105 (PO 103, 104 and 110).

Comment: Transferred from US, six in February 1973, four in April 1973, nine in September 1973. Nine have since been deleted. Employed on Coast Guard duties. All built during Second World War. Variations are visible in the mid-ships section where some have a bulwark running from the break of the forecastle to the quarterdeck. Minesweeping gear removed. All ships re-engined 1999–2002. Some carry a Pirana 26 kt motor launch armed with 40 mm grenade launchers and 7.62 mm MGs. P 103, P 104 and P 110 have had helicopter flight decks installed aft. Plans to fit flight decks in the others have been shelved. PO 102, 103, 104, 106, 108 and 113 based at Lazaro; PO 109 and 114 based at Tampico; PO 110 and 117 based at Ensenada.

48 POLARIS CLASS (COMBATBOAT 90 HMN) (PBF)

POLARIS PI 1101	**FOMALHAUT** PI 1113	**PEACOCK** PI 1125	**ELTANIN** PI 1137
SIRIUS PI 1102	**POLLUX** PI 1114	**BETELGEUSE** PI 1126	**KOCHAB** PI 1138
CAPELLA PI 1103	**RÉGULUS** PI 1115	**ADHARA** PI 1127	**ENIF** PI 1139
CANOPUS PI 1104	**ACRUX** PI 1116	**ALIOTH** PI 1128	**SCHEDAR** PI 1140
VEGA PI 1105	**SPICA** PI 1117	**RASALHAGUE** PI 1129	**MARKAB** PI 1141
ACHERNAR PI 1106	**HADAR** PI 1118	**NUNKI** PI 1130	**MEGREZ** PI 1142
RIGEL PI 1107	**SHAULA** PI 1119	**HAMAL** PI 1131	**MIZAR** PI 1143
ARCTURUS PI 1108	**MIRFAK** PI 1120	**SUHAIL** PI 1132	**PHEKDA** PI 1144
ALPHERATZ PI 1109	**ANKAA** PI 1121	**DUBHE** PI 1133	**ACAMAR** PI 1145
PROCYÓN PI 1110	**BELLATRIX** PI 1122	**DENEBOLA** PI 1134	**DIPHDA** PI 1146
AVIOR PI 1111	**ELNATH** PI 1123	**ALKAID** PI 1135	**MENKAR** PI 1147
DENEB PI 1112	**ALNILÁN** PI 1124	**ALPHECCA** PI 1136	**SABIK** PI 1148

Displacement, tonnes: 19 full load
Dimensions, metres (feet): 16.1 × 3.8 × 0.9 *(52.8 × 12.5 × 3.0)*
Speed, knots: 47
Range, n miles: 240 at 30 kt
Complement: 4
Machinery: 2 CAT 3406E diesels; 1,605 hp(m) *(1.18 MW)*; 2 waterjets
Guns: 1 Oto Melara 12.7 mm MG (PL 1141-1148). 1—12.7 mm MG (others).
Radars: Surface search: Litton Decca Bridgemaster E; I-band.

Comment: All named after stars. First 12 ordered from Dockstavarvet, Sweden, on 15 April 1999, second batch of eight on 29 July 1999 and last batch of 20 on 1 February 2000. All delivered by 2001. A further batch of eight constructed at ASTIMAR 3, Coatzacoalcos, and delivered 2004-05. These craft are in service with the Swedish and Norwegian navies and with paramilitary forces in Malaysia and China. Based at Lerma (1103, 1104); Cozumel (1143, 1144); Yucalpeten (1107, 1108); Isla Mujeres (1109, 1110); Tuxpan (1113, 1114, 1105, 1106); Chetumel (1101, 1102, 1128, 1129); Veracruz (1131, 1132); Ensenada (1111, 1112, 1123); Manzanillo (1115, 1116); Topolobampo (1118); Mazatlan (1121, 1122); Puerto Cortes (1141); Puerto Vallarta (1124, 1136); Acapulco (1126, 1127); Guaymas (1130, 1139); Puerto Penasco (1138); Isla Socorro (1135); Frontera (1142); Puerto Chiapas (1117, 1120); Los Cabos (1119, 1147); Huatulco (1133, 1134); Isla Maria Nay (1137); San Blas Nay (1145, 1146); La Paz (1148); Lazaro (1125, 1140).

SANTOS DEGOLLADO   6/2005, Mexican Navy / 1133532

DUBHE   11/2010, Chris Sattler / 1406090

536 Mexico > Patrol forces

2 DÉMOCRATA CLASS (PBO)

Name	No	Builders	Launched	Commissioned
DÉMOCRATA	PC 241 (ex-C 101)	ASTIMAR 6, Varadero	16 Oct 1997	9 Jan 1998
FRANCISCO I MADERO	PC 242	ASTIMAR 6, Varadero	May 2008	1 Jun 2009

Displacement, tonnes: 457 (PC 241), 671 (PC 242) full load
Dimensions, metres (feet): 52.5 (PC 241), 59.1 (PC 242) × 9 × 2.7 (172.2, 193.9 × 29.5 × 8.9)
Speed, knots: 30
Complement: 36 (13 officers)
Machinery: 2 MTU 20V 956 TB92 (Caterpillar 4640 (PC 242)) diesels; 6,119 hp(m) (4.5 MW); 2 shafts
Guns: 2 Bofors 40 mm/60 (twin) (PC 241). 1—25 mm (PC 242).
Radars: Surface search: Racal Decca; E/F-band.

Comment: *Madero* is a longer version of the first of class. A 50 kt Boston Whaler launch is carried at the stern. PC 241 based at Yukalpeten and PC 242 at Guaymas.

FRANCISCO I MADERO 6/2010, Mexican Navy / 1406091

2 CAPE (PGM 71) CLASS (LARGE PATROL CRAFT) (PB)

Name	No	Builders	Recommissioned
CABO CORZO (ex-*Nayarit*; ex-*Cape Hedge*)	PC 272 (ex-P 43)	CG Yard, Curtis Bay	21 Apr 1990
CABO CATOCHE (ex-*Cape Hattaras*)	PC 273 (ex-P 44)	CG Yard, Curtis Bay	18 Mar 1991

Displacement, tonnes: 100 standard; 150 full load
Dimensions, metres (feet): 28.9 × 6.2 × 1.85 (94.8 × 20.3 × 6.1)
Speed, knots: 20
Range, n miles: 2,500 at 10 kt
Complement: 14 (1 officer)
Machinery: 2 GM 16V-149TI diesels; 2,322 hp (1.73 MW) sustained; 2 shafts
Guns: 1—20 mm. 2—12.7 mm MGs.
Radars: Navigation: Raytheon SPS-64; I-band.

Comment: Built in 1953; have been re-engined and extensively modernised. Transferred under the FMS programme, having paid off from the US Coast Guard. Pennant numbers changed in 2001. PC 272 based at Puerto Vallarta and PC 273 at Isla Cozumel.

CABO CORZO 6/2005, Mexican Navy / 1133531

15 + 9 POLARIS II CLASS (COMBATBOAT 90 HMN) (PBF)

MIAPLACIDUS PI 1401	CAPH PI 1405	ALGORAB PI 1409	ALFIRK PI 1413
ALGOL PI 1402	MIRACH PI 1406	ALBIREO PI 1410	ALDERAMIN PI 1414
BEAVER PI 1403	ALHENA PI 1407	ALNITAK PI 1411	MENKALINAN PI 1415
MERAK PI 1404	SAIPH PI 1408	MINTAKA PI 1412	+9

Displacement, tonnes: 19 full load
Dimensions, metres (feet): 16 × 3.4 × 0.9 (52.5 × 11.2 × 3.0)
Speed, knots: 50. **Range, n miles:** 240 at 30 kt
Complement: 4
Machinery: 2 MAN diesels; 2,200 hp(m) (1.62 MW); 2 waterjets
Guns: 1 Oto Melara 12.7 mm MG.
Radars: Surface search: Litton Decca Bridgemaster E; I-band.

Comment: A further development of the Polaris and Acuario classes. The first of class was delivered by Dockstavarvet, Sweden, on 10 August 2005 and three further craft were completed at ASTIMAR 3 by 2007. Two further craft delivered in January 2008 and two in 2009. Based at Isla Cozumel (1401, 1402), Ciudad del Carmen (1403, 1404, 1405, 1406) and Coatzacoalcos (1407, 1408). A class of 24 is expected.

MIAPLACIDUS 6/2005, Mexican Navy / 1133524

2 POINT CLASS (LARGE PATROL CRAFT) (PB)

Name	No	Builders	Recommissioned
PUNTA MORRO (ex-*Point Verde*)	PC 281 (ex-P 60, ex-P 45)	CG Yard, Curtis Bay	19 Jul 1991
PUNTA MASTUN (ex-*Point Herron*)	PC 282 (ex-P 61, ex-P 46)	CG Yard, Curtis Bay	19 Jul 1991

Displacement, tonnes: 68 full load
Dimensions, metres (feet): 25.3 × 5.2 × 1.8 (83 × 17.1 × 5.9)
Speed, knots: 12
Range, n miles: 1,500 at 8 kt
Complement: 10
Machinery: 2 Caterpillar diesels; 1,600 hp (1.19 MW); 2 shafts
Guns: 1—20 mm. 2—12.7 mm MGs.
Radars: Surface search: Raytheon SPS-64; I-band.

Comment: Ex-US Coast Guard craft built in 1961. Steel hulls and aluminium superstructures. Speed much reduced from original 23 kt. Pennant numbers changed in 2001. Both based at Lerma.

PUNTA MASTUN 6/2005, Mexican Navy / 1133529

8 AZTECA CLASS (LARGE PATROL CRAFT) (PB)

Name	No	Builders	Commissioned
JUAN ANTONIO DE LA FUENTE (ex-*Mexica*)	PC 208 (ex-P 08)	Ailsa Shipbuilding Co Ltd, Troon	28 Dec 1975
IGNACIO RAMIREZ (ex-*Huastela*)	PC 210 (ex-P 10)	Ailsa Shipbuilding Co Ltd, Troon	1 Jun 1975
IGNACIO MARISCAL (ex-*Mazahua*)	PC 211 (ex-P 11)	Ailsa Shipbuilding Co Ltd, Troon	25 Dec 1975
FRANCISCO J MUGICA (ex-*Tarasco*)	PC 216 (ex-P 16)	Ailsa Shipbuilding Co Ltd, Troon	1 Jun 1976
JOSE NATIVIDAD MACIAS (ex-*Pimas*)	PC 220 (ex-P 20)	Lamont & Co Ltd	29 Dec 1976
YUCATAN (ex-*Tolteca*)	PC 224 (ex-P 24)	ASTIMAR 3, Coatzacoalcos	18 May 1977
TABASCO (ex-*Maya*)	PC 225 (ex-P 25)	ASTIMAR 3, Coatzacoalcos	1 Dec 1978
COCHIMIE (ex-*Veracruz*)	PC 226 (ex-P 26)	ASTIMAR 3, Coatzacoalcos	1 Dec 1978

Displacement, tonnes: 150 full load
Dimensions, metres (feet): 34.4 × 8.7 × 2.2 (112.9 × 28.5 × 7.2)
Speed, knots: 24
Range, n miles: 1,537 at 14 kt
Complement: 24 (2 officers)
Machinery: 2 Paxman 12YJCM diesels; 3,000 hp (2.24 MW) sustained; 2 shafts
Guns: 1 Bofors 40 mm/60; 300 rds/min to 12 km (6.5 n miles) anti-surface; 4 km (2.2 n miles) anti-aircraft; weight of shell 2.4 kg. 1 Oerlikon 20 mm or 1—7.62 mm MG.
Radars: Surface search: Kelvin Hughes; I-band.

Comment: Ordered by Mexico on 27 March 1973 from Associated British Machine Tool Makers Ltd to a design by TT Boat Designs, Bembridge, Isle of Wight. The first 21 were modernised in 1987 in Mexico with spare parts and equipment supplied by ABMTM Marine Division who supervised the work which included engine refurbishment and the fitting of air conditioning. Names and pennant numbers changed in 2001. Based at: Yukaltepen (PC 224, 225, 226); Puerto Chiapas (PC 220); Guaymas (PC 210); Mazatlan (PC 211); La Paz (PC 208, 216).

JUAN ANTONIO DE LA FUENTE 3/2010, Alex Pape / 1406089

Patrol forces < **Mexico** 537

8 ACUARIO CLASS (COMBATBOAT 90HMN) (PBF)

| ACUARIO PI 1301 | ARIES PI 1303 | CANCER PI 1305 | CENTAURO PI 1307 |
| AGUILA PI 1302 | AURIGA PI 1304 | CAPRICORNO PI 1306 | GEMINIS PI 1308 |

Displacement, tonnes: 19 full load
Dimensions, metres (feet): 17.1 (PI 1301–1306), 15.9 (PI 1307–1308) × 3.95 (PI 1301–1306), 3.8 (PI 1307–1308) × 2.7 *(56.1, 52.2 × 13.0, 12.5 × 8.9)*
Speed, knots: 47
Range, n miles: 240 at 30 kt
Complement: 4
Machinery: 2 CAT 3406E diesels; 1,605 hp(m) *(1.18 MW)*; 2 waterjets
Guns: 1 Sconta 50 automatic 12.7 mm MG or 1 Oto Melara 12.7 mm MG.
Radars: Surface search: Litton Decca Bridgemaster E; I-band.

Comment: A further development of the Polaris class which are based on the Swedish Combatboat 90 and built by ASTIMAR 3, Coatzacoalcos. 1301 and 1302 commissioned on 1 June 2004 and 1303–1306 on 1 September 2004. P 1307–1308 commissioned on 1 September 2004 and known as Acuario B class. Based at Puerto Peñasco (1301); El Mezquital (1302, 1303); Frontera (1304); Ciudad Madero (1305, 1306); Ensenada (1307); Manzanillo (1308). All named after stars.

ARIES   6/2010, *Mexican Navy* / 1406093

36 DEFENDER CLASS (RESPONSE BOATS) (PBF)

BR 01–036

Displacement, tonnes: 3 full load
Dimensions, metres (feet): 10.05 × 3.05 × 0.8 *(33 × 10.0 × 2.6)*
Speed, knots: 60
Complement: 4
Machinery: 3 outboard motors; 825 hp *(615 kW)*
Radars: To be announced.

Comment: High-speed inshore rescue craft of aluminium construction and foam collar built by SAFE Boats International, Port Orchard, Washington. An initial order for six craft was made in December 2007 for delivery in 2009. A further 30 craft were delivered by mid-2012.

BR 27   9/2011, *Julio Montes* / 1486294

2 + 8 (18) TENOCHTITLAN (DAMEN STAN PATROL 4207) CLASS (PB)

Name	No	Builders	Commissioned
TENOCHTITLAN	PC 331	ASTIMAR 1, Tampico	1 Jun 2012
TEOTIHUACAN	PC 332	ASTIMAR 1, Tampico	1 Jun 2012
PALENQUE	PC 333	ASTIMAR 1, Tampico	2014
MITLA	PC 334	ASTIMAR 1, Tampico	2015
UXMAL	PC 335	ASTIMAR 1, Tampico	2015

Displacement, tonnes: 208 standard
Dimensions, metres (feet): 42.8 × 7.11 × 2.52 *(140.4 × 23.3 × 8.3)*
Speed, knots: 26
Complement: 14
Machinery: 2 Caterpillar 3516B DI-TA; 5,600 hp *(4.17 MW)*; 2 cp props
Guns: 2—12.7 mm MGs. 2—7.62 mm MGs.

Comment: Construction of two Damen 4207 patrol vessels started at Tampico in July 2011. An order for a third vessel was made on 23 May 2013 for a fourth on 22 August and fifth in November 2013. Orders for a further five are expected in 2014 and an overall programme of 28 ships is projected. While details of the craft have not been formally released, they are expected to be similar to those in service in Barbados and Jamaica.

TEOTIHUACAN   6/2012, *Mexican Navy* / 1486295

© 2014 IHS

56 FAST PATROL CRAFT (PBF)

G 01–36   +20

Dimensions, metres (feet): 6.8 × 2.3 × 0.3 *(22.3 × 7.5 × 1.0)*
Speed, knots: 40
Range, n miles: 190 at 40 kt
Complement: 2
Machinery: 2 Johnson outboards; 280 hp *(209 kW)*
Guns: 1 or 2—7.62 mm MGs.

Comment: Details are for the 36 G 01–36 Piraña class. Acquired in 1993/94. An 11.6 m 50 kt Interceptor class launch is carried in *Démocrata* and modified 10.5 m versions are embarked in the Sierra, Durango and Oaxaca classes. Ten are in service. There are also ten 29 ft Mako Marine craft, with twin Mercury outboards acquired in 1995.

PIRAÑA CLASS   9/2002, *Julio Montes* / 0533285

INTERCEPTOR (old number)   7/1998, *Mexican Navy* / 0052610

INTERCEPTOR (mod)   6/2004, *Mexican Navy* / 0589767

6 MOTOR LIFEBOATS (MLB/SAR)

Displacement, tonnes: 20 full load
Dimensions, metres (feet): 14.6 × 4.4 × 1.4 *(47.9 × 14.4 × 4.6)*
Speed, knots: 25
Range, n miles: 220 at 25 kt
Complement: 4
Machinery: 2 Detroit diesels; 870 hp *(649 kW)* sustained; 2 shafts

Comment: Ordered in 2008 from Textron Marine, New Orleans. Aluminium hulls and self-righting capability. The sixth craft was delivered in 2011.

BR 101   6/2011, *Mexican Navy* / 1406760

IHS Jane's Fighting Ships 2014-2015

AMPHIBIOUS FORCES

2 PAPALOAPAN (NEWPORT) CLASS (LSTH)

Name	No	Builders	Laid down	Launched	Commissioned	Recommissioned
RIO PAPALOAPAN (ex-*Sonora*; ex-*Newport*)	A 411 (ex-04, ex-LST-1179)	Philadelphia Naval Shipyard	1 Nov 1966	3 Feb 1968	7 Jun 1969	5 Jun 2001
USUMACINTA (ex-*Frederick*)	A 412 (ex-LST-1184)	National Steel & Shipbuilding Co, San Diego	13 Apr 1968	8 Mar 1969	11 Apr 1970	1 Dec 2002

Displacement, tonnes: 5,055 standard; 8,586 full load
Dimensions, metres (feet): 159.2 × 21.2 × 5.3 *(522.3 × 69.6 × 17.4)*
Speed, knots: 20. **Range, n miles:** 14,250 at 14 kt
Complement: 257 (13 officers)

Military lift: 400 troops; 500 tons vehicles; 3 LCVPs and 1 LCPL on davits.
Machinery: 6 General Motors 16-645-E5 diesels; 16,500 hp *(12.3 MW)* sustained; 2 shafts; cp props; bow thruster
Guns: 4 USN 3 in *(76 mm)*/50 (A 411).
Radars: Surface search: Raytheon SPS-10F; G-band.
Navigation: Terma Scanter 2001; I-band.

Helicopters: Platform only.

Programmes: A-411 sold to Mexico by the US Navy on 18 January 2001. A-412 sold on 9 December 2002. Both ships employed in amphibious role rather than as transport ships as previously reported. A 411 based at Tampico and A 412 at Manzanillo.
Modernisation: A new surface search radar reported installed in both ships by 2008.
Operational: *Papaloapan* is also used as a cadet training ship.

PAPALOAPAN 8/2008, Michael Nitz / 1353235

PAPALOAPAN 8/2008, A A de Kruijf / 1353236

2 + 4 LOGISTIC SUPPORT SHIPS (AK)

Name	No	Builders	Laid down	Launched	Commissioned
MONTES AZULES	BAL 01	ASTIMAR 20, Salina Cruz	2010	30 Aug 2011	23 Nov 2011
LIBERTADOR	BAL 02	ASTIMAR 20, Salina Cruz	30 Aug 2011	8 Sep 2012	20 Nov 2012

Displacement, tonnes: 3,666 full load
Dimensions, metres (feet): 99.89 × 15.244 × 3.4 *(327.7 × 50.0 × 11.2)*
Speed, knots: 12

Machinery: 2 diesels; 3,500 hp *(2.6 MW)*; 2 shafts
Guns: 5—40 mm Bofors.
Radars: Surface search: To be announced.
Navigation: To be announced.
Helicopters: Platform for 1 medium.

Comment: Two logistic supply vessels to replace two already decommissioned and *Manzanillo*. The design is derived from US LSTs previously in service. Four further vessels are expected. Fitted with a bow ramp, helicopter deck, a hydraulic crane and two LCVPs.

SURVEY AND RESEARCH SHIPS

1 ONJUKU CLASS (SURVEY SHIP) (AGS)

Name	No	Builders	Commissioned
ONJUKU	BI 02 (ex-H 04)	Uchida Shipyard	10 Jan 1980

Displacement, tonnes: 502 full load
Dimensions, metres (feet): 36.9 × 8 × 3.5 *(121.1 × 26.2 × 11.5)*
Speed, knots: 10
Range, n miles: 5,645 at 10.5 kt
Complement: 20 (4 officers)
Machinery: 1 Yanmar 6UA-UT diesel; 700 hp(m) *(515 kW)*; 1 shaft
Radars: Navigation: Furuno; I-band.
Sonars: Furuno; hull-mounted; high frequency active.

Comment: Launched 9 December 1977 in Japan. Sonar is a fish-finder type. New pennant number in 2001. Based at Veracruz.

MONTES AZULES 8/2011, Mexican Navy / 1402472

ONJUKU 6/2005, Mexican Navy / 1133520

Survey and research ships — **Mexico** 539

2 ROBERT D CONRAD CLASS (RESEARCH SHIPS) (AGOR)

Name	No	Builders	Commissioned
ALTAIR (ex-*James M Gilliss*)	BI 03 (ex-H 05, ex-AGOR 4)	Christy Corporation, Sturgeon Bay	5 Nov 1962
ANTARES (ex-*S P Lee*)	BI 04 (ex-H 06, ex-AG 192)	Defoe SB Co, Bay City	2 Dec 1962

Displacement, tonnes: 1,392 full load
Dimensions, metres (feet): 63.7 × 12.2 × 4.7 *(209 × 40.0 × 15.4)*
Speed, knots: 13.5. **Range, n miles:** 10,500 at 10 kt
Complement: 41 (12 officers) + 15 scientists
Machinery: Diesel-electric; 2 Caterpillar diesel generators; 1,200 hp *(895 kW)*; 2 motors; 1,000 hp *(746 kW)*; 1 shaft; bow thruster
Radars: Navigation: Raytheon 1025; Raytheon R4iY; I-band.

Comment: *Altair* leased from US 14 June 1983. Refitted and modernised in Mexico. Recommissioned 23 November 1984. Primarily used for oceanography. *Antares* served as an AGI with the USN until February 1974 when she transferred on loan to the Geological Survey. Acquired by sale and recommissioned on 1 December 1992. New pennant numbers in 2001. Based at Manzanillo (BI 03) and Tampico (BI 04).

ALTAIR   6/2005, Mexican Navy / 1133519

4 ARRECIFE (EX-OLMECA II) CLASS (SURVEY CRAFT) (YGS)

ALACRAN BI 08 (ex-PR 301)   CABEZO BI 10 (ex-PR 304)
RIZO BI 09 (ex-PR 310)   ANEGAGADA DE ADENTRO BI 11 (ex-PR 309)

Displacement, tonnes: 18 full load
Dimensions, metres (feet): 16.7 × 4.4 × 1.2 *(54.8 × 14.4 × 3.9)*
Speed, knots: 20. **Range, n miles:** 460 at 10 kt
Complement: 15 (2 officers)
Machinery: 2 Detroit 8V-92TA diesels; 700 hp *(562 kW)* sustained; 2 shafts
Guns: 1 — 12.7 mm MG.
Radars: Navigation: Raytheon 1900; I-band.

Comment: Built at Acapulco and completed between 1982 and 1989. GRP hulls. Converted for inshore hydrographic duties in 2003. All have *Arrecife* in front of the names. Based at Manzanillo (BI 08, 09) and Veracruz (BI 10, 11).

ALACRAN   6/2005, Mexican Navy / 1133518

1 SURVEY SHIP (AGS)

Name	No	Builders	Commissioned
RIO HONDO (ex-*Deer Island*)	BI 06 (ex-A 26, ex-YAG 62)	Halter Marine	May 1962

Displacement, tonnes: 406 full load
Dimensions, metres (feet): 36.6 × 8.5 × 2.1 *(120.1 × 27.9 × 6.9)*
Speed, knots: 10. **Range, n miles:** 6,000 at 10 kt
Complement: 20
Machinery: 2 General Motors 7122-700 diesels; 2 shafts

Comment: Acquired from US on 1 August 1996 and adapted for a support ship role in 1997. Converted to Survey Ship in 1999. Used in US service from 1983 as an acoustic research ship to test noise reduction equipment. Started life as an oil rig supply tug. New pennant number in 2001. Based at Coatzacoalcos.

RIO HONDO   6/2004, Mexican Navy / 1484110

© 2014 IHS

1 HUMBOLDT CLASS (RESEARCH SHIP) (AGOR)

Name	No	Builders	Recommissioned
ALEJANDRO DE HUMBOLDT	BI 01 (ex-H 03)	JG Hitzler, Elbe	22 Jun 1987

Displacement, tonnes: 594 standard; 711 full load
Dimensions, metres (feet): 42.3 × 9.6 × 4.1 *(138.8 × 31.5 × 13.5)*
Speed, knots: 14
Complement: 20 (4 officers)
Machinery: 1 MAN R8V 22/30 diesel; 2 shafts
Radars: Navigation: Kelvin Hughes; I-band.

Comment: Former trawler built in Germany and launched in January 1970. Converted in 1982 to become a hydrographical and acoustic survey ship. Based at Manzanillo. New pennant number in 2001.

ALEJANDRO DE HUMBOLDT   6/2010, Mexican Navy / 1406084

1 SUPPORT SHIP (AKS)

Name	No	Builders	Commissioned
RIO SUCHIATE (ex-*Monob 1*)	BI 05 (ex-YAG 61, ex-YW 87)	Zenith Dredge Co, Duluth	11 Nov 1943

Displacement, tonnes: 1,412 full load
Dimensions, metres (feet): 59.9 × 9.9 × 2.9 *(196.5 × 32.5 × 9.5)*
Speed, knots: 9. **Range, n miles:** 2,500 at 9 kt
Complement: 21
Machinery: 1 Caterpillar D 398 diesel; 850 hp *(634 kW)*; 1 shaft

Comment: Acquired from US on 1 August 1996. The ship was converted from a water carrier to an acoustic research role in 1969, and had four laboratories in US service. Adapted to act also in support ship role in 1997. New pennant number in 2001. Based at Guaymas.

RIO SUCHIATE   6/2010, Mexican Navy / 1406083

1 SURVEY SHIP (AGS)

Name	No	Builders	Commissioned
RIO TUXPAN (ex-*Whiting*)	BI 12	Marietta Manufacturing Company, Mt Pleasant	Jul 1963

Displacement, tonnes: 922 standard
Dimensions, metres (feet): 49.7 × 10.1 × 3.7 *(163.1 × 33.1 × 12.1)*
Speed, knots: 12
Range, n miles: 5,700 at 11 kt
Complement: 30 (7 officers)
Machinery: 2 General Motors diesels; 1,600 hp *(1.2 MW)*; 2 cp props
Radars: Surface search: E/F-band.
Navigation: I-band.

Comment: Ex-US NOAA ship designed for hydrographic and bathymetric survey work. Decommissioned in May 2003 and transferred to the Mexican Navy in April 2005. Equipped (in NOAA service) with Intermediate Depth Swath Survey System (IDSSS) (36 kHz), Deep Water Echo Sounder (12 kHz), Shallow Water Echo Sounder (100 kHz), Hydrographic Survey Sounder (24 and 100 kHz), EG&G 270 Side Scan Sonar and Klein T-5000 High Speed/High Resolution Side Scan Sonar. Based at Tuxpan.

RIO TUXPAN   4/2005, NOAA / 1133517

IHS Jane's Fighting Ships 2014-2015

Complement: 17 (3 officers)
Machinery: 2 Caterpillar 3516TA diesels; 4,400 hp *(3.28 MW)* sustained; 2 shafts
Radars: Surface search: Furuno 1011; I-band.

Comment: First pair ordered in June 1989 from Australian Shipbuilding Industries. Training and support provided by Australia at Port Kolonia. Third of class negotiated with Transfield (former ASI) in 1997. Following the decision by the Australian government to extend the Pacific Patrol Boat programme to enable 30-year boat lives, *Palikir*, *Micronesia* and *Independence*, underwent half-life refits in 1998, 1999 and 2003. A life-extension refit for *Palikir* was completed in 2007 and for *Micronesia* in 2009. Refit for *Independence* is due during 2014.

MICRONESIA   9/2013*, Chris Sattler / 1525539

© 2014 IHS   *IHS Jane's Fighting Ships 2014-2015*

540 Mexico > Training ships — Auxiliaries

TRAINING SHIPS

1 SAIL TRAINING SHIP (AXS)

Name	No	Builders	Launched	Commissioned
CUAUHTÉMOC	BE 01 (ex-A 07)	Astilleros Talleres Calaya SA, Bilbao	9 Jan 1982	23 Sep 1982

Displacement, tonnes: 1,689 full load
Dimensions, metres (feet): 90.5 oa; 67.2 wl × 12 × 5.4 (296.9; 220.5 × 39.4 × 17.7)
Speed, knots: 17; 7 diesel

AUXILIARIES

Notes: There is a small auxiliary ship *Alvarado* AM 60801.

1 LOGISTIC SUPPORT SHIP (AK)

Name	No	Builders	Recommissioned
TARASCO (ex-*Rio Lerma*; ex-*Sea Point*, ex-*Tricon*, ex-*Marika*, ex-*Arneb*)	ATR 03	Solvesborg	1 Mar 1990

---

542 Montenegro > Introduction — Tugs

Montenegro

Country Overview

Montenegro was formed following a referendum on 21 May 2006 in which the people voted for independence and for the dissolution of the Federal Republic of Serbia and Montenegro which itself was the rump of the former Yugoslavia. A formal declaration of independence was made by the Montenegro Assembly on 3 June 2006. With an area of 5,333 square miles, it is located in south-eastern Europe in the Balkan Peninsula and is bordered to the north-west by Bosnia, to the east by Serbia, to the south by Albania and to the west by Croatia. It has a 158 n mile coastline with the Adriatic Sea on which Bar and Kotor are the principal ports. The capital and largest city is Podgorica. Territorial waters (12 n miles) are claimed but an EEZ has not been claimed.

The provisions of the Union Constitution were that, in the event of dissolution, the armed forces of Serbia and Montenegro would be split in such a way that each state keeps the assets in its territory. Therefore, all of the former navy of Serbia and Montenegro transferred to Montenegro in June 2006, except for the former Danuble Flotilla, which transferred to Serbian land forces. The navy is now a component of the joint Armed Forces of Montenegro.

Headquarters Appointments

Chief of General Staff:
  Vice Admiral Dragan Samardzić
Commander of the Navy:
  Captain Darko Vukovic

Personnel

2014: 351

Bases

Headquarters and Base: Bar

Organisation

Montenegrin naval forces are organised into five detachments: Patrol ships; search and rescue; maritime detachment (special forces), coastal reconnaissance, training and logistics. All naval facilities have been concentrated at Bar and the naval repair and maintenance yard at Tivat has been sold. All special forces have been consolidated under the command of the army. Former coastal defence missile systems have reportedly been sold to Egypt.

PATROL FORCES

2 KOTOR CLASS (FFGM)

Name	No	Builders	Launched	Commissioned
—	P 33	Tito Shipyard, Kraljevica	21 May 1985	Jan 1987
—	P 34	Tito Shipyard, Kraljevica	18 Dec 1986	Nov 1988

Displacement, tonnes: 1,900 full load
Dimensions, metres (feet): 96.7 × 12.8 × 4.2 (317.3 × 42.0 × 13.8)
Speed, knots: 27; 22 diesel
Range, n miles: 1,800 at 14 kt
Complement: 110
Machinery: CODAG; 1 SGW Nikolayev gas turbine; 18,000 hp(m) (13.2 MW); 2 SEMT-Pielstick 12 PA6 V 280 diesels; 9,600 hp(m) (7.1 MW) sustained; 3 shafts
Missiles: SSM: 4 SS-N-2C Styx ❶; active radar or IR homing to 83 km (45 n miles) at 0.9 Mach; warhead 513 kg; sea-skimmer at end of run.
SAM: SA-N-4 Gecko twin launcher ❷; semi-active radar homing to 15 km (8 n miles) at 2.5 Mach; height envelope 9–3,048 m (29.5–10,000 ft); warhead 50 kg.
Guns: 4 USSR 3 in (76 mm)/59 AK 726 (2 twin) (1 mounting only in 33 and 34) ❸; 90 rds/min to 16 km (8.5 n miles); weight of shell 6.8 kg. 4 USSR 30 mm/65 (2 twin) ❹; 500 rds/min to 5 km (2.7 n miles); weight of shell 5.9 kg.
A/S Mortars: 2 RBU 6000 12-barrelled trainable ❺; range 6,000 m; warhead 31 kg.
Mines: Can lay mines.
Physical countermeasures: Decoys: 2 Wallop Barricade double layer chaff launchers.
Radars: Air/surface search: Strut Curve ❻; F-band.
Navigation: Palm Frond; I-band.
Fire control: PEAB 9LV200 ❽; I-band (for 76 mm and SSM). Drum Tilt ❾; H/I-band (for 30 mm). Pop Group ❿; F/H/I-band (for SAM).
IFF: High Pole; 2 Square Head.
Sonars: Bull Nose; hull-mounted; active search and attack; medium frequency.

Programmes: Built under licence. Type name, VPB (Veliki Patrolni Brod).
Modernisation: Combat data system fitted in 2000. SS-N-2C missiles, SA-N-4 missiles and RBU 6000 rocket launchers are reported to have been removed. Following refits, which may have included installation of a flight deck aft, the ships are employed as offshore patrol vessels.
Structure: The hull is similar to the Russian Koni class.
Operational: Both ships are operational. Based at Bar.

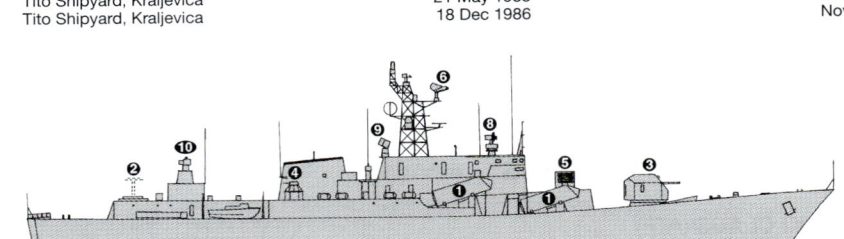

KOTOR CLASS (Scale 1 : 900), Ian Sturton / 0506341

KOTOR CLASS 6/1998, Yugoslav Navy / 0050746

TUGS

Notes: There is one coastal tug PR 41 (armed with two 20 mm guns).

PR 41
6/2007, Freivogel Collection
1167938

IHS Jane's Fighting Ships 2014-2015 © 2014 IHS

AUXILIARIES

Notes: (1) Two 22 m inshore survey vessels BH 12 and CH 1 are operated by the Naval Hydrological Institute. BH 11 has been donated to a civilian institute.
(2) There is one tender, BS 22.
(3) There are two diving tenders BRM 81 and BRM 85.

1 SAIL TRAINING SHIP (AXS)

JADRAN

Displacement, tonnes: 749 full load
Dimensions, metres (feet): 60 × 8.9 × 4.05 (196.9 × 29.2 × 13.3)
Speed, knots: 10.4
Machinery: 1 Burmeister Alpha diesel; 353 hp (263 kW)
Radars: 1 FR 2120 and 1 FR 7061; I-band.

Comment: The contract for a barquentine sail training ship was signed on 4 September 1930 with the German shipbuilding company H C Stülcken & Son of Hamburg. She was launched on 25 June 1931 and arrived in Tivat on 16 July 1933. During the Second World War, she was used by the Italian Navy under the name of *Marco Polo* before being allowed to fall into disrepair. She returned to Yugoslavia in 1946 and was reconstructed in her original form at Tivat.

JADRAN
10/2008, Camil Busquets i Vilanova
1305873

POLICE

2 MIRNA CLASS (TYPE 140) (PB)

BAR (ex-*Učka*) P 01 (ex-174)
HERCEG NOVI (ex-*Kosmaj*) P 03 (ex-178)

Displacement, tonnes: 144 full load
Dimensions, metres (feet): 32 × 6.7 × 2.3 (105 × 22.0 × 7.5)
Speed, knots: 28
Range, n miles: 400 at 20 kt
Complement: 19 (3 officers)

Machinery: 2 SEMT-Pielstick 12 PA4 200 VGDS diesels; 5,292 hp(m) (3.89 MW) sustained; 2 shafts
Guns: 1—20 mm.
Depth charges: 8 DCs.
Radars: Surface search: Racal Decca 1216C; I-band.
Sonars: Simrad SQS-3D/3F; active; high frequency.

Comment: Builders, Kraljevica Yard. Launched between June 1981 and December 1983. An unusual feature of this design is the fitting of an electric outboard motor giving a speed of up to 6 kt. One sunk possibly by a limpet mine in November 1991. Four held by Croatia, five have been sold to civilian use and two transferred from the Montenegrin Navy to the Police in 2006.

HERCEG NOVI
4/2007, Marco Ghiglino / 1167914

Morocco
MARINE ROYALE MAROCAINE

Country Overview

Formerly divided into French and Spanish protectorates, the Kingdom of Morocco gained independence in 1956. Situated in north-western Africa, it has an area of 172,414 square miles and is bordered to the east by Algeria; it occupies 80% of Western Sahara (formerly Spanish Sahara), the country to the south. Two Spanish exclaves, Ceuta and Melilla, are located on the Mediterranean coast. It has coastlines with Atlantic Ocean (756 n miles) and Mediterranean Sea (238 n miles). The capital is Rabat while Casablanca is the largest city and principal port. Other ports are at Tangier, Agadir, Kenitra, Mohammedia, and Safi. Territorial seas (12 n miles) are claimed. An EEZ (200 n mile) has also been claimed but its limits have not been fully defined.

Headquarters Appointments

Inspector of the Navy: Rear Admiral Mohamed Laghmari

Personnel

(a) 2014: 7,800 officers and ratings (including 1,500 Marines)
(b) 18 months' national service

Bases

Casablanca (HQ), Safi, Agadir, Kenitra, Tangier, Dakhla, Al Hoceima, Qsar-es-Seghir (opening 2010).

Aviation

The Ministry of Fisheries operates 11 Pilatus Britten-Norman Defender maritime surveillance aircraft.

Morocco > Frigates

FRIGATES

2 FLOREAL CLASS (FFGHM)

Name	No	Builders	Laid down	Launched	Commissioned
MOHAMMED V	611	Chantiers de L'Atlantique, St Nazaire	Jun 1999	9 Mar 2001	12 Mar 2002
HASSAN II	612	Chantiers de L'Atlantique, St Nazaire	Dec 1999	11 Feb 2002	20 Dec 2002

Displacement, tonnes: 2,997 full load
Dimensions, metres (feet): 93.5 × 14 × 4.3 (306.8 × 45.9 × 14.1)
Speed, knots: 20. **Range, n miles:** 10,000 at 15 kt
Complement: 89 (11 officers)

Machinery: CODAD; 4 SEMT-Pielstick 6 PA6 L 280 diesels; 9,600 hp(m) (7.06 MW) sustained; 2 shafts; LIPS cp props; bow thruster; 340 hp(m) (250 kW)
Missiles: SSM: 2 Aerospatiale MM 38 Exocet ❶.
SAM: 2 Matra Simbad twin launchers ❷ can replace 20 mm guns or Dagaie launcher.
Guns: 1 Otobreda 76 mm/62 Compact ❸. 2 Giat 20 F2 20 mm ❹.
Physical countermeasures: Decoys: 2 CSEE Dagaie Mk II ❺; 10-barrelled trainable launchers; chaff and IR flares.
Electronic countermeasures: ESM: Thomson-CSF ARBR 17 ❻; radar intercept.
Radars: Surface search/Fire control: Thales WM28 ❽; I/J-band.
Navigation: 2 Decca Bridgemaster E ❾; I-band (1 for helicopter control).
Weapon control systems: CSEE Najir 2000 optronic director ❼.
Helicopters: 1 Aerospatiale AS 565MA Panther ❿.

Programmes: Contract signed with Alstom on 12 July 1999. 611 delivered on 12 March 2002 and 612 on 20 December 2002.
Structure: Constructed to DNV standards. Very similar to ships in French service with 76 mm in place of 100 mm gun.

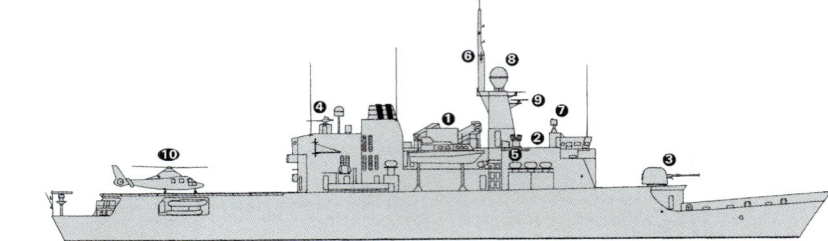

MOHAMMED V (Scale 1 : 900), Ian Sturton / 1151071

HASSAN II 3/2006, M Declerck / 1164956

HASSAN II 7/2008, M Declerck / 1353239

MOHAMMED V 10/2009, B Prézelin / 1366978

Frigates ‹ **Morocco** 545

1 MODIFIED DESCUBIERTA CLASS (FFGM)

Name	No	Builders	Laid down	Launched	Commissioned
LIEUTENANT COLONEL ERRHAMANI	501	Bazán, Cartagena	20 Mar 1979	26 Feb 1982	28 Mar 1983

Displacement, tonnes: 1,253 standard; 1,503 full load
Dimensions, metres (feet): 88.8 × 10.4 × 3.8 *(291.3 × 34.1 × 12.5)*
Speed, knots: 25.5
Range, n miles: 4,000 at 18 kt
Complement: 100

Machinery: 4 MTU-Bazán 16V 956 TB91 diesels; 15,000 hp(m) *(11 MW)* sustained; 2 shafts; cp props
Missiles: SSM: 4 Aerospatiale MM 38 Exocet ❶; inertial cruise; active radar homing to 42 km *(23 n miles)* at 0.9 Mach; warhead 165 kg; sea-skimmer. Frequently not embarked.
SAM: Selenia/Elsag Albatros octuple launcher ❷; 24 Aspide; semi-active radar homing to 13 km *(8 n miles)* at 2.5 Mach; height envelope 15–5,000 m *(49.2–16,405 ft)*; warhead 30 kg.
Guns: 1 Oto Melara 3 in *(76 mm)*/62 compact ❸; 85 rds/min to 16 km *(8.6 n miles)* anti-surface; 12 km *(6.5 n miles)* anti-aircraft; weight of shell 6 kg. 2 Breda Bofors 40 mm/70 ❹; 300 rds/min to 12.5 km *(6.7 n miles)*; weight of shell 0.96 kg.
Torpedoes: 6—324 mm Mk 32 (2 triple) tubes ❺. Honeywell Mk 46 Mod 1; anti-submarine; active/passive homing to 11 km *(5.9 n miles)* at 40 kt; warhead 44 kg.
A/S Mortars: 1 Bofors SR 375 mm twin trainable launcher ❻; range 3.6 km *(1.9 n miles)*; 24 rockets.
Physical countermeasures: Decoys: 2 CSEE Dagaie double trainable mounting; IR flares and chaff; H/J-band.
Electronic countermeasures: ESM/ECM: Elettronica ELT 715; intercept and jammer.
Radars: Air/surface search: Signaal DA05 ❼; E/F-band (see *Operational*).
Surface search: Signaal ZW06 ❽; I-band.
Fire control: Signaal WM25/41 ❾; I/J-band; range 46 km *(25 n miles)*.
Navigation: 2 Decca ❿; I-band.
Sonars: Raytheon DE 1160 B; hull-mounted; active/passive; medium range; medium frequency.
Combat data systems: Signaal SEWACO-MR action data automation. SATCOM.

Programmes: Ordered 7 June 1977.
Modernisation: New 40 mm guns fitted in 1995. Refit in Spain in 1996.
Operational: The ship is fitted to carry Exocet but the missiles are seldom embarked. The air search radar was removed in 1998 but reinstated in 1999.

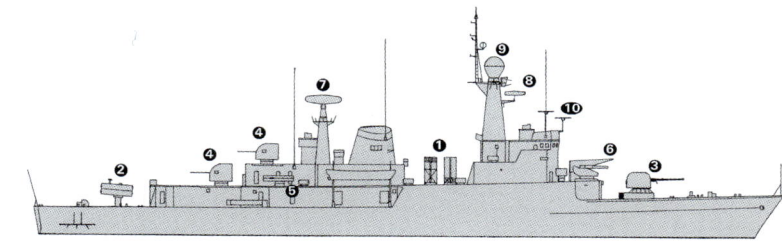

LIEUTENANT COLONEL ERRHAMANI   (Scale 1 : 900), Ian Sturton / 1151072

LIEUTENANT COLONEL ERRHAMANI   6/2006, B Prézelin / 1040673

LIEUTENANT COLONEL ERRHAMANI   6/2006, B Prézelin / 1040674

© 2014 IHS   IHS Jane's Fighting Ships 2014-2015

Morocco > Frigates — Corvettes

1 FREMM CLASS (FFGHM)

Name	No	Builders	Laid down	Launched	Commissioned
MOHAMMAD VI	701	DCNS, Lorient	2009	14 Sep 2011	30 Jan 2014

Displacement, tonnes: 4,572 standard; 6,096 full load
Dimensions, metres (feet): 142.2 oa; 137.1 wl × 19.7 × 5.4 *(466.5; 449.8 × 64.6 × 17.7)*
Speed, knots: 28
Range, n miles: 6,000 at 15 kt
Complement: 108 (22 officers) + 37 spare berths

Machinery: CODLOG; 1 Fiat/GE LM 2500+ G4 gas turbine; 47,370 hp(m) *(34.8 MW)*; 2 Jeumont motors; 2 shafts
Missiles: SAM: 16 (2 octuple) cell Sylver A43 VLS for MBDA Aster 15 ❶; inertial guidance, mid-course update and final active homing to 30 km *(16.2 n miles)* at 3 Mach.
SSM: 8 MBDA MM 40 Exocet Block III ❷; inertial cruise; active radar homing to 180 km *(100 n miles)* at 0.9 Mach; warhead 165 kg; sea-skimmer.
Guns: 1 Oto Melara 3 in *(76 mm)*/62 Super Rapid ❸; 120 rds/min to 16 km *(8.7 n miles)*; weight of shell 6 kg. 2—12.7 mm MGs.
Torpedoes: 6—324 mm (2 B 515 triple) tubes; Eurotorp Mu-90; active/passive homing to 25 km *(13.5 n miles)* at 29/50 kt.
Physical countermeasures: Decoys: 2 EADS NGDS 12-barrelled chaff, IR and anti-torpedo decoy launchers.
Electronic countermeasures: ESM: ARBR 21; intercept.
ECM: To be announced.
Radars: Air/Surface search: Thales Herakles 3-D multifunction ❹; E/F-band.
Navigation: To be announced ❺.
Fire control: Alenia Marconi NA-25XP; J-band.
Sonars: Thales TUS 4110CL; hull mounted (bow dome); active search and attack. Thales Captas UMS-4249 active/passive towed array (to be confirmed).
Combat data systems: DCN/Thales SETIS CMS.
Weapon control systems: 1 optronic FCS.

Helicopters: To be announced ❻.

Programmes: The order for one FREMM frigate was first announced on 22 October 2007 and later confirmed by French Prime Minister François Fillon on 18 April 2008. The ship, to be delivered by 2013, is the first export order for the 17-ship Franco-Italian joint programme. The ship is required to extend the patrol capabilities of the Moroccan Navy and to enable joint operations with NATO and other navies. Details are based on the ships being procured for the French Navy and may be different.

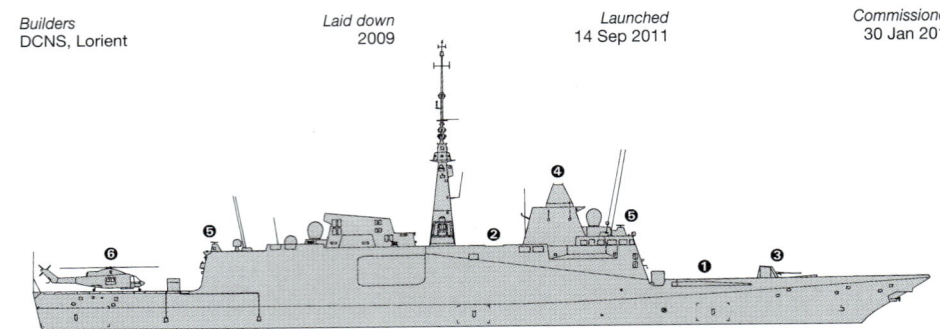

MOHAMMAD VI (Scale 1 : 1,200), Ian Sturton / 1406944

MOHAMMAD VI 1/2014*, B Prézelin / 1531001

Structure: FREMM has a conventional hull design. The main engine room contains the gas turbine and two diesel generators while the aft machinery space contains the motors. Particular attention has been paid to signature reduction. The radar signature is expected to be comparable to that of the La Fayette class while exhaust cooling measures are expected to achieve a comparatively low IR signature. Acoustic quietening is to be achieved by the rafting of engines and motors and the use of electric propulsion. The Herakles radar is housed in the foremast and communications and IFF in the after mast.
Operational: Sea trials began on 17 April 2013 and the ship was handed over on 25 November 2013.

CORVETTES

3 SIGMA CLASS (FSG)

Name	No	Builders	Laid down	Launched	Commissioned
TARIK BEN ZIYAD	613	Schelde Shipbuilding, Vlissengen	15 Apr 2008	12 Jul 2010	10 Sep 2011
SULTAN MOULAY ISMAEL	614	Schelde Shipbuilding, Vlissengen	24 Mar 2009	4 Feb 2011	23 Mar 2012
ALLAL BEN ABDALLAH	615	Schelde Shipbuilding, Vlissengen	21 Sep 2009	27 Aug 2011	10 Sep 2012

Displacement, tonnes: 2,075, 2,335 (613) full load
Dimensions, metres (feet): 97.9, 105.11 (613) × 13 × 4.0 *(321.2, 344.8 × 42.7 × 13.1)*
Speed, knots: 28
Range, n miles: 4,000 at 18 kt
Complement: 91

Machinery: 2 Pielstick 20PA6B STC diesels; 22,030 hp *(16.2 MW)*; 2 shafts; cp props
Missiles: SAM: 12 (2 sextuple) MBDA VL MICA ❶; command/inertial guidance; radar/IR homing to 20 km *(10.8 n miles)*; warhead 12 kg.
SSM: 4 MBDA MM40 Exocet Block III ❷; inertial cruise; active radar homing to 180 km *(97 n miles)* at 0.9 Mach; warhead 165 kg; sea-skimmer.
Guns: 1 Oto Melara 3 in *(76 mm)*/62 Super Rapid ❸; 120 rds/min to 16 km *(8.7 n miles)*; weight of shell 6 kg. 2 Giat 20 mm ❹.
Torpedoes: 6—324 mm (2 B 515 triple) tubes ❺; Eurotorp Mu-90; active/passive homing to 25 km *(13.5 n miles)* at 29/50 kt.
Electronic countermeasures: ESM: Thales Vigile.
ECM: Thales Scorpion.
Radars: Air/Surface search: Thales SMART-S; 3D ❻; E/F-band.
Fire control: Thales LIROD Mk 2 ❼; K-band.
Navigation: Sperry Bridgemaster; I-band.
Sonars: Thales Kingklip.
Combat data systems: Thales Tacticos.
Electro-optic systems: Thales LIROD Mk 2 optronic tracker.

Helicopters: To be announced.

Programmes: Contract for the construction of three corvettes, all to be built in the Netherlands, announced on 6 February 2008. Two (614, 615) ships are 98 m Sigma 9813 and one (613) is a lengthened 105 m Sigma 10513. First steel was cut for *Tarik Ben Ziyad* on 27 February 2008. The ships, which are to have common systems, are required to extend the patrol capabilities of the Moroccan Navy and to enable joint operations with NATO and other navies.

SULTAN MOULAY ISMAEL (Scale 1 : 900), Ian Sturton / 1528811

SULTAN MOULAY ISMAEL 3/2013*, B Prézelin / 1530922

TARIK BEN ZIAD
3/2013*, B Prézelin
1530923

SHIPBORNE AIRCRAFT

Numbers/Type: 3 Eurocopter AS 565MB Panther.
Operational speed: 165 kt *(305 km/h)*.
Service ceiling: 16,700 ft *(5,100 m)*.
Range: 483 n miles *(895 km)*.
Role/Weapon systems: Procured from France for operation from Floréal class. Sensors: Thomson-CSF Varan radar. FLIR. Weapons: 7.62 mm MG.

PANTHER (French colours)   *9/1998, M Declerck* / 0052167

PATROL FORCES

Notes: There are two patrol craft, pennant numbers 105–106, of unknown type.

106   *11/2010, Deryck Swetnam* / 1406094

4 LAZAGA CLASS (FAST ATTACK CRAFT—MISSILE) (PGG)

Name	No	Builders	Commissioned
COMMANDANT EL KHATTABI	304	Bazán, San Fernando	26 Jul 1981
COMMANDANT BOUTOUBA	305	Bazán, San Fernando	2 Aug 1982
COMMANDANT EL HARTY	306	Bazán, San Fernando	20 Nov 1981
COMMANDANT AZOUGGARH	307	Bazán, San Fernando	25 Feb 1982

Displacement, tonnes: 432 full load
Dimensions, metres (feet): 58.1 × 7.6 × 2.7 *(190.6 × 24.9 × 8.9)*
Speed, knots: 30. **Range, n miles:** 3,000 at 15 kt
Complement: 41

Machinery: 2 MTU-Bazán 16V 956 TB91 diesels; 7,500 hp(m) *(5.51 MW)* sustained; 2 shafts
Missiles: SSM: 4 Aerospatiale MM 38 Exocet; inertial cruise; active radar homing to 42 km *(23 n miles)* at 0.9 Mach; warhead 165 kg; sea-skimmer.
Guns: 1 Oto Melara 3 in *(76 mm)*/62 compact; 85 rds/min to 16 km *(8.6 n miles)* anti-surface; 12 km *(6.5 n miles)* anti-aircraft; weight of shell 6 kg. 1 Breda Bofors 40 mm/70; 300 rds/min to 12.5 km *(6.7 n miles)*; weight of shell 0.96 kg. 2 Oerlikon 20 mm/90 GAM-BO1; 800 rds/min to 2 km.
Radars: Surface search: Signaal ZW06; I-band; range 26 km *(14 n miles)*.
Fire control: Signaal WM25; I/J-band; range 46 km *(25 n miles)*.
Navigation: Furuno; I-band.
Weapon control systems: CSEE Panda optical director.

Comment: Ordered from Bazán, San Fernando (Cadiz), Spain 14 June 1977. New Bofors guns fitted aft in 1996/97. 76 mm gun removed from 305 in 1998 and also from 307. *El Harty* and *Azouggarh* refitted by Navantia, Cartagena, 2008–09.

COMMANDANT BOUTOUBA   *9/2008, Diego Quevedo* / 1353240

COMMANDANT AZOUGGARH   *11/2010, Deryck Swetnam* / 1406096

2 OKBA (PR 72) CLASS (LARGE PATROL CRAFT) (PG)

Name	No	Builders	Commissioned
OKBA	302	SFCN, Villeneuve la Garenne	16 Dec 1976
TRIKI	303	SFCN, Villeneuve la Garenne	12 Jul 1977

Displacement, tonnes: 381 standard; 452 full load
Dimensions, metres (feet): 57.5 × 7.6 × 2.1 *(188.6 × 24.9 × 6.9)*
Speed, knots: 20. **Range, n miles:** 2,500 at 16 kt
Complement: 53 (5 officers)

Machinery: 2 SACM AGO V16 ASHR diesels; 5,520 hp(m) *(4.1 MW)*; 2 shafts
Guns: 1 Oto Melara 3 in *(76 mm)*/62 compact; 85 rds/min to 16 km *(8.6 n miles)* anti-surface; 12 km *(6.5 n miles)* anti-aircraft; weight of shell 6 kg. 1 Bofors 40 mm/70; 300 rds/min to 12.5 km *(6.7 n miles)*; weight of shell 0.96 kg.
Radars: Surface search: Racal Decca 1226; I-band.
Weapon control systems: 2 CSEE Panda optical directors.

Comment: Ordered June 1973. *Okba* launched 10 October 1975, *Triki* 1 February 1976. Can be Exocet fitted (with Vega control system). *Triki* refitted at Lorient 2002–03. Modifications included installation of a funnel and removal of two diesels and two shafts. Speed reduced to 20 kt. Similar refit for *Okba* completed in early 2005.

TRIKI   *6/2003, B Prézelin* / 0589787

4 OSPREY MK II CLASS (LARGE PATROL CRAFT) (PBO)

Name	No	Builders	Commissioned
EL HAHIQ	308	Danyard A/S, Frederikshavn	11 Nov 1987
EL TAWFIQ	309	Danyard A/S, Frederikshavn	31 Jan 1988
EL HAMISS	316	Danyard A/S, Frederikshavn	9 Aug 1990
EL KARIB	317	Danyard A/S, Frederikshavn	23 Sep 1990

Displacement, tonnes: 483 full load
Dimensions, metres (feet): 54.8 × 10.5 × 2.6 *(179.8 × 34.4 × 8.5)*
Speed, knots: 22. **Range, n miles:** 4,500 at 16 kt
Complement: 15 + 20 spare berths

Machinery: 2 MAN Burmeister & Wain Alpha 12V23/30-DVO diesels; 4,440 hp(m) *(3.23 MW)* sustained; 2 water-jets
Guns: 1 Bofors 40 mm/60. 2 Oerlikon 20 mm.
Radars: Surface search: Racal Decca; I-band.
Navigation: Racal Decca; I-band.

Comment: First two ordered in September 1986; two more on 30 January 1989. There is a stern ramp with a hinged cover for launching the inspection boat. Used for Fishery Protection duties.

EL HAHIQ   *5/2010, B Prézelin* / 1366980

10 VCSM CLASS (PATROL CRAFT) (PB)

P 107–116

Displacement, tonnes: 41 full load
Dimensions, metres (feet): 20 × 5.0 × 1.5 *(65.6 × 16.4 × 4.9)*
Speed, knots: 25. **Range, n miles:** 530 at 15 kt
Complement: 5
Machinery: 2 MAN V12 diesels; 2,000 hp *(1.47 MW)*; 2 shafts
Guns: 1—7.62 mm MG.
Radars: Navigation: Furuno; I-band.

Comment: Coastal Surveillance craft ordered from Raidco Marine in 2005 (4) and 2006 (6), built at l'Herbaudière and delivered in 2006–08. Raidco RPB 20 design. GRP hull and superstructure. A 4.9 m RIB can be embarked on an inclined ramp at the stern. There are 24 similar 'Vedettes' craft in service with the French Navy.

P 107   *6/2006, B Prézelin* / 1040671

Morocco > Patrol forces

6 CORMORAN CLASS (LARGE PATROL CRAFT) (PBO)

Name	No	Builders	Launched	Commissioned
L V RABHI	310	Bazán, San Fernando	23 Sep 1987	16 Sep 1988
ERRACHIQ	311	Bazán, San Fernando	23 Sep 1987	16 Dec 1988
EL AKID	312	Bazán, San Fernando	29 Mar 1988	4 Apr 1989
EL MAHER	313	Bazán, San Fernando	29 Mar 1988	20 Jun 1989
EL MAJID	314	Bazán, San Fernando	21 Oct 1988	26 Sep 1989
EL BACHIR	315	Bazán, San Fernando	21 Oct 1988	19 Dec 1989

Displacement, tonnes: 432 full load
Dimensions, metres (feet): 58.1 × 7.6 × 2.7 (190.6 × 24.9 × 8.9)
Speed, knots: 22
Range, n miles: 6,100 at 12 kt
Complement: 36 (4 officers) + 15 spare berths

Machinery: 2 MTU-Bazán 16V 956 TB82 diesels; 8,340 hp(m) (6.13 MW) sustained; 2 shafts
Guns: 1 Bofors 40 mm/70. 2 Giat 20 mm.
Radars: Surface search: Racal Decca; I-band.
Weapon control systems: CSEE Lynx optronic director.

Comment: Three ordered from Bazán, Cadiz in October 1985 as a follow on to the Lazaga class of which these are a slower patrol version with a 10-day endurance. Option on three more taken up. Used for fishery protection. Armament removed from some. *El Akid, El Majid* and *El Bachir* refitted by Raidco Marine 2007–08.

EL BACHIR   3/2009, B Prézelin / 1366133

5 RAÏS BARGACH CLASS (TYPE OPV 64) (PSO)

Name	No	Builders	Launched	Commissioned
RAÏS BARGACH	318	Leroux & Lotz, Lorient	9 Oct 1995	14 Dec 1995
RAÏS BRITEL	319	Leroux & Lotz, Lorient	19 Mar 1996	14 May 1996
RAÏS CHARKAOUI	320	Leroux & Lotz, Lorient	25 Sep 1996	10 Dec 1996
RAÏS MAANINOU	321	Leroux & Lotz, Lorient	7 Mar 1997	21 May 1997
RAÏS AL MOUNASTIRI	322	Leroux & Lotz, Lorient	15 Oct 1997	17 Dec 1997

Displacement, tonnes: 589 full load
Dimensions, metres (feet): 64 × 11.4 × 3 (210 × 37.4 × 9.8)
Speed, knots: 24; 7 motor
Range, n miles: 4,000 at 12 kt
Complement: 24 (3 officers) + 30 spare berths
Machinery: 2 Wärtsilä Nohab 25 V16 diesels; 10,000 hp(m) (7.36 MW) sustained; 2 Leroy auxiliary motors; 326 hp(m) (240 kW); 2 shafts; cp props
Guns: 1 Bofors 40 mm/60. 1 Oerlikon 20 mm. 4—14.5 mm MGs (2 twin).
Radars: Surface search: Racal Decca Bridgemaster; I-band.

Comment: First pair ordered to a Serter design from Leroux & Lotz, Lorient in December 1993, second pair in October 1994. Option on fifth taken up in 1996. There is a stern door for launching a 7 m RIB, a water gun for firefighting and two passive stabilisation tanks. This version of the OPV 64 does not have a helicopter deck and the armament is fitted after delivery. Manned by the Navy for the Fisheries Department. Based at Agadir.

RAÏS BRITEL   4/2007, Rafael Carrera Gonzalez / 1170199

RAÏS AL MOUNASTIRI   5/2010, B Prézelin / 1366979

6 EL WACIL (P 32) CLASS (COASTAL PATROL CRAFT) (PB)

Name	No	Builders	Launched	Commissioned
EL WACIL	203	CMN, Cherbourg	12 Jun 1975	9 Oct 1975
EL JAIL	204	CMN, Cherbourg	10 Oct 1975	3 Dec 1975
EL MIKDAM	205	CMN, Cherbourg	1 Dec 1975	30 Jan 1976
EL KHAFIR	206	CMN, Cherbourg	21 Jan 1976	16 Apr 1976
EL HARIS	207	CMN, Cherbourg	31 Mar 1976	30 Jun 1976
EL ESSAHIR	208	CMN, Cherbourg	2 Jun 1976	16 Jul 1976

Displacement, tonnes: 75 standard; 90 full load
Dimensions, metres (feet): 32 × 5.4 × 1.4 (105 × 17.7 × 4.6)
Speed, knots: 28
Range, n miles: 1,500 at 15 kt
Complement: 17
Machinery: 2 SACM MGO 12V BZSHR diesels; 2,700 hp(m) (1.98 MW); 2 shafts
Guns: 1 Oerlikon 20 mm.
Radars: Surface search: Decca; I-band.

Comment: Ordered in February 1974. In July 1985 a further four of this class were ordered from the same builders but for the Customs Service. Wooden hull sheathed in plastic.

EL KHAFIR   4/2013*, Bob Fildes / 1530924

1 PATROL VESSEL (PBO)

Name	No	Builders	Commissioned
— (ex-Cygnet)	323 (ex-P 261)	R Dunston Ltd, Hessle	8 Jul 1976

Displacement, tonnes: 197 full load
Dimensions, metres (feet): 36.6 × 7.2 × 2.0 (120.1 × 23.6 × 6.6)
Speed, knots: 21
Range, n miles: 2,000 at 14 kt
Complement: 21 (4 officers)
Machinery: 2 Paxman 16YJCM diesels; 4,200 hp (3.1 MW); 2 shafts
Guns: 1—12.7 mm MG.
Radars: Navigation: I-band.

Comment: Former Royal Navy Bird-class patrol craft sold to a private buyer and delivered to Agadir on 11 April 1997. The ship was later implicated in a counter-drugs operation and the vessel was confiscated by the Moroccan authorities. It has since been operated by the Moroccan Navy.

323   3/2006, M Declerck / 1164950

1 OPV 70 CLASS (OFFSHORE PATROL VESSELS) (PSO)

Name	No	Builders	Laid down	Launched	Commissioned
BIN AN ZARAN	341	STX Lorient	2008	25 Aug 2010	23 Jun 2011

Displacement, tonnes: 800 full load
Dimensions, metres (feet): 70 × 11.3 × 3.2 (229.7 × 37.1 × 10.5)
Speed, knots: 22
Range, n miles: 4,500 at 12 kt
Complement: 64
Machinery: 2 Wärtsilä 12V26 diesels; 10,730 hp (8 MW); 2 shafts
Guns: 1 Oto Melara 3 in (76 mm)/62 Super Rapid; 120 rds/min to 16 km (8.7 n miles); weight of shell 6 kg. 1—40 mm. 2—14.5 mm. 2—12.7 mm MGs.
Radars: Surface search/navigation: Sperry BridgeMaster II; E/F/I-bands.

Comment: The order for one patrol vessel was announced on 30 May 2008. Designed by Raidco Marine, the ship has been built by STX, Lorient. Further ships are projected.

BIN AN ZARAN   7/2011, M Declerck / 1406628

10 RODMAN 101 CLASS (PB)

130–139

Displacement, tonnes: 63 full load
Dimensions, metres (feet): 30 × 5.9 × 1.3 *(98.4 × 19.4 × 4.3)*
Speed, knots: 30. **Range, n miles:** 800 at 12 kt
Complement: 9
Machinery: 2 Caterpillar 3412C diesels; 2,800 hp *(2.06 MW)*; 2 Hamilton waterjets
Radars: Navigation: I-band.

Comment: GRP hull. Built by Rodman, Vigo and delivered in 2009–10.

130   11/2010, Deryck Swetnam / 1406097

AMPHIBIOUS FORCES

3 BATRAL CLASS (LSMH)

Name	No	Builders	Commissioned
DAOUD BEN AICHA	402	Dubigeon, Normandie	28 May 1977
AHMED ES SAKALI	403	Dubigeon, Normandie	Sep 1977
ABOU ABDALLAH EL AYACHI	404	Dubigeon, Normandie	Mar 1978

Displacement, tonnes: 762 standard; 1,432 full load
Dimensions, metres (feet): 80 × 13 × 2.4 *(262.5 × 42.7 × 7.9)*
Speed, knots: 16
Range, n miles: 4,500 at 13 kt
Complement: 47 (3 officers)

Military lift: 140 troops; 12 vehicles or 300 tons
Machinery: 2 SACM Type 195 V12 CSHR diesels; 3,600 hp(m) *(2.65 MW)* sustained; 2 shafts
Guns: 2 Bofors 40 mm/70. 2—81 mm mortars. 2—12.7 mm MGs.
Radars: Surface search: Thomson-CSF DRBN 32 (Racal Decca 1226); I-band.
Helicopters: Platform only.

Comment: Two ordered on 12 March 1975. Third ordered 19 August 1975. Of same type as the French *Champlain*. Vehicle-stowage above and below decks. *Daoud Ben Aicha* was refitted in Lorient by Leroux & Lotz in 1995 and *Abou Abdallah el Ayachi* in 1997.

DAOUD BEN AICHA   10/2004, Carlos Pardo Gonzalez / 1133132

1 NEWPORT CLASS (LSTH)

Name	No	Builders	Commissioned
SIDI MOHAMMED BEN ABDALLAH (ex-*Bristol County*)	407 (ex-1198)	National Steel & Shipbuilding Co, San Diego	5 Aug 1972

Displacement, tonnes: 5,055 standard; 8,586 full load
Dimensions, metres (feet): 159.2 × 21.2 × 5.3 *(522.3 × 69.6 × 17.4)*
Speed, knots: 20
Range, n miles: 14,250 at 14 kt
Complement: 257 (13 officers)

Military lift: 400 troops (20 officers); 500 tons vehicles; 3 LCVPs and 1 LCPL on davits
Machinery: 6 ALCO 16-251 diesels; 16,500 hp *(12.3 MW)* sustained; 2 shafts; cp props; bow thruster
Guns: 1 Raytheon 20 mm 6-barrelled Vulcan Phalanx Mk 15.
Radars: Surface search: Raytheon SPS-67; G-band.
Navigation: Marconi LN66; I/J-band.
Helicopters: Platform only.

Comment: Received from the US by grant transfer on 16 August 1994. Has replaced *Arrafiq*. The ship was non-operational by late 1995 and although back in service, has so far proved to be a poor bargain. The bow ramp is supported by twin derrick arms. A ramp just forward of the superstructure connects the lower tank deck with the main deck and a vehicle passage through the superstructure provides access to the parking area amidships. A stern gate to the tank deck permits unloading of amphibious tractors into the water, or unloading of other vehicles into an LCU or on to a pier. Vehicle stowage covers 19,000 sq ft. Length over derrick arms is 562 ft *(171.3 m)*; full load draught is 11.5 ft forward and 17.5 ft aft. Based at Casablanca.

© 2014 IHS

SIDI MOHAMMED BEN ABDALLAH   7/2007, Shaun Jones / 1170198

1 CTM (LCM)

Displacement, tonnes: 60 standard; 152 full load
Dimensions, metres (feet): 23.8 × 6.4 × 1.3 *(78.1 × 21.0 × 4.3)*
Speed, knots: 9.5
Range, n miles: 380 at 8 kt
Complement: 4 + 200 embarked forces

Military lift: 90 tons (maximum); 48 tons normal
Machinery: 2 Poyaud V8520NS diesels; 450 hp *(331 kW)*; 2 shafts
Guns: 1—12.7 mm MG.
Radars: Navigation: I-band.

Comment: Ex-CTM-5 transferred from France in August 2000.

LCM   6/2006, S Dominguez Llosá / 1040668

SURVEY AND RESEARCH SHIPS

1 HYDROGRAPHIC SURVEY VESSEL (AGSC)

Name	No	Builders	Commissioned
–	H 01	Damen Shipyards	Apr 2011

Measurement, tonnes: 28 gt
Dimensions, metres (feet): 15.25 × 4.4 × 0.99 *(50 × 14.4 × 3.2)*
Speed, knots: 18
Complement: 8
Machinery: 2 MAN D2866 diesels; 592 hp *(442 kW)*; 2 shafts
Radars: Furuno 1832; I-band.

Comment: Based on the Damen Stan Tender 1504 design, the ship is equipped with Simrad EM3002 multibeamed echo-sounder, Simrad E400 single-beam echo-sounder and C-Max side-scan sonar.

H-01   6/2012, DAMEN / 1486299

IHS Jane's Fighting Ships 2014-2015

550 Morocco > Auxiliaries — Customs/coast guard/police

AUXILIARIES

Notes: (1) There is also a yacht, *Essaouira*, 60 tons, from Italy in 1967, used as a training vessel for watchkeepers.
(2) Bazán delivered a harbour pusher tug, similar to Spanish Y 171 class, in December 1993.
(3) There are two sail training craft *Al Massira* and *Boujdour*.
(4) There is a stern trawler used as a utility and diver support vessel (803 (ex-YFU 14)).

803   9/2004, S D Llosá / 1044141

1 LOGISTIC SUPPORT SHIP (AKS)

EL AIGH (ex-*Merc Nordia*) 405

Measurement, tonnes: 1,524 gt
Dimensions, metres (feet): 77 × 12.2 × 4.7 *(252.6 × 40.0 × 15.4)*
Speed, knots: 11
Complement: 25
Machinery: 1 Burmeister & Wain diesel; 1,250 hp(m) *(919 kW)*; 1 shaft
Guns: 2 — 14.5 mm MGs.

Comment: Logistic support vessel with four 5 ton cranes. Former cargo ship with ice-strengthened bow built by Fredrickshavn Vaerft in 1973 and acquired in 1981.

EL AIGH   5/1994, M Declerck / 0506199

1 DAKHLA CLASS (LOGISTIC SUPPORT SHIP) (AKS)

Name	No	Builders	Launched	Commissioned
DAKHLA	408	Leroux & Lotz, Lorient	5 Jun 1997	1 Aug 1997

Displacement, tonnes: 2,195 full load
Dimensions, metres (feet): 69 × 11.5 × 4.2 *(226.4 × 37.7 × 13.8)*
Speed, knots: 12
Range, n miles: 4,300 at 12 kt
Complement: 24 + 22 spare berths

Cargo capacity: 800 tons
Machinery: 1 Wärtsilä Nohab 8V25 diesel; 2,300 hp(m) *(1.69 MW)* sustained; 1 shaft; cp prop
Guns: 2 — 12.7 mm MGs.
Radars: Navigation: 2 Racal Decca Bridgemaster ARPA; I-band.

Comment: Ordered from Leroux & Lotz, Nantes in 1995. Side entry for vehicles. One 15 ton crane. Based at Agadir.

DAKHLA   8/1997, Leroux & Lotz / 0012789

CUSTOMS/COAST GUARD/POLICE

Notes: (1) The Coast Guard was created by Royal Decree on 9 September 1997. Responsibility for Search and Rescue conferred on the Ministère des Pêches Maritimes (MPM). Operational control is exercised from the National Rescue Service HQ at Rabat in co-ordination with the Merchant Marine HQ at Casablanca.
(2) There is a 17 m SAR craft *Al Fida* delivered in August 2002.
(3) There are four SAR craft: *Rif*, *Loukouss*, *Souss* and *Dghira*.

AL FIDA   7/2004, S D Llosá / 1044137

SOUSS   7/1995, Zamacona / 1044138

2 SAR CRAFT (SAR)

AL AMANE 2344   **AIT BAÂMRANE** 2345

Displacement, tonnes: 69 full load
Dimensions, metres (feet): 15.75 × 4.48 × 1.05 *(51.7 × 14.7 × 3.4)*
Speed, knots: 34
Complement: 4
Machinery: 2 Volvo D12; 1,300 hp *(970 kW)*; Hamilton waterjets

Comment: Constructed by Auxnaval Shipbuilders, Spain and delivered in March 2003. Aluminium hull.

AL AMANE   7/2003, Auxnaval / 1044136

3 SAR CRAFT (SAR)

HAOUZ   **ASSA**   **TARIK**

Displacement, tonnes: 41 full load
Dimensions, metres (feet): 19.4 × 4.8 × 1.3 *(63.6 × 15.7 × 4.3)*
Speed, knots: 20
Complement: 6
Machinery: 2 diesels; 1,400 hp(m) *(1.03 MW)*; 2 shafts

Comment: Rescue craft built by Schweers, Bardenfleth and delivered in 1991.

Customs/Coast guard/Police < **Morocco** — Patrol forces < **Mozambique** 551

4 SAR CRAFT (SAR)

| AL WHADA 12-64 | SEBOU 12-65 | DOUKKALA | TENSIFT |

Displacement, tonnes: 71 full load
Dimensions, metres (feet): 20.7 × 5.8 × 1.8 *(67.9 × 19.0 × 5.9)*
Speed, knots: 20
Complement: 4
Machinery: 2 MAN D2842 LE401 diesels; 2,000 hp *(1.49 MW)*; 2 shafts

Comment: Constructed by Auxnaval, Asturias, Spain and delivered in 2004.

AL WHADA   7/2004, Auxnaval / 1044139

4 ERRAID (P 32) CLASS (COASTAL PATROL CRAFT) (WPB)

Name	No	Builders	Launched	Commissioned
ERRAID	209	CMN, Cherbourg	20 Dec 1987	18 Mar 1988
ERRACED	210	CMN, Cherbourg	21 Jan 1988	15 Apr 1988
EL KACED	211	CMN, Cherbourg	10 Mar 1988	17 May 1988
ESSAID	212	CMN, Cherbourg	19 May 1988	4 Jul 1988

Displacement, tonnes: 90 full load
Dimensions, metres (feet): 32 × 5.4 × 1.4 *(105 × 17.7 × 4.6)*
Speed, knots: 28
Range, n miles: 1,500 at 15 kt
Complement: 17
Machinery: 2 SACM MGO 12V BZSHR diesels; 2,700 hp(m) *(1.98 MW)*; 2 shafts
Guns: 1 Oerlikon 20 mm.
Radars: Navigation: Decca; I-band.

Comment: Similar to the El Wacil class listed under Patrol Forces. Ordered in July 1985.

EL KACED   4/2013*, Bob Fildes / 1530925

18 ARCOR 46 CLASS (COASTAL PATROL CRAFT) (WPB)

001–018

Displacement, tonnes: 15 full load
Dimensions, metres (feet): 14.5 × 4.2 × 1.3 *(47.6 × 13.8 × 4.3)*
Speed, knots: 32
Range, n miles: 300 at 20 kt
Complement: 6
Machinery: 2 SACM UD18V8 M5D diesels; 1,010 hp(m) *(742 kW)* sustained; 2 shafts
Guns: 2 Browning 12.7 mm MGs.
Radars: Surface search: Furuno 701; I-band.

Comment: Ordered from Arcor, La Teste in June 1985. GRP hulls. Delivered in groups of three from April to September 1987. Used for patrolling the Mediterranean coastline.

ARCOR 009   11/2010, Deryck Swetnam / 1406099

15 ARCOR 53 CLASS (COASTAL PATROL CRAFT) (WPBF)

Displacement, tonnes: 17 full load
Dimensions, metres (feet): 16 × 4 × 1.2 *(52.5 × 13.1 × 3.9)*
Speed, knots: 35
Range, n miles: 300 at 20 kt
Complement: 6
Machinery: 2 Saab DSI-14 diesels; 1,250 hp(m) *(919 kW)*; 2 shafts
Guns: 1 – 12.7 mm MG.
Radars: Surface search: Furuno; I-band.

Comment: Ordered from Arcor, La Teste in 1990 for the Police Force. Delivered at one a month from October 1992.

ARCOR 53 CLASS   3/2006, M Declerck / 1164948

Mozambique

MARINHA MOÇAMBIQUE

Country Overview

The Republic of Mozambique gained independence from Portugal in 1975. Situated in south-eastern Africa, it has an area of 308,642 square miles and is bordered to the north by Tanzania, to the south by South Africa and Swaziland and to the west by Zimbabwe, Zambia, and Malawi. It has a 1,334 n mile coastline with the Mozambique Channel of the Indian Ocean. Maputo (formerly Lourenço Marques) is the capital, largest city and principal port. There is another major port at Beira. Territorial Seas (12 n miles) are claimed. A 200 n mile EEZ has also been claimed but the limits are not fully defined by boundary agreements.

All the Russian built Zhuks and Yevgenyas have sunk alongside or been sold. There are some motorboats operational on Lake Malawi.

Headquarters Appointments

Commander of the Navy:
Rear Admiral L L Menete

Personnel

2014: 200

Bases

Maputo (Naval HQ); Nacala; Beira; Pemba (Porto Amelia); Metangula (Lake Malawi); Tete (River Zambesi); Inhambane.

PATROL FORCES

Notes: (1) An order for three 42 m Ocean Eagle trimaran patrol vessels, three 32 m interceptors, and 24 fishing trawlers was reportedly made from French shipyard CMN by private Mozambiquan company Ematum in September 2013. Although the Mozambique government is believed to be guarantor, it is unclear whether the vessels are to have a formal security role. Deliveries are to be completed in 2016.

(2) A total of eight patrol craft have been reported donated by the US. Three were delivered on 21 December 2006, three on 19 March 2007 and the remaining two in late 2007.

(3) Two vessels are used for fisheries patrol: *Antilles Reefer* is a former fishing vessel seized by Mozambique authorities in July 2008 and former South African environmental patrol ship *Kuswag 1*.

(4) There is one Vredenburger-class 9 m patrol craft donated by South Africa in 2007.

(5) Four former Portuguese 20 m Jupiter class are based at Metangula (Lake Malawi). Operational status doubtful.

552 **Mozambique** > Patrol forces — **Myanmar** > Introduction

12 DEFENDER CLASS (RESPONSE BOATS) (PBF)

Displacement, tonnes: 2.7 full load
Dimensions, metres (feet): 7.6 × 2.6 × 1.0 (24.9 × 8.5 × 3.3)
Speed, knots: 46
Range, n miles: 175 at 35 kt
Complement: 4
Machinery: 2 Honda outboard motors; 450 hp (335 kW)
Guns: 1 — 12.7 mm MG.
Radars: To be announced.

Comment: High-speed inshore patrol craft of aluminium construction and foam collar built by SAFE Boats International, Port Orchard, Washington. Purchased with assistance from the United States government on 10 November 2010.

2 NAMACURRA CLASS (INSHORE PATROL CRAFT) (PB)

P 01 (ex-Y 1507)   P 02 (ex-Y 1530)

Displacement, tonnes: 5 full load
Dimensions, metres (feet): 9 × 2.7 × 0.8 (29.5 × 8.9 × 2.6)
Speed, knots: 32. **Range, n miles:** 180 at 20 kt
Complement: 4

Machinery: 2 Yamaha outboards; 380 hp(m) (2.79 kW)
Guns: 1 — 12.7 mm MG. 2 — 7.62 mm MGs.
Depth charges: 1 rack.
Radars: Surface search: Furuno; I-band.

Comment: Built in South Africa in 1980–81. Can be transported by road. Donated by South Africa in 2004.

1 CONEJERA CLASS (COASTAL PATROL CRAFT) (PB)

Name	No	Builders	Commissioned
PEBANE (ex-*Dragonera*)	P 001 (ex-P 32)	Bazán, Ferrol	31 Dec 1981

Displacement, tonnes: 85 full load
Dimensions, metres (feet): 32.2 × 5.3 × 1.4 (105.6 × 17.4 × 4.6)
Speed, knots: 13. **Range, n miles:** 1,200 at 13 kt
Complement: 12
Machinery: 2 MTU-Bazán MA 16V 362 SB80 diesels; 2,450 hp(m) (1.8 MW); 2 shafts
Guns: 1 Oerlikon 20 mm/120 Mk 10. 1 — 12.7 mm MG.
Radars: Surface search: Furuno; I-band.

Comment: Former Spanish patrol craft commissioned in 1981 and decommissioned in 2010. Transferred to Mozambique in 2013.

CONEJERA CLASS 8/2007, Marco Ghiglino / 1170222

NAMACURRA (South African colours) 8/2001, van Ginderen Collection / 0132783

Myanmar
TATMADAW YAY

Country Overview

The Union of Myanmar, also known as the Republic of Burma, gained independence in 1948. Situated in South East Asia, it has an area of 261,218 square miles, is bordered to the north-east by China, to the north-west by India and Bangladesh and to the south-east by Laos and Thailand. It has a 1,042 n mile coastline with the Andaman Sea and the Bay of Bengal. The administrative capital became Pyinmana on 6 November 2005. Rangoon (Yangon) is the commercial capital, largest city and principal port. Some 6,900 n miles of navigable inland waterways are important transport arteries. Territorial waters (12 n miles) are claimed. A 200 n mile EEZ has been claimed although the limits have only been partly defined by boundary agreements.

Headquarters Appointments

Commander in Chief:
 Rear Admiral Thura Thet Swe

Personnel

(a) 2014: 13,000 (this may include 800 naval infantry)
(b) Voluntary service

Bases

There are five regional commands with principal bases as indicated:
Ayeyarwady (Irawaddy): Monkey Point (Navy HQ), Yangon (Rangoon), Thilawa (dockyard), Great Coco Island
Taninthayi (Tenasserim): Myeik (Mergui) (Regional HQ), Zadetgyi Island (Base 58, St Matthew's Island), Kathekyun (Ketthayin), Pale Island, Thetkatan (Kadan Island)
Panmawaddy: Hainggyi Island (Regional HQ), Pathein
Mawrawady: Mawlamyine (Moulmein) (Regional HQ), Kyaikkami, Dawei (Tavoy)
Danyawady: Kyaukpyu (Regional HQ), Akyab (Base 18, Sittwe), Thandwe

The Headquarters of Training Command is at Thilawa in Rangoon. The main training depot is currently at Syriam (Thanlyin), but is to be transferred to Seikkyi, near the mouth of the Hlaing (Rangoon) River.
The Pathein base will reportedly be moved to Pyadatgyi Island, where an expanded airfield will permit the basing of air force equipment and personnel as well as navy. The Great Coco Island base has also been expanded through the construction of a large landing jetty to replace the existing small pier. It is also the site of a Chinese surveillance installation.

Organisations

Naval units are usually commanded directly from Rangoon, but operational control is occasionally delegated to regional commands.

Naval Infantry

The existence of 800 naval infantry has been previously reported but not confirmed.

IHS Jane's Fighting Ships 2014-2015 © 2014 IHS

FRIGATES

2 JIANGHU II (TYPE 053H1) CLASS (FFG)

MAHAR BANDOOLA (ex-*Anshun*) F 21 (ex-554)
MAHAR THIHA THURA (ex-*Jishou*) F 23 (ex-557)

Displacement, tonnes: 1,448 standard; 1,729 full load
Dimensions, metres (feet): 103.2 × 10.8 × 3.1 *(338.6 × 35.4 × 10.2)*
Speed, knots: 26
Range, n miles: 4,000 at 15 kt, 2,700 at 18 kt
Complement: 200 (30 officers)

Machinery: 2 Type 12E 390V diesels; 14,400 hp(m) *(10.6 MW)* sustained; 2 shafts
Missiles: SSM: 4 HY-2 (CSSC-3 Seersucker) (2 twin) launchers ❶; active radar or IR homing to 95 km *(51 n miles)* at 0.9 Mach; warhead 513 kg.
Guns: 4 China 3.9 in *(100 mm)*/56 (2 twin) ❸; 25 rds/min to 22 km *(12 n miles)*; weight of shell 15.6 kg. 4 China 37 mm/63 (2 twin) ❹; 180 rds/min to 8.5 km *(4.6 n miles)* anti-aircraft; weight of shell 1.42 kg.
A/S Mortars: 2 RBU 1200 5-tubed fixed launchers ❺; range 1,200 m; warhead 34 kg.
Mines: Can carry up to 60.
Depth charges: 2 BMB-2 projectors; 2 racks (in some).

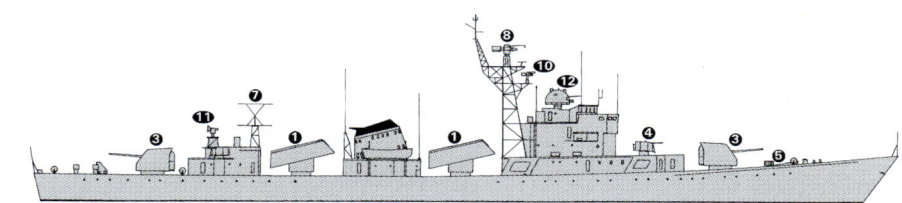

TYPE 053H1 *(Scale 1 : 900), Ian Sturton / 0130728*

Physical countermeasures: Decoys: 2 China 26-barrelled launchers.
Electronic countermeasures: ESM: Jug Pair or Watchdog; radar warning.
Radars: Air search: Type 517 Knife Rest ❼; A-band.
Air/surface search: Type 354 Eye Shield (MX 902) ❾; G-band.
Surface search/fire control: Type 352 Square Tie ❿; I-band.
Navigation: Done 2 or Fin Curve or Racal Decca; I-band.
Fire control: Type 347G Rice Bowl ⓫; I/J-band. Type 343 (Wok Won) (Wasp Head) ⓬; I-band.
IFF: High Pole A. Yard Rake or Square Head.

Sonars: Echo Type 5; hull-mounted; active search and attack; medium frequency.

Comment: Former Chinese Navy Jianghu-class frigates built in the 1980s. Transferred to Myanmar on 8 March 2012. Details reflect sensors and weapons in Chinese service and, while expected to be same, may differ.

MAHAR BANDOOLA *3/2013*, Chris Sattler / 1486547*

2 + (6) FRIGATES (FSG)

AUNG ZEYA F 11   **KYAN-SIT-THAR** F 12   +6

Displacement, tonnes: 2,500 standard (est.)
Dimensions, metres (feet): 106 × 13.5 × ? *(347.8 × 44.3 × ?)*

Machinery: To be announced.
Missiles: SSM: 4 C-602 (YJ-62); 2 twin; inertial GPS guidance and terminal active radar homing to 280 km *(151 n miles)* at 0.8 Mach; warhead 300 kg.
Guns: 1 Oto Melara 3 in *(76 mm)* Super Rapid; 120 rds/min to 16 km *(8.7 n miles)*; weight of shell 6 kg. 4—30 mm AK 630; 6 barrels; 3,000 rds/min combined to 2 km.
Physical countermeasures: To be announced.
Electronic countermeasures: To be announced.
Radars: Surface search: Bharat RAWL-02 (LW 08); D-band.
Navigation: To be announced.
Fire control: To be announced.
Sonars: To be announced.

Helicopters: Platform for 1 medium.

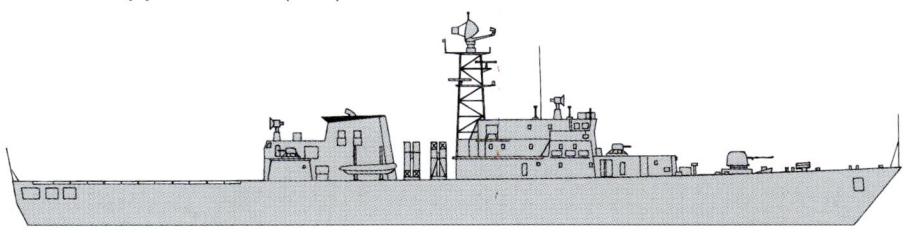

AUNG ZEYA *(Scale 1 : 900), Ian Sturton / 1406947*

Comment: Few details have emerged of these ships, which were probably built with Chinese assistance at Sinmailik Naval Dockyard, Rangoon. The design appears to be based on the Jiangwei class. It is reported that the first of class (F 11) was laid down in 2006 and commissioned in December 2011. There is a helicopter deck aft. A second ship (F 12) is reported to have been launched in 2012. Generally similar in design, the lattice mast has been replaced by a solid structure and the aft superstructure has been extended to include a hangar. A further six ships are projected.

AUNG ZEYA *6/2013* / 1530926*

CORVETTES

2 ANAWRAHTA CLASS (CORVETTES) (FSG)

Name	No	Builders	Laid down	Launched	Commissioned
ANAWRAHTA	771	Sinmalaik Shipyard, Rangoon	1998	2000	2001
BAYINTNAUNG	772	Sinmalaik Shipyard, Rangoon	1998	2001	2003

Displacement, tonnes: 1,105 full load
Dimensions, metres (feet): 77 × ? × ? *(252.6 × ? × ?)*
Complement: 101 (15 officers)

Machinery: To be announced
Missiles: SSM: 4 YJ-83 (C-802) (fitted in at least one in lieu of aft 76 mm gun); mid-course guidance and active radar homing to 120 km *(65 n miles)* at 0.9 Mach; warhead 165 kg.
Guns: 1 or 2 Oto Breda 3 in *(76 mm)*/62 compact; 85 rds/min to 16 km *(8.7 n miles)* anti-surface; 12 km *(6.5 n miles)* anti-aircraft; weight of shell 6 kg. 1—40 mm/70; 300 rds/min to 12.5 km *(6.8 n miles)*; weight of shell 0.96 kg. 4—14.5 mm (2 twin).
Physical countermeasures: To be announced.
Radars: Surface search: To be announced.
Navigation: To be announced
Fire control: To be announced
Sonars: To be announced

Helicopters: Platform for 1 medium.

Programmes: The programme to acquire ships to replace the now decommissioned PCE-827 and Admirable class corvettes was probably instituted in the 1990s. As frigates initially proved to be too expensive, three Chinese hulls are believed to have been acquired in about 1998 for fitting out at Sinmalaik Shipyard. There have been reports that Israeli electronic systems (radars and sonar) have been fitted. The details of the programme are speculative.
Operational: The first ship conducted sea trials in 2001 when the second ship was reportedly nearing completion. *Bayintnaung* participated in Exercise Milan 2008 and *Anawrahta* in Exercise Milan 2010.

ANAWRAHTA
1/2006, Indian Navy
1158724

ANAWRAHTA   2/2010, M Mazumdar / 1366449

1 OFFSHORE PATROL SHIP (PGG)

Name	No	Builders	Launched	Commissioned
—	491	Sinmalaik Shipyard, Rangoon	2012	2013

Dimensions, metres (feet): 52.4 × 7.0 × ? *(171.9 × 23.0 × ?)*
Machinery: To be announced.
Missiles: 4 C-802.
Guns: 1 Norinco 30 mm. 4—25 mm (2 twin).
Radars: Surface search: Type 348; I-band.
Surface search: To be announced.
Fire control: Type 344; I-band.
Navigation: To be announced.

Comment: The first of what is assumed to be a new class of missile-armed patrol vessels was launched in 2012 and is likely to have become operational in 2013. The ship is likely to have been built with Chinese advice and/or assistance.

OPV 491   6/2012, Mazumdar Collection / 1486302

PATROL FORCES

Notes: There is a new class of river patrol craft known as the Ngaman class. These 8 m craft are of a Boston Whaler type and are armed with a 12.5 mm gun in the bow and a twin 7.62 mm aft. Locally built, the craft have probably replaced the PBR Mk II class which have been decommissioned.

1 ADMIRABLE CLASS (FS)

Name	No	Builders	Commissioned
YAN GYI AUNG	42 (ex-MSF 356)	Willamette Iron & Steel Co, Portland	18 Dec 1945
(ex-*Creddock*)			

Displacement, tonnes: 650 standard; 945 full load
Dimensions, metres (feet): 56.2 × 10.1 × 3 *(184.4 × 33.1 × 9.8)*
Speed, knots: 14.8
Range, n miles: 4,300 at 10 kt
Complement: 73
Machinery: 2 Busch-Sulzer BS-539 diesels; 1,500 hp(m) *(1.1 MW)*; 2 shafts
Guns: 1 US 3 in *(76 mm)*/50 Mk 26; 20 rds/min to 12 km *(6.6 n miles)*; weight of shell 6 kg. 4 Bofors 40 mm/60 (2 twin). 4 Oerlikon 20 mm (2 twin).
Radars: Surface search: Raytheon SPS-5; G/H-band.

Programmes: Laid down on 10 November 1943 and launched on 22 July 1944. Transferred from USA at San Diego on 31 March 1967.
Operational: Minesweeping gear and ASW equipment removed. Following reports of being decommissioned in the 1990s, the ship was reported in service in 2013, although operational status is questionable.

YAN GYI AUNG   12/1991 / 0056640

6 HOUXIN (TYPE 037/1G) CLASS
(FAST ATTACK CRAFT—GUN) (PTG)

MAGA 471   SAITTRA 472   DUWA 473   ZEYDA 474   HANTHA 475   BANDA 476

Displacement, tonnes: 486 full load
Dimensions, metres (feet): 62.8 × 7.2 × 2.4 *(206 × 23.6 × 7.9)*
Speed, knots: 28
Range, n miles: 1,300 at 15 kt
Complement: 71

Machinery: 4 PR 230ZC diesels; 4,000 hp(m) *(2.94 MW)*; 4 shafts
Missiles: SSM: 4 YJ-8 (C-801) (2 twin); active radar homing to 40 km *(22 n miles)* at 0.9 Mach; warhead 165 kg; sea skimmer. C-802 may be fitted in due course.
Guns: 4—37 mm/63 Type 76A (2 twin); 180 rds/min to 8.5 km *(4.6 n miles)*; weight of shell 1.42 kg. 4—14.5 mm Type 69 (2 twin).
Electronic countermeasures: ESM/ECM: intercept and jammer.
Radars: Surface search: Square Tie; I-band.
Fire control: Rice Lamp; I-band.

Programmes: First pair arrived from China in December 1995, second pair in mid-1996 and last two in late 1997. The first four were wrongly reported as Hainan class.
Structure: Details given are for this class in Chinese service.
Operational: 475 damaged in a collision during sea trials in August 1996. All based at Rangoon.

ZEYDA   6/2001 / 0130747

2 OSPREY CLASS (OFFSHORE PATROL VESSELS) (PBO)

Name	No	Builders	Commissioned
INDAW	FV 55	Frederikshavn Dockyard, Frederikshavn	30 May 1980
INYA	FV 57	Frederikshavn Dockyard, Frederikshavn	25 Mar 1982

Displacement, tonnes: 391 standard; 513 full load
Dimensions, metres (feet): 50 × 10.5 × 2.8 *(164 × 34.4 × 9.2)*
Speed, knots: 20
Range, n miles: 4,500 at 16 kt
Complement: 20 (5 officers)
Machinery: 2 Burmeister and Wain Alpha diesels; 4,640 hp(m) *(3.4 MW)*; 2 shafts; cp props
Guns: 1 Bofors 40 mm/60. 2 Oerlikon 20 mm.

Comment: Operated by Burmese Navy for the People's Pearl and Fishery Department. Helicopter deck with hangar in *Indaw*. Carry David Still craft or RIBs capable of 25 kt. *Inya* reported to be in poor condition. Both based at Rangoon. A third of class, *Inma*, reported to have sunk in 1987. A similar ship is in service in Namibia.

© 2014 IHS

INYA   1980 / 0056642

6 MYANMAR (MISSILE) CLASS
(COASTAL PATROL CRAFT) (PGG)

556–559   561–562

Displacement, tonnes: 216 full load
Dimensions, metres (feet): 45 × 7 × 2.5 *(147.6 × 23.0 × 8.2)*
Speed, knots: 30 (est.)
Complement: 34 (7 officers)

Machinery: 2 Mercedes-Benz diesels; 2 shafts
Missiles: 4 YJ-83 (C-802) (2 twin) launchers; active radar homing to 120 km *(65 n miles)* at 0.9 Mach; warhead 165 kg.
Guns: 4—30 mm AK 230 (2 twin), 4—14.5 mm (1 quad).
Radars: Surface search: I-band.
Fire control: Rice Lamp; I-band.

Comment: Missile-armed variants of the remainder of the (gun-armed) Myanmar class. The vessels entered service in about 2004.

MYANMAR CLASS 556   6/2007 / 1170200

13 MYANMAR (GUN) CLASS (COASTAL PATROL CRAFT) (PGG)

551–555   560   563–565   567–570

Displacement, tonnes: 216 full load
Dimensions, metres (feet): 45 × 7 × 2.5 *(147.6 × 23.0 × 8.2)*
Speed, knots: 30 (est.)
Complement: 34 (7 officers)

Machinery: 2 Mercedes-Benz diesels; 2 shafts
Guns: 2—37 mm (1 twin) (551–555). 1—40 mm (567–570). 4—30 mm AK 230 (2 twin) (563–565). 2—30 mm CRN 91 (560). 4—25 mm (2 twin).
Radars: Surface search: I-band.
Fire control: Rice Lamp; I-band.

Comment: There are three batches of gun-armed craft. Construction of the first batch (551–555) began at Rangoon in 1991 and all had entered service by 2004. A second batch (560–565) began entering service in about 2008 and a third (567–570) in 2012. There are differences in armament and superstructure.

MYANMAR CLASS 553   6/2001 / 0130746

9 RIVER PATROL CRAFT (PBR)

RPC 11–19

Displacement, tonnes: 38 full load
Dimensions, metres (feet): 15.2 × 4.3 × 1.1 *(49.9 × 14.1 × 3.6)*
Speed, knots: 10
Range, n miles: 400 at 8 kt
Complement: 8
Machinery: 2 Thornycroft RZ 6 diesels; 250 hp *(186 kW)*; 2 shafts
Guns: 1 Oerlikon 20 mm or 2—12.7 mm MGs (twin). 1—12.7 mm MG.

Comment: Built by the Naval Engineering Depot, Rangoon. First five in mid-1980s; second batch of a modified design in 1990–91. Sometimes used by the Naval Infantry and can carry up to 35 troops. Based at Rangoon.

IHS Jane's Fighting Ships 2014-2015

556 Myanmar > Patrol forces

9 HAINAN (TYPE 037) CLASS (COASTAL PATROL CRAFT) (PC)

YAN MIN AUNG 445	YAN YE AUNG 446	YAN PAING AUNG 447
YAN WIN AUNG 448	YAN AYE AUNG 449	YAN ZWE AUNG 450
YAN MYAT AUNG 442	YAN NYEIN AUNG 443	YAN KHWIN AUNG 444

Displacement, tonnes: 381 standard; 398 full load
Dimensions, metres (feet): 58.8 × 7.2 × 2.2 *(192.9 × 23.6 × 7.2)*
Speed, knots: 30.5
Range, n miles: 1,300 at 15 kt
Complement: 69

Machinery: 4 PCR/Kolomna Type 9-D-8 diesels; 4,000 hp(m) *(2.94 MW)* sustained; 4 shafts
Guns: 4 China 57 mm/70 (2 twin); 120 rds/min to 12 km *(6.5 n miles)*; weight of shell 6.31 kg. 4 USSR 25 mm/60 (2 twin); 270 rds/min to 3 km *(1.6 n miles)* anti-aircraft; weight of shell 0.34 kg.
A/S Mortars: 4 RBU 1200 5-tubed fixed launchers; range 1,200 m; warhead 34 kg.
Mines: Rails fitted.
Depth charges: 2 BMB-2 projectors; 2 racks.
Electronic countermeasures: ESM: Intercept.
Radars: Surface search: Pot Head; I-band.
Navigation: Raytheon Pathfinder; I-band.
IFF: High Pole.
Sonars: Stag Ear; hull-mounted; active search and attack; high frequency.

Comment: First six delivered from China in January 1991, four more in mid-1993. The first six originally had double figure pennant numbers which have been changed to three figures. These ships are the later variant of this class with tripod masts. Based at Rangoon. *Yan Sit Aung* (441) reported sunk during cyclone Nargis (May 2008).

YAN WIN AUNG   9/1993 / 0056641

YAN KHWIN AUNG   3/2008 / 1353241

3 PB 90 CLASS (COASTAL PATROL CRAFT) (PB)

424–426

Displacement, tonnes: 93 full load
Dimensions, metres (feet): 27.4 × 6.6 × 2.2 *(89.9 × 21.7 × 7.2)*
Speed, knots: 32
Range, n miles: 400 at 25 kt
Complement: 17
Machinery: 3 diesels; 4,290 hp(m) *(3.15 MW)*; 3 shafts
Guns: 8—20 mm M75 (two quad). 2—128 mm launchers for illuminants.
Radars: Surface search: Decca 1226; I-band.

Comment: Built by Brodotechnika, Yugoslavia for an African country and completed in 1986–87. Laid up when the sale did not go through and shipped to Burma arriving in October 1990. All are active. Based at Rangoon.

PB 90 (Yugoslav colours)   1990, Yugoslav FDSP / 0056643

6 BURMA PGM TYPE (COASTAL PATROL CRAFT) (PB)

PGM 412–415   THIHAYARZAR I   THIHAYARZAR II

Displacement, tonnes: 171 full load
Dimensions, metres (feet): 33.5 × 6.7 × 2 *(109.9 × 22.0 × 6.6)*
Speed, knots: 16
Range, n miles: 1,400 at 14 kt
Complement: 17
Machinery: 2 Deutz SBA16MB816 LLKR diesels; 2,720 hp(m) *(2 MW)*; 2 shafts
Guns: 2 Bofors 40 mm/60.

Comment: Built by Burma Naval Dockyard modelled on the US PGM 43 type. First two completed 1983. Two more craft with different superstructure but with identical dimensions and named *Thihayarzar I* and *II* were delivered by Myanma Shipyard to the Customs on 27 June 1993. Both craft may be lightly armed.

IHS Jane's Fighting Ships 2014-2015

PGM 415   4/1993 / 0056644

THIHAYARZAR 332   2/2013 / 1168764

4 RIVER GUNBOATS (EX-TRANSPORTS) (PBR)

SAGU   SEINDA   SHWETHIDA   SINMIN

Displacement, tonnes: 100 full load
Dimensions, metres (feet): 28.8 × 6.7 × 1.4 *(94.5 × 22.0 × 4.6)*
Speed, knots: 12
Complement: 32
Machinery: 1 Crossley ERL 6-cyl diesel; 160 hp *(119 kW)*; 1 shaft
Guns: 1—40 mm/60 (*Sagu*). 1—20 mm (3 in *Sagu*).

Comment: Built in mid-1950s. *Sinmin*, *Seinda* and *Shwethida* have a roofed-in upper deck with a 20 mm gun forward of the funnel. *Sagu* has an open upper deck aft of the funnel but with a 40 mm gun forward and mountings for 20 mm aft on the upper deck and midships either side on the lower deck. Based at Moulmein and at least two are operational. Four other ships of the same type are unarmed and are listed under *Auxiliaries*.

SEINDA   8/1994 / 0056649

1 IMPROVED Y 301 CLASS (RIVER GUNBOAT) (PBR)

Y 311

Displacement, tonnes: 254 full load
Dimensions, metres (feet): 37 × 7.3 × 1.2 *(121.4 × 24.0 × 3.9)*
Speed, knots: 12
Complement: 37
Machinery: 2 MTU MB diesels; 1,000 hp(m) *(735 kW)*; 2 shafts
Guns: 2 Bofors 40 mm/60. 4 Oerlikon 20 mm.
Radars: Surface search: Raytheon; I-band.

Comment: Built at Simmilak in 1969 and based on similar Yugoslav craft which have been scrapped. Y 312 sunk during cyclone Nargis (May 2008). Based at Sittwe.

Y 311   11/2005 / 1151120

25 MICHAO CLASS (PBR)

001–025

Comment: Small craft, 52 ft *(15.8 m)* long, acquired from Yugoslavia in 1965. Also used to ferry troops and two are used as VIP launches. 1 to 7 based at Rangoon; 8 to 16 at Moulmein and 17 to 25 at Sittwe.

MICHAO CLASS   5/1995 / 0056650

© 2014 IHS

6 CARPENTARIA CLASS (RIVER PATROL CRAFT) (PBR)

112–117

Displacement, tonnes: 26 full load
Dimensions, metres (feet): 15.7 × 4.8 × 1.3 (51.5 × 15.7 × 4.3)
Speed, knots: 29. **Range, n miles:** 950 at 18 kt
Complement: 10
Machinery: 2 MTU 8V 331 TC92 diesels; 1,770 hp(m) (1.3 MW) sustained; 2 shafts
Guns: 1 Oerlikon 20 mm. 1 — 12.7 mm MG.

Comment: Built by De Havilland Marine, Sydney. First two delivered 1979, remainder in 1980. Similar to craft built for Indonesia. Based at Rangoon.

CARPENTARIA 113   1991 / 0056651

2 CGC TYPE (RIVER GUNBOATS) (PBR)

MGB 102   MGB 110

Displacement, tonnes: 50 standard; 67 full load
Dimensions, metres (feet): 25.3 × 4.9 × 1.7 (83 × 16.1 × 5.6)
Speed, knots: 11
Complement: 16
Machinery: 4 GM diesels; 800 hp (596 kW); 2 shafts
Guns: 1 Bofors 40 mm/60. 1 Oerlikon 20 mm.

Comment: Ex-USCG type cutters with new hulls built in Burma. Completed in 1960. Based at Rangoon but have not been seen recently.

MGB 110   0505966

10 Y 301 CLASS (RIVER GUNBOATS) (PBR)

Y 301–310

Displacement, tonnes: 122 full load
Dimensions, metres (feet): 32 × 7.3 × 0.9 (105 × 24.0 × 3.0)
Speed, knots: 13
Complement: 29
Machinery: 2 MTU MB diesels; 1,000 hp(m) (735 kW); 2 shafts
Guns: 2 Bofors 40 mm/60 or 1 Bofors 40 mm/60 and 1 Vickers 2-pdr.

Comment: All of these boats were completed in 1958 at the Uljanik Shipyard, Pula, Yugoslavia. Y 301, 303 and 307 based at Moulmein. The remainder at Rangoon.

Y 302   2/2013 / 1168763

3 SWIFT TYPE PGM (COASTAL PATROL CRAFT) (PB)

PGM 421–423

Displacement, tonnes: 130 full load
Dimensions, metres (feet): 31.5 × 7.2 × 3.1 (103.3 × 23.6 × 10.2)
Speed, knots: 27. **Range, n miles:** 1,800 at 18 kt
Complement: 25
Machinery: 2 MTU 12V 331 TC81 diesels; 2,450 hp(m) (1.8 MW) sustained; 2 shafts
Guns: 2 Bofors 40 mm/60. 2 Oerlikon 20 mm. 2 — 12.7 mm MGs.
Radars: Surface search: Raytheon 1500; I-band.

Comment: Swiftships construction completed between March and September 1979. Acquired 1980 through Vosper, Singapore. PGM 421 previously reported sunk in 1990s but reported to have been repaired. Based at Rangoon.

PGM   6/1991 / 0056645

6 PGM 43 TYPE (COASTAL PATROL CRAFT) (PB)

PGM 401–406

Displacement, tonnes: 143 full load
Dimensions, metres (feet): 30.8 × 6.4 × 2.3 (101 × 21.0 × 7.5)
Speed, knots: 17
Range, n miles: 1,000 at 15 kt
Complement: 17
Machinery: 8 GM 6-71 diesels; 1,392 hp (1.04 MW) sustained; 2 shafts
Guns: 1 Bofors 40 mm/60. 2 Oerlikon 20 mm (twin). 2 — 12.7 mm MGs.
Radars: Surface search: Raytheon 1500 (PGM 405–406). EDO 320 (PGM 401–404); I/J-band.

Comment: First four built by Marinette Marine in 1959; last pair by Peterson Shipbuilders in 1961. PGM 401–403 based at Moulmein and 404–405 at Rangoon. PGM 406 at Sittwe.

PGM 406   3/1992 / 0056646

AMPHIBIOUS FORCES

1 LCU

AIYAR LULIN 603

Displacement, tonnes: 366 full load
Dimensions, metres (feet): 36.3 × 10.4 × 1.8 (119.1 × 34.1 × 5.9)
Speed, knots: 10
Range, n miles: 1,200 at 8 kt
Complement: 14
Military lift: 168 tons
Machinery: 4 GM diesels; 600 hp (448 kW); 2 shafts
Guns: 1 — 12.7 mm MG.

Comment: Completed in Rangoon in 1966 to the US 1610 design. Based at Rangoon.

AIYAR LULIN   1990 / 0056654

10 LCM 3 TYPE

LCM 701–710

Displacement, tonnes: 53 full load
Dimensions, metres (feet): 15.2 × 4.3 × 1.2 (49.9 × 14.1 × 3.9)
Speed, knots: 9
Complement: 5
Machinery: 2 Gray Marine 64 HN9 diesels; 330 hp (246 kW); 2 shafts

Comment: US-built LCM type landing craft. Used as local transports for stores and personnel. Cargo capacity, 30 tons. Guns have been removed. Based at Sittwe.

LCM 704   5/1994 / 0056655

4 ABAMIN CLASS (LCU)

AIYAR MAI 604   AIYAR MAUNG 605   AIYAR MINTHAMEE 606   AIYAR MINTHAR 607

Displacement, tonnes: 254 full load
Dimensions, metres (feet): 38.3 × 9.1 × 1.4 (125.7 × 29.9 × 4.6)
Speed, knots: 10
Complement: 10
Military lift: 100 tons
Machinery: 2 Kubota diesels; 600 hp(m) (441 kW); 2 shafts
Guns: 1 — 12.7 mm MG.

Comment: All built by Yokohama Yacht in 1969. Based at Rangoon.

AIYAR MAUNG   1991 / 0056653

558 Myanmar > Amphibious forces — Auxiliaries

3 LCU

001–003

Comment: Operated by the Army. Dimensions not known.

LANDING CRAFT 003   7/1992 / 0056652

SURVEY SHIPS

Notes: Thu Tay Thi means 'survey vessel'.

1 SURVEY CRAFT (AGSC)

Name	No	Builders	Commissioned
YAY BO	807	Damen Shipyard	1958

Displacement, tonnes: 110 full load
Dimensions, metres (feet): 30 × 6.8 × 1.5 (98.4 × 22.3 × 4.9)
Speed, knots: 10
Complement: 34 (2 officers)
Machinery: 2 diesels; 2 shafts
Guns: 1 — 12.7 mm MG.

Comment: Used for river surveys. Based at Rangoon.

YAY BO   1990 / 0056656

AUXILIARIES

Notes: As well as the ships listed in this section there are two coastal tankers 603 and 615, a harbour tug, several harbour launches, and personnel carriers.

1 TRANSPORT VESSEL (AK)

AYIDAWAYA

Displacement, tonnes: 818 full load
Dimensions, metres (feet): 49.8 × 8.4 × 3.7 (163.4 × 27.6 × 12.1)
Speed, knots: 12
Complement: 30
Machinery: 1 diesel; 600 hp(m) (441 kW); 1 shaft

Comment: Built in Norway in 1975. Acquired in 1991 and used as transport for stores and personnel.

AYIDAWAYA   12/1991 / 0056658

1 BUOY TENDER (ABU)

Displacement, tonnes: 717 full load
Dimensions, metres (feet): 39.8 × 11.3 × 2.7 (130.6 × 37.1 × 8.9)
Speed, knots: 10
Complement: 23
Machinery: 2 Deutz BA8M816 diesels; 1,341 hp(m) (986 kW); 2 shafts

Comment: Built by Italthai in 1986. Operated by the Rangoon Port Authority but manned by the Navy.

HSAD DAN   5/1992 / 0056662

8 MFVS

511   520–523   901   905–906

Comment: Armed vessels of approximately 200 tons (901), 80 tons (905, 906) and 50 tons (remainder) with a 12.7 mm or 6.72 mm MG mounted above the bridge in some. All have navigational radars. Based at Rangoon.

MFV   8/1990 / 0104255

4 TRANSPORT VESSELS (AKL)

SABAN   SETHYA   SHWEPAZUN   SETYAHAT

Displacement, tonnes: 100 full load
Dimensions, metres (feet): 28.8 × 6.7 × 1.4 (94.5 × 22.0 × 4.6)
Speed, knots: 12
Complement: 30
Machinery: 1 Crossley ERL 6-cyl diesel; 160 hp (119 kW); 1 shaft

Comment: These are sister ships to the armed gunboats shown under *Patrol Forces*. It is possible that a 20 mm gun may be mounted on some occasions. Based at Rangoon.

SHWEPAZUN   1991 / 0056661

1 COASTAL CARGO SHIP (AK)

Name	No	Builders	Commissioned
PYI DAW AYE (ex-*Lone Clipper*)	601	Nordsøvaerftet, Ringkobing	1974

Displacement, tonnes: 850 full load
Dimensions, metres (feet): 49.7 × 8.34 × 3.47 (163.1 × 27.4 × 11.4)
Speed, knots: 11
Machinery: 1 Alpha diesel; 600 hp (441 kW); 1 shaft
Radars: Navigation: To be announced.

Comment: Acquired by the Myanmar Navy in about 1982. Has two cargo holds and two derricks.

PYI DAW AYE   6/1991* / 0056663

1 COASTAL TANKER (AK)

Name	No	Builders	Laid down	Launched	Commissioned
– (ex-*Seria Maru*)	609	Shimoda Dockyard	27 Aug 1973	24 Feb 1974	10 Apr 1974

Measurement, tonnes: 972 gt
Dimensions, metres (feet): 68.66 × 11.0 × 5.0 (225.3 × 36.1 × 16.4)
Speed, knots: 10
Machinery: 1 Daihatsu diesel; 1,300 hp (956 kW); 1 shaft
Radars: Navigation: To be announced.

Comment: Acquired by the Myanmar Navy in about 1986.

PRESIDENTIAL YACHT

1 TRANSPORT SHIP (YAC)

YADANABON

Comment: Built in Burma and used for VIP cruises on the Irrawaddy river and in coastal waters. Armed with 2 – 7.62 mm MGs and manned by the Navy.

PRESIDENT'S YACHT
1990
0056665

Namibia

Country Overview

Formerly South West Africa and governed by South Africa, Namibia gained independence in 1990 although South Africa continued to administer an enclave containing the principal seaport, Walvis Bay, until 1994. With an area of 318,252 square miles, it has borders to the north with Angola and to the south with South Africa. It has an 848 n mile coastline with the south Atlantic Ocean. The capital and largest city is Windhoek and there is another port at Lüderitz. Territorial seas (12 n miles) are claimed. It also claims a 200 n mile Exclusive Economic Zone (EEZ) but its limits have not been fully defined by boundary agreements.

The Maritime Wing became the Navy on 7 October 2004.

Headquarters Appointments

Head of Navy:
Commodore Peter Vilho

Bases

Walvis Bay

Personnel

2014: 350

Aviation

Five ex-US Air Force Cessna O-2A Skymaster and one Sikorsky S-61L helicopter operate in a maritime surveillance role.

PATROL FORCES

Notes: A variety of small craft were reported in course of delivery during 2013. All designed by KND, Cape Town, the vessels include: two 14 m interception craft capable of 60 kt, two 8 m swamp boats, two 6 m harbour patrol craft, five 6 m RHIBs, and five 4 m training craft.

2 RIO (MARLIM) CLASS (COASTAL PATROL CRAFT) (PB)

MÖWE BAY HPB 21   **TERRACE BAY** HPB 20

Displacement, tonnes: 31 standard; 46 full load
Dimensions, metres (feet): 20.8 × 5.5 × 1.06 *(68.2 × 18.0 × 3.5)*
Speed, knots: 27
Range, n miles: 850 at 15 kt
Complement: 9 (1 officer)
Machinery: 2 MTU 8V 2000 M92 diesels; 2,100 hp(m) *(1.54 MW)* sustained; 2 shafts
Guns: 2 – 12.7 mm MGs.
Radars: Surface search: Furuno; I-band.

Comment: Construction of two new craft began once the patrol ship *Brendan Simbwaye* was completed in 2009. Delivery was made on 25 May 2011.

MARLIM CLASS (Brazilian colours)   6/2007, L Frangetto / 1170090

1 PATROL SHIP (PBO)

Name	No	Builders	Commissioned
ORYX (ex-*S to S*)	P 01	Burmeister & Wain, Copenhagen; Abeking & Rasmussen	May 1975

Displacement, tonnes: 413 full load
Dimensions, metres (feet): 45.7 × 8.8 × 2.4 *(149.9 × 28.9 × 7.9)*
Speed, knots: 14. **Range, n miles:** 4,100 at 11 kt
Complement: 20 (6 officers)

Machinery: 2 Deutz RSBA 16M diesels; 2,000 hp(m) *(1.47 MW)*; 1 shaft; cp prop; bow thruster
Guns: 1 – 12.7 mm MG.
Radars: Surface search: Furuno ARPA FR 1525; I-band.
Navigation: Furuno FR 805D; I-band.

Comment: Built for the Nautical Investment Company, Panama and used as a yacht by the Managing Director of Fiat. Acquired in 1993 by Namibia. Replaced by *Nathanael Maxwilili* in fishery protection role and transferred to the navy as a patrol ship in 2002.

ORYX   6/1997 / 0081282

1 GRAJAÚ CLASS (LARGE PATROL CRAFT) (PBO)

Name	No	Builders	Laid down	Launched	Commissioned
BRENDAN SIMBWAYE	P 11	INACE, Fortalesa	25 Feb 2005	1 May 2008	2 Jun 2009

Displacement, tonnes: 197 standard; 217 full load
Dimensions, metres (feet): 46.5 × 7.5 × 2.3 *(152.6 × 24.6 × 7.5)*
Speed, knots: 26. **Range, n miles:** 2,200 at 12 kt
Complement: 29 (4 officers)
Machinery: 2 MTU 16V 396 TB94 diesels; 5,800 hp(m) *(4.26 MW)* sustained; 2 shafts
Guns: 1 Bofors 40 mm/70. 2 Oerlikon 20 mm.
Radars: Surface search: Racal Decca 1290A; I-band.

Comment: Following an agreement between the governments of Namibia and Brazil in November 2003, the project for a new patrol ship was conducted by EMGEPRON which contracted Inace for the construction of the vessel. The ship is similar to *Guanabara* built for the Brazilian Navy in 1999, and on which details are based.

BRENDAN SIMBWAYE   6/2010*, Felipe Salles / 1525540

1 OFFSHORE PATROL VESSEL (PBO)

Name	No	Builders	Laid down	Launched	Commissioned
ELEPHANT	P 11	Wuchang Shipyard, Wuhan	27 Jan 2011	12 Nov 2011	2012

Measurement, tonnes: 2,834 gt
Dimensions, metres (feet): 108 × 14.0 × ? *(354.3 × 45.9 × ?)*
Speed, knots: 22. **Range, n miles:** 6,000 at 12 kt

Machinery: 2 Caterpillar C280-16 diesels; 12,508 hp *(9.2 MW)*; 2 shafts
Guns: 1 – 37 mm. 2 – 14.5 mm.
Radars: Surface search: Furuno 2137S; E/F-band.
Navigation: Furuno 2127; I-band.
Fire control: Type 347G(1) Rice Bowl; I/J-band.
Helicopters: Platform for 1 medium.

Comment: Offshore patrol vessel of steel construction. The design appears to be similar to the Chinese Yuzheng-310-class fishery protection ships.

ELEPHANT   11/2012*, M Globke / 1525509

SUBMARINES

Notes: Operational analysis to establish the requirements for a future submarine capability, has been initiated although there are no firm plans to replace the Walrus class.

4 WALRUS CLASS (SSK)

Name	No	Builders	Laid down	Launched	Commissioned
WALRUS	S 802	Rotterdamse Droogdok Mij, Rotterdam	11 Oct 1979	26 Oct 1985	25 Mar 1992
ZEELEEUW	S 803	Rotterdamse Droogdok Mij, Rotterdam	24 Sep 1981	20 Jun 1987	25 Apr 1990
DOLFIJN	S 808	Rotterdamse Droogdok Mij, Rotterdam	12 Jun 1986	25 Apr 1990	29 Jan 1993
BRUINVIS	S 810	Rotterdamse Droogdok Mij, Rotterdam	14 Apr 1988	25 Apr 1992	5 Jul 1994

Displacement, tonnes: 2,505 surfaced; 2,845 dived
Dimensions, metres (feet): 67.7 × 8.4 × 7 *(222.1 × 27.6 × 23.0)*
Speed, knots: 12 surfaced; 20 dived
Range, n miles: 10,000 at 9 kt snorting
Complement: 52 (7 officers)

Machinery: Diesel-electric; 3 SEMT-Pielstick 12 PA4 200 VG diesels; 6,300 hp(m) *(4.63 MW)*; 3 alternators; 2.88 MW; 1 Holec motor; 6,910 hp(m) *(5.1 MW)*; 1 shaft
Missiles: SSM: McDonnell Douglas Sub Harpoon; active radar homing to 130 km *(70 n miles)* at 0.9 Mach; warhead 227 kg.
Torpedoes: 4—21 in *(533 mm)* tubes. Honeywell Mk 48 Mod 4; wire-guided; active/passive homing to 38 km *(20.5 n miles)* active at 55 kt; 50 km *(27 n miles)* passive at 40 kt; warhead 267 kg; 20 torpedoes or missiles carried. Mk 19 Turbine ejection pump. Mk 67 water-ram discharge.
Mines: 40 in lieu of torpedoes.
Electronic countermeasures: ESM: L3 DR 3000; radar warning.
Radars: Surface search: Signaal/Racal ZW07; I-band.
Sonars: Thomson Sintra TSM 2272 Eledone Octopus; hull-mounted; passive/active search and attack; medium frequency. GEC Avionics Type 2026; towed array; passive search; very low frequency. Thomson Sintra DUUX 5; passive ranging and intercept.
Weapon control systems: Signaal SEWACO VIII action data automation. Signaal Gipsy data system. GTHW integrated Harpoon and Torpedo FCS.

Programmes: Contract for the building of the first was signed 16 June 1979, the second was on 17 December 1979. In 1981 various changes to the design were made which resulted in a delay of one to two years. *Dolfijn* and *Bruinvis* ordered 16 August 1985; prefabrication started late 1985. Completion of *Walrus* delayed by serious fire 14 August 1986; hull undamaged but cabling and computers destroyed. *Walrus* relaunched 13 September 1989.
Modernisation: A snort exhaust diffuser was fitted to *Zeeleeuw* in 1996. The rest of the class have been similarly modified. A life-extension program for all four boats started in May 2013 for completion in 2020. The Combat Management System is to be replaced by a Guardian design based on the system in RNLN frigates. This is to integrate a new optronic mast that is to replace the navigation periscope. Upgrade of the sonar systems is to include replacement of inboard equipment and processing and the addition of L-3 Elac Nautik Scout mine and obstacle avoidance sonar. Navigation enhancements are to include WECDIS and AIS functionality while communications improvements are to include SHF SATCOM. The programme is also to include the Mk 48 mod 7AT torpedo upgrade. The HABETAS submarine escape system is also to be installed. *Zeeleeuw* started refit in 2013. She is to be followed by *Dolfijn* in 2015, *Bruinvis* in 2017, and *Walrus* in 2019.
Structure: These are improved Zwaardvis class with similar dimensions and silhouettes except for X stern. Use of H T steel increases the diving depth by some 50%. Diving depth, 300 m *(984 ft)*. Pilkington Optronics CK 24 search and CH 74 attack periscopes.
Operational: Weapon systems evaluations completed 1990–93. Sub Harpoon is not carried. *Zeeleeuw* was assigned to anti-piracy operations off Somalia September to November 2010.

BRUINVIS *12/2012, Adolfo Ortigueira Gil* / 1486304

ZEELEEUW *6/2009, Frank Findler* / 1305628

WALRUS *8/2010, A A de Kruijf* / 1366664

DOLFIJN *6/2013*, Michael Nitz* / 1530928

FRIGATES

Notes: Replacement of the M-class frigates in the mid-2020s is under consideration. The broad requirement is for the ships to have a primary ASW role with AAW self-defence capabilities, an anti-ship missile system, a medium-calibre gun and helicopter/UAV hangar. Close-range picture-compilation and self-defence are also likely to be key drivers. Initiation of a formal acquisition programme is expected 2014.

2 M CLASS (FFGHM)

Name	No	Builders	Laid down	Launched	Commissioned
VAN AMSTEL	F 831	Koninklijke Maatschappij De Schelde, Flushing	3 May 1988	19 May 1990	27 May 1993
VAN SPEIJK	F 828	Koninklijke Maatschappij De Schelde, Flushing	1 Oct 1991	26 Mar 1994	7 Sep 1995

Displacement, tonnes: 3,340 full load
Dimensions, metres (feet): 123.8 oa; 114.2 wl × 14.4 × 6.2 *(406.2; 374.7 × 47.2 × 20.3)*
Flight deck, metres: 24.5 × 14.4 *(80.4 × 47.2)*
Speed, knots: 29; 19 diesel
Range, n miles: 6,200 at 18 kt
Complement: 156 (16 officers) + 7 spare berths

Machinery: CODOG; 2 RR Spey SM1A; 34,200 hp *(25.5 MW)* sustained; 2 Stork-Wärtsilä 12SW280 diesels; 9,790 hp(m) *(7.2 MW)* sustained; 2 shafts; LIPS cp props
Missiles: SSM: 8 McDonnell Douglas Harpoon Block 1C (2 quad) launchers ❶; active radar homing to 124 km *(67 n miles)* at 0.9 Mach; warhead 227 kg.
SAM: Raytheon Sea Sparrow RIM 7P2 Mk 48 vertical launchers ❷; semi-active radar homing to 16 km *(8.5 n miles)* at 2.5 Mach; warhead 38 kg; 16 missiles. Canisters mounted on port side of hangar.
Guns: 1—3 in *(76 mm)*/62 Oto Melara compact Mk 100 ❸; 100 rds/min to 16 km *(8.6 n miles)* anti-surface; 6 km *(3.2 n miles)* anti-aircraft; weight of shell 6 kg. 1 Signaal GAU-8A Goalkeeper with General Electric 30 mm 7-barrelled ❹; 4,200 rds/min combined to 2 km. 2—12.7 mm MGs. 2—7.62 mm MGs.
Torpedoes: 4—324 mm US Mk 32 Mod 9 (2 twin) tubes (mounted inside the after superstructure) ❺. Honeywell Mk 46 Mod 5A (SW); anti-submarine; active/passive homing to 11 km *(5.9 n miles)* at 44 kt; warhead 44 kg.
Physical countermeasures: Decoys: 4 SRBOC Mk 36 IR/chaff launchers. SLQ-25 Nixie towed torpedo decoy.
Electronic countermeasures: ESM/ECM: Argo APECS II (includes AR 700 ESM); intercept and jammers.
Radars: Air/surface search: Thales SMART-S ❻; 3D; E/F-band.
Air search: Thales LW08 ❼; D-band.
Surface search: Thales Scout ❽; I-band. Thales Seastar ❾; I/J-band.
Navigation: Kelvin Hughes 1007; I-band.
Fire control: 2 Thales STIR ❿; I/J/K-band; range 140 km *(76 n miles)* for 1 m² target.
Sonars: Signaal PHS-36; hull-mounted; active search and attack; medium frequency. TNO IRLFAS; towed array; active/passive low frequency.
Combat data systems: SEWACO XI; Link 11/16. SATCOM ⓫. WSC-6 twin aerials.
Electro-optic systems: Thales Gatekeeper ⓬.

Helicopters: 1 NH90 NFH ⓭.

Programmes: Declaration of intent signed on 29 February 1984 although the contract was not signed until 29 June 1985 by which time the design had been completed. A further four ordered 10 April 1986. Names were shuffled to make the new *Van Speijk* the last of the class but she retained her allocated pennant number.
Modernisation: A mid-life modernisation is in progress 2010–14. Upgrades include modifications to operate the NH90 helicopter, replacement of the combat data system by Force Vision Guardian, addition of a Thales Seastar radar and an IR/EO Gatekeeper system, installation of a low-frequency active sonar and replacement of ESM/ECM systems. Upgrade of Harpoon from Block 1D to Block 2, Goalkeeper MLU and Links 11/16/22. Platform systems are also to be upgraded.
Structure: The VLS SAM is similar to Canadian Halifax and Greek MEKO classes. The SAM launchers have added protection and better stealth features with a flat screen. A new mast is to be designed to accommodate Seastar radar, new ESM and Gatekeeper.
Operational: NH90 helicopter from 2013.
Sales: F 832 and F 830 sold to Chile and transferred in November 2005 and mid-2006 respectively. F 827 and F 829 transferred to Belgium in March 2007 and March 2008 respectively and F 833 and F 834 to Portugal in December 2008 and January 2010 respectively.

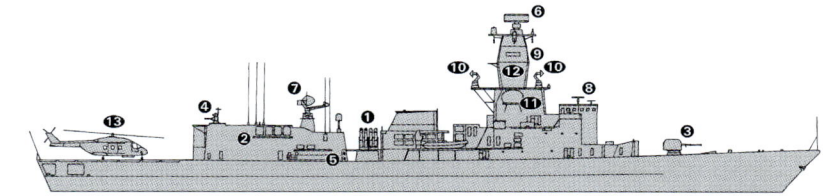

VAN SPEIJK *(Scale 1 : 1,200), Ian Sturton / 1455823*

VAN SPEIJK *7/2012, A A de Kruijf / 1486305*

VAN SPEIJK *7/2012, A A de Kruijf / 1486306*

4 DE ZEVEN PROVINCIEN CLASS (FFGHM)

Name	No	Builders	Laid down	Launched	Commissioned
DE ZEVEN PROVINCIEN	F 802	Royal Schelde, Vlissingen	1 Sep 1998	8 Apr 2000	26 Apr 2002
TROMP	F 803	Royal Schelde, Vlissingen	3 Sep 1999	7 Apr 2001	14 Mar 2003
DE RUYTER	F 804	Royal Schelde, Vlissingen	1 Sep 2000	13 Apr 2002	22 Apr 2004
EVERTSEN	F 805	Royal Schelde, Vlissingen	6 Sep 2001	19 Apr 2003	10 Jun 2005

Displacement, tonnes: 6,145 full load
Dimensions, metres (feet): 144.2 oa; 130.7 wl × 18.8 × 5.2 *(473.1; 428.8 × 61.7 × 17.1)*
Flight deck, metres: 27 × 18.8 *(88.6 × 61.7)*
Speed, knots: 28
Range, n miles: 5,000 at 18 kt
Complement: 204 (32 officers)

Machinery: CODOG; 2 RR SM1C Spey; 52,300 hp *(39 MW)* sustained; 2 Stork-Wärtsilä 16V 26 ST diesels; 13,600 hp(m) *(10 MW)*; 2 shafts; LIPS; cp props
Missiles: SSM: 8 McDonnell Douglas Harpoon Block 2; active radar homing to 124 km *(67 n miles)* at 0.9 Mach; warhead 227 kg ❶.
SAM: Mk 41 VLS (40 cells) ❷; 32 Raytheon Standard SM2-MR (Block IIIA); command/inertial guidance; semi-active radar homing to 167 km *(90 n miles)* at 2.5 Mach. 32 Evolved Sea Sparrow RIM 162B (quad pack); semi-active radar homing to 18 km *(9.7 n miles)* at 3.6 Mach; warhead 38 kg.
Guns: 1 Otobreda 5 in *(127 mm)*/54 ❸; 45 rds/min to 23 km *(12.42 n miles)* anti-surface; weight of shell 32 kg. 1 or 2 (F 805) Thales Goalkeeper 30 mm ❹; 4,200 rds/min to 2 km. 2 Browning 12.7 mm MGs. 4—7.62 mm MGs ❺.
Torpedoes: 4—323 mm (2 twin) Mk 32 Mod 9 fixed launchers ❻. Mk 46 Mod 5 torpedoes.
Physical countermeasures: Decoys: 4 SRBOC Mk 36 IR/chaff launchers; Nixie torpedo decoy.
Electronic countermeasures: ESM/ECM: Thales Sabre ❼; intercept/jammer.
Radars: Air search: Thales SMART L ⓫; 3D; D-band.
Air/surface search/fire control: Thales APAR ⓬; E/F/I-band.
Surface search: Thales Scout ⓭; I-band.
IFF: Mk XII.
Sonars: STN Atlas DSQS 24C; bow-mounted; active search and attack; medium frequency.
Combat data systems: Thales SEWACO 11; Link 11/16; UHF/SHF SATCOMS ❽. MCCIS.
Electro-optic systems: Thales Sirius IRST optronic director ❾. Thales Mirador Trainable Electro-Optical Observation System (TEOOS) ❿.
Helicopters: 1 NH90 NFH ⓮.

Programmes: Project definition awarded to Royal Schelde on 15 December 1993 with a contract for first two ships and detailed design following on 30 June 1995. Second pair ordered 5 February 1997. Shipyards in Germany (ARGE for Type 124) collaborated to achieve some commonality of design and equipment.
Modernisation: The following upgrades are planned: Modify 127 mm to fire long-range ammunition (under consideration); Goalkeeper MLU (2013). Modification of SMART-L radar with an Extended Long Range mode from 2018. This will enable a ballistic missile detect and track capability. The APAR radar has been modified to operate at F-band and I-band.
Structure: High standards of stealth and NBC protection are part of the design. DCN Samahé helicopter handling system. Space exists to retrofit an additional 8-cell Mk 41 launcher alongside the five already fitted.
Operational: All ships fitted with command facilities.

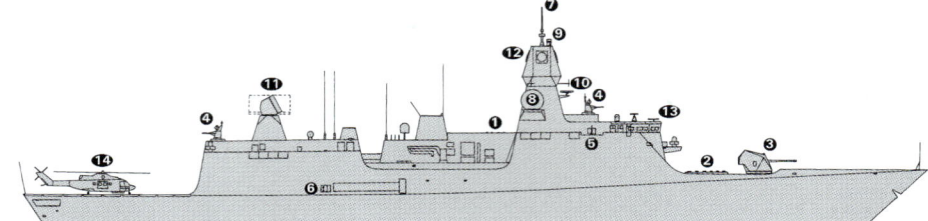

DE ZEVEN PROVINCIEN *(Scale 1 : 1,200), Ian Sturton* / 0569256

TROMP *1/2007, Chris Sattler* / 1164992

DE ZEVEN PROVINCIEN *9/2009, Shaun Jones* / 1305508

DE RUYTER *10/2013*, Michael Nitz* / 1530931

Frigates ‹ **Netherlands** 565

EVERTSEN   *6/2013*, Michael Nitz* / 1530930

DE ZEVEN PROVINCIEN   *6/2013*, Michael Winter* / 1530929

DE RUYTER   *6/2011, Michael Nitz* / 1406631

SHIPBORNE AIRCRAFT

Numbers/Type: 12/8 NHIndustries NH90 NFH/NH90 TNFH.
Operational speed: 175 kt (324 km/h).
Service ceiling: 17,000 ft (5,180 m).
Range: 621 n miles (1,150 km).
Role/Weapon systems: Twelve NH90 NFH of which first entered service on 21 April 2010 and eight maritime transport TNFH from 2013. NFH variant equipped for ASW/ASuW duties and for SAR. Sensors: Thales Oceanmaster radar, Elac Nautic HELRAS dipping sonar, FLIR and ESM. Weapons: 2 Mk 46 torpedoes. 1 – 7.62 mm MG.

NH90
6/2013*, Michael Winter
1530933

PATROL FORCES

4 HOLLAND CLASS (OFFSHORE PATROL SHIPS) (PSO)

Name	No	Builders	Laid down	Launched	Commissioned
HOLLAND	P 840	Schelde Shipbuilding, Vlissingen	8 Dec 2008	30 Oct 2009	6 Jul 2012
ZEELAND	P 841	Schelde Shipbuilding, Vlissingen	22 Sep 2009	20 Nov 2010	23 Aug 2013
FRIESLAND	P 842	Damen Shipyard, Galati	26 Nov 2009	4 Nov 2010	22 Jan 2013
GRONINGEN	P 843	Damen Shipyard, Galati	9 Apr 2010	21 Apr 2011	29 Nov 2013

Displacement, tonnes: 3,810 full load
Dimensions, metres (feet): 108.4 oa; 102.7 wl × 15.24 × 4.55 (355.6; 336.9 × 50.0 × 14.9)
Speed, knots: 22
Range, n miles: 5,000 at 16 kt
Complement: 50 + 40 spare berths
Machinery: CODOE: 2 MAN 12V 28/33D diesels; 14,480 hp (10.8 MW); 3 Caterpillar 3508B generators; 3,895 hp (2.9 MW); 2 motors; 1,070 hp (800 kW); 2 shafts; cp props; 1 Rolls Royce/ABB bow thruster; 737 hp (550 kW)
Guns: 1 Oto Melara 3 in (76 mm)/62 compact ❶; 85 rds/min to 16 km (8.7 n miles); weight of shell 6 kg. 1 Oto Melara 30 mm/70 (remotely operated) ❷; 200 rds/min; 2 Oto Melara Hitrole (remotely operated) 12.7 mm MGs ❸; 6 – 7.62 mm MGs.
Radars: Thales SMILE ❹; E/F-band.
Surface search: Thales SEASTAR ❺; I-band.
Navigation: 2 Consilium Selesmar ❺; E/F/I-bands.
Combat data systems: SEWACO CMS. Link 11/16 SATCOM.
Electro-optic systems: Thales Gatekeeper ❹; IR and TV.
Helicopters: 1 NH90 ❻.

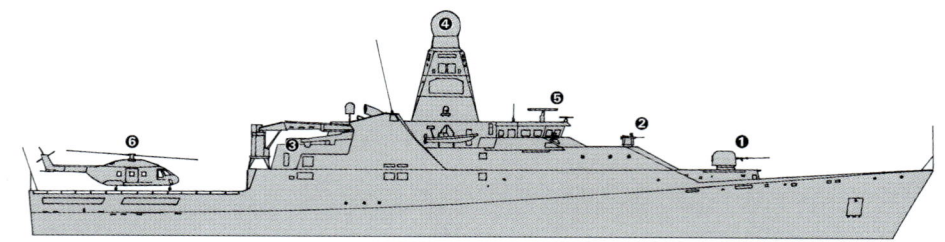

HOLLAND
(Scale 1 : 900), Ian Sturton / 1455824

Comment: Contract for the design and build of four patrol ships awarded to Damen Schelde Naval shipbuilding on 20 December 2007. The role of the ships is to conduct low-intensity military operations including maritime interdiction, counter-terrorism and humanitarian assistance. Design features include an integrated mast for sensors and communications. There is provision to accommodate additional payloads, including stowage for two 20 ft containers or pallets in a multifunction space beneath the flight deck. A 10-tonne crane is fitted on the starboard side for cargo handling. Two 12 m raiding craft (FRISC) can be embarked: one may be launched and recovered via a stern slipway, the other from a boat davit on the port side.

ZEELAND
6/2013*, Michael Winter / 1530932

AMPHIBIOUS FORCES

5 LCU MK IX (LCU)

L 9525–9529

Displacement, tonnes: 275 full load
Dimensions, metres (feet): 35.8 × 6.7 × 1.6 (117.5 × 22.0 × 5.2)
Speed, knots: 9
Range, n miles: 400 at 8 kt
Complement: 5 + 2 spare berths

Military lift: 153 troops or 65 tons of vehicles, 4 × 4 tonne or beach armoured recovery vehicle.
Machinery: Diesel-electric; 2 Caterpillar 3412C diesel generators; 1,496 hp(m) (1.1 MW); 2 Alconza D400 motors; 2 Schottel pumpjets; 2 pump jets
Guns: 1 – 12.7 mm MG; 1 – 7.62 mm MG.
Radars: Navigation: Furuno; I-band.

Comment: Ordered from Visser Dockyard, Den Helder on 19 July 1996. Steel vessels of which the first commissioned 7 April 1998. The others have been fabricated in Romania (Damen Shipyard, Galati) and fitted out at Den Helder. All five were lengthened 2003–05 to reduce draft. The craft can interlock with a 'rhino-horn' system so that they can through load while in dock. The rear ramp is only used for Ro-Ro procedures but not on the beach as originally designed. Two craft are allocated to *Rotterdam* and two to *Johan de Witt*. The remaining craft is at Texel for training and maintenance.

L 9526
7/2011, Michael Winter / 1406502

Amphibious forces < **Netherlands** 567

1 ROTTERDAM CLASS (LPD)

Name	No	Builders	Laid down	Launched	Commissioned
ROTTERDAM	L 800	Royal Schelde, Vlissingen	25 Jan 1996	22 Feb 1997	18 Apr 1998

Displacement, tonnes: 12,955 full load
Dimensions, metres (feet): 166 × 25 × 5.9 (544.6 × 82.0 × 19.4)
Flight deck, metres: 56 × 25 (183.7 × 82.0)
Speed, knots: 18
Range, n miles: 6,000 at 12 kt
Complement: 136 (16 officers)

Military lift: 538 troops (38 officers); 170 APCs or 33 MBTs. 2 LCU Mk II and 3 LCVP.
Machinery: Diesel-electric; 4 Stork Wärtsilä 12SW28 diesel generators; 14.6 MW sustained; 2 Holec motors; 16,320 hp(m) *(12 MW)*; 2 shafts; bow thruster
Guns: 2 Thales Goalkeeper 30 mm ❶. 9 – 12.7 mm MGs. 2 – 7.62 mm MGs.
Physical countermeasures: Decoys: 4 SRBOC IR/chaff launchers ❷; Nixie torpedo decoy system.
Electronic countermeasures: ESM/ECM: ARGO AR-900; intercept.
Radars: Air/surface search: Signaal DA08 ❹; E/F-band.
Surface search: Signaal Scout/Kelvin Hughes ARPA ❺; I-band.
Navigation and CCA: 2 sets; I-band.
Combat data systems: SATCOM ❸. Link 11. MCCIS.

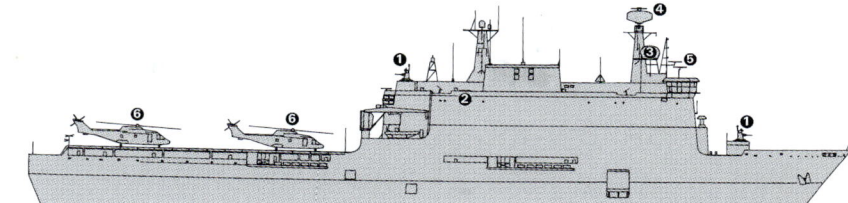

ROTTERDAM (Scale 1 : 1,500), Ian Sturton / 0534086

Weapon control systems: Signaal IRSCAN infra-red director.
Helicopters: 6 NH90 ❻ or 4 Merlin/Sea King.
Programmes: Project definition for a joint design with Spain completed in December 1993. Contract signed with Royal Schelde 25 April 1994.
Structure: Facilities to transport a Marine battalion with docking facilities for landing craft and a two spot helicopter flight deck with hangar space for six NH90. 25 ton crane for disembarkation. Full hospital facilities. Built to commercial standards with military command and control and NBCD facilities. Can carry up to 300 sonobuoys.
Operational: Alternative employment as an SAR ship for environmental and disaster relief tasks. The first landing of an Apache attack helicopter was conducted on 9 May 2012.

ROTTERDAM 9/2011, Michael Nitz / 1454768

ROTTERDAM 9/2011, Michael Nitz / 1454769

ROTTERDAM 3/2011, Shaun Jones / 1406634

© 2014 IHS   IHS Jane's Fighting Ships 2014-2015

1 JOHAN DE WITT CLASS (LPD)

Name	No	Builders	Laid down	Launched	Commissioned
JOHAN DE WITT	L 801	Royal Schelde, Vlissingen	18 Jun 2003	13 May 2006	30 Nov 2007

Displacement, tonnes: 16,948 full load
Dimensions, metres (feet): 176.4 × 29.2 × 6.7 (578.7 × 95.8 × 22.0)
Flight deck, metres: 58 × 25 (190.3 × 82.0)
Speed, knots: 17. **Range, n miles:** 10,000 at 12 kt
Complement: 146 (17 officers)

Military lift: 555 troops; 170 APCs or 33 MBTs. 4 LCVP and 2 LCU or 2 LCM
Machinery: Diesel-electric; 4 Wärtsilä 12V26A diesel generators; 19,310 hp (14.4 MW) sustained; 2 Schottel SEP 5 podded propulsors; 14,750 hp (11 MW); 2 Schottel bow thrusters; 2,400 hp (1.8 MW)
Guns: 2 Signaal Goalkeeper 30 mm ❶. 9—12.7 mm MGs. 2—7.62 mm MGs.
Physical countermeasures: Decoys: 4 SRBOC chaff/IR launchers ❷; Nixie torpedo decoy system (fitted for).
Electronic countermeasures: ESM: ARGO Systems AR-900; intercept.
Radars: Air/surface search: Thales VARIANT 2 ❹; G/I-band. Surface search: Thales/Kelvin Hughes ARPA ❺; I-band. Navigation: 1 Consilium Selesmar; I-band. 2 Consilium Selesmar ❻; E/F-band.
Combat data systems: 1 CAMS/Force Vision CMS; 2 Raytheon SHF SATCOM ❸; 2 Surcom UHF SATCOM; 1 AEHF SATCOM; Link 11 (16 and 22 planned); MCCIS.

Helicopters: 6 NH 90 ❼ or 4 Merlin.

Programmes: Contract signed with Royal Schelde 3 May 2002. The hull was constructed at the Damen-owned Galati yard in Romania and arrived at the Schelde yard on 3 December 2004 for completion. To be fitted with command and control facilities for an afloat CJTF-HQ.
Structure: Facilities to transport a fully equipped Marine battalion with docking facilities for landing craft and a two spot helicopter flight deck with hangar space for six NH90. 25 ton crane for disembarkation. Full hospital facilities. Built to commercial standards with military command and control and NBCD facilities. Can carry up to 300 sonobuoys. Modifications to carry up to 30 torpedoes may be made at a later date. Based on the L 800 design but larger and wider. The flight deck is also stronger.
Operational: Alternative employment as an SAR ship for environmental and disaster relief tasks.

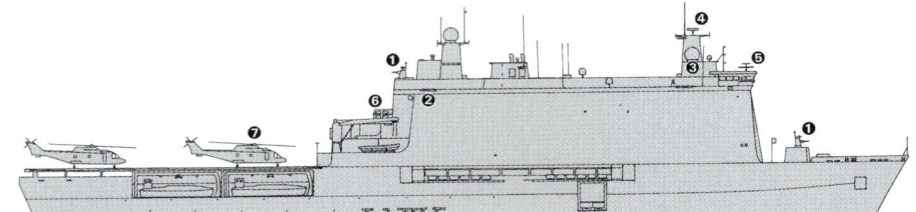

JOHAN DE WITT (Scale 1 : 1,500), Ian Sturton / 1170084

JOHAN DE WITT 10/2013*, Shaun Jones / 1530934

JOHAN DE WITT 7/2007, Michael Nitz / 1166628

JOHAN DE WITT 4/2009, A Campanera i Rovira / 1305518

Amphibious forces — Mine warfare forces < **Netherlands** 569

12 LCVP MK V (LANDING CRAFT) (LCVP)

L 9565–9576

Displacement, tonnes: 16 standard; 25 full load
Dimensions, metres (feet): 15.7 × 4.27 × 0.66 (51.5 × 14.0 × 2.2)
Speed, knots: 16
Range, n miles: 200 at 15 kt
Complement: 3

Military lift: 35 troops and equipment or 8.2 tonnes of stores or 1 vehicle and trailer.
Machinery: 2 Volvo D9 575 diesels; 2 Ultrajet 410 waterjets
Guns: 2 — 7.62 mm MGs.
Radars: Raymarine; I-band.

Comment: Contract signed on 13 December 2006 with Damen Shipyards Gorinchem for the construction and delivery of 12 Landing Craft Vehicle Personnel (LCVP) to replace the LCVP Mk II and III classes. The craft are being built by Visser Shipyard in Den Helder. The first four were delivered in 2009 and the remaining eight by 2011.

RHIBC 11 7/2012, Frank Findler / 1483637

L 9568 6/2013*, Frank Findler / 1529170

17 RIGID RAIDING CRAFT (PBF)

RHIRC 1–17

Displacement, tonnes: 8 full load
Dimensions, metres (feet): 12 × 3.3 × 0.7 (39.4 × 10.8 × 2.3)
Speed, knots: 40
Range, n miles: 200 at 30 kt
Complement: 2 + 8 embarked forces
Machinery: 2 Volvo Penta D6; 870 hp (649 kW); 2 waterjets
Guns: 1 — 7.62 mm MG.
Radars: Skydec Tactiplot; I-band.

Comment: Contract for the design and build of 48 FRISCs (Fast Raiding, Interception, and Special Forces Craft) awarded to MST Liverpool on 18 March 2009. The programme includes acquisition of three types of craft: Raiding Craft (RHIRC, Boarding Craft (RHIBC) and Support Craft Caribbean (RHISC). The 17 RHIRC are for amphibious raids, riverine operations and fleet protection worldwide. The craft were delivered in 2013.

RHIRC 1 6/2012, RNLN / 1486310

19 RIGID BOARDING CRAFT (PBF)

RHIBC 1–19

Displacement, tonnes: 8 full load
Dimensions, metres (feet): 12 × 3.3 × 1.15 (39.4 × 10.8 × 3.8)
Speed, knots: 40
Range, n miles: 300 at 35 kt
Complement: 2 + 8 embarked forces
Machinery: 2 Volvo Penta D6 diesels; 740 hp (552 kW); 2 shafts
Guns: 1 — 7.62 mm MG.
Radars: Skydec Tactiplot; I-band.

Comment: Contract for the design and build of 48 FRISCs (Fast Raiding, Interception, and Special Forces Craft) awarded to MST Liverpool on 18 March 2009. The programme includes acquisition of three types of craft: Raiding Craft (RHIRC), Boarding Craft (RHIBC) and Support Craft Caribbean (RHISC). Eight RHIBC are used for special operations and maritime counter-terrorism and 11 RHIBC are used as interceptor craft for the Holland class and for homeland defence (patrol and pursuit).

MINE WARFARE FORCES

6 ALKMAAR (TRIPARTITE) CLASS (MINEHUNTERS) (MHC)

Name	No	Laid down	Launched	Commissioned
MAKKUM	M 857	25 Feb 1983	27 Sep 1984	13 May 1985
SCHIEDAM	M 860	6 May 1984	20 Dec 1985	9 Jul 1986
URK	M 861	1 Oct 1984	2 May 1986	10 Dec 1986
ZIERIKZEE	M 862	25 Feb 1985	4 Oct 1986	7 May 1987
VLAARDINGEN	M 863	6 May 1986	4 Aug 1988	15 Mar 1989
WILLEMSTAD	M 864	3 Oct 1986	27 Jan 1989	20 Sep 1989

Displacement, tonnes: 543 standard; 660 full load
Dimensions, metres (feet): 51.5 × 8.9 × 2.6 (169 × 29.2 × 8.5)
Speed, knots: 13; 7 electric. **Range, n miles:** 3,000 at 12 kt
Complement: 38 (est.)

Machinery: 1 Stork Wärtsilä A-RUB 215X-12 diesel; 1,860 hp(m) (1.35 MW) sustained; 1 shaft; LIPS cp prop; 2 active rudders; 2 motors; 240 hp(m) (179 kW); 2 bow thrusters
Guns: 3 — 12.7 mm MGs.
Physical countermeasures: MCM: Atlas Seafox MIDS.
Radars: Navigation: Consilium Selesmar MM 950; I-band.
Sonars: Thales TSM 2022 Mk III; hull-mounted; minehunting; 100, 200 and 400 kHz and Double Eagle Mk III Mod 1 variable depth sonar.
Combat data systems: Atlas Elektronic IMCMS. SATCOM.

Programmes: The two Indonesian ships ordered in 1985 took the place of M 863 and 864 whose laying down was delayed as a result. This class is the Netherlands' part of a tripartite co-operative plan with Belgium and France for GRP hulled minehunters. The whole original class of 15 built by van der Giessen-de Noord. Ships were launched virtually ready for trials.
Modernisation: An extensive modernisation programme was completed at Den Helder 2003–11 to extend service life to 2020. Upgrades included a MCM command and control system, an Integrated Mine Countermeasures System (comprising hull-mounted and self-propelled variable-depth sonar (installed in Double Eagle Mk III Mod 1 RoV)) and a Mine-Identification and Disposal System (MIDS) based on the Atlas Seafox. Linked to the ship by a 3,000 m fibre optic tether, one variant (Seafox-C) is used for mine disposal and the other (Seafox-I) is used for identification.
Structure: A 5 ton container can be shipped, stored for varying tasks-research; patrol; extended diving; drone control.
Operational: Endurance, 15 days. MHCs are sometimes assigned to coast guard operations. M 853, 856, 858 and 859 decommissioned in 2011.
Sales: Two of a modified design to Indonesia, completed March 1988. M 850–852 decommissioned in 2000. M 854 and M 855 decommissioned in 2003. All five were sold to Latvia and transfer was completed by 2009. M 853, M 856, M 858, M 859 were decommissioned in 2011.

MAKKUM 9/2013*, Per Körnefeldt / 1530936

SCHIEDAM 6/2013*, Michael Nitz / 1530935

© 2014 IHS IHS Jane's Fighting Ships 2014-2015

SURVEY SHIPS

2 SNELLIUS CLASS (SURVEY SHIPS) (AGSH)

Name	No	Builders	Launched	Commissioned
SNELLIUS	A 802	Royal Schelde, Vlissingen	30 Apr 2003	11 Dec 2003
LUYMES	A 803	Royal Schelde, Vlissingen	22 Aug 2003	3 Jun 2004

Displacement, tonnes: 1,905 full load
Dimensions, metres (feet): 75 × 13.1 × 4 (246.1 × 43.0 × 13.1)
Speed, knots: 12
Range, n miles: 4,300 at 12 kt
Complement: 13 + 5 scientists + 24 spare berths
Machinery: Diesel electric; 3 diesel generators; 2,652 hp(m) (1.95 MW); 1 motor; 1,360 hp(m) (1 MW); 1 shaft; cp prop
Radars: Navigation: E/F- and I-band.
Sonars: Multi and single beam; high frequency; active

Comment: Designed for military and civil hydrographic surveys. Both laid down on 25 June 2002.

VAN KINSBERGEN   6/2013*, Frank Findler / 1529171

SNELLIUS   7/2011, Michael Winter / 1406503

LUYMES   1/2013, Michael Nitz / 1486307

1 SAIL TRAINING SHIP (AXS)

Name	No	Builders	Commissioned
URANIA	Y 8050	Haarlem	23 Apr 1938

Displacement, tonnes: 76 full load
Dimensions, metres (feet): 26.8 × 6.05 × 2.6 (87.9 × 19.8 × 8.5)
Speed, knots: 12; 10 diesel
Complement: 17 (14 trainees)
Machinery: 1 Caterpillar diesel; 235 hp(m) (186 kW); 1 shaft

Comment: Bermudan ketch used for training in seamanship, navigation and personal development. Refit 2001–04 included a new hull and aluminium masts. Sail number is NED 31.

TRAINING SHIPS

Notes: Two Dokkum-class minesweepers are used by Sea Cadets.

1 TRAINING SHIP (AXL)

Name	No	Builders	Commissioned
VAN KINSBERGEN	A 902	Damen Shipyards	2 Nov 1999

Displacement, tonnes: 640 full load
Dimensions, metres (feet): 41.5 × 9.2 × 3.3 (136.2 × 30.2 × 10.8)
Speed, knots: 13
Complement: 24 (3 instructors; 16 cadets)
Machinery: 2 Caterpillar 3508 BI-TA; 1,572 hp(m) (1.16 MW) sustained; 2 shafts; bow thruster; 272 hp(m) (200 kW)
Radars: Navigation: Consilium Selesmar; I-band.

Comment: Launched 30 August 1999. Replaced *Zeefakkel* as the local training ship at the naval base, Den Helder. Carries a 25 kt RIB.

URANIA   6/2013*, Michael Winter / 1530937

AUXILIARIES

Notes: (1) In addition to the vessels listed there are large numbers of non self-propelled craft with Y pennant numbers.
(2) *Thetis* (A 887) is based at Den Helder and acts as a platform for divers and underwater swimmers during harbour training.

1 AMSTERDAM CLASS (FAST COMBAT SUPPORT SHIP) (AORH)

Name	No	Builders	Laid down	Launched	Commissioned
AMSTERDAM	A 836	Merwede, Hardinxveld; Royal Schelde	25 May 1992	11 Sep 1993	2 Sep 1995

Displacement, tonnes: 17,313 full load
Dimensions, metres (feet): 166 × 22 × 8 (544.6 × 72.2 × 26.2)
Speed, knots: 20
Range, n miles: 13,440 at 20 kt
Complement: 160 (23 officers; 24 air crew) + 20 spare berths
Cargo capacity: 6,815 tons dieso; 1,660 tons aviation fuel; 290 tons solids
Machinery: 2 Bazán/Burmeister & Wain 16V 40/45 diesels; 24,000 hp(m) (17.6 MW) sustained; 1 shaft; LIPS cp prop
Guns: 1 Thales Goalkeeper 30 mm CIWS. 6 – 12.7 mm MGs. 6 – 7.62 mm MGs.
Physical countermeasures: Decoys: 4 SRBOC Mk 36 chaff/IR launchers. SLQ-25 Nixie towed torpedo decoy.
Electronic countermeasures: ESM: ARGO AR-900; intercept.
Radars: Surface search and helo control: 2 Kelvin Hughes; F-band.
Weapon control systems: Signaal IRSCAN infrared director.
Helicopters: 3 Lynx or 3 SH-3D or 3 NH90 or 2 EH 101.

Programmes: NP/SP AOR 90 replacement for *Poolster* ordered 14 October 1991. Hull built by Merwede, with fitting out by Royal Schelde from October 1993. A similar ship has been built for the Spanish Navy.
Modernisation: New communicators and command and control facilities were installed during 2011.
Structure: Close co-operation between Dutch Nevesbu and Spanish Bazán led to this design which has maintenance workshops as well as four abeam and one stern RAS/FAS station, and one Vertrep supply station. Built to merchant ship standards but with military NBC damage control.
Operational: To be decommissioned when *Karel Doorman* enters service.

AMSTERDAM   6/2010, Michael Nitz / 1366658

Auxiliaries < **Netherlands** 571

0 + 1 KAREL DOORMAN CLASS (JOINT LOGISTIC SUPPORT SHIP) (AFSH)

Name	No	Builders	Laid down	Launched	Commissioned
KAREL DOORMAN	A 833	Damen Shipyard, Galati	7 Jun 2011	17 Oct 2012	2015

Displacement, tonnes: 27,800 full load
Dimensions, metres (feet): 204.7 × 30.4 × 8.0 *(671.6 × 99.7 × 26.2)*
Speed, knots: 18
Complement: 170

Military lift: 130 troops
Machinery: Diesel-electric; 4 Rolls Royce Bergen generators; 30,575 hp *(22.8 MW)*; 1 Rolls Royce generator; 3,755 hp *(2.8 MW)*; 2 motors; 24,138 hp *(18 MW)*; 2 shafts; 2 bow thrusters; 1 stern thruster
Guns: 2 Thales Goalkeeper 30 mm. 2 Oto Melara 30 mm/70 (remote operated). 4 Oto Melara 12.7 mm Hitrole. 6−7.62 mm MGs.
Physical countermeasures: Decoys: 4 SRBOC Mk 36 chaff/IR launchers.
Radars: Air/Surface search/navigation: Thales SMILE; E/F-band.
Surface search: Thales SEASTAR; I-band.
Navigation: I-band.
Combat data systems: CAMS/Force Vision Guardian CMS. Link 11. Provision for Link 16/22. SATCOM.
Electro-optic systems: Thales Gatekeeper; IR and TV.

Helicopters: 6 NH90 or 2 Chinooks.

Programmes: A contract was let with Damen Schelde Naval Shipbuilding on 18 December 2009. The ship's hull is being built by Damen Shipyard, Galati (Romania). Primary tasks are to be maritime logistic support, strategic sealift and support of land-based forces. Secondary tasks are to be disaster relief, humanitarian aid and civil operations. It was announced in September 2013 that, as part of budget cuts, the ship would not enter Royal Netherlands Navy service. The decision was reversed a month later and the ship is to be completed as planned.
Structure: There are to be two abeam replenishment (fuel, water, solids) stations. There is to be approximately 2,000 lane metres (617 vehicle deck; 1,300 flight deck and hangar) of space for vehicles/containers, weapons for an infantry company, 8,700 cum of fuel and 125 cum/day drinking water. Embarkation/debarkation is to be facilitated by a 100 ton quarter ramp, 40 ton crane or the stern beach (mooring area for LCU). Two LCVPs are to be accommodated at davits. The ship is to be equipped with a role 3 medical facility, several workshops and a logistic support centre. The two-spot flight deck is to be capable of dual spot operations with up to Chinook-sized helicopters. An integrated mast, similar to that on the Holland class, includes the SMILE and SEASTAR radars and Gatekeeper EO system.

KAREL DOORMAN 8/2013*, A A de Kruijf / 1530939

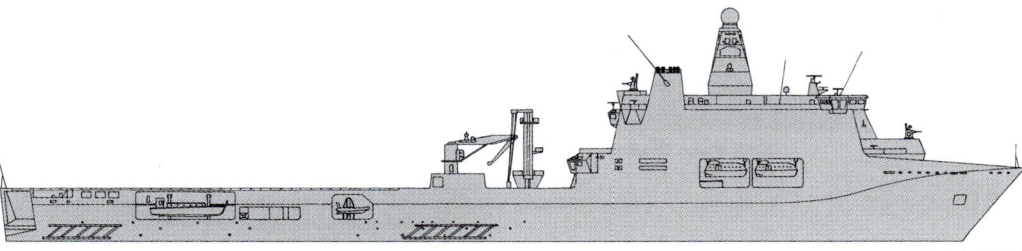

KAREL DOORMAN (Scale 1 : 1,500), Ian Sturton / 1406509

1 SUBMARINE SUPPORT SHIP AND TORPEDO TENDER (ASL/YTT)

Name	No	Builders	Commissioned
MERCUUR	A 900	Koninklijke Maatschappij de Schelde, Flushing	21 Aug 1987

Displacement, tonnes: 1,422 full load
Dimensions, metres (feet): 64.8 × 12 × 4.3 *(212.6 × 39.4 × 14.1)*
Speed, knots: 14
Complement: 39 (6 officers)

Machinery: 2 Brons 61-20/27 diesels; 1,100 hp(m) *(808 kW)*; 2 shafts; bow thruster
Guns: 2−12.7 mm MGs.
Torpedoes: 3−324 mm (triple) tubes. 1−21 in *(533 mm)* underwater tube.
Mines: Can lay mines.
Radars: Navigation: Consilium Selesmar; I-band.
Sonars: SQR-01; hull-mounted; passive search.

Comment: Replacement for previous ship of same name. Ordered 13 June 1984. Laid down 6 November 1985. Floated out 25 October 1986. Can launch training and research torpedoes above and below the waterline. Services, maintains and recovers torpedoes.

MERCUUR 6/2013*, Michael Nitz / 1530938

4 CERBERUS CLASS (DIVING TENDERS) (YDT)

Name	No	Builders	Commissioned
CERBERUS	A 851	Visser, Den Helder	28 Feb 1992
ARGUS	A 852	Visser, Den Helder	2 Jun 1992
NAUTILUS	A 853	Visser, Den Helder	18 Sep 1992
HYDRA	A 854	Visser, Den Helder	20 Nov 1992

Displacement, tonnes: 223 full load
Dimensions, metres (feet): 27.9 (A 851–852), 38.4 (A 853–854) × 8.76 × 1.5 *(91.5, 126 × 28.7 × 4.9)*
Speed, knots: 12. **Range, n miles:** 750 at 12 kt
Complement: 6 (A 851–852), 8 (A 853–854)
Machinery: 2 Volvo Penta TAMD122A diesels; 760 hp(m) *(560 kW)*; 2 shafts
Radars: Navigation: Racal Decca; I-band.

Comment: Ordered 29 November 1990. Platform for diving operations. Capable of maintaining 10 kt in Sea State 3. Can handle a 2 ton load at 4 m from the ship's side. *Hydra* lengthened by 10.5 m to provide more accommodation and recommissioned on 13 March 1998. *Nautilus* similarly refitted and re-entered service in May 2009.

NAUTILUS 4/2011, A A de Kruijf / 1406638

HYDRA 6/2008, A A de Kruijf / 1335275

1 TANKER (AOTL)

Name	No	Builders	Commissioned
PATRIA	Y 8760	De Hoop, Schiedam	9 Jun 1998

Displacement, tonnes: 692 full load
Dimensions, metres (feet): 44.4 × 6.9 × 2.8 *(145.7 × 22.6 × 9.2)*
Speed, knots: 9.5
Complement: 2
Machinery: 1 Volvo Penta TADM 122A; 381 hp(m) *(280 kW)*; 1 shaft
Radars: Navigation: Furuno RHRS-2002R; I-band

PATRIA 7/2005, A A de Kruijf / 1151128

© 2014 IHS IHS Jane's Fighting Ships 2014-2015

FRIGATES

2 ANZAC (MEKO 200) CLASS (FFHM)

Name	No	Builders	Laid down	Launched	Commissioned
TE KAHA	F 77	Transfield Defence Systems, Williamstown	19 Sep 1994	22 Jul 1995	22 Jul 1997
TE MANA	F 111	Tenix Defence Systems, Williamstown	28 Jun 1996	10 May 1997	10 Dec 1999

Displacement, tonnes: 3,759 full load
Dimensions, metres (feet): 118 oa; 109 wl × 14.8 × 4.4 *(387.1; 357.6 × 48.6 × 14.4)*
Speed, knots: 27
Range, n miles: 5,900 at 18 kt
Complement: 163

Machinery: CODOG; 1 GE LM 2500 gas turbine; 30,172 hp *(22.5 MW)* sustained; 2 MTU 12V 1163 TB93 diesels; 11,800 hp(m) *(8.8 MW)* sustained; 2 shafts; cp props
Missiles: SAM: Raytheon Sea Sparrow RIM-7P; Lockheed Martin Marietta Mk 41 Mod 5 octuple cell vertical launcher ❶; semi-active radar homing to 16 km *(8.5 n miles)* at 2.5 Mach; warhead 38 kg.
Guns: 1 FMC 5 in *(127 mm)*/54 Mk 45 Mod 2 ❷; 20 rds/min to 23 km *(12.6 n miles)*; weight of shell 32 kg. 1 Raytheon 20 mm Vulcan Phalanx 6 barrelled Mk 15 Block 1B ❸; 4,500 rds/min combined to 1.5 km. 2 Rafael Mini-Typhoon 12.7 mm remote-controlled guns ❹.
Torpedoes: 6—324 mm US Mk 32 Mod 5 (2 triple) tubes ❺; Mk 46 Mod 2; anti-submarine; active/passive homing to 11 km *(5.9 n miles)* at 40 kt; warhead 44 kg.
Physical countermeasures: Decoys: 2 Loral Hycor Mk 36 Mod 1 chaff launchers ❻. SLQ-25A torpedo decoy system.
Electronic countermeasures: ESM: DASA Maigret; Racal Centaur; intercept.
Radars: Air search: Raytheon SPS-49(V)8 ❼; C/D-band.
Air/surface search: CelsiusTech 9LV 453 TIR (Ericsson Tx/Rx) ❽; G-band.
Navigation: Atlas Elektronik 9600 ARPA ❾; I-band.
Fire control: CelsiusTech 9LV 453 ❿; G-band.
IFF: Cossor Mk XII.
Sonars: Thomson Sintra Spherion B Mod 5; hull-mounted; active search and attack; medium frequency.
Combat data systems: CelsiusTech 9LV 453 Mk 3. Link 11; GCCS-M.
Weapon control systems: CelsiusTech 9LV 453 optronic director. Raytheon CWI Mk 73 Mod 1 (for SAM).
Helicopters: 1 SH-2G (NZ) Super Seasprite ⓫.

Programmes: Contract signed with Amecon consortium on 19 November 1989 to build eight Blohm + Voss designed MEKO 200 ANZ frigates for Australia and two for New Zealand. Options on a third of class were turned down in November 1998. Modules constructed at Newcastle, Australia and Whangarei, New Zealand, and shipped to Melbourne for final assembly. The two New Zealand ships are the second and fourth of the class. First steel cut on *Te Kaha* on 11 February 1993. *Te Kaha* means Fighting Prowess. *Te Mana* means Authority.
Modernisation: Beginning with *Te Kaha* in 2009, both ships entered a Platform Systems Upgrade (PSU) including a propulsion diesel engine replacement and stability enhancement. *Te Kaha* is to complete the second phase of PSU in 2014. Upgrades include improved air conditioning and an Integrate Platform Management System. *Te Mana* will start Phase 2 in 2014. The tender process for a Frigate Systems Upgrade was completed in 2013. This upgrade, which is likely to start in 2016, is to include a new combat management system, surveillance radar, and local area air-defence (LAAD) capability. MBDA Sea Ceptor was down-selected as preferred LAAD system in 2013. As a separate project, upgrade of the Phalanx CIWS to Block 1B was completed by mid-2012.
Structure: The ships include space and weight provision for considerable enhancement including canister-launched SSM, an additional fire-control channel and ECM. Signature suppression features are incorporated in the design. All-steel construction. Fin stabilisers. McTaggert Scott Trigon 3 helicopter traversing system. Two RHIBs are carried.

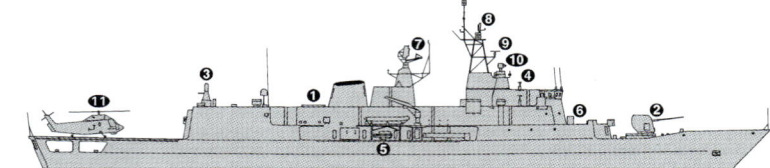

TE KAHA *(Scale 1: 1,200), Ian Sturton / 0081317*

TE KAHA *2/2012, Chris Sattler / 1486311*

TE MANA *11/2013*, Michael Nitz / 1529069*

SHIPBORNE AIRCRAFT

Notes: The current SH-2G Super Seasprite fleet is to be replaced by eight new SH2-G(NZ) aircraft. Approval for the purchase from Kaman Aerospace was given in April 2013 and the package is to include a training simulator and missiles. The replacement helicopters were originally built for the Australian Defence Force but the contract was subsequently cancelled in 2009 by the Australian government. The aircraft are for operation from the ANZAC-class frigates, Otago-class patrol vessels, and *Canterbury* and are to start entering service in 2015.

Numbers/Type: 9 NHIndustries NH90.
Operational speed: 163 kt *(300 km/h)*.
Service ceiling: 15,000 ft *(4,572 m)*.
Range: 440 n miles *(815 km)*.
Role/Weapon systems: Nine helicopters, similar to the MH 90s ordered by Australia, but configured for multirole operations. Delivery is to be completed in 2014. To be operated by the air force. Up to four may be embarked in *Canterbury*; the first landing on board was successfully completed on 29 July 2012. Sensors: Weather radar (with surface mapping capability), future growth for FLIR and dipping sonar. Weapons: 2 MAG58 7.62 mm MGs.

Numbers/Type: 5 Kaman SH-2G (NZ) Super Seasprite.
Operational speed: 130 kt *(241 km/h)*.
Service ceiling: 22,500 ft *(6,860 m)*.
Range: 400 n miles *(740 km)*.
Role/Weapon systems: Last of five delivered in February 2003. To be replaced by SH-2G(NZ) from 2015. Sensors: Litton ASN 150 C2; Telephonics APS 143 radar; AAQ 32 Safire IRDS; LR 100 ESM; ALE 47 ECM. Weapons: ASW; 2 Mk 46 torpedoes or Mk 11 depth bomb; ASV; 2 Hughes Maverick AGM 65D (NZ); 1—7.62 mm MAG58 MG.

NH90 *10/2009, RNZN / 1340403*

SUPER SEASPRITE *5/2013*, Chris Sattler / 1529068*

LAND-BASED MARITIME AIRCRAFT

Notes: Replacement of the P-3K2 fleet from about 2025 is under consideration.

Numbers/Type: 6 Lockheed P-3K2 Orion.
Operational speed: 405 kt *(750 km/h)*.
Service ceiling: 30,000 ft *(9,146 m)*.
Range: 4,000 n miles *(7,410 km)*.
Role/Weapon systems: Purchased in 1966. Long-range surveillance and reconnaissance patrol; updated 1984. Modernisation of airframes (Project Kestrel) undertaken 1995–2001 for 20 year extension. Upgrade project in progress to modernise mission avionics, sensors and communication/navigation systems. The upgrade is to include an Elta EL/M-2022(V)3 radar and Wescam MX-20 FLIR. Contract signed with L-3 communications on 4 October 2004. The first two upgraded aircraft were delivered in 2010 and programme is to be completed in 2014. A further upgrade, to restore ASW capability, is under consideration. Operated by RNZAF. Sensors: Elta EL/M-2022(V)3 radar, Wescam MX-20 FLIR, ASQ-10 MAD, acoustic processor, AYK 14 computers, IFF, ESM, SSQ 53/62 sonobuoys. Weapons: ASW; eight Mk 46 torpedoes, Mk 80 series depth bombs.

P-3K2　　　　　　　　　　　　　　　　　　　7/2004, Paul Jackson / 0589788

PATROL FORCES

2 OTAGO CLASS (OFFSHORE PATROL VESSELS) (PBO)

Name	No	Builders	Laid down	Launched	Commissioned
OTAGO	P 148	Tenix Defence Systems, Williamstown	16 Dec 2005	18 Nov 2006	18 Feb 2010
WELLINGTON	P 55	Tenix Defence Systems, Williamstown	2 Jun 2007	27 Oct 2007	6 May 2010

Displacement, tonnes: 1,626 standard
Dimensions, metres (feet): 85 × 14.0 × 3.6 *(278.9 × 45.9 × 11.8)*
Speed, knots: 22
Range, n miles: 6,000 at 15 kt
Complement: 35 + 44 spare berths

Machinery: 2 MAN Burmeister & Wain 12 RK 280 diesels; 2 shafts; cp props
Guns: 1 MSI DS 25M Autsig 25 mm. 2—12.7 mm MGs.
Radars: Navigation: I-band.
Helicopters: 1 SH-2G Super Seasprite.

Programmes: Following selection as 'Project Protector' prime contractor in April 2004, Tenix Defence awarded contract for final design and construction on 28 July 2004. The ships are to meet patrol and surveillance requirements in support of civil agencies in New Zealand's EEZ and the Southern Ocean and to assist South Pacific states to patrol their EEZs. Manufacturing of modules started at Tenix's Whangerai Shipyard in New Zealand in February 2005. Final assembly was undertaken at Williamstown, Victoria.
Structure: The design is a lengthened, helicopter-capable variant of a Kvaerner Masa Marine design in service in Ireland and Mauritius. They are ice-strengthened.
Operational: The ships also undertake the military hydrography role following the decommissioning of *Resolution* in 2012. SH-2G Seasprite flying trials were conducted on *Otago* in 2012.

WELLINGTON　　　　　　　　　　　　　　　　5/2013*, Chris Sattler / 1529067

4 LAKE CLASS (INSHORE PATROL VESSELS) (PBI)

Name	No	Builders	Laid down	Launched	Commissioned
ROTOITI	P 3569	Tenix Defence Systems, Williamstown	3 Mar 2006	4 Aug 2007	17 Apr 2009
HAWEA	P 3571	Tenix Defence Systems, Williamstown	13 Dec 2006	15 Dec 2007	1 May 2009
PUKAKI	P 3568	Tenix Defence Systems, Williamstown	21 Jun 2007	10 May 2008	14 May 2009
TAUPO	P 3570	Tenix Defence Systems, Williamstown	14 Dec 2007	23 Aug 2008	29 May 2009

Displacement, tonnes: 345 standard
Dimensions, metres (feet): 55 × 9.0 × 2.9 *(180.4 × 29.5 × 9.5)*
Speed, knots: 25
Range, n miles: 3,000 at 15 kt
Complement: 20 + 16 spare berths
Machinery: 2 MAN Burmeister & Wain 12VP 185 diesels; 2 shafts; cp props
Guns: 3—12.7 mm MGs.
Radars: Navigation: I-band.

Programmes: Following selection as 'Project Protector' prime contractor in April 2004, Tenix Defence awarded contract for final design and construction on 29 July 2004. The ships are to operate in support of civil agencies to meet patrol and surveillance requirements in New Zealand's inshore zone (out to 24 n miles), particularly around North Island, Marlborough Sounds and Tasman Bay. Manufacturing started at Tenix's Whangerai Shipyard in New Zealand in early 2005.
Structure: The Tenix design is based on the 56 m San Juan class built for the Philippines Coast Guard. Capable of operating in up to Sea State 5, they are able to launch and recover rigid hull inflatable boats in up to Sea State 4.

HAWEA　　　　　　　　　　　　　　　　　　9/2011, Chris Sattler / 1454710

SURVEY AND RESEARCH SHIPS

Notes: Following the decommissioning of *Resolution* in June 2012, acquisition of a new ship to replace her and *Manawanui* with a single vessel are under consideration. The new ship is expected to enter service in 2017–18.

1 SURVEY MOTOR BOAT (YGS)

ADVENTURE A 05

Displacement, tonnes: 9 standard
Dimensions, metres (feet): 9.7 × 3.5 × 0.7 *(31.8 × 11.5 × 2.3)*
Speed, knots: 25
Range, n miles: 1,000 at 10 kt
Complement: 3
Machinery: 2 Volvo-Penta AD31P/DP diesels; 300 hp *(223 kW)*
Radars: Navigation: I-band.

Comment: Aluminium catamaran craft built in Kumeu, North Auckland, in 1998. She is capable of independent inshore hydrographic operations and short coastal passages and is fitted with an echo sounder.

ADVENTURE　　　　　　　　　　　　　　　　4/2008, Chris Sattler / 1305309

2 RAPID ENVIRONMENTAL ASSESSMENT CRAFT (YGS)

TAKAPU A 07　　　　TARAPUNGA A 08

Displacement, tonnes: 5 standard
Dimensions, metres (feet): 10 × 2.7 × 0.6 *(32.8 × 8.9 × 2.0)*
Speed, knots: 25
Range, n miles: 150 at 25 kt
Complement: 2 + 4 personnel
Machinery: 2 Yanmar 6LY-STP diesels; 370 hp *(276 kW)*; 2 shafts
Radars: Navigation: I-band.

Comment: Aluminium catamaran craft built by Northland Spars and Rigging Ltd, Opua, New Zealand, delivered early 2013. Designed to be deployed by road, ship (Otago class and *Canterbury*) or air (C-130), the craft are capable of supporting diver, MCM, and military hydrography operations in the littoral environment.

AUXILIARIES

Notes: (1) Options for the replacement of *Endeavour* are under consideration. The successor ship is likely to be a more versatile vessel with a broader range of capabilities.
(2) In addition to vessels listed below there are three 12 m sail training craft used for seamanship training: *Paea II*, *Mako II*, *Manga II* (sail nos 6911–6913).
(3) Replacement of the diving tender *Manawanui* is under consideration.

PAEA   *2002, RNZN /* 0525919

1 CANTERBURY CLASS (MULTIROLE VESSEL) (AKRH/AX)

Name	No	Builders	Laid down	Launched	Commissioned
CANTERBURY	L 421	Merwede Shipyard	6 Sep 2005	11 Feb 2006	12 Jun 2007

Displacement, tonnes: 9,012 standard
Dimensions, metres (feet): 131.2 × 23.4 × 5.6 *(430.4 × 76.8 × 18.4)*
Speed, knots: 19. **Range, n miles:** 6,000 at 15 kt
Complement: 53

Military lift: 1 infantry company including Light Armoured Vehicles and equipment. 2 LCM.
Machinery: CODADE; 2 Wärtsilä 9L32 diesels; 12,000 hp *(9 MW)*; 3 Volvo Penta diesels; 4,916 hp *(3.67 MW)*; 2 shafts; cp props
Guns: 1 MSI DS 25M Autsig 25 mm. 2—12.7 mm MGs.
Radars: Navigation: 2 I-band.
Helicopters: 2 SH-2G Super Seasprites.

Programmes: Following selection as 'Project Protector' prime contractor in April 2004, Tenix Defence awarded contract for final design and construction on 29 July 2004. The ship was constructed in the Netherlands and fitted out by Tenix at Williamstown, Victoria.
Structure: With a design based on a commercial roll-on/roll-off vessel, the ship is built to comply with Lloyds Register of Shipping rules. The ship is ice-strengthened for operations in the Southern Ocean and the Ross Sea. Staff facilities are incorporated. Remedial maintenance work, to improve the ship's safety and stability in high seas, was undertaken in 2013. Work included relocation of alcoves used for RHIB stowage and installation of an automated accommodation ladder. Further modifications are under consideration.
Operational: The ship provides a limited tactical sealift capacity for disaster relief, humanitarian relief operations, peace support operations, military support activities and development assistance support. The ship is also used as the principal sea training platform for the RNZN.

CANTERBURY   *7/2012, A A de Kruijf /* 1486314

CANTERBURY   *11/2013*, A A de Kruijf /* 1529071

2 LANDING CRAFT (LCM)

LC 01   LC 02

Displacement, tonnes: 56 standard; 102 full load
Dimensions, metres (feet): 23 × 6.4 × ? *(75.5 × 21.0 × ?)*
Speed, knots: 9. **Range, n miles:** 250 at 9 kt
Complement: 3
Military lift: 2 armoured fighting vehicles (NZLAV)
Machinery: 2 Scania D19 44M diesels; 630 hp *(470 kW)*; 2 Veth Z-drive azimuth thrusters

Comment: Designed by Iv-Nevesbu b.v. (Papendrecht, Netherlands) and constructed by Zwijnenburg, Rotterdam, the craft are designed to be carried in *Canterbury*. Operable in Sea State 3, onload and offload can be achieved (empty) using *Canterbury's* 60 tonne crane or alternatively via the stern ramp. The LCMs are designed for beach landings and are fitted with a ballasting system to allow the safe onload and offload of cargo. They are also fitted with a kedge anchor. The stern ramp of *Canterbury* has 'marriage blocks' to facilitate correct alignment on the ramp.

LC 02   *6/2007, RNZN /* 1170065

1 REPLENISHMENT TANKER (AORH)

Name	No	Builders	Launched	Commissioned
ENDEAVOUR	A 11	Hyundai	14 Aug 1987	6 Apr 1988

Displacement, tonnes: 10,325 full load
Dimensions, metres (feet): 138.1 × 18.4 × 7.3 *(453.1 × 60.4 × 24.0)*
Speed, knots: 13.5
Range, n miles: 8,000 at 13.5 kt
Complement: 49 (10 officers)

Cargo capacity: 4,200 tons dieso; 100 tons Avcat; 4 containers
Machinery: 1 MAN-Burmeister & Wain 12V32/36 diesel; 5,780 hp(m) *(4.25 MW)* sustained; 1 shaft; LIPS cp prop
Radars: Navigation: Racal Decca 1290A/9; ARPA 1690S; I-band.
Helicopters: Platform only.

Comment: Ordered July 1986. Laid down 10 April 1987. Completion delayed by engine problems but arrived in New Zealand in May 1988. Two abeam RAS rigs (one QRC, one Probe). Fitted with Inmarsat. Standard merchant design modified on building to provide a relatively inexpensive replenishment tanker. The ship has undergone a series of modifications to comply with MARPOL regulations for single-hulled tankers, with the latest modification completed in 2013. Options for replacement of the ship with a more versatile vessel are under consideration.

ENDEAVOUR   *3/2013*, Chris Sattler /* 1529070

1 DIVING TENDER (YDT)

Name	No	Builders	Commissioned
MANAWANUI (ex-*Star Perseus*)	A 09	Cochrane, Selby	May 1979

Displacement, tonnes: 926 full load
Dimensions, metres (feet): 43.6 × 9.5 × 3.2 *(143 × 31.2 × 10.5)*
Speed, knots: 10.7
Range, n miles: 5,000 at 10 kt
Complement: 24 (2 officers)
Machinery: 2 Caterpillar D 379TA diesels; 1,130 hp *(843 kW)*; 2 shafts; cp props; bow thruster
Radars: Surface search: Racal Decca Bridgemaster 2000; I-band.
Sonars: Klein 595 Tracpoint; side scan; active high frequency.

Comment: Former North Sea Oil Rig Diving support vessel commissioned into the RNZN on 5 April 1988 and used primarily to support the RNZN operational diving team. Equipment includes two Phantom HDX remote-controlled submersibles, a decompression chamber (to 250 ft), wet diving bell and 13 ton crane. Fitted with Inmarsat. MCAIS data system, side scan sonar and GPS fitted in 1995. To be replaced by 2018.

MANAWANUI   *5/2013*, Chris Sattler /* 1529072

Nicaragua
FUERZA NAVAL-EJERCITO DE NICARAGUA

Country Overview

The Republic of Nicaragua is the largest Central American republic. After many years of civil war, a 1989 peace plan introduced a more stable period of democratic government. With an area of 50,893 square miles, it is situated between Honduras to the north and Costa Rica to the south. It has a 381 n mile coastline with the Caribbean and a 225 n mile coastline with the Pacific Ocean. Lake Nicaragua (Cocibolca), the largest lake in central America, and Lake Managua (Xolotlán) are connected by the river Tipitapa. The capital and largest city is Managua while Corinto, on the Pacific coast, is the principal port. Territorial seas (12 n miles) are claimed.

Headquarters Appointments

Head of Navy:
Rear Admiral Marvin Elias Corrales Rodriguez

Personnel

2014: 910 officers and men

Bases

Pacific: Corinto (HQ), San Juan del Sur, Puerto Sandino y Potosi
Atlantic: Bluefields (HQ), El Bluff, Puerto Cabezas, Corn Island, San Juan del Norte

---

PATROL FORCES

Notes: (1) There is an unknown number of Coastal Enforcement 30 ft Intrepid-class patrol craft. Powered by two 275 hp outboard engines, they are capable of 50 kt.
(2) Procurement of up to four Mirazh (Project 14310)-class patrol craft from Vympel Shipbuilding is reported to be under consideration.

3 DABUR CLASS (PB)

| RIO GRANDE DE MATAGALPA GC 201 | TENDERT GC 202 | RIO ESCONDIDO GC 205 |

Displacement, tonnes: 40 full load
Dimensions, metres (feet): 19.8 × 5.5 × 1.8 *(65 × 18.0 × 5.9)*
Speed, knots: 20
Range, n miles: 450 at 13 kt
Complement: 12
Machinery: 2 Caterpillar 3406 diesels; 1,500 hp *(1.1 MW)* sustained; 2 shafts
Guns: 2 — 14.5 mm MGs.
Radars: Surface search Furuno 2115; I-band.

Comment: Acquired from Israel in May 1996. All three craft re-engined: GC 205 in 2004, GC 202 in 2006 and GC 201 in 2008. All are operational on the Atlantic coast.

GC 201   4/2008, Nicaraguan Navy / 1335288

19 ASSAULT AND RIVER CRAFT (PBF)

Comment: There are approximately 36 Colombian-built Eduardoño class 10 to 13 m assault craft, capable of 50 kt, six 12 m 'Cigarette' craft capable of 45 kt and four SeaCats. These are divided between the Atlantic and Pacific.
Sixteen Zodiac RIBs with 40 hp engines were donated by the US in mid-2006 and another six in March 2009. Two Boston Whaler Justice 370 were donated by the US in July 2013.

CIGARETTE CLASS   6/2008, Nicaraguan Navy / 1335290

EDUARDOÑO CLASS   6/2008, Nicaraguan Navy / 1335291

EDUARDOÑO CLASS   6/2008, Nicaraguan Navy / 1335287

4 RODMAN 101 CLASS (PB)

| GENERAL JOSE DOLORES ESTRADA 401 | – 403 |
| CACIQUE AGATEYTE 402 | CACIQUE DIRIANGEN 404 |

Displacement, tonnes: 64 full load
Dimensions, metres (feet): 30 × 5.9 × 1.3 *(98.4 × 19.4 × 4.3)*
Speed, knots: 30
Range, n miles: 800 at 12 kt
Complement: 12
Machinery: 2 Caterpillar 3412C diesels; 2,800 hp *(2.06 MW)*; 2 Hamilton waterjets
Guns: 2 — 14.5 mm.
Radars: Navigation: Furuno FR 2115; I-band.

Comment: GRP hull. Built by Rodman, Vigo and donated by the Spanish government in 2007. Employed on fishery protection duties and operated by the navy. Two based in the Atlantic (401, 403) and two in the Pacific (402, 404).

404   6/2008, Nicaraguan Navy / 1335289

4 INTERCEPTOR CRAFT (PBF)

LI 112   +3

Displacement, tonnes: 4 full load
Dimensions, metres (feet): 13.4 × 2.75 × 0.9 *(44 × 9.0 × 3.0)*
Speed, knots: 60
Range, n miles: 600 at 25 kt
Complement: 6
Machinery: 3 Yanmar diesels; 945 hp *(704 kW)*; Bravo X drives

Comment: Manufactured by Nor-Tech, Fort Myers, FL. Composite and glass-fibre hull with V-bottomed hull. Donated by the US Southern Command in 2007. Employed on counter drugs, arms trafficking and illegal immigration duties.

INTERCEPTOR CRAFT   6/2007, US Southern Command / 1167968

578 Nicaragua > Patrol forces — Nigeria > Frigates

1 ZHUK (GRIF) CLASS (PROJECT 1400M) (PB)

RIO SEGOVIA 301

Displacement, tonnes: 40 full load
Dimensions, metres (feet): 24 × 5 × 1.2 (78.7 × 16.4 × 3.9)
Speed, knots: 30
Range, n miles: 1,100 at 15 kt
Complement: 13
Machinery: 2 Type M 401B diesels; 2,200 hp(m) (1.6 MW) sustained; 2 shafts
Guns: 4 — 14.5 mm (2 twin) MGs.
Radars: Surface search: Spin Trough; I-band.

Comment: One of seven ex-Russian craft built in the 1970s and transferred, probably via Cuba, in the 1980s. This craft returned to operational service in 2006 and a further craft may follow.

ZHUK 301   10/2006, Nicaraguan Navy / 1167883

Nigeria

Country Overview

Formerly a British protectorate, the Federal Republic of Nigeria gained full independence in 1960. With an area of 356,669 square miles, it is situated in western Africa and is bordered to the north by Niger, to the east by Chad and Cameroon and to the west by Benin. It has a 459 n mile coastline with the Gulf of Guinea. Abuja is the capital while Lagos (the capital until 1991) is the largest city, commercial centre and one of its principal ports. There are other ports at Port Harcourt, Warri, Calabar, Bonny, and Burutu. Territorial Seas (12 n miles) are claimed. An EEZ (200 n miles) has been claimed but the limits have not been defined.

The Navy has suffered from chronic lack of investment over the last ten years but an ongoing refit programme is attempting to restore a core seagoing capability for operations within the Nigerian EEZ.

Headquarters Appointments

Chief of the Naval Staff: Vice Admiral Usman Jibrin
Flag Officer Western Command: Rear Admiral S I Alade
Flag Officer Eastern Command: Rear Admiral H O Ngonadi
Flag Officer Central Command: Rear Admiral J O Olutoyin

Personnel

a) 2014: 17,864 (2,299 officers) including Coast Guard
b) Voluntary service

Bases

Western Naval Command, Apapa Lagos; NNS Beechcroft, FOB Igbokoda, FOB Badagry, Naval Base Lokoja.
Eastern Naval Command, Calabar; NNS Pathfinder, NNS Victory, NNS Jubilee, FOB Ibaka, FOB Bonny.
Central Naval Command, Yenegoa; NNS Delta, NNS Lugard, FOB Escravos, FOB Formoso.
Naval Training Command, Apapa Lagos; NNS Quorra, NNEC Sapele, NNC Onne, NNBTS Onne, Hydrographical School Port Harcourt, NNFLS Owerinta, NNSM Otta, NNSHS Offa, NNSAT Kachia, NNCETT Apapa, NPRS Makurdi.

Naval Aviation

The Nigerian Naval Air Arm operates six Agusta 109E and two Agusta Bell 206 helicopters. The A109E helicopters entered service in batches in 2002, 2004 and 2009. The Air Arm also operates and maintains AW 139 helicopters for the Nigerian Maritime Administration and Safety Agency (NIMASA), a Federal Government Agency charged with the safety and administration of seafarers. The Nigerian Navy and the Nigerian Air Force operate jointly two ATR 42 Maritime Patrol Aircraft (MPAs) in support of naval operations. The aircraft were acquired between 2009 and 2010 for EEZ patrol and are equipped with Gabbiano T-200 Surveillance Radar, Long Range Camera (LRTV) and EOST-45 Optronic suite.

Prefix to Ships' Names

NNS

Port Security Police

A separate force of 1,600 officers and men in Lagos.

Coastal Defence

The Regional Maritime Awareness Centre (RMAC) project is installed to cover the entire coastline from Lagos to Calabar and up to territorial waters. There are also remote sensing stations at DHQ, ONSA and NHQ.

FRIGATES

1 MEKO TYPE 360 H1 (FFGHM)

Name	No	Builders	Laid down	Launched	Commissioned
ARADU (ex-*Republic*)	F 89	Blohm + Voss, Hamburg	1 Dec 1978	25 Jan 1980	20 Feb 1982

Displacement, tonnes: 3,414 full load
Dimensions, metres (feet): 125.6 × 15 × 5.8 screws (412.1 × 49.2 × 19.0)
Speed, knots: 30.5
Range, n miles: 6,500 at 15 kt
Complement: 195 (26 officers)

Machinery: CODOG; 2 RR Olympus TM3B gas turbines; 50,880 hp (37.9 MW) sustained; 2 MTU 20V 956 TB92 diesels; 10,420 hp(m) (7.71 MW) sustained; 2 shafts; 2 Kamewa cp props
Missiles: SSM: 8 Oto Melara/Matra Otomat Mk 1 ❶; active radar homing to 80 km (43.2 n miles) at 0.9 Mach; warhead 210 kg.
SAM: Selenia Elsag Albatros octuple launcher ❷; 24 Aspide; semi-active radar homing to 13 km (7 n miles) at 2.5 Mach; warhead 30 kg.
Guns: 1 Oto Melara 5 in (127 mm)/54 ❸; 45 rds/min to 23 km (12.4 n miles); weight of shell 32 kg. 8 Breda Bofors 40 mm/70 (4 twin) ❹; 300 rds/min to 12.5 km (6.8 n miles) anti-surface; weight of shell 0.96 kg.
Torpedoes: 6 — 324 mm Plessey STWS-1B (2 triple) tubes ❺. 18 Whitehead A244S; anti-submarine; active/passive homing to 7 km (3.8 n miles) at 33 kt; warhead 34 kg (shaped charge).
Depth charges: 1 rack.
Physical countermeasures: Decoys: 2 Breda 105 mm SCLAR 20-tubed trainable; chaff to 5 km (2.7 n miles); illuminants to 12 km (6.6 n miles).
Electronic countermeasures: ESM: Decca RDL-2; intercept. ECM: RCM-2; jammer.
Radars: Air/surface search: Plessey AWS 5 ❻; E/F-band.
Navigation: Racal Decca 1226; I-band.
Fire control: Signaal STIR ❼; I/J/K-band. Signaal WM 25 ❽; I/J-band.
Sonars: Atlas Elektronik EA80; hull-mounted; active search and attack; medium frequency.
Combat data systems: Sewaco-BV action data automation.
Weapon control systems: M20 series GFCS. Signaal Vesta ASW.
Helicopters: 1 medium ❾.

Modernisation: Refit started at Wilmot Point, Lagos with Blohm & Voss assistance in 1991 and completed in February 1994.
Operational: Had two groundings and a major collision in 1987 and ran aground again during post refit trials in early 1994. Assessed as beyond economical repair in 1995 but managed to go to sea in early 1996, and again in 1997 when she broke down for several months in Monrovia. Back in Lagos on one engine in 1998 for further repairs. SSM system reported being refitted in 1999. Following a refit at Lagos, attended Fleet Review at Portsmouth, UK, in June 2005 and participated in fleet exercises in Brazil in January 2007. A temporary structure on the flight deck suggests that helicopters are no longer operated. A life-extension refit is planned.

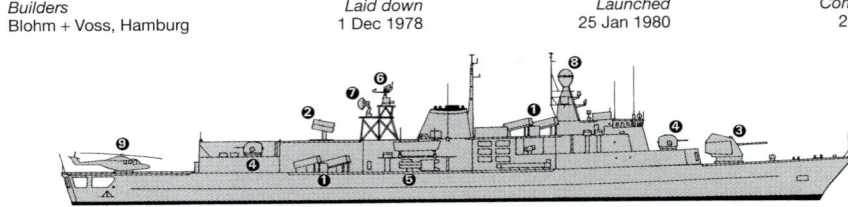

ARADU   (Scale 1 : 1,200), Ian Sturton / 0081331

ARADU   9/2007, Mario R V Carneiro / 1353243

IHS Jane's Fighting Ships 2014-2015   © 2014 IHS

Frigates — Corvettes < **Nigeria** 579

1 + 1 HAMILTON AND HERO CLASS (FFH)

Name	No
THUNDER (ex-*Chase*)	F 90 (ex-WHEC 718)
– (ex-*Gallatin*)	–

Builders	Laid down	Launched	Commissioned
Avondale Shipyards, New Orleans	27 Oct 1966	20 May 1967	1 Mar 1968
Avondale Shipyards, New Orleans	27 Feb 1967	18 Nov 1967	20 Dec 1968

Displacement, tonnes: 3,353 full load
Dimensions, metres (feet): 115.2 × 13.1 × 6.1 *(378 × 43.0 × 20.0)*
Flight deck, metres (feet): 26.8 × 12.2 *(87.9 × 40.0)*
Speed, knots: 29
Range, n miles: 9,600 at 15 kt
Complement: 162 (19 officers)

Machinery: CODOG; 2 Pratt & Whitney FT4A-6 gas turbines; 36,000 hp *(26.86 MW)*; 2 Fairbanks-Morse 38TD8-1/8-12 diesels; 7,000 hp *(5.22 MW)* sustained; 2 shafts; cp props; retractable bow propulsor; 350 hp *(261 kW)*
Guns: 1 Oto Melara 3 in *(76 mm)*/62 Mk 75 Compact; 85 rds/min to 16 km *(8.7 n miles)* anti-surface; 12 km *(6.6 n miles)* anti-aircraft; weight of shell 6 kg. 4 — 12.7 mm MGs.
Radars: Surface search: Hughes/Furuno SPS-73; E/F- and I-bands.
Tacan: URN 25.
Combat data systems: SCCS 378.

Helicopters: 1 medium.

Programmes: Twelve built of a total of 36 originally planned.
Modernisation: FRAM programme from October 1985 to October 1992. Work included standardising the engineering plants, improving the clutching systems, and replacing the Mk 56 fire-control system and 5 in/38 gun mount with the Mk 92 system and a single 76 mm Oto Melara Compact gun. The flight deck and other aircraft facilities upgraded including a telescopic hangar.

THUNDER 10/2013*, Chris Sattler / 1530943

Structure: These ships have clipper bows, twin funnels enclosing a helicopter hangar, helicopter platform aft. All are fitted with elaborate communications equipment. Superstructure is largely of aluminium construction. Bridge control of manoeuvring is by aircraft-type joystick rather than wheel.

Operational: *Thunder* transferred from the US Coast Guard on 13 May 2011 and recommissioned on 23 January 2012. She participated in the Australian Fleet Review in 2013. Ex-*Gallatin* is to be transferred in 2014.

THUNDER 11/2013*, Michael Nitz / 1530942

CORVETTES

1 MK 9 VOSPER THORNYCROFT TYPE (FSM)

Name	No	Builders	Commissioned
ENYMIRI	F 84	Vosper Thornycroft	2 May 1980

Displacement, tonnes: 691 standard; 793 full load
Dimensions, metres (feet): 69 × 9.6 × 3 *(226.4 × 31.5 × 9.8)*
Speed, knots: 27. **Range, n miles:** 2,200 at 14 kt
Complement: 90

Machinery: 4 MTU 20V 956 TB92 diesels; 22,140 hp(m) *(16.27 MW)* sustained; 2 shafts; 2 Kamewa cp props
Missiles: SAM: Short Brothers Seacat triple launcher.
Guns: 1 Oto Melara 3 in *(76 mm)*/62 Mod 6 compact; 85 rds/min to 16 km *(8.7 n miles)*; weight of shell 6 kg. 1 Breda Bofors 40 mm/70 Type 350; 300 rds/min to 12.5 km *(6.8 n miles)*; weight of shell 0.96 kg. 2 Oerlikon 20 mm.
A/S Mortars: 1 Bofors 375 mm twin launcher; range 1,600 or 3,600 m.
Electronic countermeasures: ESM: Decca Cutlass; radar warning.
Radars: Air/surface search: Plessey AWS 2; E/F-band.
Navigation: Racal Decca TM 1226; I-band.
Fire control: Signaal WM24; I/J-band; range 46 km *(25 n miles)*.
Sonars: Plessey PMS 26; lightweight; hull-mounted; active search and attack; 10 kHz.
Weapon control systems: Signaal WM20 series.

Programmes: Ordered from Vosper Thornycroft 22 April 1975.
Operational: *Enymiri* was damaged by fire in 2005 but had returned to service by early 2007 when it took part in fleet exercises. She is awaiting a docking period to renovate main machinery and generators. Sister ship *Erinomi* was decommissioned in 2011.

ENYMIRI 5/1999 / 0081333

© 2014 IHS IHS Jane's Fighting Ships 2014-2015

PATROL FORCES

Notes: (1) All the Coastal Patrol Craft belong to the Coast Guard. Some 38 craft were acquired in the mid-1980s from various shipbuilders including Simonneau (six craft P 233–238), Damen (six craft P 227–232), Swiftships (four craft P 221–224), Intermarine, Watercraft, Van Mill and Rotork. Few of these vessels have been reported at sea in recent years although some are visible, laid up ashore, and are still serviceable.
(2) A Damen 2600 Mk II patrol craft was acquired from South Africa in 2001.
(3) Four 8 m Night Cat 27, capable of 70 kt, were delivered by Intercept Boats in 2003–04.
(4) There are about 150 Nigerian Army Scorpion craft of length 10 m armed with 2 – 12.7 mm MGs.
(5) There is an unknown number of RHIBs which are also known as Jedi craft.
(6) There are plans to build a 38-m patrol ship in Nigeria.

SIKA 4   10/2010, Ann Till / 1366667

P 236 (Simonneau)   5/2002 / 0528302

SCORPION CRAFT H 224   10/2010, Ann Till / 1366665

2 EKPE (LÜRSSEN 57) CLASS (LARGE PATROL CRAFT) (PGF)

Name	No	Builders	Commissioned
EKPE	P 178	Lürssen, Bremen-Vegesack	Aug 1980
DAMISA	P 179	Lürssen, Bremen-Vegesack	Apr 1981

Displacement, tonnes: 451 full load
Dimensions, metres (feet): 58.1 × 7.6 × 3.1 (190.6 × 24.9 × 10.2)
Speed, knots: 42. **Range, n miles:** 2,000 at 10 kt
Complement: 40
Machinery: 4 MTU 16V 956 TB92 diesels; 17,700 hp(m) (13 MW) sustained; 2 shafts
Guns: 1 Oto Melara 3 in (76 mm)/62 Compact; 60 rds/min to 16 km (8.7 n miles); weight of shell 6 kg. 2 Breda 40 mm/70 (twin); 4 Emerlec 30 mm (2 twin).
Radars: Surface search: Racal Decca TM 1226; I-band.
Fire control: Signaal WM28; I/J-band.

Programmes: Ordered in 1977. Major refit in 1984 at Vegesack.
Operational: Both vessels are scheduled to be refitted. P 180 has been decommissioned.

EKPE   3/1998 / 0052656

4 CAT (BALSAM) CLASS (PBO)

Name	No	Builders	Commissioned
KYANWA (ex-*Sedge*)	A 501 (ex-WLB 402)	Marine Iron and Shipbuilding Corp, Duluth	5 Jul 1944
OLOGBO (ex-*Cowslip*)	A 502 (ex-WLB 277)	Marine Iron and Shipbuilding Corp, Duluth	17 Oct 1942
NWAMBA (ex-*Firebush*)	A 503 (ex-WLB 393)	Marine Iron and Shipbuilding Corp, Duluth	20 Jul 1944
OBULA (ex-*Sassafras*)	A 504 (ex-WLB 401)	Marine Iron and Shipbuilding Corp, Duluth	23 May 1944

Displacement, tonnes: 1,051 full load
Dimensions, metres (feet): 54.9 × 11.3 × 3.8 (180.1 × 37.1 × 12.5)
Speed, knots: 13. **Range, n miles:** 8,000 at 12 kt
Complement: 53
Machinery: Diesel electric; 2 diesels; 1,402 hp (1.06 MW); 1 motor; 1,200 hp (895 kW); 1 shaft; bow thruster
Guns: 2 – 12.7 mm MGs.
Radars: Navigation: Raytheon SPS-64(V)1.

Comment: First ship transferred from the US Coast Guard on 30 September 2002, second on 30 December 2002, third on 30 June 2003 and fourth on 30 October 2003. All four vessels are operational.

OLOGBO   10/2010, Ann Till / 1366666

15 DEFENDER CLASS (RESPONSE BOATS) (PBF)

P 313–327

Displacement, tonnes: 3 full load
Dimensions, metres (feet): 7.6 × 2.6 × 2.7 (24.9 × 8.5 × 8.9)
Speed, knots: 46. **Range, n miles:** 175 at 35 kt
Complement: 4
Machinery: 2 Honda outboard motors; 450 hp (335 kW)
Guns: 1 – 12.7 mm MG.
Radars: Navigation: Furuno; I-band.

Comment: High-speed inshore patrol craft of aluminium construction and foam collar built by SAFE Boats International, Port Orchard, Washington. An initial order for ten craft, with an option for five further craft, placed in August 2004 through USCG Foreign Military Sales programme. First four delivered on 13 December 2004 and second batch of four on 9 February 2005. Two were delivered in May 2005 and the final five on 5 July 2005.

P 316   6/2009, Nigerian Navy / 1305874

1 MODIFIED TOWN CLASS (PB)

Name	No	Builders	Laid down	Launched	Commissioned
ANDONI	P 100	Nigerian Naval Dockyard	Dec 2007	2011	1 Jun 2012

Displacement, tonnes: 150 full load
Dimensions, metres (feet): 31.3 × 6.1 × 1.4 (102.7 × 20.0 × 4.6)
Speed, knots: 20
Complement: 20 (3 officers)
Machinery: 2 Caterpillar C32 diesels; 3,200 hp (2.4 MW); 2 shafts
Guns: 1 – 20 mm. 2 – 12.7 mm MGs. 1 – 40 mm AGL.
Radars: Navigation: I-band.

Comment: Based on the Brooke Marine Town-class patrol craft, *Andoni* was laid down in December 2007 and full construction began in early 2008. However, work was subsequently suspended due to funding difficulties and it was not until 2010–11 that work re-started.

ANDONI   7/2012 / 1455797

Patrol forces < **Nigeria** 581

2 SEA EAGLE CLASS (PB)

Name	No	Builders	Commissioned
BURUTU	P 173	Nautica Nova Shipbuilding	14 Apr 2009
ZARIA	P 174	Nautica Nova Shipbuilding	14 Apr 2009

Displacement, tonnes: 190 full load
Dimensions, metres (feet): 38 × 7.0 × 1.65 (124.7 × 23.0 × 5.4)
Speed, knots: 32. **Range, n miles:** 1,500 at 15 kt
Complement: 25
Machinery: 2 MTU 16V 4000M90 diesels; 7,295 hp (5.44 MW); 2 shafts
Guns: 3—12.7 mm MGs. 1—40 mm AGL.
Radars: Surface search/navgiation: Furuno; I-band.

Comment: Procured from Suncraft International and built by Nautica Nova Shipbuilding Malaysia with funding from the Nigerian National Petroleum Corporation. Commissioned on 14 April 2009. The main role of the craft is to provide protection to offshore oil facilities.

ZARIA 6/2010, Ann Till / 1366668

3 COMBATTANTE IIIB CLASS
(FAST ATTACK CRAFT—MISSILE) (PGGF)

Name	No	Builders	Commissioned
SIRI	P 181	CMN, Cherbourg	19 Feb 1981
AYAM	P 182	CMN, Cherbourg	11 Jun 1981
EKUN	P 183	CMN, Cherbourg	18 Sep 1981

Displacement, tonnes: 391 standard; 437 full load
Dimensions, metres (feet): 56.2 × 7.6 × 2.1 (184.4 × 24.9 × 6.9)
Speed, knots: 38. **Range, n miles:** 2,000 at 15 kt
Complement: 42

Machinery: 4 MTU 16V 956 TB92 diesels; 17,700 hp(m) (13 MW) sustained; 2 shafts
Missiles: SSM: 4 Aerospatiale MM 38 Exocet; inertial cruise; active radar homing to 42 km (23 n miles) at 0.9 Mach; warhead 165 kg; sea-skimmer.
Guns: 1 Oto Melara 3 in (76 mm)/62 Compact; 60 rds/min to 16 km (8.7 n miles); weight of shell 6 kg. 2 Breda 40 mm/70 (twin); 300 rds/min to 12.5 km (6.8 n miles); weight of shell 0.96 kg. 4 Emerlec 30 mm (2 twin); 1,200 rds/min combined to 6 km (3.3 n miles); weight of shell 0.35 kg.
Electronic countermeasures: ESM: Decca RDL; radar intercept.
Radars: Air/surface search: Thomson-CSF Triton (TRS 3033); G-band.
Navigation: Racal Decca TM 1226; I-band. Fire control: Thomson-CSF Castor II (TRS 3203); I/J-band.
Weapon control systems: Thomson-CSF Vega system. 2 CSEE Panda optical directors.

Programmes: Ordered in late 1977. Finally handed over in February 1982 after delays caused by financial problems.
Modernisation: Major refit and repairs carried out at Cherbourg from March to December 1991 but the ships were delayed by financial problems.
Operational: All three ships are non-operational. Ekun was undergoing repairs in 2010 and Siri and Ayam are programmed for docking periods.

AYAM (outboard DAMISA) 5/2002 / 0528300

0 + 2 OFFSHORE PATROL VESSELS (PSO)

Name	No	Builders	Commissioned
–	F 91	Wuchang Shipyard, Wuhan	2014
–	–	Wuchang Shipyard, Wuhan	2015

Displacement, tonnes: 1,800 full load
Dimensions, metres (feet): 95 × ? × 3.5 (311.7 × ? × 11.5)
Speed, knots: 21
Complement: 20

Machinery: 2 MTU 20V 4000 diesels; 2 shafts; cp props
Guns: 1—76 mm. 2—30 mm.
Radars: Surface search/navigation: Kelvin Hughes SharpEye; E/F/I-bands.
Fire control: To be announced.
Helicopters: 1 medium.

Comment: It was announced on 29 March 2012 that Presidential approval had been given for the acquisition of two offshore patrol ships. The first ship is to be built at Wuchang Shipyard, Wuhan, China while approximately 50% of the second ship is to be built in Nigeria. Full details of the ships have not been released but the ship is expected to be armed with a 76 mm gun and design features are expected to include a flight deck and hangar. Delivery of the first ship is expected in 2014 and the second in 2015.

FUTURE PATROL SHIP 10/2013*, CSOC / 1513611

20 SUNCRAFT MANTA MK III CLASS (PB)

P 241–257   P 260–262

Displacement, tonnes: 21 full load
Dimensions, metres (feet): 17.95 × 3.6 × 0.95 (58.9 × 11.8 × 3.1)
Speed, knots: 45
Complement: 6
Machinery: 2 MTU diesels; 3,600 hp (2.7 MW); 2 surface piercing drives
Guns: 3—12.7 mm MGs. 1—40 mm AGL.
Radars: Navigation: Furuno; I-band.

Comment: Four procured from ST Marine in late 2008. A further four had been delivered by late 2010 and a further 12 by late 2013. The main role of the craft is to provide protection to offshore oil facilities.

MANTA CLASS 6/2009, Nigerian Navy / 1305875

4 SHALDAG MK II CLASS (PBF)

| ONITSHA P 247 | IKOT-ABASI P 255 | BENIN P 256 | – P 257 |

Displacement, tonnes: 59 full load
Dimensions, metres (feet): 24.8 × 6.0 × 1.2 (81.4 × 19.7 × 3.9)
Speed, knots: 45
Range, n miles: 700 at 32 kt
Complement: 10 (2 officers)
Machinery: 2 MTU diesels; 4,570 hp(m) (3.36 MW); 2 waterjets
Guns: 2—14.5 mm (twin). 2—12.7 mm MGs.
Radars: Navigation: Furuno; I-band.

Comment: A Shaldag fast patrol craft was delivered by Israel Shipyards in mid-2009 and a further two in September 2012. Delivery of a fourth craft was made on 19 November 2013.

ONITSHA 9/2010, Israel Shipyards / 1366961

2 TOWN CLASS (PB)

Name	No	Builders	Commissioned
MAKURDI	P 167	Brooke Marine, Lowestoft	14 Aug 1974
HADEJIA	P 168	Brooke Marine, Lowestoft	14 Aug 1974

Displacement, tonnes: 117 standard; 145 full load
Dimensions, metres (feet): 32.6 × 6.1 × 2.1 (107 × 20.0 × 6.9)
Speed, knots: 20
Complement: 21 (4 officers)
Machinery: 2 Ruston-Paxman Venture diesels; 2 shafts
Guns: 2 Emerlec 30 mm (twin).
Radars: Navigation: Decca; I-band.

Comment: Originally delivered in 1974 and subsequently decommissioned in the 1990s. Both craft started re-activation refits in 2009 and Makurdi became operational in 2010.

© 2014 IHS   IHS Jane's Fighting Ships 2014-2015

Nigeria > Patrol forces — Amphibious forces

3 OCEA FPB 72 MK II (PATROL CRAFT) (PB)

OKPOKU P 175   **BOMADI** P 176   **BADAGRY** P 177

Dimensions, metres (feet): 23.6 × 5.8 × 1.5 (77.4 × 19.0 × 4.9)
Speed, knots: 32. **Range, n miles:** 600 at 12 kt
Complement: 11
Machinery: 2 MTU diesels; 3,000 hp (2.24 MW); 2 shafts
Guns: 2—7.62 mm MGs.
Radars: Surface search/navigation: To be announced.

Comment: All three patrol craft were built at Lorient, launched between February and March 2012 and subsequently delivered in 2012.

OKPOKU   3/2012, B Prézelin / 1455681

2 NAUTIC SENTINEL CLASS (PB)

TORIE P 258   **EGEDE** P 259

Displacement, tonnes: 26 full load
Dimensions, metres (feet): 17 × 4.5 × 0.95 (55.8 × 14.8 × 3.1)
Speed, knots: 45
Complement: 6
Machinery: 2 MTU 8V200M84L diesels; 2,400 hp (1.79 MW); 2 surface-piercing propellers
Radars: To be announced.

Comment: Procured from Paramount Group (Nautic) in 2013.

EGEDE   6/2013*, Nigerian Navy / 1530944

2 YOLA CLASS (PB)

Name	No	Builders	Commissioned
YOLA	P 166	Abeking & Rasmussen, Lemwerder	Aug 1973
BRASS	P 169	Abeking & Rasmussen, Lemwerder	Aug 1976

Displacement, tonnes: 91 full load
Dimensions, metres (feet): 29 × 5.5 × 1.6 (95.1 × 18.0 × 5.2)
Speed, knots: 20 (P 166), 30 (P 169)
Complement: 20
Machinery: 2 Ruston-Paxman Venture (P 166) or 2 MTU 12V 331 TC71 (P 169) diesels; 2 shafts
Guns: 2 Emerlec 30 mm (twin).
Radars: Navigation: Furuno; I-band.

Comment: Originally delivered in the 1970s and subsequently decommissioned in the 1990s. Following re-activation in 2009, Brass is undergoing repairs and Yola is awaiting refit.

1 OCEA FPB 98 CLASS (PATROL CRAFT) (PB)

DORINA P 101

Displacement, tonnes: 116 full load
Dimensions, metres (feet): 35.2 × 6.8 × 1.2 (115.5 × 22.3 × 3.9)
Speed, knots: 30
Range, n miles: 1,000 at 12 kt
Complement: 12 (3 officers)
Machinery: 2 Caterpillar diesels; 3,586 hp (2.6 MW); 2 Kamewa waterjets
Guns: 1—20 mm. 2—12.7 mm MGs.
Radars: Navigation: Furuno FAR 2117; I-band.

Comment: The contract with OCEA for the construction of one patrol craft was announced in March 2011. Delivery of the vessel was made in 2012. Details similar to craft of the same class in Algerian service.

DORINA   8/2013*, B Prézelin / 1513614

FPB 98 CLASS (Benin colours)   3/2012, B Prézelin / 1455682

AMPHIBIOUS FORCES

1 FDR TYPE RO-RO 1300 (LST)

Name	No	Builders	Commissioned
AMBE	LST 1312	Howaldtswerke, Hamburg	11 May 1979

Displacement, tonnes: 1,494 standard; 1,890 full load
Dimensions, metres (feet): 87 × 14 × 2.3 (285.4 × 45.9 × 7.5)
Speed, knots: 17
Range, n miles: 5,000 at 10 kt
Complement: 56 (6 officers)

Military lift: 460 tons and 220 troops long haul; 540 troops or 1,000 troops seated short haul; can carry 5—40 ton tanks
Machinery: 2 MTU 16V 956 TB92 diesels; 8,850 hp(m) (6.5 MW) sustained; 2 shafts
Guns: 1 Breda 40 mm/70. 2 Oerlikon 20 mm.
Radars: Navigation: Racal Decca 1226; I-band.

Comment: Ordered September 1976. Built to a design prepared for the FGN. Has 19 m bow ramps and a 4 m stern ramp. Reported that bow ramps are welded shut. Second of class, Offiom, beyond repair but Ambe started repairs and machinery upgrade in 2010.

AMBE
7/1997
0012836

IHS Jane's Fighting Ships 2014-2015   © 2014 IHS

MINE WARFARE FORCES

2 LERICI CLASS (MINEHUNTERS/SWEEPERS) (MHSC)

Name	No	Builders	Commissioned
OHUE	M 371	Intermarine	28 May 1987
BARAMA	M 372	Intermarine	25 Feb 1988

Displacement, tonnes: 549 full load
Dimensions, metres (feet): 51 × 9.9 × 2.8 *(167.3 × 32.5 × 9.2)*
Speed, knots: 15.5
Range, n miles: 2,500 at 12 kt
Complement: 50 (5 officers)

Machinery: 2 MTU 12V 396 TB83 diesels; 3,120 hp(m) *(2.3 MW)* sustained; 2 waterjets
Guns: 2 Emerlec 30 mm (twin); 1,200 rds/min combined to 6 km *(3.3 n miles)*; weight of shell 0.35 kg. 2 Oerlikon 20 mm GAM-BO1.
Physical countermeasures: MCM: Fitted with 2 Pluto remote-controlled submersibles, Oropesa 'O' Mis 4 and Ibis V control system.
Radars: Navigation: Racal Decca 1226; I-band.
Sonars: Thomson Sintra TSM 2022; hull-mounted; mine detection; high frequency.

Comment: *Ohue* ordered in April 1983 and *Barama* in January 1986. *Ohue* laid down 23 July 1984 and launched 22 November 1985, *Barama* laid down 11 March 1985, launched 6 June 1986. GRP hulls but, unlike Italian and Malaysian versions they do not have separate hydraulic minehunting propulsion. Carry Galeazzi two-man decompression chambers. Both were refitted in 1999, after operations off Liberia. *Barama* reported refitted in late 2004 and both ships were undergoing refit in 2009–10. The operational effectiveness of both ships in their MCM role is doubtful.

OHUE   7/1987, *Marina Fraccaroli* / 0506063

SURVEY SHIPS

0 + 1 SURVEY SHIP (AGS)

Name	No	Builders	Commissioned
– (ex-*John McDonnell*)	– (ex-T-AGS 51)	Halter Marine	16 Dec 1991

Displacement, tonnes: 2,087 full load
Dimensions, metres (feet): 63.4 × 13.7 × 4.3 *(208 × 44.9 × 14.1)*
Speed, knots: 12
Range, n miles: 13,800 at 12 kt
Complement: 22 personnel + 11 scientists
Machinery: 1 GM EMD 12-645E6 diesel; 2,550 hp *(1.9 MW)* sustained; 1 auxiliary diesel; 230 hp *(172 kW)*; 1 shaft

Comment: Laid down on 3 August 1989 and launched on 15 August 1990. In US service, carried 34 ft survey launches for data collection in coastal regions with depths between 10 and 600 m and in deep water to 4,000 m. A small diesel is used for propulsion at towing speeds of up to 6 kt. Simrad high-frequency active hull-mounted and side scan sonars are carried. Following de-activation from US service in 2010, the ship is to be acquired by Nigeria and is to be transferred in 2014.

Ex-JOHN McDONNELL (US colours)   8/2004, *Hachiro Nakai* / 1043695

1 SURVEY SHIP (AGS)

Name	No	Builders	Launched	Commissioned
LANA	A 498	Brooke Marine, Lowestoft	4 Mar 1976	18 Jul 1976

Displacement, tonnes: 1,105 full load
Dimensions, metres (feet): 57.8 × 11.4 × 3.7 *(189.6 × 37.4 × 12.1)*
Speed, knots: 16
Range, n miles: 4,500 at 12 kt
Complement: 52 (12 officers)
Machinery: 2 Lister Blackstone diesels; 2,640 hp *(1.97 MW)*; 2 shafts
Radars: Navigation: Decca; I-band.

Comment: Similar to UK Bulldog class. Ordered in 1973. Underwent major refit in 1994 and subsequently recommissioned in July 1999. A docking period was undertaken in 2005 and a further docking period and machinery upgrade is planned to restore her to operational service.

LANA   5/1999 / 0081334

TUGS

4 COASTAL TUGS (YTB/YTL)

COMMANDER APAYI JOE A 499	DOLPHIN RIMA
DOLPHIN MIRA	COMMANDER RUDOLPH A 500

Comment: A 499 and A 500 are of 310 tons and were built in 1983. The two Dolphin tugs were built by Damen in 1986.

DOLPHIN MIRA   6/2009, *Nigerian Navy* / 1305877

GOVERNMENT MARITIME FORCES

MARITIME SAFETY ADMINISTRATION

Notes: The Nigerian Maritime Administration and Safety Agency (NIMASA) is the principal maritime agency formed on 1 August 2006 when the former National Maritime Authority and Joint Maritime Labour Industrial Council were merged. The agency is responsible to the Federal Ministry of Transport for maritime safety standards and security, shipping regulation and pollution prevention and control. Its key roles include the conduct of air and maritime surveillance, search and rescue and the safety and security of shipping. The agency operates a zonal structure to co-ordinate activities: Western Zone (HQ Lagos); Central Zone (Warri); Eastern Zone (Port Harcourt) and Northern Zone (Abuja). Since September 2012, some 25 craft have entered NIMASA service including some vessels operated by private company Global West Vessel Specialists Nigeria Limited (GWVSNL) under a partnering agreement.

0 + 2 ARESA 4200 GUARDIAN (PB)

Displacement, tonnes: 240 full load
Dimensions, metres (feet): 42 × 9.0 × 1.2 *(137.8 × 29.5 × 3.9)*
Speed, knots: 30
Range, n miles: 1,400 at 20 kt
Complement: 40
Machinery: 4 Cummins KTA50-M2 diesels; 7,500 hp *(5.6 MW)*; 4 Hamilton waterjets
Guns: 1 — 20 mm. 2 — 12.7 mm MGs.

Comment: Built by Aresa Internacional, Barcelona, the craft are expected to be delivered in 2014. Of GRP construction, the vessels are expected to conduct offshore patrol and search and rescue missions. Design features include an aft ramp for two 7.5 m RHIBs.

2 ARESA 1800 CPV FIGHTER (PB)

Displacement, tonnes: 30 full load
Dimensions, metres (feet): 17.3 × 4.5 × 0.6 *(56.8 × 14.8 × 2.0)*
Speed, knots: 43
Range, n miles: 500 at 25 kt
Complement: 12

Machinery: 2 MAN diesels; 2,000 hp *(894 kW)*; 2 shafts
Guns: 2 — 12.7 mm MGs.
Radars: Navigation: I-band.
Electro-optic systems: FLIR.

Comment: Built by Aresa Internacional, Barcelona, the craft were delivered in February 2013. Of GRP construction, the vessels are expected to conduct coastal patrol and maritime security missions.

4 ARESA 1300 SENTINEL (PB)

Displacement, tonnes: 11 full load
Dimensions, metres (feet): 13.1 × 3.33 × 0.7 *(43 × 10.9 × 2.3)*
Speed, knots: 50
Range, n miles: 350 at 25 kt
Complement: 10

Machinery: 4 Evinrude outboard motors; 1,200 hp *(894 kW)*
Guns: 1 — 12.7 mm MG.
Radars: Navigation: I-band.
Electro-optic systems: FLIR.

Comment: Built by Aresa Internacional, Barcelona, the craft were delivered in January 2013. Of GRP construction, the vessels are expected to conduct interception and maritime security missions.

ARESA 1800 FIGHTER   6/2013*, Aresa Internacional / 1486566

ARESA 1300 SENTINEL
6/2013*, Aresa Internacional
1486565

Norway

Country Overview

The Kingdom of Norway is a constitutional monarchy occupying the northwest part of the Scandinavian Peninsula. With an area of 125,016 square miles, it is bordered to the east by Sweden and to the northeast by Finland and Russia. The coastline of 11,842 n miles with the Atlantic Ocean (Norwegian Sea), Arctic Ocean (Barents Sea), North Sea and Skagerrak Strait contains numerous fjords and offshore islands. External territories in the Arctic Ocean include the Svalbard archipelago and Jan Mayen Island while the uninhabited Bouvet Island lies in the south Atlantic. Territorial claims in Antarctica include the territory known as Queen Maud Land and Peter I Island. The capital, largest city and principal port is Oslo. Other ports include Bergen, Trondheim and Stavanger. Territorial seas (12 n miles) and an EEZ (200 n miles) are claimed.

Headquarters Appointments

Chief of Naval Staff:
   Rear Admiral Lars Saunes
Deputy Chief of Naval Staff:
   Commodore Commodore Lars Johan Fleisje
Commander Coast Guard:
   To be announced
Commander Norwegian Fleet:
   Commodore H Amundsen
Acting Commander Norwegian Fleet:
   Captain Nils Stensønes

Personnel

(a) 2014: 4,140 officers and ratings
(b) 9 to 12 months' national service (up to 40% of ships complement)

Naval Special Warfare Group

The NSWG was established in 2006. It consists of the Coastal Ranger Commando, the Naval EOD Commando, the Tactical Boat Squadron and Naval Special Warfare Training Centre.

Coast Guard

Founded April 1977 with operational command held by Norwegian Defence Command. Main base at Sortland. Tasks include fishery protection, customs, police, SAR and environmental duties at sea.

Bases

Reitan (Bodø): National Operational HQ
Haakonsvern (Bergen): Main Naval Base
Laksevaag (Bergen): Submarine Repair
Ramsund: Supply, repair and maintenance
Sortland: Coast Guard Base.

Air Force Squadrons (see *Shipborne* and *Land-based Aircraft*)

Aircraft (Squadron)	Location	Duties
Sea King Mk 43 (330)	Bodø, Banak, Sola, Ørland	SAR
Orion P-3N/C (333)	Andøya	MPA
Lynx (337)	Coast Guard vessels/ Bardufoss	MP
Bell 412 (719, 339 & 720)	Bodø, Rygge, Bardufoss	Army Transport

Prefix to Ships' Names

KNM (Naval)
K/V (Coast Guard)

Strength of the Fleet

Type	Active	Building (Projected)
Submarines — Coastal	6	—
Frigates	5	—
Patrol vessels	26	1
Minesweepers/Hunters	6	—
Auxiliaries	7	1
Coast Guard Vessels	15	—
Survey Vessels	6	1

PENNANT LIST

Notes: Naval District Auxiliaries are listed on page 591.

Submarines

S 300	Ula
S 301	Utsira
S 302	Utstein
S 303	Utvaer
S 304	Uthaug
S 305	Uredd

Frigates

F 310	Fridtjof Nansen
F 311	Roald Amundsen
F 312	Otto Sverdrup
F 313	Helge Ingstad
F 314	Thor Heyerdahl

Minesweepers/Hunters

M 341	Karmøy
M 342	Måløy
M 343	Hinnøy
M 350	Alta
M 351	Otra
M 352	Rauma

Minelayers

N 50	Tyr

Patrol Forces

P 960	Skjold
P 961	Storm
P 962	Skudd
P 963	Steil
P 964	Glimt
P 965	Gnist

Auxiliaries

A 533	Norge
A 535	Valkyrien

Coast Guard

W 303	Svalbard
W 312	Ålesund
W 314	Stålbas
W 318	Harstad
W 319	Leikven
W 320	Nordkapp
W 321	Senja
W 322	Andenes
W 330	Nornen
W 331	Farm
W 332	Heimdal
W 333	Njord
W 334	Tor
W 340	Barentshav
W 341	Sortland
W 342	Bergen

SUBMARINES

Notes: Studies into the replacement of the current submarine capability from about 2020 are in progress. Options include life extension of the current inventory, a replacement programme or a combination of the two. A Request for Information was issued on 11 September 2012 to DCNS, Fincantieri, Navantia, Thyssenkrupp and Daewoo to inform the decision-making process. A recommendation for the way-ahead is expected in 2014 and a three-year project definition phase is planned to follow.

UTSTEIN *12/2011, Derek Fox / 1446502*

6 ULA CLASS (SSK)

Name	No	Builders	Laid down	Launched	Commissioned
ULA	S 300	Thyssen Nordseewerke, Emden	29 Jan 1987	28 Jul 1988	27 Apr 1989
UREDD	S 305	Thyssen Nordseewerke, Emden	23 Jun 1988	22 Sep 1989	3 May 1990
UTVAER	S 303	Thyssen Nordseewerke, Emden	8 Dec 1988	19 Apr 1990	8 Nov 1990
UTHAUG	S 304	Thyssen Nordseewerke, Emden	15 Jun 1989	18 Oct 1990	7 May 1991
UTSTEIN	S 302	Thyssen Nordseewerke, Emden	6 Dec 1989	25 Apr 1991	14 Nov 1991
UTSIRA	S 301	Thyssen Nordseewerke, Emden	15 Jun 1990	21 Nov 1991	30 Apr 1992

Displacement, tonnes: 1,057 surfaced; 1,168 dived
Dimensions, metres (feet): 59 × 5.4 × 4.6 *(193.6 × 17.7 × 15.1)*
Speed, knots: 11 surfaced; 23 dived
Complement: 21 (5 officers)

Machinery: Diesel-electric; 2 MTU 16V 396 SB83 diesels; 3,620 hp(m) *(2.7 MW)* sustained; 1 Siemens motor; 6,000 hp(m) *(4.41 MW)*; 1 shaft
Torpedoes: 8—21 in *(533 mm)* bow tubes. 14 AEG DM 2A3 Sehecht; dual purpose; wire-guided; active/passive homing to 28 km *(15 n miles)* at 23 kt; 13 km *(7 n miles)* at 35 kt; warhead 260 kg; depth to 460 m.
Electronic countermeasures: ESM: ITT ES-3701; radar intercept.
Radars: Surface search: Kelvin Hughes 1007; I-band.
Sonars: Atlas Elektronik CSU 83; active/passive intercept search and attack; medium frequency. Thomson Sintra; flank array; passive; low frequency.
Weapon control systems: Kongsberg MSI-90(U) TFCS.

Programmes: Contract signed on 30 September 1982. This was a joint West German/Norwegian effort known as Project 210 in Germany. Although final assembly was at Thyssen a number of pressure hull sections were provided by Norway.
Modernisation: MSI-90U being upgraded 2000–2005. A mid-life upgrade of all six boats is in progress 2007–15. The programme includes updates to the sonar and communications systems and a number of platform improvements.

UTHAUG *5/2009, Marco Ghiglino / 1305973*

Structure: Diving depth, 250 m *(820 ft)*. The basic command and weapon control systems are Norwegian, the attack sonar is German but the flank array, based on piezoelectric polymer antenna technology, was developed in France and substantially reduces flow noise. Calzoni Trident modular system of non-penetrating masts has been installed. Zeiss periscopes.

UTHAUG *4/2013*, Tony Roper / 1530945*

FRIGATES

5 FRIDTJOF NANSEN CLASS (FFGHM)

Name	No	Builders	Laid down	Launched	Commissioned
FRIDTJOF NANSEN	F 310	Navantia, Ferrol	9 Apr 2003	3 Jun 2004	5 Apr 2006
ROALD AMUNDSEN	F 311	Navantia, Ferrol	3 Jun 2004	25 May 2005	21 May 2007
OTTO SVERDRUP	F 312	Navantia, Ferrol	25 May 2005	28 Apr 2006	30 Apr 2008
HELGE INGSTAD	F 313	Navantia, Ferrol	28 Apr 2006	23 Nov 2007	29 Sep 2009
THOR HEYERDAHL	F 314	Navantia, Ferrol	23 Nov 2007	11 Feb 2009	18 Jan 2011

Displacement, tonnes: 5,375 full load
Dimensions, metres (feet): 133.2 × 16.8 × 4.9 *(437 × 55.1 × 16.1)*
Speed, knots: 26
Range, n miles: 4,500 at 16 kt
Complement: 120 (50 officers) + 26 spare berths

Machinery: CODAG; 1 GE LM 2500 gas turbine; 26,112 hp *(19.2 MW)*; 2 Bazán Bravo 12V diesels; 12,240 hp(m) *(9 MW)*; 2 shafts; cp props; bow thruster; 1,360 hp(m) *(1 MW)*.
Missiles: SSM: 8 Kongsberg NSM ❶; inertial, GPS and terrain mapping guidance and passive IR homing to 185 km *(100 n miles)* at 0.95 Mach; warhead 120 kg.
SAM: Mk 41 VLS (8 cells) ❷; 32 Evolved Sea Sparrow RIM 162B; semi-active radar homing to 18 km *(9.7 n miles)* at 3.6 Mach; warhead 38 kg.
Guns: 1 Oto Melara 76 mm/62 Super Rapid ❸. 120 rds/min to 15.75 km *(8.5 n miles)* anti-surface; 12 km *(6.5 n miles)* anti-aircraft; weight of shell 6 kg. 4—12.7 mm MGs. Fitted for 1—40 mm/70.
Torpedoes: 4—324 mm (2 double) tubes ❹. BAE Systems Stingray Mod 1; active/passive homing to 11 km *(5.9 n miles)* at 45 kt; warhead 35 kg shaped charge.
Physical countermeasures: Decoys: Terma SKWS chaff, IR. LOKI 130 mm acoustic decoy.
Electronic countermeasures: ESM: Condor CS-3701; intercept ❺.
Radars: Air search: Lockheed Martin SPY-1F ❼; E/F-band.
Air/surface search: Reutech RSR 210N ❾; I-band.
Surface search: Litton; E/I-band ❽.
Fire control: 2 Mk 82 (SPG-62); I/J-band ❿.
Navigation: Litton; I-band ⓫. IFF: Mk XII.
Sonars: Thomson Marconi Spherion MRS 2000 and Mk 2 CAPTAS; combined active/passive towed array.
Combat data systems: AEGIS with ASW and ASuW segments from Kongsberg; Link 11/16.
Weapon control systems: Sagem VIGY 20 optronic director ❻.
Helicopters: 1 NH90 ⓬.

Programmes: Design Definition for a new class of frigates started in March 1997. Izar (later Navantia) and Lockheed Martin selected in March 2000 and contract signed 23 June 2000. Most of the construction was undertaken by Navantia with some modules built in Norway.

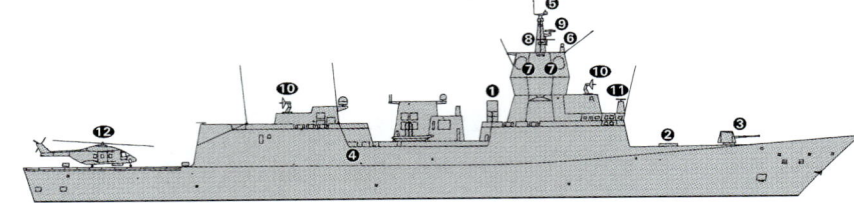

FRIDTJOF NANSEN   (Scale 1 : 1,200), Ian Sturton / 1366679

HELGE INGSTAD   9/2010, Harald Carstens / 1421704

Modernisation: Stingray torpedoes are to be upgraded or replaced in due course. Military SATCOM to be installed by 2014.
Structure: The design is based on the Alvaro de Bazan class. Damage control is automated through the Integrated Platform Management System.

Operational: *Fridtjof Nansen* successfully conducted Combat Systems Ship Qualification Trials with the US Navy in mid-2007 and participated in Operation Atalanta (anti-piracy) in 2009. *Roald Amundsen* conducted a successful firing of NSM missile on 14 October 2012.

OTTO SVERDRUP   5/2011, Frank Findler / 1406507

THOR HEYERDAHL   6/2012, M Declerck / 1486315

HELGE INGSTAD

HELGE INGSTAD

FRIDTJOF NANSEN

SHIPBORNE AIRCRAFT

Numbers/Type: 14 NHIndustries NH90 NFH.
Operational speed: 157 kt *(291 km/h).*
Service ceiling: 13,940 ft *(4,250 m).*
Range: 621 n miles *(1,150 km).*
Role/Weapon systems: A total of 14 aircraft, reconfigurable for ASW and Coast Guard missions have been ordered since 2001. However, as of late 2013, only two had been delivered and other options may be considered. Sensors: Thales ENR surveillance radar, TUS FLASH dipping sonar. Weapons: NSM missiles, Stingray Mod 1 torpedoes.

NH90   6/2008, Norwegian Navy / 1335801

Numbers/Type: 6 Westland Lynx Mk 86.
Operational speed: 125 kt *(232 km/h).*
Service ceiling: 12,500 ft *(3,810 m).*
Range: 320 n miles *(590 km).*
Role/Weapon systems: Operated by Air Force on behalf of the Coast Guard for fishery protection, offshore oil protection and SAR; embarked in CG vessels and shore-based. To be replaced by NH90 in due course. Sensors: Search radar, FLIR may be fitted, ESM. Weapons: Generally unarmed.

LYNX   6/2002, Royal Norwegian Navy / 0572608

LAND-BASED MARITIME AIRCRAFT

Notes: (1) The Air Force has a total of 57 F-16AM/BM Fighting Falcons armed with Penguin 3 ASMs.
(2) The Norwegian All Weather Search and Rescue Helicopter (NaWSarh) programme is for 16 SAR helicopters, with an option for six more, to replace the Sea King Mk 43 force. AgustaWestland AW 101 was selected in December 2013 and delivery is expected to start in 2017.

Numbers/Type: 4 Lockheed P-3C Orion.
Operational speed: 410 kt *(760 km/h).*
Service ceiling: 28,300 ft *(8,625 m).*
Range: 4,000 n miles *(7,410 km).*
Role/Weapon systems: Long-range MR and oceanic surveillance duties in peacetime, with ASW added as a war role. Updated in 1998-99 with new radars and new tactical computers. P-3Ns used by Coast Guard paid off in 1999. Sensors: APS-137(V)5 radar, ASQ-81 MAD, AQS-212 processor and computer, IFF, AAR-36 IR detection; AAR-47 ESM; ALE 47 countermeasures; sonobuoys. Weapons: ASW; 8 BAE Systems Stingray Mod 1 torpedoes, depth bombs or mines. ASV; Penguin NFT Mk 3 ASM.

P-3C   6/2001, A Sharma / 0130100

Numbers/Type: 12 Westland Sea King Mk 43B.
Operational speed: 125 kt *(232 km/h).*
Service ceiling: 10,500 ft *(3,200 m).*
Range: 630 n miles *(1,165 km).*
Role/Weapon systems: SAR, surface search and surveillance helicopter; supplemented by civil helicopters in wartime. Two 43B delivered in May 1996; remainder updated to 43B standard. Sensors: FLIR 2000 and dual Bendix radars RDR 1500 and RDR 1300. Weapons: Generally unarmed.

SEA KING 43B   2001, GKN Westland / 0051448

PATROL FORCES

Notes: There is an inshore patrol craft *Kaholmen* SHV 121 delivered in about 2005.

6 SKJOLD CLASS (PTGMF)

Name	No	Builders	Launched	Commissioned
SKJOLD	P 960	Kvaerner Mandal	22 Sep 1998	26 Apr 2013
STORM	P 961	Umoe Mandal	30 Oct 2006	9 Sep 2010
SKUDD	P 962	Umoe Mandal	3 May 2007	28 Oct 2010
STEIL	P 963	Umoe Mandal	15 Jan 2008	30 Jun 2011
GLIMT	P 964	Umoe Mandal	15 Aug 2008	30 Mar 2012
GNIST	P 965	Umoe Mandal	18 May 2009	7 Nov 2012

Displacement, tonnes: 274 full load
Dimensions, metres (feet): 47.5 × 13.5 × 0.8 *(155.8 × 44.3 × 2.6)*
Speed, knots: 60
Range, n miles: 800 at 40 kt
Complement: 21

Machinery: COGAG: 2 Pratt & Whitney ST40 gas turbines; 10,730 hp *(8.0 MW)*; 2 Pratt and Whitney ST18 gas turbines; 5,365 hp *(4 MW)*; 2 Kamewa waterjets; 2 MTU diesels (for lifting fans); 737 hp *(550 kW).*
Missiles: 8 SSM; 8 Kongsberg NSM; inertial, GPS and terrain matching guidance and passive IR homing to 185 km *(100 n miles)* at 0.95 Mach; warhead 120 kg.
SAM; Mistral; IR homing to 4 km *(2.2 n miles)* at 2.5 Mach; warhead 3 kg.
Guns: 1 Oto Melara 76 mm/62. Super Rapid; 120 rds/min to 16 km *(8.7 n miles)*; weight of shell 6 kg. 2—12.7 mm MGs.
Physical countermeasures: 1 Rheinmetall MASS-1L.
Electronic countermeasures: ESM: EDO RSS CS 3701; intercept.
Radars: Air/surface search: Thales MRR; 3D-NG; G-band.
Navigation: Litton; I-band.
Fire control: CelsiusTech Ceros 200; J-band.
Combat data systems: DCN Senit 2000; Link 11/16.
Electro-optic systems: Sagem VIGX-20 optronic director. Sofresud QPD optical sight.

Programmes: Project SMP 6081. A preproduction version (P 960) ordered 30 August 1996. This was tested by the Norwegian Navy from 1999-2001 and was evaluated by the USN and USCG in 2001-02. The Norwegian parliament decided on 23 October 2003 that five additional vessels were to be built. Contract with Skjold Prime Consortium, comprising Umoe Mandal, Armaris and Kongsberg Defence & Aerospace, was signed 28 November 2003. Ships were built at Umoe Mandal shipyard. P 960 was used for crew training and was upgraded to production standard after the other five vessels had been delivered.
Structure: FRF Sandwich SES hull with advanced stealth technology (RAS/RAM is built into the structure) including anechoic coatings. Building on experience in US trials, a more raked bow has been adopted to improve performance into sea. The foredeck structure is also to be strengthened around the gun mounting. Two quadruple SSM launchers are to be recessed aft of the bridge. These will elevate to fire and then retract.
Operational: First operational firing of NSM missile was successfully conducted from *Glimt* on 10 October 2012.

STEIL   4/2013*, Tony Roper / 1530948

GLIMT   5/2013*, Per Körnefeldt / 1530947

11 COMBATBOAT 90N (LCP)

HELLEN L 4512 (ex-KA 3)	**KOPÅS** L 4525 (ex-KA 16)
TORÅS L 4513 (ex-KA 4)	**TANGEN** L 4526 (ex-KA 17)
MØVIK L 4514 (ex-KA 5)	**ODDANE** L 4527 (ex-KA 18)
SKROLSVIK L 4520 (ex-KA 11)	**SØVIKNES** L 4531 (ex-KA 23)
KRÅKENES L 4521 (ex-KA 12)	**OSTERNES** L 4532 (ex-KA 31)
STANGNES L 4522 (ex-KA 13)	

Displacement, tonnes: 19 full load
Dimensions, metres (feet): 15.9 × 3.8 × 0.8 *(52.2 × 12.5 × 2.6)*
Speed, knots: 40 (est.)
Range, n miles: 240 at 20 kt
Complement: 3

Military lift: 2.8 tons or 20 troops
Machinery: 2 SAAB Scania DSI 14 diesels; 1,104 hp(m) *(812 kW)* or 1,251 hp(m) *(920 kW)* (KA 21–43) sustained; 2 FF 450 water-jets or 2 Kamewa FF 410 (KA 21–43)
Guns: 1 – 12.7 mm MG.
Radars: Navigation: I-band.

Comment: Ordered from Dockstavarvet, Sweden. Four Batch 1 units delivered for trials in July and October 1996. Three more of the class delivered in 1997, 13 in 1998. Four were decommissioned in 2010 and five in early 2013. Used to carry mobile light missile units and prime method of transportation for new Coastal Ranger Commando. Similar in most details to the Swedish Coastal Artillery craft. Names are mostly taken from Coastal Fortresses. Pennant numbers were changed in 2004.

KJØKØY   6/2008, Royal Norwegian Navy / 1335798

5 ALUSAFE 1290 CLASS (INSHORE PATROL CRAFT) (PB)

KAMØY L 4541	**RYPØY** L 4543	**REINØY** L 4546
BONDØY L 4542	**KJELMØY** L 4545	

Displacement, tonnes: 8 standard
Dimensions, metres (feet): 12.9 × 3.5 × 0.75 *(42.3 × 11.5 × 2.5)*
Speed, knots: 42
Complement: 15
Machinery: 2 Volvo Penta TAMD 74 EDC diesels; 900 hp *(670 kW)*; 2 Kamewa K28 waterjets
Guns: 2 – 12.7 mm MGs.

Comment: Aluminium hull. Built by Maritime Partner, Ålesund and delivered in 2002. Designed for used by the Norwegian Naval Home Guard as multifunction assault and patrol vessels by the coastal rangers. The craft are also available as tactical logistics craft and to support police, customs, environmental and fishery authorities. Two were decommissioned in early 2013.

L 4540   6/2008, Richard Scott / 1335808

4 HÅREK (ALUSAFE 1300) CLASS
(INSHORE PATROL CRAFT) (PB)

HÅREK SHV 101	**KVITSØY** SHV 104	**SLOTTERØY** SHV 105	**HALTEN** SHV 106

Displacement, tonnes: 10 standard
Dimensions, metres (feet): 13.3 × 3.65 × 0.75 *(43.6 × 12.0 × 2.5)*
Speed, knots: 40
Complement: 15
Machinery: 2 Volvo Penta TAMD 74EDC diesels; 900 hp *(670 kW)*; 2 Kamewa K28 waterjets
Guns: 2 – 12.7 mm MGs.

Comment: Aluminium hull. Built by Maritime Partner, Ålesund and delivered in 2003. Based at Stavanger, Bergen and Trondheim. Designed for use by the Norwegian Naval Home Guard as multifunction patrol vessels. The craft are also available to support police, customs, environmental and fishery authorities.

KVITSØY   6/2008, Richard Scott / 1335807

2 GYDA CLASS (INSHORE PATROL CRAFT) (PB)

HVASSER SHV 102	**HEKKINGEN** SHV 103

Displacement, tonnes: 14 full load
Dimensions, metres (feet): 13.65 × 4.1 × 0.9 *(44.8 × 13.5 × 3.0)*
Speed, knots: 42
Complement: 15
Machinery: 2 Volvo Penta TAMD 74 EDC diesels; 1,750 hp *(1.3 MW)*; 2 Kamewa K32 waterjets
Guns: 2 – 12.7 mm MGs.

Comment: Aluminium hull. Built by Henriksen Mekaniske Verksted, Tønsberg and delivered in 2003. Designed for use by the Norwegian Naval Home Guard as patrol vessels.

HVASSER   6/2006, E & M Laursen / 1040662

2 REINE CLASS (PBO)

Name	No	Builders	Commissioned
OLAV TRYGGVASON	P 380	Gryfia Shipyard, Szczecin	17 Dec 2010
MAGNUS LAGABØTE	W 335 (ex-P 381)	Gryfia Shipyard, Szczecin	Feb 2011

Displacement, tonnes: 760 full load
Dimensions, metres (feet): 49.6 × 10.3 × 3.3 *(162.7 × 33.8 × 10.8)*
Speed, knots: 16
Complement: 20
Machinery: Diesel-electric; 2 Cummins KTA 38M2 diesels; 3,042 hp *(2.24 MW)*; 2 azimuth thrusters

Comment: Contract signed in February 2008 with Remøy Management and Remøy Shipping for the construction and rental of two variants of the Coast Guard's Nornen-class patrol vessels. The contract is to extend for 15 years. The slightly longer (2.4 m) ships have the capacity to embark multipurpose craft on the aft deck, for which a crane is fitted. Operated by the Home Guard, the ships act as command and control and logistic support vessels. There is space for containerised cargo on deck. *Magnus Lagobøte* transferred to the Coast Guard on 1 January 2013.

OLAV TRYGGVASON   7/2010, L-G Nilsson / 1366788

0 + 2 IC 20 PATROL CRAFT (PB)

Displacement, tonnes: 36 full load
Dimensions, metres (feet): 20 × 4.6 × 1.1 *(65.6 × 15.1 × 3.6)*
Speed, knots: 34
Range, n miles: 500 at 25 kt
Complement: 4
Machinery: 2 MTU 183 TE92 diesels; 1,830 hp(m) *(1.35 MW)* sustained; 2 MTP 7500S or Kamewa waterjets
Radars: Navigation: 2 Kelvin Hughes 6000; I-band.

Comment: Two craft (with an option for a third) ordered from Dockstavarvet, Sweden, in December 2013. The craft are to be for SAR and general patrol duties.

MINE WARFARE FORCES

6 OKSØY/ALTA CLASS
(MINEHUNTERS/SWEEPERS) (MHCM/MSCM)

Name	No	Builders	Commissioned
KARMØY	M 341	Kvaerner Mandal	24 Oct 1994
MÅLØY	M 342	Kvaerner Mandal	24 Mar 1995
HINNØY	M 343	Kvaerner Mandal	8 Sep 1995
ALTA	M 350	Kvaerner Mandal	12 Jan 1996
OTRA	M 351	Kvaerner Mandal	8 Nov 1996
RAUMA	M 352	Kvaerner Mandal	2 Dec 1996

Displacement, tonnes: 395 full load
Dimensions, metres (feet): 55.2 × 13.6 × 0.84 *(181.1 × 44.6 × 2.8)*
Speed, knots: 20. **Range, n miles:** 1,500 at 20 kt
Complement: 37 (M 341–343), 32 (M 350–352) (13 officers (M 341–343); 10 officers (M 350–352))
Machinery: 2 MTU 12V 396 TE84 diesels; 3,700 hp(m) *(2.72 MW)* sustained; 2 Kvaerner Eureka water-jets; 2 MTU 8V 396 TE54 diesels; 1,740 hp(m) *(1.28 MW/60 Hz)* sustained; lift engines
Missiles: SAM: Matra Simbad twin launcher; Mistral; IR homing to 4 km *(2.2 n miles)*; warhead 3 kg.
Guns: 3—12.7 mm MGs.
Physical countermeasures: MCMV: 2 Pluto submersibles (minehunter); mechanical, AGATE (air gun and transducer equipment) acoustic and Elma magnetic sweep (minesweepers). Minesweeper mini torpedoes can be carried.
Radars: Navigation: Racal Decca; I-band. Sperry Bridgemaster; I-band.
Sonars: Thales TSM 2022 Mk 3; hull mounted; high frequency.

Programmes: Order placed with Kvaerner on 9 November 1989.
Modernisation: Both minehunters and minesweepers are being upgraded with new sonars (TSM 2022), new KDA tactical C2 system and new KM dynamic positioning system. The Kongsberg Simrad Hugin 1000 MR is to be installed in all minehunters. The AUV is to be used for mine reconnaissance and rapid environmental assessment.
Structure: Design developed by the Navy in Bergen with the Defence Research Institute and Norsk Veritas and uses an air cushion created by the surface effect between two hulls. The hull is built of Fibre Reinforced Plastics (FRP) in sandwich configuration. The ROVs are carried in a large hangar and are launched by either of the two hydraulic cranes. The minesweeper has an A frame aft for the sweep gear. SAM launcher mounted forward of the bridge.
Operational: SkyDec MIL-GPS with Controlled Reception Pattern Antenna (CRPA) and FUGRO Satellite Based Augmentation System (SBAS). The catamaran design is claimed to give higher transit speeds with lesser installed power than a traditional hull design. Other advantages are lower magnetic and acoustic signatures, clearer water for sonar operations and less susceptibility to shock. *Orkla* M 353 was lost after a catastrophic fire on 19 November 2002. M 354 was decommissioned in 2004 and M 340 in 2005.

HINNOY   6/2013*, Michael Nitz / 1530949

MÅLØY   9/2011, L-G Nilsson / 1486318

ALTA   6/2012, Michael Nitz / 1486319

SURVEY AND RESEARCH SHIPS

0 + 1 RESEARCH SHIP (AGE)

Measurement, tonnes: 5,000 gt
Dimensions, metres (feet): 125 × 23.0 × ? *(410.1 × 75.5 × ?)*
Machinery: 2 diesels; 2 shafts
Radars: To be announced.
Helicopters: To be announced.

Comment: It was announced on 30 November 2011 that STX OSV (changed name to Vard in 2013) had been awarded the contract for the construction of a research vessel for the Norwegian Defence Logistics Organisation. The ship is to be built at Langsten and delivery is planned for late 2014.

1 RESEARCH SHIP (AGEH)

Name	Builders	Launched	Commissioned
MARJATA	Tangen Verft, Krager	18 Dec 1992	Jul 1994

Displacement, tonnes: 7,681 full load
Dimensions, metres (feet): 81.5 × 39.9 × 6 *(267.4 × 130.9 × 19.7)*
Speed, knots: 15
Complement: 14 + 31 scientists
Machinery: Diesel-electric; 2 MTU Siemens 16V 396 TE diesels; 7,072 hp(m) *(5.2 MW)*; 2 Dresser Rand/Siemens gas-turbine generators; 9,792 hp(m) *(7.2 MW)*; 2 Siemens motors; 8,160 hp(m) *(6 MW)*; 2 Schottel 3030 thrusters. 1 Siemens motor; 2,720 hp(m); *(2 MW)*; 1 Schottel thruster (forward)
Helicopters: Platform for one medium

Comment: Ordered in February 1992 from Langsten Slip og Batbyggeri to replace the old ship of the same name. Called Project Minerva. Design developed by Ariel A/S, Horten. The three superstructure-mounted cupolas contain ELINT and SIGINT equipment. Hull-reinforced to allow operations in fringe ice. Equipment includes Sperry radars, Elac sonars, Siemens TV surveillance, and a fully equipped helicopter flight deck. The unconventional hull which gives the ship an extraordinary length to beam ratio of 2:1 is said to give great stability and dynamic qualities. To be replaced in 2015.

MARJATA   6/2000, Royal Norwegian Navy / 0105173

5 SURVEY SHIPS (AGS)

OLJEVERN 01 1978	OLJEVERN 03 1978	GEOFJORD 1958
OLJEVERN 02 1978	OLJEVERN 04 1978	

Comment: Under control of Ministry of Environment based at Stavanger. *Oljevern 01* and *03* have red hulls and work for the Pollution Control Authority.

GEOFJORD   5/2002, L-G Nilsson / 0528972

TRAINING SHIPS

2 KVARNEN CLASS (TRAINING VESSELS) (AXL)

KVARNEN P 361   NORDNES P 362

Displacement, tonnes: 25 full load
Dimensions, metres (feet): 14.99 × 4.55 × 0.8 *(49.2 × 14.9 × 2.6)*
Speed, knots: 40
Complement: 15 (15 trainees)
Machinery: 2 Caterpillar C18 diesels; 1,746 hp *(1.3 MW)*; 2 Rolls-Royce FF410S waterjets

Comment: Both craft built to a Watercat 1500 design by Marine Alutech, Finland. Christened on 27 August 2010, they were both commissioned in August 2011. Their primary role is to provide navigation training for cadets at the Royal Norwegian Naval Academy but they are also to be available as logistics vessels during exercises and operations.

NORDNES   9/2010, Royal Norwegian Navy / 1366789

AUXILIARIES

1 SUPPLY AND RESCUE VESSEL

Name	No	Builders	Commissioned
VALKYRIEN	A 535	Ulstein Hatlo	1981

Displacement, tonnes: 3,048 full load
Dimensions, metres (feet): 68 × 14.5 × 5 *(223.1 × 47.6 × 16.4)*
Speed, knots: 16
Complement: 13

Machinery: Diesel-electric; 4 diesels; 10,560 hp(m) *(7.76 MW)* sustained; 2 motors; 3.14 MW; 2 shafts; 2 bow thrusters; 1,600 hp(m) *(1.18 MW)*; 1 stern thruster; 800 hp(m) *(588 kW)*
Missiles: SAM: Mistral; IR homing to 4 km *(2.2 n miles)* at 2.5 Mach; warhead 3 kg.
Guns: 3 — 12.7 mm MGs.
Radars: Navigation: 2 Furuno; H/I-band.

Comment: Tug/supply ship acquired in 1994 for supply and SAR duties. Bollard pull 128 tons. Can carry a 700 ton deck load. Oil recovery equipment is also carried.

VALKYRIEN   6/2012, Michael Nitz / 1486320

5 COASTAL VESSELS (YPT/YDT)

Notes: Due to re-organisation of the coastal vessels, the naval districts no longer operate many of the vessels previously assigned. The following remain in service and are prefaced by two letters as follows: HM (multirole), HS (tugs), HD (diving), HP (personnel). Hitra (HP 15) is also used for training cruises. All are less than 300 tons displacement.

Name	No	Speed, knots	Commissioned	Role
VIKEN	HD 2	12	1984	Cargo (4 tons)/Passengers (40) Diving vessel
KJEØY	HM 7	10	1993	Training ship/Passengers (30)
HITRA	HP 15	—	—	Passengers (30)
SLEIPNER	HS 4	11	2002	Tug/Cargo (10 tons)
MJØLNER	HS 5	11	2002	Tug/Cargo (10 tons)

SLEIPNER   5/2008, Marco Ghiglino / 1335805

HITRA   5/2008, Marco Ghiglino / 1305970

0 + 1 LOGISTICS SUPPORT SHIP (AFS)

Name	No	Builders	Laid down	Launched	Commissioned
—	—	Daewoo, Okpo	2015	2016	2016

Displacement, tonnes: 26,200 full load
Dimensions, metres (feet): 180.7 × 25.9 × 8.6 *(592.8 × 85.0 × 28.2)*
Speed, knots: 18
Complement: 43 + 73 spare berths
Machinery: 2 Wärtsilä 6L46F diesels; 20,115 hp *(15.0 MW)*; 2 shafts
Guns: To be announced.
Radars: To be announced.

Comment: The contract for the construction of a new Logistics Support Vessel (LSV) was awarded to Daewoo Shipbuilding and Marine Engineering (DSME) on 28 June 2013. Based on the Aegir LSV design from BMT Defence Services, the ship is similar to but smaller than the design of the UK's Tide class. The ship is to have the capacity to carry 7,000 tonnes of diesel fuel, 300 tonnes of helicopter fuel, 650 tonnes of other cargo and 40 TEU containers. Replenishment facilities include a light and heavy jackstay and an astern refuelling capability. There is to be a hangar for two NH90 helicopters and the ship is also to have Role 1 hospital facilities.

© 2014 IHS

1 SUPPORT SHIP (AGDS)

Name	No	Builders	Commissioned
TYR (ex-*Standby Master*)	N 50	Mekaniske Verksted, Alesund	1981

Displacement, tonnes: 503 full load
Dimensions, metres (feet): 42.3 × 10.1 × 3.5 *(138.8 × 33.1 × 11.5)*
Speed, knots: 12
Complement: 22 (7 officers)
Machinery: 2 Deutz SBA12M816 diesels; 1,300 hp(m) *(956 kW)*; 1 shaft; cp prop; 1 MWM diesel; 150 hp(m) *(110 kW)*; bow and stern thrusters
Mines: 2 rails.
Radars: Navigation: Furuno 711 and Furuno 1011; I-band.

Comment: Former oil rig pollution control ship. Acquired in December 1993 and converted by Mjellum & Karlsen, Bergen. Recommissioned 7 March 1995 as a minelayer, and for the maintenance of controlled minefields but principal current task is to support underwater operations. Carries a ROV.

TYR   6/2008, Maritime Photographic / 1335804

ROYAL YACHTS

1 ROYAL YACHT (YAC)

Name	No	Builders	Commissioned
NORGE (ex-*Philante*)	A 533	Camper & Nicholson's Ltd, Southampton	1937

Displacement, tonnes: 1,815 full load
Dimensions, metres (feet): 80.2 × 11.6 × 4.6 *(263.1 × 38.1 × 15.1)*
Speed, knots: 17
Complement: 50 (18 officers)
Machinery: 2 Bergen KRMB-8 diesels; 4,850 hp(m) *(3.6 MW)* sustained; 2 shafts; bow thruster
Radars: Navigation: 2 Decca; I-band.

Comment: Built to the order of the late T O M Sopwith as an escort and store vessel for the yachts *Endeavour I* and *Endeavour II*. Launched on 17 February 1937. Served in the Royal Navy as an anti-submarine escort during the Second World War, after which she was purchased by the Norwegian people for King Haakon and reconditioned as a Royal Yacht at Southampton. Can accommodate about 50 people in addition to crew. Repaired after serious fire on 7 March 1985 when the ship was fitted with a bow-thruster.

NORGE   6/2009, Michael Winter / 1366138

COAST GUARD (KYSTVAKT)

1 PATROL VESSEL (WPSO)

Name	No	Builders	Commissioned
ALESUND	W 312	Myklebust Verft AS, Gursken	20 Apr 1996

Displacement, tonnes: 1,379 full load
Dimensions, metres (feet): 63 × 11.5 × 5.2 *(206.7 × 37.7 × 17.1)*
Speed, knots: 16
Complement: 26
Machinery: 1 Wärtsilä Wichmann 8V28B diesel; 3,589 hp *(2.64 MW)*; 1 shaft; cp prop; 2 thrusters
Guns: 1 — 40 mm.

Comment: On charter from Remøy Shipping until 2016. Used for miscellaneous tasks including patrol duties, oil recovery tasks, firefighting, search and rescue, towing contracts and fisheries inspection.

ALESUND   5/2009, Marco Ghiglino / 1305968

IHS Jane's Fighting Ships 2014-2015

600 Pakistan > Submarines

SUBMARINES

Notes: (1) It was announced in February 2014 that a contract to procure up to six submarines from China may be signed in 2014. The boats are expected to be of the Yuan class, modified to meet Pakistani requirements, and to feature air-independent propulsion.
(2) Reports that Pakistan may have developed, or be in the process of developing, a sea-based nuclear strike capability have not been confirmed. Although one possibility would be a submarine-launched variant of the Babur/Hatf-7 cruise missile, with a reported range of 700 km *(430 n miles)*, the operational deployment of such a missile would pose a significant technological challenge.

3 KHALID (AGOSTA 90B) CLASS (SSK)

Name	No	Builders	Laid down	Launched	Commissioned
KHALID	S 137	DCN, Cherbourg	15 Jul 1995	18 Dec 1998	6 Sep 1999
SAAD	S 138	DCN, Cherbourg; PN Dockyard	2 Dec 1999	24 Aug 2002	12 Dec 2003
HAMZA	S 139	Karachi Shipyard and Engineering Works	2000	10 Aug 2006	26 Sep 2008

Displacement, tonnes: 1,534 surfaced; 1,788 (S 137–138), 2,012 (S 139) dived
Dimensions, metres (feet): 67.2 (S 137–138), 76.2 (S 139) × 6.8 × 5.4 *(221.8, 250 × 22.3 × 17.7)*
Speed, knots: 12 surfaced; 20 dived
Range, n miles: 8,500 at 9 kt snorting; 350 at 3.5 kt dived
Complement: 36 (7 officers)

Machinery: Diesel-electric; 2 SEMT-Pielstick 16 PA4 V 185 VG diesels; 3,600 hp(m) *(2.65 MW)*; 2 Jeumont Schneider alternators; 1.7 MW; 1 Jeumont motor; 2,992 hp(m) *(2.2 MW)*; 1 cruising motor; 32 hp(m) *(23 kW)*; 1 shaft
Missiles: SSM: 4 Aerospatiale Exocet SM 39; inertial cruise; active radar homing to 50 km *(27 n miles)* at 0.9 Mach; warhead 165 kg.

Torpedoes: 4—21 in *(533 mm)* bow tubes. 16 ECAN F17P Mod 2; wire-guided; active/passive homing to 20 km *(10.8 n miles)* at 40 kt; warhead 250 kg. Total of 20 weapons.
Mines: Stonefish.
Electronic countermeasures: ESM: Thomson-CSF DR-3000U; intercept.
Radars: Surface search: KH 1007; I-band.
Sonars: Thomson Sintra TSM 2233 suite; bow cylindrical, passive ranging and intercept, and clip-on towed arrays.
Weapon control systems: Thomson Sintra SUBTICS Mk 2.

Programmes: A provisional order for a second batch of three more Agostas was reported in September 1992 and this was confirmed on 21 September 1994. First one built in France. Parts for S 138 sent to Pakistan in April 1998 and for S 139 in September 1998.

Structure: The last of the class has a 200 kW MESMA liquid oxygen AIP system, thereby extending the hull by 8.6 m. The MESMA AIP system has a power output of 200 kW which quadruples dived performance at 4 kt. The MESMA system is to be retrofitted in S 137 and S 138 during their next major refits. Work on S 137 has begun and was expected to be completed in 2013. Work on S 138 is being undertaken 2012–14. Hulls also have much improved acoustic quietening and a full integrated sonar suite including flank, intercept and towed arrays. SOPOLEM J 95 search and STS 95 attack periscopes. Sagem integrated navigation system. HLES 80 steel. Diving depth of 320 m *(1,050 ft)*.
Operational: *Khalid* completed 29 April 1999 and sailed for Pakistan in November 1999. Assigned to 5th Submarine Squadron.

HAMZA

3/2010, Guy Toremans / 1406106

SAAD

9/2003, DCN / 0562934

2 HASHMAT (AGOSTA 70) CLASS (SSK)

Name	No
HASHMAT (ex-*Astrant*)	S 135
HURMAT (ex-*Adventurous*)	S 136

Builders	Laid down	Launched	Commissioned
Dubigeon Normandie, Nantes	15 Sep 1976	14 Dec 1977	17 Feb 1979
Dubigeon Normandie, Nantes	18 Sep 1977	1 Dec 1978	18 Feb 1980

Displacement, tonnes: 1,514 surfaced; 1,768 dived
Dimensions, metres (feet): 67.6 × 6.8 × 5.4 *(221.8 × 22.3 × 17.7)*
Speed, knots: 12 surfaced; 20 dived
Range, n miles: 8,500 at 9 kt snorting; 350 at 3.5 kt dived
Complement: 59 (8 officers)

Machinery: Diesel-electric; 2 SEMT-Pielstick 16 PA4 V 185 VG diesels; 3,600 hp(m) *(2.65 MW)*; 2 Jeumont Schneider alternators; 1.7 MW; 1 motor; 4,600 hp(m) *(3.4 MW)*; 1 cruising motor; 32 hp(m) *(23 kW)*; 1 shaft
Missiles: SSM: McDonnell Douglas Sub Harpoon; active radar homing to 124 km *(67 n miles)* at 0.9 Mach; warhead 227 kg.
Torpedoes: 4—21 in *(533 mm)* bow tubes. ECAN F17P; wire-guided; active/passive homing to 20 km *(10.8 n miles)* at 40 kt; warhead 250 kg; water ram discharge gear. E14, E15 and L3 torpedoes are also available. Total of 20 torpedoes and missiles.
Mines: Stonefish.
Electronic countermeasures: ESM: DR-3000; intercept and warning.
Radars: Surface search: Thomson-CSF DRUA 33; I-band.
Sonars: Thomson Sintra TSM 2233D; passive search; medium frequency. Thomson Sintra DUUA 2B; active/passive search and attack; 8 kHz active. Thomson Sintra TSM 2933D towed array; passive; very low frequency.

Programmes: Purchased from France in mid-1978 after United Nations' ban on arms sales to South Africa. *Hashmat* arrived Karachi 31 October 1979, *Hurmat* arrived 11 August 1980.
Structure: Diving depth, 300 m *(985 ft)*. Both were modified to fire Harpoon in 1985 but may have had to acquire the missiles through a third party.
Operational: Assigned to 5th Submarine Squadron.

HURMAT 3/2008, Michael Nitz / 1305311

3 MIDGET SUBMARINES (SSW)

X 01–03

Displacement, tonnes: 120 dived
Dimensions, metres (feet): 27.8 × 5.6 × ? *(91.2 × 18.4 × ?)*
Speed, knots: 7 dived. **Range, n miles:** 2,200 surfaced; 60 dived
Complement: 16 (8 divers)
Torpedoes: 2—21 in *(533 mm)* tubes; 2 ALCATEL E 14/E 15; active homing to 12 km *(6.5 n miles)* at 25 kt; passive homing to 28 km *(15 n miles)* at 23 kt; warhead 300 kg plus either two short range active/passive homing torpedoes or two SDVs.
Mines: 12 Mk 414 Limpet type.
Sonars: Hull mounted; active/passive; high frequency.

Comment: MG 110 type built in Pakistan under supervision by Cosmos. These are enlarged SX 756 of Italian Cosmos design. Diving depth of 150 m and can carry eight swimmers with 2 tons of explosives as well as two CF2 FX 60 SDVs (swimmer delivery vehicles). Pilkington Optronics CK 39 periscopes. Reported as having a range of 1,000 n miles and an endurance of 20 days. All have been upgraded since 1995 with improved sensors and weapons. However, reports that *X 01* has been equipped with Harpoon are not considered likely. All are active.

X 03 5/2003 / 0569226

FRIGATES

Notes: Procurement of up to four further frigates from China is reported to be under consideration. Further Sword class (possibly two ships) is the most likely option but larger ships such as the Jiangkai I or II classes may also be under consideration.

4 + (2) SWORD (F-22P) CLASS (FFGH)

Name	No	Builders	Laid down	Launched	Commissioned
ZULFIQUAR	251	Hudong-Zhonghua Shipyard, Shanghai	12 Oct 2006	7 Apr 2008	30 Jul 2009
SHAMSHEER	252	Hudong-Zhonghua Shipyard, Shanghai	13 Jul 2007	31 Oct 2008	10 Feb 2010
SAIF	253	Hudong-Zhonghua Shipyard, Shanghai	4 Nov 2008	28 May 2009	11 Nov 2010
ASLAT	254	Karachi Shipyard, Karachi	10 Dec 2009	17 Jun 2011	17 Apr 2013

Displacement, tonnes: 2,980 full load
Dimensions, metres (feet): 123 × 13.2 × 3.8 *(403.5 × 43.3 × 12.5)*
Speed, knots: 29
Range, n miles: 4,000 at 18 kt
Complement: 202 (14 officers)

Machinery: CODAD: 4 SEMT Pielstick 16PA6STC diesels; 28,200 hp *(20.7 MW)*; 2 shafts
Missiles: SSM: 8 C-802 (YJ-83/CSS-N-8 Saccade) ❶; mid-course guidance and active radar homing to 120 km *(65 n miles)* at 0.9 Mach; warhead 165 kg; sea skimmer.
SAM: 1 HQ-7 (FM-90N) octuple launcher CSA-N-4 ❷; line of sight guidance to 13 km *(7 n miles)* at 2.4 Mach; warhead 14 kg.
Guns: 1 – 3 in *(76 mm)* AK 176M ❸. 2 – 30 mm Type 730B ❹; 7 barrels per mounting; 4,200 rds/min combined to 1.5 km.
Torpedoes: 6 – 324 mm (2 triple) B 515 ILAS tubes; ET52; anti-submarine; active/passive homing to 7 km *(3.8 n miles)* at 33 kt; warhead 34 kg.
A/S Mortars: 2 Type 87 6-tubed launchers.
Physical countermeasures: Decoys: To be announced.
Electronic countermeasures: ESM: RWD-8.
ECM: NJ81-3 (Type 981).

Radars: Air search: Type 517 Knife Rest ❺; A-band.
Air/surface search: Type 363 Seagull S ❻; E/F-band.
Surface search: Unknown ❼; G-band.
Fire control: Type 343G ❽; I-band (for SSM and 76 mm gun). Type 347G(2) ❾; I-band for Type 730. Type 345 ❿; I/J-band (for SAM).
Navigation: Kelvin Hughes 1007; I-band.
Sonars: Echo Type 5; hull-mounted; active search and attack; medium frequency.
Combat data systems: ZKJ-3C.
Weapon control systems: Optronic director to be announced.

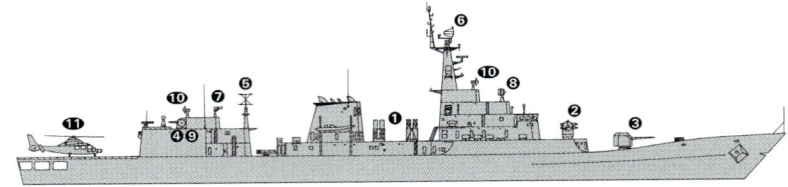

ZULFIQUAR

(Scale 1 : 1,200), Ian Sturton / 1454766

Helicopters: 1 Harbin Zhi-9EC Haitun ⓫.

Programmes: A contract to procure four frigates from China was signed on 4 April 2005. The ships, three of which have been built in Shanghai and the fourth at Karachi, look to be based on the Type 051 Jiangwei II class in service in the PLA(N). Technology transfer was a key element of the deal and the contract included the upgrade of KSEW Shipyard, training and technical assistance. Steel was first cut for the KSEW vessel on 6 March 2009. Two further ships may be built at Karachi.

SAIF

12/2011, Chris Sattler / 1454771

ASLAT

6/2013, Pakistan Navy / 1530955*

SAIF

6/2013, Pakistan Navy / 1530959*

Frigates < **Pakistan** 603

ZULFIQUAR   6/2013*, Pakistan Navy / 1530957

ASLAT   11/2013*, Shaun Jones / 1530958

SHAMSHEER   6/2013*, Pakistan Navy / 1530956

Pakistan > Frigates

5 TARIQ (AMAZON) CLASS (TYPE 21) (FFHM/FFGH)

Name	No	Builders	Laid down	Launched	Commissioned	Recommissioned
TARIQ (ex-Ambuscade)	D 181 (ex-F 172)	Yarrow Shipbuilders, Glasgow	1 Sep 1971	18 Jan 1973	5 Sep 1975	28 Jul 1993
BABUR (ex-Amazon)	D 182 (ex-F 169)	Vosper Thornycroft, Woolston	6 Nov 1969	26 Apr 1971	11 May 1974	30 Sep 1993
KHAIBAR (ex-Arrow)	D 183 (ex-F 173)	Yarrow Shipbuilders, Glasgow	28 Sep 1972	5 Feb 1974	29 Jul 1976	1 Mar 1994
TIPPU SULTAN (ex-Avenger)	D 185 (ex-F 185)	Yarrow Shipbuilders, Glasgow	30 Oct 1974	20 Nov 1975	19 Jul 1978	23 Sep 1994
SHAHJAHAN (ex-Active)	D 186 (ex-F 171)	Vosper Thornycroft, Woolston	23 Jul 1971	23 Nov 1972	17 Jun 1977	23 Sep 1994

Displacement, tonnes: 3,150 standard; 3,759 full load
Dimensions, metres (feet): 117 oa; 109.7 wl × 12.7 × 5.9 *(383.9; 359.9 × 41.7 × 19.4)*
Speed, knots: 30; 18 Tynes
Range, n miles: 4,000 at 17 kt, 1,200 at 30 kt
Complement: 221 (23 officers)

Machinery: COGOG; 2 RR Olympus TM3B gas turbines; 50,000 hp *(37.3 MW)* sustained; 2 RR Tyne RM1C gas turbines (cruising); 9,900 hp *(7.4 MW)* sustained; 2 shafts; cp props
Missiles: SSM: 4 (or 8) McDonnell Douglas Harpoon Block II (D 182, 186) ❶; active radar homing to 124 km *(67 n miles)* at 0.9 Mach; warhead 227 kg.
SAM: China LY 60N sextuple launchers ❷ (D 181, 183, 185); semi-active radar homing to 13 km *(7 n miles)* at 2.5 Mach; warhead 33 kg.
Guns: 1 Vickers 4.5 in *(114 mm)*/55 Mk 8 ❸; 25 rds/min to 22 km *(11.9 n miles)* anti-surface; 6 km *(3.3 n miles)* anti-aircraft; weight of shell 21 kg. Raytheon 20 mm Vulcan Phalanx Mk 15 Block 0 ❹; 3,000 rds/min to 1.5 km (D 181, 182, 183, 186). 2 MSI DS 30B 30 mm/75 ❺ (D 185). 4—12.7 mm MGs.
Torpedoes: 6—324 mm Plessey STWS Mk 2 (2 triple) tubes ❼ (D 186); others fitted with 2 Bofors Type 43X2 single launchers for Swedish Type 45 torpedoes.
Physical countermeasures: Decoys: Graseby Type 182; towed torpedo decoy. 2 Rheinmetall MASS launchers ❽ Mk 36 SRBOC ❾ (D 181, 182).
Electronic countermeasures: ESM: Thomson-CSF DR 3000S; intercept.
Radars: Air/surface search: Marconi Type 992R ⓫; E/F-band (D 182, 186). Signaal DA08 ⓬; F-band (D 181, 183, 185).
Surface search: Kelvin Hughes Type 1007 ⓭ or Type 1006 (D 186); I-band.
Fire control: 1 Selenia Type 912 (RTN 10X) ⓮; I/J-band (D 182, 186). 1 China LL-1 ⓯ (for LY 60N); I/J-band (D 185, 181, 183).
Sonars: Graseby Type 184P; hull-mounted; active search and attack; medium frequency. Kelvin Hughes Type 162M; hull-mounted; bottom classification; 50 kHz.
Combat data systems: CAAIS combat data system with Ferranti FM 1600B computers (D 186). CelsiusTech 9LV Mk 3 including Link Y (in remainder).
Weapon control systems: Ferranti WSA-4 digital fire-control system. CSEE Najir Mk 2 optronic director ❿ (D 182, 185, 186).

Helicopters: 1 Alouette III ⓰.

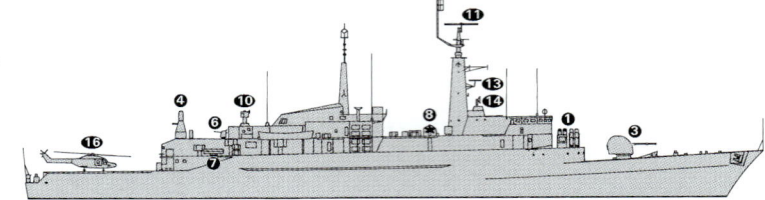

SHAHJAHAN *(Scale 1 : 1,200), Ian Sturton / 0114784*

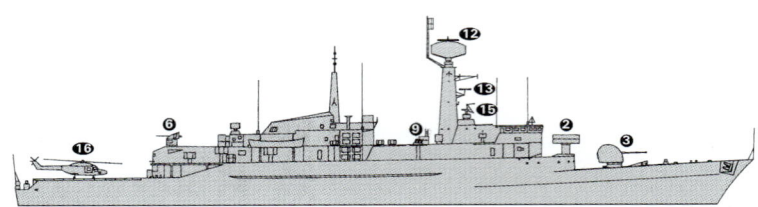

TIPPU SULTAN *(Scale 1 : 1,200), Ian Sturton / 1133556*

Programmes: Acquired from the UK in 1993–94. *Tariq* arrived in Karachi 1 November 1993 and the last pair in January 1995. These ships replaced the Garcia and Brooke classes and have been classified as destroyers.
Modernisation: Exocet, torpedo tubes and Lynx helicopter facilities were all added in RN service, but torpedo tubes were subsequently removed in all but *Badr* and *Shahjahan* and all retrofitted by Pakistan using Swedish equipment. Exocet was not transferred and the obsolete Seacat SAM system was replaced by Phalanx taken from the Gearings. Chinese LY 60N, which is a copy of Aspide, has been fitted in three of the class, Harpoon in three others. New EW equipment has been installed. There are still plans to update the hull sonars but there is no evidence that ATAS sonar has been fitted in D 183 and D 185. Other equipment upgrades include a DA08 search radar in three of the class, an optronic director, new 30 mm and 20 mm guns, SRBOC chaff launchers. An improved combat data system with a datalink to shore HQ is also fitted in four of the class.
Structure: Due to cracking in the upper deck structure large strengthening pieces have been fixed to the ships' side at the top of the steel hull as shown in the illustration. The addition of permanent ballast to improve stability has increased displacement by about 350 tons. Further hull modifications to reduce noise and vibration started in 1988 and completed in all of the class by 1992.
Operational: Form 25th Destroyer Squadron. *Badr* was decommissioned in 2013.

TARIQ *8/2013*, Shaun Jones / 1530953*

SHAHJAHAN *10/2013*, Shaun Jones / 1530954*

Frigates — Shipborne aircraft < **Pakistan** 605

1 OLIVER HAZARD PERRY CLASS

Name	No	Builders	Laid down	Launched	Commissioned
ALAMGIR (ex-*McInerney*)	260 (ex-FFG 8)	Bath Iron Works, Maine	7 Nov 1977	4 Nov 1978	15 Dec 1979

Displacement, tonnes: 2,794 standard; 4,166 full load
Dimensions, metres (feet): 138.1 × 13.7 × 4.5 hull; 7.5 sonar *(453.1 × 44.9 × 14.8; 24.6)*
Speed, knots: 29
Range, n miles: 4,500 at 20 kt
Complement: 245 (17 officers; 19 air crew)

Machinery: 2 GE LM 2500 gas turbines; 41,000 hp *(30.59 MW)* sustained; 1 shaft; cp prop 2 auxiliary retractable props; 650 hp *(484 kW)*
Guns: 1 Oto Melara 3 in *(76 mm)*/62 Mk 75 ❶; 85 rds/min to 16 km *(8.7 n miles)* anti-surface; 12 km *(6.6 n miles)* anti-aircraft; weight of shell 6 kg. 1 Raytheon 20 mm/76 6-barrelled Mk 15 Block 0 Vulcan Phalanx ❷; 3,000 rds/min combined to 1.5 km. 2 Boeing 25 mm Mk 38 guns can be fitted amidships. 4—12.7 mm MGs.
Torpedoes: 6—324 mm Mk 32 (2 triple) tubes ❸. 24 Honeywell Mk 46 Mod 5; anti-submarine; active/passive homing to 11 km *(5.9 n miles)* at 40 kt; warhead 44 kg or Alliant/Westinghouse Mk 50; active/passive homing to 15 km *(8.1 n miles)* at 50 kt; warhead 45 kg shaped charge.
Physical countermeasures: Decoys: 2 Loral Hycor SRBOC 6-barrelled fixed Mk 36 ❹; IR flares and chaff to 4 km *(2.2 n miles)*. Mk 34 launcher for Mk 53 Nulka decoys. T-Mk 6 Fanfare/SLQ-25 Nixie; torpedo decoy.
Radars: Air search: Raytheon SPS-49(V)4 ❺; C/D-band.
Surface search: Kelvin Hughes Sharpeye; E/F-band.
Fire control: Sperry Mk 92 (Signaal WM28) ❻; I/J-band.
Navigation: Furuno; I-band.
Tacan: URN 25. IFF Mk XII.
Sonars: Raytheon SQS 56; hull-mounted active search and attack; medium frequency. Gould SQR 19; passive towed array; very low frequency.
Combat data systems: NTDS, WSC-3 (UHF).
Weapon control systems: Mk 92 Mod 2.

Helicopters: To be announced.

Programmes: Transfer and re-commissioning of ex-FFG 8 was completed on 31 August 2010. The ship then undertook a refit in the United States but arrival in Pakistan was delayed by a collision with the pier while conducting basin trials on 21 January 2011. Up to three further ships of the class may be acquired by 2016.
Modernisation: It is likely that the ship will be refitted with CIWS Phalanx Block 1B and Kelvin Hughes Sharpeye surface search radar. Replacement of the Mk 13 launchers is unlikely.
Structure: The original single hangar was changed to two adjacent hangars. Provided with 19 mm Kevlar armour protection over vital spaces. 25 mm guns can be fitted for some operational deployments.

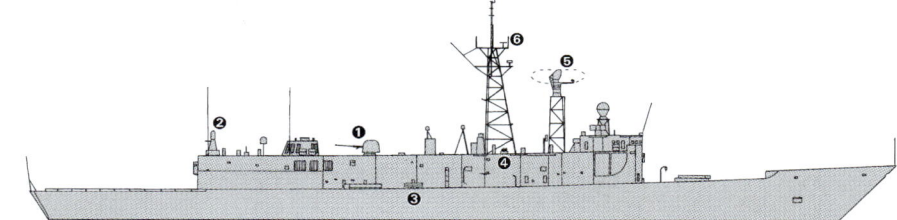

ALAMGIR *(Scale 1 : 1,200), Ian Sturton / 1528813*

ALAMGIR *1/2013*, Shaun Jones / 1530960*

SHIPBORNE AIRCRAFT

Numbers/Type: 6 Westland Sea King Mk 45/45A.
Operational speed: 125 kt *(232 km/h)*.
Service ceiling: 10,500 ft *(3,200 m)*.
Range: 630 n miles *(1,165 km)*.
Role/Weapon systems: Sensors: ARI 5955 search radar, Marconi Type 2069 dipping sonar, Star SAFIRE FLIR. AQS-928G acoustic processors. Weapons: ASW; two Mk 46 torpedoes; Mk 11 depth charges. ASV; one AM 39 Exocet missile.

Numbers/Type: 7/10 Aerospatiale SA 316 Alouette III/SA 319B Alouette III.
Operational speed: 113 kt *(210 km/h)*.
Service ceiling: 10,660 ft *(3,250 m)*.
Range: 270 n miles *(500 km)*.
Role/Weapon systems: Reconnaissance helicopter. Two SA 319B procured in mid-1970s and fitted with radar and MAD. Eight further SA 319B (eight ex-French Air Force) purchased in 2005 for delivery in 2008 after refurbishment. Four SA 316 acquired in 1994. Sensors: Weather/search radar and MAD (in two SA 319B). Weapons: ASW: Mk 11 depth charges, and MG1 A3 gun.

SEA KING *12/2009, Michael Nitz / 1406107*   ALOUETTE III *6/2003, Pakistan Navy / 0569234*

© 2014 IHS   IHS Jane's Fighting Ships 2014-2015

606 Pakistan > Shipborne aircraft — Patrol forces

Numbers/Type: 6 Hai Z-9EC.
Operational speed: 135 kt *(250 km/h)*.
Service ceiling: 15,000 ft *(4,572 m)*.
Range: 236 n miles *(437 km)*.
Role/Weapon systems: ASW helicopter procured in conjunction with Zulfiquar-class frigate programme. Chinese design based on Dauphin 2. Sensors: KLC-1 radar; ESM, DSE-1 Sonar. Weapons: up to four ET-52C torpedoes.

Z-9EC   6/2011, Royal Australian Navy / 1406644

LAND-BASED MARITIME AIRCRAFT

Notes: (1) The Maritime Security Agency operates three Britten-Norman Maritime Defenders, with Bendix RDR 1400C radars.
(2) A squadron of four indigenously produced Uqab II unmanned air vehicles was established at Mehran naval air station in 2011. The aircraft have a range of 65 n miles and can operate up to 3,000 m above sea level. The aircraft are used in a reconnaissance role and are equipped with cameras.

DEFENDER   8/1996, MSA / 0081375

Numbers/Type: 8 Lockheed P-3C Orion (Update II).
Operational speed: 410 kt *(760 km/h)*.
Service ceiling: 28,300 ft *(8,625 m)*.
Range: 4,000 n miles *(7,410 km)*.
Role/Weapon systems: Order of first two completed in 1991 but held up by the Pressler amendment, until delivery in December 1996. May be used for Elint. Eight further aircraft donated by the United States in September 2005. The first two, delivered in early 2007, are to have an avionics upgrade at a later date. The remaining six were to be upgraded before delivery by 2012. Two upgraded aircraft were destroyed in a terrorist attack on Mehran air station on 22 May 2011. Sensors: APS-115 search radar; up to 100 sonobuoys; ASQ 81 MAD; ESM. Weapons: four Whitehead A 244 torpedoes or Mk 11 depth charges for ASW; Harpoon.

P-3C   11/2013*, Michael Nitz / 1530961

Numbers/Type: 6 Fokker F27-200.
Operational speed: 250 kt *(463 km/h)*.
Service ceiling: 29,500 ft *(8,990 m)*.
Range: 2,700 n miles *(5,000 km)*.
Role/Weapon systems: Acquired in 1994–96 for maritime surveillance and limited ASW. One further aircraft acquired in 2007. Sensors: OM 100 radar, Thomson-CSF DR 3000A ESM. Star SAFIRE FLIR. Weapons: Mk II depth charge.

FOKKER F27-200   6/2001, Pakistan Navy / 0114780

Numbers/Type: 10 AMD-BA Mirage 5PA3.
Operational speed: 750 kt *(1,390 km/h)*.
Service ceiling: 59,055 ft *(18,000 m)*.
Range: 740 n miles *(1,370 km)*.
Role/Weapon systems: Operated by the Air Force, and all can be used for maritime strike. Sensors: Thomson-CSF radar. Weapons: ASV; two AM 39 Exocet or Harpoon; two 30 mm DEFA.

MIRAGE 5PA3   6/2004, Pakistan Navy / 1044171

Numbers/Type: 4 Saab 2000 AEW.
Operational speed: 250 kt *(463 km/h)*.
Service ceiling: 25,000 ft *(7,620 m)*.
Range: 570 n miles *(1,056 km)*.
Role/Weapon systems: Air Force operated early warning aircraft. Modified version of Saab 340B special mission aircraft derived from regional transport aircraft. Dorsal-mounted main radar. First aircraft rolled out in mid-June 2008 and entered final system testing in October 2009. The second aircraft was delivered on 26 April 2010 the third in late 2010 and fourth in April 2011. Sensors: Ericsson Erieye radar, ESM, ECM.

SAAB 2000 AEW   5/2008, Saab / 1330714

Numbers/Type: 2 EADS ATR 72-500.
Operational speed: 250 kt *(463 km/h)*.
Service ceiling: 25,000 ft *(7,620 m)*.
Range: 800 n miles *(1,482 km)*.
Role/Weapon systems: Both aircraft acquired in August 2013. Transport and training aircraft being converted to maritime surveillance role.

Numbers/Type: 1 Hawker Beechcraft 800XP.
Operational speed: 530 kt *(980 km/h)*.
Service ceiling: 41,000 ft *(12,497 m)*.
Range: 2,000 n miles *(3,704 km)*.
Role/Weapon systems: Military variant of business jet being converted for electronic and maritime surveillance.

PATROL FORCES

2 JALALAT CLASS (FAST ATTACK CRAFT — MISSILE) (PTG)

Name	No	Builders	Launched	Commissioned
JALALAT	1029 (ex-1022)	PN Dockyard, Karachi	16 Nov 1996	14 Aug 1997
SHUJAAT	1030	PN Dockyard, Karachi	26 Mar 1999	30 Sep 1999

Displacement, tonnes: 188 full load
Dimensions, metres (feet): 39 × 6.7 × 1.64 *(128 × 22.0 × 5.4)*
Speed, knots: 23. **Range, n miles:** 2,000 at 17 kt
Complement: 31 (3 officers)

Machinery: 2 MTU diesels; 5,984 hp(m) *(4.4 MW)* sustained; 2 shafts
Missiles: SSM: 4 China C 802 YJ-83 Saccade (2 twin); active radar homing to 120 km *(66 n miles)* at 0.9 Mach; warhead 165 kg; sea skimmer.
Guns: 2—37 mm/63 (twin); 180 rds/min to 8.5 km *(4.6 n miles)*; weight of shell 1.42 kg.
Physical countermeasures: Decoys: chaff launcher.
Electronic countermeasures: ESM: Thales DR 3000.
Radars: Surface search: Kelvin Hughes Type 756; I-band.
Fire control: Type 47G (for gun); Type TR 47G A/R (for SSM); I-band.

Comment: Designed with Chinese assistance to replace deleted Hegu class. Same hull as Larkana.

SHUJAAT   3/2012, Michael Nitz / 1455799

IHS Jane's Fighting Ships 2014-2015

2 JURRAT CLASS (FAST ATTACK CRAFT—MISSILE) (PGG)

Name	No	Builders	Launched	Commissioned
JURRAT	1023	Karachi Shipyard, Karachi	9 Sep 2004	24 Feb 2006
QUWWAT	1028	Karachi Shipyard, Karachi	13 Sep 2004	24 Feb 2006

Displacement, tonnes: 229 standard; 254 full load
Dimensions, metres (feet): 38.85 × 8.0 × 1.85 *(127.5 × 26.2 × 6.1)*
Speed, knots: 32
Range, n miles: 1,700 at 15 kt
Complement: 35 (3 officers)

Machinery: 3 MTU 16V 4000 M70; 3 shafts; ducted propellers
Missiles: SSM: 4 China C 802 YJ-83 Saccade (2 twin); active radar homing to 120 km *(66 n miles)* at 0.9 Mach; warhead 165 kg; sea skimmer.
Guns: 2—25 mm (twin).
Physical countermeasures: Decoys: chaff launcher.
Electronic countermeasures: ESM: Thales DR-3000; warning receiver.
Radars: Surface search: Type SR 47A; I-band.
Fire control: Type 47G (for gun); Type TR 47G A/R (for SSM); I-band.

Comment: Both ordered in September 2002 and laid down 4 April 2003. Built at KSEW, reportedly in co-operation with Thai company Marsun. Steel hull and aluminium superstructure.

JURRAT *3/2010, Michael Nitz / 1406109*

1 TOWN CLASS (LARGE PATROL CRAFT) (PB)

Name	No	Builders	Commissioned
RAJSHAHI	P 140	Brooke Marine, Lowestoft	1965

Displacement, tonnes: 117 standard; 145 full load
Dimensions, metres (feet): 32.6 × 6.1 × 2.1 *(107 × 20.0 × 6.9)*
Speed, knots: 24
Complement: 19
Machinery: 2 MTU 12V 538 diesels; 3,400 hp(m) *(2.5 MW)*; 2 shafts
Guns: 2 Bofors 40 mm/60. 2—12.7 mm MGs.
Radars: Surface search: Pot Head; I-band.

Comment: The last survivor in Pakistan of a class of four built by Brooke Marine in 1965. Steel hull and aluminium superstructure. Assigned to 10th Patrol Squadron.

RAJSHAHI *6/2003, Pakistan Navy / 0569233*

1 LARKANA CLASS (LARGE PATROL CRAFT) (PB)

Name	No	Builders	Commissioned
LARKANA	P 157	PN Dockyard, Karachi	6 Jun 1994

Displacement, tonnes: 183 full load
Dimensions, metres (feet): 39 × 6.7 × 1.7 *(128 × 22.0 × 5.6)*
Speed, knots: 23. **Range, n miles:** 2,000 at 17 kt
Complement: 25 (3 officers)

Machinery: 2 MTU diesels; 5,984 hp(m) *(4.4 MW)* sustained; 2 shafts
Guns: 2 Type 76A 37 mm/63 (twin). 4—25 mm/60 (2 twin).
Depth charges: 2 Mk 64 launchers.
Radars: Surface search: Kelvin Hughes Type 756; I-band.

Comment: Ordered in 1991 and started building in October 1992. Assigned to 10th Patrol Squadron.

LARKANA *9/2004 / 1133552*

4 GRIFFON 2000 TDX(M) (HOVERCRAFT) (UCAC)

211–214

Displacement, tonnes: 8 full load
Dimensions, metres (feet): 12.7 on cushion × ? *(41.7 × ?)*
Speed, knots: 35
Range, n miles: 300 at 25 kt
Complement: 2

Military lift: 25 troops or 2 tons
Machinery: 1 Deutz BF8L513 diesel; 355 hp *(265 kW)* sustained
Guns: 1—12.7 mm MG. 1—7.62 mm MG.
Radars: Navigation: I-band.

Comment: Acquired from Griffon, UK. First craft delivered in April 2004 and the last in July 2005. The first two are of a modular design to enable rapid role-change. The second two have fixed roofs.

GRIFFON 2000 *6/2005, Griffon Hovercraft / 1153502*

2 KAAN 15 (FAST INTERVENTION CRAFT) (PBF)

P 01–02

Displacement, tonnes: 21 full load
Dimensions, metres (feet): 16.7 × 4.04 × 1.2 *(54.8 × 13.3 × 3.9)*
Speed, knots: 65
Range, n miles: 350 at 35 kt
Complement: 12
Machinery: 2 MTU 12V 183 TE93 diesels; 2,300 hp(m) *(1.69 MW)*; 2 Arneson ASD 12 B1L surface drives
Guns: 2—12.7 mm MGs.

Comment: Built by Yonca Shipyard, Turkey. Advanced composites structure. The first delivered on 17 August 2004 and the second on 14 October 2004. To be operated by Special Services Group based at PNS Iqbar. Details based on those in Turkish Coast Guard service.

P 01 *7/2004, Selçuk Emre / 1044173*

2 KAAN 33 (FAST ATTACK CRAFT) (PGGF)

ZARRAR P 03   KARRAR P 04

Displacement, tonnes: 115 full load
Dimensions, metres (feet): 35.6 × 6.7 × 1.4 *(116.8 × 22.0 × 4.6)*
Speed, knots: 65; 28 diesel
Range, n miles: 970 at 15 kt
Complement: 18 (2 officers)
Machinery: CODAG: 1 Honeywell TF50 gas turbine; 2 MTU 12V 4000 M90 diesels; 7,396 hp(m) *(5.44 MW)*; 3 MJP 650/750 waterjets
Missiles: SSM: 4 McDonnell Douglas Harpoon Block 2; active radar homing to 130 km *(70 n miles)* at 0.9 Mach; warhead 227 kg.
Guns: 1—25 mm. 2—12.7 mm MGs.

Comment: Following a tendering process, two MRTP 33 fast attack craft ordered from Yonca-Onuk Shipyard, Turkey on 8 June 2006. Construction began in February 2007 and delivery of the first of class was made on 26 November 2007. The second followed in April 2008. With advanced composites structure, the craft are improved versions of those in service in the Turkish Coast Guard. The craft are to be used for patrol of littoral waters, maritime interdiction and special forces operations.

ZARRAR *6/2007, Yonca-Onuk / 1353251*

608 Pakistan > Patrol forces — Survey ships

2 AZMAT CLASS (PTG)

Name	No	Builders	Laid down	Launched	Commissioned
AZMAT	1013	Xingang Shipyard, Tianjin	1 Mar 2011	21 Sep 2011	21 Jun 2012
DEHSHAT	–	Karachi Shipyard and Engineering Works	28 Oct 2011	16 Aug 2012	2014

Displacement, tonnes: 560 standard
Dimensions, metres (feet): 63 × 8.8 × 2.4 (206.7 × 28.9 × 7.9)
Speed, knots: 30
Range, n miles: 1,000 at 18 kt
Complement: 15

Machinery: 3 diesels; 2 shafts
Missiles: SSM: 8 C-802A (YJ-83/CSS-N-8 Saccade) ❶; mid-course guidance and active radar homing to 120 km (65 n miles) at 0.9 Mach; warhead 165 kg; sea skimmer.
Guns: 2 – 37 mm (twin) ❷; 1 – 30 mm Type 630 ❸; 6 barrels per mounting.
Radars: Surface search: Type 360 ❹; E/F-band.
Fire control: Type 347G(1) ❺; I-band.
Navigation: To be announced.
Combat data systems: To be announced.
Weapon control systems: Optronic director to be announced.

Comment: A contract was signed with the Chinese government in late 2010 for the construction of two missile patrol craft. The first was built in China and the second in Pakistan. The design is based on the Houjian class on which details are based.

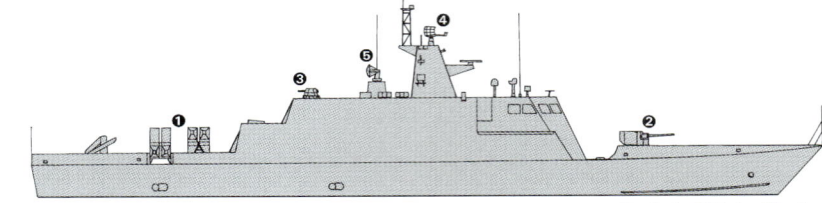

AZMAT (Scale 1 : 600), Ian Sturton / 1528814

AZMAT
6/2012, Ships of the World
1455818

4 MILITARY ASSAULT CRAFT (LCP)

111–114

Speed, knots: 30
Complement: 18
Machinery: 2 Volvo Penta IPS 600 diesels; 870 hp (648 kW); 2 shafts
Guns: 1 – 12.7 mm MG. 1 – 7.62 mm MG.

Comment: Built by Karachi Shipyard to a design by Marsun, Thailand. The first was delivered on 11 December 2004. Appearance is similar to SEAL assault craft in service with the Thai Navy.

MILITARY ASSAULT CRAFT 114 3/2007, Yonca-Onuk / 1170019

MUHAFIZ 3/2008, Guy Toremans / 1305314

SURVEY SHIPS

1 SURVEY SHIP (AGS/AGOR)

Name	No	Builders	Laid down	Launched	Commissioned
BEHR PAIMA	SV 48	Ishikawajima, Japan	16 Feb 1982	7 Jul 1982	27 Dec 1982

Measurement, tonnes: 1,202 gt
Dimensions, metres (feet): 61 × 11.8 × 3.7 (200.1 × 38.7 × 12.1)
Speed, knots: 13.7. **Range, n miles:** 5,400 at 12 kt
Complement: 84 (16 officers)
Machinery: 2 Daihatsu 6DSM-22 diesels; 2,000 hp(m) (1.47 MW); 2 shafts; cp props; bow thruster

Comment: Ordered in November 1981. Hydrographic and oceanographic research vessel. Equipped with multibeam echo-sounder, deep echo sounder and carries two survey motor boats for inshore operations.

MINE WARFARE FORCES

3 MUNSIF (ÉRIDAN) CLASS (MINEHUNTERS) (MHSC)

Name	No	Builders	Launched	Commissioned
MUNSIF (ex-Sagittaire)	M 166	Lorient Dockyard, Lorient	9 Nov 1988	27 Jul 1989
MUHAFIZ	M 163	Lorient Dockyard, Lorient	8 Jul 1995	15 May 1996
MUJAHID	M 164	Lorient, Lorient; PN Dockyard	28 Jan 1997	9 Jul 1998

Displacement, tonnes: 536 standard; 605 full load
Dimensions, metres (feet): 51.5 × 8.9 × 2.9 (169 × 29.2 × 9.5)
Speed, knots: 15; 7 motor. **Range, n miles:** 3,000 at 12 kt
Complement: 46 (5 officers)

Machinery: 1 Stork Wärtsilä A-RUB 215X-12 diesel; 1,860 hp(m) (1.37 MW) sustained; 1 shaft; LIPS cp prop; auxiliary propulsion; 2 motors; 240 hp(m) (179 kW); 2 active rudders; 2 bow thrusters
Guns: 1 GIAT 20F2 20 mm; 1 – 12.7 mm MG.
Physical countermeasures: MCM: 2 PAP 104 Mk 5 systems; mechanical sweep gear. Elesco MKR 400 acoustic sweep; MRK 960 magnetic sweep.
Radars: Navigation: Racal Decca 1229 (M 166) or Kelvin Hughes 1007; I-band.
Sonars: Thomson Sintra DUBM 21B or 21D (163 and 164); hull-mounted; active; high frequency; 100 kHz (±10 kHz). Thomson Sintra TSM 2054 MCM towed array may be included.
Combat data systems: Thomson-CSF TSM 2061 Mk 2 tactical system in the last pair.

Comment: Contract signed with France 17 January 1992. The first recommissioned into the Pakistan Navy on 24 September 1992 after active service in the Gulf with the French Navy in 1991. Sailed for Pakistan in November 1992. The second was delivered in April 1996. The last one was transferred to Karachi by transporter ship in April 1995 with a final package following in November 1995. Form 9th Auxiliary and Mine Warfare Squadron.

BEHR PAIMA 6/2003, Pakistan Navy / 0569231

IHS Jane's Fighting Ships 2014-2015 © 2014 IHS

TRAINING SHIPS

1 SAIL TRAINING SHIP (AXS)

Name	No	Builders	Laid down	Launched	Commissioned
RAH NAWARD (ex-*Prince William*)	A 24	Abeking and Rasmussen; Appledore Shipyard	Sep 1994	Sep 1995	24 Sep 2010

Measurement, tonnes: 493 gt
Dimensions, metres (feet): 59.35 × 9.9 × 4.8 *(194.7 × 32.5 × 15.7)*
Machinery: 2 MTU 8V183 TE62 diesels; 898 hp *(660 kW)*; 2 shafts; cp props

Comment: Sail Training Ship (brigantine) acquired by the Pakistan Navy on 24 September 2010. The ship was laid down at Abeking and Rasmussen in September 1994 and launched in September 1995. She was completed at Appledore Shipyard, Bideford in 2001. The total sail area is 949 m².

RAH NAWARD   6/2012, *Pakistan Navy* / 1455801

AUXILIARIES

2 COASTAL TANKERS (AOTL)

Name	No	Builders	Commissioned
GWADAR	A 49	Karachi Shipyard, Karachi	1984
KALMAT	A 21	Karachi Shipyard, Karachi	29 Aug 1992

Measurement, tonnes: 844 gt
Dimensions, metres (feet): 62.8 × 11.3 × 3 *(206 × 37.1 × 9.8)*
Speed, knots: 10
Complement: 25
Cargo capacity: 340 m³ fuel or water
Machinery: 1 Sulzer diesel; 550 hp(m) *(404 kW)*; 1 shaft
Guns: 2—7.62 mm MGs.

Comment: Assigned to 9th Auxiliary and Mine Warfare Squadron.

GWADAR   3/2008, *Guy Toremans* / 1305313

1 TANKER (AOTL)

ATTOCK A 40

Displacement, tonnes: 1,219 full load
Dimensions, metres (feet): 54 × 9.8 × 4.6 *(177.2 × 32.2 × 15.1)*
Speed, knots: 8
Complement: 18
Cargo capacity: 550 tons fuel
Machinery: 2 diesels; 800 hp(m) *(276 kW)*; 2 shafts
Guns: 2 Oerlikon 20 mm.

Comment: Built in Italy in 1957. Assigned to 9th Auxiliary and Mine Warfare Squadron.

ATTOCK   6/2004, *Pakistan Navy* / 1044169

1 FUQING CLASS (AORH)

Name	No	Builders	Commissioned
NASR (ex-*X-350*)	A 47	Dalian Shipyard	27 Aug 1987

Displacement, tonnes: 7,620 standard; 22,099 full load
Dimensions, metres (feet): 171 × 21.8 × 9.4 *(561 × 71.5 × 30.8)*
Speed, knots: 18. **Range, n miles:** 18,000 at 14 kt
Complement: 130

Cargo capacity: 10,550 tons fuel; 1,000 tons dieso; 200 tons feed water; 200 tons drinking water
Machinery: 1 Sulzer 8RLB66 diesel; 13,000 hp(m) *(9.56 MW)*; 1 shaft
Guns: 1 GE/GD Vulcan Phalanx CIWS. 4—37 mm (2 twin). 2—12.7 mm MGs.
Physical countermeasures: Decoys: SRBOC Mk 36 chaff launcher. 2 Rheinmetall MASS launchers.
Electronic countermeasures: ESM: Thales DR 3000.
Radars: Navigation: 1 Kelvin Hughes 1007; 1 SPS 66; I-band.
Helicopters: 1 Sea King.

Comment: Similar to Chinese ships of the same class. Two replenishment at sea positions on each side for liquids and one for solids. Phalanx fitted on the hangar roof in 1995. Assigned to 9th Auxiliary and Mine Warfare Squadron.

NASR   5/2009, *Guy Toremans* / 1305881

1 POOLSTER CLASS (AORH)

Name	No	Builders	Commissioned	Recommissioned
MOAWIN (ex-*Poolster*)	A 20 (ex-A 835)	Rotterdamse Droogdok Mij	10 Sep 1964	28 Jul 1994

Displacement, tonnes: 17,070 full load
Measurement, tonnes: 10,160 dwt
Dimensions, metres (feet): 168.3 × 20.3 × 8.2 *(552.2 × 66.6 × 26.9)*
Speed, knots: 21
Complement: 200 (17 officers)

Cargo capacity: 10,300 tons including 8-9,000 tons oil fuel
Machinery: 2 boilers; 2 turbines; 22,000 hp(m) *(16.2 MW)*; 1 shaft
Guns: 1 Raytheon 20 mm Vulcan Phalanx Mk 15; 3,000 rds/min to 1.5 km. 4—20 mm Oerlikon (2 twin). 2—12.7 mm MGs.
Physical countermeasures: Decoys: SRBOC Mk 36 chaff launcher.
Electronic countermeasures: ESM: SLQ-32.
Radars: Air/surface search: Racal Decca 2459; F/I-band.
Navigation: Racal Decca TM 1229C; I-band.
Sonars: Signaal CWE 10; hull-mounted; active search; medium frequency.
Helicopters: 1 Sea King.

Comment: Acquired from the Netherlands Navy. Helicopter deck aft. Funnel heightened by 4.5 m *(14.8 ft)*. Capacity for five Lynx sized helicopters. Two fuelling stations each side for underway replenishment. Phalanx to be fitted in due course. Assigned to 9th Auxiliary and Mine Warfare Squadron.

MOAWIN   10/2009, *Michael Nitz* / 1406110

1 DREDGING VESSEL

Name	No	Builders	Laid down	Commissioned
BEHR KUSHA	A 41	Zhaobao Ship Building Company	2 Feb 2004	15 Aug 2008

Displacement, tonnes: 5,118 full load
Dimensions, metres (feet): 85.3 × 8.1 × 4.1 *(279.9 × 26.6 × 13.5)*
Speed, knots: 8
Complement: 72
Machinery: 2 diesels; 4 waterjets
Radars: Navigation: I-band.

Comment: Trailer Suction Hopper Dredger formally commissioned in the Pakistan Navy after a refit. The ship's role is to conduct maintenance dredging of harbours, approaches, channels and basins.

BEHR KUSHA   6/2013*, *Pakistan Navy* / 1530962

610 Pakistan > Auxiliaries — Maritime security agency

0 + 1 FLEET TANKER (AOR)

Name	No	Builders	Laid down	Launched	Commissioned
–	–	Karachi Shipbuilding and Engineering	2013	2016	2017

Displacement, tonnes: 17,000 full load
Dimensions, metres (feet): 154.3 × 21.7 × 7.0 (506.2 × 71.2 × 23.0)
Speed, knots: 20. **Range, n miles:** 10,000 at 15 kt

Cargo capacity: 8,585 tonnes F-75; 170 tonnes JP-5; 1,000 tonnes fresh water; 100 tonnes dry cargo.
Machinery: 2 diesels; 15,556 hp (11.6 MW); 2 shafts; cp props; bow thruster
Guns: 1 Raytheon Vulcan Phalanx. 1—25 mm. 2—12.7 mm MGs.
Radars: Surface search/navigation: I-band.
Helicopters: 1 medium.

Comment: A contract between Pakistan's Ministry of Defence Production and Turkish technology and systems engineering house Savunma Teknolojileri Mühendislik ve Ticaret (STM) was signed in early 2013. The agreement is for STM to provide technical and material support for the construction of a new double-hull fleet tanker to be built by Karachi Shipyard and Engineering Works. The contract covers the detailed design of the new ship together with the supply of materiels, integrated logistic support, training test, and trials. STM is also to provide consultancy and management support for the four-year programme. The ship is to be equipped with two abeam stations for liquid and solids replenishment and one stern station for liquids and is also to be capable of supporting anti-submarine and anti-surface vessel operations through embarked helicopters. The ship is also to have a disaster relief role and is to be equipped with medical facilities. First steel was cut on 27 November 2013.

2 UTILITY SHIPS (AG)

Name	No	Builders	Laid down	Launched	Commissioned
MADADGAR	A 22	Karachi Shipyard	2007	12 Jan 2009	19 Jul 2011
RASADGAR	A 23	Karachi Shipyard	2007	28 Feb 2009	19 Jul 2011

Displacement, tonnes: 1,600 full load
Dimensions, metres (feet): 47.2 × 11.0 × 4.8 (154.9 × 36.1 × 15.7)
Speed, knots: 15
Complement: 56

Cargo capacity: 300 tonnes
Machinery: 2 Wärtsilä 9L20 diesels; 2,400 hp (1.8 MW); 2 shafts
Guns: 2—12.7 mm MGs.
Radars: Navigation: I-band.

Comment: Order placed with Karachi Shipyard and Engineering Works in May 2007. Known as Small Tanker cum Utility Ships (STUS) their roles include logistic support, SAR, minelaying and torpedo recovery.

MADADGAR 3/2012, Michael Nitz / 1454885

TUGS

Notes: Jandar and Jafakash are two pusher tugs (10 ton bollard pull) built by Karachi Shipyard and commissioned in 2000.

JANDAR and JAFAKASH 6/2003, Pakistan Navy / 1044170

7 COASTAL TUGS (YTB)

Name	No	Builders	Commissioned
BHOLU	A 44	Giessendam Shipyard	Apr 1991
GAMA	A 45	Giessendam Shipyard	Apr 1991
JANBAZ	–	Karachi Shipyard, Karachi	Sep 1990
JOSHILA	–	Karachi Shipyard, Karachi	Sep 2000
DELAIR	–	Karachi Shipyard, Karachi	Sep 2000
PURJOSH	–	Karachi Shipyard, Karachi	Mar 2013
SHERDIL	–	Karachi Shipyard, Karachi	Mar 2013

Displacement, tonnes: 269 full load
Dimensions, metres (feet): 26 × 6.8 × 2.9 (85.3 × 22.3 × 9.5)
Speed, knots: 12
Complement: 6
Machinery: 2 Cummins KTA38-M diesels; 1,836 hp (1.26 MW) sustained; 2 shafts

Comment: Details are for Bholu and Gama, built by Damen Shipyards and which entered service in 1991. Janbaz and Joshila were built by Karachi Shipyard and delivered in 1990 and 2000 respectively. Purjosh and Sherdil were delivered in March 2013.

IHS Jane's Fighting Ships 2014-2015

JOSHILA 5/2003 / 0569222

MARITIME SECURITY AGENCY

Notes: (1) All ships are painted white with a distinctive diagonal blue and red band and MSA on each side.
(2) One Britten-Norman Maritime Defender acquired in 1993, a second in 1994 and a third on 8 August 2004. Based near Karachi with 93 Squadron.
(3) Plans for new aircraft are under consideration.

1 GEARING (FRAM 1) CLASS (DD)

Name	No	Builders	Commissioned
NAZIM (ex-Tughril)	D 156 (ex-D 167)	Todd Pacific Shipyard Corporation, Seattle	4 Aug 1945

Displacement, tonnes: 2,464 standard; 3,556 full load
Dimensions, metres (feet): 119 × 12.6 × 5.8 (390.4 × 41.3 × 19.0)
Speed, knots: 32
Range, n miles: 4,500 at 16 kt
Complement: 180 (15 officers)

Machinery: 4 Babcock & Wilcox boilers; 600 psi (43.3 kg/cm^2); 850°F (454°C); 2 GE turbines; 60,000 hp (45 MW); 2 shafts
Guns: 2 US 5 in (127 mm)/38 Mk 38 (twin); 15 rds/min to 17 km (9.3 n miles) anti-surface; 11 km (5.9 n miles); anti-aircraft; weight of shell 25 kg. 4—25 mm (2 twin).
Torpedoes: 6—324 mm Mk 32 (2 triple) tubes.
Physical countermeasures: Decoys: 2 Plessey Shield 6-barrelled fixed launchers; chaff and IR flares in distraction, decoy or centroid modes.
Radars: Surface search: Raytheon/Sylvania; SPS-10; G-band.
Navigation: KH 1007; I-band.
Fire control: Western Electric Mk 25; I/J-band.
Weapon control systems: Mk 37 for 5 in guns. OE 2 SATCOM.

Comment: Transferred from the US on 30 September 1980 to the Navy. Passed on to the MSA in 1998 and renamed. This is the third Gearing to be renamed Nazim, the previous pair having been sunk as targets. All weapon systems removed except the torpedo tubes and main gun. Serves as the MSA Flagship.

NAZIM 3/2007, Paul Daly / 1170020

3 GUNS CLASS (PATROL BOATS) (PB)

GUNS MS 111   **SUR** MS 112   **MALAN** MS 113

Displacement, tonnes: 15 full load
Dimensions, metres (feet): 13 × 3.65 × 1.0 (42.7 × 12.0 × 3.3)
Speed, knots: 21
Complement: 6
Machinery: 2 Yamaha ME 730 TIL diesels; 636 hp (475 kW); 2 shafts
Guns: 1—7.62 mm MG.
Radars: Navigation: JRC 1500; I-band.

Comment: Manufactured by Karachi Shipyard and Engineering Works and commissioned in 2006. GRP construction.

MALAN 6/2006, Maritime Security Agency / 1164331

© 2014 IHS

Maritime security agency < **Pakistan** — Introduction < **Palau**

2 SHANGHAI II CLASS (FAST ATTACK CRAFT—GUN) (PB)

SABQAT P 1066 **RAFAQAT** P 1068

Displacement, tonnes: 133 full load
Dimensions, metres (feet): 38.8 × 5.4 × 1.7 *(127.3 × 17.7 × 5.6)*
Speed, knots: 30. **Range, n miles:** 700 at 16.5 kt
Complement: 34

Machinery: 2 Type L12-180 diesels; 2,400 hp(m) *(1.76 MW)* (forward); 2 Type 12-D-6 diesels; 1,820 hp(m) *(1.34 MW)* (aft); 4 shafts
Guns: 4—37 mm/63 (2 twin). 2—25 mm/80 (twin).
Mines: Fitted with mine rails for approx 10 mines.
Depth charges: 2 projectors; 8 weapons.
Radars: Surface search: Anritsu ARC-32A; I-band.

Comment: Twelve craft were originally acquired from China 1972–1976. Of these, four were transferred from the navy in 1994 and renamed *Sabqat*, *Rafaqat*, *Rahat* and *Sadaqat*. The latter two were decommissioned in 1994 and replaced by two further naval craft in about 1998. These have also since been decommissioned.

SABQAT   5/2003 / 0569223

1 HUANGFEN CLASS (PATROL BOAT) (PB)

SADAQAT (ex-*Dehshat*) P 1069 (ex-P 1026)

Displacement, tonnes: 174 standard; 208 full load
Dimensions, metres (feet): 38.6 × 7.6 × 2.7 *(126.6 × 24.9 × 8.9)*
Speed, knots: 28. **Range, n miles:** 800 at 22 kt
Complement: 28
Machinery: 3 Type 42-160 diesels; 12,000 hp(m) *(8.8 MW)* sustained; 3 shafts
Guns: 4 Norinco 25 mm/80 (2 twin); 270 rds/min to 3 km *(1.6 n miles)*; weight of shell 0.34 kg.
Radars: Surface search: Square Tie; I-band.

Comment: Originally transferred to the Pakistan Navy in April 1984. The then missile-armed craft were Chinese versions of the Soviet Osa II class. This craft was transferred to the MSA on 25 June 2005.

SADAQAT   6/2005, Maritime Security Agency / 1164330

5 RESPONSE BOAT MEDIUM (PBF)

Name	No	Builders
–	FRB 001	Marinette Marine Industries, Seattle
–	FRB 002	Marinette Marine Industries, Seattle
–	FRB 003	Marinette Marine Industries, Seattle
–	FRB 004	Marinette Marine Industries, Seattle
–	FRB 005	Marinette Marine Industries, Seattle

Displacement, tonnes: 17 full load
Dimensions, metres (feet): 13.5 × 4.45 × 0.9 *(44.3 × 14.6 × 3.0)*
Speed, knots: 42
Range, n miles: 250 at 30 kt
Complement: 4
Machinery: 2 diesels; 1,130 hp *(842 kW)*; 2 waterjets
Guns: 2—7.62 mm MGs.
Radars: Furuno; I-band.

Comment: Multimission self-righting response craft donated by the US in 2010. Built by Marinette Marine Industries, Seattle and based on the craft in US Coast Guard service.

4 BARKAT CLASS (PBO)

Name	No	Builders	Commissioned
BARKAT	1060 (ex-P 60)	China SB Corporation	29 Dec 1989
REHMAT	1061 (ex-P 61)	China SB Corporation	29 Dec 1989
NUSRAT	1062 (ex-P 62)	China SB Corporation	13 Jun 1990
VEHDAT	1063 (ex-P 63)	China SB Corporation	13 Jun 1990

Displacement, tonnes: 442 full load
Dimensions, metres (feet): 58 × 7.6 × 2.3 *(190.3 × 24.9 × 7.5)*
Speed, knots: 27. **Range, n miles:** 1,500 at 12 kt
Complement: 50 (5 officers)
Machinery: 4 MTU 16V 396 TB93 diesels; 8,720 hp(m) *(6.4 MW)* sustained; 4 shafts
Guns: 2—37 mm/63 (1 twin). 2—14.5 mm/60 (twin).
Radars: Surface search: 2 Anritsu ARC-32A; I-band.

Comment: Type P58A patrol craft built in China for the MSA. First two arrived in Karachi at the end of January 1990, second pair in August 1990. Some of this type of ship are in service with Chinese paramilitary forces.

VEHDAT   6/1994, Maritime Security Agency / 0081380

COAST GUARD

Notes: (1) Unlike the Maritime Security Agency which comes under the Defence Ministry, the official Coast Guard was set up in 1985 and is manned by the Army and answerable to the Ministry of the Interior.
(2) The Customs Service is manned by naval personnel. It operates approximately 18 craft.

1 SWALLOW CLASS (PB)

SAIF

Displacement, tonnes: 53 full load
Dimensions, metres (feet): 20 × 4.7 × 1.3 *(65.6 × 15.4 × 4.3)*
Speed, knots: 25. **Range, n miles:** 500 at 20 kt
Complement: 8
Machinery: 2 GM Detroit 12V71 T1 diesels; 2,120 hp *(1.58 MW)*; 2 shafts
Guns: 2—12.7 mm MGs

Comment: Built by Swallowcraft/Kangnam and delivered in 1986.

4 CRESTITALIA MV 55 CLASS (PBF)

SADD P 551   **SHABHAZ** P 552   **VAQAR** P 553   **BURQ** P 554

Displacement, tonnes: 23 full load
Dimensions, metres (feet): 16.5 × 5.2 × 0.9 *(54.1 × 17.1 × 3.0)*
Speed, knots: 35. **Range, n miles:** 425 at 25 kt
Complement: 5
Machinery: 2 MTU diesels; 2,200 hp *(1.64 MW)*; 2 shafts

Comment: Delivered in 1987.

SHABHAZ   5/2003 / 0569228

0 + 4 FAST PATROL CRAFT (PBF)

Displacement, tonnes: 42.5 full load
Dimensions, metres (feet): 24 × 6.6 × 1.18 *(78.7 × 21.7 × 3.9)*
Speed, knots: 60
Machinery: 4 diesels; 5,800 hp *(4.3 MW)*; 2 shafts

Comment: The contract with McTay Marine, Bromborough, for the construction of four 24 m high-speed patrol craft was announced on 4 July 2013. The aluminum craft are designed by Alucatz. Delivery is expected in 2015.

Palau

Country Overview

The Republic of Palau was a US-administered UN Trust territory from 1947 before becoming independent in 1994 when a Compact of Free Association, delegating to the US the responsibility for defence and foreign affairs, came into effect. Situated in the western Pacific Ocean, the country comprises about 200 of the Caroline Islands archipelago spread in a chain about 350 n miles long. These include Koror (the administrative centre), Babelthuap (the largest island), Arakabesan, Malakal and Peleliu. The capital is currently on Koror, but a new capital is being built in eastern Babelthuap. Territorial seas (3 n miles) are claimed. An extended fisheries zone (200 n miles) is also claimed but limits have not been fully defined.

Headquarters Appointments

Chief of Division of Marine Law Enforcement:
Captain Ellender Ngirameketii

Palau > Patrol forces — Panama > Patrol forces

PATROL FORCES

1 PACIFIC CLASS (LARGE PATROL CRAFT) (PB)

Name	No	Builders	Commissioned
PRESIDENT H I REMELIIK	001	Transfield Shipbuilding	May 1996

Displacement, tonnes: 165 full load
Dimensions, metres (feet): 31.5 × 8.1 × 2.1 *(103.3 × 26.6 × 6.9)*
Speed, knots: 20
Range, n miles: 2,500 at 12 kt
Complement: 17 (3 officers)
Machinery: 2 Caterpillar 3516TA diesels; 4,400 hp *(3.28 MW)* sustained; 2 shafts
Guns: 2—7.62 mm MGs.
Radars: Surface search: Furuno 1011; I-band.

Comment: Ordered in 1995. This was the 21st hull in the Pacific class programme. Following the decision by the Australian government to extend the Pacific Patrol Boat project, the ship underwent a half-life refit at Gladstone in 2003. A life-extension refit was completed in July 2011.

PRESIDENT H I REMELIIK
6/2004, Division of Marine Law Enforcement, Palau
1044175

Panama

SERVICIO NACIONAL AERONAVAL

Country Overview

The Republic of Panama is an independent state situated on the isthmus linking South America with Central and North America. Bordered to the west by Costa Rica and to the east by Colombia, it has an area of 29,157 square miles and a 664 n mile coastline with the north Pacific Ocean and of 370 n miles with the Caribbean. The country is bisected by the Panama Canal. A new treaty in 1977 ended US operation, maintenance and defence of the canal in 1999. The capital is Panama City while the main ports are Balboa, Cristóbal, Manzanillo, Bahía Las Minas, Vacamonte, Almirante and Puerto Armuelles. Territorial seas (12 n miles) are claimed. An Exclusive Economic Zone (EEZ) (200 n miles) has been defined by boundary agreements. Reform of the security apparatus led to the creation of the Panamanian Public Forces. The National Maritime Service was subsequently amalgamated with its air counterpart in 2008.

Headquarters Appointments

Director General National Air and Naval Service: Commissioner Belsio Giolis González Sánchez

Personnel

(a) 2014: 3,000 (including air and naval personnel)
(b) Voluntary service

Bases

Vasco Nuñez de Balboa, Cristóbal Colon, Quebrada de Piedra, Puerto Mutis, Puerto Agudulee, Puerto Mensabé, Isla Flamenco (under construction), Punta Cocos.

PATROL FORCES

Notes: The reported transfer of two former Italian Coast Guard Saettia-class patrol ships had not been completed by late 2013.

2 VOSPER TYPE (COASTAL PATROL CRAFT) (PB)

Name	No	Builders	Commissioned
PANQUIACO	P 301 (ex-GC 10)	Vosper Ltd, Portsmouth	Jul 1971
LIGIA ELENA	P 302 (ex-GC 11)	Vosper Ltd, Portsmouth	Jul 1971

Displacement, tonnes: 98 standard; 147 full load
Dimensions, metres (feet): 31.4 × 5.8 × 1.8 *(103 × 19.0 × 5.9)*
Speed, knots: 18. **Range, n miles:** 1,500 at 14 kt
Complement: 17 (3 officers)
Machinery: 2 Detroit diesels; 5,000 hp *(3.73 MW)*; 2 shafts
Guns: 2—7.62 mm MGs.
Radars: Surface search: Raytheon R-81; I-band.

Comment: *Panquiaco* launched on 22 July 1970, *Ligia Elena* on 25 August 1970. Hull of welded mild steel and upperworks of welded or buck-bolted aluminium alloy. Vosper fin stabiliser equipment. P 302 was sunk in December 1989, but subsequently recovered. Both vessels had major repairs in the Coco Solo shipyard from September 1992. This included new engines, a new radar and replacement guns. Both craft refitted at Astillero Bayano in 2009. Pacific Flotilla. Similar craft in service in MMEA Malaysia.

PANQUIACO *6/2011*, SENAN* / 1525634

1 COASTAL PATROL CRAFT (PB)

TABOGA P 306

Displacement, tonnes: 226 full load
Dimensions, metres (feet): 34.3 × 6.6 × 2.3 *(112.5 × 21.7 × 7.5)*
Speed, knots: 12
Range, n miles: 1,500 at 10 kt
Complement: 13 (5 officers)
Machinery: 2 Caterpillar diesels; 2 shafts
Guns: 2—12.7 mm MGs. 2—7.62 mm MGs.
Radars: Surface search: Raytheon R-81; I-band.

Comment: Former utility craft employed for training at US Navy's Naval Small Craft Instruction and Technical Training School (NAVSCIATTS). Transferred to Panama in March 1999. Refitted at Veracruz Shipyard in 2013.

TABOGA *6/2013*, SENAN* / 1525627

5 POINT CLASS (COASTAL PATROL CRAFT) (PB)

Name	No	Builders	Commissioned
3 DE NOVIEMBRE (ex-*Point Barrow*)	P 204 (ex-82348)	CG Yard, MD	4 Oct 1964
10 DE NOVIEMBRE (ex-*Point Huron*)	P 206 (ex-82357)	CG Yard, MD	17 Feb 1967
28 DE NOVIEMBRE (ex-*Point Frances*)	P 207 (ex-82356)	CG Yard, MD	3 Feb 1967
4 DE NOVIEMBRE (ex-*Point Winslow*)	P 208 (ex-82360)	J M Martinac, Tacoma	3 Mar 1967
5 DE NOVIEMBRE (ex-*Point Hannon*)	P 209 (ex-82355)	J M Martinac, Tacoma	23 Jan 1967

Displacement, tonnes: 70 full load
Dimensions, metres (feet): 25.3 × 5.2 × 1.8 *(83 × 17.1 × 5.9)*
Speed, knots: 18. **Range, n miles:** 1,500 at 8 kt
Complement: 10 (2 officers)
Machinery: 2 Caterpillar 3412 diesels; 1,600 hp *(1.18 MW)*; 2 shafts
Guns: 1—12.7 mm MG. 2—7.62 mm MGs.
Radars: Surface search: Raytheon R-81; I-band.

Comment: P 204 transferred from US Coast Guard 7 June 1991 and recommissioned 10 July 1991. P 206 and P 207 transferred 22 April 1999. P 208 transferred 20 September 2000 and P 209 on 11 January 2001. Carry a RIB with a 40 hp engine. Caribbean Flotilla.

28 DE NOVIEMBRE *6/2012*, SENAN* / 1525626

IHS Jane's Fighting Ships 2014-2015

Patrol forces < **Panama** 613

7 FAST PATROL BOATS (PBF)

BPC 3208   BPC 3216   BPC 3218   BPC 3227   BPC 3243–3244   BPC 3246

Dimensions, metres (feet): 10.2 × 2.3 × 0.6 *(33.5 × 7.5 × 2.0)*
Speed, knots: 30
Complement: 5
Machinery: 2 Yamaha outboards; 400 hp(m) *(294 kW)*
Guns: 1—7.62 mm MG.

Comment: Eduardoño class acquired between in 1995.

EDUARDOÑO CLASS   6/2003, Panama Maritime Service / 0587789

4 INTERCEPTOR CRAFT (PBF)

BIM 4301–4304

Dimensions, metres (feet): 13.4 × 2.75 × 0.9 *(44 × 9.0 × 3.0)*
Speed, knots: 60. **Range, n miles:** 600 at 25 kt
Complement: 5
Machinery: 3 Yanmar diesels; 945 hp *(704 kW)*; Bravo X drives

Comment: Manufactured by Nor-Tech, Fort Myers, FL. Composite and glass-fibre hull with V-bottomed hull. Donated by the US Southern Command in 2007. Employed on counter drugs, arms trafficking and illegal immigration duties.

INTERCEPT CRAFT   6/2007, US Southern Command / 1167968

1 INSHORE PATROL CRAFT (PBR)

COCLE P 814

Displacement, tonnes: 22.5 full load
Dimensions, metres (feet): 15.3 × 4.0 × 1.1 *(50.2 × 13.1 × 3.6)*
Speed, knots: 18
Complement: 5 (1 officer)
Machinery: 2 Detroit V12 diesels; 960 hp *(716 kW)*; 2 shafts
Guns: 1—12.7 mm MG. 2—7.62 mm MGs.
Radars: Surface search: Furuno 1411; I-band.

Comment: Former US Navy Swift class built by Sewart Marine in about 1967. Used for training at US Navy's Naval Small Craft Instruction and Technical Training School (NAVSCIATTS) before transfer to Panama in March 1999.

COCLE   6/2013*, SENAN / 1525630

2 INTERCEPTOR CRAFT (PBF)

BPC 3401   BPC 3402

Dimensions, metres (feet): 11.3 × 2.3 × 0.6 *(37.1 × 7.5 × 2.0)*
Speed, knots: 30. **Range, n miles:** 200 at 25 kt
Complement: 5
Machinery: 2 Yamaha outboard motors; 500 hp *(298 kW)*
Guns: 2—7.62 mm MGs.

Comment: Manufactured by Donzi Marine, Florida, and delivered in 2005. Employed on counter drugs, arms trafficking, and illegal immigration duties.

BPC 3401   6/2013*, SENAN / 1525633

2 FAST PATROL CRAFT (PBF)

BPC 4506   BPC 4508

Dimensions, metres (feet): 15.3 × 2.3 × 0.6 *(50.2 × 7.5 × 2.0)*
Speed, knots: 30. **Range, n miles:** 200 at 25 kt
Complement: 5
Machinery: 4 Yamaha outboard motors; 800 hp *(596 kW)*
Guns: 2—7.62 mm MGs.

Comment: Manufactured in Panama and delivered in 2011. Employed on counter drugs, arms trafficking, and illegal immigration duties.

1 COASTAL PATROL CRAFT (PB)

BOCAS DEL TORO P 843

Displacement, tonnes: 47 full load
Dimensions, metres (feet): 22.5 × 5.3 × 0.9 *(73.8 × 17.4 × 3.0)*
Speed, knots: 20
Complement: 4 (1 officer)
Machinery: 3 Detroit 12V 71 diesels; 1,260 hp *(940 kW)* sustained; 3 shafts
Guns: 2—12.7 mm MGs.
Radars: Surface search: Furuno 1411; I-band.

Comment: Ex-US Sea Spectre PB Mk IV Class transferred as Grant-Aid from the US in March 1998. Used for drug prevention patrols.

BOCAS DEL TORO   6/2013*, SENAN / 1525625

4 TYPE 200 PATROL CRAFT (PB)

GUILLERMO ENDARA PC 220 (ex-266)   MIREYA MOSCOSO PC 222 (ex-272)
ERNESTO PEREZ BALLADARES PC 221 (ex-270) MARTIN TORRIJOS PC 223 (ex-275)

Displacement, tonnes: 55 full load
Dimensions, metres (feet): 25 × 5.76 × 0.92 *(82 × 18.9 × 3.0)*
Speed, knots: 34. **Range, n miles:** 900 at 18 kt
Complement: 6 (2 officers)
Machinery: 3 Isotta-Fraschini diesels; 2,977 hp *(2.2 MW)*; 2 props; 1 waterjet
Guns: 1—12.7 mm MG. 2—7.62 mm MGs.
Radars: Navigation: Raytheon R-81; I-band.

Comment: Patrol craft built by Intermarine Sarzana for the Italian Coast Guard in about 2001. Aluminium hull. Donated to Panama in June 2011 and delivered in May 2012.

MARTIN TORRIJOS   6/2011* / 1525628

5 FAST PATROL CRAFT (PBF)

BPC 3803–3807

Dimensions, metres (feet): 12.6 × 2.3 × 0.6 *(41.3 × 7.5 × 2.0)*
Speed, knots: 40
Complement: 5
Machinery: 3 Yamaha outboard motors; 600 hp *(447 kW)*
Guns: 2—7.62 mm MGs.

Comment: Manufactured by Eduardoño, Cartagena, Colombia and delivered in 2011–13. Employed on counter drugs, arms trafficking, and illegal immigration duties.

BPV 3806   6/2013*, SENAN / 1525631

IHS Jane's Fighting Ships 2014-2015

LAND-BASED MARITIME AIRCRAFT

Notes: (1) In addition to the aircraft listed below, there are four CASA 212-200/300, one Embraer EMB-135 Legacy, two Cessna 208B Grand Caravan, seven T-35 Pillan, two Bell 412, seven Bell 212, six AW-139 (ordered 2010), five Bell UH-1H, three EC-145 and two Bell 407.
(2) There are plans to acquire two EADS-CASA C-235 maritime patrol aircraft.

Numbers/Type: 3 CASA C-212 Aviocar.
Operational speed: 190 kt (353 km/h).
Service ceiling: 24,000 ft (7,315 m).
Range: 1,650 n miles (3,055 km).
Role/Weapon systems: Air Force operated coastal patrol aircraft for EEZ protection and anti-smuggling duties. Sensors: APS-128 radar, limited ESM. Weapons: ASW; two Mk 44/46 torpedoes. ASV; two rocket or machine gun pods.

C-212   6/2003, Adolfo Ortigueira Gil / 0587787

Numbers/Type: 1 Pilatus Britten-Norman Islander.
Operational speed: 150 kt (280 km/h).
Service ceiling: 18,900 ft (5,760 m).
Range: 1,500 n miles (2,775 km).
Role/Weapon systems: Air Force operated coastal surveillance duties. Sensors: Search radar. Weapons: Unarmed.

AUXILIARIES

1 LANDING CRAFT (LCU)

GENERAL ESTEBAN HUERTAS A 402 (ex-YFU 81)

Displacement, tonnes: 380 full load
Dimensions, metres (feet): 38.1 × 10.9 × 2.3 (125 × 35.8 × 7.5)
Speed, knots: 8
Complement: 11 (3 officers)
Machinery: 2 Caterpillar diesels; 715 hp (533 kW); 2 shafts
Guns: 2 – 12.7 mm MGs. 2 – 7.62 mm MGs.
Radars: Navigation: Furuno R-96; I-band.

Comment: Former US Navy landing craft built by Pacific Coast Engineering in 1968 and transferred in 2004.

GENERAL ESTEBAN HUERTAS   6/2013*, SENAN / 1525632

2 UTILITY VESSELS (YAG)

LINA MARIA L 24   ISLA JICARON L 26

Displacement, tonnes: 200 full load
Dimensions, metres (feet): 34.3 × 6.6 × 2.3 (112.5 × 21.7 × 7.5)
Speed, knots: 10
Range, n miles: 1,500 at 8 kt
Complement: 8 (2 officers)
Machinery: 2 Detroit V8 S90 diesels; 2 shafts
Guns: 2 – 7.62 mm MGs.
Radars: Navigation: Raytheon R-82; I-band.

Comment: Confiscated craft taken into service and refitted in 2003. Both used for general utility tasks.

LINA MARIA   6/2013*, SENAN / 1525629

Papua New Guinea

Country Overview

Papua New Guinea lies north of Australia in the eastern half of New Guinea which it shares with the Indonesian province of Irian Jaya. An Australian-administered UN Trust territory from 1949, it became independent in 1975. Its head of state is the British sovereign, who is represented by a Governor-General. Its many island groups include the Bismarck and Louisiade Archipelagos, the Trobriand Islands, the D'Entrecasteaux Islands and Woodlark Island. Amongst other islands are Bougainville (a nine-year separatist conflict ended in 1997) and Buka. It has a 2,781 n mile coastline. The capital, principal city and port is Port Moresby. An archipelagic state, territorial seas (12 n miles) are claimed. A 200 n mile Exclusive Economic Zone (EEZ) has also been claimed but the limits have not been fully defined by boundary agreements.

Headquarters Appointments

Commander Defence Forces:
  Brigadier Francis Agwi
Director Maritime Operations:
  Commander Perry Sundie

Bases

Port Moresby (HQ PNGDF and PNGDF Landing Craft Base); Lombrum (Manus)

Naval Aviation

Two IAI Arava 201 and two Airtech CN-235M-100 transport aircraft are available for maritime surveillance.

Prefix to Ships' Names

HMPNGS

PATROL FORCES

4 PACIFIC CLASS (LARGE PATROL CRAFT) (PB)

Name	No	Builders	Commissioned
RABAUL (ex-*Tarangau*)	01	Australian Shipbuilding Industries	16 May 1987
DREGER	02	Australian Shipbuilding Industries	31 Oct 1987
SEEADLER	03	Australian Shipbuilding Industries	29 Oct 1988
MORESBY (ex-*Basilisk*)	04	Australian Shipbuilding Industries	1 Jul 1989

Displacement, tonnes: 165 full load
Dimensions, metres (feet): 31.5 × 8.1 × 2.1 (103.3 × 26.6 × 6.9)
Speed, knots: 20
Range, n miles: 2,500 at 12 kt
Complement: 17 (3 officers)
Machinery: 2 Caterpillar 3516TA diesels; 4,400 hp (3.3 MW) sustained; 2 shafts
Guns: 1 Oerlikon GAM-BO1 20 mm. 2 – 7.62 mm MGs.
Radars: Surface search: Furuno 1011; I-band.

Comment: Contract awarded in 1985 to Australian Shipbuilding Industries (Hamilton Hill, West Australia) under Australian Defence co-operation. These are the first, third, sixth and seventh of the class and some of the few to be armed. All upgraded, during half-life refits in Australia with new radars and navigation support systems in 1997/98. Following the decision by the Australian government to extend the Pacific Patrol Boat project, *Rabaul* underwent a life-extension refit at Gladstone in 2003 and *Dreger* at Townsville in 2004. Similar refits conducted for *Seeadler* and *Moresby* at Townsville in 2006 and 2007 respectively.

DREGER   10/2013*, Mick Prendergast / 1530963

AUXILIARIES

2 LANDING CRAFT (LSM)

Name	No	Builders	Commissioned
SALAMAUA	31	Walkers Ltd, Maryborough	19 Oct 1973
BUNA	32	Walkers Ltd, Maryborough	7 Dec 1973

Displacement, tonnes: 315 standard; 511 full load
Dimensions, metres (feet): 44.5 × 10.1 × 1.9 *(146 × 33.1 × 6.2)*
Speed, knots: 10
Range, n miles: 3,000 at 10 kt
Complement: 15 (2 officers)

Military lift: 160 tons
Machinery: 2 GM diesels; 2 shafts
Guns: 2 – 12.7 mm MGs.
Radars: Navigation: Racal Decca RM 916; I-band.

Comment: Transferred from Australia in 1975. Underwent extensive refits 1985–86. Both vessels reported non-operational (awaiting refits) in 2012.

SALAMAUA　　　　　　　　　　　　　　　　　　　　　　12/1990, James Goldrick / 0081510

Paraguay

ARMADA NACIONAL

Country Overview

The Republic of Paraguay is one of two landlocked countries in South America; Bolivia is the other. With an area of 157,048 square miles, it has borders to the north with Bolivia, to the east with Brazil and to the south with Argentina. There are some 1,800 n miles of internal waterways including the principal rivers, the Pilcomayo, Paraguay and Alto Paraná. Navigable by large ships for much of their length, they link the capital, largest city and principal port, Asunción, with the Rio de la Plata estuary on the Atlantic Ocean. Other ports include Ciudad del Este, Encarnación and Concepción.

Headquarters Appointments

Commander-in-Chief of the Navy:
　Admiral Silvio Salvador Guanes Solis

Personnel

2014: 1,800 including Coast Guard, marines and naval air

Organisation and Bases

Main Base: Puerto Sajonia, Asunción
There are seven Areas Navales (AN):
AN del Este
AN de Itapua
AN de Canindeyu
AN de Bahia Negra
AN de Pozo Hondo
AN de Paña Hermosa
AN de Confluencia

Training

Specialist training is done with Argentina, Bolivia, Brazil, and Uruguay (Exercise Acrux).

Marine Corps

BIM 1: Puerto Rosario
BIM 2: Puerto Vallemi
BIM 3: Asunción

Naval Aviation

Fixed Wing　Asunción International Airport
Helicopters　Puerto Sajonia

Coast Guard

Prefectura General Naval

PATROL FORCES

Notes: An unnamed 10 m patrol craft, LP 12, is based at Asuncion. There are also over 80 smaller craft of 5–6.5 m including 12 (LP 30–41) commissioned in 2007, 16 (LP 42 Series) in 2008, 15 in 2009, five in 2011 (LP 70 series) and 32 in 2012.

2 NOVATEC RIVER PATROL CRAFT (PBR)

YHAGUY P 08　　　**TEBICUARY** P 09

Displacement, tonnes: 25 standard
Dimensions, metres (feet): 16.1 × 4.5 × 0.58 *(52.8 × 14.8 × 1.9)*
Speed, knots: 40
Machinery: 2 Caterpillar 2406E diesels; 1,600 hp *(1.2 MW)*; 2 shafts
Guns: 3 – 7.62 mm MGs (fitted for).
Radars: Surface search: Furuno; I-band.

Comment: Built by Nova Marine Co Ltd, Taiwan. Commissioned 23 June 1999.

TEBICUARY　　　　　　　　　　　　　　　　　　　　　　5/2011, Hartmut Ehlers / 1406650

1 RIVER DEFENCE VESSEL (PGR)

Name	No	Builders	Commissioned
PARAGUAY	C 1	Odero, Genoa	May 1931

Displacement, tonnes: 646 standard; 879 full load
Dimensions, metres (feet): 70 × 10.7 × 1.7 *(229.7 × 35.1 × 5.6)*
Speed, knots: 17
Range, n miles: 1,700 at 16 kt
Complement: 86

Machinery: 2 boilers; 2 Parsons turbines; 3,800 hp *(2.83 MW)*; 2 shafts
Guns: 4 – 4.7 in *(120 mm)* (2 twin). 3 – 3 in *(76 mm)*. 2 – 40 mm.
Mines: 6.
Radars: Navigation: I-band.

Comment: Refitted in 1975. Has 0.5 in side armour plating and 0.3 in on deck. Based at Asunción. Gun tubs on either side of bridge can be fitted with single 20 mm guns. Plans to re-engine with diesels have not yet been implemented and the ship is probably non-operational.

PARAGUAY and TENIENTE FARINA　　　　　　　　　　4/2003, Hartmut Ehlers / 0587791

5 RIVER PATROL CRAFT (PBR)

P 107 (ex-*P 07*)　　**LP 09** (ex-*P 09*)　　**LP 11** (ex-*P 11*)
LP 08 (ex-*P 08*)　　**LP 10** (ex-*P 10*)

Displacement, tonnes: 18 full load
Dimensions, metres (feet): 14.7 × 3.1 × 0.8 *(48.2 × 10.2 × 2.6)*
Speed, knots: 12. **Range, n miles:** 240 at 12 kt
Complement: 4
Machinery: 1 GM 6-71 diesel; 340 hp *(254 kW)*; 1 shaft
Guns: 2 – 12.7 mm MGs.

Comment: Built by Arsenal de Marina, Paraguay. P 107 launched March 1989, LP 08 and LP 09 in February 1990 and LP 10–11 in October 1991. The programme was then aborted. Bases: LP 08 (Bahia Negra); P 107 (under repair); LP 09 (Isla Margarita); LP 10 (Asuncion); LP 11 (Encarnacion).

LP 10　　　　　　　　　　　　　　　　　　　　　　　　　4/2003, Hartmut Ehlers / 0567477

616 Paraguay > Patrol forces

1 ITAIPÚ CLASS (RIVER DEFENCE VESSEL) (PBR)

Name	No	Builders	Commissioned
ITAIPÚ	P 05 (ex-P 2)	Arsenal de Marinha, Rio de Janeiro	2 Apr 1985

Displacement, tonnes: 371 full load
Dimensions, metres (feet): 46.3 × 8.5 × 1.4 (151.9 × 27.9 × 4.6)
Speed, knots: 14
Range, n miles: 6,000 at 12 kt
Complement: 40 (9 officers) + 30 embarked forces

Machinery: 2 MAN V6V16/18TL diesels; 1,920 hp(m) (1.41 MW); 2 shafts
Guns: 1 Bofors 40 mm/60. 2—81 mm mortars. 6—12.7 mm MGs.
Radars: Navigation: Furuno; I-band.
Helicopters: Platform for 1 HB 350B or equivalent.

Comment: Ordered late 1982. Launched 16 March 1984. Same as Brazilian Roraima class. Has some hospital facilities. Based at Asunción. Refitted in 2010.

ITAIPÚ 5/2011, Hartmut Ehlers / 1406645

2 BOUCHARD CLASS (PATROL SHIPS) (PBR)

Name	No	Builders	Commissioned
TENIENTE FARINA (ex-Py M 10)	P 04 (ex-M 3)	Rio Santiago Naval Yard	1 Jul 1939
NANAWA (ex-Bouchard)	P 02 (ex-M 1)	Rio Santiago Naval Yard	27 Jan 1937

Displacement, tonnes: 457 standard; 660 full load
Dimensions, metres (feet): 60 × 7.3 × 2.6 (196.9 × 24.0 × 8.5)
Speed, knots: 16
Range, n miles: 6,000 at 12 kt
Complement: 70

Machinery: 2 sets MAN 2-stroke diesels; 2,000 hp(m) (1.47 MW); 2 shafts
Guns: 4 Bofors 40 mm/60 (2 twin). 2—12.7 mm MGs.
Mines: 1 rail.
Radars: Navigation: I-band.

Comment: Former Argentinian minesweepers of the Bouchard class. Transferred from the Argentine Navy to the Paraguayan Navy and recommissioned 14 March 1964 (Nanawa) and 6 May 1968 (Teniente Farina). A third ship, Capitán Meza, was scrapped between 1995–97. Based at Asunción but both probably non-operational.

BOUCHARD CLASS 6/1990, Paraguay Navy / 0081514

1 RIVER PATROL CRAFT (PBR)

Name	No	Builders	Commissioned
CAPITÁN CABRAL (ex-Triunfo)	P 01 (ex-P 1, ex-A 1)	Werf-Conrad, Haarlem	1908

Displacement, tonnes: 183 standard; 209 full load
Dimensions, metres (feet): 33.9 × 7.2 × 2.0 (111.2 × 23.6 × 6.6)
Speed, knots: 9
Complement: 25 (7 officers)
Machinery: 1 Caterpillar 3408 diesel; 365 hp (272 kW); 1 shaft
Guns: 1 Bofors 40 mm/60. 2 Oerlikon 20 mm. 2—12.7 mm MGs.
Radars: Navigation: Furuno; I-band.

Comment: Former tug. Launched in 1907. Still in excellent condition. Vickers guns were replaced and a diesel engine fitted by Arsenal de Marina in 1984. Took part in exercise NINFA XXIII in 2010. Based at Asunción.

CAPITÁN CABRAL 5/2011, Hartmut Ehlers / 1406646

2 MODIFIED HAI OU CLASS (PBF)

CAPITÁN ORTIZ P 06 TENIENTE ROBLES P 07

Displacement, tonnes: 48 full load
Dimensions, metres (feet): 21.6 × 5.5 × 1 (70.9 × 18.0 × 3.3)
Speed, knots: 36
Range, n miles: 700 at 32 kt
Complement: 10
Machinery: 2 MTU 12V 331 TC82 diesels; 2,605 hp(m) (1.92 MW) sustained; 2 shafts
Guns: 3—12.7 mm MGs. 2—7.62 mm MGs.
Radars: Surface search: I-band.

Comment: Developed by Taiwan from Dvora class hulls and presented as a gift in 1996. It is possible that these craft are the two original Dvora hulls acquired by Taiwan. Refitted in 2010–11.

TENIENTE ROBLES 5/2011, Hartmut Ehlers / 1406651

6 TYPE 701 CLASS (PBR)

MIGUEL SOTOA LP 01 (ex-P 105) – P 103
– LP 101 (ex-P 101) – LP 104 (ex-P 104)
MANUEL TRUJILLO LP 102 (ex-P 102) – P 106

Displacement, tonnes: 15 full load
Dimensions, metres (feet): 13 × 3.9 × 0.9 (42.7 × 12.8 × 3.0)
Speed, knots: 20
Complement: 7
Machinery: 2 diesels; 500 hp (373 kW); 2 shafts
Guns: 2—12.7 mm MGs.

Comment: Built by Sewart in 1970. Delivered 1967–71. Bases: LP 01 (Asuncion); LP 101 (Ayolas); LP 102 (Asuncion); P 103 (Lake Itaipu); LP 104 (Conception); P 106 (Lake Itaipu).

LP 102 5/2011, Hartmut Ehlers / 1406649

3 + (3) CROQ 15 (PATROL CRAFT) (PBR)

MARINA PELAYO PRATT GILL LP 201 – LP 202 – LP 203

Displacement, tonnes: 12 full load
Dimensions, metres (feet): 15 × 3.85 × 0.8 (49.2 × 12.6 × 2.6)
Speed, knots: 38
Complement: 2
Military lift: 18 troops
Machinery: 2 diesels; 900 hp (671 kW); 2 waterjets
Guns: 2—12.7 mm MGs.

Comment: Built by Armacraft Ltd, Australia. The craft were ordered in October 2010. The first two were delivered in September 2012 and the third is to follow in 2014. Aluminium construction with ballistic protection. The acquisition of three further craft is under consideration.

LP 201 9/2012, Paraguay Navy / 1486326

IHS Jane's Fighting Ships 2014-2015 © 2014 IHS

Land-based maritime aircraft < **Paraguay** — Introduction < **Peru** 617

LAND-BASED MARITIME AIRCRAFT

Notes: The Naval Aviation inventory includes two fixed-wing aircraft (two Cessna 150, one Cessna 210N) in operational use.

Numbers/Type: 1 Helibras HB 350B Esquilo.
Operational speed: 125 kt *(232 km/h)*.
Service ceiling: 10,000 ft *(3,050 m)*.
Range: 390 n miles *(720 km)*.
Role/Weapon systems: Support helicopter for riverine patrol craft. Delivered in July 1985.

ESQUILO 5/2000, *Hartmut Ehlers* / 0105198

AUXILIARIES

Notes: In addition to the craft listed, there are three LCVPs (EDVP 1–3), two service craft (Arsenal 1 and 2), one floating crane *(Grua Flotante)* and one floating dry dock *(Dique Flotante (ex-AFDL 26))*.

EDVP-03 5/2011, *Hartmut Ehlers* / 1406648

1 PRESIDENTIAL YACHT (MYAC)

3 DE FEBRERO (ex-*26 de Febrero*)

Displacement, tonnes: 100 full load
Dimensions, metres (feet): 28.1 × 6.0 × 1.6 *(92.2 × 19.7 × 5.2)*
Speed, knots: 11. **Range, n miles:** 1,350 at 11 kt
Complement: 6 + 8 personnel
Machinery: 2 Rolls Royce; 517 hp *(386 kW)*; 2 shafts

Comment: Built by Naval Arsenal Asunción and launched in 1972. Entered service in 1982.

3 de FEBRERO 5/2011, *Hartmut Ehlers* / 1406647

1 HYDROGRAPHIC LAUNCH (YGS)

SUBOFICIAL ROGELIO LESME LPH 01 (ex-LH 1)

Displacement, tonnes: 16 full load
Dimensions, metres (feet): 14.7 × 3.1 × 0.8 *(48.2 × 10.2 × 2.6)*
Speed, knots: 13
Complement: 5
Machinery: 1 Mercedes-Benz diesel; 100 hp *(74 kW)*; 1 shaft

Comment: Built in 1958. Refitted in 2010.

TUGS

2 TUGS (YTM/YTL)

ANGOSTURA R 5 (ex-YTL 211) **ESPERANZA** R 7

Displacement, tonnes: 71 full load
Dimensions, metres (feet): 19.8 × 5 × 2.3 *(65 × 16.4 × 7.5)*
Speed, knots: 9
Complement: 5
Machinery: 1 Scania DSI 14MO3 diesel; 360 hp *(269 kW)*; 1 shaft

Comment: *Angostura* is a harbour tug transferred under MAP. Leased in March 1965 and sold on 11 February 1977. Rebuilt by Arsenal de Marinha in 1992. *Esperanza* is a smaller 15 m vessel built by Arsenal de Marina.

Peru
ARMADA PERUANA

Country Overview

The Republic of Peru is situated in western South America. With an area of 496,225 square miles it has borders to the north with Ecuador and Colombia, to the east with Brazil and Bolivia and to the south with Chile. It has a coastline of 1,850 n miles with the Pacific Ocean. Lima is capital and largest city and is served by the port of Callao. There are further ports at Paita, Salaverry, Chimbote, Pisco, San Juan, Matarani and Ilo. Inland, Iquitos and Pucallpa are linked to the Atlantic Ocean by the Amazon River. Lake Titicaca is also an important waterway. Peru has not claimed an EEZ but is one of a few coastal states which claims a 200 n mile territorial sea.

Headquarters Appointments

Commander of the Navy:
  Admiral Carlos Tejada Mera
Chief of the Naval Staff:
  Vice Admiral Reynaldo Pizarro Antram
Inspector General:
  Vice Admiral Mauro Cacho de Armero
Commander Pacific Operations Command (Callao):
  Vice Admiral Wladimiro Giovannini y Freire
Commander Amazon Operations Command (Iquitos):
  Vice Admiral Moscoso Flores

Director General, Coast Guard:
  Vice Admiral Edmundo Deville del Campo
Commander, Surface Forces:
  Rear Admiral Manuel Vascones Morey
Commander, Submarines:
  Rear Admiral Fernando Cerdan Ruiz
Commander, Special Operations Force:
  Rear Admiral José Vía Mezarina
Commander, Naval Aviation:
  Rear Admiral Ricardo Menendez Calle
Commander, Marine Infantry Force:
  Rear Admiral Enrique de la Flor Rivero

Personnel

(a) 2014: 23,719 (2,000 officers)
(b) 2 years' voluntary military service

Organisation and Bases

2 Operational Commands: Pacific (Callao) and Amazon (Iquitos).
6 Naval Forces: Surface Fleet, Submarine, Naval Aviation, Marine Corps, Special Operations (Callao) and Naval Amazon Force (Iquitos).
5 Naval Zones: 1st (Piura), 2nd (Callao), 3rd (Arequipa), 4th (Pucallpa) and 5th (Iquitos).
Coast Guard General Directorate (Callao).
Callao: Main Naval Base, dockyard with shipbuilding capacity, one dry dock, three floating docks, training schools, Submarine Naval Station. Main Naval Air Base near Jorge Chavez International Airport.
San Lorenzo: Naval Station.
Iquitos: River base for Amazon Flotilla; small building yard, repair, facilities, floating dock.
Pucallpa: River base with logistic facilities.
San Juan de Marcona: Naval Aviation Training School and airfield.
Paita: Naval Station with logistic facilities.
Chimbote: Naval Base, dockyard for small vessels, logistic facilities.
Mollendo: Naval Station with logistic facilities
El Salto (Tumbes): Naval Station with logistics facilities
Puno: Lake Titicaca Lake Station.
La Punta (Naval Academy).
River Stations: Puero Maldonado, Pichari, El Estrecho, and Güepí.

Marines

The Peruvian Marines comprise 3,500 men whose Headquarters is at Ancón. The force includes a Marine Brigade, the Amphibious Support Group and Recon Forces. The Marine Brigade has three battalions: First Battalion – Guarnición de Marina; Second Battalion – Guardia Chalaca; Third Battalion (including Fire Support Group armed with 122 mm howitzer and 120 mm mortar and Engineer Support company) – Vencedores de Punta Melpelo. The Amphibious Support Group is composed of the Vehicles and Motor Transport battalions. Recon Forces include a Commando and anti-terrorist companies. Additionally, the Peruvian Marines have jungle battalions at Iquitos and Pucallpa (BIMSE 1 and BIMSE 2).

Special Operations

The Special Operations Command is responsible for the organisation, equipment, training and control of the operations of its subordinate Units; these Units are: the North, Central, South and Northwest Special Operations Groups, the Diving and Salvage Group, the Explosives Ordnance Unit, The Special Operations Station and the Special Operations School.

Prefix to Ships' Names

BAP (Buque Armada Peruana).

Coast Guard

A separate service set up in 1975 with a number of light forces transferred from the Navy.

PENNANT LIST

Submarines		Frigates		CF 12	Loreto	Amphibious Forces		Auxiliaries		ARA 123	Guardian Rios
		FM 52	Villavisencio	CF 13	Marañón					ARB 120	Mejia
SS 31	Angamos	FM 53	Montero	CF 14	Ucayali	DT 143	Callao	ABH 302	Morona	ARB 121	Huertas
SS 32	Antofagasta	FM 54	Mariategui	CF 15	Clavero	DT 144	Eten	ABH 303	Corrientes	ARB 124	Selendán
SS 33	Pisagua	FM 55	Aguirre	CF 16	Castilla			ABH 304	Curaray	ARB 125	Medina
SS 34	Chipana	FM 56	Palacios	CM 21	Velarde	Survey Ships		ABH 305	Pastaza	ARB 126	Dueñas
SS 35	Islay	FM 57	Bolognesi	CM 22	Santillana			ABH 306	Puno	ARB 128	Olaya
SS 36	Arica	FM 58	Quiñones	CM 23	De los Heros	AH 171	Carrasco	ACA 111	Caloyeras	ART 322	San Lorenzo
				CM 24	Herrera	AH 172	Stiglich	ACP 118	Noguera	ATP 154	Bayovar
Cruisers		Patrol Forces		CM 25	Larrea	AEH 174	La Macha	ACP 119	Gauden	ATP 155	Zorritos
CLM 81	Almirante Grau	CF 11	Amazonas	CM 26	Sanchez Carrión	AH 175	Carrillo	ALY 313	Marte	LSP 180	Mantilla
						AH 176	Melo	AMB 160	Unanue	LSP 181	Aguilar

© 2014 IHS IHS *Jane's Fighting Ships 2014-2015*

SUBMARINES

6 ANGAMOS/ISLAY (TYPE 209/1200) CLASS (SSK)

Name	No	Builders	Laid down	Launched	Commissioned
ANGAMOS (ex-Casma)	SS 31	Howaldtswerke, Kiel	15 Jul 1977	31 Aug 1979	19 Dec 1980
ANTOFAGASTA	SS 32	Howaldtswerke, Kiel	3 Oct 1977	19 Dec 1979	20 Feb 1981
PISAGUA	SS 33	Howaldtswerke, Kiel	15 Aug 1978	19 Oct 1980	12 Jul 1983
CHIPANA	SS 34	Howaldtswerke, Kiel	1 Nov 1978	19 May 1981	20 Sep 1982
ISLAY	SS 35	Howaldtswerke, Kiel	15 Mar 1971	11 Oct 1973	29 Aug 1974
ARICA	SS 36	Howaldtswerke, Kiel	1 Nov 1971	5 Apr 1974	21 Jan 1975

Displacement, tonnes: 1,204 surfaced; 1,311 dived
Dimensions, metres (feet): 56 × 6.2 × 5.5 *(183.7 × 20.3 × 18.0)*
Speed, knots: 11 surfaced; 11 snorting; 21.5 dived
Complement: 31 (SS 31–34), 35 (SS 35–36)

Machinery: Diesel-electric; 4 MTU 12V 493 AZ80 GA31L diesels; 2,400 hp(m) *(1.76 MW)* sustained; 4 Siemens alternators; 1.7 MW; 1 Siemens motor; 4,600 hp(m) *(3.38 MW)* sustained; 1 shaft
Torpedoes: 8—21 in *(533 mm)* tubes. 14 AEG SUT 264; wire-guided; active/passive homing to 12/28 km *(6.5/15 n miles)* at 35/23 kt; warhead 260 kg. Swim-out discharge.
Electronic countermeasures: ESM: Radar warning.
Radars: Surface search: Thomson-CSF Calypso; I-band.
Sonars: Atlas Elektronik CSU 3; active/passive search and attack; medium/high frequency. Thomson Sintra DUUX 2C or Atlas Elektronik PRS 3; passive ranging.
Weapon control systems: Sepa Mk 3 or Signaal Sinbad M8/24 (*Angamos* and *Antofagasta*).

Programmes: Two Type 209/1200 (SS 35–36) ordered 1969. Two further Type 209/1200 Mod boats (SS 31–32) ordered 12 August 1976. Two further Type 209/1200 Mod (SS 33–34) ordered 21 March 1977. Designed by Ingenieurkontor, Lübeck for construction by Howaldtswerke, Kiel and sale by Ferrostaal, Essen all acting as a consortium.
Modernisation: Sepa Mk 3 fire control fitted progressively from 1986. *Islay* and *Arica* modified under Delfin II programme 2007–08. Upgrade included new command and control system, new batteries, sonar improvements and SUT 264 torpedoes. The four Type 209/1200 Mod are also to be upgraded.
Structure: A single-hull design with two ballast tanks and forward and after trim tanks. Fitted with snort and remote machinery control. The single screw is slow revving, very high-capacity batteries with GRP lead-acid cells and battery cooling-by Wilh Hagen and VARTA. Fitted with two periscopes and Omega receiver. Foreplanes retract. Diving depth, 250 m *(820 ft)*.
Operational: Endurance, 50 days. Four are in service, two in refit or reserve at any one time. *Angamos* took part in multinational exercises in mid-2004 during which she achieved 156 days at sea.

ANTOFAGASTA *6/2011, Peruvian Navy / 1406652*

TYPE 209 *6/2011, Peruvian Navy / 1406653*

CRUISERS

1 DE RUYTER CLASS (CG/CLM)

Name	No	Builders	Laid down	Launched	Commissioned
ALMIRANTE GRAU (ex-De Ruyter)	CLM 81	Wilton-Fijenoord, Schiedam	5 Sep 1939	24 Dec 1944	18 Nov 1953

Displacement, tonnes: 12,360 full load
Dimensions, metres (feet): 190.3 × 17.3 × 6.7 *(624.3 × 56.8 × 22.0)*
Speed, knots: 32. **Range, n miles:** 7,000 at 12 kt
Complement: 953 (49 officers)

Machinery: 4 Werkspoor-Yarrow boilers; 2 De Schelde-Parsons turbines; 85,000 hp *(62.5 MW)*; 2 shafts
Missiles: SSM: 8 Oto Melara/Matra Otomat Mk 2 (TG 1) ❶; active radar homing to 80 km *(43.2 n miles)* at 0.9 Mach; warhead 210 kg; sea-skimmer for last 4 km *(2.2 n miles)*.
Guns: 8 Bofors 6 in *(152 mm)*/53 (4 twin) ❷; 15 rds/min to 26 km *(14 n miles)*; weight of shell 46 kg. 4 Otobreda 40 mm/70 (2 twin) ❸; 120 rds/min to 12.5 km *(6.8 n miles)*; weight of shell 0.96 kg. 4 Bofors 40 mm/70 ❹; 300 rds/min to 12 km *(6.6 n miles)*; weight of shell 0.96 kg.
Physical countermeasures: Decoys: 2 Dagaie and 1 Sagaie chaff launchers.
Radars: Air search: AN/SPS-6 ❼; D-band.
Surface search/target indication: Signaal DA08 ❽; E/F-band.
Navigation: Racal Decca 1226; I-band.
Fire control: Signaal WM25 ❾; I/J-band (for 6 in guns); range 46 km *(25 n miles)*. Signaal STIR ❿; I/J/K-band; range 140 km *(76 n miles)* for 1 m² target.
Combat data systems: Signaal Sewaco PE SATCOM ❺.
Weapon control systems: 2 Lirod 8 optronic directors ❻.

Programmes: Transferred by purchase from Netherlands 7 March 1973 and commissioned in Peruvian Navy 23 May 1973.
Modernisation: Taken in hand for a two and a half year modernisation at Amsterdam Dry Dock Co in March 1985. This was to include reconditioning of mechanical and electrical engineering systems, fitting of SSM and SAM, replacement of electronics and fitting of one CSEE Sagaie and two Dagaie launchers. In 1986 financial constraints limited the work but much had been done to update sensors and fire-control equipment. Sailed for Peru 23 January 1988 without her secondary gun armament, which was completed at Sima Yard, Callao. Sonar has been removed. AN/SPS-6 antenna reportedly removed in 2012. SATCOM fitted aft.

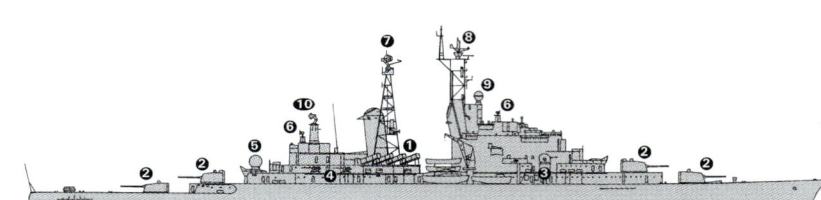
ALMIRANTE GRAU *(Scale 1 : 1,800), Ian Sturton / 0126352*

ALMIRANTE GRAU *11/2012*, Peruvian Navy / 1525635*

FRIGATES

3 CARVAJAL (MODIFIED LUPO) CLASS (FFGHM)

Name	No	Builders	Laid down	Launched	Commissioned
VILLAVISENCIO	FM 52	Fincantieri, Riva Trigoso	6 Oct 1976	7 Feb 1978	25 Jun 1979
MONTERO	FM 53	SIMA, Callao	Oct 1978	8 Oct 1982	25 Jul 1984
MARIATEGUI	FM 54	SIMA, Callao	1979	8 Oct 1984	10 Oct 1987

Displacement, tonnes: 2,243 standard; 2,540 full load
Dimensions, metres (feet): 113.2 × 11.3 × 3.7 *(371.4 × 37.1 × 12.1)*
Speed, knots: 35
Range, n miles: 3,450 at 20.5 kt
Complement: 185 (20 officers)

Machinery: CODOG; 2 GE/Fiat LM 2500 gas turbines; 50,000 hp *(37.3 MW)* sustained; 2 GMT A 230.20 M diesels; 8,000 hp(m) *(5.88 MW)* sustained; 2 shafts; LIPS cp props
Missiles: SSM: 8 Oto Melara/TESEO Mk 2 (TG 2) ❶; active radar homing to 180 km *(91.2 n miles)* at 0.9 Mach; warhead 210 kg; sea-skimmer.
SAM: Selenia Elsag Albatros octuple launcher ❷; 8 Aspide; semi-active radar homing to 13 km *(7 n miles)* at 2.5 Mach; height envelope 15–5,000 m *(49.2–16,405 ft)*; warhead 30 kg. An SA-N-10 launcher (MPG-86) may be fitted on the stern.
Guns: 1 Oto Melara 5 in *(127 mm)*/54 ❸; 45 rds/min to 16 km *(8.7 n miles)*; weight of shell 32 kg. 4 Breda 40 mm/70 *(2 twin)* ❹; 300 rds/min to 12.5 km *(6.8 n miles)*; weight of shell 0.96 kg. 2—12.7 mm MGs.
Torpedoes: 6—324 mm ILAS (2 triple) tubes ❺. Mk 44; anti-submarine; active homing to 5 km *(2.7 n miles)* at 30 kt; warhead 34 kg (shaped charge).
Physical countermeasures: Decoys: 2 Breda 105 mm SCLAR 20-barrelled trainable launchers ❻; multipurpose; chaff to 5 km *(2.7 n miles)*; illuminants to 12 km *(6.6 n miles)*; HE bombardment.
Electronic countermeasures: ESM: Elettronica Lambda; intercept.

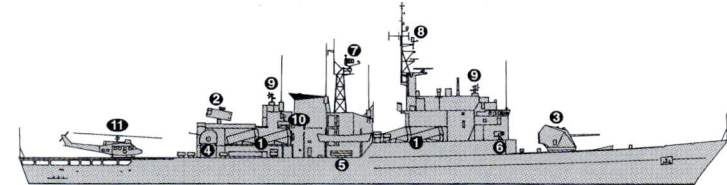

MARIATEGUI

Radars: Air search: Selenia RAN 10S ❼; E/F-band.
Surface search: Selenia RAN 11LX ❽; D/I-band.
Navigation: SMA 3 RM 20R; I-band.
Fire control: 2 RTN 10X ❾; I/J-band. 2 RTN 20X ❿; I/J-band (for Dardo).
Sonars: EDO 610E; hull-mounted; active search and attack; medium frequency.
Combat data systems: Selenia IPN-10 action data automation.
Weapon control systems: 2 Elsag Mk 10 Argo with NA-21 directors. Dardo system for 40 mm.
Helicopters: 1 Agusta AB 212ASW ⓫. 1 Agusta ASH-3D Sea King (deck only) (FM 53 and 54).

Programmes: *Montero* and *Mariategui* were the first major warships to be built on the Pacific Coast of South America, although some equipment was provided by Fincantieri.

(Scale 1 : 1,200), Ian Sturton / 0105275

Modernisation: FM 53 and FM 54 have had flight deck extensions in order to operate Sea Kings from the deck although they cannot be stowed in the hangar. Exocet MM 40 Block 3 is to replace TESEO in FM 53 and 54 2014–15. These ships are also to be equipped with indigenously developed command system Varayoc and ESM system Qhawax.
Structure: FM 52–54 differ from those built for Italian service by having a fixed hangar and higher 40 mm mounts. The SAM system is also different. FM 52 has not had a flight-deck extension and retains a step-down from the flight-deck to the stern.
Operational: Helicopter provides an over-the-horizon targeting capability for SSM. HIFR facilities fitted to FM 52 and 53 in 1989 to allow refuelling of Sea King helicopters. FM 52 is also used in a training role. FM 51 transferred to the Coast Guard in 2013.

MONTERO

10/2013, Guy Toremans* / 1525636

MARIATEGUI

2/2013, Guy Toremans* / 1486549

Peru > Frigates

4 AGUIRRE (LUPO) CLASS (FFGHM)

Name	No	Builders	Laid down	Launched	Commissioned
AGUIRRE (ex-*Orsa*)	FM 55 (ex-F 567)	Fincantieri, Muggiano	1 Aug 1977	1 Mar 1979	1 Mar 1980
PALACIOS (ex-*Lupo*)	FM 56 (ex-F 564)	Fincantieri, Riva Trigoso	11 Oct 1974	29 Jul 1976	12 Sep 1977
BOLOGNESI (ex-*Perseo*)	FM 57 (ex-F 566)	Fincantieri, Riva Trigoso	24 Feb 1977	12 Jul 1978	1 Mar 1980
QUIÑONES (ex-*Sagittario*)	FM 58 (ex-F 565)	Fincantieri, Riva Trigoso	4 Feb 1976	22 Jun 1977	18 Nov 1978

Displacement, tonnes: 2,243 standard; 2,540 full load
Dimensions, metres (feet): 113.2 × 11.3 × 3.7 *(371.4 × 37.1 × 12.1)*
Speed, knots: 35; 21 diesel
Range, n miles: 4,350 at 16 kt
Complement: 185 (20 officers)

Machinery: CODOG; 2 GE/Fiat LM 2500 gas turbines; 50,000 hp *(37.3 MW)* sustained; 2 GMT BL 230. 20M diesels; 10,000 hp(m) *(7.3 MW)* sustained; 2 shafts; LIPS cp props
Missiles: SSM: 8 Oto Melara/TESEO Mk 2 (TG 2) (FM 56, 58) ❶; active radar homing to 180 km *(91.2 n miles)* at 0.9 Mach; warhead 210 kg; sea-skimmer. 8 MBDA Exocet MM Block 3 (FM 55, 57); inertial cruise; active radar homing to 180 km *(97.2 n miles)* at 0.9 Mach; warhead 165 kg.
SAM: Raytheon NATO Sea Sparrow RIM-7M Mk 29 octuple launcher ❷; semi-active radar homing to 14.6 km *(8 n miles)* at 2.5 Mach; warhead 39 kg.
Guns: 1 Oto Melara 5 in *(127 mm)*/54 ❸; 45 rds/min to 16 km *(8.7 n miles)*; weight of shell 32 kg. 4 Breda 40 mm/70 (2 twin) ❹; 300 rds/min to 12.5 km *(6.8 n miles)*; weight of shell 0.96 kg. 2 Oerlikon 20 mm may be fitted.
Torpedoes: 6—324 mm Mk 32 (2 triple) tubes ❺. Mk 44; anti-submarine; active homing to 5 km *(2.7 n miles)* at 30 kt; warhead 34 kg (shaped charge).
Physical countermeasures: Decoys: Rheinmetall MASS. 2 Breda 105 mm SCLAR 20-barrelled trainable launchers ❻; multipurpose; chaff to 5 km *(2.7 n miles)*; illuminants to 12 km *(6.6 n miles)*; HE bombardment.
Electronic countermeasures: ESM: SLR-4; intercept. ECM: 2 SLQ-D; jammer.
Radars: Air search: Selenia SPS-774 (RAN 10S) (FM 56, 58). Selex Kronos NV 3D (FM 55, 57) ❼; E/F-band.
Surface search: SMA SPS 702 ❽; I-band. SMA SPQ-2F ❾; I-band.
Navigation: SMA SPN-748; I-band.
Fire control: Selenia SPG-70 (RTN 10X) ❿ I/J-band. 2 Selenia SPG-74 (RTN 20X) ⓫; I/J-band. 1 US Mk 95 Mod 1 ⓬; I-band.
Sonars: Raytheon DE 1160B; hull-mounted; active search and attack; medium frequency.
Combat data systems: Selenia IPN 20 (SADOC 2) action data automation (FM 56, 58). Varayoc (FM 55, 57). Link 11 (SATCOM).
Weapon control systems: 2 Elsag Mk 10 Argo with NA-21 directors. Dardo system for 40 mm.
Helicopters: 1 Agusta AB 212ASW ⓭.

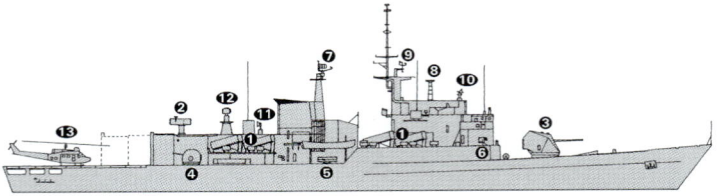

BOLOGNESI *(Scale 1 : 1,200), Ian Sturton / 1159992*

Programmes: *Palacios* and *Aguirre* formally transferred from the Italian Navy on 3 November 2004, without ammunition, torpedoes, SSM and helicopters. Following eight-month refits at Fincantieri, Muggiano they both arrived at Callao in mid-2005. A contract for the refit and transfer of two further decommissioned ships, *Sagittario* and *Perseo* was signed on 28 October 2005. Both ships were commissioned on 23 January 2006. *Bolognesi* arrived at Callao on 18 August 2006 and *Quiñones* on 20 January 2007.
Modernisation: *Bolognesi* and *Aguirre* are being upgraded with Selex Kronos 3D radar to replace RAN 10S (SPS-774), a new command system (Varayoc), eight Exocet MM40 Block 3 (to replace Teseo), Rheinmetall MASS (to replace SCLAR) decoy systems and MAGE QHAWAX ESM suite. Improvements to the Mk 95 fire-control system are also under consideration. *Palacios* and *Quiñones* are to be similarly upgraded at a later date.

AGUIRRE *4/2009, Guy Toremans / 1305528*

QUIÑONES *2/2013*, Guy Toremans / 1486550*

IHS Jane's Fighting Ships 2014-2015 © 2014 IHS

SHIPBORNE AIRCRAFT

Notes: Plans to acquire new shipborne aircraft are under consideration.

Numbers/Type: 5 Agusta AB 212ASW.
Operational speed: 106 kt (196 km/h).
Service ceiling: 14,200 ft (4,330 m).
Range: 230 n miles (425 km).
Role/Weapon systems: ASW and surface search helicopter for smaller escorts. Sensors: RDR 1700B search radar, Bendix ASQ-18 dipping sonar, ECM. Weapons: ASW; two Mk 44, Mk 46 or 244/S torpedoes or depth bombs.

AB 212   8/2006, Michael Nitz / 1335400

Numbers/Type: 3/6 Agusta-Sikorsky ASH-3D Sea King/UH-3H Sea King.
Operational speed: 120 kt (222 km/h).
Service ceiling: 12,200 ft (3,720 m).
Range: 630 n miles (1,165 km).
Role/Weapon systems: Six UH-3H utility aircraft acquired from US in 2010–11. Four are reported operational. Three ASH-3D ASW helicopters can be operated from FFGs. Sensors: Selenia search radar, Bendix ASQ-18 dipping sonar, sonobuoys. Weapons: ASW; four Mk 44, Mk 46 or 244/S torpedoes or depth bombs or mines. ASV; two AM 39 Exocet missiles.

ASH-3D   6/2013*, Peruvian Navy / 1525640

LAND-BASED MARITIME AIRCRAFT (FRONT LINE)

Notes: (1) There are two Mi-8T transport helicopters.
(2) There are three Fokker: one F-27 200, one F-27 600 and one F-27 500 is operational as an ELINT aircraft.
(3) There are two Antonov AN-32B transport aircraft.
(4) Five Beech T-34C are used for training.
(5) There is one Cessna 206 and two Cessna 150.
(6) There are three Bell 206B training helicopters.
(7) Six Enstrom F-28F training helicopters were acquired in 2008. One was lost and subsequently replaced in 2010.

Mi-8T   6/2011, Peruvian Navy / 1406656

ENSTROM F-28F   6/2011, Peruvian Navy / 1406655

Numbers/Type: 5 Beechcraft B-200CT.
Operational speed: 282 kt (523 km/h).
Service ceiling: 35,000 ft (10,670 m).
Range: 2,030 n miles (3,756 km).
Role/Weapon systems: Coastal surveillance and EEZ patrol duties. Sensors: Telephonics APS-143 Ocean Eye radar (in four aircraft), Star Safire III FLIR, cameras. Weapons: Unarmed.

B-200   10/2013*, Peruvian Navy / 1525638

Numbers/Type: 4 Fokker F-60 MPA.
Operational speed: 245 kt (453 km/h).
Service ceiling: 25,000 ft (7,620 m).
Range: 1,110 n miles (2,055 km).
Role/Weapon systems: Former Netherlands Air Force transport aircraft, two of which were converted in 2005 to a temporary MPA role for service in the Netherlands Antilles until replaced in 2007. These two were transferred in December 2010. Employed as a maritime patrol aircraft. Sensors: Selex Seaspray 7000E radar, ESM and IR sensors.

FOKKER F-60   8/2013*, Peruvian Navy / 1525639

PATROL FORCES

6 VERLARDE (PR-72P) CLASS
(FAST ATTACK CRAFT — MISSILE) (CM/PGGFM)

Name	No	Builders	Launched	Commissioned
VELARDE	CM 21	SFCN, Villeneuve la Garenne	16 Sep 1978	25 Jul 1980
SANTILLANA	CM 22	SFCN, Villeneuve la Garenne	11 Sep 1978	25 Jul 1980
DE LOS HEROS	CM 23	SFCN, Villeneuve la Garenne	20 May 1979	17 Nov 1980
HERRERA	CM 24	SFCN, Villeneuve la Garenne	16 Feb 1979	26 Feb 1981
LARREA	CM 25	SFCN, Villeneuve la Garenne	12 May 1979	16 Jun 1981
SANCHEZ CARRIÓN	CM 26	SFCN, Villeneuve la Garenne	28 Jun 1979	18 Sep 1981

Displacement, tonnes: 478 standard; 569 full load
Dimensions, metres (feet): 64 × 8.4 × 2.6 (210 × 27.6 × 8.5)
Speed, knots: 37. **Range, n miles:** 2,500 at 16 kt
Complement: 36 + 10 spare berths

Machinery: 4 SACM AGO 240 V16 M7 (CM 21, 23, 25) or 4 MTU 12V 595 (CM 22, 24, 26) diesels; 22,200 hp(m) (16.32 MW) sustained; 4 shafts
Missiles: SSM: 4 Aerospatiale MM 38 Exocet; inertial cruise; active radar homing to 42 km (23 n miles) at 0.9 Mach; warhead 165 kg; sea-skimmer.
SAM: An SA-N-10 launcher (MPG-86) may be fitted on the stern.
Guns: 1 Oto Melara 3 in (76 mm)/62 Compact; 85 rds/min to 16 km (8.7 n miles); weight of shell 6 kg. 2 Breda 40 mm/70 (twin); 300 rds/min to 12.5 km (6.8 n miles); weight of shell 0.96 kg.
Electronic countermeasures: ESM: Thomson-CSF DR 2000; intercept.
Radars: Surface search: Thomson-CSF Triton; G-band; range 33 km (18 n miles) for 2 m² target.
Navigation: Racal Decca 1226; I-band.
Fire control: Thomson-CSF/Castor II; I/J-band; range 15 km (8 n miles) for 1 m² target.
Weapon control systems: CSEE Panda director. Vega system.

Programmes: Ordered late 1976. Hulls of Velarde, De los Heros, Larrea subcontracted to Lorient Naval Yard, the others being built at Villeneuve-la-Garenne. Classified as corvettes.
Modernisation: CM 22, 24 and 26 re-engined in 2000. The other three craft are to be similarly modernised by 2014. The upgrade plan also includes fire-control systems.

DE LOS HEROS   2/2013*, Guy Toremans / 1486551

2 MARAÑON CLASS (RIVER GUNBOATS) (CF/PGR)

Name	No	Builders	Commissioned
MARAÑON	CF 13 (ex-CF 401)	John I Thornycroft & Co Ltd	Jul 1951
UCAYALI	CF 14 (ex-CF 402)	John I Thornycroft & Co Ltd	Jun 1951

Displacement, tonnes: 325 full load
Dimensions, metres (feet): 48.54 × 9.75 × 1.98 *(159.3 × 32.0 × 6.5)*
Speed, knots: 12
Range, n miles: 6,000 at 10 kt
Complement: 40 (2 officers)
Machinery: 2 Caterpillar 475 diesels (CF 13); 950 hp *(708 kW)*. 2 MTU 480 diesels (CF 14); 960 hp *(716 kW)*; 2 shafts
Guns: 2 – 3 in *(76 mm)*/50. 3 Bofors 40 mm/60. 2 Oerlikon 20 mm.

Comment: Ordered early in 1950 and launched 7 March and 23 April 1951 respectively. Employed on police duties in Upper Amazon. Superstructure of aluminium alloy. Based at Iquitos.

UCAYALI   6/2013*/ 1525637

2 AMAZONAS CLASS (RIVER GUNBOATS) (CF/PGR)

Name	No	Builders	Commissioned
AMAZONAS	CF 11 (ex-CF 403)	Electric Boat Co, Groton	1935
LORETO	CF 12 (ex-CF 404)	Electric Boat Co, Groton	1935

Displacement, tonnes: 250 standard
Dimensions, metres (feet): 43 × 8.0 × 2.34 *(141.1 × 26.2 × 7.7)*
Speed, knots: 12
Range, n miles: 4,000 at 10 kt
Complement: 33 (2 officers)
Machinery: 2 diesels; 750 hp(m) *(551 kW)*; 2 shafts
Guns: 1 – 3 in *(76 mm)*. 4 Bofors 40 mm/60.

Comment: Launched in 1934. In Upper Amazon Flotilla, based at Iquitos.

AMAZONAS   11/2009, A Sheldon-Duplaix / 1305525

1 + 1 (2) MANUEL CLAVERO CLASS
(RIVER GUNBOATS) (CF/PGR)

Name	No	Builders	Launched	Commissioned
CLAVERO	CF 15	SIMA, Iquitos	10 Jun 2008	19 Apr 2010
CASTILLA	CF 16	SIMA, Iquitos	2012	2014

Displacement, tonnes: 344 full load
Dimensions, metres (feet): 46.57 × 10.6 × 2.3 *(152.8 × 34.8 × 7.5)*
Speed, knots: 15. **Range, n miles:** 4,000 at 10 kt
Complement: 35 (2 officers)

Military lift: 20 marines
Machinery: 3 CAT C3408 diesels; 1,365 hp *(1 MW)*; 3 shafts
Guns: 1 – 20 mm. 2 – 12.7 mm MGs. 3 – 40 mm grenade launchers.
Radars: Navigation: Furuno; I-band.

Comment: *Clavero* was damaged by fire on her first operational deployment in May 2010. She returned to service in 2012. *Castilla* was laid down on 8 April 2010. A further two vessels may be ordered. Two 5.8 m Bora fast interception craft are carried.

CLAVERO   6/2013*, Peruvian Navy / 1525641

7 GRIFFON 2000 TDS HOVERCRAFT (UCAC)

UIF 101–107

Displacement, tonnes: 5 full load
Dimensions, metres (feet): 12.6 on cushion × ? *(41.3 × ?)*
Speed, knots: 35
Complement: 3
Machinery: 1 Deutz BF8L diesel; 355 hp *(265 kW)*
Guns: 2 – 7.62 mm MGs.
Radars: Furuno 1000C; I-band.

Comment: Griffon Hoverwork Ltd awarded the contract for the construction of two craft in December 2009. The vessels are used to counter drug smuggling activities along the Amazonian river bordering Peru and Colombia. They have a payload of 2 tonnes or 20 to 24 passengers. Similar craft are in service in Estonia, Finland, Lithuania, Pakistan, Sweden and UK. A further five craft were ordered in January 2012 and delivery was completed in 2012.

GRIFFON 2000   10/2013*, Guy Toremans / 1525642

AMPHIBIOUS FORCES

2 PAITA (TERREBONNE PARISH) CLASS (LSTH)

Name	No	Builders	Commissioned
CALLAO (ex-*Washoe County* LST 1165)	DT 143	Ingalls Shipbuilding, Pascagoula	30 Nov 1953
ETEN (ex-*Traverse County* LST 1160)	DT 144	Bath Iron Works, Maine	19 Dec 1953

Displacement, tonnes: 2,632 standard; 5,893 full load
Dimensions, metres (feet): 117.1 × 16.8 × 5.2 *(384.2 × 55.1 × 17.1)*
Speed, knots: 15. **Range, n miles:** 15,000 at 9 kt
Complement: 116

Military lift: 2,000 tons; 395 troops
Machinery: 4 GM 16-278A diesels; 6,000 hp *(4.48 MW)*; 2 shafts
Guns: 5 Bofors 40 mm/60 (2 twin, 1 single).
Radars: Navigation: I-band.

Comment: Four transferred from USA on loan 7 August 1984, recommissioned 4 March 1985. Have small helicopter platform. Original 3 in guns replaced by 40 mm. Lease extended by grant aid in August 1989, again in August 1994, and again in April 1999. *Paitá* and *Pisco* were decommissioned in 2012.

CALLAO   10/2013*, Guy Toremans / 1525643

0 + 2 MULTIROLE VESSEL (LPD)

Name	No	Builders	Laid down	Launched	Commissioned
–	–	SIMA, Callao	Jul 2013	2015	2016
–	–	SIMA, Callao	2015	2017	2018

Displacement, tonnes: 7,300 standard; 11,000 full load
Dimensions, metres (feet): 122 × 22.0 × 4.9 *(400.3 × 72.2 × 16.1)*
Speed, knots: 16.5. **Range, n miles:** 10,000 at 12 kt
Complement: 100

Military lift: 457 troops; 13 heavy vehicles; 2 LCVPs; 2 LCUs
Machinery: 2 MAN 9L28/32A diesels; 5,996 hp *(4.4 MW)*; 2 shafts
Guns: To be announced.
Radars: Navigation: 2 – I-band.
Helicopters: Platform for 2 Sea Kings.

Comment: A contract was signed with Dae Sun Shipbuilders and Engineering of South Korea for the construction in Peru of two multipurpose vessels. The ships are based on the Indonesian Makassar-class landing ships. Two 23 m LCUs are carried in the stern dock.

MULTIROLE VESSEL   10/2013*, Peruvian Navy / 1525650

Amphibious forces – Training ships < **Peru** 623

4 PUNTA MALPELO CLASS
(RIVER ASSAULT CRAFT) (DLS/PBF)

Name	No	Builders	Commissioned
PUNTA CAPONES	DLS 380	Construcciones Náuticas	1996
PUNTA MALPELO	DLS 381	Construcciones Náuticas	1996
PUNTA MERO	DLS 382	Construcciones Náuticas	1996
PUNTA SAL	DLS 383	Construcciones Náuticas	1996

Dimensions, metres (feet): 12.8 × 3.36 × 0.91 *(42 × 11.0 × 3.0)*
Speed, knots: 32
Range, n miles: 500 at 27 kt
Complement: 7
Machinery: 2 Diesel Volvo Penta TAMD/B; 286 hp(m) *(200 kW)*; 2 Hamilton waterjets
Guns: 1 – 40 mm AGL. 1 – 12.7 mm MG. 2 – 7.62 mm MGs.

Comment: Acquired in 1996.

SURVEY AND RESEARCH SHIPS

1 INSHORE SURVEY CRAFT (AGSC/EH)

Name	No	Builders	Commissioned
LA MACHA	AEH 174	SIMA, Chimbote	Apr 1982

Displacement, tonnes: 50 standard; 54 full load
Dimensions, metres (feet): 19.8 × 4.0 × 2.2 *(65 × 13.1 × 7.2)*
Speed, knots: 13
Complement: 8 (2 officers)
Machinery: 2 Caterpillar 3406-TA diesels; 543 hp *(400 kW)*; 2 shafts

Comment: Side scan sonar for plotting bottom contours. EH (Embarcacion Hidrográfica).

MACHA 6/2000, Peruvian Navy / 0105213

1 DOKKUM CLASS (AGSC/EH)

Name	No	Builders	Commissioned
CARRASCO (ex-*Abcoude*)	AH 171 (ex-M 810)	Smulders, Schiedam	18 May 1956

Displacement, tonnes: 379 standard; 460 full load
Dimensions, metres (feet): 46.6 × 8.8 × 2.3 *(152.9 × 28.9 × 7.5)*
Speed, knots: 12
Range, n miles: 2,500 at 10 kt
Complement: 36 (est.)
Machinery: 2 Volvo Penta diesels; 1,100 hp(m) *(820 kW)*; 2 shafts
Guns: 2 Oerlikon 20 mm/70.
Radars: Navigation: Racal Decca TM 1229C; I-band.

Comment: Service with the Netherlands Navy as a minesweeper included modernisation in the mid-1970s and a life prolonging refit in the late 1980s. *Carrasco* placed in reserve in 1993 and transferred to Peru 16 July 1994. The ship has been acquired for hydrographic duties. Two more were planned to follow in mid-1996 but the transfer was cancelled.

CARRASCO 6/2012, A Sheldon-Duplaix / 1455730

1 HUMBOLDT CLASS (AGOR)

Name	No	Builders	Laid down	Launched	Commissioned
HUMBOLDT	–	SIMA, Callao	17 Sep 1977	13 Oct 1978	25 Jan 1980

Displacement, tonnes: 1,980 standard
Dimensions, metres (feet): 76.2 × 12.6 × 4.4 *(250 × 41.3 × 14.5)*
Speed, knots: 15
Machinery: 2 B&W 10V23LU diesels; 2,898 hp *(2.13 MW)*; 1 shaft

Comment: Designed as a stern trawler and modified for oceanographic survey duties. The ship received new engines in 2010–11.

HUMBOLDT 6/2013*, INANPE / 1525655

2 VAN STRAELEN CLASS (AGSC/EH)

Name	No	Builders	Commissioned
CARRILLO (ex-*van Hamel*)	AH 175	De Vries, Amsterdam	14 Oct 1960
MELO (ex-*van der Wel*)	AH 176	De Vries, Amsterdam	6 Oct 1961

Displacement, tonnes: 172 full load
Dimensions, metres (feet): 33.1 × 5.6 × 1.6 *(108.6 × 18.4 × 5.2)*
Speed, knots: 12
Complement: 17 (2 officers)
Machinery: 2 Volvo Penta diesels; 1,100 hp(m) *(820 kW)* sustained; 2 shafts
Guns: 1 – 20 mm

Comment: Both built as inshore minesweepers. Acquired 23 February 1985 for conversion with new engines and survey equipment.

MELO 10/2013*, Guy Toremans / 1525649

1 RIVER SURVEY CRAFT (AGSC/AH)

Name	No	Builders	Commissioned
STIGLICH	AH 172	SIMA, Iquitos	3 Nov 1980

Displacement, tonnes: 234 standard; 254 full load
Dimensions, metres (feet): 34.2 × 7.9 × 0.9 *(112.2 × 25.9 × 3.0)*
Speed, knots: 9
Complement: 22 (2 officers)
Machinery: 2 Caterpillar 3304 diesels; 500 hp *(367 kW)*; 2 shafts

Comment: *Stiglich* is based at Iquitos for survey work on the Upper Amazon.

STIGLICH 6/1999, Peruvian Navy / 0081533

TRAINING SHIPS

0 + 1 SAIL TRAINING SHIP (AXS)

Name	No	Builders	Laid down	Launched	Commissioned
LA UNIÓN	–	SIMA, Callao	6 Dec 2012	Jul 2014	2015

Displacement, tonnes: 3,500 full load
Dimensions, metres (feet): 113.5 × 13.0 × ? *(372.4 × 42.7 × ?)*
Complement: 257
Machinery: To be announced.

Comment: Construction of a new sail training ship began at SIMA, Callao, in December 2012. The ship is reported to be based on a Spanish design and, when completed in 2015, will be the largest sail training vessel in South America. The ship will be four-masted with a sail area of 3,500 m^3.

624 Peru > Training ships — Auxiliaries

1 MARTE CLASS (SAIL TRAINING CRAFT) (AXS)

Name	No	Builders	Commissioned
MARTE (ex-*Neptuno*; ex-*Noah's Ark*)	ALY 313	James O Rasborough, Halifax	1974

Displacement, tonnes: 50 standard; 56 full load
Dimensions, metres (feet): 20.3 × 5.18 × 1.95 *(66.6 × 17.0 × 6.4)*
Speed, knots: 8
Complement: 26
Machinery: Two Perkins 130C diesels; 260 hp *(194 kW)*; 2 props
Radars: Surface search: Furuno; I-band.

Comment: Used for cadet instruction at the Naval Academy.

MARTE 6/2004, Peruvian Navy / 1127042

AUXILIARIES

Notes: (1) All auxiliaries may be used for commercial purposes if not required for naval use.
(2) Plans for the procurement of a replacement logistics support ship are under consideration.
(3) There are four Rio Comaina class 30 m fuel barges (ABP 336–339); *Rio Comaina*, *Rio Huazaga*, *Rio Chinganaza*, *Rio Cenepa*
(4) There are two 15 m river cargo barges (ABC 360–361).

3 HARBOUR TANKERS (FUEL/WATER) (YW/YO)

CALOYERAS ACA 111 (ex-YW 128)   **GAUDEN** ACP 119 (ex-YO 171)
NOGUERA ACP 118 (ex-YO 221)

Displacement, tonnes: 1,412 full load
Dimensions, metres (feet): 52.3 × 9.8 × 4.1 *(171.6 × 32.2 × 13.5)*
Speed, knots: 8
Complement: 23
Cargo capacity: 200,000 gallons
Machinery: 1 GM diesel; 560 hp *(418 kW)*; 1 shaft
Radars: Navigation: Raytheon; I-band.

Comment: *Noguera* (fuel) transferred from US to Peru January 1975; *Gauden* (fuel) 20 January 1981; *Caloyeras* (water) 26 January 1985.

NOGUERA 10/2013*, Guy Toremans / 1525648

1 TORPEDO RECOVERY VESSEL (YPT)

Name	No	Builders	Commissioned
SAN LORENZO	ART 322	Lürssen, Burmeister	1 Dec 1981

Displacement, tonnes: 59 standard; 66 full load
Dimensions, metres (feet): 25.2 × 5.6 × 1.7 *(82.7 × 18.4 × 5.6)*
Speed, knots: 19
Range, n miles: 500 at 15 kt
Complement: 9
Machinery: 2 MTU 8V 396 TC82 diesels; 1,740 hp(m) *(1.28 MW)* sustained; 2 shafts

Comment: Can carry four long or eight short torpedoes.

IHS Jane's Fighting Ships 2014-2015

SAN LORENZO 10/2013*, Guy Toremans / 1525646

1 MORONA CLASS (RIVER HOSPITAL CRAFT) (ABH)

Name	No	Builders	Commissioned
MORONA	ABH 302	SIMA, Iquitos	13 May 1976

Displacement, tonnes: 152 full load
Dimensions, metres (feet): 30 × 8.0 × 1.54 *(98.4 × 26.2 × 5.1)*
Speed, knots: 7
Complement: 20 (1 officer)
Machinery: 2 Caterpillar 3304 diesels; 150 hp *(112 kW)* sustained; 2 shafts

Comment: *Morona* is used as a hospital craft and has a red cross on her superstructure.

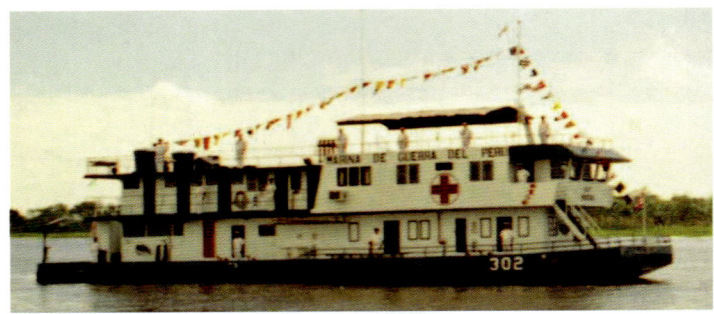

MORONA 6/2008, Peruvian Navy / 1335406

2 BAYÓVAR CLASS (TANKERS) (AOT)

Name	No	Builders	Commissioned
BAYÓVAR (ex-*Petr Shmidt*)	ATP 154	Kherson Shipyard	1987
ZORRITOS (ex-*Grigoriy Nesterenko*)	ATP 155	Kherson Shipyard	1986

Displacement, tonnes: 38,904 full load
Measurement, tonnes: 18,924 gt
Dimensions, metres (feet): 179 × 25.3 × 11.0 *(587.3 × 83.0 × 36.1)*
Speed, knots: 15
Machinery: 1 B&W 6L67GFCA diesel; 11,900 hp *(8.9 MW)*; 1 shaft

Comment: Russian-built tankers acquired on 22 December 2006 and commissioned into the Peruvian Navy at Callao on 15 April 2007 and 17 May 2007.

BAYÓVAR 6/2008, Peruvian Navy / 1335407

1 + 11 RIO NAPO CLASS (RIVER SUPPLY CRAFT)

Name	No	Builders	Commissioned
RIO NAPO	–	SIMA, Iquitos	8 Jun 2013

Displacement, tonnes: 250 full load
Dimensions, metres (feet): 42 × 7.0 × 1.8 *(137.8 × 23.0 × 5.9)*
Speed, knots: 14
Complement: 20 (1 officer)
Machinery: 2 Caterpillar diesels; 900 hp *(671 kW)*; 2 shafts

Comment: The first of a new class of 12 river support class was commissioned by the President in June 2013. The role of the ship is to provide logistic and medical services to some 7,000 people in 30 communities along the Napo River. Further vessels will serve other river-based communities.

RIO NAPO 6/2013*, Peruvian Navy / 1525644

© 2014 IHS

Auxiliaries — Coast guard < **Peru** 625

3 FLOATING DOCKS (AH)

ADF 104   ADF 106   ADF 107

Displacement, tonnes: 4,572 (ADF 104), 1,930 (ADF 106), 5,283 (ADF 107) standard

Comment: *106* (ex-US *AFDL 33*) transferred 1959; *107* (ex-US *ARD 8*) transferred 1961; *104* built at SIMA, Callao in 1991.

3 PASTAZA CLASS (RIVER HOSPITAL CRAFT) (AH)

Name	No	Builders	Commissioned
CORRIENTES	ABH 303	SIMA, Iquitos	1982
CURARAY	ABH 304	SIMA, Iquitos	1983
PASTAZA	ABH 305	SIMA, Iquitos	1981

Displacement, tonnes: 150 full load
Dimensions, metres (feet): 20.5 × 6.5 × 1.45 *(67.3 × 21.3 × 4.8)*
Speed, knots: 8
Complement: 10 (1 officer)
Machinery: 1 Caterpillar diesel; 190 hp *(142 kW)* sustained; 1 shaft

Comment: Small river-based hospital craft.

CURARAY   *6/2013*, Peruvian Navy* / 1525647

2 HARBOUR LAUNCHES (YFL)

MANTILLA LSP 180   **AGUILAR** LSP 181

Displacement, tonnes: 22 full load
Dimensions, metres (feet): 15 × 4.9 × 2.27 *(49.2 × 16.1 × 7.4)*
Speed, knots: 20
Complement: 2 + 8 personnel
Machinery: 2 Caterpillar C18 diesels; 671 hp *(500 kW)*; 2 shafts
Radars: Navigation: Furuno M-1623; I-band.

Comment: Both built at SIMA, Callao and delivered in October 2013. Based at Callao and used for personnel transfer and harbour pilot duties.

MANTILLA   *10/2013*, Guy Toremans* / 1525645

TUGS

Notes: (1) There are three river tugs *Rio Tapuina* AER 180, *Rio Gaudin* AER 186 and *Rio Zambrano* AER 187.
(2) There are also four small harbour tugs *Mejia* ARB 120, *Huertas* ARB 121, *Dueñas* ARB 126, and *Olaya* ARB 128.
(3) There is a 43 m salvage tug *Unanue* (AMB 160), first commissioned in 1944, transferred from the US in 1961.

2 HARBOUR TUGS (YTL)

SELENDÁN ARB 124   **MEDINA** ARB 125

Displacement, tonnes: 150 full load
Dimensions, metres (feet): 19.5 × 6.2 × 3.0 *(64 × 20.3 × 9.8)*
Speed, knots: 10
Complement: 3
Machinery: 2 Caterpillar 3412 diesels; 1,500 hp *(1.1 MW)*; 2 shafts
Radars: Navigation: Furuno M-1623; I-band.

Comment: Both built at SIMA, Callao and delivered in October 2012. Based at Callao.

© 2014 IHS

1 CHEROKEE CLASS (SALVAGE TUG) (ATS)

Name	No	Builders	Commissioned
GUARDIAN RIOS (ex-*Pinto* ATF 90)	ARA 123	Cramp, Philadelphia	1 Apr 1943

Displacement, tonnes: 1,666 full load
Dimensions, metres (feet): 62.5 × 11.7 × 5.2 *(205.1 × 38.4 × 17.1)*
Speed, knots: 16.5. **Range, n miles:** 6,500 at 16 kt
Complement: 99
Machinery: Diesel-electric; 4 GM 12-278 diesels; 4,400 hp *(3.28 MW)*; 4 generators; 1 motor; 3,000 hp *(2.24 MW)*; 1 shaft
Guns: 2—20 mm.

Comment: Transferred from USA on loan in 1960, sold 17 May 1974. Fitted with powerful pumps and other salvage equipment.

GUARDIAN RIOS   *6/2006, Peruvian Navy* / 1164738

COAST GUARD

1 CARVAJAL (MODIFIED LUPO) CLASS (PSO)

Name	No	Builders	Laid down	Launched	Commissioned
GUARDIAMARINA SAN MARTIN (ex-*Carvajal*)	PO 201 (ex-FM 51)	Fincantieri, Riva Trigoso	8 Aug 1974	17 Nov 1976	5 Feb 1979

Displacement, tonnes: 2,243 standard; 2,540 full load
Dimensions, metres (feet): 113.2 × 11.3 × 3.7 *(371.4 × 37.1 × 12.1)*
Speed, knots: 35. **Range, n miles:** 3,450 at 20.5 kt
Complement: 185 (20 officers)

Machinery: CODOG: 2 GE/Fiat LM 2500 gas turbines; 50,000 hp *(37.3 MW)* sustained; 2 GMT A 230.20 M diesels; 8,000 hp(m) *(5.88 MW)* sustained; 2 shafts; LIPS cp props
Guns: 1 Oto Melara 5 in *(127 mm)*/54; 45 rds/min to 16 km *(8.7 n miles)*; weight of shell 32 kg. 4 Breda 40 mm/70 (2 twin); 300 rds/min to 12.5 km *(6.8 n miles)*; weight of shell 0.96 kg. 2—12.7 mm MGs.
Electronic countermeasures: ESM: Elettronica Lambda; intercept.
Radars: Surface search: Selenia RAN 11LX; D/I-band.
Navigation: SMA 3 RM 20R; I-band.
Fire control: 2 RTN 10X; I/J-band.
Combat data systems: Selenia IPN-10 action data automation.
Weapon control systems: 2 Elsag Mk 10 Argo with NA-21 directors. Dardo system for 40 mm.
Helicopters: 1 Agusta AB 212ASW. 1 Agusta ASH-3D Sea King (deck only).

Comment: Former Peruvian Navy vessel transferred to the Coast Guard in 2013 following the removal of major weapons/sensors including SAM, SSM, torpedoes, sonar, and air search radar. The ship is capable of operating Sea King helicopters although they cannot be stowed in the hangar.

GUARDIAMARINA SAN MARTIN   *10/2013*, Guy Toremans* / 1525652

0 + 5 PATROL SHIPS (PBO)

Name	No	Builders	Commissioned
–	–	SIMA, Chimbote	2015
–	–	SIMA, Chimbote	2015
–	–	SIMA, Chimbote	2016
–	–	SIMA, Chimbote	2016
–	–	SIMA, Chimbote	2016

Displacement, tonnes: 560 full load
Dimensions, metres (feet): 54.5 × 8.5 × 2.3 *(178.8 × 27.9 × 7.5)*
Speed, knots: 22. **Range, n miles:** 3,600 at 14 kt
Complement: 25 + 14 embarked forces
Machinery: 2 Caterpillar 3516C diesels; 6,690 hp *(5.0 MW)*; 2 shafts
Guns: 1—30 mm. 2—12.7 mm MGs.

Comment: The tender for construction of five, with an option for a further five, patrol vessels was awarded to STX Offshore and Shipbuilding on 25 September 2013. The ships are likely to be based on the South Korean Coast Guard Tae Geuk class. Design features include the capability to carry two RHIBs, deployed and recovered from a stern ramp. The contract is to involve the supply of drawings, material packages and technical assistance for assembly of the ships in Peru. Construction of the first two vessels is to start at SIMA, Chimbote, in January 2014 with a view to delivery in mid-2015.

IHS Jane's Fighting Ships 2014-2015

Peru > Coast guard

5 RIO NEPEÑA CLASS (LARGE PATROL CRAFT) (WPB)

Name	No	Builders	Commissioned
RIO NEPEÑA	PM 243	SIMA, Chimbote	1 Dec 1981
RIO TAMBO	PM 244	SIMA, Chimbote	10 Mar 1982
RIO OCOÑA	PM 245	SIMA, Chimbote	14 Jul 1983
RIO HUARMEY	PM 246	SIMA, Chimbote	8 Oct 1984
RIO ZAÑA	PM 247	SIMA, Chimbote	12 Feb 1985

Displacement, tonnes: 257 standard; 301 full load
Dimensions, metres (feet): 50.9 × 7.4 × 1.7 (167 × 24.3 × 5.6)
Speed, knots: 23
Range, n miles: 3,050 at 17 kt
Complement: 39 (4 officers)
Machinery: 4 Bazán MAN V8V diesels; 5,640 hp(m) (4.15 MW); 2 shafts
Guns: 1 Oerlikon 20 mm. 2—12.7 mm MGs.
Radars: Surface search: Decca 1226; I-band.

Comment: Have aluminium alloy superstructures. The prototype craft was scrapped in 1990. *Rio Ocoña* completed refit in July 1996 and the rest of the class were refitted at one per year.

RIO HUARMEY   2/2013*, Guy Toremans / 1486555

1 PGM 71 CLASS (LARGE PATROL CRAFT) (PB)

Name	No	Builders	Commissioned
RIO CHIRA	PM 223 (ex-PGM 111)	SIMA, Callao	29 May 1972

Displacement, tonnes: 138 standard; 149 full load
Dimensions, metres (feet): 36 × 6.4 × 1.8 (118.1 × 21.0 × 5.9)
Speed, knots: 15
Range, n miles: 1,500 at 10 kt
Complement: 16 (3 officers)
Machinery: 2 Detroit GN-71 diesels; 1,450 hp (1.08 MW); 2 shafts
Guns: 1—12.7 mm MG.
Radars: Surface search: Raytheon; I-band.

Comment: Acquired from the Navy in 1975. Paid off in 1994 but back in service again in 1997, with refurbished engines.

RIO CHIRA   2000, Peruvian Coast Guard / 0105218

6 CHICAMA (DAUNTLESS) CLASS (PBR)

CHICAMA PC 216	CHORRILLOS PC 218	CAMANA PC 220
HUANCHACO PC 217	CHANCAY PC 219	CHALA PC 221

Displacement, tonnes: 14 full load
Dimensions, metres (feet): 12.2 × 4.3 × 1.3 (40 × 14.1 × 4.3)
Speed, knots: 27. **Range, n miles:** 600 at 18 kt
Complement: 5 (1 officer)
Machinery: 2 Caterpillar 3208TA diesels; 870 hp (650 kW); 2 shafts
Guns: 1—12.7 mm MG. 1—7.62 mm MG.
Radars: Surface search: Furuno 821; I-band.

Comment: Ordered in February 2000 under FMS funding. Built by SeaArk Marine, Arkansas. First pair delivered in August 2000 remainder in November 2000. Formerly river patrol craft, now operational on Pacific Coast.

CHANCAY   6/2004, Peruvian Coast Guard / 1127045

12 ZORRITOS CLASS (RIVER PATROL CRAFT) (PBR)

Name	No	Builders	Commissioned
CANCAS (ex-Zorritos)	PC 222	SIMA, Callao	23 Sep 2003
PUNTA ARENAS	PC 224	SIMA, Callao	23 Sep 2003
SANTA ROSA	PL 225	SIMA, Callao	23 Sep 2003
PACASMAYO	PC 226	SIMA, Callao	23 Sep 2003
BARRANCA	PC 227	SIMA, Callao	23 Sep 2003
COISHCO	PC 228	SIMA, Callao	Oct 2004
INDEPENDENCIA	PC 229	SIMA, Callao	Oct 2004
SAN NICOLAS	PL 230	SIMA, Callao	Oct 2004
MATARANI	PC 234	SIMA, Callao	Oct 2004
SAMA	PC 238	SIMA, Callao	Oct 2004
JULI	PL 293	SIMA, Callao	Oct 2004
MOHO	PL 294	SIMA, Callao	Oct 2004

Displacement, tonnes: 12 full load
Dimensions, metres (feet): 12.2 × 4.2 × 0.7 (40 × 13.8 × 2.3)
Speed, knots: 24
Complement: 5 (1 officer)
Machinery: 2 Caterpillar 3126 diesels; 550 hp (411 kW); 2 shafts
Guns: 1—12.7 mm MG.
Radars: Surface search: Furuno; I-band.

Comment: PL 225, 230, 293, and 294 are based at Puno, Lake Titicaca.

SAMA   10/2013*, Guy Toremans / 1525651

3 RÍO SANTA CLASS (COASTAL PATROL CRAFT) (PBR)

Name	No	Builders	Commissioned
RÍO SANTA	PP 232	Cía Nauticas, Callao	1981
RÍO MAJES	PP 233	Cía Nauticas, Callao	1982
RIO SURCO	PP 237	Cía Nauticas, Callao	1982

Displacement, tonnes: 14 full load; 15 full load
Dimensions, metres (feet): 10.5 × 3.1 × 1.9 (34.4 × 10.2 × 6.2)
Speed, knots: 20. **Range, n miles:** 86 at 20 kt
Complement: 6
Machinery: 2 Evinrude BE200CXCM outboard; 400 hp (300 kW)
Radars: Surface search: Furuno.

Comment: Built in 1981–82.

RÍO SANTA   6/2008, Peruvian Navy / 1335404

3 MÁNCORA CLASS (HARBOUR PATROL BOATS) (PBR)

Name	No	Builders	Commissioned
MÁNCORA	DCB 212	Cougar Marine, Miami	1993
HUAURA	DCB 213	Cougar Marine, Miami	1993
QUILCA	DCB 214	Cougar Marine, Miami	1993

Displacement, tonnes: 4 standard
Dimensions, metres (feet): 7.62 × 2.33 × 0.91 (25 × 7.6 × 3.0)
Speed, knots: 32
Range, n miles: 180 at 35 kt
Complement: 3
Machinery: 1 Volvo Penta AD 41B diesel; 200 hp (150 kW); 1 shaft

MÁNCORA   6/2008, Peruvian Navy / 1335402

IHS Jane's Fighting Ships 2014-2015

Coast guard < **Peru** 627

1 RÍO VIRU CLASS (COASTAL PATROL CRAFT) (PBR)

Name	No	Builders	Commissioned
RÍO VIRU	PC 235	Camcraft Inc, Louisiana	1981

Displacement, tonnes: 13 standard; 14 full load
Dimensions, metres (feet): 13.1 × 4 × 1.8 *(43 × 13.1 × 5.9)*
Speed, knots: 15. **Range, n miles:** 210 at 11 kt
Complement: 6
Machinery: 2 General Motors Detroit Diesel 6-71 diesels; 500 hp *(373 kW)*; 2 shafts
Guns: 1 – 12.7 mm MG.
Radars: Surface search: Raytheon 2800; I-band.

Comment: Aluminium hull.

RÍO VIRU 6/2008, Peruvian Navy / 1335403

RÍO VIRU 2000, Peruvian Coast Guard / 0105222

8 LA CRUZ CLASS (HARBOUR PATROL CRAFT) (PBR)

Name	No	Builders	Commissioned
LA CRUZ	DCB 350	Cougar Marine, Miami	1992
CABO BLANCO	DCB 351	Cougar Marine, Miami	1992
COLÁN	DCB 352	Cougar Marine, Miami	1992
SAMANCO	DCB 353	Cougar Marine, Miami	1992
BESIQUE	DCB 354	Cougar Marine, Miami	1992
SALINAS	DCB 355	Cougar Marine, Miami	1993
ANCÓN	DCB 356	Cougar Marine, Miami	1993
PARACAS	DCB 357	Cougar Marine, Miami	1993

Displacement, tonnes: 2 standard
Dimensions, metres (feet): 6.6 × 2.2 × 0.2 *(21.7 × 7.2 × 0.7)*
Speed, knots: 30. **Range, n miles:** 240 at 15 kt
Complement: 4
Machinery: 1 Evinrude outboard diesel; 200 hp *(150 kW)*

12 RIVER AND LAKE PATROL CRAFT (PBR)

RÍO HUALLAGA PF 260	RIO NAPO PF 264	RIO PATAYACU PF 271
RÍO SANTIAGO PF 261	RIO YAVARI PF 265	RIO ZAPOTE PF 272
RÍO PUTUMAYO PF 262	RIO MATADOR PF 266	RIO CHAMBIRA PF 273
RÍO NANAY PF 263	RIO ITAYA PF 270	RIO TAMBOPATA PF 274

Comment: PF 260–263 are 10 m craft built by Sima, Iquitos 1994–95. PF 264–266 are 8 m craft built by Sima, Iquitos 1998–99. They are employed on the Amazon River. PF 270–273 are 6 m craft built by Sima, Iquitos 1998–99. Deployed on the Amazon River. PF 274 is an aluminium-hulled 8 m craft, originally commissioned into the Peruvian Navy in 1977. All operate in the Amazon basin.

RIO HUALLAGA 6/2008, Peruvian Navy / 1335401

RIO NAPO 6/2000, Peruvian Navy / 0105221

1 LAKE HOSPITAL CRAFT (AH)

Name	No	Builders	Commissioned
PUNO (ex-*Yapura*)	ABH 306	J Watt Co; Thames Iron Works	18 May 1872

Displacement, tonnes: 508 full load
Dimensions, metres (feet): 38.13 × 6.0 × 4.0 *(125.1 × 19.7 × 13.1)*
Speed, knots: 14
Complement: 24 (1 officer)
Machinery: 1 diesel; 1 shaft

Comment: Two ships, ex-*Yapura* and *Yavari* were built in kit form in 1861–62 and after arrival at the Peruvian (now Chilean) port of Arica, transported by rail and mule to Puno on Lake Titicaca where they were subsequently assembled. Ex-*Yavari* was later lengthened and re-engined in 1914. Ex-*Yapura* continues in Peruvian Coast Guard service while *Yavari* was purchased from the Peruvian Navy in 1987 as a restoration project. She is now a museum ship at Puno and has overnight accommodation for seven guests.

PUNO 8/2009*, Peruvian Navy / 1525654

1 + 3 (8) RIVER PATROL CRAFT (PBR)

Name	No	Builders	Commissioned
–	–	SIMA, Iquitos	2013
–	–	SIMA, Iquitos	2014
–	–	SIMA, Iquitos	2014
–	–	SIMA, Iquitos	2014

Displacement, tonnes: 18 full load
Dimensions, metres (feet): 18 × 3.75 × 2.3 *(59.1 × 12.3 × 7.5)*
Speed, knots: 26
Complement: 20
Machinery: 2 Caterpillar C18 Acer diesels; 740 hp *(552 kW)*; 2 shafts

Comment: Designed and built by SIMA, Iquitos for river patrol duties. A second batch of eight vessels is under consideration.

RIVER PATROL CRAFT 10/2013*, Guy Toremans / 1525653

8 PARACHIQUE CLASS (PATROL CRAFT) (PBR)

Name	No	Builders	Commissioned
PARACHIQUE	PC 213	SIMA, Chimbote	2013
SAN ANDREAS	PC 214	SIMA, Chimbote	2013
ATICO	PC 239	SIMA, Chimbote	2012
MALABRIGO	PC 240	SIMA, Chimbote	2013
CATARINDO	PC 241	SIMA, Chimbote	2012
PUNTA PARIÑAS	PC 242	SIMA, Chimbote	2013
CASMA	PC 248	SIMA, Chimbote	2012
TORTUGAS	PC 249	SIMA, Chimbote	2012

Displacement, tonnes: 12 full load
Dimensions, metres (feet): 12.2 × 4.2 × 0.7 *(40 × 13.8 × 2.3)*
Speed, knots: 25
Complement: 5 (1 officer)
Machinery: 2 Caterpillar C7 diesels; 740 hp *(552 kW)*; 2 shafts
Guns: 1 – 12.7 mm MG.
Radars: Navigation: JRC; I-band.

Comment: Built by SIMA, Chimbote and delivered 2012–13.

PARACHIQUE 6/2013*, Peruvian Navy / 1525656

© 2014 IHS IHS Jane's Fighting Ships 2014-2015

Philippines

Country Overview

The Republic of the Philippines was formally proclaimed in 1946. Situated between Taiwan to the north and Indonesia and Malaysia to the south, the country comprises about 7,100 islands with a total coastline of 19,597 n miles with the South China, Philippine and Celebes Seas. Eleven islands, Bohol, Cebu, Leyte, Luzon, Masbate, Mindanao, Mindoro, Negros, Palawan, Panay, and Samar, contain the majority of the population. Most remaining islands are less than 1 square mile in area. The capital, principal city and port is Manila. Other important ports include Davao, Cebu and Zamboanga. An archipelagic state, territorial seas (12 n miles) are claimed. A 200 n mile EEZ has also been claimed but the limits have not been defined.

Headquarters Appointments

Flag Officer-in-Command:
  Vice Admiral Jose Luis M Alano
Commander Fleet:
  Rear Admiral Orwen J Cortez
Commandant Coast Guard:
  Rear Admiral Rodolfo Isorena
Commandant Marines:
  Major General Romeo Tanalgo

Personnel

(a) 2014: 22,000 Navy; 8,700 Marines; 6,000 Coast Guard
(b) Reserves: 17,000

Organisation

The Naval Headquarters is at Manila. The fleet is divided into functional units including the Ready Force, Patrol Force, Service Force, Assault Craft Force, Naval Air Group and Naval Special Warfare Group. There are six operational areas of responsibility: Southern Luzon; Northern Luzon; Central; West; Western Mindanao and Eastern Mindanao. The Coast Guard was transferred to the Department of Transport and Communication in 1998.

Marine Corps

Marines comprise three tactical brigades composed of 10 tactical battalions, one support regiment, a service group, a guard battalion and a reconnaissance battalion. Headquarters at Ternate, Manila Bay. Deployed in Mindanao and Palawan.

Bases

Main: Cavite.
Operational: San Vicente, Mactan, Ternate.
Stations: Cebu, Davao, Legaspi, Bonifacio, Tacloban, San Miguel, Ulugan, Balabac, Puerto Princesa, Pagasa.

Prefix to Ships' Names

BRP: Barko Republika Pilipinas

Strength of the Fleet

Type	Active	Building
Frigates	3	—
Corvettes	12	—
Fast Attack Craft	8	—
Large Patrol Craft	4	1 (3)
Coastal Patrol Craft	34	2
LPD	—	(2)
LST/LSV Transports	7	—
LCM/LCU/RUC/LCVP	44	—
Repair Ship	1	—
Tankers	4	—
Coast Guard		
Tenders	4	—
Patrol Craft	58	—

PENNANT LIST

Frigates

PF 11	Rajah Humabon
PF 15	Gregorio del Pilar
PF 16	Ramon A Alcaraz

Corvettes

PS 19	Miguel Malvar
PS 20	Magat Salamat
PS 22	Sultan Kudarat
PS 28	Cebu
PS 31	Pangasinan
PS 32	Iloilo
PS 35	Emilio Jacinto
PS 36	Apolinario Mabini
PS 37	Artemio Ricarte
PS 38	General Mariano Alvares
PS 70	Quezon
PS 74	Rizal

Patrol Forces

PG 102	Bagong Lakas
PG 104	Bagong Silang
PG 111	Bonny Serrano
PG 112	Bienvenido Salting
PG 114	Salvador Abcede
PG 115	Ramon Aguirre
PG 116	Nicolas Mahusay
PG 117	Dionisio Ojeda
PG 118	Emilio Liwanag
PG 140	Emilo Aguinaldo
PG 141	Antonio Luna
PG 370	José Andrada
PG 371	Enrique Jurado
PG 372	Alfredo Peckson
PG 374	Simeon Castro
PG 375	Carlos Albert
PG 376	Heracleo Alano
PG 377	Liberato Picar
PG 378	Hilario Ruiz
PG 379	Rafael Pargas
PG 380	Estor Reinoso
PG 381	Dioscoro Papa
PG 383	Ismael Lomibao
PG 384	Leovigildo Gantioqui
PG 385	Federico Martir
PG 386	Filipino Flojo
PG 387	Anastacio Cacayorin
PG 388	Manuel Gomez
PG 389	Teotimo Figuracion
PG 390	José Loor SR
PG 392	Juan Magluyan
PG 393	Florencio Inigo
PG 394	Alberto Navarette
PG 395	Felix Apolinario
PG 396	Abraham Campo
PG 847	Leopoldo Regis
PG 851	Apollo Tiano
PG 853	Sulpicio Hernandez

Auxiliaries

LT 87	South Cotabato
LT 501	Laguna
LT 504	Lanao Del Norte
LC 550	Bacolod City
LC 551	Dagupan City
AT 25	Ang Pangulo
AW 33	Lake Balusan
AW 34	Lake Paoay
AF 72	Lake Taal
AF 78	Lake Buhi
AC 90	Mactan
AD 617	Yakal
AT 71	Mangyan
AT 291	Subanon
AT 293	Bagobo
AT 295	Tausug
AT 296	Tagbanua
AT 297	Manobo

Coast Guard

AE 46	Cape Bojeador
PG 64	Palawan
AU 75	Bessang Pass
AE 79	Limasawa
AG 89	Kalinga
AU 100	Tirad Pass
001	San Juan
002	Esda II

FRIGATES

Notes: A programme to procure two new-build frigates was initiated in October 2013. In December 2013, four shipbuilders were selected to proceed to a second stage of bidding: Navantia, Hyundai Heavy Industries, STX Offshore and Shipbuilding, and Daewoo.

2 PILAR (HAMILTON AND HERO) CLASS (FFH)

Name	No	Builders	Laid down	Launched	Commissioned
GREGORIO DEL PILAR (ex-*Hamilton*)	PF 15 (ex-WHEC 715)	Avondale Shipyards, New Orleans	Jan 1965	18 Dec 1965	20 Feb 1967
RAMON A ALCARAZ (ex-*Dallas*)	PF 16 (ex-WHEC 716)	Avondale Shipyards, New Orleans	7 Feb 1966	1 Oct 1966	1 Oct 1967

Displacement, tonnes: 3,353 full load
Dimensions, metres (feet): 115.2 × 13.1 × 6.1 *(378 × 43.0 × 20.0)*
Flight deck, metres: 26.8 × 12.2 *(87.9 × 40.0)*
Speed, knots: 29
Range, n miles: 9,600 at 15 kt
Complement: 162 (19 officers)

Machinery: CODOG; 2 Pratt & Whitney FT4A-6 gas turbines; 36,000 hp *(26.86 MW)*; 2 Fairbanks-Morse 38TD8-1/8-12 diesels; 7,000 hp *(5.22 MW)* sustained; 2 shafts; cp props; retractable bow propulsor; 350 hp *(261 kW)*.
Guns: 1 Oto Melara 3 in *(76 mm)*/62 Mk 75 Compact; 85 rds/min to 16 km *(8.7 n miles)* anti-surface; 12 km *(6.6 n miles)* anti-aircraft; weight of shell 6 kg. 2 – 20 mm. 1 Boeing 25 mm/87 Mk 38 Bushmaster. 6 – 12.7 mm MGs.
Physical countermeasures: 2 Loral Hycor SRBOC 6-barrelled Mk 36; IR flares and chaff.
Electronic countermeasures: ESM: WLR-1C, WLR-3; intercept.
Radars: Surface search: Hughes/Furuno SPS-73; E/F- and I-bands.
Fire control: Sperry Mk 92; I/J-band.
Tacan: URN 25.
Combat data systems: SCCS 378.
Weapon control systems: Mk 92 Mod 1 GFCS.

Helicopters: 1 MBB BO 105.

Programmes: There have been 12 built for the US Coast Guard of a total of 36 originally planned. *Gregorio del Pilar* transferred from the USCG on 13 May 2011 and recommissioned on 14 December 2011. *Ramon A Alcaraz* was transferred on 22 May 2012.
Modernisation: FRAM programme was conducted from October 1985 to October 1992. Work included standardising the engineering plants, improving the clutching systems and replacing the Mk 56 fire-control system and 5 in/38 gun mount with the Mk 92 system and a single 76 mm Oto Melara Compact gun. The flight deck and other aircraft facilities upgraded including a telescopic hangar. URN 25 TACAN added. 25 mm Mk 38 guns replaced the 20 mm Mk 67. Shipboard Command and Control System (SCCS) fitted by 1996. Surface search radar replaced 1997–99. Plans to upgrade the vessels were announced on 17 December 2013. The scope of work is likely to include propulsion, weapons, and sensors.
Structure: These ships have clipper bows, twin funnels enclosing a helicopter hangar, helicopter platform aft. All are fitted with elaborate communications equipment. Superstructure is largely of aluminium construction. Bridge control of manoeuvring is by aircraft-type joystick rather than wheel.

GREGORIO DEL PILAR *7/2011, M Mazumdar* / 1406399

IHS Jane's Fighting Ships 2014-2015 © 2014 IHS

Frigates — Corvettes ‹ **Philippines**

1 CANNON CLASS (FF)

Name	No	Builders	Laid down	Launched	Commissioned
RAJAH HUMABON (ex-*Hatsuhi* DE 263; ex-*Atherton* DE 169)	PF 11 (ex-PF 78)	Norfolk Navy Yard, Portsmouth	14 Jan 1943	27 May 1943	29 Aug 1943

Displacement, tonnes: 1,412 standard; 1,778 full load
Dimensions, metres (feet): 93.3 × 11.2 × 4.3 *(306.1 × 36.7 × 14.1)*
Speed, knots: 18
Range, n miles: 6,000 at 14 kt
Complement: 165

Machinery: Diesel-electric; 2 GM EMD 16V-645E7 diesels; 5,800 hp *(4.32 MW)*; 4 generators; 2 motors; 2 shafts
Guns: 3 US 3 in *(76 mm)*/50 Mk 22; 20 rds/min to 12 km *(6.6 n miles)*; weight of shell 6 kg. 6 US/Bofors 40 mm/56 (3 twin). 4 Oerlikon 20 mm/70; 4—12.7 mm MGs.
Radars: Surface search: Raytheon SPS-5; G/H-band.
Navigation: RCA/GE Mk 26; I-band.
Sonars: SQS-17B; hull-mounted; active search and attack; medium/high frequency.
Weapon control systems: Mk 52 GFCS with Mk 41 rangefinder for 3 in guns. 3 Mk 51 Mod 2 GFCS for 40 mm.

Programmes: *Hatsuhi* originally transferred by the US to Japan 14 June 1955 and paid off June 1975 reverting to US Navy. Transferred to Philippines 23 December 1978. Towed to South Korea 1979 for overhaul and modernisation. Recommissioned 27 February 1980. A sister ship *Datu Kalantiaw* lost during Typhoon Clara 20 September 1981.
Modernisation: Upgrade plans have been suspended.
Operational: Hedgehog A/S mortars have been reported.

RAJAH HUMABON   *10/2001*, Chris Sattler / 0126280

CORVETTES

3 JACINTO (PEACOCK) CLASS (FS)

Name	No	Builders	Launched	Commissioned	Recommissioned
EMILIO JACINTO (ex-*Peacock*)	PS 35 (ex-P 239)	Hall Russell, Aberdeen	1 Dec 1982	14 Jul 1984	4 Aug 1997
APOLINARIO MABINI (ex-*Plover*)	PS 36 (ex-P 240)	Hall Russell, Aberdeen	12 Apr 1983	20 Jul 1984	4 Aug 1997
ARTEMIO RICARTE (ex-*Starling*)	PS 37 (ex-P 241)	Hall Russell, Aberdeen	11 Sep 1983	10 Aug 1984	4 Aug 1997

Displacement, tonnes: 775 full load
Dimensions, metres (feet): 62.6 × 10 × 2.7 *(205.4 × 32.8 × 8.9)*
Speed, knots: 25
Range, n miles: 2,500 at 17 kt
Complement: 31 (6 officers) + 7 spare berths

Machinery: 2 Crossley Pielstick 18 PA6 V 280 diesels; 14,000 hp(m) *(10.6 MW)* sustained; 2 shafts; 1 retractable Schottel prop; 181 hp *(135 kW)*
Guns: 1—3 in *(76 mm)*/62 Oto Melara compact; 85 rds/min to 16 km *(8.6 n miles)* anti-surface; 12 km *(6.5 n miles)* anti-aircraft; weight of shell 6 kg. 1 MSI Defence Systems 25 mm. 2—12.7 mm MGs. 2 FN 7.62 mm MGs.
Radars: Sperry Marine Bridgemaster E; E/F/I-bands.
Weapon control systems: Radamec 1500 optronic director.

Programmes: Letter of Intention to purchase from the UK signed in November 1996. Transferred 1 August 1997 after sailing from Hong Kong on 1 July 1997. Others of the class in service with the navy of the Irish Republic.
Modernisation: An upgrade programme was agreed in 2002 and a contract was signed on 6 December 2004 for phase one of the work which included overhaul of the 76 mm gun, installation of a MSI Defense Systems 25 mm mounting on the stern, replacement of Sea Archer fire-control system with a Radamec 1500 optronic director, replacement of the navigation radar with Sperry Marine Bridgemaster E and new navigation systems. Phase one was completed in September 2006. Phases two and three are to involve new propulsion and safety systems.

Structure: Fitted with telescopic cranes, loiter drive and replenishment at sea equipment. In UK service, two fast pursuit craft were carried.
Operational: These ships are the workhorses of the fleet. Based at Cavite.

EMILIO JACINTO   *12/2011*, Chris Sattler / 1406761

2 RIZAL (AUK) CLASS (FS)

Name	No	Builders	Laid down	Launched	Commissioned
RIZAL (ex-*Murrelet* MSF 372)	PS 74 (ex-PS 69)	Savannah Machine & Foundry Co, GA	24 Aug 1944	29 Dec 1944	21 Aug 1945
QUEZON (ex-*Vigilance* MSF 324)	PS 70	Associated Shipbuilders, Seattle	28 Nov 1942	5 Apr 1943	28 Feb 1944

Displacement, tonnes: 1,107 standard; 1,270 full load
Dimensions, metres (feet): 67.4 × 9.8 × 3.3 *(221.1 × 32.2 × 10.8)*
Speed, knots: 18
Range, n miles: 5,000 at 14 kt
Complement: 80 (5 officers)

Machinery: Diesel-electric; 2 GM EMD 16V-645E6 diesels; 5,800 hp *(4.32 MW)*; 2 generators; 2 motors; 2 shafts
Guns: 2 US 3 in *(76 mm)*/50 Mk 26; 20 rds/min to 12 km *(6.6 n miles)*; weight of shell 6 kg. 4 US/Bofors 40 mm/56 (2 twin); 160 rds/min to 11 km *(5.9 n miles)*; weight of shell 0.9 kg. 1 Oerlikon 20 mm (twin). 2—12.7 mm MGs.
Radars: Surface search: Raytheon SPS-5C; G/H-band.
Navigation: DAS 3; I-band.

Programmes: *Rizal* transferred from the US to the Philippines on 18 June 1965 and *Quezon* on 19 August 1967.
Modernisation: Upgrade plans have been suspended.
Structure: Upon transfer the minesweeping gear was removed and a second 3 in gun fitted aft.
Operational: Both ships were to have been deleted in 1994 but have been retained until new class of OPVs is built. Sonar equipment and depth charges have been removed.

QUEZON
12/2009,
Michael Nitz
1305976

Philippines > Corvettes — Patrol forces

1 CYCLONE CLASS (COASTAL PATROL SHIP) (PB)

Name	No	Builders	Commissioned
GENERAL MARIANO ALVARES (ex-Cyclone)	PS 38 (ex-PC 1)	Bollinger, Lockport	7 Aug 1993

Displacement, tonnes: 392 full load
Dimensions, metres (feet): 54.6 × 7.9 × 2.4 (179.1 × 25.9 × 7.9)
Speed, knots: 35
Range, n miles: 2,500 at 12 kt
Complement: 28 (4 officers) + 8 spare berths

Machinery: 4 Paxman Valenta 16RP200CM diesels; 13,400 hp (10 MW) sustained; 4 shafts
Physical countermeasures: Decoys: 2 Mk 52 sextuple and/or Wallop Super Barricade Mk 3 chaff launchers.
Electronic countermeasures: ESM: Privateer APR-39; radar warning.
Radars: Surface search: 2 Sperry RASCAR; E/F/I/J-band.
Sonars: Wesmar; hull-mounted; active; high frequency.
Weapon control systems: Marconi VISTAR IM 405 IR system.

Programmes: Transferred from the USN to the Philippines in February 2004 following refit at Bollinger. Recommissioned on 8 March 2004.
Modernisation: All armament was removed before transfer from the USN. New armament is likely to include two 25 mm guns and 12.7 mm machine guns.
Structure: Design based on Vosper Thornycroft Ramadan class modified for USN requirements including 1 in armour on superstructure. The craft has a slow speed loiter capability and has been modified to incorporate a semi-dry well, boat ramp and stern gate to facilitate deployment and recovery of a fully loaded RIB while the ship is making way.

GENERAL MARIANO ALVARES   3/2004, US Embassy, Manila / 0563762

6 PCE 827 CLASS (FS)

Name	No	Builders	Commissioned
MIGUEL MALVAR (ex-Ngoc Hoi; ex-Brattleboro PCER 852)	PS 19	Pullman Standard Car Co, Chicago	26 May 1944
MAGAT SALAMAT (ex-Chi Lang II; ex-Gayety MSF 239)	PS 20	Winslow Marine Co, Seattle	14 Jun 1944
SULTAN KUDARAT (ex-Dong Da II; ex-Crestview PCER 895)	PS 22	Willamette Iron & Steel Corp, Portland	30 Oct 1943
CEBU (ex-PCE 881)	PS 28	Albina E and M Works, Portland	31 Jul 1944
PANGASINAN (ex-PCE 891)	PS 31	Willamette Iron & Steel Corp, Portland	15 Jun 1944
ILOILO (ex-PCE 897)	PS 32	Willamette Iron & Steel Corp, Portland	6 Jan 1945

Displacement, tonnes: 650 standard; 929 full load
Dimensions, metres (feet): 56.3 × 10.1 × 2.9 (184.7 × 33.1 × 9.5)
Speed, knots: 15. **Range, n miles:** 6,600 at 11 kt
Complement: 85 (8 officers)

Machinery: 2 GM 12-278A diesels; 2,200 hp (1.64 MW); 2 shafts
Guns: 1 US 3 in (76 mm)/50; 20 rds/min to 12 km (6.6 n miles); weight of shell 6 kg. 2 to 6 US/Bofors 40 mm/56 (single or 1–3 twin); 160 rds/min to 11 km (5.9 n miles); weight of shell 0.9 kg. 2 Oerlikon 20 mm/70; 800 rds/min to 2 km.
Radars: Surface search: SPS-21D (PS 19, 28). CRM-NIA-75 (PS 29, 31, 32). SPS-53A (PS 20).
Navigation: RCA SPN-18; I/J-band.

Programmes: PS 28, 31 and 32 transferred from the US in July 1948. PS 19, 20 and 22 first transferred from the US to South Vietnam in the 1960s and then on to the Philippines in November 1975.
Modernisation: PS 19, 22, 31 and 32 refurbished in 1990–91, PS 28 in 1992 and the last one in 1996/97.
Structure: First three were originally fitted as rescue ships (PCER). A/S equipment has been removed or is inoperable. PS 20 has some minor structural differences having been built as an Admirable class MSF.
Operational: PS 23 and 29 reported decommissioned.

CEBU   5/2000, M Declerck / 0105225

LAND-BASED MARITIME AIRCRAFT

Notes: There are two Cessna 177 Cardinal transport aircraft, four Cessna T-41 trainers and one Robinson R-22 helicopter trainer.

Numbers/Type: 8 PADC (Pilatus Britten-Norman) Islander F27MP.
Operational speed: 150 kt (280 km/h).
Service ceiling: 18,900 ft (5,760 m).
Range: 1,500 n miles (2,775 km).
Role/Weapon systems: Short-range MR and SAR aircraft. First purchased in 1989. Three transferred from the Air Force. An upgrade programme, including engines, avionics and communications systems has been completed on five aircraft. The remaining three aircraft are to be similarly modernised. Sensors: Search radar, cameras. Weapons: Unarmed.

F-27MP   10/2001, Adolfo Ortigueira Gil / 0567482

Numbers/Type: 3 PADC (MBB) BO 105C.
Operational speed: 145 kt (270 km/h).
Service ceiling: 17,000 ft (5,180 m).
Range: 355 n miles (657 km).
Role/Weapon systems: Sole shipborne helicopter; some shore-based for SAR; some commando support capability. Purchased at the rate of one per year from 1986 to 1992. One lost in August 2010. Upgrade of avionics and communications is planned. Sensors: Some fitted with search radar. Weapons: Unarmed.

Numbers/Type: 3 AgustaWestland A 109 Power.
Operational speed: 152 kt (280 km/h).
Service ceiling: 16,500 ft (5,029 m).
Range: 520 n miles (827 km).
Role/Weapon systems: Contract signed in December 2012 for three aircraft to be delivered in 2014. The aircraft are to be used in maritime and internal security missions.

PATROL FORCES

4 PCF 65 (SWIFT MK 3) CLASS (COASTAL PATROL CRAFT) (PB)

PC 351–354

Displacement, tonnes: 29 standard; 38 full load
Dimensions, metres (feet): 19.8 × 4.9 × 1 (65 × 16.1 × 3.3)
Speed, knots: 25
Complement: 8
Machinery: 3 GM 12V-71TI diesels; 840 hp (616 kW) sustained; 3 shafts
Guns: 2—12.7 mm MGs.
Radars: Surface search: Koden; I-band.

Comment: Improved Swift type inshore patrol boats built by Peterson and delivered 1975–76. Aluminium construction. Some that were laid up have been returned to service. New radars fitted.

PC 354   5/1998, John Mortimer / 0081551

2 AGUINALDO CLASS (LARGE PATROL CRAFT) (PBO)

Name	No	Builders	Commissioned
EMILIO AGUINALDO	PG 140	Cavite, Sangley Point	21 Nov 1990
ANTONIO LUNA	PG 141	Cavite, Sangley Point	27 May 1999

Displacement, tonnes: 240 full load
Dimensions, metres (feet): 44 × 7.4 × 1.6 (144.4 × 24.3 × 5.2)
Speed, knots: 28
Range, n miles: 1,100 at 18 kt
Complement: 58 (6 officers)
Machinery: 2 MTU 16V-396TB94 diesels; 3,480 hp (2.59 MW) sustained; 2 shafts
Guns: 2 Bofors 40 mm/60. 2 Oerlikon 20 mm. 4—12.7 mm MGs.
Radars: Surface search: Raytheon; I-band.

Comment: Steel hulls of similar design to *Tirad Pass*. First of class launched 23 June 1984 but only completed in 1990. Second laid down 2 December 1990 and launched 23 June 1992. A third ship was laid down on 14 February 1994 and launched in April 2000. While the superstructure is 70 per cent completed, outfitting was not completed due to budget constraints.

EMILIO AGUINALDO   6/1993 / 0081540

2 POINT CLASS (PB)

Name	No	Builders	Commissioned
ALBERTO NAVARETTE (ex-Point Evans)	PG 394 (ex-82354)	CG Yard, Maryland	10 Jan 1967
ABRAHAM CAMPO (ex-Point Doran)	PG 396 (ex-82375)	CG Yard, Maryland	1 Jun 1970

Displacement, tonnes: 68 full load
Dimensions, metres (feet): 25.3 × 5.2 × 1.8 (83 × 17.1 × 5.9)
Speed, knots: 23. **Range, n miles:** 1,500 at 8 kt
Complement: 10
Machinery: 2 Caterpillar 3412 diesels; 1,600 hp (1.19 MW); 2 shafts
Guns: 2—12.7 mm MGs.
Radars: Surface search: Furuno; I-band.

Comment: PG 394 transferred from US Coast Guard 16 November 1999. Second transferred 22 March 2001. This class is in service with many other navies.

POINT CLASS (US colours)   4/1992, van Ginderen Collection / 0081549

2 KAGITINGAN CLASS (LARGE PATROL CRAFT) (PB)

Name	No	Builders	Commissioned
BAGONG LAKAS	PG 102 (ex-P 102)	Hamelin SY	9 Feb 1979
BAGONG SILANG	PG 104 (ex-P 104)	Hamelin SY	Jul 1979

Displacement, tonnes: 152 full load
Dimensions, metres (feet): 37 × 6.2 × 1.7 (121.4 × 20.3 × 5.6)
Speed, knots: 21
Complement: 30 (4 officers)
Machinery: 2 MTU 16V-538TB91 diesels; 2,500 hp(m) (1.86 MW) sustained; 2 shafts
Guns: 2 Emerlec 30 mm (twin). 4—12.7 mm MGs. 2—7.62 mm MGs.
Radars: Surface search: I-band.

Comment: Based at Cavite. P 103 paid off and used for spares. Two remain in service.

BAGONG LAKAS   1993, Philippine Navy / 0506161

7 TOMAS BATILO (SEA DOLPHIN) CLASS (FAST ATTACK CRAFT) (PBF)

BONNY SERRANO	PG 111	NICOLAS MAHUSAY	PG 116
BIENVENIDO SALTING	PG 112	DIONISIO OJEDA	PG 117
SALVADOR ABCEDE	PG 114	EMILIO LIWANAG	PG 118
RAMON AGUIRRE	PG 115		

Displacement, tonnes: 152 full load
Dimensions, metres (feet): 37 × 6.9 × 1.7 (121.4 × 22.6 × 5.6)
Speed, knots: 38
Range, n miles: 600 at 20 kt
Complement: 31 (5 officers)

Machinery: 2 MTU 20V-538TB91 diesels; 9,000 hp(m) (6.71 MW) sustained; 2 shafts
Guns: 2 Emerlec 30 mm (twin); 1,200 rds/min combined to 6 km (3.2 n miles); weight of shell 0.35 kg. 1 Bofors 40 mm/60. 2 Oerlikon 20 mm.
Radars: Surface search: Raytheon 1645; I-band.
Weapon control systems: Optical director.

Comment: Six transferred from South Korea on 15 June 1995. Part of the PKM 200 series. Different armament to South Korean ships of the same class. A further two were transferred on 7 December 2006. *Tomas Batilo* was scrapped in 2003.

BIENVENIDO SALTING   6/1996, Philippine Navy / 0506311

22 JOSÉ ANDRADA CLASS (COASTAL PATROL CRAFT) (PB)

JOSÉ ANDRADA	PG 370	ISMAEL LOMIBAO	PG 383
ENRIQUE JURADO	PG 371	LEOVIGILDO GANTIOQUI	PG 384
ALFREDO PECKSON	PG 372	FEDERICO MARTIR	PG 385
SIMEON CASTRO	PG 374	FILIPINO FLOJO	PG 386
CARLOS ALBERT	PG 375	ANASTACIO CACAYORIN	PG 387
HERACLEO ALANO	PG 376	MANUEL GOMEZ	PG 388
LIBERATO PICAR	PG 377	TEOTIMO FIGURACION	PG 389
HILARIO RUIZ	PG 378	JOSÉ LOOR SR	PG 390
RAFAEL PARGAS	PG 379	JUAN MAGLUYAN	PG 392
NESTOR REINOSO	PG 380	FLORENCIO INIGO	PG 393
DIOSCORO PAPA	PG 381	FELIX APOLINARIO	PG 395

Displacement, tonnes: 57 full load
Dimensions, metres (feet): 23.8 × 6.1 × 1.8 (78.1 × 20.0 × 5.9)
Speed, knots: 28
Range, n miles: 1,200 at 12 kt
Complement: 12 (est.) (1 officer)
Machinery: 2 Detroit 16V-92TA diesels; 1,380 hp (1.03 MW) sustained; 2 shafts
Guns: 1 Bushmaster 25 mm or Bofors 40 mm/60. 4—12.7 mm Mk 26 MGs. 2—7.62 mm M60 MGs.
Radars: Surface search: Raytheon SPS-64(V)2; I-band.

Comment: There are four batches of this class. Batch I (PCF 370–378), Batch II (PCF 379–390), Batch III (PCF 392–393) and Batch IV (PCF 395). The main difference between batches include weapons, electronics and accommodation. First four ordered from Halter Marine in August 1989 under FMS and built at Equitable Shipyards, New Orleans, as were a further four ordered in 1990. Eight more ordered in March 1993 with co-production between Halter Marine and AG&P Shipyard, Batangas. An additional three were ordered in 1995. Built to US Coast Guard standards with an aluminium hull and superstructure. The main gun may be fitted in all after some minor modifications. PG 392 delivered in March 1998, PG 393 in July 1998 and PG 395 on 10 October 2000.

TEOTIMO FIGURACION   5/2000, M Declerck / 0105226

JUAN MAGLUYAN   6/2008, Ships of the World / 1353253

632 Philippines > Patrol forces — Amphibious forces

3 CONRADO YAP (SEA HAWK/KILLER) CLASS
(COASTAL PATROL CRAFT) (PBF)

LEOPOLDO REGIS PG 847   **APOLLO TIANO** PG 851   **SULPICIO FERNANDEZ** PG 853

Displacement, tonnes: 76 full load
Dimensions, metres (feet): 25.5 × 5.4 × 1.9 *(83.7 × 17.7 × 6.2)*
Speed, knots: 38. **Range, n miles:** 290 at 20 kt
Complement: 15 (3 officers)
Machinery: 2 MTU 16V-538TB91 diesels; 5,000 hp(m) *(3.72 MW)*; 2 shafts
Guns: 1 Bofors 40 mm/60. 2 Oerlikon 20 mm (twin) Mk 16. 4—7.62 mm MGs.
Radars: Surface search: Raytheon 1645; I-band.

Comment: Type PK 181 built by Korea Tacoma and Hyundai 1975–78. Twelve craft transferred from South Korea 19 June 1993. Eight were commissioned 23 June 1993 and a further four on 23 June 1994. Nine vessels are reported to have been decommissioned.

CONRADO YAP CLASS   1993, Philippine Navy / 0506162

6 + 6 (30) MULTIPURPOSE ATTACK CRAFT (LCP)

482–487   +6

Displacement, tonnes: 19 full load
Dimensions, metres (feet): 16.5 × 4.76 × 0.8 *(54.1 × 15.6 × 2.6)*
Speed, knots: 42. **Range, n miles:** 300 at 30 kt
Complement: 4

Military lift: 16 troops plus equipment or 2.8 tons.
Machinery: 2 diesels; 2 waterjets
Guns: 1—12.7 mm MG. 2—7.62 mm MGs.
Radars: Navigation: I-band.

Comment: The design is reported to have been developed in the Philippines from the Swedish Combatboat 90 design, on which details are based. The craft, of aluminium construction, were built in Taiwan. The first batch of three was delivered in May 2009, a second batch of three in mid-2012. A further six units are expected and a class of 42 is projected.

MPAC 483   6/2012, Philippines Navy / 1486327

6 RIVERINE PATROL BOATS (PBF)

Displacement, tonnes: 11.4 full load
Dimensions, metres (feet): 12 × 3.1 × 0.6 *(39.4 × 10.2 × 2.0)*
Speed, knots: 36
Complement: 17
Machinery: 2 Yanmar 6LY2A-STP diesels; 440 hp *(328 kW)*; 2 Hamilton HJ292 waterjets
Guns: 3—12.7 mm MGs.

Comment: Six craft were donated by the United States government in September 2013. Built by Silver Ships, the craft are designed for tactical mobility and deployment of troops in a riverine environment. The aluminium hull is armoured for small arms protection and the craft are assumed to be similar to those in US service.

SURVEY AND RESEARCH SHIPS

Notes: (1) Two 52 m SWATH design hydrographic ships, *Presbitero* and *Ventura* were acquired from Spain in 1998–99 and are operated by the National Mapping and Resource Information Authority. The Coast and Geodetic Survey department operates one ship, *Atyimba* acquired in 1967. The 55 m ship *Explorer* is operated by the Mines and Geosciences Bureau and 44 m *Researcher* by the Bureau of Fisheries.
(2) Two intelligence vessels *Fort San Antonio* (AM 700) and *Fort Abad* (AM 701) were acquired in 1993.

AMPHIBIOUS FORCES

3 MARK 6 LANDING CRAFT (LCU)

SUBANON AT 291   **BAGOBO** AT 293   **TAUSUG** AT 295

Displacement, tonnes: 258 full load
Dimensions, metres (feet): 36.3 × 9.96 × 1.02 *(119.1 × 32.7 × 3.3)*
Speed, knots: 7. **Range, n miles:** 700 at 7 kt
Complement: 12

Military lift: 136 t of cargo.
Machinery: 3 Gray Marine diesels; 675 hp *(503 kW)*; 3 shafts
Guns: 2—12.7 mm MGs.
Radars: Navigation: I-band.

Comment: Former US LCT Mk 6.

2 BACOLOD CITY (FRANK S BESSON) CLASS (LSVH)

Name	No	Builders	Commissioned
BACOLOD CITY	LC 550	Moss Point Marine, Escatawpa	1 Dec 1993
DAGUPAN CITY (ex-*Cagayan De Oro City*)	LC 551	Moss Point Marine, Escatawpa	5 Apr 1994

Displacement, tonnes: 4,333 full load
Dimensions, metres (feet): 83.1 × 18.3 × 3.7 *(272.6 × 60.0 × 12.1)*
Speed, knots: 11.6. **Range, n miles:** 6,000 at 11 kt
Complement: 30 (6 officers)

Military lift: 2,280 tons (900 for amphibious operations) of vehicles, containers or cargo, plus 150 troops; 2 LCVPs on davits
Machinery: 2 GM EMD 16V-645E6 diesels; 5,800 hp *(4.32 MW)* sustained; 2 shafts; bow thruster; 250 hp *(187 kW)*
Radars: Navigation: Raytheon SPS-64(V)2; I-band.
Helicopters: Platform for 1 BO 105C.

Comment: Contract announced by Trinity Marine 3 April 1992 for two ships with an option on a third which was not taken up. Ro-ro design with 10,500 sq ft of deck space for cargo. Capable of beaching with 4 ft over the ramp on a 1 : 30 offshore gradient with a 900 ton cargo. Similar to US Army vessels but with only a bow ramp. The stern ramp space is used for accommodation for 150 troops and a helicopter platform is fitted over the stern.

BACOLOD CITY   12/2011, Chris Sattler / 1406762

DAGUPAN CITY   12/2009, Michael Nitz / 1305975

3 LST 512–1152 CLASS (TRANSPORT SHIPS) (LST)

Name	No	Builders	Commissioned
SOUTH COTABATO (ex-*Cayuga County LST 529*)	LT 87	Bethlehem Steel Corporation, Hingham	28 Feb 1944
LAGUNA (ex-*T-LST 230*)	LT 501	American Bridge, Ambridge	3 Nov 1943
BENGUET (ex-*Davies County LST 692*)	LT 507	Missouri Valley Bridge and Iron Co, Evansville	29 May 1944

Displacement, tonnes: 1,646 standard; 4,145 full load
Dimensions, metres (feet): 100 × 15.2 × 4.3 *(328.1 × 49.9 × 14.1)*
Speed, knots: 10
Complement: 110 (est.)

Military lift: 2,100 tons. 16 tanks or 10 tanks plus 200 troops
Machinery: 2 GM 12-567A diesels; 1,800 hp *(1.34 MW)*; 2 shafts
Guns: 6 US/Bofors 40 mm (2 twin, 2 single) or 4 Oerlikon 20 mm (in refitted ships).
Radars: Navigation: Raytheon SPS-64(V)2; I-band.

Programmes: Transferred from US Navy in 1976 with exception of LT 87 which was used as a light craft repair ship in South Vietnam and has retained amphibious capability (transferred to Vietnam 1970 and to Philippines 1976, acquired by purchase 5 April 1976). LT 501 commissioned in Philippine Navy 8 August 1978 and LT 507 on 18 October 1978.
Modernisation: Several have had major refits including replacement of frames and plating as well as engines and electrics and provision for four 20 mm guns to replace the 40 mm guns.
Structure: Some of the later ships have tripod masts, others have pole masts.
Operational: All are used for general cargo work in Philippine service. There were 14 deleted in 1989 and one sank in 1991. Two paid off in 1992, one in 1993 and two in 2010. *South Cotabato* was also paid off in 1993 but brought back in to service in 1994. *Lanao del Norte* grounded in the Spratly Islands on 3 November 1999 and remains there as a wreck. One further ship, *Sierra Madre* is reported to be used as an observation post in the Spratly Islands. Replacements are needed but have not been given priority.

BENGUET   1/2009 / 1366174

Amphibious forces — Auxiliaries < **Philippines** 633

0 + 2 STRATEGIC SUPPORT SHIPS (AKR)

Displacement, tonnes: 7,417 standard; 11,583 full load
Dimensions, metres (feet): 122 × 21.8 × 4.5 *(400.3 × 71.5 × 14.8)*
Speed, knots: 15
Complement: 126

Military lift: 500 troops, 12 trucks
Machinery: CODAD: 2 B&W 8L28/32A diesels; 5,250 hp *(3.9 MW)*; 2 shafts
Radars: Navigation: To be announced.
Helicopters: 3 large.

Comment: It was announced on 10 January 2014 that PT PAL, Indonesia, had received notice of award of a contract to build two strategic support ships. Details are based on the Makassar class operated by the Indonesian Navy. The design includes a docking well, capable of operating 2 LCUs, although Philippines Navy requirements may be different.

1 LANDING CRAFT (LCU)

TAGBANUA AT 296

Displacement, tonnes: 579 full load
Dimensions, metres (feet): 51.4 × 10.0 × 1.52 *(168.6 × 32.8 × 5.0)*
Speed, knots: 14
Complement: 15

Military lift: 110 t of cargo; 200 troops.
Machinery: To be announced.
Guns: 6 – 12.7 mm MGs.
Radars: Navigation: I-band.

Comment: Contract awarded on 16 March 2010 to Propmech Corporation for construction of a new landing craft in conjunction with Philippine Iron Construction and Marine Works (PICMW). Propmech was responsible for the engine and propulsion system and PICMW for construction of the hull. The ship was launched on 28 September 2011 and commissioned on 14 December 2011.

AUXILIARIES

Notes: All LSTs, LSVs, LCMs and LCUs are classified as Transports.

1 ACHELOUS CLASS (REPAIR SHIP) (ARL)

Name	No	Builders	Commissioned
YAKAL (ex-*Satyr* ARL 23; ex-*LST 852*)	AD 617 (ex-AR 517)	Chicago Bridge & Iron Co	20 Nov 1944

Displacement, tonnes: 4,412 full load
Dimensions, metres (feet): 100 × 15.2 × 4.3 *(328.1 × 49.9 × 14.1)*
Speed, knots: 11.6
Complement: 220
Machinery: 2 GM 12-567A diesels; 1,800 hp *(1.34 MW)*; 2 shafts
Guns: 4 US/Bofors 40 mm (quad). 10 Oerlikon 20 mm (5 twin).

Comment: Transferred from the US to the Philippines on 24 January 1977 by sale. (Originally to South Vietnam 30 September 1971.) Converted during construction. Extensive machine shop, spare parts stowage, and logistic support.

YAKAL 1994, Philippine Navy / 0081545

1 TRANSPORT VESSEL (AP)

Name	No	Builders	Commissioned
ANG PANGULO (ex-*Pag-Asa*; ex-*The President*, ex-*Roxas*, ex-*Lapu-Lapu*)	AT 25 (ex-TP 777)	Ishikawajima, Japan	1959

Displacement, tonnes: 2,275 standard; 2,771 full load
Dimensions, metres (feet): 78.5 × 13 × 6.4 *(257.5 × 42.7 × 21.0)*
Speed, knots: 18
Range, n miles: 6,900 at 15 kt
Complement: 81 (8 officers)
Machinery: 2 Mitsui DE642/VBF diesels; 5,000 hp(m) *(3.68 MW)*; 2 shafts
Guns: 3 Oerlikon 20 mm/70 Mk 4. 8 – 7.62 mm MGs.
Radars: Navigation: RCA CRMN-1A-75; I-band.

Comment: Built as war reparation; launched in 1958. Was used as presidential yacht and command ship with accommodation for 50 passengers. Originally named *Lapu-Lapu* after the chief who killed Magellan; renamed *Roxas* on 9 October 1962 after the late Manuel Roxas, the first President of the Philippines Republic. Renamed *The President* in 1967 and *Ang Pangulo* in 1975. In early 1987 was earmarked to transport President Marcos to Hong Kong and exile. The ship is now used as an attack transport, and still as a Presidential Yacht.

ANG PANGULO 5/1998, John Mortimer / 0081546

1 ALAMOSA CLASS (SUPPLY SHIP) (AK)

Name	No	Builders	Commissioned
MACTAN (ex-*Kukui*; ex-*Colquith*)	AC 90 (ex-TK 90)	Froemming, Milwaukee	22 Sep 1944

Displacement, tonnes: 2,540 standard; 7,691 full load
Dimensions, metres (feet): 103.2 × 15.2 × 5.5 *(338.6 × 49.9 × 18.0)*
Speed, knots: 11
Complement: 85
Machinery: 1 Nordberg diesel; 1,700 hp *(1.27 MW)*; 1 shaft
Guns: 2 – 12.7 mm MGs.

Comment: Transferred from the US Coast Guard on 1 March 1972. Used to supply military posts and lighthouses in the Philippine archipelago. Operational status doubtful.

MACTAN 4/1996, Philippine Navy / 0506312

4 FLOATING DOCKS (YFD)

YD 200 (ex-*AFDL 24*)   YD 205 (ex-*AFDL 44*)
YD 204 (ex-*AFDL 20*)   – (ex-*AFDL 40*)

Comment: Floating steel dry docks built in the USA; all are former US Navy units with YD 200 transferred in July 1948, YD 204 in October 1961 (sale 1 August 1980), YD 205 in September 1969 and AFDL 40 in 1994. Capacities: YD 205, 2,800 tons; YD 200 and YD 204, 1,000 tons. In addition there are two floating cranes, YU 206 and YU 207, built in US in 1944 and capable of lifting 30 tons.

42 LCM/LCU

Comment: Ex-US minor landing craft mostly transferred in the mid-1970s. 11 LCM 6, five LCM 8, eight LCU, 14 RUC and two LCVP. Used as transport vessels.

LCU 286 5/1998, van Ginderen Collection / 0052706

© 2014 IHS IHS Jane's Fighting Ships 2014-2015

634 Philippines > Auxiliaries — Coast guard

2 YW TYPE (WATER TANKERS) (AWT)

Name	No	Builders	Commissioned
LAKE BALUSAN	AW 33 (ex-YW 111)	Marine Iron and Shipbuilding Corp, Duluth	1 Aug 1945
LAKE PAOAY	AW 34 (ex-YW 130)	Leathem D Smith, Sturgeon Bay	28 Aug 1945

Displacement, tonnes: 1,257 full load
Dimensions, metres (feet): 53 × 10 × 4 *(173.9 × 32.8 × 13.1)*
Speed, knots: 7.5
Complement: 29
Cargo capacity: 200,000 gallons
Machinery: 2 GM 8-278A diesels; 1,500 hp *(1.12 MW)*; 2 shafts
Guns: 1 Bofors 40/60. 1 Oerlikon 20 mm.

Comment: Basically similar to YOG type but adapted to carry fresh water. Transferred from the US to the Philippines on 16 July 1975.

LAKE PAOAY *5/1998, van Ginderen Collection* / 0052708

2 YOG TYPE (TANKERS) (YO)

Name	No	Builders	Commissioned
LAKE BUHI (ex-*YOG 73*)	AF 78 (ex-YO 78)	Puget Sound, Bremerton	28 Nov 1944
LAKE TAAL (ex-*YOG*)	AF 72 (ex-YO 72)	Puget Sound, Bremerton	14 Apr 1945

Displacement, tonnes: 454 standard; 1,422 full load
Dimensions, metres (feet): 53 × 10 × 4 *(173.9 × 32.8 × 13.1)*
Speed, knots: 8
Complement: 28
Cargo capacity: 6,570 barrels dieso and gasoline
Machinery: 2 GM 8-278A diesels; 1,500 hp *(1.12 MW)*; 2 shafts
Guns: 2 Oerlikon 20 mm/70 Mk 4.

Comment: Former US Navy gasoline tankers. Transferred in July 1967 on loan and by purchase 5 March 1980.

LAKE BUHI *1993, Philippine Navy* / 0506164

TUGS

Notes: There is a 30 m YTM 764-class tug *Manobo* AT 297. There are three 20 m YTL 422 class *Igorot* YQ 222, *Ilongot* YQ 225 and *Tasaday* YQ 226.

HARBOUR TUG *5/1998, John Mortimer* / 0052709

COAST GUARD

Notes: (1) The Philippine Coast Guard (PCG) is headed by a Commandant with the rank of Admiral. First established in 1901, it was a part of the Philippine Navy from 1967 until 1998 when it became part of the Department of Transportation and Communication. With a complement of approximately 5,000, it is responsible for Maritime Search and Rescue, Maritime Safety Administration, Maritime Security and Marine Environmental Protection. There is also a 19,000-strong Philippine Coast Guard Auxiliary, a volunteer support arm established in 1972. There are 12 Coast Guard Districts, 64 stations and 236 Coast Guard detachments.
(2) The Coast Guard also operates one LCM 6 (BM 270), one LCVP (BV 182) and a River Utility Craft VU 463.
(3) Aviation: There are three Britten Norman Islanders and one Cessna Skyeagle. There are two BO-105 helicopters and the procurement of eight EC-45 helicopters is under consideration.
(4) One 82 m offshore patrol ship was ordered from France in October 2012.
(5) Ten 40-m patrol craft have been ordered from Japan. They are to be delivered in 2015.

4 SAN JUAN CLASS (WPBO)

Name	No	Builders	Commissioned
SAN JUAN	001	Tenix Defence Systems	19 Jun 2000
EDSA II (ex-*Don Emilio*)	002 (ex-419)	Tenix Defence Systems	14 Dec 2000
PAMPANGA	003	Tenix Defence Systems	30 Jan 2003
BATANGAS	004	Tenix Defence Systems	8 Aug 2003

Displacement, tonnes: 508 full load
Dimensions, metres (feet): 56 × 10.5 × 3 *(183.7 × 34.4 × 9.8)*
Speed, knots: 24.5. **Range, n miles:** 3,000 at 15 kt
Complement: 38
Machinery: 2 Caterpillar 3612 diesels; 4,800 hp(m) *(3.53 MW)* sustained; 2 shafts; cp props
Radars: Navigation: I-band
Helicopters: Platform for one light.

Comment: First reported ordered in mid-1997. Construction of first of class started in February 1999. Steel hull and aluminium superstructure. Primarily used for SAR with facilities for 300 survivors. Fire-fighting and pollution control equipment included. A contract for a further two vessels was finalised in December 2001.

SAN JUAN *6/2000, Tenix Shipbuilding* / 0105228

4 ILOCOS NORTE CLASS (PATROL CRAFT) (PB)

Name	No	Builders	Commissioned
ILOCOS NORTE	3501	Tenix Defence Systems	26 May 2003
NUEVA VIZCAYA	3502	Tenix Defence Systems	26 Aug 2003
ROMBLON	3503	Tenix Defence Systems	4 Nov 2003
DAVAO DEL NORTE	3504	Tenix Defence Systems	12 Feb 2004

Displacement, tonnes: 117 standard
Dimensions, metres (feet): 35 × 7.3 × 2.3 *(114.8 × 24.0 × 7.5)*
Speed, knots: 23
Range, n miles: 2,000 at 12 kt
Complement: 11
Machinery: 2 diesels; 2 shafts. 1 loiter waterjet
Guns: 2—30 mm (1 twin). 2—12.7 mm MGs.
Radars: Navigation: I-band

Comment: Contract on 9 December 2001 for the construction of four search and rescue vessels. An option for a further ten craft has not been taken. Based on Bay class design with steel hull and aluminium superstructure.

NUEVA VIZCAYA *8/2003, Tenix* / 0569803

1 CORREGIDOR CLASS (BUOY TENDER) (ABU)

Name	No	Builders	Commissioned
CORREGIDOR	AG 891	Niigata Engineering	2 Mar 1998

Displacement, tonnes: 1,148 full load
Dimensions, metres (feet): 56.9 × 11.0 × 3.8 *(186.7 × 36.1 × 12.5)*
Speed, knots: 13. **Range, n miles:** 4,000 at 11 kt
Complement: 37
Machinery: 2 Niigata diesels; 2 shafts
Radars: Navigation: I-band.

Comment: Lighthouse and buoy tender. Similar to *Jadayat* in service in Indonesia.

Coast guard < **Philippines** 635

1 BALSAM CLASS (TENDER) (AKLH)

Name	No	Builders	Commissioned
KALINGA (ex-*Redbud*, WAGL 398; ex-*Redbud*, T-AKL 398)	AG 89	Marine Iron and Shipbuilding Corp, Duluth	2 May 1944

Displacement, tonnes: 965 standard; 1,058 full load
Dimensions, metres (feet): 54.8 × 11.3 × 4 *(179.8 × 37.1 × 13.1)*
Speed, knots: 12
Range, n miles: 3,500 at 7 kt
Complement: 53

Machinery: Diesel-electric; 2 diesels; 1,710 hp *(1.28 MW)*; 2 generators; 1 motor; 1,200 hp *(895 kW)*; 1 shaft
Guns: 2 — 12.7 mm MGs.
Radars: Navigation: Sperry SPS-53; I/J-band.
Helicopters: Platform for 1 light.

Comment: Originally US Coast Guard buoy tender (WAGL 398). Transferred to US Navy on 25 March 1949 as AG 398 and then to the Philippine Navy 1 March 1972. One 20 ton derrick. New engines fitted.

KALINGA   *1994, Philippine Navy* / 0506201

3 BUOY TENDERS (ABU)

CAPE BOJEADOR (ex-*FS 203*) AE 46 (ex-TK 46)
LIMASAWA (ex-*Nettle* WAK 129; ex-*FS 169*) AE 79 (ex-TK 79)
MANGYAN (ex-*Nasami*; ex-*FS 408*) AT 71 (ex-AE 71, ex-AS 71)

Displacement, tonnes: 478 standard; 965 full load
Dimensions, metres (feet): 54.9 × 9.8 × 3 *(180.1 × 32.2 × 9.8)*
Speed, knots: 10
Range, n miles: 4,150 at 10 kt
Complement: 50

Cargo capacity: 400 tons
Machinery: 2 GM 6-278A diesels; 1,120 hp *(836 kW)*; 2 shafts
Guns: 1 — 12.7 mm MG can be carried.
Radars: Navigation: RCA CRMN 1A 75; I-band.

Comment: Former US Army FS 381 and FS 330 type freight and supply ships built in 1943–44. First two are employed as tenders for buoys and lighthouses. *Mangyan* transferred 24 September 1976 by sale. *Limasawa* acquired by sale 31 August 1978. One 5 ton derrick. *Cape Bojeador* paid off in 1988 but was back in service in 1991 after a major overhaul. *Mangyan* reclassified AT in 1993 and belongs to the Navy. Masts and superstructures have minor variations.

CAPE BOJEADOR   *1993, Philippine Navy* / 0506165

2 LARGE PATROL CRAFT (PB)

Name	No	Builders	Commissioned
TIRAD PASS	AU 100 (ex-SAR 100)	Sumidagawa	1974
BESSANG PASS	AU 75 (ex-SAR 99)	Sumidagawa	1974

Displacement, tonnes: 283 full load
Dimensions, metres (feet): 44 × 7.4 × 1.5 *(144.4 × 24.3 × 4.9)*
Speed, knots: 27.5
Range, n miles: 2,300 at 14 kt
Complement: 32
Machinery: 2 MTU 12V 538 TB82 diesels; 4,050 hp(m) *(2.98 MW)*; 2 shafts
Guns: 4 — 12.7 mm (2 twin) MGs.

Comment: Paid for under Japanese war reparations. Similar type as *Emilio Aguinaldo*. *Bessang Pass* grounded in 1983 but was recovered.

© 2014 IHS

TIRAD PASS   *1992, Philippine Navy* / 0081548

1 PGM-39 CLASS (LARGE PATROL CRAFT) (PB)

Name	No	Commissioned
PALAWAN (ex-*PGM 42*)	PG 64	Jun 1960

Displacement, tonnes: 126 full load
Dimensions, metres (feet): 30.6 × 5.7 × 2.1 *(100.4 × 18.7 × 6.9)*
Speed, knots: 17
Range, n miles: 1,400 at 11 kt
Complement: 30 (est.)
Machinery: 2 MTU MB 12V 493 TY57 diesels; 2,200 hp(m) *(1.6 MW)* sustained; 2 shafts
Guns: 2 Oerlikon 20 mm. 2 — 12.7 mm MGs. 1 — 81 mm mortar.
Radars: Surface search: Alpelco DFR-12; I/J-band.

Comment: Steel-hulled craft built under US military assistance programmes. Transferred upon completion. Three other craft have been decommissioned.

PGM-39 CLASS   *1994, Philippine Navy* / 0081550

10 PCF 46 CLASS (COASTAL PATROL CRAFT) (PB)

| DB 411 | DB 417 | DB 422 | DB 429 | DB 432 |
| DB 413 | DB 419 | DB 426 | DB 431 | DB 435 |

Displacement, tonnes: 21 full load
Dimensions, metres (feet): 14 × 4.4 × 1 *(45.9 × 14.4 × 3.3)*
Speed, knots: 25
Range, n miles: 1,000 at 15 kt
Complement: 8
Machinery: 2 Cummins diesels; 740 hp *(552 kW)*; 2 shafts
Guns: 2 — 12.7 mm (twin) MGs. 1 — 7.62 mm M60 MG.
Radars: Surface search: Kelvin Hughes 17; I-band.

Comment: Built by Marcelo Yard, Manila and were to have been delivered 1976–78 at the rate of two per month. By the end of 1976, 25 had been completed but a serious fire in the shipyard destroyed 12 new hulls and halted production. Some deleted.

DB 435   *1993, Philippine Navy* / 0506166

IHS Jane's Fighting Ships 2014-2015

636 Philippines > Coast guard

12 PCF 50 (SWIFT MK 1 AND MK 2) CLASS
(COASTAL PATROL CRAFT) (PB)

DF 300–303   DF 305   DF 307–313

Displacement, tonnes: 23 full load
Dimensions, metres (feet): 15.6 (DF 305, 307–313), 15.2 (DF 300–303) × 4.1 × 1.2 *(51.2, 49.9 × 13.5 × 3.9)*
Speed, knots: 28. **Range, n miles:** 685 at 16 kt
Complement: 6
Machinery: 2 GM 12-71 diesels; 680 hp *(504 kW)* sustained; 2 shafts
Guns: 2 – 12.7 mm (twin) MGs. 2 M-79 40 mm grenade launchers.
Radars: Surface search: Decca 202; I-band.

Comment: Most built in the USA. Built for US military assistance programmes and transferred in the late 1960s. Some built in 1970 in the Philippines (ferro-concrete) with enlarged superstructure. *DF 300–303* are Swift Mk 1. *DF 305* and *DF 307–313* are Swift Mk 2.

DF 308   5/1998, van Ginderen Collection / 0081552

10 PCF 65 (SWIFT MK 3) CLASS
(COASTAL PATROL CRAFT) (PB)

DF 325–332   DF 334   DF 347

Displacement, tonnes: 29 standard; 38 full load
Dimensions, metres (feet): 19.8 × 4.9 × 1 *(65 × 16.1 × 3.3)*
Speed, knots: 25
Complement: 8
Machinery: 3 GM 12V-71TI diesels; 840 hp *(616 kW)* sustained; 3 shafts
Guns: 2 – 12.7 mm MGs.
Radars: Surface search: Koden; I-band.

Comment: Improved Swift type inshore patrol boats built by Peterson and delivered 1975–76. Aluminium construction. Some that were laid up have been returned to service. New radars fitted.

DF 347   5/1998, Sattler & Steele / 0052711

4 RODMAN 38 CLASS (PB)

M 1101–1104

Displacement, tonnes: 10 full load
Dimensions, metres (feet): 11 × 3.9 × 0.7 *(36.1 × 12.8 × 2.3)*
Speed, knots: 28. **Range, n miles:** 300 at 15 kt
Complement: 4
Machinery: 2 diesels; 2 waterjets

Comment: GRP hull. Built in 2004 by Rodman, Vigo. The vessels are owned by the Bureau of Fisheries and Aquatic Resources (BFAR) and are manned jointly by BFAR and PCG personnel. The vessels are deployed to conduct surveillance in the strategic chokepoints between the Philippines major islands.

3 DE HAVILLAND CLASS (PB)

DF 321–323

Displacement, tonnes: 25 full load
Dimensions, metres (feet): 16.7 × 5 × 1.3 *(54.8 × 16.4 × 4.3)*
Speed, knots: 25. **Range, n miles:** 450 at 14 kt
Complement: 8
Machinery: 2 diesels; 740 hp *(552 kW)*; 2 shafts
Guns: 2 – 12.7 mm MGs.

Comment: Locally built in the mid-1980s. Others of this type have been paid off and numbers are uncertain.

IHS Jane's Fighting Ships 2014-2015

DF 321   5/1998, van Ginderen Collection / 0052713

11 CUTTERS (PBR)

CGC 103   CGC 110   CGC 115   CGC 128–130   CGC 132–136

Displacement, tonnes: 13 full load
Dimensions, metres (feet): 12.2 × 4.1 × 0.9 *(40 × 13.5 × 3.0)*
Speed, knots: 28
Complement: 5
Machinery: 2 Detroit diesels; 560 hp *(418 kW)*; 2 shafts
Guns: 1 – 12.7 mm MG. 1 – 7.62 mm MG.

Comment: Built at Cavite Yard from 1984. One deleted in 1994. Used for harbour patrols. There are also some small unarmed Police craft.

CGC 130   1994, Philippine Navy / 0081553

0 + 4 OCEA FPB 72 MK II (PATROL CRAFT) (PB)

Dimensions, metres (feet): 23.6 × 5.8 × 1.5 *(77.4 × 19.0 × 4.9)*
Speed, knots: 32
Range, n miles: 600 at 12 kt
Complement: 11
Machinery: 2 MTU diesels; 3,000 hp *(2.24 MW)*; 2 shafts
Guns: 2 – 7.62 mm MGs.
Radars: Surface search/navigation: To be announced.

Comment: Plans to acquire four patrol craft were announced on 31 October 2012. Delivery is expected in 2014.

FPB 72 CLASS   3/2012, B Prézelin / 1455681

10 RODMAN 101 CLASS (PB)

MCS 3001–3010

Displacement, tonnes: 63 full load
Dimensions, metres (feet): 30 × 5.9 × 1.3 *(98.4 × 19.4 × 4.3)*
Speed, knots: 28
Range, n miles: 800 at 12 kt
Complement: 9
Machinery: 2 Caterpillar 3412C diesels; 2,800 hp *(2.06 MW)*; 2 Hamilton waterjets
Guns: 4 – 12.7 mm MGs.
Radars: Navigation: I-band.

Comment: GRP hull. Built by Rodman, Vigo and delivered in 2005. The vessels are owned by the Bureau of Fisheries and Aquatic Resources (BFAR) and are manned jointly by BFAR and PCG personnel. The vessels are deployed to conduct surveillance in the strategic chokepoints between the Philippines major islands.

© 2014 IHS

Poland

MARYNARKA WOJENNA

Country Overview

The modern democratic era of the Republic of Poland began in 1989 after 42 years of communist rule. Situated in central Europe, the country has an area of 120,725 square miles and is bordered to the north by Russia (Kaliningrad), to the east by Lithuania, Belarus, and Ukraine, to the south by the Czech Republic and Slovakia and to the west by Germany. It has a 265 n mile coastline with the Baltic Sea. Warsaw is the capital and largest city while Gdansk, Szczecin and Gdynia are the principal ports. Territorial seas (12 n miles) are claimed but while it has claimed a 200 n mile EEZ, its limits have not been fully defined by boundary agreements.

Headquarters Appointments

Commander-in-Chief:
Vice Admiral Tomasz Mathea
Deputy Commander-in-Chief:
Rear Admiral Ryszard Demczuk
Commander Maritime Operations Centre:
Rear Admiral Stanislaw Zarychta
Commander 3rd Flotilla:
Rear Admiral Marian Ambroziak
Commander 8th Flotilla:
Rear Admiral Krzysztof Teryfter

Diplomatic Representation

Defence and Naval Attaché in London:
Captain Stanislaw Król

Personnel

2014   8,000

Prefix to Ships' Names

ORP, standing for *Okret Rzeczypospolitej Polskiej*

Strength of the Fleet

Type	Active	Building
Submarines—Patrol	5	—
Frigates	2	—
Corvettes	4	1
Minehunters—Coastal	20	—
LSTs	5	—
LCUs	3	—
Survey and Research Ships	2	—
AGIs	2	—
Training Ships	2	—
Salvage Ships	6	—
Tankers	4	—
Logistic Support Ship	1	—

Sea Department of the Border Guard (MOSG)

A para-naval force, subordinate to the Minister of the Interior.

Bases

Gdynia (3rd Naval Flotilla), Swinoujscie (8th Coastal Defence Flotilla), Kolobrzeg, Gdansk (Frontier Guard)

Naval Aviation

HQ at Gdynia-Babie Doly
43rd Naval Air Base (Gdynia) (An-28, W-3, SH-2G, Mi-17, Mi-2)
44th Naval Air Base (Siemirowice and Darlowo) (An-28, W-3, Mi-14, Mi-2)

Coast Defence

Two AA defence squadrons each with 12—57 mm guns.

PENNANT LIST

Submarines
291 Orzeł
294 Sokół
295 Sęp
296 Bielik
297 Kondor

Frigates
272 Generał Kazimierz Pułaski
273 Generał Tadeusz Kościuszko

Corvettes
240 Kaszub
421 Orkan
422 Piorun
423 Grom

Mine Warfare Forces
621 Flaming
623 Mewa
624 Czajka
630 Goplo
631 Gardno
632 Bukowo
633 Dabie
634 Jamno
635 Mielno
636 Wicko
637 Resko
638 Sarbsko
639 Necko
640 Naklo
641 Druzno
642 Hancza
643 Mamry
644 Wigry
645 Sniardwy
646 Wdzydze

Amphibious Forces
821 Lublin
822 Gniezno
823 Krakow
824 Poznan
825 Torun
851 —
852 —
853 —

Survey Ships and AGIs
262 Nawigator
263 Hydrograf
265 Heweliusz
266 Arctowski

Auxiliaries
251 Wodnik
253 Iskra
281 Piast
282 Lech

511 Kontradmiral X Czernicki
R 14 Zbyszko
R 15 Macko
SD 11 —
SD 13 —
Z 1 Baltyk
Z 8 —

Maritime Frontier Guard
SG 311 Kaper I
SG 312 Kaper II
SG 323 Zefir
SG 325 Tecza

SUBMARINES

Notes: A programme (Orka) to procure three new submarines was launched in early 2014. The principal contenders are likely to be the German (Thyssen Krupp) Type 214 and the French (DCNS) Scorpene class. The boats are expected to enter service in the 2020–30 timeframe.

4 SOKÓŁ (KOBBEN) (TYPE 207) CLASS (SSK)

Name	No	Builders	Laid down	Launched	Commissioned	Recommissioned
SOKÓŁ (ex-*Stord*)	294 (ex-S 308)	Rheinstahl – Nordseewerke, Emden	1 Apr 1966	2 Sep 1966	14 Feb 1967	4 Jun 2002
SEP (ex-*Skolpen*)	295 (ex-S 306)	Rheinstahl – Nordseewerke, Emden	1 Nov 1965	24 Mar 1966	17 Aug 1966	16 Aug 2002
BIELIK (ex-*Svenner*)	296 (ex-S 309)	Rheinstahl – Nordseewerke, Emden	8 Sep 1966	27 Jan 1967	12 Jun 1967	8 Sep 2003
KONDOR (ex-*Kunna*)	297 (ex-S 319)	Rheinstahl – Nordseewerke, Emden	3 Mar 1964	16 Jul 1964	29 Oct 1964	20 Oct 2004

Displacement, tonnes: 466 surfaced; 532 dived
Dimensions, metres (feet): 47.4 × 4.6 × 4.3 *(155.5 × 15.1 × 14.1)*
Speed, knots: 12 surfaced; 18 dived
Range, n miles: 5,000 at 8 kt snorting
Complement: 21 (5 officers)

Machinery: Diesel-electric; 2 MTU 12V 493 AZ80 GA31L diesels; 1,200 hp(m) *(880 kW)* sustained; 1 motor; 1,800 hp(m) *(1.32 MW)* sustained; 1 shaft
Torpedoes: 8—21 in *(533 mm)* bow tubes.
Electronic countermeasures: ESM: AR-700-S1 radar warning.
Radars: Surface search: Sperry Marine Bridgemaster; I-band.
Sonars: Atlas Elektronik CSU 83; passive search and attack; medium/high frequency.
Weapon control systems: Kongsberg MSI-70U TFCS.

Programmes: Commissioned into the Norwegian Navy from 1964, the original building cost was shared between the Norwegian and US governments. Decommissioned from the Norwegian Navy in 2001. Following announcement on 18 January 2002, four submarines transferred to the Polish Navy. A fifth, ex-*Kobben*, was transferred for spares and as a floating training base. The contract also includes provision of in-service support.
Modernisation: All modernised at Urivale Shipyard, Bergen between 1989–1992.
Structure: A development of the German Type 205 class, they have a diving depth of 650 ft *(200 m)*. Pilkington optronics CK 30 search periscope.
Operational: Based at Gdynia. These submarines are expected to be decommissioned before the first of three new submarines enters service.

KONDOR *6/2011, Michael Nitz* / 1406657

BIELIK
4/2010, E & M Laursen
1406243

Poland > Submarines — Frigates

1 KILO CLASS (PROJECT 877EM) (SSK)

Name	No	Builders	Commissioned
ORZEŁ	291	Sudomekh, Leningrad	29 Apr 1986

Displacement, tonnes: 2,496 surfaced; 3,231 dived
Dimensions, metres (feet): 72.6 × 9.9 × 6.5 *(238.2 × 32.5 × 21.3)*
Speed, knots: 10 surfaced; 9 snorting; 17 dived
Range, n miles: 6,000 at 7 kt snorting; 400 at 3 kt dived
Complement: 60 (16 officers)

Machinery: Diesel-electric; 2 DL 42M diesels; 3,650 hp(m) *(2.68 MW)*; 2 generators; 6 MW; 1 PG 141 motor; 5,900 hp(m) *(4.34 MW)*; 1 shaft; 2 auxiliary motors; 204 hp(m) *(150 kW)*; 1 economic speed motor; 130 hp *(95 kW)*
Missiles: SAM: 8 SA-N-5 (Strela 2M).
Torpedoes: 6—21 in *(533 mm)* tubes. Combination of 53–65; anti-surface; passive/wake homing to 19 km *(10.3 n miles)* at 45 kt; warhead 300 kg and TEST-71; anti-submarine; active/passive homing to 15 km *(8.1 n miles)* at 40 kt; warhead 205 kg. 53–56 WA and SET 53 M can also be carried. Total of 18 torpedoes.
Mines: 24 in lieu of torpedoes.
Electronic countermeasures: ESM: Brick Group (MRP-25); radar warning; Quad Loop HF D/F.
Radars: Surface search: Sperry Marine Bridgemaster; I-band.
Sonars: Shark Teeth (MGK-400); hull-mounted; passive search and attack (some active capability); low/medium frequency. Mouse Roar (MG 519); active mine detection; high frequency.
Weapon control systems: Murena MWU 110 TFCS.

Programmes: This was the second transfer of this class, the first being to India and others have since gone to Romania, Algeria, Iran and China. It was expected that more than one would be acquired as part of an exchange deal with the USSR for Polish-built amphibious ships, but this class was considered too large for Baltic operations and subsequent transfers were of the since deleted Foxtrot class.
Structure: Diving depth, 240 m *(787 ft)*. Has two torpedo tubes modified for wire guided anti-submarine torpedoes.
Operational: Based at Gdynia.

ORZEŁ 6/2013*, J Ciślak / 1531013

FRIGATES

2 OLIVER HAZARD PERRY CLASS (FFGHM)

Name	No	Builders	Laid down	Launched	Commissioned	Recommissioned
GENERAŁ KAZIMIERZ PUŁASKI (ex-*Clark*)	272 (ex-FFG 11)	Bath Iron Works, Maine	17 Jul 1978	24 Mar 1979	9 May 1980	15 Mar 2000
GENERAŁ TADEUSZ KOŚCIUSZKO (ex-*Wadsworth*)	273 (ex-FFG 9)	Todd Pacific Shipyard Corporation, San Pedro	13 Jul 1977	29 Jul 1978	28 Feb 1980	28 Jun 2002

Displacement, tonnes: 2,794 standard; 3,696 full load
Dimensions, metres (feet): 135.6 × 13.7 × 4.5 hull; 7.5 sonar *(444.9 × 44.9 × 14.8; 24.6)*
Speed, knots: 29. **Range, n miles:** 4,500 at 20 kt
Complement: 200 (15 officers; 19 air crew)

Machinery: 2 GE LM 2500 gas turbines; 41,000 hp *(30.59 MW)* sustained; 1 shaft; cp prop 2 auxiliary retractable props; 650 hp *(484 kW)*
Missiles: SSM: 4 McDonnell Douglas Harpoon Block 1G; active radar homing to 95 km *(51 n miles)* at 0.9 Mach; warhead 227 kg.
SAM: 36 Raytheon SM-1MR Block VI; command guidance; semi-active radar homing to 38 km *(20.5 n miles)* at 2 Mach. 1 Mk 13 Mod 4 launcher for both SSM and SAM missiles ❶.
Guns: 1 Oto Melara 3 in *(76 mm)*/62 Mk 75 ❷; 85 rds/min to 16 km *(8.7 n miles)* anti-surface; 12 km *(6.6 n miles)* anti-aircraft; weight of shell 6 kg. 1 Raytheon 20 mm/76 6-barrelled Mk 15 Vulcan Phalanx Block 1B ❸; 4,500 rds/min combined to 1.5 km. 4—12.7 mm MGs.
Torpedoes: 6—324 mm Mk 32 (2 triple) tubes ❹. 24 Eurotorp MU-90; active/passive homing to 25 km *(13.5 n miles)* at 29/50 kt.
Physical countermeasures: Decoys: 2 Loral Hycor SRBOC 6-barrelled fixed Mk 36 ❺; IR flares and chaff to 4 km *(2.2 n miles)*. T-Mk 6 Fanfare/SLQ-25 Nixie; torpedo decoy.
Electronic countermeasures: ESM/ECM: SLQ-32(V)2 ❻; radar warning.
Radars: Air search: Raytheon SPS-49(V)4 ❼; C/D-band.
Surface search: ISC Cardion SPS-55 ❽; I-band.
Fire control: Lockheed STIR (modified SPG-60) ❾; I/J-band. Sperry Mk 92 (Signaal WM28) ❿; I/J-band.
Navigation: 2 Sperry Marine Bridgemaster; I-band.
Tacan: URN 25. IFF Mk XII AIMS UPX-29.
Sonars: SQQ 89(V)2 (Raytheon SQS 56 and Gould SQR 19); hull-mounted active search and attack; medium frequency and passive towed array; very low frequency.
Combat data systems: NTDS with Link 11 and 14. SATCOM SRR-1, WSC-3 (UHF).
Weapon control systems: SWG-1 Harpoon LCS. Mk 92 (Mod 2), WCS with CAS (Combined Antenna System). The Mk 92 is the US version of the Signaal WM28 system. Mk 13 weapon direction system. 2 Mk 24 optical directors.
Helicopters: 2 Kaman SH-2G Seasprite ⓫.

Programmes: *Pułaski* approved for transfer from US by grant in 1999.
Modernisation: A contract to modernise *Pułaski* was signed with the US government in November 2013. The work is to include overhaul of the propulsion system and modernisation of the 76 mm gun, SQS-56 sonar, SM-1 launcher, and SRBOC. The work is to be completed by early 2016.
Structure: Details given are for the ship in service with the US Navy.
Operational: From 2012, the principal role of *Pułaski* is to be a training ship. Based at Gdynia.

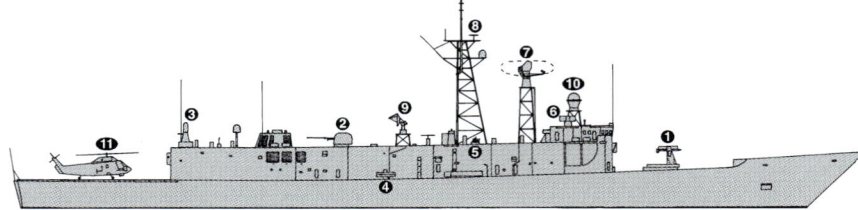

PUŁASKI (Scale 1 : 1,200), Ian Sturton / 0105229

PUŁASKI 5/2011, Per Körnefeldt / 1406660

IHS Jane's Fighting Ships 2014-2015 © 2014 IHS

Frigates — Corvettes < **Poland** 639

KOŚCIUSZKO   6/2011, L-G Nilsson / 1406658

KOŚCIUSZKO   6/2011, Michael Nitz / 1406659

CORVETTES

3 ORKAN (SASSNITZ) CLASS (PROJECT 660 (EX-151)) (FSGM)

Name	No	Builders	Launched	Commissioned
ORKAN	421	Peenewerft; Northern Shipyard	29 Sep 1990	18 Sep 1992
PIORUN	422	Peenewerft; Northern Shipyard	19 Oct 1990	11 Mar 1994
GROM (ex-*Huragan*)	423	Peenewerft; Northern Shipyard	11 Dec 1990	28 Mar 1995

Displacement, tonnes: 336 standard; 331 full load
Dimensions, metres (feet): 49.8 oa; 45 wl × 8.7 × 2.2 *(163.4; 147.6 × 28.5 × 7.2)*
Speed, knots: 38
Range, n miles: 1,600 at 14 kt
Complement: 37 (5 officers)

Machinery: 3 Type M 520T diesels; 16,000 hp(m) *(11.93 MW)* sustained; 3 shafts
Missiles: SSM; 4 launchers; Saab RBS15 Mk 3 (to be installed); active radar homing to 200 km *(108 n miles)* at 0.8 Mach; warhead 200 kg.
SAM: SA-N-5 Grail quad launcher; manual aiming; IR homing to 6 km *(3.2 n miles)* at 1.5 Mach; warhead 1.5 kg.

Guns: 1 USSR 3 in *(76 mm)*/59 AK 176; 120 rds/min to 16 km *(8.5 n miles)*; weight of shell 5.9 kg. 1—30 mm/65 AK 630; 6 barrels; 3,000 rds/min combined to 2 km. 2—12.7 mm MGs.
Electronic countermeasures: ESM: PIT Bren-R IFM wideband receiver.
Radars: Surface search: AMB Sea Giraffe; G-band.
Fire control: Thales STING; I/K-bands.
Navigation: Sperry Marine Bridgemaster; I-band.
IFF: Mk XII.
Combat data systems: Signaal TACTICOS.
Weapon control systems: Thales STING optronic director.

Programmes: Originally six of this former GDR Sassnitz class were to be built at Peenewerft for Poland. Three units were acquired and completed at Gdansk.

Modernisation: Contract with Thales Naval Nederland (TNNL) as prime contractor for upgrade of all three ships signed 29 June 2001. It embraced the following main items: RBS-15 missiles, TACTICOS combat data system, STING optronic director, AMB Sea Giraffe surveillance radar and new navigational radar, ESM equipment and improved communications. Refit of *Piorun* was completed by 2003 and the other two ships in 2006. RBS 15 Mk 3 missiles are to replace Mk 2 missiles.
Structure: Unlike the German Coast Guard vessels of the same class, these ships have retained three engines.
Operational: Based at Gdynia.

PIORUN   6/2013*, J Ciślak / 1531014

© 2014 IHS   IHS Jane's Fighting Ships 2014-2015

1 KASZUB CLASS (PROJECT 620) (FSM)

AMPHIBIOUS FORCES

5 LUBLIN CLASS (PROJECT 767)
(LST/MINELAYERS) (LST/ML)

Name	No	Builders	Launched	Commissioned
LUBLIN	821	Northern Shipyard, Gdansk	12 Jul 1988	12 Oct 1989
GNIEZNO	822	Northern Shipyard, Gdansk	7 Dec 1988	23 Feb 1990
KRAKOW	823	Northern Shipyard, Gdansk	7 Mar 1989	27 Jun 1990
POZNAN	824	Northern Shipyard, Gdansk	5 Jan 1990	8 Mar 1991
TORUŃ	825	Northern Shipyard, Gdansk	8 Jun 1990	24 May 1991

Displacement, tonnes: 1,372 standard; 1,773 full load
Dimensions, metres (feet): 95.4 × 10.8 × 2 (313 × 35.4 × 6.6)
Speed, knots: 16
Range, n miles: 1,400 at 16 kt
Complement: 50 (5 officers)

Military lift: 9 Type T-72 tanks or 9 APC or 17 medium or light trucks. 80 troops plus equipment (821–823); 125 troops plus equipment (824); 135 troops and equipment (825).
Machinery: 3 Cegielski 6ATL25D diesels; 5,390 hp(m) (3.96 MW) sustained; 3 shafts
Guns: SAM/Guns: 4 ZU-23-2MR 23 mm Wrobel II (2 twin); combined with 2 SA-N-5 missiles; IR homing to 6 km (3.2 n miles) at 1.5 Mach; warhead 1.5 kg and guns; 400 rds/min combined.
Mines: 50–134.
Depth charges: 9 throwers for counter-mining.
Physical countermeasures: Decoys: 2—12-barrelled 70 mm Derkacz chaff launchers (821 and 825). 2—12-barrelled Jastrzab chaff launchers (822–824).
Radars: Navigation: Sperry Marine Bridgemaster; E/F/I-bands.
IFF: SC10D2.

Comment: Designed with a through deck from bow to stern and can be used as minelayers as well as for amphibious landings. Folding bow and stern ramps and a stern anchor are fitted. The ship has a pressurised citadel for NBC defence and an upper deck washdown system. Mining capabilities upgraded in 1997/98. Based at Swinoujscie.

KRAKOW   6/2012*, Michael Nitz / 1531021

3 DEBA CLASS (PROJECT 716) (LCU)

Name	No	Builders	Launched	Commissioned
–	851	Naval Shipyard, Gdynia	13 Nov 1987	7 Aug 1988
–	852	Naval Shipyard, Gdynia	2 Jul 1990	2 Jan 1991
–	853	Naval Shipyard, Gdynia	26 Oct 1990	3 May 1991

Displacement, tonnes: 179 full load
Dimensions, metres (feet): 37.2 × 7.1 × 1.7 (122 × 23.3 × 5.6)
Speed, knots: 20
Range, n miles: 430 at 16 kt
Complement: 12

Military lift: 1 tank or 2 vehicles up to 20 tons and 50 troops
Machinery: 3 Type M 401A diesels; 3,000 hp(m) (2.2 MW); 3 shafts
Guns: 2 ZU-23-2M 23 mm (twin).
Radars: Surface search: Sperry Marine Bridgemaster; I-band.

Comment: The plan was to build 12 but the programme was suspended at three through lack of funds. A similar design has been assembled in Iran. Can carry up to six launchers for strung-out charges. Based at Swinoujscie.

DEBA CLASS   9/2003, J Ciślak / 0567514

MINE WARFARE FORCES

Notes: Plans to procure three Kormaran II (Project 258) minehunters were confirmed in September 2013. The details and timescale of the programme have not been announced.

3 KROGULEC CLASS (PROJECT 206FM) (MHCM)

Name	No	Builders	Commissioned
FLAMING	621	Gdynia Shipyard	11 Oct 1966
MEWA	623	Gdynia Shipyard	21 May 1967
CZAJKA	624	Gdynia Shipyard	23 Jun 1967

Displacement, tonnes: 559 full load
Dimensions, metres (feet): 58.2 × 7.7 × 2.1 (190.9 × 25.3 × 6.9)
Speed, knots: 17
Range, n miles: 2,000 at 12 kt
Complement: 52 (6 officers)

Machinery: 2 Sulzer/Cegielski 6AL 25/30 diesels; 2,203 hp(m) (1.62 MW); 2 shafts; LIPS cp props
Missiles: SAM: 2 Fasta-4M quad launchers. SA-N-5. 2 SA-N-10 (Grom) to be fitted in due course.
Guns: SAM/guns: 2 Wrobel ZU-23-2MR 23 mm (twin) with 2 SA-N-5 missiles.
Mines: 6–12 depending on type.
Depth charges: 2 racks.
Physical countermeasures: Decoys: 6–9 barrelled Jastrzab 2 launchers for chaff.
MCM: 2 Bofors MT2W mechanical, 1 TEM-PE-2MA magnetic and 1 MTA-2 acoustic sweeps. CTM Ukwial ROV with sonar, TV and charges. 10 ZHH 230 sonobuoys.
Electronic countermeasures: ECM: PIT Bren system being fitted.
Radars: Navigation: Sperry Marine Bridgemaster; I-band.
IFF: RAWAR SC-10D2
Sonars: CTM SHL-100MA hull mounted; active minehunting; high frequency; Politechnica Gdansk SHL-200 VDS.
Combat data systems: CTM Pstrokosz command support system.

Comment: All taken out of service in 1997. New armament and minehunting equipment installed. Divers recompression chamber carried. *Mewa* returned to service in May 1999, *Czajka* in May 2000 and *Flaming* in 2001. Based at Gdynia.

CZAJKA   6/2013*, J Ciślak / 1531019

4 MAMRY (NOTEC II) CLASS (PROJECT 207M)
(MINESWEEPERS/HUNTERS—COASTAL) (MHSCM)

Name	No	Builders	Launched	Commissioned
MAMRY	643	Naval Shipyard, Gdynia	20 Sep 1991	25 Sep 1992
WIGRY	644	Naval Shipyard, Gdynia	28 Nov 1992	14 May 1993
SNIARDWY	645	Naval Shipyard, Gdynia	20 Jun 1993	28 Jan 1994
WDZYDZE	646	Naval Shipyard, Gdynia	24 Jun 1994	2 Dec 1994

Displacement, tonnes: 219 full load
Dimensions, metres (feet): 38.5 × 7.4 × 1.8 (126.3 × 24.3 × 5.9)
Speed, knots: 14
Range, n miles: 865 at 14 kt
Complement: 27 (5 officers)

Machinery: 2 M 401A diesels; 1,874 hp(m) (1.38 MW); 2 shafts 2 auxiliary motors; 816 hp(m) (60 kW)
Guns: SAM/Guns: 2 ZU-23-2MR 23 mm Wrobel II (twin); combination of 2 SA-N-5 missiles; IR homing to 6 km (3.2 n miles) at 1.5 Mach; warhead 1.5 kg and guns; 400 rds/min combined to 2 km.
Mines: 6–24 depending on type.
Physical countermeasures: MCM: MMTK 1m mechanical, MTA 2 acoustic and TEM-PE 1m magnetic sweeps.
Radars: Navigation: Sperry Marine Bridgemaster; I-band.
Sonars: SHL 100/200; hull mounted/VDS; active minehunting; high frequency.

Comment: Modified version of the 207P and equipped to carry divers. All based at Gdynia.

MAMRY   6/2013*, Michael Nitz / 1531020

Mine warfare forces — Survey and research ships < **Poland** 643

13 GOPLO (NOTEC) CLASS (PROJECT 207D/207P) (MHC)

Name	No	Builders	Launched	Commissioned
GOPLO	630	Naval Shipyard, Gdynia	16 Apr 1981	13 Mar 1982
GARDNO	631	Naval Shipyard, Gdynia	23 Jun 1993	31 Mar 1984
BUKOWO	632	Naval Shipyard, Gdynia	28 Jul 1984	23 Jun 1985
DĄBIE	633	Naval Shipyard, Gdynia	21 Jun 1985	11 May 1986
JAMNO	634	Naval Shipyard, Gdynia	11 Feb 1986	11 Oct 1986
MIELNO	635	Naval Shipyard, Gdynia	27 Jun 1986	9 May 1987
WICKO	636	Naval Shipyard, Gdynia	20 Mar 1987	12 Oct 1987
RESKO	637	Naval Shipyard, Gdynia	1 Oct 1987	26 Mar 1988
SARBSKO	638	Naval Shipyard, Gdynia	10 May 1988	12 Oct 1988
NECKO	639	Naval Shipyard, Gdynia	21 Nov 1988	9 May 1989
NAKŁO	640	Naval Shipyard, Gdynia	29 May 1989	2 Mar 1990
DRUŻNO	641	Naval Shipyard, Gdynia	29 Nov 1989	21 Sep 1990
HANCZA	642	Naval Shipyard, Gdynia	9 Jul 1990	1 Mar 1991

Displacement, tonnes: 219 full load
Dimensions, metres (feet): 38.5 × 7.4 × 1.8 *(126.3 × 24.3 × 5.9)*
Speed, knots: 14
Range, n miles: 1,100 at 9 kt
Complement: 29 (6 officers)

Machinery: 2 M 401A1 diesels; 1,874 hp(m) *(1.38 MW)* sustained; 2 shafts
Guns: 2 ZU-23-2MR 23 mm (twin); 400 rds/min combined to 2 km.
Mines: 6–24
Depth charges: 24
Physical countermeasures: MCM: MMTK1 mechanical; MTA 1 acoustic and TEM-PE 1 magnetic sweeps.
Radars: Navigation: Bridgemaster; I-band.
Sonars: MG 89 or MG 79; active minehunting; high frequency.

Comment: The class is of GRP construction and named after lakes. The original 25 mm guns have been replaced by 23 mm. *Goplo* is Project 207D and is used as a trials ship based at Gdynia. The other 12 vessels are Project 207P and are based at Swinoujscie.

HANCZA — 6/2013*, Michael Nitz / 1531022

SURVEY AND RESEARCH SHIPS

2 MODIFIED FINIK 2 CLASS (PROJECT 874) (AGS)

Name	No	Builders	Launched	Commissioned
HEWELIUSZ	265	Northern Shipyard, Gdansk	11 Sep 1981	27 Nov 1982
ARCTOWSKI	266	Northern Shipyard, Gdansk	20 Nov 1981	27 Nov 1982

Displacement, tonnes: 1,153 standard; 1,238 full load
Dimensions, metres (feet): 61.6 × 10.8 × 3.3 *(202.1 × 35.4 × 10.8)*
Speed, knots: 13
Range, n miles: 5,900 at 11 kt
Complement: 49 (10 officers)
Machinery: 2 Cegielski-Sulzer 6AL25/30 diesels; 1,920 hp(m) *(1.4 MW)*; 2 auxiliary motors; 204 hp(m) *(150 kW)*; 2 shafts; cp props; bow thruster
Radars: Navigation: Sperry Marine Bridgemaster; E/F/I-bands.

Comment: Sister ships to Russian class which were built in Poland, except that *Heweliusz* and *Arctowski* have been modified and have no buoy handling equipment. Equipment includes Atlas Deso, Atlas Ralog and Atlas Dolog survey. Both ships are based at Gdynia. One sister ship, *Planeta*, is civilian operated and the other, *Zodiak*, was decommissioned in 2003.

HEWELIUSZ — 6/2011, L-G Nilsson / 1406670

ARCTOWSKI — 7/2009, B Prézelin / 1305632

2 SURVEY CRAFT (PROJECT 4234) (AGSC)

Name	Builders	Commissioned
K 10	Wisla Yard, Gdansk	6 Feb 1989
K 4	Wisla Yard, Gdansk	25 Sep 1989

Displacement, tonnes: 46 full load
Dimensions, metres (feet): 18.9 × 4.4 × 1.5 *(62 × 14.4 × 4.9)*
Speed, knots: 9
Complement: 10
Machinery: 1 Wola DM 150 diesel; 160 hp(m) *(117 kW)* sustained; 1 shaft
Radars: Navigation: SRN 207A; I-band.

Comment: Coastal survey craft based at Gdynia. There are a number of survey launches and buoy tenders listed under *Auxiliaries*.

K 10 — 5/2000, J Ciślak / 0105248

4 SURVEY CRAFT (PROJECT III/C) (AGSC)

M 35   M 38–40

Displacement, tonnes: 10 full load
Dimensions, metres (feet): 11 × 3.2 × 0.7 *(36.1 × 10.5 × 2.3)*
Speed, knots: 8
Range, n miles: 184 at 8 kt
Complement: 5
Machinery: 1 Puck Rekin SW 400/MZ diesel; 95 hp(m) *(70 kW)*; 1 shaft
Radars: Navigation: SRN 207A; I-band.

Comment: Based at Gdynia and Swinoujscie (M 35).

M 40 — 3/2003, J Ciślak / 0567515

Poland > Intelligence vessels — Auxiliaries

INTELLIGENCE VESSELS

2 MODIFIED MOMA CLASS (PROJECT 863) (AGI)

Name	No	Builders	Commissioned
NAWIGATOR	262	Northern Shipyard, Gdansk	17 Feb 1975
HYDROGRAF	263	Northern Shipyard, Gdansk	8 May 1976

Displacement, tonnes: 1,704 full load
Dimensions, metres (feet): 73.3 × 10.8 × 3.9 *(240.5 × 35.4 × 12.8)*
Speed, knots: 17. **Range, n miles:** 7,200 at 12 kt
Complement: 87 (10 officers)

Machinery: 2 Zgoda-Sulzer 6TD48 diesels; 3,300 hp(m) *(2.43 MW)* sustained; 2 shafts
Missiles: 2 Fasta-4M quad launchers. SA-N-5.
Guns: 4 — 25 mm (2 twin) (262).
Electronic countermeasures: ESM/ECM: ESM/ELINT equipment.
Radars: Navigation: 2 SRN 7453 Nogat; I-band.

Comment: Much altered in the upperworks and unrecognisable as Momas. The forecastle in *Nawigator* is longer than in *Nawigator* and one deck higher. *Hydrograf* fitted for but not with two twin 25 mm gun mountings. Forward radome replaced by a cylindrical type in *Nawigator* and after ones removed on both ships. Based at Gdynia.

NAWIGATOR — 6/2013*, J Ciślak / 1531025

HYDROGRAF — 4/2010, J Ciślak / 1406260

TRAINING SHIPS

Notes: The three masted sailing ship *Dar Mlodziezy* is civilian owned and operated but also takes naval personnel for training.

1 WODNIK CLASS (PROJECT 888) (AXTH)

Name	No	Builders	Launched	Commissioned
WODNIK	251	Northern Shipyard, Gdansk	19 Nov 1975	28 May 1976

Displacement, tonnes: 1,724 standard; 1,773 full load
Dimensions, metres (feet): 72.2 × 11.9 × 4.1 *(236.9 × 39.0 × 13.5)*
Speed, knots: 16. **Range, n miles:** 7,200 at 11 kt
Complement: 157 (24 officers; 101 midshipmen)

Machinery: 2 Zgoda-Sulzer 6TD48 diesels; 2,650 hp(m) *(1.95 MW)* sustained; 2 shafts; cp props
Guns: 4 ZU-23-2MR Wrobel 23 mm (2 twin). 2 — 30 mm AK 230 (1 twin).
Radars: Navigation: Sperry Marine Bridgemaster; E/F/I-bands.
Helicopters: Platform for 1 light.

Comment: Sister to former GDR *Wilhelm Pieck* and two Russian ships. Converted to a hospital ship (150 beds) in 1990 for deployment to the Gulf. Armament removed as part of the conversion but partially restored in 1992. Based at Gdynia. Second of class in reserve from 1999.

WODNIK — 6/2011, L-G Nilsson / 1406671

1 ISKRA CLASS (PROJECT B79) (SAIL TRAINING SHIP) (AXS)

Name	No	Builders	Launched	Commissioned
ISKRA	253	Gdansk Shipyard	6 Mar 1982	11 Aug 1982

Displacement, tonnes: 506 full load
Dimensions, metres (feet): 49 × 8.1 × 4.0 *(160.8 × 26.6 × 13.1)*
Speed, knots: 9
Complement: 64 (6 officers; 50 cadets)
Machinery: 1 Wola 75H12 diesel; 310 hp(m) *(228 kW)*; 1 auxiliary shaft; cp prop
Radars: Navigation: SRN 206; I-band.

Comment: Barquentine with 1,040 m² of sail. Used by the Naval Academy for training with a secondary survey role. Based at Gdynia.

ISKRA — 6/2013*, Michael Nitz / 1531024

AUXILIARIES

Note: Procurement of up to four Strategic Support Ships is under consideration. The broad requirement is for ships of approximately 10,000 tons with the capability of transporting about 500 troops plus some 20 vehicles and up to six helicopters.

1 PROJECT 890 CLASS
(MINE/COUNTERMEASURES COMMAND SHIP) (MCCS)

Name	No	Builders	Commissioned
KONTRADMIRAL XAWERY CZERNICKI	511	Northern Shipyard, Gdansk	1 Sep 2001

Displacement, tonnes: 2,286 full load
Dimensions, metres (feet): 73.8 × 13.8 × 4.3 *(242.1 × 45.3 × 14.1)*
Speed, knots: 14.1. **Range, n miles:** 7,000 at 12 kt
Complement: 38

Military lift: 140 troops with full individual armament or ten 20 ft containers or four 20 ft containers and six STAR 266 army trucks
Machinery: 2 Cegielski-Sulzer AL25D diesels; 2,934 hp(m) *(2.16 MW)* sustained; 2 shafts
Guns: SAM/Guns: 1 ZU 23-2MR Wrobel I/II mounts: combination of 2 Strela 2M (Grail) missiles and 2 — 23 mm guns.
Physical countermeasures: Decoys: 4 WNP81/9 9 barrelled 81 mm Jastrzab chaff launchers.
Electronic countermeasures: ESM: PIT intercept.
Radars: Navigation: Sperry Marine Bridgemaster; E/F/I-bands.
Helicopters: Platform for 1 helicopter (up to 10 ton).

Comment: Conversion from a Project 130 Degaussing Vessel to Logistic Support Ship in Northern Shipyard, Gdansk, included new upper and forward hull sections, provision of a helicopter deck and NBC protection. The ship has a 16 ton hydraulic crane and after ramp. In 2009, the ship was adapted for a new role as MCM command and control ship. She was equipped with SHF Satcom, MCCIS system, upgraded communications, an MCM command system, hyperbaric chamber and mine-disposal weapon. Based at Swinoujscie.

KONTRADMIRAL XAWERY CZERNICKI — 6/2013*, Michael Nitz / 1531023

1 BALTYK CLASS (PROJECT ZP 1200) (TANKER) (AORL)

Name	No	Builders	Commissioned
BALTYK	Z 1	Naval Shipyard, Gdynia	11 Mar 1991

Displacement, tonnes: 2,984 standard; 3,098 full load
Dimensions, metres (feet): 84.8 × 13.1 × 4.7 *(278.2 × 43.0 × 15.4)*
Speed, knots: 15
Range, n miles: 4,250 at 12 kt
Complement: 34 (4 officers)

Cargo capacity: 1,184 tons fuel, 92.7 tons lub oil
Machinery: 2 Cegielski 8 ASL 25 diesels; 4,025 hp(m) *(2.96 MW)*; 2 shafts; cp props
Guns: 4 ZU-23-2S Wrobel 23 mm (2 twin).
Radars: Navigation: Sperry Marine Bridgemaster; E/F/I-bands.

Comment: Beam replenishment stations, one each side. First of a projected class of four, of which the others were cancelled. Refitted in 2013. Based at Gdynia.

BALTYK — 6/2011, L-G Nilsson / 1406673

Auxiliaries < **Poland** 645

1 MOSKIT CLASS (PROJECT B 199) (TANKER) (AOTL)

Name	No	Builders	Launched	Commissioned
–	Z 8	Wroclaw Shipyard, Rzeczna	14 Sep 1969	21 Jul 1970

Displacement, tonnes: 1,245 full load
Dimensions, metres (feet): 58 × 9.3 × 3.3 *(190.3 × 30.5 × 10.8)*
Speed, knots: 10
Range, n miles: 1,200 at 10 kt
Complement: 21 (3 officers)

Cargo capacity: 656.5 tons
Machinery: 1 Magdeburg diesel; 965 hp(m) *(720 kW)*; 1 shaft
Guns: 4 ZU-23-2M 23 mm (2 twin).
Radars: Navigation: TRN 823; I-band.

Comment: Z 3 decommissioned in 2007. Based at Swinoujscie.

MEDUZA   7/2004, J Ciślak / 1044484

1 KORMORAN CLASS (YPT)

No	Builders	Launched	Commissioned
K 8	Naval Shipyard, Gdynia	26 Aug 1970	3 Jul 1971

Displacement, tonnes: 152 full load
Dimensions, metres (feet): 35 × 6 × 1.6 *(114.8 × 19.7 × 5.2)*
Speed, knots: 19
Range, n miles: 550 at 15 kt
Complement: 24
Machinery: 2 Type M 50F5 diesels; 2,200 hp(m) *(1.6 MW)*; 2 shafts
Guns: 2 ZU-23-2M Wrobel 23 mm (twin).
Radars: Navigation: SRN 206/301; I-band.

Comment: Armament updated in 1993. Based at Gdynia.

K 8   9/2009, Frank Findler / 1406215

2 MROWKA CLASS (PROJECT B 208) (DEGAUSSING VESSELS) (YDG)

Name	No	Builders	Commissioned
–	SD 11	Naval Shipyard, Gdynia	10 Oct 1971
–	SD 13	Naval Shipyard, Gdynia	16 Dec 1972

Displacement, tonnes: 671 full load
Dimensions, metres (feet): 44 × 8.1 × 2.9 *(144.4 × 26.6 × 9.5)*
Speed, knots: 9.5. **Range, n miles:** 2,230 at 9.5 kt
Complement: 37
Machinery: 1 6NV D36 diesel; 957 hp(m) *(704 kW)*; 1 shaft
Guns: 2—23 mm (twin) (SD 11 and 13); 2 ZU-23-2M Wrobel 23 mm (twin) (SD 12).
Radars: Navigation: Sperry Marine Bridgemaster; I-band.

Comment: SD 12 decommissioned in 2005. SD 11 based at Swinoujscie and SD 13 based at Gdynia.

SD 13   5/2008, J Ciślak / 1353269

2 PIAST CLASS (PROJECT 570M) (SALVAGE SHIPS) (ARS)

Name	No	Builders	Commissioned
PIAST	281	Northern Shipyard, Gdansk	26 Jan 1974
LECH	282	Northern Shipyard, Gdansk	30 Nov 1974

Displacement, tonnes: 1,917 full load
Dimensions, metres (feet): 72.7 × 11.9 × 4.1 *(238.5 × 39.0 × 13.5)*
Speed, knots: 15
Range, n miles: 3,000 at 12 kt
Complement: 56 (8 officers) + 12 spare berths

Machinery: 2 Zgoda-Sulzer 6TD48 diesels; 3,300 hp(m) *(2.43 MW)* sustained; 2 shafts; cp props
Missiles: SAM: 2 Fasta 4M twin launchers for SA-N-5.
Guns: 4—12.7 mm (2 twin) MGs.
Radars: Navigation: Sperry Marine Bridgemaster; E/F/I-bands.

Comment: Basically a Moma class hull with towing and firefighting capabilities. Ice-strengthened hulls. Wartime role as hospital ships. Carry three-man diving bells capable of 100 m depth and a decompression chamber. ROV added and other salvage improvements made in 1997/98. Based at Gdynia. Guns may not be carried.

LECH   6/2011, L-G Nilsson / 1406674

2 ZBYSZKO CLASS (PROJECT B 823) (SALVAGE SHIPS) (ARS)

Name	No	Builders	Commissioned
ZBYSZKO	R 14	Ustka Shipyard	8 Nov 1991
MACKO	R 15	Ustka Shipyard	20 Mar 1992

Displacement, tonnes: 386 full load
Dimensions, metres (feet): 35 × 8 × 3 *(114.8 × 26.2 × 9.8)*
Speed, knots: 11
Range, n miles: 3,000 at 10 kt
Complement: 15
Machinery: 1 Sulzer 6AL20/24D; 750 hp(m) *(551 kW)*; 1 shaft
Radars: Navigation: Sperry Marine Bridgemaster; I-band.

Comment: Type B-823 ordered 30 May 1988. Carries a decompression chamber and two divers. Mobile gantry crane on the stern. Based at Gdynia.

MACKO   6/2013*, J Ciślak / 1531027

1 TRANSPORT CRAFT (PROJECT MS-3600) (YFB)

M 1

Displacement, tonnes: 75 full load
Dimensions, metres (feet): 28.7 × 5.8 × 1.3 *(94.2 × 19.0 × 4.3)*
Speed, knots: 27
Complement: 37
Machinery: 3 M50F5 diesels; 3,600 hp(m) *(2.65 MW)*; 3 shafts
Radars: Navigation: SRN 207A; I-band.

Comment: Can be used as emergency patrol craft. Based at Gdynia as an Admirals' launch.

M1   6/2013*, J Ciślak / 1531026

6 MISCELLANEOUS HARBOUR CRAFT (YFB)

B 3   B 7   B 11   M 12   M 21   M 22

Comment: M numbers are patrol launches; B numbers are freighters and oil lighters.

M 22   6/2008, J Ciślak / 1353273

B 7   8/2009, Martin Mokrus / 1305634

3 DIVING TENDERS (PROJECT 4234) (YDT)

K 5   K 7   K 9

Displacement, tonnes: 45 full load
Dimensions, metres (feet): 18.9 × 4.4 × 1.5 (62 × 14.4 × 4.9)
Speed, knots: 9
Complement: 10
Machinery: 1 Wola DM 150 diesel; 160 hp(m) (117 kW) sustained; 1 shaft
Radars: Navigation: SRN 207A; I-band.

Comment: Identical to survey craft of same class. Built in 1989.

K 9   6/2010, J Cislak / 1406258

TUGS

2 H 960 CLASS (ATA)

H 6   H 8

Displacement, tonnes: 345 full load
Dimensions, metres (feet): 27.8 × 8 × 3.7 (91.2 × 26.2 × 12.1)
Speed, knots: 12
Range, n miles: 1,150 at 12 kt
Complement: 17 (1 officer)
Machinery: 1 Sulzer GATL 25 D diesels; 1,306 hp(m) (960 kW); 1 shaft
Radars: Navigation: SRN 401 XTA; I-band.

Comment: Built at Nauta Ship Repair Yard, Gdynia and commissioned 25 September 1992 and 19 March 1993 respectively. Based at Swinoujscie (H 6) and Gdynia (H 8).

H 8   6/2011, L-G Nilsson / 1406680

5 HARBOUR TUGS (PROJECTS H 900, H 820) (YTB/YTM)

H 4   H 5   H 7   H 9   H 10

Displacement, tonnes: 221 full load
Dimensions, metres (feet): 25.6 × 6.8 × 3.5 (84 × 22.3 × 11.5)
Speed, knots: 11
Range, n miles: 1,500 at 10 kt
Complement: 17
Machinery: 1 Cegielski-Sulzer 6AL20/24H diesel; 935 hp(m) (687 kW); 1 shaft
Radars: Navigation: SRN 206; I-band.

Comment: Details given are for H 4, 5 and 7. Completed 1979–81. Have firefighting capability except H 9–10. H 9–10 completed in 1993.

H 5   6/2011, L-G Nilsson / 1406679

MARITIME REGIONAL UNIT OF THE BORDER GUARD (MOSG)

Headquarters Appointments

Commandant MOSG:
Rear Admiral Piotr Stocki
Deputy Commandant:
Commander Wojciech Heninborch
Deputy Commandant:
Commander Roman Słowiński

Bases
Gdansk (HQ and Kaszubski Division)
Swinoujscie (Pomorski Division)

General

MOSG (Morski Oddzial Strazy Granicznej) formed on 1 August 1991 and is subordinated to the Ministry of Internal Affairs. Vessels have blue hulls with red and yellow striped insignia. Superstructures are painted white. MOSG also operates one PZL M 20 Mewa, one W-3 AM Anakonda helicopter and one M-28 Skytruck.

2 KAPER CLASS (PROJECT SKS-40)
(LARGE PATROL CRAFT) (WPB)

No	Builders	Commissioned
SG-311	Wisla Yard, Gdansk	21 Jan 1991
SG-312	Wisla Yard, Gdansk	3 Apr 1992

Displacement, tonnes: 380 full load
Dimensions, metres (feet): 42.5 × 8.38 × 2.8 (139.4 × 27.5 × 9.2)
Speed, knots: 17.5
Range, n miles: 2,800 at 14 kt
Complement: 14

Machinery: 2 Sulzer 8ATL25/30 diesels; 4,720 hp(m) (3.47 MW); 2 shafts; cp props
Guns: 2 — 7.62 mm MGs.
Radars: SRN 207; I-band.
Navigation: Racal Decca; I-band.

Comment: Used for Fishery Protection. 311 based at Gdansk and 312 at Kolobrzeg.

SG 311   5/2011, J Ciślak / 1455739

Poland (MARITIME REGIONAL UNIT OF THE BORDER GUARD (MOSG)) 647

2 BALTIC 24 CLASS (PATROL CRAFT) (WPB)

SG 111   SG 112

Displacement, tonnes: 51 full load
Dimensions, metres (feet): 23.7 × 5.65 × 1.7 *(77.8 × 18.5 × 5.6)*
Speed, knots: 21
Range, n miles: 500 at 15 kt
Complement: 6
Machinery: 2 Scania DI 16 M43 diesels; 1,475 hp *(1.1 MW)*; 2 shafts
Guns: 1—12.7 mm MG (fitted for).
Radars: Surface search/navigation: JMA 5300; I-band.

Comment: Contract with Baltic Workboats A/S, Estonia, on 9 February 2009 for the construction of two 24 m workboats. Of aluminium construction, the craft are equipped with an aft launch/recovery ramp for a 5 m RHIB. Both craft delivered on 19 May 2010.

SG 112   6/2010, MOSG / 1421638

4 IC 16 M III (PBF)

SG 213–216

Displacement, tonnes: 20 full load
Dimensions, metres (feet): 15.9 × 3.96 × 1.1 *(52.2 × 13.0 × 3.6)*
Speed, knots: 42
Range, n miles: 330 at 32 kt
Complement: 4
Machinery: 2 Scania diesels; 1,602 hp(m) *(1.19 MW)*; 2 Rolls-Royce FF 410 waterjets
Guns: 1—7.62 mm MG.
Radars: Surface search: Furuno M 1934C; I-band.

Comment: Four ordered from Dockstavarvet, Sweden and entered into service October–November 2007.

SG 215   6/2010, J Ciślak / 1406676

SG 215   5/2011, J Ciślak / 1455735

2 STRAZNIK CLASS (PROJECT SAR-1500) (WPBF)

No	Builders	Commissioned
SG-211	Damen Yard, Gdynia	29 Apr 2000
SG-212	Damen Yard, Gdynia	7 Jul 2000

Displacement, tonnes: 16 standard
Dimensions, metres (feet): 15.2 × 5.39 × 0.90 *(49.9 × 17.7 × 3.0)*
Speed, knots: 30
Range, n miles: 200 at 30 kt
Complement: 4 (1 officer)
Machinery: 2 MAN D2848 diesels; 1,360 hp *(1,000 kW)*; water jet system
Guns: 1—7.62 mm MG.
Radars: Surface search: SIMRAD; I-band.

Comment: Contract between MOSG and Damen Shipyard signed 5 October 1999. Based on Dutch SAR 1500 lifeboat. Hull and superstructure of aluminium alloy.

SG-211   10/2007, J Ciślak / 1166607

2 GRIFFON 2000 TDX CLASS (HOVERCRAFT) (UCAC)

SG 411   SG 412

Displacement, tonnes: 4 full load
Dimensions, metres (feet): 12.7 on cushion × ? *(41.7 × ?)*
Speed, knots: 30
Range, n miles: 450 at 35 kt
Complement: 3
Machinery: 1 Deutz BF 6M 1015 CP diesel; 442 hp *(330 kW)*
Radars: Navigation: SIMRAD RA 83P; I-band.

Comment: Built by Griffon Hovercraft, Southampton and delivered in 2006. Aluminium hull. Employed as patrol craft in shallow waters and rivers.

SG 412   5/2011, J Ciślak / 1455737

4 PARKER 900 PATROL CRAFT (WPB)

SG 044–047

Displacement, tonnes: 6.6 full load
Dimensions, metres (feet): 9.95 × 3.2 × 0.8 *(32.6 × 10.5 × 2.6)*
Speed, knots: 35
Complement: 2
Machinery: 2 Cummins MerCruiser QSD; 530 hp *(396 kW)*; 2 Kamewa waterjets
Radars: Surface search/navigation: Furuno; I-band.

Comment: Four Parker 900 cabin RHIBs constructed by Parker Poland, Gdansk. Delivery completed in April 2010.

SG 045   6/2010, MOSG / 1366669

648 Poland (MARITIME REGIONAL UNIT OF THE BORDER GUARD (MOSG))

1 PATROL CRAFT (PROJECT MI-6) (WPB)

SG-008

Displacement, tonnes: 16 standard
Dimensions, metres (feet): 13 × 3.7 × 1.1 *(42.7 × 12.1 × 3.6)*
Speed, knots: 11
Complement: 4
Machinery: 1 Wola; 210 hp *(154 kW)*; 1 shaft

Comment: Harbour craft built at Wisla Shipyard, Gdansk, 1989.

SG-008   3/2006, J Ciślak / 1164429

4 SPORTIS CLASS (PROJECT 7500)
(FAST INTERCEPT CRAFT) (WPBF)

SG-004   SG-005   SG-006   SG-007

Displacement, tonnes: 2 full load
Dimensions, metres (feet): 7.5 × 2.8 × 0.4 *(24.6 × 9.2 × 1.3)*
Speed, knots: 40
Complement: 2
Machinery: 1 Volvo Penta (SG 004) diesel; 230 hp *(171 kW)*; 1 Yanmar (SG 005–007) diesel; 315 hp *(235 kW)*

Comment: Built in Bojano in 1996.

SG-007   3/2006, J Ciślak / 1164430

1 SPORTIS S-8900 (WPBF)

SG 009

Displacement, tonnes: 2 full load
Dimensions, metres (feet): 8.9 × 3.5 × 0.45 *(29.2 × 11.5 × 1.5)*
Speed, knots: 50
Complement: 2
Machinery: 2 Suzuki diesels; 500 hp *(373 kW)*

Comment: Commissioned in 1994.

1 PATROL LAUNCH (PROJECT M-35) (WYFL)

SG 036

Displacement, tonnes: 29 full load
Dimensions, metres (feet): 10.49 × 4.05 × 1.72 *(34.4 × 13.3 × 5.6)*
Speed, knots: 8
Complement: 2
Machinery: 1 Mielec SW 680 diesel; 165 hp *(123 kW)*; 1 shaft

Comment: Built at Tczew River Shipyard in 1985. Similar to those in Polish naval service.

SG 036   3/2006, J Ciślak / 1164431

1 MODIFIED SPORTIS CLASS (PROJECT S-6100)
(FAST INTERCEPT CRAFT) (WPBF)

SG 066

Displacement, tonnes: 2 standard
Dimensions, metres (feet): 6.1 × 2.3 × 0.4 *(20 × 7.5 × 1.3)*
Speed, knots: 35
Complement: 2
Machinery: 2 Johnson outboard motors; 110 hp *(82 kW)*

Comment: Built at Bojano in 2001. Located at Border units along the coast.

SG 066   6/2012, MOSG / 1455740

1 MODIFIED SPORTIS S-8900 (WPBF)

SG 017

Displacement, tonnes: 2 full load
Dimensions, metres (feet): 9.2 × 3.3 × 0.56 *(30.2 × 10.8 × 1.8)*
Speed, knots: 43
Complement: 2
Machinery: 2 Yanmar diesels; 520 hp *(387 kW)*

Comment: Commissioned in 2009.

SG 017   5/2011, J Ciślak / 1455736

IHS Jane's Fighting Ships 2014-2015   © 2014 IHS

Portugal

MARINHA PORTUGUESA

Country Overview

The Republic of Portugal is situated in south-western Europe in the western portion of the Iberian Peninsula. It is bordered to the north and east by Spain and has a 967 n mile coastline with the Atlantic Ocean. The Azores and Madeira Islands in the Atlantic are integral parts of the republic, the total area of which is 35,553 square miles. Lisbon is the capital, largest city and principal port. There are further ports at Leixões (near Oporto), Setúbal, and Funchal (Madeira). Territorial seas (12 n miles) and an EEZ (200 n miles) are claimed.

Headquarters Appointments

Chief of Naval Staff:
 Admiral Luís Manuel Fourneaux Macieira Fragoso
Deputy Chief of Naval Staff:
 Vice Admiral João da Cruz de Carvalho Abreu
Naval Commander:
 Vice Admiral José Alfredo Monteiro Montenegro
Azores Maritime Zone Commander:
 Rear Admiral Fernando Manuel de Macedo Pires da Cunha
Madeira Maritime Zone Commander:
 Captain Fernando Manuel Felix Marques
Marine Corps Commander:
 Rear Admiral Luís Miguel de Matos Cortes Picciochi

Diplomatic Representation

Defence Attaché in Washington and Ottawa:
 Colonel Jorge Manuel Cabrita Alão Correia da Silva
Defence Attaché in Luanda, Kinshasa, Brazzaville, and Windhoek: Colonel Fernando Manuel Rodrigues Pereira de Albuquerque
Defence Attaché in Maputo, Lillongwe, and Harare:
 Colonel José Rui de Sousa Pacheco
Defence Attaché in Madrid, Cairo and Athens:
 Colonel Tito Augusto P Quintanilha e Mendonça
Defence Attaché in S. Tomé and Libreville:
 Colonel José Manuel Mota Lourenço da Saúde
Defence Attaché in Bissau, Conakry and Dakar:
 Captain António da Silva Campos
Defence Attaché in Brasilia, Buenos Aires, and Santiago:
 Colonel António Emidio da Silva Salgueiro
Defence Attaché in Berlin, London, The Hague, and Stockholm: Captain Jorge Manuel da Costa e Sousa
Defence Attaché in Canberra, Dili, and Jakarta:
 Captain José Manuel Ministro Ribeiro da Costa
Defence Attaché in Paris:
 Colonel Alberto Sebastião Neves Marinheiro
Defence Attaché in Rabat and Tunis:
 Colonel José António Gurreiro Martins
Defence Attaché in Praia and Abuja:
 Colonel Nuno Álvaro Pereira Matos Rocha
Defence Attaché in Algiers:
 Lieutenant Colonel António Manuel Marques da Silva

Personnel

2014: 9,029 (1,364 officers) including 1,530 marines

Marine Corps

2 battalions, 1 special operations detachment, 1 naval police unit

Bases

Main Base: Lisbon—Alfeite
Dockyard: Arsenal do Alfeite
Fleet Support: Porto, Portimão, Funchal, Ponta Delgada, Tróia
Air Base: Montijo (Lisbon)

Naval Air

The helicopter squadron was formally activated on 23 September 1993 at Montijo air force base, Lisbon. Operational and logistic procedures are similar to the air force.

Prefix to Ships' Names

NRP (Navio da República Portuguesa)

Strength of the Fleet

Type	Active (Reserve)	Building (Projected)
Submarines (Patrol)	2	—
Frigates	5	—
Corvettes	7	—
Offshore Patrol Vessels	1	1
Patrol Craft	4	—
Coastal/River Patrol Craft	11	—
LPD	—	(1)
LCU	1	—
Survey Ships and Craft	7	—
Sail Training Ships	5	—
Replenishment Tanker	1	(1)
Buoy Tenders	2	1

PENNANT LIST

Submarines
S 160 Tridente
S 161 Arpão

Frigates
F 330 Vasco da Gama
F 331 Álvares Cabral
F 332 Corte Real
F 333 Bartolomeu Dias
F 334 D Francisco de Almeida

Corvettes
F 471 António Enes
F 475 João Coutinho
F 476 Jacinto Cândido
F 477 Gen Pereira d'Eça
F 486 Baptista de Andrade
F 487 João Roby
F 488 Afonso Cerqueira

Patrol Forces
P 360 Viana do Castelo
P 361 Figueira da Foz
P 370 Rio Minho
P 1140 Cacine
P 1144 Cuanza
P 1146 Zaire
P 1150 Argos
P 1151 Dragão
P 1152 Escorpião
P 1153 Cassiopeia
P 1154 Hidra
P 1155 Centauro
P 1156 Orion
P 1157 Pégaso
P 1158 Sagitário
P 1165 Águia
P 1167 Cisne

Amphibious Forces
LDG 203 Bacamarte

Service Forces
A 520 Sagres
A 521 Schultz Xavier
A 522 D. Carlos I
A 523 Almirante Gago Coutinho
A 5203 Andrómeda
A 5204 Polar
A 5205 Auriga
A 5210 Bérrio
UAM 201 Creoula
UAM 813 Bellatrix
UAM 814 Canopus

SUBMARINES

2 TRIDENTE (TYPE 209PN) CLASS (SSK)

Name	No	Builders	Laid down	Launched	Commissioned
TRIDENTE	S 160	Howaldtswerke, Kiel	7 Mar 2005	15 Jul 2008	17 Jun 2010
ARPÃO	S 161	Howaldtswerke, Kiel	5 Jul 2006	18 Jun 2009	22 Dec 2010

Displacement, tonnes: 1,840 surfaced; 2,020 dived
Dimensions, metres (feet): 67.9 × 6.3 × 6.6 (222.8 × 20.7 × 21.7)
Speed, knots: 10 surfaced; 20 dived
Complement: 33 (7 officers)

Machinery: 2 MTU 16V 396 SE 84 diesels; 3,620 hp(m) (2.7 MW); 1 Siemens Permasyn motor; 1 shaft; 2 HDW PEM fuel cells; 240 kW
Torpedoes: 8—21 in (533 mm) bow tubes. WASS Black Shark; wire (fibre optic cable)-guided; active/passive homing to 50 km (27 n miles) at 50 kt; warhead 250 kg. 16 weapons including torpedoes and SSM.
Radars: To be announced.
Electronic countermeasures: CESM/RESM: Saab Avitronics interception suite (UME 200/CRS-8000).
Sonars: Cylindrical array with intercept passive array, passive range sonar, flank array and mine-avoidance sonar.
Weapon control systems: Atlas Elektronik ISUS 90/50.

Programmes: Contract signed on 21 April 2004 with German Submarine Consortium (GSC) for construction and delivery of two boats with option for a third. The consortium consists of Howaldtswerke-Deutsche Werft, Kiel, Nordseewerke, Emden (NSWE) and Ferrostaal, Essen.
Structure: Very similar to the Type 214 Air-Independent Propulsion (AIP) submarines built for Greece. Diving depth likely to be about 400 m (1,300 ft).
Operational: Forms 5 Squadron.

ARPÃO

9/2012, M Declerck / 1486328

FRIGATES

2 BARTOLOMEU DIAS (M) CLASS (FFGHM)

Name	No	Builders	Laid down	Launched	Commissioned
BARTOLOMEU DIAS (ex-*Van Nes*)	F 333 (ex-F 833)	Koninklijke Maatschappij De Schelde, Flushing	10 Jan 1990	16 May 1992	2 Jun 1994
D. FRANCISCO DE ALMEIDA (ex-*Van Galen*)	F 334 (ex-F 834)	Koninklijke Maatschappij De Schelde, Flushing	7 Jun 1990	21 Nov 1992	1 Dec 1994

Displacement, tonnes: 2,800 standard; 3,320 full load
Dimensions, metres (feet): 122.3 oa; 114.2 wl × 14.4 × 6.4 *(401.2; 374.7 × 47.2 × 21.0)*
Flight deck, metres: 22 × 14.4 *(72.2 × 47.2)*
Speed, knots: 29; 20 diesel
Range, n miles: 5,000 at 18 kt
Complement: 164 (20 officers) + 7 spare berths

Machinery: CODOG; 2 RR Spey SM1A; 33,800 hp *(25.2 MW)* sustained; 2 Stork-Wärtsilä 12SW280 diesels; 9,790 hp(m) *(7.2 MW)* sustained; 2 shafts; LIPS cp props
Missiles: SSM: 8 McDonnell Douglas Harpoon Block 1C (2 quad) launchers ❶; active radar homing to 130 km *(70 n miles)* at 0.9 Mach; warhead 227 kg.
SAM: Raytheon Sea Sparrow Mk 48 vertical launchers ❷; semi-active radar homing to 14.6 km *(8 n miles)* at 2.5 Mach; warhead 39 kg; 16 missiles. Canisters mounted on port side of hangar.
Guns: 1—3 in *(76 mm)*/62 Oto Melara compact Mk 100 ❸; 100 rds/min to 16 km *(8.6 n miles)* anti-surface; 12 km *(6.5 n miles)* anti-aircraft; weight of shell 6 kg. 1 Signaal SGE-30 Goalkeeper with General Electric 30 mm 7-barrelled ❹; 4,200 rds/min combined to 2 km. 2 Oerlikon 20 mm; 800 rds/min to 2 km.
Torpedoes: 4—324 mm US Mk 32 Mod 9 (2 twin) tubes (mounted inside the after superstructure) ❺. Honeywell Mk 46 Mod 5; anti-submarine; active/passive homing to 11 km *(5.9 n miles)* at 40 kt; warhead 44 kg.
Physical countermeasures: Decoys: 2 Loral Hycor SRBOC 6-tubed fixed Mk 36 quad launchers; IR flares and chaff to 4 km *(2.2 n miles)*. SLQ-25 Nixie towed torpedo decoy.
Electronic countermeasures: ESM/ECM: Argo APECS II (includes AR 700) ESM ❻; intercept and jammers.
Radars: Air/surface search: Signaal SMART ❽; 3D; F-band.
Air search: Signaal LW08 ❾; D-band.
Surface search: Signaal Scout ❿; I-band.
Navigation: Racal Decca 1226; I-band.
Fire control: 2 Signaal STIR ⓫; I/J/K-band; range 140 km *(76 n miles)* for 1 m² target.
Sonars: Signaal PHS-36; hull-mounted; active search and attack; medium frequency. Thomson Sintra Anaconda DSBV 61; towed array; passive low frequency.
Combat data systems: Signaal SEWACO VIIB action data automation; Link 11. SATCOM ❼. WSC-6 twin aerials.
Weapon control systems: Signaal IRSCAN infrared detector. Signaal VESTA helo transponder.
Helicopters: 1 Super Sea Lynx Mk 95 ⓬.

Programmes: The Declaration of Intent to purchase two ex-Netherlands frigates was announced on 1st November 2006. The ships have replaced the João Belo-class frigates. *Bartolomeu Dias* recommissioned on 16 January 2009 and *D. Francisco da Almeida* on 15 January 2010.
Modernisation: Modernisation plans have been suspended.
Structure: The VLS SAM is similar to Canadian Halifax and Greek MEKO classes. The ship is designed to reduce radar and IR signatures and has extensive NBCD arrangements. Full automation and roll stabilisation fitted. The APECS jammers are mounted starboard forward of the bridge and port aft corner of the hangar.

BARTOLOMEU DIAS *(Scale 1 : 1,200), Ian Sturton* / 1164924

D. FRANCISCO DE ALMEIDA *9/2009, Piet Cornelis* / 1406111

3 VASCO DA GAMA (MEKO 200 PN) CLASS (FFGH)

Name	No	Builders	Laid down	Launched	Commissioned
VASCO DA GAMA	F 330	Blohm + Voss, Hamburg	1 Feb 1989	26 Jun 1989	18 Jan 1991
ÁLVARES CABRAL	F 331	Howaldtswerke, Kiel	2 Jun 1989	6 Jun 1990	24 May 1991
CORTE REAL	F 332	Howaldtswerke, Kiel	24 Nov 1989	6 Jun 1990	22 Nov 1991

Displacement, tonnes: 3,180 standard; 3,353 full load
Dimensions, metres (feet): 115.9 oa; 109 pp × 14.8 × 6.1 *(380.2; 357.6 × 48.6 × 20.0)*
Speed, knots: 32; 20 diesel
Range, n miles: 4,900 at 18 kt, 9,600 at 12 kt
Complement: 198 (27 officers, 16 air crew) + 16 flag staff

Machinery: CODOG; 2 GE LM 2500 gas turbines; 53,000 hp *(39.5 MW)* sustained; 2 MTU 12V 1163 TB83 diesels; 8,840 hp(m) *(6.5 MW)*; 2 shafts; cp props
Missiles: SSM: 8 McDonnell Douglas Harpoon (2 quad) launchers ❶; active radar homing to 130 km *(70 n miles)* at 0.9 Mach; warhead 227 kg.
SAM: Raytheon Sea Sparrow Mk 29 Mod 1 octuple launcher ❷; RIM-7M/P; semi-active radar homing to 16 km *(8.5 n miles)* at 2.5 Mach; warhead 38 kg. Space left for VLS Sea Sparrow ❸.
Guns: 1 Creusot-Loire 3.9 in *(100 mm)*/55 Mod 68 CADAM ❹; 60 rds/min to 17 km *(9 n miles)* anti-surface; 8 km *(4.4 n miles)* anti-aircraft; weight of shell 13.5 kg. 1 General Electric/General Dynamics Vulcan Phalanx 20 mm Block 1B ❺; 6 barrels per mounting; 3,000 rds/min combined to 1.5 km. 2 Oerlikon 20 mm (on VLS deck) ❻ can be carried.
Torpedoes: 6—324 mm US Mk 32 (2 triple) tubes ❼. Honeywell Mk 46 Mod 5; anti-submarine; active/passive homing to 11 km *(5.9 n miles)* at 40 kt; warhead 44 kg.
Physical countermeasures: Decoys: 2 Loral Hycor Mk 36 SRBOC 6-barrelled chaff launchers ❽. Sea Gnat. SLQ-25 Nixie; towed torpedo decoy.
Electronic countermeasures: ESM/ECM: APECS II; intercept and jammer.
Radars: Air search: Signaal MW08 (derived from Smart 3D) ❿; 3D; G-band.
Air/surface search: Signaal DA08 ⓫; F-band.
Navigation: Kelvin Hughes Type 1007; I-band.
Fire control: 2 Signaal STIR ⓬; I/J/K-band; range 140 km *(76 n miles)* for 1 m² target. IFF Mk 12 Mod 4.
Sonars: Computing Devices (Canada) SQS-510(V); hull-mounted; active search and attack; medium frequency.
Combat data systems: Signaal SEWACO action data automation with STACOS tactical command; Link 11 and 14. Matra Marconi SCOT 3 SATCOM ❾ (2 sets between 3 ships).
Weapon control systems: SWG 1A(V) for SSM. Vesta Helo transponder with datalink for OTHT.
Helicopters: 2 Super Sea Lynx Mk 95 ⓭.

Programmes: The contract for all three was signed on 25 July 1986. These are Meko 200 type ordered from a consortium of builders. As well as Portugal, which provided 40 per cent of the cost, assistance was given by NATO with some missile, CIWS and torpedo systems being provided by the US.
Modernisation: Mid-life refits are planned 2015–2021. Upgrades are likely to include improvements to the combat data system, increased force protection capabilities and measures to counter asymmetric threats.
Structure: All-steel construction. Stabilisers fitted. Full RAS facilities. Space has been left for a sonar towed array and for VLS Sea Sparrow.
Operational: Designed primarily as ASW ships. SCOT SATCOM rotated between the three ships. 20 mm guns can be mounted on the VLS deck. Three year running cycles include 18 months at full readiness, three months training and six months refit.

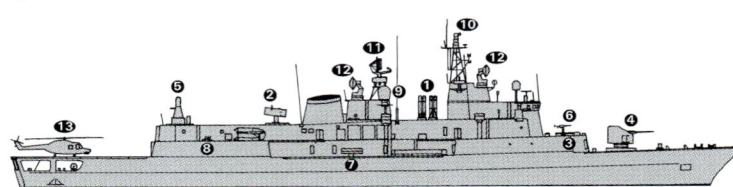

VASCO DA GAMA *(Scale 1 : 1,200), Ian Sturton* / 0567520

VASCO DA GAMA *3/2012, Martin Mokrus* / 1483643

CORTE REAL *5/2010, Michael Nitz* / 1406112

Frigates — Corvettes ‹ Portugal

ÁLVARES CABRAL
2/2009, Maritime Photographic / 1305543

CORVETTES

3 BAPTISTA DE ANDRADE CLASS (FSH)

Name	No	Builders	Laid down	Launched	Commissioned
BAPTISTA DE ANDRADE	F 486	Empresa Nacional Bazán, Cartagena	1 Sep 1972	13 Mar 1973	19 Nov 1974
JOÃO ROBY	F 487	Empresa Nacional Bazán, Cartagena	1 Dec 1972	3 Jun 1973	18 Mar 1975
AFONSO CERQUEIRA	F 488	Empresa Nacional Bazán, Cartagena	10 Mar 1973	6 Oct 1973	28 Jun 1975

Displacement, tonnes: 1,222 standard; 1,402 full load
Dimensions, metres (feet): 84.6 × 10.3 × 3.1 *(277.6 × 33.8 × 10.2)*
Speed, knots: 22. **Range, n miles:** 5,900 at 18 kt
Complement: 72 (7 officers)

Machinery: 2 OEW Pielstick 12 PC2.2 V 400 diesels; 12,000 hp(m) *(8.82 MW)* sustained; 2 shafts
Guns: 1 Creusot-Loire 3.9 in *(100 mm)*/55 Mod 1968; 80 rds/min to 17 km *(9 n miles)* anti-surface; 8 km *(4.4 n miles)* anti-aircraft; weight of shell 13.5 kg. 2 Bofors 40 mm/70; 300 rds/min to 12 km *(6.6 n miles)*; weight of shell 0.96 kg.
Radars: Navigation: 1 Racal Decca RM 316P and 1 KH 5000 Nucleos 2; I-band.

Helicopters: Platform only.

Programmes: Reclassified as corvettes.
Modernisation: Communications equipment updated 1988–91. Between 1999 and 2001 ASW and weapons control systems removed.

JOÃO ROBY
5/2010, Michael Nitz / 1406113

Operational: Class is used for Maritime Law Enforcement/SAR/Fishery Protection and for Humanitarian Operations. A fourth ship, *Oliveira E Carmo* decommissioned in 1999 and a further ship is to be decommissioned in 2014. All the class is to be deleted by 2020.

BAPTISTA DE ANDRADE
3/2009, Guy Toremans / 1305541

4 JOÃO COUTINHO CLASS (FSH)

Name	No	Builders	Laid down	Launched	Commissioned
ANTÓNIO ENES	F 471	Empresa Nacional Bazán, Cartagena	10 Apr 1968	16 Aug 1969	18 Jun 1971
JOÃO COUTINHO	F 475	Blohm + Voss, Hamburg	24 Dec 1968	2 May 1969	28 Feb 1970
JACINTO CÂNDIDO	F 476	Blohm + Voss, Hamburg	10 Feb 1969	16 Jun 1969	29 May 1970
GENERAL PEREIRA D'EÇA	F 477	Blohm + Voss, Hamburg	21 Apr 1969	26 Jul 1969	10 Oct 1970

Displacement, tonnes: 1,222 standard; 1,402 full load
Dimensions, metres (feet): 84.6 × 10.3 × 3.3 *(277.6 × 33.8 × 10.8)*
Speed, knots: 22
Range, n miles: 5,900 at 18 kt
Complement: 72 (7 officers)

Machinery: 2 OEW Pielstick 12 PC2.2 V 400 diesels; 12,000 hp(m) *(8.82 MW)* sustained; 2 shafts
Guns: 2 US 3 in *(76 mm)*/50 (twin) Mk 33; 50 rds/min to 12.8 km *(7 n miles)*; weight of shell 6 kg. 2 Bofors 40 mm/60 (twin); 300 rds/min to 12 km *(6.6 n miles)*; weight of shell 0.89 kg.
Radars: Air/surface search: Kelvin Hughes 1007; I-band. Navigation: Racal Decca RM 1226C; I-band.
Weapon control systems: Mk 51 GFCS for 40 mm.

Helicopters: Platform only.

Programmes: Reclassified as corvettes.
Modernisation: A programme for this class to include SSM and PDMS has been shelved. In 1989–91 the main radar was updated and SATCOM (INMARSAT) installed. Also fitted with SIFICAP which is a Fishery Protection data exchange system by satellite to the main database ashore.
Operational: A/S equipment no longer operational and laid apart on shore. Crew reduced as a result. Two ships, *Augusto de Castilho* and *Honório Barreto* were decommissioned in 2004 and 2003 respectively and a further two ships are to follow in 2014. The whole class is to be decommissioned by 2020.

ANTONIO ENES
12/2007, Diego Quevedo / 1335417

SHIPBORNE AIRCRAFT

Numbers/Type: 5 Westland Super Navy Lynx Mk 95.
Operational speed: 125 kt *(231 km/h)*.
Service ceiling: 12,000 ft *(3,660 m)*.
Range: 320 n miles *(593 km)*.
Role/Weapon systems: Ordered 2 November 1990 for MEKO 200 frigates; two are updated HAS 3 and three were new aircraft, all delivered in August and November 1993. There are plans to modernise the sensors. Service life extended to 2030. Sensors: Bendix 1500B radar; Bendix AQS-18V dipping sonar; Racal RNS 252 datalink. Weapons: Mk 46 torpedoes. 1 — 12.7 mm MG.

SUPER LYNX 7/2009, Guy Toremans / 1305540

LAND-BASED MARITIME AIRCRAFT

Notes: All Air Force manned.

Numbers/Type: 5/2 CASA C-212-200 Aviocar/C-212-300 Aviocar.
Operational speed: 190 kt *(353 km/h)*.
Service ceiling: 24,000 ft *(7,315 m)*.
Range: 1,650 n miles *(3,055 km)*.
Role/Weapon systems: The first five are for short-range SAR support and transport operations. The last pair were ordered in February 1993 for maritime patrol and fisheries surveillance off the Azores and Madeira. Sensors: Search radar and MAD. FLIR and datalink (last pair). Weapons: Unarmed.

CASA 212 6/2001, Adolfo Ortigueira Gil / 0529552

Numbers/Type: 5 Lockheed P3 CUP Orion.
Operational speed: 410 kt *(760 km/h)*.
Service ceiling: 28,300 ft *(8,625 m)*.
Range: 4,000 n miles *(7,410 km)*.
Role/Weapon systems: Five P-3B long-range surveillance aircraft acquired from Australia and modernised to 3P standard in 1987. These aircraft have been replaced by five P-3 CUP aircraft acquired from the Netherlands in 2005. Contract signed with Lockheed Martin on 3 January 2008 for upgrade work on all five aircraft. This is to include upgrade of the mission system and provision of improved ESM, acoustic processing, communications and sensor systems. First upgraded aircraft to be delivered late 2009 and programme to be completed by late 2012. Sensors: APS-134/137 radar, ASQ-81 MAD, AQS-901 sonobuoy processor, AQS-114 computer, IFF, ALR-66 ECM/ESM. Weapons: ASW; eight Mk 46 torpedoes, depth bombs or mines; ASV; 10 underwing stations for Harpoon.

P-3 CUP 2/2006, Portuguese Airforce / 1130518

Numbers/Type: 6/2/4 AgustaWestland EH 101 Mk 514/EH 101 Mk 515/EH 101 Mk 516.
Operational speed: 160 *(296 km/h)*.
Service ceiling: 15,000 ft *(4,572 m)*.
Range: 550 n miles *(1,019 km)*.
Role/Weapon systems: Contract in 2001 for a total of 12 utility variants of the EH 101. Six Mk 514 are configured for SAR duties, two Mk 515 for fishery protection and four Mk 516 for Combat SAR. The aircraft are designed for rapid role-change. Military lift is 28 troops and up to four tonnes underslung. Sensors: Galileo search radar, FLIR and defensive aids suite. Weapons: unarmed.

EH 101 6/2007, Portuguese Navy / 1166694

Numbers/Type: 5 EADS CASA C-295 MP Persuader.
Operational speed: 260 kt *(482 km/h)*.
Service ceiling: 13,540 ft *(4,125 m)*.
Range: 840 n miles *(1,555 km)*.
Role/Weapon systems: Five maritime surveillance aircraft ordered in February 2006. The first aircraft was delivered in late 2009. Sensors: EL/M 2022A9(V)3 surveillance radar, Star Safire IV thermal-imaging; Ericsson side scan radar; Elisra RWR and Fully Integrated Tactical System (FITS).

PATROL FORCES

1 + 1 VIANA DO CASTELO (NPO 2000) CLASS (PSOH)

Name	No	Builders	Commissioned
VIANA DO CASTELO	P 360	Viana do Castelo Shipyards	30 Dec 2010
FIGUEIRA DA FOZ	P 361	Viana do Castelo Shipyards	2014

Displacement, tonnes: 1,868 full load
Dimensions, metres (feet): 83.1 × 12.95 × 3.85 *(272.6 × 42.5 × 12.6)*
Speed, knots: 20
Range, n miles: 5,000 at 15 kt
Complement: 38 (5 officers)

Machinery: 2 Wärtsilä 12V 26 diesels; 10,460 hp *(7.8 MW)*; 2 shafts
Guns: 1 Oto Melara Marlin 30 mm. 2 — 12.7 mm MGs.
Radars: Surface search/navigation: 2 Kelvin Hughes; E/F/I-band.
Weapon control systems: Sagem optronic director.
Helicopters: Platform for one Lynx Mk 95.

Comment: Designed for EEZ patrol duties. Contract on 15 October 2002 with Viana do Castelo Shipyards for two Offshore Patrol vessels. Construction started in 2003 and the first two ships were floated out on 1 October 2005. Plans to build a further four patrol vessels were cancelled in 2012 and the order for two further modified vessels, a Buoy Tender and a Pollution Control Ship, has also been cancelled.

VIANA DO CASTELO 1/2010, Portuguese Navy / 1366175

3 CACINE CLASS (LARGE PATROL CRAFT) (PBO)

Name	No	Builders	Commissioned
CACINE	P 1140	Arsenal do Alfeite, Lisbon	May 1969
CUANZA	P 1144	Estaleiros Navais do Mondego	May 1969
ZAIRE	P 1146	Estaleiros Navais do Mondego	Nov 1970

Displacement, tonnes: 297 standard; 315 full load
Dimensions, metres (feet): 44 × 7.7 × 2.2 *(144.4 × 25.3 × 7.2)*
Speed, knots: 20
Range, n miles: 4,400 at 12 kt
Complement: 33 (3 officers)
Machinery: 2 MTU 12V 538 TB80 diesels; 3,750 hp(m) *(2.76 MW)* sustained; 2 shafts
Guns: 1 Bofors 40 mm/60. 1 Oerlikon 20 mm/65.
Radars: Surface search: Kelvin Hughes Type 1007; I/J-band.

Comment: Originally mounted a second Bofors aft but most have been removed as has the 37 mm rocket launcher. Have SIFICAP satellite data handling system for Fishery Protection duties. An RIB is carried. Two of the class are based at Madeira on a two month rotational basis. Re-engined in 1992–94. One ship is to be decommissioned in 2014.

CUANZA 3/2012*, Martin Mokrus / 1529097

ZAIRE 5/2006, A A de Kruijf / 1040764

Patrol forces — Amphibious forces < **Portugal** 653

2 ALBATROZ CLASS (RIVER PATROL CRAFT) (PBR)

Name	No	Builders	Commissioned
ÁGUIA	P 1165	Arsenal do Alfeite, Lisbon	28 Feb 1975
CISNE	P 1167	Arsenal do Alfeite, Lisbon	31 Mar 1976

Displacement, tonnes: 46 full load
Dimensions, metres (feet): 23.6 × 5.6 × 1.6 *(77.4 × 18.4 × 5.2)*
Speed, knots: 20. **Range, n miles:** 2,500 at 12 kt
Complement: 8 (1 officer)
Machinery: 2 Cummins diesels; 1,100 hp *(820 kW)*; 2 shafts
Guns: 1 Oerlikon 20 mm/65. 2 — 12.7 mm MGs
Radars: Surface search: Decca RM 316P; I-band

Comment: One other is used for harbour patrol duties. Two transferred to East Timor in 2001.

4 CENTAURO CLASS (RIVER PATROL CRAFT) (PBR)

Name	No	Builders	Commissioned
CENTAURO	P 1155	Arsenal do Alfeite, Lisbon	21 Mar 2000
ORION	P 1156	Arsenal do Alfeite, Lisbon	27 Mar 2001
PÉGASO	P 1157	Estaleiros Navais do Mondego	27 Mar 2001
SAGITARIO	P 1158	Estaleiros Navais do Mondego	27 Mar 2001

Displacement, tonnes: 90 full load
Dimensions, metres (feet): 28.4 × 5.95 × 1.4 *(93.2 × 19.5 × 4.6)*
Speed, knots: 26
Range, n miles: 640 at 20 kt
Complement: 8 (1 officer)
Machinery: 2 Cummins KTA-50-M2 diesels; 3,600 hp(m) *(2.64 MW)*; 2 shafts
Guns: 1 Oerlikon 20 mm/65.
Radars: 1 Furuno FCR-1411 MK3.

Comment: Similar to Argos class but of aluminium hull. Capable of full speed operation up to Sea State 3. Carries a semi-rigid boat with a 50 hp outboard engine. The boat is recoverable via a stern well at up to 10 kt.

ÁGUIA 8/2009, *Adolfo Ortigueira Gil* / 1305538

ORION 1/2011, *Jurg Kürsener* / 1406682

5 ARGOS CLASS (RIVER PATROL CRAFT) (PBR)

Name	No	Builders	Commissioned
ARGOS	P 1150	Arsenal do Alfeite, Lisbon	1 Jul 1991
DRAGÃO	P 1151	Arsenal do Alfeite, Lisbon	18 Oct 1991
ESCORPIÃO	P 1152	Arsenal do Alfeite, Lisbon	26 Nov 1991
CASSIOPEIA	P 1153	Conafi	11 Nov 1991
HIDRA	P 1154	Conafi	18 Dec 1991

Displacement, tonnes: 96 full load
Dimensions, metres (feet): 27.2 × 5.9 × 1.4 *(89.2 × 19.4 × 4.6)*
Speed, knots: 26. **Range, n miles:** 1,350 at 15 kt
Complement: 8 (1 officer)
Machinery: 2 MTU 12V 396 TE84 diesels; 3,700 hp(m) *(2.73 MW)* sustained; 2 shafts
Guns: 2 — 12.7 mm MGs (1150-1154).
Radars: Navigation: Furuno 1505 DA or Furuno FR 1411; I-band.

Comment: First five ordered in 1989 and 50 per cent funded by the EC. Of GRP construction, capable of full speed operation up to Sea State 3. Carries a RIB with a 37 hp outboard engine. The boat is recoverable via a stern well at up to 10 kt.

1 RIO MINHO CLASS (RIVER PATROL CRAFT) (PBR)

Name	No	Builders	Commissioned
RIO MINHO	P 370	Arsenal do Alfeite, Lisbon	1 Aug 1991

Displacement, tonnes: 73 full load
Dimensions, metres (feet): 22.4 × 6 × 0.8 *(73.5 × 19.7 × 2.6)*
Speed, knots: 9.5
Range, n miles: 420 at 7 kt
Complement: 8 (1 officer)
Machinery: 2 KHD-Deutz diesels; 664 hp(m) *(488 kW)*; 2 Schottel pumpjets
Guns: 1 — 7.62 mm MG.
Radars: Navigation: Furuno FR 1505DA; I-band.

HIDRA 9/2007, *Marco Ghiglino* / 1170064

RIO MINHO 6/2008, *Portuguese Navy* / 1335421

AMPHIBIOUS FORCES

Notes: Four new LCMs are planned as part of the LPD (NAVPOL) contract.

0 + 1 AMPHIBIOUS TRANSPORT SHIP (LPD)

Name	Builders	Laid down	Launched	Commissioned
AFONSO DE ALBUQUERQUE	Viana do Castelo Shipyards	2014	2016	2018

Displacement, tonnes: 10,668 full load
Dimensions, metres (feet): 162 × 25.0 × 5.2 *(531.5 × 82.0 × 17.1)*
Speed, knots: 19. **Range, n miles:** 6,000 at 14 kt
Complement: 150

Military lift: 650 troops; 4 LCM, 76 vehicles (including 40 light armoured vehicles), 53 light inflatable boats, 3,000 m³ of storage space
Machinery: Diesel-electric; 4 diesels; 18,775 hp *(14 MW)*; 2 shafts
Missiles: SAM: 2 RAM 21-cell Mk 49 launchers.
Guns: Medium calibre and CIWS.
Electronic countermeasures: To be announced.
Radars: Air/surface search: 3D radar to be announced.
Surface search: To be announced.
Navigation: To be announced.
Combat data systems: To be announced.
Weapon control systems: To be announced.

Helicopters: Landing spots for 4 EH-101 or 6 Lynx.

Programmes: The project is known as Navio Polivalente Logístico (NAVPOL).

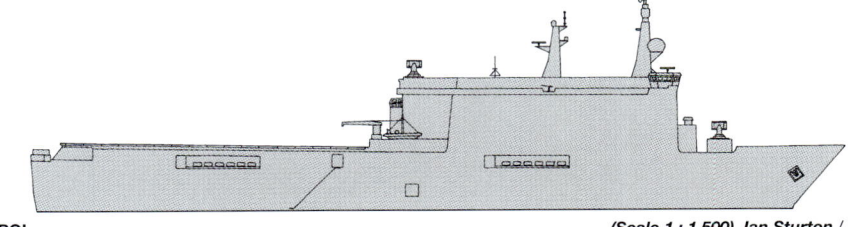

NAVPOL (Scale 1 : 1,500), *Ian Sturton* / 1153002

Structure: The design is very similar to the Schelde Enforcer 1300 and is to include a dock, flight deck, hangar, vehicle garage and hospital.
Operational: Following endorsement of the Portuguese National Defence Strategic Concept (NDSC) in 2003, the new LPD is to be designed to support worldwide joint operations of national and allied armed forces, including humanitarian aid and/or disaster relief. The ship is to be capable of projecting and supporting a battalion of troops.

© 2014 IHS IHS Jane's Fighting Ships 2014-2015

654 Portugal > Amphibious forces — Training ships

1 BOMBARDA CLASS (LCU)

Name	No	Builders	Commissioned
BACAMARTE	LDG 203	Arsenal do Alfeite, Lisbon	Aug 1985

Displacement, tonnes: 662 full load
Dimensions, metres (feet): 56.2 × 11.8 × 1.9 *(184.4 × 38.7 × 6.2)*
Speed, knots: 9.5
Range, n miles: 2,600 at 9 kt
Complement: 26 (3 officers)

Military lift: 350 tons
Machinery: 2 MTU MB diesels; 910 hp(m) *(669 kW)*; 2 shafts
Guns: 2 Oerlikon 20 mm.
Radars: Navigation: Decca RM 316P; I-band.

Comment: Similar to French EDIC. To be decommissioned in 2015.

BACAMARTE   2/2009, Adolfo Ortigueira Gil / 1305537

SURVEY SHIPS

2 D. CARLOS I (STALWART) CLASS (AGS)

Name	No	Builders	Commissioned
D. CARLOS I (ex-*Audacious*)	A 522 (ex-T-AGOS 11)	Tacoma Boatbuilding Co, Tacoma	18 Jun 1989
ALMIRANTE GAGO COUTINHO (ex-*Assurance*)	A 523 (ex-T-AGOS 5)	Tacoma Boatbuilding Co, Tacoma	1 May 1985

Displacement, tonnes: 2,322 full load
Dimensions, metres (feet): 68.3 × 13.1 × 4.6 *(224.1 × 43.0 × 15.1)*
Speed, knots: 11
Range, n miles: 4,000 at 11 kt, 6,450 at 3 kt
Complement: 34 (6 officers) + 15 scientists
Machinery: Diesel-electric; 4 Caterpillar D 398B diesel generators; 3,200 hp *(2.39 MW)*; 2 GE motors; 1,600 hp *(1.2 MW)*; 2 shafts; bow thruster; 550 hp *(410 kW)*
Radars: Navigation: 2 Raytheon; I-band.

Comment: Paid off from USN in November 1995. First one acquired 21 July 1996. Refitted to serve as a hydrographic ship, operating predominantly off the west coast of Africa. Recommissioned 9 December 1996. A second of class acquired by gift 30 September 1999, has been similarly refitted and recommissioned 26 January 2000. The Phantom S2 ROV in *D. Carlos I* has been replaced by Navajo.

D. CARLOS I   11/2006, Marco Ghiglino / 1164952

3 CORAL CLASS (YGS)

CORAL UAM 801   **ATLANTA** (ex-*Hidra*) UAM 802   **FISÁLIA** UAM 805

Comment: Craft are of 36 tons launched in 1980.

FISÁLIA   3/1992, van Ginderen Collection / 0081611

---

2 ANDRÓMEDA CLASS (AGSC)

Name	No	Builders	Commissioned
ANDRÓMEDA	A 5203	Arsenal do Alfeite, Lisbon	3 Jul 1987
AURIGA	A 5205	Arsenal do Alfeite, Lisbon	2 Mar 1988

Displacement, tonnes: 249 full load
Dimensions, metres (feet): 31.5 × 7.7 × 2.5 *(103.3 × 25.3 × 8.2)*
Speed, knots: 12. **Range, n miles:** 1,980 at 10 kt
Complement: 19 (3 officers)
Machinery: 1 MTU 12V 396 TC62 diesel; 1,200 hp(m) *(880 kW)* sustained; 1 shaft
Radars: Navigation: Koden; I-band.

Comment: Both ordered in January 1984. *Auriga* has a research submarine ROV Phantom S2 and a Klein side scan sonar. *Andrómeda* is equipped with a Navajo ROV. Mostly used for oceanography.

ANDRÓMEDA   8/1997, van Ginderen Collection / 0012932

TRAINING SHIPS

1 SAIL TRAINING YACHTS (AXS)

POLAR (ex-*Anne Linde*) A 5204

Comment: Built in Rotterdam in 1977 and commissioned in the Portuguese Navy on 21 August 1983. Sail number P-551 is displayed.

POLAR   6/2007, Portuguese Navy / 1166692

1 SAIL TRAINING SHIP (AXS)

Name	No	Builders	Commissioned
CREOULA	UAM 201	Lisbon Shipyard	1937

Displacement, tonnes: 831 standard; 1,072 full load
Dimensions, metres (feet): 67.4 × 9.9 × 4.2 *(221.1 × 32.5 × 13.8)*
Machinery: 1 MTU 8V 183 TE92 auxiliary diesel; 665 hp(m) *(490 kW)*; 1 shaft

Comment: Ex-deep sea sail fishing ship used off the coast of Newfoundland for 36 years. Bought by Fishing Department in 1976 to turn into a museum ship but because she was still seaworthy it was decided to convert her to a training ship. Recommissioned in the Navy on 28 May 1987. Refit completed in 1992 including a new engine and improved accommodation. A life-extension refit is under consideration.

CREOULA   6/2005, Portuguese Navy / 1153417

IHS Jane's Fighting Ships 2014-2015

Training ships — Auxiliaries < **Portugal** 655

1 SAIL TRAINING SHIP (AXS)

Name	No	Builders	Commissioned
SAGRES (ex-*Guanabara*; ex-*Albert Leo Schlageter*)	A 520	Blohm + Voss, Hamburg	14 Feb 1938

Displacement, tonnes: 1,753 standard; 1,971 full load
Dimensions, metres (feet): 90 oa; 70.4 wl × 12 × 5.2 *(295.3; 231 × 39.4 × 17.1)*
Speed, knots: 10.5
Range, n miles: 5,450 at 7.5 kt
Complement: 140 (9 officers)
Machinery: 2 MTU 12V 183 TE92 auxiliary diesels; 1 shaft
Radars: Navigation: 1 Racal Decca and 1 KH 1500 Nucleos 2; I-band.

Comment: Former German sail training ship launched 30 October 1937. Sister of US Coast Guard training ship *Eagle* (ex-German *Horst Wessel*) and Soviet *Tovarisch* (ex-German *Gorch Fock*). Taken by the USA as a reparation after the Second World War in 1945 and sold to Brazil in 1948. Purchased from Brazil and commissioned in the Portuguese Navy on 2 February 1962 at Rio de Janeiro and renamed *Sagres*. Sail area, 20,793 sq ft. Height of main mast, 142 ft. Phased refits 1987–88 and again in 1991–92 which included new engines, improved accommodation, hydraulic crane and updated navigation equipment. A further refit is planned 2015–17.

SAGRES *4/2009, Marco Ghiglino* / 1305536

2 SAIL TRAINING YACHTS (AXS)

BELLATRIX UAM 813   **CANOPUS** UAM 814

Displacement, tonnes: 12 (UAM 813), 10 (UAM 814) standard
Dimensions, metres (feet): 14.45 (UAM 813), 14.52 (UAM 814) × 4.4 (UAM 813), 4.23 (UAM 814) × 2.7 (UAM 813), 2.1 (UAM 814) *(47.4, 47.6 × 14.4, 13.9 × 8.9, 6.9)*
Complement: 8
Radars: Furuno; I-band.

Comment: Both attached to the naval school at Lisbon.

BELLATRIX *6/2007, Portuguese Navy* / 1166691

AUXILIARIES

Notes: (1) Two craft are employed on Pollution Control tasks. *Vazante* (UAM 687) is 14 tons and *Enchente* (UAM 688) is 65 tons.
(2) Studies for the procurement of a new AOR, to enter service in the 2020s, are in progress.

49 MISCELLANEOUS SERVICE CRAFT (YAG)

UAM 101–102	UAM 618–619	UAM 636	UAM 675	UAM 852
UAM 122	UAM 623	UAM 639–641	UAM 684	UAM 901
UAM 203	UAM 624	UAM 650–651	UAM 686–696	UAM 907
UAM 304	UAM 626	UAM 666	UAM 810–812	UAM 908
UAM 601–603	UAM 631	UAM 669	UAM 830	UAM 913
UAM 610	UAM 634	UAM 673	UAM 840	UAM 919

Displacement, tonnes: 18 full load
Dimensions, metres (feet): 14.5 × 4.3 × 0.8 *(47.6 × 14.1 × 2.6)*
Speed, knots: 27
Range, n miles: 150 at 15 kt
Machinery: 2 diesels; 640 hp *(478 kW)*; 2 waterjets

Comment: Details are for UAM 601–602 commissioned in 2007. The remaining craft are personnel and other service craft.

UAM 696 *4/2008, Marco Ghiglino* / 1335413

UAM 908 *4/2013*, Bob Fildes* / 1529073

1 BUOY TENDER (ABU)

Name	No	Builders	Commissioned
GUIA	UAM 676	S Jacinto, Aveiro	19 Aug 1983

Displacement, tonnes: 71 full load
Dimensions, metres (feet): 22 × 7.9 × 2.2 *(72.2 × 25.9 × 7.2)*
Speed, knots: 8.5; 3.5
Complement: 6
Machinery: 1 Deutz MWM SBA6M816 diesel; 465 hp(m) *(342 kW)* sustained; 1 Schottel Navigator prop

Comment: Belongs to the Lighthouse Service.

GUIA *6/2005, Portuguese Navy* / 1153416

IHS *Jane's Fighting Ships* 2014-2015

660 **Romania** > Submarines — Frigates

SUBMARINES

Notes: The Kilo-class submarine *Delfinul* 521 has not been to sea in recent years and there are no plans to refit her.

FRIGATES

2 BROADSWORD CLASS (TYPE 22) (FFHM)

Name	No	Builders	Laid down	Launched	Commissioned	Recommissioned
REGINA MARIA (ex-*London*)	F 222 (ex-F 95)	Yarrow Shipbuilders, Glasgow	7 Feb 1983	27 Oct 1984	5 Jun 1987	21 Apr 2005
REGELE FERDINAND (ex-*Coventry*)	F 221 (ex-F 98)	Swan Hunter Shipbuilders, Wallsend-on-Tyne	29 Mar 1984	8 Apr 1986	14 Oct 1988	9 Sep 2004

Displacement, tonnes: 4,166 standard; 4,877 full load
Dimensions, metres (feet): 146.5 × 14.8 × 6.4 (480.6 × 48.6 × 21.0)
Speed, knots: 30; 18 Tynes
Complement: 203

Machinery: COGOG: 2 RR Olympus TM3B gas turbines; 50,000 hp *(37.3 MW)* sustained; 2 RR Tyne RM1C gas turbines; 9,900 hp *(7.4 MW)*; 2 shafts; cp props
Guns: 1 Oto Melara 3 in *(76 mm)*/62 Super Rapid ❶; 120 rds/min to 16 km *(8.7 n miles)*; weight of shell 6 kg.
Torpedoes: 6—324 mm Plessey STWS Mk 2 (2 triple) tubes.
Physical countermeasures: Decoys: 2 Terma 130 mm DL-12 12-barrelled chaff launchers ❷.
Radars: Air/surface search: Marconi Type 967 ❹; D/E-band.
Navigation: Kelvin-Hughes Type 1007 ❺; I-band.
Sonars: Ferranti/Thomson Sintra Type 2050; hull-mounted search and attack.
Combat data systems: Ferranti CACS 1.
Weapon control systems: Radamec 2500 optronic director ❸. NAUTIS 3 fire-control system.

Helicopters: 1 medium.

Programmes: Originally successors to the UK Leander class, these ships entered RN service in 1987 but were withdrawn, half-way through their ships' lives, as a result of the 1998 UK Defence Review. Sale agreement signed on 14 January 2003 included platform overhaul, installation of reconditioned engines and combat system modernisation. Training is also included in the package. Following trials and sea training, *Regele Ferdinand* arrived in Romania on 10 December 2004 and *Regina Maria* in 2005. A 15-year through-life support contract with BAE Systems was signed in October 2005.
Modernisation: BAE Systems was prime contractor and FSL sub-contractor for reactivation and modernisation. The CACS command system was upgraded and 76 mm gun installed. A second-phase upgrade is to be undertaken in Romania although a firm timetable is yet to be announced. This is expected to include a towed-array sonar, air-defence and anti-ship weapons, an improved EW suite and small calibre guns.
Structure: Broadsword Batch 2 ships were stretched versions of Batch 1. The flight decks are capable of embarking medium helicopters.
Operational: Trials with a Puma helicopter were conducted in May 2008.

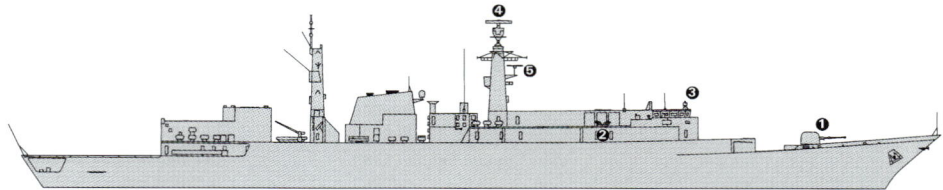

REGELE FERDINAND *(Scale 1 : 1,200), Ian Sturton / 1044184*

REGINA MARIA *6/2005, Maritime Photographic / 1133562*

REGELE FERDINAND *7/2011, C D Yaylali / 1406418*

REGELE FERDINAND *11/2004, B Sullivan / 1133560*

Frigates ◁ Romania

1 MARASESTI CLASS (FFGH)

Name	No	Builders	Laid down	Launched	Commissioned
MARASESTI (ex-*Muntenia*)	F 111	Mangalia Shipyard	7 Aug 1979	4 Jun 1981	3 Jun 1985

Displacement, tonnes: 5,883 full load
Dimensions, metres (feet): 144.6 × 14.8 × 7 *(474.4 × 48.6 × 23.0)*
Speed, knots: 27
Complement: 232 (21 officers)

Machinery: 4 diesels; 32,000 hp(m) *(23.5 MW)*; 4 shafts
Missiles: SSM: 8 SS-N-2C Styx ❶; active radar or IR homing to 83 km *(45 n miles)* at 0.9 Mach; warhead 513 kg.
Guns: 4 USSR 3 in *(76 mm)*/59 AK 726 (2 twin) ❷; 90 rds/min to 16 km *(8.5 n miles)*; weight of shell 5.9 kg. 4 – 30 mm/65 ❸; 6 barrels per mounting; 3,000 rds/min to 2 km.
Torpedoes: 6 – 21 in *(533 mm)* (2 triple) tubes ❹. Russian 53-65; passive/wake homing to 25 km *(13.5 n miles)* at 50 kt; warhead 300 kg.
A/S Mortars: 2 RBU 6000 ❺; 12-tubed trainable; range 6,000 m; warhead 31 kg.
Physical countermeasures: Decoys: 2 PK 16 chaff launchers.
Electronic countermeasures: ESM/ECM: 2 Watch Dog; intercept. Bell Clout and Bell Slam.
Radars: Air/surface search: Strut Curve ❻; F-band.
Surface search: Plank Shave ❼; E-band.
Fire control: Two Drum Tilt ❽; H/I-band. Hawk Screech ❾; I-band.
Navigation: Nayada (MR 212); Racal Decca; I-band.
IFF: High Pole B.
Sonars: Hull-mounted; active search and attack; medium frequency.

Helicopters: 2 IAR-316 Alouette III ❿.

Modernisation: Attempts have been made to modernise some of the electronic equipment. Also topweight problems have been addressed by reducing the height of the mast structures and lowering the Styx missile launchers by one deck. Two RBU 6000s have replaced the RBU 1200. Communications have been upgraded to enable NATO interoperability but there are no further modernisation plans.

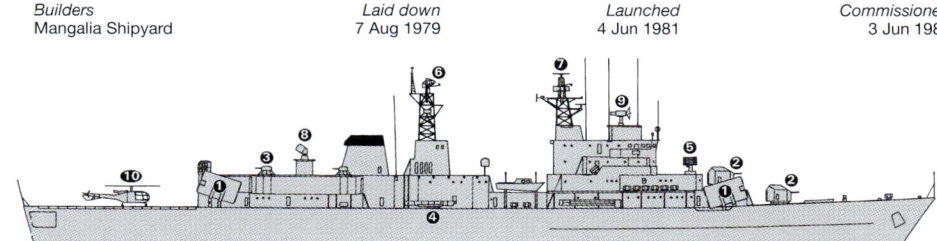

MARASESTI *(Scale 1 : 1,200), Ian Sturton / 1044186*

MARASESTI *1/2001, van Ginderen Collection / 0106855*

Structure: A distinctive Romanian design. Originally thought to be powered by gas turbines but a diesel configuration including four shafts is now confirmed.
Operational: Deactivated in June 1988 due to manpower and fuel shortages but modernisation work was done from 1990 to 1992 and sea trials started in mid-1992. Carried out a major naval exercise in September 1993, which included firing the Styx missile. Deployed to the Mediterranean in September 1994 for a short cruise, in 1995 on two occasions and again in March 1998. Reclassified as frigate in 2001. Based at Constanta.

MARASESTI *6/2004, C D Yaylali / 0589801*

MARASESTI *4/2011, C D Yaylali / 1406426*

Romania > Corvettes

CORVETTES

Notes: The multifunction corvette programme is for a class of four corvettes to replace the Tetal and Improved Tetal classes. The broad requirement is for a 2,000-ton ship that is fully compatible with NATO requirements. The ships, to be built in Romania, are likely to have ASW, ASuW and crisis-stabilisation roles.

2 IMPROVED TETAL CLASS (FSH)

Name	No	Builders	Launched	Commissioned
CONTRE ADMIRAL EUSTATIU SEBASTIAN	264	Mangalia Shipyard	12 Apr 1988	30 Dec 1989
ADMIRAL HORIA MACELARIU	265	Mangalia Shipyard	15 May 1994	29 Sep 1997

Displacement, tonnes: 1,524 full load
Dimensions, metres (feet): 92.4 × 11.7 × 3.1 *(303.1 × 38.4 × 10.2)*
Speed, knots: 24
Complement: 95
Machinery: 4 diesels; 13,000 hp(m) *(9.6 MW)*; 4 shafts
Guns: 1 USSR 3 in *(76 mm)*/59 AK 176 ❶; 120 rds/min to 15 km *(8 n miles)*; weight of shell 5.9 kg. 2–30 mm/65 AK 630 ❷; 6 barrels per mounting; 3,000 rds/min to 2 km. 2–30 mm/65 AK 306 ❸; 6 barrels per mounting; 3,000 rds/min to 2 km.
Torpedoes: 4–21 in *(533 mm)* (2 twin) tubes ❹. Russian 53-65; passive/wake homing to 25 km *(13.5 n miles)* at 50 kt; warhead 300 kg.
A/S Mortars: 2 RBU 6000 ❺; 12-tubed trainable; range 6,000 m; warhead 31 kg.
Physical countermeasures: Decoys: 2 PK 16 chaff launchers ❻.
Electronic countermeasures: ESM: 2 Watch Dog; intercept.
Radars: Air/surface search: Strut Curve ❼; F-band.
Fire control: Drum Tilt ❽; H/I-band.
Navigation: Nayada; I-band.
IFF: High Pole.
Sonars: Hull-mounted; active search and attack; medium frequency.
Helicopters: 1 IAR-316 Alouette III ❾.

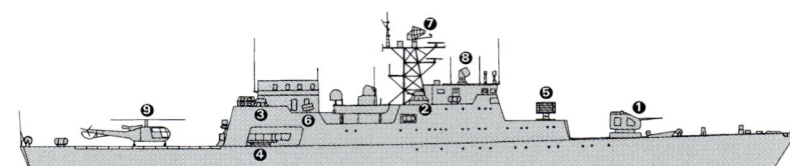

ADMIRAL HORIA MACELARIU *(Scale 1 : 900), Ian Sturton* / 1044187

Programmes: Follow on to Tetal class. Second of class was delayed when work stopped for a time in 1993–94.
Structure: As well as improved armament and a helicopter deck, there are superstructure changes from the original Tetals, but the hull and propulsion machinery are the same.
Operational: Both based at Mangalia.

ADMIRAL HORIA MACELARIU 9/2005, *Per Körnefeldt* / 1167804

ADMIRAL HORIA MACELARIU 8/2012, *C D Yaylali* / 1486334

ADMIRAL HORIA MACELARIU 11/2007, *Selim San* / 1353283

2 TETAL CLASS (FS)

Name	No	Builders	Launched	Commissioned
ADMIRAL PETRE BARBUNEANU	260	Mangalia Shipyard	23 May 1981	4 Feb 1983
VICE ADMIRAL EUGENIU ROSCA	263	Mangalia Shipyard	11 Jul 1985	23 Apr 1987

Displacement, tonnes: 1,463 full load
Dimensions, metres (feet): 92.4 × 11.7 × 3 (303.1 × 38.4 × 9.8)
Speed, knots: 24
Complement: 98

Machinery: 4 diesels; 13,000 hp(m) (9.6 MW); 4 shafts
Guns: 4 USSR 3 in (76 mm)/59 AK 726 (2 twin) ❶; 90 rds/min to 16 km (8.5 n miles); weight of shell 5.9 kg. 4 USSR 30 mm/65 (2 twin) ❷; 500 rds/min to 4 km (2.2 n miles); weight of shell 0.54 kg. 2—14.5 mm MGs ❸.
Torpedoes: 4—21 in (533 mm) (2 twin) tubes ❹. Russian 53–65; passive/wake homing to 25 km (13.5 n miles) at 50 kt; warhead 300 kg.
A/S Mortars: 2 RBU 2500 16-tubed trainable ❺; range 2,500 m; warhead 21 kg.
Physical countermeasures: Decoys: 2 PK 16 chaff launchers ❻.
Electronic countermeasures: ESM: 2 Watch Dog; intercept.

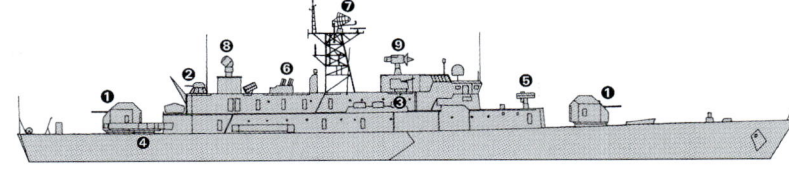

ADMIRAL PETRE BARBUNEANU   (Scale 1 : 900), Ian Sturton / 1167440

Radars: Air/surface search: Strut Curve ❼; F-band.
Fire control: Drum Tilt ❽; H/I-band. Hawk Screech ❾; I-band.
Navigation: Nayada; I-band.
IFF: High Pole.
Sonars: Hercules (MG 322); Hull-mounted; active search and attack; medium frequency.

Programmes: Building terminated in 1987 in favour of the improved design with a helicopter platform.
Structure: A modified Soviet Koni design.
Operational: Both based at Constanta. Two decommissioned in 2004.

ADMIRAL PETRE BARBUNEANU   8/2009, C D Yaylali / 1305977

SHIPBORNE AIRCRAFT

Numbers/Type: 3 IAR Brasov 330 Puma.
Operational speed: 139 kt (257 km/h).
Service ceiling: 15,750 ft (4,800 m).
Range: 297 n miles (550 km).
Role/Weapon systems: Eurocopter Puma built under license in Romania. Three aircraft procured by Romanian Navy in 2008. All upgraded to SOCAT configuration undertaken by IAR Brasov and Elbit Systems, Israel. The upgrade includes improved avionics. The helicopters are currently used in utility and SAR roles but there are plans to equip the aircraft with Thales TMS 2000 acoustic processors.

IAR 330   6/2011, Annati Collection / 1486335

LAND-BASED MARITIME AIRCRAFT

Numbers/Type: 5 Mil Mi-14PL Haze A.
Operational speed: 124 kt (230 km/h).
Service ceiling: 15,000 ft (4,570 m).
Range: 432 n miles (800 km).
Role/Weapon systems: Medium-range ASW helicopter. Sensors: Short Horn search radar, dipping sonar, MAD, sonobuoys. Weapons: ASW; internally stored torpedoes, depth mines and bombs.

HAZE PL (Polish colours)   6/2000 / 0105235

PATROL FORCES

5 BRUTAR II CLASS (RIVER MONITORS) (PGR)

Name	No	Builders	Commissioned
RAHOVA	176	Mangalia Shipyard	14 Apr 1988
OPANEZ	177	Mangalia Shipyard	24 Jul 1990
SMARDAN	178	Mangalia Shipyard	24 Jul 1990
POSADA	179	Mangalia Shipyard	14 May 1992
ROVINE	180	Mangalia Shipyard	30 Jun 1993

Displacement, tonnes: 417 full load
Dimensions, metres (feet): 45.7 × 8 × 1.5 *(149.9 × 26.2 × 4.9)*
Speed, knots: 16

Machinery: 2 diesels; 2,700 hp(m) *(2 MW)*; 2 shafts
Guns: 1—100 mm (tank turret). 2—30 mm (twin). 10—14.5 mm (2 quad, 2 single) MGs.
Rockets: 2—122 mm BM-21 rocket launchers; 40-tubed trainable.
Radars: Navigation: I-band.

Comment: Operational as patrol craft on the Danube. The first is a Brutar I. The next pair are Brutar IIs based at Tulcea and the last two are Brutar IIs based at Mangalia.

ROVINE   6/2008, Lemachko Collection / 1353281

3 KOGALNICEANU CLASS (RIVER MONITORS) (PGR)

Name	No	Builders	Recommissioned
MIKHAIL KOGALNICEANU	45	Drobeta Santierul, Turnu Severin	19 Dec 1993
I C BRATIANU	46	Drobeta Santierul, Turnu Severin	28 Dec 1994
LASCAR CATARGIU	47	Drobeta Santierul, Turnu Severin	22 Nov 1996

Displacement, tonnes: 584 full load
Dimensions, metres (feet): 52 × 9 × 1.7 *(170.6 × 29.5 × 5.6)*
Speed, knots: 18

Machinery: 2 24-H-165 RINS diesels; 4,400 hp(m) *(3.3 MW)*; 2 shafts
Guns: 2—100 mm (tank turrets). 4—30 mm (2 twin). 4—14.5 mm (2 twin).
Rockets: 2—122 mm BM-21 rocket launchers.
Radars: Navigation: I-band.

Comment: Based at Braila.

MIKHAIL KOGALNICEANU   6/2008, Lemachko Collection / 1366140

12 VD 141 CLASS (RIVER PATROL CRAFT) (PBR)

141–143	147–151	154	157	163	165

Displacement, tonnes: 99 full load
Dimensions, metres (feet): 33.3 × 4.8 × 0.9 *(109.3 × 15.7 × 3.0)*
Speed, knots: 13

Machinery: 2 diesels; 870 hp(m) *(640 kW)*; 2 shafts
Guns: 4—14.5 mm (2 twin) MGs.
Mines: 6.
Radars: Navigation: Nayada; I-band.

Comment: Built in Romania at Dobreta Severin Shipyard 1976–84. Ex river minesweepers now employed as patrol craft on the Danube.

VD 159   6/2008, Lemachko Collection / 1353279

3 ZBORUL (TARANTUL I) CLASS (PROJECT 1241 RE) (FSG)

Name	No	Builders	Commissioned
ZBORUL	188	Petrovsky Shipyard, Leningrad	Dec 1990
PESCARUSUL	189	Petrovsky Shipyard, Leningrad	Feb 1992
LASTUNUL	190	Petrovsky Shipyard, Leningrad	Feb 1992

Displacement, tonnes: 391 standard; 462 full load
Dimensions, metres (feet): 56.1 × 11.5 × 2.5 *(184.1 × 37.7 × 8.2)*
Speed, knots: 36
Range, n miles: 2,000 at 20 kt, 400 at 36 kt
Complement: 41 (5 officers)

Machinery: COGAG; 2 Type DR 77 gas turbines; 16,016 hp(m) *(11.77 MW)* sustained; 2 Nikolayev DR 76 gas turbines with reversible gearboxes; 4,993 hp(m) *(3.67 MW)* sustained; 2 shafts
Missiles: 4 SS-N-2C Styx (2 twin); active radar or IR homing to 83 km *(45 n miles)* at 0.9 Mach; warhead 513 kg.
Guns: 1 USSR 3 in *(76 mm)*/59 AK 176; 120 rds/min to 15 km *(8 n miles)*; weight of shell 5.9 kg. 2—30 mm/65 AK 630; 6 barrels per mounting; 3,000 rds/min to 2 km.
Physical countermeasures: Decoys: 2 PK 16 chaff launchers.
Electronic countermeasures: ESM: 2 Watch Dog; intercept.
Radars: Air/surface search: Plank Shave; E-band.
Fire control: Bass Tilt; H/I-band.
Navigation: Spin Trough; I-band.
IFF: Square Head. High Pole.
Weapon control systems: Hood Wink optronic director.

Comment: Built in 1985 and later transferred from the USSR. Export version similar to those built for Poland, India and Yemen. Based at Mangalia. *Pescarusul* carried out SS-N-2C firing on 28 May 2006.

LASTUNUL   6/1998, Valentino Cluru / 0052766

3 NALUCA CLASS (FAST ATTACK CRAFT—TORPEDO) (PTF)

Name	No	Builders	Commissioned
SMEUL	202	Mangalia Shipyard	25 Oct 1979
VIJELIA	204	Mangalia Shipyard	7 Feb 1980
VULCANUL	209	Mangalia Shipyard	26 Oct 1981

Displacement, tonnes: 215 full load
Dimensions, metres (feet): 36.8 × 7.6 × 1.8 *(120.7 × 24.9 × 5.9)*
Speed, knots: 36
Range, n miles: 500 at 35 kt
Complement: 22 (4 officers)

Machinery: 3 Type M 503A diesels; 8,025 hp *(5.9 MW)* sustained; 3 shafts
Guns: 4—30 mm/65 (2 twin) AK 230.
Torpedoes: 4—21 in *(533 mm)* tubes; anti-surface.
Radars: Surface search: Pot Drum; H/I-band.
Fire control: Drum Tilt; H/I-band.
IFF: High Pole A.

Comment: Based on the Osa class hull with torpedo tubes in lieu of SSMs. These craft were reported decommissioned in 2004 but have returned to service. Based at Mangalia.

VULCANUL   6/2012, Romanian Navy / 1486347

MINE WARFARE FORCES

Notes: The MCMV programme is for a class of four ships to replace the Corsar and Musca classes. The broad requirement is for a minehunter equipped with mine-detection sonar, unmanned vehicles for mine-detection and destruction and a decompression chamber for EOD teams.

1 CORSAR CLASS (MINELAYER/MCM SUPPORT SHIP) (ML/MCS)

Name	No	Builders	Commissioned
VICE ADMIRAL CONSTANTIN BALESCU	274	Mangalia Shipyard	16 Nov 1981

Displacement, tonnes: 1,473 full load
Dimensions, metres (feet): 79 × 10.6 × 3.6 *(259.2 × 34.8 × 11.8)*
Speed, knots: 19
Complement: 75

Machinery: 2 diesels; 6,400 hp(m) *(4.7 MW)*; 2 shafts
Guns: 1—57 mm/70. 4—30 mm/65 (2 twin) AK 230. 8—14.5 mm (2 quad) MGs.
A/S Mortars: 2 RBU 1200 5-tubed fixed; range 1,200 m; warhead 34 kg.
Mines: 200.
Electronic countermeasures: ESM: Watch Dog; intercept.
Radars: Air/surface search: Strut Curve; F-band.
Navigation: Don 2; I-band.
Fire control: Muff Cob; G/H-band. Drum Tilt; H/I-band.
Sonars: Tamir II; hull-mounted; active search; high frequency.

Comment: Has a large crane on the after deck. Similar to survey ship *Grigore Antipa*. Based at Constanta.

VICE ADMIRAL CONSTANTIN BALESCU 5/2012, N Sifferlinger / 1486337

4 MUSCA CLASS (MINESWEEPERS—COASTAL) (MSC)

Name	No	Builders	Commissioned
LIEUTENANT REMUS LEPRI	24	Mangalia Shipyard	23 Apr 1987
LIEUTENANT LUPU DUNESCU	25	Mangalia Shipyard	6 Jan 1989
LIEUTENANT DIMITRIE NICOLESCU	29	Mangalia Shipyard	7 Dec 1989
SUB LIEUTENANT ALEXANDRU AXENTE	30	Mangalia Shipyard	7 Dec 1989

Displacement, tonnes: 803 full load
Dimensions, metres (feet): 59.2 × 9.5 × 2.8 *(194.2 × 31.2 × 9.2)*
Speed, knots: 17
Complement: 60

Machinery: 2 diesels; 4,800 hp(m) *(3.5 MW)*; 2 shafts
Missiles: SAM: 2 quad SA-N-5 launchers.
Guns: 4—30 mm/65 (2 twin) AK 230.
A/S Mortars: 2 RBU 1200 5-tubed fixed; range 1,200 m; warhead 34 kg.
Radars: Surface search: Krivach; I-band.
Fire control: Drum Tilt; H/I-band.
Navigation: Nayada; I-band.
Sonars: Hull-mounted; active search; high frequency.

Comment: Reported as having a secondary mining capability but this is not confirmed. Based at Mangalia.

LIEUTENANT LUPU DUNESCU 5/2012, N Sifferlinger / 1486336

SURVEY AND RESEARCH SHIPS

1 CORSAR CLASS (RESEARCH SHIP) (AGOR)

Name	No	Builders	Commissioned
GRIGORE ANTIPA	75	Mangalia Shipyard	25 May 1980

Displacement, tonnes: 1,473 full load
Dimensions, metres (feet): 79 × 10.6 × 3.6 *(259.2 × 34.8 × 11.8)*
Speed, knots: 19
Complement: 75
Machinery: 2 diesels; 6,400 hp(m) *(4.7 MW)*; 2 shafts
Radars: Navigation: Nayada; I-band.

Comment: Large davits aft for launching manned submersible. Same hull as Corsar class. Used as a research ship and for diving support. Based at Constanta.

GRIGORE ANTIPA 5/1998, Diego Quevedo / 0052770

1 RESEARCH SHIP (AGS)

Name	No	Builders	Commissioned
EMIL RACOVITA	115	Drobeta Santierul, Turnu Severin	30 Oct 1977

Displacement, tonnes: 1,930 full load
Dimensions, metres (feet): 70.1 × 10 × 3.9 *(230 × 32.8 × 12.8)*
Speed, knots: 11
Complement: 80
Machinery: 1 diesel; 3,285 hp(m) *(2.4 MW)*; 1 shaft

Comment: Merchant ship design converted in the 1980s for hydrographic duties.

1 RESEARCH SHIP (AGS)

CPT. CDOR AL. CATUNEANU (ex-*Hercules*)

Displacement, tonnes: 2,450 full load
Dimensions, metres (feet): 64.8 × 14.6 × 5.5 *(212.6 × 47.9 × 18.0)*
Complement: 6
Machinery: 2 12R251 FMA diesels; 2,465 hp(m) *(1.8 MW)*; 2 shafts
Radars: Navigation: To be announced.

Comment: Originally entered service as an ocean tug in 1995. Subsequently converted to hydrographic research role. Equipped with Kongsberg EM 3000 multibeam echo-sounder, Kongsberg EM 400 bottom-profiler and Thales Sagitta positioning system. Also equipped with water velocity, tidal and bio-chemical sensors.

CPT. CDOR AL. CATUNEANU 5/2012, N Sifferlinger / 1486343

TRAINING SHIPS

1 SAIL TRAINING SHIP (AXS)

Name	No	Builders	Launched	Commissioned
MIRCEA	288	Blohm + Voss, Hamburg	29 Sep 1938	29 Mar 1939

Displacement, tonnes: 1,874 full load
Dimensions, metres (feet): 62.8 × 12 × 5.2 *(206 × 39.4 × 17.1)*
Speed, knots: 9
Range, n miles: 5,000 at 8 kt
Complement: 223 (5 officers; 140 midshipmen)
Machinery: 1 MaK 6M 451 auxiliary diesel; 1,000 hp(m) *(735 kW)*; 1 shaft
Radars: Navigation: Decca 202; I-band.

Comment: Refitted at Hamburg in 1966. Sail area, 5,739 m² *(18,830 sq ft)*, the length including the bowsprit is 81.2 m (266.4 ft). A smaller version of US Coast Guard cutter *Eagle*, German *Gorch Fock* and Portuguese *Sagres*. Based at Constanta.

MIRCEA — 5/2012, N Sifferlinger / 1486342

AUXILIARIES

(1) There are four river dredgers with pennant numbers 591, 593, 596 and 597.
(2) There are a number of river VIP, command and transport craft including: *Siret*, *Rindunica*, *Fortuna*, *Mures*, *Egreta* and *Luceafarul*.
(3) There are two river tankers 341 and 342.
(4) There are three river stations: *Dunarea* 593, *Tirnava* 597 and *Motri*.

DREDGER 597 — 10/2012, Hartmut Ehlers / 1486341

SIRET — 10/2012, Hartmut Ehlers / 1486340

1 TANKER (AOT)

TULCEA 532

Displacement, tonnes: 2,205 full load
Dimensions, metres (feet): 76.3 × 12.5 × 5 *(250.3 × 41.0 × 16.4)*
Speed, knots: 16
Cargo capacity: 1,200 tons oil
Machinery: 2 diesels; 4,800 hp(m) *(3.5 MW)*; 2 shafts
Guns: 2 – 30 mm/65 (twin). 4 – 14.5 mm (2 twin) MGs.

Comment: First one built by Tulcea Shipyard and commissioned 24 December 1992. Second of class reported in 1997 but not confirmed. Based at Constanta.

TULCEA — 2001, Romanian Navy / 0114542

1 DEGAUSSING SHIP (ADG/AGI)

Name	No	Builders	Commissioned
MAGNETICA	298	Mangalia Shipyard	18 Dec 1989

Displacement, tonnes: 440 full load
Dimensions, metres (feet): 44.7 × 8.0 × 2.2 *(146.7 × 26.2 × 7.2)*
Speed, knots: 12.5
Complement: 10
Machinery: Diesel-electric; 1 diesel generator; 600 kW; 1 shaft
Guns: 2 – 14.5 mm (twin) MGs. 2 – 12.7 mm MGs.

Comment: Built for degaussing ships up to 3,000 tons displacement.

MAGNETICA — 6/1999, Romanian Navy / 0081632

3 BRAILA CLASS (RIVER TRANSPORTS) (AG)

417–419

Displacement, tonnes: 240 full load
Dimensions, metres (feet): 38.3 × 8.6 × 1 *(125.7 × 28.2 × 3.3)*
Speed, knots: 4
Cargo capacity: 160 tons.
Machinery: 2 diesels; 300 hp(m) *(220 kW)*; 2 shafts

Comment: Used by civilian as well as naval authorities. Built at Braila Shipyard 1967–70. Danube Flotilla. Based at Turnul Magurele. Reported decommissioned in 2004 but have returned to service.

BRAILA 418 — 10/2012, Hartmut Ehlers / 1486339

Auxiliaries — Tugs < **Romania** 667

2 CROITOR CLASS (LOGISTIC SUPPORT SHIPS) (AETLMH)

Name	No	Builders	Commissioned
CONSTANTA	281	Braila Shipyard	15 Sep 1980
MIDIA	283	Braila Shipyard	26 Feb 1982

Displacement, tonnes: 2,896 standard; 3,556 full load
Dimensions, metres (feet): 108 × 13.2 × 3.8 *(354.3 × 43.3 × 12.5)*
Speed, knots: 17
Complement: 71 (7 officers)

Machinery: 2 diesels; 6,500 hp(m) *(4.8 MW)*; 2 shafts
Missiles: SAM: 2 SA-N-5 Grail quad launchers; manual aiming; IR homing to 6 km *(3.2 n miles)* at 1.5 Mach; warhead 1.5 kg.

Guns: 2—57 mm (twin). 4—30 mm/65 (2 twin). 8—14.5 mm (2 quad).
A/S Mortars: 2 RBU 1200 5-tubed fixed; range 1,200 m; warhead 34 kg.
Electronic countermeasures: ESM: 2 Watch Dog; intercept.
Radars: Air/surface search: Strut Curve; F-band.
Navigation: Krivach; I-band.
Fire control: Muff Cob; G/H-band. Drum Tilt; H/I-band.
Sonars: Tamir II; hull-mounted; active attack; high frequency.

Helicopters: 1 IAR-316 Alouette III type.

Comment: These ships are a scaled down version of Soviet Don class. Forward crane for ammunition replenishment. Some ASW escort capability. Can carry Styx missiles and torpedoes. Based at Constanta.

CONSTANTA 6/2001, Schaeffer/Marsan / 0533268

TUGS

Notes: There are a number of river tugs including pennant numbers 302, 303, 310, 327, 328, 457 and 578.

RIVER TUG 328 10/2012, Hartmut Ehlers / 1486338

1 VITEAZUL CLASS (ARS)

VITEAZUL 101

Displacement, tonnes: 552 full load
Dimensions, metres (feet): 44.5 × 9.4 × 3.3 *(146 × 30.8 × 10.8)*
Speed, knots: 12
Complement: 21
Machinery: 2 LDSR-2 diesels; 2,400 hp(m) *(1.8 MW)*; 2 shafts
Guns: 2—30 mm (twin). 2—14.5 mm (twin).
Radars: Surface search/navigation: To be announced.

Comment: Entered service in 1954. Salvage tug which is also used for patrol duties.

VITEAZUL 6/2012, Romanian Navy / 1486348

1 OCEAN TUGS (ATA)

GROZAVU 500

Displacement, tonnes: 3,658 full load
Dimensions, metres (feet): 64.8 × 14.6 × 5.5 *(212.6 × 47.9 × 18.0)*
Speed, knots: 12
Machinery: 2 diesels; 5,000 hp(m) *(3.7 MW)*; 2 shafts
Guns: 2—30 mm (twin). 8—14.5 mm (2 quad) MGs.

Comment: Built at Oltenitza Shipyard and commissioned 29 June 1993. Based at Constanta. A second ship, *Hercules* converted to hydrographic role.

GROZAVU 5/2012, N Sifferlinger / 1486346

6 DELTA IV (DELFIN) CLASS (PROJECT 667BDRM) (SSBN)

Name	No	Builders	Laid down	Launched	Commissioned
VERCHOTURE (K 51)	827	Severodvinsk Shipyard	23 Feb 1981	7 Mar 1984	29 Dec 1984
EKATERINBURG (K 84)	807	Severodvinsk Shipyard	17 Feb 1982	Sep 1984	30 Dec 1985
TULA (K 114)	805	Severodvinsk Shipyard	22 Feb 1984	22 Jan 1987	30 Oct 1987
BRIANSK (K 117)	820	Severodvinsk Shipyard	20 Apr 1985	8 Feb 1988	30 Sep 1988
KARELIA (K 18)	839	Severodvinsk Shipyard	7 Feb 1986	2 Feb 1989	10 Oct 1989
NOVOMOSKOVSK (K 407)	849	Severodvinsk Shipyard	14 Jun 1987	28 Feb 1990	27 Nov 1990

Displacement, tonnes: 10,973 surfaced; 13,717 dived
Dimensions, metres (feet): 166 oa; 158 wl × 12 × 8.7 *(544.6; 518.4 × 39.4 × 28.5)*
Speed, knots: 14 surfaced; 24 dived
Complement: 130 (40 officers)

Machinery: Nuclear; 2 VM-4 PWR; 180 MW; 2 GT3A-365 turbines; 37,400 hp(m) *(27.5 MW)*; 2 emergency motors; 612 hp(m) *(450 kW)*; 2 shafts
Missiles: SLBM: 16 Makeyev SS-N-23 (R 29RMU2 Sineva); 3-stage liquid fuel rocket; stellar inertial guidance to 8,300 km *(4,500 n miles)*; warhead nuclear 4–10 MIRV each of 100 kT; CEP 500 m. The R 29RMU2.1 (Liner) missile, with up to 12 MIRVs, is being fitted progressively from 2011. These are reported to have a range of 12,000 km.
A/S: Novator SS-N-15 Starfish; inertial flight to 45 km *(24.3 n miles)*; warhead nuclear 200 kT or Type 40 torpedo.
Torpedoes: 4 — 21 in *(533 mm)* tubes. Combination of 53 cm torpedoes. Total of 18 weapons.
Electronic countermeasures: ESM: Brick Pulp/Group; radar warning. Park Lamp D/F.
Radars: Surface search: Snoop Tray; I-band.
Sonars: Shark Gill; hull-mounted; passive/active search and attack; low/medium frequency. Shark Hide flank array; passive; low frequency. Mouse Roar; hull-mounted; active attack; high frequency. Pelamida towed array; passive search; very low frequency.

Programmes: Construction first ordered 10 December 1975. This programme completed in late 1990 and included seven boats.
Modernisation: The Sineva missile is being progressively fitted throughout the class. A successful test firing from K 114 was conducted on 11 October 2008.
Structure: A slim fitting is sited on the after fin which is reminiscent of a similar tube in one of the November class in the early 1980s. This is a dispenser for a sonar thin line towed array. The other distinguishing feature, apart from the size being greater than Delta III, is the pressure-tight fitting on the after end of the missile tube housing, which may be a TV camera to monitor communications buoy and wire retrieval operations. This is not fitted in all of the class. Brick Spit optronic mast. Diving depth, 1,300 ft *(400 m)*. The outer casing has a continuous acoustic coating and fewer free flood holes than the Delta III.
Operational: Two VLF/ELF communication buoys. Navigation systems include SATNAV, SINS, Cod Eye. Pert Spring SATCOM. Missile launch is conducted at keel depth 55 m and at a speed of 6 kt.

All operational units are part of the 12th Squadron based in the Northern Fleet at Saida Guba. Long refits have been completed as follows: K 51 (1999); K 84 (2002); K 114 (2005); K 117 (2007); K 18 (2008). The refit of K 407 started thereafter and was completed in 2011. K 51 started a life-extension refit at Zvezdochka Shipyard in August 2010 and completed in late 2012. This is likely to have included conversion to fire the Liner (a modernised version of Sineva) missile. The class is likely to remain in service until at least 2020. K 64 has been paid off as an SSBN but it is to be converted to an auxiliary submarine role. Test-firings of Liner were conducted from K 84 on 20 May 2011 and from K 114 on 29 September 2011. K 84 suffered a serious fire on 29 December 2011 while in a floating dock at Roslyakovo, near Severodvinsk. She is unlikely to return to service until 2015–16.

KARELIA and VERCHOTURE 9/2000, Lemachko Collection / 0126226

TULA 5/2010, Lemachko Collection / 1406115

Submarines < **Russian Federation** 673

3 DELTA III (KALMAR) CLASS (PROJECT 667BDR) (SSBN)

Name	No	Builders	Laid down	Launched	Commissioned
PODOLSK (K 223)	915	Severodvinsk Shipyard	19 Feb 1977	30 Apr 1979	27 Nov 1979
SYVATOY GIORGIY POBEDONOSETS (K 433)	993	Severodvinsk Shipyard	24 Aug 1978	20 Jun 1980	15 Dec 1980
RYAZAN (K 44)	862	Severodvinsk Shipyard	31 Jan 1980	19 Jan 1982	17 Sep 1982

Displacement, tonnes: 10,719 surfaced; 13,463 dived
Dimensions, metres (feet): 160 oa; 152 wl × 12 × 8.7 *(524.9; 498.7 × 39.4 × 28.5)*
Speed, knots: 14 surfaced; 24 dived
Complement: 130 (20 officers)

Machinery: Nuclear; 2 VM-4 PWR; 180 MW; 2 GT3A-635 turbines; 37,400 hp(m) *(27.5 MW)*; 2 emergency motors; 612 hp(m) *(450 kW)*; 2 shafts
Missiles: SLBM: 16 Makeyev SS-N-18 (RSM 50) Stingray (Volna); 2-stage liquid fuel rocket with post boost vehicle (PBV); stellar inertial guidance; 3 variants: Mod 1; range 6,500 km *(3,500 n miles)*; warhead nuclear 3 MIRV each of 200 kT; CEP 900 m. Mod 2; range 8,000 km *(4,320 n miles)*; warhead nuclear 450 kT; CEP 900 m. Mod 3; range 6,500 km *(3,500 n miles)*; warhead nuclear 7 MIRV 100 kT; CEP 900 m. Mods 1 and 3 were the first MIRV SLBMs in Soviet service.
Torpedoes: 4—21 in *(533 mm)* and 2—400 mm tubes. Combination of torpedoes. Total of 16 weapons.
Electronic countermeasures: ESM: Brick Pulp/Group; radar warning. Park Lamp D/F.
Radars: Surface search: Snoop Tray; I-band.
Sonars: Shark Teeth; hull-mounted; passive/active search and attack; low/medium frequency. Shark Hide flank array; passive; low frequency. Mouse Roar; hull-mounted; active attack; high frequency. Pelamida towed array; passive search; very low frequency.

Modernisation: The dispenser tube on the after fin has been fitted to most of the class. It was planned to retrofit SS-N-23 but this was shelved.
Structure: The missile casing is higher than in the decommissioned Delta I class to accommodate SS-N-18 missiles. The outer casing has a continuous 'acoustic' coating but is less streamlined and has more free flood holes than the Delta IV. Brick Spit optronic mast. Diving depth, 1,050 ft *(320 m)*.
Operational: ELF/VLF communications with floating aerial and buoy; UHF and SHF aerials. Navigation equipment includes Cod Eye radiometric sextant, SATNAV, SINS and Omega. Pert Spring SATCOM. Kremmny 2 IFF. This class consisted of 14 boats which were constructed at Severodvinsk and commissioned 1976–81. Of these, 10 have been decommissioned: K 441 and K 424 in 1996, K 449 in 1997, K 455, K 487 and K 180 in 1998–99, K 490 in 2006 and K 496 in 2008, K 211, and K 506 in 2010. The last hull of the class, K 129, was converted to a DSRV carrier in about 2003 and has been recommissioned as *Orienburg*. Following a test-firing of an SS-N-18 missile on 1 August 2008, the last remaining Northern Fleet unit, K 44, conducted an Arctic transit to join the Pacific Fleet in September 2008. It became part of the 16th Squadron based at Rybachiy (Kamchatka). The last remaining three boats are expected to decommission as units of the Dolguruky class enter service in the Pacific Fleet.

DELTA III
1/2007, Ships of the World
1167422

DELTA III *1/2007, Ships of the World* / 1167421

DELTA III *6/2011, Ships of the World* / 1406876

© 2014 IHS

IHS Jane's Fighting Ships 2014-2015

1 TYPHOON (AKULA) CLASS (PROJECT 941U) (SSBN)

Name	No	Builders	Laid down	Launched	Commissioned
DMITRIY DONSKOY (TK 208)	824	Severodvinsk Shipyard	30 Jun 1976	27 Sep 1979	29 Dec 1981

Displacement, tonnes: 18,797 surfaced; 26,925 dived
Dimensions, metres (feet): 171.5 oa; 165 wl × 24.6 × 13 *(562.7; 541.3 × 80.7 × 42.7)*
Speed, knots: 12 surfaced; 25 dived
Complement: 175 (55 officers)

Machinery: Nuclear; 2 VM-5 PWR; 380 MW; 2 GT3A turbines; 81,600 hp(m) *(60 MW)*; 2 emergency motors; 517 hp(m) *(380 kW)*; 2 shafts; shrouded props; 2 thrusters (bow and stern); 2,860 hp(m) *(1.5 MW)*
Missiles: SLBM: 1 Bulava 30 (R-30) SS-NX-32; three-stage solid fuel rocket; inertial guidance with Stellar and Glonass update to 8,300 km *(4,500 n miles)*; warhead nuclear 6–10 MIRV each of 150 kT; CEP 250 m.
SAM: SA-N-8 SAM capability when surfaced.
A/S: Novator SS-N-15 Starfish; inertial flight to 45 km *(24.3 n miles)*; warhead nuclear 200 kT or Type 40 torpedo.
Torpedoes: 6—21 in *(533 mm)* tubes. Combination of torpedoes. The weapon load includes a total of 22 torpedoes and A/S missiles.
Mines: Could be carried in lieu of torpedoes.
Physical countermeasures: Decoys: MG 34/44 tube launched decoys.
Electronic countermeasures: ESM: Rim Hat (Nakat M); radar warning. Park Lamp D/F.
Radars: Surface search: Snoop Pair (Albatros); I/J-band.
Sonars: Shark Gill; hull-mounted; passive/active search and attack; low/medium frequency. Shark Rib flank array; passive; low frequency. Mouse Roar; hull-mounted; active attack; high frequency. Pelamida towed array; passive search; very low frequency.
Weapon control systems: 3R65 data control system.

Modernisation: First of class TK 208 started refit at Severodvinsk in 1994, was relaunched on 26 June 2002 and started sea trials in August 2004. One (reported) missile tube was converted for Bulava missile test firings and the first submerged test launch took place on 21 December 2005. A series of tests, with mixed results, were completed in October 2010. Further tests were then conducted from *Yuri Dolgoruky*.
Structure: This is the largest type of submarine ever built. Two separate 7.2 m diameter hulls covered by a single outer free-flood hull with anechoic Cluster Guard tiles plus separate 6 m diameter pressure-tight compartments in the fin and fore-ends. There is a 1.2 m separation between the outer and inner hulls along the sides. The unique features of Typhoon are her enormous size and the fact that the missile tubes are mounted forward of the fin. The positioning of the launch tubes mean a fully integrated weapons area in the bow section leaving space abaft the fin for the provision of two nuclear reactors, one in each hull. The fin configuration indicates a designed capability to break through ice cover up to 3 m thick; the retractable forward hydroplanes, the rounded hull and the shape of the fin are all related to under-ice operations. Diving depth, 1,000 ft *(300 m)*.

TYPHOON CLASS 1/1997 / 0081639

Operational: Strategic targets are within range from anywhere in the world. Two VLF/ELF communication buoys are fitted. VLF navigation system for under-ice operations. Pert Spring SATCOM mast, Cod Eye radiometric sextant and Kremmny 2 IFF. Based in the Northern Fleet at Zapadnaya Litsa. Of six boats completed, the second and third, TK 202, TK 12 and TK 13 have been formally decommissioned. TK 17 and TK 20 were both decommissioned in 2013. TK 208, which was expected to decommission after the Bulava missile test programme, may be retained in service to assist in trials of new submarines in the White Sea. She remains capable of firing a Bulava missile.

Attack Submarines

Notes: Attack submarines are coated with Cluster Guard anechoic tiles. All submarines are capable of laying mines from their torpedo tubes. All SSNs are fitted with non-acoustic environmental sensors for measuring discontinuities caused by the passage of a submarine in deep water.

2 SIERRA I (BARRACUDA) CLASS (PROJECT 945) (SSN)

Name	No	Builders	Laid down	Launched	Commissioned
KARP (K 239)	622	Nizhny Novgorod; Severodvinsk Shipyard	20 Jul 1979	29 Jul 1983	29 Sep 1984
KOSTROMA (K 276) (ex-*Krab*)	648	Nizhny Novgorod; Severodvinsk Shipyard	21 Apr 1984	26 Jul 1986	27 Oct 1987

Displacement, tonnes: 7,316 surfaced; 8,230 dived
Dimensions, metres (feet): 107 × 12.5 × 8.8 *(351 × 41.0 × 28.9)*
Speed, knots: 10 surfaced; 34 dived
Complement: 61 (31 officers)

Machinery: Nuclear; 1 VM-5 PWR; 190 MW; 1 GT3A turbine; 47,500 hp(m) *(70 MW)*; 2 emergency motors; 2,004 hp(m) *(1.5 MW)*; 1 shaft; 2 spinners; 1,006 hp(m) *(740 kW)*
Missiles: SLCM: Raduga SS-N-21 Sampson (Granat) fired from 21 in *(533 mm)* tubes; land-attack; inertial/terrain-following to 2,500 km *(1,350 n miles)* at 0.7 Mach; warhead 410 kg HE or nuclear 200 kT. CEP 150 m. Probably flies at a height of about 200 m.
A/S: Novator SS-N-15 Starfish (Tsakra) fired from 53 cm tubes; inertial flight to 45 km *(24.3 n miles)*; warhead nuclear 200 kT or Type 40 torpedo. Novator SS-N-16 Stallion fired from 65 cm tubes; inertial flight to 100 km *(54 n miles)*; payload nuclear 200 kT (Vodopad) or Type 40 torpedo (Veder).
Torpedoes: 4—25.6 in *(650 mm)* and 4—21 in *(533 mm)* tubes. Combination of 65 and 53 cm torpedoes. Total of 40 weapons.
Mines: 42 in lieu of torpedoes.
Electronic countermeasures: ESM: Rim Hat/Bald Head; intercept. Park Lamp D/F.
Radars: Surface search: Snoop Pair with back-to-back ESM aerial.
Sonars: Shark Gill; hull-mounted; passive/active search and attack; low/medium frequency. Shark Rib flank array; passive; low frequency. Mouse Roar; hull-mounted; active attack; high frequency. Skat 3 towed array; passive; very low frequency.

Programmes: Launched at Gorky (Nizhny Novgorod) and transferred by river/canal to be fitted out at Severodvsinsk.
Structure: Based on design experience gained with deleted Alfa class, pressure hull constructed of titanium alloy, providing deep diving capability. Magnetic signature also reduced. Distance between hulls increases survivability and reduces radiated noise. There are six watertight compartments. The pod on the after fin is larger than that in 'Victor III'. Bulbous casing at the after end of the fin is for a towed communications buoy. Diving depth 2,460 ft *(750 m)*.

KOSTROMA 6/2002, Lemachko Collection / 0547070

Operational: Pert Spring SATCOM. Based in the Northern Fleet at Ara Guba. It is believed that K 276 was in a collision with USS *Baton Rouge* on 11 February 1992. Second of class *Karp* K 239 was laid up in 1994 but it was announced in early 2013 that both boats are to be refitted at Zvezdochka Shipyard. This is to include nuclear refuelling. *Karp* is to be refitted first, followed by *Kostroma*.

2 SIERRA II (KONDOR) CLASS (PROJECT 945B) (SSN)

Name	No	Builders	Laid down	Launched	Commissioned
PSKOV (K 336) (ex-*Okun*)	663	Nizhny Novgorod	29 Jul 1989	28 Jul 1992	14 Dec 1993
NIZHNY NOVGOROD (K534) (ex-*Zubatka*)	602	Nizhny Novgorod	15 Feb 1986	8 Jul 1989	26 Dec 1990

Displacement, tonnes: 7,722 surfaced; 9,246 dived
Dimensions, metres (feet): 111 × 14.2 × 8.8 *(364.2 × 46.6 × 28.9)*
Speed, knots: 10 surfaced; 32 dived
Complement: 61 (31 officers)

Machinery: Nuclear; 1 VM-5 PWR; 190 MW; 1 GT3A turbine; 47,500 hp(m) *(70 MW)*; 2 emergency motors; 2,004 hp(m) *(1.5 MW)*; 1 shaft; 2 spinners; 1,006 hp(m) *(740 kW)*
Missiles: SLCM: Raduga SS-N-21 Sampson (Granat) fired from 21 in *(533 mm)* tubes; land-attack; inertial/terrain-following to 2,500 km *(1,350 n miles)* at 0.7 Mach; warhead 410 kg HE or nuclear 200 kT. CEP 150 m. Flies at a height of about 200 m.
SAM: SA-N-5/8 Strela portable launcher; 12 missiles.
A/S: Novator SS-N-15 Starfish (Tsakra) fired from 53 cm tubes; inertial flight to 45 km *(24.3 n miles)*; warhead nuclear 200 kT or Type 40 torpedo. Novator SS-N-16 Stallion fired from 65 cm tubes; inertial flight to 100 km *(54 n miles)*; payload nuclear 200 kT (Vodopad) or Type 40 torpedo (Veder).
Torpedoes: 4 — 25.6 in *(650 mm)* and 4 — 21 in *(533 mm)* tubes. Combination of 65 and 53 cm torpedoes. Total of 40 weapons.
Mines: 42 in lieu of torpedoes.
Electronic countermeasures: ESM: Rim Hat; intercept. Park Lamp D/F.
Radars: Surface search: Snoop Pair with back-to-back ESM aerial.
Sonars: Shark Gill; hull-mounted; passive/active search and attack; low/medium frequency. Shark Rib flank array; passive; low frequency. Mouse Roar; hull-mounted; active attack; high frequency. Skat 3 towed array; passive; very low frequency.

Programmes: A third of class K 536 *Mars*, was scrapped before completion in July 1992.
Structure: Titanium hull. The towed communications buoy has been recessed. A 10 point environmental sensor is fitted at the front end of the fin. The standoff distance between hulls is considerable and has obvious advantages for radiated noise reduction and damage resistance. Diving depth, 2,460 ft *(750 m)*. Numbers and sizes of torpedo tubes are uncertain with different figures given by Russian sources. There are seven watertight compartments.
Operational: Based in the Northern Fleet, at Ara Guba. K 534 completed a refit/refuel in May 2008. She reportedly conducted a patrol off the US East Coast in October 2012. The operational state of K 336 is doubtful. It was reported in March 2013 that both boats are to be refitted. This work is likely to start in about 2017 when refits of the Sierra I class have been completed.

PSKOV (with KOSTROMA (Sierra I))   6/2002, Lemachko Collection / 0570928

NIZHNY NOVGOROD   3/2010, Lemachko Collection / 1406116

SIERRA II   8/1998 / 0050009

6 OSCAR II (ANTYEY) (PROJECT 949A) (SSGN)

Name	No	Builders	Laid down	Launched	Commissioned
VORONEZH (K 119)	812	Severodvinsk Shipyard	25 Feb 1986	16 Dec 1988	29 Dec 1988
SMOLENSK (K 410)	816	Severodvinsk Shipyard	9 Dec 1986	20 Jan 1990	22 Dec 1990
TVER (K 456) (ex-*Vilyachinsk*)	920	Severodvinsk Shipyard	9 Feb 1988	28 Jun 1991	18 Aug 1992
OREL (K 266) (ex-*Severodvinsk*)	847	Severodvinsk Shipyard	19 Jan 1989	22 May 1992	30 Dec 1992
OMSK (K 186)	947	Severodvinsk Shipyard	13 Jul 1989	8 May 1993	15 Dec 1993
TOMSK (K 150)	902	Severodvinsk Shipyard	27 Aug 1991	20 Jul 1996	30 Dec 1996

Displacement, tonnes: 14,123 surfaced; 18,594 dived
Dimensions, metres (feet): 154 × 18.2 × 9 *(505.2 × 59.7 × 29.5)*
Speed, knots: 15 surfaced; 28 dived
Complement: 107 (48 officers)

Machinery: Nuclear; 2 VM-5 PWR; 380 MW; 2 GT3A turbines; 98,000 hp(m) *(72 MW)*; 2 shafts; 2 spinners
Missiles: SSM: 24 Chelomey SS-N-19 Shipwreck (Granit); inertial with command update guidance; active radar homing to 450 km *(243 n miles)* at 2.5 Mach; warhead 750 kg HE or 500 kT nuclear. Novator Alfa SS-N-27 may be carried in due course.
A/S: Novator SS-N-15 Starfish (Tsakra) fired from 53 cm tubes; inertial flight to 45 km *(24.3 n miles)*; warhead nuclear 200 kT or Type 40 torpedo. Novator SS-N-16 Stallion fired from 65 cm tubes; inertial flight to 100 km *(54 n miles)*; payload nuclear 200 kT (Vodopad) or Type 40 torpedo (Veder).
Torpedoes: 4—21 in *(533 mm)* and 2—26 in *(650 mm)* tubes. Combination of 65 and 53 cm torpedoes. Total of 28 weapons including tube-launched A/S missiles.
Mines: 32 can be carried.
Electronic countermeasures: ESM: Rim Hat; intercept.
Radars: Surface search: Snoop Pair or Snoop Half; I-band.
Sonars: Shark Gill; hull-mounted; passive/active search and attack; low/medium frequency. Shark Rib flank array; passive; low frequency. Mouse Roar; hull-mounted; active attack; high frequency. Pelamida towed array; passive search; very low frequency.
Weapon control systems: Punch Bowl for third party targeting.

Programmes: Building of a class of 14 began in 1978. Two Oscar Is and 11 Oscar IIs were completed. Work on the 12th Oscar II (K 139, *Belgorod*) was thought to have stopped but it was announced by the Defence Minister on 16 July 2004 that the boat would be completed. Reports in 2013 suggest that the boat is being completed as an auxiliary SSN, capable of acting as a mother-ship for underwater operations. Construction was halted on the 13th and 14th boats.
Modernisation: Replacement of the SS-N-19 missiles is reported to be under consideration. Refit of *Voronezh* was started at Zvezdochka Shipyard in 2010 and was completed in 2012. She was followed into refit by K 410. K 150 started refit at Komsomolsk ini 2010 and is to return to service in 2014.
Structure: SSM missile tubes are in banks of 12 either side and external to the 8.5 m diameter pressure hull; they are inclined at 40° with one hatch covering each pair, the whole resulting in the very large beam. The position of the missile tubes provides a large gap of some 4 m between the outer and inner hulls. Diving depth, 1,000 ft *(300 m)* although 2,000 ft *(600 m)* is claimed. There are 10 watertight compartments.
Operational: ELF/VLF communications buoy. All have a tube on the rudder fin as in Delta IV which is used for dispensing a thin line towed sonar array. Pert Spring SATCOM. K 119, K 410 and K 266 are based at Litsa South in the Northern Fleet and K 186, K 150 and K 456 at Tarya Bay in the Pacific. In 1999 one Northern Fleet unit deployed for the first Russian SSGN patrol in the Mediterranean for 10 years. At the same time a Pacific Fleet unit sailed to the western seaboard of the United States. The two Oscar Is (K 206 and K 525) have been scrapped. K 148 and K 132 are laid up awaiting disposal and K 141 (*Kursk*) sunk as the result of an internal weapon explosion on 12 August 2000. The submarine was raised in late 2001 and broken up ashore. K 173 was decommissioned in 2009 and K 442 is laid up.

OSCAR II
11/2001, Ships of the World / 0528392

OREL
6/2005, Lemachko Collection / 1159844

Submarines < **Russian Federation** 677

OMSK 6/2010 / 1406764

OSCAR II 11/2006, Ships of the World / 1159978

TOMSK 6/2005, Lemachko Collection / 1159843

11 AKULA (SCHUKA-B) CLASS (PROJECT 971/971U/09710) (SSN)

Name	No	Builders	Laid down	Launched	Commissioned
KASHALOT (K 322)	985	Komsomolsk Shipyard	5 Sep 1986	18 Jul 1987	30 Dec 1988
BRATSK (K 391)	990	Komsomolsk Shipyard	23 Feb 1988	14 Apr 1989	29 Dec 1989
MAGADAN (K 331) (ex-*Narwhal*)	997	Komsomolsk Shipyard	28 Dec 1989	23 Jun 1990	23 Dec 1990
PANTERA (K 317)	878	Severodvinsk Shipyard	6 Nov 1986	21 May 1990	27 Dec 1990
VOLK (K 461)	867	Severodvinsk Shipyard	14 Nov 1987	11 Jun 1991	30 Dec 1991
KUZBASS (K 419) (ex-*Morzh*)	951	Komsomolsk Shipyard	28 Jul 1991	18 May 1992	31 Dec 1992
LEOPARD (K 328)	872	Severodvinsk Shipyard	26 Oct 1988	28 Jul 1992	30 Dec 1992
TIGR (K 154)	853	Severodvinsk Shipyard	10 Sep 1989	10 Jun 1993	29 Dec 1993
SAMARA (K 295) (ex-*Drakon*)	970	Komsomolsk Shipyard	7 Nov 1993	15 Jul 1994	17 Jul 1995
VEPR (K 157)	890	Severodvinsk Shipyard	13 Jul 1990	10 Dec 1994	25 Nov 1995
GEPARD (K 335)	835	Severodvinsk Shipyard	23 Sep 1991	17 Sep 1999	4 Dec 2001

Displacement, tonnes: 7,620 surfaced; 9,246 (853, 867, 872, 878, 951, 970, 985, 990, 997), 9,652 (835, 890) dived
Dimensions, metres (feet): 110 oa; 103 wl × 14 × 10.4 *(360.9; 337.9 × 45.9 × 34.1)*
Speed, knots: 10 surfaced; 28 dived
Complement: 62 (31 officers)

Machinery: Nuclear; 1 VM-5 PWR; 190 MW; 2 GT3A turbines; 47,600 hp(m) *(35 MW)*; 2 emergency propulsion motors; 750 hp(m) *(552 kW)*; 1 shaft; 2 spinners; 1,006 hp(m) *(740 kW)*
Missiles: SLCM/SSM: Reduga SS-N-21 Sampson (Granat) fired from 21 in *(533 mm)* tubes; land-attack; inertial/terrain-following to 2,500 km *(1,350 n miles)* at 0.7 Mach; warhead 410 kg HE or nuclear 200 kT. CEP 150 m. Flies at a height of about 200 m. Novator Alfa Club S SS-N-27 (3M 54 anti-ship); active radar homing to 220 km *(120 n miles)* at 0.8 Mach (cruise) and 2.9 Mach (attack); warhead 400 kg.
SAM: SA-N-5/8 Strela portable launcher. 18 missiles.
A/S: Novator SS-N-15 Starfish (Tsakra) fired from 53 cm tubes; inertial flight to 45 km *(24.3 n miles)*; warhead nuclear 200 kT or Type 40 torpedo. Novator SS-N-16 Stallion fired from 650 mm tubes; inertial flight to 100 km *(54 n miles)*; payload nuclear 200 kT (Vodopad) or Type 40 torpedo (Veder).
Torpedoes: 4—21 in *(533 mm)* and 4—25.6 in *(650 mm)* tubes. Combination of 53 and 65 cm torpedoes. Tube liners can be used to reduce the larger diameter tubes to 533 mm. Total of 40 weapons. In addition the Improved Akulas and Akula IIs have six additional 533 mm external tubes in the upper bow area.
Electronic countermeasures: ESM: Rim Hat; intercept.
Radars: Surface search: Snoop Pair or Snoop Half with back-to-back aerials on same mast as ESM.
Sonars: Shark Gill (Skat MGK 503); hull-mounted; passive/active search and attack; low/medium frequency. Mouse Roar; hull-mounted; active attack; high frequency. Skat 3 towed array; passive; very low frequency.

Programmes: Malakhit design. From K 461 onwards, the Akula Is were 'improved'. K 157 was the first Akula II to complete and she was followed by K 335. The fate of a third Akula II (K 337 Cougar) has not been confirmed but it is believed that the bow and stern pressure sections have been incorporated in the new SSBN *Yuri Dolgoruky*. Akula I K 152 *Nerpa* had been building for nearly 20 years at Komsomolsk before being launched in 2006. The submarine is being leased for 10 years to the Indian Navy from 2011. She is named *Chakra* in Indian service.
Modernisation: An overhaul programme for the whole class is in progress. K 328 is reported to have begun a refit at Zvezdochka Shipyard in 2011 and she is to be followed by K 391 and K 295 from the Pacific Fleet.
Structure: The very long fin is particularly notable. Has the same broad hull as Sierra and has reduced radiated noise levels by comparison with Victor III of which she is the traditional follow-on design. A number of prominent non-acoustic sensors appear on the fin leading-edge and on the forward casing in the later Akulas. The engineering standards around the bridge and casing are noticeably of a higher quality than other classes. The design has been incrementally improved with reduced noise levels, boundary layer suppression and active noise cancellation reported in the later units. The Improved hulls have an additional six external torpedo tubes and the two Akula IIs have been lengthened by 3.7 m to incorporate further noise reduction developments. There are six watertight compartments. Operational diving depth, 1,476 ft *(450 m)*.
Operational: Pert Spring SATCOM. K 461, K 328, K 154, K 335, K 157, and K 317 are based in the Northern Fleet at Saida Guba. K 331, K 419, K 295, and K 322 are based in the Pacific Fleet at Tarya Bay. These submarines are the core units of the Russian SSN force. *Vepr* visited Brest in September 2004, the first visit by a Russian nuclear submarine to a foreign port. K 317 *Pantera* had a serious fire in November 2006.

AKULA CLASS   6/2008 / 1406765

VEPR (Akula II)   9/2004, B Prézelin / 1042291

KUZBASS   6/2007, *Ships of the World* / 1305156

MAGADAN — 7/2009, Lemachko Collection / 1305547

1 + 5 SEVERODVINSK (YASEN) CLASS (PROJECT 885/885M) (SSN/SSGN)

Name	No	Builders	Laid down	Launched	Commissioned
SEVERODVINSK (K 329)	–	Sevmash Shipyard, Severodvinsk	21 Dec 1993	15 Jun 2010	30 Dec 2013
KAZAN (K 561)	–	Sevmash Shipyard, Severodvinsk	24 Jul 2009	2014	2015
NOVOSIBIRSK (K 573)	–	Sevmash Shipyard, Severodvinsk	26 Jul 2013	2018	2019
–	–	Sevmash Shipyard, Severodvinsk	2014	2018	2019

Displacement, tonnes: 9,500 surfaced; 11,800 dived
Dimensions, metres (feet): 133 × 11.5 × 8.4 (436.4 × 37.7 × 27.6)
Speed, knots: 17 surfaced; 28 dived
Complement: 90 (30 officers)

Machinery: Nuclear; 1 PWR; 195 MW; 2 GT3A turbines; 43,000 hp(m) (31.6 MW); 1 shaft; pump-jet propulsor; 2 spinners
Missiles: SLCM/SSM: 8 VLS launchers in after casing. Total of 24 missiles. A mixture of SS-N-27 (3M54) anti-ship missiles and SS-N-30 Kaliber (3M14) land-attack missiles.
A/S: SS-N-15 Starfish. Fired from torpedo tubes.
Torpedoes: 8—21 in (533 mm) tubes. Inclined outwards. Total of about 30 weapons.
Electronic countermeasures: ESM: Rim Hat (Nakat M); intercept.
Radars: Surface search: I-band.
Sonars: Irtysh Amfora system includes bow-mounted spherical array; passive/active search and attack; low frequency. Flank and towed arrays; passive; very low frequency.

Programmes: Malakhit design. Confirmed building in 1993. Reported plans were for seven of the class to replace the Victor III class. While it was initially reported that these were to be multipurpose SSNs derived from the Akula II class, delays in the programme suggest that there has been considerable scope for re-design and/or technical upgrade. Building of Kazan began in 2009 and a contract for four further Project 885M boats was signed in 2011.
Structure: The reactor is reported to be of a new design with core-life of 25 years.
Operational: Sea trials of Severodvinsk started in September 2011. She conducted a test firing of Kaliber land-attack missile in the White Sea in November 2012. Acceptance into service is expected in 2014, providing reported problems with propeller shaft bearings can be rectified.

SEVERODVINSK
6/2010, National Defense
1366987

4 VICTOR III (SCHUKA) CLASS (PROJECT 671 RTMK) (SSN)

Name	No	Builders	Laid down	Launched	Commissioned
PETROZAVODSK (B 388) (ex-*Snezhnogorsk*)	671	Admiralty Yard, Leningrad	8 May 1987	3 Jun 1988	30 Nov 1988
OBNINSK (B 138)	618	Admiralty Yard, Leningrad	7 Dec 1988	5 Aug 1989	10 May 1990
DANIL MOSKOVSKIY (B 414)	684	Admiralty Yard, Leningrad	1 Dec 1988	31 Aug 1990	30 Dec 1990
TAMBOV (B 448)	661	Admiralty Yard, Leningrad	31 Jan 1991	17 Oct 1991	24 Sep 1992

Displacement, tonnes: 4,928 surfaced; 6,401 dived
Dimensions, metres (feet): 107 × 10.6 × 7.4 *(351 × 34.8 × 24.3)*
Speed, knots: 10 surfaced; 30 dived
Complement: 98 (17 officers)

Machinery: Nuclear; 2 VM-4 PWR; 150 MW; 2 turbines; 31,000 hp(m) *(22.7 MW)*; 1 shaft; 2 spinners; 1,020 hp(m) *(750 kW)*
Missiles: SLCM: Raduga SS-N-21 Sampson (Granat) fired from 21 in *(533 mm)* tubes; land-attack; inertial/terrain-following to 2,500 km *(1,350 n miles)* at 0.7 Mach; warhead 410 kg HE or nuclear 200 kT. CEP 150 m or Novator Alfa SS-N-27 (3M 54 anti-ship); active radar homing to 200 km *(120 n miles)* at 0.8 Mach (cruise) and 2.9 Mach (attack); warhead 400 kg.
A/S: Novator SS-N-15 Starfish (Tsakra) fired from 53 cm tubes; inertial flight to 45 km *(24.3 n miles)*; Type 40 torpedo. Novator SS-N-16 Stallion fired from 65 cm tubes; inertial flight to 100 km *(54 n miles)*; payload nuclear 200 kT (Vodopad) or Type 40 torpedo (Veder).
Torpedoes: 4—21 in *(533 mm)* and 2—25.6 in *(650 mm)* tubes. Combination of 53 and 65 cm torpedoes. Can carry up to 24 weapons. Liners can be used to reduce 650 mm tubes to 533 mm.
Mines: Can carry 36 in lieu of torpedoes.
Electronic countermeasures: ESM: Brick Group (Brick Spit and Brick Pulp); intercept. Park Lamp D/F.
Radars: Surface search: Snoop Tray 2; I-band.
Sonars: Shark Gill; hull-mounted; passive/active search and attack; low/medium frequency. Shark Rib flank array; passive; low frequency. Mouse Roar; hull-mounted; active attack; high frequency. Scat 3 towed array; passive; very low frequency.

Programmes: The first of class was completed at Komsomolsk in 1978. With construction also being carried out at Admiralty Yard, Leningrad, there was a very rapid building programme up to the end of 1984. Construction then continued only at Leningrad and at a rate of about one per year which terminated in 1991. The last of the class of 26 boats completed sea trials in October 1992. Of these, the first 21 hulls were designated Type 671RTM. The final five hulls were designated Type 671RTMK to reflect modifications to fire cruise missiles. The last four of these are in service.

VICTOR III *6/2000, Lemachko Collection / 0126230*

Structure: The streamlined pod on the stern fin is a towed sonar array dispenser. Water environment sensors are mounted at the front of the fin and on the forward casing as in the Akula and Sierra classes. Diving depth, 1,300 ft *(400 m)*.
Operational: VLF communications buoy. VHF/UHF aerials. Navigation equipment includes SINS and SATNAV. Pert Spring SATCOM. Kremmny 2 IFF. Much improved acoustic quietening puts the radiated noise levels at the upper limits of the USN Los Angeles class. Four remaining operational units are based in the Northern Fleet at Litsa South or Ara Guba although they rarely go to sea and the operational status of B 448 and B 138 is doubtful. They are probably awaiting disposal.

DANIL MOSKOVSKIY *7/2010, Lemachko Collection / 1406117*

DANIL MOSKOVSKIY *7/2004 / 1042331*

Patrol Submarines

1 + 2 LADA CLASS (PROJECT 677) (SSK)

Name	No	Builders	Laid down	Launched	Commissioned
SANKT PETERBURG (B 585)	477	Admiralty Yard, St Petersburg	26 Dec 1997	28 Oct 2004	8 May 2010
KRONSHTADT (B 586)	–	Admiralty Yard, St Petersburg	28 Jul 2005	2014	2015
SEVASTOPOL (B 587)	–	Admiralty Yard, St Petersburg	10 Nov 2006	2014	2016

Displacement, tonnes: 1,793 surfaced; 2,693 dived
Dimensions, metres (feet): 66.8 × 7.1 × 4.4 *(219.2 × 23.3 × 14.4)*
Speed, knots: 10 surfaced; 21 dived
Range, n miles: 6,000 at 7 kt snorting
Complement: 37

Machinery: Diesel-electric; 2 diesel generators; 3,400 hp(m) *(2.5 MW)*; 1 motor; 5,576 hp(m) *(4.1 MW)*; 1 shaft
Missiles: SLCM: Novator Alfa Klub SS-N-27 (3M-54 anti-ship missiles); active radar homing to 220 km *(120 n miles)* at 0.8 Mach (cruise) and 2.9 Mach (attack); warhead 450 kg. Novator Klub SS-N-30 (3M14) land-attack; inertial guidance to 275 km *(148 n miles)* at 0.8 Mach; warhead 450 kg.
Torpedoes: 6 – 21 in *(533 mm)* tubes. 18 weapons.
Mines: In lieu of torpedoes.
Electronic countermeasures: ESM: Intercept.
Radars: Surface search: I-band.
Sonars: Conformal bow and flank arrays; active/passive; medium frequency. Towed array (low frequency).

Programmes: Work on the first of class began in 1996 and construction started in St Petersburg in 1997. Construction of the second and third of class was suspended during the sea trials of the first of class and future of the programme was uncertain for a few years. However, it was announced by the Commander of the Navy in July 2012 that building work on these two boats was to resume, although the timetable has not been published. The export version of the submarine is known as the Amur class.
Structure: The first Russian single-hulled submarine, built to a Rubin design based on the 'Amur 1650'. A fuel cell plug (for AIP) of about 12 m can be inserted to allow installation of AIP although this is unlikely in the near future. Diving depth: 820 ft *(250 m)*. A non-hull penetrating optronic periscope supplied by Elektropribor, is fitted.
Operational: Sea trials of *Saint Petersburg* started in the Baltic on 29 November 2005 and a second round of trials in August 2006. Thereafter, sea trials continued until 2012. Apparent delays in achieving operational status suggest there have been technical problems and design modifications, which are likely to be reflected in successor boats.

SANKT PETERBURG — *6/2010, Lemachko Collection* / 1366475

SANKT PETERBURG — *7/2011, Erik Laursen* / 1406767

SANKT PETERBURG — *7/2011, Maritime Photographic* / 1406766

682 Russian Federation > Submarines

18 + 6 KILO CLASS (PROJECT 877K/877M/636) (SSK)

Name	No	Builders	Laid down	Launched	Commissioned	F/S
CHITA (ex-*B 260*)	504	Komsomolsk Shipyard	22 Feb 1981	23 Aug 1981	30 Dec 1981	Pacific
VYBORG (ex-*B 227*)	469	Komsomolsk Shipyard	23 Feb 1982	16 Sep 1982	23 Feb 1983	Baltic
NOVOSIBIRSK (ex-*B 401*)	440	Nizhny Novgorod	6 Oct 1982	15 Mar 1984	30 Oct 1984	Northern
VOLOGDA (ex-*B 402*)	405	Nizhny Novgorod	24 Aug 1983	23 Sep 1984	30 Dec 1984	Northern
DIMITROV (ex-*B 806*)	487	Nizhny Novgorod	15 Oct 1984	30 Apr 1986	25 Sep 1986	Baltic
SYVATITEL NIKOLAY CHUDOTVORETS (ex-*B 445*)	531	Komsomolsk Shipyard	21 Mar 1987	26 Sep 1987	30 Jan 1988	Pacific
JAROSLAVL (ex-*B 808*)	425	Nizhny Novgorod	29 Sep 1986	30 Jul 1988	27 Dec 1988	Northern
B 394	522	Komsomolsk Shipyard	15 Apr 1988	3 Sep 1988	30 Dec 1988	Pacific
KALUGA (ex-*B 800*)	468	Nizhny Novgorod	5 Mar 1987	7 May 1989	30 Sep 1989	Northern
UST-KAMSHATS (ex-*B 464*)	547	Komsomolsk Shipyard	26 May 1989	23 Sep 1989	30 Jan 1990	Pacific
MAGNITOGORSK (ex-*B 471*)	409	Nizhny Novgorod	26 Oct 1988	22 Sep 1990	30 Dec 1990	Northern
UST-BOLSHERETSK (ex-*B 494*)	549	Komsomolsk Shipyard	5 May 1990	4 Oct 1990	30 Dec 1990	Pacific
VLADIKAVKAZ (ex-*B 459*)	431	Nizhny Novgorod	25 Feb 1988	24 Sep 1990	30 Dec 1990	Northern
ALROSA (ex-*B 871*)	554	Nizhny Novgorod	17 May 1988	Sep 1989	30 Dec 1990	Black
LIPETSK (ex-*B 177*)	429	Nizhny Novgorod	3 Nov 1989	27 Jul 1991	30 Dec 1991	Northern
B 187	529	Komsomolsk Shipyard	7 May 1991	5 Oct 1991	30 Dec 1991	Pacific
KRASNOKAMENSK (ex-*B 190*)	521	Komsomolsk Shipyard	8 May 1992	25 Sep 1992	30 Dec 1992	Pacific
MOGOCHEY (ex-*B 345*)	507	Komsomolsk Shipyard	22 Apr 1993	6 Oct 1993	22 Jan 1994	Pacific
NOVOROSSIYSK (ex-*B 261*)	–	Admiralty Shipyard, St Petersburg	20 Aug 2010	28 Nov 2013	Jul 2014	Black
ROSTOV NA DONU (ex-*B 237*)	–	Admiralty Shipyard, St Petersburg	21 Nov 2011	2014	2015	Black
STARY OSKOL (ex-*B 262*)	–	Admiralty Shipyard, St Petersburg	17 Aug 2012	2015	2016	Black
KRASNODAR	–	Admiralty Shipyard, St Petersburg	20 Feb 2014	2016	2017	Black

Displacement, tonnes: 2,362 surfaced; 3,125 dived
Dimensions, metres (feet): 72.6, 73.8 *(429, 431, 521, 529, Novorossiysk)* × 9.9 × 6.6 *(238.2, 242.1 × 32.5 × 21.7)*
Speed, knots: 10 surfaced; 9 snorting; 17 dived
Range, n miles: 6,000 at 7 kt snorting; 400 at 3 kt dived
Complement: 52 (13 officers)

Machinery: Diesel-electric; Type 4-2DL-42M 2 diesels (Type 4-2AA-42M in Project 636); 3,650 hp(m) *(2.68 MW)*; 2 generators; 1 motor; 5,900 hp(m) *(4.34 MW)*; 1 shaft; 2 auxiliary MT-168 motors; 204 hp(m) *(150 kW)*; 1 economic speed motor; 130 hp(m) *(95 kW)*
Missiles: SSM: Novator Alfa SS-N-27 may be fitted in due course.
SAM: 6–8 SA-N-5/8; IR homing from 600 to 6,000 m at 1.65 Mach; warhead 2 kg; portable launcher stowed in a well in the fin between snort and W/T masts.
Torpedoes: 6—21 in *(533 mm)* tubes. 18 combinations of 53 cm torpedoes. USET-80 is wire-guided in the 4B version (from 2 tubes).
Mines: 24 in lieu of torpedoes.
Electronic countermeasures: ESM: Squid Head or Brick Pulp; radar warning. Quad Loop D/F.
Radars: Surface search: Snoop Tray (MRP-25); I-band.
Sonars: Shark Teeth/Shark Fin (MGK-400); hull-mounted; passive/active search and attack; medium frequency. Mouse Roar; hull-mounted; active attack; high frequency.
Weapon control systems: MVU-110EM or MVU-119EM Murena torpedo fire-control system.

Programmes: Also known as the Vashavyanka class, first launched in 1979 at Komsomolsk and commissioned 12 September 1980. Subsequent construction also at Nizhny Novgorod. A total of 24 were built for Russia (1980–94) of which six were of the improved Project 636 variant. The building programme was re-activated in 2010 when a further Project 636 was laid down. Further boats were laid down in 2011 and 2012 and three additional submarines are planned to enter service by 2018.
Structure: Had a better hull form than the Tango class but was nevertheless considered fairly basic by comparison with contemporary western designs. Diving depth 790 ft *(240 m)* normal. Battery has a 9,700 kW/h capacity. The basic 'Kilo' was the Project 877; 877K has an improved fire-control system and 877M includes wire-guided torpedoes from two tubes. Project 636 is an improved design with uprated diesels, a propulsion motor rotating at half the speed (250 rpm), higher standards of noise reduction and an automated combat information system capable of providing simultaneous fire-control data on five targets. Pressure hull length is 51.8 m *(170 ft)* or 53 m *(174 ft)* for Project 636. Foreplanes on the hull are just forward of the fin. Project 636 can be identified by a vertical cut off to the after casing. B 871 has been fitted with a pump jet propulsor.
Operational: With a reserve of buoyancy of 32 per cent and a heavily compartmented pressure hull, this class is capable of being holed and still surviving. B 401, B 402, B 808, B 459, B 471, B 800, and B 177 are based in the Northern Fleet, B 260, B 445, B 494, B 190, B 345, B 187, B 464, and B 394 are based in the Pacific, B 806 and B 227 in the Baltic and B 871 in the Black Sea. The latest batch of Project 636 boats are to be based in the Black Sea. Russian made batteries have been a source of problems in warm water operations. *Alrosa* took part in the submarine rescue exercise Bold Monarch in 2011.
Sales: Exports of Project 877 have been to Poland (one), Romania (one), India (ten), Algeria (two), Iran (three) and China (two). The only exports of Project 636 have been to China (two). A further eight were ordered by China in 2002. Export versions have the letter E after the project number.

MOGOCHEY

7/2009, Lemachko Collection / 1305546

ALROSA — 9/2012, C D Yaylali / 1483532

CHITA — 7/2010, Lemachko Collection / 1406118

MAGNITOGORSK — 7/2012, Hartmut Ehlers / 1483531

Auxiliary Submarines

Notes: (1) There are a number of swimmer delivery vessels (SDV) in service including Siren (three-man) and Triton, Sever and Elbrus types.
(2) A new auxiliary submarine (SSAN) was launched at Severodvinsk Shipyard on 6 August 2003. Nicknamed 'Losharik', she is likely to be employed for scientific research and is reported to be similar to but not the same as the Uniform class. The internal construction comprises a number of linked spherical modules to increase diving performance. Her length is approximately 60 m. Known as Project 10831. It has a pennant number of AS 31 (previously reported as AS 12) and is reported to have become operational in 2007. Construction of a second Project 10831 unit began in December 2012.
(3) The Delta IV class BS 64, which was decommissioned in about 2002, is being converted (Project 09787) to an auxiliary submarine role, to replace *Orienburg* in due course. The conversion is reported to involve insertion of a 43 m (approximately) section from the decommissioned Yankee Stretch submarine to replace the missile tube section. This will increase the overall length of the boat to about 170 m. Completion of the project is expected in 2015. She is to be named *Podmoskovye*.
(4) Project 09852 is for the construction of a further research submarine. Designed by the Rubin Design Bureau, she is to be based on the Oscar-class submarine. This is reported to be the unfinished 12th Oscar II *Belgorod* K 139.

'LOSHARIK'   6/2010 / 1406768

1 DELTA III STRETCH (PROJECT 667 BDR) (SSAN)

Name	No	Builders	Laid down	Launched	Commissioned
ORIENBURG (BS 136, K 129)	656	Severodvinsk Shipyard	9 Apr 1979	15 Apr 1981	5 Nov 1981

Dimensions, metres (feet): 163 × 12 × 8.7 (534.8 × 39.4 × 28.5)
Speed, knots: 14 surfaced; 24 dived
Complement: 130 (40 officers)

Machinery: Nuclear: 2 VM-4 PWR; 180 MW; 2 GT 3A-635 turbines; 37,400 hp(m) (27.5 MW); 2 emergency motors; 612 hp(m) (450 kW); 2 shafts

Torpedoes: 4—21 in (533 mm) and 2—400 mm tubes.
Electronic countermeasures: ESM: Brick Pulp/Group; radar warning.
Radars: Surface search: Snoop Tray; I-band.
Sonars: Shark Teeth; hull mounted; active/passive search; low/medium frequency. Shark Hide; flank array; passive low frequency. Mouse Roar; hull mounted; active high frequency.

Comment: Originally launched in 1981, this former SSBN has been converted by replacing the central section with a 43 m plug, extending the overall hull length by 3 m. The submarine was reported to have returned to service in 2003 when she replaced the Yankee Stretch as the Paltus mothership. Based in the Northern Fleet.

ORIENBURG   6/2010* / 1530964

Submarines ‹ **Russian Federation** 685

2 PALTUS (PROJECT 1851) CLASS (SSAN/SSA)

AS 35   AS 21

Displacement, tonnes: 742 dived
Dimensions, metres (feet): 53 × 3.8 × 4.2
(173.9 × 12.5 × 13.8)
Speed, knots: 6 dived
Complement: 14
Machinery: Nuclear; 1 reactor; 10 MW; 1 shaft; ducted thrusters

Comment: The first was launched at Sudomekh, St Petersburg on 29 April 1991, a second of class on 29 September 1994 and a third was started but not completed. This is a follow-on to the single 44 m 520 ton X-Ray (AS 33) class which was first seen in 1984 but which has been reported non-operational. Paltus probably owes much to the USN NR 1. Paltus is associated with the Delta III Stretch SSAN which acts as a mother ship for special operations. Titanium hulled and very deep diving to 1,000 m (3,280 ft). Paltus based in the Northern Fleet at Olenya Guba.

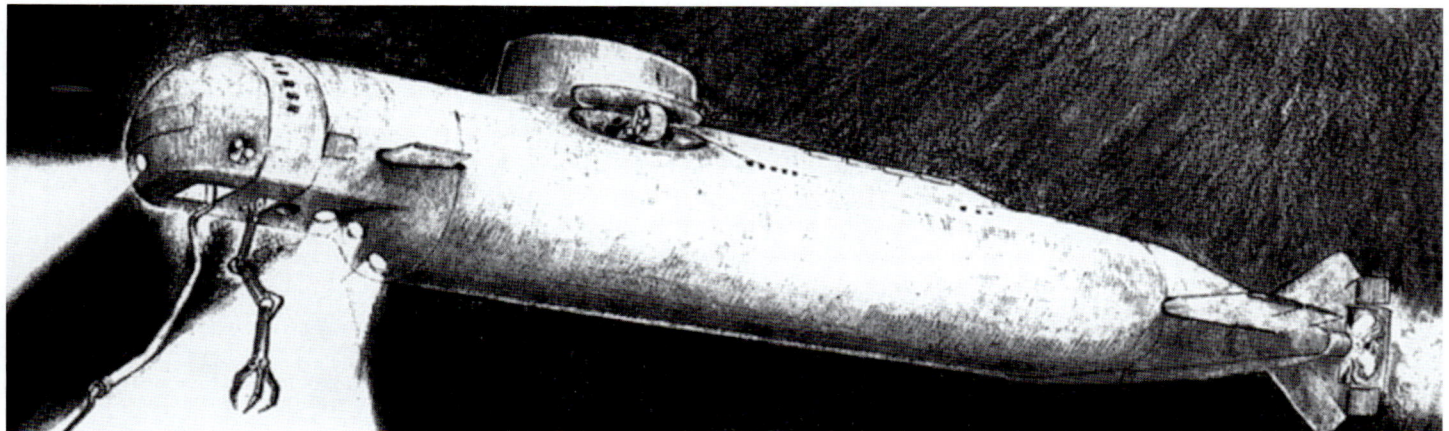

PALTUS (artist's impression)   1994 / 0506318

3 UNIFORM (KACHALOT) CLASS (PROJECT 1910) (SSAN)

Name	No	Builders	Laid down	Launched	Commissioned
AS 13	–	Sudomekh, Leningrad	20 Oct 1977	25 Nov 1982	31 Dec 1986
AS 15	–	Sudomekh, Leningrad	23 Feb 1983	29 Apr 1988	30 Dec 1991
AS 23	–	Sudomekh, St Petersburg	28 Nov 1988	23 Nov 1993	16 Dec 1994

Displacement, tonnes: 1,362 surfaced; 1,605 dived
Dimensions, metres (feet): 69 × 7.0 × 5.2
(226.4 × 23.0 × 17.1)
Speed, knots: 10 surfaced; 28 dived
Complement: 36
Machinery: Nuclear; 1 PWR; 15 MW; 2 turbines; 10,000 hp(m) (7.35 MW); 1 shaft; 2 thrusters

Radars: Navigation: Snoop Slab; I-band.

Comment: Research and development nuclear-powered submarines. Have single hulls and 'wheel' arches either side of the fin which house side thrusters. These are titanium hulled and very deep diving submarines (possibly down to 700 m (2,300 ft)), based in the Northern Fleet at Olenya Guba, and are used mainly for ocean bed operations. Plans to build more of the class were thought to have been shelved.

UNIFORM   6/2004, Lemachko Collection / 1159848

1 PROJECT 20120 EXPERIMENTAL SUBMARINE (SSA)

Name	No	Builders	Laid down	Launched	Commissioned
SAROV (B-90)	–	Nizhny Novgorod; Severodvinsk Shipyard	1989	14 Dec 2007	7 Aug 2008

Displacement, tonnes: 4,064 dived (est.)
Dimensions, metres (feet): 98.6 × 10.6 × ?
(323.5 × 34.8 × ?)
Speed, knots: 10 surfaced; 17 dived
Complement: 52

Machinery: Diesel-electric
Torpedoes: To be announced.
Mines: To be announced.

Electronic countermeasures: ESM: To be announced.
Radars: Surface search: To be announced.
Sonars: To be announced.

Comment: It is reported that the design of the submarine was developed by Rubin in about 1989, that construction was initiated at Nizhny Novgorod and that, following transfer in about 2003, the boat was completed at the Sevmash Shipyard at Severodvinsk. Although initially thought to be equipped with a small nuclear reactor to provide an air-independent capability, this is now believed unlikely. Details of the submarine have not been confirmed, and dimensions are approximate. There is a raised area on the upper casing aft of the fin and also bulges on the side of the forward casing. It is believed that the principal role of the submarine is to act as test bed for the development and testing of unmanned submersibles, weapons and underwater equipment.

SAROV   7/2009, Lemachko Collection / 1305461

© 2014 IHS   IHS Jane's Fighting Ships 2014-2015

686 Russian Federation > Aircraft carriers

AIRCRAFT CARRIERS

Notes: (1) Of the former aircraft carriers of the Kiev class, *Kiev* was sold to China for scrap in 2000; *Minsk* and *Novorossiysk* were sold to a South Korean Corporation in 1994. *Minsk* later became a tourist attraction in Shenzen, China, while *Novorossiysk* was scrapped in India. *Admiral Gorshkov* (ex-*Baku*) was sold to the Indian Navy and was commissioned as *Vikramaditya* on 16 November 2013.
(2) The requirement for a new class of four aircraft carriers was announced in mid-2005. This was re-stated by the Commander-in-Chief in 2008. While work has almost certainly begun on the project funding is proving to be an obstacle and such ships are not included in the 2011–20 Military Plan.

1 KUZNETSOV (OREL) CLASS (PROJECT 1143.5/6) (CVGM)

Name	No	Builders	Laid down	Launched	Commissioned
ADMIRAL KUZNETSOV (ex-*Tblisi*; ex-*Leonid Brezhnev*)	063	Nikolayev South	1 Apr 1982	16 Dec 1985	25 Dec 1990

Displacement, tonnes: 46,637 standard; 59,439 full load
Dimensions, feet (metres): 302.3 oa; 280 wl × 70 oa; 37 wl × 10.5 (*991.8*; *918.6* × *229.7*; *121.4* × *34.4*)
Flight deck, metres: 304.5 × 70 (*999* × *229.7*)
Speed, knots: 30
Range, n miles: 3,850 at 29 kt, 8,500 at 18 kt
Complement: 2,586 (200 officers; 626 air crew) + 40 flag staff

Machinery: 8 boilers; 4 turbines; 200,000 hp(m) (*147 MW*); 4 shafts
Missiles: SSM: 12 Chelomey SS-N-19 Shipwreck (3M-45) launchers (flush mounted) ❶; inertial guidance with command update; active radar homing to 450 km (*243 n miles*) at 2.5 Mach; warhead 500 kT nuclear or 750 kg HE.
SAM: 4 Altair SA-N-9 Gauntlet (Kinzhal) sextuple vertical launchers (192 missiles) ❷; command guidance and active radar homing to 12 km (*6.5 n miles*) at 2 Mach; warhead 15 kg. 24 magazines; 192 missiles; 4 channels of fire.
Guns: SAM/Guns: 8 Altair CADS-N-1 (Kashtan) ❸; each has a twin 30 mm Gatling combined with 8 SA-N-11 (Grisson) and Hot Flash/Hot Spot fire-control radar/optronic director. Laser beam-riding guidance for missiles to 8 km (*4.4 n miles*); warhead 9 kg; 9,000 rds/min combined to 2 km (for guns).
6 – 30 mm/65 ❹ AK 630; 6 barrels per mounting; 3,000 rds/min combined to 2 km. Probably controlled by Hot Flash/Hot Spot on CADS-N-1.
A/S Mortars: 2 RBU 12,000 ❺; range 12,000 m; warhead 80 kg. UDAV-1M; torpedo countermeasure.

Physical countermeasures: Decoys: 10 PK 10 and 4 PK 2 chaff launchers.
Electronic countermeasures: ESM/ECM: 8 Foot Ball. 4 Wine Flask (intercept). 4 Flat Track. 10 Ball Shield A and B.
Radars: Air search: Sky Watch; four Planar phased arrays ❽; 3D.
Air/surface search: Top Plate B ❾; D/E-band.
Surface search: 2 Strut Pair ❿; F-band.
Navigation: 3 Palm Frond; I-band.
Fire control: 4 Cross Sword (for SAM) ⓫; K-band. 8 Hot Flash; J-band.
Aircraft control: 2 Fly Trap B; G/H-band.
Tacan: Cake Stand ⓬.
IFF: 4 Watch Guard.
Sonars: Bull Horn and Horse Jaw; hull-mounted; active search and attack; medium/low frequency.
Weapon control systems: 3 Tin Man optronic trackers. 2 Punch Bowl SATCOM datalink ❻. 2 Low Ball SATNAV ❼. 2 Bell Crown and 2 Bell Push datalinks.

Fixed-wing aircraft: 18 Su-33 Flanker D; 4 Su-25 UTG Frogfoot.
Helicopters: 15 Ka-27 Helix. 2 Ka-31 RLD Helix AEW.

Programmes: This was a logical continuation of the deleted Kiev class. The full name of *Kuznetsov* is *Admiral Flota Sovietskogo Sojuza Kuznetsov*. The second of class, *Varyag*, was between 70 and 80 per cent complete by early 1993 at Nikolayev in the Ukraine. Building was then terminated after an unsuccessful attempt by the Navy to fund completion. Subsequently the ship was bought by Chinese interests and, having arrived at Dalian in March 2002 was completed by the Chinese Navy and commissioned in September 2012.
Modernisation: It was announced in April 2010 that a major modernisation was to be undertaken 2012–17. Work may have been delayed by the Indian carrier *Vikramaditya* and is now expected to start in 2014.
Structure: The hangar is 183 × 29.4 × 7.5 m and can hold up to 18 Flanker aircraft. There are two starboard side lifts, a ski jump of 14° and an angled deck of 7°. There are four arrester wires. The SSM system is in the centre of the flight deck forward with flush deck covers. The ship has some 16.5 m of freeboard. There is no Bass Tilt radar and the ADG guns are controlled by Kashtan fire-control system. The ship suffers from severe water distillation problems.
Operational: AEW, ASW and reconnaissance tasks undertaken by Helix helicopters. The aircraft complement listed is based on the number which might be embarked for normal operations but the Russians claim a top limit of 60. *Kuznetsov* conducted extensive flight operations throughout the second half of both 1993 and 1994, and was at sea again by September 1995 after a seven month refit. Deployed to the Mediterranean for 80 days in early 1996 before returning to the Northern Fleet. Refitted from mid-1996 to mid-1998. Operations since 2000 have been sporadic and limited in scope; exercises have been conducted in the North Atlantic and there have been three-month deployments to the Mediterranean. The most recent began on 17 December 2013.

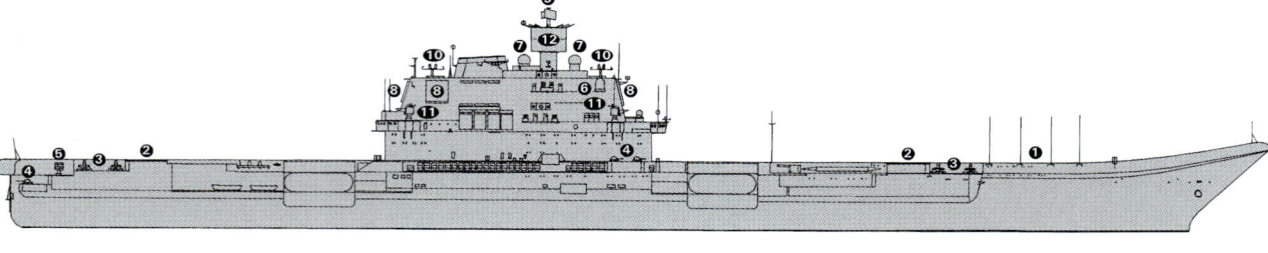

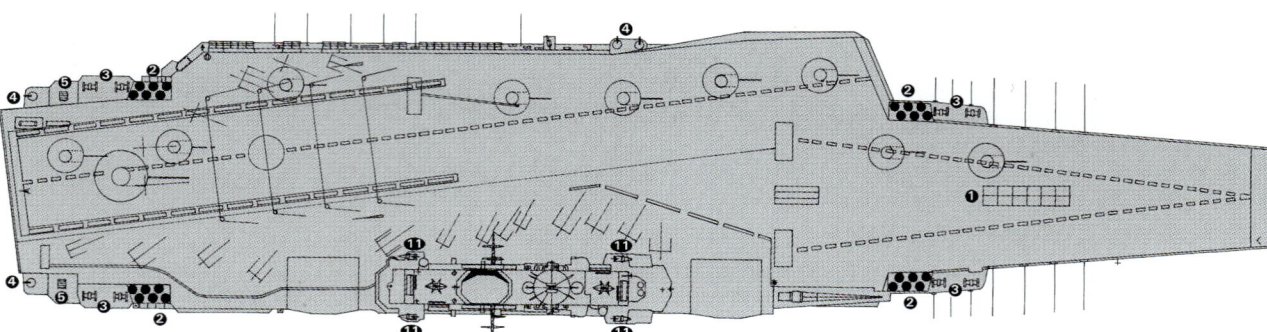

ADMIRAL KUZNETSOV (Scale 1 : 1,800), Ian Sturton / 0506078

ADMIRAL KUZNETSOV 1/2008*, *Ships of the World* / 1305318

IHS Jane's Fighting Ships 2014-2015 © 2014 IHS

ADMIRAL KUZNETSOV *1/2008*, Ships of the World* / 1305317

ADMIRAL KUZNETSOV *10/2004** / 1042330

ADMIRAL KUZNETSOV *6/2003, Lemachko Collection* / 1042294

688 Russian Federation > Battle cruisers

BATTLE CRUISERS

1 KIROV (ORLAN) CLASS (PROJECT 1144.1/1144.2) (CGHMN)

Name	No	Builders	Laid down	Launched	Commissioned
PYOTR VELIKIY (ex-*Yuri Andropov*)	099 (ex-183)	Baltic Yard 189, St Petersburg	11 Mar 1986	29 Apr 1989	9 Apr 1998

Displacement, tonnes: 19,305 standard; 24,690 full load
Dimensions, metres (feet): 252 oa; 230 wl × 28.5 × 9.1 *(826.8; 754.6 × 93.5 × 29.9)*
Speed, knots: 30
Range, n miles: 14,000 at 30 kt
Complement: 744 (82 officers; 18 air crew)

Machinery: CONAS; 2 KN-3 PWR; 300 MW; 2 oil-fired boilers; 2 GT3A-688 turbines; 140,000 hp(m) *(102.9 MW)*; 2 shafts
Missiles: SSM: 20 Chelomey SS-N-19 Shipwreck (3M 45) (P-700 Granit) (improved SS-N-12 with lower flight profile) ❶; inertial guidance with command update; active radar homing to 450 km *(243 n miles)* at 2.5 Mach; warhead 350 kT nuclear or 750 kg HE; no reloads.
SAM: 12 SA-N-20 Gargoyle (Fort-M) vertical launchers ❷; 8 rounds per launcher; command guidance; semi-active radar homing to 120 km *(64.5 n miles)*; warhead 90 kg (or nuclear?); 140 missiles. 2 SA-N-4 Gecko twin launchers ❸; semi-active radar homing to 15 km *(8 n miles)* at 2.5 Mach; warhead 50 kg; altitude 9.1–3,048 m *(30–10,000 ft)*; 40 missiles. 2 SA-N-9 Gauntlet (Kinzhal) octuple vertical launchers ❹; command guidance; active radar homing to 12 km *(6.5 n miles)* at 2 Mach; warhead 15 kg; altitude 3.4–12,192 m *(10–40,000 ft)*; 128 missiles; 4 channels of fire.
A/S: Novator SS-N-15 (Starfish); inertial flight to 45 km *(24.3 n miles)*; payload Type 40 torpedo or nuclear warhead; fired from fixed torpedo tubes behind shutters in the superstructure.
Guns: SAM/Guns: 6 CADS-N-1 (Kortik/Kashtan) ❺; each has a twin 30 mm Gatling combined with 8 SA-N-11 (Grisson) and Hot Flash/Hot Spot fire-control radar/optronic director. Laser beam-riding guidance for missiles to 8 km *(4.4 n miles)*; warhead 9 kg; 9,000 rds/min combined to 2 km (for guns).
2–130 mm/54 (twin) AK 130 ❻; 70 rds/min to 22 km *(12 n miles)*; weight of shell 33.4 kg.

Torpedoes: 10—21 in *(533 mm)* (2 quin) tubes. Combination of 53 cm torpedoes. Mounted in the hull adjacent the RBU 1000s on both quarters. Fixed tubes behind shutters can fire either SS-N-15 or Type 40 torpedoes.
A/S Mortars: 1 RBU 12,000 ❼; 10 tubes per launcher; range 12,000 m; warhead 80 kg. 2 RBU 1000 6-tubed aft ❽; range 1,000 m; warhead 55 kg. UDAV-1M; torpedo countermeasures.
Physical countermeasures: Decoys: 2 twin PK 2 150 mm chaff launchers. Towed torpedo decoy.
Electronic countermeasures: ESM/ECM: 8 Foot Ball. 4 Wine Flask (intercept). 8 Bell Bash. 4 Bell Nip. Half Cup (laser intercept).
Radars: Air search: Top Pair (Top Sail + Big Net) ⓫; 3D; C/D-band; range 366 km *(200 n miles)* for bomber, 183 km *(100 n miles)* for 2 m² target.
Air/surface search: Top Plate ⓬; 3D; D/E-band.
Navigation: 3 Palm Frond; I-band.
Fire control: Cross Sword ⓭; K-band (for SA-N-9). Top Dome for SA-N-6 ⓮; Tomb Stone J-band (for Fort M) ⓯; 2 Pop Group, F/H/I-band (for SA-N-4) ⓰. Kite Screech ⓱; H/I/K-band (for main guns). 6 Hot Flash for CADS-N-1; I/J-band.
Aircraft control: Flyscreen B; I-band.
IFF: Salt Pot A and B.
Tacan: 2 Round House B ⓲.
Sonars: Horse Jaw (Polinom); hull-mounted; active search and attack; low/medium frequency. Horse Tail; VDS; active search; medium frequency. Depth to 150–200 m *(492.1–656.2 ft)* depending on speed.
Combat data systems: Lesorub-44.
Weapon control systems: 4 Tin Man optronic trackers ❾. 2 Punch Bowl C SATCOM ❿. 4 Low Ball SATNAV. 2 Bell Crown and 2 Bell Push datalinks.
Helicopters: 3 Ka-27 Helix ⓳.

Programmes: Design work started in 1968. Type name is *atomny raketny kreyser* meaning nuclear-powered missile cruiser. A fifth of class was scrapped before being launched in 1989.
Structure: The Kirov class were the first Russian surface warships with nuclear propulsion. In addition to the nuclear plant a unique maritime combination with an auxiliary oil-fuelled system has been installed. This provides a superheat capability, boosting the normal steam output by some 50 per cent. The SS-N-19 tubes are set at an angle of about 45°. CADS-N-1 with a central fire-control radar on six mountings, each of which has two cannon and eight missile launchers. Two are mounted either side of the SS-N-19 forward and four on the after superstructure. Same A/S system as the frigate *Neustrashimy* with fixed torpedo tubes in ports behind shutters in the superstructure for firing SS-N-15 or Type 45 torpedoes. There are reported to be about 500 SAM of different types. *Velikiy*, the only operational ship, has a Tomb Stone fire-control radar instead of a forward Top Dome for SA-N-20 which is a maritime variant of SA-10C.
Operational: Based in the Northern Fleet. Over-the-horizon targeting for SS-N-19 provided by Punch Bowl SATCOM or helicopter. The first ship of the class of four, *Admiral Ushakov* (at Severodvinsk), was formally decommissioned in 2004 and the second ship, *Admiral Lazarev* (at Razboynik Bay) was decommissioned in 1998. Plans to refit the third ship, *Admiral Nakhimov*, laid up (at Severodvinsk) since 1999, have been revived. The scope of the work is substantial and includes nuclear refuelling and replacement of the SS-N-19 missile system, probably with SS-N-26 (Oniks). A contract for the upgrade was signed with Sevmash in June 2013 and the ship is planned to return to service in 2018. Plans to similarly modernise the other two ships by 2020 were announced in mid-2010 but the ships are reported to be beyond economical repair. *Pyotr Velikiy* conducted an Arctic Ocean deployments in 2012 and 2013. These are reported to include navigation as far east as the Laptev Sea.

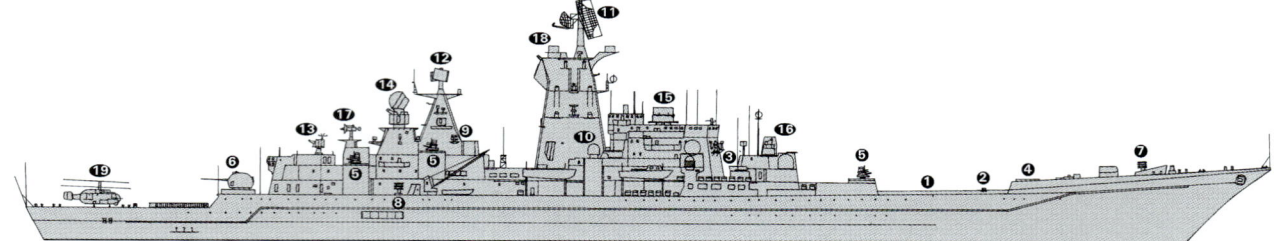

PYOTR VELIKIY *(Scale 1 : 1,500), Ian Sturton / 0528401*

PYOTR VELIKIY *9/2010, B Prézelin / 1366981*

PYOTR VELIKIY *6/2010, Ships of the World / 1366986*

IHS Jane's Fighting Ships 2014-2015

CRUISERS

3 SLAVA (ATLANT) CLASS (PROJECT 1164) (CGHM)

Name	No	Builders	Laid down	Launched	Commissioned
MOSKVA (ex-*Slava*)	121	Nikolayev North (61 Kommuna)	5 Nov 1976	27 Jul 1979	30 Dec 1982
MARSHAL USTINOV	055	Nikolayev North (61 Kommuna)	5 Oct 1978	25 Feb 1982	15 Sep 1986
VARYAG (ex-*Chervona Ukraina*)	011	Nikolayev North (61 Kommuna)	31 Jul 1979	28 Aug 1983	25 Dec 1989

Displacement, tonnes: 9,531 standard; 11,674 full load
Dimensions, metres (feet): 186.4 × 20.8 × 8.4 *(611.5 × 68.2 × 27.6)*
Speed, knots: 32
Range, n miles: 2,200 at 30 kt, 7,500 at 15 kt
Complement: 476 (62 officers)

Machinery: COGAG; 4 gas-turbines; 88,000 hp(m) *(64.68 MW)*; 2 M-70 gas-turbines; 20,000 hp(m) *(14.7 MW)*; 2 shafts
Missiles: SSM: 16 Chelomey SS-N-12 (8 twin) Sandbox (Bazalt) launchers ❶; inertial guidance with command update; active radar homing to 450 km *(243 n miles)* at 2.5 Mach; warhead nuclear 350 kT or HE 1,000 kg.
SAM: 8 SA-N-20 Gargoyle (Fort-M) vertical launchers ❷; 8 rounds per launcher; command guidance; semi-active radar homing to 120 km *(64.5 n miles)*; warhead 140 kg (or nuclear?); altitude 27,432 m *(90,000 ft)*. 64 missiles. 2 SA-N-4 Gecko twin retractable launchers ❸; semi-active radar homing to 15 km *(8 n miles)* at 2.5 Mach; altitude 9.1–3,048 m *(30–10,000 ft)*; 40 missiles.
Guns: 2 – 130 mm/54 (twin) AK 130 ❹; 70 rds/min to 22 km *(12 n miles)*; weight of shell 33.4 kg. 6 – 30 mm/65 AK 650 ❺; 6 barrels per mounting; 3,000 rds/min to 2 km.
Torpedoes: 10 – 21 in *(533 mm)* (2 quin) tubes ❻. Combination of 53 cm torpedoes.
A/S Mortars: 2 RBU 6000 12-tubed trainable ❼; range 6,000 m; warhead 31 kg.
Physical countermeasures: Decoys: 2 PK 2 chaff launchers.
Electronic countermeasures: ESM/ECM: 8 Side Globe (jammers). 4 Rum Tub (intercept).
Radars: Air search: Top Pair (Top Sail + Big Net) ❽; 3D; C/D-band; range 366 km *(200 n miles)* for bomber, 183 km *(100 n miles)* for 2 m² target.
Air/surface search: Top Steer ❾ or Top Plate (*Varyag*); 3D; D/E-band.
Navigation: 3 Palm Frond; I-band.
Fire control: Front Door ❿; F-band (for SS-N-12). Top Dome ⓫; J-band (for SA-N-6). 2 Pop Group ⓬; F/H/I-band (for SA-N-4). 3 Bass Tilt ⓭; H/I-band (for Gatlings). Kite Screech ⓮; H/I/K-band (for 130 mm).
IFF: Salt Pot A and B. 2 Long Head.
Sonars: Bull Horn and Steer Hide (Platina); hull-mounted; active search and attack; low/medium frequency.

VARYAG *9/2011, Hachiro Nakai / 1406873*

Weapon control systems: 2 Tee Plinth and 3 Tilt Pot optronic directors. 2 Punch Bowl satellite data receiving/targeting systems. 2 Bell Crown and 2 Bell Push datalinks.

Helicopters: 1 Ka-27 Helix ⓯.

Programmes: Built at the same yard as the Kara class. This is a smaller edition of the dual-purpose surface warfare/ASW *Kirov*, designed as a conventionally powered back-up for that class. The fourth of class, originally being completed for Ukraine, was transferred to Russia in July 1995 but returned to Nikolayev, Ukraine in February 1999 for completion. However, work was not finished due to lack of funds. Following the Russo-Ukrainian agreement on the basing of the Black Sea Fleet, the Ukrainian President announced in May 2010 that work on the ship is to be completed, but there have been no further reports of developments. A fifth of class was started but cancelled in October 1990.

Structure: The notable gap abaft the twin funnels (SA-N-6 area) is traversed by a large crane which stows between the funnels. The hangar is recessed below the flight deck with an inclined ramp. The torpedo tubes are behind shutters in the hull below the Top Dome radar director aft. Air conditioned citadels for NBCD. There is a bridge periscope.

Operational: Over-the-horizon targeting for SS-N-12 provided by helicopter or Punch Bowl SATCOM. *Moskva* is based in the Black Sea Fleet at Sevastopol. *Marshal Ustinov* deployed to the Northern Fleet in March 1987 and completed refit at St Petersburg in May 1995 where she remained until January 1998, when she transferred back to the Northern Fleet and is based at Severomorsk. *Varyag* transferred to Petropavlovsk in the Pacific in October 1990.

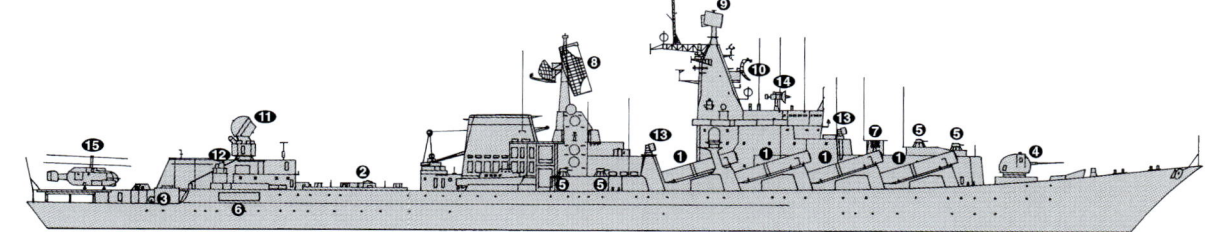

VARYAG *(Scale 1 : 1,200), Ian Sturton / 0050017*

MOSKVA *6/2010, Ships of the World / 1366984*

VARYAG *6/2010, Ships of the World / 1366983*

Russian Federation > Cruisers — Destroyers

1 KARA (BERKOT-B) CLASS (PROJECT 1134B) (CGHM)

Name	No	Builders	Laid down	Launched	Commissioned
KERCH	713 (ex-711)	Nikolayev North (61 Kommuna)	30 Apr 1971	21 Jul 1972	25 Dec 1974

Displacement, tonnes: 7,773 standard; 10,059 full load
Dimensions, metres (feet): 173.2 × 18.6 × 6.7 *(568.2 × 61.0 × 22.0)*
Speed, knots: 32. **Range, n miles:** 9,000 at 15 kt, 3,000 at 32 kt
Complement: 390 (49 officers)

Machinery: COGAG; 4 gas turbines; 108,800 hp(m) *(80 MW)*; 2 gas turbines; 13,600 hp(m) *(10 MW)*; 2 shafts
Missiles: SAM: 2 SA-N-3 Goblet twin launchers ❶; semi-active radar homing to 55 km *(30 n miles)* at 2.5 Mach; warhead 125 kg; altitude 91.4–22,860 m *(300–75,000 ft)*; 72 missiles. 2 SA-N-4 Gecko twin launchers (twin either side of mast) ❷; semi-active radar homing to 15 km *(8 n miles)* at 2.5 Mach; warhead 14.5 kg; altitude 9.1–3,048 m *(30–10,000 ft)*; 40 missiles.
A/S: 2 Raduga SS-N-14 Silex (Rastrub) quad launchers ❸; command guidance to 50 km *(27 n miles)* at 0.95 Mach; payload nuclear 5 kT or Type 40 torpedo or E53-72 torpedo. SSM version; range 35 km *(19 n miles)*; warhead 500 kg.
Guns: 4–3 in *(76 mm)*/59 AK 726 (2 twin) ❹; 90 rds/min to 16 km *(8.5 n miles)*; weight of shell 5.9 kg. 4–30 mm/65 ❺; 6 barrels per mounting; 3,000 rds/min combined to 2 km.
Torpedoes: 10–21 in *(533 mm)* (2 quin) tubes ❻. Combination of 53 cm torpedoes.
A/S Mortars: 2 RBU 6000 12-tubed trainable ❼; range 6,000 m; warhead 31 kg. 2 RBU 1000 6-tubed (aft) ❽; range 1,000 m; warhead 55 kg; torpedo countermeasures.
Physical countermeasures: Decoys: 2 PK 2 chaff launchers. 1 BAT-1 torpedo decoy.
Electronic countermeasures: ESM/ECM: 8 Side Globe (jammers). 2 Bell Slam. 2 Bell Clout. 4 Rum Tub (intercept) (fitted on mainmast).
Radars: Air search: Flat Screen ❾; E/F-band.
Air/surface search: Head Net C ❿; 3D; E-band; range 128 km *(70 n miles)*.
Navigation: 2 Don Kay; I-band. Don 2 or Palm Frond; I-band.
Fire control: 2 Head Light B/C ⓫; F/G/H-band (for SA-N-3 and SS-N-14). 2 Pop Group ⓬; F/H/I-band (for SA-N-4). 2 Owl Screech ⓭; G-band (for 76 mm). 2 Bass Tilt ⓮; H/I-band (for 30 mm).
Tacan: Fly Screen A or Fly Spike.
IFF: High Pole A. High Pole B.
Sonars: Bull Nose (Titan 2-MG 332); hull-mounted; active search and attack; low/medium frequency. Mare Tail; VDS (Vega-M 325) ⓯; active search; medium frequency.
Weapon control systems: 4 Tilt Pot optronic directors. Bell Crown, Bike Pump and Hat Box datalinks.

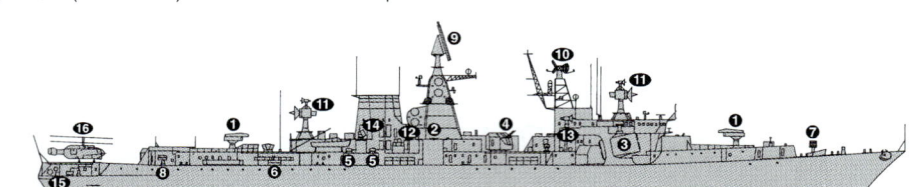

KERCH   *(Scale 1 : 1,500), Ian Sturton / 0081651*

KERCH   *7/2013*, E & M Laursen / 1530966*

Helicopters: 1 Ka-27 Helix ⓰.

Programmes: Type name is *bolshoy protivolodochny korabl*, meaning large anti-submarine ship.
Modernisation: The Flat Screen air search radar, replaced Top Sail.
Structure: The helicopter is raised to flight deck level by a lift. In addition to the 8 tubes for the SS-N-14 A/S system and the pair of twin launchers for SA-N-3 system with Goblet missiles, Kara class mounts the SA-N-4 system in 2 silos, either side of the mast. The SA-N-3 system has only 2 loading doors per launcher and a larger launching arm.
Operational: Two of the class started refits in July 1987 and have been scrapped by the Ukraine. One more was scrapped in the Pacific in 1996. *Petropavlovsk* is laid up in the Pacific and is unlikely to go to sea again. In the Black Sea, *Ochakov* was decommissioned in August 2011, *Azov* was cannibalised for spares in 1998 and *Kerch* remains in service.

DESTROYERS

Notes: While Project 21956, for a new class of destroyers, had been publicised, it is now more likely that replacement of the current inventory of Sovremenny and Udaloy classes is to be achieved with the smaller Project 22350 Admiral Gorshkov class.

1 KASHIN (PROJECT 61) CLASS (DDGM)

Name	No	Builders	Laid down	Launched	Commissioned
SMETLIVY	810	Nikolayev North	15 Jul 1966	26 Aug 1967	25 Sep 1969

Displacement, tonnes: 4,074 standard; 4,826 full load
Dimensions, metres (feet): 144 × 15.8 × 4.7 *(472.4 × 51.8 × 15.4)*
Speed, knots: 32. **Range, n miles:** 4,000 at 18 kt, 1,520 at 32 kt
Complement: 280 (25 officers)

Machinery: COGAG; 4 DE 59 gas turbines; 72,000 hp(m) *(52.9 MW)*; 2 shafts
Missiles: SSM: 8 Zvezda SS-N-25 (KH 35 Uran) Switchblade (2 quad) ❶; active radar homing to 130 km *(70.2 n miles)* at 0.9 Mach; warhead 145 kg.
SAM: 2 SA-N-1 Goa twin launchers ❷; command guidance to 31.5 km *(17 n miles)* at 2 Mach; warhead 72 kg; altitude 91.4–22,860 m *(300–75,000 ft)*; 32 missiles.
Guns: 2–3 in *(76 mm)*/59 AK 726 (1 or 2 twin) ❸; 90 rds/min to 16 km *(8.5 n miles)*; weight of shell 5.9 kg.
Torpedoes: 5–21 in *(533 mm)* (quin) tubes ❹. Combination of 53 cm torpedoes.
A/S Mortars: 2 RBU 6000 12-tubed trainable ❺; range 6,000 m; warhead 31 kg; 120 rockets.
Physical countermeasures: Decoys: PK 16 chaff launchers (modified). 2 towed torpedo decoys.
Electronic countermeasures: ESM/ECM: 2 Bell Shroud. 2 Watch Dog.
Radars: Air/surface search: Head Net C ❻; 3D; E-band. Big Net ❼; C-band.

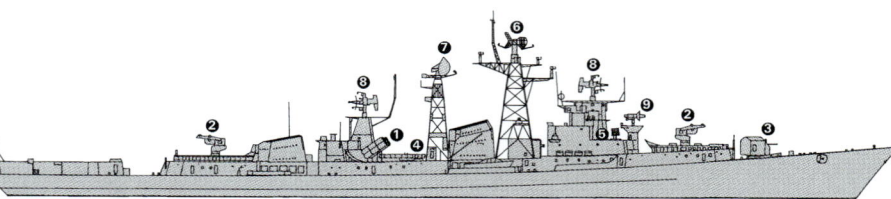

SMETLIVY   *(Scale 1 : 1,200), Ian Sturton / 0126351*

Navigation: 2 Don 2/Don Kay/Palm Frond; I-band.
Fire control: 2 Peel Group ❽; H/I-band (for SA-N-1). 1 Owl Screech ❾; G-band (for guns).
IFF: High Pole B.
Sonars: Bull Nose (MGK 336) or Wolf Paw; hull-mounted; active search and attack; medium frequency. Vega; VDS; active search; medium frequency.
Weapon control systems: 3 Tee Plinth and 4 Tilt Pot optronic directors.

Programmes: The first class of warships in the world to rely entirely on gas-turbine propulsion. Type name is *bolshoy protivolodochny korabl*, meaning large anti-submarine ship.
Modernisation: Modernised with a VDS aft, vice the after gun, and fitted for SS-N-25 in place of the RBU 1000 launchers.
Operational: Based in the Black Sea. Refitted from 1990 to 1996 but back in service in 1997. Deployed to the Indian Ocean in 2003 and remains active.
Sales: Additional ships of a modified design built for India. First transferred September 1980, the second in June 1982, the third in 1983, the fourth in August 1986 and the fifth and last in January 1988.

SMETLIVY   *7/2013*, E & M Laursen / 1530965*

Destroyers — **Russian Federation** 691

1 UDALOY II (FREGAT) CLASS (PROJECT 1155.1) (DDGHM)

Name	No	Builders	Laid down	Launched	Commissioned
ADMIRAL CHABANENKO	650 (ex-437)	Yantar Shipyard, Kaliningrad 820	15 Sep 1988	14 Dec 1992	20 Feb 1999

Displacement, tonnes: 7,824 standard; 9,043 full load
Dimensions, metres (feet): 163.5 × 19.3 × 7.5 (536.4 × 63.3 × 24.6)
Speed, knots: 28. **Range, n miles:** 4,000 at 18 kt
Complement: 249 (29 officers)

Machinery: COGAG; 2 gas turbines; 48,600 hp(m) (35.72 MW); 2 gas turbines; 24,200 hp(m) (17.79 MW); 2 shafts
Missiles: SSM: 8 Raduga SS-N-22 Sunburn (3M-82 Moskit) (2 quad) ❶; active/passive radar homing to 160 km (87 n miles) at 3.0 Mach (4.5 for attack); warhead nuclear or HE 300 kg; sea-skimmer.
SAM: 8 SA-N-9 Gauntlet (Kinzhal) vertical launchers ❷; command guidance; active radar homing to 12 km (6.5 n miles) at 2 Mach; warhead 15 kg. 64 missiles; 4 channels of fire.
A/S: Novator SS-N-15 (Starfish); inertial flight to 45 km (24.3 n miles); payload Type 40 torpedo or nuclear, fired from torpedo tubes.
Guns: SAM/Guns: 2 CADS-N-1 (Kashtan) ❸; each with twin 30 mm Gatling; combined with 8 SA-N-11 (Grisson) and Hot Flash/Hot Spot fire-control radar/optronic director. Laser beam guidance for missiles to 8 km (4.4 n miles); warhead 9 kg; 9,000 rds/min combined to 1.5 km for guns.
2—130 mm/54 (twin) AK 130 ❹; 70 rds/min to 22 km (12 n miles); weight of shell 33.4 kg.
Torpedoes: 8—21 in (533 mm) (2 quad tubes) ❺. Combination of 53 cm torpedoes. The tubes are protected by flaps in the superstructure.
A/S Mortars: 2 RBU 6000 ❻. 12-tubed trainable; range 6,000 m; warhead 31 kg.
Physical countermeasures: Decoys: 8 PK 10 and 2 PK 2 chaff launchers ❼.
Electronic countermeasures: ESM/ECM: 2 Wine Glass (intercept). 2 Bell Shroud. 2 Bell Squat. 4 Half Cup laser warner. 2 Shot Dome.
Radars: Air search: Strut Pair II ❾; F-band. Top Plate ❿; 3D; D/E-band.
Surface search: 3 Palm Frond ⓫; I-band.
Fire control: 2 Cross Swords ⓬; K-band (for SA-N-9). Kite Screech ⓭; H/I/K-band (for 100 mm gun). Band Stand (Mineral ME) ❽; D-band (for SS-N-22).
CCA: Fly Screen B ⓮.
IFF: Salt Pot B and C.
Sonars: Horse Jaw (Polinom); hull-mounted; active search and attack; medium/low frequency. Horse Tail; VDS; active search; medium frequency.
Weapon control systems: M 145 radar and optronic system. 2 Bell Crown datalink. Band Stand ❽ datalink for SS-N-22; 2 Light Bulb, 2 Round House and 1 Bell Nest datalinks.
Helicopters: 2 Ka-27 Helix A ⓯.

Programmes: A single ship follow-on class from the Udaloys. NATO designator Balcom 12. At least two more were projected with names Admiral Basisty and Admiral Kucherov; Basisty was scrapped in March 1994, and Kucherov was never started.
Structure: Similar size to the Udaloy and has the same propulsion machinery. Improved combination of weapon systems owing something to both the Sovremenny and the Neustrashimy classes. The distribution of SA-N-9 launchers may be the same as Udaloy class. The torpedo tubes are protected by a hinged flap in the superstructure.
Operational: Sea trials started on 14 September 1995 from Baltiysk. Deployed to the Northern Fleet in March 1999 when the pennant number changed. Based at Severomorsk.

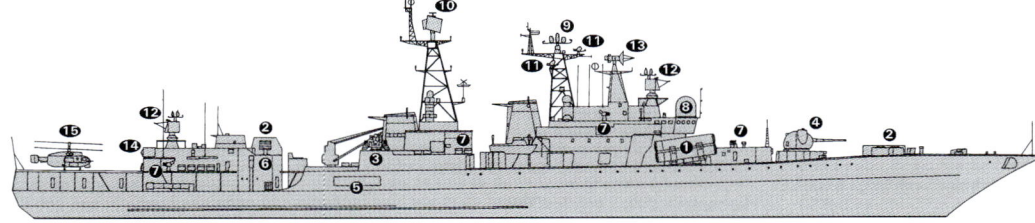

ADMIRAL CHABANENKO (Scale 1 : 1,200), Ian Sturton / 0569929

ADMIRAL CHABANENKO 8/2002 / 0528328

ADMIRAL CHABANENKO 9/2006, B Sullivan / 1164812

© 2014 IHS IHS Jane's Fighting Ships 2014-2015

FRIGATES

Notes: Construction of the Grom class (Project 1244.1) frigate *Borodino* was abandoned in 2009. The ship had been under construction at Yantar Shipyard, Kaliningrad, since 1997 but has been subject to repeated budget cuts and delays. The ship remains laid up at Kaliningrad.

2 KRIVAK (PROJECT 1135/1135M) CLASS (FFM)

Name	No	Builders	Laid down	Launched	Commissioned
LADNY	801	Kamish-Burun, Kerch	25 May 1979	7 May 1980	29 Dec 1980
PYTLIVY	808	Yantar Shipyard, Kaliningrad	27 Jun 1979	16 Apr 1981	30 Nov 1981

Displacement, tonnes: 3,150 standard; 3,709 full load
Dimensions, metres (feet): 123.5 × 14.3 × 7.3 sonar (405.2 × 46.9 × 24.0)
Speed, knots: 32. **Range, n miles:** 4,000 at 14 kt, 1,600 at 30 kt
Complement: 194 (18 officers)

Machinery: COGAG; 2 M8K gas-turbines; 55,500 hp(m) (40.8 MW); 2 M 62 gas-turbines; 13,600 hp(m) (10 MW); 2 shafts
Missiles: SAM: 2 SA-N-4 Gecko (Zif 122) twin launchers ❶; Osa-M semi-active radar homing to 15 km (8 n miles) at 2.5 Mach; warhead 50 kg; altitude 9.1–3,048 m (30–10,000 ft); 40 missiles.
A/S: Raduga SS-N-14 Silex quad launcher ❷; command guidance to 50 km (27 n miles) at 0.95 Mach; payload nuclear 5 kT or Type 40 torpedo or Type E53-72 torpedo. SSM version; range 35 km (19 n miles); warhead 500 kg.
Guns: 4–3 in (76 mm)/59 AK 726 (2 twin) (Krivak I); 90 rds/min to 16 km (8.5 n miles); weight of shell 5.9 kg. 2–3.9 in (100 mm)/70 AK 100 (Krivak II) ❸; 60 rds/min to 21.5 km (11.5 n miles); weight of shell 15.6 kg.
Torpedoes: 8–21 in (533 mm) (2 quad) tubes ❹. Combination of 53 cm torpedoes.
A/S Mortars: 2 RBU 6000 12-tubed trainable ❺; range 6,000 m; warhead 31 kg.
Mines: Capacity for 16.
Physical countermeasures: Decoys: 4 PK 16 or 10 PK 10 chaff launchers. Towed torpedo decoy.
Electronic countermeasures: ESM/ECM: 2 Bell Shroud. 2 Bell Squat. Half Cup laser warning (in some).
Radars: Air search: Head Net C ❻; 3D; E-band.
Surface search: Don Kay or Palm Frond or Don 2 or Spin Trough ❼; I-band.
Fire control: 2 Eye Bowl ❽; F-band (for SS-N-14). 2 Pop Group ❾; F/H/I-band (for SA-N-4). Owl Screech (Krivak I); G-band. Kite Screech (Krivak II) ❿; H/I/K-band. Plank Shave (Harpun B) (for SS-N-25) not fitted.
IFF: High Pole B.
Sonars: Bull Nose (MGK-335S or MG-332); hull-mounted; active search and attack; medium frequency. Mare Tail (MGK-345); VDS (MG 325) ⓫; active search; medium frequency.

Programmes: Type name was originally *bolshoy protivolodochny korabl*, meaning large anti-submarine ship. Changed in 1977–78 to *storozhevoy korabl* meaning escort ship. The naval Krivaks I and II are known as the Burevestnik class and the border guard ships Krivak III (listed separately) as Nerey class.
Structure: Krivak II class has Y-gun mounted higher than in Krivak I and the break to the quarterdeck further aft apart from other variations noted above.
Operational: Both based in Black Sea and likely to be decommissioned as Admiral Grigorovich class enter service.

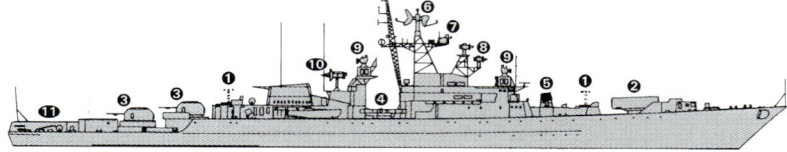

KRIVAK II (Scale 1 : 1,200), Ian Sturton / 0506084

LADNY 6/2010, A Sheldon-Duplaix / 1406120

PYTLIVY 7/2013*, E & M Laursen / 1530968

PYTLIVY 7/2013*, E & M Laursen / 1530967

Frigates ‹ Russian Federation

2 NEUSTRASHIMY (JASTREB) CLASS (PROJECT 1154) (FFHM)

Name	No	Builders	Laid down	Launched	Commissioned
NEUSTRASHIMY	712	Yantar Shipyard, Kaliningrad	27 Mar 1987	25 May 1988	24 Jan 1993
YAROSLAV MUDRYY	727	Yantar Shipyard, Kaliningrad	27 May 1988	1991	24 Jul 2009

Displacement, tonnes: 3,505 standard; 4,318 full load
Dimensions, metres (feet): 129.6 oa; 123 wl × 15.5 × 4.8 *(425.2; 403.5 × 50.9 × 15.7)*
Speed, knots: 30
Range, n miles: 4,500 at 16 kt
Complement: 210 (35 officers)

Machinery: COGAG; 2 gas turbines; 48,600 hp(m) *(35.72 MW)*; 2 gas turbines; 24,200 hp(m) *(17.79 MW)*; 2 shafts
Missiles: SSM: 8 SS-N-25 Switchblade (Kh 35E Uran) *(Yaroslav Mudryy)* ❶ (2 quad); active radar homing to 130 km *(70.2 n miles)* at 0.9 Mach; warhead 145 kg.
SAM: 4 SA-N-9 Gauntlet (Kinzhal) octuple vertical launchers ❷; command guidance; active radar homing to 12 km *(6.5 n miles)* at 2 Mach; warhead 15 kg. 32 missiles.
A/S: SS-N-15/16; inertial flight to 50 km *(27 n miles)*; payload Type 40 torpedo or nuclear warhead; fired from torpedo tubes.
Guns: SAM/Guns: 2 CADS-N-1 (Kashtan) (3M87) ❸; each has a twin 30 mm Gatling combined with 8 SA-N-11 (Grisson) and Hot Flash/Hot Spot fire-control radar/optronic director. Laser beam guidance for missiles to 8 km *(4.4 n miles)*; warhead 9 kg; 9,000 rds/min (combined) to 1.5 km (for guns).
1—3.9 in *(100 mm)*/59 A 190E ❹; 80 rds/min to 21.5 km *(11.5 n miles)*; weight of shell 15.6 kg.
Torpedoes: 6—21 in *(533 mm)* tubes combined with A/S launcher ❺; can fire SS-N-15/16 missiles with Type 40 anti-submarine torpedoes or 53 cm torpedoes.
A/S Mortars: 1 RBU 12,000 ❻; 10-tubed trainable; range 12,000 m; warhead 80 kg.
Mines: 2 rails.
Physical countermeasures: Decoys: 8 PK 10 and 2 PK 16 chaff launchers.
Electronic countermeasures: ESM/ECM: Intercept and jammers. 2 Foot Ball; 2 Half Hat; 4 Half Cup laser intercept.
Radars: Air search: Top Plate ❼; 3D; D/E-band.
Air/surface search: Cross Dome ❽; E/F-band.
Navigation: 2 Palm Frond; I-band.
Fire control: Cross Sword ❾ (for SAM); K-band. Kite Screech B ❿ (for SSM and guns); I-band.
IFF: 2 Salt Pot; 4 Box Car.
Sonars: Ox Yoke and Whale Tongue; hull-mounted; active search and attack; medium frequency. Ox Tail; VDS ⓫ or towed sonar array.
Weapon control systems: 2 Bell Crown datalink.

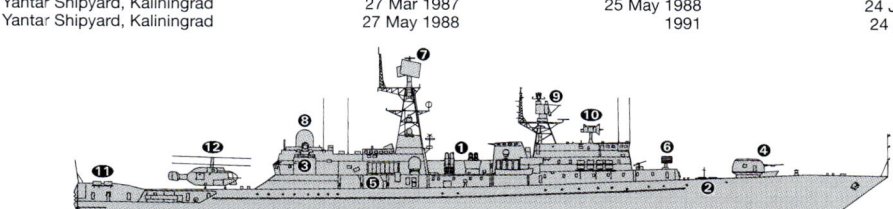

YAROSLAV MUDRYY *(Scale 1 : 1,200), Ian Sturton / 1366995*

NEUSTRASHIMY *6/2008, Michael Nitz / 1353330*

Helicopters: 1 Ka-27 Helix ⓬.

Programmes: At least four of the class were planned. The first of the class started sea trials in the Baltic in December 1990. Second of class *(Yaroslav Mudryy)* was launched in May 1991, but in October 1988 the shipyard stated that the hull would be sold for scrap. However, after several years' inaction, work recommenced in 2002 and it was confirmed in 2005 that the ship is to be completed. Sea trials began on 26 February 2009. The export version of the ship is known as 'Korsar'. The third ship *(Tuman)* was launched in July 1993 with only the hull completed and work stopped in December 1997 without any work being done. She is unlikely to be completed.

Structure: Slightly larger than the Krivak and has a helicopter which is a standard part of the armament of modern Western frigates. There are two horizontal launchers at main deck level on each side of the ship, angled at 18° from forward. These double up for A/S missiles of the SS-N-15/16 type using a 'plunge-fly-plunge' launch and flight and normal torpedoes. Similar launchers are behind shutters in the last three of the Kirov class. The helicopter deck extends across the full width of the ship. The after funnel is unusually flush decked but both funnels have been slightly extended after initial sea trials. Attempts have been made to incorporate stealth features. Main propulsion is the same as the Udaloy II class. Reported as having a basic computerised combat data system.
Operational: Based in the Baltic at Baltiysk.

NEUSTRASHIMY *11/2008, US Navy / 1353331*

YAROSLAV MUDRYY *7/2011, Erik Laursen / 1406775*

© 2014 IHS — IHS Jane's Fighting Ships 2014-2015

698 Russian Federation > Frigates — Corvettes

0 + 6 (3) ADMIRAL GRIGOROVICH (PROJECT 11356M) CLASS (FFGH)

Name	No	Builders	Laid down	Launched	Commissioned
ADMIRAL GRIGOROVICH	–	Yantar Shipyard, Kaliningrad	18 Dec 2010	14 Mar 2014	2014
ADMIRAL ESSEN	–	Yantar Shipyard, Kaliningrad	8 Jul 2011	2014	2015
ADMIRAL MAKAROV	–	Yantar Shipyard, Kaliningrad	29 Feb 2012	2015	2016
ADMIRAL BUTAKOV	–	Yantar Shipyard, Kaliningrad	12 Jul 2013	2015	2017
ADMIRAL ISTOMIN	–	Yantar Shipyard, Kaliningrad	15 Nov 2013	2016	2018
ADMIRAL KORNILOV	–	Yantar Shipyard, Kaliningrad	2014	2017	2019

Displacement, tonnes: 3,620 standard; 4,035 full load
Dimensions, metres (feet): 124.8 × 15.2 × 4.6 *(409.4 × 49.9 × 15.1)*
Speed, knots: 32. **Range, n miles:** 4,850 at 14 kt
Complement: 190 (18 officers)

Machinery: COGAG: 2 Zorya DN-59 gas turbines; 43,488 hp *(34.2 MW)*; 2 Zorya UGT 6000 gas turbines; 16,628 hp *(12.4 MW)*; 2 shafts
Missiles: SSM: 8-cell UKSK VLS for SS-N-27 (3M54) Novator Alfa Klub-N; active radar homing to 220 km *(120 n miles)* at 0.8 Mach (cruise) and 2.9 Mach (attack); warhead 400 kg.
SAM: VLS launcher for SA-N-7C (9M317); command. Semi-active radar and IR homing to 25 km *(13.5 n miles)* at 3 Mach; warhead 70 kg.
Guns: 1–3.9 in *(100 mm)*/70 A-190; 80 rds/min to 21.5 km *(11.6 n miles)*; weight of shell 15.6 kg.
SAM/Guns: 2 CADS-N-1 (Kashtan); each with twin 30 mm Gatling; combined with 8 SA-N-11 (Grisson). Laser beam guidance for missiles to 8 km *(4.4 n miles)*; warhead 9 kg; 9,000 rds/min combined to 1.5 km for guns.
Torpedoes: 4 DTA-53 21 in *(533 mm)* (2 twin) fixed launchers.
A/S Mortars: 1 RBU 6000 12-barrelled launcher.
Physical countermeasures: To be announced.
Electronic countermeasures: ECM/ESM: Pribor TK 25.
Radars: Air search: Top Plate (Fregat-M2EM); 3D; E/F-band.
Air/surface search: Cross Dome (Positiv-E); E/F-band.
Fire control: 4 Front Dome (MR-90); H/I-band (for SA-N-7). Plank Shave (Garpun B); I/J-band (for SSM). Ratep 5P-10E Puma; I-band (for 100 mm guns).
Navigation: I-band.
Sonars: Bow mounted.

Helicopters: 1 Ka-27 Helix.

Programmes: It was announced in 2010 that a contract had been awarded to Yantar Shipyard for the construction of three Project 11356 frigates. These are likely to be similar to the Talwar (Krivak IV) class that have been built for the Indian Navy. Details are based on the Talwar class and are illustrative. A contract for a second batch of three ships was announced on 14 September 2011. A third batch of three ships is likely.

0 + 6 (14) ADMIRAL GORSHKOV (PROJECT 22350) CLASS (FFGH)

Name	No	Builders	Laid down	Launched	Commissioned
ADMIRAL GORSHKOV	417	Severnaya Verf, St Petersburg	1 Feb 2006	29 Oct 2010	2014
ADMIRAL KASATONOV	–	Severnaya Verf, St Petersburg	26 Nov 2009	2014	2014
ADMIRAL GOLOVKO	–	Severnaya Verf, St Petersburg	1 Feb 2012	2015	2016
ADMIRAL ISAKOV	–	Severnaya Verf, St Petersburg	14 Nov 2013	2016	2017

Displacement, tonnes: 4,550 standard
Dimensions, metres (feet): 135 × 16.4 × 4.4 *(442.9 × 53.8 × 14.4)*
Speed, knots: 29. **Range, n miles:** 4,000 at 14 kt

Machinery: CODAG: 2 M90 gas turbines; 27,500 hp *(20.5 MW)*; 2 diesels; 5,200 hp *(3.9 MW)*; 2 shafts
Missiles: SSM: 16-cell (2 × 8) UKSK VLS ❶ for SS-N-26 (Oniks) (3M55); inertial guidance and active/passive radar homing to 300 km *(162 n miles)* at 2.6 Mach; sea skimmer in terminal phase; warhead 250 kg; VLS silo.
SAM: Redut (9M96D); 32 missiles; VLS launcher ❷; semi-active homing to 120 km *(64.8 n miles)*; warhead 24 kg. 2 Palma (CADS-N-2) CIWS; twin 30 mm 6-barrelled Gatling combined with 8 Sosna-R missiles with electro-optic director and laser/TV/IR guidance to 10 km *(5.4 n miles)*; warhead 5 kg.
A/S: Medvedka 2 (SS-N-29); inertial flight to 25 km *(13.5 n miles)*; payload Type 40 torpedo.
Guns: 1–130 mm A-192 ❸. 2–14.5 mm.
Torpedoes: 8 Paket 324 mm (2 quad) tubes.
Electronic countermeasures: ECM/ESM: Pribor TK 25.
Radars: Air search: Furke-2 ❹; E/F-band.
Air search/fire control: Poliment-K ❺; 3D; E/F-band.
Fire control: Ratep 5P-10E ❻; I-band (for 130 mm).
Navigation: To be announced.
Sonars: Zarya; bow mounted. Vinyetka LFAS towed array.
Combat data systems: To be announced.
Weapon control systems: To be announced.

Helicopters: 1 Ka-31 ❼.

Programmes: Severnaya Verf shipyard contracted on 21 October 2005 to build the lead Project 22350 frigate. The construction of a second ship was announced in June 2009 and a third ship was laid down in February 2012. A further three ships are to be built at Severnaya Verf and a class of at least 20 ships is expected. Designed by the Severnoye Design Bureau, it is the first new class of major surface combatants to be procured in 15 years. The ship may be equipped with Brahmos rather than SS-N-26 missiles. There is an export version of this ship, Project 22356.
Structure: Slightly longer and wider than the Talwar class from which the design is reported to be developed.
Operational: The first ship is likely to be based in the Baltic.

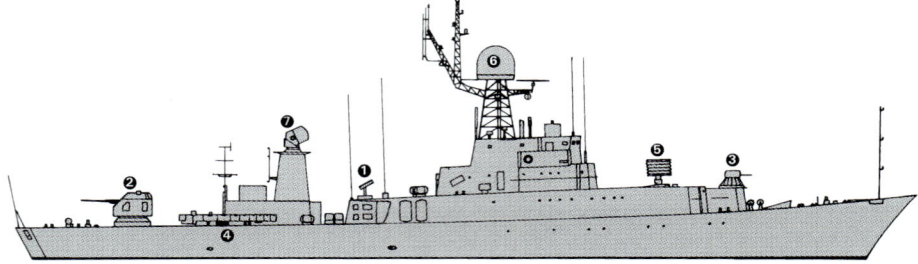

ADMIRAL GORSHKOV (Scale 1 : 1,200), Ian Sturton / 1528815

CORVETTES

7 PARCHIM II CLASS (PROJECT 1331) (FFLM)

ZELENODOLSK (ex-*MPK 99*) 308
MPK 105 245
URENGOY (ex-*MPK 192*) 304
KAZANETS (ex-*MPK 205*) 311
ALEKSIN (ex-*MPK 224*) 218
MPK 227 243
KALMYKIA (ex-*MPK 229*) 232

Displacement, tonnes: 781 standard; 975 full load
Dimensions, metres (feet): 75.2 × 9.8 × 4.4 *(246.7 × 32.2 × 14.4)*
Speed, knots: 26
Range, n miles: 2,500 at 12 kt
Complement: 70 (8 officers)

Machinery: 3 Type M 504A diesels; 10,812 hp(m) *(7.95 MW)* sustained; 3 shafts
Missiles: SAM: 2 SA-N-5 Grail quad launchers ❶; manual aiming; IR homing to 6 km *(3.2 n miles)* at 1.5 Mach; altitude to 2,500 m *(8,000 ft)*; warhead 1.5 kg.
Guns: 1–3 in *(76 mm)*/59 AK 176 ❷; 120 rds/min to 15 km *(8 n miles)*; weight of shell 5.9 kg. 1–30 mm/65 AK 630 ❸; 6 barrels; 3,000 rds/min combined to 2 km.
Torpedoes: 4–21 in *(533 mm)* (2 twin) tubes ❹. Combination of 53 cm torpedoes.
A/S Mortars: 2 RBU 6000 12-tubed trainable ❺; range 6,000 m; warhead 31 kg. 96 weapons.
Mines: Rails fitted.
Depth charges: 2 racks.
Physical countermeasures: Decoys: 2 PK 16 chaff launchers.
Electronic countermeasures: ESM: 2 Watch Dog; intercept.
Radars: Air/surface search: Cross Dome ❻; E/F-band.
Navigation: TSR 333 or Nayala or Kivach III; I-band.
Fire control: Bass Tilt ❼; H/I-band.
IFF: High Pole A.
Sonars: Bull Horn; hull-mounted; active search and attack; medium frequency. Lamb Tail; helicopter type VDS; high frequency.
Weapon control systems: Hood Wink and Odd Box optronic systems.

Programmes: Built in the GDR at Peenewerft, Wolgast for the USSR. First one commissioned 19 December 1986 and the last on 6 April 1990.
Structure: Similar design to the ex-GDR Parchim I class now serving with the Indonesian Navy but some armament differences.
Operational: All operate in the Baltic and are based at Baltiysk or Kronshtadt. All of the class refitted at Rostock in 1994–95. *Bashkortostan* (MPK 228) damaged by fire in 1999 and again in 2008. She has been decommissioned.

PARCHIM II (Scale 1 : 600), Ian Sturton / 0506204

MPK 227 7/2012, Erik Laursen / 1483539

IHS Jane's Fighting Ships 2014-2015 © 2014 IHS

Corvettes — **Russian Federation** 699

2 DERGACH (SIVUCH) (PROJECT 1239) CLASS (PGGJM)

Name	No	Builders	Launched	Commissioned
BORA (ex-MRK 27)	615	Zelenodolsk Shipyard, Kazan	1987	20 May 1997
SAMUM (ex-MRK 17)	616 (ex-575, ex-890)	Zelenodolsk Shipyard, Kazan	1992	31 Dec 1995

Displacement, tonnes: 1,067 full load
Dimensions, metres (feet): 64.5 × 17 × 3.8 *(211.6 × 55.8 × 12.5)*
Speed, knots: 53; 12 hullborne;
Range, n miles:
Complement: 67 (8 officers)

Machinery: CODOG; 2 gas turbines; 55,216 hp(m) *(40.6 MW)*; 2 diesels; 10,064 hp(m) *(7.4 MW)*; 2 hydroprops; 2 auxiliary diesels; 2 props on retractable pods
Missiles: SSM: 8 SS-N-22 (2 quad) Sunburn (3M-82 Moskit) launchers ❶; active radar homing to 160 km *(87 n miles)* at 3.0 Mach; warhead nuclear or 200 kT or HE 300 kg; sea-skimmer.
SAM: SA-N-4 Gecko twin launcher ❷; semi-active radar homing to 15 km *(8 n miles)* at 2.5 Mach; warhead 50 kg; 20 missiles.
Guns: 1—3 in *(76 mm)*/59 AK 176 ❸; 120 rds/min to 15 km *(8 n miles)*; weight of shell 5.9 kg. 2—30 mm/65 AK 630 ❹; 6 barrels per mounting; 3,000 rds/min combined to 2 km.
Electronic countermeasures: ESM/ECM: 2 Foot Ball A, 2 Half Hats.
Radars: Air/surface search: Positiv-ME1 ❼; I-band.
Fire control: Bass Tilt ❽; H/I-band (for guns). Pop Group ❾; F/H/I-band (for SAM). Band Stand (Mineral ME) ❻; D-band (for SS-N-22).
Navigation: SRN-207; I-band.
IFF: Square Head. Salt Pot.

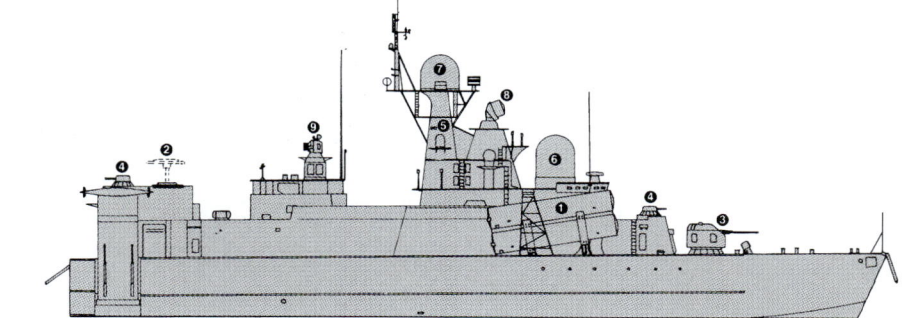

BORA *(Scale 1 : 600), Ian Sturton / 0506086*

Weapon control systems: 2 Light Bulb datalink ❺. Band Stand ❻ datalink for SS-N-22; Bell Nest.

Programmes: Almaz design approved 24 December 1980. Classified as a PGGA (Guided Missile Patrol Air Cushion Vessels). Both did trials from 1989 (Bora) and 1993 (Samum) before being accepted into service.

Structure: Twin-hulled surface effect design. The auxiliary diesels are for slow speed operations.
Operational: The design was unreliable but efforts were made in 1996/97 to restore both to an operational state. SS-N-22 missiles were test-fired in April 2003. Both ships have camouflaged hulls and are based at Sevastopol.

BORA *7/2013*, E & M Laursen / 1530972*

SAMUM *5/2011, L-G Nilsson / 1406777*

© 2014 IHS

IHS Jane's Fighting Ships 2014-2015

Russian Federation > Corvettes

25 TARANTUL (MOLNYA) (PROJECT 1241.1/1241M/1242.1) CLASS (FSGM)

Baltic
R 47 819
DIMITROVGRAD (ex-R 291) 825
R 257 833
KUZNETSK (ex-R 129) 852
ZARECHNY (ex-R 187) 855
R 2 870
MORSHANSK (ex-R 293) 874

Pacific
R 29 916
R 20 921
R 14 924
R 18 937
R 11 940
R 24 946
R 297 951
R 298 971
R 19 978
R 79 996

Black Sea
R 109 952
GROZA (ex-R 239) 953
IVANOVETS (ex-R 334) 954
BURYA (ex-R 60) 955
R 71 962

Caspian
R 32 (ex-R 5) 700 (ex-992)
GOMEL (ex-R 160) 054
STUPINETS (ex-R 101) 705

Displacement, tonnes: 391 standard; 462 full load
Dimensions, metres (feet): 56.1 × 11.5 × 2.5 *(184.1 × 37.7 × 8.2)*
Speed, knots: 36
Range, n miles: 400 at 36 kt, 1,650 at 14 kt
Complement: 34 (5 officers)

Machinery: COGAG; 2 Nikolayev Type DR 77 gas turbines; 16,016 hp(m) *(11.77 MW)* sustained; 2 Nikolayev Type DR 76 gas turbines with reversible gearboxes; 4,993 hp(m) *(3.67 MW)* sustained; 2 shafts or CODOG with 2 CM 504 diesels; 8,000 hp(m) *(5.88 MW)*, replacing second pair of gas turbines in Tarantul IIIs

Missiles: SSM: 4 Raduga SS-N-2D Styx (2 twin) launchers (Tarantul II); active radar or IR homing to 83 km *(45 n miles)* at 0.9 Mach; warhead 513 kg; sea-skimmer at end of run. 4 Raduga SS-N-22 Sunburn (3M-82 Moskit) (2 twin) launchers (Tarantul III); active radar homing to 160 km *(87 n miles)* at 3.0 Mach; warhead nuclear 200 kT or HE 300 kg; sea-skimmer. Modified version in Type 1242.1.
SAM: SA-N-5 Grail quad launcher; manual aiming; IR homing to 6 km *(3.2 n miles)* at 1.5 Mach; altitude to 2,500 m *(8,000 ft)*; warhead 1.5 kg.

Guns: 1—3 in (76 mm)/59 AK 176; 120 rds/min to 15 km *(8 n miles)*; weight of shell 5.9 kg. 2—30 mm/65; 6 barrels per mounting; 3,000 rds/min to 2 km.

Physical countermeasures: Decoys: 2 PK 16 or 4 PK 10 (Tarantul III) chaff launchers.

Electronic countermeasures: ESM: 2 Foot Ball, 2 Half Hat (in some).

Radars: Air/surface search: Plank Shave or Positiv E (Tarantul 700); I-band.
Navigation: Kivach III; I-band.
Fire control: Bass Tilt; H/I-band. Band Stand (Mineral ME); D-band (for SS-N-22).
IFF: Square Head. High Pole B.

Sonars: Foal Tail; VDS; active search; high frequency.

Weapon control systems: Hood Wink optronic director. Light Bulb datalink. Band Stand; datalink for SSM; Bell Nest.

Programmes: Tarantul II were built at Kolpino, Petrovsky, Leningrad and in the Pacific in 1980–86. Production of Taruntul IIIs then continued until 1995. One more was launched in September 1997 at Rybinsk, and a Tarantul III at Kolpino completed in December 1999 for the Baltic Fleet. Type name is *raketny kater* meaning missile cutter.

Modernisation: Tarantul III 874 served as a trials platform for a modified version of SS-N-22 with a longer range; the missile is distinguished by end caps on the launcher doors. Tarantul II 962 served as a trials platform for the CADS-N-1 point defence system in the Black Sea.

Structure: Basically same hull as Pauk class, without extension for sonar. The single Type 1242.1 has a Positiv E radar.

BURYA — 7/2013*, E & M Laursen / 1530973

ZARECHNY — 7/2012, Erik Laursen / 1483540

Sales: Tarantul I class-one to Poland 28 December 1983, second in April 1984, third in March 1988 and fourth in January 1989. One to India in April 1987, second in January 1988, third in December 1988, fourth in November 1989 and fifth in January 1990. Two to Yemen in November 1990 and January 1991. One to Romania in December 1990, two more in February 1992. One Tarantul II to Bulgaria in March 1990. Two Tarantul Is to Vietnam in 1996 and two more in 1999.

R 2 — 7/2012, Hartmut Ehlers / 1483542

13 NANUCHKA CLASS (PROJECT 1234.1/1234.7) (FSG)

North	Baltic	Pacific	Black Sea
RASSVET 520	**LIVEN** 551	**MOROZ** 409	**SHTYL** 620
AYSBERG 535	**GEYZER** 555	**RAZLIV** 450	**MIRAZH** 617
NAKAT 526	**ZYB** 560	**SMERCH** 423	
	PASSAT 570	**INEJ** 418	

Displacement, tonnes: 671 full load
Dimensions, metres (feet): 59.3 × 11.8 × 2.6 (194.6 × 38.7 × 8.5)
Speed, knots: 33
Range, n miles: 2,500 at 12 kt, 900 at 31 kt
Complement: 42 (7 officers)

Machinery: 6 M 504 diesels; 26,112 hp(m) (19.2 MW); 3 shafts
Missiles: SSM: 6 Chelomey SS-N-9 Siren (Malakhit) (2 triple) launchers ❶; command guidance and IR and active radar homing to 110 km (60 n miles) at 0.9 Mach; warhead nuclear 250 kT or HE 500 kg.
SAM: SA-N-4 Gecko twin launcher ❷; semi-active radar homing to 15 km (8 n miles) at 2.5 Mach; warhead 50 kg; altitude 9.1–3,048 m (30–10,000 ft); 20 missiles. Some anti-surface capability.
Guns: 1—3 in (76 mm)/59 AK 176 ❸; 120 rds/min to 15 km (8 n miles); weight of shell 5.9 kg. 1—30 mm/65 ❹; 6 barrels; 3,000 rds/min combined to 2 km.
Physical countermeasures: Decoys: 4 PK 10 chaff launchers ❺.
Electronic countermeasures: ESM: Foot Ball and Half Hat A and B. 4 Half Cup laser warners.
Radars: Air/surface search: Peel Pair ❼; I-band or Plank Shave; E/F-band.

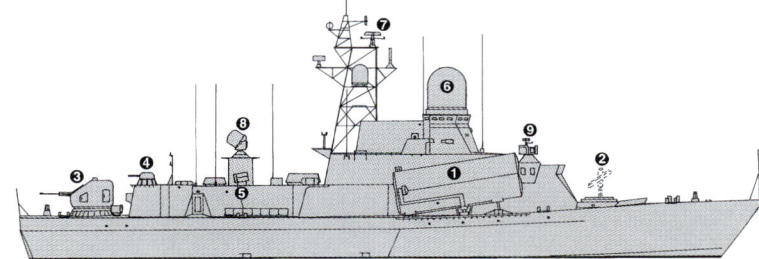

NANUCHKA III

(Scale 1 : 600), Ian Sturton / 0105552

Fire control: Bass Tilt ❽; H/I-band. Pop Group ❾; F/H/I-band (for SA-N-4). Band Stand (Mineral ME) ❻; D-band (for SS-N-9).
Navigation: Nayada; I-band.
IFF: High Pole. Square Head. Spar Stump. Salt Pot A and B.
Weapon control systems: 2 Bell Nest or Light Bulb (datalinks). Band Stand ❻ datalink for SS-N-9.

Programmes: Built from 1969 onwards at Petrovsky, Leningrad and in the Pacific. The Nanuchka III were first seen in 1978. *Nakat* was used as a trials ship for SS-NX-26 (Yakhont) in 1995. Type name is *maly raketny korabl* meaning small missile ship.
Operational: Intended for deployment in coastal waters although formerly deployed in the Mediterranean (in groups of two or three), North Sea and Pacific.

ZYB

4/2011, L-G Nilsson / 1406778

MIRAZH

7/2013*, E & M Laursen / 1530974

SHIPBORNE AIRCRAFT

Numbers/Type: 17/2 Sukhoi Su-33 Flanker D/Su-33 UB.
Operational speed: 1,345 kt *(2,500 km/h)*.
Service ceiling: 59,000 ft *(18,000 m)*.
Range: 2,160 + n miles *(4,000 km)*.
Role/Weapon systems: Fleet air defence fighter. 20 production aircraft delivered of which 2 have been lost. 10 are believed to be operational. All based in the Northern Fleet. Most training is done from a simulated flight deck ashore. Likely to be replaced by MiG 29K in due course. Sensors: Track-while-scan pulse Doppler radar, IR scanner. Weapons: One 30 mm cannon, 10 AAMs (AA-12, AA-11, AA-8).

FLANKER   2/1996 / 0506323

Numbers/Type: 5 Sukhoi Su-25UT Frogfoot UTG.
Operational speed: 526 kt *(975 km/h)*.
Service ceiling: 22,965 ft *(7,000 m)*.
Range: 675 n miles *(1,250 km)*.
Role/Weapon systems: The UTG version is the two seater ground attack aircraft used for deck training in the carrier *Kuznetsov*. About 40 more of these aircraft are Air Force. Sensors: Laser rangefinder, ESM, ECM. Weapons: One 30 mm cannon, AAMs (AA-8), rockets, bombs.

FROGFOOT   2/1996 / 0506324

Numbers/Type: 2 Kamov Ka-31 Helix RLD.
Operational speed: 119 kt *(220 km/h)*.
Service ceiling: 11,480 ft *(3,500 m)*.
Range: 162 n miles *(300 km)*.
Role/Weapon systems: AEW conversions with a solid-state radar under the fuselage. Eight sold to India. Sensors: Oko E-801 Surveillance radar, datalinks. Weapons: Unarmed.

HELIX RLD   9/1995 / 0506325

Numbers/Type: 1 Kamov KA-52K.
Operational speed: 162 kt *(300 km/h)*.
Service ceiling: 18,045 ft *(5,500 m)*.
Range: 248 n miles *(459 km)*.
Role/Weapon systems: Maritime version of KA-52 Alligator attack helicopter under development. First prototype expected 2014. Up to 16 aircraft may be deployed to Mistral-class helicopter carrier. Sensors: to be announced. Weapons: a variety of weapons may be deployed on four pylons. These include 12 Vikhr-M ASMs or 23 mm gun pods or Igla AAMs or 80 mm air-to-surface rockets.

Numbers/Type: 56/28/5 Kamov Ka-27Pl Helix A/Ka-29 Helix B/Ka-32 Helix D.
Operational speed: 135 kt *(250 km/h)*.
Service ceiling: 19,685 ft *(6,000 m)*.
Range: 432 n miles *(800 km)*.
Role/Weapon systems: ASW helicopter; three main versions — 'A' for ASW, 'B' for assault and D for SAR; deployed to surface ships and some shore stations. K-27M advanced tactical aircraft reportedly under development. Sensors: Osminog Splash Drop search radar, VGS-3 dipping sonar, sonobuoys, MAD, ESM. Weapons: ASW; three APR-2 torpedoes, nuclear or conventional S3V depth bombs or mines. Assault type: Two UV-57 rocket pods (2 × 32).

HELIX   6/2008, *Ships of the World* / 1353315

Numbers/Type: 20/4 MiG 29K Fulcrum/29 KUB.
Operational speed: 750 kt *(1,400 km/h)*.
Service ceiling: 57,000 ft *(17,400 m)*.
Range: 1,400 n miles *(2,600 km)*.
Role/Weapon systems: All-weather single-seat fighter with attack capability. An order for 24 aircraft was announced in October 2009 and delivery started in 2012. The capabilities are expected to be simialr to aircraft procured by the Indian Navy. Sensors: Phazotron-NIIR Zhuk-ME radar; jammer. Weapons: AAM; ASM; conventional bombs; 30 mm cannon.

MIG 29 KUB   6/2011, Annati Collection / 1406780

LAND-BASED MARITIME AIRCRAFT (FRONT LINE)

Numbers/Type: 25 Ilyushin Il-38 May.
Operational speed: 347 kt *(645 km/h)*.
Service ceiling: 32,800 ft *(10,000 m)*.
Range: 3,887 n miles *(7,200 km)*.
Role/Weapon systems: Long-range MR and ASW. 12 in the North, 13 in the Pacific. A contract for the modernisation of Northern Fleet aircraft to IL-38N standard was issued in May 2013. The upgrade is to include the Novella mission system, a national version of the Sea Dragon system installed in Indian IL-38SD aircraft. One IL-38N has been on operational trials since 2011. Test flights of an upgraded version started in 2002 and continued in 2003. Sensors: Wet Eye search/weather radar, MAD, sonobuoys. Weapons: ASW; internal storage for 6 tons weapons.

MAY   6/2004, Paul Tompkins / 0001128

Numbers/Type: 55 Tupolev Tu-22 M Backfire C.
Operational speed: 2.0 Mach.
Service ceiling: 60,000 ft *(18,300 m)*.
Range: 2,500 n miles *(4,630 km)*.
Role/Weapon systems: Medium-range nuclear/conventional strike and reconnaissance. Operated by the Air Force. Sensors: Down Beat search/Fan Tail attack radars, EW. Weapons: ASV; 12 tons of 'iron' bombs or standoff missiles AS-4 Kitchen (Kh 22N(A)) and AS-6 Kickback (Kh 15P). Self-defence; two 23 mm cannon.

BACKFIRE   6/2003, Paul Jackson / 0547316

Numbers/Type: 15/5 Tupolev Tu-142 Bear F/Tu-142 Bear J.
Operational speed: 500 kt *(925 km/h)*.
Service ceiling: 60,000 ft *(18,300 m)*.
Range: 6,775 n miles *(12,550 km)*.
Role/Weapon systems: Multimission long-range aircraft (ASW and communications variants). Numbers are reducing. Sensors: Wet Eye search radar, ESM; search radar, sonobuoys, EW, MAD (F), ELINT systems (J). The Bear J is reported to be equipped with VLF communications for SSBN connectivity. Weapons: ASW; various torpedoes, depth bombs and/or mines (F). Self-defence; some have two 23 mm or more cannon.

BEAR F 6/2008, *Ships of the World* / 1353316

Numbers/Type: 40 Sukhoi Su-24 Fencer D/E.
Operational speed: 1.15 Mach.
Service ceiling: 57,400 ft *(17,500 m)*.
Range: 950 n miles *(1,755 km)*.
Role/Weapon systems: Fitted for maritime reconnaissance and strike. A number of aircraft have been transferred to the Air Force and numbers are uncertain. Sensors: Radar and EW. Weapons: 30 mm Gatling gun; various ASM missiles and bombs; some have 23 mm cannon.

FENCER E 6/1999, *IHS* / 0048910

Numbers/Type: 4 Beriev Be-12 Mail.
Operational speed: 328 kt *(608 km/h)*.
Service ceiling: 37,000 ft *(11,280 m)*.
Range: 4,050 n miles *(7,500 km)*.
Role/Weapon systems: Long-range ASW/MR amphibian. Previously reported withdrawn from service but four remain operational. Sensors: Short Horn search radar, MAD, EW. Weapons: ASW; 5 tons of depth bombs, mines or torpedoes. ASV; limited missile and rocket armament.

MAIL (Russian colours) 7/1996, *J Ciślak* / 0084250

PATROL FORCES

3 BUYAN (PROJECT 21630) CLASS (PG)

ASTRAKHAN 012 (ex-101) **VOLGODONSK** (ex-*Kaspiysk*) 014 (ex-702) **MAKHACHKALA** 015

Displacement, tonnes: 528 full load
Dimensions, metres (feet): 61.45 × 9.6 × 2.1 *(201.6 × 31.5 × 6.9)*
Speed, knots: 26. **Range, n miles:** 1,500 at 15 kt

Machinery: 2 Zvezda M520 diesels; 9,900 hp *(7.35 MW)*; 2 waterjets
Missiles: SAM: SA-16 Gubka (Strelets); IR homing to 5 km *(2.7 n miles)* at 2.6 Mach; warhead 1.3 kg.
Guns: 1—3.9 in *(100 mm)* A 190; 80 rds/min to 20 km *(10.8 n miles)*; weight of shell 16 kg. 2—30 mm/65 AK 630; 6 barrels per mounting; 3,000 rds/min to 2 km. 2—14.5 mm MGs. 3—7.62 mm MGs.
Rockets: 1—122 mm UMS-73 Grad-M multibarrelled rocket launcher.
Physical countermeasures: Decoys: 2 KT 216 launchers.
Radars: Air/surface search: Positiv-ME1; I-band.
Fire control: Bass Tilt (MR-123); I/J-band.
Navigation: I-band.

Comment: Designed by Zelenodolsk Design Bureau and built by Almaz, St Petersburg. *Astrakhan* laid down on 30 January 2004 and launched on 7 October 2005. *Volgodonsk* laid down on 25 February 2005 and *Makhachkala* on 24 March 2006. Heavily armed gunboat designed for littoral operations. The design includes radar and IR signature reduction measures including below water-line exhaust. The first became operational in the Caspian Sea in mid-2007 and the second was commissioned 28 December 2011. The third was commissioned on 2 March 2013. The export version is known as Project 21632 Tornado class.

MAKHACHKALA 8/2013*, *Selim San* / 1530976

2 + 4 BUYAN-M (PROJECT 21631) CLASS (PG)

| GRAD SVIYAZHSK 021 | VELIKI USTYUG | SEPUKHOV |
| UGLICH 022 | ZELENI DOL | VISHNY VOLOCHEK |

Displacement, tonnes: 949 full load
Dimensions, metres (feet): 74.1 × 11.0 × 2.6 *(243.1 × 36.1 × 8.5)*
Speed, knots: 25

Machinery: To be announced.
Missiles: SSM: 8-cell UKSK VLS launcher for SS-N-27 Club.
SAM: 2 Gibka (Igla-1M).
Guns: 1—3.9 in *(100 mm)* A 190. 2—30 mm/65 AK 630. 2—14.5 mm. 3—7.62 mm MGs.
Radars: Air/surface search: Positiv-ME1; I-band.
Fire control: Bass Tilt; H/I-band.

Comment: The first of an enlarged variant of the Buyan class was laid down at Zelenodolsk (A M Gorky) Shipyard on 27 August 2010, a second on 22 July 2011 and a third on 27 August 2011. A further three have been ordered. The design features a (reported) carbon-fibre superstructure and an 8-cell Vertical Launch System (VLS) aft of the mainmast. The first vessel was launched on 9 March 2013 and the second on 10 April 2013. All six are likely to be based in the Caspian Sea.

3 MATKA (VEKHR) CLASS (PROJECT 206MP)
(FAST ATTACK CRAFT—MISSILE HYDROFOIL) (PGGK)

BOROVSK (ex-*R 25*) 706 **KARACHEJEVO-CHERKESSIA** 701 **BUDENOVSK** 702

Displacement, tonnes: 229 standard; 264 full load
Dimensions, feet (feet): 39.6 × 7.6 oa; 12.5 over foils × 2.1 hull, 4 over foils *(129.9 × 24.9; 41.0 × 6.9; 13.1)*
Speed, knots: 40. **Range, n miles:**
Complement: 33

Machinery: 3 Type M 504 diesels; 10,800 hp(m) *(7.94 MW)* sustained; 3 shafts
Missiles: SSM: 2 SS-N-2C/D Styx; active radar or IR homing to 83 km *(45 n miles)* at 0.9 Mach; warhead 513 kg; sea-skimmer at end of run. 8 SS-N-25 (in 966); radar homing to 130 km *(70.2 n miles)* at 0.9 Mach; warhead 145 kg; sea-skimmer.
Guns: 1—3 in *(76 mm)*/59 AK 176; 120 rds/min to 15 km *(8 n miles)*; weight of shell 5.9 kg. 1—30 mm/65 AK 630; 6 barrels per mounting; 3,000 rds/min to 2 km.
Electronic countermeasures: Clay Brick; intercept.
Radars: Air/surface search: Plank Shave; E-band.
Navigation: SRN-207; I-band.
Fire control: Bass Tilt; H/I-band.
IFF: High Pole B or Salt Pot B and Square Head.
Weapon control systems: Hood Wink optronic directors.

Programmes: In early 1978 the first of class was seen. Built at Kolpino Yard, Leningrad. Production stopped in 1983 being superseded by Tarantul class. Type name is *raketny kater* meaning missile cutter.
Structure: Similar hull to the deleted Osa class with similar single hydrofoil system to Turya class. The combination has produced a better sea-boat than the Osa class.
Operational: Based in the Caspian. Five units transferred to Ukraine in 1996.

BUDENOVSK 7/2008, *Lemachko Collection* / 1305660

1 MUKHA (SOKOL) (PROJECT 1145) CLASS
(FAST ATTACK CRAFT—PATROL HYDROFOIL) (PGK)

VLADIMIRETS (ex-*MPK 220*) 060

Displacement, tonnes: 406 full load
Dimensions, metres (feet): 50 × 8.5 × 4 *(164 × 27.9 × 13.1)*
Speed, knots: 40
Complement: 45

Machinery: CODOG; 2 Type NK-12M gas turbines; 23,046 hp(m) *(16.95 MW)* sustained; 2 diesels; 2,400 hp(m) *(1.76 MW)*; 2 shafts
Guns: 1—3 in *(76 mm)*/59 AK 176; 120 rds/min to 15 km *(8 n miles)*; weight of shell 5.9 kg. 2—30 mm/65 AK 630; 6 barrels per mounting; 3,000 rds/min combined to 2 km.
Torpedoes: 8—16 in *(406 mm)* (2 quad) tubes. SAET-40; anti-submarine; active/passive homing to 10 km *(5.4 n miles)* at 30 kt; warhead 100 kg.
Physical countermeasures: Decoys: 2 PK 16 chaff launchers.
Electronic countermeasures: ESM: Radar warning.
Radars: Surface search: Peel Cone; E-band.
Navigation: SRN 206; I-band.
Fire control: Bass Tilt; H/I-band.
Sonars: Foal Tail; VDS; active search; high frequency.

Comment Built in 1986 at Feodosuja. Features include a hydrofoil arrangement with a single fixed foil forward, large gas-turbine exhausts aft, and trainable torpedo mountings. The only ship of the class, which was used as a trials platform for the Medveka ASW guided weapon, is based in the Black Sea.

VLADIMIRETS 6/2011 / 1406782

3 + 7 PROJECT 21980 (GRACHONOK) CLASS
(DIVING TENDER) (PB/YTD)

Name	No	Builders	Laid down	Launched	Commissioned
P 104	889	Zelenodolsk Shipyard	8 Feb 2008	25 Apr 2009	Nov 2009
P 191	840	Zelenodolsk Shipyard	7 May 2010	Jul 2011	Oct 2011
P 349	841	Zelenodolsk Shipyard	6 May 2011	16 Jun 2012	14 Nov 2012
P 350	842	Zelenodolsk Shipyard	5 May 2012	Apr 2013	Aug 2013
P 377	996	Vostochnaya Verf, Vladivostok	Mar 2012	24 Jun 2013	2014
P 351	930	Zelenodolsk Shipyard	27 Jul 2012	2014	2014
–	–	Zelenodolsk Shipyard	7 May 2013	2014	2015
–	–	Zelenodolsk Shipyard	27 Jul 2013	2014	2015

Displacement, tonnes: 138 full load
Dimensions, metres (feet): 31 × 7.4 × 1.85 (101.7 × 24.3 × 6.1)
Speed, knots: 23
Complement: 8

Machinery: 2 diesels; 2 shafts
Missiles: SAM: SA-N-10 Igla quad launcher; manual aiming; IR homing to 6 km (3.2 n miles) at 1.5 Mach; warhead 1.5 kg.
Guns: 1 — 14.5 mm. 1 — 55 mm grenade launcher.
Electronic countermeasures: ESM: Pribor TK 25.
Radars: MR-231; I-band.

Comment: New class of counter-sabotage vessel. Seven craft have been or are to be built at Zelenodolsk while the first of three Pacific-based vessels was laid down at Vladivostok in 2012. The first is based in the Black Sea. The craft is designed to defend offshore installations on the continental shelf and is equipped with short-range sonar as well as diving equipment for underwater survey and inspection.

GRACHONOK CLASS — 6/2011, A Sheldon-Duplaix / 1406794

AMPHIBIOUS FORCES

15 ROPUCHA (PROJECT 775/775M) CLASS (LSTM)

North	OLENEGORSKIY GORNIAK 012	GEORGIY POBEDONOSETS 016	KONDOPOGA 027	ALEXANDER OTRAKOVSKIY 031
Baltic	KALININGRAD 102	ALEXANDER SHABALIN 110	MINSK 127	KOROLEV 130
Black	NOVOCHERKASSK 142	AZOV 151	YAMAL 156	TSESAR KUNIKOV 158
Pacific	ADMIRAL NEVELSKY 055	OSLYABYA (ex-Mukhtar Avezov) 066	PERESVET (ex-Nicolay Korsakov) 077	

Displacement, tonnes: 4,471 full load
Dimensions, metres (feet): 112.5 × 15 × 3.7 (369.1 × 49.2 × 12.1)
Speed, knots: 17.5
Range, n miles: 3,500 at 16 kt, 6,000 at 12 kt
Complement: 95 (7 officers)
Military lift: 10 MBT plus 190 troops or 24 AFVs plus 170 troops or mines
Machinery: 2 Zgoda-Sulzer 16ZVB40/48 diesels; 19,230 hp(m) (14.14 MW) sustained; 2 shafts
Missiles: SAM: 4 SA-N-5 Grail quad launchers (in at least two ships); manual aiming; IR homing to 6 km (3.2 n miles) at 1.5 Mach; altitude to 2,500 m (8,000 ft); warhead 1.5 kg; 32 missiles.
Guns: 4 — 57 mm/75 AK 725 (2 twin) ❶ (Ropucha I); 120 rds/min to 12.7 km (6.8 n miles); weight of shell 2.8 kg. 1 — 76 mm/59 AK 176 (Ropucha II); 120 rds/min to 15 km (8 n miles); weight of shell 5.9 kg. 2 — 30 mm/65 AK 630 (Ropucha II). 2 — 122 mm UMS-73 Grad-M (in some) ❷.
Rockets: 2 — 40-barrelled rocket launchers; range 9 km (5 n miles).
Mines: 92 contact type.
Radars: Air/surface search: Strut Curve ❹ (Ropucha I) or Cross Dome (Ropucha II); F-band.
Navigation: Don 2 or Nayada; I-band.
Fire control: Muff Cob ❺ (Ropucha I); G/H-band. Bass Tilt (Ropucha II); H/I-band.
IFF: 2 High Pole A or Salt Pot A.
Sonars: Mouse Tail VDS can be carried.
Weapon control systems: 2 Squeeze Box optronic directors ❸. Hood Wink and Odd Box.

Programmes: Ropucha Is completed at Northern Shipyard, Gdansk, Poland in two spells from 1974–78 (12 ships) and 1980–88. Ropucha IIs started building in 1987 with the first one commissioning in May 1990. The third and last of the class completed in January 1992. Type name is *bolshoy desantny korabl* (BDK) meaning large landing ship.

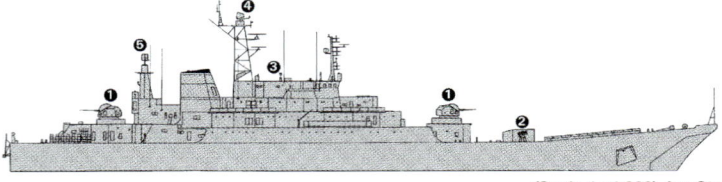

ROPUCHA I — (Scale 1 : 1,200), Ian Sturton / 0506247

AZOV — 2/2013*, C D Yaylali / 1530975

Structure: A Ro-Ro design with a tank deck running the whole length of the ship. All have very minor differences in appearance. These ships have a higher troop-to-vehicle ratio than the Alligator class. At least five of the class have rocket launchers at the after end of the forecastle. The second type have a 76 mm gun forward in place of one twin 57 mm and an ADG aft instead of the second. Radar and EW suites are also different. The after mast has been replaced by a solid extension to the superstructure.
Operational: Eleven more have been deleted so far.
Sales: One to South Yemen in 1979, returned to Russia in late 1991 for refit and was back in Aden in 1993. One to Ukraine in 1996.

KALININGRAD — 7/2013*, C D Yaylali / 1530978

Amphibious forces < **Russian Federation** 705

0 + 2 MISTRAL CLASS (AMPHIBIOUS ASSAULT SHIPS) (LHD)

Name	No	Builders	Laid down	Launched	Commissioned
VLADIVOSTOK	–	STX, Saint Nazaire	2012	15 Oct 2013	2014
SEVASTOPOL	–	STX, Saint Nazaire	2013	2014	2015

Displacement, tonnes: 16,529 standard; 21,000 full load
Dimensions, metres (feet): 199 × 32 × 8 (652.9 × 105.0 × 26.2)
Speed, knots: 19
Range, n miles: 11,000 at 15 kt
Complement: 177 (20 officers)

Military lift: 450 (up to 900 in austerity conditions) troops plus 60 armoured vehicles (approximately 1,200 tons of cargo). 4 LCU or 2 LCACs.
Machinery: Electric propulsion: 4 (3 Wärtsilä 16V32 and 1 Wärtsilä 18V200) diesel generators provide total of 20.8 MW for propulsion and services. 2 Alsthom Mermaid podded propulsors trainable through 360°; 19,040 hp(m) (14.0 MW) sustained; 1 bow thruster; 2,040 hp (1.5 MW).
Missiles: SSM: 8-cell UKSK VLS for SS-N-27 or SS-N-30.
SAM: 2 Gibka (Igla-1M).
Guns: 4—30 mm/65 AK 630.
Radars: Air/surface search: Thales MRR-3D NG; G-band.
Navigation: To be announced.
Combat data systems: To be announced.
Weapon control systems: To be announced.

Helicopters: 16 KA-52K.

Programmes: The possibility of purchasing Mistral-class amphibious vessels from France was first publicly raised in 2009. After lengthy negotiations, it was announced on 24 December 2010 that agreement had been reached to assemble two ships in France and a contract was finalised on 17 June 2011. First steel was cut at STX Marine on 1 February 2012 and work on the first Russian-built block began in August 2012 at Baltic Shipyard. Negotiations for the construction of two further ships in Russia had not been finalised by late 2012 when it was reported that plans for these two ships had been abandoned. There is (reportedly) no agreement to transfer the combat system technology, weapons or communications systems and weapons/sensors are likely to be installed in Russia.
Structure: The ships are to be largely built to commercial (Bureau Veritas) standards with fully automated platform management systems. Flight deck is 5,000 m², has six helicopter spots and is connected to the hangar by two aircraft lifts, one astern and one near island. The docking-well is capable of operating two LCACs or four LCUs. A port-side ramp facilitates loading. There is space for a 63-bed hospital. The design is to be adapted to accommodate Kamov twin rotor helicopters and the Ka-52K attack helicopter. The ships are also to be modified to operate in Arctic conditions.
Operational: Both ships are to be based at Vladivostok.

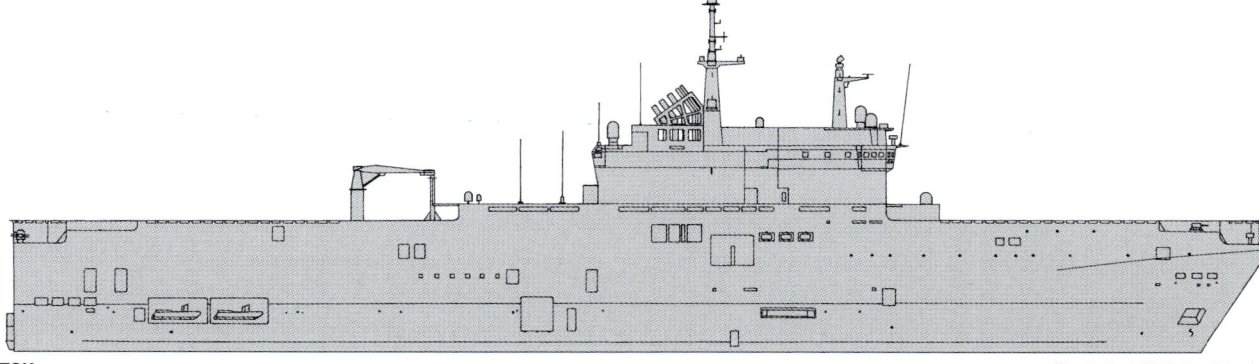

VLADIVOSTOK (Scale 1 : 1,200), Ian Sturton / 1483597

VLADIVOSTOK 3/2014*, B Prézelin / 1531228

1 LEBED (KALMAR) (PROJECT 1206) CLASS (ACV/LCUJ)

D 145 640

Displacement, tonnes: 88 full load
Dimensions, metres (feet): 24.4 on cushion × ? (80.1 × ?)
Speed, knots: 50. **Range, n miles:** 100 at 50 kt
Complement: 6 (2 officers)

Military lift: 2 light tanks or 40 tons cargo or 120 troops
Machinery: 2 Ivchenko AI-20K gas turbines for lift and propulsion; 8,000 hp(m) (5.88 MW)
Guns: 2—30 mm/65 AK 306; 6 barrels per mounting; 3,000 rds/min combined to 2 km
Radars: Navigation: Kivach; I-band

Comment: First entered service 1975. Has a bow ramp with gun on starboard side and the bridge to port. Last remaining one based in the Caspian.

9 + 4 SERNA (PROJECT 11770) CLASS (LCU)

DKA 144 575	DKA 172 645	KONTR-ADMIRAL DEMIDOV (ex-DKA 1441) –
DKA 131 631	DKA 107 650	KONTR-ADMIRAL OLENIN (ex-DKA 1442) –
DKA 56 634	DKA 67 747	+4
DKA 156 630		

Displacement, tonnes: 107 full load
Dimensions, metres (feet): 26.3 × 5.8 × 1.52 (86.3 × 19.0 × 5.0)
Speed, knots: 30. **Range, n miles:** 100 at 30 kt, 600 at 22 kt
Complement: 6
Military lift: 45 tons or 100 troops
Machinery: 2 M 503A3 diesels; 5,522 hp(m) (4.06 MW); 2 shafts

Comment: High-speed utility landing craft capable of beaching and in service in May 1995. Have an 'air-lubricated' hull. Designed for both military and civilian use by the R Alexeyev Central Design Bureau and built at Nizhny Novgorod. Can be armed. The original ship, DKA 67, was built in the 1990s and is based in the Baltic. There are six further craft of a modified design: DKA 144 is based in the Black Sea and DKA 56, 131, 156 and 172 in the Caspian. DKA 107 is in the Pacific. Four further craft are also reported under construction.

LEBED CLASS 6/2005, Lemachko Collection / 1159874

DKA 67 7/2012, Hartmut Ehlers / 1483546

© 2014 IHS IHS Jane's Fighting Ships 2014-2015

706 Russian Federation > Amphibious forces

4 ALLIGATOR (TAPIR) (PROJECT 1171) CLASS (LSTM)

SARATOV (ex-*Voronezhsky Konsomolets*) 150　　**NIKOLAY VILKOV** 081 (ex-IV)
NIKOLAY FILCHENKOV 152　　**ORSK** (ex-*Nicolay Obyenko*) 148

Displacement, tonnes: 3,455 standard; 4,775 full load
Dimensions, metres (feet): 113 × 15.5 × 4.5 *(370.7 × 50.9 × 14.8)*
Speed, knots: 18
Range, n miles: 10,000 at 15 kt
Complement: 100

Military lift: 300 troops; 1,750 tons including about 20 tanks and various trucks; 40 AFVs
Machinery: 2 diesels; 9,000 hp(m) *(6.6 MW)*; 2 shafts
Missiles: SAM: 2 or 3 SA-N-5 Grail twin launchers; manual aiming; IR homing to 6 km *(3.2 n miles)* at 1.5 Mach; altitude to 2,500 m *(8,000 ft)*; warhead 1.5 kg; 16 missiles.
Guns: 2—57 mm/75 AK 725 (twin); 120 rds/min to 12 km *(6.8 n miles)*; weight of shell 2.8 kg.
4—25 mm/80 (2 twin) (Type 4); 270 rds/min to 3 km *(1.6 n miles)*; weight of shell 0.34 kg.
Rockets: 1—122 mm UMS-72 Grad-M; 2—40-barrelled rocket launchers (in Types 3 and 4); range 9 km *(5 n miles)*.
Radars: Surface search: 2 Don 2; I-band.
Weapon control systems: 1 Squeeze Box optronic director (Types 3 and 4).

Programmes: First ship commissioned in 1966 at Kaliningrad. Last of class in service completed in 1976. Type name is *bolshoy desantny korabl* meaning large landing ship. One more Type 3 in service with Ukraine.
Structure: These ships have ramps on the bow and stern. In Type 3 the bridge structure has been raised and a forward deck house has been added to accommodate shore bombardment rocket launchers. Type 4 is similar to Type 3 with the addition of two twin 25 mm gun mountings on centreline abaft the bridge superstructure. As well as a tank deck 300 ft long stretching right across the hull there are two smaller deck areas and a hold.
Operational: In the 1980s the class operated regularly off West Africa, in the Mediterranean and in the Indian Ocean, usually with Naval Infantry units embarked. Half the class have been scrapped or laid up. Of the remainder, *Vilkov* is in the Pacific and the others in the Black Sea.
Sales: One to Ukraine in 1995.

SARATOV　　2/2013*, C D Yaylali / 1530977

0 + 1 (1) IVAN GREN (PROJECT 11711E) CLASS (LSTHM)

Name	No	Builders	Laid down	Launched	Commissioned
VITSE ADMIRAL IVAN GREN	–	Yantar Shipyard, Kaliningrad	23 Dec 2004	18 May 2012	2014

Displacement, tonnes: 5,080 full load
Dimensions, metres (feet): 120 × 16.5 × 3.6 *(393.7 × 54.1 × 11.8)*
Speed, knots: 18
Range, n miles: 3,500 at 16 kt
Complement: 100

Military lift: 300 troops; 13–60 ton tanks or 36 armoured personnel carriers
Machinery: 2 diesels; 10,000 hp(m) *(7.5 MW)*; 2 shafts
Guns: 1—3 in *(76 mm)*/60; AK-176; 120 rds/min to 15 km *(8 n miles)*; weight of shell 7 kg.
2—30 mm/65 AK-630; 6 barrels per mounting; 3,000 rds/min to 2 km.
Rockets: 2—140 mm Grad-M multilaunch rocket system.
Radars: Surface search: Positiv ME; I-band.
Helicopters: 1 Ka-29 Helix B.

Comment: A new class of amphibious ship which is a modified version of the Alligator class landing ships which were built between 1966–76. Progress in building the first of class has been slow and the future of the class is uncertain. However, it was announced in mid-2012 that a second ship was to be ordered.

IVAN GREN　　7/2012, Hartmut Ehlers / 1483544

10 ONDATRA (AKULA) (PROJECT 1176) CLASS (LCMS)

DKA 57 675	DKA 106 570	DKA 182 533	DKA 325 799	DKA 465 746
DKA 70 677	DKA 148 578	DKA 185 642	DKA 464 590	NIKOLAY RUBTSOV 555

Displacement, tonnes: 147 full load
Dimensions, metres (feet): 24 × 5 × 1.5 *(78.7 × 16.4 × 4.9)*
Speed, knots: 10
Range, n miles: 500 at 5 kt
Complement: 5
Military lift: 1 MBT
Machinery: 2 diesels; 300 hp(m) *(220 kW)*; 2 shafts

Comment: First of approximately 35 units built in 1971. Of remaining units, two (DKA 325, 465) in the Baltic, one (DKA 106) in the Black Sea, four (DKA 148, 182, 464, *Nikolay Rubtsov*) in the North, two (DKA 57 and 76) in the Pacific and one (DKA 185) in the Caspian. Tank deck of 45 × 13 ft. Two to Yemen in 1983.

ONDATRA CLASS　　7/2010, Lemachko Collection / 1406125

2 POMORNIK (ZUBR) (PROJECT 1232.2) CLASS (ACVM/LCUJM)

YEVGENIY KOCHESHKOV (ex-*MDK-118*) 770　　**MORDOVIYA MDK-94** (ex-*MDK-94*) 782

Displacement, tonnes: 559 full load
Dimensions, metres (feet): 57.6 on cushion × ? *(189 × ?)*
Speed, knots: 63
Range, n miles: 300 at 55 kt
Complement: 31 (4 officers)

Military lift: 3 MBT or 10 APC plus 230 troops (total 130 tons)
Machinery: 5 Type NK-12MV gas-turbines; 2 for lift, 23,672 hp(m) *(17.4 MW)* nominal; 3 for drive, 35,508 hp(m) *(26.1 MW)* nominal
Missiles: SAM: 2 SA-N-5 Grail quad launchers; manual aiming; IR homing to 6 km *(3.2 n miles)* at 1.5 Mach; altitude to 2,500 m *(8,000 ft)*; warhead 1.5 kg.
Guns: 2—30 mm/65 AK 630; 6 barrels per mounting; 3,000 rds/min combined to 2 km.
2—140 mm A-22 Ogon 22-barrelled rocket launchers.
Mines: 2 rails can be carried for 80.
Radars: Surface search: Curl Stone; I-band.
Fire control: Bass Tilt; H/I-band.
IFF: Salt Pot A/B. Square Head.
Weapon control systems: Quad Look (DWU-3) (modified Squeeze Box) optronic director.

Comment: First of class delivered 1986, commissioned in 1988. Last of class launched December 1994. Produced at St Petersburg and at Feodosiya. Bow and stern ramps for ro-ro working. Last survivors are based at Baltiysk and one is still operated by Ukraine. One (plus one from Ukraine) transferred and two new build for Greece by 2005. These were the first Former Soviet Union (FSU) naval platform sales to a NATO country. Four units are being built for the Chinese Navy.

MORDOVIYA　　6/2011, A Sheldon-Duplaix / 1406784

1 + 4 DYUGON (PROJECT 21820) CLASS (LCU)

ATAMAN PLAKHOV　　**DENIS DAVYDOV**　　**IVAN KARTSOV**　　+2

Displacement, tonnes: 234 full load
Dimensions, metres (feet): 45 × 7.6 × 2.2 *(147.6 × 24.9 × 7.2)*
Speed, knots: 50
Complement: 6
Military lift: 140 tons
Machinery: 2 Zvezda diesels; 7,000 hp *(5.2 MW)*; 1 Saturn M70FRU gas turbine; 2 shafts
Radars: To be announced.

Comment: Designed by Alekseeva, Nizhny Novgorod, *Ataman Plakhov*, the first of an expected five fast landing ships was laid down at the Sredne-Nevskiy Shipyard on 21 February 2006 and launched in July 2009. Of aluminium alloy construction, the vessel is a larger version of the Serna class and similarly achieves high speed through the generation of air cavities and a planing effect. The main machinery consists of two diesel engines boosted by a single gas turbine. A second unit, *Denis Davydov*, was laid down at Yaroslavl Shipyard on 18 January 2012. A further craft, named *Lieutenant Rimsky-Korsakov* was laid down on 21 June 2012 and a fourth *Midshipman Lermontov* on 18 January 2013. A Pacific unit, *Ivan Kartsov* is expected to commission in 2014.

DYUGON CLASS　　9/2009, Lemachko Collection / 1305564

Amphibious forces — Mine warfare forces < **Russian Federation** 707

2 AIST (DZHEYRAN) (PROJECT 1232.1) CLASS (ACV/LCUJ)

MDK 18 608 MDK 88 609

Displacement, tonnes: 303 full load
Dimensions, metres (feet): 47.3 on cushion × ? (155.2 × ?)
Speed, knots: 70
Range, n miles: 120 at 50 kt
Complement: 15 (3 officers)

Military lift: 80 tons or 4 light tanks plus 50 troops or 2 medium tanks plus 200 troops or 3 APCs plus 100 troops
Machinery: 2 Type NK-12M gas turbines driving 4 axial lift fans and 4 propeller units for propulsion; 19,200 hp(m) (14.1 MW) nominal
Guns: 4 – 30 mm/65 (2 twin) AK 630; 6 barrels per mounting; 3,000 rds/min combined to 2 km.
Radars: Surface search: Kivach; I-band.
Fire control: Drum Tilt; H/I-band.
IFF: High Pole B. Square Head.

Comment: First produced at Leningrad in 1970, subsequent production at rate of about six every four years. The first large hovercraft for naval use. Similar to UK SR. N4. Type name is *maly desantny korabl na vozdushnoy podushke* meaning small ACV. Modifications have been made to the original engines and some units have been reported as carrying two SA-N-5 quadruple SAM systems and chaff launchers. Based in the Caspian.

AIST CLASS   9/2000, J Cislak / 0105561

MINE WARFARE FORCES

Notes: (1) All remaining Project 1258 Yevgenya (Korond)-class MHCs were laid up by 2001, except for two which may still be used as patrol craft. These are RT 46 (201) and RT 236 (259).
(2) Some 40 to 50 craft of various dimensions, some with cable reels, some self-propelled and unmanned, some towed and unmanned are reported including the 8 m Kater and Volga unmanned mine clearance craft.

RT 236   7/2006, Lemachko Collection / 1305146

0 + 1 (3) ALEXANDRIT (PROJECT 12700) CLASS (MHSC)

Name	No	Builders	Laid down	Launched	Commissioned
BT 730	–	Sredne-Nevsky Shipyard, St Petersburg	22 Sep 2011	2014	2015

Displacement, tonnes: 620 standard; 800 full load
Dimensions, metres (feet): 51.75 × 10.2 × 2.68 (169.8 × 33.5 × 8.8)
Speed, knots: 15
Range, n miles: 1,600 at 10 kt
Complement: 41
Machinery: 1 diesel; 2,500 hp (1.86 MW); 2 shafts
Guns: 1 – 30 mm AK 306. 1 – 14.5 mm.
Radars: To be announced.

Comment: The first of a new class of MCMV was formally laid down in September 2011 although preliminary work is reported to have started in 2002. Designed by Almaz, the ships are to be of GRP construction. It is believed that the first of class was still under construction when, in October 2012, plans to start batch production were announced. It is expected that the class is to become the core of the MCM flotilla. The ships are believed to be equipped with sonars and remotely-operated underwater vehicles for mine-hunting and with traditional minesweeping gear. The machinery configuration has not been confirmed but is expected to include an electrically-powered active-rudder system for mine hunting.

© 2014 IHS

10 NATYA I (AKVAMAREN) (PROJECT 266M) CLASS
(MINESWEEPERS – OCEAN) (MSOM)

North	Pacific	Black
KOMENDOR 808	MT 265 718	VALENTIN PIKUL 770
KONTRADMIRAL VLASOV	MT 264 738	VITSEADMIRAL ZHUKOV 909
(ex-*Machinist*) 855		IVAN GOLUBETS (ex-*Radist*) 911
		TURBINIST 912
		KOVROVETS 913
		VITSE-ADMIRAL ZAKHARIN 908 (ex-611)

Displacement, tonnes: 817 full load
Dimensions, metres (feet): 61, 67 (718, 738, 770) × ? × ? (200.1, 219.8 × ? × ?)
Speed, knots: 16
Range, n miles: 3,000 at 12 kt
Complement: 67 (8 officers)

Machinery: 2 Type M 504 diesels; 5,000 hp(m) (3.67 MW) sustained; 2 shafts; cp props
Missiles: SAM: 2 SA-N-5/8 Grail quad launchers (in some); manual aiming; IR homing to 6 km (3.2 n miles) at 1.5 Mach; altitude to 2,500 m (8,000 ft); warhead 1.5 kg; 18 missiles.
Guns: 4 – 30 mm/65 (2 twin) AK 230; 500 rds/min to 6.5 km (3.5 n miles); weight of shell 0.54 kg or 2 – 30 mm/65 AK 306; 6 barrels per mounting; 3,000 rds/min combined to 2 km. 4 – 25 mm/80 (2 twin); 270 rds/min to 3 km (1.6 n miles); weight of shell 0.34 kg.
A/S Mortars: 2 RBU 1200 5-tubed fixed; range 1,200 m; warhead 34 kg.
Mines: 10.
Depth charges: 62.
Physical countermeasures: MCM: 1 or 2 GKT-2 contact sweeps; 1 AT-2 acoustic sweep; 1 TEM-3 magnetic sweep.
Radars: Surface search: Don 2 or Long Trough; I-band.
Fire control: Drum Tilt; H/I-band (not in all).
IFF: 2 Square Head. High Pole B.
Sonars: MG 79/89; hull-mounted; active minehunting; high frequency.

Programmes: First reported in 1970. Built at Kolpino and Khabarovsk. Type name is *morskoy tralshchik* meaning seagoing minesweeper. MT 264 and MT 265 were a new variant commissioned in 1989 in which AK 306 mounts replaced the twin AK 230 mounts. One further unit, known as Natya III started construction in 1994. *Valentin Pikul*, left St Petersburg for the Black Sea in July 2002. A further development of the class (known as the Agat class), *Vitseadmiral Zacharin*, was launched at the Kolpino Yard on 26 May 2006.
Structure: Some have hydraulic gantries aft. Have aluminium/steel alloy hulls. Some have Gatling 30 mm guns and a different radar configuration without Drum Tilt. The Natya IIIs are 6 m longer than earlier ships.
Operational: Usually operate in home waters but have deployed to the Mediterranean, Indian Ocean and West Africa. Sweep speed is 14 kt.
Sales: India (two in 1978, two in 1979, two in 1980, one in August 1986, two in 1987, three in 1988). Libya (two in 1981, two in February 1983, one in August 1983, one in January 1984, one in January 1985, one in October 1986). Syria (one in 1985). Yemen (one in 1991). Ethiopia (one in 1991). Some have been deleted.

KOMENDOR   5/2010, Lemachko Collection / 1406126

TURBINIST   7/2013*, E & M Laursen / 1530979

VITSE-ADMIRAL ZAKHARIN   7/2009, Lemachko Collection / 1305563

IHS Jane's Fighting Ships 2014-2015

708 Russian Federation > Mine warfare forces

710 Russian Federation > Survey and research ships

22 FINIK (PROJECT 872) CLASS (AGS/AGE/AE)

North	GS 87	GS 260	GS 278	GS 297	GS 392	GS 405
Pacific	GS 44	GS 47	GS 84	GS 272	GS 296	GS 397	GS 404
Black	GS 86	GS 402	PETR GRADOV (ex-VTR 75)
Baltic	GS 270	GS 399	GS 400	GS 403
Caspian	GS 202	ANATOLY GUZHVIN (ex-GS 301)

Displacement, tonnes: 1,219 full load
Dimensions, metres (feet): 61.3 × 10.8 × 3.3 (201.1 × 35.4 × 10.8)
Speed, knots: 13
Range, n miles: 3,000 at 13 kt
Complement: 26 (5 officers) + 9 scientists
Machinery: 2 Cegielski-Sulzer 6AL25/30 diesels; 1,920 hp(m) (1.4 MW); auxiliary propulsion; 2 motors; 204 hp(m) (150 kW); 2 shafts; cp props; bow thruster
Radars: Navigation: Kivach B; I-band.

Comment: Improved Biya class. Built at Northern Shipyard, Gdansk 1978–83. Fitted with 7 ton crane for buoy handling. Can carry two self-propelled pontoons and a boat on well-deck. Some have been used commercially. Ships of same class serve in the Polish Navy. Three transferred to Ukraine in 1997. Some may be laid up. VTR 75, originally built as a survey ship, was converted for use as an ammunition carrier in 2000.

GS 118   6/2003, Lemachko Collection / 0570901

GS 86   7/2013*, E & M Laursen / 1530982

7 ONEGA (PROJECT 1806) CLASS (AGS)

VICTOR SUBBOTIN	SFP 173	SFP 562
AKADEMIK SEMINIKHIN SFP 183	SFP 286
AKADEMIK ISANIN SFP 586	SFP 295

Displacement, tonnes: 2,185 full load
Dimensions, metres (feet): 81 × 11 × 4.2 (265.7 × 36.1 × 13.8)
Speed, knots: 20
Complement: 45
Machinery: 2 gas turbines; 8,000 hp(m) (5.88 MW); 1 shaft
Radars: Navigation: Nayada; I-band.

Comment: Built at Zelenodolsk and first seen in September 1973. Helicopter platform but no hangar in earlier ships of the class but in later hulls the space is taken up with more laboratory accommodation. Used as hydroacoustic monitoring ships. *Akademik Seminikhin* was completed in October 1992 and *Victor Subbotin* in 2006. One to Ukraine in 1997. *Victor Subbotin* based in the Baltic. *Akademik Seminikhin* in the Black Sea, *Akademik Isanin*, SFP 286, and SFP 562 in the Northern Fleet and SFP 173 and SFP 295 in the Pacific.

AKADEMIK SEMINIKHIN   7/2013*, E & M Laursen / 1530984

SFP 295   4/2012, John Mortimer / 1483552

PETR GRADOV   5/2011, L-G Nilsson / 1406787

8 BIYA (PROJECT 871) CLASS (AGS)

North	Pacific	Baltic	Caspian
GS 193	GS 200	GS 204	GS 202
	GS 210	GS 208	
	GS 269	GS 214	

Displacement, tonnes: 778 full load
Dimensions, metres (feet): 55 × 9.8 × 2.6 (180.4 × 32.2 × 8.5)
Speed, knots: 13
Range, n miles: 4,700 at 11 kt
Complement: 25
Machinery: 2 diesels; 1,200 hp(m) (882 kW); 2 shafts; cp props
Radars: Navigation: Don 2; I-band.

Comment: Built at Northern Shipyard, Gdansk 1972–76. With laboratory and one survey launch and a 5 ton crane. Two transferred to Ukraine in 1997.

2 VINOGRAD CLASS (AGOR)

GS 525   GS 526

Displacement, tonnes: 506 full load
Dimensions, metres (feet): 33 × 10.4 × 2.8 (108.3 × 34.1 × 9.2)
Speed, knots: 9
Range, n miles: 1,000 at 6 kt
Complement: 19
Machinery: Diesel-electric; 2 diesels, 2 motors; 1,200 hp(m) (882 kW); 2 trainable props

Comment: Built by Rauma-Repola, Finland, 1985–87 as hydrographic research ships. *GS 525* commissioned 12 November 1985 and *GS 526* on 17 December 1985. *525* is in the Baltic and *526* in the North. Both have side scan sonars. A similar ship has been reported operating with the Northern Fleet.

GS 269   9/2007, Lemachko Collection / 1353287

6 KAMENKA (PROJECT 870) CLASS (AGS)

GS 66   GS 113   GS 118   GS 199   GS 207   GS 211

Displacement, tonnes: 772 full load
Dimensions, metres (feet): 53.5 × 9.1 × 2.6 (175.5 × 29.9 × 8.5)
Speed, knots: 14
Range, n miles: 4,000 at 10 kt
Complement: 25
Machinery: 2 Sulzer diesels; 1,800 hp(m) (1.32 MW); 2 shafts; cp props
Radars: Navigation: Don 2; I-band.
IFF: High Pole.

Comment: Built at Northern Shipyard, Gdansk 1968–69. A 5 ton crane forward. They do not carry a survey launch but have facilities for handling and stowing buoys. Two in the Baltic and four in the Pacific. One transferred to Vietnam in 1979, one to Estonia in 1996 and one to Ukraine in 1997.

GS 526   7/2012, W Globke / 1483553

IHS Jane's Fighting Ships 2014-2015   © 2014 IHS

Survey and research ships < **Russian Federation** 711

1 MARSHAL NEDELIN (PROJECT 1914) CLASS
(MISSILE RANGE SHIP) (AGMH)

MARSHAL KRYLOV

Displacement, tonnes: 24,893 full load
Dimensions, metres (feet): 212 × 27.1 × 7.7 (695.5 × 88.9 × 25.3)
Speed, knots: 20. **Range, n miles:** 22,000 at 16 kt
Complement: 450

Machinery: 2 gas turbines; 54,000 hp(m) (40 MW); 2 shafts
Radars: Air search: Top Plate.
Navigation: 3 Palm Frond; I-band.
Helo control: Fly Screen B; I-band.
Space trackers: End Tray (balloons). Quad Leaf. 3 Quad Wedge. 4 smaller aerials.
Tacan: 2 Round House.
Helicopters: 2–4 Ka-32 Helix C.

Comment: Completed at Admiralty Yard, Leningrad 23 February 1990. Fitted with a variety of space and missile associated electronic systems. Fitted for but not with six twin 30 mm/65 ADG guns and three Bass Tilt fire-control radars. Naval subordinated, the task is monitoring missile tests with a wartime role of command ship. The Ship Globe radome is for SATCOM. Based in the Pacific but operational status doubtful. Second of class deleted.

MARSHAL KRYLOV   3/2007, Lemachko Collection / 1406130

2 PROJECT 19910 CLASS (AGS)

VAYGACH   VALERY FALEYEV

Displacement, tonnes: 1,219 full load
Dimensions, metres (feet): 56.4 × 11.17 × 2.94 (185 × 36.6 × 9.6)
Speed, knots: 12
Complement: 20
Machinery: Diesel-electric; 2 diesel generators; 2 motors; 1,475 hp (1.1 MW); 2 shafts
Radars: To be announced.

Comment: New class of hydrographic ship built by Vympel Shipyard, Rybinsk. *Vaygach* launched on 28 August 2006 and completed in November 2007, she is based in the Baltic. *Valery Faleyev* was completed at Vladivostok on 26 January 2013.

VAYGACH   9/2007, Lemachko Collection / 1305148

1 MOD SORUM (PROJECT 1454) CLASS
(RESEARCH SHIP) (AGE)

TCHUSOVOY GS 31 (ex-OS 572)

Displacement, tonnes: 1,270 standard; 1,722 full load
Dimensions, metres (feet): 59.1 × 12.6 × 4.6 (193.9 × 41.3 × 15.1)
Speed, knots: 14. **Range, n miles:** 3,500 at 13 kt
Complement: 60
Machinery: Diesel electric; 2 Type 5-2 DW2 diesel generators; 2,900 hp(m) (2.13 MW); 1 motor; 2,000 hp(m) (1.47 MW); 1 shaft
Radars: Navigation: 2 Don 2 or Nayada; I-band.

Comment: A variant of the Sorum class ocean tug design completed at Yaroslavl in 1987. The ship was originally built as a towed-array trials platform; the array and towing winch are contained in the aft superstructure. Based in the Northern Fleet, the ship is deployed on general research duties.

1 + 2 ZVEZDOCHKA (PROJECT 20180/20181/20183) CLASS
(AGE/ASR)

Name	No	Builders	Laid down	Launched	Commissioned
ZVEZDOCHKA	600	Zvezdochka Shipyard, Severodvinsk	3 Sep 2004	20 Dec 2007	24 Jul 2010
AKADEMIK KOVALEV	–	Zvezdochka Shipyard, Severodvinsk	20 Dec 2011	2014	2015
AKADEMIK ALEXSANDROV	–	Zvezdochka Shipyard, Severodvinsk	20 Dec 2012	2014	2015

Displacement, tonnes: 5,080 standard; 5,500 full load
Dimensions, metres (feet): 96, 107.6 (*Akademik Kovalev*) × 17.8 × 9.3 (315, 353 × 58.4 × 30.5)
Speed, knots: 14
Complement: 65
Machinery: Diesel electric: 4 generators; 9,011 hp (6.72 MW); 2 motors; 8,756 hp (6.53 MW); 2 azimuth propellers
Radars: To be announced.

Comment: The first multipurpose ship is capable of conducting and supporting salvage operations, underwater research and transport of ammunition. The ship is capable of operating small submersibles and is equipped with a 150 ton crane and a forward helicopter deck. The second ship (Project 20181) is an ice-class armament support ship, which is reported to be 11 m longer than the first of class. The third ship (Project 20183) is an ice-class ship for scientific testing and special-purpose cargo.

ZVEZDOCHKA   6/2012* / 1530983

3 + 2 BAKLAN (PROJECT 19920) CLASS (AGS)

Name	No	Builders	Laid down	Launched	Commissioned
–	BGK-797	Blagoveshchensk Shipyard	17 Oct 2006	14 Jul 2008	17 Jun 2009
–	BGK-2090	Vympel Shipbuilding, Rybinsk	27 Jun 2004	15 May 2008	27 Dec 2008
–	BGK-2148	Vympel Shipbuilding, Rybinsk	–	–	2013
–	–	Blagoveshchensk Shipyard	Mar 2012	–	2014
–	–	Vympel Shipbuilding, Rybinsk	–	–	2014

Displacement, tonnes: 254 full load
Dimensions, metres (feet): 36.44 × 7.9 × 1.98 (119.6 × 25.9 × 6.5)
Speed, knots: 11
Complement: 11
Machinery: 2 diesels; 2 shafts
Radars: To be announced.

Comment: Research vessels. BGK-797 laid down at Blagoveshchensk Shipyard on 17 October 2006.

BGK-797   10/2012, Chris Sattler / 1483555

0 + 1 PROJECT 22010 CLASS (AGOR)

Name	Builders	Laid down	Launched	Commissioned
YANTAR	Yantar Shipyard, Kaliningrad	2010	5 Dec 2012	2014

Displacement, tonnes: 5,314 full load
Dimensions, metres (feet): 108.1 × 17.2 × ? (354.7 × 56.4 × ?)
Machinery: To be announced
Radars: To be announced.

Comment: New research vessel announced in mid-2009. The ship may be operated by the Russian Navy or by the Russian Academy of Science. The size suggests extensive research facilities and long endurance. A gantry on the stern may be to facilitate the operation of submersibles.

YANTAR   1/2013* / 1530980

TCHUSOVOY   7/2008 / 1336045

© 2014 IHS   IHS Jane's Fighting Ships 2014-2015

712 Russian Federation > Survey and research ships — Intelligence vessels

1 + 1 PROJECT 11982 CLASS (AGOR)

Name	Builders	Laid down	Launched	Commissioned
SELIGER	Yantar Shipyard, Kaliningrad	20 Aug 2009	20 Jul 2011	25 Dec 2012
LADOGA	Pella Shipyard, St Petersburg	2014	2015	2016

Displacement, tonnes: 1,135 full load
Dimensions, metres (feet): 59.7 × 10.8 × ? *(195.9 × 35.4 × ?)*
Speed, knots: 13. **Range, n miles:** 1,000 at 12 kt
Complement: 16 + 9 scientists
Machinery: Diesel electric; 3 BA 840 diesel generators; 3,350 hp *(2.5 MW)*; 2 BA 280 diesel generators; 750 hp *(560 kW)*; 2 Rolls Royce fixed pitch thrusters; 1,880 hp *(1.4 MW)*; 2 Schottel bow thrusters; 535 hp *(400 kW)*
Radars: To be announced.

Comment: New research and oceanographic vessel designed by Almaz. A previous ship with this name was the Moma class AGI SSV 514. The ship is reported to be capable of operating a deep-diving submersible. An order for a second ship was made in December 2013.

SELIGER 6/2012 / 1483554

INTELLIGENCE VESSELS

Notes: (1) About half the AGIs are fitted with SA-N-5/8 SAM launchers.
(2) SSV in pennant numbers of some AGIs is a contraction of *sudno svyazy* meaning communications vessel.
(3) GS in pennant numbers of some AGIs is a contraction of *gidrograficheskoye sudno* meaning survey ship.

2 BALZAM (ASIA) (PROJECT 1826) CLASS (AGIM)

Name	No	Builders	Commissioned
PRIBALTIKA	SSV 080	Yantar Shipyard, Kaliningrad	Jul 1984
BELOMORE	SSV 571	Yantar Shipyard, Kaliningrad	Dec 1987

Displacement, tonnes: 4,572 full load
Dimensions, metres (feet): 105 × 15.5 × 5 *(344.5 × 50.9 × 16.4)*
Speed, knots: 20
Range, n miles: 7,000 at 16 kt
Complement: 200

Machinery: 2 diesels; 18,000 hp(m) *(13.2 MW)*; 2 shafts
Missiles: SAM: 2 SA-N-5 Grail quad launchers; manual aiming; IR homing to 6 km *(3.2 n miles)* at 1.5 Mach; altitude to 2,500 m *(8,000 ft)*; warhead 1.5 kg; 16 missiles.
Guns: 1—30 mm/65 AK 630; 6 barrels per mounting.
Radars: Surface search: Palm Frond and Don Kay; I-band.
Sonars: Lamb Tail/Mouse Tail VDS can be fitted.

Comment: Notable for twin radomes. Full EW and optronic fits. The first class of AGI to be armed. SSV 080 is based in the Pacific. *Belomore* SSV 571, based in the Northern Fleet, was reported to have been decommissioned, but remains in service. Capable of underway replenishment.

PRIBALTIKA 6/2006, Ships of the World / 1159979

5 VISHNYA (PROJECT 864) CLASS (AGIM)

Name	No	Builders	Commissioned
VIKTOR LEONOV	SSV 175	Northern Shipyard, Gdansk	Jul 1988
PRIAZOVE	SSV 201	Northern Shipyard, Gdansk	Jan 1987
KURILY	SSV 208	Northern Shipyard, Gdansk	Apr 1987
VASSILY TATISCHEV (ex-*Pelengator*)	SSV 231	Northern Shipyard, Gdansk	Apr 1989
FEODOR GOLOVIN (ex-*Meridian*)	SSV 520	Northern Shipyard, Gdansk	Jul 1986

Displacement, tonnes: 3,526 full load
Dimensions, metres (feet): 94.4 × 14.6 × 4.5 *(309.7 × 47.9 × 14.8)*
Speed, knots: 16. **Range, n miles:** 7,000 at 14 kt
Complement: 146

Machinery: 2 Zgoda 12AV25/30 diesels; 4,406 hp(m) *(3.24 MW)* sustained; 2 auxiliary electric motors; 286 hp(m) *(210 kW)*; 2 shafts; cp props
Missiles: SAM: 2 SA-N-5 Grail quad launchers; manual aiming; IR homing to 6 km *(3.2 n miles)* at 1.5 Mach; altitude to 2,500 m *(8,000 ft)*; warhead 1.5 kg.
Guns: 2—30 mm/65 AK 630; 6 barrels per mounting.
Rockets: 2—72 mm 4-tubed rocket launchers.
Radars: Surface search: 2 Nayada; I-band.
Sonars: Lamb Tail VDS can be carried.

Comment: SSV 231 and 520 based in the Baltic, SSV 201 in the Black Sea, SSV 175 in the Northern Fleet and SSV 208 in the Pacific. All have a full EW ship optronic systems and datalinks. Punch Bowl is fitted in SSV 231 and possibly in others. Some superstructure differences in all of the class. SSV 231 reported with modified mainmast in 2006. NBC pressurised citadels. Ice-strengthened hulls. All are comparatively active.

IHS Jane's Fighting Ships 2014-2015

FEODOR GOLOVIN 3/2012, Alan Irwin / 1483556

0 + 2 PROJECT 18280 CLASS (AGI)

Name	No	Builders	Laid down	Launched	Commissioned
ADMIRAL YURI IVANOV	–	Severnaya, St Petersburg	28 Dec 2004	30 Sep 2013	2014
IVAN KHURS	–	Severnaya, St Petersburg	14 Nov 2013	2017	2019

Displacement, tonnes: 4,166 standard
Dimensions, metres (feet): 95 × 16.0 × 4.0 *(311.7 × 52.5 × 13.1)*
Complement: 120
Machinery: 2 diesels; 5,440 hp *(4.06 MW)*
Guns: 1—30 mm AK 630.
Radars: To be announced.

Comment: A new class of AGI. The first is to be based in the Pacific Fleet and the second in the Northern Fleet.

3 MOMA (PROJECT 861M) CLASS (AGI/AGIM)

EKVATOR SSV 418   LIMAN SSV 824   KILDIN – (ex-mod SSV 512)

Displacement, tonnes: 1,260 standard; 1,626 full load
Dimensions, metres (feet): 73.3 × 11.2 × 3.9 *(240.5 × 36.7 × 12.8)*
Speed, knots: 17
Range, n miles: 9,000 at 11 kt
Complement: 66 + 19 scientists
Machinery: 2 Zgoda-Sulzer 6TD48 diesels; 3,300 hp(m) *(2.43 MW)* sustained; 2 shafts; cp props
Missiles: SAM: 2 SA-N-5 Grail quad launchers in some.
Radars: Surface search: 2 Don 2; I-band.

Comment: Modernised ships have a foremast in the fore well-deck and a low superstructure before the bridge. Non-modernised ships retain their cranes in the forward well-deck. Similar class operates as survey ships. Built at Gdansk, Poland between 1968–72. All based in the Black Sea. One to Ukraine in 1996.

KILDIN 2/2013*, C D Yaylali / 1530985

2 ALPINIST (PROJECT 503M/R) CLASS (AGIM)

ZHIGULEVSK GS 19   SYZRAN GS 39

Displacement, tonnes: 1,280 full load
Dimensions, metres (feet): 54 × 10.5 × 4 *(177.2 × 34.4 × 13.1)*
Speed, knots: 13
Range, n miles: 7,000 at 13 kt
Complement: 50

Machinery: 1 SKL 8 NVD 48 A2U diesel; 1,320 hp(m) *(970 kW)* sustained; 1 shaft; bow thruster
Missiles: SAM: 1 SA-N-5 Grail quad launcher (GS 39).
Electronic countermeasures: ESM: 2 Watch Dog; intercept.
Radars: Surface search: Nayada and Kivach; I-band.
Sonars: Paltus; active; high frequency.

Comment: Similar to Alpinist stern-trawlers which were built at about 10 a year at the Leninskaya Kuznitsa yard at Kiev and at the Volvograd shipyard. These AGIs were built at Kiev. In 1987 and 1988 forecastle was extended further aft and the electronics fit upgraded. Both based in the Baltic. GS 7 probably non-operational in the Pacific. A fourth of class converted for ASW training was laid up in 1997.

SYZRAN 9/2012*, Frank Findler / 1529172

© 2014 IHS

DEEP SUBMERGENCE VEHICLES

1 BESTER CLASS RESCUE SUBMERSIBLES (PROJECT 18270) (DSRV)

AS 36

Displacement, tonnes: 51 dived
Dimensions, metres (feet): 17.5 × 3.9 × 5.1 (57.4 × 12.8 × 16.7)
Speed, knots: 4
Complement: 3 + 18 casualty berths
Machinery: Battery-powered; 1 propeller; 2 vertical thrusters; 2 horizontal thrusters

Comment: Designed by the Lazurit Central Design Bureau and built at the Krasnoye Sormovo Shipyard, Nizhny Novgorod in 1994. Can mate with hulls at angles of 45° to horizontal. Endurance 4 hours. Can be carried onboard rescue ship *Alagez* or the salvage *Mikhail Rudnitsky*. Reported diving depth of over 750 m. Has an underwater manipulation system and four viewing ports. AS 36 based in Northern Fleet.

BESTER   6/2004, S Breyer / 1127289

4 PRIZ (PROJECT 1855) CLASS
(SALVAGE SUBMERSIBLES) (DSRV)

AS 26   AS 28   AS 30   AS 34

Displacement, tonnes: 59 dived
Dimensions, metres (feet): 13.5 × 3.8 × 3.9 (44.3 × 12.5 × 12.8)
Speed, knots: 3.3
Complement: 4 + 20 casualty berths

Comment: Designed by the Lazurit Central Design Bureau and built in Nizhny Novgorod 1986–89. Can be carried onboard rescue ship *Alagez* or from the salvage ship *Mikhail Rudnitsky*. Has titanium hull and reported diving depth of over 1,000 m. Endurance 2–3 hours submerged. Has an underwater manipulation system. One (possibly AS 32) was involved in the *Kursk* rescue attempt. AS 28 became trapped on the sea-bottom off the Kamchatka peninsula on 5 August 2005. It was later rescued with the help of the British submarine rescue system. AS 34 based in Northern Fleet. AS 28 in Pacific Fleet.

AS 34   6/2008, Richard Scott / 1336057

3 UNDERWATER WORKING VEHICLES (PROJECT 1839)

AS 25   AS 29   AS 32

Displacement, tonnes: 48 dived
Dimensions, metres (feet): 13.6 × 3.5 × 2.9 (44.6 × 11.5 × 9.5)
Speed, knots: 3.5
Complement: 3

Comment: Entered naval service from 1984. Designed to perform underwater technical work and to assist in submarine rescue operations in depths up to 500 m. Double-hulled.

AS 25   6/2008, Richard Scott / 1336054

4 SALVAGE SUBMERSIBLES (PROJECT 1837) (DSRV)

AS 14   AS 16   AS 18   AS 19

Displacement, tonnes: 46 dived
Dimensions, metres (feet): 12.7 × 3.5 × 3.25 (41.7 × 11.5 × 10.7)
Speed, knots: 3.6
Complement: 3 + 11 casualty berths

Comment: Designed and built by Sudomekh, St Petersburg. Can be carried onboard Kashtan class SS 750 rescue ship and Elbrus class *Alagez*. Has double hull and diving depth of 500 m. Equipped with an underwater manipulation system. Twelve reported to have been built of which some were reported to have been decommissioned in the 1990s. Operational numbers are approximate.

1 RUS (PROJECT 16810) CLASS (RESEARCH SUBMERSIBLE)

AS 37

Displacement, tonnes: 25 dived
Dimensions, metres (feet): 8 × 3.9 × 3.85 (26.2 × 12.8 × 12.6)
Speed, knots: 3
Complement: 3

Comment: Entered naval service in 2000. Designed to perform research and technical underwater work at up to 6,000 m. Titanium spherical hull. Three horizontal propulsion motors, two vertical propulsion motors and one thruster. Based in Baltic fleet.

2 POISK-2 (PROJECT 1832) CLASS (RESEARCH SUBMERSIBLES)

AS 24   AS 27

Displacement, tonnes: 66 dived
Dimensions, metres (feet): 16.3 × 2.5 × 3.3 (53.5 × 8.2 × 10.8)
Speed, knots: 3
Complement: 3

Comment: Entered naval service in 1988 and 1989. Designed to perform research and technical underwater work at up to 2,000 m.

1 KONSUL (PROJECT 16811) CLASS (DSV)

AS 39

Displacement, tonnes: 25 dived
Dimensions, metres (feet): 9 × 3.9 × 2.6 (29.5 × 12.8 × 8.5)
Speed, knots: 3
Complement: 3

Comment: Similar to Project 16810 Rus and designed by the same company. Designed to perform research and underwater technical work at up to 6,000 m. Commissioned in 2011.

AS 39   6/2011 / 1406789

TRAINING SHIPS

Notes: The Mir class sail training ships have no military connections.

10 PETRUSHKA (UK-3) CLASS (AXL)

MK 391   MK 405   MK 1277   MK 1303   MK 1407–1411   MK 1556

Displacement, tonnes: 340 full load
Dimensions, metres (feet): 39.4 × 8.4 × 2.2 (129.3 × 27.6 × 7.2)
Speed, knots: 11. **Range, n miles:** 1,000 at 11 kt
Complement: 43 (30 cadets)
Machinery: 2 Wola H12 diesels; 756 hp(m) (556 kW); 2 shafts

Comment: Training vessels built at Wisla Shipyard, Poland; first one commissioned in 1989. Very similar to the SK 620 class used as ambulance craft. Used for seamanship and navigation training and may be commercially owned. Operational status doubtful.

PETRUSHKA 164   7/2010, Lemachko Collection / 1406131

714 **Russian Federation** > Training ships — Auxiliaries

2 SMOLNY (PROJECT 887) CLASS

PEREKOP 200   **SMOLNY** 210

Displacement, tonnes: 9,297 full load
Dimensions, metres (feet): 138 × 16.2 × 6.5 (452.8 × 53.1 × 21.3)
Speed, knots: 20
Range, n miles: 9,000 at 15 kt
Complement: 467 (12 officers; 330 cadets)

Machinery: 2 Zgoda Sulzer 12ZV 40/48 diesels; 15,000 hp(m) (11 MW); 2 shafts
Guns: 4—3 in (76 mm)/60 (2 twin). 4—30 mm/65 (2 twin)
A/S Mortars: 2 RBU 2500
Electronic countermeasures: ESM: 2 Watch Dog; radar warning
Radars: Air/surface search: Head Net C; 3D; E-band; range 128 km (70 n miles).
Navigation: 4 Don 2; I-band. Don Kay (Perekop); I-band.
Fire control: Owl Screech; G-band. Drum Tilt; H/I-band.
IFF: 2 High Pole A. Square Head.
Sonars: Mouse Tail VDS; active; high frequency.

Comment: Built at Szczecin, Poland. *Smolny* completed in 1976, *Perekop* in 1977. Have considerable combatant potential. Both are active in the Baltic.

PEREKOP   6/2007, Selim San / 1170204

AUXILIARIES

Notes: Two Project 10680 (Belyanka)-class tankers *Amur* and *Pinega* are used for stowing low level radioactive waste.

2 AMGA (PROJECT 1791) CLASS
(MISSILE SUPPORT SHIPS) (AEM)

VETLUGA   **DAUGAVA**

Displacement, tonnes: 6,198 (Vetluga), 6,452 (Daugava) full load
Dimensions, metres (feet): 108 (Vetluga) × 18 (Vetluga) × 4.5 (Vetluga) (354.3 × 59.1 × 14.8)
Speed, knots: 16
Range, n miles: 4,500 at 14 kt
Complement: 210

Machinery: 2 diesels; 9,000 hp(m) (6.6 MW); 2 shafts
Guns: 4—25 mm/80 (2 twin)
Radars: Surface search: Strut Curve; F-band.
Navigation: Don 2; I-band.
IFF: High Pole B.

Comment: Built at Gorkiy. Ships with similar duties to the Lama class. Fitted with a large 55 ton crane forward and thus capable of handling much larger missiles than their predecessors. Each ship has a different length and type of crane to handle later types of missiles. Designed for servicing submarines. *Vetluga* completed in 1976 and *Daugava* (5 m longer than *Vetluga*) in 1981. Both are in the Pacific Fleet. A third of class is laid up in the North.

DAUGAVA   4/2010, Lemachko Collection / 1406132

1 VYTEGRALES II (PROJECT 596P) CLASS
(SUPPLY SHIPS) (AKH/AGF)

DAURIYA (ex-*Vyborgles*) 506

Displacement, tonnes: 6,249 full load
Dimensions, metres (feet): 122.1 × 16.8 × 6.8 (400.6 × 55.1 × 22.3)
Speed, knots: 15
Complement: 46

Machinery: 1 Burmeister & Wain 950VTBF diesel; 5,200 hp(m) (3.82 MW); 1 shaft
Radars: Navigation: Nayada or Palm Frond or Spin Trough; I-band.
CCA: Fly Screen.
Helicopters: 1 Ka-25 Hormone C.

Comment: Standard timber carriers of a class of 27. These ships were modified for naval use in 1966–68 with helicopter flight deck. Built at Zhdanov Yard, Leningrad between 1963 and 1966. *Dauriya* has a deckhouse over the aft hold. The first of class, completed in 1962, was originally *Vytegrales*, but this was later changed to *Kosmonaut Pavel Belyayev* and, with three other ships of this class, converted to Space Support Ships. The civilian-manned ships together with these naval ships are often incorrectly called Vostok or Baskunchak class. *Dauriya* is in the Black Sea, *Sevan*, *Aspheron*, and *Yamal* have been decommissioned. Two others transferred to Ukraine in 1996.

DAURIYA   7/2000, Hartmut Ehlers / 0105572

1 MALINA (PROJECT 2020) CLASS
(NUCLEAR SUBMARINE SUPPORT SHIP) (AS)

PM 63

Displacement, tonnes: 10,668 full load
Dimensions, metres (feet): 137 × 21 × 5.6 (449.5 × 68.9 × 18.4)
Speed, knots: 17
Complement: 260
Machinery: 4 gas turbines; 60,000 hp(m) (44 MW); 2 shafts
Radars: Navigation: 2 Palm Frond or 2 Nayada; I-band.

Comment: Built at Nikolayev. First deployed to Pacific in 1986. PM is an abbreviation of Plavuchaya Masterskaya (Floating workshop). A fourth of class (PM 16) launched early in 1992, was not completed. Designed to support nuclear-powered submarines and surface ships. Carry two 15 ton cranes. Based in the Northern Fleet. PM 12 and PM 74 are inactive.

PM 74   7/1996 / 0081704

10 AMUR (PROJECT 304/304M) CLASS (REPAIR SHIPS) (AR)

| PM 30 | PM 82 | PM 140 | PM 59 | PM 86 |
| PM 56 | PM 138 | PM 156 | PM 69 | PM 97 |

Displacement, tonnes: 5,588 full load
Dimensions, metres (feet): 122 × 17 × 5.1 (400.3 × 55.8 × 16.7)
Speed, knots: 12
Range, n miles: 13,000 at 8 kt
Complement: 145
Machinery: 1 Zgoda 8 TAD-48 diesel; 3,000 hp(m) (2.2 MW); 1 shaft
Radars: Navigation: Kivach or Palm Frond or Nayada; I-band.

Comment: Amur I class general purpose depot and repair ships completed 1968–83 in Szczecin, Poland. Successors to the Oskol class. Carry two 5 ton cranes and have accommodation for 200 from ships alongside. Amur II class has extra deckhouse forward of the funnel. Built at Szczecin 1983–85. Four are based in the Pacific and one in the North. Three are laid up in the Baltic and two in the Black Sea. PM 9 transferred to Ukraine.

AMUR 1 PM 56   8/2011, C D Yaylali / 1406790

AMUR II PM 86   9/2000, J Cislak / 0105571

IHS Jane's Fighting Ships 2014-2015

Auxiliaries < **Russian Federation** 715

4 BORIS CHILIKIN (PROJECT 1559V) CLASS
(REPLENISHMENT SHIPS) (AOR)

| BORIS BUTOMA | SEGEI OSIPOV (ex-*Dnestr*) |
| IVAN BUBNOV | GENRICH GASANOV |

Displacement, tonnes: 23,826 full load
Dimensions, metres (feet): 162.1 × 21.4 × 10.3 *(531.8 × 70.2 × 33.8)*
Speed, knots: 17. **Range, n miles:** 10,000 at 16 kt
Complement: 75 (est.)

Cargo capacity: 13,000 tons oil fuel and dieso; 400 tons ammunition; 400 tons spares; 400 tons victualling stores; 500 tons fresh water
Machinery: 1 diesel; 9,600 hp(m) *(7 MW)*; 1 shaft
Guns: 4 — 57 mm/80 (2 twin). Most are fitted for but not with the guns.
Radars: Air/surface search/fire control: Strut Curve (fitted for but not with). Muff Cob (fitted for but not with).
Navigation: 2 Nayada or Palm Frond; I-band.
IFF: High Pole B.

Programmes: Based on the Veliky Oktyabr merchant ship tanker design. Built at the Baltic Yard, Leningrad; *Vladimir Kolechitsky* completed in 1972, *Osipov* in 1973, *Ivan Bubnov* in 1975, *Genrich Gasanov* in 1977. Last of class *Boris Butoma* completed in 1978.
Structure: This is the only class of purpose-built underway fleet replenishment ships for the supply of both liquids and solids. Although most operate in merchant navy paint schemes, all wear naval ensigns.
Operational: Earlier ships can supply solids on both sides forward. Later ships supply solids to starboard, liquids to port forward. All can supply liquids either side aft and astern. *Osipov* and *Gasanov* are based in the North, *Bubnov* in the Black Sea, and *Butoma* in the Pacific. Most are used for commercial purposes. *Boris Chilikin* transferred to Ukraine in 1997.

SERGEI OSIPOV 6/2011, Bob Fildes / 1406791

BORIS BUTOMA 8/2012, Hachiro Nakai / 1483619

30 BOLVA (PROJECT 688/688A) CLASS
(BARRACKS SHIPS) (YPB)

Displacement, tonnes: 6,604 full load
Dimensions, metres (feet): 171 × 14 × 3 *(561 × 45.9 × 9.8)*
Cargo capacity: 350–400 tons

Comment: A total of 59 built by Valmet Oy, Helsinki between 1960 and 1984. Of the remaining 30 ships, six are Bolva 1, 16 are Bolva 2 and eight are Bolva 3. Used for accommodation of ships' companies during refit and so on. The Bolva 2 and 3 have a helicopter pad. Have accommodation facilities for about 400 people. No means of propulsion but can be steered. In addition there are several other types of Barracks Ships including five ex-Atrek class depot ships as well as converted merchant ships and large barges. At least 18 have been scrapped.

IMATRA (at Sevastopol) 3/2002, Hartmut Ehlers / 0529803

3 DUBNA CLASS (REPLENISHMENT TANKERS) (AOL/AOT)

DUBNA   PECHENGA   IRKUT

Displacement, tonnes: 11,685 full load
Dimensions, metres (feet): 130 × 20 × 7.2 *(426.5 × 65.6 × 23.6)*
Speed, knots: 16. **Range, n miles:** 7,000 at 16 kt
Complement: 70
Cargo capacity: 7,000 tons fuel; 300 tons fresh water; 1,500 tons stores
Machinery: 1 Russkiy 8DRPH23/230 diesel; 6,000 hp(m) *(4.4 MW)*; 1 shaft
Radars: Navigation: 2 Nayada; I-band.

Programmes: Completed 1974 at Rauma-Repola, Finland.
Structure: *Dubna* has 1 ton replenishment stations forward. Normally painted in merchant navy colours.
Operational: *Dubna* can refuel on either beam and astern. *Pechenga* has had RAS gear removed. One of the class transferred to Ukraine in 1997. *Irkut* and *Pechenga* are based in the Pacific and *Dubna* in the North.

© 2014 IHS

IRKUT 10/2012, Chris Sattler / 1483557

PECHENGA 11/2011, Ships of the World / 1406881

6 MOD ALTAY CLASS (PROJECT 160)
(REPLENISHMENT TANKERS) (AOL)

| PRUT | YELNYA | ILIM |
| KOLA | IZHORA | YEGORLIK |

Displacement, tonnes: 7,366 full load
Dimensions, metres (feet): 106.2 × 15.5 × 6.7 *(348.4 × 50.9 × 22.0)*
Speed, knots: 14
Range, n miles: 8,600 at 12 kt
Complement: 60
Cargo capacity: 4,400 tons oil fuel; 200 m³ solids
Machinery: 1 Burmeister & Wain BM550VTBN110 diesel; 3,200 hp(m) *(2.35 MW)*; 1 shaft
Radars: Navigation: 2 Don 2 or 2 Spin Trough; I-band.

Comment: Built from 1967–72 by Rauma-Repola, Finland. Modified for alongside replenishment. This class is part of 38 ships, being the third group of Rauma types built in Finland in 1967. *Ilim* and *Yegorlik* transferred to civilian companies in 1996/97 and operate in the Pacific with *Izhora*. *Prut* in the North, *Yelnya* and *Kola* in the Baltic.

IZHORA 4/2009, Ships of the World / 1305382

1 OLEKMA CLASS (PROJECT 6404)
(REPLENISHMENT TANKERS) (AORL)

IMAN

Displacement, tonnes: 7,417 full load
Dimensions, metres (feet): 105.1 × 14.6 × 6.7 *(344.8 × 47.9 × 22.0)*
Speed, knots: 14. **Range, n miles:** 8,000 at 14 kt
Complement: 40
Cargo capacity: 4,500 tons oil fuel; 180 m³ solids
Machinery: 1 Burmeister & Wain diesel; 2,900 hp(m) *(2.13 MW)*; 1 shaft
Radars: Navigation: Don 2 or Nayada and Spin Trough; I-band.

Comment: Built by Rauma-Repola, Finland in 1966. Modified for replenishment with refuelling rig abaft the bridge as well as astern refuelling. Based in the Baltic.

IMAN 6/2006 / 1164807

17 SK 620 CLASS (DRAKON) (TENDERS) (YH/YFL)

MK 391	PSK 382	SK 1411–1412	SN 126	SN 1318
MK 1303	PSK 405	PSK 1518	SN 128	SN 1520
MK 1407–1409	PSK 673	SN 109	SN 401	

Displacement, tonnes: 240 full load
Dimensions, metres (feet): 33 × 7.4 × 2.1 *(108.3 × 24.3 × 6.9)*
Speed, knots: 12
Range, n miles: 1,000 at 12 kt
Complement: 14 + 3 spare berths
Machinery: 2 56ANM30-H12 diesels; 620 hp(m) *(456 kW)* sustained; 2 shafts

Comment: Built at Wisla Shipyard, Poland as a smaller version of the Petrushka class training ship. PSK series serve as harbour ferries. Mostly used as hospital tenders capable of carrying 15 patients.

IHS Jane's Fighting Ships 2014-2015

716 Russian Federation > Auxiliaries

4 UDA CLASS (PROJECT 577D)
(REPLENISHMENT TANKERS) (AOL)

LENA   VISHERA   KOYDA   DUNAY

Displacement, tonnes: 5,588 standard; 7,240 full load
Dimensions, metres (feet): 122.1 × 15.8 × 6.2 *(400.6 × 51.8 × 20.3)*
Speed, knots: 17. **Range, n miles:** 4,000 at 15 kt
Complement: 85

Cargo capacity: 2,900 tons oil fuel; 100 m³ solids
Machinery: 2 diesels; 9,000 hp(m) *(6.6 MW)*; 2 shafts
Radars: Navigation: 2 Don 2 or Nayada/Palm Frond; I-band.
IFF: High Pole A.

Comment: Built between 1962 and 1967 at Vyborg Shipyard. All have a beam replenishment capability. Guns removed. *Vishera* and *Dunay* in the Pacific, *Koyda* in the Black Sea and *Lena* in the Baltic.

LENA   8/2004 / 1042325

2 MANYCH (PROJECT 1549) CLASS (WATER TANKERS) (AWT)

MANYCH   TAGIL

Displacement, tonnes: 7,824 full load
Dimensions, metres (feet): 116 × 15.7 × 7.0 *(380.6 × 51.5 × 23.0)*
Speed, knots: 18. **Range, n miles:** 7,500 at 16 kt
Complement: 90

Cargo capacity: 4,400 tons
Machinery: 2 diesels; 9,000 hp(m) *(6.6 MW)*; 2 shafts
Radars: Air/surface search: Strut Curve; E/F-band.
Navigation: Don Kay; I-band.

Comment: Distilled water carrier built at Vyborg and completed in 1972. Decommissioned and disarmed in 1996 but returned to service in 1998 after refit in Bulgaria. *Manych* re-assigned from the Black Sea to the Northern Fleet in 2011. *Tagil* based in the Pacific.

MANYCH   7/2012, W Globke / 1483558

2 KALININGRADNEFT CLASS (SUPPORT TANKER) (AORL)

VYAZMA (ex-*Katun*)   KAMA (ex-*Argun*)

Displacement, tonnes: 8,738 full load
Dimensions, metres (feet): 116 × 17 × 6.5 *(380.6 × 55.8 × 21.3)*
Speed, knots: 14
Range, n miles: 5,000 at 14 kt
Complement: 32
Cargo capacity: 5,400 tons oil fuel and other liquids
Machinery: 1 Russkiy Burmeister & Wain 5DKRP50/110-2 diesel; 3,850 hp(m) *(2.83 MW)*; 1 shaft
Radars: Navigation: Okean; I-band.

Comment: Built by Rauma-Repola, Finland in 1982. Can refuel astern. At least an additional 20 of this class operate with the fishing fleets. Operational in the Northern Fleet.

KAMA   8/2012, W Globke / 1483559

3 KHOBI CLASS (PROJECT 437M) CLASS (YO)

LOVAT   SOSHA   ORSHA

Displacement, tonnes: 1,544 full load
Dimensions, metres (feet): 67.4 × 10.1 × 3.6 *(221.1 × 33.1 × 11.8)*
Speed, knots: 13. **Range, n miles:** 2,000 at 10 kt
Complement: 30
Machinery: 1 diesel; 1,600 hp *(1.2 MW)*; 2 shafts
Radars: Navigation: Don-2; I-band.

Comment: All based in the Baltic. Used for the transport of all forms of liquids.

IHS Jane's Fighting Ships 2014-2015

LOVAT   7/2008, Hartmut Ehlers / 1353289

30 TOPLIVO CLASS (PROJECT 1844/1844D) CLASS (YO)

VTN SERIES

Displacement, tonnes: 1,199 full load
Dimensions, metres (feet): 54.3 × 7.4 × 3.2 *(178.1 × 24.3 × 10.5)*
Speed, knots: 10
Range, n miles: 1,500 at 10 kt
Complement: 20
Machinery: 1 diesel; 600 hp *(450 kW)*; 1 shaft
Radars: Navigation: Don-2; I-band.

Comment: Details given are for the Toplivo-2 class, some of which were built in Egypt but the majority in the USSR. The Toplivo-3 class, built in the USSR, are slightly larger at 1,300 tons full load. Numbers remaining in service are approximate. All the original Toplivo-1 class are believed to have been decommissioned.

VTN 30   7/2008, Hartmut Ehlers / 1353290

3 OB (PROJECT 320) CLASS (HOSPITAL SHIPS) (AHH)

YENISEI   SVIR   IRTYSH

Displacement, tonnes: 11,756 full load
Dimensions, metres (feet): 152.3 × 19.4 × 6.3 *(499.7 × 63.6 × 20.7)*
Speed, knots: 19
Range, n miles: 10,000 at 18 kt
Complement: 207 (83 medical personnel)

Machinery: 2 Zgoda-Sulzer 12ZV40/48; 15,600 hp(m) *(11.47 MW)* sustained; 2 shafts; cp props
Radars: Navigation: 3 Don 2 or 3 Nayada; I-band.
IFF: High Pole A.
Helicopters: 1 Ka-25 Hormone C.

Comment: Built at Szczecin, Poland. *Yenisei* completed 1981 and is based in the Black Sea. *Svir* completed in early 1989 and transferred to the Northern Fleet in September 1989. *Irtysh* completed in June 1990, was stationed in the Gulf in 1990–91 and is now based in the Pacific. A fourth of class is derelict and a fifth was cancelled. Have 100 beds and seven operating theatres. The first purpose-built hospital ships in the Navy, a programme which may have been prompted by the use of several merchant ships off Angola for Cuban casualties in the 'war of liberation.' NBC pressurised citadel. Ship stabilisation system. Decompression chamber. All are in use, mostly as alongside medical facilities.

IRTYSH   11/2011, Ships of the World / 1406885

0 + 3 PROJECT 23120 (LOGISTIC SUPPORT SHIPS) (AFT)

ELBRUS   MB 75

Displacement, tonnes: 9,000 full load
Dimensions, metres (feet): 90 × 22.0 × 8.65 *(295.3 × 72.2 × 28.4)*
Speed, knots: 18
Complement: 27
Machinery: To be announced.
Radars: To be announced.

Comment: The contract for the construction of three logistic support ships was placed with Severnaya Verf, St Petersburg, on 29 June 2012. The first ship was laid down in October 2012 and they are to be delivered in November 2014, November 2015 and November 2016 to the Northern, Black Sea and Pacific fleets respectively. The ships, of steel construction, are for the shipment of dry cargoes.

© 2014 IHS

Auxiliaries < **Russian Federation** 717

2 KLASMA (PROJECT 1274) CLASS (CABLE SHIPS) (ARC)

DONETS   INGURI

Displacement, tonnes: 6,096 standard; 7,011 full load
Measurement, tonnes: 5,879 gt; 3,455 dwt
Dimensions, metres (feet): 130.5 × 16 × 5.8 (428.1 × 52.5 × 19.0)
Speed, knots: 14. **Range, n miles:** 12,000 at 14 kt
Complement: 118
Machinery: Diesel-electric; 5 Wärtsilä Sulzer 624TS diesel generators (4 in Inguri); 5,000 hp(m) (3.68 MW); 2 motors; 2,150 hp(m) (1.58 MW); 2 shafts
Radars: Navigation: Spin Trough and Nayada; I-band.

Comment: Donets built at the Wärtsilä, Åbovarvet in 1968–69. Inguri completed in 1978. Both are ice strengthened and can carry 1,650 miles of cable. Donets is in the Baltic, and Inguri in the North. One to Ukraine in 1997.

INGURI   7/2012, W Globke / 1483560

4 EMBA (PROJECT 1172/1175) CLASS (CABLE SHIPS) (ARC)

SETUN   NEPRYADAVA   KEM   BIRIUSA

Displacement, tonnes: 2,083 (Nepryadava, Setun), 2,439 (Biriusa, Kem) full load
Dimensions, metres (feet): 75.9 (Nepryadava, Setun), 86.1 (Biriusa, Kem) × 12.6 × 3 (249, 282.5 × 41.3 × 9.8)
Speed, knots: 11
Complement: 40
Machinery: Diesel-electric; 2 Wärtsilä Vasa 6R22 diesel alternators; 2,350 kVA 60 Hz; 2 motors; 1,360 hp(m) (1 MW); 2 shafts (Group I) 2 Wärtsilä Vasa 8R22 diesel alternators; 3,090 kVA 60 Hz; 2 motors; 2,180 hp(m) (1.6 MW); 2 shafts (Group II) The 2 turnable propulsion units can be inclined to the ship's path giving, with a bow thruster, improved turning movement
Radars: Navigation: Kivach and Don 2; I-band.

Comment: Both Emba Is built in 1981. Designed for shallow water cable-laying. Carry 380 tons of cable. Order placed with Wärtsilä in January 1985 for two larger (Group II) ships; Kem completed on 23 October 1986. Can lay about 600 tons of cable. Designed for use off Vladivostok but also capable of operations in inland waterways. Setun is based in the Black Sea, Nepryadava in the Baltic, and Kem and Biriusa are in the Pacific. Both of the latter two were active in 2005.

SETUN   6/2003, Lemachko Collection / 0573515

4 MIKHAIL RUDNITSKY (PROJECT 05360/1) CLASS (SALVAGE AND MOORING VESSELS) (ARS)

MIKHAIL RUDNITSKY   GEORGY KOZMIN   GEORGY TITOV   SAYANY

Displacement, tonnes: 10,872 full load
Dimensions, metres (feet): 130.3 × 17.3 × 7.3 (427.5 × 56.8 × 24.0)
Speed, knots: 16. **Range, n miles:** 12,000 at 15.5 kt
Complement: 72 (10 officers)
Machinery: 1 S5DKRN62/140-3 diesel; 6,100 hp(m) (4.48 MW); 1 shaft
Radars: Navigation: Palm Frond; Nayada; I-band.
Sonars: MG 89 (Sayany).

Comment: Built at Vyborg, based on Moskva Pionier class merchant ship hull. First completed 1979, second in 1980, third in 1983 and fourth in 1984. Fly flag of Salvage and Rescue Service. Have two 40 ton and one 20 ton lift with cable fairleads forward and aft. This lift capability is adequate for handling small submersibles, such as Project 1855 Priz, one of which is carried in the centre hold. Sayany is also described as a research ship and has a high-frequency sonar. Rudnitsky and Project 1837 submersible took part in the Kursk rescue attempts in August 2000. Rudnitsky and Titov in the Northern Fleet and the other two in the Pacific.

GEORGY TITOV   8/2012, W Globke / 1483566

8 KASHTAN (PROJECT 141) CLASS (BUOY TENDERS) (ABU/AGL/ARS)

ALEXANDR PUSHKIN – (ex-KIL 926)	– KIL 498	– SS 750 (ex-KIL 140)
– KIL 143	– KIL 927	– KIL 168
– KIL 164	– KIL 158	

Displacement, tonnes: 4,674 full load
Dimensions, metres (feet): 95.5 × 17.2 × 5 (313.3 × 56.4 × 16.4)
Speed, knots: 13.5
Complement: 51 + 20 spare berths
Machinery: 4 Wärtsilä diesels; 29,000 hp(m) (2.31 MW); 2 shafts
Radars: Navigation: 2 Nayada; I-band.

Comment: Enlarged Sura class built at the Neptun Shipyard, Rostock. Ordered 29 August 1986; Alexandr Pushkin handed over in June 1988 and is classified as an AGL in the Baltic; 927 to the Pacific in July 1989; 143 to the North in July 1989; 158 to the Black Sea in November 1989; 164 to the North in January 1990; 498 to the Pacific in November 1990 and 168 to the Pacific in mid-1991. Lifting capacity: one 130 ton lifting frame, one 100 ton derrick, one 12.5 ton crane and one 10 ton derrick. All are civilian operated except SS 750 in the Baltic which is used to support Project 1837 submersibles AS 22 and AS 26.

KIL 164   8/2012, W Globke / 1483561

KIL 158   9/2013*, C D Yaylali / 1530986

6 SURA (PROJECT 419) CLASS (BUOY TENDERS) (ABU)

KIL 1–2   KIL 22   KIL 25   KIL 27   KIL 31

Displacement, tonnes: 2,408 standard; 3,201 full load
Dimensions, metres (feet): 87 × 14.8 × 5 (285.4 × 48.6 × 16.4)
Speed, knots: 12
Range, n miles: 2,000 at 11 kt
Complement: 40
Cargo capacity: 900 tons cargo; 300 tons fuel for transfer
Machinery: Diesel-electric; 4 diesel generators; 2 motors; 2,240 hp(m) (1.65 MW); 2 shafts
Radars: Navigation: 2 Don 2; I-band.

Comment: Heavy lift ships built as mooring and buoy tenders at Rostock in East Germany between 1965 and 1976. Lifting capacity: one 65 ton derrick and one 65 ton stern cage. Have been seen to carry 12 m DSRVs. KIL 1 in the Baltic, KIL 25 in the Black Sea, KIL 2, 22, and 31 in the North. Four others are laid up. One to Ukraine in 1997.

KIL 31   8/2012, W Globke / 1483565

0 + 16 PROJECT 20340 DIVING TENDERS (YTD)

PBK 762   PBK 764   PBK 767   PBK 771

Displacement, tonnes: 118 full load
Dimensions, metres (feet): 28.1 × 5.56 × 1.5 (92.2 × 18.2 × 4.9)
Speed, knots: 13.5
Complement: 8
Machinery: 2 diesels; 1,200 hp (894 kW); 2 shafts
Radars: Navigation: I-band.

Comment: The first of an expected class of 16 vessels was launched at Nizhny Novgorod on 17 September 2013. Larger versions of the Flamingo class, they are equipped with a towed sonar, an unmanned underwater vehicle and a pressure chamber. Four are to be based in the Black Sea, nine in the Baltic and three in the Caspian Flotilla.

718 Russian Federation > Auxiliaries

1 ELBRUS (OSIMOL) (PROJECT 537) CLASS
(SUBMARINE RESCUE SHIP) (ASRH)

ALAGEZ

Displacement, tonnes: 19,305 standard; 22,861 full load
Dimensions, metres (feet): 175.5 × 24.5 × 8.5 (575.8 × 80.4 × 27.9)
Speed, knots: 17. **Range, n miles:** 14,500 at 15 kt
Complement: 420
Machinery: Diesel-electric; 4 diesel generators; 2 motors; 20,000 hp(m) (14.7 MW); 2 shafts
Radars: Navigation: 2 Nayada and 2 Palm Frond; I-band.
Helicopters: 1 Ka-25 Hormone C.

Comment: Very large submarine rescue and salvage ship with icebreaking capability, possibly in view of under-ice capability of some SSBNs. Built at Nikolayev, and completed in 1982. Can carry two submersibles in store abaft the funnel which are launched from telescopic gantries. Based in the Pacific. Probably disarmed.

ALAGEZ   11/2011, Ships of the World / 1406882

12 SHELON I/II (PROJECT 1388/1388M) CLASS (YPT/YAG)

Displacement, tonnes: 274 full load
Dimensions, metres (feet): 46 × 6 × 2 (150.9 × 19.7 × 6.6)
Speed, knots: 26
Range, n miles: 1,500 at 10 kt
Complement: 14
Machinery: 2 diesels; 8,976 hp(m) (6.6 MW); 2 shafts
Radars: Navigation: Spin Trough or Kivach; I-band.

Comment: Type I built 1978–84. Built-in weapon recovery ramp aft. Type II built 1985–87. Type IIs can be used as environmental monitoring ships. One is an Admirals' yacht in the Baltic, and others are used as personnel transports.

KRKH 1668   7/2012, Erik Laursen / 1483564

TL 289   7/2009, Lemachko Collection / 1305557

20 NYRYAT 2 (PROJECT 522) CLASS (DIVING TENDERS) (YDT)

Displacement, tonnes: 57 full load
Dimensions, metres (feet): 21.5 × 3.5 × 1 (70.5 × 11.5 × 3.3)
Speed, knots: 12
Complement: 8
Machinery: 1 Type 3-D-12 diesel; 300 hp(m) (220 kW) sustained; 1 shaft
Guns: Some carry 1—12.7 mm MG on the forecastle.

Comment: Nyryat 2 are the diving tender variants of the 1950s PO 2 class workboat design widely used for both military and civilian use. Transfers: Albania, Bulgaria, Cuba, Guinea, Iraq. Many deleted.

BGK 663   7/2012, W Globke / 1483569

IHS Jane's Fighting Ships 2014-2015

48 FLAMINGO (TANYA) (PROJECT 1415) CLASS
(TENDERS) (YDT)

Displacement, tonnes: 43 full load
Dimensions, metres (feet): 22.2 × 3.9 × 1.4 (72.8 × 12.8 × 4.6)
Speed, knots: 12
Complement: 8
Machinery: 1 Type 3-D-12 diesel; 300 hp(m) (220 kW) sustained; 1 shaft

Comment: Successor to Nyryat II. There are some 28 with RVK numbers (diving tenders). There are also about 20 (PSKA numbers) assigned to the Border Guard for harbour patrol duties. These are known as the Kulik class. Other craft have BSK, RK (workboats) PRDKA (counterswimmer cutter) and BGK (inshore survey) numbers.

FLAMINGO CLASS   7/2013*, E & M Laursen / 1530987

30 NYRYAT I (PROJECT 1896) CLASS (TENDERS) (YDT)

Displacement, tonnes: 122 full load
Dimensions, metres (feet): 28.4 × 5.5 × 1.7 (93.2 × 18.0 × 5.6)
Speed, knots: 12.5
Range, n miles: 1,500 at 10 kt
Complement: 15
Machinery: 1 diesel; 450 hp(m) (331 kW); 1 shaft
Guns: 1—12.7 mm MG (in some).

Comment: Built from 1955. Can operate as patrol craft or diving tenders with recompression chamber. Similar hull and propulsion used for inshore survey craft. Some have BGK, VM or GBP (survey craft) numbers.
Transfers: Albania, Algeria, Cuba, Egypt, Iraq, North Yemen. Many deleted.

NYRYAT I   10/2008, Laursen/Jarnasen / 1353354

32 YELVA (KRAB) (PROJECT 535M) CLASS
(DIVING TENDERS) (YDT)

VM 20	VM 159	VM 270	VM 425	VM 907–911
VM 72	VM 227	VM 277	VM 429	VM 915–916
VM 143	VM 230	VM 409	VM 519	VM 919
VM 146	VM 250	VM 414	VM 725	
VM 152	VM 253	VM 415	VM 807	
VM 154	VM 268	VM 420	VM 809	

Displacement, tonnes: 300 full load
Dimensions, metres (feet): 40.9 × 8 × 2 (134.2 × 26.2 × 6.6)
Speed, knots: 12.5
Range, n miles: 1,870 at 12 kt
Complement: 30
Machinery: 2 Type 3-D-12A diesels; 630 hp(m) (463 kW) sustained; 2 shafts
Radars: Navigation: Spin Trough; I-band.

Comment: Diving tenders built 1971–83. Carry a 1 ton crane and diving bell. Some have submersible recompression chamber. Ice strengthened. One to Cuba 1973, one to Libya 1977. Some have probably been decommissioned.

VM 154   7/2013*, E & M Laursen / 1530988

© 2014 IHS

Auxiliaries < **Russian Federation**   719

1 PROJECT 11980 (DIVING TENDERS) (YDT)

VM 596

Displacement, tonnes: 335 full load
Dimensions, metres (feet): 37.1 × 7.7 × 2.5 (121.7 × 25.3 × 8.2)
Speed, knots: 12.5
Complement: 29
Machinery: 2 diesels; 525 hp (385 kW); 2 shafts

Comment: A new class of diving vessel designed by Almaz Central Design Bureau and built at Vympel Shipyard, Rybinsk. Construction started in the early 1990s but the building programme was suspended until new funds were assigned in 2002. The ship is designed to support diving and salvage operations down to a depth of 60 m and is equipped with the Falkon remote-controlled underwater equipment, which can work at depths up to 300 m. It also carries hydrological instruments and welding equipment for deep-sea work, a satellite television system and a barochamber. The lead vessel was commissioned in the Northern Fleet on 28 November 2004 and is based at Severomorsk.

VM 596 7/2008 / 1336044

1 SALVAGE LIFTING SHIP (YS)

Name	Builders	Launched	Commissioned
KOMMUNA (ex-Volkhov)	De Schelde, Vlissingen	30 Nov 1913	27 Jul 1915

Displacement, tonnes: 2,489 full load
Dimensions, metres (feet): 96 × 20.4 × 4.7 (315 × 66.9 × 15.4)
Speed, knots: 10
Range, n miles: 1,700 at 6 kt
Complement: 250
Machinery: 2 diesels; 2 shafts
Radars: Navigation: I-band.

Comment: Catamaran-hulled vessel fitted with four lifting rigs to enable sunken submarines to be lifted between the hulls. Laid down in 1912, the vessel was thought to have been decommissioned in 1978 but returned to service after a refit from 1980–84. Now based at Sevastopol to support the operation of submersibles.

KOMMUNA 7/2013*, E & M Laursen / 1530989

27 POZHARNY I (PROJECT 364) CLASS
(FIREFIGHTING CRAFT) (YTR)

PZHK 3	PZHK 36–37	PZHK 59	PZHK 79
PZHK 5	PZHK 41–47	PZHK 64	PZHK 82
PZHK 17	PZHK 49	PZHK 66	PZHK 84
PZHK 30–32	PZHK 53–55	PZHK 68	PZHK 86

Displacement, tonnes: 183 full load
Dimensions, metres (feet): 34.9 × 6.1 × 1.8 (114.5 × 20.0 × 5.9)
Speed, knots: 12
Complement: 26
Machinery: 2 Type M 50 diesels; 2,200 hp(m) (1.6 MW) sustained; 2 shafts
Guns: 4 – 12.7 mm (2 twin) MGs (in some).

Comment: Total of 84 built from mid-1950s to mid-1960s. Harbour fire boats but can be used for patrol duties. One transferred to Iraq (now deleted) and two to Ukraine.

PZHK 37 7/2013*, E & M Laursen / 1530991

© 2014 IHS

15 MORKOV (PROJECT 1461.3) CLASS (YTR)

PZHK 415	PZHK 1296	PZHK 1544–1547	PZHK 1859
PZHK 417	PZHK 1378	PZHK 1560	PZHK 2055
PZHK 900	PZHK 1514–1515	PZHK 1680	

Displacement, tonnes: 325 full load
Dimensions, metres (feet): 36.5 × 7.8 × 2.2 (119.8 × 25.6 × 7.2)
Speed, knots: 12.5
Complement: 20
Machinery: 2 diesels; 1,040 hp(m) (764 kW); 2 shafts

Comment: Carry four water monitors. Completed in 1984–86 at Rybinsk. Can be used for patrol/towage. Some are under civilian control.

PZHK 1680 7/2012, Hartmut Ehlers / 1483571

12 PELYM (PROJECT 1799) CLASS
(DEGAUSSING SHIPS) (YDG)

SR 26	SR 188	SR 267	SR 370
SR 111	SR 203	SR 280	AKADEMIK VLADIMIR KOTELNIKOV
SR 179–180	SR 233	SR 334	– (ex-SR 72)

Displacement, tonnes: 1,392 full load
Dimensions, metres (feet): 65.5 × 11.6 × 3.4 (214.9 × 38.1 × 11.2)
Speed, knots: 14
Range, n miles: 1,000 at 13 kt
Complement: 70
Machinery: 1 diesel; 1,536 hp(m) (1.13 MW); 1 shaft
Radars: Navigation: Don 2; I-band

Comment: Built from 1970 to 1987 at Khabarovsk and Gorokhovets. Earlier ships have stump mast on funnel, later ships a tripod main mast and a platform deck extending to the stern. Kotelnikov was laid down in 1991 and commissioned into the Northern Fleet in 2007. Type name is sudno razmagnichivanya meaning degaussing ship. One to Cuba 1982. Several in reserve.

AKADEMIK VLADIMIR KOTELNIKOV 7/2008, Lemachko Collection / 1305642

HARBOUR CRAFT (YFL/YFU)

Comment: There are numerous types of officers' yachts, harbour work-boats, training cutters and trials vessels in all of the major Fleet bases. Class names include P 02 (Project 376) Bryza (Project 772), Project 21270, Kronshdadt class, Urugan and Pobeda.

SERAFIM SAROVSKIY (Project 21270) 7/2012, Hartmut Ehlers / 1483570

IHS Jane's Fighting Ships 2014-2015

720 Russian Federation > Auxiliaries

15 BEREZA (PROJECT 130) CLASS
(DEGAUSSING SHIPS) (YDG)

North	Baltic	Black
SR 74	SR 28	SR 137
SR 216	SR 120	SR 541
SR 478	SR 245	SR 939
SR 548	SR 479	
SR 569	SR 570	
SR 938	SR 936	

Displacement, tonnes: 1,880 standard; 2,084 full load
Dimensions, metres (feet): 69.5 × 13.8 × 4 (228 × 45.3 × 13.1)
Speed, knots: 13. **Range, n miles:** 1,000 at 13 kt
Complement: 48
Machinery: 2 Zgoda-Sulzer 8AL25/30 diesels; 2,938 hp(m) (2.16 MW) sustained; 2 shafts; cp props
Radars: Navigation: Kivach; I-band.

Comment: First completed at Northern Shipyard, Gdansk 1984–1991. One transferred to Bulgaria in 1988. Have NBC citadels and three laboratories. Several in reserve. One to Ukraine in 1997. SR 938 converted to a logistic ship for service in the Polish Navy.

SR 541   10/2008, Laursen/Jarnasen / 1353357

0 + 1 (2) IGOR BELOUSOV (PROJECT 21300) CLASS
(SUBMARINE RESCUE SHIP) (ASRH)

Name	Builders	Laid down	Launched	Commissioned
IGOR BELOUSOV	Admiralty Yard, St Petersburg	24 Dec 2005	30 Oct 2012	2014

Displacement, tonnes: 5,385 standard
Dimensions, metres (feet): 105.1 × 17.2 × 8.1 (344.8 × 56.4 × 26.6)
Speed, knots: 15. **Range, n miles:** 3,000 at 12 kt
Complement: 98

Machinery: Diesel-electric; 6 diesel generators; 2 motors; 6,520 hp (4.86 MW); 2 shafts
Guns: To be announced.
Radars: To be announced.
Helicopters: To be announced.

Comment: Developed by the Almaz Central Marine Design Bureau. Initially, it is expected that two ships are to be built, one each for the Northern and Pacific fleets. The first of class is named after a former minister of shipbuilding of the USSR. A further two ships may be built in order to equip all four fleets. In addition to its principal submarine rescue role, it is likely to have a secondary role as a research ship. Equipment is likely to include a submergence vehicle capable of operation at a depth of down to 700 m, special-purpose deep diver equipment, and a helicopter. In addition, the ship is to be capable of deploying the British Seaeye Panther Plus Remotely Operated Vehicle (ROV). The ROV is to be fitted with sonar, an acoustic tracking system, a suite of cameras to provide rescue planners with underwater pictures of the submarine on the seabed and various cutters and manipulators. The ROV is also capable of inserting emergency life support stores into a distressed submarine and of connecting hoses and lines to a submarine's salvage connections.

IGOR BELOUSOV   8/2013*, Selim San / 1530990

2 BIRA (MP 6) CLASS (STORES SHIP) (AK)

BIRA   IRGIZ

Displacement, tonnes: 2,164 full load
Dimensions, metres (feet): 75 × 11.3 × 4.4 (246.1 × 37.1 × 14.4)
Speed, knots: 11. **Range, n miles:** 3,000 at 10 kt
Complement: 36
Machinery: 2 diesels; 1,600 hp (1.2 MW); 2 shafts
Radars: Navigation: Don-2; I-band.

Comment: Two survivors of 10 cargo (capacity 1,000 t) ships converted into MP 6-class amphibious vessels in the 1960s. Bow doors were subsequently sealed. Both ships have at times been chartered commercially but both now back in military service. Based at Baltysk.

IHS Jane's Fighting Ships 2014-2015

IRGIZ   7/2009, Hartmut Ehlers / 1305410

1 LAMA (TYPE 323/323B) CLASS
(MISSILE SUPPORT SHIP) (AEM)

GENERAL RYABIKOV

Displacement, tonnes: 4,674 full load
Dimensions, metres (feet): 112.8 × 15 × 4.4 (370.1 × 49.2 × 14.4)
Speed, knots: 14
Range, n miles: 6,000 at 10 kt
Complement: 200

Machinery: 2 diesels; 4,800 hp(m) (3 MW); 2 shafts
Missiles: SAM: 4 SA-N-5 Grail quad launchers.
Guns: 2—57 mm (twin). 4—25 mm/80.
Radars: Surface search: Strut Curve; F-band.
Navigation: Don 2; I-band.
IFF: 2 Square Head. High Pole A.

Comment: Built 1968 at Nikolayev. The engines are sited aft to allow for a very large and high hangar or hold amidships for carrying missiles or weapons' spares for submarines, surface ships and missile craft. This is about 12 ft high above the main deck. There are doors at the forward end with rails leading in and a raised turntable gantry or 20 ton travelling cranes for armament supply. The well-deck is about 40 ft long, enough for most missiles to fit horizontally before being lifted for loading. Type name is *plavuchaya masterskaya* meaning floating workshop. Based in the Black Sea and used as a troop-ship during Black Sea operations in 2008. *Voronezh* has been renamed VTR 33 and is an alongside civilian-manned support ship.

GENERAL RYABIKOV   7/2013*, E & M Laursen / 1530992

4 MUNA (TYPE 1823) CLASS (AEL)

VTR 92   VTR 94   OS 114   OS 213

Comment: Built in the 1970s and converted at Nikolayev in 1990. VTR 94 used as ammunition transport ship in the Black Sea.

VTR 94   10/2008, Laursen/Jarnasen / 1353350

2 + 3 (3) PROJECT 14157 DIVING TENDERS (YTD)

Displacement, tonnes: 82 standard
Dimensions, metres (feet): 25 × 4.8 × 1.4 (82 × 15.7 × 4.6)
Speed, knots: 12
Complement: 8
Machinery: 1 diesel; 415 hp (309 kW); 1 shaft
Radars: To be announced.

Comment: The first two craft were ordered in June 2011 and built at Blagoveshchensk Shipyard. Both were delivered in 2013. Three further vessels were ordered from Moscow Shipbuilding and Ship Repair for delivery in 2014. A class of eight is expected.

© 2014 IHS

Auxiliaries — Icebreakers ◄ **Russian Federation** 721

1 PROJECT 20360 AMMUNITION TRANSPORT SHIP (AETL)

Name	No	Builders
DUBNYAK	VTR 79	Oka Shipyard, Navashino

Displacement, tonnes: 2,017 full load
Dimensions, metres (feet): 61.5 × 15.7 × 3.1 *(201.8 × 51.5 × 10.2)*
Speed, knots: 10.5
Machinery: 2 Deutz BF8M 1015 diesels; 1 shaft
Radars: Navigation: I-band.

Comment: Built at Oka Shipyard, Navashino and entered service in 2009.

DUBNYAK   *12/2009, Lemachko Collection* / 1305978

1 POTOK CLASS (TRIALS SHIP) (AG)

OS-138

Displacement, tonnes: 864 full load
Dimensions, metres (feet): 72.1 × 9.4 × 2.5 *(236.5 × 30.8 × 8.2)*
Speed, knots: 17. **Range, n miles:** 2,500 at 12 kt
Machinery: 2 diesels; 4,000 hp(m) *(2.94 MW)*; 2 shafts
Torpedoes: 1 — 12 in *(533 mm)* tube. 1 — 16 in *(406 mm)* tube.

Comment: Last survivor of six purpose-built weapons trials ships constructed in late 1970s. Modified version of the T-58 minesweeper. Originally intended for torpedo testing, tubes formerly fitted on focsle have been removed. There is a large recovery crane aft. Based in the Black Sea.

OS-138   *7/2013*, E & M Laursen* / 1530993

1 KONDA CLASS (SUPPORT TANKER) (YO)

YAKHROMA

Displacement, tonnes: 2,012 full load
Dimensions, metres (feet): 69 × 10.1 × 4.3 *(226.4 × 33.1 × 14.1)*
Speed, knots: 12. **Range, n miles:** 2,200 at 12 kt
Complement: 36
Machinery: 1 diesel; 1,600 hp *(1.2 MW)*; 1 shaft
Radars: Navigation: Don-2; I-band.

Comment: One of four coastal tankers built at Turku, Finland, between 1955–65. Capable of stern refuelling.

YAKHROMA   *7/2008, Hartmut Ehlers* / 1305649

1 LUZA (PROJECT 1541) CLASS (TANKER) (AOS)

DON

Displacement, tonnes: 1,900 full load
Dimensions, metres (feet): 62.5 × 10.7 × 4.3 *(205.1 × 35.1 × 14.1)*
Speed, knots: 12. **Range, n miles:** 2,000 at 11 kt
Complement: 60
Machinery: 1 diesel; 1,000 hp *(746 kW)*; 1 shaft
Radars: Navigation: I-band.

Comment: Built in the 1960s at Kolpino as a Project 1541 special products (for example missile fuel) tanker. Having been reported to have been converted (Project 1852) to a salvage vessel, capable of operating a Poisk submersible, it appears now to have reverted to its original role. Based at Sevastopol.

© 2014 IHS

DON   *7/2013*, E & M Laursen* / 1530994

1 NUCLEAR FUEL CARRIER (AOSR)

IMANDRA

Measurement, tonnes: 5,806 gt
Dimensions, metres (feet): 130.5 × 17.34 × 7.02 *(428.1 × 56.9 × 23.0)*
Speed, knots: 15
Machinery: 1 BMZ 5DKRN62/140 diesel; 6,098 hp *(4.48 MW)*; 1 shaft
Radars: To be announced.

Comment: Built by Baltic Shipyard, St Petersburg, and entered service in 1980.

IMANDRA   *8/2012, W Globke* / 1483562

ICEBREAKERS

Notes: Only military icebreakers are shown in this section. Other icebreakers come under civilian management and are now used predominantly for commercial purposes. Civilian ships include the nuclear powered *50 Let Pobeda, Taymyr, Vaygach, Rossiya, S Soyuz, Yamal*, all of which are operated by the Murmansk Shipping Company. A new Project 22220 33,000-tonne nuclear-powered icebreaker was ordered from Baltic Shipyard on 3 July 2012. Construction started in 2013 and entry into service is planned for 2018. Diesel powered ships include: 20,000 tons: *Ermak, Admiral Makarov, Krasin*; 15,400 tons: *Moskva, Sankt Peterburg*; 14,600 tons: *Kapitan Sorokin, Kapitan Dranitsyn, Kapitan Nikolayev, Kapitan Khlebnikov*; 7,700 tons: *Mudyug*; 6,200 tons: *Magadan, Dikson*; 2,900 tons: *Afanasy Nikitin, Fedor Litke, Georgiy Sedov, Ivan Kruzenshtern, Ivan Moskvitin, Petr Pakhtsuvov, Semen Chelyushkin, Semen Dezhnev, Vasily Poyarkov, Vladimir Rusanov, Yuriy Lisyansky*; 2,240 tons: *Kapitan Bukayev, Kapitan Chadayev, Kapitan Chechkin, Kapitan Krutov, Kapitan Plakhin, Kapitan Zarubin*; 2,200 tons: *Kapitan Babichev, Kapitan Borodkin, Kapitan Chudinov, Kapitan Demidov, Kapitan Evdokimov, Kapitan Metsayk, Kapitan Moshkin, Kapitan Yevdokimov, Avraamiy Zavenyagin*; 2,100 tons: *Kapitan A Radzhabov, Kapitan Kosolabov, Kapitan M Izmaylov*. A Project 22600 142 m icebreaker was ordered from Baltic Shipyard in December 2011. Entry into service is planned for 2015.

50 LET POBEDA   *8/2012, W Globke* / 1483574

SANKT PETERBURG   *5/2011, Per Körnefeldt* / 1406796

IHS Jane's Fighting Ships 2014-2015

722 Russian Federation > Icebreakers — Tugs

5 DOBRYNYA NIKITICH (PROJECT 97) CLASS (AGB)

| BURAN | SADKO | VLADIMIR KAVRAISKY |
| PERESVET | ILYA MUROMETS | |

Displacement, tonnes: 3,043 full load
Measurement, tonnes: 2,290 gt; 1,136 dwt
Dimensions, metres (feet): 67.7 × 18.1 × 6.1 (222.1 × 59.4 × 20.0)
Speed, knots: 14.5
Range, n miles: 5,500 at 12 kt
Complement: 45
Machinery: Diesel-electric; 3 Type 13-D-100 or 3 Wärtsilä 6L 26 (Kruzenshtern) diesel generators; 3 motors; 5,400 hp(m) (4 MW); 3 shafts (1 fwd, 2 aft)
Guns: 2—57 mm/70 (twin). 2—37 mm/63.
Radars: Navigation: 2 Don 2; I-band.

Comment: Built at Admiralty Yard, Leningrad between 1960 and 1971. *Kavraysky* is in the Northern Fleet and *Buran* in the Baltic. Of the 18 others originally built, some have been decommissioned and others (about eight) transferred to civilian service.

BURAN 6/2004, Marco Ghiglino / 1151373

TUGS

Notes: SB means *Spasatelny Buksir* or Salvage Tug. MB means *Morskoy Buksir* or Seagoing Tug.

2 BAKLAZHAN (PROJECT 5757) CLASS
(SALVAGE TUGS) (ATS)

Name	No	Builders	Laid down
NICOLAY CHIKER	SB 131	Hollming, Rauma	1987
FOTIY KRYLOV	SB 135	Hollming, Rauma	1987

Displacement, tonnes: 7,417 standard; 8,128 full load
Dimensions, metres (feet): 99 × 19.5 × 7.1 (324.8 × 64.0 × 23.3)
Speed, knots: 18
Range, n miles: 11,000 at 16 kt
Machinery: 4 Wärtsilä diesels; 24,120 hp(m) (18.0 MW); 2 shafts
Radars: Navigation: I-band.

Comment: Both ships constructed by Hollming, Rauma, Finland. Laid down in 1987 and entered service with the Soviet Navy in 1989. Under ownership of Russian company Sovfracht, operated by Greek company Tsavliris during the 1990s before returning to Russian naval service in about 2006. Both tugs are probably still available for commercial use. Equipped with three water cannons.

FOTIY KRYLOV 2/2011, Ships of the World / 1406877

13 SORUM (PROJECT 745) CLASS (ATA)

| MB 4 | MB 28 | MB 56 | MB 61 | MB 99—100 | MB 148 |
| MB 19 | MB 37 | MB 58 | MB 76 | MB 110 | MB 304 |

Displacement, tonnes: 1,687 full load
Dimensions, metres (feet): 58 × 12.6 × 4.6 (190.3 × 41.3 × 15.1)
Speed, knots: 14
Range, n miles: 3,500 at 13 kt
Complement: 35
Machinery: Diesel-electric; 2 Type 5—2-DW2 diesel generators; 2,900 hp(m) (2.13 MW); 1 motor; 2,000 hp(m) (1.47 MW); 1 shaft
Guns: 4—30 mm/65 (2 twin) (all fitted for, but only Border Guard ships carry them).
Radars: Navigation: 2 Don 2 or Nayada; I-band.
IFF: High Pole B.

Comment: A class of ocean tugs with firefighting and diving capability. Built in Yaroslavl and Oktyabskoye from 1973 to 1989, design used for Ministry of Fisheries rescue tugs.

MB 304 7/2013*, C D Yaylali / 1530997

1 PRUT (PROJECT 527M) CLASS (RESCUE TUG) (ATS)

EPRON (ex-*MB 26*)

Displacement, tonnes: 2,154 standard; 2,845 full load
Dimensions, metres (feet): 90.2 × 14.3 × 5.5 (295.9 × 46.9 × 18.0)
Speed, knots: 20
Range, n miles: 9,000 at 16 kt
Complement: 140
Machinery: Diesel-electric; 4 diesel generators; 2 motors; 10,000 hp(m) (7.35 MW); 2 shafts
Radars: Navigation: Don-2; I-band.

Comment: Large rescue tug built at Nikolayev, Ukraine and completed in 1959. Carries two heavy-duty derricks, submersible recompression chambers, rescue chambers and bells. Last survivor of the class which is based in the Black Sea.

EPRON 9/2004, Hartmut Ehlers / 1042286

4 INGUL (PROJECT 1453) CLASS (SALVAGE TUGS) (ATS)

| PAMIR | MASHUK | ALTAY (ex-*Karabakh*) | ALATAU |

Displacement, tonnes: 4,115 full load
Dimensions, metres (feet): 92.8 × 15.4 × 5.8 (304.5 × 50.5 × 19.0)
Speed, knots: 19
Range, n miles: 9,000 at 19 kt
Complement: 71 + 18 personnel
Machinery: 2 Type 58-D-4R diesels; 6,000 hp(m) (4.4 MW); 2 shafts; cp props
Radars: Navigation: 2 Palm Frond; I-band.
IFF: High Pole. Square Head.

Comment: Built at Admiralty Yard, Leningrad in 1975–84. NATO class name the same as one of the Klasma class cable-ships. Naval-manned arctic salvage and rescue tugs. Two more, *Yaguar* (Murmansk) and *Bars* (Vladivostok), operate with the merchant fleet. Carries 18 salvage personnel, salvage pumps, diving and firefighting gear as well as a high-line for transfer of personnel. Fitted for guns but these are not carried. *Pamir* and *Altay* in the North, *Mashuk*, and *Alatau* in the Pacific.

ALTAY 8/2012, W Globke / 1483577

IHS Jane's Fighting Ships 2014-2015

Tugs — **Russian Federation** 723

7 KATUN CLASS (PROJECT 1893/1993) (SALVAGE TUGS) (ATS)

PZHS 98	PZHS 273	PZHS 282	PZHS 95
PZHS 123	PZHS 279	PZHS 92	

Displacement, tonnes: 1,021 (PZHS 123, 273, 279, 282, PZHS 98), 1,240 (PZHS 92,95) full load
Dimensions, metres (feet): 62 (PZHS 123, 273, 279, 282, PZHS 98), 65.4 (PZHS 92, 95) × 10.1 × 3.5 (PZHS 123, 273, 279, 282, PZHS 98), 3.65 (PZHS 92, 95) (203.4, 214.6 × 33.1 × 11.5, 12.0)
Speed, knots: 17
Range, n miles: 2,000 at 17 kt
Complement: 32
Machinery: 2 diesels; 5,000 hp(m) (3.68 MW); 2 shafts
Radars: Navigation: Spin Trough or Kivach (Katun II); I-band.
IFF: High Pole A.

Comment: Eight Katun I (of which six remain) built at Kolpino 1970–78. Equipped for firefighting and rescue. Two remaining Katun II PZHS 92 and 95 have an extra bridge level and lattice masts. 273 based in the Caspian; 95 in the Pacific; 92 and 98 in the North, 282 in the Baltic and 123 in the Black Sea.

PZHS 92 (KATUN II)   11/2011, Ships of the World / 1406880

10 GORYN (PROJECT 714) CLASS (ARS/ATA)

MB 15	MB 38	MB 119	SB 521–523
EVGENY KHOROV MB 35	MB 105	SB 36	SB 931

Displacement, tonnes: 2,276 standard; 2,642 full load
Dimensions, metres (feet): 63.5 × 14.3 × 5.1 (208.3 × 46.9 × 16.7)
Speed, knots: 15
Complement: 43 + 16 spare berths
Machinery: 2 Russkiy SEMT-Pielstick 6 PC2.5 L 400 diesels; 7,020 hp(m) (5.2 MW) sustained; 2 shafts; cp props; bow thruster
Radars: Navigation: 2 Don 2 or Nayada or Kivach; I-band.

Comment: Built by Rauma-Repola 1977–83. Have sick-bay. First ships have goalpost mast with 10 and 5 ton derricks and bollard pull of 35 tons. Remainder have an A-frame mast with a 15 ton crane and bollard pull of 45 tons. SB number indicates a 'rescue' tug. Three in the North, four in the Pacific, two in the Baltic and one in the Black Sea. One transferred to Ukraine in 1997.

MB 15   8/2012, W Globke / 1483575

2 NEFTEGAZ (PROJECT B-92) CLASS (ATA)

ILGA   KALAR

Displacement, tonnes: 4,077 full load
Dimensions, metres (feet): 80.3 × 16.3 × 5.0 (263.5 × 53.5 × 16.4)
Speed, knots: 15. **Range, n miles:** 5,000 at 12 kt
Complement: 23
Machinery: 2 Sulzer diesels; 7,200 hp (5.3 MW); 2 shafts; cp props
Radars: Navigation: I-band.

Comment: Large oilfield support tugs built by A Warski SY, Szczecin, Poland. Taken over for naval service; some 40 others are in civilian service. Now employed as ocean-going rescue tugs with heavy towing and firefighting capabilities. Kalar also operates in the salvage role. Capacity of 600 tons cargo on deck and 1,000 m³ of liquid cargo. Entered naval service in 1983 (Ilga) and 1990 (Kalar). Ilga based in the Northern Fleet and Kalar in the Pacific.

KALAR   12/2005, Ships of the World / 1151145

12 OKHTENSKY (PROJECT 733/733S) CLASS (ARS/ATA)

KODOR SB 4	– MB 23	– MB 166
MOSHCHNY SB 6	– MB 54	POCHETNYY MB 169
– SB 5	– MB 160	LOKSA MB 171
– MB 8	SERDITY MB 165	– MB 174

Displacement, tonnes: 963 full load
Dimensions, metres (feet): 47.6 × 10.4 × 4.1 (156.2 × 34.1 × 13.5)
Speed, knots: 13
Range, n miles: 8,000 at 7 kt, 6,000 at 13 kt
Complement: 40
Machinery: Diesel-electric; 2 BM diesel generators; 1 motor; 1,500 hp(m) (1.1 MW); 1 shaft
Guns: 2 – 57 mm/70 (twin) or 2 – 25 mm/80 (twin) (Border Guard only).
Radars: Navigation: 1 or 2 Don 2 or Spin Trough; I-band.
IFF: High Pole B.

Comment: Ocean-going salvage (MB) and rescue tugs (SB). First of a total of 62 completed 1958. Fitted with powerful pumps and other apparatus for salvage. A number of named ships are operated by the Border Guard and are armed. Two to Ukraine in 1997. Many have been scrapped.

MB 174   7/2013*, E & M Laursen / 1530998

18 PROMETEY (PROJECT 498/04983/04985) CLASS (TUGS) (YTB)

RB 1	RB 158	RB 202	RB 265	RB 360
RB 7	RB 173	RB 217	RB 296	RB 362
RB 57	RB 179	RB 239	RB 314	
RB 98	RB 201	RB 262	RB 327	

Displacement, tonnes: 366 full load
Dimensions, metres (feet): 29.4 × 8.3 × 3.2 (96.1 × 27.2 × 10.5)
Speed, knots: 11
Machinery: 2 diesels; 1,200 hp(m) (895 kW); 2 shafts

Comment: Entered service 1973–83. Bollard pull 14 tons. Later versions have more powerful engines. Based in the Northern, Pacific, Baltic and Black Sea Fleets.

RB 296   5/2011, L-G Nilsson / 1406798

1 OKHOTSK (PROJECT 22030) CLASS (TUG) (ATA)

ALEXANDER PISKUNOV MB 11

Displacement, tonnes: 1,465 full load
Dimensions, metres (feet): 56.8 × 11.8 × 3.72 (186.4 × 38.7 × 12.2)
Speed, knots: 14.5
Complement: 27
Machinery: Diesel-electric; 3 ABC 8MDZC diesel generators; 6,034 hp (4.5 MW); 2 diesel generators; 536 hp (400 kW); 2 Schorch KL6535B-AS06 motors; 4,190 hp (3.1 MW); 2 azimuth thrusters; 1 bow thruster
Radars: Navigation: I-band.

Comment: Built at Khabarovsk Shipbuilding and delivered in 2013.

2 PROJECT PE-65 TUGS (YTB)

MB 92   MB 93

Displacement, tonnes: 862 full load
Dimensions, metres (feet): 34.4 × 12.1 × 4.4 (112.9 × 39.7 × 14.4)
Speed, knots: 12
Complement: 8
Machinery: 2 Caterpillar 3512B-HD diesels; 5,070 hp (3.78 MW); 2 azimuth thrusters
Radars: Navigation: I-band.

Comment: Built at Pella Shipyard, Otradny, and delivered in September 2013. Based in the Pacific Fleet.

724 Russian Federation > Tugs

38 SIDEHOLE I AND II (PROJECT 737 K/M) CLASS (TUGS) (YTB)

RB 2	RB 43–44	RB 168	RB 240
RB 5	RB 46	RB 192–194	RB 244
RB 17	RB 49	RB 197–199	RB 246–250
RB 20	RB 48	RB 212	RB 255–256
RB 23	RB 50	RB 232–233	RB 295–296
RB 25–26	RB 52	RB 237	RB 310–311

Displacement, tonnes: 209 full load
Dimensions, metres (feet): 24.2 × 7.0 × 3.4 (79.4 × 23.0 × 11.2)
Speed, knots: 10
Machinery: 2 diesels; 900 hp(m) (670 kW); 2 shafts

Comment: Entered service 1973–83. Bollard pull 10 tons. Based in all fleets.

RB 247 7/2013*, E & M Laursen / 1530996

2 + 5 PROJECT 16609 (HARBOUR TUGS) (YTB)

DELFIN **KASATKA** +5

Displacement, tonnes: 504 full load
Dimensions, metres (feet): 29.44 × 10.14 × 3.5 (96.6 × 33.3 × 11.5)
Speed, knots: 12
Complement: 8
Machinery: 2 diesels; 3,550 hp (2.64 MW); 2 azimuth thrusters

Comment: Built at Pella Shipyard, St Petersburg, and entered service in 2012. Both based in the Northern Fleet. Five further vessels are under construction.

2 + 1 MOD SORUM (PROJECT 745MB/745MBS) CLASS (ATA)

– MB 12 **VIKTOR KOPECKY** – +1

Displacement, tonnes: 1,390 full load
Dimensions, metres (feet): 56.5 × 12.65 × 4.3 (185.4 × 41.5 × 14.1)
Speed, knots: 14
Range, n miles: 6,200 at 13 kt
Complement: 22
Machinery: Diesel-electric; 3 ABC 8MDZC diesel generators; 4,023 hp (3.0 MW); 2 diesel generators; 536 hp (400 kW); 1 Schorch KL7540B-AS12 motor; 2,720 hp (2.0 MW); 2 azimuth thrusters; 1 bow thruster

Comment: Modified variants of the Sorum-class ocean tugs. MB 12 was constructed at Zelenodolsk and entered service on 16 December 2011. She is based in the Northern Fleet. *Viktor Kopecky* built at Jaroslavl. A further unit is under construction at Zelenodolsk.

MB 12 8/2012, W Globke / 1483578

19 PROJECT 90600 CLASS (YTL)

RB 20	RB 42–43	RB 386	POMORYE
RB 27	RB 45	RB 389	RB 394–395
RB 34	RB 47–48	RB 391–392	RB 398–401

Displacement, tonnes: 417 full load
Dimensions, metres (feet): 25.4 × 9.3 × 3.3 (83.3 × 30.5 × 10.8)
Speed, knots: 12
Range, n miles: 3,000 at 12 kt
Machinery: 2 diesels; 2,030 hp(m) (1.51 MW); 2 propeller rudders

Comment: A new class of tugs, constructed at Pella Shipyard, Otradny, began entering service in 2009. At least two other vessels are under construction. *Pomorye*, RB 34, 47, 48, 386, 399, and 400 are based in the Northern Fleet, RB 43, 45, 389, 392, and 398 in the Black Sea and RB 20, 27, 42, 394, and 395 in the Baltic.

RB 48 8/2012, W Globke / 1455702

3 SLIVA (PROJECT 712) CLASS (SALVAGE TUGS) (ATS)

VIKR SB 406 **PARADOKS** SB 921 **SHAKHTER** SB 922

Displacement, tonnes: 3,099 full load
Dimensions, metres (feet): 69.2 × 15.4 × 5.1 (227 × 50.5 × 16.7)
Speed, knots: 16
Complement: 43 + 16 personnel
Machinery: 2 Russkiy SEMT-Pielstick 6 PC2.5 L 400 diesels; 7,020 hp(m) (5.2 MW) sustained; 2 shafts; cp props; bow thruster
Radars: Navigation: 2 Nayada; I-band.

Comment: Built at Rauma-Repola, Finland. Based on Goryn design. SB 406 completed 20 February 1984. Second pair ordered 1984, SB 921 completed 5 July 1985 and SB 922 on 20 December 1985. SB 922 named *Shakhter* in 1989. A fourth of class, *Iva* SB 408, was sold illegally to a Greek company in March 1993 and now flies the flag of Cyprus but is operated as a 'joint venture' with the Russian Navy. Diving facilities to 60 m and salvage party of 10 people. Bollard pull 60 tons. SB 406 based in the Northern Fleet, SB 921 in the Baltic and SB 922 in the Black Sea.

VIKR 8/2012, W Globke / 1483576

10 STIVIDOR (PROJECT 192) CLASS (TUGS) (YTB)

RB 22	RB 100	RB 109	RB 167	RB 325
RB 40	RB 108	RB 136	RB 293	RB 326

Displacement, tonnes: 584 full load
Dimensions, metres (feet): 35.7 × 9.5 × 4.6 (117.1 × 31.2 × 15.1)
Speed, knots: 12
Machinery: 2 diesels; 2,400 hp(m) (1.78 MW); 2 shafts; bow-thruster

Comment: Built in former Yugoslavia. Entered service 1980–90. Bollard pull 35 tons. Equipped with three water cannons. Based in the Northern, Pacific and Black Sea Fleets.

RB 136 7/2013*, E & M Laursen / 1530995

RUSSIAN FEDERAL BORDER GUARD SERVICE (EX MARITIME BORDER GUARD)

General

(1) The Border Guard would be integrated with naval operations in a crisis. Formerly run by the KGB, the force came under the Ministry of Defence in October 1991 and was then given to the Ministry of Interior in December 1993. It merged with the Federal Security Service on 11 March 2003.
(2) From 1993 the Border Guard started to fly its own ensign which is the St Andrews Cross with a white border on a green background. Diagonal stripes are painted on the hull which from 2004 have been painted blue.
(3) Roles include Law Enforcement, Port Security, Counter Intelligence, Counter Terrorism and Fishery Protection.

Personnel

2014: 10,000 approx

FRIGATES

3 KRIVAK III (NEREY) (PROJECT 1135MP) CLASS (FFHM)

Name	No	Builders	Laid down	Launched	Commissioned
DZERZHINSKY	158 (ex-097)	Kamish-Burun, Kerch	11 Jan 1984	2 Mar 1984	29 Dec 1984
OREL (ex-*Imeni XXVII Sezda KPSS*)	156	Kamish-Burun, Kerch	26 Sep 1983	2 Nov 1985	30 Sep 1986
VOROVSKY	160 (ex-052)	Kamish-Burun, Kerch	20 Feb 1990	28 Jul 1990	29 Dec 1990

Displacement, tonnes: 3,150 standard; 3,709 full load
Dimensions, metres (feet): 123.5 × 14.3 × 7.3 sonar *(405.2 × 46.9 × 24.0)*
Speed, knots: 32. **Range, n miles:** 4,000 at 14 kt, 1,600 at 30 kt
Complement: 194 (18 officers)

Machinery: COGAG; 2 M8K gas-turbines; 55,500 hp(m) *(40.8 MW)*; 2 M 62 gas-turbines; 13,600 hp(m) *(10 MW)*; 2 shafts
Missiles: SAM: 1 SA-N-4 Gecko (Zif 122) twin launchers ❶; Osa-M semi-active radar homing to 15 km *(8 n miles)* at 2.5 Mach; warhead 50 kg; altitude 9.1–3,048 m *(30–10,000 ft)*; 20 missiles.
Guns: 1—3.9 in *(100 mm)*/70 AK 100 ❷; 60 rds/min to 21.5 km *(11.5 n miles)*; weight of shell 15.6 kg. 2—30 mm/65 ❸; 6 barrels per mounting; 3,000 rds/min combined to 2 km.
Torpedoes: 8—21 in *(533 mm)* (2 quad) tubes ❹. Combination of 53 cm torpedoes (see table at front of section).
A/S Mortars: 2 RBU 6000 12-tubed trainable ❺; range 6,000 m; warhead 31 kg. MRG-7 55 mm grenade launcher.
Mines: Capacity for 16.
Physical countermeasures: Decoys: 4 PK 16 or 10 PK 10 chaff launchers. Towed torpedo decoy.
Electronic countermeasures: ESM/ECM: 2 Bell Shroud. 2 Bell Squat. Half Cup laser warning (in some).
Radars: Air search: Top Plate ❻; 3D; D/E-band.
Surface search: Peel Cone ❼; I-band.
Fire control: Pop Group ❽; F/H/I-band (for SA-N-4). Kite Screech ❾; H/I/K-band. Bass Tilt ❿; H/I-band.
IFF: High Pole B. Salt Pot.
Sonars: Bull Nose (MGK-335S or MG-332); hull-mounted; active search and attack; medium frequency.
Helicopters: 1 Ka-27 Helix ⓫.

KRIVAK III *(Scale 1 : 1,200), Ian Sturton / 0506085*

Programmes: Type name was originally *bolshoy protivolodochny korabl*, meaning large anti-submarine ship. Changed in 1977–78 to *storozhevoy korabl* meaning escort ship. The naval Krivaks are known as the Burevestnik class.
Structure: Krivak III class built for the former KGB but now under Border Guard Control. The removal of SS-N-14 and one SA-N-4 mounting compensates for the addition of a hangar and flight deck.
Operational: *Menzhinsky*, *Pskov*, *Anadyr* and *Kedrov* reported non-operational. Flight decks do not appear to be used for helicopter operations.

VOROVSKY
9/2010, Lemachko Collection / 1406800

VOROVSKY
8/2009, Michael Winter / 1367108

PATROL FORCES

Notes: In addition to the patrol forces listed, *Pluton* 028 and *Strelets* 025, two Yug class former research vessels, are operated as patrol craft in Arctic waters.

10 ALPINIST (PROJECT 503) CLASS (PBO)

| ANTIAS | BARS | PARELLA | ASKOLD PYNIKOV | PS 818 |
| ARGAL | PALIYA | MAGADAN | SKAT | PS 819 |

Displacement, tonnes: 1,168 full load
Dimensions, metres (feet): 53.7 × 10.5 × 4.1 *(176.2 × 34.4 × 13.5)*
Speed, knots: 12
Range, n miles: 7,000 at 12 kt
Complement: 44
Machinery: 1 diesel; 1 shaft; cp prop

Comment: Trawler design adapted for use as fishery protection role. All built at Jaroslavl 1997–2001.

SKAT — *8/2012, W Globke / 1483582*

4 KOMANDOR CLASS (PSO)

| KOMANDOR | SHKIPER GYEK | HERLUF BIDSTRUP | MANCHZHUR |

Displacement, tonnes: 2,474 full load
Dimensions, metres (feet): 88.3 × 13.6 × 4.7 *(289.7 × 44.6 × 15.4)*
Speed, knots: 20
Range, n miles: 7,000 at 19 kt
Complement: 42
Machinery: 2 Russkiy SEMT-Pielstick 6 PC2.5 L400 diesels; 7,020 hp(m) *(5.2 MW)*; 1 shaft; cp prop; bow thruster
Radars: Navigation: Furuno; I-band.
Helicopters: 2 Ka-32 Helix D for SAR.

Comment: Specialist offshore patrol vessels ordered in December 1987 from Danyard, Frederikshaven, Denmark and delivered 1989–1990. The hangar is below the helicopter deck. Transferred from the Ministry of Fisheries to the Federal Border Guard and based in the Pacific.

HERLUF BIDSTRUP — *4/2012, John Mortimer / 1483581*

5 IVAN SUSANIN (PROJECT 97P) CLASS
(PATROL SHIPS) (PGH)

| MURMANSK (ex-*Dunay*) 018 | NEVA 170 | ANADYR 173 | VOLGA 183 | DUNAY 145 |

Displacement, tonnes: 3,624 full load
Dimensions, metres (feet): 70 × 18.1 × 6.4 *(229.7 × 59.4 × 21.0)*
Speed, knots: 14.5
Range, n miles: 5,500 at 12.5 kt
Complement: 45

Machinery: Diesel-electric; 3 Type 13-D-150 diesel generators; 3 motors; 5,400 hp(m) *(4 MW)*; 3 shafts (1 fwd, 2 aft)
Guns: 2—3 in *(76 mm)*/59 AK 726 (twin); 90 rds/min to 16 km *(8.5 n miles)*; weight of shell 5.9 kg. 2—30 mm/65 AK 630 (not in all).
Radars: Surface search: Strut Curve; F-band.
Navigation: 2 Don Kay or Palm Frond; I-band.
Fire control: Hawk Screech; I-band.
Helicopters: Platform only.

Comment: Built at Admiralty Yard, Leningrad between 1974 and 1981. Generally similar to Dobrynya Nikitich class though larger with a tripod mast and different superstructure. Former icebreakers operated primarily as patrol ships. Two in the Pacific and three in the Northern Fleet. Two deleted so far.

VOLGA — *11/2007, Lemachko Collection / 1353309*

18 SORUM (PROJECT 745P) CLASS (PBO)

AMUR 043	DON 053	PRIMORYE 172
MAGADNETS 044	SAKHALIN 185	LADOGA 058
BREST 106	URAL 016	VICTOR KINGISEPP
CHUKOTKA 011	BAYKAL (ex-*Yan Berzin*) 105	(ex-*Vyatka*) 035
KARELIA 045	ZAPOLARYE 038	BUG 142
KAMCHATKA 198	ZABAYKALYE 196	
GENERAL MATROSOV 101	TVER 022	

Displacement, tonnes: 1,687 full load
Dimensions, metres (feet): 58 × 12.6 × 4.6 *(190.3 × 41.3 × 15.1)*
Speed, knots: 14
Range, n miles: 3,500 at 13 kt
Complement: 35

Machinery: Diesel-electric; 2 Type 5—2-DW2 diesel generators; 2,900 hp(m) *(2.13 MW)*; 1 motor; 2,000 hp(m) *(1.47 MW)*; 1 shaft
Guns: 4—30 mm/65 (2 twin) (all fitted for, but only Border Guard ships carry them).
Radars: Navigation: 2 Don 2 or Nayada; I-band.
IFF: High Pole B.

Comment: A class of ocean tugs armed for use as patrol vessels in the North, Pacific, Baltic and Caspian. Built in Yaroslavl and Oktyabskoye from 1973 to 1989, design used for Ministry of Fisheries rescue tugs.

URAL — *5/2010, Lemachko Collection / 1421640*

1 SPRUT (PROJECT 6457S) CLASS (PSO)

Name	No	Builders	Laid down	Launched	Commissioned
SPRUT	823	Yantar Shipyard, Kaliningrad	27 May 2002	12 Oct 2007	2008

Displacement, tonnes: 914 standard
Dimensions, metres (feet): 65.9 × 10.6 × 3.5 *(216.2 × 34.8 × 11.5)*
Speed, knots: 21.5
Range, n miles: 12,000 at 12 kt
Complement: 15 + 10 spare berths

Machinery: 1 MTU 16V 1163 diesel; 7,000 hp(m) *(5.2 MW)*; 1 shaft; fixed propeller
Radars: Surface search: I-band.
Navigation: I-band.
Helicopters: Platform for 1 light.

Comment: Specialist Fishery Protection vessel based on German Coast Guard Bad Bramstedt design. Steel hull with aluminium superstructure. Equipped with a high speed RHIB for interception.

SPRUT — *8/2009, Per Körnefeldt / 1305467*

16 PAUK I (MOLNYA) (PROJECT 12412) CLASS
(FAST ATTACK CRAFT – PATROL) (PCM)

TOLYATTI (ex-*PSKR-804*) 021	**CHEBOKSARY** (ex-*PSKR-817*) 052
NAKHODKA (ex-*PSKR-818*) 023	**SOKOL** (ex-*PSKR-812*) 063
KALININGRAD (ex-*PSKR-802*) 024	**MINSK** (ex-*PSKR-806*) 065
YAROSLAVL (ex-*PSKR-810*) 031	**KOBCHIK** (ex-*PSKR-807*) 078
YASTREB (ex-*PSKR-816*) 037	**KRECHET** (ex-*PSKR-809*) 134
SARYCH (ex-*PSKR-811*) 040	**BERKUT** (ex-*PSKR-800*) 152
GRIF (ex-*PSKR-808*) 041	**KORSHUN** (ex-*PSKR-805*) 161
ORLAN (ex-*PSKR-814*) 042	**VORON** (ex-*PSKR-801*) 163

Displacement, tonnes: 447 full load
Dimensions, metres (feet): 57.6 × 10.2 × 3.3 *(189 × 33.5 × 10.8)*
Speed, knots: 32
Range, n miles: 2,400 at 14 kt
Complement: 38

Machinery: 2 Type M 521 diesels; 16,184 hp(m) *(11.9 MW)* sustained; 2 shafts
Missiles: SAM: SA-N-5 Grail quad launcher; manual aiming; IR homing to 6 km *(3.2 n miles)* at 1.5 Mach; altitude to 2,500 m *(8,000 ft)*; warhead 1.5 kg; 8 missiles.
Guns: 1 – 3 in *(76 mm)*/59 AK 176; 120 rds/min to 15 km *(8 n miles)*; weight of shell 5.9 kg.
1 – 30 mm/65 AK 630; 6 barrels; 3,000 rds/min combined to 2 km.
Torpedoes: 4 – 16 in *(406 mm)* tubes.
A/S Mortars: 2 RBU 1200 5-tubed fixed; range 1,200 m; warhead 34 kg.
Depth charges: 2 racks (12).
Physical countermeasures: Decoys: 2 PK 16 or 4 PK 10 chaff launchers.
Electronic countermeasures: ESM: 3 Brick Plug and 2 Half Hat; radar warning.
Radars: Air/surface search: Peel Cone; E/F-band.
Surface search: Kivach or Pechora or SRN 207; I-band.
Fire control: Bass Tilt; H/I-band.
Sonars: Foal Tail; VDS (mounted on transom); active attack; high frequency.
Weapon control systems: Hood Wink optronic director.

Programmes: First laid down in 1977 and completed in 1979. In series production at Yaroslavl in the Black Sea and at Vladivostok until 1988 when the Svetlyak class took over. Type name is *maly protivolodochny korabl* meaning small anti-submarine ship. An improved version building at Kharbarovsk in 1995 was not completed.
Structure: An ASW version of the Tarantul class having the same hull form with a 1.8 m extension for dipping sonar. *Berkut, Voron* and *Kaliningrad* have a lower bridge than others. A modified version (Pauk II) with a longer superstructure, two twin 533 mm torpedo tubes and a radome similar to the Parchim class built for export.
Operational: Five in the Baltic, two in the Black Sea and the remainder in the Pacific. In addition five naval craft are laid up in the Baltic and one in the Black Sea.
Sales: One to Bulgaria in September 1989 and a second in December 1990. Two to Ukraine in 1996. A variant design built for Vietnam.

MINSK *7/2012, Erik Laursen* / 1483583

2 PAUK II (PROJECT 1241 PE) CLASS (PCM)

NOVOROSSIYSK (ex-*MPK 291*) 043   **KUBAN** (ex-*MPK 292*) 149

Displacement, tonnes: 503 full load
Dimensions, metres (feet): 58.5 × 10.2 × 3.4 *(191.9 × 33.5 × 11.2)*
Speed, knots: 32
Range, n miles: 2,400 at 14 kt
Complement: 32

Machinery: 2 Type M 521 diesels; 16,184 hp(m) *(11.9 MW)* sustained; 2 shafts
Missiles: SAM: SA-N-5 quad launcher; manual aiming, IR homing to 10 km *(5.4 n miles)* at 1.5 Mach; warhead 1.1 kg.
Guns: 1 USSR 76 mm/50 AK 176; 120 rds/min to 15 km *(8 n miles)*; weight of shell 5.9 kg.
1 – 30 mm/65 AK 630; 6 barrels; 3,000 rds/min combined to 2 km.
Torpedoes: 4 – 21 in *(533 mm)* (2 twin) fixed tubes.
A/S Mortars: 2 RBU 1200 5-tubed fixed; range 1,200 m; warhead 34 kg.
Radars: Air/surface search: Cross Dome (Positiv E); E/F-band.
Navigation: Pechora; I-band.
Fire control: Bass Tilt; H/I-band.
Sonars: Rat Tail; VDS (on transom); attack; high frequency.

Comment: Built at Yaroslav Shipyard and entered service in 1997–98 when they were transferred to the Border Guard. Originally intended for Iraq, export Pauk II variant of the type sold to India and Cuba. Has a longer superstructure than the Pauk I with a radome similar to the Parchim II class. The torpedo tubes must be trained out to launch. Both operate in the Black Sea.

NOVOROSSIYSK *5/2009, Lemachko Collection* / 1305466

© 2014 IHS

17 TERRIER (PROJECT 14170) CLASS (PB)

PSKA 450–466

Displacement, tonnes: 8 full load
Dimensions, metres (feet): 11.7 × 3.1 × 0.5 *(38.4 × 10.2 × 1.6)*
Speed, knots: 32
Range, n miles: 120 at 30 kt
Complement: 6
Machinery: 2 diesels; 2 waterjets

Comment: Built at Zelenodolsk, Sosnovka Shipyard and Khabarovsk Shipyard 2000–04.

TERRIER CLASS *5/2006, Lemachko Collection* / 1159853

2 MURAVEY (ANTARES) (PROJECT 133) CLASS (PCK)

PSKR 110 135   **PSKR 109** 136

Displacement, tonnes: 215 full load
Dimensions, metres (feet): 38.6 × 7.6 × 1.9 hull; 4.4 over foils *(126.6 × 24.9 × 6.2; 14.4)*
Speed, knots: 60
Complement: 30 (5 officers)
Machinery: 2 gas turbines; 22,600 hp(m) *(16.6 MW)*; 2 shafts
Radars: Surface search: Peel Cone; E-band.
Fire control: Bass Tilt; H/I-band.

Comment: Thirteen hydrofoil craft built at Feodosiya in the mid-1980s for the USSR Border Guard. Three transferred to Ukraine and the remainder decommissioned. Two remain in service but hydrofoils and main armament have been removed and operational status is doubtful.

PSKR 109 and 110 *9/2009, Lemachko Collection* / 1305556

3 MUSTANG (PROJECT 18623) CLASS (PBF)

PSKA 816   **PSKA 817**   **PSKA 818**

Displacement, tonnes: 36 full load
Dimensions, metres (feet): 20 × 4.5 × 1.1 *(65.6 × 14.8 × 3.6)*
Speed, knots: 45
Range, n miles: 350 at 40 kt
Complement: 6
Machinery: 2 Zvezda M-470 diesels; 2,950 hp *(2.2 MW)*; 2 Kamewa waterjets

Comment: Designed by Redan Bureau, St Petersburg and built at Yaroslavl from 2000.

MUSTANG 817 *9/2008, Lemachko Collection* / 1305646

Russian Federation (BORDER GUARD) > Patrol forces

3 + 4 (18) RUBIN (PROJECT 22460) CLASS (PATROL SHIP) (PSO)

Name	No	Builders	Laid down	Launched	Commissioned
RUBIN	050	Almaz, St Petersburg	3 Sep 2007	26 Jun 2009	14 Nov 2009
BRILLIANT	046	Almaz, St Petersburg	12 May 2010	25 Nov 2011	26 Jun 2012
ZEMCHUK	052	Almaz, St Petersburg	22 Dec 2010	21 Apr 2012	21 Sep 2012
IZUMRUD	–	Almaz, St Petersburg	21 Sep 2012	14 Aug 2013	2014
AMETIST	–	Almaz, St Petersburg	2012	2014	2015
SAPFIR	–	Vostochnaya Verf	2012	2014	2015
KORALL	–	Vostochnaya Verf	2013	2014	2015

Displacement, tonnes: 630 full load
Dimensions, metres (feet): 62.5 × 11.0 × 3.37 (205.1 × 36.1 × 11.1)
Speed, knots: 30. **Range, n miles:** 3,500 at 12 kt
Complement: 44

Machinery: 4 MTU diesels; 15,440 hp (11.5 MW); 2 shafts
Guns: 1 – 30 mm/65 AK 630; 6 barrels; 3,000 rds/min combined to 2 km. 2 – 12.7 mm MGs.
Helicopters: Platform for 1 medium.

Comment: Rubin was the first vessel of a new class of helicopter-capable patrol ships laid down at Almaz, St Petersburg on 3 September 2007 and commissioned in 2009. The ship features a launch/recovery ramp at the stern. A total of 25 vessels are planned to enter service by 2020.

RUBIN   7/2012, C D Yaylali / 1483590

17 FLAMINGO (PROJECT 1415PV) CLASS (PB)

Displacement, tonnes: 55 full load
Dimensions, metres (feet): 21.2 × 3.9 × 1.4 (69.6 × 12.8 × 4.6)
Speed, knots: 12
Range, n miles: 200 at 11 kt
Complement: 4
Machinery: 1 diesel; 300 hp (225 kW); 1 shaft
Radars: Navigation: I-band.

Comment: Harbour patrol craft built in the 1970/80s. Similar craft, known as the Kulik class, built for the Navy.

FLAMINGO CLASS   6/2006, Lemachko Collection / 1305142

2 ENFORCER II (IC 16) CLASS (PATROL CRAFT) (PB)

Displacement, tonnes: 8 full load
Dimensions, metres (feet): 11.36 × 2.94 × 0.9 (37.3 × 9.6 × 3.0)
Speed, knots: 42
Range, n miles: 200 at 30 kt
Complement: 2
Machinery: 2 Volvo Penta D9 diesels; 1,000 hp (736 kW); 2 Rolls Royce FF310 waterjets

Comment: Built by Dockstavarvet, Sweden, the craft are derived from the Combatboat 90 concept and are known as the HSPC 11.3M design. The aluminium construction craft are to be used for patrols on Russian inland waterways. Both delivered in June 2008.

ENFORCER II   6/2012, Erik Laursen / 1483589

2 MODIFIED OKEAN (PROJECT 502) CLASS (PBO)

ANTUR   DIANA

Displacement, tonnes: 912 full load
Dimensions, metres (feet): 54.2 × 9.3 × 4.0 (177.8 × 30.5 × 13.1)
Speed, knots: 13
Range, n miles: 7,900 at 11 kt
Machinery: 1 diesel; 800 hp (596 kW); 1 shaft

Comment: Former intelligence collection vessels, built in the former East Germany in the early 1960s, converted to undertake fishery protection duties. Based in the Pacific.

ANTUR   6/2009, Lemachko Collection / 1305409

0 + 1 PROJECT 22100 PATROL SHIP (PSO)

Name	No	Builders	Laid down	Launched	Commissioned
OKEAN	–	Zelenodolsk Shipyard	31 May 2012	2014	2015

Displacement, tonnes: 2,700 standard
Dimensions, metres (feet): 100 × ? × ? (328.1 × ? × ?)
Machinery: To be announced.
Guns: 1 – 76 mm. 2 – 14.5 mm.
Radars: Surface search/navigation: I-band.

Comment: The first of a new class of large offshore patrol vessels was laid down in May 2012. It is expected that the ship, the first ocean-going unit to be designed for the Russian Border Guard, is to replace the Krivak III frigates and Ivan Susanin class in due course.

1 GORNOSTAY (PROJECT 20990) CLASS (PATROL CRAFT) (PBF)

Displacement, tonnes: 17 full load
Dimensions, metres (feet): 15.57 × 3.95 × 0.66 (51.1 × 13.0 × 2.2)
Speed, knots: 52
Range, n miles: 400 at 35 kt
Complement: 3
Machinery: 2 D2842LE402 diesels; 2,200 hp (1.64 MW); 2 shafts
Guns: 1 – 7.62 mm MG.

Comment: One vessel was completed at Khabarovsk Shipyard in 2004.

Patrol forces — River patrol forces < **Russian Federation** (BORDER GUARD) 731

2 PROJECT 22120 PATROL SHIP CLASS (PSO)

Name	No	Builders	Laid down	Launched	Commissioned
PURGA	PS 824	Almaz, St Petersburg	Mar 2007	24 Dec 2009	22 Dec 2010
–	PS 825	Almaz, St Petersburg	25 Nov 2011	5 Dec 2012	24 Jun 2013

Displacement, tonnes: 1,083 full load
Dimensions, metres (feet): 71 × 10.4 × 3.44 (232.9 × 34.1 × 11.3)
Speed, knots: 24. **Range, n miles:** 6,000 at 14 kt

Complement: 25
Machinery: 2 diesels; 10,057 hp (7.5 MW); 2 shafts
Helicopters: Platform for 1 medium.

Comment: A new class of ice-reinforced helicopter-capable patrol ships. The first unit became operational in 2010 and the second in 2013.

PS 825   7/2013*, Per Körnefeldt / 1531000

0 + 1 (5) PROJECT 22160 (PATROL SHIPS) (PSO)

Displacement, tonnes: 1,300 full load
Dimensions, metres (feet): 94 × 14.0 × 3.4 (308.4 × 45.9 × 11.2)
Speed, knots: 24. **Range, n miles:** 6,000 at 12 kt
Machinery: To be announced.
Guns: 1 — 57 mm.
Helicopters: Platform for 1 medium.

Comment: Designed by Severnoye design bureau, the first of class was laid down at Zelenodolsk on 26 February 2014. The new class is designed to undertake EEZ patrol, SAR, anti-piracy and anti-smuggling activities, environmental monitoring, and coastal defence. A class of six is expected.

RIVER PATROL FORCES

Notes: Attached to Black Sea and Pacific Fleets for operations on the Danube, Amur and Usuri Rivers, and to the Caspian Flotilla.

2 YAZ (SLEPEN) (PROJECT 1208) CLASS (PGR)

BLAGOVESHCHENSK 066   **VYUGA** (ex-*Shkval*) 106

Displacement, tonnes: 447 full load
Dimensions, metres (feet): 55 × 9 × 1.5 (180.4 × 29.5 × 4.9)
Speed, knots: 24. **Range, n miles:** 1,000 at 10 kt
Complement: 32 (4 officers)

Machinery: 3 diesels; 11,400 hp(m) (8.39 MW); 3 shafts
Guns: 2 — 115 mm tank guns (TB 62) or 100 mm/56. 2 — 30 mm/65 AK 630; 6 barrels per mounting. 4 — 12.7 mm MGs (2 twin). 2 — 40 mm mortars on after deckhouse.
Radars: Surface search: Spin Trough; I-band.
Fire control: Bass Tilt; H/I-band.
IFF: High Pole B. Square Head.

Comment: First entered service in Amur Flotilla 1978. Built at Khabarovsk until 1987. All but these last two have been placed in reserve.

VYUGA   9/2010, Lemachko Collection / 1406802

8 PIYAVKA (PROJECT 1249) CLASS (PBR)

| PSKR 52 117 | PSKR 54 146 | PSKR 56 093 | PSKR 58 123 |
| PSKR 53 065 | PSKR 55 013 | PSKR 57 058 | PSKR 59 189 |

Displacement, tonnes: 233 full load
Dimensions, metres (feet): 41.6 × 6.3 × 0.9 (136.5 × 20.7 × 3.0)
Speed, knots: 17
Complement: 30 (4 officers)
Machinery: 3 diesels; 3,300 hp(m) (2.42 MW); 2 shafts
Guns: 1 — 30 mm/65 AK 630; 6 barrels. 2 — 14.5 mm (twin) MGs.
Radars: Surface search: Spin Trough; I-band.

Comment: Built at Khabarovsk 1979–84. Based in Amur Flotilla mostly for logistic support.

PSKR 54   7/2009, Lemachko Collection / 1305551

4 OGONEK (PROJECT 12130) CLASS (PBR)

PSKR 200–203

Displacement, tonnes: 100 full load
Dimensions, metres (feet): 33.4 × 4.2 × 0.8 (109.6 × 13.8 × 2.6)
Speed, knots: 25
Complement: 17 (2 officers)
Machinery: 2 diesels; 2 shafts
Guns: 2 — 30 mm AK 630.

Comment: A smaller version of the Piyavka class built at Khabarovsk from 1999. Numbers in service are uncertain.

PSKR 200   7/2009, Lemachko Collection / 1305550

732 Russian Federation (BORDER GUARD) > River patrol forces — Auxiliaries

6 SHMEL (PROJECT 1204) CLASS (PGR)

| AK 209 044 | AK 224 155 | AK 582 147 |
| AK 223 045 | AK 248 047 (ex-151) | AK 599 143 |

Displacement, tonnes: 78 full load
Dimensions, metres (feet): 27.7 × 4.3 × 1.2 *(90.9 × 14.1 × 3.9)*
Speed, knots: 25. **Range, n miles:** 600 at 12 kt
Complement: 12 (4 officers)

Machinery: 2 Type M 50 diesels; 2,200 hp(m) *(1.6 MW)* sustained; 2 shafts
Guns: 1—3 in *(76 mm)*/48 (tank turret). 1—25 mm/70 (later ships). 2—14.5 mm (twin) MGs (earlier ships). 5—7.62 mm MGs.
Rockets: 1 BP 6 rocket launcher; 18 barrels.
Mines: Can lay 9.
Radars: Surface search: Spin Trough; I-band.

Comment: Completed at Kerch and Nikolayev North (61 Kommuna) 1967–74. Some of the later ships also mount one or two multibarrelled rocket launchers amidships. The 7.62 mm guns fire through embrasures in the superstructure with one mounted on the 76 mm. Can be carried on land transport. Type name is *artillerisky kater* meaning artillery cutter. About 70 have been scrapped or laid up so far including the last naval units. These last survivors are based on the Amur River and belong to the Border Guard.
Transfers: Four to Cambodia (1984–85) (since decommissioned). Some have been taken over by Belorussian forces, and others allocated to Ukraine.

SHMEL CLASS 6/2000, Lemachko Collection / 0106875

6 VOSH (MOSKIT) (PROJECT 1248) CLASS (PGR)

| STORM 056 | KHABAROVSK 137 | 60 LETIYA POGRANVOYSK 062 |
| GROZA 057 | SHKVAL 138 | PSKR 322 198 |

Displacement, tonnes: 233 full load
Dimensions, metres (feet): 42 × 6.3 × 1 *(137.8 × 20.7 × 3.3)*
Speed, knots: 17
Complement: 34 (3 officers)

Machinery: 3 diesels; 3,300 hp(m) *(2.42 MW)*; 3 shafts
Guns: 1—3 in *(76 mm)*/48 (tank turret). 1—30 mm/65 AK 630. 2—12.7 mm (twin) MGs.
Physical countermeasures: Decoys: 1 twin barrel decoy launcher.
Radars: Surface search: Spin Trough; I-band.

Comment: Built at Sretensk on the Shilka river 1980–84. Based on Amur River. Same hull as Piyavka.

60 LETIYA POGRANVOYSK 7/2009, Lemachko Collection / 1305549

15 SAYGAK (PROJECT 14081/14081M) CLASS (PBF)

Displacement, tonnes: 12 full load
Dimensions, metres (feet): 14 × 3.5 × 0.65 *(45.9 × 11.5 × 2.1)*
Speed, knots: 38. **Range, n miles:** 135 at 35 kt
Complement: 2 + 8 personnel
Machinery: 1 Zvezda M-401B diesel; 1,000 hp *(746 kW)*; 1 waterjet
Radars: Navigation: I-band.

Comment: Built by Kama Zavod, Perm and entered service 1986–2000. Used for riverine and lake patrol. Others are used by the Customs service.

SAYGAK 069 7/2006, Lemachko Collection / 1159868

AUXILIARIES

7 NEON ANTONOV (PROJECT 1595) CLASS
(TRANSPORTS) (AK)

VYACHESLAV DENISOV 176	MIKHAIL KONOVALOV 184	NIKOLAY STARSHINOV 119
IVAN YEVTEYEV 105	SERGEY SUDETSKY 143	
IVAN LEDNEV 115	NIKOLAY SIPYAGIN 090	

Displacement, tonnes: 6,503 full load
Dimensions, metres (feet): 95 × 14.7 × 6.5 *(311.7 × 48.2 × 21.3)*
Speed, knots: 17
Range, n miles: 8,500 at 13 kt
Complement: 45

Cargo capacity: 2,500 tons
Machinery: 2 diesels; 7,000 hp(m) *(5.15 MW)*; 2 shafts
Missiles: SAM: 2 SA-N-5 Grail twin launchers; manual aiming; IR homing to 6 km *(3.2 n miles)* at 1.5 Mach; altitude to 2,500 m *(8,000 ft)*; warhead 1.5 kg.
Guns: 2—30 mm/65 (twin). 4—14.5 mm (2 twin) MGs. 4—12.7 mm MGs.
Radars: Navigation: Don Kay; Spin Trough or Palm Frond; I-band.

Comment: Ten (of which eight remain) of the class built at Nikolayev from 1975 to early 1980s. All in the Pacific. Have two small landing craft aft. Armament is not normally mounted.

IVAN YEVTEYEV 10/2012, Chris Sattler / 1483591

6 KANIN CLASS (PROJECT 16900A) (AKL)

| CHANTIJ-MANSISK | ARCHANGELSK | URENGOY |
| JURGA | KANIN | ANATOLY SHILINSKY |

Displacement, tonnes: 935 full load
Dimensions, metres (feet): 45.6 × 8.8 × 2.5 *(149.6 × 28.9 × 8.2)*
Speed, knots: 9
Range, n miles: 3,500 at 9 kt
Complement: 22
Machinery: 2 diesels; 800 hp(m) *(558 kW)*; 2 shafts

Comment: Built in the Pacific since 1996 for the Border Guard. Others may be building for commercial service. Ice reinforced bows for Arctic service. *Chantij-Mansisk* based in the Black Sea.

JURGA 10/2009, Lemachko Collection / 1366492

5 PROJECT 1481 CLASS (AO)

| BNS 180150 | BNS 181150 | BNS 182150 | BNS 183150 | BNS 711150 |

Displacement, tonnes: 440 full load
Dimensions, metres (feet): 52.43 × 8.2 × 1.12 *(172 × 26.9 × 3.7)*
Speed, knots: 10
Complement: 7
Machinery: 2 diesels; 630 hp *(470 kW)*; 1 shaft
Radars: Surface search/navigation: I-band.

Comment: Self-propelled barges constructed at Stretensky Shipyard in the 1970s. Based on the Amur River.

Auxiliaries — Amphibious forces < **Russian Federation** (BORDER GUARD) 733

1 BASKUNCHAK CLASS (PROJECT 1545) (AO)

SOVETSKIY POGRANICHNIK 102

Displacement, tonnes: 1,280 standard; 2,987 full load
Dimensions, metres (feet): 83.6 × 12.0 × 4.9 *(274.3 × 39.4 × 16.1)*
Speed, knots: 13
Range, n miles: 5,000 at 12 kt
Complement: 30
Machinery: 1 diesel; 2,000 hp *(1.49 MW)*; 1 shaft
Radars: Navigation: 1 Don-2; I-band.

Comment: Built at Zaliv Shipyard, Kerch, and completed in about 1968. Has ice-reinforced bow and is based in the Pacific.

SOVETSKIY POGRANICHNIK 6/2005, Lemachko Collection / 1159248

1 PROJECT 15010 CLASS (AO)

Name	No	Builders	Commissioned
ISHIM	139	Nikolaevskiy Shipyard, Vladivostok	29 Nov 2008

Displacement, tonnes: 5,750 full load
Dimensions, metres (feet): 94.7 × 12.0 × 3.7 *(310.7 × 39.4 × 12.1)*
Speed, knots: 14
Complement: 35
Machinery: 2 diesels; 5,000 hp *(3.7 MW)*; 1 shaft
Radars: Surface search/navigation: I-band.

Comment: Built at Komsomolsk in 1994. Carries oil, petrol, water and provisions. Based at Vladivostok.

CZILIM 6/2001, S Breyer / 0126219

2 TSAPLYA (MURENA) (PROJECT 12061) CLASS (ACV)

DK-143 659   DK-447 699

Displacement, tonnes: 151 full load
Dimensions, metres (feet): 31.6 on cushion × ? *(103.7 × ?)*
Speed, knots: 50
Range, n miles: 500 at 50 kt
Complement: 111 (3 officers)
Machinery: 2 MT-70M gas turbines for lift and propulsion; 8,000 hp *(5.88 MW)*
Guns: 2—30 mm AK 306M. 2—30 mm grenade launchers. 2—12.7 mm MGs.

Comment: Larger version of the Lebed class designed for river patrol. Built at Khabarovsk between 1987 and 1992. Operated on Amur river system.

ISHIM 10/2012, Chris Sattler / 1455741

AMPHIBIOUS FORCES

1 CZILIM (PROJECT 20910) CLASS (ACV/UCAC)

PSKA 80

Displacement, tonnes: 9 full load
Dimensions, metres (feet): 12 on cushion × ? *(39.4 × ?)*
Speed, knots: 40
Range, n miles: 300 at 30 kt
Complement: 2 + 6 personnel
Machinery: 2 Deutz BF 6M 1013 diesels; 435 hp(m) *(320 kW)* sustained; for lift and propulsion
Guns: 1—7.62 mm MG. 1—40 mm RPG.
Radars: Navigation: I-band.

Comment: Ordered from Jaroslawski Sudostroiteinyj Zawod to an Almaz design for Special Forces of the Border Guard. Laid down 24 February 1998 and in service in early 2001.

DK-143 5/2009, Lemachko Collection / 1305548

© 2014 IHS   IHS Jane's Fighting Ships 2014-2015

St Kitts and Nevis

Country Overview

The Federation of St Kitts and Nevis gained independence in 1983; the British monarch, represented by a governor-general, is the head of state. Located at the northern end of the Leeward Islands in the Lesser Antilles chain, the country comprises St Kitts (formerly Saint Christopher) (68 square miles) and, 2 n miles to the southeast, Nevis (36 square miles). The constitution allows for the secession of Nevis from the federation. The capital of St Kitts and of the federation is Basseterre; Charlestown is the capital and largest town on Nevis. Territorial seas (12 n miles) are claimed. A 200 n mile Exclusive Economic Zone (EEZ) has been claimed but the limits are not defined. The Coast Guard was part of the Police Force until 1997 when it transferred to the Regular Corps of the Defence Force.

Headquarters Appointments

Commanding Officer Coast Guard:
  Captain Anthony J Comrie

Bases
Basseterre

Personnel
2014: 33

COAST GUARD

Notes: A 920 Zodiac RHIB was donated by the US government in 2003.

1 SWIFTSHIPS 110 ft CLASS (PB)

STALWART PB 01

Displacement, tonnes: 102 full load
Dimensions, metres (feet): 35.5 × 7.6 × 2.1 *(116.5 × 24.9 × 6.9)*
Speed, knots: 18. **Range, n miles:** 1,800 at 15 kt
Complement: 14

Machinery: 4 Iveco diesels; 4 shafts
Guns: 2 – 12.7 mm MGs. 2 – 7.62 mm MGs.
Radars: Surface search: Raytheon; I-band.
Navigation: Furuno; I-band.

Comment: Built by Swiftships, Morgan City, and delivered August 1985. Aluminium alloy hull and superstructure.

STALWART   6/2010, St Kitts-Nevis CG / 1366670

2 SPECIAL PURPOSE CRAFT (PBF)

HERMITAGE BAY PB 04   OUALIE BAY PB 05

Displacement, tonnes: 7 full load
Dimensions, metres (feet): 10.1 × 3.0 × 0.7 *(33.1 × 9.8 × 2.3)*
Speed, knots: 50
Complement: 4
Machinery: 3 Mercury outboard motors; 825 hp *(615 kW)*
Guns: 2 – 7.62 mm MGs.
Radars: Navigation: Raymarine; I-band.

Comment: Built by SAFE Boats International and very similar to the craft in service with the US Coast Guard. Both craft donated by US Southern Command under the US Foreign Military Sales Program in November 2012.

1 BOSTON WHALER (PBF)

ROVER I PB 03

Displacement, tonnes: 3 full load
Dimensions, metres (feet): 6.7 × 2.3 × 0.6 *(22 × 7.5 × 2.0)*
Speed, knots: 35. **Range, n miles:** 70 at 35 kt
Complement: 2
Machinery: 1 Johnson outboard; 223 hp *(166 kW)*

Comment: Delivered in May 1988.

ROVER I   1990, St Kitts-Nevis Police / 0081727

1 DAUNTLESS CLASS (PB)

ARDENT PB 02

Displacement, tonnes: 11 full load
Dimensions, metres (feet): 12.2 × 4.3 × 1.3 *(40 × 14.1 × 4.3)*
Speed, knots: 27
Range, n miles: 600 at 18 kt
Complement: 4
Machinery: 2 Caterpillar 3208TA diesels; 870 hp *(650 kW)*; 2 shafts
Guns: 1 – 7.62 mm MG.
Radars: Surface search: Raytheon; I-band.

Comment: Built by SeaArk Marine under FMS funding and commissioned 8 August 1995. Aluminium construction.

ARDENT   6/2010, St Kitts-Nevis CG / 1366671

St Lucia

Country Overview

St Lucia gained independence in 1979; the British monarch, represented by a governor-general, is the head of state. The island (238 square miles) is one of the Windward Islands of the Lesser Antilles chain and is located between Martinique to the north and St Vincent to the south. The capital, main town and principal port is Castries, on the northwestern coast. Territorial seas (12 n miles) are claimed. Exclusive Economic Zone (EEZ) limits will not be fully defined until outstanding boundary disagreements have been resolved.

Headquarters Appointments

Coast Guard Commander:
  Superintendent Milton Desir

Bases
Castries, Vieux-Fort

Personnel
2014: 61

COAST GUARD

5 HARBOUR CRAFT (PB)

P 03   P 05–08

Comment: *P 03* is a 9 m Zodiac 920 RHIB donated by the US in 2004. *P 05* is a 35 kt Hurricane RIB acquired in June 1993, *P 06* a 45 kt Mako craft acquired in November 1995 and *P 07* a Mako 234 acquired in about 2006 and *P 08* a Brunswick 1000, capable of 48 kt, acquired in 2011.

P 08
6/2011, St Lucia CG
1406466

1 SWIFT 65 ft CLASS (PB)

DEFENDER P 02

Displacement, tonnes: 43 full load
Dimensions, metres (feet): 19.8 × 5.6 × 2 (65 × 18.4 × 6.6)
Speed, knots: 22
Range, n miles: 1,500 at 18 kt
Complement: 7
Machinery: 2 Detroit 12V-71 diesels; 680 hp (507 kW) sustained; 2 shafts
Radars: Surface search: Furuno; I-band.

Comment: Ordered from Swiftships, Morgan City in November 1983. Commissioned 3 May 1984. Similar to craft supplied to Antigua and Dominica.

DEFENDER 1/2004, St Lucia CG / 1367100

1 DAUNTLESS CLASS (PB)

PROTECTOR P 04

Displacement, tonnes: 11 full load
Dimensions, metres (feet): 12.2 × 4.3 × 1.3 (40 × 14.1 × 4.3)
Speed, knots: 27
Range, n miles: 600 at 18 kt
Complement: 4
Machinery: 2 Caterpillar 3208TA diesels; 870 hp (650 kW); 2 shafts
Radars: Surface search: Raytheon; I-band.

Comment: Ordered October 1994. Built by SeaArk Marine under FMS funding and commissioned 9 October 1995.

PROTECTOR 6/2008, St Lucia CG / 1335425

1 PLASCOA 21 M PATROL CRAFT (PB)

LESMOND REMY (ex-Karina)

Displacement, tonnes: 35 full load
Dimensions, metres (feet): 21.3 × 5.55 × 1.5 (69.9 × 18.2 × 4.9)
Speed, knots: 24
Complement: 8
Machinery: 2 GM 12V 92TA diesels; 2,190 hp (1.6 MW); 2 shafts
Radars: Navigation: Furuno 2400; I-band.

Comment: Former French customs patrol craft constructed by Plascoa in 1991. Transferred to St Lucia in December 2010.

LESMOND REMY 6/2011, St Lucia CG / 1406465

2 SPECIAL PURPOSE CRAFT (PBF)

VERONICA ADLEY PO 10   **EUSEBE LAWRENCE** PO 11

Displacement, tonnes: 7 full load
Dimensions, metres (feet): 10.1 × 3.0 × 0.7 (33.1 × 9.8 × 2.3)
Speed, knots: 50
Complement: 4
Machinery: 3 Mercury outboard motors; 825 hp (615 kW)
Guns: 2 – 7.62 mm MGs.
Radars: Navigation: Raymarine; I-band.

Comment: Built by SAFE Boats International and very similar to the craft in service with the US Coast Guard. Both craft donated by US Southern Command under the US foreign Military Sales Program in July 2012.

VERONICA ADLEY 6/2013*, St Lucia Coast Guard / 1486571

St Vincent and the Grenadines

Country Overview

St Vincent and the Grenadines gained independence in 1979; the British monarch, represented by a governor-general, is the head of state. Lying between St Lucia to the north and Grenada to the south, they form part of the Windward Islands in the Lesser Antilles chain and comprise the island of St Vincent (133 square miles) and the 32 northernmost islands and cays of the Grenadines group including (north to south): Bequia, Mustique, Canouan, Mayreau, Union Island, Palm (formerly Prune) Island, and Petit St Vincent. The capital, largest town, and principal port is Kingstown, St Vincent. An archipelagic state, territorial seas (12 n miles) are claimed. A 200 n mile Exclusive Economic Zone (EEZ) has been claimed.

Headquarters Appointments

Coast Guard Commander:
Commander Brenton Cain

Bases

Calliaqua, Canouan

Personnel

2014: 88

COAST GUARD

1 DAUNTLESS CLASS (PB)

HAIROUN SVG 04

Displacement, tonnes: 11 full load
Dimensions, metres (feet): 12.2 × 4.3 × 1.3 (40 × 14.1 × 4.3)
Speed, knots: 27
Range, n miles: 600 at 18 kt
Complement: 4
Machinery: 2 Caterpillar 3208TA diesels; 870 hp (650 kW); 2 shafts
Guns: 1 – 7.62 mm MG.
Radars: Surface search: Raytheon; I-band.

Comment: Ordered October 1984. Built by SeaArk Marine under FMS funding and commissioned 8 June 1995. Aluminium construction. The craft was refitted in late 2008.

HAIROUN
2/2013*, St Vincent
Coast Guard
1486569

736 St Vincent and the Grenadines > Coast guard — **Saudi Arabia** > Introduction

5 HARBOUR CRAFT (PB)

– SVG 02–03   – SVG 06–07   **H K TANNIS** SVG 10

Comment: *SVG 03* is a 30 kt Zodiac RIB. *SVG 06*, acquired in 2008, is a 7.5 m RHIB. *SVG 02* and *SVG 07*, acquired in 2003, are 9 m RHIBs. *H K Tannis* is a 13.5 m RHIB with waterjet propulsion. It was acquired in 2005.

2 SPECIAL PURPOSE CRAFT (PBF)

SVG 09   SVG 11

Displacement, tonnes: 7 full load
Dimensions, metres (feet): 10.1 × 3.0 × 0.7 *(33.1 × 9.8 × 2.3)*
Speed, knots: 50
Complement: 4
Machinery: 3 Mercury outboard motors; 825 hp *(615 kW)*
Guns: 2 – 7.62 mm MGs.
Radars: Navigation: Raymarine; I-band.

Comment: Built by SAFE Boats International and very similar to the craft in service with the US Coast Guard. Both craft donated by US Southern Command under the US foreign Military Sales Program in August 2012.

SVG 07   6/2008, *St Vincent Coast Guard* / 1335426

SVG 11   6/2013*, *St Vincent Coast Guard* / 1486570

H K TANNIS   6/2008, *St Vincent Coast Guard* / 1335427

Samoa

Country Overview

Samoa was a New Zealand-administered UN Trust territory until it became independent in 1962. At the same time a Treaty of Friendship delegated responsibility to New Zealand for foreign affairs. An island nation, it lies in the south Pacific Ocean, approximately midway between Hawaii and New Zealand, in the western portion of the Samoan archipelago. There are two main islands, Savai'i and Upolu, and several smaller islands, of which only two, Apolima and Manono, are inhabited. The capital and chief port is Apia on Upolu. An archipelagic state, territorial seas (12 n miles) are claimed. An Exclusive Economic Zone (EEZ) (200 n miles) is also claimed but limits have not been fully defined by boundary agreements.

Headquarters Appointments

Head of Police Maritime Division:
  Superintendent Tagaolo Iosefatu Wright

Bases
Apia

PATROL FORCES

1 PACIFIC CLASS (LARGE PATROL CRAFT) (PB)

Name	Builders	Commissioned
NAFANUA	Australian Shipbuilding Industries	5 Mar 1988

Displacement, tonnes: 168 full load
Dimensions, metres (feet): 31.5 × 8.1 × 2.1 *(103.3 × 26.6 × 6.9)*
Speed, knots: 20. **Range, n miles:** 2,500 at 12 kt
Complement: 17 (3 officers)
Machinery: 2 Caterpillar 3516TA diesels; 4,400 hp *(3.28 MW)* sustained; 2 shafts
Guns: 2 – 7.62 mm MGs.
Radars: Surface search: Furuno FR-1510; I-band.

Comment: Under the Defence Co-operation Programme Australia has provided 22 Pacific class patrol craft to Pacific islands. Training, operational and technical assistance is provided by the Royal Australian Navy. *Nafanua* ordered 3 October 1985. Refitted in 1996. Following the decision by the Australian government to extend the Pacific Patrol Boat programme, a life-extension refit was undertaken at Townsville in 2005. A further refit was completed at Cairns in May 2013.

NAFANUA   10/2009, *Peter Mansfield* / 1366488

Saudi Arabia

Country Overview

The Kingdom of Saudi Arabia occupies most of the Arabian Peninsula and is bordered to the north by Jordan, Iraq, and Kuwait, to the south by Oman and the Republic of Yemen and to the east by Qatar, Bahrain and the United Arab Emirates. With an area of 864,869 square miles, it has coastlines with the Red Sea (972 n miles) and the Gulf (454 n miles). The capital and largest city is Riyadh while the principal ports are Jiddah and Yanbu and Jizan on the Red Sea, and the major oil-exporting ports of Al Jabayl, Ad Dammam, and Ras Tanura on the Gulf. Territorial seas (12 n miles) are claimed. An EEZ has not been claimed.

Headquarters Appointments

Chief of Naval Staff: Vice Admiral Dakheel Allah Al-Wagdani
Commander Eastern Fleet: Rear Admiral Ibrahim Bin Nasser Al-Maghlooth
Commander Western Fleet: Rear Admiral Saeed Al-Zahrani
Director Frontier Force (Coast Guard): Lieutenant General Mujib bin Muhammad Al-Qahtani

Personnel

(a) 2014: 15,500 officers and men (including 3,000 marines)
(b) Voluntary service

Bases

Naval HQ: Riyadh
Main bases: Jiddah (HQ Western Fleet), Al Jubail (HQ Eastern Fleet), Aziziah (Coast Guard). Jizan (Red Sea)
Minor bases (Naval and Coast Guard): Ras Tanura, Al Dammam, Yanbou Al Bahr, Ras al-Mishab, Al Wajh, Al Qatif, Haqi, Al Sharmah, Qizan, Duba

Command and Control

The USA provided an update of command and control capabilities during the period 1991–95, including a commercial datalink to improve interoperability.

Coast Defence

Truck-mounted Otomat batteries.

Coast Guard

Part of the Frontier Force under the Minister of Interior. 5,500 officers and men. It is not always clear which ships belong to the Navy and which to the Coast Guard.

Strength of the Fleet

Type	Active	Building
Frigates	7	—
Corvettes – Missile	4	—
Fast Attack Craft – Missile	9	—
Patrol Craft	56	—
Minehunters	3	—
Minesweepers – Coastal	4	—
Replenishment Tankers	2	—

IHS Jane's Fighting Ships 2014-2015   © 2014 IHS

SUBMARINES

Notes: Reports of Saudi Arabian interest in acquiring five Type 209 submarines from Germany appeared in the German press in November 2013. These were denied by ThyssenKrupp.

FRIGATES

Notes: A programme to replace the Madina-class frigates is likely to have been delayed by plans, reported in 2013, to modernise all four ships.

4 MADINA (TYPE F 2000S) CLASS (FFGHM)

Name	No	Builders	Laid down	Launched	Commissioned
MADINA	702	Lorient (DTCN), Lorient	15 Oct 1981	23 Apr 1983	4 Jan 1985
HOFOUF	704	CNIM, Seyne-sur-Mer	14 Jun 1982	24 Jun 1983	31 Oct 1985
ABHA	706	CNIM, Seyne-sur-Mer	7 Dec 1982	23 Dec 1983	4 Apr 1986
TAIF	708	CNIM, Seyne-sur-Mer	1 Mar 1983	25 May 1984	29 Aug 1986

Displacement, tonnes: 2,032 standard; 2,916 full load
Dimensions, metres (feet): 115 × 12.5 × 4.9 sonar *(377.3 × 41.0 × 16.1)*
Speed, knots: 30
Range, n miles: 8,000 at 15 kt, 6,500 at 18 kt
Complement: 179 (15 officers)

Machinery: CODAD; 4 SEMT-Pielstick 16 PA6 280V BTC diesels; 38,400 hp(m) *(28 MW)* sustained; 2 shafts
Missiles: SSM: 8 Oto Melara/Matra Otomat Mk 2 (2 quad) ❶; active radar homing to 160 km *(86.4 n miles)* at 0.9 Mach; warhead 210 kg; sea-skimmer for last 4 km *(2.2 n miles)*. ERATO system allows mid-course guidance by ship's helicopter.
SAM: Thomson-CSF Crotale Naval octuple launcher ❷; command line of sight guidance; radar/IR homing to 13 km *(7 n miles)* at 2.4 Mach; warhead 14 kg; 26 missiles.
Guns: 1 Creusot-Loire 3.9 in *(100 mm)*/55 compact Mk 2 ❸; 20/45/90 rds/min to 17 km *(9.3 n miles)* weight of shell 13.5 kg. 4 Breda 40 mm/70 (2 twin) ❹; 300 rds/min to 12.5 km *(6.8 n miles)*; weight of shell 0.96 kg.
Torpedoes: 4—21 in *(533 mm)* tubes ❺. ECAN F17P; anti-submarine; wire-guided; active/passive homing to 20 km *(10.8 n miles)* at 40 kt; warhead 250 kg.
Physical countermeasures: Decoys: CSEE Dagaie double trainable mounting ❻; IR flares and chaff; H/J-band.
Electronic countermeasures: ESM: Thomson-CSF DR 4000; intercept; HF/DF.
ECM: Thomson-CSF Janet; jammer.

Radars: Air/surface search/IFF: Thomson-CSF Sea Tiger (DRBV 15) ❼; E/F-band; range 110 km *(60 n miles)* for 2 m² target.
Navigation: 2 Racal Decca TM 1226; I-band.
Fire control: Thomson-CSF Castor IIB/C ❽; I/J-band; range 15 km *(8 n miles)* for 1 m² target. Thomson-CSF DRBC 32 ❾; I/J-band (for SAM).
Sonars: Thomson Sintra Diodon TSM 2630; hull-mounted; active search and attack with integrated Sorel VDS ❿; 11, 12 or 13 kHz.
Combat data systems: Thomson-CSF TAVITAC action data automation; capability for Link W.
Weapon control systems: Vega system. 3 CSEE Naja optronic directors. Alcatel DLT for torpedoes.
Helicopters: 1 SA 365F Dauphin 2 ⓫.

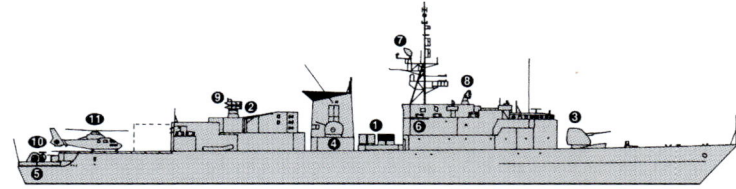

MADINA *(Scale 1 : 1,200), Ian Sturton / 0506097*

Programmes: Ordered in 1980, the major part of the Sawari I contract. Agreement for France to provide supplies and technical help.
Modernisation: The class have been upgraded by DCN Toulon, *Madina* completed in April 1997. *Hofouf* in mid-1998. *Abha* in late 1999, and *Taif* in March 2000. Improvements included updating TAVITAC, Otomat missiles, both sonars and fitting a Samahé 110 helo handling system. A contract for the refit and modernisation of all four ships was reportedly signed in mid-2013. No details have been published but DCNS, Thales, and MBDA are reported to be involved.
Structure: Fitted with Snach/Saphir folding fin stabilisers.
Operational: Navigation: CSEE Sylosat. Helicopter can provide mid-course guidance for SSM. All based at Jiddah. Only a few weeks a year are spent at sea.

HOFOUF *7/2009, B Prézelin / 1366143*

HOFOUF *3/2006 / 1167507*

3 AL RIYADH (MODIFIED LA FAYETTE) CLASS (TYPE F-3000S) (FFGHM)

Name	No	Builders	Laid down	Launched	Commissioned
AL RIYADH	812	DCN, Lorient	29 Sep 1999	1 Aug 2000	26 Jul 2002
MAKKAH	814	DCN, Lorient	25 Aug 2000	20 Jul 2001	3 Apr 2004
AL DAMMAM	816	DCN, Lorient	26 Aug 2001	7 Sep 2002	23 Oct 2004

Displacement, tonnes: 4,725 full load
Dimensions, metres (feet): 133.6 × 17.2 × 4.1 *(438.3 × 56.4 × 13.5)*
Speed, knots: 25
Range, n miles: 7,000 at 15 kt
Complement: 181 (25 officers) + 9 spare berths

Machinery: CODAD; 4 SEMT-Pielstick 16 PA6 STC diesels; 28,000 hp(m) *(20.58 MW)* sustained; 2 shafts; LIPS cp props; bow thruster
Missiles: SSM: 8 Aerospatiale MM 40 Block II Exocet ❶; inertial cruise; active radar homing to 70 km *(40 n miles)* at 0.9 Mach; warhead 165 kg; sea-skimmer.
SAM: Eurosam SAAM ❷; 2 octuple Sylver A43 VLS for Aster 15; command guidance active radar homing to 15 km *(8.1 n miles)* anti-missile, at 30 km *(16.2 n miles)* anti-aircraft. 16 missiles.
Guns: 1 Oto Melara 3 in *(76 mm)*/62 Super Rapid ❸; 120 rds/min to 16 km *(8.7 n miles)*; weight of shell 6 kg. 2 Giat 15B 20 mm ❹; 800 rds/min to 3 km; weight of shell 0.1 kg. 2−12.7 mm MGs.
Torpedoes: 4−21 in *(533 mm)* tubes; ECAN F17P; anti-submarine; wire-guided active/passive homing to 20 km *(10.8 n miles)* at 40 kt; warhead 250 kg.
Physical countermeasures: Decoys: 2 Matra Dagaie Mk 2 ❺; 10-barrelled trainable launchers; chaff and IR flares. SLAT anti-wake homing torpedoes system (when available).
Electronic countermeasures: RESM: Thomson-CSF (DR 3000-S2) ❻; intercept. Sagem Telegon 10.
CESM: Thales Altesse; intercept.
ECM: 2 Thales Salamandre; jammers.
Radars: Air search: Thales DRBV 26C Jupiter II ❼; D-band.
Surveillance/Fire control: Thomson-CSF Arabel 3D ❽; I/J-band.
Fire control: Thomson-CSF Castor II UJ ❾; J-band; range 15 km *(8 n miles)* for 1 m² target.
Navigation: 2 Racal Decca 1226 ❿; I-band. A second set fitted for helicopter control.
Sonars: Thomson Marconi CAPTAS 20; active low frequency; towed array.
Combat data systems: Thales Senit 7.
Weapon control systems: Thales Castor IIJ radar/EO tracker.

Helicopters: 1 Dauphin 2 ⓫.

Programmes: A provisional order was made on 11 June 1989, but this was not finally confirmed until 19 November 1994 when a contract for two ships was authorised under the Sawari II programme. Thomson-CSF was the prime contractor. On 25 May 1997 an order for a third ship was placed together with a substantial enhancement of the weapon systems in all three. SAM successfully tested in 816 in April 2004.
Structure: The design is a development of the French La Fayette class. Some 10 m longer, space and weight included for two more octuple SAM launchers or A50 launcher for Aster 30. Provision is made for a larger NH 90 type helicopter in the future, DCN Samahé helo handling system. STAF stabilisers. Originally planned to be fitted with a 100 mm gun, the contract was amended to incorporate a 76 mm mounting instead.
Operational: OTHT link for helicopters and Air Force F-15s. *Makkah* seriously damaged in a grounding incident 80 miles north of Jiddah in December 2004. The ship was refloated by the Tsavliris Salvage Group in early 2005 and was towed to Jiddah. Subsequently repairs were carried out, with the assistance of DCNS, and the ship returned to operational service in 2009.

AL RIYADH *(Scale 1 : 1,200), Ian Sturton / 1044496*

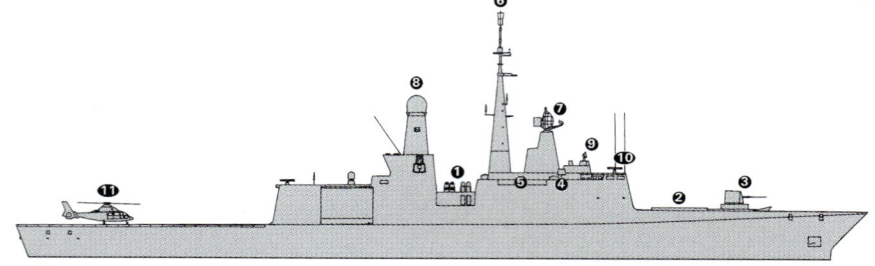

AL RIYADH *7/2009, B Prézelin / 1366142*

AL RIYADH
3/2006
1167506

AL RIYADH *9/2012, Ships of the World / 1483644*

CORVETTES

4 BADR CLASS

Name	No	Builders	Laid down	Launched	Commissioned
BADR	612	Tacoma Boatbuilding Co, Tacoma	6 Oct 1979	26 Jan 1980	30 Nov 1980
AL YARMOOK	614	Tacoma Boatbuilding Co, Tacoma	3 Jan 1980	13 May 1980	18 May 1981
HITTEEN	616	Tacoma Boatbuilding Co, Tacoma	19 May 1980	5 Sep 1980	3 Oct 1981
TABUK	618	Tacoma Boatbuilding Co, Tacoma	22 Sep 1980	18 Jun 1981	10 Jan 1983

Displacement, tonnes: 884 standard; 1,055 full load
Dimensions, metres (feet): 74.7 × 9.6 × 2.7 *(245.1 × 31.5 × 8.9)*
Speed, knots: 30; 20 diesel. **Range, n miles:** 4,000 at 20 kt
Complement: 58 (7 officers)

Machinery: CODOG; 1 GE LM 2500 gas turbine; 23,000 hp *(17.2 MW)* sustained; 2 MTU 12V 652 TB91 diesels; 3,470 hp(m) *(2.55 MW)* sustained; 2 shafts; cp props
Missiles: SSM: 8 McDonnell Douglas Harpoon (2 quad) launchers ❶; active radar homing to 130 km *(70 n miles)* at 0.9 Mach; warhead 227 kg.
Guns: 1 FMC/Oto Melara 3 in *(76 mm)*/62 Mk 75 Mod 0 ❷; 85 rds/min to 16 km *(8.7 n miles)*; weight of shell 6 kg. 1 Raytheon 20 mm 6-barrelled Vulcan Phalanx Block 0 ❸; 3,000 rds/min combined to 2 km. 2 Oerlikon 20 mm/80 ❹. 1—81 mm mortar. 2—40 mm Mk 19 grenade launchers.
Torpedoes: 6—324 mm US Mk 32 (2 triple) tubes ❺. Honeywell Mk 46; anti-submarine; active/passive homing to 11 km *(5.9 n miles)* at 40 kt; warhead 44 kg.
Physical countermeasures: Decoys: 2 Loral Hycor SRBOC 6-barrelled fixed Mk 36 ❻; IR flares and chaff to 4 km *(2.2 n miles)*.
Electronic countermeasures: ESM: SLQ-32(V)1 ❼; intercept.
Radars: Air search: Lockheed SPS-40B ❾; B-band; range 320 km *(175 n miles)*.
Surface search: ISC Cardion SPS-55 ❿; I/J-band.
Fire control: Sperry Mk 92 ⓫; I/J-band.
Sonars: Raytheon SQS-56 (DE 1164); hull-mounted; active search and attack; medium frequency.
Weapon control systems: Mk 24 optical director ❽. Mk 309 for torpedoes. Mk 92 Mod 5 GFCS. FSI Safire FLIR.

Modernisation: Refitting done in Saudi Arabia with US assistance. FLIR being fitted from 1998.
Structure: Fitted with fin stabilisers.
Operational: All based at Al Jubail on the east coast and spend little time at sea.

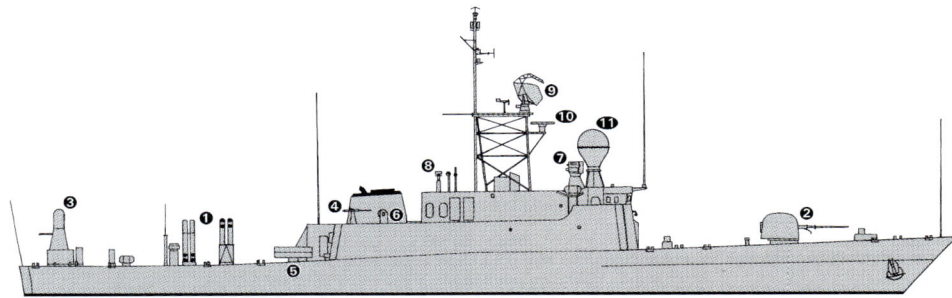

BADR *(Scale 1 : 600), Ian Sturton /* 0506250

TABUK *3/2012, Michael Nitz /* 1454884

SHIPBORNE AIRCRAFT

Notes: Procurement of a new shipborne helicopter is under consideration. Up to 10 are required for deployment to the Al Riyadh class frigates and for other tasks.

Numbers/Type: 15/6 Aerospatiale AS 565SA Dauphin 2/AS 365N Dauphin 2.
Operational speed: 140 kt *(260 km/h)*.
Service ceiling: 15,000 ft *(4,575 m)*.
Range: 410 n miles *(758 km)*.
Role/Weapon systems: AS 565SA is the ASV/ASW helicopter; procured for embarked naval aviation force; surface search/attack is the primary role. Sensors: Thomson-CSF Agrion 15 radar; Crouzet MAD. Weapons: ASV; four AS/15TT missiles. ASW; 2 Mk 46 torpedoes. AS 365N is for SAR and is operated by the Armed Forces Medical Services. Sensors: Omera DRB 32 search radar. Weapons: Unarmed.

LAND-BASED MARITIME AIRCRAFT

Notes: (1) Six P-3C Orion or CASA CN-235 patrol aircraft may be acquired in due course.
(2) Five Boeing E3-A AEW aircraft in service with Air Force.

Numbers/Type: 10 Aerospatiale AS 532SC Cougar.
Operational speed: 150 kt *(280 km/h)*.
Service ceiling: 15,090 ft *(4,600 m)*.
Range: 335 n miles *(620 km)*.
Role/Weapon systems: First pair delivered in August 1989. Total of 12 by the end of 1990. Ten reported operational. Shared with the Coast Guard. Sensors: Omera search radar Safire AAQ-22 FLIR from 1998. Weapons: ASV; Giat 20 mm cannon; AM39 Exocet or Sea Eagle ASM.

DAUPHIN 2 *4/2002, Aerospatiale /* 0093192

AS 532 Cougar *6/1990, Paul Jackson /* 0062140

Serbia

Country Overview

The Republic of Serbia was formed following a referendum on 21 May 2006 in which the people of Montenegro voted for independence and for the dissolution of the Federal Republic of Serbia and Montenegro; this itself was the rump of the former Yugoslavia. On 5 June 2006 the Serbian National Assembly decreed Serbia to be the continuing international personality of Serbia and Montenegro.

With an area of 34,116 square miles, it is located in south-eastern Europe in the Balkan Peninsula and is bordered to the west by Montenegro, Bosnia and Croatia, to the north by Hungary, to the east by Romania and Bulgaria and to the south by Albania and Macedonia. A land-locked country, the principal river is the Danube which enters the country from the north and after passing through the capital Belgrade goes on to form part of the eastern border. Other rivers include the Sova and Tisza.

The armed forces of Serbia and Montenegro were split in such a way that each state kept the assets in its territory. The former Danube Flotilla became subordinate to the Serbian land forces as the River Flotilla.

Headquarters Appointments

Commander, Riverine Flotilla:
Captain Andrija Andric

Personnel

2014: 300

Bases

Novi Sad and Belgrade.

PATROL FORCES

3 TYPE 20 BISCAYA CLASS (RIVER PATROL CRAFT) (PBR)

RPC 213   RPC 214   RPC 216

Displacement, tonnes: 53 standard; 57 full load
Dimensions, metres (feet): 21.3 × 7.5 × 1.2 (69.9 × 24.6 × 3.9)
Speed, knots: 16. **Range, n miles:** 200 at 15 kt
Complement: 10
Machinery: 2 diesels; 1,156 hp(m) (850 kW); 2 shafts
Guns: 2 M71 20 mm.
Radars: Surface search: Decca 110; I-band.

Comment: *RPC 213* built at Josip Broz Shipyard, Belgrade and RPC 214 and 216 at Brodotehnika Shipyard, Belgrade, in 1980. Steel hull with GRP superstructure. All active with the River Flotilla.

RPC 213   6/2013*, Serbian Army / 1525547

1 RIVER PATROL BOAT (PBR)

RPC 111

Displacement, tonnes: 27 standard; 29 full load
Dimensions, metres (feet): 24.1 × 4.1 × 0.9 (79.1 × 13.5 × 3.0)
Speed, knots: 18. **Range, n miles:** 720 at 17 kt
Complement: 6
Machinery: 2 diesels; 435 hp(m) (324 kW); 2 shafts
Guns: 2 M71 20 mm.

Comment: Built in 1956 for US Navy's Rhine River patrol and later transferred. Steel hull and aluminium superstructure. Capable of carrying 30 troops.

RPC 111   6/2008, Freivogel Collection / 1335423

1 RIVER PATROL BOAT (PBR)

CMP 22

Displacement, tonnes: 33 standard; 43 full load
Dimensions, metres (feet): 8 × 2.95 × 1.5 (26.2 × 9.7 × 4.9)
Speed, knots: 30
Complement: 3
Machinery: 2 diesels; 257 hp (192 kW); 2 shafts
Guns: 1—12.7 mm MG.

Comment: Built in 1979 at Greben Shipyard, Bela Luka. Transferred from the border security service to the river flotilla in 2006.

CMP 22   6/2013*, Serbian Army / 1525546

MINE WARFARE FORCES

4 NESTIN CLASS (RIVER MINESWEEPERS) (MSR)

No	Builders	Commissioned
RML 332	Brodotehnika, Belgrade	18 Dec 1976
RML 335	Brodotehnika, Belgrade	1979
RML 336	Brodotehnika, Belgrade	1980
RML 341	Brodotehnika, Belgrade	8 Jun 1996

Displacement, tonnes: 64 standard; 71 full load
Dimensions, metres (feet): 27 × 6.5 × 2.7 (88.6 × 21.3 × 8.9)
Speed, knots: 14
Range, n miles: 860 at 11 kt
Complement: 17

Machinery: 2 diesels; 512 hp(m) (382 kW); 2 shafts
Missiles: 4 (1 quad) MTU-4 launcher; Strela 2M (Grail); IR homing to 3.6 km (2 n miles); warhead 1 kg.
Guns: 4 (8 in 341) M71 20 mm (1 (2) quad). 2 M71 20 mm (332, 335, 336).
Mines: 24 can be carried.
Physical countermeasures: MCMV: Magnetic, acoustic and explosive sweeping gear.
Radars: Surface search: Racal Decca 1216R; I-band.

Comment: Some transferred to Hungary and Iraq. The next generation of the craft (*RML 341*) to a slightly modified design, was completed in 1996. The class is based at Novi Sad as part of the River Flotilla. One deleted in 1997 and a further three in 2007. Capable of transporting 100 troops.

RML 336   6/2013*, Serbian Army / 1525545

AMPHIBIOUS FORCES

5 TYPE 22 (LCU)

DJC 411 (ex-DJC 632)   DJC 413 (ex-DJC 630)   DJC 415 (ex-DJC 631)
DJC 412 (ex-DJC 625)   DJC 414 (ex-DJC 621)

Displacement, tonnes: 34 standard; 43 full load
Dimensions, metres (feet): 22.2 × 4.8 × 1.1 (72.8 × 15.7 × 3.6)
Speed, knots: 30
Range, n miles: 320 at 22 kt
Complement: 6

Military lift: 60 troops or 8 tons cargo
Machinery: 2 MWM diesels; 1,740 hp(m) (1.28 MW); 2 waterjets
Guns: 2 (1 twin) M71 20 mm. 1—30 mm grenade launcher.
Radars: Navigation: Decca 101; I-band.

Comment: Built at Greben Shipyard of polyester and glass fibre. Last one completed in 1985.

DJC 413   6/2013*, Serbian Army / 1525544

IHS Jane's Fighting Ships 2014-2015

AUXILIARIES

Notes: There is a 21 m oil supply ship (RPN-43), a floating dock RDOK-15 and six loading platforms RPP 70–75.

1 KOZARA CLASS (HEADQUARTERS SHIP) (PBR)

KOZARA (ex-*Oregon*; ex-*Kriemhild*) RPB 30

Displacement, tonnes: 559 standard; 611 full load
Dimensions, metres (feet): 67 × 9.5 × 2.7 *(219.8 × 31.2 × 8.9)*
Speed, knots: 14
Machinery: 2 Rade Koncar B6 AVJ 354-04 diesels; 680 hp *(507 kW)*; 2 shafts
Guns: 9 M71 20 mm (3 triple).
Radars: Navigation: Furuno RHRS-200S; I-band.

Comment: The ship has spent its life on the River Danube. Built for the German Navy in Linz in 1939, she was later used from 1945–46 to house US soldiers at Regensberg. Sold in 1948, she was used as a hotel-restaurant until 1953. The vessel was subsequently procured for the Yugoslavian River Flotilla in 1960 and continues to be used to house Serbian River Flotilla staff. She is capable of transporting 250 troops.

1 SABAC CLASS (DEGAUSSING VESSEL) (YDG)

SABAC RSRB 36

Displacement, tonnes: 109 standard; 118 full load
Dimensions, metres (feet): 32.2 × 7.1 × 1.2 *(105.6 × 23.3 × 3.9)*
Speed, knots: 11
Range, n miles: 3,200 at 10 kt
Complement: 20

Machinery: 2 diesels; 528 hp(m) *(388 kW)*; 2 shafts
Missiles: 4 (1 quad) MTU-4 launcher; Strela 2M (Grail); IR homing to 3.6 km *(2 n miles)*; warhead 1 kg.
Guns: 2—20 mm M71.
Radars: Navigation: Decca 1216; I-band.

Comment: Built at Brodotehnika Shipyard, Belgrade, in 1984. Used to degauss River vessels up to a length of 50 m. Capable of transporting 80 troops.

KOZARA   6/2003, Serbia Riverine Forces / 0572439

SABAC   6/2013*, Serbian Army / 1525543

Seychelles

Country Overview

A former British colony, the Republic of the Seychelles became independent in 1976. Situated in the western Indian Ocean, northeast of Madagascar, the archipelago consists of some 90 islands, disposed over 13,000 square miles in two groups. The 40 islands of the northern group include the principal islands: Mahé (the largest), Praslin, Silhouette and La Digue. The 50 or so low-lying coral islands in the south are mostly uninhabited. Victoria (Mahé) is the capital, largest town and principal port. Territorial seas (12 n miles) are claimed. A 200 n mile Exclusive Economic Zone (EEZ) has been declared but the limits have not been fully defined by boundary agreements.

Headquarters Appointments

Commander of the Coast Guard:
  Lieutenant Colonel Georges Adeline

Bases

Ile Perseverance, Mahé.

Personnel

2014: 300 including 80 air wing and 100 marines

LAND-BASED MARITIME AIRCRAFT

Numbers/Type: 1 Britten-Norman BN-2A21 Maritime Defender.
Operational speed: 150 kt *(280 km/h)*.
Service ceiling: 18,900 ft *(5,760 m)*.
Range: 1,500 n miles *(2,775 km)*.
Role/Weapon systems: Coastal surveillance and surface search aircraft delivered in 1980. Sensors: Search radar. Weapons: Provision for rockets or guns.

BN2T-4S (Irish Police colours)   8/1997 / 0016662

COAST GUARD

Notes: Three 7 m fast response craft, capable of about 30 kt, were donated by the United Arab Emirates on 4 January 2011.

2 PATROL CRAFT (PB)

ARIES   LIBRA

Displacement, tonnes: 18 full load
Dimensions, metres (feet): 13.4 × 3.8 × 1.2 *(44 × 12.5 × 3.9)*
Speed, knots: 13
Range, n miles: 200 at 11 kt
Complement: 3
Machinery: 2 General Motors Detroit 6V53 diesels; 2 shafts
Radars: Surface search: Furuno; I-band.

Comment: Former US Coast Guard lifeboats (MLB) constructed in the 1960s. Three were transported to the Seychelles onboard USS *Anchorage* in October 2000 and a further two onboard USS *Tarawa* in December 2000. Two remain in service.

MLBs   9/2003, Seychelles Coast Guard / 0568334

748 Seychelles › Coast guard

1 COASTAL PATROL CRAFT (PB)

JUNON 602

Displacement, tonnes: 41 full load
Dimensions, metres (feet): 18.3 × 5.1 × 1.8 *(60 × 16.7 × 5.9)*
Speed, knots: 20
Complement: 5
Radars: Surface search: Furuno; I-band.

Comment: Former Port and Marine Services patrol boat reintegrated into the Coast Guard in 2003.

JUNON 9/2003, Seychelles Coast Guard / 0568333

2 RODMAN 101 CLASS (PB)

LE VIGILANT P 31   **LA FLÈCHE** P 32

Displacement, tonnes: 63 full load
Dimensions, metres (feet): 30 × 5.9 × 1.3 *(98.4 × 19.4 × 4.3)*
Speed, knots: 30. **Range, n miles:** 800 at 12 kt
Complement: 9
Machinery: 2 Caterpillar 3412C diesels; 2,800 hp *(2.06 MW)*; 2 Hamilton waterjets
Guns: 2 – 12.7 mm MGs.
Radars: Navigation: I-band.

Comment: Donated to the Seychelles by the UAE on 5 January 2011 in order to boost counter-piracy capabilities. The vessels, which were acquired by UAE in about 2005, are the patrol variant of the Rodman 101 design. GRP hull. Details are based on those in Spanish service.

LE VIGILANT 11/2013*, US Navy / 1531229

1 SDB MK 5 CLASS (LARGE PATROL CRAFT) (PBO)

TOPAZ (ex-*Tarmugli*) 606 (ex-T 64)

Displacement, tonnes: 264 full load
Dimensions, metres (feet): 46 × 7.5 × 2.5 *(150.9 × 24.6 × 8.2)*
Speed, knots: 30
Range, n miles: 2,000 at 12 kt
Complement: 34 (4 officers)
Machinery: 2 MTU 16V 538 TB92 diesels; 6,820 hp(m) *(5 MW)* sustained; 2 shafts
Guns: 1 Medak 30 mm 2A42.
Radars: Surface search: Bharat 1245; I-band.

Comment: Built at Garden Reach and first commissioned in 2002. Transferred from the Indian Navy and recommissioned on 23 February 2005. The ship arrested 12 pirates in August 2009.

TOPAZ 4/2010, Seychelles Coast Guard / 1366442

1 TYNE CLASS (PB)

FORTUNE (ex-*Baltic Exchange II*)

Displacement, tonnes: 27 full load
Dimensions, metres (feet): 14 × 4.47 × 1.27 *(45.9 × 14.7 × 4.2)*
Speed, knots: 17
Range, n miles: 240 at 12 kt
Complement: 7
Machinery: 2 diesels; 2 shafts
Radars: Navigation: I-band.

Comment: Tyne-class lifeboat, formerly (1988–2010) in service with the Royal National Lifeboat Institution, donated by UK in July 2010. The role of the vessel is to conduct short-range counter-piracy patrols in the inner islands, as well as search-and-rescue and fisheries enforcement operations.

FORTUNE 9/2010, Seychelles Coast Guard / 1421690

1 TYPE FPB 42 (LARGE PATROL CRAFT) (PB)

Name	No	Builders	Commissioned
ANDROMACHE	605	Picchiotti, Viareggio	10 Jan 1983

Displacement, tonnes: 272 full load
Dimensions, metres (feet): 41.8 × 8 × 2.5 *(137.1 × 26.2 × 8.2)*
Speed, knots: 26
Range, n miles: 3,000 at 16 kt
Complement: 22 (3 officers)
Machinery: 2 Paxman Valenta 16 CM diesels; 6,650 hp *(5 MW)* sustained; 2 shafts
Guns: 1 Oerlikon 25 mm. 2 – 7.62 mm MGs.
Radars: Surface search: 2 Furuno; I-band.

Comment: Ordered from Inma, La Spezia in November 1981. Refitted 2005–06.

ANDROMACHE
3/2010, Seychelles
Coast Guard
1366441

IHS Jane's Fighting Ships 2014-2015

Introduction < **Sierra Leone** — Introduction < **Singapore** 749

Sierra Leone

Country Overview

A former British colony, Sierra Leone became independent in 1961. Located in west Africa, the country has an area of 27,699 square miles, a 217 n mile coastline with the Atlantic Ocean and is bordered to the north by Guinea and to the south by Liberia.

The capital, largest city and principal port is Freetown. Territorial seas (12 n miles) and an EEZ (200 n miles) are claimed.

Headquarters Appointments

Commander Maritime Wing:
  Captain Sal Kanu

Personnel

(a) 2014: 350 (38 officers)
(b) Voluntary service

Bases

Freetown (Murray Town) HQ and Training Base.

Freetown (Government wharf) Main Base. Forward operating bases at Yeliboya, Tombo, Bonthe (Sherbo Is) and Sulima.

PATROL FORCES

Notes: (1) Five small inshore patrol craft have been acquired to operate from Murray Town and the forward operating bases.
(2) Acquisition of a surveillance aircraft and of further patrol craft is under consideration.
(3) There are two 5 m RHIBs for inshore patrols.

1 SHANGHAI III (TYPE 062/1) CLASS (PB)

SIR MILTON PB 105

Displacement, tonnes: 173 full load
Dimensions, metres (feet): 41 × 5.3 × 1.8 *(134.5 × 17.4 × 5.9)*
Speed, knots: 25
Range, n miles: 750 at 17 kt
Complement: 43

Machinery: 4 Chinese L12-180A diesels; 4,400 hp(m) *(3.22 MW)* sustained; 4 shafts
Guns: 4 China 37 mm/63 (2 twin); 180 rds/min to 8.5 km *(4.6 n miles)*; weight of shell 1.42 kg.
  4 China 25 mm (2 twin).
Radars: Surface search: Pot Head or Anritsu 726; I-band.

Comment: Transferred from China on 10 March 2006 to replace *Alimamy Rasin*. The vessel was reported in 2013 to be non-operational, awaiting refit.

3 SEA ARK 32 FT CUTTERS (PB)

01–03

Displacement, tonnes: 6 full load
Dimensions, metres (feet): 9.8 × 3.6 × 1.0 *(32.2 × 11.8 × 3.3)*
Speed, knots: 34
Complement: 4
Machinery: 2 Yanmar 6LYAM-STP diesels; 740 hp *(550 kW)*; 2 Konrad drives
Radars: Navigation: Furuno; I-band.

Comment: Sea Ark Dauntless RAM design donated by the US on 26 May 2006. Although not permanently fitted with weapons, always patrol with light weapons. One reported non-operational in 2013.

SIR MILTON 6/2006, RSLAF (MW) / 1164270

SEA ARK 01 6/2006, RSLAF (MW) / 1164269

Singapore

Country Overview

Formerly under British rule, the Republic of Singapore became self-governing in 1959. It joined Malaysia in 1963, but separated from the Federation in 1965 to become a sovereign state. With an area of 247 square miles and a coastline of 104 n miles, the main island is separated from the southern tip of Malaysia by the narrow Johore Strait. There are 59 small adjacent islets. To the south the Singapore Strait, an important shipping channel linking the Indian Ocean with the South China Sea, separates the island from the Riau archipelago of Indonesia. Territorial seas (3 n miles) are claimed. An EEZ is not claimed.

Headquarters Appointments

Chief of the Navy: Rear Admiral Ng Chee Peng
Chief of Staff: Rear Admiral Tan Wee Beng
Fleet Commander: Rear Admiral Timothy Lo
Commander Maritime Security Task Force:
  Rear Admiral Harris Chan
Commander Police Coast Guard:
  Deputy Assistant Commissioner Hsu Sin Yun

Personnel

(a) 2014: 4,500 officers and men including 1,800 conscripts
(b) National Service: two and a half years for Corporals and above; two years for the remainder
(c) 5,000 reservists (operationally trained)

Bases

Tuas (Jurong), Changi, Sembawang

Organisation

Five Commands: Fleet, Naval Diving Unit, Naval Logistics, Maritime Security Task Force, Training.

Fleet: First Flotilla (185 Squadron, 188 Squadron).
Third Flotilla (191, 192, 193 Squadrons).
Submarines (171 Squadron).
Maritime Security Command: (182, 189, 194 Squadrons).
Coastal Command operates five unmanned Giraffe 100 air/surface surveillance radar sites at Changi, Pedra Branca, St John's Island, Sultan Shoal Lighthouse and Raffles Lighthouse. Air and surface track data is passed to HQ RSN.

Prefix to Ships' Names

RSS

Special Forces

Singapore Armed Forces special forces grouped under its Special Operations Task Force include the Naval Diving Unit and Singapore Army Special Operations Force. Police units include the Singapore Special Tactics and Rescue unit and PCG's Special Task Squadron.

Maritime Security Task Force

The Maritime Security Task Force (MSTF) was established in February 2009. It is operationally responsible to the chief of Defence Force and comprises three groups:
1. **The Comprehensive Maritime Awareness (CMA) Group** builds and maintains a comprehensive maritime situation picture through its information-sharing networks. It works closely with national agencies, international partners and the shipping community (such as ship owners, ships charterers, agents and port operators) to share maritime information.
2. **The Operations Group**, comprising operations planners from the Army, Navy and Air Force, undertakes an integrated approach towards planning and execution of all maritime security operations. It conducts daily patrols, boarding and escort operations in the Singapore Strait and Sea Lines of Communication (SLOC).
3. **The Inter-Agency Coordination Group** comprises representatives from the Police Coast Guard (PCG), the Maritime and Port Authority of Singapore (MPA), the Immigration and Checkpoints Authority (ICA), and the Singapore Customs (Customs).

Police Coast Guard

The Police Coast Guard is a unit of the Singapore Police Force and was first established in 1924. Its role is to maintain coastal security within Singapore territorial waters and to support the Singapore Armed Forces in emergencies. Its four regional commands are Brani (SE sector), Gul (SW sector), Seletar (NE sector) and Lim Chu Kang (NW sector). The PCG HQ moved to a new site at Brani (near Sentosa) on 20 March 2006. The Coastal Patrol Squadron and Special Task Squadron operate under central control. All vessels have Police Coast Guard on the superstructure and a white-red-white diagonal stripe on the hull except for Interceptor craft which have dark blue hulls with grey superstructures. Personnel numbers are about 1,000.

Strength of the Fleet

Type	Active	Building (Projected)
Submarines	4	1
Frigates	6	—
Missile Corvettes	6	—
Offshore Patrol Vessels	11	—
Inshore Patrol Craft	12	—
Minehunters	4	—
LSL/LPD	4	—
LCMs	4	—

© 2014 IHS

IHS Jane's Fighting Ships 2014-2015

SUBMARINES

Notes: It was announced on 2 December 2013 that a contract had been signed with ThyssenKrupp Marine Systems to procure two new Type 218SG submarines. The design is reported to be an adaptation of the Type 214 design in service with the navies of South Korea and Greece. Key features are to include air-independent propulsion and a combat system co-developed by Atlas Elektronik and ST Electronics. The boats are expected to replace the Challenger class and are to enter service from 2020.

2 ARCHER (VÄSTERGÖTLAND) CLASS (SSK)

Name	No	Builders	Laid down	Launched	Commissioned
ARCHER (ex-*Hälsingland*)	–	Kockums, Malmö	1 Jan 1984	31 Aug 1987	20 Oct 1988
SWORDSMAN (ex-*Västergötland*)	–	Kockums, Malmö	10 Jan 1983	17 Sep 1986	27 Nov 1987

Displacement, tonnes: 1,524 surfaced; 1,626 dived
Dimensions, metres (feet): 60.5 × 6.1 × 5.6 *(198.5 × 20.0 × 18.4)*
Speed, knots: 10 surfaced; 20 dived
Complement: 27 (5 officers)

Machinery: Diesel-Stirling-electric; 2 Hedemora V12A/15 diesels; 2,200 hp(m) *(1.62 MW)*; 2 Kockums Stirling Mk III AIP; 204 hp *(150 kW)*; 1 Jeumont Schneider motor; 1,800 hp(m) *(1.32 MW)*; 1 shaft; LIPS prop
Torpedoes: 6—21 in *(533 mm)* tubes. 12 WASS Black Shark; wire (fibre-optic cable) guided; active/passive homing to 50 km *(27 n miles)* at 50 kt; warhead 250 kg; swim-out discharge. 3—15.75 in *(400 mm)* tubes. 6 FFV Type 431/451; anti-submarine; wire-guided; active/passive homing to 20 km *(10.8 n miles)* at 25 kt; warhead 45 kg shaped charge or a small charge anti-intruder version is available.
Mines: 12 Type 47 swim-out mines in lieu of torpedoes.

Electronic countermeasures: ESM: Argo AR-700-S5; or Condor CS 3701; intercept.
Radars: Navigation: Terma; I-band.
Sonars: Atlas Elektronik CSU 83; hull-mounted; passive search and attack; medium frequency. Flank array; passive search; low frequency.
Weapon control systems: UDS SUBTICS.

Programmes: Original design contract awarded by the Swedish Navy to Kockums, Malmö on 17 April 1978. Contract for construction signed 8 December 1981. Following discussions between the governments of Sweden and Singapore in 2005, both submarines are to be transferred to the Singapore Navy as part of a package that includes modernisation refits to incorporate Air-Independent Propulsion (AIP) systems prior to delivery. On entry into Singapore service in about 2011, the boats are likely to replace two of the Challenger class, also procured from Sweden, that entered service from 2000.

Modernisation: The modernisation package is similar to that given to the Swedish Södermanland class. This included the installation of Air Independent Propulsion (Stirling Mk 3 AIP) by the insertion of a 12 m plug in the pressure hull. Other work included the installation of a pressurised diver's lock-out in the base of the sail to facilitate special forces operations and a new climate control system. The Thales Optronics CK 038 periscope was upgraded with a thermal imaging camera and an improved image intensifier.
Structure: Single hulled with an X-type rudder/after hydroplace design. Diving depth 300 m *(984 ft)*. Anechoic coating.
Operational: *Archer* was relaunched on 16 June 2009 and arrived in Singapore on 17 August 2011. She was recommissioned on 2 December 2011. *Swordsman* was relaunched on 20 October 2010 arrived in Singapore on 31 December 2012 and was recommissioned on 30 April 2013. Both submarines are assigned to 171 Squadron.

SWORDSMAN 10/2010, L-G Nilsson / 1406274

3 CHALLENGER (SJÖORMEN) CLASS (SSK)

Name	No	Builders	Laid down	Launched	Commissioned
CENTURION (ex-*Sjöormen*)	–	Kockums	1965	25 Jan 1967	31 Jul 1968
CONQUEROR (ex-*Sjölejonet*)	–	Kockums	1966	29 Jun 1967	16 Dec 1968
CHIEFTAIN (ex-*Sjohunden*)	–	Kockums	1966	21 Mar 1968	25 Jun 1969

Displacement, tonnes: 1,148 surfaced; 1,229 dived
Dimensions, metres (feet): 51 × 6.1 × 5.8 *(167.3 × 20.0 × 19.0)*
Speed, knots: 12 surfaced; 20 dived
Complement: 23 (7 officers)

Machinery: Diesel-electric; 2 Hedemora-Pielstick V12A/A2/15 diesels; 2,200 hp(m) *(1.62 MW)*; 1 ASEA motor; 1,500 hp(m) *(1.1 MW)*; 1 shaft
Torpedoes: 4—21 in *(533 mm)* bow tubes. 10 FFV Type 613; anti-surface; wire-guided; passive homing to 15 km *(8.2 n miles)* at 45 kt; warhead 250 kg. 2—16 in *(400 mm)* tubes. 4 FFV Type 431; anti-submarine; wire-guided; active/passive homing to 20 km *(10.8 n miles)* at 25 kt; warhead 45 kg shaped charge.

Mines: Minelaying capability.
Radars: Navigation: Terma; I-band.
Sonars: Plessey Hydra; hull-mounted; passive search and attack; medium frequency.
Weapon control systems: UDS SUBTICS.

Programmes: It was announced on 23 September 1995 that a submarine would be acquired from Sweden for training purposes only. Three more of the same class acquired in July 1997 for conversion plus one more for spares.
Modernisation: A contract for new periscope systems was awarded to Kollmorgen Electro Optical in January 2005. Model 90 is the most likely option.

Structure: Albacore hull. Twin-decked. Diving depth, 150 m *(492 ft)*. Air conditioning added for tropical service, together with battery cooling.
Operational: *Challenger* re-launched on 26 September 1997, *Conqueror* and *Centurion* on 28 May 1999 and *Chieftain* on 22 May 2001. *Conqueror* was recommissioned in Singapore on 24 July 2000 and *Chieftain* on 24 August 2002. *Challenger* and *Centurion* remained in Sweden to support training until January 2004 when they were transported to Singapore. Ex-*Sjohasten* was also shipped as a source of spares. *Centurion* was recommissioned on 26 June 2004 but *Challenger* was not commissioned and maintained in a training role. The three submarines are assigned to 171 squadron. Based at Changi.

CONQUEROR — 3/2000, Per Körnefeldt / 0084434

CONQUEROR — 9/2000, Sattler/Steele / 0105592

CONQUEROR — 8/2006, Jürg Kürsener / 1164545

FRIGATES

6 FORMIDABLE (PROJECT DELTA) CLASS (FFGHM)

Name	No	Builders	Laid down	Launched	Commissioned
FORMIDABLE	68	DCN, Lorient	14 Nov 2002	7 Jan 2004	5 May 2007
INTREPID	69	Singapore SB and Marine	8 Mar 2003	3 Jul 2004	5 Feb 2008
STEADFAST	70	Singapore SB and Marine	15 Nov 2003	28 Jan 2005	5 Feb 2008
TENACIOUS	71	Singapore SB and Marine	22 May 2004	15 Jul 2005	5 Feb 2008
STALWART	72	Singapore SB and Marine	12 Nov 2005	9 Dec 2005	16 Jan 2009
SUPREME	73	Singapore SB and Marine	17 May 2005	9 May 2006	16 Jan 2009

Displacement, tonnes: 3,251 full load
Dimensions, metres (feet): 114 × 16.0 × 5.0 *(374 × 52.5 × 16.4)*
Speed, knots: 27
Range, n miles: 4,000 at 15 kt
Complement: 86 (15 air crew)

Machinery: CODAD; 4 MTU 20V 8000 M90 diesels; 48,276 hp *(36 MW)*; 2 shafts; cp props; bow thruster
Missiles: SSM: 8 Boeing Harpoon ❶; active radar homing to 130 km *(70 n miles)* at 0.9 Mach; warhead 227 kg.
SAM: Eurosam SAAAM; 2 octuple Sylver A 43 VLS; 2 octuple Sylver A 50 VLS ❷ for MBDA Aster 15; command guidance active radar homing to 15 km *(8.1 n miles)* anti-missile and to 30 km *(16.2 n miles)* anti-aircraft. 32 missiles.
Guns: 1 Oto Melara 3 in *(76 mm)*/62 Super rapid ❸; 120 rds/min to 16 km *(8.7 n miles)*; weight of shell 6 kg. 2 M134D 7.62 mm miniguns.
Torpedoes: 6 — 324 mm (2 triple (recessed)) ❹ tubes. Eurotorp A 244/S Mod 3; anti-submarine; active/passive homing to 7 km *(3.8 n miles)* at 33 kt; warhead 34 kg (shaped charge).
Physical countermeasures: Decoys: 3 EADS NGDS 8-barrelled chaff ❺, IR and anti-torpedo decoy launchers.
Electronic countermeasures: ESM: RAFAEL C-PEARL-M; intercept.
Radars: Air/search: Thales Herakles 3-D radar multifunction ❼; E/F-band.
Surface search/Navigation: 2 Terma Scanter 2001 ❽; I-band.
Sonars: EDO 980 ALOFTS VDS; low frequency (2 kHz).
Combat data systems: DSTA/ST Electronics system.
Weapon control systems: 2 EADS Nagir 2000 optronic directors ❻.

Helicopters: 1 S-70B Seahawk ❾.

Programmes: Ordered from DCN International on 6 March 2000. First steel cut for hulls two and three on 2 October 2002. Prime Contractor is Singapore's Defence Science and Technology Agency (DSTA) who are also leading combat system integration in partnership with ST Electronics.
Structure: Derived from La Fayette class but there are notable differences to accommodate the weapon and sensor fit. Two of the four VLS modules are reported to be Sylver A 50, capable of launching the longer Aster 30 area-defence missile. It was reported in September 2013 that Aster 30 was to be procured as a land-based system, possibly a precursor to a sea-based system. Aster 15 successfully launched from *Intrepid* on 3 April 2008. The ships are equipped with a handling system for unmanned surface craft.
Operational: The ships form 185 Squadron based at Changi. *Steadfast* conducted trials with a Boeing Scan Eagle UAV in February 2009. *Stalwart* deployed to San Diego in 2009 for S-70B integration training.

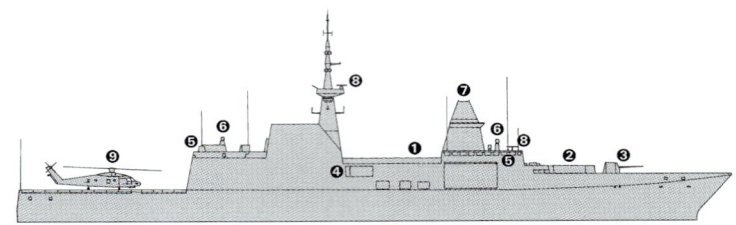
FORMIDABLE (Scale 1 : 1,200), Ian Sturton / 1366437

STALWART 9/2012, John Mortimer / 1486351

STEADFAST 3/2013*, Chris Sattler 1531029

TENACIOUS 8/2010, John Mortimer / 1406275

Corvettes < **Singapore** 753

CORVETTES

6 VICTORY CLASS (FSGM)

Name	No	Builders	Launched	Commissioned
VICTORY	P 88	Lürssen, Bremen	8 Jun 1988	18 Aug 1990
VALOUR	P 89	Singapore SB and Marine	10 Dec 1988	18 Aug 1990
VIGILANCE	P 90	Singapore SB and Marine	27 Apr 1989	18 Aug 1990
VALIANT	P 91	Singapore SB and Marine	22 Jul 1989	25 May 1991
VIGOUR	P 92	Singapore SB and Marine	1 Dec 1989	25 May 1991
VENGEANCE	P 93	Singapore SB and Marine	23 Feb 1990	25 May 1991

Displacement, tonnes: 605 full load
Dimensions, metres (feet): 62.4 oa; 58 wl × 8.5 × 3.1 *(204.7; 190.3 × 27.9 × 10.2)*
Speed, knots: 35. **Range, n miles:** 2,000 at 22 kt
Complement: 49 (8 officers)

Machinery: 4 MTU 16V 538 TB93 diesels; 15,020 hp(m) *(11 MW)* sustained; 4 shafts
Missiles: SSM: 8 McDonnell Douglas Harpoon ❶; active radar homing to 130 km *(70 n miles)* at 0.9 Mach; warhead 227 kg.
SAM: 2 Octuple IAI/Rafael Barak I ❷ radar or optical guidance to 10 km *(5.5 m)* at 2 Mach; warhead 22 kg.
Guns: 1 Oto Melara 3 in *(76 mm)*/62 Super Rapid ❸; 120 rds/min to 16 km *(8.7 n miles)*; weight of shell 6 kg. 4 CIS 50 12.7 mm MGs.
Torpedoes: 6—324 mm Whitehead B 515 (2 triple) tubes ❹. Whitehead A 244S; anti-submarine; active/passive homing to 7 km *(3.8 n miles)* at 33 kt; warhead 34 kg (shaped charge).
Physical countermeasures: Decoys: 2 Plessey Shield 9-barrelled chaff launchers ❺. 4 Rafael (2 twin) long-range chaff launchers to be fitted below the bridge wings.
Electronic countermeasures: ESM: Elisra SEWS ❻; intercept.
ECM: Rafael RAN 1101; ❼ jammer.
Radars: Surface search: Saab Sea Giraffe AMB ❿; G-band.
Navigation: Kelvin Hughes 1007; I-band.
Fire control: 2 Elta EL/M-2221(X) ⓫; I/J/K-band.
Sonars: Thomson Sintra TSM 2064; VDS ⓬; active search and attack.
Combat data systems: Elbit command system. SATCOM ❽.
Weapon control systems: Elbit MSIS optronic director ❾.

Programmes: Ordered in June 1986 to a Lürssen MGB 62 design similar to Bahrain and UAE vessels.
Modernisation: Barak launchers fitted on either side of the VDS, together with a second fire-control radar on the platform aft of the mast and an optronic director on the bridge roof. Rudder roll stabilisation retrofitted to improve sea-keeping qualities. Unidentified EW antennae have been installed below RAN 1101. A mid-life update is planned. Modifications are reported to include removal of the VDS and torpedo tubes. Replacement of the Sea Giraffe 150HC radar with Sea Giraffe AMB began in 2011.
Operational: Form 188 Squadron, part of the First Flotilla. Designated Missile Corvettes (MCV). Live Barak firing conducted from *Victory* on 23 July 2012.

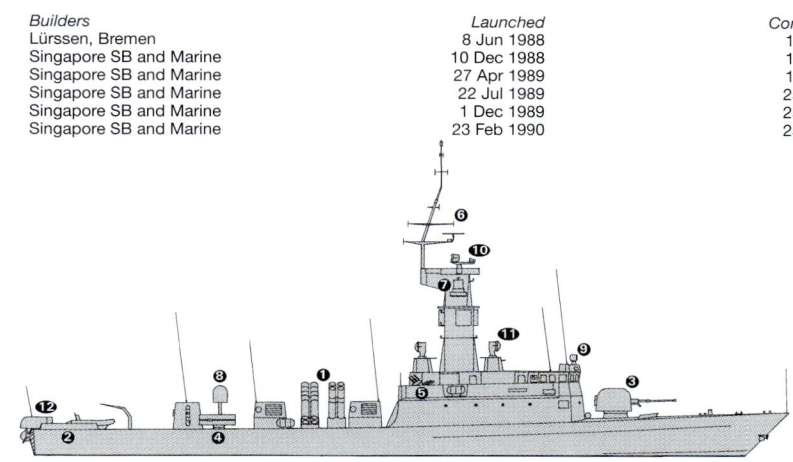

VICTORY *(Scale 1 : 600), Ian Sturton* / 0114802

VALOUR *5/2010, David Boey* / 1406277

VALIANT *9/2012, John Mortimer* / 1486352

© 2014 IHS IHS Jane's Fighting Ships 2014-2015

754 Singapore > Shipborne aircraft — Patrol forces

SHIPBORNE AIRCRAFT

Numbers/Type: 6 Sikorsky S-70B Seahawk.
Operational speed: 135 kt (250 km/h).
Service ceiling: 10,000 ft (3,050 m).
Range: 600 n miles (1,110 km).
Role/Weapon systems: Contract placed 21 January 2005 for six new helicopters for operation from Formidable class frigates. Following integration training by Stalwart at San Diego in 2009, inauguration of all six aircraft into 123 Squadron was completed on 18 January 2011. Roles ASW, ASV and surveillance. Weapons and sensors to be announced but likely to be similar to US Navy MH-60R.

S-70B   4/2013*, Chris Sattler / 1531032

LAND-BASED MARITIME AIRCRAFT

Notes: (1) The Air Force also has 40 F-16D Block 52+, 20 F-5 S/T Tiger II and 20 F-16C.
(2) There are also six CH-47D used for maritime tasks and 20 AH-64D Apache Longbow.

Numbers/Type: 5 Fokker F50 Mk 2S Enforcer.
Operational speed: 220 kt (463 km/h).
Service ceiling: 29,500 ft (8,990 m).
Range: 2,700 n miles (5,000 km).
Role/Weapon systems: In service from September 1995. Part of Air Force 121 Squadron but with mixed crews and under naval op con. One modified for Sigint. Sensors: Texas Instruments APS-134(V)7 radar; GEC FLIR; Elta ESM. Jammer fitted under wing-tip. Weapons: Harpoon ASM; mines; A-244S torpedoes.

FOKKER F 50   9/2003, David Boey / 0567532

Numbers/Type: 4 Gulfstream G550 CAEW.
Operational speed: 488 kt (904 km/h).
Service ceiling: 51,000 ft (15,545 m).
Range: 6,750 n miles (12,501 km).
Role/Weapon systems: Plans to replace E-2C Hawkeye G550 announced in April 2007. First aircraft arrived February 2009 and remainder in 2011. The aircraft achieved full operational capability in April 2012. Sensors: Elta EL/W-2085 conformal system including radar, ESM and COMINT.

PATROL FORCES

Notes: Following procurement in 2004 of a least two Rafael Protector unmanned surface vehicles (USVs), and their subsequent deployment in Resolution in 2005, the Singapore Navy has maintained its interest in the potential use of such craft. This has included participation in the US Navy's Spartan Scout technology demonstrator programme. ST Electronics first launched its propeller-driven Venus-9 USV at the IMDEX Asia 2009 trade show and subsequently a water-jet powered variant was showcased at the 2010 Singapore Air Show. The company has since expanded its Venus family of USVs to include the 16-m Venus-16 and 11-m Venus-11. Potential tasks include force protection (combining radar and electro-optical sensors with a small-calibre remote control weapon station); anti-submarine warfare (with an active dipping sonar); mine countermeasures (equipped with a synthetic aperture sonar and expendable mine neutralisation device); electronic warfare; maritime surveillance; and Precision Fire (equipped with short-range missile system).

0 + 8 LITTORAL MISSION VESSELS (PGM)

Displacement, tonnes: 1,200 full load
Dimensions, metres (feet): 80 × 12.0 × ? (262.5 × 39.4 × ?)
Speed, knots: 27. **Range, n miles:** 5,000 at 15 kt

Machinery: 2 MTU 20V 4000 M93 diesels; 11,532 hp (8.6 MW); 2 shafts
Missiles: SAM: To be announced.
Guns: To be announced.
Radars: Air/surface search: To be announced.
Navigation: To be announced.
Combat data systems: To be announced.
Electro-optic systems: To be announced.
Helicopters: Platform for 1 medium.

Comment: It was announced on 30 January 2013 that a contract had been signed with Singapore Technologies Engineering Ltd for the design and construction of eight new Littoral Mission Vessels to replace the Fearless class. Weapons are to include a medium-calibre gun and a point-defence system while sensors are to include a surveillance radar and an electro-optic system. A stern ramp is to enable launch and recovery of two interception craft. Delivery of the first of class is planned for 2016 and all eight ships are to be operational by 2020.

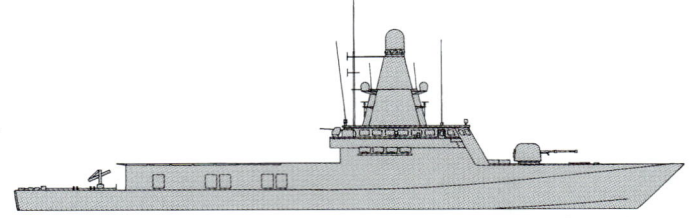

LITTORAL MISSION VESSEL   (Scale 1 : 900), Ian Sturton / 1531003

11 FEARLESS CLASS (PCM/PGM)

Name	No	Builders	Launched	Commissioned
FEARLESS	94	Singapore STEC	18 Feb 1995	5 Oct 1996
BRAVE	95	Singapore STEC	9 Sep 1995	5 Oct 1996
GALLANT	97	Singapore STEC	27 Apr 1996	3 May 1997
DARING	98	Singapore STEC	27 Apr 1996	3 May 1997
DAUNTLESS	99	Singapore STEC	23 Nov 1996	3 May 1997
RESILIENCE	82	Singapore STEC	23 Nov 1996	7 Feb 1998
UNITY	83	Singapore STEC	19 Jul 1997	7 Feb 1998
SOVEREIGNTY	84	Singapore STEC	19 Jul 1997	7 Feb 1998
JUSTICE	85	Singapore STEC	18 Oct 1997	7 Feb 1998
FREEDOM	86	Singapore STEC	18 Oct 1997	22 Aug 1998
INDEPENDENCE	87	Singapore STEC	18 Apr 1998	22 Aug 1998

Displacement, tonnes: 508 full load
Dimensions, metres (feet): 55 × 8.6 × 2.7 (180.4 × 28.2 × 8.9)
Speed, knots: 20
Range, n miles: 1,800 at 15 kt84
Complement: 32 (5 officers)

Machinery: 2 MTU 12V 595 TE90 diesels; 8,554 hp(m) (6.29 MW) sustained; 2 Kamewa water-jets
Missiles: SAM: Matra Simbad twin launcher (in some); Mistral; IR homing to 4 km (2.2 n miles); warhead 3 kg.
Guns: 1 Oto Melara 3 in (76 mm)/62 Super Rapid; 120 rds/min to 16 km (8.7 n miles); weight of shell 6 kg. 4 CIS 50 12.7 mm MGs. 1—25 mm Bushmaster (in some).
Physical countermeasures: Decoys: 2 GEC Marine Shield III 102 mm sextuple fixed chaff launchers.
Electronic countermeasures: ESM: Elisra NS-9010C; intercept.
Radars: Surface search and fire control: Elta EL/M-2228(X); I-band.
Navigation: Kelvin Hughes 1007; I-band.
Sonars: Thomson Sintra TSM 2362 Gudgeon; hull-mounted; active attack; medium frequency (94–99 only).
Weapon control systems: ST 3100 WCS. Elbit MSIS optronic director.

Programmes: Contract awarded on 27 February 1993 for 12 patrol vessels to Singapore Shipbuilding and Engineering.
Structure: First six are ASW specialist ships. All have water-jet propulsion. Second batch were to have been fitted with Gabriel II SSMs but this plan has been shelved. MSIS director being fitted. Fearless modified with new EW radome on mainmast. Simbad SAM in Resilience by 25 mm Bushmaster. Sovereignty has deck crane to facilitate special forces operations. Towed array installed temporarily in Brave has been removed.
Operational: All serve with Maritime Security Task Force. The first five form 189 Squadron and the second six 182 Squadron. Unity is to be used as a test bed for new technologies including an Indep 21 combat system. Courageous badly damaged in collision on 3 January 2003 and subsequently decommissioned. Independence damage during collision with underwater object in July 2009. Torpedoes have been removed. A 25 mm gun has replaced the Simbad SAM launcher in some. The class is double-crewed to increase availability.

SOVEREIGNTY   5/2004, David Boey / 1044510

RESILIENCE   8/2007, Bob Fildes / 1353367

BRAVE   8/2007, Edward McDonnell / 1367089

IHS Jane's Fighting Ships 2014-2015   © 2014 IHS

12 INSHORE PATROL CRAFT (PB)

FB 31–42

Displacement, tonnes: 20 full load
Dimensions, metres (feet): 14.5 × 4.2 × 1.1 (47.6 × 13.8 × 3.6)
Speed, knots: 30
Complement: 5
Machinery: 2 MAN D2848 LE 401 diesels; 1,341 hp(m) (1 MW); 2 Hamilton 362 water-jets
Guns: 1 — 40 mm grenade launcher. 1 — 12.7 mm MG. 2 — 7.62 mm MGs.
Radars: Surface search: Racal Decca; I-band.

Comment: Built by Singapore SBEC and delivered in 1990–91. Based at Tuas. Designated Fast Boats (FB). Some are kept in storage at Tuas. Similar to Police PT 1-19 class.

FB 35   5/2007, Guy Toremans / 1167801

AMPHIBIOUS FORCES

Notes: (1) The Tiger 40 hovercraft acquired in 1997 is beyond repair but the design may be used again for a repeat order.
(2) Trials of at least one hovercraft ACVI were reported in early 2005. This may have been renamed Eagle-1.

EAGLE-1   5/2010, David Boey / 1406278

4 RPL TYPE (LCU)

RPL 60–63

Displacement, tonnes: 153 standard
Dimensions, metres (feet): 36.7 × 8.5 × 1.8 (120.4 × 27.9 × 5.9)
Speed, knots: 10.7
Complement: 6
Military lift: 2 tanks or 450 troops or 110 tons cargo (fuel or stores)
Machinery: 2 MAN D2540MLE diesels; 860 hp(m) (632 kW); 2 Schottel props

Comment: First pair built at North Shipyard Point, second pair by Singapore SBEC. First two launched August 1985, next two in October 1985. Cargo deck 86.9 × 21.6 ft (26.5 × 6.6 m). Bow ramp suitable for beaching.

RPL 60   6/2001, John Mortimer / 0126301

30 LANDING CRAFT UTILITY (LCU)

300 series

Dimensions, metres (feet): 23 × 6 × 0.8 (75.5 × 19.7 × 2.6)
Speed, knots: 20
Range, n miles: 180 at 15 kt
Complement: 4
Military lift: 18 tons
Machinery: 2 MAN 2842 LZE diesels; 4,400 hp(m) (3.23 MW); 2 Kamewa water-jets
Guns: 2 — 12.7 mm MGs or 40 mm grenade launchers.

Comment: This is a larger and much faster version of the LCVPs. Construction started in 1993. Designated Fast Craft Utility (FCU).

LCU 394   12/2007, Chris Sattler / 1353370

100 LANDING CRAFT (LCVP/FCEP)

Displacement, tonnes: 4 full load
Dimensions, metres (feet): 13.6 × 3.7 × 0.6 (44.6 × 12.1 × 2.0)
Speed, knots: 20
Range, n miles: 100 at 20 kt
Complement: 3
Military lift: 4 tons or 30 troops
Machinery: 2 MAN D2866 LE diesels; 816 hp(m) (600 kW); sustained; 2 Hamilton 362 water-jets
Guns: 2 — 7.62 mm MGs.

Comment: Fast Craft, Equipment and Personnel (FCEP), built by Singapore SBEC from 1989 and are used to transport troops around the Singapore archipelago. They have a single bow ramp and can carry a rifle platoon. More than 25 are in service and the rest in storage.

FCEP 476   7/2009, David Boey / 1305476

10 DIVING SUPPORT CRAFT (YTB)

Comment: Boston Whalers used by the Naval Diving Unit. Armed with 7.62 mm MGs and 40 mm grenade launchers.

BOSTON WHALER   8/2000, David Boey / 0105601

6 FAST INTERCEPT CRAFT (HSIC)

Displacement, tonnes: 13 full load
Dimensions, metres (feet): 14.5 × 2.85 × 1.35 (47.6 × 9.4 × 4.4)
Speed, knots: 55 (est.)
Machinery: Triple Seatek diesels coupled to Trimax drives
Guns: 2 CIS 40 mm AGL. 2 CIS 50 12.7 mm MGs. 1 — 7.62 mm GPMG.
Radars: Raytheon SL 72.

Comment: Details are of craft used by Naval Diving Unit. Multistep planing hull design. At least five other planing and wave-piercing craft are reported to be in service with special forces units.

FIC 145   9/2002, David Boey / 0554729

756 Singapore > Amphibious forces

450 ASSAULT CRAFT (LCA)

Dimensions, metres (feet): 5.4 × 1.8 × 0.7 (17.7 × 5.9 × 2.3)
Speed, knots: 12
Military lift: 12 troops
Machinery: 1 outboard; 50 hp(m) (37 kW)
Guns: 1—7.62 mm MG or 40 mm grenade launcher.

Comment: Built by Singapore SBEC. Man-portable craft which can carry a section of troops in the rivers and creeks surrounding Singapore island. Numbers are approximate.

ASSAULT CRAFT
9/1995, David Boey
0080588

4 ENDURANCE CLASS (LPDM)

Name	No	Builders	Laid down	Launched	Commissioned
ENDURANCE	207	Singapore Technologies Marine, Banoi	26 Mar 1997	14 Mar 1998	18 Mar 2000
RESOLUTION	208	Singapore Technologies Marine, Banoi	22 Oct 1997	1 Aug 1998	18 Mar 2000
PERSISTENCE	209	Singapore Technologies Marine, Banoi	3 Apr 1998	13 Mar 1999	7 Apr 2001
ENDEAVOUR	210	Singapore Technologies Marine, Banoi	15 Oct 1998	12 Feb 2000	7 Apr 2001

Displacement, tonnes: 8,636 full load
Dimensions, metres (feet): 141 pp × 21 × 5 (462.6 × 68.9 × 16.4)
Speed, knots: 15
Range, n miles: 10,400 at 12 kt
Complement: 65 (8 officers)

Military lift: 350 troops; 18 tanks; 20 vehicles; 4 LCVP
Machinery: 2 Ruston 16RK 270 diesels; 12,000 hp(m) (8.82 MW); 2 shafts; Kamewa cp props; bow thruster
Missiles: SAM: 2 Matra Simbad twin launchers for Mistral ❶; IR homing to 4 km (2.2 n miles); warhead 3 kg. 2 Barak octuple launchers may be fitted in due course.
Guns: 1 Otobreda 76 mm/62 Super Rapid ❷; 120 rds/min to 16 km (8.7 n miles); weight of shell 6 kg. 2—25 mm Bushmaster (can be fitted). 5—12.7 mm MGs.
Radars: Air/surface search: Elta EL/M-2238 ❹; E/F-band.
Navigation: Kelvin Hughes Type 1007; I-band.
Weapon control systems: CS Defense NAJIR 2000 optronic director ❸.

Helicopters: 2 Super Pumas.

Programmes: Ordered in September 1994 and confirmed in mid-1996.
Structure: US drive through design with bow and stern ramps. Single intermediate deck with three hydraulic ramps. Helicopter platform aft. Indal ASIST helo handling system. Dockwell for four LCUs and davits for four LCVPs. Two 25 ton cranes. Four 36 m self-propelled pontoons can be secured to winching points on the ships' sides. Protector unmanned surface vehicles were operated from Resolution in 2005.
Operational: Endurance completed the RSN's first round-the-world deployment in late 2000. Persistence deployed to the Indian Ocean for anti-piracy patrols in 2009. Based at Changi. Form 191 Squadron.

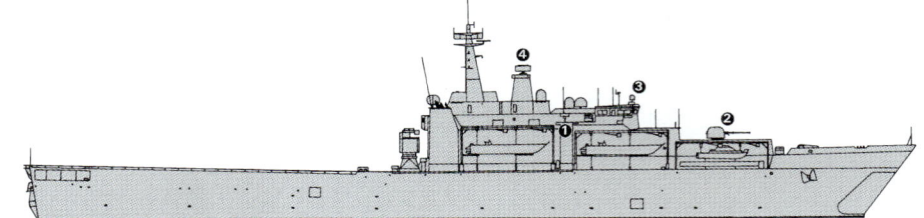

RESOLUTION (Scale 1 : 1,200), Ian Sturton / 1153491

RESOLUTION 12/2007, Chris Sattler / 1353368

ENDEAVOUR 11/2013*, Michael Nitz / 1531031

PERSISTENCE 10/2008, Michael Nitz / 1353369

IHS Jane's Fighting Ships 2014-2015

MINE WARFARE FORCES

4 BEDOK (LANDSORT) CLASS (MINEHUNTERS) (MHS)

Name	No	Builders	Launched	Commissioned
BEDOK	M 105	Kockums, Karlskrona	24 Jun 1993	7 Oct 1995
KALLANG	M 106	Singapore Shipbuilding	29 Jan 1994	7 Oct 1995
KATONG	M 107	Singapore Shipbuilding	8 Apr 1994	7 Oct 1995
PUNGGOL	M 108	Singapore Shipbuilding	16 Jul 1994	7 Oct 1995

Displacement, tonnes: 366 full load
Dimensions, metres (feet): 47.5 × 9.6 × 2.3 *(155.8 × 31.5 × 7.5)*
Speed, knots: 15. **Range, n miles:** 2,000 at 10 kt
Complement: 31 (5 officers)

Machinery: 4 Saab Scania diesels; 1,592 hp(m) *(1.17 MW)*; coupled in pairs to 2 Voith Schneider props
Guns: 1 Bofors 40 mm/70. 4 — 12.7 mm MGs.
Mines: 2 rails.
Radars: Navigation: Norcontrol DB 2000; I-band.
Sonars: Thomson-CSF TSM 2022; hull-mounted; minehunting; high frequency.
Weapon control systems: Thomson-CSF TSM 2061 Mk II minehunting and mine disposal system. Signaal WM20 director.

Programmes: Kockums/Karlskrona design ordered in February 1991. *Bedok* started trials in Sweden in December 1993, and was shipped to Singapore in early 1994 to complete. Prefabrication work done for the other three in Sweden with assembly and fitting out in Singapore at Benoi Basin.
Modernisation: DSTA Singapore awarded contract to Thales in May 2009 for Life Extension programme. Thales is to provide an integrated MCM combat system including modernisation of the hull-mounted sonar (TSM 2022), towed synthetic aperture sonar (DUBM-44) and expendable mine disposal system (reported to be ECA K-ster).
Structure: GRP hulls. Two PAP 104 Mk V ROVs embarked. Racal Precision Navigation system. Two sets of Swedish SAM minesweeping system. Magnavox GPS.
Operational: Form 194 Squadron, based at Tuas.

KATONG *5/2009, Guy Toremans* / 1367087

PUNGGOL *2/2010, M Mazumdar* / 1406279

TRAINING SHIPS

1 TRAINING TENDER (AXL)

Name	Builders
STET POLARIS	Sam Aluminium Engineering Ltd

Measurement, tonnes: 166 gt
Dimensions, metres (feet): 25 × 6.6 × 1.3 *(82 × 21.7 × 4.3)*
Speed, knots: 30
Complement: 4 (26 trainees)
Machinery: 2 Caterpillar C32 diesels; 2,230 hp *(1.39 MW)*; 2 shafts

Comment: Built by Sam Aluminium Engineering Ltd, Singapore and delivered in June 2010. Training ship owned and operated by ST Engineering and Training and leased by the Singapore Navy for navigation and seamanship training.

AUXILIARIES

Notes: There is one Floating Dock with a lift of 600 tons. *FD 2* at Changi.

1 SUBMARINE RESCUE SHIP (ASRH)

Name	Laid down	Launched	Commissioned
SWIFT RESCUE	1 Apr 2008	29 Nov 2008	30 Apr 2009

Measurement, tonnes: 4,359 gt
Dimensions, metres (feet): 83.7 × 18.3 × 4.3 *(274.6 × 60.0 × 14.1)*
Speed, knots: 12.5. **Range, n miles:** 3,000 at 12 kt
Machinery: 2 MAN 6L27/38 diesels; 5,548 hp *(4.08 MW)*; 2 shafts; cp props

Comment: Contract awarded to ST Marine Singapore on 14 March 2007 to design, build and maintain a submarine rescue ship capable of operating a submarine rescue vehicle (DSAR-6), its associated handling systems and recompression chamber. Of steel construction with a bulbous bow, the ship is equipped with a helicopter deck. The ship was launched on 29 November 2008 and is entered service in April 2009. The ship is manned by Swire Pacific Offshore; mission command and medical teams are provided by the Singapore Navy.

SWIFT RESCUE *5/2010, David Boey* / 1366445

DEEP SUBMERGENCE AND RESCUE VEHICLES

1 RESCUE SUBMERSIBLE

DSAR-6

Dimensions, metres (feet): 9.6 × 2.7 × 3.8 *(31.5 × 8.9 × 12.5)*
Speed, knots: 3
Complement: 3 (2 pilots)
Machinery: 2 electric motors; 26.8 hp *(20 kW)*; 4 tiltable side thrusters; 16 hp *(12 kW)*

Comment: James Fisher Defence Limited awarded a contract in January 2007 to deliver a new Submarine Rescue Vehicle (SRV) to the Singapore Navy. It was delivered in 2009. The submarine rescue vehicle, DSAR-6, is based on the LR5/DSAR-1, in service with the Royal Navy until replaced in 2008. Lithium based battery technology enables the underwater endurance to be increased significantly over conventional lead-acid based systems. DSAR-6 has two compartments. The forward compartment houses the pilot and co-pilot while the aft compartment carries the RCO and up to 17 rescuees. Capable of operating at a depth of 500 m in currents of up to 3 kt, rescuees may be transferred under pressure to the medical and decompression facilities onboard the submarine rescue mothership *Swift Rescue*.

DSAR-6 *6/2009, Singapore Navy* / 1421688

POLICE COAST GUARD

Notes: (1) There are plans to acquire two floating command and control centres. Each one is to be of the order of 80 × 30 m and the design is to include a four-storey accommodation/office block, a helipad and mooring points for patrol craft.
(2) A contract for six 14 m patrol craft was signed with Lung Teh Shipbuilding, Taiwan, was reported in early 2014.

2 COMMAND CRAFT (WPB)

MANTA RAY PT 20   **EAGLE RAY** PT 30

Dimensions, metres (feet): 20 × 6.0 × 1.0 *(65.6 × 19.7 × 3.3)*
Speed, knots: 30
Complement: 5
Machinery: 2 MTU 16V 2000 M90 diesels; 2 Hamilton 521 water-jets
Guns: 2 — 7.62 mm MGs.

Comment: Built by Asia-Pacific Geraldton, Singapore to a Geraldton, Australia design. These command craft are larger versions of the 18 m patrol craft.

MANTA RAY *4/2002, David Boey* / 0554731

FRIGATES

4 VALOUR CLASS (MEKO A-200 SAN) (FFGHM)

Name	No	Builders	Laid down	Launched	Commissioned
AMATOLA	F 145	Blohm + Voss, Hamburg	2 Aug 2001	6 Jun 2002	16 Feb 2006
ISANDLWANA	F 146	Howaldtswerke, Kiel	26 Oct 2001	5 Dec 2002	27 Jul 2006
SPIOENKOP	F 147	Blohm + Voss, Hamburg	28 Feb 2002	6 Jun 2003	16 Feb 2007
MENDI	F 148	Howaldtswerke, Kiel	28 Jun 2002	15 Jun 2004	20 Mar 2007

Displacement, tonnes: 3,648 full load
Dimensions, metres (feet): 121 × 16.4 × 6.2 (397 × 53.8 × 20.3)
Speed, knots: 28
Range, n miles: 7,700 at 15 kt
Complement: 100 + 20 spare berths

Machinery: CODAG; 1 GE LM 2500 gas turbine 26,820 hp(m) (20 MW); 2 MTU 16V 1163 TB93 diesels 16,102 hp(m) (11.84 MW); 2 shafts; LIPS cp props; 1 LIPS LJ210E waterjet (centreline)
Missiles: SSM: 8 MBDA Exocet MM 40 Block 2 ❶; inertial cruise; active radar homing to 70 km (40 n miles) at 0.9 Mach; warhead 165 kg.
SAM: Denel Umkhonto 32-cell VLS ❷ inertial guidance with mid-course guidance and IR homing to 12 km (6.5 n miles) at 2.4 Mach; warhead 23 kg.
Guns: 1 Otobreda 76 mm/62 compact ❸. 2 LIW DPG 35 mm (twin) ❹. 2 Oerlikon 20 mm Mk 1. 2 Reutech remote-control 12.7 mm MGs. 4 − 7.62 mm MGs.
Physical countermeasures: Decoys: 2 Super Barricade chaff launchers ❺.
Electronic countermeasures: CESM: Grintek EWASION. RESM: Avitronics/Sysdel.
Radars: Air/surface search: Thales MRR ❻ 3D; G-band.
Fire control: 2 Reutech RTS 6400 ❼; I/J-band.
Navigation/helo control: 2 Racal Bridgemaster E ❽; I-band.
Sonars: Thomson Marconi 4132 Kingklip; hull mounted, active search; medium frequency.
Combat data systems: ADS CMS.
Weapon control systems: 2 Reutech RTS 6400 optronic trackers.

Helicopters: 2 Super Lynx ❾.

Programmes: Contract for four ships, with option for one further, signed on 3 December 1999 with ESACC which includes Blohm + Voss, HDW, TRT, African Defence Systems and Thomson-CSF. Contract effective 28 April 2000. *Amatola* arrived at Simon's Town on 4 November 2003 for weapon systems integration by African Defence Systems. An option for a fifth ship is not to be exercised.
Structure: The design includes radar and IR signature reduction measures. Exhaust gases are expelled just above the waterline. Two RHIBs can be launched from davits amidships.
Modernisation: Exocet MM 40 Block 2 to be replaced by Block 3 missiles. Installation of a bow-thruster to improve low-speed manoeuvring is under consideration. The 76 mm gun may be replaced by a 127 mm or navalised 155 mm gun.
Operational: Since January 2011, the ships have been the principal units assigned to Operation Copper, an anti-piracy mission in the Mozambique Channel.

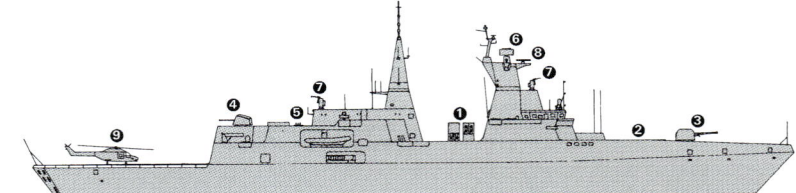
AMATOLA (Scale 1 : 1,200), Ian Sturton / 1159221

MENDI 10/2011, Ann Till / 1454718

AMATOLA 9/2009, Michael Nitz / 1305985

ISANDLWANA 9/2009, Michael Nitz / 1305986

SHIPBORNE AIRCRAFT

Numbers/Type: 4 Agusta-Westland Super Lynx 300.
Operational speed: 120 kt (222 km/h).
Service ceiling: 10,000 ft (3,048 m).
Range: 320 n miles (593 km).
Role/Weapon systems: Ordered on 14 August 2003 for delivery in 2007. Surveillance. Sensors: Telephonics APS-143 B(V)3 radar; ESM: Sea Raven 118; Cumulus Leo Mk II FLIR. Weapons: Unarmed (torpedoes and ASM may be fitted in future upgrades).

Numbers/Type: 8 Aerospatiale SA 330E/H/J Oryx.
Operational speed: 139 kt (258 km/h).
Service ceiling: 15,750 ft (4,800 m).
Range: 297 n miles (550 km).
Role/Weapon systems: Support helicopter; allocated by SAAF for naval duties and can be embarked in *Drakensberg*. Sensors: Doppler navigation with search radar. Weapons: Unarmed but can mount Armscor 30 mm Rattler.

SUPER LYNX 3/2008, Guy Toremans / 1335820

ORYX 9/2008, Michael Nitz / 1335819

LAND-BASED MARITIME AIRCRAFT

Notes: (1) Alouette utility helicopters have been replaced by Agusta A-109.
(2) The requirement to replace the Turbodaks with modern maritime patrol aircraft is under consideration.

Numbers/Type: 5 Douglas Turbodaks.
Operational speed: 161 kt *(298 km/h)*.
Service ceiling: 24,000 ft *(7,315 m)*.
Range: 1,390 n miles *(2,575 km)*.
Role/Weapon systems: A number of Dakotas has been converted for MR/SAR and other tasks. Additional fuel tanks extend the range to 2,620 n miles *(4,800 km)*. Sensors: Elta M-2022 search radar and FLIR; Sysdel ESM; sonobuoy acoustic processor. Weapons: Unarmed.

DOUGLAS DC-3 6/2003, South African Navy / 0568890

PATROL FORCES

Notes: (1) Project Biro is a programme to acquire eight 80 m offshore patrol vessels, and six 55 m inshore patrol vessels. A Request for Information was issued in mid-2011 with a view to building both types of vessel in a South African shipyard.
(2) As a result of experience in peacekeeping operations, a Maritime Reaction Squadron has been established to support the Army in peacekeeping, in other operations on lakes and rivers and in coastal security operations.

3 T CRAFT CLASS (PB)

Name	No	Builders	Commissioned
TOBIE	P 1552	T Craft International, Cape Town	18 Jul 2003
TERN	P 1553	T Craft International, Cape Town	18 Jul 2003
TEKWANE	P 1554	T Craft International, Cape Town	22 Jul 2003

Displacement, tonnes: 37 full load
Dimensions, metres (feet): 22 × 7 × 0.9 *(72.2 × 23.0 × 3.0)*
Speed, knots: 32
Range, n miles: 530 at 22 kt
Complement: 16 (1 officer)

Machinery: 2 ADE 444 TI 12V diesels; 2,000 hp *(1.5 MW)*; 2 Hamilton waterjets
Guns: 1—12.7 mm MG.
Radars: Surface search: Racal Decca; I-band.
Weapon control systems: Hesis optical director.

Comment: Twin hulled catamarans of GRP sandwich construction. Capable of carrying up to 15 people. Originally ordered in mid-1991 but not fully commissioned until 2003. Carries an RIB in the stern well. Three of this type built for Israel in 1997. *Tekwane* based at Durban, and the other two at Simon's Town.

TOBIE 9/2009, Michael Nitz / 1305983

TOBIE 9/2004, M Declerck / 1133141

23 NAMACURRA CLASS (INSHORE PATROL CRAFT) (PB)

Y 02–04 Y 08–09 Y 11–19 Y 21–29

Displacement, tonnes: 5 full load
Dimensions, metres (feet): 9 × 2.7 × 0.8 *(29.5 × 8.9 × 2.6)*
Speed, knots: 32. **Range, n miles:** 180 at 20 kt
Complement: 4

Machinery: 2 Yamaha outboards; 380 hp(m) *(279 kW)*
Guns: 1—12.7 mm MG. 2—7.62 mm MGs.
Depth charges: 1 rack.
Radars: Surface search: Furuno; I-band.

Comment: Built in South Africa in 1980–81. Can be transported by road. Two transferred to Malawi in 1988 (Y 1520) and 2008 (Y 1505) and Y 1506 has sunk at sea. Y 1501 and Y 1510 donated to Namibia on 29 November 2002 and Y 1507 and Y 1530 donated to Mozambique in 2004. Two further craft donated to the Angolan Navy in 2006. Based at Simon's Town, Durban, Cape Town, Saldanha Bay, Gordon's Bay, Port Elizabeth and East London. Three are operated on Lake Tanganyika as part of a peace-keeping force in Burundi. Five boats have been converted for riverine work under Project Xena. Pennant numbers have been changed.

NAMACURRA 2/2008, Guy Toremans / 1335816

4 WARRIOR (EX-MINISTER) CLASS (PATROL SHIP) (PG)

Name	No	Builders	Commissioned
ADAM KOK (ex-*Frederic Creswell*)	P 1563	Haifa Shipyard	6 Apr 1978
ISAAC DYOBHA (ex-*Frans Erasmus*)	P 1565	Sandock Austral, Durban	27 Jul 1979
GALESHEWE (ex-*Hendrik Mentz*)	P 1567	Sandock Austral, Durban	11 Feb 1983
MAKHANDA (ex-*Magnus Malan*)	P 1569	Sandock Austral, Durban	4 Jul 1986

Displacement, tonnes: 437 full load
Dimensions, metres (feet): 62.2 × 7.8 × 2.4 *(204.1 × 25.6 × 7.9)*
Speed, knots: 32
Range, n miles: 1,500 at 30 kt, 3,600 (est.)
Complement: 52 (7 officers)

Machinery: 4 Maybach MTU 16V 965 TB91 diesels; 15,000 hp(m) *(11 MW)* sustained; 4 shafts
Guns: 2 Oto Melara 3 in *(76 mm)*/62 compact; 85 rds/min to 16 km *(8.7 n miles)*; weight of shell 6 kg; 500 rounds per gun. 2 LIW Mk 1 20 mm. 2—12.7 mm MGs.
Physical countermeasures: Decoys: 4 ACDS launchers for chaff.
Electronic countermeasures: ESM: Delcon (ADS/Sysdel) EW system.
ECM: Rattler; jammer.
Radars: Elta EL/M 2208; E/F-band.
Fire control: Selenia RTN 10X; I/J-band.
Combat data systems: ADS Diamant (after upgrade). Mini action data automation with Link.

Programmes: Contract signed with Israel in late 1974 for this class, similar to Saar 4 class. Three built in Haifa and reached South Africa in July 1978. The ninth craft launched late March 1986. Three more improved vessels of this class were ordered but subsequently cancelled. The last of the class was finally christened in March 1992. Pennant numbers restored to the ships side and stern in 1994.
Modernisation: Ship life extension programme included a new communications refit, improvements to EW sensors, a third-generation target designation assembly, a computer-assisted action information system served by datalinks, improvements to fire control, and a new engine room monitoring system. P 1565 completed upgrade in April 1999, P 1567 in March 2000. All three craft were refurbished at South African Shipyards, Durban, 2012–13.
Operational: All are based at Simon's Town. Skerpioen missiles have been removed. *Makhanda*, decommissioned in 2007, was reactivated in 2010 and *Adam Kok* is also being re-activated at Durban. The ships are to continue in service as offshore patrol vessels until Project Biro vessels enter service.

GALESHEWE 9/2008, Michael Nitz / 1335818

5 XENA PATROL CRAFT (PB)

Displacement, tonnes: 5.6 standard
Dimensions, metres (feet): 10.5 × 2.6 × ? *(34.4 × 8.5 × ?)*
Speed, knots: 29
Range, n miles: 180 at 15 kt
Complement: 4
Machinery: 2 diesels; 2 water-jets
Guns: 1—12.7 mm MG. 1—7.62 mm MG.

Comment: The first five of a potential 16 riverine patrol craft acquired for the Maritime Reaction Squadron were delivered in 2010. The craft, of aluminium construction, were manufactured by Vee Craft Marine, Cape Town.

MINE WARFARE FORCES

Notes: Mine countermeasures capability is likely to be replaced by autonomous underwater vehicles rather than by specialist ships. The first step is to provide an interim 'offboard' system that is to form the basis on which an expanded capability can be built. The Atlas SeaOtter UUV is reported to be a contender.

3 RIVER CLASS (COASTAL MINEHUNTERS) (MHC)

Name	No	Builders	Commissioned
UMKOMAAS (ex-*Navors I*)	M 1499	Abeking & Rasmussen, Lemwerder; Sandock Austral	13 Jan 1981
UMZIMKULU (ex-*Navors III*)	M 1142	Sandock Austral	30 Oct 1981
UMHLOTI (ex-*Navors IV*)	M 1212	Sandock Austral	15 Dec 1981

Displacement, tonnes: 386 full load
Dimensions, metres (feet): 48 × 8.5 × 2.5 *(157.5 × 27.9 × 8.2)*
Speed, knots: 16. **Range, n miles:** 2,000 at 13 kt
Complement: 40 (7 officers)

Machinery: 2 MTU 12V 652 TB81 diesels; 4,515 hp(m) *(3.32 MW)*; 2 Voith Schneider props
Guns: 1 Oerlikon 20 mm GAM-BO1. 2 – 12.7 mm MGs. 2 – 7.62 mm MGs.
Physical countermeasures: MCM: 2 PAP 104 remote-controlled submersibles.
Radars: Navigation: Decca; I-band.
Sonars: Klein VDS; side scan; high frequency.

Comment: Ordered in 1978 as Research Vessels to be operated by the Navy for the Department of Transport. The lead ship *Navors I* was shipped to Durban from Germany in the heavy lift ship *Uhenfels* in June 1980 for fitting out, shortly followed by the second. The last pair were built in Durban. The vessels were painted blue with white upperworks and formed the First Research Squadron. Painted grey and renamed in 1982 but continued to fly the national flag and not the naval ensign. The prefix RV was only changed to SAS on 3 February 1988 when they were formally accepted as naval ships. Minehunting capability could be enhanced by substituting the diving container on the after deck with lightweight mechanical and acoustic sweeping gear. Carry an RIB and a decompression chamber. M1499 refitted in 2002. M 1213 placed in reserve in 2005 and unlikely to be re-activated.

UMKOMAAS   9/2008, Michael Nitz / 1335815

SURVEY AND RESEARCH SHIPS

Notes: It is planned to acquire (Project Hotel) a new hydrographic survey ship to replace *Protea*. The ship is likely to be built in South Africa, in parallel with but distinct from the offshore patrol ship project.

1 HECLA CLASS (AGSH)

Name	No	Builders	Commissioned
PROTEA	A 324	Yarrow Shipbuilders	23 May 1972

Displacement, tonnes: 2,777 full load
Dimensions, metres (feet): 79.3 × 15 × 4.7 *(260.2 × 49.2 × 15.4)*
Speed, knots: 14. **Range, n miles:** 12,000 at 11 kt
Complement: 124 (10 officers)

Machinery: Diesel-electric; 3 MTU diesels; 3,840 hp *(2.68 MW)* sustained; 3 generators; 1 motor; 2,000 hp *(1.49 MW)*; 1 shaft; cp prop; bow thruster
Guns: 2 – 12.7 mm MGs.
Radars: Navigation: Racal Decca; I-band.
Helicopters: 1 Alouette III.

Comment: Laid down 20 July 1970. Launched 14 July 1971. Equipped for hydrographic survey with limited facilities for the collection of oceanographical data and for this purpose fitted with special communications equipment, Polaris survey system, 8 m survey launches *Malgas* and *Seemeeu* and facilities for helicopter operations. Hull strengthened for navigation in ice and fitted with a passive roll stabilisation system. New engines and full overhaul in 1995–96. Carries EGNG sidescan sonar and two survey boats. Fitted for two 20 mm guns. To be replaced by a new ship under Project Hotel.

PROTEA   3/2008, Frank Findler / 1335810

1 ANTARCTIC SURVEY AND SUPPLY VESSEL (AGOBH)

Name	No	Builders	Laid down	Launched	Commissioned
S A AGULHAS II	–	STX Finland Oy, Turku	Jan 2011	21 Jul 2011	4 May 2012

Measurement, tonnes: 12,900 gt
Dimensions, metres (feet): 134 × 21.7 × 7.65 *(439.6 × 71.2 × 25.1)*
Speed, knots: 14
Complement: 45 + 100 personnel
Machinery: 4 Wärtsilä 6L32 diesels; 16,316 hp *(12 MW)*; 1 shaft
Radars: To be announced.

Comment: Contract for the construction of a new Antarctic Survey and Supply ship announced by the South African Department of Environmental Affairs on 17 November 2009. The ice-strengthened vessel was built at Rauma Shipyard. The ship is capable of carrying cargo, passengers and two Puma helicopters. Equipped with 800 m^2 of laboratories and a small hospital, the ship, operated by Smit Amandla Marine, also acts as a research base for the three South Africa stations on Marion Island, Gough Island and Antarctica. The first Antarctic deployment was conducted in 2012.

AGULHAS II   6/2012, STX / 1486354

AUXILIARIES

Notes: It is planned to acquire (Project Millennium) two Strategic Support Ships to enter service from about 2020. The primary function is to transport, land and support some 1,500 troops with up to 350 vehicles either by sealift or by a combination of helicopters and landing craft. An LHD design is likely to be required to fulfil this and wider missions of disaster relief and logistic support.

1 FLEET REPLENISHMENT SHIP (AORH)

Name	No	Builders	Launched	Commissioned
DRAKENSBERG	A 301	Sandock Austral, Durban	24 Apr 1986	11 Nov 1987

Displacement, tonnes: 6,096 standard; 12,701 full load
Dimensions, metres (feet): 147 × 19.5 × 7.9 *(482.3 × 64.0 × 25.9)*
Speed, knots: 20 (est.). **Range, n miles:** 8,000 at 15 kt
Complement: 116 (10 officers; 10 air crew) + 22 casualty berths

Cargo capacity: 5,500 tons fuel; 750 tons ammunition and dry stores; 2 Lima LCUs
Machinery: 2 diesels; 16,320 hp(m) *(12 MW)*; 1 shaft; cp prop; bow thruster
Guns: 4 Oerlikon 20 mm GAM-BO1. 8 – 12.7 mm MGs.
Helicopters: 2 SA 330H/J Oryx.

Comment: The largest ship built in South Africa and the first naval vessel to be completely designed in that country. In addition to her replenishment role she is employed on SAR, patrol and surveillance with a considerable potential for disaster relief. There are two 10-ton cranes to lower/recover LCUs or Namacurra craft. Two abeam positions and astern fuelling, jackstay and vertrep. Two helicopter landing spots, one forward and one astern. Main secondary role is the transport of consumables, but can also be used to support small craft and transport a limited number of troops. Completed a year-long refit in February 2010 and to remain in service until 2026.

DRAKENSBERG   5/2008, M Declerck / 1335813

6 LIMA CLASS (LCU)

Displacement, tonnes: 7 full load
Dimensions, metres (feet): 9.1 × 3.55 × 0.7 *(29.9 × 11.6 × 2.3)*
Speed, knots: 38
Range, n miles: 120 at 26 kt
Complement: 3
Machinery: 2 outboards; 400 hp *(298 kW)*

Comment: Built in 2003 by Stingray Marine, Cape Town. GRP construction. Capable of carrying 24 troops or 2.5 tons of cargo. Two craft can be carried in *Drakensberg*.

L 27   8/2003, Helmoed-Römer Heitman / 0530510

NAVARRA 9/2009, M Declerck / 1366012

SANTA MARIA 7/2010, Edward McDonnell / 1406138

REINA SOFIA 6/2007, John Brodie / 1166820

5 ALVARO DE BAZÁN CLASS (FFGHM)

Name	No	Builders	Laid down	Launched	Commissioned
ALVARO DE BAZÁN	F 101	Navantia, Ferrol	14 Jun 1999	31 Oct 2000	19 Sep 2002
ALMIRANTE JUAN DE BORBÓN	F 102	Navantia, Ferrol	27 Oct 2000	28 Feb 2002	3 Dec 2003
BLAS DE LEZO	F 103	Navantia, Ferrol	28 Feb 2002	16 May 2003	16 Dec 2004
MENDEZ NUÑEZ	F 104	Navantia, Ferrol	16 May 2003	12 Nov 2004	21 Mar 2006
CRISTÓBAL COLÓN	F 105	Navantia, Ferrol	20 Feb 2009	4 Nov 2010	23 Oct 2012

Displacement, tonnes: 5,947 full load
Dimensions, metres (feet): 146.4 oa; 133.1 pp × 18.6 × 7.2 *(480.3; 436.7 × 61.0 × 23.6)*
Flight deck, metres: 26.4 × 17 *(86.6 × 55.8)*
Speed, knots: 28. **Range, n miles:** 4,500 at 18 kt
Complement: 201 (21 officers)

Machinery: CODOG; 2 GE LM 2500 gas turbines; 47,328 hp(m) *(34.8 MW)* sustained; 2 Bazán/Caterpillar diesels; 12,240 hp(m) *(9 MW)* sustained; 2 shafts; LIPS cp props
Missiles: SSM: 8 Boeing Harpoon Block 2 ❶; active radar homing to 124 km *(67 n miles)* at 0.9 Mach; warhead 227 kg.
SAM: Mk 41 VLS (48 cells) ❷ 32 Raytheon SM-2MR (Block IIIB); command/inertial guidance; semi-active radar homing to 167 km *(90 n miles)* at 2.5 Mach. 64 Evolved Sea Sparrow RIM 162B (in quadpacks); semi-active radar homing to 18 km *(9.7 n miles)* at 3.6 Mach; warhead 39 kg.
Guns: 1 FMC 5 in *(127 mm)*/54 Mk 45 Mod 2 ❸ (ex-US); 20 rds/min to 23 km *(12.6 n miles)*; weight of shell 32 kg. 2 BAE 25 mm Mk 38 Mod 2 ❹. 2 Oerlikon 20 mm.
Torpedoes: 4—323 mm (2 twin) Mk 32 Mod 9 fixed launchers ❺. Honeywell Mk 46 Mod 5; anti-submarine; active/passive homing to 11 km *(5.9 n miles)* at 40 kt; warhead 44 kg.
A/S Mortars: 2 ABCAS/SSTDS launchers.
Physical countermeasures: Decoys: 4 SRBOC Mk 36 Mod 2 chaff launchers ❻. SLQ-25A Nixie torpedo decoy.
Electronic countermeasures: ESM/ECM: Regulus Mk-9500 (F 101–104) ❼ and Ceselsa Aldebaran (F 101–104) ❽. Indra Rigel (F 105).
Radars: Air/surface search: Aegis SPY-1D (SPY-1D(V) in F 105) ❿; E/F-band.
Surface search: DRS SPS-67 (RAN 12S) (F 101–104) ⓫; G-band. Indra Aries II (F 105); I-band.
Fire control: 2 Raytheon SPG-62 Mk 99 (for SAM) ⓬; I/J-band. FABA Dorna; I-band.
Navigation: 1 Raytheon SPS-73(V) ⓭; I-band.
Sonars: Raytheon DE 1160 LF; hull-mounted; active search and attack; medium frequency. Indra/Lockheed Martin LW-HP 53 (F 105); hull-mounted; active/passive search and attack. Possible ATAS active towed sonar.
Combat data systems: Lockheed Aegis Baseline 5 Phase III (DANCS); Link 11/16; Link 22 (F 105); SCOT 3, SATURN 3S; FABA-Lockheed BL S3 (F 105).
Weapon control systems: Sirius optronic director ❾; FABA DORNA radar, TV, and IR tracker. Sainsel DLT 309 TFCS. SQR-4 helo datalink.
Helicopters: 1 SH-60B Seahawk Lamps III ⓮.

Programmes: Project definition from September 1992 to July 1995, and then extended to July 1996 to incorporate Aegis. Design collaboration with German and Netherlands shipyards started 27 January 1994. Spain withdrew from the APAR air defence radar project in June 1995 and decided to incorporate Aegis SPY-1D into the design. Production order for four Flight I ships agreed on 21 October 1996 and building approved 24 January 1997. FSC in November 1997. The acquisition of a fifth ship was authorised by the Spanish government on 27 May 2005 and a contract was signed in July 2006.
Modernisation: Flight I ships had been upgraded to Baseline S-2 Standard by 2010. This included installation of SM 2 Block IIIB and Evolved Sea Sparrow Missiles and provision for the installation of Tomahawk land-attack missiles. F 105 combat system is Baseline S3.
Structure: The inclusion of SPY-1D radar increased the original size of the ship and caused major changes to the shape of the superstructure. Stealth technology incorporated. Indal RAST helicopter system. Hangar for one helicopter. All five ships can operate NH90 but the RAST system has not been adapted to the aircraft. 127 mm gun for gunfire support to land forces, taken from USN *Tarawa* class.
Operational: All based at Ferrol as the 31st Squadron. F 104 completed TBMD trials with the US Navy in 2007. Sea trials of *Cristóbal Colón* started on 14 March 2012.

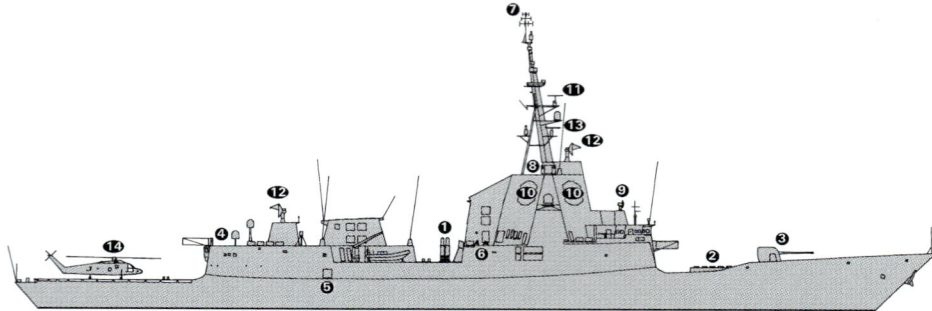

BLAS DE LEZO (Scale 1 : 1,200), Ian Sturton / 1153003

ALMIRANTE JUAN DE BOURBÓN 5/2011, Per Körnefeldt / 1454722

CRISTÓBAL COLÓN 6/2013*, Diego Quevedo / 1531033

SHIPBORNE AIRCRAFT

Notes: In January 2007, an order was placed for 45 (subsequently reduced to 38) NH90 in the TTH configuration. The first flight took place at Marignane on 17 December 2010. The first aircraft is to be delivered to the Spanish Army in 2014.

Numbers/Type: 4/12/1 BAe/McDonnell Douglas EAV-8B (Night Attack)/EAV-8B (Harrier Plus)/TAV-8B.
Operational speed: 562 kt *(1,041 km/h)*.
Service ceiling: Not available.
Range: 480 n miles *(889 km)*.
Role/Weapon systems: First batch of 12 EAV-8B delivered in 1987–88 and a further eight EAV-8B Harrier Plus in 1996–97. Five from the first batch upgraded to EAV-8B Harrier II by 2004. The remaining EAV-8B have been upgraded to AV-8B Night Attack standards. A further TAV-8B twin seat delivered in September 2000. Sensors: Targeting POD (Litening AT Block1), APG-65 (AV8B+), EW suite (RWR, ECM POD, CMDS). Weapons: Air-to-ground (GP bombs, JDAM, LGB, AGM-65E Maverick, 25 mm gun), air-to-air (AIM-9L Sidewinder, AIM-120 AMRAAM, 25 mm gun GAU12/U).

HARRIER PLUS   5/2009, A Campanera i Rovira / 1366011

Numbers/Type: 7 Sikorsky SH-3D/G/H Sea King.
Operational speed: 110 kt *(204 km/h)*.
Service ceiling: 10,000 ft *(3,050 m)*.
Range: 542 n miles *(1,005 km)*.
Role/Weapon systems: Former ASW helicopters converted to tactical transport, CSAR and special forces role, maintaining all-weather SAR capability. NVG and FLIR fitted. Converted to 3H standard in 1996–97. Weapons: 1 – 7.62 mm MG.

SH-3D   9/2009, Adolfo Ortigueira Gil / 1366008

Numbers/Type: 3 Sikorsky SH-3D Sea King AEW.
Operational speed: 110 kt *(204 km/h)*.
Service ceiling: 10,000 ft *(3,050 m)*.
Range: 542 n miles *(1,005 km)*.
Role/Weapon systems: Three Sea King helicopters were taken in hand in 1986 for conversion to AEW role to provide organic cover; first entered service August 1987. Sensors: THORN EMI Searchwater radar, ESM. Weapons: Unarmed.

SEA KING AEW   10/2008, Adolfo Ortigueira Gil / 1335864

Numbers/Type: 12 Sikorsky SH-60B Seahawk (LAMPS III).
Operational speed: 140 kt *(259 km/h)*.
Service ceiling: 10,000 ft *(3,050 m)*.
Range: 420 n miles *(778 km)*.
Role/Weapon systems: ASW and ASuW helicopter. First six delivered in 1988-89 for FFG 7 frigates. Six more Block 1 acquired in 2002 for F 100 class. There are 11 aircraft at Block 1 standard. The other one is to be upgraded in 2013. Sensors: Search radar, FLIR fitted with LRD, sonobuoys, ESM. Weapons: ASW; three Mk 46 torpedoes. ASuW; AGM-119B Penguin, AGM-114B/K Hellfire, 1 – 12.7 mm MG.

SH-60B   10/2008, Adolfo Ortigueira Gil / 1335863

Numbers/Type: 7 Agusta AB 212.
Operational speed: 100 kt *(185 km/h)*.
Service ceiling: 14,200 ft *(4,330 m)*.
Range: 230 n miles *(426 km)*.
Role/Weapon systems: Tactical transport, SAR, logistic support, maritime security operations, fire support. All seven aircraft are being upgraded to include improvements to avionics, communications, NVG and self-protection. The programme is to be completed in 2015. Weapons: 1 – 7.62 mm MG and 70 mm rocket launchers.

AB 212   5/2009, Camil Busquets i Vilanova / 1366010

Numbers/Type: 9 Hughes 500MD.
Operational speed: 110 kt *(204 km/h)*.
Service ceiling: 10,000 ft *(3,050 m)*.
Range: 203 n miles *(376 km)*.
Role/Weapon systems: Used for training, light transport and surface search. ASW role removed.

500MD   9/2009, Adolfo Ortigueira Gil / 1366009

LAND-BASED MARITIME AIRCRAFT (FRONT LINE)

Notes: (1) The Air Force F/A-18 Hornet (C.15) and Eurofighter Typhoon (C.16) can be armed with Harpoon ASM. Air Force CN-235 are not used for maritime role.
(2) Three CASA C-212/400 are operated by the Fishery Department and are based at Torrejón (Madrid), Jerez and Alicante. Two Agusta A-109C and two Dauphin N3 helicopters are based at Alicante, Jerez and Santander and Canary Islands.

Numbers/Type: 3/1 Cessna Citation II (C-550)/VII (C-650).
Operational speed: 466 kt *(862 km/h)*.
Service ceiling: 43,000 ft *(13,106 m)*.
Range: 1,100 n miles *(2,037 km)*.
Role/Weapon systems: Used for transport, training and reconnaissance.

CESSNA CITATION   6/2013*, Adolfo Ortigueira Gil / 1531034

Numbers/Type: 2 CASA C-212 Aviocar.
Operational speed: 190 kt *(353 km/h)*.
Service ceiling: 24,000 ft *(7,315 m)*.
Range: 1,040 n miles *(1,925 km)*.
Role/Weapon systems: Operated by Air Force. Primary role SAR. Based at Mallorca.

C-212   5/2002, Adolfo Ortigueira Gil / 0528933

Numbers/Type: 1 Fokker F27 Maritime.
Operational speed: 250 kt (463 km/h).
Service ceiling: 25,000 ft (7,620 m).
Range: 2,200 n miles (4,074 km).
Role/Weapon systems: Operated by Air Force in SAR role. Sensors: APS-504 search radar, cameras. Weapons: none.

F-27   6/2004, Adolfo Ortigueira Gil / 1044552

Numbers/Type: 10 Eurocopter AS 332 Super Puma.
Operational speed: 130 kt (240 km/h).
Service ceiling: 15,090 ft (4,600 m).
Range: 672 n miles (1,245 km).
Role/Weapon systems: Air Force operated for SAR/CSAR. Based at Mallorca, Las Palmas and Madrid.

AS 332   8/2013*, Adolfo Ortigueira Gil / 1531036

Numbers/Type: 2/4 Lockheed P-3A Plus Orion/P-3B Plus Orion.
Operational speed: 410 kt (760 km/h).
Service ceiling: 28,300 ft (8,625 m).
Range: 4,000 n miles (7,410 km).
Role/Weapon systems: Air Force operation for long-range MR/ASW. Original P-3A aircraft supplemented in 1988 by P-3B Orions from Norway after Lockheed modernisation. Two P-3A aircraft upgraded to P-3A plus in 1995–97. Four P-3Bs are undergoing modernisation to P3M standard with FITS data system, SPAS-16 acoustic processor, Elta 2022 radar, and AMES-C ESM system. Sensors: APS-134; FLIR; search radar, AQS-81 MAD, ALR 66 V(3) ECM/ESM, 87 sonobuoys. Weapons: ASW; eight torpedoes or depth bombs internally; 10 underwing stations. ASV; four Harpoon or 127 mm rockets.

P-3B   8/2013*, Adolfo Ortigueira Gil / 1531035

Numbers/Type: 8 EADS CASA CN-235 VIGMA.
Operational speed: 225 kt (422 km/h).
Service ceiling: 25,000 ft (7,620 m).
Range: 1,970 n miles (3,648 km).
Role/Weapon systems: Acquired in 2008 and operated by the Air Force. Primary role SAR, secondary role maritime air patrol. Based at Majorca, Gran Canaria, and Madrid. Equipped with FITS system. Sensors: FLIR/EO System Star Safire III, APS-143C(V)3 radar, H-764GU EGI, datalink system with Satcom capability.

CN-235 VIGMA   6/2012, Spanish Air Force / 1486368

PATROL FORCES

4 + (5) METEORO CLASS (PSO)

Name	No	Builders	Laid down	Launched	Commissioned
METEORO	P 41	Navantia, San Fernando	13 Mar 2009	16 Oct 2009	28 Jul 2011
RAYO	P 42	Navantia, San Fernando	3 Sep 2009	18 May 2010	26 Oct 2011
RELÁMPAGO	P 43	Navantia, San Fernando	17 Dec 2009	6 Oct 2010	7 Feb 2012
TORNADO	P 44	Navantia, San Fernando	5 May 2010	21 Mar 2011	19 Jul 2012

Displacement, tonnes: 2,675 full load
Dimensions, metres (feet): 93.9 × 14.2 × 4.3 (308.1 × 46.6 × 14.1)
Speed, knots: 20.5
Range, n miles: 8,000 at 12 kt
Complement: 46 (5 officers) + 32 personnel

Machinery: CODOE: 2 MTU 16V 1163 diesels; 12,000 hp (9 MW); 2 Siemens motors; 2,000 hp (1.5 MW); 2 cp props; bow thruster
Guns: 1 Oto Melara 3 in (76 mm)/62. 2 BAE Systems 25 mm/87 Typhoon.
Physical countermeasures: Decoys: 4 SRBOC Mk 36.
Electronic countermeasures: CESM: Indra Regulus Mk 9500.
RESM: Indra Rigel Mk 3800.
Radars: Surface search: Indra Aries; I-band.
Fire control: Dorna; K-band.
Navigation: Sperry VisionMaster; I-band.
Combat data systems: SCOMBA. LINPRO (internet-based Links 11, 16, and 22). Inmarsat. Secomsat.
Weapon control systems: Dorna optronic director.
Helicopters: Platform for 1 medium.

Comment: A programme for the procurement of a new class of up to eight multirole offshore patrol vessels known as Buques de Accion Maritima (BAM) was initiated in 2004. The modular design allows other variants of the BAM design to be capable of conducting intelligence, hydrographic and diving support tasks. Authorisation for the first batch of four patrol ships was made by the Spanish government on 20 May 2005 and a contract for their construction was signed with Navantia on 31 July 2006. The ships are capable of operating a helicopter and are equipped with two RIBs for boarding/interception operations. All four ships are based at Las Palmas (Canary Islands). A second batch of five ships (three patrol, one submarine rescue, and one hydrographic) is expected.

METEORO   9/2011, Diego Quevedo / 1454724

TORNADO   9/2012, Diego Quevedo / 1486359

RELÁMPAGO   3/2012, Diego Quevedo / 1455665

4 DESCUBIERTA CLASS (PSOH/FSGM)

Name	No	Builders	Laid down	Launched	Commissioned
INFANTA ELENA	P 76 (ex-F 33)	Bazán, Cartagena	26 Jan 1976	14 Sep 1976	12 Apr 1980
INFANTA CRISTINA	P 77 (ex-F 34)	Bazán, Cartagena	11 Sep 1976	25 Apr 1977	24 Nov 1980
CAZADORA	P 78 (ex-F 35)	Bazán, Ferrol	14 Dec 1977	17 Oct 1978	20 Jul 1982
VENCEDORA	P 79 (ex-F 36)	Bazán, Ferrol	1 Jun 1978	27 Apr 1979	18 Mar 1983

Displacement, tonnes: 1,253 standard; 1,693 full load
Dimensions, metres (feet): 88.8 × 10.4 × 3.8 *(291.3 × 34.1 × 12.5)*
Speed, knots: 25
Range, n miles: 4,000 at 18 kt, 7,500 at 12 kt
Complement: 148 (10 officers)

Machinery: 4 MTU-Bazán 16V 956 TB91 diesels; 15,000 hp(m) *(11 MW)* sustained; 2 shafts; cp props
Missiles: SSM: 4 Boeing Harpoon Block II; active radar homing to 124 km *(67 n miles)* at 0.9 Mach; warhead 227 kg.
Guns: 1 Oto Melara 3 in *(76 mm)*/62 compact; 85 rds/min to 16 km *(8.7 n miles)*; weight of shell 6 kg. 2 Oerlikon 20 mm/120.
Electronic countermeasures: ESM: Elsag Mk 1000 (part of Deneb system); or Mk 1600; intercept.
ECM: Ceselsa Canopus; or Mk 1900; jammer.
Radars: Air/surface search: Signaal DA05/2; E/F-band; range 137 km *(75 n miles)* for 2 m² target.
Surface search: Signaal ZW06; I-band.
Navigation: 2 Furuno; I-band.
Fire control: Signaal WM22/41 or WM25 system; I/J-band; range 46 km *(25 n miles)*.
Combat data systems: Tritan IV. Saturn SATCOM.
Weapon control systems: Signaal WM25; GM 101.

Programmes: Officially rated as Corvettes. *Infanta Elena* and *Infanta Cristina* are named after the daughters of King Juan Carlos. First four (of which *Diana* and *Descubierta* decommissioned) ordered 7 December 1973 (83 per cent Spanish ship construction components) and four (of which F 37 and F 38 sold to Egypt) more from Bazán, Ferrol on 25 May 1976.

Structure: Original Portuguese 'João Coutinho' design by Comodoro de Oliveira PN developed by Blohm + Voss and considerably modified by Bazán including use of Y-shaped funnel. Noise reduction measures include Masker fitted to shafts, auxiliary gas-turbine generator fitted on upper deck, all main and auxiliary diesels sound-mounted. Fully stabilised. Automatic computerised engine and alternator control; two independent engine rooms; normal running on two diesels.
Operational: P 79 converted to OPV role in 2003, P 77 and P 78 in 2004 and P 76 in 2005. P 76 and P 77 based at Cartagena and P 78 and P 79 at Las Palmas.
Sales: F 37 and F 38 sold to Egypt prior to completion. One to Morocco in 1983.

INFANTA ELENA — 9/2010, *Bob Fildes* / 1406139

INFANTA CRISTINA — 10/2005, *Camil Busquets i Vilanova* / 1167159

INFANTA ELENA — 11/2008, *M Declerck* / 1335860

4 SERVIOLA CLASS (OFFSHORE PATROL VESSELS) (PSOH)

Name	No	Builders	Laid down	Launched	Commissioned
SERVIOLA	P 71	Bazán, Ferrol	17 Oct 1989	10 May 1990	22 Mar 1991
CENTINELA	P 72	Bazán, Ferrol	12 Dec 1989	30 Mar 1990	24 Sep 1991
VIGÍA	P 73	Bazán, Ferrol	30 Oct 1990	12 Apr 1991	24 Mar 1992
ATALAYA	P 74	Bazán, Ferrol	14 Dec 1990	22 Nov 1991	29 Jun 1992

Displacement, tonnes: 1,165 full load
Dimensions, metres (feet): 68.7 oa; 63 pp × 10.4 × 3.4 *(225.4; 206.7 × 34.1 × 11.2)*
Speed, knots: 19
Range, n miles: 8,000 at 12 kt
Complement: 42 (8 officers) + 6 spare berths

Machinery: 2 MTU-Bazán 16V 956 TB91 diesels; 7,500 hp(m) *(5.5 MW)* sustained; 2 shafts; LIPS cp props
Guns: 1 US 3 in *(76 mm)*/50 Mk 27; 20 rds/min to 12 km *(6.6 n miles)*; weight of shell 6 kg. 2—12.7 mm MGs.
Electronic countermeasures: ESM: ULQ-13 (in P 71).
Radars: Surface search: Racal Decca 2459; I-band.
Navigation: Racal Decca ARPA 2690 BT; I-band.
Weapon control systems: Bazán Alcor or MSP 4000 (P 73) optronic director. Hispano mini combat system. SATCOM.

Helicopters: Platform for 1 AB 212.

Programmes: Project B215 ordered from Bazán, Ferrol in late 1988. The larger Milano design was rejected as being too expensive.
Modernisation: The guns are old stock refurbished but could be replaced by an Oto Melara 76 mm/62 or a Bofors 40 mm/70 Model 600 if funds can be found. Other equipment fits could include four Harpoon SSM, Meroka CIWS, Sea Sparrow SAM or a Bofors 375 mm ASW rocket launcher.

VIGIA   11/2008, Adolfo Ortigueira Gil / 1335859

No plans to carry out any of these improvements so far. EW equipment fitted in Serviola for training.
Structure: A modified Halcón class design similar to ships produced for Argentina and Mexico. Helicopter facilities enabling operation in up to Sea State 4 using non-retractable stabilisers. Three firefighting pumps.
Operational: For EEZ patrol. *Vigía* based at Cádiz, *Serviola* and *Atalaya* at Ferrol and *Centinela* at Las Palmas.

ATALAYA   9/2012, Adolfo Ortigueira Gil / 1486357

2 TORALLA CLASS (COASTAL PATROL CRAFT) (PB)

Name	No	Builders	Commissioned
TORALLA	P 81	Viudes, Barcelona	29 Apr 1987
FORMENTOR	P 82	Viudes, Barcelona	23 Jun 1988

Displacement, tonnes: 104 full load
Dimensions, metres (feet): 28.5 × 6.5 × 1.8 *(93.5 × 21.3 × 5.9)*
Speed, knots: 19
Range, n miles: 1,000 at 12 kt
Complement: 13

Machinery: 2 MTU-Bazán 8V 396 TB93 diesels; 2,100 hp(m) *(1.54 MW)* sustained; 2 shafts
Guns: 1 Browning 12.7 mm MG.
Radars: Surface search: Racal Decca RM 1070; I-band.
Navigation: Racal Decca RM 270; I-band.

Comment: Wooden hull with GRP sheath. Very similar to Customs Alcaravan class. *Formentor* refitted in 1996–97. Based at Cartagena.

TORALLA   10/2008, Adolfo Ortigueira Gil / 1335856

1 INSHORE/RIVER PATROL LAUNCH (PBR)

Name	No	Builders	Commissioned
CABO FRADERA	P 201	Bazán, La Carraca	11 Jan 1963

Displacement, tonnes: 21 full load
Dimensions, metres (feet): 17.8 × 4.2 × 0.9 *(58.4 × 13.8 × 3.0)*
Speed, knots: 11
Complement: 9
Machinery: 2 diesels; 280 hp(m) *(206 kW)*; 2 shafts
Guns: 1—7.62 mm MG.
Radars: Surface search: Furuno; I-band.

Comment: Based at Tuy on River Miño for border patrol with Portugal.

CABO FRADERA   4/2003, Camil Busquets i Vilanova / 0570981

Patrol forces — Amphibious forces < **Spain** 775

3 ALBORAN CLASS (OFFSHORE PATROL CRAFT) (PSOH)

Name	No	Builders	Commissioned
ALBORAN	P 62	Freire, Vigo	8 Jan 1997
ARNOMENDI	P 63	Freire, Vigo	13 Dec 2000
TARIFA	P 64	Freire, Vigo	14 Jun 2004

Displacement, tonnes: 1,995 full load
Dimensions, metres (feet): 66.5 × 11 × 4.4 (218.2 × 36.1 × 14.4)
Speed, knots: 13 (P 62), 15.8 (P 63); 3.5 diesel
Range, n miles: 20,000 at 13 kt
Complement: 37 (7 officers) + 9 spare berths

Machinery: 1 Krupp MaK 6 M 453C diesel; 2,400 hp(m) (1.76 MW) sustained (P 62); 1 Krupp MaK 8M25 diesel; 3,250 hp(m) (2.39 MW) sustained (P 63); 1 diesel generator and motor for emergency propulsion; 462 hp(m) (340 kW); 1 shaft; bow thruster; 350 hp(m) (257 kW)
Guns: 2 — 12.7 mm MGs.
Radars: Surface search: Furuno FAR-2825; I-band.
Navigation: Furuno FR-2130S; I-band.
Helicopters: Platform for 1 light.

Comment: *Alboran* launched in 1991 and purchased by the Fisheries Department to use as a Fishery Protection vessel. *Arnomendi* has a slightly larger bridge and more powerful engine. *Tarifa* is fitted with anti-pollution equipment. All based at Cartagena.

TARIFA 9/2010, Adolfo Ortigueira Gil / 1406140

ARNOMENDI 4/2002, Spanish Navy / 0105634

3 ANAGA CLASS (COASTAL PATROL CRAFT) (PB)

Name	No	Builders	Commissioned
TAGOMAGO	P 22	Bazán, La Carraca	30 Jan 1981
MEDAS	P 26	Bazán, La Carraca	16 Oct 1981
TABARCA	P 28	Bazán, La Carraca	30 Dec 1981

Displacement, tonnes: 324 full load
Dimensions, metres (feet): 44.4 × 6.6 × 2.5 (145.7 × 21.7 × 8.2)
Speed, knots: 16. **Range, n miles:** 4,000 at 13 kt
Complement: 25 (3 officers)

Machinery: 1 MTU-Bazán 16V 956 SB90 diesel; 4,000 hp(m) (2.94 MW) sustained; 1 shaft; cp prop
Guns: 1 FMC 3 in (76 mm)/50 Mk 22. 1 Oerlikon 20 mm Mk 10. 2 — 7.62 mm MGs.
Radars: Surface search: 1 Racal Decca 1226; I-band.
Navigation: Consilium Selesmar SRL MM 950; F/I-band.

Comment: Ordered from Bazán, Cádiz on 22 July 1978. For fishery and EEZ patrol duties. Rescue and firefighting capability. Speed reduced from original 20 kt. *Grosa* decommissioned in 2012.

TABARCA 10/2008, Adolfo Ortigueira Gil / 1335857

2 P 101 CLASS (PBR)

P 114	P 101 (ex-P 111)

Displacement, tonnes: 19 standard; 21 full load
Dimensions, metres (feet): 13.7 × 4.4 × 1.3 (44.9 × 14.4 × 4.3)
Speed, knots: 23.3
Range, n miles: 430 at 18 kt
Complement: 6
Machinery: 2 Baudouin-Interdiesel DNP-350; 768 hp(m) (564 kW); 2 shafts
Guns: 1 — 12.7 mm MG.
Radars: Surface search: Decca 110; I-band.

Comment: Ordered under the programme agreed 13 May 1977, funded jointly by the Navy and the Ministry of Commerce. Built to the Aresa LVC 160 design by Aresa, Arenys de Mar, Barcelona. GRP hull. Eight of the class conduct harbour auxiliary duties with Y numbers, the remainder paid off in 1993. P 101 transferred back again to patrol duties in 1996 and is based at Ayamonte (Huelva). P 114 is also used for patrol duties and is based at Ceuta.

P 101 3/2013*, Tony Roper / 1531037

AMPHIBIOUS FORCES

14 LCM (1E)

L 601–614

Displacement, tonnes: 110 full load
Dimensions, metres (feet): 23.3 × 6.4 × 1.1 (76.4 × 21.0 × 3.6)
Speed, knots: 14
Range, n miles: 160 at 12 kt
Complement: 3
Military lift: 100 tons or one main battle tank
Machinery: 2 MAN-D 2842-LE 402 diesels; 2,200 hp(m) (1.62 MW); 2 MJP-650 DD waterjets

Comment: L 601-602 built by IZAR, San Fernando, for LPDs and delivered in early 2001. Bow and stern ramps. Steel construction with wheelhouse of composites. Maximum speed in ballast is 22 kt. Based at Puntales. An order for a further 12 craft made in November 2004. First three laid down in 2005 and completed by 2007. A further nine built by Navantia, San Fernando. All delivered by early 2008.

L 604 6/2007, Spanish Navy / 1486358

40 LANDING CRAFT

Comment: Apart from those used for divers there are 14 LCM 6 (L 161–167, L 261–267), 16 LCVP and 8 LCPL. All of the LCM 6, eight of the LCVPs and most of the LCPs were built in Spanish Shipyards 1986–88. There are also two tug pontoons (mexeflotes) (L 91–L 92) completed in 1995. Most of these craft are laid up.

LCM L 162 10/1993, Diego Quevedo / 0506170

776 | **Spain** > Amphibious forces

2 GALICIA CLASS (LPD)

Name	No	Builders	Laid down	Launched	Commissioned
GALICIA	L 51	Bazán, Ferrol	31 May 1996	21 Jul 1997	30 Apr 1998
CASTILLA	L 52	Bazán, Ferrol	11 Dec 1997	14 Jun 1999	26 Jun 2000

Displacement, tonnes: 14,037 full load
Dimensions, metres (feet): 160 oa; 142 pp × 25 × 5.9 *(524.9; 465.9 × 82.0 × 19.4)*
Flight deck, metres: 60 × 25 *(196.9 × 82.0)*
Speed, knots: 20
Range, n miles: 6,000 at 12 kt
Complement: 115 (L 51), 189 (L 52)

Military lift: 543 or 404 (L 52) fully equipped troops and 72 (staff and aircrew) 6 LCVP or 4 LCM or 1 LCU and 1 LCVP. 130 APCs or 33 MBTs.
Machinery: 2 Bazán/Caterpillar 3612 diesels; 12,512 hp(m) *(9.2 MW)*; 2 shafts; LIPS cp props; bow thruster 680 hp(m) *(500 kW)*
Guns: 1 Bazán 20 mm/120 12-barrelled Meroka (fitted for) ❶; 3,600 rds/min combined to 2 km. 2 Oerlikon GAM-B01 20 mm.
Physical countermeasures: Decoys: 4 SRBOC chaff launchers.
Electronic countermeasures: ESM: Intercept.
Radars: Surface search: TRS 3D/16 (L 52) ❷; G-band. Surface search: Kelvin Hughes ARPA ❸; I-band. Navigation/helo control: I-band.
Combat data systems: SICOA (L 52); SATCOM; Link 11.
Helicopters: 6 AB 212 or 4 SH-3D Sea King ❹ or 4 Eurocopter Tiger.

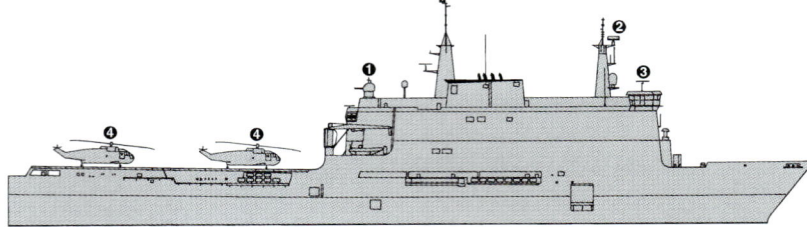

CASTILLA (Scale 1 : 1,500), Ian Sturton / 0106549

Programmes: Originally started as a national project by the Netherlands. In 1990 the ATS was seen as a possible solution to fulfil the requirements for a new LPD. Joint project definition study announced in July 1991 and completed in December 1993 and the first ship was authorised on 29 July 1994. The second of class ordered 9 May 1997.
Modernisation: L 52 C² capabilities upgraded in 2002-03 to support Flagship requirements. L 52 embarked the HQ of the Spanish High Readiness Force (Maritime) in November 2003 as part of the NATO Response Force. Both ships are to be fitted with RAM CIWS.
Structure: Able to transport a fully equipped battalion of marines providing a built-in dock for landing craft and a helicopter flight deck for debarkation in offshore conditions. Docking well is 885 m²; vehicle area 1,010 m². Access hatch on the starboard side. Hospital facilities. Built to commercial standards with military command and control and NBCD facilities. *Castilla* has improved command and control facilities with two operations centres, one for amphibious and one for a combat group.
Operational: Alternatively can also be used for a general logistic support for both military and civil operations, including environmental and disaster relief tasks. Based at Rota.

GALICIA 5/2009, A Campanera i Rovira / 1366002

CASTILLA 2/2011, Albert Campanera i Rovira / 1454725

IHS Jane's Fighting Ships 2014-2015 © 2014 IHS

1 JUAN CARLOS I CLASS (LHD)

Name	No	Builders	Laid down	Launched	Commissioned
JUAN CARLOS I	L 61	Navantia, Ferrol	20 May 2005	10 Mar 2008	30 Sep 2010

Displacement, tonnes: 27,514 full load
Dimensions, metres (feet): 230.8 × 32 × 7 (757.2 × 105.0 × 23.0)
Flight deck, metres: 202.3 × 32 (663.7 × 105.0)
Speed, knots: 21. **Range, n miles:** 9,000 at 15 kt
Complement: 296

Machinery: CODAGE; 1 GE LM 2500 gas turbine; 26,550 hp (19.8 MW); 2 MAN 324016V; 21,080 hp (15.7 MW); 2 Siemens-Schottel podded propulsors; 29,500 hp (22 MW)
Guns: 4 – 25 mm. 2 – 12.7 mm MGs.
Physical countermeasures: Decoys: Chaff launchers.
Electronic countermeasures: ESM/ECM: Indra Rigel.
Radars: Air search: Indra Lanza; D-band.
Surface search/navigation: 3 Indra Aries; I-band.
Navigation: Sperry Marine VisionMaster; I-band.
Combat data systems: SCOMBA. SICOA (ampbhibious command). Link 11, 16. SATCOM.

Helicopters: 6 landing spots for helicopter or AV-8 operations.

Programmes: Approval for the procurement of a Strategic Projection Ship was given by the Spanish Cabinet on 5 September 2003. Contract for design and construction was awarded in March 2004.
Structure: The hangar is 1,000 m². There are two 27-tonne aircraft elevators to the flight deck. Below the hangar there is a 1,400 m² lower garage, capable of carrying heavy vehicles up to the size and weight of Leopard 2 MBT. A further 975 m² of dock space is also available if the ship is not docked down. Above the lower garage is a 1,800 m light vehicle deck, which can accommodate 77 vehicles (or 144 ISO containers), and the aircraft hangar which can hold 12 NH90, eight Chinooks and up to seven Harrier. The landing dock (69.3 × 16.8 m) is to be capable of operating four LCM (1E) landing craft or one landing craft air cushion. Medical facilities include operating rooms, intensive care unit and sick bay (22 berths). There is space and weight reserved for a point-defence system.

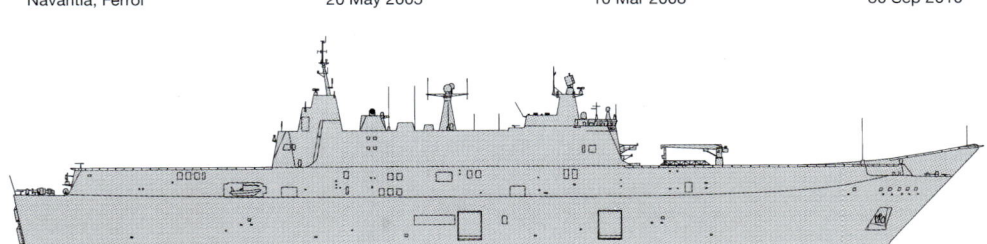

JUAN CARLOS I (Scale 1 : 1,800), Ian Sturton / 1366996

JUAN CARLOS I 4/2011, C D Yaylali / 1454726

Operational: The principal roles are amphibious, strategic projection of land forces and disaster relief. The ship is also capable of operating the fixed-wing aircraft of *Príncipe de Asturias*. Trials with Army AS 665 Tiger attack helicopters and AS 532 Cougar transport helicopters were conducted in 2012. Based at Rota.
Sales: Two similar ships are being built for the Australian Navy.

JUAN CARLOS I 5/2011, B Prézelin / 1454727

JUAN CARLOS I 6/2012, Spanish Navy / 1486360

MINE WARFARE FORCES

6 SEGURA CLASS (MINEHUNTERS) (MHC)

Name	No	Builders	Launched	Commissioned
SEGURA	M 31	Bazán, Cartagena	25 Jul 1997	27 Apr 1999
SELLA	M 32	Bazán, Cartagena	6 Jul 1998	28 May 1999
TAMBRE	M 33	Bazán, Cartagena	5 Mar 1999	18 Feb 2000
TURIA	M 34	Bazán, Cartagena	22 Nov 1999	16 Oct 2000
DUERO	M 35	Izar, Cartagena	28 Apr 2003	5 Jul 2004
TAJO	M 36	Izar, Cartagena	10 Jun 2004	10 Jan 2005

Displacement, tonnes: 539 full load
Dimensions, metres (feet): 54 oa; 51 wl × 10.7 × 2.2 (177.2; 167.3 × 35.1 × 7.2)
Speed, knots: 147 hunting. **Range, n miles:** 2,000 at 12 kt
Complement: 41 (7 officers)

Machinery: 2 MTU-Bazán 6V 396 TB83 diesels; 1,523 hp(m) (1.12 MW); 2 motors (for hunting); 200 kW; 2 Voith Schneider props; 2 side thrusters; 150 hp(m) (110 kW)
Guns: 1 Bazán/Oerlikon 20 mm GAM-BO1.
Physical countermeasures: MCM: FABA/Inisel system. 2 Gayrobot Pluto Plus ROVs.
Radars: Navigation: Kelvin-Hughes 1007; I-band.
Sonars: Raytheon/ENOSA SQQ-32 multifunction VDS mine detection; high frequency.
Combat data systems: FABA/SMYC NAUTIS.

Comment: On 4 July 1989 a technology transfer contract was signed with Vosper Thornycroft to allow Bazán to design a new MCM vessel based on the Sandown class. The order for four of the class was authorised on 7 May 1993, and an agreement signed on 26 November 1993 between DCN and Bazán provided for training in GRP technology. The first of class laid down 30 May 1995. Two more ordered on 26 January 2001. An option for two further ships is unlikely to be exercised. Sonar includes side scanning, and a towed body tracking and positioning system. M 35 and M 36 are to be fitted with the Minesniper mine disposal system. M 31–34 are to be retrofitted in due course. Form 1st MCM Squadron based at Cartagena.

TURIA 4/2012, Frank Findler / 1483645

SURVEY AND RESEARCH SHIPS

1 DARSS CLASS (RESEARCH SHIP) (AGI/AGOR)

Name	No	Builders	Commissioned
ALERTA (ex-Jasmund)	A 111	Peenewerft, Wolgast	6 Dec 1992

Displacement, tonnes: 2,329 full load
Dimensions, metres (feet): 76.3 × 12.1 × 4.2 (250.3 × 39.7 × 13.8)
Speed, knots: 11. **Range, n miles:** 1,000 at 11 kt
Complement: 60
Machinery: 1 Kolomna Type 40-DM diesel; 2,200 hp(m) (1.6 MW) sustained; 1 shaft; cp prop
Guns: Fitted for 3 twin 25 mm/70. 2–14.5 mm MGs.
Radars: Navigation: Racal Decca; I-band.

Comment: Former GDR depot ship launched on 27 February 1982 and converted to an AGI, with additional accommodation replacing much of the storage capacity. Was to have transferred to Ecuador in 1991 but the sale was cancelled. Commissioned in the Spanish Navy and sailed from Wilhelmshaven for a refit at Las Palmas prior to being based at Cartagena and used as an AGI and equipment trials ship. Saturn 35 SATCOM.

ALERTA 9/2010, Bob Fildes / 1406146

1 RESEARCH SHIP (AGOB)

Name	No	Builders	Commissioned
LAS PALMAS (ex-Somiedo)	A 52	Astilleros Atlántico, Santander	1978

Displacement, tonnes: 1,473 full load
Dimensions, metres (feet): 41 × 11.6 × 5.5 (134.5 × 38.1 × 18.0)
Speed, knots: 13
Range, n miles: 27,000 at 12 kt
Complement: 33 (8 officers) + 45 police
Machinery: 2 AESA/Sulzer 16ASV25/30 diesels; 7,744 hp(m) (5.69 MW); 2 shafts
Guns: 2–12.7 mm MGs.
Radars: Navigation: 2 Racal Decca; I-band.

Comment: Built as a tug for Compania Hispano Americana de Offshore SA. Commissioned in the Navy 30 July 1981. Converted in 1988 for Polar Research Ship duties in Antarctica with an ice strengthened bow, an enlarged bridge and two containers aft for laboratories. Based at Cartagena.

LAS PALMAS 10/2009, Diego Quevedo / 1406145

1 RESEARCH SHIP (AGOBH)

Name	No	Builders	Commissioned
HESPÉRIDES (ex-Mar Antártico)	A 33	Bazán, Cartagena	16 May 1991

Displacement, tonnes: 2,782 full load
Dimensions, metres (feet): 82.5 oa; 77.8 wl × 14.3 × 4.5 (270.7; 255.2 × 46.9 × 14.8)
Speed, knots: 15
Range, n miles: 12,000 at 13 kt
Complement: 39 (9 officers) + 30 police

Machinery: Diesel-electric; 4 MAN-Bazán 14V20/27 diesels; 6,860 hp(m) (5 MW) sustained; 4 generators; 2 AEG motors; 3,800 hp(m) (2.8 MW); 1 shaft; bow and stern thrusters; 350 hp(m) (257 kW) each
Radars: Surface search: Racal/Hispano ARPA 2690; I-band.
Navigation: Racal 2690 ACS; F-band.
Helicopters: 1 AB 212

Comment: Ordered in July 1988 from Bazán, Cartagena, by the Ministry of Education and Science. Laid down in 1989, launched 12 March 1990. Has 330 sq m of laboratories, Simbad ice sonar. Dome in keel houses several sensors. Ice-strengthened hull capable of breaking first year ice up to 45 cm at 5 kt. Based at Cartagena, the main task is to support the Spanish base at Livingston Island, Antarctica. Manned and operated by the Navy. Has a telescopic hangar. Modifications made to superstructure in 2004 to increase accommodation for scientific staff.

HESPÉRIDES 3/2011, Chris Sattler / 1454729

2 MALASPINA CLASS (SURVEY SHIPS) (AGS)

Name	No	Builders	Commissioned
MALASPINA	A 31	Bazán, La Carraca	21 Feb 1975
TOFIÑO	A 32	Bazán, La Carraca	23 Apr 1975

Displacement, tonnes: 833 standard; 1,107 full load
Dimensions, metres (feet): 57.6 × 11.7 × 3.9 (189 × 38.4 × 12.8)
Speed, knots: 15
Range, n miles: 4,000 at 12 kt, 3,140 at 14.5 kt
Complement: 63 (9 officers)
Machinery: 2 San Carlos MWM TbRHS-345-61 diesels; 3,600 hp(m) (2.64 MW); 2 shafts; LIPS cp props
Guns: 2 Oerlikon 20 mm.
Radars: Navigation: Raytheon 1220/6XB; I/J-band.

Comment: Ordered mid-1972. Both named after their immediate predecessors. Developed from British Bulldog class. Fitted with two Atlas DESO-10 AN 1021 (280–1,400 m) echo-sounders, retractable Burnett 538–2 sonar for deep sounding, Egg Mark B side scan sonar, Raydist DR-S navigation system, Hewlett Packard 2100A computer inserted into Magnavox Transit satellite navigation system, active rudder with fixed pitch auxiliary propeller. *Malaspina* used for a NATO evaluation of a Ship's Laser Inertial Navigation System (SLINS) produced by British Aerospace. Based at Cadiz.

MALASPINA 11/2008, Adolfo Ortigueira Gil / 1335850

2 LHT-130 CLASS (SURVEY MOTOR BOATS) (YGS)

Name	No	Builders	Commissioned
ASTROLABIO	A 91	Rodman, Vigo	30 Nov 2001
ESCANDALLO	A 92	Rodman, Vigo	27 Feb 2004

Displacement, tonnes: 8 full load
Dimensions, metres (feet): 12.6 × 4.2 × 0.5 (41.3 × 13.8 × 1.6)
Speed, knots: 30
Machinery: 2 diesels; 700 hp (522 kW); 2 shafts

Comment: Support craft of the Hydrographic Flotilla. Based at Puntales and transportable by road, rail, ship or aircraft.

ASTROLABIO   6/2011, Adolfo Ortigueira Gil / 1454728

1 CASTOR CLASS (SURVEY SHIPS) (AGS)

Name	No	Builders	Commissioned
ANTARES	A 23	Bazán, La Carraca	21 Nov 1974

Displacement, tonnes: 363 full load
Dimensions, metres (feet): 38.4 × 7.6 × 3.1 (126 × 24.9 × 10.2)
Speed, knots: 11.5
Range, n miles: 3,620 at 8 kt
Complement: 36 (4 officers)
Machinery: 1 Sulzer 4TD36 diesel; 720 hp(m) (530 kW); 1 shaft
Radars: Navigation: Raytheon 1620; I/J-band.

Comment: Fitted with Raydist, Omega and digital presentation of data. Based at Cadiz.

ANTARES   9/2011, Adolfo Ortigueira Gil / 1454730

TRAINING SHIPS

1 SAIL TRAINING SHIPS (AXS)

Name	No	Builders	Commissioned
JUAN SEBASTIÁN DE ELCANO	A 71	Echevarrieta, Cádiz	17 Aug 1928

Displacement, tonnes: 3,475 standard; 3,715 full load
Dimensions, metres (feet): 94.1 × 13.15 × 7.46 (308.7 × 43.1 × 24.5)
Speed, knots: 9
Range, n miles: 10,000 at 9 kt
Complement: 467 (120 trainees)
Machinery: 1 Deutz MWM KHD 6M diesel; 1,950 hp(m) (1.43 MW); 1 shaft
Guns: 2 – 37/80 mm Bazán saluting guns.
Radars: Navigation: 2 Racal Decca; I-band.

Comment: A four-masted top-sail schooner-near sister of Chilean *Esmeralda*. Named after the first circumnavigator of the world (1519–22) who succeeded to the command of the expedition led by Magellan after the latter's death. Laid down 24 November 1925. Launched on 5 March 1927. Carries 230 tons oil fuel. Engine replaced in 1992.

JUAN SEBASTIAN DE ELCANO   6/2012, Spanish Navy / 1486361

4 TRAINING CRAFT (AXL)

Name	No	Commissioned
CONTRAMAESTRE NAVARRETE (ex-*Guardiamarina Salas*)	A 82	10 May 1983
CONTRAMAESTRE SÁNCHEZ FERNÁNDEZ (ex-*Guardiamarina Godínez*)	A 83	4 Jul 1984
CONTRAMAESTRE ANTERO (ex-*Guardiamarina Rull*)	A 84	11 Jun 1984
CONTRAMAESTRE LAMADRID (ex-*Guardiamarina Chereguini*)	A 85	11 Jun 1984

Displacement, tonnes: 57 full load
Dimensions, metres (feet): 18.9 × 5.1 × 1.6 (62 × 16.7 × 5.2)
Speed, knots: 13
Complement: 15
Machinery: 2 MAN diesels; 2 shafts
Radars: Navigation: Halcon 948; I-band.

Comment: Former tenders to Naval School transferred to Naval Specialist School, Ferrol, in 2007. The craft have been assigned new names.

CONTRAMAESTRE LAMADRID   6/2007, Roberto Marin / 1170044

4 RODMAN 66 CLASS (AXT)

Name	No	Builders	Commissioned
GUARDIAMARINA BARRUTIA	A 121	Rodman, Vigo	2007
GUARDIAMARINA CHEREGUINI	A 122	Rodman, Vigo	2007
GUARDIAMARINA RULL	A 123	Rodman, Vigo	2007
GUARDIAMARINA SALAS	A 124	Rodman, Vigo	2008

Displacement, tonnes: 37 standard
Dimensions, metres (feet): 20.5 × 4.9 × 0.96 (67.3 × 16.1 × 3.1)
Speed, knots: 20
Range, n miles: 500 at 15 kt
Complement: 16
Machinery: 2 Caterpillar diesels; 1,500 hp (1.1 MW); 2 shafts
Radars: Navigation: Furuno; I-band.

Comment: GRP hull. New craft which replaced Naval School tenders in 2007/08.

GUARDIAMARINA CHEREGUINI   12/2006, Adolfo Ortigueira Gil / 1167149

7 SAIL TRAINING SHIPS (AXS)

Name	No	Builders	Commissioned
AROSA	A 72	Inglaterra	1 Apr 1981
BLANCA	A 73	–	23 Mar 2013
LA GRACIOSA (ex-*Dejá Vu*)	A 74	Inglaterra	30 Jun 1988
GIRALDA (ex-*Southern Cross*)	A 76	Morris & Mortimer, Argyll	26 Aug 1993
SISARGAS	A 75	Novo Glass, Polinya	18 May 1995
ALMANSA	A 77	–	–
PEREGRINA	A 78	–	22 Feb 2007

Comment: Based at the Naval School, Marín. A ketch (A 72) (52 tons and 22.84 m in length), a schooner (A 74) (16.8 m in length), a 90 tons ketch (A 75) launched in 1958 and formerly owned by the father of King Juan Carlos I and presented to the Naval School in 1993 and an ex-yacht (A 76). A 73 and A 77 were both built in France and are based at Marin and Ferrol respectively. A 78 was built in the Netherlands in 1988 and is based at Marin.

AUXILIARIES

Notes: There are plans to acquire a new submarine rescue ship to replace *Neptuno*. This is to be included in the second batch of BAM.

1 CANTABRIA CLASS (AORH)

Name	No	Builders	Laid down	Launched	Commissioned
CANTABRIA	A 15	Navantia, San Fernando	18 Jul 2007	21 Jul 2008	29 Jul 2010

Displacement, tonnes: 19,813 full load
Dimensions, metres (feet): 173.9 × 23.0 × 8.0 *(570.5 × 75.5 × 26.2)*
Speed, knots: 21. **Range, n miles:** 6,000 at 13 kt
Complement: 122 (19 air crew) + 20 spare berths + 10 casualty berths

Cargo capacity: 6,400 tons dieso; 1,600 tons aviation fuel
Machinery: 2 diesels; 29,200 hp *(21.8 MW)*; 1 shaft; 1 bow thruster
Guns: To be announced.
Physical countermeasures: Decoys: 6 SRBOC Mk 36 chaff launchers.
Electronic countermeasures: ESM/ECM: Indra Rigel.
Radars: Navigation: 3 Indra Aries; I-band.
Combat data systems: Scomba.
Helicopters: 2 SH-3D Sea King or 3 AB 212.

Comment: Similar in design to Patiño class with improved capabilities including double-hull, container cargo capacity, enhanced sensors and a combat data system. Two RAS stations on each side and one stern refuelling station. There is a small hospital with 10 beds. Contract for construction of the ship signed on 30 December 2004. Sea trials began in October 2009. These were followed by a full set of evaluation and certification trials. The ship deployed to Australia for much of 2013 in support for RAN operations. Based at Ferrol.

CANTABRIA   *2/2013, Chris Sattler* / 1168762

1 PATIÑO CLASS (FLEET LOGISTIC TANKER) (AORH)

Name	No	Builders	Launched	Commissioned
PATIÑO	A 14	Bazán, Ferrol	22 Jun 1994	16 Jun 1995

Displacement, tonnes: 5,854 standard; 17,319 full load
Dimensions, metres (feet): 166 × 22 × 8 *(544.6 × 72.2 × 26.2)*
Speed, knots: 20. **Range, n miles:** 13,440 at 20 kt
Complement: 165 (19 air crew) + 20 spare berths

Cargo capacity: 6,815 tons dieso; 1,660 tons aviation fuel; 500 tons solids
Machinery: 2 Bazán/Burmeister & Wain 16V40/45 diesels; 24,000 hp(m) *(17.6 MW)* sustained; 1 shaft; LIPS cp prop
Guns: 2 Bazán 20 mm/120 Meroka CIWS (fitted for). 2 Oerlikon 20 mm/90.
Physical countermeasures: Decoys: 4 SRBOC chaff launchers. Nixie torpedo decoy.
Electronic countermeasures: ESM/ECM: Aldebaran intercept and jammer.
Radars: 3 navigation/helo control; I-band.
Helicopters: 2 SH-3D Sea King or 3 AB 212.

Comment: The Bazán design AP 21 was rejected in favour of this joint Netherlands/Spain design. Ordered on 26 December 1991. Laid down 1 July 1993. Two supply stations each side for both liquids and solids. Stern refuelling. One Vertrep supply station, and workshops for aircraft maintenance. Medical facilities. Built to merchant ship standards with military NBC. Accommodation for up to 50 female crew members. SCOT 3 SATCOM to be fitted. Based at Ferrol.

PATIÑO   *4/2009, B Prézelin* / 1366145

1 TRANSPORT SHIP (APH)

Name	No	Builders	Recommissioned
CONTRAMAESTRE CASADO (ex-*Thanasis-K*; ex-*Fortuna Reefer*, ex-*Bonzo*, ex-*Bajamar*, ex-*Leeward Islands*)	A 01	Eriksberg-Göteborg	15 Dec 1982

Displacement, tonnes: 5,045 full load
Dimensions, metres (feet): 104.7 × 14.3 × 8.9 *(343.5 × 46.9 × 29.2)*
Speed, knots: 14. **Range, n miles:** 8,000 at 14 kt
Complement: 72
Machinery: 1 Burmeister & Wain diesel; 3,600 hp(m) *(2.65 MW)*; 1 shaft
Guns: 2 Oerlikon 20 mm.
Radars: Navigation: Racal Decca 1226 and 626; I-band.

Comment: Built in 1953. Impounded as smuggler. Delivered after conversion 6 December 1983. Has a helicopter deck. Since 2001, based at La Carraca (Cadiz).

CONTRAMAESTRE CASADO   *10/2008, Adolfo Ortigueira Gil* / 1335846

1 TRANSPORT SHIP (AKRH)

Name	No	Builders	Commissioned
MARTÍN POSADILLO (ex-*Rivanervión*; ex-*Cala Portals*)	A 04 (ex-ET-02)	Duro Felguera, Gijon	1973

Displacement, tonnes: 1,951 full load
Dimensions, metres (feet): 75 × 13 × 4.3 *(246.1 × 42.7 × 14.1)*
Speed, knots: 10
Complement: 18
Military lift: 42 trucks plus 25 jeeps
Machinery: 1 BMW diesel; 2,400 hp(m) *(1.77 MW)*; 1 shaft
Helicopters: Platform for 1 Chinook.

Comment: Ro-Ro ship taken on by the Army in 1990 and transferred to the Navy on 14 February 2000. Based at Cartagena.

MARTÍN POSADILLO   *3/2011, Albert Campanera i Rovira* / 1454731

HARBOUR TANKERS (YO)

No	Displacement, tons	Dimensions, metres	Cargo, tons fuel	Commissioned
Y 231	524	37.9 × 7.0 × 3.1	300	1981
Y 251	830	46.7 × 8.4 × 3.1	500	1981
Y 254	214.7	27.2 × 6.2 × 2.2	100	1981
Y 255	524	37.6 × 7 × 2.9	300	1981

Comment: All built by Bazán at Cádiz and Ferrol.

Y 251   *11/2003, Diego Quevedo* / 0570953

1 TRANSPORT SHIP (AKR)

Name	No	Builders	Commissioned
EL CAMINO ESPAÑOL (ex-*Araguary*; ex-*Cyndia*)	A 05 (ex-ET 03)	Maua, Rio de Janeiro	Oct 1984

Displacement, tonnes: 5,897 full load
Dimensions, metres (feet): 95.5 × 18.3 × 4.6 *(313.3 × 60.0 × 15.1)*
Speed, knots: 12
Complement: 64 (3 officers)
Military lift: 24 tanks plus 15 trucks and 102 jeeps
Machinery: 2 Sulzer diesels; 6,482 hp(m) *(4.76 MW)*; 2 shafts
Radars: Navigation: I-band.

Comment: Acquired by the Army in early 1999 but commissioned into the Navy on 21 September 1999. Ro-Ro design converted for military use by Bazán in Cartagena. Used for logistic support of armed forces. Has two 25 ton cranes. Based at Cartagena.

EL CAMINO ESPAÑOL   *10/2008, Adolfo Ortigueira Gil* / 1335845

18 BARGES (YO/YE)

Comment: Have Y numbers. Three in 200 series carry fuel, Five in 300 for ammunition and general stores, 10 in 400 for anti-pollution. Some floating pontoons have L numbers.

Y 221   *10/2005, Adolfo Ortigueira Gil* / 1153458

Auxiliaries — Tugs < **Spain** 781

2 LOGISTIC SUPPORT SHIPS (ATF/AGDS)

Name	No	Builders	Commissioned
MAR CARIBE (ex-*Amatista*)	A 101	Duro Felguera, Gijon	24 Mar 1975
NEPTUNO (ex-*Mar Rojo*; ex-*Amapola*)	A 20 (ex-A 102)	Duro Felguera, Gijon	24 Mar 1975

Displacement, tonnes: 1,890 full load
Dimensions, metres (feet): 53.8 × 11.8 × 4.5 *(176.5 × 38.7 × 14.8)*
Speed, knots: 12. **Range, n miles:** 6,000 at 10 kt
Complement: 44
Machinery: 2 Echevarria-Burmeister & Wain 18V23HU diesels; 4,860 hp(m) *(3.57 MW)*; 2 shafts; bow thruster

Comment: Two offshore oil rig support tugs were acquired and commissioned into the Navy 14 December 1988. Bollard pull, 80 tons. *Neptuno* converted as a diver support vessel and submarine rescue ship. She has a dynamic positioning system and carries a side scan mine detection high-frequency sonar as well as a semi-autonomous remote-controlled DSRV. The control cable restricts operations to within 75 m of an auxiliary diving unit. The DSRV is launched and recovered by a hydraulic arm. *Mar Caribe* works with Amphibious Forces and is based at Cadiz. *Neptuno* based at Cartagena.

MAR CARIBE   10/2008, *Adolfo Ortigueira Gil* / 1335847

NEPTUNO   6/2012, *Spanish Navy* / 1486363

45 HARBOUR LAUNCHES (YDT/YFL)

Y 503–509	Y 524	Y 537	Y 549–564	Y 588–592
Y 511	Y 526–529	Y 539–541	Y 581	
Y 521	Y 534	Y 545	Y 583–585	

Comment: Miscellaneous craft used for personnel transfer, pilot boats and torpedo-recovery.

Y 549   10/2007, *Adolfo Ortigueira Gil* / 1170036

Y 558   3/2011, *Albert Campanera i Rovira* / 1454733

TUGS

1 OCEAN TUG (ATA)

Name	No	Builders	Commissioned
MAHÓN (ex-*Circos*)	A 51	Astilleros Atlántico, Santander	1978

Displacement, tonnes: 1,473 full load
Dimensions, metres (feet): 41 × 11.6 × 5.5 *(134.5 × 38.1 × 18.0)*
Speed, knots: 13
Complement: 33 (8 officers) + 45 police
Machinery: 2 AESA/Sulzer 16ASV25/30 diesels; 7,744 hp(m) *(5.69 MW)*; 2 shafts
Guns: 2—12.7 mm MGs.
Radars: Navigation: 2 Racal Decca; I-band.

Comment: Built for Compania Hispano Americana de Offshore SA. Commissioned in the Navy 30 July 1981. Based at Ferrol.

MAHÓN   7/2000, *Adolfo Ortigueira Gil* / 0105651

1 OCEAN TUG (ATA)

Name	No	Builders	Commissioned
LA GRAÑA (ex-*Punta Amer*)	A 53 (ex-Y 119)	Astilleros Luzuriaga, San Sebastian	1982

Displacement, tonnes: 675 full load
Dimensions, metres (feet): 31.2 × 8.4 × 3.2 *(102.4 × 27.6 × 10.5)*
Speed, knots: 13
Range, n miles: 1,750 at 12 kt
Complement: 28
Machinery: 1 diesel; 3,240 hp(m) *(2.38 MW)*; 1 Voith Schneider prop

Comment: Former civilian tug acquired by Navy on 20 October 1987. Now designated as ocean-going. Based at Cadiz.

LA GRAÑA   3/2011, *Albert Campanera i Rovira* / 1454734

© 2014 IHS   IHS Jane's Fighting Ships 2014-2015

Spain > Tugs — Government maritime forces

COASTAL AND HARBOUR TUGS (YTB/YTM/YTL)

No	Displacement tons (full load)	HP/speed	Commissioned
Y 116	458	1,500/12	1981
Y 118	236	1,750/14	1990
Y 120 (ex-*Punta Roca*)	260	1,750/12	1988
Y 121	236	1,750/14	1990
Y 122	423	3,520/12	1999
Y 123	423	3,520/12	1999
Y 124	242	1,770/11	2001–02
Y 126	242	1,770/11	2001–02
Y 138	65	200/8	1967
Y 139	65	200/8	1967
Y 142	229	950/11	1981
Y 144	195	2,030/11	1979–80
Y 145	195	2,030/11	1979–80
Y 147	87	400/10	1987
Y 148	87	400/10	1987
Y 174	17	536/9	1999–05
Y 175	17	536/9	1999–05
Y 176	17	536/9	1999–05
Y 177	17	536/9	1999–05
Y 178	17	536/9	1999–05
Y 179	17	536/9	1999–05
Y 180	10	540/9	2005
Y 181	10	540/9	2005
Y 182	10	540/9	2005
Y 183	10	540/9	2005

Comment: There are eight tugs (Y 116, Y 118, Y 120–124, Y 126), seven harbour tugs (Y 138–139, Y 142, Y 144–145, Y 147–148) and ten submarine pushers (Y 174–183). Y 120 was built in 1973 and commissioned in the Spanish Navy in 1988.

Y 124   4/2010, Diego Quevedo / 1406148

GOVERNMENT MARITIME FORCES

POLICE (GUARDIA CIVIL — MARITIME SERVICE)

Notes: Created by Royal decree on 22 February 1991 and owned by the Ministry of Interior. Bases at Algeciras, Alicante, Almeria, Barcelona, Cadiz, Cartagena, Castellon, Ceuta, Corralejo, Gijón, Girona, Huelva, La Coruña, Las Palmas, Malaga, Motril, Palma, Pontevedra, Santander, Tarragona, Tenerife, Valencia and Vizcaya. Personnel strength 1,200 (35 officers). The force has taken over the anti-terrorist role and some general patrol duties as a peacetime paramilitary organisation coming under the Ministry of Defence in war. In addition to the craft listed there are some 42 smaller craft (under 9 m). All vessels are armed. 18 BO 105, 8 BK-117 and 31 Eurocopter EC-135 helicopters are used for coastal patrols and are based at Tenerife, Seville, Valencia, Mallorca, Huesca, Logroño, Leon and La Coruña. Two EADS/CASA 235 maritime patrol aircraft entered service in 2009.

BK 117   10/2005, Adolfo Ortigueira Gil / 1153462

C-235   6/2013*, Adolfo Ortigueira Gil / 1531038

---

1 IZAR IVP-22 CLASS (WPB)

SALEMA A 01

Displacement, tonnes: 90 full load
Dimensions, metres (feet): 24.5 × 5.96 × 1.8 (80.4 × 19.6 × 5.9)
Speed, knots: 20
Range, n miles: 400 at 12 kt
Complement: 8
Machinery: 2 MAN diesels; 2,200 hp (1,640 kW)
Guns: 1 — 12.7 mm MG.
Radars: Navigation: I-band.

Comment: Built by Bazán, San Fernando. Steel hull. Commissioned on 24 June 1999 having been procured by Agriculture and Fisheries Ministry for operation by Guardia Civil. Hull lengthened in 2003 to facilitate operation of RIB. Based at Cádiz.

SALEMA   10/2005, Adolfo Ortigueira Gil / 1153463

3 RODMAN 82 CLASS (WPB)

RIO GUADIARO (ex-*Seriola*) A 02   **RIO PISUERGA** A 03   **RIO NALON** A 04

Displacement, tonnes: 62 full load
Dimensions, metres (feet): 26 × 5.9 × 1.3 (85.3 × 19.4 × 4.3)
Speed, knots: 30
Range, n miles: 720 at 17 kt
Complement: 10
Machinery: 2 diesels; 3,600 hp (2.68 MW); 2 waterjets
Guns: 1 LAG 40 mm grenade launcher.
Radars: Navigation: I-band.

Comment: Built in 2001 by Rodman, Vigo. A 02 based at Alicante, A 03 at Cantabria and A 04 at Asturias. A 02 purchased by Fisheries department.

RIO NALON   6/2007, Camil Busquets i Vilanova / 1170025

11 RODMAN 66 CLASS (PB)

CANAL BOCAYNA M 34	**RIO PAS** M 38	**RIO GALLEGO** M 45
PICO DEL TEIDE M 35	**RIO GUADALENTIN** M 39	**RIO NACIMIENTO** M 46
RIO GUADALQUIVIR M 36	**RIO TER** M 43	**RIO NANSA** M 47
RIO TORDERA M 37	**RIO GENIL** M 44	

Displacement, tonnes: 37 full load
Dimensions, metres (feet): 20.5 × 4.9 × 0.96 (67.3 × 16.1 × 3.1)
Speed, knots: 30. **Range, n miles:** 450 at 22 kt
Complement: 6
Machinery: 2 MAN D2848 diesels; 2,200 hp (1.64 MW); 2 Hamilton HM 461 waterjets
Radars: Navigation: Furuno; I-band.

Comment: Built by Rodman to replace the Saeta class, which have been withdrawn from service. *Canal Bocayna* (Fuerteventura) delivered on 28 January 2008, *Pico del Tiede* (Tenerife) on 11 February 2008, *Rio Guadalquivir* (Cadiz) on 16 May 2008, *Rio Tordera* (Barcelona) in June 2008, *Rio Pas* (Algeciras) on 11 June 2008, *Rio Guadalentin* (Cartagena) in July 2008, *Rio Genil* (Granada) in May 2010 and *Rio Nacimiento* (Almeriá) in November 2010.

RIO TORDERA   11/2010, Camil Busquets i Vilanova / 1406153

Government maritime forces < **Spain** 783

14 RODMAN 101 CLASS (WPB)

RIO ANDARAX A 06	RIO CABRIEL A 11	RIO GUADIANA A 16
RIO GUADALOPE A 07	RIO CERVANTES A 12	RIO FRANCOLI A 17
RIO ALMANZORA A 08	RIO ARA A 13	RIO GUADALETE A 18
RIO NERVION A 09	RIO ADAJA A 14	CABALEIRO A 19
RIO GUADALAVIAR A 10	RIO DUERO A 15	

Displacement, tonnes: 70 full load
Dimensions, metres (feet): 30 × 5.9 × 1.3 (98.4 × 19.4 × 4.3)
Speed, knots: 30
Range, n miles: 800 at 12 kt
Complement: 10
Machinery: 2 Caterpillar 3412C diesels; 2,800 hp (2.06 MW); 2 Hamilton waterjets
Guns: 1 LAG 40 mm grenade launcher.
Radars: Navigation: I-band.

Comment: GRP hull. Built by Rodman, Vigo and delivered in 2003 (A 06–08), 2004 (A 09–13), 2005 (A 14), 2006 (A 15–17) and 2010 (A 18–19). A 05 A 06, A 08 and A 16 purchased by Agriculture and Fisheries Ministry. All operated by Guardia Civil.

RIO FRANCOLI 10/2010, Adolfo Ortigueira Gil / 1406150

9 RODMAN 55M CLASS (WPBF)

M 02–M 14 series

Displacement, tonnes: 17 full load
Dimensions, metres (feet): 16.5 × 3.8 × 0.7 (54.1 × 12.5 × 2.3)
Speed, knots: 35
Range, n miles: 500 at 25 kt
Complement: 6
Machinery: 2 MAN D2848-LXE diesels; 1,360 hp(m) (1 MW) sustained; 2 Hamilton water-jets
Guns: 1 — 12.7 mm MG.
Radars: Surface search: Ericsson; I-band.

Comment: GRP hulls built by Rodman, Vigo. First five in service in 1992, three in 1993, six more in 1995–96. M 01 sunk in 2002. Known as Baltic class. Two transferred to Mauritania in 2006, one to Gambia in 2007 and one to Senegal to 2006.

M 13 3/2012, Martin Mokrus / 1483646

1 RODMAN 55 CANARIAS CLASS (WPBF)

ALMIRANTE DIAZ PIMIENTA M 16

Displacement, tonnes: 19 standard
Dimensions, metres (feet): 17.4 × 3.8 × 0.8 (57.1 × 12.5 × 2.6)
Speed, knots: 48
Range, n miles: 400 at 25 kt
Complement: 6
Machinery: 2 MAN D2848 LXE406 diesels; 2,300 hp (1.71 MW); 2 Hamilton waterjets
Guns: 1 — 12.7 mm MG.
Radars: Navigation: I-band.

Comment: GRP hull built by Rodman, Vigo. Purchased in 1999 by Canary Islands Agriculture and Fishery Department. Based at Lanzarote. Same class sold to Cyprus.

ALMIRANTE DIAZ PIMIENTA 2/2008, Adolfo Ortigueira Gil / 1335842

15 RODMAN 55HJ CLASS (PB)

RIO ARBA M 17	RIO CEDENA M 22	RIO JILOCA M 27
RIO CAUDAL M 18	RIO LADRA M 23	RIO ARAGÓN M 28
RIO BERNESGA M 19	RIO CERVERA M 24	RIO SANTA EULALIA M 29
RIO MARTIN M 20	RIO JUCAR M 25	RIO ULLA M 30
RIO GUADALOBON M 21	RIO GALLO M 26	RIO MIJARES M 32

Displacement, tonnes: 20 full load
Dimensions, metres (feet): 17 × 3.8 × 0.9 (55.8 × 12.5 × 3.0)
Speed, knots: 52. **Range, n miles:** 400 at 25 kt
Complement: 5
Machinery: 2 MAN D2848 LXE406 diesels; 2,720 hp (2.02 MW); 2 Hamilton waterjets
Radars: Navigation: I-band.

Comment: GRP hull built by Rodman, Vigo. Similar to Colimbo class of Spanish Customs. M 17–24 delivered in 2004 and M 25–30 in 2005.

RIO SANTA EULALIA 12/2010, Camil Busquets i Vilanova / 1406152

1 RODMAN 58 CLASS (PB)

CORVO MARINO M 31

Displacement, tonnes: 38 full load
Dimensions, metres (feet): 18 × 4.9 × 0.9 (59.1 × 16.1 × 3.0)
Speed, knots: 34. **Range, n miles:** 450 at 25 kt
Complement: 6
Machinery: 2 diesels; 2,000 hp (1.5 MW); 2 Hamilton waterjets
Guns: 1 — 7.62 mm MG.
Radars: Navigation: Furuno; I-band.

Comment: GRP hull built by Rodman, Vigo. Purchased in 2006 for patrol duties around Cíes Islands off Vigo. Based at Naval School, Marin.

CORVO MARINO 6/2007, L M Rodriguez Garcia / 1170039

1 PATROL SHIP (PBO)

RIO MIÑO (ex-*Hoyo Maru*; ex-*Amazonas Reefer I*)

Measurement, tonnes: 605 gt
Dimensions, metres (feet): 50.8 × 8.6 × 3.3 (166.7 × 28.2 × 10.8)
Speed, knots: 12
Complement: 30
Machinery: 1 Niigata 6M28 AFTE diesel; 1,000 hp (736 kW); 1 shaft
Radars: Navigation: I-band.

Comment: Former fishing boat constructed by Kanasashi Zosen Shipyard, Japan, in 1984. Steel construction with bulbous bow. Converted to patrol boat in 2007 and recommissioned on 2 September 2007. Based at Cádiz.

RIO MIÑO 5/2012, Adolfo Ortigueira Gil / 1486364

1 PATROL SHIP (PBO)

RIO SEGRE A 20

Displacement, tonnes: 138 full load
Dimensions, metres (feet): 30 × 6.7 × 2.85 (98.4 × 22.0 × 9.4)
Speed, knots: 27. **Range, n miles:** 900 at 17 kt
Complement: 10
Machinery: 2 Caterpillar diesels; 3,700 hp (2.76 MW); 2 shafts
Guns: 1 — 12.7 mm MG. 1 LAG 40 mm grenade launcher.
Radars: Navigation: I-band.

Comment: Built by Asfibe Shipbuilders, Castellón and completed in 2010. GRP construction. A diving and rescue platform and launch/recovery system for a RHIB are located aft. Based at Valencia.

784 **Spain** › Government maritime forces

1 PATROL SHIP (PBO)

RIO SEGURA

Displacement, tonnes: 1,735 full load
Dimensions, metres (feet): 73 × 12.0 × 4.2 (239.5 × 39.4 × 13.8)
Speed, knots: 16
Complement: 30
Machinery: Diesel-electric; 2 MTU 16V4000M61 diesels; 5,438 hp (3.4 MW); 2 Schottel 1212 azimuth thrusters
Radars: Navigation: I-band.
Helicopters: Platform for 1 medium.

Comment: Astilleros Gondan selected on 17 December 2008 as successful bidder to construct a new offshore patrol ship. Delivered in December 2010 and based at Las Palmas.

RIO SEGURA   1/2011, Astilleros Gondán / 1421692

1 PATROL SHIP (PBO)

RIO TAJO (ex-*Britannia Uno*; ex-*Britannia Chieftan*, ex-*Arstertum*)

Measurement, tonnes: 682 gt
Dimensions, metres (feet): 52.86 × 11.33 × 3.44 (173.4 × 37.2 × 11.3)
Speed, knots: 12
Machinery: 2 MWM TBD440-8 diesels; 1,900 hp (1.4 MW); 2 shafts
Radars: Navigation: I-band.

Comment: Former offshore supply vessel constructed in 1973 by J G Hitzler Schiffswerft und Masch. Steel construction. Converted in 2008 to undertake patrol boat duties for Guardia Civil. Based at Las Palmas.

RIO TAJO   6/2011, B Prézelin / 1454735

3 AUXNAVAL ARMÓN 69 CLASS (PB)

RIO TAMBRE M 48   **RIO ÓRBIGO** M 49   **RIO JÁNDULA** M 50

Dimensions, metres (feet): 21.19 × 5.5 × 0.95 (69.5 × 18.0 × 3.1)
Speed, knots: 35
Complement: 5
Machinery: 2 MTU diesels; 1,450 hp (1.08 MW); 2 waterjets
Radars: Navigation: Furuno; I-band.

Comment: Delivered in 2011.

RIO TAMBRE   6/2011*, Armón / 1531039

3 AUXNAVAL ARMÓN 51 CLASS (PB)

RIO ALLER M 40   **CABO DE GATA** M 41   **RIO TORMES** M 42

Dimensions, metres (feet): 15.75 × 4.48 × 0.82 (51.7 × 14.7 × 2.7)
Speed, knots: 35
Complement: 4
Machinery: 2 MAN diesels; 1,400 hp (1.04 MW); 2 Hamilton waterjets
Radars: Navigation: Furuno; I-band.

Comment: Built by Armón Shipbuilders (Asturias) and delivered in 2008. *Rio Aller* based at Huelva, *Cabo de Gata* at Almeriá and *Rio Tormes* at Cadiz.

CABO DE GATA   6/2012* / 1531040

8 DUARRY MEGATECH 12 (PATROL CRAFT) (PB)

Displacement, tonnes: 9 full load
Dimensions, metres (feet): 12 × 3.9 × 0.72 (39.4 × 12.8 × 2.4)
Speed, knots: 40
Complement: 3
Machinery: 2 Yanmar diesels; 880 hp (656 kW); 2 Hamilton waterjets
Radars: Navigation: I-band.

Comment: Built by Duarry Shipbuilders, Barcelona and delivery began in 2010. GRP construction with Neoprene-Hypalon inflatable collar.

DUARRY MEGATECH 12   10/2012, Adolfo Ortigueira Gil / 1486365

RESEARCH SHIPS

Notes: Nine civilian research ships are owned by the Government Science and Technology Ministry and by the Agriculture Fishery and Food Ministry. Those operated by the Instituto Español de Oceanografía (IEO) are *Vizconde de Eza* (1,400 tons), *Cornide de Saavedra* (1,113 tons), *F P Navarro* (178 tons), *Odón de Buen* (64 tons), *Lura* (34 tons), *José Rioja* (32 tons), *J M Navaz* (30 tons), *Emma Bardán* (209 tons) and *Miguel Oliver* (1,200 tons). Two new 500 ton ships, *Angeles Alvariño* and *Ramon Margalef* entered service in 2011. Those operated by CSIC are *Garcia del Cid* (539 tons), *Mytilus* (170 tons) and *Sarmiento de Gamboa* (2,980 tons). The ships operate in co-operation with the Spanish Navy ship *Hespérides* and the French research ship *Thalassa*. There is a fisheries research vessel *Intermares*.

CORNIDE DE SAAVEDRA   8/2012, Adolfo Ortigueira Gil / 1486366

GARCIA DEL CID   10/2010, Adolfo Ortigueira Gil / 1406154

IHS Jane's Fighting Ships 2014-2015   © 2014 IHS

CUSTOMS

Notes: Customs service is the responsibility of the Ministry of Treasure. All carry ADUANAS on ships' sides. Some of the larger vessels are armed with machine guns. There are also three MBB-105, one MBB-117 and two AS 365 Dauphin helicopters. Six CASA C-212 patrol aircraft were transferred from the Air Force to the Inaer Group in 2011.

CASA C-212   3/2003, Adolfo Ortigueira Gil / 0570963

PATROL CRAFT (PB)

Name	Displacement, tons (full load)	HP/speed	Commissioned
ABANTO/ALBATROS/ALCA/ ALCATRAZ/ALCOTAN/ARAO/ DECIMO ANIVERSARIO/GERIFALTE I/ HALCON/PAIÑO/SACRE	87	4,732/35	2000–09
AGUILA I/AGUILA II/ AGUILA III/AGUILA IV/AGUILA V	22	2,410/45	2008–09
ALCAVARÁN I	95	4,080/23	1984
ALCAVARÁN II/ALCAVARÁN III/ ALCAVARÁN IV/ALCAVARÁN V	87	3,920/28	1984–87
ALCAUDON II	18	2,300/52	1999
COLIMBO II/COLIMBO III/COLIMBO IV	19	2,300/52	1999–03
CORMORÁN	22	2,400/60	1989
FENIX	18	2,300/50	1998
FULMAR	580	7,400/21	2006
GAVILÁN III	65	3,200/24	1983
GAVILÁN IV	70	3,920/25	1987
HALCÓN III	68	3,200/26	1983
HJ-1	19	2,300/51	1986
HJ-VIII/HJ-X	20	1,958/50	1987–88
MILANO II	15	2,300/50	1999
PETREL I	1,150	3,500/14	1993
VA II/VA IV	24	1,400/23	1985

ALCARAVAN V   5/2009, A Campanera i Rovira / 1305991

PAIÑO   11/2011, Adolfo Ortigueira Gil / 1454736

PETREL I   4/2010, Diego Quevedo / 1406157

FULMAR   5/2009, A Campanera i Rovira / 1305990

MARITIME RESCUE, SAFETY AND LOGISTIC SUPPORT

Notes: These roles are discharged by two services: SASEMAR (Sociedad Estatal de Salvamento y Seguridad Marítima) is under the direction of the Merchant Marine but may come under a Coast Guard service in due course. It operates 14 salvage and multipurpose tugs (*Don Inda, Clara Campoamor, Punta Mayor, Alonso de Chaves, Punta Salinas, María de Maeztu, Maria Pita, Maria Zambrano, Miguel de Cervantes, Gavia, Mastelero, Mesana, Luz de Mar* and *Remolcanosa Cinco*), one anti-pollution ship and 55 fast rescue craft. Aircraft assets include three EADS/CASA C-235 and 10 helicopters (AW-139 and Sikorsky S-61). All ships are painted red with a white stripe on the hull. ISM (Instituto Social de la Marina) operates two specialised medical and logistic ships for support of fishing vessels. *Esperenza del Mar* (5,000 tons) is based at Las Palmas (Canary Islands) Naval Base and *Juan de la Cosa* (1,500 tons) is based at Santander.

SALVAMAR ALBORAN   1/2005, Adolfo Ortigueira Gil / 1153456

JUAN DE LA COSA   10/2010, Adolfo Ortigueira Gil / 1406156

LUZ DE MAR   10/2010, Adolfo Ortigueira Gil / 1406155

CN-235   6/2012, Annati Collection / 1486367

Sri Lanka

Country Overview

Formerly known as Ceylon, the Democratic Socialist Republic of Sri Lanka gained independence in 1948. Situated off the southeast coast of India, from which it is separated by the Palk Strait and Gulf of Mannar, it has an area of 25,326 square miles and a coastline of 723 n miles with the Indian Ocean. The capital of Sri Lanka is Sri Jayavardhanapura (Kotte) while Colombo is the largest city and principal port. New ports are being constructed at Hambantota and Oluvil. There are further ports at Trincomalee, Kankasanthurai and Galle. Territorial waters (12 n miles) are claimed. A 200 n mile EEZ has been claimed although the limits have only been partly defined by boundary agreements.

Headquarters Appointments

Commander of the Navy:
  Vice Admiral J S K Colombage, RSP, VSV, USP
Chief of Staff:
  Rear Admiral S A M J Perera, RWP, VSV, USP
Director General, Operations:
  Rear Admiral N B J Rosayro, RSP, USP
Director General, Coast Guard:
  Rear Admiral R C Wejegunarathne, WV, RWP, RSP, USP

Area Commanders

Commander Western Naval Area:
  Rear Admiral S S Ranasinghe, WWV, RWP, USP
Commander North Central Naval Area:
  Rear Admiral S Jayakody, RSP, USP
Commander Northern Naval Area:
  Rear Admiral D M S Dissanayake, USP

Area Commanders – continued
Commander Eastern Naval Area:
  Rear Admiral R C A R Amarasinghe, RSP, USP
Commander Southern Naval Area:
  Rear Admiral D M B Wettewa, USP
Commander North Western Naval Area:
  Rear Admiral N K D Nanayakkara, RSP, USP
Commander South East Naval Area:
  Rear Admiral D E C Jayakody, RSP, USP

Personnel

(a) 2014: 41,248 (2,294 officers) regulars
(b) SLVNF: 5,344 (492 officers)
(c) Reserve force (regular): 671 (10 officers)
(d) Reserve force (volunteer): 137 (21)

Bases

Navy HQ: Colombo.
Western Command HQ: Colombo (other bases at Welisara, Uswetakeiyawa, and Training Centre at Talatuoya).
Southern Command HQ: Galle (other bases at Tangalle, Kirinda, Kirindoya, Amaduwa, and Training Centre at Boosa).
South Eastern Command HQ: Panama (other bases at Arugambay, Olluwil, and Kalmanai).
Eastern Command HQ: Trincomalee (other bases at Sampoor, Thoppoor, Vakarai, Kinniya, Nilaweli, Thiriyaya, Pulmudai, Lankapatuna, Mullaitive, and Naval Academy, Trincomalee).
Northern Command HQ: Kankasanthurai (other bases at Madagal, Karainagar, Valanai Island, Mandativu Island, Nainative Island, Pungudathive Island, and Delft Island).
North Central Command HQ: Thalaimanar (other bases at Punewa, Mannar Island, Nachchikudah, Vankalai, and Iranative Island).
North Western Command HQ: Mullikulam (other bases at Kaipitaya, Puttalam, and Silawaturai).

Pennant Numbers

Pennant numbers were reviewed in 1996, 2002 and 2010.

Prefix To Ships Names

SLNS.

Coast Guard

Plans to establish a Coast Guard, under the Ministry of Defence, were approved by parliament in July 2009. The role of the force is to assist the Sri Lankan Navy in the control of illegal activities in coastal waters. These include immigration/emigration, drug trafficking, fishing and weapons.

DELETIONS

Patrol Forces

2012 P 250, P 252 (to CG), P 253 (to CG), P 404–406 (to CG)
2013 P 403, P 440–443, P 052, P 125, P 171, P 174 (all to CG)

Auxiliaries

2012 Jetliner

PATROL FORCES

0 + 2 SANKALP CLASS (OFFSHORE PATROL VESSEL) (PSO)

Displacement, tonnes: 2,266 full load
Dimensions, metres (feet): 105 × 12.9 × 3.6 *(344.5 × 42.3 × 11.8)*
Speed, knots: 24
Range, n miles: 6,500 at 12 kt
Complement: 106 (12 officers)

Machinery: 2 SEMT-Pielstick 20 PA6B diesels; 20,900 hp(m) *(15.58 MW)*; 2 shafts; cp props
Guns: 1 – 76 mm.
Radars: Surface search: To be announced.
Navigation: To be announced.
Helicopters: 1 HAL Dhruv.

Comment: The supply of a new 105 m offshore patrol ship to Sri Lanka was sanctioned by the Indian government in 2010. Subsequently, technical discussions were conducted between Goa Shipyard and the Sri Lankan Navy in December 2010. The proposal is reported to be similar to the Sankalp-class patrol ships in service with the Indian Coast Guard and armament is expected to include a 76 mm gun. A contract was signed with Goa Shipyard in April 2013.

2 SAAR 4 CLASS (FAST ATTACK CRAFT – MISSILE) (PGG)

Name	No	Builders	Launched	Commissioned
NANDIMITHRA (ex-*Moledt*)	P 701	Israel Shipyards, Haifa	22 Mar 1979	May 1979
SURANIMILA (ex-*Komemiut*)	P 702	Israel Shipyards, Haifa	19 Jul 1978	Aug 1980

Displacement, tonnes: 422 standard; 457 full load
Dimensions, metres (feet): 58 × 7.8 × 2.4 *(190.3 × 25.6 × 7.9)*
Speed, knots: 32
Range, n miles: 1,650 at 30 kt, 4,000 at 17.5 kt
Complement: 75

Machinery: 4 MTU/Bazán 16V 956 TB91 diesels; 15,000 hp(m) *(11.03 MW)* sustained; 4 shafts
Missiles: 3 Gabriel II; radar or TV optical guidance; semi-active radar plus anti-radiation homing to 36 km *(20 n miles)* at 0.7 Mach; warhead 75 kg.
Guns: 1 Oto Melara 3 in *(76 mm)*/62 compact; 85 rds/min to 16 km *(8.7 n miles)*; weight of shell 6 kg. Adapted for shore bombardment. 1 – 40 mm. 2 Rafael Typhoon 20 mm. 2 – 20 mm. 2 – 12.7 mm MGs. 2 – 40 mm AGL.
Radars: Air/surface search: Thomson-CSF TH-D 1040 Neptune; G-band; range 33 km *(18 n miles)* for 2 m² target.
Fire control: Selenia Orion RTN 10X; I-band.

Comment: Transferred from Israel and recommissioned on 9 December 2000.

NANDIMITHRA 10/2003, Hartmut Ehlers / 0570991

1 JAYESAGARA CLASS (OFFSHORE PATROL VESSEL) (PB)

Name	No	Builders	Launched	Commissioned
JAYESAGARA	P 601	Colombo Dockyard	26 May 1983	9 Dec 1983

Displacement, tonnes: 335 full load
Dimensions, metres (feet): 39.8 × 7 × 2.1 *(130.6 × 23.0 × 6.9)*
Speed, knots: 15
Range, n miles: 3,000 at 11 kt
Complement: 52 (4 officers)
Machinery: 2 MAN 8L20/27 diesels; 2,180 hp(m) *(1.6 MW)* sustained; 2 shafts
Guns: 2 China 25 mm/80 (twin). 2 China 14.5 mm (twin) MGs. 2 – 12.7 mm MGs. 2 – 40 mm AGL. 2 – 7.62 mm MGs.
Radars: Surface search: Anritsu RA 723; I-band.

Comment: Ordered from Colombo Dockyard on 31 December 1981. Second of class sunk by Tamil forces in September 1994.

JAYESAGARA 6/2004, Sri Lanka Navy / 1044193

1 SUKANYA CLASS (OFFSHORE PATROL VESSEL) (PSOH)

Name	No	Builders	Launched	Commissioned
SAYURA (ex-*Saryu*)	P 620 (ex-54)	Hindustan Shipyard, Vishakapatnam	16 Oct 1989	8 Oct 1991

Displacement, tonnes: 1,920 full load
Dimensions, metres (feet): 101.1 oa; 96 wl × 11.5 × 4.4 *(331.7; 315 × 37.7 × 14.4)*
Speed, knots: 21
Range, n miles: 5,800 at 15 kt
Complement: 140 (15 officers)

Machinery: 2 SEMT-Pielstick 16 PA6 V 280 diesels; 12,800 hp(m) *(9.41 MW)* sustained; 2 shafts
Guns: 1 Bofors 40 mm/60. 4 China 14.5 mm (twin). 4 – 12.7 mm MGs.
Radars: Surface search: Racal Decca 2459; I-band.
Navigation: Bharat1245; I-band.

Comment: Transferred from India and recommissioned on 9 December 2000.

SAYURA 5/2009, Guy Toremans / 1367097

Patrol forces < **Sri Lanka** 787

1 RELIANCE CLASS (PSOH)

Name	No	Builders	Commissioned
SAMUDURA (ex-*Courageous*)	P 621 (ex-WMEC 622)	Coast Guard Yard, Baltimore	8 Dec 1967

Displacement, tonnes: 1,147 full load
Dimensions, metres (feet): 64.2 × 10.4 × 3.2 *(210.6 × 34.1 × 10.5)*
Speed, knots: 18
Range, n miles: 6,100 at 14 kt, 2,700 at 18 kt
Complement: 75 (12 officers)

Machinery: 2 Alco 16V-251 diesels; 6,480 hp *(4.83 MW)* sustained; 2 shafts; LIPS cp props
Guns: 1 Boeing 25 mm/87 Mk 38 Bushmaster; 200 rds/min to 6.8 km *(3.4 n miles)*. 2—12.7 mm MGs.
Radars: Surface search: Hughes/Furuno SPS-73; I-band.
Helicopters: Platform for one medium.

Comment: Transferred from USCG to Sri Lanka on 24 June 2004. During 34 years in USCG service, underwent Major Maintenance Availability (MMA) in 1989. The exhausts for main engines, ship service generators and boilers were run in a vertical funnel which reduced flight deck size. Capable of towing ships up to 10,000 tons.

SAMUDURA 2/2010, *M Mazumdar* / 1406158

5 SHANGHAI II (TYPE 062) CLASS
(FAST ATTACK CRAFT—GUN) (PB)

WEERAYA P 311 (ex-P 3141)	ABEETHA II P 316	WICKRAMA II P 318
JAGATHA P 315 (ex-P 3146)	EDITHARA II P 317	

Displacement, tonnes: 141 full load
Dimensions, metres (feet): 38.8 × 5.4 × 1.6 *(127.3 × 17.7 × 5.2)*
Speed, knots: 28
Range, n miles: 750 at 16 kt
Complement: 44

Machinery: 4 Type L12-180 diesels; 4,800 hp(m) *(3.53 MW)*; 4 shafts
Guns: 4 Royal Ordnance GCM-AO3 30 mm (2 twin) (P 316, P 317, P 318). 4—37 mm 2 (twin) (P 311, P 315). 4 China 14.5 mm (2 twin) MG. 2—7.62 mm MGs. 2—40 mm AGL (P 311, P 315).
Radars: Surface search: Koden MD 3220 Mk 2; I-band.
Navigation: Furuno 825 D; I-band.

Comment: Five transferred by China in 1971 of which four since decommissioned and *Weeraya* remains in service. Two further craft transferred in 1980 of which *Jagatha* remains in service. Three further craft (*Abeetha II*, *Edithara II* and *Wickrama II*) are modified craft with improved habitability but similar specifications. These were built at Qinxin Shipyard and commissioned on 11 June 2000.

EDITHARA II 6/2003, *Sri Lanka Navy* / 0570992

WEERAYA 6/2001, *Sri Lanka Navy* / 0130146

1 MOD SHANGHAI II CLASS
(FAST ATTACK CRAFT—GUN) (PB)

Name	No	Builders	Commissioned
RANARISI	P 322	Guijiang Shipyard	14 Jul 1992

Displacement, tonnes: 152 full load
Dimensions, metres (feet): 41 × 5.4 × 1.6 *(134.5 × 17.7 × 5.2)*
Speed, knots: 29
Range, n miles: 750 at 16 kt
Complement: 44 (4 officers)
Machinery: 4 diesels; 4,800 hp(m) *(3.53 MW)*; 4 shafts
Guns: 2 Royal Ordnance GCM-AO3 30 mm (1 twin). 2—25 mm. 4 China 14.5 mm (twin) Type 69. 2—12.7 mm MGs. 2—40 mm AGL.
Radars: Surface search: Racal Decca; I-band.

Comment: Acquired from China in September 1991. Automatic guns and improved habitability. *Ranaviru* and *Ranasuvu* destroyed by Tamil guerrillas.

RANARISI 6/2003, *Sri Lanka Navy* / 0570988

3 HAIZHUI (TYPE 062/1G) CLASS (PB)

Name	No	Builders	Commissioned
RANAJAYA	P 330	Guijiang Shipyard	22 May 1996
RANADEERA	P 331	Guijiang Shipyard	22 May 1996
RANAWICKRAMA	P 332	Guijiang Shipyard	22 May 1996

Displacement, tonnes: 173 full load
Dimensions, metres (feet): 41 × 5.3 × 1.8 *(134.5 × 17.4 × 5.9)*
Speed, knots: 21
Complement: 44
Machinery: 4 Type L12-180A diesels; 4,400 hp(m) *(3.22 MW)* sustained; 4 shafts
Guns: 2 China 37 mm/63 (1 twin). 2—30 mm GCM-A03. 4 China 25 mm/60 (2 twin). 2—12.7 mm MGs. 2—40 mm AGL.
Radars: Surface search: Anritsu 726UA; I-band.

Comment: Transferred from China by lift ship after delivery in 1995.

RANAWICKRAMA 6/2008, *Sri Lanka Navy* / 1335871

2 MOD HAIZHUI (LUSHUN) (TYPE 062/1G) CLASS
(FAST ATTACK CRAFT—GUN) (PB)

PRATHPA P 340   UDARA P 341

Displacement, tonnes: 215 full load
Dimensions, metres (feet): 45.5 × 6.4 × 1.7 *(149.3 × 21.0 × 5.6)*
Speed, knots: 28
Range, n miles: 750 at 16 kt
Complement: 44 (3 officers)
Machinery: 4 Type Z12V 190 BCJ diesels; 4,800 hp(m) *(3.53 MW)*; 4 shafts
Guns: 4 China 37 mm/63 (2 twin) Type 76. 2 China 14.5 mm (1 twin) Type 82 MGs. 2—12.7 mm MGs. 2—40 mm AGL.
Radars: Surface search: Racal Decca RM 1070A; I-band.

Comment: Built at Lushun Dockyard, Darlin. Commissioned on 2 March 1998. Larger version of Haizhui class.

UDARA 6/2005, *Sri Lanka Navy* / 1153483

© 2014 IHS   IHS Jane's Fighting Ships 2014-2015

788 Sri Lanka > Patrol forces

5 TRINITY MARINE CLASS (FAST ATTACK CRAFT—GUN) (PBF)

P 480–481   P 483–485

Displacement, tonnes: 69 full load
Dimensions, metres (feet): 24.9 × 5.4 × 1.5 *(81.7 × 17.7 × 4.9)*
Speed, knots: 47
Range, n miles: 600 at 17 kt
Complement: 20
Machinery: 2 MTU 12V 396 TE94 diesels; 4,570 hp(m) *(3.36 MW)* sustained; 2 water-jets
Guns: 2 Oerlikon 20 mm. 2—12.7 mm MGs. 2—7.62 mm MGs. 1 Grenade launcher.
Radars: Surface search: Raytheon R 1210; I-band.

Comment: All built at Equitable Shipyard, New Orleans. First three delivered in January 1997; second three in September 1997. All aluminium construction. P 482 sunk in action in 2000.

P 480   *1/1997, Sri Lanka Navy /* 0080701

26 COLOMBO MK I/II/III/IV CLASS
(FAST ATTACK CRAFT—GUN) (PBF)

P 450–451	P 497	P 417	P 432–433
P 490–492	P 410–412	P 419–424	P 435–437
P 494	P 414–415	P 430	P 439

Displacement, tonnes: 57 full load
Dimensions, metres (feet): 24.3 × 5.7 × 1.2 *(79.7 × 18.7 × 3.9)*
Speed, knots: 45
Range, n miles: 850 at 16 kt
Complement: 20

Machinery: 2 MTU 12V 396 TE94 diesels (Mk I/II) or 2 Deutz TBD 620 16V (Mk III/IV); 4,570 hp(m) *(3.36 MW)*; ASD 16 surface drives
Guns: 1 Rafael Typhoon 23 mm. 1 Oerlikon 20 mm. 2—12.7 mm MGs. 6—7.62 mm MGs. 2—40 mm AGL.
Radars: Surface search: Furuno FR 8250 or Corden Mk 2; I-band.
Weapon control systems: Elop MSIS optronic director; Typhoon GFCS.

Comment: Built by Colombo Dockyard to the Israeli Shaldag design. Deliveries of Mk I (P 450–451) began in 1996 and of Mk II (P 490–492, P 494, P 497) began in 1997. P 493 and P 496 sunk in action in 2000. Deliveries of Mk III (P 410–415, P 417, P 419–424) began in 2000 and of Mk IV (P 430, 432–439) in 2005. P 418 sunk in action in May 2006, P 438 in March 2008 and P 434 in 2009.

P 412   *3/2013*, Bob Fildes /* 1529076

COLOMBO MK III   *6/2006, Sri Lanka Navy /* 1164411

3 COASTAL PATROL CRAFT (PB)

P 240–242

Displacement, tonnes: 21 full load
Dimensions, metres (feet): 14.2 × 3.9 × 1 *(46.6 × 12.8 × 3.3)*
Speed, knots: 20
Range, n miles: 450 at 14 kt
Complement: 15 (1 officer)
Machinery: 2 Detroit 8V-71TA diesels; 460 hp *(343 kW)*; 2 shafts
Guns: 2—12.7 mm MGs.
Radars: Surface search: Furuno FR 2010; I-band.

Comment: Built by Colombo DY and commissioned in 1982 *(P 240)*, June 1986 *(P 241)* and 1993 *(P 242)*.

IHS Jane's Fighting Ships 2014-2015

P 201   *6/2003, Sri Lanka Navy /* 0570987

6 SHALDAG CLASS (FAST ATTACK CRAFT—GUN) (PBF)

P 470 (ex-P 491)   P 471 (ex-P 492)   P 472–475

Displacement, tonnes: 59 full load
Dimensions, metres (feet): 24.8 × 6 × 1.2 *(81.4 × 19.7 × 3.9)*
Speed, knots: 50
Range, n miles: 700 at 32 kt
Complement: 20

Machinery: 2 Deutz 620 TB 16V diesels; 5,000 hp(m) *(3.68 MW)*; 2 LIPS or MJP water-jets
Guns: 1 Rafael Typhoon 23 mm. 1—20 mm. 2—12.7 mm MGs. 6—7.62 mm MGs. 2—40 mm AGL.
Radars: Surface search: MD 3220 Mk II; I-band.
Weapon control systems: ELOP compass optronic director. Typhoon GFCS.

Comment: Originally launched in December 1989, first one acquired from Israel Shipyard, Haifa on 24 January 1996, second 20 July 1996 and third on 16 February 2000. Four more followed. Same hull used for the Colombo class. Also in service in Cyprus. P 476 sunk on 7 January 2006.

SHALDAG CLASS   *6/2003, Sri Lanka Navy /* 0570989

4 SUPER DVORA MK II CLASS
(FAST ATTACK CRAFT—GUN) (PBF)

P 460 (ex-P 441)   P 462 (ex-P 497)   P 464–465

Displacement, tonnes: 65 full load
Dimensions, metres (feet): 25 × 5.6 × 1.1 *(82 × 18.4 × 3.6)*
Speed, knots: 50
Range, n miles: 700 at 30 kt
Complement: 20 (1 officer)

Machinery: 2 MTU 12V 396 TE94 diesels; 4,570 hp(m) *(3.36 MW)*; ASD16 surface drives
Guns: 1 Rafael Typhoon 23 mm. 1—20 mm. 4—12.7 mm MGs. 6—7.62 mm MGs. 2—40 mm AGL.
Radars: Surface search: Koden MD 3220; I-band.
Weapon control systems: Elop MSIS optronic director; Typhoon GFCS.

Comment: First four ordered from Israel Aircraft Industries Ramta in early 1995. A slightly larger version of the Mk 1. First one delivered 5 November 1995, second 30 April 1996, third 22 June 1996 and fourth in December 1996. Two more were acquired on 9 June 1999 and 15 September 1999 respectively. The engines are an improved version of those fitted in the Israeli Navy craft. P 463 sunk in action in 2000. P 461 reported lost in action.

SUPER DVORA Mk II   *11/1999 /* 0080697

© 2014 IHS

Patrol forces < **Sri Lanka** 789

2 DVORA CLASS (FAST ATTACK CRAFT—GUN) (PBF)

P 401 (ex-P 420, ex-P 453)   P 402

Displacement, tonnes: 48 full load
Dimensions, metres (feet): 21.6 × 5.5 × 1.8 (70.9 × 18.0 × 5.9)
Speed, knots: 36
Range, n miles: 1,200 at 17 kt
Complement: 18
Machinery: 2 MTU 12V 331 TC81 diesels; 2,605 hp(m) (1.91 MW) sustained; 2 shafts
Guns: 2 Oerlikon 20 mm. 2—12.7 mm MGs. 6—7.62 mm MGs. 2—40 mm AGL.
Radars: Surface search: Anritsu 721UA; I-band.

Comment: 'Dvora' class, first pair of which transferred from Israel early 1984, next four in October 1986. Built by Israel Aircraft Industries. One sunk by Tamil forces on 29 August 1995 and second on 30 March 1996. P 403 transferred to the Coast Guard in 2013. One more deleted in late 1996. Not downgraded to patrol craft as previously reported but speed may have been reduced.

P 401   3/2013*, Bob Fildes / 1529077

4 CHEVERTON CLASS (COASTAL PATROL CRAFT) (PB)

P 243 (ex-P 221)   P 244 (ex-P 222)   P 245 (ex-P 223)   P 246 (ex-P 224)

Displacement, tonnes: 22 full load
Dimensions, metres (feet): 17 × 4.5 × 1.2 (55.8 × 14.8 × 3.9)
Speed, knots: 23
Range, n miles: 1,000 at 12 kt
Complement: 15
Machinery: 2 Detroit 8V-71TA diesels; 460 hp (343 kW); 2 shafts
Guns: 1—12.7 mm MG.
Radars: Surface search: Racal Decca 110; I-band.

Comment: Used for general patrol duties. Built by Cheverton Workboats, UK and commissioned in 1977. One paid off in 1996.

CHEVERTON CLASS   6/2004, Sri Lanka Navy / 1044192

4 INSHORE PATROL CRAFT (TYPE CME-OBM) (PBR)

P 110–113

Displacement, tonnes: 4 full load
Dimensions, metres (feet): 12.8 × 2.4 × 0.5 (42 × 7.9 × 1.6)
Speed, knots: 30
Complement: 9
Machinery: 2 outboard motors; 280 hp (209 kW)
Guns: 1—12.7 mm MG.

Comment: Built by Consolidated Marine Engineers in 1986.

INSHORE PATROL CRAFT   6/2004, Sri Lanka Navy / 1044190

26 INSHORE PATROL CRAFT (PBR)

P 106–107   P 114–124   P 126–138

Displacement, tonnes: 10 full load
Dimensions, metres (feet): 13.5 × 3 × 0.5 (44.3 × 9.8 × 1.6)
Speed, knots: 33
Range, n miles: 330 at 25 kt
Complement: 10
Machinery: 2 Cummins 6BTA5.9-M2; 584 hp (436 kW) sustained; 2 water-jets
Guns: 2—12.7 mm MGs. 4—7.62 mm MGs.
Radars: Surface search: Furuno 1941; I-band.

Comment: First pair (P 106, 107) built by TAOS Yacht Company, Colombo, and delivered in 1991. Next 25 (P 114–138) built by Blue Star Marine, Colombo and delivered between 1994 and 1998. There are minor superstructure differences between the first pair and the rest. A number were lost in action and/or have been decommissioned. P 125 transferred to the Coast Guard in 2013.

P 107   3/2013*, Bob Fildes / 1529078

33 INSHORE PATROL CRAFT (TYPE CME) (PBR)

P 140–146   P 148–161   P 163–166   P 170   P 172–173   P 175–179

Displacement, tonnes: 5 full load
Dimensions, metres (feet): 13.4 × 3 × 0.5 (44 × 9.8 × 1.6)
Speed, knots: 26
Complement: 5
Machinery: 2 Yamaha D 343 diesels; 730 hp(m) (544 kW) sustained; 2 shafts
Guns: 1—12.7 mm MG. 1—7.62 mm MG.
Radars: Surface search: Furuno FR 1941; I-band.

Comment: Built by Consolidated Marine Engineers, Sri Lanka. First nine delivered in 1988; four more in 1992 and two more in 1994. P 171 and P 174 transferred to the Coast Guard in 2013.

P 140   10/2013*, Susie Best / 1529079

1 VIKRAM CLASS (OFFSHORE PATROL VESSEL) (PSOH)

Name	No	Builders	Launched	Commissioned	Recommissioned
SAGARA	P 622	Goa Shipyard	5 Nov 1990	11 Mar 1992	25 Feb 2007
(ex-Varaha)	(ex-41)	Ltd, Goa			

Displacement, tonnes: 1,244 full load
Dimensions, metres (feet): 74.1 × 11.4 × 3.2 (243.1 × 37.4 × 10.5)
Speed, knots: 22
Range, n miles: 4,250 at 12 kt
Complement: 96 (11 officers)

Machinery: 2 SEMT-Pielstick 16 PA6V280 diesels; 12,800 hp (9.41 MW) sustained; 2 shafts; cp props
Guns: 1—30 mm. 1—23 mm.
Radars: Surface search/navigation: Furuno 2127; I-band.
Helicopters: Platform for 1 light.

Comment: Former Indian Coast Guard ships. *Sagara* leased and recommissioned on 25 February 2007. *Sayurala* loaned on 28 August 2009 and returned to Indian Coast Guard on 23 January 2011. It is likely that *Sagara* will be returned to the Indian Coast Guard but the timescale has not been announced.

SAGARA   6/2008, Sri Lanka Navy / 1335872

© 2014 IHS   IHS Jane's Fighting Ships 2014-2015

790 Sri Lanka › Patrol forces — Amphibious forces

15 MK III INSHORE PATROL CRAFT (PBF)

P 010–016   P 018–019   P 021   P 050–051   P 053–055

Displacement, tonnes: 8 full load
Dimensions, metres (feet): 14.2 × ? × ? (46.6 × ? × ?)
Speed, knots: 38
Complement: 8
Machinery: 2 Yanmar 6LY2A STP diesels; 737 hp(m) (550 kW); 2 shafts
Guns: 1—23 mm; 1—14.5 mm. 2—7.62 mm MGs.
Radars: Surface search: Furuno; I-band.

Comment: Built by Sri Lanka Navy. P 052 transferred to the Coast Guard in 2013. A further two vessels were handed over on 11 December 2013.

P 012   6/2008, Sri Lanka Navy / 1335870

168 INSHORE PATROL CRAFT (PBF)

Z 101–137   Z 139–141   Z 143–250   X 101   X 103–121

Displacement, tonnes: 3 full load
Dimensions, metres (feet): 7 × ? × ? (23 × ? × ?)
Speed, knots: 40
Complement: 8
Machinery: 2 outboard motors; 400 hp(m) (298 kW)
Guns: 1—23 mm or 1—14.5 mm. 4—7.62 mm MGs.

Comment: Built by Sri Lanka Navy.

Z 101   6/2008, Sri Lanka Navy / 1335869

6 SUPER DVORA MK III CLASS (PBF)

P 4442–4447

Displacement, tonnes: 73 full load
Dimensions, metres (feet): 27.4 × 5.7 × 1.1 (89.9 × 18.7 × 3.6)
Speed, knots: 45
Complement: 18
Machinery: 2 MTU 12V 400 M90 diesels; 2 Arneson ASD16 surface drives
Guns: 1—30 mm; 1—23 mm; 2—40 mm AGL; 4—12.7 mm MGs.
Radars: Surface search: Furuno; I-band.
Weapon control systems: To be announced.

Comment: A contract for the construction of six craft was made from IAI Ramta on 6 November 2008. Four craft were delivered by December 2009 and a further two were delivered in 2011.

DVORA III   10/2009, Sri Lanka Navy / 1367098

27 WAVE RIDER (INSHORE PATROL CRAFT) (PBF)

P 201–207   P 210–225   P 227–230

Displacement, tonnes: 10 full load
Dimensions, metres (feet): 14.57 × 3.17 × 1.67 (47.8 × 10.4 × 5.5)
Speed, knots: 31
Complement: 8
Machinery: 2 Yanmar 6LY2A STP diesels; 757 hp(m) (550 kW); 2 shafts
Guns: 2—23 mm (twin); 1—14.5 mm; 1—40 mm AGL; 3—12.7 mm MGs.

Comment: Built by Sri Lanka Navy.

WAVE RIDER   10/2009, Sri Lanka Navy / 1367096

AMPHIBIOUS FORCES

1 YUHAI (WUHU-A) (TYPE 074) CLASS (LSM)

Name	No	Builders	Commissioned
SHAKTHI	L 880	China	22 May 1996

Displacement, tonnes: 812 full load
Dimensions, metres (feet): 58.4 × 10.4 × 2.7 (191.6 × 34.1 × 8.9)
Speed, knots: 14
Range, n miles: 1,000 at 12 kt
Complement: 60
Military lift: 150 tons
Machinery: 2 MAN 8 L 20/27 diesels; 4,900 hp(m) (3.6 MW); 2 shafts
Guns: 10—14.5 mm/93 (5 twin) MGs. 6—12.7 mm MGs.
Radars: Navigation: Racal Decca; I-band.

Comment: Transferred by lift ship from China arriving 13 December 1995. A planned second of class was built but not acquired.

SHAKTHI   10/2009, Sri Lanka Navy / 1367095

2 LANDING CRAFT (LCM)

Name	No	Builders	Commissioned
RANAGAJA	L 839	Colombo Dockyard	15 Nov 1991
RANAVIJAYA	L 836	Colombo Dockyard	21 Jul 1994

Displacement, tonnes: 272 full load
Dimensions, metres (feet): 33 × 8 × 1.5 (108.3 × 26.2 × 4.9)
Speed, knots: 8
Range, n miles: 1,800 at 8 kt
Complement: 28 (2 officers)
Machinery: 2 Caterpillar diesels; 1,524 hp (1.14 MW); 2 shafts
Guns: 4 China 14.5 mm (2 twin). 2—12.7 mm MGs.
Radars: Nav Furuno FCR 1421; I-band.

Comment: Two built in 1983 and acquired in October 1985. Third of the class taken over by the Navy in September 1991 and a fourth in March 1992. *Kandula* sank in October 1992 and the hulk was salvaged in mid-December. *Pabbatha* sank in action in February 1998.

RANAVIJAYA   6/2008, Sri Lanka Navy / 1335867

Amphibious forces < **Sri Lanka** — Patrol forces < **Sri Lanka** (COAST GUARD) 791

2 YUNNAN CLASS (TYPE 067)

L 820   L 821

Displacement, tonnes: 137 full load
Dimensions, metres (feet): 28.6 × 5.4 × 1.5 *(93.8 × 17.7 × 4.9)*
Speed, knots: 12. **Range, n miles:** 500 at 10 kt
Complement: 22 (2 officers)

Military lift: 46 tons
Machinery: 2 diesels; 600 hp(m) *(441 kW)*; 2 shafts
Guns: 4 — 14.5 mm (2 twin) MGs. 2 — 7.62 mm MGs
Radars: Surface search: Fuji; I-band.

Comment: First one acquired from China in May 1991, second in May 1995.

L 821   *6/2008, Sri Lanka Navy* / 1335868

1 M 10 CLASS HOVERCRAFT (UCAC)

A 530

Displacement, tonnes: 18 full load
Dimensions, metres (feet): 20.6 on cushion × ? *(67.6 × ?)*
Speed, knots: 40. **Range, n miles:** 600 at 30 kt
Complement: 10

Military lift: 56 troops or 20 troops plus 2 vehicles
Machinery: 2 Deutz diesels; 1,050 hp(m) *(772 kW)*
Guns: 1 — 12.7 mm MG.
Radars: Navigation: Furuno; I-band.

Comment: Acquired from ABS Hovercraft/Vosper Thornycroft in April 1998 and designated a Utility Craft Air Cushion (UCAC). Has a Kevlar superstructure. More may be ordered in due course.

A 530   *6/2006, Sri Lanka Navy* / 1164409

3 FAST PERSONNEL CARRIERS (LCP)

Name	No	Builders	Commissioned
HANSAYA (ex-*Offshore Pioneer*)	A 540	Sing Koon Seng	20 Dec 1987
— (ex-*Lanka Rani*)	A 542	Kvaerner Fielistrand Ltd	2000
— (ex-*Lanka Devi*)	A 543	Kvaerner Fielistrand Ltd	2000

Displacement, tonnes: 451 full load
Dimensions, metres (feet): 40 × 10.1 × 1.8 *(131.2 × 33.1 × 5.9)*
Speed, knots: 30. **Range, n miles:** 650 at 20 kt
Complement: 30 (4 officers)

Military lift: 60 tons; 120 troops
Machinery: 2 MTU 16V 396 TE 74L; 1,800 hp(m) *(1.32 MW)*; 2 shafts
Guns: 1 Oerlikon 20 mm. 2 — 12.7 mm MGs.
Radars: Navigation: Furuno FR 1012; I-band.

Comment: A 540 acquired in January 1986 from Aluminium Shipbuilders. Catamaran hull built as oil rig tender. Now used as fast transport. A 541 decommissioned in 2002. Details are as for A 543 which was acquired from Ceylon Shipping Corporation.

A 543   *6/2006, Sri Lanka Navy* / 1164408

© 2014 IHS

AUXILIARIES

1 SUPPORT/TRAINING SHIP (AA/AX)

Name	No	Builders	Commissioned
— (ex-*Simon Keghian*)	A 521	BPKSKP, Gdynia	1972

Displacement, tonnes: 601 full load
Dimensions, metres (feet): 54 × 11.0 × 6.3 *(177.2 × 36.1 × 20.7)*
Speed, knots: 10
Range, n miles: 5,500 at 9 kt
Complement: 57 (7 officers)
Machinery: 1 DUVANT CREPELLE 8R26L diesel; 1,320 hp(m) *(970 kW)* sustained; 1 shaft
Guns: 1 — 20 mm. 10 — 12.7 mm MGs. 2 — 40 mm AGLs.
Radars: Surface search/navigation: Furuno FR 2125; I-band.

Comment: Former deep-sea fishing trawler donated by the Lorient-Matara Friendship Foundation of France and commissioned into the Sri Lanka Navy on 26 April 2005. The ship was donated on humanitarian grounds, following the tsunami of 26 December 2004, and it is understood that the vessel is used by the navy in support of fishing activities and as a training vessel.

A 521   *10/2009, Sri Lanka Navy* / 1367094

COAST GUARD

Notes: The Coast Guard was established by Act of Parliament on 9 July 2009 and became operational on 4 March 2010. Its roles include law enforcement, fishery protection, counter-piracy, -human trafficking, -narcotics, SAR, and environmental monitoring and protection.

Bases

CG HQ: Mirissa
Western Region HQ: Panadura
Southern Region HQ: Mirissa
Norther Region HQ: Kankasanturai.

PATROL FORCES

3 MK III INSHORE PATROL CRAFT (PBF)

CG 14 (ex-P 052)   CG 19   CG 22

Displacement, tonnes: 8 full load
Dimensions, metres (feet): 14.2 × ? × ? *(46.6 × ? × ?)*
Speed, knots: 34
Complement: 10
Machinery: 2 Yanmar 6LY2A STP diesels; 737 hp(m) *(550 kW)*; 2 shafts
Guns: 1 — 7.62 mm MG.
Radars: Surface search: Furuno; I-band.

Comment: Built by Sri Lanka Navy. CG 14 handed over to Coast Guard in 2012. CG 19 and 22 in December 2013.

1 INSHORE PATROL CRAFT (PBR)

CG 10 (ex-P 125)

Displacement, tonnes: 10 full load
Dimensions, metres (feet): 13.5 × 3 × 0.5 *(44.3 × 9.8 × 1.6)*
Speed, knots: 18
Range, n miles: 330 at 25 kt
Complement: 9
Machinery: 2 Cummins 6BTA5.9-M2; 584 hp *(436 kW)* sustained; 2 water-jets
Guns: 2 — 12.7 mm MGs.
Radars: Surface search: Furuno 1941; I-band.

Comment: Built by Blue Star Marine, Colombo and delivered between 1994 and 1998. Handed over to the Coast Guard in 2012.

1 DVORA CLASS (FAST ATTACK CRAFT — GUN) (PBF)

CG 46 (ex-P 403)

Displacement, tonnes: 47 full load
Dimensions, metres (feet): 21.6 × 5.5 × 1.8 *(70.9 × 18.0 × 5.9)*
Speed, knots: 32
Range, n miles: 1,200 at 17 kt
Complement: 18
Machinery: 2 MTU 12V 396 TB93 diesels; 2,605 hp(m) *(1.91 MW)* sustained; 2 shafts
Guns: 1 — 12.7 mm MG. 2 — 7.62 mm MGs.
Radars: Surface search: Furuno 2127; I-band.

Comment: Built by Israel Aircraft Industries in the 1980s. Transferred to the Coast Guard in 2012.

IHS Jane's Fighting Ships 2014-2015

792 **Sri Lanka** (COAST GUARD) > Patrol forces — **Sudan** > Introduction

2 INSHORE PATROL CRAFT (TYPE CME) (PBR)

CG 18 (ex-P 171)   – (ex-P 174)

Displacement, tonnes: 5 full load
Dimensions, metres (feet): 13.4 × 3 × 0.5 (44 × 9.8 × 1.6)
Speed, knots: 26
Complement: 5
Machinery: 2 Yamaha D 343 diesels; 730 hp(m) (544 kW) sustained; 2 shafts
Guns: 1 — 12.7 mm MG. 1 — 7.62 mm MG.
Radars: Surface search: Furuno FR 1941; I-band.

Comment: Built by Consolidated Marine Engineers, Sri Lanka 1988–94. Two craft handed over to the Coast Guard in 2012.

CG 18   6/2013*, Sri Lanka Coast Guard / 1529075

4 SUPER DVORA MK I CLASS
(FAST ATTACK CRAFT — GUN) (PBF)

– (ex-P 440)   – (ex-P 441)   – (ex-P 442)   – (ex-P 443)

Displacement, tonnes: 55 full load
Dimensions, metres (feet): 22.4 × 5.5 × 1.8 (73.5 × 18.0 × 5.9)
Speed, knots: 46
Range, n miles: 1,200 at 17 kt
Complement: 20 (1 officer)
Machinery: 2 MTU 12V 396 TB93 diesels; 3,260 hp(m) (2.4 MW) sustained; 2 shafts
Guns: 2 — 12.7 mm MGs.
Radars: Surface search: Decca 926; I-band.

Comment: Ordered from Israel Aircraft Industries in October 1986 and delivered in 1987–88. The last four of these craft were transferred from the Navy to the Coast Guard in 2012.

SUPER DVORA MK I CLASS   1995, Sri Lanka Navy / 0130147

2 SIMONNEAU CLASS (PBF)

P 252 (ex-P 412, ex-P 485)   **P 253** (ex-P 413, ex-P 486)

Displacement, tonnes: 28 full load
Dimensions, metres (feet): 17.3 × 4.9 × 1.4 (56.8 × 16.1 × 4.6)
Speed, knots: 42
Range, n miles: 500 at 35 kt
Complement: 15
Machinery: 2 MTU 12V 183 TE93 diesels; 2,300 hp(m) (1.69 MW); 2 Hamilton water-jets
Guns: 2 — 12.7 mm MGs. 2 — 7.62 mm MGs.
Radars: Surface search: Racal Decca; I-band.

Comment: Simonneau Marine Type 508 craft built in 1994–95. Two of the last remaining naval craft were transferred to the Coast Guard in 2012.

SIMONNEAU CLASS   6/2004, Sri Lanka Navy / 1044191

3 SOUTH KOREAN KILLER CLASS
(FAST ATTACK CRAFT — GUN) (PBF)

CG 40 (ex-P 404)   **CG 41** (ex-P 405)   – (ex-P 406)

Displacement, tonnes: 57 full load
Dimensions, metres (feet): 23 × 5.4 × 1.8 (75.5 × 17.7 × 5.9)
Speed, knots: 34
Complement: 18
Machinery: 2 MTU 396 TB93 diesels; 3,260 hp(m) (2.4 MW) sustained; 2 shafts
Guns: 2 — 12.7 mm MGs. 2 — 7.62 mm MGs.
Radars: Surface search: Furuno 1510; I-band.

Comment: 'South Korean Killer' class, built by Korea SB and Eng, Buson. All commissioned in naval service, February 1988, and subsequently transferred to the Coast Guard in 2012.

CG 41   6/2013*, Sri Lanka Coast Guard / 1529074

Sudan

Country Overview

The Republic of Sudan is situated in north-eastern Africa. The third largest country in Africa, it has an area of 718,723 square miles and is bordered to the north by Egypt, to the east by Eritrea and Ethiopia, to the south by South Sudan and to the west by the Central African Republic, Chad, and Libya. It has a 459 n mile coastline with the Red Sea. Khartoum is the capital and largest city and Port Sudan is the principal port. There are about 2,867 n miles of navigable waterways. Territorial waters (12 n miles) are claimed. An EEZ has not been claimed.

The country was ravaged by civil war 1983–2005 but a Comprehensive Peace Agreement was finally concluded on 9 January 2005. This allowed for the south to become a self-administering region. Following a referendum in January 2011 in which the majority voted for independence, a new country of South Sudan was established on 9 July 2011.

Naval Forces are part of the Army and have low budgetary priority.

Headquarters Appointments

Commander, Naval Forces:
  Lieutenant General Daleel al-Daw Mohamed Fadlalla

Personnel

(a) 2014: 1,300 officers and men
(b) Voluntary service

Establishment

The Navy was established in 1962 to operate on the Red Sea coast and on the River Nile.

Bases

Port Sudan (HQ). Flamingo Bay (Red Sea), Khartoum (Nile), Kosti (Nile).

IHS Jane's Fighting Ships 2014-2015   © 2014 IHS

PATROL FORCES

4 KURMUK (TYPE 15) CLASS (INSHORE PATROL CRAFT) (PBR)

| KURMUK 502 | QAYSAN 503 | RUMBEK 504 | MAYOM 505 |

Displacement, tonnes: 20 full load
Dimensions, metres (feet): 16.9 × 3.9 × 0.7 (55.4 × 12.8 × 2.3)
Speed, knots: 16
Range, n miles: 160 at 12 kt
Complement: 6
Machinery: 2 diesels; 330 hp(m) (243 kW); 2 shafts
Guns: 1 Oerlikon 20 mm; 2 — 7.62 mm MGs.

Comment: Delivered by Yugoslavia on 18 May 1989 for operations on the White Nile. All based at Flamingo Bay.

KURMUK 1989, G Jacobs / 0506101

4 SEWART CLASS (INSHORE PATROL CRAFT) (PBR)

| MAROUB 1161 | FIJAB 1162 | SALAK 1163 | HALOTE 1164 |

Displacement, tonnes: 9 full load
Dimensions, metres (feet): 12.2 × 3.7 × 1 (40 × 12.1 × 3.3)
Speed, knots: 31
Complement: 6
Machinery: 2 GM diesels; 348 hp (260 kW); 2 shafts
Guns: 1 — 12.7 mm MG.

Comment: Transferred by Iranian Coast Guard in 1975. All are based at Flamingo Bay but operational status is doubtful.

7 ASHOORA I CLASS (INSHORE PATROL CRAFT) (PBR)

Displacement, tonnes: 3 full load
Dimensions, metres (feet): 8.1 × 2.4 × 0.5 (26.6 × 7.9 × 1.6)
Speed, knots: 42
Complement: 2
Machinery: 2 Yamaha outboards; 400 hp(m) (294 kW)
Guns: 1 — 7.62 mm MG.

Comment: Acquired from Iran in 1992–94. Four based at Flamingo Bay and three at Khartoum but operational status is doubtful.

ASHOORA I 1992, IRI Marine Industries / 0080715

AUXILIARIES

Notes: (1) In addition there are two small miscellaneous support ships. *Baraka* 21 a water boat, and a Rotork 512 craft. Both restored with Iranian assistance.
(2) Five Type II LCVPs were delivered from Yugoslavia in 1991 and are based at Kosti.

2 SUPPLY SHIPS (AFL)

| SOBAT 221 | DINDER 222 |

Displacement, tonnes: 417 full load
Dimensions, metres (feet): 47.3 × 6.4 × 2.3 (155.2 × 21.0 × 7.5)
Speed, knots: 9
Complement: 15
Machinery: 3 Gray Marine diesels; 495 hp (369 kW); 3 shafts
Guns: 1 Oerlikon 20 mm. 2 — 12.7 mm MGs.

Comment: Two Yugoslav MFPD-class LCTs transferred in 1969. Used for transporting ammunition, petrol and general supplies.

Suriname

Country Overview

Formerly known as Dutch Guiana, the Republic of Suriname gained full independence in 1975. With an area of 63,037 square miles it has borders to the east with French Guiana, to the west with Guyana and to the south with Brazil; its 208 n mile coastline is on the Atlantic Ocean. The capital, largest city and chief port is Paramaribo. Territorial seas (12 n miles) and a fisheries zone (200 n miles) are claimed. There are further ports at Nieuw-Nickerie, Moengo, Paranam and Smalkalden. Territorial waters (12 n miles) are claimed. A 200 n mile Exclusive Economic Zone (EEZ) has also been claimed but the limits are not defined.

A Coast Guard was formally established on 1 September 2013. The new service is responsible to the Ministry of Internal Affairs and is staffed by personnel transferred from the Marine Section of the army.

Headquarters Appointments

Commander Coast Guard:
   Colonel Jerry Slijngaard

Personnel

2014: 240 (25 officers)

Bases

Kruktu Tere, Paramaribo

Aircraft

Two CASA C-212–400 Aviocar aircraft acquired for maritime patrol in 1998/99.

PATROL FORCES

3 RODMAN 101 CLASS (PB)

| JARABAKKA P 01 | SPARI P 02 | GRAMORGU P 03 |

Displacement, tonnes: 73 full load
Dimensions, metres (feet): 30 × 5.9 × 1.3 (98.4 × 19.4 × 4.3)
Speed, knots: 26
Range, n miles: 800 at 12 kt
Complement: 9
Machinery: 2 MTU 12V 2000 diesels; 2,900 hp (2.16 MW) sustained; 2 Hamilton 571 water-jets
Guns: 1 — 40 mm grenade launcher.
Radars: Surface search: 2 Furuno; I-band.

Comment: Ordered in December 1997, from Rodman, Vigo. First one delivered in February 1999, second and third on 3 July 1999. Carry a RIB with twin outboards. Operational status doubtful.

SPARI 3/2001, Adolfo Ortigueira Gil / 1305139

5 RODMAN 55M CLASS (PBR)

P 04–08

Displacement, tonnes: 16 full load
Dimensions, metres (feet): 17.4 × 3.9 × 0.7 (57.1 × 12.8 × 2.3)
Speed, knots: 35
Range, n miles: 500 at 25 kt
Complement: 7
Machinery: 2 MAN D2848-LXE diesels; 1,360 hp(m) (1 MW) sustained; 2 Hamilton water-jets
Guns: 1 — 12.7 mm MG.
Radars: Surface search: Furuno; I-band.

Comment: Ordered in December 1997 from Rodman, Vigo. First one delivered in October 1998, remainder in April 1999. Carry a RIB with a single outboard engine. Operational status doubtful.

P 06 4/1999, Rodman Group / 1305140

1 PROJECT 414 (PBO)

– (ex-*Karl Heinz*; ex-*Hai*) T 001

Measurement, tonnes: 205 gt
Dimensions, metres (feet): 30.6 × 8.4 × 2.5 *(100.4 × 27.6 × 8.2)*
Speed, knots: 11
Complement: 15
Machinery: 1 SKL 6VD26/20-AL-1 diesel; 721 hp *(530 kW)*; 1 shaft; cp prop
Guns: To be announced.
Radars: Navigation: To be announced.

Comment: Built at Yachtwerft, Berlin, in the former Democratic Republic of Germany, the vessel was completed in November 1990, just after unification of the two Germanys as a single state. Not required by the Federal German Navy, the ship operated as a commercial tug until laid up in the Netherlands in about 1996. After repairs, she was sold to Suriname in 2000 and subsequently employed as a patrol ship.

T 001   *6/2011* / 1406427

1 LARGE PATROL CRAFT (PB)

P 401 (ex-S 401)

Displacement, tonnes: 140 full load
Dimensions, metres (feet): 32 × 6.5 × 1.7 *(105 × 21.3 × 5.6)*
Speed, knots: 17.5. **Range, n miles:** 1,200 at 13 kt
Complement: 15
Machinery: 2 Paxman 12YHCM diesels; 2,110 hp *(1.57 MW)*; 2 shafts
Guns: 2 Bofors 40 mm. 2 – 7.62 mm MGs.
Radars: Navigation: Decca; I-band.

Comment: Surviving craft of three built by De Vries, Aalsmeer, Netherlands and commissioned in 1976–77. The fate of the other two craft is uncertain but one has been reported converted into a private yacht.

2 OCEA FPB 72 MK II (PATROL CRAFT) (PB)

P 101   P 102

Dimensions, metres (feet): 23.6 × 5.8 × 1.5 *(77.4 × 19.0 × 4.9)*
Speed, knots: 32
Complement: 11
Machinery: 2 MTU diesels; 3,000 hp *(2.24 MW)*; 2 shafts
Guns: 2 – 7.62 mm MGs.
Radars: Surface search/navigation: To be announced.

Comment: The contract with OCEA for the construction of two patrol craft was announced in late 2012. Delivery of both vessels was made in 2013.

P 101   *6/2013*, B Prézelin* / 1525554

1 OCEA FPB 98 CLASS (PATROL CRAFT) (PB)

P 201

Displacement, tonnes: 116 full load
Dimensions, metres (feet): 35.2 × 6.8 × 1.2 *(115.5 × 22.3 × 3.9)*
Speed, knots: 30
Range, n miles: 1,000 at 12 kt
Complement: 12 (3 officers)
Machinery: 2 Caterpillar diesels; 3,586 hp *(2.6 MW)*; 2 Kamewa waterjets
Guns: 1 – 20 mm. 2 – 12.7 mm MGs.
Radars: Navigation: Furuno FAR 2117; I-band.

Comment: The contract with OCEA for the construction of one patrol craft was announced in late 2012. Delivery of the first vessel was made in May 2013. Similar vessels have been sold to Algeria, Benin, and Senegal.

P 201   *6/2013*, B Prézelin* / 1525553

Sweden

SVENSKA MARINEN

Country Overview

The Kingdom of Sweden is a constitutional monarchy occupying the eastern part of the Scandinavian Peninsula. With an area of 173,730 square miles, it is bordered to the north and west by Norway and to the north-east by Finland. It has a 1,740 n mile coastline with the Gulf of Bothnia, the Baltic Sea, the Öresund, Kattegatt, and Skagerrak. The country comprises the mainland and the islands of Gotland and Öland in the Baltic Sea. The capital and largest city is Stockholm which is also a leading port. Others include Göteborg, Malmö and Norrköping. Territorial seas (12 n miles) and an EEZ (200 n miles) are claimed.

Headquarters Appointments

Chief of Naval Staff: Rear Admiral Jan Thörnquist

Diplomatic Representation

Defence Attaché in Washington and Ottawa:
  Rear Admiral Jörgen Ericsson
Defence Attaché in Tel Aviv:
  Lieutenant Colonel Björn Blomberg
Defence Attaché in Beijing and Pyongyang:
  Colonel Martin Bodin
Defence Attaché in Riga:
  Lieutenant Colonel Magnus Lundgren
Defence Attaché in Talinn:
  Commander Peter Laurin
Defence Attaché in Warsaw:
  Colonel Claes Nilsson
Defence Attaché in Vilnius:
  Lieutenant Colonel Tapani Mattus
Defence Attaché in Addis Ababa, Khartoum and Nairobi:
  Lieutenant Colonel Sven-Åke Wickström
Naval Attaché in Washington and Mexico City:
  Captain Magnus Lüning

Diplomatic Representation – continued

Defence Attaché in London: Colonel Mats Danielsson
Defence Attaché in Moscow and Minsk:
  Captain Håkan Andersson
Defence Attaché in Paris and Madrid:
  Colonel Jonas Olsson
Defence Attaché in Singapore, Bangkok and Brunei Darussalam: Captain Lennart Bengtsson
Defence Attaché in Berlin and Vienna:
  Colonel Ult Gunnehed
Defence Attaché in Belgrade and Sarajevo:
  Lieutenant Colonel Thomas Klementsson
Defence Attaché in Helsinki: Colonel Bo Stenabb
Defence Attaché in Pretoria, Luanda and Kinshasa:
  Colonel Leif Tonn-Carlsson
Defence Attaché in Ankara and Tbilisi:
  Colonel Peter Adolfsson
Defence Attaché in Budapest, Zagreb and Sofia:
  Lieutenant Colonel Laci Bonivart
Defence Attaché in Kiev and Chisinau:
  Colonel Håkan Hedström
Defence Attaché in Islamabad and Kabul:
  Colonel Anders Waldén
Defence Attaché in New Delhi:
  Colonel Mats Wigselius
Defence Attaché in Prague, Bratislava and Bucharest:
  Lieutenant Colonel Per Råstedt
Defence Attaché in Canberra:
  Captain Karl Henriksson

Organisation

The Navy consists of the Fleet and the Amphibious Battalion (ex-Coastal Artillery). The Navy is organised into one submarine flotilla, two naval warfare flotillas, one amphibious regiment, one main naval base and one naval warfare centre.

Personnel

2014:   6,070 including 1,600 officers, 350 civilians, and 1,750 reserve officers

Bases

Karlskrona, Berga (Stockholm).

Strength of the Fleet

Type	Active	Building (Planned)
Submarines—Patrol	5	2
Missile Corvettes	9	—
Inshore Patrol Craft	14	—
Minesweepers/Hunters—Coastal	5	—
Minesweepers—Inshore	4	—
LCMs	9	—
Electronic Surveillance Ship	1	—
Repair and Support Ships	9	1

DELETIONS

Patrol Forces

2011   *Tapper*

Mine Warfare Forces

2010   *Landsort, Arholma*

Auxiliaries

2011   *Grundsund*
2012   *Fårösund*

Introduction — Submarines < **Sweden**

PENNANT LIST

Corvettes		Patrol Forces		91	Munter	M 74	Kullen	A 264	Trossö
K 11	Stockholm	82	Djärv	92	Orädd	M 75	Vinga	A 320	Furusund
K 12	Malmö	83	Dristig			M 76	Ven	A 322	Heros
K 22	Gävle	84	Händig	**Mine Warfare Forces**		M 77	Ulvön	A 324	Hera
K 24	Sundsvall	85	Trygg					A 344	Loke
K 31	Visby	86	Modig	M 11	Styrsö	**Auxiliaries**		P 04	Carlskrona
K 32	Helsingborg	87	Hurtig	M 12	Spårö	A 201	Orion		
K 33	Härnösand	88	Rapp	M 13	Skaftö	A 212	Ägir	**Training ships**	
K 34	Nyköping	89	Stolt	M 14	Sturkö	A 214	Belos III	S 01	Gladan
K 35	Karlstad	90	Ärlig	M 73	Koster	A 247	Pelikanen	S 02	Falken

SUBMARINES

Notes: A prototype Swimmer Delivery Vehicle (SDV) (called Diver Group Boat) began trials in late 2008; the 10.3 m carbon-fibre craft is capable of carrying six divers and 300 kg of equipment. There are three modes of operation: surface, semi-submerged (skimmer) and submerged. A 235 kW MTU diesel driving a waterjet provides propulsive power in surfaced and skimmer modes while four Tecnadyne thrusters powered by 24 Optima batteries provides underwater propulsion. Buoyancy is adjusted using inflatable pontoons on the sides of the craft. Delivery of the first SDV was reported in 2012.

3 GOTLAND (A 19) CLASS (SSK)

Name	Builders	Laid down	Launched	Commissioned
GOTLAND	Kockums, Malmö	20 Nov 1992	2 Feb 1995	2 Sep 1996
UPPLAND	Kockums, Malmö	14 Jan 1994	9 Feb 1996	1 May 1997
HALLAND	Kockums, Malmö	21 Oct 1994	27 Sep 1996	1 Oct 1997

Displacement, tonnes: 1,518 surfaced; 1,625 dived
Dimensions, metres (feet): 60.4 × 6.2 × 5.6
(198.2 × 20.3 × 18.4)
Speed, knots: 10 surfaced; 20 dived
Complement: 27 (5 officers)

Machinery: Diesel-Stirling-electric; 2 MTU 16V 396 SE84 diesels; 3,620 hp (2.7 MW); 2 Kockums V4-275R Stirling AIP; 204 hp(m) (150 kW); 1 Jeumont Schneider motor; 1 shaft; LIPS prop
Torpedoes: 4 − 21 in (533 mm) bow tubes; 12 FFV Type 613/62; anti-surface; wire-guided; passive homing to 20 km (10.8 n miles) at 45 kt; warhead 240 kg or Bofors Type 62 (2000); wire-guided; active/passive homing to 50 km (27 n miles) at 20–50 kt; warhead 250 kg. swim-out discharge. 2 − 15.75 in (400 mm) bow tubes; 6 Swedish Ordnance Type 432/451; anti-submarine; wire-guided; active/passive homing to 20 km (10.8 n miles) at 25 kt; warhead 45 kg.
Electronic countermeasures: ESM: Racal THORN Manta S; radar warning.
Radars: Navigation: Terma Scanter; I-band.
Sonars: STN/Atlas Elektronik CSU 90-2; hull-mounted; bow, flank and intercept arrays; passive search and attack. Reson Subac; active search (from 2008).
Weapon control systems: Saab SESUB 960B.

Programmes: In October 1986 a research contract was awarded to Kockums for a design to replace the Sjöormen class. Ordered on 28 March 1990.
Modernisation: A contract for the upgrade of the combat system of all three boats was placed with TKMS (Kockums) in late 2010. Halland was completed in 2011, Uppland in 2012, and Gotland in 2013. A further mid-life upgrade contract was let to TKMS in October 2012 to address end-of-life systems, upgrade sensor systems, installation of a new non-penetrating optronic mast, and installation of a pressurised divers' lock-out to facilitate special forces operations.
Structure: The design has been developed on the basis of the Type A 17 series but this class is the first to be built with Air Independent Propulsion as part of the design. This type of AIP runs on liquid oxygen and diesel in a helium environment. Space has been reserved to fit two more V4-275R engines in due course. Single electro-optic periscope. The periscope is the only hull penetrating mast. Anechoic coatings are being applied. The four 21 in torpedo tubes are mounted over the smaller 15.75 in tubes. The smaller tubes can be tandem-loaded with two torpedoes per tube.
Operational: Reported as being able to patrol at 5 kt for several weeks without snort charging.

UPPLAND   9/2013*, Martin Mokrus / 1531004

GOTLAND   9/2011, Frank Findler / 1455819

UPPLAND   9/2013*, Guy Toremans / 1531042

Sweden > Corvettes

2 GÄVLE CLASS (FSG)

Name	No	Builders	Laid down	Launched	Commissioned
GÄVLE	K 22	Karlskronavarvet	21 Mar 1988	23 Mar 1990	1 Feb 1991
SUNDSVALL	K 24	Karlskronavarvet	20 Nov 1989	29 Nov 1991	7 Jul 1993

Displacement, tonnes: 305 standard; 405 full load
Dimensions, metres (feet): 57 × 8 × 2 (187 × 26.2 × 6.6)
Speed, knots: 30
Complement: 36 (7 officers) + 4 spare berths

Machinery: 3 MTU 16V 396 TB94 diesels; 8,700 hp(m) *(6.4 MW)* sustained; Kamewa 80562-6 water-jets; bow thrusters
Missiles: SSM: 8 Saab RBS15 Mark II (4 twin) launchers ❶; inertial guidance; active radar homing to 110 km *(59.4 n miles)* at 0.8 Mach; warhead 150 kg.
Guns: 1 Bofors 57 mm/70 Mk 3 ❷; 220 rds/min to 17 km *(9.3 n miles)*; weight of shell 2.4 kg. 1 Bofors 40 mm/70 (stealth dome) ❸; 330 rds/min to 12.5 km *(6.8 n miles)*; weight of shell 0.96 kg. 4—12.7 mm MGs.
Torpedoes: 4—15.75 in *(400 mm)* tubes can be fitted ❹. Swedish Ordnance Type 43/45; anti-submarine.
Mines: Minelaying capability.
Depth charges: On mine rails.
Physical countermeasures: Decoys: Rheinmetal MASS-1L decoy system.
Electronic countermeasures: ESM: Condor CS 3701; intercept.
ECM: Rafael Shark/RAN-1101; jammer.
Radars: Air/surface search: Ericsson Sea Giraffe 150 HC ❺; G-band.
Navigation: Terma PN 621 ❻; I-band.
Fire control: 2 Bofors Electronics 9GR 400 ❼; I/J-band.
Sonars: Hydra multisonar system ❽; hull-mounted active high-frequency plus passive towed array and active VDS. Simrad SA 950 (K 24); hull-mounted; attack. Thomson-Sintra TSM 2643 VDS.
Combat data systems: CelsiusTech 9LV Mk 3 SESYM. Link 11.
Weapon control systems: 2 Bofors Electronics 9LV 200 Mk 3 Sea Viking (K 22) or Signaal IRST (K 24) optronic directors. Bofors Electronics 9LV 450 GFCS. RC1-400 MFCS. 9AU-300 ASW control system with AQS 928G/SM sonobuoy processor. Bofors 9EW 400 EW control.

Programmes: Ordered 1 December 1985 as replacements for Spica I class.
Modernisation: *Gävle* refitted to accommodate Hydra towed array, jammer and 40 mm gun with stealth dome. Bridge wings removed and topmast modified. *Sundsvall* has been refitted with a similar 40 mm gun and modified topmast. Both ships began a mid-life upgrade programme in 2013 and are to become known as the Gävle class.
Structure: Efforts have been made to reduce radar and IR signatures. Saab 601 A/S mortars have been removed.

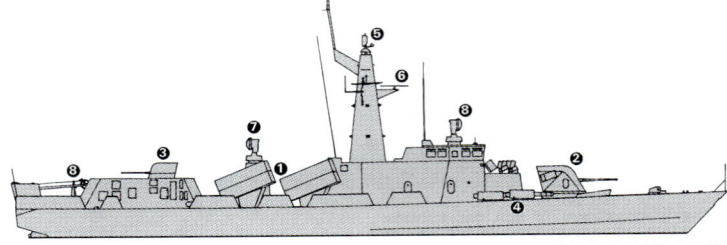

GÄVLE

(Scale 1 : 600), Ian Sturton / 1153882

SUNDSVALL

9/2011, L-G Nilsson / 1455664

GÄVLE

6/2011, Per Körnefeldt / 1406423

SUNDSVALL

4/2013*, Tony Roper / 1531045

IHS Jane's Fighting Ships 2014-2015

© 2014 IHS

Corvettes — Sweden 799

2 STOCKHOLM CLASS (FSG)

Name	No	Builders	Laid down	Launched	Commissioned
STOCKHOLM	K 11	Karlskronavarvet	1 Aug 1982	24 Aug 1984	22 Feb 1985
MALMÖ	K 12	Karlskronavarvet	14 Mar 1983	22 Mar 1985	10 May 1985

Displacement, tonnes: 356 standard; 378 full load
Dimensions, metres (feet): 50 × 7.5 × 3.3
(164 × 24.6 × 10.8)
Speed, knots: 32; 20 diesel
Complement: 33 (7 officers)

Machinery: CODAG; 1 Allied Signal TF50A gas turbine; 5,440 hp(m) (4.0 MW) sustained; 2 MTU 16V 396 TB94 diesels; 5,277 hp(m) (3.9 MW) sustained; 3 shafts; Kamewa props
Missiles: SSM: 4 Saab RBS15 Mk II (2 twin) launchers ❶; inertial guidance; active radar homing to 110 km (54 n miles) at 0.8 Mach; warhead 150 kg.
Guns: 1 Bofors 57 mm/70 Mk 2 ❷; 220 rds/min to 13.5 km (7.3 n miles); weight of shell 2.4 kg.
Torpedoes: 4 – 15.75 in (400 mm) tubes; Swedish Ordnance Type 45; anti-submarine/surface; wire guided active homing to 20 km (10.8 n miles) at 25 kt; warhead 45 kg shaped charge.
Mines: Minelaying capability.
Physical countermeasures: Decoys: Rheinmetal MASS-1L decoy system ❸.
Electronic countermeasures: ESM: Condor CS 3701; intercept and warning.
Radars: Air/surface search: Ericsson Sea Giraffe 50HC ❹; G-band.
Navigation: Terma Scanter ❺; I-band.
Fire control: Philips 9LV 200 Mk 3 ❻; J-band.
Sonars: Simrad SA 950; hull-mounted; active attack.
Thomson Sintra TSM 2642 Salmon ❼; VDS; search; medium frequency.
Combat data systems: SAAB Tech 9LV Mk 3E Cetris; datalink.
Weapon control systems: Philips 9LV 300 GFCS including a 9LV 100 optronic director and laser range-finder.

Programmes: Orders placed in September 1981. Developed from Spica II class.
Modernisation: RBS 15 missile upgraded to Mk II from 1994. Improved A/S mortar fitted in 1998-99. Extensive mid-life upgrade carried out 1999-2002. Modernisation included removal of the 21 in torpedo tubes and the aft 40 mm mounting and modification of the superstructure to reduce radar and IR signatures. The bridge wings have been removed and a pylon mast has replaced a lattice structure. Upgrades include a new propulsion system, combat data system and EW systems. The decoys are situated on either side of the gun turret. Both ships are to be fitted with CDC Hydra sonar.
Operational: Both ships are to be decommissioned in 2015 without replacement.

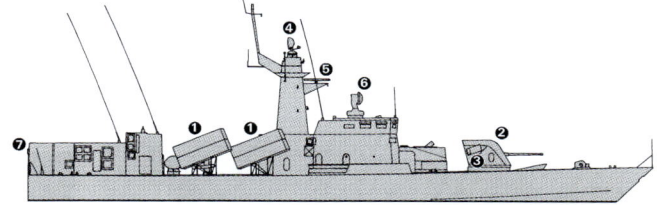

MALMÖ (Scale 1 : 600), Ian Sturton / 1153880

MALMÖ 9/2011, Per Körnefeldt / 1406422

MALMÖ 9/2013*, L-G Nilsson / 1531046

STOCKHOLM 8/2010, E & M Laursen / 1406164

© 2014 IHS IHS Jane's Fighting Ships 2014-2015

806 Sweden > Coast guard

2 KBV 590 (GRIFFON 2450 TD) CLASS (HOVERCRAFT) (UCAC)

KBV 590   KBV 592

Displacement, tonnes: 11 full load
Dimensions, metres (feet): 14.1 on cushion × ? (46.3 × ?)
Speed, knots: 50. **Range, n miles:** 400 at 35 kt
Complement: 3
Machinery: 1 Deutz BF8L diesel; 590 hp(m) (440 kW)
Radars: Navigation: Furuno 7010 D; I-band.

Comment: Built by Griffon Hovercraft, Southampton. KBV 592 delivered in 1993 and KBV 590 in 2011. KBV 590 is a Griffon 2450 TD, an enlarged version of the 2000 TDX design of KBV 592. Basing: Luleå (KBV 590); Umeå (KBV 592).

KBV 590   6/2013*, Swedish Coast Guard / 1455836

4 KBV 031 CLASS (MULTIPURPOSE VESSELS)

No	Builders	Launched	Commissioned
KBV 031	Peenewerft, Wolgast	Dec 2010	Feb 2012
KBV 032	Peenewerft, Wolgast	Aug 2011	Mar 2012
KBV 033	Peenewerft, Wolgast	Nov 2011	Jul 2012
KBV 034	Peenewerft, Wolgast	Mar 2012	Oct 2012

Displacement, tonnes: 1,270 full load
Dimensions, metres (feet): 52.1 × 10.5 × 4.0 (170.9 × 34.4 × 13.1)
Speed, knots: 16
Complement: 6
Machinery: 4 MTU 16V 2000 M60; 4,300 hp (3.2 MW); 2 shafts; cp props; 1 Schottel bow thruster; 350 kW

Comment: Contract signed on 16 July 2008 with Peene-Werft, Wolgast for the construction of four multipurpose vessels to be capable of oil-recovery, environmental control, fishery protection, patrol of territorial waters, rescue operations, towing, fire fighting and diving support. Basing: Djurö (KBV 031); Lysekil (KBV 032); Oskarshamm (KBV 033) and Helsingborg (KBV 034).

KBV 032   6/2012, Swedish Coast Guard / 1483001

KBV 032   4/2013*, Per Körnefeldt / 1455834

5 KBV 312 (BALTIC 24) PATROL CRAFT (WPB)

KBV 312–316

Displacement, tonnes: 51 full load
Dimensions, metres (feet): 23.7 × 5.65 × 1.35 (77.8 × 18.5 × 4.4)
Speed, knots: 31. **Range, n miles:** 500 at 15 kt
Complement: 6
Machinery: 3 Volvo Penta IPS 1050 diesels; 3,277 hp (2.44 MW); 2 shafts
Guns: 1 – 12.7 mm MG (fitted for).
Radars: Surface search/navigation: JMA 5300; I-band.

Comment: Contract with Baltic Workboats A/S, Estonia, in mid-2009 for the construction of five 24 m workboats. Of aluminium construction, the craft are equipped with an aft launch/recovery ramp for a 5 m RHIB. All delivered by mid-2013. Basing: Falkenberg (KBV 312); Oxelösund (KBV 313); Höllviken (KBV 314); Furusund (KBV 315); Gothenburg (KBV 316).

KBV 312   6/2012, Swedish Coast Guard / 1483002

5 POLLUTION CONTROL CRAFT (YPC)

KBV 010   KBV 047–048   KBV 050–051

Comment: KBV 010: Displacement, tons: 430. Built by Lunde in 1985. Oil spill clean-up craft. Based at Gothenburg.
KBV 047: Displacement, tons: 230. Pollution control craft built by Lunde 1980–83. Has bow ramp. Based at Slite.
KBV 048: Displacement, tons: 230. Pollution control craft built by Lunde 1980–83. Has bow ramp. Based at Simrishamn.
KBV 050: Displacement, tons: 340. Enlarged version of KBV 045 class with bow ramp. Built by Lunde in 1983. Based at Södertälje.
KBV 051: Displacement, tons: 340. Enlarged version of KBV 045 class with bow ramp. Built by Lunde in 1983. Based at Värnersborg.

KBV 047   6/2013*, Swedish Coast Guard / 1455837

KBV 010   6/2012, Swedish Coast Guard / 1483000

IHS Jane's Fighting Ships 2014-2015   © 2014 IHS

66 COAST GUARD PATROL CRAFT (SMALL) (WPB)

Comment: There are 66 miscellaneous inshore craft. In addition there are two SM hovercraft KBV 594 and 595.

KBV 474   4/2013*, Per Körnefeldt / 1455833

GOVERNMENT MARITIME FORCES

CIVILIAN SURVEY AND RESEARCH SHIPS

Notes: (1) Owned and manned (since 1 January 2002) by the National Maritime Administration.
(2) There is a 61 m research ship *Argos*. Civilian manned and owned by the National Board of Fisheries. A second civilian ship *Ocean Surveyor* belongs to the Geological Investigation but has been leased as a Support Ship on occasions.
(3) The Swedish Maritime Administration owns two buoy tenders *Scandica* and *Baltica* built in 1982 and two lighthouse tenders *Fyrbyggaren* and *Fyrbjörn*.

OCEAN SURVEYOR   5/1997, H M Steele / 0006214

SCANDICA   6/2000, Curt Borgenstam / 0106585

1 SURVEY SHIP (AGS)

JOHAN MÅNSSON (ex-*Svärten*)

Displacement, tonnes: 39 full load
Dimensions, metres (feet): 20 × 4.8 × 1.3 *(65.6 × 15.7 × 4.3)*
Speed, knots: 15
Complement: 9
Machinery: 2 Volvo Penta TAMD122 diesels; 366 hp *(269 kW)* each; 2 shafts
Radars: Navigation: Terma; I-band.

Comment: Former Ejdern class sonobuoy craft with GRP hull built in 1991 by Djupviksvarvet. Converted for survey duties in 2010.

JOHAN MÅNSSON   5/2010, Per Körnefeldt / 1454811

Switzerland

Country Overview

A landlocked western European country, the Swiss Confederation has an area of 15,940 square miles and is bordered by France, Germany, Austria, Liechtenstein and Italy. The largest city is Zurich and the capital is Bern. The principal lakes are Lake Geneva in the southwest and Lake Constance in the northeast. Others not wholly within Swiss borders are Lake Lugano and Lake Maggiore. The river Rhine, whose source is in the Swiss Alps, is navigable northwards and downstream from the port of Basel. One company of patrol boats, part of the Swiss Armed Forces, is available for operations on lakes Constance, Geneva and Maggiore.

Diplomatic Representation

Defence Attaché in London:
Colonel Hans Eberhart

SWISS ARMED FORCES

Notes: (1) There are also large numbers of flat bottomed raiding craft powered by single 60 hp outboard engines.
(2) There are a number of 6 m rescue craft equipped with a hydraulic ramp.

11 AQUARIUS CLASS (PATROUILLENBOOT 80) (PBR)

ANTARES	SATURN	PERSEUS	MARS
AQUARIUS	URANUS	SIRIUS	POLLUX
ORION	CASTOR	VENUS	

Displacement, tonnes: 7 full load
Dimensions, metres (feet): 10.7 × 3.3 × 1.1 *(35.1 × 10.8 × 3.6)*
Speed, knots: 35
Complement: 6
Machinery: 2 Volvo KAD 43 P-A diesels; 460 hp(m) *(338 kW)*; 2 shafts
Guns: 2—12.7 mm MGs.
Radars: Surface search: JFS Electronic 364; I-band.

Comment: Builders Müller AG, Spiez. GRP hulls, wooden superstructure. *Aquarius* commissioned in 1978, *Pollux* in 1984, the remainder in 1981. Re-engined with diesels which have replaced the former petrol engines.

AQUARIUS   10/1997, Swiss Army / 0019223

Syria

Country Overview

The Syrian Arab Republic was proclaimed in 1961 following brief federation with Egypt as the United Arab Republic from 1958. Situated in the Middle East, the country has an area of 71,498 square miles and is bordered to the north by Turkey, to the east by Iraq, to the south by Jordan and Israel and to the west by Lebanon. It has a 104 n mile coastline with the Mediterranean Sea. The capital and largest city is Damascus while the principal ports are Latakia and Tartus. It is the only country to claim 35 n mile Territorial seas. An EEZ is not claimed.

Headquarters Appointments

Commander-in-Chief Navy:
  Major General Taleb al-Barri

Organisation

Naval Forces come under the command of the Chief of General Staff, Commander of Land Forces.

Personnel

(a) 2014: 3,200 officers and men (2,500 reserves)
(b) 18 months' national service

Bases

Latakia, Tartous, Al-Mina-al-Bayda, Baniyas

Coast Defence

Coastal defence has been under naval control since 1984. A missile brigade is equipped with SS-C-1 Sepal and SS-C-3 Styx with sites at Tartous (2), Baniyas and Latakia. These may be upgraded to SS-N-26 Yakhont if such missiles are acquired from Russia. Two artillery battalions have a total of 36—130 mm guns and 12—100 mm guns. Coastal observation sites are manned by an Observation Battalion. There are two infantry brigades each of which is assigned to a coastal zone.

FRIGATES

2 PETYA III (PROJECT 159A) CLASS (FFL)

1–508 (ex-12)   AL HIRASA 2–508 (ex-14)

Displacement, tonnes: 965 standard; 1,199 full load
Dimensions, metres (feet): 81.8 × 9.1 × 2.9 *(268.4 × 29.9 × 9.5)*
Speed, knots: 32. **Range, n miles:** 4,870 at 10 kt, 450 at 29 kt
Complement: 98 (8 officers)

Machinery: CODAG; 2 gas turbines; 30,000 hp(m) *(22 MW)*; 1 Type 61V-3 diesel; 5,400 hp(m) *(3.97 MW)* sustained (centre shaft); 3 shafts
Guns: 4—3 in *(76 mm)*/59 AK 726 (2 twin) ❶; 90 rds/min to 16 km *(8.5 n miles)*; weight of shell 5.9 kg.
Torpedoes: 3—21 in *(533 mm)* (triple) tubes ❷. SAET-60; active/passive homing to 15 km *(8.1 n miles)* at 40 kt; warhead 100 kg.
A/S Mortars: 4 RBU 2500 16-tubed trainable ❸; range 2,500 m; warhead 21 kg.
Mines: Can carry 22.
Depth charges: 2 racks.
Radars: Surface search: Slim Net ❹; E/F-band.
  Navigation: Don 2; I-band.
  Fire control: Hawk Screech ❺; I-band.
IFF: High Pole B. 2 Square Head.
Sonars: Herkules; hull-mounted; active search and attack; high frequency.

Programmes: Transferred by the USSR in July 1975 and March 1975.
Operational: Based at Tartous. 2–508 in dock in mid-1998 to 2000. Both ships reported operational.

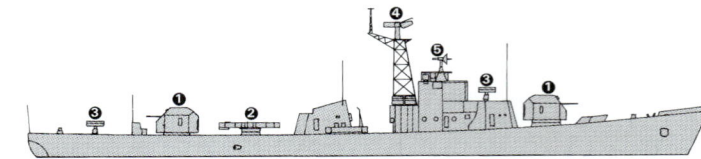

PETYA 1–508   *(Scale 1 : 900), Ian Sturton / 0506171*

AL HIRASA   6/2001 / 0121400

LAND-BASED MARITIME AIRCRAFT

Numbers/Type: 11/2 Mil Mi-14P Haze A/Mi-14P Haze C.
Operational speed: 124 kt *(230 km/h)*.
Service ceiling: 15,000 ft *(4,570 m)*.
Range: 432 n miles *(800 km)*.
Role/Weapon systems: Medium-range ASW helicopter. Sensors: Short Horn search radar, dipping sonar, MAD, sonobuoys. Weapons: ASW; internally stored torpedoes, depth mines and bombs.

Numbers/Type: 2 Kamov Ka-28 Helix.
Operational speed: 135 kt *(250 km/h)*.
Service ceiling: 19,685 ft *(6,000 m)*.
Range: 432 n miles *(800 km)*.
Role/Weapon systems: ASW helicopter. Delivered in February 1990. Sensors: Splash Drop search radar, dipping sonar, sonobuoys, MAD, ECM. Weapons: ASW; 3 torpedoes, depth bombs, mines.

PATROL FORCES

Notes: There is an unarmed 18 m diving tender *Palmyra* built by Ocea de Saint-Nazaire in 2005.

6 TIR II (IPS 18) CLASS (INSHORE PATROL CRAFT) (PTFG)

Displacement, tonnes: 29 standard
Dimensions, metres (feet): 21.1 × 5.8 × 0.9 *(69.2 × 19.0 × 3.0)*
Speed, knots: 52
Complement: 6
Missiles: SSM: 2 Noor (C-802); active radar homing to 120 km *(66 n miles)* at 0.9 Mach; warhead 165 kg.

Comment: The first three craft with missile racks, but without missiles, were delivered in mid-2006. Probably built in Iran and based on those supplied in December 2002 by North Korea. At least two had been equipped with missiles by 2012.

TIR II   6/2012 / 1486369

16 OSA (PROJECT 205) CLASS
(FAST ATTACK CRAFT — MISSILE) (PTFG)

21–26   31–40

Displacement, tonnes: 249 full load
Dimensions, metres (feet): 38.6 × 7.6 × 2.7 *(126.6 × 24.9 × 8.9)*
Speed, knots: 35 (21–26), 37 (31–40)
Range, n miles: 500 at 35 kt
Complement: 25 (3 officers)

Machinery: 3 Type M 504 (Osa II)/M 503 (Osa I) diesels; 8,025/10,800 hp(m) *(6.0/8.1 MW)* sustained; 3 shafts
Missiles: SSM: 4 SS-N-2C; active radar or IR homing to 83 km *(43 n miles)* at 0.9 Mach; warhead 513 kg; sea-skimmer at end of run.
Guns: 4—30 mm/65 (2 twin); 500 rds/min to 5 km *(2.7 n miles)*; weight of shell 0.54 kg.
Physical countermeasures: Decoys: PK 16 chaff launcher.
Radars: Surface search: Square Tie; I-band.
  Fire control: Drum Tilt; H/I-band.
IFF: 2 Square Head. High Pole A or B.

Programmes: Delivered: October 1979 (two), November 1979 (two), August 1982 (one), September 1982 (one) and May 1984 (two). Further craft acquired.
Structure: Two are modified (Nos 39 and 40).
Operational: Osa I are based at Tartous and Osa II based at Latakia. All are still fully operational and active. The Osa Is are fitted with SSN 2A/B.

OSA II 38   6/1998 / 0050214

IHS Jane's Fighting Ships 2014-2015   © 2014 IHS

8 ZHUK (GRIF) (PROJECT 1400M) CLASS
(COASTAL PATROL CRAFT) (PB)

1–8

Displacement, tonnes: 40 full load
Dimensions, metres (feet): 24 × 5 × 1.2 (78.7 × 16.4 × 3.9)
Speed, knots: 30
Range, n miles: 1,100 at 15 kt
Complement: 11 (3 officers)
Machinery: 2 Type M 401B diesels; 2,200 hp(m) (1.6 MW) sustained; 2 shafts
Guns: 4 – 14.5 mm (2 twin) MGs.
Radars: Surface search: Spin Trough; I-band.

Comment: Three transferred from USSR in August 1981, three on 25 December 1984 and two more in the late 1980s. Based at Tartous and Latakia. About half the craft are operational.

ZHUK 5–8 6/1998 / 0050215

AMPHIBIOUS FORCES

3 POLNOCHNY B CLASS (PROJECT 771) (LSM)

1-114 2-114 3-114

Displacement, tonnes: 772 standard; 847 full load
Dimensions, metres (feet): 75 × 9.6 × 2.3 (246.1 × 31.5 × 7.5)
Speed, knots: 19
Range, n miles: 1,500 at 15 kt
Complement: 40

Military lift: 180 troops; 350 tons cargo
Machinery: 2 Kolomna Type 40-D diesels; 4,400 hp(m) (3.2 MW) sustained; 2 shafts
Guns: 4 – 30 mm/65 (2 twin); 500 rds/min to 5 km (2.7 n miles); weight of shell 0.54 kg.
Rockets: 2 – 140 mm rocket launchers; 18 barrels per launcher; range 9 km (5 n miles).
Radars: Surface search: Spin Trough; I-band.
Fire control: Drum Tilt; H/I-band.

Comment: Built at Northern Shipyard, Gdansk. First transferred from USSR January 1984, two in February 1985 from Black Sea. All based at Tartous and still active.

POLNOCHNY B (Azerbaijan colours) 6/2008, Ian Sturton / 1335326

MINE WARFARE FORCES

1 SONYA (YAKHONT) (PROJECT 12650) CLASS
(COASTAL MINEHUNTER) (MHC)

532

Displacement, tonnes: 457 full load
Dimensions, metres (feet): 48 × 8.8 × 2 (157.5 × 28.9 × 6.6)
Speed, knots: 15
Range, n miles: 3,000 at 10 kt
Complement: 43 (5 officers)

Machinery: 2 Kolomna Type 9-D-8 diesels; 2,000 hp(m) (1.47 MW) sustained; 2 shafts
Missiles: SAM: 2 quad SA-N-5 launchers.
Guns: 2 – 30 mm/65 AK 630 or 2 – 30 mm/65 (twin) and 2 – 25 mm/80 (twin).
Mines: 8.
Radars: Surface search: Don 2 or Kivach or Nayada; I-band.
IFF: 2 Square Head. High Pole B.
Sonars: MG 69/79; hull-mounted; active minehunting; high frequency.

Comment: Wooden hull with GRP sheath. Transferred to Syria in 1986. Reported decommissioned in 2004 but apparently operational again in 2006.

SONYA CLASS (Russian colours) 6/2003, Guy Toremans / 0570933

1 NATYA (PROJECT 266M) CLASS (MSC/AGORM)

642

Displacement, tonnes: 817 full load
Dimensions, metres (feet): 61 × 10.2 × 3 (200.1 × 33.5 × 9.8)
Speed, knots: 16
Range, n miles: 3,000 at 12 kt
Complement: 65

Machinery: 2 Type 504 diesels; 5,000 hp(m) (3.67 MW) sustained; 2 shafts
Missiles: SAM: 2 SA-N-5 Grail quad launchers; manual aiming; IR homing to 6 km (3.2 n miles) at 1.5 Mach; altitude to 2,500 m (8,000 ft); warhead 1.5 kg; 16 missiles.
Guns: 4 – 30 mm/65 (2 twin) can be fitted.
Radars: Surface search: Don 2; I-band.
Fire control: Drum Tilt; H/I-band.

Comment: Arrived in Tartous from USSR in January 1985. Has had sweeping gear and guns removed and converted to serve as an AGOR. Painted white. Based at Latakia in reasonable condition. Reported active.

NATYA 642 6/1996 / 0080764

5 YEVGENYA (PROJECT 1258) CLASS
(MINESWEEPERS – INSHORE) (MSI/PC)

4-507 5-507 6-507 7-507 8-507

Displacement, tonnes: 78 standard; 91 full load
Dimensions, metres (feet): 24.6 × 5.5 × 1.5 (80.7 × 18.0 × 4.9)
Speed, knots: 11
Range, n miles: 300 at 10 kt
Complement: 10

Machinery: 2 Type 3-D-12 diesels; 600 hp(m) (444 kW); 2 shafts
Guns: 2 – 14.5 mm (twin) MGs (first pair). 2 – 25 mm/80 (twin) (second pair).
Radars: Surface search: Spin Trough; I-band.
IFF: High Pole.
Sonars: MG-7; stern-mounted VDS; active; high frequency.

Comment: First transferred from USSR 1978, two in 1985 and two in 1986. Second pair by Ro-flow from Baltic in February 1985 being new construction with tripod mast. Based at Tartous, at least two are operational. Both 4-507 and 5-507, thought to have been deleted, were reported operational in 2006.

YEVGENYA (Ukraine colours) 6/2003, Ships of the World / 0572652

810 Syria > Training ships — Taiwan > Introduction

TRAINING SHIPS

1 TRAINING SHIP (AX/AKR)

AL ASSAD

Displacement, tonnes: 3,556 full load
Dimensions, metres (feet): 105 × 17.2 × 4 *(344.5 × 56.4 × 13.1)*
Speed, knots: 16
Range, n miles: 4,500 at 15 kt
Complement: 196 (140 cadets)
Machinery: 2 Zgoda-Sulzer 6ZL40/48 diesels; 8,700 hp(m) *(6.4 MW)*; 2 shafts; bow thruster
Radars: Decca Seamaster; E/F- and I-band.

Comment: Built in Polnochny Shipyard, Gdansk and launched 18 February 1987. Delivered in late 1988. Ro-ro design used as a naval training ship. Unarmed but has minelaying potential. Based at Latakia and occasionally deploys on cruises.

COAST GUARD

Notes: In addition to the craft in this section, there are four Polish-built Szkwal 12 m patrol craft transferred from the Polish MOSG in 1995. Procurement of further patrol craft, possibly from Turkey, is also reported to be under consideration.

2 MAWANI CLASS (PBF)

Dimensions, metres (feet): 20.02 × 5.3 × 1.1 *(65.7 × 17.4 × 3.6)*
Speed, knots: 35
Range, n miles: 325 at 28 kt
Complement: 3 + 8 personnel
Machinery: 4 MTU diesels; 2,682 hp *(2.0 MW)*; 2 shafts
Radars: Navigation: I-band.

Comment: Acquired from Düzgit Shipyard, Turkey, in 2009.

AL ASSAD 6/2009, A Campanera i Rovira / 1366014

MAWANI CLASS 6/2012 / 1168761

AL ASSAD
5/2009, Adolfo Ortigueira Gil
1366013

Taiwan
REPUBLIC OF CHINA

Country Overview

The Republic of China was established in 1949 when the Nationalist government of China withdrew to Taiwan (Formosa) and established its headquarters. Though in practice an autonomous state, Taiwan is still formally a province of China and, as such, is claimed by the People's Republic of China. The country comprises the island of Taiwan (area 13,900 square miles), the Pescadores, or P'eng-hu Islands, the Quemoy Islands off the mainland city of Amoy (Xiamen), and the Matsu group off Fuzhou (Foochow). It has a 783 n mile coastline with East China Sea, Pacific Ocean and South China Sea. The capital and largest city of Taiwan is Taipei while Chi-lung (Keelung), Hualien, Kao-hsiung and T'ai-chung are the principal ports. Territorial seas (12 n miles) are claimed. A 200 n mile EEZ and Fishery Zone have also been claimed.

Headquarters Appointments

Commander-in-Chief:
  Admiral Chen, Yeong-Kang
Commandant of Marine Corps:
  Lieutenant General Pan, Jin-lung

Senior Flag Officers

Fleet Commander: Vice Admiral Pu, Ze-chun

Personnel

(a) 2014: 46,500 in Navy, 15,000 in Marine Corps
(b) 1 year 4 months conscript service

Bases

Tsoying: HQ First Naval District (Southern Taiwan, Pratas and Spratly). Main Base, HQ of Fleet Command, Naval Aviation Group and Marine Corps. Base of southern patrol and transport squadrons. Officers and ratings training, Naval Academy, Naval shipyard.
Kaohsiung; Naval shipyard.
Makung (Pescadores): HQ Second Naval District (Pescadores, Quemoy and Wu Ch'iu). Base for attack squadrons. Naval shipyard and training facilities.
Keelung: HQ Third Naval District (Northern Taiwan and Matsu group). Base of northern patrol and transport squadrons. Naval shipyard.
Hualien: Naval Aviation Command.
Suao: East Coast Command, submarine depot and shipyard.
Minor bases at Hualien, Tamshui, Hsinchu, Wuchi and Anping.
Building: Taitung.

Organisation

1. Fleet Command:
124th Attack squadron, based at Tsoying
142nd Support squadron, based at Kaohsiung
146th Attack squadron, based at Pescadores
151st Amphibious squadron, based at Tsoying
168th Patrol squadron, based at Suao
192nd Mine Warfare squadron, based at Tsoying
256th Submarine Unit, based at Tsoying.
2. Naval Aviation Command: There are two Groups. The fixed-wing Group based at Pingtung-North consists of two squadrons (133 and 134). The helicopter Group consists of three squadrons 501 squadron is based at Tsoying, 701 squadron at Hualien and 702 squadron at Tsoying.

Coast Defence

The land-based SSM command has six squadrons equipped with Hsiung-Feng II SSM at Tonying Island of the Matsu Group, Siyu Island of the Pescadores, Shiao Liuchiu off Kaohsiung, north of Keelung harbour, Tsoying naval base and Hualien. The ROCMC deploy eight SAM Platoons, equipped with Chaparral SAM quad-launchers, to the offshore island of Wuchiu, and Pratas islets in the South China Sea. There are also a number of 127 mm guns.

Marine Corps

Increased to three brigades in 2002 supported by one amphibious regiment and one logistics regiment. Equipped with M-116, M-733, LARC-5, LVTP5 (to be replaced by AAV-7A-IRAM/RS) personnel carriers and LVTH6 armour tractors. Based at Tsoying and in southern Taiwan. Spratly detachment provided by the Coast Guard from 1 January 2000 and Marine Corps detachment withdrawn from Pratas Islands at the same time.

Coast Guard

Formerly the Maritime Security Police but name changed on 1 January 2000. Comes under the Minister of the Interior but its numerous patrol boats are integrated with the Navy for operational purposes.

Strength of the Fleet

Type	Active (Reserve)	Building/ Transfer (Planned)
Submarines	4	(8)
Destroyers	4	—
Frigates	22	(4)
Corvettes	—	(10)
Patrol Craft (Missile)	43	—
Large Patrol Craft	8	—
Ocean Minesweepers	4	—
Coastal Minesweepers/Hunters	8	—
LSD	1	(1)
Landing Ships (LST and LSM)	13	—
LCUs	8	—
Survey Ships	1	—
Combat Support Ships	1	—
Transports	3	—
Salvage Ships	1	—
Coast Guard	23	—

IHS Jane's Fighting Ships 2014-2015   © 2014 IHS

PENNANT LIST

Submarines		1107	Tzu-I	**Amphibious Forces**		167	Yung Ren
		1108	Pan Chao			168	Yung Sui
791	Hai Shih	1109	Chang Chien	191	Chung Cheng	1301	Yung Feng
792	Hai Bao	1110	Tien Tan (bldg)	205	Chung Chien	1302	Yung Chia
793	Hai Lung	1202	Kang Ding	208	Chung Shun	1303	Yung Ting
794	Hai Hu	1203	Si Ning	216	Chung Kuang	1305	Yung Shun
		1205	Kun Ming	217	Chung Chao	1306	Yung Yang
Destroyers		1206	Di Hua	218	Chung Chi	1307	Yung Tzu
		1207	Wu Chang	221	Chung Chuan	1308	Yung Ku
1801	Kee Lung	1208	Chen Te	226	Chung Chih	1309	Yung Teh
1802	Suao			227	Chung Ming	1310	Yung Jin
1803	Tsoying			230	Chung Pang	1311	Yung An
1805	Makung	**Patrol Forces**		231	Chung Yeh		
				232	Chung Ho		
Frigates		PCL 1	Ning Hai	233	Chung Ping	**Auxiliaries and Survey Ships**	
		PCL 2	An Hai	488	Ho Shan		
932	Chin Yang	603	Jin Chiang	489	Ho Chuan	525	Wu Kang
933	Fong Yang	605	Tan Chiang	490	Ho Seng	530	Wu Yi
934	Feng Yang	606	Hsin Chiang	491	Ho Meng	552	Ta Hu
935	Lan Yang	607	Feng Chiang	492	Ho Mou	1601	Ta Kuan
936	Hae Yang	608	Tseng Chiang	493	Ho Shou		
937	Hwai Yang	609	Kao Chiang	LCC1	Kao Hsiung	**Tugs**	
938	Ning Yang	610	Jing Chiang				
939	Yi Yang	611	Hsian Chiang	**Mine Warfare Forces**		ATF 551	Ta Wan
1101	Cheng Kung	612	Tsi Chiang			ATF 553	Ta Han
1103	Cheng Ho	614	Po Chiang	158	Yung Chuan	ATF 554	Ta Kang
1105	Chi Kuang	615	Chan Chiang	162	Yung Fu	ATF 555	Ta Fung
1106	Yueh Fei	617	Chu Chiang			ATF 563	Ta Tai

SUBMARINES

Notes: Project Kwang Hua 8: Following the announcement in 2001 by the US government that it will support the acquisition of eight diesel submarines, debate has centred on how these will be procured. Northrop Grumman has reportedly offered a modernised version of the Barbel class, which dates from the 1950s. The licence of a design from a third country has proved to be problematic but commitment to the submarine programme was re-affirmed by Taiwan's Ministry of Defence on 21 February 2012. Options include the procurement of second-hand boats or an indigenous construction programme with foreign assistance, for which a four-year feasibility study was launched on 12 March 2013.

2 HAI LUNG CLASS (SSK)

Name	No	Builders	Laid down	Launched	Commissioned
HAI LUNG	793	Wilton Fijenoord	Dec 1982	6 Oct 1986	9 Oct 1987
HAI HU	794	Wilton Fijenoord	Dec 1982	20 Dec 1986	9 Apr 1988

Displacement, tonnes: 2,414 surfaced; 2,703 dived
Dimensions, metres (feet): 66.9 × 8.4 × 6.7 *(219.5 × 27.6 × 22.0)*
Speed, knots: 12 surfaced; 20 dived
Range, n miles: 10,000 at 9 kt surfaced
Complement: 67 (8 officers)

Machinery: Diesel-electric; 3 Bronswerk D-RUB 215-12 diesels; 4,050 hp(m) *(3 MW)*; 3 alternators; 2.7 MW; 1 Holec motor; 5,100 hp(m) *(3.74 MW)*; 1 shaft
Missiles: SSM: McDonnell Douglas Harpoon UGM-84L Block II; active radar homing to 124 km *(67 n miles)* at 0.9 Mach; warhead 227 kg (to be fitted).
Torpedoes: 6—21 in *(533 mm)* bow tubes. 20 AEG SUT; dual purpose; wire-guided; active/passive homing to 12 km *(6.6 n miles)* at 35 kt; warhead 250 kg.

Electronic countermeasures: ESM: Argo AR 700SF and Elbit Timnex 4CH(V)2; intercept.
Radars: Surface search: Signaal ZW06; I-band.
Sonars: Signaal SIASS-Z; hull-mounted; passive/active intercept search and attack; low/medium frequency. Fitted for but not with towed passive array.
Weapon control systems: Sinbads M TFCS.

Programmes: Order signed with Wilton Fijenoord in September 1981 for these submarines with variations from the standard Netherlands Zwaardvis design. Construction was delayed by the financial difficulties of the builders but was resumed in 1983. Sea trials of *Hai Lung* in March 1987 and *Hai Hu* in January 1988 and both submarines were shipped out on board a heavy dock vessel. The names mean *Sea Dragon* and *Sea Tiger*.

Modernisation: Plans to fit both submarines with McDonnell Douglas UGM-84L Block II Sub Harpoon were first announced in September 2005 and later confirmed on 3 October 2008. Delivery of the missiles began in December 2013.
Structure: The four horns on the forward casing are Signaal sonar intercept transducers. Torpedoes manufactured under licence in Indonesia.
Operational: Hsiung Feng II submerged launch SSMs are planned to be part of the weapons load and a torpedo tube launched version is being developed, although no recent progress has been reported. Belong to 256th Submarine Unit based at Tsoying.

HAI HU and HAI LUNG

11/2004, Ships of the World / 1044575

2 GUPPY II CLASS (SS)

Name	No	Builders	Laid down	Launched	Commissioned
HAI SHIH (ex-*Cutlass*)	791 (ex-SS 91, ex-SS 478)	Portsmouth Navy Yard	22 Jul 1944	5 Nov 1944	17 Mar 1945
HAI BAO (ex-*Tusk*)	792 (ex-SS 92, ex-SS 426)	Federal SB & DD Co, Kearney	23 Aug 1943	8 Jul 1945	11 Apr 1946

Displacement, tonnes: 1,900 surfaced; 2,459 dived
Dimensions, metres (feet): 93.7 × 8.3 × 5.5 *(307.4 × 27.2 × 18.0)*
Speed, knots: 18 surfaced; 15 dived
Range, n miles: 8,000 at 12 kt surfaced
Complement: 75 (7 officers)
Machinery: Diesel-electric; 3 Fairbanks-Morse diesels; 4,500 hp *(3.3 MW)*; 2 Elliott motors; 5,400 hp *(4 MW)*; 2 shafts

Torpedoes: 10—21 in *(533 mm)* (6 fwd, 4 aft) tubes. AEG SUT; active/passive homing to 12 km *(6.5 n miles)* at 35 kt; 28 km *(15 n miles)* at 23 kt; warhead 250 kg.
Electronic countermeasures: ESM: WLR-1/3; radar warning.
Radars: Surface search: US SS 2; I-band.
Sonars: EDO BQR 2B; hull-mounted; passive search and attack; medium frequency. Raytheon/EDO BQS 4C; adds active capability to BQR 2B. Thomson Sintra DUUG 1B; passive ranging.

Programmes: Originally fleet-type submarines of the US Navy's Tench class; extensively modernised under the Guppy II programme. *Hai Shih* transferred in April 1973 and *Hai Bao* in October the same year.
Structure: After nearly 70 years in service diving depth is very limited.
Operational: Kept in service because of difficulty in buying replacements, but operational status doubtful. Likely to have an alongside training role only. Belong to the 256th Submarine Unit based at Tsoying.

HAI BAO 11/2004, *Ships of the World* / 1044574

DESTROYERS

Notes: Acquisition of the Aegis Combat System remains an aspiration but, following the procurement of the Kidd-class DDGs the purchase of further ships is unlikely in the short-term. However, procurement of a component system, such as the SPY-1F radar, is a possibility.

4 KEELUNG (KIDD) CLASS (DDGHM)

Name	No	Builders	Laid down	Launched	Commissioned
KEELUNG (ex-*Chi Teh*; ex-*Scott*)	1801 (ex-DD 995)	Ingalls Shipbuilding, Pascagoula	12 Feb 1979	1 Mar 1980	24 Oct 1981
SUAO (ex-*Wu Teh*; ex-*Callaghan*)	1802 (ex-DD 994)	Ingalls Shipbuilding, Pascagoula	23 Oct 1978	1 Dec 1979	29 Aug 1981
TSOYING (ex-*Ming Teh*; ex-*Kidd*)	1803 (ex-DD 993)	Ingalls Shipbuilding, Pascagoula	26 Jun 1978	11 Aug 1979	27 Jun 1981
MAKUNG (ex-*Tong-Teh*; ex-*Chandler*)	1805 (ex-DD 996)	Ingalls Shipbuilding, Pascagoula	7 May 1979	24 May 1980	13 Mar 1982

Displacement, tonnes: 7,062 standard; 9,728 full load
Dimensions, metres (feet): 171.7 × 16.8 × 6.2 *(563.3 × 55.1 × 20.3)*
Speed, knots: 33. **Range, n miles:** 6,000 at 20 kt
Complement: 363 (31 officers)
Machinery: 4 GE LM 2500 gas turbines; 86,000 hp *(64.16 MW)* sustained; 2 shafts
Missiles: SSM: 4 McDonnell Douglas RGM 84L Block 2 Harpoon (1 quad) launchers ❶; active radar homing to 124 km *(67 n miles)* at 0.9 Mach; warhead 227 kg.
SAM: 64 Raytheon Standard SM-2 MR Block IIIA; command/inertial guidance; semi-active radar homing to 167 km *(90 n miles)* at 2.5 Mach. 2 twin Mk 26 launchers ❷.
Guns: 2 FMC 5 in *(127 mm)*/54 Mk 45 Mod 0 ❸; 20 rds/min to 23 km *(12.6 n miles)*; weight of shell 32 kg. 2 Raytheon 20 mm Vulcan Phalanx 6-barrelled Block 1B ❹; 4,500 rds/min (4,500 in Block 1). 4–12.7 mm MGs.
Torpedoes: 6–324 mm Mk 32 (2 triple) tubes ❺. Honeywell Mk 46 Mod 5; anti-submarine; active/passive homing to 11 km *(5.9 n miles)* at 40 kt; warhead 44 kg. Torpedoes fired from inside the hull under the hangar.
Physical countermeasures: Decoys: 4 Loral Hycor SRBOC 6-barrelled fixed Mk 36; IR flares and chaff to 4 km *(2.2 n miles)*. SLQ-25 Nixie; torpedo decoy.
Radars: Air search: ITT SPS-48E ❻; 3D; E/F-band. Raytheon SPS-49(V)5 ❼; C/D-band.

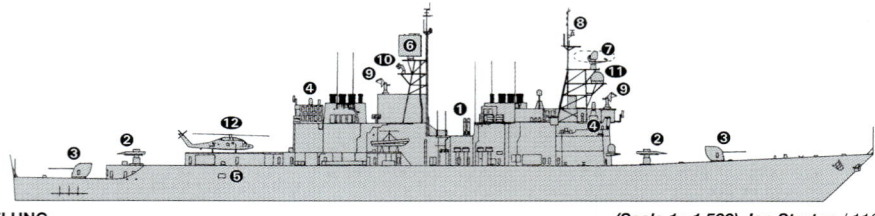

KEELUNG (Scale 1 : 1,500), Ian Sturton / 1167441

Air/surface search: ISC Cardion SPS-55 ❽; I/J-band.
Navigation: Raytheon SPS-64; I/J-band.
Fire control: 2 Raytheon SPG-51D ❾, 1 Lockheed SPG-60 ❿, 1 Lockheed SPQ-9A ⓫.
Sonars: General Electric/Hughes SQS-53D; bow-mounted; search and attack; medium frequency. Gould SQR-19 (TACTAS); passive towed array (may be fitted).
Combat data systems: ACDS Block 1 Level 1 with datalinks.
Weapon control systems: SWG-1A Harpoon LCS. 2 Mk 74 MFCS. Mk 86 Mod 5 GFCS. Mk 116 FCS for ASW. Mk 14 WDS. SYS 2(V)2 IADT. 4 SYR 3393 for SAM mid-course guidance.
Helicopters: 1 Sikorsky S-70C(M) ⓬.

Programmes: Originally ordered by the Iranian government in 1974, the contracts were taken over by the US Navy on 25 July 1979. All paid off from USN service in 1998–99. Offered to the Taiwan government, intention to buy confirmed on 2 October 2001.
Modernisation: All received major modernisation from 1988–90. Further package completed prior to transfer. ASROC has been removed.
Operational: *Keelung* and *Suao* arrived in Taiwan on 8 December 2005 and were recommissioned on 17 December 2005. *Tsoying* and *Makung* arrived in October 2006. All four ships are based at Tsoying.

TSOYING 3/2011, *Ships of the World* / 1406372

FRIGATES

Notes: The Kuang Hua 7 programme has superseded the former Kuang Hua 5 programme for the procurement of new frigates/corvettes to replace the Knox class. It is understood that there is a requirement for up to eight ships of above 2,000 tons with a main armament of Hsiung Feng-II missiles. This is likely to have been met in part by the acquisition of four Oliver Hazard Perry-class frigates from the US. These are reported to be USS *Taylor* (FFG 50), USS *Gary* (FFG 51), USS *Carr* (FFG 52) and USS *Elrod* (FFG 55). The first two are expected to be transferred in 2015.

8 + (4) CHENG KUNG CLASS (KWANG HUA 1 PROJECT) (FFGHM)

Name	No	Builders	Laid down	Launched	Commissioned
CHENG KUNG	1101	China SB Corporation, Kaohsiung	7 Jan 1990	5 Oct 1991	7 May 1993
CHENG HO	1103	China SB Corporation, Kaohsiung	21 Dec 1990	15 Oct 1992	28 Mar 1994
CHI KUANG	1105	China SB Corporation, Kaohsiung	4 Oct 1991	27 Sep 1993	4 Mar 1995
YUEH FEI	1106	China SB Corporation, Kaohsiung	5 Sep 1992	26 Aug 1994	7 Feb 1996
TZU-I	1107	China SB Corporation, Kaohsiung	7 Aug 1994	13 Jul 1995	9 Jan 1997
PAN CHAO	1108	China SB Corporation, Kaohsiung	25 Jul 1995	4 Jul 1996	16 Dec 1997
CHANG CHIEN	1109	China SB Corporation, Kaohsiung	4 Dec 1995	14 May 1997	1 Dec 1998
TIEN TAN	1110	China SB Corporation, Kaohsiung	21 Feb 2001	15 Oct 2002	11 Mar 2004

Displacement, tonnes: 2,794 standard; 4,171 full load
Dimensions, metres (feet): 138.1 × 13.7 × 4.5 hull; 7.5 sonar *(453.1 × 44.9 × 14.8; 24.6)*
Speed, knots: 29
Range, n miles: 4,500 at 20 kt
Complement: 234 (15 officers; 19 air crew)

Machinery: 2 GE LM 2500 gas turbines; 41,000 hp *(30.59 MW)* sustained; 1 shaft; cp prop 2 auxiliary retractable props; 650 hp *(484 kW)*
Missiles: SSM: 8 Hsiung Feng II/III ❶ (2 quad); inertial guidance; active radar/IR homing to 130 (200 Hsiung Feng III) km *(70 (108) n miles)* at 0.85 Mach (2 Mach); warhead 225 kg.
SAM: 40 Raytheon Standard SM1-MR Block VIA; Mk 13 launcher ❷; command guidance; semi-active radar homing to 38 km *(20.5 n miles)* at 2 Mach.
Guns: 1 Oto Melara 76 mm/62 Mk 75 ❸; 85 rds/min to 16 km *(8.7 n miles)*; weight of shell 6 kg. 2 Bofors 40 mm/70 ❹. 3—20 mm Type 75 (on hangar roof when fitted). 1 Raytheon 20 mm/76 Vulcan Phalanx 6-barrelled Block 1B ❺; 4,500 rds/min combined to 1.5 km.
Torpedoes: 6—324 mm Mk 32 (2 triple) tubes ❻. Honeywell/Alliant Mk 46 Mod 5; anti-submarine; active/passive homing to 11 km *(5.9 n miles)* at 40 kt; warhead 44 kg.
Physical countermeasures: Decoys: 4 Kung Fen 6 chaff launchers or locally produced version of RBOC *(114 mm)*. SLQ-25A Nixie; torpedo decoy.
Electronic countermeasures: ESM/ECM: Chang Feng IV (locally produced version of SLQ-32(V)2 with Sidekick); combined radar warning and jammers.
Radars: Air search: Raytheon SPS-49(V)5 or SPS-49A (1108–9) ❼; C/D-band.
Surface search: ISC Cardion SPS-55 ❽ or Raytheon Chang Bai; I/J-band.

Fire control: USN UD 417 STIR ❾; I/J-band. Unisys Mk 92 Mod 6 ❿; I/J-band.
Sonars: Raytheon SQS-56/DE 1160P; hull-mounted; active search and attack; medium frequency. SQR-18A(V)2; passive towed array or BAe/Thomson Sintra ATAS active towed array (from *Chi Kuang* onwards).
Combat data systems: Norden SYS-2(V)2 action data automation with UYK 43 computer. Ta Chen link (from *Chi Kuang* onwards and being backfitted).
Weapon control systems: Loral Mk 92 Mod 6. Mk 13 Mod 4 weapon direction system. Mk 114 ASW. 2 Mk 24 optical directors. Mk 309 TFCS.
Helicopters: 2 Sikorsky S-70C(M) ⓫ (only 1 embarked).

Programmes: First two ordered 8 May 1989. Named after Chinese generals and warriors. An eighth of class was ordered in late July 1999. Originally this ship was planned to be the first of a Flight II design, which was scrapped. Plans to procure a further four second-hand ships from the US were announced in April 2012. The first two ships are expected in 2015.
Modernisation: Hsiung Feng III supersonic missiles have been installed in a reported five ships including *Cheng Kung*. All eight ships are to be fitted. A mid-life upgrade for the class is likely to include the installation of RAM PDMS and the replacement of Standard SM-1 with SM-2. The Mk 96 direction system is also likely to be upgraded to Mod 12.
Structure: Similar to the USS *Ingraham*. RAST helicopter hauldown. The area between the masts had to be strengthened to take the Hsiung Feng II missiles. Prairie Masker hull acoustic suppression system fitted.
Operational: Form the 146th Squadron based at Makung (Pescadores).

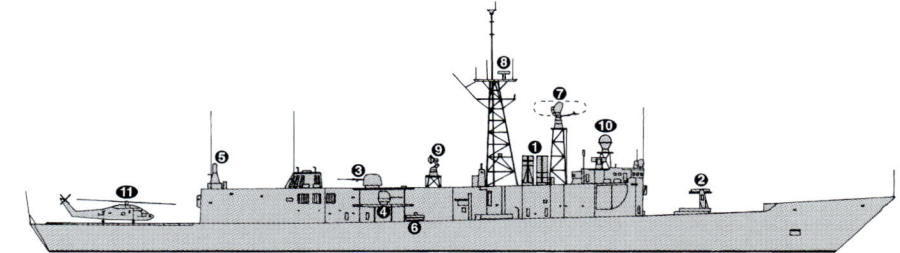

CHENG KUNG *(Scale 1 : 1,200), Ian Sturton / 0019226*

CHENG HO *4/2011, Chris Sattler / 1454737*

PAN CHAO *3/2010, Chris Sattler / 1406284*

6 KANG DING (LA FAYETTE) CLASS (KWANG HUA 2 PROJECT) (FFGHM)

Name	No	Builders	Laid down	Launched	Commissioned
KANG DING	1202	Lorient Dockyard, Lorient; Kaohsiung Shipyard	26 Aug 1993	12 Mar 1994	24 May 1996
SI NING	1203	Lorient Dockyard, Lorient; Kaohsiung Shipyard	27 Apr 1994	5 Nov 1994	15 Sep 1996
KUN MING	1205	Lorient Dockyard, Lorient; Kaohsiung Shipyard	7 Nov 1994	13 May 1995	26 Feb 1997
DI HUA	1206	Lorient Dockyard, Lorient; Kaohsiung Shipyard	1 Jul 1995	27 Nov 1995	14 Aug 1997
WU CHANG	1207	Lorient Dockyard, Lorient; Kaohsiung Shipyard	1 Jul 1995	27 Nov 1995	16 Dec 1997
CHEN TE	1208	Lorient Dockyard, Lorient; Kaohsiung Shipyard	27 Dec 1995	2 Aug 1996	16 Jan 1998

Displacement, tonnes: 3,861 full load
Dimensions, metres (feet): 124.2 × 15.4 × 5.5 screws (407.5 × 50.5 × 18.0)
Speed, knots: 25
Range, n miles: 7,000 at 15 kt
Complement: 134 (15 officers) + 25 spare berths

Machinery: CODAD; 4 SEMT-Pielstick 12 PA6 V 280 STC diesels; 23,228 hp(m) (17.08 MW); 2 shafts; LIPS cp props
Missiles: SSM: 8 Hsiung Feng II (2 quad) ❶; inertial guidance; active radar/IR homing to 130 km (70 n miles) at 0.85 Mach; warhead 225 kg.
SAM: 1 Sea Chaparral quad launcher ❷; IR homing to 3 km (1.6 n miles) supersonic; warhead 5 kg.
Guns: 1 Oto Melara 76 mm/62 Mk 75 ❸; 85 rds/min to 16 km (8.7 n miles); weight of shell 6 kg. 1 Raytheon 20 mm/76 Vulcan Phalanx Block 1B ❹; 4,500 rds/min combined to 1.5 km (0.8 n miles). 2 Bofors 40 mm/70 ❺. 2 CS 20 mm Type 75.
Torpedoes: 6—324 mm Mk 32 (2 triple) tubes ❻; Alliant Mk 46 Mod 5; active/passive homing to 11 km (5.9 n miles) at 40 kt; warhead 44 kg.
Physical countermeasures: Decoys: 2 CSEE Dagaie chaff launchers ❼.
Electronic countermeasures: ESM/ECM: Thomson-CSF DR 3000S; intercept and jammer. Chang Feng IV (1206); intercept and jammer.
Radars: Air/surface search: Thomson-CSF DRBV-26D Jupiter II (with LW08 aerial) ❾; D-band.
Surface search: Thomson-CSF Triton G ❿; G-band.
Fire control: 2 Thomson-CSF Castor IIC ⓫; I/J-band.
Navigation/helo control: 2 Racal Decca 20V90; I-band.
Sonars: BAe/Thomson Sintra ATAS (V)2; active towed array. Thomson Sintra Spherion B; bow-mounted; active search; medium frequency.
Combat data systems: Thomson-CSF TACTICOS. Link W (Ta Chen).
Weapon control systems: CSEE Najir Mk 2 optronic director ❽.

Helicopters: 1 Sikorsky S-70C(M)1 ⓬ Thunderhawk.

Programmes: Sale of up to 16 of the class authorised by the French government in August 1991. Contract for six signed with Thomson-CSF in early 1992, manufactured in France with some weapon assembly by China SB Corporation at Kaohsiung in Taiwan. First one to Taiwan in March 1996 and the last in January 1998. Names are those of Chinese cities. Second batch of 10 to be built by China SB Corporation was planned but this now seems unlikely.

Modernisation: There are plans to move Phalanx to the bridge roof and fit two 10-round RAM launchers on the hangar. A mid-life upgrade of the ships is also under consideration.

Structure: There are considerable differences with the French 'La Fayette' design in both superstructure and weapon systems. A comprehensive ASW fit has been added as well as additional gun armament. There is also no stern hatch for launching RIBs. Some of the weapons were fitted after arrival in Taiwan. DCN Samahé helicopter landing gear installed.

Operational: Form 124 Squadron based at Tsoying.

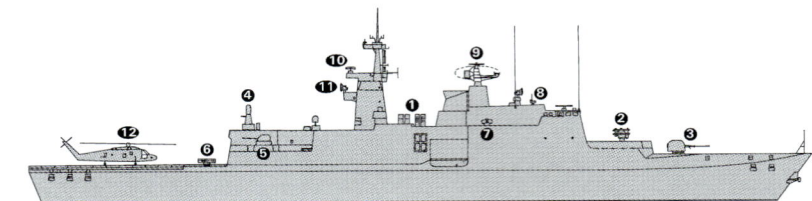

KANG DING (Scale 1 : 1,200), Ian Sturton / 0121405

KANG DING 4/2007, Chris Sattler / 1170236

CHEN TE 4/2011, Chris Sattler / 1454738

KUN MING 3/2010, Chris Sattler / 1406285

8 KNOX CLASS (FFGH)

Name	No	Builders	Laid down	Launched	Commissioned	Recommissioned
CHIN YANG (ex-*Robert E Peary*)	932 (ex-FF 1073)	Lockheed SB & Construction Co	20 Dec 1970	23 Jun 1971	23 Sep 1972	6 Oct 1993
FONG YANG (ex-*Brewton*)	933 (ex-FF 1086)	Avondale Shipyards, New Orleans	2 Oct 1970	24 Jul 1971	8 Jul 1972	6 Oct 1993
FENG YANG (ex-*Kirk*)	934 (ex-FF 1087)	Avondale Shipyards, New Orleans	4 Dec 1970	25 Sep 1971	9 Sep 1972	6 Oct 1993
LAN YANG (ex-*Joseph Hewes*)	935 (ex-FF 1078)	Avondale Shipyards, New Orleans	15 May 1969	7 Mar 1970	22 Apr 1971	4 Aug 1995
HAE YANG (ex-*Cook*)	936 (ex-FF 1083)	Avondale Shipyards, New Orleans	20 Mar 1970	23 Jan 1971	18 Dec 1971	4 Aug 1995
HWAI YANG (ex-*Barbey*)	937 (ex-FF 1088)	Avondale Shipyards, New Orleans	5 Feb 1971	4 Dec 1971	11 Nov 1972	4 Aug 1995
NING YANG (ex-*Aylwin*)	938 (ex-FF 1081)	Avondale Shipyards, New Orleans	13 Nov 1969	29 Aug 1970	18 Sep 1971	18 Oct 1999
YI YANG (ex-*Valdez*)	939 (ex-FF 1096)	Avondale Shipyards, New Orleans	30 Jun 1972	24 Mar 1973	27 Jul 1974	18 Oct 1999

Displacement, tonnes: 3,059 standard; 3,939 (932, 935), 4,328 (933–934) full load
Dimensions, metres (feet): 134 × 14.3 × 4.6 hull; 7.8 sonar *(439.6 × 46.9 × 15.1; 25.6)*
Speed, knots: 27. **Range, n miles:** 4,000 at 22 kt
Complement: 288 (17 officers)

Machinery: 2 Combustion Engineering/Babcock & Wilcox boilers; 1,200 psi *(84.4 kg/cm²)*; 950°F *(510°C)*; 1 turbine; 35,000 hp *(26 MW)*; 1 shaft
Missiles: SSM: 8 McDonnell Douglas Harpoon ❶; active radar homing to 130 km *(70 n miles)* at 0.9 Mach; warhead 227 kg.
SAM: 10 General Dynamics SM1-MR (2 triple, 2 twin) ❷; command guidance; semi-active radar homing to 46 km *(25 n miles)* at 2 Mach (fitted in all but 932 and 937).
A/S: Honeywell ASROC Mk 16 octuple launcher with reload system (has 2 cells modified to fire Harpoon) ❶; inertial guidance from 1.6–10 km *(1–5.4 n miles)*; payload Mk 46 Mod 5 Neartip.
Guns: 1 FMC 5 in *(127 mm)*/54 Mk 42 Mod 9 ❸; 20–40 rds/min to 24 km *(13 n miles)* anti-surface; 14 km *(7.7 n miles)* anti-aircraft; weight of shell 32 kg to be replaced by 1 Oto Melara 3 in *(76 mm)*/62 Mk 75; 85 rds/min to 16 km *(8.7 n miles)*; weight of shell 6 kg. 1 Raytheon 20 mm/76 6-barrelled Vulcan Phalanx Block 1B ❹; 4,500 rds/min combined to 1.5 km. 4 Type 75 20 mm.
Physical countermeasures: Decoys: 2 Loral Hycor SRBOC 6-barrelled fixed Mk 36 ❺; IR flares and chaff to 4 km *(2.2 n miles)*. T Mk 6 Fanfare/SLQ-25 Nixie; torpedo decoy. Prairie Masker hull and blade rate noise suppression.
Electronic countermeasures: ESM/ECM: SLQ-32(V)2 ❻; radar warning. Sidekick modification adds jammer and deception system.
Radars: Air search: Lockheed SPS-40B (fitted in 932, 937) ❼; B-band or Signaal DA 08; E/F-band.
Surface search: Raytheon SPS-10 or Norden SPS-67 ❽; G-band.
Navigation: Marconi LN66; I-band.
Fire control: Western Electric SPG-53A/D/F (fitted in 932, 937) ❾; or Signaal STIR; I/J-band.
Tacan: SRN 15. IFF: UPX-12.
Sonars: EDO/General Electric SQS-26CX; bow-mounted; active search and attack; medium frequency. EDO SQR-18A(V)1; passive towed array.
Combat data systems: Link 14 receive only. Link W may be fitted. FFISTS (Frigate Integrated Shipboard Tactical System). RADDS (Radar Displays and Distribution System).
Weapon control systems: SWG-1A Harpoon LCS. Mk 68 GFCS. Mk 114 ASW FCS. Mk 1 target designation system. SRQ-4 for LAMPS I.

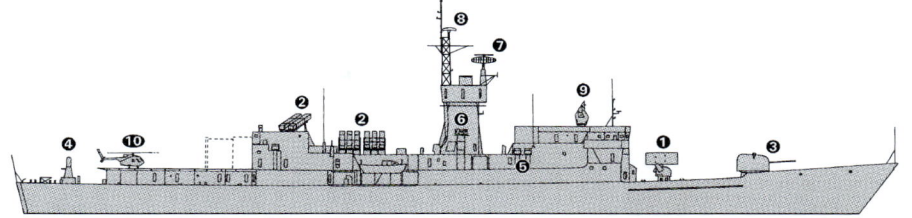

FONG YANG *(Scale 1 : 1,200), Ian Sturton / 1293480*

FONG YANG *3/2011, Ships of the World / 1406371*

Helicopters: 1 MD 500 ❿.

Programmes: *Fong Yang* leased from the US on 23 July 1992, *Chin Yang* 7 August 1992 and *Feng Yang* 6 August 1993. *Hae Yang* leased 31 May 1994; *Hwai Yang* 21 June 1994 and *Lan Yang* 30 June 1994. The second batch of three were overhauled and upgraded by Long Beach Shipyard, California. *Ning Yang* and *Yi Yang* transferred for sale on 29 April 1998, and refitted at Denton Shipyard, South Carolina. The transfer of a third (ex-*Pharris* 1094) was declined as were further offers of ex-*Whipple* (1062) and ex-*Downes* (1070) for use as spares.
Modernisation: A programme to equip all eight ships with a limited air-defence capability has been initiated. SPS-40 radar is being replaced by DA-08; SPG-53A is being replaced by STIR; 10 standard SM-1 MR (ex-Gearing class) are being installed on top of the hangar. The 127 mm gun is being replaced by the Oto Melara 76 mm/62.
Structure: ASROC-torpedo reloading capability (note slanting face of bridge structure immediately behind ASROC). Four Mk 32 torpedo tubes are fixed in the midships structure, two to a side, angled out at 45°. The arrangement provides improved loading capability over exposed triple Mk 32 torpedo tubes. A 4,000 lb lightweight anchor is fitted on the port side and an 8,000 lb anchor fits into the after section of the sonar.
Operational: Seasprite helicopters were planned to be embarked but this now seems unlikely. All of the class are assigned to 168 Patrol Squadron at Suao. *Lan Yang* is the Flagship.

SHIPBORNE AIRCRAFT

Notes: Negotiations to acquire SH-2F Seasprite helicopters for the Knox class, conducted for several years, have not been satisfactorily concluded.

Numbers/Type: 9 Hughes MD 500/ASW.
Operational speed: 110 kt *(204 km/h)*.
Service ceiling: 16,000 ft *(4,880 m)*.
Range: 203 n miles *(376 km)*.
Role/Weapon systems: Short-range ASW helicopter with limited surface search capability. 501 ASW Squadron. Sensors: Search radar, Texas Instruments ASQ 81(V)2 MAD. Weapons: ASW; one Mk 46 Mod 5 torpedo or two depth bombs. ASV; could carry machine gun pods.

Numbers/Type: 21 Sikorsky S-70C(M)1 Thunderhawks.
Operational speed: 145 kt *(269 km/h)*.
Service ceiling: 19,000 ft *(5,790 m)*.
Range: 324 n miles *(600 km)*.
Role/Weapon systems: First delivered in 1991. This is a variant of the SH-60B and became seaborne with the first Cheng Kung and Kang Ding class frigates. 701 and 702 Squadrons. Two modified for EW and Sigint role. Another 14 S-70B/C SAR and assault aircraft belong to the Air Force. Sensors: APS 128 search radar; Litton ALR 606(V)2 ESM; ARR 84 sonobuoy receiver with Litton ASN 150 datalink; Allied AQS 18(V)3 dipping sonar; ASQ 504 MAD. Ta Chen datalink to be fitted. Weapons: ASW; two Hughes Mk 46 Mod 5 torpedoes or two Mk 64 depth bombs. ASV; could carry ASM.

MD 500 *1/1995, L J Lamb / 0080778*

THUNDERHAWK *1/2000, C Chung / 0106599*

LAND-BASED MARITIME AIRCRAFT

Notes: (1) Four Grumman E-2T Hawkeye AEW aircraft were acquired by the Air Force in February 1995. These were augmented in 2005 by two further aircraft equipped with AN/APS-145 radar. The four earlier aircraft are reported to have been similarly upgraded by 2013.
(2) Plans to acquire 12 MH-53E Sea Dragon minehunting helicopters were also agreed in the 2001 agreement but is unlikely to proceed.

Numbers/Type: 3/21 Grumman S-2E/S-2T (Turbo) Trackers.
Operational speed: 130 kt (241 km/h).
Service ceiling: 25,000 ft (7,620 m).
Range: 1,350 n miles (2,500 km).
Role/Weapon systems: Patrol and ASW tasks transferred to the Navy in July 1998; 21 aircraft updated with turboprop engines and new sensors. Based at Pintung. To be augmented/replaced by P-3C as they enter service 2014–15. A total of 11 aircraft is to be transferred to the Air Force to form an ASW aviation group. Sensors: APS 504 search radar, ESM, MAD, AAS 40 FLIR, SSQ-41B, SSQ-47B sonobuoys; AQS 902F sonobuoy processor; ASN 150 datalink. Weapons: ASW; four Mk 44 torpedoes, Mk 54 depth charges or Mk 64 depth bombs or mines. ASV; Hsiung Feng II ASM; six 127 mm rockets.

Numbers/Type: 12 Lockheed P-3C Orion.
Operational speed: 410 kt (760 km/h).
Service ceiling: 28,300 ft (8,625 m).
Range: 4,000 n miles (7,410 km).
Role/Weapon systems: Plans to acquire 12 aircraft, under the 2001 US arms package, were confirmed on 16 February 2008 and, in early 2009, it was agreed that 12 refurbished aircraft were to be removed from storage in Arizona to be refurbished by Lockheed Martin at Greenville, South Carolina. Work includes structural service life extension and avionics upgrades. The first aircraft completed its first test-flight in July 2012 and was delivered on 25 September 2013. The programme is to be completed by August 2015. The aircraft are operated by the Air Force and, with 11 S-2T Trackers, are to form an ASW aviation group. Sensors and weapons to be confirmed.

TRACKER　　　　　　　　　　　6/2002, Adolfo Ortigueira Gil / 0569245

P-3C　　　　　　　　　9/2013*, ROC Ministry of National Defence / 1513534

PATROL FORCES

Notes: (1) All coastal patrol craft were transferred to the Maritime Police on 8 December 1992. The Maritime Police became the Coast Guard 1 February 2000.
(2) The Hsun Hai (Sea Swift) programme is for up to 12 corvettes to be armed with the Hsiung Feng III anti-ship missile. A contract to build a 450-tonne prototype vessel was let with Lung Teh Shipbuilding in May 2012. When completed, the catamaran design is to undergo a period of operational test and evaluation before decisions on the way-ahead are made. If the new design is suitable, it is expected that a larger 1,500-tonne variant will be developed for replacement of the Knox-class frigates.

12 JIN CHIANG CLASS (LARGE PATROL CRAFT) (PCG)

Name	No	Builders	Launched	Commissioned
JIN CHIANG	603	Lien-Ho, Kaohsiung	1 May 1994	1 Dec 1994
TAN CHIANG	605	China SB Corporation, Kaohsiung	18 Jun 1998	7 Sep 1999
HSIN CHIANG	606	China SB Corporation, Kaohsiung	14 Aug 1998	7 Sep 1999
FENG CHIANG	607	China SB Corporation, Kaohsiung	22 Oct 1998	29 Oct 1999
TSENG CHIANG	608	China SB Corporation, Kaohsiung	16 Nov 1998	29 Oct 1999
KAO CHIANG	609	China SB Corporation, Kaohsiung	15 Dec 1998	29 Oct 1999
JING CHIANG	610	China SB Corporation, Kaohsiung	13 May 1999	15 Feb 2000
HSIAN CHIANG	611	China SB Corporation, Kaohsiung	16 Jul 1999	15 Feb 2000
TSI CHIANG	612	China SB Corporation, Kaohsiung	22 Dec 1999	15 Feb 2000
PO CHIANG	614	China SB Corporation, Kaohsiung	22 Dec 1999	21 Jul 2000
CHAN CHIANG	615	China SB Corporation, Kaohsiung	21 Jan 2000	21 Jul 2000
CHU CHIANG	617	China SB Corporation, Kaohsiung	25 Feb 2000	21 Jul 2000

Displacement, tonnes: 691 full load
Dimensions, metres (feet): 61.4 × 9.5 × 2.9 (201.4 × 31.2 × 9.5)
Speed, knots: 25
Range, n miles: 4,150 at 15 kt
Complement: 50 (7 officers)

Machinery: 2 MTU 20V 1163 TB93 diesels; 20,128 hp(m) (14.79 MW); 2 shafts
Missiles: SSM: 4 Hsiung Feng I; radar or optical guidance to 36 km (19.4 n miles) at 0.9 Mach; warhead 150 kg or 4 Hsiung Feng II (606, 607); inertial guidance; active radar/IR homing to 130 km (70 n miles) at 0.85 Mach; warhead 225 kg.
Guns: 1—76 mm. 1 Bofors 40 mm/70. 1 CS 20 mm Type 75. 2—12.7 mm MGs.
Mines: 2 rails for Mk 6.
Depth charges: 2 racks.
Radars: Air/surface search: Marconi LN66; I-band.
Fire control: Hughes HR-76C5; I/J-band.
Navigation: Racal Decca Bridgemaster; I-band.
Sonars: Simrad; search and attack; high frequency.
Weapon control systems: Honeywell H 930 Mod 2 MFCS. Contraves WCS. Rafael Sea Eye FLIR; range out to 3 km.

Programmes: Kwang Hua Project 3 design by United Ship Design Centre. First one laid down 25 June 1993. Eleven more ordered 26 June 1997. Plans to procure a further 12 have probably been superseded by the new corvette programme.
Modernisation: Hsin Chiang and Feng Chiang have been upgraded with an Oto Melara 76 mm gun and four Hsiung Feng II missiles. A mast to carry the datalink radome has also been added. All ships may be similarly modified in due course and it is planned that Hsiung Feng II missiles are to be replaced with Hsiung Feng III.

HSIN CHIANG　　　　　　　　　1/2009, Ships of the World / 1305664

HSIN CHIANG
6/2006
1167508

Patrol forces — Amphibious forces < **Taiwan** 817

31 KWANG HUA 6 CLASS (PTG)

| FACG 60–66 | 68–75 | 77–84 | 86–93 |

Displacement, tonnes: 183 standard
Dimensions, metres (feet): 34.2 × 7.6 × 1.9 *(112.2 × 24.9 × 6.2)*
Speed, knots: 33
Range, n miles: 1,150 at 22 kt
Complement: 14

Machinery: 3 MTU 16V 4000 diesels; 9,600 hp *(7.2 MW)*; 3 shafts
Missiles: SSM: 4 Hsiung Feng II; inertial guidance; active radar/IR homing to 130 km *(70 n miles)* at 0.85 Mach; warhead 225 kg.
Guns: 1 CS 20 mm Type 75. 2—7.62 mm MGs.
Physical countermeasures: Decoys: Chaff launchers. ESM.
Radars: Surface search. Fire control.
Weapon control systems: Optronic director.

Comment: Funds allocated for the budget period July 1998 to June 2003 to build these craft in Taiwan to replace the Hai Ou class. Prototype (FACG 60) laid down in early 2001 and launched on 26 September 2002. Commissioned in October 2003, subsequently damaged by typhoon in September 2008 but reported repaired by December 2009. Construction of remaining craft by CSBC was started in 2007. The 5th Missile Boat Squadron was established at Tsoying in May 2010, the 1st Missile Boat Squadron at Suao on 7 April 2011 and the 2nd Missile Boat Squadron at Tsoying on 2 December 2011.

FACG 77   *4/2012, Ships of the World* / 1483648

8 NING HAI CLASS (LARGE PATROL CRAFT) (PCF)

| NING HAI PCL 1 | – PCL 3 | – PCL 6 | – PCL 8 |
| AN HAI PCL 2 | – PCL 5 | – PCL 7 | – PCL 9 |

Displacement, tonnes: 145 full load
Dimensions, metres (feet): 32 × 9 × 1.8 *(105 × 29.5 × 5.9)*
Speed, knots: 40
Complement: 18 (2 officers)

Machinery: 3 MTU 12V 396 TB93 diesels; 4,890 hp(m) *(3.6 MW)* sustained; 3 shafts
Guns: 1 Bofors 40 mm/60. 1 CS 20 mm Type 75.
Depth charges: 2 racks.
Radars: Surface search: Decca; I-band.
Sonars: Hull-mounted; active search and attack; high frequency.

Comment: Built to Vosper QAF design by China SB Corporation, Kaohsiung in 1987–90. Previously reported numbers had been exaggerated. They are used mainly for harbour defence against midget submarines and frogmen and also for Fishery protection tasks.

PCL 5   *6/2002, Ships of the World* / 0569243

0 + 1 (11) HSUN HAI CLASS (ATTACK CRAFT) (PGGF)

TUO JIANG 618

Displacement, tonnes: 500 full load
Dimensions, metres (feet): 60.4 × 14.0 × ? *(198.2 × 45.9 × ?)*
Speed, knots: 38
Complement: 41

Machinery: To be announced.
Missiles: SSM: 8 Hsiung Feng II; inertial guidance; active radar/IR homing to 130 km *(70 n miles)* at 0.85 Mach; warhead 225 kg.
Guns: 1—76 mm. 1 Raytheon 20 mm Vulcan Phalanx. 4—12.7 mm MGs.
Radars: To be announced.

Comment: A contract to build a catamaran-hulled vessel was let with Lung Teh Shipbuilding, Suao, in May 2012. The christening ceremony was conducted on 16 March 2014 and the ship is expected to enter service in 2015. The ship is of aluminium alloy construction and is expected to be the prototype for a class of 12 patrol craft under the Hsun Hai (Sea Swift) programme.

© 2014 IHS

TUO JIANG   *3/2014*, ROC Navy* / 1531196

AMPHIBIOUS FORCES

1 ANCHORAGE CLASS (LSDH)

Name	No	Builders	Commissioned
SHIU HAI	LSD 193 (ex-LSD 38)	General Dynamics, Quincy	27 Mar 1971
(ex-*Pensacola*)			

Displacement, tonnes: 8,738 standard; 13,920 full load
Dimensions, metres (feet): 168.6 × 25.6 × 6 *(553.1 × 84.0 × 19.7)*
Speed, knots: 22. **Range, n miles:** 14,800 at 12 kt
Complement: 374 (24 officers)

Military lift: 366 troops (18 officers); 2 LCU or 18 LCM 6 or 9 LCM 8 or 50 LVT; 1 LCM 6 on deck; 2 LCPL and 1 LCVP on davits. Aviation fuel, 90 tons
Machinery: 2 Foster-Wheeler boilers; 600 psi *(42.3 kg/cm^2)*; 870°F *(467°C)*; 2 De Laval turbines; 24,000 hp *(18 MW)*; 2 shafts
Guns: 2 Raytheon 20 mm/76 6-barrelled Vulcan Phalanx Mk 15; 3,000 rds/min combined to 1.5 km. 2—25 mm Mk 38. 6—12.7 mm MGs.
Physical countermeasures: Decoys: 4 Loral Hycor SRBOC 6-barrelled Mk 36; IR flares and chaff to 4 km *(2.2 n miles)*.
Electronic countermeasures: ESM: SLQ-32(V)1; intercept.
Radars: Air search: Lockheed SPS-40B; B-band.
Surface search: Raytheon SPS-10F; G-band.
Navigation: Marconi LN66; I-band.
Helicopters: Platform only.

Comment: First one acquired from US Navy 30 September 1999 and arrived in Taiwan on 2 June 2000. Transfer of ex-*Anchorage* (LSD 36) did not take place as expected in 2004 although procurement of a further amphibious ship remains a requirement. Has a docking well 131.1 × 15.2 m and two 50 ton cranes. Based at Tsoying.

SHIU HAI   *6/2000, Ships of the World* / 1167449

6 LCU 1466 CLASS (LCU)

HO SHAN (ex-*LCU 1596*) 488	HO MENG (ex-*LCU 1599*) 491
HO CHUAN (ex-*LCU 1597*) 489	HO MOU (ex-*LCU 1600*) 492
HO SENG (ex-*LCU 1598*) 490	HO SHOU (ex-*LCU 1601*) 493

Displacement, tonnes: 183 standard; 366 full load
Dimensions, metres (feet): 36.3 × 10.4 × 1.8 *(119.1 × 34.1 × 5.9)*
Speed, knots: 10
Range, n miles: 800 at 11 kt
Complement: 25 (est.)
Military lift: 167 tons or 300 troops
Machinery: 3 Gray Marine 64 YTL diesels; 675 hp *(504 kW)*; 3 shafts
Guns: 3 Oerlikon 20 mm. Some may also have 2—12.7 mm MGs.

Comment: Built by Ishikawajima Heavy Industries Co, Tokyo, Japan, for transfer to Taiwan; completed in March 1955. All originally numbered in 200 series; subsequently changed to 400 series.

HO CHUAN   *1991, DTM* / 0506105

IHS Jane's Fighting Ships 2014-2015

818 Taiwan > Amphibious forces

2 NEWPORT CLASS (LSTH)

Name	No	Builders	Commissioned
CHUNG HO (ex-*Manitowic*)	232 (ex-LST 1180)	Philadelphia Naval Shipyard	24 Jan 1970
CHUNG PING (ex-*Sumter*)	233 (ex-LST 1181)	Philadelphia Naval Shipyard	20 Jun 1970

Displacement, tonnes: 5,055 standard; 8,586 full load
Dimensions, metres (feet): 159.2 × 21.2 × 5.3 *(522.3 × 69.6 × 17.4)*
Speed, knots: 20. **Range, n miles:** 14,250 at 14 kt
Complement: 257 (13 officers)

Military lift: 400 troops; 500 tons vehicles; 3 LCVPs and 1 LCPL on davits
Machinery: 6 ALCO 16-251 diesels; 16,500 hp *(12.3 MW)* sustained; 2 shafts; cp props; bow thruster
Guns: 1 Raytheon 20 mm Vulcan Phalanx Mk 15; 3,000 rds/min combined to 1.5 km. 4 — 40 mm/60 (2 twin).
Electronic countermeasures: ESM: WD-2A (233); intercept.
ESM/ECM: Chang Feng III (232); intercept and jammer.
Radars: Surface search: Raytheon SPS-67; G-band.
Navigation: Marconi LN66; I-band.
Helicopters: Platform only.

Comment: First pair transferred from USA by lease confirmed for both ships on 1 July 1995. Refitted at Newport News and recommissioned 8 May 1997, sailing for Taiwan after a short operational work-up. Purchased outright on 29 September 2000. These ships unload by a 112 ft ramp over their bow. The ramp is supported by twin derrick arms. A ramp just forward of the superstructure connects the lower tank deck with the main deck and a vehicle passage through the superstructure provides access to the parking area amidships. A stern gate to the tank deck permits unloading of amphibious tractors into the water, or unloading of other vehicles into an LCU or on to a pier. Vehicle stowage covers 19,000 sq ft. Length over derrick arms is 562 ft *(171.3 m)*; full load draught is 11.5 ft forward and 17.5 ft aft. Bow thruster fitted to hold position offshore while unloading amphibious tractors.

CHUNG HO   6/2000, Sattler/Steele / 0106602

10 LST 1–510 AND 512–1152 CLASSES (LST)

CHUNG CHIEN (ex-*LST 716*) 205 (ex-679)
CHUNG SHUN (ex-*LST 732*) 208 (ex-624)
CHUNG KUANG (ex-*LST 503*) 216 (ex-646)
CHUNG SUO (ex-*Bradley County LST 400*) 217 (ex-667)
CHUNG CHI (ex-*LST 279*) 218
CHUNG CHUAN (ex-*LST 1030*) 221 (ex-651)
CHUNG CHIH (ex-*Sagadahoc County LST 1091*) 226 (ex-655)
CHUNG MING (ex-*Sweetwater County LST 1152*) 227 (ex-681)
CHUNG PANG (ex-*LST 578*) 230 (ex-629)
CHUNG YEH (ex-*Sublette County LST 1144*) 231 (ex-699)

Displacement, tonnes: 1,680 standard
Dimensions, metres (feet): 100 × 15.2 × 4.3 *(328.1 × 49.9 × 14.1)*
Speed, knots: 11.6. **Range, n miles:** 15,000 at 10 kt
Complement: 125 (est.)
Machinery: 2 GM 12-567A diesels; 1,800 hp *(1.34 MW)*; 2 shafts
Guns: Varies-up to 10 Bofors 40 mm/56 (2 twin, 6 single) with some modernised ships rearmed with 2 USN 3 in *(76 mm)*/50 and 6 — 40 mm (3 twin). Several Oerlikon 20 mm (twin or single).
Radars: Navigation: US SO 1, 2 or 8; I-band.

Comment: Constructed between 1943 and 1945. These ships have been rebuilt in Taiwan. Six transferred from US in 1946; two in 1947; one in 1948; eight in 1958; one in 1959; two in 1960; and one in 1961. Some have davits forward and aft. Pennant numbers have reverted to those used in the 1960s. One deleted in 1990, six more in 1993, one more in 1996 after going aground, and three more in 1998. *Chung Hai* 201 was deleted in 2010. The midships deck is occasionally used as a helicopter platform. These last 10 may be retained due to the cancellation of the programme for more locally built AKs.

CHUNG SUO   6/2000, DTM / 0126196

100 LCVPS AND ASSAULT CRAFT

Comment: Some ex-US, and some built in Taiwan. Most are armed with one or two 7.62 mm MGs. Two transferred to Indonesia in 1988. About 20 deleted in the last three years and 30 transferred to Honduras in 1996 for River operations. There are also a number of amphibious reconnaissance boats in the ARP 1000, 2000 and 3000 series. Form part of 151 Squadron.

TYPE 272   1989, DTM (Raymond Cheung) / 0506106

IHS Jane's Fighting Ships 2014-2015

2 TAIWAN TYPE LCU (LCU)

HO FONG LCU 497   **HO HU** LCU 498

Displacement, tonnes: 193 standard; 446 full load
Dimensions, metres (feet): 41.3 × 9.1 × 2.1 *(135.5 × 29.9 × 6.9)*
Speed, knots: 11. **Range, n miles:** 1,200 at 10 kt
Complement: 16
Military lift: 180 tons or 350 troops
Machinery: 4 Detroit diesels; 1,200 hp *(895 kW)*; 2 Kort nozzle props
Guns: 2 — 12.7 mm MGs.

Comment: Locally built versions of US types. Ramps at both ends.

HO FONG   6/2000, DTM / 0569237

160 LCM 6 CLASS (LCM)

Displacement, tonnes: 58 full load
Dimensions, metres (feet): 17.2 × 4.2 × 1.2 *(56.4 × 13.8 × 3.9)*
Speed, knots: 9
Military lift: 34 tons
Machinery: 2 diesels; 450 hp *(336 kW)*; 2 shafts
Guns: 1 — 12.7 mm MG.

Comment: Some built in the US, some in Taiwan. 20 were exchanged for torpedoes with Indonesia. Some 65 have been deleted. Form part of 151 Squadron.

LCM 6   7/2000, C Chung / 0106603

1 LST 512–1152 CLASS (FLAGSHIP) (AGF)

Name	No	Builders	Commissioned
KAO HSIUNG (ex-*Chung Hai*; ex-*Dukes County LST 735*)	LCC 1 (ex-219, ex-663)	Dravo Corporation, Neville Island	26 Apr 1944

Displacement, tonnes: 1,680 standard; 3,734 full load
Dimensions, metres (feet): 100 × 15.2 × 4.3 *(328.1 × 49.9 × 14.1)*
Speed, knots: 11.6. **Range, n miles:** 11,200 at 10 kt
Complement: 195

Machinery: 2 GM 12-567A diesels; 1,800 hp *(1.34 MW)*; 2 shafts
Guns: 8 Bofors 40 mm/56 (3 twin, 2 single).
Radars: Air search: Raytheon SPS 58; D-band.
Surface search: Raytheon SPS-10; G-band.

Comment: Launched on 11 March 1944. Transferred from US in May 1957 for service as an LST. Converted to a flagship for amphibious operations and renamed and redesignated (AGC) in 1964. Purchased November 1974. Note lattice mast above bridge structure, modified bridge levels, and antenna mountings on main deck. Redesignated as Command and Control Ship LCC 1.

KAO HSIUNG   6/1999 / 0080783

© 2014 IHS

MINE WARFARE FORCES

Notes: There are plans to acquire eight GRP minehunters to replace the inventory of ageing wooden-hull minesweepers.

2 OSPREY CLASS (MINEHUNTERS—COASTAL) (MHC)

Name	No	Builders	Launched	Commissioned
YUNG JIN (ex-*Oriole*)	1310 (ex-MHC 55)	Intermarine, Savannah	22 May 1993	16 Sep 1995
YUNG AN (ex-*Falcon*)	1311 (ex-MHC 59)	Intermarine, Savannah	3 Jun 1995	26 Oct 1997

Displacement, tonnes: 945 full load
Dimensions, metres (feet): 57.2 × 11 × 2.9 *(187.7 × 36.1 × 9.5)*
Speed, knots: 10. **Range, n miles:** 1,500 at 10 kt
Complement: 51 (5 officers)

Machinery: 2 Isotta Fraschini ID 36 SS 8V AM diesels; 1,600 hp(m) *(1.18 MW)* sustained; 2 Voith-Schneider props; 3 Isotta Fraschini ID 36 diesel generators; 984 kW
Guns: 2—12.7 mm MGs.
Physical countermeasures: MCM: Alliant SLQ-48 mine neutralisation system ROV (with 1,070 m cable). Degaussing DGM-4.
Radars: Surface search: Raytheon SPS-64(V)9; I-band. Navigation: R41XX; I-band.
Sonars: Raytheon/Thomson Sintra SQQ-32(V)3; VDS; active minehunting; high frequency.
Combat data systems: Unisys SYQ 13 and SYQ 109; integrated combat and machinery control system. USQ-119E(V), UHF Dama, and OTCIXS provide GCCS connectivity.

Programmes: Original design contract for Lerici-class minehunters was awarded in August 1986 to Intermarine USA which built eight of the 12 ships of the class for the US Navy. Transfer of both vessels to Taiwan was authorised in 2007 and confirmed on 29 January 2010 by the Pentagon as part of a wider arms package. Both vessels were transferred after a reactivation and overhaul work package and recommissioned in August 2012.
Structure: Construction is of monocoque GRP throughout hull, with frames eliminated. Main machinery is mounted on GRP cradles and provided with acoustic enclosures. SQQ-32 is deployed from a central well forward. Fitted with Voith cycloidal propellers which eliminate need for forward thrusters during station keeping.

YUNG AN   8/2012, *Ships of the World* / 1483647

4 YUNG FENG (MWV 50) CLASS
(MINEHUNTERS—COASTAL) (MHC)

YUNG FENG 1301   YUNG CHIA 1302   YUNG TING 1303   YUNG SHUN 1305

Displacement, tonnes: 508 full load
Dimensions, metres (feet): 49.7 × 8.7 × 3.1 *(163.1 × 28.5 × 10.2)*
Speed, knots: 14. **Range, n miles:** 3,500 at 14 kt
Complement: 45 (5 officers)

Machinery: 2 MTU 8V 396 TB93 diesels; 2,180 hp(m) *(1.6 MW)* sustained; 2 shafts
Guns: 1—20 mm. 2—12.7 mm MGs.
Radars: Navigation: I-band.
Sonars: TSM-2022; hull-mounted; active minehunting; high frequency.

Comment: Built for the Chinese Petroleum Corporation by Abeking & Rasmussen at Lemwerder, Germany. First four delivered in 1991 as offshore oil rig support ships and then converted for minehunting in Taiwan. Thomson Sintra IBIS V minehunting system is fitted and two STN Pinguin B3 ROVs are carried.

YUNG KU   6/2000, DTM / 0569242

4 ADJUTANT AND MSC 268 CLASSES
(MINESWEEPERS—COASTAL) (MSC)

YUNG CHUAN (ex-*MSC 278*) 158   YUNG REN (ex-*St Nicholas*; ex-*MSC 64*) 167
YUNG FU (ex-*Macaw*; ex-*MSC 77*) 162   YUNG SUI (ex-*Disksmude*; ex-*MSC 65*) 168

Displacement, tonnes: 381 full load
Dimensions, metres (feet): 43.9 × 8.5 × 2.4 *(144 × 27.9 × 7.9)*
Speed, knots: 13
Range, n miles: 2,500 at 12 kt
Complement: 35

Machinery: 2 GM 8-268A diesels; 880 hp *(656 kW)*; 2 shafts
Guns: 1 Oerlikon 20 mm.
Radars: Navigation: Decca 707; I-band.
Sonars: Simrad 950; hull-mounted; minehunting; high frequency.

Comment: Non-magnetic, wood-hulled minesweepers built in the US in the 1950s specifically for transfer to allied navies. All refitted 1984–86. All are in very poor condition. Several deleted so far. Two put back in service in 1996 and one in 1997 to replace three others paid off.

YUNG FENG   6/2000, DTM / 0569240

4 AGGRESSIVE CLASS (MINESWEEPERS) (MSO)

Name	No	Builders	Commissioned
YUNG YANG (ex-*Implicit*)	1306 (ex-455)	Wilmington Boat	10 Mar 1954
YUNG TZU (ex-*Conquest*)	1307 (ex-488)	Martenac, Tacoma	20 Jul 1955
YUNG KU (ex-*Gallant*)	1308 (ex-489)	Martenac, Tacoma	14 Sep 1955
YUNG TEH (ex-*Pledge*)	1309 (ex-492)	Martenac, Tacoma	20 Apr 1956

Displacement, tonnes: 732 standard; 793 full load
Dimensions, metres (feet): 52.6 × 10.7 × 4.3 *(172.6 × 35.1 × 14.1)*
Speed, knots: 14. **Range, n miles:** 3,000 at 10 kt
Complement: 86 (7 officers)

Machinery: 4 Packard ID-1700 or Waukesha diesels; 2,280 hp *(1.7 MW)*; 2 shafts; cp props
Guns: 2—12.7 mm MGs.
Radars: Navigation: Sperry SPS-53L; I-band.
Sonars: General Electric SQQ-14; VDS; active minehunting; high frequency.

Comment: Transferred by sale to Taiwan from the USN 3 August and 30 September 1994. Delivery was delayed into 1995 while replanking work was carried out in the US. All recommissioned 1 March 1995. Second batch of three planned to transfer but were subsequently scrapped after cannibalisation for spares. All are fitted with SLQ-37 mechanical acoustic and magnetic sweeps and can carry an ROV. Plans to update the class with a Unisys SYQ-12 minehunting system and Pluto ROVs have probably been overtaken by the new MCMV programme.

YUNG CHUAN   6/2000, DTM / 0569241

826 Tanzania > Patrol forces — Auxiliaries

2 HUCHUAN CLASS (FAST ATTACK CRAFT—TORPEDO) (PTK)

P 43   P 44

Displacement, tonnes: 40 standard; 47 full load
Dimensions, metres (feet): 21.8 × 6.3 × 3.6 *(71.5 × 20.7 × 11.8)*
Speed, knots: 50
Complement: 16

Machinery: 3 Type M 50 diesels; 3,300 hp(m) *(2.4 MW)* sustained; 3 shafts
Guns: 2—25 mm (twin). 2—14.5 mm (twin).
Torpedoes: 2—21 in *(533 mm)* tubes.
Radars: Surface search: Skin Head; E/F-band.

Comment: Four transferred from the People's Republic of China 1975. After a major effort in 1992, were all operational and reported to be in good condition but by 1998 two had been laid up. Present operational status is unclear but one at least appears to have had torpedo tubes removed. Based at Kigoma, Lake Tanganyika.

HUCHUAN   6/2003 / 0587794

2 VOSPER THORNYCROFT 75 ft TYPE (COASTAL PATROL CRAFT) (PB)

Displacement, tonnes: 71 full load
Dimensions, metres (feet): 22.9 × 6 × 2.4 *(75.1 × 19.7 × 7.9)*
Speed, knots: 24.5
Range, n miles: 800 at 20 kt
Complement: 11
Machinery: 2 Caterpillar D 348 diesels: 1,450 hp *(1.08 MW)* sustained; 2 shafts
Guns: 2 Oerlikon 20 mm GAM-BO1.
Radars: Surface search: Furuno; I-band.

Comment: First pair delivered 6 July 1973, second pair 1974. Used for anti-smuggling patrols off Zanzibar. Two still operational.

VOSPER 75 ft (Omani colours)   1984, N Overington / 0506066

2 PROTECTOR CLASS (PATROL CRAFT) (PB)

NGUNGURI (ex-*Vincent*) P 19   MAMBA (ex-*Vigilant*) KM 205 (ex-P 20)

Displacement, tonnes: 102 full load
Dimensions, metres (feet): 25.7 × 6.2 × 1.7 *(84.3 × 20.3 × 5.6)*
Speed, knots: 25
Complement: 4
Machinery: 2 Paxman diesels; 2,880 hp *(2.15 MW)*; 2 shafts. 1 Perkins diesel; 200 hp *(150 kW)*; 1 waterjet
Radars: Navigation: 2 Decca; I-band.

Comment: Both built for UK Customs by FBM Marine in 1998 (ex-*Vigilant*) and 1993 (ex-*Vincent*). Subsequently sold to Damen Shipyards, Netherlands, in September 2004. Following refit, entered Tanzanian service in 2005. Based at Dar Es Salaam.

MAMBA   6/2013*, Mazumdar Collection / 1531062

2 DEFENDER CLASS (PBF)

001   002

Displacement, tonnes: 2.7 full load
Dimensions, metres (feet): 7.6 × 2.6 × 2.7 *(24.9 × 8.5 × 8.9)*
Speed, knots: 46
Range, n miles: 175 at 35 kt
Complement: 4
Machinery: 2 Honda outboard motors; 450 hp *(335 kW)*
Guns: 1—12.7 mm MG.
Radars: To be announced.

Comment: High-speed inshore patrol craft of aluminium construction and foam collar built by SAFE Boats International, Port Orchard, Washington. Donated by the United States government in November 2010.

001   6/2013*, US Navy / 1531063

AUXILIARIES

Notes: There is a large tug *Karambisi*.

2 YUCH'IN (TYPE 069) CLASS (LCU)

PONO L 08   KIBUA L 09

Displacement, tonnes: 86 full load
Dimensions, metres (feet): 24.8 × 5.2 × 1.3 *(81.4 × 17.1 × 4.3)*
Speed, knots: 12
Range, n miles: 450 at 11.5 kt
Complement: 12

Military lift: 46 tons
Machinery: 2 diesels; 600 hp(m) *(441 kW)*; 2 shafts
Guns: 4—14.5 mm (2 twin) MGs.
Radars: Navigation: Fuji; I-band.

Comment: Transferred from China in 1995 probably to replace the Police Yuchai transport craft. Based at Dar-es-Salaam. *Pono* reported to be operational.

PONO   1/2001 / 0109946

1 LANDING CRAFT (LCU)

KASA

Measurement, tonnes: 473 gt
Dimensions, metres (feet): 51.2 × 12.8 × 2.0 *(168 × 42.0 × 6.6)*
Speed, knots: 10
Complement: 4
Machinery: 2 Kelvin diesels; 830 hp *(610 kW)*; 2 shafts
Radars: To be announced.

Comment: Built by Phoenix Shipbuilders, Kingston upon Hull in 1978.

KASA
6/2013*, Mazumdar Collection
1531061

Thailand

Country Overview

The Kingdom of Thailand (formerly Siam) is a constitutional monarchy in South East Asia. With an area of 198,114 square miles, it is bordered to the west by Burma, to the east by Laos and Cambodia and to the south by Malaysia. It has a 1,739 n mile coastline with the Gulf of Thailand and with the Andaman Sea. The capital, largest city and principal port (which also serves neighbouring Laos) is Bangkok. Territorial seas (12 n miles). An EEZ (200 n miles) is claimed and the limits have been partly defined by boundary agreements.

Headquarters Appointments

Commander-in-Chief of the Navy:
  Admiral Narong Pipatanasai
Deputy Commander-in-Chief:
  Admiral Jakchai Poocharoenyos
Assistant Commander-in-Chief:
  Admiral Kraisorn Chansuvanich
Chief of Staff: Admiral Tawewuth Pongsapipatt
Deputy Chiefs of Staff: Vice Admiral Yutthana Kerdduayboon
  Vice Admiral Panlop Tamisanon
  Vice Admiral Graivut Vattanatham

Senior Appointments

Commander-in-Chief, Fleet:
  Vice Admiral Pichan Dhiranetra
Deputy Commander-in-Chief, Fleet:
  Vice Admiral Vasan Changyodsuk
Chief of Staff, Fleet:
  Vice Admiral Saknarin Charoensuk

Diplomatic Representation

Naval Attaché in London:
  Captain Worawut Pruksarungruang
Naval Attaché in Washington:
  Captain Suvin Jangyodsuk
Naval Attaché in Paris:
  Captain Prachacharti Sirisawat
Naval Attaché in Canberra:
  Captain Apichai Sompolgrunk
Naval Attaché in Madrid:
  Captain Amnuay Thongrod
Naval Attaché in New Delhi:
  Captain Talerngsak Sirisawat
Naval Attaché in Singapore:
  Captain Prawin Chittinan
Naval Attaché in Kuala Lumpur:
  Captain Pogrong Monthardpalin
Naval Attaché in Beijing:
  Captain Paisan Chanopat
Naval Attaché in Rome:
  Captain Sombat Naraviroj
Naval Attaché in Manila:
  Captain Tanyakorn Senalaksna
Naval Attaché in Tokyo:
  Captain Satit Naksung
Naval Attaché in Yangon:
  Captain Apakorn Youkongkaew
Naval Attaché in Hanoi:
  Captain Prakob Suksamai
Naval Attaché in Phnom Penh:
  Captain Ronnarong Sittinan
Naval Attaché in Jakarta:
  Captain Huttakomun Sirathanapornpat
Naval Attaché in Moscow:
  Captain Chutintorn Thadtanone
Naval Attaché in Berlin:
  Captain Soonpuen Somapee

Personnel

(a) 2014: Navy, 42,000 (including Marines and Air and Coastal Command)
(b) 2 years' national service (37,000 conscripts)

Organisation

First naval area command (Upper Thai Gulf)
Second naval command (Lower Thai Gulf)
Third naval command (Andaman Sea)

Bases

Bangkok, Sattahip, Songkhla, Phang-Nga (west coast)

Naval Aviation

First air wing (U-Tapao)
Second air wing (Songkhla)
101 Sqdn MPA/ASW
102 Sqdn MPA/ASuW
103 Sqdn Utility
104 Sqdn Maritime Strike
105 Sqdn Matador
201 Sqdn Central Patrol
202 Bell Helos
203 Sikorsky Helos

Prefix to Ships' Names

HTMS

Strength of the Fleet

Type	Active	Building (Projected)
Submarines	—	(4)
Aircraft Carrier	1	—
Frigates	9	(2)
Corvettes	10	—
Fast Attack Craft (Missile)	6	—
Fast Attack Craft (Gun)	3	—
Offshore Patrol Craft	9	—
Coastal Patrol Craft	43	6 (3)
MCM Support Ship	1	—
Minehunters	4	—
MSBs	12	—
LPDs	1	—
LSTs	2	—
Hovercraft	3	—
Survey Vessels	4	—
Replenishment Ship	1	—
Tankers/Transports	7	1
Training Ships	1	—

Coast Defence

Coastal Defence Command was rapidly expanded to the 1992 two Division level after the government charged the RTN with the responsibility of defending the entire Eastern Seaboard. Ships and aircraft are rotated monthly from the Navy. Equipment includes 10 batteries of truck-mounted Exocet MM 40, 155 and 130 mm guns for coastal defence, 76, 40, 37, 20 mm guns and PL-9B SAM for air defence.

Marine Police

Acts as a Coast Guard in inshore waters with some 60 armed patrol craft and another 65 equipped with small arms only.

PENNANT LIST

Aircraft Carriers

911 Chakri Naruebet

Frigates

421 Naresuan
422 Taksin
433 Makut Rajakumarn
455 Chao Phraya
456 Bangpakong
457 Kraburi
458 Saiburi
461 Phuttha Yotfa Chulalok
462 Phuttha Loetla Naphalai

Corvettes

431 Tapi
432 Khirirat
441 Rattanakosin
442 Sukothai
511 Pattani

512 Narathiwat
531 Khamronsin
532 Thayanchon
533 Longlom
551 Krabi

Patrol Forces

311 Prabparapak
312 Hanhak Sattru
313 Suphairin
321 Ratcharit
322 Witthayakhom
323 Udomdet
331 Chon Buri
332 Songkhla
333 Phuket
521 Sattahip
522 Klongyai
523 Takbai
524 Kantang
525 Thepha
526 Taimuang
541 Hua Hin
542 Klaeng
543 Si Racha

Mine Warfare Forces

621 Thalang
631 Bang Rachan
632 Nongsarai
633 Lat Ya
634 Tha Din Daeng

Amphibious Forces

721 Sichang
722 Surin
771 Thong Kaeo
772 Thong Lang
773 Wang Nok
774 Wang Nai
781 Man Nok
782 Man Klang
783 Man Nai
784 Mattapon
785 Rawi
791 Angthong

Training Ships

413 Pin Klao

Survey and Research Ships

811 Chanthara
812 Suk
813 Pharuehaisbodi
821 Suriya

Auxiliaries

831 Chula
832 Samui
833 Prong
834 Proet
835 Samed
836 Matra (bldg)
841 Chuang
842 Chik
851 Klueng Badaan
852 Marn Vichai
853 Rin
854 Rang
855 Samaesan
856 Raet
871 Similan

SUBMARINES

Notes: The acquisition of a submarine capability is under consideration. Options include the procurement of up to four ex-German Type 206A submarines (ex-U 16, U 18, U 23, and U 24) with two additional boats (ex-U 15 and U 17) as spares. A contract for the provision of submarine technology training courses was signed with ThyssenKrupp Marine Systems on 9 January 2013. Purchase of submarines from South Korea is another possibility.

AIRCRAFT CARRIERS

1 CHAKRI NARUEBET CLASS (CVM)

Name	No	Builders	Laid down	Launched	Commissioned
CHAKRI NARUEBET	911	Bazán, Ferrol	12 Jul 1994	20 Jan 1996	27 Mar 1997

Displacement, tonnes: 11,669 full load
Dimensions, metres (feet): 182.6 oa; 164.1 wl × 30.5 oa; 22.5 wl × 6.2 *(599.1; 538.4 × 100.1; 73.8 × 20.3)*
Flight deck, metres: 174.6 × 27.5 *(572.8 × 90.2)*
Aircraft lift: 2
Speed, knots: 26; 16 diesel
Range, n miles: 10,000 at 12 kt
Complement: 601 (62 officers; 146 air crew) + 4 scientists

Machinery: CODOG; 2 GE LM 2500 gas turbines; 44,250 hp *(33 MW)* sustained; 2 MTU 16V 1163 TB83 diesels; 11,780 hp(m) *(8.67 MW)*; 2 shafts; LIPS cp props
Missiles: SAM: 1 Mk 41 LCHR 8 cell VLS launcher (fitted for but not with) ❶. 4 Matra Sadral sextuple launchers for Mistral ❷; IR homing to 4 km *(2.2 n miles)*; warhead 3 kg.
Guns: 2 — 30 mm. To be fitted.
Radars: Air search: Hughes SPS-52C ❸; E/F-band.
Surface search: SPS-64 ❹; I-band. To be fitted.
Fire control: to be fitted.
Navigation: Kelvin Hughes; I-band.
Aircraft control: Kelvin Hughes; E/F-band.
Tacan: URN 25.
Combat data systems: Tritan derivative with Unisys UYK-3 and 20 computers.

Fixed-wing aircraft: 6 AV-8S Matador (Harrier).
Helicopters: 6 S-70B-7 Seahawk; Chinook capable.

Programmes: An initial contract for a 7,800 ton vessel with Bremer Vulcan was cancelled on 22 July 1991 and replaced on 27 March 1992 with a government to government contract for a larger ship to be built by Bazán. Fabrication started in October 1993. Sea trials conducted from November 1996 to January 1997 followed by an aviation work-up at Rota from April 1997. The ship arrived in Thailand on 10 August 1997.

Structure: Similar to Spanish *Príncipe de Asturias*. 12° ski jump and two 20 ton aircraft lifts. Provision made to fit a Mk 41 VLS launcher, a surface search radar, EW systems, a hull mounted sonar and CIWS. Matra Sadral fitted in 2001. Hangar can take up 10 Sea Harrier or Seahawk aircraft.

Operational: Main tasks are SAR co-ordination and EEZ surveillance. Secondary role is air support for all maritime operations. Due to funding shortages, the ship rarely goes to sea and fixed-wing flying has been conducted from shore bases.

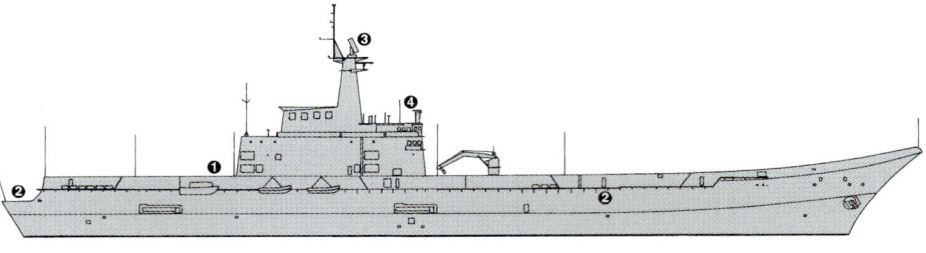

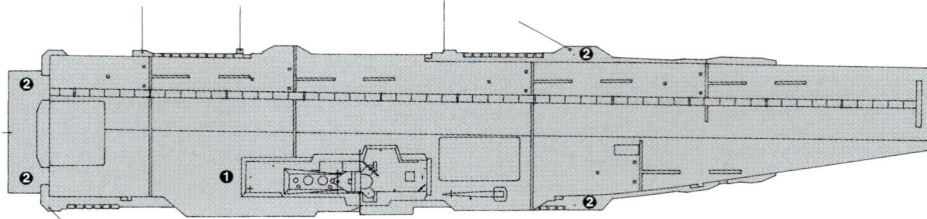

CHAKRI NARUEBET *(Scale 1 : 1,500), Ian Sturton / 0080799*

CHAKRI NARUEBET *5/1997, S G Gaya / 0019250*

CHAKRI NARUEBET *1/2004, Thai Navy League / 0589816*

CHAKRI NARUEBET *1/2004, Thai Navy League / 0589817*

FRIGATES

2 NARESUAN CLASS (TYPE 25T) (FFGHM)

Name	No	Builders	Laid down	Launched	Commissioned
NARESUAN	421 (ex-621)	Zhonghua Shipyard, Shanghai	Feb 1992	24 Jul 1993	15 Dec 1994
TAKSIN	422 (ex-622)	Zhonghua Shipyard, Shanghai	Nov 1992	14 May 1994	28 Sep 1995

Displacement, tonnes: 2,540 standard; 3,028 full load
Dimensions, metres (feet): 120 × 13 × 3.8 *(393.7 × 42.7 × 12.5)*
Speed, knots: 32
Range, n miles: 4,000 at 18 kt
Complement: 150

Machinery: CODOG; 2 GE LM 2500 gas turbines; 44,250 hp *(33 MW)* sustained; 2 MTU 20 V 1163 TB83 diesels; 11,780 hp(m) *(8.67 MW)* sustained; 2 shafts; LIPS cp props
Missiles: SSM: 8 McDonnell Douglas Harpoon (2 quad) launchers ❶; active radar homing to 130 km *(70 n miles)* at 0.9 Mach; warhead 227 kg.
SAM: Mk 41 CHR 8 cell VLS launcher ❷ Evolved Sea Sparrow RIM-162; semi-active radar homing to 18 km *(9.7 n miles)* at 3.6 Mach; warhead 38 kg (to be installed in mid-life upgrade).
Guns: 1 FMC 5 in *(127 mm)*/54 Mk 45 Mod 2 ❸; 20 rds/min to 23 km *(12.6 n miles)*; weight of shell 32 kg. 4 China 37 mm/76 (2 twin) H/PJ 76 A ❹; 180 rds/min to 8.5 km *(4.6 n miles)* anti-aircraft; weight of shell 1.42 kg.
Torpedoes: 6—324 mm Mk 32 Mod 5 (2 triple) tubes ❺. Honeywell Mk 46; active/passive homing to 11 km *(5.9 n miles)* at 40 kt; warhead 44 kg.
Physical countermeasures: Decoys: 4 China Type 945 GPJ 26-barrelled launchers ❻; chaff and IR.
Electronic countermeasures: ESM/ECM: Elettronica Newton Beta EW System: intercept and jammer.
Radars: Air search: Signaal LW08 ❽; D-band.
Surface search: China Type 360 ❾; E/F-band.
Navigation: 2 Raytheon SPS-64(V)5; I-band.
Fire control: 2 Signaal STIR ❿; I/J/K-band (for SSM and 127 mm). After one to be fitted. China 374 G ⓫ (for 37 mm).
Sonars: China SJD-7; hull-mounted; active search and attack; medium frequency.
Combat data systems: Saab, Link E. Link GADLS.
Weapon control systems: 1 JM-83H Optical Director ❼.

Helicopters: 1 Super Lynx ⓬ in due course or 1 Sikorsky S-70B-7 Seahawk.

Programmes: Contract signed 21 September 1989 for construction of two ships by the China State Shipbuilding Corporation (CSSC) with delivery in 1994. US and European weapon systems were fitted as funds became available. The first ship sailed for Bangkok without most weapon systems in January 1995 with the second following in October 1995.
Modernisation: A mid-life upgrade programme started in 2013. Evolved Sea Sparrow RIM-162 is to be installed forward of the bridge in the space allocated to an eight-cell Mk 41 launcher (which was not fitted on build as originally intended). The LW 08 and Type 360 radars are to be replaced by Saab Sea Giraffe AMB, the navigation radar by KH SharpEye, the JM-83H optical director is to be replaced by EOS 500 and the Elettronica EW system is to be replaced by CS 3701 Condor. In addition, the sonar is to be replaced, 9LV Mk 4 combat management and Ceros 200 fire-control radar are to be installed. Hull, mechanical and engineering work is also to be carried out.

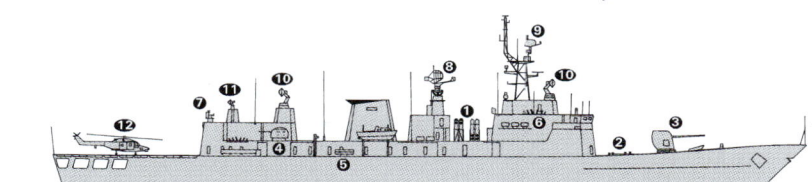

NARESUAN (Scale 1 : 1,200), Ian Sturton / 0543398

TAKSIN 8/2005, Chris Sattler / 1153917

TAKSIN 3/2011, Jurg Kürsener / 1406471

Structure: Jointly designed by the Royal Thai Navy and CSSC. This is a design incorporating much Western machinery and equipment and provides enhanced capabilities by comparison with the four Type 053 class. The anti-aircraft guns are Breda 40 mm types with 37 mm ammunition and they are controlled by a Chinese RTN-20 Dardo tracker.
Operational: *Naresuan* acted as one of the escorts for the aircraft carrier during her aviation work-up in Spanish waters in 1997.

NARESUAN 10/2008, Michael Nitz / 1353391

0 + 2 FRIGATE (FFGHM)

Displacement, tonnes: 3,700 full load
Dimensions, metres (feet): 123 × 14.4 × ? *(403.5 × 47.2 × ?)*
Speed, knots: 30
Range, n miles: 4,000 at 18 kt
Complement: 136

Machinery: 1 gas turbine; 2 diesels; 2 shafts; cp props
Missiles: SSM: 8 missiles to be announced.
SAM: 32 Raytheon RIM-162 ESSM; Mk 41 8-cell launcher.
Guns: 1 Oto Melara 76 mm. 2 MSI DS30 30 mm. 1 Vulcan Phalanx Block 1B. 2—12.7 mm MGs.
Torpedoes: 6—324 mm (2 twin).
Physical countermeasures: Nulka.
Electronic countermeasures: CESM. RESM.
Radars: Air/surface search: E/F-band.
Surface search: E/F-band.
Fire control: Thales Sting; I/J-band.
Navigation: I-band.
Sonars: Atlas Elektronik bow and towed array sonar.
Combat data systems: To be announced.
Electro-optic systems: Thales Mirador. Thales Sting optronic director.

Helicopters: 1 S-70B Seahawk or MH-60S.

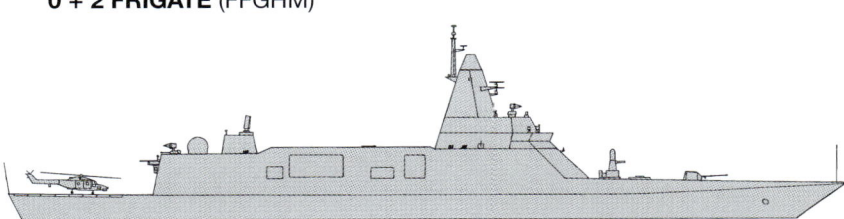

FUTURE FRIGATE (Scale 1 : 1,200), Ian Sturton / 1529099

Comment: A contract for the construction of two frigates was signed with Daewoo Shipbuilding and Marine Engineering on 7 August 2013. The first ship is to be delivered in 2016 and the second in 2018. Design features include an integrated mast for sensors and communications. Both ships are to replace the Knox class.

4 CHAO PHRAYA CLASS (TYPES 053 HT AND 053 HT (H)) (FFG/FFGH)

Name	No	Builders	Laid down	Launched	Commissioned
CHAO PHRAYA	455	Hudong SY, Shanghai	1989	24 Jun 1990	5 Apr 1991
BANGPAKONG	456	Hudong SY, Shanghai	1989	25 Jul 1990	20 Jul 1991
KRABURI	457	Hudong SY, Shanghai	1990	28 Dec 1990	16 Jan 1992
SAIBURI	458	Hudong SY, Shanghai	1990	27 Aug 1991	4 Aug 1992

Displacement, tonnes: 1,703 standard; 1,955 full load
Dimensions, metres (feet): 103.2 × 11.3 × 3.1 *(338.6 × 37.1 × 10.2)*
Speed, knots: 30
Range, n miles: 3,500 at 18 kt
Complement: 168 (22 officers)

Machinery: 4 MTU 20V 1163 TB83 diesels; 29,440 hp(m) *(21.6 MW)* sustained; 2 shafts; LIPS cp props
Missiles: SSM: 8 C 802A (YJ-83) launchers ❶; mid-course guidance and active radar homing to 120 km *(65 n miles)* at 0.9 Mach; warhead 165 kg.
SAM: 1 HQ-61 launcher for PL-9 or Matra Sadral for Mistral to be fitted.
Guns: 2 (457 and 458) or 4 China 100 mm/56 (1 or 2 twin) ❷; 25 rds/min to 22 km *(12 n miles)*; weight of shell 15.9 kg. 8 China 37 mm/76 (4 twin) H/PJ 76 A ❸; 180 rds/min to 8.5 km *(4.6 n miles)* anti-aircraft; weight of shell 1.42 kg.
A/S Mortars: 2 RBU 1200 (China Type 86) 5-tubed fixed launchers ❹; range 1,200 m.
Depth charges: 2 BMB racks.
Physical countermeasures: Decoys: 2 China Type 945 GPJ 26-barrelled chaff launchers.
Electronic countermeasures: ESM: China Type 923(1); intercept.
ECM: China Type 981(3); jammer.
Radars: Air/surface search: China Type 354 Eye Shield ❺; G-band.
Surface search/fire control: China Type 352C Square Tie ❻; I-band (for SSM).
Fire control: China Type 343 Sun Visor ❼; I-band (for 100 mm). Type 347G(1) Rice Bowl ❽; I-band (for 37 mm).
Navigation: Racal Decca 1290 A/D ARPA and Anritsu RA 71CA ❾; I-band.
IFF: Type 651.
Sonars: China Type SJD-5A; hull-mounted; active search and attack; medium frequency.
Combat data systems: China Type ZKJ-3 or STN Atlas mini COSYS action data automation.

Helicopters: Platform for 1 Bell 212 (457 and 458) ❿.

Programmes: Contract signed 18 July 1988 for four modified Jianghu-class ships to be built by the China State SB Corporation (CSSC).
Modernisation: A mini COSYS system was acquired for *Chao Phraya* and *Bangpakong* in 1999. Both ships were upgraded by 2012. C-801 missiles have been replaced by C-802, the 37 mm systems have been upgraded and EW systems modernised.
Structure: Thailand would have preferred only the hulls but China insisted on full armament. The first two ships are the Type III variant with 100 mm guns, fore and aft, and the second two are a variation with a helicopter platform replacing the after 100 mm gun. German communication equipment fitted. The EW fit is Italian designed.
Operational: On arrival in Thailand each ship was docked to make good poor shipbuilding standards and improve damage control capabilities. The ships are mostly used for rotating monthly to the Coast Guard, and for training, although *Kraburi* was part of the escort force for the aircraft carrier in Spanish waters in 1997. *Kraburi* damaged by the tsunami on 26 December 2004 but had been restored to operational service by February 2005.

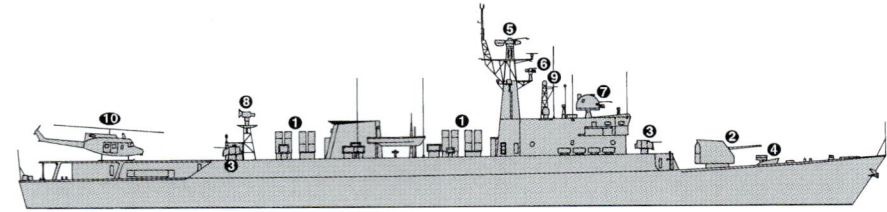

CHAO PHRAYA (Scale 1 : 900), Ian Sturton / 0080802

KRABURI (Scale 1 : 900), Ian Sturton / 0080803

SAIBURI 3/2011, Jurg Kürsener / 1406469

KRABURI 7/2010, Thai Navy League / 1406470

Frigates ◁ **Thailand** 831

2 KNOX CLASS (FFGHM)

Name	No	Builders	Laid down	Launched	Commissioned
PHUTTHA YOTFA CHULALOK (ex-*Truett*)	461 (ex-FF 1095)	Avondale Shipyards, New Orleans	27 Apr 1972	3 Feb 1973	1 Jun 1974
PHUTTHA LOETLA NAPHALAI (ex-*Ouellet*)	462 (ex-FF 1077)	Avondale Shipyards, New Orleans	15 Jan 1969	17 Jan 1970	12 Dec 1970

Displacement, tonnes: 3,059 standard; 4,328 full load
Dimensions, metres (feet): 134 × 14.3 × 4.6 hull; 7.8 sonar *(439.6 × 46.9 × 15.1; 25.6)*
Speed, knots: 27
Range, n miles: 4,000 at 22 kt
Complement: 288 (17 officers)

Machinery: 2 Combustion Engineering/Babcock & Wilcox boilers; 1,200 psi $(84.4 \ kg/cm^2)$; 950°F (510°C); 1 turbine; 35,000 hp *(26 MW)*; 1 shaft
Missiles: SSM: 8 McDonnell Douglas Harpoon; active radar homing to 130 km *(70 n miles)* at 0.9 Mach; warhead 227 kg.
A/S: Honeywell ASROC Mk 16 octuple launcher with reload system (has 2 starboard cells modified to fire Harpoon) ❶; inertial guidance to 1.6–10 km *(1–5.4 n miles)*; payload Mk 46.
Guns: 1 FMC 5 in *(127 mm)*/54 Mk 42 Mod 9 ❷; 20–40 rds/min to 24 km *(13 n miles)* anti-surface; 14 km *(7.7 n miles)* anti-aircraft; weight of shell 32 kg. 1 Raytheon 20 mm/76 6-barrelled Mk 15 Vulcan Phalanx ❸; 3,000 rds/min combined to 1.5 km.
Torpedoes: 4—324 mm Mk 32 (2 twin) fixed tubes ❹. 22 Honeywell Mk 46; anti-submarine; active/passive homing to 11 km *(5.9 n miles)* at 40 kt; warhead 44 kg.
Physical countermeasures: Decoys: 2 Loral Hycor SRBOC 6-barrelled fixed Mk 36 ❺; IR flares and chaff to 4 km *(2.2 n miles)*. T Mk-6 Fanfare/SLQ-25 Nixie; torpedo decoy. Prairie Masker hull and blade rate noise suppression.
Electronic countermeasures: ESM/ECM: SLQ-32(V)2 ❻; radar warning. Sidekick modification adds jammer and deception system.
Radars: Air search: Lockheed SPS-40B ❼; B-band; range 320 km *(175 n miles)*.
Surface search: Raytheon SPS-10 or Norden SPS-67 ❽; G-band.
Navigation: Marconi LN66; I-band.
Fire control: Western Electric SPG-53A/D/F ❾; I/J-band.
Tacan: SRN 15. IFF: UPX-12.
Sonars: EDO/General Electric SQS-26CX; bow-mounted; active search and attack; medium frequency. EDO SQR-18(V) TACTASS; passive; low frequency.
Combat data systems: Link 14 receive only.

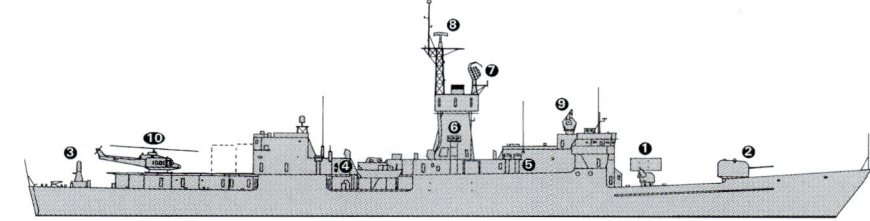
PHUTTHA YOTFA CHULALOK (Scale 1 : 1,200), Ian Sturton / 0543397

PHUTTA LOETLA NAPHALAI 8/2009, Michael Nitz / 1366017

Weapon control systems: SWG-1A Harpoon LCS. Mk 68 GFCS. Mk 114 ASW FCS. Mk 1 target designation system. MMS target acquisition sight (for mines, small craft and low-flying aircraft).
Helicopters: 1 Bell 212 ❿.

Programmes: The first ship transferred on five year lease from the USA on 30 July 1994. This was renewed by grant in 1999. The second transferred on lease 27 November 1996 and arrived in Thailand in November 1998.
Structure: Four Mk 32 torpedo tubes are fixed in the midships structure, two to a side, angled out at 45°. The arrangement provides improved loading capability over exposed triple Mk 32 torpedo tubes. A 4,000 lb lightweight anchor is fitted on the port side and an 8,000 lb anchor fits into the after section of the sonar dome.
Operational: To be replaced by new frigates in 2016 and 2018.

1 YARROW TYPE (FFH)

Name	No	Builders	Laid down	Launched	Commissioned
MAKUT RAJAKUMARN	433 (ex-7)	Yarrow Shipbuilders	11 Jan 1970	18 Nov 1971	7 May 1973

Displacement, tonnes: 1,676 standard; 1,930 full load
Dimensions, metres (feet): 97.6 × 11 × 5.5 *(320.2 × 36.1 × 18.0)*
Speed, knots: 26; 18 diesel
Range, n miles: 5,000 at 18 kt, 1,200 at 26 kt
Complement: 140 (16 officers)

Machinery: CODOG; 1 RR Olympus TM3B gas turbine; 22,500 hp *(16.8 MW)* sustained; 1 Crossley-SEMT-Pielstick 12 PC2.2 V 400 diesel; 6,000 hp(m) *(4.4 MW)* sustained; 2 shafts
Guns: 2 Vickers 4.5 in *(114 mm)*/55 Mk 8 ❶; 25 rds/min to 22 km *(12 n miles)* anti-surface; 6 km *(3.3 n miles)* anti-aircraft; weight of shell 21 kg. 2 Breda 40 mm/70 (twin) ❷; 300 rds/min to 12.5 km *(6.8 n miles)*; weight of shell 0.96 kg. 2 Oerlikon 20 mm.
Torpedoes: 6 Plessey PMW 49A tubes ❸ Mk 46; active/passive homing to 11 km *(5.9 n miles)* at 40 kt; warhead 44 kg.
Depth charges: 1 rack.
Physical countermeasures: Decoys: 2 Loral Mk 135 chaff launchers.
Electronic countermeasures: ESM/ECM: Elettronica Newton ❹; intercept and jammer. WLR-1; radar warning.
Radars: Air/surface search: Signaal DA05 ❺; E/F-band; range 137 km *(75 n miles)* for 2 m² target.
Surface search: Signaal ZW06 ❻; I-band.
Fire control: Signaal WM22/61 ❼; I/J-band; range 46 km *(25 n miles)*.
Navigation: Racal Decca; I-band.

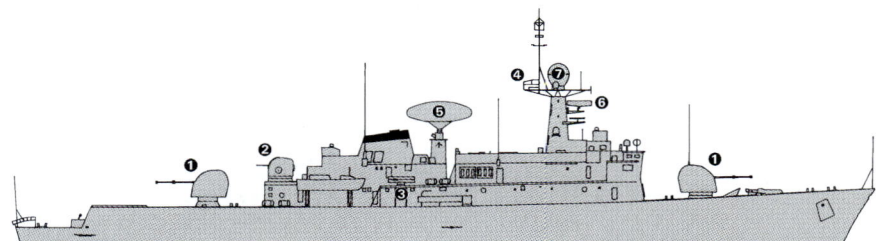
MAKUT RAJAKUMARN (Scale 1 : 900), Ian Sturton / 1167964

Sonars: Atlas Elektronik DSQS-21C; hull-mounted; active search and attack; medium frequency.
Combat data systems: Signaal Sewaco TH.

Helicopters: A small helicopter can land when the Mortar Mk 10 well is closed.

Programmes: Ordered on 21 August 1969.
Modernisation: A severe fire in February 1984 resulted in extensive work including replacement of the Olympus gas turbine, a new ER control room and central electric switchboard. Further modifications included the removal of Seacat SAM system and the installation of new EW equipment in 1993. In 1997 two Bofors 40 mm were fitted on the old Seacat mounting, and torpedo tubes replaced the old Bofors abreast the funnel. The Limbo mortar mountings have been removed.
Operational: The ship is largely automated with a consequent saving in complement, and has been most successful in service. Has lost its Flagship role to one of the Chinese-built frigates and is employed on general duties rather than as a training ship as previously reported.

MAKUT RAJAKUMARN 8/2010, Chris Sattler / 1406291

© 2014 IHS IHS Jane's Fighting Ships 2014-2015

CORVETTES

2 PATTANI CLASS (OFFSHORE PATROL VESSELS) (PBOH)

Name	No	Builders	Laid down	Launched	Commissioned
PATTANI	511	Hudong Shipyard, Shanghai	2003	19 Sep 2004	16 Dec 2005
NARATHIWAT	512	Hudong Shipyard, Shanghai	2004	Mar 2005	16 Apr 2006

Displacement, tonnes: 1,321 standard; 1,463 full load
Dimensions, metres (feet): 95.5 × 11.6 × 3.1 *(313.3 × 38.1 × 10.2)*
Speed, knots: 25
Range, n miles: 3,500 at 15 kt
Complement: 78 (18 officers)

Machinery: 2 Ruston diesels; 15,660 hp *(11.7 MW)*; 2 shafts; cp props
Guns: 1 Oto Melara 3 in *(76 mm)*/62 Super Rapid ❶; 85 rds/min to 16 km *(8.6 n miles)*. 2 — 20 mm ❷.
Radars: Air/surface search: Alenia Marconi SPS 791 (RAN-30X/I) ❸; I-band.
Surface search ❹: To be announced.
Fire control: Oerlikon/Contraves TMX ❺; I/J-band.
Navigation: I-band.
Combat data systems: COSYS.
Weapon control systems: Optronic director combined with TMX.

Helicopters: One medium.

Programmes: The contract for two Offshore Patrol Vessels was signed with China Shipbuilding Trading Company on 20 December 2002.
Structure: Space and weight provision for the addition of eight SSM, CIWS (probably Matra Sadral) and ASW capabilities at a later date.
Operational: *Pattani* arrived at Sattahip on 16 December 2005. *Narathiwat* followed on 4 May 2006.

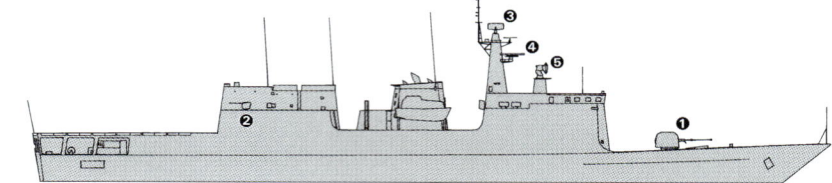

PATTANI *(Scale 1 : 900), Ian Sturton* / 1529098

NARATHIWAT *3/2007, Thai Navy League* 1353393

NARATHIWAT *11/2009, Mick Prendergast* / 1406292

PATTANI *10/2008, Michael Nitz* / 1353392

2 RATTANAKOSIN CLASS (FSGM)

Name	No	Builders	Laid down	Launched	Commissioned
RATTANAKOSIN	441 (ex-1)	Tacoma Boatbuilding Co, Tacoma	6 Feb 1984	11 Mar 1986	26 Sep 1986
SUKHOTHAI	442 (ex-2)	Tacoma Boatbuilding Co, Tacoma	26 Mar 1984	20 Jul 1986	10 Jun 1987

Displacement, tonnes: 975 full load
Dimensions, metres (feet): 76.8 × 9.6 × 2.4 *(252 × 31.5 × 7.9)*
Speed, knots: 26. **Range, n miles:** 3,000 at 16 kt
Complement: 87 (15 officers)

Machinery: 2 MTU 20V 1163 TB83 diesels; 14,730 hp(m) *(10.83 MW)* sustained; 2 shafts; Kamewa cp props
Missiles: SSM: 8 McDonnell Douglas Harpoon (2 quad) launchers ❶; active radar homing to 130 km *(70 n miles)* at 0.9 Mach; warhead 227 kg (84A) or 258 kg (84B/C).
SAM: Selenia Elsag Albatros octuple launcher ❷; 24 Aspide; semi-active radar homing to 13 km *(7 n miles)* at 2.5 Mach; height envelope 15–5,000 m *(49.2–16,405 ft)*; warhead 30 kg.
Guns: 1 Oto Melara 3 in *(76 mm)*/62 Compact ❸; 60 rds/min to 16 km *(8.7 n miles)*; weight of shell 6 kg. 2 Breda 40 mm/70 (twin) ❹; 300 rds/min to 12.5 km *(6.8 n miles)*; weight of shell 0.96 kg. 2 Rheinmetall 20 mm ❺.
Torpedoes: 6—324 mm US Mk 32 (2 triple) tubes ❻. MUSL Stingray; active/passive homing to 11 km *(5.9 n miles)* at 45 kt; warhead 35 kg (shaped charge); depth to 750 m *(2,460 ft)*.
Physical countermeasures: Decoys: CSEE Dagaie 6- or 10-tubed trainable; IR flares and chaff; H- to J-band.
Electronic countermeasures: ESM: Elettronica; intercept.
Radars: Air/surface search: Signaal DA05 ❽; E/F-band; range 137 km *(75 n miles)* for 2 m² target.
Surface search: Signaal ZW06 ❾; I-band.
Navigation: Decca 1226; I-band.
Fire control: Signaal WM25/41 ❿; I/J-band; range 46 km *(25 n miles)*.
Sonars: Atlas Elektronik DSQS-21C; hull-mounted; active search and attack; medium frequency.
Weapon control systems: Signaal Sewaco TH action data automation. Lirod 8 optronic director ❼.

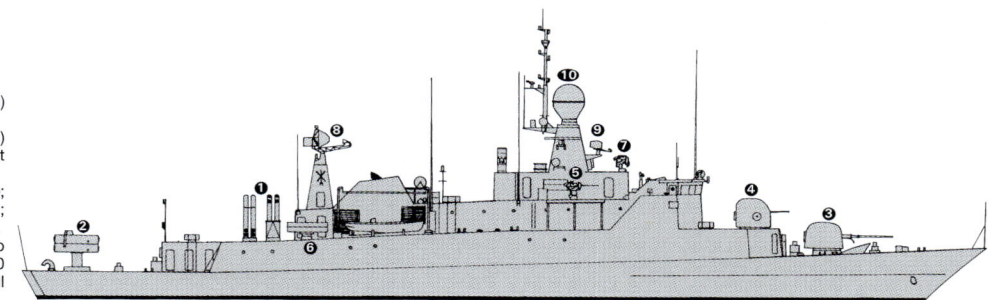

RATTANAKOSIN (Scale 1 : 600), Ian Sturton / 0506173

Programmes: Contract signed with Tacoma on 9 May 1983. Intentions to build a third were overtaken by the Vosper corvettes.
Structure: There are some similarities with the missile corvettes built for Saudi Arabia five years earlier. Space for Phalanx aft of the Harpoon launchers, but there are no plans to fit.

RATTANAKOSIN 9/2012, Chris Sattler / 1486371

3 KHAMRONSIN CLASS (FS)

Name	No	Builders	Laid down	Launched	Commissioned
KHAMRONSIN	531 (ex-1)	Ital Thai (Samutprakarn) Ltd, Bangkok	15 Mar 1988	15 Aug 1989	29 Jul 1992
THAYANCHON	532 (ex-2)	Ital Thai (Samutprakarn) Ltd, Bangkok	20 Apr 1988	7 Dec 1989	5 Sep 1992
LONGLOM	533 (ex-3)	Bangkok Naval Dockyard	15 Mar 1988	8 Aug 1989	2 Oct 1992

Displacement, tonnes: 640 full load
Dimensions, metres (feet): 62 oa; 56.7 wl × 8.2 × 2.5 *(203.4; 186 × 26.9 × 8.2)*
Speed, knots: 25
Range, n miles: 2,500 at 15 kt
Complement: 57 (6 officers)

Machinery: 2 MTU 12V 1163 TB93; 9,980 hp(m) *(7.34 MW)* sustained; 2 Kamewa cp props
Guns: 1 Oto Melara 76 mm/62 Mod 7 Compact ❶; 60 rds/min to 16 km *(8.7 n miles)*; weight of shell 6 kg. 2 Breda 30 mm/70 (twin) ❷; 800 rds/min to 12.5 km *(6.8 n miles)*; weight of shell 0.37 kg. 2—12.7 mm MGs.
Torpedoes: 6 Plessey PMW 49A (2 triple) launchers ❸; MUSL Stingray; active/passive homing to 11 km *(5.9 n miles)* at 45 kt; warhead 35 kg shaped charge.
Radars: Air/surface search: Plessey AWS 4 ❺; E/F-band.
Navigation: Racal Decca 1226; I-band.
Sonars: Atlas Elektronik DSQS-21C; hull-mounted; active search and attack; medium frequency.
Combat data systems: Plessey NAUTIS P action data automation.
Weapon control systems: British Aerospace Sea Archer 1A Mod 2 optronic GFCS ❹.

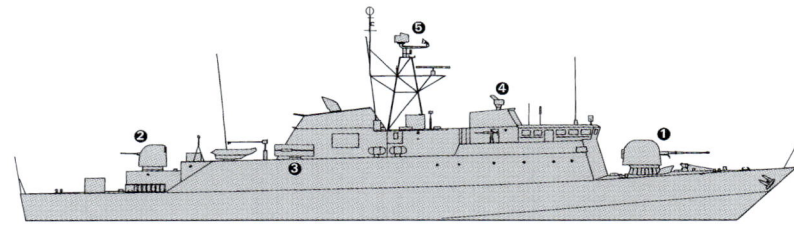

KHAMRONSIN (Scale 1 : 600), Ian Sturton / 0572649

Programmes: Contract signed on 29 September 1987 with Ital Thai Marine of Bangkok for the construction of two ASW corvettes and for technical assistance with a third to be built in Bangkok Naval Dockyard. A fourth of the class with a different superstructure and less armament was ordered by the Police in September 1989.
Structure: The vessels are based on a Vosper Thornycroft Province class 56 m design stretched by increasing the frame spacing along the whole length of the hull. Depth charge racks and mine rails may be added.

THAYANCHON 12/2007, Michael Nitz / 1353395

834 Thailand > Corvettes

2 TAPI (PF 103) CLASS (FS)

Name	No	Builders	Laid down	Launched	Commissioned
TAPI	431 (ex-5)	American SB Co, Toledo	1 Jul 1970	17 Oct 1970	19 Nov 1971
KHIRIRAT	432 (ex-6)	Norfolk SB & DD Co	18 Feb 1972	2 Jun 1973	10 Aug 1974

Displacement, tonnes: 899 standard; 1,191 full load
Dimensions, metres (feet): 83.8 × 10 × 3 hull; 4.3 sonar (274.9 × 32.8 × 9.8; 14.1)
Speed, knots: 20
Range, n miles: 2,400 at 18 kt
Complement: 135 (15 officers)

Machinery: 2 Fairbanks-Morse 38TD8-1/8-9 diesels; 5,250 hp (3.9 MW) sustained; 2 shafts
Guns: 1 Oto Melara 3 in (76 mm)/62 compact ❶; 85 rds/min to 16 km (8.7 n miles) anti-surface; 12 km (6.6 n miles) anti-aircraft; weight of shell 6 kg. 1 Bofors 40 mm/70 ❷; 300 rds/min to 12.5 km (6.8 n miles); weight of shell 0.96 kg. 2 Oerlikon 20 mm ❸. 2−12.7 mm MGs.
Torpedoes: 6−324 mm US Mk 32 (2 triple) tubes ❹. Honeywell Mk 46; anti-submarine; active/passive homing to 11 km (5.9 n miles) at 40 kt; warhead 44 kg.
Depth charges: 1 rack.
Radars: Air/surface search: Signaal LW04 ❺; D-band; range 137 km (75 n miles) for 2 m² target.
Surface search: Raytheon SPS-53E ❻; I-band.
Fire control: Signaal WM22-61 ❼; I/J-band; range 46 km (25 n miles).
IFF: UPX-23.

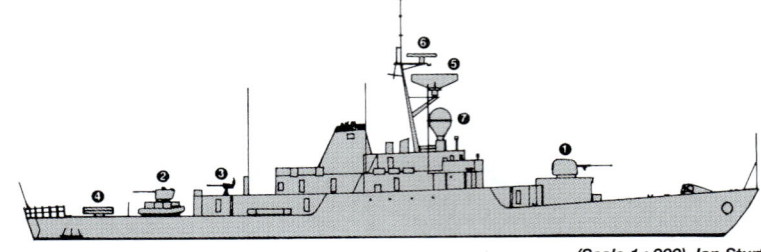

TAPI *(Scale 1 : 900), Ian Sturton* / 0506109

Sonars: Atlas Elektronik DSQS-21C; hull-mounted; active search and attack; medium frequency.
Combat data systems: Signaal Sewaco TH.

Programmes: *Tapi* was ordered on 27 June 1969. *Khirirat* was ordered on 25 June 1971.
Modernisation: *Tapi* completed 1983 and *Khirirat* in 1987. This included new gunnery and radars and a slight heightening of the funnel. Further modernisation in 1988–89 mainly to external and internal communications.
Structure: Of similar design to the Iranian ships of the Bayandor class.
Operational: Used for EEZ patrols.

TAPI 6/2001, Royal Thai Navy / 0130171

1 OFFSHORE PATROL VESSEL (PSO)

Name	No	Builders	Laid down	Launched	Commissioned
KRABI	551	Mahidol Royal Dockyard	3 Sep 2010	3 Dec 2011	26 Aug 2013

Displacement, tonnes: 2,540 full load
Dimensions, metres (feet): 90.5 × 13.5 × 3.5 (296.9 × 44.3 × 11.5)
Speed, knots: 25
Range, n miles: 3,500 at 15 kt
Complement: 39 (5 trainees) + 50 embarked forces

Machinery: 2 MAN 16RK280 diesels; 2 shafts
Guns: 1−76 mm. 2−30 mm.
Radars: Air/surface search: Thales Variant; G/I-bands.
Surface search: E/F-band.
Navigation: I-band.
Weapon control systems: Optronic director.
Helicopters: Platform for one medium.

Comment: A 90 m offshore patrol vessel based on a design provided by UK-shipbuilder BAE Systems Surface Ships. The initial requirement is for one ship with a view to up to five further ships as funds become available. The ship is to be deployed on Exclusive Economic Zone patrol duties.

KRABI
10/2013*, Chris Sattler
1529080

IHS Jane's Fighting Ships 2014-2015 © 2014 IHS

SHIPBORNE AIRCRAFT

Numbers/Type: 4 Bell 214 ST.
Operational speed: 120 kt *(228 km/h)*.
Service ceiling: 13,200 ft *(4,025 m)*.
Range: 400 n miles *(740 km)*.
Role/Weapon systems: Procured in 1987 for maritime surveillance and utility roles.

BELL 214   *6/2004, Royal Thai Navy* / 1044195

Numbers/Type: 2 AgustaWestland Super Lynx 300.
Operational speed: 125 kt *(231 km/h)*.
Service ceiling: 12,000 ft *(3,660 m)*.
Range: 340 n miles *(630 km)*.
Role/Weapon systems: Two helicopters ordered 7 August 2001 for ASW, ASV and surveillance roles. Delivered in 2005.

SUPER LYNX   *9/2004, AgustaWestland* / 0566704

Numbers/Type: 7/2 BAe/McDonnell Douglas AV-8A (Harrier)/TAV-8A (Harrier).
Operational speed: 640 kt *(1,186 km/h)*.
Service ceiling: 51,200 ft *(15,600 m)*.
Range: 800 n miles *(1,480 km)*.
Role/Weapon systems: AV-8S supplied via USA to Spain and transferred in 1996. Sensors: None. Weapons: Strike; two 30 mm Aden cannon, two AIM-9 Sidewinder or 20 mm/127 mm rockets and 'iron' bombs.

HARRIER   *1/2001, Thai Navy League* / 0130153

Numbers/Type: 6 Sikorsky S-70B7 Seahawk.
Operational speed: 135 kt *(250 km/h)*.
Service ceiling: 10,000 ft *(3,050 m)*.
Range: 600 n miles *(1,110 km)*.
Role/Weapon systems: Multimission helicopters delivered by June 1997. Plans to acquire ASW equipment have been abandoned. Sensors: Telephonics APS-143(V)3 radar; ASN 150 databus; provision for sonobuoys and dipping sonar; ALR 606(V)2 ESM. Weapons: Provision for ASM and MUSL Stingray torpedoes.

SEAHAWK   *7/2005, Thai Navy League* / 1153913

Numbers/Type: 2 Sikorsky MH-60S Seahawk.
Operational speed: 147 kt *(272 km/h)*.
Service ceiling: 10,000 ft *(3,050 m)*.
Range: 400 n miles *(741 km)*.
Role/Weapon systems: First export order for this aircraft acquired under the US FMS programme. The aircraft are employed to provide logistics support, transport, Search And Rescue (SAR) and general utility duties. The contract includes training, spares and logistics support. The aircraft entered service in August 2011 and a further four may be purchased. Sensors/weapons: to be announced.

Numbers/Type: 4 Bell 212.
Operational speed: 100 kt *(185 km/h)*.
Service ceiling: 13,200 ft *(4,025 m)*.
Range: 200 n miles *(370 km)*.
Role/Weapon systems: Commando assault and general support. At least two transferred from Army. May be sold to help pay for new shipborne helicopter. Mostly based ashore but operate from Normed class and frigates. Weapons: Pintle-mounted M60 machine guns.

BELL 212   *6/2000, Thai Navy League* / 0105842

LAND-BASED MARITIME AIRCRAFT (FRONT LINE)

Notes: There are two Embraer EMB-135 aircraft used for VIPs, transport and Medevac. They were acquired 2007–09.

Numbers/Type: 4 Sikorsky S-76B.
Operational speed: 145 kt *(269 km/h)*.
Service ceiling: 6,500 ft *(1,980 m)*.
Range: 357 n miles *(661 km)*.
Role/Weapon systems: Six originally acquired in 1996 for maritime surveillance and utility purposes. Sensors: Weather radar. Weapons: Unarmed.

S-76   *8/1996, Royal Thai Navy* / 0050241

Numbers/Type: 2/1 Lockheed P-3T Orion/UP-3T Orion.
Operational speed: 411 kt *(761 km/h)*.
Service ceiling: 28,300 ft *(8,625 m)*.
Range: 4,000 n miles *(7,410 km)*.
Role/Weapon systems: Delivered in 1996. Two for ASW and one utility. Two more are required. Sensors: APS-115 radar, ECM/ESM. Weapons: ASW; Mk 46 or Stingray torpedoes. ASV; four Harpoon.

ORION   *8/1997, Royal Thai Navy* / 0019261

Numbers/Type: 13/4 Vought A-7E Corsair II/TA-7E Corsair II.
Operational speed: 600 kt *(1,112 km/h)*.
Service ceiling: 50,000 ft *(15,240 m)*.
Range: 2,000 n miles *(3,705 km)*.
Role/Weapon systems: Delivered in 1996–97 from the US. Reconditioning programme in progress 2004. Weapons: AIM-9L Sidewinder; 1−20 mm cannon.

CORSAIR II   *8/1996, Royal Thai Navy* / 0053451

838 Thailand > Patrol forces

3 PRABPARAPAK CLASS
(FAST ATTACK CRAFT—MISSILE) (PTFG)

Name	No	Builders	Commissioned
PRABPARAPAK	311 (ex-1)	Singapore SB and Marine	28 Jul 1976
HANHAK SATTRU	312 (ex-2)	Singapore SB and Marine	6 Nov 1976
SUPHAIRIN	313 (ex-3)	Singapore SB and Marine	1 Feb 1977

Displacement, tonnes: 228 standard; 272 full load
Dimensions, metres (feet): 45.4 × 7.4 × 2.3 (149 × 24.3 × 7.5)
Speed, knots: 40
Range, n miles: 2,000 at 15 kt, 750 at 37 kt
Complement: 41 (5 officers)
Machinery: 4 MTU 16V 538 TB92 diesels; 13,640 hp(m) (10 MW) sustained; 4 shafts
Missiles: SSM: 5 IAI Gabriel I (1 triple, 2 single) launchers; radar or optical guidance; semi-active radar homing to 20 km (10.8 n miles) at 0.7 Mach; warhead 75 kg.
Guns: 1 Bofors 57 mm/70; 200 rds/min to 17 km (9.3 n miles); weight of shell 2.4 kg. 8 rocket illuminant launchers on either side of 57 mm gun. 1 Bofors 40 mm/70; 300 rds/min to 12 km (6.6 n miles); weight of shell 0.96 kg.
Electronic countermeasures: ESM: Racal RDL-2; intercept.
Radars: Surface search: Kelvin Hughes Type 17; I-band.
Fire control: Signaal WM28/5 series; I/J-band.

Programmes: Ordered June 1973. Built under licence from Lürssen. Launch dates-Prabparapak 29 July 1975, Hanhak Sattru 28 October 1975, Suphairin 20 February 1976.
Modernisation: There are plans to replace Gabriel possibly by RBS 15.
Structure: Same design as Lürssen standard 45 m class built for Singapore. Normally only three Gabriel SSM are carried.

PRABPARAPAK 6/2001, Royal Thai Navy / 0130172

3 T 81 CLASS (COASTAL PATROL CRAFT) (PB)

T 81–83

Displacement, tonnes: 122 full load
Dimensions, metres (feet): 30.1 × 6.3 × 1.7 (98.8 × 20.7 × 5.6)
Speed, knots: 25
Range, n miles: 1,300 at 15 kt
Complement: 28 (3 officers)
Machinery: 2 MTU 16V 2000 TE90 diesels; 3,600 hp(m) (2.56 MW); 2 shafts
Guns: 1 Bofors 40 mm/70. 1 Oerlikon 20 mm. 2—12.7 mm MGs.
Radars: Surface search: Sperry SM 5000; I-band.

Comment: Ordered in October 1996 from ASC Silkline in Pranburi. First one commissioned 5 August 1999, second 9 December 1999 and the third in 2000. Plans for seven more have been shelved.

T 83 3/2004, Bob Fildes / 0589810

7 PGM 71 CLASS (COASTAL PATROL CRAFT) (PB)

T 11–12   T 15–19

Displacement, tonnes: 132 standard; 149 full load
Dimensions, metres (feet): 30.8 × 6.4 × 1.9 (101 × 21.0 × 6.2)
Speed, knots: 18.5
Range, n miles: 1,500 at 10 kt
Complement: 30
Machinery: 2 GM diesels; 1,800 hp (1.34 MW); 2 shafts
Guns: 1 Bofors 40 mm/60. 1 Oerlikon 20 mm. 2—12.7 mm MGs. In some craft the 20 mm gun has been replaced by an 81 mm mortar/12.7 mm combined mounting aft.
Radars: Surface search: Decca 303 (T 11 and 12) or Decca 202 (remainder); I-band.

Comment: Built by Peterson Inc between 1966 and 1970. Transferred from US. T 13 deleted in 2011, T 14 and T 110 in 2013.

IHS Jane's Fighting Ships 2014-2015

T 16 10/1999, Royal Thai Navy / 0080816

3 T 991 CLASS (COASTAL PATROL CRAFT) (PB)

No	Builders	Laid down	Launched	Commissioned
T 991	Bangkok Naval Dockyard	9 Sep 2005	2006	30 Apr 2007
T 992	Marsun Shipyard	–	6 Sep 2007	Dec 2007
T 993	Marsun Shipyard	–	6 Sep 2007	Dec 2007

Displacement, tonnes: 189 full load
Dimensions, metres (feet): 38.7 × 6.45 × 1.8 (127 × 21.2 × 5.9)
Speed, knots: 27
Complement: 30

Machinery: 2 MTU 16V 4000 M 90 diesels; 7,400 hp (5.5 MW); 2 shafts
Guns: 2 MSI DS-30M 30 mm. 2—12.7 mm MGs.
Radars: Surface search/navigation: To be announced.
Weapon control systems: Thales Mirador optronic director.

Comment: Modified versions of the T 91 class. First vessel laid down at Naval Dockyard on 9 September 2005. Two further craft delivered by December 2007.

T 991 12/2007, M Mazumdar / 1353396

3 + (3) T 994 CLASS (COASTAL PATROL CRAFT) (PBM)

No	Builders	Laid down	Launched	Commissioned
T 994	Bangkok Naval Dockyard	21 Mar 2010	11 Jul 2011	Sep 2011
T 995	Marsun Shipbuilding	5 Apr 2010	Sep 2011	Sep 2011
T 996	Marsun Shipbuilding	5 Apr 2010	Sep 2011	Sep 2011

Displacement, tonnes: 215 full load
Dimensions, metres (feet): 41.45 × 7.2 × 1.8 (136 × 23.6 × 5.9)
Speed, knots: 27
Complement: 30

Machinery: 2 MTU 16V 4000 M 90 diesels; 7,400 hp (5.5 MW); 2 shafts
Missiles: SSM: 2 to be announced.
Guns: 2 MSI DS-30M 30 mm. 2—12.7 mm MGs.
Radars: Surface search/navigation: To be announced.
Weapon control systems: Thales Mirador optronic director.

Comment: Modified missile-armed variants of the T 991 class. A contract for three craft was signed in mid-2009. A class of six is expected.

T 994 4/2012, Royal Thai Navy / 1455676

3 SEAL ASSAULT CRAFT (LCP)

Comment: Locally built for special forces operations. Details are not known but reported to be larger and faster than PBR Mk II craft. Equipped with stern ramp.

T 242 (SEAL) 5/1997, A Sharma / 0050242

© 2014 IHS

Patrol forces < **Thailand** 839

4 SWIFT CLASS (COASTAL PATROL CRAFT) (PB)

T 21– 24

Displacement, tonnes: 22 full load
Dimensions, metres (feet): 15.2 × 4 × 1.1 *(49.9 × 13.1 × 3.6)*
Speed, knots: 25
Range, n miles: 400 at 25 kt
Complement: 8 (1 officer)
Machinery: 2 Detroit diesels; 480 hp *(358 kW)*; 2 shafts
Guns: 1—81 mm mortar. 2—12.7 mm MGs.
Radars: Surface search: Raytheon Pathfinder; I-band.

Comment: Transferred from US Navy from 1967–75. *T 25* and *T 26* deleted in 2011. *T 27–29* deleted in 2012.

T 21   *7/2008, Thai Navy League* / 1353397

13 T 213 CLASS (COASTAL PATROL CRAFT) (PB)

T 213–214   T 216–226

Displacement, tonnes: 36 standard
Dimensions, metres (feet): 19.5 × 5.3 × 1.5 *(64 × 17.4 × 4.9)*
Speed, knots: 25
Complement: 8 (1 officer)
Machinery: 2 MTU diesels; 715 hp(m) *(526 kW)*; 2 shafts
Guns: 1 Oerlikon 20 mm. 1—81 mm mortar with 12.7 mm MG.
Radars: Surface search: Racal Decca 110; I-band.

Comment: Built by Ital Thai Marine Ltd. Commissioned-*T 213–214*, 29 August 1980; *T 216–218*, 26 March 1981; *T 219–223*, 16 September 1981; *T 224*, 19 November 1982; *T 225* and *T 226*, 28 March 1984. Construction of *T 227–230* is not to have been completed. Of alloy construction. Used for fishery patrol and coastal control duties. *T 215* damaged beyond repair by tsunami on 26 December 2004 and replaced by *T 227*. Replacement by further T 227-class craft is under consideration.

T 219   *9/2003, Hartmut Ehlers* / 0572643

1 T 227 CLASS (COASTAL PATROL CRAFT) (PB)

T 227

Displacement, tonnes: 43 full load
Dimensions, metres (feet): 21.3 × 5.3 × 1.5 *(69.9 × 17.4 × 4.9)*
Speed, knots: 28
Complement: 8 (1 officer)
Machinery: 2 MTU diesels; 1,200 hp *(895 kW)*; 2 shafts
Guns: 1—20 mm. 2—12.7 mm MGs.
Radars: Surface search/navigation: I-band.

Comment: Larger variant of the T 213 class built by Marsun and launched on 1 September 2006 to replace T 215 which was lost in the tsunami of 26 December 2004.

T 227   *12/2006, Marsun* / 1190410

13 PBR MK II (RIVER PATROL CRAFT) (PBR)

Displacement, tonnes: 8 full load
Dimensions, metres (feet): 9.8 × 3.5 × 0.7 *(32.2 × 11.5 × 2.3)*
Speed, knots: 25
Range, n miles: 150 at 23 kt
Complement: 4
Machinery: 2 Detroit diesels; 430 hp *(321 kW)*; 2 Jacuzzi water-jets
Guns: 2—7.62 mm MGs. 1—60 mm mortar.
Radars: Raytheon SPS-66; I-band.

Comment: Transferred from US from 1967–73. Employed on Mekong River. Reported to be getting old, numbers are reducing and maximum speed has been virtually halved. All belong to the Riverine and SEAL Squadron.

PBR MK II   *6/2002, Thai Navy League* / 0543390

90 ASSAULT BOATS (LCP)

Displacement, tonnes: 0.4 full load
Dimensions, metres (feet): 5 × 1.9 × 0.4 *(16.4 × 6.2 × 1.3)*
Speed, knots: 24
Complement: 2
Machinery: 1 outboard; 150 hp *(110 kW)*
Guns: 1—7.62 mm MG.

Comment: Part of the Riverine Squadron with the PBRs and two PCFs. Can carry six people. Numbers uncertain.

ASSAULT BOAT   *6/2002, Thai Navy League* / 0530060

4 SEAL ASSAULT CRAFT (LCP)

P 51–54

Displacement, tonnes: 25 full load
Dimensions, metres (feet): 18.3 × 4.2 × ? *(60 × 13.8 × ?)*
Speed, knots: 42
Range, n miles: 370 at 35 kt
Complement: 22
Machinery: 2 MAN diesels; 2,600 hp *(1.94 MW)*; 2 shafts
Guns: 1—12.7 mm MG.

Comment: Designed and built by Marsun Shipyards, Thailand. Ordered in August 2008. Equipped with bow door and loading ramp aft for a RHIB.

P 51   *7/2009* / 1305391

842 Thailand > Amphibious forces — Mine warfare forces

2 LANDING CRAFT (LCU)

MATTAPON 784   **RAWI** 785

Displacement, tonnes: 145 full load
Dimensions, metres (feet): 55 × 11.0 × 1.6 *(180.4 × 36.1 × 5.2)*
Speed, knots: 12
Complement: 33 + 18 embarked forces
Military lift: Five 5-ton trucks
Machinery: 2 MAN D2842 LE 412 diesels; 1,700 hp *(1.3 MW)*; 2 shafts
Guns: 2—20 mm. 2—12.7 mm MGs.

Comment: Constructed at Marsun Shipyards, Thailand and delivered in 2010.

RAWI   6/2011, Thai Navy League / 1390485

MINE WARFARE FORCES

1 MCM SUPPORT SHIP (MCS)

Name	No	Builders	Commissioned
THALANG	621 (ex-1)	Bangkok Dock Co Ltd	4 Aug 1980

Displacement, tonnes: 1,016 standard
Dimensions, metres (feet): 55.7 × 10 × 3.1 *(182.7 × 32.8 × 10.2)*
Speed, knots: 12
Complement: 77
Machinery: 2 MTU diesels; 1,310 hp(m) *(963 kW)*; 2 shafts
Guns: 1 Bofors 40 mm/60. 2 Oerlikon 20 mm. 2—12.7 mm MGs.
Radars: Surface search: Racal Decca 1226; I-band.

Comment: Has minesweeping capability. Two 3 ton cranes provided for change of minesweeping gear in MSCs-four sets carried. Design by Ferrostaal, Essen. Has dormant minelaying capability.

THALANG   11/2001, Maritime Photographic / 0130165

2 LAT YA (GAETA) CLASS
(MINEHUNTERS/SWEEPERS) (MHSC)

Name	No	Builders	Launched	Commissioned
LAT YA	633	Intermarine, Sarzana	30 Mar 1998	18 Jun 1999
THA DIN DAENG	634	Intermarine, Sarzana	31 Oct 1998	18 Dec 1999

Displacement, tonnes: 691 full load
Dimensions, metres (feet): 52.5 × 9.9 × 2.9 *(172.2 × 32.5 × 9.5)*
Speed, knots: 14
Range, n miles: 2,000 at 12 kt
Complement: 50 (8 officers)

Machinery: 2 MTU 8V 396 TE74K diesels; 1,600 hp(m) *(1.18 MW)* sustained; 2 Voith Schneider props; auxiliary propulsion; 2 hydraulic motors
Guns: 1 MSI 30 mm.
Physical countermeasures: MCM: Atlas MWS 80-6 minehunting system. Magnetic, acoustic and mechanical sweeps; ADI Mini Dyad, Noise Maker, Bofors MS 106, 2 Pluto Plus ROVs.
Radars: Navigation: Atlas Elektronik 9600M (ARPA); I-band.
Sonars: Atlas Elektronik DSQS-11M; hull-mounted; active; high frequency.

Comment: Invitations to tender lodged by 3 April 1996. Ordered 19 September 1996. Specifications include hunting at up to 6 kt and sweeping at 10 kt. No further ships are planned.

IHS Jane's Fighting Ships 2014-2015

THA DIN DAENG   4/2004, John Mortimer / 1153909

2 BANG RACHAN CLASS
(MINEHUNTERS/SWEEPERS) (MHSC)

Name	No	Builders	Commissioned
BANG RACHAN	631 (ex-2)	Lürssen, Bremen-Vegesack	29 Apr 1987
NONGSARAI	632 (ex-3)	Lürssen, Bremen-Vegesack	17 Nov 1987

Displacement, tonnes: 451 full load
Dimensions, metres (feet): 49.1 × 9.3 × 2.5 *(161.1 × 30.5 × 8.2)*
Speed, knots: 17; 7 electric
Range, n miles: 3,100 at 12 kt
Complement: 33 (7 officers)

Machinery: 2 MTU 12V 396 TB83 diesels; 3,120 hp(m) *(2.3 MW)* sustained; 2 shafts; Kamewa cp props; auxiliary propulsion; 1 motor
Guns: 3 Oerlikon GAM-BO1 20 mm.
Physical countermeasures: MCM: MWS 80R minehunting system. Acoustic, magnetic and mechanical sweeps. 2 Gaymarine Pluto 15 remote-controlled submersibles.
Radars: Navigation: 2 Atlas Elektronik 8600 ARPA; I-band.
Sonars: Atlas Elektronik DSQS-11H; hull-mounted; minehunting; high frequency.

Comment: First ordered from Lürssen late 1984, arrived Bangkok 22 October 1987. Second ordered 5 August 1985 and arrived in Bangkok May 1988. Amagnetic steel frames and deckhouses, wooden hull. Motorola Miniranger MRS III precise navigation system. Draeger decompression chamber.

NONGSARAI   2/2005, Chris Sattler / 1153911

12 MSBS (MSR)

MLM 6–10   **MSB 11–17**

Displacement, tonnes: 25 full load
Dimensions, metres (feet): 15.3 × 4 × 0.9 *(50.2 × 13.1 × 3.0)*
Speed, knots: 8
Complement: 10
Machinery: 1 Gray Marine 64 HN9 diesel; 165 hp *(123 kW)*; 1 shaft
Guns: 2—7.62 mm MGs.

Comment: Three transferred from USA in October 1963 and two in 1964. More were built locally from 1994. Wooden hulled, converted from small motor launches. Operated on Chao Phraya river.

MLM 11   10/1995, Royal Thai Navy / 0080822

SURVEY AND RESEARCH SHIPS

Notes: There is also a civilian research vessel *Chulab Horn* which completed in 1986.

1 SURVEY SHIP (AGSH)

Name	No	Builders	Laid down	Launched	Commissioned
PHARUEHATSABODI	813	Unithai Shipyard and Engineering, Laem Chambang	25 Aug 2006	14 Feb 2008	19 Aug 2008

Dimensions, metres (feet): 66.3 × 13.2 × 3.1 *(217.5 × 43.3 × 10.2)*
Speed, knots: 12. **Range, n miles:** 3,000 at 12 kt
Complement: 13 + 58 spare berths
Machinery: Diesel-electric; 3 diesel generators; 2,652 hp(m) *(1.95 MW)*; 1 motor; 1,073 hp(m) *(800 kW)*; 2 azimuth thrusters; 1 bow thruster
Radars: Navigation: E/F- and I-band.
Sonars: Multi- and single-beam; high frequency; active.

Comment: Multipurpose hydrographic and oceanographic survey, training and mine countermeasures vessel ordered 22 December 2005 from a consortium comprising Schelde Naval Shipbuilding, Flushing, and Unithai Shipyard and Engineering, Thailand. The ship is a derivative of the Snellius class vessels built for the RNLN. The ship was built in Thailand. Hydrographic equipment includes an exploration computer system; multibeam echosounder; single-beam echosounder; side-scan sonar; Ultra-Short BaseLine (USBL); Motion and Reference Unit (MRU); draught indication system; tidal measurement system; seawater collection system; seawater measurement system; expendable bathythermograph/sound velocity meter; current flow measurement system; current meter system; sediment collection system; and oceanography equipment.

PHARUEHATSABODI   6/2011, Thai Navy League / 1406467

1 OCEANOGRAPHIC SHIP (AGOR)

Name	No	Builders	Commissioned
SUK	812	Bangkok Dock Co Ltd	3 Mar 1982

Displacement, tonnes: 1,473 standard; 1,550 full load
Dimensions, metres (feet): 62.9 × 11 × 4.1 *(206.4 × 36.1 × 13.5)*
Speed, knots: 15
Complement: 86 (20 officers)
Machinery: 2 MTU diesels; 2,400 hp(m) *(1.76 MW)*; 2 shafts
Guns: 2 Oerlikon 20 mm. 2 – 7.62 mm MGs.
Radars: Navigation: Racal Decca 1226; I-band.

Comment: Laid down 27 August 1979, launched 8 September 1981. Designed for oceanographic and survey duties.

SUK   5/1999, van Ginderen Collection / 0080828

1 SURVEY SHIP (AGS)

Name	No	Builders	Commissioned
CHANTHARA	811 (ex-AGS 11)	Lürssen	30 May 1961

Displacement, tonnes: 884 standard; 1,012 full load
Dimensions, metres (feet): 69.9 × 10.5 × 3 *(229.3 × 34.4 × 9.8)*
Speed, knots: 13.25. **Range, n miles:** 10,000 at 10 kt
Complement: 68 (8 officers)
Machinery: 2 KHD diesels; 1,090 hp(m) *(801 kW)*; 2 shafts
Guns: 2 Bofors 40 mm/60.

Comment: Laid down on 27 September 1960. Launched on 17 December 1960. Has served as a Royal Yacht.

CHANTHARA   7/2008, Thai Navy League / 1353398

1 BUOY TENDER (ABU)

Name	No	Builders	Commissioned
SURIYA	821	Bangkok Dock Co Ltd	15 Mar 1979

Displacement, tonnes: 701 full load
Dimensions, metres (feet): 54.2 × 10.2 × 3.1 *(177.8 × 33.5 × 10.2)*
Speed, knots: 12
Complement: 60 (12 officers)
Machinery: 2 MTU diesels; 1,310 hp(m) *(963 kW)*; 2 shafts; bow thruster; 135 hp(m) *(99 kW)*
Radars: Navigation: Racal Decca; I-band.

Comment: Can carry 20 mm guns.

SURIYA   3/2010*, Adolfo Ortigueira Gil / 1529082

TRAINING SHIPS

1 CANNON CLASS (FFT)

Name	No	Builders	Laid down	Launched	Commissioned
PIN KLAO (ex-*Hemminger* DE 746)	413 (ex-3, ex-1)	Western Pipe & Steel Co, San Francisco	1943	12 Sep 1943	30 May 1944

Displacement, tonnes: 1,260 standard; 1,961 full load
Dimensions, metres (feet): 93.3 × 11.2 × 4.3 *(306.1 × 36.7 × 14.1)*
Speed, knots: 20. **Range, n miles:** 10,800 at 12 kt, 6,700 at 19 kt
Complement: 192 (14 officers)
Machinery: Diesel-electric; 4 GM 16-278A diesels; 6,000 hp *(4.5 MW)*; 4 generators; 2 motors; 2 shafts
Guns: 3 USN 3 in *(76 mm)*/50 Mk 22; 20 rds/min to 12 km *(6.6 n miles)*; weight of shell 6 kg. 6 Bofors 40 mm/60 (3 twin); 120 rds/min to 10 km *(5.5 n miles)*; weight of shell 0.89 kg.
Torpedoes: 6 – 324 mm US Mk 32 (2 triple) tubes; anti-submarine.
A/S Mortars: 1 Hedgehog Mk 10 multibarrelled fixed; range 250 m; warhead 13.6 kg; 24 rockets.
Depth charges: 8 projectors; 2 racks.
Electronic countermeasures: ESM: WLR-1; radar warning.
Radars: Air/surface search: Raytheon SPS-5; G/H-band
Navigation: Raytheon SPS-21; G/H-band
Fire control: Western Electric Mk 34; I/J-band RCA/General Electric Mk 26; I/J-band
IFF: SLR 1.
Sonars: SQS-11; hull-mounted; active attack; high frequency.
Weapon control systems: Mk 52 radar GFCS for 3 in guns. Mk 63 radar GFCS for aft gun only. 2 Mk 51 optical GFCS for 40 mm.

Programmes: Transferred from US Navy at New York Navy Shipyard in July 1959 under MDAP and by sale 6 June 1975.
Modernisation: The three 21 in torpedo tubes were removed and the 20 mm guns were replaced by 40 mm. The six A/S torpedo tubes were fitted in 1966.
Operational: Used mostly as a training ship. Likely to be decommissioned in 2012.

PIN KLAO   6/1997, Royal Thai Navy / 0019254

AUXILIARIES

0 + 1 COASTAL TANKER (AOL)

Name	No	Builders	Laid down	Launched	Commissioned
MATRA	836	Bangkok Dock Co	24 May 2012	27 Feb 2013	2014

Dimensions, metres (feet): 63.5 × 12.0 × 3.92 *(208.3 × 39.4 × 12.9)*
Speed, knots: 12. **Range, n miles:** 4,000 at 12 kt
Complement: 42 (12 officers)
Machinery: 2 Wärtsilä diesels; 2,145 hp *(1.6 MW)*; 2 shafts
Guns: 1 – 20 mm. 2 – 12.7 mm MGs.
Radars: Surface search/navigation: I-band.
Helicopters: Platform for one medium.

Comment: The keel of a new oiler was laid down on 24 May 2012.

844 Thailand > Auxiliaries — Tugs

1 SIMILAN (HUDONG) CLASS (TYPE R22T)
(REPLENISHMENT SHIP) (AORH)

Name	No	Builders	Launched	Commissioned
SIMILAN	871	Hudong Shipyard, Shanghai	9 Nov 1995	12 Sep 1996

Displacement, tonnes: 23,369 full load
Dimensions, metres (feet): 171.4 × 24.6 × 9 *(562.3 × 80.7 × 29.5)*
Speed, knots: 19
Range, n miles: 10,000 at 15 kt
Complement: 157 (19 officers) + 26 spare berths

Cargo capacity: 9,000 tons fuel, water, ammunition and stores
Machinery: 2 HD-SEMT-Pielstick 16 PC2 6V400; 24,000 hp(m) *(17.64 MW)*; 2 shafts; Kamewa cp props
Radars: Air/surface search: Eye Shield (Type 354); E/F-band.
Navigation: Racal Decca 1290 ARPA; I-band.
Helicopters: 1 Seahawk type.

Comment: Contract signed with China State Shipbuilding Corporation on 29 September 1993. Fabrication started in December 1994. Two replenishment at sea positions each side and facilities for Vertrep. This ship complements the carrier and the new frigates to give the Navy a full deployment capability. Four twin 37 mm guns (Type 354) and associated Rice Lamp FC radar were not fitted.

SIMILAN   10/1998, Thai Navy League / 0050248

1 REPLENISHMENT TANKER (AORL)

Name	No	Builders	Launched
CHULA	831 (ex-2)	Singapore SEC	24 Sep 1980

Displacement, tonnes: 2,032 full load
Measurement, tonnes: 975 dwt
Dimensions, metres (feet): 67 × 9.5 × 4.4 *(219.8 × 31.2 × 14.4)*
Speed, knots: 14
Complement: 39 (7 officers)

Cargo capacity: 800 tons oil fuel
Machinery: 2 MTU 12V 396 TC62 diesels; 2,400 hp(m) *(1.76 MW)* sustained; 2 shafts
Guns: 2 Oerlikon 20 mm.
Radars: Navigation: Racal Decca 1226; I-band.

Comment: Replenishment is done by a hose handling crane boom.

CHULA   6/1998, Royal Thai Navy / 0050249

4 HARBOUR TANKERS (YO)

PRONG 833 (ex-YO 5)   **SAMED** 835 (ex-YO 10)
PROET 834 (ex-YO 9)   **CHIK** 842 (ex-YO 11)

Displacement, tonnes: 366 standard; 493 full load
Dimensions, metres (feet): 37.4 × 6 × 2.7 *(122.7 × 19.7 × 8.9)*
Speed, knots: 9
Cargo capacity: 210 tons
Machinery: 1 GM 8-268A diesel; 500 hp(m) *(368 kW)*; 1 shaft

Comment: Details are for 834, 835 and 842. Built by Bangkok Naval Dockyard. 834 commissioned 27 January 1967, remainder the same year. Details of 833 not known but reported to be approximately 180 tons.

PRONG   3/2010*, Adolfo Ortigueira Gil / 1529083

1 HARBOUR TANKER (YO)

SAMUI 832 (ex-YOG 60, ex-YO 4)

Displacement, tonnes: 1,443 full load
Dimensions, metres (feet): 53.2 × 9.7 × 4.6 *(174.5 × 31.8 × 15.1)*
Speed, knots: 8
Complement: 29

Cargo capacity: 985 tons fuel
Machinery: 1 Union diesel; 600 hp *(448 kW)*; 1 shaft
Guns: 2 Oerlikon 20 mm can be carried.
Radars: Navigation: Raytheon Pathfinder; I-band

SAMUI   12/1995 / 0506255

1 WATER TANKER (YW)

Name	No	Builders	Commissioned
CHUANG	841 (ex-YW 5)	Royal Thai Naval Dockyard, Bangkok	1965

Displacement, tonnes: 310 standard; 493 full load
Dimensions, metres (feet): 42 × 7.5 × 3.1 *(137.8 × 24.6 × 10.2)*
Speed, knots: 11
Complement: 29
Machinery: 1 GM diesel; 500 hp *(373 kW)*; 1 shaft
Guns: 1 Oerlikon 20 mm.

Comment: Launched on 14 January 1965.

CHUANG (alongside PROET)   5/1997, Maritime Photographic / 0019284

TUGS

2 COASTAL TUGS (YTB)

RIN 853 (ex-ATA 5)   **RANG** 854 (ex-ATA 6)

Displacement, tonnes: 356 standard
Dimensions, metres (feet): 32.3 × 9 × 4.6 *(106 × 29.5 × 15.1)*
Speed, knots: 12. **Range, n miles:** 1,000 at 10 kt
Complement: 19
Machinery: 1 MWM TBD441V/12K diesel; 2,100 hp(m) *(1.54 MW)*; 1 shaft

Comment: Launched 12 and 14 June 1980 at Singapore Marine Shipyard. Both commissioned 5 March 1981. Bollard pull 22 tons.

RIN   4/2007, Royal Thai Navy / 1455675

2 SAMAESAN CLASS (COASTAL TUGS) (YTR)

SAMAESAN 855   **RAET** 856

Displacement, tonnes: 305 standard
Dimensions, metres (feet): 25 × 8.5 × 2.4 (82 × 27.9 × 7.9)
Speed, knots: 10
Complement: 6
Machinery: 2 Caterpillar 3512TA diesels; 2,350 hp(m) (1.75 MW) sustained; 2 Aquamaster US 901 props

Comment: Contract signed 23 September 1992 for local construction at Thonburi Naval dockyard. Completed in December 1993. Equipped for firefighting.

RAET   5/1997, A Sharma / 0050251

2 YTL 422 CLASS (YTL)

KLUENG BADAAN 851 (ex-YTL 2)   **MARN VICHAI** 852 (ex-YTL 3)

Displacement, tonnes: 64 standard
Dimensions, metres (feet): 19.7 × 5 × 1.8 (64.6 × 16.4 × 5.9)
Speed, knots: 8
Machinery: 1 diesel; 240 hp (179 kW); 1 shaft

Comment: Built by Central Bridge Co, Trenton and bought from Canada 1953.

MARN VICHAI   3/2010*, Adolfo Ortigueira Gil / 1529084

POLICE

Notes: (1) There is also a Customs service, subordinate to the Marine Police, which operates unarmed patrol craft with CUSTOMS on the hull, and a Fishery Patrol Service also unarmed but vessels are painted blue with broad white and narrow gold diagonal stripes on the hull. Two Hydrofoil craft are on loan from the Police to the Customs service.
(2) There are large numbers of RIBs in service.

1 VOSPER THORNYCROFT TYPE (LARGE PATROL CRAFT) (PSO)

SRINAKARIN 1804

Displacement, tonnes: 640 full load
Dimensions, metres (feet): 62 × 8.2 × 2.5 (203.4 × 26.9 × 8.2)
Speed, knots: 25. **Range, n miles:** 2,500 at 15 kt
Complement: 45
Machinery: 2 Deutz MWM BV16M628 diesels; 9,524 hp(m) (7 MW) sustained; 2 shafts; Kamewa cp props
Guns: 4—30 mm (2 twin).
Radars: Surface search: Racal Decca 1226; I-band.

Comment: Ordered in September 1989 from Ital Thai Marine. Same hull as the Khamronsin class corvettes for the Navy but much more lightly armed. Delivered in April 1992.

SRINAKARIN   6/2003, Royal Thai Navy / 0572648

2 HAMELN TYPE (LARGE PATROL CRAFT) (PBO)

DAMRONG RACHANUPHAP 1802   **LOPBURI RAMES** 1803

Displacement, tonnes: 437 full load
Dimensions, metres (feet): 56.7 × 8.1 × 2.4 (186 × 26.6 × 7.9)
Speed, knots: 23
Complement: 45
Machinery: 4 MTU diesels; 4,400 hp(m) (3.23 MW); 2 shafts
Guns: 2 Oerlikon 30 mm/75 (twin). 2 Oerlikon 20 mm.
Radars: Surface search: Racal Decca 1226; I-band.

Comment: Delivered by Schiffwerft Hameln, Germany, on 3 January 1969 and 10 December 1972 respectively.

LOPBURI RAMES   6/2003, Royal Thai Navy / 0572647

2 SUMIDAGAWA TYPE (COASTAL PATROL CRAFT) (PB)

CHASANYABADEE 1101   **PHROMYOTHEE** 1103

Displacement, tonnes: 132 full load
Dimensions, metres (feet): 34 × 5.8 × 2.8 (111.5 × 19.0 × 9.2)
Speed, knots: 32
Complement: 23
Machinery: 3 Ikegai diesels; 4,050 hp(m) (2.98 MW); 3 shafts
Guns: 2—12.7 mm MGs.
Radars: Surface search: Racal Decca; I-band.

Comment: Commissioned in August 1972 and May 1973 respectively.

PHROMYOTHEE   1990, Marine Police / 0080833

1 YOKOHAMA TYPE (COASTAL PATROL CRAFT) (PB)

CHAWENGSAK SONGKRAM 1102

Displacement, tonnes: 193 full load
Dimensions, metres (feet): 35.5 × 7 × 3.5 (116.5 × 23.0 × 11.5)
Speed, knots: 32
Complement: 18
Machinery: 4 Ikegai diesels; 5,400 hp(m) (3.79 MW); 2 shafts
Guns: 2 Oerlikon 20 mm.

Comment: Commissioned 13 April 1973. A second of class operates for the Customs with the number 1201.

CHAWENGSAK SONGKRAM   1990, Marine Police / 0080834

846 Thailand > Police

1 ITAL THAI MARINE TYPE (COASTAL PATROL CRAFT) (PB)

SRIYANONT 901

Displacement, tonnes: 53 full load
Dimensions, metres (feet): 27.4 × 4.9 × 2 *(89.9 × 16.1 × 6.6)*
Speed, knots: 23
Complement: 14
Machinery: 2 Deutz BA16M816 diesels; 2,680 hp(m) *(1.97 MW)* sustained; 2 shafts
Guns: 1 Oerlikon 20 mm. 2—7.62 mm MGs.
Radars: Surface search: Racal Decca; I-band.

Comment: Commissioned 12 June 1986.

SRIYANONT   *12/2001, Thai Navy League* / 0130155

3 CUTLASS CLASS (COASTAL PATROL CRAFT) (PB)

PHRAONGKAMROP 807   **PICHARNPHOLAKIT** 808   **RAMINTHRA** 809

Displacement, tonnes: 35 full load
Dimensions, metres (feet): 19.8 × 5.2 × 2.5 *(65 × 17.1 × 8.2)*
Speed, knots: 25
Complement: 14
Machinery: 3 Detroit 12V-71TA diesels; 1,020 hp(m) *(761 kW)* sustained; 3 shafts
Guns: 1 Oerlikon 20 mm. 2—7.62 mm MGs.

Comment: Delivered by Halter Marine, New Orleans, and all commissioned on 9 March 1969. Aluminium hulls.

PHRAONGKAMROP   *4/2011, Peter Mansfield* / 1406434

3 TECHNAUTIC TYPE (COASTAL PATROL CRAFT) (PB)

810–812

Displacement, tonnes: 51 full load
Dimensions, metres (feet): 27 × 5.9 × 1.9 *(88.6 × 19.4 × 6.2)*
Speed, knots: 27
Complement: 14
Machinery: 3 Isotta Fraschini diesels; 2,500 hp(m) *(1.84 MW)*; 3 Castoldi hydrojets
Guns: 1 Oerlikon 20 mm GAM-BO1. 2—7.62 mm MGs.

Comment: Delivered by Technautic, Bangkok in 1984.

812   *1990, Marine Police* / 0080837

5 ITAL THAI MARINE TYPE (COASTAL PATROL CRAFT) (PB)

625–629

Displacement, tonnes: 43 full load
Dimensions, metres (feet): 19.5 × 5.3 × 1.5 *(64 × 17.4 × 4.9)*
Speed, knots: 27
Complement: 14
Machinery: 2 MAN D2842LE diesels; 1,350 hp(m) *(992 kW)* sustained; 2 shafts
Guns: 1—12.7 mm MG.

Comment: Built in Bangkok 1987–90. Aluminium hulls. More of the class operated by the Fishery Patrol Service.

IHS Jane's Fighting Ships 2014-2015

ITAL THAI 626   *3/2004, Bob Fildes* / 0589814

8 MARSUN TYPE (COASTAL PATROL CRAFT) (PB)

630–637

Displacement, tonnes: 39 full load
Dimensions, metres (feet): 20 × 5.6 × 1.5 *(65.6 × 18.4 × 4.9)*
Speed, knots: 25
Complement: 11
Machinery: 2 MAN D2840LXE diesels; 1,640 hp(m) *(1.2 MW)* sustained; 2 shafts
Guns: 1—12.7 mm MG.

Comment: Built by Marsun, Thailand and commissioned from 2 August 1994.

MARSUN 634   *3/2004, Bob Fildes* / 0589815

38 RIVER PATROL BOATS (PBR)

301–338

Displacement, tonnes: 5 full load
Dimensions, metres (feet): 11.3 × 3.4 × 0.7 *(37.1 × 11.2 × 2.3)*
Speed, knots: 25
Machinery: 2 diesels; 2 shafts

2 MARSUN TYPE (PB)

539–540

Displacement, tonnes: 30 full load
Dimensions, metres (feet): 17.4 × 4.9 × 0.9 *(57.1 × 16.1 × 3.0)*
Speed, knots: 25
Complement: 8
Machinery: 2 Detroit 12V-71TA diesels; 840 hp *(627 kW)* sustained; 2 shafts
Guns: 1—12.7 mm MG.

Comment: Built in Thailand. Both commissioned 26 March 1986.

MARSUN 539   *11/2001, Maritime Photographic* / 0130169

© 2014 IHS

Police < **Thailand** — Introduction < **Togo** 847

20 CAMCRAFT TYPE (RIVER PATROL CRAFT) (PBR)

415–440 series

Displacement, tonnes: 13 full load
Dimensions, metres (feet): 12.2 × 3.7 × 1 *(40 × 12.1 × 3.3)*
Speed, knots: 25
Complement: 6
Machinery: 2 Detroit diesels; 540 hp *(403 kW)*; 2 shafts

Comment: Delivered by Camcraft, Louisiana. Aluminium hulls. Numbers uncertain.

CAMCRAFT *3/2011, Jurg Kürsener* / 1406468

17 TECHNAUTIC TYPE (COASTAL PATROL CRAFT) (PB)

608–624

Displacement, tonnes: 30 full load
Dimensions, metres (feet): 18.3 × 4.9 × 0.9 *(60 × 16.1 × 3.0)*
Speed, knots: 27
Complement: 11
Machinery: 2 Isotta Fraschini ID 36 SS 8V diesels; 1,760 hp(m) *(1.29 MW)* sustained; 2 Castoldi hydrojets
Guns: 1 — 12.7 mm MG.

Comment: Built from 1983-87 in Bangkok. Operational status of some of these craft doubtful.

TECHNAUTIC 609 *11/2001, Maritime Photographic* / 0130168

26 SUMIDAGAWA TYPE (RIVER PATROL CRAFT) (PBR)

513–538

Displacement, tonnes: 18 full load
Dimensions, metres (feet): 16.5 × 3.8 × 0.7 *(54.1 × 12.5 × 2.3)*
Speed, knots: 23
Complement: 6
Machinery: 2 Cummins diesels; 800 hp *(597 kW)*; 2 shafts
Guns: 1 — 12.7 mm MG.

Comment: First 21 built by Sumidagawa, last five by Captain Co, Thailand 1978–79.

SUMIDAGAWA 526 *6/2003, Royal Thai Navy* / 0572646

SUMIDAGAWA 529 *6/1999, Marine Police* / 0080841

1 RIVER PATROL CRAFT (PBR)

339

Displacement, tonnes: 5 full load
Dimensions, metres (feet): 11.3 × 3.4 × 1.8 *(37.1 × 11.2 × 5.9)*
Speed, knots: 25
Complement: 4
Machinery: 2 diesels; 2 shafts

Comment: Built in 1990.

RIVER PATROL CRAFT 339 (alongside Technautic 609) *7/2000* / 0106613

1 PATROL CRAFT (PB)

814

Displacement, tonnes: 65 full load
Dimensions, metres (feet): 25 × 6.0 × 1.8 *(82 × 19.7 × 5.9)*
Speed, knots: 30
Range, n miles: 650 at 12 kt
Complement: 12
Machinery: 2 diesels; 2 shafts
Guns: 2 — 30 mm. 2 — 12.7 mm MGs.
Radars: Surface search/navigation: To be announced.

Comment: Built by Marsun in 2006 as a replacement for *Burespadoongkit*, which was damaged in the tsunami of 26 December 2004 and subsequently became a museum ship.

814 *4/2013*, Bob Fildes* / 1529085

Togo

Country Overview

Formerly French Togoland, the Togolese Republic gained full independence in 1960 having rejected proposals to be united with Ghana. Situated in west Africa, it has an area of 21,925 square miles and borders to the east with Benin and to the west with Ghana. Togo has a short coastline of 30 n miles with the Gulf of Guinea. Lomé is the capital, largest town and principal port. Territorial seas (12 n miles) are claimed. A 200 n mile Exclusive Economic Zone (EEZ) is also claimed but this has not been defined by boundary agreements.

Headquarters Appointments

Commanding Officer, Navy:
  Captain Yawo Atiogbé Ametsipe

Personnel

2014:
(a) 260
(b) Conscription (2 years)

Bases

Lomé

© 2014 IHS IHS Jane's Fighting Ships 2014-2015

848 Togo > Patrol forces — Trinidad and Tobago > Introduction

PATROL FORCES

1 + 1 RAIDCO RPB 33 CLASS (PATROL CRAFT) (PB)

Dimensions, metres (feet): 33.95 × 6.53 × 1.9 *(111.4 × 21.4 × 6.2)*
Speed, knots: 35. **Range, n miles:** 1,500 at 15 kt
Complement: 17
Machinery: 2 diesels; 3,916 hp *(2.9 MW)*; 2 shafts
Guns: 1—20 mm. 2—12.7 mm MGs.
Radars: Navigation: I-band.

Comment: The contract with Raidco Marine for the construction of one patrol craft was announced in May 2013. Delivery of the vessels is expected in March and July 2014. The design includes a launch and recovery platform aft for one RHIB.

2 COASTAL PATROL CRAFT (PB)

Name	No	Builders	Launched
KARA	P 761	Chantiers Navals de l'Esterel, Cannes	18 May 1976
MONO	P 762	Chantiers Navals de l'Esterel, Cannes	16 Jun 1976

Displacement, tonnes: 81 full load
Dimensions, metres (feet): 32 × 5.8 × 1.6 *(105 × 19.0 × 5.2)*
Speed, knots: 30. **Range, n miles:** 1,500 at 15 kt
Complement: 17 (1 officer)

Machinery: 2 MTU MB 12V 493 TY60 diesels; 2,000 hp(m) *(1.47 MW)* sustained; 2 shafts
Missiles: SSM: Aerospatiale SS 12M; wire-guided to 5 km *(3 n miles)* subsonic; warhead 30 kg.
Guns: 1 Bofors 40 mm/70. 1 Oerlikon 20 mm.
Radars: Surface search: Decca 916; I-band.

Comment: Both craft seagoing but missile system probably not operational.

2 DEFENDER CLASS (RESPONSE BOATS) (PBF)

Displacement, tonnes: 2.7 full load
Dimensions, metres (feet): 7.6 × 2.6 × 2.7 *(24.9 × 8.5 × 8.9)*
Speed, knots: 46
Range, n miles: 175 at 35 kt
Complement: 4
Machinery: 2 Honda outboard motors; 450 hp *(335 kW)*
Guns: 1—12.7 mm MG.
Radars: Navigation: Furuno; I-band.

Comment: High-speed inshore patrol craft of aluminium construction and foam collar built by SAFE Boats International, Port Orchard, Washington. Two craft donated by the US government on 29 January 2010.

DEFENDER CRAFT 6/2010, Togo Navy / 1366489

KARA
6/2010, Togo Navy
1366490

Tonga

Country Overview

A former British protectorate, the Kingdom of Tonga became a sovereign state in 1970. Situated in the southwestern Pacific Ocean some 1,080 n miles northeast of New Zealand, the country consists of more than 170 islands and islets running generally north-south. There are three main groups, Tongatapu, Ha'apai and Vava'u, and several outlying islands. Nuku'alofa, on Tongatapu Island, is the capital, largest town and principal port. Territorial seas (12 n miles) are claimed. An Exclusive Economic Zone (EEZ) (200 n miles) is claimed but limits have not been fully defined by boundary agreements.

Headquarters Appointments

Commanding Officer, Navy:
  Commander Sione Fifita

Personnel

2014: 125

Bases

Touliki Base, Nuku'alofa (HMNB *Masefield*)

Prefix to Ships' Names

VOEA (Vaka O Ene Afio)

PATROL FORCES

Notes: A Beech 18 aircraft was acquired in May 1995 for maritime surveillance.

3 PACIFIC CLASS (LARGE PATROL CRAFT) (PB)

Name	No	Builders	Commissioned
NEIAFU	P 201	Australian Shipbuilding Industries	28 Oct 1989
PANGAI	P 202	Australian Shipbuilding Industries	30 Jun 1990
SAVEA	P 203	Australian Shipbuilding Industries	23 Mar 1991

Displacement, tonnes: 165 full load
Dimensions, metres (feet): 31.5 × 8.1 × 2.1 *(103.3 × 26.6 × 6.9)*
Speed, knots: 20. **Range, n miles:** 2,500 at 12 kt
Complement: 17 (3 officers)
Machinery: 2 Caterpillar 3516TA diesels; 2,820 hp *(2.1 MW)* sustained; 2 shafts
Guns: 2—12.7 mm MGs.
Radars: Surface search: Furuno 1101; I-band.

Comment: Part of the Pacific Forum Australia Defence co-operation. First laid down 30 January 1989, second 2 October 1989, third February 1990. *Savea* has a hydrographic survey capability. Following half-life refits 1998–99 and the decision of the Australian government to extend the Pacific Patrol Boat programme, *Neiafu* completed a life-extension refit at Townsville in 2009 and *Pangai* in 2010. *Savea* completed life-extension refit at Townsville in 2011.

SAVEA 10/2013*, Chris Sattler / 1525555

Trinidad and Tobago

Country Overview

Trinidad and Tobago gained independence in 1962 and became a republic in 1976. The country lies at the southern end of the Lesser Antilles chain and comprises the main islands of Trinidad (1,864 square miles), Tobago (116 square miles) and 21 minor islands and rocks. Trinidad is close to the northeastern coast of Venezuela and the mouth of the Orinoco River. The capital, largest town, and principal port is Port-of-Spain, Trinidad. An archipelagic state, territorial seas (12 n miles) are claimed. While a 200 n mile Exclusive Economic Zone (EEZ) has been claimed, the limits have only been partly defined by boundary agreements.

Headquarters Appointments

Commanding Officer, Coast Guard:
  Captain Mark Williams

Aircraft

The Coast Guard operates three Cessna (Types 172, 402B and 310R) for surveillance and two C26B acquired in 1999. These aircraft can be backed by Air Division Gazelle and Sikorsky S-76 helicopters when necessary. An order for three AW 139 helicopters was made in August 2009.

Personnel

(a) 2014: 1,381 (50 officers)
(b) Voluntary service

Bases

Staubles Bay (HQ)
Hart's Cut, Tobago, Point Fortin
Piarco (Air station), Cedros
Galeota

Coast Defence

There are plans to install a coastal radar system.

Prefix to Ships' Names

TTS

IHS Jane's Fighting Ships 2014-2015

COAST GUARD

Notes: (1) It is planned to procure six interceptor craft and four helicopters.
(2) The contract for the construction of three offshore patrol vessels was cancelled by the Trinidad government in September 2010. The three 90 m ships were subsequently sold to Brazil.

1 ISLAND CLASS (PBO)

Name	No	Builders	Commissioned
NELSON (ex-*Orkney*)	CG 20 (ex-P 299)	Hall Russell, Aberdeen	25 Feb 1977

Displacement, tonnes: 940 standard; 1,280 full load
Dimensions, metres (feet): 59.5 oa; 53.7 wl × 11 × 4.5 *(195.2; 176.2 × 36.1 × 14.8)*
Speed, knots: 16.5
Range, n miles: 7,000 at 12 kt
Complement: 35 (5 officers)
Machinery: 2 Ruston 12RKC diesels; 5,640 hp *(4.21 MW)* sustained; 1 shaft; cp prop
Guns: 2—7.62 mm MGs can be carried.
Radars: Navigation: Kelvin Hughes Type 1006; I-band.

Comment: Transferred from the UK Navy on 18 December 2000 and recommissioned on 22 February 2001. Based at Port of Spain.

NELSON *6/2009, Coen Den Dekker* / 1305594

1 TYPE CG 40 (LARGE PATROL CRAFT) (PB)

Name	No	Builders	Commissioned
CASCADURA	CG 6	Karlskronavarvet	15 Jun 1980

Displacement, tonnes: 213 full load
Dimensions, metres (feet): 40.6 × 6.7 × 1.6 *(133.2 × 22.0 × 5.2)*
Speed, knots: 30
Range, n miles: 3,000 at 15 kt
Complement: 25
Machinery: 2 Paxman Valenta 16CM diesels; 6,700 hp *(5 MW)* sustained; 2 shafts
Guns: 1 Bofors 40 mm/70. 1 Oerlikon 20 mm.
Radars: Surface search: Racal Decca 1226; I-band.
Weapon control systems: Optronic GFCS.

Comment: Ordered in Sweden mid-1978. Laid down early 1979. Fitted with foam-cannon oil pollution equipment and for oceanographic and hydrographic work. Nine spare berths. The hull is similar to Swedish Spica class but with the bridge amidships. Refitted in 1989 and 1998/99. *Barracuda* decommissioned in 2005.

CASCADURA *1/1994, Maritime Photographic* / 0506207

1 AUXILIARY VESSEL

REFORM A 04

Comment: Used for Port Services and other support functions.

4 POINT CLASS (COASTAL PATROL CRAFT) (PB)

Name	No	Builders	Commissioned
COROZAL POINT (ex-*Point Heyer*)	CG 7 (ex-82369)	J M Martinac, Tacoma	3 Aug 1967
CROWN POINT (ex-*Point Bennett*)	CG 8 (ex-82351)	Coast Guard Yard, Curtis Bay	19 Dec 1966
GALERA POINT (ex-*Point Bonita*)	CG 9 (ex-82347)	J M Martinac, Tacoma	12 Sep 1966
BARCOLET POINT (ex-*Point Highland*)	CG 10 (ex-82333)	Coast Guard Yard, Curtis Bay	27 Jun 1962

Displacement, tonnes: 67 full load
Dimensions, metres (feet): 25.3 × 5.2 × 1.8 *(83 × 17.1 × 5.9)*
Speed, knots: 23
Range, n miles: 1,500 at 8 kt
Complement: 10
Machinery: 2 Caterpillar 3412 diesels; 1,600 hp *(1.19 MW)*; 2 shafts
Guns: 2—7.62 mm MGs.
Radars: Surface search: Raytheon SPS-64(V)I and Raytheon SPS 69AN; I-band.

Comment: CG 7 and CG 8 transferred from US Coast Guard 12 February 1999 and CG 9 on 14 November 2000. CG 10 transferred on 24 July 2001.

GALERNA POINT *6/2007, Trinidad and Tobago Coast Guard* / 1170216

4 SOUTER WASP 17 METRE CLASS (COASTAL PATROL CRAFT) (PB)

Name	No	Builders	Commissioned
PLYMOUTH	CG 27	WA Souter, Cowes	27 Aug 1982
CARONI	CG 28	WA Souter, Cowes	27 Aug 1982
GALEOTA	CG 29	WA Souter, Cowes	27 Aug 1982
MORUGA	CG 30	WA Souter, Cowes	27 Aug 1982

Displacement, tonnes: 20 full load
Dimensions, metres (feet): 16.8 × 4.2 × 1.4 *(55.1 × 13.8 × 4.6)*
Speed, knots: 32
Range, n miles: 500 at 18 kt
Complement: 7 (2 officers)
Machinery: 2 MANN 8V diesels; 1,470 hp *(1.1 MW)*; 2 shafts
Guns: 1—7.62 mm MG.
Radars: Surface search: Raytheon SPS 69AN; I-band.

Comment: GRP hulls. All refitted from September 1997 with new engines.

PLYMOUTH *6/2007, Trinidad and Tobago Coast Guard* / 1335303

9 INTERCEPTION CRAFT (PBF)

CG 001–002   CG 012–018

Comment: CG 001–002 are 31 ft Bowen craft acquired in May 1991. They are capable of 40 kt. CG 012–013 are Midnight Express craft. CG 014–015 are 40 ft Formula 111 craft acquired from the US in 2005. They are capable of 60 kt. CG 016–018 are 40 ft Phantom Enforcer craft manufactured in Trinidad. They are capable of 60 kt.

CG 002 *6/2007, Trinidad and Tobago Coast Guard* / 1170215

CG 012 *6/2007, Trinidad and Tobago Coast Guard* / 1170214

2 WASP 20 METRE CLASS (COASTAL PATROL CRAFT) (PB)

Name	No	Builders	Commissioned
KAIRI (ex-*Sea Bird*)	CG 31	WA Souter, Cowes	Dec 1982
MORIAH (ex-*Sea Dog*)	CG 32	WA Souter, Cowes	Dec 1982

Displacement, tonnes: 33 full load
Dimensions, metres (feet): 20.1 × 5 × 1.5 *(65.9 × 16.4 × 4.9)*
Speed, knots: 30
Range, n miles: 450 at 30 kt
Complement: 6 (2 officers)
Machinery: 2 MANN 12V diesels; 2,400 hp *(1.79 MW)*; 2 shafts
Guns: 2 — 7.62 mm MGs.
Radars: Surface search: Decca 150; I-band.

Comment: Ordered 30 September 1981. Aluminium alloy hull. Transferred from the Police in June 1989. New engines in 1999.

KAIRI *7/2001, Margaret Organ* / 0114370

2 OFFSHORE PATROL VESSELS (PBO)

GASPAR GRANDE CG 21   CHACACHACARE CG 22

Displacement, tonnes: 203 full load
Dimensions, metres (feet): 46.3 × 9.1 × 2.4 *(151.9 × 29.9 × 7.9)*
Speed, knots: 20
Range, n miles: 3,300 at 12 kt
Complement: 19

Machinery: 4 Cummins K38 diesels; 2 shafts
Guns: 2 — 30 mm. 2 — 12.7 mm MGs.
Radars: Surface search: E/F-band.
Navigation: I-band.

Comment: As part of the contract, signed with VT Shipbuilding on 5 April 2007, to build three new 90 m offshore patrol vessels (OPVs), an interim patrol capability was provided by two vessels. The order for the OPVs was subsequently cancelled in 2010. The vessels are former US oil-rig crew ships modified to undertake patrol tasks and were commissioned on 23 April 2008. They are equipped with a 39 kt Halmatic Pacific 24 RIB.

GASPAR GRANDE *6/2009, Piet Cornelis* / 1305349

1 SWORD CLASS (COASTAL PATROL CRAFT) (PB)

Name	No	Builders	Commissioned
MATELOT (ex-*Sea Skorpion*)	CG 33	SeaArk Marine, Monticello	May 1979

Displacement, tonnes: 16 full load
Dimensions, metres (feet): 13.7 × 4.1 × 1.3 *(44.9 × 13.5 × 4.3)*
Speed, knots: 28
Range, n miles: 500 at 20 kt
Complement: 6
Machinery: 2 GM diesels; 850 hp *(634 kW)*; 2 shafts
Guns: 1 — 7.62 mm MG.
Radars: Surface search: Decca 150; I-band.

Comment: Two transferred from the Police 30 June 1989, one scrapped in 1990. Refitted in 1998.

MATELOT *1/1994, Maritime Photographic* / 0506174

6 PATROL CRAFT (PB)

SCARLET IBIS CG 11	HUMMING BIRD CG 13	POUI CG 15
HIBISCUS CG 12	CHACONIA CG 14	TEAK CG 16

Measurement, tonnes: 16 dwt
Dimensions, metres (feet): 30 × 6.4 × 1.5 *(98.4 × 21.0 × 4.9)*
Speed, knots: 40
Range, n miles: 1,000 at 10 kt
Complement: 12
Machinery: 2 MTU 16V 2000 M92 diesels; 4,370 hp *(3.26 MW)*; 2 Kamewa waterjets
Guns: 1 — 20 mm. 3 — 12.7 mm MGs.
Radars: Surface search/Navigation: To be announced.

Comment: The contract with Austal Shipbuilding for the construction of six patrol craft was announced on 18 March 2008. The monohull craft, of aluminium construction, are to be used for safety of shipping, environmental protection, counter-drugs and SAR duties. The contract includes a five-year support programme and training package. All six craft were delivered by February 2010.

CG 12 *10/2009, Austal* / 1305593

CUSTOMS

Notes: Among other craft, the Customs service operate a High Speed Interception craft *Kenneth Mohammed*.

KENNETH MOHAMMED *2/2001, van Ginderen Collection* / 0114369

Tunisia

Country Overview

Formerly a French protectorate, the Tunisian Republic gained independence in 1956 and became a republic in 1957. Situated in northern Africa, it has an area of 63,170 square miles and is bordered to the west by Algeria and to the south by Libya. It has a 619 n mile coastline with the Mediterranean Sea. The capital and largest city is the seaport of Tunis. There are further ports at Bizerta, Sousse, Sfax and Gabès while as-Sukhayrah, specialises in petroleum bunkering. Territorial seas (12 n miles) are claimed. An EEZ has not been claimed.

Headquarters Appointments

Naval Chief of Staff:
  Rear Admiral Mohamed Khamassi

Personnel

(a) 2014: 4,800 officers and men (including 800 conscripts)
(b) 1 year's national service

Bases

Bizerte, Sfax, La Goulette, Kelibia

PATROL FORCES

3 COMBATTANTE III M CLASS
(FAST ATTACK CRAFT—MISSILE) (PGGF)

Name	No	Builders	Launched	Commissioned
LA GALITÉ	501	CMN, Cherbourg	16 Jun 1983	27 Feb 1985
TUNIS	502	CMN, Cherbourg	27 Oct 1983	27 Mar 1985
CARTHAGE	503	CMN, Cherbourg	24 Jan 1984	29 Apr 1985

Displacement, tonnes: 351 standard; 432 full load
Dimensions, metres (feet): 56 × 8.2 × 2.2 (183.7 × 26.9 × 7.2)
Speed, knots: 38.5. **Range, n miles:** 700 at 33 kt, 2,800 at 10 kt
Complement: 35

Machinery: 4 MTU 20V 538 TB93 diesels; 18,740 hp(m) (13.8 MW) sustained; 4 shafts
Missiles: SSM: 8 Aerospatiale MM 40 Exocet (2 quad) launchers; inertial cruise; active radar homing to 70 km (40 n miles) at 0.9 Mach; warhead 165 kg; sea-skimmer.
Guns: 1 Oto Melara 3 in (76 mm)/62 Compact; 55–65 rds/min to 16 km (8.7 n miles); weight of shell 6 kg. 2 Breda 40 mm/70 (twin); 300 rds/min to 12.5 km (6.8 n miles); weight of shell 0.96 kg. 4 Oerlikon 30 mm/75 (2 twin); 650 rds/min to 10 km (5.5 n miles); weight of shell 1 kg or 0.36 kg.
Physical countermeasures: Decoys: 1 CSEE Dagaie trainable launcher; IR flares and chaff.
Electronic countermeasures: ESM: Thomson-CSF DR 2000; intercept.
Radars: Air/surface search: Thomson-CSF Triton S; G-band; range 33 km (18 n miles) for 2 m^2 target.
Fire control: Thomson-CSF Castor II; I/J-band; range 31 km (17 n miles) for 2 m^2 target.
Combat data systems: Tavitac action data automation.
Weapon control systems: 2 CSEE Naja optronic directors for 30 mm. Thomson-CSF Vega II for SSM, 76 mm and 40 mm.

Programmes: Ordered 27 June 1981.
Operational: CSEE Sylosat navigation system. All three ships operating but reported in need of refits.

LA GALITÉ 11/2009, B Prézelin / 1366148

TUNIS 11/2009, B Prézelin / 1366147

6 COASTAL PATROL CRAFT (PB)

V 101–106

Displacement, tonnes: 39 full load
Dimensions, metres (feet): 25 × 4.8 × 1.3 (82 × 15.7 × 4.3)
Speed, knots: 23. **Range, n miles:** 900 at 15 kt
Complement: 11
Machinery: 2 Detroit 12V-71TA diesels; 840 hp (627 kW) sustained; 2 shafts; LIPS cp props
Guns: 1 Oerlikon 20 mm.
Radars: Surface search: Racal Decca 1226; I-band.

Comment: Built by Chantiers Navals de l'Esterel and commissioned in 1961–63. Two further craft of the same design (*Sabaq el Bahr* T 2 and *Jaouel el Bahr* T 1) but unarmed were transferred to the Fisheries Administration in 1971-same builders. Refitted in 1997/98. V 102 is Coast Guard.

V 105 3/2006, M Declerck / 1167532

3 MODIFIED HAIZHUI CLASS (LARGE PATROL CRAFT) (PB)

UTIQUE P 207 JERBA P 208 KURIAT P 209

Displacement, tonnes: 122 full load
Dimensions, metres (feet): 35 × 5.4 × 1.8 (114.8 × 17.7 × 5.9)
Speed, knots: 28. **Range, n miles:** 750 at 17 kt
Complement: 39
Machinery: 4 MWM TB 604 BV12 diesels; 4,400 hp(m) (3.22 MW) sustained; 4 shafts
Guns: 4 China 25 mm/80 (2 twin).
Radars: Surface search: Pot Head; I-band.

Comment: Delivered from China in March 1994. These craft resemble a smaller version of the Haizhui class in service with the Chinese Navy but with a different armament and some superstructure changes. Built to Tunisian specifications.

© 2014 IHS

KURIAT 4/1995 / 0080852

3 + 3 P 270 CLASS (PB)

JOUMHOURIA P 202   – P 203   – P 204   +3

Displacement, tonnes: 80 full load
Dimensions, metres (feet): 27.3 × 6.7 × ? (89.6 × 22.0 × ?)
Speed, knots: 38
Complement: 14
Machinery: 3 MTU 12V 4000 M84 diesel; 4,908 hp (3.66 MW); 2 shafts; 2 Rolls-Royce Kamewa 50A3 waterjets; 1 Rolls-Royce Kamewa 40A3 waterjet
Radars: Navigation: I-band.

Comment: Built by Cantieri Navale Vittoria, Adria. First delivered in 2012 and two further in 2013. Aluminium construction.

JOUMHOURIA 7/2013*, Annati Collection / 1531064

3 BIZERTE CLASS (TYPE PR 48)
(LARGE PATROL CRAFT) (PBOM)

Name	No	Builders	Commissioned
BIZERTE	P 301	SFCN, Villeneuve la Garenne	10 Jul 1970
HORRIA (ex-*Liberté*)	P 302	SFCN, Villeneuve la Garenne	Oct 1970
MONASTIR	P 304	SFCN, Villeneuve la Garenne	25 Mar 1975

Displacement, tonnes: 254 full load
Dimensions, metres (feet): 48 × 7.1 × 2.3 (157.5 × 23.3 × 7.5)
Speed, knots: 20. **Range, n miles:** 2,000 at 16 kt
Complement: 34 (4 officers)

Machinery: 2 MTU 16V 652 TB81 diesels; 4,600 hp(m) (3.4 MW) sustained; 2 shafts
Missiles: SSM: 8 Aerospatiale SS 12M; wire-guided to 5.5 km (3 n miles) subsonic; warhead 30 kg.
Guns: 4—37 mm/63 (2 twin). 2—14.5 mm MGs.
Radars: Surface search: Thomson-CSF DRBN 31; I-band.

Comment: First pair ordered in 1968, third in August 1973. Guns changed in 1994. All are active.

BIZERTE 3/2002, van Ginderen Collection / 0141859

HORRIA 10/2001 / 0533311

IHS Jane's Fighting Ships 2014-2015

8 PREVEZE (TYPE 209/1400) CLASS (SSK)

Name	No	Builders	Laid down	Launched	Commissioned
PREVEZE	S 353	Gölcük, Kocaeli	12 Sep 1989	22 Oct 1993	22 Mar 1994
SAKARYA	S 354	Gölcük, Kocaeli	1 Feb 1990	28 Jul 1994	6 Jan 1995
18 MART	S 355	Gölcük, Kocaeli	28 Jul 1994	25 Aug 1997	27 Aug 1997
ANAFARTALAR	S 356	Gölcük, Kocaeli	1 Aug 1995	1 Sep 1998	12 Oct 1998
GÜR	S 357	Gölcük, Kocaeli	21 Feb 2000	24 Feb 2003	21 Apr 2006
ÇANAKKALE	S 358	Gölcük, Kocaeli	19 Dec 2000	23 Jun 2004	22 Jun 2006
BURAKREIS	S 359	Gölcük, Kocaeli	19 Dec 2001	5 Sep 2005	1 Nov 2006
İNÖNÜ	S 360	Gölcük, Kocaeli	2 Jan 2003	24 May 2007	27 Jun 2008

Displacement, tonnes: 1,477 surfaced; 1,611 dived
Dimensions, metres (feet): 62 × 6.2 × 5.5 *(203.4 × 20.3 × 18.0)*
Speed, knots: 10 surfaced; 10 snorting; 21.5 dived
Range, n miles: 8,200 at 8 kt surfaced; 400 at 4 kt dived
Complement: 30 (8 officers)

Machinery: Diesel-electric; 4 MTU 12V 396 SB83 diesels; 5,364 hp(m) *(4.0 MW)* sustained; 4 alternators; 1 Siemens motor; 4,000 hp(m) *(3.38 MW)* sustained; 1 shaft
Missiles: McDonnell Douglas Sub Harpoon; active radar homing to 130 km *(70 n miles)* at 0.9 Mach; warhead 227 kg.
Torpedoes: 8—21 in *(533 mm)* bow tubes. GEC-Marconi Tigerfish Mk 24 Mod 2; wire-guided; active/passive homing to 13 km *(7 n miles)* at 35 kt active; 29 km *(15.7 n miles)* at 24 kt passive; warhead 134 kg or STN Atlas DM 2A4 (S 357 onwards). Total of 14 torpedoes and missiles.
Mines: In lieu of torpedoes.
Electronic countermeasures: ESM: Racal Porpoise or Racal Sealion (UAP) (S 357 onwards); intercept.
Radars: Surface search: I-band.
Sonars: Atlas Elektronik CSU 83; passive/active search and attack; medium/high frequency. STN Atlas flank array; passive low frequency.
Weapon control systems: Atlas Elektronik ISUS 83-2 TFCS. Link 11 receive only.

Programmes: Order for first two signed in Ankara on 17 November 1987 with option on two more taken up in 1993. Four more ordered 22 July 1998. All built with HDW prefabrication and assembly at Gölcük. The last four are called the Gür class.
Structure: Single hull design. Diving depth, 280 m *(820 ft)*. Kollmorgen masts. Four torpedo tubes can be used for SSM. STN Atlas flank arrays fitted in 1998/99 to the first four.
Operational: Endurance, 50 days.

18 MART — 11/2013*, Guy Toremans / 1531066

ANAFARTALAR — 5/2013*, C D Yayali / 1531067

PREVEZE — 9/2009, B Prézelin / 1366149

ANAFARTALAR — 6/2011, Diego Quevedo / 1454740

Submarines — Frigates < **Turkey** 857

0 + 6 TYPE 214 CLASS (SSK)

Displacement, tonnes: 1,845 surfaced; 2,023 dived
Dimensions, metres (feet): 65 × 6.3 × 6
(*213.3 × 20.7 × 19.7*)
Speed, knots: 12 surfaced; 20 dived
Complement: 27 (5 officers)

Machinery: 2 MTU 16V 396 SE 84 diesels; 4,243 hp (m) *(3.12 MW)*; 1 Siemens Permasyn motor; 3,875 hp(m) *(2.85 MW)*; 1 shaft; 2 HDW PEM fuel cells; 240 kW; lead acid batteries
Torpedoes: 8 – 21 in *(533 mm)* bow tubes. Raytheon Mk 48 and Atlas DM2A4. Total of 16 weapons.
Physical countermeasures: Decoys.
Electronic countermeasures: ESM.
Radars: Surface search: I-band.
Sonars: Bow, flank and towed arrays.

Weapon control systems: STN Atlas ISUS.

Programmes: Contract signed with HDW-MFI Partnership on 2 July 2009 for the delivery of material packages for six submarines. The boats are to be built at Gölcük Shipyard and delivery of the first boat is planned in 2017.
Structure: The Type 214 is a synthesis of the proven Type 209 design with AIP from the Type 212.

TYPE 214 CLASS *10/2008, Michael Nitz* / 1353402

FRIGATES

Notes: (1) The Turkish Frigate 2000 project is for a class of up to six ships to enter service from about 2018. The ships, which are to be built locally to a nationally developed or foreign design, are expected to be of the order of 6,000 tons and to have a principal role of AAW. This is to be based around the Raytheon SM-2 missile system. Options for the air-defence radar include CAFRAD, an indigenously developed multifunction phased-array radar or an off-the-shelf system. The overall aim is to maximise local content and an RFI for selected sub-systems was issued in January 2010.

(2) The development of a new intermediate class frigate is also under consideration. The TF100 class would be smaller and less capable than TF2000 and would replace the Meko class in the 2020s.

8 GABYA (OLIVER HAZARD PERRY) CLASS (FFGHM)

Name	No	Builders	Laid down	Launched	Commissioned	Recommissioned
GAZIANTEP (ex-*Clifton Sprague*)	F 490 (ex-FFG 16)	Bath Iron Works, Maine	30 Sep 1979	16 Feb 1980	21 Mar 1981	24 Jul 1998
GIRESUN (ex-*Antrim*)	F 491 (ex-FFG 20)	Todd Pacific Shipyard Corporation, Seattle	21 Jun 1978	27 Mar 1979	26 Sep 1981	24 Jul 1998
GEMLIK (ex-*Flatley*)	F 492 (ex-FFG 21)	Bath Iron Works, Maine	13 Nov 1979	15 May 1980	20 Jun 1981	24 Jul 1998
GELIBOLU (ex-*Reid*)	F 493 (ex-FFG 30)	Todd Pacific Shipyard Corporation, San Pedro	8 Oct 1980	27 Jun 1981	19 Feb 1983	22 Jul 1999
GÖKÇEADA (ex-*Mahlon S Tisdale*)	F 494 (ex-FFG 27)	Todd Pacific Shipyard Corporation, San Pedro	19 Mar 1980	7 Feb 1981	27 Nov 1982	8 Jun 2000
GEDIZ (ex-*John A Moore*)	F 495 (ex-FFG 19)	Todd Pacific Shipyard Corporation, San Pedro	19 Dec 1978	20 Oct 1979	14 Nov 1981	25 Jul 2000
GOKOVA (ex-*Samuel Eliot Morison*)	F 496 (ex-FFG 13)	Bath Iron Works, Maine	4 Aug 1978	14 Jul 1979	11 Oct 1980	11 Apr 2002
GÖKSU (ex-*Estocin*)	F 497 (ex-FFG 15)	Bath Iron Works, Maine	2 Apr 1979	3 Nov 1979	10 Jan 1981	4 Apr 2003

Displacement, tonnes: 2,794 standard; 3,696 full load
Dimensions, metres (feet): 138.1 × 13.7 × 4.5 hull; 7.5 sonar
(*453.1 × 44.9 × 14.8; 24.6*)
Speed, knots: 29
Range, n miles: 4,500 at 20 kt
Complement: 206 (13 officers; 19 air crew)

Machinery: 2 GE LM 2500 gas turbines; 41,000 hp *(30.59 MW)* sustained; 1 shaft; cp prop 2 auxiliary retractable props; 650 hp *(484 kW)*
Missiles: SSM: 4 McDonnell Douglas Harpoon Block 1B; active radar homing to 92 km *(50 n miles)* at 0.9 Mach; warhead 227 kg.
SAM: 36 Raytheon Standard SM-1MR Block VIB; command guidance; semi-active radar homing to 38 km *(20.5 n miles)* at 2 Mach. 1 Mk 13 Mod 4 launcher for both SSM and SAM missiles ❶. 32 Raytheon RIM-162 ESSM (F 495, 496 and 497) ❷; Mk 41 8-cell VLS launchers; semi-active radar homing to 18.5 km *(10 n miles)* at 3.6 Mach; warhead 227 kg.
Guns: 1 Oto Melara 3 in *(76 mm)*/62 Mk 75 ❸; 85 rds/min to 16 km *(8.7 n miles)* anti-surface; 12 km *(6.6 n miles)* anti-aircraft; weight of shell 6 kg. 1 Raytheon 20 mm/76 6-barrelled Mk 15 Vulcan Phalanx ❹; 3,000 rds/min combined to 1.5 km. 4 – 12.7 mm MGs.
Torpedoes: 6 – 324 mm Mk 32 (2 triple) tubes ❺. 24 Honeywell Mk 46 Mod 5; anti-submarine; active/passive homing to 11 km *(5.9 n miles)* at 40 kt; warhead 44 kg.
Physical countermeasures: Decoys: 2 Loral Hycor SRBOC 6-barrelled fixed Mk 36 ❻; IR flares and chaff to 4 km *(2.2 n miles)*. T-Mk-6 Fanfare/SLQ-25 Nixie; torpedo decoy.
Electronic countermeasures: ESM/ECM: SLQ-32(V)2 ❼; radar warning. Sidekick modification adds jammer and deception system.
Radars: Air search: Raytheon SPS-49(V)4 (F 490–496); C/D-band. Thales SMART-S (F 497) ❽; 3D; E/F-band.
Surface search: ISC Cardion SPS-55 ❾; I-band.
Fire control: Lockheed STIR (modified SPG-60) ❿; I/J-band.
Sperry Mk 92 (Signaal WM28) ⓫; I/J-band.
Navigation: GEM SPN-753/730; I-band.
Tacan: URN 25.
Sonars: Raytheon SQS-56; hull-mounted; active search and attack; medium frequency.
Combat data systems: NTDS with Link 11 and 14. SATCOM. Genesis (F 490–493).
Weapon control systems: SWG-1 Harpoon LCS. Mk 92 Mod 4 WCS with CAS (Combined Antenna System). The Mk 92 is the US version of the Signaal WM28. Mk 13 weapon direction system. 2 Mk 24 optical directors.

Helicopters: 1 S-70B Seahawk ⓬.

Programmes: Three approved for transfer by grant aid. Transfer delayed by Greek objections, and Turkish sailors were sent home from the US in mid-1996. Congress authorised the go-ahead again on 27 August 1997. Two more approved for transfer by sale 30 September 1998, one in February 2000, one in April 2002 and one in April 2003. At least one other *Duncan* FFG 10 for spares.
Modernisation: The combat data system has been upgraded under the 'Genesis' programme. All ships had been modernised by late 2012. MilSoft awarded a contract in August 2006 to develop a Link 11/16 datalink system. This will be completed in 2014. A programme to install an 8-cell Mk 41 VLS launcher for 32 Evolved Sea Sparrow in four ships is in progress. *Gediz* has been completed and *Gokova* followed. The associated Mk 92 fire-control radar is also to be upgraded. The arrangement is similar to the Australian FFG upgrade programme. The SPS-49 radar has been replaced by SMART-S in F 497 and F 495.
Structure: A flight deck extension programme, to enable S-70 helicopters has been completed. The work involved angling the transom as in later USN ships of the class.
Operational: Sonar towed arrays were not transferred.

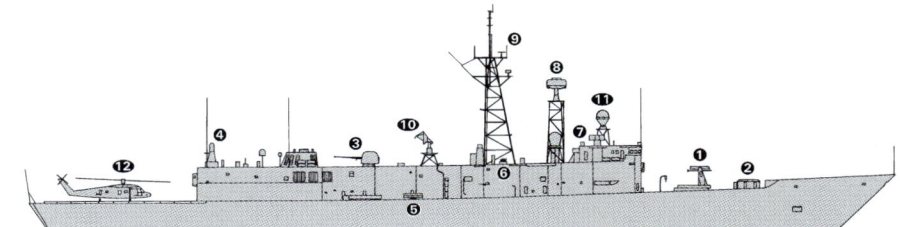

GÖKSU *(Scale 1 : 1,200), Ian Sturton* / 1455827

GEDIZ *9/2013*, C D Yaylali* / 1531072

GELIBOLU *11/2013*, Guy Toremans* / 1531071

858 Turkey > Frigates

4 BARBAROS CLASS (MEKO 200 TN II-A/B) (FFGHM)

Name	No	Builders	Laid down	Launched	Commissioned
BARBAROS	F 244	Blohm + Voss, Hamburg	18 Mar 1993	29 Sep 1993	16 Mar 1995
ORUCREIS	F 245	Gölcük, Kocaeli	15 Sep 1993	28 Jul 1994	10 May 1996
SALIHREIS	F 246	Blohm + Voss, Hamburg	24 Jul 1995	26 Sep 1997	17 Dec 1998
KEMALREIS	F 247	Gölcük, Kocaeli	4 Apr 1997	24 Jul 1998	8 Jun 2000

Displacement, tonnes: 3,434 full load
Dimensions, metres (feet): 118 × 14.8 × 4.3 hull; 6.4 sonar *(387.1 × 48.6 × 14.1; 21.0)*
Speed, knots: 32
Range, n miles: 4,100 at 18 kt
Complement: 196 (22 officers; 9 air crew) + 8 spare berths

Machinery: CODOG; 2 GE LM 2500 gas turbines; 60,000 hp *(44.76 MW)* sustained; 2 MTU 16V 1163 TB83 diesels; 11,780 hp(m) *(8.67 MW)* sustained; 2 shafts; Escher Wyss; cp props
Missiles: SSM: 8 McDonnell Douglas Harpoon (2 quad) launchers ❶; active radar homing to 130 km *(70 n miles)* at 0.9 Mach; warhead 227 kg.
SAM: Raytheon Sea Sparrow RIM-7M Mk 29 Mod 1 octuple launcher ❷ (F 244 and F 245) and VLS Mk 41 Mod 8 ❸ (F 246 and F 247); semi-active radar homing to 16 km *(8.5 n miles)* at 2.5 Mach; warhead 38 kg. RIM-162 Evolved Sea Sparrow (ESSM) in due course.
Guns: 1 FMC 5 in *(127 mm)*/54 Mk 45 Mod 1/2 ❹; 20 rds/min to 23 km *(12.6 n miles)* anti-surface; 15 km *(8.2 n miles)* anti-aircraft; weight of shell 32 kg. 3 Oerlikon-Contraves 25 mm Sea Zenith ❺; 4 barrels per mounting; 3,400 rds/min combined to 2 km.
Torpedoes: 6—324 mm Mk 32 Mod 5 (2 triple) tubes ❻. Honeywell Mk 46 Mod 5; anti-submarine; active/passive homing to 11 km *(5.9 n miles)* at 40 kt; warhead 44 kg.
Physical countermeasures: Decoys: 2 Loral Hycor 6-tubed fixed Mk 36 Mod 1 SRBOC ❼; IR flares and chaff to 4 km *(2.2 n miles)*. Nixie SLQ-25; towed torpedo decoy.
Electronic countermeasures: ESM/ECM: Racal Cutlass/Scorpion; intercept and jammer.
Radars: Air search: Siemens/Plessey AWS 9 (Type 996) (F 244, 245, 247) ❾; 3D; E/F-band. Thales SMART-S (F 246 and 247); 3D; E/F-band.
Air/surface search: Plessey/BAe AWS 6 Dolphin ❿; G-band.
Fire control: 1 or 2 (F 246-247) Signaal STIR ⓫; I/J/K-band (for SAM); range 140 km *(76 n miles)* for 1 m² target. Contraves TMX (F 244-245) ⓬; I/J-band (for SSM and 127 mm). 2 Contraves Seaguard ⓭; I/J-band (for 25 mm).
Navigation: GEM SPN-753/730; I-band.
Tacan: URN 25. IFF Mk XII Mod 4.
Sonars: Raytheon SQS-56 (DE 1160); hull-mounted; active search and attack; medium frequency.
Combat data systems: Thomson-CSF/Signaal STACOS Mod 3; Link 11. WSC 3V(7) SATCOMs. Marisat.
Weapon control systems: 2 Siemens Albis optronic directors ❽. SWG-1A for Harpoon.
Helicopters: 1 AB 212ASW ⓮ or S-70B Seahawk.

Programmes: First pair ordered 19 January 1990, second pair authorised 14 December 1992. Programme started 5 November 1991 with construction commencing in June 1992 in Germany. Completion of F 247 delayed by the Gölcük earthquake in 1999.
Structure: An improvement on the Yavuz class. Mk 29 Sea Sparrow launchers fitted in the first two, while the second pair have Mk 41 VLS aft of the funnel, which are to be retrofitted in the first two. The ships have CODOG propulsion for a higher top speed. Other differences with Yavuz include a full command system, improved radars and a citadel for NBCD protection. A bow bulwark has been added in the second pair.
Operational: The AB 212 helicopter has Sea Skua anti-ship missiles. All can be used as Flagships.

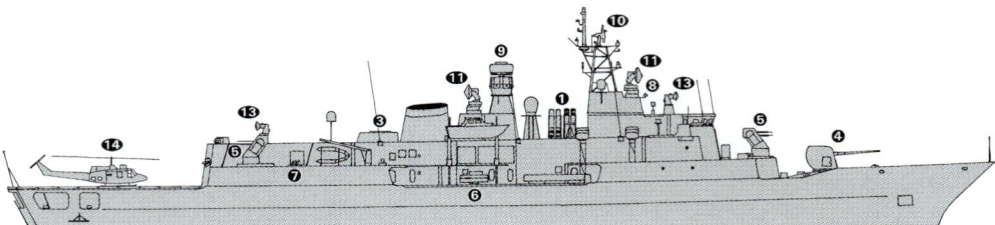
BARBAROS *(Scale 1 : 900), Ian Sturton /* 0019315

SALIHREIS *(Scale 1 : 900), Ian Sturton /* 1455826

SALIHREIS *9/2013*, Guy Toremans /* 1531069

ORUCREIS *3/2012, C D Yaylali /* 1486376

BARBAROS *11/2013*, Guy Toremans /* 1531068

IHS Jane's Fighting Ships 2014-2015 © 2014 IHS

4 YAVUZ CLASS (MEKO 200 TN) (FFGHM)

Name	No	Builders	Laid down	Launched	Commissioned
YAVUZ	F 240	Blohm + Voss, Hamburg	30 May 1985	7 Nov 1985	17 Jul 1987
TURGUTREIS (ex-*Turgut*)	F 241	Howaldtswerke, Kiel	20 May 1985	30 May 1986	4 Feb 1988
FATIH	F 242	Gölcük, Izmit	1 Jan 1986	24 Apr 1987	28 Aug 1988
YILDIRIM	F 243	Gölcük, Izmit	24 Apr 1987	22 Jul 1988	17 Nov 1989

Displacement, tonnes: 2,453 standard; 2,966 full load
Dimensions, metres (feet): 115.5 × 14.2 × 4.1 *(378.9 × 46.6 × 13.5)*
Speed, knots: 27
Range, n miles: 4,100 at 18 kt
Complement: 180 (24 officers)

Machinery: CODAD; 4 MTU 20V 1163 TB93 diesels; 29,940 hp(m) *(22 MW)* sustained; 2 shafts; cp props
Missiles: SSM: 8 McDonnell Douglas Harpoon (2 quad) launchers ❶; active radar homing to 130 km *(70 n miles)* at 0.9 Mach; warhead 227 kg.
SAM: Raytheon Sea Sparrow RIM-7M Mk 29 Mod 1 octuple launcher ❷; semi-active radar homing to 16 km *(8.5 n miles)* at 2.5 Mach; warhead 38 kg.
Guns: 1 FMC 5 in *(127 mm)*/54 Mk 45 Mod 1 ❸; 20 rds/min to 23 km *(12.6 n miles)* anti-surface; 15 km *(8.2 n miles)* anti-aircraft; weight of shell 32 kg. 3 Oerlikon-Contraves 25 mm Sea Zenith ❹; 4 barrels per mounting; 3,400 rds/min combined to 2 km.
Torpedoes: 6—324 mm Mk 32 (2 triple) tubes ❺. Honeywell Mk 46 Mod 5; anti-submarine; active/passive homing to 11 km *(5.9 n miles)* at 40 kt; warhead 44 kg.
Physical countermeasures: Decoys: 2 Loral Hycor 6-tubed fixed Mk 36 Mod 1 SRBOC ❻; IR flares and chaff to 4 km *(2.2 n miles)*. Nixie SLQ-25; towed torpedo decoy.
Electronic countermeasures: ESM/ECM: Signaal Rapids/Ramses; intercept and jammer.
Radars: Air search: Signaal DA08 ❼; F-band.
Air/surface search: Plessey AWS 6 Dolphin ❽; G-band.
Fire control: Signaal STIR ❾; I/J/K-band (for SAM); range 140 km *(76 n miles)* for 1 m² target. Signaal WM25 ❿; I/J-band (for SSM and 127 mm). 2 Contraves Seaguard ⓫; I/J-band (for 25 mm).
Navigation: GEM SPN-753/730; I-band.
Tacan: URN 25. IFF Mk XII.
Sonars: Raytheon SQS-56 (DE 1160); hull-mounted; active search and attack; medium frequency.
Combat data systems: Signaal STACOS-TU; action data automation; Link 11. WSC 3V(7) SATCOMs. Marisat.
Weapon control systems: 2 Siemens Albis optronic directors (for Sea Zenith). SWG-1A for Harpoon.

Helicopters: 1 AB 212ASW ⓬.

Programmes: Ordered 29 December 1982 with builders and Thyssen Rheinstahl Technik of Dusseldorf. Meko 200 type similar to Portuguese frigates. *Turgutreis* was renamed on 14 February 1988.
Operational: Helicopter has Sea Skua anti-ship missiles.

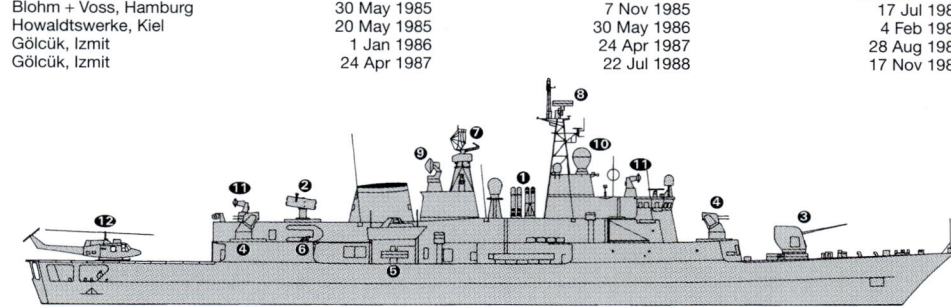

YAVUZ *(Scale 1 : 900), Ian Sturton* / 1153493

YILDIRIM *7/2013*, C D Yaylali* / 1531070

YILDIRIM *8/2010, C D Yaylali* / 1366797

YAVUZ *4/2012, C D Yaylali* / 1486377

TURGUTREIS *4/2010, C D Yaylali* / 1366799

CORVETTES

2 + 2 (4) ADA (MILGEM) CLASS (FFLG)

Name	No	Builders	Laid down	Launched	Commissioned
HEYBELIADA	F 511	Istanbul Naval Shipyard	22 Jan 2007	27 Sep 2008	27 Sep 2011
BÜYÜKADA	F 512	Istanbul Naval Shipyard	27 Sep 2008	27 Sep 2011	27 Sep 2013
BURGAZADA	F 513	Istanbul Naval Shipyard	27 Sep 2013	2014	2016
KINALIADA	F 514	Istanbul Naval Shipyard	2014	2015	2017

Displacement, tonnes: 1,524 standard; 2,032 full load
Dimensions, metres (feet): 99 × 14.4 × 3.6 *(324.8 × 47.2 × 11.8)*
Speed, knots: 29
Range, n miles: 3,500 at 15 kt
Complement: 93 + 13 spare berths

Machinery: CODAG; 2 MTU 16V 595 TE 90 diesels; 11,580 hp *(8.64 MW)*; 1 GE LM 2500 gas turbine; 30,800 hp *(23 MW)*; 2 shafts; cp props
Missiles: SSM: 8 McDonnell Douglas Harpoon (2 quadruple) ❶; active radar homing to 130 km *(70 n miles)* at 0.9 Mach; warhead 227 kg.
SAM: 1 RIM-116 RAM 21-cell Mk 49 launcher ❷.
Guns: 1—3 in *(76 mm)* Super Rapid; 120 rds/min to 16 km *(8.7 n miles)*; weight of shell 6 kg ❸. 2—12.7 mm Aselsan STAMP.
Torpedoes: 4—324 mm (2 twin) tubes ❹.
Physical countermeasures: Decoys: 2 SRBOC 6-barrelled Mk 36 ❺. Ultra Sea Sentor.
Electronic countermeasures: ESM: Aselsan ARES-2N.
Radars: Air/surface search: Thales SMART-S Mk 2; 3D ❻; E/F-band.
Fire control: Thales Sting ❼; I/K-band.
Navigation: Aselsan ALPER/Sperry Vision Master ❽; I-band.
Sonars: Active; medium frequency (under development).
Combat data systems: Genesis' derivative.
Electro-optic systems: Aselsan ASELFLIR-300D.

Helicopters: S-70B Seahawk.

Programmes: The MILGEM project was launched in 1996 for the in-country design and construction of up to 12 anti-submarine warfare and offshore patrol vessels. A contract to construct a further six ships was awarded to RMK Marine in January 2013 but, following an investigation into tender proceedings, the intention to cancel the contract was announced in September 2013. The third and fourth ships are now to be built at Istanbul Naval Shipyard and a re-tender for the subsequent four vessels is likely to be initiated.

HEYBELIADA *(Scale 1 : 900), Ian Sturton / 1455828*

BÜYÜKADA *7/2013*, C D Yaylali / 1531073*

6 BURAK (TYPE A 69) CLASS (FFLG)

Name	No	Builders	Laid down	Launched	Commissioned	Recommissioned
BOZCAADA (ex-*Commandant de Pimodan*)	F 500 (ex-F 787)	Lorient Naval Dockyard, Lorient	15 Jul 1975	7 Aug 1976	20 May 1978	22 Jun 2001
BODRUM (ex-*Drogou*)	F 501 (ex-F 783)	Lorient Naval Dockyard, Lorient	16 Oct 1973	30 Nov 1974	1 Oct 1976	18 Oct 2001
BANDIRMA (ex-*Quartier Maitre Anquetil*)	F 502 (ex-F 786)	Lorient Naval Dockyard, Lorient	1 Aug 1975	7 Aug 1976	4 Feb 1978	15 Oct 2001
BEYKOZ (ex-*d'Estienne d'Orves*)	F 503 (ex-F 781)	Lorient Naval Dockyard, Lorient	1 Sep 1972	1 Jun 1973	10 Sep 1976	18 Mar 2002
BARTIN (ex-*Amyot d'Inville*)	F 504 (ex-F 782)	Lorient Naval Dockyard, Lorient	2 Jul 1973	30 Nov 1974	13 Oct 1976	3 Jun 2002
BAFRA (ex-*Second Maitre Le Bihan*)	F 505 (ex-F 788)	Lorient Naval Dockyard, Lorient	1 Nov 1976	13 Aug 1977	7 Jul 1979	26 Jun 2002

Displacement, tonnes: 1,194 standard
Dimensions, metres (feet): 80.5 × 10.3 × 5.5 sonar *(264.1 × 33.8 × 18.0)*
Speed, knots: 23
Range, n miles: 4,500 at 15 kt
Complement: 104 (10 officers)

Machinery: 2 SEMT-Pielstick 12 PC2 V 400 diesels; 12,000 hp(m) *(8.82 MW)*; 2 shafts; LIPS cp props
Missiles: SSM: 2 Aerospatiale MM 38 Exocet ❶; inertial cruise; active radar homing to 70 km *(40 n miles)* or 42 km *(23 n miles)* at 0.9 Mach; warhead 165 kg; sea-skimmer.
Guns: 1 DCN/Creusot-Loire 3.9 in *(100 mm)*/55 Mod 68 CADAM automatic ❷; 80 rds/min to 17 km *(9 n miles)* anti-surface; 8 km *(4.4 n miles)* anti-aircraft; weight of shell 13.5 kg. 2 Giat 20 mm ❸; 720 rds/min to 10 km *(5.5 n miles)*. 4—12.7 mm MGs.
Torpedoes: 4—324 mm Mk 32 (2 twin) fixed tubes; Honeywell Mk 46 Mod 5; active/passive homing to 11 km *(5.9 n miles)* at 40 kt; warhead 44 kg. Mk 44 Mod 1; active homing to 5.5 km *(3 n miles)* at 30 kt; warhead 34 kg ❹; dual purpose.
A/S Mortars: 1 Creusot-Loire 375 mm Mk 54 6-tubed trainable launcher ❺; range 1,600 m; warhead 107 kg.
Physical countermeasures: Decoys: 2 CSEE Dagaie 10-barrelled trainable launchers ❻; chaff and IR flares; H- to J-band. Nixie torpedo decoy.
Electronic countermeasures: ESM: ARBR 16; radar warning.
Radars: Air/surface search: Thomson-CSF DRBV 51A ❼; G-band.
Navigation: Decca 20V90; I-band.
Fire control: Thomson-CSF DRBC 32E ❽; I-band.
Sonars: Thomson Sintra DUBA 25; hull-mounted; search and attack; medium frequency.
Weapon control systems: Thomson-CSF Vega system; CSEE Panda optical secondary director.

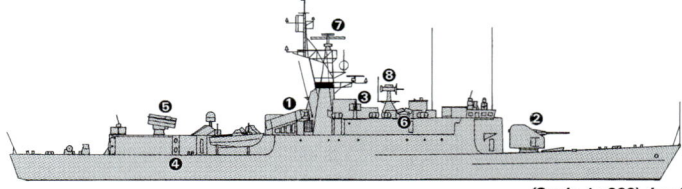

BOZCAADA *(Scale 1 : 900), Ian Sturton / 0114803*

BARTIN *11/2013*, Guy Toremans / 1531074*

Comment: Six Type A 69 class bought second-hand from France in October 2000. All, except *Bafra*, refitted at Brest. Work done on propulsion and weapons systems. Exocet MM 38 SSMs procured under separate contract. Operational use is coastal patrol duties, for which they were designed, in order to release more capable ships for front-line service.

SHIPBORNE AIRCRAFT

Numbers/Type: 13 Agusta AB 212.
Operational speed: 106 kt *(196 km/h)*.
Service ceiling: 14,200 ft *(4,330 m)*.
Range: 230 n miles *(426 km)*.
Role/Weapon systems: Multirole helicopter. Sensors: L3 AQS-18 dipping sonar (in seven aircraft); BAe Ferranti Sea Spray Mk 3 radar. Weapons: ASW; two Mk 46 torpedoes, Mk 54 Mod 1 depth charge. ASuW; M-60 MG.

AB 212   4/2010, C D Yaylali / 1366802

Numbers/Type: 23 Sikorsky S-70B Seahawk.
Operational speed: 135 kt *(250 km/h)*.
Service ceiling: 10,000 ft *(3,050 m)*.
Range: 600 n miles *(1,110 km)*.
Role/Weapon systems: Contracts placed 3 June 1998 for first four. Second contract for four further aircraft on 31 December 1998. First three delivered 26 April 2002 and second four on 24 July 2003. One aircraft lost in accident. An order for a further 17 aircraft was placed on 24 June 2005. Deliveries started in 2010. Sensors: APS-143 search radar; Helras dipping sonar, LR-100 ESM. Weapons: ASW: 2 Mk 46/54 torpedoes, Mk 54 Mod 1 depth charge. ASuW: AGM-114B Hellfire II/Penguin Mk 2 Mod 7.

S-70B   8/2009, C D Yaylali / 1366025

LAND-BASED MARITIME AIRCRAFT

Numbers/Type: 6 CASA CN-235 D/K MPA.
Operational speed: 240 kt *(445 km/h)*.
Service ceiling: 26,600 ft *(8,110 m)*.
Range: 669 n miles *(1,240 km)*.
Role/Weapon systems: Initial batch of two delivered in 2001 and a further four in 2002. First flight with mission systems took place on 18 June 2007. Thales AMACOS mission control system. Long-range maritime patrol for surface surveillance and ASW. Sensors: Ocean Master radar (SAR, ISAR, MTI and air-to-air modes); FLIR; AAR-60 missile warning; DR 3000 A ESM; MAD; Link 11. Weapons: 2 Mk-46 torpedoes.

CN-235   6/2006, Turkish Navy / 1158698

Numbers/Type: 8 Alenia ATR-72 ASW.
Operational speed: 255 kt *(472 km/h)*.
Service ceiling: 22,000 ft *(6,705 m)*.
Range: 1,200 n miles *(2,200 km)*.
Role/Weapon systems: Project Meltem-3. The contract for 10 MPA based on the ATR-72 500 aircraft was signed in July 2005. This was amended in 2013 to comprise eight aircraft (two utility, six MPA) based on the large ATR-72 600. Both utility aircraft were delivered in July and August 2013 respectively. Delivery of the first MPA is expected in early 2017. Weapons and sensors are to be integrated by the Thales AMASCOS mission system and sensors are to include a Thales Ocean Master radar, ESM and MAD. Weapons are to include Raytheon Mk 54 torpedo.

ATR-72 (model)   9/2005, C D Yaylali / 1133582

© 2014 IHS

PATROL FORCES

9 KILIÇ CLASS (FAST ATTACK CRAFT—MISSILE) (PGFG)

Name	No	Builders	Launched	Commissioned
KILIÇ	P 330	Lürssen, Bremen-Vegesack	15 Jul 1997	17 Mar 1998
KALKAN	P 331	Taşkızak, Istanbul	22 Sep 1998	22 Jul 1999
MIZRAK	P 332	Taşkızak, Istanbul	5 Apr 1999	8 Jun 2000
TUFAN	P 333	Lürssen, Bremen-Vegesack	3 Feb 2003	26 Jul 2005
MELTEM	P 334	Istanbul Naval Shipyard	1 Sep 2004	26 Jul 2005
IMBAT	P 335	Istanbul Naval Shipyard	26 Jul 2005	30 Mar 2007
ZIPKIN	P 336	Istanbul Naval Shipyard	27 Sep 2006	4 Jan 2008
ATAK	P 337	Gölcük	4 Jan 2008	14 Jan 2009
BORA	P 338	Gölcük	2009	7 Apr 2010

Displacement, tonnes: 559 full load
Dimensions, metres (feet): 62.4 × 8.3 × 2.6 *(204.7 × 27.2 × 8.5)*
Speed, knots: 38. **Range, n miles:** 1,050 at 30 kt
Complement: 46 (12 officers)

Machinery: 4 MTU 16V 956 TB91 diesels; 15,120 hp(m) *(11.1 MW)* sustained; 4 shafts
Missiles: SSM: 8 McDonnell Douglas Harpoon (2 quad) launchers; active radar homing to 130 km *(70 n miles)* at 0.9 Mach; warhead 227 kg.
Guns: 1 Otobreda 3 in *(76 mm)*/62 compact; 85 rds/min to 16 km *(8.7 n miles)* anti-surface; 12 km *(6.6 n miles)* anti-aircraft; weight of shell 6 kg. 2 Otobreda 40 mm/70 (twin); 300 rds/min to 12 km *(6.6 n miles)*; weight of shell 0.96 kg.
Physical countermeasures: Decoys: 2 Mk 36 SRBOC chaff launchers.
Electronic countermeasures: ESM: Racal Cutlass; intercept.
Radars: Surface search: Signaal MW08; G-band.
Fire control: Signaal STING; I/J-band.
Navigation: KH 1007; I-band.
Combat data systems: Signaal/Thomson-CSF STACOS.
Weapon control systems: LIROD Mk 2 optronic director.

Programmes: Contract for first three signed in May 1993 but there was a delay in confirming it. Further four ordered 19 June 2000 and a further two thereafter.
Structure: A development of the Yildiz class but with reduced radar cross-section mast and a redesigned bow to improve sea-keeping. The after gun and radars are also different to *Yildiz*. *Tufan* fitted with stealthy 76 mm gun turret and other ships are likely to be similarly upgraded.
Operational: First of class arrived in Turkey in April 1998.

IMBAT   9/2011, C D Yaylali / 1454748

KILIÇ   4/2011, C D Yaylali / 1454747

6 KARTAL CLASS (FAST ATTACK CRAFT—MISSILE) (PTFG)

Name	No	Builders	Commissioned
DENIZKUŞU	P 321 (ex-P 336)	Lürssen, Bremen-Vegesack	9 Mar 1967
ŞAHIN	P 323 (ex-P 334)	Lürssen, Bremen-Vegesack	3 Nov 1966
PELIKAN	P 326	Lürssen, Bremen-Vegesack	11 Feb 1970
ALBATROS	P 327 (ex-P 325)	Lürssen, Bremen-Vegesack	18 Mar 1970
ŞIMŞEK	P 328 (ex-P 332)	Lürssen, Bremen-Vegesack	6 Nov 1969
KASIRGA	P 329 (ex-P 338)	Lürssen, Bremen-Vegesack	25 Nov 1967

Displacement, tonnes: 163 standard; 193 full load
Dimensions, metres (feet): 42.5 × 7 × 2.4 *(139.4 × 23.0 × 7.9)*
Speed, knots: 42
Range, n miles: 500 at 40 kt
Complement: 39 (4 officers)

Machinery: 4 MTU MD 16V 538 TB90 diesels; 12,000 hp(m) *(8.82 MW)* sustained; 4 shafts
Missiles: SSM: 2 or 4 Kongsberg Penguin Mk 2; IR homing to 27 km *(14.6 n miles)* at 0.8 Mach; warhead 120 kg.
Guns: 2 Bofors 40 mm/70; 300 rds/min to 12 km *(6.6 n miles)*; weight of shell 0.96 kg.
Torpedoes: 2—21 in *(533 mm)* tubes; anti-surface.
Mines: Can carry 4.
Radars: Surface search: Racal Decca 1226; I-band.

Operational: *Meltem* sunk in collision with Soviet naval training ship *Khasan* in Bosphorus in 1985. Subsequently salvaged but beyond repair. *Atmaca* and *Kartal* decommissioned in 2013.

KARTAL CLASS   4/2009, C D Yaylali / 1366021

862 Turkey > Patrol forces

8 DOĞAN CLASS (FAST ATTACK CRAFT—MISSILE) (PGFG)

Name	No	Builders	Commissioned
DOĞAN	P 340	Lürssen, Bremen-Vegesack	23 Dec 1977
MARTI	P 341	Taşkizak Yard, Istanbul	1 Aug 1978
TAYFUN	P 342	Taşkizak Yard, Istanbul	9 Aug 1979
VOLKAN	P 343	Taşkizak Yard, Istanbul	25 Jul 1980
RÜZGAR	P 344	Taşkizak Yard, Istanbul	24 May 1985
POYRAZ	P 345	Taşkizak Yard, Istanbul	28 Aug 1986
GURBET	P 346	Taşkizak Yard, Istanbul	24 Jul 1988
FIRTINA	P 347	Taşkizak Yard, Istanbul	14 Oct 1988

Displacement, tonnes: 443 full load
Dimensions, metres (feet): 58.1 × 7.6 × 2.7 *(190.6 × 24.9 × 8.9)*
Speed, knots: 38. **Range, n miles:** 1,050 at 30 kt
Complement: 40 (5 officers)

Machinery: 4 MTU 16V 956 TB92 diesels; 17,700 hp(m) *(13 MW)* sustained; 4 shafts
Missiles: SSM: 8 McDonnell Douglas Harpoon (2 quad) launchers; active radar homing to 130 km *(70 n miles)* at 0.9 Mach; warhead 227 kg.
Guns: 1 Oto Melara 3 in *(76 mm)*/62 compact; 85 rds/min to 16 km *(8.7 n miles)* anti-surface; 12 km *(6.6 n miles)* anti-aircraft; weight of shell 6 kg. 2 Oerlikon 35 mm/90 (twin); 550 rds/min to 6 km *(3.3 n miles)*; weight of shell 1.55 kg.
Physical countermeasures: Decoys: 2 Mk 36 SRBOC chaff launchers.
Electronic countermeasures: ESM: Aselsan ARES-2N; intercept.
Radars: Surface search/navigation: GEM SPN-753/730; I-band.
Fire control: Signaal WM28/41; I/J-band.
Combat data systems: Signaal mini TACTICOS (344–347).
Weapon control systems: LIOD Mk 2 optronic director.

Programmes: First ordered 3 August 1973 to a Lürssen FPB 57 design.
Modernisation: A mid-life programme includes upgrade to the combat data system, communications and ESM. Work on the first four was completed in 2002. Work on the second four has not been confirmed.
Structure: Aluminium superstructure; steel hulls. The last pair were built with optronic directors which are being retrofitted in all, together with an improved Signaal combat data system.

POYRAZ　　　　　　　　　　　　　　　　　　　　*4/2011, C D Yaylali* / 1454746

RÜZGAR　　　　　　　　　　　　　　　　　　　　*8/2011, C D Yaylali* / 1454745

10 + 6 TUZLA CLASS (PC)

Name	No	Builders	Laid down	Launched	Commissioned
TUZLA	P 1200	Dearsan Shipyard, Istanbul	3 May 2008	9 Apr 2010	4 Jan 2011
KARABURUN	P 1201	Dearsan Shipyard, Istanbul	2008	16 Jul 2010	19 Apr 2011
KÖYCEĞIZ	P 1202	Dearsan Shipyard, Istanbul	–	11 Nov 2010	12 Aug 2011
KUMKALE	P 1203	Dearsan Shipyard, Istanbul	–	17 Mar 2011	24 Nov 2011
TARSUS	P 1204	Dearsan Shipyard, Istanbul	–	7 Jun 2011	26 Apr 2012
KARABIGA	P 1205	Dearsan Shipyard, Istanbul	–	27 Sep 2011	15 Jun 2012
KARŞIYAKA	P 1206	Dearsan Shipyard, Istanbul	–	1 Dec 2011	7 Sep 2012
TEKIRDAĞ	P 1207	Dearsan Shipyard, Istanbul	–	22 Feb 2012	4 Jan 2013
KAŞ	P 1208	Dearsan Shipyard, Istanbul	–	30 May 2012	3 May 2013
KILIMLI	P 1209	Dearsan Shipyard, Istanbul	–	5 Sep 2012	13 Aug 2013
TÜRKELI	P 1210	Dearsan Shipyard, Istanbul	–	8 Jan 2013	2014
TAŞUCU	P 1211	Dearsan Shipyard, Istanbul	–	25 Apr 2013	2014
KARATAŞ	P 1212	Dearsan Shipyard, Istanbul	–	17 Sep 2013	2014
KARPAZ	P 1213	Dearsan Shipyard, Istanbul	–	2014	2016
KOZLU	P 1214	Dearsan Shipyard, Istanbul	–	2014	2016
KUŞADASI	P 1215	Dearsan Shipyard, Istanbul	–	2014	2016

Displacement, tonnes: 406 full load
Dimensions, metres (feet): 56.9 × 8.9 × 2.5 *(186.7 × 29.2 × 8.2)*
Speed, knots: 27. **Range, n miles:** 1,000 at 14 kt
Complement: 34

Machinery: 2 MTU 16V 4000 M 90 diesels; 7,300 hp *(5.44 MW)*; 2 shafts; cp props
Guns: 2 Oto Melara 40 mm/70 (1 twin). 2—12.7 mm STAMP MGs.
A/S Mortars: 1 Aselsan 6-barrelled launcher.
Radars: Surface search/navigation: Raytheon NSC-3425/7XU.
Sonars: Simrad SP92 Mk II; hull-mounted; high frequency; 20–30 kHz.
Combat data systems: To be announced.
Electro-optic systems: Aselsan ASELFLIR 200.

Comment: Contract signed with Dearsan Shipyard, Istanbul, on 23 August 2007 for the construction of 16 anti-submarine patrol craft. Delivery of the sixteenth vessel is expected in 2014. Steel hull and superstructure. The craft are to be employed on patrol duties in the vicinity of ports and bases. Ten have been sold to Turkmenistan.

IHS Jane's Fighting Ships 2014-2015

KAŞ　　　　　　　　　　　　　　　　　　　　*11/2013*, Guy Toremans* / 1531078

TARSUS　　　　　　　　　　　　　　　　　　　*4/2013*, C D Yaylali* / 1531077

2 YILDIZ CLASS (FAST ATTACK CRAFT—MISSILE) (PGFG)

Name	No	Builders	Commissioned
YILDIZ	P 348	Taskizak Yard, Istanbul	3 Jun 1996
KARAYEL	P 349	Taskizak Yard, Istanbul	19 Sep 1996

Displacement, tonnes: 440 full load
Dimensions, metres (feet): 57.8 oa; 54.4 wl × 7.6 × 2.7 *(189.6; 178.5 × 24.9 × 8.9)*
Speed, knots: 38
Range, n miles: 1,050 at 30 kt
Complement: 45 (6 officers)

Machinery: 4 MTU 16V 956 TB91 diesels; 15,120 hp(m) *(11.1 MW)* sustained; 4 shafts
Missiles: SSM: 8 McDonnell Douglas Harpoon (2 quad) launchers; active radar homing to 130 km *(70 n miles)* at 0.9 Mach; warhead 227 kg.
Guns: 1 Oto Melara 3 in *(76 mm)*/62 compact; 85 rds/min to 16 km *(8.7 n miles)* anti-surface; 12 km *(6.6 n miles)* anti-aircraft; weight of shell 6 kg. 2 Oerlikon 35 mm/90 (twin); 550 rds/min to 6 km *(3.3 n miles)*; weight of shell 1.55 kg.
Physical countermeasures: Decoys: 2 Mk 36 SRBOC chaff launchers.
Electronic countermeasures: ESM: Racal Cutlass; intercept.
Radars: Surface search: Siemens/Plessey AW 6 Dolphin; G-band.
Fire control: Oerlikon/Contraves TMX; I/J-band.
Navigation: GEM SPN-753/730; I-band.
Combat data systems: Signaal/Thomson-CSF TACTICOS.
Weapon control systems: LIOD Mk 2 optronic director; Vesta helo datalink/transponder.

Programmes: Ordered in June 1991. *Karayel* launched 20 June 1995.
Structure: Doğan class hull with much improved weapon systems.

YILDIZ　　　　　　　　　　　　　　　　　　　*9/2008, C D Yaylali* / 1353411

2 MRTP 22 CLASS (FAST INTERVENTION CRAFT) (PBF)

SAT 1　　SAT 2

Displacement, tonnes: 48 full load
Dimensions, metres (feet): 24.08 × 5.53 × 1.3 *(79 × 18.1 × 4.3)*
Speed, knots: 54
Range, n miles: 350 at 32 kt
Complement: 5
Machinery: 2 MTU 12V 2000 M93 diesels; 4,800 hp(m) *(6.4 MW)*; 2 Arneson ASD 15 surface drives
Guns: 1—30 mm. 2—12.7 mm MGs. 2—7.62 mm MGs.

Comment: Yonuk MRTP 22 (enlarged MRTP 20) design built at Yonca Shipyard. The craft are of advanced composites construction and features a deepV hull. Delivered in 2008.

SAT 1　　　　　　　　　　　　　　　　　　　　*5/2013*, C D Yaylali* / 1531081

© 2014 IHS

3 AB 27 CLASS (LARGE PATROL CRAFT) (PC)

Name	No	Builders	Commissioned
AB 27	P 127 (ex-P 1227)	Haliç Shipyard	27 Jun 1969
AB 29	P 129 (ex-P 1229)	Haliç Shipyard	21 Feb 1969
AB 36	P 136 (ex-P 1236)	Taskizak Naval DY, Istanbul	13 Apr 1976

Displacement, tonnes: 173 full load
Dimensions, metres (feet): 40.2 × 6.4 × 1.7 *(131.9 × 21.0 × 5.6)*
Speed, knots: 22
Complement: 31 (3 officers)

Machinery: 4 SACM-AGO V16CSHR diesels; 9,600 hp(m) *(7.06 MW)* 2 cruise diesels; 300 hp(m) *(220 kW)*; 2 shafts
Guns: 1 or 2 Bofors 40 mm/70. 1 Oerlikon 20 mm (in those with 1–40 mm). 2–12.7 mm MGs.
A/S Mortars: 1 Mk 20 Mousetrap 4 rocket launcher; range 200 m; warhead 50 kg.
Depth charges: 1 rack.
Radars: Surface search: Racal Decca; I-band.
Sonars: Plessey PMS 26; hull-mounted; active search and attack; high frequency.

Comment: Pennant numbers changed in 1991. Similar to *SG 21* Coast Guard class. One to Georgia (*AB 30*) in December 1998, one to Azerbaijan (*AB 34*) in July 2000 and one to Kazakhstan (*AB 26*) in July 2001. *AB 33* decommissioned in 2005. *AB 28*, *31*, and *35* decommissioned in 2013.

AB 29   10/2013*, C D Yaylali / 1531076

4 PGM 71 CLASS (LARGE PATROL CRAFT) (PB)

PGM 104 P121 (ex-P1221)   **PGM 106** P123 (ex-P1223)
PGM 105 P122 (ex-P1222)   **PGM 107** P124 (ex-P1224)

Displacement, tonnes: 132 standard; 149 full load
Dimensions, metres (feet): 30.8 × 6.4 × 2.1 *(101 × 21.0 × 6.9)*
Speed, knots: 18.5. **Range, n miles:** 1,500 at 10 kt
Complement: 31 (3 officers)

Machinery: 8 GM diesels 2,040 hp *(1.52 MW)*; 2 shafts
Guns: 1 Bofors 40 mm/60. 4 Oerlikon 20 mm (2 twin). 1–7.62 mm MG.
A/S Mortars: 2 Mk 22 Mousetrap 8 rocket launchers; range 200 m; warhead 50 kg.
Depth charges: 2 racks (4).
Radars: Surface search: Raytheon 1500B; I-band.
Sonars: EDO SQS-17A; hull-mounted; active attack; high frequency.

Comment: Built by Peterson, Sturgeon Bay and commissioned 1967–68. Transferred from US almost immediately after completion. Pennant numbers changed in 1991.

PGM 107   11/2005, Manuel Declerck / 1153506

2 KAAN 20 CLASS (FAST INTERVENTION CRAFT) (PBF)

YUNUS 1 P 141   **YUNUS 2** P 142

Displacement, tonnes: 38 full load
Dimensions, metres (feet): 22.55 × 4.76 × 1.3 *(74 × 15.6 × 4.3)*
Speed, knots: 50. **Range, n miles:** 350 at 35 kt
Complement: 7
Machinery: 2 MTU 12V 2000 M92 diesels; 3,600 hp(m) *(4.8 MW)*; 2 Arneson ASD 12 B1L surface drives
Guns: 2–12.7 mm MGs.

Comment: Yonuk MRTP 20 (enlarged MRTP 15) design built at Yonca Shipyard. The craft are of advanced composites construction and feature a Deep V hull. Delivered in 2008 and commissioned in 2009.

YUNUS 1   11/2013*, Guy Toremans / 1531075

© 2014 IHS

AMPHIBIOUS FORCES

Notes: (1) The prefix 'Ç' for smaller amphibious vessels stands for 'Çikarma Gemisi' (landing vessel) and indicates that the craft are earmarked for national rather than NATO control.
(2) A Request for Proposals for a new landing platform dock was issued in February 2010. Capable of both military and humanitarian operations, the ship is expected to be of the order of 25,000 tons and to be equipped with at least four helicopter landing spots and a stern dock for two LCACs. It was announced on 27 December 2013 that SEDEF, partnered with Navantia, had been selected to build the ship, reportedly based on the Spanish Juan Carlos LPD design. As part of the project, four LCM, four LCVP, one RHIB, and one command boat are also to be procured. The ship is to enter service in 2017.

1 OSMAN GAZI CLASS (LSTH/ML)

Name	No	Builders	Launched	Commissioned
OSMAN GAZI	NL 125	Taskizak Yard, Istanbul	20 Jul 1990	27 Jul 1994

Displacement, tonnes: 3,834 full load
Dimensions, metres (feet): 105 × 16.1 × 4.8 *(344.5 × 52.8 × 15.7)*
Speed, knots: 17
Range, n miles: 4,000 at 15 kt

Military lift: 900 troops; 15 tanks; 4 LCVPs
Machinery: 2 MTU 12V 1163 TB73 diesels; 8,800 hp(m) *(6.47 MW)*; 2 shafts
Guns: 2 Oerlikon 35 mm/90 (twin). 1 Raytheon 20 mm Mk 15 Vulcan Phalanx 6-barrelled; 3,000 rds/min combined to 1.5 km. 2 Oerlikon 20 mm.
Electronic countermeasures: SLQ-32(V)2; intercept.
Radars: Surface search: SPS-55; I-band.
Navigation: Sperry RASCAR 2500C; I-band.
Helicopters: Platform for 1 large.

Comment: Laid down 7 July 1989. Full NBCD protection. Equipped with a support weapons co-ordination centre to control amphibious operations. The ship has about a 50% increase in military lift capacity compared with the Sarucabey class. Secondary role as minelayer. Second of class cancelled in 1991 and *Osman Gazi* took a long time to complete. Marisat fitted.

OSMAN GAZI   9/2013*, Selim San / 1531079

OSMAN GAZI   9/2012, Selim San / 1486381

1 ERTUĞRUL (TERREBONNE PARISH) CLASS (LSTH/ML)

Name	No	Builders	Commissioned	Recommissioned
ERTUĞRUL (ex-*Windham County* LST 1170)	L 401	Christy Corporation, Sturgeon Bay	15 Dec 1954	14 Dec 1973

Displacement, tonnes: 2,632 standard; 5,893 full load
Dimensions, metres (feet): 117.1 × 16.8 × 5.2 *(384.2 × 55.1 × 17.1)*
Speed, knots: 15
Complement: 163 (14 officers)

Military lift: 395 troops; 2,200 tons cargo; 4 LCVPs
Machinery: 4 GM 16-278A diesels; 6,000 hp *(4.48 MW)*; 2 shafts; cp props
Guns: 6 USN 3 in *(76 mm)*/50 (3 twin).
Radars: Surface search: Racal Decca 1226; I-band.
Fire control: 2 Western Electric Mk 34; I/J-band.
Weapon control systems: 2 Mk 63 GFCS.

Comment: Transferred by US and recommissioned 14 December 1973. Purchased outright 6 August 1987. Marisat fitted. *Serdar* decommissioned in 2013.

ERTUĞRUL   10/2009, Selim San / 1366808

IHS Jane's Fighting Ships 2014-2015

868 Turkey > Auxiliaries

1 DIVER CLASS (SALVAGE SHIP) (ARS)

Name	No	Builders	Launched	Commissioned
IŞIN (ex-*Safeguard* ARS 25)	A 589	Basalt Rock Co, Napa	20 Nov 1943	31 Oct 1944

Displacement, tonnes: 1,555 standard; 2,002 full load
Dimensions, metres (feet): 65.1 × 12.5 × 4 *(213.6 × 41.0 × 13.1)*
Speed, knots: 14.8
Complement: 110
Machinery: Diesel-electric; 4 Cooper-Bessemer GSB-8 diesels; 3,420 hp *(2.55 MW)*; 4 generators; 2 motors; 2 shafts
Guns: 2 Oerlikon 20 mm.

Comment: Transferred from US 28 September 1979 and purchased outright 6 August 1987.

IŞIN   9/2011, C D Yaylali / 1454757

1 CHANTICLEER CLASS (SUBMARINE RESCUE SHIP) (ASR)

Name	No	Builders	Launched	Commissioned
AKIN (ex-*Greenlet* ASR 10)	A 585	Moore SB & DD Co, Oakland	12 Jul 1942	29 May 1943

Displacement, tonnes: 1,680 standard; 2,358 full load
Dimensions, metres (feet): 76.7 × 13.4 × 4.9 *(251.6 × 44.0 × 16.1)*
Speed, knots: 15
Complement: 111 (9 officers)
Machinery: Diesel-electric; 4 Alco 539 diesels; 3,532 hp *(2.63 MW)*; 4 generators; 1 motor; 1 shaft
Guns: 1 Bofors 40 mm/60. 4 Oerlikon 20 mm (2 twin).
Radars: Navigation: Racal Decca BM II; I-band.

Comment: Transferred from US, recommissioned 23 December 1970 and purchased 15 February 1973. Carries a Diving Bell.

AKIN   9/2013*, C D Yaylali / 1531086

8 SMALL TRANSPORTS (YC)

ŞALOPA 12	Y 12	ŞALOPA 24	Y 24	ŞALOPA 31	Y 31
ŞALOPA 22	Y 22	ŞALOPA 27	Y 27	ŞALOPA 33	Y 33
ŞALOPA 23	Y 23	ŞALOPA 30	Y 30		

Comment: Of varying size and appearance.

ŞALOPA 23   5/2009, Selim San / 1366031

2 BARRACK SHIPS (YB)

YÜZBASI NASIT ÖNGÖREN (ex-*US APL 47*) Y 38 (ex-Y 1204)
BINBASI METIN SÜLÜS (ex-*US APL 53*) Y 39 (ex-Y 1205)

Comment: Ex-US barrack ships transferred on lease: Y 1204 in October 1972 and Y 1205 on 6 December 1974. Y 1204 based at Ereğli and Y 1205 at Gölcük. Purchased outright June 1987. Pennant numbers changed in 1991.

BINBASI METIN SÜLÜS   5/2012, C D Yaylali / 1486390

IHS Jane's Fighting Ships 2014-2015

1 BOOM DEFENCE VESSEL (ABU)

Name	No	Builders	Commissioned
AG 6 (ex-*AN 93*; ex-*Cerberus*)	P 306 (ex-*A 895*)	Bethlehem Steel Corporation, Staten Island	10 Nov 1952

Displacement, tonnes: 793 standard; 869 full load
Dimensions, metres (feet): 50.3 × 10.1 × 3 *(165 × 33.1 × 9.8)*
Speed, knots: 12.8
Range, n miles: 5,200 at 12 kt
Complement: 32 (3 officers)
Machinery: Diesel-electric; 2 GM 8-268A diesels; 880 hp *(656 kW)*; 2 generators; 1 motor; 1 shaft
Guns: 1 USN 3 in *(76 mm)*/50. 4 Oerlikon 20 mm.
Radars: Navigation: Racal Decca 1226; I-band.

Comment: Netlayer. Transferred from US to Netherlands in December 1952. Used first as a boom defence vessel and latterly as salvage and diving tender since 1961 but retained her netlaying capacity. Handed back to US Navy on 17 September 1970 but immediately turned over to the Turkish Navy under grant aid.

AG 6   7/1995, Frank Behling / 0080906

1 BOOM DEFENCE VESSEL (ABU)

Name	No	Builders	Commissioned
AG 5 (ex-*AN 104*)	P 305	Krögerwerft, Rendsburg	25 Feb 1962

Displacement, tonnes: 975 full load
Dimensions, metres (feet): 53 × 10.7 × 4.1 *(173.9 × 35.1 × 13.5)*
Speed, knots: 12. **Range, n miles:** 6,500 at 11 kt
Complement: 32 (3 officers)
Machinery: Diesel-electric; 1 MAN G7V40/60 diesel generator; 1 motor; 1,470 hp(m) *(1.08 MW)*; 1 shaft
Guns: 1 Bofors 40 mm/60. 3 Oerlikon 20 mm.
Radars: Navigation: Racal Decca 1226; I-band.

Comment: Netlayer P 305 built in US offshore programme for Turkey.

AG 5   8/2008, C D Yaylali / 1353416

3 TORPEDO RETRIEVERS (YPT)

TORPIDO TENDERI Y 95 (ex-Y 1051)   **TAKIP 1** Y 98 (ex-Y 1052)   **TAKIP 2** Y 99

Comment: Of different types.

TAKIP 2   11/2013*, Guy Toremans / 1531085

© 2014 IHS

Auxiliaries — Tugs < **Turkey** 869

2 OFFICERS' YACHTS (YAC)

GÜL NEVCIVAN

Comment: Pennant numbers not displayed.

GÜL *6/2003, Turkish Navy* / 0567541

NEVCIVAN *7/2012, C D Yaylali* / 1486391

13 FLOATING DOCKS/CRANES (YD/YFDM)

LEVENT Y 59 (ex-Y 1022)	**HAVUZ 3** Y 123 (ex-Y 1083)	**HAVUZ 10** Y 130 (ex-Y 1090)
ALGARNA 1 Y 58	**HAVUZ 4** Y 124 (ex-Y 1084)	**HAVUZ 11** Y 134
ALGARNA 3 Y 60 (ex-Y 1021)	**HAVUZ 5** Y 125 (ex-Y 1085)	**HAVUZ 13** Y 136
HAVUZ 1 Y 121 (ex-Y 1081)	**HAVUZ 8** Y 128 (ex-Y 1088)	
HAVUZ 2 Y 122 (ex-Y 1082)	**HAVUZ 9** Y 129 (ex-Y 1089)	

Comment: Algarna and *Levent* are ex-US floating cranes. The lift capability of each is as follows: *Havuz 1* (16,000 tons), *Havuz 2* (12,000 tons), *Havuz 3* (2,500 tons), *Havuz 4* (4,500 tons), *Havuz 5* (400 tons), *Havuz 8* (700 tons); *Havuz 9* (4,500 tons), *Havuz 10* (3,500 tons), *Havuz 11* (14,500 tons) and *Havuz 13* (7,500 tons).

HAVUZ 10 *8/2008, C D Yaylali* / 1353417

0 + 2 SUBMARINE RESCUE AND TOWING SHIPS (ASR)

Dimensions, metres (feet): 69 × 13.5 × 4.0 *(226.4 × 44.3 × 13.1)*
Speed, knots: 18. **Range, n miles:** 4,500 at 14 kt
Complement: 104
Machinery: To be announced.

Comment: The tendering process for the procurement of two Submarine Rescue and Towing Ships (RATSHIP) was initiated by a Request for Information on 25 July 2006. The ships are to have a pressure chamber for up to six evacuees, facilities for helicopter in-flight refuelling and vertical replenishment, a 460 m² aft deck, a 12-tonne crane and five firefighting monitors. They are also to serve as target/recovery ships for torpedo firing exercises. A dynamic positioning system is to be capable of maintaining a fixed position in a 4 kt current. A contract for the construction of the ship was let with the Istanbul Tersanesi Shipyard on 28 October 2011. Delivery is expected in late 2014.

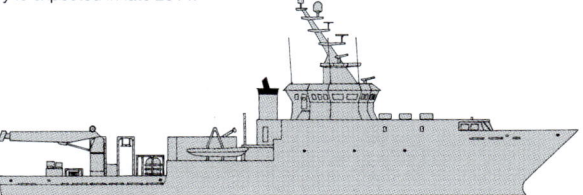

RATSHIP *(Scale 1 : 900), Ian Sturton* / 1406949

0 + 1 SUBMARINE RESCUE SHIP (ASR)

Dimensions, metres (feet): 91 × 18.5 × 5.0 *(298.6 × 60.7 × 16.4)*
Speed, knots: 18. **Range, n miles:** 4,500 at 14 kt
Complement: 131
Machinery: To be announced.

Comment: The tendering process for the procurement of a Submarine Rescue Mother Ship (MOSHIP) to replace *Akin* was initiated by a Request for Information on 25 July 2006. The ship is to be interoperable with the NATO Submarine Rescue System (NSRS) and to be capable of rescuing the crew of a distressed submarine at depths of up to 600 m. The ship is to be equipped with twin pressure chambers for up to 32 evacuees, allowing them to be transferred ashore at pressures up to 5 bar. These chambers will have docking ports for an L-type submarine rescue vehicle. The ship is also to be fitted with a stern-mounted A-frame and a 35-tonne crane for a submarine rescue diving and recompression system. Other features include a helicopter deck. A dynamic positioning system is to be capable of maintaining a fixed position in a 4 kt current. A contract for the construction of the ship was let with the Istanbul Tersanesi Shipyard on 28 October 2011. Delivery is expected in late 2014.

© 2014 IHS

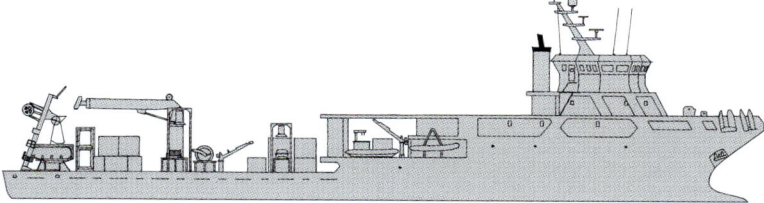

MOSHIP *(Scale 1 : 900), Ian Sturton* / 1406948

TUGS

1 POWHATAN CLASS (FLEET OCEAN TUGS) (ATF)

Name	No	Laid down	Commissioned
INEBOLU (ex-*Powhatan*)	A 590 (ex-T-ATF 166)	30 Sep 1976	15 Jun 1979

Displacement, tonnes: 2,296 full load
Dimensions, metres (feet): 68.9 × 12.8 × 4.6 *(226 × 42.0 × 15.1)*
Speed, knots: 14.5
Range, n miles: 10,000 at 13 kt
Complement: 4 + 16 personnel
Machinery: 2 GM EMD 20-645F7B diesels; 7,250 hp(m) *(5.41 MW)* sustained; 2 shafts; Kort nozzles; cp props; bow thruster; 300 hp *(224 kW)*
Guns: Space provided to fit 2 — 20 mm and 2 — 12.7 mm MGs.
Radars: Navigation: E/F/I-band.

Comment: Built at Marinette Marine Corp, Wisconsin patterned after commercial offshore supply ship design. Following de-activation from US MSC in 1999, operated on commercial lease to Don John Commercial Co until 25 February 2008. Following a refit at Detyens Shipyard, Charleston, commissioned in Turkish Navy on 15 March 2008. Equipped with 10-ton crane, two fire pumps and capable of supporting salvage operations. Bollard pull 54 tons.

INEBOLU *5/2011, Diego Quevedo* / 1406393

1 OCEAN TUG (ATR)

DARICA A 578 (ex-Y 1125)

Displacement, tonnes: 762 full load
Dimensions, metres (feet): 40.9 × 9.8 × 3.9 *(134.2 × 32.2 × 12.8)*
Speed, knots: 14
Range, n miles: 2,500 at 14 kt
Machinery: 2 ABC diesels; 4,000 hp *(2.94 MW)*; 2 shafts

Comment: Built at Taşkizak Naval Yard and commissioned 13 June 1991. Equipped for firefighting and as a torpedo tender. Pennant number changed in 1991.

DARICA *11/1994, van Ginderen Collection* / 0080910

IHS Jane's Fighting Ships 2014-2015

870 **Turkey** > Tugs — Coast guard

1 CHEROKEE CLASS (ATS)

GAZAL (ex-*Sioux*) A 587 (ex-ATF 75)

Displacement, tonnes: 1,255 standard; 1,702 full load
Dimensions, metres (feet): 62.5 × 11.7 × 5.2 *(205.1 × 38.4 × 17.1)*
Speed, knots: 16
Range, n miles: 15,000 at 8 kt
Complement: 85
Machinery: Diesel-electric; 4 GM 12-278 diesels; 4,400 hp *(3.28 MW)*; 4 generators; 1 motor; 3,000 hp *(2.24 MW)*; 1 shaft
Guns: 1—40 mm. 2 Oerlikon 20 mm.
Radars: Navigation: Racal Decca; I-band.

Comment: Originally completed on 6 December 1942. Transferred from US and commissioned 9 March 1973. Purchased 15 August 1973. Can be used for salvage.

GAZAL   6/2008, Selim San / 1353418

1 TENACE CLASS (ATA)

Name	No	Builders	Commissioned
DEĞİRMENDERE (ex-*Centaure*)	A 576 (ex-A 674)	Chantiers de la Rochelle	14 May 1974

Displacement, tonnes: 1,477 full load
Dimensions, metres (feet): 51 × 11.5 × 5.7 *(167.3 × 37.7 × 18.7)*
Speed, knots: 13
Range, n miles: 9,500 at 13 kt
Complement: 37 (3 officers)
Machinery: 2 SACM AGO 240 V12 diesels; 4,600 hp(m) *(3.38 MW)*; 1 shaft; Kort nozzle
Radars: Racal Decca RM 1226 and Racal Decca 060; I-band.

Comment: Transferred from French Navy 16 March 1999. Recommissioned after refit 22 July 1999. Bollard pull 60 tons.

DEĞİRMENDERE   1/2008, Selim San / 1353419

1 SORUM CLASS (ATS)

AKBAŞ A 586

Displacement, tonnes: 1,682 full load
Dimensions, metres (feet): 58.3 × 12.64 × 5.48 *(191.3 × 41.5 × 18.0)*
Speed, knots: 13.5
Range, n miles: 12,600 at 10 kt
Machinery: 2 diesels; 1,650 hp *(12.3 MW)*; 1 shaft
Radars: Navigation: To be announced.

Comment: Sorum class ocean tug built in St Petersburg and commissioned on 5 March 1987. Equipped for firefighting and as torpedo tender.

AKBAŞ   7/2013*, Selim San / 1531087

14 COASTAL/HARBOUR TUGS (YTB/YTM/YTL)

Name	No	Commissioned
SÖNDÜREN 2	A 1542	2000
SÖNDÜREN 3	A 1543	1999
SÖNDÜREN 4	A 1544	2000
SÖNDÜREN 1	Y 51 (ex-Y 1117)	1954
KUVVET	Y 53 (ex-Y 1122)	1962
DOĞANARSLAN	Y 52 (ex-Y 1123)	1985
ATIL	Y 55 (ex-Y 1132)	1976
PENDIK	Y 56	2000
AKSAZ (ex-*Koos*)	Y 57 (ex-Y 1651, ex-A 08)	1962
ÖNDER	Y 160	14 Apr 1999
ÖNCÜ	Y 161	2 Jul 1999
ÖZGEN	Y 162	8 Sep 1999
ÖDEV	Y 163	11 Nov 1999
ÖZGÜR	Y 164	9 Mar 2000

Comment: The displacement and speed of each are as follows: *Söndüren 2–4* (320 t/12 kt); *Söndüren 1* (128 t/12 kt); *Kuvvet* (390 t/10 kt); *Doğanarslan* (500 t/12 kt); *Atil* (300 t/10 kt); *Pendik* (238 t/10 kt); *Aksaz* (320 t/11 kt); *Önder, Öncü, Özgen, Ödev* and *Özgür* (230 t/12 kt). In addition there are 47 Katir pusher berthing tugs. *Aksaz* (ex-*Koos*) was transferred from Germany on 7 October 1996.

ÖZGEN   11/2013*, Guy Toremans / 1531088

AKSAZ   3/2010, Marco Ghiglino / 1366814

COAST GUARD (SAHIL GÜVENLIK)

Notes: (1) A Request for Proposal for the procurement of eight offshore patrol ships, to supplement the Dost class, is expected in 2013.
(2) Patrol craft based in north Cyprus include KKTCSG 101 (*Raif Denktas*), two 40 m craft (KKTCSG 01–02), two Kaan 15 class (KKTCSG 11, KKTCSG 12), two 14 m craft (KKTCSG 102–103) and a converted cabin cruiser KKTCSG 104.
(3) Four Vigilante class Boston Whalers were acquired by the Police in September 1999.

KKTCSG 104   6/2004, Selçuk Emre / 1133570

4 DOST CLASS (OFFSHORE PATROL VESSELS) (PSOH)

Name	No	Builders	Laid down	Launched	Commissioned
DOST	SG 701	RMK Marine, Tuzla	3 May 2008	9 Jun 2010	5 Apr 2013
GÜVEN	SG 702	RMK Marine, Tuzla	2009	17 Dec 2010	22 Nov 2013
UMUT	SG 703	RMK Marine, Tuzla	2009	31 May 2011	5 Apr 2013
YAŞAM	SG 704	RMK Marine, Tuzla	2009	13 Sep 2012	13 May 2014

Displacement, tonnes: 1,727 full load
Dimensions, metres (feet): 88.6 × 12.2 × 4.6 (290.7 × 40.0 × 15.1)
Speed, knots: 22
Range, n miles: 3,500 at 14 kt
Complement: 60 (5 officers)

Machinery: 2 diesels; cp props; bow thruster
Guns: 1—76 mm. 2—25 mm.
Radars: Surface search: Thales Variant; G/I-bands.
Fire control: To be announced.
Navigation: To be announced.
Weapon control systems: To be announced.
Helicopters: 1 AB 412.

Comment: Contract signed with RMK Marine on 16 January 2007 for the construction of four offshore patrol vessels to carry out SAR and EEZ patrol duties. The design, based on the Italian Sirio class, includes a telescopic hangar, two high-speed RIBs, two fire-fighting monitors and anti-pollution equipment.

UMUT   11/2013*, Guy Toremans / 1531089

18 + 1 LARGE PATROL CRAFT (WPB)

SG 80–97

Displacement, tonnes: 198 full load
Dimensions, metres (feet): 40.7 × 7.1 × 2.2 (133.5 × 23.3 × 7.2)
Speed, knots: 27
Complement: 25
Machinery: 2 diesels; 5,700 hp(m) (4.19 MW); 2 shafts
Guns: 1 Breda 40 mm/70. 2—12.7 mm MGs.
Radars: Surface search: Racal Decca; I-band.

Comment: All built at Taşkizak Shipyard except SG 89–92 which were built at Istanbul Shipyard. SG 80–82 commissioned in 1996, 83–84 in 1997, 85 in 1998, 86–87 and 89–90 in 2000, 88 in 2002, 91 in 2007, 92 in 2009, 93–94 in 2010, 95–96 in 2011, and 97 by 2013.

SG 94   4/2013*, C D Yaylali / 1531093

14 LARGE PATROL CRAFT (WPB)

SG 121–134

Displacement, tonnes: 183 full load
Dimensions, metres (feet): 40.2 (SG 121–129), 40 (SG 130–134) × 6.4 (SG 121–129), 6.5 (SG 130–134) × 1.7 (SG 121–129), 1.5 (SG 130–134) (131.9, 131.2 × 21.0, 21.3 × 5.6, 4.9)
Speed, knots: 22
Complement: 25
Machinery: 2 SACM AGO 195 V16 CSHR diesels; 4,800 hp(m) (3.53 MW) sustained 2 cruise diesels; 300 hp(m) (220 kW); 2 shafts
Guns: 1 or 2 Bofors 40 mm/60. 2—12.7 mm MGs.
Radars: Surface search: Raytheon Pathfinder; I-band.

Comment: SG 121 and 122 built by Gölcük Naval Yard, remainder by Taşkizak Naval Yard. SG 134 commissioned in 1977, remainder 1968–71. SG 130–134 have minor modifications-knuckle at bow, radar stirrup on bridge and MG on superstructure sponsons. These are similar craft to the AB 27 class listed under Patrol Forces for the Navy.

SG 134   9/2013*, C D Yaylali / 1531092

10 SAR 33 TYPE (LARGE PATROL CRAFT) (WPB)

SG 61–70

Displacement, tonnes: 183 full load
Dimensions, metres (feet): 34.6 × 8.6 × 3 (113.5 × 28.2 × 9.8)
Speed, knots: 33
Range, n miles: 450 at 24 kt, 550 at 18 kt
Complement: 24
Machinery: 3 SACM AGO 195 V16 CSHR diesels; 7,200 hp(m) (5.29 MW) sustained; 3 shafts; cp props
Guns: 1 Bofors 40 mm/60. 2—12.7 mm MGs.
Radars: Surface search: Raytheon Pathfinder; I-band.

Comment: Prototype Serter design ordered from Abeking & Rasmussen, Lemwerder in May 1976. The remainder were built at Taşkizak Naval Yard, Istanbul between 1979 and 1981. The engines have been governed back and the top speed correspondingly reduced from the original 12,000 hp and 40 kt.

SG 65   7/2013*, C D Yaylali / 1531091

4 SAR 35 TYPE (LARGE PATROL CRAFT) (WPB)

SG 71–74

Displacement, tonnes: 213 full load
Dimensions, metres (feet): 36.6 × 8.6 × 1.9 (120.1 × 28.2 × 6.2)
Speed, knots: 26
Range, n miles: 450 at 24 kt, 550 at 18 kt
Complement: 24
Machinery: 2 MTU 16V 4000 M 90 diesels; 7,398 hp (5.5 MW); 2 shafts
Guns: 1 Bofors 40 mm/60. 2—12.7 mm MGs.
Radars: Surface search: Raytheon Pathfinder; I-band.

Comment: A slightly enlarged version of the Serter designed SAR 33 Type built by Taşkizak Shipyard between 1985 and 1987. A contract was signed on 21 May 2007 with Istanbul Denizcilik Gemi Inşaa Shipyard for the modernisation of the main machinery.

SG 71   8/2000, C D Yaylali / 0106645

9 KAAN 29 CLASS (LARGE PATROL CRAFT) (WPBF)

SG 101–109

Displacement, tonnes: 95 full load
Dimensions, metres (feet): 31.7 × 6.7 × 1.4 (104 × 22.0 × 4.6)
Speed, knots: 47
Range, n miles: 750 at 20 kt
Complement: 14 (2 officers)
Machinery: 2 MTU 16V 400 M90 diesels; 7,300 hp(m) (5.44 MW); 2 MJP 753DD waterjets
Guns: 4—12.7 mm MGs.
Radars: Surface search/navigation: Raytheon Pathfinder; I-band.

Comment: All built at Yonca Shipyard. Onuk MRTP 29 design. Advanced composites structure. SG 101–103 commissioned 25 July 2001, SG 104–105 on 25 July 2002, SG 106–108 in 2003 and SG 109 in February 2004.

SG 109   6/2012, C D Yaylali / 1486395

10 ARES 42 HECTOR PATROL CRAFT (WPBF)

SAGET 26–35

Dimensions, metres (feet): 12.99 × 3.99 × 1.0 (42.6 × 13.1 × 3.3)
Speed, knots: 35
Machinery: To be announced.
Guns: To be announced.

Comment: Constructed by Ares Shipyard and delivered on 6 December 2013.

872 Turkey > Coast guard

15 KAAN 33 CLASS (LARGE PATROL CRAFT) (WPBF)

SG 301–315

Displacement, tonnes: 126 full load
Dimensions, metres (feet): 35.6 × 6.7 × 1.4 (116.8 × 22.0 × 4.6)
Speed, knots: 43
Range, n miles: 650 at 20 kt
Complement: 18 (2 officers)
Machinery: 2 MTU 16V 4000 M90 diesels; 7,396 hp(m) (5.44 MW); 2 MJP 753DD waterjets
Guns: 4 – 12.7 mm MGs.
Radars: Navigation: Raytheon Pathfinder; I-band.

Comment: All built at Yonca Shipyard. Onuk MRTP 33 design. Advanced composites structure. SG 301 commissioned in 2004, SG 302 and 303 in 2005, SG 304–307 in 2006, SG 308 and 309 in 2007, SG 310 (2008), SG 311 (2009), SG 312 (2010) and SG 313 (2011). A further nine craft (of which two have been delivered) have been ordered. ASELSAN stabilised gun (STAMP) fitted in some Kaan 33 and 29 craft.

SG 312   8/2013*, C D Yaylali / 1531090

20 KAAN 15 CLASS (FAST INTERVENTION CRAFT) (WPBF)

SG 1–18   KKTCSG 11–12

Displacement, tonnes: 20 full load
Dimensions, metres (feet): 16.7 × 4.04 × 1.2 (54.8 × 13.3 × 3.9)
Speed, knots: 54
Range, n miles: 350 at 35 kt
Complement: 12
Machinery: 2 MTU 12V 183 TE93 diesels; 2,300 hp(m) (1.69 MW); 2 Arneson ASD 12 B1L surface drives
Guns: 2 – 12.7 mm MGs.
Radars: Surface search: Raytheon; I-band.

Comment: Contract for first six with Yonca Technical Investment signed in May 1997, second order for six more in February 1999, and third for 6 more in August 2000. All built at Tuzla-Istanbul shipyard. Three delivered in 1998, seven in 1999, two in April 2000, four in July 2001 and two in July 2002. Onuk MRTP 15 design. Advanced composites structure. Two based in northern Cyprus with pennant numbers KKTCSG 11–12.

SG 5   8/2013*, C D Yaylali / 1531096

10 + 8 KAAN 19 CLASS (FAST INTERVENTION CRAFT) (WPBF)

SG 19–28   +8

Displacement, tonnes: 30 full load
Dimensions, metres (feet): 22.55 × 4.76 × 1.3 (74 × 15.6 × 4.3)
Speed, knots: 60. **Range, n miles:** 350 at 35 kt
Complement: 5
Machinery: 2 MTU 12V 2000 M 92 diesels; 3,600 hp(m) (2.7 MW); 2 MJP waterjets
Guns: 2 – 12.7 mm MGs.
Radars: Navigation: Raymarine E-120; I-band.

Comment: Yonuk MRTP 20 (enlarged MRTP 15) design built at Yonca shipyard. A total of 10 had been delivered by early 2014 and a further eight craft are to be delivered by 2017.

SG 25   7/2013*, C D Yaylali / 1531095

42 CONTROL CRAFT (PBR)

Dimensions, metres (feet): 5.8 × 2.2 × 0.8 (19 × 7.2 × 2.6)
Speed, knots: 35
Complement: 3
Machinery: 2 outboard motors; 100 hp (75 kW)

Comment: Inflatable hull craft.

IHS Jane's Fighting Ships 2014-2015

CONTROL CRAFT   6/2007, Turkish Coast Guard / 1170047

11 COASTAL PATROL CRAFT (WPB)

SG 50–58   KKTCSG 102–103

Displacement, tonnes: 29 full load
Dimensions, metres (feet): 14.6 × 4.2 × 1.1 (47.9 × 13.8 × 3.6)
Speed, knots: 15
Complement: 7
Machinery: 2 diesels; 700 hp(m) (514 kW); 2 shafts
Guns: 1 – 12.7 mm MG or 1 Oerlikon 20 mm (SG 102–103).
Radars: Surface search: Raytheon; I-band.

Comment: KKTCSG 102–103 were built for North Cyprus and have been based there since August 1990 and July 1991 respectively. Both these craft were given a heavier gun in 1992. Second batch of three completed by Taşkizak in October 1992, three more in June 1993, three more in December 1993.

SG 56   11/2012*, Selim San / 1531094

1 INSHORE PATROL CRAFT (WPBI)

RAIF DENKTAS 101 (ex-74)

Displacement, tonnes: 10 full load
Dimensions, metres (feet): 11.6 × 3.5 × 0.7 (38.1 × 11.5 × 2.3)
Speed, knots: 28. **Range, n miles:** 250 at 25 kt
Complement: 6
Machinery: 2 Volvo Aquamatic AQ200F petrol engines; 400 hp(m) (294 kW); 2 shafts
Guns: 1 – 12.7 mm MG.
Radars: Surface search: Raytheon; I-band.

Comment: Built by Protekson, Istanbul. Transferred to North Cyprus 23 September 1988. Can be equipped with a rocket launcher.

RAIF DENKTAŞ   6/2004, Selçuk Emre / 1044201

10 SECURITY AND SAFETY CRAFT (PBR)

SG 20–29

Dimensions, metres (feet): 7.75 × 2.9 × 0.5 (25.4 × 9.5 × 1.6)
Speed, knots: 35
Complement: 3
Machinery: 2 outboard motors; 180 hp (135 kW)

Comment: Rigid-inflatable hull with fibre cabin.

SG 21   4/2012, C D Yaylali / 1486393

© 2014 IHS

20 SECURITY AND SAFETY CRAFT (PBR)

SAGET 1–20

Dimensions, metres (feet): 9.5 × 3.0 × 0.5 (31.2 × 9.8 × 1.6)
Speed, knots: 40
Complement: 3
Machinery: 2 outboard motors; 350 hp (260 kW)

Comment: Rigid-inflatable hull with fibre cabin.

SAGET 5   4/2012, C D Yaylali / 1486394

10 SAFETY CRAFT (PBR)

Dimensions, metres (feet): 5.8 × 2.2 × 0.8 (19 × 7.2 × 2.6)
Speed, knots: 30
Complement: 7
Machinery: 2 outboard motors; 140 hp (105 kW)

Comment: Inflatable hull craft.

SAFETY CRAFT   6/2007, Turkish Coast Guard / 1170046

LAND-BASED MARITIME AIRCRAFT

Numbers/Type: 14 Agusta AB 412 EP.
Operational speed: 122 kt (226 km/h).
Service ceiling: 17,000 ft (5,180 m).
Range: 374 n miles (656 km).
Role/Weapon systems: Nine aircraft ordered 15 April 1999. One lost on 30 July 2005. A further six ordered in early 2005. Operated by Coast Guard/Frontier Force for patrol SAR. Sensors: Radar and FLIR. Weapons: Unarmed.

AB 412   7/2011, C D Yaylali / 1454763

Numbers/Type: 3 Casa CN-235.
Operational speed: 240 kt (445 km/h)
Service ceiling: 26,600 ft (8,110 m)
Range: 669 n miles (1,240 km)
Role/Weapon systems: Three delivered in July 2002. Long range maritime patrol for surveillance.

CN-235   6/2007, Turkish Coast Guard / 1353426

Turkmenistan

Country Overview

Formerly part of the USSR, the Republic of Turkmenistan declared its independence in 1991. Situated in Central Asia, it has an area of 188,460 square miles and is bordered to the north by Kazakhstan, to the east by Uzbekistan and Afghanistan and to the south by Iran. It has a 954 n mile coastline with the Caspian Sea. Türkmenbashi, the principal port, is linked by rail to Ashgabat, the capital and largest city. Maritime claims in the Caspian Sea are yet to be resolved. The Navy acts under the operational control of the Border Guard but is the weakest component of the Turkmen armed forces.

Personnel

2014: 700

Base

Türkmenbashi (formerly Krasnovodsk)

PATROL FORCES

Notes: One 40 m Stenka class patrol craft has been reported operational and there are also two 28 m Sobol-class patrol craft.

4 KALKAN (PROJECT 50030) M CLASS
(INSHORE PATROL CRAFT) (PBI)

Displacement, tonnes: 9 full load
Dimensions, metres (feet): 11.6 × 3.3 × 0.6 (38.1 × 10.8 × 2.0)
Speed, knots: 34
Complement: 2
Machinery: 1 Type 475K diesel; 496 hp (370 kW); 1 waterjet
Guns: 1—12.7 mm MG.

Comment: Four craft delivered during 2002. Further craft were expected but reportedly not delivered. Built by Morye Feodosiya (Ukraine) and constructed with aluminium hulls and GRP superstructure. Can be armed with 7.62 mm or 12.7 mm MGs.

KALKAN   6/2003, Morye / 0573698

5 GRIF-T CLASS (PB)

Displacement, tonnes: 40 full load
Dimensions, metres (feet): 24.4 × 5.2 × 1.57 (80.1 × 17.1 × 5.2)
Speed, knots: 40
Range, n miles: 500 at 15 kt
Complement: 13 (1 officer)
Machinery: 2 MTU 12V 2000 M 90 diesels; 2,700 hp (2 MW); 2 shafts
Guns: 1—20 mm. 1—12.7 mm MG.
Radars: Surface search: I-band.

Comment: Built by Morye Shipyard, Feodosia, Ukraine and delivered in about 2005. Modified versions of Zhuk class. Aluminium construction.

2 TARANTUL V (PROJECT 1241.8) CLASS (FSGM)

EDERMEN 828	GAYRATLY 829

Displacement, tonnes: 450 standard; 510 full load
Dimensions, metres (feet): 59.9 × 11.5 × 2.5 (196.5 × 37.7 × 8.2)
Speed, knots: 40
Range, n miles: 2,300 at 12 kt
Complement: 41 (5 officers)

Machinery: 1 Zorya GGTA M-15E gas turbine; 32,015 hp (23.54 MW); 2 shafts
Missiles: SSM: 16 (4 quad) SS-N-25 (Kh 35 Uran); active radar homing to 130 km (70.2 n miles) at 0.9 Mach; warhead 145 kg; sea skimmer.
SAM: SA-N-5 Grail quad launcher; manual aiming; IR homing to 6 km (3.2 n miles) at 1.5 Mach; warhead 1.5 kg.
Guns: 1—3 in (76 mm)/59 AK 176; 120 rds/min to 15 km (8 n miles); weight of shell 5.9 kg.
2—30 mm/65 AK 630; 6 barrels per mounting; 3,000 rds/min combined to 2 km.
Physical countermeasures: Decoys: 2 PK 16 chaff launchers.
Radars: Air/surface search: Strut Curve (Positiv-ME); I-band.
Surface search: Plank Shave; E/F-band (SS-N-25).
Fire control: Bass Tilt; H/I-band.
Navigation: Pechora; I-band.
Weapon control systems: Hood Wink optronic director.

Comment: A contract was reportedly signed in 2008 for the construction of two Project 1241.8 missile corvettes. Both ships had been delivered by late 2011 and are based at Turkmenbashi.

Turkmenistan > Patrol forces — Ukraine > Introduction

1 POINT CLASS (WPB)

Name	No	Builders	Commissioned
MERJEN (ex-Point Jackson)	PB-129 (ex-82378)	US Coast Guard Yard, Curtis Bay	3 Aug 1970

Displacement, tonnes: 67 standard; 70 full load
Dimensions, metres (feet): 25.3 × 5.2 × 1.8 (83 × 17.1 × 5.9)
Speed, knots: 23.5. **Range, n miles:** 1,200 at 8 kt
Complement: 10 (1 officer)
Machinery: 2 Caterpillar 3412 diesels; 1,600 hp (1.19 MW); 2 shafts
Guns: 2—12.7 mm MGs.
Radars: Surface search: Hughes/Furuno SPS-73; I-band.

Comment: Steel hulled craft with aluminium superstructure. Transferred from United States on 30 May 2000.

2 + 8 DEARSAN PATROL CRAFT (PC)

ARKADASH 101   BARKARAR 102   +8

Displacement, tonnes: 400 full load
Dimensions, metres (feet): 55.75 × 8.85 × 2.5 (182.9 × 29.0 × 8.2)
Speed, knots: 25
Range, n miles: 2,000 at 12 kt
Complement: 34

Machinery: 2 diesels; 2 shafts; cp props
Missiles: SSM: 4 Otomat Teseo; active radar homing to 80 km (43.2 n miles) at 0.9 Mach; warhead 210 kg.
SAM: 2 Matra Simbad twin launchers for Mistral; IR homing to 4 km (2.2 n miles); warhead 3 kg.
Guns: 2 Oto Melara 40 mm (twin). 2 Aselsan 25 mm STOP. 2 Aselsan 12.7 mm STAMP.
Radars: Surface search/navigation: Thales Variant; G/H/I-bands.
Weapon control systems: Aselflir 200 or 400.

Comment: Contract signed with Dearsan Shipyard, Istanbul, in early 2011 for the construction of two P 1200-class patrol craft, similar to those being built for the Turkish Navy. Delivery was made in 2012. Eight further craft are to be built at Turkmenbashi.

5 DEARSAN 14M PATROL CRAFT (PB)

Dimensions, metres (feet): 14.5 × 4.15 × 0.75 (47.6 × 13.6 × 2.5)
Speed, knots: 40
Range, n miles: 250 at 12 kt
Machinery: 2 MTU 8V 2000 M72 diesels; 965 hp (720 kW); 2 waterjets
Guns: 2—12.7 mm MGs.
Radars: Navigation: To be announced.

Comment: Manufactured by Dearsan and delivered in 2011.

MERJEN (inboard ship)
11/2000, Selim San
0104495

Tuvalu

Country Overview

Tuvalu, formerly the Ellice Islands, is a south Pacific island group which gained independence in 1978; the other part of the former British colony, the Gilbert Islands, became independent as Kiribati the following year. Situated some 1,600 n miles east of Papua New Guinea, the country comprises nine atolls of which Funafuti is the location of the capital, Fongafale, and home to more than 30 per cent of the population. An archipelagic state, territorial seas (12 n miles) are claimed. An Exclusive Economic Zone (EEZ) (200 n miles) is also claimed but limits have not been fully defined by boundary agreements.

Headquarters Appointments

Commander Maritime Wing: Superintendent Talafou Esekia

Bases

Funafuti

PATROL FORCES

1 PACIFIC CLASS (LARGE PATROL CRAFT) (PB)

Name	No	Builders	Commissioned
TE MATAILI	801	Transfield Shipbuilding, WA	8 Oct 1994

Displacement, tonnes: 168 full load
Dimensions, metres (feet): 31.5 × 8.1 × 2.1 (103.3 × 26.6 × 6.9)
Speed, knots: 18
Range, n miles: 2,500 at 12 kt
Complement: 18 (3 officers)
Machinery: 2 Caterpillar 3516TA diesels; 4,400 hp (3.28 MW) sustained; 2 shafts
Guns: Can carry 1—12.7 mm MG but is unarmed.
Radars: Navigation: Furuno 1011; I-band.

Comment: This is the 18th of the class to be built by the Australian Government for Exclusive Economic Zone (EEZ) patrols in the Pacific islands. The programme originally terminated at 15 but was re-opened on 19 February 1993 to include construction of five more craft for Fiji, Kiribati and Tuvalu. Training and support assistance is given by the Australian Navy. Half-life refit completed at Gladstone in 2001. Following the decision by the Australian government to extend the Pacific Patrol Boat programme, *Te Mataili* underwent a life-extension refit in 2010.

TE MATAILI   8/2011, Tuvalu Police / 1454764

Ukraine

Country Overview

Formerly part of the USSR, Ukraine declared its independence in 1991. Situated in eastern Europe, it has an area of 233,090 square miles and is bordered to the north by Belarus, to the east by Russia, to the south-west by Romania and Moldova and to the west by Hungary, Slovakia and Poland. It has a 1,501 n mile coastline with the Black Sea and the Sea of Azov. Kiev is the capital and largest city while Sevastopol, Odessa, Kerch, and Mariupol are the principal ports. Territorial Seas (12 n miles) have been claimed. An EEZ (200 n miles) has been claimed but the limits have not been defined.

Division of the former Soviet Black Sea Fleet between Russia and Ukraine had been achieved on 28 May 1997. An agreement signed on 21 April 2010 allows for the continuance of leasing port facilities until 2042 with a clause allowing for a potential five-year extension.

The order of battle reflects the situation before the Ukraine–Russia dispute of March 2014 and subsequent annexation of Crimea by Russia. As of early April 2014, most of the Ukrainian Navy had been seized by the Russian Black Sea Fleet.

Headquarters Appointments

Commander of the Navy: To be announced

Bases

Sevastopol (HQ), Novoozerne (Southern Region), Odessa (Western Region), Mikolaiv, Feodosiya, Izmail, Balaklava, Kerch

Personnel

2014: 14,800 navy

Border Guard

The Maritime Border Guard is an independent subdivision of the State Committee for Border Guards, and is not part of the Navy. It has three cutter brigades, based in Kerch, Odessa and Balaklava, to patrol the 827 mile coastline and two river brigades, which patrol a gunship squadron, a minesweeping squadron, an auxiliary ship group and a training division. Pennant numbers changed in July 1999.

PENNANT LIST

Submarines

U 01	Zaporizya

Frigates

U 130	Hetman Sagaidachny
U 205	Lutsk
U 206	Vinnitsa
U 209	Ternopil

Patrol Forces

U 120	Skadovsk
U 153	Priluki
U 154	Kahovka
U 155	Nikopol
U 156	Kremenchuk
U 207	Uzhgorod
U 208	Khmelnitsky

Mine Warfare Forces

U 310	Zhovti Vody
U 311	Cherkasy
U 330	Melitopol
U 331	Mariupol
U 360	Genichesk

Amphibious Forces

U 410	Kirovograd
U 402	Konstantin Olshansky
U 420	Donetsk
U 862	Korosten
U 904	Bilyaïvka

Auxiliaries

U 240	Feodisiya
U 510	Slavutich
U 540	Chigirin
U 541	Smila
U 542	Darnicha
U 635	Skvyra
U 700	Netisin
U 705	Kremenets
U 706	Izyaslav
U 722	Borshev
U 728	Evpatoriya
U 733	Tokmak
U 753	Kriviy Rig
U 756	Sudak
U 757	Makivka
U 759	Bahmach
U 760	Fastiv
U 782	Sokal
U 783	Illichivsk
U 803	Krasnodon
U 811	Balta
U 830	Korets
U 831	Kovel
U 852	Shostka
U 860	Kamyanka
U 891	Kherson
U 947	Krasnoperekovsk
U 953	Dubno

Survey Ships

U 511	Simferopol
U 512	Pereyaslav
U 601	Alchevsk
U 754	Dzhankoi

SUBMARINES

1 FOXTROT CLASS (PROJECT 641) (SS)

Name	No	Builders	Laid down	Launched	Commissioned
ZAPORIZYA (ex-B 435)	U 01	Sudomekh, Leningrad	24 Mar 1970	29 May 1970	20 Feb 1971

Displacement, tonnes: 1,952 surfaced; 2,475 dived
Dimensions, metres (feet): 91.3 × 7.5 × 6 (299.5 × 24.6 × 19.7)
Speed, knots: 16 surfaced; 9 snorting; 15 dived
Range, n miles: 20,000 at 8 kt surfaced; 380 at 2 kt dived
Complement: 75

Machinery: Diesel-electric; 3 Type 37-D diesels; 6,000 hp(m) (4.4 MW); 3 motors (1 × 2,700 and 2 × 1,350); 5,400 hp(m) (3.97 MW); 3 shafts; 1 auxiliary motor; 140 hp(m) (103 kW)

Torpedoes: 10—21 in (533 mm) (6 bow, 4 stern) tubes. Combination of 22–53 cm torpedoes.
Mines: 44 in lieu of torpedoes.
Electronic countermeasures: ESM: Stop Light; radar warning.
Radars: Surface search: Snoop Tray; I-band.
Sonars: Pike Jaw; hull-mounted; passive/active search and attack; high frequency.

Programmes: Transferred from Russia in August 1997 with three others of the class.

Operational: Based at Balaklava and, following a three-year refit which completed in 2000, further work was then required to restore hydraulic systems and replace batteries. The submarine was expected to return to service in 2004 but was delayed by technical difficulties and lack of funding. Following further repair work, the submarine sailed for sea trials on 25 April 2012 and formally returned to operational service on 21 January 2013.

ZAPORIZYA   7/2013*, E & M Laursen / 1531097

FRIGATES

0 + 1 (3) PROJECT 58250 CLASS (FFGHM)

Name	No	Builders	Laid down	Launched	Commissioned
–	–	Chernomorsky Shipyard, Nikolayev	18 May 2011	2014	2016

Displacement, tonnes: 2,500 full load
Dimensions, metres (feet): 112 × 13.0 × ? (367.5 × 42.7 × ?)
Speed, knots: 32. **Range, n miles:** 4,000 at 12 kt
Complement: 100 (est.)

Machinery: CODOG; 2 Gorya gas turbines; 2 Caterpillar diesels; 2 shafts
Missiles: SSM: 8 Exocet MM40 Block 3.
SAM: 16 (2 octuple) Sylver A43 VLS MBDA Aster 15.
Guns: 1 Oto Melara 76 mm/76 Super Rapid. 2 Rheinmetall Millennium 35 mm.
Torpedoes: 2 triple launchers for MU90 or A244.
Electronic countermeasures: EW System. Type 2170 SSTDS.
Radars: Air/surface search: To be announced.
Navigation: To be announced.
Fire control: To be announced.
Sonars: Gidropribor hull-mounted sonar. Thales VDS.
Combat data systems: DCNS system to be announced.
Electro-optic systems: To be announced.

Helicopters: 1 Ka-27 or NH90.

Programmes: The keel of the first of the Project 58250 class, designed by the Nikolayev Shipbuilding Centre, was laid by the Ukraine President in May 2011. The programme was originally endorsed in 2009 but the full design and production phase was delayed until funding was approved. An original requirement for 10 ships has been reduced to four, all of which are planned to have entered service by 2021. The baseline design, on which details are based, incorporated a significant number of western sensors and weapons in a quest for NATO interoperability. However, it has been reported that these aspirations may have to be de-scoped in order to reduce the overall cost of the programme. In which case, it is likely that some key systems may yet be acquired from Russia. This may in turn affect the final design. Until this has been finalised, current details remain illustrative.

PROJECT 58250   6/2009, SRDSC / 1394710

876 Ukraine > Frigates

1 KRIVAK III (NEREY) CLASS (PROJECT 1135.1) (FFHM)

Name	No	Builders	Laid down	Launched	Commissioned
HETMAN SAGAIDACHNY (ex-*Kirov*)	U 130 (ex-201)	Kamysh-Burun, Kerch	5 Oct 1990	29 Mar 1992	5 Jul 1993

Displacement, tonnes: 3,150 standard; 3,709 full load
Dimensions, metres (feet): 123.5 × 14.3 × 5 *(405.2 × 46.9 × 16.4)*
Speed, knots: 32
Range, n miles: 4,600 at 20 kt, 1,600 at 30 kt
Complement: 180 (18 officers)

Machinery: COGAG; 2 gas turbines; 55,500 hp(m) *(40.8 MW)*; 2 gas turbines; 13,600 hp(m) *(10 MW)*; 2 shafts
Missiles: SAM: 1 SA-N-4 Gecko twin launcher ❶; semi-active radar homing to 15 km *(8.1 n miles)* at 2.5 Mach; warhead 50 kg; altitude 9.1–3,048 m *(30–10,000 ft)*; 20 missiles. The launcher retracts into the mounting for stowage and protection, rising to fire and retracting to reload. The two mountings are forward of the bridge and abaft the funnel.
Guns: 1—3.9 in *(100 mm)*/70 AK 100 ❷; 60 rds/min to 21.5 km *(11.5 n miles)*; weight of shell 15.6 kg. 2—30 mm/65 ❸; 6 barrels per mounting; 3,000 rds/min combined to 2 km.
Torpedoes: 8—21 in *(533 mm)* (2 quad) tubes ❹. Combination of Russian 53 cm torpedoes.
A/S Mortars: 2 RBU 6000 12-tubed trainable ❺; range 6,000 m; warhead 31 kg.
Physical countermeasures: Decoys: 4 PK 16 chaff launchers. Towed torpedo decoy.
Electronic countermeasures: ESM: 2 Bell Shroud; intercept. ECM: 2 Bell Squat; jammers.
Radars: Air search: Top Plate ❻; 3D; D/E-band. Surface search: Spin Trough ❼; I-band. Peel Cone ❽; E-band. Fire control: Pop Group ❾; F/H/I-band (for SA-N-4). Kite Screech ❿; H/I/K-band. Bass Tilt (Krivak III) ⓫; H/I-band. Navigation: Kivach; I-band.
IFF: Salt Pot (Krivak III).
Sonars: Bull Nose (MGK 335MS); hull-mounted; active search and attack; medium frequency.

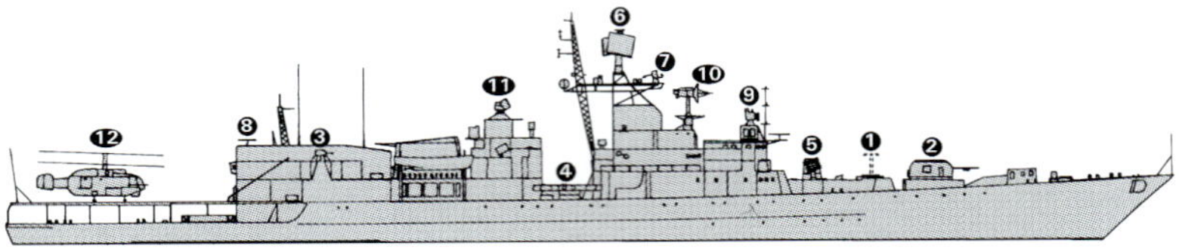

Helicopters: 1 Ka-27 Helix ⓬.

HETMAN SAGAIDACHNY (Scale 1 : 1,200), Ian Sturton / 0506208

Programmes: This is the last of the 'Krivak IIIs' originally designed for the USSR Border Guard. The seven others are based in the Russian Pacific Fleet. A ninth of class was not completed.
Operational: *Sagaidachny* has so far not been sighted with a helicopter embarked. Deployed to the Mediterranean in 1994 and late 1995, to the Indian Ocean in early 1995 and to the US in late 1996. Three further Krivak class have been decommissioned: 'Krivak II' *Sevastopol* (U 132) is probably being used for spares while 'Krivak I' *Mikolaiv* is to be scrapped. *Dnipropetrovsk* is reported to have sunk in the Black Sea in 2005.

HETMAN SAGAIDACHNY 6/2003, Ships of the World / 0572651

HETMAN SAGAIDACHNY 5/2011, L-G Nilsson / 1454812

IHS Jane's Fighting Ships 2014-2015   © 2014 IHS

Frigates — Patrol forces < **Ukraine** 877

3 GRISHA CLASS (PROJECT 1124EM/P) (FFLM)

Name	No	Builders	Laid down	Launched	Commissioned
LUTSK	U 205	Leninskaya Kuznitsa, Kiev	–	12 May 1993	12 Feb 1994
VINNITSA (ex-*Dnepr*)	U 206	Zrelenodolsk	23 Dec 1975	12 Sep 1976	31 Dec 1976
TERNOPIL	U 209	Leninskaya Kuznitsa, Kiev	–	20 Mar 2002	16 Feb 2006

Displacement, tonnes: 965 standard; 1,168 full load
Dimensions, metres (feet): 71.2 × 9.8 × 3.7 *(233.6 × 32.2 × 12.1)*
Speed, knots: 30
Range, n miles: 2,500 at 14 kt, 1,750 at 20 kt, 950 at 27 kt
Complement: 70 (5 officers)

Machinery: CODAG; 1 gas turbine; 15,000 hp(m) *(11 MW)*; 2 diesels; 16,000 hp(m) *(11.8 MW)*; 3 shafts
Missiles: SAM: SA-N-4 Gecko twin launcher ❶ *(Lutsk)*; semi-active radar homing to 15 km *(8 n miles)* at 2.5 Mach; warhead 14.5 kg; altitude 9.1–3,048 m *(30–10,000 ft)*; 20 missiles.
Guns: 1—3 in *(76 mm)*/59 AK 176 ❷; *(Lutsk and Ternopil)*; 120 rds/min to 15 km *(8 n miles)*; weight of shell 5.9 kg. 4—57 mm/75 AK 725 (twin) *(Vinnitsa)*; 120 rds/min to 12.7 km *(6.8 n miles)*; weight of shell 2.8 kg. 1—30 mm/65 ❸; *(Lutsk and Ternopil)*; 6 barrels; 3,000 rds/min combined to 2 km.
Torpedoes: 4—21 in *(533 mm)* (2 twin) tubes ❹. SAET-60; passive homing to 15 km *(8.1 n miles)* at 40 kt; warhead 400 kg.
A/S Mortars: 1 or 2 RBU 6000 12-tubed trainable ❺; range 6,000 m; warhead 31 kg.
Mines: Capacity for 18 in lieu of depth charges.
Depth charges: 2 racks (12).
Physical countermeasures: Decoys: 2 PK 16 chaff launchers.
Electronic countermeasures: ESM: 2 Watch Dog.
Radars: Air/surface search: Half Plate B ❻; *(Lutsk and Ternopil)*; E/F-band. Strut Curve *(Vinnitsa)*; F-band.
Navigation: Don 2; I-band.
Fire control: Pop Group ❼; *(Lutsk and Ternopil)*; F/H/I-band (for SA-N-4). Bass Tilt ❽; *(Lutsk and Ternopil)*; H/I-band (for 76 mm and 30 mm). Muff Cobb *(Vinnitsa)*; G/H-band.
IFF: High Pole A or B. Square Head. Salt Pot.
Sonars: Bull Nose (MGK 335MS); hull-mounted; active search and attack; high/medium frequency. Elk Tail VDS ❾; active search; high frequency.

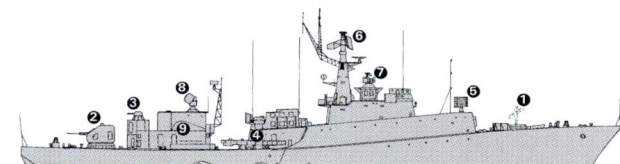

LUTSK *(Scale 1 : 900), Ian Sturton / 0506209*

VINNITSA *7/2000, Hartmut Ehlers / 0106652*

Programmes: *Lutsk* is a 'Grisha V' (Type 1124EM) launched 12 May 1993 and completed 27 November 1993. *Ternopil* is also a Grisha V and was the first new ship to join the fleet since 1994 when it commissioned in 2006. *Vinnitsa* is a 'Grisha II' (Type 1124P) ex-Russian Border Guard ship transferred in 1996. Two 'Grisha I' were also transferred but have been deleted.
Operational: All three are active. *Vinnitsa* damaged in storm on 11 November 2007.

TERNOPIL *7/2013*, E & M Laursen / 1531099*

LAND-BASED MARITIME AIRCRAFT

Notes: The Naval Aviation Force is based at Sevastopol. It comprises 12 Ka-29 assault helicopters, five Ka-27 Helix A ASW helicopter, five Mi-14 Haze, four Mi-8, one An-12 Cub, and three Be-12 Mail. The Air Force inventory includes 30 Su-24 Fencer and 140 MiG-29 Fulcrum.

PATROL FORCES

2 MATKA (VEKHR) CLASS (PROJECT 206MP)
(FAST ATTACK CRAFT—MISSILE HYDROFOIL) (PGGK)

PRILUKI (ex-*R-262*) U 153   **KAHOVKA** (ex-*R-265*) U 154

Displacement, tonnes: 229 standard; 264 full load
Dimensions, metres (feet): 39.6 × 7.6 oa; 12.5 over foils × 2.1 hull; 4 over foils *(129.9 × 24.9; 41.0 × 6.9; 13.1)*
Speed, knots: 40
Range, n miles: 600 at 35 kt; 1,500 at 14 kt
Complement: 33

Machinery: 3 Type M 504 diesels; 10,800 hp(m) *(7.94 MW)* sustained; 3 shafts
Missiles: SSM: 2 SS-N-2C/D Styx; active radar or IR homing to 83 km *(45 n miles)* at 0.9 Mach; warhead 513 kg; sea-skimmer at end of run.
Guns: 1—3 in *(76 mm)*/59 AK 176; 120 rds/min to 15 km *(8 n miles)*; weight of shell 5.9 kg. 1—30 mm/65 AK 630; 6 barrels per mounting; 3,000 rds/min to 2 km.
Electronic countermeasures: Clay Brick; intercept.
Radars: Air/surface search: Plank Shave; E-band.
Navigation: SRN-207; I-band.
Fire control: Bass Tilt; H/I-band.
IFF: High Pole B or Salt Pot B and Square Head.
Weapon control systems: Hood Wink optronic directors.

Comment: Five Russian Black Sea Fleet units transferred in 1996. Built between 1978 and 1983 with similar hulls to the Osa class. One was transferred to Georgia in 1999 and two others (*Uman* and *Tsurupinsk*) have been cannibalised for spares.

PRILUKI *7/2013*, E & M Laursen / 1531098*

0 + 2 GURZA-M (PROJECT 58155) CLASS
(INSHORE PATROL CRAFT) (PB)

Displacement, tonnes: 51 full load
Dimensions, metres (feet): 23 × 4.8 × 1.0 *(75.5 × 15.7 × 3.3)*
Speed, knots: 25
Range, n miles: 700 at 12 kt
Complement: 2
Machinery: 2 diesels; 2 shafts
Guns: 1—30 mm. 1—7.62 mm MG. 1—30 mm grenade launcher.

Comment: The first of two new river patrol craft were laid down at Lenin Shipyard, Kiev, on 25 October 2012. The craft are planned to be deployed in the Danube River Basin and in the coastal zone of the Black and Azov Seas. Overall numbers will depend on availability of funding.

© 2014 IHS   IHS Jane's Fighting Ships 2014-2015

2 PAUK I (MOLNYA) (PROJECT 1241P) CLASS (PCM)

KHMELNITSKY (ex-*MPK 116*) U 208   **UZHGOROD** (ex-*MPK 93*) U 207

Displacement, tonnes: 447 full load
Dimensions, metres (feet): 57.6 × 10.2 × 3.3 *(189 × 33.5 × 10.8)*
Speed, knots: 32
Range, n miles: 2,400 at 14 kt
Complement: 32

Machinery: 2 Type M 521 diesels; 16,184 hp(m) *(11.9 MW)* sustained; 2 shafts
Missiles: SAM: SA-N-5 Grail quad launcher; manual aiming; IR homing to 6 km *(3.2 n miles)* at 1.5 Mach; altitude to 2,500 m *(8,000 ft)*; warhead 1.5 kg; 8 missiles.
Guns: 1 – 3 in *(76 mm)*/59 AK 176; 120 rds/min to 15 km *(8 n miles)*; weight of shell 5.9 kg.
1 – 30 mm/65 AK 630; 6 barrels; 3,000 rds/min combined to 2 km.
Torpedoes: 4 – 16 in *(406 mm)*.
A/S Mortars: 2 RBU 1200 5-tubed fixed; range 1,200 m; warhead 34 kg.
Depth charges: 2 racks (12).
Physical countermeasures: Decoys: 2 PK 16 or 4 PK 10 chaff launchers.
Electronic countermeasures: ESM: 3 Brick Plug and 2 Half Hat; radar warning.
Radars: Air/surface search: Peel Cone; E/F-band.
Surface search: Kivach or Pechora; I-band.
Fire control: Bass Tilt; H/I-band.
Sonars: Foal Tail; VDS (mounted on transom); active attack; high frequency.
Weapon control systems: Hood Wink optronic director.

Programmes: Built at Yaroslavl in 1985. Transferred from Black Sea Fleet Border Guard in 1996. Others of this class are in the Ukraine Border Guard.
Structure: ASW version of the Russian Tarantul class.
Operational: Second of class *Uzhgorod* was reportedly back in service in 2006 but operational status has not been confirmed.

KHMELNITSKY   *7/2013*, E & M Laursen* / 1531102

2 TARANTUL II (MOLNYA) (PROJECT 1241.1/2) CLASS (FSGM)

PRIDNEPROVYE (ex-*Nikopol*; ex-*R-54*) U 155   **KREMENCHUK** (ex-*R 63*) U 156

Displacement, tonnes: 391 standard; 462 full load
Dimensions, metres (feet): 56.1 × 11.5 × 2.5 *(184.1 × 37.7 × 8.2)*
Speed, knots: 36. **Range, n miles:** 1,650 at 14 kt
Complement: 34 (5 officers)

Machinery: COGAG: 2 Nikolayev Type DR 77 gas turbines; 16,016 hp(m) *(11.77 MW)*;
2 Nikolayev DR 76 gas turbines with reversible gearboxes; 4,993 hp(m) *(3.67 MW)* sustained 2 shafts; cp props
Missiles: SSM: 4 Raduga SS-N-2D Styx (2 twin); active radar or IR homing to 83 km *(45 n miles)* at 0.9 Mach; warhead 513 kg; sea skimmer at end of run
SAM: 1 SA-N-5 Grail quad launcher; manual aiming; IR homing to 6 km *(3.2 n miles)* at 1.5 Mach; warhead 1.5 kg.
Guns: 1 – 3 in *(76 mm)*/59 AK 176; 120 rds/min to 15 km *(8 n miles)*; weight of shell 5.9 kg.
2 – 30 mm/65 AK-630; 6 barrels per mounting; 3,000 rds/min to 2 km.
Physical countermeasures: Decoys: 4 PK 16 chaff launchers.
Radars: Plank Shave; I-band.
Fire control: Bass Tilt; H/I-band (for guns). Band Stand (Mineral ME); D-band (for SS-N-2D).
Navigation: Kivach III; I-band.
IFF: High Pole B.
Weapon control systems: Hood Wink optronic director. Light bulb datalink. Band Stand; I-band datalink).

Programmes: Built at Kolpino. U 155 originally commissioned in 1983 and U 156 in 1985. Both transferred in 1997 and recommissioned in 2002.

PRIDNEPROVYE   *3/2002, Hartmut Ehlers* / 0529997

1 ZHUK (GRIF) CLASS (PROJECT 1400M) (PB)

SKADOVSK (ex-*AK 327*) U 120

Displacement, tonnes: 40 full load
Dimensions, metres (feet): 24 × 5 × 1.2 *(78.7 × 16.4 × 3.9)*
Speed, knots: 30. **Range, n miles:** 1,100 at 15 kt
Complement: 13
Machinery: 2 Type M 401B diesels; 2,200 hp(m) *(1.6 MW)* sustained; 2 shafts
Guns: 2 – 14.5 mm (twin). 1 – 12.7 mm MG.
Radars: Surface search: Spin Trough; I-band.

Comment: Transferred from Russia in 1997 and became operational in 2000. Others of the class are in service with the Border Guard.

SKADOVSK   *7/2013*, E & M Laursen* / 1531101

AMPHIBIOUS FORCES

Notes: Two Vydra class LCUs, *Korosten* U 862 and *Bilyaïvka* U 904, are used as trials and transport craft. There are also two non-operational Ondatra class LCM, *Svatove* U 430 and *Vil* U 537 and a T-4 LCM *Tarpan* U 538 which are laid up.

1 POMORNIK (ZUBR) (PROJECT 1232.2) CLASS (ACV/LCUJM)

DONETSK U 420

Displacement, tonnes: 559 full load
Dimensions, metres (feet): 57.6 on cushion × ? *(189 × ?)*
Speed, knots: 60. **Range, n miles:** 300 at 55 kt
Complement: 31 (4 officers)

Military lift: 3 MBT or 10 APC plus 230 troops (total 170 tons)
Machinery: 5 Type NK-12MV gas turbines; 2 for lift, 23,672 hp(m) *(17.4 MW)* nominal; 3 for drive, 35,508 hp(m) *(26.1 MW)* nominal
Missiles: SAM: 2 SA-N-5 Grail quad launchers; manual aiming; IR homing to 6 km *(3.2 n miles)* at 1.5 Mach; altitude to 2,500 m *(8,000 ft)*; warhead 1.5 kg.
Guns: 2 – 30 mm/65 AK 630; 6 barrels per mounting; 3,000 rds/min combined to 2 km.
2 retractable 122 mm rocket launchers.
Mines: 80.
Radars: Air/surface search: Cross Dome (Ekran); I-band.
Fire control: Bass Tilt MR 123; H/I-band.
IFF: Salt Pot A/B. Square Head.
Weapon control systems: Quad Look (modified Squeeze Box) (DWU 3) optronic director.

Comment: *Donetsk* was completed by Morye, Feodosiya on 20 July 1993. Sister U 421 was incomplete in 1999 when procured by Greece, delivery being made in 2001. Three further craft were transferred from Russia in 1996. Of these, U 423 (ex-*MDK 123*) was also sold to Greece, U 422 (ex-*MDK 57*) and U 424 (ex-*MDK 93*) have been decommissioned.

DONETSK   *8/2000, Lemachko Collection* / 0131164

1 POLNOCHNY C (PROJECT 773 I) CLASS (LSM)

KIROVOGRAD (ex-*SDK 123*) U 401

Displacement, tonnes: 1,138 standard; 1,168 full load
Dimensions, metres (feet): 81.3 × 9.7 × 2.4 *(266.7 × 31.8 × 7.9)*
Speed, knots: 18. **Range, n miles:** 2,000 at 12 kt
Complement: 42 (est.)

Military lift: 350 tons including 6 tanks; 180 troops
Machinery: 2 Kolomna Type 40-D diesels; 4,400 hp(m) *(3.2 MW)* sustained; 2 shafts
Missiles: 4 SA-N-5 Grail quad launchers; manual aiming; IR homing to 6 km *(3.2 n miles)* at 1.5 Mach; warhead 1.5 kg; 32 missiles.
Guns: 4 – 30 mm/65 (2 twin).
Rockets: 2 – 140 mm 18-tubed rocket launchers.
Radars: Surface search: Spin Trough; I-band.
Fire control: Drum Tilt; H/I-band (for 30 mm guns).

Comment: Built in 1970s and transferred from Russian Fleet in 1994. Reported operational again in 2001 following refit.

KIROVOGRAD   *7/2013*, E & M Laursen* / 1531100

1 ROPUCHA I (PROJECT 775) CLASS (LST)

KONSTANTIN OLSHANSKY (ex-*BDK 56*) U 402

Displacement, tonnes: 4,471 full load
Dimensions, metres (feet): 113 × 14.5 × 3.6 *(370.7 × 47.6 × 11.8)*
Speed, knots: 17.5
Range, n miles: 3,500 at 16 kt
Complement: 95 (7 officers)

Military lift: 10 MBT plus 190 troops or 24 AFVs plus 170 troops
Machinery: 2 Zgoda-Sulzer 16ZVB40/48 diesels; 19,230 hp(m) *(14.14 MW)* sustained; 2 shafts
Missiles: SAM: 4 SA-N-5 Grail quad launchers.
Guns: 4 — 57 mm/75 AK 725 (2 twin); 120 rds/min to 12.7 km *(6.8 n miles)*; weight of shell 2.8 kg.
Radars: Strut Curve; F-band.
Navigation: Don 2; I-band.
Fire control: Muff Cob; G/H-band.
IFF: High Pole B.
Weapon control systems: 2 Squeeze Box optronic directors.

Comment: Built at Gdansk, Poland in 1978 and transferred from Russia in 1996. Can be used to carry mines. Ro-Ro design with 540 m² of parking space between the stern gate and the bow doors.

KONSTANTIN OLSHANSKY   10/2008, Laursen/Jarnasen / 1353556

MINE WARFARE FORCES

2 NATYA I CLASS (PROJECT 266M) (MSO)

CHERNIGIV (ex-*Zhovti Vody*; ex-*Zenitchik*) U 310   **CHERKASY** (ex-*Razvedchik*) U 311

Displacement, tonnes: 817 full load
Dimensions, metres (feet): 61 × 10.2 × 3 *(200.1 × 33.5 × 9.8)*
Speed, knots: 16
Range, n miles: 3,000 at 12 kt
Complement: 67 (8 officers)

Machinery: 2 Type M 504 diesels; 5,000 hp(m) *(3.67 MW)* sustained; 2 shafts; cp props
Guns: 4 — 30 mm/65 (2 twin) AK 306 or 2 — 30 mm/65 AK 630; 4 — 25 mm/80 (2 twin).
A/S Mortars: 2 RBU 1200 5-tubed fixed.
Mines: 10.
Depth charges: 62.
Physical countermeasures: MCM: 1 or 2 GKT-2 contact sweeps; 1 AT-2 acoustic sweep. 1 TEM-3 magnetic sweep.
Radars: Surface search: Long Trough; E-band.
Fire control: Drum Tilt; H/I-band.
IFF: 2 Square Head. High Pole B.
Sonars: MG 79/89; hull-mounted; active minehunting; high frequency.

Comment: Built in the mid-1970s. Transferred from Russia in 1996. Both are operational.

CHERNIGIV   7/2013*, E & M Laursen / 1531103

2 SONYA (YAKHONT) (PROJECT 1265) CLASS (MHSC)

MELITOPOL (ex-*BT 79*) U 330   **MARIUPOL** (ex-*BT 126*) U 331

Displacement, tonnes: 467 full load
Dimensions, metres (feet): 48 × 8.8 × 2 *(157.5 × 28.9 × 6.6)*
Speed, knots: 15
Range, n miles: 3,000 at 10 kt
Complement: 43

Machinery: 2 Kolomna diesels; 2,000 hp(m) *(1.47 MW)* sustained; 2 shafts
Guns: 2 — 30 mm/65 (twin). 2 — 25 mm/80 (twin).
Mines: 8.
Radars: Surface search: Don 2; I-band.
IFF: Two Square Head.
Sonars: MG 69/79; hull-mounted; active; high frequency.

Comment: Built in 1978. Transferred from Russia in 1996. Wooden hull.

© 2014 IHS

MELITOPOL   10/2008, Laursen/Jarnasen / 1353555

1 YEVGENYA (KOROND) (PROJECT 1258) CLASS (MHC)

GENICHESK (ex-*RT 214*) U 360

Displacement, tonnes: 78 standard; 91 full load
Dimensions, metres (feet): 24.6 × 5.5 × 1.5 *(80.7 × 18.0 × 4.9)*
Speed, knots: 11
Range, n miles: 300 at 10 kt
Complement: 10

Machinery: 2 Type 3-D-12 diesels; 600 hp(m) *(440 kW)* sustained; 2 shafts
Guns: 2 — 14.5 mm (twin) MGs.
Mines: 8 racks.
Radars: Surface search: Spin Trough or Mius; I-band.
IFF: Salt Pot.
Sonars: A small MG-7 sonar is lifted over stern on crane; a TV system may also be used.

Comment: Transferred from Russia in 1996. Reported as being operational.

GENICHESK   6/2003, Ships of the World / 0572652

SURVEY SHIPS

Notes: (1) Also transferred in 1997 were two Muna class AGIs, *Pereyaslav* U 512 and *Dzhankoi* U 754. Both are used as transports, mostly for commercial goods.
(2) Ten former Russian civilian research ships were transferred in 1996/97. All are now in commercial service.
(3) There is an Onega class, *Severodonetsk* U 812.

1 MOMA (PROJECT 861M) CLASS (AGS)

SIMFEROPOL (ex-*Jupiter*) U 511

Displacement, tonnes: 1,626 full load
Dimensions, metres (feet): 73.3 × 11.2 × 3.9 *(240.5 × 36.7 × 12.8)*
Speed, knots: 17
Range, n miles: 9,000 at 11 kt
Complement: 56
Machinery: 2 Zgoda-Sulzer diesels; 3,300 hp(m) *(2.43 MW)* sustained; 2 shafts; cp props
Radars: Navigation: Don 2; I-band.

Comment: U 511 transferred from Russia in February 1996 and is active. A second of class U 602 has been decommissioned.

SIMFEROPOL   7/2000, Hartmut Ehlers / 0106669

IHS Jane's Fighting Ships 2014-2015

884 United Arab Emirates > Submarines — Corvettes

SUBMARINES

Notes: Submarine training has been conducted in the past but acquisition of submarines is understood to be a long-term aspiration.

SWIMMER DELIVERY VEHICLES (SDV)

Comment: Two classes of indigenously built Long Range Submersible Carriers (LRSC) have been developed by Emirates Marine Technologies. The 7.35 × 0.95 m Class 4 variant, of which approximately ten are believed to have been in service with UAE Special Forces since 1998, is capable of deploying a 200 kg payload. These are likely to be augmented by the larger 9.1 × 1.15 m Class 5 variant which can deliver 450 kg. Constructed of glass and carbon fibres, both variants are manned by two people, have a top speed of 7 kt, a range of 60 n miles at 6 kt and an operational depth of 30 m. They are equipped with depth-sounder, sonar and built-in breathing system.

CLASS 5 LRSC *2001*
Emirates Marine Technologies
0095256

CORVETTES

2 MURAY JIB (MGB 62) CLASS (FSGHM)

Name	No	Builders	Commissioned	Recommissioned
MURAY JIB	P 161 (ex-CM 01, ex-P 6501)	Lürssen, Bremen-Vegesack	Mar 1989	Nov 1990
DAS	P 162 (ex-CM 02, ex-P 6502)	Lürssen, Bremen-Vegesack	May 1989	Jan 1991

Displacement, tonnes: 640 full load
Dimensions, metres (feet): 63 × 9.3 × 2.5 (206.7 × 30.5 × 8.2)
Speed, knots: 32. **Range, n miles:** 4,000 at 16 kt
Complement: 43

Machinery: 4 MTU 16V 538 TB92 diesels; 13,640 hp(m) *(10 MW)* sustained; 4 shafts
Missiles: SSM: 8 Aerospatiale MM 40 Exocet (Block II) ❶; inertial cruise; active radar homing to 70 km *(40 n miles)* at 0.9 Mach; warhead 165 kg; sea-skimmer.
SAM: Thomson-CSF modified Crotale Navale octuple launcher ❷; radar guidance; IR homing to 13 km *(7 n miles)* at 2.4 Mach; warhead 14 kg.
Guns: 1 Oto Melara 3 in *(76 mm)*/62 Super Rapid ❸; 120 rds/min to 16 km *(8.7 n miles)*; weight of shell 6 kg. 1 Signaal Goalkeeper with GE 30 mm 7-barrelled ❹; 4,200 rds/min combined to 2 km. 2—12.7 mm MGs.
Physical countermeasures: Decoys: 2 Dagaie launchers ❺; IR flares and chaff.
Electronic countermeasures: ESM/ECM: Racal Cutlass/Cygnus ❻; intercept/jammer.
Radars: Air/surface search: Bofors Ericsson Sea Giraffe 50HC ❽; G-band.
Navigation: Racal Decca 1226; I-band.

Fire control: Bofors Electronic 9LV 223 ❾; J-band (for gun and SSM). Thomson-CSF DRBV 51C ❿; J-band (for Crotale).
Weapon control systems: CSSE Najir optronic director ❼.
Helicopters: 1 Aerospatiale Alouette SA 316 ⓫.

Programmes: Ordered in late 1986. Similar vessels to Bahrain craft. Delivery in October 1991.

Modernisation: Muray Jib completed a one-year refit at Dubai Maritime City in October 2011.
Structure: Lürssen design adapted for the particular conditions of the Gulf. This class has good air defence and a considerable anti-ship capability. The helicopter hangar is reached by flight deck lift.
Operational: Pennant numbers changed in 2002.

MURAY JIB *(Scale 1 : 600), Ian Sturton* / 0080921

DAS *2/2011, Guy Toremans* / 1454817

IHS Jane's Fighting Ships 2014-2015 © 2014 IHS

Corvettes < **United Arab Emirates** 885

1 + (1) ABU DHABI (COMANDANTE) CLASS (FSH)

Name	No	Builders	Laid down	Launched	Commissioned
ABU DHABI	P 191	Fincantieri, Riva Trigoso	2009	15 Feb 2011	8 Jan 2013

Displacement, tonnes: 1,650 full load
Dimensions, metres (feet): 88.4 × 12.2 × 4.6 screws (290 × 40.0 × 15.1)
Speed, knots: 25
Range, n miles: 3,000 at 14 kt
Complement: 60 (5 officers)

Machinery: 2 diesels; 18,780 hp(m) *(14.0 MW)*; 2 shafts; cp props; bow thruster
Missiles: SSM: 4 MBDA Exocet ❶.
Guns: 1 Otobreda 3 in *(76 mm)*/62 Super Rapid ❷. 2 Oto Melara Marlin 30 mm ❸. 2 − 12.7 mm MGs.
Torpedoes: 4 (2 twin) 324 mm tubes.
Physical countermeasures: Decoys: 2 Rheinmetall MASS. 2 WASS C 310 torpedo defence systems.
Electronic countermeasures: ESM: To be announced.
Radars: Air/surface search: SELEX KRONOS ❹; 3D; G-band. Navigation: To be announced ❺.
Fire control: SELEX NA-30S (RTN 30X) ❻; I-band.
Sonars: WASS SNA-2000 mine-avoidance; Thales Captas Nano; low-frequency active.
Combat data systems: IPNS.
Electro-optic systems: NA-30S TV and IR tracking. Medusa Mk 4B surveillance.
Helicopters: To be announced.

Programmes: Plans to procure a new anti-submarine corvette were announced in March 2009. A contract with Fincantieri (with the option for a second ship) was signed on 3 August 2009.
Structure: The ship is based on the Comandante class, in service in the Italian Navy. WASS, in partnership with Thales, designed and built the underwater system.
Modernisation: The Captas Nano may be replaced by Captas Mk 2. It is likely that the ship will be upgraded with Raytheon RIM-116 (RAM) at a later date.

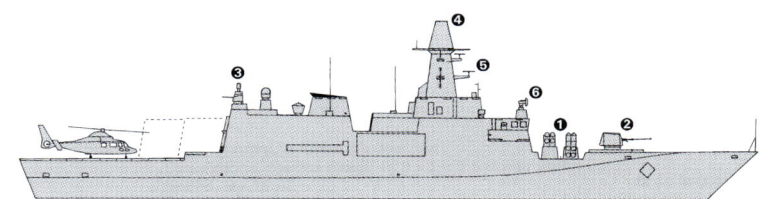

ABU DHABI *(Scale 1 : 900), Ian Sturton* / 1454773

ABU DHABI *1/2013, Giorgio Ghiglione* / 1483651

ABU DHABI *1/2013, Giorgio Ghiglione* / 1483650

© 2014 IHS IHS Jane's Fighting Ships 2014-2015

888 **United Arab Emirates** > Patrol forces — Mine warfare forces

20 RAIDING CRAFT (PBF)

Displacement, tonnes: 4 full load
Dimensions, metres (feet): 8.5 × 3 × 0.6 *(27.9 × 9.8 × 2.0)*
Speed, knots: 38
Complement: 12
Machinery: 2 outboards; 450 hp *(336 kW)*

Comment: There are eight Arctic 28 RIBs ordered from Halmatic, Southampton in June 1992 and delivered in mid-1993. GRP hulls. Speed given is fully laden. Used by Special Forces. There are also 12 Al-Shaali type ordered in 1994 and built in Dubai.

7 + 5 MODIFIED GHANNATHA CLASS (PB)

| BUTINAH P 301 | – P 303 | AL HAMRYAH P 305 | AL ZAWRAA P 307 |
| AL BAZAM P 302 | AL MUROOM P 304 | AL KHAN P 306 | +5 |

Displacement, tonnes: 49 full load
Dimensions, metres (feet): 26.5 × 5.1 × 1.1 *(86.9 × 16.7 × 3.6)*
Speed, knots: 40
Complement: 6

Machinery: 2 MTU 12V 2000 M93 diesels; 4,800 hp *(3.58 MW)*; 2 Rolls Royce FF 600 waterjets
Missiles: SSM: 4 MBDA Marte Mk 2/N.
Guns: 1 Rheinmetall 27 mm. 1 — 12.7 mm Hitrole-G.
Radars: Surface search/navigation: To be announced.
Combat data systems: Selex.
Electro-optic systems: To be announced.

Comment: The order for 12 stretched missile-armed variants of the Ghannatha-class fast troop carriers was announced on 23 February 2009. The first three vessels were built by Swede Ship Marine and the remaining nine are being built by ADSB. The first Swede Ship Marine-built craft began sea trials in late 2011. The first ADSB-built craft was launched on 11 July 2012. The last craft is to be delivered in 2014.

ARCTIC 3/1995, H M Steele / 0080926

AL BAZAM 2/2013, Guy Toremans / 1430801

MINE WARFARE FORCES

2 FRANKENTHAL CLASS (TYPE 332) (MINEHUNTERS — COASTAL) (MHC)

Name	No	Builders	Launched	Commissioned
AL MURJAN (ex-*Frankenthal*)	M 02 (ex-M 1066)	Lürssen	6 Feb 1992	16 Dec 1992
AL HASBAH (ex-*Weiden*)	M 01 (ex-M 1060)	Abeking & Rasmussen, Lemwerder	14 May 1992	30 Mar 1993

Displacement, tonnes: 660 full load
Dimensions, metres (feet): 54.5 × 9.2 × 2.6 *(178.8 × 30.2 × 8.5)*
Speed, knots: 18
Complement: 37 (5 officers)

Machinery: 2 MTU 16V 396 TB84 diesels; 5,550 hp(m) *(4.08 MW)* sustained; 2 shafts; cp props; 1 motor (minehunting)
Missiles: SAM: 2 Stinger quad launchers.
Guns: 1 Bofors 40 mm/70; being replaced by Mauser 27 mm.
Radars: Navigation: Raytheon SPS-64; I-band.
Sonars: Atlas Elektronik DSQS-11M; hull-mounted; high frequency.

Combat data systems: STN MWS 80-4.

Programmes: Originally ordered for the German Navy in September 1988 with STN Systemtechnik Nord as main contractor. *Al Murjan* laid down at Lürssen 6 December 1989. Agreement for the purchase of both ships to UAE concluded in early 2006. Following decommissioning from the German Navy and recommissioning in the UAE Navy on 28 June 2006, both ships undertook refit work at the Neue Jadewerft shipyard in Wilhelmshaven before being transported to Abu Dhabi. *Al Hasbah* arrived in August 2006 and *Al Murjan* in Autumn 2006. A training programme for the crews was conducted prior to transfer.

Modernisation: Overhaul of the main machinery and upgrade of the air-conditioning systems was completed in both ships in March 2011.
Structure: Built of amagnetic steel with same hull, similar superstructure and high standardisation as Type 333 and 352 in service with the German Navy. Equipped with two STN Systemtechnik Nord Pinguin-B3 drones with sonar, TV cameras and two countermining charges.
Operational: Weapons and sensors are as for the ships in German service and may be different.

AL MURJAN 12/2010, Harald Carstens / 1406294

IHS Jane's Fighting Ships 2014-2015 © 2014 IHS

AMPHIBIOUS FORCES

Notes: There are also four civilian LCM ships, *El Nasirah 2, Baava 1, Makasib* and *Ghagha II*. Two Serna class LCUs are also civilian owned.

4 LCT

– L 61 (ex-6401) **AL KHAZNAH** L 62 (ex-6402) – L 63 (ex-6403) – L 64 (ex-6404)

Displacement, tonnes: 864 standard (est.)
Dimensions, metres (feet): 64 × 12.0 × 2.7 *(210 × 39.4 × 8.9)*
Speed, knots: 12
Machinery: 2 Caterpillar diesels; 3,620 hp *(2.7 MW)*; 2 shafts
Guns: 2 – 12.7 mm MGs.

Comment: Built at Abu Dhabi Naval Base and completed in 1996-99. Pennant numbers changed in 2001.

L 62 *3/2006, Ships of the World* / 1127292

3 AL FEYI CLASS (LCU)

AL FEYI L 51 (ex-5401) **DAYYINAH** L 52 (ex-5402) **JANANAH** L 53 (ex-5403)

Displacement, tonnes: 660 full load
Dimensions, metres (feet): 50 × 11 × 2.8 *(164 × 36.1 × 9.2)*
Speed, knots: 11. **Range, n miles:** 1,800 at 11 kt
Complement: 10
Military lift: 4 vehicles
Machinery: 2 diesels; 1,248 hp *(931 kW)*; 2 shafts
Guns: 2 – 12.7 mm MGs.

Comment: *Al Feyi* built by Siong Huat, Singapore; completed 4 August 1987. The other pair built by Argos Shipyard, Singapore to a similar design and completed in December 1988. Used mostly as transport ships. Pennant numbers changed in 2001.

DAYYINAH (old number) *6/1996* / 0080929

3 LANDING CRAFT (LCT)

– L 65 – L 66 **ZARKOH** L 67

Displacement, tonnes: 864 standard (est.)
Dimensions, metres (feet): 64 × 12.0 × 2.7 *(210 × 39.4 × 8.9)*
Speed, knots: 11
Complement: 75
Military lift: military vehicles
Machinery: 2 Caterpillar 3508 diesels; 3,620 hp *(2.7 MW)*; 2 shafts

Comment: Fully designed in the UAE, the vessels were ordered from ADSB in November 2001 and laid down in early 2002 and delivery reportedly started in 2004. Details based on the L 61 class. Weapons are likely to include medium calibre machine guns.

L 67 *3/2008, Guy Toremans* / 1353500

2 LANDING CRAFT (LCU)

– L 41 **UMM AL NARR** L 42

Measurement, tonnes: 386 dwt
Dimensions, metres (feet): 44.4 × 10.0 × 2.2 *(145.7 × 32.8 × 7.2)*
Speed, knots: 10
Range, n miles: 1,000 at 8.5 kt
Complement: 51 (3 officers)
Military lift: Military vehicles
Machinery: 2 Caterpillar CAT 3406 TA diesels; 730 hp *(544 kW)*; 2 shafts

Comment: Fully designed in the UAE, these multimission landing craft were constructed by Abu Dhabi Shipbuilding in marine grade steel. The craft are equipped with three hydraulic deck cranes. The first ship was delivered to the navy in mid-2004 and the second to UAE Special Forces Command on 22 June 2006.

UMM AL NARR *6/2006, Abu Dhabi Shipbuilding* / 1159231

4 FAST SUPPLY VESSELS (LCP)

L 21–24

Displacement, tonnes: 54 standard; 74 full load
Dimensions, metres (feet): 26 × 5.4 × 1.3 *(85.3 × 17.7 × 4.3)*
Speed, knots: 32
Complement: 3
Military lift: 2 – 3 m containers plus 18 troops
Machinery: 2 MTU 16V 2000 M70 diesels; 2,775 hp *(2.1 MW)*; 2 Rolls Royce FF 550 waterjets
Guns: 2 – 12.7 mm MGs.

Comment: A contract for the construction of four fast supply vessels was made with Abu Dhabi Shipbuilding on 27 June 2004. The aluminium craft were built to a SwedeShip Marine design, developed in conjunction with the Swedish Defence Material Administration. The principal design features include a hydraulically operated bow door and space for storage of two fully loaded 3 m containers or a vehicle. An NBC citadel includes the wheelhouse and medical quarters. The latter can accommodate four stretcher cases or 18 fully equipped troops. All four vessels had been delivered by March 2007.

FAST SUPPLY VESSEL *2/2011, Guy Toremans* / 1406349

1 78M LANDING SHIP

SIR BU'NUER L 82

Measurement, tonnes: 1,687 gt
Dimensions, metres (feet): 78.1 × 16.0 × 3.5 *(256.2 × 52.5 × 11.5)*
Speed, knots: 12
Machinery: 2 Caterpillar 3512 diesels; 2,720 hp *(2.0 MW)*; 2 shafts

Comment: Constructed by Shin Yang Shipyard, Malaysia, and completed in October 2004.

12 GHANNATHA (TRANSPORTBÅT 2000) CLASS (LCP)

L 201–212

Displacement, tonnes: 44 full load
Dimensions, metres (feet): 24.2 × 5.1 × 1.1 (79.4 × 16.7 × 3.6)
Speed, knots: 35
Complement: 3
Military lift: 42 troops or 10 tons
Machinery: 2 MTU 12V 2000 diesels; 2,660 hp (2.0 MW); 2 Rolls Royce FF 550 waterjets
Guns: 1 Rheinmetall 27 mm. 1—12.7 mm MG. Patria NEMO 120 mm mortar.
Radars: Navigation: Terma; I-band.

Comment: Project 'Ghannatha' was for 12 amphibious transport craft based on the Transportbåt 2000 craft in service with the Royal Swedish Navy. Three craft were constructed at the Djupviks yard in Sweden while ADSB built the other nine. Details of the aluminium craft are based on those in Swedish service. Delivery was completed in 2004. All 12 craft are being converted into two variants. Six are to be modified to receive the Patria NEMO 120 mm mortar system and the other six are to receive a Rheinmetall 27 mm gun. The first converted craft, L 210, was delivered in early 2012.

L 211   2/2011, Guy Toremans / 1406348

AUXILIARIES

1 COASTAL TUG (YTB)

ANNAD A 3501

Displacement, tonnes: 808 full load
Dimensions, metres (feet): 35 × 9.8 × 4.2 (114.8 × 32.2 × 13.8)
Speed, knots: 14. **Range, n miles:** 2,500 at 14 kt
Complement: 14 (3 officers)
Machinery: 2 Caterpillar 3606TA diesels; 4,180 hp (3.12 MW) sustained; 2 shafts; cp props; bow thruster; 362 hp (266 kW)
Radars: Navigation: Racal Decca 2070; I-band.

Comment: Built by Dunston, Hessle, and completed in April 1989. Bollard pull, 55 tons. Equipped for SAR and is also used for logistic support.

AL GAFFA   3/1997 / 0019363

2 HARBOUR TUGS (YTM)

TEMSAH A 51   **UGAAB** A 52

Displacement, tonnes: 91 full load
Dimensions, metres (feet): 16.5 × 5.0 × 1.8 (54.1 × 16.4 × 5.9)
Machinery: 2 Volvo Penta TAMD-122A diesels; 760 hp (560 kW); 2 shafts

Comment: Ordered from Damen shipyard, Gorinchem in 1996 and entered service in 1998. Main role to attend Kortenaer class frigates. Equipped with fire-fighting platform abaft the mainmast.

ANNAD   6/1994 / 0080930

1 DIVING TENDER (YDT)

AL GAFFA D 1051

Displacement, tonnes: 102 full load
Dimensions, metres (feet): 31.4 × 6.9 × 1.1 (103 × 22.6 × 3.6)
Speed, knots: 26. **Range, n miles:** 390 at 24 kt
Complement: 6
Machinery: 2 MTU 12V 396 TB93 diesels; 3,260 hp(m) (2.4 MW) sustained; 2 water-jets

Comment: Ordered from Crestitalia in December 1985 for Abu Dhabi and delivered in July 1987. GRP hull. Used primarily for mine clearance but also for diving training, salvage and SAR. Fitted with a decompression chamber and diving bell. Lengthened version of Italian *Alcide Pedretti*.

TEMSAH   2/2011, Guy Toremans / 1406350

1 + 1 LOGISTIC SUPPORT SHIP (AFT)

RMAH – – A 61

Dimensions, metres (feet): 58 × 11.0 × 2.4 *(190.3 × 36.1 × 7.9)*
Machinery: To be announced.
Guns: To be announced.
Radars: Navigation: I-band.

Comment: Both built by Nobiskrug, Rendsburg, in 2013. The ships are equipped with a crane and there is space for two TEU on deck. Design features include two doors in the transom and a loading ramp aft.

RMAH   *11/2013*, Martin Mokrus* / 1531005

COAST GUARD

Notes: (1) Under control of Minister of Interior. In addition to the vessels listed below there is a number of Customs and Police launches including Barracuda craft, three Swedish Boghammar 13 m craft of the same type used by Iran and delivered in 1985, two Baglietto police launches acquired in 1988, about 10 elderly Dhafeer and Spear class of 12 and 9 m respectively, and two Halmatic Arun class Pilot craft delivered in 1990-91; some of these launches carry light machine guns.
(2) There are three 13-m Boghammar craft that entered service in 1985.

2 PROTECTOR CLASS (WPB)

101 (ex-1101)   **102** (ex-1102)

Displacement, tonnes: 183 full load
Dimensions, metres (feet): 33 × 6.7 × 2.1 *(108.3 × 22.0 × 6.9)*
Speed, knots: 33
Complement: 14

Machinery: 2 MTU 16V 396 TE94 diesels; 5,911 hp(m) *(4.35 MW)* sustained; 2 shafts; LIPS props
Guns: 1 Mauser 20 mm. 2−12.7 mm MGs.
Radars: Surface search: I-band.
Weapon control systems: 1 SAGEM optronic director.

Comment: Ordered from FBM Marine, Cowes in 1998. Aluminium hulls. First one laid down 15 June 1998. Both delivered in late 1999. More may be built by Abu Dhabi Shipbuilders. Similar to Bahamas and Chilean naval craft.

PROTECTOR 101 (old number)   *11/1999, UAE Coast Guard* / 0106675

5 CAMCRAFT 77 ft (COASTAL PATROL CRAFT) (WPB)

103 (ex-753)   **104** (ex-754)   **105** (ex-755)   **106** (ex-756)   **107** (ex-757)

Displacement, tonnes: 71 full load
Dimensions, metres (feet): 23.4 × 5.5 × 1.5 *(76.8 × 18.0 × 4.9)*
Speed, knots: 25
Complement: 8
Machinery: 2 GM 12V-71TA diesels; 840 hp *(627 kW)* sustained; 2 shafts
Guns: 2 Lawrence Scott 20 mm (not always embarked).
Radars: Surface search: Racal Decca; I-band.

Comment: Completed 1975 by Camcraft, New Orleans. Not always armed.

CAMCRAFT 105 (old number)   *6/1997* / 0019364

16 CAMCRAFT 65 ft (COASTAL PATROL CRAFT) (WPB)

108 (ex-650)	**112** (ex-654)	**116** (ex-658)	**120** (ex-662)
109 (ex-651)	**113** (ex-655)	**117** (ex-659)	**121** (ex-663)
110 (ex-652)	**114** (ex-656)	**118** (ex-660)	**122** (ex-664)
111 (ex-653)	**115** (ex-657)	**119** (ex-661)	**123** (ex-665)

Displacement, tonnes: 51 full load
Dimensions, metres (feet): 19.8 × 5.5 × 1.5 *(65 × 18.0 × 4.9)*
Speed, knots: 25
Complement: 8
Machinery: 2 MTU 6V 396 TB93 diesels; 1,630 hp(m) *(1.2 MW)* sustained; 2 shafts (in 14)
2 Detroit 8V-92TA diesels; 700 hp *(522 kW)* sustained; 2 shafts (in 2)
Guns: 1 Oerlikon 20 mm GAM-BO1.
Radars: Surface search: Racal Decca; I-band.

Comment: Built by Camcraft, New Orleans and delivered by September 1978.

CAMCRAFT 113 (old number)   *12/2001, A Sharma* / 0534118

6 BAGLIETTO GC 23 TYPE
(COASTAL PATROL CRAFT) (WPBF)

124 (ex-758)   **126** (ex-760)   **128** (ex-762)
125 (ex-759)   **127** (ex-761)   **129** (ex-763)

Displacement, tonnes: 52 full load
Dimensions, metres (feet): 24 × 5.5 × 0.9 *(78.7 × 18.0 × 3.0)*
Speed, knots: 43
Range, n miles: 700 at 20 kt
Complement: 9
Machinery: 2 MTU 12V 396 TB93 diesels; 3,260 hp(m) *(2.4 MW)* sustained; 2 Kamewa water-jets
Guns: 1 Oerlikon 20 mm. 2−7.62 mm MGs.
Radars: Surface search: I-band.

Comment: Built by Baglietto, Varazze. First two completed in March and May 1986, second pair in July 1987 and two more in 1988. All were delivered to UAE Coast Guard in Dubai.

BAGLIETTO 124 (old number)   *1987, UAE Coast Guard* / 0080934

3 BAGLIETTO 59 ft (COASTAL PATROL CRAFT) (WPBF)

501–503

Displacement, tonnes: 22 full load
Dimensions, metres (feet): 18.1 × 4.3 × 0.7 (59.4 × 14.1 × 2.3)
Speed, knots: 40
Complement: 6
Machinery: 2 MTU 12V 183 TE92 diesels; 2 shafts
Guns: 2—7.62 mm MGs.
Radars: Surface search: Racal Decca; I-band.

Comment: Ordered in 1992 and delivered in late 1993.

BAGLIETTO 503   10/1993, UAE Coast Guard / 0080935

6 WATERCRAFT 45 ft (COASTAL PATROL CRAFT) (PB)

Displacement, tonnes: 25 full load
Dimensions, metres (feet): 13.7 × 4.3 × 1.4 (44.9 × 14.1 × 4.6)
Speed, knots: 26
Range, n miles: 380 at 18 kt
Complement: 5
Machinery: 2 MAN D2542 diesels; 1,300 hp(m) (956 kW); 2 shafts
Guns: Mounts for 2—7.62 mm MGs.
Radars: Surface search: Racal Decca; I-band.

Comment: Ordered from Watercraft, UK in February 1982. Delivery in early 1983. Four deleted. Two similar craft built by Halmatic were delivered to the Dubai Port Authority in October 1997.

WATERCRAFT 45 ft   1984, UAE Coast Guard / 0506115

35 HARBOUR PATROL CRAFT (PB/YDT)

Comment: The latest are 11 Shark 33 built by Shaali Marine, Dubai and delivered in 1993–94. The remainder are a mixture of Barracuda 30 ft and FPB 22 ft classes. All are powered by twin outboard engines and most carry a 7.62 mm MG and have a Norden radar. There are also two Rotork craft used as diving tenders. Customs boats are operated separately by each of the UAE states. Some have been built for Kuwait.

BARRACUDA 271   12/2001, A Sharma / 0534117

54 SEASPRAY ASSAULT BOATS (PB)

Dimensions, metres (feet): 9.5 × 3.1 × 0.5 (31.2 × 10.2 × 1.6)
Speed, knots: 50
Range, n miles: 450 at 17 kt
Complement: 5
Machinery: 2 outboards; 500 hp (375 kW)
Radars: Navigation: I-band.

Comment: Initial batch of 24 craft delivered in September 2003. A further thirty were ordered in early 2004. Designed by Sea Spray Aluminium Boats.

SEASPRAY   2/2004, ADSB / 0563487

12 HALMATIC WORK BOATS (PB)

Displacement, tonnes: 14 standard
Dimensions, metres (feet): 16 × 4.0 × 0.7 (52.5 × 13.1 × 2.3)
Speed, knots: 24
Complement: 5
Machinery: 2 diesels; 2 waterjets
Radars: Navigation: I-band.

Comment: Construction of 12 craft started at Abu Dhabi Shipbuilding Composites in mid-2006. The first of class was completed by March 2007 but the delivery schedule for the remainder has not been confirmed. Based on the VT Halmatic Sea Keeper design with an asymmetric catamaran hull, the craft are highly manoeuvrable and are capable of carrying a 10 tonne payload.

WORK BOAT   2/2013*, Guy Toremans / 1531111

12 + 22 MRTP 16 FAST INTERVENTION CRAFT (PBF)

1801–1834

Displacement, tonnes: 22 full load
Dimensions, metres (feet): 17.75 × 4.19 × 1.2 (58.2 × 13.7 × 3.9)
Speed, knots: 60
Range, n miles: 300 at 35 kt
Complement: 4 + 8 embarked forces
Machinery: 2 MTU 8V2000 M93 diesels; 2,400 hp(m) (1.8 MW); 2 Arneson ASD 12 B1L surface drives
Guns: 2—12.7 mm MGs.

Comment: Plans to procure a fleet of 34 fast intercept craft for the UAE Critical National Infrastructure Authority were announced in March 2009. The craft are based on the Yonca-Onuk MRTP 16 design, an extended version of the Kaan 15 craft in service with the Turkish Coast Guard. The first 12 craft are being built by Yonca-Onuk at Tuzla and the remaining 22 by Abu Dhabi Shipbuilding. Delivery of the first craft was made on 10 March 2010.

1821   2/2013*, Guy Toremans / 1531110

18 AL FATTAN CLASS (PATROL BOATS) (PB)

1501 **1515** **+16**

Displacement, tonnes: 8 full load
Dimensions, metres (feet): 14.5 × 3.0 × 0.7 *(47.6 × 9.8 × 2.3)*
Speed, knots: 50
Complement: 6
Machinery: 3 Mercury Verado outboard motors; 900 hp *(671 kW)*
Guns: 2—12.7 mm MGs.

Comment: An order for 18 patrol craft was placed with Al Fattan Ship Industry on 27 February 2011. The all composite patrol craft are employed by the UAE Coast Guard to conduct coastal patrol missions, support search and rescue, and to protect the integrity of territorial waters.

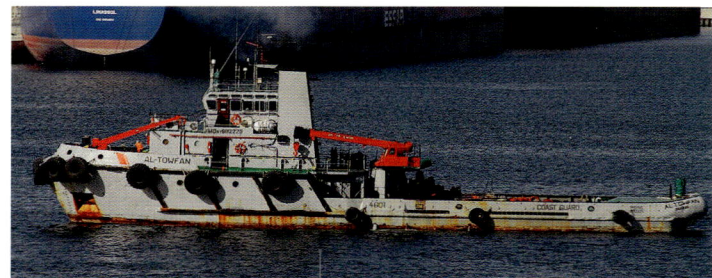

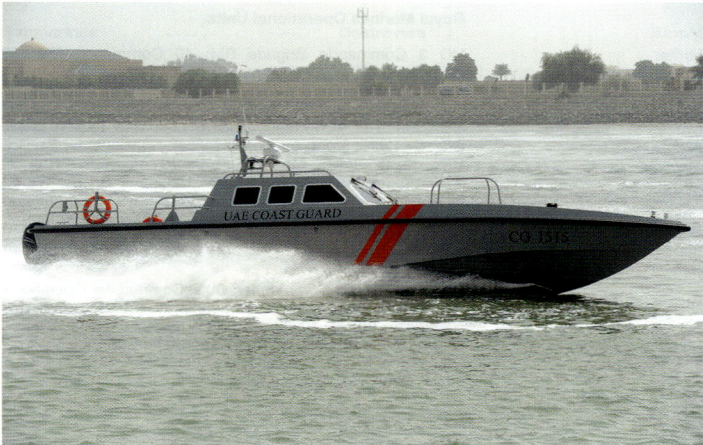

AL FATTAN 1515 ***2/2013*, Guy Toremans*** / 1531112

1 OFFSHORE PATROL VESSEL (PSO)

AL TOWFAN

Measurement, tonnes: 642 gt
Dimensions, metres (feet): 44.5 × 11.87 × 5.0 *(146 × 38.9 × 16.4)*
Speed, knots: 15
Machinery: 1 MAK 6M552AK diesel; 5,200 hp *(3.82 MW)*; 2 shafts

Comment: Built as an offshore supply vessel by Damen Shipbuilding and completed in June 1982.

AL TOWFAN ***1/2013*, Shaun Jones*** / 1565439

11 CMN DV 15 FAST INTERVENTION CRAFT (PBF)

1512 **+10**

Displacement, tonnes: 12 full load
Dimensions, metres (feet): 15.5 × 3.0 × 0.8 *(50.9 × 9.8 × 2.6)*
Speed, knots: 55
Complement: 4
Machinery: 2 Caterpillar diesels; 1,420 hp *(1.0 MW)*; 2 waterjets

Comment: Constructed by CMN Cherbourg for the UAE Critical National Infrastructure Authority. Five delivered in 2010 and six further in 2012.

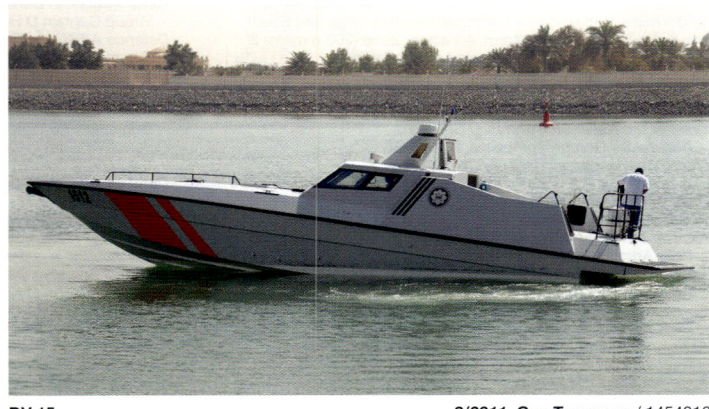

DV 15 ***2/2011, Guy Toremans*** / 1454816

1 OFFSHORE PATROL VESSEL (PSO)

AL WTAID

Measurement, tonnes: 1,700 gt
Dimensions, metres (feet): 60 × 16.0 × 4.5 *(196.9 × 52.5 × 14.8)*
Speed, knots: 13
Machinery: 2 Caterpillar 3516B diesels; 5,224 hp *(3.84 MW)*; 2 shafts

Comment: Constructed as an offshore supply vessel by Dubai Shipbuilding and completed in March 2013.

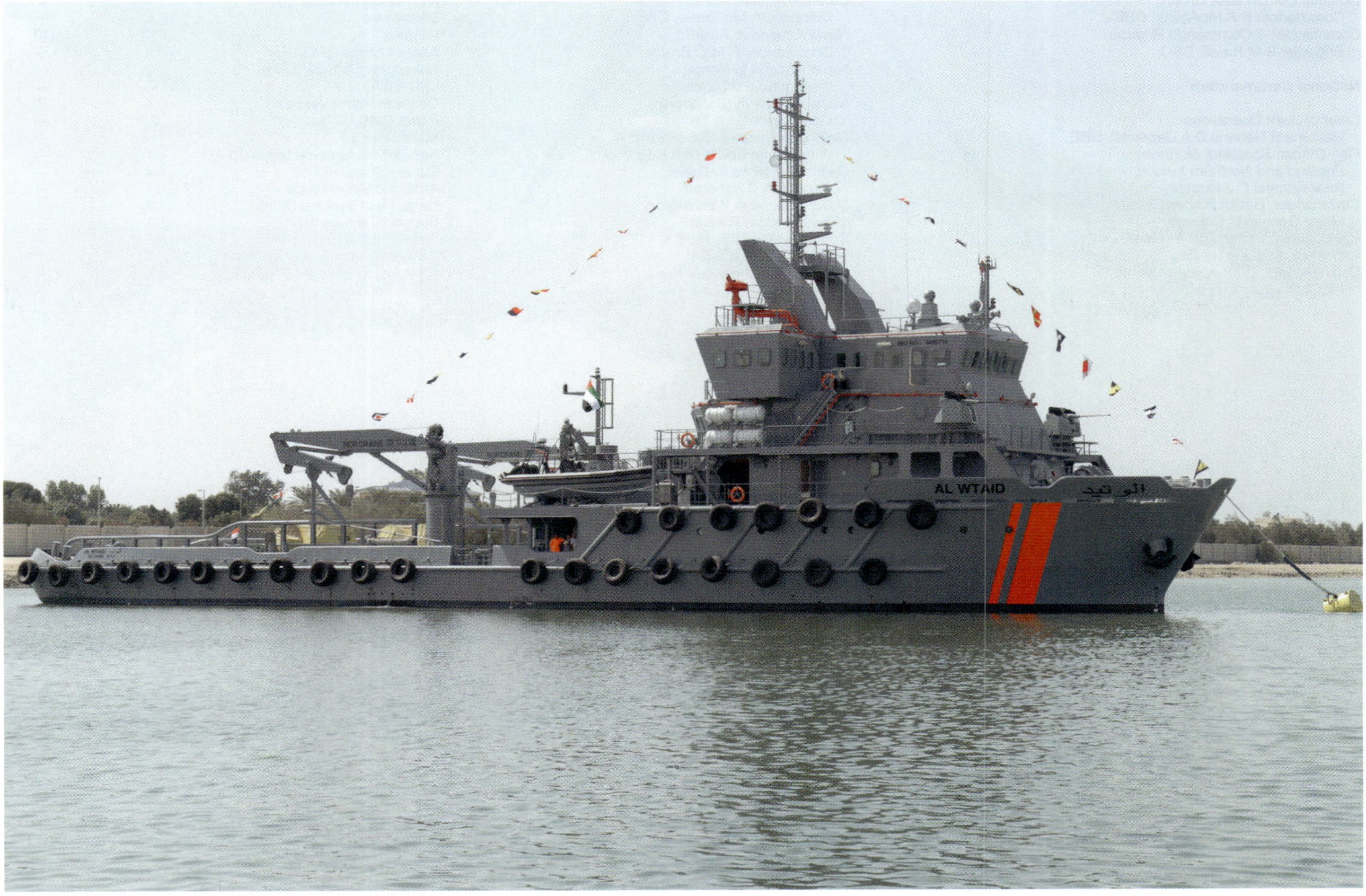

AL WTAID ***2/2013*, Guy Toremans*** / 1486557

© 2014 IHS *IHS Jane's Fighting Ships* 2014-2015

898 UK (NAVY) > Submarines

Attack Submarines

Notes: Future submarine requirements are being taken forward in a twin-track approach. In the short-term, technology advances to an extended Astute class are under consideration. Conceptual studies are also investigating requirements for a 'Maritime Underwater Future Capability' (MUFC) post 2030. Options include a development of the Astute class and linkage to development work on the future SSBN.

5 TRAFALGAR CLASS (SSN)

Name	No	Builders	Laid down	Launched	Commissioned
TIRELESS	S 88	Vickers Shipbuilding & Engineering, Barrow-in-Furness	6 Jun 1981	17 Mar 1984	5 Oct 1985
TORBAY	S 90	Vickers Shipbuilding & Engineering, Barrow-in-Furness	3 Dec 1982	8 Mar 1985	7 Feb 1987
TRENCHANT	S 91	Vickers Shipbuilding & Engineering, Barrow-in-Furness	28 Oct 1985	3 Nov 1986	14 Jan 1989
TALENT	S 92	Vickers Shipbuilding & Engineering, Barrow-in-Furness	13 May 1986	15 Apr 1988	12 May 1990
TRIUMPH	S 93	Vickers Shipbuilding & Engineering, Barrow-in-Furness	2 Feb 1987	16 Feb 1991	12 Oct 1991

Displacement, tonnes: 4,816 surfaced; 5,292 dived
Dimensions, metres (feet): 85.4 × 9.8 × 9.5 (280.2 × 32.2 × 31.2)
Speed, knots: 32 dived
Complement: 130 (18 officers)

Machinery: Nuclear; 1 RR PWR 1; 2 GEC turbines; 15,000 hp (11.2 MW); 1 shaft; pump jet propulsor; 2 WH Allen turbo generators; 3.2 MW; 2 Paxman diesel alternators; 2,800 hp (2.09 MW); 1 motor for emergency drive; 1 auxiliary retractable prop
Missiles: SLCM: Raytheon Tomahawk Block IV; TERCOM and GPS aided inertial navigation system with DSMAC to 1,600+ km (865+ n miles) at 0.7 Mach; warhead (WDU 36B) 454 kg.
Torpedoes: 5—21 in (533 mm) bow tubes. Marconi Spearfish; wire-guided; active/passive homing to 26 km (14 n miles) at 65 kt; or 31.5 km (17 n miles) at 50 kt; attack speed 55 kt; warhead 300 kg directed charge; 20 reloads.
Mines: Can be carried in lieu of torpedoes.
Physical countermeasures: Decoys: SAWCS. 2 SSE Mk 8 launchers. Type 2066 torpedo decoys.
Electronic countermeasures: RESM: Thales UAA4; passive intercept.
CESM: Eddystone.
Radars: Navigation: Kelvin Hughes Type 1007; I-band.

Sonars: TUSL 2074 LRE; hull-mounted; passive/active, search and attack; low frequency. TUSL 2046; towed array, passive search, very low frequency. Ultra Electronics 2082; active intercept and ranging. TUSL 2076 (S90-93) integrated sonar suite comprising flank array, towed array, conformal bow array, mine avoidance array. TUSL 2077; ice navigation.
Combat data systems: BAE Systems SMCS NG tactical data handling system.
Weapon control systems: BAE Systems SMCS.

Programmes: Tireless ordered 5 July 1979; Torbay 26 June 1981; Trenchant 22 March 1983; Talent 10 September 1984; Triumph 3 January 1986.
Modernisation: Tireless completed refuelling and refit with Sonar 2074, SMCS and Spearfish torpedoes in January 1999. Refuel periods for the last four boats were undertaken in parallel with a major tactical modernisation programme, the main feature of which installation of the sonar 2076 integrated sonar suite. Other upgrades included enhancements to SMCS, a new command console and improved signature reduction measures. Torbay and Trenchant were the first and second boats to complete a 2076 refit and refuel in 2003 and 2004 respectively. Talent completed her three-year refit in January 2007 and Triumph completed her refit in March 2010. Meanwhile, an ongoing programme of software replacement will continue to realise capability improvements in the last four boats. As a parallel programme, SMCS is being upgraded to SMCS NG. Tomahawk Block IV missiles started to replace Block III missiles in March 2008. Following a joint UK/US feasibility study, a Torpedo-Tube Launched (TTL) variant of the missile was developed for UK use. A series of developmental tests began in 2005, culminating in the successful completion of a 650 n mile flight of a TTL missile, fired from Trenchant, on 21 June 2007. In parallel, the Tactical Tomahawk Weapons Control (TTWC) and Tomahawk Strike Network (TSN), has been fitted. Replacement of the CESM system was initiated in 2002 and upgrade of the RESM system UAP 3 to UAA4 was started in 2012. The command system is to be replaced by BAE Systems Common Combat System in later boats.
Structure: The pressure hull and outer surfaces are covered with conformal anechoic noise reduction coatings. Retractable forward hydroplanes and strengthened fins for under ice operations. Diving depth in excess of 300 m (985 ft). Fitted with Pilkington Optronics CK 34 search and CH 84 attack periscopes.
Operational: All five boats are based at Devonport. The last three boats are to transfer to Faslane by 2017. Trafalgar decommissioned in 2009 and Turbulent in July 2012. The remainder of the class are planned to pay off as follows: Tireless 2014, Torbay 2015, Trenchant 2017, Talent 2019, and Triumph 2022.

TRENCHANT 7/2008, Ian Harris / 1353552

TIRELESS 5/2008, B Sullivan / 1353554

Submarines < **UK** (NAVY) 899

TORBAY 7/2010, J Brodie / 1406295

TORBAY 9/2008, B Sullivan / 1353553

TALENT 6/2009, B Sullivan / 1366180

© 2014 IHS *IHS Jane's Fighting Ships* 2014-2015

2 + 5 ASTUTE CLASS (SSN)

Name	No	Builders	Laid down	Launched	Commissioned
ASTUTE	S 94	BAE Systems, Barrow	31 Jan 2001	8 Jun 2007	27 Aug 2010
AMBUSH	S 96	BAE Systems, Barrow	22 Oct 2003	5 Jan 2011	1 Mar 2013
ARTFUL	S 95	BAE Systems, Barrow	11 Mar 2005	17 May 2014	2015
AUDACIOUS	S 97	BAE Systems, Barrow	24 Mar 2009	2016	2018
ANSON	S 98	BAE Systems, Barrow	13 Oct 2011	2018	2020
AGAMEMNON	S 99	BAE Systems, Barrow	18 Jul 2013	2020	2022
AJAX	S 100	BAE Systems, Barrow	2015	2022	2024

Displacement, tonnes: 6,604 surfaced; 7,519 dived
Dimensions, metres (feet): 97 × 11.27 × 10 *(318.2 × 37.0 × 32.8)*
Speed, knots: 29 dived
Complement: 98 (18 officers)

Machinery: Nuclear; 1 RR PWR 2; 2 Alsthom turbines; 27,500 hp *(20.5 MW)*; 1 shaft; pump jet propulsor; 2 turbo generators; 2 diesel alternators; 2 MTU 8V 396 SE 84 motors for emergency drive; 1 auxiliary retractable prop
Missiles: SLCM: Raytheon Tomahawk Block IV; TERCOM and GPS aided navigation with DSMAC to 1,600+ km *(865+ n miles)* at 0.7 Mach; warhead (WDU 36B) 454 kg.
Torpedoes: 6—21 in *(533 mm)* tubes Marconi Spearfish; wire-guided; active/passive homing to 26 km *(14 n miles)* at 65 kt; or 31.5 km *(17 n miles)* at 50 kt; attack speed 55 kt; warhead 500 kg directed charge. Total of 38 weapons.
Mines: In lieu of torpedoes.
Physical countermeasures: Decoys.
Electronic countermeasures: ESM:Thales UAA4; intercept.
Radars: Navigation: I-band.
Sonars: Thales 2076 integrated suite (bow, flank, fin and towed arrays). Ultra 2082 passive intercept.
Combat data systems: BAE Systems ACMS tactical data handling system (Common Combat System in S 95 and later boats). Links 11/16.

Programmes: Invitations to tender issued on 14 July 1994 to build three of the class with an option for two more. GEC-Marconi selected as prime contractor in December 1995. Contract to start building the first three placed on 17 March 1997. First steel cut late 1999 but although formal keel-laying took place in 2001, design, engineering and programme management difficulties led to a three-year delay to the first of class. This was extended to four years by a range of emergent first of class issues. There were similar delays to the second and third of class. A class of seven boats was confirmed by the 2010 Strategic Defence and Security Review.

Structure: While originally based on the Trafalgar design, the Astute class has developed into a stand-alone design that incorporates significantly increased weapon load and reduced radiated noise. The fin is slightly longer and there are two Thales Optronics CM010 non-hull-penetrating optronic masts. A more advanced variant is to be fitted to *Audacious* and subsequent boats. The boats have a dry dock hangar capability (Project Chalfont), trials of which were conducted by *Astute* in late 2012. A fully reelable towed-array handling system is incorporated. A 'Thin Flank' array is to be fitted in *Audacious* and a lighter bow array in *Anson*.

Operational: Fitted with Core H, nuclear refuelling will not be necessary in the lifetime of the submarine. All boats to be based at Faslane. Sea trials of *Astute* began on 15 November 2009 although these were interrupted by repairs necessitated by a grounding incident off Skye on 22 October 2010. Subsequently, she conducted a successful Tomahawk firing on 8 November 2011 and became operational in late 2013. *Artful* is expected to start sea trials in 2015.

ASTUTE
9/2012, Tony Roper / 1486400

ASTUTE
11/2009, BAE Systems / 1366178

AIRCRAFT CARRIERS

0 + 2 QUEEN ELIZABETH CLASS (CV)

Name	No	Builders	Laid down	Launched	Commissioned
QUEEN ELIZABETH	R 08	BAE Systems Surface Ships; Babcock International	1 Sep 2009	Jul 2014	2018
PRINCE OF WALES	–	BAE Systems Surface Ships; Babcock International	2013	2018	2020

Displacement, tonnes: 65,500 full load
Dimensions, metres (feet): 282.9 × 38.8 × 11.0 (928.1 × 127.3 × 36.1)
Flight deck, metres: 277 × 73.0 (908.8 × 239.5)
Aircraft lift: 2
Speed, knots: 26
Complement: 686 + 830 air group + 95 flag staff

Machinery: Integrated Full Electric Propulsion; 2 Rolls-Royce MT 30 gas turbine alternators; 93,870 hp (70 MW); 4 Wärtsilä diesel generators; 53,064 hp (39.6 MW); 4 induction motors; 53,640 hp (40 MW); 2 shafts
Guns: 3 Raytheon 20 mm Phalanx Block 1B ❶. 4—30 mm ❷. Miniguns.
Physical countermeasures: Torpedo defence: Type 2170 (SLQ-25A).
Radars: Air search: Thales Type 1046 (S 1850M) ❸; D-band.
Air/surface search: BAE Insyte Type 997 (ARTISAN) ❹; 3D; E/F-band.
Navigation: Sperry Marine VisionMaster ❺.
PAR: Selex Galileo SPN-720; I-band.
Combat data systems: BAE CMS-1. Link 16.
Electro-optic systems: To be announced.

Fixed-wing aircraft: Approximately 40: Up to 36 F-35B Lightning II and 4 Crowsnest. Typically a mix of between 12 and 24 F-35B and a mix of helicopters, which might include Chinook, Merlin (Mk 2 and/or Mk 4) and Apache, depending on mission.
Helicopters: Westland Merlin

Programmes: Following completion of the Demonstration Phase in 2007, it was announced on 25 July 2007 that approval (Main Gate) for the procurement of two aircraft carriers had been given. Approval for the Manufacturing Phase was announced on 20 May 2008 and contracts for construction were signed on 3 July 2008. Construction of the ships is being undertaken by the Aircraft Carrier Alliance (ACA) formed of BAE Systems Surface Ships (BAESSS), Thales UK, Babcock Marine and UK MoD as both client and participant. An Alliance Management Board, chaired by the UK MoD, leads and collectively manages the project; BAESSS is responsible for the integration of design, build, commissioning and acceptance of the ships; Thales leads the management of the Stage 1 design of platform, power and propulsion and takes responsibility for the aviation interface. Construction and assembly of the ships is as follows: Lower Blocks 3 and 4 (aft section) at BAESSS Govan; Lower Block 2 (forward midships section) and the two superstructure islands at BAESSS Portsmouth; Lower Block 1 (bow section) at Babcock Appledore, central upper blocks at A&P Tyne, flight deck at Cammell Laird and final assembly at Babcock, Rosyth. Following a delay of up to two years to the construction programme announced on 11 December 2008, first steel of Queen Elizabeth was formally cut on 7 July 2009. The construction programme has since been further delayed first by the decision of the 2010 Strategic Defence and Security Review (SDSR) to build Prince of Wales with catapults and arrestor gear to operate the F-35C and subsequently by a reversal of that decision on 10 May 2012. First steel of Prince of Wales was cut on 26 May 2011 and, while the 2010 SDSR announced that only one ship is to be operated, a decision on the operation of both ships is expected in 2015.
Structure: The systems and structural design is to Lloyds Naval Ships Rules with some specific naval standards for certain equipments. Of steel construction, the principal design features include a two-island arrangement, with flight control from the after island, two deck-edge aircraft lifts and a ski-jump to operate STOVL aircraft. The flight-deck has a single runway and ramp; five landing spots are positioned on the runway and there is a sixth spot to starboard for helicopter landings only. Other features include eight internal decks, 19 watertight sections each of which contains vertical access trunks for fire-fighting, and an integrated waste management system which is fully MARPOL compliant. Key spaces, such as the Operations Room, have been designed with reconfigurability in mind. All major machinery is controlled and monitored by an Integrated Platform Management System. Air weapons are supplied from two automated deep stores by a Highly Mechanised Weapon Handling System which uses commercial warehousing techniques. Blown Fibre Optic technology is used to connect over 850 compartments via 112 km of fibre.
Operational: To be based at Portsmouth. Sea trials of Queen Elizabeth are planned to start in 2017. These are to be followed by ship/air integration trials in 2018 leading to an Initial Operating Capability in the carrier strike role in 2020.

QUEEN ELIZABETH *7/2008, Thales UK /* 1167814

QUEEN ELIZABETH *7/2008, Thales UK /* 1353551

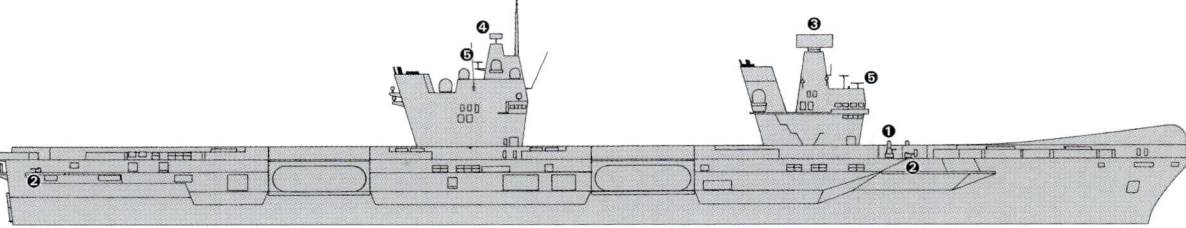

QUEEN ELIZABETH *(Scale 1 : 1,800), Ian Sturton /* 1353510

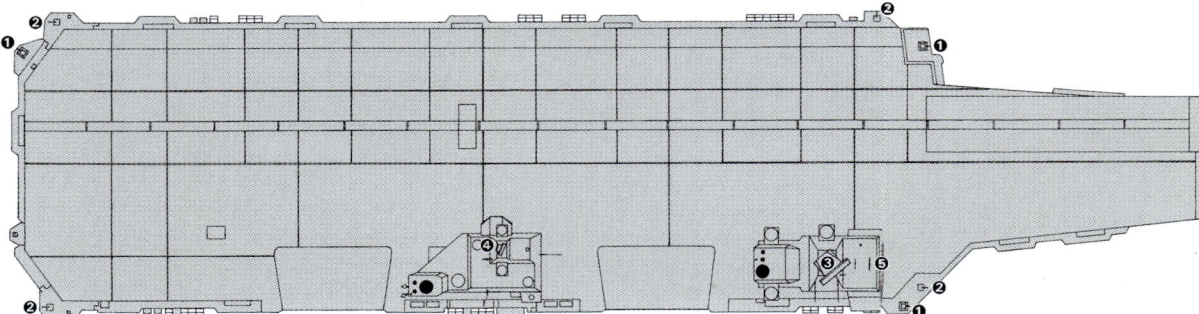

QUEEN ELIZABETH *(Scale 1 : 1,800), Ian Sturton /* 1353509

902 UK (NAVY) > Destroyers

DESTROYERS

Notes: *Bristol* (D 23) is an immobile tender used for training in Portsmouth Harbour.

6 DARING CLASS (TYPE 45) (DDGHM)

Name	No	Builders	Laid down	Launched	Commissioned
DARING	D 32	BAE Systems Marine; Vosper Thornycroft	28 Mar 2003	1 Feb 2006	20 Apr 2012
DAUNTLESS	D 33	BAE Systems Marine	26 Aug 2004	23 Jan 2007	3 Jun 2010
DIAMOND	D 34	BAE Systems Marine	25 Feb 2005	27 Nov 2007	6 May 2011
DRAGON	D 35	BAE Systems Marine	19 Dec 2005	17 Nov 2008	20 Apr 2012
DEFENDER	D 36	BAE Systems Marine	31 Jul 2006	21 Oct 2009	21 Mar 2013
DUNCAN	D 37	BAE Systems Marine	26 Jan 2007	11 Oct 2010	26 Sep 2013

Displacement, tonnes: 5,893 standard; 7,570 full load
Dimensions, metres (feet): 152.4 oa; 143.5 wl × 21.2 × 5.3 *(500; 470.8 × 69.6 × 17.4)*
Speed, knots: 31
Range, n miles: 6,500 at 18 kt
Complement: 191 + 41 spare berths

Machinery: Integrated Electric Propulsion; 2 RR WR-21 gas turbine alternators; 67,600 hp *(49.7 MW)*; 2 Wärtsilä 12V 200 diesel generators; 4 MW; 2 motors; 40 MW; 2 shafts; fixed props
Missiles: SSM: 8 Boeing Defense Harpoon Block 1C (2 quad in some ships) ❶; active radar homing to 130 km *(70 n miles)* at 0.9 Mach; warhead 227 kg.
SAM: Sea Viper (GWS 45); DCN Sylver A 50 48-cell VLS ❷; typical mix of 32 Aster 30; active pulse Doppler radar homing to 120 km *(65 n miles)* at 4.5 Mach; warhead 15 kg and 16 Aster 15; active pulse Doppler radar homing to 30 km *(16 n miles)* at 3.0 Mach.
Guns: 1 Vickers 4.5 in *(114 mm)*/55 Mk 8 Mod 1 ❸. 24 rds/min to 27.5 km *(14.8 n miles)*; weight of shell 21 kg. 2 Raytheon 20 mm Vulcan Phalanx Block 1B ❹. 2 REMSIG MSI DS 30A 30 mm/75; 650 rds/min to 10 km *(5.4 n miles)*; weight of shell 0.36 kg ❺.
Physical countermeasures: Decoys: 4 6-barrelled DLH (chaff, IR); DLF offboard decoys ❻. Type 2170 torpedo defence system.
Electronic countermeasures: ECM: To be announced.
RESM: Thales Type UAT Mod 2 ❼; intercept.
CESM: AN/SSQ-137(V) SSEE.
Radars: Air/surface search: Signaal/Marconi Type 1046 (S 1850M) ❿. D-band.
Surveillance/fire control: BAE Systems Type 1045 (Sampson) ⓫; E/F-band; multifunction.
Surface search: Raytheon Type 1048 ⓬. E/F-band.
Navigation: 2 Raytheon Type 1047 ⓭; I-band.
Sonars: Type 2091 (EDO/ULTRA MFS-7000); bow mounted; medium frequency.

Combat data systems: CMS-1 (based on DNA SSCS with additional AAW functions); Links 11, 16 STDL and 22. SATCOM ❽.
Electro-optic systems: GSA 9 with 2 EOSP sensor heads (EOGCS) (based on Radamec 2500) ❾.
Helicopters: 2 Lynx Mk HMA 8 or Merlin 1 HM.Mk 1 ⓮.

Programmes: This project has gone through many stages. Starting life as NFR 90 in the 1980s, it was taken forward via the Anglo-French Future Frigate, the tri-nation Common New Generation Frigate (Horizon) and finally, when UK withdrew from the collaborative ship programme on 25 April 1999, a national Type 45 ship project. The contract for the design and build of the first three ships (Batch 1) was placed with BAE Systems, on 20 December 2000. This was amended in late 2001 to reflect a new procurement strategy in which commitment was made to six ships. The second three ships comprise Batch 2. Final assembly of D 32 was at Scotstoun and assembly of follow-on ships is at Govan. Procurement of the missile system was pursued separately and a contract for full development and initial production of PAAMS (Sea Viper) was placed with the tri-national consortium, EUROPAAMS, in August 1999. The first ship-launch of the Sea Viper missile system was conducted by *Dauntless* on 29 September 2010.
Structure: Built to Lloyd's Naval Ship Rules. Provision for future installation of CEC, 155 mm gun or a 16-cell VLS silo, SSM and magazine-launched torpedoes. An integrated technology mast is another potential modification. The ships are designed to support and deploy at least 30 troops. OTC facilities are included. There are Tier 2 medical facilities.
Operational: *Daring*, *Dauntless*, *Diamond*, and *Dragon* have completed at least one operational deployment. *Defender* and *Duncan* are to conduct first deployments in 2014 and 2015 respectively. During 2013, *Daring* conducted a series of operational US Missile Defense Agency ballistic missile tests at the Pacific Reagan Test Site. The scope of participation was to explore potential roles in both tactical and theatre BMD.

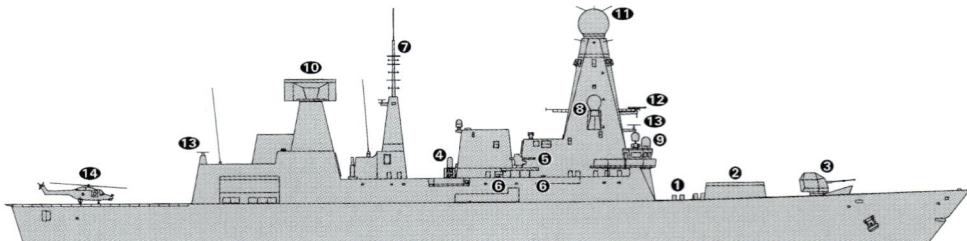

DARING *(Scale 1 : 1,200), Ian Sturton / 1353511*

DEFENDER *5/2013*, Michael Nitz / 1531117*

DARING *10/2013*, Chris Sattler / 1531116*

IHS Jane's Fighting Ships 2014-2015 © 2014 IHS

Destroyers < **UK** (NAVY) 903

DRAGON
8/2013*, Shaun Jones / 1531115

DIAMOND
4/2011, Derek Fox / 1454781

DUNCAN
7/2013*, Gordon Brodie / 1531114

© 2014 IHS
IHS Jane's Fighting Ships 2014-2015

UK (NAVY) > Frigates

FRIGATES

13 DUKE CLASS (TYPE 23) (FFGHM)

Name	No	Builders	Laid down	Launched	Commissioned
ARGYLL	F 231	Yarrow Shipbuilders, Glasgow	20 Mar 1987	8 Apr 1989	31 May 1991
LANCASTER	F 229 (ex-F 232)	Yarrow Shipbuilders, Glasgow	18 Dec 1987	24 May 1990	1 May 1992
IRON DUKE	F 234	Yarrow Shipbuilders, Glasgow	12 Dec 1988	2 Mar 1991	20 May 1993
MONMOUTH	F 235	Yarrow Shipbuilders, Glasgow	1 Jun 1989	23 Nov 1991	24 Sep 1993
MONTROSE	F 236	Yarrow Shipbuilders, Glasgow	1 Nov 1989	31 Jul 1992	2 Jun 1994
WESTMINSTER	F 237	Swan Hunter Shipbuilders, Wallsend-on-Tyne	18 Jan 1991	4 Feb 1992	13 May 1994
NORTHUMBERLAND	F 238	Swan Hunter Shipbuilders, Wallsend-on-Tyne	4 Apr 1991	4 Apr 1992	29 Nov 1994
RICHMOND	F 239	Swan Hunter Shipbuilders, Wallsend-on-Tyne	16 Feb 1992	6 Apr 1993	22 Jun 1995
SOMERSET	F 82	Yarrow Shipbuilders, Glasgow	12 Oct 1992	25 Jun 1994	20 Sep 1996
SUTHERLAND	F 81	Yarrow Shipbuilders, Glasgow	14 Oct 1993	9 Mar 1996	4 Jul 1997
KENT	F 78	Yarrow Shipbuilders, Glasgow	16 Apr 1997	27 May 1998	8 Jun 2000
PORTLAND	F 79	Yarrow Shipbuilders, Glasgow	14 Jan 1998	15 May 1999	3 May 2001
ST ALBANS	F 83	Yarrow Shipbuilders, Glasgow	18 Apr 1999	6 May 2000	6 Jun 2002

Displacement, tonnes: 3,556 standard; 4,267 full load
Dimensions, metres (feet): 133 × 16.1 × 5.5 screws; 7.3 sonar *(436.4 × 52.8 × 18.0; 24.0)*
Speed, knots: 28; 15 diesel-electric
Range, n miles: 7,800 at 15 kt
Complement: 181 (13 officers)

Machinery: CODLAG; 2 RR Spey SM1A (F 229-F 236) or SM1C (F 237 onwards) gas turbines (see *Structure*); 31,100 hp *(23.2 MW)* sustained; 4 Paxman 12CM diesels; 8,100 hp *(6 MW)*; 2 GEC motors; 4,000 hp *(3 MW)*; 2 shafts
Missiles: SSM: 8 McDonnell Douglas Harpoon Block 1C (2 quad) launchers ❶; active radar homing to 130 km *(70 n miles)* at 0.9 Mach; warhead 227 kg.
SAM: 32 British Aerospace Seawolf GWS 26 Mod 1 VLS ❷; command line of sight (CLOS) radar/TV tracking to 6 km *(3.3 n miles)* at 2.5 Mach; warhead 14 kg.
Guns: 1 Vickers 4.5 in *(114 mm)*/55 Mk 8 Mod 1 ❸; 25 rds/min to 27.5 km *(14.8 n miles)*; weight of shell 21 kg. 2 – 30 mm ASCG ❹; 650 rds/min to 10 km *(5.4 n miles)* anti-surface; 3 km *(1.6 n miles)* anti-aircraft; weight of shell 0.36 kg. 2 M 323 Mk 44 7.62 mm Miniguns. 4 – 7.62 mm MGs.
Torpedoes: 4 Cray Marine 324 mm fixed (2 twin) tubes ❺. Marconi Stingray; active/passive homing to 11 km *(5.9 n miles)* at 45 kt; warhead 35 kg (shaped charge); depth to 750 m *(2,460 ft)*.
Physical countermeasures: Decoys: Outfit DLH; 4 Sea Gnat 6-barrelled 130 mm/102 mm launchers ❻. DLF 3B offboard decoys. Type 2170 torpedo defence system.
Electronic countermeasures: RESM: Racal UAT ❼; intercept. CESM: Babcock Hammerhead.
Radars: Air/surface search: BAE Insyte Type 996(I) ❿; 3D; E/F-band. BAE Insyte Type 997 (ARTISAN) (F 234); 3D; E/F-band.
Surface search: Racal Decca Type 1008 ⓫; E/F-band.
Navigation: Kelvin Hughes Type 1007; I-band.
Fire control: 2 Marconi Type 911 ⓬; I/Ku-band.
IFF: 1010/1011 or 1018/1019.
Sonars: Ferranti/Thomson Sintra Type 2050; bow-mounted; active search and attack. Thales Type 2087 (F 78, 79, 81, 82, 83, 237, 238, 239); active low-frequency (500 Hz) towed body with passive array (100 Hz).

IRON DUKE *(Scale 1 : 1,200), Ian Sturton / 0530055*

Combat data systems: Insyte Surface Ship Command System (DNA 2); Links 11 and 16. 2 Astrium SCOT 5 SATCOMs ❽. BOWMAN. XERES. RNCSS. WECDIS.
Weapon control systems: BAe GSA 8B/GPEOD TV/IR/Laser director ❾. GWS 60 (for SSM). GWS 26 (for SAM) and ASCG for 30 mm.
Helicopters: 1 Westland Lynx HMA 8 or 1 Merlin HM 1/2 (Sonar 2087 fitted ships) ⓭.

Programmes: The first of this class was ordered from Yarrows on 29 October 1984. Further batches of three ordered in September 1986, July 1988, December 1989, January 1992 and February 1996.
Modernisation: Major improvement programmes are in progress. The Command System has been upgraded to Phase 5 and DNA(2) has been installed. The Mk 8 Mod 1 gun has been installed. Type 996 radar is being replaced by BAe Insyte Type 997 ARTISAN 3D radar 2013–15. F 234 was the first to be fitted in 2013 and is to be followed by F 231. The Seawolf system is being upgraded from 2008–2014; F 81 was the first ship to be upgraded in 2009; enhancements include improved I-band radar, an additional optronic tracker to improve low level performance and improved software. In a separate contract the Mk 4 SWELL (Seawolf Enhanced Low Level) fuze is being incorporated into existing rounds and in Block 2 missiles. The Sea Wolf system is to be replaced by Sea Ceptor from 2016 in later ships of the class. This will involve removal of the Type 911 trackers. Surface Ship Torpedo Defence, a development of Sonar 2070, can be fitted. Low Frequency Active Sonar (Type 2087) has replaced Type 2031 in eight ships (F 237, F 238, F 239, F 82, F 83, F 81, F 78 and F 79). Trial launch and recovery of a Scan Eagle UAV were conducted in March 2006. The 30 mm gun has been replaced by the MSI Automated Small Calibre Gun (ASCG). A new CESM system, Hammerhead, is to be progressively fitted (2014–21).
Structure: Incorporates stealth technology to minimise acoustic, magnetic, radar and IR signatures. The design includes a 7° slope to all vertical surfaces, rounded edges, reduction of IR emissions and a hull bubble system to reduce radiated noise. The combined diesel electric and GT propulsion system provides quiet motive power during towed sonar operations. The SM1C engines although capable of 41 MW of power combined are constrained by output into the gearbox. MacTaggart Scott Helios helo landing system. PRISM enhanced helicopter landing and handling system being fitted to all except F 231.
Operational: F 78, F 83, F 229, F 234, F 237 and F 239 are based at Portsmouth and the remainder, at Devonport. *Norfolk* and *Marlborough* decommissioned in 2005 and *Grafton* in March 2006. All three ships were sold to Chile. Further decommissionings are not planned to start until 2023, thereby extending ship-life to up to 35 years. ASW trials to test and prove Sonar 2087 and the Merlin helicopter were conducted in the Indian Ocean during 2008.

SUTHERLAND *10/2013*, Michael Nitz / 1531113*

ST ALBANS *5/2012, Michael Nitz / 1486404*

IHS Jane's Fighting Ships 2014-2015 © 2014 IHS

Frigates — Shipborne aircraft < UK (NAVY)

0 + (13) TYPE 26 CLASS (FFGHM)

Displacement, tonnes: 5,400 full load
Dimensions, metres (feet): 148 × 20.0 × 7.2 (485.6 × 65.6 × 23.6)
Speed, knots: 28. **Range, n miles:** 7,000 at 15 kt
Complement: 118 + 72 spare berths

Machinery: CODLOG; 4 MTU 20V 4000 M35B diesel generators; 16,360 hp (12.2 MW); 1 Rolls-Royce MT30 gas turbine; 46,935 hp (35.0 MW); 2 motors; 2 shafts
Missiles: SLCM: To be announced.
SSM: To be announced.
SAM: Sea Ceptor (Common Anti-air Modular Missile); up to 48 cells.
Guns: 1—127 mm. 2—30 mm. 2 Raytheon 20 mm Phalanx.
Torpedoes: To be announced.
Electronic countermeasures: EW System. Type 2170 SSTDS.
Radars: Air/surface search: Type 997 (BAE Insyte ARTISAN); 3D; E/F-band.
Navigation: To be announced.
Sonars: Type 2087 low-frequency active; Type 2050.
Combat data systems: Development of BAE Insyte DNA(2).
Electro-optic systems: To be announced.

Helicopters: 1 Merlin or Wildcat.

Programmes: The Type 26 frigate programme is for a class of 13 ships to replace the current Type 23 inventory on a one-for-one basis. There are likely to be eight ASW and five general purpose variants. The Assessment Phase (Initial Gate) was launched by a design contract awarded to BAE Ship Systems on 25 March 2010. Approval for the Demonstration and Manufacture Phase is expected in late 2014. Fabrication is expected to start in 2015. Entry into service of the first of class is planned for 2022.
Structure: A modular payload area is to be incorporated forward of the hangar. The area can accommodate up to four 12 m RHIBs or up to 10 TEUs. The flight deck is to be capable of operating a Chinook-sized aircraft and the hangar is to be large enough for a Merlin. A separate hangar space is to be provided for a maritime unmanned air vehicle. There are to be two 24-cell VLS systems, capable of launching Sea Ceptor and a mixture of anti-ship or land-attack missiles.
Sales: The Type 26 is the UK variant of what is being marketed as a 'Global Combat Ship'. Brazil was invited to be a partner in the programme in September 2010 and interest has been expressed by the governments of Australia, Canada, Malaysia, New Zealand, and Turkey.

TYPE 26 (artist's impression)   8/2012, BAE Systems / 1454256

SHIPBORNE AIRCRAFT

Notes: (1) The Crowsnest programme is to replace the current organic airborne early-warning capability provided by the Sea King ASAC Mk 7. This is to be a helicopter-based solution based in the Merlin Mk 2 and is to enter service from 2018.
(2) There are two Dauphin 2 helicopters with Royal Naval markings leased by FOST for staff transfers.
(3) A contract for the provision of a contractor-owned, contractor-operated unmanned air system has been implemented. ScanEagle RM1 is to operate from some RFA vessels and Type 23 frigates. The contract is to run until 2015.

Numbers/Type: 16/26 Westland/Agusta Merlin HM Mk 2/HM Mk 1.
Operational speed: 150 kt (277 km/h).
Service ceiling: 10,000 ft (3,048 m).
Range: 550 n miles (1,019 km).
Role/Weapon systems: Primary anti-submarine and anti-surface roles with secondary troop-carrying capability. Contract for 44 signed 9 October 1991. In service with 814, 820, 824 and 829 Squadrons. A contract for a Capability Sustainment Programme (CSP), to upgrade 30 aircraft, was awarded to Lockheed Martin UK on 21 December 2005. The first modernised Mk 2 aircraft made her maiden flight in October 2010 and conducted sea trials from *Illustrious* in 2012. The conversion to Mk 2 aircraft is to be completed in 2014. CSP features include: upgrades to the sonar, radar, aircraft and tactical management systems and other key avionics systems. Sensors: GEC-Marconi Blue Kestrel 5000 radar, Thales Flash AQS 950 dipping sonar, GEC-Marconi sonobuoy acoustic processor AQS-903A, Thales Orange Reaper ESM, ECM. Link 11. Weapons: ASW; four Stingray torpedoes or Mk 11 Mod 3 depth bombs. ASV; 12.7 mm or 7.62 mm MG and OTHT for ship-launched missiles.

DAUPHIN 2   9/2008, Ian Harris / 1353541

MERLIN HM MK 2   10/2013*, Lockheed Martin / 1531119

Numbers/Type: 3 Lockheed Martin F-35B Lightning II.
Operational speed: 1,058 kt (1,960 km/h).
Service ceiling: 50,000 ft (15,240 m).
Range: 900 n miles (1,667 km).
Role/Weapon systems: The first three of four pre-production initial operational test and evaluation (IOT&E) aircraft were handed over on 19 July 2012 (BK-1) 19 October 2012 (BK-2) and 31 May 2013 (BK-3). Long-lead items for the first production aircraft, to be operated from the Queen Elizabeth-class carriers, have been ordered. Overall a total of 48 IOT&E and production aircraft is expected although further orders may follow. Sensors: Northrop Grumman AN/APG-81 AESA radar; Lockheed Martin AN/AAQ-40 EOTS; Advanced Helmet Mounted Display System; Northrop Grumman AN/AAQ-37 DAS. Weapons: AIM-132 ASRAAM; AIM-120 AMRAAM; PAVEWAY IV.

Numbers/Type: 21/6 Westland/Agusta Merlin HC Mk 3/Merlin HC Mk 3A.
Operational speed: 150 kt (277 km/h).
Service ceiling: 13,125 ft (4,000 m).
Range: 550 n miles (1,019 km).
Role/Weapon systems: Medium-lift support helicopter. Roles include cargo and troop transport and combat SAR. Military lift is 24 troops and up to 4 tonnes underslung. Currently operated by the RAF, the entire fleet is to transfer to RN service in 2016 and 25 upgraded airframes, re-designated Merlin Mk 4, are to replace the Sea King Mk 4 in support of amphibious operations. Sensors: integrated defensive aids including Raytheon laser detection, BAE Systems Sky Guardian 2000 RWR, Doppler-based MAWS, Northrop Grumman AN/AAQ-24 Nemesis DIRCM and BAE Systems North America AN/ALE-47 chaff/flare dispensers. Weapons: machine guns.

BK-1   5/2012, Lockheed Martin / 1486403

MERLIN MK 3   6/2005, IHS/Patrick Allen / 1144750

906 UK (NAVY) > Shipborne aircraft

Numbers/Type: 15 Westland Sea King HU Mk 5.
Operational speed: 112 kt *(207 km/h)*.
Service ceiling: 10,000 ft *(3,050 m)*.
Range: 400 n miles *(740 km)*.
Role/Weapon systems: Sea King HU Mk 5 is primarily a SAR platform and utilises night vision goggles and FLIR for overland and overseas operations. These aircraft are due out of service in 2016 when a UK SAR operator will take over as contracted by the Department of Transport. Sensors: Sea Searcher radar, FLIR. Weapons: Can be fitted with 7.62 mm MGs.

SEA KING HU MK 5 *8/2007, B Sullivan / 1305160*

Numbers/Type: 30 Westland Lynx HMA Mk 8.
Operational speed: 120 kt *(222 km/h)*.
Service ceiling: 10,000 ft *(3,048 m)*.
Range: 320 n miles *(593 km)*.
Role/Weapon systems: Primary role anti-surface warfare with capability to carry a variety of weapons. Embarked in a number of RN and RFA ships. HMA Mk 8 aircraft have been upgraded with Saturn-capable radios. To be replaced from 2015 by Wildcat HMA Mk 2. Sensors: Ferranti Sea Spray Radar, Orange Crop ESM, Sea Owl PID, Missile Approach Warner, IR Jammer and flare dispenser. Weapons: Up to four Sea Skua missiles, two Stingray torpedoes, Mk 11 depth charge, M3M 0.5 in CL Cabin mounted Heavy Machine Gun.

LYNX HMA Mk 8 *7/2013*, Gordon Brodie / 1531120*

Numbers/Type: 13 Westland Sea King ASAC Mk 7.
Operational speed: 90 kt *(167 km/h)*.
Service ceiling: 10,000 ft *(3,050 m)*.
Range: 400 n miles *(740 km)*.
Role/Weapon systems: Primary role Airborne Surveillance and Control (ASaC) for maritime strike, littoral manoeuvre and force protection operations. Conversion contract awarded in October 1996 to upgrade AEW Mk 2 Fleet to ASAC Mk 7. Programme completed mid-2002. Two aircraft lost in Iraq conflict (March 2003) have been replaced by conversion of two ex-Mk 6 aircraft. Aircraft to reach end of operational lives in 2016 and subsequently are to be replaced by Merlin Mk 2 in 2018. Sensors: Thales Searchwater 2000, Racal MIR-2 'Orange Crop' ESM, IFF Mk XII, Litton 100g navigation system and JTIDS/Link 16. Weapons: None.

SEA KING ASAC MK 7 *7/2008, B Sullivan / 1353539*

Numbers/Type: 37 Westland Sea King HC Mk 4.
Operational speed: 112 kt *(208 km/h)*.
Service ceiling: 10,000 ft *(3,050 m)*.
Range: 600 n miles *(1,110 km)*.
Role/Weapon systems: Medium lift Commando support helicopter capable of carrying most Commando Force equipment either internally or underslung. To be replaced by Merlin Mk 4 from 2016. Engines have been upgraded since 2003 to improve hot and high performance. Carson blades and AW 5 bladed tail rotors were added in 2007 to provide additional lift at high density altitudes. The six HAS 6 ASW aircraft converted to Mk 6C Commando configuration were withdrawn from service in 2010. Sensors: AAR-47 ESM; IR jammer, chaff and IR flares ECM. Weapons: Can fit 7.62 mm GPMG or similar.

IHS Jane's Fighting Ships 2014-2015

SEA KING HC Mk 4 *5/2009, Maritime Photographic / 1366190*

Numbers/Type: 32/6/8/3 Boeing Chinook HC Mk 2/Chinook HC Mk 2A/Chinook HC Mk 3/Chinook HC Mk 6.
Operational speed: 140 kt *(259 km/h)*.
Service ceiling: 15,000 ft *(4,572 m)*.
Range: 651 n miles *(1,207 km)*.
Role/Weapon systems: All-weather day and night heavy-lift helicopter equivalent to CH-47D and operated by RAF. Operable from surface ships (LPH/LPD/LSL). Capable of carrying 44 fully equipped troops or 54 light fighting order troops and up to 11.3 tonnes of internal or external cargo. The 38 Mk 2 variants are currently being upgraded to Mk 4 standard, and the 8 Mk 3 variants, which entered service in 2010, are being upgraded to Mk 5 standard, with the upgrade programme due to be complete in 2016. Under the Strategic Defence and Security Review of 2010, a further 14 CH-47Fs (HC Mk 6) were ordered; three were delivered in 2013 and the remaining 11 by 2015. Sensors: defensive aids suite including missile approach warning, IR jammers and chaff/flare dispensers. Weapons: machine guns.

CHINOOK *6/2004, Royal Navy / 1043609*

Numbers/Type: 66 Westland/Boeing WAH-64D AH Mk 1.
Operational speed: 150 kt *(278 km/h)*.
Service ceiling: 21,000 ft *(6,400 m)*.
Range: 260 n miles *(480 km)*.
Role/Weapon systems: All-weather attack helicopter with day and night capability. Agusta Westland selected on 13 July 1995 to build UK AH Mk 1 based on AH-64D Longbow Boeing Apache. One squadron of six earmarked for amphibious operations. Operable from surface ships (LPH/LPD/LSL). Five aircraft undertook first operational deployment to *Ocean* in 2011 to conduct sorties over Libya. Sensors: Lockheed Martin/Northrop Grumman AN/APG-78 Longbow radar, Lockheed Martin Target Acquisition and Designation Sight (TV and direct view) and Pilot's Night Vision FLIR sensor (TADS/PNVS) (being upgraded to M-TADS/MPNVS 2008–10), Selex SAS HIDAS helicopter integrated defensive aids system, including Sky Guardian 2000 RWR, Type 1223 Laser warning receiver, Thales (Vinten) Vicon 78 Srs 455 chaff/flare dispenser, BAE Systems AN/AAR-57(V) common missile warning system (CMWS) and Lockheed Martin AN/APR-48A radar frequency interferometer. Weapons: 16 Hellfire missiles or 76 CRV-7 70 mm rockets. 1 – 30 mm chain gun.

APACHE AH MK 1 *3/2004, Royal Navy / 1153998*

Numbers/Type: 28/34 AgustaWestland AW 159 Wildcat HMA. Mk 2/AW 159 Wildcat AH. Mk 1.
Operational speed: 165 kt *(306 km/h)*.
Service ceiling: 20,000 ft *(6,096 m)*.
Range: 300 n miles *(555 km)*.
Role/Weapon systems: A total of 28 naval and 34 army variants of Wildcat are to be procured. Formal entry into service of army aircraft is to begin in 2014 and of naval aircraft in 2015. Army versions are to be operated by 1 Regt AAC and 847 Naval Air Squadron. Primary role of naval variant is anti-surface warfare. Sensors: Selex Galileo Seaspray 7400E radar; ESM, ECM. Weapons: Future Anti-Surface Guided Weapons (FASGW), 2 Stingray torpedoes; 7.62 or 12.7 mm machine guns.

WILDCAT *3/2011, AgustaWestland / 1401858*

© 2014 IHS

Amphibious forces < **UK** (NAVY) 911

2 ALBION CLASS (ASSAULT SHIPS) (LPD)

Name	No	Builders	Laid down	Launched	Commissioned
ALBION	L 14	BAE Systems, Barrow	22 May 1998	9 Mar 2001	19 Jun 2003
BULWARK	L 15	BAE Systems, Barrow	27 Jan 2000	15 Nov 2001	28 Apr 2005

Displacement, tonnes: 14,834 standard; 18,797 full load
Dimensions, metres (feet): 176 × 28.9 × 7.1 (577.4 × 94.8 × 23.3)
Speed, knots: 18
Range, n miles: 8,000 at 15 kt
Complement: 325

Military lift: 305 troops; 710 troops (including overload); 67 support vehicles; 4 LCU Mk 10 or 2 LCAC (dock); 4 LCVP Mk 5 (davits)
Machinery: Diesel-electric; 2 Wärtsilä Vasa 16V 32E diesel generators; 17,000 hp(m) *(12.5 MW)*; 2 Wärtsilä Vasa 4R 32LNE diesel generators; 4,216 hp(m) *(3.1 MW)*; 2 motors; 2 shafts; LIPS props; 1 bow thruster; 1,176 hp(m) *(865 kW)*
Guns: 2 – 20 mm ❶. 2 Signaal/General Dynamics 30 mm 7-barrelled Goalkeeper; 4,200 rds/min to 1.5 km ❷. 4 M323 Mk 44 7.62 mm Miniguns. 4 – 7.62 mm MGs.
Physical countermeasures: Decoys: 8 Sea Gnat launchers ❸ and DLH offboard decoys.
Torpedo defence: Type 2170 (SLQ-25A).
Electronic countermeasures: ESM/ECM: Thales Thorn UAT Mod 1.
Radars: Air/surface search: BAE Insyte Type 996 ❺; E/F-band.
Surface search: Racal Decca 1008; E/F-band.
Navigation/aircraft control: 2 Kelvin Hughes Type 1007 ❻; I-band.
IFF: Type 1016/1017.
Combat data systems: ADAWS 20 Ed 3.1.2. Thomson-CSF/Redifon/BAeSEMA/CS comms system. Marconi Matra SCOT 5 SATCOM ❹. BOWMAN. Link 16 SAC. Link 11.
Weapon control systems: 2 Ultra UECCS EOSS optronic directors (L 15).
Helicopters: Platform for 3 Sea King Mk 4 ❼. Chinook capable.

ALBION   (Scale 1 : 1,500), Ian Sturton / 0572733

BULWARK   3/2008, Michael Nitz / 1353535

Programmes: A decision was taken in mid-1991 to replace the then existing LPDs. Project definition studies by YARD completed in February 1994. Invitations to tender for design and build of two ships were issued to VSEL and Yarrow on 18 August 1994 with an additional tender package to Vosper Thornycroft in November 1994. In March 1995 it was announced that only VSEL would bid, conforming to the rules governing non-competitive tenders. The contract to build the ships was awarded on 18 July 1996. First steel cut 17 November 1997.
Modernisation: Davits have been replaced in both ships. BOWMAN installed to support amphibious warfare staffs. Replacement of the combat data system with Surface Ship Command System (DNA 2) is expected in 2015 and MIDAS EW System in due course. BAE Insyte Type 997 ARTISAN radar is to replace Type 996 radar.
Structure: The design includes a floodable well dock, garage (with capacity for six Challenger tanks), stern gate and side ramp access. The Flight Deck has two helicopter landing spots. A large joint operations room contains substantial command and control facilities. The ships are built to military damage control standards.
Operational: In November 2011, *Bulwark* became the 'ready' ship when *Albion* was placed at extended readiness. *Albion* is to become the 'ready' ship in 2016. Based at Devonport.

BULWARK   9/2013*, Shaun Jones / 1531127

ALBION   7/2009, Maritime Photographic / 1366193

© 2014 IHS   IHS Jane's Fighting Ships 2014-2015

912   UK (NAVY) > Amphibious forces — Mine warfare forces

FAST INTERCEPT CRAFT (HSIC)

Comment: A new class of up to four 18 m high-speed interceptor craft, reported to have been built by VT Halmatic, began to enter service in 2006. They have replaced the 16 m Very Slender Vessel craft. The new craft feature a stepped planing hull and are powered by two MAN diesels driving twin Arneson drives. Top speed is likely to be in the region of 60 kt. Capable of operating in extreme climatic conditions, they are transportable by C 130.

FAST INTERCEPTION CRAFT   11/2013*, Derek Fox / 1531128

RRC AND RIB

Comment: (1) 36 RRC Mk 3: 2.6 tons and 7.4 m *(24.2 ft)* powered by single Yamaha 220 hp *(162 kW)* diesel; 36 kt fully laden (40 light); carry 8 troops. All used by the Army. Entered service 1996–98.
(2) RIBs: Halmatic Arctic 22/Pacific 22/Arctic 28/Pacific 28. Rolling contract for all four types. Capable of carrying 10 to 15 fully laden troops at speeds of 26 to 35 kt.
(3) Offshore Raiding Craft: 9 m aluminium RIB hull with removable armour plating. Diesel powered and capable of up to 40 kt. Up to 39 craft to replace the RRC Mk 3.

RRC   4/2005, Per Körnefeldt / 1153943

ORC   6/2013*, Bram Plokker / 1531125

4 GRIFFON 2400 TD(M) CLASS (HOVERCRAFT) (LCAC)

C 21–24

Displacement, tonnes: 10.6 standard
Dimensions, metres (feet): 12.6 × 6.77
Speed, knots: 40
Range, n miles: 300 at 25 kt
Complement: 2

Military lift: 16 troops plus equipment or 2.4 tonnes
Machinery: 1 Deutz diesel; 585 hp *(440 kW)*
Guns: 1 – 7.62 mm MG.
Radars: Navigation: Raytheon; I-band.

Comment: Contract was let in June 2009 for the construction and delivery of four new Griffon 2400TD to replace the Griffon 2000 fleet that were decommissioned in 2011. The new craft offer a greater payload performance and obstacle clearance than the 2000TD. Centre sections of the cabin roof can be removed in order to carry two 1 tonne NATO pallets and the craft can be transported on a standard low loader truck or in the hold of a Hercules aircraft. They are to be operated directly from the well-deck of a LPH, LPD and LSDs. All four units delivered in 2011.

GRIFFON 2400   3/2010, Griffon / 1365906

IHS Jane's Fighting Ships 2014-2015

MINE WARFARE FORCES

Notes: (1) The long-term replacement of the current mine-countermeasures (MCM) force is under consideration. The future capability is likely to be based on the requirement to conduct MCM in support of joint expeditionary operations in littoral waters. Speed of deployment is an important consideration. Future capability is likely to be delivered by a combination of portable mission packages that could be delivered rapidly into theatre by air or from future classes of surface combatants or commercial contract shipping.
(2) Replacement of the Combined Influence Sweep (CIS), removed from the Hunt class, is under development. Future options under examination include an influence sweep system deployable from an Unmanned Surface Vehicle (USV). A technology readiness programme, known as FAST (Flexible Agile Sweeping Technology), commenced trials in 2009 with a view to continued development in a Hunt-class vessel from 2014.
(3) The Remote Control Mine Disposal Systems Mk 1 (PAP Mk 3) and 2 (PAP Mk 5) have been replaced by the Atlas Electronik Seafox mine disposal system. Stowage for 21 warshots and two surveillance vehicles is provided on each MCM platform. There are 15 training variants available to MCMVs.
(4) Autonomous underwater vehicles are now deployed for survey, environmental monitoring and mine reconnaissance in both shallow and deep waters. Ten Hydroid Remus 100 UUVs were procured in 2006 for mine reconnaissance in waters less than 30 m deep. Two Hydroid Remus 600 UUVs were delivered in 2009 for hydrographic survey and environmental monitoring in waters of 30–200 m. They can be deployed from the Hunt class, from other surface vessels or from shore.

7 SANDOWN CLASS (MINEHUNTERS) (MHC/SRMH)

Name	No	Builders	Launched	Commissioned
PENZANCE	M 106	Vosper Thornycroft, Woolston	11 Mar 1997	14 May 1998
PEMBROKE	M 107	Vosper Thornycroft, Woolston	15 Dec 1997	6 Oct 1998
GRIMSBY	M 108	Vosper Thornycroft, Woolston	10 Aug 1998	25 Sep 1999
BANGOR	M 109	Vosper Thornycroft, Woolston	16 Apr 1999	26 Jul 2000
RAMSEY	M 110	Vosper Thornycroft, Woolston	25 Nov 1999	22 Jun 2001
BLYTH	M 111	Vosper Thornycroft, Woolston	4 Jul 2000	20 Jul 2001
SHOREHAM	M 112	Vosper Thornycroft, Woolston	9 Apr 2001	2 Sep 2002

Displacement, tonnes: 546 standard; 416 full load
Dimensions, metres (feet): 52.5 × 10.5 × 2.3 *(172.2 × 34.4 × 7.5)*
Speed, knots: 13; 6.5 electric
Range, n miles: 2,500 at 12 kt
Complement: 39 (5 officers) + 6 spare berths

Machinery: 2 Paxman Valenta 6RP200E/M diesels; 1,523 hp *(1.14 MW)* sustained; Voith-Schneider propulsion; 2 Schottel bow thrusters
Guns: 1 DES/MSI DS 30B 30 mm/75; 650 rds/min to 10 km *(5.4 n miles)* anti-surface; 3 km *(1.6 n miles)* anti-aircraft; weight of shell 0.36 kg. 3 Dillon Aero M 134 7.62 mm Minigun; 6 barrels; 3,000 rds/min.
Physical countermeasures: MCM: Seafox C expendable mine-disposal system.
Radars: Navigation: Kelvin Hughes Type 1007; I-band.
Sonars: Marconi Type 2093; VDS; multifunction with 5 arrays; mine search and classification.
Combat data systems: BAE Insyte NAUTIS 4.

Programmes: A class of originally 12 designed for hunting and destroying mines and for operating in deep and exposed waters. The Sandown class complements the Hunt class. On 9 January 1984 the Vosper Thornycroft design for this class was approved. First one ordered August 1985, four more on 23 July 1987. A contract was to have been placed for a second batch in 1990 but this was deferred twice, until an order for seven more (M 106–112) was placed in July 1994.
Modernisation: Seafox C has replaced RCMDS 2. Drumgrange Precise Fixing System fitted in 2004. NAUTIS M combat system replaced by NAUTIS 4. M 134 Minigun CIWS fitted in 2006.
Structure: GRP hull. Combines vectored thrust units with bow thrusters. The sonar is deployed from a sea chest in the hull. Voith-Schneider props and a transportable manned decompression chamber.
Operational: All based at Faslane from mid-2006. *Shoreham* and *Ramsey* forward deployed in the Arabian Gulf. *Walney* decommissioned in 2010. *Cromer* became BRNC Dartmouth training ship and renamed *Hindostan* in 2001.
Sales: Three to Saudi Arabia. *Bridport*, *Sandown* and *Inverness* to Estonia 2007–08.

PEMBROKE   3/2010, Michael Nitz / 1406311

BLYTH   11/2012*, Selim San / 1531129

© 2014 IHS

Mine warfare forces — Survey ships < UK (NAVY) 913

8 HUNT CLASS (MINEHUNTERS—COASTAL) (MHC)

Name	No	Builders	Launched	Commissioned
LEDBURY	M 30	Vosper Thornycroft, Woolston	5 Dec 1979	11 Jun 1981
CATTISTOCK	M 31	Vosper Thornycroft, Woolston	22 Jan 1981	16 Jun 1982
BROCKLESBY	M 33	Vosper Thornycroft, Woolston	12 Jan 1982	3 Feb 1983
MIDDLETON	M 34	Yarrow Shipbuilders, Glasgow	27 Apr 1983	15 Aug 1984
CHIDDINGFOLD	M 37	Vosper Thornycroft, Woolston	6 Oct 1983	10 Aug 1984
ATHERSTONE	M 38	Vosper Thornycroft, Woolston	1 Mar 1986	30 Jan 1987
HURWORTH	M 39	Vosper Thornycroft, Woolston	25 Sep 1984	2 Jul 1985
QUORN	M 41	Vosper Thornycroft, Woolston	23 Jan 1988	21 Apr 1989

Displacement, tonnes: 643 standard; 752 full load
Dimensions, metres (feet): 60 oa; 57 wl × 10 × 2.9 hull; 3.4 screws (196.9; 187 × 32.8 × 9.5; 11.2)
Speed, knots: 15; 8 hydraulic
Range, n miles: 1,500 at 12 kt
Complement: 43 (5 officers)

Machinery: 2 Ruston-Paxman 9-59K Deltic diesels; 1,900 hp (1.42 MW); 1 Deltic Type 9-55B diesel for pulse generator and auxiliary drive; 780 hp (582 kW); 2 shafts; bow thruster
Guns: 1 DES/MSI DS 30B 30 mm/75; 650 rds/min to 10 km (5.4 n miles) anti-surface; 3 km (1.6 n miles) anti-aircraft; weight of shell 0.36 kg. 3 Dillon Aero M 134 7.62 mm Minigun; 6 barrels; 3,000 rds/min

Physical countermeasures: MCM: Seafox C expendable mine-disposal system.
Radars: Navigation: Kelvin Hughes Type 1007; I-band.
Sonars: Thales 2193; hull-mounted; minehunting; 100/300 kHz. Hull-mounted; active; high frequency.
Combat data systems: BAE Insyte NAUTIS 3.

Programmes: A class of originally 13 MCM Vessels with hunting capabilities.
Modernisation: Seafox C has replaced RCMDS 1. 30 mm gun has replaced the Bofors 40 mm. Drumgrange Precise Fixing System fitted 2003–04. A new minehunting sonar (Sonar 2193) and NAUTIS 3 command system fitted in all eight ships 2008–10. M 134 Minigun CIWS fitted in 2007. The influence sweeping system has been removed. A contract to replace the propulsion systems, including new engines, gearboxes, propellers and machinery control systems was let in November 2010. *Chiddingfold* was the first to be modernised in 2011 and the project is to be completed in 2016.
Structure: GRP hull.
Operational: *Brecon* was decommissioned in 2005. All eight ships based at Portsmouth. *Atherstone* and *Quorn* forward deployed to the Arabian Gulf.
Sales: *Bicester* and *Berkeley* to Greece in July 2000 and February 2001 respectively. *Cottesmore* and *Dulverton*, decommissioned in 2005, transferred to Lithuania in 2011–12.

BROCKLESBURY 1/2014*, Derek Fox / 1531130

SURVEY SHIPS

4 NESBITT CLASS (YGS)

NESBITT	9423	COOK	9425
PAT BARTON	9424	OWEN	9426

Displacement, tonnes: 11 full load
Dimensions, metres (feet): 10.6 × 2.9 × 1 (34.8 × 9.5 × 3.3)
Speed, knots: 15
Range, n miles: 300 at 8 kt
Complement: 2 + 10 spare berths
Machinery: 2 Perkins Sabre 185C diesels; 430 hp(m) (316 kW) sustained; 2 shafts

Comment: *Nesbitt*, *Pat Barton*, *Cook* and *Owen* delivered by Halmatic, Southampton by September 1996. *Pat Barton*, *Cook* and *Owen* are used as training vessels by FOSTHM, Devonport, *Nesbitt* is operated by FHMU. Fitted with C-Nav WADGPS positioning system, EM 3000 MBES (*Pathfinder*, *Pat Barton*), EM 3002 MBES (*Nesbitt*, *Cook*, *Owen*), EA 400 SBES and 2094 SSS. *Pioneer* was lost in an accident at Gibraltar in early 2010 and *Pathfinder* was withdrawn from service in late 2011.

1 SCOTT CLASS (AGSH)

Name	No	Builders	Launched	Commissioned
SCOTT	H 131	Appledore Shipbuilders, Bideford	13 Oct 1996	30 Jun 1997

Displacement, tonnes: 13,717 full load
Dimensions, metres (feet): 131.1 × 21.5 × 9 (430.1 × 70.5 × 29.5)
Speed, knots: 17.5
Complement: 71 (12 officers)
Machinery: 2 Krupp MaK 9M32 9-cyl diesels; 10,800 hp(m) (7.94 MW); 1 shaft; LIPS cp prop; retractable bow thruster
Radars: Navigation: Sperry Marine Bridgemaster; E/F/I-bands.
Helicopters: Platform for 1 light.

Comment: Designed by BAeSEMA/YARD and ordered 20 January 1995 to replace *Hecla*. Ice-strengthened bow. Foredeck strengthened for helicopter operations. The centre of the OSV surveying operations consists of an integrated navigation suite, the Sonar Array Sounding System (SASS) and data processing equipment. Additional sensors include a Sonar 2090 ocean environment sensor. The SASS IV multibeam depth-sounder is capable of gathering 361 individual depth samples concurrently over a 120° swathe, producing a three-dimensional image of the seabed. 8,000 tons of seawater ballast can be used to achieve a sonar trim. The ship operates a watch rotation manning scheme giving an at sea availability in excess of 307 days per year. Based at Devonport.

NESBITT 11/1998, John Brodie / 0053244

SCOTT 8/2010, Derek Fox / 1406217

© 2014 IHS IHS Jane's Fighting Ships 2014-2015

914 UK (NAVY) > Survey ships — Rescue vehicles

1 GLEANER CLASS (YGS)

Name	No	Builders	Launched	Commissioned
GLEANER	H 86	Emsworth Shipyard	18 Oct 1983	5 Dec 1983

Displacement, tonnes: 26 full load
Dimensions, metres (feet): 15.6 × 4.7 × 1.6 *(51.2 × 15.4 × 5.2)*
Speed, knots: 19.5
Complement: 9 (2 officers)
Machinery: 2 Volvo Penta TMD 112; 524 hp(m) *(391 kW)*; 2 shafts
Radars: Navigation: Raymarine Pathfinder; I-band.

Comment: This craft is prefixed HMSML-HM Survey Motor Launch. Primary task is conduct of high-resolution survey operations around UK ports and harbours. Integrated survey suite includes C-Nav WADGPS for positioning, POSMV 320 for inertial navigation and Simrad EM 2040 MBES, EA 400 SBES, and a 2094 SSS towed sensor.

GLEANER 11/2013*, Derek Fox / 1531132

4 SURVEY MOTOR BOATS (YGS)

JAMES CAIRD IV	SAPPHIRE
SPITFIRE	SHACKLETON

Displacement, tonnes: 9.27 full load
Dimensions, metres (feet): 10.5 × 3.38 × 0.7 *(34.4 × 11.1 × 2.3)*
Speed, knots: 22
Range, n miles: 96 at 6 kt
Complement: 10
Machinery: 2 Yanmar 6LYA-STP diesels; 630 hp *(470 kW)*; 2 Hamilton HJ274 water jets
Radars: Navigation: Furuno; I-band.

Comment: Constructed by Mustang Marine, *James Caird IV* delivered in November 2011 for operations from Antarctic patrol ship *Protector*. The vessel is modified to operate in ice and to withstand extreme weather conditions. The craft is also used for personnel transfers and as a training vessel. *Spitfire* embarked on *Enterprise* in August 2012, *Sapphire* on *Echo* in September 2012, and *Shackleton* delivered to Fleet Hydrographic Unit in April 2013. The hull is of aluminium-welded construction. Survey equipment includes a Kongsberg Marine echosounder, EM3002 MBES (*James Caird IV*), EM 2040 MBES (*Spitfire* and *Sapphire*), EA 400 SBES and a sidescan sonar system. These have all been mounted internally into the hull.

SPITFIRE 8/2012, Mustang Marine / 1486417

2 ECHO CLASS (AGSH)

Name	No	Builders	Launched	Commissioned
ECHO	H 87	Appledore Shipbuilders, Bideford	4 Mar 2002	7 Mar 2003
ENTERPRISE	H 88	Appledore Shipbuilders, Bideford	2 May 2002	17 Oct 2003

Displacement, tonnes: 3,526 full load
Dimensions, metres (feet): 90 × 16.8 × 5.5 *(295.3 × 55.1 × 18.0)*
Speed, knots: 15. **Range, n miles:** 9,000 at 12 kt
Complement: 72

Machinery: Diesel electric; 4.8 MW; 2 azimuth thrusters; 1 bow thruster
Guns: 2—20 mm. 2 Mk 44 7.62 mm. 5—7.62 mm MGs.
Radars: Navigation: E/F/I-bands.
Helicopters: Platform for VERTREP only.

Comment: The order for two multirole Hydrographic and Oceanographic Survey Vessels was placed with the prime contractor, Vosper Thornycroft Ltd, on 19 June 2000. The ships were built by Appledore Shipbuilders in Devon. The contract covers the design, build and through-life support of the ships over their 25 year service. In addition to specialist surveying tasks, the ships' operational roles include Rapid Environmental Assessment, military data gathering and Mine Countermeasures Tasking Support. The survey suite consists of EM 710 MBES, EA 600 single beam echo-sounder, towed side scan sonar, acoustic Doppler current profilers, towed undulating sensors, sub-bottom profilers and deployable tide and current meters. Survey Motor Boat *Spitfire* is embarked in *Enterprise* and *Sapphire* in *Echo*. Both ships operate a watch rotation manning scheme giving an at sea availability of 320 days per year. Both based at Devonport.

ECHO 1/2012, Shaun Jones / 1454818

RESCUE VEHICLES

1 NATO SUBMARINE RESCUE SYSTEM (DSRV)

Displacement, tonnes: 30 full load
Dimensions, metres (feet): 8.7 × 3.4 × 3.5 *(28.5 × 11.2 × 11.5)*
Speed, knots: 3.8
Complement: 3
Machinery: 2 external ZEBRA rechargeable sodium nickel chloride battery pods

Comment: The three participant nations for NSRS are UK, Norway and France with the UK Equipment and Support Organisation acting as contracting authority and host nation for project management and in-service phases. Following Invitations to Tender, a 10-year contract for the design and manufacture phase was awarded in June 2004 to a team led by Rolls-Royce Naval Marine. The core of the service is a new free-swimming Submarine Rescue Vehicle (SRV), built by Perry Slingsby, capable of accommodating 15 rescued personnel from a submarine at depths down to 600 m and at an angle of up to 60°. The SRV may be launched and recovered from suitable commercial or from military 'motherships', primarily offshore support vessels, capable of fitting the NSRS Portable Launch-And-Recovery (PLARS) installation. Battery endurance allows up to five rescue cycles without recharge but trickle charging during rescuee transfer will enable almost continuous operation. The NSRS includes an unmanned Intervention Remotely-Operated Vehicle (IROV), the Perry Slingsby Super Spartan, which can operate down to depths of 1,000 m and may be used to locate a stricken submarine, to conduct survey and rescue preparations and to resupply Emergency Life Support Stores in pressure tight pods whilst awaiting rescue. Other assets include a Transfer under Pressure system with decompression chambers for up to 72 personnel; medical treatment facilities and support equipment. The system entered service in November 2008 and is expected to remain in service unitl 2033. It is permanently maintained at HM Naval Base Clyde, Scotland, at 12 hours notice to move worldwide.

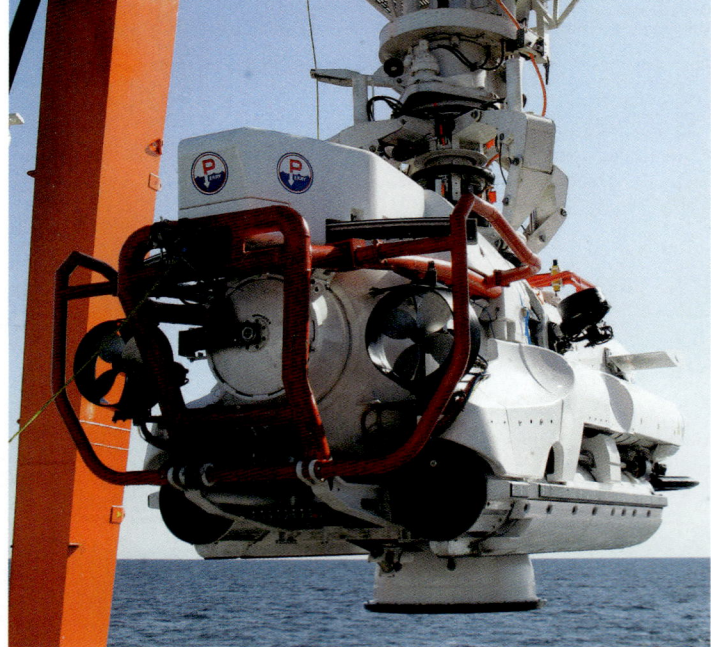

NSRS SRV 6/2008, Richard Scott / 1353524

IHS Jane's Fighting Ships 2014-2015

AUXILIARIES

General

The Royal Fleet Auxiliary Service is a civilian-manned fleet under the command of the Commander in Chief Fleet from 1 April 1993. Its main task is to supply warships at sea with fuel, food, stores and ammunition. It also provides aviation platforms, amphibious support for the Navy and Marines and sea transport for Army units. All ships take part in operational sea training. An order in council on 30 November 1989 changed the status of the RFA service to government-owned vessels on non-commercial service.

New Construction

(1) The Maritime Role 3 Medical Capability was formerly known as the Joint Casualty Treatment Ship (JCTS). The requirement for such a vessel was identified in the 1998 Strategic Defence Review. The aviation support ship *Argus* was configured as a PCRF during the 1990–91 Gulf War and the 2003 Iraq War. The contract for the Assessment Phase was awarded to BMT Ltd in February 2002 since when the key drivers have been identified as a need for eight operating tables and a 150–200 bed hospital. A two-spot flight deck and the ability to embark personnel by sea or land are also required. Development of the Systems Requirement Document (SRD) by Atkins Aviation and Defence Systems has been completed and potential solutions range from a bespoke vessel to conversion/modification of an existing military or merchant hull. The ship will be manned by RFA personnel but is unlikely to enter service before 2020. The requirement for a second ship at 12 months notice is to be met by chartering a commercial hull.

(2) The future afloat support capability is being taken forward through the Military Afloat Reach and Sustainability programme (MARS), the next phase of which is the Future Solid Support programme to replace *Fort Austin*, *Fort Rosalie*, and *Fort Victoria* from the early 2020s. Replacement of the forward repair ship *Diligence* is also under consideration.

Personnel

1 January 2014: 1,945 (663 officers)

2 WAVE CLASS (LARGE FLEET TANKERS) (AORH)

Name	No	Builders	Laid down	Launched	Commissioned
WAVE KNIGHT	A 389	BAE Systems, Barrow	22 Oct 1998	29 Sep 2000	8 Apr 2003
WAVE RULER	A 390	BAE Systems, Govan	10 Feb 2000	9 Feb 2001	27 Apr 2003

Displacement, tonnes: 32,005 full load
Measurement, tonnes: 23,294 gt
Dimensions, metres (feet): 196.3 × 30.2 × 10.0 *(644 × 99.1 × 32.8)*
Speed, knots: 18. **Range, n miles:** 10,000 at 15 kt
Complement: 80 (22 air crew)

Cargo capacity: 16,000 m³ total liquids including 3,000 m³ aviation fuel; 8–20 ft refrigerated containers plus 500 m³ solids. Capable of carrying 168 m³ of bulk Lub Oil and 47.1 m³ of drummed Lub Oil.

Machinery: Diesel-electric: 4 Wärtsilä 12V 32E/GECLM diesel generators; 25,514 hp(m) *(18.76 MW)*; 2 GECLM motors; 19,040 hp(m) *(14 MW)*; 1 shaft; Kamewa bow and stern thrusters
Guns: 2 Raytheon Vulcan Phalanx CIWS. 2—30 mm. 5—7.62 mm MGs. 2 Mk 44 7.62 mm Miniguns.
Physical countermeasures: Outfit DLJ(2).
Radars: Navigation: KH 1007; E/F/I-band.
IFF: Type 1018/9.

Helicopters: 1 Merlin HM.Mk 1.

Comment: Feasibility studies by BAeSEMA/YARD completed in early 1995. Draft invitation to tender issued 10 October 1995 followed by full tender on 26 June 1996. Contracts to build placed with VSEL (BAE Systems) on 12 March 1997. One spot flight deck with full hangar facilities for one Merlin. Enclosed bridge including bridge wings. Double hull construction. Inclined RAS gear with three rigs and two cranes.

WAVE RULER *11/2013*, Shaun Jones / 1531133*

1 APPLELEAF CLASS (SUPPORT TANKERS) (AOT)

Name	No	Builders	Launched	Commissioned
ORANGELEAF (ex-*Balder London*; ex-*Hudson Progress*)	A 110	Cammell Laird, Birkenhead	1975	2 May 1984

Displacement, tonnes: 38,353 full load
Measurement, tonnes: 18,854 gt
Dimensions, metres (feet): 170.7 × 25.9 × 11 *(560 × 85.0 × 36.1)*
Speed, knots: 15.5
Complement: 56 (19 officers)

Cargo capacity: 22,000 m³ dieso; 3,800 m³ Avcat
Machinery: 2 Pielstick 14 PC2.2 V 400 diesels; 14,000 hp(m) *(10.29 MW)* sustained; 1 shaft
Guns: 2 BMARC GAM-BO1 20 mm. 6—7.62 mm MGs. 2 Mk 44 7.62 mm Miniguns.
Radars: Navigation: Racal Decca 1226 and 1229; I-band.

Comment: Part of a four-ship order cancelled by Hudson Fuel and Shipping Co, but completed by the shipbuilders, being the only mercantile order then in hand. Major refit September 1985 to fit full RAS capability and extra accommodation. Single-hull construction. To be replaced from 2015. *Bayleaf* withdrawn from service in 2011.

ORANGELEAF *3/2009, Rene Baak / 1366198*

916 UK (NAVY) > Auxiliaries

2 ROVER CLASS (SMALL FLEET TANKERS) (AORLH)

Name	No	Builders	Launched	Commissioned
GOLD ROVER	A 271	Swan Hunter Shipbuilders, Wallsend-on-Tyne	7 Mar 1973	22 Mar 1974
BLACK ROVER	A 273	Swan Hunter Shipbuilders, Wallsend-on-Tyne	30 Oct 1973	23 Aug 1974

Displacement, tonnes: 4,775 standard; 11,707 full load
Measurement, tonnes: 7,894 gt
Dimensions, metres (feet): 140.6 × 19.3 × 7.3 *(461.3 × 63.3 × 24.0)*
Speed, knots: 19
Range, n miles: 15,000 at 15 kt
Complement: 55 (18 officers)

Cargo capacity: 3,000 m³ fuel

Machinery: 2 Pielstick 16PC2 2V 400 diesels; 15,360 hp(m) *(11.46 MW)*; 1 shaft; Kamewa cp prop; bow thruster
Guns: 2 BMARC GAM-BO1 20 mm. 4—7.62 mm MGs. 2 Mk 44 7.62 mm Miniguns.
Radars: Navigation: Racal Decca 52690 ARPA; Racal Decca 1690; I-band.

Helicopters: Platform for Westland Sea King HAS. Mk 5 or HC.Mk 4.

Comment: Single-hull construction. Small fleet tankers designed to replenish HM ships at sea with fuel, fresh water, limited dry cargo and refrigerated stores under all conditions while under way. No hangar but helicopter landing platform is served by a stores lift, to enable stores to be transferred at sea by 'vertical lift'. Capable of HIFR. Siting of SATCOM aerial varies. *Green Rover* sold in September 1992 to Indonesia. *Blue Rover* to Portugal in March 1993. *Grey Rover* decommissioned in 2006. To be replaced 2016–17.

GOLD ROVER 5/2008, B Sullivan / 1353520

1 FORT VICTORIA CLASS (FLEET REPLENISHMENT SHIPS) (AORH)

Name	No	Builders	Laid down	Launched	Commissioned
FORT VICTORIA	A 387	Harland & Wolff, Cammell Laird	4 Apr 1988	12 Jun 1990	24 Jun 1994

Displacement, tonnes: 37,167 full load
Measurement, tonnes: 28,821 gt
Dimensions, metres (feet): 204 oa; 185 wl × 30.4 × 9.8 *(669.3; 607 × 99.7 × 32.2)*
Speed, knots: 20
Complement: 264 (28 officers; 154 air crew) + 24 personnel

Cargo capacity: 12,505 m³ liquids; 3,000 m³ solids
Machinery: 2 Crossley SEMT-Pielstick 16 PC2.6 V 400 diesels; 23,904 hp(m) *(17.57 MW)* sustained; 2 shafts

Guns: 2—20 mm GAM-BO. 2 Raytheon Vulcan Phalanx 20 mm Mk 15. 2 Mk 44 7.62 mm Miniguns.
Physical countermeasures: Decoys: DLH.
Radars: Navigation: Kelvin Hughes Type 1007; I-band.
Aircraft control: Kelvin Hughes NUCLEUS; E/F-band.
Combat data systems: SCOT 5 SATCOM. NAMESIS LITE.

Helicopters: 5 Westland Sea King/Merlin helicopters.

Programmes: The requirement for this ship is to provide fuel and stores support to the Fleet at sea. Ordered 23 April 1986.

Structure: Single-hull construction. Four dual-purpose abeam replenishment rigs for simultaneous transfer of liquids and solids. Stern refuelling. Repair facilities for Merlin helicopters. The plan to fit Seawolf GWS 26 VLS was abandoned in favour of Phalanx CIWS fitted in 1998/99.
Operational: Two helicopter spots. *Fort George* withdrawn from service in 2011.

FORT VICTORIA 10/2006, B Sullivan / 1167568

IHS Jane's Fighting Ships 2014-2015 © 2014 IHS

2 FORT GRANGE CLASS (FLEET REPLENISHMENT SHIPS) (AFSH)

Name	No	Builders	Laid down	Launched	Commissioned
FORT ROSALIE (ex-Fort Grange)	A 385	Scott-Lithgow, Greenock	9 Nov 1973	9 Dec 1976	6 Apr 1978
FORT AUSTIN	A 386	Scott-Lithgow, Greenock	9 Dec 1975	9 Mar 1978	11 May 1979

Displacement, tonnes: 23,759 full load
Measurement, tonnes: 20,043 gt
Dimensions, metres (feet): 185.1 × 24.1 × 8.6 (607.3 × 79.1 × 28.2)
Speed, knots: 22
Range, n miles: 10,000 at 20 kt
Complement: 159 (31 officers; 45 air crew) + 36 personnel

Cargo capacity: 3,500 tons armament, naval and victualling stores in 4 holds of 12,800 m³

Machinery: 1 Sulzer RND90 diesel; 23,200 hp(m) (17.05 MW); 1 shaft; 2 bow thrusters
Guns: 2 BMARC GAM-BO1 20 mm. 4—7.62 mm MGs. 2 Mk 44 7.62 mm Miniguns.
Radars: Navigation: Kelvin Hughes Type 1007; I-band.

Helicopters: 4 Westland Sea King.

Comment: Ordered in November 1971. Fitted for SCOT SATCOMs but carry Marisat. Normally only one helicopter is embarked. ASW stores for helicopters carried on board. Emergency flight deck on the hangar roof. There are six cranes, three of 10 tons lift and three of 5 tons. Fort Austin completed a major refit in 2012 and returned to service in 2013. Decommissioning dates: Fort Rosalie 2022, Fort Austin 2021.

FORT AUSTIN 11/2013*, Shaun Jones / 1531134

6 TRANSPORT SHIPS (AKR)

Name	Builders	Commissioned
HURST POINT	Flensburger Schiffbau	16 Aug 2002
HARTLAND POINT	Harland & Wolff, Belfast	11 Dec 2002
EDDYSTONE	Flensburger Schiffbau	28 Nov 2002
ANVIL POINT	Harland & Wolff, Belfast	17 Jan 2003
LONGSTONE	Flensburger Schiffbau	24 Apr 2003
BEACHY HEAD	Flensburger Schiffbau	17 Apr 2003

Displacement, tonnes: 20,321 full load
Measurement, tonnes: 14,428 dwt
Dimensions, metres (feet): 193 × 26.0 × 7.4 (633.2 × 85.3 × 24.3)
Speed, knots: 21.5. **Range, n miles:** 9,200 at 21.5 kt
Complement: 18
Military lift: 2,650 linear metres of space for vehicles equating to 130 armoured vehicles plus 60 trucks and ammunition

Machinery: 2 MaK 9M43 diesels; 21,700 hp (16.2 MW); 2 cp props; bow thruster.
Radars: Navigation: I-band.

Comment: On 26 October 2000, it was announced that AWSR Ltd had been awarded the contract to provide a strategic sealift service in support of the Joint Rapid Reaction Force (JRRF) until late 2024. A key feature of the contract is that four Ro-Ro are in constant MoD use while the remaining ships are available for use by AWSR for the generation of commercial revenue. These can be called upon to support major operations and exercises. It was reported in early 2013 that two ships, possibly Longstone and Beachy Head, may be removed from the sealift force as they have seen little MoD service.

HARTLAND POINT 4/2012, Maritime Photographic / 1486412

1 STENA TYPE (FORWARD REPAIR SHIP) (ARH)

Name	No	Builders	Commissioned	Recommissioned
DILIGENCE (ex-*Stena Inspector*)	A 132	Oresundsvarvet AB, Landskrona	1981	12 Mar 1984

Displacement, tonnes: 10,938 full load
Measurement, tonnes: 8,177 gt
Dimensions, metres (feet): 112 × 20.5 × 6.8 (367.5 × 67.3 × 22.3)
Flight deck, metres: 25.4 × 25.4 (83.3 × 83.3)
Speed, knots: 12
Range, n miles: 5,000 at 12 kt
Complement: 38 (15 officers) + 202 spare berths

Cargo capacity: Long-jib crane SWL 5 tons; maximum lift, 40 tons
Machinery: Diesel-electric; 5 V16 Nohab-Polar diesel generators; 16,896 hp (12.6 MW); 4 NEBB motors; 6,000 hp(m) (4.41 MW); 1 shaft; Kamewa cp prop; 2 Kamewa bow tunnel thrusters; 3,000 hp(m) (2.2 MW); 2 azimuth thrusters (aft); 3,000 hp(m) (2.2 MW)
Guns: 2 BMARC GAM-BO1 20 mm. 4—7.62 mm MGs. 2 Mk 44 7.62 mm Miniguns.

Helicopters: Facilities for up to Boeing Chinook HC. Mk 1 (medium lift) size.

Programmes: *Stena Inspector* was designed originally as a Multipurpose Support Vessel for North Sea oil operations, and completed in January 1981. Chartered on 25 May 1982 for use as a fleet repair ship during the Falklands War. Purchased from Stena (UK) Line in October 1983, and converted for use as Forward Repair Ship in the South Atlantic (Falkland Islands). Conversion by Clyde Dock Engineering Ltd, Govan from 12 November 1983 to 29 February 1984.
Modernisation: Following items added during conversion: large workshop for hull and machinery repairs (in well-deck); accommodation for naval Junior Rates (new accommodation block); accommodation for crew of conventional submarine (in place of Saturation Diving System); extensive craneage facilities; overside supply of electrical power, water, fuel, steam, air, to ships alongside; large naval store (in place of cement tanks); armament and magazines; Naval Communications System; decompression chamber. Major refit conducted in Singapore 2005. Work included replacement/update of dynamic positioning system.
Structure: Four 5 ton anchors for four-point mooring system. Strengthened for operations in ice (Ice Class 1A). Kongsberg Albatross Positioning System has been retained in full. Uses bow and stern thrusters and main propeller to maintain a selected position to within a few metres, up to Beaufort Force 9. Controlled by Kongsberg KS 500 computers.
Operational: Principal role is operational maintenance and repair with Engineering Support Naval Party embarked. Has also been used as MCMV support ship in the Gulf and is capable of SSN support. To remain in service until 2020.

DILIGENCE   10/2011, Shaun Jones / 1454827

1 PRIMARY CASUALTY RECEIVING SHIP (APCR)

Name	No	Builders	Commissioned	Recommissioned
ARGUS (ex-*Contender Bezant*)	A 135	CNR Breda, Venice	1981	1 Jun 1988

Displacement, tonnes: 18,573 standard; 26,845 full load
Measurement, tonnes: 26,845 gt
Dimensions, metres (feet): 175.1 × 30.4 × 8.2 (574.5 × 99.7 × 26.9)
Speed, knots: 18. **Range, n miles:** 15,000 at 19 kt
Complement: 273 (22 officers; 137 air crew) + 54 personnel

Military lift: 930 tons dieso; 1,100 tons aviation fuel; 138 4 ton vehicles in lieu of aircraft
Machinery: 2 Lindholmen SEMT-Pielstick 18 PC2.5 V 400 diesels; 23,400 hp(m) (17.2 MW) sustained; 2 shafts
Guns: 2—20 mm GAM-BO. 6—7.62 mm MGs. 2 Mk 44 7.62 mm Miniguns.
Physical countermeasures: Decoys: DLJ.

Electronic countermeasures: ESM: THORN EMI Guardian; radar warning.
Radars: Surface search: Plessey Type 994; E/F-band.
Navigation: Kelvin Hughes Type 1007; I-band.
Combat data systems: Racal CANE DEB-1 data automation. Inmarsat SATCOM communications. Marisat.

Fixed-wing aircraft: Provision to transport 12 Harriers.
Helicopters: 6 Westland Sea King HAS.Mk 5/6 or similar.

Programmes: Ro-Ro container ship whose conversion to aviation training ship was begun by Harland and Wolff in March 1984 and completed on 3 March 1988. Work to convert her to PCRF role completed in 2001 and upgraded in 2007.
Structure: Uses former Ro-Ro deck as hangar with four sliding WT doors able to operate at a speed of 10 m/min. Can replenish other ships underway. One lift abaft funnel. Domestic facilities are very limited if she is to be used in the Command support role. Flight deck is 372.4 ft (113.5 m) long and has a 5 ft thick concrete layer on its lower side. First RFA to be fitted with a command system. PCRF conversion work included modification of three decks into permanent 100-bed hospital with three operating theatres. The forward lift is to be adapted for evacuation and only the aft lift remains available for aircraft.
Operational: Based at Falmouth. Operational life extended to 2020. Can conduct subsidiary role as aviation training ship.

ARGUS   6/2012, M Declerck / 1486411

Auxiliaries < **UK** (NAVY) 919

0 + 4 TIDE CLASS (FLEET TANKERS) (AORH)

Name	No	Builders	Laid down	Launched	Commissioned
TIDESPRING	–	Daewoo Shipbuilding	2013	2015	Oct 2015
TIDERACE	–	Daewoo Shipbuilding	2014	2015	Apr 2016
TIDESURGE	–	Daewoo Shipbuilding	2014	2016	Oct 2016
TIDEFORCE	–	Daewoo Shipbuilding	2015	2016	Apr 2017

Displacement, tonnes: 37,000 full load
Measurement, tonnes: 22,000 gt
Dimensions, metres (feet): 200.9 × 28.6 × 10.0
(659.1 × 93.8 × 32.8)
Speed, knots: 14.5
Complement: 63 + 26 spare berths

Machinery: 2 Wärtsilä 6L46F diesels; 20,394 hp (15.0 MW); 2 shafts
Guns: To be announced.
Physical countermeasures: To be announced.
Electronic countermeasures: To be announced.
Radars: Surface search/navigation: Kelvin Hughes SharpEye; E/F/I-bands.

Helicopters: 1 Merlin or Wildcat.

Comment: It was announced on 22 February 2012 that Daewoo Shipbuilding and Marine Engineering (DSME) had been selected as preferred bidder to construct four new double-hulled logistic support vessels. They are to be operated by the Royal Fleet Auxiliary (RFA). The ships are based on a BMT AEGIR design and are to be built in Korea. The ships are to be equipped with three abeam RAS stations for diesel oil, aviation fuel and fresh water. There is also a solid RAS reception point and a stern refuelling reel. There is storage on deck for lubricating oil and for eight 20 ft containers. Aviation facilities include a flight deck for a medium-sized helicopter and a maintenance hangar. The first vessel is to enter operational service in 2016 and the others are to follow at six-month intervals.

FUTURE LOGISTICS VESSEL

2/2012, BMT / 1454889

3 BAY CLASS LANDING SHIPS DOCK (AUXILIARY) (LSD)

Name	No	Builders	Laid down	Launched	Commissioned
LYME BAY	L 3007	Swan Hunter Shipbuilders, Wallsend-on-Tyne	22 Nov 2002	3 Sep 2005	26 Nov 2007
MOUNTS BAY	L 3008	BAE Systems Govan, Govan	25 Aug 2002	9 Apr 2004	13 Jul 2006
CARDIGAN BAY	L 3009	BAE Systems Govan, Govan	13 Oct 2003	8 Apr 2005	18 Dec 2006

Displacement, tonnes: 16,419 full load
Dimensions, metres (feet): 176.6 × 26.4 × 5.8
(579.4 × 86.6 × 19.0)
Speed, knots: 18. **Range, n miles:** 10,000 at 15 kt
Complement: 477

Military lift: 1,130 linear metres of space for vehicles equating to 24 Challenger MBTs or 150 light trucks plus 200 tons ammunition or 24 — 24 TEU containers
Machinery: Diesel-electric; 2 Wärtsilä 8L26 generators; 6,000 hp (4.5 MW); 2 Wärtsilä 12V26 generators; 9,000 hp (6.7 MW); 2 steerable propulsors; bow thruster

Guns: 2—30 mm. 2 Mk 44 7.62 Miniguns. 6—7.62 mm MGs.
Radars: Navigation: E/F/I-bands.
Combat data systems: Link 16.

Helicopters: Platform capable of operating Chinook.

Programmes: Two ships ordered from Swan Hunter on 18 December 2000. Contract for two further ships of the class, placed on 19 November 2001 with BAE Systems (Marine) at Govan. The programme was badly affected by escalating costs and delays and the whole project was passed to BAE Systems on 13 July 2006.

Modernisation: Fitted with Link 16 2013–14.
Structure: Based on the Royal Schelde Enforcer design, the LSD(A)s are designed to transport troops, vehicles, ammunition and stores in support of amphibious operations. Offload is enabled by a flight deck capable of operating heavy helicopters, an amphibious dock capable of operating one LCU Mk 10 and mexeflotes which can be hung on the ships' sides. There is no beaching capability. Davit-launched infantry landing craft (LCVPs) are not fitted but two can be carried in the dock or on deck. There are two 30 t cranes.
Operational: *Largs Bay* withdrawn from service 2011 and sold to Australia.

MOUNTS BAY

3/2009, Shaun Jones / 1366199

MOUNTS BAY

10/2013*, Shaun Jones / 1531135

MARINE SERVICES AND GOVERNMENT AGENCY SERVICES

Notes: (1) A contract was awarded to SERCo. Denholm Marine Services Ltd in January 2008 for the provision of support to naval bases, mooring maintenance and support to military training and exercises. The contract expires in 2022.
(2) A small workboat, *SD Assist*, is based at Devonport.

1 SUPPORT VESSEL (AG)

Name	Builders	Commissioned
SD NORTHERN RIVER	Myklebust Verft	1998

Measurement, tonnes: 3,612 gt
Dimensions, metres (feet): 92.8 × 18.8 × 6.34 (304.5 × 61.7 × 20.8)
Speed, knots: 13
Machinery: 2 Bergen BRM-8 diesels; 9,598 hp *(7.06 MW)*; 2 shafts; cp props; 2 bow thrusters; 671 kW; 1 stern thruster; 895 kW; 1 retractable thruster; 895 kW
Radars: Surface search/navigation: E/F-band.
Navigation: I-band.

Comment: Former oil-rig supply vessel acquired by Serco Ltd in December 2011. The vessel is an Ulstein UT-745L design of steel construction and is equipped with one 50-ton crane and one 20-ton crane. She is capable of carrying 2,000 tonnes of deck cargo. The principal task of the ship is to act as a mothership for the NATO submarine search and rescue system including the submersible and all support elements.

NORTHERN RIVER   12/2013*, Derek Fox / 1531136

1 SUPPORT SHIP (AG)

Name	Builders	Commissioned
SD VICTORIA	Damen Shipyard, Galati	Jun 2010

Displacement, tonnes: 2,540 full load
Measurement, tonnes: 864 dwt
Dimensions, metres (feet): 83 × 16.0 × 4.25 *(272.3 × 52.5 × 13.9)*
Speed, knots: 14
Complement: 16 + 72 casualty berths
Machinery: 2 Caterpillar 3526B diesels; 4,000 hp *(3.0 MW)*; 2 shafts; cp props; 1 bow thruster; 805 hp *(600 kW)*
Radars: Surface search/navigation: E/F-band.
Navigation: I-band.

Comment: Damen Support Ship 8316 design. The ship, which has replaced *Newton*, is capable of worldwide operations including military training, transport of personnel and equipment and conduct of diving support operations. Facilities include classrooms, briefing and operations rooms, workshops, extensive storage areas, a helicopter winching deck, and provision to carry and operate Rigid Inflatable Boats (RIBs). *Victoria* was launched on 27 November 2009 and delivered in June 2010. Based on the Clyde.

VICTORIA   7/2010, B Sullivan / 1366504

1 FBM CATAMARAN CLASS (YFL)

SD NORTON

Displacement, tonnes: 21 full load
Dimensions, metres (feet): 15.8 × 5.5 × 1.5 *(51.8 × 18.0 × 4.9)*
Speed, knots: 10. **Range, n miles:** 400 at 10 kt
Complement: 2
Machinery: 2 Mermaid Turbo 4 diesels; 280 hp *(209 kW)*; 2 shafts

Comment: Built by FBM Marine in 1989. Catamaran design. Can carry 30 passengers or 2 tons stores. Based at Portsmouth. To remain in service until 2022.

NORTON   2/2013*, Maritime Photographic / 1531137

2 MOORHEN CLASS (MOORING SHIPS) (ARS)

Name	Builders	Commissioned
SD MOORHEN	McTay Marine Ltd, Bromborough	26 Apr 1989
SD MOORFOWL	McTay Marine Ltd, Bromborough	30 Jun 1989

Displacement, tonnes: 539 full load
Dimensions, metres (feet): 32.3 × 11.5 × 3.8 *(106 × 37.7 × 12.5)*
Speed, knots: 8
Complement: 12 (2 officers)
Machinery: 2 Cummins KT19-M diesels; 796 hp *(594 kW)*; 2 Aquamasters; bow thruster

Comment: Classified as powered mooring lighters. The whole ship can be worked from a 'flying bridge' which is constructed over a through deck. Day mess for five divers. Both based at Kyle of Lochalsh. To remain in service until 2022.

MOORHEN   6/2009, Alistair MacDonald / 1366200

5 FELICITY CLASS (YTL)

SD FRANCES   SD FLORENCE   SD GENEVIEVE   SD HELEN   SD GEORGINA

Displacement, tonnes: 146 full load
Dimensions, metres (feet): 21.5 × 6.4 × 2.6 *(70.5 × 21.0 × 8.5)*
Speed, knots: 10. **Range, n miles:** 925 at 9 kt
Complement: 4
Machinery: 1 Mirrlees-Blackstone ESM8 diesel; 615 hp *(459 kW)*; 1 Voith-Schneider cp prop
Radars: Navigation: Raytheon; I-band.

Comment: *Frances*, *Florence* and *Genevieve* ordered early 1979 from Richard Dunston (Thorne) and completed by end 1980. Nominal bollard pull, 5.7 tons. *Helen* and *Genevieve* based at Portsmouth; *Frances*, *Florence* and *Georgina* at Devonport.

GENEVIEVE   12/2012*, Maritime Photographic / 1531138

1 RANGE SUPPORT VESSEL (YFRT)

SD WARDEN

Displacement, tonnes: 914 full load
Dimensions, metres (feet): 48.6 × 10.5 × 5.0 *(159.4 × 34.4 × 16.4)*
Speed, knots: 15. **Range, n miles:** 2,000 at 10 kt
Complement: 7
Machinery: 2 Ruston 8RKCZ diesels; 4,000 hp *(2.98 MW)*; 2 shafts; cp props
Radars: Navigation: Racal Decca RM 1250; I-band.
Sonars: Dowty 2053; high frequency.

Comment: Built by Richards, Lowestoft and completed 20 November 1989. Modified in 1998 to act, at BUTEC, as a ROV host ship and weapons launch and recovery platform. Based at Kyle of Lochalsh. To remain in service until 2022.

WARDEN   4/2009, Alistair MacDonald / 1366201

UK (MS/GAS)

6 ADEPT CLASS (COASTAL TUGS) (YTB)

| SD FORCEFUL | SD ADEPT | SD CAREFUL |
| SD POWERFUL | SD CAPABLE | SD FAITHFUL |

Displacement, tonnes: 448 standard; 549 full load
Dimensions, metres (feet): 38.8 × 9.1 × 4.0 *(127.3 × 29.9 × 13.1)*
Speed, knots: 12. **Range, n miles:** 1,500 at 10 kt
Complement: 5
Machinery: 2 Ruston 6RKC diesels; 2,575 hp *(1.92 MW)*; 2 Voith-Schneider props

Comment: 'Twin Unit Tractor Tugs' (TUTT). *Adept, Capable* and *Careful* ordered from Richard Dunston (Hessle) on 22 February 1979 and *Forceful, Powerful,* and *Faithful* on 8 February 1984. *Adept* accepted 28 October 1980, *Capable* 11 September 1981, *Careful* 12 March 1982, *Forceful* 18 March 1985, *Powerful* 30 October 1985 and *Faithful* 21 December 1985. Basing: *Adept, Careful, Forceful,* and *Faithful* at Devonport, *Powerful* at Portsmouth. *Capable* is operated by Commander British Forces Gibraltar.

POWERFUL *6/2013*, Derek Fox /* 1531139

1 TRANSPORT VESSEL (YFB)

SD EVA

Displacement, tonnes: 122 standard
Dimensions, metres (feet): 33.2 × 7.4 × 1.95 *(108.9 × 24.3 × 6.4)*
Speed, knots: 22
Complement: 4 + 34 scientists
Machinery: 2 Caterpillar C32-C diesels; 2,800 hp *(2.1 MW)*; 2 shafts
Radars: JRC 5210; I-band.

Comment: Damen FCS 3307 design. A crew transport vessel embodying a 'Sea Axe' bow. Aluminium construction. Delivered on 8 September 2009 and based on the Clyde. Replaced *Adamant*.

EVA *9/2009, Jan Oosterboer /* 1353959

2 SUBMARINE BERTHING TUGS (YTL)

Name	Builders	Commissioned
SD IMPULSE	Dunston, Hessle	11 Mar 1993
SD IMPETUS	Dunston, Hessle	28 May 1993

Displacement, tonnes: 539 full load
Dimensions, metres (feet): 32.5 × 10.0 × 5.2 *(106.6 × 32.8 × 17.1)*
Speed, knots: 12
Complement: 5
Machinery: 2 WH Allen 8S12 diesels; 3,400 hp *(2.54 MW)* sustained; 2 Aquamaster Azimuth thrusters; 1 Jastrom bow thruster

Comment: Ordered 28 January 1992 for submarine berthing duties. There are two 10 ton hydraulic winches forward and aft with break capacities of 110 tons. Bollard pull 38.6 tons ahead, 36 tons astern. Fitted with firefighting and oil pollution equipment. Designed for one-man control from the bridge with all round vision and a comprehensive Navaids fit. *Impulse* launched 10 December 1992; *Impetus* 9 February 1993. Based on the Clyde. To remain in service until 2022.

IMPETUS *4/2013*, Tony Roper /* 1531142

© 2014 IHS

9 RANGE SAFETY CRAFT (YFRT)

| SMIT STOUR | SMIT ROTHER | SMIT ROMNEY | SMIT CERNE | SMIT WEY |
| SMIT FROME | SMIT MERRION | SMIT PENALLY | SMIT NEYLAND | |

Displacement, tonnes: 6 full load
Dimensions, metres (feet): 11.3 × 3.4 × 1.2 *(37.1 × 11.2 × 3.9)*
Speed, knots: 35. **Range, n miles:** 160 at 21 kt
Complement: 2
Machinery: 2 Volvo Penta KAD 42P diesels; 680 hp *(507 kW)*; 2 — Hamilton waterjets

Comment: MP-1111 class of vessels designed (based on a fast rescue boat) and built at Maritime Partners Ltd (Norway). Aluminium alloy hull and GRP superstructure. The order for the craft followed a contract awarded to Smit International (Scotland) Ltd for the provision of Range Clearance and Safety duties in and around the various sea danger areas of UK military ranges. Three based at Dover, Portland and Pembroke Dock.

SMIT ROMNEY *7/2013*, Maritime Photographic /* 1531141

8 AIRCREW TRAINING CRAFT (YXT)

| SMIT DEE | SMIT YARE | SMIT SPEY | SMIT TAMAR |
| SMIT DON | SMIT TOWY | SMIT DART | SMIT CYMRAN |

Displacement, tonnes: 56 full load
Dimensions, metres (feet): 27.6 × 6.6 × 1.5 *(90.6 × 21.7 × 4.9)*
Speed, knots: 21. **Range, n miles:** 650 at 21 kt
Complement: 6
Machinery: 2 Cummins KTA 19M4 diesels; 1,400 hp *(1.04 MW)*; 2 shafts 1 Ultrajet 305 centreline waterjet; 305 hp *(227 kW)*
Radars: Furuno FR-2115 EPA; I-band.

Comment: Vessels built at Babcock Engineering Services, Rosyth, and FBMA Babcock Marine, Cebu, Philippines *(Yare, Towy* and *Spey)*. All delivered by 11 July 2003. Of aluminium alloy construction, the design is an adaptation of FBM Babcock Marine's Protector class patrol vessel. The order for the craft followed a contract awarded to MoD and to Smit International for provision of marine support to aircrew training, high speed marine target towing and recovery of air-sea rescue apparatus. The craft have an after docking well for a daughter craft. Based at Buckie *(Dee)*, Blyth *(Don)*, Great Yarmouth *(Yare)*, Pembroke Dock *(Towy)* and Plymouth *(Spey* and *Dart)*. *Smit Dart* is employed as a passenger craft. *Tamar* (Plymouth) and *Cymran* (Holyhead) are similar second-hand craft used for passengers.

SMIT DON *8/2010, Maritime Photographic /* 1366503

2 STORM CLASS (YFB)

Name	Builders	Commissioned
SD CAWSAND	FBM Marine, Cowes	Jul 1997
SD BOVISAND	FBM Marine, Cowes	Sep 1997

Displacement, tonnes: 99 standard
Dimensions, metres (feet): 23.9 × 11.1 × 4.95 *(78.4 × 36.4 × 16.2)*
Speed, knots: 15. **Range, n miles:** 450 at 14 kt
Complement: 5 + 75 spare berths
Machinery: 2 Caterpillar 3408TA diesels; 1,224 hp(m) *(900 kW)*; 2 shafts

Comment: Both based at Devonport. Swath design with hydraulically operated telescopic gangways. To remain in service until 2022.

CAWSAND *5/2008, Peter Ford /* 1353503

IHS Jane's Fighting Ships 2014-2015

922 UK (MS/GAS)

3 OBAN CLASS (YFL)

SD OBAN   SD ORONSAY   SD OMAGH

Displacement, tonnes: 302 full load
Dimensions, metres (feet): 27.7 × 7.3 × 3.8 (90.9 × 24.0 × 12.5)
Speed, knots: 10
Range, n miles: 1,700 at 10 kt
Complement: 4
Machinery: 2 Cummins N14M diesels; 1,050 hp(m) (785 kW); 2 Kort-Nozzles

Comment: Built by McTay Marine and completed January to July 2000. Capable of carrying 60 passengers. *Oban* based at Devonport and the other two on the Clyde. To remain in service until 2022.

ORONSAY   4/2013*, Tony Roper / 1531140

4 PADSTOW AND NEWHAVEN CLASSES (YFL)

Name	Builders
SD PADSTOW	Aluminium Shipbuilders, Fishbourne
SD NEWHAVEN	Aluminium Shipbuilders, Fishbourne
SD NUTBOURNE	Aluminium Shipbuilders, Fishbourne
SD NETLEY	Aluminium Shipbuilders, Fishbourne

Displacement, tonnes: 58 standard; 127 full load
Dimensions, metres (feet): 18.3 × 6.5 × 2.7 (60 × 21.3 × 8.9)
Speed, knots: 10. **Range, n miles:** 230 at 10 kt
Complement: 3
Machinery: 2 Cummins 6 CTA diesels; 710 hp(m) (522 kW); 2 shafts

Comment: Built by Aluminium Shipbuilders at Fishbourne, Isle of Wight and completed May to November 2000. Capable of carrying 60 passengers and based at Devonport (*Padstow* and *Newhaven*) and Portsmouth (*Nutbourne* and *Netley*). Catamaran hulls. To remain in service until 2022.

NUTBOURNE   6/2012, Derek Fox / 1455694

3 MANLY CLASS (YAG)

SD MELTON   SD MENAI   SD MEON

Displacement, tonnes: 145 full load
Dimensions, metres (feet): 24.4 × 6.4 × 3.0 (80.1 × 21.0 × 9.8)
Speed, knots: 10
Range, n miles: 700 at 10 kt
Complement: 6 (2 officers)
Machinery: 1 Lister-Blackstone ESR4 MGR diesel; 320 hp (239 kW); 1 shaft

Comment: All built by Richard Dunston, Thorne. All completed by early 1983. *Melton* is at Kyle of Lochalsh, the other two are at Devonport. To remain in service until 2022.

MEON   8/2008, Marco Ghiglino / 1353517

4 BERTHING TUGS (YTM)

SD RELIABLE   SD BOUNTIFUL   SD RESOURCEFUL   SD DEPENDABLE

Displacement, tonnes: 376 standard
Dimensions, metres (feet): 29.1 × 10.0 × 4.8 (95.5 × 32.8 × 15.7)
Speed, knots: 12
Machinery: 2 Caterpillar 3512 diesels; 4,025 hp (3.0 MW); 2 Rolls Royce US 175 thrusters
Radars: Navigation: 2 JRC 5210; I-band.

Comment: Damen Azimuth Tractor Drive (ATD) Tug 2909 design. *Reliable* (based on the Clyde) delivered on 27 November 2009, *Bountiful* (Portsmouth) on 2 April 2010, *Resourceful* (Clyde) on 28 May 2010 and *Dependable* (Clyde) on 23 July 2010.

BOUNTIFUL   5/2013*, Maritime Photographic / 1531146

1 HARBOUR WORKBOAT (YTL)

SD TILLY

Displacement, tonnes: 46 standard
Dimensions, metres (feet): 14.55 × 4.98 × 1.8 (47.7 × 16.3 × 5.9)
Speed, knots: 9
Machinery: 2 Caterpillar 3406C diesels; 600 hp (447 kW); 2 shafts; 2 Van de Giessen nozzles
Radars: Navigation: JRC JAMA-5210; I-band.

Comment: Damen Stan Tug 1405 design. General purpose inshore waters and harbour workboat delivered on 9 January 2009. Based at Devonport.

TILLY   1/2009, Damen Shipyards / 1366301

3 LARGE WORKBOATS (YTM)

SD HERCULES   SD MARS   SD JUPITER

Displacement, tonnes: 274 standard
Dimensions, metres (feet): 26.61 × 8.44 × 3.12 (87.3 × 27.7 × 10.2)
Speed, knots: 12
Complement: 9
Machinery: 2 Caterpillar 3508B diesels; 2,200 hp (1.6 MW); 2 shafts; 2 Van de Giessen nozzles
Radars: Navigation: 2 JRC 5210; I-band.

Comment: Damen Stan Tug 2608 design. *Hercules* (based at Devonport) delivered on 9 January 2009, *Mars* (Devonport) on 27 March 2009 and *Jupiter* (Clyde) on 26 June 2009.

HERCULES   7/2010, A A de Kruijf / 1366501

UK (MS/GAS) 923

4 BERTHING TUGS (YTM)

SD EILEEN   SD SUZANNE   SD CHRISTINA   SD DEBORAH

Displacement, tonnes: 249 standard
Dimensions, metres (feet): 21.2 × 9.4 × 3.6 (69.6 × 30.8 × 11.8)
Speed, knots: 11
Machinery: 2 Caterpillar 3508 diesels; 2,000 hp (1.5 MW); 2 Rolls Royce US 155 thrusters
Radars: Navigation: 2 JRC 5210; I-band.

Comment: Damen Azimuth Stern Drive Tug 2009 design. *Eileen* (based at Devonport) delivered on 21 May 2010, *Suzanne* (Portsmouth) on 30 July 2010, *Christina* (Portsmouth) on 8 October 2010 and *Deborah* (Devonport) on 17 December 2010.

CHRISTINA   5/2013*, Maritime Photographic / 1531148

2 BERTHING TUGS (YTM)

SD INDEPENDENT   SD INDULGENT

Displacement, tonnes: 351 standard
Dimensions, metres (feet): 26.09 × 9.44 × 4.3 (85.6 × 31.0 × 14.1)
Speed, knots: 13
Complement: 8
Machinery: 2 Caterpillar 3512B diesels; 3,500 hp (2.6 MW); 2 Rolls Royce US 155 thrusters; 1 bow thruster
Radars: Navigation: JRC 5210; I-band.

Comment: Damen Azimuth Stern Drive Tug 2509 design. *Independent* delivered on 16 October 2009 and *Indulgent* on 31 December 2009. Both based at Portsmouth.

INDULGENT   6/2013*, Derek Fox / 1531145

2 MULTIPURPOSE VESSELS (YAG)

SD NAVIGATOR   SD RAASAY

Displacement, tonnes: 315 standard
Dimensions, metres (feet): 26.3 × 10.64 × 2.55 (86.3 × 34.9 × 8.4)
Speed, knots: 8
Complement: 3 + 12 scientists
Machinery: 2 Caterpillar C18 diesels; 957 hp (713 kW); 2 shafts; 1 Veth-jet bow thruster
Radars: JRC JMA 5210; I-band.

Comment: Damen Multi Cat 2510 design. *Navigator* is used for buoy handling and mooring; equipped with a single crane capable of lifting up to 9 tonnes, the ship is to be capable of support diving operations. Based at Portsmouth, she was delivered on 17 July 2009. *Raasay*, is equipped with two cranes to carry out torpedo recovery, towed sonar array deployment and recovery, diving training and other trials duties. Based at Kyle of Lochalsh, she was delivered on 8 January 2010.

NAVIGATOR   6/2013*, Per Körnefeldt / 1531147

3 PERSONNEL TENDERS (YFL)

SD CLYDE SPIRIT   SD SOLENT SPIRIT   SD TAMAR SPIRIT

Measurement, tonnes: 102 gt
Dimensions, metres (feet): 19.15 × 5.3 × 1.65 (62.8 × 17.4 × 5.4)
Speed, knots: 20
Complement: 3 + 12 scientists
Machinery: 2 Caterpillar C 32 diesels; 2,200 hp (1.64 MW); 2 shafts
Radars: Navigation: JRC 5210; I-band.

Comment: Damen Stan Tender 1905 design. Steel hull with aluminium superstructure. Transport craft used for transfer of pilots, VIPs and personnel. *Clyde Spirit* (based on the Clyde) delivered on 27 June 2008, *Solent Spirit* (based at Portsmouth) on 25 July 2008 and *Tamar Spirit* (based at Devonport) on 17 October 2008.

SOLENT SPIRIT   5/2013*, Per Körnefeldt / 1531143

3 PERSONNEL TENDERS (YFL)

SD CLYDE RACER   SD SOLENT RACER   SD TAMAR RACER

Measurement, tonnes: 102 gt
Dimensions, metres (feet): 16 × 4.85 × 1.25 (52.5 × 15.9 × 4.1)
Speed, knots: 20
Complement: 3 + 10 scientists
Machinery: 2 Caterpillar 3406 diesels; 1,100 hp (820 kW); 2 shafts
Radars: Navigation: JRC 5210; I-band.

Comment: Damen Stan Tender 1505 design. Transport craft used for transfer of pilots, VIPs and personnel. *Clyde Racer* (based on the Clyde) delivered on 20 June 2008, *Solent Racer* (based at Portsmouth) on 19 September 2008 and *Tamar Racer* (based at Devonport) on 10 October 2008. Aluminium construction.

SOLENT RACER   5/2013*, Per Körnefeldt / 1531144

2 HARBOUR WORKBOATS (YTL)

SD CATHERINE   SD EMILY

Displacement, tonnes: 30 standard
Dimensions, metres (feet): 12.3 × 4.13 × 1.55 (40.4 × 13.5 × 5.1)
Speed, knots: 8
Machinery: 1 Caterpillar 3056 diesel; 165 hp (123 kW); 1 shaft
Radars: JRC JMA-5104; I-band.

Comment: Inshore waters and harbour workboats. Damen Pushy Cat 1204 design. Steel construction. *Catherine* (based at Portsmouth) delivered on 4 January 2008 and *Emily* (based at Devonport) on 8 March 2008.

CATHERINE   2/2009, Maritime Photographic / 1353723

924 UK (MS/GAS)

1 KYLE OF LOCHALSH CLASS (YTM)

SD KYLE OF LOCHALSH (ex-*Lenie*)

Displacement, tonnes: 295 full load
Dimensions, metres (feet): 25.5 × 9.0 × 2.75 *(83.7 × 29.5 × 9.0)*
Speed, knots: 12
Machinery: 2 Caterpillar 3508 TA diesels; 2,992 hp *(2.2 MW)*; 2 shafts

Comment: Brought into service by Serco Denholm in 2008. Built by David Abels Yard, Bristol, in 1997. Trials support vessel based at Kyle of Lochalsh.

KYLE OF LOCHALSH　　　　　7/2009, Alistair MacDonald / 1305390

1 MULTIPURPOSE VESSELS (YAG)

SD ENGINEER (ex-*Forth Engineer*)

Measurement, tonnes: 103 gt
Dimensions, metres (feet): 17.49 × 8.06 × ? *(57.4 × 26.4 × ?)*
Speed, knots: 8
Complement: 3
Machinery: 2 Caterpillar diesels; 600 hp *(442 kW)*; 2 shafts
Radars: Navigation: I-band.

Comment: Damen Multi Cat 1908 design. Built in 1996 and acquired by Serco Ltd in January 2008. General work and repair vessel of steel construction. Based at Devonport.

ARMY (ROYAL LOGISTIC CORPS)

Notes: (1) There are six Mk 4 LCVPs: four are based at Marchwood and two in the Falklands.
(2) 32 Combat Support Boats delivered by 2002. These are 8.2 m craft, road transportable and with a top speed of 30 kt.
(3) 17 Port and Maritime Regt, RLC is based at Marchwood, Southampton.

COMBAT SUPPORT BOAT　　　　　1/2004 / 1167573

5 RAMPED CRAFT, LOGISTIC (RCL)

Name	No	Builders	Commissioned
ANDALSNES	L 107	James and Stone, Brightlingsea	22 May 1984
AACHEN	L 110	James and Stone, Brightlingsea	12 Feb 1987
AREZZO	L 111	James and Stone, Brightlingsea	26 Mar 1987
ARROMANCHES	L 105	James and Stone, Brightlingsea	12 Jun 1987
(ex-*Agheila*)	(ex-L 112)		
AUDEMER	L 113	James and Stone, Brightlingsea	21 Aug 1987

Displacement, tonnes: 301 full load
Dimensions, metres (feet): 33.3 × 8.6 × 1.5 *(109.3 × 28.2 × 4.9)*
Speed, knots: 10
Range, n miles: 900 at 10 kt
Complement: 6
Military lift: 68 tons
Machinery: 2 Volvo Penta D9MH diesels; 700 hp *(521 kW)*; 2 shafts
Radars: Navigation: Racal Decca; I-band.

Comment: *Andalsnes* based in Cyprus, remainder at Marchwood, Southampton.

IHS Jane's Fighting Ships 2014-2015

AREZZO　　　　　6/2013*, Richard Sharpe / 1531006

4 WORK BOATS (YAG)

STORM 41　　**DIABLO** 42　　**MISTRAL** 43　　**SIROCCO** 44

Displacement, tonnes: 29 standard
Dimensions, metres (feet): 14.75 × 4.3 × ? *(48.4 × 14.1 × ?)*
Speed, knots: 10
Range, n miles: 480 at 10 kt
Complement: 4
Machinery: 2 John Deere diesels; 402 hp *(300 kW)*; 2 shafts
Radars: Navigation: Furuno; I-band.

Comment: Four new craft entered service in 2008. Capable of fire-fighting, pollution control, mexeflote operations, towed flexible barge duties, diving operations and general tug duties.

STORM　　　　　6/2013*, Derek Fox / 1531152

SCOTTISH FISHERIES PROTECTION AGENCY

Notes: (1) The Agency is responsible for the enforcement of sea fisheries regulations around the Scottish coast to a distance of 200 n miles. It has a complement of 275.
(2) There are two Cessna F-406 Caravan II aircraft with Bendix 1500 radars.

2 JURA CLASS (PSO)

JURA　　**HIRTA**

Measurement, tonnes: 2,217 gt
Dimensions, metres (feet): 84 × 13.0 × 4.5 *(275.6 × 42.7 × 14.8)*
Speed, knots: 18
Complement: 16 (8 officers)
Machinery: Diesel-electric; 3 Wärtsilä Gensets; 6,500 hp *(4.8 MW)*; 1 shaft; cp prop; 1 Brunvoll bow thruster; 1 Brunvoll stern thruster
Radars: Surface search/navigation: Sperry Marine Bridgemaster; E/F/I-bands.

Comment: *Jura* built by Ferguson Shipbuilders, Port Glasgow. Launched on 28 April 2005 and entered service in early 2006 to replace *Sulisker*. *Hirta* built by Stocznia Polnocna, Gdansk, launched on 17 August 2007 and entered service in May 2008. A third ship is no longer planned.

JURA　　　　　1/2006, SFPA / 1159412

© 2014 IHS

UK (MS/GAS) 925

1 MINNA CLASS (PBO)

MINNA

Displacement, tonnes: 869 full load
Dimensions, metres (feet): 47.7 × 10.0 × 4.5 *(156.5 × 32.8 × 14.8)*
Speed, knots: 14
Complement: 15 (6 officers)
Machinery: 2 Wärtsilä Gensets; 2,896 hp *(2.16 MW)*; 2 Indar propulsion motors; 2,145 hp *(1.6 MW)*; 2 shafts; 1 Kamewa transverse thruster *(150 kW)*

Comment: Built by Ferguson Shipbuilders, Port Glasgow. Launched in February 2003 and accepted by SFPA on 31 July 2003 as replacement for *Westra*. Procurement of a second similar ship was cancelled following a review in 2006.

MINNA — 10/2012, Tony Roper / 1486415

UK BORDER FORCE

Notes: The UK Border Force operates five cutters: four Damen 42 m craft (*Seeker*, *Searcher*, *Vigilant*, *Valiant*), that entered service 2001–04 and *Protector* (a former Finnish Border Guard vessel *Tavi* built in 2003). Their principal role is to detect prohibited and restricted goods and to prevent tax fraud. Each cutter carries a 7.4 m RHIB powered by an inboard diesel engine and waterjet.

SEARCHER — 10/2013*, Derek Fox / 1531150

PROTECTOR — 11/2013*, Derek Fox / 1531149

TRINITY HOUSE

Notes: The Corporation of Trinity House, with its HQ in London, has three responsibilities. It is the General Lighthouse Authority (GLA) for England, Wales and the Channel Islands; a Deep Sea Pilotage Authority for UK; and a major maritime charity, funded by its endowments, which supports the education, welfare and training of mariners and the promotion of safety at sea. In its GLA role, Trinity House provides nearly 600 aids to navigation including lighthouses, lightvessels, buoys, beacons, a differential global positioning service and an experimental radio-navigation service e-LORAN. Funding for these operations is by light dues levied on commercial shipping calling at UK ports. Operations are controlled from its Harwich centre while a depot at Swansea serves the west coast.

1 PATRICIA

Name	Builders	Commissioned
PATRICIA	Henry Robb Ltd., Leith	1982

Displacement, tonnes: 3,189 full load
Dimensions, metres (feet): 86.3 × 13.8 × 4.3 *(283.1 × 45.3 × 14.1)*
Speed, knots: 14
Range, n miles: 10,000 at 12 kt
Complement: 25 (8 officers)
Machinery: 4 Ruston Oil diesels; 4,285 bhp *(3.2 MW)*; connected via 4 generators to 2 motors; 3,452 hp *(2.54 MW)*; 2 shafts

Comment: Built by Henry Robb Ltd, Leith. Commissioned in May 1982.

PATRICIA — 5/2013*, Per Körnefeldt / 1531151

1 GALATEA

Name	Builders	Launched	Commissioned
GALATEA	Stocznia Remontowa SA, Gdansk	2006	2007

Displacement, tonnes: 4,024 full load
Dimensions, metres (feet): 84 × 16.5 × 4.25 *(275.6 × 54.1 × 13.9)*
Speed, knots: 13
Range, n miles: 5,250 at 12 kt
Complement: 18
Machinery: 3 Wärtsilä 8L20 diesels; 2 shafts

Comment: New multifunction tender built by Stocznia Remontowa SA shipbuilders at Gdansk, Poland. Launched on 26 July 2006, she was named by The Queen on 17 October 2007. Design features include a large working deck area and a forward helicopter flight deck. She is equipped with a dynamic positioning system. Similar to Northern Lighthouse Board vessel *Pharos*.

GALATEA — 9/2010, J Brodie / 1406314

1 ALERT

Name	Builders	Launched	Commissioned
ALERT	Stocznia Remontowa SA, Gdansk	2005	2006

Displacement, tonnes: 330 full load
Dimensions, metres (feet): 39.3 × 8.0 × 2.4 *(128.9 × 26.2 × 7.9)*
Speed, knots: 16
Range, n miles: 400 at 12 kt
Complement: 5
Machinery: 2 Caterpillar 3512 diesels; 4,023 hp *(3 MW)*; 2 shafts; cp props; bow thruster

Comment: Rapid intervention vessel built by Stocznia Remontowa SA shipbuilders at Gdansk, Poland. Launched on 11 October 2005, she was delivered in 2006. In addition to maintaining aids to navigation, the vessel provides a fast response capability and the means to carry out emergency wreck marking and hydrographic survey services. Her primary areas of operation are the Dover Strait, English Channel and Southern North Sea. The ship is equipped with a dynamic positioning system.

ALERT — 5/2006, Mark Rayner / 1167574

© 2014 IHS

IHS Jane's Fighting Ships 2014-2015

NORTHERN LIGHTHOUSE BOARD

Notes: The Northern Lighthouse Board (NLB) is the General Lighthouse Authority for Scotland and the Isle of Man. The Board provides Aids to Navigation (AtoN) including lighthouses, buoys and beacons and radio navigation aids. NLB is funded from the General Lighthouse Fund, which draws most of its income from the levy of light dues on commercial and fishing vessels calling at UK and Republic of Ireland ports. Operations are directed from its headquarters in Edinburgh.

1 PHAROS

Name	Builders	Launched	Commissioned
PHAROS	Remontowa Shipyard, Gdansk	2006	2007

Displacement, tonnes: 4,024 full load
Measurement, tonnes: 3,731 gt
Dimensions, metres (feet): 84.2 × 16.5 × 4.25 (276.2 × 54.1 × 13.9)
Speed, knots: 12
Complement: 18 (7 officers) + 12 spare berths
Machinery: Diesel-electric; 2 azimuth props; 2 bow thrusters

Comment: The contract for the construction of a new multifunction tender was signed with Remontowa Shipyard, Gdansk, Poland on 11 November 2004. The ship was launched on 3 February 2006 and delivered in March 2007. *Pharos* the tenth NLB vessel to carry the name has replaced the former vessel now used as a fishery patrol vessel by the government of South Georgia. The ship is fitted with dynamic positioning, a large aft working deck area, buoy and chain handling equipment, towing winch, integrated bridge management system, survey suite and moon pool, helicopter deck and a 30-tonne crane. Similar to Trinity House vessel *Galatea*.

PHAROS 4/2007, NLB / 1170250

1 POLE STAR

Name	Builders	Laid down	Commissioned
POLE STAR	Ferguson Shipbuilders, Port Glasgow	1999	2000

Displacement, tonnes: 1,395 full load
Dimensions, metres (feet): 51.5 × 12.0 × 3.5 (169 × 39.4 × 11.5)
Speed, knots: 12
Complement: 15 (6 officers)
Machinery: Diesel-electric; 3 Cummins Wärtsilä generators; 3,700 hp (2.8 MW); 2 motors; 2,680 hp (2 MW); 2 azimuth props; 2 bow thrusters
Radars: Sperry Marine; E/F/I-bands.

Comment: Built by Ferguson Shipbuilders, Port Glasgow. Laid down on 28 July 1999 and delivered on 15 September 2000. Principal roles are hydrographic survey and buoy handling. Equipped with dynamic positioning and an 18-tonne crane.

MARITIME & COASTGUARD AGENCY

Notes: The Maritime & Coastguard Agency is responsible for the development, promotion and enforcement of high standards of marine safety, response to maritime emergencies 24 hours a day, reduction of the risk of pollution of the marine environment from ships and, where pollution occurs, minimisation of its impact on the United Kingdom.

Response to maritime emergencies within the UK SAR region is undertaken by HM Coastguard, the MCA's Counter-pollution Response Branch and firefighting teams from the Maritime Incident Response Group. SAR and counter pollution is co-ordinated through a network of 19 Maritime Rescue Co-ordination Centres (MRCCs). Each MRCC provides continuous emergency telephone, radio and satellite communications distress watch plus safety information and radio medical advice services. The counter-pollution branch provides response to marine pollution and provides scientific and technical advice on shoreline clean up.

The MCA provides four civilian SAR helicopters (Sikorsky S-92 and Agusta Westland 139) under contract from CHC Scotia. They are based at Sumburgh, Stornoway, Lee-on-Solent and Portland. Fixed-wing aircraft include a Cessna 404 and Cessna 406 for counter-pollution surveillance. Fitted with radar, IR and UV detection equipment they are operated by PVL Group of Coventry. Additionally a Cessna 406 and two Lockheed Electra aircraft are available for dispersant spraying. It was announced in March 2013 that Bristow had been awarded a contract to operate UK SAR services from 2016.

HM Coastguard has its own corps of 3,500 volunteer Auxiliary Coastguards divided into 380 Coastguard Rescue Teams around the coast of UK. HM Coastguard also make significant use of Royal National Lifeboat Institution all-weather and inshore lifeboats and military SAR helicopters.

The MCA is also responsible for inspections and surveys of UK vessels, port state control inspections of non UK ships, the enforcement of merchant shipping legislation, the setting of ship and seafarer standards and maritime security.

A new organisation of the MCA is to be implemented by 2016. Under current proposals the 19 MRCCs would be reduced to a networked Maritime Operations Centre and nine sub-centres.

S-92 7/2007, MCA / 1170247

AW-139 7/2010, Maritime Photographic / 1406315

POLE STAR 11/2007, NLB / 1170249

United States

Country Overview

The United States of America is a federal republic which comprises 48 contiguous states (bounded to the north by Canada and to the south by Mexico) and the states of Alaska and Hawaii. External territories include Puerto Rico, American Samoa, Guam and the US Virgin Islands. With an area of 3,717,800 square miles, it occupies much of North America and has a coastline of 10,762 n miles with the Atlantic and Pacific Oceans and with the Gulf of Mexico. Washington, DC is the capital while New York, New York, is the largest city and a leading seaport. Other principal ports include New Orleans, Louisiana; Houston, Texas; Valdez, Alaska; Baton Rouge, Louisiana; Corpus Christi, Texas; Long Beach, California; Norfolk, Virginia; Tampa, Florida; Los Angeles, California; St Louis, Missouri; and Duluth, Wisconsin. There is an extensive inland waterway network, the three main components of which are the Mississippi river system (13,000 n miles long), the Great Lakes (ocean-going vessels can sail between the Great Lakes and the Atlantic Ocean via the St Lawrence Seaway (opened 1959)) and coastal waterways. Territorial seas (12 n miles) are claimed. A 200 n mile EEZ has been claimed but the limits have only been partly defined by boundary agreements.

Unified Combatant Commanders

Commander, US Strategic Command:
Admiral Cecil D Haney
Commander, US Pacific Command:
Admiral Samuel J Locklear III
Commander, US European Command:
General Philip M Breedlove
Commander, US Northern Command:
General Charles H Jacoby Jr
Commander, US Southern Command:
General John F Kelly
Commander, US Central Command:
General Lloyd J Austin III
Commander, US Africa Command:
General David M Rodriguez
Commander, US Special Operations Command:
Admiral William H McRaven

Headquarters Appointments

Chief of Naval Operations:
Admiral Jonathan W Greenert
Vice Chief of Naval Operations:
Admiral Michelle J Howard
Director, Naval Nuclear Propulsion:
Admiral John M Richardson
Chief of Naval Personnel:
Vice Admiral William F Moran
Commander, Naval Sea Systems Command:
Vice Admiral William H Hilarides

Headquarters Appointments – continued

Commander, Naval Air Systems Command:
Vice Admiral David Dunaway
Commander, Space and Naval Warfare Systems Command:
Rear Admiral Patrick H Brady

Fleet Commanders

Commander, US Fleet Forces Command:
Admiral William E Gortney
Commander, US Pacific Fleet:
Admiral Harry B Harris, Jr
Commander, US Naval Forces Europe-Africa and Commander Allied Joint Force Command, Naples: Admiral Mark E Ferguson III
Commander, Military Sealift Command:
Rear Admiral Thomas K Shannon

Flag Officers (Atlantic Area)

Commander, Naval Surface Force, Atlantic Fleet:
Rear Admiral Pete A Gumataotao
Commander, Sixth Fleet and Commander Naval Striking and Support Forces NATO and Joint Force Maritime Component Commander Europe:
Vice Admiral Philip S Davidson
Commander, US Submarine Forces, Commander Submarine Force Atlantic and Allied Submarine Command:
Vice Admiral Michael J Connor
Commander, Naval Air Force, Atlantic Fleet:
Rear Admiral Troy M Shoemaker
Commander, Navy Region Europe, Africa, Southwest Asia:
Rear Admiral John C Scorby
Commander, Naval Forces Southern Command and Fourth Fleet: Rear Admiral Sinclair M Harris

Flag Officers (Pacific Area)

Commander, Seventh Fleet:
Vice Admiral Robert L Thomas Jr
Commander, Naval Surface Force, Pacific Fleet:
Vice Admiral Thomas H Copeman
Commander, Third Fleet:
Vice Admiral Kenneth E Floyd
Commander, Naval Air Forces and Naval Air Force, Pacific Fleet: Vice Admiral David H Buss
Commander, US Naval Forces, Japan:
Rear Admiral Terry B Kraft
Commander, Submarine Force, Pacific Fleet:
Rear Admiral Philip G Sawyer
Commander, US Naval Forces, Korea:
Rear Admiral Lisa Franchetti
Commander, US Naval Forces, Marianas:
Rear Admiral Tilghman D Payne

Flag Officers (Pacific Area) – continued

Commander, Naval Mine and Anti-Submarine Warfare Command: Rear Admiral William R Merz
Commander, Fleet Cyber Command and Tenth Fleet:
Vice Admiral Jan E Tighe

Flag Officer (Central Area)

Commander, US Naval Forces, Central Command, and Fifth Fleet: Vice Admiral John W Miller

Marine Corps

Commandant:
General James F Amos
Assistant Commandant:
General John M Paxton Jr
Commander, US Marine Corps Forces Command:
Lieutenant General Richard T Tryon
Commander, US Marine Corps Forces Pacific:
Lieutenant General Terry G Robling
Commander, Marine Forces Reserve and Commander Marine Forces North:
Lieutenant General Richard P Mills
Commander US Marine Corps Forces Central Command:
Lieutenant General Robert B Neller
Commanding General I MEF:
Lieutenant General John A Toolan
Commanding General II MEF:
Major General Raymond C Fox
Commanding General III MEF and Commander, Marine Corps Bases, Japan:
Lieutenant General John L Wissler

Prefix to Ships' Names

USS (United States Ship) Warships
USNS (United States Naval Ship) Military Sealift Command

Personnel

	1 Jan 2012	1 Jan 2013	1 Jan 2014
Navy			
Officers	51,346	51,152	51,882
Warrants	1,604	1,572	1,616
Enlisted	266,215	259,876	265,161
Marine Corps			
Officers	19,842	19,744	19,249
Warrants	2,058	2,054	2,014
Enlisted	178,763	174,226	172,734

Strength of the Fleet (1 January 2014)

Type	Active (NRF) (Reserve)	Building (Projected) + Conversion/SLEP
SHIPS OF THE FLEET		
Strategic Missile Submarines		
SSBN (Ballistic Missile Submarines) (nuclear-powered)	14	(12)
Cruise Missile Submarines (SSGN) (nuclear-powered)	4	—
Attack Submarines		
SSN Submarines (nuclear-powered)	55	7
Aircraft Carriers		
CVN Multipurpose Aircraft Carriers (nuclear-powered)	10	2 (1)
Cruisers		
CG Guided Missile Cruisers	22	—
Destroyers		
DDG 1000	—	3
DDG Guided Missile Destroyers	62	4 (6)
Frigates		
FFH Frigates	6 (6)	—
LCS Littoral Combat Ships	4	13 (7)
Patrol Forces		
PC Coastal Defense Ships	13	—
Command Ships		
LCC Command Ships	2	—
Amphibious Warfare Forces		
LHA Amphibious Assault Ships (general purpose)	1	2 (3)
LHD Amphibious Assault Ships (multipurpose)	8	—
LPD Amphibious Transport Docks	10	2
LSD Dock Landing Ships	12	—
LSV Logistic Support Vessels	8	—
Mine Warfare Forces		
MCM Mine Countermeasures Ships	13	—
Research		
AGE Research	2	—
AGOR Oceanographic	6	—

Type	Active (NRF) (Reserve)	Building (Projected) + Conversion/SLEP
MILITARY SEALIFT COMMAND INVENTORY		
Naval Fleet Combat Logistics		
T-AOE Fast Combat Support	4	—
T-AKE Auxiliary Cargo and Ammunition	14	—
T-AO Oilers	15	(17)
Special Mission Ships		
T-AGM	2	—
T-AG	1	—
T-AGOS Surveillance/Patrol	5	—
T-AGS Surveying	6	1
SBX	1	—
Prepositioning		
T-MLP	1	2
T-AK	10	—
T-AKR Large, Medium-Speed, Ro-Ro	8	—
T-AG	1	—
Service Support		
T-ATF	4	—
T-ARS	4	—
AS	2	—
T-AH	2	—
T-ARC	1	—
AFSB(i)	1	—
Sealift		
T-AKR	11	—
T-AK	5	—
T-AOT	1	—
JHSV	4	6
HST	—	2
HSV	1	—
Ready Reserve Force		
T-ACS Crane Ships	6	—
T-AKR Ro-ro	37	—
T-AOT	1	—

928 **US** > Introduction

Special Notes

To provide similar information to that included in other major navies' Deployment Tables the fleet assignment (abbreviated 'F/S') status of each ship in the US Navy has been included. The assignment appears in a column immediately to the right of the commissioning date. In the case of the Floating Dry Dock section this system is not used. The following abbreviations are used to indicate fleet assignments:

AA	active Atlantic Fleet
Active	active under charter with MSC
AR	in reserve Out of Commission, Atlantic Fleet
ASA	active In Service, Atlantic Fleet
ASR	in reserve Out of Service, Atlantic Fleet
Bldg	Building
CONV	ship undergoing conversion
LOAN	ship or craft loaned to another government, or non-government agency, but US Navy retains title and the ship or craft is on the NVR
MAR	in reserve Out of Commission, Atlantic Fleet and laid up in the temporary custody of the Maritime Administration
MPR	same as 'MAR', but applies to the Pacific Fleet
NRF	assigned to the Naval Reserve Force (ships so assigned are listed in a special table for major warships and amphibious ships)
Ord	the contract for the construction of the ship has been let, but actual construction has not yet begun
PA	active Pacific Fleet
PR	in reserve Out of Commission, Pacific Fleet
Proj	the ship is scheduled for construction at some time in the immediate future
PSA	active In Service, Pacific Fleet
PSR	in reserve Out of Service, Pacific Fleet
ROS	reduced Operating Status
TAA	active Military Sealift Command, Atlantic Fleet
TAR	in Ready Reserve, Military Sealift Command, Atlantic Fleet
TPA	active Military Sealift Command, Pacific Fleet
TPR	in Ready Reserve, Military Sealift Command, Pacific Fleet
TWWR	active Military Sealift Command, Worldwide Routes

Ship Status Definitions

In Commission: as a rule any ship, except a Service Craft, that is active, is in commission. The ship has a Commanding Officer and flies a commissioning pennant. 'Commissioning date' as used in this section means the date of being 'in commission' rather than 'completion' or 'acceptance into service' as used in some other navies.

In Service: all service craft (dry docks and with classifications that start with 'Y'), with the exception of *Constitution*, that are active, are 'in service'. The ship has an Officer-in-Charge and does not fly a commissioning pennant.

Ships 'in reserve, out of commission' or 'in reserve, out of service' are put in a state of preservation for future service. Depending on the size of the ship or craft, a ship in 'mothballs' usually takes from 30 days to nearly a year to restore to full operational service.

The above status definitions do not apply to the Military Sealift Command.

Approved Fiscal Year 2014 Programme

	Appropriations (US dollars millions)
CVN 21 (CVN 79 continued funding)	917
CVN Refueling and Complex Overhaul completed (CVN 72)	1,609
2 Virginia-class submarines (SSN 792, 793 and advance procurement for future submarines)	3,881
Virginia class advance funding for future submarines	2,355
1 Arleigh Burke destroyer	1,616
Arleigh Burke Flight III advance funding	370
DDG 1000 destroyer (mission system equipment)	232
4 Littoral Combat Ships	1,793
1 Afloat Forward Staging Base	579
Joint High Speed Vessel	2,732

Proposed Fiscal Year 2015 Programme

	Appropriations (US dollars millions)
CVN 21 (CVN 78 ship construction variance)	663
CVN 21 (CVN 79 continued funding)	1,300
SSBN(X) (Ohio-class submarine replacement program)	1,200
2 Virginia-class submarines (SSN 794, 795)	3,553
Virginia class advance procurement	2,330
2 DDG 51 Arleigh Burke destroyers (DDG 119, 120)	2,805
DDG 1000 destroyers (mission system equipment)	420
3 Littoral Combat Ships	1,427
LHA Replacement advance procurement	29
Joint High Speed Vessel (completion of prior year shipbuilding)	5
2 Ship-to-Shore Connector	123
1 Moored Training Ship (converted Los Angeles-class submarine)	737

Naval Aviation

Naval Aviation had an active inventory of 3,988 aircraft as of 1 January 2014, with approximately 42% of those being operated by the US Marine Corps. The principal aviation organisations are 10 active carrier air wings and one reserve Tactical Support Wing, 12 active and two reserve maritime patrol squadrons and three active and one reserve Marine aircraft wings. Reserve squadrons fly and maintain their own aircraft. Fleet Replacement Squadrons (FRS) train winged aviators in the aircraft they will fly in fleet.

Fighter Attack: 10 Navy active squadrons, two Navy reserve squadrons and one Navy demonstration squadron with F/A-18A-D legacy Hornets. 22 Navy active Squadrons and two Navy FRS with F/A-18E-F legacy and Super Hornets. Two Navy reserve squadrons of F-5 Tiger. 11 Marine active squadrons, one reserve and one FRS with F/A-18A-D Hornets. One Marine active squadron with F-35B Lightnings. One Marine active squadron and one reserve squadron of F-5 Tigers.

Attack: Seven Marine active squadrons and one FRS with AV-8B Harriers.

Airborne Command and Control: 10 Navy active squadrons, and one FRS with E-2C Hawkeyes and E-2D Advanced Hawkeyes (FRS shared with C-2A Greyhounds)

Fleet Logistics Support: Two Navy active squadrons, and one FRS with C-2A Greyhounds (FRS shared with E-2C and E-2D); one Navy reserve squadron with C-9; four Navy reserve squadrons with C-40; five Navy reserve squadrons with C-130; three Navy reserve squadrons with C-20 (one of which also has C-37), one Marine active squadron with C-9B, one Marine squadron with C-20s

Electronic Attack: Eight Navy active squadrons and one FRS with EA-18G Growlers; three Navy active squadrons and one FRS squadron with EA-16B Prowlers. Four Marine squadrons with EA-6B.

Airborne Command Post: Two Navy active and one FRS squadron of E-6B Mercury.

Maritime Patrol: 11 Navy active, two reserve and one FRS squadrons with P-3C Orion. Two Navy active squadrons and one FRS with P-8A Poseidons.

Signals Intelligence Reconnaissance: One Navy active squadron, with EP-3E (Aries II).

Helicopter Anti-Submarine: Two Navy active squadrons, two reserve squadrons and one FRS squadron with SH-60F and HH-60H Seahawks.

Helicopter Anti-Submarine Light: Seven Navy active, one reserve and one FRS squadron with SH-60B Seahawks.

Helicopter Mine Countermeasures: Two Navy squadrons with MH-53E Sea Dragons. Active and reserve in both squadrons.

Helicopter Sea Combat: 14 Navy active squadrons, two FRS with MH-60S Knighthawks.

Helicopter Combat Support/Gunship: Six Marine active squadrons, with AH-1W Super Cobras and UH-1Y Hueys. Two Marine active squadrons with AH-1Z and UH-1Y and one FRS Squadron with all types of these aircraft.

Helicopter Transport: Three Marine active squadrons, one reserve squadron and one FRS with CH-46E Sea Knights, seven Marine active squadrons, one reserve squadron and one FRS with CH-53E Super Stallions.

Helicopter Maritime Strike/Anti-Submarine Light: 14 Navy active squadrons, one reserve squadron and two FRS with SH-60B and MH-60R Seahawks.

In-Flight Refueling: Three Marine squadrons with KC-130J and two reserve with KC-130T.

Tilt Rotor: 14 Marine squadrons and one FRS with MV-22B.

Aircraft Procurement Plan FY2013–2015

	13	*14*	*15*
F-35B	6	6	6
F-35C	4	4	2
F/A-18E/F Super Hornet	26	–	–
EA-18G Growler	12	21	–
MV-22B Osprey	17	19	19
AH-1Z/UH-1Y Super Cobra/Huey	28	21	26
MH-60S Seahawk	18	18	8
MH-60R Seahawk	19	19	29
E-2D Advanced Hawkeye	5	5	4
C-40A Clipper	–	–	–
P-8A MMA	13	16	8
T-6A/B JPATS	33	29	–
KC-130J Tanker	–	1	1
MQ-8B VTUAV	6	2	–
STUAS	5	–	–

Naval Special Warfare (NSW)

The Naval Special Warfare Command was commissioned 16 April 1987.

SEAL (Sea Air Land) teams are comprised of seven platoons per team. Four teams are attached to NSW Group One, Coronado, California, and four teams to NSW Group Two, Little Creek, Virginia. Additionally there is one reserve team per coast, attached to NSW Group Eleven. NSW Group Three, located on Ford Island, Hawaii, has one SEAL Delivery Vehicle (SDV) Team located in Pearl City, Hawaii. NSW Group Four includes three Special Boat Teams, located at Coronado, Little Creek and Stennis, Mississippi. NSW recently established Group Ten, which is comprised of two support activity commands, located on both coasts, as well as the NSW Mission Support Center at Coronado. NSW Teams are allocated to theatre commanders during operational deployments. The NSW community comprises approximately 9,600 personnel including 2,670 SEALs, 720 Special Warfare Combatant-craft Crewmen (SWCC), 760 reserve personnel, 4,200 support personnel and more than 1,150 civilian personnel.

Bases

Naval Air Stations, Air Facilities, and Naval Air Weapons Stations

Naval Air Weapons Station (NAWS) China Lake, CA; Naval Air Facility (NAF) El Centro, CA; Naval Air Station (NAS) Lemoore, CA; NAS Jacksonville, FL; NAS Key West, FL; NAS Whiting Field (Milton), FL; NAS Pensacola, FL; NAS Joint Reserve Base, New Orleans, LA; NAS Patuxent River, MD; NAS Meridian, MS; NAS Fallon, NV; NAS Corpus Christi, TX; NAS Joint Reserve Base Fort Worth, TX; NAS Kingsville, TX; NAS Oceana, VA; NAS Whidbey Island (Oak Harbor), WA; NAF Atsugi, Japan; NAF Misawa, Japan.

Naval Stations and Naval Bases

NB Coronado, CA; NB Ventura City, Point Mugu, CA; NB Point Loma (San Diego), CA; Naval Station San Diego, CA; NS Mayport, FL; NS Great Lakes, IL; NS Newport, RI; NS Norfolk, VA; NB Kitsap, WA; NS Everett, WA; NS Guantanamo Bay, Cuba; NB Guam; NS Rota, Spain.

Naval Support Activities

NSA Monterey, CA; NSA Washington, DC; NSA Orlando, FL; NSA Panama City, FL; NSA Crane IN; NSA Annapolis, MD; NSA Bethesda, MD; NSA Saratoga Springs NY; NSA Mechanicsburg, PA; NSA Mid-South (Millington), TN; NSA South Potomac Dahlgren, VA; NSA Orlando FL; NSA Souda Bay, Greece; NSA Naples, Italy; NSA Bahrain; NSA Anderson, Guam; Singapore Area Coordinator; NSA Hampton Roads, VA; NSA Souda Bay, Crete.

Naval Support Facilities and Naval Weapons Station

NSF Diego Garcia; NWS Seal Beach CA; NWS Earle Colts Neck NJ; NWS Yorktown, VA.

Other installations

PMRF Barking Sands, HI; Commander Fleet Activities (CFA) Okinawa, Japan; CFA Sasebo, Japan; CFA Chinhae, Korea; CFA Yokosuka, Japan; Naval Construction Battalion Center (NCBC) Gulfport, MS; Camp Lemonnier (Djibouti); NAVMAG Indian Island WA.

Submarine Bases

NSB Kings Bay, GA (East Coast); NSB New London, CT (East Coast).

Naval Shipyards

Norfolk Naval Shipyard, Norfolk VA; Portsmouth Naval Shipyard, Kittery, ME; Pearl Harbor Naval Shipyard, Pearl Harbor, HI; Puget Sound Naval Shipyard, Bremerton, WA.

Marine Corps Air Stations and Helicopter Facilities

MCAS: Yuma, AZ; Miramar (San Diego), CA; Camp Pendleton, CA; Cherry Point, NC; New River (Jacksonville), NC; Beaufort, SC; Futema, Okinawa, Japan; Iwakuni, Japan.

Marine Corps Bases

Marine Corps Logistic Base, Bastow, CA; Marine Corps Mountain Warfare Training Center, Bridgeport, CA; Camp Pendleton, Oceanside, CA; Marine Corps Recruit Depot, San Diego; Marine Corps Air-Ground Combat Centre, Twentynine Palms, CA; Marine Corps Barracks, Washington, DC; Marine Corps Logistic Base, Albany, GA; Camp H M Smith (Oahu), HI; Kaneohe Bay, Hi; Camp Lejeune, Jacksonville, NC; Marine Corps Recruit Depot, Parris Island, SC; Marine Corps Base, Quantico, VA; Combined Arms Training Centre, Camp Fuji, Japan; Camp Smedley D Butler, Okinawa, Japan.

Joint Service Bases

Joint Base Anacostia-Bolling, Washington, DC; Joint Base Pearl-Hickam, HI; Joint Expeditionary Base Little Creek-Fort Story, VA.

Command and control of US naval forces

Strategic and Operational Command

All US Military Forces operate under Title 10 of US Code and subsidiary Joint Force Doctrine publications. The President of the United States is the Commander-in-Chief of all US forces and exercises authority for the application of military force through the Secretary of Defense who is advised by the Chairman of the Joint Chiefs of Staff. The Unified Combatant Commanders are four-star officers who have broad geographic area of functional responsibilities. Exercising Combatant Command (COCOM), they have authority to employ forces as necessary to accomplish assigned military missions and are as follows:

Commander US European Command (Stuttgart-Vaihingen, Germany)
Commander US Africa Command (Stuttgart, Germany)
Commander US Northern Command (Peterson AFB, Colorado)
Commander US Pacific Command (Honolulu, Hawaii)
Commander US Southern Command (Miami, Florida)
Commander US Central Command (MacDill AFB, Florida)
Commander US Special Operations Command (MacDill AFB, Florida)
Commander US Transportation Command (Scott AFB, Illinois)
Commander US Strategic Command (Offutt AFB, Nebraska)

The Unified Combatant Commanders may decide to exercise Operational (OPCON) command of naval forces directly. Alternatively, they may delegate such powers to another officer who might be a subordinate Unified Commander (for example Commander, US Forces Korea), a service component commander (Army, Navy, Air Force, Marine Corps and so on), a functional component commander (air, maritime, land, special forces), a joint task force commander or a single service force commander.

US Africa Command (AFRICOM) was established in October 2007 as a sub-unified command subordinated to US European Command for a transition period of one year. It became a stand-alone unified command on 1 October 2008, with the commander reporting to the Secretary of Defense like other unified commanders. AFRICOM Headquarters is at Kelley Barracks in Stuttgart, Germany. Unlike traditional unified commands, AFRICOM focuses on war prevention rather than war-fighting. The aim is to work with African partner nations and organisations to build regional security and crisis-response capacity. AFRICOM has assumed control over existing US government programmes in Africa that had been administered by US Central, European and Pacific Commands. The US force presence in Africa includes approximately 2,000 personnel at Camp Lemonnier in Djibouti. Military advisers assigned to US embassies and diplomatic missions help to co-ordinate military-to-military activities training and other programs in support of US foreign policy. Joint Forces Command was disestablished in August 2011 and its critical functions reassigned to other combatant commands, the military services and Joint Staff in Washington, DC.

Navy force commanders have a dual chain of command. They report to the Chief of Naval Operations for administrative matters such as training and equipping of forces and are

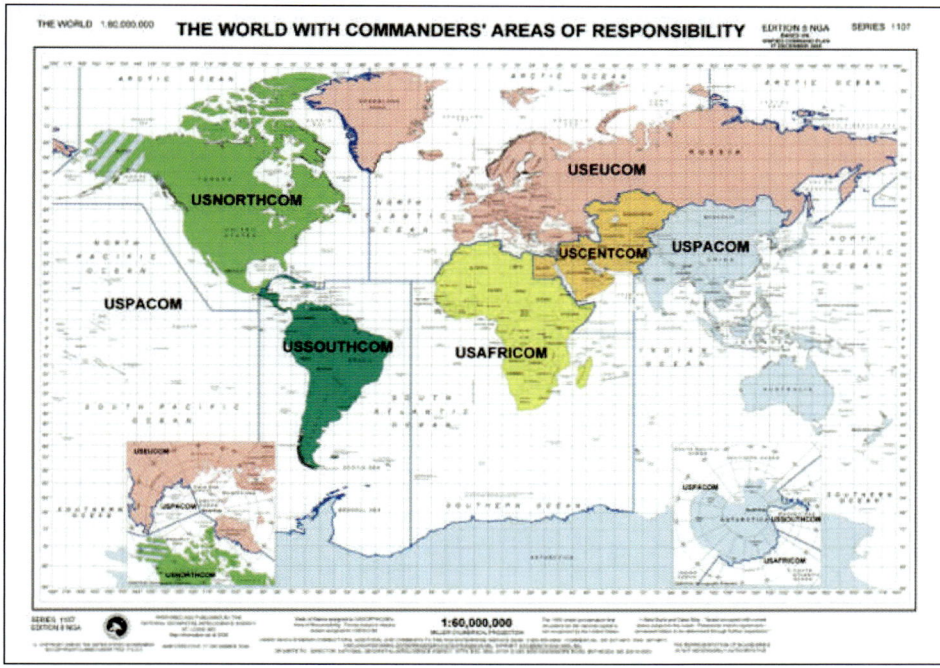

AREAS OF RESPONSIBILITY *12/2008, US DOD / 1353649*

responsible to the combatant commanders for providing forces to accomplish missions. They include the following:

Commander US Fleet Forces Command
Commander US Pacific Fleet
Commander US Naval Forces Europe
Commander US Naval Forces Central Command
Commander US Naval Forces Southern Command
Commander Navy Reserve Force
Commander Military Sealift Command

Once deployed in theatre, naval forces are operationally assigned to three-star numbered fleet commanders:

Commander US Third Fleet (Eastern Pacific)
Commander US Fourth Fleet (Caribbean, Central and South America)
Commander US Fifth Fleet (Arabian Gulf and Indian Ocean)
Commander US Sixth Fleet (Mediterranean)
Commander US Seventh Fleet (Western Pacific)
Commander US Tenth Fleet (Fort Meade, MD)

These arrangements are intended to provide a framework that provides a clear chain of command while retaining the flexibility to be adapted to the operational circumstances. For example, it is feasible for a multimission naval task group, such as a carrier strike group (CSG) (baseline composition: 1 CVN/CV, 2 CG/DDG, 1 DD/FFG, 1 SSN and 1 logistic support ship) or an expeditionary strike group (ESG) (baseline composition: 3 amphibious ships (LHD/LHA, LPD and LSD), 2 CG/DDG, 1 DD/FFG and 1 SSN) to support service, component, and other superior commanders simultaneously. The US Tenth Fleet was established on 29 January 2010. Its mission is to integrate cyberspace and information-based weapons within the US Navy. This followed similar steps taken by the Air Force and Marine Corps. Activities are co-ordinated with US Cyber Command (which became fully operational on 31 October 2010), a subordinate of US Strategic Command. US Second Fleet was disbanded on 30 September 2011 and its responsibilities for Atlantic fleet readiness and training were transferred to US Fleet Forces Command.

Tactical Command and Composite Warfare Commander

US naval task groups and forces operate under Composite Warfare Commander (CWC) doctrine. The officer in tactical command (OTC) is responsible for accomplishing the missions of his assigned forces. The CWC directs the force and controls warfare functions. The OTC may designate a subordinate commander as CWC but, in general practice, the roles are combined. The OTC/CWC is supported by Principal Warfare Commanders (PWC), Functional Warfare Commanders (FWC) and Coordinators.

PWCs include the Air Defense Commander (ADC), Strike Warfare Commander (STWC), Information Warfare Commander (IWC), Anti-submarine Warfare Commander (ASWC), and Surface Warfare Commander (SUWC). ASW and SUW areas can be combined under a Sea Combat Commander (SCC). PWCs collect and distribute information pertinent to their warfare areas and can be delegated authority to respond to threats with assigned assets.

FWCs perform duties of a scope or duration more limited than that of a PWCs. Typical FWCs include Maritime Interception Operations Commander (MIOC), Mine Warfare Commander (MIWC), Operational Deception Group Commander, Screen Commander (SC) and Underway Replenishment Group (URG) Commander.

Coordinators are responsible to the OTC/CWC for managing assets and resources. Among assigned Coordinators are the Air Resource Element Coordinator (AREC), Air Control Authority (ACA), Cryptologic Resource Coordinator (CRC), Force Over-the-horizon Track Coordinator (FOTC), Force Track Coordinator (FTC), Helicopter Element Coordinator (HEC), Submarine Operations Coordinating Authority (SOCA), TLAM Launch Area Coordinator (LAC) and TLAM Strike Coordinator (TSC).

The OTC/CWC may activate any or all of these warfare commanders and coordinators as necessary. The guiding principle of CWC doctrine is flexibility to meet operational requirements.

Multinational Operations

US naval forces regularly participate in peacetime and wartime multinational operations. Although the President always retains command authority over US forces, he may place them under control of a foreign commander as required to achieve specific military objectives. Multinational operations may be conducted under the structure of a formal alliance (such as NATO) or of an ad hoc coalition (Operation Desert Shield/Desert Storm).

Complex Naval Task Forces

Complex Task Forces usually consist of multiple CSGs and/or ESGs and may also include naval assets of allied nations. Such forces may operate together under three generic command and control structures.

In Situation A, the forces integrate, the senior officer present or officer designated by higher authority becomes the overall OTC/CWC and a new single CWC organisation is established.

In Situation B, task groups do not integrate. Unless otherwise directed, the senior OTC/CWC coordinates the tactical operations of all assigned naval forces and delegates responsibilities and TACON of specific forces to junior commanders as appropriate. The senior OTC/CWC may also designate junior commanders as sector OTC/CWCs.

In Situation C, each group retains its own OTC/CWC and its own set of warfare commanders and coordinators. The OTC/CWC of the supported force (or a common superior) draws on the assets of the entire force to achieve joint and combined force objectives.

Amphibious Operations

While the terms 'Commander Amphibious Task Force' (CATF) and 'Commander Landing Force' (CLF) no longer imply a command relationship, they are useful, descriptive terms that convey functional responsibilities. Command and Control of Amphibious Operations is established by the common superior. Generally, the Navy and Marine Corps commanders are considered co-equal for planning. As planning shifts to execution, the commander who is most responsible for the impending mission is generally seen as the supported commander with his counterpart acting as the supporting commander. This support relationship is now recognised in Joint Doctrine.

US Marine Corps Organisation

Marine Corps Structure

Title 10 directs that the Marine Corps is to consist of three divisions and three air wings with their necessary logistics support and that there is to be a similar organisation in the reserves consisting of one division, one air wing, and their respective logistical support groups. MEFs I (Camp Pendleton, CA), II (Camp Lejeune, NC) and III (Okinawa, Japan) are the three standing Marine Expeditionary Forces (MEFs).

The MEF is the USMC's principal war-fighting organisation for larger operations. Commanded by a lieutenant general, it consists of 46–50,000 personnel and includes, typically, a Marine ground division, air wing, Marine Logistics Group (MLG) and headquarters group. MEFs can conduct a broad scope of missions in any environment for 60 days and are supported by amphibious shipping and/or Maritime Prepositioning Squadrons (MPS). Because of its size, the MEF is normally committed sequentially, building on a smaller operational unit such as a Marine Expeditionary Brigade (MEB) or Marine Expeditionary Unit (MEU).

The MEB is designed as the lead element for a MEF or for small-scale contingencies. Normally commanded by a major general or brigadier, it consists of up to 16,000 Marine and Navy personnel and has thirty days sustainability. The ground combat element consists of an infantry regiment reinforced by artillery, some armour, light armoured vehicles, assault amphibian vehicles, and combat engineers. These assets can be divided into four battalion-size manoeuvre elements, supported by three to six fixed- and rotary-wing aircraft squadrons.

MEUs routinely forward deploy on Expeditionary Strike Groups (ESG). Commanded by a colonel, MEUs contain approximately 2,200 Marine and Navy personnel and can sustain operations for fifteen days. MEUs normally consist of a reinforced

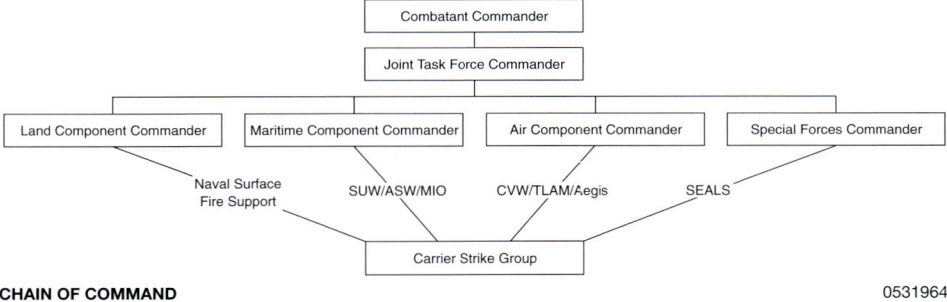

CHAIN OF COMMAND *0531964*

Composite Warfare Commander Structure

OTC/CWC

Principal Warfare Commanders

| Air Defense Commander (ADC) | Antisubmarine Warfare Commander (ASWC) | Information Warfare Commander (IWC) | Sea Combat Commander (SCC) | Strike Warfare Commander (STWC) | Surface Warfare Commander (SUWC) |

Functional Warfare Commanders

| Maritime Interception Operations Commander (MIOC) | Mine Warfare Commander (MIWC) | Operational Deception Group Commander | Screen Commander (SC) | Underway Replenishment Group (URG) Commander |

Coordinators

| Air Resource Element Coordinator (AREC) | Cryptologic Resources Coordinator (CRC) | Force Track Coordinator (FTC) | Helicopter Element Coordinator (HEC) | Launch Area Coordinator (LAC) |

| Airspace Control Authority (ACA) | Force Over-the-Horizon Coordinator (FOTC) | Submarine Operations Coordinating Authority (SOCA) | TLAM Strike Coordinator (TSC) |

CWC STRUCTURE *0531963*

US > Introduction

infantry Battalion Landing Team (BLT), a composite helicopter squadron (with air command and control and six Harriers), and a MEU Combat Logistics Battalion (CLB). Typically, such a force can act as the lead element for a larger force and/or provide shaping/engagement activities, deterrence, and limited power projection. It has the capability to conduct company to battalion-sized raids to the range limits of assigned helicopters, roughly 70–100 miles from the ESG. With the addition of the MV-22B to the operational inventory and the implementation of new tactics, the MEU is now capable of conducting raids up to three times further. An ESG typically consists of 1 LHD/LHA, 1 LPD and 1 LSD.

Marine Corps Operations

Operations are conducted by Marine Air Ground Task Forces (MAGTFs) whose size and composition will be dictated by operational circumstances. A MAGTF can be established by drawing ground, aviation, and combat service support assets from divisions, air wings, and their support groups. At the lower end of the scale, MEUs are available as immediately responsive, sea-based MAGTFs while, on a much greater scale, a full MEF might be required. This might be based on one of the standing MEFs or, as in Operation Desert Shield/Desert Storm, drawn from all three standing MEFs. A MAGTF always consists of a Command Element (CE), Ground Combat Element (GCE), Aviation Combat Element (ACE) and a Logistics Combat Element (LCE).

Embarked MEU

Marine Corps amphibious forces embarked on ESGs come under the OPCON of the naval or maritime component commander. They remain under the naval or maritime component commander throughout an amphibious operation if they will re-embark. If they transition to sustained operations ashore, they chop to either the Marine component commander or the land component commander. A Marine Corps component commander may be designated as the joint force maritime, land, or air component commander.

Communications and Data Systems

Advanced Combat Direction System (ACDS)

ACDS is a centralised, automated command and control system. An upgrade from the Naval Tactical Data System (NTDS) for aircraft carriers and large-deck amphibious ships, it provides the capability to identify and classify targets, prioritise and conduct engagements, and exchange targeting information and engagement orders within the battle group and among different service components in the joint theatre of operations. ACDS is a core Sea Shield component of non-Aegis/non-SSDS combat systems.

ACDS consists of two variants. The ACDS Block 0 system replaces obsolete NTDS computers and display consoles and incorporates new software. ACDS Block 0 is deployed on five aircraft carriers, five Wasp (LHD-1)-class amphibious assault ships, and all five Tarawa (LHA-1)-class amphibious assault ships. ACDS Block 1 is installed in one ship: *Wasp*. Following the OPEVAL failure of ACDS Block 1, it is to be replaced by the Ship Self Defense System (SSDS).

AEGIS Combat System

The AEGIS system is designed as a total weapon system, from detection to kill in the air, surface and sub-surface domains.

The SPY-1 radar system is the primary air and surface radar for the Aegis Combat System installed in the Ticonderoga (CG-47) and Arleigh Burke (DDG-51)-class warships. It is a multifunction, phased-array radar capable of search, automatic detection, transition to track, tracking of air and surface targets, and missile engagement support. The third variant of this radar, SPY-1D(V), the Littoral Warfare Radar, improves the radar's capability against low-altitude, reduced radar cross-section targets in heavy clutter environments, and in the presence of intense electronic countermeasures. The SPY-1 Series radars also demonstrated the capability to detect and track theatre ballistic missiles. AEGIS equipped platforms include Spanish F-100 and Japanese DDG ship classes.

Automated Digital Network System (ADNS)

The Automated Digital Network System is responsible for the transport of all Wide Area Network (WAN) Internet Protocol (IP) services that connect afloat units to various global shore sites. It provides ship and shore IP connectivity and promotes efficient use of available satellite and line of sight communications bandwidth. ADNS converges all voice, video, and data communications between ship and shore to an IP medium and takes advantage of all shipborne RF to transmit data efficiently. Specifically, it automates routing and switching of tactical and strategic C4I data via Transmission Control Protocol/Internet Protocol (TCP/IP) networks linking deployed battle group units with each other and with the Defense Information Systems Network (DISN) ashore. ADNS uses Commercial Off-the-Shelf (COTS) and Non-Developmental Item (NDI) Joint Tactical Architecture (JTA) – compliant hardware (routers, processors and switches), and commercial-compliant software in a standardised, scalable, shock-qualified rack design.

Commercial Broadband Satellite Program

Commercial Broadband Satellite Program (CBSP) – AN/USC-69(V), 1, 2, 3 CBSP terminals and architecture replaced Inmarsat terminals installed on unit level ships (for example, CG, DDG, LSD) and is replacing the Commercial Wideband Satellite Program (CWSP) terminal/architecture (formerly known as Challenge Athena) on force level ships (for example, LCC, CVN, LHA, LHD and LPD-17 class). Currently the CBSP terminal and architecture operate in the C-band, Ku-band and Military/Commercial X-Band. Future plans may include the military and commercial Ka-band. CBSP augments and assures bandwidth to surface combatant ships that may not be available from military satellites; that is, Wideband Global Satellite (WGS). Specific every day requirements include: Joint Service Imagery Processing System-Navy/Concentrator Architecture (JSIPS-N/JCA), Video Tele-Conferencing (VTC), Video Information Exchange system (VIXS), Video Tele-Medicine (VTM), Video Tele-Training (VTT), Automated Digital Network System (ADNS), Integrated Digital Switching Network (IDSN) for voice/telephone, Secret/Unclassified Internet Protocol Router Networks (SIPRNET/NIPRNET) and Joint Worldwide Intelligence Communications System (JWICS). The CBSP terminal uses commercial satellite connectivity and COTS/NDI Equipment. CBSP is an integral part of Navy's SATCOM architecture because of overburdened military satellite communications systems.

Navy Extremely High Frequency Satellite Communications Terminal (NESP), AN/USC-38(V)

The NESP terminal provides a highly reliable protected, (Anti-Jam, Low Probability of Interception/Detection, Anti-Scintillation) satellite communications capability to tactical fleet units. NESP allows operational units to access the family of Department of Defense communications satellites Milstar, Advanced Extremely High Frequency (AEHF) and Interim Polar Satellite System. NESP terminals provide vital survivable wartime command and control satellite communication systems for submarine, ship and shore platforms with significant terminal commonality between platform types. The Navy EHF Communications Controller is an appliqué for the NESP supports the exchange of computer-to-computer tactical data over SATCOM circuits. NESP terminals began fleet introduction in 1994. The Follow-On Terminal (FOT) replaced many of legacy NESP terminals. The NESP terminal family will be replaced by the Navy Multiband Terminal.

Navy Multiband Terminal (NMT) WSC-9, USR-10, & BSC-3

NMT is the Navy's next generation military satellite communications terminal. NMT's multiband capabilities will also enable communications over existing and emerging military EHF and SHF SATCOM systems while reducing the communications equipment footprint. NMT provides communication services via the follow Department of Defense satellites: AEHF/Wideband Global System/Interim and Enhanced Polar Systems/Defense Satellite Communications System and Milstar. All major ships, submarines and selected land based Navy communications facilities will be equipped with the NMT. It supports critical command and control communications by providing delivery of BMD targeting data over a protected EHF network, Low Probability of Detection/Low Probability of Intercept) and Anti-Jam satellite communications. Foreign Military Sales cases for an international partner variant have been signed for Canada, The Netherlands and United Kingdom.

AN/USC-61(C) Digital Modular Radio (DMR)

AN/USC-61(C) Digital Modular Radio (DMR) is the Navy's first software-defined radio to have become a communications system standard for the US military. DMR has four independent, full-duplex channels, which provide surface ships, submarines and shore commands with multiple waveforms and associated internal multilevel information security for voice and data communications. A single DMR is capable of replacing multiple existing legacy radios in the High Frequency (HF), Very High Frequency (VHF) and Ultra High Frequency (UHF), Line of Sight (LOS) and UHF Satellite Communications (SATCOM) frequency bands. DMR is software configurable and programmable with an open system architecture using Commercial Off-The-Shelf (COTS)/Non-Developmental Item (NDI) hardware. DMR achieved Full Rate Production (FRP) status on 07 May 2012.

Miniature Demand Assigned Multiple Access (Mini-DAMA)

Mini-DAMA is a communications system that supports the exchange of secure and non-secure battle group co-ordination data, tactical data and voice between base band processing equipment over UHF SATCOM, 25/5 kHz DAMA, 25/5 kHz Non-DAMA and UHF LOS. Navy has completed installations for submarines and Arleigh Burke destroyers AV(2), mine warfare ships V(2) and aircraft V(3). These Mini-DAMA radio installations provide the channel utilisation efficiencies by employing Time Division Multiple Access (TDMA) methods that have been achieved for surface warfare ships and shore stations equipped with the larger version TD-1271 DAMA multiplexer.

Battle Force Technical Network

The Battle Force Tactical Network (BFTN) provides High-Frequency Internet Protocol (HFIP) and subnet relay (SNR) to allied, coalition, and national naval and maritime units with a direct platform-to-platform tactical networking capability using legacy ultra-high-frequency (UHF) and high-frequency (HF) radios. The two technologies operate efficiently with current legacy equipment providing a cost-effective solution for achieving tactical IP networking at sea. BFTN enables warfighters on Combined Enterprise Regional Information Exchange System-Maritime (CENTRIXS-M) and Secure Internet Protocol Routing Network (SIPRNET) networks to execute and plan in a real-time tactical environment by transporting IP data directly to and from ships, submarines and aircraft. HFIP operates in the HF spectrum and is capable of data rates of 9.6 kbps in single side band and 19.2 kbps in independent side band. SNR operates in the UHF spectrum and is capable of data rates up to 64 kbps. BFTN allows surface platforms the ability to share a single SATCOM resource for reach-back capability. HFIP also supports the hardware/software upgrade requirements for battle force email (BFEM).

Common Data Link – Navy (CDL-N) Communications Data Link System (CDLS) AN/USQ-167

The CDLS terminal provides a high bandwidth, LOS datalink for exchange of signal and imagery Intelligence, Surveillance and Reconnaissance (ISR) data as well as Full Motion Video (FMV) between Navy/Joint airborne sensors and shipboard processing systems. CDLS is installed on CVNs, LHAs and LHDs. Initial versions of CDL-N began fleet introduction in 1998. CDLS has been operational since 2005. CDLS is to be replaced by Network Tactical Common Data Link (NTCDL) after 2015.

Consolidated Afloat Networks and Enterprise Services (CANES)

CANES is the technical infrastructure consolidation of existing, separately managed afloat networks including Integrated Shipboard Network Systems (ISNS), Combined Enterprise Regional Information Exchange System – Maritime (CENTRIXS-M), Sensitive Compartmented Information (SCI) Networks, and Submarine Local Area Networks (SubLAN). CANES brings Infrastructure and Platform as a Service (IaaS/PaaS), within which current and future iterations of Tasking, Collection, Processing, Exploitation, and Dissemination (TCPED) computing and storage capabilities will reside. CANES will provide complete infrastructure, inclusive of hardware, software, processing, storage, and end-user devices for unclassified, coalition, secret and SCI for all basic network services (email, web, chat, collaboration) to a wide variety of Navy surface combatants, submarines, maritime operations centers and aircraft. CANES reached Milestone C in December 2012 and Initial Operating Capability (IOC) in October 2013.

Co-operative Engagement Capability (CEC)

CEC improves battle force air defense capabilities by integrating sensor data of each co-operating ship and aircraft into a single, highly jam-resistant, real-time, fire-control-quality, composite track picture. CEC is a critical pillar of Naval Integrated Fire Control-Counter Air (NIFC-CA) capability and provides a significant contribution to the Joint Integrated Fire Control operational architecture. CEC interfaces the sensor capabilities of each CEC-equipped ship and aircraft in the strike group to support integrated engagement capability. CEC systems are installed on 133 locations including at sea in 70 ships (Aegis CGs and DDGs, carriers, and amphibious ships), 26 E-2C Hawkeye 2000 aircraft and five E-2D Hawkeye aircraft. It is also installed on eight Army aerostats, 10 USMC Composite Tracking Networks and 14 land-based test sites. Total future CEC installation are planned in approximately 269 ships, aircraft and land units. Ongoing system improvements are delivered in a new family of antennas and with reductions in system cost, weight, cooling and power. The USG-3B variant of the airborne CEC processor is in operational test in conjunction with the same for the Navy's E-2D. Navy co-ordinated with the Joint Staff, Office of the Secretary of Defense and other services to explore potential multiservice avenues for CEC capability implementation to expand sensor netting track data availability to meet a variety of warfighting requirements across various platforms. This effort has resulted in implementation of CEC into ground mobile systems including the Marine Corps' Composite Tracking Network (CTN) and the Army's Joint Land Attack Cruise Missile Defense Elevated Netted Sensor (JLENS).

Deployable Joint Command and Control (DJC2)

The DJC2 system is a transformation initiative of the Secretary of Defense and Joints Chiefs of Staff that provides a standardised, rapidly deployable, scalable and reconfigurable joint Command and Control (C2) and collaboration operations center. It can be set up anywhere in the world to support combatant commanders and their joint component commands in the rapid standup of a Joint Task Force (JTF) headquarters. DJC2 supports operations ranging in scale from first responder or small early-entry, to a full JTF combat operations center. Within 24 hours of arriving in theatre, the Joint Force Commander (JFC) and staff can securely communicate across the world, send and receive information across five different computer networks (Non-Classified Internet and secure networks NIPRnet, SIPRnet, CENTRIXS, and JWICS), participate in secure video teleconferences with remote locations and use a fully integrated Command and Control (C2)/collaborative software tool suite (including Global Command and Control – Joint (GCCS-J)) to plan and execute missions. Future upgrades to fielded systems are planned and will be delivered via technical insertion and technical refresh.

Distributed Common Ground System-Navy (DCGS-N)

DCGS-N Increment 1 is the Navy component of the Department of Defense (DoD) DCGS family of systems. DCGS-N provides integration of Intelligence, Surveillance, Reconnaissance, and Targeting (ISR&T) capabilities. DCGS-N fields to aircraft carriers (CVN), amphibious assault ships (LHA/LHD), fleet command ships (LCC), Maritime Operations Centers (MOC) and at selected shore sites. DCGS-N makes maximum use of mature Commercial-Off-The-Shelf (COTS) and Government-Off-The-Shelf (GOTS), and joint services software, tools, and standards to provide a scalable, modular, and extensible multisource capability that is interoperable with the other service and agency DCGS systems. Increment 1 includes: the Global Command and Control System-Joint Integrated Imagery and Intelligence (GCCS-I3) for intelligence analysis and processing tools and capabilities; Generic Area Limitation Environment (GALE) Lite for SIGINT analysis; Common Geopositioning Services (CGS) for imagery processing and exploitation, as well as aim-point mensuration in support of precision guided and coordinate seeking weapons; implementation of the DCGS Integration Backbone (DIB) for sharing intelligence within the DCGS family of systems; use of Net Centric Enterprise Services (NCES) standards to enhance interoperability and expose ISR data to the wider DoD audience; and exchange of ISR&T and Command and Control (C2) track information with the fielded GCCS family of systems. DCGS-N is migrating to a Common Computing Environment (CCE) construct in alignment with Consolidated Afloat Networks and Enterprise Services (CANES). As of December 2013, DCGS-N Increment 1 (Block 1) had been fielded to 24 of 34 locations. DCGS-N Increment 1 is replacing the legacy Joint Service Imagery Processing System – Navy (JSIPS-N) systems. Increment 1 will reach Full Operational Capability at the end of FY14.

DCGS-N Increment 2 builds upon the capabilities provided by DCGS-N Increment 1 and Maritime Domain Awareness (MDA) Spiral 1, converging afloat and ashore ISR into an integrated Information Dominance enterprise. Increment 2 will be a software-centric program that will support evolving fleet needs through early and frequent delivery of capabilities starting in FY2016. It will leverage the CANES, DoD and Intelligence

Community (IC) hardware and software infrastructures, including the widget construct and emerging cloud architecture, to ensure the Navy's joint C4ISR interoperability. Increment 2 will also address the remaining Navy Tasking, Collection, Processing, Exploitation, and Dissemination (TCPED) capability gaps that have been identified to include the ability to process, exploit and disseminate sensor data from emerging sensors such as the Navy's Broad Area Maritime Surveillance (BAMS) and Unmanned Carrier Launched Airborne Surveillance and Strike System (U-CLASS) unmanned sensors as well as data from the new P-8 manned aircraft. Increment 2 will greatly improve the Navy's ability to: 1) detect and identify maritime threats, 2) fuse National, Tactical and inter-theatre data for operational use, and 3) allow better DCGS FoS and IC visibility into maritime collection requirements. DCGS-N Increment 2 consists of two Releases beginning in FY16. The first Release provides an enhanced Navy ISR enterprise that: converges and builds on the DCGS-N Increment 1 and Maritime Domain Awareness Enterprise Nodes; leverages the Defense Intelligence Information Enterprise (DI2E) framework; federates ISR and TCPED workflow and production improving throughput through automation; exploits new and evolving sensors; provides Multi-INT cross-queuing and provides modular tools accessible via a web browser. The second Release enhances afloat ISR capabilities by providing: a set of software centric tools hosted on CANES providing Multi-INT fusion and analysis, behaviour prediction and intelligent knowledge management designed to operate in disconnected or denied communications environment.

Global Broadcast Service (GBS)

The Global Broadcast Service augments and interfaces with other systems to provide virtual two-way IP-networked communications to deliver a continuous, high-speed, one-way flow of high-volume information broadcast. GBS supports routine operations, training and military exercises, special activities, crisis, situational awareness, weapons targeting, intelligence, and the transition to and conduct of operations short of nuclear war. Homeland defensive operations are supported by a requirement for continental US coverage, which also provides exercise support, training and work-ups for deployment. GBS also supports military operations with US allies or coalition forces. GBS is an information technology, mission-essential, national security system providing network-centric warfare communications, but does not incorporate nuclear survivability and hardening features. GBS provides the largest bandwidth for afloat users. GBS provides capability to disseminate quickly large information products to various joint and small user platforms. With increased capacity, faster delivery of data, and near real-time receipt of imagery and data to the war fighter, it reduced reliance on MILSATCOM systems.

Global Command and Control System (GCCS)

GCCS is a comprehensive, worldwide network-centric system which provides the National Command Authority (NCA), Joint Chiefs of Staff, combatant and functional unified commands, Services, Defense Agencies, Joint Task Forces and their Service components, and others with information processing and dissemination capabilities necessary to conduct Command and Control (C2) of forces. GCCS is a means to implement the Command, Control, Communications, Computers, and Intelligence for the Warrior (C4IFTW) concept. GCCS provides the operational commanders with a near-realtime Common Operational Picture, intelligence information, collaborative joint operational planning and execution tools, and other information necessary for the execution of joint operations.

Global Command and Control System (Maritime) (GCCS-M)

GCCS-Maritime (GCCS-M) is the designated Command and Control (C2) system for the Navy and the naval implementation of the Global Command and Control System (GCCS). The evolutionary integration of previous C2 and intelligence systems, GCCS-M supports multiple war-fighting and intelligence missions for commanders at every echelon, in all afloat, ashore, and tactical naval environments, and for joint, coalition, and allied forces. GCCS-M meets the joint and service requirements for a single, integrated, scalable C2 system that receives, displays, correlates, fuses, and maintains geo-locational track information on friendly, hostile, and neutral land, sea, and air forces and integrates it with available intelligence and environmental information. GCCS-M supports evolving concepts for Network-Centric Operations by receiving, displaying, correlating, fusing, and integrating all available track, intelligence and imagery information for the warfighter. GCCS-M is implemented afloat, at ashore fixed command centers, and as the C2 portion of mobile command centers to exchange data among approximately 20,000 users for near real-time situational awareness critical to operational and tactical analysis and decision-making for controlling US, Allied and multinational forces.

The GCCS-M Program is currently structured in two acquisition increments:

- Increment 1 includes GCCS-M 3.x and 4.0.x versions. It is in the Operations and Support phase for GCCS-M 3.x and the Production and Deployment phase for GCCS-M 4.0.x. GCCS-M 4.0.3 completed Operational Testing in October 2009. This software-only maintenance build was installed on the ship's Common Computing Environment (CCE).
- Increment 2 consists of GCCS-M Version 4.1. Initial Operational Capability (afloat) in July 2010 on Patrol Coastal Craft. IOC for Force and Unit Level ships for GCCS-M 4.1 in August 2011. GCCS-M 4.1 is a software-only product that will be installed on existing Navy Afloat Enterprise Network hardware infrastructure (for example, ISNS, CANES, and so on). In addition to various other new capabilities, GCCS-M 4.1 increases track database capacity to 100,000+ and incorporates Ballistic Missile Defense planning capabilities.

Integrated Broadcast Service/Joint Tactical Terminal (IBS/JTT)

The Integrated Broadcast Service (IBS) is a system-of-systems that will migrate the Tactical Receive Equipment and Related Applications Data Dissemination System (TDDS), Tactical Information Broadcast Service (TIBS), Tactical Reconnaissance Intelligence Exchange System (TRIXS), and Near Real-Time Dissemination (NRTD) system into an integrated service with a common format. The IBS will send data via communications paths, such as UHF, SHF, EHF, GBS, and via networks. This program supports Indications Warning (I&W), surveillance, and targeting data requirements of tactical and operational commanders and targeting staffs across all warfare areas. It comprises broadcast-generation and transceiver equipment that provides intelligence data to tactical users. The Joint Tactical Terminal (JTT) will receive, decrypt, process, format, distribute, and transmit tactical data according to preset user-defined criteria across open-architecture equipment. JTT will be modular and will have the capability to receive all current tactical intelligence broadcasts (TDDS, TADIXS-B, TIBS, and TRIXS). JTT will also be interoperable with the follow-on IBS UHF broadcasts. However, the current JTT form factor does not meet space and weight constraints for a majority of the Navy and Air Force airborne platforms. Therefore, to ensure joint interoperability, the Navy and Air Force are pursuing a Special Operations Command-designed Embedded National Tactical Receiver (ENTR) for airborne platforms.

Integrated Radar Optical Surveillance and Sighting System (IROS3)

IROS3 is the Situational Awareness component of the Shipboard Protection System (SPS) Increment one. It employs COTS-based/Open Architecture products, and its key components include SPS-73 or equivalent surface search radar, electro-optical/infrared devices, an integrated surveillance system, spotlights, long range acoustic devices, and remotely operated stabilised small arms mounts. SPS Increment I is designed to detect, classify and engage real-time asymmetric threats at close-range to ships in port, at anchor and while transiting choke points or operating in restricted waters. The system provides 360° Situational Awareness (SA) and employs COTS integration to support incremental modifications as needed to tailor the system to the mission. The system has undergone extensive testing in the laboratory and a prototype is being tested at sea in Ramage. The system is to be installed in most ship classes, including surface combatants, patrol boats, amphibious and auxiliary ships, and Coast Guard cutters.

Joint Service Imagery Processing System (JSIPS-N)

JSIPS-N provides a digital imagery processing and management system, with the capability to task, process, exploit, and disseminate imagery, imagery-derived products, and imagery intelligence (IMINT) based on National, theatre, and tactical sensors. As a primary mission, JSIPS-N assists strike planners, tactical aviators, and USMC amphibious planners in the delivery of precision ordnance. JSIPS-N is installed on aircraft carriers (CVN), amphibious assault ships (LHA/LHD), fleet command ships (LCC), and at selected shore sites. A Service Life Extension Program (SLEP) was developed, integrated, and fielded by August 2010. The JSIPS-N SLEP is comprised of four subsystems: Common Geopositioning Services (CGS); Image Product Library (IPL); Imagery Exploitation Support System (IESS) Client; and the VANTAGE Shared Airborne Reconnaissance Pod (SHARP) processing and exploitation capability. The JSIPS-N system will be fully replaced by the Distributed Common Ground System-Navy (DCGS-N) by FY14.

Joint Surveillance Target Attack Radar System (JSTARS)

JSTARS is described as a 'bulletproof anti-jam datalink', utilising omnidirectional broadcast on UHF SATCOM. It receives and transmits real time MTI/FTI/SAR data via a secure uplink and downlink. It is used to demonstrate 'sensor to shooter' technology.

Joint Tactical Information Distribution System (JTIDS)

A joint program directed by the Office of the Secretary of Defense, JTIDS is a digital information-distribution system that provides rapid, crypto-secure, jam-resistant (frequency-hopping), and low-probability-of-exploitation tactical data and voice communication at a high data rate to Navy tactical aircraft and ships and Marine Corps units. JTIDS also provides capabilities for common-grid navigation and automatic communications relay. Joint and Coalition forces use JTIDS to maintain a fused, comprehensive, timely and consistent Common Tactical Picture. It has been integrated into numerous platforms and systems, including US Navy aircraft carriers, cruisers, destroyers, amphibious assault ships, E-2C Hawkeye aircraft and EP-3 Aries aircraft; US Air Force Airborne Warning and Command System (AWACS) aircraft; and US Marine Corps Tactical Air Operations Centers (TAOCs) and Tactical Air Command Centers (TACCs). Foreign country participants include Australia, Canada, France, Germany, Japan, NATO, Saudi Arabia and the United Kingdom. Additionally, JTIDS has been identified as the preferred communications link for Theatre Ballistic Missile Defense programs. JTIDS is the first implementation of the Link-16 (TDL-J) message standard supporting a near real-time, joint datalink network for information exchange among joint and combined forces for command and control of tactical operations.

Land Attack Warfare System (LAWS)

This prototype system networks all shooters (tactical air, shore artillery and seaborne fire support) into a Battle Local Area Network (Battle LAN) known as the 'Ring of Fire'. This automatically assigns fire missions to the most capable unit in the Battle LAN. LAWS controls preplanned missions, including Tomahawk, as well as time critical calls for fire from land forces. Fleet Battle Experiment ALFA was the initial test of this system.

Mark XIIA Identification Friend or Foe (IFF) Mode 5

IFF provides positive friendly identification to improve mission effectiveness, increase situational awareness, and minimise likelihood of fratricide. It supports Common Operational and Tactical Pictures. The Mark XIIA system is an upgrade, adding Mode 5, an encrypted waveform, to existing modes included in the Mark XII system to provide positive, secure, and reliable line-of-sight identification of friendly aircraft and ships, and to better support the Combat Identification Family of Systems strategy employed by United States, NATO, and other allied forces. Mode 5 is an ACAT II program that achieved a Full Rate Production decision in July 2012 and IOC in November 2012. Implementation is being accomplisehd through technology insertion Engineering Change Proposals (ECPs) to existing IFF digital interrogators and transponders aboard selected US Navy USMC and US Coast Guard aircraft, surface, and subsurface units. Other Services and some NATO nations are also fielding IFF Mode 5 capability.

Mission Data System (MDS)

This system allows planners to view Tomahawk Land Attack Missile information. MDS receives via TADIXS A or OTCIXS I digital Mission Data Updates (MDUs) from the Cruise Missile Support Activity (CMSA) and stores preplanned TLAM strike plans. Initial TLAM mission data fill is distributed via magnetic tape media provided by the CMSA.

Multifunctional Information Distribution System - Low Volume Terminal (MIDS-LVT)

MIDS-LVT is a multinational co-operative development program to design, develop and sustain a Link-16 tactical information distribution system similar in capability to Joint Tactical Information Distribution System (JTIDS) but in a lightweight, compact Link-16 terminal designed for fighter aircraft, helicopters, ships, and ground sites. MIDS-LVT is the most widely employed Link-16 terminal in the world. The United States serves as MIDS-LVT program leader, with France, Germany, Italy and Spain as full partners in all program phases. The MIDS-LVT employs the Link-16 (TADIL-J) message standard of US Navy/NATO publications. MIDS-LVT is fully interoperable with JTIDS and MIDS Joint Tactical Radios System (MIDS JTRS), as well as all other Link-16 systems in the world. MIDS-LVT was designed in response to current aircraft, surface ship, submarine, and ground-host size and weight constraints. The solution variants – MIDS-LVT (1) through MIDS-LVT (11) – support US Navy, US Marine Corps, and US Air Force aircraft; US Navy ships; US Army Patriot, THAAD, MEADS and other mobile ground-based defense systems; USAF and USMC ground-based Command and Control platforms; and potentially other tactical aircraft and ground-based systems. As of 30 September 2013, over 9,200 MIDS-LVTs have been delivered or are on contract and are integrated in 81 platforms and 33 foreign military sales customer nations.

Multifunctional Information Distribution System – Joint Tactical Radio System (MIDS JTRS)

MIDS JTRS is a Pre-Planned Product Improvement (P3I) to the MIDS-LVT that migrates the capabilities to a Joint Tactical Radio System Software Communication Architecture (SCA) compliant terminal. MIDS JTRS is fully interoperable with JTIDS and MIDS-LVT and is a form and fit replacement to the MIDS-LVT. The MIDS JTRS provides Link-16, TACAN, J-Voice and three additional channels for future growth to additional JTRS waveforms. The terminal incorporates the NSA and DoT/DoD Link-16 Information Assurance Modernization (IAM) and Frequency Remapping (FR) mandates. Additionally, MIDS JTRS Link-16 capabilities include Enhanced Throughput (ET), providing data rates up to 1.1 Mbps, and Time Slot Reallocation (TSR). Currently US Navy and US Air Force platforms have procured the MIDS JTRS terminal through Limited Production buys, but with the growth capability and the continued interest in capabilities being added to the terminal, additional platforms have shown tremendous interest in procuring the MIDS JTRS terminal.

NATO Improved Link Eleven (NILE)/Link 22 Program

This program, known as either NILE or Link 22, fulfils a North Atlantic Treaty Organization (NATO) operational staff requirement to develop a digital datalink with the aim of increasing the timeliness of the tactical information transfer even in a dense and hostile communications threat environment. The system is capable of using both fixed frequency and frequency hopping waveforms in both the UHF and HF bands. While designed to replace Link 11 on these media, and to provide a more robust Tactical Beyond Line of Sight capability, the Link 22 message set is designed to be more aligned with and to complement Link 16, easing multilink operations. Modern automated Network Management capabilities minimise the pre-planning requirements associated with Link 16 Networks. Link 22 has been developed to fulfil the operational requirement to exchange tactical data between tactical data systems (including operators) and to exchange necessary network management data. Link 22 incorporates F-series and FJ-series message standards (formats and protocols), a Dynamic Time Division Multiple Access (DTDMA) architecture, specific communications media and protocols, and specific procedures.

Ship Self-Defense System (SSDS) Mk 1 and 2

SSDS provides the integrated combat system for aircraft carriers and amphibious ships, enabling them to keep pace with the Anti-Ship Cruise Missile (ASCM) threat with a robust Anti-Air Warfare (AAW) and self-defense capability in support of the Navy's Sea Shield pillar. SSDS integrates the detection and engagement elements of the ship's combat system in a fully open-architecture distributed-processing system.

SSDS Mk 1 provides a doctrine-based Quick Reaction Combat Capability (QRCC), integrating ship's sensors for automated detection, and integrating onboard weapon systems to support multithreat engagement capability for ship's self defense. SSDS Mk 1 is the command and control system for LSD 41/49 class ships. SSDS Mk 2 (Mod 1-Mod 4) integrates with the Cooperative Engagement Capability (CEC) and Tactical Data Links, automates the weapons control doctrine (updated from SSDS Mk 1), and implements selected

© 2014 IHS IHS Jane's Fighting Ships 2014-2015

US > Introduction — Submarines

PREPOSITIONING PROGRAMME

Mobile Landing Platforms

T-MLP 1	Montford Point
T-MLP 2	John Glenn (bldg)
T-MLP 3	Lewis B Puller (bldg)

High Speed Vessels

HSV 4676	Westpac Express

Container Ships

T-AK 4396	Maj Bernard L Fisher
T-AK 4543	Lt Col John U D Page
T-AK 4544	SSGT Edward A Carter Jr

Maritime Prepositioning Ships

T-AK 3008	2nd Lt John P Bobo
T-AK 3009	PFC Dewayne T Williams
T-AK 3010	1st Lt Baldomero Lopez
T-AK 3011	1st Lt Jack Lummus
T-AK 3012	SGT William R Button
T-AK 3017	GYSGT Fred W Stockham
T-AK 323	TSGT John A Chapman

Large, Medium-Speed, Ro-Ro

T-AKR 310	Watson
T-AKR 311	Sisler
T-AKR 312	Dahl

Large, Medium-Speed, Ro-Ro – continued

T-AKR 313	Red Cloud
T-AKR 314	Charlton
T-AKR 315	Watkins
T-AKR 316	Pomeroy
T-AKR 317	Soderman

Aviation Logistic Ship

T-AG 5001	VADM K R Wheeler

SERVICE SUPPORT SHIPS

Fleet Ocean Tugs

T-ATF 168	Catawba
T-ATF 169	Navajo
T-ATF 171	Sioux
T-ATF 172	Apache

Salvage Ships

T-ARS 50	Safeguard
T-ATF 169	Navajo
T-AF 171	Sioux
T-ATF 172	Apache

Submarine Tenders

AS 39	Emory S Land
AS 40	Frank Cable

Hospital Ships

T-AH 19	Mercy
T-AH 20	Comfort

Cable Repair Ship

T-ARC 7	Zeus

SEALIFT

Large, Medium-speed Ro-Ro

T-AKR 295	Shughart
T-AKR 296	Gordon
T-AKR 297	Yano
T-AKR 298	Gilliland
T-AKR 300	Bob Hope
T-AKR 301	Fisher
T-AKR 302	Seay (PREPO)
T-AKR 303	Mendonca
T-AKR 304	Pililaau (PREPO)
T-AKR 305	Brittin
T-AKR 306	Benavidez

Vehicle Cargo Ships

T-AK 3005	SGT Matej Kocak
T-AK 3006	PFC Eugene A Obregon
T-AK 3007	MAJ Stephen W Pless

Cargo Ships

T-AK 3015	1st LT Harry L Martin
T-AK 3016	LCpl Roy M Wheat

Tankers

T-AOT 1125	Lawrence H Gianella

Joint High Speed Vessels

JHSV 1	Spearhead
JHSV 2	Choctaw County (bldg)
JHSV 3	Millinocket (bldg)
JHSV 4	Fall River (bldg)
JHSV 5	Trenton (bldg)
JHSV 6	Brunswick (bldg)
JHSV 7	Carson City (bldg)
JHSV 8	Yuma (bldg)
JHSV 9	Bismarck (bldg)
JHSV 10	Burlington (bldg)

High Speed Transport

HST 1	Guam
HST 2	Puerto Rico

READY RESERVE FORCE

(see pages 982 – 983)

SUBMARINES

Notes: (1) **Deep submergence vehicles:** The Deep Submergence Vehicles (DSV) are listed following the 'Research Ships' section.

(2) **Seal Delivery Vehicles (SDVs):** An operational requirement for SDVs has existed in the US Navy since 1965. The requirement was for a method to transport clandestinely Naval Special Warfare operators (SEALs), and their cargo, to and from assigned targets. SDVs can be transported to the general target area via small boats, ships or submarines and then transit on their own during the attack phase. Since SDVs inception, the need to increase combat effectiveness has been addressed with increasing endurance, larger payloads (to include weapons/explosives), reduction of exposure to the environment and provision of better multimission capabilities. Various vehicle iterations have evolved culminating in present day SDV Mk 8 Mod 1. This SDV is a fully flooded 'wet', combat submersible capable of carrying a pilot, co-pilot and up to four mission specialists with cargo. SDVs are propelled by an all-electric propulsion system powered by rechargeable silver-zinc batteries. It is transportable by Naval Special Warfare operators (SEALs) and can be deployed and recovered at-sea via surface ship or host submarine with dry deck shelter. It can also be launched and recovered pier side via crane or an improved boat ramp. Naval Special Warfare Command has 10 operational Mk 8 MOD 1 SDVs. One EDM SDV addresses modernisation, reliability and obsolescence issues and introduces alternative power source and other enhancements. SDV Mk 8 Mod 1 physical characteristics are: length: 21 ft, diameter 4.5 ft, dry weight is approximately 5,700 lb. It is constructed around an aluminium frame and bottom skin with acrylonitrile-butadiene-styrene (ABS) plastic bow and composite top-skin.

(3) **Unmanned Undersea Vehicles (UUVs):** Torpedo-sized and larger unmanned undersea vehicles are under development. Potential applications include underwater surveillance, mine-countermeasures, and anti-submarine warfare. Early experience was gained with the Mine Search System (MSS), operational testing of which was completed in 1993. The 35 ft long vehicle had a titanium hull and demonstrated the performance of mine-detection sonars and the ability of a UUV to survey designated areas with precise navigation. Further proof-of-concept experience was gained with the Long-Term Mine Reconnaissance System (LMRS), which was designed to be launched from the 21 inch torpedo tubes of an SSN. An engineering development system was delivered in 2002 but the programme was discontinued in favour of development of a modular UUV in which payloads can be swapped. This concept is to be demonstrated in the experimental Advanced Development UUV (ADUUV). In late 2008, the Navy cancelled phase 1 of the Mission-Reconfigurable UUV (MRUUV) due to 'technical and engineering limitations'. The vehicle had been scheduled to enter service in 2016. Development of a large diameter UUV continued. In October 2007, the first successful end-to-end submerged operation of two UUVs was demonstrated in SSN 768 Hartford. Further ahead, MRUUVs of larger size and longer endurance might be developed for launch from submarines and surface ships. Surface ship near-term programmes include the Battlespace Preparation Autonomous Underwater Vehicle (to be deployed from the Littoral Combat Ship) and the Surface Mine Countermeasure (SMCM) UUV. The SMCM UUV system will address Navy's need to reliably detect and identify undersea volume and bottom mines in high-clutter environments and areas with potential for mine case burial. The SMCM UUV will also gather environmental data to provide intelligence support for other mine warfare systems. The system will be a part of the Littoral Combat Ship Mine Warfare Mission Package and be capable of operating from a platform of opportunity. The SMCM UUV will reduce risk to personnel by operating in the minefield as an offboard sensor while the host ship stays outside the minefield boundaries.

SDV Mk VIII

10/1997, A McKaskle, USN / 0053312

Strategic Missile Submarines (SSBN)

Notes: (1) The Trident missile fitted SSBN force provides the principal US strategic deterrent under the control of US Strategic Command at Offutt Air Force Base, Nebraska. The Strategic Arms Reduction Treaty (START), implemented in December 2001, limited the combined number of SLBM and ICBM re-entry bodies (RBs) to 4,900. As part of this reduction, the first four Ohio-class submarines, no longer required for strategic service, were converted into conventionally-armed guided missile SSGNs, capable also of deploying Special Forces. The new START treaty, which further limited the number of deployed ballistic missiles to 700 and deployed warheads to 1,550, was signed on 8 April 2010 and ratified by the US Senate on 22 December 2010.

(2) An Analysis of Alternatives study for a follow-on sea-based strategic deterrent was formally launched on 19 February 2009. The Ohio Replacement Program achieved Milestone A and was approved to enter the technology development phase on 10 January 2011. In December 2012, General Dynamics Corp was awarded a five-year, cost-plus, fixed-fee contract for design, research and development of a new submarine class to replace Ohio-class vessels. The new submarine class is expected to begin construction in 2021, to be delivered from 2028 and to start deterrance patrol from 2031. The new class will consist of 12 submarines of similar dimensions to the Ohio class. They will have 16 missile tubes, and carry forward the Trident D5 strategic weapons system. The class will include a life-of-ship reactor core that will power the ship for its full 42-year service life, electric drive, and X-shaped stern control surfaces. This will be first US submarine class designed for mixed gender crew. The Ohio Replacement Program will capitalise on cost reduction and design features already proven in the Virginia-class SSNs and legacy Ohio-class SSBNs.

14 OHIO CLASS (SSBN)

Name	No	Builders	Launched	Commissioned	F/S
HENRY M JACKSON	SSBN 730	General Dynamics (Electric Boat Div)	15 Oct 1983	6 Oct 1984	PA
ALABAMA	SSBN 731	General Dynamics (Electric Boat Div)	19 May 1984	25 May 1985	PA
ALASKA	SSBN 732	General Dynamics (Electric Boat Div)	12 Jan 1985	25 Jan 1986	AA
NEVADA	SSBN 733	General Dynamics (Electric Boat Div)	14 Sep 1985	16 Aug 1986	PA
TENNESSEE	SSBN 734	General Dynamics (Electric Boat Div)	13 Dec 1986	17 Dec 1988	AA
PENNSYLVANIA	SSBN 735	General Dynamics (Electric Boat Div)	23 Apr 1988	9 Sep 1989	PA
WEST VIRGINIA	SSBN 736	General Dynamics (Electric Boat Div)	14 Oct 1989	20 Oct 1990	AA
KENTUCKY	SSBN 737	General Dynamics (Electric Boat Div)	11 Aug 1990	13 Jul 1991	PA
MARYLAND	SSBN 738	General Dynamics (Electric Boat Div)	10 Aug 1991	13 Jun 1992	AA
NEBRASKA	SSBN 739	General Dynamics (Electric Boat Div)	15 Aug 1992	10 Jul 1993	PA
RHODE ISLAND	SSBN 740	General Dynamics (Electric Boat Div)	17 Jul 1993	9 Jul 1994	AA
MAINE	SSBN 741	General Dynamics (Electric Boat Div)	16 Jul 1994	29 Jul 1995	PA
WYOMING	SSBN 742	General Dynamics (Electric Boat Div)	15 Jul 1995	13 Jul 1996	AA
LOUISIANA	SSBN 743	General Dynamics (Electric Boat Div)	27 Jul 1996	6 Sep 1997	PA

Displacement, tonnes: 17,033 surfaced; 19,000 dived
Dimensions, metres (feet): 170.7 × 12.8 × 11.1 *(560 × 42.0 × 36.4)*
Speed, knots: 24 dived
Complement: 155 (15 officers)

Machinery: Nuclear; 1 GE PWR S8G; 2 turbines; 60,000 hp *(44.8 MW)*; 1 shaft; 1 Magnetek auxiliary prop motor; 325 hp *(242 kW)*
Missiles: SLBM: 24 Lockheed Trident II; stellar inertial guidance to 12,000 km *(6,500 n miles)*; thermonuclear warheads of up to 12 MIRVs of either Mk 4 with W76 of 100 kT each, or Mk 5 with W88 of 300–475 kT each; CEP 90 m. A limit of 8 RVs was set in 1991 under the START counting rules.
Torpedoes: 4 – 21 in *(533 mm)* Mk 68 bow tubes. Raytheon Mk 48 ADCAP Mod 5/6/7; wire-guided (option); active/passive homing to 50 km *(27 n miles)*/38 km *(21 n miles)* at 40/55 kt; warhead 267 kg; depth to 800 m *(2,950 ft)*.

Physical countermeasures: Decoys: External and internal (reloadable) anti-torpedo decoy.
Electronic countermeasures: ESM: WLR-8(V)5; intercept. WLR-10; radar warning.
Radars: Surface search/navigation/fire control: AN/BPS-15J and AN/BPS-16(V)2; I/J-band.
Sonars: BQQ-6; passive search. BQS-13 spherical array for BQQ-6. BQS-15; active/passive for mine detection and obstacle avoidance; high frequency. BQR-15 (with BQQ-9 signal processor); TB-16 passive towed array. TB-23 thin line array. BQR-19; active for navigation; high frequency. ARCI Phase II.
Combat data systems: CCS Mk 2 Mod 3 with UYK 43/UYK 44 computers.
Weapon control systems: Mk 98 fire-control system.

Programmes: The size of the SSBN forces has been reduced to 14 hulls. *Ohio* completed conversion to SSGN in 2005, *Florida* and *Michigan* in 2006 and *Georgia* in 2007.

Modernisation: All Ohio class SSBNs have been converted to deploy Trident II D5 missiles. They have also been upgraded with Acoustic Rapid COTS Insertion (ARCI) sonar. CCS Mk 2 Block 1C fire-control systems has also been installed.
Structure: The size of the Trident submarine is dictated primarily by the 24 vertically launched Trident missiles and the reactor plant to drive the ship. The reactor has a nuclear core life of about 20 years. Diving depth is 244 m *(800 ft)*. Type 15 and Type 8 periscopes. Mk 19 Air Turbine Pump for torpedo discharge.
Operational: The eight Pacific Fleet units are based at Bangor, Washington, while the six Atlantic Fleet units are based at King's Bay, Georgia. Hull life of the class has been extended to 42 years and SSBN 730 is due to be decommissioned in 2027.

ALABAMA — *4/2004, Ships of the World* / 1043704

PENNSYLVANIA — *12/2005, Ships of the World* / 1154028

Cruise Missile Submarines (SSGN)

Notes: The force of Ohio-class SSGNs begins to reach the end of its operational life in the mid-2020s at about the same time as the 30-hull Virginia-class SSN programme is completed. One option for replacement of both classes is to modify the Virginia-class hull to incorporate a 21.4 m payload section, known as the Virginia Payload Module (VPM). The VPM would include four 87 inch (2.21 m) in-line missile tubes, each accommodating seven Tomahawk cruise missiles with the possibility of adding additional payloads when they become available. If this approach is adopted, VPMs could be brought into service in the Block V Virginia class, which is to start construction in 2019 and achieve initial operating capability in 2026.

4 OHIO CLASS (SSGN)

Name	No	Builders	Launched	Commissioned	F/S
OHIO	SSGN 726 (ex-SSBN 726)	General Dynamics (Electric Boat Div)	7 Apr 1979	11 Nov 1981	PA
MICHIGAN	SSGN 727 (ex-SSBN 727)	General Dynamics (Electric Boat Div)	26 Apr 1980	11 Sep 1982	PA
FLORIDA	SSGN 728 (ex-SSBN 728)	General Dynamics (Electric Boat Div)	14 Nov 1981	18 Jun 1983	AA
GEORGIA	SSGN 729 (ex-SSBN 729)	General Dynamics (Electric Boat Div)	6 Nov 1982	11 Feb 1984	AA

Displacement, tonnes: 17,033 surfaced; 19,000 dived
Dimensions, metres (feet): 170.7 × 12.8 × 11.1 *(560 × 42.0 × 36.4)*
Speed, knots: 25 (est.)
Complement: 159 (15 officers)

Machinery: Nuclear; 1 GE PWR S8G; 2 turbines; 60,000 hp *(44.8 MW)*; 1 shaft; 1 Magnetek auxiliary prop motor; 325 hp *(242 kW)*
Missiles: SLCM: Up to 154 Raytheon Tomahawk Block III and Block IV; TERCOM and GPS aided inertial navigation system with DSMAC to 1,600+ km *(865+ n miles)* at 0.7 Mach; warhead (WDU 36B) 454 kg.
Torpedoes: 4—21 in *(533 mm)* Mk 68 bow tubes. Raytheon Mk 48 ADCAP Mod 5/6/7; wire-guided (option); active/passive homing to 50 km *(27 n miles)*/38 km *(21 n miles)* at 40/55 kt; warhead 267 kg; depth to 800 m *(2,950 ft)*.
Physical countermeasures: Decoys: 8 launchers for Countermeasures Set Acoustic (CSA) and internal (reloadable) anti-torpedo decoy system.
Electronic countermeasures: ESM: BLQ-10; radar and comms intercept and analysis.
Radars: Surface search/navigation/fire control: AN/BPS 15J; I/J-band.
Sonars: AN/BQQ-10 suite; passive search (spherical array). TB-23/29; passive towed array (thin line). TB-16; passive towed array (fat line). ARCI.

Combat data systems: AN/BYG-1 Combat Control System.
Weapon control systems: AN/BYG-1.

Programmes: The 1994 nuclear posture review recommended a 14-strong SSBN force and that the remaining four Ohio class be converted to SSGN role. The SSGN would include land attack, special forces insertion and support and ISR roles. Conversion contract with General Dynamics Electric Boat in October 2002. *Ohio* started mid-life refuelling on 15 November 2002 and conversion work (at Puget Sound Naval Shipyard) on 19 November 2003. She completed conversion in December 2005. *Florida* started mid-life refuelling in August 2003 and conversion work (at Norfolk Naval Shipyard) in April 2004. She completed conversion in April 2006. *Michigan* started refuelling in March 2004 and conversion work (at Puget Sound) in October 2004. She completed conversion in November 2006. *Georgia* started refuelling in March 2005 and started conversion (at Norfolk) in October 2005. She completed conversion in November 2007.
Modernisation: Conversion work allows an SSGN to carry up to 154 Tomahawk or Tactical Tomahawk missiles by enabling seven cruise missiles to be fired from each of 22 of the current 24 Trident missile tubes. Some of these tubes are interchangeable with Special Forces stowage canisters. The remaining two tubes are permanently configured for wet/dry launch of up to 66 special operations forces. The combat system is also to be upgraded and future payloads are being developed to augment the baseline configuration. The sonar suite has been upgraded to BQQ-10 (V4) ARC-I system standard, which improves legacy sonars by the use of system architecture and improved data processing. The ARC-I programme facilitates progressive sonar upgrades. Technology refreshes of the AN/BYG-1 combat data system were completed in all four SSGNs in fiscal year 2012.
Structure: The size of the submarine was dictated primarily by the 24 missile tubes and the reactor plant to drive the ship. The reactor has a nuclear core life of about 20 years. Diving depth is 244 m *(800 ft)*. Type 8J periscope and Integrated Submarine Imaging System (ISIS). Mk 19 Air Turbine Pump for torpedo discharge.
Operational: *Georgia* played the part of an SSGN during Exercise 'Silent Hammer' in 2004. This tested procedures for strikes against time-critical targets and use of special operations forces. An onboard battle-centre tested communications and networking required to support them. All boats returned to the fleet by 2007. *Florida* launched more than 90 Tomahawk missiles during international operations in Libya in March 2011. *Ohio* and *Michigan* are based at Bangor, WA, and *Florida* and *Georgia* are based at King's Bay, GA.

FLORIDA *4/2006, US Navy* / 1167577

GEORGIA *8/2009, US Navy* / 1366212

OHIO *11/2008, US Navy* / 1353648

Attack Submarines

11 + 7 (10) VIRGINIA CLASS (SSN)

Name	No	Builders	Laid down	Launched	Commissioned	F/S
VIRGINIA	SSN 774	General Dynamics (Electric Boat Div)	15 Aug 1997	8 Aug 2003	23 Oct 2004	AA
TEXAS	SSN 775	Northrop Grumman (Newport News Shipbuilding)	26 Aug 1998	9 Apr 2005	9 Sep 2006	PA
HAWAII	SSN 776	General Dynamics (Electric Boat Div)	26 Oct 1999	28 Apr 2006	5 May 2007	PA
NORTH CAROLINA	SSN 777	Northrop Grumman (Newport News Shipbuilding)	11 Apr 2001	5 May 2007	3 May 2008	PA
NEW HAMPSHIRE	SSN 778	General Dynamics (Electric Boat Div)	9 Oct 2002	21 Feb 2008	25 Oct 2008	AA
NEW MEXICO	SSN 779	Northrop Grumman (Newport News Shipbuilding)	4 Mar 2004	17 Jan 2009	27 Mar 2010	AA
MISSOURI	SSN 780	General Dynamics (Electric Boat Div)	24 Feb 2005	20 Nov 2009	31 Jul 2010	AA
CALIFORNIA	SSN 781	Huntington Ingalls	15 Feb 2006	13 Nov 2010	29 Oct 2011	AA
MISSISSIPPI	SSN 782	General Dynamics (Electric Boat Div)	19 Feb 2007	13 Oct 2011	2 Jun 2012	AA
MINNESOTA	SSN 783	Huntington Ingalls	19 Feb 2008	3 Nov 2012	7 Sep 2013	AA
NORTH DAKOTA	SSN 784	General Dynamics (Electric Boat Div)	2 Mar 2009	2 Aug 2013	31 May 2014	AA
JOHN WARNER	SSN 785	Huntington Ingalls	2 Mar 2010	2014	2015	Bldg
ILLINOIS	SSN 786	General Dynamics (Electric Boat Div)	2 Mar 2011	2015	2016	Bldg
WASHINGTON	SSN 787	Huntington Ingalls	2 Sep 2011	2015	2017	Bldg
COLORADO	SSN 788	General Dynamics (Electric Boat Div)	2 Mar 2012	2016	2017	Bldg
INDIANA	SSN 789	Huntington Ingalls	2 Sep 2012	2016	2018	Bldg
SOUTH DAKOTA	SSN 790	General Dynamics (Electric Boat Div)	2 Mar 2013	2017	2019	Bldg
DELAWARE	SSN 791	Huntington Ingalls	2 Sep 2013	2017	2019	Bldg

Displacement, tonnes: 7,925 dived
Dimensions, metres (feet): 114.8 × 10.06 × 9.3 *(376.6 × 33.0 × 30.5)*
Speed, knots: 34 dived
Complement: 132 (15 officers)

Machinery: Nuclear; 1 GE PWR S9G; 2 turbines; 40,000 hp *(29.84 MW)*; 1 shaft; pump jet propulsor; 1 secondary propulsion submerged motor
Missiles: SLCM: Raytheon Tomahawk Block IV; land attack; TERCOM and GPS aided inertial navigation system with DSMAC to 1,600+ km *(865+ n miles)* at 0.7 Mach; warhead (WDU-36B) 454 kg. 12 VLS tubes (SSN 774–783) external to the pressure hull. 2 VPT (SSN 784–801) (six missiles per tube).
Torpedoes: 4—21 in *(533 mm)* bow tubes. Raytheon Mk 48 ADCAP Mod 5/6/7; wire-guided (option); active/passive homing to 50 km *(27 n miles)*/38 km *(21 n miles)* at 40/55 kt; warhead 267 kg; depth to 800 m *(2,625 ft)*. Air turbine pump discharge. Total of 38 including SLCM and torpedoes.
Mines: Mk 67 Mobile and Mk 60 Captor mines (until new mines are available).
Physical countermeasures: Decoys: External and internal (reloadable); anti-torpedo decoy.
Electronic countermeasures: ESM: AN/BLQ-10; radar and comms intercept and analysis.
Radars: Surface search/navigation/fire control: AN/BPS-16; I/J-band.
Sonars: Lockheed Martin BQQ-10 sonar suite including bow spherical active/passive array; wide aperture flank passive arrays; high-frequency active keel and fin arrays; TB-16 and TB-23/29A passive towed arrays; WLY-1 acoustic intercept. ARCI.
Combat data systems: AN/BYG-1.

Programmes: In February 1997, a teaming agreement was reached between Electric Boat Division of General Dynamics Corporation and Newport News Shipbuilding (now Huntington Ingalls Industries) jointly to build and deliver the Virginia class. Electric Boat is the lead design yard and prime contractor and delivers the even numbered hulls. Newport News delivers the odd numbered hulls. Construction of sub-assemblies is undertaken at the Electric Boat facilities in Groton, CT, and at Quonset Point RI and at Newport News Shipbuilding. Components are then shipped either to the Groton shipyard or to Newport News for final assembly and delivery. This division of construction responsibility takes advantage of modular design and construction and provides the most affordable approach to submarine construction at the two shipyards. Advanced funding was available in fiscal year 1996 for the first two submarines followed by advanced funding in FY 1998 and FY 2000 for the third and fourth. SSN 774–777 constitute Block I. A follow-on block buy procurement contract, signed in August 2003 for six Block II submarines (SSN 778–783), maintained the Electric Boat and Newport News teaming arrangement. This contract was modified in January 2004 to a multi-year procurement contract. This modification includes provisions to provide economic order quantity funding, allowing the bulk purchase of materials for more than one submarine at a significant overall cost saving. Another multi-year contract for the procurement of eight Block III (SSN 784–791) was signed on 22 December 2008. It funded one boat per year in FY 2009 and FY 2010 and two per year in FY 2011, 2012, and 2013. The Block IV contract for SSN 792–801, which will continue the two-per-year build rate from FY 2014–2018 was awarded on 28 April 2014. A program of at least 30 hulls is expected.
Modernisation: Periscope ESM Performance Improvement (PEPI) introduced from 2011.
Structure: Reactor core will last the life of the ship. Automated steering and diving control, using fly-by-wire technology, and can serve as host ship for Dry Deck Shelter (DDS). Integral lockout trunk and reconfigurable torpedo room to accommodate special operations forces and equipment. Block III boats (SSN 784–791) and Block IV (SSN 792–801) are to be built with a modified bow to incorporate a new Large Aperture Bow array and two 7 ft-diameter Virginia Payload Tubes (VPT) to replace 12 VLS tubes in Blocks I and II. Photonics AN/BVS-1 non-hull penetrating masts replace conventional periscopes for imaging. Test depth 488 m *(1,600 ft)*.
Operational: Optimised for coastal operations without sacrificing traditional deep-water capabilities. Designed for flexibility to change missions and perform a variety of mission areas: anti-submarine warfare, anti-surface warfare, covert intelligence/surveillance and reconnaissance, clandestine mine warfare, battle group support, covert support of special operations forces, irregular warfare and power projection/strike. SSN 774 completed the first full operational deployment of the class in April 2010. SSN 776 began a full deployment later in 2010.

MISSOURI   *8/2013*, Tony Roper* / 1531154

VIRGINIA   *8/2013*, Derek Fox* / 1531153

41 LOS ANGELES CLASS (SSN)

Name	No	Builders	Laid down	Launched	Commissioned	F/S
BREMERTON	SSN 698	General Dynamics (Electric Boat Div)	8 May 1976	22 Jul 1978	28 Mar 1981	PA
JACKSONVILLE	SSN 699	General Dynamics (Electric Boat Div)	21 Feb 1976	18 Nov 1978	16 May 1981	PA
DALLAS	SSN 700	General Dynamics (Electric Boat Div)	9 Oct 1976	28 Apr 1979	18 Jul 1981	AA
LA JOLLA	SSN 701	General Dynamics (Electric Boat Div)	16 Oct 1976	11 Aug 1979	30 Sep 1981	PA
CITY OF CORPUS CHRISTI	SSN 705	General Dynamics (Electric Boat Div)	4 Sep 1979	25 Apr 1981	8 Jan 1983	PA
ALBUQUERQUE	SSN 706	General Dynamics (Electric Boat Div)	27 Dec 1979	13 Mar 1982	21 May 1983	PA
SAN FRANCISCO	SSN 711	Newport News Shipbuilding	26 May 1977	27 Oct 1979	24 Apr 1981	PA
HOUSTON	SSN 713	Newport News Shipbuilding	29 Jan 1979	21 Mar 1981	25 Sep 1982	PA
NORFOLK	SSN 714	Newport News Shipbuilding	1 Aug 1979	31 Oct 1981	21 May 1983	AA
BUFFALO	SSN 715	Newport News Shipbuilding	25 Jan 1980	8 May 1982	5 Nov 1983	PA
OLYMPIA	SSN 717	Newport News Shipbuilding	31 Mar 1981	30 Apr 1983	17 Nov 1984	PA
PROVIDENCE	SSN 719	General Dynamics (Electric Boat Div)	14 Oct 1982	4 Aug 1984	27 Jul 1985	AA
PITTSBURGH	SSN 720	General Dynamics (Electric Boat Div)	15 Apr 1983	8 Dec 1984	23 Nov 1985	AA
CHICAGO	SSN 721	Newport News Shipbuilding	5 Jan 1983	13 Oct 1984	27 Sep 1986	PA
KEY WEST	SSN 722	Newport News Shipbuilding	6 Jul 1983	20 Jul 1985	12 Sep 1987	PA
OKLAHOMA CITY	SSN 723	Newport News Shipbuilding	4 Jan 1984	2 Nov 1985	9 Jul 1988	PA
LOUISVILLE	SSN 724	General Dynamics (Electric Boat Div)	16 Sep 1984	14 Dec 1985	8 Nov 1986	PA
HELENA	SSN 725	General Dynamics (Electric Boat Div)	28 Mar 1985	28 Jun 1986	11 Jul 1987	PA
NEWPORT NEWS	SSN 750	Newport News Shipbuilding	3 Mar 1984	15 Mar 1986	3 Jun 1989	AA
SAN JUAN	SSN 751	General Dynamics (Electric Boat Div)	16 Aug 1985	6 Dec 1986	6 Aug 1988	AA
PASADENA	SSN 752	General Dynamics (Electric Boat Div)	20 Dec 1985	12 Sep 1987	11 Feb 1989	PA
ALBANY	SSN 753	Newport News Shipbuilding	22 Apr 1985	13 Jun 1987	7 Apr 1990	AA
TOPEKA	SSN 754	General Dynamics (Electric Boat Div)	13 May 1986	23 Jan 1988	21 Oct 1989	PA
SCRANTON	SSN 756	Newport News Shipbuilding	29 Jun 1986	3 Jul 1989	26 Jan 1991	AA
ALEXANDRIA	SSN 757	General Dynamics (Electric Boat Div)	19 Jun 1987	23 Jun 1990	29 Jun 1991	AA
ASHEVILLE	SSN 758	Newport News Shipbuilding	1 Jan 1987	24 Feb 1990	28 Sep 1991	PA
JEFFERSON CITY	SSN 759	Newport News Shipbuilding	21 Sep 1987	17 Aug 1990	29 Feb 1992	PA
ANNAPOLIS	SSN 760	General Dynamics (Electric Boat Div)	15 Jun 1988	18 May 1991	11 Apr 1992	AA
SPRINGFIELD	SSN 761	General Dynamics (Electric Boat Div)	29 Jan 1990	4 Jan 1992	9 Jan 1993	AA
COLUMBUS	SSN 762	General Dynamics (Electric Boat Div)	7 Jan 1991	1 Aug 1992	24 Jul 1993	PA
SANTA FE	SSN 763	General Dynamics (Electric Boat Div)	9 Jul 1991	12 Dec 1992	8 Jan 1994	PA
BOISE	SSN 764	Newport News Shipbuilding	25 Aug 1988	23 Mar 1991	7 Nov 1992	AA
MONTPELIER	SSN 765	Newport News Shipbuilding	19 May 1989	23 Aug 1991	13 Mar 1993	AA
CHARLOTTE	SSN 766	Newport News Shipbuilding	17 Aug 1990	3 Oct 1992	16 Sep 1994	PA
HAMPTON	SSN 767	Newport News Shipbuilding	2 Mar 1990	3 Apr 1992	6 Nov 1993	PA
HARTFORD	SSN 768	General Dynamics (Electric Boat Div)	27 Apr 1992	4 Dec 1993	10 Dec 1994	AA
TOLEDO	SSN 769	Newport News Shipbuilding	6 May 1991	28 Aug 1993	24 Feb 1995	AA
TUCSON	SSN 770	Newport News Shipbuilding	15 Aug 1991	20 Mar 1994	18 Aug 1995	PA
COLUMBIA	SSN 771	General Dynamics (Electric Boat Div)	24 Apr 1993	24 Sep 1994	9 Oct 1995	PA
GREENEVILLE	SSN 772	Newport News Shipbuilding	28 Feb 1992	17 Sep 1994	16 Feb 1996	PA
CHEYENNE	SSN 773	Newport News Shipbuilding	6 Jul 1992	16 Apr 1995	13 Sep 1996	PA

Displacement, tonnes: 7,011 surfaced; 7,124 dived
Dimensions, metres (feet): 109.73 × 10.1 × 9.9 *(360 × 33.1 × 32.5)*
Speed, knots: 33 dived
Complement: 143 (16 officers)

Machinery: Nuclear; 1 GE PWR S6G; 2 turbines; 35,000 hp *(26 MW)*; 1 shaft; 1 Magnetek auxiliary prop motor; 325 hp *(242 kW)*

Missiles: SLCM: Raytheon Tomahawk Block III and Block IV; TERCOM and GPS aided navigation with DSMAC to 1,600+ km *(865+ n miles)* at 0.7 Mach; warhead (WDU 36B) 454 kg. SSN 719 and later are equipped with the Vertical Launch System, which places 12 launch tubes external to the pressure hull behind the spherical array forward.

Torpedoes: 4—21 in *(533 mm)* bow tubes. Raytheon Mk 48 ADCAP Mod 5/6/7; wire-guided (option); active/passive homing to 50 km *(27 n miles)*/38 km *(21 n miles)* at 40/55 kt; warhead 267 kg; depth to 900 m *(2,950 ft)*. Total of 26 weapons can be tube-launched, for example-12 Tomahawk, 14 torpedoes.

Mines: Mk 67 Mobile and Mk 60 Captor mines.

Physical countermeasures: Decoys: External and internal (reloadable) anti-torpedo decoy.

Electronic countermeasures: ESM: BLQ-10; radar and comms intercept ICADF.

Radars: Surface search/navigation/fire control: AN/BPS-15H; I/J-band.

Sonars: Lockheed Martin AN/BQQ-10 sonar suite including bow spherical active/passive array. TB-23/29A thin line array and TB-16/TB-34 passive fat line towed array. BQS-15 active close range including ice detection; high frequency. ARCI.

Combat data systems: CCS Mk 2. JOTS, BGIXS and TADIX-A can be fitted. USC-38 EHF. Link 11; Link 16 being fitted. AN/BYG-1 fire control.

Programmes: Various major improvement programmes and updating design changes caused programme delays in the late 1980s. From SSN 751 onwards the class is prefixed by an 'I' for 'improved'. Programme concluded at 62 hulls.

Modernisation: Mk 117 TFCS backfitted in earlier submarines of the class. EHF communications and Link 16 are being fitted. HDR antenna fitted on the majority of the class. BQQ-10 sonar suite and TB 29 fitted to the majority of the class. An ARCI programme from 1997–2006 backfitted BQQ-5 sonars with open system architecture. Three of the class (SSN 700, 701 and 715) are capable of operating with DDS.

Structure: Every effort has been made to improve sound quieting and from SSN 751 onwards the class has acoustic tile cladding to augment the 'mammalian' skin which up to then had been the standard USN outer casing coating. From SSN 751 onwards the forward hydroplanes are fitted forward instead of on the fin. The forward hydroplanes are retractable mainly for surfacing through ice. The S6G reactor is a modified version of the D2G type. The towed sonar array is stowed in a blister on the side of the casing. Diving depth is 450 m *(1,475 ft)*. Various staged design improvements have added some 220 tons to the class displacement between 688 and 773.

Operational: The Los Angeles class is the mainstay of the attack submarine force. The land-attack mission has been a notable feature of operations in Iraq, Kosovo, Afghanistan, and Libya. Under-ice operations are still a priority and several have surfaced at the North Pole. Special forces and intelligence gathering missions are also conducted. Normally additional Tomahawk missiles are carried internally (in addition to those stored externally). Weapon types/numbers vary according to mission. SSN 711 seriously damaged in collision with an undersea mountain south of Guam on 8 January 2005 and subsequently fitted with the bow section of SSN 718. SSN 755 seriously damaged by fire at Portsmouth Naval Shipyard on 23 May 2012 and is to be decommissioned in 2014 due to funding reductions. SSN 700 to be decommissioned in September 2014.

DALLAS (with DDS) 9/2006, *US Navy* / 1167579

OLYMPIA 10/2012, *Hachiro Nakai* / 1483659

Submarines < **US** 939

TUCSON 2/2012, Hachiro Nakai / 1483658

HELENA 10/2007, Michael Nitz / 1353647

SANTA FE 4/2011, Hachiro Nakai / 1454789

ASHEVILLE 2/2006, US Navy / 1167581

© 2014 IHS IHS Jane's Fighting Ships 2014-2015

3 SEAWOLF CLASS (SSN)

Name	No	Builders	Laid down	Launched	Commissioned	F/S
SEAWOLF	SSN 21	General Dynamics (Electric Boat Div)	25 Oct 1989	24 Jun 1995	19 Jul 1997	PA
CONNECTICUT	SSN 22	General Dynamics (Electric Boat Div)	14 Sep 1992	1 Sep 1997	11 Dec 1998	PA
JIMMY CARTER	SSN 23	General Dynamics (Electric Boat Div)	12 Dec 1995	13 May 2004	19 Feb 2005	PA

Displacement, tonnes: 8,189 surfaced; 9,285 (SSN 21–22), 12,353 (SSN 23) dived
Dimensions, metres (feet): 107.6 (SSN 21–22), 138.1 (SSN 23) × 12.9 × 10.9 *(353, 453.1 × 42.3 × 35.8)*
Speed, knots: 39 dived
Complement: 140 (14 officers)

Machinery: Nuclear; 1 Westinghouse PWR S6W; 2 turbines; 45,000 hp *(33.57 MW)*; 1 shaft; pumpjet propulsor; 1 (4 in SSN 23) Westinghouse secondary propulsion submerged motor(s)
Missiles: SLCM: Raytheon Tomahawk Block III and Block IV; TERCOM and GPS aided navigation with DSMAC to 1,600+ km *(865+ n miles)* at 0.7 Mach; warhead (WDU 36B) 454 kg.
Torpedoes: 8—26 in *(660 mm)* tubes (external measurement is 30 in *(762 mm)*); Raytheon Mk 48 ADCAP Mod 5/6/7; wire-guided (option); active/passive homing to 50 km *(27 n miles)*/38 km *(21 n miles)* at 40/55 kt; warhead 267 kg; depth to 800 m *(2,625 ft)*. Air turbine discharge. Total of 50 tube-launched missiles and torpedoes.
Mines: 100 in lieu of torpedoes.
Physical countermeasures: Decoys: External and internal (reloadable) anti-torpedo decoy.
Electronic countermeasures: ESM: AN/BLQ-10 radar and comms intercept.
Radars: Navigation: AN/BPS-16(V)3; I/J-band.
Sonars: BSY-2 suite with bow spherical active/passive array and wide aperture passive flank arrays; TB-16 and TB-29A surveillance and tactical towed arrays. WLY-1 system.
Combat data systems: AN-BYG-1. USC-38 EHF. JMCIS.
Weapon control systems: AN/BYG-1.

Programmes: First of class ordered on 9 January 1989; second of class on 3 May 1991 and third on 30 April 1996. Design changes to *Carter* contracted in late 1999 delayed the launch by four years.

Modernisation: The hull of SSN 23 is about 30 m longer to accommodate an hour-glass shaped Ocean Interface section and provides for additional payload to accommodate advanced technology used to carry out classified research and development, and for enhanced warfighting and special operations capabilities. *Carter* retains all of the Seawolf class's original war-fighting capability. All three boats have converted to a common open architecture and COTS Submarine Warfare Federated Tactical System (SWFTS) to establish a common submarine baseline that can be easily upgraded.
Structure: There are no external weapons. Emphasis has been put on sub-ice capabilities including retractable bow planes. Test depth 1,950 ft *(594 m)*.
Operational: A quoted 'silent' speed of 20 kt. Other operational advantages include greater manoeuvrability and space for subsequent weapon systems development. All three boats based at Bangor, WA.

JIMMY CARTER — 2/2005, Ships of the World / 1127057

CONNECTICUT — 11/2009, US Navy / 1366211

Aircraft carriers < US 941

AIRCRAFT CARRIERS

0 + 2 (1) GERALD R FORD CLASS (CVN)

Name	No	Builders	Laid down	Launched	Commissioned
GERALD R FORD	CVN 78	Huntington Ingalls, Newport News Shipbuilding	14 Nov 2009	19 Nov 2013	2016
JOHN F KENNEDY	CVN 79	Huntington Ingalls, Newport News Shipbuilding	2015	2019	2022
ENTERPRISE	CVN 80	Huntington Ingalls, Newport News Shipbuilding	2019	2023	2027

Displacement, tonnes: 101,605 full load (est.)
Dimensions, metres (feet): 332.8 × 40.8 × 12.4 (1,091.9 × 133.9 × 40.7)
Flight deck, metres: 332.8 × 78.0 (1,091.9 × 255.9)
Aircraft lift: 3
Speed, knots: 30 (est.)
Complement: 4,660

Machinery: Nuclear; 2 AIB reactors; 4 shafts
Missiles: SAM: 2 Raytheon GMLS Mk 29 launchers for Raytheon Evolved Sea Sparrow RIM-7 ❶. 2 GMLS Mk 49 RAM RIM-116 ❷.
Guns: 3 Raytheon 20 mm Vulcan Phalanx Block 1B Mk 15 6-barrelled ❸.
Physical countermeasures: Decoys: SLQ-25C.
Electronic countermeasures: ESM/ECM: SEWIP Block 2/3.
Radars: Air search: Dual Band Radar (DBR); Raytheon SPY-3 ❹; 3D; I-band and Lockheed Martin SPY-4 Volume Search Radar (VSR) ❺; 3D; E/F-band.
Navigation: SPS-73V(18).
Fire control: 4 Mk 95; I/J-band (2 per GMLS 29 launchers).
Tacan: URN-25.

Combat data systems: CEC; USG-2A; Links 4, 11, 16.
Weapon control systems: SSDS Mk II.

Fixed-wing aircraft: Composition will depend on mission but will comprise 75+ aircraft (F-35C, F/A-18E/F, EA-18G, E-2D, MH-60R/S, J-UCAS).
Helicopters: MH-60R/S

Programmes: Northrop Grumman Newport News awarded a construction preparation contract in May 2004 for detailed design, component development, long-lead procurement and advanced construction of the lead ship CVN 78. First steel cut on 11 August 2005. Detailed design and construction contract signed in September 2008. First steel of CVN 79 cut on 25 February 2011. CVN 78 programme has been subject to signifcant cost growth and delays; cost could be 20% high than projected by original construction plan. A more efficient build strategy is to be applied to CVN 79, although primary construction contract, expected in September 2013, was delayed in order to implement CVN 78 lessons learned. The construction schedule for CVN 80 has not been confirmed.

Structure: The Ford class flight deck and below deck have been optimised to increase sortie rates and improve weapons movement. This is to be accomplished with a new design and relocation of the island, three aircraft lifts and an advanced weapons elevators (AWE). Other features include four Electromagnetic Aircraft Launching Systems (EMALS), Advanced Arresting Gear (AAG) system, new SSDS open architecture combat system, fully integrated warfare system, a new nuclear power plant, and a flexible ship architecture to support the rapid insertion of future warfighting technologies. Significant habitability improvements are to be incorporated.

Operational: CVN 78 class ships will require 1,300 fewer personnel than the Nimitz class complement. Increased sortie generation rates (by 25 per cent) and reduced depot maintenance requirements will increase operational availability. New command centre to combine force networking with flexible, open system architecture to support simultaneous multiple missions, including integrated strike planning, joint/coalition operations and special warfare missions. Planned service life 50 years with one mid-life refuelling. In December 2010, four F/A-18E aircraft successfully launched from a land-based runway using EMALS.

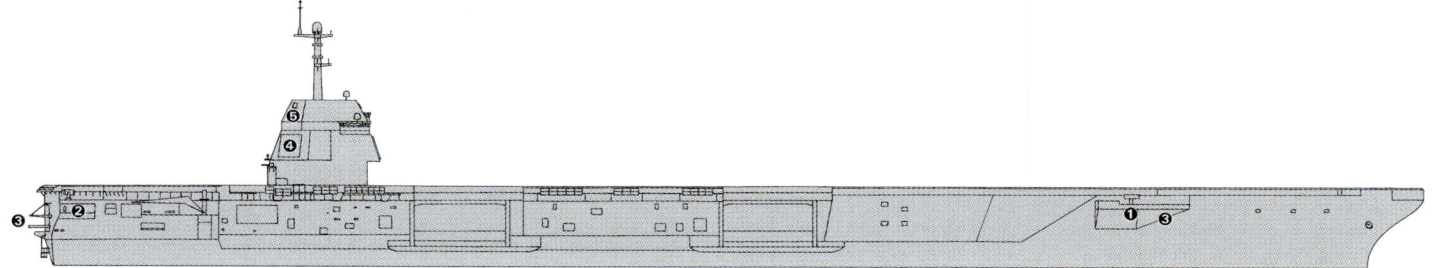

GERALD R FORD   *(Scale 1 : 1,800), Ian Sturton* / 1483678

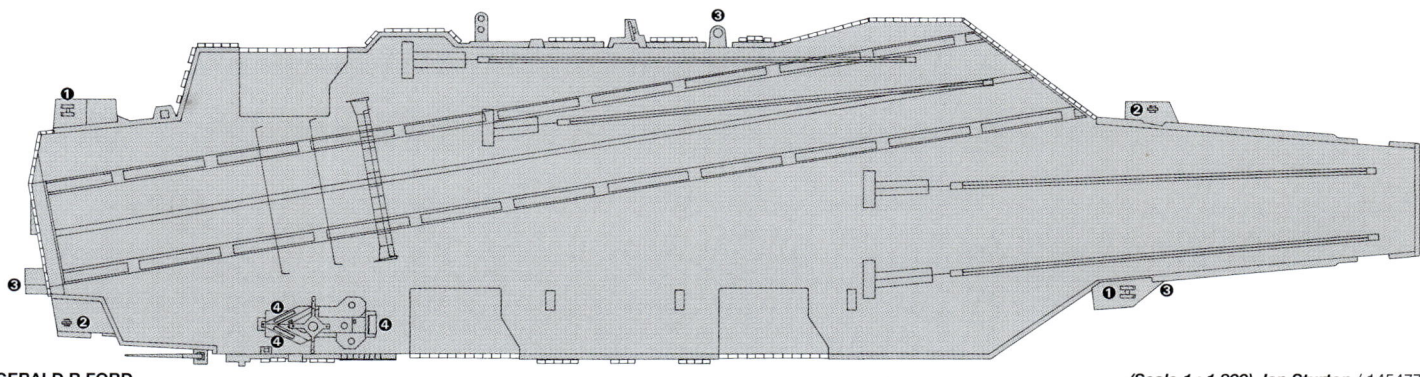

GERALD R FORD   *(Scale 1 : 1,800), Ian Sturton* / 1454777

CVN 78 (artist's impression)   4/2006, US Navy / 1159240

© 2014 IHS   IHS Jane's Fighting Ships 2014-2015

10 NIMITZ CLASS (CVNM)

Name	No	Builders	Laid down	Launched	Commissioned	F/S
NIMITZ	CVN 68	Newport News Shipbuilding	22 Jun 1968	13 May 1972	3 May 1975	PA
DWIGHT D EISENHOWER	CVN 69	Newport News Shipbuilding	15 Aug 1970	11 Oct 1975	18 Oct 1977	AA
CARL VINSON	CVN 70	Newport News Shipbuilding	11 Oct 1975	15 Mar 1980	13 Mar 1982	PA
THEODORE ROOSEVELT	CVN 71	Newport News Shipbuilding	13 Oct 1981	27 Oct 1984	25 Oct 1986	AA
ABRAHAM LINCOLN	CVN 72	Newport News Shipbuilding	3 Nov 1984	13 Feb 1988	11 Nov 1989	AA
GEORGE WASHINGTON	CVN 73	Newport News Shipbuilding	25 Aug 1986	21 Jul 1990	4 Jul 1992	PA
JOHN C STENNIS	CVN 74	Newport News Shipbuilding	13 Mar 1991	13 Nov 1993	9 Dec 1995	PA
HARRY S TRUMAN	CVN 75	Newport News Shipbuilding	29 Nov 1993	7 Sep 1996	25 Jul 1998	AA
RONALD REAGAN	CVN 76	Newport News Shipbuilding	12 Feb 1998	4 Mar 2001	12 Jul 2003	PA
GEORGE H W BUSH	CVN 77	Newport News Shipbuilding	6 Sep 2003	8 Oct 2006	10 Jan 2009	AA

Displacement, tonnes: 74,086 (CVN 68–70), 75,160 (CVN 71) standard; 92,955 (CVN 68–70), 97,933 (CVN 71), 103,637 (CVN 72–77) full load
Dimensions, metres (feet): 332.9 oa; 317 pp × 40.8 wl × 11.3 (CVN 68–70), 11.8 (CVN 71), 11.9 (CVN 72–76), 12.1 (CVN 77) *(1,092.2; 1,040 × 133.9 × 37.1, 38.7, 39.0, 39.7)*
Flight deck, metres: 332.9; 237.7 angled × 76.8 *(1,092.2; 779.9 × 252.0)*
Aircraft lift: 4
Speed, knots: 30 (est.)
Complement: 5,750 (505 officers; 2,480 air group) + 70 flag staff

Machinery: Nuclear; 2 Westinghouse/GE PWR A4W/A1G reactors; 4 turbines; 280,000 hp *(209 MW)*; 4 emergency diesels; 10,720 hp *(8 MW)*; 4 shafts
Missiles: SAM: 2 (CVN 68, 69, 70, 73, 74, 76, 77) or 3 (CVN 71, 72, 75) Raytheon GMLS Mk 29 octuple launchers ❶; NATO Sea Sparrow RIM-7P; semi-active radar homing to 16 km *(8.5 n miles)* at 2.5 Mach; warhead 38 kg. ESSM in due course. 2 GMLS Mk 49 RAM RIM-116 launchers ❷; 21 rds/launcher; passive IR/anti-radiation homing to 9.6 km *(5.2 n miles)* at 2.5 Mach; warhead 9.1 kg.
Guns: 2 (CVN 70, 71, 73) or 3 (CVN 72, 74, 75) Raytheon 20 mm Vulcan Phalanx 6-barrelled Mk 15; 4,500 rds/min combined to 1.5 km.
Physical countermeasures: Decoys: SLQ 25 Torpedo Countermeasures Transmitting Set (Nixie).
Electronic countermeasures: ESM/ECM: SLQ-32(V)4 intercept and jammers.
Radars: Air search: ITT SPS-48E ❸; 3D; E/F-band. Raytheon SPS-49(V)5 (CVN 71, 72, 75) or SPS-49A(V)1 (CVN 68, 69, 70, 73, 74, 76, 77) ❹; C/D-band. Hughes Mk 23 TAS (CVN 71, 72, 75) ❺; D-band or SPQ-9B (CVN 68–70, 73, 74, 76, 77).
Surface search: Norden SPS-67(V)1; G-band.
CCA: SPN-41, SPN-43C, 2 SPN-46; J/F/J/K-band. TPX-42A Direct Altitude and Identity Readout (DAIR).
Navigation: Raytheon SPS-64(V)9 (CVN 71, 72, 75) or SPS-73(V)12 (CVN 68, 70, 73, 74) or SPS-73(V)17 (CVN 69, 76, 77); Furuno 900; I/J-band.
Fire control: 4 Mk 95; I/J-band (2 per GMLS Mk 29 launcher).
Tacan: URN 25.

Combat data systems: ACDS Block 0 (CVN 71–72, 75) naval tactical and advanced combat direction systems; Links 4A, 11, 16 and Satellite Tadil J. GCCS (M) SATCOMS; SSR-1, WCS-3A (UHF DAMA), WSC-6 (SHF), WSC-8 (SHF), USC-38 (EHF), SSR-2A (GBS) (see Data Systems at front of section). SSDS Mk 2 (CVN 68, 69, 70, 73, 74, 76, 77). To be back-fitted in all as part of the CAPSTONE combat system upgrade.
Weapon control systems: 3 Mk 91 Mod 1 MFCS directors (part of the NSSMS Mk 57 SAM system).

Fixed-wing aircraft: Composition of air-wing depends on mission and typically includes: 44 F/A-18C/E/F Hornet; 4 EA-6B Prowler (or 5 EA-18G Growler); 4 E-2C Hawkeye.
Helicopters: A mix of SH-60F, HH-60F and MH-60S (or a mix of MH-60R and MH-60S).

Programmes: *Nimitz* was authorised in fiscal year 1967, *Dwight D Eisenhower* in FY 1970, *Carl Vinson* in FY 1974, *Theodore Roosevelt* in FY 1980 and *Abraham Lincoln* and *George Washington* in FY 1983. Construction contracts for *John C Stennis* and *Harry S Truman* were awarded in June 1988 and for *Ronald Reagan* in December 1994. Authorised in FY 1999, construction contract for *George H W Bush* awarded in January 2001.
Modernisation: CVN 68 completed a three-year Refuelling and Complex Overhaul (RCOH) in 2001. RCOH of CVN 69 started in 2001 and completed in January 2005. RCOH of CVN 70 started in November 2005 and completed in July 2009. RCOH of CVN 71 started in 2009 and was completed in 2013. The contract for the RCOH of CVN 72 was awarded on 29 March 2013. The refit to be completed in November 2016, is to include modifications to receive the F-35C Lightning II. CVN 73 is planned to start RCOH in 2017. SSDS Mk 2 Mod 0 originally installed in CVN 68 (upgraded to Mk 2 Mod 1 in 2006). This includes fitting two RAM systems and SPQ-9B radar vice Mk 23 TAS. SSDS Mk 2 Mod 1 fitted to CVN 68, 69, 76 and 77. RAM systems replace one Mk 29 and all Phalanx launchers on CVN 68 and 69. CVN 74 similarly refitted during 2005 docking but retains upgraded CIWS (Phalanx) mounts as well. CVN 73 was similarly upgraded in 2007. The SSDS upgrade package in CVN 73 and 74 is known as the CAPSTONE combat system upgrade. CAPSTONE was installed in CVN 70 during RCOH and is scheduled for CVN 71, 72, and 75 in due course.
Structure: Damage control measures include sides with system of full and empty compartments (full compartments can contain aviation fuel), approximately 2.5 in Kevlar plating over certain areas of side shell, box protection over magazine and machinery spaces. Aviation facilities include four lifts, two at the forward end of the flight deck, one to starboard abaft the island and one to port at the stern. There are four steam catapults (C13-1 (CVN 68–71), C13-2 (CVN 72–77)) and four (or three on CVN 76 and 77) Mk 7 Mod 3 arrester wires. Launch rate is one every 20 seconds. The hangar can hold less than half the full aircraft complement. Deckhead is 25.6 ft. Aviation fuel, 8,500 tons. Tactical Flag Command Centre for Flagship role. During RCOH, CVN 68 and 69 fitted with reshaped island (the mainmast has three yardarms to support more antennas). Major structural differences in CVN 76 and 77 include: a three-wire arresting system (to replace the four-wire system), an enlarged island structure which incorporates a bigger bridge, a three yardarm mainmast and the after mast (separate in previous ships) and an internal ordnance elevator. Other changes include a bulbous bow to reduce drag and a modified flight deck (angled deck increased by 0.1°) to allow the use of two catapults while aircraft land.
Operational: Multimission role of 'strike/ASW'. From CVN 70 onwards ships have an A/S control centre and A/S facilities; CVN 68 and 69 are backfitted. Endurance of 16 days for aviation fuel (steady flying) with greater than 1 million miles before nuclear reactor refuelling is required. Only one refuelling is required in the life of the ship. Ships' complements and air wings can be changed depending on the operational task. CVNs 69, 71 (to San Diego in 2015), 72, 75, and 77 based at Norfolk, VA. CVNs 70 and 76 (to Yokosuka in 2016) based at San Diego, CA, CVN 74 at Bremerton, WA, and CVN 68 at Everett, WA. CVN 73 is to be replaced at Yokosuka, Japan, by CVN 76 in 2015, when CVN 73 moves to Norfolk for RCOH preparations. First launch of X-47B UCAS-D was conducted by CVN 77 in May 2013.

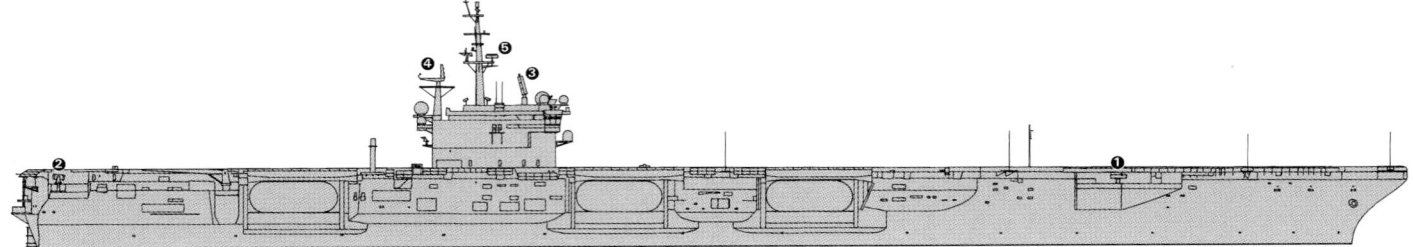

RONALD REAGAN *(Scale 1 : 1,800), Ian Sturton / 1043489*

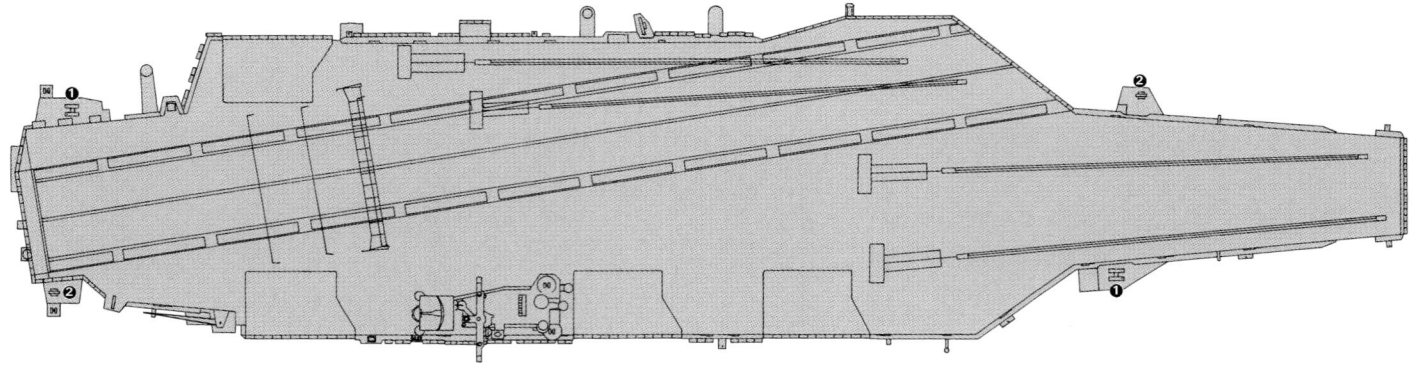

RONALD REAGAN *(Scale 1 : 1,800), Ian Sturton / 1043490*

GEORGE H W BUSH *6/2012, M Declerck / 1486418*

GEORGE H W BUSH　　　　　　　　　　　　　　　　　　　　　　　　　　　　　　　　　　　*12/2013*, US Navy* / 1531223

NIMITZ　　　　　　　　　　　　　　　　　　　　　　　　　　　　　　　　　　　　　　*12/2013*, US Navy* / 1531221

HARRY S TRUMAN　　　　　　　　　　　　　　　　　　　　　　　　　　　　　　　　　　*1/2014*, US Navy* / 1531222

CRUISERS

Notes: (1) **Ballistic Missile Defense.** In September 2009, the Obama administration announced a change in the architecture of ballistic missile defense in Europe. Under the new plan, the US is to utilise ship-based Standard SM-3 missiles, which are to be stationed in the North and Mediterranean seas from 2011, and mobile land-based SM-3 missiles in Central Europe by 2015. Therefore, much of the BMD mission is to fall on the US Navy and an expansion of the programme to convert cruisers and destroyers to the Aegis BMD capability is expected. As of late 2013, 30 ships had been equipped with BMD capability. While there are plans to increase this total to 33 ships by 2017, it is possible that all cruisers and destroyers will be required for the BMD role if the demands of regional defence worldwide are to be met.

(2) **Integrated Ship Controls.** Formerly known as Smart Ship, Integrated Ship Controls (ISC) began as Naval Research Advisory Committee recommendation in 1996 to reduce manning and workload through technology. *Yorktown* (CG 48) was selected as the Smart Ship with implementation of 47 workload-reduction initiatives tested and evaluated during a five-month deployment completed in June 1997. Core systems included: Integrated Bridge System (IBS), Integrated Condition Assessment System (ICAS), Machinery Control System (MCS), Damage Control System (DCS), Fuel Control System (FCS), fibre optic Local Area Network (LAN) and Hierarchial Yet Dynamically Reprogrammable Architecture (HYDRA). *Yorktown's* experience validated these technologies, combined with changes in policies, procedures and new watch routines, to generate substantial reductions in workload. The seven core technologies were chosen to be the ISC production system, first installed in *Monterey* (CG 61) in 2000; *Mobile Bay* (CG 53) in 2001; *Antietam* (CG 54) in 2002; *Hue City* (CG 66) in 2003; *Cape St George* (CG 71) in 2004; *San Jacinto* (CG 56) in 2006. A system upgrade and redesign was completed and installed in: *Leyte Gulf* (CG 55), *Philippine Sea* (CG 58) and *Chancellorsville* (CG 62) in 2007; *Bunker Hill* (CG 52), *Lake Champlain* (CG 57) and *Princeton* (CG 59) in 2008 and *Normandy* (CG 60), *Monterey* (CG 61), *Gettysburg* (CG 64) in 2009 and *Chosin* in 2010. As of October 2013, seven cruisers had not received the ISC upgrade.

22 TICONDEROGA CLASS (CGHM)

Name	*No*	*Builders*	*Laid down*	*Launched*	*Commissioned*	*F/S*
BUNKER HILL	CG 52	Ingalls Shipbuilding, Pascagoula	11 Jan 1984	11 Mar 1985	20 Sep 1986	PA
MOBILE BAY	CG 53	Ingalls Shipbuilding, Pascagoula	6 Jun 1984	22 Aug 1985	21 Feb 1987	PA
ANTIETAM	CG 54	Ingalls Shipbuilding, Pascagoula	15 Nov 1984	14 Feb 1986	6 Jun 1987	PA
LEYTE GULF	CG 55	Ingalls Shipbuilding, Pascagoula	18 Mar 1985	20 Jun 1986	26 Sep 1987	AA
SAN JACINTO	CG 56	Ingalls Shipbuilding, Pascagoula	24 Jul 1985	14 Nov 1986	23 Jan 1988	AA
LAKE CHAMPLAIN	CG 57	Ingalls Shipbuilding, Pascagoula	3 Mar 1986	3 Apr 1987	12 Aug 1988	PA
PHILIPPINE SEA	CG 58	Bath Iron Works, Maine	8 May 1986	12 Jul 1987	18 Mar 1989	AA
PRINCETON	CG 59	Ingalls Shipbuilding, Pascagoula	15 Oct 1986	2 Oct 1987	11 Feb 1989	PA
NORMANDY	CG 60	Bath Iron Works, Maine	7 Apr 1987	19 Mar 1988	9 Dec 1989	AA
MONTEREY	CG 61	Bath Iron Works, Maine	19 Aug 1987	23 Oct 1988	16 Jun 1990	AA
CHANCELLORSVILLE	CG 62	Ingalls Shipbuilding, Pascagoula	24 Jun 1987	15 Jul 1988	4 Nov 1989	PA
COWPENS	CG 63	Bath Iron Works, Maine	23 Dec 1987	11 Mar 1989	9 Mar 1991	PA
GETTYSBURG	CG 64	Bath Iron Works, Maine	17 Aug 1988	22 Jul 1989	22 Jun 1991	AA
CHOSIN	CG 65	Ingalls Shipbuilding, Pascagoula	22 Jul 1988	1 Sep 1989	12 Jan 1991	PA
HUE CITY	CG 66	Ingalls Shipbuilding, Pascagoula	20 Feb 1989	1 Jun 1990	14 Sep 1991	AA
SHILOH	CG 67	Bath Iron Works, Maine	1 Aug 1989	8 Sep 1990	2 Jul 1992	PA
ANZIO	CG 68	Ingalls Shipbuilding, Pascagoula	21 Aug 1989	2 Nov 1990	2 May 1992	AA
VICKSBURG	CG 69	Ingalls Shipbuilding, Pascagoula	30 May 1990	2 Aug 1991	14 Nov 1992	AA
LAKE ERIE	CG 70	Bath Iron Works, Maine	6 Mar 1990	13 Jun 1991	24 Jul 1993	PA
CAPE ST GEORGE	CG 71	Ingalls Shipbuilding, Pascagoula	19 Nov 1990	10 Jan 1992	12 Jun 1993	AA
VELLA GULF	CG 72	Ingalls Shipbuilding, Pascagoula	22 Apr 1991	13 Jun 1992	18 Sep 1993	AA
PORT ROYAL	CG 73	Ingalls Shipbuilding, Pascagoula	18 Oct 1991	20 Nov 1992	9 Jul 1994	PA

Displacement, tonnes: 10,117 full load
Dimensions, metres (feet): 172.8 × 16.8 × 9.5 sonar *(566.9 × 55.1 × 31.2)*
Speed, knots: 30 (est.)
Range, n miles: 6,000 at 20 kt
Complement: 330 (30 officers) + 47 spare berths

Machinery: 4 GE LM 2500 gas turbines; 86,000 hp *(64.16 MW)* sustained; 2 shafts; cp props
Missiles: SLCM: Raytheon Tomahawk Block III and Block IV; TERCOM and GPS aided navigation with DSMAC to 1,600+ km *(865+ n miles)* at 0.7 Mach; warhead (WDU 36B) 454 kg.
SSM: 8 McDonnell Douglas Harpoon (2 quad) ❶; active radar homing to 240 km *(130 n miles)* at 0.9 Mach; warhead 227 kg. Extended range SLAM can be fired from modified Harpoon canisters.
SAM: 122 Raytheon Standard SM-2 Block III and IVA; command/inertial guidance; semi-active radar and IR homing to 167 km *(90 n miles)* at 2.5 Mach. SAM and ASROC missiles are fired from 2 Mk 41 Mod 0 vertical launchers ❷ (61 missiles per launcher). Standard SM-3 Block 1A (in designated ships); command/inertial/GPS guidance and IR homing to 650 n miles *(1,200 km)* at 3 Mach. Raytheon RIM-162 ESSM (modernised ships); semi-active radar homing to 18.5 km *(10 n miles)* at 3.6 Mach; warhead 38 kg.
A/S: Loral ASROC VLA which has a range of 16.6 km *(9 n miles)*; inertial guidance of 1.6–10 km *(1–5.4 n miles)*; payload Mk 46 Mod 5 Neartip or Mk 50.
Guns: 2 FMC 5 in *(127 mm)*/54 Mk 45 Mod 1 ❸; 20 rds/min to 23 km *(12.6 n miles)* anti-surface; weight of shell 32 kg. 2 Raytheon 20 mm/76 Vulcan Phalanx Block 1B 6-barrelled ❹; 4,500 rds/min. 2 McDonnell Douglas 25 mm. 4—12.7 mm MGs.
Torpedoes: 6—324 mm Mk 32 (2 triple) Mod 14 tubes (fitted in the ship's side aft) ❺. 36 Honeywell Mk 46 Mod 5; anti-submarine; active/passive homing to 11 km *(5.9 n miles)* at 40 kt; warhead 44 kg or Alliant/Westinghouse Mk 50; active/passive homing to 15 km *(8.1 n miles)* at 50 kt; warhead 45 kg shaped charge.
Physical countermeasures: Decoys: Up to 8 Loral Hycor SRBOC 6-barrelled fixed Mk 36 Mod 2 ❻; IR flares and chaff. Mk 53 Mod 5 (Nulka). SLQ-25 Nixie; towed torpedo decoy.
Electronic countermeasures: ESM/ECM: Raytheon SLQ-32V(3)/SLY-2 ❼; intercept, jammers.
Radars: Air search/fire control: RCA SPY-1A phased arrays ❾; 3D; E/F-band (CG 52–58). Raytheon SPY-1B phased arrays; 3D; E/F-band (CG 59 on).
Air search: Raytheon SPS-49(V)7 or 8 ❿; C/D-band; range 457 km *(250 n miles)*.
Surface search: ISC Cardion SPS-55 ⓫; I/J-band.
Navigation: Raytheon SPS-64(V)9; I-band.
Fire control: Lockheed SPQ-9A/B ⓬; I/J-band. Four Raytheon SPG-62 ⓭; I/J-band.
Tacan: URN 25. IFF Mk XII AIMS UPX-29.
Sonars: Gould/Raytheon SQQ-89(V)5; combines hull-mounted active SQS-53B (CG 52–67) or SQS-53C (CG 68–73) and passive towed array SQR-19.
Combat data systems: CEC. NTDS with Links 4A, 11, 14. GCCS (M) and Link 16 being fitted. Link 22 in due course. SATCOM WRN-5, WSC-3 (UHF), USC-38 (EHF). UYK-7 computers (CG 52–58); UYK 43/44 (CG 59 onwards); SQQ-28 for LAMPS sonobuoy datalink ❽ (see Data Systems at front of section).
Weapon control systems: SWG-3 Tomahawk WCS. SWG-1A Harpoon LCS. Aegis Mk 7 Mod 4 multitarget tracking with Mk 99 MFCS (includes 4 Mk 80 illuminator directors); has at least 12 channels of fire. Singer Librascope Mk 116 Mod 6 (53B) or Mod 7 (53C) FCS for ASW. Lockheed Mk 86 Mod 9 GFCS (to be replaced by Mk-160 Mod 11 from 2008).

Helicopters: 2 SH-60B Seahawk LAMPS III ⓮. UAV in due course.

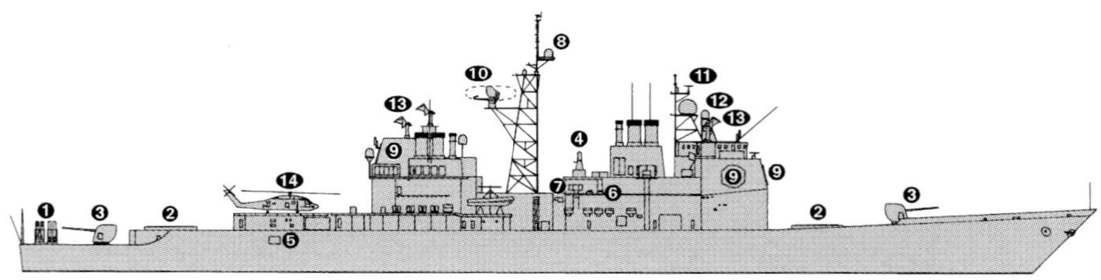

BUNKER HILL *(Scale 1 : 1,500), Ian Sturton* / 0581793

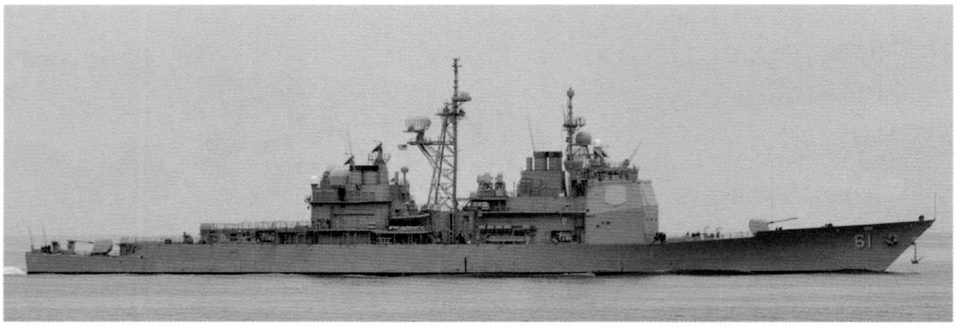

MONTEREY *6/2012, M Declerck* / 1486420

Modernisation: The Cruiser Modernisation (CG Mod) Programme is an extensive capability enhancement and service-life extension that is to be applied to all 22 ships. The principal feature is to be installation of Aegis Open Architecture (AOA) to upgrade the Aegis Weapon System (AWS), ACB 08/ACB 12 Computer Programme and associated displays and computing infrastructure. The new computer programme is to replace several existing computer programme baselines and provide improved tactical performance and functionality. The AOA upgrade is to provide capacity for future combat system growth over the life of the class. The Mk 34 Mod 4 Gun Weapon System upgrade includes the Mk 45 Mod 2 5 in/62 caliber guns, associated Mk 160 Mod 11 fire-control system and optical sights for improved land-attack capability. Additionally, several upgraded command, control, communications, computers and intelligence (C4I) systems and enhanced force-protection capabilities are to be installed. Layered defence is improved through installation of the Vulcan Phalanx Block 1B, modification of the Mk 41 VLS launchers to fire ESSM, installation of the Mk 53 Mod 5 Decoy Launch System (Nulka) and replacement of SPQ-9A with SPQ-9B radar to increase detection and engagement of surface and air threats. Modernised Baseline 3 (CG 59–64) and 4 (CG 65–73) cruisers will rejoin the fleet equipped with improved anti-submarine warfare capability through installation of the SQQ-89A(V)15 upgrade and the Multi-Function Towed Array. Baseline 2 (CG 52–58) are to retain SQQ-89(V)3. The programme will also include a significant Hull, Mechanical and Electrical (HM&E) package that features alterations in weight and movement correction, hull and deckhouse structural improvements, corrosion-control enhancements, hangar deck strengthening, distributive system enhancements and many quality-of-service upgrades. The modernisation will install the Integrated Ship Controls (ISC), or Smartship, and all-electric modifications on ships that have not yet received the alterations. Cruisers with ISC previously installed will receive system upgrades.

Complete modernisation is to be accomplished in two primary phases. The first phase involves HM&E Centric Modernisation availabilities to include ISC and all-electric modifications, in addition to stand-alone combat systems ship changes. Duration of this phase is projected to be less than six months and is to occur in ship's homeport. The second phase involves the full Combat Systems Modernisation refits which include a fully integrated combat system upgrade and those HM&E ship changes not previously completed. *Bunker Hill* (CG 52) was the first ship to complete the full upgrade in 2008. *Mobile Bay*, *Philippine Sea*, *Antietam*, *San Jacinto*, *Lake Champlain*, *Leyte Gulf*, *Chancellorsville*, and *Normandy* have completed modernisation. *Chosin* received the ISC and all-electric modifcations in 2011 and HM&E centric modernisation in 2013. *Princeton* began combat system modernisation in 2013.

Structure: The Ticonderoga class design is a modification of the Spruance class. The same basic hull is used, with the same gas-turbine propulsion plant although the overall length is slightly increased. The design includes Kevlar armour to protect vital spaces. No stabilisers. Later ships have a lighter tripod mainmast vice the square quadruped of the first two. Cracking in the superstructure was identified as a class-wide issue in 2010. Remedial work is, in most cases, being carried out during planned refit (availability) periods.

Operational: The sea-based element of the Ballistic Missile Defense Programme is known as Aegis BMD. *Lake Erie* has acted as the principal trials platform. Since the first intercept test in January 2002, a total of 20 tests were conducted by late 2008. Of these, 16 were successful. In December 2009, *Lake Erie* successfully test the Aegis BMD Version 4.0.1. Four cruisers (*Lake Erie*, *Shiloh*, *Vella Gulf* and *Monterey*) are capable of launching Standard Missile-3s (SM-3s). Plans to decommission four ships: (*Anzio*, *Vicksburg*, *Cowpens*, and *Port Royal*) in fiscal year 2013 and three further (*Chosin*, *Gettysburg*, *Hue City*) FY 2014 were delayed by Congress until at least 2015.

Cruisers ◄ **US** 945

CHOSIN
10/2013, Chris Sattler* / 1531155

ANTIETAM
10/2011, Michael Nitz / 1454832

NORMANDY
6/2012, Michael Nitz / 1486419

PORTER
10/2012, US Navy* / 1531218

DESTROYERS

28 ARLEIGH BURKE (FLIGHTS I AND II) CLASS (DDGHM)

Name	No	Builders	Laid down	Launched	Commissioned	F/S
ARLEIGH BURKE	DDG 51	Bath Iron Works, Maine	6 Dec 1988	16 Sep 1989	4 Jul 1991	AA
BARRY (ex-*John Barry*)	DDG 52	Ingalls Shipbuilding, Mississippi	26 Feb 1990	10 May 1991	12 Dec 1992	AA
JOHN PAUL JONES	DDG 53	Bath Iron Works, Maine	8 Aug 1990	26 Oct 1991	18 Dec 1993	PA
CURTIS WILBUR	DDG 54	Bath Iron Works, Maine	12 Mar 1991	16 May 1991	19 Mar 1994	PA
STOUT	DDG 55	Ingalls Shipbuilding, Mississippi	8 Aug 1991	16 Oct 1992	13 Aug 1994	AA
JOHN S MCCAIN	DDG 56	Bath Iron Works, Maine	3 Sep 1991	26 Sep 1992	2 Jul 1994	PA
		Ingalls Shipbuilding, Mississippi	12 Feb 1992	7 May 1993	10 Dec 1994	AA

34 + 4 (10) ARLEIGH BURKE (FLIGHT IIA) CLASS

Name	No	Builders	Laid down	Launched	Commissioned	F/S
OSCAR AUSTIN	DDG 79	Bath Iron Works, Maine	9 Oct 1997	7 Nov 1998	19 Aug 2000	AA
ROOSEVELT	DDG 80	Ingalls Shipbuilding, Mississippi	15 Dec 1997	10 Jan 1999	14 Oct 2000	AA
WINSTON S CHURCHILL	DDG 81	Bath Iron Works, Maine	7 May 1999	17 Apr 1999	10 Mar 2001	AA
LASSEN	DDG 82	Ingalls Shipbuilding, Mississippi	24 Aug 1998	16 Oct 1999	21 Apr 2001	PA
HOWARD	DDG 83	Bath Iron Works, Maine	9 Dec 1998	20 Nov 1999	20 Oct 2001	PA
BULKELEY	DDG 84	Ingalls Shipbuilding, Mississippi	10 May 1999	21 Jun 2000	8 Dec 2001	AA
MCCAMPBELL	DDG 85	Bath Iron Works, Maine	15 Jul 1999	2 Jul 2000	17 Aug 2002	PA
SHOUP	DDG 86	Ingalls Shipbuilding, Mississippi	13 Dec 1999	22 Nov 2000	22 Jun 2002	PA
MASON	DDG 87	Bath Iron Works, Maine	20 Jan 2000	23 Jun 2001	12 Apr 2003	AA
PREBLE	DDG 88	Northrop Grumman, Mississippi	22 Jun 2000	1 Jun 2001	9 Nov 2002	PA
MUSTIN	DDG 89	Ingalls Shipbuilding, Mississippi	15 Jan 2001	12 Dec 2001	26 Jul 2003	PA
CHAFFEE	DDG 90	Bath Iron Works, Maine	12 Apr 2001	2 Nov 2002	18 Oct 2003	PA
PINCKNEY	DDG 91	Northrop Grumman, Mississippi	16 Jul 2001	26 Jun 2002	29 May 2004	PA
MOMSEN	DDG 92	Bath Iron Works, Maine	16 Nov 2001	19 Jul 2003	28 Aug 2004	PA
CHUNG-HOON	DDG 93	Northrop Grumman, Mississippi	14 Jan 2002	15 Dec 2002	18 Sep 2004	PA
NITZE	DDG 94	Bath Iron Works, Maine	17 Sep 2002	3 Apr 2004	5 Mar 2005	AA
JAMES E WILLIAMS	DDG 95	Northrop Grumman, Mississippi	15 Jul 2002	25 Jun 2003	11 Dec 2004	AA
BAINBRIDGE	DDG 96	Bath Iron Works, Maine	7 May 2003	30 Oct 2004	12 Nov 2005	AA
HALSEY	DDG 97	Northrop Grumman, Mississippi	5 Feb 2003	9 Jan 2004	30 Jul 2005	PA
FORREST SHERMAN	DDG 98	Northrop Grumman	12 Aug 2003	30 Jun 2004	28 Jan 2006	AA
FARRAGUT	DDG 99	Bath Iron Works, Maine	7 Jan 2004	9 Jul 2005	10 Jun 2006	AA
KIDD	DDG 100	Northrop Grumman, Mississippi	1 Mar 2004	15 Dec 2004	9 Jun 2007	PA
GRIDLEY	DDG 101	Bath Iron Works, Maine	30 Jul 2004	28 Dec 2005	10 Feb 2007	PA
SAMPSON	DDG 102	Bath Iron Works, Maine	14 Mar 2005	17 Sep 2006	3 Nov 2007	PA
TRUXTUN	DDG 103	Northrop Grumman, Mississippi	11 Apr 2005	17 Apr 2007	25 Apr 2009	AA
STERETT	DDG 104	Bath Iron Works, Maine	17 Nov 2005	20 May 2007	9 Aug 2008	PA
DEWEY	DDG 105	Northrop Grumman, Mississippi	4 Oct 2006	26 Jan 2008	6 Mar 2010	PA
STOCKDALE	DDG 106	Bath Iron Works, Maine	10 Aug 2006	24 Feb 2008	18 Apr 2009	PA
GRAVELY	DDG 107	Northrop Grumman, Mississippi	26 Nov 2007	30 Mar 2009	20 Nov 2010	AA
WAYNE E MEYER	DDG 108	Bath Iron Works, Maine	17 May 2007	19 Oct 2008	10 Oct 2009	PA
JASON DUNHAM	DDG 109	Bath Iron Works, Maine	11 Apr 2008	2 Aug 2009	13 Nov 2010	AA
WILLIAM P LAWRENCE	DDG 110	Northrop Grumman, Mississippi	16 Sep 2008	15 Dec 2009	4 Jun 2011	PA
SPRUANCE	DDG 111	Bath Iron Works, Maine	14 May 2009	6 Jun 2010	1 Oct 2011	PA
MICHAEL MURPHY	DDG 112	Bath Iron Works, Maine	12 Jun 2010	8 May 2011	6 Oct 2012	PA
JOHN FINN	DDG 113	Huntington Ingalls, Mississippi	4 Nov 2013	2015	2016	Bldg
RALPH JOHNSON	DDG 114	Huntington Ingalls	2014	2015	2017	Bldg
RAFAEL PERALTA	DDG 115	Bath Iron Works, Maine	2014	2015	2016	Bldg
THOMAS HUDNER	DDG 116	Bath Iron Works, Maine	2015	2016	2017	Bldg
PAUL IGNATIUS	DDG 117	Huntington Ingalls	2015	2016	2017	Ord
DANIEL INOUYE	DDG 118	Huntington Ingalls	2016	2017	2018	Ord
–	DDG 119	Huntington Ingalls	–	–	–	Ord
–	DDG 120	Bath Iron Works, Maine	–	–	–	Ord
–	DDG 121	Huntington Ingalls	–	–	–	Ord
–	DDG 122	Bath Iron Works, Maine	–	–	–	Ord
–	DDG 123	Huntington Ingalls	–	–	–	Ord
–	DDG 124	Bath Iron Works, Maine	–	–	–	Ord
–	DDG 125	Huntington Ingalls	–	–	–	Ord
–	DDG 126	Bath Iron Works, Maine	–	–	–	Ord

Displacement, tonnes: 9,425 full load
Dimensions, metres (feet): 155.3 oa; 143.6 wl × 20.3 × 6.7 hull; 9.8 sonar *(509.5; 471.1 × 66.6 × 22.0; 32.2)*
Speed, knots: 31. **Range, n miles:** 4,300 at 20 kt
Complement: 278 (DDG 79–84), 276 (DDG 97, DDG 100–102, DDG 85–96, 98–99), 282 (DDG 103–116),
Machinery: 4 GE LM 2500-30 gas turbines; 100,000 hp *(74.6 MW)* sustained; 2 shafts; cp props
Missiles: SLCM: Raytheon Tomahawk Block III and Block IV; TERCOM and GPS aided navigation with DSMAC to 1,600+ km *(865+ n miles)* at 0.7 Mach; warhead (WDU 36B) 454 kg.
SAM: Raytheon Standard SM-2 Block III and IVA; command/inertial guidance; semi-active radar and IR homing to 167 km *(90 n miles)* at 2 Mach. 2 Lockheed Martin Mk 41 Vertical Launch Systems (VLS) for Tomahawk, Standard and ASROC VLS ❶; 2 magazines; 32 missile tubes forward, 64 aft. 32 Raytheon RIM-162 ESSM (4 quad forward, 4 quad aft); semi-active radar homing to 18.5 km *(10 n miles)* at 3.6 Mach; warhead 38 kg.
A/S: Loral ASROC VLA; inertial guidance to 1.6–16.6 km *(1–9 n miles)*; payload Mk 46 Mod 5 Neartip.
Guns: 1 BAE Systems 5 in *(127 mm)*/54 Mk 45 Mod 2 (DDG 79–80) ❷; 20 rds/min to 23 km *(12.6 n miles)*; weight of shell 32 kg. BAE Systems 5 in *(127 mm)*/62 (DDG 81 onwards); 20 or 10 rds/min; GPS guidance to 116.7 km *(63 n miles)*; warhead 72 bomblets; cep 10 m. 2 Raytheon 20 mm/76 Vulcan Phalanx Block 1B 6-barrelled ❸; 4,500 rds/min combined to 1.5 km.
Torpedoes: 6—324 mm Mk 32 Mod 14 (2 triple) tubes ❹. Alliant Mk 46 Mod 5; anti-submarine; active/passive homing to 11 km *(5.9 n miles)* at 40 kt; warhead 44 kg or Alliant/Westinghouse Mk 50; active/passive homing to 15 km *(8.1 n miles)* at 50 kt; warhead 45 kg shaped charge.
Physical countermeasures: Decoys: 2 Loral Hycor SRBOC 6-barrelled fixed Mk 36 Mod 12 ❺; Nulka decoy (DDG 91 onwards); IR flares and chaff to 4 km *(2.2 n miles)*. SLQ-25A Nixie; torpedo decoy. NATO Sea Gnat. SLQ-95 AEB. SLQ-39 chaff buoy.
Electronic countermeasures: ESM/ECM: Raytheon SLQ-32(V)3/SLY-2 ❻; intercept and jammer.
CESM: AN/SRS-1A(V) CDF (DDG 79–95); COBLU (DDG 96–104); SSEE Increment E (DDG 105–112); SSEE Increment F (DDG 113+).
Radars: Air search/fire control: Lockheed Martin SPY-1D (SPY-1D(V) DDG 91 onwards) phased arrays ❼; 3D; E/F-band.
Surface search: DRS SPS-67(V)3 (DDG 79–102), SPS-67(V)5 (DDG 103 onwards) ❽; G-band.
Navigation: Raytheon SPS-64(V)9 (DDG 79–86, 88); Sperry-Marine BME 740 (DDG 87, 89–112); I-band.
Fire control: Three Raytheon AN/SPG-62 ❾; I/J-band.
Tacan: AN/URN 25 ❿. IFF AIMS Mk XII with AN/UPX-29.
Sonars: Lockheed Martin SQQ-89(V)10 (DDG 79–84); SQQ-89(V)14 (DDG 85–90); SQQ-89(V)15 (DDG 91 and following); underwater combat system with SQS-53C; bow-mounted; active search and attack. Remote Minehunting System (DDG 91). SQR-20 Multi-Function Towed Array (MFTA) (DDG 113+).
Combat data systems: TADIX-B and TADIL-J. CEC. Links 4A, 11 and 16. (See Data Systems at front of section.) Link 22 in due course. Command and Decision (upgrade for DDG 91 and following ships).
Weapon control systems: SWG-4 or SWG-5 Tomahawk WCS. Aegis multitarget tracking with Mk 99 Mod 3 MFCS and three Mk 80 illuminators. Mk 34 GWS (consisting of Mk 160 computing system and Kollmorgen Mk 46 optronic sight). AWCS Mk 116 Mod 7 NFCS.

PINCKNEY *(Scale 1 : 1,500), Ian Sturton / 1454772*

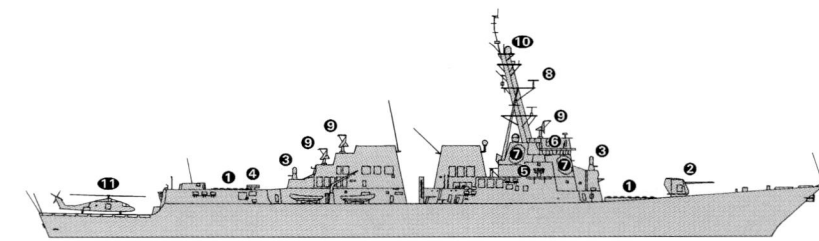

LASSEN *8/2013*, Chris Sattler / 1531156*

Helicopters: 2 LAMPS III SH-60R helicopters ⓫.

Programmes: DDG 79 was authorised in the fiscal year 94 budget. Funding for DDG 80–82 provided in FY 1995 and DDG 83–84 in FY 1996 plus partial funding for a third. Balance for DDG 85 plus DDG 86–88 in FY 1997. On 6 March 1998, multi-year contract for six ships and one option awarded to Ingalls Shipbuilding (DDGs 89, 91, 93, 95, 97, 98, 100) and contract for six ships awarded to Bath Iron Works (DDGs 90, 92, 94, 96, 99, 101, 102). On 1 August 2002, contract awarded to Bath Iron Works for the construction of DDG 102 and on 13 September 2002 a fixed-price multi-year contract awarded to Bath Iron Works (DDGs 104, 106, 108, 109, 111 and 112) and Northrop Grumman Shipbuilding (DDGs 103, 105, 107, 110) for the construction of 10 ships. Following the curtailment of the DDG 1000 programme on 23 July 2008, plans to restart the DDG 51 building programme were announced in early 2010. One Flight IIA design (DDG 113) was authorised and appropriated in FY 2010. A design and construction contract awarded to Huntington Ingalls on 15 June 2011. Construction of DDG 113, the first ship of the re-started programme, began on 10 September 2012. A contract for DDG 114 was awarded to Huntington Ingalls and for DDG 115, with an option for DDG 116, to Bath Iron Works on 26 September 2011. These ships are to incorporate all DDG 51 improvements and BMD capabilities to date. A multi-year contract for the construction of nine ships (DDG 117–125) was awarded to Huntington Ingalls (five ships) and Bath Iron Works (four ships) on 3 June 2013. An option for a tenth ship (DDG 126) was executed on 14 March 2014. The Flight III baseline (including Air and Missile Defence Radar (AMDR)) is planned as part of a future Engineering Change Proposal for the second ship in FY 2016 (DDG 124). SPS-67 radar is to be replaced by SPQ-9B in Flight III ships.

Modernisation: A mid-life modernisation programme is in progress in two distinct packages: Hull, Mechanical and Electrical (HM&E), and Combat Systems (C/S). The scope of the upgrades was included in the construction of DDG 111 and subsequent ships and is to be retrofitted in DDGs 79–110. The contract for AMDR was awarded to Raytheon in October 2013.

Structure: The upgrade from Flight II includes two hangars for embarked helicopters and an extended transom to increase the size of a dual RAST fitted flight deck. Vertical launchers are increased at each end by three cells. Other changes include the Kingfisher minehunting sonar, a reconfiguration of the SPY-1D arrays and the inclusion of a Track Initiation Processor in the Aegis radar system. Use of fibre optic technology reduced weight and improved bandwidth.

Operational: The helicopter carries Hellfire missiles. ESSM fired from DDG 86 on 24 July 2002, the first to be fired from a USN ship. Ships' lives are to be 40 years.

CHUNG-HOON   *8/2013*, Mick Prendergast* / 1531157

SPRUANCE   *1/2012*, US Navy* / 1531217

PREBLE   *7/2013*, Michael Nitz* / 1531158

950 US > Destroyers — Frigates

0 + 3 ZUMWALT (DDG 1000) CLASS (DDGH)

Name	No	Builders	Laid down	Launched	Commissioned	F/S
ZUMWALT	DDG 1000	General Dynamics Bath Iron Works	17 Nov 2011	28 Oct 2013	Jul 2015	Bldg/PA
MICHAEL MONSOOR	DDG 1001	General Dynamics Bath Iron Works	23 May 2013	Dec 2014	Feb 2017	Bldg/PA
LYNDON B JOHNSON	DDG 1002	General Dynamics Bath Iron Works	Sep 2015	Mar 2017	Mar 2019	Bldg/PA

Displacement, tonnes: 15,610 standard
Dimensions, metres (feet): 185.9 × 24.6 × 8.4 (609.9 × 80.7 × 27.6)
Speed, knots: 30. **Range, n miles:** 4,500 at 13 kt
Complement: 158 (28 air group)

Machinery: Integrated Power System (IPS); 2 Rolls Royce MT 30 Main Turbine Generators (MTG); 2 Rolls Royce RR4500 Auxiliary Turbine Generators (ATG); 2 propulsion motors; 105,270 hp (78.5 MW); 2 shafts
Missiles: 80 Advanced Vertical Launching System (AVLS) cells ❶.
SLCM: Raytheon Tomahawk Block IV; land attack; TERCOM and GPS aided inertial navigation system with DSMAC to 1,600+ km (865+ n miles) at 0.7 Mach; warhead (WDU-36B) 454 kg.
SAM: Standard SM-2 and Evolved Sea Sparrow.
A/S: Vertical launched ASROC.
Guns: 2 BAE 155 mm ❷ advanced gun systems capable of firing Long Range Land Attack Projectiles (LRLAP) at ranges over 63 n miles. 2 — 30 mm Mk 46 ❸ close-in guns.
Torpedoes: To be announced.
Physical countermeasures: Decoys: Torpedo decoys.
Radars: Air/surface search: Raytheon SPY-3 ❹; 3D; I-band.
Navigation: To be announced.
Surface search: Raytheon SPS-73(V).
Sonars: SQQ-90 tactical sonar suite comprising SQS-60 hull-mounted (mid-frequency); SQS-61 hull-mounted (high-frequency); SQR-20 multifunction towed array.
Combat data systems: Open Architecture (OA). Total Ship Computing Environment (TSCE).
Electro-optic systems: To be announced.

Helicopters: 2 MH-60R ❺ or 1 MH-60R and 3 UAVs.

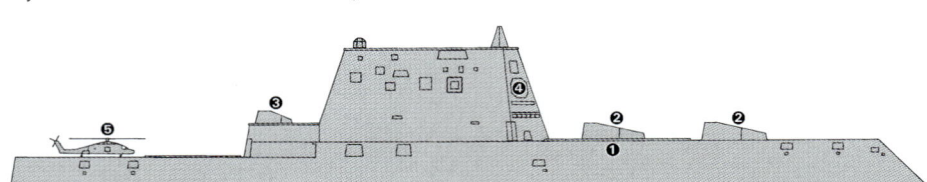

ZUMWALT *(Scale 1 : 1,500), Ian Sturton* / 1353579

ZUMWALT *10/2013*, US Navy* / 1531216

Programmes: The DDG 1000 (formerly DD(X)) programme was initiated in November 2001. Principal roles are sustained operations in the littorals and land-attack. Critical Design Review (CDR) completed September 2005 and on 23 November 2005 received approval to proceed with Milestone B, the authorised commencement of design and construction of DDG 1000 class. Under the Dual Lead Ships acquisition strategy, detailed design contracts were awarded in August 2006 to Northrop Grumman Ship Systems and General Dynamics Bath Iron Works. Raytheon is the Mission Systems Integrator and BAE Systems provides the gun systems. Northrop Grumman to build the deckhouses and hangars. Construction contracts were awarded to the shipyards for the construction of the first ship on 14 February 2008. It was announced on 23 July 2008 that the DDG 1000 programme was to be truncated to three ships. A contract for the construction of the second and third ships was awarded on 15 September 2011. All 13 Engineering Development Models (EDMs) have entered production including 155 mm Advanced Gun System and its Long Range Land Attack Projectile (LRLAP) with a range to 63 miles; the Mark 57 Advanced Vertical Launch System; composite deckhouse; the infrared suppression engine exhaust and heat suppression system; components of the Integrated Power System; the Automatic Fire Suppression System, portions of the SQQ-90 Integrated Undersea Warfare System and Dual Band Radar (DBR). Software development for the Total Ship Computing Environment is progressing while the hardware is being procured. As of October 2013, *Zumwalt* was more than 87% complete and *Michael Monsoor* more than 70%.

Structure: Features of the ship include a wave-piercing 'tumblehome' hull, optimised for stealth. Hull structure and missile cells spread impacts outward to increase survivability and reduce risk of single-hit ship loss. Integrated deckhouse and composite superstructure incorporates masts, sensors, antennas, bridge and exhaust silos. There are two shielded 155 mm Advanced Gun Systems (AGS) and an 80-cell peripheral (port and starboard) Vertical Launch System for both land attack and air defense missiles. An Integrated Power System (IPS), enables power to be distributed to any system as the tactical situation demands. IPS is designed to create sufficient reserve energy to power energy weapons in the future.

Operational: All three ships are likely to be based at San Diego.

FRIGATES

2 + 7 (3) FREEDOM CLASS LITTORAL COMBAT SHIP FLIGHT 0

Name	No	Builders	Laid down	Launched	Commissioned	F/S
FREEDOM	LCS 1	Marinette Marine, Marinette	2 Jun 2005	23 Sep 2006	8 Nov 2008	PA
FORT WORTH	LCS 3	Marinette Marine, Marinette	11 Jul 2009	4 Dec 2010	22 Sep 2012	PA
MILWAUKEE	LCS 5	Marinette Marine, Marinette	27 Oct 2011	18 Dec 2013	2015	Bldg
DETROIT	LCS 7	Marinette Marine, Marinette	8 Nov 2012	2014	2015	Bldg
LITTLE ROCK	LCS 9	Marinette Marine, Marinette	27 Jun 2013	2014	2015	Bldg
SIOUX CITY	LCS 11	Marinette Marine, Marinette	2014	2015	2015	Bldg
WICHITA	LCS 13	Marinette Marine, Marinette	2015	2016	2016	Ord
BILLINGS	LCS 15	Marinette Marine, Marinette	2016	2016	2017	Ord
INDIANAPOLIS	LCS 17	Marinette Marine, Marinette	2017	2017	2017	Ord

Displacement, tonnes: 3,354 full load
Dimensions, metres (feet): 118.8 × 17.6 × 4.3 (389.8 × 57.7 × 14.1)
Speed, knots: 40. **Range, n miles:** 3,500 at 14 kt
Complement: 50

Machinery: CODAG: 2 Rolls Royce MT-30 gas turbines; 96,550 hp (72 MW); 2 Fairbanks Morse Colt-Pielstick 16PA6B diesels; 17,160 hp (12.8 MW); 4 Rolls Royce Kamewa 153SII waterjets
Missiles: SAM: 1 Raytheon RAM RIM-116 21-cell Mk 99 launcher ❶; passive IR/anti-radiation homing to 9.6 km (5.2 n miles) at 2.5 Mach; warhead 9.1 kg.
SSM: Lockheed Martin Longbow Apache Hellfire (AGM 114L) ❷; active radar acquisition and homing to 8.3 km (4.5 n miles); warhead 9 kg.
Guns: 1 BAE Systems 57 mm/70 Mk 2 ❸; 220 rds/min to 17 km (9 n miles); weight of shell 2.4 kg. 4 — 12.7 mm MGs. 2 Mk 46 Mod 2 Bushmaster 30 mm (ASUW module) ❹; 200 rds/min to 6.8 km (3.7 n miles).
Physical countermeasures: Decoys: 2 Terma SKWS/SRBOC decoy launching systems.
Electronic countermeasures: ESM: Argon ST WBR 2000.
Radars: Air/surface search: EADS TRS-3D ❻; C-band.
Navigation: Sperry Bridgemaster ❼; E/F/I-bands.
Fire control: FABA DORNA; I-band.
Sonars: VDS (under development).
Combat data systems: COMBATSS-21.
Weapon control systems: FABA DORNA TV/IR tracker and laser range-finder ❺.

Helicopters: 2 MH-60 R/S helicopters ❽ or 1 MH-60 R/S and 3 Firescout VTUAVs.

Programmes: Two industry teams, one led by Lockheed Martin and the other by General Dynamics, were contracted in 2004 to develop designs for a fast, agile and networked surface combatant. In the original procurement programme, it was planned to build a number of each design and left open the option that both designs could proceed to series production. The keys to this approach were a fast building time of two years per ship and a relatively inexpensive cost. In April 2007, the Navy cancelled its contract with Lockheed Martin for the construction of LCS 3 after negotiations to control cost overruns failed. The second General Dynamics ship (LCS 4) was also cancelled, in November 2007, after similar coast overruns. The funding of three further ships has also been cancelled or re-allocated. In March 2009, the decision to proceed with the construction of one of each LCS variant, re-using previous hull numbers, was announced. In December 2010 it was decided that, rather than select a single LCS design for future construction, procurement of the next 20 ships is to be split between the two variants. One of each design was built from funds from fiscal year 2010 and FY 2011 and two of each from FY 2012 to FY 2015. A contract for the block-buy of 10 ships was awarded to Lockheed Martin on 29 December 2010. In 2012, the Navy Board of Inspection and Survey found numerous sustainment issues in both LCS variants. The US Comptroller General reported in July 2013 that progress had been made in addressing design and production problems but, in late 2013, the Defense Department's Director of Operational Test and Evaluation reported a number of concerns about the performance of weapon systems.

Structure: Semi-planing steel monohull design. Steel hull and aluminium superstructure. The design incorporates a large reconfigurable seaframe to allow rapidly interchangeable mission modules, a flight deck with integrated helicopter launch, recovery and handling system and the capability to launch and recover maritime vehicles (manned and unmanned) from both the stern and side. The hull of LCS 5 is 3.5 m longer than LCS 1 and 3 to improve through-water performance. The bow design has also been modified. The ship also features new ONR-designed waterjets.

Operational: The three mission packages are ASuW, ASW and MCM. The ASuW mission package includes the following key components: The missile module comprises three container units each containing 15 surface-to-surface missiles (Lockheed Martin Hellfire). The gun module comprises two Mk 46 30 mm guns and the MH-60R helicopter is capable of carrying eight Hellfire missiles and 12.7 mm and 7.62 mm machine guns. Two 11 m RHIBs are available for use by boarding teams carrying small arms. ASW mission systems are deployed from three principal off-board platforms. The ASW Unmanned Surface Vehicle (USV) is capable of carrying the USV Dipping Sonar (UDS), USV Towed Array System (UTAS) and Multistatic Offboard Source (MSOBS). It also carries a surface search radar and communications. The MH-60R helicopter carries Airborne Low-Frequency Sonar (ALFS), 25 sonobuoys and FLIR. A Fire Scout MQ-8B VTUAV is required to provide reconnaissance, surveillance and target acquisition, using an EO/IR sensor, and communications relay support. The MCM mission package comprises a variety of systems deployed by off-board systems including manned and unmanned aircraft and unmanned surface and semi-submersible systems. Detection and identification of mines is performed by the AQS-20A minehunting sonar towed by the MH-60S helicopter or the AN/WLD-1 Remote Multi-Mission Vehicle (RMMV). Mines near the surface are detected with the Airborne Laser Mine Detection System (ALMDS), also deployed in the MH-60S helicopter. Neutralisation of mines is performed by the MH-60S using Airborne Mine Neutralisation System (AMNS). Mine-sweeping is accomplished by the MH-60S or Unmanned Surface Vehicle towing the Unmanned Influence Surface Sweep System (UISS) respectively. *Freedom* conducted an eight-month deployment to Singapore in 2013 to test the viability of the modular LCS concept and of crew rotation and maintenance concepts.

IHS Jane's Fighting Ships 2014-2015 © 2014 IHS

Frigates < **US** 951

FREEDOM   *7/2008, Lockheed Martin* / 1335209

FORT WORTH   *10/2011, US Navy* / 1486429

FREEDOM   *7/2008, Lockheed Martin* / 1335208

© 2014 IHS   *IHS Jane's Fighting Ships 2014-2015*

12 OLIVER HAZARD PERRY CLASS (FFH)

Name	No	Builders	Laid down	Launched	Commissioned	F/S
HALYBURTON	FFG 40	Todd Pacific Shipyard Corporation, Seattle	26 Sep 1980	15 Oct 1981	7 Jan 1984	AA
MCCLUSKY	FFG 41	Todd Pacific Shipyard Corporation, San Pedro	21 Oct 1981	18 Sep 1982	10 Dec 1983	NRF
VANDEGRIFT	FFG 48	Todd Pacific Shipyard Corporation, Seattle	13 Oct 1981	15 Oct 1982	24 Nov 1984	PA
TAYLOR	FFG 50	Bath Iron Works, Maine	5 May 1983	5 Nov 1983	1 Dec 1984	AA
GARY	FFG 51	Todd Pacific Shipyard Corporation, San Pedro	18 Dec 1982	19 Nov 1983	17 Nov 1984	NRF
FORD	FFG 54	Todd Pacific Shipyard Corporation, San Pedro	16 Jul 1983	23 Jun 1984	29 Jun 1985	PA
ELROD	FFG 55	Bath Iron Works, Maine	21 Nov 1983	12 May 1984	6 Jun 1985	NRF
SIMPSON	FFG 56	Bath Iron Works, Maine	27 Feb 1984	21 Aug 1984	9 Nov 1985	NRF
SAMUEL B ROBERTS	FFG 58	Bath Iron Works, Maine	21 May 1984	8 Dec 1984	12 Apr 1986	NRF
KAUFFMAN	FFG 59	Bath Iron Works, Maine	8 Apr 1985	29 Mar 1986	21 Feb 1987	AA
RODNEY M DAVIS	FFG 60	Todd Pacific Shipyard Corporation, San Pedro	8 Feb 1985	11 Jan 1986	9 May 1987	NRF
INGRAHAM	FFG 61	Todd Pacific Shipyard Corporation, San Pedro	30 Mar 1987	25 Jun 1988	5 Aug 1989	PA

Displacement, tonnes: 3,010 standard; 4,166 full load
Dimensions, metres (feet): 138.1 × 13.7 × 4.5 hull; 7.5 sonar (453.1 × 44.9 × 14.8; 24.6)
Speed, knots: 29
Range, n miles: 4,500 at 20 kt
Complement: 198 (17 officers; 19 air crew)

Machinery: 2 GE LM 2500 gas turbines; 41,000 hp (30.59 MW) sustained; 1 shaft; cp prop 2 auxiliary retractable props; 650 hp (484 kW)
Guns: 1 Oto Melara 3 in (76 mm)/62 Mk 75 ❶; 85 rds/min to 16 km (8.7 n miles) anti-surface; 12 km (6.6 n miles) anti-aircraft; weight of shell 6 kg. 1 Raytheon 20 mm/76 6-barrelled Mk 15 Block 1B Vulcan Phalanx ❷; 4,500 rds/min combined to 1.5 km. 2 Boeing 25 mm Mk 38 guns can be fitted amidships. 1—25 mm Mk 38 Mod 2 (FFGs 48, 50, 51, 55, 59, 60, 61). 4—12.7 mm MGs.
Torpedoes: 6—324 mm Mk 32 (2 triple) tubes ❸. 24 Honeywell Mk 46 Mod 5; anti-submarine; active/passive homing to 11 km (5.9 n miles) at 40 kt; warhead 44 kg or Alliant/Westinghouse Mk 50; active/passive homing to 15 km (8.1 n miles) at 50 kt; warhead 45 kg shaped charge.
Physical countermeasures: Decoys: 2 Loral Hycor SRBOC 6-barrelled fixed Mk 36 ❹; IR flares and chaff to 4 km (2.2 n miles). Mk 34 launcher for Mk 53 Nulka decoys. T-Mk 6 Fanfare/SLQ-25 Nixie; torpedo decoy.
Electronic countermeasures: ESM/ECM; SLQ-32(V)2 ❺; radar warning. Sidekick modification adds jammer and deception system.
Radars: Air search: Raytheon SPS-49(V)4 or 5 (FFG 61 and during modernisation of others) ❼; C/D-band; range 457 km (250 n miles).
Surface search: ISC Cardion SPS-55 ❽; I-band.
Fire control: Sperry Mk 92 (Signaal WM28) ❾; I/J-band.
Navigation: Furuno; I-band.
Tacan: URN 25. IFF Mk XII AIMS UPX-29.
Sonars: SQQ 89(V)2 (Raytheon SQS 56 and Gould SQR 19); hull-mounted active search and attack; medium frequency and passive towed array; very low frequency.
Combat data systems: NTDS with Link 11 and 14. Link 14 only (NRF ships). SATCOM ❻ SRR-1, WSC-3 (UHF). SQQ-28 for LAMPS III datalink.

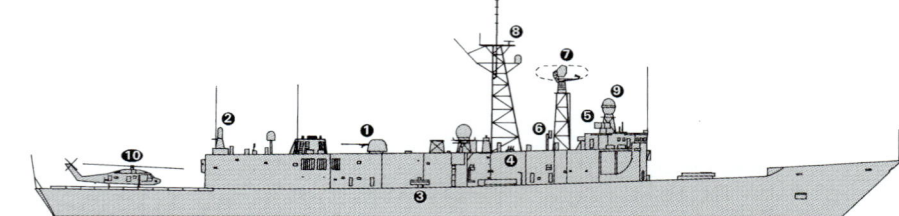

OLIVER HAZARD PERRY CLASS (Scale 1 : 1,200), Ian Sturton / 0572737

Weapon control systems: Mk 92 (Mod 4 or Mod 6 (FFG 61 and during modernisation in 11 others of the class)), WCS with CAS (Combined Antenna System). The Mk 92 is the US version of the Signaal WM28 system. SYS 2(V)2 IADT (FFG 61 and in 11 others of the class - see *Modernisation*). SRQ-4 for LAMPS III.
Helicopters: 2 SH-60B LAMPS III ❿ in Flight III/IV and certified ships.

Programmes: The lead ship was authorised in fiscal year 1973.
Modernisation: To accommodate the helicopter landing system (RAST), the overall length of the ship was increased by 8 ft (2.4 m) by increasing the angle of the ship's transom, between the waterline and the fantail, from virtually straight up to a 45° angle outwards. LAMPS III support facilities and RAST were fitted in all ships authorised from FFG 36 onwards, during construction and have been backfitted to all. FFG 61 has much improved Combat Data and Fire-Control equipment which has been retrofitted in FFG 36, 47, 48, 50–52, 54–55, 57 and 58. SQS-56 is modified for mine detection. Block 1B Phalanx fitted first in FFG 36 in October 1999. Engineering and platform improvements programme initiated in 2003. Upgrades included new diesel generators, the addition of reverse osmosis plants, COTS slewing arm davits, and self-contained breathing apparatus. Mk 13 launchers for Standard SM-1 and Harpoon missiles have been removed. Combat system improvements include the installation of Mk 53 Nulka decoys and Mk 15 Block 1B gun with surface mode capability.
Structure: The original single hangar has been changed to two adjacent hangars. Provided with 19 mm Kevlar armour protection over vital spaces. 25 mm guns can be fitted for some operational deployments.
Operational: Ships of this class were the first Navy experience in implementing a design-to-cost acquisition concept. On 14 April 1988, *Samuel B Roberts* (FFG 58), was mined in the Gulf but was subsequently repaired. Seven ships are assigned to the Combatant Naval Reserve Force.
Sales: Australia bought four (FFG 17, 18, 35, and 44) of the class and has built two more. Spain has six and Taiwan eight. Transfers include eight to Turkey plus one for spares, four to Egypt, one to Bahrain and two to Poland. FFG 8 decommissioned in August 2010 and transferred to Pakistan. FFGs 33 and 39 decommissioned in 2011. FFGs 28, 29 and 32 decommissioned in 2012. FFGs 36, 38, 42, 43, 52, 54, and 57 decommissioned in 2013 and FFGs 45, 46, 47, and 49 in 2014. All are designated for foreign sales.

RODNEY M DAVIS 6/2011, M Mazumdar / 1454839

SAMUEL B ROBERTS 3/2014*, US Navy / 1565442

IHS Jane's Fighting Ships 2014-2015

Frigates < US

2 + 6 (4) INDEPENDENCE CLASS LITTORAL COMBAT SHIP FLIGHT 0

Name	No	Builders	Laid down	Launched	Commissioned	F/S
INDEPENDENCE	LCS 2	Austal USA, Mobile	19 Jan 2006	28 Apr 2008	16 Jan 2010	PA
CORONADO	LCS 4	Austal USA, Mobile	17 Dec 2009	10 Jan 2012	5 Apr 2014	PA
JACKSON	LCS 6	Austal USA, Mobile	18 Oct 2012	14 Dec 2013	2015	Bldg
MONTGOMERY	LCS 8	Austal USA, Mobile	25 Jun 2013	2014	2015	Bldg
GABRIELLE GIFFORDS	LCS 10	Austal USA, Mobile	2014	2015	2015	Bldg
OMAHA	LCS 12	Austal USA, Mobile	2015	2015	2016	Bldg
MANCHESTER	LCS 14	Austal USA, Mobile	2015	2016	2016	Bldg
TULSA	LCS 16	Austal USA, Mobile	2016	2017	2017	Bldg

Displacement, tonnes: 2,841 full load
Dimensions, metres (feet): 127.6 × 31.6 × 4.4 *(418.6 × 103.7 × 14.4)*
Speed, knots: 40. **Range, n miles:** 3,500 at 14 kt
Complement: 40
Machinery: CODAG: 2 General Electric LM 2500 gas turbines; 52,500 hp *(39.16 MW)*; 2 MTU 20V 8000 diesels; 22,298 hp *(16.4 MW)*; 4 Wärtsilä steerable waterjets; 1 steerable thruster
Missiles: 1 Raytheon SeaRAM RIM-116B 11-cell launcher ❶; passive IR/anti-radiation homing to 9.6 km *(5.2 n miles)* at 2.5 Mach; warhead 9.1 kg.
SSM: Lockheed Martin Longbow Hellfire (AGM 114L); active radar acquisition and homing to 8.3 km *(4.5 n miles)*; warhead 9 kg.
Guns: 1 BAE Systems 57 mm/70 Mk 2 ❷; 220 rds/min to 17 km *(9 n miles)*; weight of shell 2.4 kg. 4—12.7mm MGs. 2 Mk 46 Mod 2 Bushmaster 30 mm (ASUW module); 200 rds/min to 6.8 km *(3.7 n miles)*.
Physical countermeasures: Decoys: 4 Loral/Hycor SRBOC 6-barrelled fixed launchers.
Electronic countermeasures: ESM/ECM.
Radars: Air/surface search: Saab Sea Giraffe AMB ❹; 3D; G/H-band.
Navigation: Sperry Bridgemaster; I-band.
Fire control: Seastar Safire III.
Sonars: VDS (under development).
Combat data systems: Northrop Grumman Electronic Systems Integrated Combat Management System (ICMS).
Weapon control systems: Seastar Safire III optronic director ❸.

Helicopters: 1 MH-60R/S and 3 VTUAV.

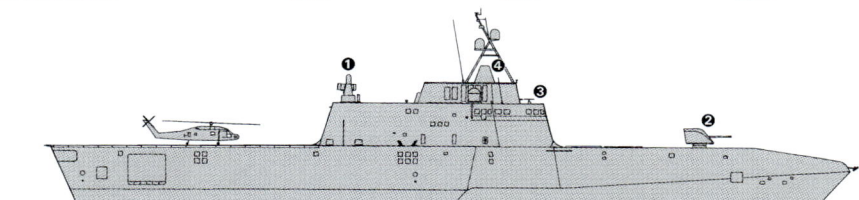

INDEPENDENCE *(Scale 1 : 1,200), Ian Sturton / 1366176*

INDEPENDENCE *7/2009, Austal Ships / 1335491*

Programmes: Two industry teams, one led by Lockheed Martin and the other by General Dynamics, were contracted in 2004 to develop designs for a fast, agile and networked surface combatant. In the original procurement programme, it was planned to build a number of each design and left open the option that both designs could proceed to series production. The keys to this approach were a fast building time of two years per ship and a relatively inexpensive cost. In April 2007, the Navy cancelled its contract with Lockheed Martin for the construction of LCS 3 after negotiations to control cost overruns failed. The second General Dynamics ship (LCS 4) was also cancelled, in November 2007, after similar cost overruns. The funding of three further ships has been cancelled or re-allocated. In March 2009, the decision to proceed with the construction of one of each LCS variant, re-using previous hull numbers, was announced. In December 2010 it was decided that, rather than select a single LCS design for future construction, procurement of the next 20 ships is to be split between the two variants. One of each design is to be built from funds from fiscal year 2010 and FY 2011 and two of each from FY 2012 to FY 2015. A contract for the block-buy of 10 ships was awarded to Austal USA on 29 December 2010. In 2012, the Navy Board of Inspection and Survey found numerous sustainment issues in both LCS variants. The US Comptroller General reported in July 2013 that progress had been made in addressing design and productionj problems but, in late 2013, the Defense Department's Director of Operational Test and Evaluation reported a number of concerns about the performance of weapon systems.

Structure: Trimaran hullform based on fast commercial ferry design for Fred Olsen Line. Aluminium construction. Large flight deck capable of operating heavy-lift helicopter. Stern launch of boats (manned and unmanned). Side-ramp Ro-Ro capability. Reconfigurable seaframe to allow rapidly interchangeable mission modules.

Operational: The three mission packages are ASuW, ASW and MCM. The ASuW mission package includes the following key components: The missile module comprises three container units each containing 15 surface-to-surface missiles (Lockheed Martin Hellfire). The gun module comprises two Mk 46 30 mm guns and the MH-60R helicopter is capable of carrying eight Hellfire missiles and 12.7 mm and 7.62 mm machine guns. Two 11 m RHIBs are available for use by boarding teams carrying small arms. ASW mission systems are deployed from three principal off-board platforms. The ASW Unmanned Surface Vehicle (USV) is capable of carrying the USV Dipping Sonar (UDS), USV Towed Array System (UTAS) and Multistatic Offboard Source (MSOBS). It also carries a surface search radar and communications. The MH-60R helicopter carries Airborne Low-Frequency Sonar (ALFS), 25 sonobuoys and FLIR. A Fire Scout MQ-8B VTUAV is required to provide reconnaissance, surveillance and target acquisition, using an EO/IR sensor, and communications relay support. The MCM mission package comprises a variety of systems deployed by off-board systems including manned and unmanned aircraft and unmanned surface and semi-submersible systems. Detection and identification of mines is performed by the AQS-20A minehunting sonar towed by the MH-60S helicopter or the AN/WLD-1 Remote Multi-Mission Vehicle (RMMV). Mines near the surface are detected with the Airborne Laser Mine Detection System (ALMDS), also deployed in the MH-60S helicopter. Neutralisation of mines is performed by the MH-60S using Airborne Mine Neutralisation System (AMNS). Mine-sweeping is accomplished by the MH-60S towing the Organic Airborne and Surface Influence Sweep (OASIS) or Unmanned Surface Vehicle towing the Unmanned Influence Surface Sweep System (UISS). The first successful launch and recovery of a RMMV was conducted in June 2011.

INDEPENDENCE *7/2009, Austal Ships / 1335492*

SHIPBORNE AIRCRAFT

Notes: (1) Numbers given are for 1 January 2014.
(2) **Joint Strike Fighter:** The JSF F-35 Lightning II is a family of next-generation strike aircraft combining stealth and enhanced sensors. The F-35C Carrier Variant (CV) will replace F/A-18A/C and complement the F/A-18E/F fleet. Marine Corps F-35B Short Take Off and Vertical Landing (STOVL) variant is to replace USMC AV-8B and F/A-18A/C/D. F-35 variants are to share a high level of commonality. The contract for the Systems Development and Demonstration (SDD) phase was awarded on 26 October 2001 to the industry team of Lockheed Martin, Northrop Grumman and BAE Systems. International participants in the SDD phase are Australia, Canada, Denmark, Italy, Netherlands, Norway, Turkey and the UK. All of these had, by early 2007, signed MOUs for the subsequent Production, Sustainment, and Follow-on Development (PSFD) phase. Security Cooperation Partnership MOUs have been established with Israel and Singapore. Engine development is being undertaken by Pratt and Whitney and by General Electric/Rolls Royce. Initial operating capability for F-35C is planned for 2014. However then-defense secretary Robert Gates stated in January 2011 that F-35B had "significant testing problems" and, as a result, the programme had been placed in a two-year probationary period. This was lifted by then-defense secretary Panetta on 20 January 2012. The F-35B conducted its first deck landing on *Wasp* on 3 October 2011. In 2012, F-35B test aircraft accomplished 396 flights and had also executed 102 vertical landings. The first of three operational F-35B were delivered in November 2012 to mark the beginning of tactical operational training at Marine Corps Air Station, Yuma. First F-35C carrier variant delivered to VFA 101 at Eglin Air Force Base, Florida, on 22 June 2013.
(3) **Tacair Integration:** Navy/Marine Corps Tactical Aviation Integration (TAI) plan was approved in 2002 to optimise combat capability and efficiencies by merging Navy and Marine Corps TACAIR at sea and ashore. As part of the TAI, Marine Corps squadrons are integrated into carrier air wings and Navy squadrons into the Marine Corps' Unit Deployment Plan (UDP).
(4) **Capabilities Based Scheduling (CBS):** This scheduling mechanism is designed to source tactical aviation (TACAIR) requirements while promoting goals of TAI. Under CBS, all Department of the Navy (DON) TACAIR squadrons are available to fill land or sea-based requirements. Objective is to fill all operational and training requirements with the most appropriate unit while balancing operational tempo across force.
(5) **Global Force Management (GFM):** This scheduling management process is designed to fill tactical aviation (TACAIR) operational and training requirements with the most appropriate unit while balancing unit operational tempo across force.
(6) **CH-53K:** Contract for the System Development and Demonstration of the CH-53K helicopter was awarded to Sikorsky in Q2 2006. The aircraft is to replace the CH-53E helicopter. USMC expects to buy 200 aircraft. The program has transitioned from design to manufacturing phase. First ground test vehicle was delivered in 2012. Flight testing is planned for 2014. IOC expected in 2018.
(7) **Air Launch and Recovery Programs:**
(a) Electromagnetic Aircraft Launch System (EMALS) is a complete carrier-based launch system designed for *Gerald R Ford* (CVN 78) and all future CVN 78-class carriers. It is designed to achieve increased sortie rates and reliability, lower operating costs, better control of launch forces and reduced wear and tear on aircraft. The system is to be capable of launching all current and future carrier air wing platforms, from lightweight unmanned aerial vehicles to heavy strike fighters. The programme entered full System Functional Demonstration (SFD) in September 2010 and in December 2010, four F/A-18E aircraft successfully test launched from land-based runway. The system was delivered to *Gerald R Ford* in 2011. As of December 2013, EMALS had successfully launched F-35C, F/A-18E, T-45C, C-2A, and E-2D aircraft.
(b) Advanced Arresting Gear (AAG) is a modular, integrated system consisting of cable shock absorbers, energy absorbers, power conditioning equipment, a thermal management system and digital controls. It is designed to replace existing Mk-7 arresting gear when landing a broad range of fixed-wing tailhook-equipped aircraft at sea. Dead-load arrestments began on the shipboard representative system in 2011.
(c) Aviation Data Management and Control System (ADMACS) is a tactical, real-time internal data management system connecting the air department, ship divisions and embarked staff who manage aircraft launch and recovery operations on CV/CVN ships. ADMACS utilises the ADMACS Local Area Network and Integrated Shipboard Network System to electronically display position and location of aircraft on flight and hangar decks. In addition, aircraft status, launch and recovery equipment, fuel, weapons types and quantity, and other aviation and ship-related information are also displayed.

Numbers/Type: 234/252 Boeing F/A-18E Super Hornet/F/A-18F Super Hornet.
Operational speed: 930 kt *(1,721 km/h)*.
Service ceiling: 50,000 ft *(15,240 m)*.
Range: 1,320 n miles *(2,376 km)*.
Role/Weapon systems: Single-seat (F/A-18E) and two-seat (F/A-18F) strike fighters for USN. First one rolled out in September 1995. First 12 production aircraft ordered in FY97. First sea trials January 1997. Entered operational service November 1999. Initial deployment to CVN 72 in July 2002. 200th aircraft delivered in August 2004. The balance of 565 aircraft to be delivered by 2016. Sensors: APG-73 radar, APG-79 AESA radar, ALR-67(V)3 RWR. ECM: ALQ-214 RFCM, towed decoys. Weapons: 11 wing stations for 8,680 kg of weapons (same weapons as C/D) including 20 mm guns.

F/A-18F *4/2013*, Hachiro Nakai* / 1531008

Numbers/Type: 50 Grumman EA-6B Prowler.
Operational speed: 566 kt *(1,048 km/h)*.
Service ceiling: 41,200 ft *(12,550 m)*.
Range: 955 n miles *(1,769 km)*.
Role/Weapon systems: Electronic Attack (EM) and Suppression of Air Defences (SEAD) aircraft to provide electronic attack in support of strikes and armed reconnaissance. Block 89A avionics/computer upgrades first delivered 2001. ICAP III receiver system upgrade first delivered in 2005. In 2010, the Marine Corps began its transition to an all ICAP III variant, scheduled for completion in 2012. Sensors: APS-130 radar; ALQ-99, ALQ-218 (ICAP III), USQ-113 communication jammer. Weapons: AGM-88 HARM, ALQ-99 jamming pods.

EA-6B *3/2012*, Tony Roper* / 1531159

Numbers/Type: 95/21/343/129 Boeing F/A-18A Hornet/F/A-18B Hornet/F/A-18C Hornet/F/A-18D Hornet.
Operational speed: 1,032 kt *(1,910 km/h)*.
Service ceiling: 50,000 ft *(15,240 m)*.
Range: 1,000 n miles *(1,850 km)*.
Role/Weapon systems: Single-seat (F/A-18A/C) and two-seat (F/A-18B/D) strike interdictor (VFA) for USN/USMC air groups. Some are used for EW support with ALQ-167 jammers. Sensors: ESM: Litton ALR 67(V)2, ALQ 165 ASPJ jammer (18C/D), ALQ-126B jammer, APG-65 or APG-73 radar, AAS-38 FLIR, ASQ-228 ATFLIR, AN/AAQ-28 Litening FLIR (USMC only), AAR-50 Nav FLIR, ASQ-173 tracker. Weapons: ASV; four Harpoon or SLAM (ER) or AGM-88 HARM missiles. AGM-65 Maverick. Strike; up to 7.7 tons of bombs (or LGM). AD; one 20 mm Vulcan cannon, nine AIM-120/AIM-7/AIM-9 missiles. Typical ASV load might include 20 mm gun, 7.7 ton bombs including AGM 154A JSOW, two AIM-9 missiles. Typical AAW load might include 20 mm gun, four AIM-7 or AIM-120, two AIM-9 missiles.

F/A-18C *6/2011, Jurg Kürsener* / 1454841

Numbers/Type: 19/82/16 Boeing/British Aerospace AV-8B Harrier II/AV-8B II Harrier II Plus/TAV-8B Harrier II.
Operational speed: 585 kt *(1,083 km/h)*.
Service ceiling: 45,000 ft *(13,716 m)*.
Range: 800 n miles *(1,480 km)*.
Role/Weapon systems: Attack and destroy surface and air targets in support of USMC. Operational since 1985. Sensors: AN/AAQ-28 LITENING targeting pod, Navigation FLIR, AN/AVS-9 night vision goggles, Angle Rate Bombing system with laser spot tracker (AV-8B II), APG-65 radar (AV-8B II Plus), ALR-67 RWR, ALQ-164 DECM pod. Weapons: Strike; 500 and 1,000 lb general purpose bombs, Paveway II LGB, Joint Direct Attack Munition, Dual Mode Guide Bomb, AGM-65 Maverick, Cluster Bomb Units, 300—25 mm rounds, 2.75 in and 5.00 in rockets. Self-defence: one GAU-12/U 25 mm cannon, four AIM-9M Sidewinder and four AIM-120 AMRAAM.

AV-8B *6/2010, M Mazumdar* / 1406325

Numbers/Type: 27 Lockheed Martin F-35B Lightning II.
Operational speed: 1,058 kt *(1,960 km/h)*.
Service ceiling: 50,000 ft *(15,240 m)*.
Range: 900 n miles *(1,667 km)*.
Role/Weapon systems: F-35B is to replace Marine Corps' F/A-18 Hornet, AV-8B Harrier and EA-6B Prowler. There are 16 F-35Bs at Marine Fighter Attack Squadron 121, Yuma, Arizona as a first step towards progression from testing and training to full tactical aviation capability. A total of 27 F-35B have been delivered including 11 to the 2nd Marine Aircraft Wing's Marine Fighter/Attack Training Squadron 501 at Eglin Air Force Base, Florida, to support pilot and maintainer training. Low-Rate Initial Production agreement reached in November 2012 to procure three more B variants. Long-range procurement plans call for eventual total buy of 340 F-35Bs. Sensors: Northrop Grumman AN/APG-81 AESA radar; Lockheed Martin AN/AAQ-40 EOTS; Northrop Grumman AN/AAQ-37 DAS. Weapons: 2 AIM-120C air-to-air missiles, two 1,000 lb GBU-32 JDAM guided bombs.

F-35B *9/2012, US Navy* / 1486431

Numbers/Type: 55 Northrop Grumman E-2C Hawkeye.
Operational speed: 300 kt *(555 km/h)*.
Service ceiling: 37,000 ft *(11,278 m)*.
Range: 1,540 n miles *(2,852 km)*.
Role/Weapon systems: Carrier-borne multimission aircraft with primary AEW, strike control, area surveillance, SAR and battle-management roles. Current configurations include 2 Group II aircraft, 27 Group II NAV upgrades, five Group II MCU/ACIS and 27 HE2000. Sensors: APS 145 radar; ALQ 217 ESM (HE 2000); Airborne tactical data system with Links 11 and 16; CEC (HE 2000). Weapons: Unarmed.

E-2C *4/2013*, Hachiro Nakai* / 1531009

Shipborne aircraft ‹ US 955

Numbers/Type: 91 Boeing EA-18G Growler.
Operational speed: 930 kt (1,721 km/h).
Service ceiling: 50,000 ft (15,240 m).
Range: 1,320 n miles (2,376 km).
Role/Weapon systems: Electronic Attack (EA) and Suppression of Enemy Air Defences (SEAD) aircraft started replacing the EA-6B Prowler from 2009. First production aircraft, built on F/A-18E/F Block II Super Hornet platform, handed over on 24 September 2007. Initial operational capability achieved September 2009 with one squadron replacing legacy EA-6B at Naval Air Station Whidbey Island, WA. Design includes ALQ-218 receivers on both wingtips compared with single ALQ-99 receiver in vertical tail fin of EA-6B. All 14 EA-6B Prowler squadrons to be converted to EA-18G by 2016. As of September 2013, eight Navy squadrons had transitioned from EA-6B to EA-18G. Sensors: APG-79 AESA radar; ALQ-218 receivers; ALQ-227 communication countermeasures. Weapons: AGM-88 HARM; AIM-120C AMRAAM.

EA-18G   4/2013*, Hachiro Nakai / 1531010

Numbers/Type: 10 Northrop Grumman E-2D Advanced Hawkeye.
Operational speed: 300 kt (555 km/h).
Service ceiling: 34,500 ft (10,515 m).
Range: 1,320 n miles (2,445 km).
Role/Weapon systems: Advanced Hawkeye uses E-2C 2000 configuration as a baseline but features a new 360° radar (with automatic air and sea surface detection) and upgraded aircraft systems. New capabilities include battle space target detection and situational awareness, support of Theatre Air and Missile Defense (TAMD) operations, and improved operational availability. Production deliveries began in 2010 and a total of 75 aircraft is planned by 2022. Sensors: ADS-18 ESA radar. ESM: ALQ-217; airborne tactical data system including Links 11 and 16 and CEC. Weapons: unarmed.

E-2D   3/2012*, Tony Roper / 1531161

Numbers/Type: 35 Northrop Grumman C-2A Greyhound.
Operational speed: 300 kt (555 km/h).
Service ceiling: 31,000 ft (9,448 m).
Range: 1,400 n miles (2,592 km).
Role/Weapon systems: Twin turbo-prop COD (Carrier Onboard Delivery) transport for high-priority cargo and passengers to and from aircraft carriers. Mission includes airlift and airdrop of special operating forces and airdrops for search and rescue. Maximum weight for payload and route support equipment is 10,000 lb and 26 passengers. Weapons: unarmed.

C-2A   4/2013*, Tony Roper / 1531160

Numbers/Type: 9 Sikorsky SH-60F Seahawk (CV).
Operational speed: 145 kt (268 km/h).
Service ceiling: 10,000 ft (3,050 m).
Range: 600 n miles (1,111 km).
Role/Weapon systems: Derivation of SH-60B that replaced SH-3H Sea King to provide close-in ASW protection to Carrier Battle Groups. First deployed in *Nimitz* 1991. To be replaced by MH-60R. Sonar: AQS-13F dipping sonar; ASQ-81 (V) MAD; UYS-2 acoustic processor; 14 sonobuoys. Weapons: ASW: three Mk 46/54 torpedoes. ASUW: One GAU 16 12.7 mm MG or one M 240 7.62 mm MG.

SH-60F   10/2006, Michael Nitz / 1305199

© 2014 IHS

P-8A POSEIDON   2/2012, US Navy / 1486433

Numbers/Type: 16 Boeing E-6B Mercury.
Operational speed: 455 kt (842 km/h).
Service ceiling: 42,000 ft (12,800 m).
Range: 6,350 n miles (11,760 m).
Role/Weapon systems: Derived from Boeing's 707 aircraft, the E-6B provides Commander, US Strategic Command with the command, control and communications capability to direct and employ strategic forces. Designed to support a flexible nuclear deterrent posture with VLF emergency communications and Airborne National Command Post (ABNCP) missions. Sensors: Radar Bendix APS-133; ALR-68(V)4 ESM; supports Trident Fleet radio communications with up to 28,000 ft of VLF trailing wire antenna. Weapons: Unarmed.

© 2014 IHS

Numbers/Type: 45 Sikorsky SH-60B Seahawk (LAMPS Mk III).
Operational speed: 145 kt (268 km/h).
Service ceiling: 10,000 ft (3,050 m).
Range: 450 n miles (833 km).
Role/Weapon systems: LAMPS Mk III is airborne platform for ASW and ASUW: operated from cruisers, destroyers and frigates. First deployed in 1984. To be replaced by MH-60R. Sensors: APS-124 search radar, AAS-44 FLIR with laser designator, ASQ-81(V) MAD, 25 sonobuoys, ALQ-142 ESM, AAR-47 MWR, ALQ-144 IRCM suppressor and ALE-39 CMDS. UYS-1 Acoustic processor. Weapons: ASW; three Mk 46 or Mk 50 torpedoes. ASUW; one 7.62 mm MG or 12.7 mm MG, four AGM-114B/K Hellfire missile.

SH-60B   10/2012, Hachiro Nakai / 1483660

Numbers/Type: 178/10 Boeing MV-22B Osprey/CV-22B Osprey.
Operational speed: 275 kt (509 km/h).
Service ceiling: 25,000 ft (7,620 m).
Range: 400 n miles (740 km).
Role/Weapon systems: Replacement for legacy assault/support helicopter (CH-46E) for Marines (MV), projected rescue and resupply for the Navy (Navy MV), and special operations for USAF SOCOM (CV). Eight active MV-22 squadrons: VNM-161, 162, 166, 261, 263, 264, 266 and 365. In addition, one squadron (HX-21) used for developmental testing, one (VMX-22) for operational testing and evaluation and one (VMMT-204) for FRS. Final operational evaluation led to full rate production decision in September 2005. Contract awarded to Boeing in March 2008 for 141 MV-22 and 26 CV-22. Supplemental funding added five CV-22. Full fleet of 360 MV-22, 50 CV-22 and 48 Navy MV-22 projected. Sensors: AAR-47 ESM; AN/ALQ-211 Suite of Integrated RF Countermeasures (SIRFC) (CV-22 only), AN/AAQ-24(V) Nemesis Directional Infra-Red Countermeasures (DIRCM) (CV-22 only); AN/AAQ-27 FLIR; APR 39A(V)2 (MV-22 only). Weapons: M-240D 7.62 mm machine gun.

MV-22B   10/2011, Michael Nitz / 1454843

Numbers/Type: 169 Sikorsky MH-60R Seahawk.
Operational speed: 145 kt (268 km/h).
Service ceiling: 10,000 ft (3,050 m).
Range: 450 n miles (833 km).
Role/Weapon systems: MH-60R is to replace the SH-60B/F fleet as the primary USW and ASUW platform operated from aircraft carriers, cruisers and destroyers. The first production aircraft was flown on 28 July 2005 and the MH-60R completed its first operational deployment in July 2009. Sensors: APS-147 long-range search radar with ISAR, ALQ-210 ESM, AQS-22 dipping sonar, acoustic processor, AAS-44 multispectral FLIR including Low Light Camera, Hawklink C-band sensor datalink, AAR-47 MWR, ALE-47 CMDS, and ALQ-144 IRCM. Weapons: ASW: three Mk 46/50 torpedoes. ASUW: four AGM-114B/K Hellfire missiles, one 7.62 mm MG or 12.7 mm MG. Pre-planned Product Improvements are in work which will upgrade the aircraft with a CDL-N Ku-band sensor datalink, a fourth weapons station for eight AGM-114s or four torpedoes and the ability to fire Mk 54 torpedoes.

MH-60R   4/2013*, Hachiro Nakai / 1531012

IHS Jane's Fighting Ships 2014-2015

MQ-8C   10/2013*, US Navy / 1531214

IHS Jane's Fighting Ships 2014-2015

956 US › Shipborne aircraft

Numbers/Type: 128/36 Bell AH-1W Super Cobra/AH-1Z Viper.
Operational speed: 132 kt (AH-1W); 142 kt (AH-1Z) (244/263 km/h).
Service ceiling: 12,200 ft (AH-1W); 20,000 ft (AH-1Z) (3,720/6,096 m).
Range: 58 n miles (AH-1W); 137 n miles (AH-1Z) (107; 254 km).
Role/Weapon systems: Close air support helicopter (HMLA), anti-armour, armed escort, armed/visual reconnaissance, and fire support coordination with own air-to-air capability. AH-1Z is upgraded aircraft with four-bladed rotor that improves speed, range and lift. Other features include glass cockpit, new engines and gearboxes. Sensors: Target Sight System (laser and FLIR targeting sensor). Weapons: Strike/assault; one triple 20 mm cannon, 8 Hellfire missiles, 14 rockets and 20 mm gun. AAW; two AIM-9M Sidewinder missiles.

Numbers/Type: 151 Sikorsky CH-53E Super Stallion.
Operational speed: 150 kt (278 km/h).
Service ceiling: 18,500 ft (5,638 m).
Range: 480 n miles (888 km).
Role/Weapon systems: Upgraded, three-engined version of Sea Stallion for USMC (two are operated by US Navy) heavy lift mission. Has seating capacity for up to 55 passengers or space for 24 litters or 36,000 lb (16,329 kg). Sensors: AN/AAQ-29A FLIR. Weapons: Up to three 12.7 mm machine guns.

958 US › Unmanned air vehicles — Patrol forces

Numbers/Type: 2 Northrop Grumman RQ-4A UAS.
Operational speed: 340 kt (630 km/h).
Service ceiling: 65,000 ft (19,810 m).
Range: 10,000 n miles (18,518 km).
Role/Weapon systems: Two production RQ-4A with maritime sensor capabilities were acquired for evaluation as the Global Hawk Maritime Demonstration (GHMD) in 2004. Delivered in 2006, GHMD conducted two years of demonstrations and experiments with the Navy and facilitated development of high altitude UAS concepts of operations and tactics, techniques and procedures for operating in conjunction with manned aircraft and surface fleet components. Re-designated as Broad Area Maritime Surveillance Demonstrator (BAMS-D) in 2009, deployed aircraft operations have completed over 7,500 combat-support flight hours. BAMS-D also performs training and system development activities on a rotational basis.

RQ-4A   6/2010, US Navy / 1486434

RQ-21A   1/2012, INSITU / 1450965

Numbers/Type: 13 AAI RQ-7 Shadow.
Operational speed: 110 kt (195 km/h).
Service ceiling: 14,000 ft (4,267 m).
Range: 67 n miles (124 km).
Role/Weapon systems: The Shadow UAS provides Reconnaissance, Surveillance and Target Acquisition (RSTA), precision laser targeting and communications relay capability to the Marine Air-Ground Task Force. Two systems are deployed. Each system is comprised of two Ground Control Stations and four Air Vehicles. An integrated weapons delivery capability is planned. Sensors: electro-optic/FLIR/infrared imaging sensor.

SHADOW 200   10/2006, US Army / 1122581

PATROL FORCES

Notes: (1) In June 2012, Navy Expeditionary Combat Command's (NECC's) Maritime Expeditionary Security Force (MESF) merged with the Riverine Force to become the Coastal Riverine Force (CRF). The CRF operates in harbours, rivers, bays, across the littorals and ashore. The CRF primary mission is to conduct maritime security operations across all phases of military operations by defending high-value assets, critical maritime infrastructure, ports, and harbours both inland and on coastal waterways against enemies and when commanded, conduct offensive combat operations.
(2) A family of modular unmanned surface vehicles (USVs), capable of carrying a variety of payloads for employment with the Littoral Combat ship classes, is under development. Flight 1 USV is the Unmanned Influence Sweep System (UISS) that is to provide an autonomous mine sweeping capability against magnetic and acoustic mines. In 2014, it is planned that up to two industry teams build Engineering Development Models with low rate initial production (LRIP) system options. Subsequent fabrication and formal testing of two LRIP units is to take place. Flight 2 USV is to add a mine-hunting capability and Flight 3 is to further add mine-neutralisation. Ultimately, USV is planned to be capable of undertaking all the mine-clearance functions. Flight 4 is envisioned to be a multifunction, multimission USV. The whole programme will be informed by experience gained in Fifth Fleet of four mine-hunting USVs (MHU), deployed in response to an urgent operational requirement for a stand-off, long-endurance, unmanned, day-night, semi-autonomous mine-hunting capability. The MHUs, the first of which was delivered in theatre in January 2014, are 11 m RHIBs fitted with AN/AQS-20A towed sonar and a remote command and control suite.

Numbers/Type: 2 Northrop Grumman X-47B UCAS-D.
Operational speed: 450 kt (833 km/h).
Service ceiling: 40,000 ft (12,192 m).
Range: 2,100 n miles (3,889 km).
Role/Weapon systems: Unmanned Combat Air System Carrier Demonstration (UCAS-D) is a computer-controlled tailless, autonomous, low-observable air vehicle built to demonstrate autonomous carrier operations capability and aerial refuelling. The aircraft flies a preprogrammed mission in response to (but not 'flown' by) mission operator and using precision GPS. The design includes space and power to accommodate weapons and intelligence, surveillance and reconnaissance sensors. First flight of AV-1 was February 2011 at Edwards Air Force Base, California. Followed by AV-2 in November 2011. Both aircraft moved to Naval Air Station, Patuxent River, Maryland in 2012 for shore-based carrier suitability testing in preparation for sea trials. In November 2012, an X-47B was hoisted on board Harry S Truman (CVN-75) for deck-handling trials. Carrier testing resumed on board George H W Bush (CVN 77) in May 2013 when carrier-based catapult launch and arrested landing were achieved. Autonomous aerial refuelling demonstrations planned for 2014.

MHU   11/2013*, US Navy / 1531212

24 SPECIAL OPERATIONS CRAFT RIVERINE (SOCR)

Displacement, tonnes: 9 full load
Dimensions, metres (feet): 10.1 × 2.7 × 0.6 (33.1 × 8.9 × 2.0)
Speed, knots: 40 (est.)
Range, n miles: 195
Complement: 4
Military lift: 8 fully equipped troops
Machinery: 2 Yanmar 6LY2A-STP diesels; 440 hp (328 kW); 2 Hamilton HJ292 waterjets
Guns: Combination of Mk 19 40 mm, 12.7 mm MG, 7.62 mm/M60, M240, GAU17 at 5 stations.

Comment: Built by United States Marine, Inc. Aluminium hull.

X-47B   5/2013*, US Navy / 1531213

Numbers/Type: 1 Insitu RQ-21A Integrator.
Operational speed: 80 kt (148 km/h).
Service ceiling: 15,000 ft (4,572 m).
Range: 50 n miles (93 km).
Role/Weapon systems: A small unmanned aircraft system designed to provide persistent reconnaissance, surveillance and target acquisition support for tactical manoeuvres and for unit-level force protection of Navy ships, Marine Corps land forces and Special Warfare Units. A follow-on to ScanEagle (uses same launcher and recovery system), RQ-21A consists of a number of air vehicles, ground control systems and multimission payloads. Operates up to 15 hours per day continuously with a short surge capability for 24 hours a day. Payloads include day or night full motion video cameras, infrared marker, laser range finder and automatic identification system receivers. Land-based operational assessment and developmental testing completed November 2012. Ship-based testing started in early 2013, low-rate initial production in late 2013 and initial operational testing in early 2014.

SOCR   2/2005, US Navy / 1043672

Patrol forces < **US** 959

13 CYCLONE CLASS (PATROL COASTAL SHIPS) (PBFM)

Name	No	Builders	Commissioned	F/S
TEMPEST	PC 2	Bollinger, Lockport	21 Aug 1993	AA
HURRICANE	PC 3	Bollinger, Lockport	15 Oct 1993	AA
MONSOON	PC 4	Bollinger, Lockport	22 Jan 1994	AA
TYPHOON	PC 5	Bollinger, Lockport	12 Feb 1994	AA
SIROCCO	PC 6	Bollinger, Lockport	11 Jun 1994	AA
SQUALL	PC 7	Bollinger, Lockport	4 Jul 1994	AA
ZEPHYR	PC 8	Bollinger, Lockport	15 Oct 1994	AA
CHINOOK	PC 9	Bollinger, Lockport	28 Jan 1995	AA
FIREBOLT	PC 10	Bollinger, Lockport	10 Jun 1995	AA
WHIRLWIND	PC 11	Bollinger, Lockport	1 Jul 1995	AA
THUNDERBOLT	PC 12	Bollinger, Lockport	7 Oct 1995	AA
SHAMAL	PC 13	Bollinger, Lockport	27 Jan 1996	AA
TORNADO	PC 14	Bollinger, Lockport	24 Jun 2000	AA

Displacement, tonnes: 385 full load
Dimensions, metres (feet): 54.6 × 7.9 × 2.4 *(179.1 × 25.9 × 7.9)*
Speed, knots: 35
Range, n miles: 2,500 at 12 kt
Complement: 48 (4 officers)

Machinery: 4 Paxman Valenta 16RP200CM diesels; 13,400 hp *(10 MW)* sustained; 4 shafts
Missiles: SSM: Raytheon Griffin; inertial guidance with semi-active terminal homing to 15 km *(8.1 n miles)*; warhead 5.9 kg (installed in Bahrain-based units).
Guns: 2 Bushmaster 25 mm Mk 38. 8—12.7 mm MGs (4 twin). 2—7.62 mm MGs. 2—40 mm Mk 19 grenade launchers (MGs and grenade launchers are interchangeable).
Electronic countermeasures: ESM: Argon ST WBR 2000.
Radars: Surface search: 2 Sperry RASCAR; E/F/I/J-band.
Sonars: Wesmar; hull-mounted; active scanning sonar; high frequency.
Weapon control systems: FLIR systems AN/KAX-1 Marflir.

Programmes: Contract awarded for eight in August 1990, five in July 1991 and one in August 1997.
Structure: Design based on Vosper Thornycroft Ramadan class modified for USN requirements including ballistic plating to protect electronics, communications and the pilot house. The craft have a slow speed loiter capability. Swimmers can be launched from a platform at the stern.
Modernisation: The ships have been modernised to incorporate an integrated bridge system, a Mk 38 stabilised weapon platform and improved communications. Raytheon Griffin SSMs have been installed in deployed units.
Operational: The ships perform maritime interdiction, homeland security, law enforcement and SAR missions. Can be operated in pairs with a maintenance team in two vans ashore. Operational control transferred from Special Operations Command to the Atlantic and Pacific Fleets on 1 October 2002. Five stern-ramp fitted ships (PC 2, PC 4, PC 8, PC 13 and PC 14) were transferred to the USCG 2004–05. Of those, two (PC 2, PC 4) returned to the Navy in FY09 and PC 8, 13 and 14 on 30 September 2011. In 2009, a sustainment programme to update communication, engineering and support systems was initiated. Work includes radar upgrades and installation of new diesel generators and air-conditioning units. Two ships are home-ported at Little Creek, Virginia, three in Mayport, Florida, and eight are forward-deployed to Manama, Bahrain: *Tempest, Typhoon, Sirocco, Squall, Chinook, Firebolt, Whirlwind,* and *Thunderbolt*. Two further ships, *Hurricane* and *Monsoon* are to transfer to Bahrain in mid-2014.
Sales: PC 1 *(Cyclone)* transferred to the Philippines Navy for counter-terrorism duties.

SHAMAL *6/2012, M Declerck* / 1486435

CHINOOK *9/2008, Shaun Jones* / 1353589

46 28 ft PATROL CRAFT (PB)

Displacement, tonnes: 6 full load
Dimensions, metres (feet): 8.5 × 3.0 × 0.96 *(27.9 × 9.8 × 3.1)*
Speed, knots: 38
Complement: 4
Machinery: 2 Mercury (or Yamaha) outboards; 500 hp *(373 kW)*
Guns: 4—12.7 mm MGs.
Radars: Navigation: Raytheon RL70; I-band.

Comment: SeaArk Marine Dauntless RAM design (Model 28V) is medium-sized harbour security and force protection boat. Heavy-duty aluminium hull. All delivered 2001–09. Original 14 powered by twin Mercury 225 Optimax. Remaining 40 boats have twin 225 Yamaha 4-stroke engines. Used primarily by Pacific Fleet. Length overall is 31 ft with aft platform.

SEA ARK 28 ft PATROL CRAFT *5/2008, Hachiro Nakai* / 1366285

16 LIGHT PATROL BOATS (PBF)

Displacement, tonnes: 1 full load
Dimensions, metres (feet): 6.8 × 2.6 × 0.5 *(22.3 × 8.5 × 1.6)*
Speed, knots: 35
Complement: 3
Machinery: 2 E-Tech outboards; 300 hp *(224 kW)*
Guns: 3—12.7 mm MGs. 1—7.62 mm MG.
Radars: Surface search: Furuno 1731; I-band

Comment: Built by Boston Whaler in 1988 for US Special Operations Command. Air transportable. Glass fibre hulls. Replacement began in 2001.

PBL-CD *1996, Boston Whaler* / 0084150

62 NSW 11 METRE RIB (RIGID INFLATABLE BOATS) (PBF)

Displacement, tonnes: 9 full load
Dimensions, metres (feet): 11 × 3.2 × 0.9 *(36.1 × 10.5 × 3.0)*
Speed, knots: 35
Range, n miles: 200 at 33 kt
Complement: 13
Machinery: 2 Caterpillar 3126 diesels; 490 hp *(365 kW)*; 2 Kamewa FF 280 water-jets
Guns: 1—12.7 mm MG, 1—7.62 mm MG or Mk 19 Mod 3 grenade launcher.

Comment: Naval Special Warfare (NSW) RIB capable of carrying nine SEALS at 35 kt. Built by USMI, New Orleans. Entered service from 1998 to 2009.

NSW RIB *1/1998, US Navy* / 0016492

NSW RIB *1/2002, M Declerck* / 0529972

© 2014 IHS IHS Jane's Fighting Ships 2014-2015

960 US > Patrol forces

118 34 ft PATROL CRAFT (PBF)

Displacement, tonnes: 9 full load
Dimensions, metres (feet): 10.36 × 3.66 × 1.3 (34 × 12.0 × 4.3)
Speed, knots: 36
Complement: 6
Machinery: 2 Cummins QSB5.9-420 GS diesels; 740 hp (550 kW); 2 Konrad 520 drives or Hamilton 292 waterjets
Guns: 4 — 12.7 mm MGs.
Radars: Navigation: Furuno; I-band.

Comment: SeaArk Marine Dauntless RAM design (model 34V) delivered from 2002. Aluminium construction transportable by aircraft. Employed on harbour and offshore installation protection tasks. First deployed with Naval Coastal Warfare Squadron-Five near San Diego. All delivered by late 2011.

CUTTER 408 10/2011, Michael Nitz / 1454849

26 27 ft PATROL CRAFT (PB)

Displacement, tonnes: 3 full load
Dimensions, metres (feet): 8.2 × 2.44 × 0.91 (26.9 × 8.0 × 3.0)
Speed, knots: 34
Complement: 4
Machinery: 2 Honda outboards; 260 hp (194 kW)
Guns: 4 — 12.7 mm MGs.
Radars: Navigation: Raytheon RL70; I-band.

Comment: SeaArk Marine Model 27VC employed on harbour security and force protection duties. All 68 delivered from 2001 to 2007. Used primarily by Atlantic Fleet. Welded aluminium hull.

12 RIVERINE ASSAULT BOATS (PB)

Displacement, tonnes: 10.25 full load
Dimensions, metres (feet): 10.1 × 2.7 × 0.6 (33.1 × 8.9 × 2.0)
Speed, knots: 40
Complement: 7
Machinery: 2 Yanmar 6LY2A-STP diesels; 440 hp (328 kW); 2 Hamilton HJ292 waterjets
Guns: 5 multipurpose mount locations capable of accepting 12.7 mm and 7.62 mm MGs and 40 mm grenade launchers.

Comment: Built by United States Marine, Inc. as variant of the Special Operations Craft—Riverine. Riverine Assault Boat (RAB) is designed for speed and manoeuvrability for combat operations in riverine environment. Aluminium hull is partially armoured for small arms protection. Used by Coastal Riverine Squadron (CRS)-2 in Portsmouth, Virginia, CRS-4 in Virginia Beach, Virginia, and CRS-3 in San Diego, California.

ASSAULT BOAT 3/2009, US Navy / 1366253

6 RIVERINE COMMAND BOATS (PBF)

Displacement, tonnes: 22.8 full load
Dimensions, metres (feet): 16.1 × 3.78 × 0.9 (52.8 × 12.4 × 3.0)
Speed, knots: 43
Complement: 8 + 10 embarked forces
Machinery: 2 Scania diesels; 850 hp (634 kW); 2 Rolls-Royce waterjets
Guns: 4 multipurpose mount locations capable of accepting 12.7 mm and 7.62 mm MGs and 40 mm grenade launchers.

Comment: Built by SAFE Boats International to provide command and control, tactical mobility and personnel transport for troops in a riverine environment. Aluminium hull with beaching plates is armoured for small arms protection.

COMMAND BOAT 9/2009, US Navy / 1366251

24 RIVERINE PATROL BOATS (PBF)

Displacement, tonnes: 11.4 full load
Dimensions, metres (feet): 12 × 3.1 × 0.6 (39.4 × 10.2 × 2.0)
Speed, knots: 38
Complement: 17
Machinery: 2 Yanmar 6LY2A-STP diesels; 440 hp (328 kW); 2 Hamilton HJ292 waterjets
Guns: 3 multipurpose mount locations capable of accepting 12.7 mm and 7.62 mm MGs and 40 mm grenade launchers.

Comment: Built by SAFE Boats International, the RPB is designed for tactical mobility and personnel transport of ground combat element in riverine environment. Aluminium hull is partially armoured for small arms protection.

PATROL BOAT 3/2009, US Navy / 1366252

0 + 6 (42) MK VI CLASS (PATROL CRAFT) (PBF)

Dimensions, metres (feet): 25.83 × 6.25 × 1.37 (84.7 × 20.5 × 4.5)
Speed, knots: 41
Range, n miles: 600 at 12 kt
Machinery: 2 MTU 16V 2000 M94 diesels; 5,200 hp (3.87 MW); 2 Hamilton HM651 waterjets
Guns: 2 — 25 mm. 1 — 12.7 mm MG.

Comment: The contract for six of a new class of patrol craft was placed with SAFE Boats International on 24 May 2012. The first is to be delivered in 2015 and a class of 48 is expected.

MK VI PATROL CRAFT (artist's impression) 6/2012, SAFE BOATS / 1451149

IHS Jane's Fighting Ships 2014-2015

COMMAND SHIPS

Notes: Plans to replace the two command ships have been abandoned in favour of extending the lives of *Blue Ridge* and *Mount Whitney* to 2039.

2 BLUE RIDGE CLASS (COMMAND SHIPS) (LCCH/AGFH)

Name	No	Builders	Laid down	Launched	Commissioned	F/S
BLUE RIDGE	LCC 19	Philadelphia Naval Shipyard	27 Feb 1967	4 Jan 1969	14 Nov 1970	PA
MOUNT WHITNEY	LCC 20	Newport News Shipbuilding	8 Jan 1969	8 Jan 1970	16 Jan 1971	AA

Displacement, tonnes: 13,287 (LCC 19), 12,635 (LCC 20) standard; 19,963 (LCC 19), 17,766 (LCC 20) full load
Dimensions, metres (feet): 193.2 × 32.9 × 7.6 *(633.9 × 107.9 × 24.9)*
Speed, knots: 23
Range, n miles: 13,000 at 16 kt
Complement: 599 (LCC 19), 562 (LCC 20) (34 officers (LCC 19); 32 officers (LCC 20))

Military lift: 700 troops; 3 LCPs; 2 LCVPs; 2 – 7 m RHIBs
Machinery: 2 Foster-Wheeler boilers; 600 psi *(42.3 kg/cm^2)*; 870°F *(467°C)*; 1 GE turbine; 22,000 hp *(16.4 MW)*; 1 shaft
Guns: 2 Raytheon 20 mm/76 6-barrelled Vulcan Phalanx Mk 15; 4,500 rds/min combined to 1.5 km. 2 – 25 mm Mk 38. 2 – 12.7 mm MGs.
Physical countermeasures: Decoys: 4 Mk 137 Mod 2 SRBOC 6-barrelled fixed launcher; IR flares and chaff to 4 km *(2.2 n miles)*. SLQ-25 Nixie; torpedo decoy.
Electronic countermeasures: ESM/ECM: SLQ-32(V)3; combined radar intercept, jammer and deception system.
Radars: Air search: Lockheed SPS-40E (LCC 19); B-band.
Surface search: Lockheed SPS-10B; G-band. Raytheon SPS-67; G-band.
Navigation: Marconi LN66 (LCC 19); Raytheon SPS-73 (LCC 19); I-band. Kelvin Hughes SharpEye; E/F/I-bands (LCC 20).
Tacan: URN 25. IFF: Mk XII AIMS UPX-29.
Combat data systems: GCCS (M) Link 4A (LCC 19), Link 11, Link 14 (LCC 19) and JTIDS (LCC 20). Theatre Battle Management Core Systems (TBMCS). CBSP (LCC 19). Wide band commercial SATCOM (LCC 20), USC-38 SATCOM, WSC-3 EHF SATCOM, NMT (LCC 19), WSC-6(V)1 and 5 (LCC 20), and WSC-6A(V)4 (LCC 20) SHF SATCOM. High Frequency Radio Group (HFRG). Mission Display System (MDS). Demand Assigned Multiple Access (DAMA QUAD). Area Air Defense Commander, Naval Fires Network, Joint Service Imagery Processing System (JSIPS-N), Common High Bandwidth Data Link, Ring Laser Gyro Network (RLGN), Joint Tactical Information Distribution System (JTIDS), Navigational Sensor System Interface (NAVSSI). (See Data Systems at front of section.)

Helicopters: Platform for 1 MH-60S Seahawk.

Programmes: Authorised in FY65 and 1966. Originally designated Amphibious Force Flagships (AGC); redesignated Command Ships (LCC) on 1 January 1969.
Modernisation: Modernisation completed FY87. 3 in guns removed in 1996/97 and Sea Sparrow missile launchers have been disembarked. Mk 23 TAS and RAM are not to be fitted. Service lives of both ships to be extended to 2039. An Extended Life Programme is to include improvement of electrical generation and distribution, HVAC upgrades, habitability and safety modifications and corrosion control.

MOUNT WHITNEY *6/2013*, Michael Nitz* / 1531164

BLUE RIDGE *7/2013*, Chris Sattler* / 1531163

Structure: General hull design and machinery arrangement are similar to the Iwo Jima class assault ships. Accommodation for 250 officers and 1,300 enlisted men.
Operational: These are large force command ships of post-Second World War design. They can provide integrated command and control facilities for sea, air and land commanders in all types of operations. *Blue Ridge* is the Seventh Fleet flagship, based at Yokosuka, Japan. *Mount Whitney* served since January 1981 as flagship Second Fleet, based at Norfolk, Virginia except during the period June to November 1999 when she served as Sixth Fleet flagship. In March 2005, *Mount Whitney* became part of MSC Special Mission programme and replaced *La Salle* as flagship Sixth Fleet, based at Gaeta, Italy. *Mount Whitney* retains US Navy status but with a 'hybrid' military/civilian crew.

AMPHIBIOUS FORCES

Notes: (1) Additional capacity is provided by the maritime pre-positioning ships (see listing under *Military Sealift Command* (MSC) section) which are either new construction or conversions of commercial ships. One squadron is maintained on station at Guam, and a second at Diego Garcia. Each squadron carries equipment to support a Marine Expeditionary Brigade.
(2) **Minesweeping:** Several of the larger amphibious ships have been used as operating bases for minesweeping helicopters.
(3) Five decommissioned LKAs and four LSTs are kept in Amphibious Lift Enhancement Program (ALEP) status. These are *Fresno* (LST 1182), *Tuscaloosa* (LST 1187), *Boulder* (LST 1190), *Racine* (LST 1191), *Charleston* (LKA 113), *Durham* (LKA 114), *Mobile* (LKA 115), *St Louis* (LKA 116) and *El Paso* (LKA 117).
(4) The LX(R) programme is planned to replace the LSD 41 class and is to start in FY19.
(5) A contract was awarded to Textron Inc, on 6 July 2012 for a new Ship-to-Shore connector craft, to replace LCAC.

8 FRANK S BESSON CLASS (LOGISTIC SUPPORT VESSELS) (LSV-ARMY)

Name	No	Builders	Commissioned
GEN FRANK S BESSON JR	LSV 1	Moss Point Marine, Escatawpa	18 Dec 1987
CW 3 HAROLD C CLINGER	LSV 2	Moss Point Marine, Escatawpa	20 Feb 1988
GEN BREHON B SOMERVELL	LSV 3	Moss Point Marine, Escatawpa	2 Apr 1988
LTG WILLIAM B BUNKER	LSV 4	Moss Point Marine, Escatawpa	18 May 1988
MG CHARLES P GROSS	LSV 5	Moss Point Marine, Escatawpa	30 Apr 1991
SP/4 JAMES A LOUX	LSV 6	Moss Point Marine, Escatawpa	16 Dec 1994
SSGT ROBERT T KURODA	LSV 7	VT Halter Marine	26 Aug 2006
MG ROBERT SMALLS	LSV 8	VT Halter Marine	15 Sep 2007

Displacement, tonnes: 4,199 full load
Dimensions, metres (feet): 83.1 (LSV 1–6), 95.7 (LSV 7–8) × 18.3 × 3.7 *(272.6, 314 × 60.0 × 12.1)*
Speed, knots: 11.5, 12.5 (LSV 7–8). **Range, n miles:** 8,200 at 11 kt
Complement: 31 (8 officers)
Military lift: 2,280 tons of vehicles including 26 M-1 tanks, containers or general cargo
Machinery: 2 GM EMD 16-645E2 diesels; 3,900 hp *(2.9 MW)* sustained; 2 shafts; Schottel bow thruster; 650 hp *(485 kW)*; 2 Caterpillar 3516B diesels (LSV 7 and 8); 4,520 hp *(3.37 MW)*; 2 shafts; Thrustmaster bow thruster 540 hp *(402 kW)*; Brunvoll stern thruster; 300 hp *(223 kW)*
Radars: Navigation: 2 Raytheon; E/F-band; I-band.

Comment: First ship approved in fiscal year 1985, second in FY 1987, remainder from Army reserve funds. Army owned ro-ro design with 10,500 sq ft of deck space for cargo. Capable of beaching with 4 ft over the ramp on a 1:30 offshore gradient. Payload is 2,000 tons of cargo. LSVs 1 and 4 are based at Fort Eustis, VA, LSVs 2, 5 and 7 are based at Pearl Harbour, HI. LSV 3 is with the Army Reserve and based at Tacoma, WA. LSV 6 is based in Kuwait. LSV 8 is based at Baltimore, MD. Two modified ships of the class built for the Philippines Navy in 1993–94.

FRANK S BESSON JR *6/2012, M Declerck* / 1486443

962 US > Amphibious forces

9 + 2 SAN ANTONIO CLASS (AMPHIBIOUS TRANSPORT DOCKS) (LPDM)

Name	No	Builders	Laid down	Launched	Commissioned	F/S
SAN ANTONIO	LPD 17	Northrop Grumman Shipbuilding	9 Dec 2000	12 Jul 2003	14 Jan 2006	AA
NEW ORLEANS	LPD 18	Northrop Grumman Shipbuilding	14 Oct 2002	11 Dec 2004	10 Mar 2007	PA
MESA VERDE	LPD 19	Northrop Grumman Shipbuilding	8 Nov 2002	19 Nov 2004	15 Dec 2007	AA
GREEN BAY	LPD 20	Northrop Grumman Shipbuilding	7 Aug 2003	11 Aug 2006	24 Jan 2009	PA
NEW YORK	LPD 21	Northrop Grumman Shipbuilding	30 Aug 2004	20 Dec 2007	7 Nov 2009	AA
SAN DIEGO	LPD 22	Huntington Ingalls Industries	23 May 2007	7 May 2010	19 May 2012	PA
ANCHORAGE	LPD 23	Huntington Ingalls Industries	24 Sep 2007	12 Feb 2011	4 May 2013	PA
ARLINGTON	LPD 24	Huntington Ingalls Industries	26 May 2008	23 Nov 2010	6 Apr 2013	AA
SOMERSET	LPD 25	Huntington Ingalls Industries	11 Dec 2009	14 Apr 2012	1 Mar 2014	PA
JOHN P MURTHA	LPD 26	Huntington Ingalls Industries	6 Jun 2012	2014	2016	Bldg
PORTLAND	LPD 27	Huntington Ingalls Industries	5 May 2013	2016	2017	Bldg

Displacement, tonnes: 24,900 full load
Dimensions, metres (feet): 208.5 × 31.9 × 7 *(684.1 × 104.7 × 23.0)*
Speed, knots: 22
Complement: 381 (29 officers) + 22 spare berths

Military lift: 699 (surge 800) troops; 2 LCACs, 14 AAVs
Machinery: 4 Colt Pielstick PC 2.5 diesels; 40,000 hp *(29.84 MW)*; 2 shafts; cp props
Missiles: SAM: 2 Raytheon RAM RIM-116 21-cell Mk 49 launchers; passive IR/anti-radiation homing to 9.6 km *(5.2 n miles)* at 2.5 Mach; warhead 9.1 kg ❶.
Guns: 2—30 mm Mk 46 ❷. 9—12.7 mm MGs.
Physical countermeasures: Decoys: 6 Mk 53 Mod 4 Nulka and chaff launcher ❸. SLQ-25A Nixie towed torpedo decoy.
Electronic countermeasures: ESM/ECM: SLQ-32A(V)2 ❹; intercept and jammer.
Radars: Air search: ITT SPS-48E ❺; 3D; E/F-band.
Surface search/navigation: Raytheon SPS-73(V)13 ❻; I-band.
Fire control: Lockheed SPQ-9B ❼; I-band.
Combat data systems: SSDS Mk 2; GCCS (M), CEC, JTIDS (Link 16), AADS (see Data Systems at front of section).

Helicopters: 1 CH-53E Sea Stallion or 2 CH-46E Sea Knight or 1 MV-22 Osprey.

Programmes: The LPD 17 programme entered technology development phase on 11 January 1993. It is the functional replacement for four classes of amphibious ships: LPD 4s, LSTs, LKAs and LSD 36s. Contract for LPD 17, with an option on two more, awarded to Avondale on 17 December 1996. Contract options for FY99 and FY00 on LPD 18 and LPD 19 were exercised in December 1998 and February 2000 respectively. A negotiated modification added the second FY00 ship, LPD 20, to the lead ship contract in May 2000. Avondale Industries was purchased by Litton Industries in 1999 and Litton Industries by Northrop Grumman in 2001. Contract awarded to Northrop Grumman Shipbuilding for LPD 21 in November 2003, for LPD 22 and 23 in June 2006 and for long-lead items for LPD 24 and 25 on 6 November 2006. Launch and commissioning dates for LPD 18-20 delayed due to shipyard damage caused by Hurricane Katrina in 2005. Northrop Grumman sold their shipbuilding division to Huntington Ingalls Industries in 2011. Construction contracts for LPD 26 and LPD 27 were let in April 2011 and July 2012 respectively.
Structure: Panama Canal-capable ships able to control and support landing forces disembarking either via surface craft such as LCACs or by VTOL aircraft, principally helicopters. The design supports a lift capability of 24,000 sq ft of deck space for vehicles, 34,000 cu ft of cargo below decks and 720 embarked Marines with surge lift capacity to 800 troops. The well-deck and stern gate arrangements are similar to those of the Wasp class; the well-deck can carry two LCACs or one LCU, or 14 Amphibious Assault Vehicles. The Flight deck can land/launch four CH-46s or two CH-53s or two MV-22s. The hangar will accommodate two CH-46s or one CH-53 or one MV-22. There is a 24-bed medical facility. There is a crane for support of boat operations and an Advanced Enclosed Mast System is being fitted in all. On 9 September 2003, salvaged steel from the World Trade Centre was cast into the bow section of USS *New York*.
Operational: The first five ships of the class have completed deployments. Material reliability and quality issues in the initial ships have been resolved and follow-on verification tests and evaluation continue.

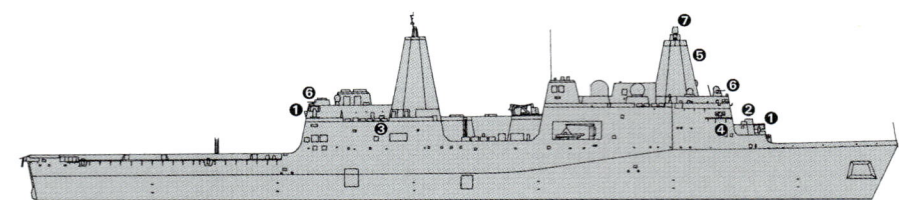

SAN ANTONIO   *(Scale 1 : 1,800), Ian Sturton / 1167439*

MESA VERDE   *4/2009, Guy Toremans / 1366229*

NEW YORK   *11/2009, US Navy / 1366230*

Amphibious forces < US   963

1 TARAWA CLASS (AMPHIBIOUS ASSAULT SHIPS) (LHAM)

Name	No	Builders	Laid down	Launched	Commissioned	F/S
PELELIU (ex-*Da Nang*)	LHA 5	Ingalls Shipbuilding, Pascagoula	12 Nov 1976	25 Nov 1978	3 May 1980	PA

Displacement, tonnes: 40,608 full load
Dimensions, metres (feet): 254.2 × 40.2 × 7.9 *(834 × 131.9 × 25.9)*
Flight deck, metres: 250 × 36 *(820.2 × 118.1)*
Speed, knots: 24
Range, n miles: 10,000 at 20 kt
Complement: 964 (56 officers)

Military lift: 1,703 troops; 4 LCU 1610 type or 2 LCU and 2 LCM 8 or 17 LCM 6 or 45 Assault Amphibian Vehicles; 1,200 tons aviation fuel. 1 LCAC may be embarked. 4 LCPL (replacement by RHIBs in progress)
Machinery: 2 Combustion Engineering boilers; 600 psi *(42.3 kg/cm²)*; 900°F *(482°C)*; 2 Westinghouse turbines; 70,000 hp *(52.2 MW)*; 2 shafts; bow thruster; 900 hp *(670 kW)*
Missiles: SAM: 2 GDC Mk 49 RAM RIM-116 ❶; 21 rounds per launcher; passive IR/anti-radiation homing to 9.6 km *(5.2 n miles)* at 2.5 Mach; warhead 9.1 kg.
Guns: 2 Raytheon 20 mm/76 6-barrelled Vulcan Phalanx Mk 15 ❷; 3,000 rds/min (4,500 in Block 1) combined to 1.5 km. 6 Mk 242 25 mm automatic cannons. 8—12.7 mm MGs.
Physical countermeasures: Decoys: 4 Loral Hycor SRBOC 6-barrelled fixed Mk 36; IR flares and chaff to 4 km *(2.2 n miles)*. SLQ-25 Nixie; acoustic torpedo decoy system. NATO Sea Gnat. SLQ-49 chaff buoys. AEB SSQ-95.
Electronic countermeasures: ESM/ECM: SLQ-32(V)3; intercept and jammers.
Radars: Air search: ITT SPS-48E ❸; E/F-band. Lockheed SPS-40E ❹; B-band. Hughes Mk 23 TAS ❺; D-band.
Surface search: Raytheon SPS-67(V)3 ❻; G-band.
Navigation: Raytheon SPS-73; I-band.
CCA: SPN-35A; SPN-43B.
Tacan: URN 25. IFF: CIS Mk XV/UPX-36.
Combat data systems: ACDS Block 0. Advanced Combat Direction System to provide computerised support in control of helicopters and aircraft, shipboard weapons and sensors, navigation, landing craft control and electronic warfare. Links 4A, 11 and 16. SATCOM SRR-1, WSC-3 (UHF), USC-38 (EHF). SMQ-11 Metsat (see Data Systems at front of section).

Fixed-wing aircraft: Harrier AV-8B VSTOL aircraft in place of some helicopters as required. MV-22 Osprey Squadron can replace CH-46D/E.
Helicopters: 19 CH-53D Sea Stallion or 26 CH-46D/E Sea Knight. UAV in due course.

Programmes: Originally intended to be a class of nine ships although only five were built. LHA 5 was authorised in FY71.
Modernisation: Two Vulcan Phalanx CIWS replaced the GMLS Mk 25 Sea Sparrow launchers. Programme completed in early 1991. RAM launchers fitted 1993–95. One launcher is above the bridge offset to port, and the other on the starboard side at the after end of the flight deck. Mk 23 TAS target acquisition radar fitted in 1992. SPS-48E replaced SPS-52D by 1994 to improve low altitude detection of missiles and aircraft. ACDS Block 0 in 1996. 5 in guns removed in 1997/98. Plans to fit SSDS have been shelved. Modifications to accommodate MV-22 Osprey operations and Collective Protection Systems completed. Fuel oil compensation system has been installed to improve damaged stability.
Structure: There are two lifts, one on the port side aft and one at the stern. Beneath the after elevator is a floodable docking well measuring 268 ft in length and 78 ft in width which is capable of accommodating four LCU 1610 type landing craft. Also included is a large garage for trucks and AFVs and troop berthing for a reinforced battalion. 33,730 sq ft available for vehicles and 116,900 cu ft for palletted stores. Extensive medical facilities including operating rooms, X-ray room, hospital ward, isolation ward, laboratories, pharmacy, dental operating room and medical store rooms.
Operational: The flight deck can operate a maximum of nine CH-53D Sea Stallion or 10 CH-46D/E Sea Knight helicopters or a mix of these and other helicopters at any one time. With some additional modifications, ships of this class can effectively operate AV-8B aircraft. The normal mix of aircraft allows for six AV-8Bs. The optimum aircraft configuration is dependent upon assigned missions. Unmanned Reconnaissance Vehicles (URVs) can be operated. LHA 3 decommissioned 28 October 2005, LHA 2 in 2007, LHA 1 in 2009 and LHA 4 in 2011. LHA 5 is expected to decommission in 2015.

PELELIU   2/2004, US Navy / 1486437

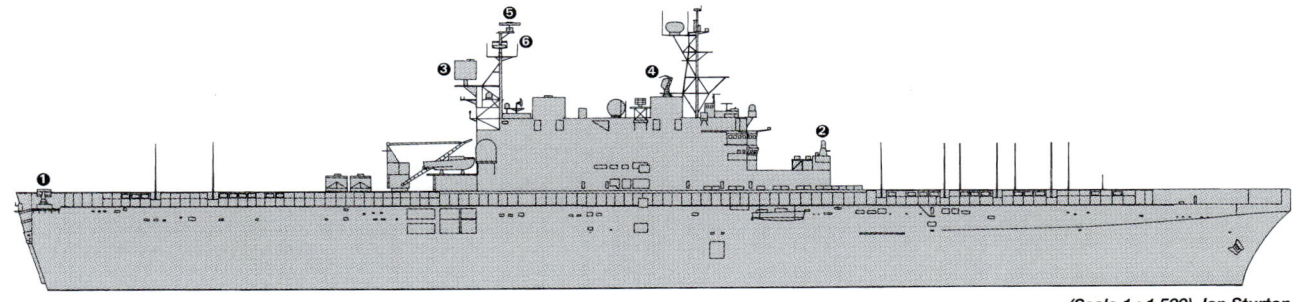

PELELIU   (Scale 1 : 1,500), Ian Sturton / 0131369

PELELIU   12/2010, US Navy / 1486436

33 MARITIME PREPOSITIONING FORCE UTILITY BOAT

Displacement, tonnes: 12 standard
Dimensions, metres (feet): 13.4 × 4.4 × ? *(44 × 14.4 × ?)*
Speed, knots: 41; 38 full load
Complement: 4
Machinery: 2 Cummins QSM11 diesels; 660 hp *(492 kW)*; 2 Hamilton 364 waterjets
Guns: 2—12.7 mm MGs.
Radars: Navigation: Furuno 1834; I-band.

Comment: Contract for the construction of 10 MPF utility craft awarded to Kvichak Marine Industries, Seattle, WA, in June 2005. The first was delivered in February 2006. Since then, there have been orders for a further 23 craft. First craft delivered in February 2006. The craft replaced the LCM-8 craft as part of the lighterage system in support of prepositioned Marine amphibious follow-on echelon missions. Aluminium construction with an articulated bow-door for beach deployment. Some are stationed at San Diego, CA, and at Norfolk, VA, and others onboard prepositioned MSC ships. Each craft can transport 30 troops and equipment.

MPF CRAFT 6/2007
Kvichak Marine
1305174

8 WASP CLASS (AMPHIBIOUS ASSAULT SHIPS) (LHDM)

Name	No	Builders	Laid down	Launched	Commissioned	F/S
WASP	LHD 1	Ingalls Shipbuilding, Pascagoula	30 May 1985	4 Aug 1987	29 Jul 1989	AA
ESSEX	LHD 2	Ingalls Shipbuilding, Pascagoula	16 Feb 1989	4 Jan 1991	17 Oct 1992	PA
KEARSARGE	LHD 3	Ingalls Shipbuilding, Pascagoula	6 Feb 1990	26 Mar 1992	16 Oct 1993	AA
BOXER	LHD 4	Ingalls Shipbuilding, Pascagoula	26 Mar 1991	13 Aug 1993	11 Feb 1995	PA
BATAAN	LHD 5	Ingalls Shipbuilding, Pascagoula	16 Mar 1994	15 Mar 1996	20 Sep 1997	AA
BONHOMME RICHARD	LHD 6	Ingalls Shipbuilding, Pascagoula	29 Mar 1995	14 Mar 1997	15 Aug 1998	PA
IWO JIMA	LHD 7	Ingalls Shipbuilding, Pascagoula	12 Dec 1997	4 Feb 2000	30 Jun 2001	AA
MAKIN ISLAND	LHD 8	Northrop Grumman Shipbuilding	14 Feb 2004	22 Sep 2006	24 Oct 2009	PA

Displacement, tonnes: 41,302 (LHD 1–4), 41,006 (LHD 5–7), 42,330 (LHD 8) full load
Dimensions, metres (feet): 258.2 oa; 240.2 wl × 42.7 oa; 32.3 wl × 8.1 *(847.1; 788.1 × 140.1; 106.0 × 26.6)*
Flight deck, metres: 249.6 × 36.0 *(818.9 × 118.1)*
Speed, knots: 22
Range, n miles: 9,500 at 20 kt
Complement: 1,123 (65 officers)

Military lift: 1,687 (plus 184 surge) troops; 12 LCM 6s or 3 LCACs; 1,232 tons aviation fuel (LHD 1–4); 1,960 tons (LHD 5–8)
Machinery: 2 Combustion Engineering boilers; 600 psi *(42.3 kg/cm²)*; 900°F *(482°C)*; 2 Westinghouse turbines; 70,000 hp *(52.2 MW)*; 2 shafts (LHD 1–7) 2 GE LM 2500+ gas turbines; 70,000 hp *(52.2 MW)*; 2 Alstom variable speed electric motors; 10,000 hp *(7.5 MW)* (LHD 8)
Missiles: SAM: 2 Raytheon GMLS Mk 29 octuple launchers ❶; 16 Sea Sparrow RIM-7P; semi-active radar homing to 16 km *(8.5 n miles)* at 2.5 Mach; warhead 38 kg. ESSM in due course. 2 GDC Mk 49 RAM RIM-116 launchers; 21 rounds per launcher ❷; passive IR/anti-radiation homing to 9.6 km *(5.2 n miles)* at 2.5 Mach; warhead 9.1 kg.
Guns: 2 Raytheon 20 mm 6-barrelled Vulcan Phalanx Mk 15 ❸; 4,500 rds/min combined to 1.5 km. 3 Boeing Bushmaster 25 mm Mk 38. 4—12.7 mm MGs.
Physical countermeasures: Decoys: 4 or 6 Loral Hycor SRBOC 6-barrelled fixed Mk 36; IR flares and chaff to 4 km *(2.2 n miles)*. SLQ-25 Nixie; acoustic torpedo decoy system. NATO Sea Gnat. SLQ-49 chaff buoys. AEB SSQ-95.
Electronic countermeasures: ESM/ECM:SLQ-32(V)3/SLY-2; intercept and jammers. Raytheon ULQ-20.

Radars: Air search: ITT SPS-48E ❺; 3D; E/F-band. Raytheon SPS-49(V)9 ❻; C/D-band. Hughes Mk 23 TAS ❼; D-band. SPQ-9B (LHD 8 on build and LHD 7 in 2007).
Surface search: Norden SPS-67 ❽; G-band.
Navigation: SPS-73; I-band.
CCA: SPN-35B (LHD 1–7), SPN-35C (LHD 8) and SPN-43C.
Fire control: 2 Mk 95; I/J-band. SPQ-9B (LHD 8).
Tacan: URN 25. IFF: CIS Mk XV UPX-29.
Combat data systems: ACDS Block 1 level 2 (LHD 1 and 7) and Block 0 (LHD 2–6). SSDS Mk 2 (LHD 8, LHD 7, and LHD 1 (2014)). Marine Tactical Amphibious C2 System (MTACCS). Links 4A, 11 (modified), 14 and 16. SATCOMS ❹ SSR-1, WSC-3 (UHF), USC-38 (EHF). SMQ-11 Metsat (see *Data Systems* at front of section). Advanced Field Artillery TDS (LHD 6–8).
Weapon control systems: 2 Mk 91 MFCS (LHD 1–6). 2 Mk 9 MFCS (LHD 7–8).

Fixed-wing aircraft: 6–8 AV-8B Harriers or up to 20 in secondary role. Capability to support 12 MV-22 Osprey in LHD 1, 3, 4, 5, 6 and 7 (LHD 2 and 8 in 2013) and F-35B (LHD 1 2014).
Helicopters: Capacity for 42 CH-46E Sea Knight but has the capability to support: AH-1W Super Cobra or AH-1Z Viper, CH-53E Super Stallion, CH-53D Sea Stallion, UH-1N Twin Huey, AH-1T Sea Cobra, and SH-60B Seahawk helicopters. UAV in due course.

Programmes: The Wasp class was a follow-on to the Tarawa clas and shares the same basic hull and engineering plant. Contract awarded to Ingalls Shipbuilding in February 1984 to build the lead ship. The same shipyard was subsequently contracted to build the other ships of the class.

Modernisation: RAM launchers retrofitted in all. All ships to be modified to accommodate MV-22 Osprey and F-35B operations.
Structure: Two aircraft elevators, one to starboard and aft of the 'island' and one to port amidships. The well-deck is 267 × 50 ft and can accommodate up to three LCACs. The flight deck has nine helicopter landing spots. Cargo capacity is 125,000 cu ft total with an additional 20,000 sq ft to accommodate vehicles. Vehicle storage is available for five M1 tanks, 25 LAVs, eight M 198 guns, 68 trucks, 10 logistic vehicles and several service vehicles. The bridge is two decks lower than that of an LHA, command, control and communication spaces having been moved inside the hull to avoid 'cheap kill' damage. Fitted with a 64 bed capacity hospital and s x operating rooms. Three 32 ft monorail trains each carrying 6,000 lbs, deliver material to the well-deck at 6.8 mph. *Iwo Jima* is likely to be the last oil-fired steam turbine ship in the USN. LHD 8 is fitted with gas turbine propulsion, electric drive, watermist fire suppression system, fibre-optic machinery control system, SPQ-9B radar and CEC.
Operational: A typical complement of aircraft is a mix of 25 helicopters and six to eight Harriers (AV-8B). In the secondary role as a sea control ship the most likely mix is 20 AV-8B Harriers and four to six SH-60B Seahawk helicopters. LHD 3 modified to provide interim Mine Countermeasures Command (MCS) capability following decommissioning of *Inchon* in June 2002. LHD 8 conducted maiden deployment in 2012. This included first operational deployment of AH-1Z Viper helicopters. LHD 1 conducted development testing of F-35B in 2013. LHDs 1, 3, 5 and 7 based at Norfolk, Virginia, and LHDs 2, 4 and 8 at San Diego, California. LHD 6 is based at Sasebo, Japan.

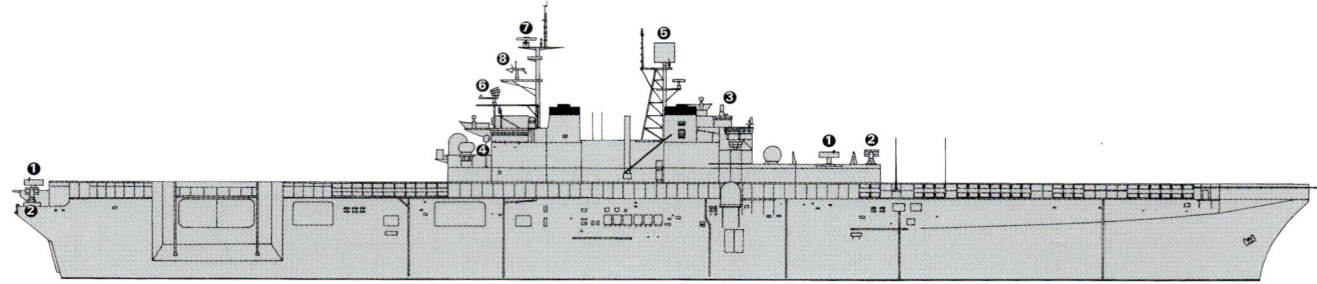

BONHOMME RICHARD *(Scale 1 : 1,500), Ian Sturton* / 0131367

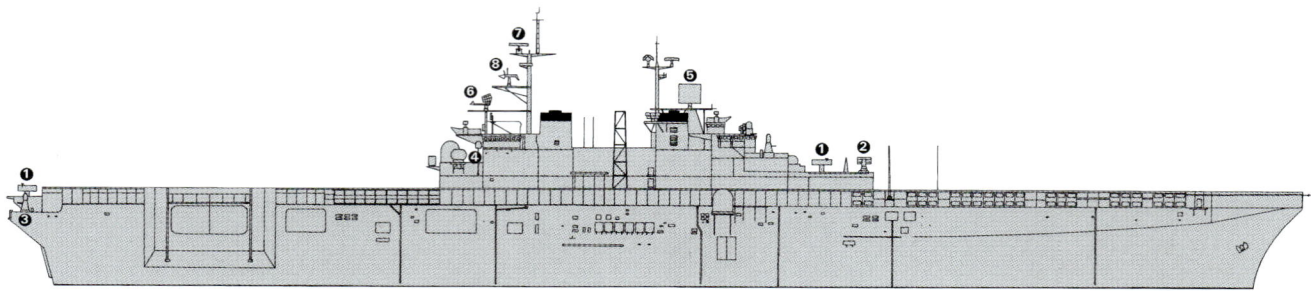

ESSEX *(Scale 1 : 1,500), Ian Sturton* / 0131368

BONHOMME RICHARD 8/2013*, Chris Sattler / 1531165

Amphibious forces < **US** 965

ESSEX   2/2012, Hachiro Nakai / 1483661

KEARSAGE   1/2012, M Declerck / 1454851

MAKIN ISLAND   10/2010, M Mazumdar / 1406329

© 2014 IHS   *IHS Jane's Fighting Ships 2014-2015*

1 AUSTIN CLASS (AMPHIBIOUS TRANSPORT DOCKS) (LPD)

Name	No	Builders	Laid down	Launched	Commissioned	F/S
DENVER	LPD 9	Lockheed SB & Construction Co	7 Feb 1964	23 Jan 1965	26 Oct 1968	PA

Displacement, tonnes: 9,277 standard; 17,521 full load
Dimensions, metres (feet): 173.8 × 30.5 × 7 *(570.2 × 100.1 × 23.0)*
Speed, knots: 21. **Range, n miles:** 7,700 at 20 kt
Complement: 420 (24 officers) + 90 flag staff

Military lift: 840 troops; 9 LCM 6s or 4 LCM 8s or 2 LCAC or 20 LVTs. 4 LCPL/LCVP
Machinery: 2 Foster-Wheeler boilers; 600 psi *(42.3 kg/cm²)*; 870°F *(467°C)*; 2 De Laval (General Electric in LPD 9 and LPD 10) turbines; 24,000 hp *(18 MW)*; 2 shafts
Guns: 2 Raytheon 20 mm/76 6-barrelled Vulcan Phalanx Mk 15 ❶; 3,000 rds/min (4,500 in Block 1) combined to 1.5 km. 2—25 mm Mk 38. 8—12.7 mm MGs.
Physical countermeasures: Decoys: 4 Loral Hycor SRBOC 6-barrelled Mk 36; IR flares and chaff to 4 km *(2.2 n miles)*.
Electronic countermeasures: ESM: SLQ-32(V)1; intercept.
Radars: Air search: Lockheed SPS-40E ❸; B-band.
Surface search: Norden SPS-67 ❹; G-band.
Navigation: Raytheon SPS-73(V)12; I-band.
Tacan: URN 25. IFF: Mk XII UPX-36.
Combat data systems: SATCOM ❷, WSC-3 (UHF), WSC-6 (SHF) (see Data Systems at front of section).

Helicopters: Up to 6 CH-46D/E Sea Knight can be carried. Hangar for only 1 light.

Programmes: LPD 9 authorised in fiscal year 1963, LPD 15 in FY 1965.
Modernisation: Modernisation carried out in normal maintenance periods from FY 1987. This included fitting two Phalanx, SPS-67 radar replacing SPS-10 and updating EW capability. 3 in guns have been removed. LPD 15 was the last LPD to receive machinery, electrical and habitability upgrades to extend life.
Structure: LPD 9 has an additional bridge and is fitted as a flagship. One small telescopic hangar. Flight deck is 168 ft *(51.2 m)* in length. Well-deck 394 × 50 ft *(120.1 × 15.2 m)*. Communications domes are not uniformly fitted.
Operational: A typical operational load might include one Seahawk, two Sea Knight, two Twin Huey, four Sea Cobra helicopters and one Cyclone patrol craft. LPD 9 based at Sasebo, Japan and LPD 15 at Norfolk. LPD 6 decommissioned in 2005, LPD 4 in 2006, LPDs 5 and 12 in 2007, LPD 10 in 2008, LPD 13 in 2009 and LPDs 7 and 8 in 2011. LPD 15 was to be decommissioned in 2012 but instead was converted to act as an Afloat Forward Staging Base-Interim (AFSB-I). LPD 9 to decommission in September 2014.
Sales: LPD 14 transferred to the Indian Navy in January 2007.

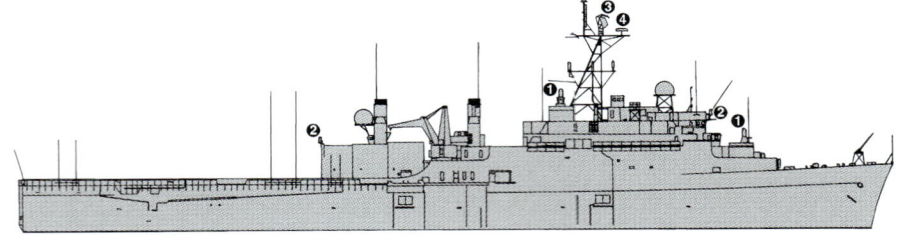

DENVER (Scale 1 : 1,500), Ian Sturton / 0016471

DENVER 7/2007, Guy Toremans / 1305194

0 + 2 (3) AMERICA CLASS (AMPHIBIOUS ASSAULT SHIP) (LHA)

Name	No	Builders	Laid down	Launched	Commissioned	F/S
AMERICA	LHA 6	Huntington Ingalls Industries, Mississippi	17 Jul 2009	4 Jun 2012	Oct 2014	PA
TRIPOLI	LHA 7	Huntington Ingalls Industries, Mississippi	2014	2017	2019	Bldg

Displacement, tonnes: 44,971 full load
Dimensions, metres (feet): 257.3 oa; 237.1 wl × 60.35 oa; 32.3 wl × 8.75 *(844.2; 777.9 × 198.0; 106.0 × 28.7)*
Flight deck, metres: 249.6 × 36.0 *(818.9 × 118.1)*
Speed, knots: 22. **Range, n miles:** 9,500 at 20 kt
Complement: 1,102 (102 officers)

Military lift: 1,687 troops (plus 184 surge)
Machinery: COGES: 2 GE LM 2500+ gas turbines; 70,000 hp *(52.2 MW)*; 2 auxiliary propulsion motors; 10,000 hp *(7.46 MW)*; 2 shafts
Missiles: SAM: 2 Raytheon GMLS Mk 29 octuple launchers; 16 Evolved Sea Sparrow RIM-162D; semi-active radar homing to 18 km *(9.7 n miles)* at 3.6 Mach; warhead 38 kg. 2 Raytheon RAM RIM-116 Mk 49 launchers; passive IR/anti-radiation homing to 9.6 km *(5.2 n miles)* at 2.5 Mach; warhead 9.1 kg.
Guns: 2 Raytheon 20 mm 6-barrelled Vulcan Phalanx Mk 15. 3—25 mm Mk 38 Mod 2 chain guns. 14—12.7 mm MGs (7 twin).
Physical countermeasures: Decoys: Mk 53 Mod 6 NULKA DLS; SLQ-25 Nixie; acoustic torpedo decoy system.
Electronic countermeasures: ESM/ECM: SLQ-32B(V)2.
Radars: Air search: ITT SPS-48E(V)10; 3D; E/F-band; Raytheon SPS-49A(V)1; SPQ-9B.
Surface search/Navigation: 2 SPS-73; I-band.
CCA: SPN-35C, SPN-41A and SPN-43C.
Tacan: URN 25. IFF: CIS UPX-29.
Combat data systems: SSDS Mk 2 Mod 4B, CEC USG-2A, Links 4A, 11 (modified), 16 and 22. SATCOMS: SSR-1, SRC-XX (UHF), USC-38 (EHF), URC-131(H)(HF), URC-139 (VHF), CBSP and 1 WSC-6C(V)9 (SHF). SMQ-11 Metsa. Advanced Field Artillery TDS.
Weapon control systems: 2 NSSMS Mk 132 Mod 2 with 2 Mk 9 MFCS.

Fixed-wing aircraft: Similar to Wasp class with improved facilities to operate and support MV-22 Osprey and up to 23 F-35B Joint Strike Fighter (JSF).

Programmes: It was announced on 6 April 2004 that the LHA Replacement design was to be a modified version of the LHD 8 design. The detailed design phase started in January 2006 following ship design approval to proceed with Milestone B. A contract for the detailed design and construction of the first of class was let on 1 June 2007 and for the second on 31 May 2012. A class of five is planned.
Structure: LHA Replacement is optimised for aviation operations and is to have additional cargo/magazine capacity in lieu of a traditional well deck. The flight deck has nine helicopter landing spots and is to be equipped with two aircraft elevators, one to starboard and aft of the island and one to port amidships; the folding capability has been removed. The aircraft hangar is larger than in previous classes and includes a significant increase in aviation parts stowage capacity. Cargo capacity is 160,000 cu ft total with an additional 11,760 sq ft to accommodate vehicle stowage. The ship is to be fitted with a 26 bed capacity hospital and two operating rooms. The bridge is two decks lower than that of an LHA 1 with the command, control and communications spaces inside the hull. The ship has gas turbine propulsion and all electric auxiliaries. The design is to be modified to include a stern dock, or well deck in the LHA 8 and subsequent vessels. LHA 8 and subsequent ships will have smaller hangars to accommodate the well-deck but modified island superstructure to increase deck area.
Operational: LHA 6 completed builder's trials off Pascagoula, Mississippi, in November 2013. To be based at San Diego.

AMERICA 11/2013*, Huntington Ingalls Industries / 1531211

Amphibious forces < US 967

12 WHIDBEY ISLAND CLASS (DOCK LANDING SHIPS) (LSD)

Name	No	Builders	Laid down	Launched	Commissioned	F/S
WHIDBEY ISLAND	LSD 41	Lockheed SB & Construction Co	4 Aug 1981	10 Jun 1983	9 Feb 1985	AA
GERMANTOWN	LSD 42	Lockheed SB & Construction Co	5 Aug 1982	29 Jun 1984	8 Feb 1986	PA
FORT MCHENRY	LSD 43	Lockheed SB & Construction Co	10 Jun 1983	1 Feb 1986	8 Aug 1987	AA
GUNSTON HALL	LSD 44	Avondale Shipyards, New Orleans	26 May 1986	27 Jun 1987	22 Apr 1989	AA
COMSTOCK	LSD 45	Avondale Shipyards, New Orleans	27 Oct 1986	16 Jan 1988	3 Feb 1990	PA
TORTUGA	LSD 46	Avondale Shipyards, New Orleans	23 Mar 1987	15 Sep 1988	17 Nov 1990	PA
RUSHMORE	LSD 47	Avondale Shipyards, New Orleans	9 Nov 1987	6 May 1989	1 Jun 1991	PA
ASHLAND	LSD 48	Avondale Shipyards, New Orleans	4 Apr 1988	11 Nov 1989	9 May 1992	AA
HARPERS FERRY	LSD 49	Avondale Shipyards, New Orleans	15 Apr 1991	16 Jan 1993	7 Jan 1995	PA
CARTER HALL	LSD 50	Avondale Shipyards, New Orleans	11 Nov 1991	2 Oct 1993	30 Sep 1995	AA
OAK HILL	LSD 51	Avondale Shipyards, New Orleans	21 Sep 1992	11 Jun 1994	8 Jun 1996	AA
PEARL HARBOR	LSD 52	Avondale Shipyards, New Orleans	27 Jan 1995	24 Feb 1996	30 May 1998	PA

Displacement, tonnes: 11,304 standard; 16,195 (LSD 41–48), 17,009 (LSD 49–52) full load
Dimensions, metres (feet): 185.8 × 25.6 × 6.3 *(609.6 × 84.0 × 20.7)*
Speed, knots: 22
Range, n miles: 8,000 at 18 kt
Complement: 413 (21 officers)

Military lift: 402 (+102 surge) troops; 2 (CV) or 4 LCACs, or 9 (CV) or 21 LCM 6, or 1 (CV) or 3 LCUs, or 64 LVTs. 2 LCPL
Cargo capacity: 5,000 cu ft for marine cargo, 12,500 sq ft for vehicles (including four preloaded LCACs in the well-deck). The cargo version' has 67,600 cu ft for marine cargo, 20,200 sq ft for vehicles but only two LCACs. Aviation fuel, 90 tons.
Machinery: 4 Colt SEMT-Pielstick 16 PC2.5 V 400 diesels; 33,000 hp(m) *(24.6 MW)* sustained; 2 shafts; cp props
Missiles: 1 GDC/Hughes Mk 49 RAM RIM-116 21-cell launcher ❶; passive IR/anti-radiation homing to 9.6 km *(5.2 n miles)* at 2.5 Mach; warhead 9.1 kg. Being fitted in all.
Guns: 2 Raytheon 20 mm/76 6-barrelled Vulcan Phalanx Mk 15 ❷; 4,500 rds/min (4,500 in Block 1) combined to 1.5 km. 2 – 25 mm Mk 38. 6 – 12.7 mm MGs.
Physical countermeasures: Decoys: 4 Loral Hycor SRBOC 6-barrelled Mk 36 and Mk 50; IR flares and chaff. SLQ-25 Nixie.
Electronic countermeasures: ESM: SLQ-32(V)1; intercept. SLQ-49.
Radars: Air search: Raytheon SPS-49(V)1 ❸; C-band.
Surface search: Norden SPS-67V ❹; G-band.
Navigation: Raytheon SPS-64(V)9 or SPS-73(V)12; I/J-band.
Tacan: URN 25. IFF: Mk XII UPX-29/UPX-36.
Combat data systems: SATCOM SRR-1, WSC-3 (UHF) (see Data Systems at front of section). SSDS Mk 1.

Helicopters: Platform only for 2 CH-53 Sea Stallion.

Programmes: Originally it was planned to construct six ships of this class as replacements for the Thomaston class LSDs. Eventually, the level of Whidbey Island class ships was established at eight, with four additional cargo-carrying variants to provide extra cargo capability. LSD 49–52 are also known as the Harper's Ferry class.
Modernisation: A Quick Reaction Combat Capability (QRCC)/Ship Self-Defense System (SSDS) was installed and successfully demonstrated in LSD 41 in 1993. During the QRCC demonstrations, the ship's SPS-49, SLQ-32, RAM and Phalanx were successfully integrated via SSDS. All ships of the class fitted with SSDS Mk 1. A mid-life upgrade package, to extend service life to 40 years, is planned for all LSD 41/49 class. The upgrade includes engine improvements, fuel efficiency measures, advanced engineering controls, increased chilled water capacity and new air compressors. The ships began their refits in 2009 and are being upgraded at the rate of two per year until the programme is completed in 2015. East and West coast ships are being upgraded at General Dynamics NASSCO.
Structure: Based on the earlier Anchorage class. One 60 (LSD 41-48) and one 20 (LSD 41-48) ton crane and one 30 ton crane (LSD 49-52). Well-deck measures 440 × 50 ft *(134.1 × 15.2 m)* in the LSD but is shorter in the Cargo Variant (CV). The cargo version is a minimum modification to the LSD 41 design. Changes in that design include additional troop magazines, air conditioning, piping and hull structure; the forward Phalanx is forward of the bridge, RAM is on the bridge roof. There is approximately 90% commonality between the two classes.
Operational: LSDs 41, 43, 44, 46, 50 and 51 are based at Little Creek, VA. LSDs 45, 47, 49, and 52 are based at San Diego, CA. LSDs 42 and 48 are based at Sasebo, Japan.

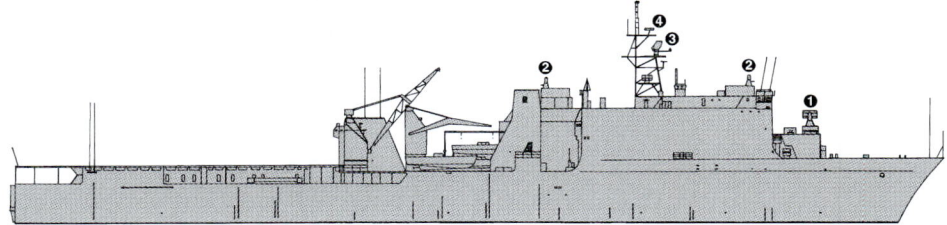

ASHLAND *(Scale 1 : 1,500), Ian Sturton / 0053362*

PEARL HARBOR *3/2012, Michael Nitz / 1486440*

OAK HILL *1/2012, M Declerck / 1454853*

WHIDBEY ISLAND *6/2012, M Declerck / 1486439*

© 2014 IHS   IHS Jane's Fighting Ships 2014-2015

US > Amphibious forces

79 LANDING CRAFT AIR CUSHION (LCAC)

Displacement, tonnes: 96 standard; 177 full load
Dimensions, metres (feet): 28 on cushion; 24.9 between hard structures × 14.7; 14.3 (91.8; 81.7 × 48.2; 46.9)
Range, n miles: 300 at 35 kt, 200 at 40 kt
Complement: 5

Military lift: 23 troops; 1 Main Battle Tank or 60–75 tons
Machinery: 4 Allied-Signal TF40B marine gas turbines for propulsion and lift; 16,000 hp (11.9 MW) sustained; 2 shrouded reversible-pitch airscrews (propulsion); 4 double-entry fans, centrifugal or mixed-flow (lift). SLEP configuration, 4 Vericor Power Systems ETF40B marine gas turbines with Full Authority Digital Engine Control (FADEC) for propulsion and lift; 19,000 hp (1.41 MW) sustained; 2 shrouded reversible-pitch airscrews (propulsion); 4 double-entry fans, centrifugal or mixed-flow (lift)
Guns: 12.7 mm and 7.62 mm MGs and 40 mm grenade launchers can be fitted.
Radars: Navigation: Marconi LN66 or Decca Bridgemaster E; I-band.

Programmes: Built by Textron Marine and Land Systems and Avondale Gulfport. A total of 90 craft delivered 1984–1997. The final craft LCAC 91 delivered in 2001 in SLEP configuration.
Modernisation: A total of 72 in-service craft are receiving Service Life Extension Programme (SLEP) from 2002–2018. The programme includes the installation of more powerful engines to provide greater lift capacity, an improved deep skirt for better handling in heavier sea states and an integrated navigation suite for precise navigation, and advanced Multimode Integrated Communications System in either normal, secure modes. Up to fiscal year 2004, 13 craft were upgraded followed by five in FY 2005, five in FY 2006, five in FY 2007, five in FY 2008, six in FY 2009, three in FY 2010 and four each in FY 2011, FY 2012, and FY 2013.
Structure: Incorporates the best attributes of the JEFF(A) and JEFF(B) learned from over five years of testing the two prototypes. Bow ramp 27.0 ft, stern ramp 14.8 ft. Cargo space capacity is 1,809 sq ft. Noise and dust levels are high and if disabled the craft is not easy to tow. 12.7 mm, 7.62 mm machine guns or 44 mm grenade launchers can be fitted.
Operational: Ship classes capable of carrying the LCAC are Wasp (three), Tarawa (one), Austin (one), Whidbey Island (four), Harpers Ferry (two), San Antonio (two), and MLP (three). A portable transport module can be carried on the cargo deck to transport up to 180 troops. Some limitations in very rough seas. Shore bases on each coast at Little Creek, VA and Camp Pendleton, CA. Forward deployed base at Yokose, Japan. Of 79 craft, 74 were operational and five undergoing SLEP as of October 2013.
Sales: Seven to Japan. One of a similar type built by South Korea.

LCAC 29   4/2012, Hachiro Nakai / 1483676

LCAC 46   6/2012, M Declerck / 1486441

35 LCU 2000 CLASS (UTILITY LANDING CRAFT) (LCU-ARMY)

RUNNYMEDE LCU 2001	**FORT DONELSON** LCU 2019
KENNESAW MOUNTAIN LCU 2002	**FORT MCHENRY** LCU 2020
MACON LCU 2003	**GREAT BRIDGE** LCU 2021
ALDIE LCU 2004	**HARPERS FERRY** LCU 2022
BRANDY STATION LCU 2005	**HOBKIRK** LCU 2023
BRISTOE STATION LCU 2006	**HOMIGUEROS** LCU 2024
BROAD RUN LCU 2007	**MALVERN HILL** LCU 2025
BUENA VISTA LCU 2008	**MATAMOROS** LCU 2026
CALABOZA LCU 2009	**MECHANICSVILLE** LCU 2027
CEDAR RUN LCU 2010	**MISSIONARY BRIDGE** LCU 2028
CHICKAHOMINY LCU 2011	**MOLINO DEL RAY** LCU 2029
CHICKASAW BAYOU LCU 2012	**MONTERREY** LCU 2030
CHURUBUSCO LCU 2013	**NEW ORLEANS** LCU 2031
COAMO LCU 2014	**PALO ALTO** LCU 2032
CONTRERAS LCU 2015	**PAULUS HOOK** LCU 2033
CORINTH LCU 2016	**PERRYVILLE** LCU 2034
EL CANEY LCU 2017	**PORT HUDSON** LCU 2035
FIVE FORKS LCU 2018	

Displacement, tonnes: 1,087 full load
Dimensions, metres (feet): 53 × 12.8 × 2.6 (173.9 × 42.0 × 8.5)
Speed, knots: 11.5
Range, n miles: 6,500 at 10 kt
Complement: 13 (2 officers)
Military lift: 350 tons
Machinery: 2 Cummins KTA50-M diesels; 2,500 hp (1.87 MW) sustained; 2 shafts; Schottel bow thruster
Radars: Navigation: 2 Raytheon; E/F-band; I-band.

Comment: Order placed with Lockheed by US Army 11 June 1986. First one completed 21 February 1990 by Moss Point Marine. These were the first LCUs built to an Army specification. There is an aft anchor to assist in retraction from beaches. Seven are active, seven in reserve, 20 prepositioned and one used for training.

MONTERREY   7/2011, M Mazumdar / 1454855

4 LCM 8 TYPE (MECHANISED LANDING CRAFT)

Displacement, tonnes: 67 standard; 129 full load
Dimensions, metres (feet): 22.5 × 6.4 × 1.6 (73.8 × 21.0 × 5.2)
Speed, knots: 12
Range, n miles: 190 at 9 kt
Complement: 4
Military lift: 67.5 tons or 1 M48/1 M60 tank or 110 fully equipped troops or 200 non-combat troops
Machinery: 2 Detroit 12V-71 diesels; 400 hp (298 kW) sustained; 2 shafts; Kort nozzles

Comment: There are 14 LCM 8s that remain in service. Of these, six are used to support the Maritime Prepositioning Force and there are four each at homeports in San Diego and Norfolk. There are 24 similar craft used by the Army.

CHICKAHOMINY   9/2012*, Michael Winter / 1531166

LCM 8   5/2003, A Sharma / 0572786

IHS Jane's Fighting Ships 2014-2015   © 2014 IHS

10 LCM 6 TYPE (MECHANISED LANDING CRAFT)

Displacement, tonnes: 65 full load
Dimensions, metres (feet): 17.1 × 4.3 × 1.2 (56.1 × 14.1 × 3.9)
Speed, knots: 9
Range, n miles: 130 at 9 kt
Complement: 5
Military lift: 34 tons or 80 troops
Machinery: 2 Detroit 6V-71 diesels; 348 hp (260 kW) sustained or 2 Detroit 8V-71 diesels; 460 hp (344 kW) sustained; 2 shafts

Comment: Welded steel construction. All 10 remaining LCMs are used to support the Offshore Petroleum Discharge System. They are no longer used as combatant landing craft.

LCU 1662 6/2012, M Declerck / 1486445

75 LANDING CRAFT PERSONNEL (LCPL)

Displacement, tonnes: 11 full load
Dimensions, metres (feet): 11 × 3.7 × 1.2 (36.1 × 12.1 × 3.9)
Speed, knots: 20. **Range, n miles:** 150 at 20 kt
Complement: 3
Military lift: 17 troops
Machinery: 1 GM 8V-71TI diesel; 425 hp (317 kW) sustained; 1 shaft
Radars: Navigation: Marconi LN66; I-band.

Comment: There are four variants of this craft: Mk 11, Mk 12, Mk 13 and 11 m LCPLs. Details given are for Mk 12 and 13. For use as control craft and carried aboard LHA, LPD and LSD classes. A total of 75 had entered service by January 2013.

LCM 6 6/1997, J W Currie / 0016482

32 LCU 1610 CLASS (UTILITY LANDING CRAFT)

Displacement, tonnes: 262 standard; 407 full load
Dimensions, metres (feet): 41.1 × 8.8 × 1.9 (134.8 × 28.9 × 6.2)
Speed, knots: 11. **Range, n miles:** 1,200 at 8 kt
Complement: 14 (2 officers)

Military lift: 127 tonnes; 2 M1A1 tanks or 400 troops
Machinery: 2 Detroit 12V-71 diesels; 900 hp (671 kW) sustained; 2 shafts; Kort nozzles
Guns: 2 — 12.7 mm MGs. 2 — 7.62 mm MGs.
Radars: Navigation: Furuno; I-band.

Comment: Steel hulled construction. Versatile craft used for a variety of tasks. Most were built between the mid-1960s and mid-1980s. A replacement craft is under consideration.

LCPL Mk 13 4/1991, Bollinger / 0084143

MINE WARFARE FORCES

Notes: (1) There are no surface minelayers. Mining is done by carrier-based aircraft, land-based aircraft and submarines. The mine inventory includes the Mk 67 submarine launched mobile mine (SLMM) and the Quickstrike series of bottom mines.
(2) MH-53E Sea Dragon helicopters can be deployed on amphibious ships (including Ponce converted to an Afloat Forward Staging Base) or transported by C-5 aircraft for mine countermeasures.
(3) Marine Mammal Systems (MMS) uses trained dolphins and sea lions for mine detection, detection of unauthorised swimmers, protection of fleet assets in port and critical infrastructure, and recovery of exercise mines and torpedoes. The dolphins can be transported by C-5 aircraft or amphibious ships. MMS is the only operational method of detecting and neutralising buried mines. This capability is to be discontinued in 2017.
(4) The AN/WLD1 Remote Minehunting System (RMS) is an off-board system that is part of the Littoral Combat Ship (LCS) mine warfare mission package. RMS is comprised of a diesel-powered semi-submersible (the Remote Multi-Mission Vehicle (RMMV)) combined with the towed AN/AQS-20A Sonar Mine Detecting Set. The vehicle tows the variable depth sensor to detect, classify and localise bottom and moored mines. System includes real-time line-of-sight datalink capability with future system increments adding an over-the-horizon communication capability, shipboard launch and recovery subsystem, and a software segment that integrates RMS into the ship's Mission Package Computing Environment allowing real-time detection and processing when using line-of-sight communications. RMMV is undergoing a three-year, three-phase Reliability Growth Programme with a view to achieving an initial operating capability in fiscal year 2015.

13 AVENGER CLASS (MINESWEEPERS/MINEHUNTERS) (MCM/MHSO)

Name	No	Builders	Laid down	Launched	Commissioned	F/S
AVENGER	MCM 1	Peterson Builders Inc, Sturgeon Bay	3 Jun 1983	15 Jun 1985	12 Sep 1987	PA
DEFENDER	MCM 2	Marinette Marine, Marinette	1 Dec 1983	4 Apr 1987	30 Sep 1989	PA
SENTRY	MCM 3	Peterson Builders Inc, Sturgeon Bay	8 Oct 1984	20 Sep 1986	2 Sep 1989	PA
CHAMPION	MCM 4	Marinette Marine, Marinette	28 Jun 1984	15 Apr 1989	27 Jul 1991	PA
DEVASTATOR	MCM 6	Peterson Builders Inc, Sturgeon Bay	9 Feb 1987	11 Jun 1988	6 Oct 1990	PA
PATRIOT	MCM 7	Marinette Marine, Marinette	31 Mar 1987	15 May 1990	18 Oct 1991	PA
SCOUT	MCM 8	Peterson Builders Inc, Sturgeon Bay	8 Jun 1987	20 May 1989	15 Dec 1990	AE
PIONEER	MCM 9	Peterson Builders Inc, Sturgeon Bay	5 Jun 1989	25 Aug 1990	7 Dec 1992	PA
WARRIOR	MCM 10	Peterson Builders Inc, Sturgeon Bay	25 Sep 1989	8 Dec 1990	3 Apr 1993	PA
GLADIATOR	MCM 11	Peterson Builders Inc, Sturgeon Bay	7 Jul 1990	29 Jun 1991	18 Sep 1993	AE
ARDENT	MCM 12	Peterson Builders Inc, Sturgeon Bay	22 Oct 1990	16 Nov 1991	18 Feb 1994	AE
DEXTROUS	MCM 13	Peterson Builders Inc, Sturgeon Bay	11 Mar 1991	20 Jun 1992	9 Jul 1994	AE
CHIEF	MCM 14	Peterson Builders Inc, Sturgeon Bay	19 Aug 1991	12 Jun 1993	5 Nov 1994	PA

Displacement, tonnes: 1,401 full load
Dimensions, metres (feet): 68.4 × 11.9 × 3.7 (224.4 × 39.0 × 12.1)
Speed, knots: 13.5. **Range, n miles:** 2,500 at 10 kt
Complement: 84 (8 officers)

Machinery: 4 Waukesha L-1616 diesels (MCM 1–2); 2,600 hp(m) (1.91 MW) or 4 Isotta Fraschini ID 36 SS 6V AM diesels (MCM 3–4, MCM 6 onwards); 2,280 hp(m) (1.68 MW) sustained; 2 Hansome Electric motors; 400 hp(m) (294 kW) for hovering; 2 shafts; cp props; 1 Omnithruster hydrojet; 350 hp (257 kW)
Guns: 2 — 12.7 mm MGs. 2 — 7.62 MGs. 2 — 40 mm grenade launchers.
Physical countermeasures: MCM: 2 SLQ-48; includes Honeywell/Hughes ROV mine neutralisation system, capable of 6 kt (1,500 m cable with cutter (MP1), and countermining charge) (MP 2). SLQ-37(V)3; magnetic/acoustic influence sweep equipment. Oropesa SLQ-38 Type 0 Size 1; mechanical sweep. SLQ-60 Seafox has replaced SLQ-48 in three ships.
Radars: Surface search: ISC Cardion SPS-55; I/J-band.
Navigation: ARPA 2525 or LN66; I-band. Both to be replaced by SPS-73.
Sonars: Raytheon/Thomson Sintra SQQ-32(V)4; VDS; active minehunting; high frequency.
Combat data systems: SATCOM SRR-1; WSC-3 (UHF). GEC/Marconi NAUTIS M in last two ships includes SSN 2 PINS command system and control. USQ-119E(V), UHF Dama and OTCIXS provide JMCIS connectivity.

Programmes: The contract for the prototype MCM was awarded in June 1982. The last three were funded in fiscal year 1990.
Modernisation: Integrated Ship Control System (ISCS) installed in all hulls. SQQ-32(V)3 is being upgraded to SQQ-32(V)4.

PATRIOT 11/2012, Hachiro Nakai / 1483662

Structure: The hull is constructed of oak, Douglas fir and Alaskan cedar, with a thin coating of glass fibre on the outside, to permit taking advantage of wood's low magnetic signature.
Operational: The entire class has been retrofitted with SQQ-32 sonar. Sentry, Devastator, Gladiator, and Dextrous are stationed at Bahrain. Warrior, Patriot, Avenger, and Defender are at Sasebo, Japan. The remaining five ships are based at San Diego. Guardian went aground on Tubbataha Reef in the Sulu Sea on 17 January 2013 and was subsequently dismantled on site. Avenger is to be decommissioned in September 2014 and used for spares.

RESEARCH SHIPS

Notes: There are many naval associated research vessels which are civilian manned and not carried on the US Naval Vessel Register. In addition civilian ships are leased for short periods to support a particular research project or trial. Some of those employed include *RSB-1* (missile booster recovery), *Acoustic Pioneer* and *Acoustic Explorer* (acoustic research).

ACOUSTIC EXPLORER 10/2007, Michael Nitz / 1353634

1 EXPERIMENTAL SWATH (AGE)

STILETTO

Displacement, tonnes: 61 full load
Dimensions, metres (feet): 26.8 × 12.2 × 0.76 (87.9 × 40.0 × 2.5)
Speed, knots: 50
Complement: 35
Machinery: 4 Caterpillar diesels; 6,600 hp (4.9 MW); 4 surface piercing propellers
Radars: Navigation: To be announced.
Helicopters: Platform for 2 SH-60R.

Programmes: Developed by the Office of Force Transformation to act as a testbed for new technologies and to evaluate the potential uses of innovative hullforms. The ship was designed by M Ship Company of San Diego, California and constructed in 15 months by Knight & Carver Yacht Center, National City, CA. The ship was delivered in 2006 and trials (including mine warfare and special operations) are expected to last several years.
Structure: Small Water Area Twin Hull (SWATH) design of lightweight all-carbon composite construction. Multiple hulls reduce drag and generate hydrodynamic lift. The ship is capable of launching/recovering an 11 m RIB via a stern ramp and can also act as a platform for UAVs. In addition, the craft includes an 'electronic keel' which enables mission planning modules to be installed and networked.
Operational: The craft took part in Exercise Trident Warrior in August 2006. Craft was shipped to Norfolk, VA area in late 2006. Homeport at Joint Expeditionary Base Little Creek, VA. Deployed to the Caribbean basin in Q3 2009 under operational control of US Naval Forces Southern Command and US 4th Fleet, and under tactical control of Joint Interagency Task Force-South (JIATF-S) to conduct counter-illicit trafficking operations. Manned by joint Army and Navy crew that included an embarked US Coast Guard Law Enforcement Detachment.

STILETTO 5/2006, US Navy / 1167633

2 ASHEVILLE CLASS (YFRT)

ATHENA (ex-*Chehalis*) **ATHENA II** (ex-*Grand Rapids*)

Displacement, tonnes: 239 full load
Dimensions, metres (feet): 50.1 × 7.3 × 2.9 (164.4 × 24.0 × 9.5)
Speed, knots: 35; 13 diesels
Range, n miles: 2,300 at 13 kt
Complement: 10 + 18 scientists
Machinery: CODOG; 1 GE LM 1500 gas-turbine; 12,500 hp (9.3 MW); 2 Cummins VT12-875 diesels; 1,450 hp (1.07 MW); 2 shafts; cp props

Comment: Work for the Naval Surface Warfare Center, at Panama City, Florida. *Athena I* and *Athena II* have seen extensive service in development of high-speed towed sensors, airborne mine countermeasures and communication systems, and in full-scale validations of model predictions for propeller stress studies. *Athena I* was built in 1969 at Tacoma Boat as a patrol craft and was later modified extensively in 1976 to act as a high-speed research vessel. *Athena II* followed in 1971. Hull and structural framework are of aluminium alloy and superstructure is glass fibre over an aluminium framework. With both diesel and gas turbine propulsion, the vessels are capable of low- or high-speed operations. To enhance ability to gather high-speed acoustic data, a Compound Air Masker (CAM) system can be fitted to reduce greatly ship's radiated noise.

ATHENA II 6/1993, Giorgio Arra / 0506179

1 EXPERIMENTAL CATAMARAN (X-CRAFT) (AGE)

SEA FIGHTER FSF-1

Displacement, tonnes: 950 standard; 1,422 full load
Dimensions, metres (feet): 83 × 22.0 × 3.5 (272.3 × 72.2 × 11.5)
Speed, knots: 50
Range, n miles: 4,000 at 20 kt
Complement: 17
Machinery: CODOG; 2 GE LM 2500 gas turbines; 67,000 hp (49.96 MW); 2 MTU 16V595 diesels; 11,585 hp (8.6 MW); four Rolls-Royce Kamewa 125 SII waterjets
Radars: Navigation: I-band.
Helicopters: Platform for 1 SH-60R.

Comment: In September 2002, the Office of Naval Research selected Titan Corporation of San Diego, California and Nigel Gee and Associates LTD of Southampton, UK to design an experimental vessel known as X-CRAFT. A contract for development and build of the vessel was awarded in February 2003. The keel was laid in June 2003 and the vessel was launched in February 2005 at Nichols Brothers Boat Builders in Whidbey Island, Washington. The vessel, an aluminium-hulled, wave-piercing catamaran, was delivered to the Navy in May 2005. Multipurpose stern ramp, with direct access to the mission bay, allows launch and recover of manned and unmanned surface and sub-surface vehicles. Flight deck has dual landing spots for two MH-60 helicopters or UAV. Between May 2005 and September 2006 the vessel was stationed in San Diego and manned by a combined Navy/Coast Guard crew to evaluate experimental manning an operational concepts. In October 2006 a civilian crew assumed operations and maintenance of the vessel. In March 2006 the homeport was changed from San Diego to Panama City, Florida where it is used as a test platform for at-sea science and technology experimentation and advanced concept demonstrations. Flight-deck extension and modifications to reduce ship signature as well as to improve hull, mechanical and electrical capabilities were completed in April 2009. A four-month docking period was completed in February 2012. Modifications included Counter Measure Washdown (CMWD) (to improve NBC defences) and installation of water-jet skirts.

SEAFIGHTER 10/2005, US Navy / 1123764

1 ELECTRIC SHIP DEMONSTRATOR

SEA JET

Displacement, tonnes: 122 full load
Dimensions, metres (feet): 40.5 × 4.3 × 1.8 (132.9 × 14.1 × 5.9)
Speed, knots: 16
Range, n miles: 200 at 8 kt
Complement: 3
Machinery: 720-cell lead-acid battery bank; 2,690 hp (2 MW); 2 motors; 2 AWJ-21 waterjets; 1 Caterpillar C9 diesel generator; 335 hp (250 kW)
Radars: Navigation: Furuno 1933C; I-band.

Comment: An Advanced Electric Ship Demonstrator (AESD) designed to test and develop electric ship and propulsor technologies. Funded by the Office of Naval Research, the craft is an approximately quarter-scale version of a destroyer-sized surface ship with tumblehome hullform. The demonstrator started trials on Lake Pend Oreille on 30 November 2005. Its first task was to test Rolls Royce Naval Marine's AWJ-21 waterjet technology. Testing of the General Dynamics RIMJET podded propulsor completed in 2009. The demonstrator was built by Dakota Creek Industries, Anacortes, WA, and is located at the Naval Surface Warfare Center Carderock Division, Acoustic Research Detachment in Bayview, Idaho.

SEA JET 6/2005, US Navy / 1116518

RESEARCH OCEANOGRAPHIC SHIPS

2 MELVILLE CLASS (AGOR)

Name	No	Builders	Commissioned	F/S
MELVILLE	AGOR 14	Defoe SB Co, Bay City	27 Aug 1969	Loan
KNORR	AGOR 15	Defoe SB Co, Bay City	14 Jan 1970	Loan

Displacement, tonnes: 2,991 full load
Dimensions, metres (feet): 85 × 14.1 × 5.0 *(278.9 × 46.3 × 16.4)*
Speed, knots: 14
Range, n miles: 10,060 at 11.7 kt
Complement: 23 (9 officers) + 38 scientists
Machinery: Diesel-electric; 3 Caterpillar 3516 diesel generators; 1 Caterpillar 3508 diesel generator; 2 motor-driven Z-drive azimuth thrusters; 3,000 hp *(2.2 MW)*; 1 bow thruster; 900 hp *(670 kW)*
Sonars: Deep-water multibeam; sub-bottom profiler; Acoustic Doppler Current Profiler.

Comment: *Melville* operated by Scripps Institution of Oceanography and *Knorr* by Woods Hole Oceanographic Institution for the Office of Naval Research. Fitted with internal wells for lowering equipment and observation ports. Problems with the propulsion system led to major modifications including electric drive (vice the original mechanical) and the insertion of a 34 ft central section increasing the displacement from the original 1,915 tons and allowing better accommodation and improved laboratory spaces. The forward thruster is retractable. These ships are highly manoeuvrable for precise position keeping.

MELVILLE *3/2003, Robert Pabst / 0572738*

3 THOMAS G THOMPSON CLASS (AGOR)

Name	No	Builders	Launched	Commissioned	F/S
THOMAS G THOMPSON	AGOR-23	Halter Marine	27 Jul 1990	8 Jul 1991	Loan
ROGER REVELLE	AGOR-24	Halter Marine	20 Apr 1995	11 Jun 1996	Loan
ATLANTIS	AGOR-25	Halter Marine	1 Feb 1996	3 Mar 1997	Loan

Displacement, tonnes: 3,455 full load
Dimensions, metres (feet): 83.5 oa; 75.2 wl × 16 × 5.6 *(274; 246.7 × 52.5 × 18.4)*
Speed, knots: 15. **Range, n miles:** 15,000 at 12 kt
Complement: 22 + 37 scientists
Machinery: Diesel-electric; 6 Caterpillar diesel generators; 6.65 MW *(3—1.5 MW and 3—715 kW)*; 2 motors; 6,000 hp *(4.48 MW)*; 2 Z-drives; bow thruster; 1,140 hp *(850 kW)*
Sonars: Various multibeam seafloor mapping sonars and sub-bottom and Acoustic Doppler Current Profiling systems.

Comment: *Thomas G Thompson* was the first of a class of oceanographic research vessels capable of operating worldwide in all seasons and suitable for use by navy laboratories, contractors and academic institutions. Dynamic positioning system enables precise station-keeping. 4,000 sq ft of laboratories. AGORs 23, 24 and 25 are operated by academic institutions for the Office of Naval Research through charter party agreements (AGOR 23-University of Washington; AGOR 24-Scripps Institution of Oceanography; AGOR 25-Woods Hole Oceanographic Institution). Ships in this series are able to meet changing oceanographic requirements for general-purpose, year-round, worldwide research. This includes launching, towing and recovering a variety of equipment. *Atlantis* is the support vessel for human-occupied research submersible, *Alvin*, operated by the National Deep Submergence Facility at Woods Hole.

THOMAS G THOMPSON *3/2009, D Brodie / 1366287*

1 AGOR-26 CLASS (AGOR)

Name	No	Builders	Commissioned
KILO MOANA	AGOR 26	Atlantic Marine, Jacksonville	3 Sep 2002

Displacement, tonnes: 2,583 full load
Dimensions, metres (feet): 56.7 × 26.8 × 7.6 *(186 × 87.9 × 24.9)*
Speed, knots: 15. **Range, n miles:** 10,000 at 11 kt
Complement: 17 + 31 scientists
Machinery: Diesel-electric; 4 Caterpillar 3508B diesel generators; 2 Westinghouse motors; 4,025 hp *(3 MW)*; 1 bow thruster 1,100 hp *(820 kW)*

Comment: Replacement for R/V *Moana Wave*. Designed to commercial standards and constructed by Atlantic Marine, Jacksonville. Launched on 17 November 2001. The ship is a small waterplane area, twin hull (SWATH) oceanographic vessel capable of performing general purpose oceanographic research in coastal and deep ocean areas. The University of Hawaii School of Ocean and Earth Science and Technology operates the ship under a charter agreement for the Office of Naval Research (ONR). The survey suite consists of a Kongsberg EM 122 multibeam echosounder (12 kHz), a Kongsberg EM 710 shallow water echo sounder (95 kHz), a Workhorse Mariner 300 kHz current profiler and an Ocean Surveyor 38 kHz current profiler.

KILO MOANA *6/2004, University of Hawaii Marine Center / 1043633*

0 + 2 ARMSTRONG CLASS (AGOR)

Name	No	Builders	Laid down	Launched	Commissioned
NEIL ARMSTRONG	AGOR 27	Dakota Creek Industries, Anacortes	17 Aug 2012	2014	2015
SALLY RIDE	AGOR 28	Dakota Creek Industries, Anacortes	17 Aug 2012	2015	2015

Displacement, tonnes: 3,024 full load
Dimensions, metres (feet): 72.5 × 15.24 × 4.6 *(237.9 × 50.0 × 15.1)*
Speed, knots: 12
Range, n miles: 11,500 at 12 kt
Complement: 20 + 24 scientists
Machinery: Integrated diesel-electric; 4 diesel generators; 16,314 hp *(12.1 MW)*; 2 motors; 2 shafts; cp props; 1 bow thruster; 1 stern tunnel thruster

Comment: Contracts for construction of two new oceanographic research ships, were awarded on 14 October 2011 and in February 2012 respectively. The ships' design includes main, wet and computer laboratories, a main crane, a stern frame and a working deck area aft. Survey equipment includes a Kongsberg EM-122 deep-water multibeam survey system, a Kongsberg EM-710 medium-water multibeam survey system, a Kongsberg EA-600 single-beam survey system, a Knudsen sub-bottom profiler and a Teledyne Acoustic Doppler Current Profiler. Both ships are to become operational in 2015 and are to be operated by the Woods Hole Oceanographic Institution *(Neil Armstrong)* and Scripps Institution of Oceanography *(Sally Ride)*.

ARMSTRONG CLASS *2/2012, US Navy / 1454888*

DEEP SUBMERGENCE VEHICLES

Notes: (1) US Navy's submarine rescue vehicles and equipment are operated by Submarine Squadron Eleven (CSS-11) and the Undersea Rescue Command (URC) in San Diego, California. URC operational assets includes a Pressurized Rescue Module (PRM), which is a remotely operated manned and tethered rescue vehicle capable of operations to depths of 2,000 ft (610 m), two Submarine Rescue Chambers (SRC) capable of rescues to depths of 850 ft (260 m), and four Atmospheric Dive System (ADS) suits capable of work supporting rescue to 2,000 ft (610 m).
(2) The Supervisor of Salvage and Diving operates four additional Navy Remotely Operated Vehicles (ROVs). They are all air-transportable and can be operated from a variety of warships and commercial vessels.
(3) The SRC is a vertical cylindrical (12 ft long, 7 ft diameter) chamber designed to mate (using a water-tight seal) with a disabled submarine (DISSUB) at depths of 850 ft (260 m) or less and is capable of rescuing personnel. SRC comprises two chambers connected by a pressure-tight hatch. The upper chamber houses two operators and six rescued personnel in a dry environment at atmospheric pressure while the lower chamber is open to the sea at the bottom. SRC descends by use of a downhaul cable attached to the SRC hatch shackle centered on the escape hatch of the DISSUB. The SRCs are air-transportable worldwide and are transported to the rescue site on board a Vessel of Opportunity (VOO). It is connected to the VOO by a backhaul cable and the DISSUB by the downhaul cable. SRC downhaul cable is mated to the DISSUB escape hatch with assistance from a diver in the ADS suit or a ROV.

US > Deep submergence vehicles — Auxiliaries

1 CUTTHROAT CLASS (DSV)

Name	No	Builders	Commissioned
CUTTHROAT	LSV-2	Newport News Shipbuilding; General Dynamics Electric Boat Division	Apr 2001

Dimensions, metres (feet): 33.8 × 3.1 × 2.7 *(110.9 × 10.2 × 8.9)*
Speed, knots: 34 dived
Machinery: Permanent Magnet electric motor; 3,000 hp(m) *(2.23 MW)*

Comment: The contract was placed with Newport News and Electric Boat in January 1999 to build *Cutthroat* LSV-2. The largest autonomous unmanned submarine in the world, it is a 1 : 3.4 scaled-down model of the Virginia-class submarine used to test advanced submarine technologies, including hydro-acoustics, hydrodynamics and manoeuvring. Its diving depth matches that of the Virginia class. The forward compartment contains 1,680 lead acid batteries and the after compartment contains the propulsion and auxiliary systems together with data recording and control systems. All appendages, including control surfaces and simulated sonar fairing, can be removed or relocated. LSV-2 is operated by the Acoustic Research Detachment at the instrumented range at Lake Pend Oreille in Bayview, Idaho. It is named after a species of trout indigenous to the lake.

CUTTHROAT *2000, Newport News* / 0105821

1 DEEP SUBMERGENCE VEHICLE: ALVIN TYPE (DSV)

Name	No	Builders	F/S
ALVIN	DSV 2	General Mills Inc, Minneapolis	PSA

Dimensions, metres (feet): 8.1 × 2.6 × ? *(26.6 × 8.5 × ?)*
Speed, knots: 2
Complement: 3 (1 pilot) + 2 personnel
Machinery: 6 brushless DC motors; 6 thrusters; 2 vertical-motion thrusters (located near the centre of gravity); 2 horizontally (near stern) (1 directed athwartships, 1 directed longitudinally); 2 on rotatable shaft near stern for vertical or longitudinal motion

Comment: *Alvin* was built for operation by the Woods Hole Oceanographic Institution for the Office of Naval Research. Named for Allyn C Vine of Woods Hole Oceanographic Institution, the original configuration had an operating depth of 6,000 ft. *Alvin* accidentally sank in 5,051 ft of water on 16 October 1968, was subsequently raised in August 1969 and refurbished in 1970–71 to its original configuration. Placed in Naval service on 1 June 1971, she was subsequently refitted with a titanium pressure sphere to provide increased depth capability. *Alvin* has an operating depth of 4,500 m *(14,764 ft)* and is powered by two banks of lead-acid batteries providing a 120 V DC system with 47 kW/h of capacity. In October 2005, after conducting dive number 4,162, *Alvin* was overhauled and provided with upgraded equipment. The National Science Foundation is funding a two-stage upgrade to *Alvin* which, when completed, will increase its operating depth to 6,500 m. Sister DSVs *Sea Cliff* and *Turtle* were placed out of service in 1998. *Turtle* resides at an aquarium in Mystic, CT.

ALVIN *5/2013*, Woods Hole Oceanographic Institution* / 1486558

1 PRESSURIZED RESCUE MODULE

Name	No	Builders	Commissioned
FALCON	PRM-1	Oceaneering International	Oct 2008

Dimensions, metres (feet): 7.4 × ? × ? *(24.3 × ? × ?)*
Speed, knots: 3.5
Complement: 3

Comment: The Submarine Rescue, Diving and Recompression System (SRDRS) is a three-phased program that replaces DSRV. Phase one was the Atmospheric Dive System 2000 (ADS 2000), a military adapted commercial diving suit that can operate to a depth of 2,000 ft *(610 m)*. The system is air-transportable worldwide and has been in service since 2006. Phase two of the SRDRS program was the Submarine Rescue System (SRS). SRS contains the Pressurized Rescue Module named *Falcon* (PRM-1) and its associated support equipment to comprise the Pressurized Rescue Module System (PRMS) in 2014. The PRM is a tethered, remotely operated submersible launched and controlled from a Vessel of Opportunity in up to Sea State 4. It has a cylindrical hull on which navigation, video, propulsion, and life support systems are mounted. The vehicle is designed for submerged transit to a depth of 2,000 ft *(610 m)* of sea water, for docking and mating (up to 45° angle) to a disabled submarine (DISSUB) and for evacuation and transfer of up to 16 rescued personnel per sortie. This system has been in service since 2008. The third and final phase of SRDRS is the Transfer Under Pressure (TUP) capability. Delivery of TUP is scheduled for 2014 and will allow for rescued submariners to remain under pressure up to 5 ata during their transfer from *Falcon* to the two specialised Submarine Decompression Chambers.

FALCON *6/2008, Richard Scott* / 1353632

AUXILIARIES

Notes: As of January 2014, the US Navy had 369 active and five inactive service craft, primarily small craft, on the US Naval Vessel Register. A majority of these vessels provide services to the fleet in various harbours and ports. Others are ocean-going ships that provide services to the fleet for research purposes. Most of the service craft are rated as 'active, in service' and some are accommodation ships.

2 CAPE FLATTERY CLASS (TORPEDO TRIALS CRAFT) (YTT)

BATTLE POINT YTT 10   **DISCOVERY BAY** YTT 11

Displacement, tonnes: 1,187 full load
Dimensions, metres (feet): 56.9 × 12.2 × 3.2 *(186.7 × 40.0 × 10.5)*
Speed, knots: 11
Range, n miles: 1,000 at 10 kt
Complement: 31 + 9 spare berths
Machinery: 1 Cummins KTA50-M diesel; 1,250 hp *(932 kW)* sustained; 1 shaft; 1 bow thruster; 400 hp *(298 kW)*; 2 stern thrusters; 600 hp *(448 kW)*

Comment: Built by McDermott Shipyard, Morgan City, and delivered in 1991–92. Fitted with two 21 in Mk 59 and three (one triple) 12.75 in Mk 32 Mod 5 torpedo tubes. Used for torpedo trials and development at Keyport, Washington. A battery is fitted for limited duration operations with the diesel shutdown. Both based at Naval Underwater Warfare Centre, Keyport, WA.

YTT *9/1999, van Ginderen Collection* / 0084162

2 DIVING TENDERS (YDT)

YDT 17   **YDT 18**

Displacement, tonnes: 279 full load
Dimensions, metres (feet): 40.2 × 8.2 × 1.8 *(131.9 × 26.9 × 5.9)*
Speed, knots: 20
Complement: 15 (7 divers)
Machinery: 2 Caterpillar diesels; 2,600 hp *(1.91 MW)*; 2 Hamilton waterjets

Comment: Tenders used to support shallow-water diving operations and are based at Panama City, FL. Ordered from Swiftships in July 1997 and delivered in April 1999.

YDT 17 *8/1999, US Navy* / 0084159

IHS Jane's Fighting Ships 2014-2015

17 PATROL CRAFT (YP)

| YP 676 | YP 683–684 | YP 694–695 | YP 700–701 |
| YP 681 | YP 686–692 | YP 697–698 | |

Displacement, tonnes: 170 full load
Dimensions, metres (feet): 32.9 × 7.3 × 2.4 *(107.9 × 24.0 × 7.9)*
Speed, knots: 13.3. **Range, n miles:** 1,500 at 12 kt
Complement: 30 (2 officers) + 24 personnel
Machinery: 2 Detroit 12V-71 diesels; 680 hp *(507 kW)* sustained; 2 shafts
Radars: Navigation: I-band.

Comment: Built in the 1980s by Peterson Builders (YP 676–682) and Marinette Marine (YP 683–700), both in Wisconsin. There are 16 based at the Naval Academy, Annapolis, and one at Naval Underwater Warfare Centre, Keyport, WA.

YP 676 CLASS   *6/2012*, Marc Piché /* 1531168

6 PATROL CRAFT (YP)

YP 703–708

Displacement, tonnes: 228 full load
Dimensions, metres (feet): 36.2 × 8.5 × 2.4 *(118.8 × 27.9 × 7.9)*
Speed, knots: 12. **Range, n miles:** 1,680 at 12 kt
Complement: 38 (2 officers; 30 trainees)
Machinery: 2 Caterpillar C-18 diesels; 715 bhp *(533 kW)* sustained; 2 shafts
Radars: Navigation: Furuno; I-band.

Comment: Built in 2010 by C&G Boatworks of Mobile, Alabama. YP 703–YP 708 are based at the Naval Academy, Annapolis.

YP 704   *5/2011*, Marc Piché /* 1531167

FLOATING DRY DOCKS

Notes: The US Navy operates a limited number of floating dry docks to supplement dry dock facilities at major naval activities. The larger floating dry docks are made sectional to facilitate movement and to render them self-docking. Some of the ARD-type docks have the forward end of their docking well closed by a structure resembling the bow of a ship to facilitate towing. Berthing facilities, repair shops and machinery are housed in sides of larger docks. None is self-propelled.

SMALL AUXILIARY FLOATING DRY DOCK (AFDL)

Name	No	Launched	F/S
DYMAMIC	AFDL 6	1944	Active, Norfolk, VA

Structure: Of steel construction and 950 tons capacity.
Sales: AFDL 1 to Dominican Republic; 4, Brazil; 5, Taiwan; 11, Kampuchea; 20, Philippines; 22, Vietnam; 24, Philippines; 26, Paraguay; 28, Mexico; 33, Peru; 34 and 36, Taiwan; 39, Brazil; 40 and 44, Philippines. AFDL 23 was sold to Gulf Copper Ship Repair in 2010.

DYNAMIC   *11/2006, US Navy /* 1305191

© 2014 IHS

MEDIUM AUXILIARY REPAIR DRY DOCKS (ARDM)

Name	No	Commissioned
SHIPPINGPORT	–	1979
–	ARDM 4	–
ARCO	ARDM 5	1986

Sales: ARD 2 to Mexico; 5, Chile; 6, Pakistan; 8, Peru; 9, Taiwan; 11, Mexico; 12, Turkey; 13, Venezuela; 14, Brazil; 15, Mexico; 17, Ecuador; 22 *(Windsor)*, Taiwan; 23, Argentina; 24, Ecuador; 25, Chile; 28, Colombia; 29, Iran; 32, Chile. ARDM 1 *(Oak Ridge)* transferred to the USCG in 2002 and in service at Baltimore. AFDM 10 sold to Todd Seattle (now Vigor Shipyards).

ARCO   *9/2008, Julio Montes /* 1353631

UNCLASSIFIED MISCELLANEOUS (IX)

Notes: (1) *Mercer* APL 39 (ex-IX 502) and *Nueces* APL 40 (ex-IX 503) are barrack ships of mid-1940s vintage.
(2) IX 516 is a decommissioned SSBN used for nuclear propulsion plant training at Charleston, South Carolina.
(3) IX 310 is an accommodation barge at Naval Undersea Warfare Center, Dresden, New York.
(4) IX 524 is Mobile At-Sea Sensor (MARSS) barge that provides advanced remote telecommunications capability during flight tests over the 42,000 square mile Pacific Missile Range off Pearl Harbor, HI.
(5) IX 527 and IX 528 are submarine test platforms.
(6) *Prevail* IX 537 is a modified Stalwart class AGOS reclassified in 2003 as a training support vessel and used in training for board, search and seizure operations. She is assigned to Commander Strike Force Training, Atlantic.
(7) *Neodesha* IX 540, a converted tug (YTB 815) is used as a mobile diving and salvage unit in Pearl Harbor, Hawaii.

PREVAIL   *6/2012, M Declerck /* 1486447

1 CONSTITUTION CLASS (AXS)

Name	Builders	Launched	Commissioned	F/S
CONSTITUTION	Edmund Hartt's Shipyard, Boston	21 Oct 1797	22 Jul 1798	AA

Displacement, tonnes: 1,900 standard (est.)
Dimensions, metres (feet): 63.1 oa; 53.3 wl × 13.26 × 6.86 *(207; 174.9 × 43.5 × 22.5)*
Speed, knots: 13 under sail
Complement: 44 (2 officers)

Comment: The oldest ship remaining on the Navy List and designated a national historic landmark in 1960 and America's "Ship of State" in 2009. One of six frigates authorised 27 March 1794 to protect American interests at sea. She first got underway on 22 July 1798 and is best remembered for her service in the war of 1812. At that time, her complement was 450. Following extensive restoration (1927–31), she went on a three-year goodwill tour around the United States (1931–34), travelling over 22,000 miles and receiving over 4.6 million visitors. Open to the public in the Charlestown Navy Yard, Boston National Historic Park, the ship receives over 500,000 visitors a year. The most recent overhaul was conducted at the Charlestown Navy Yard from 2007–10. Under fighting sails (jibs, topsails, and spanker) *Constitution* sailed for the first time in 116 years on 21 July 1997 as part of her bicentennial celebration. On 19 August 2012, *Constitution* sailed again in commemoration of her victory over HMS *Guerriere* (in which she earned the nickname 'Old Ironsides') and the 200th anniversary of the war of 1812. Armament in 1812: 24 × 32 lb carronades, 30 × 24 lb long guns, 1 × 18 lb bow chaser. Sail area in 1812: nearly 45,000 sq ft.

CONSTITUTION   *7/1997, Todd Stevens, US Navy /* 0016501

978 US > Military Sealift Command (MSC)

1 RO-RO CONTAINER: CARGO SHIP (AK)

Name	No	Builders	Commissioned	F/S
TSGT JOHN A CHAPMAN	T-AK 323	Chantiers	1978	PREPO

Displacement, tonnes: 26,801 full load
Dimensions, metres (feet): 204.1 × 26.5 × 10.5 *(669.6 × 86.9 × 34.4)*
Speed, knots: 16
Range, n miles: 23,200 at 13 kt
Complement: 19
Cargo capacity: 1,066 TEU
Machinery: Pielstick medium speed diesel; 1 shaft

Comment: The ship is owned and operated by Sealift, Inc and is under charter to Military Sealift Command. Chartered in 2010. The ship can carry more than 1,000 20 ft container equivalents of aviation munitions and serves as an NDAF ship. The ship features climate-controlled cocoons on its weather decks to allow the carriage of additional cargo.

TSGT JOHN A CHAPMAN   1/2007, Marco Ghiglino / 1167639

1 CONTAINER SHIP (AK)

Name	No	Builders	Commissioned	F/S
MAJ BERNARD F FISHER	T-AK 4396	Odense	1985	PREPO
(ex-*Sea Fox*)		Shipyard, Lindø		

Displacement, tonnes: 48,782 full load
Dimensions, metres (feet): 198.86 × 32.2 × 10.4 *(652.4 × 105.6 × 34.1)*
Speed, knots: 19
Complement: 21
Cargo capacity: 520 TEU (on deck); 1,006 TEU (under deck)
Machinery: 1 Sulzer 7RTA76 diesel; 23,030 hp *(16.9 MW)*; 1 shaft

Comment: Owned and operated by Sealift, Inc. under charter to MSC. When fully loaded the ship can carry 916 20 ft containers of various aviation munitions intended to re-supply forward-deployed fighter and attack squadrons. The ship is fitted with an extensive cocoon system enabling it to store deck-loaded munitions in an environmentally controlled atmosphere. *Fisher* is a NDAF Ship re-chartered in 2012.

MAJ BERNARD F FISHER   6/1999, US Navy / 0084185

1 + 2 MOBILE LANDING PLATFORMS (AKR)

Name	No	Builders	Laid down	Launched	Commissioned
MONTFORD POINT	T-MLP-1	General Dynamics NASSCO, San Diego	19 Jan 2012	13 Nov 2012	20 May 2013
JOHN GLENN	T-MLP-2	General Dynamics NASSCO, San Diego	17 Apr 2012	15 Sep 2013	2014
LEWIS B PULLER	T-MLP-3	General Dynamics NASSCO, San Diego	5 Nov 2013	2014	2015

Displacement, tonnes: 81,435 standard
Dimensions, metres (feet): 239.3 × 50.0 × ? *(785.1 × 164.0 × ?)*
Speed, knots: 15
Range, n miles: 9,500 at 15 kt
Complement: 33
Cargo capacity: To be announced
Machinery: Diesel-electric: 4 diesels; 2 motors; 2 shafts; bow thruster

Comment: Following the award of a preliminary design contract in February 2009, a contract for Advanced Design and Long Lead Time Material was awarded to General Dynamics NAASCO on 13 August 2010. A contract for the first two ships was awarded on 27 May 2011 and for the third in February 2012. The MLP is to be a key component of the programme to enhance the two MPSRONs and is intended to serve as a transfer station to facilitate delivery of equipment and cargo to areas where port access is limited or unavailable. The ships are on a Float On-Float Off (FLO-FLO) design. Its principal requirements are: to project a Marine combat unit and its equipment via LCACs; interface with MPF(F) LMSR, JHSV and other surface assault crafts to facilitate transfer of equipment, cargo and personnel in a sea base. There are to be three LCAC landing lanes. Consideration is also being given to modification of *Lewis B Puller* to undertake the Afloat Forward Staging Base role. The MLP is based on the Alaska class crude oil tanker. Initial Operating Capability of MLP-1 is planned for fiscal year 2015.

MONTFORD POINT   6/2013*, Military Sealift Command / 1531174

SERVICE SUPPORT

Notes: (1) MSC's 15 Service Support ships provide the Navy with towing, rescue and salvage, submarine support and cable laying and repair services, as well as a command and control platform and floating medical facilities.
(2) All Service Support ships are government-owned and are crewed by civil service mariners. USS *Frank Cable* and *Emory S Land* have hybrid crews of Navy sailors and civil service mariners.
(3) One of two Navy command ships, USS *Mount Whitney*, is in the MSC inventory of service support ships.
(4) Submarine tenders, command ship and the Afloat Forward Staging Base (Interim) have combined crews of civil service mariners and uniformed Navy personnel working under the leadership of a US Navy captain. Civil service mariners perform navigation, deck, engineering, laundry and galley service operations while military personnel on board support communications, weapons systems and security.

4 POWHATAN CLASS (FLEET OCEAN TUGS) (ATF)

Name	No	Laid down	Commissioned	F/S
CATAWBA	T-ATF 168	14 Dec 1977	28 May 1980	TPA
NAVAJO	T-ATF 169	14 Dec 1977	13 Jun 1980	TPA
SIOUX	T-ATF 171	22 Mar 1979	1 May 1981	TPA
APACHE	T-ATF 172	22 Mar 1979	30 Jul 1981	TAA

Displacement, tonnes: 2,296 full load
Dimensions, metres (feet): 68.9 × 12.8 × 4.6 *(226 × 42.0 × 15.1)*
Speed, knots: 14.5. **Range, n miles:** 10,000 at 13 kt
Complement: 4 + 16 personnel
Machinery: 2 GM EMD 20-645F7B diesels; 7,250 hp(m) *(5.41 MW)* sustained; 2 shafts; Kort nozzles; cp props; bow thruster; 300 hp *(224 kW)*
Radars: Navigation: E/F/I-band.

Comment: Built at Marinette Marine Corp, Wisconsin patterned after commercial offshore supply ship design. Originally intended as successors to the Cherokee and Abnaki class ATFs. All transferred to MSC upon completion. 10 ton capacity crane and a bollard pull of at least 54 tons. A 'deck grid' is fitted aft which contains 1 in bolt receptacles spaced 24 in apart. This allows for the bolting down of a wide variety of portable equipment in support of salvage and training evolutions worldwide. There are two fire pumps supplying three fire monitors with up to 2,200 gallons of foam per minute. A deep module can be embarked to support naval salvage teams.

SIOUX   8/2013*, L-G Nilsson / 1531173

1 ZEUS CLASS (CABLE REPAIRING SHIP) (ARC)

Name	No	Builders	Commissioned	F/S
ZEUS	T-ARC 7	National Steel & Shipbuilding Co, San Diego	19 Mar 1984	TAA

Displacement, tonnes: 8,504 standard; 15,174 full load
Dimensions, metres (feet): 156.4 × 22.3 × 7.6 *(513.1 × 73.2 × 24.9)*
Speed, knots: 15.8. **Range, n miles:** 10,000 at 15 kt
Complement: 58 personnel
Machinery: Diesel-electric; 5 GM EMD 20-645F7B diesel generators; 14.32 MW sustained; 2 motors; 10,200 hp *(7.51 MW)*; 2 shafts; bow thrusters (forward and aft)

Comment: Ordered 7 August 1979. Remotely manned engineering room controlled from the bridge. The only active cable laying/repair ship in the US Navy, she can lay up to 1,000 miles of cable in depths of 9,000 ft. The ship is also equipped for bottom mapping surveys to support cable operations.

ZEUS   8/2006, Hachiro Nakai / 1167498

IHS Jane's Fighting Ships 2014-2015   © 2014 IHS

1 AUSTIN CLASS (AFLOAT FORWARD STAGING BASE — INTERIM)

Name	No	Builders	Laid down	Launched	Commissioned	F/S
PONCE	AFSB(i) (ex-LPD 15)	Lockheed SB & Construction Co	31 Oct 1966	20 May 1970	10 Jul 1971	AA

Displacement, tonnes: 9,277 standard; 17,521 full load
Dimensions, metres (feet): 173.8 × 30.5 × 7 *(570.2 × 100.1 × 23.0)*
Speed, knots: 21. **Range, n miles:** 7,700 at 20 kt
Complement: 55 + 165 personnel

Military lift: 930 troops; 9 LCM 6s or 4 LCM 8s or 2 LCAC or 20 LVTs. 4 LCPL/LCVP
Machinery: 2 Foster-Wheeler boilers; 600 psi *(42.3 kg/cm²)*; 870°F *(467°C)*; 2 De Laval (General Electric in LPD 9 and LPD 10) turbines; 24,000 hp *(18 MW)*; 2 shafts
Guns: 2 Raytheon 20 mm/76 6-barrelled Vulcan Phalanx Mk 15 ❶; 3,000 rds/min (4,500 in Block 1) combined to 1.5 km. 2—25 mm Mk 38. 8—12.7 mm MGs.
Physical countermeasures: Decoys: 4 Loral Hycor SRBOC 6-barrelled Mk 36; IR flares and chaff to 4 km *(2.2 n miles)*.
Electronic countermeasures: ESM: SLQ-32(V)1; intercept.
Radars: Air search: Lockheed SPS-40E ❸; B-band.
Surface search: Norden SPS-67 ❹; G-band.
Navigation: Raytheon SPS-73(V)12; I-band.
Tacan: URN 25. IFF: Mk XII UPX-36.
Combat data systems: SATCOM ❷, WSC-3 (UHF), WSC-6 (SHF) (see Data Systems at front of section).

Helicopters: Up to 6 CH-46D/E Sea Knight can be carried. Hangar for only 1 light.

Programmes: LPD 15, originally authorised in fiscal year 1965, was to be decommissioned in 2012. Instead, she was reconfigured as an Afloat Forward Staging Base (Interim) to support mine countermeasure operations and provide mission support to coastal patrol ships, and aircraft operations.
Modernisation: LPD 15 was last of class to receive machinery, electrical, and habitability upgrades to extend life. In addition to LPD legacy capabilities of well-deck, flight deck, cargo storage and crane, *Ponce* has been refitted as a repair ship with a fully functional machine shop, metal shop, valve shop, hydraulic shop, filter shop, and a guest machine shop for embarks by mine countermeasures, special forces, and coalition forces. UAV Scan Eagle is a ship asset but no manned aircraft assigned. There is space to embark more than 350 personnel and there are more than 137 workstations in eight mission spaces, all with video teleconference capability. Large joint operational centre has access to six independent computer networks. *Ponce* is to become demonstration platform for solid-state laser weapon system (LaWS).

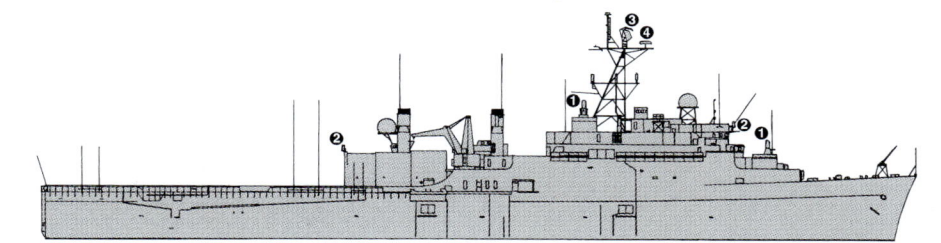

PONCE *(Scale 1 : 1,500), Ian Sturton / 0016471*

PONCE *4/2013*, US Navy / 1531209*

Structure: Flight deck is 168 ft *(51.2 m)* in length. Well-deck 394 × 50 ft *(120.1 × 15.2 m)*.

Operational: The ship is manned by a hybrid crew of Navy personnel and civilian mariners. It is based in the US Central Command area of operations.

2 EMORY S LAND CLASS (SUBMARINE TENDERS) (ASH)

Name	No	Builders	Laid down	Launched	Commissioned	F/S
EMORY S LAND	AS 39	Lockheed SB & Construction Co, Seattle	2 Mar 1976	4 May 1977	7 Jul 1979	PA
FRANK CABLE	AS 40	Lockheed SB & Construction Co, Seattle	2 Mar 1976	14 Jan 1978	29 Oct 1979	PA

Displacement, tonnes: 14,134 standard; 23,347 full load
Dimensions, metres (feet): 196.2 × 25.9 × 8.7 *(643.7 × 85.0 × 28.5)*
Speed, knots: 20. **Range, n miles:** 10,000 at 12 kt
Complement: 1,268 (AS 39), 1,270 (AS 40) (83 officers (AS 39); 81 officers (AS 40))

Machinery: 2 Combustion Engineering boilers; 620 psi *(43.6 kg/cm²)*; 860°F *(462°C)*; 1 De Laval turbine; 20,000 hp *(14.9 MW)*; 1 shaft
Guns: 4 Oerlikon 20 mm Mk 67.
Radars: Navigation: ISC Cardion SPS-55; I/J-band.

Helicopters: Platform only.

Comment: The first US submarine tenders designed specifically for servicing nuclear-propelled attack submarines. Each ship can simultaneously provide services to four submarines moored alongside. Equipped with two 30 ton cranes and one 5 ton mobile crane. There is a 23-bed sick bay. *Frank Cable* is based at Guam and *Emory S Land* at Diego Garcia. *Emory S Land* transferred to Military Sealift Command in February 2008 and *Frank Cable* in 2010. Each ship is operated by a hybrid crew of naval and civilian personnel, and remains under naval command and in commissioned status.

EMORY S LAND *4/1999, Jürg Kürsener / 0084155*

4 SAFEGUARD CLASS (SALVAGE SHIPS) (ARS)

Name	No	Builders	Commissioned	F/S
SAFEGUARD	T-ARS 50 (ex-ARS 50)	Peterson Builders Inc, Sturgeon Bay	16 Aug 1985	PA
GRASP	T-ARS 51 (ex-ARS 51)	Peterson Builders Inc, Sturgeon Bay	14 Dec 1985	AA
SALVOR	T-ARS 52 (ex-ARS 52)	Peterson Builders Inc, Sturgeon Bay	14 Jun 1986	PA
GRAPPLE	T-ARS 53 (ex-ARS 53)	Peterson Builders Inc, Sturgeon Bay	15 Nov 1986	AA

Displacement, tonnes: 3,336 full load
Dimensions, metres (feet): 77.7 × 15.5 × 5.2 *(254.9 × 50.9 × 17.1)*
Speed, knots: 14. **Range, n miles:** 8,000 at 12 kt
Complement: 52 + 26 personnel
Machinery: 4 Caterpillar diesels; 4,200 hp *(3.13 MW)*; 2 shafts; cp Kort nozzle props; bow thruster; 500 hp *(373 kW)*
Radars: Navigation: E/F/I-band.

Comment: Prototype approved in FY81, two in FY82 and one in FY83. The procurement of the fifth ARS was dropped on instructions from Congress. The design follows conventional commercial and Navy criteria. Can support surface-supplied diving operations to a depth of 58 m. Equipped with recompression chamber. Bollard pull, 65.6 tons. Using beach extraction equipment the pull increases to 360 tons. 150 ton deadlift. *Grasp* transferred from the US Navy on 19 January 2006 and *Grapple* on 13 July 2006. *Salvor* and *Safeguard* followed on 12 January 2007 and 26 September 2007 respectively.

SALVOR *12/2012*
Chris Sattler
1486451

2 MERCY CLASS (HOSPITAL SHIPS) (AHH)

Name	No	Builders	Commissioned	F/S
MERCY (ex-SS Worth)	T-AH 19	National Steel & Shipbuilding Co, San Diego	1976	ROS/TPA
COMFORT (ex-SS Rose City)	T-AH 20	National Steel & Shipbuilding Co, San Diego	1976	ROS/TAA

Displacement, tonnes: 69,552 full load
Measurement, tonnes: 55,239 gt; 36,535 net
Dimensions, metres (feet): 272.6 × 32.2 × 10 (894.4 × 105.6 × 32.8)
Speed, knots: 17. **Range, n miles:** 13,420 at 17 kt
Complement: 1,207 (820 medical personnel) + 61 personnel

Machinery: 2 boilers; 2 GE turbines; 24,500 hp (18.3 MW); 1 shaft
Radars: Navigation: E/F/I-band.
Tacan: URN 25.
Helicopters: 2 MH 60.

Comment: Converted San Clemente class tankers. *Mercy* was commissioned 19 December 1986; *Comfort* on 30 November 1987. Each ship has up to 1,000 beds, 12 operating theatres and 387 military support staff. Normally, the ships are kept in a reduced operating status in Norfolk, VA, and San Diego, CA, by a small crew of civilian mariners and active duty Navy medical and support personnel. Each ship can be fully activated and crewed within five days. In 2011, *Comfort* deployed to Latin America for 150 days for Continuing Promise in 2011. The mission stopped in nine ports, saw 67,000 patients and performed 1,100 surgeries. During Pacific Partnership 2010, *Mercy* served as a platform for humanitarian and civic assistance in Southeast Asia and treated more than 100,000 patients in Cambodia, East Timor, Indonesia and Vietnam.

MERCY   8/2008, Chris Sattler / 1353626

SEALIFT

Notes: These ships provide ocean transportation for Defense and other government agencies. In addition to the government-owned ships listed below, MSC also contracts tankers and dry cargo ships as needed on both a long- and short-term basis. As a result, these numbers vary with the operational requirement.
(2) There are three dry cargo ships and three tankers under long-term charter as of September 2012: *MV Mohegan* operates as a shuttle, carrying equipment and supplies from Singapore to Diego Garcia and is owned and operated by Sealift, Inc. *MV BBC Seattle*, owned by Teras BBC Ocean Navigation, is under charter to support Operation Enduring Freedom. A chartered tug/barge operates between Port Canaveral, Florida, and the Bahamas for the Atlantic Undersea Test and Evaluation Center (AUTEC) and is owned and operated by Nortcliffe Shipping and Trading. *MT Empire State* and *MT Evergreen State*, both owned and operated by American Petroleum Tankers LL, transport refined petroleum products between commercial refineries and storage and distribution facilities worldwide for Defense Logistics Agency, which procures and manages fuel for the US military. MSC charters additional tankers from commercial market on a short-term basis when more fuel-carrying capability is required.

7 BOB HOPE CLASS
(LARGE, MEDIUM-SPEED, RO-RO (LMSR) SHIPS) (AKR)

Name	No	Builders	Launched	Commissioned	F/S
BOB HOPE	T-AKR 300	Avondale Shipyards, New Orleans	27 Mar 1997	18 Nov 1998	TWWR
FISHER	T-AKR 301	Avondale Shipyards, New Orleans	21 Oct 1997	4 Aug 1999	TWWR
SEAY	T-AKR 302	Avondale Shipyards, New Orleans	25 Jun 1998	30 Mar 2000	PREPO
MENDONCA	T-AKR 303	Avondale Shipyards, New Orleans	25 May 1999	30 Jan 2001	TWWR
PILILAAU	T-AKR 304	Avondale Shipyards, New Orleans	18 Jan 2000	24 Jul 2001	PREPO
BRITTIN	T-AKR 305	Avondale Shipyards, New Orleans	21 Oct 2000	11 Jul 2002	TWWR
BENAVIDEZ	T-AKR 306	Avondale Shipyards, New Orleans	11 Aug 2001	10 Sep 2003	TWWR

Displacement, tonnes: 62,670 full load
Dimensions, metres (feet): 289.1 × 32.3 × 11 (948.5 × 106.0 × 36.1)
Speed, knots: 24
Range, n miles: 12,000 at 24 kt
Complement: 50 (est.)
Cargo capacity: 317,510 sq ft plus 70,152 sq ft deck cargo
Machinery: 4 Colt Pielstick 10 PC4.2 V diesels; 65,160 hp(m) (47.89 MW); 2 shafts; cp props

Comment: Contract awarded in 1993; options for additional ships exercised in 1994, 1995, 1996 and 1997. All fitted with a stern slewing ramp, side accesses and cranes for both roll-on/roll-off and lift-on/lift-off capabilities. Ramps extend to 130 ft (40 m), and two twin 55 ton cranes are installed. Ships are lay-berthed as follows: *Bob Hope* (San Diego, CA); *Mendonca* (Bremerton, WA); *Brittin* (Bremerton, WA); *Benavidez* and *Fisher* (Corpus Christi, TX). *Pililaau* lay-berthed at Corpus Christi, Texas, but expected to transfer to Prepositioning Force in 2012.

PILILAAU   1/2008, Shaun Jones / 1305187

4 + 6 JOINT HIGH SPEED VESSELS (TSV)

Name	No	Builders	Laid down	Launched	Commissioned
SPEARHEAD	JHSV 1	Austal USA, Mobile	22 Jul 2010	17 Sep 2011	5 Dec 2012
CHOCTAW COUNTY	JHSV 2	Austal USA, Mobile	13 Sep 2010	1 Oct 2012	6 Jun 2013
MILLINOCKET	JHSV 3	Austal USA, Mobile	3 May 2012	5 Jun 2013	21 Mar 2014
FALL RIVER	JHSV 4	Austal USA, Mobile	20 May 2013	16 Jan 2014	2014
TRENTON	JHSV 5	Austal USA, Mobile	8 Aug 2013	2014	2015
BRUNSWICK	JHSV 6	Austal USA, Mobile	2014	2015	2015
CARSON CITY	JHSV 7	Austal USA, Mobile	2014	2015	2016
YUMA	JHSV 8	Austal USA, Mobile	2015	2016	2016
BISMARCK	JHSV 9	Austal USA, Mobile	2015	2016	2017
BURLINGTON	JHSV 10	Austal USA, Mobile	2016	2017	2017

Displacement, tonnes: 2,400 full load
Dimensions, metres (feet): 103 × 28.5 × 3.83 (337.9 × 93.5 × 12.6)
Speed, knots: 43. **Range, n miles:** 1,200 at 35 kt
Complement: 41
Military lift: 312 troops (seated) + 545 tonnes of equipment and supplies
Machinery: 4 MTU 20V8000 M71L diesels; 48,800 hp (36.4 MW); 4 Wärtsilä WLD 1400 SR waterjets
Helicopters: Platform for one CH-53E.

Comment: The Joint High Speed Vessel (JHSV) Program is for 10 high-speed intra-theatre connector vessels. This program was initiated following signature of a Memorandum of Agreement (MOA) with the US Army which married the Army's Theater Support Vessel (TSV) program with the Navy's High Speed Connector (HSC) program. Acquisition for JHSV is under the auspices of the Navy's Program Executive Office, Ships. Original plans for the Army and Navy to own five ships each were modified by an agreement signed on 2 May 2011 that transferred all five army JHSVs to the Navy. Contracts for the preliminary design of JHSV were awarded to Austal USA, Bath Iron Works and Bollinger Shipyards (teamed with Incat) on 31 January 2008. On 13 November 2008, the Navy awarded Austal USA a fixed-price incentive contract for detailed design and construction of one vessel. The contract includes priced options for the construction of up to nine additional ships and associated shore-based spares. The contract options for the construction of JHSV 2 and 3 was let on 29 January 2010 and for JHSV 4 and 5 on 12 October 2010. The contract for JHSV 6 and 7 was awarded on 30 June 2011, for JHSV 8 and 9 on 27 February 2012 and for JHSV 10 in February 2013. The semi-SWATH catamaran design ship is of aluminium construction and has an articulated slewing stern ramp. The vessels are to be manned by Military Sealift Command with either civil service mariners or contractor crews. Trials of an electromagnetic railgun are to be conducted in JHSV 3 in 2016.

SPEARHEAD   12/2012, AUSTAL USA / 1486455

SPEARHEAD   12/2012, AUSTAL USA / 1486454

2 SHUGHART CLASS
(LARGE, MEDIUM-SPEED, RO-RO (LMSR) SHIPS) (AKR)

Name	No	Commissioned	F/S
SHUGHART (ex-*Laura Maersk*)	T-AKR 295	7 May 1996	TWWR
YANO (ex-*Leise Maersk*)	T-AKR 297	8 Feb 1997	TWWR

Measurement, tonnes: 55,169 gt
Dimensions, metres (feet): 276.4 × 32.2 × 10.5 (906.8 × 105.6 × 34.4)
Speed, knots: 24. **Range, n miles:** 12,000 at 24 kt
Complement: 44
Cargo capacity: 255,034 sq ft plus 47,023 sq ft deck cargo
Machinery: 1 Burmeister & Wain 12L90 GFCA diesel; 46,653 hp(m) (34.29 MW); 1 shaft; bow and stern thrusters
Radars: Navigation: 2 Sperry ARPA; I-band.

Comment: Both were container ships built in Denmark in 1981 and lengthened by Hyundai in 1987. Conversion contract awarded to National Steel and Shipbuilding in July 1993. Both fitted with a stern slewing ramp, side accesses and cranes for both roll-on/roll-off and lift-on/lift-off capabilities. Two twin 57 ton cranes. Conversion for *Shughart* started in June 1994; *Yano* in May 1995; *Soderman* underwent conversion to maritime prepositioning ship and renamed *Stockham*.

YANO   12/2006, Adolfo Ortigueira Gil / 1167638

Military Sealift Command (MSC) < US 981

2 GORDON CLASS
(LARGE, MEDIUM-SPEED, RO-RO (LMSR) SHIPS) (AKR)

Name	No	Commissioned	F/S
GORDON (ex-Selandia)	T-AKR 296	23 Aug 1996	TWWR
GILLILAND (ex-Jutlandia)	T-AKR 298	24 May 1997	TWWR

Measurement, tonnes: 56,311 gt
Dimensions, metres (feet): 291.4 × 32.2 × 11.9 (956 × 105.6 × 39.0)
Speed, knots: 24
Range, n miles: 12,000 at 24 kt
Complement: 50
Cargo capacity: 276,109 sq ft plus 45,722 sq ft deck cargo
Machinery: 1 Burmeister & Wain 12K84EF diesel; 26,000 hp(m) (19.11 MW); 2 Burmeister & Wain 9K84EF diesels; 39,000 hp(m) (28.66 MW); 3 shafts (centre cp prop); bow thruster
Radars: Navigation: 2 Sperry ARPA; I-band.

Comment: Built in Denmark in 1972 and lengthened by Hyundai in 1984. Conversion contract given to Newport News Shipbuilding on 30 July 1993. Both fitted with a stern slewing ramp, side accesses and improved craneage. Conversion started for both ships on 15 October 1993.

GILLILAND   2/2000, A Sharma / 0085304

3 SGT MATEJ KOCAK CLASS (VEHICLE CARGO SHIPS) (AKH)

Name	No	Builders	Commissioned	F/S
SGT MATEJ KOCAK (ex-SS John B Waterman)	T-AK 3005	Pennsylvania SB Co, Chester	14 Mar 1981	Sqn 2
PFC EUGENE A OBREGON (ex-SS Thomas Heywood)	T-AK 3006	Pennsylvania SB Co, Chester	1 Nov 1982	Sqn 1
MAJ STEPHEN W PLESS (ex-SS Charles Carroll)	T-AK 3007	General Dynamics Corp, Quincy	14 Mar 1983	Sqn 3

Displacement, tonnes: 49,536 full load
Dimensions, metres (feet): 250.2 × 32.2 × 9.8 (820.9 × 105.6 × 32.2)
Speed, knots: 20
Range, n miles: 13,000 at 20 kt
Complement: 29 + 10 scientists
Cargo capacity: Containers, 562; Ro-Ro, 152,236 sq ft; JP-5 bbls, 20,290; DF-2 bbls, 12,355; Mogas bbls, 3,717; stable water, 2,189; cranes, 2 twin 50 ton and 1—30 ton gantry
Machinery: 2 boilers; 2 GE turbines; 30,000 hp (22.4 MW); 1 shaft
Helicopters: Platform only.

Comment: Converted from three Waterman Line ships by National Steel and Shipbuilding, San Diego. Delivery dates T-AK 3005, 1 October 1984; T-AK 3006, 16 January 1985; T-AK 3007, 15 May 1985. Conversion work included the addition of 157 ft (47.9 m) amidships. All three ships purchased by MSC in 2009. All operated by Waterman SS Corp. As of October 2012, all ships used as surge sealift assets.

SGT MATEJ KOCAK   4/2012, Maritime Photographic / 1486456

1 RO-RO CONTAINER (CARGO SHIP) (AK)

Name	No	Builders	Commissioned	F/S
1ST LT HARRY L MARTIN (ex-Tarago)	T-AK 3015	Bremer Vulkan, Vegesack	20 Apr 2000	Sqn 3

Displacement, tonnes: 48,544 full load
Dimensions, metres (feet): 229.9 × 32.3 × 11 (754.3 × 106.0 × 36.1)
Speed, knots: 18. **Range, n miles:** 17,000 at 17 kt
Complement: 100 + 23 personnel
Cargo capacity: 168,547 sq ft. 735 TEU
Machinery: 1 MAN K7-SZ-90/160 diesel; 25,690 hp(m) (18.88 MW); 1 shaft

Comment: Completed in 1979. Acquired in February 1997 for conversion at Atlantic Drydock, Jacksonville. This government-owned ship carries USMC expeditionary airfield, fleet hospital package and construction equipment.

1ST LT HARRY L MARTIN   10/2010, Hachiro Nakai / 1406234

1 RO-RO CONTAINER (CARGO SHIP) (AK)

Name	No	Builders	Commissioned	F/S
L/CPL ROY M WHEAT (ex-Bazaliya)	T-AK 3016	Detyens Shipyards, Charleston	Oct 2001	Sqn 1

Displacement, tonnes: 50,905 full load
Dimensions, metres (feet): 263.3 × 30 × 10.6 (863.8 × 98.4 × 34.8)
Speed, knots: 20
Range, n miles: 12,000 at 20 kt
Complement: 30 + 100 embarked forces
Cargo capacity: 109,170 sq ft. 846 TEU
Machinery: 2 gas turbines; 47,020 hp(m) (34.56 MW); 2 shafts

Comment: Acquired in March 1997 for conversion for Maritime Prepositioning Force. The ship has been lengthened by 117 ft. Carries USMC expeditionary airfield, fleet hospital package and construction equipment.

L/CPL ROY M WHEAT   5/2010, Edward McDonnell / 1406331

1 HIGH SPEED VESSEL (HSV)

WESTPAC EXPRESS HSV 4676

Displacement, tonnes: 2,118 standard
Dimensions, metres (feet): 101 × 26.65 × 4.3 (331.4 × 87.4 × 14.1)
Speed, knots: 40
Range, n miles: 1,200 at 35 kt
Military lift: 550 tonnes of equipment and 970 personnel
Machinery: 4 Caterpillar 3618 diesels; 38,620 hp (28.8 MW); 4 Kamewa waterjets

Comment: Following trials which started in July 2001, chartered by Military Sealift Command from Austal Ships, West Australia. The charter was extended in February 2007 for up to 59 months. This was further extended to May 2014. Aluminium construction. Employed by US Marine Corps Third Expeditionary Force (III MEF) to transport equipment and troops from Okinawa for training exercises in Yokohama, Guam and other regional destinations. The benefits include reduced dependence on and cost of airlift. The vessel will retain commercial livery and markings. Based at Okinawa.

WESTPAC EXPRESS   8/2001, Mitsuhiro Kadota / 0131282

1 CHAMPION CLASS (TANKERS) (AOT)

Name	No	Builders	Commissioned
LAWRENCE H GIANELLA	T-AOT 1125	American SB Co, Tampa	22 Apr 1986

Displacement, tonnes: 40,260 full load
Dimensions, metres (feet): 187.5 × 27.4 × 10.8 (615.2 × 89.9 × 35.4)
Speed, knots: 16. **Range, n miles:** 12,000 at 16 kt
Complement: 23 (9 officers)
Cargo capacity: 238,400 barrels of oil fuel
Machinery: 1 Sulzer 5RTA76 diesel; 18,400 hp(m) (13.52 MW) sustained; 1 shaft

Comment: Built for Ocean Shipholdings Inc, Houston, Texas specifically for long-term time charter to the Military Sealift Command (20 years) as Point-to-Point fuel tanker. Purchased by the US Navy in 2003 and designated USNS. Equipped with a modular fuel delivery system to allow it to rig underway replenishment gear. Buck and Cobb deactivated in 2010 and Matthiesen in 2011.

LAWRENCE H GIANELLA   5/2009, Edward McDonnell / 1366237

982 US › Military Sealift Command (MSC) — Ready Reserve Force (RRF)

2 GUAM CLASS (HIGH SPEED TRANSPORT SHIPS) (HSV)

Name	No	Builders	Laid down	Launched	Commissioned
GUAM (ex-*Alakai*)	HST 1	Austal USA	3 Jun 2004	18 Jan 2007	14 Apr 2007
PUERTO RICO (ex-*Huakai*)	HST 2	Austal USA	3 Jun 2004	29 Sep 2008	Apr 2009

Displacement, tonnes: 1,646 full load
Measurement, tonnes: 8,235 gt
Dimensions, metres (feet): 112.3 × 23.8 × 3.65 *(368.4 × 78.1 × 12.0)*
Speed, knots: 40
Complement: 14
Machinery: 4 MTU 20V8000M71 diesels; 44,596 hp *(32.8 MW)*; 4 Rolls Royce 125 Mk II waterjets
Radars: Navigation: To be announced.

Comment: Former fast-ferries built for Hawaii Superferry. *Guam* operated in Hawaiian waters for only 11 months before being withdrawn from service due to business and environmental concerns. She and her sister ship *Puerto Rico*, which never entered revenue service, subsequently passed into the control of the US Maritime Administration. In early 2010 both ships were used to assist in relief operations following an earthquake in Haiti. In May 2012, it was announced that both ships were to be acquired by the US Navy. *Guam* is to be converted to replace the chartered *Westpac Express* in Okinawa in 2015. *Puerto Rico* is to remain laid up pending assignment.

READY RESERVE FORCE (RRF)

Notes: (1) The Ready Reserve Force was created in 1976, to support deployment and sustainment requirements and to respond to national emergencies. The RRF is designed to be made available quickly for military sealift operations. Its functions have been widened to include humanitarian and domestic security issues.
(2) On 1 January 2014 the RRF consisted of 46 ships, including 35 Ro-Ro, six auxiliary crane ships, two heavy lift barge carriers, a special mission tanker and two aviation repair ships. Eight of the Ro-Ro ships are Fast Sealift Support vessels. RRF ships are maintained in various stages of readiness and able to get underway in five or 10 days. They are located in various ports along the US East, West and Gulf coasts. Two are home-ported in National Defense Reserve Fleet's anchorage in Beaumont, TX.
(3) The Department of Transportation's Maritime Administration (MARAD) is responsible for the maintenance and administration of the ships at all times. Military Sealift Command assumes operational control only once the ships are activated for military missions.
(4) RRF ships have red, white and blue funnel stripes.
(5) The eight Algol class Fast Sealift Ships were transferred to MARAD control in October 2008.

1 PRODUCT TANKER (AOT)

PETERSBURG T-AOT 9109

Comment: *Petersburg* completed APS service in December 2007.

PETERSBURG *7/2010, M MAZUMDAR* / 1406334

6 AUXILIARY CRANE SHIPS (AK)

Name	No	Builders
KEYSTONE STATE (ex-*SS President Harrison*)	T-ACS 1	Defoe SB Co, Bay City
GEM STATE (ex-*SS President Monroe*)	T-ACS 2	Defoe SB Co, Bay City
GRAND CANYON STATE (ex-*SS President Polk*)	T-ACS 3	Dillingham SR, Portland
GOPHER STATE (ex-*Export Leader*)	T-ACS 4	Bath Iron Works, Maine
FLICKERTAIL STATE (ex-*Export Lightning*)	T-ACS 5	Bath Iron Works, Maine
CORNHUSKER STATE (ex-*Staghound*)	T-ACS 6	Bath Iron Works, Maine

Comment: Auxiliary crane ships are converted (1984–88) container ships to which have been added up to three twin boom pedestal cranes which will lift containerised or other cargo from itself or adjacent vessels and deposit it on a pier or into lighterage.

KEYSTONE STATE *10/2012, M Mazumdar* / 1486459

2 MISCELLANEOUS HEAVY LIFT SHIPS (AK/AKR)

CAPE MAY T-AKR 5063   **CAPE MOHICAN** T-AKR 5065

Comment: *Cape Ray* (T-AKR 9679) was outfitted with the Field Deployable Hydrolysis System (FDHS) and, in January 2014, deployed from Portsmouth, VA, to assist in disposal of Syrian chemical weapons stockpile. With crew of 35 civilian mariners and 64 US Army chemical specialists, *Cape Ray* was to embark the chemicals at the southern Italian port of Gioia Tauro, transport them to international waters and neutralise them using FDHS. System converts bulk amounts of mustard gas and other chemical warfare agents into compounds not usable as weapons. International disarmament effort was led by United Nations and Organisation for the Prohibition of Chemical Weapons.

IHS Jane's Fighting Ships 2014-2015

CAPE MOHICAN *3/1999, van Ginderen Collection* / 0084193

27 RO-RO SHIPS (AKR)

CAPE ISLAND (ex-*Mercury*) T-AKR 10	**CAPE EDMONT** T-AKR 5069
CAPE INTREPID (ex-*Lyra*) T-AKR 11	**CAPE INSCRIPTION** T-AKR 5076
CAPE TEXAS T-AKR 112	**CAPE KNOX** T-AKR 5082
CAPE TAYLOR (ex-*Cygnus*) T-AKR 113	**CAPE KENNEDY** T-AKR 5083
ADM WM H CALLAGHAN T-AKR 1001	**CAPE VINCENT** (ex-*Taabo Italia*) T-AKR 9666
CAPE ORLANDO (ex-*American Eagle*) T-AKR 2044	**CAPE RISE** (ex-*Saudi Riyadh*) T-AKR 9678
CAPE DUCATO T-AKR 5051	**CAPE RAY** (ex-*Saudi Makkah*) T-AKR 9679
CAPE DOUGLAS T-AKR 5052	**CAPE VICTORY** (ex-*Merzario Britania*) T-AKR 9701
CAPE DOMINGO T-AKR 5053	**CAPE TRINITY** (ex-*Santos*) T-AKR 9711
CAPE DECISION T-AKR 5054	**CAPE RACE** (ex-*G&C Admiral*) T-AKR 9960
CAPE DIAMOND T-AKR 5055	**CAPE WASHINGTON** (ex-*Hual Transporter*) T-AKR 9961
CAPE ISABEL T-AKR 5062	**CAPE WRATH** (ex-*Hual Trader*) T-AKR 9962
CAPE HUDSON T-AKR 5066	
CAPE HENRY T-AKR 5067	
CAPE HORN T-AKR 5068	

CAPE WASHINGTON *9/2007, Michael Winter* / 1305261

2 AVIATION LOGISTICS SHIPS (AVB)

Name	No	Builders	Commissioned
WRIGHT (ex-*SS Young America*)	T-AVB 3	Ingalls Shipbuilding, Pascagoula	1970
CURTISS (ex-*SS Great Republic*)	T-AVB 4	Ingalls Shipbuilding, Pascagoula	1969

Comment: To reinforce the capabilities of the Maritime Prepositioning Ship programme, conversion of two ro-ro ships into maintenance aviation support ships was approved in FY85 and FY86. *Wright* was completed 14 May 1986, *Curtiss* 18 August 1987. Both conversions took place at Todd Shipyards, Galveston, Texas. Each ship has side ports and three decks aft of the bridge superstructure and has the capability to load the vans and equipment of a Marine Aviation Intermediate Maintenance Activity. The ships' mission is to service aircraft from an afloat platform. They can then revert to a standard sealift role if required. Maritime Administration hull design is C5-S-78a. These activated RRF ships are under the operational control of the MSC but serve as part of the Army prepositioning and NDAF.

WRIGHT *1/2012, M Declerck* / 1454861

© 2014 IHS

8 ALGOL CLASS (FAST SEALIFT SHIPS) (AKRH)

Name	No	Builders	Commissioned
ALGOL	T-AKR 287	Rotterdamsche DD Mij NV, Rotterdam	7 May 1973
BELLATRIX	T-AKR 288	Rheinstahl Nordseewerke, Emden	6 Apr 1973
DENEBOLA	T-AKR 289	Rotterdamsche DD Mij NV, Rotterdam	4 Dec 1973
POLLUX	T-AKR 290	A G Weser, Bremen	20 Sep 1973
ALTAIR	T-AKR 291	Rheinstahl Nordseewerke, Emden	17 Sep 1973
REGULUS	T-AKR 292	A G Weser, Bremen	30 Mar 1973
CAPELLA	T-AKR 293	Rotterdamsche DD Mij NV, Rotterdam	4 Oct 1972
ANTARES	T-AKR 294	A G Weser, Bremen	27 Sep 1972

Displacement, tonnes: 56,243 full load
Measurement, tonnes: 25,796 net
Dimensions, metres (feet): 288.4 × 32.2 × 11.2 *(946.2 × 105.6 × 36.7)*
Speed, knots: 33
Range, n miles: 12,200 at 27 kt
Complement: 43 (est.)
Machinery: 2 Foster-Wheeler boilers; 875 psi *(61.6 kg/cm^2)*; 950°F *(510°C)*; 2 GE MST-19 steam turbines; 120,000 hp *(89.5 MW)*; 2 shafts

Helicopters: Platform only.

Comment: All were originally built as container ships for Sea-Land Services, Port Elizabeth, NJ, but used too much fuel to be cost-effective as merchant ships. Six ships of this class were approved for acquisition in FY81 and the remaining two in FY82. The purchase price included 4,000 containers and 800 container chassis for use in container ship configuration. All eight were converted to Fast Sealift Ships, which are vehicle cargo ships. Conversion included the addition of roll-on/roll-off features. The area between the forward and after superstructures allows for a helicopter flight deck. Capacities are as follows: (sq ft) 150,016 to 166,843 ro-ro; 43,407 lift-on/lift-off; and either 44 or 46 20 ft containers. In addition to one ro-ro ramp port and starboard, twin 35 ton pedestal cranes are installed between the deckhouses and twin 50 ton cranes are installed aft. Ninety-three per cent of a US Army mechanised division can be lifted using all eight ships. Seven of the class moved nearly 11 per cent of all the cargo transported between the US and Saudi Arabia during and after the Gulf War. Six were activated for the Somalian operation in December 1992 and all have been used in various operations and exercises since then. All based in Atlantic and Gulf of Mexico ports. All transferred to US Maritime Administration (MARAD) on 1 October 2008 and maintained in the Ready Reserve Force.

DENEBOLA
2/2005, Robert Pabst / 1154010

COAST GUARD

Headquarters Appointments

Commandant:
Admiral Paul F Zukunft
Vice Commandant:
Vice Admiral John P Currier
Deputy Commandant for Mission Support:
Vice Admiral Manson K Brown
Deputy Commandant for Operations:
Vice Admiral Peter V Neffenger
Commander, Atlantic Area:
Vice Admiral Robert C Parker
Commander, Pacific Area:
Vice Admiral Charles W Ray

Establishment

The United States Coast Guard was established by an Act of Congress approved 28 January 1915, which consolidated the Revenue Cutter Service (founded in 1790) and the Life Saving Service (founded in 1848). The act of establishment stated the Coast Guard 'shall be a military service and a branch of the armed forces of the USA at all times. The Coast Guard shall be a service in the Treasury Department except when operating as a service in the Navy'.

Congress further legislated that in time of national emergency or when the President so directs, the Coast Guard operates as a part of the Navy. The Coast Guard did operate as a part of the Navy during the First and Second World Wars.

The Lighthouse Service (founded in 1789) was transferred to the Coast Guard on 1 July 1939 and the Bureau of Navigation and Steamboat Inspection on 28 February 1942.

The Coast Guard was transferred from the Department of Transportation to the Department of Homeland Security on 1 March 2003.

Missions

The Coast Guard has five strategic aims:
Safety: Prevent deaths, injuries, and property damage associated with maritime transportation, fishing and recreational boating
National Defense: Defend the nation as one of the five US Armed Services. Enhance regional stability in support of the National Security Strategy specifically maritime homeland security
Maritime Security: Protect maritime borders from all intrusions by (a) halting the flow of illegal drugs, aliens, and contraband into the United States through maritime routes; (b) preventing illegal fishing; and (c) suppressing violations of federal law in the maritime arena
Mobility: Facilitate maritime commerce and eliminate interruptions and impediments to the economical movement of goods and people, while maximizing recreational access and enjoyment of the water
Protection of Natural Resources: Prevent environmental damage and natural resource degradation associated with maritime transportation, fishing, and recreational boating

Organisation

Headquarters: Washington DC
Atlantic area: Portsmouth, VA
1st District: Boston, MA
5th District: Portsmouth, VA
7th District: Miami, FL
8th District: New Orleans, LA
9th District: Cleveland, OH
Pacific area: Alameda, CA

11th District: Alameda, CA
13th District: Seattle, WA
14th District: Honolulu, HI
17th District: Juneau, AK
Each district is further sub-divided into sectors.

Personnel

2014: 6,692 officers, 1,628 warrant officers, 31,460 enlisted, 7,980 reserves

US Coast Guard Acquisition Directorate

The USCG's Aquisition Directorate is investing approximately USD27 billion 2010–35 to upgrade and replace aging aircraft, ships and missions systems — including weapons, sensors and C⁴ISR electronics. A major component of the directorate's portfolio is the former Integrated Deepwater Systems (Deepwater) programme, which assigned a Lead Systems Integrator (LSI) role in 2002 to Integrated Coast Guard Systems, a commercial partnership between Lockheed Martin and Northrop Grumman. In 2007, the USCG re-asserted its role as the LSI and brought the Deepwater projects under the management of the newly-formed Acquisition Directorate. The Directorate now manages 20 major acquisition projects, most of which are from the legacy Deepwater programme.

Cutter Strength

All Coast Guard vessels over 65 ft in length that have adequate crew accommodation are referred to as 'cutters'. All names are preceded by USCG. The first two digits of the hull number for all Coast Guard vessels under 100 ft in length indicate the approximate length overall.

Approximately 2,000 standard and non-standard boats are in service ranging in size from 11 ft skiffs to 55 ft aids-to-navigation craft. Small craft listed below are part of the Acquisition Directorate's modernisation programme.

Category/Classification		Active	Building (Projected)
Cutters			
WHEC	High Endurance Cutters	8	—
WMEC	Medium Endurance Cutters	28	—
WMSL	National Security Cutters	3	4 (1)
Patrol Forces			
WPC	Patrol Coastal	7	17 (10)
WPB	Patrol Craft	114	—
Icebreakers			
WAGB	Icebreakers	2	—
WLBB	Icebreaker	1	—
WTGB	Icebreaking Tugs	9	—
Training Cutters			
WIX	Training Cutters	1	—
Buoy Tenders			
WLB	Buoy Tenders, Seagoing	16	—
WLM	Buoy Tenders, Coastal	14	—
WLI	Buoy Tenders, Inland	4	—
WLR	Buoy Tenders, River	18	—
Construction Tenders			
WLIC	Construction Tenders, Inland	13	—
Harbour Tugs			
WYTL	Harbour Tugs, Small	11	—
Response Craft			
—	Medium	139	31
—	Small	58	392
—	LRI	1	7
—	Defender	364	—
—	Special Purpose Craft	58	392
—	Lifeboats	126	—
—	CB-OTH	14	9

DELETIONS

High Endurance Cutters

2011 Hamilton (to Philippines), Chase (to Nigeria), Acushnet
2012 Dallas (to Philippines), Jarvis

Patrol Forces

2011 Zephyr, Shamal, Tornado (all returned to US Navy)

Icebreakers

2011 Polar Sea

HIGH ENDURANCE CUTTERS

3 + 4 (1) LEGEND CLASS (NATIONAL SECURITY CUTTERS) (PSOH/WMSL)

Name	No	Builders	Laid down	Launched	Commissioned	Homeport
BERTHOLF	WMSL 750	Northrop Grumman Ingalls Shipbuilding	29 Mar 2005	29 Sep 2006	4 Aug 2008	Alameda, CA
WAESCHE	WMSL 751	Northrop Grumman Ingalls Shipbuilding	15 Sep 2006	12 Jul 2008	7 May 2010	Alameda, CA
STRATTON	WMSL 752	Huntington Ingalls	20 Jul 2009	20 Jul 2010	31 Mar 2012	Alameda, CA
HAMILTON	WMSL 753	Huntington Ingalls	5 Sep 2012	10 Aug 2013	2014	Charleston, SC
JAMES	WMSL 754	Huntington Ingalls	17 May 2013	2014	2015	Charleston, SC
MUNRO	WMSL 755	Huntington Ingalls	2014	2015	2017	Alameda, CA
KIMBALL	WMSL 756	Huntington Ingalls	2015	2016	2018	—

Displacement, tonnes: 3,257 standard; 4,178 full load
Dimensions, metres (feet): 127.4 × 16.5 × 6.4 *(418 × 54.1 × 21.0)*
Speed, knots: 28. **Range, n miles:** 12,000 at 9 kt
Complement: 108 (14 officers)

Machinery: CODAG; 1 GE LM2500 gas turbine; 29,500 hp *(22.0 MW)*; 2 MTU20V 1163 diesels; 19,310 hp *(14.4 MW)*; bow thruster; 2 shafts; cp props
Guns: 1 Bofors 57 mm/70 Mk 110; 220 rds/min to 17 km *(9.3 n miles)*; weight of shell 2.4 kg. 1 General Dynamics 20 mm Phalanx Mk 15. 4 — 12.7 mm MGs.
Physical countermeasures: Decoys: Mk 53 Mod 6 Decoy System with Nulka and SRBOC.
Electronic countermeasures: ESM/ECM: SLQ 32.
Radars: Surface search: TRS 3D/16; E/F-band.
Navigation: Hughes-Furuno SPS 73; I-band.
Fire control: SPQ-9B; I/J-band.
Tacan: AN/URN 25.
Electro-optic systems: Kollmorgen Mk 46 optronic sight.

Helicopters: 1 H-65 and two VUAV or 2 H-65.

Programmes: Contracts awarded to Northrop Grumman Ship Systems on 2 April 2003 for the design and long lead material procurement of the first of a class of National Security Cutters to replace High Endurance Cutters. Lockheed Martin providing command, control, communications and intelligence integration and hardware.

Contract for production and delivery of *Bertholf* on 28 June 2004, for *Waesche* on 18 January 2005, for *Stratton* on 8 August 2007 and for *Hamilton* on 30 November 2010. The contract for *James* was let on 12 September 2011 and for *Munro* on 1 May 2013. The programme of record is for a class of eight ships but current acquisition funding allows for only seven.

Structure: Can carry up to 11 m interceptor craft; stern ramps for rapid launch and recovery. Two helicopter hangars. The hull of the third of class was redesigned to reflect concerns about structural fatigue in the first two vessels. Modifications are to be made to the first two ships early in their lives.
Operational: Designed to deploy up to 230 days per year. *Bertholf*, *Waesche*, and *Stratton* are fully operational.

BERTHOLF 10/2011, Michael Nitz / 1454864

WAESCHE 6/2011, M Mazumdar / 1454862

8 HAMILTON CLASS (PSOH/WHEC)

Name	No	Builders	Laid down	Launched	Commissioned	Homeport	F/S
MELLON	WHEC 717	Avondale Shipyards, New Orleans	25 Jul 1966	11 Feb 1967	22 Dec 1967	Seattle, WA	PA
BOUTWELL	WHEC 719	Avondale Shipyards, New Orleans	5 Dec 1966	17 Jun 1967	14 Jun 1968	San Diego, CA	PA
SHERMAN	WHEC 720	Avondale Shipyards, New Orleans	23 Jan 1967	23 Sep 1967	23 Aug 1968	San Diego, CA	PA
GALLATIN	WHEC 721	Avondale Shipyards, New Orleans	27 Feb 1967	18 Nov 1967	20 Dec 1968	Charleston, CA	AA
MORGENTHAU	WHEC 722	Avondale Shipyards, New Orleans	17 Jul 1967	10 Feb 1968	14 Feb 1969	Honolulu, HI	PA
RUSH	WHEC 723	Avondale Shipyards, New Orleans	23 Oct 1967	16 Nov 1968	3 Jul 1969	Honolulu, HI	PA
MUNRO	WHEC 724	Avondale Shipyards, New Orleans	18 Feb 1970	5 Dec 1970	10 Sep 1971	Kodiak, AK	PA
MIDGETT	WHEC 726	Avondale Shipyards, New Orleans	5 Apr 1971	4 Sep 1971	17 Mar 1972	Seattle, WA	PA

Displacement, tonnes: 3,353 full load
Dimensions, metres (feet): 115.2 × 13.1 × 6.1 *(378 × 43.0 × 20.0)*
Flight deck, metres: 26.8 × 12.2 *(87.9 × 40.0)*
Speed, knots: 29. **Range, n miles:** 9,600 at 15 kt
Complement: 162 (19 officers)

Machinery: CODOG; 2 Pratt & Whitney FT4A-6 gas turbines; 36,000 hp *(26.86 MW)*; 2 Fairbanks-Morse 38TD8-1/8-12 diesels; 7,000 hp *(5.22 MW)* sustained; 2 shafts; cp props; retractable bow propulsor; 350 hp *(261 kW)*
Guns: 1 Oto Melara 3 in *(76 mm)*/62 Mk 75 Compact; 85 rds/min to 16 km *(8.7 n miles)* anti-surface; 12 km *(6.6 n miles)* anti-aircraft; weight of shell 6 kg. 2 Boeing 25 mm/87 Mk 38 Bushmaster. 1 Raytheon 20 mm Vulcan Phalanx 6-barrelled Mk 15; 3,000 rds/min combined to 1.5 km. 4 — 12.7 mm MGs.
Physical countermeasures: Decoys: 2 Loral Hycor SRBOC 6-barrelled fixed Mk 36; IR flares and chaff.
Electronic countermeasures: ESM: WLR-1C, WLR-3; intercept.
Radars: Air search: Lockheed SPS-40B; B-band.
Surface search: Hughes/Furuno SPS-73; E/F- and I-bands.
Fire control: Sperry Mk 92; I/J-band.
Tacan: URN 25.
Combat data systems: SCCS 378 includes OTCIXS satellite link.
Weapon control systems: Mk 92 Mod 1 GFCS.

Helicopters: 1 H-65 or 1 H-60.

Programmes: Twelve built of a total of 36 originally planned.
Modernisation: FRAM programme for all original 12 ships in this class from October 1985 to October 1992. Work included standardising the engineering plants, improving the clutching systems, replacing SPS-29 air search radar with SPS-40 radar and replacing the Mk 56 fire-control system and 5 in/38 gun mount with the Mk 92 system and a single 76 mm Oto Melara Compact gun. In addition Harpoon and Phalanx CIWS fitted to five of the class by 1992 and CIWS to all by late 1993. The flight deck and other aircraft facilities upgraded to handle a Jay Hawk helicopter including a telescopic hangar. URN 25 TACAN added along with the SQR-4 and SQR-17 sonobuoy receiving set and passive acoustic analysis systems. SRBOC chaff launchers were also fitted but not improved ESM which has been shelved. All missiles, torpedo tubes, sonar and ASW equipment removed in 1993–94. 25 mm Mk 38 guns replaced the 20 mm Mk 67. Shipboard Command and Control System (SCCS) fitted to all of the class by 1996. Surface search radar replaced 1997–99. First phase of C4ISR upgrades, including access to SIPRNET and classified networks, completed in 2004.

SHERMAN 9/2012, M Mazumdar / 1486460

Structure: These ships have clipper bows, twin funnels enclosing a helicopter hangar, helicopter platform aft. All are fitted with elaborate communications equipment. Superstructure is largely of aluminium construction. Bridge control of manoeuvring is by aircraft-type joystick rather than wheel.
Operational: Seven of the class are based in the Pacific and one on the East Coast. The removal of SSMs and all ASW equipment refocuses on Coast Guard roles. Decommissioning of the class began on 13 May 2011 with *Hamilton* and *Chase* which transferred to the Philippines and Nigeria respectively. *Dallas* was transferred to the Philippines in 2012 and *Jarvis* to Bangladesh in 2013. *Gallatin* is to be decommissioned in 2014 and transferred to Nigeria.

MEDIUM ENDURANCE CUTTERS

Notes: Medium endurance cutters of the Famous Cutter and Reliance classes are to be replaced by a new class of ships under the Offshore Patrol Cutter (OPC) programme. A Request for Proposals (RfP) was issued on 25 September 2012 and as result of which three shipyards, Bollinger Shipyards (teamed with Damen) of Lockport, LA, Eastern Shipyards (teamed with STX) of Panama City, FL, and General Dynamics Bath Iron Works (teamed with Navantia) of Bath, ME, were selected for award of Preliminary and Contract Design contracts in March 2014. One of these companies is to be down-selected in early 2016 for Detailed Design and Construction. The first ship is to be procured in FY17 and a class of 25 is planned.

13 FAMOUS CUTTER CLASS (PSOH/WMEC)

Name	No	Builders	Laid down	Launched	Commissioned	Homeport	F/S
BEAR	WMEC 901	Tacoma Boatbuilding Co, Tacoma	23 Aug 1979	25 Sep 1980	4 Feb 1983	Portsmouth, VA	AA
TAMPA	WMEC 902	Tacoma Boatbuilding Co, Tacoma	3 Apr 1980	19 Mar 1981	16 Mar 1984	Portsmouth, VA	AA
HARRIET LANE	WMEC 903	Tacoma Boatbuilding Co, Tacoma	15 Oct 1980	6 Feb 1982	20 Sep 1984	Portsmouth, VA	AA
NORTHLAND	WMEC 904	Tacoma Boatbuilding Co, Tacoma	9 Apr 1981	7 May 1982	17 Dec 1984	Portsmouth, VA	AA
SPENCER	WMEC 905	Robert E Derecktor Corp	26 Jun 1982	17 Apr 1984	28 Jun 1986	Boston, MA	AA
SENECA	WMEC 906	Robert E Derecktor Corp	16 Sep 1982	17 Apr 1984	4 May 1987	Boston, MA	AA
ESCANABA	WMEC 907	Robert E Derecktor Corp	1 Apr 1983	6 Feb 1985	27 Aug 1987	Boston, MA	AA
TAHOMA	WMEC 908	Robert E Derecktor Corp	28 Jun 1983	6 Feb 1985	6 Apr 1988	Kittery, ME	AA
CAMPBELL	WMEC 909	Robert E Derecktor Corp	10 Aug 1984	29 Apr 1986	19 Aug 1988	Kittery, ME	AA
THETIS	WMEC 910	Robert E Derecktor Corp	24 Aug 1984	29 Apr 1986	30 Jun 1989	Key West, FL	AA
FORWARD	WMEC 911	Robert E Derecktor Corp	11 Jul 1986	22 Aug 1987	4 Aug 1990	Portsmouth, VA	AA
LEGARE	WMEC 912	Robert E Derecktor Corp	11 Jul 1986	22 Aug 1987	4 Aug 1990	Portsmouth, VA	AA
MOHAWK	WMEC 913	Robert E Derecktor Corp	15 Mar 1987	5 May 1988	20 Mar 1991	Key West, FL	AA

Displacement, tonnes: 1,820 full load
Dimensions, metres (feet): 82.3 × 11.6 × 3.96 *(270 × 38.1 × 13.0)*
Speed, knots: 19.5
Range, n miles: 9,900 at 12 kt
Complement: 100 (13 officers; 5 air crew)

Machinery: 2 Alco 18V-251 diesels; 7,290 hp *(5.44 MW)* sustained; 2 shafts; cp props
Guns: 1 Oto Melara 3 in *(76 mm)*/62 Mk 75; 85 rds/min to 16 km *(8.7 n miles)* anti-surface; 12 km *(6.6 n miles)* anti-aircraft; weight of shell 6 kg. 2 — 12.7 mm MGs.
Physical countermeasures: Decoys: 2 Loral Hycor SRBOC 6-barrelled fixed Mk 36; IR flares and chaff.
Electronic countermeasures: ESM/ECM: SLQ-32(V)2; radar intercept.
Radars: Surface search: Hughes/Furuno SPS-73; E/F/I-bands.
Fire control: Sperry Mk 92 Mod 1; I/J-band.
Tacan: URN 25.
Combat data systems: SCCS-270; OTCIXS satellite link.

Helicopters: 1 HH-65 or HH-60.

Programmes: The contract for construction of WMEC 905–913 was originally awarded to Tacoma Boatbuilding Co on 29 August 1980. However, under lawsuit from the Robert E Derecktor Corp, Middletown, Rhode Island, the contract to Tacoma was determined to be invalid and was awarded to Robert E Derecktor Corp on 15 January 1981.
Modernisation: OTCIXS satellite link fitted from 1992. C4ISR upgrades completed in 2004. The work includes engineering and habitability measures. The Mission Effectiveness Project to extend service lives has resulted in the refits of 11 ships. *Mohawk* and *Forward* are to be similarly refitted.
Structure: They are the only medium endurance cutters with a helicopter hangar (which is telescopic) and the first cutters with automated command and control centre. Fin stabilisers fitted. Plans to fit SSM and/or CIWS have been abandoned as has towed array sonar and sonobuoy datalinks. New radars fitted 1997–99.
Operational: Very lively in heavy seas because the length to beam ratio is unusually small for ships required to operate in Atlantic conditions.

NORTHLAND *1/2012, M Declerck* / 1454863

14 RELIANCE CLASS (PSOH/WMEC)

Name	No	Commissioned	Homeport	F/S
RELIANCE	WMEC 615	20 Jun 1964	Kittery, ME	AA
DILIGENCE	WMEC 616	26 Aug 1964	Wilmington, NC	AA
VIGILANT	WMEC 617	3 Oct 1964	Patrick AFB, FL	AA
ACTIVE	WMEC 618	17 Sep 1966	Port Angeles, WA	PA
CONFIDENCE	WMEC 619	19 Feb 1966	Port Canaveral, FL	AA
RESOLUTE	WMEC 620	8 Dec 1966	St Petersburg, FL	AA
VALIANT	WMEC 621	3 Nov 1967	Miami, FL	AA
STEADFAST	WMEC 623	25 Sep 1968	Warrenton, OR	AA
DAUNTLESS	WMEC 624	10 Jun 1967	Galveston, TX	AA
VENTUROUS	WMEC 625	16 Sep 1968	St Petersburg, FL	AA
DEPENDABLE	WMEC 626	22 Nov 1968	Cape May, NJ	AA
VIGOROUS	WMEC 627	2 May 1969	Cape May, NJ	AA
DECISIVE	WMEC 629	23 Aug 1968	Pascagoula, MS	AA
ALERT	WMEC 630	4 Aug 1969	Warrenton, OR	PA

Displacement, tonnes: 1,110 (WMEC 618–619), 1,020 (WMEC 620–621, 623–627, 629–630) full load
Dimensions, metres (feet): 64.2 × 10.4 × 3.2 *(210.6 × 34.1 × 10.5)*
Speed, knots: 18
Range, n miles: 6,100 at 12 kt
Complement: 75 (12 officers)

Machinery: 2 Alco 16V-251 diesels; 5,000 hp *(3.72 MW)* sustained; 2 shafts; LIPS cp props
Guns: 1 McDonnell Douglas 25 mm/87 Mk 38 Bushmaster; 175 rds/min to 6.8 km *(3.4 n miles)*. 2 — 12.7 mm MGs.
Radars: Surface search: Hughes/Furuno SPS-73; E/F/I-bands.
Combat data systems: SCCS-210.

Helicopters: Platform for 1 H-65.

Modernisation: All 14 cutters underwent a Major Maintenance Availability (MMA) from 1987–94. The exhausts for main engines, ship service generators and boilers were run in a new vertical funnel which reduces flight deck size. 76 mm guns were replaced by 25 mm Mk 38. A Mission Effectiveness Project, to extend service lives, was initiated in 2005. Work, which included engineering and habitability measures, was completed in all ships by late 2010.
Structure: Designed for search and rescue duties. Design features include 360° visibility from bridge; helicopter flight deck (no hangar); and engine exhaust vent at stern which has been replaced by a funnel during MMA. Capable of towing ships up to 10,000 tons. Air conditioned throughout except engine room; high degree of habitability.
Operational: Normally operate within 500 miles of the coast. Primary roles are SAR, law enforcement homeland security and defence operations.
Sales: *Courageous* sold to Sri Lanka in 2004 and *Durable* to Colombia in 2003.

ALERT *10/2011, Michael Nitz* / 1454865

1 EDENTON CLASS (PSOH/WMEC)

Name	No	Builders	Commissioned	Homeport	F/S
ALEX HALEY (ex-*Edenton*)	WMEC 39 (ex-ATS 1)	Brooke Marine, Lowestoft	23 Jan 1971	Kodiak, AK	PA

Displacement, tonnes: 3,088 full load
Dimensions, metres (feet): 85.99 × 15.24 × 4.57 *(282.1 × 50.0 × 15.0)*
Speed, knots: 16
Range, n miles: 10,000 at 13 kt
Complement: 99 (10 officers)

Machinery: 4 Caterpillar 3516 DITAWJ diesels; 6,000 hp(m) *(4.41 MW)*; 2 shafts; cp props; bow thruster
Guns: 2 McDonnell Douglas 25 mm/87 Mk 38; 200 rds/min to 6.8 km *(3.4 n miles)*. 2 — 12.7 mm MGs.
Radars: Air/surface search: Lockheed SPS-40B; C-band.
Surface search: Hughes/Furuno SPS-73; E/F/I-bands.
Combat data systems: SCCS-282.
Helicopters: Platform for 1 H-65 or 1 H-60.

Comment: Former Navy salvage ship paid off in 1996 and taken on by the Coast Guard in November 1997 for conversion. All diving and salvage gear removed, flight deck and telescopic hangar installed, and upgraded navigation and communications. Armed with 25 mm guns. Used in the Bering Sea, Gulf of Alaska and North Pacific as a multimission cutter from 16 December 1999.

ALEX HALEY   5/2012, John Mortimer / 1486462

SHIPBORNE AIRCRAFT

Numbers/Type: 35/64 EADS MH-65C Dolphin/MH-65D Dolphin.
Operational speed: 175 kt *(324 km/h)*.
Service ceiling: 10,000 ft *(3,048 m)*.
Range: 290 n miles *(537 km)*.
Role/Weapon systems: Short-Range Recovery (SRR) helicopter and primary aircraft for USCG shipboard operations. There are 88 in operational service, 10 in depot maintenance, and one being upgraded to MH-65E configuration. Seven MH-65Ds conduct Washington DC National Capital Region Air Defence missions. All H-65 aircraft are equipped with Turbomeca Arriel 2C2-CG engines. MH-65s have airborne use-of-force upgrade, interoperable with DoD, homeland security, and local response agencies. Also configured to allow installation and removal of special AUF mission equipment. A total of 99 MH-65Cs were delivered. Upgrades to MH-65D standard continue. This includes upgraded GPS/inertial navigation and flight management systems. The MH-65E, featuring an all-glass cockpit and advanced navigation capabilities, is to enter service from fiscal year 2017. Sensors: Bendix RDR 1300 radar, FSI Talon electro-optical/infrared, Collins mission management system, fully night vision goggle compatible. Weapons: 1—M240B 7.62 mm and M107-derived .50 calibre precision fire weapon.

MH-65C   10/2012, M Mazumdar / 1486461

Numbers/Type: 42 Sikorsky MH-60T Jayhawk.
Operational speed: 180 kt *(333 km/h)*.
Service ceiling: 13,000 ft *(3,961 m)*.
Range: 700 n miles *(1,296 km)*.
Role/Weapon systems: Medium-range Recovery (MRR) helicopter. Coast Guard version of Seahawk, first flew in 1988. There are 35 in operational service; seven in depot maintenance. Airborne use-of-force modifications are complete on all aircraft. Avionics upgrade provided a Common Avionics Architecture System (CAAS) and re-designated aircraft as the MH-60T. All upgrades to be completed by 2014. Other upgrades included modern electro-optical/infrared sensor and surface search radar. Sensors: Primus 700A weather radar, FSI Talon-Electro-optical/Infrared. Weapons: 1—M240 7.62 mm MG.

MH-60T   5/2006, Takatoshi Okano / 1167495

LAND-BASED MARITIME AIRCRAFT

Notes: USCG has no Unmanned Aircraft Systems (UAS) in its aviation fleet. However, it is conducting studies to select suitable UAS candidates to augment land-based and cutter-based aircraft. USCG's Office of Research, Development, Test and Evaulation is to conduct a third and final planned demonstration of small UAS at NASA's Wallops Flight Facility in 2014. The Phase 2B demonstration is to focus on evaluation of the impact of UAS, deploying various payloads of technologically advanced sensor systems and target detection software, on National Security Cutter's (NSC) mission effectiveness. USCG has tested used of US Navy's Fire Scout and held a technical demonstration of Scan Eagle UAS on board an NSC in August 2011.

Numbers/Type: 15 EADS/CASA HC-144A (CN-235–300) 200 Ocean Sentry.
Operational speed: 236 kt *(437 km/h)*.
Service ceiling: 25,000 ft *(7,620 m)*.
Range: 1,565 n miles *(2,519 km)*.
Role/Weapon systems: Aircraft delivery to the Coast Guard began in 2007. There have been 15 aircraft acquired and options for a further three aircraft were exercised in 2013 with acceptance to occur in 2014 and 2015. Roles include SAR, law enforcement, ice patrol, and environmental protection. Aircraft is manufactured by Airbus Military while Lockheed Martin completed integration and development testing of aircraft and C4ISR mission pallet. Aircraft operate from three air stations: Coast Guard Aviation Training Center in Mobile, AL, Air Station Miami, FL, and Air Station Cape Cod, MA.

CN-235   11/2008, EADS/CASA / 1353642

Numbers/Type: 2/3/6 Dassault HU-25 D Guardian Falcon.
Operational speed: 420 kt *(774 km/h)*.
Service ceiling: 42,000 ft *(12,800 m)*.
Range: 1,500 n miles *(2,777 km)*.
Role/Weapon systems: Medium-range maritime surveillance role. The final three HU-25 D are to be withdrawn from service in 2014. Sensors: APS-127 weather/search radar (A variant). APG-66V2 air search radar (C variant); APS-143B surface search radar (D variant); Wescam 16DSI EO/IR sensor. Weapons: unarmed.

HU-25 GUARDIAN FALCON   4/2009, Jurg Kürsener / 1366241

Numbers/Type: 22/6 Lockheed HC-130H/HC-130J.
Operational speed: 325 kt *(602 km/h)*.
Service ceiling: 33,000 ft *(10,060 m)*.
Range: 4,100 n miles *(7,592 km)*; 5,500 n miles *(1,018 km)* (HC-130J).
Role/Weapon systems: Long-range maritime reconnaissance role. Sixteen Legacy HC-130Hs undergoing upgrade to deliver Deepwater requirement for Long Range Search (LRS) capability and provision of heavy air transport for Maritime Safety & Security Teams (MSSTs), Port Security Units (PSUs), and National Strike Force (NSF). When modernisation is complete, 11 of 22 HC-130H will have new avionics suites and new surface search radar. There are six HC-130J Super Hercules all of which had become operational by late 2008. A contract for three additional HC-130Js was awarded in October 2012 for delivery in 2015. Sensors (HC-130J): EDO EL/M 2022A(V)3 maritime surface search radar, mounted beneath the plane's fuselage, a nose-mounted APN-241 weather radar, electro-optical/infrared-FLIR Systems Star Safire III, DF-430 UHF/VHF Direction Finder System, and SAAB Transponder Tech AB R4A Airborne Automatic Identification System (AIS). Sensors (HC-130H): Selex Sea Spray 7500E surface search radar or APS-125 weather/search radar; DF-430 UHF/VHF direction finder; Wescam MX-20 EO/IR. Weapons: unarmed.

C-130J   4/2009, Hachiro Nakai / 1366289

Land-based maritime aircraft — Patrol forces < **US (COAST GUARD)** 987

Numbers/Type: 2 Gulfstream VC-37A.
Operational speed: 459 kt *(850 km/h)*.
Service ceiling: 51,000 ft *(15,540 m)*.
Range: 5,600 n miles *(10,370 km)*.
Role/Weapon: Military version of Gulfstream V, the first of which replaced a C-20B Gulfstream III in May 2002 and the second replaced a Bombardier Challenger 604 in 2013. Based at Air Station Washington DC. Serves as a long-range command and control aircraft for Department of Homeland Security and Coast Guard officials.

GULFSTREAM G 550   6/2003, Paul Jackson / 0568402

PIKE   5/2012, M Mazumdar / 1486466

PATROL FORCES

73 MARINE PROTECTOR CLASS (WPB)

Name	No	Commissioned	Homeport
BARRACUDA	87301	24 Feb 1998	Eureka, CA
HAMMERHEAD	87302	17 May 1998	Woods Hole, MA
MAKO	87303	28 Jun 1998	Cape May, NJ
MARLIN	87304	2 Dec 1998	Fort Meyers, FL
STINGRAY	87305	13 Jan 1999	Mobile, AL
DORADO	87306	24 Feb 1999	Crescent City, CA
OSPREY	87307	7 Apr 1999	Port Townsend, WA
CHINOOK	87308	19 May 1999	New London, CT
ALBACORE	87309	30 Jun 1999	Little Creek, VA
TARPON	87310	11 Aug 1999	Tybee Island, GA
COBIA	87311	8 Sep 1999	Mobile, AL
HAWKSBILL	87312	6 Oct 1999	Monterey, CA
CORMORANT	87313	3 Nov 1999	Fort Pierce, FL
FINBACK	87314	1 Dec 1999	Cape May, NJ
AMBERJACK	87315	29 Dec 1999	Port Isabel, TX
KITTIWAKE	87316	26 Jan 2000	Honolulu, HI
BLACKFIN	87317	23 Feb 2000	Santa Barbara, CA
BLUEFIN	87318	22 Mar 2000	Fort Pierce, FL
YELLOWFIN	87319	19 Apr 2000	Charleston, SC
MANTA	87320	17 May 2000	Freeport, TX
COHO	87321	14 Jun 2000	Pamana City, FL
KINGFISHER	87322	12 Jul 2000	Mayport, FL
SEAHAWK	87323	9 Aug 2000	Carrabelle, FL
STEELHEAD	87324	6 Sep 2000	Port Aransas, TX
BELUGA	87325	4 Oct 2000	Little Creek, VA
BLACKTIP	87326	1 Nov 2000	Oxnard, CA
PELICAN	87327	29 Nov 2000	Abbeville, LA
RIDLEY	87328	27 Dec 2000	Montauk, NY
COCHITO	87329	24 Jan 2001	Little Creek, VA
MANOWAR	87330	21 Feb 2001	Galveston, TX
MORAY	87331	21 Mar 2001	Jonesport, ME
RAZORBILL	87332	18 Apr 2001	Gulfport, MS
ADELIE	87333	16 May 2001	Port Angeles, WA
GANNET	87334	13 Jun 2001	Fort Lauderdale, FL
NARWHAL	87335	11 Jul 2001	Corona del Mar, CA
STURGEON	87336	8 Aug 2001	Grand Isle, LA
SOCKEYE	87337	5 Sep 2001	Bodega Bay, CA
iBIS	87338	3 Oct 2001	Cape May, NJ
POMPANO	87339	1 Nov 2001	Gulfport, MS
HALIBUT	87340	28 Nov 2001	Marina del Ray, CA
BONITO	87341	26 Dec 2001	Pensacola, FL
SHRIKE	87342	23 Jan 2002	Cape Canaveral, FL
TERN	87343	20 Feb 2002	San Francisco, CA
HERON	87344	20 Mar 2002	Sabine, TX
WAHOO	87345	17 Apr 2002	Port Angeles, WA
FLYINGFISH	87346	15 May 2002	Boston, MA
HADDOCK	87347	12 Jun 2002	San Diego, CA
BRANT	87348	10 Jul 2002	Corpus Christi, TX
SHEARWATER	87349	7 Aug 2002	Portsmouth, VA
PETREL	87350	4 Sep 2002	San Diego, CA
SEA LION	87352	19 Nov 2003	Bellingham, WA
SKIPJACK	87353	17 Dec 2003	Galveston, TX
DOLPHIN	87354	14 Jan 2004	Miami, FL
HAWK	87355	11 Feb 2004	St Petersburg, FL
SAILFISH	87356	10 Mar 2004	Sandy Hook, NJ
SAWFISH	87357	7 Apr 2004	Key West, FL
SWORDFISH	87358	9 Mar 2005	Port Angeles, WA
TIGER SHARK	87359	6 Apr 2005	Newport, RI
BLUE SHARK	87360	4 May 2005	Everett, WA
SEA HORSE	87361	1 Jun 2005	Portsmouth, VA
SEA OTTER	87362	29 Jun 2005	San Diego, CA
MANATEE	87363	27 Jul 2005	Ingleside, TX
AHI	87364	15 Feb 2006	Honolulu, HI
PIKE	87365	15 Jan 2006	San Francisco, CA
TERRAPIN	87366	1 Feb 2006	Bellingham, WA
SEA DRAGON	87367	14 Jan 2008	Kings Bay, GA
SEA DEVIL	87368	20 Jun 2008	Bangor, WA
CROCODILE	87369	5 Dec 2008	St Petersburg, FL
DIAMONDBACK	87370	17 Jan 2009	Miami, FL
REEF SHARK	87371	24 Mar 2009	San Juan, PR
ALLIGATOR	87372	9 Mar 2009	St Petersburg, FL
SEA DOG	87373	2 Jul 2009	Kings Bay, GA
SEA FOX	87374	18 Sep 2009	Bangor, WA

Displacement, tonnes: 92 full load
Dimensions, metres (feet): 26.5 × 5.8 × 1.6 *(86.9 × 19.0 × 5.2)*
Speed, knots: 25. **Range, n miles:** 900 at 8 kt
Complement: 10 (1 officer)
Machinery: 2 MTU 8V 396 TE94 diesels; 2,680 hp(m) *(1.97 MW)* sustained; 2 shafts
Guns: 2—12.7 mm MGs.
Radars: Navigation: I-band.

Comment: Designed by David M Cannell based on the hull of the Damen Stan Patrol 2600 which is in service with the Hong Kong police. Steel hull built by Bollinger with GRP superstructure by Halmatic. A stern ramp is used for launching a 5.5 m RIB. Following delivery of 65 vessels to the USCG and a further two to the Maltese Armed Forces, an order for a further eight was made by the USCG on 25 June 2007. Four (87367, 87368, 87373, 87374) of these latter craft were funded by the US Navy but are operated by the Coast Guard.

TERN   9/2013*, M Mazumdar / 1531175

8 + 16 (10) SENTINEL CLASS (PBO/WPC)

Name	No	Builders	Commissioned	Homeport
BERNARD C WEBBER	WPC 1101	Bollinger, Lockport	14 Apr 2012	Miami, FL
RICHARD ETHERIDGE	WPC 1102	Bollinger, Lockport	3 Aug 2012	Miami, FL
WILLIAM FLORES	WPC 1103	Bollinger, Lockport	3 Nov 2012	Miami, FL
ROBERT YERED	WPC 1104	Bollinger, Lockport	17 Feb 2013	Miami, FL
MARGARET NORVELL	WPC 1105	Bollinger, Lockport	1 Jun 2013	Miami, FL
PAUL CLARK	WPC 1106	Bollinger, Lockport	24 Aug 2013	Miami, FL
CHARLES DAVID	WPC 1107	Bollinger, Lockport	16 Nov 2013	Key West, FL
CHARLES SEXTON	WPC 1108	Bollinger, Lockport	8 Mar 2014	Key West, FL
KATHLEEN MOORE	WPC 1109	Bollinger, Lockport	2014	Key West, FL
RAYMOND EVANS	WPC 1110	Bollinger, Lockport	2014	Key West, FL
WILLIAM TRUMP	WPC 1111	Bollinger, Lockport	2014	Key West, FL
ISAAC MAYO	WPC 1112	Bollinger, Lockport	2015	Key West, FL
RICHARD DIXON	WPC 1113	Bollinger, Lockport	—	San Juan, PR
HERIBERTO HERNANDEZ	WPC 1114	Bollinger, Lockport	—	San Juan, PR

Displacement, tonnes: 359 full load
Dimensions, metres (feet): 46.7 × 7.7 × 2.6 *(153.2 × 25.3 × 8.5)*
Speed, knots: 28 (est.)
Complement: 22

Machinery: 2 MTU 20V 4000 M93L diesels; 11,520 hp *(8.6 MW)*; 2 shafts
Guns: 1 BAE Mk 38 Mod 2 25 mm. 4—12.7 mm MGs.
Radars: Surface search: Furuno XN24AF; I-band.
Navigation: Furuno XN10 A/3; I-band.

Comment: The Fast-Response Cutter was derived from the Coast Guard's Integrated Deepwater modernisation programme to replace the Island-class patrol boats. Its principal roles include fishery protection, port and coastal security, law enforcement, SAR and defence readiness. Following evaluation of bids, a contract for the first of class was awarded to Bollinger Shipyards of Lockport, LA, on 26 September 2008; contracts for further vessels were let in December 2009, September 2010 and September 2011. A follow-on construction order for a further six vessels was let on 21 September 2012 and an order for a further six vessels was made on 25 September 2013 bringing the total of vessels on contract to 24. The current programme is for 34 craft but, ultimately, a class of 58 is sought. The design is based on the Damen Stan Patrol 4708, modified to achieve 28+ kt; a stern-launch capability is incorporated; steel hull, aluminium superstructure.

BERNARD C WEBBER   2/2012, USCG / 1450984

© 2014 IHS   IHS Jane's Fighting Ships 2014-2015

988 US (COAST GUARD) > Patrol forces — Icebreakers

41 ISLAND CLASS (WPB)

Name	No	Commissioned	Homeport
FARALLON	WPB 1301	21 Feb 1986	Miami, FL
MAUI	WPB 1304	9 May 1986	Miami, FL
OCRACOKE	WPB 1307	4 Aug 1986	Miami Beach, FL
AQUIDNECK	WPB 1309	26 Sep 1986	Atlantic Beach, NC
MUSTANG	WPB 1310	3 Dec 1986	Seward, AK
NAUSHON	WPB 1311	5 Dec 1986	Ketchikan, AK
SANIBEL	WPB 1312	28 May 1987	Woods Hole, MA
EDISTO	WPB 1313	27 Mar 1987	San Diego, CA
SAPELO	WPB 1314	14 May 1987	San Juan, PR
MATINICUS	WPB 1315	19 Jun 1987	San Juan, PR
NANTUCKET	WPB 1316	10 Aug 1987	St Petersburg, FL
BARANOF	WPB 1318	25 May 1988	Miami, FL
CHANDELEUR	WPB 1319	8 Jun 1988	Ketchikan, AK
CHINCOTEAGUE	WPB 1320	8 Aug 1988	San Juan, PR
CUSHING	WPB 1321	8 Aug 1988	San Juan, PR
CUTTYHUNK	WPB 1322	5 Oct 1988	Port Angeles, WA
DRUMMOND	WPB 1323	19 Oct 1988	San Juan, PR
KEY LARGO	WPB 1324	24 Dec 1988	San Juan, PR
MONOMOY	WPB 1326	19 May 1989	Woods Hole, MA
ORCAS	WPB 1327	14 Apr 1989	Coos Bay, OR
SITKINAK	WPB 1329	31 May 1989	Miami, FL
TYBEE	WPB 1330	4 Aug 1989	Woods Hole, MA
WASHINGTON	WPB 1331	6 Oct 1989	Apra Harbor, Guam
WRANGELL	WPB 1332	15 Sep 1989	South Portland, ME
ADAK	WPB 1333	17 Nov 1989	Highlands, NJ
LIBERTY	WPB 1334	22 Sep 1989	Auke Bay, AK
ANACAPA	WPB 1335	13 Jan 1990	Petersburg, AK
KISKA	WPB 1336	21 Apr 1990	Hilo, HI
ASSATEAGUE	WPB 1337	15 Jun 1990	Apra Harbor, Guam
GRAND ISLE	WPB 1338	19 Apr 1991	Gloucester, MA
KEY BISCAYNE	WPB 1339	23 Apr 1991	Key West, FL
JEFFERSON ISLAND	WPB 1340	16 Aug 1991	South Portland, ME
KODIAK ISLAND	WPB 1341	21 Jun 1991	Key West, FL
LONG ISLAND	WPB 1342	27 Aug 1991	Valdez, AK
BAINBRIDGE ISLAND	WPB 1343	20 Sep 1991	Highlands, NJ
BLOCK ISLAND	WPB 1344	22 Nov 1991	Atlantic Beach, NC
STATEN ISLAND	WPB 1345	22 Nov 1991	Atlantic Beach, NC
ROANOKE ISLAND	WPB 1346	8 Feb 1992	Homer, AK
PEA ISLAND	WPB 1347	29 Feb 1992	Key West, FL
KNIGHT ISLAND	WPB 1348	22 Apr 1992	Key West, FL
GALVESTON ISLAND	WPB 1349	5 Jun 1992	Honolulu, HI

Displacement, tonnes: 171 (WPB 1301, 1304, 1307, 1309–1316), 156 (WPB 1318–1324, 1326–1327, 1329–1337), 136 (WPB 1338–1349) full load
Dimensions, metres (feet): 33.5 × 6.4 × 2.2 (109.9 × 21.0 × 7.2)
Speed, knots: 29. **Range, n miles:** 3,928 at 10 kt
Complement: 16 (2 officers)

Machinery: 2 Paxman Valenta 16RP 200M diesels (A and B series); 6,246 hp (4.62 MW), sustained; 2 Caterpillar 3516 DITA diesels (C series); 5,596 hp (4.17 MW) sustained; 2 shafts
Guns: 1 McDonnell Douglas 25 mm/87 Mk 38. 2 – 12.7 mm MGs.
Radars: Navigation: Hughes/Furuno SPS-73; I-band.
Combat data systems: SCCS-Lite.

Comment: All built by the Bollinger Machine Shop and Shipyard at Lockport, Louisiana. The design is based upon the 110 ft patrol craft built by Vosper Thornycroft, UK, in service in Venezuela, UAE and UK Customs, but modified to meet Coast Guard needs. Vosper Thornycroft supplied design support, stabilisers, propellers, and steering gear. Batches: A 1301–1316, B 1317–1337, C 1338–1349. Radars replaced by 1999. As part of the Deepwater programme, eight hulls were modified to include stretching of the hull to 123 ft by insertion of a 13 ft plug to enable installation of upgraded C4ISR systems, a stern launch and recovery system and various platform improvements. WPB 1303 was first to undergo conversion at Bollinger Shipyard in February 2004. She was followed by WPBs 1317, 1325 and 1328 in 2004 and 1302, 1305, 1306 and 1308 by 2007. However, following experience of significant deck cracking, hull deformation and shaft alignment problems, the conversion planned was terminated in December 2006 and all eight vessels were taken out of operational service. The remaining 110 ft cutters are to continue in service until replaced by the Fast Response Cutter. Six cutters operate from Bahrain.

EDISTO   10/2012, M Mazumdar / 1486464

ICEBREAKERS

Notes: Coast Guard budgets for fiscal year 2013 and FY 2014 includes USD9.6 million to continue polar icebreaker replacement project to preserve US surface ship presence in the Arctic.

1 HEALY CLASS (WAGBH)

Name	No	Builders	Commissioned	Homeport	F/S
HEALY	WAGB 20	Avondale Shipyards, New Orleans	29 Oct 1999	Seattle, WA	PA

Displacement, tonnes: 16,663 full load
Dimensions, metres (feet): 128 oa; 121.2 wl × 25 × 8.9 (419.9; 397.6 × 82.0 × 29.2)
Speed, knots: 17. **Range, n miles:** 16,000 at 12.5 kt
Complement: 75 (12 officers) + 45 scientists
Machinery: Diesel-electric; 4 Westinghouse/Sulzer 12ZA 40S diesels; 42,400 hp (31.16 MW); 4 Westinghouse alternators; 2 motors; 30,000 hp (22.38 MW); 2 shafts; bow thruster; 2,200 hp (1.64 MW)
Helicopters: 2 HH-65A or 1 HH-60J.

Comment: In response to the 1984 Interagency Polar Icebreaker Requirements Study and Congressional mandate, approval was given for the construction of a new icebreaker as a replacement for two Wind class which were then decommissioned in 1988. However, no action was taken to provide funds for the new ship until Congress included it in the Navy's FY91 ship construction budget and after further delays the ship was ordered 15 July 1993. Icebreaking capability of 4 ft at 3 kt. Reached North Pole in September 2005, the third by a US surface ship.

HEALY   1/2003, Bob Fildes / 0572759

1 ICEBREAKER (WLBB)

Name	No	Builders	Commissioned	Homeport
MACKINAW	WLBB 30	Manitowoc Marine, Wisconsin	10 Jun 2006	Cheboygan, MI

Displacement, tonnes: 3,556 full load
Dimensions, metres (feet): 73.1 × 17.7 × 4.8 (239.8 × 58.1 × 15.7)
Speed, knots: 15
Complement: 55 (8 officers)

Machinery: Diesel-electric; 3 diesel generators; 12,600 hp (9.4 MW); 2 podded propulsors; 6,700 hp (5 MW)
Guns: 2 – 7.62 mm MGs.
Radars: Surface search: Kongsberg Data Bridge 10.
Navigation: Kongsberg Integrated Bridge System.

Comment: Contract to build new icebreaker/buoy tender awarded 15 October 2001. Keel laid 10 February 2004. Launched in April 2005 and delivered in November 2005. Icebreaker replaced WAGB 83 and assumed the same name. In addition to breaking ice (up to 32 in thick at 3 kt ahead, 2 kt astern) for the primary shipping lanes on the Great Lakes, the new ship will service aids to navigation, as well as performing search and rescue, pollution control, homeland security, and law enforcement duties from its homeport of Cheboygan, Michigan. Principal feature is 'podded' or protected propellers that can rotate 360° for greater manoeuvrability. Other features include fully integrated bridge system, robust communications suite and 3,200 sq ft of buoy deck space. A crane of 60 ft can recover buoys weighing up to 20 tons.

MACKINAW   3/2006, Manitowoc Marine / 1154627

1 POLAR CLASS (WAGBH)

Name	No	Builders	Launched	Commissioned	Homeport	F/S
POLAR STAR	WAGB 10	Lockheed SB & Construction Co	17 Nov 1973	19 Jan 1976	Seattle, WA	PA

Displacement, tonnes: 13,402 full load
Dimensions, metres (feet): 121.6 × 25.6 × 9.8 (399 × 84.0 × 32.2)
Speed, knots: 20. **Range, n miles:** 28,275 at 13 kt
Complement: 146 (15 officers; 12 air crew) + 33 scientists

Machinery: CODOG; diesel-electric (AC/DC); 6 Alco 16V-251F/Westinghouse AC diesel generators; 21,000 hp (15.66 MW) sustained; 3 Westinghouse DC motors; 18,000 hp (13.42 MW) sustained; 3 Pratt & Whitney FT4A-12 gas turbines; 60,000 hp (44.76 MW) sustained; 3 Philadelphia 75 VMGS gears; 60,000 hp (44.76 MW) sustained; 3 shafts; cp props
Radars: Navigation: 2 Raytheon SPS-64; I-band.
Tacan: SRN 15.
Helicopters: 2 HH-65A or 1 HH-60J.

Comment: At a continuous speed of 3 kt, the ship can break ice 6 ft (1.8 m) thick, and by ramming can break 21 ft (6.4 m) pack. Conventional icebreaker hull form with 'White' cutaway bow configuration and well-rounded body sections to prevent being trapped in ice. The ice belt is 1.75 in (44.45 mm) thick supported by framing at 16 in (0.4 m) centres. Two 15 ton capacity cranes fitted aft; one 3 ton capacity crane fitted forward. Two over-the-side oceanographic winches, one over-the-stern trawl/core winch. Deck fixtures for scientific research vans, and research laboratories provided for arctic and oceanographic research. Between 1986–92, science facilities were upgraded including habitability, lab spaces and winch capabilities. Polar Star was placed in a 'special' status in 2006 and having received an extensive upgrade at Vigor Shipyards, was reactivated on 14 December 2012. She returned to operational service in 2013. Polar Sea was placed in inactive status in October 2011.

POLAR STAR   1/2014*, Chris Sattler / 1531176

IHS Jane's Fighting Ships 2014-2015   © 2014 IHS

Icebreakers — Coastal tenders < **US (COAST GUARD)** 989

9 BAY CLASS (TUGS—WTGB)

Name	No	Launched	Commissioned	Homeport
KATMAI BAY	WTGB 101	7 Nov 1977	8 Jan 1979	Sault Sainte Marie, MI
BRISTOL BAY	WTGB 102	13 Feb 1978	5 Apr 1979	Detroit, MI
MOBILE BAY	WTGB 103	13 Feb 1978	2 Sep 1979	Sturgeon Bay, WI
BISCAYNE BAY	WTGB 104	29 Aug 1978	8 Dec 1979	St Ignace, MI
NEAH BAY	WTGB 105	6 Aug 1979	18 Aug 1980	Cleveland, OH
MORRO BAY	WTGB 106	6 Aug 1979	25 Jan 1981	Cleveland, OH
PENOBSCOT BAY	WTGB 107	24 Jul 1983	4 Sep 1984	Bayonne, NJ
THUNDER BAY	WTGB 108	20 Jul 1984	29 Dec 1985	Rockland, ME
STURGEON BAY	WTGB 109	9 Jul 1986	20 Aug 1988	Bayonne, NJ

Displacement, tonnes: 673 full load
Dimensions, metres (feet): 42.7 × 11.4 × 3.8 (140.1 × 37.4 × 12.5)
Speed, knots: 14.7. **Range, n miles:** 4,000 at 12 kt
Complement: 17 (3 officers)
Machinery: Diesel-electric; 2 Fairbanks-Morse 38D8-1/8-10 diesel generators; 2.4 MW sustained; Westinghouse electric drive; 2,500 hp (1.87 MW); 1 shaft
Radars: Navigation: Raytheon SPS-73; I-band.

Comment: The size, manoeuvrability and other operational characteristics of these vessels are tailored for operations in harbours and other restricted waters and for fulfilling present and anticipated multimission requirements. All units are ice strengthened for operation on the Great Lakes, coastal waters and in rivers and can break 20 in of ice continuously and up to 8 ft by ramming. A self-contained portable bubbler van and system reduces hull friction. First six built at Tacoma Boatbuilding, Tacoma. WTGB 107–109 built in Tacoma by Bay City Marine, San Diego. *Bristol Bay* and *Mobile Bay* have had their bows reinforced to push the two aids-to-navigation barges on the Great Lakes. WTGB 106 was decommissioned in 1998 and re-activated on 4 February 2002.

MORRO BAY
4/2011*, Marc Piché
1531177

COASTAL TENDERS

14 KEEPER CLASS (BUOY TENDERS—WLM/ABU)

Name	No	Builders	Commissioned	Homeport
IDA LEWIS	WLM 551	Marinette Marine, Marinette	1 Nov 1996	Newport, RI
KATHERINE WALKER	WLM 552	Marinette Marine, Marinette	27 Jun 1997	Bayonne, NJ
ABIGAIL BURGESS	WLM 553	Marinette Marine, Marinette	19 Sep 1997	Rockland, ME
MARCUS HANNA	WLM 554	Marinette Marine, Marinette	26 Nov 1997	South Portland, ME
JAMES RANKIN	WLM 555	Marinette Marine, Marinette	26 Aug 1998	Baltimore, MD
JOSHUA APPLEBY	WLM 556	Marinette Marine, Marinette	20 Nov 1998	St Petersburg, FL
FRANK DREW	WLM 557	Marinette Marine, Marinette	17 Jun 1999	Portsmouth, VA
ANTHONY PETIT	WLM 558	Marinette Marine, Marinette	1 Jul 1999	Ketchikan, AK
BARBARA MABRITY	WLM 559	Marinette Marine, Marinette	29 Jul 1999	Mobile, AL
WILLIAM TATE	WLM 560	Marinette Marine, Marinette	16 Sep 1999	Philadelphia, PA
HARRY CLAIBORNE	WLM 561	Marinette Marine, Marinette	28 Oct 1999	Galveston, TX
MARIA BRAY	WLM 562	Marinette Marine, Marinette	6 Apr 2000	Mayport, FL
HENRY BLAKE	WLM 563	Marinette Marine, Marinette	18 May 2000	Everett, WA
GEORGE COBB	WLM 564	Marinette Marine, Marinette	22 Jun 2000	San Pedro, CA

Displacement, tonnes: 853 full load
Dimensions, metres (feet): 53.3 × 11 × 2.4 (174.9 × 36.1 × 7.9)
Speed, knots: 12. **Range, n miles:** 2,000 at 10 kt
Complement: 20 (2 officers)
Machinery: 2 Caterpillar 3508TA diesels; 1,920 hp (1.43 MW) sustained; 2 Ulstein Z-drives; bow thruster; 460 hp (343 kW)
Guns: 2—7.62 mm MGs.
Radars: Navigation: Raytheon SPS-64; I-band.

Comment: Contract awarded 22 June 1993 to Marinette Marine for first of class with an option for 13 more. Capable of breaking 9 in of ice at 3 kt or 18 in by ramming. Named after lighthouse keepers for the Lighthouse Service, one of the predecessors of the modern Coast Guard. The ship is a scaled down model of the Juniper class for coastal service. Main hoist to lift 10 tons, secondary 3.75 tons. Able to skim and recover surface oil pollution using a vessel of opportunity skimming system.

WILLIAM TATE
6/2012, M Declerck / 1461791

GEORGE COBB
10/2012, M Mazumdar / 1486467

© 2014 IHS
IHS Jane's Fighting Ships 2014-2015

SEAGOING TENDERS

16 JUNIPER CLASS (BUOY TENDERS—WLB/ABU)

Name	No	Builders	Commissioned	Homeport
JUNIPER	WLB 201	Marinette Marine, Marinette	12 Jan 1996	Newport, RI
WILLOW	WLB 202	Marinette Marine, Marinette	27 Nov 1996	Newport, RI
KUKUI	WLB 203	Marinette Marine, Marinette	9 Oct 1997	Honolulu, HI
ELM	WLB 204	Marinette Marine, Marinette	29 Jun 1998	Atlantic Beach, NC
WALNUT	WLB 205	Marinette Marine, Marinette	22 Feb 1999	Honolulu, HI
SPAR	WLB 206	Marinette Marine, Marinette	9 Mar 2001	Kodiak, AK
MAPLE	WLB 207	Marinette Marine, Marinette	21 Jun 2001	Sitka, AK
ASPEN	WLB 208	Marinette Marine, Marinette	28 Sep 2001	San Francisco, CA
SYCAMORE	WLB 209	Marinette Marine, Marinette	1 Mar 2002	Cordova, AK
CYPRESS	WLB 210	Marinette Marine, Marinette	24 Jun 2002	Pensacola, FL
OAK	WLB 211	Marinette Marine, Marinette	17 Oct 2002	Charleston, SC
HICKORY	WLB 212	Marinette Marine, Marinette	6 Mar 2003	Homer, AK
FIR	WLB 213	Marinette Marine, Marinette	27 Jun 2003	Astoria, OR
HOLLYHOCK	WLB 214	Marinette Marine, Marinette	15 Oct 2003	Port Huron, MI
SEQUOIA	WLB 215	Marinette Marine, Marinette	21 Apr 2004	Apra Harbour, Guam
ALDER	WLB 216	Marinette Marine, Marinette	2 Sep 2004	Duluth, MN

Displacement, tonnes: 2,097 full load
Dimensions, metres (feet): 68.6 × 14 × 4 *(225.1 × 45.9 × 13.1)*
Speed, knots: 15. **Range, n miles:** 6,000 at 12 kt
Complement: 42 (7 officers)
Machinery: 2 Caterpillar 3608 diesels; 6,200 hp *(4.6 MW)* sustained; 1 shaft; cp prop; bow; 460 hp *(343 kW)* and stern; 550 hp *(410 kW)* thrusters
Guns: 2—12.7 mm MGs. 2—7.62 mm MGs.
Radars: Navigation: 2 Sperry/Litton BridgeMaster E340; I-band.

Comment: Contract awarded on 18 February 1993, to Marinette Marine of Marinette, WI, to construct a new class of seagoing buoy tenders. Capable of breaking 14 in of ice at 3 kt or a minimum of 3 ft by ramming. Main hoist can lift 20 tons, secondary 5 tons. A dynamic positioning system can maintain the ship within a 10 m circle in up to 30 kt wind. The class is named after the first *Juniper*, which was built in 1940 and decommissioned in 1975.

SPAR   5/2012, John Mortimer / 1486468

BUOY TENDERS (INLAND)

2 BUOY TENDERS (WLI/ABU)

Name	No	Builders	Launched	Homeport
BLUEBELL	WLI 313 (ex-WLI 313)	Birchfield Shipyard, Tacoma	28 Sep 1944	Portland, OR
BUCKTHORN	WLI 642 (ex-WLI 642)	Mobile Ship Repair, Mobile	18 Aug 1963	Sault Sainte Marie, MI

Displacement, tonnes: 177 (WLI 313), 230 (WLI 642) full load
Dimensions, metres (feet): 30.5 × 7.3 × 1.5 *(100.1 × 24.0 × 4.9)*
Speed, knots: 10.5 (WLI 313), 11.9 (WLI 642). **Range, n miles:** 2,700 at 10 kt
Complement: 15 (1 officer)
Machinery: 2 Caterpillar diesels; 600 hp *(448 kW)*; 2 shafts

Comment: Different vintage but similar in design.

2 BUOY TENDERS (WLI/ABU)

Name	No	Builders	Homeport
BAYBERRY	WLI 65400 (ex-WLI 65400)	Reliable Shipyard, Olympia	Oak Island, NC
ELDERBERRY	WLI 65401 (ex-WLI 65401)	Reliable Shipyard, Olympia	Petersburg, AK

Displacement, tonnes: 71 full load
Dimensions, metres (feet): 19.8 × 5.2 × 1.2 *(65 × 17.1 × 3.9)*
Speed, knots: 10
Complement: 8
Machinery: 2 GM diesels; 1 or 2 shafts

Comment: Both completed in June 1954. *Blackberry* decommissioned in 2008.

BUCKTHORN   3/2000, US Coast Guard / 0084213

BAYBERRY   5/1999, Hartmut Ehlers / 0084211

Buoy tenders — Construction tenders < **US** (COAST GUARD) 991

BUOY TENDERS (RIVER)

Notes: (1) All are based on rivers of USA especially the Mississippi, Missouri, Tennessee, Cumberland and their tributaries.
(2) Two ATON (aids to navigation) barges completed in 1991–92 by Marinette Marine. For use on the Great Lakes in conjunction with icebreaker tugs *Bristol Bay* and *Mobile Bay*.

6 RIVER TENDERS (WLR)

SANGAMON WLR 65506	**SCIOTO** WLR 65504	**OSAGE** WLR 65505
OUACHITA WLR 65501	**CIMARRON** WLR 65502	**OBION** WLR 65503

Displacement, tonnes: 148 full load
Dimensions, metres (feet): 19.8 × 6.4 × 0.4 *(65 × 21.0 × 1.3)*
Speed, knots: 10. **Range, n miles:** 3,500 at 8 kt
Complement: 13
Machinery: 2 diesels; 750 hp *(560 kW)*; 2 shafts

Comment: All commissioned 1960–62. WLR 65501 and 65502 built by Platzer Shipyard, Houston, TX; 65503–65506 by Gibbs Shipyard, Jacksonville, FL. WLR push crane-equipped barges to deploy aids-to-navigation buoys on the inland river system. Some of the class have 'jetting' devices used to set and anchor buoys in sandy or muddy river beds.

SANGAMON   *8/2005, USCG* / 1353711

12 RIVER TENDERS (WLR)

WEDGE WLR 75307	**CHIPPEWA** WLR 75404	**PATOKA** WLR 75408
GASCONADE WLR 75401	**CHEYENNE** WLR 75405	**CHENA** WLR 75409
MUSKINGUM WLR 75402	**KICKAPOO** WLR 75406	**KANKAKEE** WLR 75500
WYACONDA WLR 75403	**KANAWHA** WLR 75407	**GREENBRIER** WLR 75501

Displacement, tonnes: 152 full load
Dimensions, metres (feet): 22.9 × 6.7 × 1.2 *(75.1 × 22.0 × 3.9)*
Speed, knots: 9. **Range, n miles:** 3,100 at 8 kt
Complement: 13
Machinery: 2 Caterpillar diesels; 660 hp *(492 kW)*; 2 shafts

Comment: WLR 75401–75409 built 1964–70 by four different companies. WLR 75500 and 75501 were completed in early 1990. Details given are for the WLR 75401 series, but all are much the same size. Some vessels have a pile-driving capability to support ATON equipment such as day-boards and beacons.

GASCONADE   *2/2005, USCG* / 1043673

TRAINING CUTTERS

1 EAGLE CLASS (WIX/AXS)

Name	No	Builders	Commissioned	Homeport
EAGLE (ex-*Horst Wessel*)	WIX 327	Blohm + Voss, Hamburg	15 May 1946	New London, CT

Displacement, tonnes: 1,845 full load
Dimensions, metres (feet): 89.5 oa; 70.4 wl × 12 × 4.9 *(293.6; 231 × 39.4 × 16.1)*
Speed, knots: 18; 10.5 diesel. **Range, n miles:** 5,450 at 7.5 kt
Complement: 56 (6 officers) + 150 trainees
Machinery: 1 Caterpillar D 399 auxiliary diesel; 1,125 hp *(839 kW)* sustained; 1 shaft
Radars: Navigation: SPS-73; I-band.

Comment: Former German training ship. Launched on 13 June 1936. Taken by the US as part of reparations after the Second World War for employment in US Coast Guard Practice Squadron. Taken over at Bremerhaven in January 1946; arrived at home port of New London, Connecticut, in July 1946. (Sister ship *Albert Leo Schlageter* was also taken by the USA in 1945 but was sold to Brazil in 1948 and re-sold to Portugal in 1962. Another ship of similar design, *Gorch Fock*, transferred to the USSR in 1946 and survives as *Tovarisch*.) *Eagle* was extensively overhauled 1981–82. When the Coast Guard added the orange-and-blue marking stripes to cutters in the 1960s *Eagle* was exempted because of their effect on her graceful lines; however, in early 1976 the stripes and words 'Coast Guard' were added in time for the July 1976 Operation Sail in New York harbour. During the Coast Guard's year long bicentennial celebration, which ended 4 August 1990, *Eagle* visited each of the 10 ports where the original revenue cutters were homeported: Baltimore, Maryland; New London, Connecticut; Washington, North Carolina; Savannah, Georgia; Philadelphia, Pennsylvania; Newburyport, Massachusetts; Portsmouth, New Hampshire; Charleston, South Carolina; New York, New York; and Hampton, Virginia. The cutter currently serves as a training ship for cadets and officer candidates. During 2005, *Eagle* visited Bremerhaven for the first time since leaving its original homeport in 1945.
Fore and main masts 150.3 ft *(45.8 m)*; mizzen 132 ft *(40.2 m)*; sail area, 25,351 sq ft.

EAGLE   *6/2011, Michael Nitz* / 1454868

CONSTRUCTION TENDERS (INLAND)

4 PAMLICO CLASS (WLIC)

Name	No	Builders	Homeport
PAMLICO	WLIC 800	Coast Guard Yard, Curtis Bay	New Orleans, LA
HUDSON	WLIC 801	Coast Guard Yard, Curtis Bay	Miami Beach, FL
KENNEBEC	WLIC 802	Coast Guard Yard, Curtis Bay	Portsmouth, VA
SAGINAW	WLIC 803	Coast Guard Yard, Curtis Bay	Mobile, AL

Displacement, tonnes: 466 full load
Dimensions, metres (feet): 49 × 9.1 × 1.2 *(160.8 × 29.9 × 3.9)*
Speed, knots: 11
Complement: 14 (1 officer)
Machinery: 2 Caterpillar diesels; 1,000 hp *(746 kW)*; 2 shafts
Radars: Navigation: Raytheon SPS-73; I-band.

Comment: Completed in 1976 at the Coast Guard Yard, Curtis Bay, Maryland. These ships maintain structures and buoys in bay areas along the Atlantic and Gulf coasts.

KENNEBEC   *6/2012*, Marc Piché* / 1531179

1 COSMOS CLASS (WLIC)

SMILAX WLIC 315

Displacement, tonnes: 221 full load
Dimensions, metres (feet): 30.5 × 7.3 × 1.5 *(100.1 × 24.0 × 4.9)*
Speed, knots: 10.5
Complement: 14 (1 officer)
Machinery: 2 Caterpillar D 353 diesels; 660 hp *(492 kW)* sustained; 2 shafts
Radars: Navigation: Raytheon SPS-73; I-band.

Comment: Completed in 1944. Based at Atlantic Beach, North Carolina. Primary areas of operation are intercoastal waters from Virginia to Georgia. Pushes a 70 ft construction barge equipped with a crane and other aids-to-navigation equipment. Based at Atlantic Beach, NC.

SMILAX   *5/2011*, Marc Piché* / 1531178

US (COAST GUARD) > Construction tenders — Rescue and utility craft

8 ANVIL/CLAMP CLASSES (WLIC)

ANVIL WLIC 75301	**VISE** WLIC 75305
HAMMER WLIC 75302	**CLAMP** WLIC 75306
SLEDGE WLIC 75303	**HATCHET** WLIC 75309
MALLET WLIC 75304	**AXE** WLIC 75310

Displacement, tonnes: 142 full load
Dimensions, metres (feet): 22.9 (WLIC 75301–75305), 23.2 (WLIC 75306, 75309–75310) × 6.7 × 1.2 *(75.1, 76.1 × 22.0 × 3.9)*
Speed, knots: 10
Complement: 13
Machinery: 2 Caterpillar diesels; 750 hp *(559 kW)*; 2 shafts
Radars: Navigation: SPS-73; I-band.

Comment: Completed 1962–65. Primary areas of operation are intercoastal waters from Texas to New Jersey. Push 68 ft and 84 ft construction barges equipped with cranes and other aids-to-navigation equipment.

HATCHET   10/2008, USCG / 1353709

HARBOUR TUGS

11 65 ft CLASS (WYTL)

Name	No	Homeport
CAPSTAN	WYTL 65601 (ex-WYTL 65601)	Philadelphia, PA
CHOCK	WYTL 65602 (ex-WYTL 65602)	Baltimore, MD
TACKLE	WYTL 65604 (ex-WYTL 65604)	Rockland, ME
BRIDLE	WYTL 65607 (ex-WYTL 65607)	Southwest Harbor, ME
PENDANT	WYTL 65608 (ex-WYTL 65608)	Boston, MA
SHACKLE	WYTL 65609 (ex-WYTL 65609)	South Portland, ME
HAWSER	WYTL 65610 (ex-WYTL 65610)	Bayonne, NJ
LINE	WYTL 65611 (ex-WYTL 65611)	Bayonne, NJ
WIRE	WYTL 65612 (ex-WYTL 65612)	Saugerties, NY
BOLLARD	WYTL 65614 (ex-WYTL 65614)	New Haven, CT
CLEAT	WYTL 65615 (ex-WYTL 65615)	Philadelphia, PA

Displacement, tonnes: 73 full load
Dimensions, metres (feet): 19.8 × 5.8 × 2.1 *(65 × 19.0 × 6.9)*
Speed, knots: 10
Range, n miles: 2,700 at 10 kt
Complement: 6
Machinery: 1 Caterpillar 3412TA diesel; 500 hp *(373 kW)* sustained; 1 shaft
Radars: Navigation: Raytheon SPS-73; I-band.

Comment: Built between 1961 and 1967. The multimission tugs provide icebreaking, homeland security and aids-to-navigation services to several east coast areas. Re-engined 1993–96.

HAWSER   9/2009, Marco Ghiglino / 1366248

RESCUE AND UTILITY CRAFT

Notes: Craft of several different types. All carry five or six figure numbers of which the first two figures reflect the craft's length in feet.

13 UTILITY BOATS (YAG/UTB)

Displacement, tonnes: 14 full load
Dimensions, metres (feet): 12.6 × 4.0 × 1.3 *(41.3 × 13.1 × 4.3)*
Speed, knots: 26
Range, n miles: 300 at 18 kt
Complement: 3
Machinery: 2 diesels; 626 hp *(467 kW)* sustained; 2 shafts

Comment: Built by Coast Guard Yard, Baltimore 1973–83. Aluminium hull with a towing capacity of 100 tons. Used for fast multimission response in weather conditions up to moderate.

IHS Jane's Fighting Ships 2014-2015

41484   9/2009, Marco Ghiglino / 1366245

364 DEFENDER CLASS (RESPONSE BOATS) (PBF)

Displacement, tonnes: 3 full load
Dimensions, metres (feet): 7.5 × 2.6 × 2.7 *(24.6 × 8.5 × 8.9)*
Speed, knots: 46. **Range, n miles:** 150 at 35 kt
Complement: 4
Machinery: 2 Honda outboard motors; 550 hp *(410 kW)*
Guns: 2 – 7.62 mm MGs.
Radars: Furuno; I-band.

Comment: High-speed inshore patrol craft of aluminium construction and foam collar built by SAFE Boats International, Port Orchard, Washington. First delivery in July 2003. A standardised platform for SAR, law enforcement, marine safety, and security units. Transportable in a C-130.

DEFENDER 25501   10/2011, Michael Nitz / 1454873

126 MOTOR LIFEBOATS (MLB/SAR)

Displacement, tonnes: 20 full load
Dimensions, metres (feet): 14.6 × 4.4 × 1.4 *(47.9 × 14.4 × 4.6)*
Speed, knots: 25. **Range, n miles:** 220 at 25 kt
Complement: 4
Machinery: 2 Detroit diesels; 870 hp *(649 kW)* sustained; 2 shafts

Comment: Built by Textron Marine, New Orleans, the 47 ft motor lifeboat is designed as a first response resource to conduct rescues in high seas and severe conditions. The boats are self-bailing and self-righting, and have a long cruising radius for their size. A total of 126 boats were operational by January 2014 with more being added monthly toward final inventory of 200. With aluminium hulls, and 9,000 lb bollard pull and a towing capability of 150 tons, these boats are primarily lifeboats but have multimission capability.

MLB 47267   9/2010, M Mazumdar / 1406340

58 SPECIAL PURPOSE CRAFT (PBF)

Displacement, tonnes: 7 full load
Dimensions, metres (feet): 10.1 × 3.0 × 0.7 *(33.1 × 9.8 × 2.3)*
Speed, knots: 50. **Range, n miles:** 250 at 30 kt
Complement: 4
Machinery: 3 Mercury outboard motors; 825 hp *(615 kW)*
Guns: 2 – 7.62 mm MGs.
Radars: Furuno; I-band.

Comment: Larger versions of the SAFE Boats International Defender class. Aluminium construction with foam collar. High-speed coastal craft procured for pursuit and law enforcement tasks, particularly the interception of suspicious vessels entering US territorial waters. First craft delivered in January 2006 and final craft in December 2010.

SPECIAL PURPOSE CRAFT   10/2011, Chris Sattler / 1454871

© 2014 IHS

139 + 31 RESPONSE BOAT MEDIUM (YAG/UTB)

Displacement, tonnes: 17 full load
Dimensions, metres (feet): 13.6 × 4.5 × 0.9 (44.6 × 14.8 × 3.0)
Speed, knots: 42
Range, n miles: 250 at 30 kt
Complement: 4
Machinery: 2 Detroit diesels; 825 hp (615 kW); 2 Rolls Royce FF 3753 waterjets
Guns: 2—7.62 mm MGs.
Radars: Furuno; I-band.

Comment: Multimission self-righting response craft replacing 41 ft utility boat in inland waterways and offshore up to 50 n miles. Capable of operating in seas up to 3.6 m and of towing 100 tons. First boat delivered April 2008. Built by Marinette Marine Corp., Manitowac, Wisconsin, with Kvichak Marine Industries, Seattle. Up to 170 are to be procured by 2015.

RESPONSE BOAT 45615   6/2012, M Declerck / 1486469

80 TRANSPORTABLE PORT SECURITY BOATS (YP)

Displacement, tonnes: 5.1 full load
Dimensions, metres (feet): 9.9 × 2.5 × 0.94 (32.5 × 8.2 × 3.1)
Speed, knots: 40
Complement: 4
Machinery: 2 Yanmar diesels; 315 hp (235 kW); 2 shafts
Guns: 2—12.7 mm MGs. 2—7.62 mm MGs.
Radars: Furuno; I-band.

Comment: Contract awarded in 2010 to Aluminium Chambered Boats (ACB), Bellingham, WA, for construction of 80 TPSBs to replace older boats in service. In 2010, ACB declared bankruptcy and contract was re-awarded to Kvichak Marine Industries (KMI) in Seattle, WA. Delivery began in 2011 and was completed in October 2012.

TPSB   11/2011, USCG / 1454869

14 + 8 OVER THE HORIZON CRAFT IV (PBF)

Displacement, tonnes: 4.5 standard
Dimensions, metres (feet): 7.6 × 2.74 × 0.53 (24.9 × 9.0 × 1.7)
Speed, knots: 42
Range, n miles: 250 at 30 kt
Complement: 5
Machinery: 1 Yanmar 6LP-STZP/STZP2 diesel; 315 hp (235 kW)
Guns: To be announced.
Radars: Furuno; I-band.

Comment: A first production order for six OTH-IV was placed with SAFE Boats International LLC of Bremerton, WA, on 6 February 2013; a follow-on order for 16 craft was made 6 June 2013. Under the contracts, up to 101 (of which USCG may order up to 71) OTH-IV craft may be ordered over a seven-year period. The craft is of aluminium construction and is designed to be a cutter-launched (from stern or by side davit) quick response boat whose multimission roles are to include beyond-sight pursuit and interdiction, board and search, and SAR. Design features include a radar arch, a forward-facing weapons mount and mountable ballistic protection. The craft are deployed on National Security and Fast Response Cutters.

© 2014 IHS

Rescue and utility craft< US (COAST GUARD)   993

OTH-IV   1/2013*, USCG / 1531208

58 + 392 RESPONSE BOAT SMALL (YAG/UTB)

Displacement, tonnes: 4.5 full load
Dimensions, metres (feet): 8.7 × 2.6 × 0.5 (28.5 × 8.5 × 1.6)
Speed, knots: 46
Range, n miles: 220 at 30 kt
Complement: 4
Machinery: 2 Honda outboard motors; 450 hp (335 kW)
Guns: 3—7.62 mm MGs.
Radars: Furuno; I-band.

Comment: A contract for the construction of 450 second generation Response Boat – Small (RB-S) was awarded on 14 November 2011 to Metal Shark Aluminium Boats, Jeanerette, LA. The craft are progressively to replace the Defender class from 2012. Delivery began in 2012 and is to be completed by 2018. The craft are road transportable and air-portable by C-130. An additional 20 craft may be procured by US Customs and Border Protection and by the US Navy.

RB-S   9/2011, USCG / 1454872

1 + 7 LONG RANGE INTERCEPTOR CRAFT (PBF)

Displacement, tonnes: 8 full load
Dimensions, metres (feet): 10.67 × 3.44 × 0.6 (35 × 11.3 × 2.0)
Speed, knots: 42
Range, n miles: 236 at 30 kt
Complement: 2
Machinery: 2 Cummins QSB 6.7 diesels; 960 hp (715 kW); 2 Ultra Jet 305 water-jets
Radars: To be announced.

Comment: Contract awarded to MetalCraft Marine US, Inc of Clayton, New York, on 5 June 2012, for the construction of up to 10 Long Range Interceptor (LRI) craft. Eight of the LRIs are to be operated from Coast Guard National Security Cutters and the other two by other government agencies. The roles of the craft include law enforcement, national defence, drug interdiction, migrant interdiction, environmental protection, and port security.

LONG RANGE INTERCEPTOR   1/2013, MetalCraft Marine / 1486470

IHS Jane's Fighting Ships 2014-2015

NATIONAL OCEANIC AND ATMOSPHERIC ADMINISTRATION (NOAA)

Headquarters Appointments

Under Secretary of Commerce for Oceans and Atmosphere:
Kathryn Sullivan
Director, Office of Marine and Aviation Operations and NOAA Commissioned Officer Corps:
Rear Admiral Michael S Devany
Deputy Director for Operations:
Rear Admiral David A Score

Establishment and Missions

NOAA is the largest bureau of the US Department of Commerce, with a diverse set of responsibilities in environmental sciences. NOAA components include the Office of Marine and Aviation Operations; National Ocean Service; National Weather Service; National Marine Fisheries Service; National Environmental Satellite, Data and Information Service; and the Office of Oceanic and Atmospheric Research. NOAA's research and survey vessels conduct operations in hydrography, bathymetry, oceanography, atmospheric research, fisheries assessments and research, and related programmes in marine resources. Larger research vessels operate in international waters, and smaller ones primarily in Atlantic and Pacific coastal waters, and the Gulfs of Mexico and Alaska.

NOAA conducts diving operations. It also operates fixed-wing aircraft for hurricane research and reconnaissance; oceanographic and atmospheric research; marine mammal observations; hydrologic forecasts; and aerial mapping and remote sensing.

NOAA's fleet is 17 ships, including eight former Navy ships. The T-AGOS ship *Capable*, renamed *Okeanos Explorer*, was converted to conduct ocean exploration and commissioned in 2008. It is the first US federal ship dedicated to ocean exploration. Of the remaining ex-naval ships, five are T-AGOS vessels: one has been converted for oceanographic research (*Ka'imimoana*), two for fisheries research (*Gordon Gunter*, *Oscar Elton Sette*), and two for oceanographic research (*McArthur II* and *Hi'ialakai*). *Oscar Elton Sette* replaced *Townsend Cromwell* and *McArthur II* replaced *McArthur* in 2003. *Hi'ialakai* (formerly *Vindicator*), homeported in Hawaii, was commissioned in 2004. The former naval T-AGS hydrographic survey ship *Littlehales* was transferred to NOAA in 2003 and recommissioned *Thomas Jefferson*, replacing *Whiting*. A former naval Yard Torpedo Tender (YTT) vessel was converted for coastal research and became operational in 2003 as *Nancy Foster*, replacing *Ferrel*. A newly constructed oceanographic research ship, *Ronald H Brown* (AGOR 26), was commissioned in 1997. The hydrographic survey ship *Fairweather* was deactivated in 1988, refurbished, and reactivated in 2004.

A new class of Fisheries Survey Vessels (FSV) has been designed to NOAA specifications and standards set by the International Council for the Exploration of the Sea. *Oscar Dyson*, the first of five FSVs of the same design, was commissioned in May 2005 and operates in Alaskan waters. The second FSV, *Henry B Bigelow*, was commissioned in July 2007 and operates primarily off the northeast United States. FSV 3, *Pisces*, was commissioned in November 2009. FSV 4, *Bell M Shimada*, was commissioned in August 2010 and operates along the west coast of the United States. FSVs 1–4 were built by VT Halter Marine, Moss Point, MS. A newly designed Small Waterplane Area Twin Hull (SWATH) coastal mapping vessel, *Ferdinand R Hassler*, to be used for shallow water hydrographic research, was commissioned on 8 June 2012. A new FSV, *Reuben Lasker* was built by Marinette Marine and is to be commissioned in 2014.

Ships

The following ships may be met at sea.
Oceanographic Research Ships: *Ronald H Brown*, *Ka'imimoana*, *Okeanos Explorer*.
Multipurpose Oceanographic/Coastal Research Ships: *McArthur II*, *Nancy Foster*, *Hi'ialakai*.
Hydrographic Survey Ships: *Rainier*, *Thomas Jefferson*, *Fairweather* and *Ferdinand R Hassler*.
Fisheries Research Ships: *Oregon II*, *Gordon Gunter*, *Oscar Elton Sette*, *Oscar Dyson*, *Henry B Bigelow*, *Pisces*, *Bell M Shimada*, *Reuben Lasker*.

Personnel

2014: 321 officers plus 12,000 civilians

Bases

Major: Norfolk, VA and Newport, OR.
Minor: Pascagoula, MS; Honolulu, HI; Charleston, SC; Ketchikan, AK, Kodiak, AK.

HI'IALAKAI — 6/2007, Ships of the World / 1305254

REUBEN LASKER — 12/2012*, Marc Piché — 1531180

FERDINAND R HASSLER — 6/2012*, Marc Piché / 1531181

Uruguay

Country Overview

The Oriental Republic of Uruguay is situated in south-eastern South America. With an area of 68,037 square miles it has borders to the north with Brazil and to the west with Argentina. It has a coastline of 356 n miles with the south Atlantic Ocean and River Plate. There are some 675 n miles of navigable internal waterways. The capital, largest city and principal port is Montevideo. Territorial Seas (12 n miles) and an EEZ (200 n miles) are claimed.

Headquarters Appointments

Commander-in-Chief of the Navy:
  Admiral Ricardo Giambruno

Headquarters Appointments – continued

Fleet Commander:
  Rear Admiral Daniel Héctor Nuñez Rodríguez
Commander Coast Guard:
  Captain Julio Samandu

Diplomatic Representation

Naval Attaché in London:
  Captain Carlos O Butteri Rebufello

Personnel

(a) 2014: 4,768 (609 officers) (including 480 naval infantry, 200 naval air and 1,500 Coast Guard)
(b) Voluntary service

Prefectura Nacional Naval (PNN)

Established in 1934 primarily for harbour security and coastline guard duties. In 1991 it was integrated with the Navy, although patrol craft retain Prefectura markings. There are three regions: Atlantic, Rio de la Plata, and Rio Uruguay.

Bases

Montevideo: Main naval base (*Lt Carlos Machitelli*) with two dry docks (A new naval base is under construction at Punta Lobos and will replace the current harbour facilities.)
La Paloma, Rocha: Naval station (*Lt Cdr Ernesto Motto*)
Laguna del Sauce, Maldonado: Naval air station (*Lt Cdr Carlos Curbelo*)

Marines

Cuerpo de Fusileros Navales consisting of 450 men in three rifleman companies and one combat support company.

Prefix to Ships' Names

ROU

DELETIONS

Patrol Forces

2011   *15 de Noviembre, 25 de Agosto*

FRIGATES

Notes: Following the acquisition of two João Belo-class frigates from Portugal in early 2008, there is a programme to replace both these ships. A timetable has not been announced.

2 COMANDANTE JOÃO BELO CLASS (FF)

Name	No	Builders	Laid down	Launched	Commissioned	Recommissioned
URUGUAY (ex-*João Belo*)	1 (ex-F 480)	ACB Nantes, Nantes	6 Sep 1965	22 Mar 1966	1 Jul 1967	8 Apr 2008
PEDRO CAMPBELL (ex-*Sacadura Cabral*)	2 (ex-F 483)	ACB Nantes, Nantes	18 Aug 1967	15 Mar 1968	25 Jul 1969	8 Apr 2008

Displacement, tonnes: 1,929 standard; 2,150 full load
Dimensions, metres (feet): 102.7 × 11.7 × 3.9 *(336.9 × 38.4 × 12.8)*
Speed, knots: 24. **Range, n miles:** 7,500 at 15 kt
Complement: 176 (19 officers)

Machinery: 4 SEMT-Pielstick 12 PC series diesels; 18,700 hp(m) *(13.9 MW)*; 2 shafts
Guns: 2 DCN 3.9 in *(100 mm)*/55 Mod 1964 automatic ❶; dual purpose; 60 rds/min to 17 km *(9 n miles)* anti-surface; 8 km *(4.4 n miles)* anti-aircraft; weight of shell 13.5 kg. 2 Bofors 40 mm/60 ❷; 130 rds/min to 12 km *(6.6 n miles)*; weight of shell 0.89 kg.
Torpedoes: 6—324 mm Mk 32 Mod 5 (2 triple) tubes ❸; Honeywell Mk 46 Mod 1; active/passive homing to 6.3 km *(3.4 n miles)* at 30 kt; warhead 42.7 kg.
Physical countermeasures: Decoys 2 Loral Hycor SRBOC 6-barrelled chaff launchers. SLQ-25 Nixie.
Electronic countermeasures: ESM: AR-700 (V2); intercept.
Radars: Air search: Thomson-CSF DRBV 22A ❹; D-band.
Surface search: Koden MDC-2900 ❺; E/F-band.
Navigation: Kelvin Hughes KH 1007; I-band.
Fire control: Thomson CSF DRBC 31D ❻; I-band.
Sonars: CDC SQS 510; hull mounted; medium frequency.

Weapon control systems: C T Analogique. Sagem DMAA optical director.

Comment: *Uruguay* and *Pedro Campbell* procured from Portugal and recommissioned on 8 April 2008. Based at Montevideo.

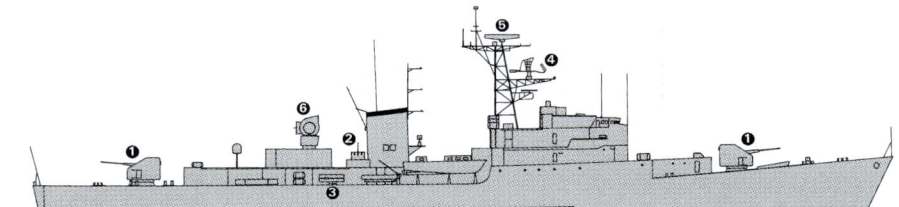

URUGUAY   *(Scale 1 : 900), Ian Sturton / 0121391*

PEDRO CAMPBELL   *2/2013*, A E Galarce / 1531183*

URUGUAY   *9/2013*, Mario R V Carneiro / 1531184*

SHIPBORNE AIRCRAFT

Numbers/Type: 1 Aerospatiale UH-13 Esquilo.
Operational speed: 121 kt (224 km/h).
Service ceiling: 11,150 ft (3,400 m).
Range: 240 n miles (445 km).
Role/Weapon systems: Utility and training aircraft transferred from Brazil in 2006. Refurbished in Brazil 2010–11. Two further aircraft may be acquired. Operated from *General Artigas*. Weapons: 2—7.62 mm MGs.

UH-13   10/2009, Uruguay Navy / 1305599

LAND-BASED MARITIME AIRCRAFT

Notes: (1) There are also two Beechcraft T-34C training aircraft.
(2) The last remaining Grumman S-2G Tracker was sold to Brazil in 2010.
(3) There are plans to acquire three multirole helicopters to operate with new offshore patrol vessels.

Numbers/Type: 2 CASA C-212-200MP.
Operational speed: 190 kt (353 km/h).
Service ceiling: 24,000 ft (7,315 m).
Range: 1,650 n miles (3,055 km).
Role/Weapon systems: Ex-Swedish Coast Guard aircraft transferred to Uruguayan Air Force on 9 September 2009. Deployed on maritime and river surveillance, SAR and border patrol roles. Sensors: Telephonics 1500B radar; FLIR Systems Star Safire III FLIR; MSS 500 mission system.

Numbers/Type: 2 Beechcraft B-200T.
Operational speed: 200 kt (372 km/h).
Service ceiling: 31,000 ft (9,448 m).
Range: 1,200 n miles (2,222 km).
Role/Weapon systems: Employed as a maritime patrol aircraft. Sensors: pod-mounted radar. A second aircraft was acquired in 2012.

B-200   2/2013*, A E Galarce / 1531182

Numbers/Type: 1 BAE Systems Jetstream T2.
Operational speed: 220 kt (409 km/h).
Service ceiling: 25,000 ft (7,619 m).
Range: 1,186 n miles (2,196 km).
Role/Weapon systems: Employed as a maritime patrol aircraft. Sensors: nose-mounted radar.

JETSTREAM   2/2012, A E Galarce / 1483670

Numbers/Type: 1 Westland Wessex HC Mk 2.
Operational speed: 90 kt (167 km/h).
Service ceiling: 10,000 ft (3,048 m).
Range: 180 n miles (333 km).
Role/Weapon systems: SAR and troop transport aircraft. Weapons: 1—7.62 mm MG.

WESSEX HC MK 2   2/2012, A E Galarce / 1483671

Numbers/Type: 6 Bolkow BO-105.
Operational speed: 113 kt (210 km/h).
Service ceiling: 9,845 ft (3,000 m).
Range: 407 n miles (754 km).
Role/Weapon systems: Coastal patrol and training aircraft acquired from Germany in August 2007.

BO-105   2/2013*, A E Galarce / 1531185

PATROL FORCES

Notes: There is a plan to acquire three offshore patrol vessels. Options include new-build vessel or second-hand.

1 WANGEROOGE CLASS (PBO/AG)

Name	No	Builders	Commissioned
MALDONADO (ex-*Norderney*)	23 (ex-A1455)	Schichau, Bremerhaven	15 Oct 1970

Displacement, tonnes: 868 standard; 1,126 full load
Dimensions, metres (feet): 52 × 11.7 × 4.1 (170.6 × 38.4 × 13.5)
Speed, knots: 14
Range, n miles: 5,000 at 10 kt
Complement: 32 (3 officers)
Machinery: Diesel-electric; 4 MWM 16-cyl diesel generators; 2 motors; 2,400 hp(m) (1.76 MW); 2 shafts; cp props
Guns: 1 Bofors 40 mm/70.
Radars: Surface search/navigation: Raytheon Pathfinder 1255; I-band.

Comment: Built as a salvage tug with ice-strengthened hull. Transferred from the German Navy on 21 November 2002. Employed as a support ship and for offshore patrol duties.

MALDONADO   11/2007, A E Galarce / 1335308

Patrol forces < **Uruguay** 997

16 VIGILANT 27 CLASS (PBR)

URUGUAY 5   6–20

Displacement, tonnes: 4 full load
Dimensions, metres (feet): 8.1 × 3.04 × 0.48 *(26.6 × 10.0 × 1.6)*
Speed, knots: 50
Complement: 3
Machinery: 2 Mercury outboards; 450 hp *(335 kW)*
Guns: 1—7.62 mm MG.

Comment: Boston Whaler craft ordered in May 2008 for deployment to Haiti as part of MINUSTAH. To be used as patrol craft on Uruguay River and lakes on return in 2014.

2 CAPE CLASS (LARGE PATROL CRAFT) (PB)

Name	No	Builders	Commissioned
COLONIA (ex-*Cape Higgon*)	10	Coast Guard Yard, Curtis Bay	14 Oct 1953
RIO NEGRO (ex-*Cape Horn*)	11	Coast Guard Yard, Curtis Bay	3 Sep 1958

Displacement, tonnes: 89 standard; 105 full load
Dimensions, metres (feet): 28.9 × 5.8 × 2 *(94.8 × 19.0 × 6.6)*
Speed, knots: 20
Range, n miles: 675 at 10 kt
Complement: 15 (3 officers)
Machinery: 2 GM 16V-149TI diesels; 2,322 hp *(1.73 MW)* sustained; 2 shafts
Guns: 2—12.7 mm MGs. 2—7.62 mm MGs.
Radars: Surface search: Sperry Marine Bridgemaster E; E/F/I-bands.

Comment: Designed for port security and search and rescue. Steel hulled. During modernisation in 1974 received new engines, electronics and deck equipment. Superstructure modified or replaced, and habitability improved. Transferred from the US Coast Guard 25 January 1990. Both based at Montevideo.

COLONIA   12/2011, A E Galarce / 1483669

1 COASTAL PATROL CRAFT (PB)

Name	No	Builders	Commissioned
PAYSANDU	12 (ex-PR 12)	Sewart Seacraft Inc, Berwick	Nov 1968

Displacement, tonnes: 61 full load
Dimensions, metres (feet): 25.5 × 5.7 × 1.85 *(83.7 × 18.7 × 6.1)*
Speed, knots: 20
Range, n miles: 600 at 10 kt
Complement: 14 (3 officers)
Machinery: 2 GM 16V-71 diesels; 811 hp *(605 kW)* sustained; 2 shafts
Guns: 1—12.7 mm MG. 2—7.62 mm MGs.
Radars: Surface search/navigation: Sperry Marine Bridgemaster E; E/F/I-bands.

Comment: Based at Montevideo.

PAYSANDU   2/2004, A E Galarce / 1044206

3 COAST GUARD PATROL CRAFT (WPB)

70–72

Displacement, tonnes: 91 full load
Dimensions, metres (feet): 22 × 5 × 1.8 *(72.2 × 16.4 × 5.9)*
Speed, knots: 12
Complement: 8
Machinery: 2 GM diesels; 400 hp *(298 kW)*; 2 shafts

Comment: Built in 1957 at Montevideo. Overhauls are planned.

PREFECTURA 70   12/2009, A E Galarce / 1454805

4 RIVER PATROL CRAFT (PBR)

URUGUAY 1–4

Displacement, tonnes: 5 full load
Dimensions, metres (feet): 11.3 × 3.25 × 0.8 *(37.1 × 10.7 × 2.6)*
Speed, knots: 32
Range, n miles: 1,500 at 24 kt
Complement: 6
Machinery: 3 Volvo AD41P 220MOP diesels
Guns: 3—7.62 mm MGs.

Comment: Built by Nuevos Ayres yacht builders. Deployed to Congo as part of UN force during 2001. Two have returned and have been operated by the Prefectura since 2013.

URUGUAY 1   2001, Uruguay Navy / 0121420

9 TYPE 44 CLASS (WPB)

441–449

Displacement, tonnes: 18 full load
Dimensions, metres (feet): 13.5 × 3.9 × 1.1 *(44.3 × 12.8 × 3.6)*
Speed, knots: 14
Range, n miles: 215 at 10 kt
Complement: 3
Machinery: 2 Detroit 6V-38 diesels; 185 hp *(136 kW)*; 2 shafts

Comment: Acquired from the US in 1999 and operated by the Coast Guard primarily as SAR craft. All nine craft are to receive overhauled engines in 2014. Basing: Punta del Este (442); Colonia del Sacramento (443); Piriápolis (444); Montevideo (448); La Paloma (449).

PREFECTURA 442   2/2013*, A E Galarce / 1531186

MINE WARFARE FORCES

3 KONDOR II CLASS (MINESWEEPERS—COASTAL) (MSC)

Name	No	Builders	Launched	Recommissioned
TEMERARIO (ex-*Riesa*)	31	Peenewerft, Wolgast	2 Oct 1972	11 Oct 1991
FORTUNA (ex-*Bernau*)	33	Peenewerft, Wolgast	3 Aug 1972	11 Oct 1991
AUDAZ (ex-*Eisleben*)	34	Peenewerft, Wolgast	2 Jan 1973	11 Oct 1991

Displacement, tonnes: 514 full load
Dimensions, metres (feet): 56.7 × 7.8 × 2.4 *(186 × 25.6 × 7.9)*
Speed, knots: 17. **Range, n miles:** 2,000 at 15 kt
Complement: 35 (7 officers)
Machinery: 2 Russki/Kolomna Type 40-DM diesels; 4,408 hp(m) *(3.24 MW)* sustained; 2 shafts; cp props
Guns: 1 Bofors 40 mm/70.
Mines: 2 rails.
Radars: Surface search/navigation: Sperry Marine Bridgemaster E; E/F/I-bands. Furuno 1621; I-band.

Comment: Belonged to the former GDR Navy. Transferred without armament. Minesweeping gear retained including MSG-3 variable depth sweep device. A fourth of class sunk after a collision with a merchant ship on 5 August 2000.

AUDAZ 10/2012, A E Galarce / 1483673

SURVEY AND RESEARCH SHIPS

1 HELGOLAND (TYPE 720B) CLASS (AGS)

Name	No	Builders	Commissioned
OYARVIDE (ex-*Helgoland*)	22 (ex-A 1457)	Unterweser, Bremerhaven	8 Mar 1966

Displacement, tonnes: 1,331 standard; 1,669 full load
Dimensions, metres (feet): 68 × 12.7 × 4.4 *(223.1 × 41.7 × 14.4)*
Speed, knots: 17. **Range, n miles:** 6,400 at 16 kt
Complement: 34
Machinery: Diesel-electric; 4 MWM 12-cyl diesel generators; 2 motors; 3,300 hp(m) *(2.43 MW)*; 2 shafts
Radars: Navigation: Raytheon; I-band.
Sonars: High definition, hull-mounted for wreck search.

Comment: Former German ocean-going tug launched on 25 November 1965. Paid off in1997 and recommissioned on 21 September 1998 after being fitted out as a survey ship. Oceanographic equipment reported to have been fitted in 2002. Ice strengthened hull. Fitted for twin 40 mm guns.

OYARVIDE 6/2002, A E Galarce / 0529549

1 INSHORE SURVEY CRAFT (AGSC)

TRIESTE

Displacement, tonnes: 12 full load
Dimensions, metres (feet): 12.1 × 3.6 × 1 *(39.7 × 11.8 × 3.3)*
Speed, knots: 16. **Range, n miles:** 500 at 16 kt
Complement: 4
Machinery: 2 Kamewa waterjets
Sonars: Elac Compact Mk II; 180 kHz. Elac LAZ 4721; 200 kHz.

Comment: Formerly owned by the Academia Maritime Internacional de Trieste. Donated by Italian government in 2000.

TRIESTE 2001, Uruguay Navy / 0121418

TRAINING SHIPS

1 SAIL TRAINING SHIP (AXS)

Name	No	Builders	Commissioned
CAPITÁN MIRANDA	20 (ex-GS 10)	SECN Matagorda, Cádiz	1930

Displacement, tonnes: 852 full load
Dimensions, metres (feet): 64 × 8 × 3.8 *(210 × 26.2 × 12.5)*
Speed, knots: 10
Complement: 49
Machinery: 1 GM diesel; 750 hp *(552 kW)*; 1 shaft
Radars: Navigation: Racal Decca TM 1226C; I-band.

Comment: Originally a diesel-driven survey ship with pronounced clipper bow. Converted for service as a three-masted schooner, commissioning as cadet training ship in 1978. Major refit by Bazán, Cadiz from June 1993 to March 1994, including a new diesel engine and a 5 m extension to the superstructure. Now has 853.4 m^2 of sail. A further major refit to extend life to 2030 began in 2013. The work is to include new engines and renovation of the hull and superstructure.

CAPITÁN MIRANDA 7/2007, Adolfo Ortigueira Gil / 1167887

1 SAIL TRAINING SHIP (AXS)

BONANZA

Displacement, tonnes: 13 full load
Dimensions, metres (feet): 15 × 4.5 × 2.8 *(49.2 × 14.8 × 9.2)*
Speed, knots: 10.4
Complement: 12

Comment: A sloop built by Green Marine, Lymington, in 1984. Donated by the US Naval Academy, Annapolis, in 1997 and used as a sail training vessel.

BONANZA 6/2013*, Uruguay Navy / 1531188

Auxiliaries — **Uruguay** 999

AUXILIARIES

Notes: (1) *Comar II* is a motor yacht used by the Commander-in-Chief.
(2) A 16 m buoy tender, *Orion* pennant number 25, built at Diques de la Armada, Montevideo, entered service in 2012. She has a catamaran hull. The new vessel is operated on the Uruguayan River.

1 BUOY TENDER (ABU)

Name	No	Builders
SIRIUS	21	Montevideo Naval Yard

Displacement, tonnes: 295 full load
Dimensions, metres (feet): 35.1 × 10 × 1.8 *(115.2 × 32.8 × 5.9)*
Speed, knots: 11
Complement: 15
Machinery: 2 Detroit 12V-71TA diesels; 840 hp *(626 kW)* sustained; 2 shafts

Comment: Buoy tender built at Montevideo Naval Yard and completed on 5 February 1988. Endurance, five days.

COMAR II   2/2007, A E Galarce / 1167932

2 LCVPS

LD 45–46

Displacement, tonnes: 15 full load
Dimensions, metres (feet): 14.1 × 3.5 × 0.8 *(46.3 × 11.5 × 2.6)*
Speed, knots: 9
Range, n miles: 580 at 9 kt
Military lift: 10 tons
Machinery: 1 GM 4-71 diesel; 115 hp *(86 kW)* sustained; 1 shaft

Comment: Built at Naval Shipyard, Montevideo and completed 1980. LD 45 operated by the Prefectura and LD 46 by the Navy.

SIRIUS   12/2012, A E Galarce / 1483675

1 LCM CLASS

LD 42

Displacement, tonnes: 24 standard; 58 full load
Dimensions, metres (feet): 17.1 × 4.3 × 1.2 *(56.1 × 14.1 × 3.9)*
Speed, knots: 9
Range, n miles: 130 at 9 kt
Complement: 5
Military lift: 30 tons
Machinery: 2 Gray Marine 64 HN9 diesels; 330 hp *(264 kW)*; 2 shafts

Comment: LD 42 built in Uruguay in 1978. LD 41 has been deleted.

LD 46   5/2011, Hartmut Ehlers / 1454876

LD 42   10/2013*, A E Galarce / 1531187

1 PIAST CLASS (PROJECT 570) (SALVAGE SHIP) (ARS)

Name	No	Builders	Commissioned
VANGUARDIA (ex-*Otto Von Guericke*)	26 (ex-A 441)	Northern Shipyard, Gdansk	29 Dec 1976

Displacement, tonnes: 1,872 full load
Dimensions, metres (feet): 73.2 × 12 × 4.2 *(240.2 × 39.4 × 13.8)*
Speed, knots: 16
Range, n miles: 6,500 at 11 kt
Complement: 61 (9 officers)
Machinery: 2 Zgoda GTD 48 diesels; 3,600 hp(m) *(2.68 MW)*; 2 shafts; cp props
Radars: Navigation: Sperry Marine Bridgemaster E; E/F/I-bands.

Comment: Acquired from Germany in October 1991 and sailed from Rostock in January 1992 after a refit at Neptun-Warnow Werft. Carries extensive towing and firefighting equipment plus a diving bell forward of the bridge.

1 LÜNEBURG CLASS (SUPPORT SHIP) (ARL)

Name	No	Builders	Commissioned
GENERAL ARTIGAS (ex-*Freiburg*)	4 (ex-A 1413)	Blohm + Voss, Hamburg	27 May 1968

Displacement, tonnes: 4,048 full load
Dimensions, metres (feet): 118.3 × 13.2 × 4.3 *(388.1 × 43.3 × 14.1)*
Speed, knots: 17. **Range, n miles:** 6,000 at 14 kt
Complement: 95 (15 officers)

Cargo capacity: 1,100 tons
Machinery: 2 MTU MD 16V 872 diesels; 5,630 hp(m) *(4.2 MW)* sustained; 2 shafts; cp props; bow thruster
Guns: 4 Bofors 40 mm/70 (2 twin).
Physical countermeasures: Decoys: 2 Breda 105 mm SCLAR launchers.
Radars: Navigation: Decca 1226/9; I-band.
Helicopters: UH-13 Esquilo or Bolkow BO-105.

Comment: Former auxiliary transferred to Uruguay on 12 April 2005. Used as a support ship for the Bremen class in German service, she was lengthened by 14.3 m in 1984 to accommodate a flight deck and a port-side crane. Her replenishment-at-sea capability provides a much needed enhancement to Uruguayan operational capability.

VANGUARDIA   12/2012, A E Galarce / 1483674

GENERAL ARTIGAS   10/2009, Uruguay Navy / 1305595

© 2014 IHS   IHS Jane's Fighting Ships 2014-2015

FRIGATES

6 MODIFIED LUPO CLASS (FFGHM)

Name	No	Builders	Laid down	Launched	Commissioned
MARISCAL SUCRE	F 21	Fincantieri, Riva Trigoso	19 Nov 1976	28 Sep 1978	10 May 1980
ALMIRANTE BRIÓN	F 22	Fincantieri, Riva Trigoso	Jun 1977	22 Feb 1979	7 Mar 1981
GENERAL URDANETA	F 23	Fincantieri, Riva Trigoso	23 Jan 1978	23 Mar 1979	8 Aug 1981
GENERAL SOUBLETTE	F 24	Fincantieri, Riva Trigoso	26 Aug 1978	4 Jan 1980	5 Dec 1981
GENERAL SALOM	F 25	Fincantieri, Riva Trigoso	7 Nov 1978	13 Jan 1980	3 Apr 1982
ALMIRANTE GARCIA (ex-*José Felix Ribas*)	F 26	Fincantieri, Riva Trigoso	21 Aug 1979	4 Oct 1980	30 Jul 1982

Displacement, tonnes: 2,243 standard; 2,560 full load
Dimensions, metres (feet): 113.2 × 11.3 × 3.7 *(371.4 × 37.1 × 12.1)*
Speed, knots: 35; 21 diesel
Range, n miles: 5,000 at 15 kt
Complement: 185

Machinery: CODOG; 2 Fiat/GE LM 2500 gas turbines; 50,000 hp *(37.3 MW)* sustained; 2 GMT A230.20M or 2 MTU 20V 1163 (F 21 and F 22) diesels; 8,000 hp(m) *(5.97 MW)* sustained; 2 shafts; LIPS cp props
Missiles: SSM: 8 Otomat Teseo Mk 2 TG1 ❶; active radar homing to 80 km *(43.2 n miles)* at 0.9 Mach; warhead 210 kg; sea-skimmer for last 4 km *(2.2 miles)*.
SAM: Selenia Elsag Albatros octuple launcher ❷; 8 Aspide; semi-active radar homing to 13 km *(7 n miles)* at 2.5 Mach; height envelope 15-5,000 m *(49.2–16,405 ft)*; warhead 30 kg.
Guns: 1 Oto Melara 5 in *(127 mm)*/54 ❸; 45 rds/min to 16 km *(8.7 n miles)*; weight of shell 32 kg. 4 Otobreda 40 mm/70 (2 twin) ❹; 300 rds/min to 12.5 km *(6.8 n miles)*; weight of shell 0.96 kg. 2 – 12.7 mm MGs.
Torpedoes: 6 – 324 mm ILAS 3 (2 triple) tubes ❺. Whitehead A244S; anti-submarine; active/passive homing to 7 km *(3.8 n miles)* at 33 kt; warhead 34 kg (shaped charge).
Physical countermeasures: Decoys: 2 Breda 105 mm SCLAR 20-barrelled trainable ❻; chaff to 5 km *(2.7 n miles)*; illuminants to 12 km *(6.6 n miles)*. Can be used for HE bombardment.
Electronic countermeasures: ESM; Elisra NS 9003/9005; intercept.
Radars: Air search: Selenia RAN 10S or Elta 2238 (F 21 and 22) ❼; E/F-band.
Air/surface search: Selenia RAN 11X; I-band.
Fire control: 2 Selenia Orion 10XP ❽; I/J-band. 2 Selenia RTN 20X ❾; I/J-band.
Navigation: SMA 3RM20; I-band.
Tacan: SRN 15A.
Sonars: EDO SQS-29 (Mod 610E) or Northrop Grumman 21 HS-7 (F 21 and F 22); hull-mounted; active search and attack; medium frequency.
Combat data systems: Selenia IPN 10. Elbit ENTCS 2000 (F 21 and F 22).
Weapon control systems: 2 Elsag NA 10 MFCS. 2 Dardo GFCS for 40 mm.
Helicopters: 1 AB 212ASW ❿.

Programmes: All ordered on 24 October 1975. Similar to ships in the Italian and Peruvian navies.
Modernisation: F 21 and F 22 were scheduled to start a refit by Ingalls Shipyard in September 1992 but contractual problems delayed start until January 1998. Refits included upgrading the gas turbines, replacing the diesels, improving the combat data system, updating sonar and ESM, and overhauling all weapon systems. The ships were redelivered in mid-2002 but further work was later required to re-install the Elta radar on a mast above the bridge. F 23 and F 24 have been upgraded by Dianca, Puerto Caballo, and returned to service in December and October 2003 respectively. Work included modernisation of the main machinery, air-conditioning and weapon systems. F 23 subsequently had a fire in about 2008 and awaits repairs. F 25 and F 26 began similar refits at Dianca in 2004 and these are expected to be completed in 2014, reportedly with Chinese assistance.
Structure: Fixed hangar means no space for Aspide reloads. Fully stabilised.
Operational: Based at Puerto Cabello.

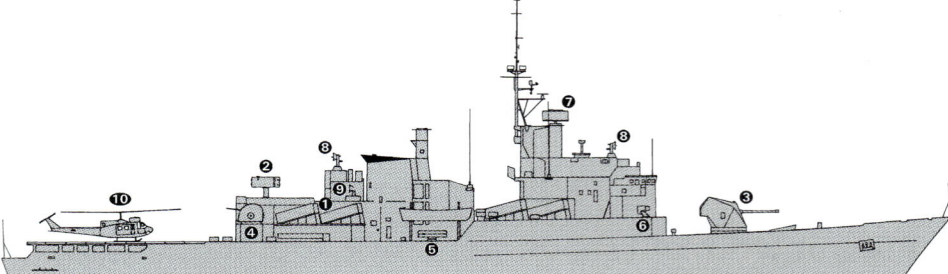
ALMIRANTE BRIÓN — *(Scale 1 : 900), Ian Sturton /* 1366436

GENERAL SOUBLETTE — *6/2009, Piet Cornelis /* 1305350

ALMIRANTE BRIÓN — *9/2007, Mario R V Carneiro /* 1353650

Shipborne aircraft — Patrol forces < **Venezuela** 1003

SHIPBORNE AIRCRAFT

Notes: (1) There are six operational Bell 412EP helicopters, equipped with radar and FLIR. Four acquired in 1999 and three more delivered in 2003 of which one has been lost. There are also one Bell 212 and one Bell 206B which is used for training. Six Mi-17 were acquired from Russia in 2007 for use by the Marines.
(2) There are plans to acquire eight helicopters for operation from the Guaiqueri- and Guaicamacuto-class patrol vessels. The first of eight Hai Z-9C helicopters is to be delivered in 2015.

Numbers/Type: 7 Agusta AB 212ASW.
Operational speed: 106 kt *(196 km/h).*
Service ceiling: 14,200 ft *(4,330 m).*
Range: 230 n miles *(426 m).*
Role/Weapon systems: ASW helicopter with secondary ASV role. Eight upgraded in Italy. One was lost in September 2010. Sensors: APS-705 search radar, Bendix AQS-18A dipping sonar. Weapons: ASW; two Mk 46 or A244/S torpedoes or depth bombs. ASV; mid-course guidance to Teseo Mk 2 missiles.

LAND-BASED MARITIME AIRCRAFT (FRONT LINE)

Notes: (1) There are also two Beech King Air and three Cessnas used for training and transport.
(2) Two CASA CN-235 maritime patrol aircraft were ordered in April 2005. The contract was signed on 29 November 2005 but cancelled in October 2006.
(3) The contract for 24 Su-30 Mk 2 Flankers was signed with the Russian government on 21 July 2006. All had been delivered by mid-2008. The aircraft are capable of carrying a variety of air-to-surface weapons.

Numbers/Type: 3/2/3 CASA C-212 S 43 Aviocar/C-212 S 200 Aviocar/C-212 S 400 Aviocar.
Operational speed: 190 kt *(353 km/h).*
Service ceiling: 24,000 ft *(7,315 m).*
Range: 1,650 n miles *(3,055 km).*
Role/Weapon systems: Medium-range MR and coastal protection aircraft; limited armed action. Acquired in 1981–82 and 1985–86. Three modernised and augmented in 1998 by S 400 type. Previous numbers have reduced. Sensors: APS-128 radar. Weapons: ASW; depth bombs. ASV; gun and rocket pods.

AB 212   6/2005, Massimo Annati / 1167652

C-212   6/2002, CASA/EADS / 0529548

PATROL FORCES

6 CONSTITUCIÓN CLASS (FAST ATTACK CRAFT — MISSILE AND GUN) (PBG/PG)

Name	No	Builders	Laid down	Launched	Commissioned
CONSTITUCIÓN	PC 11	Vosper Thornycroft	Jan 1973	1 Jun 1973	16 Aug 1974
FEDERACIÓN	PC 12	Vosper Thornycroft	Aug 1973	26 Feb 1974	25 Mar 1975
INDEPENDENCIA	PC 13	Vosper Thornycroft	Feb 1973	24 Jul 1973	20 Sep 1974
LIBERTAD	PC 14	Vosper Thornycroft	Sep 1973	5 Mar 1974	12 Jun 1975
PATRIA	PC 15	Vosper Thornycroft	Mar 1973	27 Sep 1973	9 Jan 1975
VICTORIA	PC 16	Vosper Thornycroft	Mar 1974	3 Sep 1974	22 Sep 1975

Displacement, tonnes: 173 full load
Dimensions, metres (feet): 36.9 × 7.1 × 1.8 *(121.1 × 23.3 × 5.9)*
Speed, knots: 31
Range, n miles: 1,350 at 16 kt
Complement: 20 (4 officers)
Machinery: 2 MTU MD 16V 538 TB90 diesels; 6,000 hp(m) *(4.4 MW)* sustained; 2 shafts
Missiles: SSM: 2 Oto Melara/Matra Teseo Mk 2 TG1 *(Federación, Libertad* and *Victoria);* active radar homing to 80 km *(43.2 n miles)* at 0.9 Mach; sea-skimmer for last 4 km *(2.2 n miles);* warhead 210 kg.
Guns: 1 Oto Melara 3 in *(76 mm)/62* compact *(Constitución, Independencia* and *Patria);* 85 rds/min to 16 km *(8.7 n miles);* weight of shell 6 kg. 1 Breda 30 mm/70 *(Federación, Libertad* and *Victoria);* 800 rds/min; weight of shell 0.37 kg. 2 — 12.7 mm MGs.
Radars: Surface search: SMA SPQ-2D; I-band.
Fire control: Selenia RTN 10X (in 76 mm ships); I/J-band.
Navigation: Racal; I-band.
Weapon control systems: Elsag NA 10 Mod 1 GFCS *(Constitución, Independencia* and *Patria).* Alenia Elsag Medusa optronic director *(Federación, Libertad* and *Victoria).*

Programmes: Transferred from the Navy in 1983 to the Coast Guard but now back again with Fleet Command.
Modernisation: Single Breda 30 mm guns replaced the 40 mm guns in the missile craft in 1989. All were refitted at Puerto Cabello 1992–1995 and fitted with new engines 2000–02.
Operational: Based at Punta Fijo.

LIBERTAD   6/2009, Piet Cornelis / 1305351

18 INSHORE PATROL BOATS (PBR)

CONSTANCIA	LRG 001	TENACIDAD	LRG 004	+12
PERSEVERANCIA	LRG 002	INTEGRIDAD	LRG 005	
HONESTIDAD	LRG 003	LEALTAD	LRG 006	

Displacement, tonnes: 11 full load
Dimensions, metres (feet): 12 × 2.8 × 1.7 *(39.4 × 9.2 × 5.6)*
Speed, knots: 38
Complement: 4
Machinery: 2 diesels; 640 hp(m) *(470 kW);* 2 shafts
Guns: 2 — 7.62 mm MGs.
Radars: Surface search: I-band.

Comment: First three speed boat type with GRP hulls delivered from a local shipyard in December 1991. Fourth completed in August 1993. Details given are for *Integridad* which is the first of two built at Guatire, and delivered in 1997/98. GRP construction. The twelve un-named craft are Boston Whaler Guardian class capable of 25 kt, mounting 2 — 12.7 mm and 2 — 6.72 mm MGs, and with Raytheon radars. These were donated by the US. All of these craft are used by Marines.

GUARDIAN INSHORE PATROL CRAFT   4/1999, Venezuelan Navy / 0084245

IHS Jane's Fighting Ships 2014-2015

© 2014 IHS

ESEQUIBO   3/1999 / 0084237

2 AJEERA CLASS (LCU)

Name	No	Builders	Commissioned
MARGARITA	T 71	Swiftships Inc, Morgan City	20 Jan 1984
LA ORCHILA	T 72	Swiftships Inc, Morgan City	11 May 1984

Displacement, tonnes: 435 full load
Dimensions, metres (feet): 39.6 × 11 × 1.8 *(129.9 × 36.1 × 5.9)*
Speed, knots: 13. **Range, n miles:** 1,500 at 10 kt
Complement: 26 (4 officers)

Military lift: 150 tons cargo; 100 tons fuel
Machinery: 2 Detroit 16V-149 diesels; 1,800 hp *(1.34 MW)* sustained; 2 shafts
Guns: 3 — 12.7 mm MGs.
Radars: Navigation: Raytheon 6410; I-band.

Comment: Both serve in Fluvial Command. Have a 15 ton crane.

Radars: Navigation: ARPA; I-band.

Comment: Ordered in September 1988. Developed from the Spanish Malaspina class. A multipurpose ship for oceanography, marine resource evaluation, geophysical and biological research. Equipped with Qubit hydrographic system. Carries two survey launches. EW equipment is fitted. Assigned to the OCHINA (Hydrographic department).

PUNTA BRAVA   11/2008, A A de Kruijf / 1353653

2 SURVEY CRAFT (AGSC)

Name	No	Builders	Commissioned
GABRIELA (ex-Peninsula de Araya)	LH 11	Abeking & Rasmussen, Lemwerder	5 Feb 1974
LELY (ex-Peninsula de Paraguana)	LH 12	Abeking & Rasmussen, Lemwerder	7 Feb 1974

Displacement, tonnes: 91 full load
Dimensions, metres (feet): 27 × 5.6 × 1.5 *(88.6 × 18.4 × 4.9)*
Speed, knots: 20
Complement: 9 (1 officer)
Machinery: 2 MTU diesels; 2,300 hp(m) *(1.69 MW);* 2 shafts

Comment: LH 12 laid down 28 May 1973, launched 12 December 1973 and LH 11 laid down 10 March 1973, launched 29 November 1973. Acquired in September 1986 from the Instituto de Canalizaciones. Both assigned to the Fluvial Command.

MARGARITA   6/1999, Venezuelan Navy / 0084238

1 GRIFFON 2000TD HOVERCRAFT (UCAC)

1004 Venezuela — Patrol forces — Amphibious forces

4 + (2) GUAIQUERI CLASS (OFFSHORE PATROL VESSELS) (PSOH)

Name	No	Builders	Laid down	Launched	Commissioned
GUAIQUERI (ex-*Caribe*)	PO 11 (ex-PC 21)	Navantia, Puerto Real	11 Sep 2008	24 Jun 2009	14 Apr 2011
WARAO	PO 12 (ex-PC 22)	Navantia, Puerto Real	26 May 2009	26 Oct 2009	2 Aug 2011
YECUANA	PO 13 (ex-PC 23)	Navantia, Puerto Real	22 Sep 2009	2 Mar 2010	9 Dec 2011
KARIÑA	PO 14 (ex-PC 24)	Navantia, Puerto Real	17 Feb 2010	13 Jul 2010	28 Apr 2012

Displacement, tonnes: 2,419 full load
Dimensions, metres (feet): 98.9 × 12.6 × 3.8 *(324.5 × 41.3 × 12.5)*
Speed, knots: 25. **Range, n miles:** 3,500 at 12 kt
Complement: 60 + 32 spare berths

Machinery: CODAD: 4 MTU 12V 1163 TB93 diesels; 23,600 hp *(17.6 MW)*; 2 shafts; cp props
Missiles: SAM: 8-cell VLS launcher.
Guns: 1 Oto Melara 3 in *(76 mm)* Super Rapid; 120 rds/min to 16 km *(8.7 n miles)*; weight of shell 6 kg. 1 Oerlikon Contraves Millenium 35 mm. 2—12.7 mm MGs.
Physical countermeasures: Decoys: 2—6 tubed launchers.
Electronic countermeasures: RESM: Thales Vigile.
CESM: Thales Altesse.
Radars: Surface search: Thales SMART-S; 3D; E/F-band.
Navigation: Thales Scout Mk 2; I-band.
Fire control: Thales Sting; I/J-band.
Combat data systems: Thales Tacticos. Link Y.
Electro-optic systems: Thales Sting optronic director. Thales Mirador TEOOS.

Helicopters: To be announced.

Programmes: Following agreement between the Spanish and Venezuelan governments signed on 28 November 2005, a contract for the construction of four offshore patrol vessels (POVZEE) was signed with Navantia on 26 May 2006. The ships were constructed at the Puerto Real Shipyard at Cadiz. Two further ships may be acquired.
Modernisation: There are plans to equip the class with anti-ship and/or surface-to-air missiles and sonars. There is an 8-cell VLS launcher aft of the 76 mm gun.
Operational: The vessels are employed on EEZ protection duties. *Warao* damaged by grounding incident in March 2012 and is under repair in Brazil.

GUAIQUERI (old number)   4/2011, *NAVANTIA* / 1486475

1006 Venezuela — Training ships — Coast guard

TRAINING SHIPS

1 SAIL TRAINING SHIP (AXS)

Name	No	Builders	Launched	Commissioned
SIMÓN BOLÍVAR	BE 11	AT Celaya, Bilbao	21 Nov 1979	6 Aug 1980

Displacement, tonnes: 1,280 full load
Measurement, tonnes: 949 gt
Dimensions, metres (feet): 82.5 × 10.6 × 4.4 *(270.7 × 34.8 × 14.4)*
Speed, knots: 10
Complement: 195 (17 officers; 102 trainees)
Machinery: 1 Detroit 12V-149T diesel; 875 hp *(652 kW)* sustained; 1 shaft

Comment: Ordered in 1978. Three-masted barque; similar to *Guayas* (Ecuador), *Cuauhtemoc* (Mexico) and *Gloria* (Colombia). Sail area (23 sails), 1,650 m². Highest mast, 131.2 ft *(40 m)*. Has won several international sail competitions including Cutty Sark '96. A refit is reported to have been completed in 2008.

SIMÓN BOLÍVAR   3/2010, *A E Galarce* / 1454804

AUXILIARIES

Notes: There is one 17 m navigational aids tender *Macuro* BB-11. The vessel has a catamaran hull and is armed with a 20 mm Oerlikon.

1 LOGISTIC SUPPORT SHIP (AORH)

Name	No	Builders	Commissioned
CIUDAD BOLÍVAR	T 81	Hyundai, Ulsan	2001

Displacement, tonnes: 9,906 full load
Dimensions, metres (feet): 137.7 × 18 × 6.6 *(451.8 × 59.1 × 21.7)*
Speed, knots: 18. **Range, n miles:** 4,500 at 15 kt
Complement: 104
Machinery: 2 Caterpillar 3616 diesels; 2 shafts; LIPS cp props
Guns: 2 Bofors 40 mm/70. 2—12.7 mm MGs.

Comment: Ordered from Hyundai, South Korea, in February 1999. Delivered in October 2001. Capable of carrying 4,400 tons of fuel and 900 tons of cargo. Two replenishment stations on each beam. Hangar and deck for medium size helicopter. Replenishment operations reported conducted with both French and Netherlands units. Armament is not yet fitted. Refitted at Dianca in 2012.

CIUDAD BOLÍVAR   6/2002 / 0094545

1 OCEAN TUG (ATA)

Name	No	Builders	Laid down	Launched	Commissioned
GENERAL FRANCISCO DE MIRANDA (ex-*Almirante Bruzuar*)	RA 11	Damen, Gorinchem and DIANCA, Puerto Caballo	1 Jun 2004	1 Jun 2006	19 Dec 2006

Measurement, tonnes: 1,464 gt
Dimensions, metres (feet): 60.1 × 13.6 × 5.1 *(197.2 × 44.6 × 16.7)*
Speed, knots: 12.5. **Range, n miles:** 7,000 at 10 kt
Machinery: 2 CAT 3606TA diesels; 5,400 hp *(4 MW)*; 2 shafts

GENERAL FRANCISCO DE MIRANDA   6/2010, *Venezuelan Navy* / 1406342

2 UTILITY CRAFT (YAG)

LOS TAQUES LG 11   LOS CAYOS LG 12

Displacement, tonnes: 356 full load
Dimensions, metres (feet): 26.6 × 7.1 × 1.5 *(87.3 × 23.3 × 4.9)*
Speed, knots: 8
Complement: 10
Machinery: 1 diesel; 850 hp(m) *(625 kW)*; 1 shaft
Guns: 1—12.7 mm MG.

Comment: Former trawlers. Commissioned 15 May 1981 and 17 July 1984 respectively. Used for salvage and SAR tasks.

LOS CAYOS   11/2008, *A A de Kruijf* / 1353651

COAST GUARD

Notes: (1) It is reported that a contract for the construction of one or more Damen Stan Patrol 5009 was signed in 2013. The 50 m ships are of the 'sea-axe' design and a similar vessel is in service with the Cape Verde Coast Guard. It is likely that the ship(s) would be built in Cuba and would follow-on from the Damen Stan Patrol 4207 under construction.
(2) There is a patrol vessel *San Carlos* LG 41 of approximately 35 m.

4 PETREL (POINT) CLASS (WPB)

Name	No	Builders	Commissioned
PETREL (ex-*Point Knoll*)	PG 31	US Coast Guard Yard, Curtis Bay	26 Jun 1967
ALCATRAZ (ex-*Point Judith*)	PG 32	US Coast Guard Yard, Curtis Bay	26 Jul 1966
ALBATROS (ex-*Point Franklin*)	PG 33	US Coast Guard Yard, Curtis Bay	14 Nov 1966
PELÍCANO (ex-*Point Ledge*)	PG 34	US Coast Guard Yard, Curtis Bay	18 Jul 1962

Displacement, tonnes: 69 full load
Dimensions, metres (feet): 25.3 × 5.2 × 1.8 *(83 × 17.1 × 5.9)*
Speed, knots: 23.5
Range, n miles: 1,500 at 8 kt
Complement: 10 (1 officer)
Machinery: 2 Caterpillar diesels; 1,600 hp *(1.19 MW)*; 2 shafts
Guns: 2—12.7 mm MGs.
Radars: Surface search: Raytheon SPS-64; I-band.

Comment: *Petrel* transferred from USCG on 18 November 1991 and *Alcatraz* on 15 January 1992, *Albatros* on 23 June 1998 and *Pelícano* on 3 August 1998. The transfer of four further craft is unlikely. Most of the class are believed to be operational.

12 GAVION CLASS (WPB)

GAVION PG 401	CORMORAN PG 405	NEGRON PG 409
ALCA PG 402	COLIMBO PG 406	PIGARGO PG 410
BERNACLA PG 403	FARDELA PG 407	PAGAZA PG 411
CHAMAN PG 404	FUMAREL PG 408	SERRETA PG 412

Displacement, tonnes: 46 full load
Dimensions, metres (feet): 24.4 × 5.2 × 1.5 (80.1 × 17.1 × 4.9)
Speed, knots: 25
Range, n miles: 1,000 at 12 kt
Complement: 10
Machinery: 2 Detroit 12V-92TA diesels; 2,160 hp (1.61 MW) sustained; 2 shafts
Guns: 2 – 12.7 mm MGs. 2 – 7.62 mm MGs. 1 – 40 mm Mk 19 grenade launcher.
Radars: Surface search: Raytheon R1210; I-band.

Comment: Ordered from Halter Marine 24 April 1998 and delivered from late 1999 to early 2000. Aluminium construction. Four craft refitted in 2003 and all believed to be operational.

GAVION 6/2009, Piet Cornelis / 1305352

12 PUNTA MACOLLA CLASS (PB)

PUNTA MACOLLA LSM 001	VELA DE COBO LSM 007
FARALLÓN CENTINELA LSM 002	CAYO MACEREO LSM 008
CHARAGATO LSM 003	CHUSPA LSM 009
BAJO BRITO LSM 004	ISLA DE MEDIO LSM 010
BAJO ARAYA LSM 005	MORRO DE BARCELONA LSM 011
CARECARE LSM 006	ISLA PAPAGALLO LSM 012

Displacement, tonnes: 5 full load
Dimensions, metres (feet): 12.7 × 2.8 × 2 (41.7 × 9.2 × 6.6)
Speed, knots: 30
Complement: 4
Machinery: 2 diesels; 2 shafts
Guns: 1 – 12.7 mm MG.
Radars: Surface search: Raytheon; I-band.

Comment: Built in Venezuela by Intermarine. Used by OCHINA (Hydrographic department) and for SAR. First six delivered by 1997 and last two in 2000.

BAJO ARAYA 3/1999, Venezuelan Navy / 0084246

7 POLARIS CLASS (PBF)

POLARIS LG 21	ALDEBARAN LG 24	CANOPUS LG 26
SIRIUS LG 22	ANTARES LG 25	ALTAIR LG 27
RIGEL LG 23		

Displacement, tonnes: 5 full load
Dimensions, metres (feet): 7.9 wl × 2.6 × 0.8 (25.9 × 8.5 × 2.6)
Speed, knots: 50
Complement: 4
Machinery: 1 diesel outdrive; 400 hp(m) (294 kW)
Guns: 2 – 7.62 mm MGs.
Radars: Surface search: Raytheon; I-band.

Comment: Built by Cougar Marine and delivered in 1987. Used by the Coast Guard for drug interdiction. Two more reported operational.

ALDEBARAN 4/1999, Venezuelan Navy / 0084247

© 2014 IHS

2 PROTECTOR 3612 CLASS (PB)

| CHICHIRIVICHE LG 31 | CARUANTA LG 32 |

Displacement, tonnes: 11 full load
Dimensions, metres (feet): 11.1 × 4.0 × 0.5 (36.4 × 13.1 × 1.6)

Comment: Built by SeaArk Marine, Monticello, and delivered in 1994.

1 + 5 DAMEN STAN PATROL 2606 (PB)

PÁGALO PG 51   +5

Displacement, tonnes: 75 full load
Dimensions, metres (feet): 26.5 × 5.9 × 1.9 (86.9 × 19.4 × 6.2)
Speed, knots: 22
Range, n miles: 500 at 12 kt
Complement: 12
Machinery: 2 Caterpillar 12V C32 diesels; 1,700 hp (1.36 MW); 2 shafts; cp props
Guns: 3 – 12.7 mm MGs. 3 – 7.62 mm MGs. 1 – 26.5 mm grenade launcher.
Radars: Surface search/navigation: Raymarine E-120; I-band.

Comment: Contract signed with Damen Shipyards for construction at UCOCAR, Puerto Cabello, of one Damen Stan Patrol 2606 patrol craft. Aluminium construction. Launched on 14 August 2008 and commissioned 11 November 2008. Five further craft to be built.

PÁGALO 6/2010, Venezuelan Navy / 1406341

0 + 2 DAMEN STAN PATROL 4207 (PB)

Displacement, tonnes: 208 standard
Dimensions, metres (feet): 42.8 × 7.11 × 2.52 (140.4 × 23.3 × 8.3)
Speed, knots: 26
Complement: 18 (4 officers)
Machinery: 2 Caterpillar 3516B DI-TA; 5,600 hp (4.17 MW); 2 cp props
Guns: 2 – 12.7 mm MGs.

Comment: Contract signed with Damen Shipyards for construction in Cuba, of two Damen Stan Patrol 4207. Details are based on those in Jamaican service.

3 + 1 (2) GUAICAMACUTO CLASS (PATROL VESSELS) (PSOH)

Name	No	Builders	Laid down	Launched	Commissioned
GUAICAMACUTO	GC 21	Navantia, San Fernando	17 Dec 2006	16 Oct 2008	2 Mar 2010
YAVIRE	GC 22	Navantia, San Fernando	2007	11 Mar 2009	8 Nov 2010
NAIGUATA	GC 23	Navantia, San Fernando	28 Nov 2007	24 Jun 2009	1 Feb 2011
TAMANACO	GC 24	Dianca, Puerto Cabello	3 Aug 2012	2014	2015

Displacement, tonnes: 1,453 full load
Dimensions, metres (feet): 79.9 × 11.5 × 3.7 (262.1 × 37.7 × 12.1)
Speed, knots: 22
Range, n miles: 4,000 at 12 kt
Complement: 34 + 30 spare berths

Machinery: 2 MTU 16V 1163 TB 93 diesels; 15,880 hp (11.8 MW); 2 shafts
Guns: 1 Oto Melara 3 in (76 mm)/62 Compact; 85 rds/min to 16 km (8.7 n miles); weight of shell 6 kg. 1 Oerlikon Contraves Millenium 35 mm. 2 – 12.7 mm MGs.
Electronic countermeasures: RESM: Thales Vigile.
CESM: Thales Altesse.
Radars: Surface search/navigation: Thales Variant; G/I-bands.
Fire control: Thales Sting; I/J-band.
Navigation: Kelvin Hughes; I-band.
Combat data systems: Thales Tacticos. Link Y.
Electro-optic systems: Thales Sting optronic director. Thales Mirador TEOOS.
Helicopters: Platform for one medium.

Programmes: Following agreement between the Spanish and Venezuelan governments signed on 28 November 2005, a contract for the construction of four patrol vessels (BVL) was signed with Navantia on 26 May 2006. Three ships built at the San Fernando Shipyard at Cadiz. The fourth is to be assembled in Venezuela. Five blocks were received from Navantia in July 2012 and the keel was formally laid in August 2012. Two further vessels may be acquired.
Operational: The vessels are operated by the Coast Guard.

GUAICAMACUTO 9/2009, Carlos Pardo Gonzalez / 1366290

IHS Jane's Fighting Ships 2014-2015

NATIONAL GUARD (GUARDIA NACIONAL)

Notes: (1) There are also a large number of US and Canadian built river craft of between 6 and 9 m length, which are armed with MGs.
(2) Four intercept launches were delivered in 2003; two in July 2003 and two in October 2003.
(3) Some 60 Pirana class river patrol craft were ordered. The first 15 were delivered in August 2003.

10 RIO ORINOCO II CLASS (PBF)

9903   +9

Displacement, tonnes: 30 full load
Dimensions, metres (feet): 16.5 × 4.3 × 1.4 (54.1 × 14.1 × 4.6)
Speed, knots: 36
Range, n miles: 500 at 25 kt
Complement: 5
Machinery: 2 MTU 12V 183 TE93 diesels; 2,268 hp(m) (1.67 MW) sustained; 2 shafts
Guns: 2 — 12.7 mm MGs. 2 — 7.62 mm MGs.
Radars: Surface search: Raytheon R1210; I-band.

Comment: Ordered from Halter Marine 24 April 1998. All delivered by late 1999. Aluminium construction. Some of the similar sized Orinoco I craft built in the 1970s are still in limited use.

12 PUNTA CLASS (PB)

PUNTA BARIMA A 8201	PUNTA CARDON A 8205	PUNTA UNARE A 8309
PUNTA MOSQUITO A 8202	PUNTA PLAYA A 8206	PUNTA BALLENA A 8310
PUNTA MULATOS A 8203	PUNTA MACOYA A 8307	PUNTA MACURO A 8311
PUNTA PERRET A 8204	PUNTA MORON A 8308	PUNTA MARIUSA A 8312

Displacement, tonnes: 15 full load
Dimensions, metres (feet): 13.1 × 4.1 × 1.2 (43 × 13.5 × 3.9)
Speed, knots: 34. **Range, n miles:** 390 at 25 kt
Complement: 4
Machinery: 2 MTU series 183 diesels; 1,500 hp (1.1 MW); 2 shafts
Guns: 2 — 12.7 mm MGs.
Radars: Navigation: Raytheon; I-band.

Comment: Built by Robert E Derecktor, Mamaroneck, NY. Delivered July-December 1984.

12 PROTECTOR CLASS (PB)

RIO ARAUCA II B 8421	RIO META II B 8425	RIO SINARUCO B 8429
RIO CATATUMBO II B 8422	RIO PORTUGUESA II B 8426	RIO ICABARU B 8430
RIO APURE II B 8423	RIO SARARE B 8427	RIO GUARICO II B 8431
RIO NEGRO II B 8424	RIO URIBANTE B 8428	RIO YARACUY B 8432

Displacement, tonnes: 15 full load
Dimensions, metres (feet): 13.3 × 4.5 × 1.2 (43.6 × 14.8 × 3.9)
Speed, knots: 28. **Range, n miles:** 390 at 25 kt
Complement: 4
Machinery: 2 GM diesels; 1,100 hp (810 kW); 2 shafts
Guns: 2 — 12.7 mm MGs.
Radars: Navigation: Raytheon; I-band.

Comment: Built by SeaArk Marine and completed in 1984.

ORINOCO 9903   3/2009, Marco Ghiglino / 1305600

RIO SARARE   2/1996, van Ginderen Collection / 0084249

Vietnam

Country Overview

The Socialist Republic of Vietnam was established in 1976 when the Democratic Republic of Vietnam in the north and the Republic of Vietnam in the south became one nation. The country had been divided at the 17th parallel from the end of French colonial rule in 1954 and during the ensuing Vietnam War. Located on the east coast of the Indochina peninsula, it has an area of 127,844 square miles and is bordered to the north by China and to the west by Cambodia and Laos. It has a 1,858 n mile coastline with the South China Sea. Hanoi is the capital while Ho Chi Minh City (formerly Saigon) is the largest city and a major port. There are further ports at Haiphong and Da Nang. Territorial seas (12 n miles) are claimed. An EEZ (200 n miles) has also been claimed but the limits have not been defined.

Headquarters Appointments

Chief of Naval Forces:
  Vice Admiral Nguyen Van Hien
Deputy Chief of Naval Forces:
  Captain Pham Ngoc Minh

Personnel

(a) 2014: 13,000 regulars
(b) Additional conscripts on three to four year term (about 3,000)
(c) 27,000 naval infantry

Organisation and Bases

The Vietnamese Navy is part of the People's Army of Vietnam (PAVN) and is formally known as the PAVN Navy.

The fleet is organised into four regions based on, from north to south, Haiphong (HQ), Da Nang, Nha Trang and Cân Tho. There are other bases at Cam Ranh Bay, Hue and Ha Tou.

Coast Guard

A Coast Guard was formed on 1 September 1998. It is subordinate to the Navy and may take on Customs duties.

SUBMARINES

2 YUGO CLASS (MIDGET SUBMARINES) (SSW)

Displacement, tonnes: 91 surfaced; 112 dived
Dimensions, metres (feet): 20 × 3.1 × 4.6
  (65.6 × 10.2 × 15.1)
Speed, knots: 12 surfaced; 8 dived
Range, n miles: 550 at 10 kt surfaced; 50 at 4 kt dived
Complement: 11 (est.) (7 divers (est.))
Machinery: 2 diesels; 320 hp(m) (236 kW); 1 shaft

Comment: Transferred from North Korea in 1997. May be fitted with two short torpedo tubes and a snort mast, but used primarily for diver related operations. The conning tower acts as a wet/dry diver compartment. Operational status is doubtful.

YUGO (North Korean colours)
6/1998, Ships of the World
0052525

Submarines — Frigates < **Vietnam** 1009

2 + 4 KILO CLASS (PROJECT 636) (SSK)

Name	No	Builders	Laid down	Launched	Commissioned
HANOI	HQ 182	Admiralty Yard, St Petersburg	23 Aug 2010	28 Aug 2012	3 Apr 2014
HO CHI MINH CITY	HQ 183	Admiralty Yard, St Petersburg	28 Sep 2011	28 Dec 2012	3 Apr 2014
HAIPHONG	HQ 184	Admiralty Yard, St Petersburg	2012	Aug 2013	Nov 2014
DA NANG	HQ 185	Admiralty Yard, St Petersburg	2013	2014	2015
KHANH HOA	HQ 186	Admiralty Yard, St Petersburg	2014	2015	2016
VUNG TAU	HQ 187	Admiralty Yard, St Petersburg	2014	2016	2017

Displacement, tonnes: 2,362 surfaced; 3,125 dived
Dimensions, metres (feet): 72.6 × 9.9 × 6.6 *(238.2 × 32.5 × 21.7)*
Speed, knots: 10 surfaced; 9 snorting; 17 dived
Range, n miles: 6,000 at 7 kt snorting; 400 at 3 kt dived
Complement: 52 (13 officers)

Machinery: Diesel-electric; 2 diesels; 3,650 hp(m) *(2.68 MW)*; 2 generators; 1 motor; 5,900 hp(m) *(4.34 MW)*; 1 shaft; 2 auxiliary MT-168 motors; 204 hp(m) *(150 kW)*; 1 economic speed motor; 130 hp(m) *(95 kW)*

Torpedoes: 6—21 in (533 mm) tubes. Combination of Russian TEST-71ME; anti-submarine active/passive homing to 15 km *(8.2 n miles)* at 40 kt; warhead 205 kg and 53–65; anti-surface ship passive wake homing to 19 km *(10.3 n miles)* at 45 kt; warhead 300 kg. Total of 18 weapons.
Mines: 24 in lieu of torpedoes.
Electronic countermeasures: ESM; Brick Pulp; radar warning.
Radars: Surface search: Snoop Tray; I-band.
Sonars: MGK 400 Shark Teeth/Shark Fin; hull-mounted; passive/active search and attack; medium frequency. MG 519 Mouse Roar; active attack; high frequency.

Weapon control systems: MVU 110 TFCS.

Programmes: A contract for the construction of six Project 636 boats was signed with Admiralty Shipyards in mid-2009. The first boat was laid down in early 2010 and delivery is expected to start in about 2014 at the rate of one per year. The contract is likely to include a training and maintenance package.
Structure: Diving depth, 790 ft *(240 m)*. 9,700 kWh batteries. Pressure hull 169.9 ft *(51.8 m)*.
Operational: The first of class arrived at homeport Cam Ranh on 3 January 2014 and the second on 19 March 2014.

KILO CLASS 3/1996 / 0056450

FRIGATES

Notes: The Barnegat-class frigate (ex-seaplane tender) *Pham Ngu Lao* HQ 01 has probably been decommissioned.

2 + 2 MODIFIED GEPARD (PROJECT 11661E) CLASS (FFGM)

Name	No	Builders	Laid down	Launched	Commissioned
DINH TIEN HOANG	HQ 011	Zelenodolsk Shipyard, Kazan	10 Jul 2007	12 Dec 2009	5 Mar 2011
LY THAI TO	HQ 012	Zelenodolsk Shipyard, Kazan	28 Nov 2007	16 Mar 2010	22 Aug 2011
–	–	Zelenodolsk Shipyard, Kazan	24 Sep 2013	2016	2017
–	–	Zelenodolsk Shipyard, Kazan	24 Sep 2013	2016	2017

Displacement, tonnes: 1,585 standard; 2,134 full load
Dimensions, metres (feet): 102.2 × 13.1 × 5.3 *(335.3 × 43.0 × 17.4)*
Speed, knots: 26; 18 diesel. **Range, n miles:** 5,000 at 10 kt
Complement: 103 + 28 spare berths

Machinery: CODOG; 2 gas turbines; 30,850 hp(m) *(23.0 MW)*; 1 Type 61D diesel; 7,375 hp(m) *(5.5 MW)*; 2 shafts; cp props
Missiles: SSM: 8 (2 quad) Zvezda SS-N-25 Switchblade (KH 35 Uran); IR or radar homing to 130 km *(70.2 n miles)* at 0.9 Mach; warhead 145 kg; sea-skimmer.
SAM: 1 Palma (CADS-N-2) mount; combination of 8 Sosna-R missiles; laser/TV/IR/EO guidance to 10 km *(5.4 n miles)*; warhead 5 kg.
Guns: 1—3 in *(76 mm)*/59 AK-176; 120 rds/min to 15 km *(8 n miles)*; weight of shell 5.9 kg.
2—30 mm/65 AK-630; 6 barrels per mounting; 3,000 rds/min combined to 2 km.
2—30 mm AO-18KD; 6 barrels per mounting; 10,000 rds/min to 4 km *(2.1 n miles)*
Torpedoes: 4—21 in *(533 mm)* (2 twin) tubes.
A/S Mortars: 1 RBU 6000 12-tubed trainable.
Mines: 2 rails. 48 mines.
Physical countermeasures: Decoys: 4 PK 10 chaff launchers.
Electronic countermeasures: ESM/ECM: 2 Bell Shroud. 2 Bell Squat. Intercept and jammers.
Radars: Air/surface search: Pozitiv ME 1.2; I/J-band.
Fire control: Bass Tilt; H/I-band (for guns). Band Stand (Mineral ME); D-band (for SS-N-25).
Navigation: Nayada; I-band.
IFF: 2 Square Head. 1 Salt Pot B.
Ox Tail; VDS; active search and attack; medium frequency.
Sonars: Ox Yoke; hull-mounted; active search and attack; medium frequency.
Weapon control systems: Band Stand datalink.

Helicopters: Platform for 1 medium.

Programmes: Contract signed with Rosoboronexport in late 2005 for the procurement of two Modified Gepard 3.9 class-frigates. The contract for construction was signed with Zelenodolsk Shipyard on 22 December 2006. Delivery of the first ship was made in October 2010. The second ship was loaded on the transport ship *Eide Transporter* on 25 May 2011 and was delivered to Vietnam in August 2011. The contract for the construction of two further ships at Zelenodolsk was signed in February 2013. These are reported to be Anti-Submarine Warfare (ASW) variants. Building work began in September 2013.
Operational: Details of weapons and sensors are speculative and based on those originally designated for the ships in Russian naval service.

DINH TIEN HOANG 10/2010, Lemachko Collection / 1406344

DINH TIEN HOANG 5/2011, Zelenodolsk Shipyard / 1454880

© 2014 IHS IHS Jane's Fighting Ships 2014-2015

Vietnam > Frigates — Corvettes

5 PETYA (PROJECT 159A/AE) CLASS (FFL)

HQ 09 (ex-SKR-82) HQ 11 (ex-SKR-96) HQ 13 (ex-SKR-141) HQ 15 (ex-SKR-130) HQ 17 (ex-SKR-135)

Displacement, tonnes: 965 standard; 1,199 full load
Dimensions, metres (feet): 81.8 × 9.1 × 2.9
  (268.4 × 29.9 × 9.5)
Speed, knots: 32
Range, n miles: 4,870 at 10 kt, 450 at 29 kt
Complement: 98 (8 officers)

Machinery: CODAG; 2 gas turbines; 30,000 hp(m) (22 MW); 1 Type 61V-3 diesel; 5,400 hp(m) (3.97 MW) sustained; centre shaft; 3 shafts
Guns: 4 USSR 3 in (76 mm)/59 AK 726 (2 twin); 90 rds/min to 15 km (8 n miles); weight of shell 5.9 kg. 4—37 mm (2 twin) (HQ 11). 4—23 mm (2 twin) (HQ 11, 15).
Torpedoes: 3—21 in (533 mm) (triple) tubes (Petya III). SAET-60; passive homing to 15 km (8.1 n miles) at 40 kt; warhead 400 kg. 5—16 in (406 mm) (1 quin) tubes (Petya II). SAET-40; active/passive homing to 10 km (5.5 n miles) at 30 kt; warhead 100 kg.
A/S Mortars: 4 RBU 6000 12-tubed trainable (HQ 09, 13, 17); range 6,000 m; warhead 31 kg. 4 RBU 2500 16-tubed trainable (HQ 11); range 2,500 m; warhead 21 kg.
Mines: Can carry 22.
Depth charges: 2 racks.
Electronic countermeasures: ESM: 2 Watch Dog; radar warning.
Radars: Air/surface search: Strut Curve; F-band.
  Navigation: Don 2; I-band.
  Fire control: Hawk Screech; I-band.
IFF: High Pole B. 2 Square Head.
Sonars: Vychada MG 311; hull-mounted; active attack; high frequency.

HQ 13 6/2012 / 1486476

Programmes: All built at Khabarovsk. Two Petya III (HQ 09, 11) (export version) transferred from USSR in December 1978 and three Petya IIs, (HQ 13, 15, 17); HQ 13 transferred in December 1983; HQ 15 in May 1984 and HQ 17 in December 1984.
Modernisation: Refitted and updated 1994 to 1999. The RBUs or torpedo tubes replaced by 23 mm guns in some of the class. HQ 17 completed major overhaul at Ba Son Shipyard in 2001.
Structure: The Petya IIIs have the same hulls as the Petya IIs but different armament.
Operational: Reported active between the coast and the Spratly Islands.

HQ 15 (PETYA II) 6/2008, Lemachko Collection / 1305385

0 + 2 SIGMA 9814 CLASS (FS)

Displacement, tonnes: 2,400 full load
Dimensions, metres (feet): 98 × 14.02 × ? (321.5 × 46.0 × ?)
Speed, knots: 28. **Range, n miles:** 4,000 at 18 kt
Complement: 91

Machinery: 2 diesels; 2 shafts; cp props
Missiles: SSM: To be announced.
  SAM: MBDA VL MICA.
Guns: 1 Oto Melara 3 in (76 mm).
Radars: Air/surface search: Thales SMART-S; 3D; E/F-band.
  Fire control: Thales STING; I/J-band.
  Navigation: To be announced.
Combat data systems: Thales Tacticos.
Electro-optic systems: Thales STING optronic tracker.
Helicopters: To be announced.

Programmes: It was announced by Damen Schelde Naval Shipbuilding on 22 August 2013 that agreement had been reached to deliver two Sigma 9814 frigates. While the building programme has not been announced, it is likely that one ship will be built in the Netherlands and one in a Damen yard in Vietnam. The ship is broadly similar to the two Sigma 9813 frigates built for Morocco in 2011–12. Thales Netherlands is reported to be responsible for the electronics and sensor suite.

CORVETTES

Notes: Ex-US Admirable class HQ 07 is an alongside training hulk.

1 BPS 500 (PROJECT 12418) CLASS (FSGM)

HQ 381

Displacement, tonnes: 525 full load
Dimensions, metres (feet): 62 × 11 × 2.5
  (203.4 × 36.1 × 8.2)
Speed, knots: 32. **Range, n miles:** 2,200 at 14 kt
Complement: 28

Machinery: 2 MTU diesels; 19,600 hp(m) (14.41 MW); 2 Kamewa waterjets
Missiles: SSM: 8 Zvezda SS-N-25 Switchblade (KH-35 Uran) (2 quad) ❶; active radar homing to 130 km (70.1 n miles) at 0.9 Mach; warhead 145 kg.
  SAM: SA-N-10. 24 missiles.
Guns: 1—3 in (76 mm)/59 AK 176 ❷; 120 rds/min to 15 km (8 n miles); weight of shell 5.9 kg. 1—30 mm/65 AK 630 ❸; 6 barrelled; 3,000 rds/min combined to 2 km. 2—12.7 mm MGs.
Mines: Rails fitted.
Physical countermeasures: Decoys: 2 chaff launchers ❹.
Radars: Air/surface search: Cross Dome ❻; E/F-band.
  Navigation: I-band.
  Fire control: Bass Tilt ❼; H/I-band.
Weapon control systems: Optronic director ❺.

Comment: Severnoye design (improved Pauk) ordered in 1996 and two ships subsequently delivered in kit form to Ba Son Shipyard, Ho Chi Minh City. First unit launched in June 1998 and became operational in late 2001. The second unit was not completed.

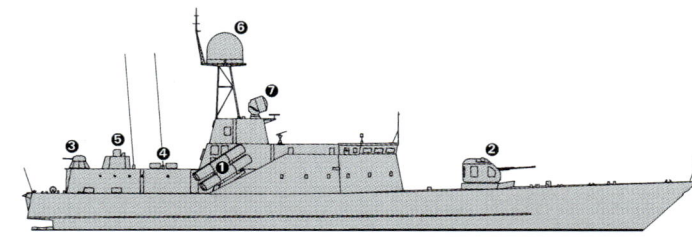

BPS 500 (not to scale), Ian Sturton / 0530054

HQ 381
6/2008
1353656

IHS Jane's Fighting Ships 2014-2015   © 2014 IHS

4 TARANTUL CLASS (PROJECT 1241RE) (FSGM)

HQ 371–372   HQ 374   HQ 378

Displacement, tonnes: 391 standard; 457 full load
Dimensions, metres (feet): 56.1 × 11.5 × 2.5 *(184.1 × 37.7 × 8.2)*
Speed, knots: 36. **Range, n miles:** 2,000 at 20 kt, 400 at 36 kt
Complement: 41 (5 officers)

Machinery: 2 Nikolayev Type DR 77 gas turbines; 16,016 hp(m) *(11.77 MW)* sustained; 2 Nikolayev Type DR 76 gas turbines with reversible gearboxes; 4,993 hp(m) *(3.67 MW)* sustained; 2 shafts
Missiles: SSM: 4 SS-N-2D Styx; IR homing to 83 km *(45 n miles)* at 0.9 Mach; warhead 513 kg; sea-skimmer at end of run.
SAM: SA-N-5 Grail quad launcher; manual aiming; IR homing to 6 km *(3.2 n miles)* at 1.5 Mach; warhead 1.5 kg.
Guns: 1—3 in *(76 mm)*/59 AK 176; 120 rds/min to 15 km *(8 n miles)*; weight of shell 5.9 kg.
2—30 mm/65 AK 630; 6 barrels per mounting; 3,000 rds/min combined to 2 km.
Physical countermeasures: Decoys: 2 PK 16 chaff launchers.
Radars: Air/surface search: Plank Shave; E-band.
Navigation: Pechora; I-band.
Fire control: Bass Tilt; H/I-band.
IFF: Salt Pot, Square Head A.
Sonars: Foal Tail; active; high frequency.
Weapon control systems: Hood Wink optronic director.

Programmes: First pair ordered in October 1994. These were new hulls exported at a favourable price and completed by 1996. Some delay in delivery because of late payments, but both were in service by April 1996. Two further vessels were ordered in 1999 and delivered in 2000.
Operational: Based at Da Nang.

HQ 376   10/2009 / 1366255

LAND-BASED MARITIME AIRCRAFT

Notes: (1) There are six Air Force Su-27 Flankers and 20 Su-22 Fitter-H that can be used for maritime surveillance.
(2) Procurement of up to six P-3C Orion maritime patrol aircraft from the United States was reportedly under consideration in 2013.

Numbers/Type: 2 PZL Mielec M-28 B1R Bryza.
Operational speed: 181 kt *(335 km/h)*.
Service ceiling: 13,770 ft *(4,200 m)*.
Range: 736 n miles *(1,365 km)*.
Role/Weapon systems: Polish-built aircraft originally based on the USSR Cash light transport. Contract in October 2003 for the procurement of up to ten aircraft configured for maritime surveillance. First two delivered in 2005 by late 2004. Sensors: MSC-400 mission system, ARS-400 radar (with SAR/ISAR modes). Weapons: to be announced.

HQ 378   6/2007, Mazumdar Collection / 1170244

M-28 (Polish colours)   6/2003, J Ciślak / 0567493

PATROL FORCES

Notes: (1) At least one (HQ 205) Shanghai II class PC may still be operational.
(2) Some of the craft listed may be transferred to the Coast Guard and Maritime Police.

6 SVETLYAK (PROJECT 1041.2) CLASS (PGM)

HQ 261   HQ 263–267

Displacement, tonnes: 371 full load
Dimensions, metres (feet): 49.5 × 9.2 × 2.4 *(162.4 × 30.2 × 7.9)*
Speed, knots: 30
Range, n miles: 2,200 at 13 kt
Complement: 28 (4 officers)

Machinery: 3 diesels; 15,900 hp(m) *(11.85 MW)* sustained; 3 shafts; cp props
Missiles: SAM: SA-N-10; shoulder launched and (manual aiming); IR homing to 5 km *(2.7 n miles)* at 1.7 Mach; warhead 1.5 kg.
Guns: 1—3 in *(76 mm)*/59 AK 176; 120 rds/min to 15 km *(8 n miles)*; weight of shell 5.9 kg.
1—30 mm/65 AK 630; 6 barrels; 3,000 rds/min combined to 2 km.
Physical countermeasures: Decoys: 2 chaff launchers.
Radars: Air/surface search: Peel Cone; E-band.
Fire control: Bass Tilt; H/I-band.
Navigation: Palm Frond B; I-band.
Weapon control systems: Hood Wink optronic director.

Comment: Contract for two craft signed with Almaz, St Petersburg in November 2001. First vessel launched on 17 July 2002 and second on 30 July 2002. Following acceptance on 17 October 2002, both vessels were shipped from St Petersburg on 14 December 2002. Two further vessels were laid down at Almaz Shipyard on 26 June 2009. The first was launched in November 2010 and the second on 22 April 2011. Two further vessels (HQ 264, 265) were built by Almaz and delivered in March 2012 and the fifth and sixth, built by Vostochnaya Shipyard, Vladivostok (HQ 266, 267) were commissioned in 2012.

HQ 371   6/2007, Mazumdar Collection / 1353655

2 + 8 TARANTUL V CLASS (PROJECT 1241.8) (FSGM)

HQ 375–376   +8

Displacement, tonnes: 391 standard; 457 full load
Dimensions, metres (feet): 59.9 × 11.5 × 2.5 *(196.5 × 37.7 × 8.2)*
Speed, knots: 36. **Range, n miles:** 2,000 at 20 kt, 400 at 36 kt
Complement: 41 (5 officers)

Machinery: 2 Nikolayev Type DR 77 gas turbines; 16,016 hp(m) *(11.77 MW)* sustained; 2 Nikolayev Type DR 76 gas turbines with reversible gearboxes; 4,993 hp(m) *(3.67 MW)* sustained; 2 shafts
Missiles: SSM: 16 (4 quad) SS-N-25 (Kh 35 Uran); active radar homing to 130 km *(70.2 n miles)* at 0.9 Mach; warhead 145 kg; sea-skimmer.
SAM: SA-N-5 Grail quad launcher; manual aiming; IR homing to 6 km *(3.2 n miles)* at 1.5 Mach; warhead 1.5 kg.
Guns: 1—3 in *(76 mm)*/59 AK 176; 120 rds/min to 15 km *(8 n miles)*; weight of shell 5.9 kg.
2—30 mm/65 AK 630; 6 barrels per mounting; 3,000 rds/min combined to 2 km.
Physical countermeasures: Decoys: 2 PK 16 chaff launchers.
Radars: Air/surface search: Strut Curve (Pozitiv-ME); I-band.
Surface search: Plank Shave; E/F-band (SS-N-25).
Fire control: Bass Tilt; H/I-band.
Navigation: Pechora; I-band.
Weapon control systems: Hood Wink optronic director.

Comment: A contract was signed in March 2004 for the supply of 10 modified Tarantul V, armed with SS-N-25 (Kh 35 Uran). Two of these, built by Vympel Shipyard, Rybinsk, were delivered in late 2007. The remaining craft are being built under licence in Vietnam. By October 2013, the first two had begun sea trials, the next two were being fitted out and the following two had begun construction.

HQ 261   9/2009 / 1366254

1012 Vietnam > Patrol forces

8 OSA II (PROJECT 205) CLASS
(FAST ATTACK CRAFT – MISSILE) (PTFG)

| HQ 354 | HQ 358–361 | HQ 384–386 |

Displacement, tonnes: 249 full load
Dimensions, metres (feet): 38.6 × 7.6 × 2.7 (126.6 × 24.9 × 8.9)
Speed, knots: 37
Range, n miles: 500 at 35 kt
Complement: 30

Machinery: 3 Type M 504 diesels; 10,800 hp(m) (7.94 MW) sustained; 3 shafts
Missiles: SSM: 4 SS-N-2B Styx; active radar or IR homing to 46 km (25 n miles) at 0.9 Mach; warhead 513 kg.
Guns: 4 USSR 30 mm/65 (2 twin); 500 rds/min to 5 km (2.7 n miles); weight of shell 0.54 kg.
Radars: Surface search: Square Tie; I-band.
Fire control: Drum Tilt; H/I-band.
IFF: High Pole. 2 Square Head.

Comment: Transferred from USSR: two in October 1979, two in September 1980, two in November 1980 and two in February 1981. All based at Da Nang. Operational status doubtful.

HQ 359   3/2010, Flor Van Otterdyk / 1406343

5 TURYA (PROJECT 206M) CLASS
(FAST ATTACK CRAFT – HYDROFOIL) (PCK)

HQ 331–335

Displacement, tonnes: 193 standard; 254 full load
Dimensions, metres (feet): 39.6 × 7.6 oa; 12.5 over foils × 1.8 hull; 4 over foils (129.9 × 24.9; 41.0 × 5.9; 13.1)
Speed, knots: 40
Range, n miles: 600 at 35 kt (foilborne); 1,450 at 14 kt (hullborne)
Complement: 30

Machinery: 3 Type M 504 diesels; 10,800 hp(m) (7.94 MW) sustained; 3 shafts
Guns: 2 USSR 57 mm/75 AK 725 (twin, aft); 120 rds/min to 12.7 km (6.8 n miles); weight of shell 2.8 kg. 2 USSR 25 mm/80 (twin, fwd); 270 rds/min to 3 km (1.6 n miles); weight of shell 0.34 kg.
Torpedoes: 4 – 21 in (533 mm) tubes (not in all).
Depth charges: 2 racks.
Radars: Surface search: Pot Drum; H/I-band.
Fire control: Muff Cob; G/H-band.
IFF: High Pole B. Square Head.
Sonars: Foal Tail (not in all); VDS; high frequency.

Comment: Transferred from USSR: two in mid-1984, one in late 1984, two in January 1986. Two more acquired from Russia. Two of the five do not have torpedo tubes or sonar. HQ 321 has been decommissioned and a few may be non-operational.

TURYA 331   5/2000, Bob Fildes / 0105741

4 SHERSHEN (PROJECT 206T) CLASS
(FAST ATTACK CRAFT) (PTFM)

| HQ 301 | HQ 305 | +2 |

Displacement, tonnes: 147 standard; 173 full load
Dimensions, metres (feet): 34.7 × 6.7 × 1.5 (113.8 × 22.0 × 4.9)
Speed, knots: 45
Range, n miles: 850 at 30 kt, 460 at 42 kt
Complement: 23

Machinery: 3 Type 503A diesels; 8,025 hp(m) (5.9 MW) sustained; 3 shafts
Missiles: SAM: 1 SA-N-5 Grail quad launcher; manual aiming; IR homing to 6 km (3.2 n miles) at 1.5 Mach; altitude to 2,500 m (8,000 ft); warhead 1.5 kg.
Guns: 4 USSR 30 mm/65 (2 twin); 500 rds/min to 5 km (2.7 n miles); weight of shell 0.54 kg.
Torpedoes: 4 – 21 in (533 mm) tubes (not in all).
Mines: Can carry 6.
Depth charges: 2 racks (12).
Radars: Surface search: Pot Drum; H/I-band.
Fire control: Drum Tilt; H/I-band.
IFF: High Pole A. Square Head.

Comment: A total of 16 transferred from USSR: two in 1973, two in April 1979 (without torpedo tubes), two in September 1979, two in August 1980, two in October 1980, two in January 1983 and four in June 1983. Most have been cannibalised for spares.

HQ 305   3/2011, Jurg Kürsener / 1454882

4 + (12) STOLKRAFT CLASS (PBR)

HQ 56–59

Displacement, tonnes: 45 full load
Dimensions, metres (feet): 22.4 × 7.5 × 1.2 (73.5 × 24.6 × 3.9)
Speed, knots: 30
Complement: 7
Machinery: 2 MTU 12V 183 TE93 diesels; 2,301 hp(m) (1.69 MW) sustained; 2 Doen waterjets 1 Volvo Penta diesel; 360 hp(m) (265 kW); 1 shaft
Guns: 1 Oerlikon 20 mm.

Comment: Four built by Oceanfast Marine, Western Australia and delivered in early 1997. Trimaran construction forward, transforming into a catamaran at the stern. Shallow draft needed for inshore and river operations. The centreline single shaft is used for loitering. The craft show the colours of the Customs department. Up to 12 more may have been built in Vietnam but this has not been confirmed.

STOLKRAFT   8/2005, Kuvel/Marsan / 1154063

4 MODIFIED ZHUK CLASS (PB)

| HQ 37 | HQ 55 | BP-29-01-01 | BP-29-98-01 |

Displacement, tonnes: 39 full load
Dimensions, metres (feet): 29 × ? × ? (95.1 × ? × ?)
Speed, knots: 30
Complement: 11 (3 officers)
Machinery: 2 Saab Scania diesels; 2,500 hp(m) (18.64 MW) sustained; 2 shafts
Guns: 2 – 12.7 mm MGs (2 twin).
Radars: Navigation: I-band.

Comment: Built in Vietnam to design based on Zhuk class.

HQ 55 (under construction)   8/2000, P Marsan / 0105744

14 ZHUK (PROJECT 1400M) CLASS (PB)

| T 864 | T 874 | T 880–881 | +10 |

Displacement, tonnes: 40 full load
Dimensions, metres (feet): 24 × 5 × 1.2 (78.7 × 16.4 × 3.9)
Speed, knots: 30. **Range, n miles:** 1,100 at 15 kt
Complement: 11 (3 officers)
Machinery: 2 Type M 401B diesels; 2,200 hp(m) (1.6 MW) sustained; 2 shafts
Guns: 4—14.5 mm (2 twin) MGs.
Radars: Surface search: Spin Trough; I-band.

Comment: Transferred: three in 1978, three in November 1979, one in November 1981, one in May 1985, three in February 1986, two in December 1989, two in January 1990, three in January 1996, two in January 1998 and two in April 1998. So far seven have been deleted but operational numbers are uncertain. Some are allocated to the Coast Guard.

7 BP-29-12-01 PATROL CRAFT (PB)

BP-06-12-01	BP-33-12-01	BP-06-12-02
BP-29-12-01	BP-06-11-02	BP-13-12-01
BP-33-11-01		

Machinery: 2 diesels; 2 shafts
Guns: 2—12.7 mm MGs.
Radars: Navigation: I-band.

Comment: Indigenously-built patrol craft of an unknown type.

BP-13-12-01   6/2013* / 1531189

RIVER PATROL CRAFT

Comment: There are large numbers of river patrol boats, mostly armed with MGs. A 14.5 m craft ordered from Singapore TSE in 1994. More are being built locally with Volvo Penta engines.

RIVER PATROL BOAT   8/2000 / 0105743

3 TT400TP CLASS (PG)

HQ 272–274

Displacement, tonnes: 420 full load
Dimensions, metres (feet): 54.16 × 9.16 × 2.7 (177.7 × 30.1 × 8.9)
Speed, knots: 32
Range, n miles: 2,500 at 15 kt
Complement: 35

Machinery: 2 MTU 16V 595 TE890 diesels; 16,856 hp (12.57 MW); 3 shafts
Guns: 1—3 in (76 mm)/59 AK 176; 120 rds/min to 15 km (8 n miles); weight of shell 5.9 kg.
2—30 mm/65 AK 630; 6 barrels per mounting; 3,000 rds/min combined to 2 km.
Radars: Surface search: To be announced.
Fire control: Bass Tilt; H/I-band.
Navigation: To be announced.
Weapon control systems: Optronic director (to be announced).

Comment: The first of a new class of patrol vessels, built at the Z713 Shipbuilding Plant, Hai Phong, was commissioned on 16 January 2012. The design appears to be based on the MTD FC54 design (known locally as TT400TP) of which four have been built for the Vietnamese Marine Police. Steel hull and aluminium superstructure. There are davits for a Rigid Hull Inflatable Boat (RHIB) on the starboard side aft. A second ship was commissioned on 31 August 2012 and a third on 28 May 2014.

© 2014 IHS

HQ 272   10/2011 / 1454881

2 TP-01 CLASS (PC)

| HQ 251 | HQ 253 |

Displacement, tonnes: 170 standard; 215 full load
Dimensions, metres (feet): 42 × 6.0 × 1.8 (137.8 × 19.7 × 5.9)
Speed, knots: 31
Range, n miles: 1,100 at 13 kt
Complement: 31
Machinery: 3 diesels; 7,500 hp (5.6 MW); 3 shafts
Guns: 2—57 mm (twin). 2—37 mm (twin).
Radars: Surface search: Pot Head; I-band.

Comment: Indigenously built at Ba Son Shipyard, Ho Chi Minh City (Saigon), both ships entered service in about 1980. Probably based on the Soviet SO-1-class submarine chaser on which details are based.

AMPHIBIOUS FORCES

Notes: There are two landing ships HQ 521 and 522 of approximately 68 m. They are known as the Nau Dinh class.

HQ 521   10/2009 / 1366257

3 POLNOCHNY (PROJECT 771) CLASS (LCM)

| HQ 511 (ex-SDK-71) | HQ 512 (ex-SDK-112) | HQ 513 (ex-SDK-74) |

Displacement, tonnes: 772 standard; 847 full load
Dimensions, metres (feet): 75 × 9.6 × 2.3 (246.1 × 31.5 × 7.5)
Speed, knots: 19
Complement: 40

Machinery: 2 Kolomna Type 40-D diesels; 4,400 hp(m) (3.2 MW) sustained; 2 shafts
Guns: 2 or 4 USSR 30 mm/65 (1 or 2 twin).
Rockets: 2—140 mm rocket launchers.
Radars: Surface search: Spin Trough; I-band.
Fire control: Drum Tilt; H/I-band.

Comment: Transfers from USSR: one in May 1979 (B), one in November 1979 (A) and one in February 1980 (B). Details are for Polnochny B class. All are reported to be in poor condition.

HQ 512 and 513   6/1995, Giorgio Arra / 0084254

IHS Jane's Fighting Ships 2014-2015

1014 Vietnam > Amphibious forces — Mine warfare forces

3 TANK LANDING SHIPS (LST)

TRAN KHANH DU (ex-*Da Nang*; ex-*Maricopa County LST 938*) HQ 501
VUNG TAU (ex-*Cochino County LST 603*) HQ 502
QUI NONH (ex-*Bulloch County LST 509*) HQ 503

Displacement, tonnes: 4,145 full load
Dimensions, metres (feet): 100 × 15.2 × 4.3 *(328.1 × 49.9 × 14.1)*
Speed, knots: 11. **Range, n miles:** 6,000 at 10 kt
Complement: 110
Machinery: 2 GM 12-567A diesels; 1,800 hp *(1.34 MW)*; 2 shafts
Guns: 8 Bofors 40 mm/60 (2 twin, 4 single). 4 Oerlikon 20 mm.

Comment: HQ 501 is LST 1-510 class and HQ 502 and 503 are LST 512–1152 class. All built in 1943–44. Transferred from US to South Vietnam in mid-1960s. Seldom seen at sea.

TRAN KHANH DU   8/2000 / 0105745

30 LANDING CRAFT (LCM AND LCU)

Comment: About five LCUs, 12 LCM 8 and LCM 6, and three LCVPs remain of the 180 minor landing craft left behind by the USA in 1975. In addition there are about 10 T4 LCUs acquired from the USSR in 1979.

LCU   6/2001 / 0131363

MINE WARFARE FORCES

2 YURKA (RUBIN) (PROJECT 266) CLASS
(MINESWEEPERS—OCEAN) (MSO)

HQ 851   HQ 885

Displacement, tonnes: 549 full load
Dimensions, metres (feet): 52.4 × 9.4 × 2.6 *(171.9 × 30.8 × 8.5)*
Speed, knots: 17. **Range, n miles:** 1,500 at 12 kt
Complement: 45

Machinery: 2 Type M 503 diesels; 5,350 hp(m) *(3.91 MW)* sustained; 2 shafts
Guns: 4 USSR 30 mm/65 (2 twin); 500 rds/min to 5 km *(2.7 n miles)*; weight of shell 0.54 kg.
Mines: 10.
Radars: Surface search: Don 2; I-band.
Fire control: Drum Tilt; H/I-band.
Sonars: Stag Ear; hull-mounted; active minehunting; high frequency.

Comment: Transferred from USSR December 1979. Steel-hulled, built in early 1970s.

YURKA (Egyptian colours)   10/1998, F Sadek / 0017818

4 SONYA (YAKHONT) (PROJECT 1265) CLASS
(MINESWEEPERS/HUNTER—COASTAL) (MHSC)

HQ 861   HQ 862 (ex-BT-228)   HQ 863 (ex-BT-296)   HQ 864 (ex-BT-212)

Displacement, tonnes: 457 full load
Dimensions, metres (feet): 48 × 8.8 × 2 *(157.5 × 28.9 × 6.6)*
Speed, knots: 15
Range, n miles: 3,000 at 10 kt
Complement: 43

Machinery: 2 Kolomna 9-D-8 diesels; 2,000 hp(m) *(1.47 MW)* sustained; 2 shafts
Guns: 2 USSR 30 mm/65 AK 630. 2—25 mm/80 (twin).
Mines: 8.
Radars: Surface search: Nayada; I-band.
Sonars: MG 69/79; active; high frequency.

Comment: First one transferred from USSR 16 February 1987, second in February 1988, third in July 1989, fourth in March 1990. Two based at Da Nang.

SONYA 862   5/2000, R Fildes / 0105746

2 YEVGENYA (KOROND) (PROJECT 1258) CLASS
(MINEHUNTERS—INSHORE) (MHI)

HQ 871   HQ 782

Displacement, tonnes: 91 full load
Dimensions, metres (feet): 24.6 × 5.5 × 1.5 *(80.7 × 18.0 × 4.9)*
Speed, knots: 11
Range, n miles: 300 at 10 kt
Complement: 10

Machinery: 2 Type 3-D-12 diesels; 600 hp(m) *(440 kW)* sustained; 2 shafts
Guns: 2 USSR 25 mm/80 (twin).
Mines: 8.
Radars: Surface search: Spin Trough; I-band.
Sonars: MG 7; active; high frequency.

Comment: First transferred from USSR in October 1979; two in December 1986. One deleted in 1990.

YEVGENYA (Ukraine colours)   6/2003, Ships of the world / 0572652

5 K 8 (PROJECT 361T) CLASS
(MINESWEEPING BOATS) (PBR)

Displacement, tonnes: 26 full load
Dimensions, metres (feet): 16.9 × 3.2 × 0.8 *(55.4 × 10.5 × 2.6)*
Speed, knots: 18
Complement: 6
Machinery: 2 Type 3-D-6 diesels; 300 hp(m) *(220 kW)* sustained; 2 shafts
Guns: 2—14.5 mm (twin) MGs.

Comment: Transferred from USSR in October 1980. Probably used as river patrol craft.

SURVEY AND RESEARCH SHIPS

1 KAMENKA (PROJECT 870) CLASS (AGS)

Displacement, tonnes: 772 full load
Dimensions, metres (feet): 53.5 × 9.1 × 2.6 (175.5 × 29.9 × 8.5)
Speed, knots: 14. **Range, n miles:** 4,000 at 10 kt
Complement: 25
Machinery: 2 Sulzer diesels; 1,800 hp(m) (1.32 MW); 2 shafts; cp props
Radars: Navigation: Don 2; I-band.

Comment: Transferred from USSR December 1979. Built at Northern Shipyard, Gdansk in the late 1960s. May be civilian manned.

KAMENKA (Russian colours) *1984 / 0506127*

1 SURVEY SHIP (AGSH)

Name	No	Builders	Laid down	Launched	Commissioned
TRAN DAI NGHIA	–	Song Thu, Danang	26 Jul 2008	Oct 2010	26 Nov 2011

Dimensions, metres (feet): 66.3 × 13.2 × 3.1 (217.5 × 43.3 × 10.2)
Speed, knots: 12. **Range, n miles:** 3,000 at 12 kt
Complement: 13 + 58 spare berths
Machinery: Diesel-electric; 3 diesel generators; 2,652 hp(m) (1.95 MW); 1 motor; 1,073 hp(m) (800 kW); 2 azimuth thrusters; 1 bow thruster
Radars: Navigation: E/F- and I-band.
Sonars: Multi and single beam; high frequency; active.

Comment: Damen 6613 design multipurpose hydrographic and oceanographic survey, training and mine countermeasures vessel. The ship is to be a derivative of the Snellius-class vessels built for the RNLN and is similar to the ship that entered Thai service in 2008. This ship was built by Song Thu Company in Danang, Vietnam from a design- and material package supplied by Damen Shipyards, Gorinchem. The hydrographical survey equipment is to be supplied by Atlas and the ship is capable of oil-spill response. The ship completed trials in May 2011 and was handed over on 26 November 2011.

DAMEN 6613 *12/2005, Damen Shipyards / 1159224*

AUXILIARIES

Notes: In addition to the vessels listed below there are two YOG 5 fuel lighters, two floating cranes, two ex-USSR unarmed Nyryat 2 diving tenders, an ocean tug HQ 950 and approximately 10 harbour tugs.

1 VODA (PROJECT 561) CLASS (WATER TANKER) (AWT)

BO 82 (ex-MVT 19)

Displacement, tonnes: 2,149 full load
Dimensions, metres (feet): 81.5 × 11.5 × 4.3 (267.4 × 37.7 × 14.1)
Speed, knots: 12. **Range, n miles:** 3,000 at 10 kt
Complement: 38
Machinery: 2 diesels; 1,600 hp (1.2 MW); 2 shafts
Radars: Navigation: Don 2; I-band.

Comment: Built by Yantar, Kaliningrad, in the 1950s. Probably an ex-Russian Pacific Fleet unit transferred in about 1996. Carries about 1,000 tons of water.

1 TRANSPORT SHIP (AP)

Name	No	Builders	Laid down	Launched	Commissioned
TRUONG SA	HQ 571	Z189 Shipyard, Haiphong	2010	5 Oct 2011	2012

Displacement, tonnes: 2,050 standard
Dimensions, metres (feet): 71 × 13.2 × ? (232.9 × 43.3 × ?)
Speed, knots: 16.
Complement: 20 + 180 spare berths
Machinery: To be announced.
Radars: To be announced.

Comment: New transport ship, said to be the largest to be designed and built in Vietnam was launched in October 2011 and delivered in 2012.

15 OFFSHORE SUPPLY VESSELS (AKL)

HQ 606–609	HQ 627–628	HQ 647	HQ 687
HQ 621–622	HQ 630–633	HQ 649	TRUONG HQ 966

Measurement, tonnes: 1,016 dwt
Dimensions, metres (feet): 70.6 × 11.8 × 4 (231.6 × 38.7 × 13.1)
Speed, knots: 12
Complement: 30
Machinery: 1 diesel; 1 shaft

Comment: Details are for HQ 966 launched at Halong Shipyard in June 1994. This is one of a group of freighters used by the Navy for coastal transport, and to service the Spratleys garrison. The ships are of various sizes and are likely to be armed with machine guns.

BD 621 (old number) *3/1997 / 0050742*

2 FLOATING DOCKS

Comment: One has a lift capacity of 8,500 tons. Transferred from USSR August 1983. Second one (Khersson) has a lift capacity of 4,500 tons and was supplied in 1988.

1 SORUM (PROJECT 745) CLASS (ATA)

HQ 957 (ex-BD 105)

Displacement, tonnes: 1,687 full load
Dimensions, metres (feet): 58 × 12.6 × 4.6 (190.3 × 41.3 × 15.1)
Speed, knots: 14
Range, n miles: 3,500 at 13 kt
Complement: 35
Machinery: Diesel-electric; 2 Type 2-DW2 diesel generators; 2,900 hp (2.13 MW); 1 motor; 2,000 hp (1.47 MW); 1 shaft
Radars: Navigation: I-band.

Comment: Ocean tug built at Yaroslavl. Transferred from Russia in 1995.

HQ 957 *6/2013* / 1531190*

COAST GUARD

Notes: The Marine Police (Cuc Canh Sat Bien) was formed in 1998 and subsequently became the Coast Guard on 12 October 2013. In addition to the craft listed there are 12 35-m TT120 patrol craft (3001–3012), 12 44-m TT200 patrol craft (2001–2008, 2010–2011, 2013–2014), four converted Shershen-class (Project 206) patrol craft (CB 5011–5014), three Damen 4207 search-and-rescue vessels (SAR 411–413) and three Damen 4612 Salvage tugs (9001–9004).

4 TT400TP CLASS (PBO)

4031–4034

Displacement, tonnes: 420 full load
Dimensions, metres (feet): 54.16 × 9.16 × 2.7 (177.7 × 30.1 × 8.9)
Speed, knots: 32
Range, n miles: 2,500 at 15 kt
Complement: 35
Machinery: 3 MTU 16V 595 TE890 diesels; 16,856 hp (12.57 MW); 3 shafts
Radars: Surface search: To be announced.
Navigation: To be announced.

Comment: Very similar to the TT400TP class built for the Vietnamese Navy. The design appears to be based on the MTD FC54 design. Steel hull and aluminium superstructure. There are davits for a RHIB on the starboard side aft.

Vietnam > Coast guard — Yemen > Introduction

1 + 3 DAMEN 9014 CLASS (PSO)

Name	No	Builders	Laid down	Launched	Commissioned
–	8001	Haiphong	17 Dec 2010	23 Oct 2012	27 Nov 2013
–	8002	Song Thu, Da Nang	20 Jun 2012	2013	2014
–	–	Ha Long Shipbuilding	15 Nov 2012	Dec 2013	2014
–	–	Ha Long Shipbuilding	15 Nov 2012	2014	2014

Displacement, tonnes: 2,400 full load
Dimensions, metres (feet): 90.5 × 14.0 × 3.4 *(296.9 × 45.9 × 11.2)*
Speed, knots: 22
Range, n miles: 4,000 at 18 kt
Complement: 40

Machinery: 4 Caterpillar 3516 diesels; 12,015 hp *(8.96 MW)*; 2 shafts; 1 bow thruster
Guns: 2—25 mm (twin).
Radars: Surface search: To be announced.
Navigation: To be announced.

Helicopters: One medium.

Comment: Four offshore patrol vessels are being built in Vietnam to the 9014 OPV (DN 2000) design provided by Damen Shipyards Group, Netherlands. The first was built by the Ministry of Defence-owned 189 Shipbuilding Company (Z189) in Hai Phong. The second ship is being built at Da Nang to a slightly modified design. Two further vessels are under construction at Ha Long.

OPV 9014 (artist's impression)   6/2012, DAMEN / 1452180

Virgin Islands (UK)

Country Overview

A British dependency, the British Virgin Islands are situated in the eastern Caribbean Sea at the northern end of the Leeward Islands in the Lesser Antilles chain. Puerto Rico lies some 52 n miles to the west. Comprising a group of 36 islands, 16 of them inhabited, and more than 20 islets and cays there are four main islands: Tortola (21 square miles); Anegada (15 square miles); Virgin Gorda (8 square miles); and Jost Van Dyke (3.5 square miles). Other inhabited islands include Peter Island, Cooper Island, Beef Island, Salt Island, and Norman Island. The capital, only town and principal port is Road Town, Tortola. Territorial seas (3 n miles) and a Fishery Zone (200 n miles) are claimed. The remainder of the Virgin Islands form a separate external territory of the US.

Headquarters Appointments
Commander Marine Unit: Chief Inspector St Clair Amory

Bases
Road Town, Tortola

POLICE

Notes: (1) There is also a 12 m Scarab, fitted with three 225 hp outboard motors, and a 10 m Mako with two 150 hp outboards.
(2) Two Dauntless class 12 m patrol boats are operated by the US Virgin Islands whose waters are also patrolled by USCG craft.

1 DAUNTLESS CLASS (PATROL CRAFT) (PB)

ST URSULA

Displacement, tonnes: 18 full load
Dimensions, metres (feet): 16.8 × 4.9 × 1.5 *(55.1 × 16.1 × 4.9)*
Speed, knots: 32.
Range, n miles: 300 at 28 kt
Complement: 4
Machinery: 2 Caterpillar C-15 diesels; 1,600 hp *(1.2 MW)*; 2 shafts
Radars: Navigation: Furuno; I-band.

Comment: Dauntless design craft constructed by SeaArk Marine, Monticello, AR, and delivered on 1 December 2006 to replace previous vessel of same name decommissioned in 2003. Aluminium construction. The craft is employed on drug interdiction, combatting illegal immigration, search and rescue and border control duties.

ST URSULA   12/2006, SeaArk Marine / 1167666

Yemen

Country Overview

The Republic of Yemen was formed in 1990 through the union of the People's Democratic Republic of Yemen and the Yemen Arab Republic. The country includes the islands of Socotra, Kamaran and Perim. With an area of 207,285 square miles, it is situated on the south-west coast of the Arabian Peninsula and is bordered to the north by Saudi Arabia and to the east by Oman. It has a 1,030 n mile coastline with the Red Sea and the Gulf of Aden, which are linked by a strategic strait, the Bab el Mandeb. The capital and largest city is Sanaa while the principal ports are Aden and Al Hudaydah. Territorial seas (12 n miles) are claimed. A 200 n mile EEZ has been claimed but the limits have only been partly defined by boundary agreements.

Headquarters Appointments

Commander, Naval Forces:
   Major General Abdullah Salem Abdullah

Headquarters Appointments – continued

Commander, Coast Guard:
   Brigadier Ali Rasa'a

Personnel

2014: 1,700 naval plus 500 marines

Bases

Main: Aden, Hodeida

Secondary: Mukalla, Perim (new base under construction), Socotra, Al Katib
Coast Defence regions: Al Ghaydah, Aden and Cameron Island

Coast Defence

Two mobile SS-C-3 Styx batteries. Some 100 mm guns installed in tank turrets at Perim Island.

PATROL FORCES

Notes: (1) In addition there are two 'Osa IIs', 122 and 124. Osa 122 is reported to be seagoing.
(2) Three 32 m Halter Marine Broadsword patrol craft have been reported: *26th of September* (141); *Sanaa* (200); *Ghamdan* (300).
(3) There are 14 11 m Sea Spirit patrol craft built by Al Yousuf, UAE.

3 HOUNAN (TYPE 021) CLASS
(FAST ATTACK CRAFT—MISSILE) (PTG)

Displacement, tonnes: 174 standard; 208 full load
Dimensions, metres (feet): 38.6 × 7.6 × 2.7 *(126.6 × 24.9 × 8.9)*
Speed, knots: 34
Range, n miles: 800 at 30 kt
Complement: 28

Machinery: 3 Type 42-160 diesels; 12,000 hp(m) *(8.8 MW)* sustained; 3 shafts
Missiles: SSM: 4 YJ-8 (Eagle Strike) (C-801); inertial cruise; active radar homing to 40 km *(22 n miles)* at 0.9 Mach; warhead 165 kg; sea-skimmer.
Guns: 4 30 mm/(2 twin AK 230); 500 rds/min to 5 km *(2.7 n miles)*.
Radars: Surface search: Square Tie; I-band.
Fire control: Rice Lamp; H/I-band.
IFF: 2 Square Head. High Pole A.

Comment: A variation of the Chinese Huangfen (Osa 1 type) class design. Delivered on 6 June 1995 at Aden having been built by the China Shipbuilding Corporation and completed in 1993. Payment was delayed by the Yemeni civil war. Based at Al Katib. One vessel (former 128) ran aground in September 1997 but was salvaged and may be operational again. One vessel (with new pennant number) is in a reasonable state of repair but is not armed with missiles.

HOUNAN 1017   4/2009, Shaun Jones / 1366258

1 TARANTUL I CLASS (PROJECT 1241) (FSGM)

124 (ex-971)

Displacement, tonnes: 391 standard; 589 full load
Dimensions, metres (feet): 56.1 × 11.5 × 2.5 *(184.1 × 37.7 × 8.2)*
Speed, knots: 36
Range, n miles: 400 at 36 kt, 2,000 at 20 kt
Complement: 50

Machinery: 2 Nikolayev Type DR 77 gas turbines; 16,016 hp(m) *(11.77 MW)* sustained; 2 Nikolayev Type DR 76 gas turbines with reversible gearboxes; 4,993 hp(m) *(3.67 MW)* sustained; 2 shafts
Missiles: SSM: 4 SS-N-2C Styx (2 twin) launchers; active radar or IR homing to 83 km *(45 n miles)* at 0.9 Mach; warhead 513 kg; sea-skimmer at end of run.
SAM: SA-N-5 Grail quad launcher; manual aiming; IR homing to 10 km *(5.4 n miles)* at 1.5 Mach; altitude to 2,500 m *(8,000 ft)*; warhead 1.1 kg.
Guns: 1—3 in *(76 mm)*/59 AK 176; 120 rds/min to 5.9 km *(3.8 n miles)*; weight of shell 7 kg.
2—30 mm/65 AK 630; 6 barrels per mounting; 3,000 rds/min to 2 km.
Physical countermeasures: Decoys: 2 PK 16 chaff launchers.
Radars: Air/surface search: Plank Shave (also for missile control); E-band.
Navigation: Spin Trough; I-band.
Fire control: Bass Tilt; H/I-band.
IFF: Square Head. High Pole.
Weapon control systems: Hood Wink optronic director.

Programmes: Two export versions of the ship originally delivered from the USSR. First one on 7 December 1990, second on 15 January 1991. One decommissioned by 2001.
Operational: Facilities for servicing missiles in Aden were destroyed in mid-1994. This remaining ship is still in a reasonable state of repair although probably without missiles.

TARANTUL 124   10/1995 / 0016612

10 AUSTAL PATROL SHIPS (PB)

P 1022–1031

Displacement, tonnes: 91 full load
Dimensions, metres (feet): 37.5 × 7.2 × 2.2 *(123 × 23.6 × 7.2)*
Speed, knots: 29
Range, n miles: 1,000 at 25 kt
Complement: 19 (3 officers)
Machinery: 2 Caterpillar 3512 diesels; 3,500 hp *(2.61 MW)*; 2 shafts
Guns: 2—14.5 mm (twin). 2—12.7 mm MGs.

Comment: Contract with Austal Ships on 9 June 2003 for a total of 10 patrol craft. All were shipped to Yemen in February 2005. Of aluminium construction, the design is based on the Bay class Australian Customs vessels. The contract included engineering and practical training for 60 Yemeni crew.

P 1022   3/2007 / 1170245

AMPHIBIOUS FORCES

Notes: Ropucha 139 is an alongside hulk.

1 NS-722 CLASS (LSMM)

BILQIS

Displacement, tonnes: 1,405 full load
Dimensions, metres (feet): 90 × 9.7 × 2.4 *(295.3 × 31.8 × 7.9)*
Speed, knots: 18
Complement: 49

Military lift: 5 T-72 tanks and 111 marines
Machinery: 2 Caterpillar diesels; 5,670 hp *(4.2 MW)*; 2 shafts
Missiles: SAM: SA-16 or ZM Mesko.
Guns: 4 ZSU-23-2MR Wrobel 23 mm (2 twin).

Comment: Ordered in late 1999 for delivery in 2002, development of the Polnochny class built by Naval Shipyard Gdynia, Poland. Shipped from Poland to Yemen on 24 May 2002. Roles include disaster relief and cadet training as well as amphibious warfare.

BILQIS   10/2001, J Ciślak / 0131343

3 DEBA CLASS (PROJECT NS-717) (LCU)

HIMYER (ex-*Dhaffar*)   SAMBA   ABDULKORI (ex-*Thamoud*)

Displacement, tonnes: 225 full load
Dimensions, metres (feet): 41 × 7.1 × 1.7 *(134.5 × 23.3 × 5.6)*
Speed, knots: 15. **Range, n miles:** 500 at 14.5 kt
Complement: 10

Military lift: 16 tons and 50 troops
Machinery: 2 Cummins diesels; 2 shafts
Guns: 2 ZU-23-2MR Wrobel 23 mm/87 (1 twin). 2—12.7 mm MGs.
Radars: Navigation: I-band.

Comment: Ordered from Poland in October 1999 and delivered in mid-2001. AK-630 CIWS may also be fitted at a later date.

ABDULKORI (on transport ship)   5/2001, J Ciślak / 0131342

MINE WARFARE FORCES

1 NATYA CLASS (PROJECT 266ME)
(MINESWEEPER—OCEAN) (MSO)

201

Displacement, tonnes: 817 full load
Dimensions, metres (feet): 61 × 10.2 × 3 *(200.1 × 33.5 × 9.8)*
Speed, knots: 16
Range, n miles: 3,000 at 12 kt
Complement: 67

Machinery: 2 Type M 504 diesels; 5,000 hp(m) *(3.67 MW)* sustained; 2 shafts; cp props
Guns: 4 – 30 mm/65 (2 twin); 500 rds/min to 5 km *(2.7 n miles)*; weight of shell 0.54 kg.
4 – 25 mm/80 (2 twin); 270 rds/min to 3 km *(1.6 n miles)*; weight of shell 0.34 kg.
A/S Mortars: 2 RBU 1200 five-tubed fixed launchers; range 1,200 m; warhead 34 kg.
Mines: 10
Physical countermeasures: Decoys: Carries contact, acoustic and magnetic sweeps.
Radars: Surface search: Don 2; I-band.
Sonars: MG 69/79; hull-mounted; active minehunting; high frequency.

Comment: Transferred from USSR in February 1991. Operational status doubtful. A second of class was delivered to Ethiopia in October 1991 but sheltered in Aden for a time in 1992.

NATYA 201   6/2002, Rahn/Globke / 0530089

AUXILIARIES

Notes: (1) A 4,500 ton Floating Dock acquired from the USSR.
(2) A 14 m Hydrographic craft acquired from Cougar Marine in 1988.
(3) An oil-pollution control craft acquired in 1999.
(4) Two Toplivo class tankers, *135* and *140*, are reported to have been decommissioned.

COAST GUARD

Notes: (1) The Yemeni Coast Guard was established in 2002 and began operating in 2003. Its tasks include counter-smuggling and immigration control duties, SAR, fishery protections, environmental protection and pollution control. Its headquarters are at Sana'a with regional headquarters at Aden (Gulf of Aden), Hodeidah (Red Sea) and Mukalla (Arabian Sea). There are plans to establish a coastal radar system with Italian assistance.
(2) In addition to the craft listed, there are reported to be a Fairey Marine Tracker II (1034), four Plascoa 15 m fast patrol craft (1501–1504) and three Geraldton 23 m patrol craft (2201–2203).

P 2202   3/2007 / 1170246

2 MARINE PROTECTOR CLASS (WPB)

SANA'A 2601   **ADEN** 2602

Displacement, tonnes: 92 full load
Dimensions, metres (feet): 26.5 × 5.8 × 1.6 *(86.9 × 19.0 × 5.2)*
Speed, knots: 25
Range, n miles: 900 at 8 kt
Complement: 10 (1 officer)
Machinery: 2 MTU 8V 396 TE94 diesels; 2,680 hp(m) *(1.97 MW)* sustained; 2 shafts
Guns: 2 – 12.7 mm MGs.
Radars: Navigation: I-band.

Comment: Designed by David M Cannell based on the hull of the Damen Stan Patrol 2600 which is in service with the Hong Kong police. Steel hull with aluminium superstructure. Contract awarded on 11 September 2009 to Bollinger Shipyard, Lockport, LA, for the construction of two patrol craft for the Yemen Coast Guard. Delivery was made in 2011. Two are in service in the Maltese Armed Forces and 73 in the USCG.

SANA'A   3/2011* / 1531191

4 ARCHANGEL CLASS (RESPONSE BOATS) (PBF)

Dimensions, metres (feet): 12.8 × 4.0 × 0.8 *(42 × 13.1 × 2.6)*
Speed, knots: 40. **Range, n miles:** 300 at 25 kt
Complement: 4
Machinery: 2 Caterpillar diesels; 2 Hamilton 322 waterjets
Guns: 1 – 12.7 mm MG.
Radars: To be announced.

Comment: High-speed fast response craft of aluminium construction and foam collar built by SAFE Boats International, Port Orchard, Washington. Donated by the US government in October 2005.

ARCHANGEL CLASS   6/2006, SAFE BOATS / 1167667

12 DEFENDER CLASS (RESPONSE BOATS) (PBF)

0801–0812

Displacement, tonnes: 3 full load
Dimensions, metres (feet): 7.6 × 2.6 × 2.7 *(24.9 × 8.5 × 8.9)*
Speed, knots: 46. **Range, n miles:** 175 at 35 kt
Complement: 4
Machinery: 2 Mercury outboard motors; 450 hp *(335 kW)*
Guns: 1 – 12.7 mm MG.
Radars: To be announced.

Comment: High-speed inshore patrol craft of aluminium construction and foam collar built by SAFE Boats International, Port Orchard, Washington. Donated by the US government in October 2005.

DEFENDER 0804   6/2006, SAFE BOATS / 1167668

8 PATROL CRAFT (PC)

1301–1308

Displacement, tonnes: 18 full load
Dimensions, metres (feet): 13.4 × 3.8 × 1.2 *(44 × 12.5 × 3.9)*
Speed, knots: 13. **Range, n miles:** 200 at 11 kt
Complement: 4
Machinery: 2 General Motors Detroit 6V53 diesels; 2 shafts

Comment: Former US Coast Guard lifeboats constructed in the 1960s. Transferred on 16 February 2004.

6 BAKLAN (CMN 15–60) CLASS (HSIC)

| BAKLAN 1201 | ZUHRAB 1203 | HUNAISH 1205 |
| SIYAN 1202 | AKISSAN 1204 | ZAKR 1206 |

Displacement, tonnes: 12 full load
Dimensions, metres (feet): 15.5 × 3 × 0.8 *(50.9 × 9.8 × 2.6)*
Speed, knots: 55
Range, n miles: 400 at 30 kt
Complement: 4
Machinery: 2 diesels; 2 surface drives
Guns: 2 — 12.7 mm MGs.
Radars: Surface search: Furuno; I-band.

Comment: Ordered from CMN Cherbourg on 3 March 1996. First five were delivered 1 August 1996 and the last one in mid-1997. Top speed in Sea States up to 3. Composite hull construction.

BAKLAN CLASS
8/1996, C M N Cherbourg
0084259

Zimbabwe

Country Overview

The Republic of Zimbabwe gained independence on 17 April 1980. Formerly the British colony of Southern Rhodesia and, between 1953 and 1963, part of the Federation of Rhodesia and Nyasaland (now Malawi), a unilateral declaration of independence on 11 November 1965 precipitated a turbulent period of guerilla war. This eventually led to a peace settlement in 1979 and elections in 1980. A landlocked country with an area of 150,873 square miles, it is situated in central southern Africa and is bordered to the north by Zambia, to the east by Mozambique, to the south by South Africa and to the west by Botswana and Namibia. It has a shoreline of approximately 350 n miles with Lake Kariba, artificially formed by the Kariba Dam, from which the country gets much of its electric power. The capital, largest city and commercial centre is Harare (formerly Salisbury). The railway system is linked to the port of Beira in Mozambique.

Bases

Kariba, Binga.

PATROL FORCES

5 RODMAN 790 CLASS (PB)

Displacement, tonnes: 2 full load
Dimensions, metres (feet): 8.1 × 2.72 × 0.7 *(26.6 × 8.9 × 2.3)*
Speed, knots: 30
Complement: 2
Machinery: 2 Volvo Penta TAMD diesels

Comment: GRP hull. Built in 1999 by Rodman, Vigo. Operated by Zimbabwe Police.

RODMAN 790 *6/1999, Rodman* / 0570999

3 RODMAN 38 CLASS (PB)

Displacement, tonnes: 10 full load
Dimensions, metres (feet): 11 × 3.9 × 0.7 *(36.1 × 12.8 × 2.3)*
Speed, knots: 28
Range, n miles: 300 at 15 kt
Complement: 4
Machinery: 2 diesels; 2 waterjets

Comment: GRP hull. Built in 1999 by Rodman, Vigo. Operated by Zimbabwe Police.

RODMAN 38 *6/1999, Rodman* / 0571000

2 RODMAN 46HJ CLASS (PB)

Displacement, tonnes: 13 full load
Dimensions, metres (feet): 14 × 3.8 × 0.6 *(45.9 × 12.5 × 2.0)*
Speed, knots: 30
Range, n miles: 350 at 18 kt
Complement: 4
Machinery: 2 Caterpillar 3280 diesels; 850 hp *(633 kW)*

Comment: GRP hull. Built in 1999 by Rodman, Vigo. Operated by Zimbabwe Police.

RODMAN 46 *6/1999, Rodman* / 0570998

INDEXES

Indexes

Country abbreviations

Alb	Albania	DR	Dominican Republic	Kwt	Kuwait	SA	South Africa
Alg	Algeria	Ecu	Ecuador	Lat	Latvia	Sam	Samoa
Ana	Anguilla	Egy	Egypt	Lby	Libya	SAr	Saudi Arabia
Ang	Angola	ElS	El Salvador	Leb	Lebanon	Sen	Senegal
Ant	Antigua and Barbuda	EqG	Equatorial Guinea	Lit	Lithuania	Ser	Serbia
Arg	Argentina	Eri	Eritrea	Mac	Macedonia, Former Yugoslav Republic of	Sey	Seychelles
Aust	Australia	Est	Estonia			Sin	Singapore
Az	Azerbaijan	ETim	East Timor	Mad	Madagascar	SL	Sierra Leone
Ban	Bangladesh	Fae	Faroe Islands	Mex	Mexico	Slo	Slovenia
Bar	Barbados	FI	Falkland Islands	MI	Marshall Islands	Sol	Solomon Islands
Bel	Belgium	Fij	Fiji	Mic	Micronesia, Federated States of	Spn	Spain
Ben	Benin	Fin	Finland	Mld	Maldives	Sri	Sri Lanka
Bhm	Bahamas	Fra	France	Mlt	Malta	StK	St Kitts and Nevis
Bhr	Bahrain	Gab	Gabon	Mlw	Malawi	StL	St Lucia
Blz	Belize	Gam	Gambia	Mly	Malaysia	StV	St Vincent and the Grenadines
Bmd	Bermuda	GB	Guinea-Bissau	Mon	Montenegro	Sud	Sudan
Bol	Bolivia	Geo	Georgia	Mor	Morocco	Sur	Suriname
Bru	Brunei	Ger	Germany	Moz	Mozambique	Swe	Sweden
Brz	Brazil	Gha	Ghana	Mtn	Mauritania	Swi	Switzerland
Bul	Bulgaria	Gn	Guinea	Mrt	Mauritius	Syr	Syria
Cam	Cameroon	Gra	Grenada	Myn	Myanmar	Tan	Tanzania
Can	Canada	Gre	Greece	Nam	Namibia	Tkm	Turkmenistan
Cay	Cayman Islands	Gua	Guatemala	NATO	NATO	Tld	Thailand
Chi	Chile	Guy	Guyana	Nic	Nicaragua	Tog	Togo
CI	Cook Islands	HT	Haiti	Nig	Nigeria	Ton	Tonga
Cmb	Cambodia	HK	Hong Kong	Nld	Netherlands	TT	Trinidad and Tobago
Col	Colombia	Hon	Honduras	Nor	Norway	Tun	Tunisia
Com	Comoros	Hun	Hungary	NZ	New Zealand	Tur	Turkey
ConB	Congo-Brazzaville	Ice	Iceland	Omn	Oman	Tuv	Tuvalu
ConD	Congo, Democratic Republic	Ind	India	Pak	Pakistan	Twn	Taiwan
CPR	China, People's Republic	Indo	Indonesia	Pal	Palau	UAE	United Arab Emirates
CpV	Cape Verde	Iran	Iran	Pan	Panama	UK	United Kingdom
CR	Costa Rica	Iraq	Iraq	Par	Paraguay	Ukr	Ukraine
Cro	Croatia	Ire	Ireland	Per	Peru	Uru	Uruguay
Ctl	Côte d'Ivoire	Isr	Israel	Plp	Philippines	US	United States
Cub	Cuba	Ita	Italy	PNG	Papua New Guinea	Van	Vanuatu
Cypr	Cyprus (Republic)	Jam	Jamaica	Pol	Poland	Ven	Venezuela
Den	Denmark	Jor	Jordan	Por	Portugal	VI	Virgin Islands, UK
Dji	Djibouti	Jpn	Japan	Qat	Qatar	Vtn	Vietnam
Dom	Dominica	Kaz	Kazakhstan	RoK	Korea, Republic of (South)	Yem	Yemen
DPRK	Korea, Democratic People's Republic (North)	Ken	Kenya	Rom	Romania	Zim	Zimbabwe
		Kir	Kiribati	Rus	Russian Federation		

Named ships

Ship	Page
1st Lt Baldomero Lopez (US)	977
1st Lt Harry L Martin (US)	981
1st Lt Jack Lummus (US)	977
2nd Lt John P Bobo (US)	977
3 De Febrero (Par)	617
3 De Noviembre (Pan)	612
4 De Noviembre (Pan)	612
5 De Noviembre (Pan)	612
5 Février 1979 (ConB)	182
6 Of October (Egy)	221
7 De Agosto (Col)	174
10 De Noviembre (Pan)	612
10 Juin 1991 (ConB)	182
11 De Noviembre (Col)	176
15 Août 1960 (ConB)	182
18 Mart (Tur)	856
18 Of June (Egy)	221
20 De Julio (Col)	174
21 Of October (Egy)	221
23 Of July (Egy)	221
25 Of April (Egy)	221
28 De Noviembre (Pan)	612
31 Juillet 1968 (ConB)	182
60 Letiya Pogranvoysk (Rus)	732

A

Ship	Page
A 33-35 (Lby)	501
A 61 (UAE)	891
A 72 (Ind)	352
A 521, 530, 542-543 (Sri)	791
A 641 (Alg)	7
A 648, 671 (Az)	42
A 702, 753, 755 (Swe)	804
A Gad (Egy)	222
A Leblanc (Can)	109
Aachen (UK)	924
Aadesh (Ind)	357
Aamchit (Leb)	497
AB 27, 29, 36 (Tur)	863
AB 2000-2005 (Aust)	37
Abad (Iran)	391
Abadejo (Arg)	23
Abalone (Can)	106
Abanto (Spn)	785
Abay (Kaz)	462
Abdul Aziz (SAr)	740, 742
Abdul Halim Perdanakusuma (Indo)	360
Abdul Rahman Al Fadel (Bhr)	48
Abdulkori (Yem)	1017
Abeetha II (Sri)	787
Abeille Bourbon (Fra)	277
Abeille Flandre (Fra)	278
Abeille Languedoc (Fra)	278
Abeille Liberté (Fra)	277
Aber-Wrach (Fra)	279
Abha (SAr)	737
Abhay (Ind)	341
Abheek (Ind)	357
Abhinav (Ind)	357
Abhiraj (Ind)	357
Abigail Burgess (US)	989
Able (US)	976
Abou Abdallah El Ayachi (Mor)	549
Aboubekr Ben Amer (Mtn)	525
Abraham Campo (Plp)	631
Abraham Lincoln (US)	942
Abrolhos (Brz)	82
Absalon (Den)	200
Abu Bakr (Ban)	54
Abu Dhabi (UAE)	885
Abu El Ghoson (Egy)	225
Abu Obaidah (SAr)	740
Abu Qir (Egy)	219
Abukuma (Jpn)	441, 456
Aby (Ctl)	184
Acamar (DR)	207
Acamar (Mex)	535
Acanthe (Fra)	274
Acheloos (Gre)	312
Achernar (Alg)	9
Achernar (Mex)	535
Achéron (Fra)	268
Achilleus (Gre)	312
Achimota (Gha)	300
Achook (Ind)	357
Aconit (Fra)	259
Acrux (DR)	207
Acrux (Mex)	535
Active (US)	985
Acuario (Mex)	537
Adak (US)	988
Adam Kok (SA)	763
Addriyah (SAr)	741
Adelaide (Aust)	31
Adelie (US)	987
Aden (Yem)	1018
ADF 104, 106-107 (Per)	625
Adhara (Mex)	535
Adhara II (Arg)	23
ADI 01-04 (Mex)	540
Adil (Mly)	516
Aditya (Ind)	352
Adm W M H Callaghan (US)	982
Admiral Butakov (Rus)	698
Admiral Chabanenko (Rus)	698
Admiral Cowan (Est)	234
Admiral Essen (Rus)	698
Admiral Golovko (Rus)	698
Admiral Gorshkov (Rus)	698
Admiral Grigorovich (Rus)	698
Admiral Horia Macelariu (Rom)	662
Admiral Isakov (Rus)	698
Admiral Istomin (Rus)	698
Admiral Kasatonov (Rus)	698
Admiral Kornilov (Rus)	698
Admiral Kuznetsov (Rus)	686
Admiral Levchenko (Rus)	692
Admiral Makarov (Rus)	698
Admiral Nevelsky (Rus)	704
Admiral Panteleyev (Rus)	692
Admiral Petre Barbuneanu (Rom)	663
Admiral Tributs (Rus)	692
Admiral Ushakov (Rus)	693
Admiral Vinogradov (Rus)	692
Admiral Vladimirskiy (Rus)	709
Admiral Yuri Ivanov (Rus)	712
Adomya (Ban)	57
Adour (Fra)	279
ADRI XXXII-LVIII (Indo)	377
Adrias (Gre)	305
Adventure (NZ)	575
Aegeon (Gre)	305
Aegir (Ice)	325
Afiat (Bru)	88
Afif (SAr)	741
Afonso Cerqueira (Por)	651
Afonso De Albuquerque (Por)	653
Africana (SA)	765
AG 5-6 (Tur)	868
Agalina (Bul)	92
Agamemnon (UK)	900
Al Agami (Egy)	227
Agathos (Cypr)	192
Ägir (Swe)	803
Agradoot (Ban)	59
Agray (Ind)	341
Agrim (Ind)	357
Aguascalientes (Mex)	541
Águia (Por)	653
Aguila (Mex)	537
Aguila I-V (Spn)	785
Aguilar (Per)	625
Aguirre (Per)	620
S A Agulhas II (SA)	764
Ahalya Bai (Ind)	356
Aheloi (Bul)	93
Ahi (US)	987
Ahmad El Fateh (Bhr)	48
Ahmad Yani (Indo)	360
Ahmadi (Kwt)	493
Al-Ahmadi (Kwt)	490
Ahmed Es Sakali (Mor)	549
Ahn Jung-Geun (RoK)	474
Al Ahweirif (Lby)	501
Aias (Gre)	312
El Aigh (Mor)	550
Aigle (Fra)	269
Ailette (Fra)	278
Ain Zaghouan (Tun)	853
Air Cushion-Tei (1-6) Gou (Jpn)	444
Airavat (Ind)	349
Aishima (Jpn)	445
Aisling (Ire)	394
Ait Baâmrane (Mor)	550
Aittitos (Gre)	308
Aiyar Lulin (Myn)	557
Aiyar Mai (Myn)	557
Aiyar Maung (Myn)	557
Aiyar Minthamee (Myn)	557
Aiyar Minthar (Myn)	557
Ajak (Indo)	366
Ajax (UK)	900
Ajay (Ind)	341
Ajeera (Bhr)	48
Ajonc (Fra)	274
Ajral (Ind)	354
AK 1-2, 6 (Ger)	296
AK 209, 223-224, 248, 582, 599 (Rus)	732
Akademik Alexsandrov (Rus)	711
Akademik Isanin (Rus)	710
Akademik Kovalev (Rus)	711
Akademik Seminikhin (Rus)	710
Akademik Vladimir Kotelnikov (Rus)	719
Akagi (Jpn)	456
Akaishi (Jpn)	452
Akamas (Cypr)	193
Akar (Tur)	866
Akbaş (Tur)	870
Akçakoca (Tur)	865
Akçay (Tur)	865
Akdu (Egy)	226
Akebono (Jpn)	436
Akhir Nahr (Alg)	8
El Akid (Mor)	548
Akigumo (Jpn)	457
Akin (Bul)	92
Akin (Tur)	868
Akissan (Yem)	1019
Akitsushima (Jpn)	451

1026 **INDEXES** > Named ships

Briz (Rus) 728
Broad Run (US) 968
Brocklesby (UK) 913
Bromo (Indo) 374
Broome (Aust) 33
Bruce C Heezen (US) 975
Bruinvis (Nld) 562
Brunei (Aust) 32
Brunswick (US) 980
Brutus (Aust) 38
BT 100, 114-115, 212, 262, 230, 232, 256 (Rus) 708
BT 730 (Rus) 707
Bu Chon (RoK) 481
Bubiyan (Kwt) 491
Buckthorn (US) 990
Budenovsk (Rus) 703
Budiman (Mly) 516
Budstikken (Den) 198
Buena Vista (US) 968
Buenaventura (Col) 178
Buenos Aires (Arg) 22, 24
Buffalo (US) 938
Buffle (Fra) 277
Bug (Rus) 726
Bukhansan (RoK) 488
Bukovina (Ukr) 882
Bukowo (Pol) 643
Bulkeley (US) 948
Bull Ray (Sin) 758
Bullarijia (Tun) 854
Bulwark (UK) 911
Bums (Ger) 295
Buna (PNG) 615
Bundaberg (Aust) 33
Bundeena (Aust) 38
Bunga Mas Enam (Mly) 515
Bunga Mas Lima (Mly) 515
Bungo (Jpn) 444
LTG William B Bunker (US) 961
Bunker Hill (US) 944
Burakreis (Tur) 856
Buran (Rus) 722
Buratti (Ita) 423
Burau (Mly) 517
Burdi (Lby) 501
Burevi (Mld) 522
Burgas (Bul) 93
Burgazada (Tur) 860
Burlington (US) 980
Burq (Pak) 611
Burujulasad (Indo) 372
Burullus (Egy) 225
Al Burullus (Egy) 226
Burutu (Nig) 581
Burya (Rus) 700
Bushehr (Iran) 391
Al Bushra (Omn) 595
Butinah (UAE) 888
Butrindi (Alb) 3
Butterfly Ray (Sin) 758
SGT William R Button (US) 977
Büyükada (Tur) 860
Buzo Sobenes (Chi) 123
Byblos (Leb) 497
Bystry (Rus) 693

C

Ç 120, 128, 132, 139-158, 305, 321-327, 329-331 (Tur) 864
C 21-24 (UK) 912
C 63, 131-138, 140-153 (Ind) 356
C 154-168, 401-436 (Ind) 357
Cabaleiro (Spn) 783
Cabezo (Mex) 539
Cabo Blanco (CR) 183
Cabo Blanco (Per) 627
Cabo Catoche (Mex) 536
Cabo Corrientes (Arg) 22
Cabo Corrientes Point Warde (Col) 179
Cabo Corzo (Mex) 536
Cabo De Gata (Spn) 784
Cabo De Hornos (Arg) 20
Cabo De Hornos (Chi) 123
Cabo De La Vella (Col) 179
Cabo Fradera (Spn) 774
Cabo Manglares Point Wells (Col) 179
Cabo Reyes (Chi) 122
Cabo San Juan (EqG) 232
Cabo Tiburon (Col) 179
Caboclo (Brz) 79
Cabrales (Chi) 121
Cacheu (GB) 318
Cacine (GB) 318
Cacine (Por) 652
Cacique Agateyte (Nic) 577
Cacique Diriangen (Nic) 577
Cagarras (Brz) 81
Caieiras (Brz) 81
Caio Duilio (Ita) 406
Cakra (Indo) 359
Calaboza (US) 968
Calabrese (Ita) 422
Calamar (Arg) 23
Calanus II (Can) 113
Calchaqui (Arg) 21
Caldas (Col) 172
Caldera (Chi) 126
Calgary (Can) 100
Calicuchima (Ecu) 213
California (US) 937
Calima (Col) 178
Adm W M H Callaghan (US) 982
Callao (Per) 622
Calmaria (Por) 656
Caloyeras (Per) 624
Camana (Per) 626
Camaron (Arg) 23
Camboriú (Brz) 80
El Camino Español (Spn) 780
Campbell (US) 985
Campo (Cam) 96
Çanakkale (Tur) 856
Canal Beagle (Arg) 20
Canal Bocayna (Spn) 782
Canal Costanero (Arg) 24
Canal De Beagle (Arg) 22
Canal Emilio Mitre (Arg) 24
Canarias (Spn) 768
Canberra (Aust) 31
Cancas (Per) 626
Cancer (Mex) 537
Çandarli (Tur) 866
El Caney (US) 968
Cangzhou (CPR) 146
Cankarso (Ind) 348
Cannanore (Ind) 350
Canonchet (US) 974
Canopus (DR) 206
Canopus (Mex) 535
Canopus (Por) 655
Canopus (Ven) 1007
Cantabria (Spn) 780
Canterbury (NZ) 576
Cap Aupaluk (Can) 112
Cap Aux Meules (Can) 112
Cap Breton (Can) 112
Cap De Rabast (Can) 112
Cap D'Espoir (Can) 112
Cap Nord (Can) 112
Cap Percé (Can) 112
Cap Rozier (Can) 112
Cap Tourmente (Can) 112
Capana (Ven) 1005
Capayán (Arg) 21
Cape Ann (Can) 112
Cape Bojeador (Plp) 635
Cape Byron (Aust) 39
Cape Calvert (Can) 112
Cape Caution (Can) 112
Cape Chaillon (Can) 112
Cape Cockburn (Can) 112
Cape Commodore (Can) 112
Cape Dauphin (Can) 112
Cape Decision (US) 982
Cape Diamond (US) 982
Cape Discovery (Can) 112
Cape Domingo (US) 982
Cape Douglas (US) 982
Cape Ducato (US) 982
Cape Dundas (Can) 112
Cape Edensaw (Can) 112
Cape Edmont (US) 982
Cape Farewell (Can) 112
Cape Fox (Can) 112
Cape Hearne (Can) 112
Cape Henry (US) 982
Cape Horn (US) 982
Cape Hudson (US) 982
Cape Hurd (Can) 110
Cape Inscription (US) 982
Cape Intrepid (US) 982
Cape Isabel (US) 982
Cape Island (US) 982
Cape Jervis (Aust) 39
Cape Kennedy (US) 982
Cape Knox (US) 982
Cape Kuper (Can) 112
Cape Lambton (Can) 112
Cape Leveque (Aust) 39
Cape Light (Can) 111
Cape May (US) 982
Cape Mckay (Can) 112
Cape Mercy (Can) 112
Cape Mohican (US) 982
Cape Mudge (Can) 112
Cape Naden (Can) 112
Cape Nelson (Aust) 39
Cape Norman (Can) 112
Cape Orlando (US) 982
Cape Palmerston (Can) 112
Cape Providence (Can) 112
Cape Race (US) 982
Cape Ray (US) 982
Cape Rescue (Can) 112
Cape Rise (US) 982
Cape Roger (Can) 108
Cape Sorell (Aust) 39
Cape Spry (Can) 112
Cape St George (Aust) 39
Cape St George (US) 944
Cape St James (Can) 112
Cape Storm (Can) 112
Cape Sutil (Can) 112
Cape Taylor (US) 982
Cape Texas (US) 982
Cape Trinity (US) 982
Cape Victory (US) 982
Cape Vincent (US) 982
Cape Washington (US) 982
Cape Wessel (Aust) 39
Cape Wrath (US) 982
Cape York (Aust) 39
Capella (DR) 206
Capella (Mex) 535
Capella (US) 983
Caph (Mex) 536
Capitaine De Frégate Awore Paul (Gab) 280
Capitaine Moulié (Fra) 275
Capitán Bretel (Bol) 68
Capitán Cabral (Par) 616
Capitan David Eyama Angue Osa (EqG) 232
Capitán De Fragata Pedro Sáinz De Baranda (Mex) 535
Capitán De Navio Blas Godinez (Mex) 533
Capitán De Navio Sebastian José Holzinger (Mex) 533
Capitán Miranda (Uru) 998
Capitán Ortiz (Par) 616
Capitán Prat (Chi) 118
Capitán Rigoberto Giraldo (Col) 173
Capitán Vladimir Valek Moure (Col) 178
Caporal Kaeble V C (Can) 109
Capotillo (DR) 206
Cappelletti (Ita) 422
Caprera (Ita) 419
Capri (Ita) 419
Capricia (Ita) 417
Capricorne (Fra) 269
Capricorno (Mex) 537
Capstan (US) 992
Captain Goddard MSM (Can) 109
Car Nicobar (Ind) 348
Carabiniere (Ita) 408
Cardiel (Arg) 24
Cardigan Bay (UK) 919
Carecare (Ven) 1007
Caribe (Ven) 1001
Caribou (Can) 104
Caribou Isle (Can) 110
Carina (Den) 198
Carl M Brashear (US) 974
Carl Vinson (US) 942
Carlo Bergamini (Ita) 408
Carlo Margottini (Ita) 408
Carlos Albert (Plp) 631
Carlos Chagas (Brz) 86
Carlos Galindo (Col) 176
Carlos Manuel De Cespedes (Cub) 191
Carlskrona (Swe) 803
Carney (US) 946
Caroly (Ita) 417
Caroni (TT) 849
Carrasco (Per) 623
Carreca (Ita) 422
Carrera (Chi) 116
Carrillo (Per) 623
Carson City (US) 980
Cartagena De Indias (Col) 178
SSGT Edward A Carter (US) 977
Carter Hall (US) 967
Carthage (Tun) 851
Caruanta (Ven) 1007
Casabianca (Fra) 248
Cascadura (TT) 849
Casma (Chi) 121
Casma (Per) 627
Casotti (Ita) 423
Cassard (Fra) 256
Cassiopea (Ita) 412
Cassiopée (Fra) 269
Cassiopeia (Por) 653
Castilla (Per) 622
Castilla (Spn) 776
Castor (Bel) 64
Castor (DR) 206
Castor (Swi) 807
Casuarina (Aust) 34
Cataguazes (Brz) 81
Catarindo (Per) 627
Catawba (US) 978
Cattistock (UK) 913
Cavatorto (Ita) 422
Cavour (Ita) 404
Cavtat (Cro) 188
Cayman Defender (Cay) 115
Cayman Guardian (Cay) 115
Cayman Protector (Cay) 115
Cayo Macereo (Ven) 1007
Cazadora (Spn) 773
CB 001-002, 101-102, 201-205, 211-217, 301 (Mly) 512
CC Karamoko Cheik Conde (Gn) 318
Ceará (Brz) 81
Cebu (Plp) 630
Cedar Run (US) 968
Cekal (Mly) 516
Celurit (Indo) 376
Centauro (Mex) 537
Centauro (Por) 653
Centenario De La Revolución (Mex) 534
Centinela (Spn) 774
Centurion (Sin) 750
Céphée (Fra) 269
Cerberus (Nld) 571
César Chávez (US) 974
Cēsis (Lat) 494
Çesme (Tur) 866
Cetina (Cro) 187
Cezayirli Gazi Hasan Paşa (Tur) 866
CG 001-002, 012-018 (TT) 849
CG 10, 14, 19, 22, 46 (Sri) 791
CG 18, 40-41 (Sri) 792
CG 121-124, 131-137 (Jam) 427
CGC 103, 110, 115, 128-130, 132-136 (Plp) 636
CHA 28-30, 34-38 (Fra) 275
Chacabuco (Chi) 122
Chacachacare (TT) 850
Chacagua (Mex) 540
Chacal (Fra) 271
Chacao (Chi) 125
Chaconia (TT) 850
Chaffee (US) 948
Chairel (Mex) 540
Chakra (Ind) 327
Chakri Naruebet (Tld) 828
Chala (Per) 626
Chaman (Ven) 1007
Chamela (Mex) 540
Chamelecon (Hon) 320
Champion (US) 969
Chan Chiang (Twn) 816
Chancay (Per) 626
Chancellorsville (US) 944
Chandeleur (US) 988
Chang Bogo (RoK) 474
Chang Chien (Twn) 813
Changbaishan (CPR) 153
Changchun (CPR) 139
Changdao (CPR) 164
Changsa (CPR) 140
Changshou (CPR) 157
Changxingdao (CPR) 162
Changzhi (CPR) 147
Changzhou (CPR) 144
Chanthara (Tld) 843
Chantij-Mansisk (Rus) 732
Chao Hu (CPR) 161
Chao Phraya (Tld) 830
TSGT John A Chapman (US) 978
Charagato (Ven) 1007
Charak (Iran) 390
Charente (Fra) 279
Charger (UK) 907
Charisma (Aust) 37
Charles David (US) 987
Charles De Gaulle (Fra) 252
Charles Drew (US) 974
Charles Sexton (US) 987
Charlie Brown (Aust) 37
Charlotte (US) 938
Charlotte Maxeke (SA) 761
Charlottetown (Can) 100
Charlton (US) 977
Chasanyabadee (Tld) 845
Châteaugay (Can) 105
J Chavez Suarez (Bol) 69
Chawengsak Songkram (Tld) 845
Che Ju (RoK) 478
Cheboksary (Rus) 727
Cheetah (Ind) 350
Cheikh Oumar Fall (Sen) 745
Cheleken (Rus) 709
Chemle (Gha) 302
Chen Te (Twn) 814
Chena (US) 991
Cheng Ho (Twn) 813
Cheng Kung (Twn) 813
Chengdu (CPR) 140
Chennai (Ind) 333
Cheong Hae Jin (RoK) 486
Cheonwangbong (RoK) 484
Cheriyam (Ind) 348
Cherkasy (Ukr) 879
Chernigiv (Ukr) 879
Chetlat (Ind) 348
Chevalier Paul (Fra) 254
Cheyenne (US) 938, 991
Chi Kuang (Twn) 813
Chicago (US) 938
Chicama (Per) 626
Chichijima (Jpn) 446
Chichiriviche (Ven) 1007
Chickahominy (US) 968
Chickasaw Bayou (US) 968
Chicoutimi (Can) 98
Chiddingfold (UK) 913
Chief (US) 969
Chieftain (Sin) 750
Chigirin (Ukr) 880
El Chihab (Alg) 5
Chihaya (Jpn) 449
Chik (Tld) 844
Chikoko I (Mlw) 505
Chikugo (Jpn) 456
Chikuma (Jpn) 441
Childers (Aust) 33
Chiloé (Chi) 126
Chimborazo (Ecu) 213
Chimera (Ita) 410
Chin Yang (Twn) 815
Chincoteague (US) 988
Chinook (US) 959, 987
Chioggia (Ita) 415
Chios (Gre) 309
Chipana (Chi) 121
Chipana (Per) 618
Chippewa (US) 991
Chiquillán (Arg) 21
Chiroo (Iran) 390
Chita (Rus) 682
Chitose (Jpn) 455
Chitra (Ban) 56
Chiyoda (Jpn) 448
Chochab (Gua) 317
Chock (US) 992
Choctaw County (US) 980
Choi Muson (RoK) 474
Choi Young (RoK) 476
Cholmsk (Rus) 728
Choluteca (Hon) 320
Chon Buri (Tld) 836
Chon Nam (RoK) 478
Chongmingdao (CPR) 162
Chorrillos (Per) 626
Chosin (US) 944
Choukai (Jpn) 434
Choules (Aust) 31
Christina (Lit) 504
Chu Chiang (Twn) 816
Chuang (Tld) 844
Chukotka (Rus) 726
Chula (Tld) 844
Chulmasan (RoK) 488
Chulupi (Arg) 21
Chun Jee (RoK) 486
Chung Chi (Twn) 818
Chung Chien (Twn) 818
Chung Chih (Twn) 818
Chung Chuan (Twn) 818
Chung Ho (Twn) 818
Chung Ju (RoK) 478, 481
Chung Kuang (Twn) 818
Chung Ming (Twn) 818
Chung Nam (RoK) 478
Chung Pang (Twn) 818
Chung Ping (Twn) 818
Chung Shun (Twn) 818
Chung Suo (Twn) 818
Chung Yeh (Twn) 818
Chung-Hoon (US) 948
Chungmugong Yi Sun-Shin (RoK) 476
Ch'ungnam (RoK) 485
Churubusco (US) 968
Chuspa (Ven) 1007
Ciara (Ire) 394
Cidade De Natal (Brz) 86

Cienaga De San Juan (Col)...178
Cimarron (US)...991
Çinar (Tur)...867
Cinus (Ita)...422
Ciorlieri (Ita)...422
Cirro (Por)...656
Cisne (Arg)...23
Cisne (Hon)...321
Cisne (Por)...653
Cisne Branco (Brz)...85
Citlaltepl (Mex)...541
City of Corpus Christi (US)...938
Ciudad Bolívar (Ven)...1006
Ciudad De Rosario (Arg)...20
Ciudad De Zarate (Arg)...20
CL 01-09, 11-171 (Jpn)...458
Clamp (US)...992
Clark's Harbour (Can)...112
Clavero (Per)...622
Cleat (US)...992
CW 3 Harold C Clinger (US)...961
Clorinda (Arg)...17
Clurit (Indo)...367
Clyde (UK)...908
CMP 22 (Ser)...746
Cnl Eduardo Avoroa Hidalgo (Bol)...69
Coamo (US)...968
Cobia (US)...987
Cochimie (Mex)...536
Cochin (Ind)...352
Cochito (US)...987
Cocle (Pan)...613
Coco-Beach (Gab)...280
Coconut Queen (Aust)...37
Coho (US)...987
Coishco (Per)...626
Colán (Per)...627
Cole (US)...946
Colhue (Arg)...24
Colhue Huapi (Arg)...24
Colimbo (Ven)...1007
Colimbo II-IV (Spn)...785
Collins (Aust)...26
Colonel Djoue-Dabany (Gab)...280
Colonia (Uru)...997
Colorado (US)...937
Columbia (US)...938
Columbus (US)...938
Comamefa (Gua)...317
Comandante Arandia (Bol)...69
Comandante Bettica (Ita)...412
Comandante Borsini (Ita)...412
Comandante Cigala Fulgosi (Ita)...412
Comandante Foscari (Ita)...412
Comandante Imperial Santana (Ang)...9
Comandante Kassanje (Ang)...9
Comandante Manhães (Brz)...83
Comandante Paiva Domingos Da Silva (Ang)...9
Comandante Toro (Chi)...121
Comandante Valodia (Ang)...9
Comandante Varella (Brz)...83
Comandatuba (Brz)...81
Comfort (US)...980
Commandant Azouggarh (Mor)...547
Commandant Birot (Fra)...260
Commandant Blaison (Fra)...260
Commandant Bouan (Fra)...260
Commandant Boutouba (Mor)...547
Commandant Ducuing (Fra)...260
Commandant El Harty (Mor)...547
Commandant El Khattabi (Mor)...547
Commandant l'Herminier (Fra)...260
Commander Apayi Joe (Nig)...583
Commander Rudolph (Nig)...583
Comodoro Carlos Castillo Bretón (Mex)...535
Comodoro Rivadavia (Arg)...19
Comstock (US)...967
Concepción Del Uruguay (Arg)...17
Conceptión (Chi)...126
Conder (Aust)...34
Conejera (Sen)...744
Confidence (US)...985
Congrio (Arg)...23
Connecticut (US)...940
Conqueror (Sin)...750
Constable Carrière (Can)...109
Constancia (Ven)...1003
Constanta (Rom)...667
Constitución (Ven)...1003
Constituição (Brz)...75
Constitution (US)...973
Contentin (Fra)...275
Contralmirante Angel Ortiz Monasterio (Mex)...535
Contramaestre Antero (Spn)...779
Contramaestre Casado (Spn)...780
Contramaestre Lamadrid (Spn)...779
Contramaestre Navarrete (Spn)...779
Contramaestre Sánchez Fernández (Spn)...779
Contre Admiral Eustatiu Sebastian (Rom)...662
Contre-Almirante Oscar Viel Toro (Chi)...123
Contreras (US)...968
Conversano (Ita)...422
Cook (UK)...913
Coquimbo (Chi)...126
Cora (Mex)...541
Cora Divh (Ind)...348
Coral (Por)...654
Coral Cod (Sin)...758
Coral Snake (Aust)...38
Coralline (Fra)...274
Corinth (US)...968
Corio Bay (Aust)...39
Cormoran (Arg)...19, 23
Cormoran (Fra)...265
Cormorán (Spn)...785
Cormoran (Ven)...1007
Cormorant (US)...987
Corner Brook (Can)...98
Cornhusker State (US)...982
Cornouailles (Fra)...275
Cornwall (Jam)...427
Coronado (US)...953
Corozal Point (TT)...849
Corporal Mclaren MMV (Can)...109
Corporal Teather C V (Can)...109
Corral (Chi)...126
Correa Falcon (Arg)...24
Corregidor (Plp)...634
Corrias (Ita)...423
Corrientes (Per)...625
Corsaro II (Ita)...417
Corte Real (Por)...650
Cortile (Ita)...423
Corvina (Arg)...23
Corvo Marino (Spn)...783
Cosby (Sin)...758
Cosme García (Spn)...768
Cotopaxi (Ecu)...213
Cotuhe (Col)...173
Cotunduba (Brz)...81
Cougar (Can)...104
Courbet (Fra)...259
Courtenay Bay (Can)...112
Cove Isle (Can)...110
Cownose Ray (Sin)...758
Cowpens (US)...944
Coyuca (Mex)...540
CP 258-259, 261-264, 301-312, 316-317, 319, 454-456, 825-862, 872-881, 884-889, 2011, 2056, 2068, 2072, 2074, 2080-2117 (Ita)...425
CP 265, 267-269, 271, 273-274, 276-278, 280-292 (Ita)...424
CP 506, 512, 514-583, 601-612, 701-726, 801-824, 863-871, 882-883, 890-892, 2201-2205 (Ita)...426
CPCIM Guillermo Londoño Vargas (Col)...174
Cpt. Cdor Al. Catuneanu (Rom)...665
Creoula (Por)...654
Cristian Reyes Holguín (Col)...174
Cristóbal Colón (Spn)...770
Crocodile (US)...987
Crocus (Bel)...64
Croix Du Sud (Fra)...269
Crotone (Ita)...415
Crown Point (TT)...849
Cruzeiro Do Sul (Brz)...84
CTCIM Jorge Moreno Salazar (Col)...174
CTM (Sen)...745
CTM 12-23, 25-31 (Fra)...268
Cuanza (Por)...652
Cuauhtémoc (Mex)...540
Çubuklu (Tur)...866
Cucut (Indo)...365
Cuddalore (Ind)...350
Cuenca (Ecu)...212
Cultrona (Ita)...423
Cundrik (Indo)...376
Curaray (Per)...625
Curaumila (Chi)...125
Curiapo (Ven)...1004
Curtis Wilbur (US)...946
Curtiss (US)...982
Cushing (US)...988
Cut Nyak Dien (Indo)...362
Cutthroat (US)...972
Cuttyhunk (US)...988
CW 3 Harold C Clinger (US)...961
Cygnus (Can)...108
Cypress (US)...990
Czajka (Pol)...642

D

D 07-08 (Dom)...205
D 7-8 (Ant)...11
D 145 (Rus)...705
D 431-433, 436-437 (Az)...41
D. Carlos I (Por)...654
D. Francisco De Almeida (Por)...650
Da Nang (Vtn)...1009
Dąbie (Pol)...643
Dae Chon (RoK)...481
Dae Chung (RoK)...486
Daejoyung (RoK)...476
Dagestan (Rus)...696
Dagma (Den)...197
Dagupan City (Plp)...632
Dahl (US)...977
Daisen (Jpn)...451
Daishandao (CPR)...163
Daiyunshan (CPR)...154
Dajlah (SAr)...743
Dakhla (Mor)...550
Dalian (CPR)...136
Dallas (US)...938
D'Amato (Ita)...422
Damavand (Iran)...380
Dame Roma Mitchell (Aust)...39
Damisa (Nig)...580
Al Dammam (SAr)...738
Damour (Leb)...498
Damrong Rachanuphap (Tld)...845
Damsah (Qat)...658
Damuan (Bru)...89
Damyat (Egy)...219
Dana (Den)...199
Danga (Mly)...517
Daniel Inouye (US)...948
Danil Moskovskiy (Rus)...680
Dannebrog (Den)...201
Dante Novaro (Ita)...425
Danxiashan (CPR)...154
Daoud Ben Aicha (Mor)...549
Daqhiliya (Egy)...224
Daqingshan (CPR)...154
Dareen (SAr)...742
Darica (Tur)...869
Darida (Ita)...422
Daring (Sin)...754
Daring (UK)...902
Darnitsya (Ukr)...882
Darshak (Ban)...57
Darshak (Ind)...351
Darulaman (Bru)...88
Darulehsan (Bru)...88
Darussalam (Bru)...88
Daruttaqwa (Bru)...88
Darwin (Aust)...27
Daryavand II (Iran)...391
Das (UAE)...884
Dasher (UK)...907
Dasman (Kwt)...492
Dastoor (Kwt)...492
Dat Assawari (Egy)...225
Datong (CPR)...148
Datteln (Ger)...293
Daugava (Rus)...714
Dauntless (Sin)...754
Dauntless (UK)...902
Dauntless (US)...985
Dauriya (Rus)...714
Davao Del Norte (Plp)...634
David Hansen (Gha)...301
Daylam (Iran)...390
Dayyinah (UAE)...889
DB 411, 413, 417, 419, 422, 426, 429, 431-432, 435 (Plp)...635
DD 119-120 (Jpn)...440
DDG 119-126 (US)...948
DDH 184 (Jpn)...432
De Falco (Ita)...422
De Grasse (Fra)...247
De Ianni (Ita)...423
De Los Heros (Per)...621
De Mist (SA)...765
De Neys (SA)...765
De Rosa (Ita)...422
De Ruyter (Nld)...564
De Santis (Ita)...423
De Zeven Provincien (Nld)...564
Debundscha (Cam)...97
Decatur (US)...946
Dechaineux (Aust)...26
Decimo Aniversario (Spn)...785
Decisive (US)...985
Al Deebel (Qat)...657
Deepak (Ind)...353
Defender (StL)...735
Defender (UK)...902
Defender (US)...969
Defensora (Brz)...75
Defiant (US)...974
Değirmendere (Tur)...870
Dehloran (Iran)...391
Dehshat (Pak)...608
Dejima (Jpn)...453
Dekanawida (US)...974
Al Dekheila (Egy)...227
Delair (Pak)...610
Delaware (US)...937
Delfin (Arg)...22
Delfin (Rus)...724
Delhi (Ind)...334
Delvar (Iran)...390
Démocrata (Mex)...536
Denden (Eri)...233
Deneb (Alg)...9
Deneb (DR)...207
Deneb (Gua)...317
Deneb (Mex)...535
Deneb Algedi (Alg)...9
Denebola (DR)...207
Denebola (Mex)...535
Denebola (US)...983
Denis Davydov (Rus)...706
Denizkuşu (Tur)...861
Denti (Fra)...270
Denver (US)...966
Dependable (US)...985
Dera'a 2, 4-8 (Bhr)...50
Dera'a 11-14 (Bhr)...51
Derafsh (Iran)...384
Derbent (Rus)...728
Des Groseilliers (Can)...107
Detroit (US)...950
Devastator (US)...969
Dewa (Jpn)...453
Dewa Kembar (Indo)...372
Dewaruci (Indo)...372
Dewey (US)...948
Dextrous (US)...969
DF 300-303, 305, 307-313, 321-323, 325-332, 334, 347 (Plp)...636
Al Dhafra (UAE)...886
Dhaleshwari (Ban)...54
Dhansiri (Ban)...56
Dheba (SAr)...742
Dheeb Al Bahar 2 (Omn)...598
Dhiraj (Ind)...353
Dhofar (Omn)...595
Di Bartolo (Ita)...422
Di Hua (Twn)...814
Diablo (UK)...924
Diamantina (Aust)...33
Diamond (UK)...902
Diamondback (US)...987
Diana (Den)...198
Diana (Rus)...730
Diaz (Chi)...125
Dili (ETim)...208
Dili (Indo)...369
Diligence (UK)...918
Diligence (US)...985
Diligente (Col)...176
Dillingen (Ger)...293
Dimitrov (Rus)...682
Dimitrovgrad (Rus)...700
Dinar (Bhr)...49
Dinder (Sud)...793
Dinh Tien Hoang (Vtn)...1009
Diomidis (Gre)...312
Dionée (Fra)...274
Dionisio Ojeda (Plp)...631
Dionysos (Cypr)...193
Diopos Antoniou (Gre)...306
Dioscoro Papa (Plp)...631
Diphda (Mex)...535
Dipikar (Cam)...96
Diponegoro (Indo)...363
Discovery Bay (US)...972
Dixmude (Fra)...266
El Djari (Alg)...6
Djärv (Swe)...800
El Djasur (Alg)...6
DJB 103-104, 107 (Cro)...187
DJC 106 (Cro)...187
DJC 411-415 (Ser)...746
Djebel Chenoua (Alg)...5
Djukas (Lit)...502
DK-143, 447 (Rus)...733
DKA 56, 67, 107, 131, 144, 156, 172 (Rus)...705
DKA 57, 70, 106, 148, 182, 185, 325, 464-465 (Rus)...706
Dmitriy Donskoy (Rus)...674
Dobrotich (Bul)...92
Dock A (Ger)...297
Dock B (Ger)...297
Doğan (Tur)...862
Doğanarslan (Tur)...870
Doğanay (Tur)...855
Al Doghas (Omn)...596
Doirani (Gre)...311
Dokdo (RoK)...483
Dolfijn (Nld)...562
Dolphin (Ana)...10
Dolphin (Iran)...390
Dolphin (Isr)...395
Dolphin (Sin)...758
Dolphin (US)...987
Dolphin Mira (Nig)...583
Dolphin Rima (Nig)...583
Dolunay (Tur)...855
Dominique (Dom)...205
Don (Rus)...721, 726
Don Lucas (Arg)...23
Don Vizo (Col)...178
Donald Cook (US)...946
Donau (Ger)...296
Donbas (Ukr)...880, 882
Donets (Rus)...717
Donetsk (Ukr)...878
Dongguan (CPR)...147
Dongtingshan (CPR)...155
Dongxiu (CPR)...165
Donuzlav (Rus)...709
Dorado (Arg)...22
Dorado (Sin)...758
Dorado (US)...987
Dordanda (Ban)...54
Dore (Indo)...370
Al Dorrar (Kwt)...492
Dorina (Nig)...582
Dost (Tur)...871
Douffine (Fra)...275
Doukkala (Mor)...551
Dourado (Brz)...78
Doutor Montenegro (Brz)...86
Dr Bernardo Houssay (Arg)...23
Dr Soeharso (Indo)...368
Dragão (Por)...653
Dragon (UK)...902
Drakensberg (SA)...764
Drazki (Bul)...90
Dreger (PNG)...614
Driade (Ita)...410
Dristig (Swe)...800
Druckdock (Dock C) (Ger)...297
Drummond (Arg)...14
Drummond (US)...988
Drużno (Pol)...643
Dsar-6 (Sin)...757
DSRV II (RoK)...485
Dubhe (Den)...198
Dubhe (DR)...207
Dubhe (Mex)...535
Dubna (Rus)...715
Dubno (Ukr)...882
Dubnyak (Rus)...721
Dubrovnik (Cro)...185
Duero (Spn)...778
Dugong (Aust)...36
Duguay-Trouin (Fra)...247
Dumbea (Fra)...279
Dumit (Can)...110
Dumont d'Urville (Fra)...267
Dunafoldvar (Hun)...325
Dunai (Ukr)...883
Dunaújváros (Hun)...325
Dunay (Rus)...716, 726
Duncan (UK)...902
Dungeness (Can)...106
Dupetit-Thouars (Fra)...247
Dupleix (Fra)...257
Dupuy De Lôme (Fra)...269
Duquesne (Fra)...247
Durango (Mex)...534
Duranta (Ban)...55
Durbar (Ban)...55
Durdam (Ban)...55

1030 **INDEXES** > Named ships

Ho Mou (Twn) 817
Ho Seng (Twn) 817
Ho Shan (Twn) 817
Ho Shou (Twn) 817
Hobart (Aust) 27
Hobkirk (US) 968
Hofouf (SAr) 737
Högsåra (Fin) 243
Holdfast Bay (Aust) 39
Holger Danske (Den) 199
Holland (Nld) 566
Hollyhock (US) 990
Homburg (Ger) 293
Homigueros (US) 968
Honduras (Hon) 319
Honestidad (Ven) 1003
Hong Daeson (RoK) 482
Hong Siuk (RoK) 482
Hongzhu (CPR) 161
Hopper (US) 946
Horacio Ugarteche (Bol) 69
Hormuz (Iran) 389
Hormuz (Omn) 597
Horobetsu (Jpn) 455
Horria (Tun) 851
El Horriya (Egy) 226
Hortensia (Fra) 274
Hotaka (Jpn) 456
Houou (Jpn) 456
Houston (US) 938
Howard (US) 948
Howard O Lorenzen (US) 976
HQ 09, 11, 13, 15, 17, 381 (Vtn) 1010
HQ 37, 55-59, 301, 305, 354, 358-361, 384-386, 331-335, (Vtn) 1012
HQ 261, 263-267, 371-372, 374-376, 378 (Vtn) 1011
HQ 272-274, 251, 253, 511-513 (Vtn) 1013
HQ 606-609, 621-622, 627-628, 630-633, 647, 649, 687, 957 (Vtn) 1015
HQ 782, 851, 861-864, 871, 885 (Vtn) 1014
Hristo Botev (Bul) 93
Hrvatska Kostajnica (Cro) 188
Hsian Chiang (Twn) 816
Hsin Bei (Twn) 821
Hsin Chiang (Twn) 816
Hsun Hu 7-9 (Twn) 822
Hua Hin (Tld) 837
Hua Luogeng (CPR) 158
Huadingshan (CPR) 154
Huaibei (CPR) 146
Huaihua (CPR) 143
Huainan (CPR) 146
Huala (Arg) 23
Hualien (Twn) 822
Huancavilca (Ecu) 209
Huanchaco (Per) 626
Huanggang (CPR) 144
Huanggangshan (CPR) 154
Huangshan (CPR) 144
Huangshi (CPR) 144
Huashan (CPR) 155
Huasteco (Mex) 540
Huaura (Per) 626
Huayin (CPR) 147
Hudson (Can) 113
Hudson (Chi) 125
Hudson (US) 991
Hue City (US) 944
Huizhou (CPR) 148
Huludao (CPR) 143
Humaita (Brz) 70
Humboldt (Per) 623
Humming Bird (TT) 850
Al Hunain (Lby) 499
Hunaish (Yem) 1019
Hunze (Nld) 572
Huon (Aust) 33
Huoqiu (CPR) 157
Huracan (Mex) 532
Huravee (Mld) 521
Hurmat (Pak) 601
Al Hurreya (Egy) 226
Al Hurreya 2-3 (Egy) 226
Hurricane (US) 959
Hurst Point (UK) 917
Hurtig (Swe) 800
Hurworth (UK) 913
Al Hussein (Jor) 460
Hussein (Jor) 461
Huveaune (Fra) 279
Huwar (Qat) 657
Hvasser (Nor) 589
Hvidbjørnen (Den) 196
Hvidsten (Den) 199
Hwa Chun (RoK) 486
Hwai Yang (Twn) 815
Hwang Dohyun (RoK) 482
Hyäne (Ger) 292
Hyangro Bong (RoK) 484
Hydra (Gre) 304
Hydra (Nld) 571
Hydrograf (Pol) 644
Hylje (Fin) 244
Hymara (Guy) 318
Hyperion (Gre) 311
Hyuga (Jpn) 431
Hyun Sihak (RoK) 482

I

I Theophilopoulos-Karavogiannos (Gre) 312
Iason (Gre) 312
Ibis (US) 987
Ibn Haritha (Lby) 500
Ibn Ouf (Lby) 500
IC 109-116, 301-310 (Ind) 354
IC 117-126 (Ind) 357
Ice Fire (Aust) 37
Ida Lewis (US) 989
Idabato (Cam) 96
El Idrissi (Alg) 7
Ieshima (Jpn) 444
Igaraparaná (Col) 173
Ignacio Allende (Mex) 529
Ignacio L Vallarta (Mex) 535
Ignacio Mariscal (Mex) 536
Ignacio Ramirez (Mex) 536
Ignesti (Ita) 422
Igor Belousov (Rus) 720
Ijhtihad (Bru) 88
Ikaria (Gre) 309
Ikazuchi (Jpn) 436
Ikhlas (Mly) 516
Ikigumo (Jpn) 457
Ikot-Abasi (Nig) 581
Ikrimah (Lby) 501
Ilam (Iran) 391
Ilarion (Cypr) 193
Ile Saint-Ours (Can) 110
Ilga (Rus) 723
Ilim (Rus) 715
Iliniza (Ecu) 213
Iliria (Alb) 3
Illichivsk (Ukr) 881
Illinois (US) 937
Illustrious (UK) 909
Ilocos Norte (Plp) 634
Iloilo (Plp) 630
Ilya Muromets (Rus) 722
Imam Gazzali (Ban) 59
Iman (Rus) 715
Iman Bonjol (Indo) 362
Imandra (Rus) 721
Imanta (Lat) 495
Imbat (Tur) 861
Impeccable (US) 975
Imperial Marinheiro (Brz) 79
Inagua (Bhm) 44
Inasa (Jpn) 456
Inazuma (Jpn) 436
Incheon (RoK) 480
Indaw (Myn) 555
Independence (Mic) 541
Independence (Sin) 754
Independence (US) 953
Independéncia (Brz) 75
Independencia (Per) 626
Independencia (Ven) 1003
Independiente (Col) 172
Indiana (US) 937
Indianapolis (US) 950
Indlovu (SA) 765
Indomable (Col) 171
Indomita (Arg) 18
Inebolu (Tur) 869
Inej (Rus) 701
Infanta Cristina (Spn) 773
Infanta Elena (Spn) 773
Ingeniero Gumucio (Bol) 69
Ingeniero Julio Krause (Arg) 20
Ingeniero Mery (Chi) 123
Ingeniero Palacios (Bol) 69
Ingeniero White (Arg) 22
Ingraham (US) 952
Inguri (Rus) 717
Inhaúma (Brz) 76
Inírida (Col) 173
Inkster (Can) 113
Inönü (Tur) 856
Integridad (Ven) 1003
Intrepid (Sin) 752
Intrépida (Arg) 18
Intrépide (Gn) 318
Intrepido (Col) 171
Inttisar (Kwt) 492
Investigator (Ind) 351
Invincible (US) 976
Inya (Myn) 555
Inzerilli (Ita) 422
Inzucchi (Ita) 422
Iquique (Chi) 126
Irakleia (Gre) 309
Irgiz (Rus) 720
Irkut (Rus) 715
Iron Duke (UK) 904
Iroquois (Can) 102
Irtysh (Rus) 716
Isaac Dyobha (SA) 763
Isaac Mayo (US) 987
Isaac Peral (Spn) 768
Isandlwana (SA) 762
Isaza (Chi) 121
Isazu (Jpn) 455
Ise (Jpn) 431
Iseshio (Jpn) 459
Iseyuki (Jpn) 458
Ishigaki (Jpn) 453
Ishikari (Jpn) 455
Ishim (Rus) 733
Işin (Tur) 868
Al Iskandarani (Egy) 227
Iskandhar (Mld) 521
Iskar (Bul) 92
Iskenderun (Tur) 866
Iskra (Pol) 644
Isku (Fin) 243
Isla Burica (CR) 183
Isla De Annobon (EqG) 232
Isla De Corisco (EqG) 232
Isla De La Plata (Ecu) 214
Isla De Medio (Ven) 1007
Isla Del Coco (CR) 183
Isla Española (Ecu) 214
Isla Fernandina (Ecu) 214
Isla Isabela (Ecu) 215
Isla Jicaron (Pan) 614
Isla La Pinta (Ecu) 216
Isla Marchena (Ecu) 216
Isla Papagallo (Ven) 1007
Isla Puná (Ecu) 214
Isla San Cristóbal (Ecu) 215
Isla San Salvador (Ecu) 214
Isla Santa Clara (Ecu) 214
Isla Santa Cruz (Ecu) 216
Isla Santa Rosa (Ecu) 214
Isla Seymour (Ecu) 214
Isla Tesoro (Col) 178
Islay (Per) 618
Ismael Lomibao (Plp) 631
Isonami (Jpn) 458
Isongo (Cam) 96
Isoshi (Jpn) 459
Isoshio (Jpn) 430
Isoyuki (Jpn) 440
Isozuki (Jpn) 457
Istiqlal (Kwt) 491
Istros (Gre) 312
Isuzu (Jpn) 456
Itaf Deme (Sen) 745
Itaipú (Par) 616
Italia (Ita) 417
Ithaki (Gre) 309
Ivan Bubnov (Rus) 715
Ivan Golubets (Rus) 707
Ivan Kartsov (Rus) 706
Ivan Khurs (Rus) 712
Ivan Lednev (Rus) 732
Ivan Yevteyev (Rus) 732
Ivanovets (Rus) 700
Iver Huitfeldt (Den) 194
Iveria (Geo) 282
Iwami (Jpn) 452, 453
Iwo Jima (US) 964
Iximche (Gua) 316
Iyonami (Jpn) 458
Izet (Kaz) 463
Izhora (Rus) 715
Izreg (Lby) 501
Iztaccihuatl (Mex) 541
Izu (Jpn) 452
Izumo (Jpn) 432
Izumrud (Rus) 730
Izunami (Jpn) 457
Izushima (Jpn) 445
Izyaslav (Ukr) 882

J

Jabanne (Cam) 96
Jaberi (Kwt) 493
Al Jabiri (Bhr) 48
Jaceguai (Brz) 76
Jacinto Cândido (Por) 651
W Jackman (Can) 112
Jackpot (Sol) 760
Jackson (US) 953
Jacksonville (US) 938
Jaco (ETim) 208
Jadayat (Indo) 376
Jadran (Mon) 542
Jae Chon (RoK) 481
Jaemin I-II (RoK) 488
Jaemin III (RoK) 489
Jaemin V-XIII (RoK) 489
Jägaren (Swe) 800
Jagatha (Sri) 787
Jaguar (Fra) 271
Jaguar (Nld) 573
Jaguar Ray (Sin) 758
Al Jahili (UAE) 886
El Jail (Mor) 548
Jaime Gómez Castro (Col) 179
Jalalat (Pak) 606
Jalanidhi (Indo) 372
Jalashwa (Ind) 349
Jamalyets (Rus) 728
Jamaran (Iran) 380
James (US) 984
James Caird IV (UK) 914
James E Williams (US) 948
James Joyce (Ire) 394
James Rankin (US) 989
Jamno (Pol) 643
Jamuna (Ban) 57
Jamuna (Ind) 351
Jananah (UAE) 889
Janbaz (Pak) 610
Janzur (Lby) 501
Jarabakka (Sur) 793
Jaradah (Bhr) 48
Jarak (Mly) 516
Al Jarim (Bhr) 48
Jaroslavl (Rus) 682
Jasiri (Ken) 464
Jasmin (Fra) 279
Jason (Fra) 278
Jason Dunham (US) 948
Al Jasrah (Bhr) 48
Al Jawf (SAr) 741
Jayesagara (Sri) 786
Jean Bart (Fra) 256
Jean De Vienne (Fra) 257
Jebat (Mly) 508
Jebel Antar (Alg) 8
Jebel Hando (Alg) 8
Jefferson City (US) 938
Jefferson Island (US) 988
Jelgava (Lat) 494
Jelyana (Lby) 501
Jeonbuk (RoK) 480
Jeongji (RoK) 474
Jepak (Mly) 520
Jerai (Mly) 514
Jerba (Tun) 851, 854
Jerong (Mly) 513
Jesus Gonzalez Ortega (Mex) 535
Ji Deokchil (RoK) 482
Jian (CPR) 147, 148
Jiangmen (CPR) 147
Jiaxin (CPR) 143
Jieyang (CPR) 148
Jif Xplorer (Fra) 278
Jimmy Carter (US) 940
Jin Chiang (Twn) 816
Jin Hae (RoK) 481
Jin Ju (RoK) 481
Jinan (CPR) 139
Jing Chiang (Twn) 816
Jinggangshan (CPR) 153
Jingjiang (CPR) 157
Jinhua (CPR) 147
Jintsu (Jpn) 441
Jiuhuashan (CPR) 154
Jiujiang (CPR) 147
Jo Cheon Hyeong (RoK) 482
João Coutinho (Por) 651
João Roby (Por) 651
Johan De Witt (Nld) 568
Johan Månsson (Swe) 807
John C Stennis (US) 942
John Ericsson (US) 975
John F Kennedy (US) 941
John Finn (US) 948
John Glenn (US) 978
John Gowlland (Aust) 34
John Lenthall (US) 975
John P Murtha (US) 962
John P Tully (Can) 113
John Paul Jones (US) 946
John S McCain (US) 946
John Warner (US) 937
Jonge Jan (Nld) 572
Jonge Prins (Nld) 572
Jonquille (Fra) 279
José Andrada (Plp) 631
José Loor Sr (Plp) 631
Jose Manuel Pando (Bol) 69
José Maria Garcia Y Toledo (Col) 179
José Maria Palas (Col) 175
Jose Natividad Macias (Mex) 536
Joshan (Iran) 384
Joshila (Pak) 610
Joshua Appleby (US) 989
Joshua Humphries (US) 975
Josué Alvarez (Col) 178
Jotvingis (Lit) 503
Al Jouf (SAr) 743
Joumhouria (Tun) 851
Jounieh (Leb) 497
Joves Fiallo (Col) 178
Joy Amedume (Gha) 301
Juan Antonio De La Fuente (Mex) 536
Juan Carlos I (Spn) 777
Juan De La Barrera (Mex) 535
Juan Lucio (Col) 176
Juan Magluyan (Plp) 631
Juan N Alvares (Mex) 535
Juan Nepomuceno Eslava (Col) 179
Juan Nepomuceno Peña (Col) 179
Juan Rafael Mora Point Chico (CR) 183
Juan Ricardo Oyola Vera (Col) 174
Juan Sebastián De Elcano (Spn) 779
Juanchaco (Col) 178
Al Jubatel (SAr) 742
Juist (Ger) 298
Jujur (Mly) 516
Juli (Per) 626
Julian Apaza (Bol) 69
Julio De Noronha (Brz) 76
Julio Olmos (Bol) 69
Jumping Marlin (Sin) 758
Jung Geungmo (RoK) 482
Jung Woon (RoK) 474
Juniper (US) 990
Junon (Sey) 748
Jupiter (Den) 198
Jura (UK) 924
Jurga (Rus) 732
Jurrat (Pak) 607
Justice (Sin) 754
Justo Sierra Mendez (Mex) 533
Jymy (Fin) 243
Jyoti (Ind) 352

K

K 4, 10 (Pol) 643
K 5, 7, 9 (Pol) 646
K 8 (Pol) 645
Kaani (Mld) 521
Kabra (Ind) 348
El Kaced (Mor) 551
El Kadessaya (Egy) 222
Kadmatt (Ind) 341
Kadmos (Gre) 312
Kagayuki (Jpn) 458
Kahovka (Ukr) 877
Kaibil Balan (Gua) 317
Kaifeng (CPR) 136
Kaimon (Jpn) 456
Kairi (TT) 850
Kaiyo (Jpn) 459
Kajava (Fin) 241
Kakap (Indo) 366
Kakinada (Ind) 350
Kala Hitam (Indo) 371
Kalaat Beni Hammad (Alg) 6
Kalaat Beni Rached (Alg) 6
Kalaat Beni-Abbes (Alg) 7
Kalakae (Indo) 366
Al Kalamoun (Leb) 498
Kalar (Rus) 723
Kalat (Iran) 390
Kaliakra (Bul) 93
Kalinga (Plp) 635
Kaliningrad (Rus) 704, 727
Kalkan (Tur) 861
Kallang (Sin) 757
Kallanpää (Fin) 244
Kalliroe (Gre) 311
Kallisto (Gre) 310
Kalmat (Pak) 609
Kalmykia (Rus) 698
Kalpeni (Ind) 348
Kaluga (Rus) 682
Kalypso (Gre) 310

Kama (Rus) 716
Kaman (Iran) 384
Kamchatka (Rus) 726
Kamenassa (ETim) 208
Kamla Devi (Ind) 355
Kamorta (Ind) 341
Kamøy (Nor) 589
Kampela 3 (Fin) 242
Kamui (Jpn) 456
Kanak Lata Barua (Ind) 355
Kanaris (Gre) 305
El Kanass (Alg) 6
Kanawha (US) 975, 991
Kang Ding (Twn) 814
Kang Jin (RoK) 485
Kang Kyeong (RoK) 485
Kangan (Iran) 390
Kangwon (RoK) 485
Kanin (Rus) 732
Kaning'a (Mlw) 505
Kaniv (Ukr) 883
Kankakee (US) 991
Kano (Jpn) 456
Kantang (Tld) 837
Kao Chiang (Twn) 816
Kao Hsiung (Twn) 818
Kaoh Chhlam (Cmb) 94
Kaoh Rong (Cmb) 94
Kaohsiung (Twn) 824
Kapak (Indo) 376
Kapas (Mly) 519
Kapatakhaya (Ban) 55
Kapitan Parvi Rang Boris Rogev (Bul) 92
Kapitan Parvi Rang Dimitar Paskalev (Bul) 92
Kapitan Parvi Rang Dimitri Dobrev (Bul) 92
Kapitan Pattimura (Indo) 362
Kapitan Vtori Rang Nikola Furnadjiev (Bul) 92
Kapitan-Leytenant Evstati Vinarov (Bul) 92
Kapitan-Leytenant Kiril Minkov (Bul) 92
Kara (Tog) 848
Karabala (Iran) 389
Karabane (Sen) 745
Karabiga (Tur) 862
Karaburun (Tur) 862
Karachejevo-Cherkessia (Rus) 703
Karamürselbey (Tur) 864
Karang Banteng (Indo) 369
Karang Pilang (Indo) 369
Karang Tekok (Indo) 369
Karang Unarang (Indo) 369
Karataş (Tur) 862
Karathanasis (Gre) 306
Karatoa (Ban) 55
Karayel (Tur) 862
Karel Doorman (Nld) 571
Karel Satsuitubun (Indo) 360
Karelia (Rus) 672, 726
El Karib (Mor) 547
Kariba (Jpn) 456
Karimata (Indo) 374
Kariña (Ven) 1004
Karkheh (Iran) 391
Karlsruhe (Ger) 288
Karlstad (Swe) 797
Karmøy (Nor) 590
Karmukh (Ind) 342
Karnaphuli (Ban) 56
Karp (Rus) 674
Karpasia (Cypr) 193
Karpaz (Tur) 862
Karrar (Pak) 607
Karşiyaka (Tur) 862
Karthala (Com) 181
Karuva (Ind) 348
Karwar (Ind) 350
Kaş (Tur) 862
Kasa (Tan) 827
Kasatka (Rus) 724
Kashalot (Rus) 678
Kashima (Jpn) 447
Kasimov (Rus) 697
Kasirga (Tur) 861
Kasos (Gre) 308
Kassir (Kwt) 492
Kasturba Gandhi (Ind) 356
Kasturi (Mly) 509
Kasungu (Mlw) 505
Kaszub (Pol) 640
Katanpää (Fin) 240
Katherine Walker (US) 989
Kathleen Moore (US) 987
Kati (Est) 235
Katmai Bay (US) 989
Katon (Indo) 366
Katong (Sin) 757
Katori (Jpn) 452
Katsonis (Gre) 303
Katsuragi (Jpn) 456
Katsuren (Jpn) 460
Kauffman (US) 952
Kaus Australe (Alg) 9
Kavak (Tur) 867
Kavaratti (Ind) 341
Kavarna (Bul) 93
Kawagiri (Jpn) 457
Kayvan (Iran) 385
Kazakhstan (Kaz) 463
Kazan (Rus) 679
Kazanets (Rus) 698
KBV 010, 031-034, 047-048, 050-051, 312-316, 590, 592 (Swe) 806
KBV 181, 201-202, 302-311 (Swe) 805
Kearsarge (US) 964
El Kechef (Alg) 6
Kedah (Mly) 510
Keelung (Twn) 812, 822
Kefallinia (Gre) 309
Kekrops (Gre) 312
Kelabang (Indo) 371
Kelantan (Mly) 510
Kelefstis Stamou (Gre) 306
Kelibia (Tun) 854
Kelso (Can) 111
Kem (Rus) 717
Kemaindera (Bru) 89
Kemalreis (Tur) 858
Kendi (Mly) 519
Kennebec (US) 991
Kennesaw Mountain (US) 968
Kenryu (Jpn) 429
Kent (UK) 904
Kentucky (US) 935
Keokuk (US) 974
Kepah (Mly) 515
Kerapu (Indo) 366
Kerch (Rus) 690
Kéréon (Fra) 276
Al Keriat (Lby) 501
Keris (Indo) 365
Kerkini (Gre) 312
Kerkira (Gre) 309
Kermeur (Fra) 274
Kernaleguen (Fra) 274
Kesari (Ind) 349
Keshet (Isr) 397
Key Biscayne (US) 988
Key Largo (US) 988
Key West (US) 938
Keystone State (US) 982
Khabarovsk (Rus) 732
Khadang (Iran) 384
Khadem (Ban) 60
El Khafir (Mor) 548
Khaibar (Pak) 604
Khaireddine (Tun) 852
Khalid (Pak) 600
Khalid (SAr) 740
Khamronsin (Tld) 833
Al Khan (UAE) 888
Khandag (Iran) 391
Khanh Hoa (Vtn) 1009
Khan Jahan Ali (Ban) 60
Khanjar (Ind) 343
Khanjar (Iran) 384
Kharg (Iran) 390
Al Kharj (SAr) 741
Khassab (Omn) 595
Al Khaznah (UAE) 889
Kherson (Ukr) 881
Khirirat (Tld) 834
Khmelnitsky (Ukr) 878
Kholmsk (Rus) 697
Khukri (Ind) 343
Khyber (Egy) 222
Al Khyber (Lby) 499
Ki Hajar Dewantara (Indo) 372
Kibua (Tan) 826
Kickapoo (US) 991
Kidd (US) 948
Kidon (Isr) 397
Kie-Ntem (EqG) 232
Kiffa Borealis (Alg) 9
Kihu (Lit) 503
Kii (Jpn) 452, 453
Kiiski 3-7 (Fin) 241
Kikau (Fij) 238
Kikuchi (Jpn) 456
KIL 1-2, 22, 25, 27, 31, 143, 158, 164, 168, 498, 927 (Rus) 717
Kilat 1-53 (Mly) 519
Kildin (Rus) 712
Kiliç (Tur) 861
Kilimli (Tur) 862
Kilo Moana (US) 971
Kiltan (Ind) 341
Kim Changhak (RoK) 482
Kim Chon (RoK) 481
Kim Jwa-Jin (RoK) 474
Kim Men (Twn) 820
Kim Po (RoK) 485
Kimanis (Mly) 517
Kimball (US) 984
Kinabalu (Mly) 514
Kinaliada (Tur) 860
Kindral Kurvits (Est) 236
Kindurong (Mly) 520
Kindy Camara (Gam) 281
King (Arg) 20
King Abdullah (Jor) 460
Kingfisher (US) 987
Kingston (Can) 103
Kinmen (Twn) 822
Kino (Mex) 540
Al Kirch (Alg) 5
Kirch (Ind) 342
Kirisame (Jpn) 436
Kirishima (Jpn) 434, 456
Kiro (Fij) 238
Kirovograd (Ukr) 878
Kirpan (Ind) 343
Kiska (US) 988
Kiso (Jpn) 452
Kitaguni (Jpn) 457
Kitakami (Jpn) 455
Kitimat II (Can) 109
Kittiwake (US) 987
Kiyozuki (Jpn) 458
Kizljar (Rus) 728
Kjelmøy (Nor) 589
Kjeøy (Nor) 591
KKTCSG 11-12, 102-103 (Tur) 872
Klaeng (Tld) 837
Klints (Lat) 495
Klongyai (Tld) 837
Klueng Badaan (Tld) 845
Knechtsand (Ger) 297
Knight Island (US) 988
Knorr (US) 971
Knud Rasmussen (Den) 197
Knurrhahn (Ger) 296
Knyaz Vladimir (Rus) 671
Ko Chang (RoK) 485
Ko Ryeong (RoK) 485
Kobchik (Rus) 727
Kobra (Indo) 366
SGT Matej Kocak (US) 981
Kochab (Mex) 535
Kochi (Ind) 333
Kodiak Island (US) 988
Kodor (Rus) 723
Kojima (Jpn) 452
Kojoon Bong (RoK) 484
Kokuryu (Jpn) 429
Kola (Rus) 715
Kolkata (Ind) 333
Kolokita (Bul) 92
Kolomna (Rus) 708
Komandor (Rus) 726
Komayuki (Jpn) 457
Kombo A Janea (Cam) 97
Komendor (Rus) 707
Kommuna (Rus) 719
Konarak (Ind) 390
Kondopoga (Rus) 704
Kondor (Pol) 637
Kondul (Ind) 348
Kong Ju (RoK) 481
Kongou (Jpn) 434
Konkan (Ind) 350
Konstantin Olshansky (Ukr) 879
Kontraadmiral Sava Ivanov (Bul) 92
Kontr-Admiral Demidov (Rus) 705
Kontr-Admiral Olenin (Rus) 705
Kontradmiral Vlasov (Rus) 707
Kontradmiral Xawery Czernicki (Pol) 644
Kopâs (Nor) 589
Kora (Ind) 342
Korall (Rus) 730
Koramshahr (Iran) 390
Korcula (Cro) 188
Korets (Rus) 697
Korets (Ukr) 882
Korolev (Rus) 704
Korshun (Rus) 727
Koshiki (Jpn) 453
Koster (Swe) 802
Kostroma (Rus) 674
Koswari (Ind) 348
Kotelnich (Rus) 708
Kotobiki (Jpn) 457
Kotonami (Jpn) 456
Kou (Est) 235
Koun (Jpn) 460
Kountouriotis (Gre) 305
Kourion (Cypr) 193
Kouya (Jpn) 456
Kovel (Ukr) 882
Kovrovets (Rus) 707
Köyceğiz (Tur) 862
Koyda (Rus) 716
Kozara (Ser) 747
Kozhikode (Ind) 350
Kozlu (Tur) 862
Krabi (Tld) 834
Kraburi (Tld) 830
Krait (Indo) 367
Krajmorje (Bul) 93
Kråkenes (Nor) 589
Krakow (Pol) 642
Kralj Dmitar Zvonimir (Cro) 186
Kralj Petar Kresimir IV (Cro) 186
Krasnodar (Rus) 682
Krasnokamensk (Rus) 682
Krasnoperekopsk (Ukr) 882
Krateos (Gre) 308
Krechet (Rus) 727
Kremenchuk (Ukr) 878
Kremenets (Ukr) 882
Kristaps (Lat) 495
Krka (Cro) 187
Kronshtadt (Rus) 681
Kronsort (Ger) 294
Kuban (Rus) 727
Kubbar (Kwt) 491
Kudaka (Jpn) 453
Kuha 23-26 (Fin) 240
Kujang (Indo) 367, 376
Kukui (US) 990
Kukulkán (Gua) 316
Kukup (Mly) 516
Kula (Fij) 238
Kulish (Ind) 342
Kullen (Swe) 802
Kum Wha (RoK) 485
Kumano (Jpn) 455
Kumataka (Jpn) 443
Kumba (Indo) 376
Kumbhir (Ind) 350
Kumejima (Jpn) 445
Kumkale (Tur) 862
Kun Ming (Twn) 814
Kunashiri (Jpn) 456
Kunigami (Jpn) 453
Kunisaki (Jpn) 443, 452
Kunlunshan (CPR) 153
Kunming (CPR) 140
Kunshan (CPR) 157
Kunta Kinteh (Gam) 281
Kupang (Indo) 369
Kurama (Jpn) 433
Kuraman (Mly) 516
Kureren (Den) 198
Kuriat (Tun) 851
Kurikoma (Jpn) 453
Kurily (Rus) 712
Kurinami (Jpn) 458
Kurmuk (Sud) 793
Kurobe (Jpn) 448
SSGT Robert T Kuroda (US) 961
Kurokami (Jpn) 455
Kurose (Jpn) 456
Kuroshima (Jpn) 445
Kuroshio (Jpn) 430
Kuršis (Lit) 502
Kurushima (Jpn) 459
Kuşadasi (Tur) 862
Kusar (Az) 40
Kusiyara (Ban) 56
Kutaisi (Geo) 283
Kuthar (Ind) 343
Kutilang (Indo) 377
Kuvvet (Tur) 870
Kuzbass (Rus) 678
Kuznetsk (Rus) 700
Kvarnen (Nor) 590
Kvitsøy (Nor) 589
Kwan Myong (RoK) 481
Kwang Yang (RoK) 486
Kwanggaeto Daewang (RoK) 477
Kyan-Sit-Thar (Myn) 553
Kyanwa (Nig) 580
Kyknos (Gre) 307
Kyong Buk (RoK) 478
Kyong Ju (RoK) 481

L

L 011-013 (Ban) 58
L 21-24, 41, 61 , 63-66 (UAE) 889
L 33, 35-39 (Ind) 349
L 201-212 (UAE) 890
L 301 (Mld) 522
L 404 (Kwt) 493
L 601-614 (Spn) 775
L 820-821 (Sri) 791
L 9092-9095 (Fra) 268
L 9525-9529 (Nld) 566
L 9565-9576 (Nld) 569
L/CPL Roy M Wheat (US) 981
L'Etoile De Mer (Fra) 274
La Argentina (Arg) 13
La Belle Poule (Fra) 271
La Capricieuse (Fra) 265
La Cruz (Per) 627
La Divette (Fra) 274
La Fayette (Fra) 259
La Flèche (Sey) 748
La Galité (Tun) 851
La Glorieuse (Fra) 265
La Gracieuse (Fra) 265
La Graciosa (Spn) 779
La Graña (Spn) 781
La Grande Hermine (Fra) 271
La Grandière (Fra) 267
La Habana (Cub) 191
La Houssaye (Fra) 276
La Jolla (US) 938
La Loude (Fra) 274
La Macha (Per) 623
La Malfa (Ita) 422
La Mitre (Fra) 275
La Moqueuse (Fra) 265
La Motte-Picquet (Fra) 258
La Orchila (Ven) 1005
La Piccirella (Ita) 423
La Plata (Arg) 22
La Sota (Ben) 67
La Spina (Ita) 422
La Unión (Per) 623
Labas (Mly) 516
Laboon (US) 946
Labuan (Aust) 32
Lacar (Arg) 24
Lachs (Ger) 292
Ladny (Rus) 694
Ladoga (Rus) 712, 726
L'Adroit (Fra) 264
Ladse (Slo) 759
Laganà (Ita) 422
Lago De Atitlan (Gua) 317
Laguna (Plp) 632
Lahav (Isr) 396
Lake Balusan (Plp) 634
Lake Buhi (Plp) 634
Lake Champlain (US) 944
Lake Erie (US) 944
Lake Paoay (Plp) 634
Lake Taal (Plp) 634
Laksamana Hang Nadim (Mly) 511
Laksamana Muhammad Amin (Mly) 511
Laksamana Tan Pusmah (Mly) 511
Laksamana Tun Abdul Jamil (Mly) 511
Lakshmi Bai (Ind) 356
Lambung Mangkurat (Indo) 362
Lampo Batang (Indo) 374
Lan Yang (Twn) 815
Lana (Nig) 583
Lancaster (UK) 904
Lang (Mly) 516
Langeness (Ger) 297
Langkawi (Mly) 517
Languedoc (Fra) 255
Lanzhou (CPR) 139
Laotieshan (CPR) 154
Lapérouse (Fra) 270
Laplace (Fra) 270
Lapwing (Ana) 10
Lara (Cypr) 193
Larak (Iran) 389
Laramie (US) 975
Lardier (Fra) 276
Larkana (Pak) 607
Larrakia (Aust) 33
Larrea (Per) 621
Las (Fra) 275
Las Palmas (Spn) 778
Lascar Catargiu (Rom) 664
Lassen (US) 948
Lastunul (Rom) 664
Lat Ya (Tld) 842
Lata (Sol) 760
Latouche-Tréville (Fra) 258
Launceston (Aust) 33
Lautaro (Chi) 125
Lautoka (Fij) 239
Lavan (Iran) 389

Lavande (Fra) 275
Lawrence H Gianella (US) 981
Lawrenceville (Can) 105
Layang (Indo) 365
Layang 2 (Mly) 519
LC 01-02 (NZ) 576
LC 244 (Ind) 350
LCM 41 (Ind) 349
LCM 701-710 (Myn) 557
LCP 1-4 (Den) 197
LCT 101-102, 104 (Ban) 57
LD 42, 45-46 (Uru) 999
Le Four (Fra) 276
Le Malin (Fra) 265
Le Téméraire (Fra) 250
Le Terrible (Fra) 250
Le Triomphant (Fra) 250
Le Vigilant (Fra) 250
Le Virgilant (Sey) 748
Lealtad (Ven) 1003
Leather Jacket (Sin) 758
Lech (Pol) 645
Leconi II (Gab) 280
Ledang (Mly) 514
Ledbury (UK) 913
Lee Eokgi (RoK) 474
Lee Jongmu (RoK) 474
Lee Sunsin (RoK) 474
Al Leeth (SAr) 742
Leeuwin (Aust) 34
Legare (US) 985
Leim (Can) 113
Lekir (Mly) 509
Lekiu (Mly) 508
Lely (Ven) 1005
Lemadang (Indo) 365
Lempira (Hon) 321
Lena (Rus) 716
Lenguado (Arg) 23
Léon (Fra) 275
Leonard C Banfield (Bar) 62
Leonard J Cowley (Can) 108
Leonardo (Ita) 416
Léopard (Fra) 271
Leopard (Rus) 678
Leopold I (Bel) 63
Leopoldo Regis (Plp) 632
Leovigildo Gantioqui (Plp) 631
Leroy Grumman (US) 975
Lesbos (Gre) 309
Lesmond Remy (StL) 735
Leticia (Col) 174
Letizia (Ita) 422
L'Étoile (Fra) 271
L'Étoile Polaire (Alg) 9
Leuser (Indo) 374
Levanzo (Ita) 420
Levent (Tur) 869
Levera (Gra) 315
Leviathan (Isr) 395
Levuka (Fij) 239
Lewis And Clark (US) 974
Lewis B Puller (US) 978
Leyte Gulf (US) 944
Leytenant Ilin (Rus) 708
LF 8701-8736 (Arg) 21
LI 112 (Nic) 577
Lianjiang (CPR) 151
Lianyungang (CPR) 143
Liaoning (CPR) 135
Liaoyang (CPR) 157
Libaran (Mly) 519
Libeccio (Ita) 409
Liberal (Brz) 75
Liberato Picar (Plp) 631
Liberta (Ant) 11
Libertad (Arg) 19
Libertad (Col) 178
Libertad (Ven) 1003
Libertador (Bol) 69
Libertador (Mex) 538
Liberty (US) 988
Libra (Ita) 412
Libra (Sey) 747
Lienchiang (Twn) 822
Lieutenant Colonel Errhamani (Mor) 545
Lieutenant Commander Georgiu (Cypr) 193
Lieutenant Commander Tsomakis (Cypr) 193
Lieutenant De Vaisseau Lavallée (Fra) 260
Lieutenant De Vaisseau Le Hénaff (Fra) 260
Lieutenant De Vaisseau Mangoye Jean-Baptiste (Gab) 280
Lieutenant Dimitrie Nicolescu (Rom) 665
Lieutenant Lupu Dunescu (Rom) 665
Lieutenant Remus Lepri (Rom) 665
Ligia Elena (Pan) 612
Ligitan (Mly) 516
Lilas (Fra) 275
Lilian (Lit) 504
Lilian Ngoyi (SA) 766
Lim Byeongrae (RoK) 482
Limam El Hadrami (Mtn) 525
Liman (Rus) 712
Limasawa (Plp) 635
Limnos (Can) 113
Limnos (Gre) 305
Lina Maria (Pan) 614
Lindsay (Can) 113
Line (US) 992
Linge (Nld) 572
Lingyanshan (CPR) 155
Linosa (Ita) 418
L'Intrépide (Ctl) 184
Linyi (CPR) 144
Lion (Fra) 271
Lipari (Ita) 419
Lipetsk (Rus) 682
Lippi (Ita) 422
Liseron (Fra) 274
Listerville (Can) 105
Lisus (Alb) 3
Litoral (EqG) 232
Little Rock (US) 950
Liugongdao (CPR) 164
Liupanshan (CPR) 155
Liuyang (CPR) 157
Liuzhou (CPR) 144
Liven (Rus) 701
Ljubomir (Ukr) 882
Lobelia (Bel) 64
Loja (Ecu) 210
Loke (Swe) 804
Loksa (Rus) 723
Lokys (Lit) 503
Lom (Bul) 93
Lombardi (Ita) 422
Lomor (Ml) 524
Long Island (US) 988
Longlom (Tld) 833
Longstone (UK) 917
Lood (Est) 234
Lopburi Rames (Tld) 845
Loreto (Per) 622
Lorraine (Fra) 255
Los Cayos (Ven) 1006
Los Frailes (Ven) 1005
Los Llanos (Ven) 1005
Los Monjes (Ven) 1005
Los Rios (Ecu) 210
Los Roques (Ven) 1005
Los Taques (Ven) 1006
Los Testigos (Ven) 1005
Louhi (Fin) 242
Louis M Lauzier (Can) 109
Louis S St Laurent (Can) 106
Louise-Marie (Bel) 63
Louisiana (US) 935
Louisville (US) 938
Loutre (Fra) 276
SP/4 James A Loux (US) 961
Lovat (Rus) 716
Loyal (US) 976
LP 01-05 (Hon) 320
LP 08-11 (Par) 615
LP 101, 104, 202-203 (Par) 616
LP 412, 414, 701-732 (Bol) 68
LP 416-417 (Bol) 69
LPM 4400-4413 (Chi) 125
LR 5 (Aust) 34
LR 7 (CPR) 159
LR 71, 73 (Cro) 189
LR 2901-2908 (Hon) 319
LR 3601-3608 (Hon) 320
LS 010, 015, 020, 025, 030, 035, 040, 060, 070, 080, 050, 129 (Gre) 313
LS 51-52, 55, 56, 65, 84-88, 95, 97, 101, 103, 106-107, 109-110, 112, 114-119, 121-123, 125-128, 134-136, 601-616 (Gre) 314
LS 137-172, 401, 413-415 (Gre) 315
LS 201, 217, 220, 222-223 (Jpn) 460
LSR 4420-4431 (Chi) 125
LTC John U D Page (US) 977
LTG William B Bunker (US) 961
Lübeck (Ger) 288
Lubéron (Fra) 276
Lublin (Pol) 642
Lubny (Ukr) 883
Lucretia Paim (Ang) 9
Ludwigshafen (Ger) 290
Luigi Dattilo (Ita) 424
Luigi Durand De La Penne (Ita) 407
Luigi Rizzo (Ita) 408
Luna (Den) 198
Lunden (Den) 202
Luoxiaoshan (CPR) 154
Luoyang (CPR) 143
Lushan (CPR) 155
Lushun (CPR) 160
Lütje Hörn (Ger) 297
Lutsk (Ukr) 877
Luxi (CPR) 157
Luymes (Nld) 570
Ly Thai To (Vtn) 1009
Lyme Bay (UK) 919
Lynch (Arg) 22
Lyndon B Johnson (US) 950
Lynx (Fra) 271
Lyø (Den) 199
Lyra (Den) 198
Lyre (Fra) 269

M

M 02-14 series (Spn) 783
M 1 (Pol) 645
M 12, 21-22 (Pol) 646
M 35, 38-40 (Pol) 643
M 325-328 (Az) 42
M 1101-1104 (Plp) 636
M Charles (Can) 109
M Fahmy (Egy) 222
M Perley (Can) 113
MA 1-3 (Ger) 296
Maanshan (CPR) 142
Mabechi (Jpn) 455
Al Mabrukah (Omn) 594
Mabul (Mly) 519
Macaé (Brz) 78
Macareu (Por) 656
Macau (Brz) 78
Machado (Chi) 125
Macham (Iran) 390
Machaon (Fra) 274
Machitis (Gre) 308
Mackinaw (US) 988
Macko (Pol) 645
Macon (US) 968
Mactan (Plp) 633
Madadgar (Pak) 610
Madeleine (Lit) 503
Madhumati (Ban) 55
Madina (SAr) 737
Madryn (Arg) 22
Maejima (Jpn) 444
Maga (Myn) 555
Magadan (Rus) 678, 726
Magadnets (Rus) 726
Magar (Ind) 349
Magat Salamat (Plp) 630
Magdalena (Mex) 540
Magdeburg (Ger) 290
Magé (Brz) 78
Magnetica (Rom) 666
Magnitogorsk (Rus) 682
Magnolia (Fra) 274
Magnus Lagabøte (Nor) 589
Magomed Gadgiev (Rus) 708
Magua (DR) 207
Mahamiru (Mly) 514
Mahan (Iran) 385
Mahan (US) 946
Mahar Bandoola (Myn) 553
Mahar Thiha Thura (Myn) 553
Maharajalela (Bru) 89
Mahawangsa (Mly) 513
El Maher (Mor) 548
Mahika (Ind) 351
El Mahir (Alg) 6
Mahish (Ind) 350
Mahnavi-Hamraz (Iran) 388
Mahnavi-Taheri (Iran) 388
Mahnavi-Vahedi (Iran) 388
Mahón (Spn) 781
Mahroos (Kwt) 492
Mahury (Fra) 279
MAI 1101-1104, 2110-2112, 3055-3059 (Rom) 668
Maimon (Kwt) 492
Main (Ger) 296
Maine (US) 935
Maistros (Gre) 311
Maitland (Aust) 33
Maïto (Fra) 276
Maj Bernard F Fisher (US) 978
Maj Stephen W Pless (US) 981
Majed (Iraq) 392
El Majid (Mor) 548
Makar (Ind) 351
Makasib (UAE) 887
Makassar (Indo) 368
Makhachkala (Rus) 703
Makhanda (SA) 763
Al Makhirah (Omn) 596
Makigumo (Jpn) 457
Makin Island (US) 964
Makinami (Jpn) 438
Makishima (Jpn) 445
Makishio (Jpn) 430
Makkah (SAr) 738
Makkum (Nld) 569
Mako (US) 987
Mako Shark (Sin) 758
Al Maks (Egy) 227
Makung (Twn) 812
Makurdi (Nig) 581
Makut Rajakumarn (Tld) 831
Malabar (Fra) 276
Malabrigo (Per) 627
Malahayati (Indo) 361
Malan (Pak) 610
Malaspina (Spn) 778
Malawali (Mly) 516
Maldonado (Uru) 996
Malevo (Arg) 24
Mallet (US) 992
Malmö (Swe) 799
Måløy (Nor) 590
Malpelo (Col) 177
Malu Baizam (Aust) 36
Malvern Hill (US) 968
Malzwin (Nld) 572
Mamba (Tan) 826
Mamilossa (Can) 112
Mamry (Pol) 642
Man Klang (Tld) 840
Man Nai (Tld) 840
Man Nok (Tld) 840
Manabi (Ecu) 210
Manacacías (Col) 173
Al Manama (Bhr) 47
Manatee (US) 987
Manaure (Ven) 1004
Manawanui (NZ) 576
Manchester (US) 953
Manchzhur (Rus) 726
Máncora (Per) 626
Mandalika (Indo) 375
Mandau (Indo) 365
Mandubi (Arg) 22
Mandume (Ang) 10
Manduruyu (Arg) 23
Mangalum (Mly) 520
Mangaratiba (Brz) 78
Mangrove Jack (Sin) 758
Mangyan (Plp) 635
Manhattan (US) 974
Manini (Fra) 276
Manjung (Mly) 516
Manø (Den) 199
Manowar (US) 987
Al Mansoor (Omn) 595
Manta (US) 987
Manta Ray (Sin) 757
Manthatisi (SA) 761
Mantilla (Arg) 21
Mantilla (Per) 625
Manuel Doblado (Mex) 535
Manuel Gomez (Plp) 631
Manuel Gutierrez Zamora (Mex) 535
Manuel Trujillo (Par) 616
Manych (Rus) 716
Maple (US) 990
Mar Caribe (Spn) 781
Mar Del Plata (Arg) 22
Mara (Ven) 1004
Maracaná (Brz) 78
Maragogipe (Brz) 78
Marajo (Brz) 86
Maranon (Ecu) 213
Marañon (Per) 622
Marasesti (Rom) 661
Marban (UAE) 887
Marcus Hanna (US) 989
Margaret Norvell (US) 987
Margarita (Ven) 1005
Maria Bray (US) 989
Maria L Pendo (Arg) 24
Mariano Abasolo (Mex) 529
Mariano Escobedo (Mex) 535
Mariano Matamoros (Mex) 535
Mariano Moreno (Arg) 23
Mariategui (Per) 619
Marie Miljø (Den) 201
Marina Pelayo Pratt Gill (Par) 616
Marine 1 (Ger) 297
Marinero Fuentealba (Chi) 121
Mario Grabar (Ita) 424
Mario Marino (Ita) 420
Mario Villegas (Col) 174
Mariscal Sucre (Ven) 1002
Mariupol (Ukr) 879
Marjata (Nor) 590
Markab (Alg) 9
Markab (Mex) 535
Markhad (Alg) 8
Marlin (Brz) 78
Marlin (Mly) 520
Marlin (US) 987
Marn Vichai (Tld) 845
Marne (Fra) 272
Maroa (Fra) 276
Maroub (Sud) 793
Marra (Ita) 423
Mars (Rus) 709
Mars (Swi) 807
Marshal Gelovani (Rus) 709
Marshal Krylov (Rus) 711
Marshal Shaposhnikov (Rus) 692
Marshal Ustinov (Rus) 689
Marsopa (Arg) 23
Marsouin (Alg) 8
Marte (Per) 624
Martha L Black (Can) 107
Marti (Tur) 862
Martin Garcia (Arg) 22
Martín Posadillo (Spn) 780
Martin Torrijos (Pan) 613
Marudu (Mly) 517
Mary Sears (US) 975
Maryborough (Aust) 33
Maryland (US) 935
Maryut (Egy) 226
Marzoug (Kwt) 493
Masan (RoK) 478
Mascardi (Arg) 24
Mash'noor (Kwt) 493
Mashtan (Bhr) 48
Mashuk (Rus) 722
Mashuu (Jpn) 448
Maskan (Kwt) 490
Mason (US) 948
Mataco (Arg) 21
Matacora (Indo) 367
Matamoros (US) 968
Matanani (Mly) 519
Matanga (Ind) 353
Matarani (Per) 626
Matelot (TT) 850
Matelot Brice Kpomasse (Ben) 67
Mateo García De Los Reyes (Spn) 768
Matias Romero (Mex) 533
Matinicus (US) 988
Matlalcueye (Mex) 541
Matra (Tld) 843
Matros Mikola Mushnirov (Ukr) 883
Matrozos (Gre) 303
Matsunami (Jpn) 458
Matsushima (Jpn) 452
Matsuura (Jpn) 456
Matsuyuki (Jpn) 440
Mattapon (Tld) 842
Matthew (Can) 113
Matthew Perry (US) 974
Mattoso Maia (Brz) 81
Maui (US) 988
Maury (Fra) 279
Maury (US) 975
Max Paredes (Bol) 69
Mayank (Ind) 351
Mayom (Sud) 793
Mayor Jaime Arias Arango (Col) 178
Mayumba (Gab) 280
Mazatenango (Gua) 317
Mazzarella (Ita) 422
Mazzei (Ita) 422
Mazzeo (Ita) 422
MB 4, 19, 28, 37, 56, 58, 61, 76, 99, 100, 110, 148, 304 (Rus) 722
MB 8, 15, 23, 38, 54, 92-93, 105, 119, 160, 166, 174 (Rus) 723
MB 12 (Rus) 724
MB 75 (Rus) 716
McCampbell (US) 948
McClusky (US) 952
McFaul (US) 946
MCS 3001-3010 (Plp) 636
MDK 18, 88 (Rus) 707
MDLC Jacques (Fra) 275
MDLC Richard (Fra) 275
MDN 94-104, 108-109, 114-117 (Ita) 414
Mechanicsville (US) 968
Mecklenburg-Vorpommern (Ger) 286

Meda (Aust) 34
Medang (Mly) 520
Medas (Spn) 775
Medgar Evers (US) 974
Medina (Per) 625
Medusa (Arg) 23
Meen (Ind) 351
Meera Behn (Ind) 356
Meerkatze (Ger) 299
Mega Bakti (Mly) 515
Meghna (Ban) 57
Megrez (Mex) 535
Meiyo (Jpn) 459
Meizhou (CPR) 148
Melbourne (Aust) 27
Melita I-II (Mlt) 523
Melitopol (Ukr) 879
Mellon (US) 984
Melo (Per) 623
Meltem (Tur) 861
Melva (Arg) 23
Melville (Aust) 34
Melville (Dom) 204
Melville (US) 971
Memmon (Mly) 520
MEN 212, 215-216 (Ita) 419
MEN 217-222, 227-228, 551 (Ita) 414
MEN 562-565 (Ita) 415
Menab (Iran) 391
Mendez Nuñez (Spn) 770
Mendi (SA) 762
Mendonca (US) 980
Mengam (Fra) 276
Mengshan (CPR) 155
Menkalinan (Mex) 536
Menkar (Mex) 535
Menominee (US) 974
Merak (Mex) 536
Mercuur (Nld) 571
Mercy (US) 980
Al-Mergheb (Lby) 501
Merikarhu (Fin) 245
Merjen (Tkm) 874
Mermaid (Aust) 34
Mero (Arg) 23
Merrickville (Can) 105
Merseet (Lby) 501
Mersey (UK) 907
Mersuji (Mly) 518
Mesa Verde (US) 962
Mesaha 1-2 (Tur) 865
Mesh (Ind) 351
Mestia (Geo) 282
Metel (Rus) 697
Meteoro (Chi) 124
Meteoro (Spn) 772
Mette Miljø (Den) 201
Meuse (Fra) 272
Mewa (Pol) 642
Mezyad (UAE) 886
MG Charles P Gross (US) 961
MG Robert Smalls (US) 961
MGB 102, 110 (Myn) 557
Mhadei (Ind) 351
Mianyang (CPR) 143
Miaplacidus (Mex) 536
Micalvi (Chi) 121
Miccoli (Ita) 422
Michael Monsoor (US) 950
Michael Murphy (US) 948
Michele Fiorillo (Ita) 424
Michele Lolini (Ita) 424
Michigan (US) 936
Michishio (Jpn) 430
Micronesia (Mic) 541
Middlesex (Jam) 427
Middleton (UK) 913
Midgett (US) 984
Midhili (Mld) 521
Midia (Rom) 667
Midnight Sun (Aust) 37
Mielno (Pol) 643
Mier (Bel) 65
Miguel Ela Edjodjomo (EqG) 231
Miguel Malvar (Plp) 630
Miguel Silva (Col) 178
Miguel Sotoa (Par) 616
El Mikdam (Mor) 548
Mikhail Kogalniceanu (Rom) 664
Mikhail Konovalov (Rus) 732
Mikhail Kutuzov (Rus) 671
Mikhail Rudnitsky (Rus) 717
Mikolaiv (Ukr) 882
Milano II (Spn) 785
Milazzo (Ita) 415
Milius (US) 946
Miljø 101-102 (Den) 201
Miljø 103 (Den) 202
Millinocket (US) 980
Milwaukee (US) 950
Minabe (Jpn) 456
Mineralny Vodi (Rus) 708
Minna (UK) 925
Minnesota (US) 937
Minoo (Jpn) 457
Minsk (Rus) 704, 727
Mintaka (Mex) 536
Mirabilis (Nam) 560
Mirach (Mex) 536
Miramar (Brz) 78
Mirazh (Rus) 701
Mircea (Rom) 666
Mireya Moscoso (Pan) 613
Mirfa (Gua) 317
Mirfak (Alg) 9
Mirfak (Mex) 535
Mirzam (Alg) 9
Missionary Bridge (US) 968
Mississippi (US) 937
Missouri (US) 937
Mistral (Fra) 266
Mistral (Spn) 767
Mistral (UK) 924
Mithun (Ind) 351
Mithuna (Indo) 376
Mitilo (Ita) 417
Mitla (Mex) 537
Mitscher (US) 946
Mittelgrund (Ger) 294
Miura (Jpn) 452
Miyajima (Jpn) 445
Mizar (Alg) 9
Mizar (Mex) 535
Mizrak (Tur) 861
Mizuho (Jpn) 451
Mizuki (Jpn) 456
Mizunami (Jpn) 458
Mjølner (Nor) 591
MK 391, 405, 1277, 1303, 1407-1411, 1556 (Rus) 713
MK 391, 1303, 1407-1409 (Rus) 715
MLM 6-10 (Tld) 842
Moawin (Pak) 609
Mobark (Kwt) 492
Mobile Bay (US) 944, 989
Mochishio (Jpn) 430
Mocovi (Arg) 21
Moctezuma II (Mex) 540
Modig (Swe) 800
Mogochey (Rus) 682
Mohammed V (Mor) 544
Mohammed VI (Mor) 546
Mohammed Brahim Rejeb (Tun) 854
Mohawk (US) 985
Moho (Per) 626
Mok Po (RoK) 481
Molino Del Ray (US) 968
Momsen (US) 948
Monastir (Tun) 851
Monção (Por) 656
Monchegorsk (Rus) 697
Moncton (Can) 103
Mondolkiri (Cmb) 94
Monge (Fra) 270
Monmouth (UK) 904
Mono (Tog) 848
Monomoy (US) 988
Monsekela (Ctl) 184
Monsoon (US) 959
Montcalm (Fra) 257
Monte Cimone (Ita) 423
Monte Sperone (Ita) 423
Monterey (US) 944
Montero (Per) 619
Monterrey (US) 968
Montes Azules (Mex) 538
Montford Point (US) 978
Montgomery (US) 953
Montpelier (US) 938
Montreal (Can) 100
Montrose (UK) 904
Mooam (Iran) 390
Moonmu Daewang (RoK) 476
Moose (Can) 104
Moran Valverde (Ecu) 210
Moray (US) 987
Morazan (Hon) 321
Morcoyán (Arg) 21
Mordoviya Mdk-94 (Rus) 706
Moresby (PNG) 614
Morgenthau (US) 984
Moriah (TT) 850
Morona (Per) 624
Moroz (Rus) 701
Morro Bay (US) 989
Morro De Barcelona (Ven) 1007
Morrosquillo (Col) 176
Morse (Fra) 276
Morshansk (Rus) 700
Moruga (TT) 849
Mosel (Ger) 296
Moshchny (Rus) 723
Moskva (Rus) 689
Mosoj Huayma (Bol) 69
Motobu (Jpn) 453
Motoura (Jpn) 455
Mou Hsing (Twn) 822
Mouanco (Cam) 96
El Mouderrib I-VII (Alg) 8
El Moukadem (Alg) 6
El Moundjid (Alg) 9
El Mounkid I-IV (Alg) 8
Mount Whitney (US) 961
El Mountassir (Alg) 6
El Mourafek (Alg) 8
El Mourakeb (Alg) 6
El Mous'if (Alg) 9
El Moussanid (Alg) 9
El Moutarid (Alg) 6
Mounts Bay (UK) 919
Mourad Rais (Alg) 4
Møvik (Nor) 589
Móvil I (Col) 178
Möwe Bay (Nam) 559
SSIM Manuel A Moyar (Col) 173
MP 15, 29, 37, 39-48 (Est) 236
MPK 82, 178, 221 (Rus) 697
MPK 105, 227 (Rus) 698
MRD 3-4 (Den) 202
MSB 11-17 (Tld) 842
MSF 1-4 (Den) 199
MSO 304 (Jpn) 446
MT 264-265 (Rus) 707
Al Mua'zzar (Omn) 593
Mubarraz (UAE) 887
Muhafiz (Pak) 608
Muhammed (Jor) 461
Al Muharraq (Bhr) 47, 49
Mujahid (Pak) 608
Mukah (Mly) 520
Mulga (Indo) 367
Mulia (Mly) 516
Mull (UK) 908
Mulnaya (Bul) 90
Multatuli (Indo) 373
Mumbai (Ind) 334
Al Munassir (Omn) 596
Munro (US) 984
Munsif (Pak) 608
Munster (Ger) 296
Munter (Swe) 800
Murakumo (Jpn) 457
Murasame (Jpn) 436
Murature (Arg) 20
Muray Jib (UAE) 884
Murban (UAE) 887
Murena (Ita) 420
Murene (Alg) 8
Al Murjan (UAE) 888
Murmansk (Rus) 726
Al Muroom (UAE) 888
Muromets (Rus) 697
Muromi (Jpn) 455
Muroto (Jpn) 450
Murozuki (Jpn) 458
Murray (Can) 113
Musa Maliya Jammeh (Gam) 281
Muskingum (US) 991
Mussandam (Omn) 595
Mustaed (Bru) 88
Mustang (US) 988
Musters (Arg) 24
Mustin (US) 948
Mutiara (Mly) 514
Mutilla (Chi) 123
Mutin (Fra) 271
Myosotis (Fra) 274
Myoukou (Jpn) 434
Mysore (Ind) 334
Mzia (Tan) 825
Mzizi (Tan) 825

N

N N O Salammbo (Tun) 852
Na Daeyong (RoK) 474
Naa Gbewaa (Gha) 300
Naantali (Fin) 240
Nacaome (Hon) 320
Nachi (Jpn) 457
Nadon (Can) 113
Nafanua (Sam) 736
Naftilos (Gre) 310
Nagashima (Jpn) 445
Naghdi (Iran) 382
Nagozuki (Jpn) 458
Nahid (Iran) 385
Nahuel Huapi (Arg) 24
Naif (Kwt) 493
Naiguata (Ven) 1007
Naiki Devi (Ind) 356
Nairal Zaurac (Alg) 9
Naiza (Kaz) 463
Naja (Den) 197
Najaden (Den) 198
Al Najah (Omn) 595
Najim Al Zaffer (Egy) 220
Najran (SAr) 743
Nakat (Rus) 701
Nakhodka (Rus) 727
Nakło (Pol) 643
Nakoura (Leb) 498
Nakul (Ind) 354
Nala (Indo) 361
Nam Won (RoK) 481
Nanaimo (Can) 103
Nanawa (Par) 616
Nandimithra (Sri) 786
Nanggala (Indo) 359
Nanhai (CPR) 151
Nanjing (CPR) 140
Nantou (Twn) 823
Nantucket (US) 988
Nanyi 09, 832 (CPR) 160
Naoshima (Jpn) 445
Narathiwat (Tld) 832
Narcis (Bel) 64
Narciso Monturiol (Spn) 768
Naresuan (Tld) 829
Nariño (Col) 173
Narushio (Jpn) 430
Narval (Arg) 23
Narwhal (US) 987
Naryan-Mar (Rus) 697
Nashak (Ind) 344
Nasir (Iraq) 392
Nasr (Pak) 609
Nasr Al Bahr (Omn) 596
Nassau (Bhm) 44
El Nasser (Egy) 220
Nastoychivy (Rus) 693
Nathanael Maxwilili (Nam) 560
Natsugiri (Jpn) 457
Natsugumo (Jpn) 457
Natsui (Jpn) 456
Natsuzuki (Jpn) 457
Naushon (US) 988
Nautilus (Nld) 571
Navajo (US) 978
Navarin (Egy) 225
Navarinon (Gre) 305
Navarra (Spn) 768
Navmachos (Gre) 308
Nawigator (Pol) 644
Nayband (Iran) 390
Nazim (Pak) 610
Neah Bay (US) 989
Nebraska (US) 935
Necko (Pol) 643
Nedjmet El Kotb (Alg) 8
Al Neemran (Omn) 596
Negron (Ven) 1007
Neiafu (Ton) 848
Neil Armstrong (US) 971
Nelson (TT) 849
Neocaligus (Can) 113
Nepryadava (Rus) 717
Neptun (Rus) 728
Neptuno (Spn) 781
Nereus (Gre) 303
Nesbitt (UK) 913
Nesebar (Bul) 94
Nestor Reinoso (Plp) 631
Nestos (Gre) 312
Netisin (Ukr) 881
Neustrashimy (Rus) 695
Neustrelitz (Ger) 298
Neva (Rus) 726
Nevada (US) 935
Nevcivan (Tur) 869
Nevelsk (Rus) 728
New Hampshire (US) 937
New Mexico (US) 937
New Orleans (US) 962, 968
New York (US) 962
Newcastle (Aust) 27
Newport News (US) 938
Neyba (DR) 207
Neyzeh (Iran) 384
Ngola Kiluange (Ang) 10
Ngunguri (Tan) 826
Niah (Mly) 520
Niamh (Ire) 393
Nichinan (Jpn) 446
Nicobar (Ind) 352
Nicolas Bravo (Mex) 530
Nicolas Mahusay (Plp) 631
Nicolas Suarez (Bol) 69
Nicolay Chiker (Rus) 722
Niedersachsen (Ger) 288
Niels Juel (Den) 194
Nikiforos (Gre) 308
Nikiforos Fokas (Gre) 305
Nikolay Filchenkov (Rus) 706
Nikolay Matusevich (Rus) 709
Nikolay Rubtsov (Rus) 706
Nikolay Sipyagin (Rus) 732
Nikolay Starshinov (Rus) 732
Nikolay Vilkov (Rus) 706
Nikopol (Bul) 93
Al Nil (Egy) 226
Al Nil (SAr) 743
Nimitz (US) 942
Nimr (Egy) 228
Ning Hai (Twn) 817
Ning Yang (Twn) 815
Ningbo (CPR) 137
Nioi (Ita) 422
Nipah (Mly) 517
Nipat (Ind) 344
Nirbhik (Ind) 344
Nirbhoy (Ban) 55
Nirdeshak (Ind) 351
Nireekshak (Ind) 352
Nirghat (Ind) 344
Nirmul (Ban) 57
Nirolhu (Mld) 521
Nirupak (Ind) 351
Nishank (Ind) 344
Nisr (Egy) 228
Niterói (Brz) 75
Nitze (US) 948
Nitzhon (Isr) 398
Niven D (Cay) 115
Nividic (Fra) 276
Nivôse (Fra) 261
Nizhny Novgorod (Rus) 675
Nizyn (Ukr) 883
Njambuur (Sen) 744
Njord (Nor) 592
Noakhali (Ban) 61
Nobaru (Jpn) 456
Noguera (Per) 624
Nongsarai (Tld) 842
Al Noketha (Kwt) 491
Noor (Iran) 379
Nooradheen (Mld) 521
Nordkapp (Nor) 592
Nordnes (Nor) 590
Nordrhein-Westfalen (Ger) 290
Nordsøen (Den) 201
Nordstrand (Ger) 297
Norfolk (US) 938
Norge (Nor) 591
Norikura (Jpn) 456
Norman (Aust) 33
Normandie (Fra) 255
Normandy (US) 944
Nornen (Nor) 592
North Carolina (US) 937
North Dakota (US) 937
Northern Warrior (Aust) 37
Northland (US) 985
Northumberland (UK) 904
Noto (Jpn) 452
Notojima (Jpn) 445
Al Nour (Egy) 222
Nova Kahovka (Ukr) 880
Novigrad (Cro) 188
Novocherkassk (Rus) 704
Novomoskovsk (Rus) 672
Novorossiysk (Rus) 682, 727
Novosibirsk (Rus) 679, 682
Ntringui (Com) 181
Nueva Vizcaya (Plp) 634
Numana (Ita) 415
Numancia (Spn) 768
Nunki (DR) 207
Nunki (Mex) 535
Nunobiki (Jpn) 457
Nur (Egy) 228
Nusa (Mly) 517
Nusa Utara (Indo) 369
Nusrat (Pak) 611
Nwamba (Nig) 580
Nyalau (Mly) 520
Nyayo (Ken) 464
Nyireh (Mly) 516
Nyköping (Swe) 797
Nymfen (Den) 198
Nymphea (Fra) 275
Nzinga Mbandi (Ang) 10

O

Oak (US) 990
Oak Hill (US) 967
Oaxaca (Mex) 534
Obion (US) 991
Obninsk (Rus) 680
Obolon (Ukr) 882

1034 **INDEXES** > Named ships

PFC Eugene A Obregon (US) 981
Observer (Dom) 205
Observer (Mrt) 526
Óbuda (Hun) 325
Obula (Nig) 580
Obzor (Bul) 93
Ocean (UK) 910
Ocean Marie (Aust) 37
Ocean Protector (Aust) 39
Ocean Shield (Aust) 36
Ocracoke (US) 988
Oddane (Nor) 589
Odessa (Ukr) 882
Odet (Fra) 279
Ödev (Tur) 870
Odisseus (Gre) 312
Odysseus (Cypr) 193
Oecussi (ETim) 208
Ohene-Kwapong (Gha) 301
O'Higgins (Chi) 116
Ohio (US) 936
Ohue (Nig) 583
Oirase (Jpn) 456
O'Kane (US) 946
Okba (Mor) 547
Okean (Rus) 730
Okeanos (Gre) 303
Oker (Ger) 295
Oki (Jpn) 452
Okigumo (Jpn) 457
Okinami (Jpn) 457
Okinawa (Jpn) 451
Okishio (Jpn) 459
Okitsu (Jpn) 456
Oklahoma City (US) 938
Okpoku (Nig) 582
Okushiri (Jpn) 456
Olav Tryggvason (Nor) 589
Oldenburg (Ger) 290
Olenegorskiy Gorniak (Rus) 704
Oljevern 01-04 (Nor) 590
Ologbo (Nig) 580
Oltramonti (Ita) 422
Olympia (US) 938
Olympias (Gre) 310
Omaha (US) 953
Omboue (Gab) 280
Omsk (Rus) 676
Ona (Arg) 21
Ona (Chi) 126
Öncü (Tur) 870
Önder (Tur) 870
Onega (Rus) 697
Ongjin (RoK) 485
Onisilos (Cypr) 193
Onitsha (Nig) 581
Onjuku (Mex) 538
Oonami (Jpn) 438
Oosumi (Jpn) 443, 451
Ootaka (Jpn) 443
Ooyodo (Jpn) 441
Opanez (Rom) 664
Opelika (US) 974
Oqbah (SAr) 740
Orädd (Swe) 800
Oral (Kaz) 462, 463
Orangeleaf (UK) 915
Orazio Corsi (Ita) 424
Orca (Arg) 23
Orca (Can) 104
Orcas (US) 988
Orel (Rus) 676, 725
Orella (Chi) 120
Oreste Cavallari (Ita) 423
Organabo (Fra) 279
Orienburg (Rus) 684
Oriku (Alb) 3
Oriole (Can) 104
Orion (Col) 178
Orion (DR) 206
Orion (Ecu) 212
Orion (Fra) 269
Orion (Gre) 311
Orion (Por) 653
Orion (Swe) 802
Orion (Swi) 807
Orione (Ita) 412
Orkan (Pol) 639
Orla (Ire) 394
Orlan (Rus) 727
Ormi (Gre) 308
El Oro (Ecu) 210
Orsa Maggiore (Ita) 417
Orsha (Rus) 716
Orsk (Rus) 706
Ortiz (Chi) 121
Orucreis (Tur) 858
Oryx (Nam) 559
Orzeł (Pol) 638
OS 114, 213 (Rus) 720
OS-138 (Rus) 721
Osage (US) 991
Oscar (Sin) 758
Oscar Austin (US) 948
Oskemen (Kaz) 462
Oslyabya (Rus) 704
Osman (Ban) 53
Osman Gazi (Tur) 863
Osprey (US) 987
Oste (Ger) 295
Östergötland (Swe) 796
Osternes (Nor) 589
Östhammar (Swe) 800
Ostria (Gre) 311
Oswaldo Cruz (Brz) 86
Oswald Siahaan (Indo) 360
Otago (NZ) 575
Otarie (Fra) 276
Otomi (Mex) 541
Otra (Nor) 590
Ottawa (Can) 100
Otter Bay (Can) 113
Ottonelli (Ita) 422
Otto Sverdrup (Nor) 586
Ouachita (US) 991
Oualie Bay (StK) 734
Oueme (Ben) 67
Ouha (Kwt) 490
Oumi (Jpn) 448, 456
Ouranos (Gre) 311
Owen (UK) 913
Owusu-Ansah (Gha) 301
Oyarvide (Uru) 998
Oyashio (Jpn) 430
Ozelot (Ger) 292
Özgen (Tur) 870
Özgür (Tur) 870

P

P 001-002 (Egy) 226
P 001, 005, 007-009, 106-107 (Geo) 283
P 01-02 (Moz) 552
P 01-02 (Pak) 607
P 01-04, 21-24, 51-52, 61 (Mlt) 523
P 01, 135 (RoK) 487
P 02-05, 14-17 (Dji) 203
P 03, 05-08 (StL) 734
P 04-08 (Sur) 793
P 05-06 (Mlt) 524
P 6-7, 18-19 (Dji) 204
P 6, 13-24, 26-38, 101-104 (Fra) 277
P 010-016, 018-019, 021, 050-051, 053-055, 201-207, 210-225, 227-230, 4442-4447 (Sri) 790
P 33-34 (Mon) 542
P 38-40 (Bhm) 44
P 42-45, 48-49, 110-113 (Bhm) 45
P 43-44 (Tan) 826
P 51-54 (Tld) 839
P 63, 084-088, 101-104, 151-153 (Eri) 233
P 101, 114 (Spn) 775
P 101-102, 201, 401 (Sur) 794
P 101-105, 301-312 (Iraq) 392
P 103-104 (Geo) 282
P 103, 106 (Par) 616
P 104, 191, 349-351, 377 (Rus) 704
P 106-107, 110-124, 126-138, 140-146, 148-161, 163-166, 170, 172-173, 175-179, 243-246, 401-402 (Sri) 789
P 107 (Par) 615
P 107-116 (Mor) 547
P 115, 207-208, 218-219 (Alb) 2
P 127-128 (RoK) 489
P 201-206 (Iraq) 393
P 203-204 (Tun) 851
P 212-216 (Az) 40
P 218-219, 222 (Az) 41
P 240-242, 410-412, 414-415, 417, 419-424, 430, 432-433, 435-437, 439, 450-451, 460, 462, 464-465, 470-475, 480-481, 483-485, 490-492, 494, 497 (Sri) 788
P 241-257, 260-262 (Nig) 581
P 252-253 (Sri) 792
P 281-282 (RoK) 488
P 303 (UAE) 888
P 313-327 (Nig) 580
P 618-622 (Ang) 9
P 1022-1031 (Yem) 1017
PA 01-02 (EIS) 230
PA 06-09 (EIS) 231
Pabna (Ban) 61
Pacasmayo (Per) 626
Pacific Marlin (Brz) 87
Pacu (Arg) 23
Padma (Ban) 57
Págalo (Ven) 1007
Pagaza (Ven) 1007
LTC John U D Page (US) 977
Pahang (Mly) 510
Paiño (Spn) 785
Palacios (Per) 620
Palan (Ind) 353
Palangrin (Fra) 274
Palawan (Plp) 635
Palenque (Mex) 537
Palikir (Mic) 541
Palinuro (Ita) 417
Paliya (Rus) 726
Palmaria (Ita) 420
Palmetto (Ant) 11
Palo Alto (US) 968
Paluma (Aust) 34
Pamir (Rus) 722
Pamlico (US) 991
Pampanga (Plp) 634
Pampeiro (Brz) 80
Panagos (Cypr) 192
Panan (Indo) 376
Panana (Indo) 366
Panarea (Ita) 418
Pancha Carrasco Point Bridge (CR) 183
Pan Chao (Twn) 813
Pandora (Gre) 312
Pandrong (Indo) 366
Pandrosos (Gre) 312
Pangai (Ton) 848
Pangasinan (Plp) 630
Panquiaco (Pan) 612
Panshih (Twn) 820
Pansio (Fin) 241
Pantelleria (Ita) 419
Panter (Nld) 573
Pantera (Rus) 678
Panthère (Fra) 271
Panyu (CPR) 151
Paolini (Ita) 422
Papanikolis (Gre) 303
Pará (Brz) 85
Paracas (Per) 627
Parachique (Per) 627
Paradoks (Rus) 724
Paraguassú (Brz) 85
Paraguay (Par) 615
Parana (Arg) 24
Parang (Indo) 376
Parati (Brz) 80
Parella (Rus) 726
Pari (Indo) 368
Pari (Mly) 513
Pari Burong (Sin) 758
Park Donghyuk (RoK) 482
Park Dongjin (RoK) 482
Parker (Arg) 15
Parksville (Can) 105
Park Wi (RoK) 474
Parnaiba (Brz) 80
Paros (Gre) 309
Parramatta (Aust) 28
Partipilo (Ita) 422
Partisan (Den) 198
Parvin (Iran) 385
Pasadena (US) 938
Passat (Rus) 701
Pastaza (Per) 625
PAT 01-15 (Indo) 376
Patagonia (Arg) 20
Pat Barton (UK) 913
Pathfinder (US) 975
Patiño (Spn) 780
Pati Unus (Indo) 362
Patoka (US) 991
Patola (Indo) 366
Patonga (Aust) 38
Patria (Nld) 571
Patria (Ven) 1003
Patricia (UK) 925
Patriot (US) 969
Patrioten (Den) 198
Pattani (Tld) 832
Patuakhali (Ban) 61
Patuca (Hon) 320
Patuxent (US) 975
Paul Bogle (Jam) 427
Paul Clark (US) 987
Paul Hamilton (US) 946
Paul Ignatius (US) 948
Paulus Hook (US) 968
Paus (Mly) 513
Pavel Derzhavin (Ukr) 882
Pavois (Fra) 278
Paysandu (Uru) 997
Pazisi (Geo) 282
PB 1-4 (HK) 322
PB 07-08 (Gra) 316
PB 211-219 (Ken) 465
PB 401-403, 405, 412-417, 421-427, 429-431, 433-435, 437-439, 441-443, 446, 451-468, 472-482, 484, 490-500, 700-707 (Col) 180
PBK 762, 764, 767, 771 (Rus) 717
PC 01-09 (EIS) 230
PC 33-38 (Jpn) 456
PC 201-232 (Sin) 759
PC 351-354 (Plp) 630
PC 1001-1003 (RoK) 487
PCL 3, 5-9 (Twn) 817
PDB 11-15, 63, 68 (Bru) 89
Peacock (Alg) 9
Peacock (Mex) 535
Pea Island (US) 988
Pearl Harbor (US) 967
Pebane (Moz) 552
Pechenga (Rus) 715
Pecos (US) 975
Pedang (Indo) 376
Pedro Campbell (Uru) 995
Pedro David Salas (Col) 178
Pedro Teixeira (Brz) 79
Pegas (Rus) 709
Pégase (Fra) 269
Pégaso (Por) 653
Pegnitz (Ger) 294
Pejerrey (Arg) 23
Peleliu (US) 963
Pelias (Gre) 312
Pelican (US) 987
Pelícano (Ven) 1006
Pelikaan (Nld) 572
Pelikan (Indo) 377
Pelikan (Tur) 861
Pelikanen (Swe) 804
Pelindung 1-5 (Mly) 519
Pelluhue (Chi) 125
Pemanggil (Mly) 516
Pembroke (UK) 912
Pemburu (Bru) 88
Penac (Can) 112
Pendant (US) 992
Pendekar (Mly) 513
Pendik (Tur) 870
Penedo (Brz) 80
Penfeld (Fra) 279
Pengaman 1 (Mly) 518
Pengawal 1-8, 11-12 (Mly) 518
Pengawal 13-49 (Mly) 520
Penggalang 1-2 (Mly) 518
Penggalang 10-27 (Mly) 519
Penggalang 30-44, 46 (Mly) 520
Penhu (Twn) 822
Peninjau 1 (Mly) 518
Pennsylvania (US) 935
Penobscot Bay (US) 989
Penyelamat 1-4 (Mly) 518
Penyelamat 5-10 (Mly) 519
Penyerang (Bru) 88
Penyu (Mly) 516
Penzance (UK) 912
Perak (Mly) 510
Perantau (Mly) 514
Perca (Arg) 23
Perdana (Mly) 513
Peregrina (Spn) 779
Perekop (Rus) 714
Peresvet (Rus) 704, 722
Peringgi (Mly) 519
Perissinotto (Ita) 423
Perkasa (Mly) 513
Perle (Fra) 248
Perryville (US) 968
Perseus (Swi) 807
Perseverancia (Ven) 1003
Persistence (Sin) 756
Perth (Aust) 28
Perwira (Bru) 88
Pescarusul (Rom) 664
Petersburg (US) 982
Peter Willemoes (Den) 194
Petir 80-85 (Mly) 518
Petrel (Arg) 23
Petrel (US) 987
Petrel (Ven) 1006
Petrel I (Spn) 785
Petr Gradov (Rus) 710
Petrozavodsk (Rus) 680
Peykan (Iran) 384
PF 01-06 (EIS) 229
PFC Dewayne T Williams (US) 977
PFC Eugene A Obregon (US) 981
PFR 1-8 (EIS) 230
PGM 104-107 (Tur) 863
PGM 401-406, 421-423 (Myn) 557
PGM 412-415 (Myn) 556
Phaéton (Fra) 274
Pharos (UK) 926
Pharuehatsabodi (Tld) 843
Phekda (Mex) 535
Philemon Quaye (Gha) 301
Philippine Sea (US) 944
Phoque (Fra) 276
Phraongkamrop (Tld) 846
Phromyothee (Tld) 845
Phuket (Tld) 836
Phuttha Loetla Naphalai (Tld) 831
Phuttha Yotfa Chulalok (Tld) 831
Piast (Pol) 645
Piccinni Leopardi (Ita) 423
Picharnpholakit (Tld) 846
Pico Del Teide (Spn) 782
Pierre Radisson (Can) 107
Pietro Venuti (Ita) 401
Pigargo (Ven) 1007
Pigasos (Gre) 307
Pijao (Col) 171
Pike (US) 987
Pikker (Est) 235
Pililaau (US) 980
Piloto Pardo (Chi) 121
Pinar 3-4, 6 (Tur) 867
Pinckney (US) 948
Pineios (Gre) 312
Pinguino (Arg) 23
Pin Klao (Tld) 843
Pintar (Mly) 516
Pioneer (US) 969
Piorun (Pol) 639
Pipinos (Gre) 303
Pirai (Guy) 318
Piraim (Brz) 85
Pirajá (Brz) 80
Piranha (Sin) 758
Piratini (Brz) 80
Pirie (Aust) 33
Pisagua (Per) 618
Piton (Indo) 367
Pittsburgh (US) 938
PL 3, 46-49, 51-56, 60-65, 85-89 (HK) 322
PL 5-21, 22, 24, 26-27, 29, 31, 40-45, 90-96 (HK) 323
PL 74-76, 82-84 (Jpn) 453
Planet (Ger) 294
Playa Blanca (Col) 178
Pleidas (Alg) 9
Maj Stephen W Pless (US) 981
Plotarchis Blessas (Gre) 307
Plotarchis Maridakis (Gre) 307
Pluton (Fra) 268
Pluvier (Fra) 265
Plymouth (TT) 849
PM 6-8, 11-12 (EIS) 229
PM 10 (EIS) 230
PM 30, 56, 59, 63, 69, 82, 86, 97, 138, 140, 156 (Rus) 714
PM 41-42 (Jpn) 456
PM 2031-2033, 2035-2038, 2040-2041, 2043-2045, 2501-2515 (Chi) 126
Pochetnyy (Rus) 723
Po Chiang (Twn) 816
Podilliya (Ukr) 882
Podolsk (Rus) 673, 728
Podor (Sen) 744
Poema (Nld) 573
Point Caveau (Can) 111
Pokagon (US) 974
Polar (Ang) 10
Polar (Por) 654
Polaris (DR) 207
Polaris (Mex) 535
Polaris (Ven) 1007
Polar Star (US) 988
Polemistis (Gre) 308
Pole Star (UK) 926
Polifemo (Ita) 421
Pollux (Bel) 64
Pollux (DR) 206
Pollux (Gua) 317
Pollux (Mex) 535
Pollux (Swi) 807
Pollux (US) 983
Poltava (Ukr) 882
Polyarny (Rus) 708
Pomeroy (US) 977
Pomorye (Rus) 724
Pompano (US) 987

Ponce (US) 979
Pono (Tan) 826
Pontos (Gre) 303
Ponza (Ita) 420
Popocateptl (Mex) 541
Poponguine (Sen) 744
Pori (Fin) 239
Porkkala (Fin) 241
Porpoise (Sin) 758
Porpora (Ita) 417
Port Cros (Fra) 276
Port Gentil (Gab) 280
Port Hudson (US) 968
Port Royal (US) 944
Porte Grande (Ban) 61
Porter (US) 946
Portete (Col) 178
Portland (UK) 904
Portland (US) 962
Porto Conte (Ita) 420
Porto Corsini (Ita) 420
Porto Empedocle (Ita) 420
Porto Ferraio (Ita) 420
Porto Fossone (Ita) 420
Porto Pisano (Ita) 420
Porto Salvo (Ita) 420
Porto Torres (Ita) 420
Porto Venere (Ita) 420
Porvoo (Fin) 240
Posada (Rom) 664
Poseidon (Cypr) 193
Poseidon (Gre) 303
Poseidon (Swe) 805
Poshak (Ind) 353
Potengi (Brz) 86
Poti (Brz) 80
Poti (Geo) 283
Poui (TT) 850
Pourquoi Pas ? (Fra) 270
Povorino (Rus) 697
Poyang Hu (CPR) 161
Poyraz (Tur) 862
Poznan (Pol) 642
PP 2001, 2003, 2005-2010, 2012-2013, 2015-2019, 2021-2023, 2025-2033, 2035-2038, 2050-2053, 2055-2056, 2058-2063, 2065-2067, 3007-3008, 3017-3018, 3535-3539, 3550, 3552-3553, 3555-3559, 3561-3563, 3565-3573, 3575-3578, 3580, 5033, 5035, 5037-5039, 5050-5053, 10025-10027 (Twn) 824
PP 6001-6002, 10017-10020, 10022-10023, 10028-10029, 10031 (Twn) 823
PP 10032-10033, 10035, 10037-10039, 10050 (Twn) 825
PR 001, 005 (Cam) 95
Prabal (Ind) 344
Prabparapak (Tld) 838
Pradhayak (Ind) 353
Prairial (Fra) 261
Pralaya (Ind) 344
Prata (Ita) 423
Prathpa (Sri) 787
Preble (US) 948
Prefecto Derbes (Arg) 21
Prefecto Fique (Arg) 21
Preia-Mar (Por) 656
Premier Maître l'Her (Fra) 260
Preserver (Can) 105
President El Hadj Omar Bongo (Gab) 281
President H I Remeliik (Pal) 612
Presidente Eloy Alfaro (Ecu) 210
Prespa (Gre) 311
Preveze (Tur) 856
PRF 300-322 (Col) 175
Priazove (Rus) 712
Pribaltika (Rus) 712
Priboy (Bul) 91
Pridneprovye (Ukr) 878
Prignitz (Ger) 299
Priluki (Ukr) 877
Primauguet (Fra) 258
Primera Dama (CR) 183
Primo Longobardo (Ita) 402
Primorye (Rus) 726
Primula (Bel) 64
Prince of Wales (UK) 901
Princeton (US) 944
Private Robertson V C (Can) 109
Priyadarshini (Ind) 355
PRM 01-04 (EIS) 230
Procida (Ita) 420
Procion (DR) 206
Procyon (Gua) 317
Procyón (Mex) 535
Proet (Tld) 844
Prometheus (Gre) 311
Prong (Tld) 844
Protea (SA) 764
Protecteur (Can) 105
Protector (StL) 735
Protector (UK) 908
Protegat (Fl) 237
Proteo (Bul) 93
Proteus (Gre) 303
Provence (Fra) 255
Providence (US) 938
Providencia (Col) 177
Provorny (Rus) 696
Prut (Rus) 715
PS 818-819 (Rus) 726
PS 825 (Rus) 731
Psara (Gre) 304
PSK 382, 405, 673, 1518 (Rus) 715
PSKA 80 (Rus) 733
PSKA 276, 282, 286, 290-291, 294-295, 297, 581, 583-585, 588-589, 595 (Rus) 729
PSKA 450-466, 816-818 (Rus) 727
Pskov (Rus) 675
PSKR 52-59, 200-203 (Rus) 731
PSKR 109-110 (Rus) 727
PSKR 274, 400-401, 500, 900, 907 (Rus) 729
PSKR 322 (Rus) 732
PSKR 632, 635, 705-706, 709, 719, 627, 629, 645, 695, 609, 670, 712, 724-725, 630, 639, 500-501 (Rus) 728
PT 14, 16-17, 22, 26, 34, 61-62 (Sin) 758
PT 71 (Cro) 188
Puelo (Arg) 24
Puerto Deseado (Arg) 19
Puerto Montt (Chi) 126
Puerto Natales (Chi) 126
Puerto Quepos (CR) 183
Puerto Rico (US) 982
Pukaki (NZ) 575
Pulau Raas (Indo) 371
Pulau Rangsang (Indo) 371
Pulau Rempang (Indo) 371
Pulau Rengat (Indo) 370
Pulau Rimau (Indo) 371
Pulau Romang (Indo) 371
Pulau Rote (Indo) 371
Pulau Rupat (Indo) 370
Pulau Rusa (Indo) 371
Puleo (Ita) 422
Puma (Ger) 292
Punai (Indo) 377
Puncher (UK) 907
Punggol (Sin) 757
Puni (Bru) 89
Puno (Per) 627
Punta Alta (Arg) 20
Punta Arenas (Chi) 126
Punta Arenas (Per) 626
Punta Ballena (Ven) 1008
Punta Barima (Ven) 1008
Punta Brava (Ven) 1005
Punta Capones (Per) 623
Punta Cardon (Ven) 1008
Punta Caxinas (Hon) 321
Punta Macolla (Ven) 1007
Punta Macoya (Ven) 1008
Punta Macuro (Ven) 1008
Punta Malpelo (Per) 623
Punta Mariusa (Ven) 1008
Punta Mastun (Mex) 536
Punta Mero (Per) 623
Punta Mogotes (Arg) 18
Punta Moron (Ven) 1008
Punta Morro (Mex) 536
Punta Mosquito (Ven) 1008
Punta Mulatos (Ven) 1008
Punta Pariñas (Per) 627
Punta Perret (Ven) 1008
Punta Playa (Ven) 1008
Punta Sal (Per) 623
Punta Unare (Ven) 1008
Purak (Ind) 353
Puran (Ind) 353
Purga (Rus) 731
Purjosh (Pak) 610
Pursuer (UK) 907
Purunpää (Fin) 240
Pusan (RoK) 478, 485
Pushpa (Ind) 353
Putian (CPR) 143
Putlos (Ger) 296
Putuoshan (CPR) 154
Puyallup (US) 974
PV 5-16, 30-37 (HK) 323
PVH 1-2 (Est) 236
PVK 003 (Est) 235
PVK 010, 012, 020-021, 025 (Est) 236
Pyhäranta (Fin) 241
Pyi Daw Aye (Myn) 558
Pyong Taek (RoK) 486
Pyotr Velikiy (Rus) 688
Pytheas (Gre) 310
Pytlivy (Rus) 694
PZHK 3, 5, 17, 30-32, 36-37, 41-47, 49, 53-55, 59, 64, 66, 68, 79, 82, 84, 86, 415, 417, 900, 1296, 1378, 1514-1515, 1544-1547, 1560, 1680, 1859, 2055 (Rus) 719
PZHS 92, 95, 98, 123, 273, 279, 282 (Rus) 723

Q

Q 42-44 (Qat) 658
Qahir Al Amwaj (Omn) 593
Al Qatar (Egy) 222
Qaysan (Sud) 793
Qeshm (Iran) 389
Qiandao Hu (CPR) 161
Qianlao (CPR) 158
Qian Sanqiang (CPR) 159
Al Qiaq (SAr) 741
Qina (Egy) 225
Qingchengshan (CPR) 154
Qingdao (CPR) 141
Qinghai Hu (CPR) 161
Qingyuan (CPR) 148
Qingzhou (CPR) 157
Qinzhou (CPR) 148
Quanzhou (CPR) 148
Quartier Maître Alfred Motto (Cam) 95
Queen Elizabeth (UK) 901
Queen Modjadji I (SA) 761
Queenston (Can) 105
Queitao (Chi) 125
Querandi (Arg) 21
Quest (Can) 104
Quezon (Plp) 629
Quilca (Per) 626
Quillen (Arg) 24
Quilotoa (Ecu) 213
Quindio (Col) 177
Quiñónes (Per) 620
Qui Nonh (Vtn) 1014
Quintero (Chi) 126
Quisquis (Ecu) 213
Quitasueño (Col) 179
Quito (Ecu) 212
Quokka (Aust) 37
Al Quonfetha (SAr) 742
Quorn (UK) 913
Quwwat (Pak) 607
Al Quysumah (SAr) 741

R

R 01-05, 118, 122, 123-124, 215-216, 225-226 (Alb) 3
R 2, 11, 14, 18-20, 24, 29, 32, 47, 71, 79, 109, 257, 297-298 (Rus) 700
R 117, 125-128, 217, 224, 227-228 (Alb) 2
Raahe (Fin) 240
Rabaul (PNG) 614
L V Rabhi (Mor) 548
Rademaker (Brz) 74
Radhwa (SAr) 742
Radhwa 1-6, 14-17 (SAr) 742
Radoom (UAE) 887
Raet (Tld) 845
Al Rafa (Egy) 222
Rafael Del Castillo Y Rada (Col) 174
Rafael Pargas (Plp) 631
Rafael Peralta (US) 948
Rafaqat (Pak) 611
Raffaele Rossetti (Ita) 416
Rahav (Isr) 395
Al-Rahmani (Omn) 594
Rah Naward (Pak) 609
Rahova (Rom) 664
Raider (UK) 907
Raif Denktas (Tur) 872
Rainier (US) 974
Rais Ali (Alg) 5
Raïs Al Mounastiri (Mor) 548
Raïs Bargach (Mor) 548
Raïs Britel (Mor) 548
Raïs Charkaoui (Mor) 548
Rais Hadj Mubarek (Alg) 4
Rais Hamidou (Alg) 5
Rais Kellich (Alg) 4
Rais Korfou (Alg) 4
Raïs Maaninou (Mor) 548
Raizan (Jpn) 456
Rajah Humabon (Plp) 629
Rajdhwaj (Ind) 357
Rajdoot (Ind) 357
Rajkamal (Ind) 357
Rajkiran (Ind) 357
Rajput (Ind) 336
Rajratan (Ind) 357
Rajshahi (Pak) 607
Rajshree (Ind) 357
Rajtarang (Ind) 357
Raju (Fin) 243
Rajveer (Ind) 357
Ralph Johnson (US) 948
Ramadan (Egy) 222
Ramage (US) 946
Raminthra (Tld) 846
Ramon A Alcaraz (Plp) 628
Ramon Aguirre (Plp) 631
Ramsey (UK) 912
Ramunia (Mly) 517
Rana (Ind) 336
Ranadeera (Sri) 787
Ranagaja (Sri) 790
Ranajaya (Sri) 787
Ranarisi (Sri) 787
Ranavijaya (Sri) 790
Ranawickrama (Sri) 787
Rancagua (Chi) 122
Rānda (Lat) 496
Rang (Tld) 844
Rangamati (Ban) 61
Ranger (UK) 907
Rani Abbakka (Ind) 355
Rani Avantibai (Ind) 355
Rani Durgavati (Ind) 355
Rani Gaidinliu (Ind) 355
Rani Rashmoni (Ind) 355
Ranjit (Ind) 336
Rankin (Aust) 26
Ranvijay (Ind) 336
Ranvir (Ind) 336
Raposo Tavares (Brz) 79
Rapp (Swe) 800
Rappahannock (US) 975
Ras Ajdir (Tun) 853
Ras Al Fulaijah (Lby) 501
Ras Al Hani (Lby) 501
Ras Al Massad (Lby) 501
Ras Al Qula (Lby) 501
Ras Bougaroni (Alg) 8
Ras Djenad (Alg) 8
Ras El Blais (Tun) 853
Ras El Edrak (Tun) 853
Ras El Hilal (Lby) 501
Ras El Manoura (Tun) 853
Ras Enghela (Tun) 853
Ras Ifrikia (Tun) 853
Ras Nouh (Alg) 8
Ras Oullis (Alg) 8
Ras Sisli (Alg) 8
Ras Tamentfoust (Alg) 8
Ras Tara (Alg) 7
Ras Tekkouch (Alg) 8
Ras Tenes (Alg) 8
Rasadgar (Pak) 610
Rasalhague (Mex) 535
Rascas (Fra) 276
Rasheed (Egy) 219
Al-Rasikh (Omn) 594
El Rassed (Alg) 6
Rassvet (Rus) 701
Rasul Gamzatov (Rus) 728
Ratanakiri (Cmb) 94
Ratcharit (Tld) 836
Rattanakosin (Tld) 833
Rauma (Fin) 240
Rauma (Nor) 590
Raven (Can) 104
Rawa (Mly) 519
Rawi (Tld) 842
Raymond Evans (US) 987
Rayo (Spn) 772
Rayyan (Kwt) 493
Razia Sultana (Ind) 355
Razliv (Rus) 701
Razorbill (US) 987
RB 01-03 (Twn) 824
RB 1, 7, 57, 98, 158, 173, 179, 201-202, 217, 239, 262, 265, 296, 314, 327, 360, 362 (Rus) 723
RB 2, 5, 17, 20, 22-23, 25-27, 34, 40, 42-50, 52, 100, 108-109, 136, 167-168, 192-194, 197-199, 212, 232-233, 237, 240, 244, 246-250, 255-256, 293, 295-296, 310-311, 325-326, 386, 389, 391-392, 394-395, 398-401 (Rus) 724
Rbigah (Qat) 658
Rebun (Jpn) 453
Recalada (Arg) 24
Redang (Mly) 519
Red Cloud (US) 977
Red Viper (Aust) 38
Reef Shark (US) 987
Reform (TT) 849
Regele Ferdinand (Rom) 660
Regge (Nld) 572
Regina (Can) 100
Regina Maria (Rom) 660
Regulus (DR) 207
Régulus (Mex) 535
Regulus (US) 983
Rehmat (Pak) 611
Rei (CpV) 114
Rei Bula Matadi (Ang) 9
Rei Ekuiki II (Ang) 9
Rei Kingury Ka Banguela (Ang) 9
Rei Mandume Ndemufayo (Ang) 9
Rei Mpanzu A Nimi (Ang) 9
Reina Sofía (Spn) 768
Reinøy (Nor) 589
Reiun (Jpn) 460
Relámpago (Spn) 772
Reliance (Aust) 38
Reliance (US) 985
Reliant (US) 974
Remora (Arg) 23
Renard (Can) 104
Renato Pennetti (Ita) 423
Rencong (Indo) 365
Renggis (Mly) 520
Rentap (Mly) 517
Requin (Alg) 8
Rescue I-II (Bmd) 67
Rescuer (Dom) 205
Rescuer (Mrt) 526
Reshitelni (Bul) 91
Resilience (Sin) 754
Resko (Pol) 643
Resolute (Can) 106
Resolute (US) 985
Resolution (Sin) 756
Retalhuleu (Gua) 317
Retriever (Mrt) 526
Rettin (Ger) 298
Revi (Fra) 273
Rēzekne (Lat) 494
RG 91-94 (Qat) 659
Rhein (Ger) 296
Rheinland-Pfalz (Ger) 290
RHIBC 1-19 (Nld) 569
RHIRC 1-17 (Nld) 569
RHISC 1-12 (Nld) 573
Rhode Island (US) 935
Rhoen (Ger) 299
Rhön (Ger) 295
Rhu (Mly) 517
Riachuelo (Brz) 70
Richard Dixon (US) 987
Richard E Byrd (US) 974
Richard Etheridge (US) 987
Richmond (UK) 904
W E Ricker (Can) 113
Ridley (US) 987
Al Riffa (Bhr) 48
Rigel (Mex) 535
Rigel (Ven) 1007
Rimau (Mly) 519
Rimini (Ita) 415
Rin (Tld) 844
Ringen (Den) 199
Rio Adaja (Spn) 783
Rio Aguarico (Ecu) 215
Rio Aller (Spn) 784
Rio Almanzora (Spn) 783
Rio Andarax (Spn) 783
Rio Angosturas (Col) 174
Rio Apure II (Ven) 1008
Rio Ara (Spn) 783
Rio Aragón (Spn) 783
Rio Arauca II (Ven) 1008
Rio Arba (Spn) 783
Rio Arica (Col) 174
Rio Babahoyo (Ecu) 215
Rio Bernesga (Spn) 783
Rio Branco (Brz) 84
Rio Bucay (Ecu) 216
Rio Bulu Bulu (Ecu) 215
Rio Cabriel (Spn) 783
Rio Catamayo (Ecu) 215
Rio Catatumbo II (Ven) 1008
Rio Caudal (Spn) 783

1036 **INDEXES** > Named ships

Rio Cedena (Spn) 783
Rio Cervantes (Spn)............. 783
Rio Cervera (Spn) 783
Rio Chambira (Per) 627
Rio Chira (Per)...................... 626
Rio Chone (Ecu).................... 215
Rio Chongon (Ecu) 216
Rio Coangos (Ecu)............... 216
Rio Coca (Ecu)...................... 216
Rio Coco (Hon) 321
Rio Conambo (Ecu) 216
Rio Cononaco (Ecu)............. 216
Rio Curaray (Ecu)................. 216
Rio Daule (Ecu) 215
Rio De La Plata (Arg) 22
Rio Deseado (Arg).................. 22
Rio Duero (Spn) 783
Rio Escondido (Nic)............. 577
Rio Esmeraldas (Ecu)........... 214
Rio Francoli (Spn) 783
Rio Gallego (Spn)................. 782
Rio Gallo (Spn)..................... 783
Rio Genil (Spn)..................... 782
Rio Grande De Matagalpa (Nic) 577
Rio Guadalaviar (Spn).......... 783
Rio Guadalentin (Spn).......... 782
Rio Guadalete (Spn) 783
Rio Guadalobon (Spn) 783
Rio Guadalope (Spn) 783
Rio Guadalquivir (Spn)......... 782
Rio Guadiana (Spn) 783
Rio Guadiaro (Spn) 782
Rio Guarico II (Ven)........... 1008
Rio Hondo (Mex).................. 539
Río Huallaga (Per) 627
Rio Huarmey (Per)................ 626
Rio Icabaru (Ven) 1008
Rio Itaya (Per) 627
Rio Jándula (Spn) 784
Rio Jiloca (Spn).................... 783
Rio Jubones (Ecu) 215
Rio Jucar (Spn) 783
Rio Jujan (Ecu)..................... 216
Rio Ladra (Spn)..................... 783
Rio Lujan (Arg) 22
Rio Macara (Ecu) 215
Río Majes (Per) 626
Rio Manta (Ecu) 216
Rio Martin (Spn).................... 783
Rio Matador (Per).................. 627
Rio Mataje (Ecu) 215
Rio Meta II (Ven) 1008
Rio Mijares (Spn) 783
Rio Minho (Por)..................... 653
Rio Miño (Spn)...................... 783
Rio Muisne (Ecu)................... 216
Rio Nacimiento (Spn)........... 782
Rio Nalon (Spn)..................... 782
Río Nanay (Per)..................... 627
Rio Nansa (Spn).................... 782
Rio Napo (Ecu)...................... 213
Rio Napo (Per) 624, 627
Rio Negro (Uru)..................... 997
Rio Negro II (Ven).............. 1008
Rio Nepeña (Per)................... 626
Rio Nervion (Spn) 783
Rio Ocoña (Per) 626
Rio Órbigo (Spn).................... 784
Rio Orellana (Ecu) 213
Rio Palora (Ecu) 216
Rio Papaloapan (Mex) 538
Rio Paraguay (Arg).................. 22
Rio Parana (Arg)...................... 22
Rio Pas (Spn)........................ 782
Rio Patayacu (Per) 627
Rio Pinillos (Col).................... 174
Rio Pisuerga (Spn)................. 782
Rio Portoviejo (Ecu) 216
Rio Portuguesa II (Ven) 1008
Río Putumayo (Per)............... 627
Rio Puyo (Ecu) 216
Rio Quequen (Arg) 22
Rio Quinindé (Ecu)................ 215
Rio San Miguel (Ecu) 215
Río Santa (Per)...................... 626
Rio Santa Eulalia (Spn) 783
Rio Santiago (Arg)................... 18
Rio Santiago (Ecu)................ 214
Río Santiago (Per) 627
Rio Sarare (Ven).................. 1008
Rio Segovia (Nic) 578
Rio Segre (Spn)..................... 783
Rio Segura (Spn)................... 784
Rio Sinaruco (Ven) 1008
Rio Suchiate (Mex) 539
Rio Surco (Per)...................... 626
Rio Tajo (Spn) 784
Rio Tambo (Per) 626
Rio Tambopata (Per)............. 627
Rio Tambre (Spn).................. 784
Rio Tangare (Ecu)................. 216
Rio Taura (Ecu)..................... 216
Rio Tena (Ecu)...................... 216
Rio Ter (Spn)......................... 782
Rio Tiputini (Ecu) 215
Rio Tordera (Spn).................. 782
Rio Tormes (Spn).................. 784
Rio Tuxpan (Mex).................. 539
Rio Ulla (Spn)........................ 783
Rio Uribante (Ven).............. 1008
Rio Uruguay (Arg) 22
Rio Valdivia (Ecu) 216
Rio Verde (Ecu) 215
Rio Vinces (Ecu).................... 216
Río Viru (Per)......................... 627
Rio Yacuambi (Ecu) 216
Rio Yaguachi (Ecu)................ 215
Rio Yaracuy (Ven)............... 1008
Rio Yasuni (Ecu).................... 216
Rio Yavari (Per) 627
Rio Zamora (Ecu) 216
Rio Zaña (Per)....................... 626
Rio Zapote (Per).................... 627
Rio Zarumilla (Ecu)............... 215
Ristna (Est) 234
River Ray (Sin) 758
Al Riyadh (SAr)............ 738, 742
Rizal (Plp).............................. 629
Rizo (Mex)............................. 539
RK 21-24 (Lat) 496
Rmah (UAE) 891
RML 332, 335-336, 341 (Ser) 746
Roald Amundsen (Nor) 586
Roanoke Island (US)............. 988
Roatan (Hon)......................... 321
Robaldo (Arg).......................... 23
Robert E Peary (US).............. 974
Robert Yered (US)................. 987
Robinson (Arg)........................ 15
Roca (Arg)............................... 24
Rocca (Ita) 423
Rodney M Davis (US) 952
Rodolfo D'Agostini (Arg) 24
Rodos (Gre) 309
Rodqm (UAE)........................ 887
Roebuck Bay (Aust) 39
Roger Revelle (US) 971
Roísín (Ire)............................ 393
Romaleos (Gre)..................... 312
Romat (Isr) 397
Romblon (Plp)....................... 634
Romeo Romei (Ita)............... 401
Romzuald Muklevitch (Rus) 709
Rona (UK) 908
Ronald Reagan (US) 942
Rondache (Fra) 278
Rondônia (Brz)........................ 80
Roosevelt (US)...................... 948
Roraima (Brz) 80
Rosales (Arg) 15
Rosati (Ita)............................. 422
Ross (US).............................. 946
Rossi (Ita).............................. 423
Rostani (Iran)......................... 390
Rostov Na Donu (Rus) 682
Rota (Den)............................. 198
Rotoiti (NZ) 575
Rotte (Nld) 572
Rotterdam (Nld) 567
Rottweil (Ger)........................ 293
Roughtail Ray (Sin) 758
Rover I (StK) 734
Rovine (Rom)........................ 664
RP 101-106, 108-116, 118-134 (Ita) 421
RPC 11-19 (Myn).................. 555
RPC 111, 213-214, 216 (Ser) 746
RPL 60-63 (Sin) 755
RT 57, 231, 233-234, 248-249, 252, 273, 278 (Rus) 708
Rubin (Rus)........................... 730
Rubis (Fra)............................. 248
Rubodh (Bhr) 48
Rudyard Lewis (Bar) 62
Rumbek (Sud)........................ 793
Rumbia (Mly)......................... 516
Runnymede (US) 968
Ruposhi Bangla (Ban)............. 61
Rupsha (Ban)........................... 60
Ruse (Bul) 93
Rush (US).............................. 984
Rushmore (US) 967
Rūsiņš (Lat)........................... 495
Russell (US).......................... 946
Ruth First (SA)...................... 766
Rüzgar (Tur) 862
Ryazan (Rus)......................... 673
Rypøy (Nor)........................... 589
Ryukyu (Jpn).......................... 451
Ryusei (Jpn).......................... 457

S

S 005-008, 11-12, 14, 18, 703 (Az) 43
S Dudka (Can) 111
S Ezzat (Egy)......................... 222
Saad (Kwt) 493
Saad (Pak) 600
Saba Al Bahr (Omn).............. 596
Sabac (Ser)........................... 747
Sabalan (Iran)........................ 380
Sabalo (Arg) 23
Sábalo (Ven)....................... 1001
Saban (Myn) 558
Sabha (Bhr)............................. 46
Sabik (Mex)........................... 535
Sabotøren (Den).................... 198
Sab Sahel (Iran) 388
Sabqat (Pak) 611
Sabre (Sen)........................... 745
Sabre (UK) 907
Sacagawea (US) 974
Sachsen (Ger)....................... 289
Sachsen-Anhalt (Ger) 290
Sacre (Spn)........................... 785
Al-Sadada (Lby)..................... 501
Sadaqat (Pak) 611
Sadd (Pak) 611
Sadh (Omn) 595
Sadko (Rus) 722
Saettia (Ita)............................ 424
Safaga (Egy) 225
Safeguard (US) 979
Saffar (Kwt)........................... 494
Al Saffar (Kwt)....................... 491
Safra 2-3 (Bhr) 50
Safwa (SAr)........................... 741
Saga (Arg)............................... 24
Sagar (Ban)............................. 58
Sagar (Ind) 355
Sagar (UAE).......................... 887
Sagara (Sri)........................... 789
Sagardhwani (Ind)................. 350
Saget 1-20 (Tur).................... 873
Saget 26-35 (Tur).................. 871
Saginaw (US)........................ 991
Sagitario (Por) 653
Sagittaire (Fra)...................... 269
Sagres (Por) 655
Sagu (Myn) 556
Sahand (Iran) 380
El Saher (Alg) 6
Sahas (Ind)............................ 353
Şahin (Tur)............................ 861
Sahyadri (Ind)........................ 338
Saiburi (Tld)........................... 830
Al Said (Omn)........................ 597
Saif (Pak)...................... 602, 611
Saif 1-4 (Bhr)........................... 51
Saif 5-10 (Bhr)......................... 50
Saikai (Jpn)............................ 460
Saikat (Ban)............................. 58
Sailfish (Sin).......................... 758
Sailfish (US).......................... 987
Saiph (Mex)........................... 536
Saire (Fra)............................. 276
Saittra (Myn) 555
Sakala (Est)........................... 234
Sakarya (Tur)......................... 856
Sakhalin (Rus)....................... 726
Sakiai (Lit)............................. 502
Sakshi (Kaz).......................... 463
Salahlah (UAE)...................... 887
Salah Rais (Alg)......................... 5
Salak (Sud)........................... 793
Salam (Ban) 56
Al Salam (Egy) 222
Salamaua (PNG)................... 615
Salamis (Cypr) 192
Salamis (Gre) 304
Salawaku (Indo) 367
CTCIM Jorge Moreno Salazar (Col) 174
Saldiray (Tur)......................... 855
Salema (Spn) 782
Salerno (Ita) 423
Salihreis (Tur) 858
Salina (Ita)............................. 418
Salinas (Chi).......................... 125
Salinas (Per).......................... 627
Sally Ride (US)...................... 971
Salmon (Arg) 23
Salmon (Sin) 758
Salone (Ita)............................ 422
Şalopa 12, 22-24, 27, 30-31, 33 (Tur)................................ 868
Salta (Arg) 13
Saltholm (Den) 198
Salvador Abcede (Plp).......... 631
Salvatore Pelosi (Ita)............ 402
Salvatore Todaro (Ita)........... 401
Salvor (US)............................ 979
Salwa (SAr)........................... 742
SAM 01-02, 04, 06-07 (Swe) 801
SAM 01-06 (Jpn) 445
Sama (Per) 626
Samaesan (Tld)..................... 845
Samanco (Per) 627
Samar (Ind) 355
Samara (Rus) 678
Samba (Yem) 1017
Sambongho (RoK)................ 489
Sambro (Can)........................ 112
Sambu (Indo) 373
Samed (Tld) 844
Samidare (Jpn) 436
Samos (Gre).......................... 309
Sampson (US) 948
Samrat (Ind) 354
Samudra Pavak (Ind) 354
Samudra Pehredar (Ind)....... 354
Samudra Prahari (Ind).......... 354
Samudura (Sri)...................... 787
Samuel Beckett (Ire) 394
Samuel B Roberts (US) 952
Samuel Risley (Can)............ 108
Samui (Tld)............................ 844
Samum (Rus)........................ 699
San Andreas (Per)................ 627
San Andres (Col).................. 175
San Andres (Mex) 540
San Antonio (Chi)................. 126
San Antonio (US) 962
San Diego (US) 962
San Francisco (US).............. 938
San Giorgio (Ita).................... 414
San Giusto (Ita) 414
San Ignacio (Mex) 540
San Jacinto (US)................... 944
San Juan (Arg) 12
San Juan (Plp) 634
San Juan (US)....................... 938
San Lorenzo (Per) 624
San Marco (Ita) 414
San Martin (Arg)...................... 24
San Nicolas (Per) 626
San Pedro (Bol)....................... 68
Sana'a (Yem)..................... 1018
Sanbe (Jpn) 456
Al Sanbouk (Kwt).................. 491
Sanca (Indo) 366
Sancaktar (Tur)..................... 864
Sanchez Carrión (Per).......... 621
Sandbar Shark (Sin)............. 758
Sandhayak (Ind).................... 351
Sangamon (US) 991
Sanges (Ita)........................... 423
Sangram (Ind) 355
Sangu (Ban)............................ 55
Sanibel (US)......................... 988
Sankalp (Ind)......................... 354
Sanket (Ban) 60
Sankt Peterburg (Rus).......... 681
Sanmenxia (CPR) 144
Sanming (CPR) 143
Sanna (Ita) 422
Sanrei (Jpn).......................... 456
Santa Cruz (Arg) 12
Santa Cruz De La Sierra (Bol) 68
Santa Fe (US)....................... 938
Santa María (Spn) 768
Santa Rosa (Per).................. 626
Santamaria (CR) 183
Santaquin (US)..................... 974
Santillana (Per)..................... 621
Santos Degollado (Mex) 535
Sanya (CPR) 144
São Paulo (Brz)...................... 72
SAP 1-11 (Gre)..................... 308
Sapanca (Tur)....................... 865
Sapelo (US).......................... 988
Sapfir (Rus)........................... 730
Saphir (Fra) 248
Sapphire (UK) 914
SAR 12-14, 17-19, 510-511, 515-516, 520 (Gre) 315
Sarafand (Leb) 497
Sarah Baartman (SA) 765
Sarandi (Arg) 13
Sarang (Ind) 355
Saratov (Rus) 706
Sarbsko (Pol) 643
Sardar (Kaz) 463
Sardasht (Iran) 388
Sargento Aldea (Chi)............ 122
Sariyer (Tur).......................... 865
Sarojini Naidu (Ind) 356
Saroma (Jpn) 456
Saros (Tur)............................ 865
Sarotama (Indo).................... 375
Sarov (Rus) 685
Sarucabey (Tur).................... 864
Sarvekshak (Ind)................... 351
Sarych (Rus) 727
Saryu (Ind) 347
Saskatoon (Can) 103
Sat 1-2 (Tur) 862
Satang (Mly).......................... 516
Satauma (Jpn) 453
Satpura (Ind) 338
Sattahip (Tld) 837
Saturn (Swi) 807
Saturno (Ita) 421
Saule (Lat)............................ 495
Savea (Ton).......................... 848
Savitri (Ind)........................... 347
Savitri Bai Phule (Ind) 356
Sawagiri (Jpn)....................... 439
Sawahil (Kwt) 494
Sawfish (US) 987
Sayany (Rus)........................ 717
Sayura (Sri) 786
Sazanami (Jpn)..................... 438
SB 5, 36, 521-523, 931 (Rus) 723
SB 10, 13-14 (Arg) 24
Sbeitla (Tun)......................... 854
Scarlet Ibis (TT).................... 850
Scarpe (Fra) 279
Scharhörn (Ger) 297
Schedar (Gua)...................... 317
Schedar (Mex) 535
Schelde (Nld) 572
Schiedam (Nld) 569
Schlei (Ger) 292
Schleswig-Holstein (Ger) 286
Schultz Xavier (Por) 656
Scimitar (UK)........................ 907
Scioto (US) 991
Scirè (Ita).............................. 401
Scirocco (Ita)........................ 409
Scott (UK) 913
Scout (US) 969
Scranton (US) 938
Sculpin (Can) 106
SD 11, 13 (Pol)..................... 645
SD Adept (UK) 921
SD Bountiful (UK)................. 922
SD Bovisand (UK)................ 921
SD Capable (UK) 921
SD Careful (UK) 921
SD Catherine (UK) 923
SD Cawsand (UK)................ 921
SD Christina (UK)................. 923
SD Clyde Racer (UK) 923
SD Clyde Spirit (UK) 923
SD Deborah (UK) 923
SD Dependable (UK) 922
SD Eileen (UK) 923
SD Emily (UK) 923
SD Engineer (UK)................. 924
SD Eva (UK)......................... 921
SD Faithful (UK) 921
SD Florence (UK) 920
SD Forceful (UK) 921
SD Frances (UK) 920
SD Genevieve (UK) 920
SD Georgina (UK) 920
SD Helen (UK)...................... 920
SD Hercules (UK)................. 922
SD Impetus (UK) 921
SD Impulse (UK) 921
SD Independent (UK)........... 923
SD Indulgent (UK)................ 923
SD Jupiter (UK).................... 922
SD Kyle of Lochalsh (UK) 924
SD Mars (UK)....................... 922
SD Melton (UK).................... 922
SD Menai (UK) 922
SD Meon (UK)...................... 922
SD Moorfowl (UK)................ 920
SD Moorhen (UK) 920
SD Navigator (UK) 923
SD Netley (UK)..................... 922
SD Newhaven (UK).............. 922
SD Northern River (UK) 920
SD Norton (UK).................... 920
SD Nutbourne (UK).............. 922
SD Oban (UK)...................... 922
SD Omagh (UK) 922
SD Oronsay (UK) 922
SD Padstow (UK)................. 922
SD Powerful (UK)................. 921
SD Raasay (UK)................... 923
SD Reliable (UK) 922
SD Resourceful (UK)........... 922
SD Solent Racer (UK) 923
SD Solent Spirit (UK) 923

SD Suzanne (UK) 923
SD Tamar Racer (UK) 923
SD Tamar Spirit (UK) 923
SD Tilly (UK) 922
SD Victoria (UK) 920
SD Warden (UK) 920
Sea Based X-band Radar-1 (US) 976
Sea Devil (US) 987
Sea Dog (US) 987
Sea Dragon (Aust) 38
Sea Dragon (US) 987
Sea Fighter (US) 970
Sea Fox (US) 987
Sea Horse (US) 987
Sea Jet (US) 970
Sea Lion (US) 987
Sea Otter (US) 987
Sea Trojan (Aust) 37
Sea Widow (Aust) 37
Sea Witch (Aust) 38
Seahawk (US) 987
Seahorse 2 (Aust) 37
Seahorse Betong (Aust) 37
Seahorse Chuditch (Aust) 37
Seahorse Horizon (Aust) 36
Seahorse Kowari (Aust) 37
Seahorse Mercator (Aust) 35
Seahorse Parma (Aust) 37
Seahorse Platypus (Aust) 37
Seahorse Quenda (Aust) 37
Seahorse Quoll (Aust) 37
Seahorse Spirit (Aust) 36
Seahorse Standard (Aust) 36
Seal (Aust) 36
Seawolf (US) 940
Seay (US) 980
Sebak (Ban) 60
Sebatik (Mly) 519
Sebo (Gha) 301
Sebou (Mor) 551
Sechelt (Can) 104
Seeadler (Ger) 299
Seeadler (PNG) 614
Seeb (Omn) 595
Seefalke (Ger) 299
Seehund 1-18 (Ger) 293
Al Seep (Kwt) 491
Sefid-Rud (Iran) 391
Segantang (Mly) 516
Segei Osipov (Rus) 715
Segura (Spn) 778
Seinda (Myn) 556
Seiun (Jpn) 460
Sejeri (Col) 178
Sejong Daewang (RoK) 475
Sekiun (Jpn) 460
Selangor (Mly) 510
Selendán (Per) 625
Seleuta (Tun) 854
Seliger (Rus) 712
Selingan (Mly) 519
Selis (Lit) 502
Sella (Spn) 778
Sembilang (Indo) 368
Semilang (Mly) 518
Seminole (US) 974
Sempadi (Mly) 516
Semser (Kaz) 463
Sendai (Jpn) 441
Seneca (US) 985
Senezh (Rus) 709
Senja (Nor) 592
Sentinella (Ita) 413
Sentry (US) 969
Seongin Bong (RoK) 484
Seoul (RoK) 478
Sep (Pol) 637
Sepukhov (Rus) 703
Sequoia (US) 990
Serang (Mly) 513
Serasa (Bru) 88
Serasan (Mly) 516
Serdity (Rus) 723
Sergey Kolbassev (Rus) 708
Sergey Sudetsky (Rus) 732
Seri (Mex) 541
Serimbum (Mly) 519
Serrano (Chi) 120
Serreta (Ven) 1007
Serviola (Spn) 774
Sethya (Myn) 558
Setia (Mly) 516
Setogiri (Jpn) 439, 458
Setoshio (Jpn) 430
Setoyuki (Jpn) 447
Settsu (Jpn) 451
Setun (Rus) 717
Setyahat (Myn) 558
Sevastopol (Rus) 681, 705
Sever (Rus) 709
Severn (UK) 907
Severodvinsk (Rus) 679
Severomorsk (Rus) 692
Sévre (Fra) 279
Sfinge (Ita) 410
SFP 173, 286, 295, 562 (Rus) 710
SG-004-009, 017, 036, 066 (Pol) 648
SG 044-047, 111-112, 213-216, 411-412 (Pol) 647
SG 1-29, 50-58, 301-315 (Tur) 872
SG 61-74, 80-97, 101-109, 121-134 (Tur) 871
SG-211-212 (Pol) 647
SG-311-312 (Pol) 646
SGT Matej Kocak (US) 981
SGT William R Button (US) 977
Shabab Oman (Omn) 596
Shabab Oman 2 (Omn) 597
Shabhaz (Pak) 611
Shabla (Bul) 93
Shackle (US) 992
Shackleton (UK) 914
Shafak (Lby) 500
Shah Amanat (Ban) 58
Shah Makhdum (Ban) 58
Shah Poran (Ban) 58
Shahayak (Ban) 60
Al Shaheed (Kwt) 492
Shaheed Aktheruddin (Ban) 56
Shaheed Ali (Mld) 522
Shaheed Daulat (Ban) 56
Shaheed Farid (Ban) 56
Shaheed Mohibullah (Ban) 56
Shaheed Ruhul Amin (Ban) 55
Shaheen (UAE) 887
Shahid Absalan (Iran) 384
Shahid Dara (Iran) 384
Shahid Golzam (Iran) 384
Shahid Hejat Zadeh (Iran) 384
Shahid Kord (Iran) 384
Shahid Marjani (Iran) 390
Shahid Mehdavi (Iran) 384
Shahid Rahisi Raisi (Iran) 384
Shahid Sahrabi (Iran) 384
Shahid Shafihi (Iran) 384
Shahid Towsali (Iran) 384
Shahjahan (Pak) 604
Shahjalal (Ban) 60
Shaibal (Ban) 58
Shakhter (Rus) 724
Shakthi (Sri) 790
Shakti (Ind) 353
Shakti Sanchar (Ban) 58
Shaladein (Egy) 226
Shalki (Ind) 328
Shamal (US) 959
Al-Shamikh (Omn) 594
Shamsheer (Pak) 602
Shamshir (Iran) 384
Shangrao (CPR) 148
Shankul (Ind) 328
Shankush (Ind) 328
Shantou (CPR) 147
Shapla (Ban) 58
Shapshal (Kaz) 463
Shaqra (SAr) 741
Sharada (Ind) 347
Shardul (Ind) 349
Shark (Aust) 36
Sharm El Sheikh (Egy) 218
Al Sharqiyah (Omn) 595
Shaula (DR) 206
Shaula (Mex) 535
Shawinigan (Can) 103
Shearwater (US) 987
Sheean (Aust) 26
Shengshan (CPR) 155
Shenyang (CPR) 136
Shenzhen (CPR) 140
Shepparton (Aust) 34
Sherdil (Pak) 610
Sherman (US) 984
Shet Gang (Ban) 61
Shibsha (Ban) 60
Shichang (CPR) 160
Shijiazhuang (CPR) 136
Shikine (Jpn) 453
Shikishima (Jpn) 451
Shiloh (US) 944
Shimagiri (Jpn) 457
Shimagumo (Jpn) 457
Shimakaze (Jpn) 436
Shimanami (Jpn) 457
Shimayuki (Jpn) 447
Shimokita (Jpn) 443, 453
Shimookh (Iraq) 392
Shinas (Omn) 595, 597
Shinonome (Jpn) 458
Shinzan (Jpn) 456
Shippingport (US) 973
Shirane (Jpn) 433
Shirase (Jpn) 450
Shirataka (Jpn) 443
Shirayuki (Jpn) 447
Shiretoko (Jpn) 453
Shishijima (Jpn) 445
Shishumar (Ind) 328
Shiu Hai (Twn) 817
Shivalik (Ind) 338
Shkiper Gyek (Rus) 726
Shkval (Bul) 91
Shkval (Rus) 732
Al Shoaie (Kwt) 491
Shoalhaven (Aust) 38
Shoreham (UK) 912
Shostka (Ukr) 881
Shoup (US) 948
Shovelnose Ray (Sin) 758
Shoyo (Jpn) 459
Shrike (US) 987
Shtyl (Rus) 701
Shughart (US) 980
Shujaa (Ken) 464
Shujaat (Pak) 606
Shulyavka (Ukr) 881
Shunde (CPR) 151
Shun Hu 1 (Twn) 822
Shun Hu 5-6 (Twn) 823
Shupavu (Ken) 464
Shwepazun (Myn) 558
Shwethida (Myn) 556
Shyri (Ecu) 209
SI 1, 4 (Arg) 24
Si Ning (Twn) 814
Si Racha (Tld) 837
Siada (Indo) 367
Siagut (Mly) 520
Siakap (Mly) 518
Siamil (Mly) 516
Siangin (Mly) 517
Sibarau (Indo) 367
Sibbald (Chi) 121
Sibenik (Cro) 187
Sibiriyakov (Rus) 709
Sichang (Tld) 840
Sicié (Fra) 276
Al Siddiq (Egy) 225
Al Siddiq (SAr) 740
Sidi Ahmed Rais (Alg) 4
Sidi Bilal (Lby) 501
Sidi Bou Said (Tun) 852
Sidi Daoud (Tun) 853
Sidi Mohammed Ben Abdallah (Mor) 549
Sidon (Leb) 497
Siegburg (Ger) 294
Sigacik (Tur) 865
Sigalu (Indo) 367
Sigma T (Can) 111
Sigurot (Indo) 367
Sikanni (Can) 104
Siktivkar (Rus) 728
Sikuati (Mly) 520
Sikuda (Indo) 367
Silanos (Ita) 422
Silas Papare (Indo) 362
Silea (Indo) 367
Silifke (Tur) 865
Siliman (Indo) 367
Silistra (Bul) 93
Silver Marlin (Sin) 758
Simeoforos Kavaloudis (Gre) 307
Simeoforos Simitzopoulos (Gre) 307
Simeoforos Starakis (Gre) 307
Simeoforos Xenos (Gre) 307
Simeon Castro (Plp) 631
Simferopol (Ukr) 879
Similan (Tld) 844
Simmonds (Can) 113
Simón Bolívar (Ven) 1006
Simpson (Chi) 117
Simpson (US) 952
Şimşek (Tur) 861
Sinai (Egy) 224
Sindhudhvaj (Ind) 329
Sindhughosh (Ind) 329
Sindhukesari (Ind) 329
Sindhukirti (Ind) 329
Sindhuraj (Ind) 329
Sindhuratna (Ind) 329
Sindhushastra (Ind) 329
Sindhuvijay (Ind) 329
Sindhuvir (Ind) 329
Singa (Indo) 366
Sin Hung (RoK) 481
Sinmin (Myn) 556
Sioux (US) 978
Sioux City (US) 950
Sipu Muin (Can) 112
Sir Bu'nuer (UAE) 889
Sir Milton (SL) 749
Sir Wilfred Grenfell (Can) 109
Sir Wilfrid Laurier (Can) 107
Sir William Alexander (Can) 107
Siri (Nig) 581
Siribua (Indo) 367
Sirik (Iran) 390
Sirio (Ita) 412
Sirius (Alg) 9
Sirius (Aust) 35
Sirius (Brz) 84
Sirius (Bul) 93
Sirius (Col) 178
Sirius (DR) 206
Sirius (Ecu) 212
Sirius (Gua) 317
Sirius (Mex) 535
Sirius (Swi) 807
Sirius (Uru) 999
Sirius (Ven) 1007
Sirjan (Iran) 390
Sirocco (UK) 924
Sirocco (US) 959
Siroco (Fra) 267
Sisargas (Spn) 779
Sisler (US) 977
Sitkinak (US) 988
Sivas (Ukr) 882
Siyan (Yem) 1019
Siyay (Can) 112
SK 1411-1412 (Rus) 715
SKA 12-14 (Den) 199
Skadovsk (Ukr) 878
Skaftö (Swe) 802
Skalvis (Lit) 502
Skat (Rus) 726
Skenandoa (US) 974
Skipjack (US) 987
Skjold (Nor) 588
Skrolsvik (Nor) 589
Skrunda (Lat) 494
Skudd (Nor) 588
Skvyra (Ukr) 881
Slamet Riyadi (Indo) 360
Slavutich (Ukr) 881
Slazak (Pol) 641
Sledge (US) 992
Sleipner (Den) 201
Sleipner (Nor) 591
Slotterøy (Nor) 589
MG Robert Smalls (US) 961
Sloug (Lby) 501
Smardan (Rom) 664
Smeli (Bul) 89
Smerch (Rus) 701
Smetlivy (Rus) 690
Smeul (Rom) 664
Smila (Ukr) 880
Smilax (US) 991
Smit Cerne (UK) 921
Smit Cymyran (UK) 921
Smit Dart (UK) 921
Smit Dee (UK) 921
Smit Don (UK) 921
Smit Frome (UK) 921
Smit Merrion (UK) 921
Smit Neyland (UK) 921
Smit Penally (UK) 921
Smit Romney (UK) 921
Smit Rother (UK) 921
Smit Spey (UK) 921
Smit Stour (UK) 921
Smit Tamar (UK) 921
Smit Towy (UK) 921
Smit Wey (UK) 921
Smit Yare (UK) 921
Smiter (UK) 907
F C G Smith (Can) 113
Smolensk (Rus) 676
Smolny (Rus) 714
SN 109, 126, 128, 401, 1318, 1520 (Rus) 715
Snellius (Nld) 570
Sneznogorsk (Rus) 697
Sniardwy (Pol) 642
Snoopy (Aust) 37
So 2 Angel Orellano Vasquez (Bol) 69
Soares De Meirelles (Brz) 86
Sobat (Sud) 793
Sochi (Rus) 728
Sockeye (US) 987
Soderman (US) 977
Södermanland (Swe) 796
Soemba (Nld) 572
Sögüt (Tur) 867
Sohag (Egy) 225
Sohn Won-IL (RoK) 474
Šokadija (Cro) 186
Sokal (Ukr) 881
Sok Cho (RoK) 481
Sokhumi (Geo) 282
Sokól (Pol) 637
Sokol (Rus) 727
Sokullu Mehmet Paşa (Tur) 866
Soldado Canave (Chi) 121
Soldado Fuentes (Chi) 122
Søløven (Den) 202
Solta (Cro) 188
Somerset (UK) 904
Somerset (US) 962
Gen Brehon B Somervell (US) 961
Somme (Fra) 272
Somudra Joy (Ban) 53
Söndüren 1-4 (Tur) 870
Songkhla (Tld) 836
Songshan (CPR) 155
Song Nam (RoK) 481
Sonora (Mex) 534
Soobrazitelny (Rus) 696
Sooke (Can) 104
Sooke Post (Can) 109
Soputan (Indo) 374
Sorachi (Jpn) 455
Sorocaima (Ven) 1004
Sorokos (Gre) 311
Sorong (Indo) 373
Soroo (Iran) 390
Sortland (Nor) 592
Sosha (Rus) 716
Sottile (Ita) 422
Soummam (Alg) 7
Al Soumood (Kwt) 494
Sour (Leb) 498
Souryu (Jpn) 429
South Cotabato (Plp) 632
South Dakota (US) 937
Souun (Jpn) 460
Sovereignty (Sin) 754
Sovershenny (Rus) 696
Sovetskaya Gavani (Rus) 697
Sovetskiy Pogranichnik (Rus) 733
Søviknes (Nor) 589
Soya (Jpn) 451
Sozopol (Bul) 94
SP/4 James A Loux (US) 961
Spar (US) 990
Spari (Sur) 793
Spårö (Swe) 802
Spearfish (Sin) 758
Spearhead (US) 980
Speditøren (Den) 199
Spencer (US) 985
Spessart (Ger) 295
Spetsai (Gre) 304
Spica (Gua) 317
Spica (Ita) 412
Spica (Mex) 535
Spiekeroog (Ger) 298
Spikefish (Sin) 758
Spindrift (Can) 112
Spioenkop (SA) 762
Spiro (Arg) 15
Spitfire (UK) 914
Spotted Ray (Sin) 758
Spray (Can) 112
Springfield (US) 938
Spruance (US) 948
Sprut (Rus) 726
Squall (US) 959
SR 26, 111, 179-180, 188, 203, 233, 267, 280, 334, 370 (Rus) 719
SR 28, 74, 120, 137, 216, 245, 478-479, 541, 548, 569-570, 936, 938-939 (Rus) 720
Sri Gaya (Mly) 514
Sri Indera Sakti (Mly) 513
Sri Johor (Mly) 512
Sri Perlis (Mly) 512
Sri Tiga (Mly) 514
Srinakarin (Tld) 845
Sriyanont (Tld) 846
SRV 300 (Ita) 418
SS 34-46, 51-79 (Jpn) 458
SS 507-509 (Jpn) 429
SS 750 (Rus) 717
SSCIM Senen Alberto Araujo (Col) 174
SSGT Edward A Carter (US) 977
SSGT Robert T Kuroda (US) 961
SSIM Julio Correa Hernández (Col) 173
SSIM Manuel A Moyar (Col) 173
St Albans (UK) 904
St John's (Can) 100
St Lykoudis (Gre) 312
St Ursula (VI) 1016

1038 **INDEXES** > Named ships

Staffetta (Ita) 413
Stalwart (StK) 734
Stalwart (Sin) 752
Stangnes (Nor) 589
Stanisci (Ita) 422
Stapa (Mly) 517
Starace (Ita) 423
Starfish Ray (Sin) 758
Stary Oskol (Rus) 682
Staten Island (US) 988
Stavropol (Rus) 728
Steadfast (Sin) 752
Steadfast (US) 985
Steelhead (US) 987
Stefan Cel Mare (Rom) 663
Stefan Karadja (Bul) 93
Steil (Nor) 583
Stella Maris (Gua) 317
Stella Polare (Ita) 417
Stephen Otu (Gha) 301
Stereden (Fra) 275
Steregushchiy (Rus) 696
Sterett (US) 948
Stern (Bel) 65
Stethem (US) 946
Stet Polaris (Sin) 757
Steve Obimpeh (Gha) 301
Stiglich (Per) 623
Stikine (Can) 104
Stiletto (US) 970
Sting Ray (Sin) 758
Stingray (US) 987
Stockdale (US) 948
GYSGT Fred W Stockham (US) 977
Stockholm (Swe) 799
Stoiky (Rus) 696
Stoker (Aust) 37
Stollergrund (Ger) 294
Stolt (Swe) 800
Storm (Nor) 588
Storm (Rus) 732
Storm (UK) 924
Storm Bay (Aust) 39
Stout (US) 946
Stratton (US) 984
Stravon (Gre) 311
Striped Marlin (Sin) 758
Stromboli (Ita) 419
Stuart (Aust) 28
Stupinets (Rus) 700
Sturgeon (US) 987
Sturgeon Bay (US) 989
Sturkö (Swe) 802
Stvor (Rus) 709
Stymfalia (Gre) 311
Styrsö (Swe) 802
Styx (Fra) 268
Suão (Por) 656
Suao (Twn) 812
Subahi (Kwt) 493
Subanon (Plp) 632
Subhadra (Ind) 347
Subhadra Kumari Chauhan (Ind) 356
Sub Lieutenant Alexandru Axente (Rom) 665
Suboficial Castillo (Arg) 17
Suboficial Rogelio Lesme (Par) 617
Subteniente Osorio Saravia (Gua) 316
Subuan (Mly) 520
Success (Aust) 35
Suchetra Kripalani (Ind) 355
Sudak (Ukr) 880
Sudarshini (Ind) 351
Suduvis (Lit) 502
El Suez (Egy) 219
Sufa (Isr) 397
Suffren (Fra) 247
Suganami (Jpn) 457
Sugashima (Jpn) 445
Sugut (Mly) 520
Suhail (Mex) 535
Suh Hoowon (RoK) 482
Sujata (Ind) 347
Suk (Tld) 843
Sukanya (Ind) 347
Sukhothai (Tld) 833
Sula (Col) 178
Al Sulayel (SAr) 741
Sulayman Jun-Kung (Gam) 281
Sulpicio Fernandez (Plp) 632
Al Sultana (Omn) 597
Sultan Hasanuddin (Indo) 363
Sultan Iskandar Muda (Indo) 363
Sultan Kudarat (Plp) 630
Sultan Moulay Ismael (Mor) 546
Sultan Nuku (Indo) 362
Sultan Thaha Syaifuddin (Indo) 362
Suluh Pari (Indo) 366
Sulzbach-Rosenberg (Ger) 293
Suma (Jpn) 446
Sumedha (Ind) 347
Sumitra (Ind) 347
Sumjinkang (RoK) 489
Summerside (Can) 103
Sumner (US) 975
Sunayna (Ind) 347
Sun Chon (RoK) 481
Sundsvall (Swe) 798
Sunjin (RoK) 486
Suou (Jpn) 450
Suphairin (Tld) 838
Supply (US) 974
Supreme (Sin) 752
Suqian (CPR) 148
Sur (Pak) 610
Sura (Indo) 366
Surabaya (Indo) 368
Suranimila (Sri) 786
Surcouf (Fra) 259
Surel (Arg) 23
Surin (Tld) 840
Suriya (Tld) 843
Surma (Ban) 57
Surovi (Ban) 58
Surrey (Jam) 427
Surubi (Arg) 23
Suruga (Jpn) 452
Sutanto (Indo) 362
Sutedi Senoputra (Indo) 362
Sutherland (UK) 904
Sutlej (Ind) 351
Suvarna (Ind) 347
Suwad (Bhr) 48
Suzdalets (Rus) 697
Suzuka (Jpn) 453
Suzunami (Jpn) 438
Suzutsuki (Jpn) 435
Svalbard (Nor) 592
SVG 02-03, 06-07, 09-11 (StV) 736
Svir (Rus) 716
Svishtov (Bul) 93
Swaraj Deep (Ind) 352
Swift Rescue (Sin) 757
Swordfish (Sin) 758
Swordfish (US) 987
Swordsman (Sin) 750
Syafaat (Bru) 88
Sycamore (US) 990
Sydney (Aust) 27
Syöksy (Fin) 243
Syounan (Jpn) 446
Syvatitel Nikolay Chudotvorets (Rus) 682
Syvatoy Giorgiy Pobedonosets (Rus) 673
Syzran (Rus) 712

T

T 001 (Sur) 794
T 4-7 (Aust) 33
T 11-12, 15-19, 81-83, 991-996 (Tld) 838
T 11-15, 201-215 (Ind) 348
T 21-24, 213-214, 216-227 (Tld) 839
T 80-84, 301-304, 401-404 (Ind) 347
T 91-99 (Tld) 837
T 710 (Az) 40
T 758-759 (Az) 42
T 864, 874, 880-881 (Vtn) 1013
Ta Fung (Twn) 821
Ta Han (Twn) 821
Ta Hu (Twn) 820
Ta Kang (Twn) 821
Ta Kuan (Twn) 820
Ta Tai (Twn) 821
Ta Wan (Twn) 821
Taape (Fra) 273
Taba (Egy) 218
Tabah (Mly) 516
Tabar (Ind) 337
Tabarca (Spn) 775
Tabarja (Leb) 498
Tabark (Tun) 854
Tabarka (Tun) 853
Tabarzin (Iran) 384
Tabasco (Mex) 536
Tabbouk (SAr) 743
Al Tabkah (Lby) 501
Taboga (Pan) 612
Tabuk (SAr) 739
Tackle (US) 992
Tae Pung Yang I, III, V, VI-VIII (RoK) 488
Tae Pung Yang II (RoK) 489
Tae Pung Yang IX-XII (RoK) 490
Tagbanua (Plp) 633
Tagil (Rus) 716
Tagomago (Spn) 775
Tagreft (Lby) 501
Taguermess (Tun) 853
Al Tahaddy (Kwt) 494
Taheri (Iran) 390
Tahoma (US) 985
Taichung (Twn) 822
Taif (SAr) 737
Tai Hu (CPR) 161
Taillat (Fra) 276
Tailor (Aust) 35
Taimuang (Tld) 837
Taina (Fra) 274
Tainan (Twn) 821
Tainha (CpV) 114
Taipei (Gam) 281
Taipei (Twn) 823
Taishan (CPR) 155, 157
Taiyuan (CPR) 140
Taizhou (CPR) 137, 147
Tajo (Spn) 778
Taka (Indo) 377
Takachiho (Jpn) 456
Takanami (Jpn) 438
Takapu (NZ) 575
Takashima (Jpn) 445
Takashio (Jpn) 430
Takataki (Jpn) 457
Takatori (Jpn) 455
Takatsuki (Jpn) 456
Takbai (Tld) 837
Taketomi (Jpn) 453
Takip 1-2 (Tur) 868
Taksin (Tld) 829
Takuyo (Jpn) 459
Talcahuano (Chi) 126
Talcauano (Chi) 123
Talent (UK) 898
Talita II (Arg) 23
Talivaldis (Lat) 495
Taliwangsa (Indo) 366
Tallashi (Ban) 57
Talwar (Ind) 337
Tamanaco (Ven) 1004, 1007
Tamanami (Jpn) 457
Tambaú (Brz) 80
Tambisan (Mly) 520
Tambora (Indo) 374
Tambov (Rus) 680
Tambre (Spn) 778
Tamjeed (Ban) 56
Tammar (Aust) 37
Tamoio (Brz) 71
Tampa (US) 985
Tana (Ken) 465
Tan Chiang (Twn) 816
Tangen (Nor) 589
Tango (Arg) 24
Tanjung Fatagar (Indo) 368
Tanjung Kambani (Indo) 370
Tanjung Nusanive (Indo) 368
Tannin (Isr) 395
H K Tannis (StV) 736
Tanu (Can) 108
Tanveer (Ban) 56
Tapajó (Brz) 71
Tapi (Tld) 834
Tara Bai (Ind) 356
Tarafdaar (Ind) 354
Tarakan (Aust) 32
Tarangini (Ind) 351
Tarapunga (NZ) 575
Tarasa (Ind) 348
Tarasco (Mex) 540
Taraz (Kaz) 462
Tareq (Iran) 379
Tarif (UAE) 887
Tarifa (Spn) 775
Tarihu (Indo) 367
Tarik (Mor) 550
Tarik Ben Ziyad (Mor) 546
Tariq (Pak) 604
Tariq (SAr) 740
Tariq Ibn Ziyad (Lby) 500
Tarkash (Ind) 337
Tarpon (US) 987
Tarshish (Isr) 397
Tarsus (Tur) 862
Taşkizak (Tur) 867
Taşucu (Tur) 862
Tasuja (Est) 234
Tatarstan (Rus) 696
Tatau (Mly) 520
Taunoa (Fra) 276
Taupo (NZ) 575
Taurus (Brz) 83
Tausug (Plp) 632
Tavolara (Ita) 420
Al Tawash (Kwt) 491
Al Taweelah (Bhr) 48
Tawfiq (Ban) 56
El Tawfiq (Mor) 547
Tawheed (Ban) 56
Tayfun (Tur) 862
Taylor (US) 952
Tayrona (Col) 171
TB 1-4 (Ban) 56
TB 35-38 (Ban) 54
Tchusovoy (Rus) 711
Te Kaha (NZ) 574
Te Kukupa (CI) 182
Te Mana (NZ) 574
Te Mataili (Tuv) 874
Teak (TT) 850
Teanoai (Kir) 465
Tebicuary (Par) 615
Tebrau (Mly) 516
Tech (Fra) 279
TECIM Jaime E Cárdenas Gomez (Col) 174
Teculapa (Mex) 540
Tecun Uman (Gua) 317
Tedong Naga (Indo) 366
Tedung Selar (Indo) 367
Teg (Ind) 337
Tegas (Mly) 516
Tegernsee (Ger) 296
Tegucigalpa (Hon) 320
Teguh Samudera (Mly) 515
Teh Hsing (Twn) 822
Tehuelche (Arg) 21
Tekirdağ (Tur) 862
Tekuma (Isr) 395
Tekwane (SA) 763
Telenn Mor (Fra) 273
Teleost (Can) 113
Telopea (Aust) 36
Teluk Amboina (Indo) 369
Teluk Banten (Indo) 369
Teluk Bone (Indo) 369
Teluk Celukan Bawang (Indo) 370
Teluk Cendrawasih (Indo) 370
Teluk Cirebon (Indo) 374
Teluk Ende (Indo) 369
Teluk Gilimanuk (Indo) 370
Teluk Hading (Indo) 370
Teluk Jakarta (Indo) 370
Teluk Lampung (Indo) 370
Teluk Manado (Indo) 370
Teluk Mandar (Indo) 369
Teluk Parigi (Indo) 370
Teluk Peleng (Indo) 370
Teluk Penyu (Indo) 369
Teluk Ratai (Indo) 369
Teluk Sabang (Indo) 374
Teluk Sampit (Indo) 369
Teluk Sangkulirang (Indo) 370
Teluk Semangka (Indo) 369
Teluk Sibolga (Indo) 370
Temerario (Uru) 998
Tempest (US) 959
Temryuk (Rus) 709
Al Temsah (Omn) 596
Temsah (UAE) 890
Tenace (Fra) 276
Tenacidad (Ven) 1003
Tenacious (Sin) 752
Tendert (Nic) 577
Tenente Boanerges (Brz) 83
Tenente Castelo (Brz) 83
Tenggol (Mly) 519
Teniente Farina (Par) 616
Teniente José Azueta (Mex) 535
Teniente Luís Bernal Baquero (Col) 178
Teniente Maximiano (Brz) 86
Teniente Olivieri (Arg) 17
Teniente Robles (Par) 616
Teniente Soliz (Bol) 68
Tennessee (US) 935
Tenochtitlan (Mex) 537
Tenryu (Jpn) 448
Tensift (Mor) 551
Tenyo (Jpn) 459
Teotihuacan (Mex) 537
Teotimo Figuracion (Plp) 631
Tepoca (Mex) 540
Teraban (Bru) 88
Terengganu (Mly) 510
Terepaima (Ven) 1004
Terminos (Mex) 540
Termoli (Ita) 415
Tern (SA) 763
Tern (US) 987
Ternopil (Ukr) 877
Terrace Bay (Nam) 559
Terrapin (US) 987
Terry Fox (Can) 107
Teruzuki (Jpn) 435
Teshio (Jpn) 455
Teuku Umar (Indo) 362
Texas (US) 937
TF R Rios V (Bol) 69
Tha Din Daeng (Tld) 842
Thafir (Kwt) 493
Thalang (Tld) 842
Thames Crespo (Bol) 69
Thar (Egy) 228
That Assuari (Qat) 658
Thayanchon (Tld) 833
The Sullivans (US) 946
The Wanderer (Aust) 37
Themistocles (Gre) 305
Theodore Roosevelt (US) 942
Thepha (Tld) 837
Theseus (Gre) 312
Thetis (Den) 196
Thétis (Fra) 270
Thetis (Gre) 312
Thetis (US) 985
Thexas (Cypr) 193
Thihayarzar I-II (Myn) 556
Thomas G Thompson (US) 971
Thomas Hudner (US) 948
Thomback Ray (Sin) 758
Thompson (Arg) 21
Thomson (Chi) 117
Thong Kaeo (Tld) 840
Thong Lang (Tld) 840
Thor (Ice) 325
Thor Heyerdahl (Nor) 586
Thread Fin (Sin) 758
Thresher Shark (Sin) 758
Thunder (Nig) 579
Thunder Bay (US) 989
Thunder Cape (Can) 112
Thunderbolt (US) 959
Tiagha (Ctl) 184
Tiantaishan (CPR) 154
Tianwangxing (CPR) 158
Tianzhushan (CPR) 154
Tiaré (Fra) 274
Tiburon (Arg) 23
Ticino (Ita) 418
Tideforce (UK) 919
Tiderace (UK) 919
Tidespring (UK) 919
Tidesurge (UK) 919
Tien Tan (Twn) 813
Tierra Bomba (Col) 178
Tiger Shark (Sin) 758
Tiger Shark (US) 987
Tighatlib (Bhr) 49
Tigr (Rus) 678
Tigre (Fra) 271
Tiira (Lat) 496
Tikuna (Brz) 70
Tillicum (Can) 105
Timbira (Brz) 71
El Tinai (Alg) 6
Tippecanoe (US) 975
Tippu Sultan (Pak) 604
Tir (Ind) 351
Tirad Pass (Plp) 635
Tiran (Iran) 385
Tirapuka (Guy) 318
Tireless (UK) 898
Tirso (Ita) 418
Tista (Ban) 56
Titan (Gre) 312
Titano (Ita) 421
Titas (Ban) 56
Tjaldrid (Fae) 238
Tjiptadi (Indo) 362
Tlaloc (Mex) 541
Tlaxcala (Mex) 541
TM 247, 340 (Bol) 69
TNR 10-12 (Bol) 69
Toba (Arg) 21
Tobias Hainyeko (Nam) 560
Tobie (SA) 763
Tobishima (Jpn) 445
Tobruk (Aust) 32
Todak (Indo) 365
Todak (Mly) 513
Todak (Sin) 758
Todendorf (Ger) 296
Todos Santos (Mex) 540
Tofiño (Spn) 778
Tokachi (Jpn) 455
Tokara (Jpn) 456
Tokinami (Jpn) 458
Tokiwa (Jpn) 449
Toledo (US) 938
Toll (Arg) 22
Tolmi (Gre) 308
Tolyatti (Rus) 727

Tom Annan (Gha) 301
Tom Thumb (Aust) 34
Tomonami (Jpn) 458
Tomsk (Rus) 676
Tonb (Iran) 389
Tone (Jpn) 441
Tonelero (Brz) 70
Tongkol (Indo) 366
Tongling (CPR) 146
Tongyeong (RoK) 486
Tonina (Arg) 22
Tonnerre (Can) 106
Tonnerre (Fra) 266
Tony Pastrana Contreras
(Col) 174
Toowoomba (Aust) 28
Topaz (Sey) 748
Topeka (US) 938
Tor (Nor) 592
Toralla (Spn) 774
Torås (Nor) 589
Torbay (UK) 898
Torie (Nig) 582
Tormenta (Mex) 532
Tornado (Cay) 115
Tornado (Spn) 772
Tornado (US) 959
Tornio (Fin) 239
Toronto (Can) 100
Torpedo Ray (Sin) 758
Torpido Tenderi (Tur) 868
Tortuga (US) 967
Tortugas (Per) 627
Tortuguero (DR) 206
Toruń (Pol) 642
Tosa (Jpn) 453
Tour Royale (Fra) 275
Tourville (Fra) 247
Toushka (Egy) 218
Towada (Jpn) 449
Al Towfan (UAE) 893
Toxotis (Gre) 307
Toyoshima (Jpn) 445
Trablous (Leb) 498
Tracker (UK) 907
Tracy (Can) 108
Traful (Arg) 24
Tramontana (Spn) 767
Tran Dai Nghia (Vtn) 1015
Tran Khanh Du (Vtn) 1014
Träskö (Fin) 243
Traverse (Can) 110
Tremiti (Ita) 419
Trenchant (UK) 898
Trenton (US) 980
Trezza (Ita) 422
Trichonis (Gre) 312
Trident (Bar) 62
Tridente (Brz) 87
Tridente (Por) 649
Tridenti (Ita) 422
Trieste (Uru) 998
Trieux (Fra) 279
Triglav (Slo) 759
Trikand (Ind) 337
Triki (Mor) 547
Trinidad (Bol) 69
Trinkat (Ind) 348
Tripoli (Leb) 497
Tripoli (US) 966
Trishul (Ind) 337
Trisula (Indo) 375
Tritão (Brz) 87
Triton (Aust) 39
Triton (Den) 196
Triton (Gre) 303
Triton (Swe) 805
Triumph (UK) 898
Triunfo (Brz) 87
Tromp (Nld) 564
Trossö (Swe) 803
Trumpeter (UK) 907
Truong (Vtn) 1015
Truong Sa (Vtn) 1015
Truxtun (US) 948
Trygg (Swe) 800
Tseng Chiang (Twn) 816
Tsesar Kunikov (Rus) 704
TSGT John A Chapman
(US) 978
Tshukudu (SA) 765
Tsibar (Bul) 91
Tsi Chiang (Twn) 816
Tsotne Dadiani (Geo) 282
Tsoying (Twn) 812
Tsugaru (Jpn) 451
Tsukuba (Jpn) 456
Tsunoshima (Jpn) 445
Tsurumi (Jpn) 455
Tsuruugi (Jpn) 456
Tsushima (Jpn) 445
Tuba (Mly) 519
Tucson (US) 938
Tucunaré (Brz) 78
Tufan (Tur) 861
Tugau (Mly) 520
Tukoro (Van) 1000
Tula (Rus) 672
Tulcea (Rom) 666
Tulsa (US) 953
Tulugaq (Den) 198
Tuna (Sin) 758
Tun Razak (Mly) 507
Tunas Samudera (Mly) 515
Tunda Satu (Mly) 515
Tunis (Tun) 851
Tunku Abdul Rahman
(Mly) 507
Tuo Jiang (Twn) 817
Tupi (Brz) 71
Turag (Ban) 55
Turaif (SAr) 743
Turbinist (Rus) 707
Turgutreis (Tur) 859
Turia (Spn) 778
Türkeli (Tur) 862
Tursas (Fin) 244
Turva (Fin) 244
Tuwaig (SAr) 742
Tuzla (Tur) 862
Tver (Rus) 676, 726
Tybee (US) 988
Tyne (UK) 907
Typhoon (US) 959
Tyr (Ice) 325
Tyr (Nor) 591
Tzacol (Gua) 317
Tzu-i (Twn) 813

U

U 31-36 (Ger) 285
U 201-211, 301-313, 601-636
(Fin) 243
U 241, 631-634, 732, 926
(Ukr) 881
U 400 series (Fin) 243
UAM 101-102, 122, 203,
304, 601-603, 610, 618-619,
623-624, 626, 631, 634, 636,
639-641, 650-651, 666, 669,
673, 675, 684, 686-696,
810-812, 830, 840, 852, 901,
907-908, 913, 919 (Por) 655
Ubaldo Diciotti (Ita) 424
Überherrn (Ger) 293
Ucayali (Per) 622
Uckermark (Ger) 299
Udara (Sri) 787
Al Udeid (Qat) 657
Udomdet (Tld) 836
Uerkouane (Tun) 854
Ugaab (UAE) 890
Ugandi (Est) 234
Uglich (Rus) 703
UIF 101-107 (Per) 622
Uisko (Fin) 244
Ukale (Dom) 205
Ukushima (Jpn) 445
Ula (Nor) 585
Al Ula (SAr) 741
Ulsan (RoK) 478
Ulua (Hon) 320
Ulvön (Swe) 802
Um Almaradim (Kwt) 490
Umalusi (SA) 765
Umar Farooq (Ban) 59
Umhloti (SA) 764
Umigiri (Jpn) 439, 458
Umitaka (Jpn) 443
Umkomaas (SA) 764
Umlus (SAr) 742
Umm Al Narr (UAE) 889
Umoja (Ken) 464
Umut (Tur) 871
Umzimkulu (SA) 764
União (Brz) 75
Unity (Sin) 754
Unryu (Jpn) 429
Untung Suropati (Indo) 362
Uppland (Swe) 795
Uraga (Jpn) 444
Ural (Rus) 726
Urania (Ita) 410
Urania (Nld) 570
Uranus (Swi) 807
Urayuki (Jpn) 457
Urazuki (Jpn) 458
Ureca (EqG) 231
Uredd (Nor) 585
Urengoy (Rus) 698, 732
Urf (Swe) 802
Uribe (Chi) 120
Urk (Nld) 569
Urso (Ita) 422
Uruguay (Uru) 995
Uruguay 1-5 (Uru) 997
Ushuaia (Arg) 22
Ust-Bolsheretsk (Rus) 682
Ust-Limsk (Rus) 697
Ust-Kamshats (Rus) 682
Usumacinta (Mex) 538
Utatlan (Gua) 316
Uthaug (Nor) 585
Utila (Hon) 321
Utique (Tun) 851, 854
Utsira (Nor) 585
Utstein (Nor) 585
Uttal (Ban) 55
Utvaer (Nor) 585
Uusimaa (Fin) 240
Uxmal (Mex) 537
Uzhgorod (Ukr) 878
Uzushio (Jpn) 430, 459

V

V 3-5, 8, 10-11, 13, 16-20
(Ger) 297
V 11-16 (Mad) 505
V 101-106 (Tun) 851
V 121-124 (ConB) 182
V 601-635 (Ita) 421
V 2000-2051, 2056-2057,
2071-2078, 3000-3007,
5000-5020, 5100, 6003-6012
(Ita) 423
V Adm Vorontsov (Rus) 709
V Gumanenko (Rus) 708
VA II, IV (Spn) 785
Vaarlahti (Fin) 242
Vaccaro (Ita) 422
VADM K R Wheeler (US) 977
Vaedderen (Den) 196
Vahterpää (Fin) 240
Vaibhav (Ind) 355
Vajra (Ind) 355
Vakta (Can) 110
Valcke (Bel) 66
Valentin Chujkin (Rus) 729
Valentin Gomez Farias
(Mex) 535
Valentin Pikul (Rus) 707, 728
Valery Faleyev (Rus) 711
Valiant (Sin) 753
Valiant (US) 974, 985
Valkyrien (Nor) 591
Valle Del Cauca (Col) 179
Valour (Sin) 753
Valparaíso (Chi) 126
Valpas (Lat) 497
Valve (Est) 235
Van Amstel (Nld) 563
Van Kinsbergen (Nld) 570
Van Speijk (Nld) 563
Vancouver (Can) 100
Vandegrift (US) 952
Vanguard (UK) 896
Vanguardia (Uru) 999
Vänö (Fin) 242
Vaqar (Pak) 611
Var (Fra) 272
Varad (Ind) 355
CPCIM Guillermo Londoño
Vargas (Col) 174
Varna (Bul) 93
Varonis (Lat) 495
Varuna (Ind) 355
Varyag (Rus) 689
Vasco Da Gama (Por) 650
Vasil Levski (Bul) 92
Vassily Tatischev (Rus) 712
Vaygach (Rus) 711
Vector (Can) 113
Vedetta (Ita) 413
Veer (Ind) 344
Vega (Gua) 317
Vega (Ita) 412
Vega (Mex) 535
Vehdat (Pak) 611
Vela De Cobo (Ven) 1007
Velarde (Per) 621
Veliki Ustyug (Rus) 703
Vella Gulf (US) 944
Ven (Swe) 802
Vencedora (Spn) 773
Vendaval (Por) 656
Vendémiaire (Fra) 261
Vengadora (Col) 176
Vengeance (Sin) 753
Vengeance (UK) 896
Ventôse (Fra) 261
Venturous (US) 985
Venus (Swi) 807
Vepr (Rus) 678
Veracruz (Mex) 534
Verchoture (Rus) 672
Verdecchia (Ita) 423
Verdon (Fra) 279
Verni (Bul) 90
Veronica Adley (StL) 735
Vertonne (Fra) 279
Vestkysten (Den) 201
Vésubie (Fra) 279
Vesuvio (Ita) 419
Vetluga (Rus) 714
Viana Do Castelo (Por) 652
Viareggio (Ita) 415
Viben (Den) 201
Vibhuti (Ind) 344
Vice Admiral Constantin
Balescu (Rom) 665
Vice Admiral Eugeniu Rosca
(Rom) 663
Vicealmirante Othón
P Blanco (Mex) 535
Vicksburg (US) 944
Victor Kingisepp (Rus) 726
Victor Subbotin (Rus) 710
Victoria (Can) 98
Victoria (Spn) 768
Victoria (Ven) 1003
Victoria Mxenge (SA) 766
Victorious (UK) 896
Victorious (US) 976
Victory (Sin) 753
Videla (Chi) 121
Vidin (Bul) 93
Vidyut (Ind) 344
Viedna (Arg) 24
Viesīte (Lat) 494
Vieste (Ita) 415
Viesturs (Lat) 495
Viesulas (Lit) 504
Vigía (Spn) 774
Vigilance (Dom) 205
Vigilance (Sin) 753
Vigilant (UK) 896
Vigilant (US) 985
Vigilante (CpV) 114
Vigilante (Fra) 271
Vigorous (US) 985
Vigour (Sin) 753
Vigraha (Ind) 355
Vijelia (Rom) 664
Vijit (Ind) 355
Viken (Nor) 591
Viking Sun (Aust) 37
Vikr (Rus) 724
Vikramaditya (Ind) 330
Vikrant (Ind) 333
Viktor Kopecky (Rus) 724
Viktor Leonov (Rus) 712
Villavisencio (Per) 619
Ville De Québec (Can) 100
Vinash (Ind) 344
Vincenzo Martellotta (Ita) 416
Vinga (Swe) 802
Vinha (Fin) 243
Vinnitsa (Ukr) 877
Viola M Davidson (Can) 111
Violette (Fra) 279
Viper (Indo) 367
Vipul (Ind) 344
Viraat (Ind) 332
Virginia (US) 937
Virginio Fasan (Ita) 408
Virsaitis (Lat) 495
Visby (Swe) 797
Vise (US) 992
Vishera (Rus) 716
Vishny Volochek (Rus) 703
Vishwast (Ind) 355
Visvaldis (Lat) 495
Vital De Oliveira (Brz) 84
Vitali (Ita) 422
Viteazul (Rom) 667
Vitse Admiral Ivan Gren
(Rus) 706
Vitse Admiral Kulakov
(Rus) 692
Vitse-Admiral Zakharin
(Rus) 707
Vitseadmiral Zhukov (Rus) ... 707
Vivek (Ind) 355
Vizir (Rus) 709
Vlaardingen (Nld) 569
Vladikavkaz (Rus) 682
Vladimir Kavraisky (Rus) 722
Vladimir Monomach (Rus) ... 671
Vladimirets (Rus) 703
Vladivostok (Rus) 705
Vladykov (Can) 113
Vlahakos (Gre) 306
VM 20, 72, 143, 146, 152, 154,
159, 227, 230, 250, 253, 268,
270, 277, 409, 414-415, 420,
425, 429, 519, 725, 807, 809,
907-911, 915-916, 919,
(Rus) 718
VM 596 (Rus) 719
Vogelsand (Ger) 297
Volga (Rus) 726
Volgodonsk (Rus) 703
Volk (Rus) 678
Volkan (Tur) 862
Vologda (Rus) 682
Voron (Rus) 727
Voronezh (Rus) 676
Vorovsky (Rus) 725
Voum-Legleita (Mtn) 525
VTN Series (Rus) 716
VTR 92, 94 (Rus) 720
Vukovar (Cro) 185
Vulcain (Fra) 268
Vulcanul (Rom) 664
Vung Tau (Vtn) 1009, 1014
Vyacheslav Denisov (Rus) ... 732
Vyazma (Rus) 716
Vyborg (Rus) 682, 728
Vyuga (Rus) 731

W

El Wacil (Mor) 548
Wadah (Kwt) 493
Al Wadeeah (SAr) 741
Waesche (US) 984
Wagio (Indo) 374
Wahoo (US) 987
Wakasa (Jpn) 446, 452
Wakataka (Jpn) 443
Wakazuki (Jpn) 457
Al Wakil (Egy) 222
Wallaby (Aust) 36
Waller (Aust) 26
Wally Schirra (US) 974
Walnut (US) 990
Walrus (Nld) 562
Walter Fachin (Ita) 423
Walter S Diehl (US) 975
Wambola (Est) 234
Wanamassa (US) 974
Wang Geon (RoK) 476
Wang Nai (Tld) 840
Wang Nok (Tld) 840
Wangerooge (Ger) 298
Wanyang-Shan (CPR) 154
Warakas (Indo) 366
Warao (Ven) 1004
Waree (Aust) 37
Warnow (Ger) 296
Warramunga (Aust) 28
Warrigal (Aust) 36
Warrior (US) 969
Washington (US) 937, 988
Washington Chambers
(US) 974
Washtucna (US) 974
Wasp (US) 964
Waters (US) 976
Watkins (US) 977
Watson (US) 977
Wattle (Aust) 36
Wave Knight (UK) 915
Wave Ruler (UK) 915
Wayne E Meyer (US) 948
Wdzydze (Pol) 642
Wedge (US) 991
Weeraya (Sri) 787
Weifang (CPR) 144
Weihai (CPR) 148
Wei Hsung (Twn) 821
Weilheim (Ger) 293
Weishan Hu (CPR) 161
Welang (Indo) 366
Weling (Indo) 367
Wellington (NZ) 575
Wenzhou (CPR) 142
Werra (Ger) 296
Wesp (Bel) 65
Westminster (UK) 904
Westpac Express (US) 981
Westport (Can) 112
West Virginia (US) 935
Westwal (Nld) 572
Al Whada (Mor) 551
L/CPL Roy M Wheat (US) 981
VADM K R Wheeler (US) 977
Whidbey Island (US) 967
Whirlwind (US) 959
Whitehorse (Can) 103
White Marlin (Sin) 758
White Shark (Sin) 758
Whitetip Shark (Sin) 758
Wichita (US) 950
Wicko (Pol) 643

Name	Page
Wickrama II (Sri)	787
Wierbalg (Nld)	572
Wiesel (Ger)	292
Wigry (Pol)	642
Wilhelm Carpelan (Fin)	242
Wilhelm Pullwer (Ger)	294
Willemstad (Nld)	569
William Flores (US)	987
William Mclean (US)	974
William P Lawrence (US)	948
William Tate (US)	989
William Trump (US)	987
PFC Dewayne T Williams (US)	977
Willow (US)	990
Windsor (Can)	98
Winnipeg (Can)	100
Winston S Churchill (US)	948
Wiratno (Indo)	362
Wire (US)	992
Wische (Ger)	297
Witthayakhom (Tld)	836
Wodnik (Pol)	644
Wolf (Can)	104
Wollongong (Aust)	33
Wombat (Aust)	36
Won Ju (RoK)	481
Won San (RoK)	485
Wrangell (US)	988
Wright (US)	982
Al Wtaid (UAE)	893
Wu Chang (Twn)	814
Wu Kang (Twn)	820
Wu Yi (Twn)	820
Wudangshan (CPR)	154
Wufengshan (CPR)	155
Wuhan (CPR)	138
Wyaconda (US)	991
Wyatt Earp (Aust)	34
Wyoming (US)	935
Wyulda (Aust)	36

X

Name	Page
X 01-03 (Pak)	601
X 101, 103-121 (Sri)	790
Xavier Pinto Telleria (Bol)	69
Xia (CPR)	128
Xian (CPR)	139
Xiangfan (CPR)	143
Xiaoyi (CPR)	157
Xinantecatl (Mex)	541
Xinhui (CPR)	151
Xuefengshan (CPR)	154
Xueshan (CPR)	155
Xuzhou (CPR)	144

Y

Name	Page
Y 01, 10 (Nam)	560
Y 02-04, 08-09, 11-19, 21-29 (SA)	763
Y 116, 118, 120-124, 126, 138-139, 142, 144-145, 147-148, 174-183 (Spn)	782
Y 231, 251, 254-255 (Spn)	780
Y 301-310 (Myn)	557
Y 311 (Myn)	556
Y 503-509, 511, 521, 524, 526-529, 534, 537, 539-541, 545, 549-564, 581, 583-585, 588-589, 590-592 (Spn)	781
Y 675-679 (Fra)	275
Y 693 (Fra)	273
Y 780-781 (Fra)	274
Y 834-835 (CPR)	160
Y 8200-8203 (Nld)	572
Yaa Asantewa (Gha)	300
Yacoub Ould Rajel (Mtn)	525
Yadanabon (Myn)	559
El Yadekh (Alg)	6
Yadryn (Rus)	708
Yaegumo (Jpn)	457
Yaeshio (Jpn)	430
Yaeyama (Jpn)	445
Yaezuki (Jpn)	457, 458
Yaffo (Isr)	397
Yahiko (Jpn)	453
Yakal (Plp)	633
Yakhroma (Rus)	721
Yakushima (Jpn)	445
Yamagiri (Jpn)	439
Yamakuni (Jpn)	456
Yamal (Rus)	704
Yamayuki (Jpn)	440
Yan Aye Aung (Myn)	556
Yan Gyi Aung (Myn)	555
Yan Khwin Aung (Myn)	556
Yan Min Aung (Myn)	556
Yan Myat Aung (Myn)	556
Yan Nyein Aung (Myn)	556
Yan Paing Aung (Myn)	556
Yan Win Aung (Myn)	556
Yan Ye Aung (Myn)	556
Yan Zwe Aung (Myn)	556
Yancheng (CPR)	144
Yandanshang (CPR)	154
Yangjiang (CPR)	151
Yangmanchun (RoK)	477
Yang Yang (RoK)	485
Yano (US)	980
Yantai (CPR)	144
Yantar (Rus)	711
Yaqui (Mex)	541
Yaracuy (Ven)	1004
Yarbay Kudret Güngör (Tur)	866
Al Yarmook (SAr)	739
Al-Yarmouk (Kwt)	490
El Yarmouk (Egy)	222
Yaroslavl (Rus)	727
Yaroslav Mudryy (Rus)	695
Yarra (Aust)	33
Yaşam (Tur)	871
Yashima (Jpn)	451
Yastreb (Rus)	727
Yavaros (Mex)	540
Yavire (Ven)	1007
Yavuz (Tur)	859
Yay Bo (Myn)	558
YD 200, 204-205 (Plp)	633
YDT 01-06 (Jpn)	448
YDT 17-18 (US)	972
Yecuana (Ven)	1004
Yee Ree (RoK)	481
Yegorlik (Rus)	715
Yehuin (Arg)	24
Yellow Elder (Bhm)	44
Yellowfin (US)	987
Yellowknife (Can)	103
Yelnya (Rus)	708, 715
Yenisei (Rus)	716
Yesil (Kaz)	463
Yevgeniy Kocheshkov (Rus)	706
YF 2121, 2124-2125, 2127-2129, 2132, 2135, 2138, 2141, 2150-2151 (Jpn)	444
YG 203-207 (Jpn)	449
Yhaguy (Par)	615
Yi Chon (RoK)	474
Yi Yang (Twn)	815
Yichang (CPR)	143
Yilan (Twn)	824
Yildiray (Tur)	855
Yildirim (Tur)	859
Yildiz (Tur)	862
Yinchuan (CPR)	140
Yingkou (CPR)	148
Yiyang (CPR)	144
YL 9-15 (Jpn)	449
YO 25-29, 31-40 (Jpn)	449
Yo Su (RoK)	481
Yodo (Jpn)	457
Yogaga (Gha)	300
Yola (Nig)	582
Yonakuni (Jpn)	453
Yong Ju (RoK)	481
Yongxingdao (CPR)	162
Yoon Young-Ha (RoK)	482
Yopito (Ven)	1004
Yos Sudarso (Indo)	360
Yoshino (Jpn)	456
Young Endeavour (Aust)	35
YP 676, 681, 683-684, 686-692, 694-695, 697-698, 700-701, 703-708 (US)	973
Ypoploiarchos Daniolos (Gre)	306
Ypoploiarchos Degiannis (Gre)	307
Ypoploiarchos Grigoropoulos (Gre)	306
Ypoploiarchos Kristallidis (Gre)	306
Ypoploiarchos Mikonios (Gre)	307
Ypoploiarchos Roussen (Gre)	306
Ypoploiarchos Troupakis (Gre)	307
Ypoploiarchos Votsis (Gre)	307
YR 01-02 (Jpn)	449
Yser (Fra)	279
YT 01-03, 58, 60-99 (Jpn)	450
YTB 37-39, 45-49, 150 (Twn)	821
YTE 12 (Jpn)	447
YTE 13 (Jpn)	448
YTL 16-17, 27-30, 32-36, 41-43 (Twn)	821
Yu (Mly)	513
Yu Seong-Ryong (RoK)	475
Yuan Wang 3-5, 6 (CPR)	159
Yubari (Jpn)	455
Yucatan (Mex)	536
Yueh Fei (Twn)	813
Yueyang (CPR)	144
Yufugiri (Jpn)	458
Yugeshima (Jpn)	445
Yukigumo (Jpn)	457
Yukon (US)	975
Yulgok Yi I (RoK)	475
Yulin (CPR)	144
Yuma (US)	980
Yun Hsing (Twn)	825
Yunbou (SAr)	741
Yuncheng (CPR)	144
Yunes (Iran)	379
Yung An (Twn)	819
Yung Chia (Twn)	819
Yung Chuan (Twn)	819
Yung Feng (Twn)	819
Yung Fu (Twn)	819
Yung Jin (Twn)	819
Yung Ku (Twn)	819
Yung Ren (Twn)	819
Yung Shun (Twn)	819
Yung Sui (Twn)	819
Yung Teh (Twn)	819
Yung Ting (Twn)	819
Yung Tzu (Twn)	819
Yung Yang (Twn)	819
Yunga (Rus)	697
Yunus 1-2 (Tur)	863
Yuntaishan (CPR)	155
Yunwashan (CPR)	154
Yuri Dolgoruky (Rus)	671
Yushan (CPR)	155
Yusoutei-Ichi-Gou (Jpn)	444
Yusoutei-Ni-Gou (Jpn)	444
Yuudachi (Jpn)	436
Yuugiri (Jpn)	439
Yuzbasi Ihsan Tulunay (Tur)	867
Yüzbasi Nasit Öngören (Tur)	868
Yuzhno-Sakhalinsk (Rus)	728
Yuzuki (Jpn)	457
YW 17-24 (Jpn)	449

Z

Name	Page
Z 8 (Pol)	645
Z 101-137, 139-141, 143-250 (Sri)	790
Zabaykalye (Rus)	726
Zaccola (Ita)	422
Zaire (Por)	652
Zakr (Yem)	1019
Zannotti (Ita)	422
Zao (Jpn)	451
Zapolarye (Rus)	726
Zaporizya (Ukr)	875
Zapoteco (Mex)	540
Zarechny (Rus)	700
Zaria (Nig)	581
Zarkoh (UAE)	889
Zarrar (Pak)	607
Al Zawraa (UAE)	888
Zborul (Rom)	664
Zbyszko (Pol)	645
Zeeland (Nld)	566
Zeeleeuw (Nld)	562
Zeemeeuw (Bel)	65
Zeffiro (Ita)	409
Zefiros (Gre)	311
Zégbéla Togba Pivi (Gn)	318
Zeleni Dol (Rus)	703
Zelenodolsk (Rus)	698
Zemaitis (Lit)	502
Zemchuk (Rus)	730
Zenobe Gramme (Bel)	65
Zephyr (US)	959
Zeus (Gre)	311
Zeus (US)	978
Zeyda (Myn)	555
Zhaik (Kaz)	464
Zhangjiagang (CPR)	157
Zhanjiang (CPR)	136
Zhaotong (CPR)	147
A Zheleznyakov (Rus)	708
Zhenghe (CPR)	159
Zhengzhou (CPR)	139
Zhenis (Kaz)	463
Zhigulevsk (Rus)	712
Zhoushan (CPR)	144
Zhuhai (CPR)	136
Zhu Kezhen (CPR)	159
Zierikzee (Nld)	569
Zigong (CPR)	147
Zijinshan (CPR)	155
Zinat Al Bihaar (Omn)	597
Zipkin (Tur)	861
Zobel (Ger)	292
Zorritos (Per)	624
Zou (Ben)	67
Zoubin (Iran)	384
Zouhel (Alg)	8
Al Zubara (Bhr)	49
Zuhrab (Yem)	1019
Zuidwal (Nld)	572
Zuiryu (Jpn)	429
Zulfiquar (Pak)	602
Zulurab (SAr)	742
Zumwalt (US)	950
Zurara (UAE)	887
Zvezdochka (Rus)	711
Zyb (Rus)	701

Named classes

2nd Lt John P Bobo (US) 977
10 De Agosto (Ecu) 215
65 ft (US) 992
300 ton (Jpn) 448
430 ton (RoK) 487
690 ton (Jpn) 446
1,200 ton (RoK) 487

A

A 26 (Swe) 796
A-125 (Rus) 729
Aadesh (Ind) 357
AB 27 (Tur) 863
Abamin (Myn) 557
Abdullah (Dauntless) (Jor) 462
Abhay (Project 1241 PE) (Pauk II) (Ind) 341
Abnaki (Mex) 541
Absalon (Den) 200
Abu Dhabi (Comandante) (UAE) 885
Abukuma (Jpn) 441
Achelous (Plp) 633
Acuario (Mex) 537
Ada (Milgem) (Tur) 860
Addriyah (MSC 322) (SAr) 741
Adelaide (Oliver Hazard Perry) (Aust) 27
Adept (UK) 921
Aditya (Ind) 352
Adjutant and MSC 268 (Twn) 819
Admirable (Myn) 555
Admiral Gorshkov (Project 22350) (Rus) 698
Admiral Grigorovich (Project 11356M) (Rus) 698
Aegir (Ice) 325
Agdlek (Den) 198
Aggressive (Twn) 819
Agor-26 (US) 971
Agosta (Spn) 767
Agosta 70 (Pak) 601
Agosta 90B (Pak) 600
Aguascalientes (Mex) 541
Águila (Mex) 533
Aguinaldo (Plp) 631
Aguirre (Lupo) (Per) 620
Ahmad El Fateh (TNC 45) (Bhr) 48
Ahmad Yani (Van Speijk) (Indo) 360
Aist (Ukr) 882
Aist (Dzheyran) (Project 1232.1) (Rus) 707
Ajeera (Bhr, Ven) 48, 1005
Akademik Krylov (Project 852/856) (Rus) 709
Akizuki (Jpn) 435, 457
Akula (Rus) 674, 706
Akula (Schuka-B) (Ind, Rus) 327, 678
Akvamaren (Rus) 707
Al Bushra (Omn) 595
Al Dhaen 12 m (Bhr) 50
Al Fattan (UAE) 893
Al Feyi (UAE) 889
Al Hussein (Hawk) (Jor) 460
Al Jarim (FPB 20) (Bhr) 48
Al Jawf (Sandown) (SAr) 741
Al Jouf (SAr) 743
Al Jubatel (SAr) 742
Al Manama (MGB 62) (Bhr) 47
Al Riffa (FPB 38) (Bhr) 48
Al Riyadh (Modified La Fayette) (SAr) 738
Al Shaheed (Kwt) 492
Al Siddiq (SAr) 740
Al Tahaddy (Kwt) 494
Alamosa (Plp) 633
Albatros (Gha, Rus, Tun) 300, 697, 852
Albatros 630 (Ecu) 216
Albatros 730 (Ecu) 216
Albatros 830 (Ecu) 216
Albatros 830 Fish-Rite (Ecu) 216
Albatros 910 (Ecu) 216
Albatros 1100 (Ecu) 215
Albatros 1100 Fish-Rite (Ecu) 216
Albatroz (Etim, Por) 208, 653
Albion (UK) 911
Alboran (Spn) 775
Alexandrit (Project 12700) (Rus) 707
Algol (US) 983
Alize (Fra) 273
Alkmaar (Tripartite) (Lat, Nld) 495, 569
Allende (Knox) (Mex) 529
Alliance (Twn) 820
Alligator (RoK, Ven) 484, 1005
Alligator (Tapir) (Project 1171) (Rus) 706
Almirante Brown (Meko 360 H2) (Arg) 13
Almirante Guilhem (Brz) 87
Almirante Padilla (Col) 172
Al-Ofouq (Omn) 596
Alpinist (Project 503) (Lit, Rus) 502, 726
Alpinist (Project 503M/R) (Rus) 712
Altair (Swe) 802
Alucat 850 (Arg) 23
Alucat 1050 (Arg) 23
Alusafe 1290 (Nor) 589
Alusafe 1300 (Nor) 589
Alvand (Vosper Mk 5) (Iran) 380
Alvaro de Bazán (Spn) 770
Älvsborg (Chi) 124
Amami (Jpn) 455
Amazon (Pak) 604
Amazonas (Per) 622
Ambassador IV (Egy) 222
America (US) 966
Amga (Project 1791) (Rus) 714
Amorim do Valle (River) (Brz) 83
Amsterdam (Nld) 570
Amur (Project 304) (Ukr) 880
Amur (Project 304/304M) (Rus) 714
AN-2 (Hun) 325
Anaga (Spn) 775
Anawrahta (Myn) 554
Anchorage (Twn) 817
Andrea Doria (Horizon) (Ita) 406
Andrómeda (Col, Por) 180, 654
Andromeda (Nasty) (Gre) 307
Angamos/Islay (Type 209/1200) (Per) 618
Ankang (CPR) 164
Antares (Mlw, Rus, Ukr) 505, 727, 883
Antarès (BRS) (Fra) 268
Antoniou (Gre) 306
Antonio Zara (Ita) 422
Antyey (Rus) 676
Anvil/Clamp (US) 992
Anwei (Type 920) (CPR) 163
Anzac (Meko 200) (Aust, NZ) 28, 574
Appleleaf (UK) 915
Aquarius (Swi) 807
Aquitaine (Fra) 255
Aragosta (Ita) 420
Aragosta (Ham) (Ita) 417
Aratu (Schütze) (Brz) 82
Arauca (Col) 174
Archangel (Alb, CpV, Chi, Com, Kaz, Ken, Sen, Yem) 3, 114, 125, 181, 463, 465, 745, 1018
Archer (UK) 907
Archer (Västergötland) (Sin) 750
Arcor 46 (Mor) 551
Arcor 53 (Mor) 551
Arctic (Nor) 592
ARD 12 (Ecu) 213
Ardhana (UAE) 887
Ares (Geo) 283
Aresa PVC-170 (Ang) 9
Argos (Por) 653
Arguin (Mtn) 525
Arihant (Ind) 327
Arleigh Burke (Flight IIA) (US) 948
Arleigh Burke (Flights I and II) (US) 946
Armatolos (Osprey 55) (Gre) 308
Armidale (Aust) 33
Armstrong (US) 971
Arrecife (ex-Olmeca II) (Mex) 539
Arrow Post (Can) 109
Artigliere (Lupo) (Ita) 410
Arun 60 (Gre) 315
Arvak (Den) 201
Asagiri (Jpn) 439
Ashdod (Eri, Isr) 233, 399
Asheville (Col, Gre, US) 179, 308, 970
Ashoora I (Sud) 793
Ashoora I (MIG-G-0800) (Iran) 387
Asia (Rus) 712
Aso (Jpn) 453
Asogiri (Jpn) 458
Assad (Mly) 511
Astute (UK) 900
Asuka (Jpn) 447
Atago (Jpn) 432
Atilay (Type 209/1200) (Tur) 855
Atlant (Rus) 689
Attack (Indo) 367
Auk (Mex, Plp) 535, 629
Austin (Ind, US) 349, 966, 979
Auxnaval Armón 51 (Spn) 784
Auxnaval Armón 69 (Spn) 784
Avenger (US) 969
Axios (Lüneburg) (Type 701) (Gre) 311
Aydin (Tur) 865
Azmat (Pak) 608
Azteca (Mex) 536

B

Bacolod City (Frank S Besson) (Plp) 632
Badau (Waspada) (Indo) 367
Bad Bramstedt (Ger) 298
Baden-Württemberg (Type 125) (Ger) 290
Badr (SAr) 739
Baglietto Mangusta (Alg) 8
Baglietto Type 20 (Alg) 8
Bahamas (Bhm) 44
Bakassi (Type P 48S) (Cam) 95
Baklan (CMN 15-60) (Yem) 1019
Baklan (Project 19920) (Rus) 711
Baklazhan (Project 5757) (Rus) 722
Baldur (Ice) 325
Balsam (Col, DR, Gha, Nig, Plp) 175, 205, 301, 580, 635
Baltic 24 (Pol) 647
Baltyk (Pol) 644
Balzam (Asia) (Project 1826) (Rus) 712
Bambuk (Project 12884) (Ukr) 881
Bangaram (Ind) 348
Bang Rachan (Tld) 842
Ban Yas (TNC 45) (UAE) 887
Baptista De Andrade (Por) 651
Baradero (Dabur) (Arg) 17
Barbaros (Tur) 858
Barentshav (VS 794) (Nor) 592
Barkat (Pak) 611
Barracuda (Fra, Rus) 247, 674
Barroso (Brz) 76
Bartolomeu Dias (M) (Por) 650
Barzan (Vita) (Qat) 657
Basento (Ecu) 213
Baskunchak (Rus) 733
Batral (Chi, Mor) 122, 549
Bay (Aust, UK, US) 31, 39, 919, 989
Bayandor (PF 103) (Iran) 382
Baynunah (UAE) 886
Bayóvar (Per) 624
Beautemps-Beaupré (Fra) 269
Bedok (Landsort) (Sin) 757
Bélier (Fra) 277
Bellatrix (DR) 206
Bendeharu (Bru) 89
Bereza (Ukr) 880
Bereza (Project 130) (Bul, Rus) 92, 720
Bergamini (Ita) 408
Berkot-B (Rus) 690
Berlin (Type 702) (Ger) 295
Bérrio (Rover) (Por) 656
Bester (Rus) 713
Bigliani (Ita, Lby) 422, 501
Bira (MP 6) (Rus) 720
Biya (Project 870) (Ukr) 880
Biya (Project 871) (Cub, Rus) 191, 710
Bizerte (Tun) 851
Blanco Encalada (M) (Chi) 117
Blue Ridge (US) 961
Bob Hope (US) 980
Bogomol (Project 02061) (Rus) 729
Bolva (Project 688/688A) (Rus) 715
Bombarda (Por) 654
Boraida (Mod Durance) (SAr) 741
Borey (Rus) 671
Boris Chilikin (Project 1559V) (Rus) 715
Bormida (Ita) 418
Boston Whaler Justice (ElS) 231
Botica (Mac) 504
Bouchard (Par) 616
BPS 500 (Project 12418) (Vtn) 1010
Bracui (River) (Brz) 80
Brahmaputra (Ind) 340
Braila (Rom) 666
Brandenburg (Ger) 286
Braunschweig (K130) (Ger) 290
Bravo (Bronstein) (Mex) 530
Bredstedt (Ger) 298
Bremen (Ger) 288
Bremse (Tun) 854
Bristol (Ind) 354
Briz (Sonya) (Project 12650) (Bul) 91
Broadsword (Brz, Chi, Gua, Rom) 74, 118, 316, 660
Bronstein (Mex) 530
Brutar II (Rom) 664
Bryza (Project 722) (Az) 41
Bukhansan (RoK) 488
Burak (Type A 69) (Tur) 860
Buratti (Lby) 501
Buratti (N 23) (Ita) 423
Burya (Alg) 5
Al Bushra (Omn) 595
Buyan (Project 21630) (Rus) 703
Buyan-M (Project 21631) (Rus) 703

C

C 14 (Iran) 386
C 62 (Alg) 5
Ç 120 (Tur) 864
Ç139 (Tur) 864
Cacine (Por) 652
Cakra Type 209/1300 (Indo) 359
Calmaria (Por) 656
Canberra (Aust) 31
Cannon (Plp, Tld) 629, 843
Canopus (Swiftships 110 ft) (DR) 206
Cantabria (Spn) 780
Canterbury (NZ) 576
Cantieri del Golfo (Dji) 204
Capana (Alligator) (Ven) 1005
Cape (Aust, Can, Iran, Uru) 39, 112, 385, 997
Cape Flattery (US) 972
Cape Goéland (Can) 112
Cape Hurd (Can) 110
Cape Light (Can) 111
Cape (PGM 71) (Mex) 536
Cape Roger (Can) 108
Carlskrona (Swe) 803
Car Nicobar (Ind, Mrt) 348, 527
Carpentaria (Myn) 557
Carvajal (Modified Lupo) (Per) 619, 625
Casma (Saar 4) (Chi) 121
Cassard (Fra) 256
Cassiopea (Ita) 412
Castle (Ban) 54
Castor (Spn) 779
Cat (Balsam) (Nig) 580
Cavour (Ita) 404
CDIC (Chi, Fra) 121, 268
Ceará (Thomaston) (Brz) 81
Centauro (Por) 653
Cerberus (Nld) 571
Cetina (Silba) (Cro) 187
Chaho (DPRK) 471
Chakri Naruebet (Tld) 828
Challenger (Bhm) 44
Challenger (Sjöormen) (Sin) 750
Chamo (Eri) 233
Chamois (Fra) 273
Champion (US) 981
Chamsuri (RoK) 482
Chang Bogo (Type 209/1200) (RoK) 474
Chanticleer (Tur) 868
Chao Phraya (Tld) 830
Charles de Gaulle (Fra) 252
Cheng Kung (Twn) 813
Cheong Hae Jin (RoK) 486
Cheoy Lee (HK) 323
Cherokee (Arg, Ecu, Per, Twn, Tur) 17, 213, 625, 821, 870
Cheverton (Mld, Sri) 522, 789
Chicama (Dauntless) (Per) 626
Chihaya (Jpn) 449
Chinese P4 (Ban) 56
Chinese 27 metre (Ben, CpV) 67, 114
Chios (Jason) (Gre) 309
Chiyoda (Jpn) 448
Chon Buri (Tld) 836
Chong-Jin (DPRK) 471
Chong-Ju (DPRK) 471
Christina (Griffon 2000 TD) (Lit) 504
Chui-E (Alg) 8
Chun Jee (RoK) 486
Circé (Tur) 865
Clemenceau (Brz) 72
Cochrane (Chi) 119
Collins (Aust) 26
Colombo MK I/II/III/IV (Sri) 788
Comandante (Ita, UAE) 412, 885
Comandante João Belo (Uru) 995
Combattante I (Kwt) 490
Combattante II (Iran) 384
Combattante II G (Lby) 500

1042 **INDEXES** > Named classes

Combattante IIIB (Nig) ... 581
Combattante III M (Qat, Tun) ... 658, 851
Command Boat 450 (Swe) ... 801
Commander (Jor) ... 461
Conafi 55 (Por) ... 656
Conejera (Mtn, Moz, Sen) ... 525, 552, 744
Conrado Yap (Sea Hawk/Killer) (Plp) ... 632
Constitución (Ven) ... 1003
Constitution (US) ... 973
Converted Stalwart (US) ... 976
Converted Trawler (Cub) ... 191
Coral (Por) ... 654
Cormoran (Mor) ... 548
Corregidor (Plp) ... 634
Corrubia (Ita) ... 422
Corsar (Rom) ... 665
Cosmos (US) ... 991
Costa Sur (Arg) ... 20
Cougar (Iran) ... 388
Cougar Enforcer 40 (Kwt) ... 493
County (Damen Stan Patrol 4207) (Jam) ... 427
Cove (Tur) ... 865
Cove Island (Can) ... 110
CP 256 (Ita) ... 425
CP 259 (Ita) ... 425
CP 261 (Ita) ... 425
CP 265 (Ita) ... 424
CP 301 (Ita) ... 425
CP 313 (Ita) ... 425
CP 314 (Ita) ... 425
CP 401 (Ita) ... 423
CP 405 (Ita) ... 424
CP 409 (Ita) ... 424
CP 454 (Ita) ... 425
CP 506 (Ita) ... 426
CP 601 (Ita) ... 426
CP 701 (Ita) ... 426
CP 713 (Ita) ... 426
CP 801 (Ita) ... 426
CP 814 (Ita) ... 426
CP 825 (Ita) ... 425
CP 2001 (Ita) ... 425
CP 2084 (Ita) ... 425
CP 2201 (Ita) ... 426
CP 2301 (Ita) ... 425
Crestitalia MV-45 (Qat) ... 659
Crestitalia MV 55 (Pak) ... 611
Crestitalia MV 70 (Egy) ... 228
Croitor (Rom) ... 667
Cutlass (Gua, Tld) ... 317, 846
Cutthroat (US) ... 972
Cyclone (Plp, US) ... 630, 959
Czilim (Project 20910) (Rus) ... 733

D

D. Carlos I (Stalwart) (Por) ... 654
Dabur (Arg, Chi, Isr, Nic) ... 17, 125, 398, 577
Dachou (Type 917A) (CPR) ... 165
Dadie (Type 814A) (CPR) ... 158
Dadong (Type 946A) (CPR) ... 161
Dagger (Indo) ... 365
Dahua (Type 909) (CPR) ... 158
Dajiang (Type 925) (CPR) ... 162
Dakhla (Mor) ... 550
Dalang (Type 922 II/III) (CPR) ... 161
Dalao (Type 926) (CPR) ... 164
Damen 6210 (Ang) ... 10
Damen 9014 (Vtn) ... 1016
Damen Cougartek (HK) ... 322
Damsah (Combattante III M) (Qat) ... 658
Dandao (Type 917) (CPR) ... 162
Danlin (CPR) ... 162
Danyao (Type 904A) (CPR) ... 165
Daozha (CPR) ... 165
Daphne (EqG) ... 231
Daring (UK) ... 902
Darss (Spn) ... 778
Darussalam (PV 80) (Bru) ... 88
Dauntless (Ant, Bhm, Bar, Cay, Dom, Geo, Gra, Gua, Jam, Jor, Kaz, Per, StK, StL, StV, VI) ... 11, 45, 62, 115, 205, 283, 315, 316, 427, 462, 626, 734, 735, 1016
Daxin (Alg) ... 7
Daxin (Type 795) (CPR) ... 159
Dayun (Type 904) (CPR) ... 162
Dazhou (Type 946) (CPR) ... 162
De Havilland (Plp) ... 636
De La Penne (ex-Animoso) (Ita) ... 407
De Ruyter (Per) ... 618
De Zeven Provincien (Nld) ... 564
Deba (Pol, Yem) ... 642, 1017
Deepak (Ind) ... 353
Defender (Az, Ban, Cam, Chi, CR, ElS, Gab, Gha, Isr, Ken, Mex, Moz, Nig, Tan Tog, US, Yem) ... 43, 55, 61, 96, 126, 183, 231, 280, 301, 399, 465, 537, 552, 580, 826, 848, 992, 1018
Defender 25 (Mrt) ... 527
Defender 27 (Ben, Mrt) ... 67, 526
Defender 2710 (Iraq) ... 392
Delfin (Col, Rus) ... 180, 672
Delhi (Ind) ... 334
Delta III (Kalmar) (Rus) ... 673
Delta III Stretch (Rus) ... 684
Delta IV (Delfin) (Rus) ... 672
Delvar (Iran) ... 390
Démocrata (Mex) ... 536
Depoli (Ita) ... 419
Dergach (Sivuch) (Project 1239) (Rus) ... 699
Descubierta (Egy, Spn) ... 219, 773
D'Estienne d'Orves (Type A 69) (Fra) ... 260
Al Dhaen 12 m (Bhr) ... 50
Dhofar (Province) (Omn) ... 595
Diana (SF MK II) (Den) ... 198
Dilos (Geo, Gre) ... 282, 313
Diponegoro (Sigma) (Indo) ... 363
Diver (Tur) ... 868
Djebel Chenoua (C 58) (Alg) ... 5
Dobrynya Nikitich (Project 97) (Rus) ... 722
Doğan (Tur) ... 862
Dokkum (Per) ... 623
Dolgoruky (Borey) (Rus) ... 671
Dolphin (Ban) ... 61
Dolphin (Type 800) (Isr) ... 395
Dongdiao (Type 815) (CPR) ... 158
Dong Hae (Col) ... 173
Dost (Tur) ... 871
Drummond (Type A 69) (Arg) ... 14
Dubna (Rus) ... 715
Duke (UK) ... 904
Dumit (Can) ... 110
Durance (Arg, Aust, Fra) ... 20, 35, 272
Durango (Mex) ... 534
Durbar (Hegu) (Ban) ... 55
Durdharsha (Huangfen) (Ban) ... 54
Durjoy (Hainan) (Ban) ... 55
Dvora (Sri) ... 789, 791
Dyugon (Project 21820) (Rus) ... 706
Dzheyran (Rus) ... 707

E

Eagle (US) ... 991
Echo (UK) ... 914
Eckaloo (Can) ... 110
EDA-R (Fra) ... 268
Edenton (RoK, US) ... 486, 986
Edic 700 (Dji, Sen) ... 204, 745
Edsall (Mex) ... 540
Eduardono (Hon) ... 320
EDVM 25 (Brz) ... 81
Eilat (Saar 5) (Isr) ... 396
Eithne (Ire) ... 393
Ekpe (Lürssen 57) (Nig) ... 580
El Mouderrib (Chui-E) (Alg) ... 8
El Mounkid (Alg) ... 8
El Wacil (P 32) (Mor) ... 548
Elbe (Ger) ... 296
Elbrus (Osimol) (Project 537) (Rus) ... 718
Eleuthera (Keith Nelson) (Bhm) ... 44
Elli (Kortenaer) (Gre) ... 305
Emba (Project 1172/1175) (Rus) ... 717
Emory S land (US) ... 979
Endurance (Sin, Tld) ... 756, 841
Enforcer II (IC 16) (Rus) ... 730
Engin (Circé) (Tur) ... 865
Enoshima (Jpn) ... 446
Ensdorf (Ger) ... 294
Éridan (Pak) ... 608
Éridan (Tripartite) (Fra) ... 269
Erraid (P 32) (Mor) ... 551
Ertuğrul (Terrebonne Parish) (Tur) ... 863
Esmeraldas (Ecu) ... 210
Espada (Ecu) ... 214
Espadarte (CpV) ... 114
Esploratore (Ita) ... 413
Espora (Meko 140 A16) (Arg) ... 15
Esterel (Type RPC 50) (Fra) ... 276
Etna (Gre, Ita) ... 311, 418
Evniki (Osprey) (Gre) ... 310
Evropi (Hunt) (Gre) ... 310
Express Shark Cat (Aust) ... 38

F

Fabian Wrede (Fin) ... 242
Fairey Sword (Bhr) ... 51
Faisal (Commander) (Jor) ... 461
Falaj 2 (UAE) ... 887
Famous cutter (US) ... 985
Fatahillah (Indo) ... 361
Fateh (Iran) ... 379
Fateh (Saettia Mk IV) (Iraq) ... 392
Al Fattan (UAE) ... 893
Faysal (Jor) ... 461
FB 55SC (HK) ... 323
FBM catamaran (UK) ... 920
FB RIB 42SC (HK) ... 324
FC-19 (Kaz) ... 463
Fearless (Sin) ... 754
Felenk (Cove) (Tur) ... 865
Felicity (UK) ... 920
Al Feyi (UAE) ... 889
Finik (Project 872) (Rus) ... 710
Fire (Can) ... 106
Fish (Aust) ... 35
Flamant (OPV 54) (Cam, Fra) ... 96, 265
Flamingo (Project 1415PV) (Rus) ... 730
Flamingo (Tanya) (Project 1415) (Az, Rus) ... 42, 718
Floréal (Fra, Mor) ... 261, 544
Flower (Bel) ... 64
Flower (Tripartite) (Bul) ... 91
Flyvefisken (Den, Lit) ... 202, 502
Forbin (Horizon) (Fra) ... 254
Formidable (Project Delta) (Sin) ... 752
Fort Grange (UK) ... 917
Fort Victoria (UK) ... 916
Foudre (Chi, Fra) ... 122, 267
Foxtrot (Lby, Ukr) ... 499, 875
FPB 28 (Indo) ... 377
Frankenthal (Ger, UAE) ... 293, 888
Frank S Besson (Plp, US) ... 632, 961
Freedom (US) ... 950
Fregat (Rus) ... 691, 692
Fréhel (Fra) ... 276
Fremm (Mor) ... 546
French Edic 700 (Leb) ... 498
Fridtjof Nansen (Nor) ... 586
Frigate (Myn, Tld) ... 553, 829
Frosch I (Indo) ... 370
Frosch II (Indo) ... 374
Fuchi (Type 903) (CPR) ... 161
Fulin (CPR) ... 163
Fulmar (Fra) ... 265
Fuqing (Pak) ... 609
Fuqing (Type 905) (CPR) ... 161
Furusund (Swe) ... 803
Futami (Jpn) ... 446
Fuzhou (CPR) ... 164

G

Gabes (Tun) ... 854
Gabya (Oliver Hazard Perry) (Tur) ... 857
Gaeta (Aust, Tld) ... 33, 842
Gagah (Mly) ... 516
Gahjae (Iran) ... 386
Galana (Ken) ... 465
Galerna (Agosta) (S 70) (Spn) ... 767
Galicia (Spn) ... 776
Ganzhu (CPR) ... 158
Garibaldi (Ita) ... 413
Gavion (Ven) ... 1007
Gävle (Swe) ... 798
Gearing (Fram 1) (Pak) ... 610
Georges Leygues (Fra) ... 257
Gepard (Ger) ... 292
Gepard (Project 11661K) (Rus) ... 696
Gerald R Ford (US) ... 941
Geranium (Fra) ... 279
Ghaem (MIG-S-1800) (Iran) ... 385
Ghannatha (Transportbåt 2000) (UAE) ... 890
Glavkos (Gre) ... 303
Gleaner (UK) ... 914
Glycine (Fra) ... 271
Godavari (Ind) ... 339
Golf (CPR) ... 134
Golok (Indo) ... 376
Goplo (Notec) (Pol) ... 643
Gordon (US) ... 981
Gordon Reid (Can) ... 111
Gornostay (Project 20990) (Rus) ... 730
Gorya (Type 12660) (Rus) ... 708
Goryn (Project 714) (Rus) ... 723
Gotland (A 19) (Swe) ... 795
Gowind (Mly) ... 512
GPB-480 (Project 1896) (Rus) ... 709
Grachonok (Rus) ... 704
Grajaú (Brz, Nam) ... 79, 559
Granby (Can) ... 104
Grif (Az, Cub, EqG, Nic, Rus, Syr, Ukr) ... 41, 192, 231, 578, 728, 809, 878
Griffon (Can) ... 107
Griffon 2000 TD (Lit) ... 504
Griffon 2000 TDX (Pol) ... 647
Griffon 2400 TD(M) (UK) ... 912
Griffon 8000 TD(M) (Ind, SAr) ... 356, 743
Griffon 8100TD (Type 392) (Swe) ... 800
Grif-T (Tkm) ... 873
Grisha (Ukr) ... 877
Grisha (Albatros) (Project 1124/1124M) (Rus) ... 697
Gromovoy (CPR) ... 166
Grumete diaz (Dabur) (Chi) ... 125
Guaicamacuto (Ven) ... 1007
Guaiqueri (Ven) ... 1004
Guam (US) ... 982
Guangzhou (CPR) ... 164
Guardian (Hon) ... 320
Gulf (Can) ... 106
Gumdoksuri (RoK) ... 482
Gun (Myn) ... 555
Guns (Pak) ... 610
Gunsan (RoK) ... 490
Guppy II (Twn) ... 812
Gurza-M (Project 58155) (Ukr) ... 877
Gyda (Nor) ... 589

H

H 960 (Pol) ... 646
Hae Uri (RoK) ... 489
Hai Cheng (Twn) ... 825
Hai Lung (Twn) ... 811
Hai Ou (Gam) ... 281
Hai Ying (Twn) ... 825
Haijiu (Type 037/1) (CPR) ... 152
Hainan (Ban, Egy, DPRK) ... 55, 222, 470
Hainan (Type 037) (CPR, Myn) ... 152, 556
Haiqing (Type 037/1S) (CPR) ... 152
Haixun 01 (CPR) ... 166
Haixun 11 (CPR) ... 166
Haixun 21 (CPR) ... 166
Haixun 22 (CPR) ... 166
Haixun 31 (CPR) ... 166
Haixun 051 (CPR) ... 166
Haizhui/Shanghai III (Type 062/1) (CPR) ... 152
Haizhui (Type 062/1) (Ban) ... 56
Haizhui (Type 062/1G) (Sri) ... 787
Halcon (Type B 119) (Arg) ... 21
Halifax (Can) ... 100
Halmatic 20 metre (Bhr) ... 50
Halmatic 160 (Bhr) ... 50
Halmatic M160 (Ana, Qat) ... 10, 659
Ham (Ita) ... 417
Hamashio (Jpn) ... 459
Hämeenmaa (Fin) ... 240
Hamilton (US) ... 984
Hamilton and Hero (Ban, Nig, Plp) ... 53, 579, 628
Hamina (Fin) ... 239
Han (CPR) ... 130
Hanchon (DPRK) ... 472
Handalan (SPICA-M) (Mly) ... 513
Hang Tuah (Type 41/61) (Mly) ... 514
Hantae (DPRK) ... 472
Hårek (Alusafe 1300) (Nor) ... 589
Harushio (Jpn) ... 430
Hashidate (Jpn) ... 449
Hashmat (Agosta 70) (Pak) ... 601
Hashim (Rotork) (Jor) ... 461
Hatakaze (Jpn) ... 436
Hateruma (Jpn) ... 453
Hatsuyuki (Jpn) ... 440
Hauki (Fin) ... 242
Hawk (Jor) ... 460
Hayabusa (Jpn) ... 443
Hayagumo (Jpn) ... 457
Hayanami (Jpn) ... 458
Healy (US) ... 988
Hecla (Indo, SA) ... 372, 764
Hegu (Ban, Egy) ... 55, 223
Helgoland (Ger) ... 297
Helgoland (Type 720B) (Uru) ... 998
Hellenic 56 (Gre) ... 308
Helsinki (Cro) ... 185
Hendijan (Iran) ... 390
Hengam (Iran) ... 389
Henry J Kaiser (US) ... 975
Hercules (Type 42) (Arg) ... 18
Hermes (Ind) ... 332
Hero (Jam) ... 427
Hero (Damen Stan Patrol 4207) (Can) ... 109
Heroine (Type 209/1400 MOD (SA)) (SA) ... 761
Hetz (Saar 4.5) (Isr) ... 397
Hibiki (Jpn) ... 446
Hida (Jpn) ... 452
Hila (Fin) ... 243
Hirashima (Jpn) ... 445
Hiuchi (Jpn) ... 450
Hobart (Aust) ... 27

Named classes < INDEXES

Ho Hsing (Twn)	821
Holland (Nld)	566
Holm (Den)	198
Holzinger (Águila) (Mex)	533
Hongqi (CPR)	163
Hood (Can)	111
Horizon (Fra, Ita)	254, 406
Houbei (Type 022) (CPR)	151
Houdong (Iran)	384
Houjian (or Huang) (Type 037/2) (CPR)	151
Hounan (Type 021) (Yem)	1017
Houxin (Type 037/1G) (CPR, Myn)	151, 555
Hsun Hai (Twn)	817
Hua Hin (Tld)	837
Huangfen (Ban, Pak)	54, 611
Huangpu (Mtn)	525
Huasteco (Mex)	540
Huchuan (Ban, Tan)	54, 826
Hudong (Tld)	844
Hujiu (Ban, CPR)	60, 165
Humboldt (Mex, Per)	539, 623
Hungnam (DPRK)	472
Hunt (Gre, Lit, UK)	310, 502, 913
Huon (Gaeta) (Aust)	33
Huracan (Saar 4.5) (Mex)	532
Al Hussein (Hawk) (Jor)	460
Hydra (Gre)	304
Hyuga (Jpn)	431

I

Igor Belousov (Project 21300) (Rus)	720
Ijhtihad (FPB 41) (Bru)	88
Ilocos Norte (Plp)	634
Impeccable (US)	975
Imperial Marinheiro (Brz)	79
Improved Akizuki (Jpn)	440
Improved Osprey 55 (Sen)	744
Improved Romeo (Egy)	217
Improved Sauro (Ita)	402
Improved Tetal (Rom)	662
Improved Tursas (Fin)	245
Improved Y 301 (Myn)	556
Incheon (RoK)	480
Independence (US)	953
Indigenous Aircraft Carrier (Ind)	333
Ingul (Project 1453) (Rus)	722
Inhaúma (Brz)	76
Intrépida (Arg)	18
Inttisar (OPV 310) (Kwt)	492
Invincible (UK)	909
Iran Hormuz 21 (Iran)	389
Iran Hormuz 24 (Iran)	388
Iroquois (Can)	102
Isaac Peral (S 80A) (Spn)	768
Iskandhar (Mld)	521
Iskra (Pol)	644
Isla Fernandina (Vigilante) (Ecu)	214
Island (Ban, TT, UK, US)	55, 849, 908, 988
Itaipú (Par)	616
Ivan Gren (Project 11711E) (Rus)	706
Ivan Susanin (Project 97P) (Rus)	726
Iver Huitfeldt (Den)	194
Iwami (Jpn)	453
Izar IVP-22 (Spn)	782
Izu (Jpn)	452
Izumo (Jpn)	432

J

Jacinto (Peacock) (Plp)	629
Jacob van Heemskerck (Chi)	118
Jaco (Shanghai II) (ETim)	208
Jägaren (Swe)	800
Jaime Gómez (MK III PB) (Col)	179
Jalalat (Pak)	606
Al Jarim (FPB 20) (Bhr)	48
Jason (Fra, Gre)	278, 309
Jastreb (Rus)	695
Al Jawf (Sandown) (SAr)	741
Jayesagara (Sri)	786
Jebel Antar (Alg)	8
Jerong (Mly)	513
Jianghu (Type 056) (CPR)	148
Jianghu (CPR)	167
Jianghu I (Ban, Egy)	53, 220
Jianghu I/II/V (Type 053H/053H1/053H1G) (CPR)	147
Jianghu II (Type 053H1) (Myn)	553
Jianghu III (Type 053 H2) (Ban, CPR)	54, 146
Jianghu IV (Type 053HTH) (CPR)	160
Jiangkai I (Type 054) (CPR)	142
Jiangkai II (Type 054A) (CPR)	144
Jiangwei I (Type 053 H2G) (CPR)	146
Jiangwei II (Type 053H3) (CPR)	143
Jif Xplorer (Fra)	278
Jin (CPR)	128
Jin Chiang (Twn)	816
Jingsha II (CPR)	155
Jinyou (CPR)	163
João Coutinho (Por)	651
Johan de Witt (Nld)	568
José Andrada (Plp)	631
José Maria Palas (Swift 110) (Col)	175
Al Jouf (SAr)	743
Juan Carlos I (Spn)	777
Al Jubatel (SAr)	742
Juniper (US)	990
Jura (UK)	924
Jurmo (Fin)	243
Jurrat (Pak)	607
Jyoti (Ind)	352

K

K 8 (Project 361T) (Vtn)	1014
Kaan 15 (Tur)	872
Kaan 19 (Tur)	872
Kaan 20 (Egy, Geo, Tur)	222, 283, 863
Kaan 29 (Tur)	871
Kaan 33 (Qat, Tur)	658, 872
Kachalot (Rus)	685
Kagitingan (Plp)	631
Kajami (Iran)	385
Kakap (PB 57) (Indo)	366
KAL-40 (Indo)	367
Kaliningradneft (Rus)	716
Kalkan (Project 50030) M (EqG, Tkm, Ukr)	231, 873, 883
Kalmar (Rus)	673, 705
Kaman (Combattante II) (Iran)	384
Kamenka (Project 870) (Rus, Vtn)	710, 1015
Kamorta (Ind)	341
Kampela (Fin)	242
Kan (CPR)	158
Kang Ding (La Fayette) (Twn)	814
Kanin (Rus)	732
Kaoh (Cmb)	94
Kaper (Pol)	646
Kapitan Pattimura (Parchim I) (Indo)	362
Kara (Berkot-B) (Rus)	690
Karbala (MIG-S-3700) (Iran)	389
Karel Doorman (Nld)	571
Karnaphuli (Kraljevica) (Ban)	56
Kartal (Tur)	861
Kashdom II (Iran)	387
Kashima (Jpn)	447
Kashin II (Ind)	336
Kashin (Project 61) (Rus)	690
Kashtan (Project 141) (Rus)	717
Kasos (Hellenic 56) (Gre)	308
Kasturi (Type FS 1500) (Mly)	509
Kaszub (Pol)	640
Katanpää (Fin)	240
Katran (Project 20970) (Kaz)	463
Katun (Rus)	723
Kayvan (Cape) (Iran)	385
Kazakhstan (Kaz)	463
KBV 001 (Swe)	805
KBV 031 (Swe)	806
KBV 041 (Lit)	503
KBV 101 (Lit)	504
KBV 181 (Swe)	805
KBV 201 (Swe)	805
KBV 236 (Est, Lat)	235, 495
KBV 301 (Swe)	805
KBV 590 (Griffon 2450 TD) (Swe)	806
KCR-40 (Indo)	367
KCR-60M (Indo)	368
KDX-2 (RoK)	476
Kebir (Alg)	6
Kedah (Meko 100 RMN) (Mly)	510
Keelung (Kidd) (Twn)	812
Keeper (US)	989
Kefallinia (Zubr) (Gre)	309
Keith Nelson (Bhm)	44
Keka (HK)	322
Kelso (Can)	111
Khalid (Agosta 90B) (Pak)	600
Khamronsin (Tld)	833
Khobi (Rus)	716
Khukri (Ind)	343
Kidd (Twn)	812
Ki Hajar Dewantara (Indo)	372
Kiiski (Fin)	241
Kilat (Mly)	519
Kiliç (Tur)	861
Kilo (Alg, CPR, Ind, Iran, Pol, Rus, Vtn)	4, 132, 329, 379, 638, 682, 1009
King (Arg)	20
Kingston (Can)	103
Kirov (Orlan) (Rus)	688
Klasma (Project 1274) (Rus)	717
Klewang (Indo)	368
Knox (Egy, Mex, Twn, Tld)	219, 529, 815, 831
Knud Rasmussen (Den)	197
Knurrhahn (Ger)	296
Kobben (Pol)	637
Kogalniceanu (Rom)	664
Kojima (Jpn)	452
Kolkata (Project 15A) (Ind)	333
Komandor (Rus)	726
Komar (DPRK)	469
Končar (Type R-02) (Cro)	187
Konda (Rus)	721
Kondor (Rus)	675
Kondor I (CpV, Tun)	114, 853
Kondor II (Uru)	998
Kondor II (Type 89) (Indo)	371
Kongbang (DPRK)	472
Kongou (Jpn)	434
Koni (Alg, Bul)	4, 89
Koni (Project 1159) (Lby)	499
Konsul (Project 16811) (Rus)	713
Kora (Ind)	342
Kormoran (Pol)	645
Korond (Ukr, Vtn)	879, 1014
Kortenaer (Gre)	305
Koster (Swe)	802
Kotonami (Jpn)	456
Kotor (Mon)	542
Kowan (DPRK)	472
Kozara (Ser)	747
Krab (Rus)	718
Krait (Indo)	367
Kraljevica (Ban)	56
Kralj (Type R-03) (Cro)	186
Krivak (Project 1135/1135M) (Rus)	694
Krivak III (Nerey) (Ukr)	876
Krivak III (Nerey) (Project 1135MP) (Rus)	725
Krogulec (Pol)	642
Kronshtadt (Project 122) (Alb)	2
KSS-2 (Type 214) (RoK)	474
Kuha (Fin)	240
Kujang (Indo)	376
Kulmbach (Ger)	293
Kunigami (Jpn)	453
Kupang (Indo)	369
Kurmuk (Type 15) (Sud)	793
Kurobe (Jpn)	448
Ku Song, Sin Hung and Mod Sin Hung (DPRK)	471
Kutter (Lit)	503
Kuznetsov (Orel) (Rus)	686
Kuznetsov (Orel) (Project 1143.5/6) (CPR)	135
Kvarnen (Nor)	590
Kwang Hua 6 (Twn)	817
Kwanggaeto Daewang (KDX-1) (RoK)	477
Kyle of Lochalsh (UK)	924

L

La Belle Poule (Fra)	271
La Combattante II (Mly)	513
La Combattante IIA (Gre)	307
La Combattante III (Gre)	307
La Cruz (Per)	627
La Fayette (Fra, Twn)	259, 814
Lada (Rus)	681
Lake (NZ)	575
Laksamana (Assad) (Mly)	511
Lama (Type 323/323B) (Rus)	720
Landsort (Sin)	757
Langkawi (Mly)	517
Lapérouse (Fra)	265, 270
Lapérouse (BH2) (Fra)	270
Larkana (Pak)	607
Laskos (La Combattante III) (Gre)	307
Latorre (Jacob van Heemskerck) (Chi)	118
Lat Ya (Gaeta) (Tld)	842
Lazaga (Mor)	547
LCM (Uru)	999
LCM 6 (SAr, Twn)	742, 818
LCM 8 (Aust, EIS, RoK)	37, 231, 484
LCM 2000 (Aust)	37
LCMs (Chi)	122
LCP (Den)	197
LCU 1466 (Bhr, Twn)	48, 817
LCU 1512 (Ban)	58
LCU 1600 (DR)	207
LCU 1610 (Brz, SAr, US)	80, 741, 969
LCU 2000 (US)	968
Le Malin (Fra)	265
Le Triomphant (Fra)	250
Leander (Ecu)	210
Lebed (Kalmar) (Project 1206) (Rus)	705
Leeuwin (Aust)	34
Legend (US)	984
Lekiu (Mly)	508
Leonard J Cowley (Can)	108
Léopard (Fra)	271
Lerici (Mly, Nig)	514, 583
Lerici/Gaeta (Ita)	415
Lewis and Clark (US)	974
LHT-130 (Spn)	779
Lida (Sapfir) (Project 10750) (Rus)	708
Lima (SA)	764
Lindau (Type 331) (Lit)	502
Lindormen (Est)	234
Lokki (Fin, Lat, Lit)	241, 496, 503
Los Angeles (US)	938
Louhi (Fin)	242
Louisbourg (Can)	109
LPR-40 (Brz, Col)	79, 174
LS 51 (Gre)	314
LST 1–510 and 512–1152 (Twn)	818
LST 1–511 and 512–1152 (Indo)	369
LST 512–1152 (Plp, Twn)	632, 818
Lublin (Pol)	642
Luda (Type 051DT/051G/051G II) (CPR)	136
Luga (Az)	40
Luhai (CPR)	140
Luhu (Type 052A) (CPR)	141
Luneburg (Col)	178
Lüneburg (Egy, Gre, Uru)	226, 311, 999
Lupo (Ita, Per)	410, 620
Lürssen 45 (Ecu)	212
Lürssen 57 (Nig)	580
Lürssen FPB 45 (Gha)	301
Lürssen PB 57 (Gha)	300
Lürssen VSV 15 (Indo)	375
Lushun (Sri)	787
Luyang I (Type 052B) (CPR)	138
Luyang II (Type 052C) (CPR)	139
Luyang III (Type 052D) (CPR)	140
Luza (Project 1541) (Rus)	721
Luzhou (CPR)	136
Lynch (Arg)	22

M

M (Bel, Chi, Nld, Por)	63, 117, 563, 650
M 10 (Sri)	791
M36 (Tld)	840
Macaé (Napa 500) (Brz)	78
Machitis (Gre)	308
Madhumati (Sea Dragon) (Ban)	55
Madina (Type F 2000s) (SAr)	737
Maestrale (Ita)	409
Magar (Ind)	349
Mahamiru (Lerici) (Mly)	514
Mahar 31 (Bhr)	51
Maipo (Batral) (Chi)	122
Maïto (Fra)	276
Makar (Ind)	351
Malakhit (Rus)	708
Malaspina (Spn)	778
Malawali (Mly)	516
Malina (Project 2020) (Rus)	714
Mamry (Notec II) (Pol)	642
Máncora (Per)	626
Mandalika (Indo)	375
Mandovi (Mtn)	525
Mandume (Ang)	10
Mangust (Project 12150) (Rus)	729
Manly (UK)	922
Man Nok (Tld)	840
Al Manama (MGB 62) (Bhr)	47
Manta (Ecu, Kwt)	214, 493
Manuel Azueta (Edsall) (Mex)	540
Manuel Clavero (Per)	622
Manych (Project 1549) (Rus)	716
Marañon (Per)	622
Marasesti (Rom)	661
Mar del Plata (Z-28) (Arg)	22
Marine Protector (Mlt, US, Yem)	523, 987, 1018
Marlim (Nam)	559
Marlin (Mly)	520
Marlin (Meatini) (Brz)	78
Marshal Nedelin (Project 1914) (Rus)	711
Marte (Per)	624
Martha L Black (Can)	107
Mashuu (Jpn)	448
Matka (Vekhr) (Rus, Ukr)	703, 877
Matsunami (Jpn)	458

1044 **INDEXES** > Named classes

Mawani (Syr) 810
Mazinger (RoK) 487
Mazzei (Ita) 422
MCC 1101 (Ita) 418
Meatini (Brz, Ita) 78, 422
Meghna (Ban) 57
Meiyo (Jpn) 459
Meko A-200 (Alg) 5
Meko Type 360 H1 (Nig) 578
Melville (US) 971
Men 212 (Ita) 419
Men 215 (Ita) 419
Mercy (US) 980
Meriuisko (Fin) 243
Metal Shark Defiant 38 (Ban) 61
Meteoro (Spn) 772
MHV 800 (Den) 198
MHV 850 (Den) 198
MHV 900 (Den) 199
Micalvi (Chi) 121
Michao (Myn) 556
Midget submarines (RoK, Pak) 473, 601
Mihashi and Raizan (Jpn) 456
Mikhail Rudnitsky (Project 05360/1) (Rus) 717
MIL 40 (Iran) 388
MIL 55 (Iran) 388
Milgem (Tur) 860
Minerva (Ita) 410
Ming (CPR) 130
Minna (UK) 925
Mirazh (Project 14310) (Rus) 729
Mirna (Mon) 543
Mirna (Type 140) (Cro) 188
Missile (Myn) 555
Mistral (Fra, Rus) 266, 705
Miura (Jpn) 452
Mizuho (Jpn) 451
Mk V (Kwt) 491
Mk VI (US) 960
Mod Altay (Rus) 715
Mod Durance (SAr) 741
Mod Haizhui (Lushun) (Type 062/1G) (Sri) 787
Mod Roisín (PV 90) (Ire) 394
Mod Shanghai II (Sri) 787
Mod Sorum (Project 745MB/745MBS) (Rus) 724
Mod Sorum (Project 1454) (Rus) 711
Mod Yanha (CPR) 169
Modified Descubierta (Mor) 545
Modified Finik 2 (Pol) 643
Modified Georges Leygues (Fra) 258
Modified Gepard (Project 11661E) (Vtn) 1009
Modified Ghannatha (UAE) 888
Modified Hai Ou (Par) 616
Modified Haizhui (Tun) 851
Modified Hiryu (Jpn) 459
Modified Kiev (Ind) 330
Modified Lupo (Per, Ven) 619, 625, 1002
Modified Moma (Pol) 644
Modified MTM 217 (Ita) 415
Modified Niterói (Brz) 84
Modified Okean (Project 502) (Rus) 730
Modified Patra (Cypr) 192
Modified R (Can) 107
Modified River (UK) 908
Modified Scorpene (Brz) 70
Modified Sportis (Pol) 648
Modified Stenka (Cmb) 94
Modified Town (Nig) 580
Modified Ulsan (Ban) 52
Modified Zhuk (Vtn) 1012
Molnya (Rus, Ukr) 700, 727, 878, 882
Moma (Project 861) (Cro, Rus) 188, 709
Moma (Project 861M) (Rus, Ukr) 712, 879
Moorhen (UK) 920
Morkov (Project 1461.3) (Rus) 719
Morona (Per) 624
Morrosquillo (LCU 1466A) (Col) 176
Moskit (Pol, Rus) 645, 732
El Mouderrib (Chui-E) (Alg) 8
El Mounkid (Alg) 8
Mourad Rais (Koni) (Alg) 4
MPMB (Cro) 188
Mrowka (Pol) 645
MRTP 22 (Tur) 862
MSF (Den) 199
MTC 1011 (Ita) 419
MTM 217 (Ita) 414
MTP 96 (Ita) 414
Mubarraz (UAE) 887
Mukha (Sokol) (Project 1145) (Rus) 703
Muna (Type 1823) (Rus) 720
Munsif (Eridan) (Pak) 608
Murakumo (Jpn) 458
Murasame (Jpn) 436
Muravey (Antares) (Ukr) 883
Muravey (Antares) (Project 133) (Rus) 727
Muray Jib (MGB 62) (UAE) 884
Murce (MIG-G-0900) (Iran) 387
Murena (Rus) 733
Murena E (RoK) 484
Musca (Rom) 665
Mustaed (Bru) 88
Mustang (Project 18623) (Rus) 727
Myanmar (Gun) (Myn) 555
Myanmar (Missile) (Myn) 555

N

Nachshol (Stingray Interceptor) (Isr) 399
Nahang (Iran) 378
Najin (DPRK) 468
Naluca (Rom) 664
Namacurra (Ang, Mlw, Moz, Nam, SA) 10, 505, 552, 560, 763
Nampo (DPRK) 472
Nanuchka (Rus) 701
Nanuchka II (Burya) (Alg) 5
Nanuchka II (Project 1234) (Lby) 500
Nanyun (CPR) 161
Naresuan (Tld) 829
Nascimento (Brz) 85
Naser (Iran) 391
Nasty (Gre) 307
Natsugiri (Indo, Jpn) 377, 457
Natya (Yem) 1018
Natya I (Ind, Ukr) 350, 879
Natya I (Akvamaren) (Project 266M) (Rus) 707
Natya (Project 266M) (Syr) 809
Natya (Project 266ME) (Lby) 501
Neftegaz (Project B-92) (Rus) 723
Neon Antonov (Project 1595) (Rus) 732
Nerey (Rus, Ukr) 725, 876
Nesbitt (UK) 913
Nestin (Hun, Ser) 325, 746
Neustadt (Bul, Rom) 94, 668
Neustrashimy (Jastreb) (Rus) 695
Newport (Brz, Mex, Mor, Twn) 81, 538, 549, 818
NFI (Indo) 374
Nichinan (Jpn) 446
Nicobar (Ind) 352
Niijima (Jpn) 444
Nimitz (US) 942
Ning Hai (Twn) 817
Nisr (Egy) 228
Niterói (Brz) 75
Nodriza (Col) 174
Nojima (Jpn) 452
Nordkapp (Nor) 592
Normed (Tld) 840
Nornen (Nor) 592
Notec (Pol) 643
Notec II (Pol) 642
NS-722 (Yem) 1017
Nusa 22 (Mly) 517
Nusa 28 (Mly) 520
Nyayo (Ken) 464
Nyryat I (Project 522) (Egy) 226
Nyryat I (Project 1896) (Alb, Rus) 2, 718
Nyryat 2 (Project 522) (Rus) 718

O

Oaxaca (Mex) 534
Ob (Project 320) (Rus) 716
Oban (UK) 922
OCEA FPB 98 (Alg, Ben, Nig, Sur) 9, 67, 582, 794
October (Egy) 223
Ogonek (Project 12130) (Rus) 731
Ohio (US) 935, 936
Ohre (Ger) 297
Ojika (Jpn) 453
Okba (PR 72) (Mor) 547
Okhotsk (Project 22030) (Rus) 723
Okhtensky (Egy) 227
Okhtensky (Project 733/733S) (Rus) 723
Oksøy/Alta (Nor) 590
OL 44 (Gre) 314
Olekma (Rus) 715
Oliver Hazard Perry (Aust, Bhr, Egy, Pak, Pol, Tur, US) 27, 46, 218, 605, 638, 857, 952
Olivieri (Arg) 17
Olya (Malakhit) (Project 1259) (Rus) 708
Olya (Project 1259) (Bul) 92
Ona (Chi) 126
Ondatra (Akula) (Project 1176) (Rus) 706
Onega (Project 1806) (Rus) 710
Onjuku (Mex) 538
Onuk MRTP 16 (Qat) 658
Oosumi (Jpn) 443
OPV 54 (Cam, Fra, Mtn) 96, 265, 525
OPV 62 (EqG) 232
OPV 70 (Mor) 548
Orca (Can) 104
Orel (CPR, Rus) 135, 686
Orkan (Sassnitz) (Pol) 639
Orlan (Rus, Ukr) 688, 883
Osa (DPRK) 470
Osa (Project 205) (Syr) 808
Osa I (Project 205) (Egy) 223
Osa II (Alg, Cub) 6, 190
Osa II (Project 205) (Az, Vtn) 43, 1012
Oscar II (Antyey) (Rus) 676
Osimol (Rus) 718
Osman (Jianghu I) (Ban) 53
Osman Gazi (Tur) 863
Osprey (Egy, Gre, Myn, Twn) 225, 310, 555, 819
Osprey 55 (Gre) 308
Osprey FV 710 (Nam) 560
Osprey Mk II (Mor) 547
Oste (Ger) 295
Otago (NZ) 575
Ouranos (Gre) 311
Outrage 25 ft (Hon) 320
Outrage 29 ft (Hon) 319
Oyashio (Jpn) 430

P

P 21 (Ire) 394
P 41 Peacock (Ire) 394
P 101 (Spn) 775
P 127 (RoK) 489
P 270 (Tun) 851
P 350 (Tun) 854
P 400 (Fra, Gab, Ken) 265, 280, 465
P-2000 (Mrt) 526
Pabna (Ban) 61
Pacific (CI, Fij, Kir, MI, Mic, Pal, PNG, Sam, Sol, Ton, Tuv, Van) 182, 238, 465, 524, 541, 612, 614, 736, 760, 848, 874, 1000
Padstow and Newhaven (UK) 922
PAF-L (Col) 176
Paita (Terrebonne Parish) (Per) 622
Paltus (Project 1851) (Rus) 685
Paluma (Aust) 34
Pamlico (US) 991
Pansio (Fin) 241
Papaloapan (Newport) (Mex) 538
Papanikolis (Type 214) (Gre) 303
Pará (Brz) 85
Parachique (Per) 627
Parchim I (Indo) 362
Parchim II (Rus) 698
Parnaiba (Brz) 80
Parvin (PGM-71) (Iran) 385
Pashe (MIG-G-1900) (Iran) 385
Pastaza (Per) 625
Pathfinder (US) 975
Patiño (Spn) 780
Patra (Ctl, Fra, Gab) 184, 278, 280
Pattani (Tld) 832
Pauk I (Bul) 91
Pauk I (Molnya) (Ukr) 882
Pauk I (Molnya) (Project 1241P) (Ukr) 878
Pauk I (Molnya) (Project 12412) (Rus) 727
Pauk II (Cub, Ind) 190, 341
Pauk II (Project 1241 PE) (Rus) 727
PB 90 (Myn) 556
PC-43 (Indo) 368
PCE 827 (Plp) 630
PCF 46 (Plp) 635
PCF 50 (Swift Mk 1 and Mk 2) (Plp) 636
PCF 65 (Swift Mk 3) (Plp) 630, 636
Peacock (Plp) 629
Pedretti (Ita) 420
Pedro Teixeira (Brz) 79
Pegasus (SAr) 742
Pelindung (Mly) 519
Pelym (Project 1799) (Cub, Rus) 191, 719
Pengaman (Mly) 518
Pengawal (Mly) 518
Pengawal 13 (Mly) 520
Pengawal 23 (Mly) 520
Penggalang 16 (Mly) 520
Penggalang 18 (Mly) 518
Peninjau (Mly) 518
Penyelamat 15 (Mly) 518
Penyelamat 20 (Mly) 519
Perdana (La Combattante II) (Mly) 513
Perwira (Bru) 88
Peterson MK 4 (Gam) 281
Petir 10 (Mly) 517
Petir Open (Mly) 518
Petrel (Point) (Ven) 1006
Petrushka (UK-3) (Az, Rus, Ukr) 40, 713, 880
Petya (Project 159A/AE) (Vtn) 1010
Petya II (Project 159A) (Az) 40
Petya III (Project 159A) (Syr) 808
Peykaap I (IPS 16) (Iran) 385
Peykaap II (IPS 16 Mod) (Iran) 386
Peykaap III (IPS 16 Mod) (Iran) 386
PGM-39 (Plp) 635
PGM 71 (Ecu, Mex, Per, Tld, Tur) 215, 536, 626, 838, 863
Phaéton (Fra) 274
Piast (Pol, Uru) 645, 999
Pijao (Type 209/1200) (Col) 171
Pikker (Est) 235
Pilar (Hamilton and Hero) (Plp) 628
Piloto Pardo (Chi) 121
Piranha (EIS, Hon) 229, 320
Piratini (Brz) 80
Piyavka (Project 1249) (Rus) 731
Po Hang (RoK) 481
Point (Arg, Az, Col, CR, DR, Ecu, EIS, Geo, Mex, Pan, Plp, TT, Tkm, Ven) 18, 43, 179, 183, 206, 214, 229, 282, 536, 612, 631, 849, 874, 1006
Poisk-2 (Project 1832) (Rus) 713
Polar (US) 988
Polaris (Mex, Ven) 535, 1007
Polaris II (Mex) 536
Polnochny A (Project 770) (Az, Egy) 41, 224
Polnochny B (Alg, Az, Syr) 7, 41, 809
Polnochny C (Project 773 I) (Ukr) 878
Polnochny D (Ind) 350
Polnochny (Project 771) (Vtn) 1013
Poluchat 1 (Egy) 226
Poluchat I (Alg) 7
Poluchat (Project 368) (Az) 41
Pomornik (Zubr) (Project 1232.2) (CPR, Rus, Ukr) 156, 706, 878
Pondicherry (Natya I) (Ind) 350
Ponza (Ita) 420
Poolster (Pak) 609
Poseidon (Cypr) 193
Post (Can) 109
Potok (Rus) 721
Pourquoi Pas ? (Fra) 270
Powhatan (Tur, US) 869, 978
Pozharny I (Project 364) (Rus) 719
Pozharny (Project 364) (Ukr) 881
PR 48 (Sen) 744
PR 72M (Sen) 744
Prabparapak (Tld) 838
Predator (Iraq) 392
Preveze (Type 209/1400) (Tur) 856
Prignitz (Ger) 299
Priyadarshini (Ind) 355
Priz (Project 1855) (Rus) 713
Project 73 (Az) 42
Project 890 (Pol) 644
Project 1481 (Rus) 732
Project 1496/1496M (Rus) 729
Project 11982 (Rus) 712
Project 1398B (Aist) (Ukr) 882
Project 15010 (Rus) 733
Project 18280 (Rus) 712
Project 19910 (Rus) 711
Project 21980 (Grachonok) (Rus) 704
Project 22010 (Rus) 711
Project 22120 Patrol Ship (Rus) 731
Project 58250 (Ukr) 875
Project 90600 (Rus) 724
Project V 820 (Az) 42
Prometey (Project 498/04983/04985) (Rus) 723
Prometheus (Etna) (Gre) 311
Protecteur (Can) 105

Named classes < INDEXES 1045

Protector (Bhm, Chi, EIS, Tan UAE, Ven) 44, 126, 230, 826, 891, 1008
Protector (ASI 315) (HK) 322
Protector 3612 (Ven)...................1007
Providencia (Col) 177
Province (Omn)..............................595
Provo Wallis (Can) 108
Prut (Project 527M) (Rus)...............722
PS 700 (Lby)..................................500
Pulau Rengat (Tripartite) (Indo) 370
Punta (Ven) 1008
Punta Macolla (Ven).....................1007
Punta Malpelo (Per) 623
PV08 (Fin) 245
PV50M (EqG)..................................232
PV88 (EqG).....................................232

Q

Qahir (Omn) 593
Qing (CPR).......................................134
Qiongsha (CPR)...............................160
Queen Elizabeth (UK) 901
Quito (Lürssen 45) (Ecu) 212

R

R (Can)..107
R-2 Mala (Cro) 185
Rafael del Castillo Y Rada (Swift 105) (Col) 174
Raidco Opv 45 (Sen) 745
Raidco RPB 18 (Mtn).......................525
Raidco RPB 20 (Lby)........................501
Raidco RPB 33 (Sen, Tog)......744, 848
Raïs Bargach (Mor).........................548
Rajput (Kashin II) (Ind).....................336
Rajshree (Ind)..................................357
Ramadan (Egy) 222
Ramunia (Mly)..................................517
Rani Abbakka (Ind) 355
Ratcharit (Tld) 836
Rattanakosin (Tld)............................833
Rauma (Fin) 240
Red (Arg)..20
Reine (Nor).......................................589
Reliance (Col, Sri, US)....179, 787, 985
Reshef (Saar 4) (Isr) 398
Reshitelni (Pauk I) (Project 1241P) (Bul).....................91
Rhein (Tur).......................................866
Rhu (Mly) 517
Al Riffa (FPB 38) (Bhr).......................48
Rihtniemi (Est)..................................234
Rinker (Ecu) 215
Rio (Col)...175
Rio Coca (Ecu).................................216
Rio (Marlim) (Nam)...........................559
Rio Minho (Por)................................653
Rio Napo (Per) 624
Rio Nepeña (Per)..............................626
Rio Orinoco II (Ven)........................1008
Rio Puyango (Ecu) 215
Río Santa (Per) 626
Río Viru (Per) 627
Riquelme (Tiger) (Chi) 120
River (Ban, Brz, Guy, SA, UK).........58, 80, 83, 318, 764, 907
Al Riyadh (Modified La Fayette) (SAr) 738
Rizal (Auk) (Plp)...............................629
Robert D Conrad (Mex, Tun)...........................539, 852
Rodman 20 m (Bhr) 51
Rodman 38 (Plp, Por, Tun, Zim)...........636, 656, 854, 1019
Rodman 46 (Cam) 96
Rodman 46HJ (Zim) 1019
Rodman 55 Canarias (Spn) 783
Rodman 55HJ (Cypr, Spn)......192, 783
Rodman 55M (Spn, Sur).........783, 793
Rodman 58 (Omn, Spn)..........598, 783
Rodman 66 (Gab, Spn)......280, 779, 782
Rodman 82 (Spn).............................782
Rodman 101 (Mor, Nic, Omn, Plp, Sey, Spn, Sur) 549, 577, 597, 636, 748, 783, 793
Rodman 790 (Zim).........................1019
Rodman 800 (Chi)............................126
Roebuck (Ban)...................................59
Roísín (Ire)..393
Romeo (Project 033) (DPRK) 466
Ropucha (Project 775/775M) (Rus) 704
Ropucha I (Project 775) (Ukr) 879
Roraima (Brz).....................................80
Roslavl (CPR)...................................166
Rotork (Jor, Mlw) 461, 505
Rotterdam (Nld) 567
Roussen (Super Vita) (Gre) 306
Rover (Indo, Por, UK)......373, 656, 916
Rubin (Vtn)....................................1014
Rubin (Project 22460) (Rus) 730
Rubis Améthyste (Fra) 248
Ruposhi Bangla (Ban).......................61
Rus (Project 16810) (Rus)...............713

S

Saar 4 (Chi, Gre, Isr, Sri) 121, 313, 398, 786
Saar 4.5 (Isr, Mex)..................397, 532
Saar 5 (Isr) 396
Sabac (Ser) 747
Sábalo (Ven)...................................1001
Sachsen (Ger)..................................289
Saettia (Ita).......................................424
Saettia Mk III (Mlt)...........................523
Saettia Mk IV (Iraq) 392
Safe Boats Defender 38 (Cay)........115
Safeboats Defender 310X (Mlt) 524
Safeguard (US) 979
Sagar (T 43) (Ban).............................58
Sagardhwani (Ind)............................350
Salisbury (Ban) 59
Salta (Type 209/1200) (Arg) 13
Sam (Jpn, Swe) 445, 801
Samaesan (Tld)................................845
Samar (Ind) 355
Sambongho (RoK) 489
Samudra (UT 517) (Ind)...................354
Samuel Risley (Can).........................108
San Antonio (US) 962
San Giorgio (Ita)...............................414
San Juan (Plp) 634
Sandhayak (Ind)...............................351
Sandown (Est, SAr, UK) 234, 741, 912
Sang-O (DPRK)................................467
Sang-O II (K-300) (DPRK)...............467
Sankalp (Ind, Sri) 354, 786
Santa Cruz (Bol).................................68
Santa Cruz (TR 1700) (Arg)...............12
Santa María (Spn)............................768
Sapfir (Rus)......................................708
Sardar (Project 22180) (Kaz)...........463
Sariwon (DPRK)...............................468
Sarojini Naidu (Ind) 356
Sarucabey (Tur)................................864
Sarych (Rus) 693
Saryu (Ind) 347
Sassnitz (Ger, Pol) 298, 639
Sattahip (PSMM Mk 5) (Tld)...........837
Sav (Den) 202
Sawahil (Kwt)..........................492, 494
Saygak (Project 14081/ 14081M) (Rus) 732
Schuka (Rus) 680
Schuka-B (Ind, Rus) 327, 678
Schütze (Brz) 82
Schwedeneck (Ger) 294
Scimitar (UK)....................................907
Scorpene (Chi, Ind, Mly)..............................116, 328, 507
Scott (UK) 913
SDB Mk 3 (Mrt) 526
SDB MK 5 (Mld, Sey).......................521
SDB Mk 5 (Bangaram) (Ind)............348
Sea Dolphin (Ban, ETim, Kaz, Plp) 56, 208, 463, 631
Sea Dolphin/Wildcat (Gha) 301
Sea Dragon (Ban) 55
Sea Eagle (Nig) 581
Sea Guard (SAr)...............................742
Sea Hawk/Killer (Plp) 632
Sea Spectre PB Mk III (Egy) 228
Seaspray (HK)........................322, 323
Seawolf (US)....................................940
Sechelt (Can) 104
Seeb (Vosper 25) (Omn)..................595
Segura (Spn)....................................778
Sejong Daewang (KDX-3) (RoK) 475
Semilang (Mly) 518
Sentinel (US)....................................987
Serasa (Bru) 88
Serna (Project 11770) (Rus)............705
Serviola (Spn) 774
Severodvinsk (Yasen) (Rus) 679
Sewart (Gua, Iran, Sud)............................316, 388, 793
Seydi (Tur)..865
SGT Matej Kocak (US).....................981
Al Shaheed (Kwt) 492
Shaheed (Shanghai II) (Type 062) (Ban)..............................56
Shaldag (Cypr, Isr, Sri)........193, 397, 788
Shaldag Mk II (EqG, Nig)........232, 581
Shang (CPR) 129
Shanghai II (Alb, Ban, ETim, Egy, DPRK, Pak, Tan).......................2, 56, 208, 224, 470, 611, 825
Shanghai II (Type 062) (ConD, Sri)..........................181, 787
Shanghai III (Type 062/1) (SL).........749
Shapla (River) (Ban) 58
Shark Cat 800 (Aust) 38
Shark (Damen Stan Patrol 3507) (Sin) 758
Shelon (Project 1388M) (Az) 40
Shelon I/II (Project 1388/1388M) (Rus) 718
Shengli (CPR) 163
Shershen (Egy, DPRK)............224, 472
Shershen (Project 206T) (Cmb, Vtn) 94, 1012
Shichang (CPR) 160
Shikishima (Jpn) 451
Shimagiri (Jpn).................................457
Shimayuki (Jpn) 447
Shirane (Jpn)....................................433
Shiretoko (Jpn) 452
Shishumar (Type 209/1500) (Ind) 328
Shivalik (Project 17) (Ind) 338
Shmel (Ukr)......................................883
Shmel (Project 1204) (Rus) 732
Shoyo (Jpn) 459
Shughart (US)...................................980
Shupang (Type 636A) (CPR).......159, 169
Shupavu (Ken) 464
Shyri (Type 209/1300) (Ecu)............209
Sibarau (Attack) (Indo) 367
Sibiriyakov (Project 865) (Rus)........709
Al Siddiq (SAr) 740
Sidehole I and II (Project 737 K/M) (Rus) 724
Sierra (Mex).....................................533
Sierra I (Barracuda) (Rus)................674
Sierra II (Kondor) (Rus)....................675
Sigma (Indo, Mor)..................363, 546
Sigma 9814 (Vtn) 1010
Sigma 10514 (Indo) 364
Sigma T (Can) 111
Silas Bent (Tur).................................866
Silba (Cro)..187
Silmä (Est)..235
Silver Ships 48 ft (Az)........................43
Simeto (Ita, Tun).....................418, 853
Similan (Hudong) (Tld) 844
Simonneau (Sri) 792
Sindhughosh (Kilo) (Project 877EM/8773) (Ind)..........329
Singa (PB 57) (Indo)........................366
Sinpo (DPRK)...................................469
Sipadan (Mly)...................................516
Sipadan Kayu (Mly).........................519
Sipadan Steel (Mly)..........................519
Sir Bedivere (Brz) 82
Sir Galahad (Brz)...............................81
Sirius (Aust, Brz) 35, 84
Sivuch (Rus).....................................699
Sjöormen (Sin) 750
SK 620 (Rus, Ukr)...................715, 881
Skjold (Nor)......................................588
Skrunda (Lat) 494
Slava (Atlant) (Rus) 689
Slazak (Meko A 100) (Pol)...............641
Slepen (Rus) 731
Sliva (Project 712) (Rus)..................724
Smolny (Project 887) (Rus) 714
Snake (Gha).....................................302
Snellius (Nld)....................................570
SNR-17 (Rom) 668
SO 1 (DPRK)....................................470
Sobol (Project 12200) (Kaz, Rus) 463, 729
Södermanland (A 17) (Swe)............795
Soho (DPRK)....................................468
Soju (DPRK).....................................469
Sokol (Rus) 703
Sokół (Kobben) (Type 207) (Pol)......637
Sokzhoi (Rus)...................................728
Song (CPR).......................................131
Sonya (Bul, Cub)........................91, 191
Sonya (Yakhont) (Project 1265) (Az, Ukr, Vtn)...............42, 879, 1014
Sonya (Yakhont) (Project 12650) (Syr) 809
Sonya (Yakhont) (Project 12650/1265M) (Rus)....................708
Sorum (Tur) 870
Sorum (Project 745) (Rus, Vtn)..........................722, 1015
Sorum (Project 745P) (Rus) 726
Sotoyomo (Arg)..................................17
Souryu (Jpn) 429
Souter 20 metre (Bhr) 50
Souter Wasp 17 metre (TT)..............849
South Korean Killer (Sri) 792
Sovremenny (CPR) 137
Sovremenny (Sarych) (Rus) 693
Soya (Jpn)..451
Spasilac (Cro) 188
Sportis (Pol) 648
Sprut (Project 6457S) (Rus) 726
SSV-10 (Ukr)....................................883
Stalwart (Por) 654
Steber (Aust)......................................38
Stenka (Project 205P) (Az) 43
Stenka (Tarantul) (Cub, Ukr) 192, 882
Stenka (Tarantul) (Project 205P) (Rus)......................728
Steregushchiy (Rus)........................696
Stingray Interceptor (Isr) 399
Stividor (Project 192) (Rus).............724
Stockholm (Swe) 799
Stolkraft (Vtn)..................................1012
Stollergrund (Ger, Isr).............294, 399
Storm (Lit, UK).........................502, 921
Straznik (Pol)....................................647
Stromboli (Ita) 419
Styrsö (Swe) 802
Subahi (Kwt) 493
Suffren (Barracuda) (Fra)................247
Sugashima (Jpn)..............................445
Sukanya (Ind, Sri) 347, 786
Suma (Jpn) 446
Suncraft Manta Mk III (Nig).............581
Sunkar (Project 14081) (Kaz) 462
Super Dvora (Eri) 233
Super Dvora Mk I (Sri) 792
Super Dvora Mk I and Mk II (Isr).....398
Super Dvora Mk II (Ind, Sri).......347, 788
Super Dvora Mk III (Isr, Sri).....399, 790
Super Vita (Gre) 306
Supervittoria 800 (Mlt) 523
Supply (US)......................................974
Sura (Project 145) (Ukr)...................881
Sura (Project 419) (Rus) 717
Svetlyak (Project 1041.2) (Vtn)......1011
Svetlyak (Project 10412) (Slo).........759
Svetlyak (Project 1041Z) (Rus)........728
Swallow (RoK, Pak) 485, 611
Swallow 65 (Ind) 356
Swift (Tld)...839
Swift 36 ft (CR) 183
Swift 42 ft (CR) 183
Swift 65 ft (Ant, CR, Dom, Hon, StL) 11, 183, 204, 320, 735
Swift 85 ft (Hon)...............................320
Swift 105 ft (CR, Hon)............183, 319
Swift PBR (Cam).................................95
Swiftships (Col, Ecu)...............175, 214
Swiftships 35 m (DR, Iraq)......206, 392
Swiftships 65 ft (EIS).......................230
Swiftships 77 ft (EIS, Gn)........229, 318
Swiftships 93 ft (Egy).......................227
Swiftships 105 ft (Eri).......................233
Swiftships 110 ft (DR, StK).......206, 734
Swiftships Protector (Egy) 228
Sword (TT) 850
Sword (F-22P) (Pak).........................602
Syounan (Jpn)...................................446

T

T-4 (Project 1785) (Az) 41
T 43 (CPR, Egy) 157, 224
T 81 (Tld)...838
T 91 (Tld)...837
T 213 (Tld)..839
T 227 (Tld)..839
T 991 (Tld)..838
T 994 (Tld)..838
T Craft (SA) 763
Taechong I (DPRK) 470
Taechong II (DPRK) 470
Tae Geuk (RoK)................................490
Al Tahaddy (Kwt)..............................494
Takanami (Jpn) 438
Takatori (Jpn) 455
Takatsuki (Jpn).................................456
Takuyo (Jpn) 459
Talwar (Project 1135.6) (Ind) 337
Tanu (Can).......................................108
Tanya (Az, Rus).......................42, 718
Tapir (Rus)..706
Tapi (PF 103) (Tld)...........................834
Tapper (Swe)....................................800
Tara Bai (Ind)....................................356
Tarantul (Cub, Rus, Ukr, Vtn)............192, 728, 882, 1011
Tarantul I (Ind, Rom, Yem).........344, 664, 1017
Tarantul II (Bul)..................................90
Tarantul II (Molnya) (Project 1241.1/2) (Ukr) 878
Tarantul V (Vtn)..............................1011
Tarantul V (Project 1241.8) (Tkm) 873

INDEXES > Named classes — Aircraft by countries

Name	Page
Tarantul (Molnya) (Project 1241.1/1241M/1242.1) (Rus)	700
Tarawa (US)	963
Tariq (Amazon) (Pak)	604
Tarlan (Iran)	386
TB 11PA and 10 TB 40A (DPRK)	471
Teh Hsing (Twn)	822
Tenace (Tur)	870
Tenerife (Col)	175
Tenochtitlan (Damen Stan Patrol 4207) (Mex)	537
Tenryu (Jpn)	448
Tenyo (Jpn)	459
Terrebonne Parish (Per, Tur)	622, 863
Terrier (Project 14170) (Rus)	727
Terry Fox (Can)	107
Teshio (Jpn)	455
Tetal (Rom)	663
Thetis (Den)	196
Thomas G Thompson (US)	971
Thomaston (Brz)	81
Thomson (Type 209/1400) (Chi)	117
Thondor (Houdong) (Iran)	384
Thong Kaeo (Tld)	840
Thor (Ulstein UT 512L) (Ice)	325
Ticonderoga (US)	944
Tide (UK)	919
Tiger (Chi, Egy)	120, 221
Tikuna (Type 209/1450) (Brz)	70
Timblo (Ind)	357
Timsah (Egy)	227
Tir (Ind)	351
Tir (IPS 18) (Iran)	383
Tir II (IPS 18) (Syr)	808
Tisza (Indo)	374
Todak (PB 57) (Indo)	365
Todaro (Type 212A) (Ita)	401
Tokara (Jpn)	456
Toledo (Col)	179
Tolmi (Asheville) (Gre)	308
Tomas Batilo (Sea Dolphin) (Plp)	631
Tondar (Iran)	389
Toplivo (Rus)	716
Toplivo 2 (Egy)	226
Toralla (Spn)	774
Tornado (Ban, Geo)	61, 283
Toro (Arg)	21
Towada (Jpn)	449
Town (Nig, Pak)	581, 607
TP-01 (Vtn)	1013
Tracker II (Mld)	521
Tracker Mk 2 (Leb)	497
Tracy (Can)	108
Trafalgar (UK)	898
Tral (DPRK)	469
Transportbåt 2000 (UAE)	890
Traverse (Can)	110
Tridente (Type 209PN) (Por)	649
Trinity Marine (Sri)	788
Tripartite (Bul, Fra, Indo, Lat, Nld)	91, 269, 370, 495, 569
Tritão (Brz)	87
Trossö (Swe)	803
Tsaplya (Murena E) (Project 12061) (RoK)	484
Tsaplya (Murena) (Project 12061) (Rus)	733
Tsuruugi (Jpn)	456
TT400TP (Vtn)	1013, 1015
Tugau (Mly)	520
Tupi (Type 209/1400) (Brz)	71
Tuqiang (CPR)	165
Turk (AB 25) (Az, Geo, Kaz)	40, 283, 462
Tursas (Fin)	244
Turya (Project 206M) (Vtn)	1012
Tuzhong (CPR)	167
Tuzla (Tur)	862
Tyne (Sey)	748
Type 032 (Qing) (CPR)	134
Type 20 Biscaya (Ser)	746
Type 26 (UK)	905
Type 44 (Chi, EIS, Guy, Hon, Uru)	125, 230, 318, 321, 997
Type 83 (Egy)	227
Type 123K (Chinese P4) (Ban)	56
Type 206A (Col)	171
Type 209/1400 (Indo)	358
Type 214 (Tur)	857
Type 246 (Alb)	3
Type 701 (Par)	616
Type 800 (ETim)	208
Type 1200 (Chi)	123
Type P22 (Ger)	298
Typhoon (Ban)	61
Typhoon (Akula) (Rus)	674
Tzira (Defender) (Isr)	399

U

Name	Page
U 700 (Fin)	244
Uda (Rus)	716
Udaloy (Fregat) (Rus)	692
Udaloy II (Fregat) (Rus)	691
Ula (Nor)	585
Ulsan (RoK)	478
Ulstein UT 507 (Fra)	278
Ulstein UT 711 (Fra)	278
Um Almaradim (Combattante I) (Kwt)	490
Uniform (Kachalot) (Rus)	685
Upholder (Can)	98
Uraga (Jpn)	444
Uribe (Mex)	535
US Mk II (Iran)	387
US Mk III (Iran)	384
US 3812-VCF (HT)	319
UVL 10 (Fin)	244
Uwajima (Jpn)	445

V

Name	Page
V 600 Falco (Ita)	421
V 2000 (N 61) (Ita)	423
V 3000 (Ita)	423
V 5000 (Ita)	423
V 6000 (Ita)	423
Vakta (Can)	110
Valas (Fin)	242
Valle (Auk) (Mex)	535
Valour (SA)	762
Valpas (Lat)	497
Vanguard (UK)	896
Van Speijk (Indo)	360
Van Straelen (Per)	623
Vasco Da Gama (Meko 200 PN) (Por)	650
Västergötland (Sin)	750
VCSM (Cam, Gab, Gn, Mor, Sen)	97, 280, 318, 547, 744
VD 141 (Rom)	664
Veer (Tarantul I) (Ind)	344
Vekhr (Rus, Ukr)	703, 877
Veritas (Chi)	125
Verlarde (PR-72P) (Per)	621
Viana Do Castelo (NPO 2000) (Por)	652
Victor III (Schuka) (Rus)	680
Victoria (Upholder) (Can)	98
Victorious (US)	976
Victory (Sin)	753
Victory Team P 46 (Kwt)	492
Vidar (Lat, Lit)	495, 503
Vigilant 27 (Uru)	997
Vigilante (Ecu, Gua)	214, 317
Vihuri (Fin)	243
Vikhr (IVA) (Project B-99) (Az)	43
Vikram (Ind, Mrt, Sri)	355, 527, 789
Vinograd (Rus)	710
Virginia (US)	937
Visby (Swe)	797
Vishnya (Project 864) (Rus)	712
Vishwast (Ind)	355
Vita (Qat)	657
Viteazul (Rom)	667
Vittoria (Cypr)	193
Voda (Project 561) (Ukr, Vtn)	880, 1015
Vosh (Moskit) (Project 1248) (Rus)	732
Vosper 25 (Omn)	595
Votsis (La Combattante IIA) (Type 148) (Gre)	307
VTP (Fra)	274
Vydra (Egy)	224
Vydra (Project 106) (Az)	41
Vydra (Project 106K) (Bul)	93
Vytegrales II (Project 596P) (Rus)	714

W

Name	Page
El Wacil (P 32) (Mor)	548
Walchensee (Ger)	296
Walrus (Nld)	562
Wangerooge (Ger, Uru)	298, 996
Warrior (ex-Minister) (SA)	763
Wasp (US)	964
Wasp 20 metre (TT)	850
Wasp 30 metre (Bhr)	49
Waspada (Indo)	367
Water (Ecu)	213
Waters (US)	976
Watson (US)	977
Wattle (Aust)	36
Wave (UK)	915
Wellington (BH.7) (Iran)	389
Westerwald (Egy)	226
Whidbey Island (US)	967
White Sumac (DR, Tun)	206, 853
Wielingen (Bul)	90
Wilkes (Tun)	852
Wochi (Type 081A) (CPR)	157
Wodnik (Pol)	644
Wolei (Type 918) (CPR)	157
Won San (RoK)	485
Wosao (Type 082) (CPR)	156
Wozang (CPR)	157
Wu Kang (Twn)	820
Wuhu-A (CPR, Sri)	154, 790

X

Name	Page
Xia (CPR)	128

Y

Name	Page
Y 301 (Myn)	557
Yaeyama (Jpn)	445
Yakhont (Az, Rus, Syr, Ukr, Vtn)	42, 708, 809, 879, 1014
Yamayuri (Com)	181
Yanbing (Mod Yanha) (CPR)	169
Yanci (CPR)	164
Yang Yang (RoK)	485
Yanha (CPR)	165
Yannan (CPR)	164
Yantai (Type 073) (CPR)	162
Yarrow type (Tld)	831
Yasen (Rus)	679
Yavuz (Tur)	859
Yaz (Slepen) (Project 1208) (Rus)	731
Yelva (Krab) (Project 535M) (Rus)	718
Yelva (Project 535M) (Ukr)	881
Yenlai (Type 635) (CPR)	159
Yen Pai (Yanci) (CPR)	164
Yevgenya (Az, Cub)	42, 191
Yevgenya (Korond) (Project 1258) (Ukr, Vtn)	879, 1014
Yevgenya (Project 1258) (Syr)	809
YF 2150 (Jpn)	444
Yilan (Twn)	824
Yildiz (Tur)	862
Yodo (Jpn)	457
Yola (Nig)	582
Yono (IS 120) (Iran)	380
Yono (P4) (DPRK)	466
Youdian (Type 991) (CPR)	163
YT 802 (US)	974
YTL 422 (Tld)	845
Yuan (CPR)	133
Yubei (Type 074A) (CPR)	155
Yuch'in (Ban)	57
Yuch'in (Type 068/069) (CPR)	156
Yuch'in (Type 069) (Tan)	826
Yudeng (Type 073 III) (CPR)	154
Yug (Project 862) (Rus)	709
Yugo (DPRK, Vtn)	467, 1008
Yuhai (Type 074) (Wuhu-A) (CPR)	154
Yuhai (Wuhu-A) (Type 074) (Sri)	790
Yukan (Type 072) (CPR)	155
Yukto (DPRK)	472
Yuliang (Type 079) (CPR)	154
Yulin (CPR)	163
Yung Feng (MWV 50) (Twn)	819
Yun Hsing (Twn)	825
Yunnan (Cam, CPR, Sri)	97, 155, 791
Yunshu (Type 073 IV) (CPR)	155
Yurka (Egy)	225
Yurka (Rubin) (Project 266) (Vtn)	1014
Yusoutei (Jpn)	444
Yuting I (Type 072 II) (CPR)	154
Yuting II (Type 072 III) (CPR)	154
Yuyi (Type 726) (CPR)	156
Yuzhao (Type 071) (CPR)	153

Z

Name	Page
Z (Egy)	225
Zborul (Tarantul I) (Rom)	664
Zbyszko (Pol)	645
Zeus (US)	978
Zhuk (Grif) (Az, Cub, EqG, Nic, Ukr)	41, 192, 231, 578, 878, 882
Zhuk (Grif) (Project 1400/1400M) (Rus)	728
Zhuk (Grif) (Project 1400M) (Syr)	809
Zhuk (Project 1400) (Geo, Kaz)	282, 462
Zhuk (Project 1400M) (Bul, Vtn)	94, 1013
Zhuk (Type 1400M) (Mrt)	526
Zorritos (Per)	626
Zubr (CPR, Gre, Rus, Ukr)	156, 309, 706, 878
Zumwalt (DDG 1000) (US)	950
Zvezdochka (Project 20180/20181/20183) (Rus)	711

Aircraft by countries

Country	Page
Algeria	5-6
Argentina	16, 24-25
Australia	30-31
Bahrain	47
Bangladesh	54
Belgium	64
Brazil	77-78
Brunei	88
Bulgaria	91
Canada	103
Chile	119-120
China	149-150
Colombia	173
Croatia	185
Cyprus	193
Denmark	197
Dominican Republic	207
Ecuador	211-212
Egypt	221
Finland	241
France	262-264
Germany	291-292
Greece	306
Hong Kong	324
India	345-346
Indonesia	364-365
Iran	383
Ireland	394
Israel	397
Italy	411
Japan	441-442, 453-455
Korea, South	481-482
Latvia	496
Libya	500
Malaysia	512, 521
Malta	524
Mauritania	525
Mexico	531-532
Morocco	547
Netherlands	566
New Zealand	574-575
Norway	588
Oman	595
Pakistan	605-606
Panama	614
Paraguay	617
Peru	621
Philippines	630
Poland	641
Portugal	652
Romania	663
Russian Federation	702-703
Saudi Arabia	739
Senegal	745
Seychelles	747
Singapore	754
South Africa	762-763
Spain	771-772
Sweden	800
Syria	808
Taiwan	815-816
Thailand	835-836
Turkey	861, 873
Ukraine	877
United Arab Emirates	886
United Kingdom	905-907
United States	954-957, 986-987
Uruguay	996
Venezuela	1003
Vietnam	1011

NOTES

NOTES